Metals and Their Compounds in the Environment

Edited by Ernest Merian

© VCH Verlagsgesellschaft mbH, D-6940 Weinheim (Federal Republic of Germany), 1991

Distribution
VCH, P.O. Box 101161, D-6940 Weinheim (Federal Republic of Germany)
Switzerland: VCH, P.O. Box, CH-4020 Basel (Switzerland)
United Kingdom and Ireland: VCH (UK) Ltd., 8 Wellington Court, Cambridge CB1 1HZ (England)
USA and Canada: VCH, Suite 909, 220 East 23rd Street, New York, NY 10010–4606 (USA)

ISBN 3-527-26521-X (VCH, Weinheim)   ISBN 0-89573-562-8 (VCH, New York)

Metals and Their Compounds in the Environment

Occurrence, Analysis, and Biological Relevance

Edited by Ernest Merian

in Cooperation with
Thomas W. Clarkson, Lawrence Fishbein,
Marika Geldmacher-von Mallinckrodt, Magnus Piscator,
Hans-Werner Schlipköter, Markus Stoeppler,
Werner Stumm, and F. William Sunderman, Jr.

VCH   Weinheim · New York · Basel · Cambridge

Dr. Ernest Merian
Im Kirsgarten 22
CH-4106 Therwil
Switzerland

This book was carefully produced. Nevertheless, authors, editor and publisher do not warrant the information contained therein to be free of errors. Readers are advised to keep in mind that statements, data, illustrations, procedural details or other items may inadvertently be inaccurate.

Published jointly by
VCH Verlagsgesellschaft mbH, Weinheim (Federal Republic of Germany)
VCH Publishers, Inc., New York, NY (USA)

Editorial Director: Dr. Hans F. Ebel
Editorial Manager: Christa Maria Schultz
Production Manager: Claudia Grössl

Library of Congress Card No: 90-12615

British Library Cataloguing-in-Publication Data:
Metals and their compounds in the environment:
occurrence, analysis, and biological relevance
1. Metals. Toxic effects. Environmental aspects
I. Merian, E.
574.5222
ISBN 3-527-26521-X

Deutsche Bibliothek Cataloguing-in-Publication Data:
Metals and their compounds in the environment:
occurrence, analysis and biological relevance / ed. by Ernest Merian
in cooperation with Thomas W. Clarkson. – Weinheim;
New York; Basel; Cambridge: VCH, 1991
ISBN 3-527-26521-X (Weinheim ...)
ISBN 0-89573-562-8 (New York ...)
NE: Merian, Ernest [Hrsg.]

© VCH Verlagsgesellschaft mbH, D-6940 Weinheim (Federal Republic of Germany), 1991

Printed on acid-free paper

All rights reserved (including those of translation into other languages). No part of this book may be reproduced in any form – by photoprint, microfilm, or any other means – nor transmitted or translated into a machine language without written permission from the publishers. Registered names, trademarks, etc. used in this book, even when not specifically marked as such, are not to be considered unprotected by law.

Composition: K+V Fotosatz GmbH, D-6124 Beerfelden
Printing: Zechnersche Buchdruckerei GmbH, D-6720 Speyer
Bookbinding: Konrad Triltsch, Druck- und Verlagsanstalt, D-8700 Würzburg
Printed in the Federal Republic of Germany

Editorial Advisory Board

Prof. Dr. THOMAS W. CLARKSON
Division of Toxicology
University of Rochester
School of Medicine
P.O. Box RBB
Rochester, New York 14642, USA

Prof. Dr. LAWRENCE FISHBEIN
International Life Sciences Institute
Risk Science Institute
1126 Sixteenth St. NW
Washington, DC 20036, USA

Prof. Dr. Dr. MARIKA
GELDMACHER-VON MALLINCKRODT
Senatskommission der Deutschen
Forschungsgemeinschaft
für Klinisch-Toxikologische Analytik
Institut für Rechtsmedizin
Universitätsstraße 22
D-8520 Erlangen

Prof. Dr. MAGNUS PISCATOR
The Karolinska Institute
Department of Environmental Hygiene
P.O. Box 60400
S-10401 Stockholm

Prof. Dr. HANS-WERNER SCHLIPKÖTER
Medizinisches Institut
für Umwelthygiene
der Universität Düsseldorf
Auf'm Hennekamp 50
D-4000 Düsseldorf 1

Dr. MARKUS STOEPPLER
Institut für Chemie
Forschungszentrum Jülich GmbH
Institut 4, Angewandte Physikalische
Chemie
Postfach 1913
D-5170 Jülich

Prof. Dr. WERNER STUMM
Eidgenössische Technische
Hochschulen EAWAG
(Eidgen. Anstalt für Wasserversorgung,
Abwasserreinigung und Gewässerschutz)
CH-8600 Dübendorf

Prof. Dr. F. WILLIAM SUNDERMAN, Jr.
Departments of Laboratory Medicine
and Pharmacology
University of Connecticut
School of Medicine
263 Farmington Avenue
Farmington, Connecticut 06032, USA

Preface

Metals and metalloid elements are ubiquitous. They occur naturally as ions, compounds and complexes (speciation) and – to an increasingly relevant degree – in the anthroposphere in a variety of forms. Many of them are extracted, purified and processed for industrial use and then released to the environment again. Even non-scientists are aware of the problems involved. As a result of global and regional redistribution (cycles) and conversion, living organisms can be exposed to them (in excess of their natural levels) locally through respiration, skin contact and consumption. Certain metal compounds are, besides negative effects in high doses, essential for some organisms in small doses. Interactions with other substances in complex dynamic systems also have to be considered. The discussion of these important environmental issues is currently underway and will continue in the future. This is also true for the disposal of wastes, besides the fact that some metals are exhaustible raw materials that can only be partially recycled. It is hoped that this book will stimulate people to use metals sparingly so that pollution can be reduced at its source.

Among the ever-increasing flood of specialized technical literature on metal compounds most reviews approach the subject from the viewpoint of human toxicology only. Thus the need arises for a handbook that deals with environmental and analytical chemistry of metal compounds in addition to their biological effects. The success of the first edition of this handbook, written in German and published in 1984, prompted the decision to put out a new edition in English. This completely revised second edition contains updated information and fills some noticeable gaps found in the first edition.

Scientists from numerous fields with widely differing goals are concerned with the effects of metal compounds: geologists, metallurgists, material scientists, chemists, analytical chemists, environmental engineers (including security and waste disposal experts and administrators), biochemists, biologists, toxicologists, ecotoxicologists, veterinarians, industrial hygienists, and physicians, but also decision makers, administrators, and teachers at all levels. The book aims to provide all these specialists with a balanced presentation of the topic "Metals in the Environment" based on up-to-date international scientific knowledge. For this purpose it was necessary to compile, examine and critically evaluate the most recent comprehensive and widely distributed literature, and present the results in a unified and well structured form. In addition, the conclusions reached at many symposia have been included – for example at the Hans Wolfgang Nürnberg Memorial IAEAC Workshops on Toxic Metal Compounds (Interrelation between Chemistry and Biology) in Villars-sur-Ollon in 1986 and in Follonica in 1988.

Eighty-six scientific experts carried out an interdisciplinary evaluation, incorporating important results from all five continents; previous publications very often only concentrated either on North American or on European literature. Planning and coordination were the responsibility of the editor-in-chief in association with the eight members of the Editorial Advisory Board.

Unfortunately, our revered colleague Prof. Dr. Hans Wolfgang Nürnberg, whose contribution to the success of the first IAEAC Workshop in Geneva in 1983 and the first edition was especially important, died unexpectedly on May 12, 1985. We remember him with respect and affection.

In *Part I* of this book, "General Aspects", the fundamentals of the subject are systematically reviewed. In 20 chapters some of which consist of several subchapters, the following general problems are discussed: analytical chemistry, distribution and transformation in waters, the atmosphere and soils, emissions, bioavailability, biological effects, and evaluation of risks. *Part II*, "Selected Metals and Metalloids, and Their Ions and Compounds", presents detailed reviews of the 37 elements and their derivatives in alphabetical order. The length of the chapters and their contents vary according to the current assessment of their ecochemical and ecotoxicological relevance.

The chapters of Part II are organized according to the following scheme: The physical and chemical properties of the metal or metalloid and its compounds, including analytical methods, occurrence, production, uses (environmental balance), and wastes; resorption, metabolism and excretion in microorganisms, plants, animals and humans; and finally, acute and chronic adverse biological effects on populations, groups at risk, individuals and organs of living species in the ecosphere. It is important to take account of the bioavailability, transport and transformation processes, individual resistance mechanisms and tolerance at cellular levels when assessing the possible consequences of exposure to metal compounds (including catalytic effects). However, the equally important field of environmental radioactivity was excluded.

An attempt has been made to assess the ecological and toxicological risks on the basis of national and international requirements – for example, recommended exposure limits. In addition to environmental evidence, the discussion is also based on experience from allied fields, such as industrial medicine and clinical chemistry. In controversial cases, in which certain questions have not yet been resolved, diverging opinions have been equally considered as far as possible. In each chapter great emphasis has been placed on the cited literature. The publications and documents selected by a multidisciplinary group are intended to serve as additional sources of information and as a guide to more intensive study of specific questions.

The *Glossary* provides useful explanations of the specialized terms. The *Index* is intended to provide complete access to all topics in the book, even to information about a metal mentioned in a chapter about another element. For reasons of logic and consistency, the use of additional terms and subentries has been avoided. Unfortunately, it has become accepted to speak of lead, cadmium and mercury instead of the actually occurring compound, i.e., the appropriate chemical species. Since the co-authors did use a wide number of quasi-synonyms, it was necessary to restrict the index to a selection of terms (with cross-references). The reader may thus find different terms on the indicated pages, but with some imagination, he can collect much more information from the handbook. The index refers also to newest additional literature, which allows to find further updated data. References in bold print and those followed by "f." or "ff." provide important information about key terms and are thus useful initial guidance. The clear arrangement of the book will readily lead the reader to the main topics.

My thanks are due to all the contributing authors for their patience and compliance with the concepts of the editors and the publishers, and to my colleagues in the editorial board for their intensive and expert advice. Furthermore, I am particularly grateful to Priv.-Doz. Dr. Ulrich Ewers, Düsseldorf, Prof. Dr. Ulrich Förstner, Hamburg-Harburg, and Prof. Dr. Bernhard Ulrich, Göttingen, for their cooperation and valuable input that went beyond the chapters they contributed. Dr. Hans F. Ebel, Editorial Director of VCH, who advocated the publication of this reference book, deserves my special thanks. I am equally indebted to Mrs. Christa Schultz of the editorial department of VCH whose involvement in all the editorial work and invaluable criticism and assistance made an important contribution to the book's success. Mrs. Elizabeth Mole, Weinheim, deserves special mention for improving the arrangement and the definitions in the glossary and contributing to its high quality. Mrs. Mole and Mrs. Ann L. Gray, Leverkusen, maintained high standards in translating or revising the contributions and, in addition, made very good factual suggestions. Without the support I received, the preparation of this handbook would not have been possible.

Therwil, autumn 1990   Ernest Merian

Contents

Part I General Aspects

I.1 The Composition of the Upper Earth's Crust and the Natural Cycles of Selected Metals. Metals in Natural Raw Materials. Natural Resources 3
KARL HANS WEDEPOHL

I.2 Production and Processing of Metals: Their Disposal and Future Risks 19
ROGER C. WILMOTH, S. JACKSON HUBBARD, JOHN O. BURCKLE, and JOHN F. MARTIN

I.3 Chemical Processes in the Environment, Relevance of Chemical Speciation 67
JAMES J. MORGAN and WERNER STUMM

I.4a Analytical Chemistry of Metals and Metal Compounds 105
MARKUS STOEPPLER

I.4b Metal Concentrations in Human Body Fluids and Tissues 207
ULRICH EWERS and ARTHUR BROCKHAUS

I.4c Bioindicators for Monitoring Heavy Metals in the Environment ... 221
JÜRG HERTZ

I.5 Metals and Metal Compounds in Waters 233
BISERKA RASPOR

I.6a Metal Compounds in the Atmosphere 257
HANS PUXBAUM

I.6b Indoor Environments: The Role of Metals 287
LAWRENCE FISHBEIN

I.7a Heavy Metal Compounds in the Soil 311
HERBERT W. SCHMITT and HANS STICHER

I.7b Dumping of Wastes 333
ULRICH FÖRSTNER, CARLO COLOMBI, and RAINER KISTLER

I.7c Heavy Metals in Sewage Sludge and Town Waste Compost 357
HEINZ HÄNI

I.7d Deposition of Acids and Metal Compounds 369
BERNHARD ULRICH

I.7e Mobilization of Metals from Sediments 379
ULRICH FÖRSTNER and WILLEM SALOMONS

I.7f Uptake, Distribution, and Effects of Metal Compounds on Plants .. 399
MICHAEL L. BERROW and JOHN C. BURRIDGE

I.7g Metal Tolerance in Higher Plants 411
CAROLINE L. SCHULTZ and THOMAS C. HUTCHINSON

I.7h Bacteria, Fungi, and Blue-Green Algae 419
BETTY H. OLSON and ASHOK K. PANIGRAHI

I.8a Metal Loads of Food of Vegetable Origin Including Mushrooms ... 449
PETER WEIGERT

I.8b Metal Accumulation in the Food Chain and Load of Feed and Food   469
HANS-JÜRGEN HAPKE

I.9   Metals as Essential Trace Elements for Plants, Animals, and
      Humans 481
FELIX KIEFFER

I.10 The Significance of Interactions in Metal Essentiality and Toxicity .. 491
DETMAR BEYERSMANN

I.11 Cell Biochemistry and Transmembrane Transport of Some Metals .. 511
STEPHEN G. GEORGE

I.12 Metallothioneins 523
ANDREAS SCHÄFFER and JEREMIAS H. R. KÄGI

I.13a Effects of Metals on Domestic Animals 531
HANS-JÜRGEN HAPKE

I.13b Extrapolation of Animal Experiments 547
MAGNUS PISCATOR

I.14a Metal and Ceramic Implants 557
       ROLF MICHEL

I.14b Treatment Using Metal Ions and Complexes 565
       JOHN R. DUFFIELD and DAVID R. WILLIAMS

I.15   Intake, Distribution, and Excretion of Metals and Metal
       Compounds in Humans and Animals 571
       ULRICH EWERS and HANS-WERNER SCHLIPKÖTER

I.16   Acute Metal Toxicity in Humans 585
       MARIKA GELDMACHER-VON MALLINCKRODT

I.17a Chronic Toxicity of Metals and Metal Compounds 591
       ULRICH EWERS and HANS-WERNER SCHLIPKÖTER

I.17b Metal Compounds and Immunotoxicology 605
       BADRUL ALAM CHOWDHURY and RANJIT KUMAR CHANDRA

I.18   Mutagenicity, Carcinogenicity, Teratogenicity 617
       ERICH GEBHART and TOBY G. ROSSMAN

I.19   Ecogenetics 641
       MARIKA GELDMACHER-VON MALLINCKRODT

I.20a Standard Setting and Risk Assessment in Human Exposure to
       Metals and Their Compounds 651
       REINIER L. ZIELHUIS

I.20b Standards, Guidelines, and Legislative Regulations Concerning
       Metals and Their Compounds 687
       ULRICH EWERS

Part II   Metals and Metalloids, and Their Ions and Compounds

II.1   Aluminum 715
       JOHN SAVORY and MICHAEL R. WILLS

II.2   Antimony 743
       BRUCE A. FOWLER and PETER L. GOERING

II.3   Arsenic 751
       ALAIN LÉONARD

II.4	Beryllium .... WALLACE R. GRIFFITTS and DAVID N. SKILLETER	775
II.5	Bismuth .... DAVID W. THOMAS	789
II.6	Cadmium .... MARKUS STOEPPLER	803
II.7	Chromium .... JOHANNES GAUGLHOFER and VERA BIANCHI	853
II.8	Cobalt .... GERHARD N. SCHRAUZER	879
II.9	Copper .... I. HERBERT SCHEINBERG	893
II.10	Gallium .... PETER L. GOERING and BRUCE A. FOWLER	909
II.11	Germanium .... BENJAMIN R. FISHER, PETER L. GOERING, and BRUCE A. FOWLER	921
II.12	Gold .... C. FRANK SHAW III	931
II.13	Indium .... BRUCE A. FOWLER and PETER L. GOERING	939
II.14	Iron .... HELMUT A. HUEBERS	945
II.15	The Lanthanides .... PETER L. GOERING, BENJAMIN R. FISHER, and BRUCE A. FOWLER	959
II.16	Lead .... ULRICH EWERS and HANS-WERNER SCHLIPKÖTER	971
II.17	Lithium .... BARTHOLOMÉ RIBAS	1015
II.18	Magnesium .... JERRY K. AIKAWA	1025

II.19 Manganese 1035
RAINER SCHIELE

II.20 Mercury 1045
RUDY VON BURG and MICHAEL R. GREENWOOD

II.21 Molybdenum 1089
GEORGE K. DAVIS

II.22 Nickel 1101
F. WILLIAM SUNDERMAN, JR. and AGNETA OSKARSSON

II.23 Niobium 1127
PETER L. GOERING and BRUCE A. FOWLER

II.24 Platinum-Group Metals 1135
HERMANN RENNER and GABRIELLA SCHMUCKLER

II.25 Selenium 1153
LAWRENCE FISHBEIN

II.26 Silver 1191
H. G. PETERING and C. J. MCCLAIN

II.27 Tantalum 1203
PETER L. GOERING and BRUCE A. FOWLER

II.28 Tellurium 1211
LAWRENCE FISHBEIN

II.29 Thallium 1227
FRITZ H. KEMPER and HANS P. BERTRAM

II.30 Tin 1243
ERIC J. BULTEN and HARRY A. MEINEMA

II.31 Titanium 1261
JACK WHITEHEAD

II.32 Tungsten 1269
MICHAEL HARTUNG

II.33 Uranium, Thorium, and Decay Products 1275
WERNER BURKART

II.34 Vanadium 1289
RICHARD U. BYERRUM

II.35 Yttrium 1299
ROGER DEUBER and THOMAS HEIM

II.36 Zinc 1309
FRIEDRICH KARL OHNESORGE and MICHAEL WILHELM

II.37 Zirconium 1343
KARL-HEINZ SCHALLER

Glossary 1349

Index 1373

List of Contributors

Prof. Dr. JERRY K. AIKAWA
University of Colorado
Health Sciences Center
School of Medicine
4200 East Ninth Avenue
Denver, Colorado 80262, USA

Dr. MICHAEL L. BERROW
The Macaulay Land Use Research
Institute
Craigiebuckler, Aberdeen AB9 2QJ
Scotland, United Kingdom

Dr. HANS P. BERTRAM
Institut für Pharmakologie und
Toxikologie der Universität Münster
Domagkstraße 12
D-4400 Münster

Prof. Dr. DETMAR BEYERSMANN
Fachbereich 2 – Biochemie
Universität Bremen
Postfach 33 04 40
D-2800 Bremen

Dr. VERA BIANCHI
Università degli Studi di Padova
Dipartimento di Biologia
Via Trieste 75
I-35121 Padova

Dr. ARTHUR BROCKHAUS
Medizinisches Institut für
Umwelthygiene an der Universität
Düsseldorf
Auf'm Hennekamp 50
D-4000 Düsseldorf 1

Dr. ERIC J. BULTEN
TNO Division of Technology for
Society
Head Department of Chemistry
P. O. Box 108
NL-3700 AC Zeist

JOHN O. BURCKLE
United States Environmental
Protection Agency
Office of Research and Development
Cincinnati, Ohio 45268, USA

Dr. RUDY VON BURG
P. O. Box 432
Pinole, California 94564, USA

Dr. WERNER BURKART
Paul Scherrer Institut
Abteilung Strahlenhygiene
CH-5234 Villigen

JOHN C. BURRIDGE
The Macaulay Land Use Research
Institute
Craigiebuckler, Aberdeen AB9 2QJ,
Scotland, United Kingdom

Prof. Dr. RICHARD U. BYERRUM
Michigan State University
Department of Biochemistry
East Lansing, Michigan 48824, USA

Prof. Dr. RANJIT K. CHANDRA
Memorial University of Newfoundland
Departments of Pediatrics, Medicine
and Biochemistry
St. John's, Newfoundland A1A 1R8
Canada

Dr. BADRUL A. CHOWDHURY
3737 Beaubien
Detroit, Michigan 48201, USA

Dipl.-Ing. geol. CARLO COLOMBI
Colombi Schmutz Dorthe AG
Kirchstraße 22
CH-3097 Bern/Liebefeld

Prof. Dr. GEORGE K. DAVIS
2903 S.W. Second Court
Gainesville, Florida 32601, USA

ROGER DEUBER
Carbotech AG
Eulerstraße 68
CH-4051 Basel

Dr. JOHN R. DUFFIELD
University of Wales
Institute of Science and Technology
Department of Applied Chemistry
Redwood Building
P.O. Box 13
Cardiff CF1 3XF
Wales, United Kingdom

Priv.-Doz. Dr. ULRICH EWERS
Medizinisches Institut für
Umwelthygiene an der
Universität Düsseldorf
Auf'm Hennekamp 50
D-4000 Düsseldorf

Prof. Dr. LAWRENCE FISHBEIN
International Life Sciences Institute
Risk Science Institute
1126 Sixteenth Street N.W.
Washington, DC 20036, USA

Dr. BENJAMIN R. FISHER
U.S. Food and Drug Administration
Department of Health and Human
Services
Center for Devices and Radiological
Health
12709 Twinbrook Parkway
Rockville, Maryland 20857, USA

Prof. Dr. ULRICH FÖRSTNER
Technische Universität Hamburg-
Harburg
Arbeitsbereich Umweltschutztechnik
Postfach 901403
D-2100 Hamburg 90

Dr. BRUCE A. FOWLER
University of Maryland
School of Medicine
Toxicology Program
660 West Redwood Street
Baltimore, Maryland 21201, USA

Dr. JOHANNES GAUGLHOFER
Chapfhaldenstraße 21
CH-9032 Engelburg, St. Gallen

Prof. Dr. ERICH GEBHART
Institut für Humangenetik und
Anthropologie der Universität
Erlangen-Nürnberg
Schwabenanlage 10
D-8520 Erlangen

Prof. Dr. Dr. MARIKA
GELDMACHER- VON MALLINCKRODT
Senatskommission der Deutschen
Forschungsgemeinschaft für
Klinisch-Toxikologische Analytik
Institut für Rechtsmedizin
Universitätsstraße 22
D-8520 Erlangen

List of Contributors

Dr. STEPHEN G. GEORGE
NERC Unit of Aquatic Biochemistry
School of Molecular and Biological
Sciences
University of Stirling
Stirling FK9 4LA
Scotland, United Kingdom

Dr. PETER L. GOERING
U.S. Food and Drug Administration
Department of Health and Human
Services
Center for Devices and Radiological
Health
12709 Twinbrook Parkway
Rockville, Maryland 20857, USA

Dr. MICHAEL R. GREENWOOD
Xerox Corporation
Joseph C. Wilson Center for
Technology
Webster, New York 14580, USA

Dr. WALLACE R. GRIFFITTS
United States Department of the
Interior Denver Federal Center
Geological Survey
Branch of Geochemistry
P.O. Box 25046
Denver, Colorado 80225, USA

Dr. HEINZ HÄNI
Eidgenössische Forschungsanstalt
für Agrikulturchemie und
Umwelthygiene
CH-3097 Liebefeld-Bern

Prof. Dr. HANS-JÜRGEN HAPKE
Institut für Pharmakologie, Toxikologie
und Pharmazie der Tierärztlichen
Hochschule Hannover
Bünteweg 17
D-3000 Hannover 71

Priv.-Doz. Dr. MICHAEL HARTUNG
Institut für Arbeits- und Sozialmedizin
der Universität Erlangen-Nürnberg
Schillerstraße 25/29
D-8520 Erlangen

Dr. THOMAS HEIM
Carbotech AG
Eulerstraße 68
CH-4051 Basel

Dr. JÜRG HERTZ
Swiss Federal Institute for Forest,
Snow and Landscape Research
Department of Ecology
Zürcherstraße 111
CH-8903 Birmendorf

S. JACKSON HUBBARD
United States Environmental
Protection Agency
Office of Research and Development
Cincinnati, Ohio 45268, USA

Prof. Dr. Dr. HELMUT A. HUEBERS
United States Department of
Agriculture
Children's Nutrition Research Center
at Baylor College of Medicine
1100 Bates Street
Houston, Texas 77030, USA

Prof. Dr. THOMAS C. HUTCHINSON
Institute for Environmental Studies
Department of Botany
University of Toronto
Toronto, Ontario M5S 1A4
Canada

Prof. Dr. JEREMIAS H.R. KÄGI
Biochemisches Institut der Universität
Zürich
Winterthurerstraße 190
CH-8057 Zürich

Prof. Dr. Fritz H. Kemper
Institut für Pharmakologie und
Toxikologie der Universität Münster
Domagkstraße 12
D-4400 Münster

Dr. Felix Kieffer
Wander AG
Ernährung Schweiz
Postfach 2747
CH-3001 Bern

Dr. Rainer Kistler
Zentralstelle für Umweltschutz
Bahnhofstraße 27
CH-6300 Zug

Prof. Dr. Alain Léonard
Université Catolique de Louvain
Avenue E. Mounier, 72
B-1200 Bruxelles

John F. Martin
United States Environmental
Protection Agency
Office of Research and Development
Cincinnati, Ohio 45268, USA

Prof. Dr. Craig J. McClain
University of Kentucky
Medical School
Lexington, Kentucky 40503, USA

Dr. Harry A. Meinema
TNO Division of Technology for
Society
Head Department of Chemistry
P.O. Box 108
NL-3700 AC Zeist

Prof. Dr. Rolf Michel
Zentraleinrichtung für Strahlenschutz
der Universität Hannover
Am kleinen Felde 30
D-3000 Hannover 1

Prof. Dr. James J. Morgan
California Institute of Technology
Environmental Engineering Science
Pasadena, California 91125, USA

Prof. Dr. Friedrich Karl
Ohnesorge
Institut für Toxikologie
der Universität Düsseldorf
Moorenstraße 5
D-4000 Düsseldorf 1

Prof. Dr. Betty H. Olson
Program in Social Ecology
University of California, Irvine
Irvine, California 92717, USA

Dr. Agneta Oskarsson
The National Food Administration
Toxicology Laboratory
P.O. Box 622
S-75126 Uppsala

Dr. Ashok K. Panigrahi
Environmental Toxicology Laboratory
Department of Botany
Berhampur University
Berhampur − 760007, Orissa
India

Prof. Dr. Harold G. Petering
2484 Heather Way
Lexington, Kentucky 40503, USA

Prof. Dr. Magnus Piscator
The Karolinska Institute
Department of Environmental Hygiene
P.O. Box 60400
S-10401 Stockholm

Dr. Hans Puxbaum
Institut für Analytische Chemie
der Technischen Universität Wien
Abteilung für Umweltanalytik
Getreidemarkt 9/151
A-1060 Wien

Dr. BISERKA RASPOR
Ruder Bošković Institute
Center for Marine Research
YU-41001 Zagreb, Yugoslavia

Dr. HERMANN RENNER
DEGUSSA AG
Weißfrauenstraße 9
D-6000 Frankfurt/Main 1

Prof. Dr. BARTHOLOMÉ RIBAS
National Centre for
Environmental Health
National Institute
of Health Carlos III
Carretera Majadahonda
a Pozuelo km 2
E-28220 Madrid

Prof. Dr. TOBY G. ROSSMAN
New York University
Medical Center
Institute of Environmental Medicine
550 First Avenue
New York, NY 10016, USA

Dr. WILLEM SALOMONS
Institute for Soil Fertility
P.O. Box 30003
NL-9750 RA Haren

Prof. Dr. JOHN SAVORY
University of Virginia
Health Science Center
Clinical Chemistry and Toxicology
Laboratories
P.O. Box 168
Charlottesville, Virginia 22908, USA

Dr. ANDREAS SCHÄFFER
Biochemisches Institut der Universität
Zürich
Winterthurerstraße 190
CH-8057 Zürich

Dipl.-Ing. KARL-HEINZ SCHALLER
Institut für Arbeits- und Sozialmedizin
der Universität Erlangen-Nürnberg
Schillerstraße 25/29
D-8520 Erlangen

Prof. Dr. I. HERBERT SCHEINBERG
Albert Einstein College of Medicine
Yeshiva University
1300 Morris Park Avenue
Bronx, New York 10461, USA

Prof. Dr. RAINER SCHIELE
Institut für Arbeits- und Sozial-
Medizin der Universität
Erlangen-Nürnberg
Schillerstraße 25/29
D-8520 Erlangen

Prof. Dr. HANS-WERNER SCHLIPKÖTER
Medizinisches Institut für
Umwelthygiene an der
Universität Düsseldorf
Auf'm Hennekamp 50
D-4000 Düsseldorf 1

Dr. HERBERT W. SCHMITT
Eidgenössische Technische Hochschule
Zürich
Institut für Lebensmittelwissenschaft
ETH-Zentrum
Universitätstraße 2
CH-8092 Zürich

Prof. Dr. GABRIELLA SCHMUCKLER
Technion Israel
Institute of Technology
Technion City
Haifa 3200, Israel

Prof. Dr. GERHARD N. SCHRAUZER
University of California, San Diego
Department of Chemistry, B-014
La Jolla, California 92093, USA

CAROLINE L. SCHULTZ
Institute for Environmental Studies
Department of Botany
University of Toronto
Toronto, Ontario M5S 1A4, Canada

Prof. Dr. C. FRANK SHAW III
University of Wisconsin
Department of Chemistry
P.O. Box 413
Milwaukee, Wisconsin 53201, USA

Dr. DAVID N. SKILLETER
Medical Research Council
MRC Toxicology Unit
Woodmansterne Road
Carshalton, Surrey SM5 4EF
United Kingdom

Prf. Dr. HANS STICHER
Eidgenössische Technische Hochschule
Zürich
Institut für Lebensmittelwissenschaft
ETH-Zentrum
Universitätstraße 2
CH-8092 Zürich

Dr. MARKUS STOEPPLER
Institut für Chemie
Forschungszentrum Jülich GmbH
Postfach 1913
D-5170 Jülich

Prof. Dr. Dr. WERNER STUMM
EAWAG Eidgenössische Anstalt für
Wasserversorgung, Abwasserreinigung
und Gewässerschutz
CH-8600 Dübendorf/Zürich

Prof. Dr. F. WILLIAM SUNDERMAN, Jr.
Departments of Laboratory Medicine
and Pharmacology
University of Connecticut
School of Medicine
263 Farmington Avenue
Farmington, Connecticut 06032, USA

Dr. DAVID W. THOMAS
Institute of Medical and Veterinary
Science
Division of Clinical Chemistry
Frome Road
Adelaide, South Australia 5000

Prof. Dr. BERNHARD ULRICH
Institut für Bodenkunde und
Waldernährung der Universität
Göttingen
Büsgenweg 2
D-3400 Göttingen

Prof. Dr. KARL HANS WEDEPOHL
Geochemisches Institut der Universität
Göttingen
Goldschmidtstraße 1
D-3400 Göttingen

Dr. PETER WEIGERT
Bundesgesundheitsamt
Zentrale Erfassungs- und
Bewertungsstelle für
Umweltchemikalien
Postfach 33 00 13
D-1000 Berlin 33

B. Sc. JACK WHITEHEAD
10, Dunsmore Close
Maltby, Middlesbrough
Cleveland TS8 0BS
United Kingdom

Dr. MICHAEL WILHELM
Institut für Toxikologie
der Universität Düsseldorf
Moorenstraße 5
D-4000 Düsseldorf 1

Prof. Dr. DAVID R. WILLIAMS
University of Wales
Institute of Science and Technology
Department of Applied Chemistry
Redwood Building
P. O. Box 13
Cardiff CF1 3XF
Wales, United Kingdom

Dr. MICHAEL R. WILLS
University of Virginia
Health Science Center
Clinical Chemistry and Toxicology
Laboratories
P. O. Box 168
Charlottesville, Virginia 22908, USA

Dr. ROGER C. WILMOTH
United States Environmental
Protection Agency
Office of Research and Development
Risk Reduction Engineering
Laboratory
Water and Hazardous Waste Treatment
Research Division
Cincinnati, Ohio 45268, USA

Prof. Dr. REINIER L. ZIELHUIS
Louise de Colignylaan 4
NL-2341 CJ Oegstgeest

Part I General Aspects

I.1 The Composition of the Upper Earth's Crust and the Natural Cycles of Selected Metals. Metals in Natural Raw Materials. Natural Resources
===

KARL HANS WEDEPOHL, Göttingen, Federal Republic of Germany

1 Formation of the Earth's Crust

The present natural abundance of the chemical elements at and close to the earth's surface is a function of a sequence of processes:

a) Syntheses of the nuclei of the elements in stars, condensation of primitive compounds from solar nebula, aggregation of particles of primitive compounds to form planets.
b) Separation of the earth's crust and atmosphere from the earth's mantle during the geologic history.
c) Transformation of the earth's crust through reactions between rocks, waters, and atmosphere under internal (radiogenic) and external (solar) influences of heat.

The Earth and the planetary system were formed 4.6 billion years ago. Certain meteorites as fragments from small planets have preserved a primitive cosmic composition and contain records of the early history of the solar system. Because of the lack of an atmosphere the lunar surface has not been reworked and still exhibits the craters from the impact of large planetesimals which were abundant in space at the stage of planet formation. The oldest rocks on earth have an age of 3.5 to 4 billion years. Their masses represent the nuclei of the continents which grew from magmatic melts during the geologic history. These magmas originate from more than 50 km depth. The crust has undergone transformations due to weathering, mass transport, and increase of temperature and pressure from its growing thickness and from heat of deeper layers.

The crust is the thin skin of our planet. In the continents it has an average thickness of 40 km and underneath the oceans of 7 km. Its mass of 2×10^{19} t contributes only a proportion of 0.4% of the total earth. The crust covers the earth's mantle which represents 68% of the earth's mass. It consists of magnesium-iron silicates and oxides and reflects the large cosmic abundance of O, Si, Mg, and Fe. The most abundant isotopes of these elements have even numbers of protons and neutrons (Oddo-Harkin rule) indicating their considerable stability in the stellar synthesis of the nuclei. The selection of the elements in the earth's crust is controlled by their behavior during partial melting of the earth's mantle. The magmas, which are partial melting products of the mantle, preferentially transport volatile and so-called incompatible elements into the crust which do not fit well in the crystal structures of the mantle minerals (because of size or valency). This process of element selection is comparable to zone melting. The flux of magmas occurs only partly in a subaerial volcanism. Submarine basaltic magmatism is quantitatively far more important. It is mainly restricted to large meridional ridges in which the ocean crust grows at a rate of about 3×10^{10} t per year through magmatism. At this rate the ocean crust would double its mass within $<10^9$ years if the growth is not compensated by consumption. At the continental margins around the Pacific Ocean the ocean crust is subducted under the continental crust and finally digested by the earth's mantle. The growing crust of the Atlantic and Indian Ocean pushes the bordering continents into opposite directions. The engine of this dynamic behavior of fragments (plates) of the earth's crust is the upwelling of radiogenic heat in the earth's mantle. Large-scale convective transport of heat and matter has lateral branches on which the crustal plates can float. Under the high pressure of the mantle solid matter has a ductile behavior which allows a slow plastic flow in a convective system. The continental and ocean crustal plates with their upper mantle base of more than 50 km thickness move at velocities of a few centimeters per year. The continental crust is growing in a narrow volcanic belt close to the deep trenches caused by the subduction of ocean crust. Dehydration of the subducted slab of ocean crust triggers the andesitic and basaltic volcanism from which the continents grow. This source is responsible for the average chemical composition of the continental crust which is close to that of the magmatic rock andesite (named after the Andes).

Beside the important process in which the ocean and continental crusts grow from basaltic and andesitic melts originating in the upper mantle, granitic magmas can be formed by partial melting within the continental crust. This causes a major remobilization and vertical transport of matter.

2 Alteration of the Earth's Crust

The focal areas of this book have relatively minor relationships to the primary formation of the earth's crust which has caused a certain distribution of the chemical elements. They mainly deal with products of the alteration of the crust in geologic processes. We can presently still observe the weathering of solid rocks, the erosion of mountain ridges, and the transport of eroded materials as suspended and dissolved constituents in river and rain water, in ice and wind. In-situ weathering forms soils. Soils are the base of food production for the human nutrition. Therefore, soils need special protection against the impact of toxic substances (see Chapter I.7).

In connection with magma production from the earth's mantle and its eruption in volcanic processes, the outer shell of the earth permanently delivers steam and also the more aggressive gases CO_2, SO_2, H_2S, HCl, HF, etc. to the atmosphere and the surface waters. During earth's history oceans, lakes, ice caps, rivers, groundwater, and interstitial waters of sediments have grown to a reservoir of 2×10^{18} t water. The existence of this water reservoir at and close to the surface of the planet causes a difference between the Earth and its planetary neighbors Venus and Mars. After weathering and transport the reactive constituents of the magmatic degassing become finally fixed in the large masses of hydroxide bearing sediments, in limestones, gypsum, and salt deposits. The mentioned chemical sediments ($CaCO_3$, $CaSO_4 \cdot 2H_2O$, $NaCl$, etc.) have a proportion of 10 to 20% in the sediment shell of the earth. The major proportion of the sediments consists of the weathering products of preexisting rocks. This mass contains detrital and newly formed minerals after transport and grain size separation in suspension and sedimentation. After diagenetic reconstitution and consolidation they form clays (mudstones, shales) and sandstones. Greywackes are special sandstones formed from detrital materials which have only undergone minor chemical alteration and separation. Gravity is the major engine for the transport of weathering products at the earth's surface. Crustal masses which were folded and fractured into mountains and lifted above sea level by continent moving forces are the object of erosion. After decomposition and transport their matter will be collected in sea basins, mainly at continental margins. Rates of erosion in mountain ridges and the related rates of deposition in near-shore basins scale in the range of fractions of millimeters to millimeters per year. Shales and greywackes which represent important rock masses decomposed without major chemical fractionation approach the average chemical composition of the upper continental earth's crust (Table I.1-1).

Continuous sedimentation moves a certain layer of deposition down under increasing load which is connected with increasing pressure and temperature. The average gradient of rising pressure and temperature with depth is 0.3 kb per km and 20°C per km, respectively. Only minor chemical reactions occur in the majority of sediments up to temperatures of 200°C. In the range of 300 to 800°C former sediments approach new equilibria by various reactions. The process is called regional metamorphism. Water in the porous volume of rocks activates these material reactions. If the metamorphosed sediments which still contain 1 to 2 percent water attain temperatures of about 650 to 700°C they form granitic partial melts. These magmas rise diapirically if their proportion in the rock exceeds 15 percent. They transport the low-melting fraction from the lower into the upper crust. The lower crust rarely attains temperatures of 650 to 700°C except heat is advected from the upper mantle by convection or intrusion of large masses of mafic magmas.

Ore deposits are crustal units of low abundance but of economic importance. They are formed from magmatic melts, hot brines, or from sea water. The hot brines of geothermally heated sea water or formation water abundantly extract metals from the subsurface country rocks in their conduits and precipitate metal sulfides and other compounds, if they mix with surface waters. The total masses of metals in ore deposits are small compared with their large but highly diluted reservoir in common rock species (cf. Table I.1-1). The continental crust contains 1.6×10^{15} t Zn, 3.8×10^{14} t Cu, and 3.1×10^{14} t Pb dispersed in its normal rocks. The estimated reserves of the same metals accumulated in ore deposits are: 1.2×10^8 t Zn, 3.4×10^8 t Cu, and 9.3×10^7 t Pb. Even if the hypothetical and speculated resources of Zn, Cu, and Pb exceed ten to hundred times the known reserves, the total amount of accumulated metals in the crust is small relative to their dispersed mass. High concentrations of toxic metals in near-surface ore deposits can cause environmental problems because of the solubility of their oxidation products and their potential transport in weathering solutions. Dumps of metal mines and mills can contaminate

Table I.1-1. Concentrations of 25 Elements in the Continental and Oceanic Earth's Crust and in Abundant Rock Species (in ppm)

	Be	Mg	Ti	V	Cr	Mn	Fe	Co	Ni	Cu	Zn	As
Shales	3	16000	4600	130	90	850	48000	19	68	45	95	10
Greywackes	3	13000	3800	67	50	750	38000	20	40	45	105	8
Limestones	[0.5]a	26000	400	20	11	700	15000	2	15	4	23	2.5
Granitic rocks	5.5	6000	3000	94	12	325	20000	4	7	13	50	1.5
Gneisses, mica schists	3.8	13000	3870	60	76	600	33000	13	26	23	65	4.3
Basaltic and gabbroic rocks	0.6	37000	9700	251	168	1390	86000	48	134	90	100	1.5
Granulites	2.1	14000	3520	73	88	895	38000	15	33	27	65	[4]a
Continental crust	2.9	16000	4680	109	88	800	42000	19	45	35	69	3.4
Oceanic crust (Ocean ridge basalt)	1	50000	7000	252	317	1200	70000	45	144	81	78	[1.5]a

	Se	Zr	Mo	Ag	Cd	Sn	Te	Pt	Au	Hg	Tl	Pb	Bi
Shales	0.5	160	1.3	0.07	0.13	2.5	?	[0.01]a	0.0025	0.45	0.68	22	0.13
Greywackes	0.1	450	0.7	0.1	0.09	[3]	?	[0.01]a	0.003	0.11	0.20	14	0.07
Limestones	0.19	19	0.4	0.0xb	0.16	[0.x]b	?	[0.001]a	0.002	0.03	0.05	5	0.02
Granitic rocks	0.04	145	1.8	0.12	0.09	3.5	0.01	0.005	0.0024	0.03	1.1	32	0.19
Gneisses, mica schists	0.08	168	[1.5]a	0.08	0.10	2.5	[0.02]a	[0.01]	0.003	0.02	0.65	16	0.10
Basaltic and gabbroic rocks	0.09	137	1	0.11	0.10	1.5	0.008	0.03	0.004	0.02	0.08	3.5	0.04
Granulites	[0.08]a	153	[1.5]a	0.09	0.10	2.5	[0.02]a	[0.01]a	0.0015	[0.02]a	0.28	9.8	0.04
Continental crust	0.077	152	1.5	0.10	0.10	2.5	[0.02]a	0.013	0.0025	0.02	0.49	15	0.08
Oceanic crust (Ocean ridge basalt)	0.17	85	[0.8]	0.03	0.13	0.9	?	[0.03]a	0.002	[0.02]a	0.013	0.89	0.006

[a] Estimated concentration
[b] x = order of magnitude

ground- and surface waters so that reservoirs for drinking water have to avoid such mining areas (see Chapter I.2).

The elements beryllium, magnesium, titanium, vanadium, chromium, manganese, iron, cobalt, nickel, zinc, zirconium, molybdenum, silver, cadmium, tin, thallium, lead, and bismuth mainly occur in the crystal structures of rock-forming silicates and oxides of the common rocks in the earth's crust. These elements with the exception of magnesium, iron, titanium, chromium, and zirconium are trace elements in the minerals. They follow certain rules as reported by GOLDSCHMIDT (1954) in their tendency to enter specific crystal structures. Beside their occurrence in silicates, copper, arsenic, and selenium have an affinity to sulfides and might be mobilized under weathering conditions. Platinum and gold can occur as native metals in common rocks. Magnesium, iron, titanium, chromium, and zirconium also form their own minerals in abundant rock species. Black shales which are often characterized by higher concentrations of iron, vanadium, nickel, zinc, cobalt, molybdenum, silver, cadmium, thallium, lead, and bismuth than in normal shales contain several of the listed elements in their sulfide or carbonaceous fraction. These bituminous shales as potential source rocks of crude oils will probably become objects of future mining which might cause environmental problems.

3 Average Abundance of 25 Elements in Sedimentary, Magmatic and Metamorphic Rock Species

Data on average concentrations of 25 elements (covered by this book) in the most important sedimentary, magmatic and metamorphic rocks have been compiled in Table I.1-1 mainly from sources and reports listed by WEDEPOHL (1969–1978, 1968, 1975, 1981) and by HEINRICHS et al. (1980). These data can be used to estimate the mean abundances of the considered elements in the continental earth's crust which consist of 22% granitic rocks, 23% gneisses and mica schists, 17% gabbros, amphibolites and basalts, and 37% granulites. Granites, gabbros, and basalts are formed by the consolidation of magmatic melts. Gneisses, mica schists, amphibolites, and granulites are metamorphic products of former sedimentary and magmatic rocks which had to adjust to different temperature and pressure conditions. Granulites are the most common rock species of the lower continental earth's crust whereas the remainder of the listed rock types represent the upper crust. Sediments cover large areas of the metamorphic and magmatic crustal rocks. If spread equally over the whole earth they would have a thickness of about one kilometer. Shales (and partly greywackes) contain the weathering products of large crustal units. Therefore, typical concentrations of our 25 elements in shales are almost equal to the abundances of the respective elements in the average continental crust. Major deviations from this balance are restricted to the volatile elements As, Se, and Hg which are especially accumulated in certain clay sediments. The fact that the average concentrations in abundant crustal rocks and their major weathering products are almost equal for more than 20 of the selected elements confirms the representativeness of the compiled data.

Table I.1-1 also contains data for average concentrations of the 25 selected elements in the oceanic crust. As a first approximation we have assumed for this part of the compilation that the oceanic crust mainly consists of the so-called ocean ridge basalt. Large volumes of this basaltic ocean crust have undergone hydrothermal alteration connected with a gain of H_2O, CO_2, Na, Mg, and S from heated sea water and losses of Si, Ca, Fe, Mn, etc. from the altered basalt to the ocean water reservoir.

Data from Table I.1-1 can be easily used for estimates of the natural background in processes where elements from natural and anthropogenic sources are mixed. Many soils, which produce a large proportion of our food, have developed on shales. Therefore, they contain a natural background in the selected 25 elements which is comparable to the values for shales listed in Table I.1-1. LANTZY and MACKENZIE (1979) have confirmed that resemblance of many soils with shales. Natural dust, the suspended proportion in river water and the detrital silicate fraction in coal is often comparable to shales (and greywackes) in chemical composition. In addition to the silicate fraction coal has accumulated several elements through the physiological action of the preexisting plants (Mg, Zn, Mo, etc.) and through diagenetic precipitation of sulfides (Fe, Ag, Cu, Cd, etc.). Average concentrations of the 25 elements in coal are listed in Table I.1-3. Brick is mainly produced from clays which are comparable to shales in chemical composition (see also Sect. 5). Readily volatile elements as Hg, Bi, Cd, Tl, Pb, Zn, As, and Se are partly lost during firing in brickworks (BRUMSACK, 1977).

Cement is produced from a mixture of limestone and shale or clay. Therefore, the composition of its starting material can be estimated from data listed in Table I.1-1.

4 Concentrations and Transport of 25 Elements in Natural Waters

The most important agents of transport at the earth's surface are waters and wind. Beside matter in suspension rivers and rain carry large amounts of dissolved compounds from continental rock weathering to the oceans. Rain also moves salt spray in opposite direction from ocean to land. Rivers carry yearly about 3.6×10^{13} t water, 8.9×10^9 t suspended matter and 3.6×10^9 t dissolved compounds to the oceans. Wind transported rain compensates its more than ten times smaller concentration of dissolved constituents by its higher speed of travel. The amount of Mn, Co, Cr, Ni, Ag, and V extracted yearly from the atmosphere by rain exceeds the rate of river transport by a factor which ranges from 2 to 10 (LANTZY and MACKENZIE, 1979). For the elements As, Cd, Cu, Mo, and Zn the respective factor ranges from 20 to 100, and in case of the more volatile elements Pb, Hg, and Se it ranges from 110 to 790.

The dissolved concentrations of the 25 selected elements in ocean deep water are controlled by natural processes. This is not principally the case for river water and rain. The data on river water listed in Table I.1-2 (according to TUREKIAN, 1969; WEDEPOHL, 1969–1978, and MARTIN and MEYBECK, 1979) are mainly from rivers

I.1 The Composition of the Upper Earth's Crust

Table I.1-2. Concentrations of 25 Elements in Sea Water[a] (Deep Water) and River Water (in µg/kg)

	Be	Mg	Ti	V	Cr	Mn	Fe	Co	Ni	Cu	Zn	As
River water	0.1	4100	3	0.9	1	4	40	0.2	0.3	2	7	1.7
Sea water	0.0002[b]	1.3×10^6	1	1.9	0.2	0.01[f]	0.1	0.002[c]	0.6	0.25	0.6	1.6

	Se	Zr	Mo	Ag	Cd	Sn	Te	Pt	Au	Hg	Tl	Pb	Bi
River water	0.2	2.6	1	0.3	0.02	0.006[d]	?	?	0.002	0.07	0.04	0.3	0.05
Sea water	0.09	0.03	10	0.002[e]	0.1	0.0006[d]	?	?	0.01	0.002	0.01	0.003	0.000003[g]

[a] For several elements exists a large difference between surface and deep water. Only deep water concentrations are listed. Elements with biologically caused depletion in surface water are, e.g., Ni, Cu, Zn, and Cd.
[b] MEASURES and EDMOND (1982), [c] KNAUER et al. (1982), [d] BYRD and ANDREAE (1982), [e] MARTIN et al. (1983), [f] LANDING and BRULAND (1980), [g] LEE et al. (1985/86)

without major contamination from industrialized areas. Suspended clay materials in the rivers have a high capacitiy to adsorb organic residues and metals from anthropogenic and natural sources (sewage, industrial immissions, soil extraction by acid rain water, etc.). They keep the level of dissolved metals reasonably low. The proportion of anthropogenic and natural emissions of 18 of the 25 selected elements in the worldwide atmospheric transport has been estimated by LANTZY and MACKENZIE (1979).

Sampling and analysis of trace elements in sea water have been improved tremendously within the last decade. The investigations since the mid-seventies have demonstrated that the majority of the older data were too high due to contamination and unreliable procedures and had to be discarded. Values in Table I.1-2 on Mg, Ti, V, Cr, Se, Zr, Mo, and Au are from TUREKIAN (1969). New data on Be, Mn, Fe, Co, Ni, Cu, Zn, As, Ag, Cd, Sn, Hg, Tl, Pb, and Bi have been reported by MEASURES and EDMOND (1982), BRULAND (1983), WONG et al. (1983) and by the authors listed in the footnotes of Table I.1-2. With the exception of Mg, V, and Mo, concentrations in deep ocean water are lower than in river water due to the consumption by organisms, the precipitation of authigenic minerals (e.g., MnO_2) etc. in the sea. Mg, V, As, and Mo occur in almost equal concentrations in deep and surface sea water. Several elements, such as Ni, Zn, Cd, and Ba, are highly depleted in the surface layers of the oceans because of their consumption by organisms. The factor of depletion relative to deep water ranges from 500 (Cd) to 5 (Ni, Ba). Vertical distributions of Mn, Co, Sn, and Pb are characterized by surface maxima. The higher lead concentration is caused by anthropogenic contamination from fuel additives. An increasing extraction of Mn and Co from soils by acid rain water could have caused an additional transport of these elements from the continents to the oceans. The elements Mn, Co, Sn, and Pb have very low absolute concentrations in sea water, which allows the observation of surface contamination.

Assuming a steady state system in the continental run-off and the deposition of minerals in the ocean, the residence time of the elements in sea water can be estimated from the yearly rate of transport in the rivers and their average concentration in ocean water. For elements such as Zn, Cu, Ni, and Ti the mean residence time in sea water before incorporation in solid phases is of the order of 10^3 years.

5 Average Abundance of 25 Metals in Natural Raw Materials

The combustion of fossil fuels for the production of energy introduces numerous metals into the atmosphere and subsequently into soils, rivers, and oceans. Coal contains the degraded matter of fossil plants. Crude oil is a thermal product of the kerogen and lipid fraction in residues from microorganisms preserved in sediments. The latter is usually perfectly separated from the silicates of the host rocks whereas coal still contains a minor fraction of the interlayered sediments.

Average trace metal concentrations in coal and crude oil are listed in Table I.1-3. The ranges of variation of these metals in oil are appreciably larger than in coal. Co and Hg scatter over four orders of magnitude.

Table I.1-3. Concentrations of 25 Elements in Brown Coal, Hard Coal, and Crude Oil (in ppm). In the averages reported in this table equal statistical weight is given to mean values from different authors

	Be	Mg	Ti	V	Cr	Mn	Fe	Co	Ni	Cu	Zn	As
Brown coal	0.7[a,b]	3000[a]	140[a,b]	9.5[a,b]	9[a,b]	92[a,b]	5400[a]	2.6[a,b]	5[a,b]	2.5[a,b]	11[a]	2.2[a]
Hard coal	2.6[a,b]	1300[a]	465[a,b]	32[a,b]	13[a,b]	156[a,b]	9000[a]	8[a,b]	22[a,b]	16[a,b]	48[a,b]	21[a,b]
Crude oil	0.0004[c]		0.1[c]	39[c,d,e]	0.12[c,d,e,f]	0.5[c,d,e]	6.5[c,d,f]	0.5[c,d,e,f]	11[c,d,f]	0.7[c,e]	8.0[c,d,e,f]	0.13[c,d,e]

	Se	Zr	Mo	Ag	Cd	Sn	Te	Pt	Au	Hg	Tl	Pb	Bi
Brown coal	0.47[a]	10[b]	2.8[b]	0.01[a]	0.07[a]	1.0[a,b]	?	?	0.0x	0.26[a]	0.03[a]	9[a,b]	0.02[a]
Hard coal	2[a]	28[b]	4[b]	0.44[a]	1.8[a]	2.6[a,b]	?	?	0.0x	0.36[a]	0.62[a]	46[a,b]	0.15[a]
Crude oil	0.20[c,d,e,f]	?	10[c]	0.05[c,f]	0.01[c]	0.01[c]	?	?	0.0009[d,e]	3.4[c,d,e,f]	?	0.3[c]	?

[a] BRUMSACK et al. (1984) reported averages of dried brown coal with 18% ash and hard coal with 8.7 and with 13.9% ash mainly fired in West-German power plants
[b] YUDOVICH et al. (1972) compiled data on large numbers of samples from industrial countries of Europe, USA, and USSR
[c] BERTINE and GOLDBERG (1971) compiled data on crude oil from worldwide sampling
[d] HITCHON et al. (1975) reported on crude oils from Alberta (Canada)
[e] SHAH et al. (1970) reported on crude oils from Libya, California, Louisiana, and Wyoming
[f] ELLRICH et al. (1985) reported on crude oils from South Germany

A large proportion of the Mg, Ti, Cr, Mn, Fe, and Zr concentrations of hard coal listed in Table I.1-3 belongs to the detrital sediment material which is contained in coal. This can be concluded from the respective concentrations of the listed elements in shales and greywackes (cf. Table I.1-1). Sulfur in coal is produced by bacterial sulfate reduction and diagenetic precipitation as iron sulfide close to the carbonaceous material. Several metals which form sulfides, selenides, and arsenides of very low solubility (such as Cu, Cd, Ag, etc.) are accumulated from the waters in which the plant material was deposited. Some metals (such as Mg, Ni, Cu, Zn, etc.) had a physiological function in the preexisting plants and were, therefore, concentrated by the living material.

The only elements which are specifically high in crude oil relative to coal are V, Mo, and Hg. The former two elements are known for their special accumulation in black shales, which are black from the high proportion of organic residues. The living organisms of the primary production in the surface layers of the oceans do not contain extraordinary concentrations of V and Mo. Therefore, a diagenetic origin of the accumulation of these elements by a scavenging complex formation has to be assumed (see Chapters II.21 and II.34). The vanadium and nickel concentrations in crude oils vary with their origin (TISSOT and WELTE, 1984), and therefore, it is possible to estimate their origin from V/Ni analysis. For instance, crude oils from Venezuela, Angola, Columbia, Ecuador, and California are rich in vanadium, those from Indonesia, Libya, and Western Africa contain very little, whereas crude oils from Angola, Columbia, Ecuador, and California are rich in nickel and those from Libya and Tunisia contain little nickel (TISSOT and WELTE, 1984). Arabian and Canadian crude oils contain medium amounts of 10–50 ppm V and 3–20 ppm Ni. Mercury as the most volatile metal can be easily transported in the thermal gradient of the upper earth's crust and can be trapped by the organic residues on this way.

Of the elements listed in Table I.1-3 Zn, As, Se, Ag, Cd, Tl, Pb, and Bi are accumulated to high levels in the finest fraction of particulate aerosols which leave the stacks of coal firing power plants, and Hg is the only metal which is predominantly emitted as gas during coal firing (BRUMSACK et al., 1984). According to balance computations on the behavior of numerous metals during the combustion of hard coal in West-German power plants BRUMSACK et al. (1984) concluded that the fly-ash from the stacks contains about 1000 to 1400 ppm Ni, 3000 to 6000 ppm Zn, 800 to 2000 ppm As, 200 to 300 ppm Se, 20 to 50 ppm Ag, 100 to 200 ppm Cd, 5 to 10 ppm Hg, 10 to 60 ppm Tl, 3000 to 9000 ppm Pb, and 10 to 20 ppm Bi. The concentrations of As, Se, Ag, Pb, and Bi in the finest particulate fraction exceed those in natural dust (cf. shale in Table I.1-1) by factors of several hundreds. The level of the very toxic cadmium in the finest fly-ash is more than thousand times higher than the cadmium abundance in natural dust. The volatility of the reported elements is caused by the relatively high concentration of chlorine in the firing process (0.6 to 1.2% Cl in stack fly-ash). Large amounts of volatile elements in particulate aerosols are transported by wind over large distances. Their impact on soils and river water can be estimated from the yearly combustion of coal and oil in the world. The consumption of these raw materials for energy is about 3×10^9 t each. The yearly firing of coal probably causes the worldwide emission of 2.4×10^4 t Pb, 1.6×10^4 t Zn, 6×10^3 t As, 10^3 t Se, 8×10^2 t Hg, 5×10^2 t Cd, and 1.5×10^2 t Tl, if the data computed by BRUMSACK et al. (1984) on emissions in West Germany are extrapolated. If the yearly run-off from the continents to the oceans of 3.6×10^{13} t H_2O could extract from soils the total amount of the yearly immissions from coal firing, the following concentrations should be observed: 0.7 ppb Pb, 0.4 ppb Zn, 0.17 ppb As, 0.03 ppb Se, 0.02 ppb Hg, 0.015 ppb Cd, and 0.004 ppb Tl. The estimated concentrations from coal emissions attain the natural soluble concentrations in rivers (cf. Table I.1-2) only in the case of Pb and Cd. Admittedly, the total extraction of the soil immissions from coal firing is an extreme assumption.

In Sect. 3 it is mentioned that brick is mainly produced from clays and shales for which average concentrations of the 25 metals are summarized in Table I.1-1. The heating of clay materials up to about 1000°C causes a partial volatilization of Zn, Pb, Cd, Tl, and Bi which ranges from about 40 to 80% of the original metal concentration (BRUMSACK, 1977). Because of the primitive technology of this process, emission is usually not reduced by the installation of filters. Cement is also produced from natural raw materials. The common starting mixture contains 3 parts limestone and 1 part clay or shale with average compositions as listed in Table I.1-1. Trace metal concentrations are usually low in limestone relative to shale. The high temperatures required for partial melting of the material are in the range of 1400 to 1500°C. Electrostatic filters reduce the emission of volatilized elements. Because of the accumulation of toxic metals like Tl in the filtered particulates their recycling must be avoided.

6 Natural Resources

Human beings made systematic and organized use of natural raw materials since the Neolithic Revolution. With the onset of the industrialization the amount of matter mined and extracted from its natural occurrence has increased to a size which is comparable with the masses transported in natural cycles. The volume of the yearly water consumption for irrigation and technical purposes has passed the level of ten percent of the total river discharge to oceans. The food consumption has increased to a size of one permill of the total organic production.

The metal mineral mining started a little more than 4000 years ago. But the size of the operations was small until fossil energy became available for an exponential growth of mineral mining and processing during the last century. For the formation of an exploitable ore deposit metals have to be accumulated locally to a high degree relative to the average crustal abundance of the elements. Factors of accumulation range from hundred to several thousands with the exception of iron and mercury. Iron as a major constituent of the earth's crust only requires a ten- to fifteenfold higher concentration to form an ore deposit. Therefore, small iron ore deposits are abundant. Mercury as a very rare element needs abnormal high degrees of accumulation. But even processes for the average degree of ore concentration are statistically scarce because they need a combination of several not very abundant conditions. In many cases hot water is the transporting agent of the metals. Because of the rarity of favorable conditions, the geographic distribution of mineral resources on a worldwide basis is very uneven. No technically advanced country is currently self-sufficient. More than 80 percent of the world production for Cr and Pt is supplied by one country, and a large industrial producer like the United States has to import more than 70 percent of the demands of Cr, Pt, Ta, Al, Mn, Sn, and Ni metal or ore.

The exploitation of an ore deposit has to be preceded by prospection and exploration. Prospection is guided by experience on ore genesis and on relations between specific ores and certain geologic structures. Exploration has to investigate the size and exploitability of ore bodies. Both are related to the economic situation on the world market or in a specific country. The easily exploitable ore deposits of the world are almost exhausted. Therefore, a steady increase of the expenses of prospection, exploration, and exploitation of metal ore deposits is to be expected. The grade of the mined ores has steadily decreased. This decrease requires an increase of energy for mining and mineral processing. The price of energy certainly controls the minimum grade of an ore to be exploitable. Because of the rising expenses for prospection and exploration mining companies restrict the size of reserves. Therefore, the knowledge about the world resources in metals is not very good. The required degree of metal accumulation in ore deposits high above the level of the crustal abundance allows to foresee that the world resources of some metals will be exhausted within the time of a few generations.

Information on consumption, reserves, and potential resources for the 25 metals selected for this chapter is compiled in Table I.1-4 on a worldwide basis. An identified resource is called a reserve. Potential resources require new mining technology or processing methods. Iron is by far the most important metal for the human civilization and therefore, has the highest consumption (6.3×10^8 t/a). It is succeeded by manganese (2.2×10^7 t/a). The production of Sn, Mg, Ni, Cr, Ti, Pb, Zn, and Cu is in the range of 10^5 to 10^7 t/a. Next in consumption follows the group of Ag, Be, Cd, Co, V, As, Zr, and Mo (10^4 to 10^5 t/a). The latter two classes contain the metals Ni, Cr, Co, V, and Mo, which are mined principally to be added to iron to give it more desirable properties of strength and resistance against corrosion. Manganese is another essential additive for steels. A large proportion of the Sn, Zn, and Cd production is used as a protective coating for iron. Seven of the 25 metals are produced in quantities of 35 to 5800 t/a. If the current production is maintained instead of being increased, the lifetime of the reserves listed in Table I.1-4 will range from 11 to 760 years. The reserves of Cd, Ag, Bi, Au, Zn, Pb, Ti, Hg, and Sn ores will allow mining operations for a time no longer than between 11 and 50 years.

Three quarters of the metal minerals are processed and mostly consumed in the relatively small highly industrialized countries of a quarter of the world population. The aerial concentration of processing and consumption causes environmental problems. The risk for contamination of soils, rivers, and air by toxic trace elements is high in the industrialized countries. Beside the firing of coal and oil, processing of ores and the technical use of several metals is a major source of such contaminations.

Table I.1-4. Annual Consumption, Reserves, and Potential Resources of 25 Metals in Tons on a Worldwide Basis. The majority of data on yearly consumption according to GOCHT (1974) and HAEFS (1986). The latter refers to 1984 (Mg, V, Cr, Mn, Fe, Ni, Cu, Zn, Mo, Ag, Cd, Sn, Pt, Au, Hg, Pb). Information on reserves from GLOBAL 2000 (1980), GOCHT (1974), and HAEFS (1986). Data on potential resources according to GLOBAL 2000 (1980)

	Be	Mg	Ti	V	Cr	Mn	Fe	Co	Ni	Cu	Zn	As
Annual consumption	$>1.3 \times 10^4$	3.3×10^5	3.8×10^6	2.9×10^4	2.5×10^6	2.2×10^7	6.3×10^8	2.3×10^4	7.8×10^5	8.2×10^6	6.5×10^6	4.0×10^4
Reserves in ore deposits		large		1.6×10^8	2.2×10^7	7.8×10^8	1.8×10^9	5.1×10^6	5.4×10^7	6.0×10^8	1.6×10^8	1.3×10^7
Potential resources					6.0×10^9	1.1×10^9	1.4×10^{11}		1.0×10^8	1.8×10^9	4.0×10^9	

	Se	Zr	Mo	Ag	Cd	Sn	Te	Pt	Au	Hg	Tl	Pb	Bi
Annual consumption	1.6×10^3	5.2×10^4	6.2×10^4	1.3×10^4	2.0×10^4	2.2×10^5	2.0×10^2	1.0×10^2	1.4×10^3	5.8×10^3	35	5.4×10^6	4.8×10^3
Reserves in ore deposits	1.1×10^5	3.7×10^7	8.8×10^6	2.0×10^5	2.2×10^5	1.0×10^7	2.2×10^4	6.5×10^3	3.0×10^4	2.5×10^5	1.4×10^3	1.9×10^8	1.0×10^5
Potential resources				2.0×10^6		2.7×10^7		1.6×10^4		4.0×10^5		1.3×10^9	

7 Concluding Remarks

Human beings depend on soils uncontaminated by toxic elements to produce their food. They need clean drinking water, which is often derived from river water, and they need clean air for breathing. Anthropogenic contamination of soils, waters, and air can only be discovered if the natural background in the abundance of trace metals is known. Therefore, knowledge of the natural cycles of these elements and of the size of the natural reservoirs is required for a better understanding of numerous environmental problems.

References

Bertine, K. K., and Goldberg, E. D. (1971), *Fossil Fuel Combustion and the Major Sedimentary Cycle. Science 173*, 233–235.
Bruland, K. W. (1983), *Trace Elements in Sea Water. Chem. Oceanogr. 8*, 157–220.
Brumsack, H. J. (1977), *Potential Metal Pollution in Grass and Soil Samples around Brickworks. Environ. Geol. 2*, 33–41.
Brumsack, H. J., Heinrichs, H., and Lange, H. (1984), *West German Coal Power Plants as Sources of Potentially Toxic Emissions. Environ. Technol. Lett. 5*, 7–22.
Byrd, J. T., and Andreae, M. O. (1982), *Tin and Methyltin Species in Seawater: Concentrations and Fluxes. Science 218*, 565–569.
Ellrich, J., Hirner, A., and Stärk, H. (1985), *Distribution of Trace Elements in Crude Oils from Southern Germany. Chem. Geol. 48*, 313–323.
GLOBAL 2000 (1980), *Report to the President*, US Council on Environmental Quality and US Foreign Department, G. O. Barney (ed.). US Government Printing Office, Washington, DC.
Gocht, W. (1974), *Handbuch der Metallmärkte*. Springer Verlag, Berlin-Heidelberg-New York.
Goldschmidt, V. M. (1954), *Geochemistry*. Clarendon Press, Oxford.
Haefs, H. (1986), *Der Fischer-Weltalmanach 1987*. Fischer, Frankfurt am Main.
Heinrichs, H., Schulz-Dobrick, B., and Wedepohl, K. H. (1980), *Terrestrial Geochemistry of Cd, Bi, Tl, Pb, Zn, and Rb. Geochim. Cosmochim. Acta 44*, 1519–1533.
Hitchon, B., Filby, R. H., and Shah, K. R. (1975), in: Yen, T. F. (ed.): *The Role of Trace Elements in Petroleum*. Ann Arbor Sci. Publ., Ann Arbor, Michigan.
Knauer, G. A., Martin, J. H., and Gordon, R. M. (1982), *Cobalt in North-east Pacific Waters. Nature 297*, 49–51.
Landing, W. M., and Bruland, K. W. (1980), *Manganese in the North Pacific. Earth Planet. Sci. Lett 49*, 45–56.
Lantzy, R. J., and MacKenzie, F. T. (1979), *Atmospheric Trace Metals: Global Cycles and Assessment of Man's Impact. Geochim. Cosmochim. Acta 43*, 511–525.
Lee, D. S., Edmond, J. M., and Bruland, K. W. (1985/86), *Bismuth in the Atlantic and North Pacific: A Natural Analogue to Plutonium and Lead? Earth Planet. Sci. Lett. 76*, 254–262.
Martin, J. H., and Meybeck, M. (1979), *Elemental Mass-balance of Material Carried by Major World Rivers. Mar. Chem. 7*, 173–206.
Martin, J. H., Knauer, G. A., and Gordon, R. M. (1983), *Silver Distribution and Fluxes in North-east Pacific Waters. Nature 305*, 306–309.
Measures, C. I., and Edmond, J. M. (1982), *Beryllium in the Water Column of the Central North Pacific. Nature 297*, 51–53.

Shah, K. R., Filby, R. H., and Haller, W. A. (1970), *Determination of Trace Elements in Petroleum by Neutron Activation. J. Radioanal. Chem. 6*, 185–192, 413–422.

Tissot, B. P., and Welte, D. H. (1984), *Petroleum Formation and Occurrence*, 2nd Ed. Springer Verlag, Berlin-Heidelberg-New York-Tokyo.

Turekian, K. K. (1969), *The Oceans, Streams and Atmosphere, in: Handbook of Geochemistry*, Vol. 1, Chap. 10. Springer Verlag, Berlin-Heidelberg-New York.

Wedepohl, K. H. (1968), *Chemical Fractionation in the Sedimentary Environment,* in: Ahrens, L. H. (ed.): *Origin and Distribution of the Elements*, pp. 999–1016. Pergamon Press, Oxford-New York.

Wedepohl, K. H. (1975), *The Contribution of Chemical Data to Assumptions about the Origin of Magmas from the Mantle. Fortschr. Miner. 52*(2), 141–172.

Wedepohl, K. H. (ed.) (1969–1978), *Handbook of Geochemistry*, Vol. I, II-1 through II-5. Springer Verlag, Berlin-Heidelberg-New York.

Wedepohl, K. H. (1981), *Tholeiitic Basalts from Spreading Ocean Ridges. Naturwissenschaften 68*, 110–119.

Wong, C. S., Kremling, K., Riley, J. P., Johnson, W. K., Stukas, V., Berrang, P. G., Erickson, P., Thomas, D., Petersen, H., and Imber, B. (1983), in: Wong, C. S., et al. (eds.): *Trace Metals in Sea Water.* Plenum Press, New York.

Yudovich, Y. E., Korycheva, A. A., Obrucknikov, A. S., and Stepanov, Y. V. (1972), *Mean Trace-Element Contents in Coal. Geochem. Int. 9*, 712–720.

Additional Recommended Literature

Trueb, L. (1989), *Shortage of Resources and Environmental Problems in the Age of New Materials* (in German). *Neue Zürcher Zeitung, Forschung und Technik, No. 236*, p. 73 (11 October), Zürich.

I.2 Production and Processing of Metals: Their Disposal and Future Risks[1]

ROGER C. WILMOTH, S. JACKSON HUBBARD, JOHN O. BURCKLE, and JOHN F. MARTIN, Cincinnati, Ohio, USA[2]

1 Metals Production – Mining and Milling

The non-fuel mining industry is an integral and significant part of the United States and world economy. It provides a diversity of products, including lead used in storage batteries, ammunition, and pigments; copper for electrical equipment and supplies; iron for the construction and transportation industries; zinc for galvanizing and other uses; silver for photographic materials; gold for electronic equipment, jewelry, and medicinal use; and uranium used by electric utilities. This industrial sector also produces nonmetallic minerals such as phosphates used to produce industrial chemicals and fertilizers (MINERALS YEARBOOK, 1983). The total metal ore production in the United States alone was worth more than $5.8 billion, and the total value of raw non-fuel minerals was more than $21 billion in 1983 (MINERALS YEARBOOK, 1983). This value accounted for 1% of the U.S. Gross National Product (GNP), while products made from these raw materials account for appoximately 9% of the GNP annually (U.S. INDUSTRIAL OUTLOOK, 1985).

1.1 Non-fuel Mining Segments

There were 580 metal mines and 12117 nonmetal mines active in the United States in 1980 (U.S. INDUSTRIAL OUTLOOK, 1985). In general, the number of mines in operation has decreased over the past several years. However, a reasonable estimate for 1983 indicates that between 400 and 500 metal mines operated in the segments covered here; nearly half this number were gold and uranium mines.

Although mines are classified on the basis of their predominant product, they may also produce large quantities of other materials as coproducts. For example, in 1978, U.S. zinc mines produced 72% of all zinc; 100% of all cadmium, germanium, indium, and thallium; and 3.1, 4.1, and 6.1% of all gold, silver, and lead mined in the United States, respectively. In the same year, copper mines produced over 30% of the silver, 35% of the gold, and 100% of the rhenium, selenium, palladium, tellurium, and platinum mined in this country (MINERALS YEARBOOK, 1981).

---

[1] Information regarding the disposal of metal containing industrial and consumer's goods is to be found in Chapter I.7b and in Part II.
[2] U. Förstner, Hamburg-Harburg, has contributed valuable additional information from European sources.

Table 1.2-1. Mines in the Industry Segments in 1982, by Volume of Material Handled[a,b] (MINERALS YEARBOOK, 1981)

Mining Industry Segment	Total Number of Mines	Less than 1000 tons	1000 to 10000 tons	10000 to 100000 tons	100000 to 1000000 tons	1000000 to 10000000 tons	More than 10000000 tons
Metals							
Bauxite (aluminum)	8	–	1	5	2	–	–
Copper	32	3	1	5	1	15	7
Gold[c]	101	41	28	11	14	6	1
Iron ore	26	–	2	4	6	8	6
Lead	17	7	1	–	2	7	–
Silver	63	32	14	6	10	1	–
Titanium	5	–	–	–	1	4	–
Tungsten	23	18	2	2	1	–	–
Uranium	128	16	34	52	24	2	–
Zinc	14	–	1	2	9	2	–
Other metals[d]	21	5	7	2	2	5	–
Subtotal	438	122	91	89	72	50	14
Nonmetals							
Asbestos	3	–	–	3	–	–	–
Phosphate rock	33	–	1	–	4	23	5
Subtotal	36	–	1	3	4	23	5
Total	474	122	92	92	76	73	19

[a] Includes product and waste, but excludes wells, ponds, and pumping operations
[b] These data are reported in short tons; one short ton equals 1.1 metric tons
[c] Excludes placer operations
[d] Includes antimony, beryllium, mercury, molybdenum, nickel, rare-earth metals, and vanadium

In most mining segments, a few large mines produce most of the products. Table I.2-1 shows the number of mines in each segment, categorized by volume of material handled. This volume includes the amount of earth and rock that must be removed to reach the ore. 213 small mines handled only 10% of the material handled by the 14 largest mines.

1.2 Mining and Beneficiation Wastes

In the nonfuel mining industry, the valuable portion of the crude ore is a small fraction of the total volume of material that must be handled to obtain it (Table I.2-2). For example, over 6900 units of material must be handled to obtain one marketable unit of uranium. The high ratio of "material handled" to "marketable product" is

Table I.2-2. Ratio of Material Handled to Units of Marketable Metal and Estimated Percentage of Metals in Ore (MINERALS YEARBOOK, 1983)

Mining Industry Segment	Ratio of Material Handled to Units of Marketable Metal[a]	Typical Percentage of Metal in Ore
Copper	420:1	0.6
Gold	350000:1	0.0004
Iron ore	6:1	33.0
Lead	19:1	5.0
Mercury	NA	0.5
Molybdenum	NA	0.2
Silver	7500:1	0.03
Tungsten	NA	0.5
Uranium	6900:1	0.15
Zinc	27:1	3.7

NA indicates not available
[a] Excludes material from development and exploration activities

due primarily to the low percentage of metal in the ore and to the mining methods and processes that must be employed. As shown in Table I.2-2, no metal exceeds 5% of the crude ore in which it is embedded, except iron. Aluminum in metallurgical bauxite presents a similar picture. As high-grade ore reserves continue to dwindle, these percentages are likely to become even smaller. The fact that the materials handled consist largely of waste or unusable materials distinguishes these mining industry segments from many other process industries where waste materials make up a relatively small portion of the materials processed to produce a final product.

Several stages may be distinguished: Overburden and waste rock must be removed to expose the ore. The ores are then extracted (mined) and transported to a nearby mill, where they are beneficiated (concentrated or dressed). Mining and beneficiation processes thus generate four categories of large-volume waste: mine waste, tailings, dump and heap leach waste, and mine water.

The vast majority of nonfuel ores in the USA are mined by surface techniques. Only antimony, lead, and zinc mining are solely underground operations. The industry segments that employ both methods handled more ore in surface mines than in below-ground mines (with the exception of silver) in 1982.

Surface mining generates more waste than underground mining (US EPA, 1982a) (Table I.2-3; reliable data were not available for iron ore). As shown, the volume of waste as a percentage of the total amount of crude ore ranges from 9 to 27% for underground mines. In surface mining, the amount of waste ranges from 2 to 10 times the total volume of crude ore. The particle size of mine waste ranges from small clay-size particles (0.002 mm diameter) to boulders (0.3 m diameter). Mine waste piles cover areas ranging from 2 to 240 hectares, with a mean area of 51 ha (1 hectare equals 2.471 acres), according to U.S. BUREAU OF MINES survey of 456 waste piles in the copper, lead, zinc, gold, silver, and phosphate industry segments (MOUNTAIN STATES RESEARCH AND DEVELOPMENT, INC., 1981).

Table I.2-3. Material Handled at Surface and Underground Mines in 1982, for Selected Industry Segments (in thousands of metric tons) (MINERALS YEARBOOK, 1983)

Mining Industry Segment	Surface			Underground		
	Crude Ore	Waste	Waste/Crude Ore Ratio	Crude Ore	Waste	Waste/Crude Ore Ratio
Copper	156 004	321 985	2.06	22 040	1 968	0.09
Gold	21 768	48 797	2.24	1 896	369	0.19
Silver	2 186	19 319	8.84	3 891	584	0.15
Uranium	6 848	72 197	10.54	3 111	848	0.27

After the ore is mined, the first step in beneficiation is generally grinding and crushing. The crushed ores are then concentrated to free the valuable minerals and metal particles (termed values) from the matrix of less valuable rock (called gangue). Beneficiation processes include physical/chemical separation techniques such as gravity concentration, magnetic separation, electrostatic separation, flotation, ion exchange, solvent extraction, electrowinning, precipitation, and amalgamation. The choice of beneficiation process depends on properties of the metal or mineral ore and the gangue, the properties of other minerals or metals in the same ore, and the relative costs of alternative methods. All processes generate tailings, another type of waste.

Tailings generally leave the mill as a slurry, consisting of 50 to 70% (by weight) liquid mill effluent and 30 to 50% solids (clay, silt, and sand-sized particles). More than half of all mine tailings are disposed of in tailings ponds. Some copper tailings ponds in the American southwest cover 240 to 400 hectares (one exceeds 2000 ha), while some small lead/zinc tailings ponds cover less than 1 ha. Based on a U.S. BUREAU OF MINES (1981) survey of 145 tailings ponds in the copper, lead, zinc, gold, silver, and phosphate industries, the average size of these ponds in approximately 200 hectares. Many facilities use several ponds in series, which improves treatment efficiency. Multi-pond systems offer other advantages as well, as the tailings themselves are often used to construct dams and dikes.

Technological advances since the turn of the century have made it economically feasible to beneficiate ore taken from lower-grade ore deposits (MARTIN and MILLS, 1975). For example, froth flotation beneficiation processes have had a tremendous effect on mine production and on the amount and type of mine waste generated. The tailings from froth flotation operations are generally alkaline, because the froth flotation process is most efficient at a higher pH. The metals in the alkaline tailing solids are therefore often immobile.

Dump leaching, heap leaching, and in situ leaching are other processes used to extract metals from low-grade ore. In dump leaching, the material to be leached is placed directly on the ground. Acid is applied, generally by spraying, although many sulfide ores will generate acid during wetting. As the liquid percolates through the ore, it leaches out metals, a process that may take years or decades. The leachate, "pregnant" with the valuable metals, is collected at the base of the pile and subjected to further processing to recover the metal. Dump leach piles often cover hundreds of hectares, rise to 60 meters or more, and contain tens of millions of metric tons of low-grade ore (overburden), which becomes waste after leaching. The dump leach site is often selected to take advantage of naturally impermeable surfaces and to utilize the natural slope of ridges and valleys for the collection of pregnant leach solutions. Loss of leach solution is kept to a minimum in order to maximize metal recovery.

Heap leaching operations are much smaller than dump leach operations, generally employ a relatively impermeable pad under the leach material to maximize recovery of the leachate, and usually take place over a period of months rather than years. Heap leaching is generally used for ores of higher grade or value. For gold ore, a cyanide solution is used as a leaching solution, rather than acid. When leaching no longer produces economically attractive quantities of valuable metals and the sites are no longer in use, the spent ore is often left in place or nearby without further treatment.

In situ leaching is employed in shattered or broken ore bodies on the surface or in old underground workings. Leach solution is applied either by piping or by percolation through overburden. Leach solution is then pumped form collection sumps to a metal recovery or precipitation facility. In situ leaching is most economical when the ore body is surrounded by an impervious layer, which minimizes loss of leach solutions. However, when water is sufficient as a leach solution, in situ leaching is economical even in pervious strata.

Leaching processes are used most often in gold (cyanide leach), uranium (water leach in situ), and copper operations (sulfuric acid).

The final waste type, mine water, is water that infiltrates a mine and must be removed to facilitate mining. The quantity and quality of the mine water handled varies from mine to mine. Quantities may range from zero to thousands of liters per ton of ore mined. The number of mine water ponds at mine sites in the industry segments covered in this chapter is usually between one and six (PEDCo ENVIRONMENTAL, INC., 1984).

1.3 Waste Quantities from Mining and Milling

Table I.2-4 presents an estimate of the cumulative amount of tailings and mine waste generated by the U.S. mining and beneficiation of metallic ores, phosphate rock, and asbestos from 1910 through 1981. As shown, nearly 19 billion metric tons of waste have been generated by the mining and beneficiation of eight metals and two nonmetals. Copper, iron ore, and phosphate rock have produced over 85% of the total volume of waste.

Mining and beneficiating nonfuel ores and minerals in the U.S. generated approximately 2000 million metric tons of waste in 1980 (US EPA, 1985a). Also in other countries similar problems are known which are associated with the large quantities of these materials that are to be handled (SALOMONS and FÖRSTNER, 1988a, b). The waste handled by the U.S. mining industry declined to 1300 million metric tons for

Table I.2-4. Estimated Cumulative Mine Waste and Tailings 1910 through 1981 (millions of metric tons) in the United States (US EPA, 1985b)

Mining Industry Segment	Tailings	Mine Waste	Total Waste
Metals			
Copper	6900	17000	23900
Gold	350	400	750
Iron ore	3000	8500	11500
Lead	480	50	530
Molybdenum	500	370	870
Silver	50	30	80
Uranium	180	2000	2180
Zinc	730	70	800
Nonmetals			
Phosphate rock	2200	5500	7700
Asbestos	40	30	70
Total	14430	33950	48380

the industry as a whole in 1982 (MINERALS YEARBOOK, 1984). The industry segments covered in this chapter are responsible for more than 90% of this nonfuel mining waste. The copper mining segment alone generates approximately half of the waste produced by the metal mining segments.

The gold segment generated less than 2% of all leaching waste in 1980, but this increased to more than 5% in 1982, since the heap leaching method is of growing importance.

The wastes generated by the nonfuel mining industry in the U.S.A. are generally disposed of onsite, and thus the geographic distribution of active mining waste management sites corresponds closely to the distribution of mine sites. Accordingly, the principal mining states, i.e., Arizona (copper), Minnesota (iron ore), New Mexico and Wyoming (uranium), and Florida (phosphate rock), are the states that produce the majority of all mining waste, in the U.S.A. (US EPA, 1985b). Similar information is available from Canada, Latin American and European countries (including Eastern countries), Australia and developing countries (see also SALOMONS and FÖRSTNER, 1988a, b).

1.4 Management Practices for Mining Wastes

1.4.1 Overview of the Mining Waste Management Process

Mine waste, tailings, heap and dump leach waste, and mine water can be managed in a variety of ways. Fig. I.2-1 provides an overview of the mining waste management process. As shown in the figure, mine waste may be used on- or offsite, disposed of

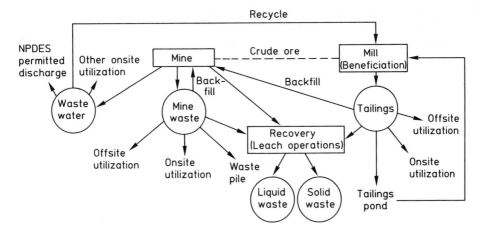

Fig. I.2-1. The mining waste management process.

in mine waste piles, or used in leach operations to recover additional valuable constituents from the ore. Similarly, tailings may be used on- or offsite, diposed of in tailings ponds, or used in leach operations to recover valuable constituents in the tailings that are still present after milling processes have been completed. Tailings also may contain residues of the reagents used in flotation processes. These reagents include forms of cyanide (used in the leaching of gold and silver and in the separation of sulfide minerals), sulfuric acid used and formed in copper dump leaching, and various organic and inorganic compounds used in copper, lead, and zinc flotation (GREBER et al., 1979).

Mine water may be discharged to surface streams (often after treatment) regulated outfalls, used as milling process makeup water (recycled), or used onsite for other purposes (e.g., dust control, drilling fluids, sluicing solids back to the mine as backfill, etc.).

The recovery of valuable constituents from mine water (e.g., ion exchange treatment for uranium), from mill process solids, or from extraction from dump leach liquors could possibly be considered to be waste treatment processes, in that such recovery extracts metals or constituents that would otherwise be potentially hazardous constituents of waste prior to disposal. However, the mining industry considers these processes to be extraction or beneficiation processes because they recover valuable products from materials that have metal concentrations below those in ore of a grade suitable or economical for milling and smelting.

Table I.2-5 presents the volumes and percentages of mine waste and tailings that are currently managed according to the various practices shown in Fig. I.2-1 and mentioned above. The table shows that more than half of all mine waste and tailings is disposed of in the U.S.A. in piles and ponds, respectively (US EPA, 1985b). Most onsite utilization of mine waste and tailings involves the dump leaching of copper mine waste and the use of sand tailings to build tailings impoundment dams in all industry segments.

Table I.2-5. Current Waste Management Practices in the U.S.A. (US EPA, 1984a)

Waste Type	Management Practice	Volume (in millions of metric tons per year)	Percent of Waste Generated
Mine waste	Pile	569	56
	Onsite utilization	313[a]	31
	Backfill	86	9
	Offsite utilization	43	4
	Total	1011	100
Tailings	Ponds	267	61
	Onsite utilization	141[b]	32
	Backfill	21	5
	Offsite utilization	8[c]	2
	Total	437	100

[a] Includes dump leach operations and starter dams for tailings impoundments
[b] Includes the sand fraction used in building tailings impoundment dams
[c] Includes 4 million metric tons of Tennessee zinc tailings sold as construction materials or soil supplements

1.4.2 Waste Management Practices

Waste management practices include process modifications for waste or potential hazard minimization, recovery operations, treatment prior to land disposal, onsite use of mine water, and offsite use of mine waste and mill tailings. Each of these practices is discussed below.

Process Modifications for Waste Minimization. Although there are no practical means of reducing the volume of solid waste produced by mining and beneficiation operations, some changes in beneficiation processes can lead to changes in the chemical composition of the tailings released into tailings impoundments. For example, pilot studies have been conducted in which less toxic sodium sulfide and bisulfide were substituted for cyanide compounds in the beneficiation of copper ores. Similarly, alkalinity in the beneficiation circuits can be maintained by lime instead of ammonia.

The thickened discharge method of tailings management involves partially dewatering the tailings slurry and discharging it from a single point. This results in a gently sloping, cone-shaped deposit. The water removed from the tailings can be treated and discharged or returned to the milling circuit. The dewatering costs associated with this method are offset by reduced earthwork costs. A disadvantage of the thickened discharge technique in some circumstances is that no water is stored with the tailings, which may mean that the dewatered slurry piles become sources of fugitive dust (VICK, 1981; GOODSON and ASSOCIATES, 1981).

Biological acid leaching, a new process under development in Canada, may be a feasible substitute for current dump leaching practices. Unlike dump leaching operations, the new process does not convert the sulfur in the ore to sulfuric acid; instead, it converts it to elemental sulfur which is both less hazardous to the environment and potentially salable. The process is still in the pilot development stage (CURTIN, 1983).

Recovery Operations. Leaching is a process used to recover metal values from low-grade ore or tailings, and is a common practice in some mining segments (i.e., copper, gold, silver, and uranium), including leaching of phosphate rock and uranium wastes (SEITER and HUNT, 1982).

In the copper, gold, and silver industries, technical efficiency and economic factors have made the recovery of mineral values by leaching processes economically feasible. Overburden, tailings, and other wastes will continue to be "remined" in the future, if extraction efficiencies continue to improve and if product prices exceed extraction costs.

Onsite Use of Mine Water. Water generated by mine dewatering may be used in the milling process as makeup water (treatment may or may not be required), or used onsite for dust control, sluicing solids to the mine as backfill or in cooling or drilling fluids. In some cases, all of the water required by the mill operation is obtained from mine drainage, which eliminates the need for wells and a mine water treatment system, or greatly reduces the volume of mine water discharged.

Waste utilization practices include agricultural lime replacement, road and building construction, and the production of bricks, ceramics, and wallboard. These methods are discussed below and summarized in Table I.2-6.

The most widespread use for these wastes is in the production of concrete and bituminous aggregates for road construction. Other applications in road construction include the use of these wastes in road bases, as embankments, and to make antiskid surfaces. Approximately 50% of the zinc tailings in Tennessee are sold for aggregate production.

The most important constraints on the use of mining wastes are imposed by energy, economic, and logistic considerations (SEITTER and HUNT, 1982). Mining wastes, therefore, are competitive only when they can be marketed or used in the geographical area close to the originating mine.

Uses of mining wastes do not and will not keep pace with the approximately 1 to 2 billion metric tons of these wastes that may be generated each year. However, research on the cost-effective utilization of mining wastes is justified, because any use that becomes widely practiced will help reduce the magnitude of the mining waste disposal problem.

1.4.3 Waste Siting and Disposal Methods

For technical reasons, most mining waste is finally disposed of on land.

Location and Siting. The topography, geography, and hydrogeology and, in some cases, meteorology, as well as population density of the geographical area in which

Table I.2-6. Uses of Mine Waste and Tailings (Based on SEITTER and HUNT, 1982)

Use	Asbestos	Copper	Gold and Silver	Iron Ore/ Taconite	Lead	Molybdenum	Phosphate	Uranium	Zinc
Material Use									
Soil supplement								1	
Wall board production	3								
Brick/block production	1	1	1	1					
Ceramic products							1		
Antiskid aggregate				3			1		
Embankments		3	3	3	3				
General aggregate			3	3	3				
Fill or pavement base		3	3	3	3				
Asphalt aggregate	2			3	3	3		1	3
Concrete aggregate			3	3	3		1		3
Development Stage									

1. Bench-scale research project
2. Full-scale demonstration project
3. Full-scale, sporadically practiced

a mine is located, affect the siting of the waste disposal area, the extent to which mitigative practices are required, and the types of mitigative systems that can be selected. The extent of the ore body, the quantity of waste to be generated, and the method of mining are also considered when siting a disposal area.

Waste Disposal Methods for Tailings. Waste disposal methods for tailings include tailings ponds, stope back-, below-grade disposal, and offshore disposal. As shown in Table I.2-5, more than half of the tailings are disposed of in tailings ponds.

(1) Tailings Impoundments. Tailings impoundments have been used at ore mills in the United States since the early 1900s. In recent years, they have become increasingly important and may account for as much as 20% of the construction cost of a mine/mill project (VICK, 1981). Some ore bodies may not be exploited, because suitable sites for tailings disposal are not available within a practical distance. Tailings impoundments retain water, making it available for recycling to the mill flotation circuits and other processes requiring water. They act as equalization basins, and protect the quality of surface waterways by preventing the release of suspended solids and dissolved chemicals.

(2) Stope Backfilling. This method, also referred to as sandfilling, involves converting a portion of the coarse fraction of tailings into a slurry and then injecting the slurry into the mined-out portions of stopes. Stope backfilling is currently practiced or is being considered as a method of disposing of such diverse materials as copper

tailings, spent shale from oil shale retorts, and tailings from Wyoming trona (sodium carbonate) mines (VICK, 1981).

(3) Below-grade Disposal. This method of tailings disposal consists of placing tailings in an excavated pit (in lieu of above-grade impoundments) so that at closure, the entire deposit is below the level of the original land surface. This method currently is unique to the uranium industry, which uses it to reduce the likelihood of erosion. This disposal method is costly unless mined-out pits can be used (US EPA, 1982a; VICK, 1981).

(4) Offshore Disposal. In the past, offshore disposal has been a euphenism for dumping tailings into a large lake or the ocean without regard for environmental consequences. Recently, more responsible proposals have shown that if the tailings are chemically innocuous, are sufficiently coarse to settle rapidly with a minimum amount of turbidity, and are piped to deep-water areas to avoid the most biologically productive nearshore zones, offshore disposal may have reasonably small environmental impacts in certain specific cases. Even so, offshore disposal is not a widely accepted alternative within regulatory agencies in the United States, and few mines have been located near the ocean in the past. Technical arguments notwithstanding, recent experience indicates that most developed countries will not approve offshore disposal of tailings (VICK, 1981).

A review including case histories of environmental problems with coastal and marine mines has been given by ELLIS (1988). In order to keep ports and channels accesible to shipping, sediments have to be removed regularly by dredging. For coastal areas 108 responses from 37 countries on a questionnaire of the International Association of Ports and Harbors (IAPH, 1981) indicate that in 1979 there were 580 million tons of dredging, about one-fourth of it is ocean-dumped and another two-thirds are deposited in wetlands and nearshore. In the river mouth to the North Sea, for example, approximately 20 million cubic meters have to be dredged from Rhine/Meuse (Rotterdam harbor). The total quantity of sediment which is dredged in the Netherlands, Belgium, and West Germany amounts to about twelve times the total suspended matter supply from the Rhine (VAN DRIEL et la., 1984). The sediments from Rotterdam or Hamburg harbor have a wide spectrum of composition depending on disposal options and local mixing (NIJSSEN, 1988; ROTTERDAM MUNICIPALITY, 1986; CHRISTIANSEN et al., 1982; TENT, 1982; SALOMONS and FÖRSTNER, 1984), and the highest rates of enrichment compared to natural background values are found for cadmium (up to 60-fold) and mercury (up to 30-fold); see Table I.2-7.

Waste Disposal Methods for Mine Waste. As shown in Table I.2-5, an estimated 56% of the mine waste removed to gain access to an ore body is disposed of in mine waste piles near or adjacent to the mine. The overburden from open pit mines is usually discarded on the outside slopes of the pit; the trend in the mining industry is, however, toward immediate backfilling. Some underground mining methods include cut-and-fill stoping and square-set stoping and provide structural stability to the mined areas, in addition to serving as a means of waste disposal (PEDCO ENVIRONMENTAL, INC., 1984).

Table I.2-7. Composition of Dredged Sludge from Rotterdam and Hamburg Harbors (ROTTERDAM MUNICIPALITY, 1986; CHRISTIANSEN et al., 1982; TENT, 1982)

	Rotterdam[a]	Hamburg[b]	(Natural Background)[c]
Arsenic	ca. 40	28 – 95	(13)
Lead	80 – 230	112 – 438	(20)
Cadmium	3 – 16	6 – 20	(0.3)
Chromium	97 – 187	94 – 244	(60)
Copper	39 – 142	170 – 897	(50)
Nickel	24 – 44	50 – 100	(35)
Mercury	0.8 – 7	3 – 12	(0.4)
Zinc	256 – 1016	1020 – 2450	(100)

All data in µg/g dry matter
[a] Europoort (smaller values)/Waalhaven (higher values) for 1984
[b] 80% range of samples with more than 10% loss of ignition
[c] fine-grained shallow water sediment (SALOMONS and FÖRSTNER, 1984)
This table has been made available by U. FÖRSTNER

Waste Disposal Methods for Dump Leach/Heap Leach Material. Whether or not active dump leach and heap leach operations are considered to be process operations rather than solid waste disposal practices, solid waste material remains after the completion of these operations. The current practice is to transport overburden and low-grade copper ore for dump leach processes (or waste and low-grade precious metal ore for heap leach operations) to leaching beds, where the dumped material is spread by bulldozers. Equipment travel on the leach dump compacts the top layer of the material; this layer is then scarified to facilitate infiltration of the leach solution. This process of layering and subsequent scarifying of the leach dump may continue for 50 years or more (GOODSON and ASSOCIATES, 1982).

Waste Disposal Methods for Mine Water. Water produced from mine dewatering may be discharged directly or indirectly (after treatment such as settling) to a surface stream, used in the milling process as makeup water (treatment may or may not be required), pumped to a tailings pond, and/or used onsite for dust control, cooling, or as drilling fluids, etc.

Treatment of mine water in onsite impoundments is the management practice used when discharge or total recycling are not possible. Such treatments include simple settling, precipitation, the addition of coagulants and flocculants, or the removal of certain species (e.g., radium-226 removal by coprecipitation with barium chloride in mine water ponds in the uranium industry).

1.5 Mitigative Measures for Land Disposal Sites

Even if greater use is made of waste utilization and alternative waste disposal methods, the greatest portion of mining wastes will still be disposed of in land disposal facilities such as waste piles, tailings ponds, and settlings impoundments.

However, various measures are available to detect or mitigate the problems associated with the land disposal of mining wastes. These measures may be classified into four general types:

1. Detection and inspection measures determine whether problems are developing. These activities include ground-water monitoring and visual inspection of other systems, erosion control, dam stability, and runoff control.
2. Liquid control measures control the potential for liquid to come into contact with mining waste, and thus minimize surface water pollution and the amount of liquid available for leachate formation.
3. Containment systems prevent leachate from entering the ground water and posing a threat to human health and the environment. Two types of containment systems are considered here: containment systems designed to prevent leachate from entering the ground water (such as liners and systems designed to control plumes of contaminated ground water) and corrective action measures.
4. Security systems prevent entry to the waste management area by animals or by unauthorized persons. These systems protect the general public and prevent activities that might damage onsite control systems.

The waste management measures that are most relevant to individual waste management sites depend, in part, on the operational phase of the waste management site. Three operational phases are distinguished here:

1. Active site life is the period during which waste is being added to the disposal site. A disposal site may be closed even though the mine itself remains active.
2. Closure is the period immediately following active site life, in which various activities are undertaken to ensure adequate protection of human health and the environment during the post-closure phase, and to minimize maintenance activities in the post-closure phase.
3. Post-closure is the period following closure during which there are no further additions of waste to the site. The main post-closure activities are the monitoring of the site for leaks and the maintenance of liquid control, containment, and security systems established during site life or at the time of closure.

Corrective action occurs after a plume of contaminated ground water or another environmental hazard is discovered. This may occur during active site life, at the time of closure, or during the post-closure phase.

Various mitigative measures are appropriate to the management of mining waste (Table I.2-8).

2 Processing of Metals (Smelting and Refining)

In general, the economically important mineral ores occur as metallic oxides or sulfides. Practically all of the metals found in nature are utilized to some extent in commerce. Those produced in major quantities, and consequently of major econom-

Table I.2-8. Mitigative Measures by Stage of Site Life (MERRIDAN RESEARCH, INC., 1985)

Stage of Site Life	Mitigative Measure	Purpose
Active site life	Hydrogeologic evaluation and ground-water monitoring	Detection of contaminants
	Runon/runoff control	Liquid control
	Liners	Containment
	Cutoff walls	Containment
	Leachate collection, removal, and treatment systems	Liquid control
	Security systems	Security of control systems and protection of public health
Closure	Continue measures initiated during active site life	All purposes mentioned above
	Wastewater treatment	Liquid control
	Pond sediment removal	Waste removal
	Dike stabilization	Liquid control
	Waste stabilization	Liquid control
	Installation of leachate collection, removal and treatment systems at surface impoundments	Liquid control
	Final cover	Liquid control
Post-closure	Ground-water monitoring	Detection of contaminants
	Inspect/maintain all existing systems	All purposes mentioned above
Corrective action	Interceptor wells	Containment
	Hydraulic barriers	Containment
	Grouting	Containment
	Cutoff walls	Containment
	Collection	Treatment

ic significance, are: iron, aluminum, copper, lead, manganese, and zinc (see also Chapter I.1 and MICHAELIS, 1984). These are referred to here as "the major metals", with the remaining loosely termed "the minor metals". This discussion will treat the minor metals only briefly because of the diversity and volume of information needed for the detailed description of the processes used for the winning of the minor metals from their ores (see also Part II).

The principal ores of iron and aluminum are oxidic although some iron is produced from important sulfide deposits. The principal ores of copper, lead, and zinc are sulfidic; some 10 to 15% of copper produced is won from oxidic ores via a heap leaching process conducted in conjunction with the ore mining operation. The sulfidic ores of the major metals may occur in combination with each other and generally are accompanied by one or more of the minor metals in sufficient quantity to warrant by-product recovery. Over two-fifths of the minor metals are produced as by-products (see also MICHAELIS, 1984). Unless the ore grade is of a relatively high content, it is necessary to beneficiate the ore as mentioned above to produce a concentrate suitable for the metal winning processes. A large variety of processes

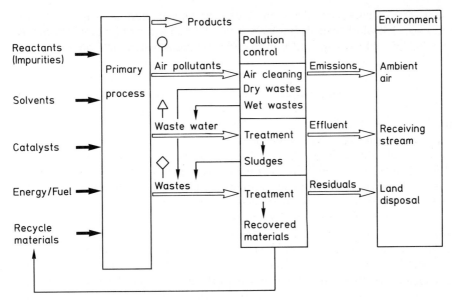

Fig. I.2-2. Generic process flow diagram.

have been developed to produce metals and metallic compounds from their natural ores (primary metals) and from recycled metallic wastes (secondary metals) in recycling and recovery operations. The processes are based primarily upon pyrometallurgical (including smelting) and hydrometallurgical technologies augmented by various chemical and electrochemical operations for purification and refining. A generic process flow diagram is given in Fig. I.2-2.

In the smelting process, the rock and gangue materials are melted, a silicaceous slag is formed (with addition of silica and limestone), and the impurities are removed by volatilization and slagging. The blast furnace is the process of choice for smelting iron and lead, and had been used for copper until the widespread use of ore concentrate feedstocks necessitated a change to a hearth-type smelting furnace to prevent excessive losses of the finely ground concentrate. There are a few still in use for processing higher-grade lump copper sulfide ores. In this process the metal sulfides, except for copper, are converted to oxides and reduced by carbon monoxide. In copper smelting, the oxidation is controlled to produce a copper matte, a mixture of copper and iron sulfides.

2.1 Iron and Steel

(see also Chapter II.14)

The iron and steel industry is composed of a variety of operations and processes for the conversion of iron ore into iron and steel (KATARI and GERSTLE, 1977). The major operations are (1) ore preparation, (2) coke production, (3) coke by-product recovery, (4) pig iron production, (5) ferro-alloy production, and (6) steel production.

The process of choice for the winning of metallic iron from its oxide ores is the blast furnace. The blast furnace is a large vertical, or shaft, furnace in which the oxide ores are reduced to elemental iron by reducing gases generated from the partial combustion of coke with oxygen enriched air. The coke, fluxes, iron ore, and recycle materials are charged from the top of the furnace in specific portions of coke/flux/iron. The furnace charge moves downward countercurrently to the air, is smelted, and the molten iron accumulates under the molten slag on the furnace hearth. The slag and metal are periodically removed from the furnace. The slag is discarded and the metal is cast into "pigs".

The air is introduced along the sides of the furnace shaft above the hearth area at sufficient velocity to achieve the temperatures required for smelting the charge. Because the velocity of this is great enough to entrain fine materials, the iron ore must be agglomerated to prevent it from being blown from the top of the furnace. The agglomeration is achieved by sintering, pelleting, nodulizing, or briquetting. All of these processes involve the use of heat (see also MICHAELIS, 1984) and generate particulate, sulfur dioxide, and, in some cases, hydrocarbon emissions.

The steel industry has been a leader in the employment of oxygen in smelting. The open hearth furnace, once the mainstay of the industry, has largely been replaced by the basic oxygen furnace which provides more than a five-fold increase in production rate. The electric furnace, which may be either of the arc or induction design, is used to produce high alloy and specialty steels. Steel making involves the combination of molten iron, ferrous scrap, alloying material, and fluxes in a furnace to achieve the removal of impurities through oxidation, volatilization, and slagging. Oxygen is used to increase the reaction rates and thermal intensity. This results in higher production rates than otherwise possible. The molten steel is then cast into ingots or continuously cast, and cast material then processed in the rolling mill to produce the desired end product (sheets, coils, etc.). The hot rolling of steel causes surface oxidation which must be removed before the product is shipped. The cleaning is achieved by treating in a "pickling bath" containing strong concentrations of sulfuric or hydrochloric acid. This operation generates large quantities of waste pickle liquor and acidic rinse water. Both must be treated before discharge as a wastewater or disposal to the land as a waste.

2.2 Lead
(see also Chapter II.16)

Lead concentrate is smelted in a blast furnace (US EPA, 1980a) in a manner very similar to iron. The ore concentrate is sintered to produce a feed material suitable for the blast furnace. The concentrate is blended with sinter recycle, flue dusts, and fluxes and then pelletized. The pellets are then spread over a horizontal metal belt and moved through the sintering machine where a hot gas passes over the pellets. The sulfur in the concentrate is oxidized, releasing heat which converts the sulfides to oxides and sulfates. Low melting silicates form and bind the pellets together to form a sinter. The sinter is then broken and screened to remove fines which are recycled to the blending step.

The sinter is fed into the top of the blast furnace with coke, limestone, and other fluxing materials. The carbon monoxide formed by the partial oxidation of the coke reduces the metallic oxides to metal. A slag composed mainly of iron and calcium silicates forms a layer above the metals on the furnace hearth. If copper is present, it forms a matte layer. A spiess (arsenides and antimonides of iron and other metals) layer will form if the charge is high in arsenic or antimony. The crude lead bullion is refined to remove and recover other metals such as copper, gold, silver, antimony, arsenic, tin, etc., using a variety of pyrometallurgical operations.

The matte and spiess are sent to other smelting operations for metal recovery. The slag is treated in a slag fuming furnace to recover zinc and lead. Germanium, cadmium, chlorine, and fluorine may also be generated in the slag fuming operation. The presence of chlorine or fluorine requires a leaching step to avoid accumulation of these impurities in recycling operations.

2.3 Zinc

(see also Chapter II.36)

Zinc is produced from sulfide ores by a pyrometallurgical process or by a roast, leach, electrowin process (US EPA, 1980b). The pyrometallurgical route to zinc involves three steps: roasting to remove practically all the sulfur and to convert the zinc sulfide to zinc oxide; sintering the calcine with recycled sinter, coke, and silica; and reduction of metallic zinc in a retort furnace. The zinc oxide in the sinter is reduced to form zinc metal vapor which is condensed and collected. If zinc oxide is the desired product, an oxidizing furnace is used in which the zinc vapor is oxidized back to zinc oxide. All of the mercury, most of the sulfur, and part of the cadmium present in the feed will be eliminated by roasting. The lead and remaining cadmium impurities are volatilized from the sintering machine.

Today, the preferred routes to zinc involve both pyro- and hydrometallurgical processing steps. The zinc concentrate is roasted to form a calcine containing the oxide and some sulfate. The calcine is then leached in recycled electrolyte (sulfuric acid solution) from the electrowinning process. The leach liquor is purified to remove impurities such as iron, arsenic, copper, cadmium, antimony, cobalt, germanium, nickel, and thallium, which interfere with electrowinning. Finally the zinc is won from the liquor by electrolysis. The zinc sheets are stripped from the cathodes, melted, and cast into ingots or slabs.

Cadmium is a principal by-product of the zinc smelter. The fumes and dusts from zinc roasting and sintering operations, dusts from the smelting of copper-lead-zinc ores and lead-zinc ores, and the purfication sludge from the electrolytic zinc plant, are leached to extract soluble cadmium. The cadmium is precipitated by the addition of zinc powder to form a sponge. In the pyrometallurgical process, the sponge is mixed with coke and lime and processed in a conventional horizontal retort furnace. In the electrolytic process, the sponge is redissolved in sulfuric acid and electrowon on cadmium cathodes. Recovery of cadmium is incomplete and is limited to about 96% of the sponge cadmium. The recovery is limited to prevent the deposition of the thallium-cadmium alloy. Thallium can be precipitated from the spent electrolysis solution with hydrogen sulfide and the precipitates redissolved and electrowon.

2.4 Copper

(see also Section 2.8, Chapter II.9, MICHAELIS, 1984, and Table I.2-10)

There are a relatively large number of process routes to copper (US EPA, 1980c). The reverberatory furnace succeeded the use of the blast furnace in the late nineteenth century. However, the reverberatory furnace, while offering significant advantages of operational flexibility and capability to process concentrates having high impurity levels, is energy-intensive and has a significantly lower specific smelting capacity than the more modern oxygen-based smelting processes introduced in the mid-twentieth century. These new smelting furnaces, employing oxygen and oxygen-enriched air, are capable of two to four times the copper production capacity per unit of hearth area. These flash smelters and the Noranda and Mitsubishi furnace designs also produce an off-gas of sufficient sulfur dioxide concentration to permit affordable control by conventional acid plant designs. These factors, taken together, provide a strong economic incentive for the use of oxygen smelting furnaces for new smelters and also for furnace conversion at existing smelters where higher capital costs are more attractive than the accumulation of higher operating costs over the remaining life of the smelter.

2.5 Aluminum

(see also Section 2.9 and Chapter II.1)

Aluminum is the most abundant metallic element in the earth's crust, occurring principally in the oxide form. The principal ore is bauxite, a mixture of hydrated aluminum oxides. The Bayer process is used to refine bauxite into alumina suitable for electrolytic reduction to aluminum metal (PARSONS, 1977). The major waste generated by bauxite refining is the "red mud", consisting of the solid residues remaining after the hydroxide leaching of the aluminum oxides. This water is best managed by disposal into closed ponds, which permits recovery of water and hydroxide content for reuse in the refining process. The calcining furnaces also emit dust, which can be effectively controlled by electrostatic precipitators or fabric filters.

The Hall-Heroult process (PARSONS, 1977) is the process used to electrolytically reduce bauxite to aluminum metal. In this process, the aluminum is dissolved in molten cryolite, a double fluoride salt of sodium and aluminum (Na_3AlF_6) in an electrolytic cell. The cell consists of a steel "pot" lined with refractory brick and an inner liner of carbon bricks which acts as the cathode. The anode is an arrangement of consumable carbon electrodes formed from petroleum coke and pitch. The molten aluminum is deposited at the cathode and oxygen burns the carbon anode, releasing carbon dioxide and small quantities of carbon monoxide. The molten bath is covered by a blanket of aluminum, which dissolves in the cryolite as the electrolysis proceeds.

Aluminum production is thus very energy-intensive (about 280 GJ/t are needed; MICHAELIS, 1984).

2.6 Recycling of Metals

The recovery and reprocessing of metals from waste materials is an integral part of the economies of most industrial nations. The metals are derived from a variety of sources including discarded wastes (old scrap), production scrap (new scrap), and products which are new, but unused and obsolete (obsolete scrap). Drosses, skimmings, and other metallurgical wastes also are recycled (see Sects. 2.9 and 3.2 and Table I.2-14). In many countries, the recycling constitutes a substantial, if not a majority, (in the cases of aluminum and copper about one third; in the cases of lead and iron more) contribution to the annual production of many metals, particularly the major segments of the market (see also MICHAELIS, 1984).

Metals may be recycled into either primary or secondary production processes. As mentioned earlier, the term "primary" metal is used for a metal derived principally from a mined ore. The term "secondary" metal connotes a metal derived from metal scrap wastes. In practice many of the primary producers use large quantities of recycled scrap, including that derived from discarded wastes. Secondary producers utilize much the same wastes as those recycled to primary producers.

Scrap is a commodity which is bought and sold, usually on an open market basis. The sale may be made on a spot price basis or involve a longer term contract. The price is influenced by competition between the primary and secondary producers. Smaller operations tend to be economically marginal and go in and out of production based upon the local market condition.

The secondary industry can be viewed as composed of six major segments (see also Part II):

1. Ferrous
2. Aluminum
3. Copper
4. Lead
5. Zinc
6. Minor metals (all others)

They recover metal from wastes (e.g., PATTERSON, 1987) and convert scrap metal to either metal products that are marketed for use or bulk metal forms sold to other metal fabricators. Processes have much in common with those used for winning metals from ores and may be classified into three basic categories of operation: (1) preparation of scrap for smelting or refining, (2) smelting/refining, and (3) casting into product form. Each segment of the secondary industry has a degree of commonality in the raw materials consumed, the process operations employed, the products produced, the pollutants (wastes) generated, the pollution control equipment used, and environmental control problems remaining unsolved.

A look at the data listed in Table I.2-14 suggests that recovery of metals such as copper and nickel from electroplating residues waste could well compete with natural resources of these elements. The same is valid for lead, zinc, and silver from certain fractions of waste combustion products. For instance, BALL et al. (1987) have evaluated the economic feasibility of a state-wide hydrometallurgical recovery facility for metal bearing wastes in Missouri (see also Chapter I.7b). With an estimated charge of approximately 200 US$/ton for the studied sludges from metal finishing industries, such a facility would compete with 100–150 US$/ton of disposal costs. However, since the facility would have substantial environmental merits, it was recommended that the State of Missouri should take initiatives to encourage the development in three directions: (1) to determine the treatability of wastes and "delisting" potential of the residuals generated by the process, (2) to monitor the regulatory environment, especially concerning the possible restrictions on land disposal of metal-bearing wastes, and (3) to review possible subsidies or incentive scenarios which would make the facility cost-competitive with landfills.

2.7 Metal Finishing

Various types of finishes are applied to metal products to improve market acceptibility. The two principal reasons for applying finishes to products are to either provide protection from the environment in which the product is intended for use or to enhance the appearance or performance of the product. There are a variety of metal finishing operations employed including electroplating, anodizing, painting, and galvanizing (US EPA, 1983b). Of these, electroplating is of the greatest importance with regard to metal consumption. Briefly, electroplating involves the deposition of a thin coating of one metal upon the surface of another metal from solution by electrodeposition. The process involves the immersion of the part to be plated into a tank containing the plating solution. The plating solution is an aqueous solution, which may be acidic, neutral, or alkaline, containing additives and a high ionic concentration of the metal to be plated. An electric circuit is arranged between an inert or sacrifical anode and the part to be plated, which is the cathode. When the current is impressed, the metal ions deposit upon the cathode surfaces from the plating solution.

2.8 Environmental Management for Processing and Manufacturing

The most noticeable form of pollution from the metals production industry is the discharge of emissions to the atmosphere. A tall stack to disperse kilning gas streams, production curtailment, and removal of pollutants before discharge by means of a gas cleaning technology are the methods used for reducing the ambient pollutant concentrations in the vicinity of a smelter.

The tall stack discharges pollutants at such heights that the gaseous emissions are sufficiently diluted when dispersed into the lower atmosphere to meet air quality requirements. It is possible to add preheated air into the stack to achieve additional dispersion and dilution.

The second method for controlling concentrations at ground level when adverse weather conditions prevail is production curtailment. This method has been referred to as "closed loop control" or as Supplementary Control System (SCS). It is based on the monitoring or pollutant concentrations at ground level at various sites in the areas surrounding the smelter and using this information to control the smelter operating rate.

The third method employs an appropriate gas cleaning technology for removing the pollutants of concern before the exhaust gas is discharged into the atmosphere. This is the method preferred as it prevents the discharge of pollutants into the environment and is effective under all meteorological conditions. In the smelting of metals, fabric filters, electrostatic precipitators, and high-energy wet scrubbers are used for control of particulate pollutants while acid plants, sulfur dioxide liquefaction, and various scrubbing processes are used for control of sulfur dioxide (US EPA, 1984b).

Control of ambient pollution concentrations by dispersion from tall stacks and production curtailment are only applicable to gaseous discharges. Particulate matter must be effectively removed prior to discharge. If this is not done, particulate pollutants will settle out of the air onto the surrounding land and surface waters resulting in heavy metal pollution. Many of the metals contained in the ore concentrates form metal compounds, particularly oxides, that are volatilized at smelting temperatures. Volatilization is one of the principal mechanisms of removing impurities from the smelting charge. The final particulate control device installed before discharge must be designed and operated at a temperature which is sufficiently below the condensation temperature of the most volatile metal compound present to provide effective control. Alternatively some volatile metallic species are controlled by chemical removal systems.

Control of air pollutant emissions from production of non-ferrous metals from sulfidic ores is focused on SO_2 control (US EPA, 1984b). Because processes for control of SO_2 require the removal of particulate contaminants, effective control of total particulate and trace element emissions is accomplished as a "by-product" of SO_2 control. The most economical approach to SO_2 control is the application of metallurgical acid plants (US EPA, 1984b; WIESENBERG et al., 1980, 1981).

The relative increase in sulfur containment (i.e., percentage of sulfur in the concentrate which is not emitted to the atmosphere) is given in Table I.2-9 for a variety of copper smelting process alternatives which represent the improvements realized in moving from reverberatory smelting to oxygen-based smelting to hydrometallurgical processes (BURCKLE and WORRELL, 1981). Smelter improvements have been achieved for zinc and are in the works for lead in the form of the QSL process.

The sulfur by-product market is extremely important to the selection of an economic route to SO_2 control.

There are a number of techniques which are technically feasible for controlling the sources of weak sulfur dioxide emissions (WIESENBERG et al., 1980, 1981). These techniques, based upon process modifications and add-on control technologies are:
1. Gas blending where off-gas strength is increased by actions resulting in reduced furnace air consumption and infiltration combined with oxygen enrichment.
2. Removal and SO_2 through an absorption process producing a "throw-away" residual.

Table I.2-9. Sulfur Containment Scenarios in Various Copper Smelting Processes

Scenario	Estimated Sulfur Containment (%)
• Conventional reverb system — green charge or calcine charge using multihearth roaster producing weak gas	50 – 55
• Conventional reverb system (as above) with flue gas scrubbing to control the weak streams	90 – 92
• Upgrading conventional reverb system using roaster to eliminate sulfur and produce strong stream for acid plant	86 – 93
• Upgrade conventional reverb to produce strong stream for acid plant through oxygen enrichment	90 – 94
• Replace reverb system with new matte smelting process technology producing only strong streams controlled by acid plants — Flash furnace — Noranda reactor	94 – 95
• Replace reverb system with new continuous system process technology producing only strong streams controlled by acid plants — Mitsubishi — QS — WORCRA	98 – 99+
• Replace reverb system with hydrometallurgical system employing roasting to produce strong stream controlled by acid plant	99+

3. Concentration of the SO_2 and control in an acid plant through a regenerable FGD process.

A review of a number of alternative processes for primary copper smelting reveals newer smelting technologies, particularly those based upon oxygen smelting, offer attractive economic benefits, particularly where sufficient capital expenditures are involved. These systems offer not only the prospect of high levels of sulfur containment but significant improvements in energy consumption and productivity. Upgrading of existing furnace operations to strengthen sulfur dioxide content can be an effective approach to SO_2 control when coupled with FGD systems (US EPA, 1985c). Estimates for representative processes of the attributes of sulfur containment, energy consumption, and capital costs for copper are compared in Table I.2-10 (BURCKLE and WORRELL, 1981).

Hydrometallurgical systems appear to offer the highest sulfur containment. However, these processes are the least developed and the quality of information in this area is more uncertain. Because of heavy reliance on electricity, the generation of air pollutants may simply be moved to the source of electrical generation.

The principal environmental problem at aluminum smelters has traditionally been the air emissions from the reduction cells. The pollutants of major concern are calcium and aluminum compounds containing fluorides and hydrogen fluoride. The fluorides can cause vegetative and material damage, and the particulate form is especially a danger to foraging animals and livestock. Effective control of such emis-

Table I.2-10. Energy, Capital Cost, and Sulfur Containment Data for Alternate Technologies of Copper Smelting

Technology	Estimated Maximum Sulfur Containment	Capital Cost Percent of Base Case	Energy Requirement Percent of Base Case (quantity MM Btu/ton of anode)		
			PITT and WADSWORTH	KELLOG and HENDERSON	US EPA (1983c)
Green feed reverb	52	90[a]	116 (30.5)	118 (18.5)	106 (26.5)
Calcine feed reverb (base case)	86–93	100	100 (26.2)	100 (15.6)	100 (25.1)
Electric furnace	94–95	100[a]	106 (27.8)	156 (24.3)	112 (28.2)
OUTOKUMPU flash furnace (O_2)	94–95	70–80[a]	62 (16.3)	79 (12.3)	56 (14.0)
INCO flash furnace	94–95	70–80[a]	67 (17.7)	64 (9.9)	
Noranda	94–95	80[a]	74 (19.3)	79 (12.3)	53 (13.3)
KIVCET	94–95	70–80[19]			78 (19.7)[a]
Mitsubishi	98–99+	70–80[18]	78 (20.4)	90 (14.0)	
QS	98–99+	70–80[a]	68 (17.9)		
WORCRA	98–99+	70–80[19]			78 (19.7)[a]

[a] Estimated

Sulfur containment range estimates based upon: 5 to 6% total sulfur entering smelter lost as fugitives during slag and metal tapping and converter operation; single absorption acid plant efficiency of 98% and double absorption acid plant efficiency at 99.7%; calcine feed reverb range is affected by assumption of amount of sulfur expelled in the furnace, i.e., from 0 to 7%

sions has evolved through improved cell designs and more effective operation of cell hoods which capture emissions, permitting their effective removal by wet scrubbers or dry scrubbing (actually filtration) devices (PARSONS, 1977). The latter are preferred where impurity build-up can be controlled to acceptable levels, as these systems capture and recycle the fluorides, including the hydrogen fluoride, and aluminum values to the cells. This greatly reduces the costs for disposal of wet scrubber waste sludges. Also of concern are the air emissions from anode baking. These emissions include sulfur dioxide, tars and oils from the pyrolysis of the anode carbon, and fluorides from recycled anode materials. Electrostatic precipitators are not effective for fluoride control and may cause burning of the oils and tars. Fabric filters are blinded by the tars. A few smelters have tried dry scrubbing systems using aluminum, and at least one smelter has used calcined fluid petroleum coke (PARSONS, 1977). These systems would have the advantage of returning the aluminum and fluoride values to the cells where organics would be burned-off. Wet electrostatic precipitators and wet scrubbers are effective, but produce a waste water containing oils and fluorides requiring control and result in production of a waste which is probably hazardous.

A review of the background documents (US EPA, 1983a) supporting the development of regulations for metallurgical waste waters in the U.S. suggests that the waste waters can be classified into seven groups, each group relating to a specific type of generic pollutant removal problem. These are treatment of acidic metal-bearing

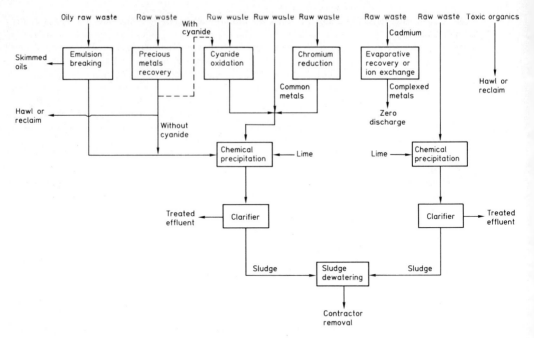

Fig. I.2-3. Treatment scheme for acidic metal-bearing waste waters.

waste waters, recovery of precious metals, destruction of cyanide, control of hexavalent chromium, treatment of complexed metals, removal of oil and grease, and control of toxic organics (Fig. I.2-3). While a broad range of alternative technologies exists for treating the waste waters from these processes, treatments most widely practiced today are based on the following technologies.

Common acidic, metal-bearing waste waters are usually chemically treated, usually with lime or alkali-earth hydroxide to achieve adjustment of pH and precipitation of metals. The precipitates are separated from the liquid phase by flocculation and clarification, eventually the effectiveness is enhanced by coagulants or flocculants. Further separation may be achieved by filtration, where required. A large range of alternative treatment technologies, which provide wastes which are more environmentally stable than hydroxide sludges or which provide for removal and capability for recycling, are also available (Table I.2-11; COLEMAN et al., 1980). A synopsis of performance and estimated cost is presented in Table I.2-12 (COLEMAN et al., 1980).

The recovery of precious metals, the reduction of hexavalent chromium, the removal of oily wastes, and the destruction of cyanide must be accomplished prior to metal removal. Oils and greases are removed by gravity separation and skimming of free oils followed by chemical and emulsion breaking and subsequent skimming for the removal of emulsified oils.

Cyanide bearing wastes are treated with an oxidizing agent (ozone or chlorine) to destroy the cyanide in the waste water. Cyanide, as well as being a highly toxic

Table I.2-11. Treatment Methods of Metal Containing Waste Waters

Treatment Method	Status of Development	Comment on Application/Applicability
1. Hydroxide precipitation	Commerically available	Applied as either a batch or continuous process for waste water treatment. Widely applicable. Limitation posed by metal complexing, does not remove alkaline and alkaline-earth metals and certain anions. Hexavalent chromium must be reduced before treatment
2. Sulfide precipitation	Installed at Boliden's copper smelter in Sweden to control smelter effluents. Also in operation at Furukawa's Ashio smelter in Japan	Process has the advantage of controlling heavy metals in acidic solutions. The sulfide sludge can be recycled to smelting furnaces
3. Ion exchange	Commercially available	Process can be used to control cations and anions. Process is not suitable as an end-of-pipe treatment. Additional treatment of secondary waste stream is required. It is suitable for concentrating wastes of high volume and low pollutant concentration and as a polishing process
4. Evaporation ponds	Commerically available	Application is limited to arid regions. End products are easily resolubilized. Large land requirements are necessary
5. Reverse osmosis	Commerically available	Pretreatment of waste water is necessary to avoid membrane damage. Membrane scaling is a problem if insoluble compounds are present. Suitable for concentrating dilute waste streams. A concentrated reject stream is produced and requires further treatment. Process is not suitable as an end-of-pipe treatment
6. Activated carbon	Commercially available	Primarily suitable for removal of organics. Secondary removal of inorganics has been observed. It is unlikely that this process would be applied to treatment of inorganics alone. Disposal of exhausted carbon can result in the gradual release of the absorbed compounds
7. Ferrite coprecipitation	Limited commerical use in Japan treating scrubber blowdown from a municipal incinerator. Also used by Nippon Electric Co.	Very good removal efficiency is achieved. This process appears to be an applicable treatment technology. However, it has only been applied on a commercial scale in Japan. The solid waste generated may be either recycled to smelting furnaces or landfilled

Table 1.2-11 (continued)

Treatment Method	Status of Development	Comment on Application/Applicability
8. Starch complexing	Applied in pilot studies as polishing process	Good removal efficiency is obtained. Metal compounds are released when starch decomposes. Starch manufacture requires carbon disulfide which is dangerous. Process is still in developmental stage but warrants further study
9. Sodium borohydride reduction	Limited commercial use	Process application is limited to reducing semi-noble and noble metals to the elemental state. Instability of sodium borohydrate in acidic solutions necessitates operation at pH 8.5 to 9.0. Process may be useful for small waste streams containing either valuable or difficult to remove metals, e.g., mercury. Not likely to be widely used because of high cost relative to hydroxide precipitation

pollutant, will complex metals such as copper, cadmium, and zinc and prevent efficient removal of these metals. Waste waters containing hexavalent chromium are treated with a reducing agent to reduce the chromium to the trivalent form which can then be precipitated from solution by hydroxide.

Chelating agents are often used in electroplating operations. When metal chelating agents are present, the waste water streams containing complexed metals must be segregated and separately treated. Chelating agents react with dissolved metal ions to form "chelate complexes", which are usually quite soluble in neutral or slightly alkaline solutions. A waste stream containing metal complexes may then be separately treated by adjusting the pH to 11.6 to 12.5 with lime to break the complex. Other alternatives may be used such as sulfide precipitation, ion exchange, or starch xanthate precipitation.

Rinsing operations are the major source of water pollution in the conventional electroplating process. Reduction of waste water flow with recovery and recycle of plating bath chemicals is the technique preferred, not only for enhanced control of environmental pollution, but also for the significant economic advantages offered (US EPA, 1982b; BURCKLE, 1987). Application of controls at the electroplating bath to concentrate plating chemicals from rinse waters for recycle to the bath such as evaporation, ion exchange, or electrolyte recovery is a preferred approach. Alternatively techniques to reduce drag-out, minimize rinse water usage, and recycle rinse water directly to the plating bath can be recommended, also because of significant cost savings in bath chemicals, process water, waste water treatment, sewage, and waste disposal.

Where chemical processes are used for pollutant reduction, greater quantities of waste water treatment chemicals are required per unit of pollutant removed as the waste water concentration decreases. This results in proportionally greater sludge

Table I.2-12. Performance of Treatment Technologies

Treatment Method	Cost	Control Effectiveness			Energy Requirement	Residues Produced
		Element	Inlet (mg/L)	Outlet (mg/L)		
1. Hydroxide precipitation	**Capital cost:**[a] $0.343/liter/day ($1.3/gallon/day) **Operating and maintenance cost:** $700/day for 3785×10^6 liters/day (1 million gallons/day) capacity and 1 gram per liter lime dosage rate	Arsenic	0.2–0.5	0.83	Energy users are: lime feeders, wastewater pumps, clarifier agitator drive, sludge rake drive, and vacuum filter. Energy consumption is 8 kWh per 3785 liters (1000 gallons) treated and 1 mg/L lime dosage. 14 kWh per 3785 liters (1000 gallons) treated and 1500 mg/L lime dosage	Solid metal hydroxides Compounds will redissolve in acidic environment
		Cadmium	1300	0.0007		
		Chromium(III)	1300	0.06		
		Copper	204–385	0.2–2.3		
		Lead	0.5–25	0.01–0.03		
		Manganese		0.5		
		Nickel	5	0.15		
		Zinc	16.1	0.02–0.23		
2. Sulfide precipitation	Not given in literature Should be similar to hydroxide precipitation	Arsenic	0.8–132	0.05–24.4	Unkown Energy requirement should be similar to those given for hydroxide precipitation	Metal sulfides are produced as end products Most sulfides are insoluble in acidic solutions They can be recycled to sulfide processing smelters. Solubilization is possible upon oxidation to sulfates
		Cadmium	0.4–1.0	0.008		
		Copper	50–115	0.5		
		Mercury	0.3–50	0.01–0.12		
		Nickel	0.13–0.29	0.09–0.51		
		Zinc	42	1.16		

Table I.2-12 (continued)

Treatment Method	Cost	Control Effectiveness			Energy Requirement	Residues Produced
		Element	Inlet (mg/L)	Outlet (mg/L)		
3. Ion exchange	Depends on capacity and solids concentration **Installed capital costs:** $0.063 to $0.719/liter/day ($0.24 to $2.72/gallon/day). **Operating costs:** $0.00018 to $0.00067/liter ($0.00067 to $0.00253/gal)	Cadmium Chromium(VI) Copper Lead Manganese Mercury Nickel Zinc	0.1 8–10.7 1.02 0.055–144.8 50 1.4–2.8 150–900 6	0.001 0–0.9 0.03 0.0015–0.53 0.046 0.005 0.01 0.4	Pumps are only energy consumers 1.1 kWh for 3785 liters (1000 gallons) treated	Regenerant waste stream will contain high metals concentration. Low volume of the waste stream may make further treatment more economical
4. Evaporation ponds	Depends on location and land cost. **Capital cost:** $1.49/liter/day ($5.6/gallon/day) assuming 76.2 cm (30 inches) evaporation a year and $1000 per 4046 m² (acre)	Complete control if liner is used. Seepage of salts into groundwater is possible if liner breaks			Only energy consumers are pumps. 0.5 kWh per 3785 liters (1000 gallons) of wastewater is reported	Process produces salts which redissolve upon contact with water

Treatment Method	Cost	Control Effectiveness			Energy Requirement	Residues Produced
		Element	Inlet (mg/L)	Outlet (mg/L)		
5. Reverse osmosis	Depends on feed characteristics, pretreatment, and desired conversion. **Installed capital costs:** $0.193 to $0.528/liter/day ($0.73 to $2.00/gal/day) capacity. **Operating costs:** $0.00017 to $0.00058/liter ($0.00066 to $0.00218/gal)	Calcium	260	0.4	Pumps are major consumers. 8–10 kWh per 3785 liters (1000 gallons) of water for a 3 785 000 liter per day (100 000 GPD[b]) unit	Process produces a concentrated aqueous effluent stream containing practically all incoming contaminants
		Magnesium	170	0.3		
		Iron	77	0.4		
		Aluminum	12	0.2		
		Sulfate	1340	0.9		
		Manganese	43	0.5		
6. Activated carbon	Data needed	Data needed			Data needed	
7. Ferrite coprecipitation	Not available, should be similar to hydroxide precipitation	Mercury	7.4	0.001	Not available, should be similar to hydroxide precipitation	Solid ferrite with heavy metals coprecipitated
		Cadmium	240	0.008		
		Copper	10	0.010		
		Zinc	18	0.016		
		Chromium	10	0.010		
		Nickel	1000	0.2		
		Manganese	12	0.007		
		Iron	600	0.060		
		Bismuth	240	0.100		
		Lead	475	0.010		
		pH				

Table I.2-12 (continued)

Treatment Method	Cost	Control Effectiveness			Energy Requirement	Residues Produced
		Element	Inlet (mg/L)	Outlet (mg/L)		
8. Starch complexing	Not available	Copper	31.8	0.007	Not available, should be higher than those for hydroxide precipitation due to the high volume of the precipitate	Concentrated acidic wastewater stream, if starch is regenerated. Exhausted starch, which decomposes if not properly stored
		Nickel	29.4	0.019		
		Cadmium	56.2	0.009		
		Lead	104	0.025		
		Chromium(III)	26	0.003		
		Silver	54	0.245		
		Zinc	33	0.046		
		Iron	27.9	Not dectectable		
		Manganese	27.5	1.630		
		Mercury	100	0.003		
9. Sodium borohydride	**Capital cost:** $0.18/liter/day ($0.67/gallon/day). **Operating costs:** $0.002 to $0.007 liter ($0.007 to $0.026/gallon). Basis: 57 000 liters/day (15 000 gallons/day) plant. (PA-322)	Removes 0.45 to 1.81 kg TDS/day (1.0 to 4.0 lbs/day). Reported effluent concentration is: Mercury 0.010			Not available	Elemental metals

[a] 1 gallon = 3.785 liters. [b] GPD – gallons per day

Table I.2-13. Techniques for Waste Water Minimization in Electroplating

Minimizing Drag-out

- Minimize viscosity and surface tension by reducing concentration, increasing temperature, and adding wetting agents
- Design and maintain racks and barrels to: promote good drainage; cover with a non-wetting coating; maintain integrity of coating; perform rack stripping in a separate line
- Withdraw the workpiece slowly from the bath
- Position the workpiece to: consolidate drainage and remove by quickest route; avoid dripping from higher to lower piece; avoid horizontal surfaces and solution products; minimize the workpiece dimension protruding vertically into the bath; minimize the surface area of the workpiece in contact with the bath surface as the piece is withdrawn

Drag-out Recovery

- Use a drain board to collect drainage between tanks
- Use a drip tank to collect drainage for return to bath
- Use spray rinse or air knife to reduce the quantity of rinse water contaminated in rinsing

Rinse Water Management

- Ensure adequate turbulence and contact time
- Use a sufficient volume of rinse water to achieve rinsing
- Control make-up water volumes
- Reuse rinse water

Multiple Rinse Stages

- Utilize multiple rinse stages – parallel or countercurrent
- Incorporate a drag-out or drag-in/drag-out tank arrangement

formation. Inefficient design and operation, therefore, significantly affects the interrelated factors of materials consumption, pollution control, and costs.

In-plant changes (Table I.2-13) involve techniques for minimizing the drag-out transferred to the rinse tanks, the amount of waste water produced in the rinsing operation, and the amount of electroplating chemicals lost as drag-out. The overall effect is a reduction of:

1. chemical purchases, resulting in lower raw material costs;
2. waste consumption, resulting in lower water and sewerage costs;
3. waste water treatment needs and residual waste disposal costs.

2.9 Waste Management

Metal concentrations in industrial waste vary over a wide spectrum, depending on the specific production process. Extreme examples with respect to metal contents are given in Table I.2-14 for residues of metal working industries, e.g., from recycling of aluminum, and for filter cakes from waste water treatment facilities, e.g., from electroplating industries.

Table I.2-14. Heavy Metals (μg/g) in Wastes from Aluminum Recycling and Filter Cake from Electroplating Industry (BRUNNER and ZIMMERLI, 1986; made available by C. COLOMBI and U. FÖRSTNER)

	Aluminum Recycling	Electroplating Filter Cake
Copper	90	65 000
Chromium	160	2 400
Nickel	40	82 000
Zinc	200	7 000

Table I.2-15. Potential Sources of Hazardous Wastes in the Production of Metals

	Cu	Pb	Zn	Al	I&S
Acid plant blowdown	+	+	+	−	−
Waste water treatment sludge	+	+	+	+	+
Bleed electrolyte	+	+	+	−	−
Direct contact cooling water	+	+	+	nd	nd
Slags	+	+	+	+[a]	nd
Spent furnace brick	+	+	+	+[b]	nd
Air pollution control dusts	+	+	+	+[c]	+[d]
Leach residues	−	−	+	−	nd
Red mud				+[e]	
Waste pickle liquor	−	−	−	−	+

+ indicates present; − not present; nd, not determined

[a] Dross and skims from casting
[b] Also spent potliner material, a source of fluoride and cyanide
[c] Includes baghouse bags used in dry fluoride control systems
[d] Electric furnace dust contains lead, cadmium, and zinc from scrap recycled
[e] If not properly pretreated, possibly high in caustic content

The major wastes generated by the smelting and refining processes which may potentially constitute a hazardous waste include both process wastes and residuals from air pollution control and waste water treatment systems (Table I.2-15). These wastes are considered hazardous if they exhibit corrosive or toxic characteristics such as an extreme in pH or leachability of an ionic species (heavy metal, fluoride, chloride, cyanide, etc.). Currently, aqueous wastes and sludges are managed by disposal in unlined ponds. The waters which separate from the solids may be recycled (as in bauxite refining), removed by evaporation in arid areas, or treated and discharged in areas of net positive rainfall. The slags are either hot dumped or granulated with water and deposited onto unlined disposal areas. Granulated slag is often sold for aggregate, railway ballast, or manufacture of mineral wool. Although not all slags exhibit toxic characteristics, specific slags should be periodically monitored to ensure that heavy metals are not being returned to the environment in a form which will result in the formation of pollution and resulting ecological damage or adverse human health effects.

3 Hazardous Waste Disposal

3.1 U.S. Definition of Hazardous Waste

The HAZARDOUS and SOLID WASTE AMENDMENTS (1984) (Public Law 98-616) are strongly oriented toward prohibition of continued land disposal of hazardous waste unless it can be determined that such disposal protects human health and the environment. These amendments further strengthen and define provisions of the Resource Conservation and Recovery Act (RCRA) of 1976 (Public Law 94-580) which authorizes the U.S. Environmental Protection Agency (EPA) to regulate the treatment, storage, and transportation of hazardous industrial wastes. Under these Acts and accompanying regulations, many metal-containing wastes are defined as hazardous. Specific industrial sources of hazardous waste include iron and steel manufacturing, secondary lead production, petroleum refining, and manufacture of inorganic chemicals and pigments. Non-specific metal-containing sources listed as hazardous include: waste water treatment sludges from electroplating operations; spent cyanide plating bath solutions and other spent quenching and stripping solutions, and sludges from plating baths, quenching baths, and quenching bath treatment. Additionally, several metal salt and cyanide complexes are defined by the Acts as hazardous waste because of their acute toxicity (GROSSE, 1986).

For solid wastes that are not specifically noted and named as above, there are a series of tests to define their hazardous nature based on four properties:

- Ignitability is determined by measuring the flash point of a substance. The flash point is the lowest temperature at which a substance gives off flammable vapors which, in contact with a spark or flame, will ignite. Substances with a flash point of 60 °C (140 F) or lower are considered ignitable.
- Corrosivity refers to the capacity of a waste to extract or solubilize toxic contaminants from other wastes. A waste is considered corrosive if it has a pH below 2 or above 12.5, or if it corrodes steel following a test developed by the National Association of Corrosion Engineers.
- Reactivity refers to the tendency of a waste to explode, autopolymerize, create a vigorous reaction with air or water, or exhibit thermal instability with regard to shock or to the generation of toxic gases.
- Toxicity refers to the release, through improper disposal, of toxic materials in sufficient amounts to pose a substantial hazard to human health or to the environment. In the EPA's toxicity test (called the Extraction Procedure (EP) Toxicity Test), soluble materials are extracted from the waste at a pH of 5 over a 24-h period. If the test extract exceeds established limits for certain contaminants, the waste is considered toxic and therefore hazardous (US EPA, 1988).

Among the 14 contaminants currently specified in the EP Toxicity Test, eight are metals: arsenic, barium, cadmium, chromium, lead, mercury, selenium, and silver. The United States found itself in a dilema, however, when applying the EP Toxicity Test to wastes classified as "Large Volume".

Large volume wastes as described earlier, are those wastes which result from the extraction and processing of minerals, including metallic and nonmetallic ores and fossil fuels. These wastes have been defined as non-hazardous, even though they contain a low level of polluting components which can cause selected samples to fail the EP Toxicity Tests. The decision to regulate such wastes as non-hazardous was based upon the analysis of risk posed by such wastes to human health and the environment. Other related wastes from processing operations, particularly specific wastes produced in metal winning operations, have been found to consistently produce high concentrations of toxic pollutants which pose significant environmental risk. Such wastes have been designated hazardous and are regulated as such.

3.1.1 Characteristics of Metal-bearing Wastes

Solids. Slag and scrap metal are obvious metal waste components found virtually everywhere. These waste sources, however, usually contain metallic forms which are not problematic except for some leaching and migration of oxidation products. Of greater concern in the solids area are those plots of ground or stream beds which have become polluted with metal species through commercial production or recovery operations. Soil contamination by metals is reported at 41% of the NATIONAL PRIORITIES LIST (NPL) (1980) sites in the U.S. (those sites which have been identified as targets for long-term "remedial action" under the "Superfund" law, the Comprehensive Environmental Response, Compensation and Liability Act (CERCLA) of 1980), and 14% of the NPL sites are reported to be contaminated by metals alone. Lead, chromium, arsenic, cadmium, copper, zinc, mercury, and nickel in a variety of species and salts are most often cited as problem contaminants. Metal pollution at the NPL sites results principally from municipal landfills, metal finishing operations, the chemical industry, and battery recycling activities.

Liquids. Metal-bearing liquid wastes may generally be categorized as aqueous process streams, non-aqueous wastes such as oils and solvents, and non-point aqueous sources such as groundwater and leachate plumes. These latter sources may be the result of naturally occurring or human generated pollution of the land. Waste waters from the metal finishing industry (see Sect. 2) make up a large portion of the aqueous, metal-bearing waste streams which may require treatment for neutralization of acid, recovery of precious metals, destruction of cyanide, control of hexavalent chromium, removal of complexed metals, and separation and control of oil, grease, and toxic organics. Each of these processes will, in general, produce an effluent for discharge or further treatment, and often a sludge requiring dewatering and disposal. This area, too, affords some of the greatest opportunities for waste minimization and metal recovery operations through waste water handling and sludge processing.

Sludges. For the purpose of this discussion, metal-bearing sludges are considered to be the solid waste generated through waste water and solvent treatment techniques. In the past, general practice called for disposal of these residues to a landfill, land farm, or surface impoundment. From these resting places, as recognized by RCRA and CERCLA, metals often re-entered the environment as pollutants. In the U.S., land disposal of this type is now giving way to techniques such as metal recovery from sludges, solidification/stabilization of sludges, and recycle/reuse options to minimize waste production.

3.2 Trends Toward Waste Minimization

To use the metal finishing industry as an example, it is evident that minimization efforts are becoming more important as disposal costs rise and landfill sites become less available (SPEAROT, 1987). Improved precipitation techniques, evaporation, recovery of plating and rinse waters, and application of separation technologies such as ion exchange, reverse osmosis, and electrodialysis are beginning to see increased use for reducing the waste water stream and for recovery of chemicals, thereby also reducing the load on waste water treatment systems and the need for sludge disposal.

On the other side of the issue, however, waste minimization techniques must provide for continued production of a high quality product as the recovery loop is tightened (SPEAROT, 1987). It is important that recycled products (water, metals, and other chemicals) be of high quality and purity so that they are compatible with the process (see also Sect. 2.6). Secondly, as an operation moves toward 100% recovery it is vitally important to remove impurities that begin to build up in the process stream. A third item of concern is that in some cases waste minimization may suggest altering a process to eliminate a troublesome constituent such as cadmium. The operator must be careful in these cases, however, to verify that the modified process gives comparable product quality. Finally, it must be recognized that many new and emerging technologies for waste minimization, recovery, and treatment are complicated to run and require a higher level of technical expertise for operation of the system.

3.3 Treatment and Recovery Techniques for Metal-bearing Sludges

Basic waste water treatment technology is generally available for all manufacturing segments producing metal-bearing wastes. Currently, the U.S. is defining Best Demonstrated Available Technology (BDAT) for a group of waste streams listed under RCRA. Sludges and concentrated wastes from these water treatment units may, however, continue to present disposal problems. The following three technologies may be applied to assist in sludge treatment and improved recycle of metals:

3.3.1 Electro-kinetics

The term "electro-kinetics" refers to processes associated with the flow of direct current electricity through a system. In this case, the sludge-water-electrolyte system may be driven by an electric current to cause migration of metal ions toward the electrodes. In addition to its applicability for recovery of metals from water and sludges, electro-kinetics may be applicable to treatment of contaminated soils (HORNG et al.,

1987). The EPA is investigating the use of electro-kinetic technology for clean-up of chromium-contaminated soils at the 0.6 hectare site of a hard chrome plating facility (BANERJEE, 1987). Laboratory experiments indicate that the electro-kinetic treatment of saturated soils from the site will accelerate leaching and movement of the chromium ions.

3.3.2 Phosphate Precipitation

Another method for treating metal-bearing sludges for metal value recovery is to acid-leach the metal ions from the sludges and then go through a selective precipitation operation to reclaim specific metals. It appears that sulfuric acid leaching, followed by combinations of precipitation of metals as phosphates, will effectively and selectively remove iron, chromium, copper, zinc, and nickel. The phosphate precipitates also lend themselves to filtration in that they are spherical in shape. Pilot-scale test work sponsored by the EPA has shown that recovered metals are of sufficient purity to serve as feedstock for commercial uses, and that the residue from the acid leaching and the low-value ferric phosphate generated are normally not classified as hazardous wastes under U.S. regulations (DAHNKE et al., 1986; TWIDWELL and DAHNKE, 1988).

3.3.3 Stabilization/Solidification

Stabilization processes and solidification processes have different goals. Stabilization systems attempt to reduce the solubility or chemical reactivity of a waste by changing its chemical state or by physical entrapment (microencapsulation). Solidification systems attempt to convert the waste into an easily handled mass with reduced hazard from volatilization, leaching, or spillage. The two are often considered together because they have the common purpose of improving the containment of potential pollutants in treated wastes. Combined processes are often termed "waste fixation" or "encapsulation".

Waste fixation systems that have potential for application to metal-bearing waste include sorption, lime-fly ash pozzolan processes, pozzolan-portland cement systems, thermoplastic microencapsulation, and microencapsulation. The sorption systems involve adding a solid material to a liquid or sludge in order to soak up any liquid present. Pozzolanic processes depend on forming a concrete matrix around the waste material to microencapsulate it and to form an easily handled block. These processes differ from thermoplastic encapsulation in that the latter uses a hot-mix cementing agent such as asphalt to drive off liquid and volatile phases of the waste and to set into a hardened block. Microencapsulation simply refers to isolating large masses of waste with some type of jacketing material (CULLINANE et al., 1986). Other stabilization techniques aimed at the immobilization of metal-containing wastes are based on additions of cement, water glass (alkali silicate), coal fly ash, lime or gypsum (CALMANO, 1988). Generally, maintenance of a pH of neutrality or slightly beyond favors adsorption or precipitation of soluble metals (GAMBRELL et al., 1983).

"Geochemical and biological engineering" emphasize the increasing efforts of using natural resources available at the disposal site for reducing negative environmental effects of all types of waste material, in particular of acid mine wastes. Practical examples for improvement of storage quality of metal-containing waste, including measures for recultivation of old and recent mining waste disposal sites and physicochemical methods to water processing, have been reviewed by several authors (SALOMONS and FÖRSTNER, 1988a, b).

Incorporation in naturally formed minerals, which remain stable over geological times, allows environmentally safe and economic immobilization of potentially toxic metals in large-volume waste materials. Metal sulfides have particularly low solubility compared to the respective carbonates, phosphates, and oxides (KERSTEN, 1988). Marine conditions, which are favorable due to the higher production of sulfide ions, in addition, seem to repress the formation of mono-methyl mercury (CRAIG and MORETON, 1984). This type of waste deposition under stable anoxic conditions, where large masses of polluted materials are covered with inert sediment, became known as "subsediment-deposit". The first example was planned for highly contaminated sludges from Stamford Harbor in the Central Long Island Sound following intensive discussions in the U.S. Congress (MORTON, 1980). Other advantages of near-shore capped mound deposits (KESTER et al., 1983) include the protection of groundwater resources, since the underlying water is saline, and enhanced degradation of organic priority pollutants (KERSTEN, 1988).

Integrated biological systems for effluent treatment from mine and mill tailings include (VON MICHAELIS, 1988): (1) metal polishing by a natural marsh; (2) treatment in artificial meanders; (3) passive treatment over limestone rocks, in ponds, or artificial bog; (4) oxidation of ferrous iron by contact with Sphagnum moss; and (5) removal of organics and metals by algae, woodchips, and agrowastes.

Of course, future efforts should not only be aimed at chemically stabilizing critical compounds in waste materials. Recycling of valuable components is a much better strategy (see Section 2.6 and PATTERSON, 1987).

4 Long-term Risks from Metal-bearing Wastes

It is technically possible now to significantly reduce waterborne discharges of metals from *active* industrial sources and to control airborne emissions from these sources, thus minimizing risks from current metals production and finishing operations; however, each control technology produces a metal-bearing waste, which in itself is a potential long-term risk. The control scenarios for active metal production operations have been previously discussed. This section on long-term risks will relate more to the metal-bearing wastes from those mining, milling, and manufacturing processes. These metal-bearing wastes will be with us for the centuries to come, and may impact life as we know it on this planet.

To recap, in the U.S. many metal-bearing wastes result from mining, beneficiation, and/or primary metals production and are categorized as large volume wastes.

These large volume wastes are particularly difficult to handle because of their size (see also SALOMONS and FÖRSTNER, 1988a, b). For example, it is not unusual for a tailings pile from a copper mine to be 100 meters high and 1000 meters long. Invariably, virtually all of these large volume wastes are potential air and water pollutant problems. Because of their size, shape, and composition, most have difficulty sustaining vegetative growth and suffer from both wind and water erosion. Rainfall on the extensive surface areas of these waste piles permeates and forms (1) seeps to surface drainage at the toe of the pile and (2) recharges the groundwater supply from the middle of the pile. If the waste contains sulfides, it may form acidic drainage which will in turn solubilize metals and transfer these to the surface seeps and groundwater via intrusion from the pile.

4.1 Airborne

One of the most serious long-term airborne environmental concerns is the evolution of potentially carcinogenic radon gas. This problem is particularly well-documented in the wastes from the phosphate fertilizer industry. Of lesser health concern is fugitive dust from the metal-bearing waste piles. Though undesirable, fugitive dust is not normally hazardous, other than in the immediate vicinity. An exception to this occurs when the waste contains asbestos, a silicate mineral, which may contain metallic forms. Asbestos is a carcinogen, and fugitive emissions from asbestos-containing wastes pose serious immediate and long-term threats to human health.

4.2 Surface Waters

Discharges to surface waters from metal bearing wastes impact: TDS (total disolved solids) levels, toxicity, acidity, and cyanides.

High TDS levels in receiving streams constitute a risk to aquatic life and also require more sophisticated and costly treatment to elevate the water to quality suitable for industrial and for potable use.

Metal toxicity is a most important long-term effect in surface waters on the viability of aquatic life and its delicate ecosystem balance. Examples of mercury accumulation within fish and associated food chains are numerous and serve as warning beacons to the longer-term, more chronic effects of metal pollution bio-accumulation in the food chain.

Acidity, produced by the oxidation of pyritic (iron-bearing) materials does double damage. First, the low pH will seriously damage aquatic life by drastically altering the ecosystem (flora and fauna), inhibiting reproduction, and killing fish. Second, the low pH dissolves metals and mobilizes them, increasing possibility for uptake by the flora and fauna, thus increasing toxicity.

Cyanides in surface waters are always associated with metals because of the great chelating characteristics of cyanides. These highly toxic cyanometallic compounds can have disasterous effect's on the life of a stream and the apparent toxicity is highly pH dependent. Photodecomposition of ferro- and ferricyanides produces high fish mortality due to the evolution of HCN.

4.3 Groundwater

Long-term groundwater problems generally relate to TDS, acidity, and toxic metals.

As in the case of surface water, the TDS impact of metals is more of an economic/water use problem than a health effect's one. Intrusion of metal-bearing seepage into a borderline aquifer can preclude its consideration as a water supply unless extensive, expensive treatment options are employed. Such a degradation can stifle the economic/industrial and agricultural growth of a region.

Acidity is perhaps the most serious long-term threat from metal-bearing wastes because acidity solubilizes and mobilizes the toxic metals. The acidity production is a latent characteristic and can develop many years after disposal of a metal-bearing waste, when the neutralizing or buffering capacity in a pyrite-containing waste is exceeded. The toxic metals mobilized under acidic conditions include barium, mercury, selenium, chromium, gold, silver, copper, lead, zinc, cobalt, arsenic, and cadmium. Consumption and use of these waters must be severely restricted to protect public health.

Problems especially arise from the oxidation of sulfidic components in wastes, which mobilize toxic metals in high concentrations. For example, cases of mine tailings deposited in less well buffered catchment areas are known from Poland, Wales, Harz/Germany, New Lead Belt and Coeur d'Alène regions in the U.S.A., Canada, Australia, New Guinea, Philippines, and South Africa (FÖRSTNER, 1981).

Typically elevated concentrations of trace elements in groundwaters have been found in the vicinity of industrial waste deposits. For instance, BALKE et al. (1973) studied a case at Nievenheim in the lower reaches of the river Rhine, where a zinc processing plant had infiltrated waste water into the substratum. In the groundwater the concentrations of arsenic surpassed 50 mg/L, and maximum concentrations have been measured for cadmium of 600 µg/L, for thallium of 800 µg/L, for mercury of 50 µg/L, and for zinc of 40 mg/L. Eighteen months after the percolation had been stopped, these contaminations were only insignificantly lower.

A simple chemical mechanism could not explain the rapid production of acidic mine drainage by the oxidation of Mn, Fe, Zn, Pb, Cu, and Cd sulfides, and thus the inhancing of their mobility. According to SINGER and STUMM (1970) ferric iron is the major oxidant of pyrite in the complex natural sequence, and it is mainly *Thiobacillus ferrooxidans*, an iron-oxidizing acidophilic bacterium (see also Chapter I.7h), which accelerates metal sulfide oxidations 10^6 times over the abiotic rate.

High concentration factors have been determined in inland waters affected by acidic mine effluents, where the levels of dissolved Ni, Mn, and Fe exceed the normal surface water concentrations by factors of more than 10000, values of Zn and Cr are increased 1000-fold, and for Pb and Cd enrichments were higher than 100-fold (Table I.2-16; for references see FÖRSTNER, 1981). There are many examples of deleterious effects of mine effluents on freshwater ecosystems: The Itai-Itai catastrophe at the Jintsu River in Japan has been connected with cadmium-rich effluents from an abandoned lead-zinc mine (KOBAYASHI, 1972; see also Chapter II.6). In streams in Wales enrichments of Pb, Cu, and Zn leached from the outcrops of mineralized zone and spoil heaps of abandoned mines still cause an increased mortality rate in fish and other organisms (ABDULLAH and ROYLE, 1972). Oysters

Table I.2-16. Metal Concentrations (in µg/L) in Inland Waters Affected by Acidic Mine Effluents (After FÖRSTNER, 1981)

Metal	Cornwall (SW England)	Silesia (Poland)	Siberia[f] (USSR)	Colorado[g] (USA)	Philippines[h]	Tasmania[i]
Arsenic	250[a]	–	499	70	–	–
Cadmium	–	1 325[d]	207	70	–	6 100
Cobalt	–	15[e]	368	–	–	–
Chromium	–	17[e]	–	–	120	–
Copper	1 160[b]	62[e]	20 710	3 900	953	1 350
Iron	23 000[b]	3 185[e]	–	213 000	176 100	20 500
Manganese	2 400[b]	315[e]	1 624	8 000	–	22 500
Nickel	–	14[e]	900	460	80	–
Lead	530[c]	23[e]	2 071	300	443	–
Zinc	10 000[b]	43 100[d]	5 770	17 000	1 280	105 000

[a] Tamar River (ASTON et al., 1975); [b] Carnon River; [c] Gannel River (ASTON et al., 1974); [d] Graniczna Woda, inflow to; [e] Mala Panew (PASTERNAK, 1974); [f] maximum values from up to 4500 samples (UDODOV and PARILOV, 1961); [g] HILL (1973); [h] Baguio Mining District, Agno River (LESACA, 1977); [i] Storys Creek in South Esk catchment (TYLER and BUCKNEY, 1973)

placed in the heavy contaminated Restronguet Creek of Southwest England contained initially 250 mg/kg Cu in dry matter, rising to 1500 mg/kg after one month and 6000 mg/kg after six months (THORNTON, 1977). From the Upper Silesian mining areas of Poland it is suggested that the process of self-purification in some of the receiving waters is inhibited by the high concentrations of several metals; during many years no fish has appeared in the Sztola River (PASTERNAK, 1974). Large-scale destruction of organic life has been found around Brunswick mines in Canada (HARVEY, 1976), from Lepanto copper mines of the Philippines, Rum Jungle mines in Northern Australia (WATSON, 1975) and the South Esk River system of Tasmania (TYLER and BUCKNEY, 1973). Contamination of rice fields from mine drains is still found in Japan, e.g., in the Ichi and Maruyama River basins (ASAMI, 1981). Renal dysfunction has been observed in inhabitants from the upper section of the Ichi River basin, and it has been stressed by several authors that the allowable limit for unpolished rice (1.0 mg Cd/kg) is probably too high (ASAMI, 1988).

Early diagenetic changes and element mobilization from *dredged* material in a man-made estuarine marsh has been investigated by DARBY et al. (1986); the results of this study are described in Chapter I.7e. Another adverse effect of dredged materials after land disposal relates to contamination of crops. High pH, high clay and organic matter contents reduce the plant availability of most metals (VAN DRIEL and NIJSSEN, 1988). Dredged materials often have high silt and organic matter contents (corresponding to soils), and in the long run, the organic matter will be partly decomposed with a corresponding increase in bioavailability. Dredged materials containing higher amounts of oxidizable sulfides are also less buffered than soils, which effect has also an influence on plant crops growth (HERMS and TENT, 1982).

4.4 Damage Cases in the United States

In the United States, the Resource Conservation and Recovery Act and Superfund legislation (CERCLA) empowers the Environmental Protection Agency to develop a list of hazardous wastes sites that pose significant environmental or health risks (NATIONAL PRIORITIES LIST, 1986). As previously noted, this list, the National

Table I.2-17. Summary of Metals in National Priority (Superfund) Sites

Summary of metal problems at NPL sites

952 total NPL sites analyzed (703 listed, 249 proposed)
41% (389) report metal problems
26% (244) report metals with organics
14% (133) report metals, but not organics
 1% (12) unclear if organics are present
29% (113) of sites reporting metals, report only one metal
71% (276) of sites reporting metals, report multiple metals

Metal/ # of sites

Metal	# of sites
Pb	133
Cr	118
As	77
Cd	65
Cu	49
Zn	40
Hg	32
Ni	24
Ba	10
Ag	10
Fe, Ra, U, Th, Mn, Se	48

# of sites	Industry	Metals most often reported
154	landfill/chemical waste dump	As Pb Cr Cd Ba Zn Mn Ni
43	metal finishing/plating/electronics	Cr Pb Ni Zn Cu Cd Fe As
35	chemical/pharmaceutical	Pb Cr Cd Hg As Cu
28	mining/ore processing/smelting	Pb As Cr Cd Cu Zn Fe Ag
21	federal (DOD, DOE)	Pb Cd Cr Ni Zn Hg As
19	battery recycle	Pb Cd Ni Cu Zn
18	wood treating	Cr Cu As
16	oil and solvent recycle	Pb Zn Cr As
13	nuclear processing/equipment	Ra Th U
5	pesticide	As
5	vehicle and drum cleaning	As
3	paint	Pb Cr Cd Hg
29	other	As Pb Hg Cr
389	total	

Priorities List (NPL), is ranked according to risk and includes such ranking factors as size, population affected, toxicity, mobility, etc. In 1986, the National Priorities List added 84 sites as imminent hazards. Of these, 47 contained metals as a primary contaminant. At least 22 of 97 sites added to the list as hazardous during mid-1987 were listed because of metals contamination.

A brief analysis was made (BATES, 1988) to estimate the extent of metals problems at NPL sites. Descriptions of 952 sites (703 listed plus 249 proposed) were reviewed to determine the extent of metals contamination problems. Results of this analysis are presented in Table I.2-17. As noted previously, 41% of these sites (389) reported metals problems. The majority of these (244) reported metals combined with organics, but a significant number (133) reported only metals or metals with inorganics. Of the 389 sites reporting metals problems, most (276) reported problems from two or more metals, but nearly 29% (113) reported only one metal as a problem. Although sites reporting metals problems are generally distributed fairly uniformly across the NPL priority groups, 7 of the top 10 sites and 30 of the top 50 have metals problems.

Although some of the NPL sites descriptions simply identified heavy metals as a problem, many of the site descriptions specified the metals. The metals most often cited as a problem are lead, chromium, arsenic, and cadmium, each of which is cited as a problem at over 50 sites. Copper, zinc, mercury, and nickel are cited as problems at more than 20 sites each, while all other metals are cited as problems at less than 10 sites each. Only a few industries or activities account for most of the NPL sites with metals problems. Not surprisingly landfills account for a high percentage (154 sites, 40%) as chemical wastes were often dumped in municipal landfills, as well as specific chemical waste dumps. Some sites such as landfills, chemical, oil recycle and wood treating usually have problems with organic compounds, as well as metals.

In conclusion, analysis indicates that problems are associated with only a few specific metals and have resulted mostly from just a few industrial activities. The experience to date in the U.S. with the frequency of metal-bearing wastes from mining, milling, processing, and manufacturing appearing as hazardous waste sites on the Superfund NPL list, in combination with the intercontinental nature of hazardous air pollutants from these processes, can be extrapolated both geographically and on a time basis as a forewarning of long-term worldwide risks posed by these wastes.

References

Abdullah, M. I., and Royle, L. G. (1972), *Heavy Metal Content of Some Rivers and Lakes in Wales.* Nature *238*, 229–230.

Asami, T. (1981), *The Ichi and Maruyama River Basins: Soil Pollution by Cadmium, Zinc, Lead and Copper Discharged from Ikuno Mine,* in: Kitagishi, K., and Yamane, I. (eds.): *Heavy Metal Pollution in Soils of Japan,* pp. 227–236. Japanese Soil Science Society, Tokyo.

Asami, T. (1988), *Soil Pollution by Metals from Mining and Smelting Activities,* in: Salomons, W., and Förstner, U. (eds.): *Chemistry and Biology of Solid Waste: Dredged Materials and Mine Tailings,* pp. 143–169. Springer-Verlag, Berlin.

Aston, S. R., Thornton, I., Webb, J. S., Purves, J. B., and Milford, B. L. (1974), *Stream Sediment Composition, an Aid to Water Quality Assessment.* Water Air Soil Pollut. *3*, 321–325.

Aston, S. R., Thornton, I., Webb, J. S., Milford, B. L., and Purves, J. B. (1975), *Arsenic in Stream Sediments and Waters of South-West England.* Sci. Total Environ. *4*, 347–358.

Balke, K. D., Kussmaul, H., and Siebert, G. (1973), *Chemische und thermische Kontamination des Grundwassers durch Industriewässer.* Z. Dtsch. Geol. Ges. *127*, 447–460.

Ball, R. O., Verret, G. P., Buckingham, P. L., and Mahfood, S. (1987), *Economic Feasibility of a State-wide Hydrometallurgical Recovery Facility,* in: Patterson, J. W., and Passino, R. (eds.): *Metals Speciation, Separation and Recovery,* pp. 690–709. Lewis Publ., Chelsea, Michigan.

Banerjee, S. (1987), *Electro-Decontamination of Chrome-Contaminated Soils,* in: *Land Disposal, Remedial Action, Incineration and Treatment of Hazardous Waste Proceedings of the Thirteenth Annual Research Symposium.* EPA/600/9-87/015, pp. 193–200. U.S. Environmental Protection Agency, Cincinnati, Ohio.

Bates, E. (1988), *Personal Communication,* February 1988. U.S. Environmental Protection Agency, Cincinnati, Ohio.

Brunner, P. H., and Zimmerli, R. (1986), *EAWAG Bericht Nr. 30-4721,* Dübendorf, Switzerland.

Burckle, J. O. (1987), *Control de la Contaminación en Galvanoplastia-Cambios en el Interior de las Plantas,* in *Proceedings of Control de la Contaminacion en Origen en la Industria de Galvanotecnia,* 23–25 September 1987. Servicio Central de Publicaciones-Gobierno Vasco, Vitoria Gasteiz.

Burckle, J. O., and Worrell, A. C. (1981), *Comparison of Environmental Aspects of Selected Nonferrous Metals Production Technologies,* in: Chatwin, T. D., and Kikumoto, N. (eds.): *Sulfur Dioxide Control in Pyrometallurgy.* The Metallurgical Society of the AIME.

Calmano, W. (1988), *Stabilization of Dredged Mud,* in: Salomons, W., and Förstner, U. (eds.): *Environmental Management of Solid Waste – Dredged Material and Mine Tailings,* pp. 80–98. Springer-Verlag, Berlin.

Christiansen, H., Öhlmann, G., and Tent, L. (1982), *Probleme im Zusammenhang mit dem Anfall von Baggergut im Hamburger Hafen.* Wasserwirtschaft *72*, 385–389.

Coleman, R. T., et al. (1980), *Sources and Treatment of Wastewater in the Nonferrous Metals Industry.* EPA 600/2-80/074. U.S. Environmental Protection Agency, Cincinnati, Ohio.

Craig, P. J., and Moreton, P. A. (1984), *The Role of Sulphide in the Formation of Dimethylmercury in River and Estuary Sediments.* Mar. Pollut. Bull. *15*, 406–408.

Cullinane, M. J., Jones, L. W., and Malone, P. G. (1986), *Handbook for Stabilization/Solidification of Hazardous Waste.* EPA/504/2-86/001. U.S. Environmental Protection Agency, Cincinnati, Ohio.

Curtin, M. E. (1983), *Microbial Mining and Metal Recovery.* Bio/Technology (May), 229–235.

Dahnke, D. R., Twidwell, L. G., and Robins, R. G. (1986), *Selective Iron Removal from Process Solutions by Phosphate Precipitation,* in: *Iron Control in Hydrometallurgy,* pp. 477–503. John Wiley & Sons, New York.

Darby, D. A., Adams, D. D., and Nivens, W. T. (1986), *Early Sediment Changes and Element Mobilization in a Man-made Estuarine Marsh,* in: Sly, P. G. (ed.): *Sediment and Water Interactions,* pp. 343–351. Springer-Verlag, New York.

Ellis, D. V. (1988), *Case Histories of Coastal and Marine Mines,* in: Salomons, W., and Förstner, U. (eds.): *Chemistry and Biology of Solid Waste – Dredged Material and Mine Tailings,* pp. 73–100. Springer-Verlag, Berlin.

Förstner, U. (1981), *Trace Metals in Fresh Waters (with particular reference to mine effluents),* in: Wolf, K. H. (ed.): *Handbook of Strata-Bound and Stratiform Ore Deposits,* Volume 9, pp. 271–303. Elsevier Publ., Amsterdam.

Gambrell, R. P., Reddy, C. N., and Khalid, R. A. (1983), *Characterization of Trace and Toxic Materials in Sediments of a Lake Being Restored.* J. Water Pollut. Control Fed. *55*, 1271–1279.

Goodson & Associates (1982), *Development of Systematic Waste Disposal Plans for Metal and Nonmetal Mines.* Bureau of Mines Open File Report 183–82. U.S. Department of the Interior, U.S. Bureau of Mines, Washington, DC.

Greber, J. S., Patel, V. P., Pfetzing, E. A., Amick, R. S., and Toftner, R. O. (1979), *Assessment of Environmental Impact of the Mineral Mining Industry.* EPA-600/2-79-107. U.S. Environmental Protection Agency, Office of Research and Development, Industrial Environmental Research Laboratory, Cincinnati, Ohio.

Grosse, D. W. (1986), *A Review of Alternative Treatment Processes for Metal-Bearing Hazardous Waste Streams. J. Air Pollut. Control Ass.* 36, 603–614.

Harvey, H. H. (1976), *Aquatic Environmental Quality: Problems and Proposals. J. Fish. Res. Board Can.* 33, 2634–2670.

Hazardous and Solid Waste Amendment (1984), Public Law 98-616. *United States Code, Congressional and Administrative News*, Vol. 5. West Publishing, St. Paul, Minnesota.

Herms, U., and Tent, L. (1982), *Schwermetallgehalte im Hafenschlick sowie in landwirtschaftlich genutzten Hafenschlick-Spülfeldern im Raum Hamburg. Geol. Jahrb. F12*, 3–11.

Hill, R. D. (1973), *Control and Prevention of Mine Drainage*, in: *Cycling and Control of Metals*, pp. 91–94. U.S. Environmental Protection Agency, National Environmental Research Center, Cincinnati, Ohio.

Horng, J., Banerjee, S., and Herrmann, J. (1987), *Evaluating Electro-kinetics as a Remedial Action Technique.* in: *Proceedings of the Second International Conference on New Frontiers for Hazardous Waste Management.* EPA/600/9-87/018F, pp. 65–78. U.S. Environmental Protection Agency, Cincinnati, Ohio.

IAPH (International Association of Ports and Harbors) (1981), *A Survey of World Port Practices in the Ocean Disposal of Dredged Material as Related to the London Dumping Convention.* Report of the Ad Hoc Dredging Committee (A. J. Tozzoli, Chairman). Port Authority of New York and New Jersey, 38 pp. International Association of Ports and Harbors, New York.

Katari, V. S., and Gerstle, R. W. (1977), *Industrial Process Profiles for Environmental Use,* Chapter 24: *The Iron and Steel Industry.* EPA600/2-77-023X. U.S. Environmental Protection Agency, Cincinnati, Ohio.

Kellogg, H., and Henderson, J. (1976), *Use in Sulfide Smelting,* Chapter 19, *Extractive Metallurgy of Copper,* Yannopoulos, J. C., et al. (eds.) AIME.

Kersten, M. (1988), *Geochemistry of Priority Pollutants in Anoxic Sludges: Cadmium, Arsenic, Methyl Mercury, and Chlorinated Organics,* in: Salomons, W., and Förstner, U. (eds.): *Chemistry and Biology of Solid Waste – Dredged Material and Mine Tailings*, pp. 170–213. Springer-Verlag, Berlin.

Kester, D. R., Ketchum, B. H., Duedall, I. W., and Park, P. K. (eds.) (1983), *Wastes in the Ocean.* Vol. 2: *Dredged-Material Disposal in the Ocean.* Wiley, New York.

Kobayashi, J. (1972), *Relation between the "Itai-Itai" Disease and the Pollution of River Water by Cadmium from a Mine. Adv. Water Pollut. Res., Proc. 5th Int. Conf. San Francisco and Hawaii I-25*, pp. 1–7.

Lesaca, R. M. (1977), *Monitoring of Heavy Metals in Philippine Rivers, Bay Waters and Lakes.* Proc. Int. Conf. Heavy Metals in the Environment, Toronto 1975, Vol. II/1, pp. 285–307. University of Toronto.

Martin, H. W., and Mills, W. R., Jr. (1975), *Water Pollution Caused by Inactive Ore and Mineral Mines.* Report prepared for the Industrial Environmental Research Laboratory, Office of Research and Development, U.S. Environmental Protection Agency, by Toups Corporation. Contract No. 68-03-2212. US EPA, Industrial Environmental Research Laboratory, Cincinnati, Ohio.

Michaelis, H. (1984), *Production, Processing, Waste Water Management, and Recycling of Metals* (in German), in: Merian, E. (ed.): *Metalle in der Umwelt*, pp. 11–20. Verlag Chemie, Weinheim-Deerfield Beach/Florida-Basel.

Minerals Yearbook (1981), U.S. Department of the Interior, U.S. Bureau of Mines, Washington, DC.

Minerals Yearbook (1983), U.S. Department of the Interior, U.S. Bureau of Mines, Washington, DC.

Minerals Yearbook (1984), U.S. Department of the Interior, U.S. Bureau of Mines, Washington, DC.

Morton, R. W. (1980), *"Capping" Procedures as an Alternative Technique to Isolate Contaminated Dredged Material in the Marine Environment,* in: *Dredge Spoil Disposal and PCB Contamination: Hearings before the Commitee on Merchant Marine and Fisheries.* House of Representatives, Ninety-sixth Congress, 2nd Session, on Exploring the Various Aspects of Dumping of Dredged Spoil Material in the Ocean and the PCB Contamination Issue, March 14, May 21, 1980. USGPO Ser. No. 96–43, pp. 623–652. Washington, DC.

Mountain States Research and Development, Inc. (1981), *Inventory of Waste Enbankments of Surface and Underground Openings: Metal and Nonmetal Active Mines.* Bureau of Mines Open File Report 110-82. Report prepared for the U.S. Bureau of Mines, Contract No. J0199054. U.S. Department of the Interior, U.S. Bureau of Mines, Washington, DC.

National Priorities List Fact Book (1986), HW 7.3., 94 pp. U.S. Environmental Protection Agency, Washington, DC.

Nijssen, J. P. J. (1988), *Rotterdam Dredged Material: Approach to Handling,* in: Salomons, W., and Förstner, U. (eds.): *Environmental Management of Solid Waste – Dredged Material and Mine Tailings,* pp. 243–281. Springer-Verlag, Berlin.

Parsons, T. (1977), *Industrial Process Profiles for Environmental Use,* Chapter 25: *Primary Aluminum Industry.* EPA600/2-77-023Y. U.S. Environmental Protection Agency, Cincinnati, Ohio.

Pasternak, K. (1974), *The Influence of Pollution of a Zinc Plant at Miasteczki Slaskie on the Content of Micro-elements in the Environment of Surface Waters.* Acta Hydrobiol. 16, 273–297.

Patterson, J. W. (1987), Metals Separation and Recovery, in: Patterson, J. W., and Passino, R. (eds.): *Metals Speciation, Separation and Recovery,* pp. 63–93. Lewis Publ., Chelsea, Michigan.

PEDCo Environmental, Inc. (1984), *Evaluation of Management Practices for Mine Solid Waste Storage, Disposal, and Treatment.* 3 Vols. Prepared for the Industrial Environmental Research Laboratory, U.S. Environmental Protection Agency. Contract No. 68-03-2900. PEDCo Environmental, Inc., Cincinnati, Ohio.

Pitt, C. H., Wadsworth, M. E. (1980), *An Assessment of Energy Requirements in Proven and New Copper Processes.* U.S. Department of Energy; Final Report for Contract EM-78-S-07-1743.

Rotterdam Municipality (1984), *Grootschalige locatie voor de berging van baggerspecie uit het benedenrivierengebied.* Project Report/Environmental Compatibility Study, Oct. 1984, 334 p. Municipality of Rotterdam/Rijkswaterstaat, The Netherlands.

Salomons, W., and Förstner, U. (1984), *Metals in the Hydrocycle.* Springer-Verlag, Berlin.

Salomons, W., and Förstner, U. (eds.) (1988a), *Chemistry and Biology of Solid Waste: Dredged Materials and Mine Tailings.* Springer-Verlag, Berlin.

Salomons, W., and Förstner, U. (eds.) (1988b), *Environmental Management of Solid Waste: Dredged Materials and Mine Tailings.* Springer-Verlag, Berlin.

Seitter, L. E., and Hunt, R. G. (1982), *Industrial Resource Recovery Practices: Mining Industries* (SIC Division B). Prepared for the Office of Solid Waste, U.S. Environmental Protection Agency by Franklin Associates, Ltd. Contract No. 68-01-6000. Franklin Associates, Ltd., Prairie Village, Kansas.

Singer, P. C., and Stumm, W. (1970), *Acidic Mine Drainage: The Rate-determining Step.* Science 167, 1171–1173.

Spearot, R. M. (1987), *Waste Minimization in the Metal Finishing Industry,* in: *Summary Proceedings of the Workshop on Hazardous Waste Minimization,* unpublished, U.S. Environmental Protection Agency, Cincinnati, Ohio.

Tent, L. (1982), *Auswirkungen der Schwermetallbelastung von Tidegewässern am Beispiel der Elbe.* Wasserwirtschaft 72, 60–62.

Thornton, I. (1977), *Some Aspects of Environmental Geochemistry in Britain.* Proc. Int. Conf. Heavy Metals in the Environment, Toronto, 1975, Vol. II/1, pp. 17–38. University of Toronto.

Twidwell, L. G., and Dahnke, D. R. (1988), *Metal Value Recovery from Metal Hydroxide Sludges. Removal of Iron and Recovery of Chromium,* 203 pp. U.S. Environmental Protection Agency, Cincinnati, Ohio.

Tyler, P. A., and Buckney, R. T. (1973), *Pollution of a Tasmanian River by Mine Effluents. I. Chemical Evidence. Int. Rev. Ges. Hydrobiol.* 58, 873–883.

Udodov, P. A., and Parilov, Y. U. S. (1961), *Certain Regularities of Migration of Metals in Natural Waters. Geochemistry* 8, 763–769.

U.S. Bureau of Mines (1981), *Mine Waste Disposal Technology.* Proceedings Bureau of Mines Technology Transfer Workshop, Denver, Co., July 16, 1981. Information Circular 8857. U.S. Department of the Interior, U.S. Bureau of Mines, Washington, DC.

U.S. Industrial Outlook (1985), U.S. Department of Commerce, Washington, DC.

US EPA (Environmental Protection Agency) (1980a), *Industrial Process Profiles for Environmental Use*, Chapter 27, *Primary Lead Industry.* EPA 600/2-80-168. U.S. Environmental Protection Agency, Cincinnati, Ohio.

US EPA (Environmental Protection Agency) (1980b), *Industrial Process Profiles for Environmental Use,* Chapter 28, *Primary Zinc Industry.* EPA 600/2-80-169. U.S. Environmental Protection Agency, Cincinnati, Ohio.

US EPA (Environmental Protection Agency) (1980c), *Industrial Process Profiles for Environmental Use,* Chapter 29, *Primary Copper Industry.* EPA 600/2-80-170, U.S. Environmental Protection Agency, Cincinnati, Ohio.

US EPA (Environmental Protection Agency) (1982a), *Development Document for Final Effluent Limitations Guidelines and New Source Performance Standards for the Ore Mining and Dressing Point.* Source Category. 440/1-82/061. U.S. Environmental Protection Agency, Washington, DC.

US EPA (Environmental Protection Agency) (1982b), *Summary Report – Control Treatment Technology for the Metal Finishing Industry: In-Plant Changes.* EPA 600/8-82-008. U.S. Environmental Protection Agency, Cincinnati, Ohio.

US EPA (Environmental Protection Agency) (1983a), *Development Document for Effluent Limitations Guidelines and Standards for the Metal Finishing Point Source Category.* EPA400/1-83/091. U.S. Environmental Protection Agency, Cincinnati, Ohio.

US EPA (Environmental Protection Agency) (1983b), *Development Document for Electroplating/MF.* EPA440/1-83-091. NTIS PB83-102004. U.S. Environmental Protection Agency, Cincinnati, Ohio.

US EPA (Environmental Protection Agency) (1983c), *Effluent Limitations Guidelines and Standards for the Nonferrous Metal Point Source Category.* EPA 440/1-83/0196. U. S. Environmental Protection Agency, Cincinnati, Ohio.

US EPA (Environmental Protection Agency) (1984a), *State Regulations and Management Practices in the U.S. Mining Industry.* Interim Report prepared for the U.S. Environmental Protection Agency, Division of Solid Waste Management. Charles River Associates.

US EPA (Environmental Protection Agency) (1984b), *Review of New Source Performance Standards for Primary Copper Smelters.* EPA450/3-83-018A. U.S. Environmental Protection Agency, Cincinnati, Ohio.

US EPA (Environmental Protection Agency) (1985a), *State Regulations and Management Practices in the U.S. Mining Industry.* Interim Report prepared for the U.S. Environmental Protection Agency, Division of Solid Waste Management. Charles River Associates.

US EPA (Environmental Protection Agency) (1985b), *Estimated Costs to the U.S. Mining Industry for Management of Hazardous Solid Wastes.* Final Report. Prepared for the U.S. Environmental Protection Agency, Division of Solid Waste Management. Charles River Associates.

US EPA (Environmental Protection Agency) (1985c), *Cost Comparisons of Selected Technologies for the Control of Sulfur Dioxide from Copper Smelters.* NTIS PB 85-215 705/HS. U.S. Environmental Protection Agency, Cincinnati, Ohio.

US EPA (Environmental Protection Agency) (1988), *Identification and Listing of Hazardous Waste.* Code of Federal Regulations, Title 40, Part 261.

Van Driel, W., and Nijssen, J. P. J. (1988), *Development of Dredged Material Disposal Site: Implications for Soil, Flora and Food Quality,* in: Salomons, W., and Förstner, U. (eds.): *Chemistry and Biology of Solid Waste – Dredged Material and Mine Tailings,* pp. 101–126. Springer-Verlag, Berlin.

Van Driel, W., Kerdijk, H. N., and Salomons, W. (1984), *Use and Disposal of Contaminated Dredged Material. Land Water Int. 53*, 13–18.
Vick, S. G. (1981), *Siting and Design of Tailings Impoundments. Min. Eng.* (June), 653–657.
Von Michaelis, H. (1988), *Integrated Biological Systems for Effluent Treatment from Mine and Mill Tailings*, in: Salomons, W., and Förstner, U. (eds.): *Environmental Management of Solid Waste – Dredged Material and Mine Tailings*, pp. 99–113. Springer-Verlag, Berlin.
Watson, G. M. (1975), *Rum Jungle Environmental Studies*. Summary Report. *Aust. At. Energy Comm. E 366*, 21 p.
Wiesenberg, I. J., et al. (1980), *Fertility of Primary Copper Smelter Leak Sulfur Dioxide Streams Control.* EPA600/2-80-152. U.S. Environmental Protection Agency, Cincinnati, Ohio.
Wiesenberg, I. J., et al. (1981), *Weak SO_2 Stream Control from Copper Smelters*, in: Chatwin, T. D., and Kikumoto, N. (eds.): *Sulfur Dioxide Control in Pyrometallurgy*. The Metallurgical Society of the AIME.

Additional Recommended Literature

Allen, H. E., and Garrison, A. W. (1989), *Metal Speciation and Transport in Groundwater. Proceedings EPA/ACS Workshop, Jekyll Island, Georgia.* Lewis Publishers, Chelsea, Michigan, in press.
Baccini, P. (1989a), *The Control of Heavy Metal Fluxes from the Anthroposphere to the Environments. Proceedings International Conference Heavy Metals in the Environment, Geneva*, Vol. 1, pp. 13–23. CEP Consultants Ltd., Edinburgh.
Baccini, P. (1989b), *Lecture Notes in Earth Sci. 20, The Landfill (Reactor and Final Storage).* Springer-Verlag, Berlin-Heidelberg-New York-London-Paris-Tokyo.
Envitec '89 (1989), *Soil – a Bottleneck in Environmental Capacity*, Summaries of Papers, Düsseldorfer Messegesellschaft mbH "Nowea", see also *Swiss Chem. 11*(7–8), 49–58.
Fleischer, G. (1989), *Waste Reduction in the Metal Industry 1*. EF-Verlag für Energie- und Umwelttechnik GmbH, Berlin.
Thomé-Kozmiensky, K. J. (1989), *Proceedings 6th International Recycling Congress and Exhibition* (see also *Abfallwirtschaftsjournal 1*(11), 1–128 (1990), and Report in *Swiss Chem. 12*(5), in press); *Incineration and Environment 3; Landfilling (Dumps of Wastes) 3; Site Remediation 3.* EF-Verlag für Energie- und Umwelttechnik GmbH, Berlin.
Trüb, L. (1989a), *Two-hundred Years Uranium* (in German). *Neue Zürcher Zeitung, Forschung und Technik No. 224*, p. 91 (27 September), Zürich.
Trüb, L. (1989b), *Shortage of Resources and Environmental Problems in the Age of New Materials* (in German). *Neue Zürcher Zeitung, Forschung und Technik No. 236*, pp. 73–74 (11 October), Zürich.
Trüb, L. (1989c), *Red Dog, Alaska – the Largest Lead-Zinc-Mine in the Western World* (in German). *Neue Zürcher Zeitung, Forschung und Technik No. 242*, pp. 77–78 (18 October), Zürich.
Trüb, L. (1990), *The Gold in the Black Hills of South Dakota – Precious Metal Mining* (in German). *Neue Zürcher Zeitung, Forschung und Technik No. 49*, p. 77 (28 February), Zürich.
Yakowitz, H. (1989), *Encouraging Waste Reduction/Minimization and Recycling: a Policy Oriented Overview.* OECD Paris, Proceedings IPRE (International Professional Association for Environmental Affairs) Brussels, in press.

I.3 Chemical Processes in the Environment, Relevance of Chemical Speciation

JAMES J. MORGAN, Pasadena, California, USA
WERNER STUMM, Zürich, Switzerland

1 Introduction

Metal cycles on a regional and global basis have been profoundly modified by human activity. Comparative flux data for land, atmosphere, river and ocean transfers of metals allow an identification of the most potentially hazardous water and soil pollutants. These metals, which existed in trace amounts in water (<1 µmol L^{-1}) under earlier, natural conditions, have been released to a greater extent to aquatic and terrestrial systems. There is a close linkage of the metal cycles to the biological processes, and the question arises to what extent the elevated levels of metals constitute disturbance of ecosystems and a threat to human health.

It is useful to know some of the basic physical and chemical properties of metal ions, especially some of their elementary electronic structures in order to understand chemical behavior and the form of occurrence (speciation), and, in turn, the cycling and fate of a metal in the natural environment, to appreciate the biochemical action of a metal as a nutrient or toxicant (see also Chapters I.5, I.6, I.7, and Part II, Sections 2.1 and 4 in the Chapters).

1.1 Definitions

In the following, any combination of cations with molecules or anions containing free pairs of electrons (bases) is called coordination (or complex formation) and can be either electrostatic or covalent, or a mixture of both. The metal cation will be called the central atom, and the anions or molecules with which it forms a coordination compound will be referred to as ligands.

If the ligand is composed of several atoms, the one responsible for the basic or nucleophilic nature of the ligand is called the ligand atom, and if there are several such atoms which can occupy more than one coordination position in the complex, it is referred to as a multidentate complex former. Ligands occupying one, two, three, etc., positions are referred to as unidentate, bidentate, tridentate, etc. Typical examples are oxalate and ethylenediamine as bidentate ligands, citrate as a tridentate ligand, ethylenediamine tetraacetate (EDTA) as hexadentate ligand. Complex formation with multidentate ligands is called chelation, and the complexes are called chelates. The most obvious feature of a chelate is the formation of a ring. For example, in the reaction between the bidentate ethylenediamine and $Cu \cdot aq^{2+}$, a chelate

```
        OH₂
     H   |   H
H₂C—N\  |  /N
     \  Cu  /  \CH₂
H₂C—N/  |  \N—CH₂
     H  |   H
        OH₂
```

with two rings, each of five members, is formed. O- and N- are the donor atoms. Complexes with more than one metal atom (central atom) are called multi- or polynuclear complexes.

There is a phenomenological similarity between the "neutralization" of H^+ with bases and that of metal ions with complex formers. The bases, molecules, or ions that can "neutralize" H^+ or metal ions possess free pairs of electrons. Acids are proton donors according to BRØNSTED. LEWIS, on the other hand, has proposed a much more generalized definition of an acid in the sense that he does not attribute acidity to a particular element but to a unique electronic arrangement: the availability of an empty orbital for the acceptance of a pair of electrons. Such acidic or acid-analogue properties are possessed by H^+, metal ions and other Lewis acids such as $SOCl_2$, $AlCl_3$, SO_2, BF_3. In aqueous solutions protons and metal ions compete with each other for the available bases.

1.2 Chemical Speciation

Fig. I.3-1 illustrates the various forms of occurrence of Cu(II) in natural water systems (STUMM and MORGAN, 1981). One would wish to have analytical possibilities to distinguish between the various solute and "adsorbed" species or to identify the solid or surface sites (organic surface, iron(III) oxide, aluminum silicates) in which the metal ion is present or bound to. Usually, the evidence for a particular form of occurrence is circumstantial and is based on complementary evaluations together with kinetic and thermodynamic considerations.

The Need for Chemical Speciation. The chemical behavior of any element in the environment depends on the nature of its components. Historically, however, the limnologists', oceanographers', earth scientists' and biologists' interests lay primarily in determining collective parameters and the elemental composition. This information alone is often inadequate for identifying the mechanisms that control processes in the environment and the composition of natural waters and to understand their perturbation.

Analytical Possibilities (see also Chapter I.4a). In order to understand the factors that control the concentrations of heavy metals in natural waters, their chemical reactivity and affinity sequence to the binding agents, their bioavailability, their ultimate fate, we need to know the form of occurrence of the metal. Although we can make many chemical equilibrium models that predict the existence of complexes in natural waters, analytically one encounters difficulties in identifying unequivocally the various solute species and in distinguishing between dissolved and particulate

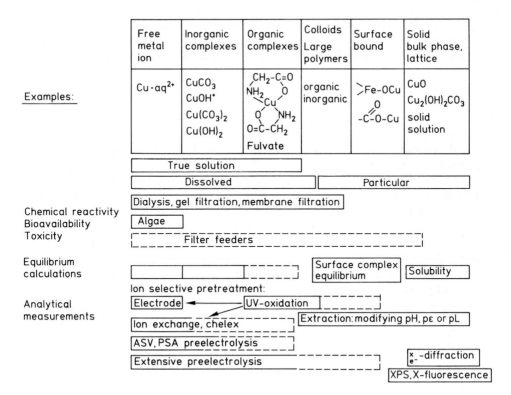

Fig. I.3-1. The forms of occurrence of Cu(II) in natural water systems. — Usually only a very small fraction of the total Cu(II) present in a natural water occurs as "free" $Cu \cdot aq^{2+}$; typically carbonato complexes, depending on pH, and also some hydroxo complexes are more prevalent in inorganic species. Organic ligands, above all fulvate and humate, tie up Cu(II) as organic complexes. Since some of Cu(II) is adsorbed or bound to the surface of colloids which may have diameters smaller than 0.1 μm, filtration techniques cannot properly distinguish between dissolved and particular Cu(II). As discussed in the text, the toxicity depends on speciation; the ecological and toxicological effects with regard to algae are related to the activity of the free aquo Cu^{2+} ion. Some of the operational procedures for the quantification of selected groups of Cu(II) species are indicated. Bioassays with algae can be used to measure free Cu^{2+} ion activity.

concentrations. The analytical task is rendered very difficult because the individual chemical species are often present at nano- and picomolar concentrations. The ion selective electrode (ISE), if it were sufficiently sensitive, would permit the measurement of free metal ion activity. Otherwise, no single simple method permits unequivocal identification of species. Although voltammetry, especially differential pulse polarography (DPP) or anodic stripping voltammetry (ASV), is an extremely sensitive technique, one must be aware that it measures the electrochemically labile, i.e., the electrochemically available, metal ion concentration which in the case of Cu(II), for example, includes the carbonate and glycine complexes. At low pH, the ASV technique gives values that may be representative of total soluble metal ion concentrations.

In assessing metal ion complexation, much attention is paid to the influence of soluble complex formers (including fulvic or humic acids) in regulating the concentration of free metal ions. But particle surfaces rival solute complex formers in tying up metal ions. Actually, particles are usually more important than soluble complex formers in regulating the "residual" concentrations of metal ions.

Because there is usually no direct way to determine the actual metal species present, it becomes highly desirable to characterize in a given natural water sample the metal ion and its coordinative environment (particles and soluble complex formers) as well as possible, in order to derive some evidence on the plausible form of occurrence. Often, one determines total "soluble" metal ion concentration in filtered and subsequently acidified samples. One then uses equilibrium calculations to estimate the free metal ion concentration that might have been present at the original pH of the water. Similarly, many workers titrate natural waters directly with trace metals by voltammetric detection methods to determine the so-called complexing capacity of the water. This capacity includes complexation by soluble complex formers (humic and fulvic acids) and by particle surfaces (oxides and organic and organically coated particles) (SIGG et al., 1984; GONÇALVES et al., 1985). Because of the very low trace metal ion concentrations in natural waters, the lack of sufficiently selective analytic sensors, and the complicated features in the interaction of the organic material with metal ions, there are many limitations involved in this approach (VARNEY et al., 1984; BUFFLE et al., 1984).

Because metal ions are typically sorbed to particles many of which are of colloidal size and are often smaller than 0.1 µm, membrane filtration usually does not permit the analytical differentiation between dissolved and particulate concentrations.

Speciation of Solid Phases. Unfortunately, no simple analytical recipes are available to determine with sufficient selectivity the site of the metal ion in the adsorbed or solid phase. It is extremely difficult to identify the "site of the adsorbed" species. The use of extractants to operationally fractionate the forms of metals in the solid or adsorbed phase has not resolved the actual molecular state in which these metals are present. Such sequential leaching procedures may give some insight, however, into the potential mobilization of metals, e.g., from sediments.

Surface speciation is of extreme importance to understand the "buffering" of metal ions by the surface phase and to appreciate the role of surface species for kinetic processes, e.g., the dissolution of oxides and minerals, nucleation of new solid phases on these heterogeneous surfaces and coagulation or dispersion of colloidal particles. There is an urgent need to develop methods, e.g., X-ray spectroscopic techniques, to determine surface speciation in atmospheric and aquatic particles (see also Chapter I.6a).

2 Coordination of Metals

2.1 Electronic Structure and Chemical Behavior

From a very general viewpoint, in all chemical reactions the molecules and ions tend to improve the electron stabilities of their valence electrons. In *coordinative* reactions, e.g., those involving Lewis acids and Lewis bases, electron pairs are donated by the bases [Lewis bases (donors) to the Lewis acids (receptors)]

$$A + :B \rightleftarrows A:B$$

Brønsted acid-base behavior in water, e.g.,

$$H^+ + B \rightleftarrows HB^+$$

and

$$HB_1 + HB_2 \rightleftarrows HB_2 + HB_1$$

and metal-ligand complex formation in solution and as well as in heterogeneous reactions,

$$M + L \rightleftarrows ML$$

and

$$M_1L_1 + M_2L_1 \rightleftarrows M_2L_1 + M_1L_2$$

are Lewis acid/Lewis base reactions. In redox processes, electrons are completely transferred between oxidant-reductant pairs,

$$Ox_1 + Red_1 \rightleftarrows Red_2 + Ox_2$$

e.g.,

$$Cr^{3+} + Cu^+ \rightleftarrows Cr^{2+} + Cu^{2+}$$

or, in a more complex reaction,

$$2\ Cu^{2+} + HSO_3^- + H_2O \rightleftarrows 2\ Cu^+ + SO_4^{2-} + 3\ H^+ \ .$$

Lewis acid-base reactions and redox reactions are examples of "closed-shell/closed-shell" interactions between species.

It is worth noting that two major pollution problems in the world today – acid pollution and metal pollution – are both manifestations of *acid-base* disturbances in the most general sense. Deposition of acidity (see also Chapter I.7d) results from intense oxidative production of Brønsted acids (H_2SO_4, HNO_3), causing increased free and/or combined base-neutralizing capacity in ecosystems (MORGAN, 1986). Metal pollution introduces excessive quantities of certain *Lewis acid* metals to ecosystems. The biological consequences of metal pollution strongly depend on the resulting chemical speciation, which is a function of the kinds and amounts of Lewis bases, the redox status (pE) and the acidity-alkalinity (pH) characteristics of particular environments. In the biota, the key Lewis bases include ligands containing oxygen (O) donors (e.g., OH), nitrogen (N) donors (e.g., NH), or sulfur (S) donors (e.g., SH), which are coordinated to essential trace metals, e.g., Mg, Mn, Fe, Co, Zn. In biochemistry, geochemistry, environmental chemistry, and in chemistry of metal emission control it is important to recognize and make use of general patterns of chemical affinities between Lewis acids and bases, i.e., the energetics of equilibria for all reactions of interest of metal coordination

$$M_i + L_j \rightleftarrows M_i L_j$$

including aqueous, solid and surface complex species and extending to multiple metal (e.g., polynuclear) or multiple ligand (e.g., mixed ligand) forms. The aqueous proton is one of the most influential M_i and the hydroxide ion one of the more influential ligands. Redox states of the electron acceptors, M_i, and of the donors, L_j, *strongly* affect the affinities (e.g., Fe^{3+} vs Fe^{2+} and SO_4^{2-} vs S^{2-}).

A comparison of typical *essential* metals for life with the potentially *hazardous* metals just mentioned is of interest (metalloids are underlined):
Essential metals: Na, K, Mg, Ca, Cr, Mn, Fe, Co, Ni, Cu, Zn, Mo.
Hazardous metals: Cr, Cu, Zn, As, Se, Ag, Cd, In, Sn, Sb, Hg, Tl, Pb, Bi.

2.2 Periodic Classification of Metals

The Periodic Table is shown in Table I.3-1. The metals and metalloids (semi-metals) comprise all of the elements except the noble gases (Group 0) and H, B, C, N, O, F, P, S, Cl, Br, I, and At. The metalloids are Si, Ge, As, Se, Sb, and Te. Hydrogen exhibits metallic properties as the Lewis acid H^+ and non-metallic properties as the Lewis base H^-. Groups Ia and IIa, the "s block" metals, form monovalent cations (alkali metal cations) and bivalent cations (earth alkali cations), respectively, each of these metal ions achieving the "noble gas structure". Groups IIIb through VIb contain the "p block" metals; among the "p block" metal ions are Al (III oxidation state), Pb (II and IV oxidation states), and Tl (I and II oxidation states). The Al^{3+} ion has the noble gas electron configuration of Ne.

The noble gas configuration (d^0) for metal ions is associated with high spherical symmetry and electron sheaths which are not readily deformed by electric fields, i.e., their polarizabilities are low, or, in more descriptive language, they are "hard" spheres.

Those metal cations that have an electron number corresponding to that of Ni^0, Pd^0, and Pt^0 ($g = 10$ or 12) have electron clouds more readily deformable by the electric fields of other species, i.e., they have many valence electrons and higher polarizabilities, and may be visualized as "soft sphere" ions. In general, higher polarizabilities are found to have increased strength of covalent bonding.

The transition elements have between 0 and 10 d electrons. The three transition series of the Periodic Table, in which the 3d, 4d, and 5d electronic orbitals are being successively filled, occupy rows 4, 5, and 6 of the Periodic Table. The first transition series runs from Sc (21) through Zn (30); the second from Y (39) through Cd (48); the third from La (57) to Hg (80). The orbital energies and configuration d electrons of the first and subsequent transition series result in a wide variety of oxidation states (except for the Group IIIb metals, which have only the III oxidation state).

Electrons in s, p, or d orbitals may be paired or unpaired, resulting in different energies for the complex electron configuration of the metal. The energy levels of the d orbitals are altered by the electric field of the electron pair donors in ligands, e.g., O, N, F, S, etc. The pairing or unpairing of electrons in octahedral transition metal complexes depends upon the number of valence electrons and the electron energies of the ligands in relation to the acceptor metal. Basically, the five d electron orbitals of the central metal are split into two higher-energy orbitals (e_g) and three lower-energy orbitals (t_{2g}), separated by an energy, Δ_0,

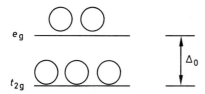

If a ligand is "weak", Δ_0 is small; for "strong" ligands Δ_0 is greater. The resulting set of electron orbitals in an octahedral complex of a metal with d^4, d^5, d^6, or d^7 configurations may have predominantly paired electrons in the orbitals, in a "low spin" complex; or it may have several unpaired electrons, in a "high spin" complex. The resulting stability of a ML complex, as well as its kinetic behavior, either "labile" or "inert", is thus a direct consequence of the weak or strong interaction of the transition metal d orbitals with ligand orbitals (for more than 3 d electrons). For example, the valence electron configuration of Mn^{2+}(aq) is $3d^5$. The complex is known to be $Mn(H_2O)_6^{2+}$, and magnetic properties show that all five electrons are unpaired, thus

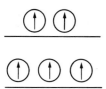

and the complex is high spin. Water is only a moderate-strength ligand. As a result, Mn^{2+}(aq) is labile, i.e., it readily releases the H_2O ligands. An example of a low spin transition metal complex is $Fe(CN)_6^{2-}$, with Fe(II) as the central metal. There are 6 d electrons. The stable (lowest energy) electron arrangement for large Δ_0 is t_{2g}^6, or t_{2g} ⇅ ⇅ ⇅.

The relative strengths of ligands with respect to energy of d orbital splitting, Δ_0, is known as the "spectrochemical" series:

$F^- < H_2O < SCN^- < NH_3 < NO_2^- < CN^-$
(smallest Δ_0)   (highest Δ_0) .

As the first transition series is crossed, the 5d configuration of Mn^{2+} represents a relatively weak tendency toward covalent bonding, as does Zn^{2+} at the end of the

Table I.3-1. Periodic Table of the Elements

Period	Group Ia	Group IIa	Group IIIa	Group IVa	Group Va	Group VIa	Group VIIa	Group VIII			Group Ib	Group IIb	Group IIIb	Group IVb	Group Vb	Group VIb	Group VIIb	Group 0
1 1s	1 H																1 H	2 He
2 2s2p	3 Li	4 Be											5 B	6 C	7 N	8 O	9 F	10 Ne
3 3s3p	11 Na	12 Mg											13 Al	14 Si	15 P	16 S	17 Cl	18 Ar
4 4s3d 4p	19 K	20 Ca	21 Sc	22 Ti	23 V	24 Cr	25 Mn	26 Fe	27 Co	28 Ni	29 Cu	30 Zn	31 Ga	32 Ge	33 As	34 Se	35 Br	36 Kr
5 5s4d 5p	37 Rb	38 Sr	39 Y	40 Zr	41 Nb	42 Mo	43 Tc	44 Ru	45 Rh	46 Pd	47 Ag	48 Cd	49 In	50 Sn	51 Sb	52 Te	53 I	54 Xe
6 6s (4f) 5d 6p	55 Cs	56 Ba	57* La	72 Hf	73 Ta	74 W	75 Re	76 Os	77 Ir	78 Pt	79 Au	80 Hg	81 Tl	82 Pb	83 Bi	84 Po	85 At	86 Rn
7 7s (5f) 6d	87 Fr	88 Ra	89** Ac															

*Lanthanide series 4f	58 Ce	59 Pr	60 Nd	61 Pm	62 Sm	63 Eu	64 Gd	65 Tb	66 Dy	67 Ho	68 Er	69 Tm	70 Yb	71 Lu
**Actinide series 5f	90 Th	91 Pa	92 U	93 Np	94 Pu	95 Am	96 Cm	97 Bk	98 Cf	99 Es	100 Fm	101 Md	102 No	103 Lr

Heavy boundary divides metals and metalloids (dashed boundary) from non-metals.
Type-B metals are marked by ×, borderline metals are marked by diagonal lines (those closer to type A by \ and those closer to type B by /).

series. In general, covalent character of bonding tends to increase from left to right for the other transition metals.

The lanthanides (see also Chapters II.15 and II.35), elements 58 through 71, constitute a so-called inner transition series, as to the actinides, elements 90 through 103. Scandium (21) and yttrium (39), together with the lanthanides are traditionally referred to as the rare earth elements. The lanthanides, with 3+ ions and decreasing radii, show strong ionic bonding and weaker covalent bonding characteristics. As discussed below the lanthanides tend to exhibit "hard sphere" or A-type behavior in their coordination compounds.

2.3 A and B Behavior of Metal Cations
(Hard and soft acid-base rules)

Inorganic and organic ligands contain the following possible donor atoms in the 4th, 5th, 6th, and 7th vertical column of the Periodic Table:

C	N	O	F
	P	S	Cl
	As	Se	Br
		Te	I .

In water, the halogens are effective complexing agents only as anions, but not if bound to carbon. For special reasons the cyanide ion is a particularly strong complex former. The more important donor atoms include nitrogen, oxygen, and sulfur.

According to AHRLAND et al. (1958) and SCHWARZENBACH's (1961) analysis of metal-ligand complex stability constants data in aqueous solutions, e.g., for $Ag^+ + F^- \rightleftarrows AgF(aq)$ vs $Ag^+ + Cl^- \rightleftarrows AgCl(aq)$ (giving the difference between Cl^- and F^- as Lewis bases in aqueous solution), or for $Mg^{2+} + OH^- \rightleftarrows MgOH^+(aq)$ vs $Mg^{2+} + F^- \rightleftarrows MgF^+(aq)$ giving relative strengths of F^- and OH^- as electron donors, or for $Cu^+ + Cl^- \rightleftarrows CuCl(aq)$ vs $Ag^+ + Cl^- \rightleftarrows AgCl(aq)$ giving electron accepting strengths of Ag^+ and Cu^+ as Lewis acid metal ions, there is an evident classification of the metal ions into "A" or "hard", and "B" or "soft" type metals (cf. Table I.3-1).

As Table I.3-2 shows, this classification into A- and B-type metal cations is governed by the number of electrons in the outer shell. Type-A metal cations having the inert gas type (d^0) electron configuration correspond to those which were classified above as "hard sphere" cations. These ions may be visualized as being of spherical symmetry; their electron sheaths are not readily deformed under the influence of electronic fields, such as those produced by adjacent charged ions. Type-B metal cations have a more readily deformable electron sheath (higher polarizability) than type-A metals and were characterized as "soft sphere" cations.

Metal cations of type A form complexes preferentially with the fluoride ion and ligands having oxygen as donor atom. Water is more strongly attracted to these metals than are ammonia or cyanide. No sulfides (precipitates of complexes) are formed by these ions in aqueous solution, since OH^- ions readily displace HS^- or S^{2-}. Chloro or iodo complexes are weak and occur most readily in acid solutions

Table I.3-2. Classification of Metal Ions

Type-A Metal Cations	Transition-Metal Cations	Type-B Metal Cations
Electron configuration of inert gas; low polarizability; "hard spheres"; (H^+), Li^+, Na^+, K^+, Be^{2+}, Mg^{2+}, Ca^{2+}, Sr^{2+}, Al^{3+}, Sc^{3+}, La^{3+}, Si^{4+}, Ti^{4+}, Zr^{4+}, Th^{4+}	One to nine outer shell electrons; not spherically symmetric; V^{2+}, Cr^{2+}, Mn^{2+}, Fe^{2+}, Co^{2+}, Ni^{2+}, Cu^{2+}, Ti^{3+}, V^{3+}, Cr^{3+}, Mn^{3+}, Fe^{3+}, Co^{3+}	Electron number corresponds to Ni^0, Pd^0 and Pt^0 (10 or 12 outer shell electrons); low electronegativity; high polarizability; "soft spheres"; Cu^+, Ag^+, Au^+, Tl^+, Ga^+, Zn^{2+}, Cd^{2+}, Hg^{2+}, Pb^{2+}, Sn^{2+}, Tl^{3+}, Au^{3+}, In^{3+}, Bi^{3+}

According to PEARSON's (1963) Hard and Soft Acids

Hard Acids	Borderline	Soft Acids
All type-A metal cations plus Cr^{3+}, Mn^{3+}, Fe^{3+}, Co^{3+}, UO^{2+}, VO^{2+}	All bivalent transition metal cations plus Zn^{2+}, Pb^{2+}, Bi^{3+}	All type-B metal cations minus Zn^{2+}, Pb^{2+}, Bi^{3+}
Also species such as BF_3, BCl_3, SO_3, RSO_2^+, RPO_2^+, CO_2, RCO^+, R_3C^+	SO_2, NO^+, $B(CH_3)_3$	All metal atoms, bulk metals I_2, Br_2, ICN, I^+, Br^+
Preference for ligand atom: $N \gg P$ $O \gg S$ $F \gg Cl$		$P \gg N$ $S \gg O$ $I \gg F$

Qualitative generalizations on stability sequence:

Cations: Stability ∝ (charge/radius)	Cations: Irving-Williams order: $Mn^{2+} < Fe^{2+} < Co^{2+} < Ni^{2+} < Cu^{2+} > Zn^{2+}$	
Ligands: $F > O > N = Cl > Br > I > S$ $OH^- > RO^- > RCO_2^-$ $CO_3^{2-} \gg NO_3^-$ $PO_4^{3-} \gg SO_4^{2-} \gg ClO_4^-$		Ligands: $S > I > Br > Cl = N > O > F$

under which conditions competition with OH^- is minimal. The univalent alkali ions form only relatively unstable ion pairs with some anions; some weak complexes of Li^+ and Na^+ with chelating agents, macrocyclic ligands, and polyphosphates are known. Chelating agents containing only nitrogen of sulfur as ligand atoms do not coordinate with type-A cations to form complexes of appreciable stability. Type-A metal cations tend to form difficultly soluble precipitates with OH^-, CO_3^{2-} and PO_4^{3-}; no reaction occurs with sulfur and nitrogen donors (addition of NH_3, alkali sulfides, or alkali cyanides produces solid hydroxides). Some stability sequences are indicated in Table I.3-2.

In contrast, type-B metal ions coordinate preferentially with bases containing I, S, or N as donor atoms. Thus, metal ions in this class may bind ammonia more strongly than water, CN^- in preference to OH^-, and form more stable I^- or Cl^- complexes than F^- complexes. These metal cations, as well as transition metal cations, form insoluble sulfides and soluble complexes with S^{2-} and HS^-.

Non-colored components often yield a colored compound (charge transfer bands), thus indicating a significant deformation of the electron orbital overlap. Hence, in addition to coulombic forces, types of interactions other than simple electrostatic forces must be considered. These other types of interactions can be interpreted in terms of quantum mechanics, and in a somewhat oversimplified picture the bond is regarded as resulting from the sharing of an electron pair between the central atom and the ligand (covalent bond). The tendency toward complex formation increases with the capability of the cation to take up electrons (increasing ionization potential of the metal) and with decreasing electronegativity of the ligand, (increasing tendency of the ligand to donate electrons). In the series F, O, N, Cl, Br, I, S, the electronegativity decreases from left to right, whereas the stability of complexes with type-B cations increases. TURNER et al. (1981) have applied:

$$\Delta \beta = \log \beta^0_{MF} - \log \beta^0_{MCl},$$

i.e., the difference in stability between the fluoro and chloro complexes of a particular element as a guide to its propensity to form covalent bonds. Elements for which $\Delta\beta > 2$ are type-A cations, and elements for which $\Delta\beta < -2$ are type-B cations and form strong complexes that are largely covalently bound.

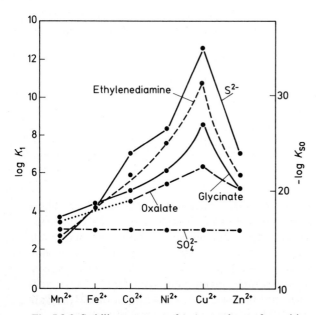

Fig. I.3-2. Stability constants of 1:1 complexes of transition metals and solubility products of their sulfides (Irving-Williams series).

For transition metal cations, a reasonably well-established rule on the sequence of complex stability, the Irving-Williams order, is valid. According to this rule, the stability of complexes increases in the series

$$Mn^{2+} < Fe^{2+} < Co^{2+} < Ni^{2+} < Cu^{2+} > Zn^{2+} .$$

An example is given in Fig. I.3-2.

WILLIAMS and HALE (1966) have proposed the following expected gradation of bivalent metals from (a) to (b) character:

(a) Be > Mg > Ca, Ba, Sr > Sn > Pb > Zn > Cd > Hg (b) .

For electron donors, a simple attempt at classifying "hard" and "soft" bases is illustrated in Table I.3-2. Such classification schemes reveal, not surprisingly, that "hard" and "soft" are not absolute, but gradually varying qualities. PEARSON (1963) has proposed, in generalizing "hardness" and "softness" properties for Lewis acids and bases for many kinds of systems, the so-called "HSAB" (hard and soft acid-base) rules.

Rule 1: *Equilibrium*. Hard acids prefer to associate with hard bases and soft acids with soft bases.

Rule 2: *Kinetics*. Hard acids react readily with hard bases and soft acids with soft bases.

2.4 Hydrolysis of Metal Ions

OH^- is one of the most influential ligands, especially for type-A metal cations. More than 30 years ago, BRØNSTED postulated that multivalent metal ions participate in a series of consecutive proton transfers:

$$Fe(H_2O)_6^{3+} = Fe(H_2O)_5OH^{2+} + H^+ = Fe(H_2O)_4(OH)_2^+ + 2H^+$$
$$= Fe(OH)_3(H_2O)_3(s) + 3H^+ = Fe(OH)_4(H_2O)_2^- + 4H^+ .$$

In the case of Fe(III), hydrolysis can go beyond the uncharged species $Fe(OH_3(H_2O)_3(s)$ to form anions such as the ferrate(III) ion, probably $[Fe(OH)_4 \cdot 2H_2O]^-$. All hydrated ions can, in principle, donate a larger number of protons than that corresponding to their charge and can form anionic hydroxo-metal complexes, but, because of the limited pH range of aqueous solutions, not all elements can exist as anionic hydroxo or oxo complexes.

The acidity of aquo-metal ions is expected to increase with decrease of the radius and an increase of charge of the central ion. BAES and MESMER (1976) have correlated the first constant for metal hydrolysis

$$M^{z+} + H_2O \rightleftharpoons MOH^{z-1} + H^+; \ ^*K_1 ,$$

with (z/d), where z is charge and d is the M-O distance in the complex. Fig. I.3-3 shows their result: different cation groups can be distinguished. The cation group farthest to the right contains cations that are most likely to form ionic M-O bonds, whereas the cation groups on the left which have a stronger tendency to hydrolyze

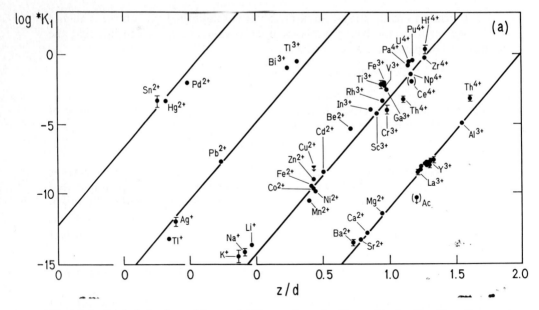

Fig. I.3-3a. Hydrolysis of metal ions; solubility product of oxides and hydroxides (from Baes and Mesmer, 1976). The linear dependence of the first hydrolysis constant $^*K_1 = \{MOH^{(z-1)+}\}\{H^+\}/\{M^{z+}\}$ on the ratio of the charge to the M-O distance (z/d) for four groups of cations (25 °C).

for their size and charge, tend to form bonds of a more covalent character (Baes and Mesmer, 1976).

2.5 Kinetics of Complex Formation

Understanding the quantitative aspects of metals chemistry in the natural environment and in biological and technological systems must ultimately rest on an understanding of the kinetics of slow vs. fast reactions. But metals' chemical kinetics can be extremely diverse and complex. Biological catalysis, e.g., in metal oxidations and metal oxide reductions, is possible in soils, sediments, column operations, sludges, and other suitable environments. For processes in lakes, Imboden and Schwarzenbach (1985) have developed a very useful comparison for physical vs. chemical time scales. A number of other kinetics plus equilibrium modeling approaches have been successfully applied to complex reaction systems in water chemistry and atmospheric chemistry (e.g., the fog acidification model of Jacob and Hoffmann, 1983).

In many cases complexation of metals by ligands is rapid relative to other processes of interest (e.g., transport, mixing, precipitation, ion exchange), and equilibrium models provide then a good representation of complexation phenomena in natural systems.

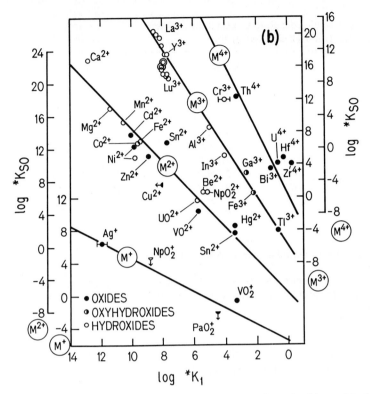

Fig. I.3-3 b. Hydrolysis of metal ions; solubility product of oxides and hydroxides (from BAES and MESMER, 1976). The correlation of the solubility product $*K_{SO}$ with the first hydrolysis constant $*K_1$ for M^+, M^{2+}, M^{3+}, and M^{4+} cations. The lines have slopes of -1, -2, -3, and -4 (25 °C). This relationship results because the equilibrium constant for the reaction

$$M(OH)_z(s) + (z-1)M^{z+} = z\, MOH^{(z-1)+}; \quad K = (*K_1)^z K_{SO}; \quad K = \{MOH^{(z-1)+}\}^z/\{M^{z+}\}^{(z-1)},$$

is often close to $10^{-5.6}$. This relatively low value reflects the general tendency of cations to precipitate shortly after hydrolysis begins unless $[M^{z+}]$ is quite low. The strong tendency for solutions to supersaturate often allows hydrolysis to proceed much further in solution than would be expected from the value of K.

For most metals of interest in aquatic chemistry, a simplified description of the exchange of hydration water for a ligand involves the formation of an outersphere complex (ion pair) between the hydrated metal and the ligand, followed by expulsion of a water molecule (SUTIN, 1966; MOREL, 1983), e.g.,

$$M(H_2O)_6^{2+} + L^- \xrightleftharpoons[k_{-L}]{k_L} M(H_2O)_6 \cdot L^+$$

$$M(H_2O)_6 \cdot L^+ \xrightleftharpoons[k_w]{k_{-w}} M(H_2O)_5 L^+ + H_2O \ .$$

Table I.3-3. Characteristic Second-order Rate Constants for Water Exchange in the Primary Solvation Shell of Metal Ions[a,b]

	K_w (mol^{-1} L^{-1} s^{-1})
Cr^{3+}	3.6×10^{-5}
Mn^{2+}	3.4×10^6
Fe^{3+}	3.3×10^2
Fe^{2+}	3.5×10^5
Co^{3+}	$\leq 10^2$
Co^{2+}	1.2×10^5
Ni^{2+}	2.9×10^3
Cu^{2+}	$\geq 9 \times 10^8$
Zn^{2+}	5×10^6
Cd^{2+}	9×10^7
Hg^{2+}	4×10^8

[a] Adapted from Sutin (1966) with permission from Annual Reviews Inc.
[b] Note: The second-order rate constant for water exchange is equal to 6/55 times the first-order rate constant (s^{-1}) for replacement of a particular water molecule

The first reaction is fast ($k_L \cong 10^9$ to 10^{10} mol^{-1} L^{-1} s^{-1}; $k_{-L} \cong 10^7$ to 10^{10} s^{-1} so that the forward rate constant for the overall reaction

$$M(H_2O)_6^{2+} + L^- \underset{k_b}{\overset{k_f}{\rightleftharpoons}} M(H_2O)_5 L^+ + H_2O$$

is given by $k_f = k_{-w} \dfrac{k_L}{k_{-L}} = k_{-w} K$

where K is the stability constant of the ion pair. K depends primarily on electrostatic interaction, i.e., on the charge number of the ions, not on their chemical nature.

This general result shows that the loss of a water molecule from the primary hydration sphere of a metal ion is often the rate determining step. This is convenient for estimating complex formation reaction rates, even in the absence of specific information (MOREL, 1983; see also Table I.3-3). Fig. I.3-4 gives water exchange rate constants and illustrates, for example, that the second-order rate constants for the formation of most copper complexes are orders of magnitude larger than those for Ni(II). Thus, the medium residence time of a H_2O molecule ($\tau = k^{-1}$) coordinated to Al^{3+} is about 1 s. This is very long in comparison to the aquo ions of Ca^{2+} and Hg^{2+} ($\tau \cong 10^{-8}$ to 10^{-9} s), but short in comparison to H_2O coordinated to Cr^{3+} ($\tau \approx 10^6$ s) (SCHNEIDER, 1968). In the case of chelate formation, more than one water molecule must be replaced by a reacting ligand group, and the complexation reaction can be represented by a stepwise water-ligand exchange. Often the presence of the first coordinated ligand facilitates the replacement of the remaining water molecules and the second reaction is relatively fast. Thus, in many instances the exchange of the first coordinated water controls the overall complexation rate. Ligand

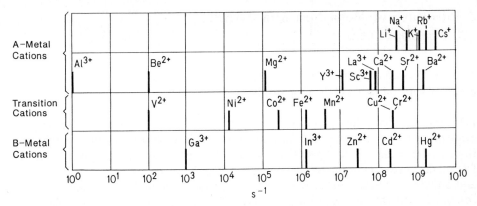

Fig. I.3-4. Exchange rate of water from aquo ions expressed as first-order constants k, i.e., $d[H_2O]_{ex}/dt = k\,[M^{z+}_{aq}]$ where $[H_2O]_{ex}$ is exchanged water from the coordination sphere of an aquo ion M^{z+}_{aq}. (Modified from SCHNEIDER, 1968).

substitution rates of aqueous metal species and their hydroxo complexes, e.g., Fe^{3+} and $FeOH^{2+}$ differ by several orders of magnitude; hence strong pH-influences on substitution rates are observed.

There is a need for more adequate experimental kinetic data. In the absence of these data, the elementary kinetic principles outlined above can provide a preliminary useful framework for identifying reactions that *might* be relatively slow.

2.6 Metal-Carbon Compounds
(see also Chapters I.7h, II.16, II.20, and II.30)

The stability of the metal-carbon bound in aqueous solution decreases from germanium to lead. Besides forming a variety of inorganic complexes, Hg has very high stability constants with organic ligands and can form stable bonds to carbon and therefore true organometallic compounds (ANDREAE, 1986).

Metals and metalloids that form alkyl compounds (e.g., methylmercury) deserve special concern, because these compounds are volatile and accumulate in cells; they are poisonous to the central nervous system of higher organisms. Methylmercury species (or other metal alkyls) may be produced at a rate faster than they are degraded by other organisms. They may accumulate in organisms. Metals and metalloids which have been reported to be biomethylated in minor amounts include Ge, Sn, As, Se, Te, Pd, Pt, Au, Hg, Tl. There is still a controversy about the biological methylation of Pb. Anthropogenic pollution by alkylated metal ions is certainly more significant than biomethylation. Methylated tin species have been observed in near-shore polluted waters. For data on As, Sb, Se, and Te see the recent review by ANDREAE (1986).

3 Interaction of Metal Ions with Particles

(see also Chapter I.6a)

3.1 The Coordination Chemistry of the Solid-Water Interface as a Major Control in Aquatic Geochemistry

Particles – through their surfaces – are scavengers for metal ions and often reactive elements in their transport from land to rivers and lakes and from continents to the floor of the oceans. TUREKIAN (1977) writes "The great particle conspiracy is active from land to sea to dominate the behavior of the dissolved species". Hydrous oxide and aluminum silicate surfaces, as well as organically coated and organic surfaces contain functional surface groups ($\equiv$MOH, $\equiv$ROH, R-COOH) that are able to act as coordinating sites of the surface.

The functional groups at the surface undergo acid-base and other coordinative interactions; thus their coordination properties are similar to those of their counterparts in soluble compounds (Fig. I.3-5). The concentration of metals is typically much larger in the solid phase than in the solution phase. Thus the buffering of metals is much higher in the presence of particles than in their absence (SIGG et al., 1984). More than 95% of the heavy metals that are transported from land to sea occur in the form of particulate matter (MARTIN and WHITFIELD, 1983).

Adsorption of H^+ and OH^- ions is thus based on protonation and deprotonation of surface hydrolysis:

$$S\text{-}OH + H^+ \rightleftarrows S\text{-}OH_2^+$$

$$S\text{-}OH \;(+OH^-) \rightleftarrows S\text{-}O^- + H^+ \;(+H_2O) \;.$$

Deprotonated surface hydroxyls exhibit Lewis base behavior. Adsorption of metal ions is therefore understood as competitive complex formation involving one or two surface hydroxyl groups:

$$S\text{-}OH + M^{z+} \rightleftarrows S\text{-}OM^{(z-1)+} + H^+$$

$$2\;S\text{-}OH + M^{z+} \rightleftarrows (S\text{-}O)_2 M^{(z-2)+} + 2\;H^+ \;.$$

The main mechanism for anion adsorption is ligand exchange, again involving one or two surface hydroxyls:

$$S\text{-}OH + L \rightleftarrows S\text{-}L^+ + OH^-$$

$$2\;S\text{-}OH + L \rightleftarrows S_2 L^{2+} + 2\;OH^- \;.$$

3.2 The Structure of the Hydrous Oxide-Water Interface

There is now increasing support from spectroscopy (MOTSCHI, 1986) elucidating the structure of surface complexes with in situ extended X-ray adsorption, fine structure studies (EXAFS-measurements). HAYES et al. (1987) have documented that the selenite ion is bound directly to the goethite (α-FeOOH) surface (for absorption

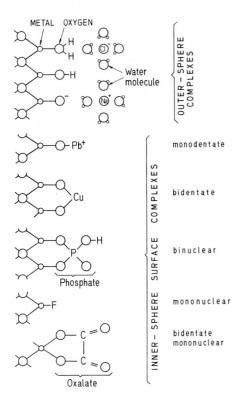

Fig. I.3-5. Surface complexation. – The surface of naturally occurring solids is characterized by functional groups, e.g., OH-groups on the surface of hydrous oxides or on organic surfaces (modified from Sposito, 1984). Specific adsorption of metal ions and of ligands (anions or weak acids) occurs through coordinative interactions. Two broad categories of surface complexes can be distinguished: inner-sphere complexes and outer-sphere complexes. Outer-sphere surface complexes involve electrostatic bonding mechanisms and are less stable than inner-sphere surface complexes which involve some covalent bonding.

isotherm see Fig. I.3-6) in a bidentate binuclear fashion with two iron atoms 338 pm from the selenium atom. The inner sphere nature of complexes of oxy-anions with Fe(III)(hydr)oxide surfaces was also indicated by SIGG and STUMM (1981) and more recently by BALISTRIERI and CHAO (1987) and GOLDBERG (1985) by exemplifying that the tendency to form surface complexes is correlated with the tendency to form (inner-sphere) complexes in solution:

$$>\text{FeOH} + H_2A \rightleftarrows\, >\text{FeHA} + H_2O$$

$$\text{FeOH}^{2+}(\text{aq}) + H_2A \rightleftarrows \text{FeHA}^{2+} + H_2O \;.$$

Studies on the kinetics of adsorption of metal ions on Al_2O_3 (see also Chapter II.1) have shown that the adsorption rate constants correlate significantly with the rate constants for the release of water molecules from the hydrated metal ions (STUMM

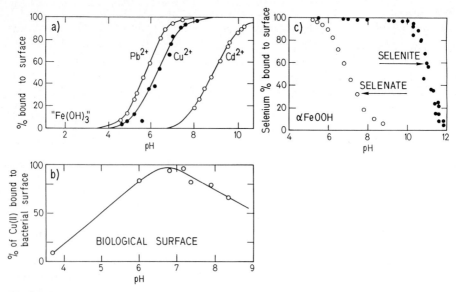

Fig. I.3-6. pH-Dependence of specific adsorption (surface complex formation) (a) of metal ions to oxide surfaces (from WANG and STUMM, 1987); (b) of Cu(II) to a biological surface (from GONÇALVES et al., 1987); (c) of oxyanions (selenite and selenate) to goethite (α-FeOOH) (from HAYES et al., 1987). This dependence is in accordance with mass law equations describing the surface coordination equilibria.

Experimental conditions: (a) 60 µg/L Fe(OH)$_3$, [M$_T$] = 10^{-5} mol L^{-1}, I = 10^{-2} mol L^{-1}; (b) 6 mg/L bacterial cells (*Klebsiella pneumoniae*) Cu(II)$_T$ = 3×10^{-6} mol L^{-1} (the decrease in adsorption at higher pH-values is due to cellular exudates in solution); (c) 30 g/L α-FeOOH, [Se$_T$] = 10^{-4} mol L^{-1}, I = 10^{-1} mol L^{-1}.

et al., 1987; HACHIYA et al., 1984). This observation is in accordance with the concept of formation of inner-sphere complexes.

Equilibrium Constants. The concept of surface complexation permits adsorption equilibria to be handled in the same way as equilibria in solution. The extent of metal binding surface coordination and its pH-dependence can be quantified by mass action equations (surface equilibrium constants) and can be explained by considering the affinity of the surface sites for metal ions or ligands and the pH-dependence of the activity of surface sites and ligands (Fig. I.3-6). The tendency to form surface complexes may be compared with that to form corresponding solute complexes (SCHINDLER and STUMM, 1987).

3.3 Redox Catalysis

(see also Chapters II.14, II.19, and II.34)

Oxidation of transition metal ions by oxygen can be catalyzed by particle surfaces. It was shown that adsorption of oxide surfaces can accelerate oxygenation of Fe(II),

Mn(II), and VO^{2+} in a similar way as hydrolysis, i.e., the specific adsorptive interaction of a metal ion with the surface hydroxo groups of a hydrous oxide surface has similar effects on oxygenation as hydroxo complex formation in solution (WEHRLI and STUMM, 1988). In the case of VO^{2+}, the complex formation with (hydr)oxide surfaces, e.g., VOOM<, enhances the reactivity of vanadyl towards oxidation by O_2 by more than three orders of magnitude and in a comparable way as complex formation with hydroxides in solution.

4 Cycling of Metals

The dispersion of metals into the atmosphere appears to rival, and sometimes exceed, natural mobilizations. Metals are released into the atmosphere, both as particles and as vapors, as a result not only of fossil fuel (coal, oil, natural gas) combustion but also of cement production and extractive metallurgy. Elements are termed atmophile when their mass transport through the atmosphere is greater than that in

Table I.3-4. Natural and Anthropogenic Sources of Atmospheric Emissions[a]

Element	Continental Dust Flux	Volcanic Dust Flux	Volcanic Gas Flux	Industrial Particulate Emissions	Fossil Fuel Flux	Total Emissions, Industrial Plus Fossil Fuel	Atmospheric Interference Factor (%)[b]
Al	356 500	132 750	8.4	40 000	32 000	72 000	15
Ti	23 000	12 000	–	3 600	1 600	5 200	15
Sm	32	9	–	7	5	12	29
Fe	190 000	87 750	3.7	75 000	32 000	107 000	39
Mn	4 250	1 800	2.1	3 000	160	3 160	52
Co	40	30	0.04	24	20	44	63
Cr	500	84	0.005	650	290	940	161
V	500	150	0.05	1 000	1 100	2 100	323
Ni	200	83	0.0009	600	380	980	346
Sn	50	2.4	0.005	400	30	430	821
Cu	100	93	0.012	2 200	430	2 630	1 363
Cd	2.5	0.4	0.001	40	15	55	1 897
Zn	250	108	0.14	7 000	1 400	8 400	2 346
As	25	3	0.1	620	160	780	2 786
Se	3	1	0.13	50	90	140	3 390
Sb	9.5	0.3	0.013	200	180	380	3 878
Mo	10	1.4	0.02	100	410	510	4 474
Ag	0.5	0.1	0.0006	40	10	50	8 333
Hg	0.3	0.1	0.001	50	60	110	27 500
Pb	50	8.7	0.012	16 000	4 300	20 300	34 583

[a] From LANTZY and MACKENZIE (1979). See original paper for assumptions and discussion of sources of data. All fluxes are in units of 10^8 g per year
[b] Atmospheric interference factor = [total emissions ÷ (continental + volcanic fluxes)] × 100

Table I.3-5. Perturbations of the Geochemical Cycles of Metals by Society. The Elements are Grouped According to the Scale for Which Such Perturbations Can Be Documented (ANDREAE et al., 1985)

Element	1 Scale of Perturbation			2 Most Diagnostic Environments	3 Mobility	4 Health Concern	5 Critical Pathway
	Global	Regional	Local				
Pb	+	+	+	A, Sd, I, W, H, So	v, a	+	F, A^c
V	+	+, c	+	A	d	(+)	A?
As	(+)	+	+	A, Sd, So, W	v, s, a	+	A, W
Sn	(+)	+	+	A, Sd, W	v, a	+^{a}	F
Zn	(−)	+	+	A, Sd, W, So	v, s	E	F
Cd	(−)	+	+	A, Sd, So, W	v, s	+	F
Hg	(−)	+	+	A, Sd, Fish, So	v, a	+^{a}	F, (A)
Sb	(−)	+	+	A, Sd	v, s	(+)	F, W, A?
Cu	(−)	+	+	A, Sd, W, So	v, s	E	F?
Ag	(−)	+	+	A, Sd, W	(v)	(+)	?
Se	(−)	(+)	+	A	v, s, a	E	F
Ge	?	+	+	A, So, W?	v, s, a	(+)^a	?
Ni	−	+	+	A, Sd	−	E	F, W, A?
Cr	−	+	+	A, Sd, W, Gw	s, v^b	E	W, F
B	−	(+)	+	A, Sd, Gw	v, s	E	W
K	−	(+)	+	A	s	E	F
Pt	?	?	+	A, Sd	s	(+)	?
Pd	?	?	+	Sd	s	(+)	?
Mo	?	?	+	A, W, So, Sd	s	E	F, W
Tl	?	?	+	Em, So	v, s	(+)	A, F?
In	?	?	+	A, So, Em	v	(+)	?
Bi	?	?	+	A, So, Em	v	(+)	?
Be	?	?	+	A, So, Em	−	(+)	A
Ga	?	?	+	Em	v	(+)	?
Te	?	?	(+)	So	v, a?	(+)	?
Mn	−	c, +	+	A, Sd, W	r	E	A
Fe	−	c	+	A, Sd, W	r	E	F, A, W
Al	−	c	+	A, Sd	−	(+)	W?
Si	−	c	+	A	−	(E)	−
Ti	−	c	+	A, Sd, So	−	−	−
Co	−	c	(+)	Sd	r	E	F?
Na	−	c, +	+	W, A, So	s	E	F, W
Mg, Ca	−	c	+	A, Sd, Em	(s)	E	
Ba	−	c	+	A, Sd, Em		(+)	F, W
U	−	c	+	A, So, Gw	s	(+)	A, W
Th	−	c	+	A, So, Gw	−	(+)	A
Zr	−	c	+	A	−	−	−
Y	−	c	(+)	Sd	−	−	−
Au	(−)	(−)	+	Sd	−	+	−
W	(−)	(−)	+	A	s?		−
Li, Rb, Cs	−	c	c	Sd	s	+	−
Rare Earths	−	c	c	Sd	−	−	−
Ta, Hf, Sc, Sr	−	c	c	Sd	−	−	−
Os, Ir, Ru, Rh	?	?	?	−	−	−	−

streams. Many atmophile elements are volatile and have metal oxides of relatively low boiling points. It is also known that some of these metals, especially the type-B metals, i.e., Hg, As, Se, Sn, and Pb, can be methylated and/or released into the atmosphere as vapors and that Hg and probably As and Se are released as inorganic vapor from the burning of coal. In contrast, the type-A metals, i.e., Al, Ti, Sn, Mn, Co, Cr, V, and Ni, are termed lithophile because their mass transport to the oceans by streams exceeds their transport through the atmosphere.

Computations of anthropogenic fluxes have been provided by BERTINE and GOLDBERG (1971) and by MACKENZIE and WOLLAST (1977). Table I.3-4 (from LANTZY and MACKENZIE, 1979) shows the magnitude of fluxes derived for 20 trace metals. The importance of the anthropogenic flux for any metal is given by the interference factor, IF, which is calculated as (total anthropogenic emissions ÷ total natural emissions) × 100. It can be seen from the table that, for the so-called atmophile elements Sn, Cu, Cd, Zn, As, Se, Sb, Mo, Ag, Hg, and Pb, IF values are considerably higher than 100%.

LANTZY and MACKENZIE, however, point out that further documentation is needed to interpret the data and to establish global balances for atmophile elements. Indeed, there are arguments inferring that for Hg, As, and Se, and perhaps other atmophile metals, there are significant fluxes from the sea surface to the atmosphere. If such fluxes are considered in the interference calculations for these metals, the factors may be reduced to levels of no interference. Another evaluation of metal transfer rates was made by SPOSITO (1986) on the basis of the data compiled by BUAT-MENARD (1985). Despite many uncertainties the conclusion emerges from SPOSITO's review (1986) that the trace metals Cu, Zn, Ag, Sb, Sn, Hg, and Pb are the most potentially hazardous on a global or regional scale. Lead is of acute concern on the global scale, because of its prominent showing in all the enrichment factors and transfer rates considered. These conclusions are in accordance with those of ANDREAE et al. (1984) in a recent Dahlem report (Table I.3-5). In this table, the geographic scale was defined "global" when the effect of perturbation can be demonstrated at least in large parts of the northern hemisphere.

Thus, the type-B metals, i.e., the soft Lewis acids, are not only enriched in the natural environment, but they are also, because of their toxic effect, i.e., their tenden-

---

(1) + significant perturbation; (+) possible perturbation; (−) enriched relative to crustal abundances, but the enrichment may not by anthropogenic; − no perturbation; ? not enough information; c enhanced due to mobilization of crustal materials (soil, dust).
(2) A air; Sd sediments (coastal, lake); So soils; I ice cores; W surface waters; Gw groundwaters; H humans; Em emission studies (only listed when little geochemical information is available).
(3) v volatile; s soluble; r soluble only under reducing conditions; a mobile as alkylated organometallic species; − not mobile.
(4) + toxic in excess; (+) toxic, but little data available; E essential, but toxic in excess.
(5) F food; W water; A air; − no significant exposure likely.

[a] organometallic forms only
[b] hexavalent form volatile and toxic, trivalent form essential
[c] exposure through hand-to-mouth activity is critical for lead in children
[d] enriched relative to crustal abundance from fuel oil combustion (vanadium porphyrins)

cy to react with soft bases, e.g., to SH- and NH-groups in enzymes, potentially hazardous to ecology and human health.

4.1 Technology

(see also Chapters I.2, I.7b, and Part II)

The major challenge in pollution control, in a broad perspective, must be to devise improved *chemical* processes to intercept the waste products of mining, agriculture, fossil fuel, and solid waste combustion and manufacture which are now mobilizing excessive flows of potentially toxic "atmophile" and "lithophile" metals into the environment and seriously disturb the evolved delicate balance within ecosystems and in human metabolism.

Chemical technology must find more sophisticated ways to exploit those very same principles of *affinity* and *reactivity* in coordination chemistry and redox chemistry − coupled with phase separations and other physical-chemical processes − that regulate metals in rocks, the seas, and the biota. In other words, our technology must be based on the same fundamental chemical patterns for metals that we find all around us in the natural world, properly adapted and optimized for the physical and chemical conditions and time scales of our control processes (MORGAN, 1986).

5 Regulating the Metal Ion Concentrations in Oceans and Lakes

As we have seen, knowledge concerning metal distribution among the different chemical forms is critical. In natural waters those cations (Ca^{2+}, Mg^{2+}, Na^+) present at relatively high concentrations ($>10^{-4}$ mol L^{-1}) are typically type-A cations; those present at trace concentrations, i.e., the minor or trace metals, are mostly type-B metals or transition cations (Fig. I.3-6). The ligands typically encountered in fresh waters are also given in Fig. I.3-7. Some of the organic ligands are present as humic and fulvic acids. They are able to form metal complexes having different molecular scale bonding characteristics (BUFFLE, 1988).

5.1 Equilibrium Models in Natural Waters

In order to understand the chemical properties of aquatic systems one would wish to characterize different systems in terms of their dominant variables, major ions, oxidation-reduction status, acid-base components, minor ions, complexing components, and adsorbing surfaces. A systematic approach − from the abundant ions and their "simple" interactions with the rest of the system to the less abundant ions and their complex interactions with the rest of the system − should prove valuable in defining the relative importance of different variables in determining the stability

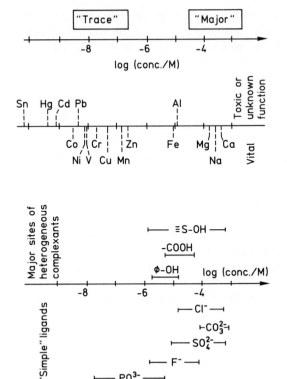

Fig. I.3-7. Average values of total concentration of metals and ranges of concentrations of ligands or complexing sites in natural fresh waters. ≡S-OH, inorganic solid surface sites. −COOH and ⌀-OH, total concentration of −COOH and phenolic sites of natural organic matter, respectively. Adapted from BUFFLE and ALTMANN (1987).

of aquatic chemical systems. From the computation of chemical equilibrium models, it is in principle possible to decide, for example, in what complex form a given metal ion is likely to be found in a body of water containing a certain set of reacting ligands and surfaces and competing metals. A particularly convenient computer method that permits one to evaluate multimetal − multiligand models − including heterogeneous equilibria such as surface complex formation, has been developed by WESTALL (1979). One has to be aware, however, that some of the heterogeneous complexing agents, especially the humic acids, possess complex qualities which impose constraints on the type of information to be gained from simplified models.

General conclusions from such computer models which attempt to simulate in a simplified way natural waters give the following tentative conclusions. At the concentrations of complex-forming organic ligands typically encountered in open surface waters, the complex-forming donor groups are bound mostly to Ca and Mg. Thus, their tendency to interact with trace metals is somewhat mitigated. For Cu(II), however, − which is somewhat intermediate between type-A and type-B behavior −

organic ligands, even at very small concentrations, are able to compete successfully with inorganic ligands for the available coordinative sites of the Cu ion. Because algal growth is strongly dependent upon free Cu^{2+}, organic matter, either incipiently present in the water or formed from exudates of algae, may indeed exert a pronounced influence on the physiological response of algae by regulating most sensitively free Cu. Natural organic matter, i.e., humic and fulvic acids, should not be considered simply as detrital metabolic waste but rather as a basic structure of nature that is essential for the maintenance of life. As has been shown by BUFFLE and ALTMANN (1987) these natural complexants are buffers for Cu(II) and some other metal ions regulating pCu and pM over wide ranges of pH and concentrations.

5.2 The Role of Particles in Regulating the Metal Ion Concentration in Oceans and Lakes

Settling particles, especially bioorganic particles, play a dominant role in binding heavy metals and transferring them to the deeper portion of oceans and lakes where they are partially mineralized and transformed into the sediments. The particle cycle in oceans and lakes is primarily driven by the photosynthetic production of plank-

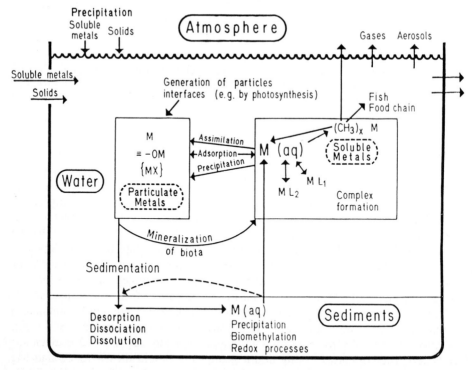

Fig. I.3-8. The residual metal ion concentrations in oceans and lakes are regulated by the adsorption of metal ions to the settling particles and by the tendency of ligands to form soluble complexes.

ton. Biological surfaces are especially efficient scavengers – probably better ones than the mineral surfaces – for heavy metals (MOREL and HUDSON, 1985; SIGG, 1987; GONÇALVES et al., 1987; XUE et al., 1988). Autochthonous particles (clays, aluminum silicates, humus) may occasionally also be of some importance in lakes.

The partitioning of metals between the particles and the water is influenced on one hand by the affinity of the metal for the particle surface, on the other hand by the chemical speciation of the metal in solution; i.e., its affinity for the solute ligands (Fig. I.3-8). In lakes and the ocean, the effects of particles on the regulation of metal ions are very pronounced, because the continuously settling particles act as a conveyor belt in transporting reactive elements. Indeed, the partitioning of metals and other reactive elements between particles and water is the key parameter in establishing the residence times and the residual concentrations of these elements in the ocean and in lakes (WHITFIELD and TURNER, 1987; MOREL and HUDSON, 1985; SIGG et al., 1987).

Lakes, despite being polluted with metal ions ten to hundred times as much as oceans from riverine and atmospheric inputs, are often nearly as much depleted of these trace metals as are the oceans. Therefore, the elimination mechanisms in lakes must be more efficient than those in oceans. Larger productivities and higher sedimentation rates for particles are primarily responsible for the more efficient scavenging in lakes through adsorption of metals on phytoplankton and to a lesser extent by other particles. The input of phosphorus into a lake, influencing the production of biogenic particles, is a major factor controlling the sedimentation rate of biogenic particles and in removing "biophile" heavy metals (SIGG et al., 1987).

In Fig. I.3-9 the depth profile of the Pb-concentrations of the Central Pacific is compared with that of Lake Constance. In either case, the Pb concentrations of the

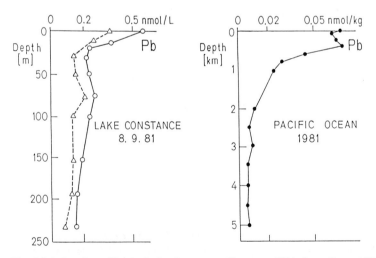

Fig. I.3-9. Lead profiles in Lake Constance (Summer 1981 data: SIGG, 1985) and in the Pacific Ocean (1981 data: SCHAULE and PATTERSON, 1981). The similar shape of these profiles, despite the difference in length scales (kilometers for the ocean and meters for the lake), illustrates the influence of the atmospheric deposition on the upper layers and the scavenging of Pb(II) by the settling particles. Modified from SIGG (1985).

surface waters are higher than in the deep water; thus, atmospheric transport plays in both cases a significant role in supplying Pb to the surface water. The decrease in the concentration of Pb with depth occurs by particles that scavenge Pb(II) most efficiently. PATTERSON (e.g., SETTLE and PATTERSON, 1980) has used data on the memory record of sediments to compare prehistoric and present-day eolian inputs. These data suggest that the present Pb(II) input is two orders of magnitude larger than that of prehistoric time. What does this mean with regard to the ecology? We really do not know.

Other potential scavenging and metal regeneration cycles operate near the sediment-water interface. Subsequent to early epidiagenesis in the partially anoxic sediments, iron(II) and manganese(II), and other elements depending on redox conditions, are released by diffusion from the sediments to the overlying water, where

Element type	Mean oceanic residence time (τ_i)	Concentration[a] range (mol L^{-1})	$\dfrac{[X]_{\text{deep Atlantic}}}{[X]_{\text{deep Pacific}}}$	Profile type
Accumulated	$> 10^5$	$10^{-8} - 10^{-1}$	1	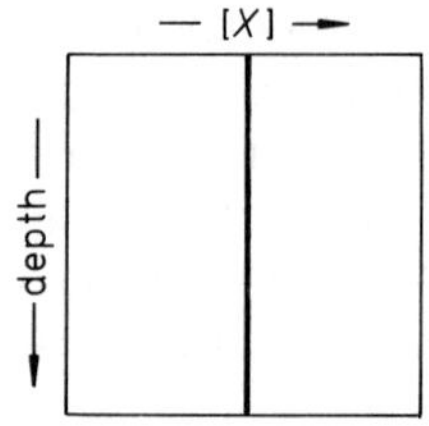 (vertical line profile)
Recycled	$10^3 - 10^5$	$10^{-11} - 10^{-5}$	<1	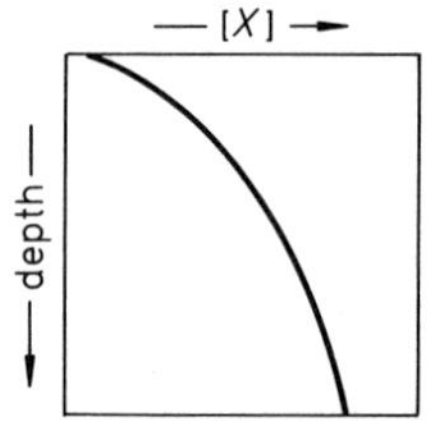 (nutrient-like profile)
Scavenged	$< 10^3$	$10^{-14} - 10^{-11}$	>1	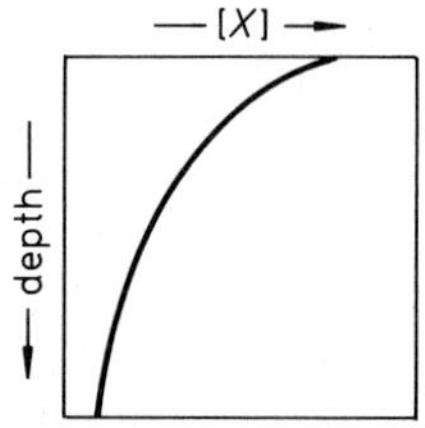 (scavenged profile)

Fig. I.3-10. Schematic deep ocean profiles for elements. This figure is based on a classification of elements according to their oceanic profiles given by WHITFIELD and TURNER (1987). Uptake of some of the elements, especially the recycled ones, occurs somewhat analogously as that of nutrients. There are some elements such as Cd that are non-essential but may be taken up (perhaps because they mimick essential elements) the same way as nutrients.

[a] These ranges show significant overlap, since the concentrations of the elements also depend on crustal abundance.

iron and manganese are oxidized to insoluble iron(III) and manganese(III, IV) oxides. These oxides are also important conveyors of heavy metals near the sediment surface.

WHITFIELD (1979) and WHITFIELD and TURNER (1987) have shown that the elements in the ocean can be classified according to their oceanic residence times, τ_i:

$$\tau_i = \frac{\text{total number of moles of i in ocean}}{\text{rate of addition or removal}}$$

which are, in turn, a measure of the intensity of their particle-water interaction. Thus the elements that show the strongest interactions with the particulate phase have very short residence times; those elements that interact little with particles are characterized by long residence times (Fig. I.3-10).

6 Bioavailability and Toxicity

(see also Chapters I.7, I.13, I.15, and Part II)

Metals are partitioned in the biota by different acid-base affinities; by their kinetics; by spatial partitioning, e.g., by membranes and compartments, and by temporal partitioning (WILLIAMS, 1981). One aspect of metal toxicity, but by no means the only one, is the chemical combination of metals and ligands (Lewis acids and bases) in organisms. The cellular bases are almost exclusively sulfur, nitrogen, and oxygen donor groups (including H_2O and solute bases, e.g., HCO_3^-, HPO_4^{2-}, OH^-, etc.). The acids are H^+, the essential metal cations (e.g., Na, K, Mg, Ca, Cr, Mn, Fe, Co, Ni, Cu, Zn, Mo), potentially hazardous metals, if present (e.g., Hg^{2+}, CH_3Hg^+, Pb^{2+}, Cd^{2+}, Cr). Among the sulfur-seeking ("soft-soft") type-B metals are, of course, the toxicants Hg, Pb, Tl, as well as the essential protein and enzyme metals Fe, Cu, and Zn. The proton, H^+, has high affinity for all donors, S, N, and O, thus the key role of pH in metal binding in organisms. Mg, Ca, Be, Al, Sn, Ge, and the lanthanides are oxygen-seeking ("hard-hard").

6.1 The Relevance of Free Metal Ion Activity

The chemical behavior of any element in the environment depends on the nature of its compounds and species. Physiological, ecological, and toxicological effects of a metal are usually strongly structure-specific, i.e., they depend on the species. Thus, for example, the effect of Cu on the growth of algae depends on whether Cu(II) is present as free Cu^{2+} ion or is present as a carbonato- or an organic complex. This has been exemplified in some studies on the relationship of Cu^{2+} ion activity and the toxicity of copper to phytoplankton. In these experiments cupric ion activity was altered independently of total Cu(II) concentration by varying the complex-forming concentration and the pH; i.e., by using pCu buffers (SUNDA and GUILLARD, 1976; JACKSON and MORGAN, 1978; MOREL, 1983; MOREL and HUDSON, 1985).

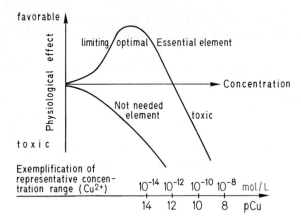

Fig. I.3-11. Relation between the concentration of an element and its physiological effect. In order to give an idea on the "reactive" concentration range for the interaction with algae, a plausible scale for pCu (= $-\log [Cu]$) is given. Note, however, that the concentration range depends on the type of algae and on the presence of other (competing) metal ions and differs for different metal ions.

Our discussion concentrates primarily on single cell organisms. They are special in the sense that they are, solely separated by a membrane, directly in contact with the aqueous medium.

Fig. I.3-11 illustrates in a simplified way the physiological dependence (growth or biomass production) on the concentration of an essential element. The lowest concentration range limits growth; in the next concentration range an optimum in growth is obtained; and at higher concentrations toxic effects (reduction in growth) are observed. As this figure shows, non-essential elements (such as Cd, Pb) cause a negative physiological effect with increasing concentration.

To demonstrate that the free ion activity of a metal is a key parameter which determines its biological effects, it is necessary to measure the same physiological responses for the same free metal ion activities, obtained by several combinations of chelating agents and ligand concentrations (MOREL, 1983). For example, Fig. I.3-12 shows how different combinations of complex formers and total metals in algal culture media result in a unique physiological response to the activity of Cu^{2+}. Thus, in designing experiments to study the biological effects of trace metals one needs first to calculate the trace metal speciation in the medium. This often involves equilibrium calculations for multi-metal multi-ligand systems.

These concepts can, with certain limitations, be also extended to higher organisms. ANDREWS et al. (1977) have shown that the toxicity of Cd to grass shrimp is related to the free Cd^{2+} ion activity. ANDREWS et al. have demonstrated that the toxicity of Cu(II) to a freshwater crustacean is directly connected with free cupric ion activity, but not with the activities of Cu-complexes or particulate forms of copper.

SANDERS et al. (1983) have also shown that the effects of Cu(II) on the growth of crab larvae and on their metallothionein with copper chelate buffer systems must be interpreted on the basis of free Cu^{2+} ion activity. The data obtained reveal

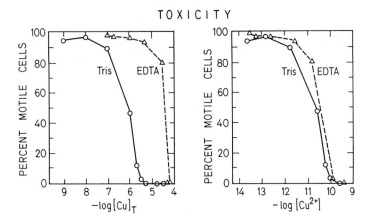

Fig. I.3-12. Experiments with different chelators and a wide range of trace metal concentrations demonstrate that trace metal toxicity and deficiency are determined by metal ion activities and not total concentrations. Motility data of the dinoflagellate *Gonyaulax tamarensis* as a function of total copper $[Cu]_T$ and cupric ion activity $[Cu^{2+}]$ for two chelators, Tris and EDTA. (After ANDERSON and MOREL, 1978).

predictable relations between $[Cu^{2+}]$ in seawater and processes at the cellular and organismic levels. Similarly, the uptake of metal ions by plants, e.g., of aluminum, is usually related to free metal ion activity. Others have shown that the chelation of a variety of metals reduces the toxicity of metals to organisms, e.g., a reduction in the uptake of mercury by fish in the presence of EDTA and cysteine (PAGENKOPF, 1986) a reduction in copper and/or zinc toxicity to fish by NTA, glycine, and humic acids, and a reduction in copper toxicity to clams by EDTA.

As SANDERS et al. (1983) pointed out, the dependencies of metal toxicity or availability to aquatic organisms on free metal ion concentrations in solution may in fact be a rather widespread phenomenon. The existence of such dependencies should allow one to predict changes in the response of an organism to a particular metal through knowledge of the variations in the aqueous chemistry of the metal. Variables such as the total concentrations of the metal in question, pH, alkalinity, the concentration of natural chelators, the concentration of competing metals, and the presence of adsorptive surfaces all can affect the concentration of free metal ion and thus affect the response of an organism to that metal.

6.2 Metal Uptake by Algae

The uptake of metal ions by phytoplankton appears in most cases to be a two-stage process involving the binding of a relatively large pool of metal on the outside surface of the cell either by biologically released ligands or by surface functional ligand groups. Subsequent to this surface complex formation, the metals are carried through the membrane – usually by porter molecules – to the inside of the cell. The uptake model depicted here (Fig. I.3-13a) is much influenced by R. J. P.

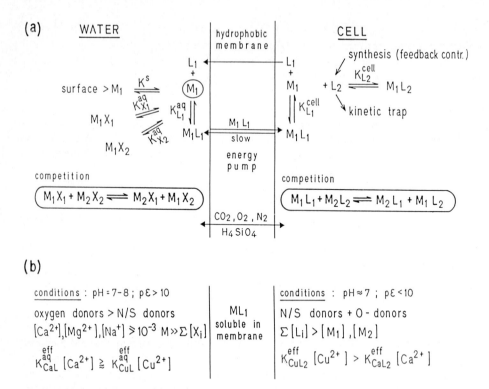

Fig. I.3-13. Schematic model of metal ion uptake through a membrane of a phytoplankton cell. (a) The metal ion is bound to the outside surface of the cell either by biologically released ligands or by surface functional ligand groups subsequently to the surface complex formation. The metals are carried – usually by portermolecules – to the inside of the cell. If the transport into the cell is slow in comparison to the preequilibration process on the solution side, then the uptake of the metal ion of the cell depends on the free metal ion activity.
(b) Solution variables outside and inside the cell.

WILLIAMS' (WILLIAMS and HALE, 1966; WILLIAMS, 1981) research on structural aspects of metal toxicity and on the discussions by MOREL and HUDSON (1985), WOOD and WANG (1985), and SIGG (1987). On the water side, an equilibrium exists between the various complexes of solutes and surface ligands with the free metal ion. A competition between the different ligands for H^+ and different metal ions establishes multidimensional equilibria. Since these equilibria are interdependent, no element is free from the label "toxic" at certain dose levels (WILLIAMS, 1981). The kinetic aspects include the rates of the ligand exchange reactions, the rate of reaction and release of the carrier ligand L_1 and the rate of transport of ML_1 across the membrane (usually by active transport). If the transport into the cell is slow in comparison to the pre-equilibration processes on the solution side, then the uptake of the metal ion by the cell depends on the free metal ion concentration. The production and release of L_1, L_2... are related to the growth rate of the algae. The selectivity of certain metal ions is given by the selectivity of the ligand L_1. Carrier molecules are often proteins, and the stability of their complexes with different metal ions corresponds to the Irwing-Williams stability order. Steric factors are also involved in the selectivity. The concentration inside the cell depends on the availability of the ligand L_1, on the rate of transport across the membrane, and possibly on the rate of transport out of the cell.

The central importance of free metal ion activities in controlling the biological effects does not necessarily mean that the free aquo metal ions are actually the chemical species taken up by aquatic organisms. It reflects the fact that the chemical reactivity of a metal is measured by the free metal ion activity and that the physiological effects of a metal are mediated by chemical reactions between the metal and the various cellular ligands (MOREL, 1983).

The metal ions in the cell are used in biochemical processes or become trapped in inactive forms, e.g., as complexes of metallothionein, as a detoxification mechanism. Toxic effects are observed in algae when the cellular concentration of toxic metal ions reaches some critical level approaching the minimum cellular concentration of essential trace metals. The system becomes "overflooded" with toxic metal ions that then *react with critical enzymes.*

In Fig. I.3-13b the solution variables outside and inside the cell (pH, redox potential, pε, and type of ligands and cations) are compared. In the surrounding water oxygen donors (hard bases such as OH^-, HCO_3^-, carboxylates) prevail over soft bases (NH_3, HS^-). The hard earth alkali and alkali cations typically exceed manifold the concentrations of trace metals and ligands; thus, the tendency to form Ca- and Mg-complexes exceeds the tendency to form trace metal complexes. On the other hand, on the inside of the cell reducing conditions prevail, and, therefore, the concentrations of soft bases (NH_3, HS^-) become preponderant in addition to oxygen donor ligands. The concentrations of the ligands may now exceed the concentrations of the trace metals, and the tendency to form complexes with Cu^{2+} exceeds that of Ca^{2+}.

Metal-Carbon-Compounds. Methylation (or alkylation) of metal ions may increase the bioavailability of non-essential elements. For example, the methylation or alkylation of Hg(II) results in an increase of toxicity. Hg-alkyl compounds – typically methylmercury is present in seawater as the non-charged species CH_3HgCl and in freshwater as CH_3HgOH – are taken up faster and permeate the membranes more easily than inorganic Hg(II), because they are more lipid soluble. On the other hand, organo-arsenic species are less toxic in the food chain and for man than inorganic arsenics.

References

Ahrland, S., Chatt, J., and Davies, N.R. (1958), *The Relative Affinities of Ligand Atoms for Acceptor Molecules and Ions. Q. Rev. Chem. Soc. 12,* 265.
Anderson, D.A., and Morel, F.M.M. (1978), *Copper Sensitivity of Gonyaulax tamarensis. Limnol. Oceanogr. 23,* 283–295.

Andreae, M. O. (1986), *Chemical Species in Seawater and Marine Particulates,* in: Bernhard, M., Brinkman, F. E., and Sadler, P. J. (eds.): *The Importance of Chemical "Speciation" in Environmental Processes,* Dahlem Konferenzen, pp. 301–335. Springer Verlag, Berlin.

Andreae, M. O., et al. (1985), *Changing Geochemical Cycles,* in: Nriagu, J. O. (ed.): *Changing Metal Cycles and Human Health,* Dahlem Konferenzen, pp. 359–373. Springer Verlag, Berlin.

Andrew, R. W., Biesinger, K. E., and Glass, G. E. (1977), *Effects of Inorganic Complexing on the Toxicity of Copper to Daphnia magna. Water Res. 11,* 309–315.

Baes, C. F., and Mesmer, R. E. (1976), *The Hydrolysis of Cations.* Wiley-Interscience, New York.

Balistrieri, L. S., and Chao, T. T. (1987), *Selenium Adsorption by Goethite. Soil Sci. Soc. Am. J. 51,* 1145–1151.

Bertine, K. K., and Goldberg, E. D. (1971), *Fossil Fuel Combustion and the Major Sedimentary Cycles. Science 173,* 233.

Buat-Menard, P. E. (1985), *Fluxes of Metals Through the Atmosphere and Oceans,* in: Nriagu, J. O. (ed.): *Changing Metal Cycles and Human Health,* Dahlem Konferenzen, pp. 43–69. Springer Verlag, Berlin.

Buffle, J. (1988), *Complexation Reactions in Aquatic Systems. An Analytical Approach.* Ellis Horwood, Chichester, UK.

Buffle, J., and Altmann, R. S. (1987), *Interpretation of Metal Complexation by Heterogeneous Complexants,* in: Stumm, W. (ed.): *Aquatic Surface Chemistry,* pp. 351–383. Wiley-Interscience, New York.

Buffle, J., Tessier, A., and Haerdi, W. (1984), *Interpretation of Trace Metal Complexation by Aquatic Organic Matter,* in: Kramer, C. J. M., and Duinker, J. C. (eds.): *Complexation of Trace Metals in Natural Waters.* M. Nijhoff/Dr. W. Junk Publishers, The Hague.

Goldberg, S. (1985), *Chemical Modelling of Anion Competition on Goethite Using the Constant Capacitance Model. Soil Sci. Soc. Am. J. 49,* 851–856.

Gonçalves, M. L. S., Sigg, L., and Stumm, W. (1985), *Voltammetric Methods for Distinguishing Between Dissolved and Particulate Metal Ion Concentrations in the Presence of Hydrous Oxides; A Case Study on Lead(II). Environ. Sci. Technol. 19.,* 141–146.

Gonçalves, M. L. S., Sigg, L., Reutlinger, M., and Stumm, W. (1987), *Metal Ion Binding by Biological Surfaces; Voltammetric Assessment in Presence of Bacteria. W. Nürnberg Spec. Issue, Sci. Total Environ. 60,* 105–119.

Hachiya, K., Sasaki, M., Ikeda, T., Mikami, N., and Yasunaga, T. (1984), *Static and Kinetic Studies of Adsorption Desorption of Metal Ions on a γ-Al_2O_3 Surface. J. Phys. Chem. 88,* 27.

Hayes, K. F., Lawrence Roe, A., Brown, G. E., Jr., Hodgson, K. O., Leckie, J. O., and Parks, G. A. (1987), *In situ X-Ray Absorption Study of Surface Complexes: Selenium Oxyanions on α-FeOOH. Science 238,* 783–786.

Imboden, D. M., and Schwarzenbach, R. P. (1985), *Spatial and Temporal Distribution of Chemical Substances in Lakes: Modelling Concepts,* in: Stumm, W. (ed.): *Chemical Processes in Lakes,* pp. 1–30. Wiley-Interscience, New York.

Jackson, G. A., and Morgan, J. J. (1978), *Trace Metal-Chelator Interactions and Phytoplankton Growth in Seawater Media: Theoretical Analysis and Comparison with Reported Observations. Limnol. Oceanogr. 23,* 268.

Jacob, D. J., and Hoffmann, M. R. (1983), *A Dynamic Model for the Production of H^+, NO_3^-, and SO_4^{2-} in Urban Fog. J. Geophys. Res. 88,* 6611.

Lantzy, R. J., and Mackenzie, F. T. (1979), *Atmospheric Trace Metals: Global Cycle and Assessment of Man's Impact. Geochim. Cosmochim. Acta 43,* 511.

Mackenzie, F. T., and Wollast, R. (1977), in: Goldberg, E. D. (ed.): *The Sea,* Vol. 6. Wiley-Interscience, New York.

Martin, J.-M., and Meybeck, M. (1979), *Elemental Mass-balance of Material Carried by Major World Rivers. Mar. Chem. 7,* 173–206.

Martin, J.-M., and Whitfield, M. (1983), *The Significance of the River Input of Chemical Elements to the Ocean,* in: Wong, C. S., Boyle, E., Bruland, K. W., Burton, J. D., Goldberg, E. D. (eds.): *Trace Metals in Seawater.* Plenum Press, New York.

Morel, F. M. M. (1983), *Principles of Aquatic Chemistry*. Wiley-Interscience, New York.
Morel, F. M. M., and Hudson, R. J. M. (1985), *The Geobiological Cycle of Trace Elements in Aquatic Systems: Redfield Revisited,* in: Stumm, W. (ed.): *Chemical Processes in Lakes,* pp. 251–281. Wiley-Interscience, New York.
Morgan, J. J. (1986), *Proceedings International Symposium on Metal Speciation, Separation and Recovery, Chicago,* July 1986.
Motschi, H. (1987), *Aspects of the Molecular Structure in Surface Complexes; Spectroscopic Investigations,* in: Stumm, W. (ed.): *Aquatic Surface Chemistry,* pp. 111–125. Wiley-Interscience, New York.
Pagenkopf, G. K. (1986), *Metal Ion Speciation and Toxicity in Aquatic Systems,* in: Sigel, H. (ed.): *Metal Ions in Biological Systems.* M. Dekker, New York.
Pearson, R. G. (1963), *Hard and Soft Acids and Bases. J. Am. Chem. Soc. 85,* 3533.
Sanders, B. M., Jenkins, K. D., Sunda, W. G., and Costlow, J. D. (1983), *Free Cupric Ion Activity in Seawater: Effects on Metalllothionein and Growth in Crab Larvae. Science 222,* 53–55.
Schaule, B. K., and Patterson, C. C. (1981), *Lead Concentrations in the Northeast Pacific: Evidence for Global Anthropogenic Perturbations. Earth Planet. Sci. Lett. 54,* 97.
Schindler, P. W., and Stumm, W. (1987), *The Surface Chemistry of Oxides, Hydroxides and Oxide Minerals,* in: Stumm, W. (ed.): *Aquatic Surface Chemistry,* pp. 83–110 . Wiley-Interscience, New York.
Schneider, W. (1968), *Einführung in die Koordinations-Chemie,* Springer Verlag, Berlin.
Schwarzenbach, G. (1961), *The General, Selective and Specific Formation of Complexes by Metal Cations. Adv. Inorg. Chem. Radiochem. 3,* 257.
Schwarzenbach, G. (1970), *Electrostatic and Non-electrostatic Contributions to Ion Association in Solution. Pure Appl. Chem. 24,* 307–334.
Settle, D. M., and Patterson, C. C. (1980), *Lead in Albacore: Guide to Lead Pollution in Americans. Science 207,* 1167–1176.
Sigg, L. (1985), *Metal Transfer Mechanisms in Lakes; The Role of Settling Particles,* in: Stumm, W. (ed.): *Chemical Processes in Lakes,* pp. 283–307. Wiley-Interscience, New York.
Sigg, L. (1987), *Surface Chemical Aspects of the Distribution and Fate of Metal Ions in Lakes,* in: Stumm, W. (ed.): *Aquatic Surface Chemistry,* pp. 319–350. Wiley-Interscience, New York.
Sigg, L., and Stumm, W. (1981), *The Interactions of Anions and Weak Acids with the Hydrous Goethite (α-FeOOH) Surface. Colloids Surf. 2,* 101–117.
Sigg, L., Stumm, W., and Zinder, B. (1984), *Chemical Processes at the Particle Water Interface,* in: Kramer, C. J. M., and Duinker, J. C. (eds.): *Complexation of Trace Metals in Natural Waters.* M. Nijhoff, Dr. W. Junk Publishers, The Hague.
Sigg, L., Sturm, M., and Kistler, D. (1987), *Vertical Transport of Heavy Metals by Settling Particles in Lake Zurich. Limnol. Oceanogr. 32*(1), 112–130.
Sposito, G. (1983), *On the Surface Complexation Model of the Oxide-Aqueous Solution Interface. J. Colloid Interf. Sci. 91,* 329–340.
Sposito, G. (1984), *The Surface Chemistry of Soils.* Oxford University Press, New York.
Sposito, G. (1986), *The Distribution of Potentially Hazardous Trace Metals,* in: Sigel, H. (ed.): *Metal Ions in Biological Systems.* M. Dekker Inc., New York.
Stumm, W., and Morgan, J. J. (1981), *Aquatic Chemistry,* 2nd Ed. Wiley-Interscience, New York.
Stumm, W., Wehrli, B., and Wieland, E. (1987), *Surface Complexation and its Impact on Geochemical Kinetics. Croat. Chim. Acta 60,* 429–456.
Sunda, W. G., and Guillard, R. R. L. (1976), *The Relationship between Cupric Ion Activity and the Toxicity of Copper to Phytoplankton. J. Mar. Res. 34,* 511.
Sutin, N. (1966), *Annu. Rev. Phys. Chem. 17,* 119–172.
Turekian, K. K. (1977), *The Fate of Metals in the Oceans. Geochim. Cosmochim. Acta 41,* 1139.
Turner, D. R., Whitfield, M., and Dickson, A. G. (1981), *The Euqilibrium Speciation of Dissolved Components in Freshwater and Seawater at 25 °C and 1 atm Pressure. Geochim. Cosmochim. Acta 45,* 855.

Varney, M. S., Turner, D. R., Whitfield, M., and Mantoura, R. F. C. (1984), in: Kramer, C. J. M., and Duinker, J. C. (eds.): *Complexation of Trace Metals in Natural Waters.* M. Nijhoff/Dr. W. Junk Publishers, The Hague.

Wang, Z.-J., and Stumm, W. (1987), *Heavy Metal Complexation by Surfaces and Humic Acids: A Brief Discourse on Assessment by Acidimetric Titration.* Neth. J. Agric. Sci. 35, 231–240.

Wehrli, B., and Stumm, W. (1988), *Oxygenation of Vanadyl (IV): Effect of Coordinated Surface-Hydroxyl Groups on OH^-.* Langmuir 4, 753–758.

Westall, J. (1979) *MICROQL: I. A Chemical Equilibrium Program in BASIC. II. Computation of Adsorption Equilibria in BASIC.* EAWAG, Duebendorf, Switzerland.

Whitfield, M. (1979), *The Mean Oceanic Residence Time (Mort) Concept – A Rationalisation.* Mar. Chem. 8, 101–123.

Whitfield, M., and Turner, D. R. (1987), *The Role of Particles in Regulating the Composition of Natural Waters,* in: Stumm, W. (ed.): *Aquatic Surface Chemistry,* pp. 457–493. Wiley-Interscience, New York.

Williams, R. J. P. (1981), *Physico-chemical Aspects of Inorganic Element Transfer through Membranes.* Philos. Trans. R. Soc. London B 294, 57–74.

Williams, R. J. P., and Hale, J. D. (1966), *Classification of Acceptors and Donors in Inorganic Reactions.* Struct. Bond. 1, 249.

Wood, J. M., and Wang, H. K. (1985), *Strategies for Microbial Resistance to Heavy Metals,* in: Stumm, W. (ed.): *Chemical Processes in Lakes,* pp. 81–98. Wiley-Interscience, New York.

Xue, H., Sigg, L., and Stumm, W. (1988), *The Binding of Heavy Metals to Algae Surface.* Water Res. 22 (7), 917–926.

Literature Recommended by the Editors for Further Reading

Baccini, P. (1989), *The Control of Heavy Metal Fluxes from the Anthroposphere to the Environment.* Proceedings International Conference Heavy Metals in the Environment, Geneva, Vol. 1, pp. 13–23. CEP Consultants Ltd., Edinburgh.

Bro-Rasmussen, F. (1988), *Proceedings 1st European SECOTOX Conference on Ecotoxicology, Copenhagen,* in press.

Bruland, K. W. (1983), *Trace Elements in Sea-water.* Chem. Oceanogr. 8, 157–220.

CEP Consultants Ltd., Edinburgh.
  Proceedings (1977) First International Conference on Heavy Metals in the Environment, Toronto.
  Proceedings (1979) Second International Conference on Heavy Metals in the Environment, London.
  Proceedings (1981) Third International Conference on Heavy Metals in the Environment, Amsterdam.
  Proceedings (1983) Fourth International Conference on Heavy Metals in the Environment, Heidelberg.
  Proceedings (1985) Fifth International Conference on Heavy Metals in the Environment, Athens.
  Proceedings (1987) Sixth International Conference on Heavy Metals in the Environment, New Orleans.
  Proceedings (1989) Seventh International Conference on Heavy Metals in the Environment, Geneva.

Hutchinson, T. C., and Meema, K. M. (1987), *SCOPE 31: Lead, Mercury, Cadmium and Arsenic in the Environment.* Wiley & Sons, Chichester-New York-Brisbane-Toronto-Singapore.

Hutzinger, O. (1980–1985), *The Handbook of Environmental Chemistry,* Volume 1, *The Natural Environment and the Biogeochemical Cycles*, Parts A, B, C, and D. Springer Verlag, Berlin-Heidelberg.
Hutzinger, O. (1980–1988), *The Handbook of Environmental Chemistry,* Volume 2, *Reactions and Processes*, Parts A, B, C, and D. Springer Verlag, Berlin-Heidelberg.
Kramer, J.R., and Allen, H.E. (1988), *Metal Speciation I (Theory, Analysis and Application).* Lewis Publishers, Chelsea, Michigan.
Merian, E., Frei, R.W., Härdi, W., and Schlatter, C. (1985), *Carcinogenic and Mutagenic Metal Compounds 1 (Speciation, Interrelation between Chemistry and Biology). Proceedings of an IAEAC Workshop in Geneva.* Gordon & Breach Science Publishers, New York.
Merian, E., Frei, R.W., Härdi, W., and Schlatter, C. (1988), *Carcinogenic and Mutagenic Metal Compounds 2 (Speciation, Interrelation between Chemistry and Biology). Proceedings of an IAEAC Workshop in Villars.* Gordon & Breach Science Publishers, New York.
Merian, E., Bronzetti, G., and Härdi, W. (1989), *Carcinogenic and Mutagenic Metal Compounds 3 (Speciation, Interrelation between Chemistry and Biology). Proceedings of an IAEAC Workshop in Follonica.* Gordon & Breach Science Publishers, New York, in press.
Rondia, D. (1986), *Belgian Research on Metal Cycling in the Environment, Proceedings.* Presses Universitaires Liège.
Salomons, W., Bayne, B.L., Duutsma, E.K., and Förstner, U. (1988), *Pollution of the North Sea.* Springer Verlag, Berlin-Heidelberg.
Salomons, W., and Förstner, U. (1984), *Metals in the Hydrocycle.* Springer Verlag, Berlin-Heidelberg.

I.4a Analytical Chemistry of Metals and Metal Compounds

MARKUS STOEPPLER, Jülich, Federal Republic of Germany

This chapter presents a systematic overview on general analytical aspects being the prerequisites for a quantitative environmental chemistry of metals and metalloids. The chapter contains the following sections:

1 Introduction
2 Analytical Tasks
3 Planning
4 Sampling and Sample Storage
   4.1 Biological and Environmental Materials
   4.2 Aqueous Matrices
5 Sample Preparation
   5.1 Biological and Environmental Materials
   5.2 Aqueous Samples
6 Analytical Methods
   6.1 Optical Atomic Spectrometry
      6.1.1 Atomic Absorption Spectrometry (AAS)
      6.1.2 Atomic Fluorescence Spectrometry (AFS)
      6.1.3 Atomic Emission Spectrometry (AES)
   6.2 Electrochemical Methods
   6.3 Activation Analysis and Application of Radiotracers
   6.4 X-Ray Fluorescence (XRF)
   6.5 Mass Spectrometry (MS)
   6.6 Inductively Coupled Plasma-Mass Spectrometry (ICP-MS)
   6.7 Spectrophotometry (Colorimetry)
   6.8 Chromatographic Methods
   6.9 Future Prospects
7 Separation, Identification, and Quantification of Chemical Compounds (Speciation)
   7.1 Introduction
   7.2 Aqueous Samples
   7.3 Biological and Other Fluids, Solid Materials, and Air
8 Economic Aspects (Cost-Benefit Relations)
9 Quality Assurance
10 Data Evaluation
   References

1 Introduction

This book deals with about forty metals and metalloids of particular environmental significance. Commonly they appear in different environmental compartments in trace concentrations, ranging from the higher mg/kg (ppm) level through the µg/kg (ppb) level down to ultratrace contents at the ng/kg (ppt) level or even below.

Reliable trace analysis thus is of vital importance for ecological chemistry and metal toxicology but also for other branches such as food control or occupational medicine. Analytical results provide the quantitative basis for detailed knowledge about the behavior and fate of metals in the environment, the observance of regulations (see Chapters I.20a and I.20b) and for decisions on the introduction or alteration of measures for environmental protection (MERIAN, 1986). Therefore, it is evident that wrong or unreliable analytical results can lead to severe errors in the judgement of ecological, economical, and health impacts of trace elements.

The determination of *total contents* for the estimation of dose relationships and threshold levels is the most common approach. However, there are increasing efforts and methods under development to identify, characterize and quantify the various chemical compounds of metals and metalloids present in the environment (see Sect. 7). This is very important for deeper insights into the fate of trace elements, i.e., their transport, action and metabolization in various environmental and biological compartments.

Despite significant methodological progress during the last few decades, we still face the situation that the reliability of analytical data is far from being satisfactory, as is obvious from many recent interlaboratory studies and comparison of published metal contents for distinct materials (TÖLG, 1978; VERSIECK and CORNELIS, 1980; PARR, 1980; KOSTA and BYRNE, 1982; BRIX et al., 1983; MÜLLER and KALLISCHNIGG, 1983; KNÖCHEL and PETERSEN, 1983; VERSIECK, 1984, 1985; ANALYTICAL QUALITY CONTROL (HARMONISED MONITORING) COMMITTEE, 1985; TAYLOR and BRIGGS, 1986; MUNTAU, 1986; FISHBEIN, 1987; LORING and RANTALA, 1988; HERTZ, 1988; WAGEMANN and ARMSTRONG, 1988; SCHELENZ et al., 1989).

The reasons for this are just as varied as the structure of trace analytical procedures. Despite significant improvements in the instrumental performance of trace analytical methods, possibly indeed due to just this progress, the complex error sources from sampling to sample preparation have gained influence. In discussing the potential of current trace analysis this has to be considered in order to arrive at realistic judgements. This chapter thus commences with the definition of different tasks for the analysis of environmental and biological materials by increasing difficulty and subsequently deals with the various steps and methodological approaches in a critical manner. Finally, also the frequently very important cost-benefit relations will be mentioned. A − not at all comprehensive, but typical − selection of references based on the author's experience should draw the reader's attention as far as possible to review papers providing more detailed information in the mentioned area. However, since the intention was also to cope with current progress and knowledge, quite a number of original publications are included as well. Quotation of promising new techniques and results in a field still expanding at a remarkable speed made the list of references somewhat more extensive compared with the first German edition of this chapter (STOEPPLER and NÜRNBERG, 1984) but the author expects that a useful introduction to the goals, prospects, and pitfalls of applied analytical chemistry has still been produced. The special field of the analysis of metal compounds in the air, including analysis of particulates, is outlined in Chapter I.6 of this book; special requirements, however, will be dealt with in this chapter as well. Also in the Sections 2 of Part II one finds information related to specific analytical chemistry of the described metal compounds. Additional information on analytical methods and speciation has also been presented at the I.A.E.A.C. Workshops on Carcinogenic and Mutagenic Metal Compounds in Geneva, Villars and Follonica (MERIAN et al., 1985, 1988 and 1990).

2 Analytical Tasks

Fig. I.4a-1 presents a general scheme for trace metal analysis. The actual analytical procedure covers steps III to VII — from sampling to data evaluation.

Depending on effort and the practical experience of an analytical laboratory, different categories in increasing order of difficulty can be defined as is given below. A detailed, comparative discussion of principles and properties of the instrumental methods applied most frequently today can be found in Sect. 6.

Difficulties Low. Rapid analysis, e.g., in case of an acute intoxication for identification and quantitation of the element under consideration. Methods for this purpose have to be quick and specific, but false negative — sometimes also false positive — results should be minimized as far as the state of the art allows.

The same is the case for analysis aiming at the identification of significantly increased trace metal values possibly exceeding distinct threshold limits, e.g., for cadmium, lead, copper, chromium, mercury, nickel, and zinc in sewage sludge. This would preclude such a material from being used as a fertilizer. Analyses of this type can be performed with comparatively inexpensive and simple and/or less sensitive methods as colorimetry, flame atomic absorption spectrometry, polarography, emission spectrometry with plasma (mainly ICP) excitation and X-ray fluorescence (XRF). Due to concentration levels in the mg/kg (ppm) range sample preparation is commonly simple and contamination risks are negligible.

Difficulties Moderate. Determination of concentrations and — predominantly — concentration ratios at the higher µg/kg (ppb) to mg/kg (ppm) level. Most important tasks are to differentiate between polluted and non-polluted areas (samples: plants, soil, small indicator animals) or to identify typical element distributions ("fingerprints") in metal-accumulating species. Methods applied to this type of investigation should attain an appreciable precision since the main aim is a relative comparison between the same species so that absolute values with utmost accuracy are commonly not required. The concentration level does not need careful contamination control in this case either. For these tasks many analytical methods, depending upon availability, can be used.

Difficulties Slightly Increased. Biological and environmental monitoring in cases of occupational or environmental exposure, however, in materials with endogenously

I	**Analytical task** General association (degree of difficulty), extent (number of samples, frequency of sampling, urgency)
II	**Planning** Orientation of task on the potential of the performing laboratory (instrumentation and staff; organizational structure and time schedule)
III	**Sampling** Minimization of contamination and losses, including conservation of properties if speciation is required
IV	**Sample storage** Short-term, medium-term or long-term storage under as far as possible controlled conditions in some cases after homogenization; long-term storage, e.g., in environmental specimen banks, under cryogenic conditions for retrospective analysis
V	**Preparation (pretreatment)** (For direct analysis already to be considered at step III), homogenization, aliquotation, decomposition, matrix or compound separation, enrichment, etc.
VI	**Analysis** Application of a method selected according to step II for routine determination Quality control by independent or reference method(s), application of control, reference or certified reference materials, (internal quality control), interlaboratory comparisons for validation of results in terms of accuracy and precision
VII	**Data evaluation** Numerical values with limits of accuracy and precision
VIII	**Interpretation of data** Ecochemical, ecological, geological, oceanographical, ecotoxicological, toxicological, occupational, epidemiological, etc.

Fig. I.4a-1. General scheme of a trace analytical procedure.

low metal levels. Examples: determination of lead, cadmium, nickel, chromium, etc. in urine and blood of occupationally exposed collectives, heavy metals in waste water, some plants, and in foodstuff. The tasks usually are the precise and accurate analysis of metal contents at different µg/kg levels (body fluids, waste water, environmental materials) or to ascertain that distinct metals are present significantly below or above legal threshold levels (food control). This requires some experience on the part of the staff and appropriately sensitive methods, e.g., graphite furnace AAS, voltammetry and also other, recently introduced methods such as total reflection X-ray fluorescence spectrometry (TXRF) and recently also ICP-MS.

Difficulties High. Determination and definition of average (normal) trace metal contents in biological fluids, tissues, and different environmental materials within base line studies. This includes the analysis of normal and slightly elevated levels of trace metals in precipitation (rain and snow), in potable water and inland water but also of higher metal levels and of various chemical forms in estuaries and indicator organisms. Metal concentrations range from the lower µg/kg (ppb) down to the higher ng/kg (ppt) level. Here significant risks for contamination, losses, problems of detection power and selectivity of the applied methods can be expected. Thus, reliable data are only attainable if the laboratory staff is already experienced in this field and continuous efforts in analytical quality control are made. Due to their detection power the methods predominantly applied are the most sensitive modes of atomic absorption spectrometry (graphite furnace, hydride and cold vapor AAS), frequently also in combination with chromatographic techniques, of voltammetry (differential pulse and adsorption voltammetry), the latter also because of its potency in discriminating between different chemical compounds (mainly in aqueous media), and – increasingly – of TXRF and ICP-MS.

Difficulties Extreme. Trace metal contents at the ng/kg (ppt) level and even below, concentration differences at natural levels due to e.g. behavioral, nutritional and geochemical influences. Examples are the determination of extremely low levels of trace metals in open oceans, in less polluted estuarine and off-shore waters, but also in rivers, formerly considered to be significantly polluted due to false analyses (MART et al., 1982b; BRULAND, 1983; BLOOM and CRECELIUS, 1983; MAY and STOEPPLER, 1983; MART and NÜRNBERG, 1984, 1985; JICKELS and BURTON, 1988; HUYNH-NGOC et al., 1988; DORTEN et al., 1989). Further examples are the determination of ubiquitous elements such as aluminum (SLAVIN 1986b; CEDERGREN and FRECH, 1987) and chromium (VERSIECK, 1985; OTTAWAY and FELL, 1986) posing severe contamination and/or determination problems in biological materials. Similar problems have to be faced for, e.g., the determination of lead, cadmium, and nickel in blood plasma or serum (EVERSON and PATTERSON, 1980; STOEPPLER, 1986; SUNDERMAN et al., 1986) as well as for the determination of cadmium and lead in human milk (DABEKA et al., 1986; MÜLLER, 1987). Trace metal levels (Hg, Cd, Pb, Co, Ni, etc.) are also very low in some foodstuffs with contents at the ng/kg level (MAY and STOEPPLER, 1983; NARRES et al., 1984b, 1985; DABEKA and McKENZIE, 1987; VOS et al., 1986a, b, 1988). The same is the case for methylmercury in various environmental and biological materials (MAY et al., 1985; AHMED et al., 1987; HORVAT et al., 1988).

Without particular experience and staff training, clean room techniques, use of the most powerful – sometimes not yet commercially available – methods and rigid quality control reliable data cannot be achieved. Thus this area will doubtless remain the domain of highly specialized and well equipped laboratories.

3 Planning

Planning for the performance of an analytical program has to be oriented towards the particular task and its urgency but also towards the laboratory equipment and staff experience. The latter might be prohibitive for a reliable performance of the analytical program if experience is poor.

If the program is limited in time but urgently required it is commonly less advisable to introduce a new procedure in order to speed up sample throughput if time to become familiar with this new procedure appears to be longer than the gain in time compared to the previous procedure, because by introducing a new procedure one has commonly to consider unexpected disturbances and delays.

Similar, but often much more severe problems might be faced if a completely new method has to be introduced. In such a case, but also if only procedural changes are made, it is strongly recommended to still make use of the earlier method (or procedure) in a transition phase, if it was quite reliable as a reference method compared to the newly introduced one so that the staff can still use a method with which they are familiar and in which they already have confidence. If the new method (procedure) performs better in direct comparison this also helps to convince and motivate them about the methodological change. This approach, of course, is only absolutely necessary if an independent method is not already at hand to check the newly introduced one. In this phase also supply of appropriate reference materials and/or participation in external quality control programs is extremely valuable (see also Sect. 9).

4 Sampling and Sample Storage

A sampling plan (or protocol) must properly depend on the particular analytical task. It is obvious that the amount taken for an individual sample and all measures to minimize contamination and losses have to be defined with respect to the element(s) to be determined, the expected concentration range, type and homogeneity of the matrix and the foreseen method(s).

Since there are – despite some general similarities – distinct differences between biological and (solid) environmental materials on the one hand and aqueous samples on the other, these are discussed separately under Sects. 4.1 and 4.2.

4.1 Biological and Environmental Materials

Human biological material shall be divided into two groups:
1. Body fluids like blood, blood plasma or serum, milk and urine. Other liquids, e.g., cerebrospinal fluid, bile, gastric juice, pancreatic juice, prostatic fluid, seminal fluid, and sweat are less frequently analyzed.
2. Solid materials like bones, feces, glands, hair, kidney, liver, muscle, nails, skin, spleen, stomach, teeth, uterus.

A meaningful selection of biological specimens requires a detailed knowledge of metabolic processes for the element(s) under consideration (SANSONI and IYENGAR,

1978; IYENGAR and KOLLMER, 1986), particularly if not only biopsy but also autopsy materials have to be analyzed. For the latter, postmortem influences will occur and can cause significant elemental changes by autolysis, internal losses, and contamination (IYENGAR, 1981, 1986; AALBERS et al., 1987).

Trace metal levels in body fluids are frequently very low for many elements (VERSIECK, 1985; SUNDERMAN et al., 1986; CLARKSON et al., 1988), thus utmost contamination precautions are necessary during sampling. Sampling vessels should either be already cleaned during manufacture – possible to some extent for commercial sampling systems for whole blood already containing anticoagulating agents (heparine, citrate, K-EDTA) or, e.g., for urine carefully cleaned prior to sampling (EVERSON and PATTERSON, 1980; BEHNE, 1981; STOEPPLER, 1983b; AITIO and JÄRVISALO, 1984; SUNDERMAN et al., 1986). The accurate determination of normal levels of nickel, cobalt, and chromium in whole blood or serum requires replacement of stainless steel canules by those made of PTFE or polypropylene (SUNDERMAN, 1980; BEHNE, 1981; SUNDERMAN et al., 1986; OTTAWAY and FELL, 1986). However, it has been shown that using appropriate steel needles, reliable results can also be obtained, provided the first 5 mL of blood are discarded (BRO et al., 1988). Sometimes, as described for aluminum, even sampling under clean room conditions might be necessary to attain accurate analytical results (FRECH et al., 1982). For urine acidification to a pH ≤2 after collection it is necessary to avoid adsorption of metal ions onto particulates. Acidification, depending on the subsequent analytical purpose, is done either with nitric, hydrochloric, or acetic acid (STOEPPLER, 1983a; AITIO and JÄRVISALO, 1984).

Sampling of blood and urine also has to be performed in accordance with physiological facts requiring the analyst's cooperation with physicians to exclude false diagnoses. Collection of, e.g., blood should be performed at distinct times of the day and from a defined position of the patient's body (BEHNE, 1981). Urine sampling with utmost contamination precautions – human beings are often a severe contamination source, particularly in occupational exposure in which case a bath or shower prior to urine voiding is mandatory – is most reliable if a 24 h collection is performed. Since in many cases this is not practicable also well defined spot urines (e.g., the 2nd morning voiding) and normalization of metal contents to creatinine or specific gravity yield quite reliable results. Practicable methods for biological monitoring have been described by ANGERER and SCHALLER (1985, 1988).

Hair is a biopsy as well as an autopsy material. Due to the difficulty in distinguishing between endogenous and exogenous influences for some trace metals use of hair for epidemiological studies is limited (CHITTLEBOROUGH, 1980). However, for pollution estimation and retrospective studies, e.g., in cases of poisoning hair is a valuable material. If sampling and sample preparation is carefully performed, longitudinal analysis of hair strands makes quite reliable estimations possible about the time of poisoning or intoxication and sometimes also about biological half-life (VALKOVIC, 1977; HOPPS, 1977; HAGEDORN-GÖTZ et al., 1977; YUKAWA, 1984). An analytical advantage of hair in comparison to body fluids is the relative enrichment of numerous elements in this matrix compared to whole blood, serum and urine, which is commonly in the order of a factor of 100 or even higher so that generally contamination during collection is not significant and determination does not require methods of utmost detection power. Nails behave similarly, however, due to the much smaller surface with a less pronounced tendency to reflect exogenous contamination.

Metal contents in feces, organs (lung, kidney, liver), tissues (such as muscle, placenta, skin) bone and teeth are in general – however, with differences – commonly somewhat to significantly higher than in body fluids (IYENGAR et al., 1978; O'CONNOR et al., 1980; VANOETEREN et al., 1986; GERHARDSSON et al., 1986). Therefore, contamination problems for these materials and a number of elements are less severe. Teeth, being a biopsy as well as an autopsy material possess a particular importance also because of their very rigid structure with low moisture content (EWERS et al., 1979; STEENHOUT and POURTOIS, 1981; HAAVIKO et al., 1984; PURCHASE and FERGUSSON, 1986).

As far as accurate analytical values are concerned, the definition of the *weight basis* is of great importance also for the comparison of data from different authors. This is still problematic for each material with a considerable moisture content. The so-called fresh weight very often cannot be determined or even defined exactly. Hence frequently the dry weight is taken as a basis for sound normalization of data. Dry weight determination in practice is performed using different techniques, e.g., freeze drying, careful drying in a desiccator, oven drying at various temperatures, etc. (PANDE, 1974). However, even if the same method is applied different results for moisture percentage are common as is known from intercomparisons, certification campaigns, etc. and are a distinct and frequent error source. A further and even more complex area is discrimination between bound and free moisture and its determination (PYPER, 1985). However, if other – still predominating – analytical error sources are taken into account, in very many cases, particularly for routine analysis, this error plays only a minor role.

Material that is heterogeneous if trace metal distribution is considered, as is the case for, e.g., lung, liver, or kidney, commonly has to be homogenized prior to analysis. For the determination of trace metals besides classical techniques using ball mills made of different materials after drying, the so-called brittle fracture technique – cryogenic homogenization of dried or fresh materials at liquid nitrogen temperature using a PTFE device (IYENGAR and KASPEREK, 1977) and similar techniques particularly for the preparation of fresh materials for long-term storage, e.g., in Environmental Specimen Banking (LANGLAND et al., 1983; KLUSSMANN et al., 1985; SCHLADOT and BACKHAUS, 1988) are promising approaches.

Since drying and/or homogenization may lead to contamination or losses in many cases experiments with materials of similar but well known elemental composition or radioisotopically labeled substances (KRIVAN, 1986, 1987) are highly recommended.

For *environmental materials* also division into two groups is advisable
1. Ecological samples as soils, sediments, plants, wild animals.
2. Samples of anthropogenic origin as foodstuff, sewage sludge, dust, technical materials.

For (1) in principle the same applies as for solid human biological materials if sample collection is considered. Due to numerous environmental influences the proper selection of sampling site, number of samples (for statistical reasons in order to reliably reflect biological or local concentration variations), individual sample size and sometimes also sampling frequency as well as homogenization measures are important (e.g., HOFFMANN and SCHWEIGER, 1981). Planning has to be performed in close cooperation between analysts, ecologists, biologists, geologists, etc. in order to reach meaningful data, see Fig. I.4a-1 and Sect. 2.

For soils and undisturbed sediment samples sampling of cores is common in order to obtain retrospective informations about anthropogenic or geological influences by stratified analysis (FÖRSTNER and WITTMANN, 1981; MARC, 1985). These sediment or soil cores were either separated in slices immediately after sampling or stored under cryogenic conditions as complete core. Tree rings are treated similarly.

For (2) it should be noted that material of anthropogenic origin predominantly originating from technological processes is frequently homogeneous, which is also the case for dust filters (KLOCKOW, 1987). Sometimes, however, larger amounts of, e.g., food such as coarse sausage, meat, bread and cakes, or sewage sludge, have to be collected for a representative sample after prior homogenization.

Some basic foods such as flour, sugar, muscle meat, fish fillet, milk and dairy products, fruit, and various beverages (juices, wine, beer) contain – as already mentioned – see Sect. 2, Chapter 8, and Part II – very low levels of ecotoxic trace metals (MAY and STOEPPLER, 1983; NARRES et al., 1984b, 1985; STOEPPLER et al., 1985; DABEKA and MCKENZIE, 1987; VOS et al., 1986a, b, 1988). Therefore, sample collection for these materials has to be performed very carefully to minimize contamination. For solid material or pastes in cans or boxes subsamples should also be taken from inside, if mixing is not possible, because boundary surfaces often show higher trace metal levels due to adsorption that may lead to erroneous values.

Short-, medium- or long-term *sample storage* is often necessary for various reasons: in cases where immediate analysis is not possible, reference samples have to be kept (forensic medicine, food control, environmental and basic research) or if surplus sample material is required for calibration or reference during a longer period. However, during this – predominantly at low temperatures – storage, changes in the composition of the original materials are possible that may negatively influence the analytical results. These changes can mainly cause structural alterations which disturb or even make impossible later analysis of chemical species. Also elemental losses by adsorption on the walls of the storage vessels or contamination from these, if not carefully checked prior to use, are possible. In the case of mercury, if storage vessels exclusively of glass or quartz are not used, reduction to metallic mercury in solution and diffusion through plastic materials as well as from outside may lead to significant concentration changes (MAY et al., 1980). Diffusion, however, is very slow at low temperatures and becomes negligible $\leq -80\,°C$. If whole blood has to be stored at lower temperatures hemolysis is common. For subsequent analysis of chemical species in urine deep frozen storage, predominantly in the dark, is needed. If compounds which may decompose easily have to be determined, utmost care in evaluating appropriate storage conditions is mandatory if detailed information is not already available (STOEPPLER, 1983a).

A particular approach for long-term storage of biological and environmental materials is preservation of selected specimens at cryogenic temperatures in Environmental Specimen Banks (ESBs). These ESBs aim at providing representative samples for retrospective analysis of hazardous chemical substances not known or not determinable at the date of collection, trend surveillance, control of legislative measures and reference materials for future analysis with improved methods (BERLIN et al., 1979; LUEPKE, 1979; LEWIS et al., 1984, 1990). After a preliminary phase of about a decade with pilot programs in the United States of America (ZEISLER et al., 1983) and the Federal Republic of Germany (KAYSER et al., 1982; STRUPP et al., 1985; NÜRNBERG et al., 1986; BMFT, 1988) permanent banking with chronological sampling campaigns was started 1985 in both countries. Storage temperatures are around $-80°C$ for human samples in the FRG stored in Münster, around and below $-140°C$ for all specimens stored in the ESB of the United States as well as for terrestrial and marine samples in the FRG stored in Jülich (BOEHRINGER and HERTEL, 1987). Similar, but less comprehensive programs also exist in Japan, Canada, and Sweden. There is a current exchange of data and of ideas at international meetings between all scientists engaged in this field of research (WISE and ZEISLER, 1985; WISE et al., 1988).

4.2 Aqueous Matrices

Before the development of suitable sample collection procedures in the analysis of trace metals in water, *contamination* during sampling was one of the most severe and most frequent sources of error. Despite some progress noted for the last few decades, sample collection still is one of the most crucial steps as far as accurate analytical results are concerned.

Rain and *snow* collection can be performed with so-called "bulk samplers" for simultaneous collection of dry and wet deposition or wet-only samplers that open only during rain- or snowfall (see recent comprehensive review and discussion of possibilities and limitations of all types of collection systems (KLOCKOW, 1987; BARTH and KLOCKOW, 1989; LARJAVA et al., 1990; PUXBAUM et al., 1988).

Since there are some problems in attaining consistent data with bulk samplers for the estimation of total trace metal deposition, the use of wet-only samplers is still common. Quite reliable systems for this purpose are e.g. those developed, continuously improved and successfully applied to the monitoring of trace metal and acid deposition by NÜRNBERG and coworkers in the FRG (NÜRNBERG et al., 1980, 1984; VALENTA et al., 1986). These samplers are equipped with an electric heater controlled by the outside temperature and are able to operate down to $-30°C$. The precleaned collection vessel is made of polyethylene. In polar and alpine regions manual collection of fresh snow samples with very clean plastic containers under utmost contamination precautions, e.g., by wearing plastic gloves, sampling opposite to wind direction, has also led to accurate results even at the ng/kg level (MART, 1983). However, reliable retrospective trace metal analyses in ice and snow cores require extreme care and special additional equipment (e.g., BOUTRON, 1979, 1986; MART et al., 1982a; MART, 1983).

Sample collection techniques for inland and sea water depend on the water depth. Surface water − 0.5 to 1 m below surface − can be manually collected by wearing plastic gloves from a boat, the bank, or shore, with 0.5 to 2 L bottles of linear polyethylene for all metals. For mercury, however, sampling in plastic vessels can be accepted, but if longer storage is foreseen, glass and quartz bottles should be used (MART, 1979b, 1982; STOEPPLER, 1989a). For automated analysis systems used for continuous water monitoring from the bank line sampling can be performed by pumps of plastics and appropriate plastic tubes. If, however, samples are taken in lakes, rivers and the open ocean with the aid of larger vessels it must be remembered that there is always a zone of water contaminated with trace metals around these vessels. Thus the use of e.g. a rubber boat and its operation at a distance of at least 500 to 1000 m from the vessel is absolutely necessary (MART, 1979b). Sampling has to be performed before the bow of the boat and against the wind direction as well as outside the sector contaminated by emissions from the funnel of the vessel. Contamination risks are very high in non-polluted waters (open ocean, distinct lakes) (ASHTON and CHAN, 1987) compared to polluted coastal waters, estuaries and rivers as well as to precipitation in Europe and Northern America.

If samples down to a depth of approximately 200 m have to be collected, so-called GoFLO samplers (General Oceanics, Miami, FL), or modifications of these, exclusively constructed from plastics (PTFE etc.) and carefully precleaned are the most reliable systems (BRULAND et al., 1979; MART et al., 1983; FREIMANN et al., 1983). If samples have to be collected contamination-free from depths down to 6000 m quite sophisticated systems, mainly based on the concept of the CIT sampler (PATTERSON and SETTLE, 1976; SCHAULE and PATTERSON, 1980), are required (HAAS and MART, 1980; MART et al., 1983; BRÜGMANN et al., 1987). They made possible accurate analysis of trace metals in open oceans down to the ng/L level in contrast to former conventional systems that frequently led to unreliable results for trace metals. For the collection of water samples very close to the sea bottom (1–2 m) a free-fall sampler has also been constructed and applied (SIPOS et al., 1980). The only remaining problem is the contamination-free storage of these sampling systems on the deck of research vessels.

If *potable water* has to be collected, first a volume of approx. 20 L should be taken and discarded in order to avoid or at least to minimize contamination from the supply line (KLAHRE et al., 1978). In water works automated analysis, e.g., by voltammetric systems, can be completed directly via a bypass to the main water line (BODEWIG and VALENTA, 1985).

Waste water commonly poses no contamination problems because of higher metal contents. Sampling is performed manually with polyethylene bottles from 0.5 to 2 L.

5 Sample Preparation

Sample preparation includes the analytical steps that are necessary immediately prior to the determination of an element or its compounds. It includes, e.g., sample irradiation in appropriate facilities for subsequent neutron activation analysis

(NAA), direct introduction of solid or liquid subsamples into systems for optical atomic spectroscopy, application of partial or complete decomposition-extraction and mere extraction procedures. As already done in Sect. 4, this section will also be divided into biological and environmental materials on the one hand and water samples on the other.

5.1 Biological and Environmental Materials

Methods such as neutron activation analysis (NAA), X-ray fluorescence (XRF), spark source mass spectrometry (SSMS), optical atomic spectrometry (OAS) and in some cases also voltammetry allow the direct analysis of liquid or even solid samples. For activation analysis the sample has to be irradiated first in appropriate facilities such as nuclear reactors or accelerators. In this method, the selection of irradiation vessels and careful blank control is very important because of possible contamination from trace metals occurring in the vessel material (see Sect. 6.3). Apart from this, direct sample introduction into an analytical system has important advantages in minimizing or even avoiding contamination and losses, increasing detection power and providing quick results. The latter is advantageous in cases where quick information is urgently needed, e.g., for the control of threshold limits in environmental and food surveillance, in occupational exposure and metal poisoning where immediate action is mandatory.

In optical atomic spectroscopy the excitation of atoms occurs at comparatively high temperatures (flames, plasmas, and electrothermal systems) that can decompose completely or to a sufficient degree interfering organic matter and thus often favors direct analysis. The influence of interfering inorganic constituents can be minimized by dilution and/or addition of matrix or element modifiers (see Sect. 6.1). However, these additional steps can lead to contamination and − less frequently − to losses. This in principle is also the case for drying and milling prior to direct sample introduction procedures if the material cannot be used without any pretreatment. In general these error sources can be controlled if careful blank determinations, e.g., with appropriate reference materials, are performed (GRETZINGER et al., 1982). However, dilution and addition of modifying chemicals can further negatively influence practically attainable determination limits.

Due to the still ongoing development and improvement of detection power and selectivity in instrumental analysis the number of direct determinations also has an increasing tendency. However, the partial or complete decomposition of interfering organic matter during element determination is still necessary in many cases in order to attain a suitable analyte solution for subsequent application of spectroscopic and voltammetric methods. A complete decomposition is unavoidable, e.g., for solid − sometimes also liquid − biological and environmental samples for voltammetry (NÜRNBERG, 1983a; WÜRFELS et al., 1987, 1989a−c; OSTAPCZUK et al., 1988) and isotope dilution mass spectrometry (IDMS) (HEUMANN, 1988a). For the development of less expensive and time consuming direct methods in optical atomic spectroscopy, mainly in atomic absorption spectroscopy (AAS), due to the lack of appropriate reference materials and during the evaluation of these (see Sect. 9) multi-stage, easy to calibrate reference methods are still necessary based as far as possible on optimized decomposition procedures (TÖLG, 1979; BROWN et al., 1981; GÖTZ and HEUMANN, 1988a). If radiochemical neutron analysis (RCNAA) to enhance detection power for certain elements after prior irradiation is applied the required selective separation of individual elements and groups of elements needs decomposition procedures for irradiated samples (GUINN and HOSTE, 1980; HEYDORN, 1984; KRIVAN, 1985; ERDTMANN and PETRI, 1986; VERSIECK et al., 1988a).

For the decomposition of various materials a wide selection of methods has been developed. Many of them are currently still being used – including some straightforward recently introduced approaches – that possess distinct advantages but also limitations (GORSUCH, 1970; SANSONI and IYENGAR, 1978; BOCK, 1979; TSCHÖPEL, 1980, 1990; SANSONI and PANDAY, 1983; JACKWERTH and GOMISCEK, 1984; KNAPP, 1984; KNAPP and GRILLO, 1986; KINGSTON and JASSIE, 1986; SCHRAMEL et al., 1987; SEILER et al., 1988; TSCHÖPEL, 1990). A critical overview of the most frequently applied techniques for the materials of concern in this chapter is given in Table I.4a-1.

Since decomposition can cause severe analytical errors if not properly performed and any contamination or loss deteriorates detection power and accuracy some remarks will be made about the methods listed in Table I.4a-1.

The comparatively simple and inexpensive *dry ashing* in muffle furnaces is generally not suitable for ashing samples with rather low trace metal contents, if special furnaces, e.g., equipped with quartz walls and a very accurate temperature control, are not used and special precautions are not taken (see, e.g., FARIWAR-MOHSENI and NEEB, 1979).

If *wet digestions* are performed in open vessels, increased consumption of reagents – i.e., higher blanks – requires the use of especially purified reagents, e.g., by subboiling distillation. However, apart from the advantage in being less expensive and providing a high sample throughput, the disadvantages of open systems are losses of volatile elements, which nevertheless can be minimized by reflux systems, and an always incomplete decomposition with varying residues even under practically identical conditions (WÜRFELS and JACKWERTH, 1985).

Decomposition in closed systems minimizes consumption of reagents and avoids loss of volatile elements. Since ultrapure or purified reagents can be used, blanks predominantly depend on the material of the decomposition vessels and their surface area. Thus decomposition systems with low trace metal contents of extremely pure quartz and PTFE as well as compact design are very promising. It is, however, necessary to carefully purify decomposition vessels prior to use in ultrapure acids or acid vapor (TSCHÖPEL et al., 1980) and to determine elemental blanks for each individual vessel.

Under normal conditions metal concentrations in solutions from decomposition are at least one order of magnitude lower than in the initial material. This can lead upon storage, additional manipulations, e.g., pipetting, and if further dilution is necessary prior to the determination step to erroneous results due to contamination and losses (TSCHÖPEL et al., 1980). This step in an analytical procedure should thus be controlled in a suitable manner, preferentially by use of appropriate reference materials.

Table I.4a-1. The at Present Most Frequently Applied Decomposition Methods in Trace and Ultratrace Analysis of Biological and Environmental Materials

Method	Risk of Contamination	Complete Decomposition of Organic Matter	Costs	Sample Throughput	Remarks
Dry ashing high temperatures	+++	+++	+	+++	Usually less applicable for low contents and high precision analysis
Low temperature ashing, oxygen plasma	+	++	+++	++	Often duration of decomposition long due to formation of residues if no special systems are used (e.g. KNAPP, 1984)
Burning in oxygen stream	+	++/+++	+++	++	Recent improvements sample throughput and technical performance to note (KNAPP, 1984)
Wet ashing, open systems, glass or quartz	++	++/+++	++	++/+++	Automated systems with considerable costs allow high sample throughput. Also reflux systems. Various acids and acid mixtures. Complete decomposition only possible at higher temperatures and by addition of, e.g., $HClO_4$ and/or H_2SO_4
Wet ashing in closed systems under pressure ≤180°C with PTFE vessels	+	++	++	+++	Difficult for mercury analysis. Mainly HNO_3 and mixtures with HNO_3, also with HF. Subsequent voltammetric analysis requires posttreatment with $HClO_4$
Wet ashing in closed systems under pressure with microwave energy PTFE etc.	+	++	++	+++	Rapid method with appreciable effectivity, however, in principle at present similar effects like HNO_3 pressure decomposition. Security precautions still somewhat difficult in routine
Wet ashing in closed systems at higher temperatures quartz vessels, temp. up to 300°C	+	+++	+++	++	Particularly useful for subsequent voltammetric analysis and mercury systems. Still very expensive (KNAPP, 1984; SCHRAMEL et al., 1987)

Scores: + low; ++ medium; +++ high

Problems also exist, if for enrichment or further separation from interfering elements chelate extraction with distribution coefficients in the order of $\leq 10^3$ is applied (BÄCHMANN, 1981; MIZUIKE, 1983). This approach is also used for the analysis of trace metals in liquid samples prior to, e.g., atomic spectroscopic methods, spark-source mass spectrometry and X-ray fluorescence; the latter particularly in the total reflection mode.

Liquid/liquid extractions and ion exchange separations are common in radiochemical activation analysis; they are, however, not influenced by contamination.

Direct extraction, e.g., from body fluids like urine and environmental materials as soils and sediments, allows the evaluation of rapid and contamination-controlled routine procedures under properly defined conditions. A limitation of these procedures, however, is that extraction yield can be influenced by various matrix constituents. The use of radiotracers in the control of those procedures with real samples is thus recommended.

To achieve better phase separation in liquid/liquid extraction centrifugation is frequently performed. Centrifugation, however, can also lead to contamination by abrasion of metal parts during centrifugation. This can be minimized if centrifuge tubes are sealed during centrifugation and subsequent opening is carefully performed.

Direct separation using extraction procedures without decomposition is unavoidable if organometallic compounds or valence states of a metal have to be quantified. This is the case for, e.g., As(III)/As(V) and organometallic compounds of arsenic and mercury (MAY et al., 1987; HORVAT et al., 1988). Contamination usually for species analysis is no problem, but the quantitative separation from the matrix or from other valence states or compounds of the same metal occurring with constant yield is difficult to achieve (see Sect. 7).

5.2 Aqueous Samples

Water samples usually consist of an aqueous phase and a particulate phase. Hence in many cases the first preparatory step is separation of particulate matter in order to distinguish between dissolved and particulate-bound metals. This is performed by filtration through membrane filters with a pore size of 0.45 µm. Though colloidal particles and some bacteria can pass through these filters, due to a convention in aquatic chemistry all trace metals occurring in the filtrate are considered to be dissolved. In order to avoid contamination, at this stage again extremely important, all filtration systems have to be cleaned with utmost care and checked for elemental blanks prior to use. Also conditioning with water from the sampling site can be useful. To further exclude contamination from outside, dust filtration is commonly performed in closed systems with the aid of nitrogen at slight overpressure (see, e.g., MART, 1979a, 1982; MART et al., 1980). The filtrate should then be acidified to approx. pH 2 with hydrochloric or nitric acid, depending on the metals to be analyzed (STANDING COMMITTEE OF ANALYSTS, 1980; STOEPPLER, 1989a).

If mercury has to be determined, it is recommended that appropriate subsamples should be transferred into glass or quartz bottles. For methylmercury determination, storage in the dark is mandatory because UV light rapidly decomposes this compound. Acidification inhibits bacterial activity and avoids wall adsorption. Even for speciation studies acidification can be useful as reported, e.g., for selenium species (CHEAM and AGEMIAN, 1980) but should be investigated for each particular case if no reliable information from the literature is available. As already mentioned, long-term storage should be undertaken at lower temperatures, if possible below $-20\,°C$ after shock freezing at liquid nitrogen temperature ($-196\,°C$) (SCHEUERMANN and HARTKAMP, 1983) which also preserves different chemical forms of metals.

Filters with particulates should also be stored at low temperatures in precleaned plastic boxes, preferably of polyethylene (MART, 1979a, 1982). Prior to the determination of trace metals these filters have to be decomposed as described for biological and environmental materials in Sect. 5.1 (see also Table I.4a-1).

There are several trace analysis methods suitable for the fairly rapid determination of trace metals in various water types ranging from rain water to fresh water and sea water. Detection power (see Table I.4a-2 and Sect. 6) and vulnerability to matrix interferences, however, are often limiting factors for simple direct application (STOEPPLER, 1989a). Hence, for the use of *graphite furnace AAS* (GFAAS) in sea water and for extremely low metal contents (mainly also the case for sea water) at the ng/L level in general, preconcentration and separation techniques using solvent extraction with different chelators, column techniques or precipitation prior to analysis are mandatory (DANIELSSON et al., 1978; BRULAND et al., 1979; STATHAM, 1985; CLARK, 1986; BURBA and WILLMER, 1987; CHAKRABORTI et al., 1987; SU-CHENG, 1988; SU-CHENG et al., 1988; NAKASHIMA et al., 1988a, b). The same applies to *inductively coupled plasma-atomic emission spectroscopy* (ICP-AES) where mainly chelating resins are used for preconcentration (THOMPSON et al., 1982; NOJIRI et al., 1985; CHENG et al., 1987; VAN BERKEL et al., 1988), *plasma source-mass spectrometry* (PS-MS) (HIEFTJE and VICKERS, 1989), and *total reflection X-ray fluorescence* (TXRF) (STÖSSEL and PRANGE, 1985; PRANGE et al., 1985, 1987). Column preconcentration techniques can be also automated for higher sample throughput (KNAPP et al., 1987).

The potential of *voltammetry* differs to some extent from the methods mentioned above for the analysis of metals in aqueous samples. Voltammetry provides, due to a contamination-free in-situ preconcentration on mercury and various solid electrodes (also with the aid of the recently increasingly studied adsorption techniques), a detection power down to the ng/L level for a number of ecologically important metals and metalloids (see Sect. 6.2). However, unbiased results can only be achieved if the water matrix does not contain perceptible amounts of *dissolved organic matter* (DOM). DOM forms electrochemically inert complexes with a distinct amount of the trace metals present thus leading to wrong – i.e., too low – analytical results. Therefore, with the exception of potable water, open sea water, deep sea water, and rain water from rural regions which contain only negligible amounts of organic matter, in most water samples DOM has to be decomposed prior to the application of electrochemical analytical methods. This decomposition, however, can be performed successfully and with minimal risk of contamination by UV irradiation in carefully precleaned vessels made of ultrapure quartz (BATLEY and FARRAR, 1978; MART, 1982; BLAZKA and PROCHAZKOVA, 1983; DORTEN et al., 1984). Even in the filtrate of highly polluted waste water containing large amounts of DOM a few hours of irradiation makes possible interference-free voltammetric metal determinations (PIHLAR et al., 1980).

6 Analytical Methods

There are a number of analytical methods available in principle for the determination of trace metals, discussed in various recent publications (e.g., TÖLG, 1978, 1987; IAEA, 1980; MARR and CRESSER, 1983; STOEPPLER, 1983b, 1988; STOEPPLER and NÜRNBERG, 1984; WOLF and HARNLY, 1984; SANSONI, 1985; FISHBEIN, 1987; IYENGAR, 1989; CREASER and DAVIES, 1989; BROEKAERT and SCHRADER, 1989; FRESENIUS and LUDERWALD, 1989). In fact, however, there are only a few instrumental methods with sufficiently high detection power currently applied in routine analysis.

In these methods one can distinguish between single element, oligo element and multielement methods (see also Table I.4a-3 in Sect. 8) that can reach detection limits close to or even below typical metal levels in environmental and biological materials. Therefore, in many cases, provided that sampling and sample preparation do not introduce significant bias, these methods, if properly applied, promise fairly accurate results.

The methods most frequently used at present in routine and reference tasks are various modes of the (still) single element method of atomic absorption spectrometry, the multi-element method of plasma-induced atomic emission spectrometry, and various versions of electrochemical oligo-element approaches. Further promising multielement methods, increasingly used despite their high investment costs, are plasma-source (mainly ICP) mass spectrometry with impressive detection power and element coverage, a new version of X-ray fluorescence, total reflection XRF (TXRF), isotope dilution mass spectrometry, and still neutron activation analysis in the instrumental as well as in the radiochemical mode. The remaining instrumental analytical methods also mentioned in this section have at present only limited yet sometimes important significance in distinct tasks.

Table I.4a-2 shows relative detection limits for the above-mentioned methods and for most of the metals and metalloids treated in this book. The given values should be considered as typical of distinct instrumental approaches rather than as easily attainable in all cases. However, the intention for this table was predominantly to provide a comparative overview and an aid for the selection of a suitable method for a particular task.

It has to be further pointed out that Table I.4a-2 summarizes *detection limits* for all these methods in an *analyte solution*. In cases where prior to determination digestion or dilution procedures are necessary the detection limit for the matrix − not to be confused with the *determination limit* (commonly at least a factor of three higher, see Sect. 10) − has to be multiplied by *at least a factor of ten*. This is, however, not the case if any preconcentration step can be applied in an analytical procedure.

Table I.4a-2. Typical Relative Detection Limits for 38 Elements for Atomic Spectrometric Methods, Voltammetric Methods, Inductively Coupled Plasma-Mass Spectrometry, Total Reflection X-Ray Fluorescence, and Neutron Activation Analysis for Aqueous Solutions or Non-interfering Matrix for Neutron Activation Analysis, respectively. Values are given in μg/L or μg/kg. The detection limit is defined as three times the standard deviation of the respective background noise or blank (3 s), tabulated values given as 2 s are converted to the more realistic 3 s values

Element	ICP-AES[a]		Atomic Absorption Spectrometry[b]			Voltammetry[c]	ICP-MS[d]	TXRF[e]	NAA[f]	
			Flame AAS	GFAAS (Zeeman)	Hydride/Cold Vapor AAS				Det. Limit	Radionuclide
Ag	1.7	(1.8)	1.5	0.015		0.1	0.03	0.4	2	[110m]Ag
Al	4.2	(7)	45	0.15		0.03*	0.16	–	4	[28]Al
As	17	(11)	30	0.6	<0.03	≤0.2	0.04	0.2	0.05	[76]As
Au	2	(3)	10	0.3		10	0.06	0.2	0.005	[198]Au
Be	0.08	(0.2)	3	0.015		–	(0.1)	–	–	–
Bi	35	(11)	30	0.3	0.03	<0.1	(0.04)	0.2	20	[210]Bi*
Cd	0.7	(1)	0.7	0.006		≤0.0002	0.03	0.4	1.5	[111m]Cd
Co	1.5	(3)	9	0.15		≤0.005*	0.01	0.1	0.03	[60m]Co
Cr	1.6	(1.5)	3	0.06		0.02*	0.01	0.4	20	[51]Cr
Cu	0.8	(1.5)	1.5	0.06		0.002	0.02	0.1	0.1	[64]Cu
Fe	1.1	(1.5)	7.5	0.06		≤0.04*	(0.2)	0.2	6	[59]Fe
Ga	17	(11)	(150)	0.2		0.004*	(0.08)	–	0.06	[72]Ga
Ge	7	(9)	(300)	0.45		–	0.02	–	0.5	[75]Ge
Hg	3	(7)	300	1.2	0.005	0.005	0.02	0.2	0.03	[197]Hg
In	14	(25)	(30)	0.15		–	0.06	–	0.0006	[116m]In
La	1.5	(2.3)	(3000)	–		–	0.01	–	0.06	[140]La
Li	3	(1.2)	0.7	0.06		–	0.1	–	–	–
Mg	0.04	(0.1)	0.15	0.012		–	0.13	0.1	25	[27]Mg
Mn	0.15	(0.3)	1.5	0.03		40	0.03	0.2	0.003	[56]Mn
Mo	3.6	(2)	45	0.12		100	0.04	–	10	[101]Mo
Nb	2	(8)	(3000)	–		–	(0.02)	–	200	[94]Nb
Ni	2	(4)	6	0.3		0.001*	0.04	0.1	15	[65]Ni
Pb	11	(15)	15	0.15		0.001	0.01	0.2	3000	[207m]Pb
Pd	7	(9)	(30)	0.75		0.02*	(0.06)	–	0.12	[109]Pd
Pt	10	(7)	60	1.5		≤0.0001*	(0.08)	0.2	1	[197]Pt

Sb	8	(11)	45	0.6	0.15	0.1	0.06	0.5	0.2	^{122}Sb
Se	15	(20)	150	0.6	<0.03	0.02*	1.0	0.2	1.1	^{77m}Se
Sn	4	(10)	30	0.6	0.8	≤0.03*	0.06	0.5	2	^{123m}Sn
Ta	3.4	(8)	(1500)	–			(0.02)	–	1.3	^{182}Ta
Te	15	(15)	30	0.3	0.03	0.06*	0.08	1.0	0.5	^{131}Te
Th	5	(8)				3	0.02	–	0.16	^{233}Th
Ti	0.5	(0.6)	75	1.2		100	0.32	0.4	10	^{51}Ti
Tl	15	(18)	15	0.15		0.015	(0.05)	0.2	40	^{204}Tl*
U	25	(26)	(45000)	–		0.03*	0.01	0.02	–	^{239}U
V	0.6	(1.2)	60	0.6		100	(0.03)	0.2	0.15	^{52}V
W	6.4	(8)	(1500)	–		1000	0.05	–	0.045	^{187}W
Zn	0.5	(0.75)	1.5	0.03		0.02	0.01	0.1	2.5	^{69m}Zn
Zr	1	(3.3)	(1500)	–		–	(0.03)	0.3	100	^{97}Zr

[a] First set of data for ICP-AES is converted to 3 s and is based on determined values for the most sensitive lines with a 50 MHz conventional argon ICP at 15 pm bandwidth (BOUMANS and VRAKKING, 1987b), second set of data (in parentheses) are the values reported by Jobin Yvon (1988) for the instruments JY 38 PLUS and JY 70 PLUS

[b] Data for AAS are converted to 3 s; they are based mainly on values tabulated or reported (in parentheses) by WELZ (1985), and a recent compilation (SLAVIN et al., 1988) for Zeeman GFAAS. For work with graphite tube furnaces a realistic injected volume of 50 µL is assumed. However, the recently introduced L'Vov platforms frequently do not allow injected volumes above 30 µL so that eventually the given value has to be multiplied with a factor of approx. 1.7 on the one hand; on the other hand, some analyte solutions do allow multiple injections before a single firing, thus compensation might be possible in particular cases (see Sect. 6.1.1). For hydride and cold vapor AAS a sample volume of 20 mL is assumed, for cold vapor preconcentration on noble metal nets and wool. Increase of sample volume, which is easily possible and additional improvements in preconcentration techniques and instrumental arrangements led to detection limits that are approximately one order of magnitude lower; see text

[c] These data stem partly from previous work, performed by the late H. W. NÜRNBERG and coworkers (STOEPPLER and NÜRNBERG, 1984) and from recently published work, mainly on adsorption voltammetry (see Sect. 6.2). Values only attainable by adsorption voltammetry are marked with an asterix

[d] Most values stem from a recent compilation from literature data (HIEFTJE and VICKERS, 1989), the remaining ones, given in parentheses, can be found in a table distributed by SCIEX (1988) for elan ICP/MS Elemental Analysis System (10 s integration)

[e] Detection limits for TXRF have been provided by MICHAELIS (1989) and are based for comparative purposes also on a sample volume of 50 µL

[f] INAA data are based on a sample weight of 500 mg (non-interfering matrix) and the following irradiation and counting conditions: Thermal neutron flux 1×10^{13} n·cm^{-2}·s^{-1}, irradiation time 5 h (maximum); counting with a 40 cm Ge(Li) detector, with a sample-to-detector distance of 2 cm; zero decay time before start of count and maximum counting time = 100 min. If β-counting after radiochemical separation is applied, Bi and Tl can also be determined after the same irradiation conditions (marked with an asterix). Some remarks on improvements by radiochemical procedures etc. can be found in Sect. 6.3

6.1 Optical Atomic Spectrometry

Atomic spectroscopy is based on the generation of free atoms, predominantly by the supply of thermal energy. The atoms thus generated can absorb or emit radiation due to defined transitions of the valence electrons of the outer shell of the atom. Thus a specific element identification based on the atomic structure is feasible in the wavelength range from somewhat below 200 nm to approx. 850 nm. For this purpose, depending on the particular method, more or less sophisticated systems for free atom production, separation, and quantification of radiation are required. These typical applications and current progress are amply treated in the literature for atomic spectroscopy in general (e.g., HARNLY and WOLF, 1984; SANSONI, 1985; GREENFIELD et al., 1986; DAWSON, 1986; SLAVIN, 1986a; METCALF, 1987; BROEKAERT and TÖLG, 1987; and yearly atomic spectrometry literature updates: e.g., CRESSER et al., 1988; BROWN et al., 1988; SHARP, B. H. et al., 1988) for atomic absorption spectrometry (e.g., KIRKBRIGHT, 1980; VAN LOON, 1980; CANTLE, 1982; TSALEV and ZAPRIANOV, 1983; SUZUKU and OHTA, 1983; SLAVIN, 1984; TSALEV, 1984; WELZ, 1985; LANGMYHR and WIBETOE, 1985; MARSHALL et al., 1986; STURGEON, 1986; SUBRAMANIAN, 1986, 1988b; SLAVIN et al., 1988; L'VOV, 1988) and for atomic emission spectrometry (e.g., BARNES, 1978; HAAS and FASSEL, 1980; MARSHALL et al., 1983; WINGE et al., 1984; FASSEL, 1986; MATUSIEWICZ, 1986; BOUMANS, 1987; BOUMANS and VRAKKING, 1987b; BROEKAERT, 1987; THOMSON and WALSH, 1988; LITTLEJOHN, 1988).

6.1.1 Atomic Absorption Spectrometry (AAS)

The basic principle that an atom absorbs radiation at the same wavelength at which it emits radiation has only been practically used for the analysis of elements since 1955 (WELZ, 1985).

The principle of an atomic absorption spectrophotometer is simple. The sample is atomized in the light path of a specific radiation source. The extent of radiation absorption is directly proportional to the number of atoms, i.e., the elemental concentration. It is mathematically described by the law of Lambert-Beer:

$$A = \log \frac{I_0}{I_t},$$

where A is the absorbance, I_0 the initital intensity of radiation, and I_t the intensity of radiation after passing through the sample.

Since the resonance lines, which only occur from atoms in the ground state, are much less than all emitted lines, an absorption spectrum is significantly simpler than an emission spectrum. Hence the requirements for its spectral dispersion (monochromator) are not very high. However, this principle and the relatively weak radiation sources limit the dynamic range. Thus calibration graphs for individual elements are only linear in the lower concentration region. With increasing concentration the slopes bend more or less. This is the most important reason why AAS under normal instrumental conditions has no promising potential for multielement determinations and is still predominantly used as a single element method.

Until 1969, except for the so-called cold vapor method in use for mercury (WELZ, 1985), only AAS instruments with flame atomization were commercially available. These comparatively cheap devices had rather simple radiation sources, burners with direct supply of the analyte solution via a mixing chamber (consumption 2 to 5 mL/min), and single beam direct current optics. Compared to colorimetric procedures quick element determinations − a single measurement could be completed in ten seconds − without any prior chemical separation, in many cases even with superior detection power for many elements, could be performed. Therefore, until the early seventies flame AAS remained the predominant trace analysis method (KIRKBRIGHT, 1980; WELZ, 1985).

However, quick expansion of this apparently simple and reliable method brought AAS instruments into many laboratories with no or only minimal trace analysis experience and hence in most cases predominantly no access to other analytical methods for comparison of results. Thus, problems with physical (radiation) and chemical interferences, baseline drift and non-specific losses of radiation (Rayleigh scattering on particles, broad-band absorption by molecules) (WELZ, 1985) often led to severe analytical errors. Therefore, scientists and manufacturers tried to eliminate or at least to correct as far as possible these error sources. Influences of emission lines from other elements and baseline drift could be eliminated by improvements of the optical systems in introducing alternating current and double-beam alternating current systems with sample and reference beam (WELZ, 1985). These improvements, with additional mirrors and lenses, however, required more powerful radiation sources, e.g., electrodeless discharge and high intensity lamps and sophisticated electronics.

Moreover, a correction for non-specific losses of radiation in flame AAS was feasible, if the absorption of a close non-absorbing element line could be measured and subtracted from the total absorption at the element line. Another, more promising, possibility was comparison with background absorption and electronic correction using suitable continuum radiation (deuterium and hydrogen) sources (WELZ, 1985). Chemical interferences could be overcome by matrix matched calibration graphs, analyte addition or admixture of chemical compounds (e.g., lanthanum salts) that enhance atomization efficiency. In very difficult cases also separation procedures such as solvent extraction and column techniques were used. The flame technique now benefits from PC-operated automated devices and very effective atomization systems such as slotted quartz tubes (KEIL, 1984; BROWN and TAYLOR, 1985) and atom trapping (FRASER et al., 1986; BROWN et al., 1987; WEST, 1988) that enhance detection power, flexibility and application range significantly. Coupling with flow-injection systems is also promising (SIMONSEN et al., 1986; BURGUERA et al., 1986; APPLETON and TYSON, 1986; NYGREN et al., 1988b; FANG and WELZ, 1989; FANG et al., 1989). Flame atomic absorption techniques are now considered to be very accurate and precise. Thus they have attained in some cases official status and in a few cases also reference method status for major and minor elements (IHNAT, 1987).

Flame AAS has an element coverage of around 70 elements. Under optimal conditions day-to-day precision is around, sometimes even below, 2% (RSD). However, modern fully equipped flame AAS systems are not cheap and there is increasing competition from low-cost sequential ICP-AES systems that reach approximately the same detection limits for most typical AAS elements and significantly lower for refractory ones with the important advantage of multielement potential (see Sect. 6.1.3 and Tables I.4a-2 and I.4a-3). It should be noted, however, that early in 1989 a commercially available multielement flame AAS system allowing the simultaneous determination of up to 12 elements was presented (SECKLER and DUNGS, 1989). Since this instrument is sold at a fair price it will certainly find successful application in many branches where only a limited multielement potential is sufficient.

The introduction of the first commercially available graphite tube furnace based on pioneering work by L'VOV and MASSMANN (WELZ, 1985) added a new dimension to AAS. Compared with flame AAS detection limits for many elements were significantly lower, for some metals by several orders of magnitude (see Table I.4a-2). The reason were the nearly complete atomization, a high local atom concentration and comparatively long residence times in the light path. All these parameters are less favorable in the flame.

The price one had to pay for these advantages, however, was the time needed for a single measurement, an initially extreme vulnerability to matrix interferences, and a lower day-to-day precision, typically around 5%.

In a graphite furnace system the measuring cycle starts with the automated injection of liquid samples (analyte solutions from various digestions, extracts, aqueous samples and body fluids) from approx. 5 to approx. 50 µL into the graphite tube or up to approx. 30 µL onto the platform or probe (details for these see below). In particular devices, discussed below, solid samples can also be introduced. Hereafter a temperature program is initiated including drying (only for liquids), thermal treatment ("charring") to partly or completely decompose organic matrix constituents, sometimes also to remove interfering inorganic matter, atomization (with temperatures up to 3000°C), and an additional heating step ("cleaning") to remove remaining matrix constituents from the furnace. The temperature program has to be adjusted to element(s) to be determined and the matrix of the sample. During this cycle the furnace is flushed with an inert purge gas stream (mainly argon or nitrogen) in order to protect hot graphite parts against oxygen attack and to remove volatile reaction products. The duration of this cycle is typically around 100 s, i.e., ten times a regular measurement with flame AAS. However, reductions in the analysis time to around 60 s including the time for automatic sampling without a significant decrease in accuracy and precision could be achieved by careful program evaluation for, e.g., Pb, Cr, and Cu in biological and environmental samples (HALLS, 1984; HALLS et al., 1987). This, most probably, is at present the optimal gain in time that can be expected under favorable conditions.

The extreme sensitivity of the graphite furnace also enhanced physical and chemical interferences already known from flame AAS but there were also various sources of difficulties and interferences not observed hitherto. In the furnace, for example, the ratio analyte to matrix is significantly lower than in flame AAS and other methods because of the much lower analyte levels. Moreover, the rapid appearance of the analyte signal and a sometimes structured background posed problems for background correction and the initially used recorder. Additionally there are reactions between the graphite surface, the analyte and sample constituents sometimes leading to the formation of gas phase molecular species rather than atoms and many influences of the temperature profile of graphite tubes. All this made the situation complex. It must be noted, however, that in many cases it has been possible to identify error sources even in the seventies due to the continuous efforts of numerous workers. Hence, only a few papers can be mentioned here (e.g., MASSMANN and GÜCER, 1974; MASSMANN, 1976; OTTAWAY, 1976, 1984; VOLLAND et al., 1977; L'VOV, 1978; CZOBIK and MATOUSEK, 1978; MATOUSEK, 1981; SLAVIN and MANNING, 1982; WELZ, 1986a; L'VOV, 1987; L'VOV et al., 1987; BERNDT and SOPCZAK, 1987; WELZ et al., 1988a; BRUMBAUGH and KOIRTYOHANN, 1988; AJAYI et al., 1988; BASS and HOLCOMBE, 1988).

The application of radiotracers in various investigations also proved to be very useful in studies of interferences and mechanisms (e.g., VEILLON et al., 1980; SCHMID and KRIVAN, 1985; WHITLEY et al., 1988).

Based on the pioneering work of the researchers mentioned above and many others, an analytical approach which, though extremely sensitive, was for some time not considered very reliable, underwent a tremendous methodological and technical development and improvement. This will be summarized and a few outstanding examples will be given.

The addition of particular chemical compounds ("modifiers") often changed the behavior of matrix and analyte in such a manner that the appearance of the analyte signal could be delayed so that higher charring temperatures are applicable. This led in many cases to a *decreased* volatility of the analyte and an *increased* volatility of interfering matrix constituents (EDIGER, 1975). This principle was systematically applied and developed further and combined with technical improvements such as automated addition of modifiers (COOKSEY and BARNETT, 1979) and their use together with graphite tubes, coated with pyrocarbon and the L'Vov platform, completely made from pyrocarbon as a practical approach to the so-called "constant temperature furnace". Here the platform inserted in a graphite tube is predominantly heated by radiation from the tube wall. Thus vaporization occurs into an atmosphere that is at a higher temperature than the platform and additionally at a nearly constant temperature, considered as an absolutely necessary prerequisite for the effective reduction of interferences (HINDERBERGER et al., 1981; SLAVIN et al., 1981; MAY and BRUMBAUGH, 1982; MANNING and SLAVIN, 1983). An alternative method for high temperature vaporization in a Massmann-type furnace is the use of a probe. This approach, now also commercially available, separates the heating of the probe in time from the heating of the graphite tube (MARSHALL et al., 1985; BROWN and RIEDEL, 1987; LITTLEJOHN, 1989). However, another important improvement became possible by introducing a new background correction technique using the Zeeman effect. Here a magnetic field is applied either to the radiation source (direct Zeeman effect) or to the atomization system (inverse Zeeman effect). The magnetic field splits the emission lines of the radiation source so that the background radiation can be measured slightly off (direct Zeeman effect) or directly at the analyte wavelength (inverse Zeeman effect). Zeeman systems thus correct over the entire wavelength range, which cannot be achieved with any continuum source. Correction efficiency is very effective for high background levels up to around 2 A and particularly for structured background, a severe problem for correction by continuum radiation sources (STEPHENS, 1980; YASUDA et al., 1980; FERNANDEZ et al., 1981; SLAVIN et al., 1983; SLAVIN and CARNRICK, 1985; WELZ, 1985; KNOWLES and FRARY, 1988). However, it must be noted that also this principle can sometimes be prone to errors by e.g. Zeeman splitting of molecules (WIBETOE and LANGMYHR, 1987).

These improvements in conjunction with the generation of so-called "integrated absorbance" values (i.e., $A \cdot s$, which is an area) made it possible to compare so-called "characteristic mass" values (i.e., mass of element in pg that produces an integrated absorbance signal equivalent to 0.004 $A \cdot s$) for various elements in different matrices (SLAVIN and CARNRICK, 1984). It could thus be shown that for a number of metals under carefully evaluated working conditions characteristic mass data differed on average by only around 20%. If higher accuracy is required, however, combined application of — if necessary matrix-adapted — aqueous reference solutions with the analyte addition technique certainly obtains better results. It should be noted, however, that the analyte addition (formerly called standard addition) technique can only correct for *multiplicative* but not for *additive* systematic errors. Moreover, the added analyte element and the element present in the sample must behave in an analytically identical manner (WELZ, 1986b).

A continuous improvement of accessories and procedures together with increased commercial introduction of AAS instruments with Zeeman background correction (BROEKAERT, 1982; EGILA et al., 1986; STOEPPLER, 1989b) led to further progress.

- The matrix modification concept was broadened by new modifiers thus increasingly permitting direct analysis in various biological and environmental matrices formerly considered difficult to analyze (e.g., SUBRAMANIAN, 1988a; PINEL et al., 1988; XIAO-QUAN et al., 1988, 1989). From the present view, however, matrix modifiers on the basis of palladium and its compounds show promising potential for a number of elements and matrices (XIAO-QUAN et al., 1984; SCHLEMMER and WELZ, 1986; VOTH-BEACH and SCHRADER, 1987; WELZ et al., 1988a, b; KNOWLES and BRODIE, 1988; HINDS et al., 1988; SAMPSON, 1988).
- Introduction of alternate gases, now also commercially available in several GFAAS systems (STOEPPLER, 1989b), offered further possibilities. The addition of *hydrogen* was found to eliminate matrix interferences (FRECH and CEDERGREN, 1976; NOVAK and STOEPPLER, 1986; NOVAK, 1989). Similar influences were observed for metal (e.g., tungsten) tube atomizers (OHTA and MIZUNO, 1989) that provide, particularly for refractory elements but also for some other elements, a comparatively high detection power but will not be discussed here in detail (SUZUKI and OHTA, 1983; OHTA and YANG SU, 1987). *Oxygen* was used for in-situ ashing of biological materials (BEATY et al., 1980; EATON and HOLCOMBE, 1983) especially for the analysis of materials with extremely low metal contents. Examples are the determination of cadmium in crude oil and oil products (NARRES et al., 1984a) and cadmium and lead in retail and human milk (NARRES et al., 1985). *Freon* was reported to reduce the memory effect in the determination of refractory elements like molybdenum (WELZ and SCHLEMMER, 1988b). Similar effects also effective for platinum group elements

have been observed for *methane* by the same authors (WELZ and SCHLEMMER, 1988a). Methane was formerly used almost exclusively for the manufacture of layers of pyrolytic graphite on electrographite tubes (WELZ, 1985).
– Zeeman background correction, to some extent Smith-Hieftje background correction which also uses radiation from the same source (WELZ, 1985), were increasingly applied to direct trace metal determination in solid samples and aqueous slurries of powdered materials in the application of various instrumental concepts (e.g., VÖLLKOPF et al., 1985, 1987; CARNRICK et al., 1986; OLAYINKA et al., 1986; SCHLEMMER and WELZ, 1987; LÜCKER et al., 1987; LINDBERG et al., 1988; EPSTEIN et al., 1989). Since calibration by aqueous solutions might be prone to errors in this type of analysis, synthetic reference materials have been proposed for calibration in solid sampling GFAAS (AKATSUKA and ATSUYA, 1989). Matrix modifiers (HINDS et al., 1988; HERBER, 1989) and oxygen ashing (MOHL et al., 1986) were also applied in solid sampling GFAAS in order to improve calibration reliability and to facilitate higher sample intakes in atomizer systems less suited for intakes above 1 mg.

An AAS instrument, using the direct Zeeman effect (i.e., magnetic field at the radiation source) particularly suited for solid sampling GFAAS was designed and commercially introduced. It had a special, comparatively large furnace and an appropriate sample boat that could take sample amounts up to 5 mg or even more with properties like a L'Vov platform (KURFÜRST, 1983a, b, 1984; ROSOPULO et al., 1984; KURFÜRST et al., 1984). This instrument, now available in an improved, PC-operated, 2nd version (a third version is ready for introduction), despite some remaining technical problems (mainly for the special radiation sources required for this concept) shows in general promising performance. Therefore it might, besides other approaches, provide the starting point for further rapid metal analysis and homogeneity studies in solids (GROBECKER et al., 1987).

Some reviews about this technique have already appeared (L'VOV, 1976; LANGMYHR, 1979; LANGMYHR and WIBETOE, 1985). In addition, most of the papers presented at three international meetings on solid sampling AAS conducted since 1984 appeared or will appear as special issues of *Fresenius Z. Anal. Chem.* (STOEPPLER, 1985; LANGMYHR, 1987; HERBER, 1989).

These improvements, together with computerization mainly using PCs and a flexible software to operate accessories such as furnaces and autosamplers and procedures, created instruments (STOEPPLER, 1989b) and methods for furnace AAS analyses that compete favorably as far as costs, accuracy, and precision are compared with other trace analysis methods (SLAVIN et al., 1988). Therefore, this method will still play an important role in many scientific branches. It is obvious that the use of instruments with Zeeman background correction will increase, though not necessarily for all purposes, but offering more versatility at a cost level not significantly higher than for comparable continuum source corrected systems. From research performed during the last decade, some progress in graphite furnace, probably also metal tube, AAS can be predicted (L'VOV, 1988). Future improvements in commercial systems should include a more comprehensive use of computerization, e.g., for signal analysis, and better monitoring and control of furnace temperatures with the goal of further gains in accuracy and precision (HERBER and PIETERS, 1988; VANDECASTEELE et al., 1988b). Technical improvements and modifications to achieve higher detection power are required in many cases. The multiple injection technique, often combined with an appropriate in-situ treatment of samples, is already available in several AAS systems (e.g., NARRES et al., 1984a). Other promising approaches are computer summation of time-resolved signals (BERNDT et al., 1987) and low-pressure vaporization, applicable to liquids as well as to solids (HASSELL et al., 1988). Also promising might be changes in the internal shape of atomization tubes (DUFFIELD et al., 1988) and a further evaluation of probe techniques (LITTLEJOHN, 1989).

At the beginning of this section it was mentioned that the method under normal conditions does not allow an extension of dynamic ranges. This is the case for the design of nearly all commercial systems available at present. However, there is a possibility of overcoming this by applying staircase modulation wave form continuum radiation source AA spectrometry (SIMAAC) proposed first by HARNLY et al. (HARNLY, 1986). An extended range background correction was recently described (O'HAVER et al., 1988). The staircase modulation wave form makes possible the computation of two absorbances (calibration curves) of different sensitivity for every atomization and thus covers 4–6 orders of concentration. This, of course, is also applicable to flame AAS but still not commercially available. Practical application of this system showed a potential for the analysis of as many as 16 elements simultaneously but also limitations in wavelength range because of the properties of the applied radiation source. This might be overcome in the future by using laser sources with extended wavelength ranges.

In 1987 a commercial Zeeman-AAS system with a graphite furnace was introduced able to measure automatically four elements during one firing of the furnace and eight elements using only two firings. The instrument contains eight conventional hollow cathode lamps as energy sources but employs the change in peak width as a function of concentration, somewhat similar to the SIMAAC system, to extend the linear dynamic range (RETZIK and BASS, 1988). This is the first commerical approach to at least an oligoelement graphite furnace AAS system.

Since neither flame nor furnace AAS initially provided a particular detection power for mercury- and hydride-forming metalloids like antimony, arsenic, selenium, tin, etc., special accessories to measure these elements after volatilization in the gas phase were developed and are commercially available from various manufacturers (STOEPPLER, 1989b). Mercury, being the first element analyzed by AAS (WELZ, 1985), is commonly reduced after digestion of solid samples or directly in liquids to elemental mercury by the addition of stannous chloride or sodium tetrahydroborate and analyzed at ambient or slightly increased temperature with detection limits at the low µg/L level (WELZ, 1985). These reducing agents act in a different manner, and thus are also a means of speciation. Stannous chloride only reduces ionic mercury compounds, while sodium tetrahydroborate is also able to transfer methylmercury into elemental mercury (see also Sect. 7). If preconcentration of the evolved elemental mercury on noble metal (gold, silver) wire or gauze and volatilization from these by rapid heating to around 500 °C is performed, sharp peaks are produced in the recording system. This leads to detection limits ≤ 100 pg absolute that are necessary for the reliable determination of the very low mercury contents in some food products and natural waters (ODA and INGLE, 1981; FREIMANN and SCHMIDT, 1982; BLOOM and CRECELIUS, 1983; STOEPPLER, 1983c; WELZ and SCHUBERT-JACOBS, 1986). Also direct reduction/volatilization of mercury in solid samples via pyrolysis has been reported. DUMAREY et al. (1980) used a special instrument and preconcentration on gold, FLECKENSTEIN (1985) a graphite furnace AAS instrument with direct Zeeman background correction. For solid sampling the use of appropriate reference materials is necessary.

Inorganic compounds of hydride-forming elements, if the total content is not determined by graphite furnace with matrix (mainly Pd) modifier, are transformed into volatile compounds by the addition of sodium tetrahydroborate to the analyte solution in appropriate, also commercially available devices, and atomized either by introduction into the flame or into electrically heated quartz tubes in the light path of AAS instruments (WELZ, 1985). The latter at present is more frequently used. Flow injection is also very promising for hydride and cold vapor techniques, and commercial systems are already available. The hydride method attains even lower detection limits than the graphite furnace but also interference effects from matrix constituents were observed with the aid of radiotracers. Thus accurate results can only be achieved if proper, matrix-dependent working conditions are maintained, already described in detail for antimony, arsenic, selenium, and tin (PIWONKA et al., 1985; KRIVAN et al., 1985; WELZ and SCHUBERT-JACOBS, 1986; WELZ and MELCHER, 1983, 1984; PETRICK and KRIVAN, 1987; ITOH et al., 1988). An interlaboratory trial for selenium in body fluids showed the significance of suitable digestion procedures prior to hydride AAS (WELZ et al., 1987).

Picogram levels of selenium and arsenic can be determined using hydride preconcentration at cryogenic temperatures (PIWONKA et al., 1985; ALT et al., 1987; ARENAS et al., 1988).

Another approach for hydride generation – AAS determination of e.g. antimony and selenium in aqueous samples – is the generation of SbH_3 or SeH_2 using sodium borohydride trapping in a graphite furnace at elevated temperatures and subsequent atomization with absolute detection limits of 200 pg for antimony (STURGEON et al., 1985) and 70 pg for selenium (WILLIE et al., 1986).

If the potential of all AAS modes available at present is considered, it becomes evident that the method has significantly gained in versatility and reliability and thus will certainly remain an important analytical tool for the foreseeable future despite many competing approaches.

6.1.2 Atomic Fluorescence Spectrometry (AFS)

The principle of atomic fluorescence spectrometry (AFS) introduced by WINEFORDNER and VICKERS (1964) is the use of fluorescent radiation emitted from atoms that have initially absorbed radiation of a suitable wavelength. The intensity of the emitted radiation is not only proportional to the number of atoms present but also to the light intensity of the radiation source. Thus this approach combines the simplicity of atomic absorption spectra with an extraordinary detection power for several elements, and the wide dynamic range of atomic emission spectrometry (VAN LOON, 1981).

There are a number of papers demonstrating the potential of the method with flame atomization for the determination of cadmium in blood and urine (MICHEL et al., 1979; OTTAWAY, 1983; EKANEM et al., 1986). Applying Zeeman background correction, cadmium, mercury, and zinc were determined in various environmental materials (NARANJIT et al., 1984), and cadmium, copper, iron, magnesium, and manganese in aqueous solutions (TIE-ZHENG and STEPHENS, 1986). Applying the hydride technique, e.g. tin (NAKAHARA and WASA, 1986) as well as arsenic and selenium (EBDON and WILKINSON, 1987c) were determined.

The most promising systems, however, with a remarkable detection power are those with vaporization in graphite tubes and laser excitation (FALK and TILCH, 1987; GOFORTH and WINEFORDNER, 1987). Several authors have shown this for lead (BOLSHOV et al., 1981; DITTRICH and STÄRK, 1987; OMENETTO et al., 1988; APATIN et al., 1989) and other elements such as aluminum, cadmium, cobalt, copper, indium, iridium, manganese, silver, thallium, and vanadium (DITTRICH and STÄRK, 1987; DOUGHERTY et al., 1987a, b, 1989; OMENETTO et al., 1988; GARDEN et al., 1988). There is, however, at present no commerical instrument available for this technique despite its outstanding potential. It appears that progress in AAS and plasma AES has until now prevented the commercial development of appropriate systems.

The multielement potential of AFS was also investigated and a commercial instrument comprising hollow cathode lamps for excitation and an inductively coupled plasma (ICP) as sample cell introduced. Detection limits for 32 diverse elements studied were comparable to flame AAS and ICP-AES. Dynamic ranges were 4 to 5 orders of magnitude. Spectral line interferences were negligible and matrix effects not very significant. However, poor detection limits were observed for refractory elements (DEMERS and ALLEMAND, 1981; GREENFIELD, 1984; DEMERS and JANSEN, 1985; SANZOLONE, 1986). Experiments were also performed with a direct current (DC) plasma as an excitation source for AFS (HENDRICK et al., 1986). Obviously owing to the strong competition of ICP-AES and also flame AAS the multielement approach has not been very successful in the last couple of years. Thus a significant growth of multielement AFS is questionable, since there are no particular advantages compared with ICP-AES and flame AAS, recently also introduced in a multielement mode. More promising for the future due to its superior sensitivity for a number of environmental and toxicologically important elements is undoubtedly laser-excited furnace AFS.

6.1.3 Atomic Emission Spectrometry (AES)

The principle of atomic emission spectrometry (also called optical emission spectrometry, OES) is the generation of radiation (excitation via sparks, arcs, flames, or plasmas) by a multitude of transitions. This leads, particularly for heavier elements, to very complex spectra with a dependence of the number of lines on the excitation temperature. Owing to the multitude of lines and thus the possibility of interferences with lines of other elements the requirements for proper optical resolution are very high. Since the radiation intensity is directly proportional to the atom concentration the dynamic range covers several orders of magnitude. This makes the method well suited for multielement determinations.

AES is basically a classical instrumental method. With excitation by sparks and arcs for solid samples and with flames for liquids, it played a leading role in trace element analysis in the past, besides colorimetry (spectrophotometry), since for a comparatively long time span it was the only commercially available multielement method. Thus, one had to accept the high instrumental costs and its, at least for many biological and environmental materials, only fair precision and accuracy. With the appearance of neutron activation analysis (NAA) and, more drastically, of AAS, modern electroanalytical methods and for higher contents of X-ray fluorescence (XRF), the application fields of AES stagnated or even decreased.

This situation, however, changed with the introduction of new excitation sources (plasmas) mainly inductively coupled plasmas in the early 1960s that reached temperatures above 5000 °C (GREENFIELD et al., 1964; WENDT and FASSEL, 1965). This led, with a general potential for determining up to 70 elements, to a significant enhancement of detection power for about 40 elements of the Periodic Table and due to the achievable high temperature to a drastic reduction of – predominantly chemical – matrix interferences (BARNES, 1978; BOUMANS, 1987).

Since instrumental developments, owing to the complexity of the task, required some time, commercial instruments were introduced from 1974, equipped mainly with ICP excitation sources and optical dispersion systems such as classical gratings or echelle systems. These instruments offered sequential as well as simultaneous multielement potential.

Due to the need for multielement determinations in many research areas ICP-AES has become established as a major technique for elemental analysis and is used today in numerous laboratories achieving in the optimal concentration region a typical day-to-day precision of about 2%.

Prices range from the level of a complete, fully equipped and PC-operated AAS setup for all AAS modes for a simple sequential system to approximately three times this basic price for fully computerized ICP-AES instruments allowing simultaneous and sequential operation with an effective background correction for a wide selection of elements (see Sect. 8 and Table I.4a-3).

Compared to single element flame AAS, plasma AES has the important advantage of rapid multielement determinations with superior detection limits for some mainly refractory elements (see Table I.4a-2). However, running costs for ICP-AES are much higher because of the quite high argon consumption. Thus it is interesting to observe acceptance and progress of the recently introduced and – compared to a low-cost ICP-AES system – cheap multielement flame AAS instrument, allowing for the simultaneous determination of up to 12 elements. No doubt, it might favorably compete with low-cost ICP-AES systems if requirements for a limited number of particular elements are met.

Compared to neutron activation analysis and classical X-ray fluorescence much faster or simpler determinations with plasma AES are important advantages in routine multielement analysis of biological and environmental materials if concentration levels are appropriate.

In the following a selection of examples from the ample recent literature on plasma AES are given as an introduction to the fields of application with main emphasis on the progress achieved within the last years (BROEKAERT, 1987; MEYER, 1987; EBDON and EVANS, 1988). The efficiency of ICP-AES and DCP-AES for trace element determination was critically discussed from the point of view of practical experience also in comparison to other methods, by SCHRAMEL et al. (1982b) and SCHRAMEL (1987).

The by far most frequently used excitation principle, ICP, is still supplemented for special applications by direct current plasma (DCP) (LEIS et al., 1989) and microwave-induced plasma (MIP). The latter at present, however, cannot be used for the direct analysis of solutions after pneumatic nebulization but has excellent performance for dry analyte vapors (BROEKAERT, 1987; ABDILLAHI, 1988). An outstanding example is the very sensitive determination of mercury after preconcentration on gold using helium at atmospheric pressure (NOJIRI et al., 1986) with a detection limit of 0.01 ng/L.

The wide application range of plasma AES in various materials shows the state of the art already attained. This includes analysis of natural waters by ICP (JÄGER, 1983; SPEER et al., 1986) and DCP (URASA, 1984) simultaneous determination of numerous elements in human organs (SUBRAMANIAN and MÉRANGER, 1982) and of ytterbium in biological samples by ICP (MOLINERO et al., 1988) as well as of copper in blood serum by DCP and ICP (BAMIRO et al., 1988). Analysis of 9 environmentally important trace elements by a low-cost sequential ICP-AES in 11 certified reference materials (MOHL and STOEPPLER, 1989) and of up to 36 elements simultaneously by ICP-AES in certified reference, biological and environmental materials (QUE HEE and BOYLE, 1988) demonstrated the reliability of the method if appropriate digestion/dilution and instrumental parameters are chosen. 14 elements were determined by simultaneous ICP-AES using certified reference material for calibration (SCHRAMEL and LI-QIANG, 1983). 16 elements were determined by an ICP-AES system allowing simultaneous as well as sequential operation within rapid screening tasks for sewage sludge and soil samples (SCHRAMEL et al., 1982a). Sequential ICP-AES with a computer-controlled rapid scanning echelle monochromator was used to determine 17 major, minor, and trace elements in certified reference sediments and soils without any corrections for line overlap interferences and good agreement with certified values found (KANDA and TAIRA, 1988). 28 elements were determined in heavy slag and electrofilter ash from a large-scale city waste incineration plant by a sequential ICP-AES instrument (KARSTENSEN and LUND, 1989).

There is also some progress to be noted for the optimization of nebulizer systems which is very important for reliable routine applications (NIXON and SMITH, 1986; LUFFER and SALIN, 1986), particularly for the analysis of microsamples (ROUTH et al., 1987).

Further progress in sample introduction can be expected from the growing application of flow injection techniques (RUZICKA, 1986, see also Sect. 6.9) particularly if preconcentration can be used (ANDERSON and MCLEOD, 1988).

There are also attempts to reduce instrument and running costs in order to expand the use of ICP further by e.g. the design of low-power ICP and the introduction of effective cooling systems for lower argon consumption (DE GALAN and VAN DER PLAS, 1986; CORR et al., 1988).

Since the analytical sensitivities of elements such as antimony, arsenic, bismuth, selenium, and tin are comparatively poor in conventional ICP-AES, the hydride generation technique sometimes with particular generators can be combined with plasma – mainly DCP-AES. By this approach detection limits ≤ 1 µg/L were obtained in different materials (OLIVEIRA et al., 1983; EK and HULDEN, 1987; PERÄMÄKI and LAJUNEN, 1988; WATLING and COLLIER, 1988; LI et al., 1988). Recently a hydride generation method for tin using DCP-AES was reported, eliminating interferences from a number of transition metals by the addition of L-cystine, thus achieving a limit of 20 pg/mL for a 5 mL sample (BRINDLE and LE, 1988).

There is also appreciable experience in direct sample insertion, introduction by thermal vaporization and direct analysis of slurries of solid materials. Direct sample insertion is performed by e.g. introduction of microliter volumes of analyte solution utilizing a graphite cup (ABDULLAH et al., 1984), a graphite probe (BAXTER et al., 1986), or a wire loop and drying of the solution (SING and SALIN, 1989) sometimes with a remarkable gain in detection limits. Sample introduction by electrothermal vaporization (MITCHELL and SNEDDON, 1987) is also a promising technique. Direct powder injection is also possible in ICP-AES systems (NG et al., 1984).

Aqueous, chemically stabilized suspensions (slurries) of finely ground materials – particle size ≤ 30 µm – can be sprayed into a conventional plasma torch if special high-solids type nebulizers are used (SPARKES and EBDON, 1986, 1988; EBDON and WILKINSON, 1987a, b; EBDON and COLLIER, 1988). The precision and sometimes also accuracy of this approach are not comparable with that for the analysis of analyte solutions. Solid sampling ICP-AES, however, has some practical advantages because of its simple and quick preparation technique for screening and homogeneity studies.

Thermal vaporization approaches for ICP-AES, including also solid sampling and future possibilities including laser applications, were comprehensively reviewed by MATUSIEWICZ (1986). Examples of solid sample analysis with plasma AES can also be found in the already mentioned solid sampling AAS review (LANGMYHR and WIBETOE, 1985).

For the determination of lower concentrations liquid-liquid extraction is very effective frequently combined with the merit of signal enhancement (KUMAMARU et al., 1987).

Interferences of different origins affecting accuracy and precision are still a problem for plasma AES in complex matrices (RAMSAY et al., 1987; RAMSAY and THOMPSON, 1986, 1987). Thus there are many and effective ways to improve the reliability of data ranging from separation from interfering elements by e.g. liquid chromatography for rare earths (TIELROOY et al., 1988), different modes of internal standardization (SCHMIDT and SLAVIN, 1982; RAMSAY and THOMPSON, 1984, 1987) also applying flow injection (GINÉ et al., 1988) to the so-called simplex optimization (EBDON and CARPENTER, 1988) and Fourier transform spectrometry (SNOOK, 1988).

Finally, carbon furnace atomic emission spectrometry should be mentioned. There are many contributions by the late JOHN OTTAWAY and his coworkers to this approach. Though not commercially available it might be of significance for the future development of simple and versatile multielement methods (MARSHALL et al., 1983; LUNDBERG et al., 1986; LITTLEJOHN, 1988).

6.2 Electrochemical Methods

Electrochemical methods belong to the classical trace analytical methods. Polarography was introduced by HEYROVSKY in the early 1920s. From its beginning this method has undergone steady, sometimes slower, but recently faster, progress reflected in many books and reviews (e.g., HEYROVSKY and ZUMAN, 1959; MEITES, 1965; HEYROVSKY and KUTA, 1965; NÜRNBERG and KASTENING, 1974; BRAININA, 1974; BOND, 1980; NEEB, 1980; NÜRNBERG, 1981; WANG, 1985, 1988; OSTERYOUNG and OSTERYOUNG, 1985; HENZE and NEEB, 1986).
For metal ions commonly the following electrode process takes place:

$$M^{n+} + ne^- \underset{\text{oxidation}}{\overset{\text{reduction}}{\rightleftarrows}} M \ .$$

In some cases, e.g., during the reduction of Cr(VI) to Cr(III), the electrode process consists only of a change in valence state.

According to the terminology recommended by IUPAC (International Union of Pure and Applied Chemistry) the original method applying a dropping mercury electrode is called *polarography* whereas *voltammetry* comprises all approaches that evaluate current-potential relations, i.e., also polarography. Thus, in the following only the term voltammetry will be used for these techniques.

Voltammetry is based on Faraday's law that 1 mol of a compound transformed in an electrode process is equivalent to the very high electric charge of $n \times 96500$ Coulomb (where n is the number of electrons transferred in the elementary step of the electrode process). This is the reason for the detection power and wide dynamic range for all metals accessible to voltammetry.

Voltammetry requires that the metals to be determined are present completely dissolved in a non-interfering analyte solution. Since even traces of residual organic matter can severely interfere leading to erroneous results, decomposition procedures must be of utmost effectivity for biological and environmental materials prior to voltammetric determinations to achieve accurate data (see Sect. 5).

It is further necessary that the analyte solution has sufficient conductivity. This is realized by the addition of an inert electrolyte, e.g., the added acid (hydrochloric or perchloric acid) or buffer for pH adjustment. If necessary, alkali salts can be added as supporting electrolytes. In all cases, especially if very low levels are to be determined, contamination from these electrolytes has to be avoided or at least minimized as far as possible.

Each metal has a certain redox potential at a certain working electrode. This frequently allows the sequential simultaneous determination of several metals, e.g., of copper, lead, cadmium, zinc, and selenium(IV) at the mercury electrode. For trace analysis of mercury, copper, chromium(VI), and arsenic(III) the gold electrode plays an important role (BODEWIG et al., 1982; WANG, 1985; HENZE and NEEB, 1986).

Since dissolved oxygen interferes, the analyte solution has to be deaerated with ultrapure nitrogen. The given term is always the potential E and the measured term the current i resulting during the electrode process. Material consumption is negligible, since it occurs only in close proximity to the working electrode. The measured current i, which is proportional to the concentration of the metal to be analyzed in the solution, consists of two components: $i = i_F + i_C$; i_F is the Faradaic current linked with the electron transfer during the electrode process and i_c the capacitive current due to the reload of the phase boundary that corresponds to a condenser if the potential changes. If a high detection power is required i_F/i_C should be $\gg 1$. The theoretical voltammetric detection limit lies at $i_F/i_C = 1$. However, this cannot be achieved even at extremely low concentrations. The limiting factor is always the achievable blank.

In contemporary voltammetric trace analysis *differential pulse methods* have gained considerable importance due to a favorable signal to noise ratio (NEEB, 1980; NÜRNBERG, 1981; WANG, 1985; HENZE and NEEB, 1986). In differential pulse voltammetry E consists of a series of rectangular pulses superimposed on a voltage ramp. This allows $i_F/i_C \gg 1$ to be maintained down to extremely low levels around 10^{-12} mol/L (0.1 ng/L).

For metal levels in the analyte solution above 50 to 500 µg/L, direct determination with the dropping mercury electrode (DME) is commonly applied. This method is termed differential pulse polarography (DPP). In the 1960s it constituted the commencement of modern and powerful polarography and voltammetry (NÜRNBERG, 1981).

If the metal levels are lower, an in-situ enrichment is necessary using the so-called stripping technique (WANG, 1985). Metals that form an amalgam with mercury such as cadmium, copper, lead, zinc, etc. are accumulated by cathodic deposition in a (drop or film) mercury electrode. Commonly the solution is agitated or the electrode (film) rotated at this stage. Subsequently the amalgam is anodically oxidized in the determination stage. The mercury drop electrode can be applied down to approx. 0.1 µg/L. For ultratrace determinations, however, the use of the mercury film electrode (MFE) is necessary achieving detection limits at the ng/L level. In this technique, mercury(II) nitrate is added to the analyte and the mercury electrode is electrolytically formed by simultaneous in-situ deposition of a few hundred Å thick mercury film during the cathodic enrichment step for the metals to be determined. Alternatively metals such as mercury, bismuth, and arsenic are deposited as an elemental monomolecular film at a solid gold or graphite electrode. Subsequently the film is also anodically oxidized. This approach is termed *anodic stripping voltammetry* (ASV). Metals forming insoluble compounds with mercury such as selenium are accumulated by anodic oxidation and the compound is subsequently cathodically reduced. This approach is termed *cathodic stripping voltammetry* (CSV).

The development of microprocessors and minicomputers made possible the recent commercial implementation of *square wave voltammetry* (SWV). Its principle had already been proposed many yeary ago by BARKER. The square wave voltammetric wave form combines a large amplitude square wave modulation with a staircase wave form.

The resulting net current, a true differential signal, can be obtained at high effective scan rates. The voltammograms obtained display excellent sensitivity and rejection of background currents. The main advantage of this methodological version compared to common stripping techniques, however, is the lowering of analysis time without losing the properties of DPSV (BUCHANAN and SOLETA, 1983; OSTERYOUNG and OSTERYOUNG, 1985; OSTAPCZUK et al., 1986, 1988).

Another very promising approach successfully applied during the last years for an increasing number of elements is *adsorptive stripping voltammetry* (ASV) (WANG, 1986). Here, metals forming a complex with large organic ligands are accumulated by adsorption of the complex at the surface of, mainly, a hanging mercury drop electrode at a suitable potential. Subsequently the complex is reduced as already described for CSV (WANG, 1985, 1986; VAN DEN BERG, 1986, 1988).

Initial applications were for nickel and cobalt based on their dimethylglyoxime (DMG) chelates (GOLIMOWSKI et al., 1980; FLORA and NIEBOER, 1980; PIHLAR et al., 1981). Up to now many experiments have been conducted to determine the optimal ligands and working conditions for a number of elements such as cobalt and nickel in sea water applying nioxime as a much more effective ligand than DMG (DONAT and BRULAND, 1988), aluminum (WANG et al., 1985; VAN DEN BERG et al., 1986), chromium (GOLIMOWSKI et al., 1985), iron (WANG and MAHMOUD, 1987; HUA et al., 1988), gallium (WANG and ZANDEII, 1986b), manganese (WANG and MAHMOUD, 1986), molybdenum (VAN DEN BERG, 1985), palladium (WANG and VARUGHESE, 1987), platinum (VAN DEN BERG and JACINTO, 1988), selenium (BREYER and GILBERT, 1987), tin (WANG and ZADEII, 1987b), thorium (ZHAO et al., 1986), uranium (VAN DEN BERG and HUANG, 1984b; VAN DEN BERG and NIMMO, 1987; WANG and ZADEII, 1987a), vanadium (VAN DEN BERG and HUANG, 1984a), and yttrium together with some heavy rare earths (WANG and ZADEII, 1986a). At present most of these methods are applied predominantly to aqueous samples due to interferences in digestion solutions from solid materials. It is to be expected, however, that further systematic studies, also with new ligands and improved decomposition methods, will result in a more universal use of adsorptive stripping for at least some of the metals mentioned above.

Another possibility for the determination of low concentrations of elements that do not form amalgams is the use of catalytic effects occurring during polarographic reduction (FERRI and BULDINI, 1981; GOLIMOWSKI, 1989).

It could be shown that the combined application of classical polarography, different normal stripping modes, and adsorption voltammetry allows the sequential determination of up to eight elements in the same analyte solution with an average analysis time of 25 min for each element (ADELOJU et al., 1985).

Commonly voltammetric cells are made of glass or quartz. For determinations at ultratrace levels, however, PTFE cells are recommended because of negligible blanks. These cells have compartments for the counter − and the reference − electrode closed by membrane diaphragms.

In order to achieve reliable results, concentration evaluation in voltammetry has to be performed by the analyte addition technique with a typical precision around and below 5%. Commonly two additions are sufficient. The time needed for a complete determination ranges from 15 to 40 min for normal voltammetry including deaeration, but is significantly lower for the square wave mode as already mentioned above. If, however, the potential for the sequential determination of several elements in the same subsample and the use of parallel cells, which allows deaeration before the start of the analytical cycle, are considered, the sample throughput is comparable to that achieved in single element atomic absorption spectrometry with the graphite furnace that has similar detection power for many elements.

From Table I.4a-2 it is obvious that voltammetric stripping approaches achieve the lowest detection limits compared to other techniques for a considerable number of toxicologically and environmentally significant elements. The extraordinary progress of adsorptive stripping voltammetry during the last years contributed greatly to this situation. A further methodological advantage is in-situ enrichment directly from the analyte solution. In contrast to other enrichment techniques blanks are negligible for this technique. Therefore, voltammetry has proved to be an ideal method for the direct and very accurate determination of extremely low metal contents in sea water, inland water, and precipitation (MART, 1983; MART and NÜRNBERG, 1984, 1985; GOLIMOWSKI et al., 1985; VAN DEN BERG, 1986; DONAT and BRULAND, 1988). For some water types, however, UV digestion is required introducing, if properly performed, only spurious contamination (DORTEN et al., 1984).

A favorable technical property of voltammetric instrumentation besides the comparatively low investment and running costs (see Sect. 8) is its ruggedness. Thus also from this aspect it is excellently suited for field missions and use on board research vessels (MART, 1982; MART et al., 1983).

For body fluids as well as for solid biological and environmental materials the complete digestion necessary prior to voltammetric determination is the limiting factor for detection power. Blanks that are typically in the range of ≤1 to a few ng per digestion vessel – intakes commonly vary from approx. 100 to 400 mg dry weight – and element in most cases prevent the utilization of the extraordinary detection power of voltammetry. In this case practically attainable determination limits (definition see Sect. 10) are in the disappointing order of µg/kg. Compared to GFAAS, which allows the direct analysis of trace metals in body fluids and solid materials, this is somewhat inferior.

All voltammetric modes, provided a proper digestion is applied, are excellently suited for single and oligoelement analysis from the µg/kg level with appreciable accuracy and precision in routine and analytical quality assessment (NÜRNBERG, 1983a; NARRES et al., 1984b; ADELOJU et al., 1985; OSTAPCZUK et al., 1988). It has to be mentioned, however, that the potential range over which most metals are oxidized is relatively narrow. Therefore, in some cases it is difficult to resolve the responses from electroactive metals that are very close to each other so that overlap might occur. This, however, varies for different types of electrodes. Several approaches, e.g., chemical (masking), mathematical, and instrumental, have been used to solve the problem. Sometimes also a separation step prior to voltammetric determination may be necessary, e.g., for the separation of tin from lead (WANG, 1985).

A further and very promising property of voltammetry is that it is specific for distinct substances and thus can be successfully used in the important and growing research field dealing with the speciation of metals in natural waters (see Sect. 7.2).

Finally a comparatively new method, potentiometric stripping analysis (PSA), will be discussed. It does not attain the extraordinary detection power of voltammetry, however, it is an instrumentally simpler approach (JAGNER and GRANELI,

1976; JAGNER, 1978). The method is based on the potentiostatic reduction and amalgamation of metals and the subsequent registration of the potential-time curve when the reduced metals are reoxidized by means of mercury(II) ions. Simultaneously a mercury film electrode is formed by deposition of mercury on glassy carbon or fiber electrodes from different materials. The reoxidation reaction used for the determination of the preconcentrated metals occurs by simply disconnecting the potentiostatic circuitry and thus without an external current. Electrochemical reoxidation is performed with the nobler Hg^{2+} ions, the oxygen dissolved in the analyte solution, and other added oxidizing substances. The potential-time relation is recorded. The metals deposited as amalgams can be sequentially and selectively reoxidized based upon their individual electrochemical redox potential. The reoxidation time is proportional to the concentration in the analyte solution.

The advantages of PSA are that deaeration to remove dissolved oxygen from the analyte solution frequently can be avoided, and that the method is less sensitive to traces of organic substances which severely interfere in voltammetry (OSTAPCZUK et al., 1989). This allows rapid direct determinations of trace metals in, e.g., whole blood and serum (JAGNER et al., 1981). The recent introduction of a more expensive computerized PSA and improved, mainly fiber, electrodes (BARANSKI and QUON, 1986; HUILIANG et al., 1987a) significantly extended the working range down to the low µg/L level for various applications such as cadmium and lead in whole blood (ALMESTRAND et al., 1987), cadmium, lead, and copper in milk and milk powder (ALMESTRAND et al., 1986), mercury in tap water (HUILIANG et al., 1987b), and arsenic in sea water and urine (HUILIANG et al., 1988). The already mentioned properties together with fiber electrodes make PSA well suited for flow systems (ESKILLSSON et al., 1985; FRENZEL and SCHULZE, 1987; NEWTON and VAN DEN BERG, 1987; HUILIANG et al., 1987c; ESKILSSON and HARALDSSON, 1987). PSA also has potential for flow injection analysis (ALMESTRAND et al., 1988; SCHULZE et al., 1989). From the present state PSA can certainly be considered as a promising extension of electrochemical methods with the possibility of a further gain in importance and detection power.

Because of their particular potential, electroanalytical approaches can be considered to be complementary to rather than competing with other instrumental methods (NÜRNBERG, 1982; BERSIER, 1987).

6.3 Activation Analysis and Application of Radiotracers

The principle of activation analysis is alteration of the atomic nucleus by irradiation with particles such as neutrons, protons, helium nuclei, or high energy photons in appropriate irradiation facilities in nuclear reactors, accelerators, etc. (ERDTMANN and NÜRNBERG, 1973; BOWEN, 1980; GUINN and HOSTE, 1980; HEYDORN, 1984; CORNELIS, 1985; KRIVAN, 1985; ERDTMANN and PETRI, 1986; HOSTE, 1986; IAEA, 1986; KOSTA, 1986). During this reaction the atomic nucleus is excited either by the energy supply or by absorption of a particle. In the case of absorption of a particle, initially either an isotope of the target element (neutron absorption) or of another element (proton, helium nucleus) is formed.

The number of excited atoms formed depends on the intensity of the radiation, the irradiation time, and on the probability of absorption of radiation by the nucleus (cross section), and is directly proportional to the number of atoms present.

The atomic nucleus excited in this way is labile in most cases, i.e., it tends to reach an energetically lower level. This is always accompanied by emission of α- or β-particles and/or gamma-quanta. The emitted gamma-radiation is characteristic of each isotope and each distinct energy level and nearly monoenergetic (sharp lines in the electromagnetic spectrum). This is called *radioactive decay*.

Since the radiation intensity only depends on the initial concentration of the excited nuclei, the radioactive decay follows the rule of a 1st order reaction. An excited nucleus is thus statistically characterized by the decay constant $\lambda = \ln 2/T$ and the energy of the emitted line(s). In this equation T is the half-life, i.e., the time after which half of the excited atoms have "decayed", which means that they have reached the lower energy level. Half-lives range from fractions of seconds up to years. Identification and quantification is performed by measuring the energy (typically from a few keV up to the MeV region) possibly the half-life, and the radiation intensity (decays per unit of time) of the isotope representing the element to be determined.

High intensity of the exciting radiation and large cross section combined with favorable half-lives lead to a high sensitivity for many elements. The multitude of lines and half-lives, however, make activation analysis a logistically complex task.

After World War II, nuclear analytical methods, in the first instance mainly neutron activation analysis (NAA), constituted for more than two decades the most reliable and most powerful multielement analytical method.

The introduction of germanium-(lithium) detectors that achieved excellent energy resolution for gamma-radiation allowed, in combination with computer evaluation, the simultaneous instrumental determination of a multitude of elements. In the case of extremely complex spectra and/or high radiation background from the matrix and also for use of the whole methodological potential, radiochemical separations for groups of elements or single elements were performed. This version is called radiochemical activation analysis in contrast to the solely instrumental approach.

Because of these properties the merit of all modes of activation analysis is the gain achieved for insights into bioinorganic relations for essential and non-essential trace metals in medicine, biology, and ecology. This was also the case for better knowledge of normal levels of trace metals in biological systems (BOWEN, 1966; IYENGAR et al., 1978; VERSIECK and CORNELIS, 1980; CORNELIS, 1985; VERSIECK, 1985).

Despite the tremendous progress in the development and application of AAS, ICP-AES, and other modern instrumental analytical techniques, activation analysis, mainly NAA, remains a very useful multielement method for fingerprint studies in many biological and environmental materials. Since it is, if properly applied (CORNELIS et al., 1982), less prone to contamination and – particularly in the radiochemical mode – quite reliable, it is still of paramount importance for quality assessment studies in comparison to other methods (see e.g., IAEA, 1985; ESPRIT et al., 1986; DE GOEIJ, 1988; ZEISLER et al., 1983, 1988; XILEI et al., 1988).

Typical detection limits for the metals discussed in this book and for instrumental NAA are given in Table I.4a-2.

Instrumental neutron activation analysis achieves for different aliquots of a sample material and under optimal conditions on average a precision between 5 and 10% which is somewhat inferior to that of modern GFAAS. Investment costs for a complete instrumental gamma-spectroscopic setup are comparable to a fully computerized AAS instrument (see Sect. 8 and Table I.4a-3). A broad routine application of nuclear analytical methods, however, is hampered by dependence on only a few nuclear research reactors – in some industrialized countries with decreasing tendency – or other irradiation facilities such as linear accelerators, cyclotrons, etc. Of additional significance are the relatively high permanent costs for irradiation and storage and the sometimes cumbersome radiochemical separations in lead-shielded analysis cells. The comparatively long "cooling" (storage) times often necessary for the decay of short-lived interfering radionuclides from matrix constituents should also be mentioned. Moreover, expensive security precautions and the need for specially trained and very experienced staff must also be considered if these methods are compared with other less demanding ones. This makes activation analysis techniques, despite their invaluable properties, time-consuming (i.e., slow), expensive, and to some extent cumbersome and thus not very frequently applied.

However, there are still fields where new versions of activation analysis techniques will be of benefit. An example is the in vivo elemental analysis of cadmium by the use of a collimated beam of fast neutrons from a neutron (e.g., ^{238}Pu-Be) source for the determination of this element in the liver and kidneys of exposed persons by measuring the generated prompt gamma-rays. The technique can also be used in principle for some other elements (SCOTT and CHETTLE, 1986). Another example is the use of "cold" neutron beams to generate short-lived isotopes by prompt gamma-neutron activation that is also expected to provide new possibilities for the sensitive determination of elements that are less accessible under normal reactor conditions and for the analysis of thin samples. Due to the fact that only extremely short-lived radioisotopes are produced by this technique, these samples can also be used after analysis by cold neutrons for other trace analysis methods (LINDSTROM et al., 1987; ROSSBACH et al., 1988).

In laboratories that are appropriately equipped, the *use of radioactive isotopes* in methodological development as well as in biochemical and environmental studies is a very promising approach (MITCHELL, 1982). The application of radioisotopes has proved to be extremely valuable in studies on reaction mechanisms and about accuracy and precision in various analytical procedures (SCHMID and KRIVAN, 1985; KRIVAN et al., 1985; KRIVAN, 1986, 1987; ARPADJAN and KRIVAN, 1988; KRIVAN and HAAS, 1988) but also for radiotracer experiments, e.g., in marine biological systems (IAEA, 1975).

6.4 X-Ray Fluorescence (XRF)

If atoms are subjected to radiation of a suitable energy, this leads to the elimination of electrons from the inner (K, L, M...) atomic shells. Following this, electrons from outer shells drop, due to distinct laws, into free positions with the emission of comparatively simple electromagnetic radiation in the energy range from 0.6 to 120 keV (wavelengths 2 to 0.01 nm) that is unique and can be exactly computed for each excited element. The frequency of the emitted lines is proportional to the square of the nuclear charge number Z of the respective element. Therefore, this radiation can be used after wavelength or energy dispersion for qualitative detection and quantitative determination of all elements with $Z>6$. These rays were called X-rays by their discoverer RÖNTGEN.

X-rays can be induced by X-ray tubes, direct X-ray radiation, gamma-radiation from radionuclide sources but also by particles such as electrons, protons, α- and heavier particles (BERTIN, 1971; JENKINS, 1975; WINEFORDNER, 1976; KATSANOS, 1980; LIEBHAFSKY et al., 1986; WHISTON, 1987). The latter is termed PIXE. PIXE is particularly suited for thin targets and elements with lower nuclear charge numbers and has remarkable detection power down to the µg/kg-level in particular applications (VALKOVIC, 1977; GONSIOR and ROTH, 1983; MAENHAUT, 1987; KLOCKENKÄMPER et al., 1987; JOHANSSON and CAMPBELL, 1988).

Both versions of X-ray fluorescence, the wavelength and energy dispersive modes, are perfect multielement methods with high sample throughput and are excellently suited for fingerprint studies in a number of appropriate materials.

With typical absolute detection limits in the order of 10^{-6} g, however, conventional XRF techniques can only be used for heavy metal determinations at higher contents from approx. 5 mg/kg (TALBOT and CHANG, 1987), i.e., in metal polluted biota, sewage sludge, soils, and dust filters. If lower contents have to be determined preconcentration techniques are usually necessary (see Sect. 5.1). Precision is quite good with values $\leq 2\%$ under optimal conditions. This makes the method well suited for homogeneity studies, e.g. during the development of certified reference materials.

As far as accuracy is concerned, the matrix dependence of each analytical line from each element present with $Z>20$ can cause erroneous data due to irregular stray effects of the exciting radiation (LIEBHAFSKY et al., 1986; HELSEN and VREBOS, 1986). Thus for XRF a proper sample preparation (superfine milling, fusion) and use of certified reference materials with a matrix composition as close as possible to the material to be analyzed is mandatory. If these materials are not available, accuracy has to be assessed by comparative analyses with independent methods to arrive at reliable results. If high elemental levels have to be analyzed, dilution with a matrix with low Z, e.g. cellulose, decreases those effects and is favorable for the production of reference samples.

If the exciting radiation is collimated by special mirrors on a very small area (microprobe techniques) very low detection limits in the order of 10^{-9} g can be achieved with e.g. PIXE techniques (VALKOVIC, 1977; GONSIOR and ROTH, 1983; JOHANSSON and CAMPBELL, 1988). If a synchrotron beam is used absolute quantities below 10^{-12} g can be measured with a beam spot size of less than 10×10 µm for the elements Zn ($Z = 30$) to K ($Z = 19$) (GIAUQUE et al., 1988). This is a promising technique for metal distribution studies in biological structures such as tissue, hair, and sediments (see Chapter I.7e, Sect. 3.2).

A comparatively new technique, total-reflection X-ray fluorescence analysis (TXRF), has gained importance during the last decade. The principle is a significant improvement of the signal-to-background ratio by total reflection of the exciting beam from conventional radiation sources at a flat sample support with absolute detection limits that are 2–3 orders of magnitude lower than those of conventional XRF (PRANGE, 1989). This method, if proper sample preparation is performed, including various preconcentration, digestion and separation procedures for interfering matrix constituents (PRANGE and SCHWENKE, 1989), has been increasingly and successfully used for multielement analysis in a number of matrices ranging from rain water and sea water to numerous solid materials (MICHAELIS and PRANGE, 1988). It is also applicable to the direct analysis of solid samples (BOHLEN et al., 1987). Because of the outstanding analytical potential of TXRF its detection limits have been included in Table I.4a-2.

A further increase in detection power and spatial resolution can be expected if excitation in TXRF is performed via synchrotron radiation with absolute detection limits down to the femtogram range (IIDA et al., 1986; PELLA and DOBBYN, 1988; JONES and GORDON, 1989).

Finally it should be noted that in vivo elemental analysis is also feasible for XRF, mainly used for lead in bones but also feasible for cadmium in kidney, mercury, strontium, and platinum (WIELOPOLSKI et al., 1983; SCOTT and CHETTLE, 1986).

6.5 Mass Spectrometry (MS)

The principle of mass spectrometry of metals is the generation of gaseous ions. This is performed thermally, by a spark of high potential, or by an electron – or ion – current. Subsequently, the generated ions or molecular fragments are separated by energy-mass focusing in strong magnetic fields. The qualitative detection or quantitative determination of charged particles was performed in classical mass spectrometry by photoplates. Nowadays, however, this is achieved by direct-reading electronic systems. A unique advantage is that all elements of the Periodic Table are accessible to the method with typical absolute detection limits in the range of 10^{-9} to 10^{-12} g (AHEARN, 1972; MORRISON, 1980; FACCHETTI, 1982; DAVIS and FREARSON, 1987; ADAMS et al., 1988; HEUMANN, 1988b). Simultaneous multielement analysis is predominantly performed by spark-source mass spectrometry (SSMS). This method requires transformation into a conducting form (target), if not primarily a conducting material, e.g., metal is taken, by mixing with ultrapure graphite. Thus practically attainable detection limits depend predominantly upon target preparation and purity of the graphite. Precision under optimal conditions is around 10% while accuracy is generally 15–30% with calibration or within an order of magnitude (BACON and URE, 1984). These properties, combined with a very expensive instrumentation, have led to a diminished use of this method which is less suited for flexible tasks in biological and environmental materials.

Excellent precision and accuracy can be achieved if isotope dilution mass spectrometry (IDMS) is applied. The method, however, requires complete decomposition of the sample. Prior to or after (the latter is less recommendable and cannot be applied in all cases) decomposition, an exactly known amount (spike) of an appropriate isotope of the element to be determined is added to the sample and subsequently the mass ratio is measured. The method, of course, is only applicable to elements with at least two stable isotopes or to those that possess radioisotopes with half-lives of more than 10^6 years. Therefore, a number of elements are not accessible for this method. These are: sodium, scandium, yttrium, cobalt, rhodium, gold, phosphorus, arsenic, fluorine and the rare earth elements, praseodymium, terbium, holmium, and thulium. The elements cesium, beryllium, niobium, manganese, aluminum, bismuth, iodine, and thorium possess long-living radioactive isotopes that do allow in some cases (e.g., for cesium, beryllium, and bismuth) application of IDMS (HEUMANN, 1988a; DE BIÈVRE, 1989).

IDMS is applicable to most ionization modes, e.g., thermal ionization (HEUMANN, 1980), spark source (MOODY and PAULSEN, 1988; JOCHUM et al., 1988b) electron impact (HEUMANN, 1988a), and field desorption (SCHULTEN et al., 1984; HEUMANN, 1988a).

Since the determination of mass ratios is much more precise than that of mass/charge ratios, this method attains excellent precision and accuracy under optimal experimental conditions. It needs, however, in most cases chemical pretreatment, i.e., group or single element separation and preconcentration procedures prior to determination (HILPERT and WAIDMANN, 1986, 1988; HEUMANN, 1988a). Thus IDMS requires utmost contamination precautions and control for reliable application down to the µg/kg or even ng/kg level, and a very experienced staff as well as contamination-minimized and technically well equipped laboratories, adapted to the particular task (e.g., SCHAULE and PATTERSON, 1981; STUKAS and WONG, 1983; MICHIELS and DE BIÈVRE, 1986; VÖLKENING and HEUMANN, 1988a). The attainable blanks during sample preparation and the perfect transformation of the added spikes into the chemical forms of the elements (isoformation) to be analyzed are of utmost importance, since the separation procedures are not usually quantitative. The latter, however, provided complete isoformation and dissolution can be achieved, is not necessary for this approach (HEUMANN, 1988a).

Due to the high costs for instrumentation and laboratory equipment and the time required for a complete set of analyses (samples and blanks) that is in the order of one week for one material and a few elements, IDMS cannot be considered as a routine method but is the method of choice if extreme requirements for accuracy and precision, e.g., for certification of reference materials and control of other methods have to be met (BARNES et al., 1982; DE BIÈVRE et al., 1988; GÖTZ and HEUMANN, 1988). The extreme precision in determination of isotope ratios makes MS a very reliable method for the identification of e.g. anthropogenic and geogenic lead sources in the environment (SCHLADOT et al., 1980; TRINCHERINI and FACCHETTI, 1983) and for metabolic studies using stable isotopes in man and experimental animals (RABINOWITZ et al., 1973; TURNLUND, 1983; TERA et al., 1985).

Mass analysis termed LAMMA, applying laser beams for ionization (FEIGL et al., 1984), or mass spectrometers coupled with ion microprobes – termed secondary ion mass spectrometry (SIMS) – are means to determine trace element distributions on surfaces of solid materials, in biological structures and in extremely small samples (MORRISON, 1980; ODOM et al., 1988; BENNINGHOVEN et al., 1987; JOCHUM et al., 1988a; ADAMS et al., 1988). Absolute detection limits are extremely low and thus allow the study of local effects and metal distribution at very low levels. The systems, however, are very expensive and cannot be used for exact quantitation due to calibration problems.

Other, very sensitive and comparatively new mass spectrometric methods are, e.g., fast atom bombardment MS (SELF et al., 1987), resonance ionization MS using thermal ionization and lasers (MOORE et al., 1984), and so-called direct loading IDMS (NAKAMURA et al., 1989). DL-IDMS permits precise concentration determinations of rare earths, alkaline earth metals, and alkali metals directly in acid digests of minerals without previous chemical separation down to approx. 10^{-15} mol.

Recently there has also been a renaissance of glow discharge mass spectrometry, already used for early mass spectrographs 50 years ago, for the direct bulk analysis of solids (HARRISON et al., 1986; JAKUBOWSKI et al., 1988).

6.6 Inductively Coupled Plasma-Mass Spectrometry (ICP-MS)

ICP-MS is a new analytical approach. Commercially introduced 1983 (HOUK, 1986) despite some residual problems, the method has been widely accepted and thus is already used in many different areas with growing tendency. At present there are three commercial instruments available. There are some detailed and critical reviews about performance and potential (e.g., DOUGLAS and HOUK, 1985; SELBY and HIEFTJE, 1987; DATE and GRAY, 1988; HIEFTJE and VICKERS, 1989). From the theoretical point of view the excellent ionization efficiency of an inductively coupled plasma is an ideal source for inorganic mass spectrometry. In combination with simple spectra, wide dynamic range, high detection power, and isotope-ratio capability of mass spectrometry a considerable analytical potential allowing high sample throughput can be expected. In particular, detection power and element coverage are significantly superior to ICP-AES (see Table I.4a-2). This, however, can also lead to severe blank problems for a number of elements.

Initial problems encountered with the interface connecting the plasma to the mass spectrometer ("pinch effect") have been substantially reduced (DOUGLAS and FRENCH, 1986). This represents an important technical improvement. Still remaining problems are, e.g., poorer precision compared to ICP-AES if isotope dilution is not used and various matrix influences that range from signal suppression at higher salt contents − only minimized by dilution of the analyte solution − to molecular ion interferences (MUNRO et al., 1986; PICKFORD and BROWN, 1986; LUCK and SIEWERS, 1988; GILLSON et al., 1988). If digestion is performed, nitric acid is very well suited due to only small or negligible interferences, while other acids (HCl, H_2SO_4, H_3PO_4) lead to interferences (MUNRO et al., 1986). Signal suppression due to alkali salts can be corrected by an internal standard (VANDECASTEELE et al., 1988a) or by an on-line sample treatment to separate interfering alkali and alkaline earth elements and anions from the elements to be determined by adsorption on appropriate columns and elution prior to ICP-MS (PLANTZ et al., 1989). This method was also of particular benefit for the determination of trace metals in sea water (MCLAREN et al., 1985) and in an open ocean water reference material (BEAUCHEMIN et al., 1988a). ICP-MS is thus excellently suited for the rapid direct analysis of a large number of elements in inland water (DIETZ, 1986; SANSONI et al., 1988; HENSHAW et al., 1989) and in ultrapure acids (PAULSEN et al., 1988).

Finally some examples are given from the exponentially growing literature on applications of ICP-MS. These include determination of trace metals in organic based materials (HUTTON, 1986), milk and blood, also isotope ratios, (DEAN et al., 1987a; DELVES and CAMPBELL, 1988), foods (MUNRO et al., 1986; SATZGER, 1988), different environmental and reference materials (WARD, 1987), marine sediments (MCLAREN et al., 1987), methylmercury and trace metals in marine biological reference materials and sea water applying standard additions, isotope dilution isotope ratios for chloride interference and flow injection (BEAUCHEMIN et al., 1988a, c, d; RIDOUT et al., 1988), and of platinum metals and tin in geological and environmental materials (GREGOIRE, 1988; BRZEZINSKA-PAUDYN and VAN LOON, 1988).

Feasibility of solid sample introduction by slurry nebulization of certified reference materials and industrial catalysts was demonstrated for ICP-MS (WILLIAMS et al., 1987). A further gain in detection power was shown for a number of metals in various materials using plasma source MS coupled with electrothermal sample introduction (PARK et al., 1987; GREGOIRE, 1988). Pneumatic nebulization and hydride generation was applied to isotopic analysis of selenium in human metabolic experiments (JANGHORBANI and TING, 1989).

6.7 Spectrophotometry (Colorimetry)

(UV/Vis-)Spectrophotometry is the optical spectroscopy of molecules. In contrast to atomic spectroscopic methods that are characterized by sharp lines, absorption and emission, processes of molecules are much more complex since electron transfer in molecules is split into numerous terms of oscillation and rotation (WINEFORDNER, 1976; KNOWLES and BURGESS, 1984; LEVER, 1984; NOWICKI-JANKOWSKA et al., 1986; BURGESS and MIELENZ, 1987).

In the gas phase with minimal interactions between molecules the different energy terms occur as series of lines (bands) in absorption and emission that might cause interferences in atomic absorption and atomic fluorescence spectrometry. In solution, however, due to strong interactions, i.e., collisions, between the analyte molecules and molecules of the solvent, absorption continua occur with characteristic relatively broad absorption or fluorescence maxima. Therefore, a distinct colored metal compound can hardly be reliably determined in a complex sample containing other metals. Colorimetric trace metal determinations are thus based, commonly after sample decomposition, on selective separations from interfering ions (e.g., ABBASI et al., 1988). The latter are commonly performed by ion exchange or chelate formation and solvent-extraction. Suitable chelates can also be used directly for determination after separation from the matrix.

The concept of spectrophotometric instruments is similar to that for atomic absorption. The commonly occurring broadband spectra, however, do not require the same high quality optics. Modern spectrophotometers, predominantly offered as double-beam instruments, provide high stability, low noise, and the advantages of computerized background control, area integration, sample supply, derivative spectroscopy (MELGAREJO et al., 1989), diode array (SANZ et al., 1988) and Fourier transform techniques, etc. Precision on average is below 5% (e.g., KOMAIA and ITOH, 1988; GARCIA et al., 1988; MELGAREJO et al., 1989). The most sophisticated systems, sometimes with special cuvettes, allow very precise element determinations down to the µg/L level. Highly sensitive spectrophotometric determinations are possible if a catalytic reaction is performed prior to the determination step (MÜLLER, 1982).

Spectrophotometers are used with growing tendency in biochemistry, organic and clinical chemistry and as detectors for e.g. HPLC (THRELFALL, 1988). For total metal determinations in environmental chemistry and in occupational exposure, however, the method is less frequently applied and has been replaced in many cases by atomic spectroscopic and electrochemical methods with some exceptions, e.g. if high elemental levels have to be determined, where it still serves as a reliable routine or reference method (e.g., KOELLING et al., 1988).

6.8 Chromatographic Methods

Analytical chromatography is based on the fact that different compounds travel under distinct conditions at different speed through a system consisting of a mobile and a stationary phase. The stationary phase commonly consists of a solid carrier material filled into suitable tubes (columns) made of quartz, glass, or metals and is impregnated with appropriate chemicals. For capillary gas chromatography, the solid phase is a thin layer or film on the inner surface of the column. During analysis the liquid phase passes the column and separation of different components of a mixture injected at the top of the column occurs by continuous adsorption and desorption depending on the properties of column and analytes over the whole length of the column. This allows the separation of numerous compounds in the same analyte solution. The chromatographic procedure leads to distinct fractions in the eluent containing several components and producing more or less sharp peaks in the time/concentration evaluation. Identification and quantification is performed, depending on the particular class of compounds, with various detectors that are able to differentiate between background (mobile phase) and the compound to be analyzed (DEYL et al., 1975).

Thus chromatographic methods consist of a combination of a compound-specific separation technique with a compound- or element-specific detection system and are also well suited for speciation analysis, treated in Sect. 7.

In *gas chromatography* (GC) the mobile phase is an inert gas (nitrogen, hydrogen, argon). Volatilization of the compound to be determined is performed by heat. This could cause problems with thermally less stable compounds.

Modern gas chromatographs have reached a high technical performance. Instruments with automated pressure and temperature control and sample changers are offered at modest prices (see Table I.4a-3). Commonly used are conductivity (CD), flame ionization (FID) and electron capture (ECD) detectors. For metal determinations in various liquids atomic spectrometric detectors (plasmas, AAS, and AFS) are directly coupled with gas chromatographs (EBDON et al., 1986).

Since inorganic compounds are not accessible to gas chromatographic techniques they have to be transformed into appropriate organometallic compounds as described for numerous elements (e.g., Co, Tl, Cd, Cu, Ni, Pb, Fe, In, Pd, Sb, Hg, Bi, Zn, Cr(VI), Au, Pt, Rh, Ir, and Os) via di(trifluorethyl)dithiocarbamato chelates using capillary GC and either an FID or an ECD, the latter is more sensitive (SCHALLER and NEEB, 1986). This technique allows very sensitive multielement determinations. Aluminum in sea water has been determined down to 0.6 nmol/L by the use of GC with ECD and transformation of aluminum into the 1,1,1-trifluoro-2,4-pentanedione derivative (MEASURES and EDMOND, 1989). An on-column hydride generation method has been described for the production of volatile hydrides of Sn, As, and Sb that was also used to determine the full range of butyltin species in water (CLARK et al., 1987).

In *liquid chromatography*, today mainly applied as high performance liquid chromatography (HPLC), the mobile phase is always a liquid. Therefore, the working temperatures are comparatively low. This makes the determination of thermally less stable compounds possible (KNOX et al., 1978; DOLAN et al., 1987).

Costs for HPLC systems are significantly higher than for gas chromatographic instruments. The reason is that HPLC operates under elevated pressure requiring technically more sophisticated systems (see Table I.4a-3). Detection can be performed by spectrophotometry, electrochemically, also for metal chelates (BOND et al., 1986) and especially for metals by atomic spectrometric methods (EBDOU et al., 1987a; NYGREN et al., 1988a). Examples are the determination of Cu, Ni, and Pb in urine after transformation into dithiocarbamate complexes (BOND et al., 1986), of Cd, Co, Cu, Hg, and Ni in water using bis(ethoxyethyl)dithiocarbamate (MUNDER and BALLSCHMITER, 1986), of Zr(IV), Hf(VI), Fe(III), Nb(V), Al(III), and Sb(III) as chelates at N-methylfluorohydroxamic acid (PALIERI and FRITZ, 1987) on-line preconcentration and separation of Cd, Pb, Hg, Cu, Co, Ni, and Bi as dithiocarbamate complexes by reversed-phase liquid chromatography and determination with an UV-VIS-diode array detector (IRTH et al., 1987), and multielement analysis by HPLC (Cu, Be, Al, Ga, Pd, and Fe) after solvent extraction with acetylacetone (ICHINOKI et al., 1988).

Another versatile analytical approach which has become remarkably popular since its introduction in 1975 (SMALL et al., 1975; FREI et al. 1988, 1990; PUXBAUM and TSITOURIDOU, 1987) is *ion chromatography* using various chelating cation exchange resins. Due to these properties, usually no pre-column derivatization is necessary. Separations, however, are performed as in anion exchange by elution with appropriate solutions and sometimes post-chromatographic derivatization for quantitative determination. The multielement potential and the low detection limits achievable with various detectors e.g., UV/VIS, conductivity and pulsed amperometry, are remarkable (RUBIN and HEBERLING, 1987; WEISS, 1987; YAN and SCHWEDT, 1987). This approach is not only complementary to atomic spectrometry and voltammetry in that it combines total metal and metal species determination, it also allows, at least in aqueous matrices, the simultaneous analysis of cations and anions (JONES and TARTER, 1985). In this context a newly published comprehensive handbook of ion exchange resins and their analytical application has to be mentioned (KORKISCH, 1988; see also recommended literature on p. 206).

The *ring-oven technique*, a special mode of spot analysis, for which papers published since 1976 have been reviewed (WEISZ, 1987), provides many useful approaches for the semi-quantitative determination of numerous metal ions. Metal chelates can also be separated and determined by *thin-layer chromatography* (SCHUSTER, 1986).

An outstanding example of the potential of analytical chromatography is the progress achieved in separation and determination of the lanthanides (ROBARDS et al., 1988). As for other trace and ultratrace methods, contamination precautions are unavoidable to achieve reliable results.

6.9 Future Prospects

Though this section is far from being comprehensive, a few, in the opinion of the author, important trends and prospects shall be mentioned here in the order of the methodological sections.

In general it can be said that further extended *computer application*, computer evaluation and increased use of *robotics* in the laboratory up to complex *expert systems* appear to be unavoidable for the proper solution of many problems already faced in analytical chemistry and those that might appear in the near future (SALIT and CIRILLO, 1986; LOCHMÜLLER and LUNG, 1986; KLAESSENS and KATEMAN, 1987; BRAITHWAITE, 1987; BETTERIDGE, 1987; SHARP, R. L., et al., 1988). Techniques that will stimulate and complement many analytical methods are the extended use of *flow injection systems* for separation, preconcentration and introduction into various analysis instruments (RUZICKA, 1986; RUZICKA and HANSEN, 1986, 1988; PACEY, 1988; DREWS et al., 1988; BURGUERA, 1989; FRESENIUS and LUDERWALD, 1989), the broad application of *lasers* and of combined methods that use lasers and allow either direct analysis of solids or decomposition, separation from bulk elements and determination in a closed on-line system thus minimizing contamination and enhancing detection ability (OMENETTO and HUMAN, 1984; WINKLER and WRIGHT, 1988; TÖLG, 1988a, b, 1989).

In *sample preparation* further evaluation of the high pressure asher system and microwave decomposition techniques can be expected. The latter can possibly be improved towards higher decomposition temperatures and safer operation. Research in polymer chemistry could provide plastic and inorganic materials with extremely smooth surfaces and excellent chemical and thermal stability for these systems in the near future.

In *atomic absorption spectrometry* the oligo- and multielement instruments already introduced, improvements in sample transportation and preconcentration (RUZICKA and ARNDAL, 1989) for flame AAS by FIA and improved nebulization (BERNDT, 1988) and progress, e.g. automation, in solid sampling systems (KURFÜRST et al., 1989) will certainly open up new application fields. Technical improvements of furnaces and of direct analysis in graphite furnaces by further optimization of matrix modification and the use of external gases can also be expected. Promising developments during the last decade might also lead in the near future to new modifications and in some cases to commercial introduction of e.g. ICP as an atomizer for AAS (LIANG and BLADES, 1988), the multielement methods *coherent forward scattering* (HERMANN, 1988) and *furnace atomic nonthermal ex-*

citation spectrometry (FANES) (FALK et al., 1984, 1986; FALK, 1988). Another path for multielement AAS could be the use of a modulated Grimm-type glow discharge plasma as primary light source (OHLS et al., 1988).

Atomic fluorescence spectrometry with laser excitation in graphite furnaces is a very attractive method and attains extremely low detection limits in the picogram to subpicogram range (GOFORTH and WINEFORDNER, 1986, 1987; DOUGHERTY et al., 1987a; PRELI et al., 1987). If laser-induced double resonance fluorescence of lead with graphite tube atomization is concerned detection limits at the femtogram level can be achieved (LEONG et al., 1988). Despite these documentations of an impressive potential, however, some doubts may be expressed about broader use, particularly about commercial instruments in the near future.

Atomic emission spectrometry will certainly benefit from progress in FIA methods (RUZICKA and HANSEN, 1986, 1988) and more effective nebulizer systems (BERNDT, 1988) as well as from intelligent applications of solid sample analysis, e.g., by vaporization into the plasma source. Another example of the use of electrothermal sample introduction is the extremely sensitive determination of arsenic, selenium, and antimony using metastable transfer emission spectrometry (HOOD and NIEMCZYK, 1987). Of particular interest for difficult elements like rare earths might be laser-excited ionic fluorescence spectrometry in an ICP (TREMBLAY et al., 1987). Fourier transform spectrometers provide a much higher resolution and complete spectral record in comparison to scanning monochromators for AES in the visible and UV regions (FAIRES, 1986, 1987; THORNE, 1987). Thus progress and improvements of existing instrumentation towards high resolution Fourier transform atomic spectrometers can be additionally predicted.

Electrochemical methods have gained detection power in the last decade due to permanent progress in adsorption voltammetry and potentiometry for which further and systematic studies can be expected, particularly for extension to environmental and biological materials. This progress will doubtless continue in combination with a further improved instrumentation. Commercial introduction of computerized PSA systems is expected for numerous direct applications in biology and medicine but also for environmental samples such as natural and soil pore waters. Problems observed hitherto in sample decomposition have now been overcome by the HPA system (WÜRFELS et al., 1987, 1989a–c). This will greatly extend the application range of voltammetric methods in the near future.

Neutron activation analysis will still play an important role in accuracy assessment for many purposes. Due to its freedom from contamination it might be also of value for the analysis of some important pollutants by radiochemical approaches. Cold neutrons are being explored for their usefulness in the detection of elements previously not accessible for NAA and for the characterization of surfaces and surface layers. The continuing use of radiotracers for methodological studies and the production of new ones certainly is of permanent or even increasing importance.

In *X-ray fluorescence* the progress of the total reflection mode will most probably continue, possibly also triggered by the HPA decomposition method. Application of synchrotron radiation sources will allow local analysis with high detection power and further improve TXRF, however, at much greater expense.

For *mass spectrometry* and *inductively coupled plasma mass spectrometry* improvements towards simpler and cheaper but for many analytical purposes still sufficient quadrupole mass spectrometers in IDMS have already begun (HEUMANN et al., 1985). Further progress is to be expected in ICP-MS with more competing manufacturers, further improved plasmas and interfaces and also the introduction of (however, much more expensive) instruments with a better resolution for further minimization of ion interferences.

Spectrophotometric approaches, mainly in combination with *chromatography* and FIA systems with laser excitation (MORIN et al., 1989), chemiluminescence detection (SAKAMATO-ARNOLD and JOHNSON, 1987; BOYLE et al., 1987; JONES et al., 1989) split-beam optics (BISHOP, 1989) and diode array techniques, the latter is also useful for various other applications (KENDALL-TOBIAS, 1989), can attain extremely low detection limits that are of importance for future studies in speciation as well as in biological and environmental systems with very low metal contents.

Photoacoustic spectroscopy, a comparatively new method, can be improved and its range extended if laser excitation is applied (GUO et al., 1988). At the moment, however, the method might only be of value for basic studies with solids and films to investigate its analytical potential.

7 Separation, Identification, and Quantification of Chemical Compounds (Speciation)

7.1 Introduction

The numerous and quite different analytical methods applied today allow on the one hand in principle total analyses for the majority of elements and materials if properly performed with acceptable accuracy. This, whatever "acceptable accuracy" means, applies to some difficult elements and lower levels or very complex matrices only if the analyses are performed in experienced laboratories. Even for them, however, problems still exist if utmost accuracy and precision are concerned (see also Sect. 9).

On the other hand, it is now well recognized that the essential or toxic, i.e., physicochemical and biochemical action of an element depends on its chemical form (compound and/or valence state) and – if there are several forms present – on the ratio of these (BERNHARD et al., 1986). Speciation in general, of course, is even more complex and more cumbersome to perform than total elemental analysis and thus is one of the greatest challenges for analytical chemists. In general successful work has already been done to approach some important tasks, but this can only be regarded as a preliminary step into a new field of progress as well as of pitfalls (BERNHARD et al., 1986; LANDNER, 1987).

The methodologies already available provide tools for many studies in this area: compound specific methods such as chromatography, voltammetry, and spectrophotometry on the one hand, and combination of element specific ones (e.g., AAS, ICP-AES, NAA) with compound specific, on the other hand. In the following sections first sea and inland waters shall be treated where some useful approaches are already available. The much more difficult tasks in biological and solid materials are treated in Sect. 7.3.

7.2 Aqueous Samples

(see also Chapters I.3 and I.5; KRAMER and ALLEN, 1988)

For speciation analysis of an element in the aqueous environment, in order to understand its biological and/or geochemical fate, its amounts in particulate matter and dissolved forms such as inorganic and organometallic species, complexes with organic compounds and adsorbed on various types of colloidal particles have to be considered (CHAU, 1986a, b; ANDREAE, 1986; DUINKER, 1986; CRAIG, 1986). All these forms might be present in a thermodynamic equilibrium or not (LAXEN and HARRISON, 1981; FLORENCE, 1986; KESTER, 1986).

Collection of water samples prior to the application of analytical methods including solvent extraction, chromatography, and electroanalytical techniques requires utmost care, as described in Sect. 4.2, particularly if contamination is concerned. Apart from the often performed filtration (0.45 µm), necessary if only dissolved species have to be investigated, further treatment and addition of acids should be strictly avoided. Long-term storage is also not acceptable since even at lower temperatures around $-20\,°C$ wall adsorption and bacterial activity occurs that might change the ratio of species significantly. The influence of extremely low temperatures, e.g. $<-80\,°C$, has not yet been investigated in detail but is an important task for the near future. For subsequent application of voltammetric methods in inland waters with low conductivity, however, the addition of a non-complexing buffer to attain an ionic strength of ≥ 0.01 mol/L is necessary and can be accepted in most cases (FLORENCE, 1986).

For the separation and quantification of metal and organometallic species in aqueous solution, besides the very promising electrochemical approach for metals and complexes treated below in detail, numerous combinations of methods have been described in the literature. Examples of *organometallic compounds* are the determination of methylmercury down to the ng/L level, after preconcentration at an anion exchanger or a sulfhydryl cotton fiber, by cold vapor AAS (MAY et al., 1987) or gas chromatography (LEE, 1987), of lead and methyllead ions in water after in situ ethylation and AAS determination (RAPSOMANIKIS et al., 1986), of various organotin compounds with HPLC followed by AAS (KADOKAMI et al., 1988), of arsenic compounds by hydride cold trapping-quartz furnace AAS (VAN CLEUVENBERGEN, 1988), of selenium compounds by various methods (ROBBERECHT and VAN GRIEKEN, 1982; ITOH et al., 1989), and of various organometallic compounds after GC preconcentration and derivation with atomic spectrometric methods (CHAU, 1986b).

The separation and determination of *inorganic species and complexes* have been described for inorganic selenium and antimony species by hydride AAS (APTE and HOWARD, 1986a, b), for copper, cadmium, and zinc by an automated two-column ion exchange system and flame AAS detection (LIU and INGLE, 1989), for determination of stability constants and concentration of copper chelators in sea water by solvent extraction and in fresh waters by a cation exchange and a chelating resin, in both cases followed by GFAAS determination (MOFFET and ZIKA, 1987; MIWA et al., 1989), and for the determination of various complexes of aluminum with ion chromatography post column determination and spectrophotometric detection (BERTSCH and ANDERSON, 1989). Different oxidation states of metals can also be determined by FIA with spectrophotometric detection, e.g., for iron(II/III) and chromium(III/IV) (BUBNIS et al., 1983). A method for higher concentrations of complexed and uncomplexed metals (Al, Ca, K, Mg, Mn, and Na) in soil pore water has been described using sequential anion and cation separations followed by NAA on the separate solutions (DUFFY et al., 1988).

Electrochemistry, however, compared to all other trace element speciation methods available at present, still appears to be the most systematic experimental approach (STOEPPLER and NÜRNBERG, 1984; FLORENCE, 1986; NIMMO et al., 1989). Thus speciation techniques based on electrochemical methods will be discussed in some detail also mentioning more general questions in this field (VALENTA, 1983; NÜRNBERG, 1983b; NÜRNBERG and VALENTA, 1983; STOEPPLER and NÜRNBERG, 1984).

As a consequence of the above mentioned neglect of optimized sample preparation for voltammetry, a decrease of detection power occurs. However, with the use of the most powerful modes, e.g., the differential pulse, differential pulse stripping, and adsorptive stripping techniques, for a variety of trace metals concentrations down to the 10^{-9} mol/L level are accessible. This still allows work under realistic conditions. In this context it is worthwhile mentioning that physicochemical parameters of the dissolved metal species such as stoichiometric stability constants are practically unaffected by the total concentration of the investigated metal as long as the concentration of the relevant ligands is significantly higher. Due to possible unknown influences on the distribution of metal species, however, there is always a possibility of significant errors so that the total concentration should not exceed values above 10^{-7} mol/L in speciation studies.

Because of the complex composition of natural waters, detailed studies can only be performed stepwise for a single inorganic or organic ligand and a certain metal. Also simultaneous studies with several metals are possible, provided their redox potentials are significantly different. From these studies detailed informations can be gained.

Labile metal complexes, formed with inorganic ligands typical of sea and inland water such as chloride, hydroxide, carbonate, hydrogen carbonate, and sulfate, but also with distinct organic ligands show a uniform and reversible voltammetric peak for the complexed as well as for the non-complexed part of the metal content. Its potential position depends on the concentration [X] of the ligand. The potential shift of this peak due to the ligand concentration yields the stoichiometric stability constants β_j and the ligand numbers j and thus the identity of the complex MX_j of the studied metals which exist in distinct regions of the ligand concentration. If these data have been determined stepwise and in separate studies, the species distribution of the total concentrations of the investigated metals expressed as percentage can be computed for a distinct water type (NÜRNBERG and VALENTA, 1983).

Stable metal complexes are formed with many organic ligands and chelating agents which predominantly occur in inland and coastal waters. They produce separate irreversible peaks which occur at a significantly more negative potential than the corresponding reversible peaks. For a distinct metal M this portion can be written ML. If the decrease of the reversible peak corresponding to MX_j with increasing ligand concentration L is followed, this also yields the speciation data. The ligand concentrations [L] are very important for the studied water type in order to attain a distinct degree of complexation (NÜRNBERG and RASPOR, 1981; NÜRNBERG and VALENTA, 1983).

It is also possible to measure the *velocity constant of the complex formation* which takes place in a second order reaction. This is performed by measuring the time-dependence of the decrease of the reversible peak representing the non-complexed metal portion by adjusting a ligand concentration [L] that corresponds to the metal concentration [M]. This results in additional basic information, e.g., on the type of reaction mechanism for the formation of stable inert complexes with organic ligands in natural waters (RASPOR et al., 1980). Fundamental and generalizable prognostic conclusions can be achieved by model studies with the ligands nitrilotriacetic acid (NTA) and ethylenediamine tetraacetic acid (EDTA), and cadmium, lead, and zinc in sea water and hard inland waters (NÜRNBERG and RASPOR, 1981).

An important empirical diagnostic parameter is the *global complexing capacity* of a water type for a certain heavy metal. To achieve this a voltammetric titration of the water with the metal being studied is performed. The end point yields the total concentration $\Sigma[L]$ of inert and stable complexes forming ligands (NÜRNBERG and VALENTA, 1983). These ligands do not occur as free compounds but form complexes with the alkaline earths calcium and magnesium frequently present in excess. Thus the value of $\Sigma[L]$ depends of the complexing ability of the heavy metal selected as titrant. Frequently copper(II) is used for this purpose but investigations have also been carried out with lead(II) and cadmium(II). The complexing capacity for these heavy metals is in the order copper > lead > cadmium. Studies previously performed in oceans and inland waters have shown that there are regional correlations between complexing capacity and biological productivity, i.e., concentration of dissolved organic matter (DOM).

As discussed in Sect. 6.2, voltammetry is based on electrode processes at the phase boundary electrode/solution, i.e., on redox reactions. This property makes voltammetry well suited for the determination of different *oxidation states* of metals. In practical application two possibilities exist:

1. Voltammetric signals can be obtained at significantly different electrode potentials for different oxidation states of a metal, example: vanadium.
2. In the utilizable potential range only one electrode process can be obtained for a distinct oxidation state. Examples are the selectivity of voltammetry for arsenic(III), selenium(IV), and chromium(VI). This selectivity for distinct oxidation states can also be used for media other than water if a suitable sample preparation for voltammetry is possible (NÜRNBERG, 1982, 1983a; VALENTA, 1983).

7.3 Biological and Other Fluids, Solid Materials, and Air

For the collection and preparation of *biological fluids* and the other materials mentioned in this section, for speciation in principle the same applies as already mentioned in Sect. 7.2 for water, i.e., utmost contamination precautions and analysis as soon as possible after sampling.

Metals are present in body fluids (whole blood, blood serum and blood plasma, saliva, milk, etc.) under normal physiological conditions either predominantly bound to blood corpuscles and various proteins or accompanied by numerous organic compounds (urine). Thus, under these conditions, and if no particular chromatographic separations were applied, electrochemical methods are less suited for analysis. In cases of, e.g., metal intoxications, however, leading to high concentrations of distinct metal species such as arsenic(III) in urine, serum, and gastric juice, voltammetry can favorably be used for the identification of toxic species and rapid screening procedures in clinical toxicology (FRANKE and DE ZEEUW, 1976; WANG, 1988).

Suitable separations of metal species, occurring as ions, as distinct organometallic compounds, or bound to proteins can be achieved by various chromatographic techniques. Quantitations, in some cases down to the ng/L level, frequently with spectroscopic and other appropriate methods, recently reviewed (KRULL, 1986; WOLF, 1986; GARDINER, 1987, 1988), have been performed in numerous studies. Examples of these studies are the determination of trimethyllead salts in whole blood by high-resolution GC-GFAAS coupling (NYGREN and NILSSON, 1987), of methylmercury in whole blood and other samples after a clean-up with hydrochloric acid by anion exchange chromatography followed by cold vapor AAS (MAY et al., 1985, 1987), of different inorganic and organic selenium compounds in urine and serum by anion exchange chromatography, followed by neutron activation analysis (BLOTCKY et al., 1988), and separation as well as speciation of toxic from non-toxic, from seafood originating arsenic species (mainly arsenobetaine) in urine by HPLC in combination with hydride AAS (CHANA and SMITH, 1987). A rapid routine method to distinguish between the more or less toxic arsenic(III), arsenic(V), their inorganic metabolites methylarsonic acid (MMA) and dimethylarsinic acid (DMA) on the one hand, and chemically very stable non-toxic arsenic compounds from seafood on the other hand, still in use today (e.g., STOEPPLER and APEL, 1984) was reported some years ago (NORIN and VAHTER, 1981). The principle is simple: addition of sodium tetrahydroborate to urine which only transforms inorganic arsenic and metabolites into volatile hydrides determined by AAS, arsenobetaine and similar compounds do not react under these conditions.

An indirect speciation method is based on the fact that only hexavalent chromium passes the membrane of red blood cells. Thus chromium content of RBCs is an indicator of exposure to chromate (WIEGAND et al., 1988). Examples of chromatographic studies on the distribution of metals in protein fractions by AAS were those performed for beryllium in blood serum (STIEFEL et al., 1980), zinc in erythrocytes (GARDINER et al., 1984) and aluminum in serum (KEIRSSE et al., 1987). General methodological studies for protein bound metals were performed for, e.g., liquid chromatography-flame AAS coupling (EBDON et al., 1987b) and HPLC-ICP-MS coupling (DEAN et al., 1987b). A very specific, and for basic studies still important, approach is the use of radiotracers with very high specific radioactivity for investigations with laboratory animals (SABBIONI et al., 1981; VAHTER et al., 1982) that can be used for optimization of methods as well (see Sect. 6.3).

Similar techniques can be applied to other liquids, examples are the direct determination of organolead compounds by HPLC with electrochemical detection in gasoline (BOND and MCLACHLAN, 1986), the determination of traces of nickel and its chemical speciation in coffee, tea, and red wine by a combination of solvent extraction, chromatography, and spectroscopic methods (WEBER and SCHWEDT, 1982) of iron(III) in wines by adsorptive stripping voltammetry (WANG and MANNINO, 1989) and of tin in different lemon juices with the result that few differences existed between tin species (WEBER, 1987).

Speciation in solid materials is analytically more challenging than in liquids. Thus it cannot attain perfect results in all tackled areas. Too many questions are posed ranging from the quantification of numerous species including valence states in numerous very different matrices, their occurrence and distribution in structures and at interfaces up to the attempt to evaluate their behavior and action under distinct environmental conditions (BERNHARD et al., 1986).

Organometallic compounds can commonly be extracted prior to determination with appreciable yields from solid materials by appropriate clean-up procedures. For these, the sample commonly has to be homogenized, sometimes also freeze-dried and milled, but in most cases fresh or deep frozen material is used.

In order to avoid deterioration by light (UV) or air (oxygen), for some compounds work in the dark (e.g., for methylmercury, MAY et al., 1987) and use of an inert gas (N_2) might be necessary. The latter, however, is more important if valence states have to be quantified.

Some compounds can be easily separated from the matrix by volatilization, e.g., methylarsines, dimethylselenide (IRGOLIC, 1987) and methylmercury, the latter after acidification in soils and sediments (HORVAT et al., 1988). For other organometallic compounds the clean-up procedure consists of a mild dissolution of tissues and other biological materials without breaking metal-carbon bonds using enzymes (CHAU, 1988), or tetramethyl ammonium hydroxide (TMAH) (STOEPPLER and APEL, 1984; CHAU, 1988). A weak NaOH treatment was used to release selenite and selenate from biogenic particles and sediments (CUTTER, 1985). Acidification to break bonds between organometallic compounds and SH-groups of proteins was applied to organomercury and organotin compounds (e.g., MAY et al., 1987; STEPHENSON and SMITH, 1988; MARTIN-LANDA et al., 1989). Also direct extraction by using various solvents and internal standardization was described for arsenic (BEAUCHEMIN et al., 1988b), organotin (EPLER et al., 1988; SULLIVAN et al., 1988), and organotin as well as organolead compounds (FORSTER and HOWARD, 1989). Following these clean-up procedures chromatographic separations and different quantitation methods were applied (IRGOLIC, 1987; HORVAT et al., 1988; EBDON et al., 1988; CLARK and CRAIG, 1988).

To some extent successful attempts were made to arrive at a rough quantitation of the bioavailability of essential and toxic elements in foods by, e.g., ion-exchange, dialysis, ion retardation, and different spectrographic and electrochemical methods and also to contribute by speciation approaches, as far as metals are involved, to the long-term storage stability of foods for Ca, Mg, and Zn (SCHWEIZER and SCHWEDT, 1988). For some plants the analysis of lead distribution within and upon contaminated plants was studied (SCHWEDT and JAHN, 1987). Distribution of cadmium and thallium in protein fractions of the dissolvable part of native rape plants was investigated (GÜNTER and UMLAND, 1988).

General methods for speciation in foods and for the elements Pb, Ca, Mg, and Cu were recently reviewed in some detail (SCHWEDT, 1988). An important field of research is the influence of chemical species on absorption and physiological fate of trace elements from the diet (MILLS, 1986). Model studies about factors affecting the solubilization of iron, copper, zinc, and cadmium during gastro-intestinal digestion of food and their association with protein fractions were recently performed (MASSEY et al., 1986). In vivo speciation in human nutrition studies with zinc was investigated by administering the stable isotope ^{67}Zn and measuring the ^{64}Zn/^{67}Zn isotope ratio by mass spectrometry (EAGLES et al., 1989).

Localization, depth profiling, and speciation of metal compounds in organic and inorganic matrices, including filter materials, are a domain of various beam techniques (GRASSERBAUER, 1987; NEBESNY et al., 1989). Examples applying the laser microprobe mass analyzer (LAMMA) are speciation of inorganic salts and of fly-ash particles from an incineration plant (DENNEMONT et al., 1985), investigations on the localization of cadmium in the renal cortex of rats after long-term exposure to this element (SCHMIDT et al., 1986), and the discrimination between coprecipitated and adsorbed lead on individual calcite particles (WOUTERS et al., 1988).

Many efforts are being made, still imperfect, if comprehensive answers are expected, at elucidating the behavior of metals and their compounds as far as environmental sinks (soils, sediments, sludges, etc.) are concerned. Metals are transported via dry and wet deposition and by aqueous effluents stemming from geogenic and anthropogenic sources to these sinks. They can then be mobilized again under the influence of precipitation, running waters (pH), competing compounds, complexing agents, and plants (BRÜMMER, 1986; LECKIE, 1986; CRAIG, 1986; FÖRSTNER, 1986).

The general approach for soils and sediments (see also ANGEHRN-BETTINAZZI et al., 1990, and Chapter I.7e, Sect. 3.2) is after sampling of representative samples, for soils mainly from different horizons, and homogenization, air or furnace drying (the latter 24 h at 60°C) and separate collection of soil solutions (pore waters). For sediments separation into various (up to eight) distinct fractions in the 850 µm to <1 µm size range is common (TESSIER et al., 1982) because of important relationships between these and biota (TESSIER et al., 1984). Then, for identification of the reactivity of binding forms of various species, successive *extractions* are performed to gain informations about the following metal species: (1) water-soluble, (2) exchangeable, (3) organically bound, (4) occluded in iron and manganese oxides, (5) definite compounds (e.g., heavy metal carbonates, phosphates, and sulfides), (6) residue (metals structurally bound in silicates). The soluble fraction is commonly obtained by extraction with, e.g., deionized water, while for fractions (2) to (5) different solutions have been recommended (TESSIER et al., 1979; BRÜMMER, 1986; SCOKART et al., 1987; SAMANIDOU and FYTIANOS, 1987), fraction (6) is always digested by an acid mixture that contains hydrofluoric acid. If particular elements are of interest, other extraction schemes have been used. An example is the sole use of acidic solutions of different pH and composition for the determination of arsenic, antimony, bismuth, and selenium in geological materials (KULDVERE, 1989). An approach still under development, yielding some useful information, also applicable to plant material, is *thermal analysis*. Here, stepwise heating of the sample and determination of the vaporized metals and non-metals can lead to useful information about volatilizable compounds and their binding forms (BAUER and NATUSCH, 1981). Especially advantageous is the comparison with reference samples of known composition and structure.

All these procedures, however, suffer from a certain lack of selectivity and thus can only provide some important though limited informations (TESSIER and CAMPBELL, 1988; AUALIITIA and PICKERING, 1988) that have to be complemented by additional investigations and analyses like those mentioned above.

Of further importance for the characterization of soils are the determination of adsorption capacity of different soil components, the evaluation of adsorption mechanisms and adsorption/desorption, and of precipitation/dissolution processes which can greatly depend on pH values again influencing metal contents and speciation in the liquid phase (BRÜMMER, 1986). The latter properties are also important for the mobilization of metal species from sediments, e.g., the pH-dependent release of arsenic from river sediment (MOK and WAI, 1989). The availability of heavy metals for plants is an important factor, and because uptake by roots only occurs from the liquid phase (soil solution) the metal content in this solution is of prime importance. Determination is commonly done in the water saturation extract. The same is possible if extraction is performed with dilute salt solutions such as 0.1 mol/L $CaCl_2$ or $Ca(NO_3)_2$ (BRÜMMER, 1986).

Similar extraction procedures as for soils and sediments are applied to characterize the metal species in sewage sludge (RUDD et al., 1988). This can be performed by either a batch method (CAMPANELLA et al., 1987a) or by utilizing a column (CAMPANELLA et al., 1987b). In a study of digested and undigested sludge, metal partitioning differences were found by applying this technique (GIBBS and ANGELIDIS, 1988). An analytical procedure based on a very similar sequential extraction scheme for the chemical associations of Fe, Pb, Cu, Mn, Cr, and Cd was applied to suspended particulates in urban air (OBIOLS et al., 1986). Various extraction procedures different from those described above were applied to separation and determination of a number of elements using chromatography and AAS (HANSEN et al., 1984) and a column method followed by multielement analysis with INAA (WARREN and DUDAS, 1988) in fly ashes.

Volatile species of lead, tin, selenium, tellurium, arsenic, and mercury in air have been separated and determined by different methods. Examples are the detection of alkylselenide, alkyllead and alkyltin compounds by GC and atomic fluorescence (D'ULIVO and PAPOFF, 1986), of various selenium and tellurium compounds by adsorption on gold-coated quartz beds, leaching with different solvents, anion exchange and GFAAS determination (MUANGNOICHAROEN et al., 1988), of methylarsenic compounds in airborne particular matter by acid extraction hydride-generation GC and AAS detection (MUKAI et al., 1986), of organic forms of mercury and arsenic in precipitation by extraction-gas chromatography and detection by AAS

Table I.4a-3. Examples of Typical Costs of Commercial Analytical Instruments for Decomposition and Analysis of Metals, Metalloids and Organometallic Compounds as of Spring 1989, Prices are Given in 10^3 DM Without Tax

Method	Instrumentation	Price Range (approx.)	Properties	Remarks
Decomposition	High pressure asher (HPA)	70	Programmable decomposition of up to 7 samples in quartz vessels	Temperature up to 300°C, pressure up to 100 bar; complete decomposition of organic matter
	Microwave decomposition	40–60	Programmable multisample digestion in PTFE vessels	Rapid decomposition with inside temperatures of up to 180°C, decomposition incomplete; safety precautions sometimes still poor
	Decomposition in closed systems	10–50	Programmable single and multiple units with PTFE vessels	Up to 160°–170°C decomposition incomplete
	Decomposition in open vessels or reflux systems	30–60	Programmable multisample decomposition in quartz and glass tubes	Decomp. depends upon the acids used and is complete at higher temperatures
	Low temperature ashing systems; also reflux possible	40–60	Ashing in an O_2-plasma (reduced pressure), quartz vessels	Decomp. depends upon the matrix and is complete for pure organic materials
	Burning in oxygen, open, but with reflux	30–40	Burning in O_2, glass and quartz vessels	Decomp. depends upon the matrix and is complete in many cases
AAS	Basic instruments with D_2 and Zeeman background correction	40–120	All possibilities: flame, graphite furnace, PC operation	Single element mode
	Multielement flame system	60–80	PC operation, up to 12 elements simultaneously	D_2-background correction
	Oligoelement graphite furnace system	140	4 elements simultaneously	Zeeman background correction; 8 elements sequentially
	Accessories: furnaces autosamplers, hydride and cold vapor systems	20–40		Different systems also from special manufactures, standards high
FIA	Basic system for automated operation	≥40	For combination with AAS and other methods	Promising approach for further development
ICP-AES	Sequential and/or simultaneous systems	100–350	Total measuring time 1–2 min for sequential multielement analysis	Argon consumption high, thus advantageous only if ≥4 elements are determined in a sample
Voltammetry	Basic systems with recorder	20–40	Oligoelement determination with all useful modes possible	More expensive systems allow semiautomated operation

Method	Description			
	Fully computerized systems	>50		Determination of 4 elements in the same solution requires up to 20 min
Potentiometry	Basic systems with recorder	20–25	Like voltammetric system, but not all modes possible	Still not completely evaluated and thus less flexible than voltammetry
INAA	Complete with 4096-channel function, PC and detector	100–150	Multielement mode. Counting time varies	Reactor facilities or access mandatory
	Sample changers	50		
	Liquid scintillation counters	30–100	Up to three channels, PC operation etc.	For radiochemical NAA and radiotracer studies, high sample throughput
XRF	Energy and wavelength dispersive for sequential and simultaneous measurements, PC operated	100–300	Time required: 10 min to 1 h for simultaneous determination	Special preparation and/or reference material necessary for accurate results
	Total reflection XRF complete system	300–400	Absolute detection limits at the pg level	Analyte addition and special preparation required for accurate results
MS	Spark-source MS	≥800	Up to 75 elements; simultaneous determination up to 1 h	High precision and detection power but poor accuracy. For accurate results isotope dilution is required
	Thermal ionization MS	300–800	Some elements simultaneously, measuring time ≥1 h	Ideal for isotope ratios and isotope dilution analysis; if properly performed, very accurate results can be obtained
ICP-MS	With low resolution MS	400–500	Potential and properties as for MS; sample throughput high	High detection power, but susceptible to interferences, accuracy appreciable if appropriate measures are taken, e.g., separation of matrix constituents, analyte addition, and isotope dilution
Spectrophotometry	Complete with double beam optics, PC evaluation, etc.	30–100	Also higher orders and other advanced techniques possible	Increasingly used as detection for FIA and chromatographic systems
GC	Complete with automated sample injection, detectors, and PC	40–80	Several elements and compounds simultaneously; measuring time variable: min to h	Useful for compounds that are stable at higher temperatures
HPLC	Complete with automated sample injection, various detectors and PC	>80	Like for GC	For compounds less stable at higher temperatures

(BRZEZINSKA et al., 1983; PAUDYN and VAN LOON, 1986). Elemental mercury, mercuric chloride, methylmercuric chloride and dimethylmercury in air were determined by applying different absorbers for these species and mercury determination by resonance atomic fluorescence spectrometry (SCHROEDER and JACKSON, 1985).

8 Economic Aspects (Cost-Benefit Relations)

In Table I.4a-3 approximate costs of analytical instruments (as of spring 1989) are listed. Frequent fluctuations in exchange rates would make a computation difficult, so the prices are basically given in DM. They include typical ranges from simple to the most sophisticated systems, if applicable, in the order of sections in this chapter. Where necessary prices for accessories are also included. For some methods an estimate is given for the average time needed for a single measurement or a single run if oligo- or multielement methods are concerned. For the latter an average of 15 elements is assumed. If for single or oligoelement methods analyte addition is used, the total time needed for determination can be found by multiplying the time for a single measurement by the number of additions plus one. There are, however, no data included about the real throughput of samples. This is only possible in a more general manner and thus in the following a few typical examples are given from the author's experience.

In urine samples a quick decision is required if a thallium intoxication has occurred. For such a yes-or-no decision, and if a reliable method is available, commonly two samples and the performance of two measurements in each sample are sufficient. Provided appropriate standard solutions are already at hand, and flame AAS can be directly applied to the sample, after measurement of a control urine with appropriate thallium concentration (three readings requiring a total measuring time of $3 \cdot 10\,s = 30\,s$) for the four measurements in the urine samples additionally $4 \cdot 10\,s = 40\,s$ are required. If a modern PC-controlled AAS system with already stored conditions for Tl can be used, instrumental and lamp parameters are adjusted automatically, and very quick information is feasible. This, however, is not the case if an older instrument is used. If lower thallium concentrations, e.g. during medical treatment of a poisoned patient, have to be analyzed, the graphite furnace must be applied. In this case, the sole measuring times have to be multiplied by approximately a factor of ten and additional time calculated for analyte additions or a complete calibration graph. It is obvious that the total time consumption for single samples due to preparatory steps and the necessary calibration graph or analyte addition is much longer than for series of samples, for which often a measurement against a single calibration graph or a few matrix matched samples are sufficient. Thus proper organization of analytical work is an important factor for cost optimization.

If samples with metal contents at very low levels have to be analyzed numerous measurements of blank values in digests or reagent solutions, depending on the required reliability of the task, are necessary which can significantly influence total costs.

A few general remarks, however, should be made, concerning the selection of instruments and methods. For large series of very similar samples, despite the high investment, mechanized or even completely computerized systems are recommendable, since total expenses for laboratory staff are very high and rise much more rapidly than costs for instruments. The latter, for some methods, even show a decreasing tendency. The decision for a routine method depends on the task but also on experience and level of knowledge of the laboratory staff. Another and sometimes decisive factor is the service capacity and ability of the instrument manufacturer.

For laboratories with high sample throughput in ecology, medicine, food control, and industrial research, and great significance of the data obtained, besides the routine method the additional use of a very reliable method for data validation is recommendable. Despite the high investment and also running costs of such a fairly complex instrumentation (e.g., isotope dilution mass spectrometry) this can finally save running costs and manpower in avoiding extensive repetitive analyses. For special laboratories acting as reference laboratories within the scope of their own tasks and capabilities as far as possible complete instrumentation and continuous training of staff is mandatory. At least two independent analytical principles (e.g., atomic spectrometry and XRF or voltammetry) should always be applied with sufficient expertise.

9 Quality Assurance

Accuracy problems recognized in trace analysis have already been addressed in Sect. 1 of this chapter. Since in environmental protection, occupational medicine, and food control analytical data provide the quantitative basis for the estimation of exposures, the introduction of legal measures, and their supervision, it is obvious that elimination of error sources and continuous protection of the quality of these data is extremely important. Thus, in the following the means of controlling and minimizing systematic errors during *sample collection, sample preparation,* and *analytical determination* will be discussed.

Errors during *sample collection* are of importance if one has to deal with materials containing only spurious amounts of the element(s) to be determined. Due to improper manipulations, thus the sample can be easily contaminated, sometimes amounting to several orders of magnitude. This is particularly the case for ubiquitously distributed elements such as lead (PATTERSON and SETTLE, 1976; MART, 1979a, b), aluminum (FRECH et al., 1982; VERSIECK, 1985), and chromium (VERSIECK, 1985; OTTAWAY and FELL, 1986). Control of this error source is largely incomplete, since it requires the accurate application of the most powerful analytical methods as well as utmost contamination minimization in the laboratory (PATTERSON and SETTLE, 1976; EVERSON and PATTERSON, 1980).

At present there are only a limited number of laboratories with very competent staff able to perform this task appropriately. Since a direct control practically does not exist, only the comparison of the results of different laboratories analyzing the same matrix is of assistance; e.g., serum or whole blood (VERSIECK, 1985; VERSIECK et al., 1988b) or sea water taken independently from the same site (WONG et al., 1983) preferably by different methods. To some extent participation in laboratory intercomparisons is also useful for this purpose in order to compare correct performance of sampling, sample preparation, and analysis. In such a trial the obtained results should agree in the participating laboratories within the typical precision of the applied method(s), e.g., for dissolved cadmium, copper, and lead in deep sea water samples at the nanogram per liter level (BRULAND et al., 1985) or for cadmium in blood at the 0.1 to 0.2 µg/L level (STOEPPLER, 1983c). If these results are compared with real analytical data after appropriate sampling procedures in the same matrix an estimation of sampling performance appears to be possible.

Errors during *sampling preparation* can occur in every laboratory. They can be identified in an objective manner to some extent by comparative analyses with homogeneous materials, the laboratory intercomparisons already mentioned, or a continuous internal and external quality control with appropriate control or reference materials (PITTWELL, 1988). An objective approach for the general improvement of the performance of an analytical laboratory is represented by good laboratory practice regulations that comprehensively deal with all aspects of analytical work including computer application, i.e. (1) organization and staff responsibilities, (2) acquisition, (3) project control and organization, (4) records and documentation (including writing of standard operating procedures, SOPs), and (5) operational code of practice (security, training, assessment) (WELLER, 1988; GARNER and BARGE, 1988).

In order to identify whether the most important errors are in sample preparation and/or analytical determination it is common in laboratory intercomparisons, besides several liquid or solid control materials, to also distribute an aqueous solution with appropriate analyte concentration. The result then allows conclusions on specific error sources to be drawn.

Since there are numerous error sources in sample preparation it is necessary, in addition to the external quality assessment programs currently performed in many branches e.g. for cadmium and lead in blood (LIND et al., 1987), for trace elements in foods (KUMPULAINEN and PAAKKI, 1987), and for aluminum in biological materials (TAYLOR, 1988), to make use of reference and certified reference materials as well. These materials, covering a wide range of environmental and biological materials, are produced and commercially available from various national and supranational agencies and producers with a steadily increasing selection of elements and materials (VEILLON, 1986; PARR et al., 1987; WOLF, 1985; WOLF and STOEPPLER, 1987, 1988). Prices for these materials are moderate, particularly if the benefit from their use is considered. They are available as finely ground and very homogeneous solids, liquids or lyophilized liquids (whole blood, blood serum, urine). The latter can be easily reconstituted by the addition of distilled water. Reference and certified reference materials (RMs and CRMs) should be used for method evaluation as well as in particular cases for simultaneous analysis together with real samples for current assessment of the methodological approach. If the reliability of a sample preparation procedure is to be assessed, in principle it is sufficient on a routine basis to carry through the whole analytical procedure surplus (control) material of the same or a matrix very similar to that of the samples. If appropriate certified reference materials are available, these should be used comparatively in the analytical characterization of laboratory control materials.

It has to be mentioned, however, that despite the increasing range of certified reference materials – for example, a new type of CRM (second generation CRM) (VERSIECK et al., 1988b) and the first certified values for methylmercury in a series of marine reference materials have recently been published (BERMAN et al., 1989) – these still do not cover by far the range of matrices and elements necessary for comprehensive use. This led, and sometimes still leads, to misuse of CRMs in that a material with completely different trace metal levels and matrix composition was taken for the control of a trace analytical procedure. An example might be the use of a liver CRM with 270 µg/kg cadmium (dry weight) for control of cadmium analysis in muscle tissue with less than 10 µg/kg cadmium. In such a case the use of an internally characterized similar control material, if possible with at least two independent analytical methods, is much more reliable (see also below). If contamination has to be faced with biological materials, the use of materials that behave similarly in digestion procedures but have extremely low trace metal contents (e.g., skim milk powder, refined sugar, native bovine blood) could be successfully used for the minimization of blanks in decomposition systems (STOEPPLER, 1983a). There are also various contamination risks in the last preparation step prior to determination, depending on the respective metal and its level in the analyte solution. Contamination can be introduced from external sources, mainly laboratory dust but also from the analyst. Thus at lower elemental concentrations the use of laminar flow zones or boxes with filtered air and use of gloves and of scrupulously cleaned laboratory ware is absolutely necessary for accurate results (ZIEF and SPEIGHTS, 1972; TSCHÖPEL et al., 1980; PATTERSON and SETTLE, 1976; EVERSON and PATTERSON, 1980; MART, 1982; TÖLG, 1987).

Errors during *analytical determination* range from simple technical errors like improper analyte solutions to chemical and physical interferences typical of the methods treated in Sect. 6 in some detail. Though for trace analytical methods interferences are known today in principle, difficult matrices or extraordinarily high contents of specifically interfering elements can still lead to a significant bias. This is not acceptable, if highly accurate data are required (see Sect. 2). This can occur if one tries to substitute a cumbersome or contamination-susceptible sample pretreatment prior to analysis by a direct method.

For the improvement of existing or the development of new analytical methods certified reference materials (CRMs) are always invaluable, if available. For many matrices and for quite a number of elements treated in this book, however, no reference material exists. Thus other approaches are required.

Atomic spectrometric techniques, for instance, can be used in combined procedures after decomposition and decomposition/extraction as well as with liquid and solid samples in the direct mode. Thus, if the blank of such a combined procedure can be controlled – which can best be approached if all steps are performed on-line, i.e., in a closed analytical system (TÖLG, 1979, 1987, 1988a; KNAPP, 1984) – this procedure might be used to validate any direct method. The results of such an approach, however, are more valid if different determination principles are applied

(LIEM et al., 1984; KAISER and TÖLG, 1986; MEYER et al., 1987). This is also important for analysis in a liquid matrix (sea water) using different extraction procedures/AAS and voltammetry as the direct approach which in this case is the simpler and, if samples are properly prepared, probably the most reliable approach (BRÜGMANN et al., 1983; BRULAND et al., 1985). Promising for method validation and also the characterization and certification of reference materials is isotope dilution mass spectrometry, which can in principle also be applied in the ICP-MS mode (HEUMANN, 1980, 1988a; HILPERT and WAIDMANN, 1988; DATE and GRAY, 1988).

In instrumentally well-equipped laboratories that can use various independent analytical methods, e.g., atomic spectrometry, voltammetry, mass spectrometry or advanced modes of X-ray fluorescence and, possibly also some nuclear methods, it is possible to achieve internal characterization in new materials for a variety of elements within the competence of the laboratory staff. In combination with rigid quality control including sample exchange with external laboratories, and permanent use of appropriate RMs and CRMs and GLP application, an appreciable state of reliability can be attained even if quite difficult materials have to be analyzed.

10 Data Evaluation

First, instrument readings obtained so far for a single analysis or an analytical series have to be evaluated from a purely analytical viewpoint. The thus generated data must then be judged for their individual or general importance.

Second, the *units* to be reported are important and thus have to be mentioned. During the last couple of years it has become more and more common to report trace metal contents in molar units, reflecting the true number of atoms involved. This certainly is more correct than reporting weight units. Many chemists, however, educated in the weight-based analytical chemistry, are still puzzled by the molar scale so that, as a compromise and for a time of transition, the simultaneous use of both units (with weight units in parentheses) is recommended for reporting results.

For data evaluation two approaches must be distinguished: (1) the formal generation of single values from calibration graphs, by analyte addition or by direct comparison with control samples, and (2) the establishment of the reliability, i.e., general precision and accuracy due to GLPs and quality assurance for the whole analytical approach based on typical and particular data generated in the laboratory involved in the study under question (TAYLOR, 1981; GARFIELD, 1984; TAYLOR and STANLEY, 1985), often also requiring careful statistical treatment (e.g. SNEDECOR and COCHRAN, 1973; GREEN and MARGERISON, 1977; DOERFFEL and ECKSCHLAGER, 1981). This is necessary since in the next phase a more general judgement is often performed in a dialogue with experts from branches other than analytical chemistry such as biologists, toxicologists, ecologists, and possibly also politicians and lawyers. Thus, if necessary, particular measures for quality control should be mentioned in the final report without too many details. In this report, and if these terms have to be addressed, the different "weights" of *limit of detection (LOD)*, i.e., the concentration at which there is certainty that the analyte is present even if it cannot be quantified, mainly assumed as 3 s (see Table I.4a-2) and *limit of quantification (or determination) (LOQ)*, i.e., the concentration at which the analyte can be quantified with an agreed degree of confidence, frequently assumed as 10 s (KEITH et al., 1983) must also be discussed and clarified if the obtained data require this.

Information given in the analytical report about a distinct uncertainty of the value assumed as true, allows the range to be clearly defined in which this value can be found. Sometimes distinct time or technical reasons do not allow extended studies for an exact confirmation of a range of uncertainty. This is acceptable if only yes-or-no decisions are wanted, but it should be clearly stated.

If analytical data with a reliable comment on their true information value are judged, often already during the first rough inspection, it becomes clear whether the task can be performed or not. If it appears that there is still too large an uncertainty in the obtained results, discussion between all institutions or laboratories involved can lead to additional measures. That could include an increase in the number of samples and/or the number of analytical determinations, or the introduction of a more accurate or more precise method to obtain statistically more certain data. For this case, however, it should be mentioned that this situation should be the exception not the rule, since already during the planning phase one should take the time and trouble to obtain all relevant information to finally arrive at trustworthy data. The judgement of the gained data from the view of biology, toxicology, ecology or legal decision(s) cannot be treated in this context. It is often done on a multidisciplinary basis either transversally, i.e., in comparison to typical normal or exposure values of statistically relevant collectives (human beings, animals, other environmental species) or longitudinally, i.e., time-dependent for the studied collective using an appropriate, computer-aided statistical evaluation.

References

Aalbers, T. H. G., Houtman, J. P. W., and Makkink, B. (1987), *Trace Element Concentrations in Human Autopsy Tissue. Clin. Chem. 32*, 2057–2064.

Abbasi, S. A., Hameed, A. S., Nipaney, P. S., and Soni, R. (1988), *Microdetermination of Vanadium, in Environmental Samples as its Ternary Complex with N-p-Aminophenyl-2-phenylacrylohydroxamic Acid and 3-(o-Carboxyphenyl)-1-phenyltriazine-N-oxide. Analyst 113*, 1561–1565.

Abdillahi, M. M. (1988), *Microwave Induced Plasmas for Trace Elemental Analysis. Int. Lab. 18*, 6, 16–24.

Abdullah, M., Fuwa, K., and Haraguchi, H. (1984), *Simultaneous Multielement Analysis of Microliter Volumes of Solution Samples by Inductively Coupled Plasma Atomic Emission Spectrometry Utilizing a Graphite Cup Direct Insertion Technique. Spectrochim. Acta 39B*, 1129–1139.

Adams, F., Gijbels, R., and van Grieken, R. (eds.) (1988), *Inorganic Mass Spectrometry.* John Wiley & Sons, New York.

Adeloju, S. B., Bond, A. M., and Briggs, M. H. (1985), *Multielement Determination in Biological Materials by Differential Pulse Voltammetry. Anal. Chem. 57*, 1286–1390.

Ahearn, A. J. (ed.) (1972), *Trace Analysis by Mass Spectrometry*. Academic Press, New York.
Ahmed, R., May, K., and Stoeppler, M. (1987), *Wet Deposition of Mercury and Methylmercury from the Atmosphere. Sci. Total Environ. 60*, 249–261.
Aitio, A., and Järvisalo, J. (1984), *Biological Monitoring of Occupational Exposure to Toxic Chemicals. Collection, Processing and Storage of Specimens. Pure Appl. Chem. 56*, 549–566.
Ajayi, O. D., Littlejohn, D., and Boss, C. B. (1988), *Comparison of Vapour-phase Temperatures and Chemical Interferences for Wall, Platform and Probe Atomisation in Electrothermal Atomisation Atomic Absorption Spectrometry. Anal. Proc. 25*, 75–77.
Ajlec, R., and Stupar, J. (1989), *Determination of Iron Species in Wine by Ion-Exchange Chromatography-Flame Atomic Absorption Spectrometry. Analyst 114*, 137–142.
Akatsuka, K., and Atsuya, I. (1989), *Synthetic Reference Material for Direct Analysis of Solid Biological Samples by Electrothermal Atomic Absorption Spectrometry. Anal. Chem. 61*, 216–220.
Almestrand, L., Jagner, D., and Renman, L. (1986), *Determination of Cadmium, Lead and Copper in Milk and Milk Powder by Means of Flow Potentiometric Stripping Analysis. Talanta 33*, 991–995.
Almestrand, L., Jagner, D., and Renman, L. (1987), *Automated Determination of Cadmium and Lead in Whole Blood by Computerized Flow Potentiometric Stripping with Carbon Fibre Electrodes. Anal. Chim. Acta 193*, 71–79.
Almestrand, L., Betti, M., Hua, C., Jagner, D., and Renman, L. (1988), *Determination of Lead in Whole Blood with a Simple Flow-Injection System and Computerized Stripping Potentiometry. Anal. Chim. Acta 209*, 329–334.
Alt, F., Messerschmidt, J., and Tölg, G. (1987), *A Contribution towards the Improvement of Se-Determination in the pg-Region by Hydride AAS. Fresenius Z. Anal. Chem. 327*, 233–234.
Analytical Quality Control (Harmonised Monitoring) Committee (1985), *Accuracy of Determination of Total Mercury in River Waters: Analytical Quality Control in the Harmonised Monitoring Scheme. Analyst 110*, 103–111.
Anderson, D. R., and McLeod, C. W. (1988), *On-line Pre-concentration with a Boron Specific Resin in Flow Injection-inductively Coupled Plasma Emission Spectrometry. Anal. Proc. 25*, 67–69.
Andreae, M. O. (1986), *Chemical Species in Seawater and Marine Particulates,* in: Bernhard, M., Brinckmann, F. E., and Sadler, P. J. (eds.): *The Importance of Chemical "Speciation" in Environmental Processes*, pp. 301–335. Springer, Berlin-Heidelberg-New York-London-Paris-Tokyo.
Angehrn-Bettinazzi, C., Hertz, J., and Stöckli, H. (1990), *Factors Affecting the Investigation of Heavy Metal Speciation in Forest Soils Using Thin-channel Ultrafiltration; Distribution of Heavy Metals in Various Litter Horizons and Forest Soils. Special Issue First IAEAC Soil Residue Analysis Workshop. Int. J. Environ. Anal. Chem.*, in press.
Angerer, J., and Schaller, K. H. (1985, 1988), *DFG Analyses of Hazardous Substances in Biological Materials (Methods for Biological Monitoring)*, Vols. 1 and 2. VCH Verlagsgesellschaft, Weinheim-Basel-Cambridge-New York.
Apatin, V. M., Arkhangel'skii, B. V., Bol'Shov, M. A., Ermolov, V. V., Koloshnikov, V. G., Kompanetz, O. N., Kuznetsov, N. I., Mikhailov, E. L., Shishkovskii, V. S., and Boutron, C. F. (1989), *Automated Laser Excited Atomic Fluorescence Spectrometer for Determination of Trace Concentrations of Elements in Liquid and Solid Samples. Spectrochim. Acta*, in press.
Appleton, J. M. H., and Tyson, J. F. (1986), *Flow Injection Atomic Absorption Spectrometry: The Kinetics of Instrument Response. J. Anal. At. Spectrom. 1*, 63–74.
Apte, S. C., and Howard, A. G. (1986a), *Determination of Dissolved Inorganic Antimony(V) and Antimony(III) Species in Natural Waters by Hydride Generation Atomic Absorption Spectrometry. J. Anal. At. Spectrom. 1*, 221–225.
Apte, S. C., and Howard, A. G. (1986b), *Determination of Dissolved Inorganic Selenium(IV) and Selenium(VI) Species in Natural Waters by Hydride Generation Atomic Absorption Spectrometry. J. Anal. At. Spectrom. 1*, 379–382.

Arenas, V., Stoepler, M., and Bergerhoff, G. (1988), *Arsenic Determination in the Ultratrace Range by Atomic Absorption Spectrometry after Preconcentration of the Hydride.* Fresenius Z. Anal. Chem. *332*, 447–452.

Arpadjan, S., and Krivan, V. (1988), *Behaviour of Chromium in the Graphite Furnace during the Performance of the Flameless Atomic Absorption Spectrometry.* Fresenius Z. Anal. Chem. *329*, 745–749.

Ashton, A., and Chan, R. (1987), *Monitoring of Microgram per Litre Concentrations of Trace Metals in Sea Water: The Choice of Methodology for Sampling and Analysis.* Analyst *112*, 841–844.

Aualiitia, T. U., and Pickering, W. F. (1988), *Sediment Analysis – Lability of Selectivity Extracted Fractions.* Talanta *35*, 559–566.

Bächmann, K. (1981), *Multielement Concentration for Trace Elements.* CRC Crit. Rev. Anal. Chem. *12*, 1–67.

Bacon, J. R., and Ure, A. M. (1984), *Spark-source Mass Spectrometry: Recent Developments and Applications – A Review.* Analyst *109*, 1229–1254.

Bamiro, F. O., Littlejohn, D., and Marshall, J. (1988), *Determination of Copper in Blood Serum by Direct Current Plasma and Inductively Coupled Plasma Atomic Emission Spectrometry.* J. Anal. At. Spectrom. *3*, 379–384.

Baranski, A. S., and Quon, H. (1986), *Potentiometric Stripping Determination of Heavy Metals with Carbon Fiber and Gold Microelectrodes.* Anal. Chem. *58*, 407–412.

Barnes, R. M. (1978), *Recent Advances in Emission Spectroscopy: Inductively Coupled Plasma Discharges for Spectrochemical Analysis.* Crit. Rev. Anal. Chem. *7*, 203–296.

Barnes, I. L., Murphy, T. J., and Michiels, E. A. I. (1982), *Certification of Lead Concentration in Standard Reference Materials by Isotope Dilution Mass Spectrometry.* J. Assoc. Off. Anal. Chem. *65*, 953–956.

Barth, S., and Klockow, D. (1989), *A Contribution to the Experimental Quantification of Dry Deposition to the Canopy of Coniferous Trees,* in: Georgii, H. W. (ed.): *Mechanism and Effect of Pollutant Transfer into Forests.* Kluwer Academic Publishers, Lancaster, U.K.

Bass, D. A., and Holcombe, J. A. (1988), *Mechanisms of Lead Vaporization from an Oxygenated Graphite Surface Using Mass Spectrometry and Atomic Absorption.* Anal. Chem. *60*, 578–582.

Batley, G. E., and Farrar, Y. J. (1978), *Irradiation Techniques for the Release of Bound Heavy Metals in Natural Waters and Blood.* Anal. Chim. Acta *99*, 283–292.

Bauer, C. F., and Natusch, D. F. S. (1981), *Speciation at Trace Levels by Helium Microwave-induced Plasma Emission Spectrometry.* Anal. Chem. *53*, 2020–2027.

Baxter, D. C., Littlejohn, D., Ottaway, J. M., Fell, G. S., and Halls, D. J. (1986), *Determination of Chromium in Urine by Probe Electrothermal Atomisation Atomic Emission Spectrometry Using a Low-Resolution Monochromator.* J. Anal. At. Spectrom. *1*, 35–39.

Beaty, M., Barnett, W., and Grobenski, Z. (1980), *Techniques for Analyzing Difficult Samples with the HGA Graphite Furnace.* Atom. Spectrosc. *1*, 72–77.

Beauchemin, D., McLaren, J. W., Willie, S. N., and Berman, S. S. (1988a), *Determination of Trace Metals in Marine Biological Reference Materials by Inductively Coupled Plasma Mass Spectrometry.* Anal. Chem. *60*, 687–691.

Beauchemin, D., Bednas, M. E., Berman, S. S., McLaren, J. W., Siu, K. W. M., and Sturgeon, R. E. (1988b), *Identification and Quantification of Arsenic Species in a Dogfish Muscle Reference Material for Trace Elements.* Anal. Chem. *60*, 2209–2212.

Beauchemin, D., Siu, K. W. M., and Berman, S. S. (1988c), *Determination of Organomercury in Biological Reference Materials by Inductively Coupled Plasma Mass Spectrometry Using Flow Injection Analysis.* Anal. Chem. *60*, 2587–2590.

Beauchemin, D., McLaren, J. W., Mykytiuk, A. P., and Berman, S. S (1988d), *Determination of Trace Metals in an Open Ocean Water Reference Material by Inductively Coupled Plasma Mass Spectrometry.* J. Anal. At. Spectrom. *3*, 305–308.

Behne, D. (1981), *Sources of Error in Sampling and Sample Preparation for Trace Element Analysis in Medicine.* J. Clin. Chem. Clin. Biochem. *19*, 115–120.

Benninghoven, A., Rudenauer, F. G., and Werner, H. W. (1987), *Secondary Ion Mass Spectrometry: Basic Concepts, Instrumental Aspects, Applications and Trends.* John Wiley & Sons, New York.

Berlin, A., Wolff, A. H., and Hasegawa, Y. (eds.) (1979), *The Use of Biological Specimens for the Assessment of Human Exposure to Environmental Pollutants.* Martinus Nijhoff Publishers, The Hague.

Berman, S. S., Siu, K. W. M., Maxwell, P. S., and Beauchemin, D., and Clancy, V. P. (1989), *Marine Biological Reference Materials for Methylmercury: Analytical Methodologies used in Certification.* Fresenius Z. Anal. Chem. *333*, 641–644.

Berndt, H. (1988), *High Pressure Nebulization: A New Way of Sample Introduction for Atomic Spectroscopy.* Fresenius Z. Anal. Chem. *331*, 321–323.

Berndt, H., and Sopczak, D. (1987), *Error Recognition and Program Optimisation in ET-AAS by Time-resolved Signals (Direct Determination of Element Traces in Urine: Cd, Co, Cr, Ni, Tl, Pb).* Fresenius Z. Anal. Chem. *329*, 18–26.

Berndt, H., Schaldach, G., and Klockenkämper, R. (1987), *Improvement of the Detection Power in Electrothermal Atomic Absorption Spectrometry by Summation of Signals. Determination of Traces of Metals in Drinking Water and Urine.* Anal. Chim. Acta *200*, 573–579.

Bernhard, M., Brinckmann, F. E., and Sadler, P. J. (eds.) (1986), *The Importance of Chemical "Speciation" in Environmental Processes.* Springer, Berlin-Heidelberg-New York-London-Paris-Tokyo.

Bersier, P. M. (1987), *Do Polarography and Voltammetry Deserve Wider Recognition in Official and Recommended Methods?* Anal. Proc. *24*, 44–49.

Bertin, E. P. (1971), *Principles and Practices of X-Ray Spectrometric Analysis.* Plenum Press, New York.

Bertsch, P. M., and Anderson, M. A. (1989), *Speciation of Aluminium in Aqueous Solutions Using Ion Chromatography.* Anal. Chem. *61*, 535–539.

Betteridge, D. (1987), *Towards Intelligent Automation.* Anal. Proc. *24*, 106–108.

Bishop, J. (1989), *An Innovative UV-VIS Spectrophotometer Design.* Int. Lab. *19/4*, 58–63.

Blazka, P., and Prochazkova, L. (1983), *Mineralization of Organic Matter in Water by U.V. Radiation.* Water Res. *17*, 355–364.

Bloom, N. S., and Crecelius, E. A. (1983), *Determination of Mercury in Seawater of Sub-nanogram per Liter Levels.* Marine Chem. *14*, 49–59.

Blotcky, A. J., Ebrahim, A., and Rack, E. P. (1988), *Determination of Selenium Metabolites in Biological Fluids Using Instrumental and Molecular Neutron Activation Analysis.* Anal. Chem. *60*, 2734–2737.

BMFT (Bundesministerium für Forschung und Technologie) (ed.) (1988), *Umweltprobenbank – Bericht und Bewertung der Pilotphase.* Springer-Verlag, Berlin-Heidelberg-New York-London-Paris-Tokyo.

Bock, R., translated and revised by Marr, L. (1979), *A Handbook of Decomposition Methods in Analytical Chemistry.* International Textbook Company Ltd., Bishopbriggs-Glasgow-London.

Bodewig, F. G., and Valenta, P. (1985), *Vollautomatisierte voltammetrische Analyse mit dem Voltammat.* Gewässerschutz Wasser Abwasser *77*, 173–179.

Bodewig, F. G., Valenta, P., and Nürnberg, H. W. (1982), *Trace Determination of As(III) and As(V) in Natural Waters by Differential Pulse Anodic Stripping Voltammetry.* Fresenius Z. Anal. Chem. *311*, 187–191.

Boehringer, U. R., and Hertel, W. (1987), *Aufbau einer Umweltprobenbank als Vorsorgeinstrument: Umweltproben unzersetzt langfristig lagern.* Umwelt *1–2*, 21–23.

Bohlen, A. V., Eller, R., Klockenkämper, R., and Tölg, G. (1987), *Microanalysis of Solid Samples by Total-reflection X-Ray Fluorescence Spectrometry.* Anal. Chem. *59*, 2551–2555.

Bolshov, M. A., Zybin, A. V., Koloshnikov, V. G., and Vasnetsov, M. V. (1981), *Detection of Extremely Low Lead Concentrations by Laser Atomic Fluorescence Spectrometry.* Spectrochim. Acta *36B*, 345–350.

Bond, A. M. (1980), *Modern Polarographic Methods in Analytical Chemistry.* Marcel Dekker, New York.

Bond, A. M., and McLachlan, N. M. (1986), *Direct Determination of Tetraethyllead and Tetramethyllead in Gasoline by High-performance Liquid Chromatography with Electrochemical Detection at Mercury Electrodes.* Anal. Chem. 54, 756–758.
Bond, A. M., Knight, R. W., Reust, J. B., Tucker, D. J., and Wallace, G. G. (1986), *Determination of Metals in Urine by Direct Injection of Sample, High-performance Liquid Chromatography and Electrochemical or Spectrophotometric Detection.* Anal. Chim. Acta 182, 47–59.
Boumans, P. W. J. M. (ed.) (1987), *Inductively Coupled Plasma Emission Spectroscopy, Part 1, Methodology, Instrumentation and Performance, Part 2, Applications and Fundamentals.* John Wiley & Sons, New York-Chichester-Brisbane-Toronto-Singapore.
Boumans, P. W. J. M., and Vrakking, J. J. A. M. (1987a), *Inductively Coupled Plasmas: Line Width and Shapes, Detection Limits and Spectral Interferences. An Integrated Picture.* J. Anal. At. Spectrom. 2, 513–525.
Boumans, P. W. J. M., and Vrakking, J. J. A. M. (1987b), *Detection Limits of about 350 Prominent Lines of 65 Elements Observed in 50 and 27 MHz Inductively Coupled Plasma (ICP): Effects of Source Characteristics, Noise and Spectral Bandwidth-"Standard" Values for the 50 MHz ICP.* Spectrochim. Acta 42B, 553–579.
Boutron, C. F. (1979), *Reduction of Contamination Problems in Sampling of Antarctic Snows for Trace Element Analysis.* Anal. Chim. Acta 106, 127–130.
Boutron, C. F. (1986), *Atmospheric Toxic Metals and Metalloids in the Snow and Ice Layers Deposited in Greenland and Antarctica from Prehistoric Times to Present.* Chapter 15 in: Nriagu, J. O., and Davidson, C. I. (eds.): *Toxic Metals in the Atmosphere (Advances in Environmental Science and Technology Series),* pp. 467–505. John Wiley & Sons, New York.
Bowen, H. J. M. (1966), *Trace Elements in Biochemistry.* Academic Press, London.
Bowen, H. J. M. (1980), *Application of Activation Techniques to Biological Analysis.* CRC Crit. Rev. Anal. Chem. 10, 127–174.
Boyle, E. A., Handy, B., and Vangeen, A. (1987), *Cobalt Determination in Natural Waters Using Cation-Exchange Liquid Chromatography with Luminol Chemiluminescence Detection.* Anal. Chem. 59, 1499–1503.
Brainina, K. Z. (1974), *Stripping Voltammetry in Chemical Analysis.* J. Wiley & Sons, New York.
Braithwaite, A. (1987), *An Overview of the Criteria for Introducing LIMS into a Laboratory or Company Complex.* Anal. Proc. 24, 125–126.
Breyer, R., and Gilbert, B. P. (1987), *Determination of Very Low Levels of Selenium(IV) in Sea Water by Differential-pulse Cathodic Stripping Voltammetry after Extraction of the 3,3'-Diaminobenzidine Piazselenol.* Anal. Chim. Acta 201, 33–41.
Brindle, I. D., and Le, X. (1988), *Determination of Trace Amounts of Tin by Hydride Generation Direct Current Plasma Atomic Emission Spectrometry: Interference Reduction by L-Cystine.* Analyst 113, 1377–1381.
Brix, H., Lyngby, J. E., and Schierup, H.-E. (1983), *The Reproducibility in the Determination of Heavy Metals in Marine Plant Material – An Interlaboratory Calibration.* Marine Chem. 12, 69–85.
Bro, S., Jørgensen, O. J., Christensen, J. M., and Hørder, M. (1988), *Concentration of Nickel and Chromium in Serum: Influence of Blood Sampling Technique.* J. Trace Elem. Electrolytes Health Dis. 2, 31–35.
Broekaert, J. A. C. (1982), *Zeeman Atomic Absorption Instrumentation.* Spectrochim. Acta 37B, 65–69.
Broekaert, J. A. C. (1987), *Trends in Optical Spectrochemical Trace Analysis with Plasma Sources.* Anal. Chim. Acta 196, 1–21.
Broekaert, J. A. C., and Tölg, G. (1987), *Recent Developments in Atomic Spectrometry Methods for Elemental Trace Determinations.* Fresenius Z. Anal. Chem. 326, 495–509.
Broekaert, J. A. C., and Schrader, B. (1989), *Sonderheft Spektroskopie.* Nachr. Chem. Tech. Lab. 37.
Brown, A. A., and Riedel, U. (1987), *An Introduction to Graphite Probe Atomization in Graphite Furnace AAS.* Int. Lab. 17, 42–50.

Brown, A. A., and Taylor, A. (1985), *Applications of a Slotted Quartz Tube and Flame Atomic-Absorption Spectrometry to the Analysis of Biological Samples.* Analyst 110, 579–582.

Brown, A. A., Roberts, D. J., and Kahokola, K. V. (1987), *Methods for Improving the Sensitivity in Flame Atomic Absorption Spectrometry.* J. Anal. At. Spectrom. 2, 201–204.

Brown, A. A., Halls, D. J., and Taylor, A. (1988), *Atomic Spectrometry Update – Clinical and Biological Materials, Foods and Beverages.* J. Anal. At. Spectrom. 3, 45R–78R.

Brown, S. S., Nomoto, S., Stoeppler, M., and Sunderman, F. W., Jr. (1981), *IUPAC Reference Method for Analysis of Nickel in Serum and Urine by Electrothermal Atomic Absorption Spectrometry.* Pure Appl. Chem. 53, 773–781.

Brügmann, L., Danielsson, L.-G., Magnusson, B., and Westerlund, S. (1983), *Intercomparison of Different Methods for the Determination of Trace Metals in Sea Water.* Marine Chem. 13, 327–339.

Brügmann, L., Geyer, E., and Kay, R. (1987), *A New Teflon Sampler for Trace Metal Studies in Seawater – "WATES".* Marine Chem. 21, 91–99.

Brümmer, G. W. (1986), *Heavy Metal Species, Mobility and Availability in Soils,* in: Bernhard, M., Brinckmann, F. E., and Sadler, P. J. (eds.): *The Importance of Chemical "Speciation" in Environmental Processes,* pp. 169–192. Springer, Berlin-Heidelberg-New York-London-Paris-Tokyo.

Bruland, K. W. (1983), *Trace Elements in Sea-water,* in: Wong, C. S., Boyle, E., Bruland, K. W., and Goldberg, E. D. (eds.): *Trace Metals in Sea Water,* pp. 157–220. Plenum Press, New York-London.

Bruland, K. W., Franks, R. P., Knauer, G. A., and Martin, J. H. (1979), *Sampling and Analytical Methods for the Determination of Copper, Cadmium, Zinc, and Nickel at the Nanogram per Liter Level in Sea Water.* Anal. Chim. Acta 105, 233–245.

Bruland, K. W., Coale, K. H., and Mart, L. (1985), *Analysis of Sea Water for Dissolved Cadmium, Copper and Lead: An Intercomparison of Voltammetric and Atomic Absorption Methods.* Marine Chem. 17, 285–300.

Brumbaugh, W. G., and Koirtyohann, S. R. (1988), *Effects of Surface on the Atomization of Lead by Graphite Furnace.* Anal. Chem. 60, 1051–1055.

Brzezinska-Paudyn, A., and van Loon, J. C. (1988), *Determination of Tin in Environmental Samples by Graphite Furnace Atomic Absorption and Inductively Coupled Plasma-Mass Spectrometry.* Fresenius Z. Anal. Chem. 331, 707–712.

Brzezinska, A., van Loon, J., Williams, D., Oguma, K., Fuwa, K., and Haraguchi, I. H. (1983), *A Study of the Determination of Dimethylmercury and Methylmercury Chloride in Air.* Spectrochim. Acta 38B, 1339–1346.

Bubnis, B. P., Straka, M. R., and Pacey, G. E. (1983), *Metal Speciation by Flow-injection Analysis.* Talanta 30, 841–844.

Buchanan, E. G., and Soleta, D. D. (1983), *Optimization of Instrumental Parameters for Square-wave Anodic Stripping Voltammetry.* Talanta 30, 459–464.

Burba, P., and Willmer, P. G. (1987), *Multielement-preconcentration for Atomic Spectroscopy by Sorption of Dithiocarbamate-metal Complexes (e.g. HMDC) on Cellulose Collectors.* Fresenius Z. Anal. Chem. 329, 539–545.

Burgess, C., and Mielenz, K. D. (eds.) (1987), *Advances in Standards and Methodology in Spectrophotometry.* Elsevier, Amsterdam.

Burguera, J. L. (1989), *Flow Injection Atomic Spectroscopy.* Marcel Dekker, New York.

Burguera, J. L., Burguera, M., and Alarcon, O. M. (1986), *Determination of Sodium, Potassium, Calcium, Magnesium, Iron, Copper and Zinc in Cerebrospinal Fluid by Flow Injection Atomic Absorption Spectrometry.* J. Anal. At. Spectrom. 1, 79–83.

Campanella, L., Cardarelli, E., Ferri, T., Petronio, B. M., and Pupella, A. (1987a), *Evaluation of Heavy Metals Speciation in an Urban Sludge I. Batch Method.* Sci. Total Environ. 61, 217–228.

Campanella, L., Cardarelli, E., Ferri, T., Petronio, B. M., and Pupella, A. (1987b), *Evaluation of Heavy Metals Speciation in an Urban Sludge II. Column Method.* Sci. Total Environ. 61, 229–234.

Cantle, J. E. (ed.) (1982), *Atomic Absorption Spectrometry,* Vol. 5 of *Techniques and Instrumentation in Analytical Chemistry.* Elsevier, Amsterdam-Oxford-New York.

Carnrick, G. R., Lumas, B. K., and Barnett, W. B. (1986), *Analysis of Solid Samples by Graphite Furnace Atomic Absorption Spectrometry Using Zeeman Background Correction. J. Anal. At. Spectrom. 1,* 443–447.

Cedergren, A., and Frech, W. (1987), *Determination of Aluminium in Biological Materials by Graphite Furnace Atomic Absorption Spectrometry (GFAAS). Pure Appl. Chem. 59,* 221–228.

Chakraborti, D., Adams, F., van Mol, W., and Irgolic, K. J. (1987), *Determination of Trace Metals in Natural Waters at Nanogram per Liter Levels by Electrothermal Atomic Absorption Spectrometry after Extraction with Sodium Diethyldithiocarbamate. Anal. Chim. Acta 196,* 23–31.

Chana, B. S., and Smith, N. J. (1987), *Urinary Arsenic Speciation by High-performance Liquid Chromatography/Atomic Absorption Spectrometry for Monitoring Occupational Exposure to Inorganic Arsenic. Anal. Chim. Acta 197,* 177–186.

Chau, Y. K. (1986a), *Analytical Aspects of Organometallic Species Determination in Freshwater Systems,* in: Bernhard, M., Brinckmann, F. E., and Sadler, P. J. (eds.): *The Importance of Chemical "Speciation" in Environmental Processes,* pp. 149–167. Springer, Berlin-Heidelberg-New York-London-Paris-Tokyo.

Chau, Y. K. (1986b), *Occurrence and Speciation of Organometallic Compounds in Fresh Water Systems. Sci. Total Environ. 49,* 305–323.

Chau, Y. K. (1988), *Speciation of Molecular and Ionic Organometallic Compounds in Biological Samples. Sci. Total Environ. 71,* 57–58.

Cheam, V., and Agemian, H. (1980), *Preservation and Stability of Inorganic Selenium Compounds at ppb Levels in Water Samples. Anal. Chim. Acta 113,* 237–245.

Cheng, C. J., Akagi, T., and Haraguchi, H. (1987), *Simultaneous Multi-element Analysis for Trace Metals in Sea Water by Inductively-coupled Plasma/Atomic Emission Spectrometry after Batch Preconcentration on a Chelating Resin. Anal. Chim. Acta 198,* 173–181.

Chittleborough, G. (1980), *A Chemist's View of the Analysis of Human Hair for Trace Elements. Sci. Total Environ. 14,* 53–75.

Clark, J. R. (1986), *Electrothermal Atomisation Atomic Absorption Conditions and Matrix Modifications for Determining Antimony, Arsenic, Bismuth, Cadmium, Gallium, Gold, Indium, Lead, Molybdenium, Palladium, Platinum, Selenium, Silver, Tellurium, Thallium and Tin Following Back-extraction of Organic Aminohalide Extracts. J. Anal. At. Spectrom. 1,* 301–308.

Clark, S., Ashby, J., and Craig, P. J. (1987), *On-column Hydride Generation Method for the Production of Volatile Hydrides of Tin, Arsenic and Antimony for the Gas-chromatographic Analysis of Dilute Solutions. Analyst 112,* 1781–1782.

Clark, S., and Craig, P. J. (1988), *The Analysis of Inorganic and Organometallic Antimony, Arsenic and Tin Compounds Using an On-column Hydride Generation Method. Appl. Organomet. Chem. 2,* 33–46.

Clarkson, T. W., Friberg, L., Nordberg, G. F., and Sager, P. F. (eds.) (1988), *Biological Monitoring of Toxic Metals.* Plenum Press, New York-London.

Cooksey, M., and Barnett, W. B. (1979), *Matrix Modification and the Method of Additions in Flameless Atomic Absorption. Atom. Spectrosc. 5,* 101–105.

Cornelis, R. (1985), *Trace Element Studies in the Biosphere with Neutron Activation Analysis. J. Trace Microprobe Tech. 2,* 237–265.

Cornelis, R., Hoste, J., and Versieck (1982), *Potential Interferences Inherent in Neutron-activation Analysis of Trace Elements in Biological Materials. Talanta 29,* 1029–1034.

Corr, S. P., Hall, D. H., Littlejohn, D., and Perkins, C. V. (1988), *Novel Instrumentation for Inductively Coupled Plasma Atomic Emission Spectrometry. Anal. Proc. 25,* 226–229.

Craig, P. J. (1986), *Chemical Species in Industrial Discharges and Effluents* in: Bernhard, M., Brinckmann, F. E., and Sadler, P. J. (eds.): *The Importance of Chemical Speciation in Environmental Processes,* pp. 443–464. Springer, Berlin-Heidelberg-New York-London-Paris-Tokyo.

Creaser, C. S., and Davies, A. M. C. (1989), *Analytical Applications of Spectroscopy*. Royal Society of Chemistry, Cambridge.

Cresser, M. S., Ebdon, L., and Dean, J. R. (1988), *Atomic Spectrometry Update – Environmental Analysis. J. Anal. At. Spectrom. 3*, 1R–28R.

Cutter, G. A. (1985), *Determination of Selenium Speciation in Biogenic Particles and Sediments. Anal. Chem. 57*, 2951–2955.

Czobik, E. J., and Matousek, J. P. (1978), *Interference Effects in Furnace AAS. Anal. Chem. 50*, 2–10.

Dabeka, R. W., and McKenzie, A. D. (1987), *Lead, Cadmium, and Fluoride Levels in Market Milk and Infant Formulas in Canada. J. Assoc. Off. Anal. Chem. 70*, 753–757.

Dabeka, R. W., Kardinski, K. F., McKenzie, A. D., and Bajdik, C. D. (1986), *Lead, Cadmium and Fluoride in Human Milk and Correlation of Levels with Environmental and Food Factors. Food Chem. Toxicol. 24*, 913–921.

Danielsson, L.-G., Magnusson, B., and Westerlund, S. (1978), *An Improved Metal Extraction Procedure for the Determination of Trace Metals in Sea Water by Atomic Absorption Spectrometry with Electrothermal Atomization. Anal. Chim. Acta 98*, 47–57.

Date, A. R., and Gray, A. L. (eds.) (1988), *Applications of Inductively Coupled Plasma Mass Spectrometry*. Blackie & Son Ltd., Glasgow.

Davis, R., and Frearson, M. (1987), *Mass Spectrometry*. John Wiley & Sons, New York.

Dawson, J. B. (1986), *Analytical Atomic Spectroscopy in Biology and Medicine. Fresenius Z. Anal. Chem. 324*, 463–471.

Dean, J. R., Ebdon, L., and Massey, R. (1987a), *Selection of Mode for the Measurement of Lead Isotope Ratios by Inductively Coupled Plasma Mass Spectrometry and its Application to Milk Powder Analysis. J. Anal. At. Spectrom. 2*, 369–374.

Dean, J. R., Munro, S., Ebdon, L., Crews, H. M., and Massey, R. C. (1987b), *Studies of Metalloprotein Species by Directly Coupled High-performance Liquid Chromatography Inductively Coupled Plasma Mass Spectrometry. J. Anal. At. Spectrom. 2*, 607–610.

de Bièvre, P. (1989), *Isotope Dilution Mass Spectrometry: What Can it Contribute to Accuracy in Trace Analysis? Fresenius Z. Anal. Chem.*, in press.

de Bièvre, P., Savory, J., Lamberty, A., and Savory, G. (1988), *Meeting the Need for Reference Measurements. Fresenius Z. Anal. Chem. 332*, 718–721.

de Galan, L., and van der Plas, P. S. C. (1986), *Low Power ICP-Physical Principles and Analytical Performance. Fresenius Z. Anal. Chem. 324*, 472–478.

de Goeij, J. J. M. (1988), *Radiochemical Neutron Activation Analysis: Status in Chemical Analysis. Trans. Am. Nucl. Soc. 56*, 194–195.

Delves, H. T., and Campbell, M. J. (1988), *Measurements of Total Lead Concentrations and of Lead Isotope Ratios in Whole Blood by Use of Inductively Coupled Plasma Source Mass Spectrometry. J. Anal. At. Spectrom. 3*, 343–348.

Demers, D. R., and Allemand, C. D. (1981), *Atomic Fluorescence Spectrometry with an Inductively Coupled Plasma as Atomization Cell and Pulsed Hollow Cathode Lamps for Excitation. Anal. Chem. 53*, 1915–1921.

Demers, D. R., and Jansen, E. B. M. (1985), *Recent Developments and Applications in Hollow Cathode Lamp-excited ICP Atomic Fluorescence Spectrometry,* in: Sansoni, B. (ed.): *Instrumentelle Multielementanalyse*, pp. 397–410. VCH Verlagsgesellschaft, Weinheim-Deerfield Beach/Florida-Basel.

Dennemont, J., Jaccard, J., and Landry, J.-C. (1985), *Speciation of Elements in the Environment by Laser Microprobe Mass Analysis (LAMMA). Int. J. Environ. Anal. Chem. 21*, 115–127.

Deyl, Z., Macek, K., and Janak, J. (eds.) (1975), *Liquid Column Chromatography. Journal of Chromatography Library*, Vol. 3. Elsevier, Amsterdam.

Dietz, F. (1986), *Experiences with ICP Mass Spectrometry in Water Analysis. Fresenius Z. Anal. Chem. 324*, 212–223.

Dittrich, K., and Stärk, H. J. (1986), *Laser-excited Atomic Fluorescence Spectrometry as a Practical Analytical Method, Part 1. Design of a Graphite Tube Atomiser for the Determination of Trace Amounts of Lead. J. Anal. At. Spectrom. 1*, 237–241.

Dittrich, K., and Stärk, H. J. (1987), *Laser Excited Atomic Fluorescence Spectrometry as a Practical Analytical Method, Part 2. Evaluation of a Graphite Tube Atomiser for the Determination of Trace Amounts of Indium, Gallium, Aluminium, Vanadium and Iridium by LAFS. J. Anal. At. Spectrom. 2*, 63–66.

Doerffel, K., and Eckschlager, K. (1981), *Optimale Strategien in der Analytik*. Harri Deutsch, Thun, Switzerland.

Dolan, J. W., Snyder, L. R., and Quarry, M. A. (1987), *HPLC Method Development and Column Reproducibility. Int. Lab. 1718*, 66–71.

Donat, J. R., and Bruland, K. W. (1988), *Direct Determination of Dissolved Cobalt and Nickel in Seawater by Differential Pulse Cathodic Stripping Voltammetry Preceded by Adsorptive Collection of Cyclohexane-1-2-Dione Dioxime Complexes. Anal. Chem. 60*, 240–244.

Dorten, W., Valenta, P., and Nürnberg, H. W. (1984), *A New Photodigestion Device to Decompose Organic Matter in Water. Fresenius Z. Anal. Chem. 317*, 367–371.

Dorten, W. S., Elbaz-Pulichet, F., Mart, L. R., and Martin, J.-M. (1989), *Reassessment of the River Input of Trace Metals into the Mediterranean Sea. Ambio*, in press.

Dougherty, J. P., Preli, F. R., Jr., McCaffrey, J. T., Seltzner, M. D., and Michel, R. G. (1987a), *Instrumentation for Zeeman Electrothermal Atomizer Laser Excited Atomic Fluorescence Spectrometry. Anal. Chem. 59*, 1112–1119.

Dougherty, J. P., Preli, F. R., and Michel, R. G. (1987b), *Laser-excited Atomic-Fluorescence Spectrometry in an Electrothermal Atomizer with Zeeman Background Correction. Talanta 36*, 151–159.

Dougherty, J. P., Preli, F. R., and Michel, R. G. (1989), *Laser-excited Atomic-Fluorescence Spectrometry in an Atomic Absorption Graphite Tube Furnace. J. Anal. At. Spectrom. 4*, 429–434.

Douglas, D. J., and Houk, R. S. (1985), *Inductively-coupled Plasma Mass Spectrometry (ICP-MS). Prog. Anal. At. Spectrosc. 8*, 1–18.

Douglas, D. J., and French, J. B. (1986), *An improved Interface for Inductively Coupled Plasma-mass Spectrometry (ICP-MS). Spectrochim. Acta 41B*, 197–204.

Drews, W., Weber, G., and Tölg, G. (1988), *Flow-injection System for the Determination of Nickel by Means of MIP-OES after Conversion to $Ni(CO)_4$. Fresenius Z. Anal. Chem. 332*, 862–865.

Duffield, R. J., Dawson, J. B., Saffar, M. H., Ellis, D. J., and Fisher, G. W. (1988), *Examination of the Effect of Furnace Design on Sensitivity in Graphite Furnace Atomic Absorption Spectrometry. J. Anal. At. Spectrom. 3*, 139–144.

Duffy, S. J., Hay, G. W., Mickledwaithe, R. K., and van Loon, G. W. (1988), *A Method for Determining Metal Species in Soil Pore Water. Sci. Total Environ. 76*, 203–215.

Duinker, J. C. (1986), *Formation and Transformation of Element Species in Estuaries*, in: Bernhard, M., Brinckmann, F. E., and Sadler, P. J. (eds.): *The Importance of Chemical "Speciation" in Environmental Processes*, pp. 365–384. Springer, Berlin-Heidelberg-New York-London-Paris-Tokyo.

D'Ulivo, A., and Papoff, P. (1986), *Simultaneous Detection of Alkylselenide, Alkyllead and Alkyltin Compounds by Gas Chromatography Using a Multi-channel Non-dispersive Atomic Fluorescence Spectrometric Detector and a Miniature Flame as the Atomiser. J. Anal. At. Spectrom. 1*, 479–484.

Dumarey, R., Heindryckx, R., and Dams, R. (1980), *Determination of Mercury in Environmental Standard Reference Materials by Pyrolysis. Anal. Chim. Acta 118*, 381–383.

Eagles, J. Fairweather-Tait, J., Portwood, D. E., Self, R., Götz, A., and Heumann, K. G. (1989), *Comparison of Fast Atom Bombardment Mass Spectrometry for the Measurement of Zinc Absorption in Human Nutrition Studies. Anal. Chem. 61*, 1023–1025.

Eaton, D. K., and Holcombe, J. A. (1983), *Oxygen Ashing and Matrix Modifiers in Graphite Furnace Atomic Absorption Spectrometric Determination of Lead in Whole Blood. Anal. Chem. 55*, 946–950.

Ebdon, L., and Carpenter, R. (1988), *Multi-element Simplex Optimisation for Inductively-coupled Plasma/Atomic Emission Spectrometry with a Plasma Torch Fitted with a Wide-bore Injector Tube.* Anal. Chim. Acta 209, 135–145.

Ebdon, L., and Collier, A. R. (1988), *Direct Atomic Spectrometric Analysis by Slurry Atomisation. Part 5. Analysis of Kaolin Using Inductively Coupled Plasma Atomic Emission Spectrometry.* J. Anal. At. Spectrom. 3, 557–561.

Ebdon, L., and Evans, E. H. (1988), *The Inductively Coupled Plasma – An Analytical Tool with Growing Application.* Eur. Spectrosc. News 79, 9–16.

Ebdon, L., and Parry, H. G. M. (1987), *Direct Atomic Spectrometric Analysis by Slurry Atomisation. Part 2. Elimination of Interferences in the Determination of Arsenic in Whole Coal by Electrothermal Atomisation Atomic Absorption Spectrometry.* J. Anal. At. Spectrom. 2, 131–134.

Ebdon, L., and Wilkinson, J. R. (1987a), *Direct Atomic Spectrometric Analysis by Slurry Atomisation. Part 1. Optimisation of Whole Coal Analysis by Inductively Coupled Plasma Atomic Emission Spectrometry.* J. Anal. At. Spectrom. 2, 39–44.

Ebdon, L., and Wilkinson, J. R., (1987b), *Direct Atomic Spectrometric Analysis by Slurry Atomisation. Part 3. Whole Coal Analysis by Inductively Coupled Plasma Atomic Emission Spectrometry.* J. Anal. At. Spectrom. 2, 325–328.

Ebdon, L., and Wilkinson, J. R. (1987c), *The Determination of Arsenic and Selenium in Coal by Continuous Flow Hydride-Generation Atomic Absorption Spectrometry and Atomic Fluorescence Spectrometry.* Anal. Chim. Acta 914, 177–187.

Ebdon, L., Hill, S., and Ward, R. W. (1986), *Directly Coupled Chromatography-Atomic Spectroscopy. Part 1. Directly Coupled Gas Chromatography-Atomic Spectroscopy.* Analyst 111, 1113–1138.

Ebdon, L., Hill, S., and Ward, R. W. (1987a), *Directly Coupled Chromatography-Atomic Spectroscopy. Part 2. Directly Coupled Liquid Chromatography-Atomic Spectroscopy.* Analyst 112, 1–16.

Ebdon, L., Hill, S., and Jones, P. (1987b), *Application of Directly Coupled Flame Atomic Absorption Spectrometry – Fast Protein Liquid Chromatography to the Determination of Protein-bound Metals.* Analyst 112, 437–440.

Ebdon, L., Hill, S., Walton, A. P., and Ward, R. W. (1988), *Coupled Chromatography-Atomic Spectrometry for Arsenic Speciation – A Comparative Study.* Analyst 113, 1159–1165.

Ediger, R. E. (1975), *Atomic Absorption Analysis with the Graphite Furnace Using Matrix Modification.* At. Absorpt. Newslett. 14, 127–130.

Egila, J. N., Littlejohn, D., Ottaway, J. M., and Xiao-Quan, S. (1986), *Clinical Applications of Electrothermal Atomic Absorption Spectrometry with Zeeman-Effect Background Correction.* Anal. Proc. 23, 426–429.

Ek, P., and Hulden, S.-G. (1987), *A Continuous Hydride-generation System for Direct Current Plasma Atomic-emission Spectrometry (DCP-AES).* Talanta 34, 495–502.

Ekanem, E. J., Barnard, C. L. R., and Ottaway, J. M. (1986), *Improved Determination of Cadmium in Blood by Flame Atomic Fluorescence Spectrometry.* J. Anal. At. Spectrom. 1, 349–353.

Epler, K. S., O'Haver, T. C., Turk, G. C., and MacCrehan, W. A. (1988), *Laser-enhanced Ionization as a Selective Detector for the Liquid Chromatographic Determination of Alkyltins in Sediment.* Anal. Chem. 60, 2062–3066.

Epstein, M. S., Carnrick, G. R., Slavin, W., and Miller-Ihli, N. J. (1988), *Automated Slurry Sample Introduction for Analysis of a River Sediment by Graphite Furnace Atomic Absorption Spectrometry.* Anal. Chem. 61, 1414–1419.

Erdtmann, G., and Nürnberg, H. W. (1973), *Activation Analysis of Organic Substances and Materials,* in: Korte, F. (ed.): *Methodicum Chimicum,* Vol. 1, Part 2, Chap. 10.4, pp. 735–791. G. Thieme, Stuttgart.

Erdtmann, G., and Petri, H. (1986), *Nuclear Activation Analysis: Fundamentals and Techniques,* in: Elving, P. J. (ed.): *Treatise on Analytical Chemistry,* 2nd Ed., Part 1, Vol. 14, Chap. 7, pp. 419–643. John Wiley & Sons, New York.

Eskilsson, H., and Haraldsson, C. (1987), *Reductive Stripping Chronopotentiometry for Selenium in Biological Materials with a Flow System*. Anal. Chim. Acta 198, 231–237.

Eskilsson, H., Haraldsson, C., and Jagner, D. (1985), *Determination of Nickel and Cobalt in Natural Waters and Biological Material by Reductive Chronopotentiometric Stripping Analysis in a Flow System without Sample Deoxygenation*. Anal. Chim. Acta 175, 79–88.

Esprit, M., Vandecasteele, C., and Hoste, J. (1986), *Determination of Cadmium in Environmental Materials by Fast Neutron Activation Analysis*. Anal. Chim. Acta 185, 307–313.

Everson, J., and Patterson, C. C. (1980), *"Ultra-clean" Isotope Dilution/Mass Spectrometric Analyses for Lead in Human Blood Plasma Indicate that Most Reported Values are Artificially High*. Clin. Chem. 26, 1603–1607.

Ewers, U., Brockhaus, A., Genter, E., Idel, H., and Schürmann, E. A. (1979), *Tooth Lead Concentration in School Children from Different Environments of North Western Germany*. Int. Arch. Occup. Environ. Health 44, 65–80.

Facchetti, S. (ed.) (1982), *Applications of Mass Spectrometry to Trace Analysis*. Elsevier, Amsterdam-Oxford-New York.

Faires, L. M. (1986), *Fourier Transforms for Analytical Atomic Spectroscopy*. Anal. Chem. 58, 1023A–1034A.

Faires, L. M. (1987), *Inductively Coupled Plasma Fourier Transform Spectrometry: A New Analytical Technique? Potentials and Problems*. J. Anal. At. Spectrom. 2, 585–590.

Falk, H. (1988), *Some Results of Joint Research Activities on the FANES Technique between the Academy of Sciences, Berlin, and Strathclyde University, Glasgow*. Anal. Proc. 25, 250–251.

Falk, H., and Tilch, J. (1987), *Atomisation Efficiency and Over-all Performance of Electrothermal Atomisers in Atomic Absorption, Furnace Atomisation Non-thermal Excitation and Laser Excited Atomic Fluorescence Spectrometry*. J. Anal. At. Spectrom. 2, 527–531.

Falk, H., Hoffmann, E., and Ludke, C. (1984), *A Comparison of Furnace Atomic Nonthermal Excitation Spectrometry (FANES) with Other Atomic Spectroscopic Techniques*. Spectrochim. Acta 39B, 283–294.

Falk, H., Hoffmann, E., Ludke, C., Ottaway, J. M., and Littlejohn, D. (1986), *Studies on the Determination of Cadmium in Blood by Furnace Atomic Non-thermal Excitation Spectrometry*. Analyst 111, 285–290.

Fang, Z., and Welz, B. (1989), *Optimisation of Experimental Parameters for Flow Injection Flame Atomic Absorption Spectrometry*. J. Anal. At. Spectrom. 4, 83–89.

Fang, Z., Welz, B., and Schlemmer, G. (1989), *Analysis of Samples with High Dissolved Solids Content Using Flow Injection Flame Atomic Absorption Spectrometry*. J. Anal. At. Spectrom. 4, 91–95.

Fariwar-Mohseni, M., and Neeb, R. (1979), *On the Simultaneous Inverse-voltammetric Determination of Cadmium, Lead and Copper in Milk Powder*. Fresenius Z. Anal. Chem. 196, 156–158.

Fassel, V. A. (1986), *Analytical Inductively Coupled Plasma Spectroscopies – Past, Present, and Future*. Fresenius Z. Anal. Chem. 324, 511–518.

Feigl, P. K. D., Krueger, F. R., and Schueler, B. (1984), *Determination of Organic and Inorganic Compounds in the Femtogram Range by Laser Microprobe Mass Spectrometry*. Microchim. Acta 1984 II, 85–96.

Fernandez, F. J., Bohler, W., Beaty, M. M., and Barnett, W. B. (1981), *Correction for High Background Levels Using the Zeeman Effect*. At. Spectrosc. 3, 73–80.

Ferri, D., and Buldini, P. L. (1981), *Differential Pulse Polarographic Determination of Traces of Iron in Solar-grade Silicon*. Anal. Chim. Acta 126, 247–251.

Fishbein, L. (1987), *Perspectives of Analysis of Carcinogenic and Mutagenic Metals in Biological Samples*. Int. J. Environ. Anal. Chem. 28, 21–69.

Fleckenstein, J. (1985), *Direct Measurement of Mercury in Solid Biological Samples by Zeeman Atomic Absorption Spectrometry in the Graphite Furnace*. Fresenius Z. Anal. Chem. 322, 704–707.

Flora, C. J., and Nieboer, E. (1980), *Determination of Nickel by Differential Pulse Polarography at a Dropping Mercury Electrode*. Anal. Chem. 52, 1013–1020.

Florence, T. M. (1986), *Electrochemical Approaches to Trace Element Speciation in Waters – A Review. Analyst 111*, 489–505.

Forster, R. C., and Howard, A. G. (1989), *The Capillary Gas Chromatography – Atomic Absorption Spectrometry of Organotin and Organolead Compounds. Anal. Proc. 26*, 34–36.

Förstner, U. (1986), *Chemical Forms and Environmental Effects of Critical Elements in Solid-Waste Materials: Combustion Residues,* in: Bernhard, M., Brinckmann, F. E., and Sadler, P. J. (eds.): *The Importance of Chemical "Speciation" in Environmental Processes,* pp. 465–491. Springer, Berlin-Heidelberg-New York-London-Paris-Tokyo.

Förstner, U., and Wittmann, G. T. W. (1981), *Metal Pollution in the Aquatic Environment,* 2nd revised Ed. Springer, Berlin.

Franke, J., and de Zeeuw (1976), *Differential Pulse Anodic Stripping Voltammetry as a Rapid Screening Technique for Heavy Metal Intoxications. Arch. Toxicol. 37*, 47–55.

Fraser, S. M., Ure, A. M., Mitchell, M. C., and West, T. S. (1986), *Determination of Cadmium in Calcium Chloride Extracts of Soils by Atom-trapping Atomic Absorption Spectrometry. J. Anal. At. Spectrom. 1*, 19–21.

Frech, W., and Cedergren, A. (1976), *Investigations of Reactions Involved in Flameless AA Procedures. Part II, An Experimental Study of the Role of Hydrogen in Eliminating the Interference from Chlorine in the Determination of Pb in Steel. Anal. Chim. Acta 82*, 93–102.

Frech, W., Cedergren, A., Cederberg, C., and Vessman, J. (1982), *Evaluation of Some Critical Factors Affecting Determination of Aluminium in Blood, Plasma or Serum by Electrothermal Atomic Absorption Spectroscopy. Clin. Chem. 28*, 2259–2263.

Frei, R. W., et al. (1988, 1990), *Proceedings of the 5th and 6th Sils-Maria IAEAC Symposia on Ion-Chromatography. J. Chromatogr. 439*(1), 1–170, and *482*(2), 263–425.

Freimann, P., and Schmidt, D. (1982), *Determination of Mercury in Seawater by Cold Vapor Atomic Absorption Spectrophotometry. Fresenius Z. Anal. Chem. 313*, 200–202.

Freimann, P., Schmidt, D., and Schomaker, K. (1983), *MERCOS – A Simple Teflon Sampler for Ultratrace Metal Analysis in Seawater. Marine Chem. 14*, 43–48.

Frenzel, W., and Schulze, G. (1987), *Open Thin-layer Cell – A Flow-through Electrode for Potentiometric Stripping. Analyst 112*, 133–136.

Fresenius, W., and Lüderwald, I. (1989), *Presentations at the 11th International Symposium on Microchemical Techniques, Wiesbaden. Fresenius Z. Anal. Chem. 334*(7), 601–721.

Garcia, I. L., Navarro, P., and Cordoba, H. (1988), *Manual and FIA Methods for the Determination of Cadmium with Malachite Green and Iodide. Talanta 35*, 885–889.

Garden, L. M., Littlejohn, D., Dittrich, K., and Stärk, H.-J. (1988), *Laser-excited Atomic and Molecular Fluorescence – Preliminary Investigations for Thallium. Anal. Proc. 25*, 230–232.

Gardiner, P. H. E. (1987), *Species Identification for Trace Inorganic Elements in Biological Materials. Top. Curr. Chem. 141*, 145–174.

Gardiner, P. H. E. (1988), *Chemical Speciation in Biology and Medicine: The Role of Atomic Spectrometric Techniques. J. Anal. At. Spectrom. 3*, 163–168.

Gardiner, P. H. E., Gessner, H., Brätter, P., Stoeppler, M., and Nürnberg, H. W. (1984), *The Distribution of Zinc in Human Erythrocytes. J. Clin. Chem. Clin. Biochem. 22*, 159–163.

Garner, W. Y., and Barge, M. (1988), *Good Laboratory Practices – An Agrochemical Perspective.* ACS Symposium Series No. 369. American Chemical Society, Washington, DC.

Garfield, F. M. (1984), *Quality Assurance Principles for Analytical Laboratories.* AOAC, Virginia.

Gerhardsson, L., Brune, D., Nordberg, G. F., and Wester, P. O. (1986), *Distribution of Cadmium, Lead and Zinc in Lung, Liver and Kidney in Long-term Exposed Smelter Workers. Sci. Total Environ. 50*, 65–85.

Giauque, R. D., Thompson, A. C., Underwood, J. H., Wu, Y., Jones, K. W., and Rivers, M. L. (1988), *Measurement of Femtogram Quantities of Trace Elements Using an X-Ray Microprobe. Anal. Chem. 60*, 855–858.

Gibbs, R. J., and Angelidis, M. (1988), *Metal Chemistry Differences between Digested and Undigested Sludges. J. WPCF 60*, 113–118.

Gillson, G. R., Douglas, D. J., Fulford, J. E., Halligan, K. W., and Tanner, S. D. (1988), *Nonspectroscopic Interelement Interferences in Inductively Coupled Plasma Mass Spectrometry. Anal. Chem. 60*, 1472–1474.

Giné, M. F., Krug, F. J., Filho, H. G., dos Reis, B. F., and Zagatto, E. A. G. (1988), *Flow Injection Calibration of Inductively Coupled Plasma Atomic Emission Spectrometry Using the Generalised Standard Additions Method. J. Anal. At. Spectrom. 3*, 673–678.

Goforth, D., and Winefordner, J. D. (1986), *Laser-excited Atomic Fluorescence of Atoms Produced in a Graphite Furnace. Anal. Chem. 58*, 2598–2602.

Goforth, D., and Winefordner, J. D. (1987), *A Graphite-tube Furnace for Use in Laser-excited Atomic-fluorescence Spectrometry. Talanta 34*, 290–292.

Golimowski, J. (1989), *Trace Analysis of Iron in Environmental Water and Snow Samples from Poland. Anal. Lett. 22*, 481–492.

Golimowski, J., Nürnberg, H. W., and Valenta, P. (1980), *Die voltametrische Bestimmung toxischer Schwermetalle im Wein. Lebensm. Chem. Gerichtl. Chem. 34*, 116–120.

Golimowski, J., Valenta, P., and Nürnberg, H. W. (1985), *Trace Determination of Chromium in Various Water Types by Adsorption Differential Pulse Voltammetry. Fresenius Z. Anal. Chem. 322*, 315–322.

Gonsior, B., and Roth, M. (1983), *Trace Element Analysis by Particle and Photon-induced X-Ray Emission Spectroscopy. Talanta 30*, 385–400.

Gorsuch, T. T. (1970), *The Destruction of Organic Matter.* Pergamon Press, Oxford-New York-Toronto-Sydney-Braunschweig.

Götz, A., and Heumann, G. (1988), *Determination of Heavy Metals (Pb, Cd, Cu, Zn, Cr) in Sedimentary Reference Materials with IDMS: Total Concentration and Aqua Regia Soluble Portion. Fresenius Z. Anal. Chem. 332*, 640–644.

Grasserbauer, M. (1987), *Analysis of Inorganic Materials by Beam Techniques – The Challenge of High Technology. Anal. Chim. Acta 915*, 1–32.

Green, J. R., and Margerison, D. (1977), *Statistical Treatment of Experimental Data.* Elsevier, Amsterdam.

Greenfield, S. (1984), *Inductively Coupled Plasmas in Atomic Fluorescence Spectrometry. Anal. Proc. 21*, 61–63.

Greenfield, S., Jones, I. L., and Berry, C. T. (1964), *High-pressure Plasmas as Spectroscopic Emission Sources. Analyst 89*, 713–720.

Greenfield, S., Hieftje, G. M., Omenetto, N., Scheeline, A., and Slavin, W. (1986), *Twenty-five Years of Analytical Atomic Spectroscopy. Anal. Chim. Acta 180*, 69–98.

Gregoire, D. C. (1988), *Determination of Platinum, Palladium, Ruthenium and Iridium in Geological Materials by Inductively Coupled Plasma Mass Spectrometry with Sample Introduction by Electrothermal Vaporisation. J. Anal. At. Spectrom. 3*, 309–314.

Gretzinger, K., Kotz, L., Tschöpel, P., and Tölg, G. (1982), *Causes and Elimination of Systematic Errors in the Determination of Iron and Cobalt in Aqueous Solutions in the ng/ml and pg/ml Range. Talanta 29*, 1011–1018.

Grobecker, K. H., Mohl, C., and Stoeppler, M. (1987), *Solid Sampling Zeeman GFAAS – A Versatile Tool for Rapid and Direct Metal Analysis and Homogeneity Studies,* in: Lindberg, S. E., and Hutchinson, T. C. (eds.): *Proceedings International Conference on Heavy Metals in the Environment,* Vol. 1, pp. 486–488. CEP Consultants Ltd., Edinburgh.

Guinn, V. P., and Hoste, J. (1980), *Neutron Activation Analysis,* in: *IAEA Elemental Analysis of Biological Materials, Tech. Rep. Ser. 197*, 105–140.

Guo, R., Chen, N., and Lai, E. P. C. (1988), *Extraction-photoacoustic Spectroscopic Determination of Antimony, Cadmium, Cobalt, Nickel, Palladium and Thallium. Analyst 113*, 595–598.

Günther, K., and Umland, F. (1988), *Speciation of Cadmium and Thallium in Native Rape (Brassica rapus). Fresenius Z. Anal. Chem. 331*, 302–309.

Haas, W. J., and Fassel, V. A. (1980), *Inductively Coupled Plasma Atomic Emission Spectroscopy,* in: *IAEA, Elemental Analysis of Biological Materials,* pp. 167–199. Tech. Rep. Ser. 197, Vienna.

Haas, H., and Mart, L. (1980), *Konstruktion und Funktionsweise eines automatischen Wasserprobennehmers für den Tiefsee-Einsatz.* Ber. KFA Jülich, Jül. 1689.

Haavikko, K., Anttila, A., Helle, A., and Vuori, E. (1984), *Lead Concentrations of Enamel and Dentine of Deciduous Teeth of Children from two Finnish Towns.* Arch. Environ. Health 39, 78–84.

Hagedorn-Götz, H., Küppers, G., and Stoeppler, M. (1977), *On Nickel Contents in Urine and Hair in a Case of Exposure to Nickel Carbonyl.* Arch. Toxicol. 38, 275–285.

Halls, D. J. (1984), *Speeding up Determinations by Electrothermal Atomic-absorption Spectrometry.* Analyst 109, 1081–1084.

Halls, D. J., Mohl, C., and Stoeppler, M. (1987), *Application of Rapid Furnace Programmes in Atomic Absorption Spectrometry to the Determination of Lead, Chromium and Copper in Digests of Plant Materials.* Analyst 112, 185–189.

Hansen, L. D., Silberman, D., Fisher, G. L., and Eatough, D. J. (1984), *Chemical Speciation of Elements in Stack-collected, Respirable-size Coal Fly Ash.* Environ. Sci. Technol. 18, 181–186.

Harnly, J. M. (1986), *Multielement Atomic Absorption with a Continuum Source.* Anal. Chem. 58, 933A–943A.

Harnly, J. M., and Wolf, W. R. (1984), *Atomic Spectrometry for Inorganic Elements in Foods,* in: Charalambous, G. (ed.): *Analysis of Foods and Beverages: Modern Techniques,* pp. 451–481. Academic Press, Orlando-San Diego-San Francisco-New York-London-Toronto-Montreal-Sydney-Tokyo-Sao Paulo.

Harrison, W. W., Hess, K. R., Marcus, R. K., and King, F. L. (1986), *Glow Discharge Mass Spectrometry.* Anal. Chem. 58, 341A–356A.

Hassell, D. C., Rettberg, T. M., Fort, F. A., and Holcombe, J. A. (1988), *Low Pressure Vaporization for Graphite Furnace Atomic Absorption Spectrometry.* Anal. Chem. 60, 2680–2683.

Helsen, J. A., and Vrebos, B. A. R. (1986), *Quantitative X-Ray Fluorescence Analysis: Limits of Precision and Accuracy.* Int. Lab. 16/10, 66–71.

Hendrick, M. S., Goliber, P. A., and Michel, R. G. (1986), *Direct Current Plasma as an Excitation Source for Flame Atomic Fluorescence Spectrometry – Some Applications.* J. Anal. At. Spectrom. 1, 45–50.

Henshaw, J. M., Heithmar, E. M., and Hinners, T. A. (1989), *Inductively Coupled Plasma Mass Spectrometry Determination of Trace Elements in Surface Waters Subject to Acidic Deposition.* Anal. Chem. 61, 335–342.

Henze, G., and Neeb, R. (1986), *Elektrochemische Analytik.* Springer, Berlin-Heidelberg-New York-Tokyo.

Herber, R. F. M. (ed.) (1989), *Proceedings 3rd International Colloquium Solid Sampling with Optical Atomic Spectroscopy, Wetzlar, Oct. 10–12, 1988.* Fresenius Z. Anal. Chem., in press.

Herber, R. F. M., and Pieters, H. J. (1988), *Computer Aided Signal Analysis in Electrothermal Atomization-Atomic Absorption Spectrometry – Accuracy and Precision: Peak Height vs Peak Area.* Spectrochim. Acta 43B, 149–158.

Hermann, G. H. (1988), *Coherent Forward Scattering Spectroscopy (CFS): Present Status and Future Perspectives.* CRC Crit. Rev. Anal. Chem. 19, 323–377.

Hertz, H. S. (1988), *Are Quality and Productivity Compatible in the Analytical Laboratory?* Anal. Chem. 60, 75A–80A.

Heumann, K. G. (1980), *Mass Spectrometric Isotope Dilution Analysis for Accurate Determination of Elements in Environmental Samples.* Toxicol. Environ. Chem. Rev. 3, 111–129.

Heumann, K. G. (1988a), *Isotope Dilution Mass Spectrometry,* in: Adams, F., Gijbels, R., and van Grieken, R. (eds.): *Inorganic Mass Spectrometry,* pp. 301–376. Wiley, New York.

Heumann, K. G. (ed.) (1988b), *Element Trace Analysis by Mass Spectrometry.* Fresenius Z. Anal. Chem. 331(2), 103–222.

Heumann, K. G., Schindlmeier, W., Zeininger, H., and Schmidt, M. (1985), *Application of an Economical and Small Thermal Ionization Mass Spectrometer for Accurate Anion Trace Analysis.* Fresenius Z. Anal. Chem. 320, 457–462.

Heydorn, K. (1984), *Neutron Activation Analysis for Clinical Trace Element Research,* 2 Volumes. CRC Press Inc. Boca Raton, Florida.

Heyrovsky, J., and Kuta, J. (1965), *Grundlagen der Polarographie.* Akademie-Verlag, Berlin.

Heyrovsky, J., and Zuman, P. (1959), *Einführung in die praktische Polarographie.* Verlag Technik, Berlin.

Hieftje, G.M., and Vickers, G.H. (1989), *Developments in Plasma Source/Mass Spectrometry.* Anal. Chim. Acta 216, 1–24.

Hilpert, K., and Waidmann, E. (1986), *Multi-element Determination in Environmental Samples by Mass Spectrometric Isotope Dilution Analysis Using Thermal Ionization.* Fresenius Z. Anal. Chem. 325, 141–145.

Hilpert, K., and Waidmann, E. (1988), *Multi-element Determination in Environmental Samples by Mass Spectrometric Isotope Dilution Analysis Using Thermal Ionization. Part II. Oyster Tissue.* Fresenius Z. Anal. Chem. 331, 111–113.

Hinderberger, E.J., Kaiser, M.L., and Koirtyohann, S.R. (1981), *Furnace Atomic Absorption Analysis of Biological Samples Using the L'Vov Platform and Matrix Modification.* At. Spectrosc. 1, 1–7.

Hinds, M.W., Katyal, M., and Jackson, K.W. (1988), *Effectiveness of Palladium plus Magnesium as a Matrix Modifier for the Determination of Lead in Solutions and Soil Slurries by Electrothermal Atomisation Atomic Absorption Spectrometry.* J. Anal. At. Spectrom. 3, 83–87.

Hoffmann, G., and Schweiger, P. (1981), *Taking Soil and Plant Samples for the Analysis of Heavy Metals.* Staub Reinhalt. Luft 41, 443–444.

Hood, W.H., and Niemczyk, T.M. (1987), *Determination of Arsenic, Selenium, and Antimony Using Metastable Transfer Emission Spectrometry.* Anal. Chem. 59, 2468–2472.

Hopps, H.C. (1977), *The Biological Bases for Using Hair and Nail for Analyses of Trace Elements.* Sci. Total Environ. 7, 71–89.

Horvat, M., May, K., Stoeppler, M., and Byrne, A.R. (1988), *Comparative Studies of Methylmercury Determination in Biological and Environmental Samples.* Appl. Organomet. Chem. 2, 515–524.

Hoste, J. (1986), *Applications of Nuclear Activation Analysis,* in: Elving, P.I. (ed.): *Treatise on Analytical Chemistry,* 2nd Ed., Part I, Vol. 14, pp. 645–777. John Wiley & Sons, New York.

Houk, R.S. (1986), *Mass Spectrometry of Inductively Coupled Plasmas.* Anal. Chem. 58, 97A–105A.

Hua, C., Jagner, D., and Renman, L. (1988), *Constant-current Stripping Analysis for Iron(III) by Adsorptive Accumulation of its Solochrome Violet RS Complex on a Carbon-fibre Electrode.* Talanta 35, 597–600.

Huiliang, H., Hua, C., Jagner, D., and Renman, L. (1987a), *Carbon Fibre Electrode in Flow Potentiometric Stripping Analysis.* Anal. Chim. Acta 193, 61–69.

Huiliang, H., Jagner, D., and Renman, L. (1987b), *Flow Potentiometric and Constant-current Stripping Analysis for Mercury(II) with Gold, Platinum and Carbon Fibre Working Electrodes. Application to the Analysis of Tap Water.* Anal. Chim. Acta 201, 1–9.

Huiliang, H., Jagner, D., and Renman, L. (1987c), *Simultaneous Determination of Mercury(II), Copper(II) and Bismuth(III) in Urine by Flow Constant-current Stripping Analysis with a Gold Fibre Electrode.* Anal. Chim. Acta 202, 117–122.

Huiliang, H., Jagner, D., and Renman, L. (1988), *Flow Potentiometric and Constant-current Stripping Analysis for Arsenic(V) without Prior Chemical Reduction to Arsenic(III). Application to the Determination of Total Arsenic in Seawater and Urine.* Anal. Chim. Acta 207, 37–46.

Hutton, R.C. (1986), *Application of Inductively Coupled Plasma Source Mass Spectrometry (ICP-MS) to the Determination of Trace Metals in Organics.* J. Anal. At. Spectrom. 1, 259–263.

Huynh-Ngoc, L., Whitehead, N.E., and Oregioni, B. (1988), *Low Levels of Copper and Lead in a Highly Industrialized River.* Toxicol. Environ. Chem. 17, 223–246.

IAEA (International Atomic Energy Agency) (1975), *Design of Radiotracer Experiments in Marine Biological Systems.* Tech. Rep. Ser. No. 167. IAEA, Vienna.

IAEA (International Atomic Energy Agency) (1980), *Elemental Analysis of Biological Materials. Current Problems and Techniques with Special Reference to Trace Elements.* Tech. Rep. Ser. No. 197. IAEA, Vienna.
IAEA (International Atomic Energy Agency) (1985), *Chemical Aspects of Nuclear Methods of Analysis.* IAEA-TECTOD-350. IAEA, Vienna.
IAEA (International Atomic Energy Agency) (1986), *Nuclear Techniques for Analysis of Environmental Samples.* IAEA/RL/135. IAEA, Vienna.
Ichinoki, S., Hongo, N., and Yamazaki, M. (1988), *Multielement Analysis by High-performance Liquid Chromatography Following Solvent Extraction with Acetylacetone.* Anal. Chem. 60, 2099–2104.
Ihnat, M. (1987), *High Reliability Atomic Absorption Spectrometry of Major and Minor Elements in Biological Materials.* Fresenius Z. Anal. Chem. 326, 739–741.
Iida, A., Yoshaga, A., Sakurai, K., and Gohshi, Y. (1986), *Synchroton Radiation Excited X-Ray Fluorescence Analysis Uing Total Reflection of X-Rays.* Anal. Chem. 58, 394–397.
Irgolic, K. J. (1987), *Analytical Procedures for the Determination of Organic Compounds of Metals and Metalloids in Environmental Samples.* Sci. Total Environ. 64, 61–73.
Irth, H., de Jong, G. J., Brinkman, U. A. T., and Frei, R. W. (1987), *Trace Enrichment and Separation of Metal Ions as Dithiocarbamate Complexes by Liquid Chromatography.* Anal. Chem. 59, 98–101.
Itoh, K., Chikuma, M., and Tanaka, H. (1988), *Determination of Selenium in Sediments by Hydride Generation Atomic Absorption Spectrometry. Elimination of Interferences.* Fresenius Z. Anal. Chem. 330, 600–604.
Itoh, K., Chikuma, M., Nishimura, M., Tanaka, T., Tanaka, M., Nakayama, M., and Tanaka, H. (1989), *Determination of Selenium(IV) and Other Forms of Selenium Dissolved in Sea Water by Anion-Exchange Resin Loaded with Sulfonic Acid Derivative of Bismuthiol-II and Hydride Generation Atomic-Absorption Spectrometry.* Fresenius Z. Anal. Chem. 333, 102–107.
Iyengar, G. V. (1981), *Autopsy Sampling and Elemental Analysis: Errors Arising from Post Mortem Changes.* J. Pathol. 134, 173–180.
Iyengar, G. V. (1986), *Presampling Factors.* J. Res. Natl. Bur. Stand. 91, 67–74.
Iyengar, G. V. (1989), *Elemental Analysis of Biological Systems, Vol. I: Biological, Medical, Environmental, Compositional, and Methodological Aspects.* CRC Press, Boca Raton, Florida.
Iyengar, G. V., and Kasperek, K. (1977), *Application of the Brittle Fracture Technique (BFT) to Homogenise Biological Samples and Some Observations Regarding the Distribution Behaviour of the Trace Elements at Different Concentration Levels in a Biological Matrix.* J. Radioanal. Chem. 39, 301–316.
Iyengar, G. V., and Kollmer, W. E. (1986), *Some Aspects of Sample Procurement from Human Subjects for Biomedical Trace Element Research.* Trace Elem. Med. 3, 25–33.
Iyengar, G. V., Kollmer, W. E., and Bowen, H. J. M. (1978), *The Elemental Composition of Human Tissues and Body Fluids.* Verlag Chemie, Weinheim-New York.
Jackwerth, E., and Gomiscek, S. (1984), *General Aspects of Trace Analytical Methods – VI. Acid Pressure Decomposition in Trace Element Analysis.* Pure Appl. Chem. 56, 479–489.
Jäger, W. (1983), *Practical Experiences with Plasma Emission Techniques during the Routine Determination of Sewage Waste, Sludge and Soil Samples.* Z. Wasser Abwasser Forsch. 16, 231–233.
Jagner, D. (1978), *Instrumental Approach to Potentiometric Stripping Analysis of Some Heavy Metals.* Anal. Chem. 50, 1924–1929.
Jagner, D., and Graneli, A. (1976), *Potentiometric Stripping Analysis.* Anal. Chim. Acta 83, 19–26.
Jagner, D., Josefson, M., Westerlund, S., and Arén, K. (1981), *Simultaneous Determination of Cadmium and Lead in Whole Blood and in Serum by Computerized Potentiometric Stripping Analysis.* Anal. Chem. 53, 1406–1410.
Jakubowski, N., Stuewer, D., and Vieth, W. (1988), *Glow Discharge Mass Spectrometry with Low Resolution – Principles, Properties and Problems.* Fresenius Z. Anal. Chem. 331, 145–149.

Janghorbani, M., and Ting, B. T. G. (1989), *Comparison of Pneumatic Nebulization and Hydride Generation Inductively Coupled Plasma Mass Spectrometry for Isotopic Analysis of Selenium.* Anal. Chem. *61*, 701–708.

Jenkins, R. (1975), *An Introduction to X-Ray Spectrometry.* Heyden & Sons, London.

Jickels, T. D., and Burton, J. D. (1988), *Cobalt, Copper, Manganese and Nickel in the Sargasso Sea.* Marine Chem. *23*, 131–144.

Jobin Yvon (1988), *ICP Spectrometer of ISA JOBIN-YVON Longjumeau Cedex (France).*

Jochum, K. P., Matus, L., and Seufert, H. M. (1988a), *Trace Element Analysis by Laser Plasma Mass Spectrometry.* Fresenius Z. Anal. Chem. *331*, 136–139.

Jochum, K. P., Seufert, H. M., Midinet-Best, S., Rettmann, E., Schönberger, K., and Zimmer, M. (1988b), *Multi-element Analysis by Isotope Dilution-Spark Source Mass Spectrometry (ID-SSMS).* Fresenius Z. Anal. Chem. *331*, 104–110.

Johansson, S. A. E., and Campbell, J. L. (1988), *PIXE: A Novel Technique for Elemental Analysis.* John Wiley & Sons, Chichester-New York-Brisbane-Toronto-Singapore.

Jones, K. W., and Gordon, B. M. (1989), *Trace Element Determinations with Synchrotron-induced X-Ray Emission.* Anal. Chem. *61*, 341A–358A.

Jones, V. K., and Tarter, J. G. (1985), *Simultaneous Analysis of Anions and Cations in Water Samples Using Ion Chromatography.* Int. Lab. *15/9*, 36–39.

Jones, P., Williams, T., and Ebdon, L. (1989), *Determination of Cobalt at Picogram Levels by High-Performance Liquid Chromatography with Chemiluminescence Detection.* Anal. Chim. Acta *217*, 157–163.

Kadokami, K., Uehiro, T., Murita, M., and Fuwa, K. (1988), *Determination of Organotin Compounds in Water by Bonded-phase Extraction and High-performance Liquid Chromatography with Long-tube Atomic Absorption Spectrometric Detection.* J. Anal. At. Spectrom. *3*, 187–191.

Kaiser, G., and Tölg, G. (1986), *Reliable Determination of Elemental Traces in the ng/g Range in Biotic Materials and in Coal by Inverse Voltammetric and Atomic Absorption Spectrometry after Combustion of the Sample in a Stream of Oxygen.* Fresenius Z. Anal. Chem. *325*, 32–40.

Kanda, Y., and Taira, M. (1988), *Sequential Multi-element Analysis of Sediments and Soils by Inductively-coupled Plasma/Atomic Emission Spectrometry with a Computer-controlled Rapid Scanning Echelle Monochromator.* Anal. Chim. Acta *207*, 269–281.

Karstensen, K. H., and Lund, W. (1989), *Multielement Analysis of City Waste Incineration Ash and Slag by Inductively Coupled Plasma Atomic Emission Spectrometry.* Sci. Total Environ. *79*, 179–189.

Katsanos, A. A. (1980), *X-Ray Methods,* in: *IAEA, Tech. Rep. Ser. No. 197*, pp. 231–251. IAEA, Vienna.

Kayser, D., Boehringer, U., and Schmidt-Bleek, F. (1982), *The Environmental Specimen Banking Project of the Federal Republic of Germany.* Environ. Monit. Assess. *1*, 241–255.

Keil, R. (1984), *Sensitivity Enhancement in Flame Atomic Absorption Analysis Using a Slotted Quartz Tube in the Atomisation Chamber.* Fresenius Z. Anal. Chem. *319*, 391–394.

Keirsse, K., Smeyers-Verbecke, J., Verbeelen, D., and Massart, D. L. (1987), *Critical Study of the Speciation of Aluminium in Biological Fluids by Size-Exclusion Chromatography and Electrothermal Atomic Absorption Spectrometry.* Anal. Chim. Acta *196*, 103–114.

Keith, L. H., Crummet, W., Deegan, J. Jr., Libby, R. A., Taylor, J. K., and Wentler, G. (1983), *Principles of Environmental Analysis.* Anal. Chem. *55*, 2210–2218.

Kendall-Tobias, M. (1989), *Diode Array Spectroscopy.* Int. Lab. *19/4*, 64–69.

Kester, D. R. (1986), *Equilibrium Models in Seawater: Applications and Limitations,* in: Bernhard, M., Brinckmann, F. E., and Sadler, P. J. (eds.): *The Importance of Chemical "Speciation" in Environmental Processes,* pp. 337–363. Springer, Berlin-Heidelberg-New York-London-Paris-Tokyo.

Kingston, H. M., and Jassie, L. B. (1986), *Microwave Energy for Acid Decomposition at Elevated Temperatures and Pressures Using Biological and Botanical Samples.* Anal. Chem. *58*, 2534–2541.

Kirkbright, G. F. (1980), *Atomic Absorption Spectroscopy,* in: *Elemental Analysis of Biological Materials,* pp. 141–165. *Tech. Rep. Ser.* No. *197.* IAEA, Vienna.

Klaessens, J., and Kateman, G. (1987), *Problem Solving by Expert Systems in Analytical Chemistry. Fresenius Z. Anal. Chem. 326,* 203–213.

Klahre, P., Valenta, P., and Nürnberg, H. W. (1978), *Ein normiertes puls-polarographisches Verfahren zur Prüfung des Trinkwassers auf toxische Metalle. Vom Wasser 51,* 199–219.

Klockenkämper, R., Raith, B., Divoux, S., Gonsior, B., Brüggerhoff, S., and Jackwerth, E. (1987), *Comparison of Different Excitation Methods for X-Ray Spectral Analysis. Fresenius Z. Anal. Chem. 326,* 105–117.

Klockow, D. (1987), *The Present State of Sampling of Trace Substances in Ambient Air. Fresenius Z. Anal. Chem. 326,* 5–24.

Klussmann, U., Strupp, D., and Ebing, W. (1985), *Development of an Apparatus for Homogenization of Deep-frozen Plant Samples. Fresenius Z. Anal. Chem. 322,* 456–461.

Knapp, G. (1984), *Routes to Powerful Methods of Elemental Trace Analysis in Environmental Samples. Fresenius Z. Anal. Chem. 317,* 213–219.

Knapp, G., and Grillo, A. (1986), *A High Pressure Asher for Trace Analysis. Int. Lab. 16/3,* 76–79.

Knapp, G., Müller, K., Strunz, M., and Wegscheider, W. (1987), *Automation in Element Pre-Concentration with Chelating Ion Exchangers. J. Anal. At. Spectrom. (JAAS) 2,* 611–614.

Knöchel, A., and Petersen, W. (1983), *Results of an Interlaboratory Test for Heavy Metals in Elbe Water. Fresenius Z. Anal. Chem. 314,* 105–113.

Knowles, M. B., and Brodie, K. G. (1988), *Determination of Selenium in Blood by Zeeman Graphite Furnace Atomic Absorption Spectrometry Using a Palladium-Ascorbic Acid Chemical Modifier. J. Anal. At. Spectrom. 3,* 511–516.

Knowles, A., and Burgess, C. (eds.) (1984), *Practical Absorption Spectrometry.* Chapman & Hall, London.

Knowles, M. B., and Frary, B. D. (1988), *Zeeman AAS Applied to Element Determinations in Complex Matrices. Int. Lab. 18,* 3, 52–64.

Knox, J. H., Done, J. N., Gilbert, M. T., Pryde, A., and Wall, R. A. (1978), *High Performance Liquid Chromatography.* University Press, Edinburgh.

Koelling, S., Kunze, J., and Tauber, C. (1988), *Determination of As, Cd, Co, Cr, Cu, Fe, Hg, Mn, Ni, Pb, V and Zn in Coals, Additives, Fly Ashes and Impactor Filter Ashes from Coal-fired Power Plants. Fresenius Z. Anal. Chem. 332,* 776–789.

Komata, M., and Itoh, J.-I. (1988), *A Highly Sensitive Spectrophotometric Determination of Cadmium with α,β,γ,σ-tetrakis(4-N-trimethylaminophenyl)-porphyrine. Talanta 35,* 723–724.

Korkisch, J. (1988), *CRC Handbook of Ion Exchange Resins: Their Application to Inorganic Analytical Chemistry.* 6 Volumes. CRC Press, Boca Raton, Florida.

Kosta, L. (1986), *Radiochemical Analysis – A Survey of its Present State. Fresenius Z. Anal. Chem. 324,* 649–654.

Kosta, L., and Byrne, A. R. (1982), *Analytical Evaluation of Comparative Data on Trace Elements in Biological Materials. J. Radioanal. Chem. 69,* 117–129.

Kramer, J. R., and Allen, H. E. (1988), *Metal Speciation, Theory, Analysis and Application.* Lewis Publishers, Chelsea, Michigan.

Krivan, V. (1985), *Neutronenaktivierungsanalyse* in: *Analytiker Taschenbuch,* Vol. 5, pp. 35–68. Springer, Heidelberg.

Krivan, V. (1986), *Application of Radiotracers to Methodological Studies in Trace Element Analysis,* in: Elving, P. I. (ed.): *Treatise on Analytical Chemistry,* 2nd Ed., Part I, Vol. 14, pp. 339–417. John Wiley & Sons, New York.

Krivan, V. (1987), *Radiotracers for the Determination of the Accuracy of Trace Element Analyses. Sci. Total Environ. 64,* 21–40.

Krivan, V., and Haas, H. F. (1988), *Prevention of Loss of Mercury(II) during Storage of Dilute Solutions in Various Containers. Fresenius Z. Anal. Chem. 332,* 1–6.

Krivan, V., Petrick, K., Welz, B., and Melcher, M. (1985), *Radiotracer Error-Diagnostic Investigation of Selenium Determination by Hydride-generation Atomic Absorption Spectrometry Involving Treatment with Hydrogen Peroxide and Hydrochloric Acid.* Anal. Chem. 57, 1703–1706.

Krull, I. S. (1986), *Analysis of Inorganic Species by Ion Chromatography and Liquid Chromatography* in: Bernhard, M., Brinckmann, F. E., and Sadler, D. J. (eds.): *The Importance of Chemical "Speciation" in Environmental Processes,* pp. 579–611. Springer, Berlin-Heidelberg-New York-London-Paris-Tokyo.

Kuldvere, A. (1989), *Extraction of Geological Materials with Mineral Acids for the Determination of Arsenic, Antimony, Bismuth and Selenium by Hydride Generation Atomic Absorption Spectrometry.* Analyst 114, 125–131.

Kumamaru, T., Okamoto, Y., Yamamoto, Y., Nakata, F., Nitta, Y., and Matsuo, H. (1987), *Enhancement Effect of Organic Solvents in Inductively Coupled Plasma Atomic Emission Spectrometry.* Fresenius Z. Anal. Chem. 327, 777–781.

Kumpulainen, J., and Paaki, M. (1987), *Analytical Quality Program Used by the Trace Elements in Foods and Diets Sub-network of the FAO European Cooperative Network on Trace Elements.* Fresenius Z. Anal. Chem. 326, 684–689.

Kurfürst, U. (1983a), *Investigations on the Analysis of Heavy Metals in Solids by Direct Zeeman Atomic Absorption Spectroscopy, II. Theory, Properties and Efficiency of Direct Zeeman AAS.* Fresenius Z. Anal. Chem. 315, 304–320.

Kurfürst, U. (1983b), *Investigations on the Analysis of Heavy Metals in Solids by Direct Zeeman Atomic Absorption Spectroscopy, III. Importance of the Graphite Boat Technique for the Analysis of Solids.* Fresenius Z. Anal. Chem. 316, 1–7.

Kurfürst, U. (1984), *Untersuchungen über die Schwermetallanalyse in Feststoffen mit der direkten Zeeman-Atom-Absorptions-Spektrometrie.* Doctoral Thesis, Universität Bremen.

Kurfürst, U., Grobecker, K. H., and Stoeppler, M. (1984), *Homogeneity Studies in Biological, Standard Reference and Control Materials with Solid Sampling and Direct Zeeman AAS,* in: Brätter, P., and Schramel, P. (eds.): *Trace Element Analytical Chemistry in Medicine and Biology,* pp. 539–558. Walter de Gruyter, Berlin-New York.

Kurfürst, U., Kempeness, M., Stoeppler, M., and Schnider, O. (1989), *An Automated Solid Sample Analysis System.* Fresenius Z. Anal. Chem., in press.

Landner, L. (ed.) (1987), *Speciation of Metals in Water, Sediment and Soil Systems.* Lecture notes in *Earth Sciences,* Vol. 11. Springer, Berlin-Heidelberg-New York-London-Paris-Tokyo-Hongkong.

Langland, J. K., Harrison, S. H., Kratochvil, B., and Zeisler, R. (1983), *Cryogenic Homogenization of Biological Tissues,* in: Zeisler, R., Harrison, S. H., and Wise, S. A. (eds.): *The Pilot National Environmental Specimen Bank – Analysis of Human Liver Specimens,* pp. 21–34. NBS Special Publication 656, U.S. Department of Commerce, Washington, D.C.

Langmyhr, F. J. (1979), *Direct Analysis of Solids by Atomic Absorption Spectrophotometry.* Analyst 104, 993–1016.

Langmyhr, F. J. (ed.) (1987), *Proceedings 2nd International Colloquium Solid Sampling with Atomic Spectroscopic Methods, Wetzlar,* Oct. 13–15, 1986. Fresenius Z. Anal. Chem. 328, 315–418.

Langmyhr, F. J., and Wibetoe, G. (1985), *Direct Analysis of Solids by Atomic Absorption Spectrophotometry.* Prog. Anal. At. Spectrosc. 8, 193–256.

Larjava, K., Reith, J., and Klockow, D. (1990), *Development and Laboratory Investigations of a Denuder Sampling System for the Determination of Heavy Metal Species in Flue Gases at Elevated Temperatures.* Int. J. Environ. Anal. Chem. 38(1), 31–45.

Laxen, D. P. H., and Harrison, R. M. (1981), *A Scheme of the Physico-chemical Speciation of Trace Metals in Freshwater Samples.* Sci. Total Environ. 19, 59–82.

Leckie, J. O. (1986), *Adsorption and Transformation of Trace Element Species at Sediment/Water Interfaces,* in: Bernhard, M., Brinckmann, F. E., and Sadler, P. J. (eds.): *The Importance of Chemical "Speciation" in Environmental Processes,* pp. 237–254. Springer, Berlin-Heidelberg-New York-London-Paris-Tokyo.

Lee, Y. H. (1987), *Determination of Methyl- and Ethylmercury in Natural Waters at Sub-nanogram per Liter Using SCF-absorbent Preconcentration Procedure. Int. J. Environ. Anal. Chem. 29*, 263–276.

Leis, F., Broekaert, J. A. C., and Waechter, H. (1989), *A Three-electrode Direct Current Plasma as Compared to an Inductively Coupled Argon Plasma. Fresenius Z. Anal. Chem. 333*, 2–5.

Leong, M., Vera, J., Smith, B. W., Omenetto, N., and Winefordner, J. D. (1988), *Laser-induced Double Resonance Fluorescence of Lead with Graphite Tube Atomization. Anal. Chem. 60*, 1605–1610.

Lever, A. B. P. (1984), *Inorganic Electronic Spectroscopy.* Elsevier, Amsterdam.

Lewis, R. A., Stein, N., and Lewis, C. W. (eds.) (1984), *Environmental Specimen Banking and Monitoring as Related to Banking.* Martinus Nijhoff Publishers, Boston.

Lewis, R. A., Klein, B., Paulus, M., and Horras, P. (1990), *Environmental Specimen Banking*, in: Stoeppler, M. (ed.): *Hazardous Metals in the Environment. Evaluation of Analytical Methods in Biological Systems*, in preparation. Elsevier, Amsterdam.

Li, Z., Xiao-Quan, S., and Zhe-Ming, N. (1988), *An Oblique Section Hydride Generator for the Simultaneous Determination of Arsenic, Antimony and Bismuth in Geological Samples by Inductively Coupled Plasma-Atomic Emission Spectrometry. Fresenius Z. Anal. Chem. 332*, 764–768.

Liang, D. C., and Blades, M. W. (1988), *Atmospheric Pressure Capacitively Coupled Plasma Atomizer for Atomic Absorption Spectrometry. Anal. Chem. 60*, 27–31.

Liebhafsky, H. W., Schweikert, E. A., and Meyers, E. A. (1986), *The Nature of X-Rays, Spectrochemical Analysis of Conventional X-Ray Methods and Neutron Diffraction and Absorption*, in: Elving, P. I. (ed.): *Treatise on Analytical Chemistry*, 2nd Ed., Part I, Vol. 8, pp. 209–309. John Wiley & Sons, New York.

Liem, I., Kaiser, G., Sager, M., and Tölg, G. (1984), *The Determination of Thallium in Rocks and Biological Materials at $ng \cdot g^{-1}$ Levels by Differential-pulse Anodic Stripping Voltammetry and Electrothermal Atomic Absorption Spectrometry. Anal. Chim. Acta 158*, 179–197.

Lind, B., Elinder, C.-G., Friberg, L., Nilsson, B., Svartengren, M., and Vahter, M. (1987), *Quality Control in the Analysis of Lead and Cadmium in Blood. Fresenius Z. Anal. Chem. 326*, 647–655.

Lindberg, I., Lundberg, E., Arkhammer, P., and Berggen, P.-O. (1988), *Direct Determination of Selenium in Solid Biological Materials by Graphite Furnace Atomic Absorption Spectrometry. J. Anal. At. Spectrom. 3*, 497–501.

Lindstrom, R. M., Zeisler, R., and Rossbach, M. (1987), *Activation Analysis Opportunities Using Cold Neutron Beams. Radioanal. Nucl. Chem. 112*, 321–330.

Littlejohn, D. (1988), *Graphite Furnace Atomic Emission Spectrometry – The Rediscovery of a Technique. Anal. Proc. 25*, 217–220.

Littlejohn, D. (1989), *Becoming Absorbed and Excited in Atomic Spectrometry. Anal. Proc. 26*, 92–95.

Liu, Y., and Ingle, J. D., Jr. (1989), *Automated Two-Column Ion Exchange System for Determination of the Speciation of Trace Metals in Natural Waters. Anal. Chem. 61*, 525–529.

Lochmüller, C. H., and Lung, K. R. (1986), *Applications of Laboratory Robotics in Spectrophotometric Sample Preparation and Experimental Optimization. Anal. Chim. Acta 183*, 257–262.

Loring, D. H., and Rantala, R. T. T. (1988), *An Intercalibration Exercise of Trace Metals in Marine Sediments. Mar. Chem. 24*, 13–28.

Luck, J., and Siewers, U. (1988), *Progress in Analytical Application of an Inductively Coupled Plasma/Mass Spectrometer System. Fresenius Z. Anal. Chem. 331*, 129–132.

Lücker, E., Rosopulo, A., Koberstein, S., and Kreuzer, W. (1987), *The Determination of Heavy Metals in Fresh Renal Matrix by Means of Solid Sampling and Atomic-Absorption Spectrometry. Part II. The Practicability of the Determination of Lead and Cadmium in Fresh Renal Matrix by Means of Solid Sampling and Zeeman-AAS. Fresenius Z. Anal. Chem. 329*, 31–34.

Luepke, N.-P. (ed.) (1979), *Monitoring Environmental Materials and Specimen Banking*. Martinus Nijhoff Publishers, The Hague.

Luffer, D. R., and Salin, E. D. (1986), *Rapid Throughput Nebulizer-spray Chamber System for Inductively Coupled Plasma Atomic Emission Spectrometry*. Anal. Chem. 58, 654–656.

Lundberg, E., Baxter, D. C., and Frech, W. (1986), *Constant-temperature Atomiser-computer Controlled Echelle Spectrometer System for Graphite Furnace Atomic Emission Spectrometry*. J. Anal. At. Spectrom. 1, 105–113.

L'Vov, B. (1976), *Trace Characterization of Powders by Atomic-Absorption Spectrometry*. Talanta 23, 109–118.

L'Vov, B. (1978), *Electrothermal Atomization – The Way Toward Absolute Methods of Atomic Absorption Analysis*. Spectrochim. Acta 33B, 153–193.

L'Vov, B. (1987), *Recent Advances in the Theory of Atomisation in Graphite Furnace Atomic Absorption Spectrometry: The Oxygen-Carbon Alternative*. Analyst 112, 355–364.

L'Vov, B. (1988), *Graphite Furnace Atomic Absorption Spectrometry – On the Way to Absolute Analysis*. J. Anal. At. Spectrom. (JAAS) 3, 9–12.

L'Vov, B., Polzik, L. K., and Yatsenko, L. F. (1987), *The Effect of Thermal Sample Pretreatment on the Absorption Signal in Graphite Furnace AAS*. Talanta 34, 141–145.

Maenhaut, W. (1987), *Particle-induced X-Ray Emission Spectrometry: An Accurate Technique in the Analysis of Biological, Environmental and Geological Samples*. Anal. Chim. Acta 195, 125–140.

Manning, D. C., and Slavin, W. (1983), *The Determination of Trace Elements in Natural Waters Using the Stabilized Temperature Platform Furnace*. Appl. Spectrosc. 37, 1–11.

MARC (Monitoring and Assessment Research Centre) (1985), *Historical Monitoring*. Report No. 31. University of London.

Marr, I. L., and Cresser, M. S. (1983), *Environmental Chemical Analysis*. International Textbook Company, Glasgow and London.

Marshall, J., Littlejohn, D., Ottaway, J. M., Harnly, J. M., Miller-Ihli, N. J., and O'Haver, T. C. (1983), *Simultaneous Multi-element Analysis by Carbon Furnace Atomic-Emission Spectrometry*. Analyst 108, 178–188.

Marshall, J., Baxter, D. C., Carroll, J., Cook, S., Corr, S. P., Giri, S. K., Durie, D., Littlejohn, D., Ottaway, J. M., Stephen, S. C., and Wright, S. (1985), *The Probe Furnace in Atomic Spectrometry*. Anal. Proc. 22, 371–373.

Marshall, J., Ottaway, B. J., Ottaway, J. M., and Littlejohn, D. (1986), *Continuum-source Atomic Absorption Spectrometry – New Lamps for Old?* Anal. Chim. Acta 180, 357–371.

Mart, L. (1979a), *Prevention of Contamination and Other Accuracy Risks in Voltammetric Trace Metal Analysis of Natural Waters. I. Preparatory Steps, Filtration and Storage of Water Samples*. Fresenius Z. Anal. Chem. 296, 350–357.

Mart, L. (1979b), *Prevention of Contamination and Other Accuracy Risks in Voltammetric Trace Metal Analysis of Natural Waters. II. Collection of Surface Water Samples*. Fresenius Z. Anal. Chem. 299, 97–102.

Mart, L. (1982), *Minimization of Accuracy Risks in Voltammetric Ultratrace Determination of Heavy Metals in Natural Waters*. Talanta 29, 1035–1040.

Mart, L. (1983), *Seasonal Variation of Cd, Pb, Cu and Ni Levels in Snow from the Eastern Arctic Ocean*. Tellus 35B, 131–141.

Mart, L., and Nürnberg, H. W. (1984), *Trace Metal Levels in the Eastern Arctic Ocean*. Sci. Total Environ. 39, 1–14.

Mart, L., and Nürnberg, H. W. (1985), *Cd, Pb, Cu, Ni and Co Distribution in the German Bight*. Mar. Chem. 18, 197–213.

Mart, L., Nürnberg, H. W., and Valenta, P. (1980), *Prevention of Contamination and Other Accuracy Risks in Voltammetric Trace Metal Analysis of Natural Waters. III. Voltammetric Ultratrace Analysis with a Multicell-System Designed for Clean Bench Working*. Fresenius Z. Anal. Chem. 300, 350–362.

Mart, L., Nürnberg, H.W., and Gravenhorst, G. (1982a), *Probennahme von Firnproben zur Schwermetallanalyse (Pb, Cd, Ni, Cu) an der Atka-Bucht.* Ber. Polarforsch. *6*, 68–69.

Mart, L., Rützel, H. Klahre, P., Sipos, L., Platzek, U., Valenta, P., and Nürnberg, H.W. (1982b), *Comparative Studies on the Distribution of Heavy Metals in the Oceans and Coastal Waters.* Sci. Total Environ. *26*, 1–17.

Mart, L., Nürnberg, H.W., and Dyrssen, D. (1983), *Low Level Determination of Trace Metals in Arctic Seawater and Snow by Differential Pulse Anodic Stripping Voltammetry,* in: Wong, C.S., Boyle, E., Bruland, K.W., and Goldberg, E.D. (eds): *Trace Metals in Sea Water,* pp. 113–130. Plenum Press, New York-London.

Martin-Landa, I., de Pablos, F., and Marr, I.L. (1989), *Determination of Organotins in Fish and Sediments by Gas Chromatography with Flame Photometric Detection.* Anal. Proc. *26*, 16–18.

Massey, R.C., Burell, J.A., McWeeny, D.J., and Crews, H. (1986), *Speciation Studies on Simulated Gastro-intestinal Digests of Foodstuffs.* Toxicol. Environ. Chem. *13*, 85–93.

Massmann, H. (1976), *State of the Development of AA and Atomic Fluorescence Spectrometry with Furnaces.* Proc. Anal. Div. Chem. Soc. *13*, 258–264.

Massmann, H., and Gücer, S. (1974), *Physical and Chemical Processes in AAS with Graphite Furnaces.* Spectrochim. Acta *39B*, 283–300.

Matousek, J.P. (1981), *Interferences in Electrothermal Atomic Absorption Spectrometry, Their Elimination and Control.* Prog. Anal. At. Spectrosc. *4*, 247–310.

Matusiewicz, H. (1986), *Thermal Vaporization for Inductively Coupled Plasma Optical Emission Spectrometry – A Review.* J. Anal. At. Spectrom. *1*, 171–184.

May, K., Reisinger, K., Flucht, R., and Stoeppler, M. (1980), *Radiochemical Studies on the Behaviour of Mercury- and Methylmercury Chloride in Fresh and Sea Water.* Vom Wasser *55*, 63–76.

May, K., and Stoeppler, M. (1983), *Studies on the Biogeochemical Cycle of Mercury, I. Mercury in Sea and Inland Water and Food Products,* in: Proceedings International Conference Heavy Metals in the Environment, Vol. 1, pp. 241–244. CEP Consultants Ltd., Edinburgh.

May, K., Ahmed, R., Reisinger, K., Torres, B., and Stoeppler, M. (1985), *Methylmercury Contents in Specimens of the Environmental Specimen Bank and Other Materials,* in: Proceedings International Conference Heavy Metals in the Environment, Vol. 2, pp. 513–515. CEP Consultants Ltd., Edinburgh.

May, K., Stoeppler, M., and Reisinger K. (1987), *Studies on the Ratio Total Mercury/Methylmercury in the Aquatic Food Chain.* Toxicol. Environ. Chem. *13*, 153–159.

May, T.W., and Brumbaugh, W.G. (1982), *Matrix Modifier and L'Vov Platform for Elimination of Matrix Interferences in the Analysis of Fish Tissues for Lead by Graphite Furnace Atomic Absorption Spectrometry.* Anal. Chem. *54*, 1032–1037.

McLaren, J.W., Mykytiuk, A.P., Willie, S.N., and Berman, S.S. (1985), *Determination of Trace Metals in Seawater by Inductively Coupled Plasma Mass Spectrometry with Preconcentration on Silica-immobilized 8-hydroxyquinoline.* Anal. Chem. *57*, 2907–2911.

McLaren, J.W., Beauchemin, D., and Berman, S.S. (1987), *Application of Isotope Dilution Inductively Coupled Plasma Mass Spectrometry to the Analysis of Marine Sediments.* Anal. Chem. *59*, 610–613.

Measures, C.I., and Edmond, J.M. (1989), *Shipboard Determination of Aluminium in Seawater at the Nanomolar Level by Electron Capture Detection Gas Chromatography.* Anal. Chem. *61*, 544–547.

Meites, L. (1965), *Polarographic Techniques,* 2nd Ed. Wiley Interscience, New York.

Melgarejo, A.G., Céspedes, A.G., and Pavon, J.M.C. (1989), *Simultaneous Determination of Nickel, Zinc and Copper by Second-derivative Spectrophotometry Using 1-(2-pyridylazo)-2-naphthol as Reagent.* Analyst *114*, 109–111.

Merian, E. (1986), *The Role of Modern Analytical Chemistry in Environmental Pollution Studies.* Kem. Kemi *13*, 959–969.

Merian, E., Frei, R.W., Haerdi, W., Schlatter, C., and Bronzetti, G.L. (1985, 1988, 1990), *Carcinogenic and Mutagenic Metal Compounds.* Vols. 1, 2 and 3, Proceedings. Gordon & Breach Science Publishers, New York-London.

Metcalf, E. (1987), *Atomic Absorption and Emission Spectroscopy – Analytical Chemistry by Open Learning.* John Wiley & Sons, Chichester-New York-Brisbane-Toronto-Singapore.

Meyer, G. A. (1987), *ICP – Still the Panacea for Trace Metal Analysis? Anal. Chem.* 59, 1345A–1354A.

Meyer, A., de la Chevallerie-Haaf, U., and Henze, G. (1987), *Determination of Zinc, Cadmium, Lead and Copper in Soils and Sewage Sludges by Microprocessor-controlled Voltammetry in Comparison with Atomic Absorption Spectrometry. Fresenius Z. Anal. Chem.* 328, 565–568.

Michaelis, W. (1989), *Private communication.*

Michaelis, W., and Prange, A. (1988), *Trace Analysis of Geological and Environmental Samples by Total Reflection X-Ray Fluorescence Spectrometry. Nucl. Geophys. Int. J. Radiat. Appl. Instrum. E 2*, 231–245.

Michel, R. G., Hall, M. L., Ottaway, J. M., and Fell, G. S. (1979), *Determination of Cadmium in Blood and Urine by Flame Atomic-fluorescence Spectrometry. Analyst* 104, 491–504.

Michiels, E., and de Bievre, P. (1986), *Determination of Cadmium in Whole Blood by Isotope Dilution Mass Spectrometry (IDMS)*, in: O'Neill, I. K., Schuller, P., and Fishbein, L. (eds.): *Environmental Carcinogens, Selected Methods of Analysis.* IARC Scientific Publication No. 71, pp. 443–450. IARC, Lyon.

Mills, C. F. (1986), *The Influence of Chemical Species on the Absorption and Physiological Utilization of Trace Elements from the Diet or Environment,* in: Bernhard, M., Brinckmann, F. E., and Sadler, P. J. (eds.): *The Importance of Chemical "Speciation" in Environmental Processes,* pp. 71–83. Springer, Berlin-Heidelberg-New York-London-Paris-Tokyo.

Mitchell, J. W. (1982), *Radioisotope Techniques in Trace Analysis and Ultrapurification. Int. Lab.* 11, 1/2, 12–25.

Mitchell, P. G., and Sneddon, J. (1987), *Direct Determination of Metals in Milligram Masses and Microlitre Volumes by Direct-current Argon-plasma Emission Spectrometry with Sample Introduction by Electrothermal Vaporization. Talanta* 34, 849–856.

Miwa, T., Murakami, M., and Mizuika, A. (1989), *Speciation of Copper in Fresh Waters. Anal. Chim. Acta* 219, 1–8.

Mizuike, A. (1983), *Enrichment Techniques for Inorganic Trace Analysis,* Vol. 19, in: *Chemical Laboratory Practice.* Springer, Berlin-Heidelberg-New York.

Moffet, J. W., and Zika, R. G. (1987), *Solvent Extraction of Copper Acetylacetonate in Studies of Copper(II) Speciation in Seawater. Mar. Chem.* 21, 301–313.

Mohl, C., and Stoeppler, M. (1989), *Multielement Determination in Biological Materials with ICP-AES,* in: Welz, B. (ed.): *5. Colloquium Atomspektrometrische Spurenanalytik.* Bodenseewerk Perkin-Elmer, in press.

Mohl, C., Narres, H. D., and Stoeppler, M. (1986), *Oxygen Ashing for the Determination of Lead and Cadmium in Difficult Materials with Graphite Tube Furnace AAS,* in: Welz, B. (ed.): *Fortschritte in der atomspektrometrischen Spurenanalytik,* Vol. 2, pp. 439–446. VCH Verlagsgesellschaft, Weinheim-Deerfield Beach/Florida-Basel.

Mok, W. H., and Wai, C. M. (1989), *Distribution and Mobilization of Arsenic Species in the Creeks around the Blackbird Mining District, Idaho. Water Res.* 23, 7–13.

Molinero, A. L., Castillo, J. R., and de Vega, A. (1988), *Determination of Ytterbium by AES-ICP, Application to Samples of Biological Origin. Fresenius Z. Anal. Chem.* 331, 721–724.

Moody, J. R., and Paulsen, P. J. (1988), *Isotope Dilution Spark-source Mass Spectrometric Determination of Total Mercury in Botanical and Biological Samples. Analyst* 113, 923–927.

Moore, L. J., Fassett, J. D., and Travis, J. C. (1984), *Systematics of Multielement Determination with Resonance Ionization Mass Spectrometry and Thermal Atomization. Anal. Chem.* 56, 2270–2775.

Morin, M., Bador, R., and Dechaud, H. (1989), *Detection of Europium(III) and Samarium(III) by Chelation and Laser-excited Time-resolved Fluorimetry. Anal. Chim. Acta* 219, 67–77.

Morrison, G. H. (1980), *Mass Spectrometry,* in: *Elemental Analysis of Biological Materials. Tech. Rep. Ser. No. 197,* pp. 201–229. IAEA, Vienna.

Muangnoicharoen, S., Chiou, K.Y., and Manuel, O.K. (1988), *Determination of Selenium and Tellurium in Air by Atomic-Absorption Spectrometry.* Talanta 35, 679–683.

Mukai, H., Ambe, Y., Muku, T., Taketshita, K., and Fukuma, T. (1986), *Seasonal Variation of Methylarsenic Compounds in Airborne Particulate Matter.* Nature 324, 239–241.

Müller, C. (1987), *Cadmium Content of Human Milk.* Trace Elem. Med. 4, 4–7.

Müller, H. (1982), *Catalymetric Methods of Analysis.* Crit. Rev. Anal. Chem. 13, 313–372.

Müller, J., and Kallischnigg, G. (1983), *Ergebnisse eines Ringversuches: Blei, Cadmium und Quecksilber in biologischem Material.* ZEBS-Berichte 1/1983, Dietrich Reimer, Berlin.

Munder, A., and Ballschmiter, K. (1986), *Chromatography of Metal Chelates XI. Trace Analysis of Cadmium, Cobalt, Copper, Mercury and Nickel in Water Using Bis(ethoxyethyl)dithiocarbamate as Reagent for RP C_{18}-HPLC and Photometric Detection.* Fresenius Z. Anal. Chem. 323, 869–874.

Munro, S., Ebdon, L., and McWeeny, D. (1986), *Application of Inductively Coupled Plasma Mass Spectrometry (ICP-MS) for Trace Metal Determination in Foods.* J. Anal. At. Spectrom. 1, 211–219.

Muntau, H. (1986), *The Problem of Accuracy in Environmental Analysis.* Fresenius Z. Anal. Chem. 324, 678–682.

Nakahara, T., and Wasa, T. (1986), *Determination of Tin by Non-dispersive Atomic Fluorescence Spectrometry Coupled with a Hydride Generation Technique.* J. Anal. At. Spectrom. 1, 473–477.

Nakamura, N., Yamamoto, K., Noda, S., Nishikawa, Y., Komi, H., Nagamato, H., Nakayama, T., and Misawa, K. (1989), *Determination of Picogram Quantities of Rare-Earth Elements in Meteoritic Materials by Direct-loading Thermal Ionization Mass Spectrometry.* Anal. Chem. 61, 755–762.

Nakashima, S., Sturgeon, R.E., Willie, S.N., and Berman, S.S. (1988a), *Determination of Trace Metals in Seawater by Graphite Furnace Atomic Absorption Spectrometry with Preconcentration on Silica-immobilized 8-hydroxyquinoline in a Flow System.* Fresenius Z. Anal. Chem. 330, 592–595.

Nakashima, S., Sturgeon, R.E., Willie, S.N., and Berman, S.S. (1988b), *Determination of Trace Elements in Sea Water by Graphite Furnace Zeeman Atomic Absorption Spectrometry after Preconcentration by Tetrahydroborate Reductive Precipitation.* Anal. Chim. Acta 207, 291–299.

Naranjit, D.A., Radziuk, B.H., and van Loon, J.C. (1984), *A Zeeman-effect Based Scatter Correction System for Non-dispersive Atomic Fluorescence Spectrometry.* Spectrochim. Acta 38B, 969–977.

Narres, H.D., Mohl, C., and Stoeppler, M. (1984a), *Metal Analysis in Difficult Materials with Platform Furnace Zeeman Atomic Absorption Spectrometry. I. Direct Determination of Cadmium in Crude Oil and Oil Products.* Int. J. Environ. Anal. Chem. 18, 267–279.

Narres, H.D., Valenta, P., and Nürnberg, H.W. (1984b), *Voltammetric Determination of Heavy Metals in Meat and Internal Organs of Slaughtered Bullocks.* Z. Lebensm. Unters. Forsch. 179, 440–446.

Narres, H.D., Mohl, C., and Stoeppler, M. (1985), *Metal Analysis in Difficult Materials with Platform Furnace Zeeman-Atomic-Absorption-Spectrometry. 2. Direct Determination of Cadmium and Lead in Milk.* Z. Lebensm. Unters. Forsch. 181, 111–116.

Nebesny, K.W., Maschhoff, B.L., and Armstrong, N.R. (1989), *Quantitation of Auger and X-Ray Photoelectron Spectroscopies.* Anal. Chem. 61, 469A–481A.

Neeb, R. (1980), *Electrochemical Methods,* in: Elemental Analysis of Biological Materials. Tech. Rep. Ser. No. 197, pp. 281–299. IAEA, Vienna.

Newton, M.P., and van den Berg, C.M.G. (1987), *Determination of Nickel, Cobalt, Copper, and Uranium in Water by Cathodic Stripping Chronopotentiometry with Continuous Flow.* Anal. Chim. Acta 199, 59–76.

Ng, K.C., Zerezghi, M., and Caruso, J.A. (1984), *Direct Powder Injection of NBS Coal Fly Ash in Inductively Coupled Plasma Atomic Emission Spectrometry with Rapid Scanning Spectrometric Detection.* Anal. Chem. 56, 417–421.

Nimmo, M., van den Berg, C. M. G., and Brown, J. (1989), *The Chemical Speciation of Dissolved Nickel, Copper, Vanadium and Iron in Liverpool Bay, Irish Sea. Estuarine Coastal Shelf Sci.* 29, 57–74.

Nixon, D. E., and Smith, G. E. (1986), *Comparison of Jarrell-Ash, Perkin-Elmer, and Modified Perkin-Elmer Nebulizers for Inductively Coupled Plasma Analysis. Anal. Chem.* 58, 2886–2888.

Nojiri, Y., Kawai, T., Otsuki, A., and Fuwa, K. (1985), *Simultaneous Multielement Determination of Trace Metals in Lake Waters by ICP Emission Spectrometry with Preconcentration and their Background Levels in Japan. Water Res.* 4, 503–509.

Nojiri, Y., Otsuki, A., and Fuwa, K. (1986), *Determination of Sub-nanogram-per-liter Levels of Mercury in Lake Water with Atmospheric Pressure Helium Microwave Induced Plasma Emission Spectrometry. Anal. Chem.* 58, 544–547.

Norin, H., and Vahter, M. (1981), *A Rapid Method for the Selective Analysis of Total Urinary Metabolites of Inorganic Arsenic. Scand. J. Work. Environ. Health* 7, 38–44.

Novak, L. (1989), *Über den Einfluß anorganischer Probenbestandteile auf die atomabsorptionsspektrometrische Bestimmung von Blei.* Doctoral Thesis, Universität Bonn, FRG.

Novak, L., and Stoeppler, M. (1986), *Use of Hydrogen for the Elimination of Matrix Interferences in the Determination of Lead by Graphite Furnace AAS. Fresenius Z. Anal. Chem.* 323, 737–741.

Nowicki-Jankowska, T., Gorkzynska, K., Michalik, A., and Wietecka, E. (1986), *Analytical Visible and Ultraviolet Spectroscopy,* in: Svehla, G. (ed.): *Comprehensive Analytical Chemistry,* Vol. 19. Elsevier, Amsterdam.

Nürnberg, H. W. (1981), *Differentielle Pulspolarographie, Pulsvoltammetrie und Pulsinversvoltammetrie,* in: *Analytiker-Taschenbuch,* Vol. 2, pp. 211–230. Springer-Verlag, Berlin-Heidelberg-New York.

Nürnberg, H. W. (1982), *Voltammetric Trace Analysis in Ecological Chemistry of Toxic Metals. Pure Appl. Chem.* 54, 853–878.

Nürnberg, H. W. (1983a), *Potentialities and Applications of Voltammetry in the Analysis of Toxic Trace Metals in Body Fluids,* in: Facchetti, S. (ed.): *Analytical Techniques for Heavy Metals in Biological Fluids,* pp. 209–232. Elsevier, Amsterdam.

Nürnberg, H. W. (1983b), *Voltammetric Studies on Trace Metal Speciation in Natural Waters, Part II: Application and Conclusions for Chemical Oceanography and Chemical Limnology* in: Leppard, G. G. (ed.): *Trace Element Speciation in Surface Waters and its Ecological Implications,* pp. 211–230. Plenum Press, New York-London.

Nürnberg, H. W., and Kastening, B. (1974), *Polarographic and Voltammetric Techniques,* in: Korte, F. (ed.): *Methodicum Chimicum,* Vol. I, Part A, pp. 548–607. Academic Press, New York-San Francisco-London.

Nürnberg, H. W., and Raspor, B. (1981), *Applications of Voltammetry in Studies of the Speciation of Heavy Metals by Organic Chelators in Sea Water. Environ. Technol. Lett.* 2, 457–483.

Nürnberg, H. W., and Valenta, P. (1983), *Potentialities and Applications of Voltammetry in Chemical Speciation of Trace Metals in the Sea,* in: Wong, C. S., Boyle, E., Bruland, K. W., Burton, D., and Goldberg, E. D. (eds.): *Trace Metals in Sea Water,* pp. 671–697. Plenum Press, New York-London.

Nürnberg, H. W., Valenta, P., and Nguyen, V. D. (1980), *Ein neuer Weg zur Messung toxischer Metalle im Regen. Jahresbericht 1979/80 der KFA Jülich,* pp. 47–54.

Nürnberg, H. W., Valenta, P., Nguyen, V. D., Gödde, M., and Urano de Carvalho, E. (1984), *Studies on the Deposition of Acid and of Ecotoxic Heavy Metals with Precipitation from the Atmosphere. Fresenius Z. Anal. Chem.* 317, 314–323.

Nürnberg, H. W., Stoeppler, M., and Dürbeck, H. W. (1986), *Environmental Specimen Banking – Our Link with the Future,* in: Hulanicki, A. (ed.): *Euroanalysis, V. Reviews on Analytical Chemistry,* pp. 33–51. Akademiai Kiado, Budapest.

Nygren, O., and Nilsson, C.-A. (1987), *Determination of Trimethyllead Salts in Blood Using High-resolution Gas-chromatography-graphite Furnace Atomic Absorption Spectrometry. J. Anal. At. Spectrom.* 2, 805–808.

Nygren, O., Nilsson, C. A., and Frech, W. (1988 a), *On-line Interfacing of a Liquid Chromatograph to a Continuously Heated Graphite Furnace Atomic Absorption Spectrophotometer for Element-specific Detection.* Anal. Chem. 60, 2204–2208.

Nygren, O., Nilsson, C. A., and Gustavsson, A. (1988 b), *Determination of Lead in Blood Using Flow Injection and a Nebulizer Interface for Flame Atomic Absorption Spectrometry.* Analyst 113, 591–594.

Obiols, J., Devesa, R., and Sol, A. (1986), *Speciation of Heavy Metals in Suspended Particulates in Urban Air.* Toxicol. Environ. Chem. 13, 121–128.

O'Connor, P., Kerrigan, G. C., Taylor, K. R., Morris, D. D., and Wright, C. R. (1980), *Levels and Temporal Trends of Trace Element Concentrations in Vertebral Bone.* Arch. Environ. Health 35, 21–28.

Oda, C. E., and Ingle, J. D., Jr. (1981), *Continuous Flow Cold Vapor Atomic Absorption Determination of Mercury.* Anal. Chem. 53, 2030–2033.

Odom, R. W., Lux, G., Fleming, R. H., Chu, P. K., Niemeyer, I. C., and Blattner, R. J. (1988), *Quantitative Trace Element Analysis of Microdroplet Residues by Secondary Ion Mass Spectrometry.* Anal. Chem. 60, 2070–2075.

O'Haver, T. C., Carroll, J., Nichol, R., and Littlejohn, D. (1988), *Extended Range Background Correction in Continuum Source Atomic Absorption Spectrometry.* J. Anal. At. Spectrom. 3, 155–157.

Ohls, K., Flock, J., and Loepp, H. (1988), *Application of a Modulated Grimm-type Glow Discharge Plasma as Primary Light Source for Simultaneous Reading Atomic Absorption Spectrometry.* Fresenius Z. Anal. Chem. 332, 456–463.

Ohta, K., and Mizuno, T. (1987), *Effect of Hydrogen on Atomic Absorption of Iron. Cobalt, and Nickel.* Anal. Chim. Acta 217, 377–382.

Ohta, K., and Yang Su, S. (1987), *Electrothermal Atomic Absorption Spectrometry with Improved Tungsten Tube Atomizer.* Anal. Chem. 59, 539–540.

Olayinka, K. O., Haswell, S. J., and Grzeskowiak, R. (1986), *Development of a Slurry Technique for the Determination of Cadmium in Dried Foods by Electrothermal Atomisation Atomic Absorption Spectrometry.* J. Anal. At. Spectrom. 1, 297–300.

Oliviera, E., de, McLaren, J. W., and Berman, S. S. (1983), *Simultaneous Determination of Arsenic, Antimony, and Selenium in Marine Samples by Inductively Coupled Plasma Atomic Emission Spectrometry.* Anal. Chem. 55, 2047–2050.

Omenetto, N., and Human, H. G. C. (1984), *Laser Excited Analytical Atomic and Ionic Fluorescence in Flames, Furnaces and Inductively Coupled Plasmas – I. General Considerations.* Spectrochim. Acta 39B, 1333–1343.

Omenetto, N., Cavalli, P., Broglia, M., Qi, P., and Rossi, G. (1988), *Laser-induced Single-resonance and Double-resonance Atomic Fluorescence Spectrometry in a Graphite Tube Atomiser.* J. Anal. At. Spectrom. 3, 231–235.

Ostapczuk, P., Valenta, P., and Nürnberg, H. W. (1986), *Square Wave Voltammetry – A Rapid and Reliable Determination Method of Zn, Cd, Pb, Cu, Ni and Co in Biological and Environmental Samples.* J. Electroanal. Chem. 214, 51–64.

Ostapczuk, P., Stoeppler, M., and Dürbeck, H. W. (1988), *Present Potential of Electrochemical Methods for Metal Determinations in Reference Materials.* Fresenius Z. Anal. Chem. 322, 662–665.

Ostapczuk, P., Stoeppler, M., and Dürbeck, H. W. (1989), *Potentiometric Stripping Determination of Cadmium in Environmental and Biological Samples.* Toxicol. Environ. Chem., in press.

Osteryoung, J. G., and Osteryoung, R. A. (1985), *Square Wave Voltammetry.* Anal. Chem. 57, 101A–110A.

Ottaway, J. M. (1976), *Atom Formation and Interferences in Flame and Carbon Furnace Atomic Spectrometry.* Proc. Anal. Div. Chem. Soc. 13, 185–192.

Ottaway, J. M. (1983), *Heavy Metal Determinations by Atomic Absorption and Emission Spectrometry,* in: Facchetti, S. (ed.): *Analytical Techniques for Heavy Metals in Biological Fluids,* pp. 171–208. Elsevier, Amsterdam.

Ottaway, J. M. (1984), *Recent Innovations in Electrothermal Atomisation. Anal. Proc. 21*, 55–59.
Ottaway, J. M., and Fell, G. S. (1986), *Determination of Chromium in Biological Materials. Pure Appl. Chem. 58*, 1707–1720.
Pacey, G. E. (ed.) (1988), *Proceedings 4th International Conference on Flow Analysis,* Las Vegas, April 17–20. *Anal. Chim. Acta, Special Issue 214*, 1–466.
Palmieri, M. D., and Fritz, J. D. (1987), *Determination of Metal Ions by High-performance Liquid Chromatographic Separation of their Hydroxamic Acid Chelates. Anal. Chem. 59*, 2226–2231.
Pande, A. (1974), *Handbook of Moisture Determination and Control.* Marcel Dekker, New York.
Park, C. J., van Loon, J. C., Arrowsmith, P., and French, J. B. (1987), *Sample Analysis Using Plasma Source Mass Spectrometry with Electrothermal Sample Introduction. Anal. Chem. 59*, 2191–2196.
Parr, R. M. (1980), *The Reliability of Trace Element Analysis as Revealed by Analytical Reference Materials,* in: Brätter, P., and Schramel, P. (eds.): *Trace Element Analytical Chemistry in Medicine and Biology,* Vol. 1, pp. 631–651. W. de Gruyter, Berlin-New York.
Parr, R. M., Muramatsu, Y., and Clements, S. A. (1987), *Survey and Evaluation of Available Biological Reference Materials for Trace Element Analysis. Fresenius Z. Anal. Chem. 326*, 601–608.
Patterson, C. C., and Settle, D. M. (1976), *The Reduction of Order of Magnitude Errors in Lead Analyses of Biological Materials and Natural Waters by Evaluating and Controlling the Extent and Sources of Industrial Lead Contamination introduced during Sample Collecting and Analysis,* in: Lafleur, P. (ed.): *Accuracy in Trace Analysis: Sampling, Sample Handling, Analysis,* pp. 321–351, NBS Special Publication 422, U.S. Dept. of Commerce, Washington, D.C.
Paudyn, A., and van Loon, J. C. (1986), *Determination of Organic Forms of Mercury and Arsenic in Water and Atmospheric Samples by Gas Chromatography-Atomic Absorption. Fresenius Z. Anal. Chem. 325*, 369–376.
Paulsen, P. J., Beary, E. S., Bushee, D. S., and Moddy, J. R. (1988), *Inductively Coupled Plasma Mass Spectrometry Analysis of Ultrapure Acids. Anal. Chem. 60*, 971–975.
Pella, P. A., and Dobbyn, R. C. (1988), *Total Reflection Energy-dispersive X-Ray Fluorescence Spectrometry Using Monochromatic Synchrotron Radiation: Application to Selenium in Blood Serum. Anal. Chem. 60*, 684–687.
Perämäki, P., and Lajunen, H. J. (1988), *Determination of Antimony in Geological Samples Using Hydride Generation and Direct Current Plasma Atomic Emission Spectrometry. Analyst 113*, 1567–1570.
Petrick, K., and Krivan, V. (1987), *Interferences of Hydride Forming Elements and of Mercury in the Determination of Antimony, Arsenic, Selenium and Tin by Hydride-generation AAS. Fresenius Z. Anal. Chem. 327*, 338–342.
Pickford, C. J., and Brown, R. M. (1986), *Comparison of ICP-MS with ICP-ES: Detection Power and Interference Effects Experienced with Complex Matrices. Spectrochim. Acta 41B*, 183–187.
Pihlar, B., Valenta, P., Golimowski, J., and Nürnberg, H. W. (1980), *Die voltammetrische Bestimmung toxischer Spurenmetalle in kommunalen Abwässern und im Ablauf biologischer Kläranlagen. Z. Wasser Abwasser Forsch. 13*, 130–138.
Pihlar, B., Valenta, P., and Nürnberg, H. W. (1981), *New High-Performance Analytical Procedure for the Voltammetric Determination of Nickel in Routine Analysis of Waters, Biological Materials and Food. Fresenius Z. Anal. Chem. 307*, 337–346.
Pinel, R., Benabdallah, M. Z., Astruc, A., and Astruc, M. (1988), *Determination of Trace Amounts of Total Tin in Water Using Extraction Followed by Graphite Furnace Atomic Absorption Spectrometry with an Oxidising Matrix Modifier. J. Anal. At. Spectrom. 3*, 475–477.
Pittwell, L. R. (1988), *Quality Assurance in the Environment. Anal. Proc. 25*, 192–194.
Piwonka, J., Kaiser, G., and Tölg, G. (1985), *Determination of Selenium at ng/g- and pg/g Levels by Hydride Generation-Atomic Absorption Spectrometry in Biotic Materials. Fresenius Z. Anal. Chem. 321*, 225–234.

Plantz, M. R., Fritz, J. R., Smith, F. G., and Houk, R. S. (1989), *Separation of Trace Metal Complexes for Analysis of Samples of High Salt Content by Inductively Coupled Plasma Mass Spectrometry.* Anal. Chem. *61*, 149 – 153.

Prange, A. (1989), *Total Reflection X-Ray Spectrometry: Method and Applications (Review).* Spectrochim. Acta *44B*, in press.

Prange, A., and Schwenke, H. (1989), *Sample Treatment for TXRF-requirements and Prospects.* Adv. X-Ray Anal. *32*, 209 – 218.

Prange, A., Knöchel, A., and Michaelis, W. (1985), *Multi-element Determination of Dissolved Heavy Metal Traces in Sea Water by Total-reflection X-Ray Fluorescence Spectrometry.* Anal. Chim. Acta *172*, 79 – 100.

Prange, A., Knoth, J., Stössel, R. P., Böddeker, H., and Kramer, K. (1987), *Determination of Trace Elements in the Water Cycle by Total Reflection X-Ray Fluorescence Spectrometry.* Anal. Chim. Acta *195*, 275 – 287.

Preli, F. R., Dougherty, J. P., and Michel, R. G. (1987), *Laser-excited Atomic Fluorescence Spectrometry with a Laboratory-constructed Tube Electrothermal Atomizer.* Anal. Chem. *59*, 1784 – 1789.

Purchase, N. G., and Fergusson, J. E. (1986), *Lead in Teeth: The Influence of the Tooth Type and the Sample within a Tooth on Lead Levels.* Sci. Total Environ. *52*, 239 – 250.

Puxbaum, H., and Tsitouridou, R. (1987), *Application of a Portable Ion Chromatograph for Field Site Measurements of the Composition of Fog Water and Atmospheric Aerosols.* Int. J. Environ. Anal. Chem. *31*, 11 – 22.

Puxbaum, H., Rosenberg, C., Winiwarter, W., Gregori, M., Pech, G., and Casensky, V. (1988), *Determination of Inorganic and Organic Volatile Acids in Ambient Air with an Annual Diffusion Denuder System.* Fresenius Z. Anal. Chem. *331*, 1 – 7.

Pyper, J. W. (1985), *The Determination of Moisture in Solids – A Selected Review.* Anal. Chim. Acta *170*, 159 – 175.

Que Hee, S. S., and Boyle, J. R. (1988), *Simultaneous Multielement Analysis of Some Environmental and Biological Samples by Inductively Coupled Plasma Atomic Emission Spectrometry.* Anal. Chem. *60*, 1033 – 1042.

Rabinowitz, M. B., Wetherill, G. W., and Kapple, J. D. (1973), *Lead Metabolism in the Normal Human: Stable Isotope Studies.* Science *182*, 725 – 727.

Ramsay, M. H., and Thompson, M. (1984), *Improved Precision in Inductively Coupled Plasma Atomic-Emission Spectrometry by a Parameter-related Internal Standard Method.* Analyst *109*, 1625 – 1626.

Ramsay, M. H., and Thompson, M. (1986), *A Predictive Model of Plasma Matrix Effects in Inductively Coupled Plasma Atomic Emission Spectrometry.* J. Anal. At. Spectrom. *1*, 185 – 193.

Ramsay, M. H., and Thompson, M. (1987), *High-accuracy Analysis by Inductively Coupled Plasma Atomic Emission Spectrometry Using the Parameter-related Internal Standard Method.* J. Anal. At. Spectrom. *2*, 497 – 502.

Ramsay, M. H., Thompson, M., and Banerjee, E. K. (1987), *Realistic Assessment of Analytical Data Quality from Inductively Coupled Plasma Atomic Emission Spectrometry.* Anal. Proc. *24*, 260 – 265.

Rapsomanikis, S., Donard, O. F. X., and Weber, J. H. (1986), *Speciation of Lead and Methyllead Ions in Water by Chromatography/Atomic Absorption Spectrometry after Ethylation with Sodium Tetraethylborate.* Anal. Chem. *58*, 35 – 38.

Raspor, B., Nürnberg, H. W., Valenta, P., and Branica, M. (1980), *Kinetics and Mechanism of Trace Metal Chelation in Sea Water.* J. Electroanal. Chem. *115*, 293 – 308.

Retzik, M., and Bass, D. (1988), *Concept and Design of a Simultaneous Multielement GFAAS.* Int. Lab. *18/8*, 49 – 56.

Ridout, P. S., Jones, H. R., and Williams, J. G. (1988), *Determination of Trace Elements in a Marine Reference Material of Lobster Hepatopancreas (TORT-1) Using Inductively Coupled Plasma Mass Spectrometry.* Analyst *113*, 1383 – 1386.

Robards, K., Clarke, S., and Patsalides, E. (1988), *Advances in the Analytical Chromatography of the Lanthanides – A Review. Analyst 113*, 1757–1779.

Robberecht, H., and van Grieken, R. (1982), *Selenium in Environmental Waters: Determination, Speciation and Concentration Levels. Talanta 29*, 823–844.

Rosopulo, A., Grobecker, K.-H., and Kurfürst, U. (1984), *Investigations on the Analysis of Heavy Metals in Solids by Direct Zeeman Atomic Absorption Spectroscopy. IV. Methodology of the Direct Solid Sample Analysis for Biological Materials. Fresenius Z. Anal. Chem. 319*, 540–546.

Rossbach, M., Schärpf, O., Kaiser, W., Graf, W. Schirmer, A., Faber, W., Duppich, J., and Zeisler, R. (1988), *The Use of Focusing Super Mirror Neutron Guides to Enhance Cold Neutron Fluence Rates. Nucl. Instrum. Methods Phys. Res. B35*, 181–190.

Routh, M. W., Goulter, J. E., Tasker, D. B., and Arellano, S. D. (1987), *Investigation of ICP-AES Nebulization Techniques for the Analysis of Microsamples. Int. Lab. 17*, March, 54–61.

Rubin, R. B., and Heberling, S. S. (1987), *Metal Determinations by Ion Chromatography: A Complement to Atomic Spectroscopy. Int. Lab. 17/9*, 54–60.

Rudd, T., Lake, D. L., Mehrotra, I., Sterritt, R. M., Kirk, P. W. W., Campbell, J. A., and Lester, J. N. (1988), *Characterisation of Metal Forms in Sewage Sludge by Chemical Extraction and Progressive Acidification. Sci. Total Environ. 74*, 149–175.

Ruzicka, J. (1986), *Flow Injection Analysis – A Survey of its Potential for Spectroscopy. Fresenius Z. Anal. Chem. 324*, 745–749.

Ruzicka, J., and Arndal, A. (1989), *Solvent Extraction in Flow Injection Analysis and its Application to Enhancement of Atomic Spectrometry. Anal. Chim. Acta 216*, 243–255.

Ruzicka, J., and Hansen, E. H. (1986), *The First Decade of Flow Injection Analysis: From Serial Assay to Diagnostic Tool. Anal. Chim. Acta 179*, 1–58.

Ruzicka, J., and Hansen, E. H. (1988), *Homogeneous and Heterogeneous Systems. Flow Injection Analysis Today and Tomorrow. Anal. Chim. Acta 214*, 1–27.

Sabbioni, E., Goetz, L., Birattari, C., and Bonardi, M. (1981), *Environmental Biochemistry of Current Environmental Levels of Heavy Metals; Preparation of Radiotracers with very High Specific Radioactivity for Metallo-Biochemical Experiments on Laboratory Animals. Sci. Total Environ. 17*, 257–276.

Sakamoto-Arnold, C. M., and Johnson, K. S. (1987), *Determination of Picomolar Levels of Cobalt in Seawater by Flow Injection Analysis with Chemiluminescence Detection. Anal. Chem. 59*, 1789–1794.

Salit, M. L., and Cirillo, M. G. (1986), *The Decision to Buy a Laboratory Robot. Int. Lab. 16*, 11, 72–87.

Samanidou, V., and Fytianos, K. (1987), *Partitioning of Heavy Metals into Selective Chemical Fractions in Sediments from Rivers in Northern Greece. Sci. Total Environ. 67*, 279–285.

Sampson, B. (1988), *Determination of Cobalt in Plasma by Electrothermal Atomisation Atomic Absorption Spectrometry Using Palladium Matrix Modification. Anal. Proc. 25*, 229–230.

Sansoni, B. (ed.) (1985), *Instrumental Multi-element Analysis.* VCH Verlagsgesellschaft, Weinheim-Deerfield Beach/Florida-Basel.

Sansoni, B., and Iyengar, V. (1978), *Sampling and Sample Preparation Methods for the Analysis of Trace Elements in Biological Materials.* Ber. KFA Jülich, Jül-Spez-13.

Sansoni, B., and Panday, V. K. (1983), *Ashing in Trace Element Analysis of Biological Material,* in: Facchetti, S. (ed.): *Analytical Techniques for Heavy Metals in Biological Fluids*, pp. 91–131. Elsevier, Amsterdam.

Sansoni, B., Brunner, W., Wolff, G., Ruppert, H., and Dittrich, R. (1988), *Comparative Instrumental Multi-element Analysis I: Comparison of ICP Source Mass Spectrometry with ICP Atomic Emission Spectrometry, ICP Atomic Fluorescence Spectrometry and Atomic Absorption Spectrometry for the Analysis of Natural Waters from a Granite Region. Fresenius Z. Anal. Chem. 331*, 154–169.

Sanz, J., Gallarta, F., Galban, J., and Castillo, J. R. (1988), *Determination of Selenium by Hydride Generation Ultraviolet-visible Molecular Absorption Spectrometry with Diode-array Detection. Analyst 113*, 1387–1391.

Sanzolone, R. F. (1986), *Inductively Coupled Plasma Atomic Fluorescence Spectrometric Determination of Cadmium, Copper, Iron, Lead, Manganese and Zinc. J. Anal. At. Spectrom. 1*, 343–347.

Satzger, R. D. (1988), *Evaluation of Inductively Coupled Plasma Mass Spectrometry for the Determination of Trace Elements in Foods. Anal. Chem. 60*, 2500–2504.

Schaller, H., and Neeb, R. (1986), *Gas-chromatographic Elemental Analysis via Di(trifluoroethyl)dithiocarbamato-chelates IX. Optimization of the Capillary-gas-chromatographic Determination of Trace Metals. Fresenius Z. Anal. Chem. 323*, 473–476.

Schaule, B. K., and Patterson, C. C. (1980), *The Occurrence of Lead in the Northeast Pacific and the Effects of Anthropogenic Inputs*, in: Branica, M., and Konrad, Z. (eds.): *Lead in the Marine Environment*, pp. 31–43. Pergamon Press, Oxford-New York-Toronto-Sydney-Paris-Frankfurt.

Schaule, B. K., and Patterson, C. C. (1981), *Lead Concentrations in the Northeast Pacific: Evidence for Global Anthropogenic Pertuberations. Earth Planet Sci. Lett. 54*, 97–116.

Schelenz, R., Parr, R. M., Zeiller, E., and Clements, S. (1989), *Chromium in Biological Materials – IAEA Intercomparison Results. Fresenius Z. Anal. Chem.*, in press.

Scheuermann, H., and Hartkamp, H. (1983), *Stabilization of Water Samples for the Analysis of Traces of Heavy Metals. Fresenius Z. Anal. Chem. 315*, 430–433.

Schladot, J. D., and Backhaus, F. W. (1988), *Preparation of Sample Material for Environmental Specimen Banking Purposes – Milling and Homogenization at Cryogenic Temperatures*, in: Wise, S. A., Zeisler, R., and Goldstein, G. M. (eds.): *Progress in Environmental Specimen Banking*, pp. 184–193, NBS Special Publication 740, U.S. Department of Commerce, Washington, D.C.

Schladot, J. D., Hilpert, K., and Nürnberg, H. W. (1980), *Mass Spectrometric Determination of Lead Isotopes in West German Pit Coal. Adv. Mass Spectrom. 8*, 325–329.

Schlemmer, G., and Welz, B. (1986), *Palladium and Magnesium Nitrates, a More Universal Modifier for Graphite Furnace Atomic Absorption Spectrometry. Spectrochim. Acta 41B*, 1157–1165.

Schlemmer, G., and Welz, B. (1987), *Determination of Heavy Metals in Environmental Reference Materials Using Solid Sampling Graphite Furnace AAS. Fresenius Z. Anal. Chem. 328*, 405–409.

Schmid, W., and Krivan, V. (1985), *Radiotracer Study of the Preatomization Behavior of Lead in the Graphite Furnace. Anal. Chem. 57*, 30–34.

Schmidt, G. J., and Slavin, W. (1982), *Inductively Coupled Plasma Emission Spectrometry with Internal Standardization and Subtraction of Plasma Background Fluctuations. Anal. Chem. 54*, 2491–2495.

Schmidt, P. F., Barckhaus, R., and Kleimeier, W. (1986), *Laser Microprobe Mass Analyzer (LAMMA) Investigations on the Localization of Cadmium in Renal Cortex of Rats after Long-term Exposure to Cadmium. Trace Elem. Med. 3*, 36–39.

Schramel, P. (1987), *ICP- and DCP-Emission Spectrometry for Trace Element Analysis in Biomedical and Environmental Samples. Proceedings Ispra Course Reliability in Environmental Trace Analysis*, May 4–8. Joint Research Centre, Ispra, Italy.

Schramel, P., and Li-Qiang, X. (1983), *Determination of 14 Elements in Botanical Samples by Simultaneous Inductively Coupled Plasma Atomic Emission Spectrometry Using Standard Reference Material as a Multielement Standard. Fresenius Z. Anal. Chem. 314*, 671–677.

Schramel, P., Li-Qiang, X., Wolf, A., and Hasse, S. (1982a), *ICP-Emission Spectroscopy: An Analytical Method for Routine Supervision of Sludge and Soil. Fresenius Z. Anal. Chem. 313*, 213–216.

Schramel, P., Klose, B.-J., and Hasse, S. (1982b), *Efficiency of ICP Emission Spectroscopy for the Determinations of Trace Elements in Bio-medical and Environmental Samples. Fresenius Z. Anal. Chem. 310*, 209–216.

Schramel, P., Hasse, S., and Knapp, G. (1987), *Application of the High-pressure Ashing System HPA According to Knapp for Voltammetric Determination of Trace Elements in Biological Material. Fresenius Z. Anal. Chem. 326*, 142–145.

Schroeder, H. W., and Jackson, R. A. (1985), *An Instrumental Analytical Technique for Speciation of Atmospheric Mercury.* Int. J. Environ. Anal. Chem. **22**, 1–18.

Schulten, H.-R., Bahr, U., and Palavinskas, R. (1984), *New Method for Mass Spectrometric Trace Analyses of Metals in Biology and Medicine.* Fresenius Z. Anal. Chem. **317**, 497–511.

Schulze, G., Han, E., and Frenzel, W. (1989), *Comparison of Potentiometric Stripping Analysis and Anodic Stripping Voltammetry in Flow Injection Analysis. Interferences in the Determination of Lead.* Fresenius Z. Anal. Chem. **332**, 844–848.

Schuster, M. (1986), *The Chromatography of Metal Chelates. Part XVI. TLC of N,N-dialkyl-N'-thiobenzoylthiourea Chelates.* Fresenius Z. Anal. Chem. **324**, 127–129.

Schwedt, G. (1988), *Mineralstoffe und Spurenelemente in Lebensmitteln und ihre Bindungspartner – Methoden und Aufgaben der Elementspezies-Analytik.* Lebensm. Chem. Gerichtl. Chem. **42**, 36–39.

Schwedt, G., and Jahn, G. (1987), *Scheme for the Analysis of Lead Distribution within and upon Contaminated Plants.* Fresenius Z. Anal. Chem. **328**, 85–88.

Schweizer, A., and Schwedt, G. (1988), *Characterization and Quantification of Mineral Bindings in Different Foodstuffs.* Fresenius Z. Anal. Chem. **330**, 518–523.

SCIEX (1988), *Commercial Information from Manufacturer* (prospect information).

Scokart, P. O., Meeus-Verdinne, K., and de Borger, R. (1987), *Speciation of Heavy Metals in Polluted Soils by Sequential Extraction and ICP Spectrometry.* Int. J. Environ. Anal. Chem. **29**, 305–315.

Scott, M. C., and Chettle O. R. (1986), *In vivo Elemental Analysis in Occupational Medicine.* Scand. J. Work Environ. Health **12**, 81–96.

Seckler, M., and Dungs, K. (1989), *Mehrelementbestimmungsverfahren in der Atomabsorptionsspektrometrie,* in: Welz, B. (ed.): *Colloquium Atomspektrometrische Spurenanalytik.* Bodenseewerk Perkin-Elmer, in press.

Seiler, H., Schaller, K. H., Angerer, J., Fleischer, M., Machata, G., Pilz, W., Stoeppler, M., and Zorn, H. (1988), *Digestion Procedures for the Determination of Metals in Biological Material,* in: Angerer, J., and Schaller, K. H.: *Analyses of Hazardous Substances in Biological Materials, Methods for Biological Monitoring,* Vol. 2, pp. 1–30. VCH Verlagsgesellschaft, Weinheim-Basel-Cambridge-New York.

Selby, M., and Hieftje, G. M. (1987), *Inductively Coupled Plasma-Mass Spectrometry: A Status Report.* Int. Lab. **17**, 8, 28–38.

Self, R., Eagles, J., Fairweather-Tait, S. J., and Portwood, D. E. (1987), *Fast Atom Bombardment Mass Spectrometry (FABMS) of Mineral Nutrients in Human Nutrition Studies.* Anal. Proc. **24**, 366–367.

Sharp, B. L., Barnett, N. W., Burridge, J. C., Littlejohn, D., and Tyson, J. F. (1988), *Atomic Spectrometry Update – Atomisation and Excitation.* J. Anal. At. Spectrom. **3**, 133R–154R.

Sharp, R. L., Whitfield, R. G., and Fox, L. E. (1988), *Robots in the Laboratory: A General Approach.* Anal. Chem. **60**, 1056A–1062A.

Simonsen, K. W., Nielsen, B., Jensen, A., and Andersen, J. R. (1986), *Direct Microcomputer Controlled Determination of Zinc in Human Serum by Flow Injection Atomic Absorption Spectrometry.* J. Anal. At. Spectrom. **1**, 453–456.

Sing, R. L. A., and Salin, E. D. (1989), *Introduction of Liquid Samples into the Inductively Coupled Plasma by Direct Insertion on a Wire Loop.* Anal. Chem. **61**, 163–169.

Sipos, L., Rützel, H., and Thijssen, T. H. P. (1980), *Performance of a New Device for Sampling Sea Water from the Sea Bottom.* Thalassia Jugosl. **16**, 89–94.

Slavin, W. (1984), *Graphite Furnace AAS – A Source Book.* Perkin-Elmer Corp. Norwalk.

Slavin, W. (1986a), *Flames, Furnaces, Plasmas – How do we Choose?* Anal. Chem. **58**, 589A–597A.

Slavin, W. (1986b), *An Overview of Recent Development in the Determination of Aluminium in Serum by Furnace Atomic Absorption Spectrometry.* J. Anal. At. Spectrom. **1**, 281–285.

Slavin, W., and Carnrick, G. R. (1984), *The Possibility of Standardless Furnace Atomic Absorption Spectroscopy.* Spectrochim. Acta **39B**, 271–282.

Slavin, W., and Carnrick, G. R. (1985), *A Survey of Applications of the Stabilized Temperature Platform Furnace and Zeeman Correction.* At. Spectrosc. 6, 157–160.

Slavin, W., and Manning, D. C. (1982), *Graphite Furnace Interferences, a Guide to the Literature.* Prog. Anal. At. Spectrosc. 5, 243–340.

Slavin, W., Manning, D. C., and Carnrick, G. R. (1981), *The Stabilized Temperature Platform Furnace.* At. Spectrosc. 2, 127–145.

Slavin, W., Carnrick, G. R., Manning, D. C., and Pruszkowska, E. (1983), *Recent Experiences with the Stabilized Temperature Platform Furnace and Zeeman Background Correction.* At. Spectrosc. 4, 69–86.

Slavin, W., Manning, D. C., and Carnrick, G. R. (1988), *Graphite Furnace Technology, where are we?* J. Anal. At. Spectrom. 3, 13–19.

Small, H., Stevens, T. S., and Bauman, W. C. (1975), *Novel Ion Exchange Chromatographic Method Using Conductimetric Detection.* Anal. Chem. 47, 1801–1809.

Snedecor, G. W., and Cochran, W. G. (1973), *Statistical Methods,* 6th Ed. The IOWA State University Press, Ames, Iowa.

Snook, R. D. (1988), *Inductively Coupled Plasma Fourier Transform Spectrometry.* Anal. Proc. 25, 354–355.

Sparkes, S. T., and Ebdon, L. (1986), *Slurry Atomisation for Agricultural Samples by Plasma Emission Spectrometry.* Anal. Proc. 23, 410–412.

Sparkes, S. T., and Ebdon, L. (1988), *Direct Atomic Spectrometric Analysis by Slurry Atomisation. Part 6. Simplex Optimisation of a Direct Current Plasma for Kaolin Analysis.* J. Anal. At. Spectrom. 3, 563–569.

Speer, R., Hoffmann, P., and Lieser, K. H. (1986), *Problems in Measuring Water Samples of Different Salt Content by ICP-AES.* Fresenius Z. Anal. Chem. 325, 558.

Standing Committee of Analysts (1980), *General Principles of Sampling and Accuracy of Results,* in: *Methods for the Examination of Waters and Associated Materials.* Her Majesty's Stationary Office, London.

Statham, P. J. (1985), *The Determination of Dissolved Manganese and Cadmium in Sea Water at low $nmol \cdot l^{-1}$ Concentrations by Chelation and Extraction followed by Electrothermal Atomic Absorption Spectrometry.* Anal. Chim. Acta 169, 149–159.

Steenhout, A., and Pourtois, M. (1981), *Lead Accumulation in Teeth as a Function of Age with Different Exposures.* Br. J. Ind. Med. 38, 297–303.

Stephens, R. (1980), *Zeeman Modulated Atomic Absorption Spectroscopy.* Crit. Rev. Anal. Chem. 9, 167–195.

Stephenson, M. D., and Smith, D. R. (1988), *Determination of Tributyltin in Tissues and Sediments by Graphite Furnace Atomic Absorption Spectrometry.* Anal. Chem. 60, 696–698.

Stiefel, T. H., Schulze, K., and Tölg, G. (1980), *Analysis of Trace Elements Distributed in Blood – Determination of Beryllium Concentrations ≥ 0.01 ng/g in Human and Animal Blood Components by Preparative Electrophoresis and Flameless Atomic Absorption Spectrometry.* Fresenius Z. Anal. Chem. 300, 189–196.

Stoeppler, M. (1983a), *Analytical Aspects of Sample Collection, Sample Storage and Sample Treatment,* in: Brätter, P., and Schramel, P. (eds.): *Trace Element Analytical Chemistry in Medicine and Biology,* Vol. 2, pp. 909–928. Walter de Gruyter, Berlin-New York.

Stoeppler, M. (1983b), *Strategies for the Reliable Analysis of Heavy Metals in Man and His Environment,* in: *Proceedings International Conference Heavy Metals in Man and His Environment,* Heidelberg, pp. 70–77. CEP Consultants Ltd., Edinburgh.

Stoeppler, M. (1983c), *Atomic Absorption Spectrometry – A Valuable Tool for Trace and Ultratrace Determinations of Metals and Metalloids in Biological Materials.* Spectrochim. Acta 38B, 1559–1568.

Stoeppler, M. (ed.) (1985), *Proceedings Colloquium on Present Status and Perspectives of the Analysis of Solid Samples by AAS,* Wetzlar, Oct. 8–10, 1984.

Stoeppler, M. (1986), *Recent Methodological Progress in Cadmium Analysis.* Int. J. Environ. Anal. Chem. 27, 231–239.

Stoeppler, M. (1988), *Analytical Methods and Quality Control for Trace Metal Determinations: A Critical Review of the State of the Art,* in: Clarkson, T. W., Friberg, L., Nordberg, G. F., and Sager, P. R. (eds.): *Biological Monitoring of Toxic Metals,* pp. 481–492. Plenum Publishing Corp., New York-London.

Stoeppler, M. (1989a), *Problems and Analytical Methods for the Determination of Trace Metals and Metalloids in Polluted and Nonpolluted Freshwater Ecosystems,* Chap. 4, in: Boudou, A., and Ribeyre, F. (eds.): *Aquatic Ecotoxicology: Fundamental Concepts and Methodologies,* Vol. 1, pp. 77–96. CRC Press Inc., Boca Raton, Florida.

Stoeppler, M. (1989b), *Marktübersicht: Atomabsorptionsspektrometrie,* in Broeckaert, J. A. C., and Schrader, B.: *Sonderheft Spektroskopie. Nachr. Chem. Tech. Lab.* 37, 29–55.

Stoeppler, M., and Apel, M. (1984), *Determination of Arsenic Species in Liquid and Solid Materials from the Environment and Food and Body Fluids. Fresenius Z. Anal. Chem.* 317, 226–227.

Stoeppler, M., and Nürnberg, H. W. (1984), Chapter I.4a, *Analytik von Metallen und ihren Verbindungen* in: Merian, E. (ed.): *Metalle in der Umwelt,* pp. 45–104. Verlag Chemie, Weinheim-Deerfield Beach/Florida-Basel.

Stoeppler, M., Apel, M., Bagschik, U., May, K., Mohl, C., Ostapczuk, P., Enkelmann, R., and Eschnauer, H. (1985), *New Results on Trace Element Contents in German and Foreign Wines. Lebensmittelchem. Gerichtl. Chem.* 39, 60–61.

Stössel, R. P., and Prange, A. (1985), *Determination of Trace Elements in Rain Water by Total-reflection X-Ray Fluorescence. Anal. Chem.* 57, 2880–2885.

Strupp, D., Klussmann, U., and Ebing, W. (1985), *Environmental Specimen Bank for the Plant-soil-system: Opportunity to Store Contaminated Samples Unchanged over Many Years. Results of a Pilot Study. Fresenius Z. Anal. Chem.* 322, 747–751.

Stukas, V. J., and Wong, C. S. (1983), *Accurate and Precise Analysis of Trace Levels of Cu, Cd, Pb, Zn, Fe and Ni in Sea Water by Isotope Dilution Mass Spectrometry,* in: Wong, C. S., Boyle, E., Bruland, K. W., Burton, J. D., and Goldberg, E. D. (eds.): *Trace Metals in Sea Water,* pp. 513–536. Plenum Publishing Corp., New York.

Sturgeon, R. (1986), *Graphite Furnace Atomic Absorption Spectrometry: Fact and Fiction. Fresenius Z. Anal. Chem.* 324, 807–818.

Sturgeon, R. E., Willie, S. N., and Berman, S. S. (1985), *Hydride Generation-Atomic Absorption Determination of Antimony in Seawater with in situ Concentration in a Graphite Furnace. Anal. Chem.* 57, 2311–2314.

Subramanian, K. S. (1986), *Determination of Trace Metals in Human Blood by Graphite Furnace Atomic Absorption Spectrometry. Prog. Anal. Spectrosc.* 9, 237–334.

Subramanian, K. S. (1988a), *Determination of Arsenic in Whole Blood by Graphite Platform-in-Furnace Atomic Absorption Spectrometry with Nitric Acid Deproteinisation and Nickel Fortification. J. Anal. At. Spectrom.* 3, 111–114.

Subramanian, K. S. (1988b), *Determination of Trace Elements in Biological Fluids Other than Blood by Graphite Furnace Atomic Absorption Spectrometry. Prog. Anal. Spectrosc.* 11, 511–608.

Subramanian, K. S., and Méranger, J. C. (1982), *Simultaneous Determination of 20 Elements in Some Human Kidney and Liver Autopsy Samples by Inductively-coupled Plasma Atomic Emission Spectrometry. Sci. Total Environ.* 24, 147–157.

Su-Cheng, P. (1988), *Pre-concentration Efficiency of Chelex-100 Resin for Heavy Metals in Seawater. Part 2. Distribution of Heavy Metals on a Chelex-100 Column and Optimization of the Column Efficiency by a Plate Simulation Method. Anal. Chim. Acta* 211, 271–280.

Su-Cheng, P., Pai-Yee, W., and Ruei-Lung, L. (1988), *Pre-concentration Efficiency of Chelex-100 Resin for Heavy Metals in Seawater. Part 1. Effects of pH and Salts on the Distribution Ratios of Heavy Metals. Anal. Chim. Acta* 211, 257–270.

Sullivan, J. S., Torkelson, J. D., Wekell, M. M., Hollingworth, T. A., Saxton, W. L., Miller, G. A., Panaro, K. W., and Uhler, A. D. (1988), *Determination of Tri-n-butyltin and Di-n-butyltin in Fish as Hydride Derivatives by Reaction Gas Chromatography. Anal. Chem.* 60, 626–630.

Sunderman, F. W., Jr. (1980), *Analytical Biochemistry of Nickel. Pure Appl. Chem.* 52, 527–544.

Sunderman, F.W., Jr., Aitio, A., Morgan, L.G., and Norseth, T. (1986), *Biological Monitoring of Nickel. Toxicol. Ind. Health 2*, 17–78.

Suzuki, M., and Ohta, K. (1983), *Electrothermal Atomic Absorption Spectrometry with Metal Atomizers. Prog. Anal. At. Spectrosc. 6*, 49–162.

Talbot, V., and Chang, W.-J. (1987), *Rapid Multielement Analysis of Oyster and Cockle Tissue Using X-Ray Fluorescence Spectrometry, with Application to Reconnaissance Marine Pollution Investigations. Sci. Total Environ. 60*, 213–223.

Taylor, A. (1988), *Reference Materials for Measurement of Aluminium in Biological Samples. Fresenius Z. Anal. Chem. 332*, 616–619.

Taylor, A., and Briggs, R.J. (1986), *An External Quality Assessment Scheme for Trace Elements in Biological Fluids. J. Anal. At. Spectrom. 1*, 391–395.

Taylor, J.K. (1981), *Quality Assurance of Chemical Measurements. Anal. Chem. 53*, 1588A–1595A.

Taylor, J.K., and Stanley, T.W. (1985), *Quality Assurance for Environmental Measurements. ASTM Special Technical Publication 867.* ASTM, Philadelphia.

Tera, O., Schwartzman, D.W., and Watkins, T.R. (1985), *Identification of Gasoline Lead in Children's Blood Using Isotopic Analysis. Anal. Environ. Health 40*, 120–123.

Tessier, A., and Campbell, P.G.C. (1988), *Comments on the Testing of the Accuracy of an Extraction Procedure for Determining the Partitioning of Trace Metals in Sediments. Anal. Chem. 60*, 1475–1476.

Tessier, A., Campbell, P.G.C., and Bisson, M. (1979), *Sequential Extraction Procedure for the Speciation of Particulate Trace Metals. Anal. Chem. 51*, 844–850.

Tessier, A., Campbell, P.G.C., and Bisson, M. (1982), *Particulate Trace Metal Speciation in Stream Sediments and Relationships with Grain Size: Implications for Geochemical Exploration. J. Geochim. Explor. 16*, 77–104.

Tessier, A., Campbell, P.G.C., Auclair, J.C., and Bisson, M. (1984), *Relationships Between the Partitioning of Trace Metals in Sediments and Their Accumulation in the Tissues of the Freshwater Mollusc Elipitio complanata in a Mining Area. Can. J. Fish. Aquat. Sci. 41*, 1463–1472.

Thomson, M., and Walsh, J.N. (1988), *Handbook of Inductively Coupled Plasma Spectrometry*, 2nd Ed. Blackie & Son Ltd., Glasgow.

Thompson, M., Ramsay, M.H., and Pahlavanpour, B. (1982), *Water Analysis by Inductively Coupled Plasma Atomic Emission Spectrometry after a Rapid Pre-concentration. Analyst 107*, 1330–1334.

Thorne, A. (1987), *High Resolution Fourier Transform Atomic Spectrometry. J. Anal. At. Spectrom. 2*, 227–232.

Threlfall, T.L. (1988), *Developments in UV/VIS Absorption Spectrometry. Eur. Spectrosc. News 78*, 8–17.

Tieh-Zheng, G., and Stephens, R. (1986), *Behavior of Zeeman Corrected Atomic Fluorescence by High Source Currents. J. Anal. At. Spectrom. 1*, 355–358.

Tielrooy, J.A., Vleeschhouwer, P.H.M., Kraak, J.C., and Maessen, F.S.M.J. (1988), *Determination of Rare-Earth Elements by High-performance Liquid Chromatography/Inductively Coupled Plasma/Atomic Emission Spectrometry. Anal. Chim. Acta 207*, 149–159.

Tölg, G. (1978), *Problems, Limitations, and Future Trends in the Analytical Characterization of High-Purity Materials. Pure Appl. Chem. 50*, 1075–1090.

Tölg, G. (1979), *New Ways for the Analytical Characterization of High-purity Materials. Fresenius Z. Anal. Chem. 294*, 1–15.

Tölg, G. (1987), *Extreme Trace Analysis of the Elements – The State of the Art Today and Tomorrow. Analyst 112*, 365–376.

Tölg, G. (1988a), *Where is Analysis of Trace Elements in Biotic Matrices Going to? Fresenius Z. Anal. Chem. 331*, 226–235.

Tölg, G. (1988b), *Must Analytical Chemistry Become ever more Sensitive? Fresenius Z. Anal. Chem. 329*, 735–736.

Tölg, G. (1989), *Personal communication.*
Tremblay, M. E., Smith, B. W., and Winefordner, J. D. *(1987), Laser-excited Ionic Fluorescence Spectrometry of Rare-Earth Elements in the Inductively-coupled Plasma. Anal. Chim. Acta 199,* 111–118.
Trincherini, P. R., and Facchetti, S. (1983), *Isotope Dilution Mass Spectrometry Applied to Lead Determination* in: Facchetti, S. (ed.): *Analytical Techniques for Heavy Metals in Biological Fluids,* pp. 255–272. Elsevier, Amsterdam-Oxford-New York.
Tsalev, D. L. (1984), *Atomic Absorption Spectrometry in Occupational and Environmental Health Practice,* Vol. II, *Determination of Individual Elements.* CRC Press, Boca Raton, Florida.
Tsalev, D. L., and Zaprianov, Z. K. (1983), *Atomic Absorption Spectrometry in Occupational and Environmental Health Practice,* Vol. I, *Analytical Aspects and Health Significance.* CRC Press, Boca Raton, Florida.
Tschöpel, P. (1980), *Aufschlußmethoden,* in: *Ullmanns Encyklopädie der technischen Chemie,* 4th Ed., Vol. 5, pp. 27–40. Verlag Chemie, Weinheim-New York.
Tschöpel, P. (1990), *Sample Preparation,* in: Stoeppler, M. (ed.): *Hazardous Metals in the Environment. Evaluation of Analytical Methods in Biological Systems,* in press. Elsevier Science Publishers, Amsterdam.
Tschöpel, P., Kotz, L., Schulz, W., Veber, M., and Tölg, G. (1980), *Causes and Elimination of Systematic Errors in the Determination of Elements in Aqueous Solutions in the ng/ml and pg/ml Range. Fresenius Z. Anal. Chem. 302,* 1–14.
Turnlund, J. R. (1983), *Use of Stable Isotopes to Determine Bioavailability of Trace Elements in Humans. Sci. Total Environ. 28,* 385–392.
Urasa, I. T. (1984), *Determination of Arsenic, Boron, Carbon, Phosphorus, Selenium, and Silicon in Natural Waters by Direct Current Plasma Atomic Emission Spectrometry. Anal. Chem. 56,* 904–908.
Vahter, M., Marafante, E., Lindgren, A., and Denker, L. (1982), *Tissue Distribution and Subcellular Binding of Arsenic in Marmoset Monkeys after Injection of ^{74}As-Arsenite. Arch. Toxicol. 51,* 65–77.
Valenta, P. (1983), *Voltammetric Studies on Trace Metal Speciation in Natural Waters. Part I. Methods,* in: Leppard, G. G.: *Trace Element Speciation in Surface Waters and its Ecological Implications,* pp. 49–69. Plenum Publishing Corp., New York.
Valenta, P., Nguyen, V. D., and Nürnberg, H. W. (1986), *Acid and Heavy Metal Pollution by Wet Deposition. Sci Total Environ. 55,* 311–320.
Valkovic, V. (1977), *Trace Elements in Human Hair.* Garland STPM Press, New York.
van Berkel, W. W., Overbosch, A. W., Feenstra, G., and Maessen, J. M. J. (1988), *Enrichment of Artificial Sea Water. A Critical Examination of Chelex-100 for Group-wise Analyte Preconcentration and Matrix Separation. J. Anal. At. Spectrom. (JAAS) 3,* 249–257.
van Cleuvenbergen, R. J. A., van Mol, W. E., and Adams, F. C. (1988), *Arsenic Speciation in Water by Hydride Cold Trapping – Quartz Furnace Atomic Absorption Spectrometry: An Evaluation. J. Anal. At. Spectrom. 3,* 169–176.
Vandecasteele, C., Nagels, M., Vanhoe, H., and Dams, R. (1988a), *Suppression of Analyte Signal in Inductively-coupled Plasma/Mass Spectrometry and the Use of an Internal Standard. Anal. Chim. Acta 211,* 91–98.
Vandecasteele, C., Windels, G., Desmet, B., Deruck, A., and Dams, R. (1988b), *Use of a Personal Computer for Signal Treatment in Atomic Absorption Spectrometry with Electrothermal Atomization Using Deuterium Arc Background Correction. Analyst 113,* 1691–1694.
van den Berg, C. M. G. (1985), *Direct Determination of Molybdenum in Sea Water by Cathodic Stripping Voltammetry. Anal. Chem. 57,* 1532–1536.
van den Berg, C. M. G. (1986), *The Determination of Trace Metals in Sea Water Using Cathodic Stripping Voltammetry. Sci. Total Environ. 49,* 89–99.
van den Berg, C. M. G. (1988), *Adsorptive Cathodic Stripping Voltammetry and Chronopotentiometry of Trace Metals in Sea Water. Anal. Proc. 25,* 265–266.

van den Berg, C. M. G., and Huang, Z. Q. (1984a), *Direct Electrochemical Determination of Dissolved Vanadium in Sea Water Using Cathodic Stripping Voltammetry with the Hanging Mercury Drop Electrode.* Anal. Chem. *56*, 2383–2386.

van den Berg, C. M. G., and Huang, Z. Q. (1984b), *Determination of Uranium(VI) in Sea Water by Cathodic Stripping Voltammetry of Complexes with Catechol.* Anal. Chim. Acta *164*, 209–222.

van den Berg, C. M. G., and Jacinto, G. S. (1988), *The Determination of Platinum in Sea Water by Adsorptive Cathodic Stripping Voltammetry.* Anal. Chim. Acta *211*, 129–139.

van den Berg, C. M. G., and Nimmo, M. (1987), *Direct Determination of Uranium in Water by Cathodic Stripping Voltammetry.* Anal. Chem. *59*, 924–928.

van den Berg, C. M. G., Murphy, K., and Riley, J. P. (1986), *The Determination of Aluminium in Sea Water and Fresh Water by Cathodic Stripping Voltammetry.* Anal. Chim. Acta *188*, 177–185.

van Loon, J. C. C. (1980), *Analytical Atomic Absorption Spectrometry – Selected Methods.* Academic Press, New York-London-Toronto-Sidney-San Francisco.

van Loon, J. C. C. (1981), *Atomic Fluorescence Spectrometry – Present Status and Future Prospects.* Anal. Chem. *53*, 332A–361A.

Vanoeteren, C., Cornelis, R., and Sabbioni, E. (1986), *Critical Evaluation of Normal Levels of Major and Trace Elements in Human Lung Tissue. EUR 10440 EN.*

Veillon, C. (1986), *Trace Element Analysis of Biological Samples – Problems and Precautions.* Anal. Chem. *58*, 851A–866A.

Veillon, C., Guthrie, B. E., and Wolf, W. R. (1980), *Retention of Cr by Graphite Furnace Tubes.* Anal. Chem. *52*, 457–459.

Versieck, J. (1984), *Trace Element Analysis – A Plea for Accuracy.* Trace Elem. Med. *1*, 2–12.

Versieck, J. (1985), *Trace Elements in Human Body Fluids and Tissues.* CRC Crit. Rev. Clin. Lab. Sci. *22*, 97–184.

Versieck, J., and Cornelis, R. (1980), *Normal Levels of Trace Elements in Human Blood Plasma or Serum.* Anal. Chim. Acta *116*, 217–254.

Versieck, J., Vanballenberghe, L., and de Kesel, A. (1988a), *Determination of Cadmium in Serum and Packed Blood Cells by Neutron Activation Analysis,* in: Safe, S., and Hutzinger, O. (eds.-in-chief): *Environmental Toxin Series,* Vol. 2, Stoeppler, M., and Piscator, M. (eds.): *Cadmium,* pp. 205–211. Springer, Berlin-Heidelberg-New York-London-Paris-Tokyo.

Versieck, J., Vanballenberghe, L., de Kesel, A., Hoste, J., Wallaeys, B., Vandenhaute, J., Baeck, N., Steyaert, H., Byrne, A. R., and Sunderman, F. W., Jr. (1988b), *Certification of a Second-Generation Biological Reference Material (Freeze Dried Human Serum) for Trace Element Determinations.* Anal. Chim. Acta *204*, 63–75.

Völkening, J., and Heumann, K. G. (1988), *Determination of Heavy Metals at the pg/g Level in Antarctic Snow with DPASV and IDMS.* Fresenius Z. Anal. Chem. *331*, 174–181.

Volland, G., Kölblin, G., Tschöpel, P., and Tölg, G. (1977), *Some Sources of Systemic Errors in the Determination of Elements in the ng and pg Range by AAS with Flameless Excitation in the Graphite Oven.* Fresenius Z. Anal. Chem. *284*, 1–12.

Völlkopf, U., Grobenski, Z., Tamm, R., and Welz, B. (1985), *Solid Sampling in Graphite Furnace Atomic-Absorption Spectrometry Using the Cup-in-tube Technique.* Analyst *110*, 573–577.

Völlkopf, U., Lehmann, R., and Weber, D. (1987), *Determination of Cadmium, Copper, Manganese and Rubidium in Plastic Materials by Graphite Furnace Atomic Absorption Spectrometry Using Solid Sampling.* J. Anal. At. Spectrom. *2*, 455–458.

Vos, G., Hovens, J. P. C., and Hagel, P. (1986a), *Chromium, Nickel, Copper, Zinc, Arsenic, Selenium, Cadmium, Mercury and Lead in Dutch Fishery Products 1977–1989.* Sci. Total Environ. *52*, 25–40.

Vos, G., Teenwen, J. J. H. M., and van Delft, W. (1986b), *Arsenic, Cadmium, Lead and Mercury in Meat, Liver and Kidneys of Swine Slaughtered in the Netherlands during the Period 1980–1985.* Z. Lebensm. Unters. Forsch. *183*, 297–401.

Vos, G., Lammers, H., and van Delft, W. (1988), *Arsenic, Cadmium, Lead and Mercury in Meat, Livers and Kidneys of Sheep Slaughtered in the Netherlands. Z. Lebensm. Unters. Forsch. 187*, 1−7.

Voth-Beach, L. M., and Schrader, D. E. (1987), *Investigations of a Reduced Palladium Chemical Modifier for Graphite Furnace Atomic Adsorption Spectrometry. J. Anal. At. Spectrom. 2*, 45−50.

Wagemann, R., and Armstrong, F. A. J. (1988), *Trace Metal Determination in Animal Tissues: An Interlaboratory Comparison. Talanta 35*, 545−551.

Wang, J. (1985), *Stripping Analysis, Principles, Instrumentation and Applications*. VCH Verlagsgesellschaft, Weinheim-Deerfield Beach/Florida-Basel.

Wang, J. (1986), *Absorptive Stripping Voltammetry. Int. Lab. 16*, 11/12, 50−59.

Wang, J. (1988), *Electroanalytical Techniques in Clinical Chemistry and Laboratory Medicine*. VCH Verlagsgesellschaft, Weinheim-Deerfield Beach/Florida-Basel.

Wang, J., and Mahmoud, J. S. (1986), *Stripping Voltammetry of Manganese Based on Chelate Adsorption at the Hanging Mercury Drop Electrode. Anal. Chim. Acta 182*, 147−155.

Wang, J., and Mahmoud J. S. (1987), *Chelate Adsorption for Trace Voltammetric Measurements of Iron(III). Fresenius Z. Anal. Chem. 327*, 789−793.

Wang, J., and Mannino, S. (1989), *Application of Adsorptive Stripping Voltammetry to the Speciation and Determination of Iron(III) and Total Iron in Wines. Analyst 114*, 643−645.

Wang, J., and Varughese, K. (1987), *Determination of Traces of Palladium by Adsorptive Stripping Voltammetry of the Dimethylglyoxime Complex. Anal. Chim. Acta 199*, 185−189.

Wang, J., and Zadeii, J. M. (1986a), *Trace Determination of Yttrium and Some Heavy Rare-Earths by Adsorptive Stripping Voltammetry. Talanta 33*, 321−324.

Wang, J., and Zadeii, J. M. (1986b), *Determination of Traces of Gallium Based on Stripping Voltammetry with Adsorptive Accumulation. Anal. Chim. Acta 185*, 229−238.

Wang, J., and Zadeii, J. M. (1987a), *Adsorptive Stripping Voltammetry Measurements of Trace Levels of Uranium following Chelation with Mordant Blue. Talanta 34*, 247−251.

Wang, J., and Zadeii, J. M. (1987b), *Ultrasensitive and Selective Measurements of Tin by Adsorptive Stripping Voltammetry of the Tin-Tropolone Complex. Talanta 34*, 909−914.

Wang, J., Farias, P. A. M., and Mahmoud, J. S. (1985), *Stripping Voltammetry of Aluminium Based on Adsorptive Accumulation of its Solochrome Violet RS Complex at the Static Mercury Drop Electrode. Anal. Chim. Acta 172*, 57−64.

Ward, N. I. (1987), *Inductively Coupled Plasma-source Mass Spectrometry and the Multielement Analysis of Environmental Matrices*, in: Proceedings International Conference Heavy Metals in the Environment, New Orleans, Vol. 2, pp. 23−28. CEP Consultants Ltd., Edinburgh.

Warren, C. J., and Dudas, M. J. (1988), *Leaching Behavior of Selected Trace Elements in Chemically Weathered Alkaline Fly Ash. Sci. Total Environ. 76*, 229−246.

Watling, R. J., and Collier, A. R. (1988), *Continuous Hydride Generation in Inductively Coupled Plasma Atomic Emission Spectrometry Using a Dual Platinum Grid Nebuliser System. Analyst 113*, 345−346.

Weber, G. (1987), *Speciation of Tin in Lemon Juice: An Example of Trace Metal Speciation in Food. Anal. Chim. Acta 200*, 79−88.

Weber, G., and Schwedt, G. (1982), *Trace Determination of Traces of Nickel and its Chemical Speciation in Coffee, Tea and Red Wine by Chromatographic and Spectroscopic Methods. Anal. Chim. Acta 134*, 81−92.

Weiss, J. (1987), *Ion Chromatography − A Review of Recent Developments. Fresenius Z. Anal. Chem. 327*, 451−455.

Weisz, H. (1987), *Recent Applications of the Ring-oven Technique: a Brief Review. Anal. Chim. Acta 202*, 25−34.

Weller, D. L. M. (1988), *Good Laboratory Practice and Computerisation in the Analytical Laboratory. Anal. Proc. 25*, 199−200.

Welz, B. (1985), *Atomic Absorption Spectrometry*, 2nd Ed. VCH Verlagsgesellschaft, Weinheim-Deerfield Beach/Florida-Basel.

Welz, B. (ed.) (1986a), *Lectures, Post Colloquium Spectroscopicum Internationale Symposium XXIV "Selected Topics from Graphite Furnace and Hydride-Generation AAS".* Fresenius Z. Anal. Chem. 323, 673–799.

Welz, B. (1986b), *Abuse of the Analyte Addition Technique in Atomic Absorption Spectrometry.* Fresenius Z. Anal. Chem. 325, 95–101.

Welz, B., and Melcher, M. (1983), *Investigations on Atomisation Mechanisms of Volatile Hydride-forming Elements in a Heated Quartz Cell.* Analyst 108, 213–224.

Welz, B., and Melcher, M. (1984), *Mechanisms of Transition Metal Interferences in Hydride Generation Atomic Absorption Spectrometry, Part 3. Releasing Effect of Iron(III) on Nickel Interference on Arsenic and Selenium.* Analyst 109, 577–579.

Welz, B., and Schlemmer, G. (1988a), *The Use of Methane as an Alternate Gas in Graphite Furnace Atomic Absorption Spectrometry.* At. Spectrom. 9, 76–80.

Welz, B., and Schlemmer, G. (1988b), *The Use of Freon as an Alternate Gas in Graphite Furnace Atomic Absorption Spectrometry.* At. Spectrom. 9, 81–83.

Welz, B., and Schubert-Jacobs, M. (1986), *Mechanisms of Transition Metal Interferences in Hydride Generation Atomic Absorption Spectrometry, Part. 4. Influence of Acid and Tetrahydroborate Concentrations on Interferences in Arsenic and Selenium Determinations.* J. Anal. At. Spectrom. 1, 23–27.

Welz, B., Wolynetz, M.S., and Verlinden, M. (1987), *Interlaboratory Trial on the Determination of Selenium in Lyophilized Human Serum, Blood and Urine Using Hydride Generation Atomic Absorption Spectrometry.* Pure Appl. Chem. 59, 927–936.

Welz, B., Sperling, M., Schlemmer, G., Wenzel, N., and Marowsky, G. (1988a), *Spatially and Temporally Gas Phase Temperature Measurements in a Massmann-type Graphite Tube Furnace Using Coherent Anti-stokes Raman Scattering.* Spectrochim. Acta 43B, 1187–1207.

Welz, B., Schlemmer, G., and Mudakavi, J.R. (1988b), *Palladium Nitrate-Magnesium Nitrate Modifier for Graphite Furnace Atomic Absorption Spectrometry.* J. Anal. At. Spectrom. 3, 695–701.

Wendt, R.H., and Fassel, V.A. (1965), *Induction-coupled Plasma Spectrometric Excitation Source.* Anal. Chem. 37, 920–922.

West, T.S. (1988), *In situ Pre-concentration in Flame Atomic Spectrometry.* Anal. Proc. 25, 240–244.

Whiston, C. (1987), *X-Ray Methods.* Wiley, Chichester.

Whitley, J.E., Hannah, R., and Littlejohn, D. (1988), *Radioanalytical Studies of Electrothermal Atomisation Atomic Absorption Spectrometry.* Anal. Proc. 25, 246–248.

Wibetoe, G., and Langmyhr, F.J. (1987), *Interferences in Inverse Zeeman-corrected Atomic Absorption Spectrometry Caused by Zeeman Splitting of Molecules.* Anal. Chim. Acta 1987, 81–86.

Wiegand, H.J., Ottenwälder, H., and Bolt, H.M. (1988), *Recent Advances in Biological Monitoring of Hexavalent Chromium Compounds.* Sci. Total Environ. 71, 309–315.

Wielopolski, L., Rosen, J.F., Slatkin, D.N., Vartsky, D., Ellis, K.J., and Cohn, S.H. (1983), *Feasibility of Noninvasive Analysis of Lead in the Human Tibia by Soft X-Ray Fluorescence.* Med. Phys. 10, 248–251.

Williams, J.G., Gray, A.L., Norman, P., and Ebdon, L. (1987), *Feasibility of Solid Sample Introduction by Slurry Nebulisation for Inductively Coupled Plasma Mass Spectrometry.* J. Anal. At. Spectrom. 2, 469–472.

Willie, S.N., Sturgeon, R.E., and Berman, S.S. (1986), *Hydride Generation Atomic Absorption Determination of Selenium in Marine Sediments, Tissues and Sea Water with in situ Concentration in a Graphite Furnace.* Anal. Chem. 58, 1140–1143.

Winefordner, J.D. (ed.) (1976), *Trace Analysis Spectroscopic Methods for Elements, Chemical Analysis,* Vol. 4b. Wiley Interscience, New York-London-Sidney-Toronto.

Winefordner, J.D., and Vickers, T.J. (1964), *Atomic Fluorescence Spectrometry as a Means of Chemical Analysis.* Anal. Chem. 36, 161–168.

Winge, R. K., Fassel, V. A., Peterson, V. J., and Floyd, M. A. (1984), *Inductively Coupled Plasma Atomic Emission Spectroscopy. An Atlas of Spectral Information.* Elsevier, Amsterdam.

Winkler, B. K., and Wright, J. C. (1988), *Nonlinear Atomic Spectroscopy of Flames.* Anal. Chem. *60*, 2599–2608.

Wise, S. A., and Zeisler, R. (eds.) (1985), *International Review of Environmental Specimen Banking.* NBS Special Publication 706. U.S. Department of Commerce, Washington, D.C.

Wise, S. A., Zeisler, R., and Goldstein, G. M. (eds.) (1988), *Progress in Environmental Specimen Banking.* NBS Special Publication 740, U.S. Department of Commerce, Washington, D.C.

Wolf, W. R. (1985), *Biological Reference Materials: Availability, Uses, and Need for Validation of Nutrient Measurement.* John Wiley & Sons, New York-Chichester-Brisbane-Toronto-Singapore.

Wolf, W. R. (1986), *Approaches to the Determination of Chemical Species in Biological Materials,* in: Bernhard, M., Brinckmann, F. E., and Sadler, P. J. (eds.): *The Importance of Chemical "Speciation" in Environmental Processes,* pp. 39–58. Springer, Berlin-Heidelberg-New York-London-Paris-Tokyo.

Wolf, W. R., and Harnly, J. R. (1984), *Trace Element Analysis,* in: King, R. D. (ed).: *Developments in Food Analysis Techniques-3,* pp. 69–97. Elsevier, London-New York.

Wolf, W. R., and Stoeppler, M. (eds.) (1987), *Proceedings Second International Symposium on Biological Reference Materials.* Fresenius Z. Anal. Chem. *326*, 597–745.

Wolf, W. R., and Stoeppler, M. (eds.) (1988), *Proceedings Third International Symposium on Biological Reference Materials.* Fresenius Z. Anal. Chem. *332*, 517–744.

Wong, C. S., Boyle, E., Bruland, K. W., Burton, J. D., and Goldberg, E. D. (eds.) (1983), *Trace Metals in Sea Water.* Plenum Press, New York.

Wouters, L. C., van Grieken, R. E., Linton, R. W., and Bauer, C. F. (1988), *Discrimination between Coprecipitated and Adsorbed Lead on Individual Calcite Particles Using Laser Microprobe Mass Analysis.* Anal. Chem. *60*, 2218–2220.

Würfels, M., and Jackwerth, E. (1985), *Investigations on the Carbon Balance in Decomposition of Biological Materials with Nitric Acid.* Fresenius Z. Anal. Chem. *322*, 354–358.

Würfels, M., Jackwerth, E., and Stoeppler, M. (1987), *Pretreatment Studies with Biological and Environmental Materials. V. About the Problem of Disturbances of Inverse Voltammetric Trace Analysis after Pressure Decomposition of Biological Samples.* Fresenius Z. Anal. Chem. *329*, 459–461.

Würfels, M., Jackwerth, E., and Stoeppler, M. (1989a), *On the Residue of Biological Materials after Pressure Decomposition with Nitric Acid I. Balancing the Decomposition Reaction.* Anal. Chim. Acta, in press.

Würfels, M., Jackwerth, E., and Stoeppler, M. (1989b), *On the Residue of Biological Materials after Pressure Decomposition with Nitric Acid II. Identifying the Reaction Products.* Anal. Chim. Acta, in press.

Würfels, M., Jackwerth, E., and Stoeppler, M. (1989c), *On the Residue of Biological Materials after Pressure Decomposition with Nitric Acid III. Influence of Reaction Products on Inverse Voltammetric Element Determination.* Anal. Chim. Acta, in press.

Xiao-Quan, S., Zhe-Ming, N., and Zhang, L. (1984), *Application of Matrix-modification in Determination of Thallium in Waste Water by Graphite-Furnace Atomic-Absorption Spectrometry.* Talanta *31*, 150–152.

Xiao-Quan, S., Shen, L., and Zhe-Ming, N. (1988), *Determination of Aluminium in Human Blood and Serum by Graphite Furnace Atomic Absorption Spectrometry Using Potassium Dichromate Matrix Modification.* J. Anal. At. Spectrom. *3*, 99–103.

Xiao-Quan, S., Yian, Z., and Zhe-Ming, N. (1989), *Determination of Beryllium in Urine by Graphite-Furnace Atomic Absorption Spectrometry.* Anal. Chim. Acta *217*, 271–280.

Xilei, L., van Renterghem, D., Cornelis, R., and Mees, L. (1988), *Radiochemical Neutron Activation Analysis for Thirteen Trace Metals in Human Blood Serum by Using Inorganic Ion-exchangers.* Anal. Chim. Acta *211*, 231–241.

Yan, D., and Schwedt (1987), *Optimization of Ion Chromatographic Trace Analysis for Heavy and Earth Alkali Metals by Post-chromatographic Derivatization.* Fresenius Z. Anal. Chem. 327, 503–508.
Yasuda, K., Koizumi, H., Ohishi, K., and Noda, T. (1980), *Zeeman Effect Atomic Absorption.* Prog. Anal. At. Spectrosc. 3, 299–368.
Yukawa, M. (1984), *The Variation of Trace Element Concentration in Human Hair: The Trace Element Profile in Human Long Hair by Sectional Analysis Using Neutron Activation Analysis.* Sci. Total Environ. 38, 41–54.
Zeisler, R., Harrison, S. H., and Wise, S. A. (1983), *The Pilot National Environmental Specimen Bank – Analysis of Human Liver Specimen.* NBS Special Publication 656, U.S. Department of Commerce, Washington, D.C.
Zeisler, R., Stone, S. F., and Sanders, R. W. (1988), *Sequential Determination of Biological and Pollutant Elements in Marine Bivalves.* Anal. Chem. 60, 2760–2765.
Zhao, Z., Cai, X., Li, P., and Yang, H. (1986), *Polarographic Determination of Trace Amounts of Thorium.* Talanta 33, 623–625.
Zief, M., and Speigths, R. (1972), *Ultrapurity – Methods and Techniques.* Marcel Dekker, New York.

Additional Recommended Literature

Chambaz, D., and Haerdi, W. (1989), *On-line Preconcentration and Elution of Trace Metals by Ion Chromatography.* J. Chromatogr. 482, 335–342.
Feret, F. R., and Sokolowski, J. (1989), *Effect of Sample Surface Integrity on X-ray Fluorescence Analysis of Aluminum Alloys.* Spectrosc. Int. 2(1), 34–39.
Hopkin, S. P. (1989), *Ecophysiology of Metals in Terrestrial Invertebrates* (with information on analysis in biological materials), 306 pages. Elsevier Applied Science, London-New York.
Klockenkämper, R. (1990), *Total-reflection X-ray Fluorescence Spectrometry: Principles and Applications* (with a comprehensive TXRF bibliography). Spectrosc. Int. 2(2), 26–37.
Leroy, M. J. F., Merian, E., et al. (1990), *Proceedings 20th International IAEAC Symposium of Environmental Analytical Chemistry.* Fresenius Z. Anal. Chem., in press.
Mottram, P. (1989), Bright Future for New Lanthanide Chelates in Time-resolved Fluorometry. Eur. Clin. Lab. News, 4 (November).
Nygard, D. D., and Bulman, F. (1990), *Analysis of Water for Arsenic, Lead, Selenium, and Thallium by Inductively Coupled Plasma Atomic Emission Spectrometry at Contract Laboratory Program Levels.* Spectrosc. Int. 2(2), 44–47.
Sarasin, H. A. (1990), *New Possibilities in Transition and Heavy Metal Analytical Chemistry: Sample Preparation for Ion Chromatography and Determination of Transition and Heavy Metals by Ion Chromatography* (in German), 42 pages. Henry A. Sarasin AG, Basel, Switzerland.

# I.4b   Metal Concentrations in Human Body Fluids and Tissues

ULRICH EWERS and ARTHUR BROCKHAUS
Düsseldorf, Federal Republic of Germany

1 Introduction

Metals occurring at low levels in human body fluids and tissues (concentration ranges: nanograms or micrograms per gram or per milliliter) generally are referred to a group of elements called trace elements. The term "trace element" stems from previous times when it was difficult to measure and quantify the minute amounts of certain elements in biological matrices. It is still in use although extremely sensitive and specific methods have been developed to exactly quantify even very low element concentrations in biological matrices.

Trace elements are best classified in two groups: the essential and the non-essential. In the absence of an essential element, a deficiency syndrome develops (see Chapter I.9). Supplementation of this specific element reverses the symptoms. According to the present knowledge the list of essential trace elements includes (in alphabetical order) at least chromium, cobalt, copper, fluorine, iodine, iron, manganese, molybdenum, nickel, selenium, silicon, tin, vanadium, and zinc (see also Chapter I.9). The group of non-essential elements includes all others which are more or less constantly found in variable concentrations in living matter but for which no proof of essentiality has been given. It should be noted, however, that some of the elements with currently unknown physiological functions will be found to participate in biochemical processes and thus will turn out to be essential.

In this context, it must be mentioned that, above a defined level, all the elements are toxic to man (see also Chapter I.16). A simplified schematic representation of the relationships between uptake, tissue element concentration, and health effects is shown in Fig. I.4b-1. Curve I represents an essential trace element showing that detrimental effects to health may occur either under conditions of deficiency or of excessive exposure. Curve II represents a non-essential trace element, which may lead to a toxic response under conditions of excessive exposure.

During the last 50 years enormous efforts have been made in investigating the role of trace elements in nutrition and in biochemical and physiological processes that take place in the body of man, both in health and disease.

The literature comprises a vast number of publications which cannot be considered here. For more detailed informations on the metabolism and function of trace elements, their importance in clinical medicine, estimates of nutritional requirements, and symptoms due to deficiency or excess, the reader is referred to the publications of UNDERWOOD (1977), PRASAD (1978), MERTZ (1981), HAMILTON

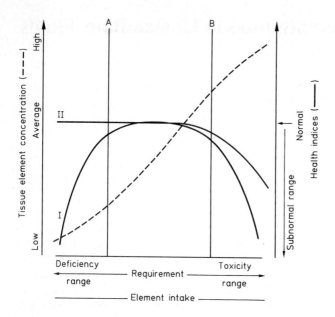

Fig. I.4b-1. Schematic representation of the relationships between element intake, tissue element concentration, and health indices. Curve I represents an essential trace element, which may produce adverse health effects in conditions of deficiency or excessive exposure. Curve II represents a non-essential trace element, which may produce adverse health effects only under conditions of excessive exposure. Intakes A and B represent intakes which produce minimal statistically significant changes from normal values for one or more health indices due to deficiency or toxicity, respectively. Modified from SPIVEY FOX and TAO (1981).

(1981), HURLEY (1981), NEWBERNE (1981), ZUMKLEY (1983), FRIBERG et al. (1986), and other chapters in this book. Reference is also made to the proceedings of several international workshops on trace elements in medicine and biology (BRÄTTER and SCHRAMEL, 1980, 1983, 1984, 1987, 1988; HEMPHILL, 1968–1989; GLADTKE et al., 1985; WHO/CEC/US EPA, 1987a, b). The interactions of trace elements in humans and animals were discussed by the TASK GROUP OF METAL INTERACTION (1978), by MILLS (1981), NORDBERG et al. (1986), and FISHBEIN (1987).

This chapter deals with the levels of trace and ultra-trace metals in human body fluids and tissues. Accurate data on the metal concentrations in these matrices are essential for the following purposes:

– establishing of normal criteria for metal concentrations in body fluids and tissues of individuals and population groups;
– detection of "abnormal" states, e.g., trace element deficiency or excessive exposures;
– retrospective or prospective studies relating trace element status to health and disease;
– establishing recommendations for trace element intake.

2 Early Studies of the Mineral Composition of Human Body Fluids and Tissues and Literature Surveys

One of the first systematic approaches to investigate the mineral composition of human body fluids and tissues was performed by SCHROEDER, TIPTON and coworkers in the 1950s and 1960s. In a program for spectrographic analysis of human tissues, which was carried out at the University of Tennessee (USA) and the Health Physics Division of Oak Ridge National Laboratory, tissues from some 250 autopsied persons from the United States and some 200 persons from outside continental United States were analyzed for a number of trace elements (TIPTON and COOK, 1963; TIPTON et al., 1963, 1965). In another series of publications SCHROEDER and coworkers presented a large amount of data on the concentrations of "essential" and "abnormal" trace metals in human body fluids and tissues as well as in non-human tissues, food, and beverages (SCHROEDER, 1965a, b; SCHROEDER and BALASSA, 1961a, b, 1965, 1966a, b, 1967; SCHROEDER et al. 1962a, b, 1963a, b, 1964, 1966a, b, 1967a, b, 1968, 1970a–c, 1972).

A compilation of published data concerning the elemental composition of human body fluids and tissues was presented by IYENGAR et al. (1978). This work covers the literature up to 1976 and exclusively deals with values for adults. Reference is also made to the review of MORRISON (1979). Reviews of published data for selected metals such as lead and cadmium were presented by BARRY (1978) and VAHTER (1982). A critical evaluation of the normal levels of trace elements in human blood plasma and serum was made by VERSIECK and CORNELIS (1980). The literature on the levels of major and trace elements in human lung tissue was reviewed by VANOETEREN et al. (1986). Probably the most extensive critical review of the literature on trace elements in human body fluids and tissues has been published by VERSIECK (1985).

3 Analytical Methods to Measure Metal Concentrations in Human Tissues and Body Fluids
(see also Chapter I.4a; ANGERER and SCHALLER, 1985, 1988)

The most frequently used analytical techniques to measure metal concentrations in human tissues and body fluids include atomic absorption spectrometry (AAS) applying different atomization methods (flame and electrothermal), emission spectroscopic methods based on flames, arcs, and sparks, neutron activation analysis (NAA), electrochemical methods, isotope dilution mass spectrometry, and atomic and X-ray fluorescence spectroscopy. Of the single element techniques AAS and electrochemical methods are most frequently applied. The most often used multielement techniques are NAA and emission spectroscopy. In particular, the inductively coupled plasma atomic emission spectrometry (ICP-AES) represents a promising technique for the simultaneous analysis of major and trace elements in biological and other materials. However, there is still a need for improved sensitivities and detection limits for certain elements of interest. A general review of the above mentioned analytical methods is given in Chapter I.4a of this book. WARD (1987) compared NAA and ICP-MS in liver standard reference materials for 66 trace elements.

The need for simultaneous localization and determination of chemical elements at the cellular and subcellular level has led to the development of combined microscopic/microanalytical methods for histochemical analysis. A summary of the characteristics of elemental localization techniques now available, though not fully developed, is shown in Table I.4b-1. With the advancement of new technologies and increasingly sophisticated instrumentation such as electron or ion microscope/microprobe, laser microprobe mass analysis (LAMMA) and proton-induced X-ray emission (PIXE) analyses with detection limits comparable to the more traditional analytical methods mentioned above and with spatial resolutions equal to or surpassing the light microscope, can be determined.

Another point of interest is that there are techniques now available for the in vivo measurement of certain trace elements in human tissues. These include the measurement of liver and kidney cadmium by means of neutron capture γ-ray analysis and the measurement of tooth and bone lead by X-ray fluorescence analysis.

The determination of kidney and liver cadmium by neutron capture γ-ray analysis is based on the specific properties of ^{113}Cd, a naturally occurring stable isotope (12.2% abundance). This isotope has a high probability (20000 barns) of capturing thermal neutrons. In the capture process, excited ^{114}Cd is produced

Table I.4b-1. Characteristics of Elemental Localization Techniques (Source: STIKA and MORRISON, 1981).

Excitation Source	Signal Detected	Technique Name	Comments/Advantages and Limitations
Electrons	Transmitted electrons	Electron energy loss spectrometry (EELS)	High sensitivity (two orders of magnitude greater than EPA). All elements Z>2, spatial resolution >1 nm, technique under recent development
	X-Rays	Electron probe analyses (EPA)	Analysis volume $<10^{-18}$ liter; sensitivity greater for heavier atomic numbers
	Auger electrons	Auger electrons spectroscopy (AES)	High vacuum requirements limit usefulness for biological systems; Z>9
Ions	X-Rays	Ion or proton-induced X-ray emission (PIXE)	Vaccum not required, H-detection, early stages of development, need accelerator, 1 µm spatial resolution
	Ions	Secondary ion mass spectrometry (SIMS) or ion microscopy	High sensitivity, especially for low atomic numbers; isotopic analysis, possible to do elemental analyses with depths 1 µm spatial resolution
Photons	Ions	Laser microprobe mass analyses (LAMMA)	Volume $10^{-15}-10^{-14}$ liter, isotopic analysis, high sensitivity, high-powered laser beams required

which, in turn, promptly decays to the ground state. This de-excitation is accomplished by the emission of a cascade of gamma-rays, which are detected externally to the body. As the gamma-rays are emitted promptly after neutron capture, the subject must be irridiated and counted simultaneously. Prior to the irridiation, the organs must be accurately located by means of an ultrasonic scanner. The partial body neutron activation (PBNA) technique is non-invasive and carries a minimal risk. It has been applied to the in vivo measurement of kidney and liver cadmium in cadmium-exposed workers and in smokers and non-smokers (ELLIS et al., 1979, 1984; ROELS et al., 1981).

In vivo X-ray fluorescence analyses of bone and tooth lead levels have been reported by SHAPIRO et al. (1978) and PRICE et al. (1984).

4 Evaluation of Published Data Pertaining to the Mineral Composition of Human Tissue

Published analytical data for the levels of trace elements in human tissues and body fluids cannot be accepted a priori as accurate, unless all details of the analytical procedure are critically evaluated, including those of sample collection, storage, preparation, and digestion. The validity of a reported result should be confirmed by simultaneous analysis by a second, independent method, by analysis of certified standard reference materials, and/or by agreement among collaborating laboratories.

An evaluation of reported trace element concentrations in human tissues and body fluids shows that widely divergent values were measured by different reseachers. For a considerable time, the wide variability of the measured values was attributed to biological sources of variation such as age, sex, dietary and smoking habits, physiological conditions, environmental exposures, geographical circumstances, and other influences. Although there is no doubt that some of these factors influence trace element concentrations in human tissues and body fluids, there is now increasing evidence indicating that a considerable number of previous studies must have suffered heavily from methodological deficiencies including inadequate sample collection and handling and gross analytical inaccuracies. As a consequence, great caution should be exercised when using and interpreting published data. Apart from these methodological deficiencies many studies are inadequate because the samples were not precisely defined nor were the donors adequately characterized in terms of state of health, history of occupation, smoking, alcoholism, medication, environmental exposures, and relevant factors.

The problems of analytical inaccuracy and sample contamination as a source of error in trace element determinations have been discussed extensively by VERSIECK (1985). The author concludes that "it may now be safely said that much of what has been published on the levels of aluminum, vanadium, chromium, manganese, cobalt, nickel, arsenic, molybdenum, cadmium, mercury, and other trace elements in biological matrices is in error". In fact, there are only a few elements such as iron, copper, zinc, and – perhaps – selenium for which fairly consistent values were measured by most investigators. However, even for these elements, problems are far from being definitively settled.

In order to ascertain the accuracy of analytical results, it is strongly recommended to analyze standard reference materials of identical or similar matrix simultaneously with the samples. It must be noted, however, that the concentration of most trace elements in currently available reference materials is markedly higher than in a number of matrices that come to analysis. Thus, it has to be taken into account that the errors for real samples are significantly higher than revealed by the results obtained in reference or intercomparison materials with high metal concentrations.

Sample contamination may result from (a) sample contamination by collection devices, (b) sample contamination by collection and storage vessels, and (c) sample contamination by ambient air, chemical reagents, laboratory instruments, and combinations of hazards. Inadequate sample collection and handling may drastically distort the intrinsic trace element content of biological matrices and may be a cause of more serious errors than any other step of the analytical procedure. It cannot be emphasized too strongly that the analysis of a contaminated sample is a futile exercise: no amount of care during the analysis may yield reliable information if an error has been introduced earlier.

Another important problem in the evaluation and interpretation of published trace element concentrations in biological matrices is their being expressed in many different and, sometimes, unfamiliar units. Numerous investigators reporting their values in dried or ashed samples do not mention the ratio of the dry or ash weight to the fresh (wet, original) weight. In these cases conversion factors based on the water content of organs and tissues and on the amount of ash in percent of wet or dry weight must be applied to calculate the element concentrations as µg/g or ng/g fresh weight or µg/L. Such conversion factors may be taken from the report of the TASK GROUP ON REFERENCE MAN (1972) or from TIPTON and COOK (1963) and TIPTON et al. (1965). Of course, this only permits an approximation of the real value because of considerable individual and physiological variations.

5 Selected Reference Values for Trace Elements in Human Tissues and Body Fluids

In a concluding comment to his review article VERSIECK (1985) states that, "despite significant advances in the course of the last 25 years, much has remained chaotic in the current status and approach to the subject of trace elements. When compared to the wealth of published data on their levels in human body fluids and tissues, results that may be recommended as reference values are remarkably scarce". Despite these difficulties the author made an attempt to set forth "selected reference values" for a number of elements in some blood fluids and tissues (plasma or serum, packed blood cells, urine, lung, liver, kidney, and skeletal muscle tissue) of adult individuals. These reference values were selected from the literature on the basis of one or more of the following criteria: (1) explicit information on the subjects included in the study; (2) detailed description of sample collection and preparation methods; (3) documented accuracy and precision of the measurement procedures; (4) proper reporting of error evaluations and adequate use of statistical methods; (5) agreement with values published by other authors; and (6) experience and low-level manipulation skill of the investigators. For details the reader is referred to the review of VERSIECK (1985). As mentioned above it should be noted that the levels of some trace metals, e.g., cadmium in blood, urine, kidney and liver and lead in blood, bones and teeth, are influenced by different factors such as age, sex, smoking, and geographic environment (CLAEYS-THOREAU et al., 1987; VAHTER, 1982). In applying any reference values these factors must be taken into account.

Quite a few "normal" concentrations of trace elements in body fluids and human tissues are given in Part II of this book. As explained in Section 8 of this chapter some precaution is needed for the interpretation of these figures given in most cases for healthy adults.

6 Trace Element Concentrations in Human Hair

In recent years, the determination of metal concentrations in hair has attracted much attention. Metal concentrations in hair have been used for forensic purposes, as an index of the nutritional status, as a diagnostic sign of disease, or as an indicator of exposure to certain pollutants. Undoubtedly, hair is an attractive material particularly because of the ease of sample collection and preservation. Concentrations of trace elements in hair are often considerably higher than in blood or other biological samples, and subsequent analysis should, therefore, be easier and more precise. Because of the physiology of hair formation, analysis of total length should give an integrated assessment of exposure over a known period of time, while examination of longitudinal sections should provide a diary or calendar of exposure (HOPPS, 1977). Newly developed methods, e.g., inductively-coupled plasma atomic emission spectrometry, allow for the simultaneous determination of up to approximately 30 elements in hair following a simple acid digestion. These features, together with an increased public awareness of matters of diet and health, have led to a growing number of commercial organizations offering a hair analysis service. Such companies often make extravagant claims, based on limited scientific data, for what can be revealed by such analysis. Frequently, physicians, toxicologists, and biochemists are asked for guidance on the interpretation of results and the follow-up of anxious and frightened patients whose 'trace element imbalance' has been discovered by hair testing.

The use and misuse of trace element analyses of human hair has been evaluated critically by SKY-PECK and JOSEPH (1983), RIVLIN (1983), BARETT (1985), TAYLOR (1986), KRAUSE and CHUTSCH (1987) and SCHREINER et al. (1987). After thorough study of the literature TAYLOR (1986) came to the following conclusions:

1. Modern analytical procedures enable sensitive, accurate, and precise measurements of trace element concentrations in human hair. Since interlaboratory com-

parisons indicate that the results are not always reliable, hair reference materials should be used throughout the analytical process to assure accurate analytical data.
2. The interpretation of the analytical data represents a complex problem since trace element concentrations of human hair are influenced by numerous factors including age and sex of the subject, color and growth site. Therefore, it is vital that reference levels should be selected carefully so that these factors are taken into consideration. Unfortunately, much of the experimental and investigative work has failed to do so, and many studies are of doubtful validity. In addition, the contribution of external contamination of the trace element content of hair is very variable, and it is difficult or even impossible to control this factor.
3. Even where it has been possible to control for these influences the most reliable experimental data indicate that the trace element content of hair does not correlate with the trace element concentrations in metabolically important tissues. Such large and variable discrepancies were found that it is difficult to accept how the elemental concentrations in hair could reflect the trace element status of a subject. Those occasions in which the concentration of trace elements in hair can be shown to reflect either body status or exposure are essentially extreme situations, usually with significantly increased concentrations and evidence of toxicity. In these situations parameters other than hair concentrations are more informative.
4. With the few exceptions mentioned above trace element analysis of hair is not a useful procedure; in many instances it provides data that may be misleading. The activities of laboratories which advertise and provide such analyses on a commercial basis can only be viewed with scepticism.

Despite these difficulties and problems it should be noted that hair analysis, on a group basis, may provide reasonable informations on exposure differences in differently exposed population groups, as was demonstrated by WIBOWO et al. (1986), WILHELM et al. (1987), and other investigators.

7 Trace Element Concentrations in Teeth and Bones

Deciduous and permanent teeth represent an interesting and relatively easily accessible matrix for trace elements with a strong affinity to calcified tissues. Presently, most data are available for lead in teeth (FERGUSSON and PURCHASE, 1987), but there are also some studies reporting on the levels of copper, zinc, cadmium, manganese, and other trace elements in human teeth (ATTRAMADAL and JONSEN, 1976; FOSSE and BERG JUSTESEN, 1978; SHARON, 1988).

Determination of lead in shed deciduous teeth has become a useful measure of past lead exposure in children. The formation of deciduous teeth begins in prenatal life, and the average lead level in the shed teeth may be taken as an integrated measure of the total exposure during early life. The lead concentration in whole deciduous as well as in permanent teeth increases with age, and the slope of the tooth lead vs. age regression line can be used as an index of a population's exposure to lead

(STEENHOUT and POURTOIS, 1981). It should be noted, however, that the lead concentration varies between different tooth types decreasing from the medial incisors to the premolars and molars, respectively (DELVES et al., 1982). Thus, the type of tooth that has been analyzed should be recorded in each study.

A significant variation in lead concentrations has also been found in relation to different tooth tissues (enamel, primary and secondary dentine; see also SHARON, 1988). The highest lead concentrations are found in the inner and outer surfaces of the tooth, i.e., the outer layer of enamel and the circumpulpal or secondary dentine. Enamel appears to change very little after its formation, and the lead concentrations are independent of age. On the other hand, the circumpulpal dentine which is formed after the eruption of the tooth continues to accumulate lead until the tooth is shed. Therefore, secondary dentine contains a much higher lead concentration than primary dentine.

8 Recommendations for Further Research

After thorough evaluation of the current state of knowledge concerning the importance and role of trace elements in biology and medicine the participants of a workshop on "research needed to improve data on mineral content of human tissues" (28–30 May 1980, University of Maryland, USA) came to the following conclusions (HOPPS and O'DELL, 1981):

1. There is urgent need for an adequate data base of the mineral and trace element content of human tissues; the present data base is grossly inadequate.
2. Such information will provide the means for important positive contributions to the health of man because it will greatly increase our knowledge of nutritional requirements as well as our ability to detect and correct early malnutrition.
3. The proposed data base will provide information of much value in the prevention and control of many diseases that, although they do not fall clearly into the category of malnutrition, are causally related to deficiency, excess, or imbalance of various trace elements.
4. To produce an adequate data base of the mineral and trace element content of human tissues will require clearly defined protocols, including standards for quality control with respect to (a) experimental design including donor and sample selection; (b) techniques for sample collection and storage; (c) analytical methods; and (d) data handling.

Detailed recommendations with respect to subject and sample selection and sample preservation and storage have been worked out by HAMILTON (1981), KOIRTYOHANN and HOPPS (1981), IMREY (1981), and more recently by IYENGAR (1986). Concerning the analytical part the following recommendations were made (SMITH et al. (1981): (1) Whenever available, certified standard reference materials of identical or similar matrix should be analyzed simultaneously with samples to ascertain accuracy of the results. This practice should be reported in resulting publications.

(2) Existing biological reference materials should be certified for additional trace elements of biological interest, and new reference materials with low concentrations of trace elements (blood plasma or serum and urine) should be certified for trace elements of biological interest. (3) Concentrations of trace elements should be reported as micrograms per gram of fresh tissue. If concentrations are reported on other bases, exact information should be given as to moisture content, ash content, and other factors so that the data could be converted to wet (fresh) tissue weight.

VERSIECK (1985) places special emphasis on the following aspects:

1. For the collection of the samples, materials should be used with the lowest possible impurities, All container materials and laboratory instruments must be checked and, if necessary, thoroughly cleaned before use. To determine trace elements in blood that occur at very low levels the blood samples should be collected only after 20 to 40 mL of blood have flushed the catheter or cannula.
2. In general, commercially available high-purity chemical reagents should be used, but it must be noted, that, in many instances, they are not sufficiently pure for the most demanding work so that additional measures (e.g., subboiling distillation of reagent acids) must be taken to reduce the level of impurities.
3. To determine trace element levels in the nanogram or subnanogram per gram range (e.g., chromium, cobalt, manganese, and vanadium in serum) working under clean room conditions plays a vital role in obtaining reliable figures on low-level trace element matrices.
4. The investigator should critically examine his own habits and be constantly aware of his own effect on the blank. There are many problem areas, involving jewelry, soaps, shampoos, cosmetics, detergents, smoking in the laboratory, which can lead to elevated levels of specific trace elements.
5. Special attention should also be directed to phenomena such as absorption of ions and compounds on the walls of the containers, losses by volatilization, evaporation or transpiration, segregation, colloidal formation, and, for autopsy samples, alterations in body fluids and tissues after death.

References

Angerer, J., and Schaller, K. H. (1985, 1988), *Analysis of Hazardous Substances in Biological Materials*, Vols. 1 and 2. VCH, Weinheim-Basel-Cambridge-New York.
Attramadal, A., and Jonsen, J. (1976), *The Content of Lead, Cadmium, Zinc and Copper in Deciduous and Permanent Human Teeth.* Acta Odontol. Scand. *34*, 127–131.
Barrett, S. (1985), *Commercial Hair Analysis.* J. Am. Med. Assoc. *254*, 1041–1045.
Barry, P. S. I. (1978), *Distribution and Storage of Lead in Human Tissues*, in: Nriagu, J. O. (ed.): *The Biogeochemistry of Lead in the Environment*, Part B, pp. 97–150. Elsevier, Amsterdam-New York-Oxford.
Brätter, P., and Schramel, P. (eds.) (1980), *Trace Element Analytical Chemistry in Medicine and Biology.* Proceedings of the First International Workshop, Neuherberg (FRG), April 1980. W. de Gruyter, Berlin-New York.

Brätter, P., and Schramel, P. (eds.) (1983), *Trace Element Analytical Chemistry in Medicine and Biology*, Vol. 2. *Proceedings of the Second International Workshop, Neuherberg (FRG)*, April 1982. W. de Gruyter, Berlin-New York.

Brätter, P., and Schramel, P. (eds.) (1984), *Trace Element Analytical Chemistry in Medicine and Biology*, Vol. 3. *Proceedings of the Third International Workshop, Neuherberg (FRG)*, April 1984. W. de Gruyter, Berlin-New York.

Brätter, P., and Schramel, P. (1987), *Trace Element Analytical Chemistry in Medicine and Biology*, Vol. 4. *Proceedings of the Fourth International Workshop, Neuherberg (FRG)*, April 1986. W. de Gruyter, Berlin-New York.

Brätter, P., and Schramel, P. (1988), *Trace Element Analytical Chemistry in Medicine and Biology*, Vol. 5. *Proceedings of the Fifth International Workshop, Neuherberg (FRG)*, April 1988. W. de Gruyter, Berlin-New York.

Claeys-Thoreau, F., Thiessen, L., Bruaux, P., Ducoffe, G., and Verduyn, G. (1987), *Assessment and Comparison of Human Exposure to Lead Between Belgium, Malta, Mexico and Sweden. Int. Arch. Occup. Environ. Health* 59, 31–41.

Delves, H. T., Clayton, B. E., Carmichael, A., Bubear, M., and Smith, M. (1982), *An Appraisal of the Analytical Significance of Tooth-lead Measurements as Possible Indices of Environmental Exposure of Children to Lead. Ann. Clin. Biochem.* 19, 329–337.

Ellis, K. J., Varisky, D., Zanzi, I., Cohn, S. H., and Yasumura, S. (1979), *Cadmium: In vivo Measurement in Smokers and Nonsmokers. Science* 205, 323–325.

Ellis, K. J., Yuen, K., Yasumura, S., and Cohn, S. H. (1984), *Dose-response Analysis of Cadmium in Man: Body Burden vs Kidney Dysfunction. Environ. Res.* 33, 216–226.

Fergusson, J. E., and Purchase, N. G. (1987), *The Analysis and Levels of Lead in Human Teeth: A Review. Environ. Pollut.* 46, 11–44.

Fishbein, L. (1987), *Trace and Ultra-trace Elements in Nutrition: An Overview (Zinc, Copper, Chromium, Vanadium, and Nickel). Toxicol. Environ. Chem.* 14, 73–99.

Fosse, G., and Berg Justesen, N. P. (1978), *Zinc and Copper in Dediciduous Teeth of Norwegian Children. Int. J. Environ. Stud.* 13, 19–34.

Friberg, L., Nordberg, G. F., and Vouk, V. B. (eds.) (1986), *Handbook on the Toxicology of Metals*, Vols. I and II. Elsevier, Amsterdam-New York-Oxford.

Gladtke, E., Heimann, G., Lombeck, I., and Eckert, I. (eds.) (1985), *Spurenelemente – Stoffwechsel, Ernährung, Imbalancen, Ultra-Trace-Elemente*. Symposium, Konstanz, 1984. G. Thieme Verlag, Stuttgart-New York.

Hamilton, E. I. (1981), *An Overview: The Chemical Elements, Nutrition, Disease, and the Health of Man. Fed. Proc. Fed. Am. Soc. Exp. Biol.* 40, 2126–2130.

Hemphill, D. D. (ed.) (1968–1989), *Trace Substances in Environmental Health. Proceedings I-XXII*, University of Missouri, Columbia.

Hopps, H. C. (1977), *The Biological Basis for Using Hair and Nail for Analyses of Trace Metals. Sci. Total Environ.* 7, 71–89.

Hopps, H. C., and O'Dell, B. L. (1981), *Introductions and Conclusions Presented at the Workshop "Research Needed to Improve Data on Mineral Content of Human Tissues,"* May 28–30, 1980, University of Maryland (USA). *Fed. Proc. Fed. Am. Soc. Exp. Biol.* 40, 2112–2114.

Hurley, L. S. (1981), *Trace Metals in Mammalian Development. John Hopkins Med. J.* 148, 1–10.

Imrey, P. B. (1981), *Design of a Tissue Mineral Data Bank. Fed. Proc. Fed. Am. Soc. Exp. Biol.* 40, 2148–2153.

Iyengar, G. V. (1986), in: O'Neill, I. K., Schuller, P., and Fishbein, L. (eds.): *Environmental Carcinogens – Selected Methods of Analysis. IARC Sci. Publ. No. 71*, pp. 141–158. IARC, Lyon.

Iyengar, G. V., Kollmer, W. E., and Bowen, H. J. M. (1978), *The Elemental Composition of Human Tissues and Body Fluids*. Verlag Chemie, Weinheim-New York.

Koirtyohann, S. R., and Hopps, H. C. (1981), *Sample Selection, Collection, Preservation and Storage for a Data Bank on Trace Elements in Human Tissues. Fed. Proc. Fed. Am. Soc. Exp. Biol.* 40, 2143–2148.

Krause, C., and Chutsch, M. (1987), *Haaranalyse in Medizin und Umwelt. Schriftenr. Ver. Wasser Boden Lufthyg. 71.*
Mertz, W. (1981), *The Essential Trace Elements.* Science *213*, 1332–1338.
Mills, C. F. (1981), *Interaction Between Elements in Tissues: Studies in Animal Models. Fed. Proc. Fed. Am. Soc. Exp. Biol. 40,* 2138–2143.
Morrison, G. M. (1979), *Elemental Trace Analysis of Biological Materials. Crit. Rev. Anal. Chem. 8,* 287–320.
Newberne, P. M. (1981), *Disease States and Tissue Mineral Elements in Man. Fed. Proc. Fed. Am. Soc. Exp. Biol. 40,* 2134–2138.
Nordberg, G. F., Parizek, J., Pershagen, G., and Gerhardsson, L. (1986), *Factors Influencing Effects and Dose-Response Relationship of Metals,* in: Friberg, L., Nordberg, G. F., and Vouk, V. B. (eds.): Handbook on the Toxicology of Metals, Vol. I, pp. 175–205. Elsevier, Amsterdam-New York-Oxford.
Prasad, A. S. (1978), *Trace Elements and Iron in Human Metabolism.* John Wiley & Sons, Chichester-New York-Brisbane-Toronto.
Price, J., Baddeley, H., Kenardy, J. A., Thomas, B. J., and Thomas, B. W. (1984), *In vivo X-ray Fluorescence Estimation of Bone Lead Concentrations in Queensland Adults. Br. J. Radiol. 57,* 29–33.
Rivlin, R. S. (1983), *Misuse of Hair Analysis for Nutritional Assessment. Am. J. Med. 75,* 489–493.
Roels, H. A., Lauwerys, R. R., Buchet, J. P., Bernard, A., Chettle, D. R., Harvey, T. C., and Al-Haddad, I. K. (1981), *In vivo Measurement of Liver and Kidney Cadmium in Workers Exposed to this Metal: Its Significance with Respect to Cadmium in Blood and Urine. Environ. Res. 26,* 217–240.
Schreiner, S., Kruse, H., and Berg, C. (1987), *Die Haaranalyse. Pharm. Ztg. 132,* 3201–3240.
Schroeder, H. A. (1965a), *Cadmium as a Factor in Hypertension. J. Chronic Dis. 18,* 217–228.
Schroeder, H. A. (1965b), *J. Chronic Dis. 18,* 647–656.
Schroeder, H. A., and Balassa, J. J. (1961a), *Abnormal Trace Elements in Man: Cadmium. J. Chronic Dis. 14,* 236–258.
Schroeder, H. A., and Balassa, J. J. (1961b), *Abnormal Trace Elements in Man: Lead. J. Chronic Dis. 14,* 408–425.
Schroeder, H. A., and Balassa, J. J. (1965), *Abnormal Trace Metals in Man: Niobium. J. Chronic Dis. 18,* 229–241.
Schroeder, H. A., and Balassa, J. J. (1966a), *Abnormal Trace Metals in Man: Arsenic. J. Chronic Dis. 19,* 85–106.
Schroeder, H. A., and Balassa, J. J. (1966b), *Abnormal Trace Metals in Man: Zirconium. J. Chronic Dis. 19,* 573–586.
Schroeder, H. A., and Balassa, J. J. (1967), *Abnormal Trace Metals in Man: Germanium. J. Chronic Dis. 20,* 211–224.
Schroeder, H. A., Balassa, J. J., and Tipton, I. J. (1962a), *Abnormal Trace Metals in Man: Nickel. J. Chronic Dis. 15,* 51–65.
Schroeder, H. A., Balassa, J. J., and Tipton, I. J. (1962b), *Abnormal Trace Metals in Man: Chromium. J. Chronic Dis. 15,* 941–964.
Schroeder, H. A., Balassa, J. J., and Tipton, I. J. (1963a), *Abnormal Trace Metals in Man: Titanium. J. Chronic Dis. 16,* 55–69.
Schroeder, H. A., Balassa, J. J., and Tipton, I. J. (1963b), *Abnormal Trace Metals in Man: Vanadium. J. Chronic Dis. 16,* 1047–1071.
Schroeder, H. A., Balassa, J. J., and Tipton, I. J. (1964), *Abnormal Trace Metals in Man: Tin. J. Chronic Dis. 17,* 483–502.
Schroeder, H. A., Balassa, J. J., and Tipton, I. H. (1966a), *Essential Trace Metals in Man: Manganese. A Study in Homeostasis. J. Chronic Dis. 19,* 545–571.
Schroeder, H. A., Nason, A. P., Tipton, I. J., and Balassa, J. J. (1966b), *Essential Trace Metals in Man: Copper. J. Chronic Dis. 19,* 1007–1034.

Schroeder, H. A., Buckman, J., and Balassa, J. J. (1967a), *Abnormal Trace Elements in Man: Tellurium*. J. Chronic Dis. 20, 147–161.

Schroeder, H. A., Nason, A. P., Tipton, I. H., and Balassa, J. J. (1967b), *Essential Trace Metals in Man: Zinc. Relation to Environmental Cadmium*. J. Chronic Dis. 20, 179–210.

Schroeder, H. A., Nason, A. P., and Tipton, I. H. (1967c), *Essential Trace Metals in Man: Cobalt*. J. Chronic Dis. 20, 869–890.

Schroeder, H. A., Nason, A. P., and Tipton, I. J. (1968), *Essential Metals in Man: Magnesium*. J. Chronic Dis. 21, 815–841.

Schroeder, H. A., Nason, A. P., and Tipton, I. H. (1970a), *Chromium Deficiency as a Factor in Arteriosclerosis*. J. Chronic Dis. 23, 123–142.

Schroeder, H. A., Frost, D. V., and Balassa, J. J. (1970b), *Essential Trace Metals in Man: Selenium*. J. Chronic Dis. 23, 227–243.

Schroeder, H. A., Balassa, J. J., and Tipton, I. H. (1970c), *Essential Trace Metals in Man: Molybdenum*. J. Chronic Dis. 23, 481–499.

Schroeder, H. A., Tipton, I. H., and Nason, A. P. (1972), *Trace Metals in Man: Strontium and Barium*. J. Chronic Dis. 25, 491–517.

Shapiro, I. M., Burke, A., Mitchell, G., and Bloch, P. (1978), *X-ray Fluorescence Analysis of Lead in Teeth of Urban Children in situ: Correlation between the Tooth Lead Level and the Concentration of Blood Lead and Free Erythroporphyrins*. Environ. Res. 17, 46–52.

Sharon, I. M. (1988), *Human and Animal Teeth: Biological Monitors of Pollutants*. 3rd International Conference on Environmental Contamination, Venice. Proceedings, pp. 286–288. CEP Consultants Ltd., Edinburgh.

Sky-Peck, H. H., and Joseph, B. J. (1983), *Chemical Toxicology and Clinical Chemistry of Metals*, pp. 159–163, Academic Press, London.

Smith, J. C., Anderson, R. A., Ferretti, R., Levander, O. A., Morris, E. R., Roginski, E. E., Veillon, C., Wolf, W. R., Anderson, J. J., and Mertz, W. (1981), *Evaluation of Published Data Pertaining to Mineral Composition of Human Tissue*. Fed. Proc. Fed. Am. Soc. Exp. Biol. 40, 2120–2125.

Spivey Fox, M. R., and Tao, S. H. (1981), *Mineral Content of Human Tissues from a Nutrition Perspective*. Fed. Proc. Fed. Am. Soc. Exp. Biol. 40, 2130–2134.

Steenhout, A., and Pourtois, M. (1981), *Lead Accumulation in Teeth as a Function of Age with Different Exposures*. Br. J. Ind. Med. 38, 297–303.

Stika, K. M., and Morrison, G. H. (1981), *Analytical Methods for the Mineral Content of Human Tissues*. Fed. Proc. Fed. Am. Soc. Exp. Biol. 40, 2115–2120.

Task Group on Metal Interaction (1978), *Factors Influencing Metabolism and Toxicity of Metals: A Consensus Report*. Environ. Health Perspect. 25, 3–41.

Task Group on Reference Man (1975), ICRP Publ. No. 23. Pergamon Press, Oxford.

Taylor, A. (1986), *Usefulness of Measurements of Trace Elements in Hair*. Ann. Clin. Biochem. 23, 364–378.

Tipton, I. H., and Cook, M. J. (1963), *Trace Elements in Human Tissue. Part II. Adult Subjects from the United States*. Health Phys. 9, 103–145.

Tipton, I. H., Cook, M. J., Steiner, R. L., Boye, C. A., Perry, H. M., and Schroeder, H. A. (1963), *Trace Elements in Human Tissue. Part I. Methods*. Health Phys. 9, 89–101.

Tipton, I. H., Schroeder, H. A., Perry, H. H., and Cook, M. J. (1965), *Trace Elements in Human Tissue. Part III. Subjects from Africa, the Near and Far East and Europe*. Health Phys. 11, 403–451.

Underwood, E. J. (1977), *Trace Elements in Human and Animal Nutrition*. Academic Press, New York.

Vahter, M. (ed.) (1982), *Assessment of Human Exposure to Lead and Cadmium Through Biological Monitoring*. Prepared for United Nations Environment Program and World Health Organization. National Swedish Institute of Environmental Hygiene and Department of Environmental Hygiene, Karolinska Institute, Stockholm.

Vanoeteren, C., Cornelis, R., and Sabbioni, E. (1986), *Critical Evaluation of Normal Levels of Major and Trace Elements in Human Lung Tissue*. Report prepared for the Commission of the

European Communities, Directorate General for Science Research and Development. Brussels-Luxembourg.
Versieck, J. (1985), *Trace Elements in Human Body Fluids and Tissues. Crit. Rev. Clin. Lab. Sci.* 22, 97–184.
Versieck, J., and Cornelis, R. (1980), *Normal Levels of Trace Elements in Human Blood Plasma of Serum. Anal. Chim. Acta 116,* 217–254.
Ward, N. I. (1987), *The Future of Multi-(ultra-)Trace Element Analysis in Assessing Human Health and Disease: A Comparison of NAA and ICPMS. 2nd Nordic Symposium on Trace Elements in Human Health and Disease, Odense. Environ. Health 20,* 118–123. WHO Regional Office for Europe, Copenhagen.
WHO/CEC/US EPA (1987a), *Trace Elements in Human Health and Disease: Extended Abstracts from the 2nd Nordic Symposium, Odense (DK),* 17.–21. August. *Environ. Health Ser. 20.* WHO Regional Office for Europe, Copenhagen.
WHO/CEC/US EPA (1987b), *Trace Elements in Human Health and Disease: Report on the 2nd Nordic Symposium. Odense (DK),* 17.–21. August. *Environ. Health Ser. 26.* WHO Regional Office for Europe, Copenhagen.
Wibowo, A. A. E., Herber, R. F. M., Das, H. A., Roeleveld, N., and Zielhuis, R. L. (1986), *Levels of Metals in Hair of Young Children as an Indicator of Environmental Pollution. Environ. Res.* 40, 346–356.
Wilhelm, M., Hafner, D., Lombeck, I., and Ohnesorge, F. K. (1987), *Variables Influencing Cadmium Concentrations in Hair of Pre-school Children Living in Different Areas of the Federal Republic of Germany. Int. Arch. Occup. Environ. Health 60,* 43–50.
Zumkley, H. (ed.) (1983), *Spurenelemente.* G. Thieme Verlag, Stuttgart-New York.

I.4c Bioindicators for Monitoring Heavy Metals in the Environment

JÜRG HERTZ, Birmensdorf, Switzerland

1 Introduction
(see also Chapter I.7h, Section 4)

Indicators are widely used in monitoring various social, political, financial, hygienic, and environmental factors. Environmental indicators are suitable for the observation of long-term developments in an ecosystem, as well as for planning and controlling effects of anthropogenic activities. In recent years biological indicators have become widespread for air and water quality assessment.

2 Definitions and Classification

Fig. I.4c-1 shows a proposal for the classification of bioindication. According to STÖCKER (1980) *bioindication* means the time-dependent, sensitive response of measurable quantities of biological objects and systems to anthropogenic influences on the environment. Bioindicators are thus objects, compartments, and ecosystems which exhibit bioindication. A bioindicator can be of optical, chemical, or physical-biochemical nature (HASELOFF, 1982). Optical indicators include the reduced growth of plant species, decoloration of leaves or changes in population and behavior of any kind of living beings. Chemical indicators are characterized by the accumulation of the substance under consideration, whereas physical-biochemical indicators show, for example, reduced enzymatic activity or derangement of physiological functions.

In general a distinction can be made between bioindication as a qualitative method for the detection of the presence of pollutants and biomonitoring as a more quantitative method for the determination of the effects of the pollutants present (TONNEIJK and POSTHUMUS, 1987). BICK (1982) gave a definition of the latter without using the expression "biomonitor": "Bioindicators (i.e., biomonitors) are organisms which can be used for the recognition and quantitative determination of anthropogenically induced environmental factors". For the detection and recognition of water or air pollution, biological organisms which respond sensitively and specifically to the pollutants under consideration can be used. In addition, organisms that readily amass the polluting components without changing their

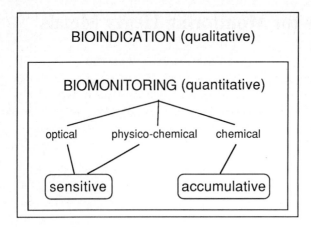

Fig. I.4c-1. Classification of bioindication.

chemical nature may be used as accumulators (POSTHUMUS, 1984). This classification into sensitive and accumulative biomonitors was first given by STEUBING (1976) and STÖCKER (1980), and is now well-accepted terminology.

3 Sensitive Biomonitors

Sensitive biomonitors are used in *aquatic ecosystems* as integrators of the pollutional stresses caused by contaminants in order to provide early warning systems (CAIRNS and VAN DER SCHALIE, 1979). Biomonitoring systems can be divided into two categories: *ecological surveys* and *toxicity testing* (ROOP and HUNSAKER, 1985).

Ecological surveys may use indicator species or assessments based on the composition of biological communities, numerical diversity, and indexes. By making comparisons between affected and control areas, ecological surveys can indicate the health of a water body exposed to pollutant loadings.

Toxicity testing is used to gain basic information about the general toxicity of effluents which are expected to be introduced into an ecosystem. By far the greatest number of toxicity tests have been performed on adult fish to answer such questions as (BUIKEMA, 1982):

– Is the material lethal to the test organism and at what concentration?
– What are the effects on an organism exposed to sublethal concentrations of toxicant for part or all of its life cycle?
– Which organism is most sensitive?
– Under which conditions are wastes most toxic?
– Does the toxicity of the material change when it enters the environment?
– Does the waste or pure compound pass or fail some regulatory standard?
– How much of the receiving system is affected?
– What are the short-term effects of episodic spills?

Various toxicity tests exist to answer these questions. The investigations can be used for the prediction of environmental effects of a waste, for the comparison of toxicants on animals or test conditions, or for the regulation of effluent discharge.

Sensitive biomonitors may equally be used for *air quality control*. Observing frequency, density, vitality, and damage to plant species yields information about the overall air quality (HAWKSWORTH, 1971). Lichens are especially sensitive to air quality because they form a sensitive symbiosis between algae and fungi, exhibit an unshielded surface against air from which they gain the main part of toxic and essential trace elements, and are long-living plants with photosynthetic activity over the whole year.

An "Index of Atmospheric Purity (*IAP*)" can be calculated from the observable quantities (LEBLANC and DESLOOVER, 1970), and hence a non-specific summation of information about air quality at the sites under consideration can be evaluated.

$$IAP = \sum_{i=1}^{n} \frac{Q \cdot f}{10}$$

n is the number of lichen species at one site; Q the toxitolerance factor (for resistant species: small); f frequency and density (1 seldom, low density; 5 frequent, high density).

The quantitative interpretation of *IAP*-values has been performed by developing regression models which include *IAP* as well as various chemical air quality data (HERZIG et al., 1987). Nevertheless, all such regression models depend on the completeness of the analytical air quality data used, but the calculation of single species concentrations on the basis of *IAP*-values is not possible.

4 Accumulation Biomonitors

4.1 Bioaccumulation

Among the many possible chemical species which could be considered, the bioaccumulation of heavy metals has been studied most extensively. The most frequently cited ones are Cu, an element which displays high accumulation efficiency in many biological systems, and Cd, Hg, Pb, and Zn, examples of important polluting elements (ROUSH et al., 1985, and literature cited therein).

Accumulation of heavy metals has been observed in many compartments of an ecosystem. High concentrations of Cd and Pb are found in certain arthropods, and Hg accumulates in the hair of mammals and feathers of birds; mussels also exhibit high concentrations of many elements.

TYLER (1972) showed that dead organic matter, such as litter and the underlying soil horizons, as well as low-level plants (especially lichens and mosses) accumulate high amounts of heavy metals. The main reason for this accumulation is the high stability of the chemical complexes formed between heavy metal ions and the negatively charged organic groups. Furthermore, the charges on unprotected surfaces provide ion-exchange facilities which allow the effective adsorption of metal ions to occur. Because their cuticle-covered epidermis is not very permeable to metal ions, higher plants accumulate metals much less effectively.

4.2 Accumulation Organisms

4.2.1 Biomonitors for Terrestrial Ecosystems

The most sensitive biological instruments for measuring the deposition of heavy metals in terrestrial ecosystems are certain mosses. Bryophytes, as low level plants without real roots, acquire water mainly through their surfaces. This kind of water uptake is additionally favored by the high surface to volume ratio and the lack of a cuticle in most moss species (BROWN, 1982). Water and, hence, metal uptake in mosses is thus mainly independent of substrate composition and contamination.

In addition, the transport of heavy metals to the plant surface in the form of dust originating from the surrounding soil is negligible compared to the contribution by atmospheric deposition (RÜHLING and TYLER, 1969). Mosses exhibit ion exchange properties; the ion exchange equilibria in, for example, *Hylocomium splendens* (Hedw.) Schimp. favor the following preferential uptake of metals (RÜHLING and TYLER, 1970): Pb^{2+}, $Cu^{2+} > Ni^{2+} > Co^{2+} > Zn^{2+}$, Mn^{2+}. Even in the presence of high concentrations of competing ions such as Mg^{2+}, Ca^{2+}, K^+, and Na^+, Cu^{2+} and Pb^{2+} were completely absorbed, thus leading to the conclusion that they are absorbed from rain water to a large extent. The concentrations in *Hylocomium splendens* in a contaminated area may be as high as 0.2 mg g^{-1} for Pb, 0.1 mg g^{-1} for Cu, 2 µg g^{-1} for Cd, and 0.4 mg g^{-1} for Zn (FOLKESON, 1979). Even higher concentrations were found in *Bryum argenteum* Hedw. in a city. The respective maximum contents for Pb, Cu, Cd, and Zn (given relatively to dry weight) are 5 mg g^{-1}, 2 mg g^{-1}, 60 µg g^{-1}, and 5.8 mg g^{-1} (THÖNI et al., 1986). It has been shown that Pb, Cd, and Zn are readily liberated from *Bryum argenteum* by extraction with dilute nitric acid, whereas Cu is bound more strongly (HERTZ et al., 1984). The wide variation in morphology, growth form and cation exchange capacity shown by different species of moss is largely responsible for the equally large variation in concentrations of heavy metals found in different species of mosses at the same sampling site (FOLKESON, 1979). Nevertheless, FOLKESON postulated that different species could be used in a monitoring survey if an interspecies relationship was made. *Hylocomium splendens, Hypnum cupressiforme* Hedw., and *Pleurozium schreberi* (brid.) Mitt. are postulated to be suitable for heavy metal monitoring (TYLER, 1976), but in cities in central Europe only *Bryum argenteum* is abundant enough to be of practical use for environmental surveys (THÖNI and HERTZ, 1987).

Another group of low level plants, lichens, are proposed as accumulating biomonitors for heavy metals in the air (BURTON, 1986). Lichens have been extensively employed as sensitive biomonitors because of their susceptibility to gaseous pollutants such as fluoride or sulfur dioxide. Their use for monitoring airborne particulate matter and atmospheric deposition of heavy metals has been discussed (e.g., JENKINS and DAVIES, 1966), but it has not become particularly widespread

(MARTIN and CAUGHTREY, 1982). The reason for this lies mainly in the inherent impossibility of simultaneously applying an organism as both a sensitive and an accumulative monitor. Other reasons for the preference of mosses over lichens include the enhanced accumulation ability of mosses (FOLKESON, 1979) and the fact that certain epiphytic lichens seem to accumulate metals not only from the atmosphere but from the inner part of the bark of trees as well (DEBRUIN and HACKENITZ, 1986).

There are various suggestions in the literature that the analysis of fungal fruit bodies may provide useful material for monitoring purposes. In contrast to mosses and lichens, *fungi* can reflect the heavy metal content of the substrate soil. The mycelium penetrates huge soil compartments during the whole year whereas the fruit body is only exposed to the air for a short time. Any direct contamination by polluted air is thus negligible. It has long been known that fruit bodies accumulate certain elements (BERTRAND and BERTRAND, 1947). Because it is easy to cultivate, *Agaricus* has been well studied. The fruit bodies of this genus are highly susceptible to the accumulation of Hg and Cd (STIJVE and BESSON, 1976, SEEGER et al., 1978). Nevertheless, the accumulation efficiency of heavy metals in the body is highly dependent on the specific type of mushroom under consideration (DIETL, 1986) and does not satisfactorily reflect the spatial variation of the heavy metal content of the underlying soil (TYLER, 1982). Thus only few applications of mushrooms of this type as monitoring organisms have been reported.

The application of *animals* as biomonitors is under intensive discussion (ELLENBERG, 1984; BICK and NEUMANN, 1982). The heavy metal content of different specimens in the same population of vertebrates may vary by a factor of 10 to 100. This is due to specific differences in diet and variation in uptake properties according to the sex and age of an individual. Some criteria for the choice of appropriate species for biomonitoring have thus to be fulfilled (ELLENBERG et al., 1985):

- Sedentary home range behavior.
- Known food composition.
- Uniform distribution.
- Small oscillation of population size, etc.

Certain *birds* are among the few vertebrate species, which meet these prerequisites; the heavy metal content of feathers has been especially investigated (GOEDE and DEBRUIN, 1986). For biomonitoring purposes in Central Europe, blackbirds (WEYERS et al., 1985), magpies (HAHN et al., 1985) and goshawks (ELLENBERG et al., 1985) are examples of birds studied.

For monitoring the soil status, the application of other organisms, such as invertebrates, has been discussed (EIJSACKERS, 1983). Earthworms were found to exhibit fairly high concentrations of heavy metals which depended on the source of the ingested material rather than local soil concentration (BENGTSSON et al., 1983). Arthropods seem to be the only invertebrates suitable for biomonitoring soil metal content in forest ecosystems (FUNKE, 1987).

4.2.2 Biomonitors for Aqueous Ecosystems

In marine aquatic ecosystems, molluscs and algae are the most frequently used biomonitors for heavy metals (PHILLIPS, 1980). *Macroalgae* fulfil many indicator requirements; they are sessile, hardy enough to allow laboratory studies, able to survive in highly polluted areas, and large enough for easy analysis. Certain species are widespread, especially *Fucus* in temperate water and *Sargassum* in subtropical and tropical areas. Species of *Ulva* are particularly resistant and are often used in areas which are severely polluted by sewage effluents. The major variables which affect the levels of trace metals in macroalgae are season, shore position, and part of the plant sampled. Using a standardized procedure, many of these variables can be eliminated. Nevertheless, certain effects such as regulation of the uptake of certain metals, metal-metal interactions, and growth inhibition by pollutants cannot be excluded and thereby limit the ability of macroalgae as well as other plants to be indicators.

Molluscs are the most commonly used monitors for metals in sea water. BRYAN (1976) stated "... it has been generally found that concentrations of heavy metals in bivalve molluscs change with those in the surrounding environment ...". Bivalves exhibit little evidence for the metabolic regulation of metal uptake, and thus the proportionality between water concentration and organism content is linear. *Mytilus edulis* seems to be especially suited as a biomonitor and was proposed for monitoring programs in sea water (SCHULZ-BALDES, 1982). Recently molluscs have also been studied in a freshwater environment (ZADORY, 1984) thus allowing direct comparison of data to be made.

Aquatic *mosses* have been suggested to be as effective accumulators as their terrestrial counterparts (MOUVET and CORDEBAR, 1986), but the applications have not yet become widespread.

4.3 Biomonitoring Methods with Plants

Bioaccumulation monitoring is now well established in the observation of air pollution by heavy metals, and thus several methods have been developed for this purpose.

A distinction can be made between passive and active biomonitoring. Active biomonitoring includes the exposure of well-defined plant species to controlled conditions, whereas passive monitoring refers to the observation or chemical analysis of indigenous plants.

HASELOFF (1982) divided the active biomonitoring into transplantation, test plant, and test chamber method. In the transplantation procedure, suitable organisms (mainly mosses and lichens) are transported from an unpolluted control area to the polluted sites under consideration. The exposure time is thus well defined but the change in uptake efficiency due to the climatic change is not fully known. The trial plants have to be fenced in for an extended length of time, and thus expensive and lengthy observation periods are the consequence. The same is true for the test plant procedure. The test plants grow under controlled conditions in a greenhouse and are then transferred to the sites under investigation. Standardized grass cultures are used for the continuous monitoring of heavy metals and fluoride

(KOSTKA-RICK and ARNDT, 1987). On the other hand, test plants may be exposed to unfiltered and filtered ambient air in closed test chambers to provide direct information about effects of air contamination. This method is usually applied to testing either synergistic or single species effects of pollutants on sensitive biomonitors.

The application of passive accumulation monitoring in air pollution studies has several advantages compared with physico-chemical measurements as it is covered by various governmental regulations. Collection of one set of samples yields information about an extended period of time, thus allowing characterization of the site with minor cost. Passive biomonitoring is a low-expense method for the evaluation of air quality thus allowing studies with high information density, i.e., about 1 sample km^{-2} to be performed (THOMAS, 1983).

Although a large number of studies exist, there is still a need for a quantitative comparison of the concentration of environmental chemicals which have accumulated in biomonitors with air pollution measurements at the same site (THOMAS, 1986); such a comparison would enable the calibration of the biomonitoring system to be made. The prerequisites for the quantitative determination of airborne heavy metal pollution using plants are the choice of suitable plant species, knowledge about the effects of genetic variation and microclimate on the accumulation efficiency of the monitoring plants, and the estimation of the accumulation time (THÖNI and HERTZ, 1987). The analytical procedure may introduce uncertainties as well and has to be evaluated carefully (WOLF et al., 1984).

4.4 Selected Applications of Mosses as Biomonitors

Today, only mosses enjoy widespread popularity for biomonitoring heavy metals in the environment.

Summaries concerning the application of mosses as biomonitors for heavy metals have been published by MASCHKE (1981) and MARTIN and COUGHTREY, 1982). Mosses have been used for evaluating the background heavy metal pollution in large areas where pleurocarp forest mosses are used (RÜHLING and TYLER, 1969). The time dependence of moss metal contents reflects the implementation of governmental regulations. It could, for example, be shown that the Cd content fell drastically within 10 years after the restriction of its use (RÜHLING and TYLER, 1984). Long-range transport of heavy metals from the British Isles to the South coast of Norway could be detected by analyzing mosses (STEINNES, 1984). Small area studies can be performed to evaluate heavy metal point sources (e.g., CAMERON and NICKLESS, 1977). Urban heavy metal air pollution was studied using the moss *Bryum argenteum* (THÖNI et al., 1986); in this study it was possible to evaluate the main emission sources by the application of statistical methods.

The field is expanding, and it can be expected that mosses will be used extensively for controlling regulatory environmental standards in the future.

References

Bengtsson, G., Nordstroem, S., and Rundgren, S. (1983), *Population Density and Tissue Metal Concentration of Lumbricids in Forest Soils near a Brass Mill. Environ. Pollut. Ser. A 30*, 897–1108.

Bertrand, G., and Bertrand, D. (1947), *Rubidium in Cryptogams. Ann. Inst. Pasteur 73*, 797–803.

Bick, H. (1982), *Bioindikatoren und Umweltschutz. Decheniana Beih. 62*, 2–5.

Bick, H., and Neumann, D. (eds.) (1982), *Bioindikatoren: Ergebnisse des Symposiums: Tiere als Indikatoren für Umweltbelastungen. Decheniana Beih. 62.*

Brown, D. H. (1982), *Mineral Nutrition*, in: Smith, A. J. E. (ed.): *Bryophyte Ecology.* Chapman & Hall, London-New York.

Bryan, G. W. (1976), in: Johnston, R. (ed.): *Marine Pollution,* pp. 185–302. Academic Press, London-New York.

Buikema, A. L., Niederlehner, B. R., and Cairns, J. (1982), *Biological Monitoring. Part IV. Toxicity Testing. Water Res. 16*, 239–262.

Burton, M. A. S. (1986), *Biological Monitoring of Environmental Contaminants (Plants).* Marc-Report No. 32. Monitoring and Assessment Research Center, London.

Cairns, J., and Van der Schalie, W. H. (1979), *Biological Monitoring in Water Pollution. Water Res. 14*, 261–266.

Cameron, A. J., and Nickless, G. (1977), *Use of Mosses as Collectors of Heavy Metals near a Smelting Complex. Water Air Soil Pollut. 7*, 117–125.

DeBruin, M, and Hackenitz, E. (1986), *Trace Element Concentrations in Epiphytic Lichens and Bark Substrate. Environ. Pollut. Ser. B 11*, 153–160.

Dietl, G. (1986), *Großpilze als Bioindikatoren – Möglichkeiten und Beispiele, Ulmer Pilzflora I*, pp. 142–151. Arbeitsgemeinschaft Mykologie, Ulm, FRG.

Eijsackers, H. (1983), *Soil Fauna and Soil Microflora as Possible Indicators of Soil Pollution*, in: Best, E. P. H., and Haeck, J. (eds.): *Ecological Indicators for the Assessment of the Quality of Air, Water, Soil and Ecosystems,* pp. 307–316. Reidel, Dordrecht-Boston.

Ellenberg, H. (ed.) (1984), *Symposium: Das freilebende Tier als Indikator für den Funktionszustand der Umwelt,* Wien, Austria.

Ellenberg, H., Dietrich, J., Stoeppler, M., and Nürnberg, H. W. (1985), *Environmental Monitoring of Heavy Metals with Birds as Pollution Integrating Biomonitors. I. Introduction, Definitions and Practical Examples for Goshawk (Accipiter gentilis)*, in: Lekkas, D. T. (ed.): *Heavy Metals in the Environment,* pp. 724–726. CEP Consultants Ltd., Edinburgh.

Folkeson, L. (1979), *Interspecies Calibration of Heavy Metal Concentrations in nine Mosses and Lichens: Applicability to Deposition Measurements. Water Air Soil Pollut. 11*, 253–260.

Funke, W. (1987), *Invertebrates as Biological Indicators in Forests. VDI Ber. 609*, 133–176.

Goede, A. A, and DeBruin, M. (1986), *The Use of Bird Feathers for Indicating Heavy Metal Pollution. Environ. Monit. Assess. 7*, 249–256.

Hahn, E., Ostapczuk, P., Ellenberg, H., and Stoeppler, M. (1985), *Environmental Monitoring of Heavy Metals with Birds as Pollution Integrating Biomonitors. II. Cadmium, Lead and Copper in Magpie (Pica pica) Feathers from a Heavily Polluted and a Control Area*, in: Lekkas, D. T. (ed.): *Heavy Metals in the Environment,* pp. 721–723. CEP Consultants Ltd., Edinburgh.

Haseloff, H. P. (1982), *Bioindikatoren und Bioindikation. Biol. Unserer Zeit 12*, 20–26.

Hawksworth, D. L. (1971), *Lichens as Litmus for Air Pollution: A Historical Review. Int. J. Environ Stud. 1*, 281–296.

Hertz, J., Schmid, I., and Thöni, L. (1984), *Monitoring of Heavy Metals in Airborne Particles by Using Mosses Collected from the City of Zurich. Int. J. Environ. Anal. Chem. 17*, 1–12.

Herzig, R., Liebendörfer, L., and Urech, M. (1987), *Lichens as Biological Indicators for Air Pollution in Switzerland: Method Evaluation and Calibration of Important Air Pollutants. VDI Ber. 609*, 619–639.

Jenkins, D. A., and Davies, R. I. (1966), *Trace Element Contents of Organic Accumulations. Nature* 210, 1296–1297.

Kostka-Rick, R., and Arndt, U. (1987), *Systematical Research to Optimize the Procedure of Standardized Grass Cultures. VDI Ber.* 609, 301–316.

LeBlanc, F., and DeSloover, J. (1970), *Relation between Industrialization and the Distribution and Growth of Epiphytic Lichens and Mosses in Montreal. Can. J. Bot.* 48, 1485–1496.

Martin, M. H., and Coughtrey, P. J. (1982), *Biological Monitoring of Heavy Metal Pollution.* Applied Science Publishers, London-New York.

Maschke, J. (1981), *Moose als Bioindikatoren von Schwermetall-Immissionen,* A. R. Gantner Verlag, Vaduz, Liechtenstein.

Mouvet, C., Pattee, E., and Cordebar, P. (1986), *Use of Aquatic Bryophytes for the Identification and precise Localization of Sources of Multiform Metallic Pollution. Acta Oecol. Oecol. Appl.* 7, 77–91.

Phillips, D. J. H. (1980), *Quantitative Aquatic Biological Indicators: Their Use to Monitor Trace Metals and Organochlorine Pollution.* Applied Science Publishers, London.

Posthumus, A. C. (1984), in: Treshow, M. (ed.): *Air Pollution and Plant Life,* pp. 73–95. J. Wiley & Sons, Chichester, UK.

Roop, R. D., and Hunsaker, C. T. (1985), *Biomonitoring for Toxics Control in NPDES Permitting. J. Water Pollut. Control Fed.* 57, 271–277.

Roush, T. H., et al. (1985), *Effects of Pollution on Freshwater Organisms. J. Water Pollut. Control Fed.* 57, 667.

Rühling, A., and Tyler, G. (1969), *Ecology of Heavy Metals – a Regional and Historical Study. Bot. Not.* 122, 248–259.

Rühling, A., and Tyler, G. (1970), *Sorption and Retention of Heavy Metals in the Woodland Moss Hylocomium splendens (Hedw.) Br. et Sch. Oikos* 21, 92–97.

Rühling, A., and Tyler, G. (1984), *Recent Changes in the Deposition of Heavy Metals in Northern Europe. Water Air Soil Pollut.* 22, 173–180.

Schulz-Baldes, M. (1982), *Decheniana Beih.* 26, 43–54.

Seeger, R., Nuetzel, R., and Dill, U. (1978), *Cadmium in Mushrooms. Z. Lebensm. Untersuch. Forsch.* 166, 23–24.

Steinnes, E. (1984), *Heavy Metal Pollution of Natural Surface Soils due to Long-distance Atmospheric Transport. Ecol. Stud.* 47, 115–122.

Steubing, L. (1976), *Niedere und höhere Pflanzen als Indikatoren für Immissionsbelastungen. Daten Dok. Umweltschutz* 19, 13–27.

Stijve, T., and Besson, R. (1976), *Mercury, Cadmium, Lead and Selenium Content of Mushroom Species Belonging to the Genus Agaricus. Chemosphere* 5, 151–158.

Stöcker, G. (1980), in: Schubert, R., and Schuh, J. (eds.): *Methodische und theoretische Grundlagen der Bioindikation (Bioindikation 1),* pp. 10–21. Martin-Luther-Universität, Halle (Saale), GDR.

Thomas, W. (1983), *Über die Verwendung von Pflanzen zur Analyse räumlicher Spurensubstanz-Immissionsmuster. Staub Reinhalt. Luft* 43, 141–148.

Thomas, W. (1986), *Representativity of Mosses as Biomonitor Organisms for the Accumulation of Environmental Chemicals in Plants and Soils. Ecotoxol. Environ. Safe.* 11, 339–346.

Thöni, L., and Hertz, J. (1987), *Moose als Biomonitoren für die flächenhafte Abschätzung der Schwermetallbelastung in der Schweiz. VDI Ber.* 609, 755–763.

Thöni, L., Schmid, I., Hertz, J., and Urmi, E. (1986), *Kartierung der Blei-, Kupfer-, Zink- und Cadmium-Belastung im Raume Zürich anhand des Mooses Bryum argenteum Hedw. als Biomonitor. Staub Reinhalt. Luft* 46, 295–299.

Tonneijk, A. E. G., and Posthumus, A. C. (1987), *Use of Indicator Plants for Biological Monitoring of Effects of Air Pollution: the Dutch Approach. VDI Ber.* 609, 205–216.

Tyler, G. (1972), *Heavy Metals Pollute Nature, May Reduce Productivity. Ambio* 1, 52–59.

Tyler, G. (1976), *Metal Concentrations in Moss, Leaves and other Indicators of Metal Exposure in the Environment*, in: Proceedings International Conference Environmental Sensing Assessment, pp. 2–4. IEEE, New York.

Tyler, G. (1982), *Accumulation and Exclusion of Metals in Collybia peronata and Amanita rubescens*. Trans. Br. Mycol. Soc. *79*, 239–245.

Weyers, B., Glück, E., and Stoeppler, M. (1985), *Environmental Monitoring of Heavy Metals with Birds as Pollution Integrating Biomonitors. III. Fate and Content of Trace Metals in Blackbirds Food, Organs and Feathers for a Highly Polluted and a Control Area*, in: Lekkas, D.T. (ed.): Heavy Metals in the Environment, pp. 718–720. CEP Consultants Ltd., Edinburgh.

Wolf, A., Schramel, P., Lill, G., and Hohn, H. (1984), *Bestimmung von Spurenelementen in Moos- und Bodenproben zur Untersuchung der Eignung als Indikatoren für Umweltbelastungen*. Fresenius Z. Anal. Chem. *317*, 512–519.

Zadory, L. (1984), *Freshwater Molluscs as Accumulation Indicators for Monitoring Heavy Metal Pollution*. Fresenius Z. Anal. Chem. *317*, 375–379.

Additional Recommended Literature

Angehrn-Bettinazzi, C., Thöni, L., and Hertz, J. (1989), *An Attempt to Evaluate Some Factors Affecting the Heavy Metal Accumulation in a Forest Stand*. IAEAC Workshop Follonica. Int. J. Environ. Anal. Chem. *35*(2), 69–79.

Boehringer, U.R., and Hertel, W. (1987), *Environmental Specimen Banking* (in German), Umwelt *1–2*, 21–23.

Brooks, R.R. (1989), *Phytoarcheology*. Endeavour New Ser. *13*(3), 129–134.

DeBruin, M. (1987), *Some Aspects of the Use of Epiphytic Lichens as Biological Monitors for Heavy Metal Air Pollution*. Proceedings 6th International Conference Heavy Metals in the Environment, New Orleans, Vol. 2, pp. 130–133. CEP Consultants Ltd., Edinburgh.

Doi, R., Ohno, H., and Harada, M. (1984), *Mercury in Feathers of Wild Birds from the Mercury-polluted Area along the Shore of the Shiranui Sea, Japan*. Sci. Total Environ. *40*, 155–167.

Ellenberg, H., Dietrich, J., Stoeppler, M., and Nürnberg, H.W. (1986), *Environmental Monitoring of Heavy Metals with Feathers of Molting Goshawks as Pollution Integrating Monitors*. Allg. Forst Z. *1/2* (in German); Birds of Prey Bull. *3*, 207–211.

Fowle III, J.R., Waters, M., and Erinoff, L. (1986), *Scientific Criteria for Biomarkers*, US Environmental Protection Agency Report, Cincinnati, Ohio.

Frahm, J.P., and Kürschner, H. (1989), *Mosses on Trees in the Rain Forest* (in German). Mitteilungen der Deutschen Forschungsgemeinschaft (DFG) *4/89*, 18–22. VCH, Weinheim-Basel-Cambridge-New York.

Glooschenko, W.A., and Arafat, N. (1988), *Atmospheric Deposition of Arsenic and Selenium across Canada using Sphagnum Moss as Biomonitor*. Sci. Total Environ. *73*, 269–275.

Goldstein, B.D., and Upton, A.C. (1984), *Human Health and the Environment: Some Research Needs*. 3rd Report (including chapters related to markers of chemical exposure and to biological monitoring), NIH Publication No. 86–1277. National Institute of Environmental Health Sciences, Research Triangle Park, North Carolina.

Herzig, R., Liebendörfer, L., Urech, M., Ammann, K., Guecheva, M., and Landolt, M. (1989), *Lichens as Biological Indicators of Air Pollution in Switzerland: Passive Biomonitoring as a Part of an Integrated Biological Measuring System for Monitoring Air Pollutants*. IAEAC Workshop, Follonica. Int. J. Environ. Anal. Chem. *35*(1), 43–57.

Hopkin, S.P. (1989), *Ecophysiology of Metals in Terrestrial Invertebrates*. Elsevier Applied Sciences, London-New York.

Levine, M. B., Taylor, D. H., Levine, D. A., and Barrett, G. W. (1987), *Effects of Sewage Sludge on the Density, Heavy Metal Accumulation, and Locomotor Activity of Lumbricid Earthworms.* Proceedings 6th International Conference Heavy Metals in the Environment, New Orleans, Vol. 1, pp. 390–392. CEP Consultants Ltd., Edinburgh.

Nash T. H., and Wirth, V. (eds.) (1988), *Lichens, Bryophytes, and Air Quality, Bibliotheca Lichenologica*, Vol. 30. J. Cramer, Berlin-Stuttgart.

Nürnberg, H. W., Stoeppler, M., and Dürbeck, H. W. (1986), *Environmental Specimen Banking, Euroanalysis V Rev. Anal. Chem. Budapest*, pp. 34–51.

Ramseier, S., Martin, M., Haerdi, W., Honsberger, P., Cuendet, G., and Tarradellas, J. (1989), *Bioaccumulation of Cadmium by Lumbricid Earthworms.* IAEAC Workshop, Follonica. Toxicol. Environ. Chem. 22(1–4), 189–196.

Rossel, D., and Liebendörfer, L. (1988), *Early Detection of Environmental Pollutants* (in German). SAGUF and Swiss Council of Sciences.

Simmers, J. W., Rhett, R. G., Crawley, D. K., and Marquenie, J. M. (1985), *Prediction of the Bioavailability of Heavy Metals to Animals Colonizing an Intertidal Wetland Created with Contaminated Dredged Material/Bioavailability of Heavy Metals in Relation to Potential Use of Dredged Material for Large-scale Acid Mine Soil Restauration.* Proceedings 5th International Conference Heavy Metals in the Environment, Athens, Vol. 2, pp. 208–210, 214–216. CEP Consultants Ltd., Edinburgh.

Simmers, J. W., Rhett, R. G., Brown, C. P., and Stafford, E. A. (1987), *Long-term Prediction of Contaminant Mobility from Dredged Material.* Proceedings Trace Substances in Environmental Health XXI, St. Louis, pp. 64–73 (ed. by D. D. Hemphill, University of Missouri).

Steinnes, E., Bølviken, and Hvatum, O. Ø (1987), Columbia, *Regional Differences and Temporal Trends in Heavy Metal Deposition from the Atmosphere Studied by Analysis of Ombrotrophic Peat/Heavy Metal Contamination of Natural Surface Soils in Norway from Long-range Atmospheric Transport: Further Evidence from Analysis of Different Soil Horizons.* Proceedings 7th International Conference Heavy Metals in the Environment, New Orleans, Vol. 1, pp. 201–203, 291–293. CEP Consultants Ltd., Edinburgh.

van Straalen, N. M., and Denneman, C. A. J. (1988), *Ecotoxicological Evaluation of Reference Values for Contaminants in a Clean Soil.* Proceedings 1st European SECOTOX Conference on Ecotoxicology, Copenhagen (ed. by F. Bro-Rasmussen), pp. 404–405. Technical University, Lyngby, Denmark.

Streit, B., Krüger, C., Lahner, G., Kirsch, S., Hauser, G., and Diehl, B. (1990), *Uptake and Storage of Heavy Metals by Earthworms in Various Soils* (in German). Z. Umweltchem. Oekotoxikol. 2(1), 10–13. ecomed, Landsberg-Zürich.

Thomas, R., and Wagener, D. (1987), *Biomarker Project.* National Academy of Sciences, Washington D.C.

van Gestel, C. A. M., van Dis, W. A., and van Breemen, P. M. (1988), *Development of a Standardised Reproduction Toxicity Test with Earth-worm Species.* Proceedings 1st European SECOTOX Conference on Ecotoxicology, Copenhagen (ed. by F. Bro-Rasmussen), pp. 175–177. Technical University, Lyngby, Denmark.

Wise, S. A., and Zeisler, R. (1985), *International Review of Environmental Specimen Banking,* NBS/SP-706. National Bureau of Standards, Gaithersburg, Maryland.

I.5 Metals and Metal Compounds in Waters

BISERKA RASPOR, Zagreb, Yugoslavia

1 Introduction
(see also Chapter I.3)

For many years only total metal concentrations and metal distribution in different environmental compartments, including the aquatic, have been determined. In the recent years, reliable measurements of ecotoxicologically significant metal compounds have been performed in natural waters (BRULAND, 1983; KESTER et al., 1986). This is especially true for the trace metals of prime environmental concern, such as Cd, Pb, Cu, Zn, and Hg in seawater, in view of their very low natural concentrations. The lower the actual dissolved metal concentration in natural water, the more critical for the ecosystem are even small additions of the same metal ions into the aquatic environment (NÜRNBERG, 1983).

According to MART et al. (1982) and BRULAND (1983) concentration levels of Cd, Pb, Cu, Ni, Co, and other metal ions differ significantly between surface and deep-seawater layers in oceans (Pacific, Atlantic, Arctic Sea, Mediterranean and North Sea) and coastal waters (affected by anthropogenic input). As already mentioned older data published may be several orders of magnitude too high (NÜRNBERG and MART, 1985). Elements with a high ratio deep/surface water levels (for instance, cadmium and zinc) are called nutrient-type elements, because their concentrations are correlated with those of nutrients such as phosphate (BRULAND, 1983; NÜRNBERG and MART, 1985). To avoid analytical errors sampling needs special care (see, e.g., MART et al., 1982, 1985).

MART et al. (1985) and MART and NÜRNBERG (1986a) studied cadmium, lead, copper, nickel, and cobalt distribution in the tidal River Elbe and in the German Bight and differentiated especially between dissolved metal ions and metal compounds in suspended particles. Mixing with the main water body and resuspension play an important role (see also Chapter I.7e). In the case of lead, total amounts are practically determined by that part bound to particles, whereas natural background levels of dissolved lead in river waters are only of the order of 10–50 ng/L, and even in highly polluted waste water plumes seldomly exceed 400 ng/L (MART et al., 1985; MART and NÜRNBERG, 1986a). Dissolved copper levels are the highest for all trace metals considered, the natural background levels in river waters are of the order of 50–200 ng/L, the mean levels in the tidal rivers Elbe or Weser may rise to more than 2000 ng/L (MART et al., 1985; MART and NÜRNBERG, 1986a). Since cadmium is mostly dissolved, differences between total and dissolved cadmium concentrations are smaller. In the nutrient-depleted central regions of the oceans cadmium concentrations are below 5 ng/L, 5−20 ng dissolved Cd per liter is the probable natural background level in river waters, 15−75 ng/L Cd are found in areas of upwelling in the Pacific and Atlantic Oceans and in the Weddell Sea, whereas in the tidal rivers Elbe and Weser dissolved levels may exceed 100 ng/L Cd (MART and NÜRNBERG, 1986a, b; see also Chapter II.6).

Information on the chemical forms of particular metals and not only the total concentrations is needed for understanding the kinetics of metal transformation and the attainment of the respective equilibria in natural waters. Therefore, investigation of natural waters should consist of the following steps (HORNE, 1969):

− identification and quantitative determination of existing chemical forms,
− elucidation of the chemical equilibria,
− examination of the kinetics and mechanism of chemical processes.

The fourth step should include the development of the equilibria and kinetic models, based on the existing chemical data. There is an iterative link between these four steps. The better the knowledge on the critical, important processes and their rates in the aquatic environment, the more realistic and relevant will be the predictions obtained by such models.

The occurrence and distribution of metal compounds in natural waters has to be described and can be understood only in terms of the physical chemistry of natural aquatic systems, including colloids, solids and biota at various pHs, assuming that the applied analytical methods are sound and based on the data quality control. Numerous chemical forms of metals exist in fresh water, estuarine, and seawater (BRULAND, 1983). In order to understand why certain types of metal compounds are usually found in certain water types, it is useful to consider the electron configurations, the Periodic Table of Elements, and the empirically defined affinities of metals and ligands in the formation of metal complexes, known as hard acid-hard base and soft acid-soft base principle (see also Chapter I.3).

1.1 Periodic Table of Elements

(see also Chapter I.3)

The properties of chemical elements are periodic functions of their atomic number (MASTERTON et al., 1986). As one moves across a period or down a group of the Periodic Table, the physical properties of elements change in a smooth, regular fashion. Within a given group, the elements show very similar chemical properties, because they have the same outer-electron configuration. Elements may thus be classified as follows:

− main-group elements in the Periodic Table are confined to the two groups at the far left and the six groups at the righthand side of the table, assigned as groups IA to VIIIA;
− transition elements are those in the center of the Periodic Table, in the gap between the IIA and IIIA main-group elements and are assigned as groups IB to VIIIB;

– lanthanides (atomic numbers 57 to 71) and actinides (atomic numbers 89 to 103) are listed separately at the bottom of the table (see also Section 5, Chapters II.15, and II.33).

According to the physical properties, elements are classified as metals, nonmetals, and metalloids.

1.2 Metals

Of 108 elements known until now, 84 belong to the group of metals, 17 to nonmetals, and 7 to the metalloids. The predominance of metals over other classes of elements is also reflected in nature. Of the ten most abundant elements in the earth's crust, seven are metals: aluminum, iron, calcium, sodium, potassium, magnesium, and titanium (Chapter I.1; GIDDINGS, 1973).

Metals have low ionization energy, i.e., they easily lose the outermost electron(s) and therefore have relatively free electrons to move about.

Due to the loss of valence electrons, positive ions are smaller than the metal atoms from which they are formed. The sodium atom has a radius of 0.186 nm while the sodium ion has a radius of 0.095 nm. The difference in radii between atom and cation is due to the excess of protons in the ion, which draws the outer electrons closer to the nucleus (MASTERTON et al., 1986).

Along the same period the ionization energy increases from left to right and in the same chemical group decreases down the group. This means that the ability of elements to form cations changes in the opposite manner. The successive alkali metals have a minimal ionization energy, which indicates that these metals form cations very easily.

All transition metals form cations of $+1$, $+2$ and $+3$ oxidation state by loss of successive s and d electrons which have energies of the same order of magnitude. These metals often have more than one oxidation state and hence more than one set of compounds, e.g., Cu^+/Cu^{2+}, Fe^{2+}/Fe^{3+}, Co^{2+}/Co^{3+}. In contrast to the transition metals, the metals of the main groups IA and IIA are present in only one oxidation state, $+1$ and $+2$, respectively (MASTERTON et al., 1986).

In general, transition metals have somewhat higher ionization energies than the main-group elements, and thus are generally less reactive since they oxidize less readily. Potassium and calcium vigorously react with water while of the transition metals in the first series only scandium reacts rapidly with water and manganese reacts slowly.

1.3 Nonmetals

Elements on the righthand side of the diagonal which consists of B, Si, As, Se, Te, and Po are classified as nonmetals. They have high ionization energies and therefore do not lose electrons in order to achieve the stable electron configuration of the noble gas, but form ions by accepting the electrons. This is the reason why nonmetals have no free electrons which could serve for conducting electricity and heat. Nonmetals are usually present in water as anions, such as O^{2-}, F^-, Cl^-. Due to the gain of electrons and the increased repulsion of the outer electrons, negative ions are larger than nonmetal atoms, from which they are formed. The radius of the chlorine atom is 0.099 nm while that of the chloride ion is 0.181 nm (MASTERTON et al., 1986).

1.4 Metalloids

On the righthand side of the Periodic Table, between metals and nonmetals, seven elements exist which are difficult to classify as metals or nonmetals according to their physical properties. They have properties in between those of elements in the two other classes. In particular their electronic configuration is intermediate between that of metals and nonmetals. These elements are boron, silicon, germanium, arsenic, antimony, tellurium, and selenium. They are often called metalloids (see Chapters II.2, II.3, II.11, II.25, and II.28).

1.5 Ionic Metal Compounds

Generally speaking, within the aqueous phase metal ions might undergo the following reactions: complexation, precipitation, and changes of the oxidation state. Metal ions in natural water systems can interact with the inorganic and organic types of ligands in the water phase and/or at the surface of the solid phase (see Chapter I.3). Ionic bonds result from electrostatic interaction of the oppositely charged ions (PYTKOWICZ, 1983). Ionic compounds result typically from the reaction of the elements of low ionization energy (usually the metals of the I A and II A groups) with the elements of high ionization energy (the nonmetals of the VI A and VII A groups).

In an aqueous electrolytic solution, ions of opposite charge are held together by electrostatic forces within the critical distance, forming ion-pairs. These forces decrease with $1/r^2$ where r is the interionic distance (PYTKOWICZ, 1983). Ion triplets may also occur as is the case for $[CaMg(CO_3)]^{2+}$ (PYTKOWICZ and HAWLEY, 1974). When the ion pair is formed, the metal ion or the ligand or both retain coordination water, so that cation and anion are separated by one or more water molecules (STUMM and MORGAN, 1981). Ion pairs are also called outer-sphere complexes.

Estimates of the stability constants of ion pairs can be made on the basis of a simple electrostatic model which considers coulombic interactions between the ions (STUMM and MORGAN, 1981). The areas of particular importance of application of the ionic model to aquatic chemistry are the hydration energies of cations and complex formation constants (WHITFIELD and TURNER, 1983).

1.6 Covalent Metal Compounds

In molecules atoms are held together by strong forces called covalent bonds (LEWIS, 1916). They are formed when a metal as an electron-acceptor (Lewis acid) reacts with

Table I.5-1. Electronegativity Values of the Elements (Listed in Groups of the Periodic Table) (PAULING, 1960)

IA	IIA	IB	IIB	IIIB	IVB	VB	VIB	VIIB	VIIIB			IIIA	IVA	VA	VIA	VIIA	VIIIA
H 2.1																	He –
Li 1.0	Be 1.5											B 2.0	C 2.5	N 3.0	O 3.5	F 4.0	Ne –
Na 0.9	Mg 1.2											Al 1.5	Si 1.8	P 2.1	S 2.5	Cl 3.0	Ar –
K 0.8	Ca 1.0	Sc 1.3	Ti 1.5	V 1.6	Cr 1.6	Mn 1.5	Fe 1.8	Co 1.8	Ni 1.8	Cu 1.8	Zn 1.6	Ga 1.6	Ge 1.8	As 2.0	Se 2.4	Br 2.8	Kr –
Rb 0.8	Sr 1.0	Y 1.2	Zr 1.4	Nb 1.6	Mo 1.8	Tc 1.9	Ru 2.2	Rh 2.2	Pd 2.2	Ag 1.9	Cd 1.7	In 1.7	Sn 1.8	Sb 1.9	Te 2.1	I 2.5	Xe –
Cs 0.7	Ba 0.9	57–71 1.1–1.2	Hf 1.3	Ta 1.5	W 1.7	Re 1.9	Os 2.2	Ir 2.2	Pt 2.2	Au 2.4	Hg 1.9	Tl 1.8	Pb 1.8	Bi 1.9	Po 2.0	At 2.2	Rn –

an electron-pair donor (Lewis base). Metals that form coordinate bonds most readily are small, and highly charged with empty orbitals, e.g., the transition metals (PYTKOWICZ, 1983).

Atoms of two different elements always differ at least slightly in their affinity for the electrons. Hence, covalent bonds between unlike atoms are always unsymmetrical resp. polar (MASTERTON et al., 1986). The greater the electronegativity of an atom, the greater the affinity for bonding electrons. PAULING (1960) used bond energies to calculate relative electronegativity values for the various elements, arbitrarily defining the most electronegative element, i.e., fluorine with the value 4.0. The assigned electronegativity values for the Periodic Table of Elements are presented in Table I.5-1. It would be helpful to know the electronegativity value for each oxidation state of an element and for each individual valence orbital (WHITFIELD and TURNER, 1983).

The greater the difference of electronegativities between two elements, the more ionic is the bond between them. A difference of 1.7 units corresponds to a bond with 50% ionic character. Electronegativity differences less than 1.7 units imply that the bonding is mainly covalent (MASTERTON et al., 1986). Bonds are stronger and the bond energy is higher for a multiple rather than for a single bond between the same two atoms.

The term "complex ion" indicates a charged species in which a metal ion is joined by coordinate covalent bond(s) to neutral molecules and/or anions (MASTERTON et al., 1986; see also Sect. 3).

Since the predominant ligands in the aquatic environment are the water molecules, the consideration of the occurrence and distribution of metal complexes should begin with the hydration of ions.

2 Structure of Liquid Water and Hydration of Ions

In a fundamental sense, metal ions dissolved in water are already complexed – they have formed aqua-ions. Thus it is logical to begin the discussion of the formation and stability of complex ions in aqueous solution with the structure of liquid water and aqua-ions themselves (COTTON and WILKINSON, 1980).

The structure of liquid water is the subject of intense study and controversy. The polar nature of the water molecule and its ability to form strong intermolecular hydrogen bonds result in the cooperative association of multimolecular aggregates (HORNE, 1969).

Hydrogen bonding is the specific association of the hydrogen atom of one molecule with the lone pair electrons of another. Hydrogen bonding is responsible for many of the extraordinary physical properties of water. Each water molecule has approximately 4.4 neighbors in the first coordination shell. Thus, liquid water is a highly structured liquid in which the tetrahedral coordination observed in ice is still evident (WESTALL and STUMM, 1980).

An understanding of ionic hydration is a prerequisite for understanding chemistry of water. The hydration atmosphere of an ion in solution has a complex internal structure and its outer boundary is difficult to establish. An ion in solution is represented as being surrounded by two zones. An inner layer can be equated to what is often called the "primary" hydration shell, which is composed of dense, electrostricted and immobilized water molecules strongly bound by the coulombic field on the ion. Furthermore, there is a region of comparative randomness, of disrupted water organization, of broken structure. At some further distance from the ion, the water structure is "normal", although the molecules may be slightly polarized by the ubiquitous charge field. FRANK and EVANS (1945) suggested that the structure-enhanced zone is present and intact in all ions, while the particular characteristics of different types of ions arise from the variability of the structure – the broken zone (HORNE, 1969).

For a metal ion in an aqua complex we wish to know the coordination number and the manner in which the water molecules are arranged around the metal ion, or according to TAUBE (1954) "formulas" of the ion-water complexes. Some experimental methods measure only the most tightly bound water molecules, whereas other methods measure the loosely bound water molecules as well. Therefore, various methods yield different hydration numbers (HORNE, 1969).

Generally speaking, cations are more hydrated than the anions of the same negative charge, and the greater the charge of the ion, the more heavily hydrated is the ion. The protons in water are hydrated as well, and although they are usually written as H^+ or H_3O^+ (hydronium ion) the best available evidence points strongly to the existence of $H_9O_4^+$, which is the prevailing form (HORNE, 1969).

The primary hydration number of Mg^{2+} is higher than that of Li^+, even though these differently charged cations have nearly the same crystal radius. In a given charge type, the smaller the crystal radius of the ion, the heavier is the hydration.

Ultrasonic velocity measurements are convenient for measuring hydration numbers from ion compressibilities (PADOVA, 1964). For the di- and trivalent cations of the first transition series, the aqua ions are octahedral $[M(H_2O)_6]^{2+}$ or $^{3+}$, although in Cr(II), Mn(II) and Cu(II) definite distortions of the octahedra are present (COTTON and WILKINSON, 1980).

The crystals of the first series of the transition metals are colored, e.g., Ti(III), V(II), V(III), Mn(II), Fe(II), Fe(III), Co(II), Ni(II). The hexaaquo-salts of these metals dissolve in water without changing the color. The absorption spectra of crystals and solutions for these transition metals are perfectly in agreement, so that there is no doubt about the octahedral coordination of these transition metals in water (SCHNEIDER, 1968).

The hydration of ions can also be conceptualized in terms of the residence time of water molecules near an ion. If an average water molecule is in a position near an ion for a longer time than it would be at some greater distance from the ion, then the ion is positively hydrated. However, if the water molecule is more mobile near the ion than it would be at some distance from the ion, the term "negative hydration" is used. The residence time concept of hydration is complementary to the concept of the hydration number (HORNE, 1969).

There are vast differences in the average length of time that a water molecule spends in the coordination sphere. For Cr(III) and Rh(III) the residence time is so long that when a solution of $Cr(H_2O)_6^{3+}$ in ordinary water is mixed with water enriched in ^{18}O, many hours are required for complete equilibration of the enriched solvent water with the coordinated water (COTTON and WILKINSON, 1980). TAUBE (1954) measured the kinetics of water exchange in the solution of Cr(III). The half time of this reaction is 2×10^6 seconds (MORGAN and STONE, 1985). For Rh(III) the reaction of water exchange is even slower. It can be concluded that Cr(III) and Rh(III) show a clear inert behavior with respect to the exchange of water molecules in their hydration sphere. For most other aqua-ions the exchange of water molecule(s) occurs too rapidly to permit the same type of measurement (SCHNEIDER, 1968).

3 Complex Formation

When covalent or inner-sphere types of complexes are formed, kinetically speaking, a dehydration step must precede the association reaction. Covalent association is accompanied by changes in the absorbance of visible light and is indicative of complex formation reactions, whereas the formation of ion pairs may be accompanied by changes in the ultraviolet region (STUMM and MORGAN, 1981).

The metal cation in a complex is called the central ion, the molecules or anions bound directly to it are called ligands, and the number of bonds formed by the central ion is its coordination number.

Based on the exchange rate of hydration water and comparing the coordination number of covalent complexes with other ligands than water it is obvious that the complex formation represents ligand exchange of an equivalent number of water molecules from the hydration shell. For example, each molecule of ammonia or pyridine as monodentate ligands displaces one molecule of water, while a molecule of bidentate ligands such as diamine or dithizone displaces two water molecules, etc. (IRVING and WILLIAMS, 1948). In many instances the exchange of the first coordinated water molecule controls the overall rate of complexation (see Chapter I.3).

3.1 Labile Complexes
(see also Section 6.3)

Complex ions that exchange ligands almost instantaneously are referred to as labile. Typically, they exchange ligands in water solution with a half time of a minute or less. The $[Cu(H_2O)_6]^{2+}$ and $[Al(H_2O)_6]^{3+}$ are aqua-complexes of this type. The half time for the former complex is 1×10^{-10} seconds, which corresponds to diffusion controlled reactions and for the latter it amounts to 6 seconds. These ions represent the labile type of complexes (MORGAN and STONE, 1985). In general, the half time of the water exchange reactions from the primary shell of the ion covers the range of about fifteen orders of magnitude (SCHNEIDER, 1968).

3.2 Inert Complexes
(see also Section 7)

In contrast to the labile complexes, in non-labile or inert complexes the hydration water is slowly exchanged with the added ligand. An example for the slow rate of water exchange is the aqua-complex $[Cr(H_2O)_6]^{3+}$. An additional illustration of labile and inert type of complexes, which reflects different bonding strength and therefore different electron configuration is given by the complexes of the same metal atom but at two different oxidation states, e.g., Fe(II) and Fe(III). The complex $[Fe(CN)_6]^{4-}$ is inert while the complex $[Fe(CN)_6]^{3-}$ is labile (MASTERTON et al., 1986).

For each reaction of complex formation in an aquatic system, the thermodynamic stability has to be defined, besides the kinetic stability of the particular chemical form, which refers to the rate of transformation leading to the attainment of equilibrium. Theoretically, the change of the free energy of complex formation indicates whether the observed reaction is thermodynamically possible to occur. As an additional practical parameter, a kinetic factor defines the height of the energy barrier E_a, which chemical reactants have to overcome. It determines the rate of the reaction and the measurable amount of the product of a chemical reaction.

Most simple ionic equilibria in aqueous solutions tend to be very rapid, their rates often being controlled by diffusion. Further it is probably correct to assume that most equilibria in the dissolved phase are reached rapidly (HORNE, 1969). Rates of precipitation and even more of dissolution are usually slower (WESTALL and STUMM, 1980).

4 Chemical Models and Equilibrium Constants
(see also Chapter I.3)

Information about the stability of metal compounds and the rate of their formation in the range of minutes to years is important for the understanding of environmental processes. The reactions that reach equilibrium in a shorter time can be described and modelled in terms of equilibrium conditions (KESTER et al., 1986). For example, equilibrium models of seawater proposed by SILLEN (1967) are based on the assumption of the rapid attainment of equilibria in seawater. This model represents the approach of a physical chemist to the elucidation of oceanographic processes, which lead to the constant composition of seawater with respect to the macrocomponents, or to the nearly constant pH of seawater. Equilibrium models of chemical forms have been extensively developed in marine and estuarine systems (see also Chapter I.7e). They have provided quantitative predictions for marine chemistry of inorganic species. However, more reliable data bases for chemical constants under marine, estuarine, and freshwater conditions are needed (KESTER et al., 1986).

Three types of equilibrium constants are in common use (STUMM and MORGAN, 1981):

1. Thermodynamic constants based on activities, the activity scale being based on the infinite dilution reference state.
2. Apparent or stoichiometric equilibrium constants expressed as concentration quotients and valid for a medium of given ionic strength.
3. Conditional constants that hold only under specified experimental conditions (e.g., at a given pH).

In dilute solutions ($I < 10^{-2}$ mol L^{-1}), that is in fresh waters, calculations are usually based on the infinite dilution activity convention and thermodynamic constants. In these dilute electrolyte mixtures, deviations from ideal behavior are primarily caused by long-range electrostatic interactions. The Debye-Hückel equation or one of its extended forms is assumed to give an adequate description of these interactions and to define the properties of the ions. Correspondingly, individual ion activities are estimated by means of individual ion activity coefficients calculated with the help of the Güntelberg or Davies equation (see Table 3.3 in STUMM and MORGAN, 1981).

In more concentrated solutions ($I > 0.5$ mol L^{-1}), calculations are usually based on the ionic medium scale, i.e., for the evaluation of the concentration of coexistent species at equilibrium, the apparent equilibrium constants (valid for the medium under consideration) are needed. By using "medium-bound" constants, we have bypassed activity coefficients in stoichiometric calculations. In this case, equilibrium constants valid for seawater or the medium of interest have to be determined. If such constants are not known, thermodynamic constants and estimated activity coefficients must be used (STUMM and MORGAN, 1981). Alternatively, apparent equilibrium constants valid for the medium of particular (or closely similar) interest or constants that have been corrected for the medium under consideration can be used in conjunction with concentrations (STUMM and MORGAN, 1981).

For electrolytic solutions of high ionic strength, such as seawater ($I \approx 0.7$ mol L^{-1}), for which most activity coefficient equations are not valid, or in the case when the presence of several electrolytes affects the activity coefficients of the others, it is best to use apparent constants measured directly in the medium of interest rather than to use thermodynamic constants. The apparent equilibrium constants determined in one ionic medium can be used in another if the *effective* ionic strength of the two solutions is the same. In such cases, the apparent equilibrium constants can be considered to be thermodynamic ones in the ionic medium under study, while the activity coefficients of free metal and ligand as well as that of the complex are essentially invariant (PYTKOWICZ, 1983).

In a solution of hydrated metal ions M and unidentate ligands L, if only soluble mononuclear complexes are formed, the system at equilibrium is described with the following equation and the apparent equilibrium constant (COTTON and WILKINSON, 1980):

$$ML_{N-1} + L = ML_N \qquad K_N = \frac{[ML_N]}{[ML_{N-1}][L]} ,$$

where N represents the maximum coordination number of the metal ion M for the ligand L. K_N is the stepwise apparent formation or stability constant. With only few exceptions, there is generally a slowly descending progression in the values of K_N in any particular system. Thus typically, as a ligand is added to the solution of a metal ion, ML is first formed more rapidly and has a higher stability than any other complex of the series. There are several reasons for a steady decrease in K_N values as the number of ligands increases: (1) statistical factors, (2) increased steric hindrance as the number of ligands increases if they are bulkier than the H_2O molecules they replace, (3) coulombic factors, mainly in complexes with charged ligands (COTTON and WILKINSON, 1980).

Another way of expressing the equilibrium relations of hydrated metal ions M and unidentate ligands L is the following:

$$M + NL = ML_N \qquad \beta_N = \frac{[ML_N]}{[M][L]^N} .$$

The term β_N is called the overall apparent formation or stability constant. The interrelationship between the overall and stepwise apparent stability constants is the following:

$$\beta_N = K_1 \cdot K_2 \cdot K_3 \cdot K_4 \ldots K_N .$$

In aqueous solutions besides the mononuclear, the polynuclear complexes might exist as well. In a polynuclear complex two or more metal atoms $M_M L_N$ are bound with the ligands to a complex. The fraction of polynuclear complexes in a solution decreases on dilution (STUMM and MORGAN, 1981).

Metal complexes which consist of two different ligands are called mixed ligand complexes. Statistically speaking, if a metal is able to form complexes MCl_2 and $M(OH)_2$ with stability constants $\beta_2(Cl)$ and $\beta_2(OH)$, then a mixed ligand complex of the form MOHCl should also be formed with an effective stability constant defined by $\beta(OH, Cl) = 2 [\beta_2(Cl) \beta_2(OH)]^{1/2}$. Although such complexes have been considered in seawater models (DYRSSEN and WEDBORG, 1975; WHITFIELD and TURNER, 1980), where they usually represent less than 12% of the total dissolved metal, little direct experimental evidence has been presented for their existence in seawater (WHITFIELD and TURNER, 1986).

5 Hard and Soft Acceptors and Donors

On the basis of experimental evidence, AHRLAND et al. (1958) and PEARSON (1963) classified the electron acceptors (metals) and electron donors (ligands) into "hard" and "soft" categories, according to the stability of the complexes of particular type of metal with the particular type of ligand (see Chapter I.3). The stability of metal complexes formed by any ligand with a series of metals may be expected to increase with electronegativity of the metal concerned (IRVING and WILLIAMS, 1948).

The separation of metals into distinct groups was based on empirical thermodynamic data, namely, trends in the magnitude of equilibrium constants that describe the formation of metal-ion/ligand complexes. On the basis of these criteria, metal ions can be divided into three groups, hard, soft, and borderline. The partition of a particular ion in each group is shown in Fig. I.5-1 (NIEBOER and RICHARDSON, 1980; see also Chapter I.3, Table I.3-2).

Hard cations include ions of the alkali metals, alkaline earth metals, lanthanides, actinides, and aluminum. They bind mainly via electrostatic interactions and form strongest complexes with electron donors from VIA and VIIA main groups in the Periodic Table. Their ligand preference has the sequence

$N \gg P > As < Sb : O \gg S > Se > Te : F^- \gg Cl^- > Br^- > I^-$ with $F^- > O > N$.

Hard or class A cations preferentially bind to hard bases, i.e., to oxygen sites of inorganic anions (O_2^{2-}, OH^-, $H_2PO_4^-$, CO_3^{2-}, SO_4^{2-}) and organic molecules provided with oxygen containing groups (carboxylate, carbonyl, alcohol, ester). Considering organic types of ligands, hard cations will also bind to nitrogen sites, although less strongly.

Soft or class B cations include transition metals from the triangle with Cu^+ at its apex (note Cu^{2+} is borderline) and Ir^{3+} and Bi^{3+} at its base. They bind mainly via covalent interactions, forming their strongest complexes with electron donors in the following sequence

$N \ll P > As > Sb : O \ll S \simeq Se \simeq Te : F^- \ll Cl^- < Br^- < I^-$ with $S > N > O > F^-$.

Soft cations preferentially bind to soft bases, i.e., with inorganic anions I^- and CN^-, while in organic molecules they preferentially bind to sulfur (sulfhydryl, disulfide, thioether) and nitrogen sites (amino, imidazole, histidine, nucleotide base).

The borderline cations comprise the first row of transition metals, in their common oxidation states, as well as Ga^{2+}, In^{3+}, Cd^{2+}, Sn^{2+} and Pb^{2+}. The hydrogen ion and the metalloid ions As(III) and Sb(III) are also included in this category

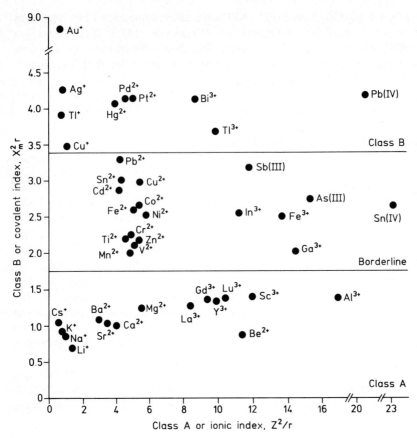

Fig. I.5-1. A separation of metal ions and metalloid ions, As(III) and Sb(III), into three categories: class A, borderline, and class B ions. The class B index $X_m^2 r$ is plotted for each ion against the class A index Z^2/r. In these expressions X_m is the metal-ion electronegativity, r its ionic radius, and Z its formal charge. Oxidation states given by Roman numerals imply that simple, hydrated cations do not exist even in acidic solutions. (From NIEBOER and RICHARDSON, 1980, reproduced with permission).

(NIEBOER and RICHARDSON, 1980). They are able to form stable complexes with numerous ligands, and preference for a given donor group will be determined by factors including the degree of hard-character, in part by the soft-character, in part by the relative availability of ligand(s) in the system, and in part by the steric environment of the reaction site(s) themselves (WHITFIELD and TURNER, 1983). Thus, according to Fig. I.5-1, among borderline metal ions, class B, i.e., soft character, increases in the order of $Mn^{2+} < Zn^{2+} < Ni^{2+} < Fe^{2+} \simeq Co^{2+} < Cd^{2+} < Cu^{2+} < Pb^{2+}$ (NIEBOER and RICHARDSON, 1980).

It should be noted to Fig. I.5-1 that Cd^{2+} belongs to the borderline metals rather than to the class B group where it has been placed by PEARSON (1963, 1969). In addition, since it is not possible to determine an ionic radius for H^+, its position is not indicated in Fig. I.5-1. However, the chemistry and chemical reactivity calculations clearly show that H^+ should be regarded as a borderline ion (KLOPMAN, 1968; EVANS and HUHEEY, 1970), although this is in contrast to the traditional view that a proton is a pure class A cation (PEARSON, 1963, 1969). It should be stressed that pH is an important factor regulating the acid-base equilibrium of inorganic and organic types of ligands, since the proton directly competes with metal ions and frequently metal displaces protons from the binding site or *vice versa*, depending on the pH of the aqueous solution.

An examination of trends in the magnitude of metal-ligand equilibrium constants determined in aqueous solution reveals some interesting features. Soft or class B cations, in spite of their own preference for soft bases, when reacting with hard bases form complexes that are more stable than those with hard cations of comparable Z^2/r values (see Fig. I.5-1). The same observation holds for borderline cations relative to hard cations. Presumably, this feature signifies that in addition to the largely ionic interactions observed for class A ion, borderline and class B ions of comparable size and charge make significant covalent contributions to the overall interaction energy (NIEBOER and RICHARDSON, 1980).

Another observation of interest is that ions with values of Z^2/r greater than 8 (Fig. I.5-1), with few exceptions, tend to hydrolyze, forming metal hydroxides and oxo anions in mildly acidic and some even in quite acidic solutions (BAES and MESMER, 1976; HUHEEY, 1978).

Ions with intermediate values of $X_m^2 r$ and concurrent large Z^2/r values, form water-soluble organometallic cations which involve metal-carbon bonds; for example, $(CH_3)_2Pb^{2+}$, $(CH_3)_2Tl^+$, $(CH_3)_3Sn^+$, CH_3Hg^+, and the volatile $(CH_3)_3As$, corresponding to the complexes of Pb^{4+}, Tl^{3+}, Sn^{4+}, Hg^{2+} and As^{3+}, respectively. Water resistent alkyl derivatives of Au^{3+} and Pt^{4+} are also known, namely, $(CH_3)_2Au^+$ and $(CH_3)_3Pt^+$ (THAYER, 1974).

6 Experimental Approaches to Study the Behavior of Metal Ions and Compounds in Water
(Especially Potentiometric and Voltammetric Techniques)

According to the modified approach of BURTON (1979), applying experimental techniques, it is possible to differentiate within the total metal concentration:

- *chemical species* or metal compounds, which are described in terms of a well defined chemical stoichiometry (e.g., ions, molecules, complexes);
- *chemical fractions*, which represent the group of chemical forms operationally defined with respect to the applied physical separation techniques. The total metal concentration is distributed between the dissolved, particulate, and colloidal fractions. Measurements of the chemical forms are usually more accessible than detailed speciation studies (WHITFIELD and TURNER, 1986).

    The most direct evidence concerning the occurrence of metal compounds is provided by chemical sensing methods (see also Chapter I.4a), which may be divided into surface and bulk techniques. Amperometric, potentiometric, and bioassay techniques are surface sensing chemical techniques, while bulk sensing is performed by spectroscopic techniques. Surface sensing chemical techniques (potentiometric and amperometric) are non-destructive (WHITFIELD and TURNER, 1986). While choosing the sensing technique and the respective working conditions, the kinetic characteristics of the studied metal compound should be known or at least judged from the rate of water exchange reaction, since it determines whether the existing distribution of metal compounds in natural water will be retained until the analyses or not. If the distribution reaction of metal complexes is rapid, special attention must be paid to minimize the redistribution of species during sampling and analysis. If the redistribution reaction is slow, time-consuming physical techniques, such as photooxidation, partial leaching, and derivatization can be used to separate or preconcentrate various metal compounds prior to analysis (WHITFIELD and TURNER, 1986).

6.1 Potentiometric Technique
(see also Chapter I.4a)

In potentiometric analysis, the ion-specific electrode (ISE) is a sensor of the activity of "free", hydrated metal ions. Its applicable concentration range is limited to a concentration range where a linear relationship between the activity of metal ions and the electrode potential, according to the Nernstian equation, exists. In reality, the lowest metal concentration which could be determined by ISE is at best 1×10^{-6} mol of metal per liter of water solution. The deviation from linearity is caused by increasing interferences of other metal ions of the same charge, as the concentration of the investigated metal decreases. No electrode is completely specific to only one metal ion. The interferences will be especially significant in seawater, which is a multicomponent mixture of metals of various concentrations. The sensitivity of ion-specific electrodes is far from being sufficient for the analysis and study of trace metal speciation in natural waters, while the actual concentration of trace metals of ecotoxicological significance is several orders of magnitude lower. It has already been pointed out that the existing and prevailing metal compounds in natural waters depend on the level of total metal concentration, while the competitive reactions of other water components significantly increase with the decrease of studied metal concentration. Therefore, ion-specific electrodes are not sensitive enough for obtaining relevant experimental data for understanding the processes and mechanisms of dissolved trace metal complexation in natural waters. Besides that, ion-specific electrodes measure only the activity of hydrated metal ions, e.g., $[Cu(H_2O)_6]^{2+}$, so that even simple, labile complexes such as $CuCl^+$, $Cu(OH)_2$, and $CuCO_3$ are not sensed by the electrode and, therefore, these labile types of complexes are included in the complexed part of total copper concentration. Such a situation is misleading while in the complexed part of the total metal concentration at least two kinetically different classes of metal complexes exist and in fact, just oppositely, these two classes, labile and inert, should be experimentally differentiated. This is the reason why the values of the complexing capacity for the same water type determined by Cu(II) titration, using an ion-specific electrode for the end-point detection, are much higher than those from the anodic stripping voltammetric (ASV) measurements (FLORENCE and BATLEY, 1980; HART, 1981).

6.2 Voltammetric Technique
(see also Chapter I.4a)

The only technique which permits quantitative and selective determination of natural trace concentrations of Cd, Cu, Co, Ni, Hg, Pb, Tl, Zn, Bi, In, As(III) and As(VI), Cr(III) and Cr(VI) in different water types is the voltammetric technique. It has the advantage that the measurements can be performed directly in the water sample without any pretreatment. Inherent high sensitivity of this technique is based on Faraday's law, which defines an enormous electric charge of 96500 coulombs for one mole equivalent of reduced substance, during the deposition step (NÜRNBERG, 1977). With respect to sensitivity, low accuracy risks, and low cost requirements, voltammetry is of all present techniques the most powerful analytical method for the direct and comparative determination of the total metal concentration and, respectively, at the natural pH of water, for the trace metal speciation (BRANICA et al., 1969; VALENTA, 1983).

In contrast to exclusively element-sensitive methods, such as atomic absorption spectroscopy (AAS), atomic emission spectroscopy (AES), X-ray fluorescence (RFA), or neutron activation analysis (NAA), voltammetry is a *species-sensitive* method (NÜRNBERG, 1977). Therefore, voltammetric methods are successfully applied to the determination of the existing metal compounds, of their redox state, of the apparent stability constants, and the reaction kinetics (homogeneous and heterogeneous) in the given water type (BRANICA et al., 1969).

Of particular importance for the information obtained on the lability of the studied metal complex is the time scale of the measuring technique, because it defines the kinetic window, i.e., the lability of the analyzed metal complex. For both polarography and voltammetry the lability of the metal complex depends not only on the dissociation rate constant, but also on the effective measuring time, which for constant electrochemical parameters (deposition potential, duration of the excitation voltage step) depends on the residence time of complex molecules in the diffusion layer, which can be deliberately changed by varying the stirring rate or the electrode rotation rate (FLORENCE, 1986).

A method of very efficient electrolyte mixing, with a thin reaction layer at the electrode surface, was reported by MAGJER and BRANICA (1977). With the optimized geometry of the vibrating disc and the conical hole, higher sensitivity than for stirring or rotation is obtained. This method was applied in a study on copper complexing capacity of seawater (PLAVŠIĆ et al., 1982) permitting the differentiation between labile and inert types of copper complexes.

6.3 Labile Complexes
(see also Section 3.1)

As mentioned above, in an aquatic system hard acids, i.e., metal ions, are present as ionic pairs. Their existence as (Cat)CO_3, (Cat)HCO_3, (Cat)SO_4 and (Cat)Cl, where (Cat) represents the cations Na^+, K^+, Ca^{2+}, Mg^{2+} in aqueous media of high salinity, has been verified by Raman spectroscopy, sound attenuation and potentiometry (see references in SIPOS et al., 1980a). At low, natural concentrations, the soft and boundary cations form labile, mononuclear complexes with the inorganic ligands (OH^-, CO_3^{2-}, HCO_3^-, SO_4^{2-}, Cl^-) in natural waters. The existence and the stability constants of labile metal complexes can be determined with the pseudo-polarographic method. From a series of curves recorded in solutions of variable ligand concentration, the shift of the half-wave or the peak potential is related to the free ligand concentration, and the stability constants of the consecutive metal complexes, as well as the number of ligands bound to the central metal ion in the complex are obtained. Applying the pseudo-polarographic approach, labile metal complexes can be reliably characterized in model solutions providing for the observed type of reaction the most important characteristics of a real water system. Comparing the shifts of the peak potential in the model solution with that in a natural water sample of known composition with respect to the macrocomponents, agreement should be achieved if the model solution was correctly postulated (BUBIĆ and BRANICA, 1973; SIPOS et al., 1980b).

In seawater, the existence of chloro complexes of cadmium has been studied, applying the described procedure of pseudo-polarography (BARIĆ and BRANICA, 1967; BUBIĆ and BRANICA, 1973; BRANICA et al., 1977). The results indicate that the predominant chloro species in the model solution of seawater, at a chloride concentration higher than 0.26 mol L^{-1} is the $CdCl_2^0$ complex. In seawater, 97% of the total cadmium concentration is in the form of chloro complexes. The mono-, di- and tri-chloro complexes occur in comparable quantities, although there is a slight preference for the neutral species, $CdCl_2^0$. The carbonato and bicarbonato complexes of cadmium are present only in trace quantities (see Table V in SIPOS et al., 1980a).

Applying the pseudo-polarographic method, the speciation of lead by carbonate anions in seawater has been elucidated (SIPOS et al., 1980b). The results indicate that in seawater the most abundant species are $Pb(OH)^+$, $Pb(CO_3)$, followed by $PbCl_2$ and $PbCl^+$. In seawater, despite the high chloride-ion concentration, the contribution of chloro complexes, in comparison with cadmium, is relatively small, while the formation of bicarbonato complexes of lead was not detected (SIPOS et al., 1980b). The theoretical shift in the case of speciation of lead with inorganic ligands in seawater was in complete agreement with the experimentally determined values (SIPOS et al., 1980b). This indicates that the model represents the actual situation with respect to the extent of formed labile metal complexes in seawater.

From the seawater model according to PYTKOWICZ (1983) it can be concluded that alkali ions are predominantly present as hydrated ions, while 50% of the total concentration of alkaline earth cations (Ca^{2+} and Mg^{2+}) is bound with SO_4^{2-} and CO_3^{2-} as ion pairs. Therefore, for lead, which is predominantly present as carbonato and hydroxo complex in seawater, its distribution of chemical species will be affected while the available ligand concentration is reduced for the amount bound to the ion pairs.

BILINSKI et al. (1976) concluded that lead(II) and copper(II) form $MeCO_3^0$ and $[Me(CO_3)_2]^{2-}$ complexes of similar stability, which suggests that if only the formation of complexes with the inorganic type of ligands is considered, metal-carbonato complexes will be the predominant lead and copper species in natural waters.

Cadmium, which in seawater predominantly exists as chloro complex, will not be significantly affected by the correction for the amount of chloride ions bound to ion pairs, since in seawater 83% of the total chloride concentration is in the hydrated form (SIPOS et al., 1980a). Cadmium(II) and zinc(II) form the complex $MeCO_3^0$ of much lower stability than that of lead and copper and, therefore, the carbonato complexes of zinc and cadmium will not be significant chemical forms of these metals in natural waters.

The method of pseudo-polarography can be applied to the irreversible electrode reaction as well (ĆOSOVIĆ et al., 1982). In that case, the method is applicable *only* if the coefficient a of electron transfer does not change in the investigated range of ligand concentration. In contrast to the tedious measurements which had previously been performed by manual changes of the deposition potential and the treatment of the experimental data after the pseudo-polarographic analysis, more recently, a computerized system for pseudo-polarographic measurement has been developed (BRANICA et al., 1986).

Although labile metal complexes are formed mainly with major inorganic anions of water, some organic types of ligands may also form labile complexes. Examples are the complexes of Cd and Zn with the anionic form of the amino acids glycine (SIMÕES GONÇALVES and VALENTA, 1982; SIMÕES GONÇALVES et al., 1983) and L-aspartic acid (VALENTA et al., 1984), the amino acids which are regarded as possible components of the dissolved organic matter (DOM). The conclusion follows that the ligand level required for any significant complex formation with Cd(II) and Zn(II) (higher than 5%) is well above the concentrations of these ligands expected in seawater. Besides amino acids in seawater there are possibly other DOM components which can form inert types of complexes or chelates, with specific functional groups.

7 Inert Complexes
(see also Section 3.2)

In seawater and lake water the interaction of the trace metals Cd, Pb and Zn with the well defined ligands NTA (nitrilotriacetic acid) and EDTA (ethylenediamine tetraacetic acid) of synthetic origin has been studied as a model for better understanding of the formation mechanism of inert types of complexes (chelates) with organic ligands of natural origin (RASPOR, 1980).

7.1 Kinetic Approach

In seawater, slow exchange of the EDTA ligand between calcium and cadmium occurs, a reaction being promoted by a several orders of magnitude higher stability constant of Cd(II)-EDTA in comparison with Ca(II)-EDTA chelate (MALJKOVIĆ and BRANICA, 1971). The authors observed the reduction of the rate of Cd(II)-EDTA chelate formation in seawater, and have concluded that calcium and chloride ions are the constitutents of seawater which exert the predominant effect on the degree of Cd(II)-EDTA chelate formation. This reaction mechanism has a general significance and can be applied to similar chelate formation reactions occurring between trace metals and various organic ligands in natural waters. The scheme reflects the fact that both alkaline earth cations, Ca^{2+} and Mg^{2+}, influence the rate of Cd(II)-EDTA chelate formation (RASPOR et al., 1977b). The rate determining step of EDTA exchange occurs between Ca(II)-EDTA chelate and Cd^{2+}, due to a higher stability constant of Cd(II)-EDTA. The same exchange reaction mechanism was determined by measuring the operation rate constants of EDTA chelate formation with lead and zinc in seawater (RASPOR et al., 1980a). The higher the stability constant of the chelate as in the case of Pb(II)-EDTA, the higher is the formation rate constant $(2 \times 10^3 \text{ mol}^{-1} \text{ L}^{-1} \text{ s}^{-1})$, while for Cd(II)- and Zn(II)-EDTA, which have similar stability constants, the formation rate constants in seawater medium are also similar $(2.5 \times 10^2 \text{ mol}^{-1} \text{ L}^{-1} \text{ s}^{-1})$.

7.2 The Equilibrium Approach
(see also Section 4)

Total dissolved cadmium concentrations were titrated with increasing NTA or EDTA concentrations, in model solutions containing only one of the major cations of seawater, then in artificial, and finally in real seawater (RASPOR and BRANICA, 1973, 1975; RASPOR et al., 1977a). After the equilibration the extent of metal chelation is followed by measuring the decrease of the labile, unchelated metal concentration, while during the short measuring time, the concentration distribution of metal between the labile and inert complexed forms does not change (VALENTA, 1983).

The same type of measurements was performed with lead and zinc in seawater and lake water, with the model ligands NTA and EDTA (RASPOR et al., 1980b, 1981). Combining the results of equilibrium and the kinetic measurements of the homogeneous rate of chelate formation (RASPOR et al., 1977b, 1980a) it has been clearly shown that competing reactions of calcium and magnesium for the NTA ligand play a significant role and determine the rate of Cd(II)-NTA chelate formation. This example allows to notice the influence of the total concentration level of metal on the relevant competing reactions. At a concentration of 10^{-7} mol L^{-1} of total cadmium, the salinity components of seawater which influence the degree of Cd(II)-NTA formation are chloride, calcium, and magnesium. If the concentration of total cadmium is decreased to 3×10^{-9} mol L^{-1}, in addition to the previously mentioned ions, the concentration of carbonate, bicarbonate, sulfate as well as sodium and zinc ions has to be taken into account. To obtain the apparent stability

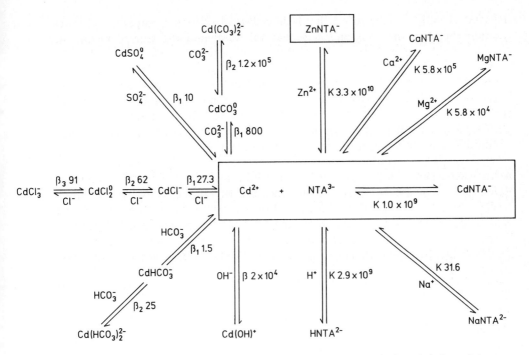

Fig. I.5-2. Complex equilibrium reactions taking place in seawater during chelation of the trace metal Cd(II) with the ligand NTA of synthetic, anthropogenic origin. (From NÜRNBERG and RASPOR, 1981, reproduced with permission).

constant of Cd(II)-NTA chelate, all these equilibria of *side reactions* have to be considered.

As an example, equilibrium reactions of the metal Cd(II) and the ligand NTA, which are relevant in seawater, are presented in Fig. I.5-2. It should be noted that the distribution of all specific chemical forms of the metal within the labile and inert complexes has to be known and the apparent stability constants have to refer to the same ionic strength, in this case of seawater.

The competitive reactions were further experimentally confirmed by voltammetric titration of Cd, Pb, and Zn with humic substances, isolated from marine and estuarine sediments, from various parts of the sea (RASPOR et al., 1984a, b). These experimental findings confirm that in the water of open sea areas, the existing dissolved humate concentrations are usually less than 0.1 mg/L. According to our experimental results this concentration of ligands is too low to significantly contribute to the complexation of Cd, Pb, and Zn. Metal ions with a different speciation pattern, such as copper and mercury, might be more significantly complexed by dissolved humic substances, especially in fresh waters, due to higher stability constants of humate complexes of these metals (MANTOURA et al., 1978; WHITFIELD and TURNER, 1980).

From these examples it can be concluded that the chemical behavior and therefore the speciation of each of these metals is different, even if they belong to the same class, i.e., to the soft or borderline class of cations. Competitive chemical reactions between the components of natural water and the studied metals and ligands have also been observed. It is highly recommended to make a detailed speciation study for each metal of ecotoxicological interest, which considers all the possible labile and inert types of complexes in a given water type, at an as low as possible concentration level. Voltammetric (and polarographic) techniques are the only species-specific techniques which make the determination of relevant metal compounds close to the natural concentration level possible, under pH and ionic strength conditions corresponding to those in natural waters.

Acknowledgement. This study is part of the joint research project "Environmental Research in Aquatic Systems" of the Institute of Applied Physical Chemistry, Nuclear Research Center (KFA) Jülich, and the Center for Marine Research Zagreb, Rudjer Boskovic Institute, Zagreb, in the bilateral German-Yugoslav Agreement. Financial support by the International Bureau of KFA, Jülich, and the Self-Managed Authority for Scientific Research, SR Croatia, Yugoslavia, is gratefully acknowledged.

References

Ahrland, S., Chatt, J., and Davies, N.R. (1958), *Q. Rev. Chem. Soc. 12*, 265–276.
Baes, C.E.F., and Mesmer, R.E. (1976), *The Hydrolysis of Cations.* Wiley-Interscience, New York.
Barić, A., and Branica, M. (1967), *J. Polarogr. Soc. 13*, 4–8.
Bilinski, H., Huston, R., and Stumm, W. (1976), *Anal. Chim. Acta 84*, 157–164.
Branica, M., Petek, M., Barić, A., and Jeftić, L. (1969), *Rapp. Comm. Int. Mer. Medit. 19*(5), 929–933.
Branica, M., Novak, D.M., and Bubić, S. (1977), *Croat. Chem. Acta 49*, 539–547.
Branica, M., Pižeta, I., and Marić, I. (1986), *J. Electroanal. Chem. 214*, 95–102.
Bruland, K.W. (1983), *Trace Elements in Seawater. Chem. Oceanogr. 8*, 157–220.
Bubić, S., and Branica, M. (1973), *Thalassia Jugosl. 9*, 47–53.
Burton, J.D. (1979), *Philos. Trans. R. Soc. London B286*, 443–456.
Cotton, F.A., and Wilkinson, G. (1980), *Advanced Inorganic Chemistry,* 4th Ed., Chap. 3, pp. 61–106. John Wiley & Sons, New York.
Ćosović, B., Degobbis, D., Bilinski, H., and Branica, M. (1982), *Geochim. Cosmochim. Acta 46*, 151–158.
Dyrssen, B., and Wedborg, M. (1975), in: Goldberg, E.D. (ed.): *The Sea*, Vol. 5, pp. 181–195. John Wiley & Sons, New York.
Evans, R.S., and Huheey, J.E. (1970), *J. Inorg. Nucl. Chem. 32*, 777–793.
Florence, T.M. (1986), *Analyst 111*, 489–505.
Florence, T.M., and Batley, G.E. (1980), *CRC Crit. Rev. Anal. Chem. 9*, 219–296.
Frank, H.S., and Evans, M.W. (1945), *J. Chem. Phys. 13*, 507–532.
Giddings, J.C. (1973), *Chemistry, Man and Environmental Change, An Integrated Approach.* Canfield Press, San Francisco.
Hart, B.T. (1981), *Environ. Technol. Lett. 2*, 95–110.
Horne, R.A. (1969), *Marine Chemistry,* pp. 11–53. John Wiley & Sons, New York.
Huheey, J.E. (1978), *Inorganic Chemistry,* 2nd Ed. Harper & Row, New York.
Irving, H., and Williams, R.J.P. (1948), *Nature 162*, 746–747.

Kester et al. Group Report (1986), in: Bernhard, M., Brinckmann, F. E., and Sadler, P. J. (eds.): *Importance of Chemical "Speciation" in Environmental Processes*, pp. 275–299. Dahlem Konferenzen, Springer-Verlag, Berlin.
Klopman, G. (1968), *J. Am. Chem. Soc. 90*, 223–234.
Lewis, G. N. (1916), *J. Am. Chem. Soc. 38*, 762–788.
Magjer, T., and Branica, M. (1977), *Croat. Chem. Acta 49*, L1–L5.
Maljković, D., and Branica, M. (1971), *Limnol. Oceanogr. 16*, 779–785.
Mantoura, R. F. C., Dickson, A., and Riley, J. P. (1978), *Estuarine Coastal Mar. Sci. 6*, 387–408.
Mart, L., and Nürnberg, H. W. (1986a), *Cd, Pb, Cu, Ni and Co Distribution in the German Bight. Mar. Chem. 18*, 197–213.
Mart, L., and Nürnberg, H. W. (1986b), *The Distribution of Cadmium in the Sea. Experientia Suppl. 50*, 28–40.
Mart, L., Rützel, H., Klahre, P., Sipos, L., Platzek, U., Valenta, P., and Nürnberg, H. W. (1982), *Comparative Studies on the Distribution of Heavy Metals in the Oceans and Coastal Waters. Sci. Total Environ. 26*, 1–17.
Mart, L., Nürnberg, H. W., and Rützel, H. (1985), *Levels of Heavy Metals in the Tidal Elbe and its Estuary and the Heavy Metal Input into the Sea. Sci. Total Environ. 44*, 35–49.
Masterton, W. L., Slowinski, E. J., and Stanitski, C. L. (1986), *Chemical Principles*, 6th Ed., pp. 224–287. College Publishing, Philadelphia.
Morgan, J. J., and Stone, A. T. (1985), in: Stumm, W. (ed.): *Chemical Processes in Lakes*, pp. 389–426. John Wiley & Sons, New York.
Nieboer, E., and Richardson, D. H. .S. (1980), *Environ. Pollut. Ser. B 1*, 3–26.
Nürnberg, H. W. (1977), *Electrochim. Acta 22*, 935–949.
Nürnberg, H. W. (1983), in: Leppard, G. G. (ed.): *Trace Element Speciation in Surface Waters and its Ecological Implications*, pp. 211–229. Plenum Publ. Corp., New York-London.
Nürnberg, H. W., and Mart, L. (1985), *Distribution and Fate of Heavy Metals in the Oceans. Proceedings Fifth International Conference on Heavy Metals in the Environment, Athens*, Vol. 2, pp. 340–342. CEP Consultants Ltd., Edinburgh.
Nürnberg, H. W., and Raspor, B. (1981), *Environ. Technol. Lett. 2*, 457–483.
Padova, J. (1964), *J. Chem. Phys. 40*, 691–694.
Pauling, L. (1960), *The Nature of the Chemical Bond*. Cornell University Press, Ithaca, New York.
Pearson, R. G. (1963), *J. Am. Chem. Soc. 85*, 3533–3539.
Pearson, R. G. (1969), *Surv. Prog. Chem. 5*, 1–52.
Plavšić, M., Krznarić, D., and Branica, M. (1982), *Mar. Chem. 11*, 17–31.
Pytkowicz, R. M. (1983), *Equilibria, Nonequilibria and Natural Waters*, Vol. II, Chap. 6, pp. 189–234. John Wiley & Sons, New York.
Pytkowicz, R. M., and Hawley, J. E. (1974), *Limnol. Oceanogr. 19*, 223–234.
Raspor, B. (1980), in: Nriagu, J. O. (ed.): *Cadmium in the Environment*, Chap. 6, Part I, pp. 147–236. John Wiley & Sons, New York.
Raspor, B., and Branica, M. (1973), *J. Electroanal. Chem. 45*, 79–88.
Raspor, B., and Branica, M. (1975), *J. Electroanal. Chem. 59*, 99–109.
Raspor, B., Valenta, P., Nürnberg, H. W., and Branica, M. (1977a), *Sci. Total Environ. 2*, 87–109.
Raspor, B., Valenta, P., Nürnberg, H. W., and Branica, M. (1977b), *Thalassia Jugosl. 13*, 79–91.
Raspor, B., Nürnberg, H. W., Valenta, P., and Branica, M. (1980a), *J. Electroanal. Chem. 115*, 293–308.
Raspor, B., Nürnberg, H. W., Valenta, P., and Branica, M. (1980b), in: Branica, M., and Konrad, Z. (eds.): *Lead in the Marine Environment*, pp. 181–195. Pergamon Press, Oxford.
Raspor, B., Nürnberg, H. W., Valenta, P., and Branica, M. (1981), *Limnol. Oceanogr. 26*, 54–66.
Raspor, B., Nürnberg, H. W., Valenta, P., and Branica, M. (1984a), *Mar. Chem. 15*, 217–230.
Raspor, B., Nürnberg, H. W., Valenta, P., and Branica, M. (1984b), *Mar. Chem. 15*, 231–249.
Schneider, W. (1968), *Einführung in die Koordinationschemie*, Chap. 4, pp. 35–46. Springer Verlag, Berlin.
Sillen, L. G. (1967), *Science 156*, 1189–1197.

Simões Gonçalves, M. L. S., and Valenta, P. (1982), *J. Electronanal. Chem. 132*, 357–375.
Simões Gonçalves, M. L. S., Valenta, P., and Nürnberg, H. W. (1983), *J. Electroanal. Chem. 149*, 249–262.
Sipos, L., Raspor, B., Nürnberg, H. W., and Pytkowicz, R. M. (1980a), *Mar. Chem. 9*, 37–47.
Sipos, L., Valenta, P., Nürnberg, H. W., and Branica, M. (1980b), in: Branica, M., and Konrad, Z. (eds.): *Lead in the Marine Environment*, pp. 61–76. Pergamon Press, Oxford.
Stumm, W., and Morgan, J. J. (1981), *Aquatic Chemistry*, 2nd Ed., Chap. 6, pp. 323–417. John Wiley & Sons, New York.
Taube, H. (1954), *J. Phys. Chem. 58*, 523–528.
Thayer, J. S. (1974), *J. Organomet. Chem. 76*, 265–295.
Valenta, P. (1983), in: Leppard, G. G. (ed.): *Trace Element Speciation in Surface Waters and its Ecological Implications*, pp. 49–69. Plenum Publ. Corp., New York-London.
Valenta, P., Simões Gonçalves, M. L. S., and Sugawara, M. (1984), in: Kramer, C. J. M., and Duinker, C. J. (eds.): *Complexation of Trace Metals in Natural Waters*, pp. 357–366. Martinus Nijhoff/W. Junk Publ., The Hague.
Westall, J., and Stumm, W. (1980), *The Hydrosphere*, in: Hutzinger, O. (ed.): *The Handbook of Environmental Chemistry*, Vol. 1, Part A, pp. 17–48. Springer Verlag, Berlin.
Whitfield, M., and Turner, D. R. (1980), in: Branica, M., and Konrad, Z. (eds.): *Lead in the Marine Environment*, pp. 109–148. Pergamon Press, Oxford.
Whitfield, M., and Turner, D. R. (1983), in: Wong, C. S., Boyle, K., Bruland, K. W., Burton, J. R., and Goldberg, E. G. (eds.): *Trace Metals in Sea Water*, pp. 719–750. Plenum Press, New York.
Whitfield, M., and Turner, D. R. (1986), *Sci. Total Environ. 58*, 9–35.

Additional Recommended Literature

Ahrland, S. (1988), *Trace Metal Complexation by Inorganic Ligands in Sea Water*, in: West, T. S., and the late Nürnberg, H. W. (eds): *The Determination of Trace Metals in Natural Waters*, Chapter 7, pp. 223–252. Blackwell Scientific Publications, Oxford.
Allen, H. E., and Garrison, A. W. (1989), *Metal Speciation and Transport in Groundwater, Proceedings Workshop Jekyll Island, Georgia*. Lewis Publishers, Chelsea, Michigan.
Aten, C. F., Bourke, J. B., and Walton, J. C. (1989), *Ion-content of Streams in the Cold R. and Chubb. R. Watersheds 1983–1986. Toxicol. Environ. Chem.*, in press.
Baretta, J. W., and Ruardij, P. (1988), *Tidal Flat Estuaries*, 270 pages. Springer Verlag, Berlin.
Bowen, H. J. M. (1985), *The Cycles of Copper, Silver and Gold*, in: Hutzinger, O. (ed.): *The Handbook of Environmental Chemistry*, Vol. 1, Part D, pp. 1–27. Springer Verlag, Berlin.
Bro-Rasmussen, F. Løkke, H., and Tyle, H. (1988), *Proceedings 1st European SECOTOX Conference on Ecotoxicology, Copenhagen*. Technical University of Denmark, Lingby.
Brown, D. S., and Allison, J. D. (1987), *MINTEQ A1, An Equilibrium Metal Speciation Model: User's Manual*, EPA/600/3-87/012 (92 pages). Environmental Research Laboratory, Athens, Georgia, USA.
Brown, D. S., Carlton, R. E., and Mulkey, L. A. (1986), *Development of Land Disposal Decisions for Metals Using MINTEQ Sensitivity Analyses*, EPE/600/S3-86/030 (8 pages). Environmental Research Laboratory, Athens, Georgia, USA.
Buffle, J. (1988), *Complexation Reactions in Aquatic Systems. An Analytical Approach*, 692 pages. Ellis Horwood, Chichester.
Buffle, J. (1988), *Electroanalytical Measurement of Trace Metals Complexation*, in: West, T. S., and the late Nürnberg, H. W. (eds.): *The Determination of Trace Metals in Natural Waters*, Chapter 6, pp. 179–230. Blackwell Scientific Publications, Oxford.

Bulman, A., and Cooper, J. R. (1986), *Speciation of Fission and Activation Products in the Environment*, 437 pages. Elsevier Applied Science Publishers, London-New York.

Burckle, J., et al. (1988), *US-EPA Report on Waste Water Treatment Technology*. EPA, Cincinnati, Ohio, and Radion Corp.

Calmano, W., Förstner, U., and Kersten, M. (1986), *Metal Associations in Anoxic Sediments and Changes Following Upland Disposal. Toxicol. Environ. Chem.* 12(3+4), 313–321.

Collins, H.-J. (1986), *Water and Pollutants Balances of Waste Disposal Sites and Influence on Waters* (in German). DFG Publication, 339 pages. VCH Verlagsgesellschaft, Weinheim-Deerfield Beach/Florida-Basel.

Daum, K. A., and Newland, L. W. (1982), *Complexing Effects on Behavior of Some Metals*, in: Hutzinger, O. (ed.): *The Handbook of Environmental Chemistry*, Vol. 2, Part B, pp. 129–139. Springer Verlag, Berlin.

Dobbs, J. C., Susetyo, M., Knight, F. E., Castles, M. A., Carreira, L. A., and Azarraga, L. V. (1989), *A Novel Approach to Metal-humic Complexation Studies by Lanthanide Ion Probe Spectroscopy. Int. J. Environ. Anal. Chem.*, in press.

Dobbs, J. C., Susetyo, W., Knight, F. E., Castles, M. A., Carreira, L. A., and Azarraga, L. V. (1989), *Characterization of Metal Binding Sites in Fulvic Acids by Lanthanide Ion Probe Spectroscopy. Anal. Chem.* 61(5), 483–488.

Dronkers, J., and van Leussen, W. (1988), *Physical Processes in Estuaries*, 570 pages. Springer Verlag, Berlin.

Gray, J. S., and Rumohr, H. (1984), *The Ecology of Marine Sediments*, 193 pages. Springer Verlag, Berlin.

Head, P. C. (ed.) (1985), *Handbook on Practical Estuarine Chemistry*, 337 pages. Cambridge University Press, Cambridge, UK.

Heck, Jr. K. L. (1987), *Ecological Studies in the Middle Reach of Chesapeake Bay*, 287 pages. Springer Verlag, Berlin.

Hellmann, H. (1987), *Analysis of Surface Waters*, 275 pages. Ellis Horwood, Chichester, UK.

Houriet, J.-P. (1989), *Heavy Metal Compounds in Swiss River Sediments* (in German). *Umweltschutz in der Schweiz 4/89*, 31–43. Swiss Federal Office for Environmental Protection.

Kramer, J. R., and Allen, H. E. (1988), *Metal Speciation*, 357 pages. Lewis Publishers, Chelsea, Michigan.

Kranck, K. (1980), *Sedimentation Processes in the Sea*, in: Hutzinger, O. (ed.): *The Handbook of Environmental Chemistry*, Vol. 2, Part A, pp. 61–75. Springer Verlag, Berlin.

Landner, L. (1987), *Speciation of Metals in Water, Sediment, and Soil Systems*. Springer Verlag, Berlin.

Lin, Y., Bailey, G. W., and Lynch, A. T. (1988), *Metal Interactions at Sulfide Mineral Surfaces. Part II. Adsorption and Desorption of Lanthanum.* Proceedings International Symposium Environment, Life Elements and Health, Beijing, China, in press.

Lin, Y., Bailey, G. W., and Lynch, A. T. (1988), *Metal Interactions at Sulfide Mineral Surfaces. Part III. Metal Affinities in Single and Multiple Ion Adsorption Reactions.* Proceedings International Symposium Environment, Life Elements and Health, Beijing, China, in press.

Lion, L. W. (1984), *The Surface of the Ocean*, in: Hutzinger, O. (ed.): *The Handbook of Environmental Chemistry*, Vol. 1, Part C, pp. 79–104. Springer Verlag, Berlin.

Loux, N. T., Garrison, A. W., and Chafin, C. R. (1989), *Acquisition and Analysis of Groundwater/Aquifer Samples: Current Technology and the Trade-off between Quality Assurance and Practical Considerations. Int. J. Environ. Anal. Chem.*, in press.

Loux, N. T., Brown, D. S., Chafin, Cl. R., Allison, J. D., and Hassan, S. M. (1989), *Chemical Speciation and Competitive Cationic Partitioning on a Sandy Aquifer Material, Chemical Speciation and Bioavailability*, in press.

Lum, K. R., and Nriagu, J. O. (1988), *Proceedings International Conference Trace Metals in Lakes*, Hamilton, Canada. *Sci. Total Environ.*, in press.

McBride, A. (1987), *The American Water Pollution Control Program.* Proceedings 17th Annual IAEAC Symposium Analytical Chemistry of Pollutants, Jekyll Island, Georgia.

Menzel, D. B. (1986), *Drinking Water and Health,* Vol. 6, 457 pages. National Academy Press, Washington, DC.

Moore, J. W., and Ramamoorthy, S. (1984), *Heavy Metals in Natural Waters,* 268 pages. Springer Verlag, Berlin.

Moore, R. V., and Noble, M. A. (1987), *Distribution Coefficients for Heavy Metal Ions with and without Complexing Agents. Proceedings 17th Annual IAEAC Symposium Analytical Chemistry of Pollutants, Jekyll Island, Georgia.*

Nriagu, J. O., and Pacyna, J. M. (1988), *Quantitative Assessment of Worldwide Contamination of Air, Water and Soils by Trace Metals. Nature 333,* 134–139.

Nriagu, J. O., and Sprague, J. B. (eds.) (1987), *Cadmium in the Aquatic Environment,* 272 pages. Wiley-Interscience, New York.

Nürnberg, H. W. (ed.) (1985), *Pollutants and Their Ecotoxicological Significance,* 515 pages. John Wiley & Sons, Chichester, UK.

Nygard, D. D., and Bulman, F. (1990), *Analysis of Water for Arsenic, Lead, Selenium, and Thallium by Inductively Coupled Plasma Atomic Emission Spectrometry at Contract Laboratory Program Levels. Spectrosc. Int. 2*(2), 44–47.

Öschger, H., and Langway, C. C., Jr. (eds.) (1989), *The Environmental Record in Glaciers and Ice Sheets. Dahlem Workshop Reports.* Wiley & Sons, Chichester-New York.

Parker, W. R. (ed.) (1988), *Developments in Coastal and Estuarine Pollution,* 299 pages. *IAWPRC, Water Sci. Technol. 20,* No. 6/7. Pergamon Press, Oxford.

rivm (1989), *Concern for Tomorrow* (with informations on concentrations in rivers, e.g., the Rhine). National Institute of Public Health and Environmental Protection, Bilthoven, The Netherlands.

Sahrhage, D. (1988), *Antarctic Ocean and Resources Variability,* 320 pages. Springer Verlag, Berlin.

Salomons, W., and Förstner, U. (1984), *Metals in the Hydrocycle,* 349 pages. Springer Verlag, Berlin.

Salomons, W., Bayne, B. L., Duursma, E. K., and Förstner, U. (1988), *Pollution of the North Sea,* 600 pages. Springer Verlag, Berlin.

Schmist, N. W. (1988), *Toxic Contamination in Large Lakes,* 4 Volumes (especially Volume III on *Sources, Fate, and Controls of Toxic Contaminants).* Lewis Publishers, Chelsea, Michigan.

Sigleo, A. C., and Hatton, A. (eds.) (1985), *Marine and Estuarine Geochemistry,* 331 pages. Lewis Publishers, Chelsea, Michigan.

Stumm, W. (ed.) (1985), *Chemical Processes in Lakes,* 435 pages. John Wiley & Sons, New York-Chichester.

Valenta, P., Breder, R., Mart, L., Rützel, H., and Merks, A. G. A. (1987), *Distribution of Cadmium and Lead Between Dissolved and Particulate Phases in Estuaries. Toxicol. Environ. Chem. 14*(1+2), 129–141.

Waldichuk, M. (1988), *Exchange of Pollutants and Other Substances Between the Atmosphere and the Oceans,* in: Hutzinger, O. (ed.), *The Handbook of Environmental Chemistry,* Vol. 2, Part D, pp. 113–151. Springer Verlag, Berlin.

Wangersky, P. J. (1980), *Chemical Oceanography,* in: Hutzinger, O. (ed.), *The Handbook of Environmental Chemistry,* Vol. 1, Part A, pp. 51–67. Springer Verlag, Berlin.

Westall, J. C. (1986), *MICROQL – A Chemical Equilibrium Program in BASIC, Report 86-02, MICROQL Version 2 for PCs,* 47 pages. Department of Chemistry, Oregon State University, Corvallis, Oregon.

Westall, J. C. (1987), *Transformation Related to Redox Chemistry and to Metal Speciation* (in the Field and in the Laboratory). *Proceedings 17th Annual IAEAC Symposium on Analytical Chemistry of Pollutants,* Jekyll Island, Georgia.

White, H. H. (1984), *Concepts in Marine Pollution Measurements,* 743 pages. Maryland Sea Grant Publication, Maryland.

I.6a Metal Compounds in the Atmosphere

HANS PUXBAUM, Wien, Austria

1 Introduction

Metal compounds are found in a great variety in the atmosphere. With some exceptions the major form of occurrence is in the particulate phase. Emissions of metal compounds into the atmosphere originate from natural (terrestrial, marine, volcanic, biogenic) as well as from anthropogenic (combustion, industrial, automobile) sources. Of great importance is the question: to what extent have anthropogenic emissions modified the natural geochemical cycles of metals (WOOD and GOLDBERG, 1977) and what is the role of natural and anthropogenic aerosol particles in the global tropospheric chemistry? There is evidence that the atmospheric input of anthropogenic toxic trace metals leads to the enrichment of the respective components even in remote environments such as the high Alpine regions and the Arctic. Finally, there is concern about the ecological impact and about health hazards by inhalation, especially in densely populated areas and at work places exposed to increased levels of metal containing vapors or particles.

In this chapter, the sources, the atmospheric occurrence and behavior, the analytical aspects and the methods to assess the contributions of different sources to the aerosol composition are discussed. Among the voluminous literature available on metals in the atmosphere, the monograph by NRIAGU and DAVIDSON (1986) and the extensive review by SCHROEDER et al. (1987) are recommended which cover the literature until 1983. The discussion will be focused on the most important trace metals and be limited to the particulate phase. For the special case of mercury, which occurs predominantly in the gaseous phase in the atmosphere (KAISER and TÖLG, 1980) the reader is referred to Chapter II.20. Also other metals such as lead, tin, arsenic, selenium, and tellurium may occur in a gaseous state in the atmosphere to some extent. The atmospheric fate and the possible mobilization of these compounds via biomethylation have been reviewed by CRAIG (1980), see also Chapter I.7h.

2 Sources of Metal Compounds in the Atmosphere

The atmospheric occurrence of trace metals is determined by the source strength, the atmospheric dispersion, and the deposition processes.

Table I.6a-1. Worldwide Atmospheric Emission of Trace Elements from Natural and Anthropogenic Sources. Compiled from Reviews by PACYNA (1986b) (A) and SALOMONS (1986) (B). Estimated Annual Emissions in 10^6 kg

Source		As	Cd	Co	Cu	Cr	Hg	Mn	Ni	Pb	Se	V	Zn
Natural	A	8	1	5	19	9	0.2	520	26	19	0.4	66	45
	B	21	0.3	7	58	19	25	610	28	4	3	65	36
Manmade	A	24	7	–	56	–	–	–	47	450	1	–	310
	B	78	6	5	94	260	11	320	98	400	14	210	840

The source strength for the atmospheric emission of a metal compound is calculated from emission factors available for the different emitting processes. While most of the national emission inventories are focused on SO_2, NO_x, and total particulate matter, data on the emission of metal compounds are relatively sparse. PACYNA (1986a) has reviewed the available trace element emission factors for natural and anthropogenic sources.

On the global scale compilations have been made for metal emissions by NRIAGU (1979), LANTZY and MACKENZIE (1979), and WEISEL (1981). The divergence of the data reflects the uncertainties in estimating global emissions from very sparse data sets on natural and anthropogenic sources. From SALOMONS' (1986) and PACYNA'S (1986b) reviews the figures in Table I.6a-1 have been compiled. These data indicate a clear dominance of anthropogenic emissions for the most important trace elements such as As, Cd, Cu, Cr, Pb, and Zn on the global scale.

Long-range transport of Saharan and Asian dust has been identified as the dominant source of mineralic particles over the Atlantic, the Arctic, and the Pacific (SCOPE, 1979; RAHN et al., 1979; DUCE et al., 1980; UEMATSU et al., 1983; PARRINGTON and ZOLLER, 1984). Other important sources of naturally emitted metal compounds are volcanoes (ZOLLER, 1983), forest fires which may be of natural occurrence as well as originate from anthropogenic activities and exudations from vegetation (PACYNA, 1986a, b).

Sea spray and gaseous emissions from the oceans contribute only a minor fraction of trace metal compounds to the atmosphere on a global basis (PACYNA, 1986b). Volcanoes are thought to be the main source of naturally emitted As and Cd and are important sources of Pb, Se, Zn, and Hg (Table I.6a-2). Forest fires are likely important emitters of Hg whereas plant exudations contribute markedly to the flux of naturally emitted As, Zn, and Cd compounds.

Globally anthropogenic emissions of metals (Table I.6a-3) already exceed the emissions of several trace elements from natural sources. On a regional scale in densely populated areas anthropogenic emissions of metals are by far the dominant contributors as compared to natural sources. This is reflected by the fact that ambient concentrations of trace elements in source regions are some orders of magnitude higher than in remote regions. Table I.6a-3 shows that anthropogenic emissions of Be, Co, Mo, Sb, and Se are mainly derived from coal combustion, Ni and V are released predominantly from oil firing, As, Cd, Cu, and Zn are emitted

Table I.6a-2. Worldwide Emissions of Trace Elements from Natural Sources, After a Compilation by PACYNA (1986b). Annual Emissions in 10^6 kg

Source	As	Cd	Co	Cu	Cr	Mn	Ni	Pb	Se	V	Zn	Hg
Atmospheric dust	0.24	0.25	4	12	50	425	20	10	0.3	50	25	0.03
Volcanoes	7	0.5	1.4	4	3.9	82	3.8	6.4	0.1	6.9	10	0.03
Forest fires	0.16	0.01	–	0.3	–	–	0.6	0.5	–	–	0.5	0.1
Vegetation	0.26	0.2	–	2.5	–	5	1.6	1.6	–	0.2	10	–
Sea spray	0.14	0.002	–	0.1	–	4	0.04	0.1	–	9	0.02	0.003

Table I.6a-3. Worldwide Anthropogenic Emissions of Trace Metals During 1975 in 10^6 kg

Source	As[b]	Cd[a]	Cu[a]	Ni[a]	Pb[a]	Se[c]	Zn[a]
Mining, nonferrous metals	0.013	0.002	0.8	–	8.2	0.005	1.6
Primary nonferrous metal production	15.2	4.71	20.8	9.4	76.5	0.28	107
Secondary nonferrous metal production	–	0.60	0.33	0.2	0.8	–	9.5
Iron and steel production	4.2	0.07	5.9	1.2	50	0.01	35
Industrial applications	0.02	0.05	4.9	1.9	7.4	0.06	26
Coal combustion	0.55	0.06	4.7	0.7	14	0.68	15
Oil combustion (includ. gasoline)	0.004	0.003	0.74	27	273	0.06	0.1
Wood combustion	0.60	0.20	12	3.0	4.5	–	75
Waste incineration	0.43	1.40	5.3	3.4	8.9	–	37
Manufacture phosphate fertilizer	2.66	0.21	0.6	0.6	0.05	–	1.8
Miscellaneous	–	–	–	–	5.9	–	6.7
Total	23.6	7.3	56	47	449	1.1	314

[a] Data after NRIAGU (1979)
[b] WALSH et al. (1979)
[c] NAS (1979)

mainly from non-ferrous smelters and secondary production plants (see also Chapter I.2). Cr and Mn are released in large amounts from iron and steel industries, and for Pb the automotive traffic is the predominant source.

Emission data for trace metals on a national scale have been compiled recently for European countries by PACYNA (1984) for 1979 and have been revised for several elements (PACYNA, 1987). The major emission areas in Europe are (1) the Soviet Union, (2) Poland and Czechoslovakia, and (3) the Benelux countries and the Western part of the Federal Republic of Germany. For Zn and V significant emission sources are also located in the UK, Spain, and Italy. While Zn and V emissions are distributed relatively evenly in Europe, emissions of As, Cd, Cu, Cr, Mn, and Be are rather concentrated in Central and Eastern Europe (PACYNA, 1986b). A detailed trace metal emission inventory is also available for the Los Angeles area (CASS and MCRAE, 1983, 1986). Regional and local emission inventories are generally used to assess the impact of sources to air quality, pollutant deposition (VANDERBORGHT et al., 1983), or human exposure (BENNETT, 1981) and to derive control strategies. Recently, the European emission inventory has been used to model long-range transport of trace metals to the Arctic (PACYNA et al., 1985).

3 Atmospheric Occurrence

A large number of data are available concerning atmospheric concentrations of trace elements associated with particulate matter. Comprehensive surveys have been compiled by SCHROEDER et al. (1987) and by WIERSMA and DAVIDSON (1986). Atmospheric concentration ranges for remote, rural, and urban locations are given for 14 elements in Table I.6a-4. The large differences for the remote sites reflect the different regions such as maritime and continental, northern and southern hemispheres.

Lowest concentrations of trace elements have been found in the Antarctic (CUNNINGHAM and ZOLLER, 1981) and very low levels in the maritime atmosphere over the Pacific ocean (GORDON et al., 1978; DUCE et al., 1983; PARRINGTON and

Table I.6a-4. Concentration Ranges of Various Elements Associated with Particulate Matter in the Atmosphere (ng/m^3). (After SCHROEDER et al., 1987)

Location	As	Cd	Ni	Pb	V	Zn
Remote	0.007 – 1.9	0.003 – 1.1	0.01 – 30	0.007 – 64	0.001 – 14	0.03 – 110
Rural	1.0 – 28	0.4 – 1000	0.6 – 78	2 – 1700	2.7 – 97	11 – 403
Urban						
Canada	7.7 – 626	2 – 103	4 – 371	353 – 3416	10 – 130	55 – 1390
USA	2 – 2320	0.2 – 7000	1 – 328	30 – 96270	0.4 – 1460	15 – 8328
Europe	5 – 330	0.4 – 260	0.3 – 1400	10 – 9000	11 – 73	160 – 8340
Other	20 – 85	0.6 – 177	2.3 – 158	1.3 – 11020	1.7 – 180	110 – 2700

Co	Cr	Cu	Fe	Mn	Se	Sb
0.001 – 0.9	0.005 – 11.2	0.029 – 12	0.62 – 4160	0.01 – 16.7	0.0056 – 0.19	0.0008 – 1.19
0.08 – 10.1	1.1 – 44	3 – 280	55 – 14530	3.7 – 99	0.01 – 3.0	0.6 – 7
1 – 7.9	4 – 26	17 – 500	700 – 5400	20 – 270	na	13 – 125
0.2 – 83	2.2 – 124	3 – 5140	130 – 13800	4 – 488	0.2 – 30	0.5 – 171
0.4 – 18.3	3.7 – 227	13 – 2760	294 – 13000	23 – 850	0.01 – 127	2 – 470
0.3 – 10	tr – 277	2.0 – 6810	21 – 32820	1.7 – 590	na	7 – 36

na = not available, tr = traces

ZOLLER, 1984). Remote sites at some distance to the high emission regions of the East USA, Europe, and USSR, such as Central USA (MOYERS et al., 1977) or the Arctic (HEIDAM, 1981; BARRIE et al., 1981; HEINTZENBERG et al., 1981), receive polluted air masses in an episodic mode. At such sites trace metal concentrations may rise to levels found in rural areas in densely populated countries. Well documented is the transport of Eurasian aerosols to the Arctic which occurs particularly in winter (RAHN, 1985). Due to meteorological effects, the Arctic atmosphere behaves as a reservoir for long-range transported pollutants during the winter months and is cleaned during spring and summer (HEIDAM, 1986).

High concentrations of trace elements are observed in densely populated and industrial areas (Table I.6a-4; SCHROEDER et al., 1987). The transport pathways of the polluted air masses can be reconstructed by trajectory analysis and elemental tracer techniques (HUSAIN, 1986; RAHN and LOWENTHAL, 1985; CHEN and DUCE, 1983; LOWENTHAL and RAHN, 1985).

Elemental balances for atmospheric particles indicate that trace metals comprise only a small fraction of the total aerosol mass. Even in a highly industrialized city such as Linz (Austria) the relative contribution of trace metal compounds (Cd, Cr, Cu, Mn, V, Zn, and Pb compounds) to the total suspended particles (TSP) was found to be about 1%, Fe-compounds comprised 1–8% of the TSP mass (PUXBAUM et al., 1985). The major part of the TSP is formed by electrolytes (Na^+, K^+, NH_4^+, Cl^-, NO_3^-, SO_4^{2-}), (25–35%), carbonaceous material (8–11%), and mineralic components (Ca, Mg, Si, Al compounds) (16–18%).

4 Size Distributions of Atmospheric Particles and Trace Metals

Atmospheric particles occur in a wide range of sizes. The size distribution can be expressed as a number-, surface-, or mass-density function. If the mass-size distribution is presented differentially versus the logarithm of the particle size ($dM/dlg\,AD$ vs. $lg\,AD$; M, particle mass; AD, aerodynamic diameter) for a typical urban aerosol a trimodal size distribution function (Fig. I.6a-1) is obtained (WHITBY, 1978). The finest mode (*nucleation mode*) (0.005–0.1 µm AD) is formed by particles from gas to particle conversion reactions as well as by particles from high energy combustion processes. The particles found in the *accumulation mode* size range (0.1–2.5 µm AD) originate from coagulation and condensation processes between and on nucleation mode particles. The interrelations between nucleation and accumulation mode particles can be described by growth laws derived from particle dynamics (MCMURRY and WILSON, 1982, 1983).

The particles in the *coarse mode* (2.5–100 µm AD) are of entirely different origin. They are formed during mechanic processes such as erosion or abrasion and during combustion of ash containing fuels. Fly ash particles are generally found in the lower size range of the coarse mode.

Particle size is a governing factor for the deposition in the human respiratory system. According to the deposition model of the TASK GROUP ON LUNG DYNAMICS

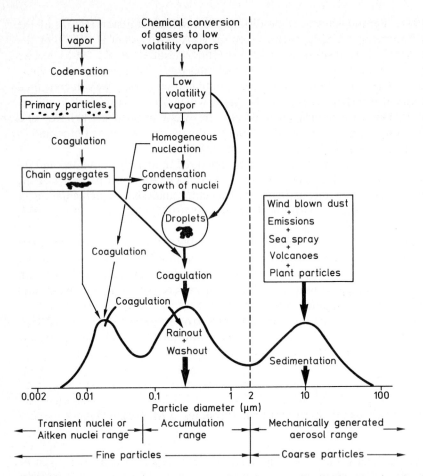

Fig. I.6a-1. Scheme of an atmospheric aerosol surface area distribution showing the three modes, the main source of mass for each mode, the principal processes involved in inserting mass into each mode, and the principal removal mechanisms. (From WHITBY, 1978).

(1966), particles larger than 10 µm preferentially deposit in the nasopharyngeal compartment, whereas smaller particles penetrate into the bronchial and alveolar compartments (*thoracic fraction*) according to the deposition functions given in Fig. I.6a-2. For ambient air quality control collection methods for particles < 10 µm *AD* have been promulgated as "PM$_{10}$ standard" (ISO, 1981; PURDUE, 1986). Using the deposition functions of the lung dynamics model or newer deposition data (STAHLHOFEN, 1986) and data on the solubility of some trace metals, the bioavailability of metals via respiratory uptake can be modeled (BENNETT, 1981; US EPA, 1982; DAVIDSON and OSBORN, 1986). In work place atmospheres, especially in the case of insoluble fibrogenous particles such as quartz and asbestos, the lung penetrating (alveolar) fraction is of major interest. Currently several standards exist to define the alveolarly "respirable" fraction. Most European countries follow the

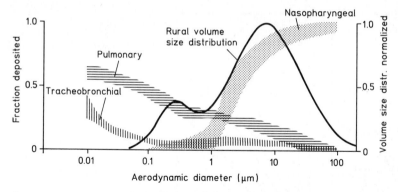

Fig. I.6a-2. Collection efficiency (fraction deposited) of various parts of the human breathing system and a normalized volume size distribution of the rural aerosol. The ordinate is given in a linear scale. (From JAENICKE, 1986).

Johannesburg convention which defines an AD of 5 μm. In the USA the AD_{50} for respirable particles is 3.5 μm.

Particle size is also a governing factor for the atmospheric lifetime of a particle. Various trace metals are found in different ranges of the size spectra of atmospheric particles. During combustion or industrial processes metal compounds may exist in the gaseous state as well as contained in various forms of fly ashes or dusts. After passing control devices such as fabric filters, electrostatic precipitators or wet scrubbers, certain fractions of the compounds will be retained, whereas a remaining part will be released to the atmosphere. Electrostatic precipitators for dust removal do not remove gaseous emissions and have a decreasing collection efficiency for smaller particles. Many trace metals, however, tend to be enriched in the small particle fraction, which results in a significantly lower removal efficiency of the control device as compared to the bulk dust or fly ash composition.

The formation of fine particles during combustion processes has been described by FLAGAN and FRIEDLANDER (1978). The result of the "vaporization and condensation model" is a bimodal size distribution with a submicrometer mode containing predominantly the volatile components and a large particle mode composed of the mineralic compounds of the fuel. Emission tests support this theory. Larger particles emitted from coal fired furnaces are primarily oxides of Al, Si, Ca, Fe, Na, Mg, and K while smaller particles are highly enriched in volatile trace elements such as As, Sb, Se, Cd, Pb, and Zn. An intermediate behavior is found for Ba, Be, Cr, Sn, Ni, Sr, U, and V showing a slight enrichment in the fine particle fraction (COLES et al., 1979). Hg is emitted from coal fired power plants predominantly (92–99%) in the gas phase (LINDBERG, 1980). The multimodal shape of trace element size distributions is also found in the ambient atmosphere. Pb, Cd, and Zn are present predominantly in the accumulation mode (AD 0.3–0.8 μm), Ca, Mg, and Al generally follow the shape of the coarse mode ($AD > 3$ μm) whereas Mn, V, Cu, and Cr exhibit an intermediate behavior with ADs of about 1–5 μm, according to the compilation by DAVIDSON and OSBORN (1986) and data from M. MAYER (1981). Near combustion

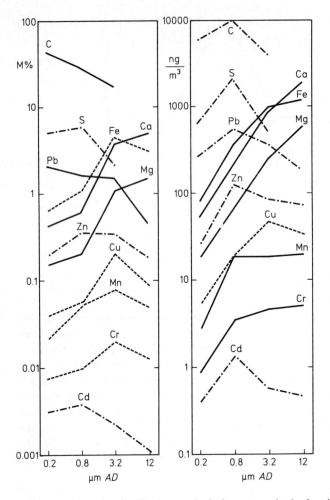

Fig. I.6a-3. Mass-size distributions and relative content in the fractional aerosol mass of major and minor elements. Urban aerosol Vienna (Getreidemarkt), average from 20 samples from Oct. 1979, sampled with a Berner 4-stage impactor (mean deposition diameters of the 4 stages: 0.2, 0.8, 3.2 and 12 µm AD). (From M. MAYER, 1981).

and metallurgical sources the respective metals are also found in the larger size ranges. The mass size distributions and the relative content in the fractional aerosol mass of some trace metals in an urban environment (Vienna) is shown in Fig. I.6a-3. According to PUXBAUM and WOPENKA (1984) Ca, Mg, Sr, Fe, and Cu have comparable relative concentrations in nucleation as well as accumulation mode particles, whereas Pb is significantly enriched in the nucleation mode as compared to the accumulation mode size fraction in an urban aerosol.

5 Chemical Speciation

The understanding of the chemical nature of metal compounds in the atmosphere is important for three reasons. First, the toxicity of a metal is dependent upon the chemical form. Second, environmental pathways in the global cycles depend on the chemical forms of the respective metal, and third, trace metal compounds have an impact on the atmospheric chemistry which is related to catalytically active species.

For ambient atmospheric aerosols, speciation is hampered by the fact that trace metal components comprise only a minor fraction of the aerosol particles in mass. Most of the speciation work for the ambient atmospheric metals has been focused upon gaseous metal compounds, e.g., of mercury, lead (reviewed by HARRISON, 1986) and selenium (JIANG et al., 1983), or coarse particles which can be analyzed by single particle methods or contain sufficient mineralic phases for XRD analysis. Advanced single particle analysis methods such as STEM (scanning transmission electron microscopy) and SAED (selected area electron diffraction) are capable of analyzing particles as small as 0.02 µm (MCCRONE and DELLY, 1973). However, these techniques are applicable only if the particle is a singular compound and is, therefore, restricted to a very small fraction of particles of the fine part of the atmospheric aerosol. STEM and SAED techniques have been successfully applied to identify submicron asbestos particles (compiled by CHATFIELD, 1984). The major fraction of the submicrometer particles in polluted atmospheres is a mixture of carbonaceous compounds and inorganic salts, mainly sulfates, nitrates, and chlorides, the main cation being ammonium (e.g., STELSON and SEINFELD, 1981). Due to the hygroscopic properties of the given mixture in many cases (at higher relative humidities or lower temperatures), the electrolytes will be present in liquid state (PILINIS and SEINFELD, 1987). Soluble trace metal compounds attached to such particles will dissolve and the metals be present in the ionic state. After an eventual drying of the particle the metals are expected to appear as sulfates or mixed salts.

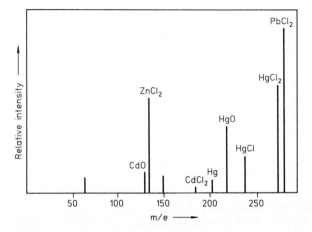

Fig. I.6a-4. Mass spectrum of gaseous emissions of a municipal incinerator (Vienna) (V & F, 1988).

Coarse particles collected by filters or by sampling from dusty surfaces such as roadways have been shown to contain various forms of Pb and Zn compounds and different minerals (BIGGINS and HARRISON, 1979a,b; O'CONNOR and JAKLEVIC, 1981; DAVIS, 1981; BLOCH et al., 1980; reviewed by HARRISON, 1986). HODGE and STALLARD (1986) recently identified Pt and Pd in roadside dust at levels of 0.7 and 0.2 ppm, respectively, on a mass basis.

Some work has been performed on speciation of particulate emissions from automobiles, combustion sources, and various industrial processes (see also Sect. 8). Recently, it has been found that a large fraction of the mercury emission from waste incinerators is present in a water-soluble form. It could be shown by mass spectroscopy that $HgCl_2$ and HgO are the major constituents of the emissions (V & F, 1988). Halides of Pb, Zn, and Cd could also be identified in the gaseous state (Fig. I.6a-4).

For coal fly ashes a differentiation of matrix versus surface enriched elements has been performed via leaching techniques (e.g., HANSEN and FISCHER, 1980) or surface analytical methods (e.g., LINTON et al., 1976, 1977). In comparative studies of compound forms of elements in oil and coal fly ashes (HENRY and KNAPP, 1980) various forms of V, Fe, Ni, Al, Si, Ca, and Mg have been identified.

6 Deposition

(see also Chapter I.4a, therein Sections 4.2 and 5.2, and Chapter I.7d)

Metals can deposit either by wet or dry deposition. For dry deposition particle size, several meteorological parameters, and the surface structure of the receptor are important parameters determining the deposition mechanism. For wet deposition rainfall rate and the concentration of the component of interest in the liquid phase determine the wet deposition flux. A simple model to express the concentration of a metal component in the liquid phase (C_L) as a function of the atmospheric concentration (C_A), the liquid water content of the cloud (L), the density of water (d), and the scavenging efficiency (E_s) has been described by JUNGE (1963):

$$C_L = C_A \cdot E_s \cdot \frac{d}{L}.$$

There is an ongoing discussion about the separate treatment of "in cloud" (rainout) and "below cloud" (washout) scavenging processes (HICKS, 1986; HALES, 1989; SEINFELD, 1986). In most cases the concentration of an aerosol constituent in wet deposition (rain water) is lower than the concentration of the respective component in cloud water (WEATHERS et al., 1988). Wet deposition can be measured relatively easily using wet deposition samplers. Contaminations from sampling instrumentation and vessels should be avoided. Dry deposits are either sampled with surrogate surface samplers or calculated from atmospheric concentrations and experimentally derived deposition velocities. Surrogate surface sampling techniques

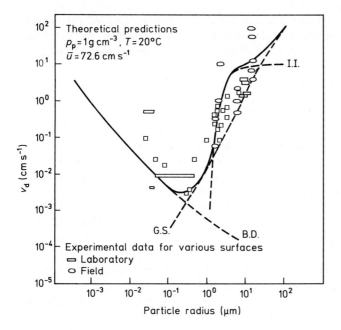

Fig. I.6a-5. Theoretically and experimentally derived deposition velocities as a function of the particle diameter. Contributing processes: B. D., Brownian diffusion; G. S., gravitational settlement; I. I., inertial impaction (NCAR, 1982).

use smooth surfaces for aerosol collection and yield non-representative results for dry deposition on rough structures such as vegetation. A different approach to measure the dry deposition flux to vegetative systems is the throughfall method (MAYER and ULRICH, 1982). This method uses statistically distributed bulk deposition samplers under trees. The difference between the integrated throughfall deposition (a mixture of wet deposits, dry deposits washed from the leaves during rainfall, and other deposits caused by fog or cloud interception as well as deposits by rime frost) and the open field wet deposition is attributed to the "dry deposition" fraction.

It is important to indicate the method used for dry deposition estimates when comparing individual results, because the uncertainties for each of the methods are very high. For example, for the flux method using averaged deposition velocities the values may differ by a factor of three (GALLOWAY et al., 1980). More detailed discussions of the dry deposition process of atmospheric particles have been reported (SLINN, 1982; SEHMEL, 1980; DAVIDSON and OSBORN, 1986).

Since particle size is a major factor influencing the deposition velocity, the relative contribution of dry to total deposition is higher near the sources. The size dependence of the deposition velocity of particles is shown in Fig. I.6a-5. The impact of atmospheric fluxes of trace metals into terrestrial and aquatic ecosystems has recently been reviewed by SALOMONS (1986). Forest areas which have been studied in detail include the Walker Branch Watershed in the USA (LINDBERG et al., 1979; LINDBERG and TURNER, 1983), the Solling Forest in the FRG (R. MAYER, 1981) and the Vienna Woods in Austria (KAZDA and GLATZEL, 1984). Detailed records on metal concentrations in atmospheric deposition are also available for the FRG (NÜRNBERG et al., 1983; VALENTA et al., 1986). Top soils under forest stands show a strong enrichment of anthropogenic trace metals.

However, a part of this enrichment is due to the leaching of trace metals (LINDBERG and HARRISS, 1981). Atmospheric dry and wet deposition contribute significantly to the metal fluxes to forest soils (LINDBERG et al., 1982).

A strong north to south gradient of Pb, As, and Sb in top soils with concentration differences of a factor of 10 in Norway is indicative for long-range transport of trace metals from Central Europe to the northern terrestrial ecosystems (RAMBAEK and STEINNES, 1980). This conclusion has been supported by similar findings in a study of trace element profiles in ombrotrophic bogs from different parts of Norway (HVATUM et al., 1983) as well as from trace element distribution patterns in snow deposits in the Arctic region (ROSS and GRANAT, 1986). Atmospheric deposition is also a major source of metal input into many aquatic ecosystems (SALOMONS, 1986).

For lakes in industrialized areas, e.g., Lake Michigan, the atmospheric load is especially important for lead (60%) and zinc (33%), while for Co, Cd, and Mn the atmospheric flux has been estimated to be 11–13% of the total input. For Al, Fe, and Co the atmosphere is a minor source (EISENREICH, 1980). For acid sensitive lakes, metal concentrations tend to increase with decreasing pH (DICKSON, 1980; BORG, 1983). This effect can be explained by a higher tendency of the metals to remain in solution at lower pH levels and by a possible solubilization of metals from the sediment.

In remote softwater lakes sediment profiles may be used to evaluate enrichment trends due to anthropogenic activities. Trace metal profiles in sediments of 10 lakes in Ontario (Algonquin Provincial Park) indicate a 2-fold enrichment of Ni, Cu, Zn, and Cd and a 25-fold increase of Pb during the past 100 years (WONG et al., 1984).

Enrichment of lead is even found in sediments from the Atlantic and Pacific oceans (SCHAULE and PATTERSON, 1981, 1983). Present-day fluxes of Pb appear to be around an order of magnitude higher (68 ng/cm^2 per year) in the Pacific as compared to pre-industrial levels (1–7.5 ng/cm^2). The present-day fluxes of Pb into the Atlantic are estimated to range from 170 to 330 ng/cm^2 per year. In coastal seawater such as the Western Mediterranean basin soil derived particles originated from arid areas (in this case the Sahara). The atmospheric flux of anthropogenic trace metals, however, was dominated by aerosols from industrialized regions of Western Europe. Volcanic activity (Mount Etna) contributes selenium. The atmospheric input of Cr, Hg, Pb, and Zn into the Western Mediterranean basin is of the same order of magnitude as the riverine and coastal inputs of these components (ARNOLD et al., 1983). For the southern bight of the North Sea estimates even indicate a predominance of the atmospheric input of Cu, Zn, Ag, and Pb (which occurs mainly via wet deposition) as compared to the input by the Scheldt river (DEDEURWAERDER et al., 1985).

7 Historical Trends

The longest historical record about the air chemistry of the atmosphere is found in the 3000 m deep ice layers in Greenland and the Antarctic. The longest core drilled in the antarctic ice contains the accumulated material from the past 160000 years (LEGRAND et al., 1988). It is highly interesting to assess whether the human emissions of metals have changed the atmospheric aerosol composition on a global or hemispheric scale. Available data compiled by BOUTRON (1986) indicate that while Greenland is clearly affected by some anthropogenic metals, the Antarctic shows no significant increase in the metal content of the recent snow layers as compared to some hundred years old ice layers.

According to data from MUROZUMI et al. (1969) and NG and PATTERSON (1981), the lead concentrations in Greenland ice have increased about 200-fold from prehistoric times to the present (Fig. I.6a-6). From the beginning of this century up to now, the increase has been about 4-fold. Also for zinc a 3- to 4-fold increase during the last century has been derived (HERRON et al., 1977).

The concentrations of trace metals in the arctic ice layers are, however, very low (for Pb and Zn in the range of $1-400$ ng/kg and $20-300$ ng/kg, respectively) so that highly specialized procedures for sample handling and analysis are required.

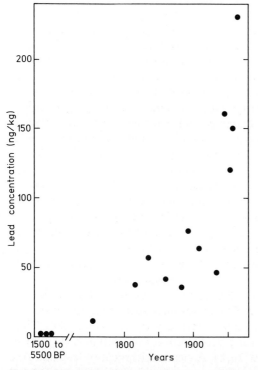

Fig. I.6a-6. Depth profiles of lead in ice cores from Northwest Greenland. (From BOUTRON, 1986; after MUROZUMI et al., 1969; NG and PATTERSON, 1981).

Table I.6a-5. Concentration Ranges and "Typical Concentrations" (in parentheses) for Trace Elements in Recent Snow from Alpine, Greenland and Antarctic Snow. Compiled from BATIFOL and BOUTRON (1984) and BOUTRON (1986). Concentration Ranges are Ranges from Averaged Results Obtained for Different Sites (in 10^{-9} g/g)

	Pb	Cd	Cu	Zn	Ag
Alps (recent snow, >3000 m altitude)	1–12 (2)	0.1–0.4 (0.2)	0.3–4 (1.5)	2–11 (2.5)	0.01–0.03 (0.02)
Greenland (recent snow)	0.1–0.9 (0.3)	0.001–0.030 (0.01)	0.03–0.1 (0.07)	0.1–0.4 (0.3)	0.007–0.01 (0.008)
Antarctic (recent snow)	0.001–0.005 (0.003)	0.001–0.01 (0.005)	0.015–0.06 (0.03)	– (0.05)	0.002–0.01 (0.005)

Due to large uncertainties regarding the contaminations involved in ice core analysis no reliable data exist on possible trends of other metals in the Greenland ice layers (see BOUTRON, 1986).

Aged ice layers can also be found at high elevation sites at mid latitudes. BRIAT (1978) analyzed the trace metal trends in an ice core from the Mont Blanc massiv covering the years 1948–1974. The author concludes that the levels of Pd, V, and Cd have increased by a factor of two during the observation period, whereas no statistically significant trends were found for Mn, Cu, and Zn. For the elements Pb, Cd, Cu, Zn, and Ag the concentrations in high alpine snow from Mont Blanc are roughly one order of magnitude higher than those in surface snow from central Greenland and two orders of magnitude higher than those measured in recent snow in the high central plateau areas of the Antarctic (BATIFOL and BOUTRON, 1984) (Table I.6a-5).

Several other approaches concerning retrospective studies of trace metal enrichment in lake sediments, ombrotrophic peatlands, and tree-rings have been proposed in the last years. The use of lake sediments for long-term retrospective trace metal monitoring has been questioned because of the strong pH dependence of trace metal mobility in lakes (NRIAGU and WONG, 1986; ARAFAT and NRIAGU, 1986). Lake sediments can, however, be used to study the impact of local metal emission sources on the lake ecology (NRIAGU and RAO, 1987). Tree rings may also reveal information about trace metal accumulation due to local anthropogenic emission sources (BAES and McLAUGHLIN, 1984); their use as long-term monitors of trace metal pollution is not conclusive at present (BERISH and RAGSDALE, 1985).

More promising for long-term retrospective deposition monitoring are ecosystems for which the major source of nutrients is atmospheric deposition. Such conditions are found in ombrotrophic bogs and peatlands (GLOOSCHENKO, 1986). Bog vegetation has been shown to be a good biomonitoring substrate for surveys of regional trace metal deposition (PAKARINEN, 1981; GLOOSCHENKO, 1986). Time resolved studies have been performed using depth profiles in peatlands (LIVETT et al., 1979; PAKARINEN and TOLONEN, 1977a, b). A clear enrichment of lead and zinc in the more recent layers dated around 1900 as compared to older layers (dated 1400–1800) has been found at a Finnish and a British site. Interpretation of the results, however, is problematic due to dating problems and trace metal mobilities in the more recent part of the substrate (GLOOSCHENKO, 1986).

8 Atmospheric Aerosol Sampling and Analysis
(for Biomonitoring see Chapter I.4c)

For the analytical characterization of airborne particles the combined use of special sampling methods and chemical and physical analytical methods is required (HEINRICH, 1980; LIOY et al., 1980; MALISSA, 1976; MCCRONE and DELLY, 1973; LIU et al., 1980).

8.1 Sampling of Airborne Particles with Not Classifying Methods

The sampling method most widely used for ambient particles is the "high volume sampler". This instrument has been standardized in many countries (FRG: VDI, 1972; USA: US EPA, 1971) and collects total suspended particles with sampling rates of 40–100 m^3/h. During a sampling interval of 24 h 0.x–x g of particles are retained which are evaluated gravimetrically and can be subjected to a trace metal analysis. Advantages of high volume samplers are: The large filter size allows to be cut to several aliquots that can be used for (a) various types of analysis, (b) replicate analysis, (c) sharing the sample with other groups, (d) storage of aliquots as "backup" samples. Disadvantages are: High power consumption, high noise production and in some cases contamination of the particulate samples by abrasion products of the pump have been reported (COUNTESS, 1974).

Since a number of years preseparating inlet systems have been available for high volume samplers which exclude the particles larger than 10 µm from penetrating the filter. Such systems fulfill the PM$_{10}$ standard established by the US EPA in 1984 (MCFARLAND et al., 1984). Preseparators for the *respirable particle* fraction are also available for "mid" and "low volume" samplers which are small apparatuses for filter sampling (ASTM, 1965; HRUDEY, 1977; WEDDING, 1982).

In automated air quality measurement stations semicontinuously registering instruments (β-radiation monitors, tape samplers) are used for sampling and measuring the particle concentrations. Under certain precautions the deposits on the filter tapes can be subjected to chemical analysis of selected components.

8.2 Sampling with Classifying Methods

The selection of a method for size separating sampling of particulate matter is dependent on the size range and the amount of sampled particles required for the analysis. A survey of size classifying sampling methods has been given by IAEA (1978), HOCHRAINER (1978), LBL (1975), and HIDY (1986). The most widely used principle of size classifying sampling of particles is the multistage impaction. Cascade impactors with 4–9 impaction stages in the size range of 0.02–20 µm aerodynamic diameter and sampling rates of 0.005–70 m^3/h are commercially available and are used for emission and ambient sampling (MARPLE and WILLEKE, 1976; HIDY, 1986; PUXBAUM, 1979; LODGE and CHAN, 1987). Very favorable calibration results in the size range of 0.06–10 µm AD have been obtained for a low-pressure impactor designed by BERNER (1978) (WANG and JOHN, 1988).

For routine sampling of two size fractions of atmospheric particles the "dichotomous virtual impactor" is used, which can be equipped with a sample changer for up to 36 samples (GOULDING et al., 1978). Further possibilities of dichotomous sampling are offered by the "tandem filter" technique (CAHILL et al., 1977) and by the use of cyclone preseparators (LIPPMANN, 1976; JOHN and REISCHL 1980).

8.3 Special Sampling Techniques

For the analytical characterization of single particles the deposition of the particle on very flat surfaces is required (e.g., "Nuclepore" filters or organic foils) (MCCRONE and DELLY, 1973; GRASSERBAUER, 1978a; SPURNY et al., 1979; CHATFIELD, 1984).

8.4 Diffusion Controlled Separation of Aerosols

Mixtures of gases and particles can be separated by diffusion denuders combined with filter samplers. Chemically reactive gases are collected on the surface of selectively coated diffusion denuder tubes, whereas particulate components are collected on subsequently placed filters (STEVENS et al., 1978; KLOCKOW, 1982). Until now this method has been exclusively used for non-metallic components. However, it can be foreseen that denuder techniques may also be used for the separation of gaseous and particulate metallic components.

8.5 Bulk Analysis of the Elements

For the bulk analysis of metallic elements in atmospheric particles spectroscopic methods (e.g., AAS, ICP, XRF, PIXE, SSMS) are widely used (Chapter I.4a). A crucial step is the sample preparation. The analyst has to decide whether the total sample should be digested by a thermal procedure or whether digestion in acids is sufficient. In special cases electrochemical methods can be used for the analysis. For large-scale monitoring non-destructive multielement techniques are preferred such as instrumental neutron activation analysis (INAA) (ONDOV et al., 1979; RAGAINI, 1978) or energy dispersive X-ray fluorescence analysis (EDXRF) (LEYDEN, 1978). For the analysis of very small samples electron beam microanalysis (EPMA) (GRASSERBAUER, 1978a, b) and more recently PIXE (CAHILL, 1975) have been used.

8.6 Compound Specific Analysis

Due to the complex composition of ambient particles the compound specific analysis of metallic components is a difficult task. As a consequence most studies of metallic compounds have been performed in emission samples where the compounds are in an enriched state and the composition is generally less complex. Examples for compound specific analysis in source samples are, e.g., lead compounds in automotive emissions (HABIBI, 1973), vanadium compounds in fly ashes (HENRY and KNAPP, 1980), or manganese compounds in emissions from turbines (HARKER et al., 1975). A speciation of Fe compounds in car exhaust gas and street dust has been performed by Moessbauer spectroscopy (EYMERY et al., 1978; ISMAIL et al., 1985).

Potential methods for the analysis of compounds in ambient particulate samples are X-ray diffraction (McCRONE and DELLY, 1973), infrared spectroscopy (KELLNER, 1978), AES (Auger electron spectroscopy), ESCA (electron spectroscopy for chemical analysis), SAED (selected area electron diffraction), and SIMS (secondary ion mass spectroscopy) (KEYSER et al., 1978) (see also Sect. 5 on speciation).

8.7 Single Particle Analysis and Surface Characterization of Airborne Particles

The determination of size, shape, surface structure, and chemical composition are key steps for the characterization of atmospheric single particles. Table I.6a-6 gives a survey of the methods used for single particle analysis: Light microscopy (LM), electron beam microanalysis by electron probes (EPMA) or scanning electron microscopy (SEM), ion probe microanalysis (IPMA), transmission electron microscopy (TEM), and scanning transmission electron microscopy (STEM). The selection criteria for the use of the listed methods are the size of the particles and the elements to be analyzed (McCRONE and DELLY, 1973; GRASSERBAUER, 1978a,b; HEINRICH, 1980; POST and BUSECK, 1984). Automated single particle analytical methods such as CCSEM (computer controlled scanning electron microscopy) are used for source apportionment studies of aerosols via the PCB (particle class balance) technique (KIM and HOPKE, 1988).

For particulate emissions particularly in fly ash samples it has been found that certain elements – antimony, arsenic, lead, cadmium, chromium, cobalt, manganese, nickel, zinc, sulfur, selenium, thallium, and vanadium – appear to be enriched on the surface of the particles in a water-soluble form (HANSEN and FISHER, 1980; KEYSER et al., 1978).

9 Source Analysis

Results for metallic elements in aerosol samples which are obtained by multielement techniques form data sets from which information about the sources of the components can be extracted (GORDON, 1980). Such methods which make use of data

Table I.6a-6. Survey of Analytical Methods for the Characterization of Individual Airborne Particles. (After GRASSERBAUER, 1978a)

Analytical Method	Reagent	Signal	Analytical Information	Relative Sensitivity	Lower Limit of Particle Diameter (μm)
LM	Light	Reflected, transmitted light	Type of compounds (species, structure), size, shape, morphology	Only pure species can be identified	0.5
EPMA, SEM	Electrons	X-Ray spectrum	Type of elements and their concentration. Number of particles of a specific composition	>0.1%	0.1
		SE	Shape, size, morphology. Number of particles of a specific composition		0.01
		BSE, AE	Shape, size, morphology. Number of particles of a specific composition		0.1
IPMA	Ions (O_2^+, O^-, Ar^+)	Secondary ions	Type of elements and their concentration	ppm	0.5
STEM, TEM	Electrons	X-Ray spectrum	Type of elements and their concentration	Major and minor compounds	0.02
		Secondary electrons	Shape, size, morphology	>0.1%	0.005
		Transmitted electrons	Size, shape		0.001
		Diffracted electrons	Structure and lattice parameters	Pure species	0.02
		Energy spectrum of	Type of elements	Major components	0.01

obtained at receptor points are called receptor models. The most important receptor models are chemical mass balances (CMB), enrichment factors, time series correlation, multivariate models and spatial models (COOPER and WATSON, 1980; GORDON, 1988). Recently dispersion modelling has also been used to explain the occurrence of trace metal concentrations at a remote site in the Arctic (PACYNA et al., 1985).

The two most widely used receptor models in industrialized regions are the chemical mass balance (CMB) and various forms of factor analysis (HOPKE, 1986). The CMB model requires information about the composition of the contributing aerosol sources in the model region. A great achievement of the CMB method was the identification of road dust as a major contributor to the urban aerosol mass

(COOPER, 1980) as well as the identification of wood combustion as an important aerosol source (CORE et al., 1981). Major drawbacks of the CMB method arise when reactive aerosol components are involved (e.g., nitrates, ammonium compounds, organic constituents), when the regional background has a significant impact on the local aerosol mass concentration, and when several sources with similar elemental profiles contribute to the aerosol composition (LOWENTHAL et al., 1987). In a simulation study it was shown that for applications on the local scale CMB models yield "acceptable" accuracy and precision (JAVITZ et al., 1988). Multivariate statistical methods, e.g., factor analysis (FA) and target transformation factor analysis (TTFA) do not require preinformation about the composition of possibly contributing particulate emissions. However, their resolving power to discriminate various source contributions is limited to 4–6 main factors influencing the aerosol mass (LOWENTHAL and RAHN, 1987). On regional aerosols satisfactory results are obtained for only 1–2 factors concerning the pollution elements.

The main advantage of FA and TTFA is to identify unusual sources that may not have previously considered and to find the major contributing source classes (HOPKE, 1988). More recently microscopic methods have also been applied to relate individually analyzed particles to suspected sources (JOHNSON and MCINTYRE, 1983; VAN BORM and ADAMS, 1988; KIM and HOPKE, 1988). This approach seems to be applicable especially in regions where emissions of metal components contribute significantly to the aerosol mass.

In remote areas CMB methods are not applicable due to the mixed influence of numberless sources. FA generally tends to uncover obviously influencing sources such as maritime, crustal, and mixed anthropogenic ones (HEIDAM, 1981). More promising for such regions is the use of enrichment factors (ZOLLER et al., 1974) and of tracer systems (RAHN, 1985). The enrichment factor EF_{crust} relates the concentration of a given element in air to the concentration (X) of a crustal element (Al, Ti, Sc, or Fe) in air, normalized to the ratio of the given element concentration in the crust related to the reference element in the crust:

$$EF_{crust} = \frac{X_{air}/Al_{air}}{X_{crust}/Al_{crust}} .$$

EFs near unity suggest that crustal erosion is the primary source of the observed element in the atmosphere. EFs greater than unity indicate that other sources are the main contributors to the concentrations of the observed element. A drawback of this method is the unability to discriminate between anthropogenic contributions and other natural processes which produce an enrichment of the abundancy of the observed metal as compared to crustal abundancies. A delicate problem when using EFs, especially in polluted regions, is the similarity of the matrix of major elements in coal fly ash and the crustal composition. A more sophisticated tool is the use of elemental ratios in a multitracer system (RAHN and LOWENTHAL, 1984). With the use of a 7-element tracer system, LOWENTHAL and RAHN (1985) came to the conclusion that roughly 70% of the tracer elements observed in Alaska came from the USSR, 25% from Europe and the remainder from North America. Support for the tracer sets used for Central Europe has been found recently by BORBELY-KISS et al. (1988).

There is an ongoing search for new tracer sets forming signatures of respective sources ("true" tracers would be emitted exclusively from single source types; while this is rather uncommon, elemental ratios are used as signatures from different sources). However, no simple solutions have arrived until now. Iridium is emitted from volcanoes (ZOLLER et al., 1983) but is also used as tracer for meteoritic material (TUNCEL and ZOLLER, 1987). Selenium is emitted from volcanoes (TUNCEL and ZOLLER, 1987) and is also an important tracer for coal emissions (DUTKIEWICZ and HUSAIN, 1988). Gaseous boron has been ascribed to be an important tracer for coal; however, a marine influence has also to be taken into account (RAHN and FOGG, 1983). Finally, lanthanides have been used as tracers for petrochemical and oil refining activities (OLMEZ and GORDON, 1985). With the further development of receptor models it can be expected that their use will be extended to calibration and testing of dispersion models (HOPKE, 1986).

Acknowledgement. Helpful comments from K. A. Rahn for the sections on sampling and source analysis are gratefully acknowledged.

References

Adams, F.C., Van Craen, M.J., and Van Espen, P.J. (1980), *Enrichment of Trace Elements in Remote Aerosols*. Environ. Sci. Technol. 14, 1002–1005.

Arafat, N., and Nriagu, J.O. (1986), *Simulated Mobilization of Metals from Sediments in Response to Lake Acidification*. Water Air Soil Pollut. 31(3–4), 991–998.

Arnold, M., Seghaier, A., Martin, D., Buat-Menard, P., and Chesselet, R. (1983), *Geochemistry of Marine Aerosol above the Western Mediterranean Sea*. Journ. Etud. Pollut. Mar. Mediterr. 6th, (Meeting Date 1982), Vol. 6, pp. 27–37. Commission Internationale pour l'Exploration Scientifique de la Mer Mediterranée, Monaco.

ASTM (American Society for Testing and Materials) (1965) ASTM D 2009–2065, Philadelphia.

Baes, III C.F., and McLaughlin, S.B. (1984), *Trace Elements in Tree Rings: Evidence of Recent and Historical Air Pollution*. Science 224, 494–497.

Barrie, L.A., Hoff, R.M., and Daggupaty, S.M. (1981), *The Influence of Mid-latitudinal Pollution Sources on Haze in the Canadian Arctic*. Atmos. Environ. 15, 1407–1419.

Batifol, F.M., and Boutron C.F. (1984), *Atmospheric Heavy Metals in the High Altitude Surface Snows from Mt. Blanc, French Alps*. Atmos. Environ. 18, 2507–2515.

Bennett, B.G. (1981), *Exposure Commitment Assessment of Environmental Pollutants*, Vol. 1, No. 1, MARC Report 23. Monitoring and Assessment Research Center, Chelsea College, London.

Berish, C.W., and Ragsdale, H.L. (1985), *Chronological Sequence of Element Concentrations in Wood of Carya spp. in the Southern Appalachian Mountains*. Can. J. For. Res. 15, 477–483.

Berner, A. (1978), *A Five Stage Cascade Impactor for Measurement of Mass-Size Distributions of Aerosols* (in German). Chem. Ing. Tech. 50, 399.

Biggins, P.D.E., and Harrison, R.M. (1979a), *Atmospheric Chemistry of Automotive Lead*. Environ. Sci. Technol. 13, 558–565.

Biggins, P.D.E., and Harrison, R.M. (1979b), *The Identification of Specific Chemical Compounds in Size-fractionated Atmospheric Particulates Collected at Roadside Sites*. Atmos. Environ. 13, 1213–1216.

Bloch, P., Adams, F., Van Landuyt, J., and Van Goethem, L. (1980), *Morphological and Chemical Characterization of Individual Aerosol Particles in the Atmosphere*, in: Versino, B. (ed.): *Proceedings Symposium Physico-Chemical Behaviour of Atmospheric Pollutants*. EUR 6621, ECSC-EEC-EAEC, Brussels, Luxembourg.

Borbely-Kiss, I., Haszpra, L., Koltay, E., Laszlo, S., Meszaros, A., Meszaros, E., and Szabo, G. (1988), *Elemental Concentrations and Regional Signatures in Atmospheric Aerosols over Hungary. Phys. Scr. 37*, 299–304.

Borg, H. (1983), *Trace Metals in Swedish Natural Waters. Hydrobiologia 101*, 27–34.

Boutron, C. F. (1986), *Atmospheric Toxic Metals and Metalloxides in the Snow and Ice Layers Deposited in Greenland and Antarctic from Prehistoric Times to Present*, in: Nriagu, J. O., and Davidson, C. I. (eds.): *Toxic Metals in the Atmosphere*. Wiley, New York.

Briat, M. (1978), *Evaluation of Levels of Pb, V, Cd, Zn, and Cu in the Snow of Mt. Blanc During the Last 25 Years. Atmospheric Pollution*, Proceedings 13th Int. Coll. Paris, in: Benarie, M. M. (ed.): *Studies in Environmental Sciences*, Vol. 1. Elsevier, Amsterdam.

Cahill, T. A. (1975), *Ion Ecxited X-ray Analysis of Environmental Samples*; in: Ziegler, J. (ed.): *New Uses of Ion Accelerators*, pp. 1–72. Plenum Press, New York.

Cahill, T. A., Ashbaugh, L. L., Barone, J. B., Eldred, R. A., Ferney, P. J., Flocchini, R. G., Goodart, Ch., Shaddoan, D. J., and Wolfe F. W. (1977), *Analysis of Respirable Fractions in Atmospheric Particulates Via Sequential Filtration. J. Air Pollut. Control Assoc. 27*, 675–678.

Cass, G. R., and McRae, G. J. (1983), *Source-Receptor Reconciliation of Routine Air Monitoring Data for Trace Metals: An Emission Inventory Assisted Approach. Environ. Sci. Technol. 17*, 129–139.

Cass, G. R., and McRae, G. J. (1986), *Emission and Air Quality Relationships for Atmospheric Trace Metals*, in: Nriagu, J. O., and Davidson, C. I. (eds.): *Toxic Metals in the Atmosphere*. Wiley, New York.

Chatfield, E. J. (1984), *Determination of Asbestos Fibers in Air and Water*. ISO-TC-1774 Report.

Chen, L., and Duce, R. A. (1983), *The Sources of SO_4^{2-}, V and Mineral Matter in Aerosol Particles over Bermuda. Atmos. Environ. 17*, 2055–2064.

Coles, D. G., Ragaini, R. C., Ondov, J. M., Fisher, G. L., Silberman, D., and Prentice, B. (1979), *Chemical Studies of Stack Fly Ash from a Coal Fired Power Plant. Environ. Sci. Technol. 13*, 455–459.

Cooper, J. A. (1980), *J. Air Pollut. Control Assoc. 30*, 855–860.

Cooper, J. A., and Watson, J. G. Jr. (1980), *Receptor Oriented Methods of Air Particulate Source Apportionment. J. Air Pollut. Control Assoc. 30*, 1116–1125.

Core, J. E., Hanrahan, P. L., and Cooper, J. A. (1981), *Air Particulate Control Strategy Development – A. New Approach Using Chemical Mass Balance Methods*, in: Macias, E. S., and Hopke, P. K. (eds.): *Atmospheric Aerosols: Source/Air Quality Relationships*. ACS Symposium Series 167, Washington, DC.

Countess, R. J. (1974), *J. Air Pollut. Control Assoc. 24*, 605.

Craig, P. J. (1980), *Metal Cycles and Biological Methylation*, in: Hutzinger, O. (ed.): *The Handbook of Environmental Chemistry*, Vol.1, Part A, *The Natural Environment and the Biogeochemical Cycles*, pp. 169–227. Springer, Berlin-New York.

Cunningham, W. C., and Zoller, W. H. (1981), *The Chemical Composition of Remote Area Aerosols. J. Aerosol. Sci. 12*, 367–384.

Davidson, C. I., and Osborn, J. F. (1986), *The Sizes of Airborne Trace Metal Containing Particles*, in: Nriagu, J. O., and Davidson, C. I. (eds.): *Toxic Metals in the Atmosphere*. Wiley, New York.

Davis, B. L. (1981), *Quantitative Analysis of Crystalline and Amorphous Airborne Particulates in the Provo-Orem Vicinity, Utah. Atmos. Environ. 15*, 613–618.

Dedeurwaerder, H. L., Baeyens, W. F., and Dehairs, F. A. (1985), *Estimates of Dry and Wet Deposition of Several Trace Metals in the Southern Bight of the North Sea*, in: Lekkas, T. D. (ed.): *Heavy Metals in the Environment, 5th International Conference*, Vol. 1, pp. 135–137. CEP Consultants Ltd., Edinburgh.

Dickson, W. (1980), *Properties of Acidified Waters*, in: Drablos, D., and Tollan, A. (eds.): *Ecological Impact of Acid Precipitation*. Proceedings International Conference, Sandefjord, Norway, March 11–14.

Duce, R. A., Unni, C. K., Ray, B. J., Prospero, J. M., and Merrill, J. T. (1980), *Long Range Atmospheric Transport of Soil Dust from Asia to the Tropical North Pacific: Temporal Variability*. Science *209*, 1522–1524.

Duce, R. A., Arimoto, R., Ray, B. J., Unni, C. K., and Harder, P. J. (1983), *Atmospheric Trace Elements at Enewetak Atoll: 1, Concentrations, Sources, and Temporal Variability*. J. Geophys. Res. *88*, 5321–5342.

Dutkiewicz, V. A., and Husain, L. (1988), *Spatial Pattern of Non-Urban Se Concentrations in the Northeastern US and its Pollution Source Implications*. Atmos. Environ. *22*, 2223–2228.

Eisenreich, S. J. (1980), *Atmospheric Input of Trace Metals to Lake Michigan*. Water Air Soil Pollut. *13*, 287–301.

Eymery, J. P., Rajn, S. B., and Moine, P. (1978), J. Phys. *D11*, 2147–2149.

Flagan, R. C., and Friedlander, S. K. (1978), *Particle Formation in Pulverized Coal Combustion – A Review*, in: Shaw, D. T. (ed.): *Recent Developments in Aerosol Science*. Wiley, New York.

Galloway, J. N., Eisenreich, S. J., and Scott, B. C. (1980), *Toxic Substances in Atmospheric Deposition: A Review and Assessment*, National Atmospheric Deposition Program Report NC-141. U.S. Environmental Protection Agency Report EPA-560/5-80-001, 146 p. Washington, DC.

Germani, M. S., Small, M., Zoller, W. H., and Moyers, J. L. (1981), *Fractionation of Elements During Copper Smelting*, Environ. Sci. Technol. *15*, 299–305.

Glooschenko, W. A. (1986), *Monitoring the Atmospheric Deposition of Metals by Use of Bog Vegetation and Peat Profiles*, in: Nriagu, J. O., and Davidson, C. I. (eds.): *Toxic Metals in the Atmosphere*. Wiley, New York.

Gordon, G. E. (1980), *Receptor Models*. Environ. Sci. Technol. *14*, 792–800.

Gordon, G. E. (1988), *Receptor Models*. Environ. Sci. Technol. *22*, 1132–1142.

Gordon, G. E., Moyers, J. L., Rahn, K. A., Gatz, D. F., Dzubay, T. G., Zoller, W. H., and Corrin, M. H. (1978), *Atmospheric Trace Elements: Cycles and Measurements*. Report of the National Science Foundation Atmospheric Chemistry Workshop. Panel on Trace Elements, National Center for Atmospheric Research, Boulder, Colorado.

Goulding, F. S., Jaklevic, J. M., and Loo, B. W. (1978), US EPA Report No. EPA-600/4/78-034.

Grasserbauer, M. (1978a), in: Malissa, H. (ed.): *Analysis of Airborne Particles by Physical Methods*, Chap. 8. CRC Press, West Palm Beach, Florida.

Grasserbauer, M. (1978b), *The Present State of Local Analysis*. Mikrochim. Acta (Wien) 1978 I, 329–350.

Greenberg, R. R., Gordon, G. E., Zoller, W. H., Jacko, R. B., Neuendorf, D. W., and Yost, K. J. (1978), *Composition of Particles Emitted from the Nicosia Municipal Incinerator*. Environ. Sci. Technol. *12*, 1329–1232.

Habibi, K. (1973), *Characterization of Particulate Matter in Vehicle Exhaust*. Environ. Sci. Technol. *7*, 223.

Hales, J. M. (1989), *A Generalized Multidimensional Model for Precipitation Scavenging and Atmospheric Chemistry*. Atmos. Environ. *23*, 2017–2031.

Hansen, L. D., and Fisher, G. L. (1980), *Elemental Distribution in Coal Fly Ash Particles*. Environ. Sci. Technol. *14*, 1111–1117.

Harker, A. B., Pagni, P. J., Novakov, T., and Hughes, L. (1975), Chemosphere *6*, 339.

Harrison, R. M. (1986), *Chemical Speciation and Reaction Pathways of Metals in the Atmosphere*, in: Nriagu, J. O., and Davidson, C. I. (eds.): *Toxic Metals in the Atmosphere*. Wiley, New York.

Heidam, N. Z. (1981), *On the Origin of the Arctic Aerosol: A Statistical Approach*. Atmos. Environ. *15*, 1421–1427.

Heidam, N. Z. (1986), *Trace Metals in the Arctic*, in: Nriagu, J. O., and Davidson, C. I. (eds.): *Toxic Metals in the Atmosphere*. Wiley, New York.

Heinrich, K. F. J. (ed.) (1980), *Characterization of Particles*. NBS Spec. Pub. 533. US Government Printing Office, Washington, D.C.

Heintzenberg, J., Hansson, H. C., and Lannefors, H. (1981), *The Chemical Composition of Arctic Haze at Ny Alesund, Spitsbergen. Tellus 33*, 162–171.

Henry, W. M., and Knapp, K. T. (1980), *Compound Forms of Fossil Fuel Fly Ash Emissions. Environ. Sci. Technol. 14*, 450–456.

Herron, M. M., Langneay, C. C., Weiss, H. V., and Cragin, J. H. (1977), *Atmospheric Trace Metals and Sulfate in the Greenland Ice Sheet. Geochim. Cosmochim. Acta 41*, 915–920.

Hicks, B. B. (1986), *Differences in Wet and Dry Particle Deposition Parameters Between North America and Europe*, in: Lee, S. D., Schneiders, T., Grant, L. D., and Verkerk, P. J. (eds.): *Aerosols*. Lewis Publishers, Chelsea, Michigan.

Hidy, G. M. (1986), *Definition and Characterization of Suspended Particles in Ambient Air*, in: Lee, S. D., Schneider, T., Grant, L. D., and Verkerk, P. J. (eds.): *Aerosols*. Lewis Publishers, Chelsea, Michigan.

Hochrainer, D. (1978), in: Malissa, H. (ed.): *Analysis of Airborne Particles by Physical Methods*, Chap. 2. CRC Press, West Palm Beach, Florida.

Hodge, V. F., and Stallard, M. O. (1986), *Platinum and Palladium in Roadside Dust. Environ. Sci. Technol. 20*, 1058–1060.

Hopke, P. K. (1986), *Quantitative Source Attribution of Metals in the Air Using Receptor Models*, in: Nriagu, J. O., and Davidson, C. I. (eds.): *Toxic Metals in the Atmosphere*. Wiley, New York.

Hopke, P. K. (1988), *Target Transformation Factor Analysis as an Aerosol Mass Apportionment Method: A Review and Sensitivity Study. Atmos. Environ. 22*, 1777–1792.

Hrudey, S. E. (1977), in: Perry, R., and Young, R. J. (eds.): *Handbook of Air Pollution Analysis*, Chap. 1. Chapman & Hall, London.

Husain, L. (1986), *Chemical Elements as Tracers of Pollutant Transport to a Rural Area*, in: Nriagu, J. O., and Davidson, C. I. (eds.): *Toxic Metals in the Atmosphere*. Wiley, New York.

Hvatum, O. O., Bolviken, B., and Steinnes, E. (1983), *Heavy Metals in Norwegian Ombrotrophic Bogs. Ecol. Bull. 35*, 351–356.

IAEA (International Atomic Energy Agency) (1978), *Particle Size Analysis in Estimating the Significance of Airborne Contamination. Tech. Rep. Ser.* No. 179. IAEA, Vienna.

Ismail, S. S., Grass, F., Wiesinger, G., and Mostafa, A. G. (1985), *Different Sources of Contamination Detected by Nuclear Methods*. WMO No. 647, pp. 563–589. Proc. TECOMAC.

ISO TC 146 (1981), *Am. Ind. Hyg. Assoc. J. 42*, A64–A68.

Jaenicke, R. (1986), *Physical Characterization of Aerosols*, in: Lee, S. D., Schneider, T., Grant, L. D., and Verkerk, P. J. (eds.): *Aerosols*. Lewis Publishers, Chelsea, Michigan.

Javitz, H. S., Watson, J. G., and Robinson, N. (1988), *Performance of the Chemical Mass Balance Model with Simulated Local Scale Aerosols. Atmos. Environ. 22*, 2309–2322.

Jiang, S., Robberecht, H., and Adams, F. (1983), *Identification and Determination of Alkylselenide Compounds in Environmental Air. Atmos. Environ. 17*, 111–114.

John, W., and Reischl, G. (1980), *A Cyclone for Size Selective Sampling of Ambient Air. J. Air Pollut. Control Assoc. 30*, 872–876.

Johnson, D. J., and McIntyre, B. L. (1983), *A Particle Class Balance Receptor Model for Aerosol Apportionment in Syracuse, N. Y.*, in: Dattner, S. L., and Hopke, P. K. (eds.): *Receptor Models Applied to Contemporary Pollution Problems*. Air Pollution Control Association, Pittsburgh, Pennsylvania.

Junge, C. E. (1963), *Air Chemistry and Radioactivity*. Academic Press, New York.

Kaiser, G., and Tölg, G. (1980), *Mercury*, in: Hutzinger, O. (ed.): *The Handbook of Environmental Chemistry*, Vol. 3, Part A, *Anthropogenic Compounds*. Springer, Berlin-New York.

Kazda, M., and Glatzel, G. (1984), *Heavy Metal Enrichment and Mobility in the Infiltration Zone of Stemflow from Beeches in the Vienna Woods* (in German). *Z. Pflanzenernähr. Bodenkd. 147*, 743–752.

Kellner, R. (1978), in: Malissa, H. (ed.): *Analysis of Airborne Particles by Physical Methods*, Chap. 22. CRC Press, West Palm Beach, Florida.

Keyser, T. R., Natusch, D. F. S., Evans, C. A. Jr., and Linton, R. W. (1978), *Characterizing the Surfaces of Environmental Particles. Environ. Sci. Technol. 12*, 768–773.

Kim, D., and Hopke, P. K. (1988), *Source Apportionment of the El Paso Aerosol by Particle Class Balance Analysis. Aerosol Sci. Technol. 9*, 221–235.

Klockow, D. (1982), *Analytical Chemistry of the Atmospheric Aerosol*, in: Georgii, H. W., and Jaeschke, W. (eds.): *Chemistry of the Unpolluted and Polluted Troposphere*. D. Reidel Publ., Dordrecht.

Lantzy, R. J., and MacKenzie, F. T. (1979), *Atmospheric Trace Metals: Global Cycles and Assessment of Man's Impact. Geochim. Cosmochim. Acta 43*, 511–525.

LBL (Lawrence Berkeley Laboratory) (1975), *Instrumentation for Environmental Monitoring – Air. Lawrence Berkeley Lab. Rep. LBL. 1*, Vol. 1, Part 2, 1st Ed. Environmental Instrumentation Group, LBL, University of California, Berkeley.

Legrand, M. R., Lorius, C., Barkov, N. I., and Petrov, V. N. (1988), *Vostok (Antarctic) Ice Core: Atmospheric Chemistry Changes over the Last Climatic Cycle (160 000 years). Atmos. Environ. 22*, 317–331.

Leyden, D. E. (1978), in: Malissa, H. (ed.): *Analysis of Airborne Particles by Physical Methods*, Chap. 3. CRC Press, West Palm Beach, Florida.

Lindberg, S. E. (1980), *Mercury Partitioning in a Power Plant Plume and its Influence on Atmospheric Removal Mechanisms. Atmos. Environ. 14*, 227–231.

Lindberg, S. E., and Harriss, R. C. (1981), *The Role of Atmospheric Deposition of an Eastern US Deciduous Forest. Water Air Soil Pollut. 16*, 13–31.

Lindberg, S. E., and Turner, R. R. (1983), *Trace Metals in Rain at Forested Sites in the Eastern United States. Proceedings International Conference on Heavy Metals in the Environment*, Heidelberg. CEP Consultants Ltd., Edinburgh.

Lindberg, S. E., Harriss, R. C., Turner, R. R., Shriner, D. S., and Huff, D. D. (1979), *Mechanisms and Rates of Atmospheric Deposition of Selected Trace Elements and Sulfate to a Deciduous Forest Watershed*. ORNL/TM-6674, 514 pp. Oak Ridge National Laboratory, Oak Ridge, Tennessee.

Lindberg, S. E., Harriss, R. C., and Turner, R. R. (1982), *Reports: Atmospheric Deposition of Metals to Forest Vegetation. Science 215*, 1609–1611.

Linton, R. W., Loh, A., Natusch, D. F. S., Evans, C. A., and Williams, P. (1976), *Surface Predominance of Trace Elements in Airborne Particles. Science 191*, 852–854.

Linton, R. W., Williams, P., Evans, C. A., and Natusch, D. F. S. (1977), *Determination of the Surface Predominance of Toxic Elements in Airborne Particles by IMMS and AES. Anal. Chem. 49*, 1514–1521.

Lioy, P. J., Watson, J. G., and Spengler, J. D. (1980), *APCA Specialty Conference Workshop on Baseline Data for Inhalable Particulate Matter. J. Air Pollut. Control Assoc. 30*, 1126–1130.

Lippmann, M. (1976), *Size Selective Sampling for Inhalation Hazard Evaluations*, in: Liu, B. Y. H. (ed.): *Fine Particles*. Academic Press, New York.

Liu, B. Y. H., Raabe, O. G., Smith, W. B., Spencer, H. W., and Kuykendal, W. B. (1980), *Advances in Particle Sampling and Measurement. Environ. Sci. Technol. 14*, 392–397.

Livett, E. A., Lee, J. A., and Tallis, J. H. (1979), *Lead, Zinc and Copper Analyses of British Blanket Peats. J. Ecol. 67*, 865–891.

Lodge, Jr. J. P., and Chan, T. L. (eds.) (1987), *L 87028, Cascade Impactor Sampling and Data Analysis*. American Industrial Hygienists Association, Akron, Ohio.

Lowenthal, D. H., and Rahn, K. A. (1985), *Regional Sources of Pollution Aerosol at Barrow, Alaska, During Winter 1979/80 as Deduced from Elemental Tracers. Atmos. Environ. 19* (12), 2011–2024.

Lowenthal, D. H., and Rahn, K. A. (1987), *Application of Factor-Analysis Receptor Model to Simulated Urban- and Regional-Scale Data Sets. Atmos. Environ. 21* (9), 2005–2013.

Lowenthal, D. H., Hanumara, R. C., Rahn, K. A., and Currie, L. A. (1987), *Effects of Systematic Error, Estimates and Uncertainties in Chemical Mass Balance Apportionments: Quail Roost II Revisited. Atmos. Environ. 21* (3), 501–510.

Malissa, H. (1976), *Integrated Dust Analysis. Angew. Chem. Int. Ed. Engl. 15*, 141–149.

Marple, V. A., and Willeke, K. (1976), *Inertial Impactors: Theory, Design, Use*, in: Liu, B. Y. H. (ed.): *Fine Particles*. Academic Press, New York.

Mayer, M. (1981), *Contributions for Determining the Size Distribution of Trace Metals in Atmospheric Particles* (in German). Diploma Thesis, Technische Universität, Wien.

Mayer, R. (1981), *Natürliche and anthropogene Komponenten des Schwermetallhaushalts von Waldökosystemen. Gott. Bodenkdl. Ber. 70*, 1–152.

Mayer, R., and Ulrich B. (1982), *Calculation of Deposition Rates from the Flux Balance and Ecological Effects of Atmospheric Deposition upon Forest Ecosystems*, in: Georgii, H. W., and Pankrath, J. (eds.): *Deposition of Atmospheric Pollutants*, pp. 195–200. Reidel Publ., Dordrecht, The Netherlands.

McCrone, W. C., and Delly, J. G. (1973), *The Particle Atlas*, Vol. 1–3, 2nd Ed. Ann Arbor Science, Ann Arbor, Michigan.

McFarland, A. R., Ortiz, C. A., and Bertch, Jr. R. W. (1984), *J. Air Pollut. Control Assoc. 34*, 544–547.

McMurry, P. H., and Wilson, J. C. (1982), *Growth Laws for the Formation of Secondary Ambient Aerosols: Implications for Chemical Conversion Mechanisms. Atmos. Environ. 16*, 121–134.

McMurry, P. H., and Wilson, J. C. (1983), *J. Geophys. Res. 88*, 5101–5108.

Miller, F. J., Gardner, D. E., Graham, J. A., Lee, R. E. Jr., Wilson, W. E., and Bachmann, J. D. (1979), *Size Considerations for Establishing a Standard for Inhalable Particles. J. Air Pollut. Control Assoc. 29*, 610–615.

Moyers, J. L., Ranweiler, L. E., Hopf, S. B., and Korte, N. E. (1977), *Evaluation of Particulate Trace Species in Southwest Desert Atmosphere. Environ. Sci. Technol. 11*, 789–795.

Murozumi, M., Chow, T. J., and Patterson, C. C. (1969), *Chemical Concentrations of Pollutant Lead Aerosols, Terrestrial Dusts and Sea Salts in Greenland and Antarctic Snow Strata. Geochim. Cosmochim. Acta 33*, 1247–1294.

NAS (National Academy of Sciences) (1976), *Selenium*. PB 251 Subcommittee on Selenium, Committee on Medical and Biological Effects of Environmental Pollutants. National Research Council, Washington, DC.

Natusch, D. F. S., and Wallace, J. R. (1974), *Urban Aersosol Toxicity: The Influence of Particle Size. Science 186*, 695–699.

NCAR (National Center for Atmospheric Research) (1982), *Regional Acid Deposition: Models and Physical Processes*. NCAR, Boulder, Colorado.

Ng, A., and Patterson, C. C. (1981), *Natural Concentrations of Lead in Ancient Arctic and Antarctic Ice. Geochim. Cosmochim. Acta 45*, 2109–2121.

Nriagu, J. O. (1979), *Global Inventory of Natural and Anthropogenic Emissions of Trace Metals into the Atmosphere. Nature 279*, 409–411.

Nriagu, J. O., and Davidson, C. I. (1986), *Toxic Metals in the Atmosphere*. Wiley, New York.

Nriagu, J. O., and Rao, S. S. (1987), *Response of Lake Sediments to Changes in Trace Metal Emission from the Smelters at Sudbury, Ontario. Environ. Pollut. 44* (3), 211–218.

Nriagu, J. O., and Wong, H. K. T. (1986), *What Fraction of the Total Metal Flux into Lakes is Retained in the Sediments? Water Air Soil Pollut. 31* (3–4), 999–1006.

Nürnberg, H. W., Valenta, P., and Nguyen, V. D. (1983), *The Wet Deposition of Heavy Metals from the Atmosphere in the Federal Republic of Germany. Proceedings International Conference on Heavy Metals in the Environment Heidelberg*, Vol. I, pp. 115–123. CEP Consultants Ltd., Edinburgh.

O'Connor, B. H., and Jaklevic, J. M. (1981), *Characterization of Ambient Aerosol Particulate Samples from the St. Louis Area by X-Ray Powder Diffractometry. Atmos. Environ. 15*, 1681–1690.

Olmez, I., and Gordon, G. E. (1985), *Science 229*, 966–968.

Ondov, J. M., Ragaini, R. C., and Biermann, A. H. (1979), *Environ. Sci. Technol. 13*, 598–607.

Pacyna, J. M. (1984), *Estimation of the Atmospheric Emissions of Trace Elements from Anthropogenic Sources in Europe. Atmos. Environ. 18*, 41–50.

Pacyna, J. M. (1986a), *Emission Factors of Atmospheric Elements*, in: Nriagu, J. O., and Davidson, C. I. (eds.): *Toxic Metals in the Atmosphere*, Wiley, New York.

Pacyna, J. M. (1986b), *Atmospheric Trace Elements from Natural and Anthropogenic Sources*, in: Nriagu, J. O., and Davidson, C. I. (eds.): *Toxic Metals in the Atmosphere*. Wiley, New York.

Pacyna, J. M. (1987), *Long Range Transport of Heavy Metals – Modelling and Measurements*, in: Preprints *16th International Technical Meeting on Air Pollution Modelling and Applications*. Committee on Challenges of Modern Society, Bruxelles.

Pacyna, J. M., Ottar, B., Tomza, U., and Maenhaut, W. (1985), *Long-Range Transport of Trace Elements to Ny Alesund, Spitsbergen*. Atmos. Environ. 19, 857–865.

Pakarinen. P. (1981), *Metal Content of Ombrotrophic Sphagnum Mosses in NW Europe*. Ann. Bot. Fenn. 18, 281–292.

Pakarinen, P., and Tolonen, K. (1977a), *Distribution of Lead in Sphagnum fuscum Profiles in Finland*. Oikos 28, 69–73.

Pakarinen, P., and Tolonen, K. (1977b), *Vertical Distributions of N, P, K, Zn and Pb in Sphagnum peat*. Suo 28, 95–102.

Parrington, J. R., and Zoller, W. H. (1984), *Diurnal and Longer Term Temporal Changes in the Composition of Atmospheric Particles at Manna Loa, Hawai*. J. Geophys. Res. 89 (D2), 2522–2534.

Pilinis, C., and Seinfeld, J. H. (1987), *Continued Development of a General Equilibrium Model for Inorganic Multicomponent Aerosol*. Atmos. Environ. 21, 2453–2466.

Post, J. E., and Buseck, P. R. (1984), *Characterization of Individual Particles in the Phoenix Urban Aerosol Using Electron-Beam Instruments*. Environ. Sci. Technol. 18, 35–42.

Purdue, L. J. (1986), *US EPA PM_{10} Methodology Review*, in: Lee, S. D., Schneider, T., Grant, L. D., and Verkerk, P. J. (eds.): *Aerosols*. Lewis Publishers, Chelsea, Michigan.

Puxbaum, H. (1979), *Sampling of Inhalable and Lung Penetrating Particles for "Integrated Aerosol Analysis"* (in German). Fresenius Z. Anal. Chem. 298, 110–128.

Puxbaum, H., and Wopenka, B. (1984), *Chemical Composition of Nucleation and Accumulation Mode Particles Collected in Vienna, Austria*. Atmos. Environ. 18, 573–580.

Puxbaum, H., Quintana, E., and Pimminger, M. (1985), *Spatial Distributions of Atmospheric Aerosol Constituents in Linz (Austria)*. Fresenius Z. Anal. Chem. 322, 205–212.

Ragaini, R. C. (1978), in: Malissa, H. (ed.): *Analysis of Airborne Particles by Physical Methods*, Chap. 7. CRC Press, West Palm Beach, Florida.

Rahn, K. A. (1985), *Progress in Arctic Air Chemistry*. Atmos. Environ. 19, 1987–1994.

Rahn, K. A., and Fogg, T. R. (1983), *Boron as a Tracer of Aerosol from Combustion of Coal. Final Technical Report*, DOE/PC/51260-4; Order No. DE84004708, 30 pp. Avail. NTIS, from: Energy Res. Abstr. 1984 9(7), Abstr. No. 11452.

Rahn, K. A., and Lowenthal, D. H. (1984), *Elemental Tracers of Distant Regional Pollution Aerosols*. Science 223, 132–139.

Rahn, K. A., and Lowenthal, D. H. (1985), *Pollution Aerosol in the Northeast: Northeastern-Midwestern Contributions*. Science 228 (4697), 275–284.

Rahn, K. A., Borys, R. D., Shaw, G. E., Schütz, L., and Jaenicke, R. (1979), *Long Range Impact of Desert Aerosol and Atmospheric Chemistry: Two Examples*, in: *Saharan Dust. SCOPE 14*, Wiley, New York.

Rambaek, J. P., and Steinnes, E. (1980), *Atmospheric Deposition of Heavy Metals Studied by Analysis of Moss Samples Using Neutron Activation Analysis and Atomic Absorption Spectrometry*. Nuclear Methods Environmental Energy Research, USDOE CONF-800433, pp. 175–180.

Ross, H. B., and Granat, L. (1986), *Deposition of Atmospheric Trace Metals in Northern Sweden as Measured in the Snowpack*. Tellus 38B (1), 27–43.

Salomons, W. (1986), *Impact of Atmospheric Inputs on the Hydrospheric Trace Metal Cycle*, in: Nriagu, J. O., and Davidson, C. I. (eds.), *Toxic Metals in the Atmosphere*. Wiley, New York.

Schaule, B. K., and Patterson, C. C. (1981), *Lead Concentrations in the Northeast Pacific: Evidence for Global Anthropogenic Perturbations*. Earth Planet Sci. Lett. 54, 97–116.

Schaule, B. K., and Patterson, C. C. (1983), *Perturbations of the Natural Depth Profile in the Sargasso Sea by Industrial Lead.* Proceedings of NATO Advanced Research Workshop of Trace Metals in Seawater, Erice, Italy, 1981, pp. 407–504. Plenum Press, New York.

Schroeder, W. H., Dobson, M., Kane, D. M., and Johnson, N. D. (1987), *Toxic Trace Elements Associated with Airborne Particulate Matter: A Review. J. Air Pollut. Control Assoc.* 37, 1267–1285.

SCOPE (1979), *Saharan Dust*, SCOPE 14. Wiley, New York.

Sehmel, G. A. (1980), *Particle and Gas Dry Deposition: A Review. Atmos. Environ.* 14, 983–1012.

Seinfeld, J. H. (1986), *Atmospheric Chemistry and Physics of Air Pollution.* Wiley, New York.

Slinn, W. G. N. (1982), *Atmos. Environ.* 7, 1785–1794.

Smith, R. D., Campbell, J. A., and Nielson, K. K. (1979), *Concentration Dependence upon Particle Size of Volatilized Elements in Fly Ash. Environ. Sci. Technol.* 13, 553–558.

Spurny, K. R., Lodge, J. P. Jr., Frank, E. R., and Sheelsley, D. C. (1979), *Aerosol Filtration by Means of Nuclepore Filters: Structural and Filtration Properties. Environ. Sci. Technol.* 3, 453.

Stahlhofen, W. (1986), *Regional Deposition of Inhalable Particles in Humans*, in: Lee, S. D., Schneider, T., Grant, L. D., and Verkerk, P. V. (eds.): *Aerosols.* Lewis Publishers, Chelsea, Michigan.

Stahlhofen, W., Gebhart, J., and Heyder, J. (1980), *Am. Ind. Hyg. Assoc. J.* 41, 385–398.

Stelson, A. W., and Seinfeld, J. H. (1981), *Chemical Mass Accounting of Urban Aerosol. Environ. Sci. Technol.* 15, 671–679.

Stevens, R. K., Dzubay, T. G., Russworm, G., and Rickel, D. (1978), *Atmos. Environ.* 12, 55–68.

Task Group on Lung Dynamics (1966), *Health Phys.* 12, 173–207.

Tuncel, G., and Zoller, W. H. (1987), *Atmospheric Iridium at the South Pole as a Measure of the Meteoritic Component. Nature* (London) 329 (6141), 703–705.

Uematsu, M., Duce, R. A., Prospero, J. M., Chen, L., Merrill, J. T., and McDonald, R. L. (1983), *J. Geophys. Res.* 88, 5343–5352.

US EPA (Environmental Protection Agency) (1971), *Reference Methods for the Determination of Suspended Particulates in the Atmosphere (High Volume Method). U.S. Fed. Reg.* 36, No. 84.

US EPA (Environmental Protection Agency) (1982), *Air Quality Criteria for Particulate Matter and Sulfur Oxides*, Vol. III, EPA 600/8-82-092C.

Valenta, P., Nguyen, V. D., and Nürnberg, H. W. (1986), *Acid and Heavy Metal Pollution by Wet Deposition. Sci. Total Environ.* 55, 311–320.

Van Borm, W. A., and Adams, F. C. (1988), *Cluster Analysis of Electron Microprobe Analysis Data of Individual Particles for Source Apportionment of Air Particulate Matter. Atmos. Environ.* 22, 2297–2307.

Vanderborght, B., Mertens, I., and Kretzschmar, J. (1983), *Comparing the Calculated and Measured Aerosol Concentrations and Depositions Around a Metallurgic Plant. Atmos. Environ.* 17, 1687–1701.

VDI (Verein Deutscher Ingenieure) (1972), *Measurement of Particles in Ambient Air* (in German). *VDI-Richtlinie 2463.* VDI-Verlag, Düsseldorf.

V & F (1988), *Application Notes for the CI-MS 500 Real Time Gas Analyzer.* V & F Analyse- und Meßtechnik, Absams, Tyrol, Austria.

Walsh, P. R., Duce, R. A., and Fasching, J. L. (1979), *Considerations of the Enrichment, Sources and Flux of Arsenic in the Troposphere. J. Geophys. Res.* 84, 1719–1726.

Wang, H. C., and John, W. (1988), *Characteristics of the Berner Impactor for Sampling Inorganic Ions. Aerosol Sci. Technol.* 8, 157–172.

Weathers, K. C., Likens, G. E., Bormann, F. H., Bicknell, S. H., Bormann, B. T., Daube, B. C., Eaton, J. S., Galloway, J. N., Keene, W. C., Kimball, K. D., McDowell, W. H., Siccama, T. G., Smiley, D., and Tarrant, R. A. (1988), *Cloudwater Chemistry from Ten Sites in North America. Environ. Sci. Technol.* 22, 1018–1026.

Wedding, J. B. (1982), *Ambient Aerosol Sampling. History, Present Thinking and a Proposed Inlet for Invaluable Particles. Environ. Sci. Technol.* 16, 154–161.

Weisel, C. P. (1981), *The Atmospheric Flux of Elements from the Ocean*. Ph. D. Dissertation, University of Rhode Island, R. I. Kingston.

Whitby, K. T. (1978), *The Physical Characterization of Sulfur Aerosols*. Atmos. Environ. 12, 135–159.

Wiersma, G. B., and Davidson, C. I. (1986), *Trace Metals in the Atmosphere of Rural and Remote Areas*, in: Nriagu, J. O., and Davidson, C. I. (eds.): *Toxic Metals in the Atmosphere*. Wiley, New York.

Wong, H. K. T., Nriagu, J. O., and Coker, R. D. (1984), *Atmospheric Input of Heavy Metals Chronicled in Lake Sediments of the Algonquin Provincial Park, Ontario, Canada*. Chem. Geol. 44 (1–3), 187–201.

Wood, J. M., and Goldberg, E. D. (1977), *Impact of Metals on the Biosphere*, in: Stumm, W. (ed.): *Global Chemical Cycles and their Alteration by Man*. Dahlem Konferenzen, Abakon Verlag, Berlin.

Zoller, W. H. (1983), *Anthropogenic Perturbation of Metal Fluxes into the Atmosphere*, in: Nriagu, J. O. (ed.): *Changing Metal Cycles and Human Health*. Dahlem Konferenzen, Springer, Berlin.

Zoller, W. H., Gladney, E. S., and Duce, R. A. (1974), *Atmospheric Concentrations and Sources of Trace Metals at the South Pole*, Science 183, 198.

Zoller, W. H., Parrington, J. R., and Kotra, J. M. P. (1983), *Iridium Enrichment in Airborne Particles from Kilauea Volcano: January 1983*. Science 222 (4628), 1118–1121.

Additional Recommended Literature

Becker, K. H., and Löbel, J. (1985), *Atmospheric Trace Substances and their Physicochemical Behavior* (in German), 264 p. Springer Verlag, Berlin.

Bølviken, B., and Steinnes, E. (1987), *Heavy Metal Contamination of Natural Surface Soils from Long-range Atmospheric Transport*. Proceedings International Conference on Heavy Metals in the Environment, New Orleans, Vol. 1, pp. 291–293. CEP Consultants Ltd., Edinburgh.

Bowen, H. J. M. (1982), *The Cycles of Copper, Silver and Gold*, in: Hutzinger, O. (ed.): *The Handbook of Environmental Chemistry*, Vol. 1, Part D, pp. 1–27. Springer Verlag, Berlin.

Brimblecombe, P. (1985), *Air, Composition and Chemistry*. Cambridge University Press, Cambridge, U.K.

Craig, P. J. (1980), *Metal Cycles and Biological Methylation*, in: Hutzinger, O. (ed.): *The Handbook of Environmental Chemistry*, Vol. 1, Part A, pp. 169–227. Springer Verlag, Berlin.

Esman, N. A., and Mehlmann, M. A. (1984), *Occupational and Industrial Hygiene, Concepts and Methods*. Princeton Scientific Publishers, Princeton, New Jersey.

Ewers, U. (1988), *WHO Air Quality Guidelines for Europe* (in German). Öff. Gesundh.-Wes. 50, 626–629.

Hertz, J., Schmid, I., and Thöni, L. (1984), *Monitoring of Heavy Metals in Airborne Particles by Using Mosses Collected from the City of Zurich*. Int. J. Environ. Anal. Chem. 17 (1), 1–12.

Hutchinson, T. C., and Meema, K. M. (1987), *Lead, Mercury, Cadmium and Arsenic in the Environment* (SCOPE 31). John Wiley & Sons, Chichester, U.K.

Hvatum, O. Ø., Steinnes, E., and Bølviken, B. (1987), *Regional Differences and Temporal Trends in Heavy Metal Deposition from the Atmosphere* (studied by analyses of ombrophobic peat). Proceedings International Conference, on Heavy Metals in the Environment, New Orleans, Vol. 1, pp. 201–203. CEP Consultants Ltd., Edinburgh.

Klockow, D., and Kaiser, R. D. (1985), *Generation of Various Types of Nickel-containing Aerosols and Detection of Nickel Species in these Aerosols*, in: Brown, S. S., and Sunderman, F. W., Jr. (eds.): *Progress in Nickel Toxicology*, p. 234ff. Blackwells, Oxford.

Larjava, K., Reith, J., and Klockow, D. (1989), *Development and Laboratory Investigations of a Denuder Sampling System for the Determination of Heavy Metal Species in Flue Gases at Elevated Temperatures. Int. J. Environ. Anal. Chem. 38* (1), 31–45.

Macias, E. S., and Hopke, P. K. (1981), *Atmospheric Aerosol Source/Air Quality Relationships. ACS Symp. Ser. 167.* American Chemical Society, Washington, D. C.

Merian, E. (1984), *Introduction on Environmental Chemistry and Global Cycles of Chromium, Nickel, Cobalt, Beryllium, Arsenic, Cadmium and Selenium, and their Derivatives. Toxicol. Environ. Chem. 8* (1), 9–38.

Mislin, H., and Ravera, O. (1986), *Cadmium in the Environment. Experientia Suppl. 50.* Birkhäuser Verlag, Basel.

Nriagu, J. O., and Pacyna, J. M. (1988), *Quantitative Assessment of Worldwide Contamination of Air, Water and Soils by Trace Metals. Nature 333*, 134–139.

Nürnberg, H. W. (1985), *Pollutants and Their Ecotoxicological Significance.* John Wiley & Sons, Chichester, U.K.

Obiols, J., Devesa, R., and Sol, A. (1986), *Speciation of Heavy Metals in Suspended Particulates in Urban Air. Toxicol. Environ. Chem. 13* (1+2), 121–128.

Purves, D. (1985), *Trace Element Contamination of the Environment*, revised 2nd Ed. Elsevier Science Publishers, Amsterdam.

rivm (1989), *Concern for Tomorrow, A National Environmental Survey* (with data on metal concentrations in air and depositions). National Institute of Public Health and Environmental Protection, Bilthoven, The Netherlands.

Rondia, D. (1986), *(Belgian Research on) Metal Cycling in the Environment.* Presses Universitaires de Liège, Belgium.

Schindler, P. W., and Gälli, B. (1988), *Particle Size Distribution of Airborne Trace Metals. 3rd IAEAC Workshop on Toxic Metal Compounds* (Interrelation between Chemistry and Biology). *Int. J. Environ. Anal. Chem. 35* (2), 111–118.

Schulte-Hostede, S. (1988), *Air Pollution and Plant Metabolism.* Elsevier Applied Science, London.

Spurni, Kv. R. (1986), *Physical and Chemical Characterization of Individual Airborne Particulates* (Analytical Chemical Series). Ellis Horwood, Chichester, U.K.

Steinnes, E. (1983), *Contamination of Surface Soils by Heavy Metals from Air Pollution: Significance of Long-range Atmospheric Transport.* Proceedings International Conference on Heavy Metals in the Environment, Heidelberg, Vol. 2, pp. 1170–1172. CEP Consultants Ltd., Edinburgh.

Steinnes, E., Gjengedal, E., Johansen, O., Rambaek, J. P., and Hanssen, J. E. (1987), *Experience from the Use of Mosses as Indicators of Metal Deposition from the Atmosphere.* Proceedings International Conference on Heavy Metals in the Environment, New Orleans, Vol. 2, 57 p. CEP Consultants Ltd., Edinburgh.

Steinnes, E. (1988), *Cadmium in the Terrestrial Environment: Impact of Long-range Atmospheric Transport. 3rd IAEAC Workshop on Toxic Metal Compounds* (Interrelation between Chemistry and Biology). *Toxicol. Environ. Chem. 19* (3+4), 139–145.

Thomé-Kozmiensky, K. J. (1989), *Incineration 3.* Proceedings 6th International Recycling Congress, 914 pages. EF-Verlag für Energie- und Umwelttechnik GmbH, Berlin.

Valerio, F., Brescianini, C., Lazzarotto, A., and Balducci, D. (1986), *Metals in Airborne Particulate as Polynuclear Aromatic Hydrocarbons Source Indicators. Toxicol. Environ. Chem. 13* (1+2), 113–120.

Valerio, F. (1988), *Airborne Metals in Urban Areas. 3rd IAEAC Workshop on Toxic Metal Compounds* (Interrelation between Chemistry and Biology). *Int. J. Environ. Anal. Chem. 35* (2), 101–110.

Van Dop, H. (1986), *Atmospheric Distribution of Pollutants and Modelling of Air Pollution Dispersion*, in: Hutzinger, O. (ed.): *The Handbook of Environmental Chemistry,* Vol. 4, Part A, pp. 107–147. Springer Verlag, Berlin.

Waldichuk, M. (1988), *Exchange of Pollutants and Other Substances between the Atmosphere and the Oceans*, in Hutzinger, O. (ed.): *The Handbook of Environmental Chemistry*, Vol. 2, Part D, pp. 113–151. Springer Verlag, Berlin.

Winchester, J. W. (1980), *Transport Processes in Air*, in: Hutzinger, O. (ed.): *Handbook of Environmental Chemistry*, Vol. 2, Part A, pp. 19–30. Springer Verlag, Berlin.

I.6b Indoor Environments: The Role of Metals

LAWRENCE FISHBEIN, Washington, DC, USA

1 Introduction

There is increasing recognition that the indoor environment may play a critical role in regard to the scope of exposure of an individual to a broad spectrum of constituents, many of which have major toxicological significance. Indoor concentrations of total suspended particles and respirable particulates often exceed outdoor concentrations and the agents cause both specific illnesses and the minor complaints which constitute the "sick building" syndrome (WHO, 1983). Indoor air may be polluted by genotoxins produced indoors and, depending on particle size and air-exchange rate, by particles infiltrating from outdoors (ALZONA et al., 1979; COHEN and COHEN, 1980; DOCKERY and SPENGLER, 1981 a, b; VAN HOUDT et al., 1986). As a consequence of smoking and cooking, indoor levels of genotoxic compounds can often exceed outdoor levels (LÖFROTH et al., 1983; VAN HOUDT et al., 1984). Wood combustion (RUDLING et al., 1981; RAMDAHL et al., 1984; ALFHEIM and RAMDAHL, 1984; VAN HOUDT et al., 1986) and emissions of kerosene heaters (YAMANAKA and MURUOKA, 1984) have also been shown to produce mutagenic compounds. Considering potential interactions it must also be taken into account that absorption of organic compounds on the surface of indoor airborne particles occurs to a larger extent than on outdoor particles (WESCHLER, 1984).

The agents often found in indoor environments are mostly known to be hazardous in high concentration, but the lower limits of their dose-response relationships are poorly defined. Additionally, synergistic toxicological relationships are believed to occur as a result of exposure to agents such as asbestos and radon with cigarette smoke in indoor environments. Hence, although the magnitude of indoor health hazards is not now known, mounting evidence suggests that both identification of agents and the measurement of indoor exposures are critical for a more realistic assessment of the effect of air pollution on health (NAS, 1981; SPENGLER and SEXTON, 1983; SEXTON and WESOLOWSKI, 1985; GAMMAGA and KAYE, 1985; WHO, 1983). This facet is underscored by summaries of human activity pattern studies which indicate that individuals spend the majority of their time indoors or indoors at home. Typical figures range from 60–75% (SZALAI, 1972; CHAPIN, 1974; SEXTON et al., 1984) to nearly 90% for employed men and 95% for homemakers (NAS, 1981; MOSCHANDREAS, 1981; DOCKERY and SPENGLER, 1981 a, b; OTT, 1980; SPENGLER and SEXTON, 1983). Further research is necessary to better understand which sources contribute to which extent to indoor atmospheric pollution and which is the ratio of total individual exposure contributed by indoor atmospheric concentrations.

In contrast to a variety of volatile synthetic organic chemicals (VOS), gases and inorganic compounds and moieties and asbestos frequently reported in the air of homes and offices (EPA, 1978a, b; NAS, 1981; SWEDISH COUNCIL FOR BUILDING RESEARCH, 1984), there is a relative paucity of data concerning aspects of metals and metalloids in indoor environments. Many experts believe that the indoor exposure to metals and their compounds is less significant than the indoor exposure to formaldehyde, hydrocarbons, asbestos, and radon.

The major sources of occurrence of a number of metallic elements and their derivatives include: cigarette smoke (main and side streams), street dusts, fuel consumption (wood, fossil fuels, biomass), and consumer products (principally contained in aerosol dispensers). The major elements of consideration in terms of their toxicological potential in indoor air pollution are lead, nickel, cadmium, and arsenic.

This chapter considers the sources and bioavailability of a number of metals and metalloids such as lead, cadmium, nickel, zinc, copper, selenium, and arsenic found in indoor environments. A number of these elements and their compounds (Pb, Cd, Ni, As) have been shown to be carcinogenic and/or mutagenic (FRIBERG et al., 1986; SUNDERMAN, Jr., 1986; IARC, 1976, 1980, 1982, 1986a, b; FISHBEIN, 1989) while others (Zn, Cu, and Se) are known to be essential micro- and trace substances.

In this chapter emphasis is directed to indoor atmospheric environments. For information on potential oral uptake, e.g., from food, beverages, cooking pots, cups, cutlery, and water pipes (still a risk for lead, copper, or cadmium exposure) the reader is referred to the elemental chapters in Part II and to Chapters I.5 and I.8. Also for information on indoor occupational atmospheric exposure the appropriate elemental chapters in Part II should be considered.

2 Sources

The sources of chemical indoor pollution can be divided into two main groups, namely intermittent and continuous sources, and the quantification of the contribution of the various sources to the pollution of an indoor environment is still evolving (SEIFERT, 1984).

In many cases indoor concentration profiles are related in some measure to outdoor levels as well as to usage patterns of unvented or improperly vented combustion appliances or usages, to air infiltration and ventilation, life styles, and to chemical reactions, and adsorption or absorption of the constituents or particulate elements on indoor surfaces (YOCUM, 1982).

Thus, numerous studies have suggested that the indoor environment can have different pollutant levels than outdoors and that infiltration losses, settling, and internal sources can result in indoor particle levels that are less than or greater than outdoor levels (MOSCHANDREAS et al., 1979; THOMPSON et al., 1973). It has been reported that indoor/outdoor mass concentration of respirable size particles are approximately equivalent in non-smoking homes, but can be two to three times higher inside smokers' homes (SPENGLER et al., 1985).

2.1 Cigarette Smoke

Smoking is considered to be a major source of indoor particles and indoor concentrations of tobacco-smoke compounds which may well exceed concentrations found outdoors (IARC, 1986a; NAS, 1981). Although for many individuals (primarily children) (BONHAM and WILSON, 1981; RONA et al., 1981), the main or sole exposure to numerous gaseous and particulate compounds results from passive exposure to tobacco smoke, the specific contribution of tobacco combustion products to personal exposures has not been well documented (NAS, 1981).

Tobacco contains minerals and other inorganic constituents derived from soil, fertilizers, mulch agricultural sprays, and polluted rainfall. Table I.6b-1 lists the concentrations of 24 metals found in cigarette smoke (IARC, 1986a). BACHE et al. (1985) reported on the levels of cadmium and nickel in tobacco smoke particles of

Table I.6b-1. Concentrations of Metals in Cigarette Smoke (IARC, 1986a)

Element	µg/Cigarette	References[a]
Na	1.3	(3)
K	70	(3)
Cs	0.0002	(3)
Mg	0.070	(3)
Sc	0.0014	(3)
La	0.0018	(3)
Cr	0.004 – 0.069	(1, 3)
Mn	0.003	(3)
Fe	0.042	(3)
Co	0.0002	(3)
Ni	0.0 – 0.51	(1, 2, 3, 4, 5)
Cu	0.19	(3)
Ag	0.0012	(3)
Au	0.00002	(3)
Zn	0.12 – 1.21	(1, 3, 4)
Cd	0.007 – 0.35	(1, 3, 4, 5)
Hg	0.004	(3)
Al	0.22	(3)
Pb	0.017 – 0.98	(1, 3, 4)
As	0.012 – 0.022	(1, 3)
Sb	0.052	(3)
Bi	0.004	(3)
Se	0.001 – 0.063	(1, 3)
Te	0.006	(3)

[a] References: (1) JENKINS et al. (1986), (2) NAS (1975), (3) NORMAN (1977), (4) PERINELLI and CARUGNO (1978), (5) BACHE et al. (1986)

tobacco grown on a low-cadmium soil/sludge mixture. Sludge-grown and control (soil-grown) tobaccos contained 5.33 and 1.87 ppb Cd and 1.15 and 0.64 ppm Ni, respectively.

Additionally, tobacco and tobacco smoke contain radioactive elements including ^{226}Ra and ^{228}Th; however, 99% of the alpha-activity is derived from ^{210}Po (COHEN et al., 1979) while ^{210}Pb and ^{210}Po in tobacco originate from phosphate fertilizers (TSO, 1966) and/or from airborne ^{210}Pb-containing aerosol particles that are trapped by the trichomes of tobacco leaves (MARTELL, 1974).

According to the results of a study of MUSSALO-RAUHAMAA et al. (1986), a one-pack-a-day smoker inhales about 0.9 µg cadmium daily during smoking, of which approximately 5–10% is retained (MORGAN, 1979). It should be noted that the amount of Cd obtained from food in Finland is about equally great (KOIVISTOINEN, 1980). Cadmium is a major element of toxicological consideration with its content in cigarette smoke believed to be due in part to the presence of Cd in phosphate fertilizers (FRIBERG et al., 1985) and in sewage sludge. Analysis of kidney, liver, and lung specimens in both smokers and non-smokers revealed a statistically significant relationship between the body burden of Cd in smokers and the number of past years of smoking history with heavy smokers accumulating more than twice the body burden of cadmium as non-smokers (LEWIS et al., 1972; IARC, 1986a). KALLIO-MÄKI et al. (1989) and ANTTILA et al. (1989) determined cadmium and chromium in human lung tissue in relation to smoking and lung cancer. Cadmium concentrations were related to the exposure time (ex-smokers after 10–15 years had again significantly lower cadmium concentrations) and to severe emphysema. Pulmonary contents of lung cancer patients were about 6 µg Cr/g d.w. and 3 µg Cd/g d.w. (controls about 4 µg Cr/g d.w. and 2 µg Cd/g d.w.), but the variations were quite high. In non-lung cancer patients a significant positive correlation between the amount of smoking and pulmonary chromium and cadmium was found. Smokers had also more magnesium and titanium in their lung tissue samples.

A number of studies have attempted to assess the fate of heavy metals including cadmium and nickel during the smoking of a cigarettes (NANDI et al., 1969; MENDEN et al., 1972; SZADKOWSKI et al., 1969; see also Table I.6b-2). While NANDI et al. (1969) assumed that 69% of the Cd in the unsmoked cigarette entered the mainstream, SZADKOWSKI et al. (1969) and LEPPANEN and LUKKONEN-LILJA (1982) reported that only 13% (and even less from filter cigarettes) of the Cd in the unsmoked cigarette was found in the inhalable mainstream smoke. Studies of MENDEN et al. (1972) on wet- and dry-ashed cigarettes, cigars, and pipe tobacco indicated values of 1.56–1.96 µg Cd and 4.25–7.55 µg Ni per cigarette (see also Table I.6b-3) and 0.93–1.86 µg Cd per gram of cigar and pipe tobacco. Only 7–10% of the Cd and 0.4–2.4% of the Ni in the smoked portion of the cigarettes appeared in the particulate phase of the mainstream smoke with the remainder present in the ash, trapped in the butt or lost in the sidestream (see also Table I.6b-2). These data suggest that 38–50% of the Cd and 11–33% of the Ni in the smoked portion of cigarettes are present in the sidestream. The tobacco smoke condensate (TSC) data in the study of MENDEN et al. (1972) suggest that a pack of 20 cigarettes would provide about 2 µg of Cd and from 0.4 to 1.6 µg of Ni in the TSC which is inhaled by the smoker while the sidestream smoke would put into the general vicinity of a non-smoker

Table I.6b-2. Concentration of Cadmium, Nickel, and Zinc in Non-filter Cigarette Mainstream Smoke (MS) and the Ratio of Their Relative Distribution in Sidestream Smoke (SS) (IARC, 1986a)

Metal	MS (ng/cigarette)	SS:MS
Cadmium	100	3.6 – 7.2
Nickel	20 – 80	0.2 – 3
Zinc	60	0.2 – 6.7

Table I.6b-3. Varying Mean Contents of Trace and Heavy Metals in Filter Cigarette Tobacco ($\pm$ SD) from 1920 to 1984 (MUSSALO-RAUHAMAA et al. (1986)

Sampling Time and Place	No. of Different Brands[a]	Cd (µg/g)	Zn (µg/g)	Cu (µg/g)	Pb (µg/g)	Mg (µg/g)	Fe (µg/g)
1920s Finland	2	1.4 ± 0.4	212 ± 22	22.6 ± 0.1	3.4 ± 1.1	9.5 ± 0.3	0.44 ± 0.12
1940s Finland[b]	1	1.0	217	11.9	2.2	10.3	0.62
1966 – 67 Finland	5	1.7 ± 0.4	54 ± 13	17.6 ± 3.0	3.3 ± 1.5	6.7 ± 1.2	0.61 ± 0.22
1978 Finland	1	1.3	38	11.8	2.2	5.1	0.41
1980 – 82 Finland	4	1.9 ± 0.6	50 ± 8	15.6 ± 4.2	1.7 ± 0.6	6.8 ± 0.7	0.46 ± 0.05
1984 Finland	1	0.7	39	9.2	1.3	2.4	0.48
1950s England	11		65	65.7	5.3	6.8	0.77
1957 USA	3		29 ± 5	24.7 ± 10.8	46 ± 31	4.9	0.52
1975 Yugoslavia	10	2.2 ± 1.0	51 ± 14	18.9 ± 8.2			1.15 ± 0.67
1977 East Germany	15	1.7 ± 0.8	28 ± 8	18.8 ± 11.2	3.3 ± 1.2		
1978 West Germany	15	1.5 ± 0.3	45 ± 8	21.7 ± 8.8	3.3 ± 0.6		0.43 ± 0.09
1979 – 80 Finland	44	1.4 ± 0.4			0.8 ± 0.4		

[a] Duplicate determinations were made from separate digests of each brand
[b] Pipe tobacco sample (contains also other than tobacco leaves)
The mean contents in filter cigarette tobacco samples were Cd, 1.7 µg/g; Cu, 15.6 µg/g; Pb, 2.4 µg/g; Mg, 6.4 µg/g; and Fe, 0.5 µg/g, in 1960 – 1980.

8.6 – 12.8 µg of Cd and 12.4 – 20.6 µg of Ni per pack. These amounts of Cd and Ni were considered by MENDEN et al. (1972) to constitute a considerable greater environmental contamination of these elements than is found in ambient air.

It is important to note that the peak temperature in the burning cone of a cigarette during puffing reaches 800 – 900 °C (WYNDER and HOFFMANN, 1967) but between puff reaches only about 600 °C (HOFFMANN et al., 1983). It has been suggested by mathematical models (BAKER, 1982) that most of the vapors leaving the pyrolysis/distillation region behind the cone diffuses out of the cigarette condensing to form sidestream smoke particles. The conditions prevailing during sidestream smoke generation favor formation of aerosol particles of smaller size (0.01 – 0.1 µm) than those occurring in the mainstream smoke (0.1 – 1.0 µm) (BAKER, 1982), which is an additional factor to be noted regarding the toxic effects of inhaled sidestream smoke (IARC, 1986a). About 300 – 400 of the more than 3800 individual compounds identified in tobacco smoke have thus been identified quantitatively in mainstream and sidestream smoke.

Additionally, aluminum and silicon concentrations increase in the lungs of smokers with increased cigarette smoking. The alveolar macrophages from cigarette smokers contain inorganic aluminium silicate particles with an elemental composition and crystalline features characteristic of kaolinite (BRODY and CRAIGHEAD, 1975; CHOX et al. 1978; VALLYATHAN and HAHN, 1985).

The analyses of cigarette tips indicated an effective accumulation of metals by commercial filters during smoking (MUSSALO-RAUHAMAA, 1986). The trace and heavy metal contents in tips after the smoking procedure were in many cases increased tenfold. About 22 – 50% of the Cd was found in the ashes and stump of the smoked cigarettes which indicated that Cd passes in large measure to sidestream smoke while the other metals determined (Zn, Cu, Pb, Mg, Fe) remained mostly in the remnants of the smoked cigarette noted above.

2.2 Wood Combustion

It is generally acknowledged that rising energy costs and uncertain availability of petroleum and natural gas following the 1973 oil embargo have led to a greatly increased use of alternative fuels for residential space heating in North America and Europe. For example, from 1973 to 1980, the number of U.S. homes using wood as a primary residential heat source increased 900% with 7.5 million American households having at least one working wood stove in 1980 (SKOG and WATTERSON, 1983).

It should also be noted that roughly half the world's households cook daily with wood, crop residues, or animal dung used in simple stoves made of rock or clay. Most of these households are in rural areas of South America, Africa, and Asia (DAVIDSON et al., 1986; HUGHART, 1979). Although emissions from these stoves can be significant (BUTCHER et al., 1984), our current understanding of the pollutants and associated human exposures is extremely limited (DAVIDSON et al., 1986).

The continuing upsurge in residential wood consumption (RWC) has raised concern about potential adverse effects on ambient air quality (SEXTON et al., 1984, 1985, 1986; TRAVIS et al., 1985) since wood burning stoves, furnaces and fireplaces have been shown to emit significant quantities of particles, carbon monoxide, nitrogen oxides, sulfur oxides, and over 50 organic species including many polycyclic aromatic hydrocarbons (see also DE ANGELIS et la., 1980; LIPFERT and DUNGAN, 1983; BUDIANSKY, 1980). Smoke from the burning of wood (or biomass) can be emitted directly into the living space from improperly installed, maintained, or operated stoves and fireplaces (SEXTON et al., 1984, 1985, 1986; DAVIDSON et al., 1986). Alternatively, woodsmoke of outdoor origin emitted from chimneys can penetrate indoors as a consequence of natural or forced ventilation.

It has been reported that almost all particle emissions from residential wood combustion are in the respirable size fraction ($d_a < 3.5$ µg), and highest concentrations are likely to occur in residential neighborhoods (SEXTON et al., 1984, 1985, 1986).

Table I.6b-4. Summary Statistics for Indoor/Outdoor Comparisons (Sexton et al., 1986)

Parameter	Concentrations Indoor				Indoor minus Outdoor				Ratios Indoor:Outdoor			
	$\bar{X}$		SD		$\bar{X}$		SD		$\bar{X}$		SD	
	FF[d]	CF[e]	FF	CF	FF	CF	FF	CF	FF	CF	FF	CF
Elements[a] (ng m^{-3})												
Si	52	892	41	464	−0.1	337	99	664	3.6	4.5	4.7	5.8
S	1016	133	771	79	−690	39	405	67	0.6	1.5	0.3	0.7
K	358	479	239	453	−16	336	180	463	0.9	5.4	0.4	4.9
Ca	152	1007	136	1086	106	903	124	1067	3.8	5.9	2.9	5.1
Fe	38	191	35	105	23	9	37	163	3.2	1.8	3.6	1.5
Cu	133	39	98	20	151	16	101	29	>10	2.8	—	3.4
Zn	15	24	9	17	6	19	6	16	2.4	5.3	1.3	3.0
Br	12	3	9	2	−12	−4	4	2	0.5	0.4	0.2	0.2
Pb	53	13	31	6	−52	−9	25	13	0.5	0.7	0.2	0.4
PAH[b] (ng m^{-3})												
BZAA	0.3		0.3		−0.9		1.2		0.7		0.9	
CHRY	0.5		0.4		−1.3		2.0		0.8		1.0	
BZFA	1.7		1.8		−1.1		3.9		1.2		1.5	
BZEP	0.6		0.5		−0.4		1.3		1.1		1.2	
BZAP	0.8		0.9		−0.03		1.1		1.5		2.1	
Particle-Phase Carbon[c] (μg m^{-3})												
Organic	10.9		7.9		1.6		9.7		1.4		1.3	
Elemental	4.0		1.4		0.2		1.9		1.1		0.5	

[a] N = 14 (5 homes); [b] N = 14 (6 homes); [c] N = 8 (5 homes); [d] Fine fraction ($d_a < 2.5$ μg); [e] Coarse fraction (d_a 2.5 – 15 μm)
$\bar{X}$ arithmetic mean, SD arithmetic standard deviation, PAH polycyclic aromatic hydrocarbons, BZAA benzo(a)anthracene, CHRY chrysene, BZFA benzofluoanthenes, BZEP benzo(e)pyrene, BZAP benzo(a)pyrene

Relatively few studies have noted the elemental composition of wood combustion (DAVIDSON et al., 1986; MOSCHANDREAS et al., 1980; COLOME and SPENGLER, 1982; COLOME et al., 1981; SEXTON et al., 1985; STEVENS et al., 1984; EPA, 1978a; WALLACE et al., 1984). WALLACE et al. (1984) identified seven elements (Fe, As, P, S, Si, Zn, Cr in decreasing frequency) on 50–100% of five filters and eight elements (K, Ca, Cd, Mg, As, Cu, Ni, V) on 10–40% of five filters in a U.S. EPA initiated indoor air monitoring program in 1982 involving five indoor locations. Potassium seems to be an indicator for indoor wood burning (SEXTON et al., 1985; EPA, 1978a). SEXTON et al. (1986) characterized also indoor air quality in wood-burning residences by mean and standard deviations for indoor, indoor–outdoor and indoor:outdoor concentrations of nine selected elements (Si, S, K, Ca, Fe, Cu, Zn, Br, Pb), five individual polycyclic aromatic hydrocarbons and particle-phase carbon (Table I.6b-4). Enhanced indoor concentrations of Ca, Fe, Cu, and Zn were commonly measured in five fraction particles (e.g., i−o>0 and i/o>1) indicating that indoor sources are primarily reasonable for these emissions. Only Br and Pb were higher outdoors for particles in the coarse fraction.

Air pollutant concentrations have been measured in residences in the Himalayas of Nepal where biomass fuels are used (in addition to charcoal and wood) for cooking and heating. Levels of total suspended particles (TSP) were in the range of 3–42 mg/m^3, with respirable suspended particles (RSP) in the range of 1–14 mg/m^3 in the houses sampled. High concentrations of several trace elements have also been measured as illustrated in Table I.6b-5 which compares airborne concentrations of particulate mass and several chemical species at indoor and outdoor locations in Nepal (DAVIDSON et al., 1986). Trace elements are characterized as crustal or enriched depending on the value of the crustal enrichment factor (ZOLLER et al., 1974). Few data are available on the probably smaller indoor concentrations of trace elements in developing countries for comparison with Table I.6b-5. A pilot project conducted in Kumjung in 1979 involved measurement of indoor and outdoor Al, Mg, Cu, and Pb (DAVIDSON et al., 1981), and concentrations of these elements were similar to levels shown in Table I.6b-5 (DAVIDSON et al., 1986).

2.3 Coal Combustion
(see also COLOME and SPENGLER, 1982)

Although the combustion of coal (containing all naturally occurring elements) has long been recognized as a major outdoor elemental pollution source, there are relatively few studies delineating the contribution of coal burning to indoor pollution either as a consequence of migration of particulate matter or the combustion of coal in indoor environments *per se.*

Apparently many of the elements and their compounds are volatile, absorb on particulates, and have abverse effects on human health (KLEIN et al., 1975; NATUSH and WALLACE, 1974). Many of the elements are enriched 100 to 1000 fold over their natural abundance (depending on local geological and geographical factors; see also Chapter I.1) in the earth's crust with the major portion of this enriched particulate

Table I.6b-5. Airborne Concentrations of Particulate Mass and Several Chemical Species at Indoor and Outdoor Locations in Nepal[a] (DAVIDSON et al., 1986). (TSP total suspended particles, RSP respirable suspended particles)

		Indoor Concn.	Outdoor Concn.	
		$N_{TSP} = 18$ and $N_{RSP} = 17$	Hotel Everest View, N = 3	Katmandu, N = 4
Mass	TSP	8800 (1.8)[c]		280 (1.2)
	RSP	4700 (1.9)[d]		
Crustal Elements (enrichment factor <10)				
Al	TSP	33 (3.1)	0.87 (1.1)	4.6 (1.4)
	RSP	2.8 (3.0)		
Ca	TSP	45 (2.7)	0.84 (1.4)	6.9 (1.3)
	RSP	3.4 (2.7)		
Fe	TSP	62 (4.1)	0.90 (1.3)	6.1 (1.1)
	RSP	8.3 (2.8)		
K	TSP	64 (3.1)	0.71 (1.1)	6.1 (1.2)
	RSP	32 (2.9)		
Mg	TSP	11 (4.1)	0.14 (1.2)	1.1 (1.1)
	RSP	0.78 (2.6)		
Mn	TSP	2.7 (3.3)	0.023 (1.1)	0.18 (1.1)
	RSP	0.21 (2.3)		
Na	TSP	3.7 (2.6)	0.19 (1.2)	1.2 (1.1)
	RSP	0.62 (4.5)		
Enriched Elements				
Ag	TSP	<0.004	0.000086 (1.4)	0.0011 (1.2)
	RSP	<0.004		
Cd	TSP	0.0083 (2.8)	<0.00016	0.0011 (1.1)
	RSP	0.0081 (2.1)		
Cu	TSP	0.22 (3.7)	[b]	0.035 (1.04)
	RSP	0.068 (2.3)		
Pb	TSP	0.047 (1.9)	0.0022 (1.4)	0.40 (1.2)
	RSP	0.029 (3.2)		
Zn	TSP		0.014 (1.8)	0.39 (1.2)
Other Species (TSP), N = 2 for each entry				
Sulfate		6.5 (2.3)	0.90 (1.06)	1.7 (1.6)
Nitrate		31 (4.2)	1.8 (2.4)	7.4 (1.3)
C (organic)		1600 (2.4)	4.4 (1.2)	44 (1.07)
C (elemental)		310 (2.6)	<3.0	17 (1.3)

[a] Values given are geometric mean concentrations in µg/m³ TSP, with geometric standard deviations shown in parentheses. The number of filters in each date set is denoted by N
[b] Outdoor concentrations of Cu at Hotel Everest View and indoor concentrations of Zn are highly variable and have been excluded
[c] Range: 2900 – 42 000
[d] Range: 870 – 14 000

Table I.6b-6. Trace Metals Emitted in Fly Ash (LEE and VON LEHMDEN, 1973)

Element[b]	Concentration (ppm)[a]				
Particle Size	25 µm	12.5 µm	10 µm	3.5 µm	1.5 µm
Al	67000	54300	57300	63600	59300
B	300	500	500	500	500
Be	2	1	2	2	2
Cd	≤5	≤5	≤5	≤5	100
Cr	130	130	130	300	300
Cu	150	150	200	200	200
Fe	40000	59000	43500	35500	32300
Mn	200	240	290	390	500
Ni	3000	200	200	300	300
Pb	300	200	300	300	500
V	200	200	200	200	200

[a] Concentrations in particles of different sizes
[b] Sample collected from a coal-fired steam power plant and analyzed by neutron activation and spark source mass spectrometry

mass occurring in the 0.5 to 10.0 micron particle diameter range which are commonly inhaled and deposited in the human respiratory system (TORREY, 1978). Submicron particles are particularly hazardous since about 50% of the inhaled particles between 0.01 and 0.1 µm deposit in the alveoli of the lungs where most trace elements are readily absorbed in the blood stream (TORREY, 1978).

The inorganic constituents of ash are those typical of rocks and soils (primarily Si, Al, Fe, and Ca, and the oxides of these four elements comprise 95 to 99% of the composition of ash which also contains smaller amounts (0.5 to 3.5%) of Mg, Ti, S, Na, and K as well as ppm quantities of from 20 to 50 elements).

For certain trace components there is a very definite partitioning between the bottom ash and fly ash, which for some elements such as selenium can be an order of magnitude (TORREY, 1978). Many studies also suggest that certain potentially hazardous elements in coal (e.g., As, Be, Cd, Pb, and Se) are concentrated in or on the small fly-ash particles while certain others, such as Hg, are emitted primarily as vapors. Table I.6b-6 lists 11 trace metals emitted in fly ash arranged in terms of five particle size concentrations (LEE and VON LEHMDEN, 1973; note also the enrichment compared to the figures in Chapter I.1!).

2.4 Personal Monitoring

As noted earlier, there have been comparatively few studies involving a comparison of metals in indoor and outdoor environments, especially involving personal monitoring. In one such study referred to as part of the Harvard Six Cities Studies, indoor, outdoor, and personal monitoring were carried out at 24 homes in Topeka, Kansas (TOSTESON et al., 1981). Eighteen non-smoking individuals took part in the personal monitoring portion of the study. The collected respirable particulate matter (RPM) was analyzed. The results below illustrate the mean indoor, outdoor, and personal monitoring results for Fe, Pb, and Al found in particulate matter in this study.

Monitoring Location	Concentrations ($\mu g/m^3$)			
	RPM	Fe	Pb	Al
Personal	29.6	0.0499	0.1605	0.0276
Indoor	30.5	0.0328	0.0919	0.0305
Outdoor	12.2	0.0349	0.1242	0.0122
Indoor/Outdoor Ratios	1.5	0.94	0.74	1.08

These results suggest that there are significant indoor sources of respirable particulate matter, and while there are some indoor sources of Al, there are no significant indoor sources of lead.

2.5 House Dust
(see also KRAUSE et al., 1987; LAXEN et al., 1987; FISHBEIN, 1989)

It is increasingly recognized that house dust which can arise from external sources including dustfall from aerosol, soil and street dust as well as dust generated within the house is a source of a broad spectrum of metals. Wide qualitative and quantitative variations in the chemical composition of the house dust can be expected as the proportion of elements arising from the above contributing sources may vary considerably, temporally and from house to house (FERGUSSON et al., 1986; FERGUSSON and SCHROEDER, 1985; SOLOMON and HARTFORD, 1976).

FERGUSSON et al. (1986) determined the concentrations of 26 elements in house dust, street dust, and soil collected in Christchurch, New Zealand (Table I.6b-7). There is a degree of uniformity in the concentrations over the 11 house dust samples. The trend in variability of the results is an increase in the order: soil < street dust < house dust.

The elements Hf, Th, Sc, Sm, Ce, La, Mn, Na, K, V, Al, and Fe were considered to be soil-based and contribute about 45 – 50% to the house dust and 87% to street dust. The elements which are enriched (>3 times) in the dusts relative to the levels found in local soils are the "pollution" type elements Br, Cu, Cl, Pb, Zn, Cr, Ca, Co, As, and Sb in house dust and Zn, Cr, Cu, and Pb in street dust.

Except for lead (e.g., from lead paints) and zinc there were no significant differences in the levels of the elements in the house dust with respect to the age of the house or construction material (FERGUSSON and RYAN, 1984; KOWALCZYK et al., 1982; FRIEDLANDER, 1983). Soils and street dust are major contributors to house dust, and the levels of most elements in the soil are relatively constant throughout the city. In fact, for house dust 45 – 50% of the dust arises from soil and street dust,

Table I.6b-7. Concentration[a] of Elements in House Dust, Street Dust, and Soil (FERGUSSON et al., 1986)

Element	House Dust[b]			Street Dust[c]			Soil[c]		
	Mean[a]	SD	Range[a]	Mean[a]	SD	Range[a]	Mean[a]	SD	Range[a]
Sm	1.22	0.41	0.56–2.11	3.10	0.38	2.40–3.64	3.90	0.41	3.16–4.39
Hf	2.12	1.10	0.71–4.71	3.45	0.81	2.06–4.46	4.12	0.74	3.21–5.40
Sc	2.90	0.84	1.35–4.64	5.45	0.69	4.54–6.95	6.84	0.62	6.06–7.89
Th	3.40	1.28	1.83–6.77	6.73	1.08	5.61–9.28	8.15	1.15	7.30–11.0
La	11.9	6.18	6.67–18.7	22.0	5.34	12.6–32.8	27.4	3.67	21.3–32.6
Ce	23.6	10.4	9.26–41.9	43.6	6.40	34.5–529	52.3	5.05	44.9–61.2
I	<8.0		<7.0–<12.0	<9.0		<8.0–<9.0	<11.0		<7.0–16.1
Br	37.1	18.2	12.3–63.6	43.0	35.5	21.9–145	<26.0		<17.0–44.8
Cl	4006	2087	1306–7497	381	116	245–655	<340		154–623
Au	0.42	0.26	0.09–0.87	0.05	0.04	<0.001–0.12	0.04		<0.001–0.06
Co	8.59	3.32	4.75–15.7	6.41	1.52	3.45–9.00	6.34	1.21	4.58–8.47
Sb	10.0	9.64	1.83–30.6	4.69	0.99	2.61–6.78	5.94	1.06	4.33–7.95
As	15.8	4.24	7.95–23.4	14.5	7.43	1.24–29.5	10.2	8.86	4.26–34.0
V	30.4	8.45	11.7–42.9	58.3	8.91	40.7–71.7	66.0	11.1	49.7–79.0
Cu	<230	91.1	<167–384	90.8	63.7	14.9–195	17.5	12.0	9.0–54.0
Cr	103	48.3	62.5–238	103	126	32.9–466	36.9	17.6	22.5–85.3
Mn	207	67.3	102–349	313	47.5	246–405	325	57.0	235–458
Zn	845	186	574–1189	716	473	153–1738	251	135	133–653
Pb	734	398	287–1408	1223	900	175–2794	202	379	35–1395
Ti	2041	995	680–4317	2117	353	1508–2729	2530	323	2168–3356
Mg	5931	4458	1032–9025	4728	1668	2167–7751	165	3317	125–10873
Ca	1.49	0.78	0.73–3.55	0.82	0.20	0.50–1.28	0.72	0.24	0.15–0.97
Na	1.18	0.40	0.57–2.05	1.73	0.19	1.38–1.99	1.86	0.12	1.71–2.04
K	1.26	0.26	1.04–1.76	1.83	0.32	1.37–2.42	2.50	0.46	1.83–3.24
Fe	1.02	0.32	0.53–1.60	2.09	0.26	1.70–2.45	2.00	0.18	1.73–2.37
Al	2.39	0.79	0.91–4.05	4.84	0.51	4.23–5.72	5.56	0.54	5.36–6.09
Organic	42.9	13.4	25.7–56.5	9.0	5.4	2.2–16.5	10.1	2.1	6.9–13.9

[a] Concentration in µg g^{-1} except for Ca, Na, K, Fe, Al, and organic which are in %
[b] Eleven samples
[c] Twelve samples

about 2–3% from tire wear, cement and car emissions and 1% from salt. The remainder of the material is presumed to be organic.

SOLOMON and HARTFORD (1976) measured lead and cadmium in and around well-kept homes painted with low-level paints situated in a small urban community (Champaign-Urbana, Illinois, population 100000). Lead levels in dusts within homes average 600 ppm and 680 µg/m^2. In non-residential interiors, dusts averaged 1400 ppm Pb and 2040 µg/m^2. Unexpectedly high cadmium accumulations were found, especially in carpet dusts from rubber-backed carpets. For example, Cd contents as high as 105 ppm and 219 µg/m^2 were measured in some homes. Large amounts of cadmium (e.g., 44 µg/m^2 on the average) were found on rugs with rubber backing in one sample assaying 3000 ppm Cd. Abrasion could cause the Cd presence in dust. Implications of the high levels of Pb and Cd found both indoors and outdoors can have potentially serious implications for a young child or infant. For example, a child might be expected to have 10 mg of dust on its fingers (LEPOW et al., 1974). If the child spent most of its time indoors and placed its fingers in its mouth ten times, an average of 60 µg Pb and 2 µg Cd per day might be ingested. If appreciable times were spent outdoors or in a non-residential area with high Pb and Cd levels (due to both high soil and automotive origins), these values could be doubled or tripled (SOLOMON and HARTFORD, 1976).

Lead in dust is in fact considered by many investigators to be a principal pathway of the element to man (DUGGAN and WILLIAMS, 1977; ROYAL COMMISSION ON ENVIRONMENTAL POLLUTION, 1983; CHARNEY et al., 1980, 1983; MILAR and COONEY, 1982; DUGGAN, 1980, 1983; MIELKE et al., 1983; YAFFE et al., 1983; ELWOOD, 1983), either by inhalation of resuspended dust or ingestion of settled material (see also Chapter II.16).

Most studies suggest that lead (atmospheric fallout of petrol lead, soil lead, and street dust lead) originates in the outdoor atmosphere (e.g., the National Ambient Air Quality Standard (NAAQS) in the U.S. for Pb is 1.5 µg/m^3 averaged over a three-month period) and that energy conservation measures provide additional isolation from outdoor exposures. Indoor/outdoor ratios for Pb are usually less than 1.0 and values in the range of 0.6 to 0.8 would probably encompass most values from typical non-airconditioned homes (YOCUM, 1982).

2.6 Consumer Products

It has become increasingly recognized that consumer products, primarily resulting from the expanded use of aerosol containers, can be an additional potential source of both organic and inorganic compounds in indoor environments (NAS, 1981). One area of potential concern involves the use of aerosolized antiperspirants. Earlier formulations contained a variety of zirconium salts (including Zr-Al complexes) and the U.S. Food and Drug Administration ruled in 1975 that these aerosol antiperspirants or any aerosolized drug or cosmetic product containing zirconium salts posed an unacceptable risk because of suspicion of chronic inhalation toxicity (SCHMIDT, 1975).

The most common active ingredient in aerosol underarm antiperspirants is generally a formulation of aluminum chlorohydrate. WARD et al. (1977) postulated that lesions observed in human lungs might have been the result of inhaling aerosol underarm deodorants. Inhalation studies with alchlor (a propyleneglycol complex of aluminum chloridehydroxide) resulted in increased lung weights and formation of granulomatous lesions in the respiratory bronchioles of hamsters exposed for six hours/day at a concentration of 5 mg/m^3 (DREW et al., 1974).

More recent studies of tissue deposition patterns after chronic inhalation exposure of rats and guinea pigs to aluminum chlorohydrate (STONE et al., 1979) showed that exposure to 0.25, 2.5, or 25 mg/m^3 aluminum chlorohydrate for periods of up to 24 months resulted in aluminum being primarily contained in the lungs and in other organs such as peribronchial lymph nodes and adrenal glands of higher dosed animals. Although it was suggested that it was not possible to make direct comparisons between the limited "bathroom type" exposure of users of aluminum chlorohydrate and the type of exposures studied in experimental animals, the inhalation of aerosolized aluminum formulations by humans cannot be excluded. To this potential route of exposure of aluminum must be added other sources and routes of exposure of the element. In indoor environments these could include Al from tobacco smoke, and aluminum oxide and aluminum silicate dusts from cooking utensils and other Al-containing items. For other sources of aluminum (e.g., food and remedies) see Chapter II.1, EPSTEIN (1985), KRUEGER et al. (1984), and CRAPPER et al. (1973, 1976) who also discussed effects.

The total body burden of aluminum in normal healthy individuals has been variously estimated at 30–60 mg (JONES and BENNETT, 1986). Although the ingestion pathway is a more significant route of transfer of aluminum from the environment to man than the inhalation pathway (JONES and BENNETT, 1986), it should be noted that the highest aluminum concentrations in the normal healthy human body are in the lungs with levels ranging up to more than 100 µg/g dry wt (ALFREY, 1984; ALFREY et al., 1980).

3 Bioavailability
(see also Chapters I.8 and I.15)

It is useful to briefly consider several germane aspects of the bioavailability of metals since the absorption (primarily via inhalation) of metals in indoor environments from sources described above may impact on homeostasis and the body burden of both trace essential and toxic metals and their compounds.

The operative aspects governing the term "bioavailability" of metals and their compounds are known to be complex and are influenced by many factors including physical and chemical forms (speciation), route of exposure, the host, absorption and retention in the body, diet, factors affecting and/or altering transport and permeability of biological membranes and other interactions including those with other elements as well as with proteins and other macromolecules. It should be particularly noted that exposure to trace contaminants including metals generally occurs in mixtures in both outdoor and indoor environments (e.g., combustion of diverse

Table I.6b-8. Representative Inhalation and Consumption Rates[a]

		Adult	Child (2 yr old)
Inhalation	Air	20 m^3/d	6 m^3/d
Consumption	Foods	1200 g/d	750 g/d
	Fish	25 g/d	
	Milk	0.3 L/d	0.5 L/d
	Water	1.2 L/d	1.3 L/d
	Dirt		20 mg/d

[a] ICRP (1975)

Table I.6b-9. Approximate Mean Concentrations of Metals in Environmental Media[a]

Metal	Air (ng/m^3)	Water (µg/L)	Food (µg/g)	
			Fish	Other Foods
Arsenic	20	1.2	3.6	0.03
Cadmium	30	–	–	0.025
Chromium	7	0.2	–	0.05
Lead	2000	10	–	0.09
Mercury	7	–	0.4	0.004
Nickel	20	8	–	0.13
Selenium	3	0.2	–	0.06
Tin	30	0.04	–	0.2
Zinc	300	40	–	10

[a] BENNETT (1981 a, b, 1982)

Table I.6b-10. Intake Rates of Metals via Ingestion and Inhalation (µg/d)[a]

Metal	Fish	Other Foods	Water	Total Ingestion	Inhalation
Arsenic	90	40	1.4	130	0.4
Cadmium	–	30	–	30	0.6
Chromium	–	60	0.2	60	0.1
Lead (adult)	–	110	12	180[b]	40
Lead (child)	–	70	13	170[c]	6
Mercury	10	5	–	15	0.14
Nickel	–	160	10	170	0.4
Selenium	–	70	0.2	70	0.06
Tin	–	200	–	200	0.6
Zinc	–	10000	50	10000	6

[a] BENNETT (1981 a, b, 1982)
[b] Includes 60 µg/d from surface contamination of food
[c] Includes 40 µg/d from surface contamination of food and 48 µg/d from dirt on hands

Table I.6b-11. Fractional Absorption of Ingested and Inhaled Metals by Blood[a]

Metals	Absorption	
	Following Ingestion	Following Inhalation
Arsenic	0.9	0.3
Cadmium	0.05	0.25
Chromium	0.01	0.2
Lead (adult)	0.1	0.2
Lead (child)	0.25	0.4
Mercury (organic)	1.0	–
Mercury (inorganic)	0.05	0.7
Nickel	0.05	0.3
Selenium	0.8	0.2
Tin	0.2	0.2
Zinc	0.3	0.2

[a] BENNETT (1981a,b, 1982)

Table I.6b-12. Distribution and Retention of Metals in Man[a]

Metals	Uptake Rate (µg/d)	Distribution	Mass (kg)	Retention Time (d)	Concentration (µg/g)
Arsenic	81[b]	1.0 (body)	70	4	
	37[c]	1.0 (body)	70	8	0.009[d]
Cadmium	1.7	0.3 (kidneys)	0.31	10950	18
Chromium	0.6	0.05 (bone)	5	1400	0.008
		0.65 (body)	65	50	0.0003
Lead (adult)	26	1.0 (blood)	5.2	23	0.12
(child)	45	1.0 (blood)	2	?	?
Mercury	10[b]	1.0 (body)	70	100	
	0.4[c]	1.0 (body)	70	60	0.015[d]
Nickel	8.6	0.3 (body)	70	200	0.007
Selenium	56	0.9 (body)	70	140	0.1
Tin	4.1	0.35 (bone)	5	350	0.1
		0.15 (body)	65	350	0.003
Zinc	3000	1.0 (body)	70	380	16

[a] BENNETT (1981a,b, 1982)
[b] From intake of organic forms
[c] From intake of inorganic forms
[d] From combined intake of organic and inorganic forms

fuels, house dust, cigarette smoke, etc.), and transfer rates and amounts are determined by factors noted above. Many of these metals are essential at certain levels (e.g., Fe, Zn, Mn, Cu, Cr, Se, Mo) and toxic at elevated levels (e.g., Se) while other metals and metalloids such as As, Be, Cd, Hg, Ni, and Pb possess in the main no defined essentiality or are clearly toxic (BENNETT, 1981a,b, 1982, 1986; FRIBERG et al., 1986; NAS, 1972, 1973, 1974a,b, 1975, 1976, 1977a–c, 1978; WHO, 1973).

Transfer of metals to the body following inhalation depends on deposition in the lungs of respirable particles of generally less than 10 µm in diameter, and absorption into blood with deposition governed by the physical characteristics of the contaminant and absorption primarily by the chemical properties (BENNETT, 1986; see also Table I.6b-6). Metals absorbed into blood are distributed to organs and tissues which vary from one metal to another and depend on many factors including similarities to essential elements and filtration or storage locations.

Except for tobacco smoke, both mainstream and sidestream (IARC, 1986), data are meager for individual exposures of metallic contaminants in indoor environments compared to measurements of airborne contaminants in occupational settings or measurements in general ambient settings.

BENNETT (1981a, b, 1982, 1986) has recently compared representative inhalation and consumption rates in adults and a two-year-old child (Table I.6b-8), approximate mean concentrations of nine metals (As, Cd, Cr, Pb, Hg, Ni, Se, Sn, and Zn) in air, water, and food (Table I.6b-9), their intake rates via ingestion and inhalation (Table I.6b-10), the fractional absorption of these metals by blood via ingestion and inhalation (Table I.6b-11), and their distribution and retention in man (Table I.6b-12). It was stressed by BENNETT (1986) that the estimates of fractional absorption following inhalation and the mean retention times in the body (Table I.6b-11) are particularly uncertain. The assessments concern the total amounts of each metal, except for As and Hg for which, due to differences in absorption and retention of inorganic and organic forms, a distinction is made.

References

Alfheim, I., and Ramdahl, T. (1984), *Contribution of Wood Combustion to Indoor Air Pollution as Measured by Mutagenicity in Salmonella and Polycyclic Aromatic Hydrocarbon Concentration.* Environ. Mutagen. 6, 121–130.

Alfrey, A.C. (1984), *Aluminum Metabolism in Uremia.* Neurotoxicology 1, 43–54.

Alfrey, A.C., Hegg, A., and Craswell, P. (1980), *Metabolism and Toxicity of Aluminum in Renal Failure.* Am. J. Clin. Nutr. 33, 1509–1516.

Alzona, J., Cohen, B.L., Rudolph, H., Jow, H.N., and Frolicher, J.O. (1979), *Indoor-Outdoor Relationships for Airborne Particulate Matter of Outdoor Origin.* Atmos. Environ. 13, 55–60.

Anttila, S., Kokkonen, P., Pääkkö, P., and Kalliomäki, P.-L. (1989), *Cadmium and Chromium in Human Lung Tissue in Relation to Smoking and Lung Cancer,* Toxicol. Environ. Chem. 22 (1–4), 91–95.

Bache, C.A., Lisk, D.J., Doss, G.J., Hoffmann, D., and Adams, J.D. (1985), *Cadmium and Nickel in Mainstream Particulates of Cigarettes Containing Tobacco Grown on a Low-Cadmium Soil-Sludge Mixture.* Environ. Health 16, 547–552.

Baker, R.R. (1982), *Variation of Sidestream Gas Formation During the Smoking Cycle.* Beitr. Tabakforsch. 11, 181–193.

Bennett, B.G. (1981a), *Exposure Commitment Assessments of Environmental Pollutants. Summary Exposure Assessments for Lead, Cadmium, Arsenic.* MARC Report No. 23, Vol. 1, No. 1. Monitoring and Assessment Research Centre, London.

Bennett, B.G. (1981b), *Exposure Commitment Assessments of Environmental Pollutants. Summary Exposure Assessments for Mercury, Nickel, Tin.* MARC Report No. 25, Vol. 1, No. 2. Monitoring and Assessment Research Centre, London.

Bennett, B. G. (1982), *Exposure Commitment Assessments of Environmental Pollutants. Summary Exposure Assessments for PCBs, Selenium, Chromium. MARC Report No. 28,* Vol. 2. Monitoring and Assessment Research Centre, London.

Bennett, B. G. (1986), *Exposure Assessment for Metals in Carcinogenesis,* in: Environmental Carcinogens – Selected Methods of Analysis, Vol. 8, *Some Metals: As, Be, Cd, Cr, Ni, Pb, Se, Zn,* pp. 115–127. International Agency for Research on Cancer, Lyon.

Bonham, G. S., and Wilson, R. W. (1981), *Childrens' Health in Families with Cigarette Smokers. Am. J. Publ. Health 71,* 290–293.

Bremner, I. (1979), *The Toxicity of Calcium, Zinc, and Molybdenum and Their Effects on Copper Metabolism. Proc. Nutr. Soc. 38,* 235–242.

Brody, A. R., and Craighead, J. E. (1975), *Cytoplasmic Inclusions in Pulmonary Macrophages of Cigarette Smokers. Lab. Invest. 32,* 125–132.

Budiansky, S. (1980), *Bioenergy: The Lesson of Wood Burning? Environ. Sci. Technol. 14,* 769–771.

Butcher, S. S., Rao, U., Smith, K. R., Osborn, J. F., Azuma, P., and Fields, H. (1984), *Emission Factors and Efficiencies for Small-scale Open Biomass Combustion: Toward Standard Measurement Techniques.* Annual Meeting, Fuel Chemistry Division, American Chemical Society, Philadelphia, August 26–31.

Chapin, F. S. (1974), *Human-Activity Patterns in the City.* Wiley-Interscience, New York.

Charney, E., Sayre, J., and Coulter, M. (1980), *Increased Lead Absorption in Inner City Children. Where Does the Lead Come from? Pediatrics 65,* 226–231.

Charney, E., Kersler, B., Farfel, M., and Jackson, D. (1983), *Childhood Lead Poisoning. N. Engl. J. Med. 309,* 1089–1093.

Chox, R., Pantrat, G., Viallat, J., Farisse, P., and Boutin, C. (1978), *Inorganic Cytoplasmic Inclusions in Alveolar Macrophages. Arch. Pathol. Lab. Med. 102,* 79–83.

Cohen, A. F., and Cohen, B. L. (1980), *Protection from Being Indoors Against Inhalation of Suspended Particulate Matter of Outdoor Origin. Atmos. Environ. 14,* 183–184.

Cohen, G. M., Mehta, R., and Meredith-Brown, M. (1979), *Large Interindividual Variations in Metabolism of Benzo(a)pyrene by Peripheral Lung Tissue from Lung Cancer Patients. Int. J. Cancer 24,* 129–133.

Colome, S. D., and Spengler, J. D. (1982), *Residential Indoor and Matched Outdoor Pollutant Measurements with Special Considerations of Woodburning Homes,* in: Cooper, J. A., and Maler, D. (eds.): *Residential Solid Fuels: Environmental Impacts and Solutions.* Oregon Graduate Center, Beaverton, Oregon.

Colome, S. D., McCarthy, S. M., and Spengler, J. D. (1981), *Residential Indoor and Ambient Outdoor Comparison of Gaseous and Particulate Air Pollutants in two Cities.* Paper No. 81–57.3, presented at *74th Annual Meeting of Air Pollution Control Association,* Philadelphia.

Crapper, D. R., Krishman, S. S., and Dalton, A. J. (1973), *Brain Aluminum Distribution in Alzheimer's Disease and Experimental Neurofibrillary Degeneration. Science 180,* 511.

Crapper, D. R., Karlik, S., and DeBoni, U. (1976), *Aluminum and Other Metals in Senile (Alzheimer's) Dementia. Ageing 7,* 471.

Davidson, C. I., Grimm, T. C., and Nasta, M. A. (1981), *Airborne Lead and Other Elements Derived from Local Fires in the Himalaya. Science 214,* 1344–1346.

Davidson, C. I., Lin, S. W., Osborn, J. F., Mandey, M. R., Rasmussen, R. A., and Khalli, M. A. K. (1986), *Indoor and Outdoor Air Pollution in the Himalaya. Environ. Sci. Technol. 20,* 561–567.

De Angelis, D. G., Ruffin, D. S., Peters, J. A., and Reznik, R. B. (1980), *Source Assessment: Residential Combustion of Wood.* EPA-600/2-80-042b. US Environmental Protection Agency, Research Triangle Park, North Carolina.

Dockery, D. W., and Spengler, J. D. (1981a), *J. Air Pollut. Control Assoc. 31,* 153–159.

Dockery, D. W., and Spengler, J. D. (1981b), *Indoor-Outdoor Relationship for Respirable Sulfates and Particles. Atmos. Environ. 15,* 335–343.

Drew, R. T., Bupta, B. N., Bend, J. R., and Hook, G. E. R. (1974), *Inhalation Studies with Glycol Complex of Aluminum-chloride-hydroxide. Arch. Environ. Health 28,* 321–326.

Duggan, M. J. (1980), *Lead in Urban Dust: An Assessment. Water Air Soil Pollut. 14,* 309–321.
Duggan, M. (1983), *Lead in Dust as a Source of Childrens' Body Lead,* in: Rutter, M., and Russel-Jones, R. (eds.): *Lead Versus Health,* pp. 115–139. Wiley, New York.
Duggan, M. J., and Williams, C. (1977), *Lead-in-dust in City Streets. Sci. Total Environ. 7,* 91–97.
Elwood, P. C. (1983), *The Lead Debate,* pp. 1–6. Congress of the Institution of Environmental Health Officers, Brighton, U.K.
EPA (1978a), *Indoor Air Pollution in the Residential Environment,* Vol. 1, *Data Collection, Analysis, and Interpretation.* EPA-600/7-78-229a. US Environmental Protection Agency, Research Triangle Park, North Carolina, December.
EPA (1978b), *Indoor Air Pollution in the Residential Environment,* Vol. 2, *Field Monitoring Protocol, Indoor Episodic Pollutant Release Experiments and Numerical Analyses.* EPA-600/7-78-229b. US Environmental Protection Agency, Research Triangle Park, North Carolina, December.
Epstein, S. G. (1985), *Aluminum in Nature, in the Body, and its Relationships to Human Health. Trace Elem. Med. 2,* 14–18.
Fergusson, J. E., and Ryan, D. E. (1984), *The Elemental Composition of Street Dust from Large and Small Urban Areas Relative to City Type, Source, and Particle Size. Sci. Total Environ. 34,* 101–116.
Fergusson, J. E., and Schroeder, R. J. (1985), *Lead in House Dust of Christchurch, New Zealand: Sampling Levels and Sources. Sci. Total Environ. 46,* 61–72.
Fergusson, J. E., Forbes, E. A., Schroeder, R. J., and Ryan, D. E. (1986), *The Elemental Composition and Sources of Home Dust and Street Dust. Sci. Total Environ. 50,* 217–221.
Fishbein, L. (1989), *Metals in the Indoor Environment. Toxicol. Environ. Chem. 22* (1–4), 1–7.
Friberg, L., Piscator, M., Nordberg, L. F., and Kjellstrom, T. (1985), *Cadmium in the Environment,* 2nd Ed., pp. 46–47, CRC Press, Cleveland, Ohio.
Friberg, L., Nordberg, G. F., and Vouk, V. B. (eds.) (1986), *Handbook on the Toxicology of Metals,* 2nd Ed., Vol. I, pp. 85–127, 128–148, 175–205. Elsevier North Holland Biomedical Press, Amsterdam.
Friedlander, S. K. (1983), *Chemical Elements Balances and Identification of Air Pollution Sources. Environ. Sci. Technol. 7,* 235–240.
Gammage, R. B., and Kaye, S. V. (eds.) (1985), *Indoor Air and Human Health.* Lewis Publ., Chelsea, Michigan.
Hoffmann, D., Haley, N. J., Brunnemann, K. D., Adams, J. D., and Wynder, E. L. (1983), *Cigarette Sidestream Smoke: Formation, Analysis, and Model Studies on the Uptake by Non-smokers.* Presented at the *US-Japan Meeting on New Etiology of Lung Cancer,* Honolulu, Hawaii, March 21–23.
Hughart, D. (1979), *Prospects for Traditional and Non-conventional Energy Sources in Developing Countries. World Bank Working Paper No. 346,* Washington, DC.
IARC (1976), *IARC Monographs on the Evolution of the Carcinogenic Risk of Chemicals to Humans,* Volume 11, *Cadmium, Nickel, Some Epoxides, Miscellaneous Industrial Chemicals, and General Considerations on Volatile Anaesthetics.* International Agency for Research on Cancer, Lyon.
IARC (1980), *IARC Monographs on the Evolution of the Carcinogenic Risk of Chemicals to Humans,* Vol. 23, *Some Metals and Metallic Compounds.* International Agency for Research on Cancer, Lyon.
IARC (1982), *IARC Monographs on the Evolution of the Carcinogenic Risk of Chemicals to Humans,* Supplement 4, *Chemicals, Industrial Processes and Industries Associated with Cancer in Humans.* International Agency for Research on Cancer, Lyon.
IARC (1986a), *IARC Monographs on the Evolution of the Carcinogenic Risk of Chemicals to Humans,* Vol. 38, *Tobacco Smoking.* International Agency for Research on Cancer, Lyon.
IARC (1986b), *Environmental Carcinogens – Selected Methods for Analysis,* Vol. 8, *Some Metals.* International Agency for Research on Cancer, Lyon.

ICRP (International Commission on Radiological Protection) (1975), *Report on the Task Group on Reference Man,* ICRP Publ. 23. Pergamon Press, Oxford.

Jenkins, R. A., O'Neill, I. K., Schuller, P., and Fishbein, L. (eds). (1986), *Occurrence of Selected Metals in Cigarette Tobaccos and Smoke,* in: *Environmental Carcinogens, Selected Methods of Analysis,* Vol. 8, *Some Metals: As, Be, Cu, Cl, Cr, Ni, Pb, Se, Zn.* IARC Scientific Publications No. 71, pp. 129–138. International Agency for Research on Cancer, Lyon.

Jones, K. C., and Bennett, B. G. (1986), *Exposure of Man to Environmental Aluminum – An Exposure Commitment Assessment.* Sci. Total Environ. *52,* 65–82.

Kalliomäki, P. L., Kokkonen, P., Päällö, P., Anttila, S, and Kalliomäki, K. (1989), *Elemental Analyses in Human Lung Tissue Correlated with Smoking, Emphysema and Lung Cancer.* Toxicol. Environ. Chem. *19* (1+2), 1–6.

Klein, D. N. (1975), *Pathways of Thirty-seven Trace Elements Through Coal-fired Power Plants.* Environ. Sci. Technol. *9,* 973–979.

Koivistoinen, P. (ed.) (1980), *Mineral Elemental Composition of Finnish Food.* Acta Agric. Scand. (Suppl. 22).

Kowalczyk, G. S., Gordon, G. E., and Rheingrover, S. W. (1982), *Identification of Atmosphere Particulate Sources in Washington, DC Using Chemical Element Balance.* Environ. Sci. Technol. *16,* 79–90.

Krause, C., Dube, P., Neumayr, U., Schulz, C., and Wolter, R. (1987), *Metal Concentrations in Indoor Dust Samples from German Homes,* in *Proceedings of the 4th International Conference on Indoor Air Quality and Climate,* Berlin, 17–21 August, pp. 509–514.

Krueger, G. L., Morris, T. K., Suskind, R. R., and Widner, E. M. (1984), *The Health Effects of Aluminum Compounds in Mammals.* CRC Crit. Rev. Toxicol. *13,* 1–24.

Laxen, D. D. H., Raab, G. M., and Fulton, M. (1987), *Children's Blood Lead and Exposure to Lead in Household Dusts and Water (A Basis for an Environmental Standard for Lead in Dust).* Sci. Total Environ. *66,* 235–244.

Lee, R. E., Jr., and von Lehmden D. J. (1973), *Trace Metal Pollution in the Environment.* J. Air Pollut. Control Assoc. *23,* 853–1857.

Leppanen, A., and Lukkonen-Lilja, H. (1982), *Analysis of Tobacco Smoke. Summary of Studies during 1977–1980.* Technical Research Centre of Finland, Research Report No. 86.

Lepow, M., Brickman, L., Robino, R. A., Markowitz, S., Gillette, M., and Kpaish, J. (1974), *Environ. Health Perspect.* 7, 99–102.

Lewis, G. P., Jusko, W., Coughly, L., and Hartz, S. (1972), *Contribution of Cigarette Smoking to Cadmium Accumulation in Man.* Lancet *1,* 291–292.

Lipfert, F. W., and Dungan, J. L. (1983), *Residential Firewood Use in the United States.* Science *219,* 1425–1427.

Löfroth, G., Nilsson, L., and Alfheim, I. (1983), *Passive Smoking and Urban Air Pollution: Salmonella Microsome Mutagenicity Assay of Simultaneously Collected Indoor and Outdoor Particulate Matter,* in: Waters, M., et al. (eds.): *Short-Term Bioassays in the Analysis of Complex Environmental Mixtures,* pp. 515–525. Plenum, New York.

Martell, E. A. (1974), *Radioactivity of Tobacco Trichomes and Unsoluble Cigarette Smoke Particles.* Nature *249,* 215–217.

Menden, E. E., Elia, V. J., Michael, L. W., and Petering, H. G. (1972), *Distribution of Cadmium and Nickel in Tobacco During Cigarette Smoking.* Environ. Sci. Technol. *6,* 830–832.

Mielke, H. W., Anderson, J. C., Berry, K. J., Mielke, P. W., Chaney, R. L., and Leech, M. (1983), *Lead Concentrations in Inner City Soils as a Factor in the Child Lead Problem.* Am. J. Publ. Health 73, 1366–1369.

Milar, I. B., and Cooney, P. A. (1982), *Urban Lead – A Study of Environmental Lead and Its Significance to School Children in the Vicinity of a Major Trunk Road.* Atmos. Environ. *16,* 615–620.

Morgan, W. D. (1979), *New Ways of Measuring Cadmium in Man.* Nature *282,* 673–674.

Moschandreas, D. J. (1981), *Exposure to Pollutants and Daily Time Budgets of People.* Bull. N.Y. Acad. Med. *57,* 845–859.

Moschandreas, D. J., Winchester, J. W., and Nelson, J. A. (1979), *Fine Particle Residential Indoor Air Pollution. Atmos. Environ. 13*, 10.

Moschandreas, D. J., Zabransky, J., and Rector, H. E. (1980), *The Effects of Woodburning on the Indoor Residential Air Quality. Environ. Int. 4*, 463–468.

Mussalo-Rauhamaa, H., Salmela, S. S., Leppanen, A., and Pyysalo, H. (1986), *Cigarettes as a Source of Some Trace and Heavy Metals and Pesticides in Man. Arch. Environ. 41*, 49–55.

Nandi, M., Slone, D., Jick, H., Shapiro, S., and Lewis, G. P. (1969), *Lancet 2*, 1329–1330.

NAS (1972), *Lead – Airborne Lead in Perspective.* National Academy of Sciences, Washington, DC.

NAS (1973), *Manganese.* National Academy of Sciences, Washington, DC.

NAS (1974a), *Chromium.* National Academy of Sciences, Washington, DC.

NAS (1974b), *Vanadium.* National Academy of Sciences, Washington, DC.

NAS (1975), *Nickel.* National Academy of Sciences, Washington, DC.

NAS (1976), *Selenium.* National Academy of Sciences, Washington, DC.

NAS (1977a), *Copper.* National Academy of Sciences, Washington, DC.

NAS (1977b), *Arsenic.* National Academy of Sciences, Washington, DC.

NAS (1977c), *Iron.* National Academy of Sciences, Washington, DC.

NAS (1978), *Zinc.* National Academy of Sciences, Washington, DC.

NAS (1981), *Indoor Pollutants.* National Academy of Sciences, Washington, DC.

Natush, P. F. S., and Wallace, J. R. (1974), *Toxic Trace Elements: Preferential Concentration in Respirable Particles. Science 183*, 202–204.

Norman, V. (1977), *An Overview of the Vapor Phase, Semivolatile and Non-volatile Components of Cigarette Smoke. Recent Adv. Tob. Sci. 3*, 25–58.

Ott, W. R. (1980), *Models of Human Exposure to Air Pollution,* Technical Report 32. SIAM Institute for Mathematics and Society. Stanford University, Stanford, California.

Perinelli, M. A., and Carugno, N. (1978), *Determination of Trace Metals in Cigarette Smoke by Flameless Atomic Absorption Spectrometry. Beitr. Tabakforsch. 9*, 214–217.

Ramdahl, T., Schjoldager, J., Currie, L. A., Hanssen, J. E., Moller, M., Klouda, G. A., and Alfheim, I. (1984), *Ambient Impact of Residential Wood Combustion in Elvirum, Norway. Sci. Total Environ. 36*, 81–90.

Rona, R. J., Florey, C., Clarke, G. C., and Chinn, S. (1981), *Parental Smoking at Home and Height of Children. Br. Med. J. 283*, 1363.

Royal Commission on Environmental Pollution (1983), Ninth Report, *Lead in the Environment.* HMSO, London.

Rudling, K., Ahling, B., and Lofroth, G. (1981), *Chemical and Biological Characterization of Emissions for Combustion of Wood and Wood Chips in Small Furnaces and Stoves.* International Conference Research on Solid Fuels, June 1–5.

Schmidt, A. M. (1975), *Aerosol Drug and Cosmetic Products Containing Zirconium. Fed. Regist. 40*, 24328–24344.

Seifert, B. (1984), in: Bergund, B., Bergund, V., Lindvall, T., and Sundell, J. (eds.): *Indoor Air,* Vol. 6, *Evaluations and Conclusions for Health, Sciences and Technology,* pp. 11–13. Swedish Council for Building Research, Stockholm.

Sexton, K., and Wesolowski, J. J. (1985), *Safeguarding Indoor Air Quality. Environ. Sci. Technol. 19*, 305–309.

Sexton, K., Spengler, J. D., and Treitman, R. D. (1984), *Effects of Residential Wood Combustion on Indoor Air Quality: A Case Study in Waterbury, Vermont. Atmos. Environ. 18*, 1371–1382.

Sexton, K., Liu, K. S., Hayward, S. B., and Spengler, J. D. (1985), *Characterization and Source Apportionment of Wintertime Aerosol in a Woodburning Community. Atmos. Environ. 19*, 1225–1236.

Sexton, K., Liu, K. S., Treitman, R. D., Spengler, J. D., and Turner, W. A. (1986), *Characterization of Indoor Air Quality in Wood-burning Residences. Environ. Int. 12*, 265–278.

Skog, K. E., and Watterson, I. A. (1983), *Residential Fuelwood Use in the United States: 1980–1981.* Survey Completion Report, US Forest Service, Forest Products Laboratory, Madison, Wisconsin.

Solomon, R. L., and Hartford, J. W. (1976), *Lead and Cadmium in Dusts and Soils in a Small Urban Community.* Environ. Sci. Technol. *10*, 773–777.

Spengler, J. D., and Sexton, K. (1983), *Indoor Air Pollution: A Public Health Perspective.* Science *221*, 9–17.

Spengler, J. D., Dockery, D. W., Turner, W. A., Solfson, J. M., and Ferris, B. J., Jr. (1985), *Long-term Measurements of Respirable Sulfate and Particles Inside and Outside Homes.* Atmos. Environ. *15*, 22.

Stevens, R. K., Dzubay, T. G., Lewis, C. W., and Shaw, R. W., Jr. (1984), *Source Apportionment Methods Applied to the Determination of the Origin of Ambient Aerosols that Affect Visibility in Forested Aereas.* Atmos. Environ. *18*, 261–272.

Stone, C. J., McLaurin, D. A., Steinhagen, W. H., Cavender, F. C., and Haseman, J. K. (1979), *Tissue Disposition Patterns after Chronic Inhalation Exposure of Rats and Guinea Pigs to Aluminum Chlorohydrate.* Toxicol. Appl. Pharmacol. *49*, 71.

Sunderman, F. W. Jr. (1986), *Carcinogenicity and Mutagenicity of Some Metals and Their Compounds,* in IARC (1986b), pp. 17–43.

Swedish Council and Building Research (1984), *Indoor Air, Proceedings of the 3rd International Conference on Indoor Air Quality and Climate,* Vols. 1–5, Stockholm, August 20–24.

Szadkowski, D., Schultze, H., Schaller, K. H., and Lehnert, G. (1969), *Ecological Significance and Heavy Metal Content in Cigarettes – Assay of Pb, Cd, and Ni and Analysis of Gas Phases and Particle Phases in Tobacco.* Arch. Hyg. *153*, 1–8.

Szalai, A. (ed.) (1972), *The Use of Time: Daily Activities of Urban and Suburban Populations in Twelve Countries.* Mouton Publ., The Hague, The Netherlands.

Thompson, C. R., Hensel, E. G., and Katz, G. (1973), *Outdoor-Indoor Levels of Six Air Pollutants.* J. Air Pollut. Control Assoc. *23*, 10.

Torrey, S. (ed.) (1978), *Trace Contaminants from Coal.* Noyes Data Corp., Park Ridge, New Jersey.

Tosteson, J. D., Spengler, J. D., and Weker, R. A. (1981), *Aluminum, Iron, and Lead Concentrations of Personal, Indoor and Outdoor Respirable Particles,* presented at *International Symposium on Indoor Air Pollution, Health and Energy Conservation,* University of Massachusetts, Amherst, MA, October.

Travis, C. C., Etnier, E. L., and Meyer, E. R. (1985), *Health Risks of Residential Wood Heat.* Environ. Manage. *9*, 209–215.

Tso, T. C. (1966), *Micro- and Secondary Elements in Tobacco.* Acad. Sin. Inst. Bot. Bull. *7*, 28–63.

Vallyathan, V., and Hahn, L. H. (1985), *Cigarette Smoking and Inorganic Dust in Human Lungs.* Arch. Environ. Health *40*, 69–73.

van Houdt, J. J., Jongen, W. M. F., Alink, G. M., and Boleij, J. S. M. (1984), *Mutagenic Activity of Airborne Particles Inside and Outside Homes.* Environ. Mutagen. *6*, 861–869.

van Houdt, J. J., Daenan, C. M. J., Boleij, J. S. M., and Alink, G. M. (1986), *Contribution of Wood Stoves and Fire Places to Mutagenic Activity of Airborne Particulate Matter Inside Homes.* Mutat. Res. *171*, 91–98.

Wallace, L., Bromberg, S., Pellizzari, E., Hartwell, T., Zelon, H., and Heldon, L. (1984), *Plan and Preliminary Results of the U.S. Environmental Protection Agency's Indoor Air Monitoring Program: 1982,* in: *Proceedings of the 3rd International Conference on Indoor Air Quality and Climate,* Stockholm, August 20–24.

Ward, G. W. (1977), *Lung Changes to Inhalation of Underarm Aerosol Deodorants,* presented at the *Annual Meeting of American Thoracic Society.*

Weschler, C. J. (1984), *Indoor-Outdoor Relationships for Non-Polar Organic Constituents of Aerosol Particles.* Environ. Sci. Technol. *18*, 648–652.

WHO (1983), *Indoor Air Pollutants. Exposure and Health Effects Assessment.* Euro Reports and Studies Working Group Report No. 78. World Health Organization, Copenhagen.

Wynder, E. L., and Hoffmann, D. (1967), *Tobacco and Tobacco Smoke. Studies in Experimental Carcinogenesis.* Academic Press, New York.

Yaffe, Y., Flessel, J. J., Wesolowsk, A. et al. (1983), *Identification of Lead Sources in California Children Using Stable Isotope Ratio Technique.* Arch. Environ. Health 38, 237–245.

Yamanaka, S., and Muruoka, S. (1984), *Mutagenicity of the Extract Recovered from Airborne Particles Outside and Inside a Home with an Unvented Kerosene Heater.* Atmos. Environ. 7, 1485–1487.

Yocum, J. E. (1982), *Indoor-Outdoor Air Quality Relationships.* J. Air Pollut. Control Assoc. 32, 500–520.

Zoller, W. H., Gladney, E. S., and Duce, R. A. (1974), *Science 83,* 198–200.

I.7a Heavy Metal Compounds in the Soil

HERBERT W. SCHMITT and HANS STICHER, Zürich, Switzerland

1 Introduction
(see also Chapters I.1 and I.3)

Heavy metal ions included in the parent soil material are set free in the process of soil formation in correspondence to the rate of weathering. The further fate of the ions depends on pedological factors such as pH, humus content, redox potential as well as on external factors such as temperature, precipitations, erosion, land use practice etc. Accordingly, some elements are accumulated in the top soil whereas others are leached out.

In addition to the native content varying amounts of heavy metals are supplied to the soil by dry and wet atmospheric deposition and by agronomic practices (fertilizers, sewage sludge). For certain metals this anthropogenic proportion constitutes, globally considered, only a fraction of the natural amounts (ADRIANO, 1986). In other cases it can be the exclusive source. As emissions very often originate from point sources, the enrichment in their vicinity can easily exceed the natural state.

Usually metals are present in the soil
- as part of the soil parent material or soil minerals of secondary origin;
- precipitated with other compounds of the soil;
- sorbed on exchange sites. Metal oxides or hydroxides, clay minerals and organic matter can serve as exchangers;
- dissolved in the soil solution, either in the aqua-form or complexed with inorganic or organic ligands;
- embodied in microorganisms, plants or animals.

Their degree of mobility, activity, and bioavailability is influenced by many factors, particularly pH, temperature, redox potential, cation exchange capacity of the solid phase, competition with other metal ions, ligation by anions, composition and quantity of the soil solution. Since these soil properties may vary already within short distances, the metal content is subject to spatial variability (TRANGMAR et al., 1985; WOPEREIS et al., 1988).

2 The Soil System

2.1 Interactions between the Solid and the Liquid Phase

The solid and the liquid phase of the soil are in close contact. The polar solvent water interacts with the surface structure of the solid and is capable of enveloping charged sites. This has important consequences:

- water can penetrate into the solid and finally cause its disintegration;
- the pH of water affects the charge pattern of the surface;
- water acts as a transport medium for ions in exchange and adsorption processes.

The solid phase is essentially composed of clays, metal oxides and organic matter and in most cases carries an excess of negative charge. This leads to a surface potential and determines the distribution of cations and anions in the liquid phase. Two opposing forces, electrical attraction and diffusion back into the bulk solution, act on the cations which have to counterbalance the negative surface charge.

Several surface potential models have been developed. Their value for the description of adsorption processes has been reviewed by BARROW (1985). The more complex models assume several planes of adsorption which form in their entity a diffuse layer. In the adjacent bulk solution the effect of the surface potential has petered out. Formulas for the calculation of ion concentrations, swelling pressures and spatial extensions of diffuse layers have been published (KEMPER and QUIRK, 1970).

In the model of BOWDEN et al. (1977) it is postulated that ions not only differ in their affinity to the surface but also in the mean position of their adsorption plane.

2.2 Redox Potential

The water and the oxygen content of the soil influence the redox status of the soil which can change even within short distances. The redox potential may reach 700 mV in oxidized and -400 mV in strongly reduced soils (GAMBRELL and PATRICK, 1978) and is pH-sensitive (about -59 mV per pH unit). Using the pE value (SPOSITO, 1981), the negative common logarithm of the aqueous free electron activity, then positive pE values favor oxidized species whereas small values of pE favor reduced species. At pH 7.0 oxidized soils have $+7 < pE < +13.5$, moderately reduced soils have $+2 < pE < +7$, reduced soils have $-2 < pE < +2$, highly reduced soils have $-6.8 < pE < -2$.

The status of several metal ions is affected by the redox conditions, particularly for Fe and Mn, but also for Cr, Cu, As, Hg, and Pb. The redox potential can change directly the oxidation state of a heavy metal. Indirectly the chemical form of a metal ion can be changed through a change in the oxidation state of a ligand atom such as C, N, O, S.

Fe^{3+} is only stable under strongly acidic and oxidizing conditions and is otherwise reduced to the soluble Fe^{2+}:

Table I.7a-1. pE Value and Reduction/Oxidation Change (SPOSITO, 1983)

Oxidation States	pE(crit)	
	pH 5	pH 8
Mn(III)/Mn(II)	15.9	6.9
Mn(IV)/Mn(II)	14.7	8.7
Cr(VI)/Cr(III)	13.0	8.0
Hg(II)/Hg(I)	11.8	5.8
Fe(III)/Fe(II)	8.9	−0.1
Cu(II)/Cu(I)	8.3	2.3
Ag(I)/Ag(0)	5.5	5.5
Mo(VI)Mo(IV) (assuming 1 mmol L^{-1} SO$_4^{2-}$)	0.5	−3.9

$$FeOOH + e^- + 3H^+ = Fe^{2+} + 2H_2O \ .$$

Under strongly reducing conditions the formation of Fe(OH)$_2$, FeCO$_3$, FeS, and FeS$_2$ may occur. The formation of insoluble sulfides under similar conditions is also observed for Cd, Zn, Ni, Co, Cu, Pb, and Sn.

Manganese solubility increases with increasing acidity and reducing conditions, and the Mn^{2+} ion is the only significant soluble ion as Mn^{4+} is very insoluble and Mn^{3+} disproportionates into Mn^{4+} and Mn^{2+}. The exposure of birnessite (γ−MnO$_2$ to reducing agents (e.g., Fe(II), nitrite, but even sewage sludge) resulted in the presence of Mn(II) on the oxide surface. This divalent Mn was susceptible to displacement during Cu sorption (TRAINA and DONER, 1985).

Organic matter, Fe(II) minerals, and other reducing agents in soil are known to reduce Cr(VI) (CrO$_4^{2-}$) to Cr(III) under acidic conditions and heat.

Diagrams relating pH and redox status for various ions in several inorganic compounds have been published (GARRELS and CHRIST, 1965; POURBAIX, 1966). Critical pE values for redox speciation of heavy metals are given in Table I.7a-1.

3 Adsorption

3.1 General Considerations

Charged constituents of the soil can be divided into two classes:

- fixed charge constituents;
- variable charge constituents.

Prominent members of the first group involve clay minerals that develop negative charge as a result of lattice substitutions. The second class is composed of constituents whose charge varies with the pH of the soil solution and/or with the amount of reactions that have occurred with ions. Members are the oxides of iron, manganese, aluminum, titanium, but also organic matter.

For variable charge constituents two concurrent properties determine the degree of adsorption, the dissociation of the soil adsorbent

$$HL = L^- + H^+$$

and the acidity of the metal ion (first hydrolysis constant)

$$M(H_2O)_m^{n+} = M(H_2O)_{m-1}(OH)^{(n-1)+} + H^+ \ .$$

The "adsorbing units" of all these variable charge exchangers are O^{2-}, OH^- and O^- either as a matrix with embedded cations (Si^{4+}, Al^{3+}, etc.) in minerals or part (by covalent linking) of organic matter. All interactions between a metal ion M^{n+} in the soil solution and surfaces of soil solids can be formulated as a hydroxyoxide complex of uncertain composition and charge (BOHN and BOHN, 1986; SCHULTHESS and SPARKS, 1988)

$$MO_p(OH)_r^{(n-2p-r)+} \ .$$

3.2 Factors of Adsorption

The amount of metal adsorbed is dependent on the type of soil component. It appears that clay minerals adsorb far smaller quantities than other sorbents such as oxides and organic material (Table I.7a-2).

Anions may affect the adsorption of metals. At moderate levels of phosphate adsorption on allophane, some enhancement of sorption has been noted (CLARK, 1983) due to a higher negative charge density on the surface. On the other hand, high levels of phosphate suppressed adsorption on aluminum hydroxide. Specific adsorption of Cu^{2+} to AlOH groups was blocked as these sites were occupied by phosphate (MCBRIDE, 1985).

Table I.7a-2. Comparison of Cobalt Sorption Isotherm Gradients for Individual Soil Materials (MCLAREN et al., 1986)

Sorbent	Isotherm Gradient ($cm^3 g^{-1}$)
Kaolinite	5
Illite	60
Montmorillonite	592 – 758
Fulvic acid	981
Humic acid	2117
Soil birnessite	1090 – 22422
Soil oxide (ferro-manganese concretions)	5733 – 43460

Measuring the adsorption of metal ions on goethite at different pHs, MCKENZIE (1980) found the following decreasing sequence of adsorption (for hematite the order of Pb^{2+} and Cu^{2+} was reversed)

$$Cu^{2+} > Pb^{2+} > Zn^{2+} > Co^{2+} > Ni^{2+} > Mn^{2+} \ .$$

This sequence correlates with the first hydrolysis constant of the metal ions (GRIMME, 1968). Obviously the adsorption is related to metal hydrolysis. For the same pK value adsorption increases with increasing ionic radius (TILLER et al. 1984a, b; BRUEMMER et al., 1986).

In general, with increasing pH the capacity of soils for most metal ions is increased but the mobility of the cations is decreased. The relative mobility, dependent on the pH, of some trace elements in soils has been investigated by FULLER (1977):

pH 4.2 – 6.6: Cd, Hg, Ni, Zn: relatively mobile
               As, Be, Cr: moderately mobile
               Cu, Pb, Se: slowly mobile
pH 6.7 – 8.8: As, Cr: relatively mobile
               Be, Cd, Hg, Zn: moderately mobile
               Cu, Pb, Ni: slowly mobile.

Adsorption studies are often conducted using metal concentrations higher than found in natural systems. If the solubility product is exceeded (e.g., for hydroxide or carbonate formation), the amount of adsorption can be falsified by precipitation. This was observed for the adsorption of Cu and Zn on montmorillonite above pH 6 (BINGHAM et al., 1964).

3.3 Adsorption Isotherms

The adsorption of a solute from the soil solution can be described by adsorption isotherms. The term implies that temperature is the only variable affecting adsorption, an assumption which is often questionable (BARROW, 1978).

The amounts in solution (abscissa) are plotted against the adsorbed amounts (ordinate) in the state of equilibrium. Usually reversibility of the processes is assumed as is a constant capacity of the exchanger although the activity is dependent on ionic strength and pH.

The two isotherms which are widely used to describe the adsorption behavior, the equations by Langmuir and by Freundlich, belong to the equilibrium models. In addition, several other isotherms (SANDERS, 1980) including kinetic models have been developed (AMACHER et al., 1988). In a study about retention/release of CrO_4^{2-}, Cd^{2+}, Hg^{2+} in a variety of soils with different properties the kinetic models could not describe the adsorption data in contrast to the Freundlich and the two-site Langmuir equations.

If there is not a maximum adsorptivity, e.g., when multilayers are formed, then the data can often be fitted to a Freundlich equation

$$q = Kc^\beta \ ,$$

where q is the adsorbed amount, c is the equilibrium concentration and K and β are constants.

If saturation occurs then the Langmuir equation is used:

$$q = bKc/(1+Kc) .$$

The constant b equals the saturation level of the adsorbent (assuming a monomolecular layer), q, c, and K have the same meaning as above. Extensions for multiple sites are easily made by adding additional terms of the same kind with new constants b_i, K_i.

An adsorption equation for variable charge adsorbents has been developed by BOWDEN et al. (1977). The equation takes separate account of all the individual ion species present in solution and includes a term Φ_a for the electrostatic potential (which is pH-dependent)

$$q = \frac{b\Sigma K_i c_i \exp(-z_i\Phi_a F/RT)}{1+\Sigma K_i c_i \exp(-z_i\Phi_a F/RT)}$$

b, K have the same significance as in the Langmuir equation; F is the Faraday constant; R the gas constant; T the temperature (Kelvin); z the charge of the adsorbed ion.

The estimation of the potential term poses problems and requires the solution of several equations. A simplification for adsorption at constant pH has been presented (POSNER and BARROW, 1982). The equations have been successfully applied to the pH-dependent adsorption of anions (phosphate) and cations (Cu^{2+}, Pb^{2+}) to goethite (BARROW et al., 1980, 1981). Other equations relating to solution concentration and surface potential have been published (NIR, 1986).

Under certain circumstances a sigmoidal adsorption curve can be observed, e.g., when the complexation of a metal ion at low concentrations with soluble ligands is preferred to sorption to the solid phase (NEAL and SPOSITO, 1986) or when an adsorption process and a precipitation reaction occur synchronously. The following formula, an extension of the Langmuir equation with an added sigmoidicity factor R, was used to describe this behavior (SCHMITT and STICHER, 1986a)

$$q = bKc/(1+Kc+R/c) .$$

Adsorption and desorption processes may differ and directly affect the plant availability of metal ions. The amounts of copper, e.g., desorbed from humic acids, soil oxide and montmorillonite, studied with a dialysis equilibrium technique, were very small compared to the sorbed quantity (MCLAREN et al., 1983).

3.4 Competitive Adsorption

If several species compete for the same adsorption sites, modifications of the most commonly used adsorption isotherms are necessary. Usually, instantaneous equilibrium between solution and adsorbed ions is assumed and for Langmuir-type processes expressed in the form (MURALI and AYLMORE, 1983a,b,c)

$$q_i = \frac{bK_i c_i}{1 + \Sigma K_j c_j} .$$

A few simplifications for binary systems can be introduced:

1. Very low effective concentrations and surface coverage, i.e.,

$\Sigma K_j c_j \ll 1$
$b \gg \Sigma q_j$

render competition effects insignificant. Thus

$$q_i = b_i K_i c_i .$$

2. For high concentrations and near total surface coverage

$\Sigma K_j c_j \gg 1$

and it follows for species 1

$$q_1 = \frac{bK_1 c_1}{K_1 c_1 + K_2 c_2} = \frac{b(K_1/K_2)(c_1/c_2)}{1 + (K_1/K_2)(c_1/c_2)}$$

which is an ordinary Langmuir equation with $K = K_1/K_2$ and $c = c_1/c_2$.

This form has been used by GRIFFIN and AU (1977) in a study of the adsorption of Pb^{2+} to montmorillonite. Compared with a pure Pb^{2+} solution, a markedly lower adsorption was observed in the presence of Ca^{2+} and found to be dependent on the Pb^{2+}/Ca^{2+} ratio in solution. The results were attributed to competition for cation exchange sites on the clay.

Competition can also be described (HENDRICKSON and COREY, 1981) through the use of a competitive selectivity coefficient (N is any competing cation):

NL + M = ML + N

$$K_{MN} = \frac{[ML][N]}{[NL][M]} .$$

Re-evaluating published data the authors found that the selectivity values increased substantially as the concentration of trace metals decreased with respect to the concentration of other competing cations.

Competing Ca^{2+} shifted the adsorption equilibrium for Cd^{2+} drastically, but had less effect on Cu^{2+} (CAVALLARO and MCBRIDE, 1978). This suggests that ion exchange is responsible for Cd^{2+} adsorption, but Cu^{2+} may be bonded more specifically.

The competitive adsorption of Cd, Cu, Pb, and Zn onto four soils with differing chemical properties was investigated by ELLIOTT et al. (1986). For two mineral soils, adsorption under acidic conditions (pH 5.0) followed the sequence Pb > Cu > Zn > Cd, which corresponds to the order of increasing pK for the first hydrolysis product. For two soils with high organic matter content the order was Pb > Cu > Cd > Zn.

Table I.7a-3. Calculated Saturation Levels for Cd, Pb, Cu in an Orthic Luvisol under Various Competition Conditions (Molar Ratios) (SCHMITT and STICHER, 1986b)

Cd	3.055 mg/g
Cd (Cd/Pb = 1:1)	1.127 mg/g
Cd (Cd/Pb = 1:18)	0.086 mg/g
Cd (Cd/Cu = 1:59)	0.033 mg/g
Cd (Cd/Pb/Cu = 1:18:59)	0.026 mg/g
Pb	6.54 mg/g
Pb (Pb/Cd = 1:18)	6.21 mg/g
Pb (Pb/Cd = 1:1)	4.68 mg/g
Pb (Pb/Cu/Cd = 18:59:1)	1.72 mg/g
Cu	2.08 mg/g
Cu (Cu/Cd = 59:1)	2.04 mg/g
Cu (Cu/Cd = 5:1)	1.77 mg/g
Cu (Cu/Pb/Cd = 59:18:1)	1.55 mg/g

The competition of cadmium with lead and copper was measured in an orthic luvisol (SCHMITT and STICHER, 1986a, b). The sorption of cadmium decreased to a greater degree than the competition ratio would predict while lead and copper were affected to a lesser degree by the competitive conditions (Table I.7a-3). The increased availability of cadmium in the soil also resulted in a higher plant uptake (GSPONER et al., 1986) corroborating findings that soil Pb increases both Cd concentration and Cd uptake in plants (MILLER et al., 1977).

4 Cation Exchange

4.1 General Considerations

The cation exchange capacity (CEC) denotes the total sum of exchangeable cations and is commonly expressed in meq/100 g or cmol of charge per kg. Clay, oxides, and organic matter contribute to the CEC of soils. The CEC of the top layers originates from 25–90% from organic matter and is much more strongly influenced by the pH than is the CEC of clay. Multiple regression analysis can be used to obtain pH-CEC relationships for organic matter and clay (HELLING et al., 1964; MCLEAN and OWEN, 1969). Table I.7a-4 lists the cation exchange capacity of various soil components.

4.2 Cation Exchange Equations

Several equations can be used to describe an equilibrium between the composition of a soil solution and the exchangeable cations on the surface of the solid phase.

Table I.7a-4. Cation Exchange Capacity of Soil Constituents (cmol/kg)

Kaolinite	3 – 15
Illite	10 – 40
Vermiculite	100 – 150
Montmorillonite	80 – 150
Bentonite[a]	450
Amorphous Al-oxide[a]	50
Amorphous Fe-oxide[a]	160
Birnessite[a]	230
Humic acid	480 – 870
Fulvic acid	640 – 1420
Humus	200

[a] At pH 7.6 (from BRUEMMER and HERMS, 1983)

Usually exchange reactions are very rapid, stoichiometric, and reversible and therefore the mass action law can be employed. The exchange reaction (ex = exchangeable, A and B are cations with the same valence)

$$A(ex) + B = A + B(ex)$$

can be described by the following equation (using activities)

$$\frac{a_{B(ex)}}{a_{A(ex)}} = K_{ex} \frac{a_B}{a_A} .$$

Since activities of the adsorbed cations cannot be measured or calculated easily, concentrations are often used in place of activities (Kerr equation). This simplification diminishes the value of the equation particularly for divalent ions. VANSELOW modified the Kerr equation assuming that the activity of an exchangeable cation is linearly proportional to its mole fraction. If activity coefficients on the soil matrix are omitted, the Vanselow selectivity coefficent K_V is defined:

$$K_V = \frac{a_B M_{A(ex)}}{a_A M_{B(ex)}} , \text{ where}$$

$$M_{A(ex)} = \frac{[A(ex)]}{[A(ex)] + [B(ex)]}$$

$$M_{B(ex)} = \frac{[B(ex)]}{[A(ex)] + [B(ex)]} .$$

For the exchange between monovalent and higher charged cations the modification of GAPON is often used. For the exchange reaction

$$D(ex) + 2M^+ = 2M(ex) + D^{2+} ,$$

where D denotes a divalent, M a monovalent ion, the Gapon equation reads

$$\frac{[M(ex)]}{[D(ex)_{1/2}]} = K_{ex} \frac{[M^+]}{[D^{2+}]^{1/2}} .$$

It corresponds to an exchange equation with equivalent amounts of ions but suffers from the same drawbacks as the Kerr equation because activities are omitted. Nevertheless, the equation holds reasonably true for the essential concentration range of many soils and can be interrelated to the Vanselow equation (EVANGELOU and PHILLIPS, 1988). Studying the exchange behavior of Pb^{2+} with Al^{3+} and Ca^{2+}, LAGERWERFF and BROWER (1973) used the Gapon exchange coefficient to compare different soil types.

Frequently, however, equations of the Langmuir and Freundlich type are likewise used for the description of exchange phenomena.

Cation exchange equilibria have also been considered to be a composite of two half complexation reactions (SHAVIV and MATTIGOD, 1985). These complexation reactions are assumed to occur between the exchanging cations and the negatively charged surfaces as hypothetical free ligands. Calculating ionic equilibria with the SOILCHEM program (SPOSITO and COVES, 1978), the distribution and the exchange of ions can be predicted.

4.3 Kinetics of Exchange Reactions

Rate laws are determined by experimentation and cannot be inferred by simply examining the overall chemical reaction equations. They help to reveal a reaction mechanism and to predict how quickly an equilibrium can be reached. Several kinetic equations have been employed (for a review see SPARKS, 1985), but even modified Freundlich equations (sorption versus time) have been used to describe the kinetics of arsenite adsorption (ELKHATIB et al., 1984).

For an ion exchange process

$$A^+ + B\text{-Soil} = B^+ + A\text{-Soil}$$

several steps can be rate controlling (SPARKS, 1985):

1. Film diffusion: diffusion of A^+ (with the counterion X^-) through the solution film surrounding the soil particle.
2. Intraparticle diffusion: diffusion through a hydrated interlayer space of the soil particle.
3. Chemical exchange of A^+ for B^+ from the surface of the solid phase.
4. Intraparticle diffusion of B^+ in reversed order, followed by a film diffusion (with the counterion X^-) back into the bulk solution.

The overall exchange process is mainly diffusion-controlled. Several factors determine whether the kinetic process is governed by film or intraparticle diffusion, e.g., flow velocity, hydrodynamic film thickness, particle size, activation energy. A detailed study on the kinetics of ion exchange on clay minerals and soils has been presented by OGWADA and SPARKS (1986a, b).

In a study on the adsorption of Ni^{2+} and Cu^{2+}, kinetic analysis revealed three separate adsorption events (HARTER and LEHMANN, 1983). The first event, roughly terminated after the first minute, was attributed to a very rapid, diffusion controlled cation exchange reaction (against Ca^{2+}, Mg^{2+}, K^+, H^+). Equilibrium was attained after about half an hour.

Equilibria as well as the rates of adsorption and desorption of Pb^{2+}, Cu^{2+}, Cd^{2+}, Zn^{2+}, Ca^{2+} by peat organic matter were investigated by BUNZL et al. (1976). The distribution coefficient with respect to H^+ as the competitive ion showed that the order of the selectivity and of the absolute rates of the metal adsorption by peat is

$$Pb^{2+} > Cu^{2+} > Cd^{2+} \geq Zn^{2+} > Ca^{2+}$$

in the pH range of 3.5 to 4.5. The adsorption of the metal ion was coupled to an exchange of two protons. The half-lives of adsorption (the time at which half of the equilibrium uptake is reached) were in the range of 5 to 15 s and decreased with increasing amounts of metal ions present in the solution. The results imply that the higher selectivity also causes a faster initial rate of adsorption. This will be observed for any ion-exchange process, as long as the diffusion coefficients of the ions are not too different.

Reactions in soil often fail to fit to single reaction models suggesting that multi-reaction models may fit the data. Several approaches involving one or more slow reversible and one or more irreversible reactions of an nth order have been presented (AMACHER et al., 1988) and can eventually provide an understanding of metal retention/release processes (NYFFELER et al., 1984).

5 Diffusion of Metals into Soil Minerals

Cations bound to the surface of minerals can diffuse into the interior of the solid phase. The relative diffusion rates depend on the ionic diameters and the pH conditions as has been shown by BRUEMMER et al. (1988) in a series of experiments with goethite. With increasing pH the affinity of the oxide surface is increased up to the point where the formation of hydroxocomplexes inhibits the access to the surface. With increasing ionic diameter the diffusion rate decreases, e.g., Cd^{2+} (0.97 nm) $< Zn^{2+}$ (0.74 nm) $< Ni^{2+}$ (0.69 nm). In the interior these ions can neutralize negative charges and remain fixed in appropriate positions. Also for other minerals such as manganese oxide (MCKENZIE, 1980) and illites and smectites (GERTH, 1985) similar diffusion has been observed. Such processes are often described as irreversible adsorption and cause heavy metals (e.g., distributed by pollution into soils) to become immobilized and unavailable with time.

6 Complexation

6.1 General Considerations

The concentrations of heavy metals in the soil solution are usually very low (10^{-8} mol L^{-1}).

Frequently the metal ions are coordinated to organic substances, mainly humic and fulvic acids forming oligodentate complexes and chelates. Complex formation arises when water ligands are replaced by other molecules.

Chelation reactions have the following effects:
- metal ions are prevented from being precipitated;
- complexing agents can act as carriers for metals in the liquid phase;
- the toxicity of the free aqua form is often reduced by complexation;
- chelation plays an important role in the chemical weathering of minerals and rocks.

6.2 Stability Constants

According to the Lewis concept of acids and bases, metal ions and ligands are acids and bases, respectively, and their reactions to form metal complexes are acid-base reactions. A dual parameter scale including an additional term for the ligand has been proposed to represent the Lewis acidity of metal ions (MISONO et al., 1967) and to give an indication of the stability of a complex with a certain ligand.

Several approaches have been used to examine and evaluate stability constants of metal complexes with humic substances (FITCH and STEVENSON, 1984)

$$mM + nA = M_mA_n \qquad K = \frac{[M_mA_n]}{[M]^m[A]^n} .$$

Using two different methods SCHNITZER and HANSEN (1970) determined stability constants of metal-fulvic acid complexes at pH 3 and 5 and for different ionic strengths (Table I.7 a-5).

Stability constants increase with increasing pH and decreasing ionic strength. For Cu complexes BUFFLE (1980) observed a linear relationship for log K values between pH = 3 and pH = 5. This concept was extended to other metal ions and used to estimate log K values in soils with a pH around 4 (BOURG and VÉDY, 1986).

At pH = 8 the log K values follow approximately the IRVING and WILLIAMS (1948) stability sequence (MANTOURA et al., 1978; NAND RAM and RAMAN, 1984)

$$Cu^{2+} > Ni^{2+} = Zn^{2+} > Co^{2+} > Mn^{2+} = Cd^{2+} > Ca^{2+} > Mg^{2+} .$$

In experiments with peat (BLOOM and McBRIDE, 1979), the large preference for Cu^{2+} has been explained by the Jahn-Teller distortion of the ligand field. This leads to an exchange of one or two aqua ligands with organic oxygen ligands (carboxylic acids) as indicated by ESR measurements. Most of the other divalent ions do not coordinate directly to functional oxygens but maintain a hydration sphere.

Table I.7a-5. Log K Values of Metal-Fulvic Acid Complexes. Effect of pH and Ionic Strength I (SCHNITZER and HANSEN, 1970)

	pH 3.0			pH 5.0
	$I = 0.00$	$I = 0.10$	$I = 0.15$	$I = 0.10$
Cu^{2+}	4.7	3.3	2.6	4.0
Ni^{2+}	4.5	3.2	2.4	4.2
Co^{2+}	4.3	2.9	2.2	4.1
Pb^{2+}	3.6	2.7	2.1	4.0
Ca^{2+}	3.6	2.7	2.1	3.3
Zn^{2+}	3.2	2.3	2.0	3.6
Mn^{2+}	2.9	2.2	1.7	3.7
Mg^{2+}	2.7	1.9	1.6	2.1

Results of the interaction of 11 different metal ions (Hg, Fe, Pb, Al, Cr, Cu, Cd, Zn, Ni, Co, Mn) with humic acids at different pHs (2.4, 3.7, 4.7, 5.8) have been reported (KERNDORFF and SCHNITZER, 1979; SCHNITZER, 1986) and indicate that Hg(II) and Fe(III) are always adsorbed most readily while Co and Mn are adsorbed least readily. TURNER et al. (1981) showed that the stability constants correlate linearly with the first hydrolysis constant of the metal ions.

Using organic substances extracted from acidic forest soils metal ions can be grouped into four categories (KOENIG, 1985; DIETZE and KOENIG, 1988). To the first group belong Fe, Al, and Pb which are almost completely complexed to substances which themselves bind strongly to mineral surfaces. Cr and Cu, the second group, are complexed to a large degree as well, but to substances which are loosely bound to mineral surfaces. The third group (Ni, Co, Cd) forms hardly any or only very weak complexes, the distribution between soil and soil solution, however, is pH dependent. For the members of the fourth group (Zn, Mn, together with Ca and Mg) there is no evidence of complexation with natural organic substances. The following stability series summarizes the results:

$$Cr > Fe > Al > Pb \gg Cu > Ni > Co \gg Cd > Zn \gg Mn = Ca = Mg \ .$$

6.3 Speciation

With the aid of specific extraction reagents the proportion of a metal within different phases can be determined:

1. in the solution phase;
2. in the exchangeable phase, either attracted to sites on clays, organic matter, oxides (hydroxides) or complexed/chelated especially to organic matter;
3. in the precipitation phase as, e.g., sulfide, carbonate, hydroxide, etc.;
4. in the lattice phase of minerals or occluded in oxides;
5. in the biophase.

Table I.7a-6. Principal Chemical Species of Heavy Metals in Acid and Alkaline Soil Solutions under Oxic Conditions (Florence, 1977; Sposito, 1983)

Mn(II)	acid:	Mn^{2+}, $MnSO_4^0$, Org
	alk.:	Mn^{2+}, $MnSO_4^0$, $MnCO_3^0$, $MnHCO_3^+$, $MnB(OH)_4^+$
Fe(II)	acid:	Fe^{2+}, $FeSO_4^0$, $FeH_2PO_4^+$
	alk.:	$FeCO_3^0$, Fe^{2+}, $FeHCO_3^+$, $FeSO_4^0$
Ni(II)	acid:	Ni^{2+}, $NiSO_4^0$, $NiHCO_3^+$, Org
	alk.:	$NiCO_3^0$, $NiHCO_3^+$, Ni^{2+}, $NiB(OH)_4^+$
Cu(II)	acid:	Org, Cu^{2+}
	alk.:	$CuCO_3^0$, Org, $CuB(OH)_4^+$, $Cu(B(OH)_4)_2^0$
Zn(II)	acid:	Zn^{2+}, $ZnSO_4^0$
	alk.:	$ZnHCO_3^+$, $ZnCO_3^0$, Zn^{2+}, $ZnSO_4^0$, $ZnB(OH)_4^+$
Cd(II)	acid:	Cd^{2+}, $CdSO_4^0$, $CdCl^+$
	alk.:	Cd^{2+}, $CdCl^+$, $CdSO_4^0$, $CdHCO_3^+$
Pb(II)	acid:	Pb^{2+}, Org, $PbSO_4^0$, $PbHCO_3^+$
	alk.:	$PbCO_3^0$, $PbHCO_3^+$, $Pb(CO_3)_2^{2-}$, $PbOH^+$

Org = organic complexes, e.g., with fulvic acid

McLaren and Crawford (1973) determined the copper fractionation in 24 soils. On average only a few percent of copper were present in the soil solution or in easily exchangeable form (representing the available form for plant uptake), about 30% were specifically sorbed on clay or organic matter, 15% sorbed on or occluded by oxides, 50% remained in the lattice. Similar studies were performed for other metals (e.g., Schwertmann et al., 1982a,b; Hickey and Kittrick, 1984; Sims, 1986; Levesque and Mathur, 1986, 1988). Obviously the type and fate of soil and the natural content of the metal in a soil are decisive and any generalization has to be avoided (King, 1988).

Concerning the speciation of heavy metal ions in the soil solution analytical methods, such as anodic stripping voltammetry, removal by resins, destruction by UV radiation, filtration, dialysis, ion selective electrodes, allow to distinguish between adsorbed or solid species, soluble species, free metal ion, labile or non-labile organic or inorganic complexes (Florence, 1977). These criteria have been applied to speciate some heavy metal ions in soil solutions (Table I.7a-6).

In the case of anoxic conditions, sulfate and nitrate complexes are replaced by sulfide and ammine complexes.

The concentration of Zn, Cd, Cu, and Pb has been measured in equilibrium solutions of samples from Ap horizons of different soils. The solubility isotherms of a series of definite inorganic metal compounds has been calculated as a function of the pH. The results provide insight into the stability of inorganic metal complexes and the relative mobility of metal ions (Herms, 1982; Herms and Bruemmer, 1984; Bruemmer et al., 1986).

The following compilation is a summary of what has been reported in the literature about the speciation of metal ions in soil solution where organic complexation agents are present (Hodgson et al., 1965, 1966; Olomu et al., 1973; Berrow

Table I.7a-7. Speciation (%) of Chromium and Nickel in the Gravitational Water of a Serpentinitic Cambisol. (JUCHLER and STICHER, 1987)

Species	O (pH 5.4)	B (pH 6.6)	C (pH 7.2)
$CrCO_3^+$	29.0		
$Cr(CO_3)_2^-$	24.1	66.6	11.2
$Cr(CO_3)_3^{3-}$		33.3	88.8
$CrOH^{2+}$	2.0		
$Cr(OH)_2^+$	1.0		
CrFulv	43.8		
Ni^{2+}	78.2	80.8	77.5
$NiHCO_3^+$		1.7	5.6
$NiCO_3^0$			2.2
$NiSO_4^0$	0.2	0.2	0.3
$NiOH^+$			0.2
NiFulv	21.5	17.3	14.2

O, B, C denote different soil horizons; Fulv = fulvic acid

and MITCHELL, 1980; CAMERLYNCK and KIEKENS, 1982; FUJII et al., 1982; SANDERS, 1982, 1983; STEVENSON and FITCH, 1986):

- mainly free cations: Co, Mn, Cd
- intermediate: Zn, Ni
- mainly complexed: Cu, Pb, Fe .

With the aid of programs such as GEOCHEM (SPOSITO and MATTIGOD, 1980), ADSORP (BOURG, 1982; BOURG and VÉDY, 1986), or SOILCHEM (SPOSITO and COVES, 1988) the speciation of metal ions in a soil solution can be calculated. A set of thermodynamic equilibrium constants for each soil component is incorporated and ionic strength and pH are taken into account. The approach is restricted to systems which have been characterized with respect to cationic and anionic components. Table I.7a-7 shows predictions (with the GEOCHEM program) for a serpentinitic acid cambisol (JUCHLER and STICHER, 1987).

The speciation of cadmium in the soil matrix has been investigated by solid state ^{113}Cd NMR-spectroscopy (BANK et al., 1989).

7 Translocation

The adsorption selectivity influences the composition of the soil solution and the amount sorbed on the soil matrix. The water flux and other parameters such as weathering which mobilizes many elements, change this pattern and translocate ions into other soil regions, eventually into the groundwater (CAMPBELL and BECKETT, 1988).

Models have been developed to predict ion movement depending on transport and chemical reactions.

These models combine
- important soil parameters, e.g., adsorptivity, water content, density;
- the composition of the soil solution;
- the water flux through the soil;
- the occurrence of many coupled equilibria like precipitation/dissolution and ion exchange.

Two types of models have been used. Segmentation or piston flow models divide soil and influx into segments with specified initial concentrations and neglect the effects of hydraulic flow velocity and dispersion. The water and the soil segment react as a completely mixed reactor. (CHRISTENSEN, 1981; FLUEHLER and JURY, 1983).

If hydraulic and dispersive effects are to be considered, then convection-diffusion equations are used which are usually solved with finite difference schemes (SELIM and MANSELL, 1976; MAYER, 1978; FLUEHLER and JURY, 1983; FOERSTER, 1986; UTERMANN and RICHTER, 1988). For both model types the chemical reactions taking place may be described by equilibrium equations or by a kinetic approach (PRENZEL, 1986).

SCHMITT and STICHER (1986a, b) measured the competitive adsorption of Pb, Cu, and Cd (in ratios according to the Swiss sewage sludge regulation) and expressed the results in the form of modified Langmuir equations. They used both models to predict the consequences of a regular load of these metals (e.g., in the form of sewage sludge) onto a soil during the next hundred years taking into account the annual precipitation rate.

Although mass flow and hydrodynamic dispersion are neglected in the equilibrium model, the predictions of both models are essentially the same and confirm similar findings of MAYER (1978). After about 30 years all of the cadmium input in the topsoil can no longer be adsorbed (in contrast to copper and lead) and is leaching into deeper horizons.

References

Adriano, D. C. (1986), *Trace Elements in the Terrestrial Environment*. Springer Verlag, New York-Berlin-Heidelberg-Tokyo.
Amacher, M. C., Selim, H. M., and Iskandar, I. K. (1988), *Kinetics of Chromium(VI) and Cadmium Retention in Soils; A Nonlinear Multireaction Model. Soil Sci. Soc. Am. J. 52*, 398 – 408.
Bank, S., Bank, J. F., Marchetti, P. S., and Ellis, P. D. (1989), *Solid State Cadmium-113 Nuclear Magnetic Resonance Study of Cadmium Speciation in Environmentally Contaminated Sediments. J. Environ. Qual. 18*, 25 – 30.
Barrow, J. J. (1978), *The Description of Phosphate Adsorption Curves. J. Soil Sci. 29*, 447 – 462.

Barrow, N. J. (1985), *Reaction of Anions and Cations with Variable-charge Soils. Adv. Agron. 38*, 183–230.
Barrow, N. J., Bowden, J. W., Posner, A. M., and Quirk, J. P. (1980), *An Objective Method for Fitting Models of Ion Adsorption on Variable Charge Surfaces. Aust. J. Soil Res. 18*, 395–404.
Barrow, N. J., Bowden, J. W., Posner, A. M., and Quirk, J. P. (1981), *Describing the Adsorption of Copper, Zinc and Lead on a Variable Charge Mineral Surface. Aust. J. Soil Res. 19*, 309–321.
Berrow, M. L., and Mitchell, R. L. (1980), *Location of Trace Elements in Soil Profiles: Total and Extractable Contents of Individual Horizons. Trans. R. Soc. Edinburgh Earth Sci. 71*, 103–121.
Bingham, F. T., Page, A. L., and Simms, J. R. (1964), *Retention of Cu and Zn by H-Montmorillonite. Soil Sci. Soc. Am. Proc. 28*, 351–354.
Bloom, P. R., and McBride, M. B. (1979), *Metal Ion Binding and Exchange with Hydrogen Ions and Acid Washed Peat. Soil Sci. Soc. Am. J. 43*, 687–692.
Bohn, H. L., and Bohn, R. K. (1986), *Solid Activity Coefficients of Soil Components. Geoderma 38*, 3–18.
Bourg, A. C. M. (1982), *ADSORP, a Chemical Equilibria Computer Program Accounting for Adsorption Processes in Aquatic Systems. Environ. Technol. Lett. 3*, 305–310.
Bourg, A. C. M., and Védy, J. C. (1986), *Expected Speciation of Dissolved Trace Metals in Gravitational Water of Acid Profiles. Geoderma 38*, 279–292.
Bowden, J. W., Posner, A. M., and Quirk, J. P. (1977), *Ionic Adsorption on Variable Charge Mineral Surfaces. Theoretical-Charge Development and Titration Curves. Aust. J. Soil. Res. 15*, 121–136.
Brümmer, G. W., and Herms, U. (1983), *Heavy Metal Species, Mobility and Availability in Soils*, in: Bernard, M., Brinckman, F. E., and Sadler, P. (eds.): *The Importance of Chemical Speciation in Environmental Processes. Dahlem Konferenzen.* Springer Verlag, Berlin-Heidelberg-New York.
Brümmer, G. W., Gerth, J., and Thiller, K. G. (1988), *Reaction Kinetics of the Adsorption and Desorption of Nickel, Zink and Cadmium by Goethite. I. Adsorption and Diffusion of Metals. J. Soil Sci. 39*, 37–52.
Buffle, J. J. (1980), *A Critical Comparison of Studies of Complex Formation between Copper(II) and Fulvic Substances of Natural Waters. Anal. Chim. Acta 118*, 29–44.
Bunzl, K., Schmidt, W., and Sansoni, B. (1976), *Kinetics of Ion Exchange in Soil Organic Matter. IV. Adsorption and Desorption of Pb^{2+}, Cu^{2+}, Cd^{2+}, Zn^{2+} and Ca^{2+} by Peat. J. Soil Sci. 27*, 32–41.
Cavallaro, N., and McBride, M. B. (1978), *Copper and Cadmium Adsorption Characteristics of Selected Acid and Calcareous Soils. Soil Sci. Soc. Am. J. 42*, 550–556.
Camerlynck, R., and Kiekens, L. (1982), *Speciation of Heavy Metals in Soils Based on Charge Separation. Plant Soil 68*, 331–339.
Campbell, D. J., and Beckett, P. H. T. (1988), *The Soil Solution in a Soil Treated with Digested Sewage Sludge. J. Soil Sci. 39*, 283–298.
Christensen, T. H. (1981), *A Model for Low-range Cadmium Migration in Soils: Principles and Verification. Proceedings International Conference Heavy Metals in the Environment.* pp. 214–217. CEP Consultants Ltd., Edinburgh.
Clark, C. J. (1983), *Surface Charge and Trace Metal Adsorption Characteristics of Allophane and Imogolite.* Ph. D. Thesis, Cornell University, New York (Diss. Abstr. 82-10799).
Dietze, G., and König, N. (1988), *Metal Speciation in Soil Solution by Dialysis and Ion Exchange Resin Procedures. Z. Pflanzenernähr. Bodenkd. 151*, 243–250.
Elliott, H. A., Liberati, M. R., and Huang, C. P. (1986), *Competitive Adsorption of Heavy Metals by Soils. J. Environ. Qual. 15*, 214–219.
Elkhatib, E. A., Bennett, O. L., and Wright, R. J. (1984), *Kinetics of Arsenite Sorption in Soils. Soil Sci. Soc. Am. J. 48*, 758–762.
Evangelou, P. V., and Phillips, R. E. (1988), *Comparison between the Gapon and the Vanselow Exchange Selectivity Coefficients. Soil Sci. Soc. Am. J. 52*, 379–382.

Fitch, A., and Stevenson, F. J. (1984), *Comparison of Models for Determining Stability Constants of Metal Complexes with Humic Substances.* Soil Sci. Soc. Am. J. *48*, 1044–1050.
Florence, T. M. (1977), *Trace Metal Species in Fresh Waters.* Water Res. *11*, 681–687.
Flühler, H., and Jury, W. A. (1983), *Estimating Solute Transport Using Nonlinear, Rate Dependent, Two-Site Adsorption Models.* Eidg. Anst. Forstl. Versuchswes. Ber. *245*, 1–48.
Förster, R. (1986), *A Multicomponent Transport Model.* Geoderma *38*, 261–278.
Fujii, R., Hendrickson, L. L., and Corey, R. B. (1982), *Activities of Trace Metals in Sludge-amended Soils,* in: Jenne, E. A., and Wildung, R. E. (eds.): *Science of the Total Environment, Proceedings 21st Hanford Life Science Symposium,* Richland, WA, pp. 179–190.
Fuller, W. H. (1977), *Movement of Selected Metals, Asbestos and Cyanide in Soil, Application to Waste Disposal Problem.* EPA 600/2-77-020. Solid and Hazardous Waste Research Division, US-EPA, Cincinnati, OH.
Gambrell, R. P., and Patrick, W. H. (1978), in: Hook, D. D., and Crawford, R. M. M. (eds.): *Plant Life in Anaerobic Environments,* pp. 375–423. Ann Arbor Science Publishers, Ann Arbor, Michigan.
Garrels, R. M., and Christ, C. L. (1965), *Minerals, Solutions and Equilibrium.* Harper, New York.
Gerth, J. (1985), *Untersuchungen zur Adsorption von Nickel, Zink and Cadmium durch Bodentonfraktionen unterschiedlichen Stoffbestandes und verschiedener Bodenkomponenten.* Dissertation, Universität Kiel, FRG.
Griffin, R. A., and Au, A. K. (1977), *Lead Adsorption by Montmorillonite Using a Competitive Langmuir Equation.* Soil Sci. Soc. Am. J. *41*, 880–882.
Grimme, H. (1968), *Die Adsorption von Mn, Co, Cu und Zn durch Goethit aus verdünnten Lösungen.* Z. Pflanzenernähr. Bodenkd. *121*, 58–65.
Gsponer, R., Schmitt, H. W., and Sticher, H. (1986), *Einfluß der Schwermetallkompetition im Boden auf die Aufnahme durch Pflanzen.* BGS Bull. *10*, 45–50.
Harter, R. D., and Lehmann, R. G. (1983), *Use of Kinetics for the Study of Exchange Reaction in Soils.* Soil Sci. Soc. Am. J. *47*, 666–669.
Helling, C. S., Chester, G., and Corey, R. B. (1964), *Contribution of Organic Matter and Clay to Soil Cation Exchange Capacity as Affected by the pH of the Saturation Solution.* Soil Sci. Soc. Am. Proc. *28*, 517–520.
Hendrickson, L. L., and Corey, R. B. (1981), *Effect of Equilibrium Metal Concentrations on Apparent Selectivity Coefficients of Soil Complexes.* Soil Sci. *131*, 163–171.
Herms, U. (1982), *Untersuchungen zur Schwermetalllöslichkeit in kontaminierten Böden und kompostierten Siedlungsabfällen in Abhängigkeit von Bodenreaktion, Redoxbedingungen und Stoffbestand.* Dissertation, Unversität Kiel, FRG.
Herms, U., and Bruemmer, G. W. (1984), *Einflußgrößen der Schwermetalllöslichkeit und -bindung in Böden.* Z. Pflanzenernähr. Bodenkd. *147*, 400–424.
Hickey, M. G., and Kittrick, J. A. (1984), *Chemical Partitioning of Cadmium, Copper, Nickel and Zinc in Soils and Sediments Containing High Levels of Heavy Metals.* J. Environ. Qual. *13*, 372–376.
Hodgson, J. F., Geering, H. R., and Norvell, W. A. (1965), *Micronutrient Cation Complexes in Soil Solution: Partition between Complexed and Uncomplexed Forms by Solvent Extraction.* Soil Sci. Soc. Am. Proc. *29*, 665–669.
Hodgson, J. F., Lindsay, W. L., and Trierweiler, J. F. (1966), *Micronutrient Cation Complexing in Soil Solution: II. Complexing of Zinc and Copper in Displaced Solution from Calcareous Soils.* Soil Sci. Soc. Am. Proc. *30*, 723–726.
Irving, H. M. N. H., and Williams, R. J. P. (1948), *Order of Stability of Metal Complexes.* Nature *162*, 746–747.
Juchler, S., and Sticher, H. (1987), *Chrom- und Nickel-Dynamik in Serpentinböden.* BGS Bull. *11*, 11–17.
Kemper, W. D., and Quirk, J. P. (1970), *Graphic Presentation of a Mathematical Solution for Interacting Double Layers.* Soil Sci. Soc. Am. Proc. *34*, 347–351.

Kerndorff, H., and Schnitzer, M. (1979), *Humic and Fulvic Acids as Indicators of Soil and Water Pollution.* Water Air Soil Pollut. *12,* 319–329.

King, L. D. (1988), *Retention of Metals by Several Soils of the Southeastern United States.* J. Environ. Qual. *17,* 239–246.

König, N. (1985), *Molekülgrößenverteilung, Komplexierungs- und Adsorptionsverhalten natürlicher organischer Substanzen eines sauren Waldbodens und ihr Einfluß auf die Schwermetallkonzentrationen in der Bodenlösung.* Mitt. Dtsch. Bodenkundl. Ges. *43/I,* 383–385.

Lagerwerff, J. V., and Brower, D. L. (1973), *Exchange Adsorption or Precipitation of Lead in Soils Treated with Chlorides of Aluminum, Calcium and Sodium.* Soil Sci. Soc. Am. Proc. *37,* 11–13.

Levesque, M. P., and Mathur, S. P. (1986), *Soil Tests for Copper, Iron, Manganese and Zinc in Histosols 1. The Influence of Soil Properties, Iron, Manganese and Zinc on the Level and Distribution of Copper.* Soil Sci. *142,* 153–163.

Levesque, M. P., and Mathur, S. P. (1988), *Soil Tests for Copper, Iron, Manganese and Zinc in Histosols 3. A Comparison of Eight Extractants for Measuring Active and Reserve Forms of the Elements.* Soil Sci. *145,* 215–221.

Mantoura, R. F. C., Dickson, A., and Riley, J. P. (1978), *The Complexation of Metals with Humic Materials in Natural Waters.* Estuarine Coastal Mar. Sci. *6,* 387–408.

Mayer, R. (1978), *Adsorption Isotherms as Regulators Controlling Heavy Metal Transport in Soils.* Z. Pflanzenernähr. Bodenkd. *141,* 11–28.

McBride, M. B. (1985), *Sorption of Copper(II) on Aluminium Hydroxide as Affected by Phosphate.* Soil Sci. Soc. Am. J. *49,* 843–846.

McKenzie, R. M. (1980), *The Adsorption of Lead and Other Heavy Metals on Oxides of Manganese and Iron.* Aust. J. Soil Res. *18,* 61–73.

McLaren, R. G., and Crawford, D. V. (1973), *Studies on Soil Copper I. Fractionation of Copper in Soils.* J. Soil Sci. *24,* 172–181.

McLaren, R. G., Williams, J. G., and Swift, R. S. (1983), *Some Observations on the Desorption and Distribution Behaviour of Copper with Soil Components.* J. Soil Sci. *34,* 325–331.

McLaren, R. G., Lawson, D. M., and Swift, R. S. (1986), *Sorption and Desorption of Cobalt by Soils and Soil Components.* J. Soil Sci. *37,* 413–426.

McLean, E. O., and Owen, E. J. (1969), *Effects of pH on the Contribution of Organic Matter and Clay to Soil Cation Exchange Capacity.* Soil Sci. Soc. Am. Proc. *33,* 855–858.

Miller, J. E., Hassett, J. J., and Koeppe, D. E. (1977), *Interactions of Lead and Cadmium on Metal Uptake and Growth of Corn Plants.* J. Environ. Qual. *6,* 18–20.

Misono, M., Ochiai, E., Saito, Y., and Yoneda, Y. (1967), *A New Dual Parameter Scale for the Strength Lewis Acids and Bases with the Evaluation of their Softness.* J. Inorg. Nucl. Chem. *29,* 2685–2691.

Murali, V., and Aylmore, L. A. G. (1983a), *Competitive Adsorption During Solute Transport in Soils: Mathematic Models.* Soil Sci. *135,* 143–150.

Murali, V., and Aylmore, L. A. G. (1983b), *Competitive Adsorption During Transport in Soils. 2. Simulations of Competitive Adsorption.* Soil Sci. *135,* 203–213.

Murali, V., and Aylmore, L. A. G. (1983c), *Competitive Adsorption During Solute Transport in Soils. 3. A Review of Experimental Evidence of Competitive Adsorption and Evaluation of Simple Competitive Models.* Soil Sci. *136,* 279–290.

Nand Ram, and Raman, K. V. (1984), *Stability Constants of Complexes of Metals with Humic and Fulvic Acids under Non-acid Conditions.* Z. Pflanzenernähr. Bodenkd. *147,* 171–176.

Neal, R. H., and Sposito, G. (1986), *Effects of Soluble Organic Matter and Sewage Sludge Amendments on Cadmium Sorption by Soils at Low Cadmium Concentration.* Soil Sci. *142,* 164–172.

Nir, S. (1986), *Specific and Nonspecific Cation Adsorption to Clays: Solution Concentrations and Surface Potentials.* Soil Sci. Soc. Am. J. *50,* 52–57.

Nyffeler, U. P., Li, Y., and Santschi, P. H. (1984), *A Kinetic Approach to Describe Trace-Element Distribution between Particles and Solution in Natural Aquatic Systems.* Geochim. Cosmochim. Acta *48,* 1513–1522.

Ogwada, R. A., and Sparks, D. L. (1986a), *Kinetics of Ion Exchange on Clay Minerals and Soil: I. Evaluation of Methods.* Soil Sci. Soc. Am. J. *50*, 1158–1162.

Ogwada, R. A., and Sparks, D. L. (1986b), *Kinetics of Ion Exchange on Clay Minerals and Soil: II. Elucidation of Rate-limiting Steps.* Soil Sci. Soc. Am. J. *50*, 1162–1164.

Olomu, M. O., Racz, G. J., and Cho, C. M. (1973), *Effect of Flooding on the Eh, pH and Concentrations of Fe and Mn in Several Manitoba Soils.* Soil Sci. Soc. Am. Proc. *37*, 220–224.

Posner, A. M., and Barrow, N. J. (1982), *Simplification of a Model for Ion Adsorption on Oxide Surfaces.* J. Soil Sci. *33*, 211–217.

Pourbaix, M. (1966), *Atlas of Electrochemical Equilibrium in Aqueous Solutions.* Pergamon, Oxford.

Prenzel, J. (1986), *A Numerical Scheme for the Calculation of Coupled Chemical Equilibria.* Geoderma *38*, 31–39.

Sanders, J. R. (1980), *The Use of Adsorption Equations to Describe Copper Complexing by Humified Organic Matter.* J. Soil Sci. *31*, 633–641.

Sanders, J. R. (1982), *The Effect of pH upon the Copper and Cupric Ion Concentrations in Soil Solutions.* J. Soil Sci. *33*, 679–689.

Sanders, J. R. (1983), *The Effect of pH on the Total and Free Ionic Concentrations of Manganese, Zinc and Cobalt in Soil Solutions.* J. Soil Sci. *34*, 315–323.

Schmitt, H. W., and Sticher, H. (1986a), *Prediction of Heavy Metal Contents and Displacement in Soils.* Z. Pflanzenernähr. Bodenkd. *149*, 157–171.

Schmitt, H. W., and Sticher, H. (1986b), *Long-term Trend Analysis of Heavy Metal Content and Translocation in Soils.* Geoderma *38*, 195–207.

Schnitzer, M., and Hansen, E. H. (1970), *Organo-metallic Interactions in Soils: 8. An Evaluation of Methods for the Determination of Stability Constants of Metal-Fulvic Acid Complexes.* Soil Sci. *109*, 333–340.

Schnitzer, M. (1986), *Binding of Humic Substances by Soil Mineral Colloids,* in: Huang, P. M., and Schnitzer, M. (eds.): *Interactions of Soil Minerals with Natural Organics and Microbes,* pp. 77–101. SSSA Special Publication Number 17, Soil Science Society of America.

Schulthess, C. P., and Sparks, D. L. (1988), *A Critical Assessment of Surface Adsorption Models.* Soil Sci. Soc. Am. J. *52*, 92–97.

Schwertmann, U., Fischer, W. R., and Fechter, H. (1982a), *Trace Elements in Pedosequences II. Two Braunerde-Podsol-Sequences on Slate.* Z. Pflanzenernähr. Bodenkd. *145*, 161–180.

Schwertmann, U., Fischer, W. R., and Fechter, H. (1982b), *Trace Elements in Pedosequences I. Two Pararendzina-Pseudogley-Sequences on Loess.* Z. Pflanzenernähr. Bodenkd. *145*, 181–196.

Selim, H. M., and Mansell, R. S. (1976), *Analytical Solution of the Equation for Transport of Reactive Solutes through Soils.* Water Resour. Res. *12*, 528–532.

Shaviv, A., and Mattigod, S. V. (1985), *Cation Exchange Equilibria in Soils Expressed as Cation-Ligand Complex Formation.* Soil Sci. Soc. Am. J. *49*, 569–573.

Sims, J. T. (1986), *Soil pH Effects on the Distribution and Plant Availability of Manganese, Copper and Zinc.* Soil Sci. Soc. Am. J. *50*, 367–373.

Sparks, D. L. (1985), *Kinetics of Ionic Reactions in Clay Minerals and Soils.* Adv. Agron. *38*, 231–265.

Sposito, G. (1981), *The Thermodynamics of Soil Solution.* Oxford University Press, Oxford-New York.

Sposito, G. (1983), *The Chemical Forms of Trace Metals in Soils,* in: Thornton, I. (ed.): *Applied Environmental Geochemistry,* pp. 123–170, Academic Press, London.

Sposito, G., and Coves, J. (1988), *SOILCHEM,* Campus Software Office, Berkeley, California.

Sposito, G., and Mattigod, S. V. (1980), *GEOCHEM: A Computer Program for the Calculation of Chemical Equilibria in Soil Solution and Other Natural Water Systems.* Department of Soil and Environmental Sciences, Unversity of California, Riverside.

Stevenson, F. J., and Fitch, A. (1986), *Chemistry of Complexation of Metal Ions with Soil Solution Organics,* in: Huang, P. M., and Schnitzer, M. (eds.): *Interactions of Soil Minerals with Natural*

Organics and Microbes, pp. 29 – 58. SSSA Special Publication Number 17. Soil Science Society of America.

Tiller, K. G., Gerth, J., and Brümmer, G. (1984a), *The Sorption of Cd, Zn, and Ni by Soil Clay Fractions: Procedures for Partition of Bound Forms and their Interpretation.* Geoderma 34, 1 – 16.

Tiller, K. G., Gerth, J., and Brümmer, G. (1984b), *The Relative Affinities of Cd, Ni and Zn for Different Soil Clay Fractions and Goethite.* Geoderma 34, 17 – 35.

Traina, S. J., and Doner, H. E. (1985), *Copper-Manganese(II) Exchange on a Chemically Reduced Birnessite.* Soil Sci. Soc. Am. J. 49, 307 – 313.

Trangmar, B. B., Yost, R. S., and Uehara, G. (1985), *Application of Geostatistics to Spatial Studies of Soil Properties.* Adv. Agron. 38, 45 – 94.

Turner, D. R., Whitfield, M., and Dickson, A. G. (1981), *The Equilibrium Speciation of Dissolved Components in Freshwater and Seawater at 25 °C and 1 atm. Pressure.* Geochim. Cosmochim. Acta 45, 855 – 881.

Utermann, J., and Richter, J. (1988), *Leaching of Physically Interactive Ions in Soils – Model Development and Calibration.* Z. Pflanzenernähr. Bodenkd. 151, 165 – 170.

Wopereis, M. C., Gascuel-Odoux, C., Bourrie, G., and Soignet, G. (1988), *Spatial Variability of Heavy Metals in Soil on a One-hectare Scale.* Soil Sci. 146, 113 – 118.

Additional Recommended Literature

Envitec '89 (1989), *Soil – a Bottleneck in Environmental Capacity.* Summaries of Papers, Düsseldorfer Messegesellschaft mbH "Nowea", see also *Swiss Chem.* 11 (7 – 8), 49 – 58.

Hopkin, S. P. (1989), *Ecophysiology of Metals in Terrestrial Invertebrates.* Elsevier Applied Science, London-New York.

Karlaganis, G. (1989), *Address of the Swiss Federal Office of Environment, Forests and Landscape to the 7th International Conference on Heavy Metals in the Environment, Geneva*, discussing especially Swiss soil observations and definitions of soil fertility; see also *Swiss Chem.* 12(3), 27 – 34 (1990).

Ramseier, S., Martin, M., Haerdi, W., Honsberger, P., Cuendet, G., and Tarradellas, J. (1989), *Bioaccumulation of Cadmium by Lumbricus terrestris.* Toxicol. Environ. Chem. 22(1 – 4), 189 – 196.

Streit, B., Krüger, C., Lahner, G., Kirsch, S., Hauser, G., and Diehl, B. (1990), *Uptake and Storage of Heavy Metals by Earthworms in Various Soils* (in German). Z. Umweltchem. Oekotoxikol. 2(1), 10 – 13.

I.7 b   Dumping of Wastes

ULRICH FÖRSTNER, Hamburg, Federal Republic of Germany
CARLO COLOMBI, Liebefeld, Switzerland
RAINER KISTLER, Zug, Switzerland

1 Introduction

The widespread use of metals for different kinds of applications (e.g., pigments, coatings, alloys, electronic equipment, etc.) leads to the fact that some of the utilized metals (or their compounds) end up in wastes. Metals in wastes can cause severe environmental impacts.

Metal containing waste materials include municipal solid wastes, industrial by-products, sewage sludge, dredged material (see Chapter I.2), wastes from mining and smelting operations (see Chapter I.2), filter residues from waste water treatment and atmospheric emission control, ashes and slags from burning of coal and oil, and from incineration of municipal refuse and sewage sludge. All these wastes pose most challenging problems, which are not only technical in nature, but attention must also be given to the social and financial aspects.

Specific concern for land disposal of contaminated material and for abandoned waste disposal sites includes the transport of heavy metals to both surface and ground waters. In an inventory of 700 cases of soil pollution in the Netherlands, for example, the following types of contamination were subdivided on the basis of main contaminants (DE KREUK, 1986): solvents − 120 sites; oil and oil-like products − 150 sites; gas works (aromatics and tar) − 70 sites; biocides (hexachlorocyclohexane) − 40 sites; heavy metals − 200 sites. With regard to loss of ground water resources, ascribable to contamination, a United States Library of Congress Report (ANONYMOUS, 1980) listed 1360 well closings in a 30 year time span, of which 619 cases were related to metal contamination.

The main problems associated with the prediction of the behavior of landfills are the slow reaction rate and the complexity of the involved processes. Laboratory or pilot scale experiments can only give some indications, but never simulate reality (see also ENVITEC '89).

2 Types of Waste Materials and Their Metal Contents

2.1 Municipal Solid Wastes

In the industrialized Western countries "production" of municipal solid waste is approximately half a ton per capita and year. It consists of containers, packages, non-

Table I-7b-1. Heavy Metals (µg/g dry mass) in Municipal Solid Wastes (EHRIG, 1989)

	(1)	(2)	(3)	(4)	(5)	(6)
Cadmium	3.5	40–50	11	3.5	2–14	3–9
Copper	238	411–532	400	400	120–210	31–345
Lead	399	210–370	400	210	110–330	294–545
Mercury	0.6	0.3–0.4	4	1.1	1–14	?
Zinc	521	588–742	1200	1200	300–1000	310–956

(1) GREINER (1983); (2) FRESENIUS et al. (cited in EHRIG, 1989); (3) BELEVI and BACCINI (1987); (4) NIELSEN (1978); (5) VOGL (1978); (6) BILITEWSKI (1989); concentrations in wastes with 30% humidity, such wastes also contain about 76–108 µg/g chromium and 13 µg/g nickel

Table I-7b-2. Heavy Metals in Sewage Sludge (KISTLER et al., 1987)

	Raw Sludge (mg/kg d.m.)	Digested Sludge (mg/kg d.m.)
Cr	250	210
Ni	130	82
Cu	380	400
Zn	1800	2400
Cd	3.0	6.8
Hg	1.9	2.9
Pb	170	260

d.m. = dry matter

durable goods and durable goods (FISHBEIN, 1989). Under moderate wet climate fresh dumped refuse has an average water content of approximately 30%; the storage capacity is much higher (450–600 L/t dry refuse = 39–46%), but decreases with age of the landfill. Elemental composition is rather variable and as a consequence of the inhomogeneous structure of refuse it is difficult to get comparable values (EHRIG, 1989). Table I.7b-1 lists some analyses of bulk composition for selected trace metals (see also TABASARAN, 1985; BRUNNER and ERNST, 1986; VELLER, 1989; YAKOWITZ, 1989).

Sorting out certain fractions of waste components may lead to significant reduction of metal concentrations both in deposited and incinerated municipal waste. For instance, 84% of the cadmium content has been found in the plastic portion (BILITEWSKI, 1989; Table I-7b-1).

Per capita release of sewage sludge ranges at approximately 1000 L per year; with 2–5% solid matter this contributes additional 10% of the municipal solid waste production. A certain proportion (ca. 35% in West Germany, about 50% in Switzerland) of less contaminated sewage sludge is still used in agriculture (see Chapter I.7c and Tables I.7b-2 and I.7b-3); due to decreasing acceptance in this sector higher percentages will be deposited in landfills or be incinerated.

Table I.7b-3. Typical Metal Concentrations in an Average Sample of Sewage Sludge from a Large City in North Rhine-Westphalia. (Measurements in mg/kg Dry Weight, Made Available by M. STOEPPLER, Jülich 1988)

Element	Measured	Maximum Allowed Concentration
Cadmium	17	20
Chromium[a]	570	1200
Copper	690	1200
Lead	1600	1200
Mercury	5	25
Nickel	180	200
Zinc	2300	3000

[a] In one of nine continuously examined purification plants between 1986 and 1988 chromium concentrations were always between 1500 and 2000 mg/kg

2.2 Incineration Ashes

(see also Chapter I.1, Section 5, and FISHBEIN, 1989)

In the near future, residues from coal-fired power plants may constitute one of the largest tonnages of solid waste material. World production of collected residues (bottom/boiler slag and fly ash) has been estimated at 300 million tons/year, of which approximately 70% is discharged to landfills (ADRIANO et al., 1980). Recycling rates (for soil amendment and construction materials) are less than 10% in the U.S. and 40–50% in some major European countries. Increasing tonnages of flue-gas desulfurization sludge can be expected from the equipment of power plants with lime/limestone desulfurization systems. With the increasing demand for better air quality, filter residues also play an increasing role in all incineration processes (see, for instance, UMWELT, 1987, BACCINI, 1989b).

Elevated factors of enrichment between coal and slag have been noted in the range of ten for Al, Ca, Ce, Cr, Fe, K, Mg, Mn, Sc, and Si (i.e., for the "lithogenic" elements), whereas the enrichment between slag and fly ash – relevant for the present considerations – is significant for Se (>300), followed by Sb ($\times 20$), Pb ($\times 13$), Cd, Cu, Zn (all $\times 7$), and As ($\times 6$) (KLEIN et al., 1975). Elemental concentrations strongly depend on surface enrichment processes, i.e., on temperature and grain-size (FÖRSTNER, 1986).

Incineration of municipal solid wastes (MSW) leads to a reduction of the volume by approximately 80%, and 60–70% by weight; reuse of slag reduces the original volume of MSW by 95%. Application of waste incineration varies widely; in West Germany, for example, approximately 30% of the MSW is incinerated, whereas in Switzerland the percentage is as high as 80%. Waste incineration ashes usually exhibit relatively high concentrations of trace metals – Zn and Pb up to the percent

Table I.7b-4. Concentrations of Elements Observed in Alexandria Municipal Incinerator Fly Ash and Suspended Particles (GREENBERG et al., 1978)

Element	Concentration (µg/g, unless % indicated)			
	Fly Ash[a]		Suspended Particles[b]	
	AV ± S.D.	Range	AV ± S.D	Range
Chromium	1330 ± 170	1070 – 1900	490 ± 350	122 – 1800
Arsenic	40 ± 13	9.4 – 74	210 ± 100	81 – 510
Selenium	3.4 ± 1.9	1.4 – 11	23 ± 21	7 – 110
Cadmium	42 ± 24	<5 – 125	1100 ± 400	520 – 2100
Antimony	270 ± 140	139 – 760	2400 ± 2400	610 – 12600
Lead (%)	0.40 ± 0.13	0.18 – 0.54	9.7 ± 2.6	5.5 – 15.5

[a] 33 samples; [b] 26 samples

range –, and particularly strong enrichment factors compared to natural contents have been observed for these elements and for cadmium and silver (BRUNNER and ZOBRIST, 1983; CARLSON, 1986). In Table I.7b-4 fly ash and suspended particles rich in chromium, cadmium, and antimony from another incinerator have been summarized.

2.3 Mine Wastes and Dredged Materials
(see also Chapter I.2)

Problems associated with these materials are the large quantities to be handled (SALOMONS and FÖRSTNER, 1988a, b) and the oxidation of sulfidic components in these wastes, which mobilizes toxic metals in high concentrations. These wastes have been and are dumped in coastal areas and in rivers (see also DE KOCK et al., 1986). In order to keep ports and channels accessible to shipping, sediments have to be removed regularly by dredging. A compilation of typical metal concentrations in sediments from Rotterdam Harbor (mixing of less contaminated marine sediments and highly polluted fluvial material gives a wide spectrum of composition, to which different disposal options are being applied; NIJSSEN, 1988) and from Hamburg Harbor is presented in Chapter I.2 (Table I.2-7). Highest rates of enrichment compared to natural background values have been found for cadmium (up to 60-fold) and mercury (up to 30-fold).

2.4 Metal Processing, Metal Recycling, and Industrial Wastes
(see also Chapter I.2 and Part II)

Metal concentrations in industrial waste vary over a wide spectrum, depending on the specific production process. FISHBEIN (1989) has summarized the main sources

Table I.7b-5. Sources of Industrial Waste (FISHBEIN, 1989)[a]

Industry	No. of Facilities	Amount Managed (million metric wet tons)	Description
Chemical	700	218	Contaminated waste waters, spent solvent residuals, still bottoms, spent catalyst, treatment sludges, filter cakes
Fabricated metals	200	4	Electroplating wastes, sludges contaminated with metals and cyanides, degreasing solvents
Electrical equipment	240	1	Degreasing solvents
Petroleum refinery	100	20	Leaded tank bottoms, slop oil emulsion solids
Primary metals	150	4	Pickle liquor, sludge with metal contaminates
Transportation equipment	150	3	Degreasing solvents, metal sludges
National security	100	1	All types of waste
Other	13360	24	
Total	3000	275	All types of waste

[a] Source: National Screening Survey of Hazardous Waste Treatment, Storage, Disposal and Recycling Facilities, U.S. Environmental Protection Agency, Office of Solid Waste, Office of Policy Planning and Information, 1986

in Table I.7b-5. Extreme examples with respect to metal contents are given in Chapter I.2 (Table I.2-14) (BRUNNER and ZIMMERLI, 1986) for residues of metal working industries, e.g., from recycling of aluminum, and for heavily polluted filter cakes from waste water treatment facilities, e.g, from electroplating industries. In the chemical industries heavy metals are commonly used as catalysts, fungicides, for chloralkali electrolysis, etc., and may end up as waste. Speciation of the metals in these wastes ranges from soluble salts to insoluble forms. For catalysts based on valuable metals the recycling rate is high.

3 Mobilization of Metals from Solid Waste Materials
(Fig. I.7b-1)

3.1 Municipal Solid Wastes

Over a long period of time landfilling has been done by the "try and error" method. The processes in the landfills were handled with a "black-box view". Only within the last few years has dumping become a defined technology. Therefore, pollution cases associated with landfills have led to negative publicity. Studies of metal fluxes of

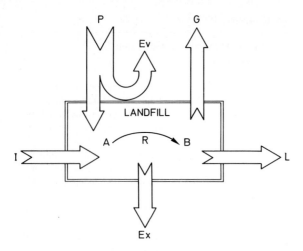

Fig. I.7b-1. Fluxes in landfills leading to emissions. P, precipitation; Ev, evaporation; G, gas; I, infiltration from ground water; Ex, exfiltration into ground water (uncontrolled); L, leachate (controlled); R, reaction within landfill; A, dumped waste; B, product of reaction R.

MSW landfills by BACCINI et al. (1987) show that metals (and their derivatives) are emitted mainly by the aquatic phase. The authors found, however, that the total mass flux of metals in the leachate is rather low compared to the total metal content in the landfill. The solubility of metals and their compounds in the leachate of landfills is controlled by the parameters pH, redox potential, solubility of the deposited metal species, concentration of complexing agents (NH_3/NH_4^+, fatty acids), and ion strength (activity coefficient). Inside the landfills metals and their compounds can undergo different transformation processes. These reactions do not lead in all cases to metal emissions, but they can cause other toxic or hazardous emissions. The variety of the different forms of metals in wastes renders a large number of reactions possible.

Subsequent to landfilling, the raw waste compounds undergo a variety of early diagenetic processes accompanying microbially mediated degradation of the organic compounds (ARAGNO, 1989). The metabolic intermediates of organic matter decay (e.g., HCO_3^-, HPO_4^{2-}, carbohydrates and other low-molecular organic acids) and those of the coupled inorganic reduction processes (e.g., Fe^{2+}, Mn^{2+}, S^{2-}, NH_4^+) accumulate in the interstitial water until concentrations are limited by physical convection/dispersion, by subsequent microbial utilization, or by diagenetic formation of secondary ("authigenic") minerals such as metal sulfides. This secondary inventory of a reactor landfill is critical both in buffering leachate water chemistry (STUMM and MORGAN, 1981) and in affecting transport of pollutants to underlying ground water aquifers. A landfill with sewage sludge only is a less complex and more homogeneous "reactor" than a municipal solid waste landfill (LICHTENSTEIGER and BRUNNER, 1987). By comparison of their long-term evolution with similar natural sediments (peat, organic soils) and their diagenesis it has been indicated that the transformation of organic material will last for geological time scales (10^3 to 10^7 years).

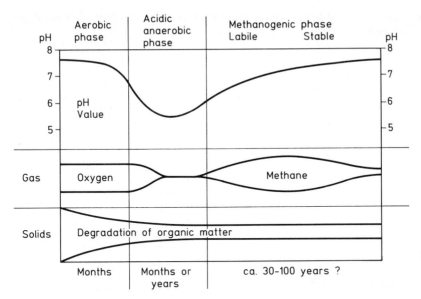

Fig. I.7b-2. Scheme of chemical evolution of municipal solid waste landfills.

In municipal solid waste landfills initial conditions are characterized by the presence of oxygen and pH values between 7 and 8 (Fig. I.7b-2). During the subsequent "acetic phase", pH values as low as 5 were measured due to the formation of organic acids in a more and more reducing milieu; concentrations of organic substances in the leachate are high. In a transition time of 1 to 2 years chemistry of the landfill changes from acetic to methanogenic conditions; the methanogenic phase is characterized by higher pH values and a significant drop of BOD_5 (biochemical oxygen demand) values from more than 5000–40000 mg/L in the acetic phase to 20–500 mg/L. Long-term evolution of a "reactor landfill", subsequent to the methanogenic phase, is still an open question (FÖRSTNER et al., 1989).

Metal concentrations in leachates from municipal landfills have been analyzed since the beginning of the 1970s. Studies performed by QUASIM and BURCHINAL (1970), WALKER (1973), MEYER (1973), and HUGHES (1975) on ground water pollution from sanitary landfill leachate and areas treated with waste compost and sewage sludge indicated that deeper fills pose less pollution problems than the shallower fills, which may leach the bulk of pollution in a shorter period of time, thereby exceeding the dilutional capacity of the moving ground water. More recent publications have focussed on the behavior of heavy metals in the various ground water zones (anaerobic, anoxic, aerobic) downstream from a landfill (compilation by EHRIG, 1983; NICHOLSON et al., 1983). Typical differences have been found for iron, manganese, and zinc in leachates between the "acetic phase" (with high organic loadings and low pH values) and the "methanogenic phase" (with low biodegradable organics and higher pH values), whereas for other trace elements such differences have not been established (Table I.7b-6). The latter finding may be related to difficulties in sampling and chemical analysis since similar effects, for example, of pH

Table I.7b-6. Concentrations of Trace Elements (µg/L) in Leachates from Municipal Solid Waste Landfills (Review by EHRIG, 1989)

Element	"Acetic Phase" Average	"Acetic Phase" Range	Average	Range	"Methanogenic Phase" Average	"Methanogenic Phase" Range
Iron	780	20 – 2100			15	3 – 280
Manganese	25	0.3 – 65			0.7	0.03 – 45
Zinc	5	0.1 – 120			0.6	0.03 – 4
Arsenic			160	5 – 1600		
Cadmium			6	0.5 – 140		
Chromium			300	30 – 1600		
Copper			80	4 – 1400		
Lead			90	8 – 1020		
Mercury			10	0.2 – 50		
Nickel			200	20 – 2050		

Table I.7b-7. Influence of Waste Disposal on Ground Water Quality (ARNETH et al., 1989). Statistical Values (in µg/L) of 33 Sites (Up-/Downstream Data)

Element or Compound	Uncontaminated Ground Water				Contaminated Ground Water				CF
	min.	mean	max.	n	min.	mean	max.	n	
Ammonium	120	190	250	2	<50	11800	15300	27	62.1
Arsenic	<2	3.2	25	17	<2	109	1930	70	34.1
Boron	<20	<20	27	18	<20	1270	31200	90	>63.5
Cadmium	<0.2	<0.2	0.6	16	<0.2	1.3	19	74	>6.5
Copper	<5	5.8	43	18	<5	27	373	90	4.7
Chromium	<5	<5	8	18	<5	25	1710	90	>5.0
Fluorine	<100	<100	120	3	<100	260	730	42	>2.6
Iron	546	6990	24200	17	<10	10400	172000	81	1.5
Lead	<2	3.4	27	17	<2	17	336	76	5.0
Manganese	19	498	1350	18	<10	2020	24700	89	4.1
Nickel	<5	9.4	16	17	<5	28	330	89	3.0
Nitrate	<500	6600	31400	18	<500	173000	$11.5 \cdot 10^6$	90	26.2
Strontium	<40	147	430	18	<40	465	6350	90	3.2
Zinc	10	36	95	17	<10	170	2620	82	4.7

n = number of measurements; CF = contamination factor

on zinc mobility can be expected for other related elements such as Cd, Ni, Pb, and Cu.

A ranking of pollutants with respect to their mobility in municipal solid waste landfills was established by CHRISTENSEN et al. (1987). Chloride was very mobile, sodium, ammonium, potassium, and magnesium were moderately mobile, and zinc, cadmium, iron and in most cases also manganese were only partly mobile; in particular the heavy metals zinc and cadmium showed very restricted mobility in the anaerobic zone of even very coarse aquifer materials (see, however, also BROWN et al., 1986, who discussed mobility of As, Ba, Cd, Pb, Ni and Tl). Comparison of inorganic ground water constituents upstream and downstream from 33 waste disposal sites in West Germany (Table I.7b-7, from ARNETH et al., 1989) indicates characteristic differences in pollutant mobilities. High contamination factors (CF values: contaminated mean/uncontaminated mean) have been found for boron, ammonium, nitrate, and arsenic; the latter element may pose problems during initial phases of landfill operations (BLAKEY, 1984). Under anaerobic conditions soluble metals precipitate as insoluble sulfides (Table I.7b-8) in the landfill. Sulfide is produced by microbiological reduction of sulfates or by the decomposition of organic compounds containing sulfur. However, the metals are only fixed in the landfill as long as the pH remains above neutral, which is the case in anaerobic phases due to biological activity.

Table I.7b-8. Solubility Products of Some Heavy Metals

Sulfide	pK_{so}
FeS	18.4
NiS	23.9
CuS	44.1
Cu_2S	46.7
ZnS	22.9
CdS	28.4
HgS	49.0
PbS	27.5

3.2 Incineration Ashes

Among the effects which coal fly ash has on the ground water environment and particularly on the release of metals are changes in pH. Typical changes of metal species occur during ash-pond disposal and in landfill leachates, which could affect ground water quality (FÖRSTNER, 1986). In actual and laboratory-simulated disposal situations it has been shown that elements with anionic aqueous speciation such as S, As, Se, and Mo are appreciably solubilized in ash sluicing/ponding systems that are neutral to alkaline in pH (TURNER et al., 1982). Obviously there is a correspondence between those elements exhibiting elevated concentrations in the effluent waters and those most extractable in water (Table I.7b-9). In contrast to literature information, recent investigations by DE GROOT et al. (1987) indicate limited solubility of anions at high pH (>11). The metals Pb, Cu, Cd, and Zn show a minimum solubility at high pH whereas major elements, such as Al and Si, show two minima in the pH range 7 to 9 and at pH values higher than 11. The latter was related to the formation of new mineral phases (ettringite).

It has been stressed by TURNER (1981) from data on fly ash deposits that, since trivalent arsenic is likely to be the predominant As species in ash pore water and

Table I.7b-9. Trace Element Concentrations in Ash-Pond Effluent Water Relative to Cooling Lake Intake and Outlet (DREESEN et al., 1977)

Concentration Ratio	Ash-Pond Effluent/ Cooling Lake Intake	Ash-Pond Effluent/ Cooling Lake Outlet
>50	B, F, Mo, Se, V	Se
10–50	As	B
2–10	–	As, Cr, F, Mo, V
<2	Cd, Cu, Zn	Cd, Cu, Zn

ground water and is also the more toxic form in water, interactions of As(III) with soils and landfill-liner material should be more closely examined to ensure the protection of drinking water aquifers in critical areas.

With respect to municipal waste ashes it has been demonstrated by BACCINI and BRUNNER (1985), that even if chlorine from incineration plants only contributes to a minor degree to the contamination of this element in surface waters (approx. 4%), there are several problems associated with this element in municipal waste incineration systems; as a (1) carrier of protons, which give rise to acidification of the atmosphere; (2) carrier of metals, affecting higher vapor pressure and solubility of certain metal chlorides; and (3) chlorination agent for aromatic hydrocarbons. It is expected that in particular elements such as cadmium, silver, and mercury, which form strong complexes with chlorine, may be mobilized during interactions with these landfill leachates. Sequential extractions performed by WADGE and HUTTON (1987) indicate that about 20% of total Cd and 1% of total Pb in coal fly ash was in the exchangeable fraction; in contrast, the single largest fractions of Cd and Pb in refuse ash, at 72% and 41%, respectively, were present in the exchangeable, relatively labile chemical form.

KISTLER et al. (1987) investigated the leachability of some heavy metals in pyrolyzed sewage sludge. They found that the metals were highly immobile due to the well-buffered neutral to alkaline properties of the char. Therefore, it was concluded that this char is well suited for disposal in landfills.

3.3 Mine Tailings and Dredged Material
(see Chapter I.2)

Acidic drainage from coal and ore mines has been recognized as a serious environmental pollution problem since long. Transformation of sulfides and a shift to more acid conditions particularly enhances the mobility of elements such as Mn, Fe, Zn, Pb, Cu, and Cd. There are many examples of deleterious effects of mine effluents on freshwater ecosystems: The Itai-Itai catastrophe at the Jintsu River in Japan has been connected with cadmium-rich effluents from an abandoned lead-zinc mine (KOBAYASHI, 1981). In streams of Wales enrichment of Pb, Cu, and Zn leached from the outcrops of mineralized zone and spoil heaps of disused mines still cause a high mortality rate in fish and other organisms (ABDULLAH and ROYLE, 1972). For other examples see Chapter I.2, especially Sect. 4.3.

Early diagenetic changes and element mobilization from dredged material in a man-made estuarine marsh has been investigated by DARBY et al. (1986); the results of this study have been described in Chapter I.7e. Another adverse effect of dredged materials after land disposal relates to contamination of crops.

3.4 Other Reactions of Metals Within Landfills Including Transfer to the Gas Phase

The transfer of metals or metal compounds to the gas phase is controlled by their vapor pressure. Only few of them (e.g., mercury and dimethyl mercury, $AsCl_3$, organic lead, etc.) have a vapor pressure which may lead to a considerable metal concentration in the gas phase at ambient temperatures (Table I.7b-10). Special attention should be focussed on mercury in landfills. Wastes containing mercury in inorganic form are, e.g., batteries (alkaline cells contain up to 1% Hg). Table I.7b-11 demonstrates, however, that the concentrations of some heavy metals in the gases from a hazardous waste material landfill were below the detection limits. At this specific site about 1000 t of batteries containing approximately 3000 kg of mercury were deposited. Further studies are necessary and are planned also in Germany.

Table I.7b-10. Vapor Pressure and Resulting Concentration of Metals in the Gas Phase

	T (K)	p (Pa)	c (mg/m^3)
Hg	273	0.025	2.2
	283	0.065	5.6
	293	0.16	13.2
	303	0.37	29.5
$AsCl_3$	296	1333	98000
$(CH_3)_4Pb$	303	5333	565000
$(C_2H_5)_4Pb$	311	133	16640

Table I.7b-11. Concentration of Heavy Metals in the Gases from a Hazardous Waste Landfill

	c (mg/m^3 dry gas)
Cadmium	<0.004
Mercury	<0.003
Zinc	<0.13
Lead	<0.26

Values could not be measured by absorption of gas in a solution of $KMnO_4$ in nitric acid because concentrations were too low

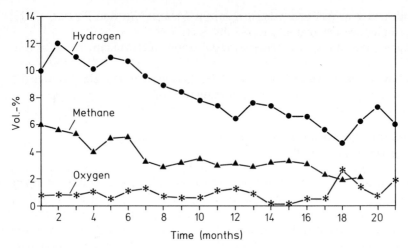

Fig. I.7b-3. Composition of hazardous waste landfill gas. (In this specific landfill more than 2000 t of elemental aluminum were disposed of.)

Organic mercury is used in fungicides (phenylmercury salts) and can therefore be contained in wastes of the chemical industries (agrochemical production).

Many experts have studied the potential of microbiological oxidation and methylation of mercury (see Chapters I.7h and II.20). Due to the limited available oxygen in landfills, the oxidation is the limiting step in the methylation to di- and monomethyl mercury. Furthermore, in the presence of sulfide ions mercury precipitates as insoluble mercury sulfide, for which the methylation rate is even slower than for other bivalent forms of mercury.

Only few other reactions of metals within landfills are known, except for "normal" oxidation (rusting) of iron, etc. (e.g. of barrels containing toxic wastes). Corrosion of the containers of toxic wastes should therefore be taken into account when predictions about long-term behavior are requested.

Another interesting example of a reaction in which metals are involved has been observed, but remains to be clarified further: Measurements of the gas composition of a hazardous material landfill showed a high concentration of hydrogen and a low concentration of oxygen (Fig. I.7b-3). The presence of methane indicates anaerobic microbiological activity (see also Fig. I.7b-2), but the biological production of hydrogen by degradation of fatty acids is endergonic (the reaction is prevented at a hydrogen partial pressure above 10^{-1} Pascal). According to this microbiological formation of hydrogen at the observed concentrations does not seem very reasonable.

At this specific landfill about 15 000 t of waste from aluminum recycling industry (see Table I.2-14) and about 30 000 t of residues (bottom ash and ESP-dust) of MSW incinerators, containing totally more than 2000 t of elemental aluminum were dumped. Due to the low redox potential of aluminum ($E_0 = -1.67$ Volt) it reacts with water to form aluminum hydroxide and hydrogen.

The formed layer of hydroxide normally stops the corrosion attack. But in the presence of transition metal ions such as Pb^{2+}, Cu^{2+}, Zn^{2+}, or at elevated pH the protecting layer seems to be destroyed (KLUGE et al., 1979).

4 Prognosis of Potential Metal Releases from Waste Dumps

Initial estimation of potential release of contaminants from solid wastes is mostly based on elutriate tests. Information on the behavior of metal pollutants is limited due to the generally low solubility of these constituents and lack of realistic environmental conditions. Common single reagent leachate tests, e.g., US EPA, German DIN, ASTM, and IAEA, use either distilled water or acetic acid (THEIS and PADGETT, 1983). Best results with respect to the estimation of short-term effects can be attained by "cascade" test procedures at variable solution/solid ratios such as the procedure of the US EPA designed for studies on the leachability of solid wastes (using artificial landfill leachates; HAM et al., 1979) and the test method developed by the Netherland Energy Research Center for studies on combustion residues (VAN DER SLOOT et al., 1984). A review of physical and chemical methods – the latter mainly elutriate test procedures – for characterization of hazardous wastes designed for landfills in the USA has been given by FRANCIS et al. (1989). HARTLEN (1987) elaborated a handbook on dumping, dealing also with the elution of heavy metals from disposed incineration residues.

The procedures described so far are restricted with regard to prediction of long-term effects in waste deposits, since these concepts neither involve mechanistic nor kinetic considerations and therefore do not allow calculations of release periods. Long-term release of metals from waste materials into ground water has been studied: (1) with an in-situ method by inserting dialysis bags with typical solid phases into ground water aquifers (FÖRSTNER and CARSTENS, 1988), and (2) with a laboratory circulating system including an ion-exchanger unit, where a "quick motion" effect can be attained by controlled significative intensivation of the relevant parameters pH value, redox potential, and temperature (SCHOER and FÖRSTNER, 1987). In both procedures, by means of sequential extraction of the contaminated solids before and after exposure, a qualitative and quantitative assessment of mobilization-relevant "pools" can be reached, by which the release can be predicted beyond the experimentally simulated time.

BROWN et al. (1986) developed a metal speciation modelling approach for evaluating potential mobilities of arsenic, barium, cadmium, chromium, lead, mercury, nickel, selenium, silver, and thallium. With the model MINTEQ they were able to understand leachate contamination from a failed land disposal facility, differentiating between "mobile" and "relatively immobile" metals.

5 Remedial Measures Including Pretreatment of Metal-rich Wastes

Within the conceptual perspectives of future landfill operations, two major alternatives have been extrapolated (BRUNNER and ZIMMERLI, 1986; BACCINI, 1989a): In "reactor landfills" these conditions are widely variable as a result of biochemical reactions, while "final storage quality" implies less variations of chemical interactions. The long-term geochemical implications can be evaluated from a comparison of the inventory of either "reactor landfill" or the "final storage material" with the earth crust composition and the processes taking place in the different sites (Table I.7b-12).

Table I.7b-12. Comparison of Inventories of Chemical Components in the Two Landfill Alternatives and in the Earth's Crust

Reactor Landfill	Final Storage	Earth's Crust
Major solid constituents		
Solid "inert" waste	Silicates, oxides	Quarz, Fe-oxide clay, carbonates
Putrefactive waste	[Gypsum, NaCl][a]	(Gypsum, NaCl)
Grease trap waste	(Char)[b]	Kerogenic compounds
Minor solid constituents		
Organic micropollutants	Organic micropollutants	–
Metals in reactive chemical forms	Metal-bearing minerals, mainly oxides	Metals mainly in inert forms
Dissolved constituents		
Protons, electrons	(Protons)	(pH: acid rain)
Organic compounds	(Organic residues)	(Humic acids)
Dissolved salts	[Dissolved salts][a]	(Dissolved salts)

[a] Partial extraction during pretreatment
[b] Minor constituent

5.1 Municipal Solid Waste Landfill as Reactor

From a geochemical point of view the "reactor landfill" is characterized by labile conditions during the initial aerobic and acid anaerobic phases, the former mainly due to uncontrolled interactions with organic solutes. Particular problems occur when leachate collection pipes are plugging during the acidic decomposition period (HAM et al. in BACCINI, 1989a). It has been inferred that oxidation of sulfidic minerals by intruding rain water may mobilize trace metals, and the impact on the underlying ground water could be even higher if a chromatography-like process, involving continuous dissolution and reprecipitation during passage of oxidized water through the deposit, would preconcentrate critical elements prior to final release with the leachate (FÖRSTNER et al., 1989). Sludge-only landfills have lower permeabilities and interparticular porosities than municipal solid waste (MSW) landfills. Thus, landfill gas and leachate control are more difficult and less effective than in MSW landfills. Dissolved substances such as nitrogen and phosphorus reach the supernatant water which consequently has to be treated; heavy metal concentrations in the surface water of the studied sludge pond complied with the quality requirements for surface waters to be upgraded to drinking water (LICHTENSTEIGER and BRUNNER, 1987).

Studies by BACCINI et al. (1987) on metal fluxes in municipal solid waste landfills have shown that most metals are emitted predominantly in the aqueous phase; of the studied elements, only mercury was emitted to a significant extent via the gas phase. In particular the pathway of the latter element merits further attention, due to its high concentrations – mainly from batteries – at specific sites and the enhanced transformation to volatile species at initial stages of landfill evolution.

Results among others by MCLELLON et al. (1974), ATWELL (1976), KELLY (1976), and SHUSTER (1976) indicated that the trace element contents of surface drainage water from landfills are mainly influenced by the adsorption capacities of the specific soil. FULLER and KORTE (1977), GRIFFIN et al. (1977) and FARQUHAR and CONSTABLE (1978) in their studies on the attenuation of metal pollutants in municipal landfill leachate by clay minerals demonstrated that the potential usefulness of clay materials as liners for waste disposal sites depends to a large extent on the pH of the leachate solution; removal of heavy metal cations takes place through precipitation as hydroxides and carbonates or by coprecipitation with Fe and Mn hydrous oxides. Exchange of resident ions is also active in removing cations with alkaline earth metals being released in most cases and possibly being discharged from the liner (review by FARQUHAR and PARKER, 1989). For metal-finishing sludges, involving acid solutions, disposal in limestone-lined segregated landfills may be a safe and economical alternative (REGAN and DRAPER, 1987).

Liquid effluents from waste materials include direct and indirect discharges of landfill leachates and process waters from cleaning-up of contaminated soil. In principle the techniques used for the purification of domestic and industrial waste water can also be used for the treatment of polluted ground water. For metal-rich effluents

Table I.7b-13. Practical Examples of Ground Water Treatment (VAN LUIN, 1988)

	Treatment System	Capacity (m^3/d)	Substance	Influent ($\mu g/L$)	Effluent ($\mu g/L$)
Thermal galvanizing plant	Flocculation tank, ultrafiltration, sludge belt press	400	Zinc Cadmium	7000–60000 2.7	50 0.1
Electroplating industry	Oxidation pond, neutralization and flocculation tank, plate separator, sand and carbon filter, thickener, sludge filter press	600	Zinc Chromium Nickel Copper	8000 3000 3000 1400	100 10 300 300

reasonable results are generally achieved by physicochemical methods (Table I.7b-13 after VAN LUIN, 1988). If the heavy metals are adsorbed on suspended matter, they can be removed to a significant extent by sedimentation. Problems may arise in precipitation-based systems when large fluctuations occur in the composition of the influent. Tests carried out at a ground water treatment plant on the site of a former electroplating factory provided better removal efficiency with sulfide precipitation than with alkaline precipitation or co-precipitation with Fe(III). If environmental standards are higher it is necessary to use a sand filter subsequent to precipitation. Costs range between 2.2 and 2.5 US $ per m^3 treated ground water (VAN LUIN, 1988).

5.2 Final Storage: Municipal Waste Incineration Ashes
(see also BACCINI, 1989b)

Final storage quality, which is defined by the composition of earth crust material, in most cases is not attained by simply incinerating municipal waste, i.e., by reduction of organic fractions only. There is, in particular, the problem of easily soluble minerals, such as sodium chloride. Measures before incineration include the separate collection of (organic) kitchen and garden wastes (containing some chlorine and sulfur), which can be transferred into compost. A major decrease of chlorine content, however, would require a significant reduction of PVC in municipal solid waste. After incineration washing of the residues can be performed either with neutral or acidified water. For example, material washed at pH 4 in closed cycles can directly be deposited, since acid precipitation – as the major influencing factor of the geochemistry on inorganic residue landfills – could not have any further effect on the mobility of metals (HÄMMERLI-WIRTH, 1987). However, it has to be considered that part of the sorption sites, for example, iron oxides, are extracted by this procedure as well. By the addition of alkaline binders such as cement, the initial leaching of some elements can also be decreased. Care must be taken not to apply too much alkaline material in order to prevent the dissolution of Pb, Zn, and other metals at high pH values, due to the formation of soluble hydroxo complexes (BRUNNER and BACCINI, 1987).

As mentioned above (KISTLER et al., 1987) metals in the pyrolysis residues ("char") are not likely to be mobilized by acidic leachates due to the highly buffered neutral to alkaline properties of the residue. Nevertheless, since these metals might be mobilized by acids combined with organic ligands, they should be desposited onto landfills where organic material is excluded ("monolandfill").

Processes for the treatment of large quantities of residues, such as mine tailings and dredged materials, will generally be built up to several unit operations combined in such a way that optimum results are realized at a minimum of operation costs. In a generalized scheme, the scenario for the treatment of residues may be visualized as indicated in Fig. I.7b-4, where two categories of techniques can be distinguished (VAN GEMERT et al., 1988).

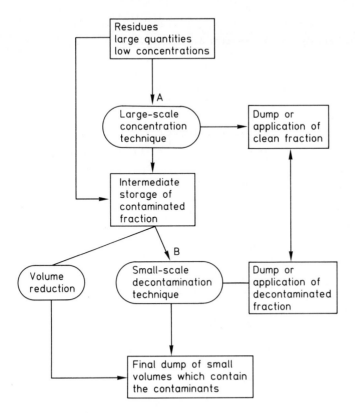

Fig. I.7b-4. Scenario for the treatment of large quantities of residues (VAN GEMERT et al., 1988).

A. Large-scale concentration techniques. These techniques are characterized by large-scale applicability, low costs per unit of residue to be treated and a low sensitivity to variations in circumstances. It is advantageous that these techniques may be constructed in mobile or transportable plants. A-techniques include methods such as hydrocyclonage, flotation, and high-gradient magnetic separation.

B. Decontamination or concentration techniques which are especially designed for relatively small scale operation. These techniques are generally suited for the treatment of residues which contain higher concentrations of contaminants; they involve higher operating costs per unit of residue to be treated; furthermore, they are more complicated, require specific experience of operators and are suitably constructed in stationary or semi-mobile plants. B-techniques include biological treatment, acid leaching of inorganic compounds, ion exchange methods, and solvent extraction of organic compounds.

As an example of A-techniques, the classification of harbor sludge from Hamburg with a highly effective combination of hydrocyclonage and elutriator as designed by WERTHER (1988) is shown in Fig. I.7b-5. In the hydrocyclone the separation of the

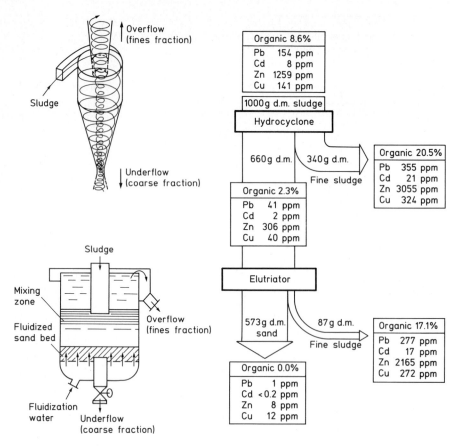

Fig. I.7b-5. Schematic view of the hydrocyclone and elutriator used for sludge classification in Hamburg Harbor (*left*). Mass balance and distribution of heavy metals (*right*). After HILLIGARDT et al. (1986). d.m. = dry material.

coarse fraction (relatively clean sand) from the highly polluted fines is effected by the action of centrifugal forces. The coarse fraction leaves the cyclone in the underflow, while the fines are contained in the overflow. The advantage of the hydrocyclone is its simplicity and its ability to handle large throughputs. A disadvantage is that the sharpness of the separation is fairly low. The elutriator, which follows in the classification scheme, effects a much better sharpness of separation. The basic principle is here separation according to the settling velocity of the particles in an upflowing water stream. There are virtually no fines found in the underflow of the elutriator. The efficiency of the separation procedure is illustrated in Fig. I.7b-5 (right) which demonstrates that the sand thus separated from the sludge has a heavy metal content which is of the same order of magnitude found in naturally occurring sandstones.

An example for B-techniques is the acid leaching method described by MÜLLER and RIETHMAYER (1982), which comprises three steps for extraction of metals from solid wastes, such as sewage sludge and dredged materials: (1) Extraction of metals with acid at pH 0.5 – 1.0; (2) separation of purified sludge from the acid solution followed by rewashing with decanting centrifuges; (3) precipitation of metals from the solution through pH increase with lime followed by carbonate precipitation with the CO_2 released with acid leaching. According to VAN GEMERT et al. (1988) the application potential of this method is determined by ecological and economic aspects. With respect to the former aspect, it is necessary to discharge waste water with large amounts of salts; it is also uncertain, whether undesirable compounds are formed when organic matter is present in the sludge. Economic aspects include costs of alternative techniques or storage and the availability of a sufficient amount of HCl at a reasonable price and quality.

5.3 Stabilization of Mine Tailings and Dredged Material

"Geochemical and biological engineering" emphasizes the increasing efforts of using natural resources available at the disposal site for reducing negative environmental effects of all types of waste material, in particular of acid mine wastes. Practical examples for improvement of storage quality of metal-containing waste, including measures for recultivation of old and recent mining waste disposal sites and physicochemical methods of water processing, have been reviewed by several authors (SALOMONS and FÖRSTNER, 1988 a, b; FÖRSTNER, 1988). Incorporation in naturally formed minerals, integrated biological systems, and solidification/stabilization techniques has been described in Chapter I.2, Section 3.3 (see also BURCKLE et al., 1988).

5.4 Hydrometallurgical Extraction from Industrial Wastes

Future efforts should not only be aimed at chemically stabilizing critical compounds in waste materials but – in particular – at recycling valuable components (e.g., PATTERSON, 1987). Recovery of metals such as copper and nickel from electroplating residues waste could well compete with natural resources of these elements (see Chapter I.2, Section 2.6; and Part II). The same is valid for lead, zinc, and silver from certain fractions of waste combustion products.

References

Abdullah, M. I., and Royle, L. G. (1972), *Heavy Metal Content of Some Rivers and Lakes in Wales.* Nature **238**, 329 – 330.
Adriano, D. C., Page, A. L., Elseewi, A. A., Chang, A. C., and Straughan, I. (1980), *Utilization and Disposal of Fly Ash and Other Coal Residues in Terrestrial Ecosystems: A Review. J. Environ. Qual.* **9**, 333 – 344.
Anonymous (1980), *Groundwater Strategies. Environ. Sci. Technol.* **14**, 1030 – 1035.
Aragno, M. (1989), *The Landfill Ecosystem: A Microbiologist's Look Inside a "Black Box",* in: Baccini, P. (ed.): *The Landfill – Reactor and Final Storage. Lecture Notes in Earth Sciences,* Vol. 20, pp. 15 – 38. Springer-Verlag, Berlin.

Arneth, J.-D., Milde, G., Kerndorff, H., and Schleyer, R. (1989), *Waste Deposit Influences on Ground Water Quality as a Tool for Waste Type and Site Selection for Final Storage Quality*, in: Baccini, P. (ed.): *The Landfill – Reactor and Final Storage. Lecture Notes in Earth Sciences*, Vol. 20, pp. 399–416. Springer-Verlag, Berlin.

Atwell, J. S. (1976), *Identifying and Correcting Groundwater Contamination at a Land Disposal Site. Proceedings 4th National Congress Waste Management*. Technol. Res. Rep. SW-Sp. 278–298. U.S. Environmental Protection Agency, Cincinnati, Ohio.

Baccini, P. (ed.) (1989a), *The Landfill – Reactor and Final Storage. Lecture Notes in Earth Sciences*, Vol. 20. Springer, Berlin.

Baccini, P. (1989b), *The Control of Heavy Metal Fluxes from the Anthroposphere to the Environment. Proceedings International Conference Heavy Metals in the Environment, Geneva*, Vol. 1, pp. 13–23. CEP Consultants Ltd., Edinburgh.

Baccini, P., and Brunner, P. H. (1985), *Behandlung und Endlagerung von Reststoffen aus Kehrichtverbrennungsanlagen. Gas Waser Abwasser 65*, 403–409.

Baccini, P., Henseler, G., Figi, R., and Belevi, H. (1987), *Water and Element Balances of Municipal Solid Waste Landfills. Waste Manage. Res. 5*, 483–499.

Belevi, H., and Baccini, P. (1987), *Water and Element Fluxes from Sanitary Landfills*, in: *Process Technology and Environmental Impact of Sanitary Landfills, Cagliari/Sardinia*, October 19–23, 1987.

Bilitewski, B. (1989), *Opportunities for Fuel from Municipal Solid Waste* (in German). *Entsorgungs-Praxis 3*, 74–78.

Blakey, N. C. (1984), *Behavior of Arsenical Wastes Co-Disposed with Domestic Solid Wastes. J. Water Pollut. Control Fed. 56*, 69–75.

Brown, D. S., Carlton, R. E., and Mulkey, L. A. (1986), *Development of Land Disposal Decisions for Metals Using MINTEQ Sensitivity Analyses*. EPA, Athens, Georgia, /600/S3-86/030, Project Summary.

Brunner, P. H., and Baccini, P. (1987), *The Generation of Hazardous Waste by MSW-Incineration Calls for New Concepts in Thermal Waste Treatment*, in: *Second International Conference on New Frontiers for Hazardous Waste Management, Pittsburgh, Pennsylvania*, Sept. 27–30, 1987.

Brunner, P. H., and Ernst, W. R. (1986), *Alternative Methods for the Analysis of Municipal Solid Waste. Waste Manage. Res. 4*, 147–160.

Brunner, P. H., and Zimmerli, R. (1986), *Beitrag zur Methodik der Zuordnung von Abfallstoffen zu Deponietypen*. EAWAG Bericht Nr. 30-4721.

Brunner, P. H., and Zobrist, J. (1983), *Die Müllverbrennung als Quelle von Metallen in der Umwelt. Müll Abfall 9*, 221–227.

Burckle, J., Christiansen, A., Hubbard, J., Martin, J., and Mayo, F. (1988), *Wastewater Treatment Technology*. Radian Corporation and EPA, Cincinnati, Ohio.

Carlsson, K. (1986), *Heavy Metals from "Energy from Waste" Plants – Comparison of Gas Cleaning Systems. Waste Manage. Res. 4*, 15–20.

Christensen, T. H., Kjeldsen, P., Lyngkilde, J., and Tjell, J. C. (1987), *Behavior of Leachate Pollutants in Groundwater*, in: *Proceedings International Symposium on Process, Technology and Environmental Impact of Sanitary Landfill, Cagliari, Italy*, October 1987, Paper No. XXXVIII.

Darby, D. A., Adams, D. D., and Nivens, W. T. (1986), *Early Sediment Changes and Element Mobilization in a Man-made Estuarine Marsh*, in: Sly, P. G. (ed.): *Sediment and Water Interactions*, pp. 343–351. Springer-Verlag, New York.

De Groot, G. A., Wijkstra, J., Hoede, D., and Van der Sloot, H. A. (1987), *Leaching Characteristics of Hazardous Elements from Coal Fly Ash as a Function of the Acidity of the Contact Solution and the Liquid/Solid Ratio*. Presentation at 4th International Hazardous Waste Symposium on Environmental Aspects of Stabilization/Solidification of Hazardous and Radioactive Wastes, Atlanta, Georgia, May 3–6, 1987.

De Kock, W. C., Kuypers, J. W. M., Nijssen, J. P. J., and Scholten, M. (1986), *A Prognosis on Ecotoxicological Effects in a North Sea Benthic System Related to Confined Disposal of Con-*

taminated Sediment and Resulting Seepage. Proceedings 2nd International Conference on Environmental Contamination, p. 36. CEP Consultants Ltd., Edinburgh (complete text in press).

De Kreuk, J. F. (1986), *Microbiological Decontamination of Excavated Soil,* in: Assink, J. W., and Van Den Brink, W. J. (eds.): *Contaminated Soil,* pp. 669–678. Martinus Nijhoff Publ., Dordrecht, The Netherlands.

Dreesen, D. R., Gladney, E. S., Owens, J. W., Perkins, B. L., Wienke, C. L., and Wangen, L. E. (1977), *Comparison of Levels of Trace Elements Extracted from Fly Ash and Levels Found in Effluent Waters from a Coal-fired Power Plant. Environ. Sci. Technol. 11,* 1017–1019.

Ehrig, H.-J. (1983), *Quality and Quantity of Sanitary Landfill Leachate. Waste Manage. Res. 1,* 53–68.

Ehrig, H.-J. (1989), *Water and Element Balances of Landfills,* in: Baccini, P. (ed.): *The Landfill – Reactor and Final Storage. Lecture Notes in Earth Sciences,* Vol. 20, pp. 83–116. Springer, Berlin.

ENVITEC '89 (1989), *Soil – a Bottleneck in Environmental Capacity.* Summaries of Papers, Düsseldorfer Messegesellschaft mbH "Nowea", see also report in *Swiss Chem. 11*(7–8), 49–58.

Farquhar, G. J., and Constable, T. W. (1978), *Leachate Contaminant Attenuation in Soil.* Waterloo Research Institute, Project No. 2123, University of Waterloo, Canada.

Farquhar, G. J., and Parker, W. (1989), *Interactions of Leachates with Natural and Synthetic Envelopes,* in: Baccini, P. (ed.): *The Landfill – Reactor and Final Storage. Lecture Notes in Earth Sciences,* Vol. 20, pp. 175–200. Springer, Berlin.

Fishbein, L. (1989), *Potential Metal Toxicity from Hazardous Waste Incineration. Toxicol. Environ. Chem. 18*(4). 287–309.

Förstner, U. (1986), *Chemical Forms and Environmental Effects of Critical Elements in Solid-Waste Materials – Combustion Residues,* in: Bernhard, M., Brinckman, F. E., and Sadler, P. J. (eds.): *The Importance of Chemical "Speciation" in Environmental Processes.* Dahlem Konferenzen, *Life Sci. Res. Rep. 33,* 465–491.

Förstner, U. (1988), *Geochemical Processes in Waste Deposits* (in German). *Geowissenschaften 6(10),* 302–306.

Förstner, U., and Carstens, A. (1988), *In-situ-Versuche zur Veränderung von festen Schwermetallphasen in aeroben und anaeroben Grundwasserleitern. Vom Wasser 71,* 113–123.

Förstner, U., Kersten, M., and Wienberg, R. (1989), *Geochemical Processes in Landfills,* in: Baccini, P. (ed.): *The Landfill – Reactor and Final Storage. Lecture Notes in Earth Sciences,* Vol. 20, pp. 39–82. Springer-Verlag, Berlin.

Francis, C. W., Maskarinec, M. P., and Lee, D. W. (1989), *Physical and Chemical Methods for the Characterization of Hazardous Wastes,* in: Baccini, P. (ed.): *The Landfill – Reactor and Final Storage. Lecture Notes in Earth Sciences,* Vol. 20, pp. 371–399. Springer-Verlag, Berlin.

Fuller, W. H., and Korte, N. (1976), *Attenuation Mechanisms Through Soil,* in: Genetelli, E. J., and Cirello, J. (eds.): *Gas and Leachate from Landfills.* Report EPA-600/9-76-004. U.S. Environmental Protection Agency, Cincinnati, Ohio.

Greenberg, R. R., Zoller, W. H., and Gordon, G. E. (1978), *Composition and Size Distribution of Particles Released in Refuse Incineration. Environ. Sci. Technol. 12,* 566–573.

Greiner, B. (1983), *Chemisch-physikalische Analyse von Hausmüll. Umweltbundesamt Berichte 7/83.* Erich Schmidt Verlag, Berlin.

Griffin, R. A., Frost, R. R., Au, A. K., Robinson, G. D., and Shimp, N. F. (1977), *Attenuation of Pollutants in Municipal Landfill Leachate by Clay Minerals. II. Heavy Metal Adsorption.* Illinois State Geological Survey, *Environ. Geol. Notes 79,* 47 p.

Ham, R. K., Anderson, M. A., Stanforth, R., and Stegmann, R. (1979), *Background Study on the Development of a Standard Leaching Test.* EPA-600/2-79-109. U.S. Environmental Protection Agency, Cincinnati, Ohio.

Hämmerli-Wirth, H. (1987), *Die Behandlung und Ablagerung von Kehrichtschlacken und Filteraschen. Phoenix International 6/87,* 11–17.

Hartlén, J. (1987), *Elution of Heavy Metals from Disposed Incineration Residues. Proceedings 337th DECHEMA Colloquium, Frankfurt.*

Hilligardt, R., Kalck, U., Kröning, H., Weber, J., and Werther, J. (1986), *Classification and Dewatering of Dredged River Sediments Contaminated with Heavy Metals.* Presentation 4th World Filtration Congress, Ostend/Belgium, April 22–25.

Hughes, J. L. (1975), *Evaluation of Ground-Water Degradation Resulting from Waste Disposal near Barstow, California.* U.S. Geol. Surv. Prof. Pap. *878*, 33 p.

Kelly, W. E. (1976), *Ground-water Pollution near a Landfill.* J. Environ. Eng. Div. Am. Soc. Civ. Eng. *102*, 1189–1193.

Kistler, R. C., Widmer, F., and Brunner, P. H. (1987), *Behavior of Chromium, Nickel, Copper, Zinc, Cadmium, Mercury and Lead During the Pyrolysis of Sewage Sludge.* Environ. Sci. Technol. *21*, 704–708.

Klein, D. H., Andren, A. W., Carter, J. A., Emery, J. F., Feldman, C., Fulkerson, W., Lyon, W. S., Ogle, J. C., Talmi, Y., Van Hook, R. I., and Bolton, N. (1975), *Pathways of Thirty-seven Trace Elements Through Coal-fired Power Plant.* Environ. Sci. Technol. *10*, 973–979.

Kluge, G., Saalfeld, H., and Dannecker, W. (1979), *Forschungsbericht Nr. 10303006.* Umweltbundesamt, Berlin.

Kobayashi, J. (1981), *Relation between the "Itai-Itai" Disease and the Pollution of River Water by Cadmium from a Mine.* Proceedings 5th International Conference on Advanced Water Pollution Research, San Francisco and Hawaii, Vol. I-25, pp. 1–7.

Lichtensteiger, T., Brunner, P. H., and Langmeier, M. (1988), *EAWAG-Projekt Nr. 30-681, Klärschlamm in Deponien.* Cost-681-Forschungsprojekt.

McLellon, W. M., Vickers, D. H., Charba, J. F., and Bengstrom, G. I. (1974), *Environmental Impact Assessment of a Sanitary Landfill in a Higher Water Table Area.* Proceedings 29th Indiana Waste Conference, Purdue University, Ext. Ser. *145*, 94–102.

Meyer, C. F. (1973), *Polluted Groundwater, Some Causes, Effects, Controls and Monitoring.* EPA Report 600/4/73-oob. U.S. Environmental Protection Agency, Cincinnati, Ohio.

Müller, G., and Riethmayer, S. (1982), *Chemische Entgiftung: das alternative Konzept zur problemlosen und endgültigen Entsorgung schwermetallbelasteter Baggerschlämme.* Chem. Ztg. *106*, 289–292.

Nicholson, R. V., Cherry, J. A., and Reardon, E. J. (1983), *Migration of Contaminants in Groundwater at a Landfill: A Case Story. 6. Hydrochemistry.* J. Hydrol. *63*, Special Issue, 131–167.

Nielsen, R. (1978), *Report on Cadmium.* Swedish Environmental Board.

Nijssen, J. P. J. (1988), *Rotterdam Dredged Material: Approach to Handling*, in: Salomons, W., and Förstner, U. (eds.): *Environmental Management of Solid Waste – Dredged Material and Mine Tailings*, pp. 243–281. Springer-Verlag, Berlin.

Patterson, J. W. (1987), *Metals Separation and Recovery*, in: Patterson, J. W., and Passino, R. (eds.): *Metal Speciation, Separation, and Recovery*, pp. 63–93. Lewis Publ., Chelsea, Michigan.

Quasim, S. R., and Burchinal, J. C. (1970), *Leaching from Simulated Landfills.* J. Water Pollut. Control Fed. *42*, 371–379.

Regan, R. W., and Draper, C. E. (1987), *Segregated Landfilling of Metal Finishing Sludge: Concept Evaluation Studies.* Proceedings International Conference Heavy Metals in the Environment, New Orleans, pp. 245–247. CEP Consultants Ltd., Edinburgh.

Salomons, W., and Förstner, U. (eds.) (1988a), *Chemistry and Biology of Solid Waste: Dredged Materials and Mine Tailings.* Springer-Verlag, Berlin.

Salomons, W., and Förstner, U. (eds.) (1988b), *Environmental Management of Solid Waste: Dredged Materials and Mine Tailings.* Springer-Verlag, Berlin.

Schoer, J., and Förstner, U. (1987), *Abschätzung der Langzeitbelastung von Grundwasser durch die Ablagerung metallhaltiger Feststoffe.* Vom Wasser *69*, 23–32.

Shuster, K. A. (1976), *Leachate Damage Assessment: Case Studies.* EPA/530 – SW-509, SW-514, SW-517. Environmental Protection Agency, Cincinnati, Ohio.

Stumm, W., and Morgan, J. J. (1981), *Aquatic Chemistry.* Wiley, New York.

Tabasaran, O. (1985), *Separierung schwermetallhaltiger Hausmüll-Komponenten durch Absieben.* Müll Abfall *16*, 15–22.

Theis, T. L., and Padgett, L. E. (1983), *Factors Affecting the Release of Trace Metals from Municipal Sludge Ashes. J. Water Pollut. Control Fed.* 55, 1271–1279.

Turner, R. R. (1981), *Oxidation State of Arsenic in Coal Ash Leachate. Environ. Sci. Technol.* 15, 1062–1066.

Turner, R. R., Lowry, P., Levin, M., Lindberg, S. E., and Tamura, T. (1982), *Leachability and Aqueous Speciation of Selected Trace Constitutents of Coal Fly Ash.* Final Report, Research Project 1061-1/EA-2588. Electric Power Research Institute, Palo Alto, California.

Umwelt (1987), *Reduction of Heavy Metal Dust Contents in Flue Gas during the Incineration of Hazardous Waste* (in German), No. 6, pp. 224–225 of 27th October (there is a Final Report No. 1059 published by the Umweltbundesamt Berlin).

Van der Sloot, H. A., Piepers, O. and Kok, A. (1984), *A Standard Leaching Test for Combustion Residues.* Studiegroep Ontwikkeling Standaard Uitloogtesten Verbrandingsresiduen (SOSUV). BEOP-31, June 1984.

Van Gemert, W. J. Th., Quakernaat, J., and Van Veen, H. J. (1988), *Methods for the Treatment of Contaminated Dredged Sediments,* in: Salomons, W., and Förstner, U. (eds.): *Environmental Management of Solid Waste – Dredged Material and Mine Tailings,* pp. 44–64. Springer-Verlag, Berlin.

Van Luin, A. B. (1988), *The Treatment of Polluted Groundwater from the Clean-up of Contaminated Soil,* in: Wolf, K., Van Den Brink, W. J., and Colon, F. J. (eds.): *Contaminated Soil '88,* pp. 1167–1174. Kluwer Academic Publ., Dordrecht, The Netherlands.

Veller, G. (1989), *Waste Reduction in Packaging,* Van Leer B. V., *Proceedings IPRE* (International Professional Association for Environmental Affairs) Brussels, in press.

Vogl, J. (1978), *Umwelteinflüsse von Abfalldeponien. Berichte aus Wassergütewirtschaft und Gesundheitsingenieurwesen Nr. 19,* Technische Universität München.

Wadge, A., and Hutton, M. (1987), *The Leachability and Chemical Speciation of Selected Trace Elements in Fly Ash from Coal Combustion and Refuse Incineration. Environ. Pollut.* 48, 85–99.

Walker, W. (1973), *Where Have all the Toxic Chemicals Gone? Groundwater* 11, 11–14.

Werther, J. (1988), *Classification and Dewatering of Sludges,* in: Salomons, W., and Förstner, U. (eds.): *Environmental Management of Solid Waste – Dredged Materal and Mine Tailings,* pp. 65–79. Springer-Verlag Berlin.

Yakowitz, H. (1989), *Encouraging Waste Reduction/Minimization and Recycling: A Policy Oriented Overview.* OECD Paris, *Proceedings IPRE* (International Professional Association for Environmental Affairs) Brussels, in press.

Additional Recommended Literature

Allen, H. E., and Garrison, A. W. (1989), *Metal Speciation and Transport in Groundwater. Proceedings EPA/ACS Workshop, Jekyll Island, Georgia.* Lewis Publishers, Chelsea, Michigan, in press.

Thomé-Kozmiensky, K. J. (1989), *6th International Recycling Congress and Exhibition, Berlin. Abfallwirtschafts-Journal 11,* 1–128. (EF Verlag für Energie- und Umwelttechnik GmbH, Berlin.) Report in *Swiss Chem. 12* (1990, in press); *Site Remediation 3,* 718 pages, EF Verlag für Energie- und Umwelttechnik GmbH, Berlin. *Landfilling – Disposal of Wastes 3,* 732 pages, EF Verlag für Energie- und Umwelttechnik GmbH, Berlin.

US EPA (Environmental Protection Agency) (1988), *Identification and Listing of Hazardous Waste. Code of Federal Regulations,* Title 40, Part 261.

I.7c Heavy Metals in Sewage Sludge and Town Waste Compost

HEINZ HÄNI, Bern-Liebefeld, Switzerland

1 Introduction

Sewage sludge and town waste compost and the mixtures thereof are municipal refuses. Sewage sludges are anaerobically digested or aerated end products of waste water treatment and purification, while town waste composts are prepared in urban composting plants. By the utilization of these products in agriculture an important amount of nutrients and organic matter can be recycled. In 1980 75% of the sludges produced in Switzerland were still utilized in agriculture. Due to different reasons, many of which may be psychological in nature, there has been a decrease in their use in agriculture which means that in 1987 no more than 50% of the 4 million m^3 of sludge corresponding to 125 000 tons of dry matter were disposed onto agricultural land (FAC, internal communication). Concerning the town waste composts, out of the 2.3 million tons produced in Switzerland in 1980 only 3% were composted (SCHRIFTENREIHE UMWELTSCHUTZ, 1984).

Sewage sludges, mainly those from plants which remove phosphate from waste waters are known to be very good phosphate fertilizers (FURRER, 1980). The importance of sewage sludge as a source of organic matter, however, is relatively small. This is in contrast to town waste composts. In these latter products the plant nutrients are much less important than the organic matter. Therefore, town waste compost can be classified as a soil improver which is mainly used in viniculture. The positive effects are seen in a favorable influence on the soil structure which improves aeration of the soil, water movement, field capacity, soil tillage, crumb stability, and resistance against erosion (FURRER, 1981).

It must be emphasized that among these positive aspects there are negative ones which become more important as the quality of the waste product declines. In this respect the presence of toxic substances plays a major part but as the title of the book implies only the problem of heavy metals is dealt with in this chapter. Some of these heavy metals, such as copper and zinc, are essential for man, animals, and plants, whereas other elements, e.g., lead, cadmium, some chromium species, and mercury show no positive effect at all. Their addition to soils which are used for food and feed production can therefore do nothing but harm. In high concentrations all of these elements whether essential or not are potentially toxic to plants and/or warm-blooded animals and man.

There are different pathways by which the heavy metals can reach the waste product. From an investigation of a municipal sewage sludge plant in New York it is evident that galvanic firms are the major source of the extremely high contents of chromium and nickel in sludge (FÖRSTNER and STIEFEL, 1978). Large amounts of the total input of cadmium, copper, and zinc can be ascribed to domestic effluents. For zinc the amount in rainwater is a major source in these effluents. A Swiss investigation comes to similar conclusions for copper and zinc (AMMANN et al., 1980). In this case a great part of cadmium is of industrial origin, namely from emissions of a foundry.

On an international scale it is evident that cadmium, zinc, copper, and nickel are of primary concern when ecological problems have to be studied. In addition to these four elements chromium and lead are included in Table I.7c-1, in which their contents in sewage sludge and town waste compost are compared with those in soil. From the figures it becomes clear that the agricultural use of these waste products must lead to an increase of the heavy metal contents in soil. As the content of zinc in soils differs most widely from its content in waste products, this element is often used for estimating the degree of accumulation. In practice cases are well known where the zinc content in soil increased by a factor of ten or more after the use of waste products (FURRER et al., 1980; FAC, FAW, RAC, 1980). The absolute figures amounted to about 1000 ppm which is very high compared to natural and tolerated contents in soil (see Table I.7c-1).

The use of town waste compost normally leads to a greater accumulation of heavy metals in soil than the spreading of sewage sludge. The reason for this is the higher application rate of compost: 30 tons dry matter per ha per year compared to 2.5 tons per ha per year (see also Sect. 3). However, it is obvious from the Swiss mean value of lead in compost that no more than 1.7 tons per ha can be applied if the maximum annual load of 2500 g of lead per ha according to the SWISS ORDINANCE FOR SEWAGE SLUDGE (1981) is considered. On the basis of the normal application rate of compost in agriculture BRUNNER and BACCINI (1981) calculated that a period of 1 to 6 years would be required for the heavy metal content in soil to be doubled, whereas at least 30 to 120 years would be needed to reach the same accumulation of sludge.

In this connection, occasionally the view is held that the heavy metals in garbage and, as a consequence, in composted refuse occur mainly in elementary form in which they are much less harmful than the more soluble metal compounds present in sewage sludge. Chemical analysis with ammonium acetate (pH 4.8) and a mixture of ammonium acetate-ethylenediamine tetraacetic acid (EDTA) (pH 4.65) showed approximately the same heavy metal solubility for waste compost and sewage sludge. Therefore, there is no reason to judge the heavy metals in these two waste products differently (FAC, FAW, RAC, 1980). The comparatively low mean values of heavy metals in waste compost from the Federal Republic of Germany are striking. It has been suggested that this may be due to differences in analytical methodology, whereas the correspondence of these higher values between Switzerland and the UK is quite good (see Table I.7c-1).

In the case of cadmium it has been shown that waste products are not the only reason for soil pollution. German and Swiss investigations agree with the finding that the atmosphere is quantitatively the most important source (see Table I.7c-2).

Table I.7c-1. Typical Heavy Metal Contents in Soil, Sewage Sludge and Composted Refuse (in mg/kg of dry matter)

Element	Soil[a]		Sewage Sludge						Composted Refuse				
	Frequent	Tolerable	USA (Michigan)[b]		England and Wales[c]		Sweden[d]		Switzerland[e]		GB[f] Mean	FRG[g] Mean	Switzerl.[h] Mean
			Mean	Median	Mean	Median	Mean	Median	Mean	Median			
Lead	0.1 to 20	100	1380	480	820	700	281	180	533	355	1200	230	1460
Cadmium	0.1 to 1	3	74	12	–	–	13	6.7	15	7	–	4	11
Chromium	2 to 50	100	2031	380	980	250	872	86	392	140	120	–	170
Copper	1 to 20	100	1024	700	970	800	791	560	506	400	800	270	715
Nickel	2 to 50	50	371	52	510	80	121	51	94	52	120	–	90
Zinc	3 to 50	300	3315	2200	4100	3000	2055	1567	2269	2022	2000	1000	2200

[a] Kloke (1980a), [b] Blakeslee (1973), [c] Berrow and Webber (1972), [d] Berggren and Svante (1972), [e] FAC (1979), [f] Berrow and Burridge (1979), [g] Hoffmann (1980), [h] Furrer (1981)

Table I.7c-2. Cadmium Pollution of Soil from Different Sources

Source of Contamination	FRG[a] (tons/year)	Switzerland[b] (tons/year)
Precipitation	90[c]	16
Farm fertilizers	–	2
Commercial fertilizers	65	2
Sewage sludge	13	1.7
Composted refuse	–	0.5
Other sources	10	–

[a] Umweltbundesamt (1980), [b] Cadmium in Switzerland (1984)
[c] Metal industries 50 t/a, furnaces 35 t/a, waste incinerations 5 t/a

The re-use of waste products in agriculture leads to local accumulations, whereas the atmospheric emissions tend to be evenly deposited over the land and are widespread. In Switzerland the application of composted refuse leads to a surface accumulation of cadmium of 20–30 mg per m^2 per year compared to 0.4 mg per m^2 per year from atmospheric emissions (CADMIUM IN SWITZERLAND, 1984).

There has been a continuous accumulation of heavy metals in soils, because their industrial consumption and dissipation is still increasing. Their contents in soil, glacial ice, sediments and dust are greater by a factor of 10 in comparison with the last century. Since these accumulations are irreversible this problem has to be taken seriously. In contrast to air and water, the soil cannot be purified from heavy metals. In addition, heavy metals of anthropogenic origin are more soluble than those which are naturally present in soil (FAC, FAW, RAC, 1980). Therefore, the risk of negative effects increases: biological processes in soil can be harmed (WALTER and STADELMANN, 1979), plants with enhanced heavy metal contents are produced with reduction in quality and consequently danger for the consumer or yields of crops may be reduced.

2 Determination of the Tolerable Heavy Metal Concentration in Soil

From the many pot and field trials which have been described in the literature it follows that the heavy metal contents in plants depend mainly upon the amount applied, the composition of the sludge or compost, and the pH value of the soil (HORAK, 1980, see also Chapters I.7 f and g). Since composted refuse is essentially used in viniculture with soil pH above 7 the plant uptake of heavy metals due to their decreased solubility is drastically reduced compared to acid soils. For this reason the following review concentrates upon experiments with sewage sludge.

At the moment of sludge application, it is the sludge itself which essentially determines the heavy metal bond. Only in the course of time do heavy metals react with the soil, mainly as a result of the degradation of sludge organic matter, in an equilibrium with the soil solution. This suggests that short-term experiments are not appropriate for conclusions to be drawn regarding longer-term effects. For this reason it is questionable whether a thirty-day pot experiment with soybean shoots and a given sludge application rate can be taken as a generally valid proof that the cation exchange capacity (CEC) of a soil has no influence on heavy metal uptake by plants. In this experiment cadmium, zinc, copper, chromium, and lead were studied, and the CEC of the soil (silt loam of pH 6.6) was adjusted by dilution with sand (LATTERELL et al., 1976). However, the exchange capacity of soils has been found to be, apart from pH, the most important factor governing the extent of heavy metal fixation (GUPTA et al., 1980).

Different findings have been reported concerning the temporal change of the heavy metal solubility after the application of sewage sludge. In field experiments BERROW and BURRIDGE (1979) found a decrease in solubility of nickel and zinc in acetic acid over the course of ten years. Similar results were obtained by KELLING et al. (1977) who used diethylenetriamine pentaacetic acid (DTPA) as the extractant. On the other hand, SCHAUER et al. (1980) found in the second year of a field trial carried out over several years that there was an increased DTP solubility together with increased contents of cadmium, copper, nickel, and zinc in lettuce plants used as a test crop. This behavior can be explained by the degradation of the organic matter of sewage sludge followed by liberation of heavy metals into the soil solution.

At the beginning of this section the considerable decrease of heavy metal uptake in alkaline soils compared to acid soils has been mentioned. Limits of accumulation can be reached at quite low concentrations, e.g., in the acid, humus-deficient soils of Berlin (KLOKE, 1980b). On the other hand, there are results which show a greater sensitivity of plants to heavy metals in alkaline soils (MITCHELL et al., 1978) where sludge artificially enriched with heavy metals was used. A possible explanation may be the fact that in alkaline soils the heavy metals function as essential micronutrients and are less soluble. Thus under these conditions a situation of deficiency could appear for one or more of these nutrients. Plants may also become more sensitive to toxic heavy metals but more extensive investigations in support of this hypothesis are necessary.

From different publications it follows that cadmium and zinc (to a lesser extent also nickel) are easily taken up by plants (FURR et al., 1976; GAYNOR and HALSTEAD, 1976; KELLING et al., 1977; BURRIDGE and BERROW, 1984). COTTENIE (1981) ascribes the behavior of these elements to their comparatively low phytotoxicity, whereas copper and chromium have a negative effect on plants at much lower contents. Thus the possibility of accumulation in the plant is less important. On the other hand, BERROW and BURRIDGE (1979, 1984, and personal communication) have reported the accumulation of copper, chromium, and lead in roots and their restricted transport to above-ground parts of the plant. However, a greater phytotoxicity of chromium and copper compared with cadmium, zinc, and nickel is not accepted by BERROW.

Plants grown on fields with high doses of sewage sludge (dry matter up to 700 tons per ha in 50 years) show, as might be expected, higher absolute heavy metal contents than plants grown on less polluted soils. However, on analyzing edible plant parts grown under these rather extreme conditions for cadmium only, values near the limit or slightly beyond the normal range of 0−0.5 ppm cited in literature (DAVIS and CARLTON-SMITH, 1980) were found: 1.3 ppm in wheat grain, 0.42 ppm in grains of oats, and 0.54 ppm in potato tubers (DIEZ and ROSOPULO, 1976). It has to be added that the 180 ppm of the sludge used in this experiment exceeded the normal range considerably. The vegetative plant parts accumulated heavy metals strongly and high contents were found in stems and leaves, e.g., 6.6 ppm cadmium in maize. This content is critical in situations where the plant parts, as in case of maize silage, are used as fodder. Finally it must be noted that different species of plants react differently to increased heavy metal contents, e.g., dicotyledons will take up more heavy metals from the same soil than monocotyledons (GUPTA and HÄNI, 1981).

Similar results were obtained in the use of composted refuse and its mixture with sewage sludge in viniculture and vegetable gardening (FAC, FAW, RAC, 1980; GYSI, 1980; KLOKE, 1980b). Accordingly the content in grapes (and therefore also in wine) was not increased, but increased contents were found in the leaves together with a distinct reduction in the yield of cut timber. This is regarded as a sign of reduced soil fertility (mainly due to cadmium, zinc, and lead). Under the conditions of these experiments the amount of 8 m^3/ha/year of composted refuse would lead to intolerable accumulations of heavy metals within 5 – 10 years. Instead of using composted refuse in viniculture to counteract erosion, a continued terracing and green manuring should be recommended. From the FRG, enhanced values in leafy vegetables and node necrosis have been reported (KLOKE, 1980b).

Thus the heavy metal input into soils has to be limited to a level which protects even the most sensitive plants from undesirable heavy metal contents. The greatest attention has to be paid to elements like cadmium and zinc which are relatively easily assimilated by plants. Cadmium is critical because of its great toxicity for warm-blooded animals and man (zootoxicity), whereas zinc is known for its phytotoxicity, mainly observed in sugar beet crops on highly polluted soils (SAUERBECK and STYPEREK, 1986).

Are the numerous trials with sewage sludge satisfactory for drawing conclusions about the approximate degree of soil pollution which can be tolerated? At first sight this seems to be a question almost impossible to answer because there are great differences in the experimental lay-outs and factors such as amount of sludge applied, single or repeated dose, kind of sludge (possibly artificially enriched with heavy metals), and especially differences in experiments carried out in the field or in a glass-house. DE VRIES and TILLER (1978) showed the difficulty in translating effects found in pot experiments to those in the field. They confirmed that in plants grown in pots treated with heavy metals their contents were greatly increased compared with the control pots, whereas the metal uptakes of plants grown with the same metal doses in field or miniplots were much less pronounced.

However, with an extractant which indicates these differences, i.e., a greater metal solubility in the soil contained in a pot than in the field it should be possible to predict the plant reaction from soil analysis. If this were the case the confusions raised by differences in the experimental design and technique would be significant.
One is therefore confronted with the problem that the same total content of heavy metals in soil can lead to different contents in plants. This is due to differences in soil properties such as pH and CEC which are the most important factors in the fixation of heavy metals. The extractant used should dissolve the amount related to the plant content (the biological relevant concentration) independently of soil properties.

Experiments carried out for the determination of trace element deficiency disorders in the field and where extractants such as EDTA and DTPA were used have influenced work in the field of heavy metal excess. However, there are some indications in the literature that their use for assessing the degree of toxicity in different soils may be less appropriate (MITCHELL et al., 1978; SOMMER, 1978). It has been argued that the easily soluble metal fractions could represent a better means for estimating toxicity. Our own experiments with cadmium, copper, nickel, and zinc confirm this suggestion (HÄNI and GUPTA, 1986; GUPTA et al., 1987). Thus a more or less straight-line relationship was found between the amounts of heavy metals extracted from different soils by sodium nitrate (0.1 mol/L) and plant contents. It is also the view of SAUERBECK and STYPEREK (1986) that the fraction soluble in calcium chloride (0.05 mol/L) gives the most reliable indication of availability. According to SAUERBECK and STYPEREK the extraction procedure with neutral salt solutions should be tested in long-term trials in which plant and soil contents are well known.

The values for tolerable heavy metal contents in soil (see Table I.7c-1) are based upon total contents which, according to the preceding paragraph, give no indication of the biologically active portion. Thus such values alone are never sufficient to define limits for soil pollution. These total contents are certainly a good method of laying down load restrictions, but in unfavorable cases, e.g., acid sandy soils, these values may be too high. In such a situation an additional control based on the easily soluble metal fraction is necessary. This fraction should never reach values which are toxic for sensitive plants or microbiological soil processes. In soils with a higher metal-binding capacity the easily soluble part should also be controlled because metal availability may change with time due to lowering of pH (leaching of lime, acid precipitations) or degradation of humus.

3 Measures and Outlook

Several countries have laid down rules for the agricultural use of sewage sludge and composted refuse in order to limit the heavy metal input to soils. Accordingly such guidelines should only be used in agriculture if the heavy metal contents do not exceed certain limits. For comparison the limiting values for sewage sludge allowed in some countries are summarized in Table I.7c-3 (WEBBER et al., 1984).

It can be shown that the heavy metal contents lie considerably below these limiting values at present in most of the countries. TJELL (1986) has reported that it is possible to reduce sludge concentrations considerably by implementing existing control on industries and waste water authorities. Therefore, in Switzerland efforts are made towards a drastical reduction of the existing values (e.g., by a factor of three for cadmium). The figures of the decrease for Switzerland between 1978 and 1983 are as follows: The average contents of lead, cadmium, copper, nickel, and zinc decreased from 564 to 331, 15.3 to 6.5, 514 to 404, 95 to 77, 2305 to 1708, respectively (SCHRIFTENREIHE UMWELTSCHUTZ, 1985). This tendency of decreasing heavy metal contents (mainly of lead, cadmium, and zinc) has continued after 1983 (CANDINAS et al., 1989).

In order to obtain the best control an additional maximum annual application rate must be laid down. In Switzerland this limit is 2.5 tons per ha (as sewage sludge dry matter) as an average value per year (SWISS ORDINANCE FOR SEWAGE SLUDGE, 1981). Based upon the Swiss limiting values for heavy metals in sewage sludge, within hundred years, there will be an increase in content in the upper soil layer (0–20 cm) which corresponds to the values for tolerable heavy metal contents in soils proposed by KLOKE (1980a).

Since 1986 a SWISS ORDINANCE FOR POLLUTANTS IN SOIL has been vigorously applied. In this ordinance the total contents correspond approximately to half the

Table I.7c-3. Maximum Permissible Concentrations for some Heavy Metal (mg/kg dry wt.) in Sludges Considered to be Acceptable for Use on Agricultural Land (WEBBER et al., 1984)

Element	Belgium	Canada	Denmark	Finland	France	FRG	Netherlands	Norway	Sweden	Switzerl.	All Countries Range	Median	CEC Directive R	M
Lead	300	500	400	1200	800	1200	500	300	300	1000	300–1200	500	750	1000
Cadmium	10	20	8	30	20	20	10	10	15	30	8–30	7	20	40
Chromium	500			1000	1000	1200	500	200	1000	1000	200–1200	1000	750	
Copper	500			3000	3000	1200	600	1500	3000	1000	500–3000	1100	1000	1500
Nickel	100	180	30	500	200	200	100	100	500	200	30–500	200	300	400
Zinc	2000	1850		5000	3000	3000	2000	3000	10000	3000	1000–10000	3000	2500	3000

CEC Directive: recommended (R) maximum and mandatory (M) not to be exceeded concentrations

Table I.7c-4. Indicative Values for Heavy Metals in Soil (SWISS ORDINANCE FOR POLLUTANTS IN SOIL, 1986)

Heavy Metal	Heavy Metal Content in Air-dried Mineral Soil (mg/kg)	
	Total (extracted by HNO_3)	Soluble (in $NaNO_3$, 0.1 mol/L)
Lead	50	1.0
Cadmium	0.8	0.03
Chromium	75	–
Cobalt	25	–
Copper	50	0.7
Molybdenum	5	–
Nickel	50	0.2
Mercury	0.8	–
Thallium	1	–
Zinc	200	0.5

tolerable values quoted by KLOKE (1980a) (see Table I.7c-4). This low level was set with the intention of avoiding any possible risk. In addition there are indicative values for easily soluble compounds (sodium nitrate 0.1 mol/L). Where such soluble values are missing there are insufficient data on plant uptake and other soil biological measurements to allow the adoption of indicative values at present time.

If the same standards controlling soil pollution are to be set for composted refuse as for composted sewage sludge the limiting values for heavy metals in these products must be lower by a factor which allows for their higher rate of application. Bearing in mind these limitations the contents of the composts of all the composting plants in Switzerland were too high (FAC, FAW, RAC, 1980). In this connection BRUNNER and BACCINI (1981) have stated that up to 20% of the net input into the domestic wastes contain the major portion of the relatively high heavy metal contents in composts.

The future of composting lies in the reduction of heavy metal contents at the source. A Swiss study has shown that the separation of plant and animal wastes from the refuse allows the manufacture of a product which meets the requirements of high quality compost (SCHRIFTENREIHE UMWELTSCHUTZ, 1985). Respective limiting values have been laid down in the SWISS ORDINANCE OF POLLUTING SUBSTANCES (1986).

The above-mentioned controls are useful for the estimation of the time in which no damage should appear provided the valid norms are respected. It would, however, be a mistake to remain inactive during this period of time since it cannot be our aim to load the soils with heavy metals to the maximum possible amount. This would also mean a misinterpretation of the indicative soil values. Once the necessity of protecting the soils by such legal means is accepted the indicative values entail the following objectives:

a) far-reaching elimination of heavy metals at the source,
b) substitution of heavy metals in different industrial processes and products,
c) increasing efforts towards purifying the air.

Only a permanent improvement in the quality of a waste product will allow its long-term agricultural use. Critical situations cannot be corrected subsequently by limiting or restricting plant production. Also the disposal of wastes (landfill, incineration, etc.) or other alternatives (extraction of fat and protein, building material, electricity generation, fuel production) instead of their use in agriculture do not represent a genuine solution to the problem at present.

References

Ammann, P., Schweizer, C., and Wyss, C. (1980), *Bilan de métaux lourds dans le bassin versant d'une station d'épuration,* in: Hermite P. L., and Ott, H. (eds.): *Proceedings of the Second European Symposium on Characterization, Treatment and Use of Sewage Sludge,* pp. 274–283. D. Reidel Publishing Company, Dordrecht-Boston-London.

Berggren, B., and Svante, O. (1972), *Analysresultat Rorande Fungmetaller Och Klorerade Kolväten I Rötslem Fran Svenska Reningsverk* 1968–71. Insitutionen für Markvetenskap Lantbrukskögskolan.
Berrow, M. L., and Burridge, J. C. (1979), in: *Proceedings of the International Conference on Management and Control of Heavy Metals in the Environment*, London, pp. 304–311. CEP Consultants Ltd., Edinburgh.
Berrow, M. L., and Burridge, J. C. (1984), *Persistence of Metals in Sewage Sludge Treated Soils.* in: Hermite, P. L., and Ott, H. (eds.): *Proceedings of the Third International Symposium on Processing and Use of Sewage Sludge*, pp. 418–422. D. Reidel, Dordrecht-Boston-Lancaster.
Berrow, M. L., and Webber, J. (1972), *Trace Elements in Sewage Sludges. J. Sci. Food Agric.* 23, 93–100.
Blakeslee, P. A. (1973), *Monitoring Considerations for Municipal Wastewater Effluent and Sludge Application to the Land.* U.S. Environmental Protection Agency, U.S. Dept. of Agriculture, Universities Workshop, Champaign, Urbana, Illinois, July 9–13.
Brunner, P., and Baccini, P. (1981), *Die Schwermetalle, Sorgenkinder der Entsorgung? Neue Zürcher Zeitung, Forschung und Technik 70*, p. 65 (25 March), Zürich.
Burridge, J. C., and Berrow, M. L. (1984), in: *Proceedings of the First International Conference on Environmental Contamination*, London, pp. 215–224. CEP Consultants Ltd., Edinburgh.
Cadmium in Switzerland (1984), BUS, *Report of a Study Group of the Federal Government.*
Candinas, A., Gupta, S. K., Zaugg, W., Besson, J. M., and Lischer, P. (1989), *Changes in the Heavy Metal Contents of Swiss Sludge and Future Perspectives. Proceedings Conference on Alternative Uses of Sewage Sludge, University of York.* WRC and CEC, in press.
Cottenie, A. (1981), *Sludge Treatment and Disposal in Relation to Heavy Metals,* in: *Proceedings of the International Conference on Heavy Metals in the Environment*, Amsterdam, pp. 167–175. CEP Consultants Ltd., Edinburgh.
Davis, R. D., and Carlton-Smith, C. (1980), *Tech. Rep. TR 140*, 44. Stevenage Laboratory, Water Research Centre.
De Vries, M. P. C., and Tiller, K. G. (1978), *Sewage Sludge as a Soil Amendment – Comparison of Results from Experiments Conducted Inside and Outside a Glasshouse. Environ. Pollut.* 16, 231–240.
Diez, T., and Rosopulo, A. (1976), *Schwermetallgehalte in Böden und Pflanzen nach extrem hohen Klärschlammgaben. Landw. Forsch. Sonderh. 33/1*, 236–248.
FAC (1979), *Klärschlammkontrolle: Erfahrungen nach einem Jahr*, p. 19. Untersuchungsbericht.
FAC, FAW, RAC (1980), *Schwermetallgehalte der Müll- und Müllklärschlammkomposte in der Schweiz; Beurteilung und Konsequenzen, vor allem für den Rebbau.* Untersuchungsbericht, p. 38.
Förstner, U., and Stiefel, R. (1978), *Umweltprobleme durch Metallanreicherungen in kommunalen Abwässern. Chem. Ztg. 102*, 161–167.
Furr, A. K., Kelly, W. C., Bache, C. A., Gutenmann, W. H., and Lisk, D. J. (1976), *Multielement Absorption by Crops Grown in Pots on Municipal Sludge-Amended Soil. J. Agric. Food Chem.* 24, 889–892.
Furrer, O. J. (1980), in: *Landwirtschaftlicher Wert des Klärschlamms.* EAS-Seminar, Basel, 24–28 September.
Furrer, O. J. (1981), in: *Siedlungskomposte in der Schweiz – positive und negative Aspekte.* XVII. Vortragstagung, Deutsche Gesellschaft für Qualitätsforschung. Speyer, 26–27 März.
Furrer, O. J., Keller, P., Häni, H., and Gupta, S. K. (1980), in: *Schadstoffgrenzwerte – Entstehung und Notwendigkeit.* EAS-Seminar, Basel, 24–28 September.
Gaynor, J. D., and Halstead, R. L. (1976), *Chemical and Plant Extractability of Heavy Metals and Plant Growth on Soils Amended with Sludge. Can. J. Soil Sci.* 56, 1–8.
Gupta, S. K., and Häni, H. (1981), *Easily Extractable Cd-Content of a Soil – its Extraction, its Relationship with the Growth and Root Characteristics of Test Plants, and its Effect on Some of the Soil Microbiological Parameters,* in: 'Hermite, P. L., and Ott, H. (eds.): *Proceedings of*

the Second European Symposium on Characterization, Treatment and Use of Sewage Sludge, pp. 665–676. D. Reidel Publishing Company, Dordrecht-Boston-London.

Gupta, S. K., Häni, H., and Schindler, P. (1980), *Mobilisierung und Immobilisierung von Metallen durch die organische Bodensubstanz.* FAC, interner Forschungsbericht, p. 48.

Gupta, S. K., Häni, H., Santschi, E., and Stadelmann, F. X. (1987), *The Effect of Graded Doses of Nickel on the Yield, the Nickel Content of Lettuce and the Soil Respiration. Toxicol. Environ. Chem. 14*, 1–10.

Gysi, C. (1980), in: *Verwendung von Müll- und Müllklärschlammkomposten in der Landwirtschaft*, pp. 29–43. Informationstagung, Gottlieb-Duttweiler-Institut, Rüschlikon/Zürich.

Häni, H., and Gupta, S. K. (1986), *Chemical Methods for the Biological Characterization of Metals in Sludge and Soil,* in: 'Hermite, P. L. (ed.): *Proceedings of the Fourth International Symposium on Processing and Use of Organic Sludge and Liquid Agricultural Wastes,* pp. 157–167. D. Reidel Publishing Company, Dordrecht-Boston-Lancaster-Tokyo.

Hoffmann, G. (1980), in: *Verwendung von Müll- und Müllklärschlammkomposten in der Landwirtschaft*, pp. 12–28. Informationstagung, Gottlieb-Duttweiler-Institut, Rüschlikon/Zürich.

Horak, O. (1980), *Untersuchung der Schwermetallbelastung von Pflanzen durch Klärschlammgaben im Gefäßversuch. Bodenkultur 31*, 172–181.

Kelling, K. A., Keeney, D. R., Walsh, L. M., and Ryan, J. A. (1977), *A Field Study of the Agricultural Use of Sewage Sludge: III. Effect on Uptake and Extractability of Sludge-Borne Metals. J. Environ. Qual. 6*, 352–358.

Kloke, A. (1980a), *Mitt. VDLUFA 1–3*, 9–11.

Kloke, A. (1980b), in: *Verwendung von Müll- und Müllklärschlammkompost in der Landwirtschaft*, pp. 58–87. Informationstagung, Gottlieb-Duttweiler-Institut, Rüschlikon/Zürich.

Latterell, J. J., Dowdy, R. H., and Ham G. E. (1976), *Sludge-Borne Uptake by Soybeans as a Function of Soil Cation Exchange Capacity. Commun. Soil Sci. Plant Anal. 7*, 465–476.

Mitchell, G. A., Bingham, F. T., and Page, A. L. (1978), *Yield and Metal Composition of Lettuce and Wheat Grown on Soils Amended with Sewage Sludge Enriched with Cadmium, Copper, Nickel and Zinc. J. Environ. Qual. 7*, 165–171.

Sauerbeck, D. R., and Styperek, P. (1986), *Long-Term Effects of Contaminants,* in: 'Hermite, P. L. (ed.): *Proceedings of the Fourth International Symposium on Processing and Use of Organic Sludge and Liquid Agricultural Wastes,* pp. 318–335. D. Reidel Publishing Company, Dordrecht-Boston-Lancaster-Tokyo.

Schauer, P. S., Wright, W. R., and Pelchat, J. (1980), *Sludge-Borne Heavy Metal Availability by Vegetable Crops Under Field Conditions. J. Environ. Qual. 9*, 69–73.

Schriftenreihe Umweltschutz (1984), BUS, Nr. 26, *Kompostierung.*

Schriftenreihe Umweltschutz (1985), BUS, Nr. 45, *Kompostieren in regionalen Anlagen.*

Schriftenreihe Umweltschutz (1985), BUS, Nr. 46, *Gewässerschutzstatistik.*

Sommer, G. (1978), *Gefäßversuche zur Ermittlung der Schadgrenzen von Cadmium, Kupfer, Blei und Zink im Hinblick auf den Einsatz von Abfallstoffen in der Landwirtschaft. Landw. Forsch. Sonderh. 35*, 350–364.

Swiss Ordinance for Sewage Sludge (1981), Federal Government.

Swiss Ordinance for Pollutants in Soil (1986), Federal Government.

Swiss Ordinance of Polluting Substances (1986), Federal Government.

Tjell, J. C. (1986), *Trace Metal Regulations for Sludge Utilisation in Agriculture; a Critical Review,* in: 'Hermite, P. L. (ed.): *Proceedings of the Fourth International Symposium on Processing and Use of Organic Sludge and Liquid Agricultural Waste,* pp. 348–361. D. Reidel Publishing Company, Dordrecht-Boston-Lancaster-Tokyo.

Walter, C., and Stadelmann, F. X. (1979), *Influence due zinc et du cadmium sur les microorganismes ainsi que sur quelques processus biochimiques du sol. Rech. Agron. Sui. 18* (4), 311–324.

Webber, M. D., Kloke, A., and Tjell, J. C. (1984), *A Review of Current Sludge Use Guidelines for the Control of Heavy Metal Contamination in Soils,* in: 'Hermite, P. L., and Ott, H. (eds.): *Proceedings of the Third International Symposium on Processing and Use of Sewage Sludge,* pp. 371–386. D. Reidel Publishing Company, Dordrecht-Boston-Lancaster-Tokyo.

Umweltbundesamt (1980), *Cadmium-Bericht,* Texte 1/81, Berlin, p. 32.

Additional Recommended Literature

Brunner, P. H., and Gaicy, D. (1989), *Changes and Significance of Metal Loads in Sewage Sludge* (in German). Mitteilung No. 28 of the EAWAG, Dübendorf, Switzerland, pp. 5–8.

Coosemans, J., and Uyttebroek, P. (1983), *Comparison of Different Extraction Methods for Heavy Metals from "Compost" Substrates and from Tomato Leaves and Fruits Grown on these Substrates.* Acta Hortic. *133,* 165–172.

Karlaganis, G. (1989), *Definitions of Soil Fertility.* Proceedings (Address) *7th International Conference on Heavy Metals in the Environment, Geneva.* CEP Consultants Ltd., Edinburgh.

Kistler, R., Widmer, F., and Brunner, P. (1986), *Pyrolysis of Sewage Sludge – Distribution of Heavy Metals* (in German). Swiss Chem. *8* (2a), 45–49.

van Assche, C., and Uyttebroek, P. (1982), *Demand, Supply and Application Possibilities of Domestic Waste Compost in Agriculture and Horticulture.* Agric. Wastes *4,* 203–212.

I.7d Deposition of Acids and Metal Compounds

BERNHARD ULRICH, Göttingen, Federal Republic of Germany

1 Definition of Acids and Bases in Ecosystems
(see also Chapters I.3 and I.5)

In the following, the definition of Brønsted is used: bases are substances which can take up protons from H_2O (proton acceptors), acids are substances which can deliver protons to H_2O (proton donors). In ecosystems endergonic and irreversible reactions, catalyzed by enzymes, are possible, and substances can react as acids or bases which will not behave in this way in aqueous equilibrium systems. Since many substances can react as bases and as acids within the pH range of ecosystems (8.5 to 2.5), the following terminology is used: Substances are called bases if they react within this pH range almost exclusively as proton acceptors. Substances which react within this pH range as proton donors are called acids.

In the frame of this terminology, the following bases can be deposited on, transported within, or exist in ecosystems:

In gas phase:

NH_3 ($NH_3 + H_2O \rightarrow NH_4^+ + OH^-$)

In solution phase:

OH^- ($OH^- + H^+ \rightarrow H_2O$)
HCO_3^- ($HCO_3^- + H^+ \rightarrow CO_2 + H_2O$) (only at pH > 5.0, alkalinity)
NO_3^- ($NO_3^- + H^+ \rightsquigarrow$ organically bound N, NO_x, N_2)
SO_4^{2-} ($SO_4^{2-} + 2\,H^+ \rightsquigarrow$ organically bound S)

In solid phase:

carbonates
silicates
alkali and earth alkali cations (shortly: M_b cations), bound ± exchangeable on
   weak acidic groups of mineral and organic matter.

The character of NO_3^- and SO_4^{2-} as bases becomes apparent only if these ions in combination with protons are transferred by organisms into uncharged molecules (proteins, N_2O, N_2, H_2S). Without biological activity these ions do not behave as bases. From the N/S ratio of proteins (15/1) it becomes obvious that the uptake of sulfate by organisms is only a small fraction of the uptake of NO_3^- and can therefore be usually neglected.

The following acids can be deposited on, transported within, or accumulated in ecosystems:

In gas phase:

SO_2 ($SO_2 + H_2O \to H_2SO_3$; $SO_2 + \frac{1}{2}O_2 + H_2O \to H_2SO_4$)
NO_x ($NO_x + H_2O$ ($+O_2$) $\rightsquigarrow HNO_3$)
H_2S (negligible)

In solution phase:

oxonium ion (H_3O^+, often the term proton, H^+, is used)
$CO_2 \cdot H_2O$ (carbonic acid)
NH_4^+ ($NH_4^+ \to NH_3 + H^+$; $NH_4^+ \rightsquigarrow N_{org} + H^+$)
cations forming weak hydroxides: Mn, Al, Fe, heavy metals (shortly: M_a cations)
organic acids

In solid phase:

sulfides (rapidly oxidized and therefore not present under oxidizing conditions)
undissociated acidic groups on clay minerals (pH dependent charge) and organic matter
exchangeable and fixed NH_4^+
M_a cations, bound ± exchangeable on acidic groups of minerals and organic matter
aluminum hydroxosulfates (PRENZEL, 1983) and sulfate adsorbed on aluminum hydroxides
organically bound N ($N_{org} \rightsquigarrow HNO_3$)
organically bound S ($S_{org} \rightsquigarrow H_2SO_4$)

2 Physical Interactions between Air Pollutants and Ecosystems

The physical interactions between air pollutants and ecosystems can be represented by the material fluxes. Fig. I.7d-1 (from SCHULTZ, 1987) shows the fluxes from the canopy of a forest ecosystem. The total atmospheric deposition (TD) consists of precipitational deposition (PD, due to gravity, independent of receptor surface) and interceptional deposition (ID, due to physical interaction between air pollutants and receptor surface). The material deposited can be adsorbed at surfaces (AD) or assimilated by the plant (AS). During rainfall part of dry deposited materials can be washed off and leave the canopy with throughfall and stemflow (IP). This flux can contain materials leached from the canopy which are part of the internal cycling of the ecosystem (LL). Part of the materials are transferred to soil by litter fall (LF). This flux can also consist of deposited materials and materials from the biological

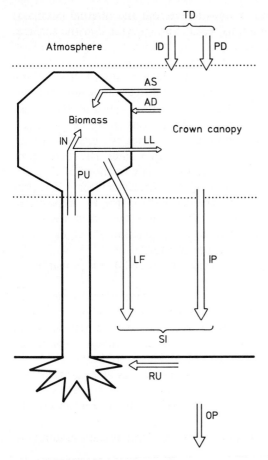

Fig. I.7d-1. Scheme of material fluxes in a forest ecosystem. *TD* total atmospheric deposition, *ID* interceptional deposition, *PD* precipitational deposition, *PU* plant uptake, *LF* litter fall, *IP* internal precipitation, *LL* leaching of leaves, *AS* assimilation, *AD* adsorption, *IN* incorporation of biomass, *SI* soil input, *RU* root uptake, *OP* output from the ecosystem. (From Schultz, 1987).

cycle. Part of the internal cycle is plant uptake (PU) and the incorporation in biomass (IN). The input into soil (SI) is given by the sum of LF and IP.

3 Deposition Pathways, and Rates of Acids and Bases

(Ulrich, 1985; Matzner, 1984; Gehrmann, 1987; Bredemeier, 1988)

Acidity can be deposited in the form of SO_2, NO, H_3O^+, carbonic acid, NH_4^+, M_a cations, and organic acids.

Sulfur dioxide (SO_2) dissolves in water films at external and internal (stomata) surfaces. The protons (forming oxonium ions, H_3O^+) are generated at the surfaces after the dissolution. If the surface contains bases, buffer reactions can take place. Mass action law calculations show that a water film will act as an infinite sink for SO_2 as long as the pH remains above 3.8. Only if pH reaches values below 3.8, the undissociated $SO_2 \cdot H_2O$ molecule has a chance to pass through cell membranes into the cytoplasm. This is a precondition for cell damage. A direct effect of SO_2 to leaf cells is thus bound to the condition that the buffer reactions in the cell wall (exchangeable bound Ca ions) and in the cytoplasm are exhausted. Balance studies show that almost all of the sulfate formed is leached with rain water. The counterions are H^+, K^+, Ca^{2+}, Mg^{2+}, and Mn^{2+} (MATZNER, 1984). The fractions of the various cations depend on the site conditions, especially the chemical soil state. The proton buffering in the leaves can vary between zero and 87% of the H^+ deposition. High percentages are reached on soils of higher base saturation and in areas with higher rates of NH_3 deposition. The acidity due to dry deposition of SO_2 can reach high values, especially in coniferous forests (up to 3 kmol $H^+ \cdot ha^{-1} \cdot a^{-1}$).

No data exist about the deposition rate of nitrogen oxide (NO), it is probably low compared to SO_2.

The wet (precipitation) deposition of *protons* varies in large closed forest areas between 0.6 and 1.1 kmol $H^+ \cdot ha^{-1} \cdot a^{-1}$. Typical are values of about 0.8 kmol, yielding a pH around 4.2 in rain water. The pH of preindustrial rain exceeded 5. In the presence of local emitters of basic dust (e.g., in rural areas, along roads, in the vicinity of cities), the wet proton deposition can be as low as 0.2 kmol.

An important pathway of deposition of protons into forests at exposed sites is the interception of cloud droplets by the forest canopy. Deposition rates between zero and 3 kmol $H^+ \cdot ha^{-1} \cdot a^{-1}$ have been reported. The pH of cloud droplets may be well below 3.

At pH values below 5, *carbonic acid* contributes only insignificant amounts to acid deposition.

Ammonium (NH_4^+) is, however, of great importance. Deposition rates between 0.4 and 3.9 kmol $\cdot ha^{-1} \cdot a^{-1}$) have been reported. In most ecosystems, the deposited NH_4^+ acts as proton donor and contributes thus immediately to soil acidification.

The deposition rates of M_a *cations* are usually low compared to other acids and may therefore be neglected. The same is true for *organic acids*.

Values for total deposition of acidity in forests vary between 1.2 and >8 kmol $\cdot ha^{-1} \cdot a^{-1}$. The few data available on deposition in small vegetation such as agricultural crops and grassland indicate that the rate may be <1.2 kmol $\cdot ha^{-1} \cdot a^{-1}$. In respect to the bases, only NO_3^- shows significant rates of deposition (0.5 to 1 kmol $\cdot ha^{-1} \cdot a^{-1}$). It may be leached, however, with the seepage water, without having reacted as a proton acceptor.

The effective deposition of acidity in an ecosystem can therefore only be assessed from the input/output balance of the canopy and of the whole ecosystem.

4 Reaction of Deposited Acidity in Soils

The buffer reactions of acids in soils can be grouped into buffer ranges as shown in Table I.7d-1 (from ULRICH, 1986). The pH values given apply to chemical equilibria which may often not be reached, especially if Al compounds are involved and leached. The weathering of silicates is the only process in which protons are consumed and nutrient cations such as Ca, Mg, K (M_b cations) appear in solution. In the exchanger buffer range the exchangeable M_b cations are replaced by M_a cations. The M_a cations are released from the silicate lattice or from oxides with proton consumption. If the base saturation (% M_b cations of cation exchange capacity) reaches low values (<15–10%) and pH drops below 4.2, the dissolution of

Table. I.7d-1. Buffer Systems and their pH Ranges in Soils

Buffer Substance	pH Range	Main Reaction Product of Lower Acid Neutralization Capacity (Chemical Change in Soil)
Carbonate buffer range $CaCO_3$	8.6>pH>6.2	$Ca(HCO_3)_2$ in solution (leaching of Ca and basicity)
Silicate buffer range Primary silicates	Whole pH scale (dominating buffer reaction in carbonate-free soils pH>5)	Clay minerals (increase of CEC)[a]
Exchange buffer range Clay minerals	5>pH>4.2	Non-exchangeable $n[Al(OH)_x^{(3-x)+}]$ (blockage of permanent charge, reduction of CEC)[a]
Mn-oxides		Exchangeable Mn^{2+} (reduction of base saturation)
Clay minerals		Exchangeable Al^{3+} (reduction of base saturation)
Interlayer Al $n[Al(OH)_x^{(3-x)+}]$		Al-hydroxosulfate (accumulation of acid in case of input of H_2SO_4)
Aluminum buffer range Interlayer Al Al-hydroxosulfate	4.2>pH	Al^{3+} in solution (Al displacement, reduction of permanent charge)
Aluminum/iron buffer range e.g., Al buffer range in addition: "Soil $Fe(OH)_3$"	3.8>pH	Organic Fe complexes (Fe displacement, bleaching)
Iron buffer range Ferrihydrite	3.2>pH	Fe^{3+} (Fe displacement, bleaching, clay destruction)

[a] CEC cation exchange capacity

polymeric Al hydroxo cations and sulfates starts, resulting in high Al^{3+} concentrations in the soil solution. If these substances are used up or missing, pH can drop below 3.8. Below pH 3.8 iron oxides dissolve in the presence of reducing and chelating organic acids. Below pH 3.2 the solubility of iron oxides becomes high enough to act as proton buffer substances.

Heavy metals participate in the buffer reactions according to the pH dependence of the solubility of their binding forms. The general tendency is that their solubility increases with decreasing pH.

5 Deposition of Alkali and Earth Alkali Metals

The sources of dissolved alkali and earth alkali ions in particles, rain, and snow are sea spray (Na, Mg), soil dust (Ca, K), and plant surfaces (K). The deposition of Na and Mg with rain and snow decreases therefore in Germany from the coastal area to the South (Table I.7d-2). The deposition of K and Ca shows no north/south gradient, but a tendency to extreme high values due to local emission sources. In a mature forest the deposition rates may be twice as high as the precipitational deposition, due to a capture of particles and cloud droplets by the forest canopy. Taking this in account, a comparison with the rates of uptake and of accumulation in the wood increment of a beech forest (see Table I.7d-2) indicates that the deposition of these nutrients can be of significance for the growth of unfertilized natural vegetation. The deposition rates of Na, Mg and K should be independent of anthropogenic air pollution. This is different for Ca, due to the dissolution of solid bound Ca in acid particles.

Table I.7d-2. Wet Deposition of Natrium, Potassium, Magnesium, and Calcium on a North/South Gradient in Germany. (Bulk samplers, data from various authors)

Region	($kg \cdot ha^{-1} \cdot a^{-1}$)			
	Na	Mg	Ca	K
Wingst (Cuxhaven)	23 – 34	2.7 – 4.1	4 – 6	4 – 6
Hamburg North	11 – 21	1.7 – 3.0	5 – 7	2 – 3
Hamburg South	9 – 14	1.4 – 2.0	4 – 6	2 – 3
Lüneburger Heide	10 – 26	1.9 – 3.4	3 – 7	2 – 6
Harz	7 – 13	1.3 – 2.6	6 – 9	3 – 5
Solling	5 – 12	1.3 – 3.9	6 – 22	2 – 5
North-Rhine Westfalia	8 – 13	1.0 – 2.0	7 – 11	3 – 10
Göttingen	3 – 8	1.0 – 2.1	8 – 16	2 – 9
Baden-Württemberg	1 – 6	0.6 – 1.7	2 – 7	2 – 11
Bavaria	–	0.9 – 2.2	5 – 15	2 – 20
Southern Black Forest	3 – 4	0.5 – 0.7	2 – 3	2 – 3
Uptake by natural vegetation (beech forest)	–	~4	~30	~40
Accumulation in the increment of beech forest	–	~2	~8	~7

6 Deposition of Heavy Metals

(see also Chapters I.4c and I.6a; ANGEHRN-BETTINAZZI et al., 1990; BARTH and KLOCKOW, 1989; LARJAVA et al., 1989; PUXBAUM et al., 1987, 1988a, b, 1989; VALENTA and NGUYEN, 1986)

As an example, the fluxes of Cr, Co, Ni, Cu, Zn, Cd, and Pb in a spruce stand (*Picea abies*) in the Solling mountains (Germany) are shown in Fig. I.7d-2 (from SCHULTZ, 1987). In the following discussion on the behavior of heavy metals, additional data from other forests on acid soils are taken into consideration.

Deposited *chromium* is in general accumulated in the ecosystem. The intercepted Cr is deposited mainly as water insoluble compounds in aerosols (SCHMIDT et al., 1985). A small fraction of the deposited Cr is washed off, leaching is negligible. The main fraction of deposited Cr is adsorbed in the canopy and reaches the soil with litter fall; however, part of it accumulates in the canopy. The output of Cr from the ecosystem with seepage water is small on loamy soils, but reaches the level of deposition on sandy soils.

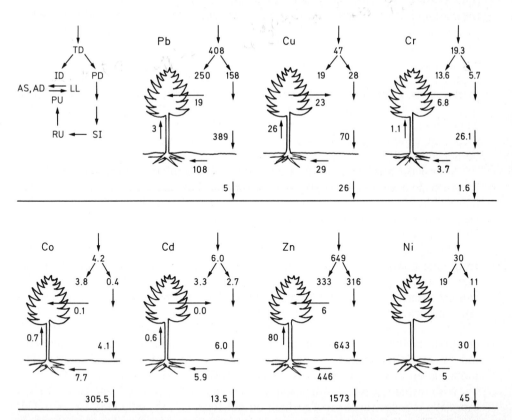

Fig. I.7d-2. Fluxes of chromium, cobalt, nickel, copper, zinc, cadmium, and lead in a spruce stand (*Picea abies*) in the Solling mountains (Germany) (g·ha^{-1}·a^{-1}). (From SCHULTZ, 1987). For abbreviations see Fig. I.7d-1.

Cobalt shows a negative ecosystem balance in most forests on acid soils. Deposition is small and contributes only little to the turnover in the ecosystem. The soils lose Co continuously with the seepage water. In the forest shown in Fig. I.7d-2, the mobilizable Co storage in soil covers the net loss from soil only for a period of 16 years.

Deposited *nickel* is absorbed in the canopy and accumulates in the bark. The balance of the soil is usually negative (output exceeds input). Compared to the turnover, however, the loss is small.

In all forests investigated, *copper* accumulates in the mineral soil and the top organic layer. The accumulation in biomass is small, most of the Cu taken up is either leached or reaches the soil with litter fall. Whereas the Cu content in spruce needles approaches the level of deficiency, the Cu content in the top organic layer of the soil approaches the level of toxicity.

The fluxes of *zinc* vary considerably between different forests. Whereas the balance for the spruce forest in Fig. I.7d-2 is strongly negative, other forest ecosystems show positive balances. The deposition increases with rainfall and therefore with altitude. The interceptional deposition seems to depend strongly on the exposure of the forest. The content in needles is generally low and approaches the level of deficiency. The balance of the soil is generally negative. On the long term, soil acidification may thus lead to Zn deficiency in the vegetation.

In acidic forest soil *cadmium* is mobilized and leached. The input/output balances of the ecosystems as a whole as well as of the soil are negative. Spruce forests show a considerable rate of Cd uptake from soil in contrast to deciduous forests. The different uptake behavior is reflected in different Cd contents in the wood and may be related to higher Cd concentrations in the soil solution. Probably the Cd now mobilized in the soil stems from deposition and has been accumulated before soil acidification has reached the Al buffer range.

Almost the total *lead* which is deposited is accumulated within the ecosystem. The seepage output of Pb is always very low and shows no trend (for the ecosystem depicted in Fig. I.7d-2 seepage outputs have been measured since 1972). Most of the lead accumulates in the top organic layer. The deposition of Pb has decreased since 1974, according to the decrease of lead content in fuel.

The deposition rate of Pb, Cd, Cu and Zn increases with increasing precipitation. In general the interceptional deposition is of the same order of magnitude as the precipitational deposition. Deposited Cd, Zn, and Co are washed off from the canopy, whereas Pb, Cu, and Cr are retained. This is connected with increased contents in bark.

Pb, Cr, As, Hg, and Ag form stable complexes with organic matter and thus accumulate in the top organic layer of forest soils. A mobilization and downward movement of these elements is bound to the mobilization of soil organic matter. The mobilization of organic matter is a characteristic feature of podsolic soils.

With increasing soil acidification the mobility of heavy metals increases. This is especially true for those metals which do not form strong complexes, e.g., Co, Cd, and Be. Pb and Cr are to a high degree fixed in the root cortex. The mobile metals Cd, Co, and Zn show a relationship between concentration in soil solution and root uptake.

The data available until now allow the conclusion that direct adverse effects of deposited heavy metals on leaves are not to be expected. Effects on soil organisms in the top organic layer of forest soils and on roots, however, cannot be excluded.

References

Angehrn-Bettinazzi, C., Hertz, J., and Stöckli, H. (1990), *Factors Affecting the Investigation of Heavy Metal Speciation in Forest Soils Using Thin-Channel Ultrafiltration; Distribution of Heavy Metals in Various Litter Horizons and Forest Soils.* Special Issue *First IAEAC Soil Residue Analysis Workshop. Int. J. Environ. Anal. Chem. 39*(1), 81–89.

Barth, S., and Klockow, D. (1989), *A Contribution to the Experimental Quantification of Dry Deposition to the Canopy of Coniferous Trees*, in: Georgii, H. W. (ed.): *Mechanism and Effect of Pollutant Transfer into Forests.* Kluwer Academic Publishers, Hingham, Massachusetts, in press.

Bredemeier, M. (1988), *Forest Canopy Transformation of Atmospheric Deposition. Water Air Soil 40*, 121–138.

Gehrmann, J. (1987), *Derzeitiger Stand der Belastung von Waldökosystemen in Nordrhein-Westfalen durch Deposition von Luftverunreinigungen. Forst Holzwirt 42*, 141–145.

Larjava, K., Reith, J., and Klockow, D. (1989), *Development and Laboratory Investigations of a Denuder Sampling System for the Determination of Heavy Metal Species in Flue Gases at Elevated Temperatures. Int. Environ. Anal. Chem. 38*(1), 31–45 (1990).

Matzner, E. (1984), *Deposition und Umsatz chemischer Elemente im Kronenraum von Waldbeständen. Ber. Forschungszentrum Waldökosysteme Univ. Göttingen 2*, 61–87.

Prenzel, J. (1983), *A Mechanism for Storage and Retrieval of Acid in Acid Soils,* in: Ulrich, B., and Pankrath, J. (eds.): *Effects on Accumulation of Air Pollutants in Forest Ecosystems,* pp. 157–170. Reidel, Dordrecht.

Puxbaum, H., and Tsitouridou, R. (1987), *Application of a Portable Ion Chromatograph for Field Site Measurements of the Ionic Composition of Fog Water and Atmospheric Aerosols. Int. J. Environ. Anal. Chem. 31*, 11–22.

Puxbaum, H., and Winiwarter, W. (1988), *Memorandum und Bibliography on the 6th Workshop Heterogeneous Atmospheric Chemistry, Vienna.* Report 16/B/88 of the Institute for Analytical Chemistry, Technical University, Vienna.

Puxbaum, H., Vitovec, W., and Kovar, A. (1988a), *Chemical Composition of Wet Deposition in the Eastern Alpine Region,* in: Unsworth, M. H., and Fowler, D. (eds.): *Acid Deposition at High Elevation Sites,* pp. 419–430. Kluwer Academic Publishers, Hingham, Massachusetts.

Puxbaum, H., Rosenberg, C., Winiwarter, W., Gregori, M., Pech, G., and Casensky, V. (1988b), *Determination of Inorganic and Organic Volatile Acids in Ambient Air with an Annular Diffusion Denuder System. Fresenius Z. Anal. Chem. 331*, 1–7.

Puxbaum, H., Rosenberg, C., Ober, E., and Gregori, M. (1989), *Occurrence and Deposition of Acid Components in Forest Ecosystems* (in German). *Staub Reinhalt. Luft 49*, 169–174.

Schmidt, M., Mayer, R., and Georgi, B. (1985), *Beeinflussung der Interceptionsdeposition (trockene Deposition) durch die Aerosolkonzentration und -größenverteilung in einem Buchenbestand (Solling). VDI Ber. 560*, 423–438.

Schultz, R. (1987), *Vergleichende Betrachtung des Schwermetallhaushalts verschiedener Waldökosysteme Norddeutschlands.* Dissertation Fachbereich Stadt- u. Landschaftsplanung, Universität Kassel. (This paper contains an updated literature review).

Ulrich, B. (1985), *Interaction of Indirect and Direct Effects of Air Pollutants in Forests,* in: Troyanowski, C. (ed.): *Air Pollution and Plants,* pp. 149–181. VCH Verlagsgesellschaft, Weinheim-Deerfield Beach/Florida-Basel.

Ulrich, B. (1986), *Natural and Anthropogenic Components of Soil Acidification. Z. Pflanzenernähr. Bodenkd. 149,* 702–717.

Valenta, P., and Nguyen, V. D. (1986), *Acid and Heavy Metal Pollution by Wet Deposition. Sci. Total Environ. 55,* 311–320; *Atmospheric Pollutants in Forest Areas* (ed. by Georgii, H.-W.), pp. 69–78. D. Reidel, Dordrecht.

Additional Recommended Literature

Adriano, D. C., Havas, M., Johnson, A. H., Lindberg, S. E., Page, A. L., and Norton, S. A. (1989), *Acidic Precipitation,* 4 Volumes (especially Vol. 2 deals with biological and ecological effects, and Vol. 3 with sources, deposition and canopy interactions). Springer-Verlag, New York-Berlin-Heidelberg-London-Paris-Tokyo-Hong Kong.

Andreae, H., and Mayer, R. (1989), *Distribution of Heavy Metals in a Landscape Subject to Acid Atmospheric Deposition. Proceedings 7th International Conference on Heavy Metals in the Environment Geneva,* Vol. 2, pp. 44–47. CEP Consultants Ltd., Edinburgh; see also *Swiss Chem. 12*(3), 27–34 (1990).

Mayer, R. (1989), *The Impact of Atmospheric Acid Deposition on Soil and Vegetation. Proceedings 7th International Conference on Heavy Metals in the Environment,* Vol. 2, pp. 19–27. CEP Consultants Ltd., Edinburgh; see also *Swiss Chem. 12*(3), 27–34 (1990).

Prinz, B., et al. (1983), *Acid Precipitation – Origin and Effects. Proceedings VDI Conference Lindau*; see also reports in *Chemosphere 13*(4), N11–N25 (1984) and *TRAC 3*(4), XXI (April 1984).

Prinz, B., et al. (1990), *Effects of Atmospheric Pollutants on Soils. Proceedings VDI Conference Lindau,* in press.

Schulze, E.-D., Lange, O. L., and Oren, R. (1989), *Forest Decline and Air Pollution* (a Study of Spruce on Acid Soils). Springer-Verlag, Berlin-Heidelberg-New York-London-Paris-Tokyo-Hong Kong.

Smith, W. H. (1990), *Air Pollution and Forests, Interaction Between Air Contaminants and Forest Ecosystems,* 2nd Ed. Springer-Verlag, New York-Berlin-Heidelberg-London-Paris-Tokyo.

Troyanowsky, C., et al. (1974), *Air Pollutants Affecting Plants* (including Indirect Effects). *Proceedings 2nd European FECS and GDCh Conference Lindau*; see also *Chemosphere 14*(2), N1–N12 (1985).

I.7e Mobilization of Metals from Sediments

ULRICH FÖRSTNER, Hamburg, Federal Republic of Germany
WILLEM SALOMONS, Haren, The Netherlands

1 Introduction

Modern research on sediment-bound contaminants presumably originated with the idea that these deposits reflect the biological, chemical, and physical conditions of a waterbody (ZÜLLIG, 1956). Sediment analyses are used for selection of critical sites for routine water sampling for non-soluble contaminants that are rapidly adsorbed by particulate matter and may thus escape detection by water analysis (FÖRSTNER and MÜLLER, 1973). Sediment data are involved in environmental forensic investigations (MEIGGS, 1980), where short-term or past-pollution events are not reflected in water analyses.

Sediments are increasingly recognized as both a carrier and a possible *source of contaminants* in aquatic systems; furthermore, there are potential biological effects from polluted solid materials disposed on agricultural land. Metals, for example, are not necessarily fixed permanently by the sediment, but may be recycled via biological and chemical agents both within the sedimentary compartment and the water column. This is especially true for "dredged materials", which threaten not only organisms, but also the quality of potable water (FÖRSTNER et al., 1986a; see also SALOMONS and FÖRSTNER, 1988; and Chapters I.2 and I.7b).

Among the criteria to assess which element or elemental species may be of major concern, one question deserves primary attention (NRIAGU, 1984): Is the element *mobile in geochemical processes* because of either its volatility or its solubility in natural water, so that the effect of geochemical perturbations can propagate through the environment? In this sense mobility and availability of trace metals for metabolic processes are closely related to their chemical species both in solution (see Chapters I.3, I.4a and I.5) and in particulate matter.

Problems of "speciation" become particularly complex in *heterogeneous systems*, e.g., in soils, sediments, and aerosol particles. The greater the instability of the polluted ("stressed") system the more difficulty is in sample handling and storage prior to analysis (WOOD et al., 1986): Many of the analytical techniques are handicapped by disruptive preparation techniques which may alter the chemical speciation of inorganic components or lead to loss of analyte before analysis, e.g., freezing, lyophilization, evaporation, oxidation, changes in pH, light catalyzed reactions, reactions with the sample container, time delays before analysis with biologically active samples, and sample contamination. On the other hand, it is just the "stressed" system, where action is immediately needed and where for an assessment or prognosis of possible adverse effects the species and their transformations of pollutants have to be evaluated. The following questions have been raised with respect to the mobility and bioavailability of potentially toxic metals in contaminated systems:

1. How reactive are the metals introduced with *solid materials* from anthropogenic activities (hazardous waste, sewage sludge, atmospheric fallout) in comparison to the natural compounds?
2. Are the *interactions* of critical metals between solution and solid phases comparable for natural and contaminated systems?
3. What are the factors and *processes of remobilization* to become particularly effective, when either the solid inputs or the solid/solution interactions lead to weaker bonding of certain metal species in contaminated compared to natural systems?

In this chapter emphasis will be given to the latter aspect – remobilization of potentially toxic elements from polluted sediment in both aquatic systems and land deposits.

2 Dissolved versus Solid Metal Fractions in Aquatic Systems

Mobility of an element in the terrestrial and aquatic environment is reflected by the ratio of dissolved and solid fractions; these ratios are firstly influenced by the respective *inputs* and subsequently by the *interactions* taking place within the different environmental compartments.

For *rivers* it seems that the dissolved fractions of metals in polluted waters are significantly higher than in the less polluted systems. Example of cadmium: Rhine River: 70% (HEINRICHS, 1975; see also BREDER, 1981, 1988); Mississippi River: 10% (TREFRY, 1977). Relative increases of dissolved metals (Co, Ni, and Cu) in the Susquehanna River during Dec./Jan. and July (CARPENTER et al., 1975) have been interpreted as an effect of decaying organic matter which is abundant in the river water during these periods (TROUP and BRICKER, 1975). For *lakes* and *oceans* it is suggested that metals from decomposed plankton are partially returned to the solution, whereas the allochthonous particles are deposited on the bottom relatively unchanged (BACCINI, 1976; BRULAND, 1980).

The situation in rivers should be considered in more detail both for practical and scientific reasons. Where is the more effective remedial action – for dissolved or solid input? Can the mobility of a certain element be described by a distribution coefficient, K_D, between dissolved and solid metal concentrations?

The question to what extent interactions between dissolved and solid metal species are affected by either *sorption or precipitation* processes is still being discussed (BRÜMMER et al., 1986). In natural systems, significant differences have been observed for the individual elements, particular with respect to the parameters which are contributing to the limitations of metal concentrations. Data measured by WOLLAST (1982) on River Meuse well correspond to calculated values for lead, copper, and zinc, whereas for cadmium the measured concentrations are approximately

Table I.7e-1. Limitation of Dissolved Metal Concentrations in River Meuse at Tailfer (after WOLLAST, 1982)

Element	Measured (µg/L)	Calculated (µg/L)	From Compound:
Lead	6	3	$Pb_3(CO_3)_2(OH)_2$
Copper	16	14	$Cu_2(CO_3)(OH)_2$
Zinc	39	20	$ZnSiO_4$
Cadmium	0.8	10	$CdCO_3$

one order of magnitude lower than predicted from the stability of cadmium carbonate (Table I.7e-1).

With respect to the effects of metals in *pore waters* (see next section) it has been stressed by SALOMONS (1985) that from an *impact* point of view it is important to know whether the concentrations were determined by *adsorption/desorption* processes or by *precipitation/dissolution* processes. If the latter is the case the concentrations in the pore waters of pollutants are independent of the concentrations in the solid phase (Fig. I.7e-1). There is strong direct and indirect evidence that the concentrations of copper, cadmium, and zinc in sulfidic pore waters are determined by precipitation-dissolution processes. The concentrations of arsenic and chromium in pore waters are probably controlled by adsorption-desorption processes and mainly depend on the concentrations in the solid phase (LU and CHEN, 1977; see also Chapters I.5 and II.6).

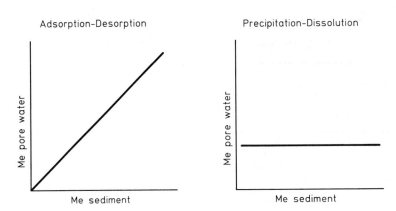

Fig. I.7e-1. Pore water concentrations for absorption-desorption versus precipitation-dissolution controlled processes (SALOMONS, 1985).

3 Assessing Metal Mobility from Pore Water Composition and Solid Speciation

3.1 Pore Water Chemistry

The composition of interstitial water in sediments is perhaps the most sensitive indicator of the types and the extent of reactions that take place between trace metal

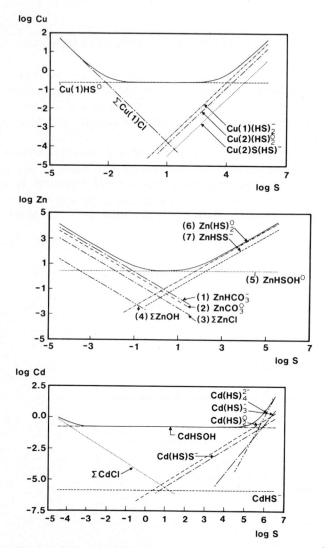

Fig. I.7e-2. The solubility of copper, zinc, and cadmium ($\mu g\,L^{-1}$) as a function of the total free sulfide concentration ($\mu g\,L^{-1}$); pH 7.0; chlorinity, 5000 mg L^{-1}; alkalinity, 50 meq L^{-1}; Eh, 200 mV (SALOMONS et al., 1987).

loaded sediment particles and the aqueous phase which contacts them. The large surface area of fine-grained sediment in relation to the small volume of its entrapped interstitial water ensures that minor reactions with the solid phases will be indicated by major changes in the composition of the aqueous phase (BATLEY and GILES, 1980). Significant enrichment of trace metals in pore waters has been found in anoxic sediment samples and has been explained by effects of *complexation* by organic substances. On the other hand, iron in pore solutions shows a typical temporal evolution, which seems to be controlled by *precipitation* reactions rather than by complexation: In a large-scale experiment on 80×30×6 m pits in the Rhine estuary, filled with dredged sediments, iron concentrations in the pore water reached a maximum after 40–50 days and then decreased as a result of precipitation of iron sulfide (marine conditions) or iron carbonate (FÖRSTNER and SALOMONS, 1983).

Conversion of the reducible iron to siderite in freshwater environments acts as a buffer against a decrease of the pH by decreasing the alkalinity. Minor element chemistry in anoxic sediments is mainly controlled by the presence of solid sulfides; this evidence is based on (SALOMONS et al., 1987): (1) comparison between calculated equilibrium concentrations and measured ones; (2) comparison between pore water concentrations and sediment concentrations; and (3) adsorption experiments with and without sulfide present.

Results of the calculations by SALOMONS et al. (1987) on the influence of the sulfide concentration on the speciation and solubility of Zn, Cd, and Cu are given in Fig. I.7e-2 and I.7e-3. The data in Fig. I.7e-2 show that cadmium is almost in-

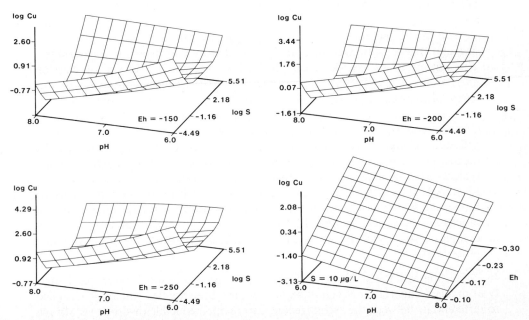

Fig. I.7e-3. The solubility of copper in sulfidic systems as a function of the redox potential (SALOMONS et al., 1987).

variant to changes in sulfide concentrations. This may also explain the nearly constant levels of cadmium concentrations in pore waters reported in the literature (SALOMONS et al., 1987). Zinc and especially copper show larger changes in their solubility by changes in sulfide concentrations. The situation for copper is even more complex because of its occurrence as either Cu^+ or Cu^{2+} depending on the redox potential. Calculations predict an increase in solubility with a decrease in redox potential (Fig. I.7e-3). These calculations, however, depend to a large extent on the assumed speciation in solution and on the thermodynamic data.

In pore waters (bi)sulfide ions compete with the dissolved organic matter for complexation of the metals. In a number of experiments performed by SALOMONS et al. (1987) the overall κ value for dissolved organic complexes in pore waters from estuarine sediments was determined. The κ value was about two orders of magnitude higher than observed in surface waters. However, calculations showed that these κ values were still too low to make organic complexation a significant factor in sulfide pore waters.

3.2 Solid Speciation

Basically there are three *methodological concepts* for determining the distribution of an element within or among small particles (KEYSER et al., 1978; FÖRSTNER, 1985):

- Analysis of *single particles* by X-ray fluorescence using either a scanning electron microscope (SEM) or an electron microprobe can identify differences in the matrix composition between individual particles. The total concentration of the element can be determined as a function of particle size. Other physical fractionation and preconcentration methods include density and magnetic separations.
- The *surface of the particles* can be studied directly by the use of electron microprobe X-ray emission spectrometry (EMP), electron spectroscopy for chemical analysis (ESCA), Auger electron spectroscopy (AES), and secondary ion-mass spectrometry. Depth-profile analysis determines the variation of chemical composition below the original surface.
- *Solvent leaching* – apart from the characterization of the reactivity of specific metals – can provide information on the behavior of metal pollutants under typical environmental conditions. Common single reagent leachate tests, e.g., US EPA, ASTM, IAEA, ICES, and German Water Chemistry Group (Deutsche Einheitsverfahren) use either distilled water or acetic acid. A large number of test procedures have been designed particularly for soil studies; these partly used organic chelators such as EDTA, DTPA, both as single extractants or in sequential procedures.

In connection with the problems arising from the disposal of *solid wastes*, particularly of dredged materials, *extraction sequences* have been applied which are designed to differentiate between the exchangeable, carbonatic, reducible (hydrous Fe/Mn oxides), oxidizable (sulfides and organic phases), and residual fractions (ENGLER et al., 1977; TESSIER et al., 1979).

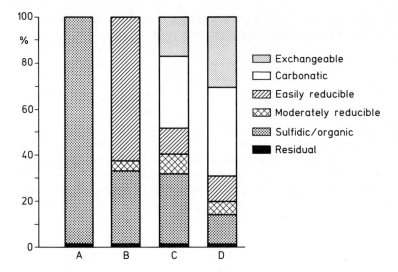

Fig. I.7e-4. Partition of cadmium in anoxic mud from Hamburg harbor in relation to the pretreatment procedures: **A** control sample extracted as received under oxygen-free conditions; **B** after treatment with the US EPA/CE elutriate test; **C** freeze-dried; **D** oven-dried (60°C) (after KERSTEN and FÖRSTNER, 1986).

A simple but impressive experiment on the *effect of oxidation* in regulating the chemical form of cadmium and other trace metals has been performed on an anoxic sediment sample from Hamburg harbor; the sample was divided into four series under an argon flushed glove box in order to study the effect of various sample pretreatments including aeration and dehydration on the chemical forms of cadmium (Fig. I.7e-4; KERSTEN and FÖRSTNER, 1986):

A. Manipulations of the first series were all done under the *inert atmosphere* to serve as a control.
B. The second series was treated by the *Elutriate Test* modified for air bubbling (LEE et al., 1976). This test was initially designed by the U.S. Environmental Protection Agency to detect any short-term release of chemical contaminants from polluted material during dredging manipulations and disposal. This test involves the mixing of one volume of the harbor sludge with four volumes of the dredging or disposal site water for a 30-min shaking period.
C. The third subsample series was *freeze-dried*, and
D. the fourth series was dried under air at *60 °C*.

Subsequent to the preservation and pretreatment measures the subsamples were extracted by a *six-step sequential leaching technique* (modification of Tessier's method). The significant differences as shown in Fig. I.7e-4 can be ascribed to the contact of the sediment with air and by dehydration rather than to experimental artifacts such as inhomogeneity of the sediments or variations in the extraction protocol. Indeed, no differences were obtained for oxic suspended matter (KERSTEN and

FÖRSTNER, 1985), and the sum of the metal concentrations in the individual fractions of all four series of each sample agreed within 10%.

From these findings it seems that one of the major applications of solid speciation data includes an estimation of the *long-term* changes of metal mobilities under variable environmental conditions by analyzing characteristic *re-arrangements of chemical forms* in these materials. Although the individual "fractions" received from the chemical leaching experiments rarely reflect specific metal "phases", but rather are defined by the selection of the extracting medium and by the experimental conditions ("operational phases"), the elution medium is designed to simulate certain — mostly extreme — environmental conditions, such as the interaction with saline waters in estuaries or reducing conditions during land disposal of sludge materials. The overall most significant effects on the remobilization of heavy metals can be expected from acidification, either from atmospheric emissions or from oxidation of sulfidic compounds in anoxic waste materials.

4 Remobilization of Metals from Polluted Sediments

Solubility, mobility, and bioavailability of sediment-bound metals can be increased by *four major factors* in terrestrial and aquatic environments:

- *lowering of pH*, locally from mining effluents, regionally from acid precipitation;
- *increasing salt concentrations*, by the effect of competition on sorption sites on solid surfaces and by the formation of soluble chloro-complexes with some trace metals;
- increasing occurrence of natural and synthetic *complexing agents*, which can form soluble metal complexes with trace metals that are otherwise adsorbed to solid matter; and
- *changing redox conditions*, e.g., after land deposition of polluted anoxic dredged materials.

In some cases, which will be described here, mobilization is a change in the chemical environment affecting lower rates of precipitation or adsorption — compared to "natural" conditions — rather than active release from contaminated solid materials.

4.1 Acidity

Acidity imposes problems in all aspects of metal mobilization in the environment: toxicity of drinking water, growth and reproduction of aquatic organisms, increased leaching of nutrients from the soil and the ensuing reduction of soil fertility, increased availability and toxicity of metals, and the undesirable acceleration of mercury methylation in sediments (FAGERSTRÖM and JERNELÖV, 1972). On a regional scale, acid precipitation is probably the prime factor affecting metal mobility in surface waters.

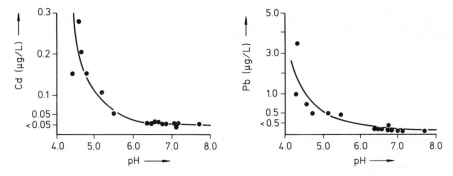

Fig. I.7e-5. Dissolved metal concentrations (in µg/L) in relation to pH values in 16 lakes at the West coast of Sweden (after DICKSON, 1980).

In Swedish lakes a pronounced correlation was observed between dissolved metal levels and pH (Fig. I.7e-5, after DICKSON, 1980); this phenomenon is probably due to the *combined effects* of

1. changing solid/dissolved equilibria in the atmospheric precipitation,
2. washout processes on soils and rocks in the catchment area,
3. enhancing groundwater mobility of metals, and
4. active remobilization from aquatic sediments.

There is a significant decrease of pH values in less buffered, low-carbonate dredged sludges − such as those from Hamburg harbor − after some time interval (months to several years) subsequent to land spreading. This can be explained by the ability of certain bacteria (*Thiobacillus thiooxidans* and *T. ferrooxidans*) to oxidize sulfur and ferrous iron; while decreasing the pH from 4−5 to about 2.0, the process of metal dissolution from dredged sludge is enhanced. In a laboratory system acidification with sulfurous acid to pH 4.0 and subsequent bacterial leaching solubilized the following metal percentages from the dredged sediments of Hamburg harbor (CALMANO et al., 1983): Cd and Co, 98%; Mn, 91%; Cu, 84%; Ni, 66%; Cr, 45%; Fe, 27%; and Pb, 17%. Despite this, the original objective to "detoxify" the material to the quality standard required for agricultural application (e.g., MÜLLER and RIETHMAYER, 1982) could not be reached, even by the combined method of acid/bacterial leaching.

4.2 Salinity

Riverborne metals entering an estuarine environment can be affected by a change in pH, chlorinity, turbidity maximum, and formation of new particulate matter (SALOMONS, 1980). Significant changes occur even in the low-salinity region of an estuary; the removal of riverborne iron and, in some estuaries, of manganese at low salinities is well established (DUINKER, 1980). However, apart from the flux of riverine material into an estuarine environment, the deposited particulates may pro-

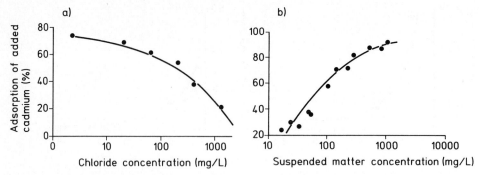

Fig. I.7e-6. Adsorption of cadmium on sediments from the Ijsselmeer in relation to chloride concentration (a) and suspended solid content (b) (after SALOMONS and MOOK, 1980).

vide a source of dissolved and newly formed particulate components. As a result of biological or biochemical pumping, the intertidal flats in many estuaries act as a source of dissolved metals (MORRIS et al., 1982).

Release of trace metals from suspended particulates and sediments has been reported from several estuaries, e.g., Scheldt estuary, Gironde estuary, Elbe and Weser estuaries in Northern Germany, and Savannah/Ogeechee estuaries (SALOMONS and FÖRSTNER, 1984). The release has been interpreted by oxidation processes and by intensive breakdown of organic matter (both mediated by microorganisms; PRAUSE et al., 1985), whereafter the released metals become complexed with chloride and/or ligands from the decomposing organic matter in the water. In this way the uptake by or precipitation on the suspended matter may be inhibited; in addition, it has been suggested by MILLWARD and MOORE (1982) that the major cations, magnesium and calcium, are probably co-adsorbed; competition from these species for adsorption sites increases with increasing salinity.

According to experimental data given by SALOMONS and MOOK (1980) these effects can even be found in salt-polluted inland waters (Fig. I.7e-6a): at chloride contents of 200 mg/L (example: Lower Rhine River) the "normal" adsorption rate of cadmium would be reduced by approx. 20%; at 1000 mg/L Cl^- (example: Weser River) this rate would be only half compared to the adsorption of Cd under natural salt concentrations. Suspended matter concentrations act opposite to the salinity effects (Fig. I.7e-6b).

4.3 Natural and Synthetic Complexing Agents
(see also FÖRSTNER et al., 1984)

It seems that naturally dissolved organic matter, such as humic material or amino acids, has little effect upon the speciation of dissolved Cd, Pb, or Zn (NÜRNBERG, 1983); only the distribution of Cu species is affected markedly by natural organic chelators (STUMM and MORGAN, 1981). Significant effects, however, on most heavy metals can be expected in inland waters from strong synthetic chelators, such as

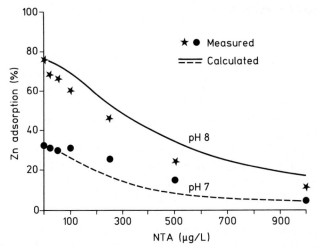

Fig. I.7e-7. Effect of NTA on the adsorption of zinc on suspended sediments (SALOMONS, 1983).

nitrilotriacetic acid (NTA), which is already used in some countries in detergents to substitute polyphosphates.

The extent of these effects depends on the concentration of the complexing agent, its pH value, the mode of occurrence of trace metals in the sediments and on competition by other cations (SALOMONS and VAN PAGEE, 1981). The influence is two-fold:

1. The complexing agent may actively desorb heavy metals from the suspended matter and deposited sediment. Similar processes may occur within the deposited sediments (increased mobilization of trace metals to the interstitial water).
2. When both NTA (or any other complexing agent) and metals are discharged into a river system, the complexing agent may negatively influence the natural adsorption processes.

The active remobilization has been studied by many investigators and seems to exhibit reliable results at NTA-concentrations above approx. 1 – 2 mg/L (FÖRSTNER et al., 1984). Such concentrations of NTA could rarely be expected in normal river waters, but may occur at even higher levels in sewage treatment plants. The "passive" effects of NTA on the sorption of metals on solid surfaces are starting at lower values (SALOMONS and VAN PAGEE, 1981), for most metals already at NTA concentrations of 200 to 500 µg/L. Under high-pH conditions (pH 8) it has been shown by SALOMONS (1983) that zinc adsorption on suspended matter is already significantly affected at NTA concentrations of 20 – 50 µg/L (Fig. I.7e-7).

4.4 Redox Changes

Much of the dredged material removed during harbor and channel maintenance dredging is high in organic matter and clay, and is both biologically and chemically active (FÖRSTNER et al., 1986a).

4.4.1 Changes of Metal Species Distribution

Our data from sequential leaching procedures on Hamburg harbor sediments (Fig. I.7e-4) have shown that oxidation drastically changes the proportion of chemical forms in the originally anoxic material: Following the application of the elutriate test, part of the oxidizable sulfidic/organic forms of Cd is now found in the easily reducible fraction. After freeze- and oven-drying of the initially anoxic samples, cadmium proportions were found even in the most mobile operationally defined carbonatic and exchangeable fractions.

One of the most obvious features is the *transformation of oxidizable fraction*. Unfortunately, this fraction is still less well-defined because it comprises both labile and more refractory organic substances in combination with sulfidic metal associations. With respect to the practical uses of these investigations, there is much information available from the percentage of oxidizable phases. In Fig. I.7e-8 bar graphs show the average percentages of peroxide-oxidizable metal fractions of the respective non-residual bound metals as obtained from the sequential extraction of the Hamburg harbor mud samples. It is shown that the concentration of metal in this fraction increase in the order $Mn < Fe < Cr = Ni < Zn < Pb < Cd < Cu$, which widely corresponds to the Irving-Williams order for the sequence of complex stability, but also to the stability of sulfidic metal compounds. Following application of the elutriate test, decreased portions of the sulfidic/organic fractions of Pb (42%) and Cu (21%) are chiefly found in the moderately reducible fraction. Transformation of oxidizable Ni (11%), Zn (62%), and Cd (67%) mainly results in increased percentages of easily reducible fractions.

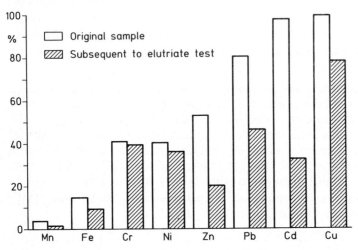

Fig. I.7e-8. Changes of oxidizable metal fractions in a sediment sample from Hamburg harbor during elutriate test procedure (KERSTEN and FÖRSTNER, 1987).

4.4.2 Species Transformations in Tidal Sediments

An example of oxidative remobilization of cadmium and other heavy metals has been studied in a *tidal freshwater flat* in the upper Elbe estuary near Hamburg (KERSTEN et al., 1986). This mudflat — diurnal tidal water fluctuations in the range of 3 m are affecting this productive site — is colonialized by dense monodominant reed stands providing an effective trap for heavy metal loaded suspended matter from upstreams. Examination of sediment cores taken at this site showed distinct patterns of redox potential and heavy metal fractionation profiles (Fig. I.7e-9): While in the anoxic zone approximately 60 to 80% of Cd is found in the oxidizable fraction, high percentages of Na-acetate extractable forms are found in the oxic and post-oxic zones of the sediment cores. The higher amounts of labile cadmium forms are accompanied with a marked depletion in the total content of the toxic metal compared to that in the anoxic sediment zone. Comparison of the fractionation patterns and total contents of other diagenetically less mobile metal examples indicates that a significant proportion of cadmium is leached from the surface sediment by a process of *"oxidative pumping"* by tidal water drainage in this high-energetic environment. This could result in migration of the remobilized metal into either the deeper anoxic zone, where it can precipitate again to contribute to the enhanced oxidizable sulfidic/organic fraction, or to the surface water, from where it can be exported into the outer estuary. It could, however, also contribute to bioavailable cadmium portions such as indicated by the enhanced macrophyte cadmium concentrations.

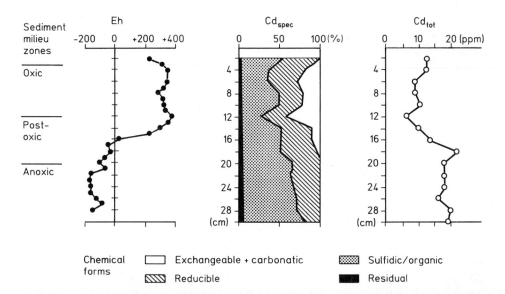

Fig. I.7e-9. Core sediments from the Heukenlock intertidal flat in the Elbe River near Hamburg (after KERSTEN et al., 1986). *Left:* Sediment milieu zones/Eh conditions; *middle:* chemical forms of cadmium in sediment; *right:* bulk cadmium distribution.

4.4.3 Pore Water Release of Metals During Dredging

Early sediment changes and element mobilization from pore water in a man-made estuarine marsh have been investigated by DARBY et al. (1986) (Table I.7e-2): Compared to the river water concentration, the channel sediment pore water is enriched by a factor of 200 for Fe and Mn, 30–50 for Ni and Pb, approx. 10 for Cd and Hg, and 2–3 for Cu and Zn. When the expected concentrations of metal following hydraulic dredging, which were calculated from a rate of pore water to river water of about 1:4, were compared with the actual measurements at the pipe exiting the dredging device, typical removal from solution was observed for iron and to a lesser extent for manganese; on the other hand, remobilization was found for zinc at the highest rates, followed by copper, lead, and cadmium. According to DARBY et al. (1986) the levels of heavy metal mobilization were higher at this time than at any time in the subsequent two years of marsh maturation. The amount of metal mobilization detected at the effluent pipe of the disposal area during dredging could be accounted for by the release of relatively small amounts of those elements bound to labile sediment phases; while only 3–5% of the labile phase Cu and Ni would account for the measured increases reported in Table I.7e-2, up to 36% of the labile phase Pb and Zn was required to account for the higher than expected concentrations of these elements.

These data demonstrate the problematic effect of dispersing anoxic waste materials in *ecologically productive*, high-energy nearshore, estuarine, and inlet zones (KHALID, 1980). This may also pertain to procedures such as "sludge-harrowing" as it is occassionally performed in the cold season in some sections of Hamburg harbor. Most critical is the dispersion of anoxic mud on land surface, rather than on a heap deposit, due to stronger interactions of the former material with the atmosphere.

In the *marine environment* several observations have been made suggesting similar – but significantly slower – oxidation processes on the release of metals from polluted sediments. Field evidence for changing cadmium and zinc mobilities

Table I.7e-2. Mobilization of Metals and Nutrients During Dredging (after DARBY et al., 1986). All Concentrations are in mg/L, Except for Hg

Metal	Channel Sediment Pore Water (a)	River Water Concn. (b)	(a/b)	Effluent at Man-made Marsh		
				Expected Concn.	Measured Concn.	% Change
Mn	6.94	0.03	230	1.34	1.19	−11
Fe	57.3	0.26	220	11.12	6.01	−46
Ni	0.054	0.001	54	0.011	0.035	+218
Pb	0.077	0.002	38	0.016	0.142	+788
Hg (µg/L)	3.2	0.26	12	0.82	2.0	+144
Cd	0.009	0.001	9	0.0025	0.019	+660
Cu	0.012	0.004	3	0.0055	0.051	+827
Zn	0.12	0.052	2	0.065	5.30	+8069

was reported by HOLMES et al. (1974) from Corpus Christi Bay harbor: During summer when the harbor water is stagnant Cd and Zn are precipitated as CdS and ZnS at the sediment-water interface; in the winter months, the intrusion of oxygen-rich water into the bay results in the release of some of the precipitated metals. Pore water analyses in shallow water sediments of Puget Sound estuary indicate higher concentrations of metals in the upper oxidized centimeters of sediment covering reduced, H_2S-containing deposits (EMERSON et al., 1984). The authors suggested a greatly enhanced flux of metals to the bottom waters due to a process of biologically mediated ventilation of the upper sediment layers with oxic overlying water. These processes have been shown to be more effective during spring and summer than during the winter months (HINES et al., 1984). From enclosure experiments in the Narragansett Bay it has been estimated that through mechanisms such as oxidation of organic and sulfidic material the anthropogenic proportion of Cd in marine sediments is released to the water within approx. three years time; for remobilization of Cu and Pb approx. 40 and 440 years is needed, respectively, according to these extrapolations (HUNT and SMITH, 1983). Mobilization during early diagenesis in the surficial sediment layer is proposed to explain the elevated dissolved concentrations of Cd, Cu, Zn, and other metals in Northeast Pacific continental slope water (JONES and MURRAY, 1984), waters of the Gulf of St. Lawrence (WESTERLUND et al., 1986), the Mediterranean Villefranche Bay (GAILLARD et al., 1986), and the North Sea shelf water (KREMLING, 1985; NOLTING, 1986).

5 Predicting Potential Metal Mobilization – Remedial Measures

Species differentiations can be used for the estimation on the remobilization of metals under *changing environmental conditions:*

– In the *estuarine environment* the "exchangeable fraction" might be affected in particular; however, changes of pH and redox potential could also influence other easily extractable phases, e.g., carbonate and manganese oxides.
– Among the factors enhancing metal mobility *acid interactions* deserve particular attention due to the fact that ionic species predominate, which are readily available for biological uptake. Lowering of pH will affect, according to its strength, the "exchangeable", then the "easily reducible" and in case part of the "moderately reducible" fraction, the latter consisting of Fe-oxyhydrates in less crystallized forms.
– The effects on *organically bound metals* are more complex; however, it has been argued that this fraction is highly susceptible to environmental changes, especially during early diagenetic reactions, where recycling of mineralized organic matter and pore fluid transfer processes are controlling the dynamics of pollutants and nutrients (see also FÖRSTNER, 1984).

Various *remedial measures* (see also SALOMONS and FÖRSTNER, 1988) against excessive release of toxic metals into the environment may be considered, giving priority to the technologies applied near the source of pollution. If metals have been dispersed, liming of soils and waters as well as mechanical and chemical *stabilization* of solid waste such as encapsulation, use of impermeable base liners and surface covers could reduce fluxes and biological availability of toxic metals (RULKENS et al., 1985; CALMANO, 1988).

Generally, maintenance of a *pH of neutrality* or slightly higher (by application of lime or limestone) favors adsorption or precipitation of soluble metals (GAMBRELL et al., 1983; CALMANO et al., 1986). For *mercury contaminated sediments* isolating from the water body by means of physical barriers, such as polymer film overlays, blanket plugs of waste wool, sand and gravel overlays, has been proposed (JERNELÖV and LANN, 1973; WOLERY and WALTERS, 1974; REIMERS et al., 1975).

The best strategy for disposing contaminated sediments is to isolate them in a permanently reducing environment (KESTER et al., 1983), e.g., in capped mound deposits above the prevailing sea-floor or (capped) disposal in subaqueous depressions; from a geochemical view the marine sulfidic environment is favorable due to the high stability of metal sulfides, particularly of mercury (disproportionation of monomethyl-mercury), and more efficient degradation of organic matter (FÖRSTNER et al., 1986b).

Acknowledgement. We would like to thank Mr. Michael Kersten for providing material on redox effects on metal speciation of sediments.

References

Baccini, P. (1976), *Untersuchungen über den Stoffhaushalt in Seen*. Z. Hydrol. 38, 121–158.
Batley, G. E., and Giles, M. S. (1980), *A Solvent Displacement Technique for the Separation of Sediment Interstitial Waters,* in: Baker, R. A. (ed.): *Contaminants and Sediments,* Vol. 2, pp. 101–117. Ann Arbor Science Publishers. Ann Arbor, Michigan.
Breder, R. (1981), *The River Rhine Pollution with Toxic Metals* (in German). Ph D Thesis, Universität Bonn.
Breder, R. (1988), in: Hutzinger, O., Safe, S., Stoeppler, M., and Piscator, M. (eds.): *Cadmium in European Inland Waters, in Environmental Toxins.* Vol. 2, pp. 159–169. Springer-Verlag, Berlin-Heidelberg.
Bruland, K. W. (1980), *Oceanographic Distribution of Cadmium, Zinc, Nickel and Copper in the North Pacific.* Earth Planet. Sci. Lett. 47, 176–198.
Brümmer, G., Gerth, J., and Herms, U. (1986), *Heavy Metal Species, Mobility and Availability in Soils.* Z. Pflanzenernaehr. Bodenkd. 149, 382–398.
Calmano, W. (1988), *Stabilization of Dredged Mud,* in: Salomons, W., and Förstner, U. (eds.): *Environmental Management of Solid Waste – Dredged Material and Mine Tailings,* pp. 80–98. Springer-Verlag, Berlin.
Calmano, W., Ahlf, W., and Förstner, U. (1983), *Heavy Metal Removal from Contaminated Sludges with Dissolved Sulfur Dioxide in Combination with Bacterial Leaching.* Proceedings International Conference Heavy Metals in the Environment, September, Heidelberg, pp. 952–955. CEP Consultants Ltd., Edinburgh.
Calmano, W., Förstner, U., Kersten, M., and Krause, D. (1986), *Behaviour of Dredged Mud after Stabilization with Different Additives,* in: Assink, J. W., and Van der Brink, W. J. (eds.): *Contaminated Soil,* pp. 737–746. Martinus Nijhoff Publishers, Dordrecht.

Carpenter, J. H., Bradford, W. L., and Grant, V. (1975), *Processes Affecting the Composition of Estuarine Water (H_2CO_3, Fe, Mn, Zn, Cu, Ni, Cr, and Cd)*, in: Cronin, L. E. (ed.): *Estuarine Research*, Vol. 1, pp. 137–152. Academic Press, London.

Darby, D. A., Adams, D. D., and Nivens, W. T. (1986), *Early Sediment Changes and Element Mobilization in a Man-made Estuarine Marsh*, in: Sly, P. G. (ed.): *Sediment and Water Interactions*, pp. 343–351. Springer-Verlag, New York.

Dickson, W. (1980), *Properties of Acidified Waters*, in: Drablos, D., and Tollan, A. (eds.): *Ecological Impact of Acid Precipitation*, pp. 75–83. SNSF-Project, Oslo-Aas.

Duinker, J. C. (1980), *Suspended Matter in Estuaries: Adsorption and Desorption Processes*, in: Olausson, E., and Cato, I. (eds.): *Chemistry and Biogeochemistry of Estuaries*, pp. 121–153. Wiley, Chichester, U. K.

Emerson, S., Jahnke, R., and Heggie, D. (1984), *Sediment-water Exchange in Shallow Water Estuarine Sediments. J. Mar. Res.* **42**, 709–730.

Engler, R. M., Brannon, J. M., Rose, J., and Bigham, G. (1977), *A Practical Selective Extraction Procedure for Sediment Characterization*, in: Yen, T. F. (ed.): *Chemistry of Marine Sediments*, pp. 161–171. Ann Arbor Science Publishers, Ann Arbor, Michigan.

Fagerström, T., and Jernelöv, A. (1972), *Aspects of the Quantitative Ecology of Mercury. Water Res.* **6**, 1193–1202.

Förstner, U. (1984), *Effects of Salinity of the Metal Sorption onto Organic Particulate Matter*, in: *Proceedings 4th International Wadden Sea Symposium: The Role of Organic Matter in the Wadden Sea*. Netherlands Institute for Sea Research, Publ. Ser. No. 10-1984, pp. 195–209.

Förstner, U. (1985), *Chemical Forms and Reactivities of Metals in Sediments*, in: Leschber, R., Davis, R. D., and L'Hermite, P. (eds.): *Chemical Methods for Assessing Bio-available Metals in Sludges and Soils*, pp. 1–30. Elsevier Applied Science Publishers, London.

Förstner, U., and Müller, G. (1973), *Heavy Metal Accumulations in River Sediments: A Response to Environmental Pollution. Geoforum* **14**, 53–61.

Förstner, U., and Salomons, W. (1983), *Trace Element Speciation in Surface Waters: Interactions with Particulate Matter*, in: Leppard, G. G. (ed.): *Trace Element Speciation in Surface Waters and Its Ecological Implications*, pp. 245–273. Plenum Press, New York.

Förstner, U., Hennes, E.-C., and Schöttler, U., et al. (1984), *Einflußnahme von NTA auf die Wechselwirkungen im Gewässer mit Sedimenten, Schwebstoffen und dem Untergrund (Uferfiltration und künstliche Grundwasseranreicherung)*, in: Bernhardt, H. (ed.): *Die aquatische Umweltverträglichkeit von Nitrilotriacetat (NTA)*, pp. 181–208. Verlag Hans Richarz, St. Augustin/FRG.

Förstner, U., Ahlf, W., Calmano, W., Kersten, M., and Salomons, W. (1986a), *Mobility of Heavy Metals in Dredged Harbour Sediments*, in: Sly, P. G. (ed.): *Sediments and Water Interactions*, pp. 371–380. Springer-Verlag, New York.

Förstner, U., Ahlf, W., Calmano, W., and Kersten, M. (1986b), *Mobility of Pollutants in Dredged Materials – Implications for Selecting Disposal Options*, in: Kullenberg, G. (ed.): *Role of the Ocean as a Waste Disposal Option*, pp. 597–615. D. Reidel Publishers, Dordrecht.

Gaillard, J. F., Jeandel, C., Michard, G., Nicolas, E., and Renard, D. (1986), *Interstitial Water Chemistry of Villefranche Bay Sediments: Trace Metal Diagenesis. Mar. Chem.* **18**, 233–247.

Gambrell, R. P., Reedy, C. N., and Khalid, R. A. (1983), *Characterization of Trace and Toxic Materials in Sediments of a Lake Being Restored. J. Water Pollut. Control Fed.* **55**, 1201–1213.

Heinrichs, H. (1975), *Die Untersuchung von Gesteinen und Gewässern auf Cd, Sb, Hg, Tl, Pb, und Bi mit der flammenlosen Atomabsorptions-Spektralphotometrie.* Doctoral Dissertation, Universität Göttingen, FRG.

Hines, M. E., Berry Lyons, W. M., Armstrong, P. B., Orem, W. H., Spencer, M. J., and Gaudette, H. E. (1984), *Seasonal Metal Remobilization in the Sediments of Great Bay, New Hampshire. Mar. Chem.* **15**, 173–187.

Holmes, C. W., Slade, E. A., and McLerran, C. J. (1974), *Migration and Redistribution of Zinc and Cadmium in Marine Estuarine Systems. Environ. Sci. Technol.* **8**, 255–259.

Hunt, C. D., and Smith, D. L. (1983), *Remobilization of Metals from Polluted Marine Sediments.* Can. J. Fish. Aquat. Sci. *40*, 132–142.

Jernelöv, A., and Lann, H. (1973), *Studies in Sweden on Feasibility of Some Methods for Restoration of Mercury-contaminated Bodies of Water.* Environ. Sci. Technol. *7*, 712–718.

Jones, C. J., and Murray, J. W. (1984), *Nickel, Cadmium and Copper in the Northeast Pacific off the Coast of Washington.* Limnol. Oceanogr. *29*, 711–720.

Kersten, M., and Förstner, U. (1985), *Trace Metal Partitioning in Suspended Matter with Special Reference to Pollution in the Southeastern North Sea,* in: Degens, E. T., Kempe, S., and Herrera, R. (eds.): *Transport of Carbon and Minerals in Major World Rivers,* Vol. 3. Mitt. Geol.-Paläontol. Inst. Univ. Hamburg, SCOPE/UNEP Spec. Vol. 58, pp. 631–645.

Kersten, M., and Förstner, U. (1986), *Chemical Fractionation on Heavy Metals in Anoxic Estuarine and Coastal Sediments.* Water Sci. Technol. *18*, 121–130.

Kersten, M., and Förstner, U. (1987), *Effect of Sample Pretreatment on the Reliability of Solid Speciation Data on Heavy Metals – Implications for the Study of Early Diagenetic Processes.* Mar. Chem. *22*, 299–312.

Kersten, M., Förstner, U., Kerner, M., and Kausch, H. (1986), *Remobilization of Cd and N from Periodically Inundated Soil,* in Trans. XIII. Congr. Int. Soc. Soil Sci., Hamburg, Vol. II. pp. 348–349.

Kester, D. R., Ketchum, B. H., Duedall, I. W., and Park, P. K. (1983) (eds.): *Dredged-Material Disposal in the Ocean.* Series *Wastes in the Ocean,* Vol. 2. Wiley, New York.

Keyser, T. R., Natusch, D. F. S., Evans, C. A., Jr., and Linton, R. W. (1978), *Characterizing the Surface of Environmental Particles.* Environ Sci. Technol. *12*, 768–773.

Kahlid, R. A. (1980), *Chemical Mobility of Cadmium in Sediment-water Systems,* in: Nriagu, J. O. (ed.): *Cadmium in the Environment,* Part I: *Ecological Cycling,* pp. 257–304. Wiley, New York.

Kremling, K. (1985), *The Distribution of Cadmium, Copper, Nickel, Manganese and Aluminum in Surface Waters of the Open Atlantic and European Shelf Area.* Deep-Sea Res. *32*, 531–555.

Lee, G. F., Lopez, J. M., and Piwoni, M. D. (1976), *Evaluation of the Factors Influencing the Result of the Elutriate Test for Dredging Material Disposal Criteria,* in: Krenkel, P. A., Harrison, J., and Burdick, J. C. III. (eds.): *Dredging and Its Environmental Effects. Proceedings Special Conference American Society Civil Engineers,* pp. 253–288. ASCE, New York.

Lu, C. S. J., and Chen, K. Y. (1977), *Migration of Trace Metals in Interfaces of Seawater and Polluted Surficial Sediments.* Environ. Sci. Technol. *11*, 174–182.

Meiggs, T. O. (1980), *The Use of Sediment Analysis in Forensic Investigations and Procedural Requirements for such Studies,* in: Baker, R. A. (ed.): *Contaminants and Sediments,* Vol. 1, pp. 297–308. Ann Arbor Science Publishers, Ann Arbor, Michigan.

Millward, G. E., and Moore, R. M. (1982), *The Absorption of Cu, Mn, and Zn by Iron Oxyhydrate in Model Estuarine Solutions.* Water Res. *16*, 981–985.

Morris, A. W., Bale, A. J., and Howland, R. J. M. (1982), *The Dynamics of Estuarine Manganese Cycling.* Estuarine Coastal Shelf Sci. *13*, 175–192.

Müller, G., and Riethmayer, S. (1982), *Chemische Entgiftung – das alternative Konzept zur problemlosen und endgültigen Entsorgung schwermetallbelasteter Baggerschlämme.* Chem. Ztg. *106*, 289–292.

Nolting, R. F. (1986), *Copper, Zinc, Cadmium, Nickel, Iron and Manganese in the Southern Bight of the North Sea.* Mar. Pollut. Bull. *17*, 113–117.

Nriagu J. O. (ed.) (1984), *Changing Metal Cycles and Human Health.* Dahlem Konferenzen. Life Sci. Res. Rep. *28*, 445 pp. Springer-Verlag, Berlin.

Nürnberg, H. W. (1983), *Investigations of Heavy Metal Speciation in Natural Waters by Voltammetric Procedures.* Fresenius Z. Anal. Chem. *316*, 557–565.

Prause, B., Rehm, E., and Schulz-Baldes, M. (1985), *The Remobilization of Pb and Cd from Contaminated Dredge Spoil after Dumping in the Marine Environment.* Environ. Technol. Lett. *6*, 261–266.

Reimers, R. S., Krenkel, P. A., Eagle, M., and Tragift, G. (1975), *Sorption Phenomenon in the Organics of Bottom Sediments,* in: Krenkel, P. A. (ed.): *Heavy Metals in the Aquatic Environment,* pp. 117–136. Pergamon Press, Oxford.

Rulkens, W. H., Assink, J. W., and Van Gemert, W. J. T. (1985), *On-site Processing of Contaminated Soil,* in: Smith, M. A. (ed.): *Contaminated Land – Reclamation and Treatment,* pp. 37 – 90. Plenum Press, New York.
Salomons, W. (1980), *Adsorption Processes and Hydrodynamic Conditions in Estuaries. Environ. Technol. Lett. 1,* 356 – 365.
Salomons, W. (1983), *Trace Metals in the Rhine, their Past and Present (1920 – 1983), Influence on Aquatic and Terrestrial Ecosystems,* in: *Proceedings International Conference Heavy Metals in the Environment,* Heidelberg, Sept. 6 – 9, pp. 764 – 771. CEP Consultants Ltd., Edinburgh.
Salomons, W. (1985), *Sediment and Water Quality. Environ. Technol. Lett. 6,* 315 – 368.
Salomons, W., and Förstner, U. (1984), *Metals in the Hydrocycle.* Springer-Verlag, Berlin.
Salomons, W., and Förstner, U. (eds.) (1988), *Management of Mine Waste and Dredged Material,* 396 pp. Springer-Verlag, Berlin.
Salomons, W., and Mook, W. G. (1980), *Biogeochemical Processes Affecting Metal Concentrations in Lake Sediments (IJsselmeer, The Netherlands). Sci. Total Environ. 16,* 217 – 229.
Salomons, W., and Van Pagee, J. A. (1981), *Prediction of NTA Levels in River Systems and their Effects on Metal Concentrations,* in: *Proceedings International Conference Heavy Metals in the Environment,* Amsterdam, Sept. 14 – 18, pp. 694 – 697. CEP Consultants Ltd., Edingburgh.
Salomons, W., De Rooij, N. M., Kerdijk, H., and Bril, J. (1987), *Sediments as a Source for Contaminants?* In: Thomas, R. L. et al. (eds.): *Ecological Effects of in-situ Sediment Contaminants. Hydrobiologia 149,* 13 – 30. Dr. W. Junk Publ., Dordrecht.
Stumm, W., and Morgan, J. J. (1981), *Aquatic Chemistry.* John Wiley & Sons, New York.
Tessier, A., Campbell, P. G. C., and Bisson, M. (1979), *Sequential Extraction Procedure for the Speciation of Particulate Trace Metals. Anal. Chem. 51,* 844 – 851.
Trefry, J. H. (1977), *The Transport of Heavy Metals by the Mississippi River and their Fate in the Gulf of Mexico.* PhD Thesis, Texas A&M University, Dallas.
Troup, B. N., and Bricker, O. P. (1975), *Processes Affecting the Transport of Materials from Continents to Oceans,* in: Church, T. M. (ed.): *Marine Chemistry in the Coastal Environment. Am. Chem. Soc. Symp. Ser. 18,* 133 – 151.
Westerlund, S. G., Anderson, L. G., Hall, P. O. J., Iverfeldt, A., Rutgers van der Loeff, M. M., and Sundby, B. (1986), *Benthic Fluxes of Cadmium, Copper, Nickel, Zinc, and Lead in the Coastal Environment. Geochim. Cosmochim. Acta 50,* 1289 – 1296.
Wolery, T. J., and Walters, L. J. (1974), *Pollutant Mercury and Sedimentation in the Western Basin of Lake Erie,* in: *Proceedings 17th Conference Great Lakes Research,* pp. 235 – 249. International Association of Great Lakes Research.
Wollast, R. (1982), *Methodology of Research in Micropollutants – Heavy Metals. Water Sci. Technol. 14,* 107 – 125.
Wood, J. M., Chakrabarty, A. M., et al. (1986), *Speciation in Systems Under Stress,* in: Bernhard, M., Brinkman, F. E., and Sadler, P. S. (eds.): *The Importance of Chemical Speciation in Environmental Processes.* Dahlem Konferenzen. *Life Sci. Res. Rep. 33,* 425 – 441. Springer-Verlag, Berlin.
Züllig, H. (1956), *Sedimente als Ausdruck des Zustandes eines Gewässers. Schweiz. Z. Hydrol. 18,* 7 – 143.

Additional Recommended Literature

Allan, R. J. (1986), *The Role of Particulate Matter in the Fate of Contaminants in Aquatic Ecosystems.* National Water Research Institute, Scientific Series No. 142. 128 pp. Canada Centre for Inland Waters, Burlington/Ontario.
Allen, H. E. (ed.) (1984), *Micropollutants in River Sediments.* Report on a WHO Working Group Meeting, Trier/FRG, August 5 – 8, 1980. Euro Reports and Studies No. 61, 85 p. WHO Regional Office for Europe, Copenhagen.

Baccini, P. (ed.) (1989), *The Landfill*. Springer-Verlag, Berlin-Heidelberg-New York-London-Paris-Tokyo.

Baker, R. A. (ed.) (1980), *Contaminants and Sediments*. 2 Volumes. Ann Arbor Science Publishers, Ann Arbor, Michigan.

Calmano, W., Förstner, U, and Kersten, M. (1986), *Metal Associations in Anoxic Sediments and Changes, Following Upland Disposal*. Toxicol. Environ. Chem. *12*(3+4), 313–321.

Goldberg, E. D. (ed.) (1978), *Biogeochemistry of Estuarine Sediments*. Proceedings UNESCO/SCOR Workshop, Melreux/Belgium, Nov. 29–Dec. 3, 1976, 293 pp. UNESCO, Paris.

Golterman, H. L. (ed.) (1977), *Interactions between Sediments and Fresh Water*. Proc. 1st Intern. Symp. Amsterdam, Sept. 6–10, 1976. 472 pp. Junk/The Hague and PUDOC/Wageningen.

Golterman, H. L., Sly, P. G., and Thomas, R. L. (eds.) (1983), *Study of the Relationship between Water Quality and Sediment Transport*. A Guide for the Collection and Interpretation of Sediment Quality Data. 231 pp. UNESCO, Paris.

Gunn, A. M., Winnard, D. A., and Hunt, D. T. E. (1988), *Trace Metal Speciation in Sediments and Soils*, in: Kramer, J. R., and Allen, H. E. (eds.): *Metal Speciation*, pp. 261–294. Lewis Publishers, Chelsea, Michigan.

Hart, B. T., and Sly, P. G. (eds.) (1989), *Sediments and Freshwater Interactions*. Proc. 4th Intern. Symp., Melbourne, Febr. 17–20, 1987. Dr. W. Junk, Publ., The Hague.

Houriet, J.-P. (1989), *Heavy Metals in Sediments of Swiss Rivers* (in German), *Bulletin Umweltschutz in der Schweiz 4/89*, 31–43. Swiss Federal Office for Environment, Forest and Landscape, Berne.

Kavanaugh, M. C., and Leckie, J. O. (eds.) (1980), *Particulates in Water – Characterization, Fate, Effects, and Removal*. Advances in Chemistry Ser. *189*, 401 pp. American Chemical Society, Washington D.C.

Landner, L. (ed.) (1987), *Speciation of Metals in Water, Sediment and Soil Systems*. Lecture Notes in Earth Sciences No. 11, 190 pp. Springer-Verlag, Berlin.

Leckie, J. O. (1988), *Coordination Chemistry at the Solid/Solution Interface*, in: Kramer, J. R., and Allen, H. E., (eds.): *Metal Speciation*, pp. 41–68. Lewis Publishers, Forest Grove, Oregon.

Loll, U. (1989), *Recycling of Sludge*. EF Verlag für Energie- und Umwelttechnik GmbH, Berlin.

Postma, H. (ed.) (1981), *Sediment and Pollution Exchange in Shallow Seas*. Rapp. P.-v. Réun, Cons. Intern. Explor. Mer (ICES), Vol. 181, Copenhagen.

Rauret, G., Rubio, R., Jopez-Sanchez, J. F., and Casassas, E. (1989), *Specific Procedure for Metal Solid Speciation in Heavily Polluted River Sediments*. Toxicol. Environ. Chem., in press.

Salomons, W., and Förstner, U. (eds.) (1988), *Chemistry and Biology of Solid Waste – Dredged Material and Mine Tailings*, 305 pp. Springer-Verlag, Berlin.

Shear, H., and Watson, A. E. P. (eds.) (1977), *The Fluvial Transport of Sediment-Associated Nutrients and Contaminants*. Proc. Intern. Workshop Kitchener/Ontario, Oct. 20–22, 1976, 309 pp. PLUARG, International Joint Commission, Windsor, Ontario.

Sly, P. (ed.) (1982), *Sediment/Freshwater Interaction*. Proc. 2nd Intern. Symp. Kingston, Ontario, June 15–18, 1981, 701 pp. Dr. W. Junk Publ., The Hague (*Hydrobiologia* Vols. *91/92*).

Sly, P. (ed.) (1986), *Sediment and Water Interactions*. Proc. 3rd Intern. Symp. Geneva, August 27–31, 1984, 521 pp. Springer-Verlag, New York.

Tessier, A., and Campbell, P. G. C. (1988), *Partitioning of Trace Metals in Sediments*, in: Kramer, J. R., and Allen, H. E. (eds.): *Metal Speciation*, pp. 183–199. Lewis Publishers, Chelsea, Michigan.

Thomas, R. L., et al. (eds.) (1987), *Ecological Effects of in-situ Sediment Contaminants*. Proc. Intern. Workshop Aberystwyth, Wales, August 21–24, 1984, 272 pp. Dr. W. Junk Publ., Dordrecht (*Hydrobiologia* Vol. *149*).

Thomé-Kozmiensky, K. J., et al. (1989), *6th International Recycling Congress and Exhibition, Abfallwirtschafts-Journal 11*, and especially *Landfilling 3*. EF Verlag für Energie- und Umwelttechnik GmbH, Berlin.

Valenta, P., Breder, R., Mart, L., Rützel, H., and Merks, A. G. A. (1987), *Distribution of Cadmium and Lead between Dissolved and Particulate Phases in Estuaries*. Toxicol. Environ. Chem. *14* (1+2), 129–141.

I.7f   Uptake, Distribution, and Effects of Metal Compounds on Plants

MICHAEL L. BERROW and JOHN C. BURRIDGE
Aberdeen, Scotland, UK

1 Uptake of Metals from the Soil

The uptake of metals from soil by plants through their roots to their above-ground parts or under-ground storage organs depends on (1) the total amount present in the soil, (2) the proportion of the total that is accessible to the plant roots, and (3) the ability of the plants to transfer the metals across the soil-root interface. These factors are not independent, but interact, for example, when the uptake affects plant growth because a deficient or toxic level of a metal exists. Such interactions are not discussed in this chapter, but must be kept in mind when evaluating plant uptake in the course of environmental studies.

The total amount in the soil is derived in the first place from natural sources, but may be increased substantially by man's industrial and agricultural activities (BERROW and BURRIDGE, 1979). This is discussed elsewhere in this book, for instance, in Chapter I.2, the other subchapters of Chapter I.7, and Part II; it is sufficient at this point to emphasize that in soils there are generally major differences between the chemical forms of metals derived from rocks by the natural processes of pedological weathering and the forms arising from environmental pollution (see also Chapter I.1).

The accessibility of an element to plants in any given soil is determined by its chemical form and its location within the soil. The most readily available elements are those present in the so-called soil solution in the ionic state or as soluble organic-matter complexes; the least available are those firmly bound within the structure of solids, for instance, within the crystal lattice of primary rock minerals. Between these extremes the most important pool of available material is associated with charged sites on surfaces of very small particles such as clay and silt, and on organic matter, which together comprise what may be termed the "exchange-complex". These sites are characterized by their ability to release one ion in exchange for another, for example, calcium may exchange with magnesium, potassium or with hydrogen. Such conditions as acidity, organic matter content and drainage status are among many factors that affect the chemical forms of metals and thus their availability to plants (MITCHELL, 1964).

The soil-root interface is not a passive, inert sieve. The root surface is an active boundary with characteristics varying with plant species and dependent on the particular element. Cation exchange capacity, for instance, is a property of roots that can be reproducibly measured (CROOKE, 1964) and which is generally greater for dicots than for monocots. Moreover, the soil environment immediately adjacent to the roots can be strongly influenced by root exudates (LINEHAN et al., 1985; MERCKX et al., 1986a, b), so that apart from biochemical processes of transfer across cell walls within the roots, chemical processes of dissolution, chelation and precipitation outside the root also occur. Microbial activity in the rhizosphere is an additional factor that must be taken into account. Elements can accumulate on plant roots, for example, Al, Cu, and Fe, sometimes without any measurable transfer to the above-ground tissues even when poor growth has occurred.

2 Assessing Plant-available Metals in Soil

At present there are no methods by which the amount of a metal in a soil, accessible to plant roots, can be quantitatively established by direct measurement. Analysis of plant tissues can establish uptake after it has taken place, but this has only a limited application to predicting future performance. The interpretation of plant analyses is further complicated by such factors as differences between plant species in their ability to extract a given element from a particular soil, and by variations in plant content with stage of growth and plant part (MITCHELL and BURRIDGE, 1980).

Relations between plant uptake and soil total contents are seldom close. Generally, in freely drained soils the amount available to plant roots is only a small fraction of the total, and the available pool comprises several different forms. Although analyzing the naturally occurring soil solution would seem best, difficulties in taking a well defined sample for analysis, together with the very low levels present in the soil solution, have hitherto made its analysis impractical. In practice, it is usual to attempt to establish, for the soil type in question, empirical relations between plant contents and the amount that can be extracted from the soil in the laboratory by specific reagents.

A very wide range of reagents has been used by soil-scientists in seeking to establish empirical relations between plant uptake and soil content (COX and KAMPRATH, 1972). On the one hand, relatively aggressive reagents such as 1 mol/L HCl and acidified ammonium oxalate have been used; and on the other hand, a less vigorous attack by ammonium acetate at pH 4.8 (HÄNI, 1977) or the mild action of water alone, has been employed. In each situation a compromise has to be made between attempting to simulate the extracting power of the roots, and overcoming limitations set by the sensitivity of the analytical technique. For research purposes, it is preferable to use separately several reagents that discriminate between the different forms in which an element is held in the soil, whereas for pollution monitoring a single extractant may be satisfactory.

The three extractants 0.43 mol/L acetic acid, 1 mol/L ammonium acetate at pH 7.0 and 0.05 mol/L ethylenediaminetetraacetic acid (EDTA) at pH 7.0 have been extensively used in Scotland, where the soils are pedologically young as a consequence of relatively recent glacial action and a subsequent cool, temperate climate. These reagents differentiate between the amounts that can be displaced from the soil by the

H$^+$ ion, readily exchanged by the NH$_4^+$ ion, or chelated by EDTA in competition with naturally occurring organic ligands (BERROW and BURRIDGE, 1980).

Good relationships between soil and plant contents obtained with these reagents are shown in Fig. I.7f-1. Soil drainage status had no effect on the relationships for Cu or Mo, but for Co the relationships of freely and poorly drained soils were significantly different. The diagnostic use of such relationships must always take into account any limitations that may be imposed by the specific soil types and plant species on which the relationships are based. Soil-plant relationships for Cd, Cu, Mn, Pb, and Zn were discussed by PICKERING (1981) in a review of selective chemical extraction methods.

In different situations, for example, with calcareous soils, other reagents may be more effective. Extraction with diethylenetriamine pentaacetic acid (DTPA) was developed by LINDSAY and NORVELL (1978), for instance, to identify near-neutral and calcareous soils which had insufficient available Cu, Fe, Mn, or Zn for maximum crop yields. The 0.005 mol/L DTPA buffered at pH 7.3 was tested against the response by corn (*Zea mays*) and sorghum (*Sorghum bicolor*) to fertilization with Fe and Zn, for which elements critical extractable levels of 4.5 and 0.8 mg/kg, respectively, were established.

At the toxicity end of the extractable scale, where Cd is probably the metal of greatest environmental concern, exchangeable or dilute-acid extractable Cd has been found useful for identifying polluted soils. In greenhouse trials using 46 Ontario soils, HAQ et al. (1980) found that for Swiss chard (*Beta vulgaris*) acetic acid was the best of nine extractants for the prediction of plant-available Cd and Ni when soil pH was included in the equation. Acetic acid- and EDTA-extractable Ni and Zn have both been found to show linear relationships with the contents of pasture herbage species growing on two sewage-sludge contaminated soils under field conditions (BERROW and BURRIDGE, 1981). SAUERBECK and STYPEREK (1983) have proposed 0.05 mol/L CaCl$_2$ as an extractant for assessing Cd availability in soils.

The effects of sewage sludge on the Cd and Zn contents of crops have been reviewed by a task force of 25 scientists assembled by the Council for Agricultural Science and Technology (CAST, 1980). In addition, PAGE (1981) concluded from a study of Cd in food crops that the concentration of Cd in food crops depends upon a wide variety of interacting soil and plant factors, the most important soil factors being the total soil Cd content and the soil pH.

A fuller understanding of the processes by which metals are released from soil minerals by weathering is still required. More information is also needed about the forms, available to plants, in which they are held in the soil. One technique for studying metal species in soil is the use of sequential extraction with different reagents. Such an approach for studying sewage-sludge amended soils and also sediments has been reviewed by several authors (LAKE et al., 1984; LESCHBER et al., 1984; PICKERING, 1986).

Root samples are difficult to study not only because of the need to remove adhering soil particles, but also because most root systems are not easily separated from the soil under field conditions. The relative composition of soil and root usually determines whether the analysis for any particular element is capable of producing meaningful results.

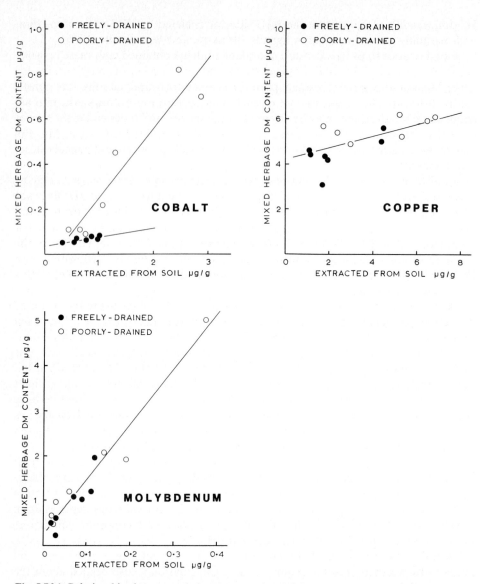

Fig. I.7f-1. Relationships between cobalt, copper, and molybdenum contents of mixed herbage dry matter (DM) from temporary grass/clover leys and the amounts of these elements extracted from the related soils by acetic acid for cobalt, EDTA for copper, and ammonium acetate for molybdenum. Each point is the average of 9 plant and 9 soil analyses, obtained by sampling 3 times each year for 3 successive years at farms on 7 different soils in northeast Scotland.

3 Soil pH, Drainage Status, and Organic Matter

The influence of these soil factors on the uptake of trace elements by plants can be clearly demonstrated by researches carried out in Scotland during the past 40 years. Similar general effects have been reported throughout the world by many other workers, but detailed variations arise from geological, climatic, agricultural, and cropping differences.

Soil reaction (pH) has a major effect on the uptake of many elements: some become more available to plants as pH decreases (e.g., Co, Mn, and Ni), others as pH increases (e.g., Mo and Se), whereas some may be only slightly affected (e.g., Cu). Table I.7f-1 shows how the grain content of oats (*Avena sativa*) was altered on a particular soil, a sandy loam, by increasing the natural pH with additions of lime and decreasing it with additions of iron and aluminum sulfates. The uptake pattern in this table is also typical for the leaf content of most plant species.

Under Scottish conditions, the drainage status of a soil is an important characteristic affecting the availability of naturally-occurring trace elements to plants. Reducing conditions in poorly-drained, gleyed soils increase the rate of weathering of ferromagnesian minerals, releasing, for instance, Co, Ni, and V (BERROW and MITCHELL, 1980) so that the readily extractable proportion may reach 20 percent of the total in comparison with 1 – 5 percent in a freely drained soil. Table I.7f-2 shows how

Table I.7f-1. Effect of Change in Soil pH on the Trace Element Content of Mature Grain of Field-grown Oats (*Avena sativa*); Results Expressed as mg/kg in Dry Matter

pH	Cu	Mn	Mo	Ni	Zn
4.5	2.4	109	0.21	4.4	55
5.0	2.6	86	0.35	3.2	62
5.5	2.8	57	0.41	2.0	48
6.0	3.3	24	0.66	0.8	36
6.5	3.2	11	0.68	0.5	31

Table I.7f-2. Effect of Drainage Status on Trace Element Availability and Plant Uptake; Results Expressed as mg/kg in Dry Matter. Cu and Mn Extracted by 0.05 mol/L EDTA, Mo by 1 mol/L Ammonium Acetate and Ni by 0.43 mol/L Acetic Acid

	Copper		Manganese		Molybdenum		Nickel	
	FD	PD	FD	PD	FD	PD	FD	PD
Soil extract	4.3	9.8	64	180	0.2	6.4	1.1	10
Mixed herbage	6.9	7.4	65	210	7.2	31	0.9	4.1
Ryegrass *Lolium* spp.	3.2	3.2	79	160	6.7	9.1	0.8	2.4
Cocksfoot *Dactylis glomerata*	4.1	4.4	123	190	6.0	23	1.0	2.3
Timothy *Phleum pratense*	3.7	4.8	66	170	3.5	15	0.6	1.4
Clover *Trifolium* spp.	8.9	10	31	65	4.6	48	1.3	6.2

FD freely drained, PD poorly drained

two adjacent fields with similar total soil contents can differ in their ability to supply trace elements to plants, as a consequence of their contrasting drainage status. The plant contents of Mn, Mo, and Ni are clearly associated with the amounts readily extracted from the soil. Soil drainage status can also affect the quantitative relationship between soil-extractable and plant content, as seen for Co in Fig. I.7f-1.

Effects of the organic matter content of a soil on plant uptake are complex and often indirect. Among important inter-related soil characteristics that can be altered when organic matter such as farmyard manure or sewage sludge is regularly added to soil are:

- its water-holding capacity,
- microbial activity, which is also strongly influenced by the quality of the organic matter, as measured by its C:N ratio,
- cation and anion exchange capacity,
- its ability to supply chelating ligands.

Poor drainage also favors the accumulation of organic matter in the surface horizons of the soil profile.

The relative significance of the soil factors pH, drainage status, and organic matter for the plant's uptake of particular metals depends on the form in which the root-accessible fraction is most likely to occur in the soil. Cation exchange processes are more important than organic chelation for Mn, while the opposite holds for Cu (BERROW and MITCHELL, 1980).

4 Diversity of Plant Metal Content

Plant species vary in their ability to extract metals from soil, as seen in Table I.7f-2: some species are notable accumulators of specific elements, for example, Co by *Crotalaria cobalticola*, Ni by *Alyssum* spp., and Se by *Astragalus* spp. Of more direct importance for agriculture and horticulture are the distribution of an element among different plant parts and variation of content with season and stage of growth

Table I.7f-3. Seasonal Distribution of Copper and Zinc in Oats (*Avena sativa*); Results Expressed as mg/kg in Dry Matter

	28 May		16 June		15 July		20 Aug		4 Sept	
	Cu	Zn	Cu	Zn	Cu	Zn	Cu	Zn	Cu	Zn
Whole tiller	6.4	41	3.0	25	1.5	15	1.3	16	1.5	16
Seed head	–	–			2.0	28	1.6	26	1.7	24
Stem	–	–	2.0	24	1.1	11	1.0	5	1.1	4
Leaf sheath	–	–			1.1	12	2.0	7	1.2	6
Leaf blade	6.4	41	4.0	27	2.5	14	1.4	20	1.7	15

(MITCHELL and BURRIDGE, 1980), since an appreciation of these factors is essential for the correct interpretation of diagnostic plant analysis. An illustration of such variations in a farm crop of oats (*Avena sativa*) is given in Table I.7f-3 for Cu and Zn. The much higher levels in young leaves and enhanced levels in the seed head are typical.

5 Plant Uptake of Metals from the Air

In plant uptake studies, one must always be aware that plant sample contents may not only be derived through the root system, but may also arise from various sources of superficial contamination. Above-ground parts are not only liable to become contaminated with soil during sampling, but are also exposed to atmospheric deposits throughout the whole period of growth. Deposition of metals from point-source emitters, such as smelters, can locally completely obscure the true uptake from the soil, while the diffuse dispersal of fine materials from motor vehicle exhausts, for example, can have very widespread effects (MUROZUMI et al., 1969). The nature and condition of plant surfaces, which vary with species, plant part, and stage of growth, have a marked effect on the retention of atmospheric deposits. Hair-covered or senescent leaves retain far more than smooth leaves with an intact waxy cuticle.

Much of the deposited particulate matter can be washed from the leaf surface and does not appear to enter the plant tissues. Any water-soluble forms deposited on leaves, however, may enter the plant. The well known efficacy of foliar applications by sprays containing water-soluble salts of Cu or Mn in correcting deficiencies of these elements in crops proves that they can be taken up through the leaf and become physiologically active. The average water-solubility of elements in deposition at seven sampling stations recorded by CAWSE (1980) varied considerably depending on the element, ranging from over 80 percent for Br, Se, and Zn to less than 25 percent for Al and Fe.

Elements associated with the very small particles that are typical of combustion sources (below 2 µm diameter) would be more susceptible than soil or clay particles to dissolution in acid rain water and could contribute to the true leaf contents. The elements As, Cd, Cu, Se, and Zn fall in this category. Pollutants such as F and S are readily taken up by plant leaves from the gaseous phase, for example, as HF or SO_2. Movement is mainly in the transpiration stream, so that there is little transfer to storage organs. It is possible that As, Hg, and Se, derived from gaseous atmospheric contaminants, would behave in a similar way, but clear experimental evidence for this is not yet available.

In a study of global atmospheric cycling of trace elements, LANTZY and MACKENZIE (1979) concluded that for As, Hg, and Se, and possibly some other atmophile metals, there were significant fluxes from the ocean surface to the atmosphere. Contributions from other natural sources to the atmosphere including those from exposed soil and rock surfaces, volcanism, and burning vegetation were also evaluated, as well as the principal anthropogenic emissions such as industrial activities and fossil fuel burning.

6 Essential and/or Toxic Metals

According to BOWEN (1979), a number of elements such as As(III), as well as Al, B, Be, Cd, Co, Cr(VI), Cu, I, Mo, Ni, Se(IV), and Tl can be harmful to crops, even at quite low concentration. Nevertheless, many of these elements are also essential for good growth. Mechanisms of toxicity may operate by altering the permeability of cell membranes by forming antimetabolites, by reacting with essential metabolites, or by substituting in part for other essential ions.

The elements considered in Part II of this book can be classified by their known essential functions or occasional toxic effects in plants and animals. For a number of these metals, for instance, Ag, Au, Be, Bi, Pt-metals, Sb, Sn, Te, Ti, Tl, and Zr, the levels present in uncontaminated soils in forms available to plants are so low that the absence of suitable analytical techniques has limited studies of soil-plant relationships. Their behavior can only be anticipated, in a general way, by comparison with the most chemically similar elements that have been more intensively studied. Caution must be exercised in making such comparisons, however, because the elements that have received most study are naturally those having a biological function. This must be borne in mind when comparing, for instance, the behavior of Cd with Zn or of Ag with Cu. A summary of current information on essentiality and toxicity is given in Table I.7f-4.

Further information on tolerable soil levels and on the need to differentiate between soil total contents and the soluble fraction is given in Chapters I.7c, I.7d, I.7g, and Part II; see also HÄNI (1977).

Table I.7f-4. Classification of Elements Discussed in Part II of this Book

	E_p	E_a	T_p	T_a[a]		E_p	E_a	T_p	T_a[a]
Aluminum			*		Manganese	*	*	*	*
Antimony					Mercury			*	*
Arsenic			*	*	Molybdenum	*	*	*	*
Beryllium			*	*	Nickel			*	*
Bismuth					Niobium				
Cadmium			*	*	Pt-metals			*	
Chromium		*			Selenium		*	*	*
Cobalt		*			Silver			*	
Copper	*	*	*	*	Tantalum				
Gallium				*	Tellurium				
Germanium					Thallium			*	*
Gold					Tin			*	*
Indium				*	Titanium				
Iron	*	*			Tungsten			*	*
Lanthanum					Uranium				*
Lead			*	*	Vanadium	*	*	*	*
Lithium				*	Zinc	*	*	*	*
Magnesium	*	*			Zirconium				*

[a] E_p essential to plants, E_a essential to animals, T_p toxic to plants, T_a toxic to animals

7 Secondary Actions of Plant Uptake

Most metals when present at abnormally high available levels in soil can cause visible injury to plants, inhibit plant growth by damaging the roots or lead to crop failure. Unfortunately, healthy plants can sometimes contain metals at levels within the plant's range of tolerance, but at levels potentially deleterious to grazing animals or to human beings in the case of crops grown as foodstuff. A good example of such hidden risks occurs with Mo (SUTTLE, 1975) because contents of 5–10 mg Mo/kg dry matter in herbage can induce Cu-deficiency in cattle, although the plants will tolerate at least ten times as much. Likewise, Cd can enter plants from fertilizer sources such as phosphates and sewage sludge and, while having no effect on plant growth, can significantly increase animal intake of this harmful element.

The same problem also arises with human food (cereal products, vegetables, fungi such as mushrooms, fruit, fruit-juices). For this reason, limiting values for heavy metals in certain food-stuffs are generally laid down by national regulations. Detailed information for various elements is given in other chapters of this book. In the case of vines, Cd for instance, although present in the leaves is at very low levels in the grapes. In fact fruit, fruit-juices, and wine are not at very high risk (MOHR, 1980; GYSI, 1980; KLOKE, 1980).

References

Berrow, M. L., and Burridge, J. C. (1979), *Sources and Distribution of Trace Elements in Soils and Related Crops,* in: *Proceedings of the International Conference on Management and Control of Heavy Metals in the Environment, London,* pp. 304–311. CEP Consultants Ltd., Edinburgh.

Berrow, M. L., and Burridge, J. C. (1980), *Trace Element Levels in Soils: Effects of Sewage Sludge,* in: *Inorganic Pollution and Agriculture, Proceedings of Conference, London, 1977.* MAFF/ADAS Reference Book 326, pp. 159–183. HMSO, London.

Berrow, M. L., and Burridge, J. C. (1981), *Persistence of Metals in Available Form in Sewage Sludge Treated Soils under Field Conditions,* in: *Proceedings of International Conference on Heavy Metals in the Environment, Amsterdam,* pp. 202–205. CEP Consultants Ltd., Edinburgh.

Berrow, M. L., and Mitchell, R. L. (1980), *Location of Trace Elements in Soil Profiles: Total and Extractable Contents of Individual Horizons. Trans. R. Soc. Edinburgh Earth Sci.* **71**, 103–121.

Bowen, H. J. M. (1979), in: *Environmental Chemistry of the Elements,* p. 333. Academic Press, London.

CAST (Council for Agricultural Science and Technology) (1980), *Effects of Sewage Sludge on the Cadmium and Zinc Content of Crops.* Report No. 83, p. 77. Ames, Iowa, ISSN 0194-4088.

Cawse, P. A. (1980), *Deposition of Trace Elements from the Atmosphere in the UK,* in: *Inorganic Pollution and Agriculture, Proceedings of Conference, London, 1977.* MAFF/ADAS Reference Book 326, pp. 22–46. HMSO, London.

Cox, F. R., and Kamprath, E. J. (1972), *Micronutrient Soil Tests,* in: Mortvedt, J. J., Giordano, P. M., and Lindsay, W. L. (eds.): *Micronutrients in Agriculture. Proceedings Symposium at Muscle Shoals, Alabama,* pp. 289–317. Soil Science Society of America, Madison, Wisconsin.

Crooke, W. M. (1964), *The Measurement of Cation-exchange Capacity of Plant Roots. Plant Soil* **21**, 43–49.

Gysi, C. (1980), in: *GDI-Proceedings: Die Verwendung von Müll- und Müllklärschlamm-Komposten in der Landwirtschaft,* pp. 29–43. Verlag GDI-Institut, Rüschlikon-Zürich.

Häni, E. (1977), Lecture at the *Informationstagung Klärschlamm-Verwertung in der Landwirtschaft, Zollikofen,* Switzerland.

Haq, A. U., Bates, T. E., and Soon, Y. K. (1980), *Comparison of Extractants for Plant-available Zinc, Cadmium, Nickel and Copper in Contaminated Soils. Soil Sci. Soc. Am. J. 44,* 772–777.

Kloke, A. (1980), in: *GDI-Proceedings: Die Verwendung von Müll- und Müllklärschlamm-Komposten in der Landwirtschaft,* pp. 58–87. Verlag GDI-Institut, Rüschlikon-Zürich.

Lake, D. L., Kirk, P. W. W., and Lester, J. N. (1984), *Fractionation, Characterization and Speciation of Heavy Metals in Sewage Sludge and Sludge-amended Soils* (A review). *J. Environ. Qual. 13,* 175–183.

Lantzy, R. J., and Mackenzie, F. T. (1979), *Atmospheric Trace Metals: Global Cycles and Assessment of Man's Impact. Geochim. Cosmochim. Acta 43,* 511–525.

Leschber, R., Davis, R. D., and L'Hermite, P. (eds.) (1984), *Chemical Methods for Assessing Bioavailable Metals in Sludges and Soils. Proceedings CEC Seminar, Münster,* April 1984. Elsevier, London, 96 pages.

Lindsay, W. L., and Norvell, W. A. (1978), *Development of a DTPA Soil Test for Zinc, Iron, Manganese and Copper. Soil Sci. Soc. Am. J. 42,* 421–428.

Linehan, D. J., Sinclair, A. H., and Mitchell, M. C. (1985), *Mobilisation of Cu, Mn and Zn in the Soil Solutions of Barley Rhizospheres. Plant Soil 86,* 147–149.

Merckx, R., van Ginkel, J. H., Sinnaeve, J., and Cremers, A. (1986a), *Plant-induced Changes in the Rhizosphere of Maize and Wheat I. Production and Turnover of Root-derived Material in the Rhizosphere of Maize and Wheat. Plant Soil 96,* 85–93.

Merckx, R., van Ginkel, J. H., Sinnaeve, J., and Cremers, A. (1986b), *Plant-induced Changes in the Rhizosphere of Maize and Wheat II. Complexation of Cobalt, Zinc and Manganese in the Rhizosphere of Maize and Wheat. Plant Soil 96,* 95–107.

Mitchell, R. E. (1964), *Trace Elements in Soils,* in: Bear, F. E. (ed.): *Chemistry of the Soil,* 2nd Ed. *ACS Monograph 160,* 320–368. Reinhold, New York.

Mitchell, R. L., and Burridge, J. C. (1980), *Trace Elements in Soils and Crops,* in: *Environmental Geochemistry and Health. A Royal Society Discussion, London,* March 1978, pp. 15–24. The Royal Society, London.

Mohr, H. D. (1980), *Heavy Metal Content of Roots and Shoots of Vines (Vitis vinifera L.) after Fertilization with Garbage-Sewage-Sludge Compost. Z. Pflanzenernähr. Bodenkd. 143,* 129–139.

Murozumi, M., Chow, T. J., and Patterson, C. (1969), *Chemical Concentrations of Pollutant Lead Aerosols, Terrestrial Dust and Sea Salts in Greenland and Antarctic Snow Strata. Geochim. Cosmochim. Acta 33,* 1247–1294.

Page, A. L. (1981), *Cadmium in Soils and its Accumulation by Food Crops,* in: *Proceedings of International Conference on Heavy Metals in the Environment, Amsterdam,* pp. 206–213. CEP Consultants Ltd., Edinburgh.

Pickering, W. F. (1981), *Selective Chemical Extraction of Soil Components and Bound Metal Species. CRC Rev. Anal. Chem. 12,* 233–266.

Pickering, W. F. (1986), *Metal Ion Speciation – Soils and Sediments* (a Review). *Ore Geol. Rev. 1,* 83–146.

Sauerbeck, D. R., and Styperek, P. (1983), *Predicting the Cadmium Availability from Different Soils by $CaCl_2$ Extraction,* in: L'Hermite, P., and Ott, H. (eds.): *Processing and Use of Sewage Sludge. Proceedings 3rd International Symposium, Brighton,* September, pp. 431–434. D. Reidel, Dordrecht.

Suttle, N. F. (1975), *Trace Element Interactions in Animals,* in: Nicholas, D. J. D., and Egan, A. R. (eds.): *Trace Elements in Soil-Plant-Animal Systems. Proceedings Symposium at Waite Institute, South Australia,* pp. 271–289. Academic Press, New York.

Additional Recommended Literature

Angehrn-Bettinazzi, C., Thöni, L., and Hertz, J. (1989), *An Attempt to Evaluate Some Factors Affecting the Heavy Metal Accumulation in a Forest Stand*. Environ. Anal. Chem. 35 (2), 69–79.

Baroccio, A., and Dottori, A. (1984), *Tolerance of Cereal Cultivars to Heavy Metals*. Proceedings VIth International Colloquium Plant Nutrition, p. 35.

Berrow, M. L., and Burridge, J. C. (1985), *A Comparison of Long-term Effects of Similar Sewage Sludge Treatments on two Different Soils*. Proceedings International Conference on Heavy Metals in the Environment, Athens, Vol. 1., p. 345. CEP Consultants Ltd., Edinburgh.

Berrow, M. L., and Ure, A. M. (1986), *Trace Element Distribution and Mobilisation in Scottish Soils with Particular Reference to Cobalt, Copper and Molybdenum*. Environ. Geochem. Health 8, 19.

Brown R. M., Pickford, C. J., and Davison, W. L. (1984), *Speciation of Metals in Soils*. Environ. Anal. Chem. 18 (3), 135–141.

Burton, M. A. S. (1986), *Biological Monitoring of Environmental Contaminants (Plants)*. MARC Report No. 32, London.

Casler, M. D., Collins, M., and Reich, J. M. (1987), *Location, Year, Maturity, and Alfalfa Competition Effects on Mineral Element Concentrations in Smooth Bromegrass*. Agron. J. 79, 774.

Cataldo, D. A., Wildung, R. E., and Garland, T. R. (1987), *Speciation of Trace Inorganic Contaminants in Plants and Bioavailability to Animals: an Overview*. J. Environ. Qual. 16, 289.

Chaney, R. L. (1988), *Metal Speciation and Interactions among Elements Affect Trace Element Transfer in Agricultural and Environmental Food-Chains*, in: Kramer, J. R., and Allen, H. E. (eds.): Metal Speciation, pp. 219–260. Lewis Publishers, Chelsey, Michigan.

Chang, A. C., Page, A. L., and Warneke, J. E. (1987), *Long-term Sludge Applications on Cadmium and Zinc Accumulation in Swiss Chard and Radish*. J. Environ. Qual. 16, 217.

Coquery, M., and Stokes, P. M. (1989), *Effect of Sediment Chemistry on the Bioavailability of Trace Metals to Aquatic Microphytes*. Proceedings 7th International Conference on Heavy Metals in the Environment, Geneva, Vol. 2, pp. 11–14. CEP Consultants Ltd., Edinburgh.

Dressler, R. L., Storm, G. L., Tzilkowski, W. M., and Sopper, W. E. (1986), *Heavy Metals in Cottontail Rabbits on Mined Lands Treated with Sewage Sludge*. J. Environ. Qual. 15, 278.

Driel, W. van, Smilde, K. W., and Luit, B., van (1985), *Comparison of the Heavy-Metal Uptake of Cyperus esculentes and of Agronomic Plants Grown on Contaminated Dutch Sediments*. Misc. Paper, US Army Corps of Engineers, 67 pages.

Godbold, D. L., Jentschke, G., Kettner, C., and Hüttermann, A. (1989), *Element Distribution in Non-mycorrhizal and Mycorrhizal Spruce Roots: Influence of Heavy Metals*. Proceedings 7th International Conference on Heavy Metals in the Environment, Geneva, Vol. 2, pp. 170–173. CEP Consultants Ltd., Edinburgh.

Gupta, S. K., Häni, H., Santschi, E., and Stadelmann, F. X. (1987), *The Effect of Graded Doses of Nickel on the Yield, the Nickel Content of Lettuce and the Soil Respiration*. Toxicol. Environ. Chem. 14 (1+2), 1–9.

Heckman, J. R., Angle, J. S., and Chaney, R. L. (1987), *Residual Effects of Sewage Sludge on Soybean: I. Accumulation of Heavy Metals*. J. Environ. Qual. 16, 113.

Herzig, R., Liebendörfer, L., Urech, M., Ammann, Kl., Cuecheva, M., and Landolt, W. (1989), *Passive Monitoring with Lichens as a Part of an Integrated Biological Measuring System for Monitoring Air Pollution in Switzerland*. Environ. Anal. Chem. 35 (1), 43–57.

Iyengar, S. S., Martens, D. C., and Miller, W. P. (1981), *Distribution and Plant Availability of Soil Zinc Fractions*. Soil Sci. Soc. Am. J. 45, 735.

Kopp, P., Oestling, O., and Burkart, W. (1989), *Availability to and Uptake by Plants of Radionuclides under Different Environmental Conditions*. Toxicol. Environ. Chem. 23 (1–4), 53–63.

Lamersdorf, N. P. (1989), *The Behavior of Lead and Cadmium in the Intensive Rooting Zone of Acid Spruce Forest Soil*. Toxicol. Environ. Chem. 18 (4), 239–247.

Lehn, H., and Bopp, M. (1987), *Prediction of Heavy-Metal Concentrations in Mature Plants by Chemical Analysis of Seedlings*. Plant Soil *101*, 9.

Liebendörfer, L., Herzig, R., Urech, M., and Ammann, K. (1988), *Evaluation and Calibration of the Swiss Lichen Indicated Method with Important Air Pollutants* (in German). Staub Reinhalt. Luft *48*, 233–238.

Linehan, D. J., Sinclair, A. H., and Mitchell, M. C. (1989), *Seasonal Changes in Cu, Mn, Zn and Co Concentrations in Soil Solution in the Root Zone of Barley (Hordeum vulgare) L)*. J. Soil Sci. *40*, 103.

Miller, J. E., Hassett, J. J., and Koeppe, D. E. (1977), *Interactions of Lead and Cadmium on Metal Uptake and Growth of Corn Plants*. J. Environ. Qual. *6*, 18.

Puddu, A., Pettine, M., La Noce, T., And Pagnotta, R. (1989), *Toxic Effects of Organotin Compounds on Marine Phytoplankton*. Proceedings 7th International Conference on Heavy Metals in the Environment, Geneva, Vol. 2, pp.166–169. CEP Consultants Ltd., Edinburgh.

Reith, J. W. S., Burridge, J. C., Berrow, M. L., and Caldwell, K. S. (1984), *Effects of Fertilisers on the Contents of Copper and Molybdenum in Herbage Cut for Conservation*. J. Sci. Food Agric. *35*, 245.

Rüegg, J. (1989), *A New Biotest with Root Fungi* (in German). Neue Zürcher Zeitung, Forschung und Technik, No. 117, p. 77 (24 May), Zürich.

Shariatpanahi, M., and Anderson, A. C. (1987), *Accumulation of Cadmium, Mercury and Lead by Vegetables Following Long-term Land Application of Waste-water*. Sci. Total Environ. *52*, 41.

Sieghardt, H. (1987), *Heavy Metal- and Nutrient Contents of Plants and Soil Samples from Metalliferous Waste Dumps in Bleiberg (Austria), I. Herbaceous Plants*. Z. Pflanzenernähr. Bodenkd. *150*, 129.

Sikora, F. J., and Wolt, J. (1986), *Effect of Cadmium- and Zinc-treated Sludge on Yield and Cadmium-Zinc Uptake of Corn*. J. Environ. Qual. *15*, 341.

Sims, J. T. (1986), *Soil pH Effects on the Distribution and Plant Availability of Manganese, Copper and Zinc*. Soil Sci. Soc. Am. J. *50*, 367.

Smith, St., Peterson, P. J., and Kwan, K. H. M. (1989), *Chromium Accumulation, Transport and Toxicity in Plants*. Toxicol. Environ. Chem., in press.

Tiller, K. G. (1989), *Heavy Metals in Soils and their Environmental Significance*. Adv. Soil Sci. *9*, 113.

Valdares, J. M., Gal, M., Mingelgrin, U., and Page, A. L. (1983), *Some Heavy Metals in Soils Treated with Sewage Sludge: their Effect on Yield and their Uptake by Plants*. J. Environ. Qual. *12*, 49.

Wassermann, J. C. (1989), *Zn, Cu, Fe, and Mo Concentrations in Cell Wall from Eelgrass (Zostera noltii)*. Proceedings 7th International Conference on Heavy Metals in the Environment, Geneva, Vol. 2, pp 5–10. CEP Consultants Ltd., Edinburgh.

Wiersma, D., van Goor, B. J., and van der Veen, N. G. (1986), *Cadmium, Lead, Mercury and Arsenic Concentrations in Crops and Corresponding Soils in the Netherlands*. J. Agric. Food Chem. *34*, 1067.

Yang, J. S. (1989), *The Cooperative Chemistries of Platinum Group Metals and Their Periodic Neighbours in Marine Microphytes*. Proceedings 7th International Conference on Heavy Metals in the Environment, Geneva, Vol. 2, pp. 1–4. CEP Consultants Ltd., Edinburgh.

I.7g Metal Tolerance in Higher Plants

CAROLINE L. SCHULTZ and THOMAS C. HUTCHINSON
Toronto, Ontario, Canada

1 Occurrence of Vegetation on "Toxic" Sites

Elevated levels of metal and metalloid compounds such as those of lead, copper, nickel, chromium, zinc, cobalt, cadmium, arsenic, and selenium often occur naturally in soils overlying ore bodies and mineralizations. They also occur in the wastes from metal extraction industries, especially around mining sites, and in surface soils and lake sediments around base-metal smelters. The concentrations of these potentially phytotoxic elements are at levels which affect most higher plants. The long-term poisoning of soils such that most or all plant species are unable to grow is apparent in places such as Ducktown in Tennessee, where vegetation has been virtually absent on mine deposits for the past 70–80 years because of very high copper levels in surface deposits. Where vegetation is present, plant cover and species diversity are considerably less than on adjacent uncontaminated soils, even on long-abandoned sites such as old Roman lead mine workings in Britain, abandoned for 1500 years, and old lead, zinc and copper mines in Germany, Yugoslavia, Czechoslovakia, and Britain worked over 300 years ago.

Around mines the soil contamination is generally high, with an abrupt spatial transition to low metal levels at the edge of the mine wastes. This rapid spatial change in metal contamination is often reflected in the sharp boundary between metal-tolerant and non-tolerant populations of plant species, suggestive of strong selection pressure. BRADSHAW (1975) noted that "the localization of metal tolerance to contaminated areas is very remarkable. Tolerance is found at a very high frequency in populations growing on metalliferous wastes and other metal contaminated soils, even when the population is very small and surrounded by non-tolerant populations growing on ordinary soil". He reports that the grass *Anthoxanthum odoratum* growing on spoil at a zinc mine in Trelogan, Wales, occurs continuously across the mine-agricultural boundary, but that the degree of zinc tolerance changes dramatically over a few meters from a fully tolerant to a fully non-tolerant population (ANTONOVICS and BRADSHAW, 1970).

The genetic basis of tolerance appears to be normal. A number of genes with additive and dominance effects appear to be involved (GARTSIDE and MCNEILLY, 1974). While the degree of dominance is variable, tolerance has generally been found to be dominant to non-tolerance. It should be noted, however, that we are dealing in comparisons rather than in absolutes when we speak of tolerance. Tolerance is, after all, dependent upon specific experimental conditions and on the way in which it is measured. The degree of tolerance a particular population of a plant species shows to a metal is also a direct function of the chemical speciation of the metal in the environment and its bioavailability (see Chapter I.7c by HÄNI on availability of metals from sewage sludge). In soils, critical factors influencing bioavailability include pH, redox potential, organic matter content, physical composition of the soil, occurrence of iron and manganese compounds, and the availability of base cations which compete for uptake sites at the root surface, e.g., calcium and magnesium.

2 Examples of Extreme Tolerances

Despite the limitations and need for qualifications in expressing degrees of tolerance, some examples obviously demonstrate remarkable extremes. For example, at a vegetated bog BOYLE (1971) found copper concentrations of up to 10% dry weight in surface peat samples at a site called the Tantramar Copper Swamp in New Brunswick, Canada. The source of contamination was artesian spring wells having high copper concentrations.

The labiate (mint family) *Becium homblei* occurs on copper deposits in Zaire, Zimbabwe, and Zambia. CANNON (1960) reported that this species was responsible for the discovery of copper deposits in Zambia and Zimbabwe, where its occurrence was believed confined to soils with >1000 ppm copper. REILLY (1967) and REILLY and REILLY (1973) described *B. homblei* as a cuprophile, tolerant to greater than 70000 ppm (d.w.) copper in soil, and accumulating up to 17% of copper in the leaves, organically bound to the cell walls.

Copper mosses were originally described from Scandinavia and later from Alaska, where they were used in mineral prospecting (PERSSON 1948; SHACKLETTE, 1965). Indeed, many plant species with high but specific metal tolerances have been used successfully as indicators of ore deposits, especially in the USSR, USA, South Africa, Africa, and Scandinavia. The metals for which good "indicator" vegetation occurs include copper, nickel, lead, chromium, zinc, selenium, mercury, arsenic, cobalt, and uranium. Botanical prospecting and biogeochemical prospecting are a direct application of knowledge of specific metal tolerances in native floras. The field violet *Viola calaminaria* is largely confined to sites of high zinc and lead in the European alps, growing well on soil with levels in excess of 10000 ppm Zn. Nickel tolerance in its most extreme form is shown by the nickel accumulators *Sibertia acuminata, Psychotria dovarrei*, and *Hybanthus austrocaledonicus* which are found in nickel-chromium rich serpentine soils in New Caledonia, where they have leaf nickel concentrations in excess of 10000 ppm dry weight. *S. acuminata* sap (latex) contained an astonishing 167000 ppm (17%) of nickel (LEE et al., 1978).

In some cases these metal accumulator species can concentrate metals in aboveground parts of the plant from soils with either high or low metal concentrations. Some of the selenium-accumulator species of the genus *Astragalus*, a legume family, do this in quite low soil selenium areas, posing a danger to livestock which graze them in the USA and Canadian prairies. PORTER and PETERSON (1975) reported

Table I.7g-1. Heavy Metal Content (μmol/g dry weight) of a Naturally Metalliferous Soil and of the Leaves of Heavy Metal Tolerant Plants at the End of the Vegetation Period. Locality: Breinig (Germany). (From ERNST, 1975)

	Zn	Pb	Cd	Cu	Total
Concentration in the Soil	383.0	96.50	15.52	1.38	496.40
Concentration in the Leaves of:					
Thlaspi alpestre	159.0	8.21	4.83	1.26	173.30
Minuartia verna	151.3	6.52	0.65	1.74	160.21
Arneria calaminaria	112.8	11.60	1.10	0.79	126.29
Silene cucubalus	40.8	0.29	0.02	0.32	41.43
Lotus corniculatus	30.4	0.05	0.02	0.27	30.74
Festuca ovina	28.3	0.97	0.13	0.19	29.59
Campanula rotundifolia	24.9	4.82	0.95	0.40	31.07
Thymus serpyllum	24.5	4.05	0.30	0.63	29.48
Rumex acetosa	21.4	2.12	0.16	0.32	24.00
Cladonia rangiformis	19.9	7.96	0.41	1.58	29.85
Euphrasia stricta	16.8	0.83	0.40	0.27	18.30
Achillea millefolium	14.4	1.38	0.02	0.40	16.20
Viola calaminaria	8.9	0.19	0.02	0.16	9.27

arsenic levels in the herb *Jasione montana* of 6640 ppm, in the heather *Calluna vulgaris* of 4130 ppm As, and in the grass *Agrostis tenuis* of 3470 ppm, at old arsenic mine tips in Cornwall and Devon in England. PETERSON (1971) also reported on a chromium accumulator *Pimelea suteri* containing 2.6% Cr in its foliage.

It is apparent that plant species differ greatly in both the extent to which they can grow and reproduce in the presence of elevated metal concentrations, and in the degree to which, on the same site, they take up and translocate the metals from the soil. Table I.7g-1 from the work of ERNST (1975) illustrates this point very clearly. It shows that overall metal levels in the leaves differed by almost 20 fold between the species growing at a naturally contaminated site. It also seems apparent that some species, such as the zinc-tolerant *Viola calaminaria*, are effective excluders of zinc, lead, and cadmium from translocation to their foliage despite the high soil concentrations. BAKER (1981) points out, however, that often this exclusion mechanism breaks down once a critical soil metal level has been reached, and unrestricted transport of the metal then occurs.

3 The Physiological Basis for Tolerance

Earlier views considered metal tolerance in plants (and animals) to be specific, such that tolerance to one metal would not confer tolerance to another (ANTONOVICS et al. 1971). However, more recent evidence shows that, in some instances, selection for tolerance to one metal enhances tolerance to another. Populations of the zinc-tolerant grass *Agrostis tenuis* in Wales were also nickel-tolerant despite the absence of nickel from the mine soil (GREGORY and BRADSHAW, 1965). Similar co-tolerances have been found in copper- and nickel-tolerant *Deschampsis cespitosa* from Sudbury, Canada (COX and HUTCHINSON, 1981), and for zinc and lead in copper-tolerant *Mimulus guttatus* (ALLEN and SHEPPARD, 1971). Indeed, in the case of the Sudbury-area population of the grass *D. cespitosa*, a remarkable degree of multiple and co-tolerances was reported. The plants were simultaneously tolerant to copper, nickel, cobalt, aluminum, arsenic, and silver, all of which are elevated in the soil. They were also tolerant to zinc, lead, and cadmium which are not elevated (COX and HUTCHINSON, 1981).

As there is no evidence for the exclusion of metals from metal-tolerant ecotypes (ERNST, 1975) and tissue metal levels are generally considerably elevated compared to those on normal soils, there must be physiological and biochemical mechanisms operating within the plant which prevents or reduces metal toxicity. Tolerance mechanisms which have been proposed include binding of metal to cell wall material (TURNER and MARSHALL, 1971, 1972), possession of metal-tolerant enzymes (MATHYS, 1975; COX and THURMAN, 1978), complexation with organic acids with subsequent removal to the vacuole (MATHYS, 1977; GODBOLD et al., 1983, 1984), and binding to specialized thiol-rich proteins or phytochelatins (RAUSER, 1984; LOLKEMA et al., 1984; GRILL et al., 1987). Although there is some evidence for each of these mechanisms, no one single has been found to account fully for tolerance (THURMAN and COLLINS, 1983). It is likely, therefore, that tolerance is achieved through the operation of a number of these mechanisms.

4 The Potential for Evolution of Metal Tolerance
(see also Chapter I.7h, Section 5)

To colonize metalliferous soils, species must generally evolve metal-tolerant races. Metal-tolerant populations must originate from populations on normal soil which have not been subject to metal toxicity. WALLEY et al. (1974) found that when seed from a non-tolerant pasture population of *Agrostis tenuis* was sown on varying proportions of soil from a copper mine diluted with potting compost, the number of survivors decreased with increasing copper contamination. After a six-month period on slightly ameliorated soil about 0.4 percent of the seedlings survived.

The low species diversity on mine and smelter wastes suggests that not all species are able to evolve metal-tolerant ecotypes. The evolution of tolerant populations is dependent on the presence of suitable variation upon which natural selection can act. By screening seed of normal populations of nine species on copper mine soil diluted with potting compost, GARTSIDE and MCNEILLY (1974) demonstrated that only two species, *Agrostis tenuis* and *Dactylis glomerata*, were capable of producing fully tolerant survivors after 16 weeks (tolerance was compared to that of *Agrostis tenuis* from a copper mine site) (GARTSIDE and MCNEILLY, 1974). Four other species, *Arrhenatherum elatius, Poa trivialis, Lolium perenne,* and *Cynosurus cristatus*, had survivors with intermediate tolerance, while the remaining three species *Plantago lanceolata, Trifolium repens,* and *Anthoxanthum odoratum* left no survivors. Selection occurred at both the seedling and the germination stages in all cases. Only *A. tenuis* is a known colonizer of copper contaminated wastes. *A. odoratum* and *P. lanceolata* are frequent colonizers of lead and zinc wastes, indicating that there is no relationship between tolerance to these metals and copper in these species. The general absence of *D. glomerata* from mine spoils in the U.K. is clearly not related to its capacity to evolve copper tolerance, but must be attributed to some other factor such as its high nutrient requirement. Mine spoils are almost inevitably low in nitrogen and phosphorus supply and are also very drought prone. Legumes are generally not frequent colonizers of metal-enriched soils in Europe, which may be due to insufficient genetic variability or to sensitivity of the *Rhizobium*/legume symbiosis. There is considerable interest in finding legume species which can grow on metalliferous wastes and fix nitrogen effectively (DAY and LUDECKE, 1981). However, two legumes, *Lupinus bicolor* and *Lotus purshianus*, both capable of fixing nitrogen, have been found on copper mine waste in California (WU and KRUCKENBERG, 1985).

The variability in the ability to evolve metal tolerance which has been demonstrated among species may also be population dependent. Screening of seed from nine populations of *Agrostis capillaris*, a known colonizer of lead, zinc, and copper wastes, on solutions containing cadmium, lead, copper, nickel, or zinc, showed that certain populations contained high frequencies of individuals capable of tolerating certain metals while other had no tolerant individuals (SYMEONIDIS et al., 1985). In small isolated populations the genes for tolerance may have been lost through random genetic drift or may not have been present in the founding individuals. Thus, not all populations of a species have the same evolutionary potential, and colonization by a species in one situation does not necessarily imply that this is a universal phenomenon for the species. *Agrostis tenuis* which is frequently found on lead mines, does not colonize lead mines around Tyndrum in Scotland even though it is common in the surrounding pastures. The mines are colonized exclusively by *A. canina* (CRAIG, 1976). It is nonetheless the case that both in Europe and North America members of the genera *Agrostis, Deschampsia, Mimulus, Silene, Vicia, Thlaspi,* and *Viola* occur in metal-contaminated mine and smelter sites, suggesting intrinsic genetic features for metal tolerance.

Selection for metal tolerance occurs at the germination, seedling, and adult stages but might also occur at the gametophyte stage. SEARCY and MULCAHY (1985a) found that selection appeared to occur during pollen development in *Silene dioica* and *Mimulus guttatus*. Although pollen tube growth was not affected, in terms of the number of fertilizations and of viable seed, the relative success of pollen from non-tolerant parents was reduced in the presence of copper or zinc (SEARCY and MULCAHY, 1985b). Although there is no published evidence, it is conceivable that some species which are capable of producing metal-tolerant individuals do not colonize mine wastes due to an adverse effect of metals on reproductive processes.

Naturally metalliferous soils provide situations in which evolution has continued uninterrupted over long periods of time. A great number of endemic metallophytes are found on extensive metalliferous soils in Africa (WILD and BRADSHAW, 1977). The copper flower, *Becium homblei*, which occurs almost exclusively on cupriferous soils, has a close relative, *Becium obovatum*, which occurs on uncontaminated soils close by. True endemics, whose closest relatives are a considerable distance away, occur only in very large and physically complex areas such as the serpentine area in Zimbabwe and the Katanga copper area in Zaire. This suggests that either geographic isolation has resulted in speciation or that the size and complexity of such areas has provided local environments into which the newly formed endemic species can retreat in times of environmental change (WILD and BRADSHAW, 1977).

The endemic heavy metal and serpentine species in South Central Africa must have begun their evolution soon after the angiosperms emerged as a group. In Europe most of the species on metalliferous soils are ecotypes or subspecies of species growing on normal soils. Only *Viola calaminaria* and *Thlaspi calaminare* (previously called *Thlaspi alpestre* ssp. *calaminare*) are distinct species which grow exclusively on calamine soils.

For the most part metal-tolerant populations are the result of natural selection acting on normal populations resulting in ecotypic differentiation. However, certain cases have been reported where populations occurring on metal-contaminated soils do not have a greater tolerance than populations on uncontaminated soils. Species which appear to have this "constitutional" tolerance are *Andropogon virginicus* (GIBSON and RISSER, 1982), *Typha latifolia* (MCNAUGHTON et al., 1974; TAYLOR and CROWDER, 1984), *Cynodon dactylon* (WU and ANTONOVICS, 1976), and *Thlaspi goesingense* (REEVES and BAKER, 1984). Although this absence of local ecotypic differentiation may be due to reduced metal uptake because of physical and chemical barriers, as may be the case in *T. latifolia* where submerged roots grow in a reducing environment, in other species, such as *A. virginicus*, high metal levels have been found in plant tissues, suggesting that the populations were at least subject to the potential toxic effects of the metals.

In conclusion, it seems that the occurrence of natural or anthropogenic sites of high toxic metal concentrations has acted as a powerful selective factor. While most species have been unable to tolerate the physiological and biochemical stresses imposed, a few have responded with the evolution of metal-tolerant populations. In some cases the tolerance is of an entire species. Metal tolerance appears to be genetically controlled in a fairly simple Mendelian manner, though the specific mechanisms involved may be several and additive.

References

Allen, W.R., and Sheppard, P.M. (1971), *Proc. R. Soc. London B 177*, 177–196.
Antonovics, J., and Bradshaw, A.D. (1970), *Heredity 25*, 349–362.
Antonovics, J., Bradshaw, A.D., and Turner, R.G. (1971), *Adv. Ecol. Res. 7*, 1–85.
Baker, A.J.M. (1981), *J. Plant Nutr. 3*, 643–654.
Boyle, R.W. (1971), *Can. Min. Metall. Bull.*, 46–50.
Bradshaw, A.D. (1975), in: Hutchinson, T.C. (ed.), *International Conference on Heavy Metals in the Environment*, Vol. 2, Part 2, pp. 599–622. Publisher: University Toronto.
Cannon, H (1960), *Science 132*, 591–598.
Cox, R.M., and Hutchinson, T.C. (1981), *J. Plant Nutr. 3*, 731–741.

Cox, R.M., and Thurman, D. (1978), *New Phytol. 80*, 17–22.
Craig, G.C. (1976), in: Antonovics, J. (1975), in: Hutchinson, T.C. (ed.), *International Conference on Heavy Metals in the Environment*, Vol. 2, Part 1, pp. 169–186. Publisher: University Toronto.
Day, A.D., and Ludcke, L. (1981), *J. Environ. Qual. 2*, 314–315.
Ernst, W.H.O. (1975), in: Hutchinson, T.C. (ed.), *International Conference on Heavy Metals in the Environment*, Vol. 2, Part 1, pp. 121–136. Publisher: University Toronto.
Gartside, D.W., and McNeilly, T. (1974), *Heredity 32*, 287–297.
Gibson, D.J., and Risser, P.G. (1982), *New Phytol. 92*, 589–599.
Godbold, D.L., Horst, W.J., Marschner, H., Collins, J.C., and Thurman, D.A. (1983), *Z. Pflanzenphysiol. 112*, 315–324.
Godbold, D.L., Horst, W.J., Collins, J.C., Thurman, D.A., and Marschner, H. (1984), *J. Plant Physiol. 116*, 59–69.
Gregory, R.P.G., and Bradshaw, A.D. (1965), *New Phytol. 64*, 131–143.
Grill E., Winnacker, E.L., and Zenk, M.H. (1987), *Proc. Natl. Acad. Sci. USA 84*, 439–443.
Lee, J., Reeves, R.D., Brooks, R.R., and Jaffré, T. (1978), *Phytochemistry 17*, 1033–1035.
Lolkema, P.C., Donker, M.H., Schouten, A.J., and Ernst, W.H.O. (1984), *Planta 162*, 174–179.
Mathys, W. (1975), *Physiol. Plant 33*, 161–165.
Mathys, W. (1977), *Physiol. Plant 40*, 130–136.
McNaughton, S.J., Folsom, T.C., Lee, T., Park, F., Price, C., Roeder, D., Schmity, J., and Stockwell, C. (1974), *Ecology 55*, 1163–1165.
Persson, H. (1948), *Rev. Bryol. Lichenol. 17*, 75–78.
Peterson, P.J. (1971), *Sci. Prog. Oxford 59*, 505–526.
Porter, E.K., and Peterson, P.J. (1975), *J. Sci. Total Environ. 4*, 365–371.
Rauser, W.E. (1984), *Plant Sci. Lett. 33*, 239–247.
Reeves, R.D., and Baker, A.J. (1984), *New Phytol. 98*, 191–204.
Reilly, C. (1967), *Nature 215*, 666–669.
Reilly, A., and Reilly, C. (1973), *Plant Soil 38*, 671–674.
Searcy, K.B., and Mulcahy, D.L. (1985a), *Am. J. Bot. 72*, 1700–1706.
Searcy, K.B., and Mulcahy, D.L. (1985b), *Am. J. Bot. 72*, 1695–1699.
Shacklette, H.T. (1965), *Geol. Surv. Bull. US*, 1198C.
Symeonidis, L., McNeilly, T., Bradshaw, A.D. (1985), *New Phytol. 101*, 309–315.
Taylor, G.J., and Crowder, A.A. (1984), *Can. J. Bot. 62*, 1304–1308.
Thurman, D., and Collins, J.C. (1983), in: *Proceedings International Conference on Heavy Metals in the Environment, Heidelberg*, pp. 599–622. CEP Consultants Ltd., Edinburgh.
Turner, R.G., and Marshall, C. (1971), *New Phytol. 70*, 539–545.
Turner, R.G., and Marshall, C. (1972), *New Phytol. 71*, 671–676.
Walley, K.A., Khan, M.S., and Bradshaw, A.D. (1974), *Heredity 32*, 309–319.
Wild, H., and Bradshaw, A.D. (1977), *Evolution 31*, 282–293.
Wu, L., and Antonovics, J. (1976), *Ecology 57*, 205–208.
Wu, L., and Kruckenberg, A.L. (1985), *New Phytol. 99*, 565–570.

Additional Recommended Literature

Chaney, R.L. (1987), *Metal Speciation Affecting Trace Element Transfer in Agricultural Food Chains*. Lecture Jekyll Island, Georgia.
Ernst, W.H.O. (1986), *Longterm Pollution and Selection. Proceedings 2nd Conference on Environmental Contamination, Amsterdam*, pp. 10–15. CEP Consultants Ltd., Edinburgh.

Grill, E., and Zenk, M. H. (1989), *How Do Plants Protect Themselves Against Heavy Metals?* (in German) *Chem. Unserer Zeit* 23(6), 193–199.

Harmens, H., Verkleij, J. A. C., Koevots, P., and Ernst, W. H. O. (1989), *The Role of Organic Acids and Phytochelatins in the Mechanism of Zinc Tolerance in Silene vulgaris. Proceedings International Conference on Heavy Metals in the Environment, Geneva*, Vol. 2, pp. 178–181. CEP Consultants Ltd., Edinburgh.

Twiss, M. R., Nalewajko, C., and Stokes, P. M. (1989), *The Influence of Phosphorus Nutrition on Copper Tolerance in Scenedesmus spp. Proceedings International Conference on Heavy Metals in the Environment, Geneva*, Vol. 2, pp. 174–177. CEP Consultants Ltd., Edinburgh.

von Frenckell, B. A. K., and Hutchinson, T. C. (1989), *Co-tolerance to Metals in the Grass Deschampsia cespitosa. Proceedings International Conference on Heavy Metals in the Environment, Geneva*, Vol. 2, pp. 182–185. CEP Consultants Ltd., Edinburgh.

I.7h Bacteria, Fungi, and Blue-Green Algae

BETTY H. OLSON, Irvine, California, USA
ASHOK K. PANIGRAHI, Berhampur, India

1 Introduction

Microorganisms are generally the first to be affected by discharges of heavy metals into the environment (WOOD, 1989). Microbial communities can be affected by changes in total viable counts, shifts in the balance of species present or alteration of the metabolic characteristics of the community (STERRIT and LESTER, 1980). Further, changes in the growth and the nitrogen fixing capacity have been observed for heterocystous cyanobacteria (RATH et al., 1983, 1985, 1986a,b; SAHU et al., 1987). Similarly, microbial ecosystems can also drastically alter the fate of metals entering aquatic or soil environments.

Bacteria, cyanobacteria, and fungi can alter the form of occurrence of metals through methylation, chelation, complexation, catalysis, or absorption. Thus, microorganisms affect the bioavailability of metals in both aquatic and soil systems, and these processes affect the movement of the metal up the food chain. Factors which affect a microorganism's ability to alter the chemical form, e.g., $Hg^{2+} \rightarrow Hg^0$, of a metal are the bioavailable concentration of the metal, the types and numbers of microorganisms present, the time period of the organism's exposure to the metal, and the physicochemical parameters of the environment. The presence of chemicals such as natural inorganic and organic materials, biological agents, and physical parameters such as pH, temperature, redox potential, sunlight, water and composition of the soil or sediment are also important mediating variables in the metal transformation process (SAXENA and HOWARD, 1977).

Biotransformation depends also on environmental metal toxicity which is influenced by a wide number of chemical, physical, and biological factors. Organisms which produce hydrogen sulfide often exhibit a tolerance to heavy metals because this compound binds with the metal to form insoluble sulfides. This mechanism may also afford protection to non-hydrogen sulfide producing microorganisms in the surrounding environment. The production of other extracellular products such as glutathione may also reduce metal toxicity. The production of extracellular nitrogenous compounds by the heterocystous cyanobacteria reduces the toxicity of mercury (RATH et al., 1986a). The extracellular products of cyanobacteria act as chelating agents, and in some cases complex organics are produced by the interaction of ionic, elemental, or organic mercury with the extracellular by-products. These play a vital role in altering the environmental conditions.

Certain yeasts in soil environments decrease metal toxicity by the production of chelating agents such as citric acid or oxalic acid or by producing methionine (GADD and GRIFFITHS, 1978). Chelation of metals by humic acids, amino acids, polypeptides, and ethylenediamine tetraacetic acid alter their bioavailability and thus toxicity to microorganisms (BITTON and FREIHOFER, 1978; RATH et al., 1986a; SAHU, 1987). ALBRIGHT et al. (1972) also investigated the effect of Ag^+, Cu^{2+}, Hg^{2+}, and Zn^{2+} ions on heterotrophic bacterial populations and concluded that Ag^+ had a greater or equal toxicity to Cu^{2+}, $Ni^{2+} > Ba^{2+}$, Cr^{2+}, $Hg^{2+} > Zn^{2+}$, Na^+, and Cd^{2+}.

A metal in a soil or any environmental system undergoes a number of physical and chemical interactions as well as interactions with microorganisms. The physical and chemical properties of the environment often mediate the bioavailability of the metal (DOELMAN, 1978). Thus it is important to emphasize that the toxicity of any metal to an organism in pure culture is markedly different than in natural habitats.

Sampling and analytical errors must also be considered (see Chapter I.4a). PICKET and DEAN (1976a) reported that 97% of Hg^{2+} was lost from an open flask within seven days. SHAW (1987) and SAHU (1987) confirmed the volatilization of mercury from experimental flasks and observed that mercury adheres to the inner walls of the glass vessel and the outer lining of the cell wall of the cyanobacterium *Westiellopsis prolifica*. Lead was not volatilized from any medium tested, and approximately 20% was bound to the inner walls of the vessel. Cadmium was not volatilized nor absorbed to the walls of flasks in the concentration range of 0 to 180 µg g^{-1}.

2 Uptake, Interactions, Transformations, and Resistance to Heavy Metals

(see also Chapters I.2, Section 3.3; I.7g, and I.10, and Part II)

2.1 Interaction of Metals with Components in Media

Combined actions or reactions of microorganisms which in the environment may be synergistic or antagonistic are often lost in laboratory studies. The interaction between metal ions and the growth medium used is critical in determining the bioavailability of the metal. Thus, the metal content in a medium may be drastically different from that which was initially added. The metal may exist as a free ion, an inorganic complex, or an organic complex which affect its availability to the microorganisms present in that medium. RAMAMOORTHY and KUSHNER (1975) demonstrated that media have a dramatic effect on Hg, Pb, Cu, and Cd. Utilizing a medium developed by FOOT and TAYLOR (1949), they found little free Hg^{2+} ion available in solution.

Foot and Taylor medium resulted in only 30% being available in the ionic form, and trypticase soy broth as well as Nelson's broth (NELSON et al., 1973) strongly bound Cd. The biochemical constituents of the media tested by RAMAMOORTHY

Table I.7h-1. Interaction Between Biochemical Ingredients in Media and Metals. (Adapted from RAMAMOORTHY and KUSHNER, 1975)

Metal	Casamino Acid	Protease Peptone	Yeast Extract	Bacto Tryptone	Bacto Peptone
Mercury	+ + + + +	+ + + +	+ + +	+ +	+
Lead	+ + + + +	+	+ + + +	+ + +	+ +
Copper	+ + + + +	+ +	+ + + +	+ + +	+
Cadmium	+ + + +	+ + +	+	+ +	−

− no binding capacity; + weakest binding capacity; + + + + + strongest binding capacity

and KUSHNER (1975) and their relative capacity to bind with the above-mentioned metals are shown in Table I.7h-1.

The amount of magnesium in a medium has been shown to be critical in determining metal toxicity. The toxicity of several metals appears to be inversely related to the concentration of Mg in the medium. This may be due to the fact that a variety of metals (Sn, Zn, Cd, Co, and Ni) replace Mg in a number of enzyme and co-enzyme systems that involve phosphate transfer (ABELSON and ALDOUS, 1950).

The biosynthesis of intercellular traps for the removal of metal ions from solution represents a temporary measure adopted by cells to prevent metals from reaching toxic levels. The strategy adopted for the biosynthesis of intracellular traps fits quite closely the predicted petitioning for elements in organic or inorganic matrices. For example, Na, K, Mg, Cu, Al, P, Si, and B prefer to react with an oxygen donor matrix, but Cu, Zn, Fe, Ni, Co, Mo, Cd, and Hg prefer a nitrogen and sulfur donor matrix. WOOD (1974) discovered that nickel tolerant mutants of the cyanobacterium *Synechococcus* sp. could synthesize an intracellular polymer that effectively removed nickel from solution. The author reported that the same mutant was resistant to copper. Mutants with intracellular trapping mechanisms tend to bioconcentrate the toxic metal intracellularly approximately 200 times over the external concentration. Intracellular concentrations of metal ions can be controlled by deposition on a solid surface, and the concentration of free metal ions can also be controlled by the biosynthesis of ligands in the form of small molecules with high stability constants.

2.2 Uptake of Metals by Microorganisms

The binding of metals to microorganisms can be divided into (1) a non-specific adsorption of the metal to surface slime layers or extracellular matrices and (2) metabolically dependent intracellular uptake. The non-specific binding of the metal to a cell surface or products will occur for viable and non-viable organisms while those uptakes which are mediated by metabolic uptake processes are only observed in viable cells. Certain organisms utilize metabolic processes to exclude metal ions from the cell thereby establishing resistance. Cadmium resistance in *Staphylococcus aureus* is an example of this phenomenon (CHOPRA, 1975); mercury resistance in *Westiellopsis prolifica* is another example (PANIGRAHI and MISRA, 1989).

Microorganisms, including the algae, synthesize extracellular ligands which complex metals and prevent their cellular uptake. Cyanobacteria were found to be more sensitive to nickel toxicity than green algae, which points out differences in transport mechanisms for prokaryotes versus eukaryotes. Several bacteria have been found that precipitate silver as Ag_2S at the cell surface, and certain fungi are very efficient at recovering uranium (SEIGEL, 1983).

Once the metal is inside the cell, a number of transformation processes are possible. Volatilization of mercury and selenium have been postulated as well as methylation of Hg, Pb, and Sn. These processes are carried out by a wide variety of bacteria, fungi, yeasts, and blue-green algae.

2.3 Undefined and Non-energy-dependent Metal Uptake Mechanisms

Cadmium uptake is temperature-dependent in *Staphylococcus aureus* (TYNECKA et al., 1975), and exclusion from the cell in *S. aureus* and other bacteria is plasmid (extra-chromosomal DNA) mediated. However, cadmium is accumulated by an energy-dependent uptake in *Saccharomyces cerevisiae*, although the mechanism of uptake at this time is unknown. Two hypotheses have been suggested by NORRIS et al. (1980). One is based upon the ability of cadmium to form complexes similar to Zn, and the other is based on the similar ionic radii of calcium and cadmium. Either process enables the cation to substitute for other ions in transport systems. Lead, on the other hand, can reach similar cellular concentrations as Co and Cd in *S. cerevisiae* but does not require an energy source.

Metal cations readily adsorb to negatively charged sites at the surface of microorganisms. This process is rapid, reversible, and independent of temperature and energy metabolism. Anionic ligands at the cell surface responsible for cation adsorption include phosphoryl, carboxyl, sulfhydryl and hydroxyl groups of proteins and lipids in the membrane and cell wall components, such as peptidoglycans and associated polymers. Evidence indicated, however, that *S. cerevisiae* cell walls bound their own weight of Hg^{2+} at high affinity sites, and this metal retention could not be correlated with available wall protein or phosphate binding sites.

2.4 Metabolically Mediated Mechanisms

Divalent cation transport systems have been shown to be important in translocation of metals, a phenomenon characteristic of several bacterial genera.

Metals can replace Mg in several enzyme and co-enzyme systems involving phosphate transfer. The metals thus far identified in this process are Mn, Zn, Cd, Co, and Ni (ABELSON and ALDOUS, 1950). In the Gram positive bacterium *Bacillus subtilis* metal transport of divalent forms of Co, Mn, and Ni occurs via formation of a citrate divalent cation complex.

Divalent cation transport systems for microorganisms, other than bacteria, are less well understood. In the filamentous fungus *Neocosmospora vasinifecta* Zn^{2+} uptake is inhibited by Mg^{2+}, except at high Zn^{2+} concentrations. Some inhibition was noted for the uptake of the divalent cation of Zn by divalent forms of Co, Ni, and Cu. Divalent Co uptake by *Saccharomyces cerevisiae* is inhibited most strongly by cations with similar ionic radii. Furthermore, uptake decreases across the series Zn > Ni > Mg > Mn > Cd > Ca.

The observation that there is high intracellular concentration of potassium in microorganisms has encouraged study on *monovalent cation uptake*. Four kinetically and genetically distinct K transport systems have been reported for *Escherichia coli*. The characteristic uptake of titanium, which may be present in the effluents of certain mining processes, suggests that it should be included close to potassium in the affinity series.

Anionic complexes occur in aquatic environments and may be important in metal transport into microorganisms. The most completely described *anion transport* system involves sulfate permease. Energy-dependent uptake of chromate has been described in the fungus *Neurospora crassa*. Anion transport was mediated in *E. coli* in diffusion rather than an active uptake mechanism. Selenate resistant cells did not incorporate selenium as readily into cell proteins and produced a sulfate permease that was resistant to selenate inhibition. Selenium granules have been observed in the cytoplasm of *E. coli* and other bacteria.

Arsenate uptake in *Streptococcus faecalis* is energy-dependent, requiring ATP and is stimulated by K^+ and other permeate cations. Further, a neutral to basic cytoplasmic pH is important in the uptake of this compound.

2.5 Concentration of Metals by Microorganisms

TORNABENE and EDWARDS (1972) demonstrated that *Micrococcus luteus* and *Azotobacter* grown in the presence of lead bromide concentrated high quantities of Pb^{2+} on a dry weight basis in the cells. *Azotobacter* sp. was more effective at immobilizing Pb^{2+}, which may be related to the production of capsular material that has the ability to accumulate biological and non-biological substances. Lead associated with the cytoplasm represented less than one percent of total cellular Pb in both of these species. Immobilization of lead by microbial systems may be important in its transfer up the food chain and suggests passive transfer.

PATRICK and LOUTIT (1976) have demonstrated that freshwater sediment bacterial isolates were able to concentrate Cr, Cu, Mg, Pb, and Zn. The concentration of these metals in bacterial populations may explain metal transfer in the food chain to tubificids which feed on these bacteria.

RATH et al. (1983) reported accumulation of mercury in the cyanobacterium *Westiellopsis prolifica* when exposed to mercuric chloride. SAHU (1987) recorded the accumulation of mercury in *W. prolifica*, exposed to phenylmercuric acetate. SHAW (1987) observed accumulation of mercury in cyanobacteria exposed to mercury containing industrial effluents and waste of a chlor-alkali industry. PANIGRAHI and MISRA (1989) suggested probable biological magnification of mercury in an aquatic and terrestrial food chain. Algae have been shown to concentrate heavy metals under agricultural field conditions (TROLLOPE and EVANS, 1976). The mechanism, by which mercury was picked up and concentrated in the tissues, is likely to be concentration/dose-dependent (SAHU, 1987).

2.6 Biomethylation of Metals and Their Transport

The microbial synthesis of organometallic compounds from inorganic precursors is not yet completely understood in both terrestrial environment and in the sea. Perhaps chemical reactions with biologically produced alkylating compounds do also occur. Mechanisms for vitamin B_{12}-dependent synthesis of metal alkyls have been stipulated for the metals Hg, Pb, Ti, Pd, Pt, Au, Sn, Cr, and for the metalloids As and Se. Pathways for the synthesis of organoarsenic compounds have also been shown to occur by a mechanism involving S-adenosylmethionine as the methylating coenzyme (McBride et al., 1978). To date two different mechanisms have been determined for methyl transfer from methylcobalamin B_{12} to heavy metals: (1) electrophilic attack by the attacking metals on the Co-bond of methyl B_{12} and (2) methyl-radical transfer to an ion pair between the attacking metal ion and the corrin macrocycle. Metal ions which displace the methyl group by electrophilic attack are Hg(II), Pb(IV), Ti(III), and Pd(II). Examples of free-radical transfer are Pt(II)/Pt(IV), Sn(II), Cr(II), and Au(III). Vitamin B_{12}-dependent strains are capable of methylating Hg(II) salts to CH_3Hg^+ (see also Sect. 7), whereas the B_{12}-independent strain is incapable of catalyzing this reaction. Both strains transport Hg(II) into the cells at the same rate, but the independent strain is inhibited by at least a 40-fold lower concentration of Hg(II) than the dependent strain. It could thus be that *Clostridium cochlearium* uses biomethylation as a mechanism for detoxification giving the organism a clear advantage in mercury contaminated systems. This biomethylation capability was shown to be plasmid-mediated (Pan-Hou and Imura, 1982). After the release of methylmercury from the microbial system, it enters food chains as a consequence of its rapid diffusion rate.

Biotransformation of selenium in aquatic environments, soil, sediment, and sewage have been reviewed by Doran (1982) (see also Sect. 10). Biomethylation mechanisms for selenium have also been reviewed, and it has been shown that the resulting dimethylselenide is less toxic than inorganic selenium species (NRC, 1976). Likely mechanisms of bacterial resistance to selenite include export, methylation or volatilization, and precipitation of insoluble selenium metal (Foster, 1983; Summers and Silver, 1978; Trevors et al., 1985). Burton et al. (1987) have shown clearly that the capability to reduce selenite to selenium metal is widespread among heterotrophic bacteria isolated from the Kesterson National Wildlife Reserve, an area highly contaminated with Se. In addition, most of the selenite resistance isolates produced a garlic odor, indicative of the production of the methylated forms of selenium (Doran, 1982). The results described by Burton et al. (1987) have revealed significant bacterial resistance to selenite in a selenium-contaminated aquatic system. The same authors, in addition, reported that such resistance has developed in light of the extensive documentation of the selection for bacterial resistance to metals and other antibacterial agents in the environment with a high concentration of these toxic agents (Foster, 1983; Summers and Jacoby, 1977; Wood and Wang, 1983). The ability of heterotrophic bacteria to convert inorganic selenium to volatile methylated selenium species may also contribute to the transport of selenium through the food chain, analogous to biological cycling of mercury that is initiated by bacterial reduction and methylation of the mercuric ion (Wood and Wang, 1983), see also Sect. 7.3.

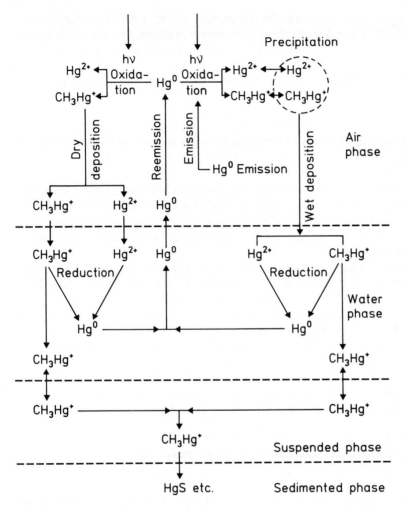

Fig. I.7h-1. Mercury cycle in the environment.

$$CH_3Hg^+ \rightleftarrows Hg^{2+} \rightleftarrows Hg^0$$
methyl mercury   ionic   elemental

Microorganisms have been isolated which catalyze the reactions described above for both the forward reaction to Hg^0 and the back reaction to methylmercury$^+$ (CH_3Hg^+) (Fig. I.7h-1).

The enzymes which carry out the forward reaction are coded by DNA associated with bacterial plasmids and transposons (SILVER, 1983; JACKSON and SUMMERS, 1982). Recently, it has been reported that chromosomally occurring Hg reductase systems have been identified (SILVER, 1989). A wide variety of bacteria have been isolated from a variety of ecosystems such as soil, water, and marine sediments which exhibits mercury detoxification mechanisms. The reduction of ionic mercury to elemental mercury, which is volatile, represents a very effective detoxification mechanism. Much less is known about the reverse reaction, i.e., the oxidation of elemental mercury to ionic mercury (Hg^0 to Hg^{2+}). However, an enzyme which is critical to this reaction is catalase. Microbial methylation of the mercuric ion is also widespread. Biomethylation has been shown to occur in sediments and in human feces (Fig. I.7h-1).

2.7 Synergistic and Antagonistic Effects between Metal Ions and Microorganisms
(see also Section 8)

BALDRY et al. (1977) were able to demonstrate slight toxicity of *Klebsiella aerogenes* to chromic chloride, but potassium chromate and cupric sulfate were far more toxic. The addition of Cd or Zn to cells in contact with Cu increased the lethal action of Cu. Similarly, zinc and chromate acted synergistically while cadmium chromate acted antagonistically.

Microorganisms' ability to adapt to exposure to a metal ion thus may be dependent on exposure to other metal ions. For example, the work of PICKETT and DEAN (1976b) indicated that *Klebsiella aerogenes* became more resistant to Cd if pre-exposed to Zn. However, the reverse was not observed. Their work also showed that chelating agents reduced Zn and Cd toxicity. Further, sulfate and magnesium (PICKETT and DEAN, 1976a) were reported to be important in establishing metal resistance in this species.

RATH et al. (1983) reported that the cyanobacterium *Westiellopsis prolifica* was extremely tolerant to inorganic mercury and the algae continued to grow and accumulate in concentrations less than 0.04 mg L^{-1}.

2.8 Heavy Metal Resistance and Antibiotic Resistance
(see also Sections 5 and 7.2, and Chapter II.6, Section 4.2.3)

Large populations of *Bacillus* sp. resistant to 20 μg g^{-1} Hg were observed in the New York Bight sediments contaminated by sewage disposal wastes. *Bacillus* populations isolated from contaminated sediments showed a greater resistance to Hg and other heavy metals than strains of the same genus isolated from non-contaminated sediments. It was found that 24% of the *Bacillus* sp. isolated from the contaminated area were resistant to Hg, Cd, and Zn (TIMONEY et al., 1978). Twenty-four percent were resistant to Hg and Zn, while 20% were resistant only to Zn and 14% only to Hg. The criterion used to determine resistance was growth in the presence of 5 μg g^{-1} Hg, 11 μg g^{-1} Cd, and 65 μg g^{-1} Zn.

Chemostat studies of *Klebsiella (Enterobacter) aerogenes* indicated that sulfate and magnesium affected resistance to both Cd and Zn. Glucose, ammonium, and phosphate limited organisms were sensitive to Cd, while glucose and potassium limited organisms were sensitive to Zn (PICKETT and DEAN, 1976a).

Utilizing *Klebsiella aerogenes* strains which were resistant to 50 μg mL^{-1} Cd and 16 μg mL^{-1} Zn, in a medium containing β-glycerophosphate as a sole phosphorus source, glucose-6-phosphate dehydrogenase activity decreased, while phosphoglucose isomerase activity increased. This suggested an increased flow of carbon through the Embden-Meyerhof pathway compared to the pentose phosphate pathway. However, conversion of glucose into biomass was largely unaffected. Glucose phosphoenolpyruvate phosphotransferase activity was also lower in resistant strains (PICKETT et al., 1976).

The resistance of microbial cells to the toxic elements arsenic, antimony, and cadmium has been shown to occur through the evolution of cellular exclusion mechanisms. An operon-like structure induced resistance to arsenate, arsenite, and antimony(III) in *Escherichia coli* (SILVER et al., 1981). In the case of cadmium, resistance has been shown to be mediated by a plasmid. Two separate plasmid genes have been shown to be responsible for Cd^{2+} resistance. These genes also prevent Zn^{2+} toxicity. The resistance to toxicity by these microorganisms resides in their ability to selectively remove these elements by energy-driven efflux systems.

Bacillus sp. isolated from sediments near the sewage sludge dump site were more resistant to ampicillin. It was concluded that heavy metal contamination of an ecosystem can result in selective pressure for antibiotic resistant bacteria in that system (TIMONEY et al., 1978). PICKETT and DEAN (1976b) have shown that strains of *Klebsiella aerogenes*, resistant to Cd and Zn, also demonstrated multiple antibiotic resistance patterns.

Copper resistance was found to be associated with resistance to chloramphenicol and gentamycin in *Klebsiella aerogenes*. However, increased sensitivity to streptomycin and no change in sensitivity to nalidixic acid were noted. In chromate resistant organisms, an associated resistance to nalidixic acid was found but none to the other three antibiotics tested (BALDRY et al., 1977).

Heavy metal tolerant bacteria isolated from surface waters and sediments in Chesapeake Bay and its tributaries also showed multiple resistance to antibiotics suggesting metal-drug resistance linkages. Most of the bacterial strains resistant to Co, Pb, Hg, or Cd were also resistant to ampicillin and chloramphenicol. TIMONEY et al. (1978) found that multiple and single metal resistance in *Bacillus* spp. was associated with ampicillin resistance in isolates from a New York Bight sewage sludge disposal. Thus, metal and drug resistance factors are common in the environment, especially in hospital environments, and are genetically linked on plasmids in certain cases (DYKE et al., 1970). The penicillinase plasmids of *Staphylococcus aureus* carry, for instance, determinants for resistance to mercury as well as for arsenate, lead, cadmium, and bismuth ions, all grouped in one region (KONDO et al., 1974). The reasons for a drug metal genetic association from an environmental viewpoint is far less clear.

3 Microbial Mediation of Biogeochemical Cycling of Metals

The interaction of microbes with the formation and alteration of minerals can be traced with certainty to proterozoic and possibly to archean times. The earliest interrelationships between metals and microorganisms were viewed as changes in the redox potential of mineral complexes or their associated free ions in aquatic environments (BOWEN, 1966). Many of these elements are essential to organisms and thus circulate in the biosphere, moving from organisms to environment in a continuous cycle. The basic phenomena that drive many of the biogeochemical cycles are microbial energy and carbon requirements. The major microorganisms involved in these processes are bacteria, cyanobacteria, green algae, fungi, and protozoa.

Autotrophic bacteria are those bacteria using inorganic carbon as a carbon source, whereas heterotrophic bacteria obtain both cellular carbon and energy from preformed organic complexes; the importance of this latter grouping mineral cycling has only recently been fully recognized.

For three reasons microorganisms are important in the cycling of minerals: (1) microorganisms comprise most of the earth's biomass and have rapid generation times; (2) microorganisms occupy the widest diversity of habitats; and (3) microorganisms colonized the earth 4–5 times earlier, in time, than higher forms of life. Thus, the cycling of metals by microorganisms is hardly a new phenomenon, although our understanding of the subject is continuously increasing.

Microorganisms affect the redistribution of metals by oxidation and reduction or by metal binding. Metals may be solubilized by oxidation, as in the case of uranium, or by reduction as in the case of iron and manganese. Some metal oxidations, as well as reductions, are microbially mediated. In some metal oxidations, the metal furnishes the only source of electrons and therefore of energy for the microbes. Both oxidations and reductions are fundamental in the redistribution of metals (EHRLICH, 1978).

Bacterial leaching of sulfide minerals, for instance, has been known in Falun (Sweden) since 1687 (HALLBERG and PICKARD, 1981): *Thiobacillus ferrooxidans* is an iron- and sulfur-oxidizing chemoautotroph with the ability to oxidize $CuFeS_2$, obtaining its carbon from carbon dioxide. Cu^{2+} ions, Fe^{2+} ions, and sulfuric acid are formed.

The concept of biogeochemical cycles was expanded dramatically during the 1970s. The finding that Hg could be methylated to a volatile and highly toxic form (JENSEN and JERNELÖV, 1969) renewed interest in metal cycles and their biological mediation. During this period of intense research, information concerning volatile metal complexes became widely known, and it became clear that changes in valence were no longer adequate for accurate description of the role of microorganisms in biogeochemical cycles.

4 Biological Indicators and Monitors

(see also Chapter I.4c)

Both terms are frequently used in environmental studies. "Indicator" is the ability of an organism involved to simply indicate the presence or absence of the heavy metals in this context. CLEMENTS (1920) wrote: "Every plant is a measure of the conditions under which it grows. To this extent it is an index of soil and climate, and consequently an indicator of the behavior of other plants and animals in the same spot." "Monitors" have two distinguishing criteria: a) the provision for regular surveillance; and b) quantification of how much pollutant is present. Biological monitors must occur in some abundance throughout the area of study, both in the polluted and non-polluted parts. They should accumulate and retain the pollutant progressively during the exposure period. Because biological materials react ecotoxicologically to contamination levels, they may translate chemical dosage situations into a biological response which in turn may be a convenient substitute for man as a target organism (GOODMAN, 1974). REAY (1979) neatly defined the purpose of monitoring as the need to observe, explain, and control both emissions of pollutants and effects of pollutants.

Often the biomonitor will be part of a biological system as implicated by CAIRNS (1980). PANIGRAHI (1980) and PANIGRAHI and MISRA (1989) described that many toxic materials accumulate among food webs. In these situations the detritivore-decomposer levels are frequently the first to show changes, since the organic matter and the soils are the ultimate sink for most, including air-borne, heavy metal contamination. Not all biological materials are suitable for use as monitoring tools for heavy metals. They must be very easy to collect during the entire year. Repeatability is essential, and cost of collection and analysis should be acceptable.

Two types of *indicator species* are recognized, "universal" and "local" ones. Local indicators are the species associated with metal bearing substrates in certain geographical areas but which also grow elsewhere in non-mineralized areas. The disadvantages with the universal indicators are: a) these species tend to be very rare; b) often have very restricted geographical distribution ranges; and c) many are endemic to relatively small. Some 122 species have been listed by CANNON (1971) in a comprehensive list of plant indicators of mineral deposits. The elements indicated by those species are Al, B, Co, Cu with Mo, C (as diamonds), Au, Fe, Pb, Li, Mn, Hg, Ni, P, Se with uranium, Ag, Sr, Sn, V, and Zn.

Biogeochemical prospecting uses the chemical analysis of plant samples, with the assumption that high concentrations of metal within the aerial parts will signify abnormally high concentrations of the metal in the soil and underlying substrate. BAKER (1981) described three responses of plants to increasing heavy metal concentrations in the soil on the basis of the concentration of metals within the plant (i.e., excluders, indicators, and accumulators, see also PETERSON, 1971).

The use of plants in both geobotanical and biogeochemical prospecting provides examples of the use of plants as indicators. Biogeochemical methods for plant analysis are more useful and more widely used than geobotanical methods. However, both geobotanical (indicating by presence) and biogeochemical prospecting are not applicable in microbiological studies. Intensive and extensive work in this line may help in the opening of a new line of research. SHAW (1987) and SAHU (1987) used blue-green algae as an indicator of contamination or pollution and found higher concentrations of residual mercury and calculated the body burden and soil/sediment/solid waste ratio of mercury.

5 Genetic Control of Metal Transformation in Microbial Populations
(see also Section 2.8 and Chapters I.7g and I.19)

Some microorganisms have inherited the ability to resist high concentrations of toxic elements through their evolution in extreme environmental conditions (BROCK, 1978). Some microorganisms have acquired a transferred resistance to the polluted environment relatively recently through the acquisition of extrachromosomal DNA molecules, i.e., plasmids (SILVER, 1983).

Many bacteria possess extrachromosomal DNA, as well as the chromosomal DNA. Those genetic elements that are totally independent of the chromosome are called plasmids. Plasmids are usually referred to by function; for example, resistance to antibiotics is referred to as drug resistance (R) factors and to metals as metal resistance factors.

Resistance in *Staphylococcus aureus* appears to be genetically controlled because growth in the absence of cadmium results in appearance of susceptible cells (WEISS et al., 1978). NOVICK and ROTH (1968) demonstrated that plasmid resistance to cadmium in *S. aureus* also conferred resistance to zinc. In addition, mutants that became susceptible to Cd simultaneously became sensitive to Zn, suggesting linkage of Cd and Zn resistance on the same plasmid. The same authors demonstrated that the plasmid that controlled penicillin resistance also determined metal resistance to arsenate, arsenite, Pb, Cd, Hg, Bi, Sb, and Zn.

The patterns of metal resistance appeared to be strain specific. Multiple metal resistance may be carried on penicillinase plasmids, but it certainly does not imply that all penicillinase plasmids in *S. aureus* confer multiple metal resistance.

The remainder of this chapter will focus on five of the most widely investigated metals: Pb, Hg, Cd, As, and Se. The chemical properties of an element are extremely important in determining its ultimate fate in the environment. Cadmium and mercury are limited in their oxidation states, obtaining valences no higher than two and are not regarded as transition elements, being more electro-positive and having lower melting points. These elements readily form complexes with ammonia, amines, halides, and cyanide, and react with sulfur and phosphorus (VALLEE and ULMER, 1972). Cadmium, a relatively rare earth element, has a much smaller concentration in the earth's crust than Hg or Pb. Cadmium contamination in the environment has increased in the last twenty years from anthropogenic activities, while the distribution of mercury is approximately evenly divided between industrial activities and natural sources.

Lead, cadmium, and mercury all have a strong affinity for ligands such as phosphate, cysteinyl and histidyl side chains of proteins, purines, pyrimidines and porphyrins (VALLEE and ULMER, 1972). Therefore, each of these elements has the ability to interact with a large number of molecules which affect their interactions with microorganisms.

6 Lead Compounds

Among the Group IV elements, lead is the most electro-positive and has the most stable divalent state. It is fairly reactive chemically and dissolves in a number of acids. Lead is widely distributed in the environment and is also readily identified in most plant and animal tissues. Most importantly, the distribution, metabolism, and toxicity of Pb appears to be significantly influenced by other elements. Excess amounts of organic Pb and Hg compounds and inorganic compounds of Pb, Hg, and Cd are toxic to microorganisms (VALLEE and ULMER, 1972).

Lead carbonates, oxides, and sulfates are released into the environment through mining and combustion activities while elemental lead is emitted through smelting operations (CRAIG, 1980). The chemistry of lead in soils is largely controlled by absorption to soil mineral surfaces, by the formation of stable organo-lead complex ions, and insoluble organo-lead chelates, particulates, and residues. The effect of Pb on microbial populations is usually greater in sandy soils than in clay or peat soils (DOELMAN, 1978).

BEWLEY and CAMPBELL (1978) showed that the phyllosphere microflora is abundant in the presence of high Cd, Zn, and Pb concentrations. They examined fungal populations present on the oak leaf surfaces of *Quercus robor* L. which contained approximately 1200 $\mu g \, g^{-1}$ Cd.

The effect of Zn, Pb, and Cd on fungi colonizing hawthorn leaves caused by aerial fallout from a Zn-Pb smelter near Bristol, England, showed that *Sporobolomyces roseus* was absent from the most heavily contaminated leaves. The occurrence of *Aureobasidium pullulans* was significantly positively correlated with Pb but was inhibited by Zn and/or Cd. The numbers of bacterial colonies were only slightly reduced by the combined effects of all three metals, but total numbers of bacteria were negatively correlated with Pb. Filamentous fungi isolated by leaf washing were only slightly inhibited by all three metals, and mycelial proliferation on senescent leaves was only slightly affected by heavy metal pollution (BEWLEY and CAMPBELL, 1980).

In Pb amended soils Gram negative rods became increasingly dominant in soil bacterial populations over coryne forms. DOELMAN and HAANSTRA (1979) concluded that the dominance of Gram negative organisms may be explained by the fact that they were contained in the interior of soil aggregates and therefore were protected against exposure (DOELMAN and HAANSTRA, 1979). In this case lead resistance would not be a characteristic of the bacterial population, but rather a passive environmental phenomenon.

Certainly, not all workers would agree with this assumption. TROYER et al. (1981) have demonstrated lead resistance in bacterial populations from soils with anomalous concentrations of Pb, Cd, and Zn in England. VARMA et al. (1976) have shown lead resistant bacteria isolated from domestic sewage. Further, simultaneous resistance to antibiotics was observed and the process appeared to be plasmid mediated either on separate plasmids or resistance transfer factors.

COLE (1979) demonstrated that several isolated heterotrophic soil bacteria were capable of solubilizing lead sulfide. The organisms were also capable of solubilizing zinc sulfide, cadmium sulfide, and calcium sulfide, but no activity was observed against lead sulfate and lead oxides. The isolates produced neither 2-ketogluconic acid nor siderochromes suggesting a previously undescribed solubilizing agent. The organisms were believed to belong to the genera *Pseudomonas* or *Xanthomonas*. The ability of one isolate to solubilize zinc sulfide and lead sulfide was increased by the addition of glucose to the medium.

The data suggest that autotrophic bacteria are not the only microorganisms capable of solubilizing metal sulfides. Although thiobacilli are able to solubilize much larger quantities of metal sulfides (SILVER and TORMA, 1974) than heterotrophic isolates, the existence of heterotrophic solubilizers provides a possible explanation for increased concentrations of soluble metals observed under conditions where *Thiobacillus* would be inactive or present in low numbers. Probably pH is also critical. It seems, however, that not all metal sulfides are solubilized. This could be due to the formation of chelating agents specific for certain metals by selection during the early cellular growth phase of certain cell strains, or by the formation of different solubilizing agents by cells grown with different metals. MELLOR and MALLEY (1948) demonstrated that the stability constants of heavy metal cations with chelating agents were of the order of $Cu > Pb > Zn \simeq Cd$, whereas the amount of metal sulfide solubilized by isolates did not follow this order.

It is not clear whether the methylation of lead in aquatic environments is carried out by both biotic and abiotic processes. WONG et al. (1975) demonstrated that in the presence of *Pseudomonas, Alcaligenes, Acinetobacter, Flavobacterium*, and *Aeromonas* isolated from lake sediments certain inorganic and organic lead compounds were transformed into volatile tetramethyl-lead, $(Me)_4Pb$. (The trimethyl-lead ion, $(Me_3)_3 Pb^+$, was most readily methylated to form $(Me)_4 Pb$.) But the synthesis of $(Me)_4 Pb$ takes place rather by disproportionation, since $(Me)_3Pb^+$ itself is a methylating agent, $(Et)_3Pb^+$ is not methylated and the methyl group of H_3C-cobalamin cannot be transferred to $(Me)_3Pb^+$ (REISINGER, 1987).

Occasionally inorganic lead in the form of lead nitrate, $Pb(NO_3)_2$, or lead chloride, $PbCl_2$, was methylated to form tetramethyl-lead. Laboratory studies by BAKER et al. (1976) of acidic oligotrophic Lake Ontario sediments indicated, however, that over the pH range of 3.5 – 7.5 the rate of transformation was very small varying from 0 – 0.009% of the total added lead. This very limited rate of transformation may explain some of the controversy surrounding the question of biologically mediated methylation of lead.

JARVIE et al. (1975) and CRAIG (1980) demonstrated already that the methylation of $(Me)_3PbOAc$ may be an abiological process. REISINGER et al. (1981) could not demonstrate biological methylation of inorganic Pb or $(Me)_3PbOAc$, $(Et)_3PbCl$ in river sediments, sewage sludge or with the bacterium *Methanosarcina barkeri*.

Because of the labile nature of alkyllead compounds in the environment and the low rate of transformation of Pb into organolead compounds, the importance of microbial populations in this pathway of the Pb-cycle is far less significant than anthropogenic sources.

7 Mercury Compounds
(see also Section 2.6 and Chapter II.20)

7.1 Overview

The chemical properties of mercury described earlier enhance its availability to microorganisms in aquatic or terrestrial environments. Microorganisms, namely bacteria (*Pseudomonas*), can reduce Hg(II) species or degrade and reduce organomercurials into elemental mercury (SPANGLER et al., 1973a; MAGOS et al., 1964; FURUKAWA et al., 1969). The mercury in this form is easily volatilized from the soil or the sediment entering the atmosphere or the water column. The formation of elemental Hg is probably very important in the global cycle of Hg.

Simultaneously, other groups of bacteria and certain fungi can mediate the methylation of inorganic mercury forming mono- or eventually also dimethylmercury (HAMDY and NOYES, 1975; YAMADA and TONOMURA, 1972; VONK and SIJPERSTEIJN, 1973). The ability to methylate Hg is not confined to a limited number of species of microorganisms; thus conditions which promote growth can promote methylation. In a natural environment bacteria are present which carry out methylation and demethylation of mercury. This view is not generally accepted; WONG (1981) suggested that divalent Hg is methylated in freshwater sediments at increased pH values with a minimum yield and no dimethylation occurs. The data of SPANGLER et al. (1973b) showed methylation to occur in freshwater sediments. Demethylation, volatilization and methylation appear to be greatest under aerobic conditions in laboratory tests (OLSON and COOPER, 1976; SPANGLER et al., 1973b; HAMDY and NOYES, 1975; VONK and SIJPERSTEIJN, 1973). In environmental experiments methylation is particularly enhanced under anaerobic conditions (WOOD, 1975). Under anaerobic conditions the rate of volatilization by mercury reducing strains, (which are likely to also be demethylating strains), has been shown to be reduced, while methylation, though affected, is not inhibited to the same degree (BARKAY et al., 1979; OLSON, 1978). Therefore, in natural aquatic environments, methylmercury produced under aerobic conditions will be rapidly degraded by demethylating organisms, the result being that net methylmercury accumulation is demonstrated in environmental samples when conditions are anoxic and demethylation is inhibited (OLSON and COOPER, 1976; SPANGLER et al., 1973b). The genetic basis of demethylation has been shown to be plasmid mediated for a number of hospital strains of bacteria and several environmental strains (SUMMERS and SILVER, 1972; OLSON et al., 1979).

Bacteria isolated from mercury contaminated soils and sediments displayed the major form of resistance which was associated with the Tn21-like transposon (OLSON and FORD, 1986; OLSON et al., 1987). The genus that accounted for the major amount of resistance was *Pseudomonas* sp., followed by *Flavobacterium*.

Elemental Hg^0 and HgS are oxidized in an aerobic environment and Hg vapor in H_2O is oxidized to Hg^{2+} (WALLACE et al., 1971). Fifty percent of methoxyethylmercury is degraded to inorganic Hg within three days in most environments. Inorganic Hg or complexed Hg may be adsorbed or absorbed on sediment particles and can then be reduced or methylated; some goes into biological tissues as CH_3Hg^+. Other forms enter the sediment/water interface as HgO which may then be reoxidized to the divalent state and returned to the sediment. Last, there is downward movement of Hg to the inorganic mineralized zones (JENSEN and JERNELÖV, 1972).

After the removal of a plasmid (extrachromosomal DNA) from a bacterial strain of *Clostridium cochlearum* it no longer demethylated methylmercury but was capable of methylating mercury (PAN-HOU et al., 1980). These authors concluded that methylation and demethylation of mercury were controlled by one plasmid. Environmental bacterial isolates examined by BARKAY et al. (1979) also showed ability to methylate or demethylate mercury depending on experimental conditions.

7.2 Incidence of the Mercury Resistance Phenotypes
(see also Section 7.4)

Bacteria have developed a variety of means, generally referred to as resistance mechanisms, for dealing with the many forms of mercury. Available data (BARKAY et al., 1985; BLAGHEN et al., 1983; BOOTH and WILLIAMS, 1984; CLARK et al., 1977; NAKAHARA et al., 1977a,b,c; NELSON et al., 1973; PICKUP et al., 1983; PORTER et al., 1982; SUMMERS et al., 1978; TIMONEY et al., 1978; WALKER and COLWELL, 1974) indicate that plasmid encoded resistance to mercury (see also Sect. 7.5) is as common as the antibiotic resistance and can easily be found in a wide variety of genera among the eubacteria (IZAKI, 1977; KONDO et al., 1974; NAKAHARA et al., 1977a,b,c; NOVICK and ROTH, 1968; NOVICK et al., 1979; SPANGLER et al., 1973a,b; SUMMERS et al., 1978; SUMMERS and SILVER, 1978). However, the incidence of mercury resistance can be increased by the presence of mercury compounds in either industrial or clinical environments. Two general classes of plasmid-determined mercury resistance are known: reduction of Hg^{2+} to metallic mercury Hg^0, providing resistance to several organomercurials including merbromin and fluorescein mercuric acetate; and the reductive biotransformation of phenylmercuric acetate and methylmercury ions (SCHOTTEL et al., 1974).

$$\begin{array}{l} \text{Narrow spectrum resistance} \left[Hg^{2+} \xrightarrow[\text{NADPH, R-SH}]{\text{Hg(II) reductase (FAD)}} Hg^0 \text{ (vapor)} \right. \\ \text{Broad spectrum resistance} \left[R-Hg^+ \xrightarrow[\text{R-SH}]{\text{Organomercurial lyase}} Hg^{2+} + R-H \right. \end{array}$$

The classes of plasmid-encoded mercury resistance (see also Section 7.5)

In addition to plasmid determined reductive biotransformations, two other modes of mercurial biotransformation have been observed. Many bacteria produce hydrogen sulfide, with which mercury reacts to form the very insoluble HgS (see also Sect. 7.3, last paragraph). A different type of transformation is effected by methylcobalamin, which is produced by certain bacteria to serve as catalyst in biosynthesis (PAN-HOU and IMURA, 1982). Methylcobalamin methylates the mercuric ion in a non-enzymatic process that yields volatile monomethyl or dimethyl mercury, which readily dissipates from the microbial environment (WOOD, 1974; WOOD et al., 1978). Some bacteria can apparently produce both hydrogen sulfide and methylmercury (COMPEAU and BARTHA, 1985).

The presence of Hg compounds in the environment has received a great deal of attention due to their highly toxic nature and translocation through the food chain. In Sweden the use of phenylmercuric acetate (PMA) and methylmercury in fungicidal agents in seed dressings resulted in a significant decrease in the populations of seed-feeding birds (FOSTER and NAKAHARA, 1979). High levels of methylmercury have also been detected in fish from the Great Lakes region of North America.

Methylmercury is 100 times more toxic than inorganic mercury and has been found to be mutagenic under experimental conditions (FRIBERG and VOSTAL, 1972). The solubility of inorganic and organic mercury compounds in lipids as well as their binding to sulfhydryl groups of proteins in membranes and enzymes account for the cytotoxicity.

The biological cycle of mercury in the environment has received a great deal of study to determine the role of microorganisms (SUMMERS and SILVER, 1978). A positive correlation between the distribution of mercury compounds and that of resistant microorganisms in metal contaminated sediments has been reported. In addition, there is a strong positive correlation between antibiotic resistance and heavy metal resistance among both clinical and environmental isolates (NELSON et al., 1973; SCHOTTEL et al., 1974; SUMMERS et al., 1974; SUMMERS and SILVER, 1978).

Mercury- and organomercurial-resistant bacteria were first isolated from mercury contaminated soil in Japan (TANAKA et al., 1983; TEZUKA and TONOMURA, 1978). Mercury resistant isolates have also been obtained from the heavily polluted Chesapeake Bay area and from the sediments of the New York Bight where high concentrations of mercury exist (see also Sect. 2.8).

Two mechanisms for the elimination of metals from the growth medium have been proposed: (1) the synthesis of thiols that bind the mercury compound, thereby reducing its toxicity to the cell; and (2) the existence of a permeability barrier that would limit access of the mercury to the cell (SUMMERS and SILVER, 1978). The distinction between the sensitive and resistant strains is absolute with respect to conversion of Hg^{2+} to a volatile form (see also Fig. I.7h-1 and SUMMERS and LEWIS, 1973). Sensitive plasmidless strains show no detectable loss of $^{203}HgCl_2$ or ^{203}Hg-PMA, whereas Hg^{2+} resistant strains volatilize added $HgCl_2$, and strains resistant to both Hg^{2+} and PMA are capable of volatilizing both (SCHOTTEL et al., 1974). Several mercury-resistant *Staphylococcus aureus* strains originally thought to be nonvolatilizing exhibited an increased uptake and binding of Hg^{2+}.

Subsequently these strains were found to be able to volatilize mercury (KONDO et al., 1974). The discrepancy is believed to be due to poor induction conditions. Mercury resistant volatilizing strains are also *Escherichia coli* and *Pseudomonas* sp. (SUMMERS and LEWIS, 1973; WEISS et al., 1977). In most but not all cases, it has been demonstrated that metallic mercury is the volatile end product of mercury detoxification, such that mercury resistance and volatilization of elemental mercury are essentially synonymous.

$$R\,Hg^+ \xrightarrow{\text{NADPH} \quad (1) \quad \text{NADP}^+} R\,H$$

$$Hg^{2+} \underset{(2)}{\overset{\text{NADPH}}{\rightleftharpoons}} Hg^0 + \text{NADP}^+$$

Detoxification of mercury by organomercurial lyase (1) and mercuric reductase (2) enzymes

The most convincing evidence that metallic mercury is the volatile form of mercury produced upon the detoxification of Hg^{2+} to date comes from the combined gas chromatography-mass spectroscopy analysis of the volatilized mercury. Speciation of the volatilized mercury has been examined in a limited number of strains (ROBINSON and TUOVINEN, 1984).

7.3 Mechanisms of Methylation of Mercury
(see also Section 2.6)

The methylation of mercury by microorganisms from soil, sediments, rotting fish, and even the human intestinal tract, but also by cell-free extracts of methanogenic bacteria, has been reported (JENSEN and JERNELÖV, 1969; JERNELÖV and MARTIN, 1975; WOOD, 1974; YAMADA and TOMOMURA, 1972; HAMDY and NOYES, 1975; OLSON et al., 1982; ROWLAND et al., 1975; VONK and SIJPERSTEIJN, 1973). The toxicity of methylmercury to microorganisms depends on its residence time and stability in the water system. Biological methylation of mercury by microorganisms is believed to play a role in the formation of methylmercury in aquatic organisms and sediments and may represent an important link in the mercury cycle. Although methylmercury is more toxic than inorganic mercury, it is more volatile, and therefore methylation may actually be a detoxification mechanism.

Three pathways involving the methylation of mercury have received attention: (1) abiotic or photochemical methylation of Hg^{2+}; (2) the methylation of mercuric ions in sediments by bacteria that excrete methylcobalamin which can act as a methyl donor; and (3) the methylation of mercury by bacterial flora in aquatic organisms, which may also utilize methylcobalamin (RIDLEY et al., 1977).

The mechanism of methylation of mercury remains unclear but appears to involve the non-enzymatic transfer of methyl groups from methylcobalamin to Hg^{2+} (BERTILSSON and NEUJAHR, 1971; DESIMONE et al., 1973; IMURA et al., 1971; RIDLEY et al., 1977). Three major coenzymes are known to be involved in the methyl transfer: (1) N5-methyltetrahydrofolate derivatives; (2) S-adenosylmethionine; and (3) vitamin B_{12} (methylcobalamin). Methylcobalamin is believed to be responsible for the methylation of inorganic Hg^{2+} salts because it is the only agent capable of transferring carbanion methyl groups (BERTILSSON and NEUJAHR, 1971;

DeSimone et al., 1973). The methylation reaction is believed to proceed via electrophilic attack of the mercuric ion on the carbanion species which is stabilized by the cobalt atom. The overall reaction proceeds as follows:

$$Hg^{2+} \xrightarrow{CH_3-B_{12}} CH_3Hg^+ \xrightarrow{CH_3-B_{12}} (CH_3)_2 Hg$$

The first methylation reaction proceeds 6000 times faster than the second (Stokes and Hall, 1985).

Enzymatic transfer of methyl groups to mercury has also been proposed but has not been clearly demonstrated. Although Hg^{2+} appears to be the most likely direct precursor of methylmercury, soil and aquatic microorganisms have been reported (Matsumura et al., 1971) to be capable of producing dimethylmercury from PMA (phenylmercuric acetate).

Bottom sediments from freshwater aquaria and putrescent homogenates of fish have been shown to produce methylmercury from Hg^{2+} and dimethylmercury from methylmercury, respectively (Jensen and Jernelöv, 1969).

The formation of mono- and dimethylmercury from Hg^{2+} under anaerobic conditions has been demonstrated in cell extracts of methanogenic bacteria isolated from canal mud in Delft, Holland (Yamada and Tonomura, 1972). Methylcobalamin was present at substrate concentrations, and the reaction was shown to require ATP and hydrogen as the source of electrons. The enzymatic transfer of methyl groups from Co^{2+} to Hg^{2+} was proposed as the mechanism of methylation of mercury. However, rapid methylation at higher Hg^{2+} levels suggested that the methyl transfer from Co^{2+} to Hg^{2+} may also proceed via a non-enzymatic pathway. Methylmercury is formed from $HgCl_2$, HgI_2, HgO, $Hg(NO_3)_2$, $HgSO_4$, and $Hg(CH_3COO)_2$ but not from HgS by the anaerobic bacterium *Clostridium cochlearium*. The formation of methylmercury was confirmed by thin layer chromatography and by the degradation of the product by *Pseudomonas* sp.

Methylmercury is said to be produced in aerobic sediments and by pure cultures of aerobic microorganisms (Hamdy and Noyes, 1975; Landner, 1971; Ramamoorthy et al., 1982; Spangler et al., 1973a, b). *Pseudomonas* sp. was found to be involved in the production of methylmercury, when $HgCl_2$ was added to lake sediments, incubated under aerobic conditions. According to Sijpersteijn and Vonk (1974) *Neurospora crassa* produces $^+Hg-S-CH_3$, and not $^+Hg-CH_3$, a reaction which is in any case more likely. A mercury-resistant strain of *Enterobacter aerogenes* isolated from river sediments was capable of methylating mercury, but it was unable to reduce Hg^{2+} to metallic mercury (Hamdy and Noyes, 1975). Pure cultures of *Escherichia coli, Streptococcus, Staphylococcus, Lactobacillus, Bacteroides, Bifidobacterium,* and yeast were examined for the ability to form methylmercury from $HgCl_2$.

In natural environments hydrogen sulfide may be evolved in anoxic sulfur-containing sediments. Mercuric sulfide is formed when divalent mercury ions and sulfide ions are simultaneously present due to the extremely low solubility in water (Craig and Bartlett, 1978; Fagerstrom and Jernelöv, 1971; Gavis and Ferguson, 1972). Therefore, the question of the availability of mercury for methylation in sediments is of interest. Methylmercury is formed from mercuric sulfide by aerobic organic sediments but at much lower rates than those observed for $HgCl_2$. No methylmercury was formed under anaerobic conditions, presumably because of the low redox potential. It is also of interest to note that hydrogen sulfide aids the volatilization of mercury (ROWLAND et al., 1977).

7.4 Detoxification of Organomercury Compounds

The detoxification of organomercurials is believed to result from the cleavage of the carbon-mercury linkage by the organomercurial lyase enzyme followed by the reduction of Hg^{2+} to Hg^0 by the mercuric reductase enzyme. The decomposition of organomercury compounds by mercury-resistant bacteria was first detected in *Pseudomonas* spK62 soil isolate. About 70% of the radiolabeled compounds was shown to disappear from the medium within 2 hours.

Broad spectrum mercury resistant plasmids of *Pseudomonas aeruginosa* encoded resistance to phenylmercury acetate (PMA), methylmercury chloride (MMC), ethylmercury chloride (EMC), thiomersal, merbromin, FMA, and pHMB, but the bacterium is capable of degrading only the first four compounds. In contrast, the broad-spectrum mercury resistance plasmids of *Escherichia coli* encode volatilization of mercury from PMA and thiomersal. The broad-spectrum strains of *Escherichia coli* are sensitive to EMC, MMC, and pHMB, but they are able to volatilize mercury from these compounds at a very slow rate. In the above cases the decomposition was measured by the volatilization of mercury.

The inducible nature of the mercury resistance systems has been well documented. Mercury resistance systems have been described for *Escherichia coli*, *Staphylococcus aureus*, and *Pseudomonas* spp. , and over 100 strains have been shown to possess inducible enzyme systems (BENNETT et al., 1978; FURUKAWA et al., 1969; FURUKAWA and TONOMURA, 1972; KLECKNER et al., 1977; KOMURA and IZAKI, 1971; SCHOTTEL, 1978; SUMMERS and SUGARMAN, 1974). Cross-induction of the mercury resistance system appears to show that induction by an organomercurial or Hg^{2+}, to which the strain is resistant, will induce resistance to other Hg compounds.

ROBINSON and TUOVINEN (1984) suggested that the resistant population was homogeneous and that mercury resistance was not being selected from a heterologous population. Previous growth in the presence of Hg^{2+} or an organomercurial compound at subinhibitory levels eliminates the lag phase observed with mercury resistant strains. Hg^r and Hg^s strains can be distinguished in that the sensitive strains are not capable of producing either one of the enzymes responsible for the detoxification of mercury compounds. Because detoxification of Hg^{2+} or organomercurials ultimately results in the volatilization of metallic mercury from the growth medium, the measurement of the enzyme levels of volatilization activity or both in response to added ^{203}Hg compounds may be used to confirm the inducible nature of the system (CLARK et al., 1977; FOSTER et al., 1979; FOSTER and NAKAHARA, 1979; FURUKAWA and TONOMURA, 1972; SCHOTTEL, 1978; SCHOTTEL et al., 1974; SUMMERS and SILVER, 1972). It is interesting that in both induced Hg^r strains of *Escherichia coli* and *Pseudomonas aeruginosa*, rates of volatilization of mercury from organomercurials were lower than those observed for Hg^{2+} under comparable conditions.

Separate enzymes appear to be involved in the detoxification of mercury and organomercury compounds. The organomercurial lyase enzyme hydrolyzes the organomercurials such as PMA to produce benzene and Hg^{2+}, which is slowed by the reduction of Hg^{2+} to Hg^0 by mercuric reductase.

7.5 The Gram Negative Systems

The most studied examples of plasmid determined mercury resistance (*mer*) are the narrow spectrum loci encoded by Tn501 (BENNETT et al., 1978; BROWN et al., 1980, 1983; BROWN and GODDETTE, 1984; FOX and WALSH, 1982, 1983; LUND et al., 1986; MISRA et al., 1984, 1985) and Tn21 (BARRINEAU et al., 1984; DE LA CRUZ and GRINSTED, 1982; IZAKI et al., 1974; JACKSON and SUMMERS, 1982; MISRA et al., 1984, 1985; RINDERLE et al., 1982) and the broad spectrum locus encoded by the IncM plasmid R831b (OGAWA et al., 1984; SCHOTTEL, 1978). Each of these loci encodes a Hg(II) reductase (HR) enzyme and a Hg(II) uptake system whose expression is induced by mercury compounds. Plasmid R831b also encodes an inducible enzyme, the organomercurial lyase (OL), which splits the covalent carbon-mercury bonds in such compounds as phenylmercuric acetate (PMA) (SCHOTTEL, 1978). The broad spectrum locus of R831b is not known to transpose; however, transposable resistance to PMA has been reported (RADFORD et al., 1981). Although the mechanism of detoxification appears to be the same, the range of substrates and inducers varies from strain to strain with patterns emerging and conserved along species lines. The mercury resistance genes, as for other resistance genes carried by the plasmids, often occur on transposons (BENNETT et al., 1978; STANISICH et al., 1977). These transposable elements are specific DNA sequences that can insert more or less randomly into other DNA sequences in the absence of host cell-mediated recombination functions. It is generally accepted that these elements play an important role in evolution.

About 10% of Gram negative bacteria are resistant to inorganic mercury and are also able to degrade phenylmercuric acetate and a variety of other aryl and alkyl mercurials. The enzyme responsible for this biotransformation is the organomercurial lyase, previously given the genotype designation *merB* in genetic studies with *Staphylococcus* plasmid. The OL reductively removes the organic moiety from certain organomercurials, yielding in the case of PMA benzene and mercuric acetate.

7.6 The Structural Genes of the *mer* Loci, Hg(II) Reductase

The best understood of the *mer* genes is a *merA*, which encodes for HR. This inducible, NADPH-dependent enzyme is found in the cytosol (SUMMERS and SUGARMAN, 1974). Recent studies indicate that the HR reaction mechanism involves the formation of a charge-transfer complex between the thiolate of Cys 140 and FAD (SCHULTZ et al., 1985). The resting enzyme (Eox) must first be activated by reduction with NADPH to an EH2 form, which binds Hg(II) but cannot reduce it. Hg(II) reduction requires another NADPH and the formation of EH2-NADPH. HR is also subject to inhibition by exogenous thiols which in excess prevent the formation of the catalytically viable monodentate Hg(II) ligand with the enzyme (MILLER et al., 1985; SCHULTZ et al., 1985).

At the DNA level, the lowest overall homology in the *mer* operons is found in *merA*, though there are regions of very high homology within the *HR* structural gene (e.g., the active site). Codon-by-codon comparison of *HR* genes of Tn21 and Tn501 (PERLER et al., 1980) highlights regions of conservation and divergence. While the enzymatic activity of HR has been found exclusively in the cytoplasm (SUMMERS and SUGARMAN, 1974), small amounts of the protein appear to partition with the cytoplasmic membrane (JACKSON and SUMMERS, 1982). This suggests that HR may interact directly with the membrane bound elements of the Hg(II) uptake system.

The early findings that HR is cytosolic rather than periplasmic provoked the hypothesis that there is a system for bringing Hg(II) through the cell wall to the enzyme (SUMMERS and SUGARMAN, 1974). Strains carrying certain subcloned fragments of certain insertional mutations of the operon were more sensitive to Hg(II) than strains that had no plasmid at all. This mercury supersensitive phenotype (Hg_{ss}) was associated with a complete loss of HR activity and the presence of an inducible Hg(II) uptake system.

8 Cadmium Compounds

(see also Sections 2.7 and 2.8)

Clays and other minerals absorb metal ions such as Cd(II)-ions and can therefore reduce their concentration in soil water and their bioavailability – also for microorganisms. Montmorillonite and kaolinite protect bacteria, actinomycetes, and filamentous fungi from Cd toxicity. BABICH and STOTZKY (1977) suggested that the availability of Cd is closely linked to the cation exchange capacity of the environment. Both Zn and Mg have been shown to reduce Cd toxicity to bacteria.

Oak saplings were sprayed with Zn(II), Pb(II), and Cd(II) to simulate smelter effects which showed little effect on viable counts of bacteria, yeast and filamentous fungi. However, fewer microorganisms were isolated from oak trees near the smelter area than from control groupings of oak trees (BEWLEY, 1979). Examination of bacteria, yeast, and filamentous fungi resistance or tolerance to Zn(II), Pb(II), and Cd(II) in vitro and in vivo indicated that bacteria proved to be less tolerant in vitro but were found in high numbers under field conditions at all metal concentrations. Yeasts proved to be very tolerant under both in vitro and in vivo conditions, while filamentous fungi appeared to be more resistant than bacteria in vitro but were decreased by Pb contamination (BEWLEY, 1979).

HOUBA and REMACLE (1980) showed that in heavily Cd(II) polluted environments 16% of the Gram negative strains of bacteria could grow on a medium containing 100 $\mu g\ mL^{-1}$ Cd. This medium contained bactotryptone, yeast extract, sodium chloride, glucose, agar, and water. Thus, the comments in the media section suggest that major binding constituents were avoided and true resistance was demonstrated for these strains.

Cadmium(II)-ions have a significantly greater effect on microorganisms than Cd complexed as $Cd(CN_4)^{2-}$ (CENCI and MOROZZI, 1977). Complexing cadmium with cyanide appeared to decrease inhibitory effects on microbial growth in activated sludge systems. Further Cd^{2+} caused a greater lag in mixed microbial population from activated sludge systems than $Cd(CN_4)^{2-}$. However, there were no differences between total substrate utilization (STERRITT and LESTER, 1980). At this time it appears that Cd bioaccumulates through transport routes similar to those for Zn(II) and Mn(II) (CRAIG, 1980).

9 Arsenic Compounds
(see also Chapter II.3 and CHALLENGER, 1978)

The reduction and methylation of arsenate by *Methanobacterium* under anaerobic conditions have been reported. Initially, arsenate is probably reduced to arsenite and then methylated to methylarsines. WOOD (1977) studied the synthesis of dimethylarsines from arsenate in a reaction that requires methylcobalamin and methane synthetase. Methylarsine, dimethylarsine, arsine, and methane were produced from methyl(aquo)cobaloxamine-As2O3 DTE in water. A reaction between As^{3+} and CH_3^- (from the cobaloxime) to produce CH_3As^{2+} has been suggested. However, because the existence of As^{3+} cations in aqueous solution is very unlikely, a displacement reaction of the following type appears to be more likely:

$$CH_3^- + As(OH)_3 \dashrightarrow CH_3As(OH)_2 + OH^-$$

Methanearsonic acid is a herbicide for some grass species. Very little is known about the molecular interaction of this acid or its salts with biologically important compounds. The formation of this compound merits serious consideration, as the trivalent arsenic interacts with protein thiol groups, inactivating, for instance, enzymes. Pentavalent arsenic compounds are also expected to react with thiol groups in biologic systems. If the conversion of methanearsonic acid to methylbis(alkylthio)arsine, (carried out in a tube) occurs in a cell, disturbance of enzyme activities is very likely. The mechanism of this type of reduction is not known in detail.

Dimethylarsinic acid (cacodylic acid) and its salts find widespread use as postemergence contact herbicides. Identical reactions with that of methylarsonic acid have been reported. The arsenic-carbon bonds are very stable, but are cleaved by heating with solid sodium hydroxide or chromium trioxide. It has been reported that cacodylic acid reacts with $HSCH_2CONH_2$ and $HSCH_2CH_2(NH)COOH$ to produce the trivalent arsenic derivatives R_2As-SR. Alkylarsines and dialkylarsines have been detected as products formed by the reduction and methylation of inorganic and methylarsenic acids. All these compounds easily condense with the sulfhydryl groups to form alkylbis(organylthio)arsines.

$$RAsX_2 + 2HSR^1 ---\rightarrow RAs(SR^1)_2 + 2HX$$

If thiol groups are present in enzymes, trivalent arsenic compounds can form stable bonds with them, thus preventing the enzymes from functioning properly. The likely reaction takes place between lipoic acid, a building block of the enzyme pyruvate oxidase, and a trivalent alkyldihaloarsine. Dialkylhaloarsines and dialkylhydroxyarsines react similarly with thiols, but cannot form the stable neutral ring compounds with dithiols.

Soils are usually contaminated with arsenic compounds through the use of pesticides, although some contamination occurs from smelting operations, burning of cotton wastes, and fallout from the burning of fuel (Fig. I.7h-2).

Most arsenic in water is added through industrial discharges. The highest concentrations, other than those occurring naturally in spring waters, are usually in areas of high industrial activity. In the soil environment, insoluble or slightly soluble compounds are constantly being resolubilized and made available to the life systems (Fig. I.7h-2).

Arsenic is ubiquitous in the plant kingdom. In green algae, the amount of arsenic varies inversely with the apparent chlorophyll content, from 0.05 to 5.0 ppm on a dry weight basis. For brown algae, values of about 30 ppm have been reported.

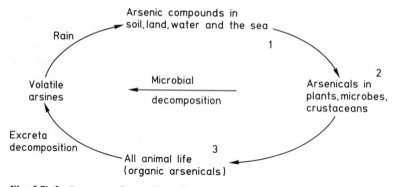

Fig. I.7h-2. A proposed arsenic cycle. *1* The cycle in nature, *2* marine algae may contain large amounts of arsenic, *3* edible tissues of food animals contain average amounts of arsenic.

10 Selenium Compounds

(see also Section 2.6 and Chapter II.25)

Selenium plays a considerable role in biochemical systems. It is essential to the health of animals and human beings, but at the same time may be poisonous in higher concentrations (GISSEL-NIELSON and GISSEL-NIELSON, 1973).

The fact that selenium compounds volatilize from the soil has been demonstrated by the use of ^{75}Se in the form of sodium selenite as a soil additive (ZIEVE and

PETERSON, 1981). The rate of selenium volatilization from soil depends on many factors: temperature, moisture, time, season of the year, microbiological activity, and concentration of water-soluble selenium.

Material balance calculations show that the amount of soil microbiological volatilization of selenium approaches the anthropogenic flow of selenium into the atmosphere.

Compounds of selenium and of sulfur are biochemically competitive in some microorganisms. A suggested mechanism for the reduction of selenite in certain microorganisms involves attachment to a carrier protein and the transformation from Se(IV) to Se^0 (WILBER, 1980). Details of the oxidation of selenium in microorganisms are not known. The same author pointed out that certain species of molds methylate inorganic compounds of selenium. The first step is the formation of methaneselenonic acid followed by the production of the ion of methane seleninic acid, which in the presence of free methane forms a dimethyl selenide. The mechanism by which some microorganisms adapt to increased levels of environmental selenium is obscure. SARATHCHANDRA et al. (1981) reported that elemental selenium can be oxidized to form selenite by the action of the microorganism *Bacillus megaterium*. The concentrations of selenium in various species of mushrooms growing wild in Norway has been described by many workers. An ecological magnification factor was calculated for the different species: ^{75}Se in organism/^{75}Se in ambient water (NASSOS et al., 1980). Some interesting facts on the movement of selenium through aquatic ecosystems have been reported recently (e.g., DE PINTO et al., 1981).

Mechanism(s) of resistance to Group V and VI elements, including arsenic, selenium, and tellurium, is poorly understood. Selenium resistance in certain bacterial genera is well known (see Sect. 2.6) and the incorporation of selenium salts into selective media for the isolation of salmonellae and certain bacilli is a standard practice (LAPAGE and BASCOMB, 1968).

Biotransformations of selenium in aquatic environments, soil, sediment, and sewage have been reviewed by DORAN (1982). Addition of selenate or selenite to lake water or addition of selenite or elemental selenium to soil or sewage sludge leads to the emission of volatile methylated selenium species as well as the production of insoluble selenium metal (CHAU et al., 1976; REAMER and ZOLLER, 1980). Biomethylation mechanisms for selenium have been reviewed, and it has been shown that the resulting dimethylselenide is less toxic than inorganic selenium species (NRC, 1976).

References

Abelson, P.H., and Aldous, E., (1950), *J. Bacteriol. 60*, 401–413.
Albright, L.J., Wentworth, J.W., and Wilson, E.N. (1972), *Water Res. 65*, 1589–1596.
Babich, H., and Stotzky, G. (1977), *Appl. Environ. Microbiol. 33*, 696–705.
Baker, A.J.M. (1981), *J. Plant Nutr. 3*, 643–654.

Baker, D. E., Amacher, M. C., and Doty, W. D. (1976), *Monitoring Sewage Sludges, Soils and Crops for Zinc and Cadmium,* in *Proceedings Eighth Cornell Agricultural Waste Management Conference, Land as a Waste Management Alternative,* pp. 261–281. Ann Arbor Science Publishers, Ann Arbor, Michigan.
Baldry, M. G. C., Hogarth, D. S., and Dean, A. C. R. (1977), *Microbios Lett. 4,* 7–16.
Barkay, T., Olson, B. H., and Colwell, R. R. (1979), in *Management and Control of Heavy Metals in the Environment,* Vol. 7, pp. 356–363. CEP Consultants Ltd., Edinburgh.
Barkay, T., Fouts, D. L., and Olson, B. H. (1985), *Appl. Environ. Microbiol. 49,* 686–692.
Barrineau, P., Gilbert, P., Jackson, W. J., Jones, C. S., Summers, A. O., and Wisdom, S. (1984), *J. Mol. Appl. Genet. 2,* 601–619.
Bennett, P., Grinsted, M. J., Choi, C. L., and Richmond, M. H. (1978), *Mol. Gen. Genet. 159,* 101–106.
Bertilsson, L., and Neujahr, H. Y. (1971), *Biochemistry 10,* 2805–2808.
Bewley, R. J. F. (1979), *J. Gen. Microbiol. 110,* 247–254.
Bewley, R. J. F., and Campbell, R. (1978), *Trans. Br. Mycol. Soc. 71,* 508–511.
Bewley, R. J. F., and Campbell, R. (1980), *Microb. Ecol. 6,* 227–240.
Bitton, G., and Freihofer, J. (1978), *J. Microbiol. Ecol. 4,* 119–125.
Blaghen, M., Lett, M. C., and Vidon, D. J. M. (1983), *FEMS Microbiol. Lett. 19,* 93–96.
Booth, J. E., and Williams, J. W. (1984), *J. Gen. Microbiol. 130,* 725–730.
Bowen, H. J. M. (1966), *Trace Elements in Biochemistry.* Academic Press, New York.
Brock, I. T. D. (1978), *Thermophilic Microorganisms and Life at High Temperatures.* Springer-Verlag, New York.
Brown, N. L., and Goddette, D. (1984), in: Bray, R. C., Engel, P. C., and Mayhew, S. G. (eds.), pp. 165–168. de Gruyter, Berlin.
Brown, N. L., Choi, C. L., Grinsted, J., Richmond, M. H., and Whitehead, P. R. (1980), *Nucleic Acids Res. 8,* 1933–1945.
Brown, N. L., Ford, S. J., Pridmore, R. D., and Fritzinger, D. C. (1983), *Biochemistry 22,* 4089–4095.
Burton, G. A., Giddings, J., DeBrine, T. H., and Fall, R. (1987), *Appl. Environ. Microbiol. 53*(1), 185–188.
Cairns, J. (1980), *Scenarios on Alternative Futures for Biological Monitoring, 1978–1985,* in: Worf, C. D. L. (ed.): *Biological Monitoring for Environmental Effects,* pp. 11–12. Lexington Books, Lexington, Massachusetts.
Cannon, H. L. (1971), *Taxonomy 20,* 227–256.
Cenci, G., and Morozzi, G. (1977), *Sci. Total Environ. 7,* 131–143.
Challenger, F. (1978), *Biosynthesis of Organometallic and Organometalloid Compounds,* in: Brinckman, F., and Bellema, J. M. (eds.): *Organometals and Organometalloids. ASC Symp. Ser. 82,* 1. American Chemical Society, Washington, DC.
Chau, Y. K., Wong, P. T. S., Silverberg, B. A., Luxon, P. L., and Bengert, G. A. (1976), *Science 192,* 1130–1131.
Chopra, I. (1975), *Antimicrob. Agents Chemother. 7,* 8–14.
Clark, D. L., Weiss, A. A., and Silver, S. (1977), *J. Bacteriol. 132,* 186–196.
Clements, F. E. (1920), *Plant Indicators. The Relation of Plant Communities to Process and Practice.* Carnegie Institute of Washington, Publ. No. 290. Washington, DC.
Cole, M. A. (1979), *Soil Sci. 127,* 313–317.
Compeau, G. C., and Bartha, R. (1985), *Appl. Environ. Microbiol. 50* (2), 498–502.
Craig, P. J. (1980), in: Hutzinger, O. (ed.): *The Handbook of Environmental Chemistry,* Vol. 1, Part A, pp. 169–227. Springer-Verlag, Berlin.
Craig, P. J., and Bartlett, P. D. (1978), *Nature 275,* 635–637.
De La Cruz, F., and Grinsted, J. (1982), *J. Bacteriol. 151,* 222–228.
De Pinto, J. W., Young, T. C., and Martin, S. C. (1981), *J. Water Pollut. Contam. Fed. 53*(6), 999–1007.

De Simone, R. E., Penley, M. W., Charbonneau, L., Smith, S. G., Wood, J. M., Hill, H. A. O., Pratt, J. M., Ridsdale, S., and Williams, R. J. P. (1973), *Biochim. Biophys. Acta 304*, 851–863.
Doelman, P. (1978), in: Nriagu, J. O. (ed.): *The Biogeochemistry of Lead in the Environment*, pp. 343–353. Elsevier/North-Holland Biomedical Press, Amsterdam.
Doelman, P., and Haanstra, L. (1979), *Soil Biol. Biochem. 11*, 487–491.
Doran, J. W. (1982), *Adv. Microb. Ecol. 6*, 17–32.
Dyke, K. G. H., Parker, M. T., and Richmond, M. H. (1970), *J. Med. Microbiol. 3*, 125–136.
Ehrlich, H. L. (1978), *Geomicrobiol. J. 1*, 69–87.
Fagerstrom, T., and Jernelöv, A. (1971), *Water Res. 5*, 121–122.
Fagerstrom, T., and Jernelöv, A. (1972), *Water Res. 6*, 1193–1202.
Faust, S. D., and Aly, O. M. (1981), *Chemstry of Natural Waters*. Ann Arbor Science Publishers, Ann Arbor, Michigan.
Foot, C. H., and Taylor, C. B. (1949), *Proc. Soc. Appl. Bacteriol. 1*, 11–13.
Foster, T. J. (1983), *Microbiol. Rev. 47*, 361–409.
Foster, T. J., and Nakahara, H. (1979), *J. Bacteriol. 140*, 301–305.
Foster, T. J., Nakahara, H., Weiss, A. A., and Silver, S. (1979), *J. Bacteriol. 140*, 167–181.
Fox, B., and Walsh, C. T. (1982), *J. Biol. Chem. 257*, 2498–2503.
Fox, B., and Walsh, C. T. (1983), *Biochemistry 22*, 4082–4088.
Friberg, L., and Vostal, J. (eds.) (1972), *Mercury in the Environment*. CRC Press, Boca Raton, Florida.
Friello, D. A., and Chakrabarty, A. M. (1980), *Transposable Mercury Resistance in Pseudomonas putida*, in: Stuttard, C., and Rozee, K. R., (eds.): *Plasmids and Transposons: Environmental Effects and Maintenance Mechanisms*, pp. 249–260. Academic Press, New York.
Furukawa, K., and Tonomura, K. (1972), *Agric. Biol. Chem. 36*, 2441–2448.
Furukawa, K. T., Suzuki, T., and Tonomura, K. (1969), *Agric. Biol. Chem. 33*, 128–130.
Gadd, G. M., and Griffiths, A. J. (1978), *Microb. Ecol. 4*, 303–317.
Gavis, J., and Ferguson, J. F. (1972), *Water Res. 6*, 989–1008.
Gissel-Nielson, G., and Gissel-Nielson, M. (1973), *AMBIO 2,*, 114–117.
Goodman, G. T. (1974), *Proc. R. Soc. London 185B*, 127–148.
Hallberg, R. O., and Pickard, D. T. (1981), *The Small "Miners" of Falun. Proceedings 5th International Symposium on Environmental Biogeochemistry, Stockholm*.
Hamdy, M. K., and Noyes, O. R. (1975), *Appl. Microbiol. 30*, 424–432.
Houba, C., and Remacle, J. (1980), *Microb. Ecol. 6*, 55–69.
Imura, N., Sukegawa, E., Pan, S., Nagao, K., Kim, J., Kwan, T., and Ukita, T. (1971), *Science 172*, 1248–1249.
Izaki, K. (1977), *J. Bacteriol. 131*, 696–698.
Izaki, K., Tashiro, Y., and Funaba, T. (1974), *J. Biochem. 75*, 591–599.
Jackson, W. J., and Summers, A. O. (1982), *J. Bacteriol. 151*, 962–970.
Jarvie, A. W. P., Markall, R. N., and Potter, H. R. (1975), *Nature 255*, 17.
Jensen, S., and Jernelöv, A. (1969), *Nature 223*, 753–753.
Jensen, S., and Jernelöv, A. (1972), *Behavior of Hg in the Environment*, in: *Mercury Contamination in Man and His Environment*, JAEA (International Atomic Energy Agency), *Tech. Rep. Ser. 137*, 43.
Jernelöv, A., and Martin, A. L. (1975), *Annu. Rev. Microbiol. 29*, 61–77.
Kleckner, N., Roth, J., and Batstein, D. (1977), *Genetic Engineering in vivo Using Translocatable Drug Resistance Elements. Mol. Biol. 116*, 125ff.
Komura, I., and Izaki, K. (1971), *J. Biochem. (Tokyo) 70*, 885–893.
Komura, I., Funaba, T., and Izaki, K. (1971), *J. Biochem. (Tokyo) 70*, 895–901.
Kondo, I., Ishikawa, T., and Nakahara, H. (1974), *J. Bacteriol. 117*, 1–7.
Landner, L. (1971), *Nature 230*, 452–454.
Lapage, S. P., and Bascomb, S. (1968), *J. Appl. Bacteriol. 31*, 568–580.
Lund, P. A., Ford, S. J., and Brown, N. L. (1986), *J. Gen. Microbiol. 132*, 465–480.
Magos, L. A., Tuffery, A., and Clarkson, T. W. (1964), *Br. J. Ind. Med. 21*, 294–298.

Matsumura, F., Gotoh, Y., and Boush, G.M. (1971), *Science 173*, 49–51.
McBride, B.C., Merilees, H., Cullen, W.R., and Pickett, W. (1978), *ACS Monogr. 82*, p. 94.
Mellor, D.P., and Malley, L. (1948), *Nature 161*, 436–437.
Miller, J.H., et al. (1985), *Mutagenic Specificity. J. Mol. Biol. 182*, 45–68.
Misra, T.K., Brown, N.L., Fritzinger, D.C., Pridmore, R.D., Barnes, W.M., et al. (1984), *Proc. Natl. Acad. Sci. USA 81*, 5975–5979.
Misra, T.K., Brown, N.L., Haberstroh, L., Schmidt, A., Goddette, D., and Silver, S. (1985), *Gene 34*, 253–262.
Nakahara, H., Ishikawa, T., Sarai, Y., Kondo, I., Kozukue, H., and Mitsuhashi, S. (1977a), *Antimicrob. Agents Chemother. 11*, 999–1003.
Nakahara, H., Ishikawa, T., Sarai, Y., Kondo, I., Kozukue, H., and Silver, S. (1977b), *Appl. Environ. Microbiol. 33*, 975–976.
Nakahara, H., Ishikawa, T., Sarai, Y., Kondo, I., and Mitsuhashi, S. (1977c), *Nature 266*, 165–167.
Nassos, P.A., Coats, J.R., Metcalf, R.L., Brown, D.D., and Hanson, L.G. (1980), *Bull. Environ. Contam. Toxicol. 24*(5), 752–758.
Nelson, J.D., Blair, W., Brinckman, F.E., Colwell, R.R., and Iverson, W. (1973), *Appl. Microbiol. 26*, 321–326.
Norris, P.R., Brierley, J.A., and Kelly, D.P. (1980), *Physiological Characteristic of Two Facultatively Thermophilic Mineral-oxidizing Bacteria. FEMS Microbiol. Lett. 7*(2), 119–122.
Novick, R.P., and Roth, C. (1968), *J. Bacteriol. 95*, 1335–1342.
Novick, R.P., Edelman, I., Schwesinger, M.D., Gruss, A.D., Swanson, E.C., and Pattee, P.A. (1979), *Proc. Natl. Acad. Sci. USA 76*, 400–404.
NRC (National Research Council) (1976), *Selenium*. Committee on Medical and Biologic Effects of Environmental Pollutants. National Academy of Sciences, Washington, DC.
Ogawa, H.I., Tolle, C.L., and Summers, A.O. (1984), *Gene 32*, 311–320.
Olson, B.H. (1978), in: Loutit, M.W., and Miles, J.A.R. (eds.): *Microbiol Ecology*, pp. 416–422. Springer-Verlag, Berlin.
Olson, B.H., and Cooper, R.C. (1976), *Water Res. 10*, 113–116.
Olson, B.H., and Ford, S. (1986), *Implications of Quantification of Bacterial Genotypes in Polluted Environments*, in: Lester, J.N., Perry, R., and Sterritt, R.M. (Eds.): *Chemicals in the Environment*, pp. 151–158. Selper Ltd., London.
Olson, B.H., Barkay, T., and Colwell, R.R. (1979), *Appl. Environ. Microbiol. 20*, 478–485.
Olson, G.J., Porter, F.D., Rubinstein, J., and Silver, S. (1982), *J. Bacteriol. 151*, 1230–1236.
Olson, B.H., Ford, S., and Lester, J. (1987), *The Occurrence of MerR and MerC Gene Sequences among Mercury Resistant Determinants in River Sediments Containing Elevated Levels of Mercury. Oceans*, 173–179.
Panigrahi, A.K. (1980), *Eco-physiological Effects of Mercurial Compounds on Some Freshwater Fishes*. Ph.D. Thesis, Berhampur University, Orissa, India.
Panigrahi, A.K., and Misra, B.N. (1989), *Mercury in the Environment*. Naya Prakash Publications, Calcutta, India.
Pan-Hou, H.S.K., and Imura, N. (1982), *Arch. Microbiol. 131*, 176–177.
Pan-Hou, H.S.K., Hosono, M., and Imura, N. (1980), *Appl. Environ. Microbiol. 40*(5), 1007–1011.
Patrick, F.M., and Loutit, M. (1976), *Water Res. 10*, 333–335.
Perler, F., Efstratiadis, A., Lomedico, P., Gilbert, W., Kolodoer, R., and Dodgson, J. (1980), *The Evolution of Genes: The Chicken Preproinsulin Gene. Cell 20*, 555–556.
Peterson, P.J. (1971), *Sci. Prog. Oxford 59*, 505–526.
Pickett, A.W., and Dean, A.C.R. (1976a), *Microbios 15*, 79–91.
Pickett, A.W., and Dean, A.C.R. (1976b), *Microbios Lett. 1*, 165–167.
Pickett, A.W., Carte, I.S., and Dean, A.C.R. (1976), *Microbios 15*, 105–111.
Pickup, R.W., Lewis, R.J., and Williams, P.J. (1983), *J. Bacteriol. 129*, 153–158.
Porter, F.D., Silver, S., Ong, C., and Nakahara, H. (1982), *Antimicrob. Agents Chemother. 22*, 852–858.

Radford, A. J., Oliver, J., Kelly, W. J., and Reanney, D. C. (1981), *J. Bacteriol. 147*, 1110–1112.
Ramamoorthy, S., and Kushner, D. J. (1975), *Microb. Ecol. 2*, 162–176.
Ramamoorthy, S., Cheng, T. C., and Kushner, D. J. (1982), *Bull. Environ. Contam. Toxicol. 29*, 167–173.
Rath, P., Panigrahi, B. K., and Misra, B. N. (1983), *Microbios Lett. 29*, 25–29.
Rath, P., Panigrahi, A. K., and Misra, B. N. (1985), *J. Environ. Biol. 4*, 103–109.
Rath, P., Panigrahi, A. K., and Misra, B. N. (1986a), *Environ. Pollut. 42*, 143–149.
Rath, P., Panigrahi, A. K., and Misra, B. N. (1986b), *Microbios Lett. 31*, 15–20.
Reamer, D. C., and Zoller, W. H. (1980), *Science 208*, 500–502.
Reay, J. S. S. (1979), *Philos. Trans. R. Soc. London 290A*, 609–623.
Reisinger, K., Stoeppler, M., and Nürnberg, H. W. (1981), *Nature 291*, 228–230. *Proceedings International Conference on Heavy Metals in the Environment, Amsterdam,* pp. 649–652. CEP Consultants Ltd., Edinburgh.
Reisinger, K. (1987), *Ph. D. Thesis,* Technische Universität Aachen, FRG.
Ridley, W. P., Dizikes, L. J., and Wood, J. W. (1977), *Science 197*, 329–332.
Rinderle, S. J., Booth, J. E., and Williams, J. W. (1982), *Biochemistry 22*, 869–876.
Robinson, J. B., and Tuovinen, O. H. (1984), *Microbiol. Rev. 48*, 95–124.
Rowland, I. R., Grasso, P., and Davies, M. J. (1975), *Experientia 31*, 1064–1065.
Rowland, I. R., Davies, M. J., and Grasso, P. (1977), *Nature 265*, 718–719.
Sahu, A. (1987), *Toxicological Effects of Pesticide on a Blue-green Alga.* Ph. D. Thesis, Berhampur University, Orissa, India.
Sahu, A., Shaw, A. P., Panigrahi, B. K., and Misra, B. N. (1987), *Microbios Lett. 33*, 45–50.
Sarathchandra, S. U., et al. (1981), *Science 211*, 600–601.
Saxena, J., and Howard, P. H. (1977), *Adv. Appl. Microbiol.*, 185–226.
Schottel, J. L. (1978), *J. Biol. Chem. 253*, 4341–4349.
Schottel, J., Mandal, A., Clark, D., Silver, S., and Hedges, R. W. (1974), *Nature 251*, 335–337.
Schultz, A. M., Copeland, T. D., Mark, G. E., Rapp, U. R., and Oroszlan, S. (1985), *Detection of Myristylated GAG-RAF Transforming Protein with RAF-specific Antipeptide Sera. Virology 146* (1), 78–89.
Seigel, B. Z. (1983), *Science 219*, 255.
Shaw, B. P. (1987), *Eco-physiological Studies on the Effluent from a Chloralkali Factory on Biosystems.* Ph. D. Thesis, Berhampur University, Orissa, India.
Sijpersteijn, A. H., and Vonk, S. W. (1974), *Antonie von Leeuwenhoek J. Microbiol. Serol. 40*, 393.
Silver, S. (1983), in: Westbroek, C. P., and de Jong, E. W. (eds.): *Bacterial Interactions with Mineral Cations and Anions Good Ions and Bad, in Biomineralization and Biological Metal Accumulation.* Reidel, Dordrecht.
Silver, S. (1989), *Biogeochemical Cycles of Metals and Human Health.* Dahlem Konferenzen, Berlin, in press.
Silver, M., and Torma, A. E. (1974), *Can. J. Microbiol. 20*, 141–147.
Silver, S., Budd, K., Leahy, K. M., Shaw, W. V., Hammond, D., Novick, R. P., Willsky, G. R., Malamy, M. H., and Rosenberg, H. (1981), *J. Bacteriol. 146*, 983.
Spangler, W. J., Spigarelli, J. L., Rose, J. M., Flippin, R. S., and Miller, H. H. (1973a), *Degradation of Methylmercury by Bacteria Isolated from Environmental Samples. Appl. Microbiol. 25*, 488–493.
Spangler, W. J., Spigarelli, J. L., Rose, J. M., and Miller, H. H. (1973b), *Methylmercury: Bacterial Degradation in Lake Sediments. Science 180*, 192–193.
Stanisich, V. A., Bennett, P. M., and Richmond, M. H. (1977), *Characterization of a Translocation Unit Encoding Resistance to Mercuric Ions that Occurs on a Nonconjugative Plasmid in Pseudomonas aeruginosa. J. Bacteriol. 129*, 1227–1233.
Sterritt, R. M., and Lester, J. N. (1980), *Sci. Total Environ. 14*, 5–17.
Stokes, H. W., and Hall, B. G. (1985), *Sequence of the ebgR Gene of Escherichia coli – Evidence that the ebg and lac Operons are Descended from a Common Ancestor. Mol. Biol. Evol. 2* (6), 478–483.

Summers, A.O., and Jacoby, G.A. (1977), *J. Bacteriol. 129*, 276–281.
Summers, A.O., and Lewis, E. (1973), *J. Bacteriol. 113*, 1070–1072.
Summers, A.O., and Silver, S. (1972), *J. Bacteriol. 112*, 1228–1236.
Summers, A.O., and Silver, S. (1978), *Annu. Rev. Microbiol. 32*, 637–672.
Summers, A.O., and Sugarman, L.I. (1974), *J. Bacteriol. 119*, 242–249.
Summers, A.O., Schottel, J., Clark, D., and Silver, S. (1974), *Plasmid-borne Hg(II) and Organomercurial Resistance,* in: Schlessinger, D. (ed.): *Microbiology – 1974,* pp. 219–226. American Society for Microbiology, Washington, DC.
Summers, A.O., Jacoby, G.A., Swartz, M.N., McHugh, G., and Sutton, L. (1978), *Metal Cation and Oxyanion Resistances in Plasmids of Gram Negative Bacteria,* in: Schlessinger, D. (ed.): *Microbiology – 1978,* pp. 128–131. American Society for Microbiology, Washington, DC.
Summers, A.O., Weiss, R.B., and Jacoby, G.A. (1980), *Plasmid 3*, 35–47.
Tanaka, M., Yamamoto, T., and Sawai, T. (1983), *J. Bacteriol. 153*, 1432–1438.
Tezuka, T., and Tonomura, K. (1978), *J. Bacteriol. 135*, 138–143.
Timoney, J.F., Port, J., Giles, J., and Spanier, J. (1978), *Appl. Environ. Microbiol. 36*, 465–572.
Tonumura, K., and Kanzaki, F. (1969), *Biochim. Biophys. Acta 184*, 227–229.
Tornabene, T.G., and Edwards, H.W. (1972), *Science 76*, 1334–1335.
Trevors, J.T., Oddie, K.M., and Belliveau, B.H. (1985), *Metal Resistance in Bacteria. FEMS Microbiol. Rev. 32*, 39–54.
Trollope, D.R., and Evans, B. (1976), *Environ. Pollut. 11*, 109.
Troyer, L.S., Olson, B.H., Thornton, I., and Matthews, H. (1981), in: Hemphill, D.D. (ed.): *Trace Substances in Environmental Health XIV.* University of Missouri, Columbia.
Tynecka, Z., Zajac, J., and Gos, Z. (1975), *Acta Microbiol. Pol. 7*(24), 11–20.
Vallee, B.L., and Ulmer, D.D. (1972), *Annu. Rev. Biochem. 41*, 91–128.
Varma, M.M., Thomas, W.A., and Prasad, C. (1976), *J. Appl. Bacteriol. 41*, 347–349.
Vonk, S.W., and Sijpersteijn, A.K. (1973), *Antonie van Leeuwenhoek J. Microbiol. Serol. 39*, 505–513.
Walker, J.D., and Colwell, R.R. (1974), *Appl. Microbiol. 27*, 285–287.
Wallace, R.A., Fulkerson, W., Shutts, W.D., and Lyon, S.W. (1971), *Mercury in the Environment – The Human Element,* uRNL-NSF-EP-1, Oak Ridge National Laboratory, Oak Ridge, Tennessee.
Weiss, A.A., Murphy, S.D., and Silver, S. (1977), *J. Bacteriol. 132*, 197–208.
Weiss, A.A., Schottel, J.L., Clark, D.L., Beller, R.G., and Silver, S. (1978), *Mercury and Organomercurial Resistance with Enteric, Staphylococcal, and Pseudomonad Plasmids,* in: Schlessinger, D. (ed.): *Microbiology – 1978,* pp. 121–124. American Society for Microbiology, Washington, DC.
Wilber, C.G. (1980), *Toxicology of Selenium: A Review. Clin. Toxicol. 17*(2), 171–230.
Wong, P.T.S. (1981), *Proceedings International Conference on Heavy Metals in the Environment, Amsterdam,* pp. 641–648. CEP Consultants Ltd., Edinburgh.
Wong, P.T.S., Chau, Y.K., and Luxon, P.L. (1975), *Nature 253*, 263–264.
Wood, J.M. (1974), *Science 183*, 1049–1052.
Wood, J.M. (1975), *Naturwissenschaften 62*, 357.
Wood, J.M. (1977), *Science 197*, 329–332.
Wood, J.M. (1989), *Transport, Bioaccumulation and Toxicity of Elements in Microorganisms under Environmental Stress. Proceedings International Conference on Heavy Metals in the Environment, Geneva,* Vol. 1, pp. 1–12. CEP Consultants Ltd., Edinburgh.
Wood, J.M., and Wang, H.K. (1983), *Environ. Sci. Technol. 17*, 582A–590A.
Wood, J.M., Kennedy, F.S., and Rosen, C.G. (1968), *Nature 220*, 173–174.
Wood, J.M., Cheh, A., Dizikes, L.J., Ridley, W.P., Rakow, S., and Lakowicz, J.R. (1978), *Fed. Proc. 37*, 16–21.
Yamada, M., and Tonomura, K. (1972), *J. Ferment. Technol. 50*, 159–166; 893–909.
Zieve, R., and Peterson, P.J. (1981), *Sci. Total Environ. 19*(3), 277–284.

I.8 a Metal Loads of Food of Vegetable Origin Including Mushrooms

PETER WEIGERT, Berlin (West), Germany

1 Introduction

Metals and other stable elements occur in varying amounts as natural ingredients of the earth's crust. In principle, they can, therefore, also be determined in variable concentrations in all natural plants and animals and consequently also in food and feed. If, in certain cases, it is impossible to ascertain their exact content this is entirely due to the analytical methods used, which are not sufficiently sensitive in that particular range of concentration. On the other hand, discovery of, e.g., arsenic, cadmium, lead, mercury, or thallium in food and feed is not necessarily a consequence of increasing pollution of the environment per se; in most cases, it is the result of the increasingly efficient chemical and/or physical analysis which permits detection of such elements even in the lowest concentration ranges (see, e.g., SHACKLETTE, 1980; FISHBEIN, 1987; or NRIAGU and PACYNA, 1988). Normally only total metal concentrations were measured, and little is known on speciation.

The concentration of the various elements in plant food and feed is influenced by different factors: the genetic properties of plants, the nature of the soil on which the plants are grown, the climate, and the degree of maturity of the plant at the time of harvesting have a considerable influence on the content of elements in the crop (see also Chapter I.7).

The influence of genetics is demonstrated by the difference in the accumulation of cadmium in rye and wheat grown on the same cultivation area. Appropriate investigations have shown that rye always absorbs distinctly less cadmium from the soil than wheat.

In contrast, rye always exhibits higher concentrations of lead than wheat grown at the same location. The reasons for these discrepancies have not yet been fully clarified. Preliminary indications suggest that these differences are related to the ability of the finest branches of the roots to release tiny quantities of acid. As heavy metals can only be absorbed in solution through the roots and as cadmium is only soluble at a pH of 5 or lower, wheat plants seem to be able to create the appropriate conditions in the proximity of their roots more easily than rye plants.

However, these differences between individual food and feed crops offer a possibility, also in the long term, of exploiting land which is no longer suitable for agricultural purposes. By selecting the most suitable crop, excessive accumulation of heavy metals in food and feed can be avoided.

Apart from genetic factors, the nature of the soil has the most important influence on the content of elements in plant food and feed. Depending on the existing geological conditions of the particular region the element content of the soil can vary by several degrees of magnitude. The concentration of elements in plants is largely correlated to the corresponding concentrations in the soil on which they grow, though differences in the acid content, the humus layer and structure of the soil itself will, to a lesser degree, be detectable in certain plants. Besides the existing heavy metal content of the soil, the elemental content of the soil can be very detrimentally affected by anthropogenic influence, such as the use of contaminated mineral fertilizers, or spreading sewage sludge contaminated with heavy metals or sediments dredged from polluted rivers. A large number of investigations has demonstrated this influence, especially of contaminated sewage sludge, on the heavy metal concentration in food and feed crops. In certain polluted areas the heavy metal content can reach levels several times higher than normal, so that the plants grown there can be used neither as feed nor food according to the applicable regulations.

Factors such as climate or the ripeness of crops can lead to seasonal dependency of the elemental concentration. However, these are generally not as significant as the genetic and soil specific factors.

An additional aspect of the assessment of heavy metals in food is of importance. Not all foods are marketed in their natural state, some have been treated before they become commercially available. The elemental content, especially that adhering to the surface of the plant, can be influenced by means of the normal food technological processing or culinary preparation. Thus, the original content of lead or thallium can be reduced by 80% in certain plants by means of intensive washing (KLEIN, 1982). Such high decontamination factors are not possible for cadmium, as it is predominantly absorbed by the roots and is relatively firmly bound to the tissue structures. In this case a reduction of only 20% compared to the original content is possible as a result of intensive cleaning. However, reduction in the concentration of possibly undesirable elements is not the only result of treatment of the raw goods. On the contrary, the concentration, especially of lead, iron, zinc, and tin, can be distinctly higher in canned goods than in the original raw goods as a result of outdated packaging material, such as uncoated soldered tin cans. However, in keeping with the current developments in food and packaging technology such secondary contamination is no longer defensible, so that these undesirable effects will hopefully be avoided in the near future. A further secondary contamination of prepared foods, which is actually avoidable, is due to the use of drinking water containing lead from out of date household installations (lead pipes) or seasoning containing heavy metals, especially cooking salt.

In the preparation of this chapter about the specific elements the most recent literature (from 1980 onwards) was preferably cited. Furthermore, the analytical methods presented in the publications were critically evaluated with respect to their correctness or comparability with other results, as numerous investigations are known to contain results which are not reproducible due to inadequate analysis (see also Chapter I.4a).

2 Aluminum

There is little reliable information about the content of aluminum in food and feed. Some values are summarized in Table I.8a-1. According to the available data, rye and wheat contain approximately 5.0 mg/kg of aluminum, and rice about 1.5 mg/kg. Concentrations of 1.2–2.5 mg/kg were determined in potatoes.

The content in fruit and vegetables is very variable and ranges from 0.1 to 5.0 mg/kg. Approximately 4.0 mg Al per kg fresh material were found in mushrooms.

The aluminum content in cocoa and coffee beans amounts to 30–45 mg/kg, while it can reach up to 128 mg/kg in tea leaves. It should, however, be noted that only a small amount of aluminum remains in the brewed or percolated beverage. According to recent American investigations 60% of the daily intake of aluminum is supplied by cereals and cereal products (GREGER, 1985). Of course, it must be considered that aluminum in vegetable food may also originate from cooking pots (see also Chapter II.1).

Table I.8a-1. Aluminum Content in Food (mg/kg)

Food	Mean Value (arithm.)	Minimum Value	Maximum Value
Cereals			
Wheat[a]	5.4	–	–
Rye[a]	4.8	–	–
Rice[a]	1.2–1.7	–	–
Potatoes[b]	1.2–2.5	0.1	–
Vegetables			
Lettuce[a]	0.6	–	–
Cabbage[b]	0.1	–	–
Asparagus[a]	5.0	1.0	9.9
Tomatoes[b]	0.1	–	–
Fruit			
Peaches[a]	0.4	–	–
Apples[a]	0.5–2.8	0.05	7.0
Bananas[a]	0.4	–	–
Oranges[a]	0.4	–	–
Mushrooms[a]	4.0	–	–
Cocoa[a]	45.0	–	–
Tea[a]	128.0	–	–
Coffee[a]	(0–30)	–	–

[a] SCHLETTWEIN-GSELL and MOMMSEN-STRAUB (1973)
[b] GREGER (1985)

3 Arsenic

In contrast to marine animal organisms food and feed plants contain only minute amounts of arsenic. Cereals and potatoes exhibit medium amounts of 0.05 mg As per kg, fruit and vegetables between 0.02 – 0.04 mg/kg fresh weight. The concentration in tea leaves, cocoa and coffee beans is about 0.1 mg/kg fresh material (see Table I.8a-2). The highest amounts were found in certain wild mushroom species with a concentration of 1.27 mg arsenic per fresh material. The arsenic concentration in the most important basic human nutrients is of no relevance to health according to current toxicological knowledge. However, it must be admitted that no evaluable data about the kind of chemical bond and the oxidation state of arsenic in foods are available at present. All the available results refer to the total arsenic content. Moreover, it can be concluded from the more recent investigations that most of the above mentioned concentrations are clearly too high due to inefficient analytical methods; realistic values will presumably be no higher than half as much.

Table I.8a-2. Arsenic Content in Food (mg/kg)

Food	Mean Value (arithm.)	Minimum Value	Maximum Value
Cereals[a]	0.05	0.001	0.610
Potatoes[a]	0.07	0.005	0.200
Vegetables[a]	0.04	0.001	0.43
Lettuce, cabbage, tomatoes[b]	0.03	0.01	0.14
Fruit			
Apples[b]	0.02	0.002	0.05
Peaches, cherries[b]	0.04	0.005	0.07
Citrus fruits[b]	0.03	0.005	0.05
Mushrooms[b]	0.14	0.01	1.27
Cocoa[a]	0.1	0.006	1.09
Tea[a]	0.13	ND	0.4
Coffee[a]	0.22	0.01	0.8

ND not detectable
[a] WEIGERT et al. (1984)
[b] BUNDESGESUNDHEITSAMT (1986)

4 Cadmium

As a natural ingredient of the earth's crust, cadmium is contained in all naturally occurring materials and is thus present in all foods, at least in trace amounts. As the content of cadmium in the soil already shows considerable variation under natural conditions, it is not surprising that relatively high cadmium concentrations can be determined in areas due to additional anthropogenic influences. Besides the pollution of the soil by industrial gas emission, deposition of waste including sediments dredged from rivers and especially the disposal of sewage sludge from municipal and commercial sewage treatment plants plays an important role. In addition, the use of cadmium-containing phosphate fertilizer can lead to accumulation of cadmium, especially in plant foods, but also in feed.

The average cadmium content of the cereals greatly varies; thus, the average values of 0.016 mg Cd per fresh weight are found in rye. On the other hand, wheat contains an average of 0.056 mg/kg and rice 0.075 mg/kg fresh material (OCKER, 1985; WEIGERT et al., 1984; see also Chapter I.8b). The reasons for these differences have, however, not yet been clarified.

In contrast to lead or thallium, cadmium is predominantly absorbed through the roots and only slightly through the surface of leaves. Thus, there must be other additional factors which cause its accumulation in the edible parts of the plants. The type of soil, the pH value, the type of bond, and the pre-existing cadmium contamination of the soil are certainly of importance for the availability of cadmium from the soil. In addition to these properties of the soil, there must be plant specific factors, otherwise the differences between wheat and rye grown on the same ground would be inexplicable. The same is true of celery on the one hand and radish on the other. Though these differences in behavior of some nutrient plants pose many questions, they also offer the possibility, from the point of view of minimizing the human intake of cadmium, of choosing to cultivate crops appropriate to the nature of the soil. In the coming years it will certainly prove necessary to reduce the consumer's total exposure to heavy metals by means of selective recommendations of what to cultivate in problematic regions.

A general increase in the cadmium content of cereals, which has been repeatedly postulated by various authors, could not be confirmed, at least not in the Federal Republic of Germany. Thus, a comparative evaluation of the cadmium content, ascertained within the context of the "Special Examination of the Harvest" since 1974, with grain samples from former years revealed that no increase in the cadmium concentration was detectable, neither during the period from 1974 till 1984 nor when very old wheat samples were compared with those of recent years.

The average cadmium concentration in potatoes is 0.03 to 0.05 mg/kg fresh material, that in fruit 0.009 mg/kg fresh material. In contrast, there is considerable variation in the cadmium content of various kinds of vegetables; the average values range between 0.015 and 0.670 mg/kg fresh material. The higher concentrations were found in a few vegetables such as spinach, celery, and occasionally in carrots. These vegetables seem to be able to absorb especially large amounts of cadmium from the soil. This should be taken into account in their cultivation and corresponding fertilization measures. Fodder maize cultivated on ground spread with sewage or dredged sludge can exhibit considerable cadmium concentrations.

A reduction of the cadmium content in the edible part of plant foods by means of industrial or culinary processing or preparation is, in contrast to lead, possible only to a very limited extent (KLEIN, 1982).

Table I.8a-3. Cadmium Content in Food (mg/kg)

Food	Mean Value (arithm.)	Minimum Value	Maximum Value
Cereals			
Wheat[a]	0.056	0.001	1.27
Rye[a]	0.016	0.003	0.26
Rice[b]	0.075	0.002	0.09
Potatoes[b]	0.03 – 0.05	0.001	0.32
Vegetables			
Lettuce[b]	0.05	0.001	1.6
Tomatoes, cucumber[b]	0.04	ND	0.50
Spinach[b]	0.23	0.019	2.76
Celery[b]	0.67	0.01	2.0
Fruit			
Apples[b]	0.011	ND	0.17
Peaches, cherries[b]	0.008	ND	0.11
Citrus fruits[b]	0.008	ND	0.046
Mushrooms[c]	0.46	0.001	27.6
Cocoa[b]	0.4	ND	4.3
Tea[b]	0.21	ND	26.2
Coffee[b]	0.017	0.003	0.09

ND not detectable
[a] OCKER (1985)
[b] WEIGERT et al. (1984)
[c] BUNDESGESUNDHEITSAMT (1986)

Extremely high cadmium concentrations of 27.9 mg/kg fresh material have been found in wild mushrooms; the average content is about 0.5 mg/kg fresh material.

The mean cadmium content of coffee beans was given as 0.017 mg/kg fresh material. Tea leaves contain an average of 0.2 mg/kg fresh material, whereby maximum values reached 26 mg/kg fresh material. However, no sound investigation to determine how much of it is transferred to the beverage or infusion has been carried out. In contrast, the relatively high content of cadmium of about 0.4 mg/kg fresh material in cocoa leads to distinctly higher cadmium concentrations in its products, especially the use of cocoa in bitter chocolate.

A summary of cadmium values found in foodstuffs is compiled in Table I.8a-3.

5 Chromium

This heavy metal, which is used in many industrial fields, is found in food and feed in concentrations between 0.03 and 2.4 mg/kg fresh material (see Table I.8a-4). The

Table I.8a-4. Chromium Content in Food (mg/kg)

Food	Mean Value (arithm.)	Minimum Value	Maximum Value
Cereals[a]	0.13 – 0.4	0.02	9.2
Potatoes[b]	0.05 – 0.3	ND	0.65
Vegetables[c]	0.11	ND	2.3
Fruit[c]	0.07	ND	1.0
Mushrooms[d]	0.06 – 0.135	–	–
Cocoa[c]	1.05	ND	2.46
Tea[c]	0.95	0.1	2.44
Coffee[c]	0.025	ND	0.05

ND not detectable
[a] FARRE and LAGARDA (1986); BUNDESGESUNDHEITSAMT (1986)
[b] SOUCI et al. (1986)
[c] BUNDESGESUNDHEITSAMT (1986)
[d] SCHLETTWEIN-GSELL and MOMMSEN-STRAUB (1973)

cereals of importance to the human diet generally contain 0.03 to 0.4 mg/kg fresh material. Potatoes, fruit, and vegetables show average contents of between 0.05 and 0.1 mg/kg fresh material. Similar values occur in wild mushrooms. According to recent Spanish investigations grain and grain products, meat, poultry, and fish make the largest contribution to the dietary intake of chromium. These studies revealed an average of 0.135 mg chromium per kg in cereals and cereal products (FARRE and LAGARDA, 1986).

In contrast, the average chromium content of cocoa beans and tea leaves is approximately 1 mg/kg fresh material; in coffee beans, however, only 0.025 mg/kg fresh material was determined. The causes of these differences are still unknown; nor is there much reliable information about the type of chemical bond and the oxidation state of chromium in food of plant origin. It is, however, increasingly necessary to differentiate between chromium(III) and chromium(VI).

SMITH et al. (1989) have investigated chromium accumulation, transport, and toxicity in plants, and found that chromium bioavailability, exchangeability, and uptake are very low. Even near smelters plants take up five times less than nickel; and cadmium, copper, and thallium are also 50 times more phytotoxic. These authors made experiments with barley seeds and lemma. Chromium(VI) has a higher transport index and seems to be more toxic than chromium(III). In the roots chromium(III) concentrations may be higher, but in the shoots chromium(VI) dominates. Oxalate-chromate complexes play a great role. Chromium concentrations in plant cell mitochondria and nuclei are, however, rather low.

6 Cobalt

According to the sparse information available about this element in plant foods, the average cobalt content of cereals, potatoes, fruit, and mushrooms ranges from 0.01 to 0.1 mg/kg fresh material. In vegetables, cocoa, tea leaves, and roasted coffee average values of 0.1 – 1.0 mg/kg fresh weight are found. Cocoa beans are reported to have cobalt concentrations exceeding 1 mg/kg (Table I.8a-5). There are as yet no reports about the influence of environmental factors on cobalt content; future monitoring investigations could possibly provide information about this.

Table I.8a-5. Cobalt Content in Food (mg/kg)

Food	Mean Value (arithm.)	Minimum Value	Maximum Value
Cereals			
Wheat[a]	0.02	0.005	0.09
Rye[a]	0.03	0.007	0.18
Potatoes[a]	0.013	0.008	0.016
Vegetables			
Lettuce, cabbage[b]	0.05 – 1.0	–	–
Tomatoes, cucumber[b]	0.05 – 0.06	–	–
Fruit			
Apples, pears[b]	0.09 – 0.32	–	–
Peaches, cherries[b]	0.02 – 0.14	–	–
Mushrooms[b]	0.17	–	–
Cocoa[b]	0.3 – 1.7	–	–
Tea[b]	0.6 – 1.0	–	–
Coffee roasted[b]	0.34 – 0.88	–	–

[a] SOUCI et al. (1986)
[b] SCHLETTWEIN-GSELL and MOMMSEN-STRAUB (1973)

7 Copper

There is a relatively large amount of data available about copper and its compounds, which should be reliable on account of the analytical situation (see Table I.8a-6). An average content of 2 to 4 mg Cu per kg fresh weight is found in bread grain, while 1.5 mg/kg fresh material is estimated in potatoes. The average copper content in fruit and vegetables amounts to 0.6 to 1.0 mg/kg fresh material, whereby maximum values of over 20 mg/kg fresh material have been measured. In Canada a copper content of 8 to 22 mg Cu per kg was determined in radish roots (VAN NETTEN and MORLEY,

Table I.8a-6. Copper Content in Food (mg/kg)

Food	Mean Value (arithm.)	Minimum Value	Maximum Value
Cereals			
Wheat[a]	2.75	0.58	7.8
Rye[b]	5.0	4.0	6.0
Rice[b]	2.4	–	–
Potatoes	1.57	0.55	5.3
Vegetables in general[a]	1.0	0.01	73.7
Radish[c]	8–22	–	–
Fruit[a]	0.6	ND	24.2
Mushrooms[a]	4.9	4.2	5.6
Cocoa[a]	27.0	19.9	37.6
Tea[a]	21.8	ND	127
Coffee[a]	13.3	0.2	19.8

ND not detectable
[a] BUNDESGESUNDHEITSAMT (1986)
[b] SOUCI et al. (1986)
[c] VAN NETTEN and MORLEY (1983)

1983). The copper concentration in wild mushrooms is approximately 5 mg/kg fresh material. Distinctly higher concentrations are detectable in coffee and cocoa beans, on average 20 mg/kg. The concentrations in tea leaves, however, show particularly large variations; thus, tea leaves bought in Europe contained between 5 and 127 mg Cu per kg fresh material (in this case, however, practically dry matter), in contrast, 6.8–26 mg Cu per kg fresh material was reported in a more recent American investigation, which could also show that only 0.14–0.19 mg Cu per kg (liter) was transferred to the beverage (KENNEY and THIMAYA, 1983).

8 Iron

Although iron is a ubiquitous element and essential for human and animal nutrition, there are very little current data about it. According to the available data (Table I.8a-7), iron is present in cereals in average concentrations of between 26.0 and 46.0 mg per kg fresh material. In potatoes about 8 mg/kg of Fe are found on an average. The corresponding amounts for fruit and vegetables are about 4.0 to 7.0 mg/kg fresh material, while 125 and 170 mg/kg fresh material were determined in cocoa and coffee beans and tea leaves. No reliable information is available about the iron content of wild mushrooms. More recent investigations were aimed at establishing the type of chemical bond iron exhibits in foods (SCHWEDT, 1984). However, no general

Table I.8a-7. Iron Content in Food (mg/kg)

Food	Mean Value (arithm.)	Minimum Value	Maximum Value
Cereals			
Wheat[a]	33.0	31.0	35.0
Rye[a]	46.0	35.0	100.0
Rice[a]	26.0	20.0	31.0
Potatoes[a]	8.0	4.4	15.0
Vegetables[b]	14.7	0.1	425.5
Fruit[b]	5.3	2.0	11.0
Cocoa[a]	125	116.0	150.0
Tea[b]	172	120.0	565.0
Coffee[a]	168	47.0	289.0

[a] Souci et al. (1986)
[b] Bundesgesundheitsamt (1986)

conclusion can be drawn from these results due to the high variability. For confirmation further research teams using additional investigation material of different origin would be necessary.

9 Lead

The lead content of plant foods is very variable and is a consequence, with certain exceptions, of exogenous influences. Both the lead uptake from the soil through the roots and lead deposition on parts of the plant above the ground are important. Lead from emission, but also from stirred up dust is retained and partly incorporated especially by rough, rugged, or sticky surfaces. Despite intensive cleaning, lead can only be partly removed even when it only adheres to the surface. Examples of this are certain kinds of cabbage which exhibited lead concentrations of up to 6.2 mg Pb per kg fresh material near industrial sites or roads. The average lead content in bread grain and potatoes lies in approximately the same range of between 0.03 and 0.09 mg/kg fresh material, whereby maximum values of up to 1.4 mg/kg fresh material could be determined in rye or potatoes, respectively. Wheat reached a maximum content of only 0.4 to 0.5 mg Pb per kg fresh material even when grown on the same site (OCKER, 1985). Lead concentrations similar to those in cereals were measured in fruit. According to current knowledge the differences between the individual sorts of fruit are not very marked. In contrast, vegetables exhibit very variable concentrations. These differences are not only caused by various growth and fertilization conditions, but also depend on plant genetic factors in most cases. Thus, mean lead concentrations of 0.8 or 0.6 mg/kg fresh material were measured in green cabbage, but

Table I.8a-8. Lead Content in Food (mg/kg)

Food	Mean Value (arithm.)	Minimum Value	Maximum Value
Cereals			
Wheat	0.035	0.004	0.42
Rye	0.07	0.01	1.46
Rice	0.18	0.01	1.0
Potatoes	0.09	0.005	1.9
Vegetables			
Lettuce, cabbage	0.2	0.001	6.1
Tomatoes, cucumber	0.07	0.005	1.9
Fruit			
Apples	0.06	0.001	0.47
Peaches, cherries	0.12	0.003	1.15
Citrus fruits	0.06	0.001	1.2
Mushrooms	0.3	ND	7.5
Cocoa	0.27	ND	1.8
Tea	0.73	ND	6.47
Coffee	0.20	0.01	1.09

ND not detectable
All values from BUNDESGESUNDHEITSAMT (1986)

also in several seasoning herbs; maximum values of over 6.0 mg Pb per kg fresh material were determined. Although these examples of vegetable species, which accumulate particularly high lead concentrations, are striking, most other sorts of vegetables, especially fruit and shoot vegetables exhibit relatively low average lead concentrations of 0.05 to 0.08 mg/kg fresh material.

In this context it must be emphasized that even these concentrations could be reduced on an average by means of selective agricultural measures, if, e.g., vegetables were no longer grown on sewage sludge or in the immediate vicinity of pertinent industrial plants or roads carrying a heavy volume of traffic. Lead contamination of food, which is completely avoidable, can be caused by certain conservation packaging methods, when cans are sealed with lead containing solder. As a result of extremely fine lead particles in the interior of the can, the effect of preparation and processing, which considerably reduces the original lead content, cannot only be reversed, but in many cases lead concentrations in the contents are higher than in the fresh material before it was cleaned. This secondary contamination by means of out-of-date packaging technology can no longer be tolerated today and should be banned in the near future, if necessary by official measures.

Wild mushrooms exhibit average lead concentrations of 0.1 to 0.3 mg/kg fresh material, whereby maximum values of 7.5 mg/kg fresh material are detectable (WEIGERT et al., 1984). The average lead concentration in cocoa and coffee beans is 0.2 mg/kg fresh material, that in tea leaves 0.7 mg/kg fresh material. How much of the lead content remains in the beverage has, however, not been sufficiently investigated. In Table I.8a-8 some lead values are compiled.

10 Manganese

The concentration of this industrially much used element in plant foods cannot be reliably stated at present. The values differ by multiples of each other depending on which author is consulted. The content of 11.0 to 34.0 mg Mn per kg fresh weight in cereals published by SOUCI et al. (1986) can be regarded as a reference point. However, the validity of the average value of 734.4 mg Mn per kg fresh material in tea leaves reported there will have to be confirmed by further investigation in the near future. For available data see Table I.8a-9.

Table I.8a-9. Manganese Content in Food (mg/kg)

Food	Mean Value (arithm.)	Minimum Value	Maximum Value
Cereals			
Wheat[a]	24.0	19.0	29.0
Rye[a]	34.0	24.0	43.0
Rice[a]	11.0	–	–
Potatoes[a]	1.5	1.0	2.5
Vegetables[b]	2.2 – 20	–	–
Fruit[b]	0.5 – 2.0	–	–
Mushrooms[b]	0.3 – 7.4	–	–
Cocoa[b]	25 – 35	–	–
Tea[a]	734	–	–
Coffee[b]	15.0	–	–

[a] SOUCI et al. (1986)
[b] SCHLETTWEIN-GSELL and MOMMSEN-STRAUB (1973)

11 Mercury

For years increased attention has been paid to the content of mercury in plant food and feed, as it is one of the most toxic elements for humans and animals. Its content in these foods originates from both the soil, which depends on the geological mercury concentration, and from emission, whereby the latter contamination is thought

Table I.8a-10. Mercury Content in Food (mg/kg)

Food	Mean Value (arithm.)	Minimum Value	Maximum Value
Cereals	0.006 – 0.03	0.002	0.12
Potatoes	0.006	ND	0.02
Vegetables			
Lettuce, cabbage	0.014	ND	0.143
Tomatoes, cucumber	0.005	ND	0.040
Fruit			
Apples, pears	0.008	ND	0.040
Peaches, cherries	0.004	ND	0.019
Citrus fruits	0.013	ND	0.080
Mushrooms	0.340	ND	8.80
Cocoa	0.005	ND	0.014
Tea	0.025	ND	0.60
Coffee	0.040	0.002	0.08

ND not detectable
All values from BUNDESGESUNDHEITSAMT (1986)

to surpass the former according to current knowledge. The influence of cultivating food and feed plants in areas fertilized with sewage sludge or sediment dredged from rivers on the mercury concentration of plants is a controversial issue.

In the Federal Republic of Germany grain, potatoes, vegetables, and fruit contain average mercury concentrations of 0.005 to 0.05 mg Hg per kg fresh material. Up to now maximum values of 0.15 mg/kg fresh material have been found (BUNDESGE-SUNDHEITSAMT, 1986). In contrast, wild mushrooms can exhibit maximum mercury concentrations of up to 10 mg/kg, with average values of around 0.3 mg/kg fresh material. An average of 0.005 to 0.02 mg Hg per kg fresh material was found in cocoa beans and tea leaves. A mean content of 0.04 mg Hg per kg fresh material was reported for coffee beans. All the data (Table I.8a-10) refer to the total amount of mercury; reliable results about the type of the chemical bond of mercury in plant foods, especially about the proportion of organic bound mercury, are hardly available. The findings of recent cross-section investigations in the Federal Republic of Germany are noteworthy (WEIGERT, 1987). It was shown that, although mercury occurs in most plant foods only in minute traces, when it is found in appreciable concentrations it is always accompanied by measurable concentrations of selenium. In contrast to food of animal origin, the selenium content is always lower than the mercury content.

12 Molybdenum

There is very little reliable information about this element in plant foods. According to the available data (Table I.8a-11), the average molybdenum content varies between 0.01 and 0.2 mg/kg fresh material in most plant foods.

Table I.8a-11. Molybdenum Content in Food (mg/kg)

Food	Mean Value (arithm.)	Minimum Value	Maximum Value
Cereals			
Wheat[a]	–	0.2	0.8
Rye[a]	–	0.07	0.62
Rice[a]	–	0.5	1.0
Potatoes[a]	–	0.03	0.86
Vegetables[b]	0.01 – 0.04	–	–
Fruit[b]	0.002 – 0.160	–	–
Mushrooms[b]	0.032	–	–
Cocoa[b]	0.730	–	–
Tea[b]	0.130	–	–
Coffee[b]	–	ND	ND

ND not detectable
[a] Souci et al. (1986)
[b] Schlettwein-Gsell and Mommsen-Straub (1973)

13 Nickel

This element is predominantly absorbed by the plant from the soil; conjecture that nickel-containing dust from industry influences the nickel content of cereals could not be confirmed in the Federal Republic of Germany (Ocker, 1985). Regardless of the region, the mean values are 0.1 – 0.2 mg Ni per kg fresh material for cereals; the maximum values seldom exceed 0.8 mg/kg. The corresponding values in potatoes, fruit, and vegetables are usually lower. In contrast, considerably higher average concentrations of 5 to 15 mg/kg fresh material are measured in cocoa beans and tea leaves (see also Chapter II.22 and, e.g., Nielsen, 1989). In coffee and tea beverage, however, only 0.014 or 0.022 mg Ni/kg (liter), respectively, were determined. It is difficult to judge the reliability of the analysis with this complicated matrix in the case of the maximum values of more than 50 mg Ni per kg fresh weight found in cocoa beans (see also Table I.8a-12).

Table I.8a-12. Nickel Content in Food (mg/kg)

Food	Mean Value (arithm.)	Minimum Value	Maximum Value
Cereals			
Wheat[a]	0.19 – 0.34	0.05	0.89
Rye[b]	0.12	0.05	2.7
Potatoes[b]	0.05 – 0.26	ND	0.56
Vegetables[c]	0.05	ND	67.9
Fruit[c]	0.08	ND	0.58
Cocoa[c]	15.1	7.4	54.7
Tea[b]	5.25	0.25	11.2
Coffee[b]	0.77	0.58	0.98

ND not detectable
[a] OCKER (1985)
[b] SOUCI et al. (1986)
[c] BUNDESGESUNDHEITSAMT (1986)

14 Selenium

The content of this element, which is essential for the human diet, in plant food and feed depends almost entirely on the selenium content of the soil on which the plants grow. Thus, the selenium content of the individual plants can vary accordingly. Many of the more recent concentration results given here are based on analyses of foods which were marketed and investigated in the Federal Republic of Germany. This explains the differences, e.g., the maximum selenium content in cereals, which SOUCI et al. (1986) gave as 1.3 mg/kg fresh weight, while the corresponding German results show a maximum value of 0.26 mg/kg fresh weight. Moreover, it can be assumed from recent knowledge that almost all data obtained before 1980 in the range below 5 mg/kg (ppm) are no longer valid from the analytical point of view. These values are included in Table I.8a-13 only to provide a complete picture.

The mean selenium concentration in grain ranges between 0.025 and 0.07 mg/kg fresh weight in Germany, whereas values of between 0.24 to 0.38 mg/kg are reported for the USA (TREPTOW et al., 1978). In selenium-rich areas in Venezuela concentrations of up to 118.8 mg/kg were determined. The selenium values for potatoes, fruit, and vegetables are on an average about 0.01 mg/kg fresh weight, whereby the variation in these foods is distinctly less than in cereals.

Wild mushrooms exhibit an average concentration of about 0.01 mg Se per kg fresh material in the Federal Republic of Germany. In Switzerland distinctly higher concentrations were determined in a comprehensive investigation; thus, in certain cases 20 mg Se per kg dry matter were measured (STIJVE, 1977). Due to the age of

Table I.8a-13. Selenium Content in Food (mg/kg)

Food	Mean Value (arithm.)	Minimum Value	Maximum Value
Cereals	0.025 – 0.068[a]	0.01[a]	0.26[a]
	0.24 – 0.38[b]	–	–
	–	–	18.8[c]
Wheat[d]	–	0.007	1.300
Rye[d]	–	0.002	0.08
Rice[d]	–	0.13	0.71
Potatoes[a]	0.025	ND	0.05
Vegetables[a]	0.027	ND	0.120
Fruit[a]	0.023	ND	0.05
Mushrooms	0.013[a]	0.01[a]	0.02[a]
	0.03 – 13.0[e,f]	ND	20.0[e,f]
Tea[a]	0.04	ND	0.07
Coffee[a]	0.034	ND	0.09

ND not detectable
[a] Bundesgesundheitsamt (1986)
[b] Morris and Levander (1970)
[c] Jaffee et al. (1978)
[d] Souci et al. (1986)
[e] Stijve (1977)
[f] Results in mg Se per kg dry matter

their mycelium, wild mushrooms generally contain up to 10 times the selenium content of cultured mushrooms. On the other hand, tea leaves and coffee beans contain 0.03 to 0.04 mg/kg fresh material.

15 Thallium

Under natural conditions the concentration of this rare element in food is generally very low. As in the case of most of the other elements, the thallium concentration of plants directly depends on the thallium content of the soil. In addition, industrial emission causes particularly high concentrations in edible plants in some cases (see Chapter II.29). According to the available data, the cabbage species (Brassicaceae) accumulate especially high amounts of thallium. The mean thallium content of cereals, fruit, and vegetables is about 0.05 – 0.1 mg/kg fresh weight. In potatoes an average of 0.02 mg/kg fresh weight was determined. In contrast, mushrooms can exhibit extremely high thallium concentrations; depending on the area where they grew, up to 1.2 mg Tl per kg fresh material was determined (Seeger and Gross, 1981) (see also Table I.8a-14).

Table I.8a-14. Thallium Content in Food (mg/kg)

Food	Mean Value (arithm.)	Minimum Value	Maximum Value
Cereals[a]	0.02 – 0.08	ND	0.250
Potatoes[a]	0.025	ND	0.025
Vegetables[a, d]	0.05 – 0.13	ND	1.9
Lettuce[b]	0.005	0.003	0.012
Fruit[a, d]	0.05 – 0.14	ND	0.34
Mushrooms[c]	0.048	0.014	1.22

ND not detectable
[a] BUNDESGESUNDHEITSAMT (1986)
[b] SHERLOCK and SMART (1986)
[c] SEEGER and GROSS (1981)
[d] Data from polluted areas

16 Vanadium

Very few reliable measurements are available about the amount of vanadium in food. According to these results the main factor influencing regional differences is obviously the special geogenic conditions. On account of insufficient analysis in almost all cases, the average vanadium content of 0.01 to 1.0 mg/kg, given in Table I.8a-15, in the majority of foods should be interpreted with great care. Even when similar values for that particular food are found in different publications it does not mean they are

Table I.8a-15. Vanadium Content in Food (mg/kg)

Food	Mean Value (arithm.)
Cereals	
Wheat	0.08[a]
	2.3[b]
Rice	0.230[a]
	0.820[a]
Potatoes	0.01 – 0.02[a]
Vegetables	0.02 – 1.0[a]
Fruit	0.008 – 0.06[a]
Mushrooms	0.001[a]

[a] SCHLETTWEIN-GSELL and MOMMSEN-STRAUB (1983)
[b] SOUCI et al. (1986)

more reliable, as they can usually be traced back to the same author. According to more recent findings vanadium concentrations of over 0.1 mg/kg fresh material seem to be based on experimental error.

17 Zinc

Zinc as a ubiquitous element is to be found in almost all foods. It is an essential dietary element for humans and animals. Zinc deficiency symptoms have rarely been described in adults, whereas zinc deficiency can cause serious damage to the skin and the enzyme system of infants and young children. The mean zinc content in cereals is about 26 mg/kg fresh weight, whereby wheat grains with 41 mg/kg contain distinctly higher amounts of zinc than rye or rice. In potatoes and vegetables the average zinc values of 3.5 mg/kg or 4.3 mg/kg fresh material are distinctly lower. Fruit contains even less zinc with 1.6 mg/kg. In mushrooms mean zinc concentrations ranging from 9.7 mg/kg fresh material to 280 mg/kg were found, whereby the latter data are based on older literature reports (see also Table I.8a-16). Cocoa and tea leaves with values of 35.0 mg Zn per kg fresh material exhibit high concentrations; coffee beans, on the other hand, contain only 6.7 mg/kg fresh material on an average. Extremely high zinc concentrations, which also lead to poisoning, can be determined when acidic food is stored or prepared in containers galvanized with zinc.

Table I.8a-16. Zinc Content in Food (mg/kg)

Food	Mean Value (arithm.)	Minimum Value	Maximum Value
Cereals			
Wheat[a]	41.0	22.0	100.0
Rye[a]	13.0	–	–
Rice[a]	–	8.0	20.0
Potatoes[b]	3.51	1.1	17.9
Vegetables[b]	4.31	ND	99.4
Fruit[b]	1.66	ND	66.0
Mushrooms[b]	9.7	2.6	21.5
	40.0 – 280.0[c]	–	–
Cocoa[a]	35.0	20.0	69.6
Tea[b]	35.0	18.3	63.9
Coffee[b]	6.7	6.1	8.0

ND not detectable
[a] SOUCI et al. (1986)
[b] BUNDESGESUNDHEITSAMT (1986)
[c] SCHLETTWEIN-GSELL and MOMMSEN-STRAUB (1973); data from 1962

References

Bundesgesundheitsamt (1986), *ZEBS*, unpublished evaluations.
Farre, R., and Lagarda, M. J. (1986), *Chromium Content of Foods and Diets in a Spanish Population*. *J. Micronutr. Anal. 2*, 297–304.
Fishbein, L. (1987), *Trace and Ultra Trace Elements in Nutrition: An Overview: Zinc, Copper, Chromium, Vanadium, and Nickel. Toxicol. Environ. Chem. 14*, 73–99.
Greger, J. L. (1985), *Aluminum Content of the American Diet, Food Technol. 39*, 73–80.
Jaffee, W. G., Chavez, J. F., and de Mondragon, M. C. (1978), cited in: Treptow, H., Bielig, H. J., and Askar, A. (1978).
Klein, H. (1982), *Einfluß von Herstellungs- und Zubereitungsverfahren auf den Arsen-, Blei-, Cadmium- und Quecksilbergehalt von Lebensmitteln. ZEBS Ber. 3/1982*, Bundesgesundheitsamt, Berlin.
Morris, V. C., and Levander, O. A. (1970), *Selenium Content of Foods, J. Nutr. 100*, 1383.
Netten, van, C., and Morley, D. R. (1983), *Uptake of Uranium, Molybdenum, Copper, and Selenium by the Radish from Uranium-rich Soils. Arch. Environ. Health 38*, 172–175.
Nielsen, G. D. (1989), *Oral Challenge of Nickel Allergic Patients with Hand Eczema*, in: Nieboer, E., and Aitio, A. (eds.): *Nickel and Human Health, Current Perspectives. Adv. Environ. Sci. Technol.*, in press.
Nriagu, J. O., and Pacyna, J. M. (1988), *Quantitative Assessment of Worldwide Contamination of Air, Water and Soils by Trace Metals. Nature 333*, 134–139.
Ocker, H.-D. (1985), *Pflanzenschutzmittelrückstände und Schwermetallgehalte in der deutschen Brotgetreideernte.* Statistische Erhebungen aus Volldruschen der besonderen Ernteermittlung. Bundesforschungsanstalt für Getreide- und Kartoffelverarbeitung, Institut für Biochemie und Analytik, Detmold.
Schlettwein-Gsell, D., and Mommsen-Straub, S. (1973), *Spurenelemente in Lebensmitteln, Int. Z. Vitamin Ernährungsforsch. Beih. 13.*
Schwedt, G. (1984), *Analytik von Metall-Bindungsformen in Lebensmitteln. GIT Fachz. Lab. 11*, 1013–1021.
Seeger, R., and Gross, M. (1981), *Thallium in Higher Fungi, Z. Lebensm. Unters. Forsch. 173*, 9–15.
Shacklette, H. T. (1980), *Elements in Fruits and Vegetables from Areas of Commercial Production in the Conterminous United States. Geological Survey Professional Paper 1178.* US Governmental Printing Office, Washington, D.C.
Sherlock, J. C., and Smart, G. A. (1986), *Thallium in Foods and the Diet. Food Addit. Contam. 3*, 363–370.
Smith, S., Peterson, P. J., and Kwan, K. H. M. (1989), *Chromium Accumulation, Transport and Toxicity in Plants.* Presentation at the 3rd IAEAC Workshop on Toxic Metal Compounds – Interrelation between Chemistry and Biology, Follonica, Italy. *Toxicol. Environ. Chem. 24*(4), 241–251.
Souci-Fachmann-Kraut (1986), *Die Zusammensetzung der Lebensmittel Nährwert-Tabellen 1986/1987*, 3rd Ed. Wissenschaftliche Verlagsgesellschaft, Stuttgart.
Stijve, T. (1977), *Selenium Content of Mushrooms. Z. Lebensm. Unters. Forsch. 164*, 201–203.
Treptow, H., Bielig, H. J., and Askar, A. (1978), *Selen in Lebensmitteln. Alimenta 17*, 15–20.
Weigert, P. (1987), *The Correlation between Selenium/Mercury and Zinc/Cadmium in Identical Food Samples.* International Conference on Heavy Metals in the Environment, New Orleans, 15th–18th Sept. CEP Consultants Ltd., Edinburgh.
Weigert, P., Müller, J., Klein, H., Zufelde, K.-P., and Hillebrand, J. (1984), *Arsen, Blei, Cadmium und Quecksilber in und auf Lebensmitteln, ZEBS Ber. 1/1984*, Bundesgesundheitsamt, Berlin.

Additional Recommended Literature

Joint FAO/WHO Food Contamination Monitoring Program, Summary of 1980–1983 (1986), Monitoring Data (WHO/EHE/FOS/86.2). World Health Organization, Geneva.

Kenney, M. A., and Thimaya, S. (1983), *Copper Content of Tea, J. Am. Diet. Assoc. 82*, 509–510.

Kirleis, Allen, W., Sommers, Lee, E., and Nelson, D. W. (1984), *Yield, Heavy Metal Content, and Milling and Baking Properties of Soft Red Winter Wheat Grown on Soils Amended with Sewage Sludge. Cereal Chem. 61*(6), 518–522.

Lorenz, H., Ocker, H.-D., Brüggemann, J., Weigert, P., and Sonneborn, M. (1986), *Cadmiumgehalte in Getreideproben der Vergangenheit – Vergleich zur Gegenwart. Z. Lebensm. Unters. Forsch. 183*, 402–405.

Meisinger, V., Pospischil, E., and Jahn, O. (1987), *Contents of Trace Elements in Alcoholic Drinks. Proceedings of Second Nordic Symposium on Trace Elements in Human Health and Disease, Odense.* W.H.O., Copenhagen.

Pieczonka, K., and Rosopulo, A. (1985), *Distribution of Cadmium, Copper, and Zinc in the Caryopsis of Wheat (Triticum aestivum L.). Fresenius Z. Anal. Chem. 322*, 697–699.

Sattar, A., and Khalid, Z. M. (1979), *Nutritional Significance of Dietary Essential Trace Elements. Lebensm. Wiss. Technol. 12*, 303–307.

Summary and Assessment of Data Received from the FAO/WHO Collaborating Centers for Food Contamination Monitoring (1982). World Health Organization, Geneva.

Waldbott, G. L. (1978), *Health Effects of Environmental Pollutants,* 2nd Ed. C.V. Mosby Comp., St. Louis, Missouri.

I.8 b Metal Accumulation in the Food Chain and Load of Feed and Food

HANS-JÜRGEN HAPKE, Hannover, Federal Republic of Germany

1 Introduction

Due to their ubiquitous occurrence, heavy metals are present in all foodstuffs. The metal load of feed and food depends also on the conditions under which feed and food are produced and processed. Many of these metals have nutritional importance and are essential. Others, such as lead, cadmium, mercury, arsenic, and thallium, have no nutritional importance, and their presence can lead to health problems.

Table I.8b-1. Recommended or Legally Established Threshold Limits (in µg/L) for Metal Content of Drinking Water (from FÖRSTNER and WITTMANN, 1979, and other sources)

	As	Pb	Cd	Cr	Cu	Hg	Se	Ag	Zn
WHO 1971 (World Health Organization International)	50	100[a]	10[b]	–[c]	50[d]	1	10	–	5000
WHO 1970 (World Health Organization European)	50	100	10	50	50	–	10	–	5000
Australia 1973	50	50	10	50	10000	–	10	50	5000
FRG 1975	40	40	6	50	–	4	8	–	2000
Japan 1968	50	100	–	50	10000	1	–	–	100
Switzerland 1980	50	50	5	20	1500	3	–	200	1500
SABS 1971 (South African Bureau of Standards)	50	50	50	50	1000	–	–	–	5000
USSR 1970	50	100	10	100	100	5	1	–	1000
USPHS 1962 (U.S. Public Health Service)	10	50	10	50	1000	–	10	50	5000
US-NAS 1972 (National Academy of Sciences)	100	50	10	50	1000	2	10	–	5000
US-EPA 1975 (Environmental Protection Agency)	50	50	10	50	–	2	10	50	–

The WHO Guidelines 1982 for Drinking Water Quality EFP/82.39 provide a number of recent recommendations:
[a] pp. 68/69 for lead 50 µg/L
[b] p. 67 for cadmium 50 µg/L
[c] p. 67 for total chromium 50 µg/L (chlorination stimulates the release of chromium(VI))
[d] pp. 93/94 for copper 1000 µg/L (not for reasons of its primary toxicity but of adverse effects in households)

Essential elements produce their intended effects within a certain concentration, or dosage, range. Values both below and above this range can lead to health problems. Non-essential components are only hazardous when the concentration exceeds certain limits (see Chapter 1.9) (cf. Table I.8b-1).

Heavy metals can reach the food chain by way of water, soil, plants, animals, and humans (EDWARD and DOOLEY, 1981). Accumulation takes place in certain target tissues, the extent being determined by the duration of exposure and the existing concentration of the metal in the organism's environment. "Dilution" may occur by the same means. For example, arsenic accumulates in marine organisms, but not in terrestrial animals (the concentration of arsenic in animal feed is only one-tenth of that in soil, and in beef and milk the concentration is only one-tenth of that in feed). Lead shows a similar behavior under normal conditions, although this can differ for specific organs (DER RAT VON SACHVERSTÄNDIGEN FÜR UMWELTFRAGEN, 1978). Since individual heavy metals and their compounds exhibit different behavior, a differentiated description of these processes is necessary. Little is known about the origin of these differences, but a possible explanation is that the differing reactivities of the heavy metals promote depot formation through the binding of ionized metals to storage structures. Absorption and distribution depend considerably upon the lipid solubility of the individual compound. Corresponding to their ability to penetrate lipid membranes, organic compounds are more likely to enter and accumulate in an organism than are inorganic ones.

2 Antimony

Accumulation occurs in the liver of domestic animals. Concentrations of 0.002 mg/kg have been reported and may be higher in older animals. Antimony compounds have been used as chemotherapeutical drugs in animals without problems (DEUTSCHE FORSCHUNGSGEMEINSCHAFT, 1975a; HAPKE, 1987).

3 Arsenic

With respect to foodstuffs, the transfer of arsenic from soil to plants is of little importance. Normal vegetation has an arsenic content of about 0.1 to 1.0 mg per kg of dry matter. Vegetation from soils that have been contaminated by industrial or agricultural use (i.e., arsenic-containing pesticides and fertilizers) may contain from 1.0 up to 20 mg/kg. Arsenic is widely transported into agricultural areas in Central Europe, and it can accumulate in fungi and certain grasses. The importance of arsenic in vegetation as an environmental hazard is limited (DEUTSCHE FORSCHUNGSGEMEINSCHAFT, 1975b; HAPKE, 1987; see also Chapters I.7 and I.8a).

In contrast, arsenic is continuously absorbed by marine organisms, where it accumulates. A concentration of 0.05 mg per liter of water leads to values of 0.1 to

6.0 mg/kg (mean value 2.6 mg/kg) in these organisms, mainly in the form of organic complexes. Concentrations of 10 to 100 mg arsenic per kg have been found in estuaries and offshore areas (DEUTSCHE FORSCHUNGSGEMEINSCHAFT, 1978; DEUTSCHE GESELLSCHAFT FÜR ERNÄHRUNG, 1976, 1988; see also Chapter II.3). By way of this accumulation in plankton, small and medium-sized fish, and therefore fishmeal, animal feed can contain a mean arsenic concentration of 0.4 mg/kg. Thus, arsenic residues are to be expected in the tissues of all domestic animals that are fed fishmeal (e.g., poultry and swine) (DEUTSCHE FORSCHUNGSGEMEINSCHAFT, 1975a, 1978).

Concentrations of about 0.1 mg/kg were found in pork, 0.5 mg/kg in pork kidney, and 0.05 to 5 mg/kg in pork liver. Beef, however, contains less than 0.01 mg/kg and poultry about 0.2 mg/kg (SAHL, 1982). The liver and kidneys of cattle from unpolluted areas show concentrations of less than 0.1 and 0.5 mg/kg, respectively (see also WARD, 1987). After exposure, these values may increase up to 10 or 15 mg/kg in both organs (HOLM and DEMUTH, 1988; HOLM, 1980). The arsenic content in cow's milk is small, resulting in concentrations of 0.03 to 0.06 mg/kg in milk and 0.003 mg/kg in milk products (SAHL, 1982).

Farm animals are exposed to arsenic almost exclusively via their feed. Since excretion is very slow, accumulation is to be expected, although preferentially in such non-edible tissue as bone and hair. Besides feed, arsenic uptake is also possible through the use of organic arsenic compounds in veterinary medicine (such as the "arsenicals" used to treat swine dysentery). Little is known about the oxidation state, As(III) or As(V) or the binding form of arsenic in food. In summary, arsenic accumulation in tissues is not expected if an animal's feed and drinking water does not contain more than 2 mg/kg and 0.2 mg/liter arsenic, respectively (FÖRSTNER and WITTMANN, 1979; HAPKE, 1987).

4 Cadmium

Relatively mobile cadmium enters the food chain beginning with soil (see also Chapters I.7, I.8a, and II.6). Different soils show different concentrations of this heavy metal. These levels can be increased considerably through the application of phosphate fertilizers as well as sewage sludge or compost on cultivated fields. Uptake of cadmium by plant roots differs depending on the pH of the soil. Due to a short vegetation period, however, only small-scale accumulation is observed in plants (EDWARD and DOOLEY, 1981; DEUTSCHE FORSCHUNGSGEMEINSCHAFT, 1980). Cadmium-containing dust can also increase the concentration of cadmium in plants because resorption via the leaves is possible in some cases (UMWELTBUNDESAMT, 1977). Thus the cadmium contents of vegetable feeds vary from 0.005 to 0.1 mg/kg, and in certain cases up to 1.0 mg/kg. These levels may increase to 50 mg/kg in cases of atmospheric cadmium contamination by industry. Levels of up to 10 mg/kg have been observed after application of cadmium-containing phosphate fertilizers. In rice, concentrations of up to 1.0 mg/kg cadmium have been found, originating from the water used for cultivation (HOLM, 1980; HAPKE, 1987; Chapters I.8a and II.6).

Due to facilitated resorption, leafy vegetables, with their large surface areas, and root crops such as potatoes, carrots, and radishes, contain higher cadmium levels than foods from other parts of the plant. Fruits such as grapes barely take up cadmium. Some species of edible mushrooms concentrate the metal to a considerable extent in their mycelia and surficial parts resulting in cadmium concentrations of about 0.4 mg/kg (DEUTSCHE FORSCHUNGSGEMEINSCHAFT, 1975a; DEUTSCHE GESELLSCHAFT FÜR ERNÄHRUNG, 1976, 1988).

As a result of contaminated animal feed, cadmium is ubiquitous in animal product foodstuffs. Since the rate of oral uptake exceeds excretion, accumulation results in some animal tissues. Apparently, the constant presence of cadmium in foodstuffs is sufficient to cause a slowly increasing concentration in the kidneys. Slaughtered animals more than 10 years old showed concentrations of up to 40 mg/kg, whereas young animals had only trace amounts of about 0.5 mg/kg. Besides the kidneys (especially the renal cortex), the other target organ is the liver, with levels of 0.08 mg/kg in young animals and from 0.3 to 1.0 mg/kg in older ones (see also WARD, 1987). Values of only 0.001 to 0.002 mg/kg have been measured in meat (corresponding to those amounts found in plants), less in milk, but more in fish, especially oysters (FRIBERG et al., 1986). As with lead, cadmium concentrations are higher in wild boars than in domestic pigs (UNGLAUB et al., 1987).

An accumulation of cadmium in the meat and milk of domestic animals is unlikely. A critical concentration in the kidneys should not occur before the age of three years if the animal's feed contains less than 0.5 mg cadmium per kg dry matter. On the basis of this level in feed, however, the kidneys of animals older than three years may contain 1 mg/kg and more. Cadmium concentrations in swine kidney may reach the same values as in feed, in liver about half that. The same applies to sheep, for which a concentration of 6 mg/kg in feed can lead to 10 mg/kg in the renal cortex. Accumulation can be avoided if the feed concentration does not exceed 0.6 mg/kg, and drinking water should not contain more than 0.05 mg/liter (HAPKE, 1987). Kidneys used for human consumption should not contain more than 1 mg/kg cadmium.

In living tissue, cadmium is bound to certain digestible protein fragments. In organic forms, cadmium is biologically disposable. Among marine organisms, concentrations ranging from 0.02 mg/kg in fish from unpolluted waters to 2.0 mg/kg in oysters and 12 mg/kg in crabs have been found. Levels of 10 to 100 mg/kg have been reported in squid. Cadmium apparently accumulates in marine organisms, but information is still controversial (WACHS, 1981). Fresh water fish and shellfish from the Elbe estuary in Germany show cadmium levels of about 0.03 mg/kg due to their contaminated environment. Liver of fish from polluted water may contain 10 to 50 mg cadmium/kg, whereas normal fish liver contains about 1 mg/kg (HOLM, 1980).

In humans, about one-third of the total cadmium burden originates from animal products and two-thirds from plant products. Taken together, cadmium intake from these sources approaches the toxicological limit of 0.525 mg per week as established by the World Health Organization. Actual weekly uptake should be about 0.1 to 0.25 mg cadmium, in extreme cases 0.5 mg (see also FRIBERG et al., 1986). These amounts do not result from meat, dairy products, calves kidneys or plants which are produced far away from industrially contaminated areas, but may be found in leafy vegetables, root crops, kidney and liver, and uncultivated mushrooms. Care should be taken to avoid all additional cadmium burden. This also applies to drinking water, since ground water may already contain concentrations of 4 µg/liter due to remobilization from sediment (in the Federal Republic of Germany the threshold value is 5 µg/liter) (DEUTSCHE FORSCHUNGSGEMEINSCHAFT, 1975a; DEUTSCHE GESELLSCHAFT FÜR ERNÄHRUNG, 1976, 1988).

5 Chromium

Chromium is an essential trace element that does not accumulate in organisms. Concentrations of 0.05 to 0.2 mg/kg have been measured in meat, and these levels do not increase in liver (0.4 mg/kg) or kidney (0.2 mg/kg). WARD (1987) found only about 0.1 mg/kg in bovine liver. The chromium content of animal drinking water should not exceed 1.0 mg/L (DEUTSCHE FORSCHUNGSGEMEINSCHAFT, 1975a; DEUTSCHE GESELLSCHAFT FÜR ERNÄHRUNG, 1976, 1988).

6 Copper

This essential element accumulates in the liver resulting in concentrations of 20 to 100 mg/kg (the mean is about 50 mg/kg). WARD (1987) found as much as 190 mg/kg copper in bovine liver. Kidney normally contains about 3 mg/kg copper, and meat from 0.3 to 5.0 mg/kg. Oysters, which are supposedly „storage animals" for copper, can contain up to 137 mg/kg. In fact, aquatic organisms accumulate copper far more easily than mercury, lead, or cadmium (WACHS, 1981). Foodstuffs normally contain 2 to 4 mg copper/kg. Levels in animal feed should not exceed 10 mg/kg, in animal drinking water no more than 0.5 mg/L (DEUTSCHE FORSCHUNGSGEMEINSCHAFT, 1987; HOLM and DEMUTH, 1988).

7 Lead

Lead found in food and animal feed comes mainly from external sources, for example, lead-containing dust can adhere to the surfaces of edible plants. Transfer of lead from the soil to plants takes place only when the lead concentration of the soil is extremely high (more than 5000 mg/kg of dry matter) (EDWARD and DOOLEY, 1981; DEUTSCHE FORSCHUNGSGEMEINSCHAFT, 1975a; UMWELTBUNDESAMT, 1976).

Plant feed and food have lead contents ranging from less than 0.1 mg/kg up to 5 or 10 mg/kg of dry matter. Industrial pollution can cause values of between 100 and 1000 mg/kg in certain areas. In general, plant foodstuffs contain more lead than foodstuffs of animal origin (DEUTSCHE FORSCHUNGSGEMEINSCHAFT, 1975a, 1980; DEUTSCHE GESELLSCHAFT FÜR ERNÄHRUNG, 1976, 1988).

The lead content of animal tissues is caused mainly by the uptake of contaminated feed. About 10% of the total burden is caused by the inhalation of lead-containing dust, even though this has a higher resorption rate. Lead shows different accumulation rates in tissue. No more than 0.1 mg/kg lead is found in the skeletal muscle of farm animals, even if the animals have been exposed to massive levels. In liver, however, concentrations can reach more than 0.1 mg/kg, and in exceptional cases even 10 mg/kg in animals from polluted areas. Similar concentrations are found in the kidneys. Milk and blood both show very low lead concentrations of between 0.002 and 0.006 mg/L (DEUTSCHE FORSCHUNGSGEMEINSCHAFT, 1975b). Lead concentrations in the muscle, liver, and kidney of wild boars are higher than those of domestic pigs, which corresponds to the more polluted feed of wild animals (HECHT, 1987). To avoid accumulation, animal drinking water should not contain more than 0.1 mg lead per liter.

Within an organism, lead is most likely present in the form of organic compounds. It is exchangeably bound to certain protein fractions of the protoplasm in liver and kidney cells. Highest accumulation takes place in bone. Domestic animal feed with a lead content of no more than 30 mg/kg dry matter will result in liver levels of less than 1 mg/kg, although swine feed should not contain more than 5 mg/kg lead (HAPKE, 1987).

Lead does not play a major role in aquatic food chains. Lead levels in fish depend upon the amount of lead pollution in the environmental water. Levels of 0.01–0.03 and 0.04–0.15 mg/kg have been determined in fish muscle and liver, respectively, but these values may increase to 0.08 and 0.09 mg/kg in fish from polluted water. There is remarkable variation in the investigative data, in particular among different species of fish (FÖRSTNER and WITTMANN, 1979; HAPKE, 1987; HOLM, 1980). Fish can contain up to 0.5 mg/kg lead, oysters up to 1 mg/kg. Accumulation in the aquatic food chain is limited (WACHS, 1981).

Human intake of lead through food is about 0.3 to 1.0 mg lead per week, with single cases of up to 2.5 mg. A weekly uptake of 3 to 4 mg is accepted as toxicologically harmless because resorption is also low. About half of human lead intake is through food, of which more than half originates from plants. The normal food chain from soil to plant to animal to man causes a dilution, rather than an accumulation, of the metal. The dilution factor is about 1000, and in polluted industrial areas it is still about 100. No acute hazard from lead in the food chain has been determined, although it should be kept in mind that lead content is higher in the liver and kidneys of animals kept near industrial emission sources. If these organs contain more than 3 mg/kg, they are considered unfit for human consumption (see also Chapter II.16; DER RAT VON SACHVERSTÄNDIGEN FÜR UMWELTFRAGEN, 1975; TSUCHIYA, 1986).

8 Mercury

Mercury present in soil originates mainly from mineral decay and is supplemented regionally through industrial or agricultural contamination. Sewage sludge can add to the normal soil content of 0.03 to 0.06 mg/kg, resulting in values of up to 3 or 4 mg/kg. Plants growing in such soil absorb only a small amount of mercury through their roots; most mercury in plants, which is in general a small amount usually in the form of inorganic compounds, results from surface contamination from mercury-containing aerosols. Food and animal feed derived from plants usually have mercury contents between 0.001 and 0.03 mg/kg (DEUTSCHE FORSCHUNGSGEMEINSCHAFT, 1975a; DEUTSCHE GESELLSCHAFT FÜR ERNÄHRUNG, 1976, 1988).

Through industrial usage, mercury is sometimes introduced into bodies of water where the metal exhibits a behavior different to that in soil which may lead to accumulation in the food chain. As is the case for arsenic, various inorganic and organic species of mercury must be distinguished. Marine organisms are especially able to transform inorganic mercury compounds into organic ones. As organic compounds, mercury is more easily transferred through the aquatic food chain (Chapter I.7h; DEUTSCHE FORSCHUNGSGEMEINSCHAFT, 1978).

As a result, marine organisms can show mercury levels of up to 5 mg/kg bound as methylmercury, the accumulation level being dependent on species and age. Accumulation factors in the marine food chain are 100 to 1000 (see also WACHS, 1981), compared to only 2 to 5 in the terrestrial food chain. Older predatory fish in particular can accumulate large quantities of mercury (0.5 to more than 1 mg/kg). Sea food is thus a particular source of mercury burden in man and animals. Fresh water and estuarine fish such as mussels and eels show higher mercury contents (up to 3 mg/kg) than fish caught in the open sea (0.05 to 0.1 mg/kg). In pike, an accumulation factor of 3000 has been established between the fish and its environment.

Mercury also passes to domestic animals and animal-derived food products via fish meal. Swine liver may contain up to 0.1 mg/kg and pork meat 0.05 mg/kg. Beef products contain only very small amounts due to a demethylation of organic mercury compounds by microorganisms in the rumen. The inorganic compounds resulting from this process are not very fat-soluble, and the enteric absorption rate is thus depressed. Beef thus contains between 0.001 and 0.02 mg mercury per kg, and milk about 0.01 mg/kg.

Poultry (0.04 mg/kg) and eggs (0.03 mg/kg) show higher values because of the widespread use of fish meal as feed. The nutritional mercury burden of man derives mainly from the consumption of fish and other aquatic animals. The Federal German Food Law has thus established a limit of 1 mg/kg mercury for all fish and shellfish destined for consumption (VERORDNUNG ÜBER HÖCHSTMENGEN AN QUECKSILBER IN FISCHEN, KRUSTEN-, SCHALEN- UND WEICHTIEREN, 1975).

A weekly uptake of 0.35 mg mercury is actually only achieved through occupational exposure and/or by the consumption of extraordinary amounts of fish or organ meats (such as liver and kidney). Otherwise, weekly uptake through food remains at between 0.03 and 0.05 mg mercury. This situation has not changed during the last decades and is not expected to in the near future.

9 Molybdenum

Molybdenum is an essential metal for plants and animals. Plants growing in molybdenum-enriched soil absorb large amounts of this metal, resulting in concentrations of up to 250 mg/kg. Plants growing in normal soil contain from 0.1 to 3 mg/kg at the most (see also Chapter I.8a). Due to rapid excretion, the use of hay with a molybdenum content of 0.2 to 2 mg/kg as animal feed does not result in an accumulation in the edible tissues. Molybdenum overload in animals results in the symptoms of copper deficiency. Animal liver normally contains 1 mg/kg molybdenum (whereas WARD, 1987, found about 3 mg/kg in bovine liver), kidney about 0.3 mg/kg (DEUTSCHE GESELLSCHAFT FÜR ERNÄHRUNG, 1976, 1988; HAPKE, 1987).

10 Nickel

Food of plant origin shows normal nickel concentrations of about 0.3 mg/kg (see also Chapters I.7 and I.8a). Uptake into plants occurs mainly through the soil, but is also possible from the atmosphere. Animals show a very slow absorption, but very high excretion rate for nickel which together result in zero-accumulation in animals, with the exception of some marine organisms (HAPKE, 1987). WARD (1987) found about 0.1 mg nickel/kg in bovine liver.

11 Selenium

This essential trace element is selectively absorbed from the soil and stored by certain plants. Animals that consume these plants (e.g., *Astragalus* spp.) can show clinical symptoms of selenium poisoning. The minimum selenium requirement in food is 0.04 mg/kg, 0.1 mg/kg being optimal. Concentrations exceeding 4 mg/kg may lead to toxic symptoms; however, one must distinguish between the effects of the various chemical species (e.g., selenites and selenates; see Chapter II.25).

The normal selenium level in plants is 0.01 to 0.1 mg/kg (see also Chapter I.8a). Soil concentrations of 100 mg/kg result in plant levels of up to 50 or even 100 mg/kg (DEUTSCHE FORSCHUNGSGEMEINSCHAFT, 1975a; DEUTSCHE GESELLSCHAFT FÜR ERNÄHRUNG, 1976, 1988). Farm animals receive selenium exclusively through their feed, and fish meal can have concentrations of 0.5 to 2.0 mg/kg. Concentrations in meat range from 0.2 to 0.5 mg/kg (sometimes more than 1 mg/kg), in liver 0.5 to 1.0 mg/kg, and in kidney 2 to 3 mg/kg. Milk concentrations rarely exceed 0.1 mg/L. If animal feed contains less than 1 mg/kg selenium and drinking water less than 0.05 mg/L, an accumulation in animals is not expected. One-third of human selenium uptake is due to animal products, the rest coming from plant sources. Health hazards resulting from nutritional selenium have not been recognized (see also Chapter II.25; DEUTSCHE FORSCHUNGSGEMEINSCHAFT, 1975a; HAPKE, 1987; OEHME, 1978).

12 Thallium

At this time, this heavy metal is only of local significance due to industrial emissions, such as from the cement industry (see Chapter II.29). An accumulation of thallium in the food chain occurs only if the vegetation is contaminated and some plants are able to absorb thallium from the soil.

A thallium concentration of 1 mg/kg in animal feed results in accumulation in animals. All tissues are affected, although liver and kidney are preferred. Feed containing less than 0.5 mg/kg can result in thallium concentrations of more than 1.0 mg/kg only in the kidneys of test animals that were fed over a period of several months (HAPKE, 1987; OEHME, 1978).

13 Vanadium

The entry of vanadium into soils and plants is caused mainly by industrial emissions, especially combustion processes. Vanadium is probably essential for both plants and animals. Plants absorb it from the soil and animals from their feed. Since the ingested amounts are eliminated quickly from the animal, an accumulation in food derived from plants or terrestrial animals has not been observed. Significantly higher vanadium concentrations have been found in oils with a high linoleic acid content (such as sunflower oil), vegetables, liver, and marine organisms. The vanadium content of bovine liver is 0.02 to 0.2 mg/kg and of kidneys 0.03 to 0.12 mg/kg (HAPKE, 1987).

14 Zinc

This essential metal is absorbed by plants through the soil or from industrial dusts. Zinc can be phytotoxic. Normal values in plants are between 1 and 40 mg/kg of dry matter. Cereals sometimes show concentrations of over 100 mg/kg (DEUTSCHE FORSCHUNGSGEMEINSCHAFT, 1975a).

Farm animals continuously receive zinc with their feed, and the excess is immediately excreted. An accumulation in animal tissues does not occur. Normal levels in meat, fish, and poultry are within 10 and 200 mg/kg with a mean of 25 mg/kg. Milk products show levels of about 5 mg/kg, liver 100 to 150 mg/kg, and kidney 50 to 100 mg/kg. (DEUTSCHE FORSCHUNGSGEMEINSCHAFT, 1975a; DEUTSCHE GESELLSCHAFT FÜR ERNÄHRUNG, 1976, 1988; HAPKE, 1987).

Since zinc does not accumulate it is of no health significance as an industrial pollutant in foodstuffs. Zinc contents of 200 to 500 mg/kg in animal feed and 25 mg/L in drinking water are of no significance. Concentrations of up to 1500 mg zinc/kg have been found in oysters from estuarine areas (HOLM, 1980).

References

Der Rat von Sachverständigen für Umweltfragen (1978), *Umweltgutachten.* Verlag W. Kohlhammer, Stuttgart und Mainz.
Deutsche Forschungsgemeinschaft (DFG) (1975a), Forschungsbericht *Rückstände in Fleisch und Fleischerzeugnissen.*
Deutsche Forschungsgemeinschaft (DFG) (1975b), *Rückstände von Bioziden und Umweltchemikalien in der Milch.* Mitteilung I der Kommission zur Prüfung von Rückständen in Lebensmitteln.
Deutsche Forschungsgemeinschaft (DFG) (1978), *Rückstände in Fischen*, Mitteilung VII der Kommission zur Prüfung von Rückständen in Lebensmitteln.
Deutsche Forschungsgemeinschaft (DFG) (1980), *Bewertung von Rückständen in Getreide*, Mitteilung VIII der Kommission zur Prüfung von Rückständen in Lebensmitteln.
Deutsche Gesellschaft für Ernährung e. V. (1976), *Ernährungsbericht.*
Deutsche Gesellschaft für Ernährung e. V. (1988), *Ernährungsbericht.*
Edward, W. C., and Dooley, A. L. (1981), *Heavy and Trace Metal Determinations in Cattle Grazing Pastures Fertilized with Treated Raffinate. Vet. Hum. Toxicol. 22*, 309–311.
Förstner, U., and Wittmann, G. T. W. (1979), *Metal Pollution in the Aquatic Environment.* Springer Verlag, Berlin.
Friberg, L., Kjellström, T., and Nordberg, G. F. (1986), Chapter 7 *Cadmium*, in: Friberg, L., Nordberg, G. F., and Vouk, V. G. (eds.): *Handbook on the Toxicology of Metals*, pp. 135–136. Elsevier Science Publishers, Amsterdam.
Hapke, H.-J. (1987), *Toxikologie für Veterinärmediziner,* 2nd Ed. Enke-Verlag, Stuttgart.
Hecht, H. (1987), *Unterschiede im Schwermetallgehalt bei Haus- und Wildschweinen und ihre Ursachen. Fleischwirtschaft 67,* 1511–1518.
Holm, J. (1980), *Blei-, Cadmium-, Arsen- und Zinkgehalte von Fischen aus unbelasteten und belasteten Binnengewässern, Fleischwirtschaft 60,* 1076–1083.
Holm, J., and Demuth, B. (1988), *Vergleichende Untersuchungen von Elementgehalten in tierischen Lebensmitteln 68,* 1472–1478.
Oehme, F. W. (1978), *Toxicity of Heavy Metals in the Environment,* Part 1. Marcel Dekker, New York – Basel.
Sahl, B. (1982), *Arsenic Concentrations in Cattle Liver, Kidney and Milk. Vet. Hum. Toxicol. 24,* 173–174.
Tsuchiya, K. (1986), Chapter 14 *Lead*, in: Friberg, L., Nordberg, G. F., and Vouk, V. G. (eds.): *Handbook on the Toxicology of Metals*, Vol. II, pp. 302–305, 309–311. Elsevier Science Publishers, Amsterdam.
Umweltbundesamt (1976), *Luftqualitätskriterien für Blei*, Berichte 3/7.
Umweltbundesamt (1977), *Luftqualitätskriterien für Cadmium,* Berichte 4/7.

Unglaub, W., Hahn, E., and Promberger, N. (1987), *Neuere Erkenntnisse zu Rückständen im Fleisch. Tierärztl. Umsch. 42*, 557–563.
Verordnung über Höchstmengen an Quecksilber in Fischen, Krusten-, Schalen- und Weichtieren (1975), *Lebensmittelrecht-Bundesgesetze und -verordnungen über Lebensmittel und Bedarfsgegenstände. Bundesgesetzblatt I*, p. 485 (6. Februar 1975). Becksche Textausgabe, 10th Ed. Ergänzungsstand Oktober 1979. C. H. Becksche Verlagsbuchhandlung, München.
Wachs, B. (1981), *Bioaccumulation of Heavy Metals in Aquatic Organisms. Sicherheit in Chemie und Umwelt 1, Nr. 3, 113–115*.
Ward, N. I. (1987), *The Future of Multi-Element Analysis. Environ. Health 20*, 118–123. 2nd Nordic Symposium Odense. WHO Regional Office for Europe, Copenhagen.

Additional Recommended Literature
From U. Ewers, Düsseldorf

Boyer, K. W., and Horwitz, W. (1986), *Special Considerations in Trace Element Analysis of Foods and Biological Materials*, in: O'Neill, I. K., Schuller, P., and Fishbein, L. (eds.): *Environmental Carcinogens – Selected Methods of Analysis*, pp. 191–220. IARC Scientific Publications No. 71, Lyon (France).
Chamberlain, A. C. (1983), *Fallout of Lead and Uptake by Crops. Atmos. Environ. 17*, 693–706.
Deutsche Gesellschaft für Ernährung (1984), *Ernährungsbericht 1984*, pp. 81–87. Deutsche Gesellschaft für Ernährung e.V., Frankfurt a. M.
Gallacher, J. E. J., Elwood, P. C., Phillips, K. M., Davies, B. E., Ginnever, R. C., Toothill, C., and Jones, D. T. (1984), *Vegetable Consumption and Blood Lead Concentrations. J. Epidemiol. Commun. Health 38*, 173–176.
Gladke, E., Heimann, G., Lombeck, I., and Eckert, I. (eds.) (1985), *Spurenelemente – Stoffwechsel, Ernährung, Imbalancen, Ultra-Trace-Elemente*. G. Thieme Verlag, Stuttgart-New York.
Page, A. L., El-Amamy, M. M., and Chang, A. C. (1986), *Cadmium in the Environment and its Entry into Terrestrial Food Chain Crops*, in: Foulkes (ed.): *Cadmium*, pp. 33–74. Springer-Verlag, Berlin-Heidelberg-New York-Toronto.
Peterson, P. J. (1978), *Lead and Vegetation*, in: Nriagu, J. O. (ed.): *The Biogeochemistry of Lead in the Environment*, Part B, pp. 355–384. Elsevier, Amsterdam-New York-Oxford.
Ryan, J. A., Pahren, H. R., and Lucas, J. B. (1982), *Controlling Cadmium in the Food Chain: A Review and Rationale Based on Health Effects. Environ. Res. 28*, 251–302.
Sachverständigenrat für Umweltfragen (1987), *Umweltgutachten 1987*, pp. 344–382. Verlag W. Kohlhammer, Stuttgart-Mainz.
Schmidt, E. H. F., and Hildebrandt, A. G. (eds.) (1983), *Health Evaluation of Heavy Metals in Infant Formula and Junior Food*. Springer-Verlag, Berlin-Heidelberg-New York.
Weigert, P., Müller, J., Klein, H., Zufelde, K. P., and Hillebrand, J. (1984), *Arsen, Blei, Cadmium und Quecksilber in und auf Lebensmitteln*. ZEBS-Hefte 1/1984.

I.9 Metals as Essential Trace Elements for Plants, Animals, and Humans

FELIX KIEFFER, Bern, Switzerland

1 Introduction

In the discussion about the pollution and protection of the environment the toxic heavy metals, such as arsenic, cadmium, lead, and mercury, always occupy a predominant position. As a consequence of the modern technological processes coupled with human negligence, they have indeed been spread throughout the whole biosphere to an unprecedented degree. This discussion, however, could easily make one forget that, starting with the monocellular microorganisms through the plant and animal world up to man, all living species are absolutely dependent on the presence of metal salts in their food in order to maintain their life and reproductive capacity. The reason for this resides in the evolutionary development of the organisms. The first living cells probably developed in a "primeval soup" which contained, in dissolved state, all elements present in the earth's crust at that period.

The chemical composition of sea water is well known today (see also Chapter I.1). It still contains more or less the same elements as in early times, namely (in descending order or concentration): chloride, sodium, magnesium, sulfur (as SO_4^{2-}), calcium, potassium, bromine, boron (as BO_3^{3-}), strontium, silicon (as SiO_4^{4-}), fluorine, lithium, rubidium, phosphorus (as PO_4^{3-}), iodine, barium, aluminum, iron, molybdenum, selenium, uranium, zinc, copper, tin, manganese, vanadium, lead, nickel, cobalt, mercury, chromium, cadmium, gold, and others. The overall concentration of solid matter reaches 35 grams/liter. Exact figures have been collected and published elsewhere (KIEFFER, 1979; BRULAND, 1983; see also Part II).

Up to now, modern research has shown that at least 24 elements are essential to life. Reference to the Periodic Table indicates that they form three coherent blocks which almost touch each other. The *non-metals* hydrogen, carbon, nitrogen, oxygen, fluorine, silicon, phosphorus, and sulfur are the main building blocks of all organic compounds such as proteins, carbohydrates, fats, nucleic acids, and vitamins. The *metals*, on the other hand, are mainly responsible for the correct functioning of innumerable enzymatic and metabolic reactions. Sodium, magnesium, potassium, calcium, phosphorus, and chlorine are often called minerals or electrolytes, whereas vanadium, chromium, manganese, iron, cobalt, nickel, copper, zinc, selenium, molybdenum, tin, and iodine are called trace elements (citations are in the order of increasing atomic number).

2 The Elemental Composition of Man

Elemental analyses of human bodies have been published at different occasions (see also Chapter I.4b and Part II). In spite of the fact that there are considerable variations in such analyses, the order of magnitude of the values given in Table I.9-1 are probably rather representative (KIEFFER, 1979).

The third column of Table I.9-1 gives the values for the approximate amounts of the different elements present in an average adult body. For many of the essential trace elements, such as vanadium, chromium, manganese, cobalt, nickel, copper, selenium, molybdenum, tin, and iodine, the quantities are between 3 and 100 mg per 70 kg body weight. Here the question arises whether such small amounts are at all capable of performing biological functions. It is easier to answer this if the weights are expressed in molar quantities (column 4 of the table). From these figures it is possible to calculate that the body contains at least 10^{19} ions of each of these trace elements. If we assume that the whole body contains approximately 10^{14} cells, a value which is given in many textbooks, and that the elements are evenly distributed among all these cells, an assumption which is obviously not correct, then each cell contains between 10^5 and 10^6 ions of these trace elements. Metabolically active cells will even contain a much larger quantity, whereas in the case of fat, cartilage, and bone cells the reverse is true. This rather simple but logical calculation reveals the true order of magnitude in which the common and well-known trace elements participate in physiological processes. On the basis of the amounts present in the body, all elements listed in Table I.9-1, even the most rare ones, are capable of exerting a physiological effect in the body. This assumption is not unrealistic, indeed, because all these elements were already present in the environment when life began on earth about three billion years ago, and they have been present in the food of all organisms ever since then.

Similar assumptions can be made for plants and animals. Even monocellular organisms, such as bacteria, need tiny amounts of a long list of trace elements. These are practically the same as those needed by human cells. Deeper research in the field of ultra-trace elements is hindered by the fact that it is still impossible to produce a culture medium or a test diet for animals which is absolutely free of one single element or metal. The following explanation makes this statement intelligible. It is common knowledge that many modern analytical methods have a limit of detection not better than about 0.1 ppb. This sensitivity allows the detection of the presence of 1 gram of a foreign substance (e.g., a trace element) in 10000 tons of a raw material.

At lower contaminations the analysts would have to declare the product as totally pure, because the sought after substance seems to be absent. Nevertheless, 100 grams of such a "pure", recrystallized substance can contain up to 10^{14} unrecognized foreign atoms of many different elements, even though the researchers firmly believe that their experimental diets are free of the element under investigation. Hence, it is impossible to establish the essentiality of a trace element simply by manipulating the composition of the diet, as it is quite commonly practiced in vitamin research.

Table I.9-1. Approximate Elemental Composition of the Human Body (70 kg body weight)

Element		Relative Atomic Mass	Grams per 70 kg b.w.	Quantity Expressed in mol per 70 kg b.w.	Number of Atoms in the Body	Number of Atoms per Cell
H	Hydrogen	1	7000	3500	4.2×10^{27}	4.2×10^{13}
B	Boron	10.8	0.01	0.00092	5.5×10^{10}	5.5×10^{6}
C	Carbon	12	12600	1050	6.4×10^{26}	6.4×10^{12}
N	Nitrogen	14	2100	75	9.1×10^{25}	9.1×10^{11}
O	Oxygen	16	45500	1425	1.7×10^{27}	1.7×10^{13}
F	Fluorine	19	0.8	0.021	2.6×10^{22}	2.6×10^{8}
Na	Sodium	23	105	4.6	2.8×10^{24}	2.8×10^{10}
Mg	Magnesium	24.3	35	1.44	8.7×10^{23}	8.7×10^{9}
Al	Aluminum	27	0.1	0.0037	2.2×10^{21}	2.2×10^{7}
Si	Silicon	28	1.4	0.05	3.0×10^{22}	3.0×10^{8}
P	Phosphorus	31	700	22.5	1.4×10^{25}	1.4×10^{11}
S	Sulfur	32	175	5.5	3.3×10^{24}	3.3×10^{10}
Cl	Chlorine	35.5	105	2.96	1.8×10^{24}	1.8×10^{10}
K	Potassium	39.1	140	3.58	2.2×10^{24}	2.2×10^{10}
Ca	Calcium	40.1	1050	26.2	1.6×10^{25}	1.6×10^{11}
Ti	Titanium	47.9	0.01	0.00021	1.3×10^{20}	1.3×10^{6}
V	Vanadium	50.9	0.02	0.00039	2.4×10^{20}	2.4×10^{6}
Cr	Chromium	52	0.005	0.0001	0.6×10^{20}	6.0×10^{5}
Mn	Manganese	55	0.02	0.00036	2.2×10^{20}	2.2×10^{6}
Fe	Iron	56	4.2	0.075	4.5×10^{22}	4.5×10^{8}
Co	Cobalt	59	0.003	0.00005	0.3×10^{20}	3.0×10^{5}
Ni	Nickel	58.7	0.01	0.00017	1.0×10^{20}	1.0×10^{6}
Cu	Copper	63.5	0.11	0.0016	1.0×10^{21}	1.0×10^{7}
Zn	Zinc	65.4	2.33	0.036	2.2×10^{22}	2.2×10^{8}
As	Arsenic	74.9	0.014	0.00019	1.1×10^{20}	1.1×10^{6}
Se	Selenium	78.9	0.02	0.00025	1.5×10^{20}	1.5×10^{6}
Rb	Rubidium	85.5	1.1	0.013	7.9×10^{21}	7.9×10^{7}
Sr	Strontium	87.6	0.14	0.0016	1.0×10^{21}	1.0×10^{7}
Zr	Zirconium	91.2	0.3	0.0033	2.0×10^{21}	2.0×10^{7}
Nb	Niobium	92.9	0.1	0.0011	7.0×10^{20}	7.0×10^{6}
Mo	Molybdenum	95.9	0.005	0.00005	0.32×10^{20}	3.2×10^{5}
Cd	Cadmium	112.4	0.03	0.00027	1.6×10^{20}	1.6×10^{6}
Sn	Tin	118.7	0.03	0.00025	1.5×10^{20}	1.5×10^{6}
Sb	Antimony	121.7	0.07	0.00057	3.5×10^{20}	3.5×10^{6}
I	Iodine	126.9	0.03	0.00024	1.5×10^{20}	1.5×10^{6}
Ba	Barium	137.3	0.016	0.00012	0.73×10^{20}	7.3×10^{5}
Pb	Lead	207.2	0.08	0.00038	2.3×10^{20}	2.3×10^{6}

3 The Elements that are Essential for Life

With the exception of iron and iodine our knowledge about the essential trace elements is less than 100 years old. A review of the history of their discovery has been published elsewhere (KIEFFER, 1979). Table I.9-1 shows that 98% of the body mass of man is made up of 9 non-metallic elements. Another 1.89% are formed by the 4 main electrolytes sodium, magnesium, potassium, and calcium. The 11 typical trace elements occupy just a tiny 0.012% or 8.6 grams of the body weight. But this small fraction exerts a tremendous influence on all body functions that could never have been imagined in earlier times. Therefore, it is increasingly important to improve our knowledge about the positive and negative effects of the heavy metals, which are present in the environment and in our food, on the human metabolism.

Whether an element is essential for life or not depends on its participation in one or several biochemical reactions. The term "essential" is borrowed from the early amino acid and protein chemistry. According to the recommendations of a WHO expert committee (WHO, 1973), this expression is not optimal when it is applied to trace elements because of the mentioned difficulties in establishing an absolute essentiality. According to a new, improved definition (MERTZ and CORNATZER, 1971) an element is already useful to the organism and to the maintenance of health when a measurable deficit in the diet reduces the growth and vitality of humans, animals, or plants to a reproducible degree. If we start from this definition it becomes plausible that even well-known toxic elements, such as arsenic, silver, cadmium, and lead, and possibly aluminum, are needed in minute quantities for the normal functioning of cell metabolism, although they are recognized as toxic in higher concentrations.

The most important biological functions of the known trace elements, taken from many different literature sources, are listed in Table I.9-2.

In most cases, the ions of trace elements act as coordination centers for building up or stabilizing the structure of enzymes and proteins. Exceptions are iron, which is the central atom of heme in cytochromes and hemoglobin, cobalt in the center of vitamin B_{12}, and iodine, which is a constituent of the hormone thyroxine. Chromium seems to be essential for the biosynthesis of the glucose tolerance factor in man (VINSON and HSIAO, 1985) which itself seems to be chromium-free (HAYLOCK, 1983).

About 50 zinc enzymes, and as many reactions, are known (MERTZ, 1987). Regarding copper deficiency it has also been observed that there may be correlations with mitochondrial changes and with a reduction of pancreatic weight and protein concentration (MYLROIE et al., 1987), with a reduction of cardiovascular integrity (SMITH, 1987), and with a reduction of disease resistance (for instance, of ewes; mountain sheep are more resistant; JONES and SUTTLE, 1987).

Food and drinking water must contain all the essential mineral elements in adequate quantities, otherwise the continuous losses with urine, feces, and sweat would produce severe deficiency syndromes within a relatively short period of time. In some countries the desirable ranges of intakes for each element are established and published in regular intervals by semi-official national nutrition boards. The values are

Table I.9-2. Main Functions of Trace Elements and Consequences of Their Absence

Element	Constituent of	Causes of Total Absence
Chromium	Needed for the biosynthesis of the glucose tolerance factor	Diabetes type II
Cobalt	Cobalamin (vitamin B_{12}), which is a cofactor of 4 enzymes	Anemia; arrest of nucleic acid synthesis etc.
Copper	Cytochrome oxidase	Blocking of all respiration
	Ceruloplasmin = laccase	Lack of copper transport
	Ascorbic acid oxidase	Disturbed redox-reactions
	Lysine oxidase	Stop of cartilage formation
	Tyrosinase	Albinism, no pigmentation
	Superoxide dismutase, SOD	Cell destruction by superoxide radicals
	Monoamine oxidase	Lack of neurotransmitter synthesis
	Uricase (not in man)	Lack of uric acid oxidation
Iodine	Thyroxine (T_3, T_4 hormones)	Defective cell differentiation, multiple malfunctions, goiter
Iron	Cytochromes a, b, c, f, P_{450}, cytocrome c reductase	Blocking of oxidative reactions, blocking of cell respiration and energy production
	Catalase, peroxidases	Multiple lipid peroxidation, hemolysis
	Hemoglobin	Anemia, no oxygen transport, asphyxia
Manganese	Arginase	Urea synthesis impossible (ammonia intoxication)
	Malic enzyme, pyruvate carboxylase	Blocking of citric acid cycle
	Superoxide dismutase, SOD	Destruction of mitochondria by oxygen radicals
Molybdenum	Aldehyde oxidase	Blocking of fatty acid biosynthesis from carbohydrates
	Xanthine oxidase	Uric acid formation impossible
Nickel	Urease (not in man), glucose tolerance	Nitrogen fixation in soil inhibited
Selenium	Glutathion peroxidase	Accumulation of lipid peroxides, hemolysis
Tin	Gastrin (?)	Impaired digestion and growth
Zinc	Alkaline phosphatase	Whole energy metabolism compromised
	Alcohol dehydrogenase	Alcohol intoxication
	Carboanhydrase	Acidosis
	Carboxypeptidase	Protein biosynthesis blocked, infertility
	Glutamate dehydrogenase	Transamination reactions stopped
	Lactate dehydrogenase	Lactate accumulation, acidosis in muscles
	Superoxide dismutase, SOD	Cell destruction by superoxide radicals

called "Recommended Dietary Allowances" or RDAs (see Part II). The best-known tables of this kind are those published in the USA. Table I.9-3 indicates the values for the electrolytes and trace elements for adults (NAS/NRC, 1980). These values consider the fact that only a certain percentage of the metal ions present in the food can be absorbed in the intestine (see also Chapter I.15).

The absorption rates differ remarkably between the elements. Table I.9-4 shows these rates for seven trace elements (see also Part II, and FRIBERG et al., 1986). The

Table I.9-3. Recommended Daily Intakes of Mineral Elements

Element	Recommended Daily Intake (in mg)
Potassium	1850 – 5500
Sodium	1100 – 3300
Calcium	800 – 1200
Phosphorus	800 – 1200
Magnesium	350 – 400
Iron	men: 10; women: 18
Zinc	15
Manganese	2.5 – 5 [a]
Copper	2 – 3 [a]
Molybdenum	0.15 – 0.5 [a]
Chromium	0.05 – 0.2 [a]
Selenium	0.05 – 0.2 [a]
Iodine	0.15

[a] Tentative ranges

values permit a calculation of the amounts which are actually needed by the organism, e.g., in parenteral nutrition.

In natural foodstuffs the trace elements are present in very different concentrations and various chemical forms (see also Chapters I.8a, I.8b, and Part II), the characterization of which is a domain only at its beginnings. Sparse data only for a few elements exist in this latter respect (SOLOMONS and ROSENBERG, 1984). The bioavailability of an element is in fact strongly influenced by its chemical form. Apart from this the intestinal mucosa contains unspecific and specific transport mechanisms for the different metals. Some metals may share the same carrier. Thus, if they are present in unbalanced concentrations in a meal, observations suggest that the host can suffer from malabsorption due to interactions, as outlined in Table I.9-5.

These interdependencies demonstrate the importance of an equilibrium in the overall trace element content of a healthy food supply. There is still an unresolved problem in making available the complete set of essential trace elements in optimal amounts at a more or less continuous rate for all people. One of the recurring problems is the widespread use of refined foodstuffs, such as white sugar, corn syrups of all kinds, isomerized corn syrup (HFCS), pure glucose, fructose, and sugar, alcohols, refined (white) flours and refined fats and oils. For instance, iron absorption changes in many ways, baked rolls may decrease it, meat, and/or vitamin C may increase it (MERTZ, 1987). The interaction of ascorbic acid with the daily intake of "absorbable" iron is even quantifiable (MERTZ, 1987). Fasting, fiber content of the diet, and fruit juices (compared to aqueous solutions) have also an influence on the uptake of iron and of other trace elements (MERTZ et al., 1986). Cardiovascular integrity of male rats is especially impaired in case of copper deficiency, when sucrose or fructose (instead of starch) is administered with the diet (female rats and human subjects are less sensitive; SMITH, 1987). Latent deficiencies are more frequent than

Table I.9-4. Absorption Rates of Minerals and Trace Elements

Element in Food	Absorption Rate (in %)[a] (Ingestion)
Sodium	90 – 95
Chlorine	95 – 100
Potassium	90 – 95
Molybdenum	70 – 80 or less
Selenium	50 – 80 or less or more
Phosphate	60 – 70
Calcium	25 – 40
Zinc	20 – 40 or more
Magnesium	30 – 35
Copper	10 – 30 or more
Iron	7 – 15 or more
Manganese	3 – 5 or less
Chromium(III)	0.5 – 1.0[b]

[a] Depending on the dose, speciation (especially in the cases of selenium, copper, iron, and chromium, but also of zinc) and transformation, health status, and eventually interactions (see also Part II and FRIBERG et al., 1986)
[b] Chromate absorption may be somewhat higher, especially in patients in whom Cr(VI) is not reduced by gastric juice

Table I.9-5. Interactions in Trace Element Absorption

An Excess of ...	Produces a Deficit of ...
Cadmium	Selenium, zinc
Calcium	Zinc
Iron	Copper, zinc
Manganese	Magnesium
Molybdenum	Copper
Zinc	Copper, iron

one would believe, and they are difficult to identify for different reasons. Blood analyses are expensive and not all that helpful. Further, the consequences of the many possible combined deficiencies are rather difficult to recognize.

Nutrients are taken up from water and soil by plants. The minimal quantity of most trace elements which accumulate is a consequence of their genetically determined cell structure and function which can vary somewhat between the species. However, trace metals which are accidentally present in the soil for geological reasons or by human deposition as, for example, with lead from gasoline along the highways, can accumulate in grass and food crops. They represent unforeseeable risks for the consumer.

The minimum biological needs for trace elements of plants, animals, and man are only proportionally, but not fundamentally different. Therefore, no distinction has been made in this respect within this chapter. Even boron makes no exception, in spite of the fact that boron is generally considered to be essential only for plants. Recent findings demonstrate that the hydroxylases which are responsible for the transformation of the vitamins D_2 and D_3 into their corresponding 1,25-dihydroxy-derivatives, and cholesterol into estradiol, strictly depend on borate ions. Hence, boron may be a potent protective factor against post-menopausal osteoporosis in women (NIELSEN, 1987, 1988). With this final remark we intend to demonstrate how important each single element eventually may be for life, even if nowadays textbooks and scientists still declare some of them as futile. It also makes clear that our present knowledge of the biological functions of many trace elements is still rather incomplete.

References

Bruland, K. W. (1983), *Trace Elements in Sea Water. Chem. Oceanogr. 8,* 158–220.
Friberg, L., Nordberg, G. F., and Vouk V. B. (1986), *Handbook on the Toxicology of Metals,* 2nd Ed., Vol. II. Elsevier, Science Publishers, Amsterdam.
Haylock, S. J. et al. (1983), *The Relationship of Chromium to the Glucose Tolerance Factor, II. J. Inorg. Biochem. 19,* 105–117.
Jones, D. G., and Suttle, N. F. (1987), *Copper and Disease Resistance. Trace Subst. Environ. Health 21,* 514–525.
Kieffer, F. (1979), *Trace Elements Govern Our Health. SANDOZ-Bulletin No. 51–53.* (Reprints in English or French or German from the author).
Mertz, W. (1987), *The Practical Importance of Interactions of Trace Elements. Trace Subst. Environ. Health 21,* 526–532.
Mertz, W., and Cornatzer, W. E. (1971), *Newer Trace Elements in Nutrition.* Marcel Dekker, New York.
Mertz, W., Hansen, C., and Hartmuth-Hoene, A. E. (1986), *Biological Reference Materials and Bioavailability. Proceedings 4th International Workshop on Trace Element Analytical Chemistry in Medicine and Biology, Neuherberg,* edited by Brätter, P., and Schramel, P. W. de Gruyter, Berlin-New York.
Mylroie, A. A., Tucker, C., Umbles, C., and Kyle, J. (1987), *Effects of Dietary Copper Deficiency on Rat Pancreas. Trace Subst. Environ. Health 21,* 44–53.
NAS/NRC (1980), *Recommended Dietary Allowances,* 9th Ed. National Academy of Sciences, National Research Council, Washington, D.C.
Nielsen, F. H. (1987), *Dietary Boron Affects Mineral, Estrogen and Testosterone Metabolism in Post-menopausal Women. FASEB J. 1,* 394–397.
Nielsen, F. H. (1988), *Boron, an Overlooked Element of Potential Nutritional Importance. Nutr. Today* (Jan./Feb.) 4–7.
Smith, J. C. (1987), *Copper Nutrition and Cardiovascular Integrity. Trace Subst. Environ. Health 21,* 499–513.
Solomons, N. W., and Rosenberg, I. H. (1984), *Absorption and Malabsorption of Mineral Nutrients.* Alan R. Liss, New York.
Vinson, J. A., and Hsiao, K.-H. (1985), *Comparative Effect of Various Forms of Chromium on Serum Glucose: An Assay for Biologically Active Chromium. Nutr. Rep. Int. 32,* 1–7.
WHO (1973), *Trace Elements in Human Nutrition.* Tech. Rep. Ser. No. 532 (Geneva).

Additional Recommended Literature

Anke, M., Brückner, C., Gürtler, H., and Grün, M. (1987, 1988), *Proceedings of Workshops on Minerals and Trace Elements, Leipzig* (in German), Part 1 1987, 255 pages; Part 2 1987, 250 pages; Part 1 1988, 243 pages; Part 2 1988, 217 pages. Sektion Tierproduktion und Veterinärmedizin, Universität Jena, GDR.

Anke, M., Baumann, W., Bräunlich, H., Brückner, C., Groppel, B., and Grün, M. (1989), *Proceedings 6th International Trace Element Symposium*, 5 Vols. Karl-Marx-Universität, Leipzig, and Friedrich-Schiller-Universität, Jena.

Bowen, H. J. M. (1979), *Environmental Chemistry of the Elements*. Academic Press, London.

Brätter, R., and Schramel, P. (1980, 1983, 1984, 1987, 1988), *Trace Element Analytical Chemistry in Medicine and Biology. Proceedings Workshop Neuherberg*, 5 Vols. Walter de Gruyter, Berlin-New York.

Fishbein, L. (1987), *Trace and Ultratrace Elements in Nutrition: An Overview (Zinc, Copper, Chromium, Vanadium and Nickel), Toxicological and Environmental Chemistry* (Gordon and Breach), Vol. 14 (1+2), p. 73–99.

Hemphill, D. D. (1967–1989), *Trace Substances in Environmental Health*. 23 Proceedings. University of Missouri, Columbia.

Iyengar, G. V., Kollmer, W. E., and Bowen, H. J. M. (1978), *The Elemental Composition of Human Tissues and Body Fluids*. Verlag Chemie, Weinheim-New York.

Kharash, N. (1979), *Trace Metals in Health and Disease*. Raven Press, New York.

Kiefer, W. (1987), *Selenium – a Medically Significant Trace Element* (in German). Ars Medici 2/87, 60–74.

Kiefer, W. (1988), *Vitamins, Minerals, and Trace Elements Controlling Body Functions* (in German). Wander, Bern.

Kiefer, W. (1988), *Effects of Food and Environments on Human Metabolism and Health* (in German). *Medizin + Ernährung* 3/88.

Nève, I., and Favier, A. (1988), *Selenium in Medicine and Biology*. Walter de Gruyter, Berlin-New York.

Prasad, A. S. (1976), *Trace Elements in Human Health and Disease*, 2 Vols. Academic Press, London.

Prasad, A. S. (1988), *Essential and Toxic Trace Elements in Human Health and Disease, Current Topics in Nutrition and Disease*, Vol. 18, 704 p., Alan R. Liss, New York.

Schroeder, H. A. (1972), *The Poisons Around Us – Toxic Metals in Food, Air and Water*. Indiana University Press, Bloomington-London.

Schroeder, H. A. (1976), *Trace Elements and Nutrition*. Faber & Faber, London.

Trueb, L. (1988), *Osteoporosis – a Nutritional Problem* (in German). *Neue Zürcher Zeitung, Forschung und Technik*, No. 9, p. 79 (13 January), Zürich.

Underwood, E. J. (1976), *Trace Elements in Human and Animal Nutrition*. Academic Press, London.

WHO (1987a), *Selenium. Environ. Health Crit. 58* (Geneva).

WHO (1987b), *2nd Nordic Symposium. Trace Elements in Human Health and Disease*, Odense. *Environ. Health* 20 and 26. WHO Regional Office for Europe, Copenhagen.

I.10 The Significance of Interactions in Metal Essentiality and Toxicity

DETMAR BEYERSMANN, Bremen, Federal Republic of Germany

1 Introduction

In 1977, ARTHUR FURST stated: "Metals: We know so much and we know so little". In particular, he felt that we knew a lot about metal chemistry and not enough about metal biochemistry. Metal biochemistry and toxicology have developed considerably since that time, but have investigated mainly single interactions with living matter. In a modified sense, the motto of A. FURST still holds true: we know so much about individual metal biochemistry and medicine, but we know so little about combined effects of metals in nutrition and toxicity. This statement is in marked contrast to our scientific and practical needs: interactions of inorganic compounds in the abiotic and in the living world are not the exception but the rule. Metal compounds interact at the levels of uptake and bioavailability, of toxic reactions, of metabolism, of signal transduction, of immune and tolerance reactions and of genetic and related effects.

In this chapter no comprehensive review will be attempted, but the main aim will be to reach an understanding of the biochemical mechanisms involved in metal interactions. The emphasis will be on human exposure and on those interactions in the microbial, the plant and the animal realms that affect the human situation by food chains or as models for the human conditions. Since we have only very limited data on interaction at the human level derived from hospital or epidemiological data, we have to extrapolate many conclusions about combined effects of metals in nutrition or toxicity from animal experiments. Further aspects of metal interactions in biological systems have been covered in recent reviews from different points of view. The reader is referred to BABICH et al. (1985) for interactions of metals with physiochemical factors in microbes, to NORDBERG (1976), NORDBERG and PERSHAGEN (1984), NORDBERG et al. (1986), FURST (1987a) and to GEBHART and ROSSMAN (this book, Chapter I.18) for medical and cancer aspects of metal interactions, and to FISHBEIN (1987) for bioavailability and interactions, especially with respect to selenium. Other examples of interactions are described in many chapters of this book, particularly in Part II, but also in Chapters I.9, Sect. 3; I.13b, Sect. 5; I.20a, Sect. 2.8, and II.9, Sect. 5.

First, it is important to have an agreement about what is meant by the term "interaction". A task group of metal interaction has discussed and formulated a definition as follows: "Interaction is a process by which metals in their various forms change the critical concentration or a critical effect of a metal under consideration" (NORDBERG, 1976). Such changes are regarded as interactions only, if they deviate from simple additivity. To use unambiguous terms, interactions should be called positive, if they cause overadditive effects, and negative, if they cause less than additive results. The term synergistic is not used here since it is ambiguous in that it covers both additive and overadditive effects (in which sense it is used in other chapters of this book). The term potentiating is used neither, since it suggests some kind of exponential relation.

Second, a further limitation has to be mentioned. This chapter is confined to chemical and biochemical factors that interact with metals in the environment. Other factors that are of equal importance with respect to the human condition are, for instance, age, sex, inherited or acquired metabolic disturbances, physical and psychic stress. It is highly desirable to acquire more knowledge about these influences, which cannot be covered in this chapter. This section will discuss interactions with essential and toxic roles of metal and metalloid compounds.

The effect described will strongly depend on the actual concentration of a metal within the scale from deficiency to essentiality and toxicity. If the supply of a metal compound is close to the lower limit of the essential range, interactions that lower the bioavailability, will produce symptoms of deficiency. If the burden of a metal is close to the upper limit of the essential range, interactions that increase its availability will move the dose into the toxic range. These relations hold true for biochemical as well as for abiotic interactions which will be discussed first.

2 Abiotic Combinations

Prior to their uptake by living matter, inorganic substances interact in the abiotic environment and modify their biological effects. Such interactions may change the solubility, the valence state and charge, the chemical ligands or chelation, which all influence the bioavailability. With regard to the food chain ending in human nutrition, the aqueous environment and the soil are of great importance, whereas in inhalative exposure the air contaminations may constitute a higher risk. Aerosols have very complex compositions, and interactions at surfaces and in pores must be considered. In water, oxidation of metal sulfides in sediments and dissolution of hydroxides and carbonates by acidification through acid rain mobilize transition or heavy metals whereas hard water precipitates some metal carbonates (FÖRSTNER, this book, Chapter I.7). A classic example of toxification by abiotic interaction is the so-called plumbosolvency. A high contamination of drinking water by lead has been observed in areas of Belgium and Scotland where the water is of low hardness and low pH and where it was supplied through lead tubes and/or tubes connected by lead-rich soldering materials. An extraordinary case of mobilization by oxidation and immobilization through reduction is chromium. Whereas the oxidation of Cr(III) to Cr(VI) generates the bioavailable chromate, the reduction of Cr(VI) to Cr(III) produces a form that is excluded by biomembranes and that precipitates at neutral pH (BARTLETT and JAMES, 1979; JAMES and BARTLETT, 1983). Humic acids act as ion exchangers which influence the availability of essential and toxic metal ions

(REASHID, 1971). Chelators from industrial waste waters such as EDTA or NTA mobilize heavy metal ions from river sediments. The resulting complexes may enhance the uptake of toxic metals such as cadmium, mercury, copper, and chromium through cell membranes (NOLEN et al., 1972; SCHARPF et al., 1972; MURAMOTO, 1980; GENTILE et al., 1981; VENIER et al., 1987).

In industrial hygiene, air contaminants are of major concern. The complex constitution of industrial emissions gives rise to complex interactions. In stainless steel welding fumes, compounds of chromium, nickel, and other metals interact in a manner not well understood up to date (see Sect. 9). Interactions of Zn, Cd, Sb, As, and Pb have to be considered in the exhausts of waste incinerators (FISHBEIN, 1989).

In a strict sense, also the mobilization of insoluble metal compounds in the acid milieu of the stomach is an abiotic interaction, since it occurs prior to the absorption through the intestinal membranes. Abiotic interactions are equally important for toxicity (uptake of heavy metals) as for nutrition (uptake of essential metals). Further aspects of environmental interactions of inorganic and organic compounds have been discussed in the Handbook of Environmental Chemistry, Volume 2 "Reactions and Processes" (HUTZINGER, 1980, 1982, 1985, 1988) and by SMITH (1990).

3 Interactions in Uptake

The administration routes and the doses (which are not always mentioned in the literature) have of course a great influence, and in vitro and in vivo data may also be different. The uptake of metal compounds through biomembranes of skin, lung, or intestine involves various mechanisms of penetration. A general factor of interaction is the pH, since it influences the charge and thus the available form of a metal. Whereas organic substances in general penetrate better in their uncharged forms, inorganics commonly use cation or anion carriers of cell membranes. The situation becomes even more complicated, if one considers not only uptake but also excretion. An example is the lead homeostasis in red cells where the influx is by facilitated diffusion, and the efflux is by active transport (SIMONS, 1988). Metal-metal interactions in uptake have been subject of intensive research, since they contribute to human health and disease and they are important in animal nutrition (MILLS, 1985). Some data have been collected on human exposure to combinations of inorganic substances, either by epidemiological methods or from patients in hospitals (FOX and JACOBS, 1986; EWERS and SCHLIPKÖTER, this book, Chapter I.15). Amazingly, much less information is available about the biochemical mechanism of trace element transport through cell membranes, and still less about the biochemistry of interactions. Since only limited data are available on interactions at the actual penetration stage, much of the evidence regarding competition or enhancement in uptake is derived from studies on toxic symptoms. Some examples are discussed below, and a survey of interactions of inorganic substances in animals is listed in Table I.10-1.

Table I.10-1. Dietary and Metabolic Interactions of Inorganic Substances in Animals. (Modified from MILLS, 1989)

	Inorganic Dietary Components Provoking or Exacerbating the Syndrome
(A) Deficiency Syndrome	
Co (Anorexia, anemia, neurological defects)	$CaCO_3$: +
Cu (Defective melanogenesis, skeletal and connective tissue defects)	Mo: +, S: +, Fe: +, Cd: +, Zn: +, Ag: +
Fe (Anemia)	PO_4^{3-}: +, Cu: ±, Zn: +, Cd: +
Mn (Skeletal defects, reproductive failure)	Ca: +, PO_4^{3-}: +
Mg (Neuromuscular defects)	K: +
Mo (Defective keratogenesis)	W: +
Zn (Anorexia, parakeratosis, defective immune function)	Ca: +, Cu: +, P as phytate: +
Se (Myopathy, myoglobinuria, liver necrosis, defective leukocyte function)	S: +, Cu: +, Ag: +
(B) Toxicity Syndrome in Chronic Exposure	
Cd (Growth retardation)	Fe: −, Cu: −
Cu (Growth retardation, tissue damage)	Fe: −, Zn: −
Mo (Secondary induction of Cu deficiency, delayed puberty and conception)	Cu: −, S: +
Pb (Neurological lesions, anemia)	Ca: −, Fe: −, S: −
Zn (Pancreatic and renal damage, anemia, skeletal and tissue lesions, perinatal mortality)	Cu: −, Fe: −

+ increased dietary concentrations exacerbate toxic manifestations
− low concentrations exacerbate
± conflicting observations

Cadmium and Calcium. In areas of Japan highly contaminated by industrial use of cadmium, an epidemy of Itai-Itai (pain) disease was observed due to excessive ingestion of contaminated rice and water (see also STOEPPLER, this book, Chapter II.6). Patients had taken up 300−480 µg of cadmium per day. The symptoms of the disease were severe renal tubular damage and pronounced osteomalacia causing great pain in back and legs. These effects were observed mainly in women, which had low daily intakes of calcium and protein. Very large doses of vitamin D were needed to alleviate the symptoms. The interpretation of the disease was aided by observations in experimental animals. Cadmium primarily caused renal tubular damage, which impaired the reabsorption of calcium and phosphate. The resulting mineral deficiency induced a depletion of calcium from bones. In calcium-deficient animals the retention of cadmium and the extent of osteoporosis was even more pronounced than in the case of calcium deficiency alone. A further factor augmenting the symptoms of osteomalacia may be the disturbance by cadmium of vitamin D hydroxylation to its active form which was observed in chick kidney (for a review of cadmium and calcium interactions see FRIBERG et al., 1986).

Calcium deficiency has also been shown to increase the absorption of cadmium and lead in the gut of experimental animals and human patients. Calcium-rich diets, on the other hand, protected from toxicity including carcinogenicity of cadmium, lead, and nickel in rats (NORDBERG et al., 1986).

Cadmium and Zinc. In rats, cadmium and zinc ions compete in uptake. Zinc protects against cadmium injury to rat testes, against cadmium-induced growth depression in chicks, against teratogenic effects of cadmium in hamsters and chicken, and it protects against cadmium-induced fetal growth retardation in mice (NORDBERG et al., 1986).

Pretreatment with low doses of cadmium also protected rats against testicular destruction by subsequent treatment with higher doses of cadmium (NORDBERG, 1971). This finding was explained by the induction of metallothionein synthesis (see also Sect. 7 and KÄGI, this book, Chapter I.12). Protection against cadmium injury by zinc may occur at two levels: (1) competition for uptake, (2) induction by zinc of metallothionein synthesis, which then binds and thus detoxifies cadmium. In experimental carcinogenicity studies zinc ions inhibited the induction of tumors by cadmium. These protective effects may not be efficient when cadmium exposure occurs in a situation of borderline or deficient zinc supplementation in food. In this case the tolerable dose of cadmium will be lower than in adequate zinc nutrition (MILLS, 1985). The similar chemical properties of zinc and cadmium explain the competition of their ions for biological binding sites in many zinc enzymes (BERTINI et al., 1986).

Cadmium and Copper. Dietary supplementation of cadmium induces a redistribution of copper resulting in copper deficiency in animals. Since a supplementation with excess zinc produces the same signs of copper deficiency these findings have been discussed by MILLS (1985) in terms of induction of metallothionein which subsequently absorbs copper. This example demonstrates that interactions at the level of uptake may occur not only with ion transport proteins of the cell membranes but also indirectly, if an intracellular sink is created like that of metallothionein.

Iron and Cadmium. Cadmium injections decrease the absorption of iron from the gut and result in anemic symptoms in rabbits (BERLIN and FRIBERG, 1960). In a reciprocal experiment FLANAGAN et al. (1978) demonstrated that iron deficiency in mice caused an increased absorption of cadmium and lead from the gut.

Interactions with Anions and Chelating Agents. Whereas an acid milieu in the environment including the stomach will enhance the absorption and toxicity of metal ions, several anions have a depressing effect on the utilization of metal ions. High sulfide concentrations abolish part of the copper, mercury, and lead effects: high carbonate or hydrogen carbonate levels lower the availability of calcium, magnesium, and zinc, and high phosphate concentrations diminish the utilization of iron (MILLS, 1986).

In animal nutrition, calcium is involved in the adsorption of essential and toxic ions to phytate (myo-inositol-hexakis-dihydrogen-phosphate). Phytate, which is a significant component of many diets rich in fibers, forms insoluble complexes with Zn, Cu, Cd, and Pb ions and restricts the bioavailability of zinc in farm animals (MILLS, 1986). A high calcium concentration increases the occlusion of zinc in a Ca-Zn-phytate precipitate. This observation may also be relevant for human nutrition involving a fiber-rich diet. Interactions of lead with calcium in regulatory functions are discussed in Sect. 5.

In general chelating agents lower the available free metal ion concentration, but the complex may have greater permeability into, e.g., the brain. Cadmium uptake may serve as an example: EDTA protected rats from cadmium-induced hypertension, and EDTA, penicillamine, diethylenetriamine pentaacetate, and dimercaptosuccinate reduced the toxicity of lethal cadmium doses (CANTILENA and KLAASSEN, 1981).

Influence of Chemical Factors on the Toxicity of Metal Compounds to Microbes. Metal effects on microorganisms are not without relevance for humans. Interactions of metals at the level of bioavailability to bacteria, fungi, and protozoa may affect the food chain up to human nutrition. The reader is referred to the review of BABICH et al. (1985), who discussed the mediation of heavy metal toxicity by inorganic factors. Table I.10-2 lists the influences of many inorganic and a few organic substances on the toxicity of some metals to microorganisms. It is surprising but understandable that most additional factors decrease the toxicity of metal compounds to microbes.

Table I.10-2. Influence of Some Chemical Factors on the Toxicity of Some Transition and Heavy Metals to Microorganisms. (Modified from BABICH et al., 1985)

Factor	Effects on Toxicity to Bacteria, Yeast and Fungi
pH increased	Ni: −, Cd: −, Zn: +
Mg(II)	Ni: −
Ca(II)	Cd: −
Cl⁻ increased	Cd: −, Hg: −
Salinity increased	Cd: −, Ni: −
Sulfide	Cu: −, Ni: −
Water hardness increased	Ni: −, Cd: −, Zn: −
EDTA	Ni: −, Cd: −, Cu: −
NTA	Ni: −, Zn: −
Citrate	Ni: −
Clay minerals	Ni: −, Cd: −
Humic acids	Ni: −, Pb: −, Cd: −

− reduced
+ increased toxicity

4 Interactions in Biochemical Redox Processes

Biochemical redox processes often generate reactive oxygen species as toxic by-products. Molecular oxygen may be reduced by single-electron steps to superoxide ($HO_2\cdot$), hydrogen peroxide (H_2O_2), hydroxy radical ($HO\cdot$), and water. Reactive oxygen species cause the formation of peroxide or hydroxy derivatives with lipids, proteins, and nucleic acids. At the level of the biomembranes, lipid peroxidation is a major destructive reaction. Unsaturated fatty acids are transformed into radical and hydroperoxy products which in turn promote free radical chain reactions with other cell constituents. Detoxifying enzymes such as superoxide dismutase, catalase, and peroxidases catalyze the dismutation and reduction of reactive oxygen intermediates. Sulfhydryl compounds, above all others the tripeptide glutathione, serve as reductants and radical scavengers.

Destructive Effects. Toxic metal compounds may interact with redox processes or may even directly generate reactive oxygen compounds. HALLIWELL and GUTTERIDGE (1984) and CHRISTIE and COSTA (1984) have discussed multiple mechanisms of interaction of metals with redox processes:

a) Metal ions capable of single electron transfers such as Fe(II) or Cu(I) generate hydroxy radicals from H_2O_2 in Fenton-type reactions.
b) Metal ions such as Cu(II) or chromate(VI) directly oxidize thiol groups in biomembranes to disulfide bridges.
c) Metal ions may stimulate lipid peroxidation indirectly, e.g., Pb(II), which causes conformational shifts in hemoglobin and thus facilitates methemoglobin and superoxide formation.
d) Metal ions such as Cd(II), Hg(II), Pb(II), and Zn(II) form very stable complexes with glutathione and thus lower the level of available antioxidant in cells.

In addition, a mechanism discussed by FISHBEIN (1987) has to be taken into account:

e) Inhibition or depletion of glutathione peroxidase by metal ions. Ag(I), Cd(II), and Hg(II) directly inhibit this enzyme, whereas a depletion of selenium from the diet causes a diminution of the enzyme protein (glutathione peroxidase may even serve as an indicator for selenium bioavailability in humans). Increased levels of lead or sulfur in the diet of rats counteract the absorption of selenium and its utilization for glutathione peroxidase.

Reactive oxygen species such as H_2O_2 and free radical oxygen compounds are not only destructive to membrane lipids and enzyme proteins, but also cause DNA breaks and DNA base hydroxylations if they are generated close enough to DNA to reach it within their life time. A review of genotoxic reactions of reduced oxygen species has been published recently (VUILLAUME, 1987, see also Sect. 8).

Protective Effects of Selenium (see also FISHBEIN, 1986, and this book, Chapter II.25). Thiol compounds such as cysteine and glutathione act as antioxidants and radical scavengers. The major protective compound in vivo is glutathione which occurs at cellular concentrations of 2–4 mmol/L. Glutathione protects against toxic effects of cadmium and mercury in rodents. Selenite plays an important role in the formation and utilization of glutathione. It induces an enzyme of the glutathione biosynthetic pathway (γ-glutamyl-cysteinyl synthetase), and it induces the formation of glutathione reductase, which reduces the oxidized form of glutathione, the disulfide, to the reduced form (CHUNG and MAINES, 1981). Furthermore, selenium is an essential constituent of the selenoenzyme glutathione peroxidase which catalyzes the reduction of H_2O_2 by glutathione. Supplementation of the diet of experimental animals with selenite at non-toxic doses results in a protection against toxic effects of Ag(I), Cd(II), Hg(II), and methylmercury (reviewed by NORDBERG et al., 1986, and FISHBEIN, 1987). Selenite especially protects against cadmium-induced damage of the testes and from mercury-induced renal damage, both in rats. The biochemical mechanisms are complex, since the administration of selenite to rats results in a marked increase of cadmium and mercury in blood and some organs. Hence the major protective mechanism of selenium seems to be the protection against oxidative damage and free radical actions via gluthathione peroxidase (CHOW, 1979). The same mechanisms may be relevant for the anticarcinogenic effects of selenium (see Sect. 8) and the augmentation of the immune response by selenite (see Sect. 9).

5 Interactions of Calcium and Lead in Regulatory Processes

An important biochemical interaction is that of Pb^{2+} with Ca^{2+} in calcium-dependent processes. Calcium transport and binding to regulatory sites control numerous cell functions, e.g., the response to electric stimuli in the nervous system and the response to hormones and growth factors in the control of cellular functions. More than 70 proteins are regulated by calcium, ranging from control of enzymic functions to the control of the cell cycle. POUNDS (1984) reviewed the "effect of lead intoxication on calcium homeostasis and calcium-mediated cell function". He discussed three levels of Pb/Ca interactions: (a) direct interference of Pb^{2+} with Ca^{2+} transport proteins and calcium storage, (b) substitution of Ca^{2+} at important cellular binding sites, e.g., calmodulin, and (c) indirect effects of Pb^{2+} on calcium-mediated functions, e.g., changes in energy production or membrane permeability. POUNDS emphasized the striking similarity between the uptake and distribution kinetics in cells and tissues of Ca^{2+} and Pb^{2+}. A central motif is that animals and humans adapted to a low calcium diet and increased calcium utilization also exhibit enhanced absorption and retention of lead. This interaction is manifested in the increased susceptibility of pregnant women and fetuses to lead compared with the adult situation. It has not been possible to correlate lead/calcium interactions to a single Ca^{2+} carrier or gate, since cell membranes contain at least four different calcium exchange proteins. These include active, ATP-dependent processes and a passive, voltage-dependent Ca^{2+}-gate. The latter is important in neurochemistry, since lead affects the synaptosomal calcium transport in a complex, biphasic way. The block of ganglionic impulse transmission by Pb^{2+} is relieved by Ca^{2+}, indicating a direct competition. Recently, the specific inhibition by micromolecular concentrations of lead of a voltage-activated calcium channel in an animal model (*Aplysia*) has been identified (BÜSSELBERG, 1989).

The substiution of Ca^{2+} by Pb^{2+} at functional binding sites not necessarily results in inhibition. Lead binds to calcium-sites in the calcium-mediating protein calmodulin and activates the calmodulin-dependent phosphodiesterase. Recently, the stimulation of the calcium-dependent proteinkinase C from rat brain by lead ions was described by MARKOVAC and GOLDSTEIN (1988). On the other hand, Pb^{2+} decreases the velocity of human blood clotting, a calcium-dependent process (ASENI et al., 1979).

An example of indirect effects of lead on calcium-mediated functions is the lead-induced decrease of the levels of calcium and cholecalciferol (the active form of vitamin D) in the serum of children, which results from an inhibition by lead of the calcium transport in gastrointestinal epithelial cells. On the other hand, lead intoxication of experimental animals causes an increase in tissue levels of calcium. In summary, the generalization of Pb/Ca interactions in terms of a competition for binding sites is an oversimplification and has to be complemented by indirect interactions like those at the level of energy production (POUNDS, 1984).

6 Effects of Metals on the Metabolism of Organic Xenobiotics

Metal compounds also affect the toxicity of organic xenobiotics. They may do so either by direct interaction with metabolizing enzymes or other biomolecules or by affecting the biosynthesis of enzyme proteins and other substances. The effects of toxic metal ions on the hepatic and lung drug metabolizing systems have been investigated intensely. As early as 1967, SUNDERMAN detected the inhibition of induction of benzopyrene hydroxylase (a cytochrome P 450 enzyme) in rat liver by nickel carbonyl.

In the seventies, the inhibition of hepatic aryl hydrocarbon hydroxylase by cadmium and other heavy metal ions was studied, and recently the inhibition of rat lung hydrocarbon hydroxylase was detected (for a review see FURST, 1987a). Besides the hydroxylases, also enzymes of glutathione metabolism (see Sect. 4) and conjugation with organic compounds are inhibited by heavy metals. Glutathione-S-transferase from rat liver is inhibited by Hg(II), Cu(II), and Cd(II) (DIERICKX, 1982).

In addition to direct interactions of toxic metals with drug metabolizing enzymes, the biosynthesis of enzymes or their prosthetic groups may be affected by metals. Pb(II) strongly inhibits the synthesis of heme at the levels of the zinc enzyme aminolevulinate dehydratase and the enzyme inserting Fe(II) into porphyrin, i.e., ferrochelatase. The result is a decrease in heme synthesis and in the level of cytochrome P 450 which is required for aryl hydroxylation (TEPHLY et al., 1979). The depletion and inactivation of aryl hydrocarbon hydroxylase causes a reduced biotransforma-

Table I.10-3. Effects of Metals on Organic Carcinogens. (Modified from FURST, 1987a, and from KASPRZAK and WAALKES, 1986)

Applied Metal	Organic Compound	Effect
Al(III)	4-Nitroquinoline-N-oxide	−
Al(III)	Dimethylnitrosamine	−
Ca(II)	Dimethylbenzanthracene	+
Cd(II)	Methylnitrosourea	+
Co(II)	3-Methylcholanthrene	±
Co(II)	Diethylnitrosamine	○
Cu(II)	3-Methyl-4-dimethylaminoazobenzene	−
Cu(II)	Benzo(a)pyrene	○
Cu(II)	Diethylnitrosamine	○
Cu(II)	Dimethylbenzanthracene	−
Cu(II)	2-Acetylaminofluorene	−
Cu(II)	Dimethylnitrosamine	−
Mg(II)	Dimethylbenzanthracene	−
Mg(II)	Diethylnitrosamine	+
Mn(II)	Benzo(a)pyrene	−
Ni_3S_2	Benzo(a)pyrene	+
Pb(II)	2-Acetylaminofluorene	○
Zn(II)	Dimethylbenzanthracene	−
Zn(II)	Methylbenzylnitrosamine	+
Zn(II)	Ethylnitrosourea	+
Zn(II)	4-Nitroquinoline-N-oxide	+

+ metal increased incidence of tumors
− metal decreased incidence
± conflicting results
○ no effect

tion of aromatic xenobiotics. Depending on whether the metabolism of these compounds results in detoxification or toxification, the inhibition of hydroxylases may augment or diminish the toxicity of the aromatic drugs. Table I.10-3 summarizes some effects of metal compounds on the carcinogenicity of organic substances. These results are interpreted in terms of metal ion modulated biotransformation of the organic compounds (see also Sect. 8).

7 Alteration of Gene Expression and Induction of Protective Proteins

Metal ions such as Mg(II) and Zn(II) are involved in the transcription of the genetic message. Toxic metal ions may affect the activity of RNA polymerase and/or induce the expression of specific proteins which in turn change the susceptibility of cells to metals. Mn(II) increases the activity of isolated RNA polymerase and augments the error frequency of transcription (EICHHORN et al., 1986). Cr(III) injected i.p. in rats binds to liver DNA, interacts with non-histone proteins, and induces the synthesis of a chromium-binding protein (OKADA et al., 1989). Pb(II), if administered to rats in the drinking water, induces the synthesis of a renal cytosolic protein which serves as a receptor for lead and is transformed to lead inclusion bodies (FOWLER, 1989). Ni(II) addition to the culture medium of mouse cells induces the formation of a nickel-resistant cell line which tolerates ten times as much nickel, i.e., 1 mmol/L, as the parent strain and is also slightly more resistant to cadmium and lead (IMBRA et al., 1989). Hg(II) activates the transcription of a mercury-resistant gene (*mer R*) in *Pseudomonas* bacteria (O'HALLORAN and WALSH, 1987). Cd(II) induces the synthesis of a membrane bound Cd-ATPase in *Staphylococcus aureus*, which pumps out cadmium (SILVER and MISRA, 1988).

The most thoroughly investigated protective reaction is the induction of the protein metallothionein in animals, plants, and microorganisms. Metallothionein is a small protein consisting of about 60 amino acids, of which about one third are cysteines. The protein has a particularly high metal-binding capacity (KÄGI et al., 1984; KÄGI, this book, Chapter I.12). This protein is induced in mammalian systems by many metal ions, namely Cd(II), Zn(II), Cu(II), Hg(II), Ag(I), by glucocorticoid hormones and growth factors. A most striking effect is the diminution of cadmium toxicity for animals which have been pretreated by low doses of cadmium or by zinc. This protective effect, however, is not due to a decreased body burden of cadmium. Administration of low doses of dietary cadmium in experimental animals caused an accumulation of cadmium, zinc, copper, mercury, and silver in the form of tissue metallothionein (NORDBERG et al., 1986). The binding of metals to thionein may have adverse effects, when patients or animals receive large oral doses of zinc. Thus, they synthesize excessive thionein and suffer from depletion of copper bound to this protein (MILLS, 1985). This type of interaction may be a major mechanism of zinc toxicity.

Besides thioneins, stress proteins are induced by certain metal compounds, e.g., those of Cu(II), Cd(II), and As(III). This type of proteins is better known as heat-shock proteins which are induced also by short treatment of animal, plant, or yeast cells at elevated temperature (about 42 °C). In animal cell cultures, the induction of stress proteins by moderate concentrations of Cu(II), Cd(II), or As(III) induced a limited resistance to these metals (NOVER, 1984). The mechanism of induction of protective proteins by metal compounds is still in debate. The finding that both metals and glucocorticoid hormones stimulate the same human metallothionein gene, suggests a connection between inducing metal ions and the transcription activating factors (KARIN et al., 1988). A recent hypothesis published by SUNDERMAN proposes that some carcinogenic metal ions substitute for zinc in finger-loop proteins that regulate the expression of oncogenes (SUNDERMAN and BARBER, 1988). In summary, the induction of transcription of specific genes by metal ions may be double-faced: it provides the cell with metallothionein acting as a sink for toxic metals but it may also cause the activation of oncogenes and growth factors which promote tumor growth.

8 Interactions in Carcinogenesis and Mutagenesis
(Table I.10-4)

Carcinogenesis. Metal compounds have been known to be involved in carcinogenesis for decades, either from epidemiological or from animal studies. The International Agency for Research on Cancer (IARC, 1973, 1976, 1980) has classified the compounds of arsenic, hexavalent chromium, and nickel as established human carcinogens. The Deutsche Forschungsgemeinschaft (DFG, 1988) each year publishes an updated list of carcinogens derived from human and animal data which additionally includes beryllium and cadmium and their compounds, antimony trioxide, and cobalt and its salts. For the time being IARC (1987) has classified beryllium and cadmium and their compounds only as probable carcinogens in humans (Group 2A). The mechanisms of metal carcinogenesis have been investigated in detail since the late seventies (reviews by FLESSEL et al., 1980; SWIERENGA et al., 1987; FURST, 1987b; GEBHART and ROSSMAN, this book, Chapter I.18; COSTA, 1988; MERIAN et al., 1985, 1988, 1989). A common motif of metal carcinogens compared to organics is the very limited concordance between carcinogenicity and short-term mutagenicity results. It should also be mentioned that in human beings occupational exposure to some metal compounds has caused lung cancer, but there are no data on cancer caused by systemic absorption; skin cancer caused by arsenic may be the exception. With

Table I.10-4. Metal-Metal Interactions in Experimental Carcinogenesis. (Modified from FURST, 1987b)

Carcinogen	Addition	Target	Effect	Reference
Cd(II)	Zn(II)	Muscle	–	GUNN et al. (1963)
	Zn(II)	Muscle	O	FURST and CASSETA (1972)
	Mg(II)	Muscle	–	POIRIER et al. (1983)
		Testes	O	
	Ca(II)	Muscle	O	POIRIER et al. (1983)
		Testes	O	
Ni₃S₂	Mn(II)	Muscle	–	SUNDERMAN et al. (1974)
	Cr(III)	Muscle	O	SUNDERMAN et al. (1974)
	Cu(II)	Kidney	–	SUNDERMAN et al. (1974)
	Ca(II)	Muscle	O	KASPRZAK et al. (1985a)
	Mg(II)	Muscle	–	KASPRZAK et al. (1985a)
Ni(II)	Ca(II)	Lung	–	POIRIER et al. (1984)
	Mg(II)	Lung	–	POIRIER et al. (1984)
Pb(II)	Mg(II)	Lung	–	POIRIER et al. (1984)
	Ca(II)	Lung	–	POIRIER et al. (1984)
	Ca(II)	Kidney	+	KASPRZAK et al. (1985b)

– inhibitory effect of added metal
+ enhancing effect
O no effect

respect to interactions, the picture is still blurred by inconsistent and sometimes conflicting findings (recent reviews by NORDBERG and PERSHAGEN, 1984; NORDBERG et al., 1986; FURST, 1987a). Since GEBHART and ROSSMAN in this book (Chapter I.18) also review some aspects of carcinogenic metal interactions the following discussion will be confined to general aspects and some additional views.

As in general metal toxicology, most important in metal carcinogenesis is speciation, i.e., the interaction with ligands, especially chelators. The metals themselves are not active unless they are solubilized by oxidation and chelation in living systems. The importance of chelation for the bioavailability of metals has been discussed in Sect. 3.

Some metal compounds are more potent cocarcinogens than carcinogens as such. Table I.10-3 summarizes data about interactions of metal compounds with organic carcinogens. The carcinogens $CdCl_2$ and Ni_3S_2 enhanced the tumorigenic effects of some organics, the weak carcinogen $CoCl_2$ showed conflicting results in combination with an aromatic carcinogen, and out of the non-carcinogens tested, Ca(II), Mg(II), Zn(II) enhanced the tumorigenic effects of some organic compounds. With regard to mechanisms of the cocarcinogenic effects, multiple levels of interaction are discussed:

a) Effect of metal ions on the drug metabolizing enzymes, which activate procarcinogens to the ultimate carcinogens (see Sect. 6).
b) Inhibition of DNA repair processes, by interaction either with enzymes or with DNA.
c) Activation of transcription of oncogenes and other cell proliferation genes or interference with signal transducing systems which control cell growth.
d) Alternatively, interactions in the generation of reactive reduced oxygen species such as superoxide and hydroxy radicals have to be considered.

Iron promotes the formation of OH radicals (VUILLAUME, 1987) whereas selenium deactivates by providing a higher level of glutathione peroxidase (SHAMBERGER, 1985). Up to date, selenite has been tested or been effective in the diminution of carcinogenicity with organics only (FISHBEIN, 1986; FISHBEIN, this book, Chapter II.25).

Mutagenesis. GEBHART and ROSSMAN (this book, Chapter I.18) discuss general mechanisms of mutagenicity and DNA damaging effects of metal compounds. An important finding is the small overlap between carcinogenic and mutagenic properties of metals. Chromate(VI) is the only metal carcinogen that is active in nearly all short-term mutation assays. One interpretation considers the limited bioavailability as a cause for the limited genotoxicity of metals. Whereas trivalent chromium in general is not mutagenic due to its extremely low uptake into cells, it is rendered bioavailable and mutagenic by chelation with o-phenanthroline or bipyridine (WARREN et al., 1981). Lead chromate, which is not mutagenic due to its very low solubility, releases the mutagenic chromate ion, if the lead ion is complexed with nitrilotriacetate (VENIER et al., 1987).

A further important finding is that several metal compounds such as As(III), Ni(II), and Cd(II) are weak mutagens but potent comutagens, which interfere with the repair of DNA damage caused by other agents (ROSSMAN et al., 1988; HARTWIG and BEYERSMANN, 1989). Also depression of mutagenicity is observed: selenite, a well-known anticarcinogenic substance, is also antimutagenic in combination with organic mutagens (SHAMBERGER, 1985), and $CoCl_2$ has been found to be antimutagenic in bacteria (INOUE et al., 1981). Whereas in experimental settings and mechanistic studies predominantly single agents are investigated in necessarily reduced systems, the situation of humans in the environment is characterized by multiple exposures. Interactions between metals and between metals and organic substances are the rule and single exposures are the exception. A major challenge to metal toxicologists is the understanding of the interaction processes in work environments. Welding fumes are an example of a complex inorganic pollutant affecting millions of workers. Welding fumes may contain carcinogenic chromium and nickel oxides. The mutagenicity and cell transforming activities of the whole fume samples exceed the sum of the effects expected from the Cr(VI) and Ni(II) contents, a finding which still awaits understanding (STERN, 1985).

9 Open Issues and Research Needs

Several health aspects of metal interactions could not been treated in this review partially due to limits in space and partially because no satisfying results are available. One such field is the reproductive toxicity of metals. As, Cd, Pb, and Hg are systemic toxicants and affect many enzymatic and regulatory processes in parallel. It is difficult to assess whether the teratogenic effects of lead and mercury are due to specific interactions or to indirect actions. Whereas several studies on heavy metal teratogenicity have been published, there is very little information about effects on the male reproductive system (CLARKSON et al., 1983). In contrast, low doses (1–3 ppm) of selenite fed to rodents augmented or restored (after selenium deficiency) all observed immunological functions including the rejection of implanted tumors (KIREMIDJIAN-SCHUMACHER and STOTZKY, 1987). After long-term exposure of male mice to lead, a decrease in the number of pregnancies in females mated with the exposed males was observed (JOHANSSON and WIDE, 1986).

Another field where sufficient knowledge is lacking is the interaction of metal compounds with the immune system. Iron or zinc deficiencies caused an impairment of multiple immunological functions in humans and experimental animals (SCHMIDT and BAYER, 1987). Also excessive intake of metals in toxic doses may affect immune functions. Excessive oral supply of copper or manganese (BEACH et al., 1982), inhalation but not oral uptake of lead (HILLAM and OKZAM, 1986), and inhalation of nickel compounds (SPIEGELBERG et al., 1984) impair the immune response in rodents. A suppression of killer cell functions has been observed after application of toxic doses of nickel in mice (SMIALOWICZ et al., 1987) and of cadmium salts in rats (STACEY et al., 1988). FURST (1987a) discussed effects of metal compounds on the immune (and interferon) response as indicators for an enhanced susceptibility to tumorigenesis, but he emphasized that these connections are far from being elucidated.

After a period of intense research on exposures to and effects of single metal compounds and some initiating studies of metal interactions in nutrition and toxicity, it is highly desirable to extend the research on combined exposures as a general subject in future work. Even when single factors are studied, the composition of the environment or the culture media should be carefully monitored. There is no single exposure in the environment or in industry, and the basic research on combined exposures will have a great impact on the setting of standards and thresholds for a better protection of humans and their environment. Of similar importance is a balanced diet with respect to trace elements in human nutrition and in agriculture. A metal taken up at a normally non-toxic dose may become toxic if the diet lacks an antagonist, and a trace element present at a normally sufficient dose may become deficient when it is accompanied by an excess of an antagonist.

References

Aseni, P., Bossa, R., Galatulas, I., and Perrone, G. (1979), *Lead Effect on Blood Coagulation: Interaction with Calcium.* IRCS Med. Sci. 7, 235.

Babich, H., Devanas, M. A., and Stotzky, G. (1985), *The Mediation of Mutagenicity and Clastogenicity of Heavy Metals by Physicochemical Factors.* Environ. Res. 37, 253–286.

Bartlett, R., and James, B. R. (1979), *Behavior of Chromium in Soils. III. Oxidation.* J. Environ. Qual. 8, 31–35.

Beach, R. S., Gershwin, M. E., and Hurley, L. S. (1982), *Zinc, Copper and Manganese in Immune Function and Experimental Oncogenesis.* Nutr. Cancer 3, 172–191.

Berlin, M., and Friberg, L. (1960), *Bone Marrow Activity and Erythrocyte Destruction in Chronic Cadmium Poisoning.* Arch. Environ. Health 1, 478–486.

Bertini, I., Lucchinat, C., Maret, W., and Zeppezauer, M. (eds.) (1986), *Zinc Enzymes.* Birkhäuser, Boston-Basel-Stuttgart.

Büsselberg, D. (1989), *Einfluß von Schwermetallionen auf einen spannungsaktivierten Calciumstrom von Nervenzellen der Meeresschnecke Aplysia californica.* Dissertation, Universität Hohenheim, FRG.

Cantilena, L. R., and Klaassen, C. D. (1981), *Comparison of the Effectiveness of Several Chelators after Single Administration on the Toxicity, Excretion and Distribution of Cadmium.* Toxicol. Appl. Pharmacol. 58, 452–460.

Chow, C. K. (1979), *Nutritional Influences on Cellular Antioxidant Defense Systems.* Am. J. Clin. Nutr. 32, 1006–1081.

Christie, N. T., and Costa, M. (1984), *In vitro Assessment of the Toxicity of Metal Compounds. IV. Disposition of Metals in Cells: Interactions with Membranes, Glutathione, Metallothionein and DNA.* Biol. Trace Elem. Res. 6, 139–158.

Chung, A. S., and Maines, M. D. (1981), *Effect of Selenium on Glutathione Metabolism: Induction of γ-Glutamyl Cysteine Synthetase and Glutathione Reductase in Rat Liver.* Biochem. Pharmacol. 30, 3217.

Clarkson, T. W., Nordberg, G. F., and Sager, P. R. (eds.) (1983), *Reproductive and Developmental Toxicity of Metals.* Plenum Press, New York-London.

Costa, M. (1988), *The Role of Trace Metals in Carcinogenesis, Oncology Overview.* International Cancer Research Data Bank, Bethesda, Maryland.

DFG (Deutsche Forschungsgemeinschaft) (1988), *Maximale Arbeitsplatzkonzentrationen und biologische Arbeitsstoff-Toleranzwerte.* VCH Verlagsgesellschaft, Weinheim-Basel-Cambridge-New York.

Dierickx, P. J. (1982), *In vitro Inhibition of Soluble Glutathione-S-transferases from Rat Liver by Heavy Metals. Enzyme 27*, 25 – 32.

Eichhorn, G. L., Clark, P., Shin, Y. A., Butzow, J. J., Rifkin, J. M., Pillai, R. P., Chuknyiski, P. P., and Waysbort, D. (1986), *The Influence of Metal Ion-Nucleic Acid Interactions on Genetic Information Transfer,* in Xavier, A. V. (ed.): *Frontiers in Bioinorganic Chemistry,* pp. 80 – 83. VCH Verlagsgesellschaft, Weinheim-Deerfield Beach/Florida-Basel.

Fishbein, L. (1986), *Perspectives of Selenium Anticarcinogenicity. Toxicol. Environ. Chem. 12,* 1 – 30.

Fishbein, L. (1987), *Bioavailability and Interactions of Metals: General Consideration,* in: Fishbein, L., Furst, A., and Mehlman, M. A. (eds.), *Genotoxic and Carcinogenic Metals: Environmental and Occupational Occurrence and Exposure,* pp. 245 – 264. Princeton Scientific Publishing Co., Princeton, New Jersey.

Fishbein, L. (1989), *Potential Toxicity from Hazardous Waste Interaction. Toxicol. Environ. Chem. 18,* 287 – 309.

Flanagan, P. R., Mc Lellan, J. S., Haist, J., Cherian, M. G., Chamberlain, M. J., and Valberg, L. S. (1978), *Increased Dietary Cadmium Absorption in Mice and Human Subjects with Iron Deficiency. Gastroenterology 74,* 841 – 846.

Flessel, C. P., Furst, A., and Radding, S. B. (1980), *A Comparison of Carcinogenic Metals,* in: Sigel, H. (ed.), *Metal Ions in Biological Systems,* Vol. 10, pp. 23 – 54. Marcel Dekker, New York.

Fowler, B. A. (1989), *Roles of Nuclear/Cytosolic Receptor-like Lead Binding Proteins in Mediating Lead-induced Alterations in Renal Gene Expression. Abstracts 1st International Meeting on Molecular Mechanisms of Metal Toxicity and Carcinogenicity, Urbino, Italy, 1988.*

Fox, M. R. S., and Jacobs, R. M. (1986), *Human Nutrition and Metal Toxicity,* in: Sigel, H. (ed.), *Metal Ions in Biological Systems,* Vol. 20, pp. 201 – 228. Marcel Dekker, New York.

Friberg, L., Kjellström, T., Nordberg, G., and Piscator, M. (1986), *Cadmium,* in: Friberg, L., Nordberg, G. F., and Vouk, V. B. (eds.), *Handbook on the Toxicity of Metals,* pp. 355 – 381. Elsevier, Amsterdam.

Furst, A. (1977), *Metals: we know so much and we know so little,* in, Drucker, H., and Wildung, R. W. (eds.), *Biological Implications of Metals in the Environment. Proceedings 15th Annual Hanford Life Science Symposium,* pp. 426 – 440. Battelle, N. W. Laboratories, Richland.

Furst, A. (1987a), *Metal Interactions in Carcinogenesis,* in: Fishbein, L., Furst, A., and Mehlman, M. A. (eds.), *Genotoxic and Carcinogenic Metals: Environmental and Occupational Occurrence and Exposure,* pp. 279 – 293. Scientific Publishing Co., Princeton, New Jersey.

Furst, A. (1987b), *Towards Mechanisms of Metal Carcinogenesis,* in: Fishbein, L., Furst, A., and Mehlman, M. A. (eds.), *Genotoxic and Carcinogenic Metals: Environmental and Occupational Occurrence and Exposure,* pp. 295 – 327. Scientific Publishing Co., Princeton, New Jersey.

Furst, A., and Cassetta, D. (1972), *Failure of Zinc to Negate Cadmium Carcinogenesis. Proc. Am. Assoc. Cancer Res. 13,* 62.

Gentile, J. M., Hyde, K., and Schubert, J. (1981), *Chromium Genotoxicity as Influenced by Complexation and Rate Effects. Toxicol. Lett. 7,* 439 – 448.

Gunn, S. A., Gould, T. C., and Anderson, W. A. D. (1963), *Cadmium-induced Interstitial Cell Tumors in Rats and Mice and their Prevention by Zinc. J. Natl. Cancer Inst. 31,* 745 – 749.

Halliwell, B., and Gutteridge, M. C. (1984), *Oxygen Toxicity, Oxygen Radicals, Transition Metals and Disease. Biochem. J. 219,* 1 – 14.

Hartwig, A., and Beyersmann, D. (1989), *Enhancement of UV-induced Mutagenesis and Sister-Chromatid Exchanges by Nickel Ions in V79-Cells: Evidence for Inhibition of DNA Repair. Mutat. Res. 217,* 65 – 73.

Hillam, R. P., and Ozkam, A. N. (1986), *Comparison of Local and Systemic Immunity after Intratracheal, Intraperitoneal and Intravenous Immunization of Mice Exposed to either Aerosolized or Ingested Lead. Environ. Res. 39,* 265 – 297.

Hutzinger, O. (1980, 1982, 1985, 1988), *The Handbook of Environmental Chemistry,* Vol. 2, *Reactions and Processes,* Parts A, B, C, and D with various scientific contributions on abiotic interactions. Springer-Verlag Berlin-Heidelberg-New York.

IARC (International Agency for Research on Cancer) (1973), *IARC Monographs on the Evaluation of Carcinogenic Risk to Man,* Vol. 2, *Some Inorganic and Organometallic Compounds.* IARC, Lyon.
IARC (1976), *IARC Monographs on the Evaluation of Carcinogenic Risk to Man,* Vol. 11, *Cadmium, Nickel, some Epoxides, Miscellaneous Industrial Chemical and General Considerations on Volatile Anaesthetics.* IARC, Lyon.
IARC (1980), *IARC Monographs on the Evaluation of Carcinogenic Risk to Man,* Vol. 23, *Some Metals and Metallic Compounds.* IARC, Lyon.
IARC (1987), *IARC Monographs on the Evaluation of Carcinogenicity, Supplement 7* (IARC differenciates between carcinogens in humans – Group 1 –, probable carcinogens in humans – Group 2A –, and possible carcinogens in humans – Group 2B).
Imbra, R. J., Wong, X.-W., and Costa, M. (1989), *Characterization of a Nickel Chloride Resistant Mouse Cell Line. Biol. Trace Elem. Res., 21,* 97–103.
Inoue, T., Ohta, Y., Sadaie, Y., and Kada, T. (1981), *Effect of Cobaltous Chloride on Spontaneous Mutation Induction in a Bacillus subtilis Mutator Strain. Mutat. Res. 91,* 41–45.
James, B. R., and Bartlett, R. J. (1983), *Behavior of Chromium in Soils. VI. Interactions between Oxidation, Reduction and Organic Complexation. J. Environ. Qual. 12,* 173–176.
Johansson, L., and Wide, M. (1986), *Long-term Exposure of the Male Mouse to Lead: Effects on Fertility. Environ. Res. 41,* 481–487.
Kägi, J. H. R., Vasak, M., Lerch, K., Gilg, D. F. O., Hunziker, P., Bernhard, W. R., and Good, M. (1984), *Structure of Mammalian Metallothionein. Environ. Health Perspect. 54,* 93–103.
Karin, M., Imagawa, M., Bodner, M., Chiu, R., Lefevre, C., Imbra, R., Dana, S., and Herrlich, P. (1988), *Regulation of Human Metallothionein and Growth Hormone Gene Expression by Steroid Hormone Receptors and Other Transacting Factors. UCLA Symp. Mol. Cell. Biol., New Ser. 75,* 177–184.
Kasprzak, K. S., and Waalkes, M. P. (1986), *The Role of Calcium, Magnesium, and Zinc in Carcinogenesis,* in: Poirier, L. A., Newberne, P. M., and Pariza, W., (eds.), *Essential Nutritients in Carcinogenesis,* pp. 497–515. Plenum Publishing Co., New York.
Kasprzak, K. S., Quander, R. V., and Poirier, L. A. (1985a), *Effects of Calcium and Magnesium Salts on Nickel Subsulfide Carcinogenicity in Fisher Rats. Carcinogenesis 6,* 1161–1166.
Kasprzak, K. S., Hoover, K. L., and Poirier, L. A. (1985b), *Effects of Dietary Calcium Acetate on Lead Subacetate Carcinogenicity in Kidneys of Male Sprague-Dawley Rats. Carcinogenesis 6,* 279–282.
Kiremidjian-Schumacher, L., and Stotzky, G. (1987), *Selenium and Immune Responses. Environ. Res. 42,* 277–308.
Markovac, J., and Goldstein, G. W. (1988), *Picomolar Concentrations of Lead Stimulate Brain Protein Kinase C. Nature 334,* 71–73.
Merian, E., Bronzetti, G. L., Frei, R. W., Haerdi, W., and Schlatter, C. (1985, 1988, 1989), *Carcinogenic and Mutagenic Metal Compounds (Interrelation between Chemistry and Biology),* Vols. 1, 2, and 3 (informing also on interactions). Gordon & Breach Science Publishers, New York-London-Paris-Montreux-Tokyo-Melbourne.
Mills, C. F. (1985), *Dietary Interactions Involving the Trace Elements. Annu. Rev. Nutr. 5,* 173–193.
Mills, C. F. (1986), *Interactions Involving Inorganic Nutrients,* in: Taylor, T. G., and Jenkins, N. K. (eds.), *Proceedings XIII. International Congress of Nutrition 1985,* pp. 532–536. John Libbey, London-Paris.
Mills, C. F. (1989), *Geochemistry and Animal Health,* In: Bowie, S. H. U., and Thornton, I. (eds.), *Environmental Geochemistry and Health,* pp. 59–76. Reidel Publishing Co., Dordrecht-Boston-Lancaster.
Muramoto, S. (1980), *Effect of Complexants on the Exposure to High Concentrations of Cadmium, Copper, Zinc and Lead. Bull. Environ. Contam. Toxicol. 25,* 941–946.

Nolen, G. A., Buchler, E. V., Geil, R. G., and Goldenthal, E. I. (1972), *Effects of Trisodium Nitrilotriacetate on Cadmium and Methyl Mercury Toxicity and Teratogenicity in Rats.* Toxicol. Appl. Pharmacol. *23*, 222–237.

Nordberg, G. F. (1971), *Effects of Acute and Chronic Exposure on the Testicles of Mice.* Environ. Physiol. Biochem. *1*, 171–187.

Nordberg, G. F. (1976), *Effects and Dose Response Relationships of Toxic Metals.* Elsevier, Amsterdam.

Nordberg, G. F., and Pershagen, G. (1984), *Metal Interactions in Carcinogenesis.* Toxicol. Environ. Chem. *9*, 63–78.

Nordberg, G. F., Parizek, J., Pershagen, G., and Gerhardsson, L. (1986), *Factors Influencing Effects and Dose-Response Relationships of Metals*, in: Friberg, L., Nordberg, G. F., and Vouk, V. (eds.), Handbook on the Toxicity of Metals, pp. 175–205. Elsevier, Amsterdam.

Nover, L. (ed.) (1984), *Heat Shock Response of Eukaryotic Cells.* Springer-Verlag, Berlin-Heidelberg-New York.

O'Halloran, T., and Walsh, C. (1987), *Metalloregulating DNA-binding Protein Encoded by the merR Gene: Isolation and Characterization.* Science *235*, 211–214.

Okada, S., Tsukada, H., and Tezuka, M. (1989), *Effect of Chromium(III) on Nucleolar RNA Synthesis.* Biol. Trace Elem. Res. *21*, 35–39.

Poirier, L. A., Kasprzak, K. S., Hoover, K. L., and Wenk, M. L. (1983), *Effects of Calcium and Magnesium Acetates on the Carcinogenicity of Cadmium Chloride in Wistar Rats.* Cancer Res. *43*, 4575–4581.

Poirier, L. A., Theiss, J. C., Arnold, L. J., and Shimkin, M. B. (1984), *Inhibition by Magnesium and Calcium Acetates of Lead Subacetate and Nickel Acetate Induced Lung Tumors in Strain A Mice.* Cancer Res. *44*, 1520–1522.

Pounds, J. G. (1984), *Effect of Lead Intoxication on Calcium Homeostasis and Calcium-mediated Cell Function: a Review.* Neurotoxicology *5*, 295–332.

Reashid, M. A. (1971), *Role of Humic Acids of Marine Origin and their Different Molecular Weight Fractions in Complexing Di- and trivalent Metals.* Soil Sci. *111*, 298–306.

Rossman, T. G., Zelikoff, J. T., Agarwal, S., and Kneip, T. J. (1988), *Genetic Toxicology of Metal Compounds: an Examination of Appropriate Cellular Models.* Toxicol. Environ. Chem. *14*, 251–262.

Scharpf, L. G., Hill, I. D., Wright, P. L., Plank, J. B., Keplinger, M. L., and Chandra, J. C. (1972), *Effect of Sodium Nitrilotriacetate on Toxicity, Teratogenicity and Tissue Distribution of Cadmium.* Nature *239*, 231–234.

Schmidt, K. H., and Bayer, W. (1987), *Interactions of Minerals and Trace Elements with Host Defence Mechanisms – a Review Article.* Trace Elem. Med. *4*, 35–41.

Shamberger, R. J. (1985), *The Genotoxicity of Selenium.* Mutat. Res. *154*, 29–48.

Silver, S., and Misra, T. K. (1988), *Plasmid-mediated Heavy Metal Resistances.* Annu. Rev. Microbiol. *42*, 717–743.

Simons, T. J. B. (1988), *Active Transport of Lead by the Calcium Pump in Human Red Cell Hosts.* J. Physiol. *405*, 105–113.

Smialowicz, R. J., Rogers, R. R., Riddle, M. M., Luebke, R. W., Fogelson, L. D., and Rowe, D. G. (1987), *Effects of Manganese, Calcium, Magnesium and Zinc on Nickel-induced Supression of Murine Natural Killer Cell Activity.* J. Toxicol. Environ. Health *20*, 67–80.

Smith, W. H. (1990), *Air Pollution and Forests, Interaction between Air Contaminants and Forest Ecosystems.* 2nd Ed. Springer-Verlag, New York-Berlin-Heidelberg-London-Paris-Tokyo.

Spiegelberg, T., Koerdel, W., and Hochrainer, D. (1984), *Effects of Nickel(II) Chloride Inhalation on Alveolar Macrophages and the Humoral Immune Systems of Rats.* Eotoxicol. Environ. Saf. *8*, 516–525.

Stacey, N. H., Craig, G., and Muller, L. (1988), *Effects of Cadmium on Natural Killer and Killer Cell Functions in vivo.* Environ. Res. *45*, 71–77.

Stern, R. M. (1985), *Nickel and Chromium Compounds and Welding Fumes in Mammalian Cell Transformation Bioassay in vitro.* Publication No. 85.37, The Danish Welding Institute, Copenhagen.

Sunderman, F. W., Jr. (1967), *Inhibition of Induction of Benzopyrene Hydroxylase by Nickel Carbonyl. Cancer Res. 27*, 950–955.

Sunderman, F. W., Jr., Lan, T. J., and Cralley, L. J. (1974), *Inhibitory Effect of Manganese upon Muscle Tumorigenesis by Nickel Subsulfide. Cancer Res. 34*, 92–95.

Sunderman, F. W., Jr., and Barber, A. M. (1988), *Finger-loops, Oncogenes and Metals. Ann. Clin. Lab. 18*, 267–288.

Swierenga, S. H. H., Gilman, J. P. W., and McLean, J. R. (1987), *Cancer Risk from Inorganics. Cancer Metastasis Rev. 6*, 113–154.

Tephly, T. R., Wagner, G., Sedman, R., and Piper, W. (1979), *Effects of Metals on Heme Biosynthesis and Metabolism. Fed. Proc. 38*, 346.

Venier, P., Gava, C., Zordan, M., Bianchi, V., Levis, A. G., DeFlora, S., Bennicelli, C., and Camoirano, A. (1987), *Interactions of Chromium with Nitrilotriacetic Acid (NTA) in the Induction of Genetic Effects in Bacteria. Toxicol. Environ. Chem. 14*, 201–218.

Vuillaume, M. (1987), *Reduced Oxygen-species, Mutation Induction and Cancer Initiation. Mutat. Res. 186*, 43–72.

Warren, G., Schultz, P., Bancroft, D., Bennett, K., Abbott, E. H., and Rogers, S. (1981), *Mutagenicity of a Series of Hexacoordinate Chromium(III) Compounds. Mutat. Res. 90*, 111–118.

I.11 Cell Biochemistry and Transmembrane Transport of Some Metals

STEPHEN G. GEORGE, Stirling, Scotland

1 Introduction

In this chapter, nutritional imbalances (deficiencies and toxicities) and the biochemistry of metal-activated and metalloenzymes will not be covered, emphasis will be placed upon uptake processes and intracellular regulation of some of the more important essential and pollutant metals. Many general principles involved in uptake, disposition, and toxicity of metals in animals have been considered by WILLIAMS (1981 a, b, 1985).

High affinity transport systems for uptake of non-alkali and alkaline earth metals have evolved because of their low bioavailability, due either to their occurrence in trace levels, complexation (e.g., Cu^{2+}), or insolubility (e.g., Fe^{3+}). Since many metals are highly reactive, cellular homeostatic mechanisms have evolved for control of free intracellular metal concentrations. The similar ligand binding properties of metals (e.g., S-coordination of Cu, Cd, Hg, and Zn) result in many competitive interactions for uptake and metabolism (see Chapter I.10), as well as toxicity by destruction of normal function (enzyme activity, membrane structure, etc.).

1.1 General Uptake Processes and Variables

Little is known of the transported chemical species of metals, and there are innumerable potential metal-binding ligands in the intestinal lumen. Many nutritional studies have demonstrated marked effects on uptake rates by synthetic chelating agents (e.g., EDTA, NTA) or naturally occurring metal-binding molecules (e.g., citrate, picolinate, lactoferrin). Binding to intestinal secretions (e.g., pancreatic juice, mucus) may facilitate uptake whereas binding to the brush border glycocalyx and intestinal cell membrane may reduce availability (BREMNER and MILLS, 1981; LONNERDAHL et al., 1985; MENARD and COUSINS, 1983; QUARTERMAN, 1985). Uptake of Ca, Cu, Fe, and Zn appears to be regulated by feedback mechanisms since there is increased uptake by deficient animals. Parallel effects on uptake of other metals are also frequently observed, e.g., Cd and Pb in Ca-deficiency, Zn in Fe deficiency, however, this does not imply common carriers. In suckling animals there is pinocytic uptake of proteins and metal-protein adjuncts in the distal jejunum and ileum (see, e.g., KELLER and DOHERTY, 1980). Many studies have indicated that paracellular uptake of metals across the intestine may also be important (MILBURN et al., 1981), and there is increasing evidence to suggest that some metals may cross cellular membranes by an uncontrolled, non-selective diffusion process as their neutral lipid soluble complexes; examples of such complexes are $CdCl_2^0$ and methyl-HgCl. Paracellular uptake across the intestinal epithelium also appears to occur for many oxy-anions, including AsO_4^{3-}, CrO_4^{2-}, PO_4^{3-}, SeO_4^{2-}, TeO_4^{2-} and VO_4^{3-} (BELL et al., 1983).

2 Cadmium
(see also Chapter I.12)

Cadmium is a non-essential metal whose metabolism is closely related to that of copper and zinc due to its preference for $-SH$ binding and tetrahedral coordination geometry similar to Zn, e.g., metallothionein can bind 7 g atoms of Cd or Zn although the relative binding affinity differs ... $Hg > Cu > Cd > Zn$.

2.1 Uptake

Intestinal absorption of cadmium is low, this is partly explained by the high degree of binding to the gut contents and the mucosal surface together with a low rate of transport from the mucosal cell to plasma.

Uptake of cadmium by gut segments perfused with Cd-containing glucose-saline is rapid and shows saturation kinetics (V_{max} 0.01 µmol/g/min, K_m 0.1 mmol L^{-1}). Entry may occur via the calcium transport system, Cd-uptake is strongly inhibited by Ca (FOULKES, 1980) and is related to Ca-status. However, studies with chicken intestine did not reveal a correlation between vitamin D-dependent Ca-binding protein concentration and Cd absorption (KOO et al., 1978). Cadmium accumulates in the intestinal mucosal cell as Cd-MT. Mucosal to serosal transfer of Cd is slow, 50 pmol/h at 10^{-6} mol L^{-1} Cd in the lumen.

2.2 Tissue Distribution and Metabolism

Organs and tissues which are able to synthesize thionein accumulate the highest concentrations of Cd. When Cd^{2+} is injected parenterally or intravenously it is rapidly taken up by the liver. There is a lag period (3–4 h) before synthesis of thionein during which Cd is distributed between the subcellular organelles and high molecular weight cytosol proteins, thereafter it becomes bound to the new high affinity sites of thionein and thus detoxified. Some cadmium is excreted from the liver via the bile, however, the majority is transported (probably as Cd-MT) and stored in the kidneys. After intravenous injection, Cd-MT is very rapidly taken up by the kidney (40% in 5 min), probably by endocytosis into lysosomes, and less than 2% is excreted (NORDBERG and NORDBERG, 1975). Excretion of kidney-Cd is minimal, and the metal is very strongly retained ($t_{1/2}$ ca. 30 years in man).

3 Chromium

Chromium is an essential trace metal required as an integral component of the glucose tolerance factor. Reactivity is dependent upon the valence state. Cr(III) has a low solubility, consequently a low bioavailability, but forms stable complexes and is hence toxic. Cr(VI), however, forms a stable and highly soluble oxyanion (CrO_4^{2-}), which is readily taken up but is only moderately reactive.

3.1 Uptake and Metabolism

Very small amounts of Cr(III) can be taken up by pinocytosis (the cell membrane invaginates, pinches off, and forms an intracellular vesicle containing the metal) or as hydrophobic complexes. Conversely, Cr(VI) is readily taken up and transported by the sulfate and phosphate transporting systems. Plasma ascorbic acid can spontaneously reduce up to 2 ppm Cr(VI), whereupon the Cr(III) becomes bound to Tf. At concentrations above 2 ppm Cr, there is significant accumulation in the liver and kidney, however, the highest concentrations are found in the red blood cells. After penetration of Cr(VI) into the red cell it is reduced by intracellular glutathione and the Cr(III) becomes bound to hemoglobin and thus has a long biological half-life. Microsomal cytochrome P450 reductase, cytosolic glutathione, ascorbic acid, and the enzyme DT diaphorase are capable of reduction of Cr(VI) in other tissues, particularly liver, often resulting in toxicity. Genotoxicity may result from intranuclear free radical production or formation of condensation polynucleotides by binding of Cr(III) to guanine and the phosphate of the sugar backbone of DNA. Recent results have demonstrated Cr(III)-induced synthesis of a 70 kDa nucleolar protein which may be involved in cell proliferation (OKADO and TSUKAKA, 1985).

4 Copper

Copper is an essential component of several metalloenzymes, both for maintenance of structure and catalysis of redox reactions involving oxygen as electron acceptor. Cu(II) binds very strongly to protein – SH groups and also catalyzes lipid peroxidation, hence excesses are extremely toxic and control of intracellular free Cu levels are therefore extremely important.

4.1 Uptake

Transepithelial transport of copper appears to occur by two processes: (1) a rapid low capacity active system saturable at low Cu-concentrations and (2) a slower higher capacity system. Cu-uptake may be regulated by Cu-MT in the intestinal mucosa, however, this has not been confirmed experimentally, and significant mucosal cell Cu-MT concentrations are only apparent after ingestion of high dietary Cu-levels (BREMNER et al., 1978).

4.2 Interorgan Transport and Metabolism

Enterohepatic transport of Cu is thought primarily to involve albumin, which in most species has an NH_3 terminal Cu-binding site. Kinetic studies indicate that histidine mobilizes Cu(II) from albumin by competition and that the His_2-Cu(II) complex interacts with a specific membrane transport protein. Transport of the free Cu(II) ion occurs by a facilitated, passive mechanism (DARWISH et al., 1984). Hepatic Cu is either stored temporarily (as Cu-MT), released into the bile, or incorporated into the glycoprotein, ceruloplasmin (Cp) and released into the plasma for transport to extrahepatic tissues. Cp is a glycoprotein, mol. wt. 151 kDa, which binds 6–7 atoms Cu/mol. Evidence for the function of Cp in an analogous manner to transferrin is not as yet definitive, however, activities of the Cu-enzymes, superoxide dismutase, lysyl oxidase and cytochrome c oxidase appear to be related to plasma Cp levels, and preliminary evidence for Cp receptors in aorta and erythrocytes have been reported (HARRIS and STEVENS, 1985).

4.3 Storage and Elimination of Copper

Under normal physiological conditions small amounts of Cu are stored as Cu-MT, and excretion is via bile, either as a Cu-glutathione complex or as Cu-MT. After exposure to excess Cu (and in Wilson's disease) Cu accumulates in hepatic lysosomes as an insoluble autoxidized and polymerized form of metallothionein, cytosolic Cu-MT levels do not appear to be elevated (BREMNER et al., 1978).

5 Iron

In biological systems iron shows a preference for *O*- and *N*-coordination and is utilized for its redox properties in several metalloenzymes and respiratory proteins. Due to its low bioavailability it is retained with great avidity, and excretion is almost entirely fecal from exfoliation of intestinal cells.

5.1 Uptake

At a redox potential of about +0.5 V and pH 7 iron is present as $Fe(OH)_3$, since this has a low solubility product (2.5×10^{-18} mol L^{-1}) high affinity iron-binding and transport mechanisms have developed to increase uptake efficiency. The mechanisms of transepithelial transport of Fe in non-bacterial systems have not been elucidated. In vivo and in vitro studies with isolated gut segments and luminal membranes have shown that (1) saturation kinetics occur and an active transport component is present, (2) bioavailability is dependent upon the free Fe-concentration (thus can be modified by chelating agents, including citrate and lactoferrin from milk), (3) it is subject to adaptive regulation (therefore is presumably mediated by a controllable carrier mechanism) (FORTH and RUMMEL, 1973; SIMPSON and PETERS, 1986). In animals fed an iron-rich diet, iron accumulates in the intestinal mucosa in ferritin, and it has been suggested that the degree of Fe-saturation of ferritin in the mucosa may play a role in the regulation of Fe-absorption (GRANICK, 1946), although other studies have suggested control by plasma transferrin saturation (see, e.g., FLETCHER and HUEHNS, 1968).

5.2 Interorgan Transport and Metabolism

In plasma, Fe-transport is mediated by transferrin (Tf), a protein of 80 kDa mol. wt., capable of binding two Fe^{3+} (K_{eq} 10^{24}). Tf also binds Al, Cr, and V. Tf binds to a specific cell surface receptor; the receptor, with a molecular weight of 170–200 kDa can bind 2 Tfs, cellular uptake is by formation of coated pits, and subsequent transfer to the cell is by endocytosis of the vesicles (NEWMAN et al., 1982).

5.3 Storage

The bulk of intracellular iron is stored as a readily available form in ferritin, a 460–480 kDa protein whose hollow core can accommodate up to 4500 atoms Fe as ferric hydroxide/phosphate micelles. Vertebrates do not appear to have any mechanism for excretion of excess Fe, and excesses accumulate as hemosiderin, a partial degradation product of ferritin containing additional Fe. In some tissues, insoluble Fe-containing peroxidized pigments, lipofuchsins, accumulate.

6 Lead

Pb is a non-essential metal whose toxicity is largely due to the inhibition of enzymes responsible for heme synthesis and competition for calcium in Ca-dependent reactions.

6.1 Uptake

In vitro studies have demonstrated a massive binding of Pb^{2+} to the mucosal surface, rapid uptake into the tissue, and a slow passive mucosal/serosal transport

(50 pmol/h at 10^{-6} mol L^{-1} which is not saturable between 10^{-7} and 10^{-2} mol L^{-1}). When chelated lead (e.g., citrate, EDTA) is used, the tissue uptake is much reduced and serosal transport is increased, possibly by paracellular uptake (HILBURN et al., 1981). Calcium depresses intestinal absorption of lead, particularly in the presence of phytate or phosphate. There are also interactions between Pb- and Fe-uptake although the mechanisms are unclear, thus Pb-uptake is reduced in Fe-deficiency states, and lactoferrin in milk increases Pb-uptake (QUARTERMAN, 1985).

6.2 Tissue Distribution, Metabolism

Plasma Pb-concentrations are very low and Pb accumulates in the erythrocytes. It is taken up by passive diffusion, and accumulation is a balance between an inward leak and an active extrusion, possibly mediated by the Ca-pump (SIMONS, 1984). Once inside the erythrocyte, the metal either precipitates as the phosphate or becomes bound to a 12 kDa binding protein. In other tissues, particularly the liver, lead accumulates in intranuclear inclusion bodies where it is bound to a 27.5 kDa protein. The toxicity of Pb is largely due to the inhibition of δ-Ala dehydratase and ferrochelatase, and its neurotoxicity. The latter may be due to competition with calcium at important sites for uptake, storage, and release of neurotransmitters. Indeed, there appear to be many substitutions of Pb for Ca, particularly where phosphate is involved. It is deposited in bone and in many molluscs where intracellular Ca-phosphate inclusions are found; deposits of Pb-phosphate are found after Pb-exposure (COOMBS and GEORGE, 1978).

7 Mercury

Mercury is a non-essential metal which is extremely toxic. It inhibits many enzymes by displacement of the essential metal of metalloenzymes, both Hg and methyl-Hg (MeHg) complex with the essential cysteine residue of $-$SH dependent enzymes, e.g., those involved in glycolysis. MeHg is extremely neurotoxic and causes structural abnormalities, brain protein synthesis being particularly sensitive, possibly by inhibition of tRNA acylases (CHEUNG and VERITY, 1985), consequently cytoskeletal abnormalities will be apparent. Another structural effect of low concentrations of MeHgCl (SEGALL and WOOD, 1974) is the catalysis of hydration and hydrolysis of the vinyl ether linkages of membrane plasmalogens (1-alkenyl-2-acyl-glycerolphospholipids), thus causing destruction of the myelin structure. In the natural environment mercury is present as the insoluble sulfide (cinnabar), however, it may be methylated by microbes and accumulates in the food chain as MeHg.

7.1 Uptake

There is extensive absorption of inorganic Hg to the intestinal cell surface, and uptake efficiency is low (3–14%). In contrast, MeHg is taken up with a high efficiency

(95–100%). Experimental studies with artificial lipid bilayer membranes have shown that both inorganic Hg and MeHg form neutral chloro-complexes which rapidly diffuse through membranes. The permeability of HgCl (1.3×10^{-2} cm^{-1}; GUTKNECHT, 1981) is lower than that of methylHg and arylHg (LAKOWITZ and ANDERSON, 1980). Chloro-complex formation is probably important in MeHg transport from ocean water to plankton. However, in higher trophic levels (pelagic fish → carnivorous fish → man) MeHg is tightly complexed to protein −SH groups, the chemical form and the uptake mechanism are not yet known.

7.2 Metabolism and Excretion of Methylmercury

About 5% of the body MeHg is found in blood, and it is bound to cysteine residues of serum proteins and glutathione in erythrocytes. It is rapidly distributed throughout the tissues ($t_{1/2}$ in blood 7.6 h, KERSHAW et al., 1980). The rapid transport of MeHg in the body is probably explained by the rapid exchange reaction which HgS$_x$ undergoes with serum chloride to form a stable MeHg chloro-complex. In mammals there are not striking differences in concentrations between tissues, however, in fish 80–95% accumulates in muscle where >95% is present as a MeHg-cysteine protein complex (WESTOO, 1966). MeHg is retained in the body for relatively long periods, $t_{1/2}$ 70 days (MEITTENEN, 1973), the major excretory route being via the bile as a MeHg-glutathione complex (BALLATORI and CLARKSON, 1982).

7.3 Metabolism and Excretion of Inorganic Mercury

In contrast to MeHg, inorganic Hg induces synthesis of metallothionein and is bound as HgMT in the liver and kidney. The half-life of inorganic Hg in man is shorter than that of MeHg and is approximately 40 days. A small proportion is retained considerably longer, probably due to formation of insoluble Hg-selenide in the liver (which accounts for the major proportion of Hg in the livers of marine mammals, MARTOJA and VIALLE, 1977). Interactions between Se and Hg are important factors in body distribution and toxicity although the findings so far do not give a clear picture of the processes involved (for reviews see SHAMBERGER, 1983; PIOTROWSKI and INSKIP, 1981).

8 Zinc

The catalytic activity of zinc, an essential component of over 150 enzymes, is dependent upon its property to form a Lewis acid. It is also believed to have an important function in the stabilization of cell membrane structure (WILLIAMS, 1985).

8.1 Uptake

Transepithelial uptake is probably carrier-mediated (i.e., it is saturable, K_m with brush border vesicles 0.39 mmol L^{-1}) and the zinc either becomes integrated into the mucosal Zn-pool or is rapidly (15 min) transported into the plasma (COUSINS, 1979; DAVIES, 1980; MENARD and COUSINS, 1983). Absorption appears to be homeostatically regulated, the sequestration of Zn as Zn-metallothionein (Zn-MT) within the enterocyte has been postulated as a mechanism of regulation (RICHARDS and COUSINS, 1975), however, significant increases in mucosal Zn-MT only occur under conditions of dietary excess; although known inducers of thionein synthesis, such as hydrocortisone, increase Zn-absorption and -retention by 50% (SAS and BREMNER, 1979).

Bioavailability can be modified by a number of dietary factors. Identification of citrate as the major Zn-complex in human milk was proposed as the explanation for the higher bioavailability of zinc in human compared with bovine milk, however, recent results have indicated that complexation by casein is also different (LONNERDAHL et al., 1985). Complexation of zinc with dietary constituents, such as phytate, lowers Zn-uptake dramatically by formation of a stable Zn-phytate complex. The modifying effects of complexing agents such as citrate, picolinate, casein, lactoferrin, etc., on Zn-uptake appear to differ between different methods of study viz. in vitro experiments, everted sac perfusions and experiments with brush border vesicles, thus the role of chelation in Zn-uptake remains unanswered (MENARD and COUSINS, 1983).

8.2 Interorgan Transport and Storage

Zinc is primarily associated with albumin; during systemic transport and in vascularly perfused intestine 94% of newly absorbed Zn is bound to albumin (SMITH et al., 1978).

As discussed for copper later, although quantitatively minor species, complexes of zinc with amino acids such as His and Tau may be important in transmembrane transport. Zinc is rapidly taken up by isolated hepatocytes into (1) a labile pool (K_{app} 29 µm L^{-1}) which changes rapidly in size, and (2) it slowly equilibrates with pre-existing cellular pools (K_m 9.5 µm L^{-1}, V_{max} 9.9 pmol Zn/min/mg protein). Zn-storage and intracellular homeostasis are primarily mediated by metallothionein, a 6 kDa cysteine-rich protein. Metallothionein synthesis is influenced by a variety of parameters including metal concentrations (Zn, Cu, Cd, Hg) as well as being subject to hormonal controls, e.g., glucocorticoids, glucagon, endotoxin, interleukin I (COUSINS, 1985). The control of metallothionein gene translation, synthesis and properties of the protein are discussed in Chapter I.12 and in recent reviews by KARIN (1985) and HAMER (1986).

8.3 Detoxication, Storage, and Elimination of Excess Zinc

After exposure to excess Zn, Zn-MT levels in gastric mucosa and liver are increased, and thus free intracellular Zn-concentrations are regulated. In marine molluscs where there appears to be little control over metal uptake, excess Zn accumulates in tertiary lysosomes, particularly in the kidney. Studies of Zn- and Cd-binding to isolated lysosomes have shown that a proportion of the metal is exchangeable (i.e., they probably function as a non-specific "chelation sink"), and a large proportion of the metal is totally insoluble due to the formation of highly cross-linked polymerized lipoprotein peroxidation products (lipofuchsins). Metals, particularly Cu, Fe, and Zn (but also Cd, etc.) become trapped in the high S-containing matrix (GEORGE and VIARENGO, 1985). Excretion of zinc is almost entirely fecal, the major proportion being derived from pancreatic Zn-metalloenzymes.

References

Ballatori, N., and Clarkson, T. W. (1982), *Developmental Changes in the Biliary Excretion of Methylmercury and Glutathione. Science 216*, 61–62.
Bell, M. V., Kelly, K. F., and Sargent, J. R. (1983), *The Transport of Orthovanadate and Similar Oxyanions in Relation to Salt and Water Transport across the Isolated Intestine of the Common Eel, Anguilla anguilla. J. Exp. Biol. 102*, 295–305.
Bremner, I., and Mills, C. F. (1981), *Absorption, Transport and Storage of Essential Trace Elements. Philos. Trans. R. Soc. London B294*, 75–94.
Bremner, I., Hoekstra, W. G., Davies, N. T., and Young, B. W. (1978), *Effect of Zinc Status of Rats on the Synthesis and Degradation of Copper-induced Metallothionein. Biochem. J. 174*, 883–892.
Cheung, M. K., and Verity, M. A. (1985), *Experimental Methylmercury Neurotoxicity: Locus of Mercurial Inhibition of Brain Protein Synthesis in vivo and in vitro. J. Neurochem. 44*, 1799–1808.
Coombs, T. L., and George, S. G. (1978), *Uptake and Storage Mechanisms of Heavy Metals in Marine Organisms*, in: McLusky, D. S. and Berry, A. J. (eds.): *Physiology and Behaviour of Marine Organisms*, pp. 179–187. Pergamon Press, Oxford.
Cousins, R. J. (1979), *Regulatory Aspects of Zinc Metabolism in Liver and Intestine. Nutr. Rev. 37*, 97–103.
Cousins, R. J. (1985), *Hormonal Regulation of Zinc Metabolism in Liver Cells*, in: Mills et al., (1985), pp. 384–390.
Darwish, H. M., Cheney, J. C., Schmitt, R. C., and Ettinger, M. J. (1984), *Mobilisation of Cu(II) from Plasma Components and Mechanism of Hepatic Copper Transport. Am. J. Physiol. 246*, G72–G79.
Davies, N. T. (1980), *Studies on the Absorption of Zinc by Rat Intestine. Br. J. Nutr. 43*, 189–203.
Fletcher, J., and Huehns, E. R. (1968), *Function of Transferrin. Nature 218*, 1211.
Forth, W., and Rummel, W. (1973), *Iron Absorption. Physiol. Rev. 53*, 724–729.
Foulkes, E. C. (1980), *Some Determinants of Intestinal Cadmium Transport in the Rat. J. Environ. Pathol. Toxicol. 3*, 471–481.
George, S. G., and Viarengo, A. (1985), *A Model for Heavy Metal Homeostasis in the Mussel*, in: Vernberg, F. J., Thurberg, F. P., Calabrese, A., and Vernberg, W. (eds.): *Marine Pollution and Physiology; Recent Advances*, pp. 125–144. University of South Carolina Press, Columbia.
Granick, S. D. (1946), *Protein Apoferritin and Ferritin in Fe Feeding and Absorption. J. Biol. Chem. 164*, 737.
Gutknecht, J. (1981), *Inorganic Mercury (Hg^{++}) Transport Through Lipid Bilayer Membranes. J. Membr. Biol. 61*, 61–66.
Hamer, D. (1986), *Metallothionein. Annu. Rev. Biochem. 55*, 913–951.
Harris, E. D., and Stevens, M. D. (1985), *Receptor for Ceruloplasmin in Aortic Cell Membranes*, in: Mills et al. (1985), pp. 320–323.

Hilburn, M. E., Blair, J. A., Coogan, M. J., Porter, J. L., and Walters, J. R. F. (1981), *The Transport of Heavy Metals across the Mammalian Small Intestine*, in: Proceedings of the International Conference on Heavy Metals in the Environment, pp. 490–494. CEP Consultants, Ltd. Edinburgh.

Karin, M. (1985), *Metallothioneins: Proteins in Search of Function*. Cell 41, 9–10.

Keller, S. A., and Doherty, R. A. (1980), Am. J. Physiol. 339, G114–G122.

Kershaw, T. G., Clarkson, T. W., and Dhahir, P. H. (1980), *The Relationship between Blood Levels and Dose of Methylmercury in Man*. Arch. Environ. Health 35, 28–36.

Koo, S. I., Fullmer, C. S., and Wasserman, R. H. (1978), *Intestinal Absorption and Retention of ^{109}Cd: Effects of Cholecalciferol, Calcium Status and Other Variables*. J. Nutr. 108, 1812–1872.

Lakowitz, J. R., and Anderson, C. J. (1980), Chem. Biol. Interact. 30, 309–323.

Lonnerdahl, B., Keen, C. L., Bell, J. B., and Hurley, L. S. (1985), *Zinc Uptake from Chelates and Milk Fractions*, in: Mills et al. (1985), pp. 427–430.

Martoja, R., and Vialle, D. (1977), *Accumulation de granules de selenium mercurique dans le foie d'Odontocetes (Mammiferes, Cetaces): un mechanisme possible de detoxication du methyl-mercure par le selenium*. C.R. Seances Acad. Sci. Paris Ser. D. 285, 109–112.

Meittenen, J. K. (1973), *The Accumulation and Excretion of Heavy Metals in Organisms*, in: Krenkel, P. A. (ed.): Heavy Metals in the Aquatic Environment, pp. 155–165. Pergamon Press, Oxford.

Menard, M. I., and Cousins, R. J. (1983), *Zinc Transport by Brush Border Membrane Vesicles from Rat Intestine*. J. Nutr. 113, 1434–1442.

Mills, C. F., Bremner, I., and Chesters, J. K. (eds.) (1985), *Trace Elements in Man and Animals*, (TEMA 5) Commonwealth Agricultural Bureaux, Slough, U.K.

Newman, R., Schneider, C., Sutherland, R., Vodinelich, L., and Greaves, M. (1982), *The Transferrin Receptor*. Trends Biochem. Sci. 7, 397–400.

Nordberg, G. F., and Nordberg, M. (1975), *Comparative Toxicity of Cadmium-metallothionein and Cadmium Chloride on Mouse Kidney*. Environ. Health Perspect. 12, 103–108.

Okado, S., and Tsukaka, H. (1985), *Nucleolar Chromium-bound Protein Enhancing Nucleolar RNA Synthesis in Regenerating Rat Liver*, in: Mills et al. (1985), pp. 133–139.

Piotrowski, J. K., and Inskip, M. J. (1981), *Health Effects of Methylmercury*, MARC Report No. 24. Monitoring and Assessment Research Centre, Chelsea College, University of London.

Quarterman, J. (1985), *The Role of Intestinal Mucus on Metal Absorption*, in: Mills et al. (1985), pp. 400–401.

Richards, M. P., and Cousins, R. J. (1975), *Mammalian Zinc Homeostasis: Requirements for RNA and Metallothionein Synthesis*. Biochem. Biophys. Res. Commun. 64, 1215–1223.

Sas, B., and Bremner, I. (1979), *Effects of Acute Stress on the Absorption of Zinc by the Chick*. J. Inorg. Biochem. 11, 67–76.

Segall, H. J., and Wood, J. M. (1974), *Reaction of Methyl Mercury with Plasmalogens Suggests a Mechanism for Neurotoxicity of Metal-alkyls*. Nature 248, 456–458.

Shamberger, R. J. (1983), *Biochemistry of Selenium*, in: Frieden, E. (ed.): Biochemistry of the Elements, Vol 2, pp. 149–151. Plenum Press, New York.

Simons, T. J. B. (1984), *Active Transport of Lead by Human Red Blood Cells*. FEBS Lett. 172, 250–254.

Simpson, R. J., and Peters, T. J. (1986), *Mouse Intestinal Fe^{3+} Uptake Kinetics in vivo. The Significance of Brush Border Membrane Vesicle Transport in the Mechanism of Mucosal Fe^{3+} Uptake*. Biochim. Biophys. Acta 856, 115–122.

Smith, K. T., Failla, M. L., and Cousins, R. J. (1978), *Identification of the Plasma Carrier for Zinc Absorption by Perfused Rat Intestine*. cBiochem. J. 184, 627–633.

Westoo, G. (1966), *Determination of Methylmercury Compounds in Foodstuffs. I. Methylmercury Compounds in Fish, Identification and Determination*. Acta Chem. Scand. 20, 2131.

Williams, R. J. P. (1981a), *The Natural Selection of the Chemical Elements*. Proc. R. Soc. London B 213, 361–397.

Williams, R. J. P. (1981 b), *Physico-chemical Aspects of Inorganic Element Transfer through Membranes*. Philos. Trans. R. Soc. London B*294*, 57–74.

Williams, R. J. P. (1985), *Homeostasis: An Outline of the Problems*, in: Mills et al., (1985), pp. 300–306.

Additional Recommended Literature

Chambers, P. L., Chambers, C. M., and Greim H. (eds.) (1989), *Biological Monitoring of Exposure and the Response at the Subcellular Level to Toxic Substances. Proceedings European Society of Toxicology.* Arch. Toxicol. Suppl. *13*.

Costa, M. (1989), *Molecular Mechanisms of Metal Toxicity and Carcinogenicity. Proceedings of a Meeting in Urbino (Italy).* Biol. Trace Elem. Res., in press.

Cousins, R. J. (1985), *Absorption, Transport and Hepatic Metabolism of Copper and Zinc: Special Reference to Metallothionein and Ceruloplasmin.* Physiol. Rev. *65*, 238–309.

Hirayama, K. (1980), *Effects of Amino Acids on Brain Uptake of Methylmercury.* Toxicol. Appl. Pharmacol. *55*, 318–323.

Hirayama, K. (1985), *Effects of Combined Administration of Thiol Compounds and Methylmercury Chloride on Mercury Distribution in Rats.* Biochem. Pharmacol. *34*, 2030–2032.

Jollow, D., Snyder, R., and Sipes, J. G. (1990), *Molecular and Cellular Effects and Their Impact on Human Health. Proceedings 4th International Symposium,* College of Pharmacy, University of Arizona, Tucson, in press.

I.12 Metallothioneins

ANDREAS SCHÄFFER and JEREMIAS H. R. KÄGI, Zürich, Switzerland

While focusing their interest on the biological role of cadmium, in 1957 MARGOSHES and VALLEE isolated from equine kidney a low molecular weight protein with highly unusual properties. In its native form it contains Zn(II), Cd(II), and Cu(I) in varying proportions. It is devoid of aromatic amino acids, is remarkably stable as a holoprotein with regard to proteolytic digestion and thermal denaturation and shows no detectable enzymatic activity. The name of this protein, "metallothionein" (MT), reflects the extremely high sulfur and metal content, both of the order of 10% (w/w) (KÄGI and VALLEE, 1960). Lately there has been a growing interest in the study of this protein and research groups in fields as diverse as inorganic chemistry, biochemistry, molecular biology, physiology, toxicology, and medicine are investigating its properties in a complementary manner.

MTs occurring in vertebrates, invertebrates, plants, eukaryotic microorganisms and in some prokaryotes (NORDBERG and KOJIMA, 1979; HAMER, 1986; ROBINSON and JACKSON, 1986), have recently been subdivided into three classes (FOWLER et al., 1987). Class I MTs are defined to include polypeptides related in primary structure to equine renal MT while those of class II display none or only very distant evolutionary correspondence to mammalian forms. Class III MTs are polypeptides from plants composed of atypical γ-glutamylcysteinyl units. The isolation procedures include solvent fractionation, gel filtration and ion exchange or covalent chromatography; moreover, reverse phase HPLC may serve as a powerful technique for this purpose (KÄGI and KOJIMA, 1987). In most mammals the predominant metallic constituent of the hepatic forms is zinc with minor amounts of cadmium and copper making up the balance. Kidney MTs tend to have a higher Cd content. The basal concentration of the protein in different tissues varies widely. There are also striking species differences; human liver and kidney are particularly rich in this protein, with quantities ranging from 0.01% to 0.1% of the dry weight of the tissue.

The MT content of tissue is greatly increased by the administration of Cd (PISCATOR, 1964) and some other metals salts, for example, those of Zn, Cu or Hg (NORDBERG and KOJIMA, 1979), which induce MT synthesis (BREMNER, 1987). In cell cultures, the optimum concentration for induction approaches a level just below that which causes cytotoxicity. In addition, as summarized in Table I.12-1, induction is effected also by many other chemicals including glucocorticoid hormones (KARIN and HERSCHMAN, 1979), bacterial endotoxin (DURNAM et al., 1984), interleukin I (KARIN et al., 1985), α-interferon (FRIEDMAN and STARK, 1985), carbon tetrachloride (OH et al., 1978), as well as by a variety of stress conditions. Although the mechanism of this induction phenomenon is only partially understood, it occurs

Table I.12-1. Factors Which Induce Metallothionein Synthesis in Cultured Cells or in vivo (for citations see PALMITER, 1987, and BREMNER, 1987)

Metal ions:	Streptozotocin
Cd, Zn, Cu, Hg	Isopropanol
Ag, Co, Ni, Bi, Au	Ethanol
	Ethionine
Glucocorticoids	Alkylating agents
Progesterone	Chloroform
Estrogen	Carbon tetrachloride
Glucagon	
Catecholamines	Starvation
Interleukin I	Infection
Interferon	Inflammation
	Laparatomy
Butyrate	Physical stress
Retinoate	X-irradiation
Phorbol esters	High O_2 tension
Endotoxin	
Carrageenan	
Dextran	

mainly at the level of transcription initiation (HAMER, 1986). In fact MT gene promoters have been identified and have been fused artificially to a variety of other genes as a means to regulate gene expression by the above stimuli (PALMITER et al., 1983; PALMITER, 1987). The corresponding yeast Cu-MT expression system may have promising applications in the biotechnological and biopharmaceutical syntheses of proteins in yeast (BUTT, 1987).

Mammalian MTs consist of a single polypeptide chain of usually 61 amino acids (Fig. I.12-1) with a strong preference for β-turn conformation. Hydrodynamic and structural analyses (see below) indicate a non-globular shape of the molecule. Class

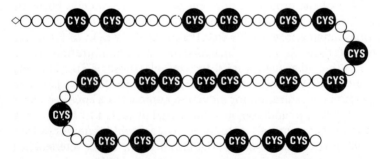

Fig. I.12-1. Schematic amino acid sequence of mammalian metallothionein (adapted from KOJIMA and KÄGI, 1978). The small circles represent amino acid residues other than Cys. The diamond signifies the amino terminal acetyl group.

Table I.12-2. Spectroscopic Evidence for Metal-Thiolate Clusters in Metallothionein

Method	Metal Derivative Studied	Reference
Ultraviolet absorption spectroscopy	Cd(II), Zn(II), Hg(II) Co(II) Fe(II)	Cited in KÄGI and KOJIMA (1987) VAŠÁK and KÄGI (1981) GOOD and VAŠÁK (1986)
Circular dichroism spectroscopy	Cd(II), Zn(II), Hg(II)	Cited in KÄGI and KOJIMA (1987)
Magnetic circular dichroism spectroscopy	Cd(II), Zn(II), Hg(II) Fe(II), Co(II)	Cited in KÄGI and KOJIMA (1987)
Luminescence spectroscopy	Cu(I)	BELTRAMINI et al. (1987)
Nuclear magnetic resonance spectroscopy (NMR)	^{113}Cd(II)	OTVOS and ARMITAGE (1980)
Electron paramagnetic resonance spectroscopy (EPR)	Co(II)	VAŠÁK and KÄGI (1981)
Extended X-ray absorption fine structure measurements (EXAFS)	Zn(II) Cu(I)	ABRAHAMS et al. (1986) SMITH et al. (1986)
Mössbauer spectroscopy	^{57}Fe(II)	DING et al. (1987)

I MTs can be divided into two subgroups, MT-1, and MT-2, carrying two and three negative charges at neutral pH, respectively. In some species these isoproteins, separable from each other by ion exchange chromatography, still represent mixtures of several isoforms whose occurrence in tissues can differ markedly. Today, the primary structure of 52 MTs is known completely or in part (KÄGI and KOJIMA, 1987).

The positions of the 20 cysteine residues (Cys) in the polypeptide (Fig. I.12-1) are highly conserved (NORDBERG and KOJIMA, 1979). Their arrangement in the prevailing Cys-X-Cys and Cys-Cys sequences (X = amino acid residue other than Cys) renders the protein a potent metal chelator. In fact, all Cys are involved in the binding of bivalent metal ions (usually seven) giving rise to a Cys/metal ratio of about 3. This unusual stoichiometry indicates that the metal complexes are part of clusters in which neighboring metals are linked by bridging thiolate ligands. The existence of such clustered structures has also become evident from a large variety of spectroscopic features of MT homogeneously substituted with particular metal ions (Table I.12-2). Cross-irradiation studies of ^{113}Cd-containing MT have moreover revealed that the protein harbors two topologically separate metal-thiolate clusters (OTVOS and ARMITAGE, 1980).

Recently, the spatial solution structure of two mammalian MTs has been elucidated by two-dimensional NMR spectroscopy (ARSENIEV et al., 1988; VAŠÁK et al., 1987; SCHULTZE et al., 1988). A stereo view of rat liver Cd$_7$-MT-2 (Fig. I.12-2) shows that the protein consists of two equally sized globular domains, made up of the amino terminal and carboxyl terminal half-chains, respectively, each holding in

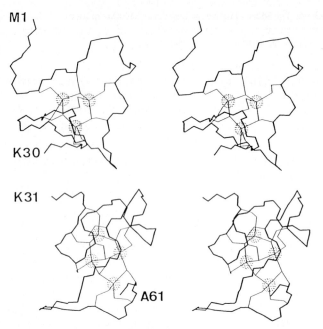

Fig. I.12-2. Two-dimensional NMR solution structure of the amino terminal (top) and carboxyl terminal (bottom) domains of rat liver Cd_7-metallothionein-2. Stereo view of polypeptide backbone (thick line), Cys side chains (thin lines) and metal positions (dotted spheres of radius 90 pm). The letters and numbers denote the terminal amino acid residues of the domains. The figures are prepared from the coordinates of SCHULTZE et al. (1988) (courtesy of P. SCHULTZE).

the interior of their helical fold a metal-thiolate cluster. The amino terminal domain made up of residues 1–30 is a right-handed spiral containing a Cd_3Cys_9 cluster with a cyclohexane-like structure. The carboxyl terminal domain harbors a Cd_4Cys_{11} cluster with a butterfly-shaped bicyclo[3.3.1]nonane-like structure. The domains are connected by a hinge region consisting of a Lys-Lys segment in the middle of the polypeptide chain. However, owing to the paucity of clearly recognizable interdomain contacts, the mutual orientation of the domains has not as yet been defined. This failure probably reflects appreciable flexibility about the hinge region.

The thermodynamic stabilities of the metal-thiolate clusters follow the order of thiolate model complexes, i.e., $Zn(II) < Cd(II) < Cu(I)$, $Ag(I)$, $Hg(II)$, $Bi(III)$ (KÄGI and KOJIMA, 1987). There is evidence that the carboxyl terminal cluster is somewhat more stable. Upon acidification, $Zn(II)$ and $Cd(II)$ are released from the protein moiety. At neutral pH, average apparent stability constants for $Zn(II)$ and $Cd(II)$ have been calculated to be of the order of 10^{12} L mol^{-1} and 10^{16} L mol^{-1}, respectively. The 10000 fold higher affinity for $Cd(II)$ accounts for the tendency of this environmentally exiguous ion to be accumulated in MT.

The physiological and toxicological significance of MT has been a topic of discussion ever since its first isolation and has as yet not been defined unambiguously. The preponderance of zinc in most preparations and the responsiveness of MT-bound Zn(II) and Cu(I) to the dietary supplies of these essential nutrients are arguments for a role in their metabolism (BREMNER, 1987). As a ubiquitous fast-responding homeostatic mediator, MT has been suggested to regulate the flow of these essential metal ions through cells and to participate as a metal donor and acceptor in the biosynthesis and degradation in the respective metalloenzymes and metalloproteins (BELTRAMINI and LERCH, 1982; BRADY, 1982). There is also considerable support for a protective role of MT in metal poisoning, especially Cd(II). Through intracellular sequestration, this protein provides for effective attenuation of Cd toxicity (WEBB, 1987). However, paradoxically the production of excessive amounts of Cd-containing MT is also believed to be a causative factor in bringing about kidney damage in chronic Cd poisoning (NORDBERG and KOJIMA, 1979). A protective role against toxic metals is also discussed for Class III MTs which were discovered in metal-tolerant plants (RAUSER and CURVETTO, 1980). A more general fast-responding homeostatic role in regulating the flow of essential metals through the organism and in providing an intracellular pathway for such metals in biosynthesis and degradation of metalloenzymes has also been proposed for these ubiquitously occurring compounds. The fact that cultured mammalian cells that overexpress MT are unusually resistant to X-ray damage (BAKKA and WEBB, 1981) has led some investigators to propose that MT may function as scavenger for free radicals (THORNALLEY and VAŠÁK, 1985). Glutathione, however, with a much higher overall abundance seems a more available candidate for this purpose (BREMNER, 1987).

There are also a number of interesting medical aspects emerging. Certain inherited disease states associated with abnormal trace metal metabolism have been suggested to involve directly or indirectly changes in MT synthesis. This includes acrodermatitis enteropathica, biliary cirrhosis, Wilson's and Menkes' disease (BREMNER, 1987). Urinary MT levels have been used to measure Cd exposure of industrial workers (ROELS et al., 1983). Also the side effects of certain metal-containing drugs, such as gold-containing antirheumatics (GLENNÅS and RUGSTAD, 1985) and platinum-containing anticancer agents can be mitigated by the metal-chelating properties of the protein. Thus, MT induction by $Bi(NO_3)_3$ administration was recently shown to decrease the renal and lethal toxicity of cis-$Pt(NH_3)_2Cl_2$ without compromising its antitumor activity (NAGANUMA et al., 1987). These experiments point to a potential utility of stimulating MT synthesis in cancer therapy.

References

Abrahams, I. L., Bremner, I., Diakun, G. P., Garner, C. D., Hasnain, S. S., Ross, I., and Vašák, M. (1986), *Structural Study of the Copper and Zinc Sites in Metallothioneins by Using Extended X-ray-absorption Fine Structure. Biochem. J.* **236**, 585–589.

Arseniev, A., Schultze, P., Wörgötter, E., Braun, W., Wagner, G., Vašák, M., Kägi, J. H. R., and Wüthrich, K. (1988), *Three-dimensional Structure of Rabbit Liver [Cd$_7$]Metallothionein-2a in Aqueous Solution Determined by Nuclear Magnetic Resonance. J. Mol. Biol.* **201**, 637–657.

Bakka, A., and Webb, M. (1981), *Metabolism of Zinc and Copper in the Neonate: Changes in the Concentrations and Contents of Thionein-bound Zn and Cu with Age in the Livers of the Newborns of Various Mammalian Species. Biochem. Pharmacol.* **30**, 721–725.

Beltramini, M., and Lerch, K. (1982), *Copper Transfer between Neurospora Copper Metallothionein and Type 3 Copper Apoproteins.* FEBS Lett. *142*, 219–222.

Beltramini, M., Münger, K., Germann, U. A., and Lerch, K. (1987), *Luminescence Emission from the Cu(I)-Thiolate Complex in Metallothioneins*, in: Kägi, J. H. R., and Kojima, Y. (eds.): *Metallothionein II*, pp. 237–241. Experientia Suppl. *52*. Birkhäuser Verlag, Basel.

Brady, F. O. (1982), *The Physiological Function of Metallothionein.* Trends Biochem. Sci. *7*, 143–145.

Bremner, I. (1987), *Nutritional and Physiological Significance of Metallothionein*, in: Kägi, J. H. R., and Kojima, Y. (eds.), *Metallothionein II*, pp. 81–107. Experientia Suppl. *52*. Birkhäuser Verlag, Basel.

Butt, T. R., and Ecker, D. J. (1987), *Yeast Metallothionein and Applications in Biotechnology.* Microbiol. Rev. *51*, 351–364.

Ding, X., Bill, E., Good, M., Trautwein, A. X., and Vašák, M. (1988), *Mössbauer Studies on the Metal-Thiolate Cluster Formation in Fe(II)-Metallothionein.* Eur. J. Biochem. *171*, 711–714.

Durnam, D. M., and Palmiter, R. D. (1981), *Transcriptional Regulation of the Mouse Metallothionein-I Gene by Heavy Metals.* J. Biol. Chem. *256*, 5712–5716.

Durnam, D. M., Hoffman, J. S., Quaife, C. J., Benditt, E. P., Chen, H. Y., Brinster, R. L., and Palmiter, R. D. (1984), *Induction of Mouse Metallothionein-I mRNA by Bacterial Endotoxin is Independent of Metals and Glucocorticoid Hormones.* Proc. Natl. Acad. Sci. USA *81*, 1053–1056.

Fowler, B. A., Hildebrand, C. E., Kojima, Y., and Webb, M. (1987), *Nomenclature of Metallothionein*, in: Kägi, J. H. R., and Kojima, Y. (eds.), *Metallothionein II*, pp. 19–22. Experientia Suppl. *52*. Birkhäuser Verlag, Basel.

Friedman, R. L., and Stark, G. R. (1985), *α-Interferon-induced Transcription of HLA and Metallothionein Genes Containing Homologous Upstream Sequences.* Nature *314*, 637–639.

Glennås, A., and Rugstad, H. E. (1985), *Acquired Resistance to Auranofin in Cultured Human Cells.* Scand. J. Rheumatol. *14*, 230–238.

Good, M., and Vašák, M. (1986), *Iron(II)-Substituted Metallothionein: Evidence for the Existence of Iron-Thiolate Clusters.* Biochemistry *25*, 8353–8356.

Hamer, D. H. (1986), *Metallothionein.* Annu. Rev. Biochem. *55*, 913–951.

Kägi, J. H. R., and Kojima, Y. (1987), *Chemistry and Biochemistry of Metallothionein*, in: Kägi, J. H. R., and Kojima, Y. (eds.), *Metallothionein II*, pp. 25–61. Experientia Suppl. *52*. Birkhäuser Verlag, Basel.

Kägi, J.-H. R, and Vallee, B. L. (1960), *Metallothionein: A Cadmium- and Zinc-containing Protein from Equine Renal Cortex.* J. Biol. Chem. *235*, 3460–3465.

Karin, M., and Herschman, H. R. (1979), *Dexamethasone Stimulation of Metallothionein Synthesis in HeLa Cell Cultures.* Science *204*, 176–177.

Karin, M., Imbra, R. J., Heguy, A., and Wong, G. (1985), *Interleukin I Regulates Human Metallothionein Gene Expression.* Mol. Cell. Biol. *5*, 2866–2869.

Kojima, Y., and Kägi, J. H. R. (1978), *Metallothionein.* Trends Biochem. Sci. *3*, 90–93.

Margoshes, M., and Vallee, B. L. (1957), *A Cadmium Protein from Equine Kidney Cortex.* J. Am. Chem. Soc. *79*, 4813–4814.

Naganuma, A., Satoh, M., and Imura, N. (1987), *Prevention of Lethal and Renal Toxicity of cis-Diaminedichloroplatinum(II) by Induction of Metallothionein Synthesis without Compromising its Antitumor Activity in Mice.* Cancer Res. *47*, 983–987.

Nordberg, M., and Kojima, Y. (1979), *Metallothionein and Other Low Molecular Weight Metalbinding Proteins*, in: Kägi, J. H. R., and Nordberg, M., (eds.), *Metallothionein*, pp. 41–121. Experientia Suppl. *34*. Birkhäuser Verlag, Basel.

Oh, S. H., Deagen, J. T., Whanger, P. D., and Weswig, P. H. (1978), *Biological Function of Metallothionein. V. Its Induction in Rats by Various Stresses.* Am. J. Physiol. *234*, E282–E285.

Otvos, J. D., and Armitage, I. M. (1980), *Structure of the Metal Clusters in Rabbit Liver Metallothionein.* Proc. Natl. Acad. Sci. USA *77*, 7094–7098.

Palmiter, R. D. (1987), *Molecular Biology of Metallothionein Gene Expression*, in: Kägi, J. H. R., and Kojima, Y. (eds.), *Metallothionein II*, pp. 63–80. *Experientia Suppl. 52*. Birkhäuser Verlag, Basel.
Palmiter, R. D., Norstedt, G., Gelinas, R. E., Hammer, R. E., and Brinster, R. L. (1983), *Metallothionein-Human GH Fusion Genes Stimulate Growth of Mice*. *Science 222*, 809–814.
Piscator, M. (1964), *On Cadmium in Normal Human Kidneys Together with a Report on the Isolation of Metallothionein from Livers of Cadmium-exposed Rabbits* (in Swedish). *Nord. Hyg. Tidskr. 45*, 76–82.
Rauser, W. E., and Curvetto, N. R. (1980), *Metallothionein Occurs in Roots of Agrostis Tolerant to Excess Copper*. *Nature 287*, 563–564.
Robinson, N. J., and Jackson, P. J. (1986), *"Metallothionein-like" Metal Complexes in Angiosperms; their Structure and Function*. *Physiol. Plant. 67*, 499–506.
Roels, H., Lauwerys, R., Buchet, J. P., Bernard, A., Garvey, J. S., and Linton, H. J. (1983), *Significance of Urinary Metallothionein in Workers Exposed to Cadmium*. *Int. Arch. Occup. Environ. Health 52*, 159–166.
Schultze, P., Wörgötter, E., Braun, W., Wagner, G., Vašák, M., Kägi, J.H.R., and Wüthrich, K. (1988), *Conformation of [Cd_7]-Metallothionein-2 from Rat Liver in Aqueous Solution Determined by Nuclear Magnetic Resonance Spectroscopy*. *J. Mol. Biol. 203*, 251–268.
Smith, T. A., Lerch, K., and Hodgson, K. O. (1986), *Structural Study of the Cu Sites in Metallothionein from Neurospora crassa*. *Inorg. Chem. 25*, 4677–4680.
Thornalley, P. J., and Vašák, M. (1985), *Possible Role for Metallothionein in Protection Against Radiation-induced Oxidative Stress. Kinetics and Mechanism of its Reaction with Superoxide and Hydroxyl Radicals*. *Biochim. Biophys. Acta 827*, 36–44.
Vašák, M., and Kägi, J.H.R. (1981), *Metal Thiolate Clusters in Cobalt(II)-Metallothionein*. *Proc. Natl. Acad. Sci. USA 78*, 6709–6713.
Vašák, M., Wörgötter, E., Wagner, G., Kägi, J.H.R., and Wüthrich, K. (1987), *Metal Coordination in Rat Liver Metallothionein-2 Prepared With or Without Reconstitution of the Metal Clusters, and Comparison With Rabbit Liver Metallothionein-2*. *J. Mol. Biol. 196*, 711–719.
Webb, M. (1987), *Toxicological Significance of Metallothionein*, in: Kägi, J.H.R., and Kojima, Y. (eds.), *Metallothionein II*, pp. 109–134. *Experientia Suppl. 52*. Birkhäuser Verlag, Basel.

Additional Recommended Literature

Ebdon, L. (1990), *Trace Metal Speciation Using HPLC-ICP-MS. Proceedings 20th IAEAC Annual Symposium on Environmental Analytical Chemistry, Strasbourg;* Special Issue *Fresenius Z. Anal. Chem.*, in press.
Fürst, P., Hu, S., Hackett, R., and Hamer, D. (1988), *Copper Activates Metallothionein Gene Transcription by Altering the Conformation of a Specific DNA Binding Protein*. *Cell 55*, 705–717.
Kägi, J.H.R., and Kojima, Y. (eds.) (1987), *Metallothionein II*. *Experientia Suppl. 52*. Birkhäuser Verlag, Basel.
Ramseier, S., Deshusses, J., and Haerdi, W. (1990), *Detection and Purification of Cadmium-binding Proteins Found in the Intestine of Earthworms. Proceedings 20th IAEAC Annual Symposium on Environmental Analytical Chemistry, Strasbourg;* Special Issue *Fresenius Z. Anal. Chem.*, in press.
Willner, H., Vašák, M., and Kägi, J.H.R. (1987), *Cadmium-thiolate Clusters in Metallothionein: Spectrophotometric and Spectropolarimetric Features*. *Biochemistry 26*, 6287–6292.

I.13a   Effects of Metals on Domestic Animals

HANS-JÜRGEN HAPKE, Hannover, Federal Republic of Germany

1 Introduction

Animals come into contact with metals either directly, through metal compounds naturally present in the environment, or indirectly, through metals which are present as a result of technology. Many metals are nutritionally essential for animals and are utilized by them within a certain concentration range (see also ANKE, 1984, 1987, who described deficiency in many case studies). Only amounts exceeding this range produce adverse effects. Non-essential metals produce adverse effects only when amounts higher than the threshold value are ingested. These limiting doses are known only for the most common domestic animals and only for a few substances and even then are dependent on various factors such as animal species, condition, age, and interrelations between the elements. Since knowledge is incomplete, the significance of only some metal compounds with respect to domestic animals, especially those animals which are used for human food, can be discussed here. Speciation has a great influence on bioavailability, uptake, and effects, and by "metals" both metal compounds and/or ions are meant.

Effects of metals on laboratory animals will not be discussed in this chapter (see Chapter I.13b and the elemental chapters of Part II). Effects on wildlife are only discussed in cases where data are available (see also HOPKIN, 1989). Interactions with microorganisms are described in Chapter I.7h. Thus, this chapter will deal only with specific situations in veterinary toxicology.

Free-living fish, in contrast to those raised in fish-farms, show different sensitivity to heavy metals. This is due to the metals' chemical form in water since there are different routes of absorption (i.e., oral ingestion, through the gills, or through the skin). The extent of absorption depends considerably on the solubility of the heavy metal compound in water and the physico-chemical conditions of the water such as pH, temperature, and the presence of other elements which may induce or inhibit absorption.

Effects depend on the actual concentration of the element in the target organs. The pharmacokinetic behavior of metals is influenced by their lipid- or water-solubility and their relations to one another. In general, organic compounds of metals are more lipid soluble and are able to permeate biological membranes. The distribution of such compounds in the body, starting with enteral absorption and ending in the cell membrane, is more rapid and nearly complete. These organic metal compounds are of higher toxicity than lipid-insoluble compounds. Compounds which are soluble in water do not penetrate the intestinal mucosa and are not absorbed after oral ingestion. They result in only local effects upon the gastrointestinal tract, not in systemic ones. A well-known example is mercury. Alkyl- or arylmercury compounds are very toxic compared to mercury chloride. Other examples have been suggested, but not well investigated, although they must be taken into account for estimation of threshold levels. In most cases of intoxication the speciation of the metal is unknown.

2 Aluminum

Intoxication in animals, acute as well as chronic, is very rare. It may, however, occur after ingestion of industrial waste water or after drug application. Sometimes feed of animal origin contains aluminum concentrations of up to 300 mg/kg, and in the case of feather meal even 450 mg/kg has been reported. Metallic aluminum is practically non-toxic.

In cattle, an uptake of 200 g of aluminum results in toxicity symptoms, and 500 g are lethal. Provided that vomiting induced by the aluminum is prevented, 35 to 50 g are fatal to dogs, 5 to 10 g for cats. Chronic aluminum intoxication occurs in animals when feed containing more than 1000 mg/kg is ingested over a long period of time. Normally animals take up only small fractions of the relatively large amounts of aluminum present in vegetable feed, perhaps because the aluminum may be precipitated. Rainbow trout, which are very sensitive, are affected by water concentrations of above 2 mg/L (see also Chapters II.1 and I.7 d). Aluminum hydroxide is practically non-toxic because it is insoluble (HAPKE, 1987; MAYER, 1983; MEYER-JONES et al. 1977; OEHME, 1978).

Oral uptake of high doses causes primarily local irritation due to the astringent and caustic effects of aluminum salts. Gastroenteritis is the cause of vomiting and diarrhea.

Long-term uptake of feed with aluminum contents of 1400 mg/kg results in hematological effects and bone damage by reducing the uptake of phosphorus, as well as producing rachitis in poultry. The chronic effect of aluminum is the disturbance of the phosphorus metabolism (MEYER-JONES et al., 1977; OEHME, 1978). Specific treatment for aluminium poisoning is not known. Animal feed should not contain more than 1000 mg per kg of dry matter for cattle and sheep and probably no more than 200 mg/kg for swine, poultry, horses, and rabbits.

3 Antimony

Antimony intoxication is possible as a result of contact with industrial dusts, contaminated rain water or the intake of antimony-containing drugs. Especially reactive are compounds of oxidation state III such as those used in colored pigments. Dogs show effects after oral uptake of 4 mg potassium antimony tartrate per kilogram of body weight, cats after 10 mg/kg. In rabbits, doses of more than 100 mg/kg are lethal. Horses, cows, and sheep are more resistant (BROWNING, 1969; HAPKE, 1987; LUCKEY and VENUGOPAL, 1977; LUCKEY et al., 1975; OEHME, 1978).

Contact with high concentrations of antimony salts causes irritation of the skin and mucous membranes. Gastrointestinal disturbances such as vomiting and diarrhea may occur. Accumulation in the body does not take place because of delayed enteral absorption and rapid elimination. Paralysis of peripheral blood vessels may cause circulatory collapse. Chronic intoxication results in metabolic disturbances and liver degeneration. Antimony intoxication is treated with dimercaprol (HAPKE, 1987).

4 Arsenic

Various inorganic and organic arsenic compounds used for technical purposes (such as preservatives) are found in the industrial wastes and waste waters of certain factories which may lead to contamination (ANKE, 1984; ANKE et al., 1987; BROWNING, 1969). In animals, poisoning from such compounds is very rare (SELBY et al., 1977). Dogs are more frequently affected than other animals, perhaps as a result of overdosage of veterinary drugs containing arsenic.

The toxicity of arsenic depends considerably upon the oxidation state of the element in the specific compounds (LUCKEY et al., 1975; LUCKEY and VENUGOPAL, 1977). Arsenic in oxidation state III (as in arsenic trioxide) is lethal after an oral intake of 30 to 100 mg per kilogram of body weight, without particular species differences. Arsenite is lethal after as little as 10 mg/kg (DECKERT et al., 1983). It is not easy to define a dose for chronic arsenic intoxication since habituation occurs after repeated uptake (HAPKE, 1987). Nevertheless, animal feed should not contain more than 2 mg/kg and drinking water no more than 0.2 mg/L in order to avoid accumulation of arsenic in edible tissues. Intoxication does not occur if feed concentrations are less than 50 mg arsenic per kg of dry matter.

Acute arsenic poisoning leads to circulatory failure and plasma extravasation after general vasodilation. Disturbance of the gastrointestinal mucosa occurs with inflammation followed by vomiting and hemorrhagic diarrhea. This is a result of the inhibition of essential cell enzymes and a resulting blockage of cell metabolism. Chronic poisoning in animals is not known. In animal experiments, however, chronic arsenic poisoning leads to metabolic disturbances, loss of appetite, weight loss, gastrointestinal disorders, infertility, and parenchymic degeneration (RIVIERE et al., 1981).

Acute arsenic poisoning is symptomatically treated. To fix unbound arsenic, sodium thiosulfate and dithioglycerol are administered simultaneously. An inactive arsenic-containing complex forms and is excreted. The easiest way to avoid arsenic poisoning in animals is the safe disposal of industrial wastes (OEHME, 1978).

5 Beryllium

Beryllium poisoning may be caused locally by industrial contamination. Toxic doses depend upon speciation and the different metal salts, but there seem to be no reports on intoxication in domestic animals. Only acute toxic doses after parenteral intake are known from experiments. Toxic doses from oral intake or inhalation are not known for farm animals. Poultry can tolerate a beryllium concentration in feed of 1000 mg/kg over a period of several months. Fish, however, are harmed by concentrations of more than 1 mg per liter water (HAPKE, 1987).

It is known from experiments that there is a very low absorption rate for beryllium, followed by deposition in bones, liver, and kidneys. This leads to degenerative lesions of the parenchyma and to impairment of hematopoiesis. An inhibition of bone metabolism causes osteoporotic lesions. In experiments with dogs, inhalation of beryllium-containing dust causes lung tumors.

Acute poisoning is treated with cortisone and adrenocorticotropic hormone (ACTH) which inhibits lymphatic reactions. Free beryllium can be bound and excreted with sodium salicylate (HAPKE, 1987; HAPKE and GRAHWIT, 1985).

6 Bismuth

In rare cases, industrial emissions of bismuth may affect farm animals, but actual cases of intoxication are not known. Bismuth shows a very slow enteral absorption rate, and thus no toxic concentrations can accumulate in the organism.

In animal experiments stomatitis and renal damage (a result of the excretion process) with polyuria have been observed, followed by anuria, methemoglobinemia, and albuminuria. Administration of dithioglycerine lessens symptoms of intoxication by chemically inactivating the bismuth. It can be suggested that feed concentrations of less than 400 mg bismuth per kilogram of dry matter do not lead to intoxication in animals (HAPKE, 1987).

7 Cadmium

Cadmium poisoning does not often occur in farm animals. The main source of chronic intoxication is the ingestion of plants that have been contaminated with industrial dust containing cadmium. Intoxication through inhalation of cadmium in the air is even rarer. The intestinal absorption rate of this element is low, but excretion is very slow. Thus accumulation occurs which may lead to slow poisoning. The speciation of cadmium compounds is not important with regard to absorption or toxicity (ANKE, 1984; ANKE et al., 1987).

In general, oral uptake of cadmium causes chronic poisoning, although a single oral dose of 5 to 10 mg/kg body weight is lethal. Concentrations of more than 30 to 60 mg cadmium per kg lead to poisoning and death. Feed concentrations of 25 mg/kg cause bone metabolism disorders in dogs after six months. A concentration of more than 0.01 mg cadmium per liter of water is toxic for fish, and mortality occurs at levels of over 10 mg/L. Disarrangement of enzymatic activity occurs already at 0.1 mg/kg.

Ingestion of relatively high amounts of cadmium (5 mg/kg body weight in cats) causes irritation of the gastric mucosa followed by vomiting. Long-term uptake of smaller doses results in the symptoms of zinc, copper, and iron deficiency. Zinc deficiency in swine leads to parakeratosis. In mammals, necrosis of the male gonads occurs, with inhibition of spermatogenesis as a result. In poultry, however, ovarial atresia is observed, which leads to inhibition of egg production. Chronic poisoning is accompanied by general growth inhibition, renal necroses (normally a tubular cell degeneration, but in sheep and rabbits as glomerular amyloidosis), hypocalcaemia, and hypercalciuria as well as proteinuria and glucosuria. Treatment of cadmium poisoning is not possible (HAPKE, 1987; HAPKE and GRAHWIT, 1985). In fish, a cadmium concentration of 2 mg/L water causes severe hematological changes such as hypochromic anemia. A concentration of 0.005 mg/L is considered safe.

Poisoning can be avoided by reducing the amount of cadmium in animal feed. For poultry, the cadmium concentration should be restricted to 0.5 mg/kg feed (dry matter). When considering toxic effects, mammals can tolerate higher amounts (10 or 30 mg/kg). In the long term, however, cadmium contents of 0.5 mg/kg feed result in accumulation in the foodstuffs derived from these animals, with higher levels found in the kidneys in particular, although not in skeletal muscle.

8 Chromium

Chromium poisoning in animals may result from the emission of industrial dusts containing chromium oxide or from waste waters polluted with chromium compounds (BROWNING, 1969; OEHME, 1978). Chromium(III) compounds are only poorly absorbed after oral ingestion, although chromium(III) is an essential element for animals. Compounds of oxidation state VI are more toxic. An oral dose of 15 grams potassium dichromate is lethal for a horse, 3 grams for a dog. Chronic intoxication results from the repeated administration of 30 mg chromate per kg body weight daily in all animals. Poultry is more resistant, and concentrations of up to 100 mg/kg feed do not produce effects. Poisoning is also not to be expected with chromium concentrations of less than 1 mg per liter drinking water. For fish, however, concentrations of more than 0.2 mg/L are hazardous, but this varies with different compounds (HAPKE, 1987).

Acute effects of chromium poisoning are characterized by local irritation of the affected tissues such as gastroenteritis and damage to liver and kidney parenchyma. No accumulation is observed; chromium is only a passing poison (HAPKE and

GRAHWIT, 1985; KIRCHGESSNER, 1987; LUCKEY et al., 1975; LUCKEY and VENUGOPAL, 1977).

9 Cobalt

Under normal conditions, cobalt intoxication in farm animals has not been reported and is not expected, with the exception of severe accidental circumstances (LUCKEY and VENUGOPAL, 1977). Cobalt is, however, an essential metal and is involved in different enzymatic systems. Anemia, due to cobalt deficiency, is the more common hazard (BROWNING, 1969; KIRCHGESSNER, 1987).

A concentration of 1000 mg cobalt per kg of feed is toxic for most farm animals. Sheep, however, are more sensitive, and feed concentrations of 400 mg/kg (a total burden of 5 g) are lethal. A daily dose of 1 mg/kg body weight results in polycytothemia due to stimulation of the hematopoietic system. At 5 mg/kg daily no toxic effects are observed, although hemoglobin levels are raised. Fish should not be exposed to concentrations higher than 0.1 mg cobalt per liter of water (HAPKE, 1987; HAPKE and GRAHWIT, 1985).

Following oral ingestion of large amounts of cobalt, local irritations such as gastroenteritis, vomiting, and diarrhea occur, followed by damage to the liver and kidneys. Since cobalt is excreted completely within two days, no accumulation in the body takes place. Acute poisoning can be treated with the chelating agent EDTA. Feed containing more than 10 mg cobalt per kg dry matter should not be used (OEHME, 1978; SCHRAMEL et al., 1982, 1984).

10 Copper

Copper is an essential element for animals, and thus deficiency symptoms are much more common than those resulting from overdosage. Copper plays an important role in different enzymatic systems and some redox processes (GOONERATNE and HOWELL, 1980). High concentrations of copper in feed due to mineral supplementation may lead to intoxication in sheep. Some individual Bedlington terriers become intoxicated at normal concentrations as their excretory systems lack the ability to eliminate it (FIEBINGER, 1981).

The toxic threshold for copper depends on the species, age, and nutritional condition of the animal as well as on the presence of molybdenum and zinc, and the specific copper compound. In sheep, the most sensitive animal species, a feed content of 45 mg/kg body weight leads to lethal liver damage, and 15 mg/kg feed are toxic. Daily doses of more than 20 mg/kg body weight may produce intoxication. In adult cattle, however, symptoms of intoxication were observed only after doses of 50 to 100 mg per kg of body weight, while 200 mg/kg were lethal. Other species such as swine and poultry are affected when concentrations exceed 300 mg/kg feed, and death results at levels of 1000 mg/kg (BANTON et al., 1987; GOONERATNE and HOWELL, 1980; GRÜNDER, 1982;, HAPKE, 1987; HAPKE and GRAHWIT, 1985; LUCKEY and VENUGOPAL, 1977; OEHME, 1978).

Lethal copper concentrations for fish are 0.02 mg/L in soft and acidic water, but 0.52 mg/L in hard and alkaline water. To avoid mortality in fish, the copper concentration of water should not exceed 0.1 mg/L (FÖRSTNER and WITTMANN, 1981).

In mammals, oral uptake of large amounts of copper causes local irritation and gastroenteritis, which leads to vomiting. Symptoms of acute poisoning are weakness, anorexia, dyspnoea, renal and muscular damage, and hemolytic anemia. In addition, liver damage, accompanied by hemochromatosis and skin changes, were observed after repeated doses. Slight effects can be inhibited by selenium and vitamin E (SUTTLE and MILLS, 1966). Acute poisoning must be treated symptomatically using calcium-EDTA, dimercaprol and penicillamine. Feed for sheep should not contain more than 10 mg copper per kg dry matter, and drinking water no more than 0.5 mg/L. Calves tolerate concentrations of up to 100 mg/kg dry matter, pigs and poultry, 250 and 300 mg/kg, respectively. Chronic copper intoxication can largely be avoided by increasing the zinc, iron and molybdenum concentrations in feed and by using protein supplements (HAPKE, 1987).

11 Iron

Iron is an essential element for all animals. It is incorporated as a complex into several important enzymes and in hemoglobin. Iron deficiency is more common in animals than damage resulting from iron surplus. Industrially induced intoxication has not been reported, but may happen because waste waters sometimes contain toxic concentrations. A common cause of iron poisoning in pigs is the therapeutic use of iron complex compounds; piglets especially are very sensitive. Doses of more than 150 mg/kg body weight are toxic. For horses, an oral dose of 250 g iron sulfate is lethal; dogs show toxicity symptoms after 2 g, and 8 g administered orally are lethal. Adult pigs are affected when their feed concentration exceeds 3000 mg iron per kg body weight (0.3%), depending on the supply of phosphorus. If the feed contains large amounts of phosphorus, even 0.5% iron can be tolerated. For fish, tolerable concentrations vary from 1 to 100 mg/L, depending on the species of fish and iron compound (HAPKE, 1987; HAPKE and GRAHWIT, 1985).

Oral ingestion of very large amounts of iron(II) compounds has cauterizing effects, but iron(III) compounds show local irritation at lower concentrations. This leads to vomiting, diarrhea, and perhaps fatal shock. A transient storage in the reticulo-endothelial system is observed. Chronic poisoning is characterized by hemosiderosis, i.e., high accumulation of iron in the system. In pigs, chronic iron poisoning leads to weight loss and the symptoms of phosphorus deficiency. Serum phosphorus content can be used as a measure. Treatment is symptomatic and can be supplemented by desferrioxamine, which is able to bind free iron (OEHME, 1978).

12 Lead

Because of an inhalative absorption rate of 10 to 50 percent, the inhalation of lead-containing dust can lead to effects in animals. But the primary source of contamination is the ingestion of vegetation burdened with lead-containing dust (UMWELTBUNDESAMT, 1976). As a consequence, lead poisoning mainly occurs in herbivores living near lead-emitting industries (ANKE, 1984; BROWNING, 1969). The most sensitive animals are ruminants (cattle, sheep, goats, red and roe deer), and young horses (BURROWS and BORCHARD, 1982). Poultry, pigs, dogs, and rabbits are much more resistant.

The enteral absorption rate of lead varies among species from 2 to about 20 percent. The rate also depends upon the nutritional composition of the feed (e.g., the calcium, phosphorus, and sulfur content). The absorbed lead is deposited temporarily in liver and kidney cells and thereafter in bone tissue. Under certain conditions (such as lactation, starvation, or other cases of increased calcium demand) lead may be released from these deposits and become reactive in the body.

Lethal poisoning in cattle results from the ingestion of about 10 g lead (as acetate). The same results occur when cattle receive feed with lead concentrations of 300 to 450 mg/kg dry matter for a period of over two weeks. Symptoms of intoxication in ruminants can be expected in the case of feed concentrations that exceed 50 mg lead per kg dry matter for several months. Concentrations of 250 mg/kg result in clinical intoxication symptoms after two to four weeks. A daily dose of 2 mg/kg body weight decreases milk production and impairs fertility. A daily dosage of 0.2 mg/kg body weight or ingestion of feed with less than 10 mg lead per kg dry matter does not result in measurable symptoms in ruminants. Young animals are more sensitive than adults (GRÜNDER, 1982; HAPKE, 1987; HAPKE and GRAHWIT, 1985).

For pigs, dogs, and other species, these doses must increase about five-fold. Ponies are poisoned by daily doses of more than 2 mg/kg body weight (BURROWS and BORCHARD, 1982). In poultry, a daily intake of lead-containing feed which leads to body concentrations of 0.02 mg/kg body weight results in the inhibition of δ-amino-levulinic acid dehydrogenase activity. This effect can be used for diagnostic purposes (VOGT et al., 1986). Visible toxic effects, however, are reached with lead concentrations of 1000 mg/kg feed. This corresponds to a daily dose of about 50 mg/kg body weight. Pigeons die after receiving a daily dose of 250 mg lead per kg body weight for 3 to 5 weeks. In fish, for example rainbow trout, a decrease in hemoglobin concentration is observed after exposure to a lead concentration of more than 1 mg/L for several weeks. Concentrations of less than 0.05 mg/L do not result in harmful effects for fish or in accumulation in edible tissues (HAPKE, 1987; HAPKE and GRAHWIT, 1985; LUCKEY et al., 1978; LUCKEY and VENUGOPAL, 1977; OEHME, 1978).

In cattle, the symptoms of acute lead poisoning are irritation of the central nervous system, or paresis and intestinal disorders. Chronic intoxication, which from a practical standpoint is more significant, causes weight loss, anemia, disorders in porphyrin synthesis, and paralysis of the peripheral nerves. Impairment of bone metabolism is also possible. All symptoms are the result of the inhibition of enzymes involved in porphyrin and heme synthesis or carbohydrate metabolism. The detection of δ-amino-levulinic acid dehydrogenase activity can be used for diagnostic purposes in ruminants (RUHR, 1984).

Lead poisoning in domestic animals can be prevented by using feed with less than 50 mg per kg dry matter. This concentration does not lead to accumulation in edible tissues. Since normal vegetation shows maximum lead concentrations of 10 mg/kg, poisoning is only possible in local areas with a high lead burden.

13 Manganese

Contamination of animals with manganese may occur as a result of pollution of vegetation with industrial dusts. More often, however, poisoning is caused by the administration of manganese-containing drugs such as potassium permanganate or manganic dioxide, both of which are toxicologically important. Since manganese is an essential element and a component of several enzymes, deficiency is more common than intoxication. For example, growth retardation occurs when food contains less than 10 mg manganese per kg (LUCKEY and VENUGOPAL, 1977).

Toxic symptoms appear only after feed concentrations have reached more than 2000 mg/kg. A dose of 10 grams potassium permanganate is lethal for horses. Fish, with the exception of some exotic species, can tolerate water concentrations of up to 1 g/L (HAPKE, 1987).

Oral ingestion of manganese causes mucosal irritation along with gastroenteritis and stimulation of the central nervous system. Further symptoms are anemia, degenerative alterations of the spinal cord and – as a result of the inhalation of manganic dust – bronchitis. Short-term accumulation takes place in the liver and kidneys, but excretion is rapid. Treatment of manganese poisoning is symptomatic using EDTA. To prevent intoxication, contamination from industrial dust should be avoided. If animal feed contains more than 1000 (cattle), 400 (swine, horses), or 2000 (poultry) mg manganese per kilogram of dry matter, it is considered unfit for consumption (HAPKE, 1987).

14 Mercury

Water, aquatic animals, or mercury-treated seed grain are possible sources of mercury contamination (BROWNING, 1969). At one time, poisoning occurred as a result of the application of inorganic mercury compounds as disinfectants or through the use of organic compounds as diuretics. Such instances are rare today due to the use of other materials for these purposes. Poisoning from mercury vapors can occur only in closed rooms. Mercury poisoning in animals as a result of industrial emissions is rare (BROWNING, 1969; KIRCHGESSNER, 1987; RIVIERE et al., 1981). In dogs, the inhalation of air containing 15 mg/m^3 metallic mercury vapor is lethal; concentrations of 3 mg/m^3 cause symptoms of severe toxicity; 0.1 mg/m^3 causes short-term symptoms.

Toxic dosages of mercury depend considerably on the type of compound (LUCKEY et al., 1975; LUCKEY and VENUGOPAL, 1977). Damaging effects are obtained with very small amounts of short-chained alkylmercury compounds. A single dose of 5 to 10 mg ethylmercury per kg of body weight is lethal; calves die after a daily dose of 0.5 mg/kg. Horses, swine, and poultry are more resistant. The production of normal eggs is altered when hens are fed feed containing 5 mg Hg (in the form of methylmercury chloride) per kg dry matter. Eggshell stability is reduced. In fish, a concentration of more than 0.01 mg/L methylmercury has biochemical effects. Higher concentrations (up to 0.1 mg/L) are still tolerated, but increasing accumulation occurs, and 0.2 mg/L is toxic for rainbow trout. Crab mortality occurs at concentrations of 0.006 mg/L ($=LC_{50}$, 96 h) (FÖRSTNER and WITTMANN, 1981; HAPKE, 1987; HAPKE and GRAHWIT, 1985).

The symptoms of mercury poisoning are unspecific. Decreased motility, psychomotoric disorders, blindness, and paralysis of the peripheral nerves (especially the muscles of the larynx and pharynx) are all observed. Acute poisoning is associated with respiratory disorders, gastroenteritis, stomatitis, and renal damage, resulting in vomiting, diarrhea, and at first polyuria, followed by oliguria and anuria (ROBERTS et al., 1982).

Acute mercury poisoning in animals can be treated with dimercaprol (dithioglycerine). The feeding of animals with seed grain that has been treated with mercury must be avoided. Care must be taken to keep wild poultry away from fields sown with treated seed grain. To avoid an accumulation of mercury in edible tissues (for example, through the use of fish meal), domestic animal feed should contain less than 2 mg mercury per kg of dry matter.

15 Molybdenum

This element is essential and a component of several enzymes. Its effects in organisms are strongly associated with the level of copper, iron, and zinc, as well as methionine metabolism (PARADA, 1981). Molybdenum poisoning may occur in herbivores which have fed on plants from contaminated pastures that contain large quantities of molybdenum (up to 250 mg/kg dry matter instead of the normal value of 3 mg/kg). Contamination of plants can result from industrial emissions or uptake from the soil, but molybdenum intoxications are rare and of only local importance (HAPKE, 1987).

The toxicity of molybdenum depends on the copper supply and protein content of the feed. Repeated ingestion of feed with a molybdenum content of more than 25 mg/kg dry matter results in toxicity, but provided enough copper, zinc, and protein are available, even a concentration of 100 mg/kg can be tolerated without effects. Chronic intoxication has only been reported in cattle, sheep, horses, and swine. The symptoms are growth retardation, anemia, anorexia and coordination disorders. In cattle, a discoloration of the hair is observed, along with enteritis and diarrhea.

Molybdenum intoxication can be avoided by using feed with less than 10 mg/kg for cattle, 20 mg/kg for pigs, 5 mg/kg for horses, or 100 mg/kg for poultry, and providing sources of copper and protein. Animal contact with soil high in molybdenum, or plants grown in such soil, should be avoided.

16 Nickel

Nickel is an essential element and is normally present in food. A potential source of contamination for animals is industrial dust containing nickel oxide (ANKE, 1984; KIRCHGESSNER, 1987), but in fact, intoxication has not been reported, and toxicity values are known only from experiments (BROWNING, 1969; LUCKEY and VENUGOPAL, 1977). Toxic doses depend on the nickel compound. Due to their lipid solubility, organic compounds show higher absorption rates than do water-soluble inorganic compounds. In dogs, an oral dose of 5 to 10 mg per kilogram of body weight has an acute lethal effect. The same effect is achieved with 25 to 50 mg nickel sulfate per kg body weight. No effects are expected in dogs if their feed contains 10 mg/kg or less nickel. Noxious effects are seen in swine when the nickel content of feed is more than 250 mg/kg (300 mg/kg for poultry). Feed containing more than 100 mg nickel as chloride or more than 500 mg as carbonate per kilogram of dry matter inhibits weight gain in young cattle. Adult cows tolerate half these doses. Effects on horses, sheep, and goats are not known, and in general the knowledge of nickel toxicity in farm animals is poor. No accumulation occurs in the body. Oral ingestion results in mucosal irritation followed by vomiting. Localized caustic effects of nickel compounds are also observed. Tremors and ataxia are reported after absorption of high doses. Long-term uptake causes weight loss and metabolic inhibition by disturbing carbohydrate metabolism in particular. Liver and kidney degeneration can also occur.

Treatment of nickel poisoning is symptomatic, the preferred method being administration of protein-enriched food. To avoid intoxication, the nickel concentration of animal feed should not exceed 50 mg/kg (ruminants and horses), 100 mg/kg (swine), and 300 mg/kg (poultry) (HAPKE, 1987; HAPKE and GRAHWIT, 1985).

17 Selenium

A possible source of contamination is the ingestion of selenium-enriched plant matter grown in selenium-rich soil. A selenium content of soil of 100 mg/kg can result in plant concentrations of up to 1000 or 1500 mg/kg (normal values in plants are 1 to 50 mg/kg). Selenium is essential for animals, and its deficiency results in growth retardation (BROWNING, 1969; KIRCHGESSNER, 1987).

Elemental selenium is not absorbed by the intestine, and intoxication is thus always due to compounds such as selenites, selenates, and organic selenium compounds bound in plants. An oral dose of 10 mg selenite per kg body weight is lethal for pigs, dogs, and cats, whereas 4 mg/kg metallic selenium is tolerated without effects. No specific sensitivities are known.

An oral intake (through feed) of 5 to 15 mg selenium (as selenate) per kg of body weight results in chronic intoxication. Ingestion of plants containing 45 mg/kg organic bound selenium (corresponding to a daily dose of 0.33 mg/kg body weight) is fatal to sheep. The same amount of selenium taken as selenite does not cause poisoning. Horses may reveal symptoms after grazing for three weeks on selenium-enriched pasture land. In cattle and sheep, a daily dose of 1 or 0.5 mg/kg body weight, respectively, causes poisoning after a few days (HAPKE, 1987). Fish are very sensitive to selenium. The LC_{50} (96 hours) for carp is 35 mg/L, and for crabs 1 mg/L (FÖRSTNER and WITTMANN, 1981).

Selenium causes metabolic disorders, and its effects are enhanced by the presence of other metals such as cadmium, zinc, nickel, and cobalt. Acute poisoning is characterized by local irritation, gastroenteritis, vomiting, diarrhea, and collapse ("blind staggering") and appears after ingesting feed with a selenium content of more than 20 mg/kg for four weeks. Further symptoms are loss of appetite, central nervous paralysis, ataxia, emaciation, psycho-depression, and myopathy in lambs and calves. In horses, hair loss in the mane and tail are observed, and cattle lose the tips of their tails. Detachment of the hooves occurs in horses. Selenium is teratogenic in sheep. The main symptom of chronic poisoning ("alkali disease") is emaciation, which appears when feed contains more than 10 mg/kg selenium. To avoid intoxication, animal feed should not contain more than 2 mg/kg selenium. The adverse effects of selenium can be reduced by using protein-rich feed (HAPKE, 1987).

18 Thallium

Thallium is a non-essential element for animals. Ingestion of contaminated vegetation is a source of thallium intoxication. Contamination may occur in the vicinity of cement factories which use thallium-containing pyrites (see also Chapter II.29). Intoxication of herbivores is therefore limited to certain local areas. More often, poisoning of pigs and dogs occurs when these animals unintentionally ingest baits containing thallium salts that were intended for rodent control (BROWNING, 1969; HAPKE, 1987; OEHME, 1978).

The lethal thallium dose for dogs is 10 to 20 mg per kg body weight (HAPKE, 1987; HAPKE and GRAHWIT, 1985; LUCKEY et al., 1975; LUCKEY and VENUGOPAL, 1977). No remarkable sensitivity differences among species are known. Even rodents show similar sensitivity. A total amount of 30 mg/kg body weight is tolerated when ingested as a single dose or in repeated fractions (thallium is accumulated in the body). A long-term uptake of 0.1 mg/kg body weight in sheep and young cattle results in damaging effects after several weeks or months. Feed containing 15 mg thallium per kg dry matter produces gastric ulcer in chicken, 40 mg/kg results in a decrease in weight gain, laying rate, feed efficiency and eggshell thickness (GRÜNDER, 1982; HAPKE, 1987; UEBERSCHAER et al., 1986; VOGT et al., 1986).

Acute poisoning causes local irritation of the intestinal mucosae with vomiting and diarrhea. Typical symptoms are hair loss and tremors as well as inflammation at body orifices and skin furunculosis. The same signs appear after chronic intoxication, particularly in dogs. Cattle show unspecific symptoms after chronic exposure. Treatment of acute poisoning is unsuccessful in animals in contrast to humans. Dithiozone or Prussian blue may be given to bind intestinal thallium and remove it from repeated absorption. Symptomatic therapy under clinical conditions is also necessary. A threshold value of 0.5 mg thallium per kg dry matter in feed should not be exceeded. Ingestion of protein-rich feed reduces the adverse effects (COYLE, 1980).

19 Tin

Occasionally, industrial dust contains tin in such amounts that it can deposit on vegetation and lead to the contamination of herbivores. Tin ingested in metallic form or as inorganic compounds (oxide, chloride), is not absorbed by the body and is therefore less toxic. Inorganic tin compounds are toxic only after very high doses. In dogs, an oral dose of 200 to 300 mg tin chloride per kg is toxic.

Animal feed containing more than 20 mg/kg tin (as an organic compound) causes poisoning in mammals. Poultry, however, can tolerate a concentration of 160 mg/kg in feed without effects, and fish do not react to levels of 2 mg tin per liter of water (HAPKE, 1987).

Organic tin compounds show increased absorption rates after oral intake. Due to their better absorption into and distribution within an organism, these compounds cause paralysis of the central nervous system and irritation of the skin and mucosae, resulting in skin changes and gastroenteritis. The toxic potency of organic tin compounds depends on the length of the alkyl chain. The shorter the chain (such as methyl and ethyl compounds), the more distinct the effect (see also Chapter II.30). Aryl compounds hardly exhibit toxic effects. Treatment with dimercaprol reduces damaging effects through chemical inactivation of the tin compound (HAPKE, 1987).

20 Vanadium

Industrial emissions of vanadium compounds (vanadium pentoxide, vanadium sulfate and vanadate) occur occasionally, but have not been reported as sources of animal poisoning. Knowledge of vanadium toxicology is therefore based only on animal experiments. Daily doses of about 10 mg/kg body weight (corresponding to a feed concentration of 310 to 350 mg/kg dry matter) produced poisoning in sheep. A single dose of 40 mg/kg is lethal after several days (ANKE, 1984; BROWNING, 1969; LUCKEY and VENUGOPAL, 1977).

Drinking water containing 50 mg vanadium per liter causes unspecific symptoms in cattle after several weeks. Feed concentrations of more than 20 mg/kg lead to chronic poisoning in poultry with a reduction in laying rate and a decreased hatching rate. Acute effects of high doses of vanadium are pronounced vasodilation and circulatory collapse in combination with central nervous convulsions. Oral ingestion results in local mucosal irritation, followed by gastroenteritis, vomiting, and diarrhea. Loss of appetite and emaciation are observed in cases of chronic vanadium poisoning (HANSARD et al., 1982). No treatment is available. Farm animals should be kept away from areas contaminated with vanadium. Concentrations in animal feed and drinking water should be less than 10 mg/kg and 0.1 mg/L, respectively (HAPKE, 1987).

21 Zinc

In certain areas, industrial dust from steel works and other manufacturing plants may contain high amounts of zinc, which can result in an increased zinc burden for vegetation (KIRCHGESSNER, 1987; LANTZSCH and SCHENKEL, 1978) that is a source of contamination of herbivores. As an essential element, zinc participates in a variety of biochemical processes. After oral ingestion, enteral absorption is very slow, and elimination is rapid, and no accumulation occurs in the body. Whether zinc effects vary among different species of domestic animals is not yet known, but suggestions can be made as a result of experiments on laboratory animals (ANKE, 1984; BROWNING, 1969; GRÜNDER, 1982).

Zinc poisoning is very rare, and acute cases are possible only after ingestion of very large doses. This is the case in cattle, sheep, pigs, horses, and poultry when the concentration in feed is about 1000 mg/kg dry matter. A prerequisite for this is the continuous ingestion of plants contaminated with adherent zinc dust. Feed concentrations of 200 to 500 mg/kg, or water concentrations of 25 mg/L, are tolerated by all animal species. After ingestion, zinc has astringent effects, resulting in gastroenteritis with vomiting and diarrhea (LUCKEY et al., 1975). Weariness, paralysis of the hind legs, growth retardation and anemia are all observed. Damage to the kidneys and pancreas may also occur. The balance and interaction of zinc and other essential trace elements (such as copper and iron) are important with regard to the toxicity of zinc. Pregnant animals are more sensitive to zinc intoxication. Treatment of intoxication is unnecessary since replacement of food with a low zinc content results in a lessening of the symptoms. Animal feed should not contain more than the following zinc concentrations: for cattle, 500 mg/kg; for sheep, 300 mg/kg; for swine, 1000 mg/kg; and for poultry, 1000 mg/kg (HAPKE, 1987).

Trout is damaged by water concentrations of 5 to 50 mg/L, depending upon the period of exposure (GRÜNDER, 1982). Fish are indeed very sensitive; damage to the gills and hepatic and/or renal degeneration may be observed.

References

Anke, M. (1984), *Significance of Newer Essential Trace Elements (such as Nickel, Arsenic, Lithium, Vanadium, Lead, Cadmium, and Tungsten) for the Nutrition of Animals (Goats)*, Proceedings 3rd International Workshop on Trace Elements, Neuherberg/München. W. de Gruyter, Berlin.

Anke, M., et al. (1987), *Effects of Deficiency and Essentiality in Ruminants, Trace Substances in Environmental Health XXI*, pp. 533–550, 556–566. University of Missouri; see also *Mengen und Spurenelemente,* two volumes proceedings 1987, Karl-Marx-Universität, Leipzig.

Banton, M. J., Nicholson, S. S., Jowett, P. L. H., Brantley, M. B., and Bondreaux, C. L. (1987), *Copper Toxicosis in Cattle Fed Chicken Litter. J. Am. Vet. Med. Assoc. 191,* 827–828.

Browning, E. C. (1969), *Toxicity of Industrial Metals.* Butterworths, London.

Burrows, G. F., and Borchard, R. E. (1982), *Experimental Lead Toxicosis in Ponies: Comparison of the Effects of Smelter Efficient-contaminated Hay and Lead Acetate, Am. J. Vet. Res. 43,* 2129–2133.

Coyle, V. (1980), *Diagnosis and Treatment of Thallium Toxicosis in a Dog, J. Small Anim. Pract. 21,* 391–397.

Deckert, W., Georgi, K., Kahl, H., and Klötzer, H.-H. (1983), *Akute Arsenvergiftungen bei Weidetieren, Monatsh. Veterinärmed. 38,* 650–652.

Fiebiger, J. (1981), *Kupferintoxikationen beim Bedlington-Terrier, Tierärztl. Umsch. 39,* 202–205.

Förstner, U., and Wittmann, G. T. W. (1981), *Metal Pollution in the Aquatic Environment.* Springer Verlag, Berlin-Heidelberg-New York.

Gooneratne, S. R., and Howell, J. M. (1980), *Creatine Kinase Release and Muscle Changes in Chronic Copper Poisoning in Sheep. Res. Vet. Sci. 28,* 351–361.

Gründer, H.-D. (1982), *Belastungsgrenzen für Schwermetalle bei Haustierwiederkäuern, Landwirtsch. Forsch. Sonderh. 39,* 60–93.

Hansard, S. L., Ammerman, C. B., Henry, P. R., and Simpson, C. F. (1982), *Vanadium Metabolism in Sheep. I. Comparative and Acute Toxicity of Vanadium Compounds in Sheep, J. Anim. Sci. 55,* 344–349; *II. Effect of Dietary Vanadium on Performance, Vanadium Excretion and Bone Deposition in Sheep, J. Anim. Sci. 55,* 350–356.

Hapke, H.-J. (1987), *Toxikologie für Veterinärmediziner,* 2nd Ed. Ferdinand Enke-Verlag, Stuttgart.

Hapke, H.-J., and Grahwit, G. (1985), *WHO-Manual: Ambient Air Pollutants from Industrial Sources,* pp. 149–160. Elsevier Science Publishers, Amsterdam.

Hopkin, S. P. (1989), *Ecophysiology of Metals in Terrestrial Invertebrates.* Elsevier Applied Science, London-New York.

Kirchgessner, M. (1987), *Tierernährung.* DLG-Verlag, Frankfurt/Main.

Lantzsch, H.-J., and Schenkel, H. (1978), *Effect of Specific Nutrient Toxicities in Animal and Man: Zinc,* in: Rechcial M. (ed.): *CRC Handbook Series in Nutrition and Food,* Sect. E. Vol. I. CRC Press Inc., Boca Raton, Florida.

Luckey, T. D., and Venugopal, B. (1977), *Metal Toxicity in Mammals,* Vols. 1 and 2. Plenum Press, New York-London.

Luckey, T. D., Venugopal, B., and Hutcheson, D. (1975), *Heavy Metal Toxicity, Safety, and Homology.* Suppl. Vol. 1 of *Environmental Quality and Safety.* Georg Thieme, Stuttgart.

Mayer, H. (1983), *Vorkommen und Bedeutung von Aluminium in Tiermehl und anderen von Tieren stammenden Eiweißfuttermitteln, Tierärztl. Umsch. 38,* 875–882.

Meyer-Jones, L., Booth, N. H., and McDonald, L. E. (1977), *Veterinary Pharmacology and Therapeutics.* 4th Ed. The Iowa State University Press, Ames, Iowa.

Oehme, F. W. (1978), *Toxicity of Heavy Metals in the Environment,* Part 1 and 2. Marcel Dekker, New York-Basel.

Parada, R. (1981), *Zinc Deficiency and Molybdenum Poisoned Cattle, Vet. Hum. Toxicol. 23,* 16–21.

Riviere, J. E., Boosinger, T. R., and Everson, R. J. (1981), *Inorganic Arsenic Toxicosis in Cattle,* Mod. Vet. Pract., 209–211.

Roberts, M., Seawright, A. A., and Norman, P. D. (1982), *Some Effects of Chronic Mercuric Chloride Intoxication on Renal Function in a Horse,* Vet. Hum. Toxicol. 24, 415–420.

Ruf, M. (1981), *Gewässerbelastung aus fischereilicher Sicht,* (Cd, Cu, Pb, Hg and Zn are discussed), in: *Wasser Berlin '81,* Kongreßberichte, pp. 415–428. Colloquium Verlag Otto H. Hess, Berlin.

Ruhr, L. P. (1984), *Blood Lead, Delta-Aminolevulinic Acid Dehydratase and Free Erythrocyte Porphyrines in Normal Cattle,* Vet. Hum. Toxicol. 26, 105–107.

Schramel, P., et al. (1982, 1984), *2nd and 3rd International Workshop on Trace Element Analytical Chemistry in Medicine and Biology,* Neuherberg/München, Proceedings. W. de Gruyter, Berlin.

Selby, L. A., Case, A. C., Osweiler, G. D., and Hayes, H. M. (1977), *Epidemiology and Toxicology of Arsenic Poisoning in Domestic Animals,* Environ. Health Perspect. 19, 183–189.

Suttle, N. F., and Mills, C. F. (1966), *Studies of the Toxicity of Copper to Pigs. 1. Effects of Oral Supplements of Zinc and Iron Salts on the Development of Copper Toxicosis,* Br. J. Nutr. 20, 135–148; 2. *Effect of Protein Source and Other Dietary Components on the Response to High and Moderate Intakes of Copper,* Br. J. Nutr. 20, 149–161.

Ueberschaer, K.-H., Matthes, S., and Vogt, H. (1986), *Einfluß von Thalliumzusätzen zum Broiler- und Legehennenfutter.* 5. Spurenelementsymposium, pp. 1233–1240. Karl-Marx-Universität, Leipzig.

Umweltbundesamt (1976), *Luftreinheitskriterien für Blei.* Bericht 3/76, Berlin.

Umweltbundesamt (1977), *Luftreinheitskriterien für Cadmium.* Bericht 4/77, Berlin.

Umweltbundesamt (1980), *Umwelt- und Gesundheitskriterien für Quecksilber.* Bericht 5/80, Berlin.

Vogt, H., Ueberschaer, K.-H., and Matthes, S. (1986), *Der Einfluß toxischer Schwermetalle auf die Eiqualität.* 7. Europäische Geflügelkonferenz, Proceedings Vol. 1, pp. 570–573.

I.13 b   Extrapolation of Animal Experiments

MAGNUS PISCATOR, Stockholm, Sweden

1 Introduction

Every year many new organic chemicals are synthesized under laboratory conditions in search of, e.g., new drugs and pesticides. It is obvious that such compounds must be subjected to extensive testing before they are released. It is not long ago that DDT was introduced as a pesticide at a time when proper testing was not performed and the effects are now seen in the ecosystem.

New compounds are now studied according to certain protocols with some minimum requirements. In addition to testing the general toxicity in animals, there must also be tests on teratogenicity and mutagenicity (WHO, 1978, 1984, 1986, 1987). However, the large number of new chemicals makes it virtually impossible to test all substances in long-term studies.

There is also a need to get more knowledge about substances already in use, which were introduced when there was no or very little testing before their release into the environment. This leads every year to banning or restrictions in use.

Unlike most organic chemicals metals cannot be banned. The use of certain synthetic metal compounds can be banned or restricted, e.g., organic lead in gasoline, but methylmercury cannot be banned, since it is naturally synthesized in certain parts of the ecosystem.

Metals have been released into the environment since the dawn of time, and during the last thousand years human beings have substantially increased that release to build up our present technology.

Today the effects of most common metals which occur in industrial environments as well as in the general environment have been extensively studied in human beings. Studies on exposed workers have given detailed knowledge of effects that may be caused by metals. It was early recognized that exposure to lead could cause severe disturbances in the synthesis of hemoglobin as well as damage to the nervous system. A continuous research on lead has today also contributed to a much better understanding of the effects of members of the general population, especially children. Some other metals or metal compounds have been only recently recognized as causing health problems. The syndrome of chronic cadmium poisoning was first described after world war II (FRIBERG, 1950), and in the 1960s it was shown that cadmium exposure could also cause effects in the general population (FRIBERG et al., 1974).

At present there are standards for maximum exposure to metals in working environments, where inhalation is the most common exposure route. With regard to the general population there are recommendations for maximum concentrations in air, food, water, or in body fluids (see Chapter I.20 and Part II).

The data basis for such standards, recommendations, or guidelines is mainly from studies on human beings. In spite of the large number of animal studies data from such studies have been of limited value for the evaluation of health risks to humans. Thus documents on metals produced by WHO, US EPA or other international or national organizations may contain many pages on effects seen in experimental animals, but in the final summary or conclusion the emphasis is on results from human studies. Also the use of LD_{50} values is of limited usefulness. Even with regard to cancer the animal experiments have given less information than the experience gained from studies on human beings. That arsenic may cause skin cancer was recognized in the last century, but skin cancer has not been induced in animals given arsenic. Arsenic can also cause lung cancer in workers, but it has been very difficult to induce lung cancer in experimental animals (PERSHAGEN et al., 1984).

However, the mechanisms behind the effects seen in humans are often difficult to study in human populations, and for information on metabolism of metals animal experiments have given much valuable information. BONNER and PARKE (1984) have discussed the potential and limitations of molecular biochemical studies and of mathematical modelling, but have warned of oversimplifications.

In the following sections an attempt is made to point out some of the problems encountered in the extrapolation of results from studies on animals to the human situation.

2 Exposure Route and Dosage

Humans are generally exposed via air, food, and water. In children hand-to-mouth transfer of dust and soil constitutes an additional risk for exposure to metals, especially lead (US EPA, 1984). An unusual exposure is via the parenteral route, which may occur in people with severe diseases treated by dialysis, total parenteral nutrition or other special methods. In the working environment some metal compounds may be absorbed via the skin, e.g., hexavalent chromium.

Even if the main exposure route is by inhalation in the working environment, inhaled metals may be transported by the mucociliary system from the respiratory tract to the gastrointestinal tract. In dusty environments there may be a considerable oral exposure from contaminated hands, cigarettes, etc. Furthermore, exposure may occur from volatilization of metals in contaminated tobacco. These factors have been discussed with regard to cadmium by ADAMSSON et al. (1979), with regard to cadmium and arsenic by ROELS et al. (1982), and with regard to lead by CHAVALIT-NITIKUL et al. (1984). Personal hygiene is obviously a factor of importance, which of course is difficult to study in animal experiments.

Among members of the general population one exposure route may be wholly dominating, e.g., oral exposure to methylmercury, but for some metals inhalation may be as important as oral exposure. Typical examples are cadmium and lead. With regard to cadmium smokers absorb as much cadmium from inhaled tobacco smoke as they do from food (FRIBERG et al., 1974). In some areas with heavy traffic lead may be absorbed in similar amounts from the lungs and the gut.

It is obvious that it is difficult to reproduce the complicated exposure situations of human beings in animals experiments. Long-term inhalation studies are expensive as are long-term feeding experiments, whereas single or short-term injection experiments are less expensive and often give more dramatic results. It even happens that metal compounds are injected directly into organs, e.g., kidney, prostate or brain. Such injections may result in extensive damage and highly significant findings, but it must be understood that such exposures are far away from what we may experience in real life.

Ideally the doses given to animals should be of the same order of magnitude as those that humans are exposed to. This means that in a study on cadmium the feed should contain from about 10 up to 200 µg/kg, i.e., from 10 to 200 ppb.

Due to the long biological half-time of cadmium such an exposure would result in relatively low cadmium concentrations in the kidneys and in other organs and any effect studies would be meaningless. Furthermore, metabolic factors such as low absorption rates play a role, as will be discussed later. It has been shown that rats can tolerate continuous exposure to 50 mg Cd per kg feed for two years (LÖSER, 1980).

It is therefore not surprising that most of the positive findings with regard to cadmium are from injection studies. For metal compounds with shorter half-times and high absorption it is easier to reproduce the human situation. One example is methylmercury. ALBANUS et al. (1972) exposed cats to methylmercury accumulated in fish and thus got a dosage which was realistic and a natural exposure medium.

An example of what a non-physiological exposure route and high doses may lead to is a study by AXELSSON and PISCATOR (1966a, b). Rabbits were given subcutaneous injections of cadmium as the chloride (0.25 mg Cd/kg three times weekly) for periods up to 26 weeks. It was intended to study the renal effects of cadmium, and renal damage could be shown to occur after several weeks of exposure. It was also found that the rabbits got hemolytic anemia and ahaptoglobinemia, which then necessitated further studies on the hemolytic component. Even if it seemed that the cadmium accumulation in the kidneys had caused the renal damage, the hemoglobin released from the red cells and reabsorbed in the kidneys might have contributed to the effect on the kidneys. This necessitated extensive discussions on the possible role of hemolysis in the pathogenesis of renal damage caused by cadmium. It also initiated haptoglobin determinations in cadmium workers, and signs of slight hemolysis were found in these workers (PISCATOR, 1971; FRIBERG et al., 1974). This example shows that high doses may cause multiple effects which may obscure the main effect to be studied. However, it also indicates that a study with a theoretically bad design may give valuable information that may lead to further studies. In the present case regrettably the slight hemolysis that may be caused by cadmium has not been studied further. It may also be of interest with regard to the transport of cadmium by metallothionein (see Chapter I.12).

3 Species Differences in Metabolism of Metals

With regard to metabolism there is the difference between human beings and experimental animals, and the interspecies differences between animals, sometimes also intraspecies differences, must be taken into account.

Within the human population there may be differences. One factor are inherited diseases (see Chapter I.19), which result in abnormal handling of essential elements such as copper and zinc, another factor may be the large variations in intake of nutrients, including essential metals.

The rat is the most common test animal, and there is an abundance of data on the fate of metals in that species. However, some metals are handled differently by the rat compared to man and other experimental animals, and metabolic differences may cause differences in effects. The most striking example is the fate of inorganic arsenic. In man and other mammals absorbed arsenic has a relatively short biological half-time, whereas it is considerably longer in the rat. The reason for this difference is the accumulation of arsenic in the red cells (WHO, 1981). On the other hand, in vitro and in vivo systems using liver cells from the rat seem to give valuable information on the mechanisms for methylation of inorganic arsenic (BUCHET and LAUWERYS, 1985, 1987). OBERDÖRSTER (1989) has evaluated the differences in environmental cadmium inhalation between rats and humans.

Other mammals also seem to be similar to humans with regard to cellular metabolism of arsenic. A striking exception is the marmoset monkey, which does not have the capacity to methylate arsenic (VAHTER et al., 1982; VAHTER and MARAFANTE, 1985).

The most important health hazards to human beings is cancer of the skin or lung after exposure to inorganic arsenic via water and air, respectively. Any animal model for the carcinogenic action of arsenic should be based on experiments using animals with a metabolism as similar as possible to that in humans. It has been extremely difficult to induce cancer in animals exposed to arsenic, the hamster seems to be a suitable species (PERSHAGEN et al., 1984).

Fortunately, data are now available also on metabolism of arsenic in humans, and such data will together with the animal data give sufficient information on metabolism of arsenic in the long run.

In the following a few comments will be made on some factors of interest in the different steps of the metabolism of metals.

3.1 Absorption
(see also Part II)

Animal experiments are generally performed with commercial diets, which contain all essential nutrients, often more than is optimal. Human beings have often marginal or deficient intakes. As will be noted in the specific chapters on metals in Part II, the absorption of many metals is generally quite low in experimental animals, being higher in fasting animals. In humans it generally is higher. The reason for this difference is that humans often have low intakes of essential elements such as calcium and iron, and there is plenty of evidence from animal studies that the absorption of metals such as cadmium and lead is influenced by the status of these essential elements. If rats are put on diets deficient in calcium, the absorption of cadmium will increase (LARSSON and PISCATOR, 1971, see also Chapter II.6).

Nutritional factors were studied by BARLTROP and KHOO (1975), who gave rats lead in synthetic diets with different amounts of fat, fibers, minerals, and protein. Very large differences in absorption were noted. There is also an age factor: Young animals will absorb higher amounts of essential as well as non-essential metals (ENGSTRÖM and NORDBERG, 1979a), especially sucklings may have a high absorption. This is in accordance with the limited data on metal absorption in young humans.

In animal experiments different metal compounds may be used, and in a study of rats the absorption rate of lead varied within a wide range depending on the compound (BARLTROP and MEEK, 1975).

3.2 Transport, Distribution and Storage

With regard to transport of metals animal studies have given valuable information. Some transport proteins are similar in animals and in humans, e.g., albumin, transferrin, ceruloplasmin, and metallothionein. In animal experiments the doses given are often high, which may sometimes give misleading results. As an example, beryllium given in small amounts will appear in the plasma in a diffusible form, whereas after large doses beryllium will be found in a phosphate complex (VACHER and STONER, 1968). Arsenic is generally transported in plasma, but as mentioned earlier, the rat stores arsenic in the red cells, which results in a long half-time in blood. The storage organs are otherwise similar in man and animals, lead is stored in the kidneys, mercury vapor is stored after oxidation as divalent mercury in the brain. There are also some dose-dependent factors with regard to storage. Thus the ratio between cadmium in liver and in kidney will increase with the dose during short-term intensive exposure (ENGSTRÖM and NORDBERG, 1979b). With time cadmium will be transferred via metallothionein from liver to kidney. The biological half-time may also be dose-dependent.

3.3 Excretion

The main excretion routes are via the kidneys and the gastrointestinal tract. In animal experiments large species differences have been seen with regard to biliary excretion, in the rat this route is much more important than in the rabbit or the dog.

There are few data on biliary excretion of metals in humans. ISHIHARA and MATSUSHIRO (1986) studied the urinary and biliary excretion of 15 metals in human beings and found that biliary excretion seemed to be more important for arsenic, iron, and lead, whereas for cadmium, inorganic mercury, and copper the opposite was found. Methylmercury is known to be excreted mainly via bile, as first shown in studies on rats.

At present it is difficult to estimate how relevant some of the data on excretion obtained in animal studies are with regard to excretion in humans. There are anatomical and functional species differences in, e.g., kidneys which may make it difficult to extrapolate to humans. On the other hand, the mechanisms in humans are sometimes difficult to study, which makes animal experiments valuable.

4 Are Effects Seen in Experimental Animals Relevant to Evaluation of Effects on Human Beings?

As mentioned in Sect. 2, many animal experiments have been performed with high injected doses of metal compounds. This has resulted in a large number of reported effects, which seldom appear in humans. Injection of large amounts of cadmium salts may cause severe testicular damage, a dramatic effect which has got a lot of attention, but is of little interest with regard to long-term effects of cadmium in humans. In the body cadmium is bound to metallothionein, and NORDBERG (1972) could show that injection of cadmium bound to metallothionein did not cause testicular damage but renal damage, which is more consistent with what we know about effects of cadmium in human beings.

There are some metals for which no or slight systemic effects have been noted in humans, but animal data show a large variety of symptoms. In Chapter II.22 many interesting findings have been noted in animals receiving injections of nickel salts, but it is doubtful if such data are of relevance to human beings. Hemodialysis with nickel contaminated water caused acute reactions, e.g., nausea and headache but the symptoms disappeared rapidly. Of greatest interest to the human situation are effects of nickel on the respiratory system and skin sensitization; the latter may affect about 10% of the female population, but are very difficult to study in animal experiments. Long-term feeding studies, even with large doses of nickel, have not shown any effect (NORSETH, 1986).

Some metals may cause changes in the nervous system, e.g., mercury vapor and lead. To detect, for instance, behavioral changes or changes in I.Q. extensive psychological test protocols are needed. There is an abundance of papers on effects of lead in experimental animals, but it is only through well-planned studies on children that the risk from "low-level" lead exposure has been recognized during recent years (see Chapter II.16).

It can be concluded that for some metals such as arsenic, cadmium, chromium, lead, manganese, mercury, and nickel studies on humans have given the best information. However, there are metals which only recently have come into use or the exposure of which may be obscured by the presence of other metals. For such metals there are virtually no human data available and in such cases the animal data must be taken into account. Germanium and indium belong to that group.

5 Interactions
(see also Chapter I.10)

As mentioned in Sect. 3, there may be large variations in the intake of essential as well as non-essential elements in human populations. In working environments simultaneous exposure may occur to a variety of metals. How these metals influence each other is of great interest, and animal data may be of great value for evaluating such interactions. A striking example is the Itai-itai disease in Japan, which mainly affected women exposed to cadmium via rice (see also Chapter II.6). Calcium deficiency was common in that group, and it could be shown in animal experiments that in rats on a calcium-deficient diet cadmium absorption increased considerably (LARSSON and PISCATOR, 1971). The relationship between zinc and cadmium is also of interest, since these two metals occur together in nature and also occur together in metallothionein.

Since human intake of zinc often is marginal it was thought that excessive cadmium exposure may cause changes in the distribution of zinc in the body. It was found that in the human kidney the increase in cadmium levels was accompanied by a parallel increase in zinc levels (PISCATOR and LIND, 1971). An evaluation of animal data revealed that in experimental animals the relationship between zinc and cadmium seemed to differ from what was found in humans (ELINDER and PISCATOR, 1978). However, the horse seemed to give the best information in this case. The horse is not an experimental animal, but during the life-time this animal accumulates large amounts of cadmium, which makes it ideal for a long-term study. Horses were shown to have similar increases in kidney zinc as humans, and by determination of metallothionein it was possible to show how total zinc and zinc bound to metallothionein was influenced by cadmium (NORDBERG et al., 1979).

In experimental animals the intake of zinc is more than optimal if commercial diets are given. In a rat study it was shown that cadmium given to rats with an optimal intake of zinc did not cause any effect related to zinc deficiency, but when the intake of zinc was made suboptimal a decrease in the zinc content of the testes was noted (PETERING et al., 1971). The explanation may be that cadmium exposure will cause retention of zinc in metallothionein in kidneys and liver, which will make it less available for other tissues. These data also helped to explain effects on the fetus. Occupationally cadmium exposed women have been reported to get children with low birth weights, which is a sign of zinc deficiency. Since only small amounts of cadmium can pass the placenta, it is conceivable that this is a result of cadmium causing redistribution of zinc in the mother and making less zinc available to the fetus.

6 Conclusions

Animal experiments may give valuable information on toxicity mechanisms and may predict effects in human beings. In the interpretation of animal data the exposure route, dosage, nutritional status, metabolic factors specific for the species or strain, organ sensitivity etc. must be taken into account. The rat may be suitable for some types of studies, but often another animal may be the best one.

Extrapolation generally means that animal data are used to obtain a no-effect level (NOEL) or a lowest observed adverse effect level (LOAEL), and then a "safety" factor is applied, usually 100. This is at present the best approach, e.g., to assess food additives (WHO, 1987) when no human data are available. The approach has sometimes been tried both for essential metals and non-essential metals but does not work. For instance, a NOEL or LOAEL obtained for zinc in animal experiments cannot be divided by 100, the resulting ADI for humans would be below the requirement. The tolerable weekly intake of methylmercury, for example, was obtained by using a "safety" factor of 10 for the blood levels found in cases of methylmercury poisoning in Japan. For metals such as cadmium and lead a "safety" factor of 10, however, cannot be used, since the gap between present "normal" levels and critical levels is much less.

As pointed out in the introduction human data are often available, and any extrapolations should first be made from such data. Extrapolation from animal data must be made when human data are scarce. A recent example is germanium, which was found to be added in Sweden to "health foods", and the Swedish Food Administration had to react since animal experiments indicated some effects of large oral doses of germanium. From Japan it has been reported that a person died after ingestion of high doses of an organic germanium compound.

References

Adamsson, E., Piscator, M., and Nogawa, K. (1979), *Pulmonary and Gastrointestinal Exposure to Cadmium Oxide Dust in a Battery Factory. Environ. Health Perspect.* 28, 219–222.
Axelsson, B., and Piscator, M. (1966a), *Serum Proteins in Cadmium Poisoned Rabbits. Arch. Environ. Health* 12, 374–381.
Axelsson, B., and Piscator, M. (1966b), *Renal Damage after Prolonged Exposure to Cadmium. Arch. Environ. Health* 12, 360–373.
Albanus, L., Frankenberg, L., Grant, C., von Haartman, U., Jernelöv, A., Nordberg, G., Rydälv, M., Schütz, A., and Skerfving, S. (1972), *Toxicity for Cats of Methylmercury in Contaminated Fish from Swedish Lakes and Methylmercury Hydroxide Added to Fish. Environ. Res.* 5, 425–442.
Barltrop, D., and Khoo, H. E. (1975), *The Influence of Nutritional Factors on Lead Absorption. Postgrad. Med. J.* 51, 795–800.
Barltrop, D., and Meek, F. (1975), *Absorption of Different Lead Compounds. Postgrad. Med. J.* 51, 805–809.
Bonner, F. W., and Parke, D. V. (1984), *Transferability of Test Results from Animal Experiments to Humans* (in German), in: Merian, E. (ed.): Metalle in der Umwelt, pp. 195–207 (with additional references). Verlag Chemie, Weinheim-Deerfield Beach/Florida-Basel.
Buchet, J. P., and Lauwerys, R. (1985), *Study of Inorganic Arsenic Methylation by Rat Liver in vitro. Relevance for the Interpretation of Observations in Man. Arch. Toxicol.* 57, 125–129.
Buchet, J. P., and Lauwerys, R. (1987), *Study of Factors Influencing the in vitro Methylation of Inorganic Arsenics in Rats. Toxicol. Appl. Pharmacol.* 91, 65–74.
Chavalitnitikul, C., Levin, L., and Chen, L. C. (1984), *Study and Models of Total Lead Exposures of Battery Workers. Am. Ind. Hyg. Assoc. J.* 45, 802–808.

Elinder, C.-G., and Piscator, M. (1978), *Cadmium and Zinc Relationships. Environ. Health Perspect. 25*, 129–132.

Engström, B., and Nordberg, G. F. (1979a), *Dose Dependence of Gastrointestinal Absorption and Biological Half-time of Cadmium in Mice. Toxicology 13*, 215–222.

Engström, B., and Nordberg, G. F. (1979b), *Factors Influencing Absorption and Retention of Oral ^{109}Cd in Mice: Age, Pretreatment and Subsequent Treatment with Non-Radioactive Cadmium. Acta Pharmacol. Toxicol. 45*, 315–324.

Friberg, L. (1950), *Health Hazards in the Manufacture of Alkaline Accumulators with Special Reference to Chronic Cadmium Poisoning. A Clinical and Experimental Study. Acta Med. Scand. Suppl. 240, 138*, 1–124.

Friberg, L., Piscator, M., Nordberg, G. F., and Kjellström, T. (1974), *Cadmium in the Environment*, 2nd Ed. CRC Press, Cleveland, Ohio.

Ishihara, N., and Matsushiro, T. (1986), *Biliary and Urinary Excretion of Metals in Humans. Arch. Environ. Health 41*, 324–330.

Larsson, S.-E., and Piscator, M. (1971), *Effect of Cadmium on Skeletal Tissue in Normal and Calcium-deficient Rats. Isr. J. Med. Sci. 7*, 495–497.

Löser, E. (1980), *A Two-year Oral Carcinogenicity Study with Cadmium in Rats. Cancer Lett. 9*, 191–198.

Nordberg, G. F. (1972), *Cadmium Metabolism and Toxicity. Environ. Physiol. Biochem. 2*, 7–36.

Nordberg, M., Elinder, C.-G., and Rahnster, B. (1979), *Cadmium, Zinc and Copper in Horse Kidney Metallothionein. Environ. Res. 20*, 341–350.

Norseth, T. (1986), *Nickel*, in: Friberg, L., Nordberg, G. F., and Vouk, V. (eds.): *Handbook on the Toxicology of Metals*, 2nd Ed., pp. 462–481. Elsevier, Amsterdam.

Oberdörster, G. (1989), *Assessment of Lung Cancer Risk from Inhaled Environmental Cadmium. Toxicol. Environ. Chem. 23* (1–4), 41–51.

Pershagen, G., Nordberg, G., and Björklund, N.-E. (1984), *Carcinomas of the Respiratory Tract in Hamsters Given Arsenic Trioxide and/or Benzo(a)pyrene by the Pulmonary Route. Environ. Res. 34*, 227–241.

Petering, H., Johnson, M. A., and Stemmer, K. L. (1971), *Studies on Zinc Metabolism in the Rat. Arch. Environ. Health 23*, 93.

Piscator, M. (1971), *Proteinuria as an Index of Renal Injury with Special Reference to Chronic Cadmium Poisoning*, in: Holden, H., and Kazantzis, G. (eds.): *Proceedings of the Symposium of the Early Detection of Occupational Hazards*. Society of Occupational Medicine, London.

Piscator, M., and Lind, B. (1972), *Cadmium, Zinc, Copper and Lead in Human Renal Cortex. Arch. Environ. Health 24*, 426–431.

Roels, H., Buchet, J. P., Truc, J., Croquet, F., and Lauwerys, R. (1982), *The Possible Role of Direct Ingestion on the Overall Absorption of Cadmium or Arsenic in Workers Exposed to CdO or As_2O_3 Dust. Am. J. Ind. Med. 3*, 53–65.

US EPA (1984), *Air Quality Criteria for Lead*, Vol I. EPA-600/8-83/028aF. US Environmental Protection Agency, Cincinnati, Ohio.

Vacher, J., and Stoner, H. F. (1968), *The Transport of Beryllium in Rat Blood. Biochem. Pharmacol. 17*, 93–107.

Vahter, M., and Marafante, E. (1985), *Reduction and Binding of Arsenate in Marmoset Monkeys. Arch. Toxicol. 57*, 119–124.

Vahter, M., Marafante, E., Lindgren, A., and Dencker, L. (1982), *Tissue Distribution and Subcellular Binding of Arsenic in Marmoset Monkeys after Injection of ^{74}As-arsenite. Arch. Toxicol. 51*, 65–77.

WHO (1978), *Principles and Methods for Evaluating the Toxicity of Chemicals*, Part I. Environ. Health Crit. 6. WHO, Geneva.

WHO (1981), *Arsenic. Environ. Health Crit. 18*. WHO, Geneva.

WHO (1984), *Principles for Evaluating Health Risks to Progeny Associated with Exposure to Chemicals During Pregnancy. Environ. Health Crit. 30*. WHO, Geneva.

WHO (1986), *Principles of Toxicokinetic Studies. Environ. Health Crit. 57*. WHO, Geneva.

WHO (1987), *Principles for the Safety Assessment of Food Additives and Contaminants in Food. Environ. Health Crit. 70*. WHO, Geneva.

I.14a   Metal and Ceramic Implants

ROLF MICHEL, Hannover, Federal Republic of Germany

1 Compositions and Applications

Medical implants are used extensively with outstanding therapeutic success for dental restorations, dental prosthetics, orthopedic surgery, restorative and replacement surgery, as well as for artificial organs and other internal devices. There is a strong development toward improvement of technical performance as well as further extension of applications.

Metallic and ceramic biomaterials make up the bulk of modern implants. The individual compositions of biomaterials vary considerably so that for each class only ranges can be given. Dental amalgams have compositions of 33–42% Ag, 0–35% Cu, 10–32% Sn, 40–62% Hg, and 0–5% Zn. Other dental materials are alloys based on gold (3–50% Ag, 10–92% Au, 0–1% Co, 8–21% Cu, 0–1% Ge, 1–23% Pd, 0–1% Pt, 0–2% Zn), silver-palladium based materials (35–90% Ag, 1–20% Au, 0–15% Cu, 0–1% Ga, 0–1% In, 3–50% Pd, 0–1% Pt, 0–2% Zn), and dental cements (17–30% Ag, 0–1% B, 0–9% Ba, 0–12% Bi, 0–1% Mg, 0–1% Na, 0–12% Pb, 27–55% Zn).

If there is special need for mechanical strength, as in maxillo-facial and orthopedic surgery, osteosyntheses or alloarthroplastics, stronger materials have to be used. For nails, screws and plates, iron-based alloys (0–2% Al, 17–20% Cr, 64–75% Fe, 0–3% Mn, 0–3% Mo, 7–14% Ni, 0–1% Ti) are frequently used, while for total joint replacement cobalt-based alloys (0–1% Be, 32–65% Co, 18–32% Cr, 0–4% Cu, 0–40% Fe, 0–3% Mn, 1–10% Mo, 0–37% Ni, 0–2% Si, 0–4% Ti, 0–16% W) dominate. Since even these extremely strong and corrosion resistant materials are not completely satisfactory, alternative titanium-based alloys (5–7% Al, 0–2% Fe, 88–100% Ti, 4–5% V) and pure metals as niobium and tantalum are used. Nickel-based alloys (0–4% Al, 0–3% B, 0–2% Be, 0–15% Co, 11–15% Cr, 0–2% Cu, 0–2% Fe, 0–4% Mn, 2–10% Mo, 54–79% Ni, 0–5% Nb, 0–3% Sn, 0–45% Ti, 0–1% W) are also applied because they have shape memory capabilities, i.e., they "remember" a preprogrammed shape and revert to this shape in response to an increase in temperature.

Ceramic implants cover materials such as dental porcelain (4–9% Al, 2–3% B, 1% Ca, 3–4% Na, 29–32% Si), bioglass (14–25% Ca, 0–3% K, 2–3% Mg, 4–22% Na, 18–23% Si), calciumphosphate ceramics (38–40% Ca) and carbon-based ceramics (88–92% C, 8–12% Si). For orthopedic applications Al_2O_3 ceramics are used increasingly for total joint replacement.

Among the more than 35 elements used in these biomaterials, there are those which are major constituent elements in the human body. But the majority are present in the body only in trace or ultra-trace concentrations. Due to their interaction with body fluids, biomaterials cause a direct internal exposure to their constituent elements. The particular feature of exposure to medical implants is their direct contact to the tissues and body fluids without any metabolic barrier. For dental materials, the ingestion and inhalative exposure is also of importance. Therefore, the application of amalgams is still controversial (ESPEVIK and MJOER, 1982), though the exposure to mercury is considered to be more problematic for dentists and dental personnel than for patients (KESSEL et al., 1980).

2 Corrosion and Destruction

All implant materials are heavily attacked and damaged by the highly aggressive in-body environment. Implants are exposed to extreme mechanical stress, resulting in wear and fatigue, and all types of chemical interactions. As of now, there is no material known which is inert in the human body. Metals and alloys all undergo diverse types of (electrochemical) corrosion, while the term destruction is used for the various processes occurring with ceramic implants. The combination of mechanical and electrochemical stress can even cause the failure of the implant, e.g., the damage of plates, screws, nails, or prostheses.

Variability within the body environment, with regard to acidity and oxygen availability, results in the presence of anodic as well as of cathodic corrosion. The combination of different materials even results in the presence of electrochemical processes. While for internal devices such combinations are today widely regarded as obsolete, in dental medicine various metals and alloys are often combined without considering the enhancement of corrosion by these combinations.

For all implant materials, the question of a possible passivation and actual corrosion rates is essential. Surface analytical techniques show that for functioning implants made of Fe-based or Co-based alloys, passivation may occur (OHNSORGE and HOLM, 1975, 1978). Model experiments using radiotracer methods resulted in corrosion rates for non-loaded implants made of these materials between 10 and 10^{-5} mg per cm^2 per year (HOFMANN et al., 1981, 1982). Depending on the actual case, other types of corrosion, including wear, can further increase the material output and be a cause of massive material deposition out of the implant into the surrounding tissues. Thus corrosion and wear affect the metal/tissue or ceramics/tissue interface as well as the total organism, the implants represent pools of trace elements which are set free by these processes.

The knowledge about the interactions of implant materials with the organism is still sparse, and much has to be done in the future. While more data exist for metallic implants, for ceramics investigations are only beginning.

3 Biomedical Response and Biocompatibility

The ingestive and inhalative exposure of man to corrosion and destruction products of dental implants is similar to metal exposure in other fields of today's modern civilized life. The decisive difference determining the biomedical response to implant materials originates from the direct contact of the embedded material and the resulting exposure to corrosion and destruction products. The action of implants in living tissues is multifactorial. Basically, the biomedical response is a foreign body reaction by which the organism answers to implantation trauma and to the presence of the implants. In the course of wound healing the body responds to the implanted foreign body, and the implant is covered by a more or less prominent layer of connective tissue which encapsulates it. The thickness of connective tissue capsules around non-loaded implants (in animal experiments) is a measure of the reaction of the organism to the implant (LAING et al., 1967). The degree of tissue reaction is found to be proportional to the amount of corrosion products released. It can be classified according to pathological appearances distinguishing loose vascularized fibrous tissues, dense non-vascularized ones without dead cells, and sterile abscesses (STEINEMANN, 1980).

For loaded implants, much stronger corrosion and destruction processes usually occur. Severe dissolution and spread of crystallites and wear products lead to severe tissue reactions. Massive inflammatory reactions, depending on size and type of implant and material released, occur and are classically known as metallosis (CONTZEN and BROGHAMMER, 1964; SCHUSTER, 1972; ZILKENS, 1981). Detailed descriptions of the histological findings in the surroundings of implants have been published (WILLERT and SEMLITSCH, 1980).

Beside local responses, more general ones are reported, namely allergic and to a certain degree carcinogenic ones. Allergic reactions to implants are contact allergies (COOMBS and GELL, 1975) for which the removal of the allergen, i.e., the implants, induces health reestablishment. Whether metal sensitivity and bone necrosis, the consequence of which is loosening of the prostheses in total joint replacements, are directly connected is still controversial (DRAENERT and DRAENERT, 1984).

The knowledge of the possible carcinogenic action of metal implants up to now is based on approximately ten clinical cases (see MICHEL, 1987, for detailed references) which exhibited various types of tumors at the implant sites. While the carcinogenic action of some constituent elements is well established (SUNDERMANN, 1977), the relatively advanced ages of patients receiving implants, the long expression times expected, and the high level of normal cancer incidences do not allow for a decision on the basis of epidemiological data.

Today, the biocompatibility of implant materials is judged mainly on the basis of histological appearances of connective tissues from the area surrounding implants. It does not consider local or systemic chemical effects caused by corrosion and implant destruction. The question whether changes of trace element concentrations in tissues and body fluids resulting from the mobilization of the implant trace element pools by corrosion and destruction are merely deviations from "normal" well tolerated by the organism, or whether they are adverse effects (WHO, 1978) with possibly hazardous consequences at low incidence rates is not yet answered.

4 Chemical Effects in Tissues and Body Fluids

In spite of the fact that trace element investigations of tissues and body fluids after application of implants have been performed for more than 25 years, only a few materials were investigated in detail. A detailed review of all such studies can be found elsewhere (MICHEL, 1987). Chemical trace analysis still plays a minor role in biocompatibility testing and is not part of testing and standardization programs for the development of new implant materials. Massive changes in trace element levels due to implant corrosion and destruction have been observed. They are, however, not completely understood. Particular problems arise from the fact that there is not yet a general consensus about normal trace element concentrations in man (VERSIECK, 1985) and that there is still rapid development in the field of trace element analysis of biological matrices. Thus investigation of the biochemical consequences of implantation still is in the descriptive phase. No connection of elemental concentrations to medical occurrences has been made successfully, and no forecasts of case developments on the basis of trace element data can be made at present. On the other hand, local as well as systemic effects have now been established for a number of implant materials.

4.1 Local Effects

Most investigations involving trace element analyses to evaluate the fate of corrosion products in the human body were directed at tissues from the surroundings of the implant. For all materials investigated so far local effects have been observed. If statements such as "no effect by this material" were made, this was usually the result of a lack of sensitivity on the part of the analytical technique. For some elements the analytical situation is relatively good (Fe, Co, Cr, Ni, Ta), while for others (Ti, Nb) either no method exists or no adequate investigation was performed up to now. Neutron activation analysis was the most often successfully used technique. Its extremely low blank values and a minimum of possible interferences compared to other analytical techniques make it particularly well suited (see also Chapter I.4a).

Local effects, which were investigated most intensively for iron- and cobalt-based alloys and for tantalum, are summarized as follows: The concentrations of the alloy's constituent elements are elevated by several orders of magnitude (LUX and ZEISLER, 1974; MICHEL and ZILKENS, 1978; MICHEL et al., 1984; REICH, 1987). Also those elements not specified and even minor impurities (e.g., 0.1% Co in Fe-based alloys) can cause enrichments of up to one order of magnitude (MICHEL and ZILKENS, 1978). The actual concentrations of corrosion products are distributed in a logarithmic normal fashion ranging from practically "normal" to the most extreme enrichments, while the distribution of normal trace element concentrations is distinctly different. For the tissues in direct contact with the implants other essential elements not contained in the implant materials are also affected. Such effects were observed as inverse correlations of the concentrations of Zn and Cr and of K and Ni for Fe-based alloys (LUX et al., 1976; ZILKENS et al., 1981). For Ta-implants, increasing selenium concentrations accompany high tantalum tissue burdening (REICH, 1987). Local effects extend up to about 10 cm from the implant, and are strongest in the area nearest the implant, thus exhibiting the existence of transport processes for the corrosion products. The connective tissue capsule, which is taken as an indication of biocompatibility, manifests its barrier function for corrosion products as seen in its trace elements content (LUX et al., 1976). The analysis of correlations among corrosion product concentrations in the tissues exhibits for some applications the prominence of passivation accompanied by selective solution of the implant constituents (MICHEL and ZILKENS, 1978). Others show dissolution according to the implant composition (LUX and ZEISLER, 1974). For Co-based alloys used for metal-polyethylene total endoprostheses of the hip joint, the observed concentrations in the tissues can only be explained by the simultaneous action of passivation and selective solution and the different physiological mobilities of the corrosion products (MICHEL et al., 1984). Besides alloy constituents, metals such as Ba, Zr and Hf contribute to the local effects arround cement fixated prostheses. They originate from X-ray contrast additives in the PMMA bone cements, which also undergo destruction and transport (LOEER et al., 1983).

Local effects of ceramic devices (mostly Al_2O_3 ceramics) have been investigated only to a minor degree. Histological studies (GRISS and HEIMKE, 1981) exhibited the existence of massive wear debris in the local tissues accompanied by transport of phagocytized particulate matter. Chemical analysis indicated massive Al enrichment as a local effect of total endoprostheses of the hip joint using an Al_2O_3 ceramics head and acetabulum and a Co-based shaft. For such prostheses only relatively small amounts of the corrosion products of Co-based materials were observed in the tissues. This effect, however, was counterbalanced by a high concentration of Al in local tissues (LOEER et al., 1986). No correlation of the tissue burdening with pathological effects and duration of implantation was observed.

4.2 Systemic Effects

(see also Chapters II.8 and II.20)

In order for systemic changes of trace element levels to occur, the corrosion-caused uptake must be large enough to counterbalance the excretion or clearance of the respective element (SMITH and BLACK, 1977). From model experiments the observed corrosion rates cover such wide ranges that both the existence and absence of systemic effects are possible, depending on the actual corrosion rates which cannot be predicted. Systemic effects can be analyzed in vivo in blood, serum, and urine, and in vitro in organs from autopsies. Analysis of hair and nails has been found to be problematic because of many extraneous influences. Systemic effects, up until now, have only been investigated for dental amalgams, for Fe- and Co-based alloys, and for Ta.

Recent investigations of dental amalgams, which have been discussed controversially in the past, indicate that an early stage of mercury burdening might be detectable on a cellular level by proton microprobe before macroscopic blood Hg-levels change (JOHANNSON and LINDH, 1987). For Fe- and Co-based alloys, recent investigations of serum and plasma demonstrated definitive systemic enrichments of cobalt, chromium, and nickel (HILDEBRAND et al., 1988; MICHEL et al., 1987) for a variety of applications. In earlier studies (OHNSORGE et al., 1978), systemic effects were observed only for (McKee-Farrar) metal-metal total endoprostheses (TEPs) of the hip joint as a consequence of high blank values and/or lack of sensitivity. Such effects have also now been observed after application of Fe-based plates and Co-based metal-polyethylene TEPs of the hip joint. Studies of the time dependence of cobalt burdening for the latter implant types showed cases exhibiting no enrichment while others showed continuous increase post operation over periods of 90 days resulting in cobalt serum levels an order of magnitude above normal (MICHEL et al., 1986).

As of now, there are just three cases of analyses of human organs of deceased implant bearers with cobalt-based TEPs which show enrichments in cobalt, chromium, and nickel in various organs, mainly those related to the reticuloendothelial system (MINSKI and DOBBS, 1980; MICHEL et al., 1986). Animal experiments on Fe-based implants (MICHEL et al., 1980) and on tantalum centromedullary nails (REICH, 1987) also give evidence of such effects in liver and lymphatic tissue. No reports of such investigations have been made for ceramic implants.

Considering the large number of applications, which for TEPs is estimated to be more than 100 000 per year, and the decreasing age of patients at implantation, together with an increasing implantation time and the possible risks of the exposure to metals, the status of knowledge about the action of metal and ceramic implants in the human body is not satisfactory. Future needs are a comprehensive analytical data base of (time-dependent) local and systemic effects, model experiments of transport, excretion and storage of corrosion products, and a connection of the analytical findings with the medical consequences of the implanted devices in order to develop a measure of the possible risk introduced by implanted devices.

References

Contzen, H., and Broghammer, H. (1964), *Korrosion und Metallose. Bruns Beitr. Klin. Chir. 208*, 75 – 84.
Coombs, R. R., and Gell, P. G. H. (1975), *Classification of Allergic Reactions Responsible for Clinical Hypersensitivity and Disease.* in: Gell, P. G. H., Coombs, R. R., and Lachmann, P. J. (eds.), *Clinical Aspects of Immunology,* p. 761. Blackwell Scientific Publications, Oxford.
Draenert, K., and Draenert, Y. (1984), *Die entzündliche Reaktion des Gewebes auf Implantate – Histomorphologische Untersuchungen langjährig implantierter Prothesen und Osteosynthesen. Hefte Unfallheilk. 164,* 490 – 498.
Espevik, S., and Mjoer, I. (1982), *Corrosion and Toxicology of Dental Amalgams,* in: Smith, D. C., and Williams, D. F. (eds.), *Biocompatibility of Dental Materials,* Vol. III, pp. 1 – 18. CRC Press, Boca Raton, Florida.
Griss, P., and Heimke, G. (1981), *Biocompatibility of High Density Alumina and its Applications in Orthopaedic Surgery,* in: Williams, D. F. (ed.), *Biocompatibility of Implant Materials,* CRC Series in Biocompatibility, pp. 155 – 198. CRC Press, Boca Raton, Florida.

Hildebrand, H. F., Ostapczuk, P., Mercier, J. F., Stoeppler, M., Roumazeille, B., and Decozlx, J. (1988), *Orthopaedic Implants and Corrosion Products – Ultrastructural and Analytical Studies of 65 Patients,* in: Hildebrand, H. F., and Champy, M. (eds.), *Biocompatibility of Co-Cr-Ni-alloys,* NATO-ASI Series. Plenum Publ. Comp., London.

Hofmann, J., Michel, R., Holm, R., and Zilkens, J. (1981), *Corrosion Behaviour of Stainless Steel Implants in Biological Media. Surf. Interfaces Anal. 3,* 110–117.

Hofmann, J., Wiehl, N., Michel, R., Loeer, F., and Zilkens, J. (1982), *Neutron Activation Studies of the In-body Corrosion of Hip-joint Prostheses Made of Co-Cr-alloys. J. Radioanal. Chem. 70,* 85–107.

Johansson, E., and Lindh, U. (1987), *Mercury in Blood Cells – Altered Elemental Profiles Toxic Events in Human Exposure. Biol. Trace Elem. Res. 12,* 309–321.

Kessel, R., Bencze, K., Hamm, M., and Sonnabend, E. (1980), *Dtsch. Zahnaerztl. Z. 35,* 457.

Laing, P. G., Ferguson, A. B., and Hodge, E. S. (1967), *Tissue Reaction in Rabbit Muscle Exposed to Metallic Implants. J. Biomed. Mater. Res. 1,* 135–149.

Loeer, F., Zilkens, J., Michel, R., Freisem-Broda, G., and Bigalke, K. H. (1983), *Gewebebelastung mit körperfremden Spurenelementen durch Röntgenkontrastmittel der Knochenzemente. Z. Orthop. 121,* 255–259.

Loeer, F., Zilkens, J., Schmidt, E., Nolte, M., Reich, M., and Michel, R. (1986), *Release of Trace Elements from Ceramic-metal Endoprostheses of the Hip Joint.* in: Christel, P., Meunier, A., and Lee, A. J. C. (eds.), *Biological and Biomechanical Performances of Biomaterials,* pp. 501–506. Elsevier, Amsterdam.

Lux, F., and Zeisler, R. (1974), *Investigations of the Corrosive Deposition of Components of Metal Implants and of the Behaviour of Biological Trace Elements in Metallosis Tissue by Means of Instrumental Multi-element Activation Analysis. J. Radioanal. Chem. 19,* 289–297.

Lux, F., Schuster, J., and Zeisler, R. (1976), *A Mechanistic Model for the Metabolism of Corrosion Products and of Biological Trace Elements in Metallosis Tissue Based on Results Obtained by Activation Analysis. J. Radioanal. Chem. 32,* 229–239.

Michel, R. (1987), *CRC Crit. Rev. Biocompat. 3* (3), 235–317.

Michel, R., and Zilkens, J. (1978), *Untersuchungen zum Verhalten von Metallspuren im umgebenden Gewebe von AO-Winkelplatten mit Hilfe der Neutronenaktivierungsanalyse. Z. Orthop. 116,* 666–674.

Michel, R., Hofmann, J., and Zilkens, J. (1980), *Trace Element Behaviour of Human and Mammalian Tissues During Excessive Supply of Metals,* in: Braetter, P., and Schramel, P. (eds.), *Trace Element Analytical Chemistry in Medicine and Biology,* pp. 137–157. Walter de Gruyter, Berlin.

Michel, R., Hofmann, J., Loeer, F., and Zilkens, J. (1984), *Trace Element Burdening of Human Tissues due to the Corrosion of Hip-joint Prostheses Made of Cobalt-chromium Alloys. Arch. Orthop. Trauma. Surg. 103,* 85–95.

Michel, R., Loeer, F., Nolte, M., Reich, M., and Zilkens, J. (1986), *Neutron Activation Analysis of Human Tissues, Organs and Body Fluids to Describe the Interaction of Orthopaedic Implants Made of Cobalt-chromium Alloys with the Patients Organism, Proceedings 7th Modern Trends in Activation Analysis,* June 23–27, Copenhagen, pp. 495–503; see also *J. Radioanal. Nucl. Chem. 113* (1987), 83–95.

Michel, R., Loeer, F., Nolte, M., Reich, M., and Zilkens, J. (1988), *Phenomenology of the Trace Element Burdening of the Human Organism by the In-body Corrosion of Co-Cr-Ni-alloys as Revealed by Neutron Activation Analysis,* in: Hildebrand, H. F., and Champy, M. (eds.), *Biocompatibility of Co-Cr-Ni-alloys.* NATO-ASI Series. Plenum Press, London, in press.

Minski, M. J., and Dobbs, H. S. (1980), *Neutron Activation Techniques Applied to Biomedical Samples in Particular Tissues Contaminated by Stainless Steel Implants,* in: Braetter, P., and Schramel, P. (eds.), *Trace Element Analytical Chemistry in Medicine and Biology,* pp. 339–350. Walter de Gruyter, Berlin.

Ohnsorge, J., and Holm, R. (1975), *ESCA-Untersuchungen der Passivschicht von Metallimplantaten. Z. Orthop. 113,* 770–772.

Ohnsorge, J., and Holm, R. (1978), *Surface Investigations of Oxide Layers on Cobalt-chromium-alloyed Orthopaedic Implants Using ESCA Technique.* Med. Progr. Tech. *5*, 171–177.

Ohnsorge, J., Abeln, M., and Zilkens, J. (1978), *Spurenelementkonzentration verschiedener Gewebe, nachgewiesen mit Hilfe der Neutronenaktivierungsanalyse.* Z. Orthop. *116*, 607–608.

Reich, M. (1987), *Zur Belastung menschlicher und tierischer Gewebe und Organe mit Verschleiss- und Korrosionsprodukten keramischer und metallischer Implantate.* Ph D Thesis, Universität Köln.

Schuster, J. (1972), *Die Metallose.* Chirurg *43*, 114.

Smith, G. K., and Black, J. (1977), *Models for Systemic Effects of Metallic Implants,* in: Weinstein, A., Horowitz, E., and Ruff, A. W. (eds.), *Retrieval and Analysis of Orthopaedic Implants*, pp. 23–30. NBS Spec. Publ. 472, U.S. Department of Commerce, Washington, DC.

Steinemann, S. G. (1980), *Corrosion of Surgical Implants – in vivo and in vitro Tests,* in: Winter, G. D., Leray, J. L., and de Groot, K. (eds.), *Evaluation of Biomaterials*, pp. 1–34. John Wiley & Sons, New York.

Sundermann, Jr., F. W. (1977), *Metal Carcinogenesis,* in: Goyer, R. A., and Mehlmann, M. A. (eds.), *Toxicology of the Trace Elements*, pp. 257–295. John Wiley & Sons, New York.

Versieck, J. (1985), *Trace Elements in Human Body Fluids and Tissues.* CRC Crit. Rev. Clin. Lab. Sci. *22*, 97–184.

WHO (World Health Organization) (1978), *Principles and Methods for Evaluating the Toxicity of Chemicals, Part I.* Environ. Health Crit. *6*, 19.

Willert, H. G., and Semlitsch, M. (1980), *Biomaterialien und orthopaedische Implantate,* in: Witt, A. N., Rettig, H., Schlegel, K. F., Hackenbroich, M., and Hupfauer, W. (eds.), *Orthopaedie in Praxis und Klinik*, Vol. II, pp. 221–253. Georg Thieme Verlag, Stuttgart.

Zilkens, J. (1981), *Metallose – Eine erweiterte Definition.* Georg Thieme Verlag, Stuttgart.

Zilkens, J., Loeer, F., Michel, R., and Hofmann, J. (1981), *Tierexperimentelle Untersuchungen zur Frage der Gewebebelastung mit Korrosionsprodukten bei Verwendung von V4A-Implantaten.* Z. Orthop. *119*, 760–763.

I.14b Treatment Using Metal Ions and Complexes

JOHN R. DUFFIELD and DAVID R. WILLIAMS, Cardiff, Wales

1 Introduction

Metals in one form or another have been used for centuries in the treatment of disease (BERMAN, 1980). Only in recent times, however, has such treatment been placed on a scientific footing with regard to optimum chemical form and mode of action of a given metal ion and its associated complexes. Due to limitations of space, it is impossible to give an exhaustive account of the applications of individual metals and compounds and so only the basic concepts will be discussed here.

2 Basic Principles

Life evolved on the surface of the earth, and thus the chemical compositions of organisms with respect to trace metals are a reflection of the abundance of those metals at the surface of the earth coupled with their chemical availabilities and properties (see also Chapter I.9). These metals tend to be the lighter elements of the Periodic Table (WILLIAMS and HALSTEAD, 1983).

In general, metals fall into two categories in relation to life: those which are essential and those which are poisonous. Between these extremes come the beneficial and detrimental metals in their various shades and guises. The average 70 kg person contains less than 10 g of the essential trace metals without which life could not exist in its present form (WILLIAMS, 1971). In addition, the concentrations of these elements, both absolute and relative, are crucial to the maintenance of a healthy existence. This is because all elicit a *biphasic response* such that too little of an essential element precludes normal biological activity whereas an excess will be toxic. Trace metals also exert stimulatory and/or antagonistic effects upon one another giving rise to a 'plateau of health' which will be obtained under steady state conditions (FIABANE and WILLIAMS, 1977).

Metals which are detrimental or toxic to life operate by challenging the roles of the essential metals and blocking metabolic pathways, i.e., the toxicity is a function of chemical combination (WILLIAMS, 1984). Although such elements have been recognized by man for thousands of years, they are relative newcomers to the biosphere being, in general, the heavier elements of the Periodic Table. These heavy metals have only gained a wide-scale ascendancy following industrialization. In addition, industrial activity can alter the natural cycling of metals, increasing their chemical and biological availabilities.

The above preamble, although brief, illustrates the extremely complicated and conflicting processes involved in trace metal metabolism and healthy life. How can this knowledge and the underlying principles involved be used in disease therapy and health care using metals and their complexes? Central to this question is the concept of *chemical speciation* (see also Chapter I.4a).

The term 'chemical speciation' is used to encompass all of the chemical forms of an element in a given system and includes a breakdown of the concentrations of protein-bound, low molecular mass and aquated metal ions. The importance of speciation will become apparent in the following, but it should be noted that in the majority of treatments involving metals it is not the total metal concentration which is crucial but the concentrations of one or more specific chemical forms of that element. This is because successful therapy often involves selectivity, both from the viewpoint of bioavailability and toxicity.

In biological systems, speciation of elements is not an easily quantifiable concept as metals are generally present at low concentrations and exist in complicated labile equilibria among a plethora of competing ligands; thus, computer simulation techniques involving thermodynamic formation constants for all metal ligand interactions are used (DUFFIELD and WILLIAMS, 1986).

3 Treatment Using the Essential Trace Metals

Treatment of this type involves manipulation of the biphasic response curve of a given organism (see also Chapter I.9).

3.1 Trace Metal Deficiency

Trace element deficiency can be caused in a number of ways:

1. Dietary factors can lead to a reduction in the absorption of a metal ion from the gut or to increased excretion of an element. For example, the formation of insoluble complexes of zinc with calcium and phytate has been shown to markedly reduce the bioavailability of zinc to humans (SOLOMONS, 1982).
2. Disease and trauma can either prevent trace metal absorption or increase the rate of depletion. An example of the former is found in acrodermatitis enteropathica (A.E.) which is a rare autosomal recessively inherited disease characterized by skin lesions, diarrhea, and alopecia (NELDNER et al., 1974). The basic molecular defect in A.E. appears to be closely related to zinc metabolism and is connected to a partial impairment of zinc absorption from the intestinal tract resulting in a severe zinc deficiency state, explaining the clinical manifestations of the disease (HAMBIDGE et al., 1978).

3. Trace metal depletion may also be brought about by existing drug therapy or by intravenous feeding (BERTHON et al., 1980).

The majority of trace element deficiency syndromes, such as those alluded to above, can be recognized and analyzed by assessing the speciation of the trace metals in the system of interest. Zinc can be speciated in human milk and its bioavailability determined as a percentage of the total zinc bound in neutral species (MAY et al., 1982). In addition, the metal-dependent side-effects of chelating agents and other therapeuticals in blood plasma and other biofluids can be investigated and possible mobilization and excretion quantified in terms of species distributions and the Plasma Mobilizing Index (MAY and WILLIAMS, 1977).

Once the cause of the deficiency has been established, treatment can then take the form of trace element supplementation in which the depleted metal is either added to the diet, injected, infused intravenously, or included in drug therapy. The route and regimen of supplementation will once again be speciation-dependent.

3.2 Therapy and Healthcare

Therapy and health care uses of essential trace metals will also be a function of the biphasic response and, to a certain extent, selective toxicity. In such systems, the biphasic response need not refer to the activity or metabolism of the organism as a whole but may refer to trace metal imbalances in certain tissues or biofluids. Such a state of affairs is thought to occur with respect to the role of copper administration in rheumatoid arthritis (R.A.).

The etiology of R.A. is not well characterized, but it appears to stem from a breakdown in the patient's autoimmune responses arising from an inadequate supply of copper. In nature, copper is widespread and nutritional deficiencies rare. There is, however, much evidence for localized copper imbalances in R.A. (SORENSON, 1982).

The majority of evidence from screening data of copper preparations is that reduction in inflammation in R.A. is proportional to the copper administered into the tissue, and it is the increased availability of copper ions per se which affords protection against inflammation. In general, the ligand complexed to the copper, in the administered form, does not influence the screening result, provided that it is labile in blood plasma. Whatever the mechanism through which copper acts, inflamed tissues are then able to acquire the metal ion from the labile equilibrium system in plasma, i.e., the copper must be bioavailable from the original complex.

A second use of essential metal ions in therapy comes in metal chelation in antibacterial and antiviral activity (PERRIN and STÜNZI, 1982).

In 1944, it was found that 8-hydroxyquinoline (oxine) exerts its antibacterial action via chelation and that this is not due to a reduction in the bioavailability of essential elements to the bacteria (ALBERT, 1973). Concentration quenching experiments showed that oxine exerts its toxic action by forming lethal metal complexes particularly with iron(III). Paradoxically, it is not the neutral 1:3 iron-oxine species which is lethal, as might be expected from increased iron uptake by passive diffusion leading to overload, but the charged 1:1 and 1:2 iron(III)-oxine complexes.

The antibacterial properties of metal complexes have also been utilized in the design and manufacture of health care products. Zinc salts have been shown to inhibit the growth of the plaque-forming bacteria, *Streptococcus mutans*, both in vivo and in vitro (HARRAP et al., 1984), and zinc citrate is now included in many dentifrice preparations. Although these findings are relatively recent, they represent a classic example of the combined use of the biphasic response and selective toxicity. In the concentrations used, the zinc is harmless to man but lethal to the bacteria. Here, too, speciation plays its part, although the bioactive zinc citrate complex(es) still awaits identification.

4 Treatment Using Exogenous Metal Ions

In this context, exogenous metals are taken to be those metals which are non-essential for life. They can thus be described as beneficial, relatively innocuous, detrimental or toxic to health. Such metals and their complexes have been, and are being, used in a wide range of medical applications, including cancer chemotherapy, ulcer therapy, radiolabeling for tumor diagnosis, stimulation of immune response and drug targeting. Although it is not possible to detail all potential uses of metals in these categories, two examples will be discussed here.

4.1 Ulcer Therapy Using Bismuth Compounds

Bismuth compounds are used in the treatment of disorders of the alimentary system. In the United Kingdom, one such medicine is 'De-Nol', a colloidal bismuth citrate solution, which is used successfully in the treatment of gastric ulcers. The mechanism of action in this preparation has been examined and elucidated by computer simulation of the bismuth speciation in the gastro-intestinal tract (WILLIAMS, 1977). It is found that the change in speciation produced by the acid stomach juices causes the bismuth to move from a charged and soluble $Bi(citrate)_2^{3-}$ complex to insoluble $Bi\text{-}citrate^0$ and bismuth hydroxychloride ($Bi \cdot OH \cdot Cl^0$) species. It is thought that these bismuth species precipitate in the stomach, coating and isolating the ulcerated areas from the digestive juices. In this way, healing is promoted and allowed to progress without interference.

4.2 Chelation in Anticancer Activity

An important discovery made during the last 20 years is that coordination compounds of platinum possess anticancer activity (ROSENBERG, 1971). In 1979, cisplatin (*cis*-diammin-chloroplatinum) was introduced clinically in the UK for use in oncology.

The anti-cancer activity of cisplatin and other platinum containing compounds is due to the inhibition of cancer cell DNA synthesis. All the complexes are neutral and contain at least two reactive ligands, such as chloride. In addition, molecular orbital considerations have established that their structures are planar, so that *cis-trans* isomerism is possible. It is the *cis* stereochemical property which is critical since the *trans* isomers are inactive.

The mode of action of cisplatin and its analogues is thought to involve passage of the neutral complex into cancer cells where the reactive ligands are released. The platinum is thus allowed to cross-link between the nitrogens of two DNA purines and pyrimidines, thus preventing replication.

5 Conclusions

The use of metals in therapy represents a promising and exciting field of endeavor. Of necessity, this chapter has touched only upon the fringes of this effort (but see, for instance, also Chapters I.10, II.1, II.3, and II.15).

In general, our understanding of the chemistry involved in designing therapeuticals is good. However, much further research is needed to elucidate mechanisms involved in the regulation of trace metal metabolism with respect to both onset of disease and subsequent treatment. In this, chemical speciation studies will play a key role.

References

Albert, A. (1973), *Selective Toxicity*, 5th Ed. Chapman & Hall, London.
Berman, E. (1980), *Toxic Metals and their Analysis*. Heyden, London.
Berthon, G., Matuchansky, C., and May, P. M. (1980), *Computer Simulation of Metal Ion Equilibria in Biofluids. Part 3: Trace Metal Supplementation in Total Parenteral Nutrition*. J. Inorg. Biochem. *11*, 63–73.
Duffield, J. R., and Williams, D. R. (1986), *The Environmental Chemistry of Radioactive Waste Disposal*. Chem. Soc. Rev. *15*, 291–307.
Fiabane, A. M., and Williams, D. R. (1977), *The Principles of Bioinorganic Chemistry*. The Chemical Society, London.
Hambidge, K. M., Neldner K. H., Walravens, P. A., Weston, W. L., Silverman, A., Sabol, J. L., and Brown, R. M. (1978), in: Hambidge, K. M., and Nichols, B. L. (eds.), *Zinc and Copper in Clinical Medicine*, pp. 81–98. Spectrum Publications, New York.
Harrap, G. J., Best, J. S., and Saxton, C. A. (1984), *Human Oral Retention of Zinc from Mouthwashes Containing Zinc Salts*. Arch. Oral Biol. *29*, 87–91.
May, P. M., and Williams, D. R. (1977), *Computer Simulation of Chelation Therapy*. FEBS Lett. *78*, 134–137.
May, P. M., Smith, G. L., and Williams, D. R. (1982), *Computer Calculation of Zinc Complexes Distributed in Milk*. J. Nutr. *112*, 1990–1993.

Neldner, K., Hagler, L., Wise, W. R., Stifel, F. B., Lufkin, E. G., and Herman, R. H. (1974), *Acrodermatitis Enteropathica. A Clinical and Biochemical Survey. Arch. Dermatol. 110*, 711.

Perrin, D. D., and Stünzi, H. (1982), in: Sigel, H. (ed.), *Metal Ions in Biological Systems*, Vol. 14, pp. 207–241. Marcel Dekker, New York.

Rosenberg, B. (1971), *Biological Effects of Platinum Compounds. New Agents for the Control of Tumors. Plat. Met. Rev. 15*, 42.

Solomons, N. W. (1982), *Biological Availability of Zinc in Humans. Am. J. Clin. Nutr. 35*, 1048.

Sorenson, R. J. (ed.) (1982), *Inflammatory Diseases and Copper*. Humana Press, Clifton, New Jersey.

Williams, D. R. (1971), *The Metals of Life*. Van Nostrand, London.

Williams, D. R. (1977), *Analytical and Computer Simulation Studies of a Bismuth Ammonium Citrate System. J. Inorg. Nucl. Chem. 39*, 711–714.

Williams, D. R., and Halstead, B. W. (1982/83), *Chelating Agents in Medicine. J. Toxicol. Clin. Toxicol. 19*, 1081–1115.

Williams, R. J. P. (1984), in: Nriagu, J. O. (ed.), *Changing Metal Cycles and Human Health*, pp. 251–263. Springer Verlag, Berlin.

I.15 Intake, Distribution, and Excretion of Metals and Metal Compounds in Humans and Animals

ULRICH EWERS and HANS-WERNER SCHLIPKÖTER
Düsseldorf, Federal Republic of Germany

1 Introduction

Metals and metal compounds may be absorbed by means of inhalation, ingestion, and absorption through the skin. In the general environment human exposure to metal compounds is predominantly through food, drinking water, and beverages, whereas in the work environment absorption following inhalation is of primary importance. Examples of industrial exposures are: exposure to mercury vapor in chloralkali plants, exposure to dust and fume containing arsenic, cadmium, iron, lead, manganese, and nickel in metal smelters and refineries, and exposure to lead in storage battery factories. In the general environment metal containing aerosols may originate from a variety of sources such as coal-fired power plants, iron and steel production, non-ferrous smelters and cement plants (VOUK and PIVER, 1983; SABBIONI et al., 1984). The combustion of leaded gasoline represents the major source of airborne lead, particularly in urban areas and along highways. In general, however, the contribution of inhalation is low compared with exposure through food and beverages. Absorption through the skin may be of importance in the case of skin contact with lipophilic organic metal compounds, but has also been observed for some inorganic metal compounds such as mercury and thallium compounds.

2 Ingestion

Metals and metal compounds taken up orally with food and beverages or by ingestion of contaminated soil and dust are mainly absorbed in the intestinal tract. Compared with absorption from the gut, absorption in the mouth and stomach is of minor importance. The mechanisms of intestinal absorption of metal ions and compounds have not yet been clarified precisely. It is assumed that absorption mainly occurs by means of diffusion processes following the concentration gradient between the intestinal lumen, mucosa cells, and blood. The strong blood circulation in the intestinal tract contributes to the maintenance of a high concentration gradient between intestinal lumen and blood. Binding of metal ions or metal compounds to plasma proteins and blood cells, rapid diffusion into various organs and tissues as well as rapid elimination tend to enhance the rate of absorption since the free concentration in the blood, which determines the diffusion gradient, is lowered.

In addition to passive diffusion there exist special transport mechanisms for certain essential metal ions which must be absorbed when needed more quickly than by passive diffusion. For example, in both animals and humans iron absorption from the gut is adjusted to a fine homeostasis with low iron stores resulting in enhanced absorption and, alternately, sufficient iron stores resulting in decreased absorption. A selective passive transport mechanism is also known for calcium ions which can pass the membranes of the mucosa cells by means of special "carrier" proteins. Active transport, in which molecule or ion movements occur against an increasing concentration gradient by means of metabolic energy, is effective for potassium and sodium ions.

The rates of absorption vary from one metal to another and may also differ from one animal species to another. Moreover, the chemical form, in which the metal occurs, is of great importance. In general, compounds with low solubility in water are less readily absorbed than soluble compounds. Lipophilic organic metal compounds are usually absorbed better than ionic metal compounds. Methylmercury and phenylmercury compounds, for example, are almost completely absorbed, whereas inorganic mercury compounds are absorbed to the extent of 10% or less. Table I.15-1

Table I.15-1. Gastrointestinal Absorption and Secretion of Metals in Humans and Animals[a]

Metal	Physicochemical Form	Gastrointestinal Absorption (%)	Gastro-intestinal Secretion	Factors Influencing Gastro-intestinal Absorption
Aluminum		<5		Increased absorption by addition of acids to the diet
Antimony		~15		
Arsenic	Inorganic As(III)- and As(V) compounds	>90		
	Organic As compounds in seafood	<80		
Barium	Insoluble Ba salts	<1		
	Soluble Ba salts	10–30	+	
Beryllium	Soluble Be salts	20		
Cadmium		4–7		Increased absorption in conditions of calcium, iron, and protein deficiency
Chromium	Cr(III) compounds	<1		
	Cr(VI) compounds	2–6	+	
Cobalt	Co oxide	<0.5	+	
	$CoCl_2$	~30		
Copper	Soluble Cu salts	50–70	+	Absorption is regulated by the copper status in the body. Increased absorption in conditions of copper deficiency

Table I.15-1 (continued)

Metal	Physicochemical Form	Gastrointestinal Absorption (%)	Gastro-intestinal Secretion	Factors Influencing Gastro-intestinal Absorption
Germanium	GeO_2, $Ge(Et)_4$	>90	+	
Indium	In salts	<2		
Iron	Fe salts	2–20	+	Absorption is regulated by the iron status in the body. Increased absorption in conditions of iron deficiency
Lead	Water-soluble Pb salts	~10	+	Absorption in infants up to 50%. Increased absorption in conditions of calcium, vitamin D, and iron deficiency
Manganese	Mn(II) salts	3–5	+	Increased absorption in conditions of iron deficiency and in young animals/infants
Mercury	Elemental Hg	<0.5		
	$HgCl_2$	<10	+	
	Organic Hg compounds	>80	+	
Molybdenum	Mo(VI) compounds	40–80	+	
Nickel	Ni salts	~10	+	
Selenium	Water-soluble Se compounds	>80		
Silver	Water-soluble Ag compounds	10–20	+	
Tellurium	Tellurites	10–25	+	
Thallium	Water-soluble Tl salts	>90	+	
Tin	Inorganic Sn(II)- and Sn(IV) compounds	<5	+	
	Monoethyltin, dibutyltin, tributyltin, triphenyltin	~10	+	
	Triethyltin	>90	+	
Titanium	TiO_2	<5	+	
Uranium	Water-soluble U compounds	10–30		
Vanadium	Water-soluble V compounds	~2		
Zinc	Water-soluble Zn salts	50–80	+	

[a] The data have been compiled from the "Handbook on the Toxicology of Metals" (L. Friberg, G.F. Nordberg, V.B. Vouk, eds.), Elsevier, Amsterdam-New York-Oxford 1986, Volume II

summarizes data on the gastrointestinal absorption of some metals and metal compounds.

The rate of absorption of metal compounds is affected by a number of dietary and constitutional factors. Dietary deficiencies of calcium, copper, iron, zinc, protein, and vitamin D have been shown to enhance the intestinal absorption of lead and cadmium (ROSEN and SORELL, 1978; BREMNER, 1979). The gastrointestinal absorption of lead, manganese, and mercury(II) is greater in infants and young children than in adults. Sufficient stores of iron tend to decrease iron absorption. A detailed discussion of the factors influencing the metabolism and toxicity of metals has been presented by the TASK GROUP ON METAL INTERACTION (1978) and by NORDBERG et al. (1986).

3 Inhalation

Metals and metal compounds may be inhaled in the form of vapors or in the form of particles with different particle size. Because of their low volatility most metals and metal compounds occur in the atmosphere as constituents of suspended particles (aerosols). At room temperature only mercury as well as a number of volatile inorganic and organic metal compounds such as tetraethyllead, arsine and nickel tetracarbonyl may be present in vapor form. Aerosols usually contain particles with many different sizes, shapes and densities. With regard to inhalation particles with a diameter <20 µm are of particular importance. Particles >20 µm rapidly sediment and thus are rarely found in the atmosphere when there is little air movement. With regard to the physical and chemical characteristics and properties of aerosols reference is made to MERCER (1973) and PHALEN (1984).

3.1 Absorption of Vapors and Gases

Agents occurring in vapor form may be inhaled into the lower airways and alveolar parts of the lungs (alveoli). Because of the large surface area of the alveoli conditions are favorable for the gas exchange between alveolar air and blood and thus for the passing of gases and vapors into the blood. Whether gases and vapors reach the lower airways and to what extent this occurs depends on their solubility in water. Whereas water-soluble gases and vapors are readily dissolved in the mucous membranes of the nose, mouth, throat, and bronchi (nasopharyngeal and tracheobronchial region), less soluble gases and vapors reach the terminal airways and the alveoli (pulmonal region). Here they are dissolved in the fluid film coating the alveoli ("surfactant"), and then pass into the blood or lymph stream. Although there are very few data available relating to special agents it is assumed that about 80% of metals or metal compounds inhaled in vapor form are absorbed (TASK GROUP ON METAL ACCUMULATION, 1973; KLAASSEN, 1986).

3.2 Deposition and Absorption of Particles

Deposition of inhaled aerosol particles in the airways mainly occurs by impaction, sedimentation, and diffusion. Impaction occurs when aerosol particles, which tend to move along their original pathway, are impacted on airway surfaces as a result of an air stream bending such as at an airway bifurcation. Gravitational sedimentation is an important mechanism for deposition in the smaller bronchi, the bronchioles, and the alveolar spaces where the airways are small and the velocity of the airstream is low. Sedimentation is effective mainly for particles with an aerodynamic diameter of 1–5 µm. This diameter is equal to the diameter of a spherical particle of unit density (1 g/mL) with the same settling velocity in air as the particle in question. Diffusion is an important deposition mechanism in the small airways and alveoli for particles below 1 µm. These particles undergo a random motion caused by discrete collisions of gas molecules (Brownian motion), which increases with decreasing particle size.

The deposition of particles in the respiratory tract is determined by a combination of the physical forces that remove particles from the air stream and the anatomy of the respiratory tract. The site of deposition affects (1) the severity of the consequences of tissue damage to the respiratory tract, (2) the degree of absorption of systemic toxicants, and (3) the clearance mechanisms available for the removal of the particles from the respiratory tract. Particles having an aerodynamic diameter of 5 to 30 µm are largely deposited in the nasopharyngeal region by impaction. Particles with an aerodynamic diameter of 1 to 5 µm are deposited in the nasopharyngeal and tracheobronchial region mainly by sedimentation. The smallest particles with an aerodynamic diameter of less than 1 µm (submicron particles) are deposited in the alveolar region primarily by diffusion.

The deposition rates of inhaled particles of a given mass median aerodynamic diameter in the nasopharyngeal, tracheobronchial, and pulmonal region are shown in Fig. I.15-1. With regard to submicron particles, 5–30% are deposited in the tracheobronchial region and 30–60% in the pulmonal region. Larger particles are more or less efficiently deposited in the nasopharyngeal and tracheobronchial region. The non-deposited particles remain suspended in the inhaled air and leave the lungs with exhaled air.

Particles that have been deposited in the respiratory tract are removed by mechanisms that vary depending on the site of deposition. The speed and efficiency of these clearance mechanisms are critical factors in the assessment of the toxic potential of the deposited particles, since rapid removal decreases the time available to cause critical damage to the pulmonary tissues and to permit systemic absorption of toxic agents.

Clearance from the ciliated surfaces of the respiratory tract, which extend from the terminal bronchioles to the nose, is primarily by mucociliary transport. The upper airways cilia drive mucus backwards and downwards to the pharynx and the lower airways-cilia drive mucus upwards to the pharynx whereafter it is swallowed. In healthy subjects the tracheal mucus transport rate has been estimated to be in the range of about 4 to 20 mm/min. The velocity is lower in the peripheral airways. Particles deposited in the most peripheral ciliated airways are eliminated from the lung

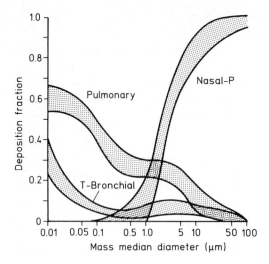

Fig. I.15-1. Deposition of inhaled particles in the respiratory tract in relation to the mass median (aerodynamic) diameter. Each of the shaded areas indicates the variability of deposition for a given mass medium diameter in each compartment when the distribution parameter (geometric standard deviation characterizing the log-normal distribution of the partical diameters) varies from 1.2 to 4.5 and the tidal volume is 1450 mL (from TASK GROUP ON LUNG DYNAMICS, 1966).

usually within 24 h. The efficiency of the mucociliary clearance depends on constitutional factors and can be affected by different conditions such as tobacco smoking, air pollution, and diseases of the respiratory tract (CAMNER et al., 1986; WOLFF, 1986).

Particles deposited in the alveolar region can be removed by several mechanisms: (1) transport to the ciliated part of the tracheobronchial tree and further elimination via the mucociliary escalator; (2) removal via the lymphatic drainage, and (3) dissolution of the particles and removal of the dissolved substances with the blood stream. For most materials deposited in the alveoli dissolution and removal of the solute are probably the most important mechanisms by which material is removed from the lung. Solubility in vivo is important even for substances that are usually regarded as insoluble in water. The absorption of toxicants in particles deposited in the pulmonal region can, therefore, be assumed to be about 100%. Insoluble particles are phagocytized by alveolar macrophages, which – possibly by ameboid motion – can move to the ciliated surface of the respiratory bronchioles, from where they are eliminated via the mucociliary escalator. A number of studies show, however, that this elimination process is extremely slow with a half-time of more than 100 days. This situation may lead to an accumulation of insoluble metal compounds in lung tissue (KOLLMEIER et al., 1988).

Data presented here refer mainly to humans, but there are great quantitative differences between species (especially between rodents and humans) in uptake and elimination (BRAIN, CUDDIHY, DAHL, MAUDERLY, 1987, and especially OBERDÖRSTER, 1989, who compared pulmonary retention of cadmium in rats and humans). Also smokers and non-smokers behave differently.

Regarding a more detailed discussion of the deposition and clearance of particles in the respiratory tract reference is made to TASK GROUP ON LUNG DYNAMICS (1966), BRAIN and VAHLBERG (1979), STUART (1984), BOWDEN (1984), CAMNER et al. (1986), and MENZEL and AMDUR (1986).

3.3 Relationship between Metal Concentrations in Air and in Biological Media

Inhalation exposure to metal containing aerosols results in increased metal concentrations in biological media such as blood, urine, and tissues. Tracer studies show no differences in the distribution of metals whether taken up from the lung or gut. The relationship between metal concentration in air and in biological media has been investigated in many experimental and epidemiological studies (WHO, 1980; SNEE, 1981; CHAMBERLAIN, 1983; VAHTER et al., 1986; ANGERER et al., 1989). It should be noted, however, that the levels of metals in biological media are influenced by a number of factors and reach a steady state only after continuous exposure for a longer time.

4 Transport and Distribution

Once absorbed in the gastrointestinal or respiratory tract or through the skin metal ions and metal compounds enter the blood stream and thereby are distributed throughout the body. Distribution usually occurs rapidly. The rate of distribution to the tissues of each organ is determined by the blood flow through the organ and the ease, with which the agents cross the capillary walls and penetrate into the cells of the particular tissue. Metal ions (especially those not bound to blood cells and/or plasma proteins, or those dissociated from them again; see the next two paragraphs) and hydrophilic molecules mainly diffuse through aqueous pores in the capillary walls and cell membranes. Liver, spleen, kidneys, red bone marrow, and intestinal mucosa as well as different glands are equipped with blood vessels exhibiting a great number of such pores. This is, however, not the case for the brain and spinal cord. The blood vessels of these organs have a particularly low permeability for hydrophilic molecules and ions (blood-brain barrier, see below) and can only be crossed by lipid-soluble molecules.

When entering the blood stream most metal ions are bound to blood cells and plasma proteins. For example, more than 90% of lead, cadmium, and methylmercury in blood are bound to erythrocytes. The protein-bound fraction of cadmium and mercury in plasma is about 99%. Iron in serum is bound to transferrin, a β_1-globulin, which represents an important transport vehicle of iron in the body. A further metal binding protein in plasma is ceruloplasmin which carries most of the copper in serum. Lipid-soluble organic metal compounds may also enter the blood cells, but are mainly bound to plasma proteins such as albumin, the most abundant protein in serum, and α- and β-lipoprotein, which are important transport proteins for lipid-soluble compounds such as vitamins and steroid hormones.

Metal ions or metal compounds bound to blood cells or proteins are not immediately available for distribution into the extravascular spaces, since high molecular weight proteins and blood cells cannot cross the capillary walls. The transfer of metal ions or metal compounds between intravascular, interstitial and intracellular fluid, therefore, depends on the 'diffusible fraction', i.e., the fraction not bound to blood cells or proteins. Usually the binding to blood cells and plasma proteins is a rapidly reversible process. As unbound metal ions or metal compounds diffuse from the capillaries, bound ions or molecules dissociate from blood cells and proteins until the concentration of free ions and molecules in the extravascular space equilibrates with the concentration of diffusible ions and molecules in the intravascular space.

4.1 Storage of Metals in Tissues

Liver and kidneys have a high capacity to bind metals and other xenobiotics. This might be related to the fact that the blood flow through these organs is very high and that the blood vessels of these organs are easily permeable to hydrophilic molecules and ions. Intracellular binding proteins may be of additional importance in concentrating metals such as lead, cadmium, and mercury within the liver and kidneys. One of the best characterized binding proteins for metals is metallothionein, a low-molecular weight protein (mol. wt. 6500) with a high cysteine content, which can bind cadmium, copper, cobalt, mercury, gold, and zinc ions (see Chapter I.12).

Bones and other mineralizing tissues such as teeth can also serve as storage organs for metals such as barium, beryllium, lead, strontium, lanthanum, and yttrium. The skeletal uptake of these metals can be considered to be a surface chemistry phenomenon, in which exchange takes place between the bone surface and the surrounding extracellular fluid. An important factor in this process seems to be that the physico-chemical behavior of these divalent metal ions is similar to that of calcium ions. About 90% of lead in the body is found in the skeleton (BARRY, 1978). Deposition and storage occurs throughout the whole life so that the lead levels in bones and teeth increase during life (STEENHOUT and POURTOIS, 1981; DRASCH et al., 1987). Lead is not toxic to the bone, but conditions of bone demineralization, e.g., during pregnancy, or conditions of osteoporosis may cause a mobilization of lead from bones, which will be reflected by increased lead levels in blood (BUCHET et al., 1977; SILBERGELD et al., 1988). The deposition and storage of radioactive strontium in bone is known to increase the risk of osteosarcoma and other neoplasmas.

4.2 Blood-Brain Barrier

There are three anatomic and physiologic reasons why metal ions have a reduced capacity for entering the central nervous system (CNS). First, the capillary endothelial cells of the CNS are tightly joined, leaving only few or no pores between the cells. Second, the capillaries of the CNS are largely surrounded by glial cell processes (astrocytes), and third, the protein content of the interstitial fluid of the CNS is much lower than in other organs. Metal ions, therefore, have difficulties to gain access to the interstitial fluid of the brain unless there are special transport mechanisms such as for sodium, potassium, and calcium ions. These features act together as a protective mechanism to decrease the distribution of toxic metals to the CNS. Easily lipid-soluble metal compounds such as tetraethyllead and methylmercury can, however, readily enter the CNS, which has a special affinity for these compounds and, therefore represents the target organ of toxicity. Additionally, it should be noted that elemental mercury can also easily cross the blood-brain barrier.

The blood-brain barrier is not completely developed at birth, and this is one reason why newborns and infants are more susceptible to neurotoxic metals than adults. Lead probably represents the best-known example: effects of lead poisoning on the CNS, which are manifested as encephalopathy, have been observed more frequently and at lower lead levels in blood in young children than in adults. Subclinical neuropsychological deficits have been observed in young children with only moderately elevated lead levels in blood or tooth (RUTTER and JONES, 1983; BORNSCHEIN and RABINOWITZ, 1985).

4.3 Placental Transfer

A number of metals including arsenic, cadmium, lead, nickel, and methylmercury have been shown to cross the placenta and to enter the fetal blood circulation. Evidence originates from studies in humans and from animal experiments (SAGER et al., 1986). The major mechanism by which metal ions and metal compounds pass through the placenta seems to be passive diffusion. Lipid-soluble substances traverse more rapidly and attain a maternal-fetal equilibrium more rapidly than polar substances. Methylmercury, which easily crosses the placenta, is a good example. There is a highly significant correlation between mercury in maternal and newborn blood. Usually the mercury levels in newborn blood are higher than in maternal blood (TAKIZAWA, 1979; LAUWERYS et al., 1978). With regard to lead there also is a good correlation between the lead levels in maternal and cord blood, but the lead concentrations in cord blood are generally lower than in maternal blood indicating that the placenta exerts a certain barrier function towards lead (LAUWERYS et al., 1978). With regard to cadmium the placental transfer seems to be lower than for lead since the cadmium levels in newborn blood are, on the average, about 50% lower than in maternal blood (LAUWERYS et al., 1978; SCHIELE et al., 1985). Compared with maternal blood the placenta concentrates cadmium about 10-fold. Smokers have significantly higher cadmium levels in the placenta than non-smokers (BUCHET et al., 1978; LAUWERYS et al., 1978; ROELS et al., 1978).

5 Biotransformation

For a number of metals biotransformation including changes of the oxidation state, methylation processes, and cleavage of metal-carbon bonds plays an important role.

Examples of changes of the valence state are mercury and arsenic. Following inhalation, mercury vapor is transported from the lungs to the brain, where it penetrates the blood-brain barrier. In the brain it is oxidized to divalent mercury ions (Hg^{2+}), which are responsible for the neurotoxic effects of mercury vapor.

Trivalent inorganic arsenic (arsenite) is oxidized in vivo to pentavalent arsenic (arsenate) as can be shown by the finding of arsenate in urine. Also the opposite reaction, i.e., the reduction of arsenate to arsenite has been demonstrated in animals. Both arsenite and arsenate are methylated in vivo. In humans the urinary excretion of arsenic consists of about 20% inorganic arsenic, 20% methylarsonic acid, and 60% dimethylarsonic acid.

Demethylation is an important detoxification mechanism for methylmercury. Both in humans and animals ingested methylmercury is excreted mainly in inorganic form. Probably, the microflora of the gut plays a key role in the demethylation process. Since the resulting Hg^{2+} ions are not well absorbed in the gut the feces represent the predominant medium of excretion.

Dealkylation also plays an important role in the biotransformation of organotin and organolead compounds. Dealkylation is carried out in liver microsomes, probably by a cytochrome P-450 dependent monooxygenase. The resulting trialkyl lead compounds (R_3Pb^+) are relatively stable and are responsible for the major toxic effects of tetraalkyllead compounds (GRANDJEAN, 1984).

6 Excretion

As discussed in Section 1, many metals are not completely absorbed in the gut. The non-absorbed fraction is eliminated with the feces. Regarding the amount of metals which is absorbed in the gut or in the lungs, gastrointestinal and urinary excretion represents the most important pathways of excretion. Other routes of excretion are salivary excretion, exhalation, lactation, and loss of hair, nails, and teeth. These processes may be of special interest, e.g., with regard to exposure assessment, but they do not contribute significantly to the excretion of metals from the body. Exhalation only plays a role for volatile metal compounds and elemental mercury circulating in the blood.

6.1 Urinary Excretion

Urinary excretion is probably the most important excretory route for metals. The main excretory mechanisms in the kidney are glomerular filtration and active or passive tubular secretion. Compared with glomerular filtration, tubular secretion of metals seems to play only a minor role. For example, the urinary excretion of beryllium is considered to take place via tubular secretion.

The glomerular capillaries have large pores, which are permeable for molecules and ions ranging in size up to plasma albumin (mol. wt. 60000). Therefore, a compound will be filtered at the glomerulus unless its molecular weight exceeds about 60 000. Glomerular filtration is thus restricted to the 'diffusible fraction' in plasma and to metals bound to low molecular weight proteins such as metallothionein (mol. wt. 6500). Metal ions and metal compounds bound to blood cells and large proteins are not filtered.

Once filtered at the glomeruli, metals may be excreted with urine or may be passively reabsorbed into the tubular cells and into the blood stream. Passive reabsorption is of particular importance for lipid-soluble agents, whereas polar compounds and ions, unless there are special transport systems, are unable to diffuse across the tubular membranes and, therefore, will be excreted into the urine. Cadmium bound to metallothionein is also very efficiently reabsorbed in the renal tubules, and this is the reason for the efficient retention and accumulation of this metal in the body.

6.2 Gastrointestinal Excretion

Gastrointestinal excretion may occur via the gastrointestinal mucosa and also via salivary, biliary, and pancreatic secretion. In humans, the stomach and the intestine each excrete 3 liter of fluid per day. Foreign substances can be excreted with these fluids into the gastrointestinal tract. Experimental studies show that particularly biliary excretion is of significant importance for a number of metals. Since the cells of the intestinal mucosa have a rapid turnover, passive loss of metals bound to these cells may also represent a significant route of excretion when the cells are shed and excreted in feces.

Metals excreted via bile may be reabsorbed in the gut and consequently can become available for re-excretion in the bile (enterohepatic circulation). When enterohepatic circulation occurs – e.g., in the case of methylmercury or thallium – it is possible to increase the net gastrointestinal excretion of the metal by stopping intestinal reabsorption through binding of the metal excreted in bile to a compound that is not absorbable.
Resins binding methylmercury and Prussian blue binding thallium ions have been used in this way in the treatment of human poisoning (KAZANTZIS, 1986). Biliary excretion has been demonstrated for both inorganic and organic compounds of metals as shown in Table I.15-1.

References

Angerer, J., Heinrich-Ramm, R., and Lehnert G. (1989), *Occupational Exposure to Cobalt and Nickel – Biological Monitoring, Int. J. Environ. Anal. Chem.* **35** (2), 81–88.
Barry, P. S. I. (1978), *Distribution and Storage of Lead in Human Tissues*, in: Nriagu J. O. (ed.): *The Biogeochemistry of Lead in the Environment*, Vol. II, pp. 97–150. Elsevier/North-Holland Biomedical Press, Amsterdam-New York-Oxford.

Bornschein, R. L., and Rabinowitz, M. B. (eds.) (1985), *Second International Conference on Prospective Studies on Lead, Cincinnati, Ohio,* April 9–11, 1984. Environ. Res. 33, 1–210.
Bowden, D. H. (1984), *The Alveolar Macrophage.* Environ. Health Perspect. 55, 327–342.
Brain, J. D., and Vahlberg, P. A. (1979), *Deposition of Aerosols in the Respiratory Tract.* Am. Rev. Respir. Dis. 120, 1325–1373.
Brain, J. D., Cuddihy, R. G., Dahl, A. R., Mauderly, J. L., Oberdörster, G., and others (1987), *Proceedings Design and Interpretation of Inhalation Studies and Their Use in Risk Assessment.* Universität Hannover, FRG, and ILSI, Washington, D.C.
Bremner, I. (1979), *Mammalian Absorption, Transport and Excretion of Cadmium,* in: Nriagu, J. O. (ed.): *The Biogeochemistry of Cadmium in the Environment,* pp. 175–194. Elsevier/North-Holland Biomedical Press, Amsterdam-New York-Oxford.
Buchet, J. P., Lauwerys, R., Roels, H., and Hubermont, G. (1977), *Mobilization of Lead During Pregnancy in Rats.* Int. Arch. Environ, Health 40, 33–36.
Buchet, J. D., Roels, H., Hubermont, G., and Lauwerys, R. (1978), *Placental Transfer of Lead, Mercury, Cadmium, and Carbon Monoxide in Women. II. Influence of Some Epidemiological Factors on the Frequency Distributions of the Biological Indices in Maternal and Umbilical Cord Blood.* Environ. Res. 15, 494.
Camner, P., Clarkson, T. W., and Nordberg, G. F. (1986), *Routes of Exposure, Dose and Metabolism of Metals,* in: Friberg, L., Nordberg, G. F., and Vouk, V. B. (eds.): *Handbook on the Toxicology of Metals,* Vol. I, pp. 85–127. Elsevier, Amsterdam-New York-Oxford.
Chamberlain, A. C. (1983) *Effect of Airborne Lead on Blood Lead.* Atmos. Environ. 17, 677–692.
Drasch, G., Böhm, J., and Baur, C. (1987), *Lead in Human Bones. Investigations on an Occupationally Non-exposed Population in Southern Bavaria.* Sci. Total Environ. 64, 303–315.
Grandjean, P. (ed.) (1984), *Biological Effects of Organolead Compounds.* CRC Press, Boca Raton, Florida.
Kazantzis, G. (1986), *Diagnosis and Treatment of Metal Poisoning-General Aspects,* in: Friberg, L., Nordberg, G. F., and Vouk, V. B. (eds.): *Handbook on the Toxicology of Metals,* Vol. I, pp. 294–318. Elsevier Amsterdam-New York-Oxford.
Klaassen, C. O. (1986), *Distribution, Excretion, and Absorption of Toxicants,* in: Klaassen, C. O., Amdur, M. O., and Doull, J. (eds.): *Casarett and Doull's Toxicology – The Basic Science of Poisoning,* 3rd Ed., pp. 33–63. Macmillan Publishing Company, New York.
Kollmeier, H., Seemann, J., Müller, K.-M., Schejbal, V., Rothe, G., Wittig, P., and Hummelsheim, G. (1988), *Assoziationen zwischen lungengeweblichen Chrom- und Nickelgehalten und Karzinomen der Lunge.* Prax. Klin. Pneumol. 42, 142–148.
Lauwerys, R., Buchet, J. P., Roels, H., and Hubermont, G. (1978), *Placental Transfer of Lead, Mercury, Cadmium, and Carbon Monoxide in Women. I. Comparison of the Frequency Distributions of the Biological Indices in Maternal and Umbilical Cord Blood.* Environ. Res. 15, 278–289.
Menzel, D. B., and Amdur, M. O. (1986), *Toxic Responses to the Respiratory System,* in: Klaassen, C. O., Amdur, M. O., and Doull, J. (eds.): *Casarett and Doull's Toxicology – The Basic Science of Poisoning,* 3rd Ed., pp. 330–358. Macmillan Publishing Company, New York.
Mercer, T. T. (1973), *Aerosol Technology in Hazard Evaluation.* Academic Press, New York-London.
Nordberg, G. F., Parizek, J., Pershagen, G., and Gerhardsson, L. (1986), *Factors Influencing Effects and Dose-response Relationships of Metals,* in: Friberg, L., Nordberg, G. F., and Vouk, V. B. (eds.): *Handbook on the Toxicology of Metals,* Vol. I, pp. 175–205. Elsevier, Amsterdam-New York-Oxford.
Oberdörster, G. (1989), *Assessment of Lung Cancer Risk from Inhaled Environmental Cadmium,* Toxicol. Environ. Chem. 23 (1–4), 41–51.
Phalen, R. F. (1984), *Inhalation Studies: Foundations and Techniques.* CRC Press, Boca Raton, Florida.
Roels, H., Hubermont, G., Buchet, J. P., and Lauwerys, R. (1978), *Placental Transfer of Lead, Mercury, and Carbon Monoxide in Women. III. Factors Influencing the Accumulation of Heavy*

Metals in the Placenta and the Relationship between Metal Concentrations in Placenta and in Maternal and in Cord Blood. Environ. Res. 86, 236.

Rosen, J. F., and Sorell, M. (1978), *The Metabolism and Subclinical Effects of Lead in Children,* in: Nriagu, J. O. (ed.): *The Biogeochemistry of Lead in the Environment,* Vol. II, pp. 151–172. Elsevier/North Holland Biomedical Press, Amsterdam-New York-Oxford.

Rutter, M., and Jones, R. R. (eds.) (1983), *Lead versus Health – Sources and Effects of Low Level Lead Exposure.* J. Wiley & Sons, Chichester-New York-Brisbane-Toronto-Singapore.

Sabbioni, E., Goetz, L., and Bignoli, G. (1984), *Health and Environmental Implications of Trace Metals Released from Coal-fired Power Plants: An Assessment Study of the Situation in the European Community. Sci. Total Environ. 40*, 141–154.

Sager, P. R., Clarkson, T. W., and Nordberg, G. F. (1986), *Reproductive and Developmental Toxicity of Metals,* in: Friberg, L., Nordberg, G. F., and Vouk, V. B. (eds.): *Handbook on the Toxicology of Metals,* Vol. I, pp. 391–434. Elsevier, Amsterdam-New York-Oxford.

Schiele, R., Glatzel, I., and Schaller, K. H. (1985), *Die usuelle Cadmium-Belastung von Müttern und ihren Neugeborenen in verschiedenen Regionen Bayerns. Zentralbl. Bakteriol. Mikrobiol. Hyg. I. Abt. Orig. B 181,* 295–308.

Silbergeld, E. K., Schwartz, J., and Mahaffey, K. (1988), *Lead and Osteoporosis: Mobilization of Lead from Bone in Postmenopausal Women. Environ. Res. 47,* 79–94.

Snee, R. D. (1981), *Evaluation of Studies of the Relationship Between Blood Lead and Air Lead. Int. Arch. Occup. Environ. Health 48,* 219–242.

Steenhout, A., and Pourtois, M. (1981), *Lead Accumulation in Teeth as a Function of Age with Different Exposures. Br. J. Ind. Med. 38,* 297–303.

Stuart, B. O. (1984), *Deposition and Clearance of Inhaled Particles. Environ. Health. Perspect. 55,* 369–390.

Takizawa, Y. (1979), *Epidemiology of Mercury Poisoning,* in: Nriagu, J. O. (ed.): *The Biogeochemistry of Mercury in the Environment,* pp. 325–366. Elsevier, North-Holland Biomedical Press, Amsterdam-New York-Oxford.

Task Group on Lung Dynamics (1966), *Deposition and Retention Models for Internal Dosimetry of the Human Respiratory Tract. Health Phys. 12,* 173–207.

Task Group on Metal Accumulation (1973), *Accumulation of Toxic Metals with Special Reference to their Absorption, Excretion and Biological Half-times. Environ. Physiol. Biochem. 3,* 65–107.

Task Group on Metal Interaction (1978), *Factors Influencing Metabolism and Toxicity of Metals: A Consensus Report. Environ. Health Perspect. 25,* 3–41.

Vahter, M., Friberg, L., Rahnster, B., Nygren, A., and Nolinder, P. (1986), *Airborne Arsenic and Urinary Excretion of Metabolites of Inorganic Arsenic among Smelter Workers. Int. Arch. Occup. Environ. Health 57,* 79–91.

Vouk, V. B., and Piver, W. T. (1983), *Metallic Elements in Fossil Fuel Combustion Products: Amounts and Form of Emissions and Evaluation of Carcinogenicity and Mutagenicity. Environ. Health Perspect. 47,* 201–225.

WHO (World Health Organization) (1980), *Recommended Health-based Limits in Occupational Exposure to Heavy Metals. Tech. Rep. Ser. 647.* WHO, Geneva.

Wolff, R. K. (1986), *Effects of Airborne Pollutants on Mucociliary Clearance. Environ. Health Perspect. 66,* 223–237.

I.16 Acute Metal Toxicity in Humans

MARIKA GELDMACHER-VON MALLINCKRODT
Erlangen, Federal Republic of Germany

1 General
(see also Chapter I.20a)

Acute metal poisoning is rarely observed in clinical practice. However, in the case of an unknown illness, differential diagnosis must take it into account. Metal poisoning produces no specific symptoms and only rarely are tests done to detect toxic metals. The result is that metal poisoning is repeatedly not recognized and the actual number of cases may be far higher than those reported.

The majority of metal poisonings are usually the result of suicide attempts. Occasionally, errors, such as in the laboratory or clinic, contaminated food, and in rare cases, medical treatment, can lead to intoxication.

The acute toxic effects of metals cannot be considered as isolated phenomena, but rather as a part of the complete spectrum of activity and/or dose-activity relationship of a metal in a biological system. WILLIAMS (1981 a) demonstrated this for an essential and a non-essential metal (Fig. I.16-1). From this, one can see that for essential metals, too little intake, as well as too much, can lead to illness and death.

Despite fluid boundaries between sub-toxic and toxic levels, acute poisonings due to the intake of a single high dose show in practice most striking symptoms which develop suddenly. If the poison cannot be removed or inactivated quickly, irreversible organ and systemic changes, which can be fatal, usually occur. In contrast, chronic poisoning develops gradually as a result of repeated intake of relatively small, but still toxic, doses. This can lead to irreversible damage and death as well.

The symptoms of acute and chronic metal poisoning can be completely different. For example, acute mercury poisoning (see also Chapter II.20) through oral ingestion of large amounts of a mercury salt such as mercury isocyanate leads to intense nausea, vomiting and diarrhea, and possibly death from shock within the first 24 to 36 hours. Chronic poisoning caused by the same compound, however, damages primarily the nervous system, followed by the kidneys.

The physical or chemical form (the speciation) of the element also affects toxicity (see also Chapter I.20a). There is also a linguistic problem when speaking of metal toxicity, since these elements practically never exist in the metallic form in living organisms. In fact, we rather discuss effects of metal ions, metal complexes, or metal compounds. For instance, metallic mercury, inorganic mercury ions, and organic mercury compounds show completely different spectra of effects (see Chapter II.20). Orally ingested metallic mercury is largely non-toxic. In contrast, a one-time inhalation of a high concentration of mercury vapor leads to lethargy, followed by

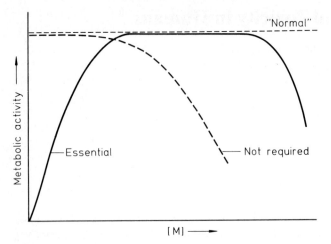

Fig. I.16-1. Effect of an essential and a non-essential metal on the metabolic activity of a cell. "Normal" indicates that the cell is functioning in its physiological range (WILLIAMS, 1981a). [M] is the metal concentration.

restlessness, nausea, diarrhea, a metallic taste in the mouth, coughing, tachypnea and possibly respiratory arrest. Histological signs are eroded bronchial tubes, bronchiolitis accompanied by interstitial pneumonia, gastroenteritis, colitis, and kidney damage. These symptoms can be accompanied by disturbances of the central nervous system (CNS) such as tremor and increased excitability.

2 Uptake and Distribution

Acutely toxic amounts of metals and their compounds are usually taken in through the mouth or lungs. In addition to the dose, the method of uptake also determines the intensity as well as the duration of toxic effects and can lead to very different symptoms (see the above example for mercury and its compounds). The method of transport of the metal compound through membranes into the blood or cell interior can also influence the toxic effects (see Chapter I.11).

Oral uptake often results in vomiting, which reduces the amount of toxin that can be absorbed. Metal compounds can react with either the acid in the stomach or the alkaline environment of the intestine. This can decidedly influence solubility. Before being distributed throughout the body, the metals pass the liver, where often detoxification processes begin.

Inhalation of fine particles can result in a direct and rapid transfer of soluble metal compounds into the blood. This leads to a rapid distribution and onset of symptoms. Toxic quantities of metals can also be absorbed through the skin and mucous membranes. Injection has been reported very rarely.

3 Biological Mechanisms Responsible for Effects

While some groups of toxins, such as organophosphates or digitalis-type drugs, have specific and clearly defined targets in the organism, metals in an acutely toxic amount possess no uniform pattern of action (see also WEBB, 1977).

WILLIAMS (1981 b) provided a general overview of the role of metals in biological systems, from which much can be drawn about the toxicity dynamics of metals. Biological systems can utilize solar energy to alter the relations of the different metals in their bodies so that these are different from the relations present in the earth's crust. The reason for this is the pressure of natural selection, which allows only those species with highly effective and optimized biochemistry to survive. Natural selection has led to an almost optimal utilization of the various metals. Specialization of metal functions was possible in as far as it occurred at the same time as the evolution of proteins. The movement of elements was determined by specific electromechanical and electrochemical events.

WILLIAMS (1982, 1983, 1985) has extensively discussed the chemical selectivity of protein side-chains and small inorganic molecules for metal ions. For example, electrostatic interactions, which are dependent upon ion size and charge, play a significant role. Also important is the electronic affinity of the ion, or in other words, the acidity of the corresponding Lewis acids. In general, one can say that cations of the Group IA or IIIA elements bind only to oxygen donors, while increasingly in the direction of the transition metals and B-Subgroups, binding to nitrogen or sulfur donating groups takes place (see also Chapter I.3). Polarizability leads to stereo-chemical consequences.

Anions behave reciprocally to cations, and the degree of hydrogen bonding becomes important. The regulation mechanisms of biological systems for maintenance of homeostasis are able to adjust to small changes in the concentration of metal compounds for short periods of time. Selection and adaptation to concentration changes occurring over longer periods of time also take place. However, sudden, significant concentration changes, and their results, caused by acute intake of an overdose of metal ion cannot be counteracted and lead to possibly fatal disturbances in the organism. Symptoms of acute poisoning appear as a result of changes in the molecular structure of proteins, breaking of hydrogen bonds, inhibition of enzymes, changes in potential, and so forth. Not only those metal compounds which are "foreign" to the system can have toxic effects, but also those which are essential, if present in large quantities (see Fig. I.16-1).

Despite these general considerations, up until now the unusual diversity of symptoms observed in the intact organism could not be explained adequately (see Part II). The effects of heavy metals manifest themselves in very different tissues, partially with a rather high specificity for certain metals and their compounds. These organ-specific effects could not be explained.

4 Quantitative Assessment of the Toxicity of Individual Metal Compounds

(see also NORDBERG, 1976; FRIBERG et al., 1986)

The toxicity of a compound can be characterized by its LD_{50}, i.e., the dose which is fatal to one-half of a population of experimental animals. In order to compare the LD_{50} values of two metals, additional information must be available. Important factors are the chemical form in which the element is present (oxidation state, inorganic or organic bonds), the way of uptake (oral, intravenous, intraperitoneal, inhalation), the type of animal, age, stage of development, and the time interval between intake and death. The basic conditions must be comparable. LUCKEY et al. (1975) indicated that comparison of the toxicity of metal compounds is more useful when the LD_{50} value is expressed in mmol/kg instead of the usual mg/kg. Metals with small differences in atomic mass can show large variances in specific gravity, which influences the toxicity. For example, tungsten and metavanadate are equally toxic if the LD_{50} is expressed in mmol/kg, but vanadate is three times as toxic as tungsten when the LD_{50} is expressed in mg/kg.

If one considers the toxicity of a metal ion with regard to its location in the Periodic System, a pattern can be seen (LUCKEY and VENUGOPAL, 1977). Toxicity decreases with an increase in the stability of the electron configuration. Metal ions of the Group IA and IIA elements are highly electropositive. These metal ions appear in the biological environment primarily as free cations. The toxicity of the Subgroups IA and IIA increases with increasing atomic number:

IA: $Na < K < Rb < Cs$

IIA: $Mg < Ca < Sr < Ba$

The lighter elements lithium and beryllium are less electropositive, but, nevertheless, more toxic than the other members of their groups. They have a smaller ionic radius and a higher charge to mass ratio.

Also in the Subgroups IB, IIB, IIIA, the acute toxicity of the metal ion increases with the electropositivity:

IB: $Cu < Ag < Au$

IIB: $Zn < Cd < Hg$

IIIA: $Al < Ga < In < Tl$

This increase in toxicity can be explained by the increasing affinity of these elements for amino, imino, and sulhydryl groups, which form the active centers of a number of enzymes. The metals of the sixth Period and their compounds are potentially the most toxic elements of the Periodic System. The generally poor water-solubility of their salts, however, often masks their inherent high degree of toxicity. This toxicity becomes apparent in those lead, mercury, and thallium salts that are relatively soluble. The metallic ions of the fourth Period form mostly covalent bonds and complexes with biological ligands and some form hydroxyacids in which the metal is part of the anion.

In addition to the electrochemical character and the solubility of a metal and its compounds, which influence its bioavailability, the various oxidation states of an element are also important. For example, manganese(VII) compounds such as permanganate are more toxic than manganese(II) compounds, and arsenic(III) oxide is more toxic than arsenic(V) oxide.

The following additional factors are significant for the toxicity of a compound (LUCKEY et al., 1975):

a) The extent of resorption, e.g., from the gastrointestinal or respiratory system.
b) The particle size of the metal or metal compound (especially important for inhalation toxicity).
c) The distribution through the blood to the various organs.
d) The extent of and the route of excretion, as well as the influence of metabolism and detoxification processes.
e) Storage in the cells in the form of harmless particles.
f) The efficiency of the mechanisms that control the absorption, excretion, distribution, and retention of toxic metals or compounds.
g) The concentration of metal compounds in the organs, which is influenced by the physical form in which the metal is present (i.e., as an ion, colloid, or hydrate).
h) The influence of the pH of body fluids and organs on the hydrolysis of heavy metal salts as well as their solubility, reactivity, and the toxicity of the hydrolysis products.
i) The ability of the metal to chelate ligands of biological macromolecules and other tissue components, as well as the stability of these chelates.
j) The ability of the toxic metal to react with other metals, or to suppress or activate essential metals.
k) The ability of other metals or body compounds to increase or reduce the toxicity of a metal.

5 Symptoms of Acute Metal Intoxication

The most common symptoms of acute metal poisoning can be classified as follows (KAZANTZIS, 1979):

a) Gastrointestinal symptoms: oral ingestion of large quantities of soluble metal salts leads relatively quickly to gastroenteritis (irritation of the stomach and intestinal mucosa). The result is nausea, vomiting, abdominal pain, diarrhea, and possibly shock due to dehydration and loss of electrolytes. An example is arsenic poisoning.
b) Damage to the respiratory tract: inhalation of metals or metal compounds can lead to pulmonary edema (swelling of the lungs through accumulation of fluid). One must also distinguish between dusts, smoke, and metallic chlorides that lead to production of hydrochloric acid.
c) Cardiovascular effects: arrhythmia (irregular heartbeat), low blood pressure, and shock.

d) Effects on the central nervous system: cramps, coma, death.
e) Kidney damage with oliguria (lowered urine production). Anuria (lack of urine production) is often the result of tubule necrosis (severe kidney damage).
f) Damage to the blood or blood-producing organs: hemolytic anemia after inhalation of arsenic hydride or ingestion of copper salts.
g) Metal fume fever as an immunological reaction to the inhalation of metallic oxide aerosols (e.g., zinc oxide smoke).

The symptoms of individual metal poisonings can be found in standard reference books, for example, LUCKEY and VENUGOPAL (1977), VENUGOPAL and LUCKEY (1978), FRIBERG et al. (1979, 1986), BROWN and KODAMA (1987), and SEILER et al. (1988), as well as in the individual chapters of this book.

References

Brown, S.S., and Kodama, Y. (eds.) (1987), *Toxicology of Metals*. John Wiley, New York-Chichester-Brisbane-Toronto.
Friberg, L., Nordberg, G.F., and Vouk, V.B. (eds.) (1979), *Handbook on the Toxicology of Metals*. Elsevier/North Holland, Amsterdam-New York-Oxford.
Friberg, L., Nordberg, G.F., and Vouk, V.B. (eds.) (1986), *Handbook on the Toxicology of Metals*, 2nd rev. Ed. Elsevier/North Holland, Amsterdam-New York-Oxford.
Kazantzis, G. (1979), in: Friberg, L., Nordberg, G.F., and Vouk, V.B. (eds.), *Handbook on the Toxicology of Metals*, p. 219. Elsevier/North Holland, Amsterdam-New York-Oxford.
Luckey, T.D., and Venugopal, B. (1977), *Metal Toxicity in Mammals*, Vol. 1, pp. 7, 105. Plenum Press, New York-London.
Luckey, T.D., Venugopal, B., and Hutcheson, D. (1975), *Heavy Metal Toxicity, Safety, Homology*, pp. 61, 62. Thieme, Stuttgart; Academic Press, New York- San Francisco-London.
Nordberg, G.F. (ed.) (1976), *Effects and Dose-Response Relationships of Toxic Metals*. Elsevier, Amsterdam-Oxford-New York.
Seiler, H.G., Sigel, H., and Sigel, A. (eds.) (1988), *Handbook on Toxicity of Inorganic Compounds*. Marcel Dekker, New York-Basel.
Venugopal, B., and Luckey, T.D. (1978), *Metal Toxicity in Mammals*, Vol. 2. Plenum Press, New York-London.
Webb, M. (1977), in: Brown, S. (ed.), *Clinical Chemistry and Chemical Toxicology of Metals*, p. 51. Elsevier/North Holland, Amsterdam-New York-Oxford.
Williams, R.J.P. (1981a), *Physico-chemical Aspects of Inorganic Element Transfer through Membranes*. Philos. Trans. R. Soc. London B *294*, 57–74.
Williams, R.J.P. (1981b), *Natural Selection of the Chemical Elements*. Proc. R. Soc. London B *213*, 361–397.
Williams, R.J.P. (1982), *Metal Ions in Biological Catalysts*. Pure Appl. Chem. *54*, 1889–1904.
Williams, R.J.P. (1983), *Inorganic Elements in Biological Space and Time*. Pure Appl. Chem. *55*, 1089–1100.
Williams, R.J.P. (1985), *Homeostasis: An Outline of the Problems*. TEMA *5*, 300–306.

I.17a Chronic Toxicity of Metals and Metal Compounds

ULRICH EWERS and HANS-WERNER SCHLIPKÖTER
Düsseldorf, Federal Republic of Germany

1 Introduction

Human exposure to metals and metal compounds in the general environment is predominantly through food, drinking water, and beverages. For smokers inhalation to metal compounds that occur in tabacco smoke is of additional importance. Other forms of uptake, by inhalation or via skin contact, may occur under special circumstances (see Part II). Every day and generally over the major part of the life-span, small doses, usually in the microgram range, are taken up, each of which is insufficient to produce signs of overt toxicity in the individual. However, over a long period of time, adverse effects may occur as a result of long-term, low-level exposure.

In the work environment, human exposure to metals and metal compounds is mainly by inhalation. Volatile metals and metal compounds such as mercury and tetraethyl lead, occur in vapor form, whereas less volatile metals and metal compounds usually are present as constituents of aerosols and suspended dust in the workroom air. Similar to the general environment, the normal every-day exposure usually does not cause detectable effects in the individual, but prolonged exposure over longer periods of time may produce a variety of toxic effects including damage of specific organs, cancer or allergic reactions.

As pointed out in other chapters, especially in Part II, speciation is important and has an influence on bioavailability, distribution in the organism, metabolism, and effects. As in the cases of arsenic, mercury, and chromium it must thus be made clear in which oxidation state a metal occurs, when toxicity is discussed. It is also necessary to distinguish between inorganic forms, organometallic compounds, and organometallic ions, e.g., tetraethyl tin, tetramethyl lead, trimethyl lead ions, dimethyl mercury, and methyl mercury ions.

Since low-level, long-term exposure is the most common way of exposure to environmental chemicals, in the general environment as well as in the work environment, the possible chronic effects of these agents are of great concern to toxicologists, regulatory officials, and the general public. This chapter tries to present an overview on the chronic toxicity of metals and metal compounds.

2 General Principles Underlying Chronic Toxicity
(see also Section 5)

Toxic effects may be produced by acute or chronic exposure to chemical agents. Acute exposure is defined as a single or multiple exposure occurring within a short time (24 hours or less). Chronic exposure is defined as daily or otherwise repeated exposure over long periods of time, e.g., over the working life-time or the entire life-span. For many agents the toxic effects of acute exposure are quite different from those produced by chronic exposure. In general, acute exposure to agents that are rapidly absorbed produces immediate toxic effects. In some cases, however, acute exposure can also produce delayed toxicity, i.e., toxic effects occurring several hours or days after exposure. Long-term, low-level exposure usually does not produce immediate toxic effects, but after a certain time signs of toxicity may become apparent.

In general, chronic toxic effects occur when the agent accumulates in the biological system (absorption exceeds metabolism and/or excretion) or when an agent produces irreversible toxic effects or when there is insufficient time for the system to recover from the toxic effect within the exposure frequency interval. When the elimination rate is less than the absorption rate the agent usually does not accumulate indefinitely, but reaches a steady state where the rate of elimination equals the rate of absorption.

The way, by which metals and metal compounds exert chronic toxic effects, is only partially understood. The most important mechanisms appear to be those of accumulation and those resulting from irreversible toxic effects.

An example of chronic toxic effects produced by accumulation of a metal in a specific organ is the nephrotoxicity of inorganic cadmium and mercury ions, which accumulate in the kidneys. The accumulation may be without effect on the functional status of the organ, but when a certain concentration is reached or exceeded the functional status or capacity of the organ is affected (kidney damage by mercury, however, may also be produced via the immune system). The concentration, at which adverse functional changes, reversible or irreversible, occur, generally is called the "critical organ concentration".

Accumulation of metals in tissues does not necessarily imply the occurrence of toxic effects. In the case of some metals, inactive complexes or storage depots are formed. Lead, for example, is stored in bones and teeth in an inert form. As discussed in Chapter I.12, cadmium and some other metals bind to a cysteine-rich, low-molecular-weight protein, metallothionein, which likewise appears to represent an inert storage form. It is supposed that kidney damage is prevented as long as the tubular cells can produce enough metallothionein to bind toxic cadmium ions. If this capacity is exhausted, free cadmium ions occur in the cells and become available for toxic actions.

In the case of some other metals, chronic toxicity seems to result from irreversible toxic effects (e.g., the action of methyl mercury in the central nervous system). Although the underlying biological mechanisms are not yet understood, the carcinogenic action of some metals and metal compounds appears also to result from such effects. Other examples of apparently irreversible effects are the toxic effects of low-level lead exposure on the immature brain and the causation of allergic reactions resulting from previous sensitization to certain metals or metal compounds, e.g., chromates, nickel, and hexachloroplatinates. For other prenatal effects of metal compounds see, e.g., DAVIS and SVENDSGAARD (1987), MARSH et al. (1981, 1987), Chapter I.18, and specific Chapters in Part II.

3 Spectrum of Chronic Toxic Effects

The toxic actions of metals and metal compounds normally involve a multiplicity of target organs and systems. In no case, the multiple manifestations of toxicity can be assigned to a particular biochemical or biophysical mechanism such as inhibition of one or several enzymes or interaction with specific receptors or ligands. However, it must be admitted that we know little about earliest effects of low doses on molecular targets, and it is, for instance, believed that α-lipoic acid may be a unique primary target for trivalent arsenic.

According to the site of action toxic effects can be divided into two groups: local and systemic effects. *Local effects* refer to those that occur at the site of first contact between the biological system and the toxic agent. Depending on the route and circumstances of exposure the gastrointestinal tract, the respiratory organs, the skin, and the eyes can be affected. Gastrointestinal effects such as anorexia, nausea, vomiting, and diarrhea followed by constipation may occur as a result of repeated ingestion of toxic metal compounds over a period of time. Intestinal colics have been observed in industrial workers and in children with relatively low-level lead exposure. Chronic pulmonary disorders such as toxic and allergic pulmonary diseases and respiratory cancer result from inhalation exposure to metals or metal compounds (e.g., arsenic and beryllium compounds, cadmium oxide, chromates, nickel and nickel compounds, hexachloroplatinates). Other local toxic effects include allergic skin reactions induced, e.g., by nickel and chromates.

Systemic effects require absorption and distribution of the toxic agent to a site distant from its entry point. Metal compounds may produce a variety of systemic effects at different sites of the organism. Usually, the major toxic effects are found to occur in one or two organs. These organs generally are called the *target organs of toxicity* for that chemical. The most prominent target organs of metals and their compounds include the nervous system (see, for instance TILSON and FOULKES, 1987; WHO, 1986; BONDY and PRASAD, 1987), the hemopoietic system, and the kidneys. The target organ of toxicity is not always the site of the highest concentration of the metal. For example, lead is concentrated in the bones, but its chronic toxicity is mainly directed to the hemopoietic and the central nervous system.

For some substances both local and systemic toxic effects can be observed. For example, tetraethyl lead produces effects on the skin at the site of absorption and is then transported systemically to produce its typical effects mainly on the central nervous system.

4 Critical Concentration, Critical Organ, and Critical Effect

In preventive medicine and in setting environmental and occupational health standards it is of particular importance to identify early effects. If these can be prevented, later, perhaps more severe, effects can be avoided. On this background, the terms critical concentration, critical organ, and critical effect have been established, especially based on experiences with the nephrotoxic action of cadmium (FRIBERG and KJELLSTRÖM, 1981; KJELLSTRÖM et al., 1984; FRIBERG, 1986). The terms critical concentration, critical organ, and critical effect are defined as follows:

- *Critical concentration of a cell:* the concentration at which adverse functional changes, reversible or irreversible, occur in the cell.
- *Critical organ concentration:* the mean concentration in the organ at the time any of its cells reaches the critical concentration.
- *Critical organ:* the particular organ which first attains the critical concentration of a metal under specified circumstances of exposure and for a given population.
- *Critical effect:* an adverse functional change which occurs when the critical concentration is reached in the critical organ.

The critical concentration varies between individuals and also appears to depend on age. The dose-response relationship, expressing the occurrence rate (response) of a particular effect as a function of metal concentration in the critical organ thus displays the frequency distribution of individual critical concentrations. In order to give an indication of this dose-response relationship, FRIBERG and KJELLSTRÖM (1981) suggested the introduction of the term *"population critical concentration"* (PCC) when evaluating population-based data. A PCC-50 would be the concentration in the critical organ at which 50% of the population exhibits the critical effect; i.e., they have exceeded their individual critical concentrations. The concept of PCC has been developed, in particular, with respect to the nephrotoxic effects of cadmium (KJELLSTRÖM et al., 1984). Based on a reanalysis of published data the PCC-10 (10% response rate) for cadmium in kidney cortex was estimated to be in the range of 180–220 µg/g and the PCC-50 to be about 250 µg/g. The concept of critical concentration has been used in risk estimations to derive "acceptable" metal concentrations in the critical organ. Using metabolic models some authors have tried to calculate the daily doses, which would lead to the critical concentration within a certain period of time (KJELLSTRÖM and NORDBERG, 1978). On the basis of these data acceptable concentrations of metals in food and water may be derived.

Since the metal concentration in the critical organ is difficult to measure in vivo, several studies have also tried to measure the critical effects in relation to the metal concentrations in indicator media such as blood, urine, and hair. For example, the critical concentration of cadmium in urine, at which the occurrence rate of tubular proteinuria starts to increase, has been estimated to be about 10 µg/g creatinine in middle-aged workers (ROELS et al., 1981b). In aged persons, this critical concentration seems to be much lower (ROELS et al., 1981a). In the case of methyl-mercury poisoning, the critical concentration of mercury in blood, at which the occurrence rate of paresthesia starts to increase, was found to be about 300 µg/L (WHO, 1976).

In such studies the term "earliest effect-level" has also been used instead of "critical concentration".

5 Evaluation of Chronic Toxicity

There are two main principles to evaluate the chronic toxicity of chemicals to humans: (1) descriptive animal toxicity testing and (2) toxicity studies in humans on the basis of controlled human studies, case reports, and epidemiological studies.

Descriptive animal toxicity testing usually consists of three steps: acute, subchronic, and chronic toxicity testing (WHO, 1978).

Acute toxicity has been defined as the adverse effects resulting from the administration of a single dose or multiple doses given within a short time (24 h or less). Acute toxicity tests are conducted to evaluate the relative toxicity of a compound, to investigate its mode of action and its specific toxic effects, and to determine the existence of species differences.

Subchronic toxicity tests generally involve daily or otherwise repeated exposure to a substance over a period of about 90 days. Such tests are performed to get information on the major toxic effects of the test substance, its toxicokinetic behavior, the target organs affected, and the reversibility of the observed effects.

The long-term or *chronic exposure studies* are performed similarly to the subchronic exposure studies except that the period of exposure is longer. Chronic exposure studies are often conducted with the aim of establishing "no-observed-effect levels" that may be used in setting acceptable daily intakes (ADIs), tolerance limits for chemicals in food or water, or occupational health standards. Moreover, such studies are used to determine the carcinogenic potential of a chemical.

The principles of the standard toxicity testing protocol (experimental design, evaluation and interpretation of test results) cannot be discussed in this chapter. Guidelines for toxicity testing of chemicals (including ecotoxicity, and short-term and long-term toxicity testing) have been recommended by the Organization for Economic Co-operation and Development (OECD, 1977) and the World Health Organization (WHO, 1978, 1986). Likewise, guidelines for good laboratory practice (GLP) have been developed (OECD, 1977).

The chronic toxicity of a chemical in humans can be evaluated from three types of study: (1) case reports, including a history of exposure to the toxic agent to be studied; (2) descriptive epidemiological studies, in which the incidence of various diseases in human populations is found to vary (spatially or temporally) with exposure to the agent; (3) analytical epidemiological studies (e.g., case-control or cohort studies), in which individual exposure to the agent is found to be associated with an increased risk for certain diseases. Since man usually is exposed simultaneously to several chemicals and numerous other risk factors, including genetic factors, it is frequently difficult to establish unequivocally the degree of hazard associated with any single chemical. This is particularly true with respect to the identification of the effects of low-level, long-term exposures. Considering the varying incidences of various diseases in the population, only significantly elevated relative risks can be identified by epidemiological studies. Another problem is that many of these effects, e.g., cancer, renal diseases and cardiovascular diseases per se are associated with increasing age. Many of these diseases are multicausal in nature, and exposure to toxic agents, both natural or synthetic, in the environment is only one of several potential risk factors.

6 Occurrence of Chronic Intoxications and Other Signs of Chronic Toxicity

Signs of chronic toxicity of metals and metal compounds are most frequently observed as a result of long-term *occupational exposures* to toxic metals, e.g., in ore mining, primary and secondary smelting, pigment production, and manufacturing and handling of specific chemicals such as arsenic containing insecticides. In many countries diseases that are related to occupational metal exposures belong to the compensable occupational diseases entitling the victim for compensation. In the Federal Republic of Germany, e.g., the list of occupational diseases includes diseases caused by lead, mercury, chromium, cadmium, manganese, thallium, vanadium, arsenic, beryllium, and their toxic compounds. In 1925, the International Labour Office (ILO) established an international list of occupational diseases. The first list covered only three occupational diseases and was amended several times. The present list (amended list of ILO Convention No. 121 (1980)) contains 29 groups of occupational diseases including diseases caused by beryllium, cadmium, chromium, manganese, arsenic, mercury, and lead, and their toxic compounds (ILO, 1983).

Signs of chronic toxicity resulting from *environmental exposure* to metals and toxic metal compounds were observed in relatively large population groups in a number of countries. The predominant pathways of exposure were by ingestion of contaminated food or drinking water. In some of these episodes several thousand people were affected.

In Japan, excessive exposure to cadmium (see also Chapter II.6) of the general population occurred by ingestion of contaminated rice and other foodstuffs. In the areas concerned (parts of the Toyama prefecture) a high incidence of a painful type of osteomalacia with multiple fractures appearing together with kidney dysfunction (involving proteinuria, glucosuria, amino aciduria, and phosphaturia) was noted, particularly among elderly women (*Itai-Itai disease*). Although the pathogenesis of the disease is still under dispute (see, for instance, NOMIYAMA et al., 1981), it is generally accepted that cadmium was a necessary etiological factor (TSUCHIYA, 1978). Autopsy studies of Itai-Itai patients showed that liver cadmium levels were very high, whereas renal cortex cadmium levels were low indicating a significant loss of cadmium from the kidney as a result of renal dysfunction (NOGAWA et al., 1986). The occurrence of the Itai-Itai disease in certain areas of Japan seems to constitute only the top of an iceberg (FRIBERG, 1984). The cadmium levels in human blood and tissue among Japanese are considered to be the highest in the world (KJELL-

STRÖM, 1979; VAHTER, 1982), and an increased prevalence of proteinuria was found in a number of areas where exposure to cadmium through consumption of rice was also high. The health significance of cadmium-induced proteinuria is still unclear, and the question remains whether this renal disturbance must be considered as an adverse health effect. On the other hand, there is evidence that it is an irreversible effect (KIDO et al., 1988), and from the standpoint of preventive medicine it is certainly advisable to prevent the occurrence of such an effect.

Two episodes of community poisoning by mercury compounds (see also Chapter II.20) were observed in Japan (Minamata bay area, 1953 – 1960, and Niigata area, 1965). The so-called *Minamata disease* has been defined as neuropathia arising from intake of large quantities of fish and shellfish containing high concentrations of methyl mercury. The disease mainly resulted from the accumulation of mercury in the food chain, during which it was transformed to the highly neurotoxic methyl mercury. Since the victims of the Minamata disease were investigated very thoroughly, the Japanese experience allows an analysis of health risks due to methyl-mercury poisoning (TAKIZAWA, 1979; for both adult and prenatal poisoning see also MARSH et al., 1981, 1987). Another mercury poisoning incident took place in Iraq in 1972. The primary cause of the poisoning was traced to be the ingestion of seed wheat treated with mercury-containing fungicides, which was used for making flour. Victims were identified only in provinces where the inhabitants ate homemade bread, and not a single case of poisoning was reported in cities, where flour was supplied from government-inspected flour mills. The initial symptoms of the poisoning occurring after an average latency period of 16 to 38 days consisted of numbness and paresthesia in the extremities. In total several thousand persons were affected, of whom 459 died (BAKIR et al., 1973; TAKIZAWA, 1979, and Chapter II.20).

Although severe intoxications of relatively large population groups that occurred under specific circumstances in certain areas are the most dramatic manifestations of the chronic toxicity of metals and their compounds, the subclinical effects of long-term, low-level metal exposure in the general population are also of importance in terms of human health. The data presently available suggest that a variety of subclinical effects may be associated with long-term, low-level metal exposure. Many of these effects are influenced by several other factors so that the effect that can be attributed to metal exposure is difficult to evaluate.

One example of such effects is the probably causal relationship between elevated lead exposure (see also Chapter II.16) and increased blood pressure that could be demonstrated in some recent large-scale general population studies as well as in a number of studies on lead workers (HARLAN et al., 1985; PIRKLE et al., 1985; DE KORT et al., 1987; WEISS et al., 1986). None of these observational studies can be stated as definitely establishing causal linkages between lead exposure and increased blood pressure or hypertension, and it should be noted that negative findings have also been reported (POCOCK et al., 1985). However, the plausibility of the observed associations is supported by the consistency of findings of several independent investigations as well as by extensive toxicological data, which clearly demonstrate increases in blood pressure for animal models under well-controlled experimental conditions (VICTERY et al., 1982). Uncertainty still exists regarding the lead exposure levels, at which significant elevations in blood pressure occur. Moreover, it should be noted that many factors, including hereditary traits and nutritional factors, contribute to the development of increased blood pressure. The contribution of lead appears to be relatively small, usually not accounting for more than 1 – 2% of the variation explained by the models employed when other significant factors are controlled for in the analyses.

Similar to lead, exposure to cadmium has been found to be associated with increased blood pressure (VIVOLI et al., 1985). The causal linkage between cadmium and increased blood pressure has been demonstrated convincingly in animal studies (KOPP et al., 1981).

Another example of subclinical chronic toxicity is the nephrotoxicity of long-term, low-level cadmium exposure (see also Chapter II.6). According to one study (ROELS et al., 1981a) there is suggestive evidence that environmental pollution by cadmium as found in some industrialized areas of Europe may exacerbate the age-related decline of renal function in population groups non-occupationally exposed to heavy metals. Similar effects were noted by ALT et al. (1983), whereas another study (EWERS et al., 1985) could not find a relationship between cadmium exposure and renal dysfunction in population groups living in two cadmium polluted areas of West Germany. A retrospective mortality study conducted in Liège (Belgium), a cadmium-polluted area, showed a higher mortality from nephritis and nephrosis in that area when compared with other areas of Belgium (LAUWERYS and DE WALS, 1981).

Another issue of concern is the neurotoxicity of lead in young children (see also Chapter II.16). A number of independent studies clearly demonstrate that increased lead exposure of young children, as indicated by increased blood lead or tooth lead levels, is associated with certain performance-deficits and behavioral alterations as measured by various psychometric and neuropsychological tests (NEEDLEMAN et al., 1979; WINNEKE et al., 1983a,b; RUTTER and JONES, 1983; BORNSCHEIN and RABINOWITZ, 1985; DAVIS and SVENDSGAARD, 1987). Animal experiments clearly indicate that there is a causal relationship between prenatal and early postnatal lead exposure and certain cognitive and performance deficits (WINNEKE, 1986).

During the last decade the possible role of aluminum (see also Chapter II.1) as an etiological factor of a number of diseases has found increasing interest. For a long time, the general opinion was that Al^{3+} is very atoxic causing intoxications and diseases only in some rare and artificial situations, e.g., chronic hemodialysis and heavy occupational exposure to Al-fumes (welding) or dust of soluble Al^{3+} compounds. In the last years, however, considerable evidence has accumulated that at least two kinds of diseases may be caused by aluminum. Both occurred in patients treated by hemodialysis, which may contribute substantially to the accumulation of aluminum in the body, if the water used for dialysis contains significant amounts of aluminum. As a result of aluminum accumulation in the body, particularly in the brain, severe neurotoxic and other effects may occur. The diseases have been named 'dialysis encephalopathy' or 'dialysis dementia' and 'dialysis osteomalacia' (for review see GANROT, 1986). Moreover, a number of other diseases such as senile dementia of Alzheimer's type, endemic amyotropic lateral sclerosis, and parkinsonism-dementia have been connected with aluminum (GANROT, 1986), but further studies are necessary to prove evidence.

In a broad sense, the carcinogenic effects of certain metals and metal compounds, which are discussed in more detail in Chapter I.18 and in Part II of this book, can also be considered as chronic toxic effects. The causation of lung cancer as a result of long-term inhalation exposure to chromates and arsenic, cadmium, cobalt, and nickel and their compounds is well known (IARC, 1980; TAKENAKA et al., 1983; THUN et al., 1985). These kinds of effects are essentially confined to workers with heavy occupational exposure to these agents, but it should be mentioned that recent studies suggest that populations living in the vicinity of large copper and zinc smelters also have an increased risk of lung cancer (MATANOSKI et al., 1981; CORDIER et al., 1983; BROWN et al., 1984; PERSHAGEN, 1985). Although no firm conclusions can be drawn on the cause of the excess lung cancer risk in these areas, it seems plausible that air emissions from the smelters, e.g., of arsenic, have played a role. In an area of Taiwan, exposure to very high levels of arsenic in drinking water has been found to be associated with an increased prevalence rate of skin cancer, hyperpigmentation, keratosis, and a peripheral vascular disorder called "blackfoot disease" (TSENG et al., 1968; TSENG, 1977). Similar associations have been reported by some other authors (for review see IARC, 1980).

7 Diagnosis of Chronic Intoxications

As can be seen from the preceding paragraphs, metal poisoning may give rise to a wide variety of effects in various organs of the body. Frequently, the presenting features, e.g., anemia, proteinuria, gastrointestinal and neurologic disorders, and allergic reactions, are non-specific. Thus, the clinical examination often does not give a lead on the cause of the illness, and it can be assumed that many cases of chronic metal intoxications remain undiagnosed. Of course, it may also be true that effects attributed to metal compounds can have other causes. The difficulties of disclosing the role of metals in the etiology of chronic diseases are even more pronounced, since many of these diseases, e.g., hypertension, renal dysfunction, and neoplasia, are multicausal in nature. Exposure to metals may represent only one of several etiologic factors, and it is usually impossible to evaluate the relative contribution of this factor.

In order to arrive at a correct diagnosis of chronic metal intoxications the following principles may be considered:

History of exposure: In most cases, the history of exposure to a toxic metal will give important clues. In an industrial situation there may be a clear history of exposure, which can be obtained from the patient, his/her relatives or co-workers, or from the employer and the company's health and safety representatives. The physician should not fail to take a full and accurate occupational history where a poisoning of occupational origin is suspected. In the domestic environment, the origin of metal exposure may be more difficult to trace. Household chemicals, medicaments, smoking, contaminated drinking water and food as well as hobby activities have to be considered as sources of exposure, and it may be of great importance to keep suspicious materials for chemical analysis. Moreover, homicidal and suicidal instances have to be taken into account.

Toxicological analysis (see also Chapter I.4a): The best way of confirming the diagnosis of a chronic metal intoxication is the detection of increased levels of the suspected metal in biological specimens such as blood, urine, feces, hair, or nails. The examinations of blood and urine samples for metal concentrations can be considered a routine procedure, but they should be performed in laboratories that are equipped for this purpose and participate in external quality control. In interpreting the results of such analyses the toxicokinetic behavior of the metals in the body should be born in mind. Some metals, e.g. arsenic and thallium, disappear rapidly from the blood and others are sparingly excreted in urine. When the metal absorption has ceased some time previously and the blood and urine levels have returned to normal, the analysis of hair or nails may be useful. It must be considered, however, that hair and nail analysis is difficult and contamination may be a serious problem.

Medical examinations: General clinical, biochemical, physiological, and neurological examinations may, in some cases, be helpful for diagnosis, and, particularly, to judge the severity of an intoxication. The biochemical abnormalities associated with increased metal absorption, e.g., anemia, decreased δ-aminolevulinic acid dehydratase activity in blood and increased δ-aminolevulinic acid levels in urine in the case of lead intoxication, increased urinary excretion of low-molecular-weight proteins as a result of cadmium- or mercury-induced renal dysfunction, are described in the chapters on the different metals in Part II of this book. Although these abnormalities are characteristic for the above named metals, such tests do not *prove* an intoxication. Due to the non-specifity of the symptoms and effects, such findings can only be interpreted in relation to the results of toxicological analyses.

The principles of treatment of chronic metal intoxications and of preventing metal poisoning cannot be discussed here. For these aspects the reader should consult other presentations, e.g., MOESCHLIN (1980), KAZANTZIS (1986), and HERNBERG (1986).

References

Alt, J.M., Maywald, A., Raguse-Degener, G., and Rühling, U. (1983), *Proteinurie bei einer Cadmium-exponierten Bevölkerungsgruppe (Oker 1980). Staub Reinhalt. Luft 43*, 294–297.
Bakir, F., Damluji, S.F., Amin-Zaki, A., Murtadha, M., Khalidi, A., Al-Rawi, S., Tikriti, S., and Dhahir, H.I. (1973), *Methylmercury Poisoning in Iraq. Science 181*, 230–241.
Bondy, S.C., and Prasad, K.N. (1987), *Metal Neurotoxicity.* CRC Press: Wolfe Medical Publications Ltd., London.
Bornschein, R.L., and Rabinowitz, M.B. (eds.) (1985), *The Second International Conference on Prospective Studies of Lead, Cincinnati (Ohio),* April 1984. *Environ. Res. 39*, 1–210.
Brown, L.M., Pottern, L.M., and Blot, W.J. (1984), *Lung Cancer in Relation to Environmetal Pollutants Emitted from Industrial Sources. Environ. Res. 34*, 250–261.

Cordier, S., Theriault, G., and Iturra, H. (1983), *Mortality Patterns in a Population Living Near a Copper Smelter. Environ. Res. 31*, 311–322.
Davis, J. M., and Svendsgaard, D. J. (1987), *Lead and Child Development. Nature 329,* 297–300.
de Kort, W. L., Verschoor, M. A., Wibowo, A. A., and van Hemmen, J. J. (1987), *Occupational Exposure to Lead and Blood Pressure: A Study in 105 Workers. Am. J. Ind. Med. 11,* 145–156.
Ewers, U., Brockhaus, A., Dolgner, R., Freier, I., Jermann, E., Bernard, A., Stiller-Winkler, R., and Manojlovic, N. (1985), *Environmental Exposure to Cadmium and Renal Function of Elderly Women Living in Cadmium-polluted Areas of the Federal Republic of Germany. Int. Arch. Occup. Environ. Health 55,* 217–223.
Friberg, L. (1984), *Cadmium and the Kidney. Environ. Health Perspect. 54,* 1–11.
Friberg, L. (1986), *Risk Assessment,* in: Friberg, L., Nordberg, G. F., and Vouk, V. B. (eds.): *Handbook on the Toxicology of Metals,* 2nd Ed., Vol. I, Chap. 12, pp. 269–293. Elsevier/North Holland, Amsterdam-New York-Oxford.
Friberg, L., and Kjellström, T. (1981), *Toxic Metals – Pitfalls in Risk Estimation,* in: *International Conference on Heavy Metals in the Environment, Amsterdam,* September 1981, pp. 1–11. CEP Consultants Ltd., Edinburgh.
Ganrot, P.O. (1986), *Metabolism and Possible Health Effects of Aluminium. Environ. Health Perspect. 65,* 363–441.
Harlan, W. R., Landis, J. R., Schmouder, R. L., Goldstein, N. G., and Harlan, L. C. (1985), *Blood Lead and Blood Pressure. J. Am. Med. Assoc. 253,* 530–534.
Hernberg, S. (1986), *General Aspects of the Prevention of Metal Poisoning,* in: Friberger L., Nordberg, G. F., and Vouk, V. B. (eds.): *Handbook on the Toxicology of Metals,* 2nd Ed., Vol. I, Chap. 10, pp. 231–252. Elsevier/North-Holland, Amsterdam-New York-Oxford.
IARC (International Agency for Research on Cancer) (1980), *IARC Monographs on the Evaluation of the Carcinogenic Risk of Chemicals to Humans,* Vol. 22, *Some Metals and Metallic Compounds.* IARC, Lyon.
ILO (International Labour Office) (1983), *Encyclopedia of Occupational Health and Safety,* 3rd (revised) Ed. International Labour Office, Geneva.
Kazantzis, G. (1986), *Diagnosis and Treatment of Metal Poisoning – General Aspects,* in: Friberg, L., Nordberg, G. F., and Vouk, V. B. (eds.): *Handbook on the Toxicology of Metals,* 2nd Ed., Vol. I, Chap. 13, pp. 294–318. Elsevier/North-Holland, Amsterdam-New York-Oxford.
Kido, T., Honda, R., Tsuritani, I., Yamaha, H., Ishihaki, M., Yamada, Y., and Nogawa, K. (1988), *Progress of Renal Dysfunction in Inhabitants Environmentally Exposed to Cadmium. Arch. Environ. Health 43,* 213–217.
Kjellström, T. (1979), *Exposure and Accumulation of Cadmium in Populations from Japan, the United States, and Sweden. Environ. Health Perspect. 28,* 169–197.
Kjellström, T., and Nordberg, G. F. (1978), *A Kinetic Model of Cadmium Metabolism in the Human Lung. Environ. Res. 16,* 248–269.
Kjellström, T., Elinder, C. G., and Friberg, L. (1984), *Conceptual Problems in Establishing the Critical Concentration of Cadmium in Human Kidney Cortex. Environ. Res. 33,* 284–295.
Kopp, S. J., Glonek, T., Perry, H. M., Erlanger, M. E., and Perry, E. F. (1982), *Cardiovascular Actions of Cadmium at Environmental Exposure Levels. Science 217,* 837–839.
Lauwerys, R. R., and De Wals, P. (1981), *Environmental Pollution by Cadmium and Mortality from Renal Diseases. Lancet I,* 383.
Marsh et al. (1981), *Dose-response Relationship for Human Fetal Exposure to Methylmercury. Clin. Toxicol. 18,* 1311–1318.
Marsh et al. (1987), *Fetal Methylmercury Poisoning. Arch. Neurol. 44,* 1017–1022.
Matanoski, G., Landau, G., Tonascia, J., Lazar, C., Elliott, E. A., McEnroe, W., and King, K. (1981), *Cancer Mortality in an Industrial Area of Baltimore. Environ. Res. 25,* 8–28.
Moeschlin, S. (1980), *Klinik und Therapie der Vergiftungen,* 6th Ed. G. Thieme Verlag, Stuttgart-New York.

Needleman, H. E., Gunnoe, C., Leviton, A., Reed, R., Peresie, H., Maher, C., and Barrett, P. (1979), *Deficits in Psychologic and Classroom Performance of Children with Elevated Dentine Lead Levels. N. Engl. J. Med.* **300**, 689–695.
Nogawa, K., Honda, R., Yamada, Y., Kido, T., Tsuritani, I., Ishizaki, M., and Yamaha, H. (1986), *Critical Concentration of Cadmium in Kidney Cortex of Humans Exposed to Environmental Cadmium. Environ. Res.* **40**, 251–260.
Nomiyama, K., Nomiyama, H., Akahori, F., and Masaoka, T. (1981), *Recent Studies on Health Effects of Cadmium in Japan*, pp. 59–104. Environmental Agency, Tokyo.
Pershagen, G. (1985), *Lung Cancer Mortality Among Men Living Near an Arsenic Emitting Smelter. Am. J. Epidemiol.* **122**, 684–694.
Pirkle, J. L., Schwartz, J., Landis, J. R., and Harlan, W. R. (1985), *The Relationship Between Blood Lead Levels and Blood Pressure and its Cardiovascular Risk Implications. Am. J. Epidemiol.* **121**, 246–258.
Pocock, S. J., Shaper, A. G., Ashby, D., and Delves, T. (1985), *Blood Lead and Blood Pressure in Middle-aged Men*, in: Lekkas, T. D. (ed.): *International Conference Heavy Metals in the Environment, Athens*, September 1985, Vol. I, pp. 303–305. CEP Consultants Ltd., Edinburgh.
OECD (Organization for Economic Co-operation and Development) (1977) *Guidelines in Respect to Procedures and Requirements for Anticipating the Effects of Chemicals on Man and in the Environment.* OECD, Paris.
Repko, J., and Corum, C. (1979), *Critical Review and Evaluation of the Neurological and Behavioral Sequelae of Inorganic Lead Absorption. CRC Rev. Toxicol.* **6**, 135–187.
Roels, H. A., Lauwerys, R. R., Buchet, J. P., and Bernhard, A. (1981), *Environmental Exposure to Cadmium and Renal Function of Aged Women in Three Areas of Belgium. Environ. Res.* **24**, 117–130.
Roels, H. A., Lauwerys, R. R., Buchet, J. P., Bernard, A., Chettle, D. R., Harvey, T. C., and Al-Haddad, I. K. (1981), *In vivo Measurement of Liver and Kidney Cadmium in Workers Exposed to this Metal: Its Significance with Respect to Cadmium in Blood and Urine. Environ. Res.* **26**, 217–240.
Rutter, M., and Jones, R. R. (eds.) (1983), *Lead versus Health – Sources and Effects of Low Level Lead Exposure.* J. Wiley & Sons, Chichester-New York-Brisbane-Toronto-Singapore.
Takenaka, S., Oldiges, H., König, H., Hochrainer, D., and Oberdörster, G. (1983), *Carcinogenicity of Cadmium Chloride Aerosols in W Rats. J. Natl. Cancer Inst. US* **70**, 367–373.
Takizawa, K. (1979), in: Nriagu, J. O. (ed.): *The Biogeochemistry of Mercury in the Environment*, Chap. 14. Elsevier/North-Holland, Amsterdam-New York-Oxford.
Thun, M. J., Schnorr, T. M., Smith, A. B., Halperin, W. E., and Lemen, R. A. (1985), *Mortality among a Cohort of U.S. Cadmium Production Workers – an Update. J. Natl. Cancer Inst. US* **74**, 325–333.
Tilson, H. A., and Foulkes, E. C. (1987), *Symposium on Neurotoxicology of Heavy Metals. Fundam. Appl. Toxicol.* **9**, 599–615.
Tseng, W. P. (1977), *Effects and Dose-response Relationships of Skin Cancer and Blackfoot Disease with Arsenic. Environ. Health Perspect.* **19**, 109–119.
Tseng, W. P., Chu, H. M., How, S. W., Fong, J. M., Lin, C. S., and Yeh, S. (1968), *Prevalence of Skin Cancer in an Endemic Area of Chronic Arsenicism in Taiwan. J. Natl. Cancer Inst. US* **40**, 453–463.
Tsuchiya, K. (1978), *Cadmium Studies in Japan: A Review.* Elsevier/North-Holland, Amsterdam-New York-Oxford.
Vahter, M. (ed.) (1982), *Assessment of Human Exposure to Lead and Cadmium Through Biological Monitoring.* Prepared for United Nations Environment Program and World Health Organization. National Swedish Institute of Environmental Medicine and Department of Environmental Hygiene, Karolinska Institute, Stockholm.
Victery, W., Vander, A. J., Shulak, J. M., Schoeps, P., and Julius, S. (1982), *Lead, Hypertension, and the Renin-Angiotensin System in Rats. J. Lab. Clin. Med.* **99**, 354–362.

Vivoli, G., Bergmoi, M., Borella, P., Fantuzzi, G., Caselgrandi, E., Fusco, A., and Tamburi, M. (1985), *Cadmium Distribution in Some Biological Matrices in Hypertension: A Case-control Study*, in: *International Conference Heavy Metals in the Environment, Athens*, September 1985, Vol. II, pp. 67–69. CEP Consultants Ltd., Edinburgh.

Weiss, S. T., Munoz, A., Stein, A., Sparrow, D., and Speizer, F. E. (1986), *The Relationship of Blood Lead to Blood Pressure in a Longitudinal Study of Working Men. Am. J. Epidemiol. 123*, 800–808.

WHO (World Health Organization) (1976), *Environmental Health Criteria 1, Mercury.* WHO, Geneva.

WHO (World Health Organization) (1978), *Environmental Health Criteria 6, Principles and Methods for Evaluating the Toxicity of Chemicals*, Part I. WHO, Geneva.

WHO (1986), *Principles and Methods for the Assessment of Neurotoxicity Associated with Exposures to Chemicals. Environmental Health Criteria 60.* World Health Organization, Geneva.

Winneke, G. (1986), *Animal Studies*, in: Lansdown, R., and Yule, W. (eds.): *The Lead Debate – The Environment, Toxicology, and Child Health*, pp. 217–234. Croom Helm, London – Sidney.

Winneke, G., Hrdina, K. G., and Brockhaus, A. (1983a), *Neuropsychological Studies in Children with Elevated Tooth-lead Concentrations. I. Pilot Study. Int. Arch. Occup. Environ. Health 51*, 169–183.

Winneke, G., Krämer, U., Brockhaus, A., Ewers, U., Kujanek, G., Lechner, H., and Jahnke, W. (1983b), *Neuropsychological Studies in Children with Elevated Tooth-lead Concentrations. II. Extended Study. Int. Arch. Occup. Environ. Health 51*, 231–252.

I.17b Metal Compounds and Immunotoxicology

BADRUL ALAM CHOWDHURY, Detroit, Michigan, USA
RANJIT KUMAR CHANDRA, St. John's, Newfoundland, Canada

1 Introduction
(see also DESCOTES, 1986; MENNÉ and NIEBOER, 1989)

Human civilization and concomitant increase in industrial activity has gradually redistributed many toxic metals from the earth's crust to the environment and increased the possibility of human exposure. Among the toxic metals arsenic, cadmium, lead, and mercury are especially prevalent in nature due to their high industrial use, and are also present in human and animal tissues in considerable amounts. These metals serve no essential biological function, and studies have shown that they are toxic to various organ systems of the body. In addition, some metals such as copper, selenium, and zinc, which are essential to life may also be toxic when present in excess. Although the toxic properties of metals were known for over a century, their effects on the immune system have only being studied for the last two decades. The studies have shown that these metals consistently impair immune functions at large doses and some even selectively affect the immune cells at doses lower than those affecting other organs of the body. Although most toxicological studies involving metals have been done on laboratory animals, the findings probably apply to man as the biological functions of different mammalian species are mostly similar. Knowledge recently acquired on the immune system effects of the biologically important metals and their relevance to human health is discussed in the following sections. Emphasis is on immunosuppression, however, allergic reactions (especially of mercury and nickel compounds) and autoimmunity are also important.

2 Arsenic

Data on immune system effects of arsenic are limited. In rodents arsenic is reported to increase the susceptibility to viral challenge (GAINER and PRY, 1972) and to decrease antibody titer and synthesis (BLAKLEY et al., 1980). However, in vitro studies using bovine and human lymphocytes show that the proliferation of T cells to mitogens is increased by arsenic (MCCABE et al., 1983). The antibody forming cell number is also seen to be increased by low doses of arsenic when stimulated with heterologous antigen in vitro. But at higher doses, the response is suppressed. The apparent enhancement of immune responses by low doses of arsenic is due to cytotoxicity of arsenic against suppressor T cells, and not an augmentation of lymphocyte function (YOSHIDA et al., 1987). The effect of arsenic on immune cells thus seems to be dose dependent like that of cadmium.

3 Cadmium

Cadmium is toxic to almost every organ system of the body including the immune system. However, the immunotoxic effects of cadmium are rather ambiguous. Primary and secondary immune response against heterologous antigens or antibody titers against infections are suppressed in experimental animals treated with large oral doses of cadmium, or following cadmium injection or inhalation (KOLLER, 1973; BOZELKA et al., 1978; GRAHAM et al., 1978; KRZYSTYNIAK et al., 1987). In contrast, low oral doses of cadmium increase the number of antibody forming cells (MALAVE and DERUFFINO, 1984; CHOWDHURY, et al., 1987). Also a single parenteral dose of cadmium before antigen challenge increases antibody production (KOLLER et al., 1976). In keeping with the animal observations, in vitro studies show that large amounts of cadmium decrease, whereas small amounts increase the primary immune response (FUJIMAKI et al., 1982).

Cadmium also affects the cell mediated immunity. Delayed type hypersensitivity skin reaction in mice is suppressed by cadmium (MÜLLER et al., 1979). Proliferation response of T cells to mitogens has been reported to be increased (MÜLLER et al., 1979; MALAVE and DERUFFINO, 1984), unchanged (BLAKLEY, 1985), or decreased (GAWORSKI and SHARMA, 1978), whereas that of B cells is generally increased (MÜLLER et al., 1979; BLAKLEY, 1985). These apparent conflicting results are explained by the observation that cadmium at low doses selectively affects the suppressor cells and thereby increases the ratio of helper to suppressor cells (Table I.17b-1). Thus, at low doses cadmium is seen to increase antibody production and

Table 1.17b-1. T-Lymphocyte Subsets in the Spleen

Treatment	CD4 Cells (L3T4$^+$ Helper) (%)	CD8 Cells (Lyt-2$^+$ Suppressor) (%)	Helper/Suppressor Ratio
None	34.90 ± 6.63	24.96 ± 3.55	1.40 ± 0.21
Cd 50 ppm for 3 wk	32.35 ± 5.98	17.24 ± 2.74[a]	1.89 ± 0.32[a]
Cd 50 ppm with Zn 500 ppm for 3 wk	31.90 ± 6.00	22.19 ± 2.79	1.44 ± 0.19
Zn 500 ppm for 3 wk	33.40 ± 6.28	23.37 ± 2.04	1.40 ± 0.21

Values are expressed as mean ± SD of eight observations
[a] Significantly different at P (probability) < 0.01 from other values in the same column by Duncan's multiple-range test

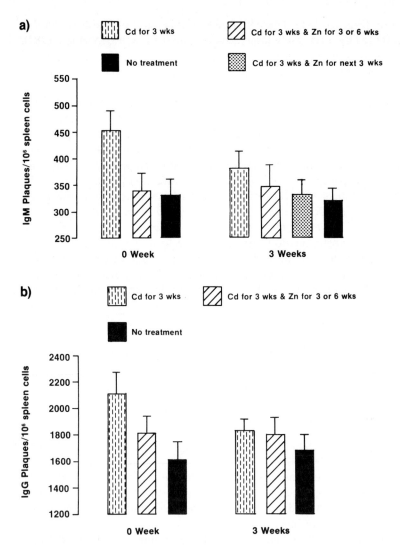

Fig. I.17b-1. Antibody producing cell response after 3 weeks of cadmium or cadmium with zinc treatment (0 week), and 3 weeks after cessation of cadmium treatment (3 weeks). (a) IgM plaques, (b) IgG plaques. Bars represent mean and SD of 10 observations. Cadmium treatment alone resulted in a greater number of antibody forming cells at 0 week than in untreated controls ($P<0.05$). Simultaneous treatment of cadmium and zinc did not result in a greater number of antibody forming cells. Three weeks later there were no significant differences between any of the groups.

mitogen stimulation (Fig. I.17b-1). These immunological effects of low doses of cadmium can be prevented by a relatively large dose of zinc (CHOWDHURY et al., 1987). At larger doses of cadmium, all cell types are probably affected resulting in general impairment of immune responses. In addition, cadmium also affects the maturation of lymphocytes and results in a shift to immature cell types (BURCHIEL et al., 1987).

Other immune cells such as the macrophages and polymorphonuclear lymphocytes are also adversely affected by cadmium. Phagocytic capability of these cells is reduced (LOOSE et al., 1978), and production of lymphokines by macrophages is suppressed in cadmium treated animals (KIREMIDJIAN-SCHUMACHER et al., 1981). As a result of generalized immunosuppression, cadmium treated animals show increased susceptibility to bacterial, viral, and protozoal infection (COOK et al., 1975; GAINER, 1977; EXON et al., 1979). Incidence of cancer in experimental animals and also in man is increased in a dose-dependent fashion by cadmium (TAKENAKA et al., 1983; THUN et al., 1985). Development of cancer is probably multifactorial, but cadmium induced immunosuppression may contribute to its pathogenesis.

The implications of cadmium immunotoxicity on human health are presently unknown. However, animal observations suggest that cadmium may well be immunotoxic in man. The observation that human lymphocytes exposed in vitro to cadmium accumulate intracellular concentrations of cadmium at a level that is 3000-fold greater than in the culture medium further supports this assumption (HILDEBRAND and CRAM, 1979). The reason for such a selective uptake of the metal is not known. Thus, in addition to other toxicological tests, immunological monitoring may be useful in populations at risk of excess cadmium exposure.

4 Lead

Very little is known about the immunotoxicologic effects of organic lead compounds. Thus, in this section, the inorganic group is mainly discussed as inorganic lead compounds are present both in the industrial and general environment, and most data on lead immunotoxicity involve inorganic lead.

In experimental animals various parameters of the immune system are affected by lead. Changes in humoral immunity are more pronounced and have been studied more extensively than others. In animals exposed to lead compounds, the level of circulating antibodies and antibody production upon sensitization to antigens is impaired. Experiments on rats, mice, and rabbits show that the number of antibody forming cells in spleen and serum antibody levels are lower in lead treated animals as compared to untreated controls (KOLLER and KOVACIC, 1974; BLAKLEY et al., 1980; LUSTER et al., 1978). The effect is seen both after oral and parenteral lead exposure, and against antigens like heterologous red blood cells, bacteria, and viruses. The developing immune system is also susceptible to lead toxicity. Antibody response is suppressed in mice exposed to lead in the pre- and immediate post-natal period, but cell mediated immunity was not found to be affected in these studies (TALCOTT and KOLLER, 1983).

Other data suggest that cellular immune function is also affected by lead. Delayed cutaneous hypersensitivity reaction which is a classical in vivo test for cell mediated immunity is impaired by lead in various rodent species (MÜLLER et al., 1977; FAITH et al., 1979). However, the in vitro tests of cell mediated immunity do not show uniform effects on lead exposure. Lymphocyte transformation in vitro to mitogens such as PHA (phytohemagglutinin), ConA (concanavalin A), and PWM (pokeweed mitogen), and to heterologous lymphocytes in mixed lymphocyte culture are the commonly employed assays of cells mediated immunity. The plant lectin PHA and ConA activate T cells, and PWM activates both T and B cells. In mixed lymphocyte culture, proliferation of lymphocytes initiated by contact with cells bearing different major histocompatibility proteins is measured. In some studies, lymphocytes from animals exposed to lead have been shown to have reduced proliferative response to mitogens (GAWORSKI and SHARMA, 1978; FAITH et al., 1979), whereas in others no changes were observed (KOLLER et al., 1979). In mixed lymphocyte culture no changes have been seen following lead treatment (KOLLER, 1980). These conflicting results may be due to differential effects of lead on various immune cell types depending on dose, duration, and route of lead exposure. It is possible that the B cells or the subset of T cells that help B cell function are affected early in lead treatment, whereas other cell types including those which proliferate in response to mitogens and heterologous antigens are relatively resistant. One study, in which temporal changes in immune functions were studied in mice following high parenteral lead exposure, reported that lead suppressed B and T cell functions at different times after treatment. Also lead caused a shift in immunocompetent cells to immature types that are functionally less competent (BURCHIEL et al., 1987). Furthermore, lead is known to possess mitogenic properties and also decreases the viability of lymphocytes in culture (SHENKER et al., 1977; GALLAGHER et al., 1979). This would further confound the results of lymphocyte proliferation response in cultures. However, taking the various studies in consideration it can be concluded that lead suppresses cell mediated immune functions which generally correlate with humoral immunity data.

Non-specific immune functions are also affected by lead. The phagocytic activity of polymorphonuclear leukocytes and macrophages is suppressed on acute lead exposure (TREJO et al., 1972; WARD et al., 1975). However, on chronic exposure macrophages are actually activated as enumerated by stimulated phagocytosis and increased acid phosphatase levels (KOLLER and ROAN, 1977). Cell mediated antibody dependent cellular cytotoxicity is reduced on chronic lead exposure in mice, but natural killer cells are not affected (NEILAN et al., 1983).

Altered immune functions by lead is expected to increase the incidence of infections. Studies have shown that oral or parenteral lead increases susceptibility of laboratory animals to bacteria and viruses (COOK et al., 1975; GAINER, 1977; EXON et al., 1979). In addition, lead also appears to promote the growth of viral and chemically induced neoplasms (GAINER, 1973; HINTON et al., 1979). It is possible that altered immune functions contribute to the development of cancers in such situations.

These animal experiments show that lead suppresses the immune system, particularly the humoral responses. Some of these effects are observed at low doses, and, therefore, may be detrimental to the health of man by mechanisms other than the typical well-documented toxicity which occurs at larger doses. Also the effects are consistent and, therefore, tests of immune functions may be of value in biomonitoring lead exposure.

5 Mercury

Among the inorganic mercury compounds, mercury vapor and mercury salts are the forms commonly used in industry. Organic mercury exists in different forms of which methylmercury is highly toxic.

Mercury compounds are toxic to various components of the immune system. Both inorganic and organic mercury enhances mortality of animals challenged with viruses (KOLLER, 1975; GAINER, 1977). These animals also produce low amounts of specific neutralizing antibodies (KOLLER et al., 1977). Primary and secondary antibody response is suppressed in animals exposed to methylmercury during the prenatal and postnatal period (BLAKLEY et al., 1980). This suggests that organic mercury compounds are not only immunotoxic but also affect the immunological ontogeny. Lymphocyte transformation response to mitogens is also reduced by mercury chloride (GAWORSKI and SHARMA, 1978). Methylmercury reduces the latent period and survival time for chemically induced carcinoma in rats (NIXON et al., 1979). Although data on mercury immunotoxicity at dose ranges comparable to possible human exposure are not available, animal results suggest that mercury compounds are probably immunotoxic for man.

6 Nickel

Nickel is an important inhalable pollutant that exists in many forms including metallic nickel, nickel subsulfide, nickel chloride, nickel oxide, and nickel carbonyl. The majority of nickel associated lesions described in man is caused by inhalation exposure, the most prominent of which being cancer of the nasal cavity and lung, and allergic reactions (see Chapter II.20, Sect. 6.5). Animal experiments also show that nickel is toxic to the immune system. Acute and chronic inhalation exposure of rabbits and mice to nickel results in deficient macrophage function as indicated by decreased phagocytic and bactericidal capacity of pulmonary alveolar macrophages (ADKINS et al., 1979; WIERNIK et al., 1983). Inhalation or parenteral exposure of mice to nickel suppresses primary antibody responses to T cell dependent antigens. Mitogen response of lymphocytes to T and B cell mitogens is also decreased (GRAHAM et al., 1978; SMIALOWICZ et al., 1984), as well as the development of delayed hypersensitivity reaction to recall antigens (PARKER and TURK, 1978). In mice, parenteral nickel suppresses the systemic natural killer cell function (SMIALO-

WICZ et al., 1985). However, in monkeys after intratracheal installation of nickel, natural killer activity of bronchial lavage cells was found to be increased, but macrophage phagocytic activity was decreased. The primary antibody forming cell number was unchanged (HALEY et al., 1987). These data suggest that the pulmonary alveolar macrophages represent the immune cell population that is most susceptible to the toxic effects of nickel compounds.

7 Zinc

Unlike other metals so far discussed zinc is an essential dietary element. It is required for the activity of many enzymes and also maintains the three-dimensional structural configuration of various proteins including the nuclear proteins. Zinc also plays an important role in the development and maintenance of the immune system. Various animal and human studies have shown the deleterious effects of zinc deficiency on the cellular immune functions (FRAKER et al., 1986). However, recent studies have shown that large doses of zinc are also toxic to the immune system. In mice toxic doses of zinc increase susceptibility to some infectious agents and inhibit the phagocytic capacity of peritoneal macrophages (SOBOCINSKI et al., 1977; KARL et

Table I.17b-2. Plasma Zinc Levels and Polymorphonuclear (PMN) Leukocyte Function[a]

Time of Sample	Plasma Zinc, μg/dL	Chemotactic Migration,[b] PMNs per 10 HPF[c]	Ingestion of Bacteria, No./100 PMNs	Bactericidal Capacity, % Viable Bacteria
Baseline	83.0 ± 9.2[d]	632 ± 101[d]	493 ± 47[d]	3.1 ± 1.1[f]
During zinc administration, wk				
2	101.1 ± 12.7[f]	585 ± 99[f]	386 ± 51[e]	3.9 ± 1.4[f]
4	181.5 ± 21.1[d]	355 ± 86[d]	272 ± 36[d]	4.6 ± 1.9[f]
6	199.7 ± 18.5[d]	292 ± 57[d]	251 ± 40[d]	3.8 ± 2.1[f]
Follow-up after cessation of zinc supplement, wk				
2	167.4 ± 20.3[d]	401 ± 71[d]	198 ± 43[d]	4.5 ± 2.3[f]
10	90.3 ± 10.0[d]	576 ± 114[d]	459 ± 56[d]	2.7 ± 1.2[f]

[a] Values are given as mean ± SE
[b] HPF indicates high-power field
[c] Results are shown for migration at 120 minutes. Similar differences were observed at 60 and 180 minutes
[d] Probability $P<.01$ for differences between baseline or follow-up figures and values obtained during zinc administration
[e] $0.5<P<0.1$
[f] NS, $P>0.1$

al., 1973). In human volunteers, a moderately large amount of oral zinc supplement, 150 mg twice a day, has been shown to reduce lymphocyte transformation response to T cell mitogen PHA and to reduce chemotactic migration and ingestion of bacteria by neutrophils (Table I.17b-2). Also cultured lymphocytes stimulated with PHA show lower prereplicative calcium accumulation in the presence of zinc. Altered serum lipid profiles and changes in cell membrane structural lipid composition by zinc have been postulated to be the mechanism by which large doses of zinc alter the function of immunocompetent cells (CHANDRA, 1984). These studies showed that the immune functions are affected both in deficiency and excess of zinc. Thus, monitoring the immune status may be useful to evaluate the functional consequence of both extremes of zinc nutrition.

8 Conclusions

Immunotoxicological potentials of metals and their compounds are well known. Arsenic, cadmium, lead, mercury, and nickel are uniformly toxic to the immune system, although the ultimate effect depends on the species of animals studied and the route and mode of administration. Recent advances in the field of immunology with the development of new techniques have clarified many of the cellular and biochemical bases of their toxicity. These investigations have probably only scratched the surface of what could be serious implications for human health. Immunotoxicological properties of large doses of some essential metals such as zinc are also of considerable concern as many people consume large amounts of zinc as dietary supplement. The same may also be true for other essential metals such as copper and selenium, as large doses of these are known to have toxic potentials. Some of these metals are also used in industries and are present in the atmosphere. An important feature of metal toxicity is the observation that in animals some metals affect the immune system at doses that are unaccompanied by other clinical manifestations of toxicity. A similar effect probably also occurs in man. Such an effect will increase the susceptibility of the host to disease processes like infection and malignancy. Altered function of immunoregulatory cells may also increase the incidence of autoimmune and atopic diseases in the exposed population. Thus immunological biomonitoring may be a valuable guide to gauge the extent of this subclinical toxicity and may lead to the prevention of these potential complications.

References

Adkins, B., Jr., Richards, J. H., and Gardner, D. E. (1979), *Enhancement of Experimental Respiratory Infection Following Nickel Inhalation.* Environ. Res. 20, 33 – 42.
Blakley, B. R., Sisodia, C. S., and Mukkur, T. K. (1980), *The Effect of Methylmercury, Tetraethyl Lead, and Sodium Arsenite on the Humoral Immune Response in Mice.* Toxicol. Appl. Pharmacol. 52, 245 – 254.

Blakley, B. R. (1985), *The Effect of Cadmium Chloride on the Immune Response in Mice.* Can. J. Comp. Med. **49**, 104–108.

Bozelka, B. E., Burkholder, P. M., and Chang, L. W. (1978), *Cadmium: A Metallic Inhibitor of Antibody-mediated Immunity in Mice.* Environ. Res. **17**, 390–402.

Burchiel, S. W., Hadley, W. M., Cameron, C. L., Fincher, R. H., Lim, T.-W., Elias, L., and Steward, C. C. (1987), *Analysis of Heavy Metal Immunotoxicity by Multiparameter Flow Cytometry: Correlation of Flow Cytometry and Immune Function Data in B6CF1 Mice.* Int. J. Immunopharmacol. **9**, 597–610.

Chandra, R. K. (1984), *Excessive Intake of Zinc Impairs Immune Responses.* J. Am. Med. Assoc. **252**, 1443–1446.

Chowdhury, B. A., Friel, J. K., and Chandra, R. K. (1987), *Cadmium-induced Immunopathology is Prevented by Zinc Administration in Mice.* J. Nutr. **117**, 1788–1794.

Cook, J. A., Hoffmann, E. O., and DiLuzio, N. R. (1975), *Influence of Lead and Cadmium on the Susceptibility of Rats to Bacterial Challenge.* Proc. Soc. Exp. Biol. Med. **150**, 741–747.

Descotes, J. (1986), *Immunotoxicology of Drugs and Chemicals.* Elsevier Science Publishers, Amsterdam.

Exon, J. H., Koller, L. D., and Kerkvliet, N. I. (1979), *Lead-Cadmium Interaction: Effects on Viral-induced Mortality and Tissue Residues in Mice.* Arch. Environ. Health **34**, 469–474.

Faith, R. E., Luster, M. I., and Kimmel, C. A. (1979), *Effect of Chronic Developmental Lead Exposure on Cell-mediated Immune Function.* Clin. Exp. Immunol. **35**, 413–420.

Fraker, P. J., Gershwin, M. E., Good, R. A., and Prasad, A. S. (1986), *Interrelationship Between Zinc and Immune Function.* Fed. Proc. **45**, 1474–1479.

Fujimaki, H., Murakami, M., and Kubota, K. (1982), *In vitro Evaluation of Cadmium-induced Augmentation of the Antibody Response.* Toxicol. Appl. Phamacol. **62**, 288–293.

Gainer, J. H. (1973), *Activation of the Rauscher Leukemia Virus by Metals.* J. Natl. Cancer Inst. U.S. **51**, 609–613.

Gainer, J. H. (1977), *Effects of Heavy Metals and of Deficiency of Zinc on Mortality Rates in Mice Infected with Encephalomyocarditis Virus.* Am. J. Vet. Res. **38**, 869–872.

Gainer, J. H., and Pry, T. W. (1972), *Effects of Arsenicals on Viral Infections in Mice.* Am. J. Vet. Res. **33**, 2299–2307.

Gallagher, K., Matarazzo, W. J., and Gray, I. (1979), *Trace Metal Modification of Immunocompetence. II. Effect of Pb^{2+}, Cd^{2+}, and Cr^{3+} on RNA Turnover, Hexokinase Activity, and Blastogenesis during B-lymphocyte Transformation in vitro.* Clin. Immunol. Immunopathol. **13**, 369–377.

Gaworski, C. L., and Sharma, R. P. (1978), *The Effects of Heavy Metals on [^{3}H]thymidine Uptake in Lymphocytes.* Toxicol. Appl. Pharmacol. **46**, 305–313.

Graham, J. A., Miller, F. J., Daniels, M. J., Payne, E. A., and Gardner, D. E. (1978), *Influence of Cadmium, Nickel, and Chromium on Primary Immunity in Mice.* Environ. Res. **16**, 77–87.

Haley, P. J., Bice, D. E., Muggenburg, B. A., Hahn, F. F., and Benjamin, S. A. (1987), *Immunopathologic Effects of Nickel Subsulfide on the Primate Pulmonary Immune System.* Toxicol. Appl. Pharmacol. **88**, 1–12.

Hildebrand, C. E., and Cram, L. S. (1979), *Distribution of Cadmium in Human Blood Cultured in Low Levels of $CdCl_2$: Accumulation of Cd in Lymphocytes and Preferential Binding to Metallothionein.* Proc. Soc. Exp. Biol. Med. **161**, 438–443.

Hinton, D. E., Lipsky, M. M., Heatfield, B. M., and Trump, B. F. (1979), *Opposite Effects of Lead on Chemical Carcinogenesis in Kidney and Liver of Rats.* Bull. Environ. Contam. Toxicol. **23**, 464–469.

Karl, L., Chvapil, M., and Zukoski, C. F. (1973), *Effect of Zinc on the Viability and Phagocytic Capacity of Peritoneal Macrophages.* Proc. Soc. Exp. Biol. Med. **142**, 1123–1127.

Kiremidjian-Schumacher, L., Stotzky, G., Likhite, V., Schwartz, J., and Dickstein, R. A. (1981), *Influence of Cadmium, Lead, and Zinc on the Ability of Sensitized Guinea Pig Lymphocytes to Interact with Specific Antigen and to Produce Lymphokine.* Environ. Res. **24**, 96–105.

Koller, L. D. (1973), *Immunosuppression Produced by Lead, Cadmium, and Mercury. Am. J. Vet. Res. 34*, 1457–1458.
Koller, L. D. (1975), *Methylmercury: Effect on Oncogenic and Non-oncogenic Viruses in Mice. Am. J. Vet. Res. 36*, 1501–1504.
Koller, L. D. (1980), *Immunotoxicology of Heavy Metals. Int. J. Immunopharmacol. 2*, 269–279.
Koller, L. D., and Kovacic, S. (1974), *Decreased Antibody Formation in Mice Exposed to Lead. Nature 250*, 148–150.
Koller, L. D., and Roan, J. G. (1977), *Effects of Lead and Cadmium on Mouse Peritoneal Macrophages. J. Reticuloendothel. Soc. 21*, 7–12.
Koller, L. D., Exon, J. H., and Roan, J. G. (1976), *Humoral Antibody Response in Mice After Single Dose Exposure to Lead or Cadmium. Proc. Soc. Exp. Biol. Med. 151*, 339–342.
Koller, L. D., Exon, J. H., and Arbogast, B. (1977), *Methylmercury: Effect on Serum Enzymes and Humoral Antibody. J. Toxicol. Environ. Health 2*, 1115–1123.
Koller, L. D., Roan, J. G., and Kerkvliet, N. I. (1979), *Mitogen Stimulation of Lymphocytes in CBA Mice Exposed to Lead and Cadmium. Environ. Res. 19*, 177–188.
Krzystyniak, K., Fournier, M., Trottier, B., Nadeau, D., and Chevalier, G. (1987), *Immunosuppression in Mice after Inhalation of Cadmium Aerosol. Toxicol. Lett. 38*, 1–12.
Loose, L. D., Silkworth, J. B., and Warrington, D. (1978), *Cadmium-induced Phagocyte Cytotoxicity. Bull. Environ. Contam. Toxicol. 20*, 582–588.
Luster, M. I., Faith, R. E., and Kimmel, C. A. (1978), *Depression of Humoral Immunity in Rats Following Chronic Developmental Lead Exposure. J. Environ. Pathol. Toxicol. 1*, 397–402.
Malave, I., and DeRuffino, D. T. (1984), *Altered Immune Response During Cadmium Administration in Mice. Toxicol. Appl. Pharmacol. 74*, 46–56.
McCabe, M., Maguire, D., and Nowak, M. (1983), *The Effects of Arsenic Compounds on Human and Bovine Lymphocyte Mitogenesis in vitro. Environ. Res. 31*, 323–331.
Menné, T., and Nieboer, E. (1989), *Metal Contact Dermatitis: A Common and Potentially Debilitating Disease. Endeavour 13* (3), 117–122.
Müller, S., Gillert, K. E., Krause, C., Gross, U., and Age-Schehr, J. L., Diamantstein, T. (1977), *Suppression of Delayed Type Hypersensitivity of Mice by Lead. Experientia 33*, 667–668.
Müller, S., Gillert, K. E., Krause, C., Jautzke, G., Gross, U., and Diamantstein, T. (1979), *Effects of Cadmium on the Immune System of Mice. Experientia 35*, 909–910.
Neilan, B. A., O'Neill, K., and Handwerger, B. S. (1983), *Effect of Low-level Lead Exposure on Antibody-dependent and Natural Killer Cell-mediated Cytotoxicity. Toxicol. Appl. Pharmacol. 69*, 272–275.
Nixon, J. E, Koller, L. D., and Exon, J. H. (1979), *Effect of Methylmercury Chloride on Transplacental Tumors Induced by Sodium Nitrite and Ethylurea in Rats. J. Natl. Cancer Inst. U.S. 63*, 1057–1063.
Parker, D., and Turk, J. L. (1978), *Delay in the Development of the Allergic Response to Metals Following Intratracheal Installation. Int. Arch. Allergy Appl. Immunol. 57*, 289–293.
Shenker, B. J., Matarazzo, W. J., Hirsch, R. L., and Gray, I. (1977), *Trace Metal Modification of Immunocompetence. I. Effect of Trace Metals in the Cultures on in vitro Transformation of B Lymphocytes. Cell. Immunol. 34*, 19–24.
Smialowicz, R. J., Rogers, R. R., Riddle, M. M., and Stott, G. A. (1984), *Immunologic Effects of Nickel. I. Suppression of Cellular and Humoral Immunity. Environ. Res. 33*, 413–427.
Smialowicz, R. J., Rogers, R. R., Riddle, M. M., Garner, R. J., Lowe, D. G., and Luebke, R. W. (1985), *Immunologic Effects of Nickel. II. Suppression of Natural Killer Cell Activity. Environ. Res. 36*, 56–66.
Sobocinski, P. Z., Cantebury, W. J., Jr., and Powanda, M. C. (1977), *Differential Effect of Parenteral Zinc on Course of Various Bacterial Infections. Proc. Soc. Exp. Biol. Med. 56*, 334–339.
Takenaka, S., Oldiges, H., König, H., Hochrainer, D., and Oberdörster, G. (1983), *Carcinogenicity of Cadmium Chloride Aerosols in W Rats. J. Natl. Cancer Inst. U.S. 70*, 367–373.

Talcott, P. A., and Koller, L. D. (1983), *The Effect of Inorganic Lead and/or a Polychlorinated Biphenyl on the Developing Immune System of Mice. J. Toxicol. Environ. Health* 12, 337–352.

Thun, M. J., Schnorr, T. M., Smith, A. B., Halperin, W. E., and Lemen, R. A. (1985), *Mortality Among a Cohort of U.S. Cadmium Production Workers: An Update. J. Natl. Cancer Inst. U.S.* 74, 325–333.

Trejo, R. A., DiLuzio, N. R., Loose, L. D., and Hoffman, E. (1972), *Reticuloendothelial and Hepatic Functional Alterations Following Lead Acetate Administration. Exp. Mol. Pathol.* 17, 145–158.

Ward, P. A., Goldschmidt, P., and Greene, N. D. (1975), *Suppressive Effects of Metal Salts on Leukocyte and Fibroblastic Function. J. Reticuloendothel. Soc.* 18, 313–321.

Wiernik, A., Johansson, A., Jarstrand, C., and Camner, P. (1983), *Rabbit Lung After Inhalation of Soluble Nickel. I. Effects on Alveolar Macrophages. Environ. Res.* 30, 129–141.

Yoshida, T., Shimamura, T., and Shigeta, S. (1987), *Enhancement of the Immune Response in vitro by Arsenic. Int. J. Immunopharmacol.* 9, 411–415.

Note Added in Proof

Trüb, L. (1990), has recently written an important review on constitutional eczemas (causes and therapy of neurodermitis constitutionalis atopica), a hereditary disposition for allergic diseases (in German), in *Neue Zürcher Zeitung, Forschung und Technik,* No. 19, p. 79 (24 January), based on lectures of S. Borelli, H. Düngemann, and W. Gühring, Davos, Switzerland, 1989.

I.18 Mutagenicity, Carcinogenicity, Teratogenicity

ERICH GEBHART, Erlangen, Federal Republic of Germany
TOBY G. ROSSMAN, New York, USA

1 Introduction

The induction of changes in genetic information (mutations) by environmental agents is one of the important aspects of modern environmental research. Mutations are events of considerable consequences for the individual but also for the afflicted population. The reaction of chemical agents with nucleic acids, the chemical carriers of genetic information, yields a rather broad and variable spectrum of consequences via complex and, in part, unsolved mechanisms: Beside lethal damage, molecular mutations (gene mutations) or structural chromosome changes (chromosome mutations) may be produced, all of which can cause more or less marked changes of the phenotype. In addition, numerical chromosome changes (genome mutations) arising from disturbances of the mitotic process may be induced by chemical mutagens. In contrast to other environmental chemicals, several metals are essential constituents of vital molecules. At the wrong concentration or at the wrong site, however, they can cause grave consequences for the afflicted cell or the afflicted individual. This is particularly true with respect to the genetic information.

The most recent data of molecular cancer genetics and cytogenetics have provided strong evidence that mutational events in somatic cells can be the initial step of a malignant process. Thus, the evidence of a clear interrelationship of mutagenesis and carcinogenesis has been strengthened by the research of the last years.

The induction of malformations by the action of metals on embryos or fetuses in utero is a problem of their toxic rather than of their mutagenic activity. Teratogenicity, therefore, is discussed in a separate section of this chapter.

2 Mutagenicity
(see also Part II)

A large range of reliable methods for testing the mutagenicity of chemicals is now available. In order to relate the results of a mutagenicity assay to man, the metabolism of the test agent in the test system should at least be approximately comparable to its actual metabolism in the human body. For a thorough analysis, there should be attempts to detect the whole spectrum of known types of mutations by the applied test battery. A catalog of mutagenicity test procedures has been published by KILBEY et al. (1984).

2.1 Reactions with Nucleic Acids

The demonstration of direct reactions of an agent with DNA is of fundamental importance in understanding its mutagenicity. The numerous nucleophilic centers in nucleic acids are favorite sites for binding of metals. The type and localization of binding apparently depends on the respective metal. The most serious reaction of metals with DNA, from a genetic point of view, is the induction of cross-links between both DNA strands. Cross-linking (see also Part II) has been shown to occur with Cu(II), Zn(II), Co(II), and Mn(II) (EICHHORN, 1979), and also with *cis*-diammin-dichloro-platinum (ROBERTS and THOMSON, 1979).

As reviewed by LEVIS and BIANCHI (1982) Cr(VI) has been clearly demonstrated to bind to purified DNA, resulting in the induction of cross-links (see also Chapter II.7). In the presence of a microsomal activating system Cr(VI) is reduced to Cr(III) which preferentially binds to guanine and results in DNA-protein cross-links (TSAPAKOS and WETTERHAHN, 1983; BEYERSMANN and KÖSTER, 1987). Recently, the induction of DNA-protein cross-links in cells treated with chromate has been shown to involve specific nuclear proteins (MILLER and COSTA, 1988). Such cross-links are apparently important in mediating the long-term toxic and carcinogenic effects of chromates (SUGIYAMA et al., 1986). The latter authors, in addition, found induction of single-strand breaks in the DNA of three cell lines of different origin. Other reactions such as chelation or formation of complexes between DNA and metals, above all by the so-called "soft" ions, e.g., Ag(II), Hg(II), at nucleophilic DNA sites could also be detected (e.g., JACOBSON and TURNER, 1980). Besides the differing affinity to DNA of different metals, the steric structure of the DNA bases is of crucial importance for those reactions. "Hard" ions, such as Na(I), Mg(II), and Ca(II), bind nearly exclusively to the phosphate groups, apparently stabilizing the DNA helix. "Borderline" ions, e.g., Co(II), Ni(II), Zn(II), Mn(II), Cd(II), Cu(II) are able to react with the phosphate groups as well as with the bases, the ionic strength evidently influencing their respective affinity. These latter ions also can form mixed chelates between the phosphates and the bases (JACOBSON and TURNER, 1980, for details). Similar reactions are also to be expected between metals and RNA, thereby affecting the specific functions of the latter (EICHHORN, 1979; JACOBSON and TURNER, 1980).

Several metal ions can react with the amino and sulfhydryl groups of proteins. Metal binding domains are especially important for proteins involved in nucleic acid metabolism (BERG, 1986). Considering the importance of these groups for enzyme activity, reactions with metal ions frequently result in an inhibition or at least a considerable misfunction of various enzymatic systems, particularly those involving nucleic acid metabolism. Such reactions are known to occur, for instance, with arsenic (LÉONARD and LAUWERYS, 1980a; see also Chapter II.3), but other metals can also inhibit or stimulate certain enzymatic functions (SIROVER and LOEB, 1976; JACOBSON and TURNER, 1980). When the enzymes are those involved in DNA

Table I.18-1. Metal Compounds Giving Positive Results in Some in vitro Assays

Infidelity of DNA Synthesis[1]	rec Assay[2]	Prophage Induction[3]	Mutagenesis		In vitro Transformation[6]
			Bacterial[4]	Mammalian[5]	
Ag(I)	As(III), As(V)	Cr(III), Cr(VI)	Cd(II)[7]	Be(II)	As(III), As(V)
Be(II)	Be(II)	Fe(II), Fe(III)	Cr(VI)	Cd(II)	Be(II)
Cd(II)	Cd(II)	Mo(VI)	Mn(II)	Cr(VI)	Cd(II)
Co(II)	Co(II)	Mn(II)	Mo(II)	Mn(II)	Cr(VI)
Cr(III)	Cr(VI)	Ni(II)	Pt (complex)	Mo(VI)	Ni(II)
Cu(II)	Co(I)	Pb(II)	Se(VI)	Ni(II)	Pb(II)
Mn(II)	Hg(II)	Pt (complex)	Rh(III)	Pb(II)	Ti(III)
Ni(II)	Mo(VI)	Se(VI)	Te(IV)	Pt (complex)	
Pb(II)	Os(VIII)	Sn(II)		W(VI)	
	Pt (complex)	W(VI)		Zn(II)	
	Rb(I), Rb(II)	Zn(II)			
	Se(IV)				
	Sb(III), Sb(V)				
	Te(IV)				
	Tl(I)				
	V(V)				

General references: ROSSMAN (1981); SIROVER (1981); VOUK and PIVER (1983); HANSEN and STERN (1984)

Specific references (selected):
[1] ZAKOUR et al. (1981); [2] KANEMATSU et al. (1980); [3] ROSSMAN et al. (1984), TAMARO et al. (1979); [4] HECK and COSTA (1982a); [5] HECK and COSTA (1982a); ZELIKOFF et al. (1986); [6] HECK and COSTA (1982b); [7] MANDEL and RYSER (1984)

replication, metals can contribute in an indirect way to alterations in genetic information by affecting replication fidelity. Table I.18-1 summarizes data on the induction of infidelity of DNA synthesis as well as other aspects of DNA interactions.

It must be pointed out, however, that the fact that a metal compound causes infidelity in DNA replication does not imply that the metal necessarily affects the DNA polymerase enzyme. For example, it has recently been found that Mn(II) at low concentration appears to cause errors in replication via interactions with DNA rather than with the polymerase (BECKMAN et al., 1985).

Metal compounds can also cause DNA strand breaks indirectly. For example, Cu(II), Ni(II), and Cr(VI) were able to cause depurination of DNA, an event which can have mutagenic consequences (SCHAAPER et al., 1987). In the cell, apurinic sites can lead to strand breaks via cleavage by AP endonucleases; in vitro, such sites are detectable as alkali-labile sites. Another indirect mechanism by which metals can damage DNA is via their involvement in generating oxygen free radicals (see review by AUST et al., 1985). In one study, Cd(II), Mg(II), Mn(II), Cr(VI), Zn(II), and selenite were all shown to induce strand breaks in human fibroblasts. In each case except selenite, the DNA damage could be blocked by hydroxyl radical scavengers (SNYDER, 1988). Cr(VI) has been shown to react with H_2O_2 to produce more highly active oxygen species (KAWANISHI et al., 1986).

There are a number of short-term in vitro assays which give indirect evidence for damage to DNA. The rec assay, for example, measures the toxicity of compounds in repair competent (rec^+) and incompetent (rec^-) strains of *Bacillus subtilis* (KANEMATSU et al., 1980). The theory behind this assay is that if a compound damages DNA, it should be more toxic to cells which lack the ability to perform DNA repair. A large number of metal compounds give a positive response in this assay (Table I.18-1). (Positive responses or results in this chapter means that a metal compound causes effects (visible interactions or damages) in a laboratory test.) It should be emphasized that results in this assay are only suggestive (and not proof) of DNA damage. A similar argument can be made for the pol test, which utilizes repair competent ($pol\ A^+$) and repair incompetent ($pol\ A^-$) strains of *Escherichia coli*. As(III), Cd (II), and Cr(VI) give a positive pol test (DE FLORA et al., 1984).

Another assay which indirectly measures DNA damage (or halting of the replication fork) is the induction of prophage in the Microscreen assay (ROSSMAN et al., 1984, 1986a). This assay measures the ability of an agent to induce the SOS system in *E. coli*. Compounds which cause any type of DNA lesion (strand breaks, cross-links, intercalation, alkylation, etc.) will cause the induction of this system, resulting in the release of free phage from its integrated (prophage) form. Metals capable of inducing prophage are listed in Table I.18-1. This assay also does not prove that DNA was damaged, since agents which halt the replication fork by inhibiting proteins required for DNA replication can also induce the SOS system.

The induction of sister chromatid exchanges (SCE) is also thought to be a reflection of DNA damage and/or repair (GEBHART, 1981). Few metal compounds have been studied using this assay system. Positive results have been reported for Cr(VI), Ni(II), As(III), and As(V). (See reviews in VOUK and PIVER, 1983, and CHRISTIE and COSTA, 1983, as well as more recent work on chromium and nickel by SEN and COSTA, 1986.)

Finally, indirect evidence of DNA damage can be obtained by assaying the ability of a compound to elicit DNA repair replication. In mammalian cells, this can be demonstrated by measuring "unscheduled DNA synthesis" (UDS) which measures the incorporation of radiolabeled precursors into DNA under conditions where the normal replication of DNA is inhibited. Very few metal compounds have been studied in this assay. A positive response has been seen for Se(IV) and Se(VI) (WHITING et al., 1980a), Cr(VI) (WHITING et al., 1979) and a number of platinum complexes (PLOOY and LOHMAN, 1980).

2.2 Mutagenesis and in vitro Transformation

Bacterial mutagenicity assays have played a major role in the preliminary assessment of genetic risk from chemicals. These assays are simple, fast and inexpensive, especially compared with whole animal carcinogenicity tests. The Ames test, which is the bacterial mutagenicity assay used most often, measures reversion from histidine-requiring (his^-) to histidine independence (his^+) in *Salmonella typhimurium*. The *Eschericia coli* WP2 system similarly measures reversion at a tryptophan locus. For a review on various short-term assays, see HOLLSTEIN and

McCann (1979). Reversion assays depend upon mutations of a specific type at specific sites in the DNA, and require different tester strains for each type of reversion. In general, a base pair substitution mutation can only be reverted by a second base pair substitution (not necessarily at the same site), and a frameshift mutation can only be reverted by a second frameshift which restores the reading frame. In addition, separate strains are required to detect mutagens which cause reversions predominantly at A-T sites or at G-C sites.

Bacterial mutagenicity systems are notoriously insensitive at detecting metal compounds. Table I.18-1 lists eight metals whose compounds are reported to be mutagenic in bacteria. Chromate is the easiest carcinogenic metal compound to detect in bacterial mutagenesis assays, and is especially active in a strain that detects base pair substitutions at A-T sites (Bennicelli et al., 1983). Although compounds of nickel, cadmium, and arsenic are considered to be human carcinogens, neither cause reversions in bacterial systems (see references in Table I.18-1). Carcinostatic platinum complexes cause cross-links in DNA and are usually positive in short-term assays (Table I.18-1). Some of the other metals listed in Table I.18-1 have been cited in only one report and have not been confirmed.

The reasons for the insensitivity of bacterial assays in detecting genotoxic metal compounds have been discussed (Rossman et al., 1984, 1987). In brief, one or more of the following reasons may apply to any specific metal: (1) Some metal ions may not be able to enter bacterial cells. This may be true for metal compounds which do not produce toxicity at even very high concentrations (Rossman et al., 1984). (2) Bacterial enzymes may be more sensitive to metal ion inhibition, resulting in toxicity which masks the mutagenicity. (3) Insoluble metal compounds may be more genotoxic and may enter mammalian cells by phagocytosis. Bacteria are unable to phagocytize (see Rossman et al., 1987). (4) Some metal compounds may be co-mutagens rather than mutagens (see Sect. 2.4). (5) The genetic end point being measured (reversion) is inappropriate for compounds which cause deletions, rearrangements, aneuploidy, or gene amplification. (6) Bacterial DNA repair systems may be very efficient at handling damage due to metal compounds.

The mutagenicity of metal compounds in mammalian cells has been assessed using two systems. In one, a line of mouse lymphoma cells is used, and forward mutation at the thymidine kinase (*tk*) locus is measured by scoring resistance to trifluorothymidine (Amacher and Paillet, 1980; Oberly et al., 1982). In the other system, Chinese hamster V79 cells are used, and forward mutation at the hypoxanthine-guanine-phosphoribosyl transferase (*hprt*) locus is measured by scoring resistance to 6-thioguanine (Miyaki et al., 1979). In both of these assays, forward mutation of a non-essential gene is being measured and, therefore, base pair substitutions, frameshifts, deletions, and inactivating rearrangements should all be measurable.

It is necessary, when analyzing the literature on metal mutagenicity, to go back to the primary sources. Often, a metal compound is reported to be mutagenic even if there is only a doubling of the mutation frequency at very low levels of survival. This "positive" result is then cited in every review article. Even very high concentrations of NaCl can induce a "positive" mutagenic response (Moore and Brock, 1988). The metals listed in Table I.18-1 include only those which appear to cause significant (>2×) increases in mutation frequency at high survival levels (around 50% or higher). However, as in the case of bacterial mutagenesis, not all of these results have been confirmed. AMACHER and PAILLET (1980) found that $NiCl_2$ and trans-$Pt(NH_3)_2Cl_2$ caused a greater than 2-fold increase in mutagenesis in the mouse lymphoma system, but not $CdCl_2$, $Pb(OOCCH_3)_2$, $MgCl_2$ or Na_2HAsO_4. Using the same system, OBERLY et al. (1982) obtained positive responses with $CdSO_4$ (4× increase at 55% survival), $MnCl_2$ (3.3× increase at 48% survival) and chromate compounds which gave a clear dose-response relationship.

Positive results with chromates have been shown in many laboratories (see HECK and COSTA, 1982b). $Pb(NO_3)_2$ also appeared to give a positive response, which the authors suggest might be due to the precipitate formed. Both soluble and insoluble lead compounds are mutagenic in V79 cells (ZELIKOFF et al., 1988). Results with $HgCl_2$, $NaAsO_2$, and Na_2HAsO_4 in the mouse lymphoma system are questionable (no effects at >50% survival). Although $AlCl_3$ did induce a doubling of mutation frequency at high survival, the pH of the medium was affected by this compound. Low pH treatment is itself mutagenic in this system (BRUSICK, 1986). In the Chinese hamster V79 cell system, positive results were obtained with $BeCl_2$ (6× increase at 50% survival) (MIYAKI et al., 1979). The concentrations of $MnCl_2$ and $CoCl_2$ were too high to evaluate properly. $NiCl_2$ caused no increase at high survival levels. Recently, ROSSMAN's laboratory (ZELIKOFF et al., 1986, 1988, and unpubl.) evaluated a number of metal salts in the V79 cell system, using chronic exposure and high survival conditions. No increase in mutagenesis was seen with $NaAsO_2$, $BaCl_2$, $CuCl_2$, $HgCl_2$, $MgCl_2$, or NaCl. A greater than 3-fold increase at high (>70%) survival levels was seen for K_2CrO_4, $Ni(OOCCH_3)_2$, $ZnCl_2$, Na_2MoO_4, $KMnO_4$ and Na_2WO_4.

The lack of a clear mutagenic response in the hprt assay does not imply the absence of a genotoxic effect. The hprt gene is X-linked, and therefore less sensitive to the detection of multilocus deletions (which might be lethal), compared with hemizygous autosomal loci (e.g., tk in the mouse lymphoma assay). However, a recent advance in mammalian mutagenicity studies involved the use of stably integrated bacterial genes as targets for mutagenesis. One such line is g12, a derivative of Chinese hamster V79 cells which carries a single copy of the E. coli gpt gene (KLEIN and ROSSMAN, manuscript in preparation). Using this line, a very high level (60×) of mutagenicity could be demonstrated for insoluble nickel sulfide, while a much weaker (2×) mutagenic response was seen for soluble nickel sulfate (CHRISTIE et al., 1989). It will be of interest to see if other metal compounds are also more mutagenic when assayed in this system. Recently, it was found that $NiCl_2$ increased mutagenesis in viral DNA intergrated in mouse cells (BIGGART and MURPHY, 1988).

In vitro cell transformation systems measure the ability of an agent to convert a non-tumorigenic cell into a tumorigenic cell. What is actually measured are changes in the growth pattern (e.g., ability to form a colony in soft agar) which correlate with tumorigenicity. Most of the studies of metal-induced transformation have been carried out using the Syrian hamster embryo (SHE) system (e.g., COSTA et al., 1980). The known human metal carcinogens are all able to cause transformation in vitro (Table I.18-1). In addition, suspect carcinogens such as Be(II) and Pb(II), as well as Ti(III), are positive in this assay, (see review by HECK and COSTA, 1982).

Cell transformation can apparently occur via a number of different mechanisms and should not be viewed as merely another mutational assay. For example, compounds of arsenic are able to transform SHE cells, but do not cause mutations at the gene level (ROSSMAN et al., 1980). However, it has recently been shown that arsenite can cause gene amplification (LEE et al., 1988), an effect which will not be detected in a standard mutagenesis assay.

One additional factor needs to be mentioned with regard to studying metal induced mutations or transformation. It is quite possible that the emphasis on studying soluble metal compounds might be an error. For example, phagocytosis of insoluble (precipitated) compounds of Pb(II), Ba(II), Be(II), Ni(II), and Mn(II) may be an important route of entry into mammalian cells (ROSSMAN et al., 1987). Insoluble nickel sulfide was much more mutagenic than soluble nickel sulfate at the *gpt* locus in the V79 transfectant line g12 (CHRISTIE et al., 1989).

2.3 Induction of Chromosome and Genome Mutations

The phenomena seen in clincial cytogenetics illustrate the grave pathologic consequences of structural and numerical chromosome aberrations (e.g., SCHINZEL, 1984). In addition, chromosomal changes in single somatic cells are now thought to be one route in the process of malignant transformation, as documented by the vast data of tumor cytogenetics (e.g., SANDBERG, 1980; MITELMAN, 1988; WOLMAN, 1986).

The demonstration of chromosome and genome mutations, therefore, has great significance in modern mutagenicity studies. Utilizing mammalian in vivo test systems guarantees data of the highest experimentally available level of relevance for man. Data from cytogenetic test systems are collected in Table I.18-2. Of particular interest, however, are chromosome studies on somatic cells of humans exposed to metallic compounds (reviewed by GEBHART, 1985). One of the first heavy metals under intense study was lead; chromsome studies have been performed for many years on various exposed populations (see GEBHART, 1985, for references). From the very beginning of such studies conflicting data on the chromosome damaging activity of lead have been obtained (GERBER et al., 1980). Reasons for these discrepancies may lie, at least in part, in the technical peculiarities of the different studies. In addition to the route of exposure, variations exist in the size of the exposed population, the number of analyzed metaphases, as well as culture time of the lymphocytes. Since some groups (e.g., SCHWANITZ et al., 1975), who reported increased rates of chromosome damage in exposed individuals, could not demonstrate a significant correlation between individual aberration rates and individual levels of lead exposure (e.g., expressed as urinary excretion of aminolevulinic acid), there must be other reasons for their findings. Several authors were able to show that the simultaneous influence of other heavy metals (cadmium, zinc) or half metals (arsenic) apparently influence the frequency and types of chromosomal aberrations found in lead-exposed individuals (NORDENSON et al., 1978; NORDENSON and BECKMAN, 1982; DEKNUDT and LÉONARD, 1975; BAUCHINGER et al., 1976; OBE et al., 1984).

Two further facts of apparent general validity should be pointed out: Several authors reported evidence of a weakly mutagenic (HOPKIN and EVANS, 1980;

Table I.18-2. Metals Inducing Chromosome Damage

Chromosome Mutations in Plants/Insects	Mammals Cell Culture	In vivo	Humans Cell Culture	In vivo	Genome Mutations in Plants	Cell Culture
Al[1]	As[1,21]	Al[1]	As[1,6,21]	As[6,11,21]	As[1,6]	As[1,6]
As[1]	Be[1]	As[1,21]	Cd[1]	(Cd)[9,12]	Au[16]	Cd[17]
Ba[1]	Cd[1]	Cd[1]	Cr[1,5]	Cr[1,5]	Ba[16]	Co[18]
Be[1]	Cr[1,5]	Cr[1,5]	Hg[1,4]	(Hg)[4,13]	Be[1,16]	Cr[1]
Cd[1,2]	(Fe)[1]	Hg[1,4]	(Li)[8]	(Li)[8]	(Bi)[1]	Hg[1,4]
(Ce)[1]	Hg[1,4]	Li[7]	Ni[20]	Ni[20]	Cd[16]	
Co[2]	Ni[1,20]	Ni[20]	(Pb)[1]	(Pb)[1,14,15]	Co[16]	
Cr[1,2,5]	Pb[1]	Pb[1]	Pt[1]	(Zn)[9,10]	Cr[16]	
(Cs)[1]	Pt[1]	Sb[1]	Se[19]		Cu[16]	
Cu[1,2]	Se[1,19]	(Se)[1,19]	Zn[9,10]		Hg[3,4]	
Fe[2]	Te[1]	Zn[1]			Pa[16]	
Hg[3,4]	Tl[1]	Pt[22]			Rb[16]	
Mn[1,2]	Zn[1]				Tl[16]	
Nd[1]						
Ni[1,2]						
Pt[1]						
Th[1]						
Tl[1]						
Zn[2]						

[1] HANSEN and STERN, 1984; [2] GLÄSS, 1956; [3] GEBHART, 1969; [4] LÉONARD et al., 1983; [5] LEVIS and BIANCHI, 1982; [6] LÉONARD and LAUWERYS, 1980a; [7] KING et al., 1979; [8] DE LA TORRE and KROMPOTIC, 1976; [9] DEKNUDT and LÉONARD, 1975; [10] PILINSKAYA, 1970, 1974; [11] NORDENSON et al., 1978, 1979; [12] BAUCHINGER et al., 1976; [13] POPESCU et al., 1979; [14] FORNI et al., 1980; [15]SCHWANITZ et al. 1975; [16] LEVAN, 1945; [17] RÖHR and BAUCHINGER et al., 1976; [18] DE SOUZA NAZARETH, 1976; [19] NORPPA et al., 1980; SHAMBERGER, 1985; NEWTON and LILLY, 1986; [20] LÉONARD et al., 1981; WAKSVIK and BOYSEN, 1982; [21] JACOBSON-KRAM and MONTALBANO, 1985; DEKNUDT et al., 1986; [22] LEVINE et al., 1980; TANDON and SODHI, 1985

HUSGAFVEL-PURSIAINEN et al., 1980; OBE et al., 1984), and also comutagenic (MÄKI-PAAKKANEN et al., 1981; WATANABE et al., 1983) action of heavy smoking. Therefore, smoking habits of test and control persons also could influence the outcome of cytogenetic studies on individuals exposed to heavy metals.

A very important factor in judging cytogenetic data from mutagen-exposed populations is the rather high interindividual variation of induced aberration frequency, as shown, e.g., by studies of GEBHART et al. (1982) on patients undergoing cytostatic therapy. This phenomenon, or problem of heterogeneity, however, reflects an important reality which must not be neglected if mutagenicity of heavy metals in man is discussed.

The limited number of cytogenetic studies on occupational exposure to cadmium revealed no distinct chromosome damaging or SCE inducing activity on man by this metal (O'RIORDAN et al., 1978; FLEIG et al., 1983; NOGAWA et al., 1986). DEGRAEVE (1981), reviewing the literature, regarded cadmium as a comutagen rather than a mutagen by itself. Increased aberration rates, as compared with non-exposed controls, however, were described from workers of a zinc smelting plant who had increased blood levels of lead and cadmium (BAUCHINGER et al., 1976).

In workers exposed to lead and zinc, a higher frequency of chromosome aberrations was observed than in those exposed to lead and cadmium; the chromosome damage recorded in the latter group, however, was of more serious type than in the former. Apparently, though not clearly clastogenic by themselves, lead, cadmium, and zinc may constitute a potential genetic hazard to man by their combined action or by increasing the weak mutagenic effect of smoking.

In fishermen who had consumed mercury-contaminated fish, an increase of chromosome damage was reported (KATO, 1976; SKERFVING et al., 1974), and there was some correlation between the observed blood levels of mercury and the extent of chromosome damage. Occupational load with mercury or some organic compounds under certain circumstances may induce chromosome damage (POPESCU et al., 1979), although this action could be due to other environmental agents. VERSCHAEVE et al. (1979) found no significant increase in aberration rates in 28 workers exposed to metallic mercury at a chloralkali plant, although urine mercury levels were higher in the exposed workers than in controls. MABILLE et al. (1984) also could not confirm a chromosome damaging action of mercury exposure. In addition, suitable precautions taken for toxicological reasons may reduce the aberration-inducing activity of mercury and its compounds to the control level. Segregational errors of chromosomes in human lymphocytes and in Indian muntjak cells, however, were reported (VERSCHAEVE et al., 1984).

Considering the well-documented chromosome damaging activity of hexavalent chromium compounds in experimental studies on cell cultures and test animals (LÉONARD and LAUWERYS, 1980b; LEVIS and BIANCHI, 1982), it is not surprising that chromium and certain of its compounds were also found distinctly clastogenic in exposed individuals (BIGALIEV et al., 1977; IMREH and RADULESCU, 1982). The only study performed to date on nickel-exposed workers revealed a significant increase of achromatic lesions but not of true breaks (WAKSVIK and BOYSEN, 1982).

Data on the chromosome damaging activity of arsenic have been reported from persons exposed to it for therapeutic and for occupational reasons. PETRES et al. (1977) found a significant increase of structural chromosome aberrations in persons after therapy with arsenic, while NORDENSON and co-workers (1978, 1982) reported the same from workers occupationally exposed to arsenic in a smelting plant in Northern Sweden.

Platinum compounds, particularly cis-Pt-diammine dichloride, increasingly have acquired an important role in the chemotherapy of certain forms of cancer. These compounds have been shown to act as strong clastogens in experimental systems (TURNBULL et al., 1979; WIENCKE et al., 1979). On the basis of past experience with clastogenic activity of other cytostatic drugs in man, it may well be assumed that these platinum compounds will also act as human clastogens, although so far only very preliminary data of cytogenetic studies on therapeutically exposed individuals have come to the authors' knowledge. A significant increase of SCE and chromatid-type aberrations was observed in peripheral lymphocytes of five patients receiving a carboplatin therapy (SHINKAI et al., 1988).

The chromosome damaging action of radioactive metals (plutonium, thorium, uranium) or of radioactive isotopes certainly may be attributed to the radiation emitted by them, and therefore are not considered here.

The significant increase of certain types of chromosome damage found in 18 gold miners suffering from progressive systemic sclerosis (BERNSTEIN et al., 1980) may be due to their illness rather than to their occupational exposure.

2.4 Modulating Effects of Metals on in vitro Systems

Although some metal compounds have been shown to be mutagenic or clastogenic, in most cases the carcinogenicity of metal compounds cannot easily be explained by their weak or absent genetic effects (e.g., arsenite, cadmium compounds). This has led to recent interest in experiments aimed at detecting indirect genotoxic effects of metal compounds by looking for synergism with known genotoxic agents. The effects of metal compounds on in vitro end-points are listed in Table I.18-3.

In theory, synergism or antagonism by metal compounds can occur at a variety of levels:

(1) Reaction of the metal compound with the genotoxicant or its metabolites might occur, resulting in a stronger or weaker agent. For example, Co(II) appears to form mutagenic complexes with some compounds (OGAWA et al., 1986). Cu(II), Fe(II), and Mn(II), in the presence of ascorbate, cause the formation of active oxygen species which are genotoxic (STICH et al., 1979). It has long been known that H_2O_2 causes single-strand breaks in cellular DNA, but this effect is now thought to be dependent upon a metal-catalyzed Fenton reaction in which hydroxyl radical is formed (reviewed by AUST et al., 1985). The comutagenic effect of Cu(II) with UV might be explainable on this basis (ROSSMAN and MOLINA, 1986; ROSSMAN et al., 1989).

(2) Metal compounds can affect the metabolic activation of compounds, resulting in alterations in the amount or spectrum of metabolites. This effect can be at the level of the mixed-function oxidases, epoxide hydratase, or formation of sulfate esters or glutathione conjugates. It is thought that some of the antimutagenic/anticlastogenic effects of selenium compounds might be due to effects on metabolic activation (SHAMBERGER, 1985).

(3) It is theoretically possible that some metal compounds could affect the binding of chemicals to DNA. This area has not been studied.

(4) It is also theoretically possible that some metal compounds could affect the integration of tumor viral DNA into the chromosome. This may be an explanation for the enhancement of SA7 virus transformation by many metal compounds (CASTO et al., 1979).

(5) A number of metal compounds have been shown to inhibit DNA repair, usually by indirect means. If a metal compound is comutagenic with (enhances mutagenicity of) another agent in a repair-proficient cell type, but not in a repair-deficient cell type, it is considered evidence that the metal is probably inhibiting DNA repair. Arsenite, for example, is comutagenic with ultraviolet (UV) light in repair-proficient *Escherichia coli*, but not in strains carrying the *uvrA* or *polA* muta-

Table I.18-3. Modulating Effects of Metals in vitro

Metal	Effect	Agents Affected	References[a]
Ag(II)	enhanced transformation	SA7 virus	1
As(III)	comutagenic (*Escherichia coli*)	UV	2
	comutagenic (V79)	UV, MMS, MNU, crosslinkers	3, 4, 5
	coclastogenic	UV, crosslinkers, EMS	3, 6
	enhanced transformation	SA7 virus	1
Be(II)	enhanced transformation	SA7 virus	1
Cd(II)	comutagenic (*Streptomyces typhimurium*)	nitrosamines	7
	enhanced transformation	SA7 virus	1
	promotion	aromatics	8
Co(II)	antimutagenic (*S. typhimurium*)	Trp-P-2	9
	comutagenic (*S. typhimurium*)	heteroaromatics	10
Cr(VI)	comutagenic (*S. typhimurium*)	sodium azide	11
	comutagenic (*S. typhimurium*)	9 amino acridine	12
	enhanced transformation	SA7 virus	1
	promotion	aromatics	8
Cu(II)	comutagenic (*E. coli*)	UV	13
	coclastogenic (CHO)	ascorbate, isoniazid	14, 15
	enhanced transformation	SA7 virus	1
Fe(II)	coclastogenic (CHO)	ascorbate, isoniazid	14, 15
	enhanced transformation	SA7 virus	1
Ge(II)	antimutagenic (*S. typhimurium*)	Trp-P-1	16
Hg(II)	enhanced transformation	SA7 virus	1
Mn(II)	comutagenic (*E. coli*)	UV	13
	coclastogenic (CHO)	ascorbate, isoniazid	14, 15
	enhanced transformation	SA7	1
Mo(VI)	comutagenic (*E. coli*)	UV	13
Ni(II)	comutagenic (*E. coli* and *S. typhimurium*)	MMS, EMS	17
	enhanced transformation	SA7 virus	1
	promotion	aromatics	8
	comutagenic (V79)	UV	21, 22
Pb(II)	enhanced transformation	SA7 virus	1
Pt(IV)	enhanced transformation	SA7 virus	1
Se(IV)	antimutagenic (*S. typhimurium*)	benzo(*a*)pyrene, AAF	18, 23
	anticlastogenic	clastogens MMS, AAF	19
	anti-SCe		20
Sb(III)	enhanced transformation	SA7 virus	1
Tl(I)	enhanced transformation	SA7 virus	1
W(VI)	enhanced transformation	SA7 virus	1
Zn(II)	enhanced transformation	SA7 virus	1

[a] 1 CASTO et al. (1979); 2 ROSSMAN (1981); 3 LEE et al. (1985, 1986); 4 OKUI and FUJIWARA (1986a,b); 5 ROSSMAN and LI (1989); 6 JAN et al. (1986); 7 MANDEL and RYSER (1984); 8 RIVEDAL and SANNER (1980); 9 MOCHIZUKI and KADA (1982); 10 OGAWA et al. (1986); 11 LA VELLE and WITMER (1984); 12 LA VELLE (1986); 13 ROSSMAN and MOLINA (1986); 14 STICH et al. (1979); 15 WHITING et al. (1980b); 16 KADA et al. (1984); 17 DUBINS and LA VELLE (1986); 18 TEEL (1984); 19 NORDENSON and BECKMAN (1985); 20 RAY et al. (1978); 21 HARTWIG and BEYERSMANN (1987); 22 CHRISTIE et al. (1989); 23 JACOBS et al. (1977a)

tions (ROSSMAN et al., 1986b). On the other hand, Ni(II) is comutagenic with the methylating agent MMS in repair-proficient and *uvrA* strains of *E. coli* but not in *polA*, *lexA*, or *recA* strains (DUBINS and LA VELLE, 1987). Ni(II) was not comutagenic with UV in *E. coli* (ROSSMAN and MOLINA, 1986) but was comutagenic with both UV and chromate in Chinese hamster cells (HARTWIG and BEYERSMANN, 1987). Cr(VI) and Cd(II) are also comutagenic with UV in these cells (HARTWIG and BEYERSMANN, 1989). DNA lesions caused by UV and MMS are recognized by different repair enzymes, and these enzymes differ in prokaryotes and eukaryotes. Thus, it is not surprising that metal compounds will show agent and species specificity in their comutagenic effects. Evidence for effects by arsenite on the repair of UV damage in mammalian cells has also been reported (LEE et al., 1985). Arsenite is the first example where the evidence for inhibition of DNA repair (of pyrimidine dimers in this case) is also supported by direct measurements of lesions (OKUI and FUJIWARA, 1986).

More recently, 15 metal salts were assayed for their abilities to inhibit excision of pyrimidine dimers from DNA. Hg(II), As(III), Cu(II), Ni(II), Co(II), and Cd(II) all showed inhibitory effects in a dose-dependent fashion. When the same metal compounds were tested for their abilities to inhibit repair of X-ray induced strand breaks, Hg(II), As(III), Ni(II), Ga(II), Zn(II), and Mo(VI) were inhibitory (SNYDER et al., 1989). Cd(II) was also found to reduce the amount of unscheduled DNA synthesis after UV-irradiation, an effect associated with an increase in the accumulation of DNA strand breaks (NOCENTINI, 1987). Using the nucleoid sedimentation technique, Ni(II) was shown to inhibit the repair of UV-induced DNA damage (HARTWIG and BEYERSMANN, 1989).

Metal compounds also inhibit the repair of alkylation damage of DNA. Cd(II), Zn(II), Hg(II), and Pb(II) were very effective inhibitors (and Al(III) and Fe(III) at higher concentrations) of human O^6-methylguanine-DNA-methyltransferase (BHATTACHARYYA et al., 1988). Arsenite is comutagenic with MNU in Chinese hamster cells, an effect which appears to be associated with prolonged strand breaks, indicating inhibition by arsenite of a late stage in repair of methylation damage (LI and ROSSMAN, 1989).

(6) Finally, by affecting the process of DNA replication, metal compounds can alter the mutagenic response of other agents. Metal-induced infidelity of DNA-replication (discussed in Sect. 2.1) might be of greater importance when the replication involves a damaged template. Although there is no proof for this mechanism, it has been hypothesized to explain why the weakly mutagenic Mn(II) shows such a strong comutagenic effect with UV (ROSSMAN and MOLINA, 1986).

3 Carcinogenicity
(see also Part II)

Recent developments in the field of genetics and cytogenetics of human neoplasia, especially in the area of oncogenes, clearly emphasize the close relationship between mutational changes of the genetic material and malignant transformation. Experimental studies on the mutagenic and carcinogenic action of radiation and chemicals have long suggested such a relationship. From an epidemiologic standpoint, the potential carcinogenic consequences due to occupational exposure to cadmium oxide, chromates, hematite, nickel and arsenic seem most important (see IARC, 1973, 1976, 1980; SUNDERMAN, 1981; FISHBEIN, 1979; LÉONARD and LAUWERYS, 1980a,b; NELSON, 1985). In this context it must be pointed out, however, that the working place concentrations of these metals reported in former studies were much higher than those allowed today. The findings of the reaction of certain metals or metal compounds with DNA and polynucleotides, infidelity of DNA synthesis, the induction of gene and chromosome mutations, as presented above, are all possible mechanisms for the carcinogenic actions of arsenic (LÉONARD, 1985), beryllium (SKILLETER, 1985), cadmium (KAZANTZIS, 1985), chromium (NELSON, 1985), and nickel (SUNDERMAN, 1985a,b). The very large number of experimental studies on the carcinogenic effects of these metals has been reviewed and evaluated by extensive review papers like those cited before. Tumors could be induced in animals by experimental application of beryllium, cadmium, chromates, cobalt, lead, nickel, zinc, and carbohydrates of iron. Ni_3S_2 displayed the highest carcinogenic activity. Those experimental studies on animals support and extend the epidemiological findings. A weak carcinogenic effect also was reported from pertinent experiments with lead subacetate, manganese sulfate, molybdenum oxide, and nickel acetate by STONER et al. (1976) as well as with lead chromate by FÜRST et al. (1976).

Summarizing the risk estimation from all data on the carcinogenicity of metals and metal compounds (which of course is incomplete) may lead to the assumption of carcinogenesis in man by beryllium, cadmium, chromium(VI), and nickel compounds if the exposure to them is intense enough. An estimation of the risk for the general population, however, is very difficult (FISHBEIN, 1976); IARC 1973, 1976, 1980). Some evidence pertaining to that problem may be derived from studies on populations living in areas of high metal pollution by industry (BLOT and FRAUMENI, 1975; NORDSTRÖM et al., 1979). The situation is further complicated by the multiple exposure of humans. A cocarcinogenic activity in experiments was reported, for instance, for chromium and nickel (LANE and MASS, 1977; RIVEDAL and SANNER, 1980), while an anticarcinogenic effect was found with selenium compounds (GRIFFIN and JACOBS, 1977; JACOBS et al., 1977b).

4 Teratogenicity
(see also Part II)

The induction of damage by environmental influences in embryonic (or fetal) development, causing malformations or embryonic death, is a problem of the toxicity rather than of the mutagenicity of the inducing agent. Teratologic effects, therefore, are not heritable, as in most cases the genetic material remains unaffected by the

Table I.18-4. Metals Found Positive in Teratogenicity Studies (From SCHARDEIN, 1985)

Compounds of	Teratogenicity in					
	Rat	Mouse	Hamster	Rabbit	Other Animals	Humans
Cadmium	+			−		−
Chromium		+	+			
Copper			+			
Gallium		+	−			
Indium			+			
Lead	±	±	+	±	−	−(+)
Manganese	−	+	−			
Mercury	±	±	+	±	±	+
Molybdenum	−		−		+	
Nickel	±	+	+			−
Selenium	−	−	−	−	+	
Strontium	±			+		
Tellurium	+					
Ytterbium			±			
Zinc (complex)		+				

+ positive; − negative outcome; ± contradictory results

teratogen (but see also Chapter I.19). In general, mechanisms of teratogenicity are different from those of mutagenicity. The statement whether a metal compound is teratogenic depends, however, on the defintion of teratogenicity. If one considers teratogens as agents causing abnormal development (structural or functional) lead and methyl mercury qualify also as teratogens. If the term is restricted to malformations and embryonic death, methyl mercury has not been shown to be teratogenic in humans. It affects the cyto-architecture of the developing brain and, in severe cases, leads to small head size, but not actual anatomic malformations have been described.

The teratogenic action of metals has been tested comprehensively on animals (see SCHARDEIN, 1985, for references) and some metals displayed some adverse effects in various test systems (Table I.18-4). In addition to these classical studies, recently organ culture systems have been used for more sophisticated examinations (JACQUET, 1985).

Of particular interest, of course are studies on human exposure: While existent data demonstrate some reproductive hazard from lead (see also Chapter II.16), no general convincing data related to birth defects per se have been presented so far (SCHARDEIN, 1985). As pointed out by the same author, only two mercury compounds other than methyl mercury have been studied in relation to congenital abnormality in the human: Mercuric chloride has been associated with abortion in several reports, and phenyl mercuric acetate was found not to have any relationship to congenital malformation among 889 pregnancies studied by HEINONEN et al. (1977). Methyl mercury, on the other hand, is well known to be associated with detrimental effects described as the so-called Minimata disease (see SCHARDEIN, 1985, for references). The fetal type of this disease is of confirmed true teratogenic nature. This "methyl mercury fetopathy" has now been reported from different parts of the world.

There have been no substantiated reports to indicate that heavy metals, other than lead and organic mercury compounds, have any causal relationship to the induction of birth defects in humans (SCHARDEIN, 1985). The teratogenic potency of some of these metals or their compounds in experimental systems, however, should prompt us to carefully examine the extent of the real hazard to man.

The use of lithium as an antidepressant during the first trimester, for instance, may be related to an increased incidence of congenital defects, particularly of the cardiovascular system (reviewed by BRIGGS et al., 1983). "The risk must be approximately 10% if it is accurately determined from the published number of reported malformed cases to those exposed" (SCHARDEIN, 1985).

5 Concluding Remarks

Any discussion of the mutagenic, carcinogenic, and teratogenic action of metals and their compounds certainly has to consider their toxicity. In many if not most cases it well may overwhelm the potential mutagenicity in man demonstrated for rather few metals (arsenic, chromates, nickel, and platinum) by this very limited review. Concerning the carcinogenicity of occupational exposure, more detailed epidemiologic data point to a possible hazard from arsenic, beryllium, cadmium, chromium, and nickel, at least in certain oxidation states. Teratogenic activity in man, so far, could definitely be demonstrated for methyl mercury only. Some teratogenic potential was also suggested for lithium.

The rather unexamined area of interactive effects, however, presents a large number of unsolved problems concerning comutagenic, cocarcinogenic, and coteratogenic activities, which urgently need to be clarified in the near future. It might well be that information on the interactive effects of metals with each other as well as with other agents could be of great practical importance.

In addition, several positive results from short-term tests on non-mammalian organisms presented above should stimulate further experimental testing on the more relevant mammalian in vivo systems as well as extensive epidemiological examinations.

Information on mutagenicity, carcinogenicity, and teratogenicity of metals and their compounds, which is available at present or expected from future research, can be expected to contribute to the safety and health of individuals exposed to these compounds as well as to that of future generations. For further reading, a list of very recent reviews of a more specialized nature follows the list of cited references.

References

Amacher, D. E., and Paillet, S. C. (1980), *Induction of Trifluorothymidine Resistant Mutants by Metal Ions in L5178Y/TK$^{+/-}$ Cells*. Mutat. Res. 78, 279–288.

Aust, S. D., Morehouse, L. A., and Thomas, C. E. (1985), *Role of Metals in Oxygen Radical Reactions*. J. Free Radicals Biol. Med. 1, 3–25.

Bauchinger, M., Schmid, E., Einbrodt, H. J., and Dresp, J. (1976), *Chromosome Aberrations in Lymphocytes after Occupational Exposure to Lead and Cadmium*. Mutat. Res. 40, 57–62.

Beckman, R. A., Mildvan, A. S., and Loeb, L. A. (1985), *On the Fidelity of DNA Replication: Manganese Mutagenesis in vitro*. Biochemistry 24, 5810–5817.

Bennicelli, C., Camoirano, A., Petruzzelli, S., Zanacchi, P., and De Flora, S. (1983), *High Sensitivity of Salmonella TA102 in Detecting Hexavalent Chromium Mutagenicity and its Reversal by Liver and Lung Preparations*. Mutat. Res. 122, 1–5.

Berg, J. M. (1986), *Potential Metal-binding Domains in Nucleic Acid Binding Proteins*. Science 232, 485–487.

Bernstein, R., Prinsloo, I., Zwi, S., Andrew, M. J. A., Dawson, B., and Jenkins, T. (1980), *Chromosomal Aberrations in Occupation Associated Progressive Systematic Sclerosis*. S. Afr. Med. J. 58, 235–237.

Beyersmann, D., and Köster, A. (1987), *On the Role of Trivalent Chromium and Chromium Genotoxicity*. Toxicol. Environ. Chem. 14, 11–22.

Bhattacharyya, D., Boulden, A. M., Foote, R. S., and Mitra, S. (1988), *Effect of Polyvalent Metal Ions on the Reactivity of Human O^6-methylguanine-DNA Methyltransferase*. Carcinogenesis 9, 683–685.

Bigaliev, A. B., Turbaev, M. N., Bigalieva, R. K., and Elemesova, M. S. (1977), *Cytogenetic Examination of Workers Engaged in Chrome Production*. Genetika 13/3, 545–547.

Biggart, N. W., and Murphy, E. C., Jr. (1988), *Analysis of Metal-induced Mutations Altering the Expression or Structure of a Retroviral Gene in a Mammalian Cell Line*. Mutat. Res. 198, 115–129.

Blot, W. J., and Fraumeni, J. F. (1975), *Arsenical Air Pollution and Lung Cancer*. Lancet 2, 142–144.

Briggs, G. G., Bodendorfer, T. M., Freeman, R. K., Yaffe, S. J. (1983), *Drugs in Pregnancy and Lactation*. Williams & Wilkins, Baltimore-London.

Brusick, D. J. (1986), *Genotoxic Effects in Cultured Mammalian Cells Produced by Low pH Treatment Conditions and Increased Ion Concentrations*. Environ. Mutagen. 8, 879–886.

Casto B. C., Meyers, J., and Di Paolo, J. A. (1979), *Enhancement of Viral Transformation for Evaluation of the Carcinogenic or Mutagenic Potential of Inorganic Metal Salts*. Cancer Res. 39, 193–198.

Christie, N. T., and Costa, M. (1983), *In vitro Assessment of the Toxicity of Metal Compounds. III. Effects of Metals on DNA Structure and Function in Intact Cells*. Biol. Trace Elem. Res. 5, 55–71.

Christie, N. T., Tummolo, D. M., Klein, C. B., and Rossman, T. G. (1989), *The Role of Ni(II) in Mutation*, in: Nieboer, E., and Antio, A. (eds.): Nickel and Human Health: Current Perspectives. John Wiley & Sons, New York (in press).

Costa, M., Jones, M., and Lindberg, O. (1980), *Metal Carcinogenesis in Tissue Culture Systems*. In: Inorganic Chemistry in Biology and Medicine. ACS Symp. Ser. 40, 45–74.

De Flora, S., Zanacchi, P., Camoirano, A., Bennicelli, C., and Badolati, G. S. (1984), *Genotoxic Activity and Potency of 135 Compounds in the Ames Reversion Test and in a Bacterial DNA-repair Test*. Mutat. Res. 133, 161–198.

Degraeve, N. (1981), *Carcinogenic, Teratogenic and Mutagenic Effects of Cadmium*. Mutat. Res. 86, 114–135.

Deknudt, G., and Léonard, A. (1975), *Cytogenetic Investigations and Leucocytes of Workers from a Cadmium Plant*. Environ. Physiol. Biochem. 5, 319–327.

Deknudt, G., Léonard, A., Arany, J., Jenar-Du Buisson, G., and Delavignette, E. (1986), *In vivo Studies in Male Mice on the Mutagenic Effects of Inorganic Arsenic*. Mutagenesis 1, 33–34.
De la Torre, R., and Krompotic, E. (1976), *The in vivo and in vitro Effects of Lithium on Human Chromosomes and Cell Replication*. Teratology 13, 131–138.
De Souza Nazareth, R. (1976), *Cobalt Chloride Effects on Nondisjunction Preliminary Results*. Cienc. Cult. 28, 1472–1475.
Dubins, J. S., and La Velle, J. M. (1986), *Nickel(II) Genotoxicity: Potentiation of Mutagenesis of Simple Alkylating Agents*. Mutat. Res. 162, 187–199.
Eichhorn, G. L. (1979), *Aging, Genetics, and the Environment: Potential of Errors Introduced into Genetic Information Transfer by Metal Ions*. Mech. Ageing Dev. 9, 291–301.
Fishbein, L. (1976), *Environmental Metallic Carcinogens: An Overview of Exposure Levels*. J. Toxicol. Environ. Health 2, 77–109.
Fishbein, L. (1979), *Potential Industrial Carcinogens and Mutagens*. Elsevier, Amsterdam-Oxford-New York.
Fleig, I., Rieth, H., Stocker, W. G., and Thiess, A. M. (1983), *Chromosome Investigations of Workers Exposed to Cadmium in the Manufacturing of Cadmium Stabilizers and Pigments*. Ecotoxicol. Environ. Saf. 7, 106–110.
Forni, A., Sciame, A., Bertazzi, B. A., and Alessio, L. (1980), *Chromosome and Biochemical Studies in Women Occupationally Exposed to Lead*. Arch. Environ. Health 35, 139–146.
Fürst, A., Schlauder, M., and Sasmore, D. P. (1976), *Tumorigenic Activity and Lead Chromate*. Cancer Res. 36, 1779–1783.
Gebhart, E. (1969), *Untersuchungen über die cytogenetische Wirkung einiger Hauptwirkstoffe von Vaginal-Anticoncipientien*. Arzneim. Forsch. (Drug Res.) 19, 364–372.
Gebhart, E. (1981), *Sister Chromatid Exchange (SCE) and Structural Chromosome Aberration in Mutagenicity Testing*. Human Genet. 58, 235–254.
Gebhart, E. (1985), *Chromosome Damage in Individuals Exposed to Heavy Metals*, in: Merian, E., Frei, R. W., Härdi, W., and Schlatter, C. (eds.): *Carcinogenic and Mutagenic Metal Compounds*, pp. 213–225. Gordon & Breach Sci., Publ., New York-London-Paris-Montreux-Tokyo.
Gebhart, E., Lösing, J., Mueller, R. L., and Windolph, B. (1982), *Interindividual Variation of Cytogenetic Damage by Cytostatic Therapy*, in: Sorsa, M., and Vainio, H. (eds.): *Mutagens in Our Environment*, pp. 89–98. Liss, New York.
Gerber, G. B., Léonard, A., and Jacquet, P. (1980), *Toxicity, Mutagenicity and Teratogenicity of Lead*. Mutat. Res. 76,'115–141.
Gläss, E. (1956), *Die Verteilung von Fragmentationen und Achromatischen Stellen auf den Chromosomen von Vicia faba nach Behandlung mit Schwermetallsalzen*. Chromosoma 8, 260–284.
Griffin, A. C., and Jacobs, M. M. (1977), *Effects of Selenium on Azo Dye Hepatocarcinogenesis*. Cancer Lett. 3, 177–181.
Hansen, K., and Stern, R. M. (1984), *A Survey of Metal-induced Mutagenicity in vitro and in vivo*. J. Am. Coll. Toxicol. 3, 381–430.
Hartwig, A., and Beyersmann, D. (1987), *Enhancement of UV and Chromate Mutagenesis by Nickel Ions in the Chinese Hamster hgprt Assay*. Toxicol. Environ. Chem. 14, 33–42.
Hartwig, A., and Beyersmann, D. (1989), *Comutagenicity and Inhibition of DNA Repair by Metal Ions in Mammalian Cells*. Biol. Trace Elem. Res. (in press).
Heck, J. D., and Costa, M. (1982a), *In vivo Assessment of the Toxicity of Metal Compounds. I. Mammalian Cell Transformation*. Biol. Trace Elem. Res. 4, 71–82.
Heck, J. D., and Costa, M. (1982b), *In vitro Assessment of the Toxicity of Metal Compounds. II. Mutagenesis*. Biol. Trace Elem. Res. 4, 319–330.
Heck, J. D., and Costa, M. (1982c), *Surface Reduction of Amorphous NiS Particles Potentiates Their Phagocytosis and Subsequent Induction of Morphological Transformation in Syrian Hamster Embryo Cells*. Cancer Lett. 15, 19–26.
Heinonen, O. P., Slone, D., and Shapiro, S. (1977), *Birth Defects and Drugs in Pregnancy*. PSG, Publishing Comp. Inc., Littleton, Massuchesetts.

Hollstein, M., and McCann, J. (1979), *Short-term Tests for Carcinogens and Mutagens. Mutat. Res. 65*, 133–226.
Hopkin, J. M., and Evans, H. J. (1980), *Cigarette Smoke-induced DNA Damage and Lung Cancer Risks. Nature 283*, 388–390.
Husgafvel-Pursiainen, K., Mäki-Paakkanen, J., Norppa, H., and Sorsa, M. (1980), *Smoking and Sister Chromatid Exchange. Hereditas 92*, 247–250.
IARC (International Agency for Research on Cancer) (1973) *IARC Monographs on the Evaluation of Carcinogenic Risk of Chemicals to Man.* Vol. 2, *Some Inorganic and Organometallic Compounds.* IARC, Lyon.
IARC (International Agency for Research on Cancer) (1976), *IARC Monographs on the Evaluation of Carcinogenic Risk of Chemicals to Man.* Vol. 11, *Cadmium, Nickel, Some Epoxides, Miscellaneous Industrial Chemicals and General Considerations on Volatile Anaesthetics.* IARC, Lyon.
IARC (International Agency for Research on Cancer) (1980), *IARC Monographs on the Evaluation of Carcinogenic Risk of Chemicals to Man.* Vol. 23, *Some Metals and Metallic Compounds.* IARC, Lyon.
Imreh, S., and Radulescu, D. (1982), *Cytogenetic Effects of Chromium in vivo and in vitro. Mutat. Res. 97*, 192–193.
Jacobs, M. M., Matney, T. S., and Griffin, A. C. (1977a), *Inhibitory Effects of Selenium on the Mutagenicity of 2-Acetylaminofluorene (AAF) and AFF Derivatives. Cancer Lett. 2*, 319–322.
Jacobs, M. M., Jansson, G., and Griffin, A. C. (1977b), *Inhibitory Effects of Selenium on 1,2-Dimethylhydrazine and Methylazoxymethanol Acetate Induction of Colon Tumors. Cancer Lett. 2*, 133–138.
Jacobson, K. B., and Turner, J. E. (1980), *The Interaction of Cadmium and Certain Other Metal Ions with Proteins and Nucleic Acids. Toxicology 16*, 1–37.
Jacobson-Kram, D., and Montalbano, D. (1985), *The Reproductive Effects Assessment Group's Report on the Mutagenicity of Inorganic Arsenic. Environ. Mutagen. 7*, 787–804.
Jacquet, P. (1985), *The Use of Embryo-culture Techniques in the Research on the Teratogenic and Mutagenic Properties of Metals,* in: Merian, E., Frei, R. W., Härdi, W., and Schlatter, C. (eds.): *Carcinogenic and Mutagenic Metal Compounds,* pp. 227–243. Gordon & Breach Sci. Publ., New York–London–Paris–Montreux–Tokyo.
Jan, K. Y., Huang, R. Y., and Lee, T. C. (1986), *Different Modes of Action of Sodium Arsenite, 3-Aminobenzamide, and Caffeine on the Enhancement of Ethyl Methanesulfonate. Cytogenet. Cell Genet. 41*, 202–208.
Kada, T., Mochizuki, H., and Miyao, K. (1984), *Antimutagenic Effects of Germanium Oxide on Trp-P-2-induced Frameshift Mutations in Salmonella typhimurium TA98 and TA1538. Mutat. Res. 125*, 145–151.
Kanematsu, N., Hara, M., and Kada, T. (1980), *Rec-assay and Mutagenicity Studies on Metal Compounds. Mutat. Res. 77*, 109–116.
Kato, R. (1976), *Chromosome Breakage Associated with Organic Mercury in Chinese Hamster Cells in vitro. Mutat. Res. 38*, 340–341.
Kawanishi, S., Inoue, S., and Sano, S. (1986), *Mechanism of DNA Cleavage Induced by Sodium Chromate(VI) in the Presence of Hydrogen Peroxide. J. Biol. Chem. 261*, 5952–5958.
Kazantzis, G. (1985), *Mutagenic and Carcinogenic Effects of Cadmium,* in: Merian, E., Frei, R. W., Härdi, W., and Schlatter, C. (eds.): *Carcinogenic and Mutagenic Metal Compounds,* pp. 387–398. Gordon & Breach, New York–London–Paris–Montreux–Tokyo.
Kilbey, B. J., Legator, M., Nichols, W., and Ramel, C. (1984), *Handbook of Mutagenicity Test Procedures.* Elsevier, Amsterdam–New York–Oxford.
King, M. T., Beikrich, H., Eckhardt, K., Gocke, E., and Wild, S. (1979), *Mutagenicity Studies with X-ray-contrast Media, Analgetics, Antipyretics, Antirheumatics and Some Other Pharmaceutical Drugs in Bacterial, Drosophila and Mammalian Test Systems. Mutat. Res. 66*, 33–43.
Lane, P. B., and Mass, M. J. (1977), *Carcinogenicity and Cocarcinogenicity of Chromium Carbonyl in Heterotopic Tracheal Grafts. Cancer Res. 37*, 1476–1479.

La Velle, J. M., and Witmer, C. M. (1984), *Chromium(VI) Potentiates Mutagenesis by Sodium Azide but not Ethylmethanesulfonate.* Environ. Mutagen. **3**, 311–320.

La Velle, J. M. (1986), *Potassium Chromate Potentiates Frameshift Mutagenesis in E. coli and S. typhimurium.* Mutat. Res. **171**, 1–10.

Lee, T. C., Huang, R. Y., and Jan, K. Y. (1985), *Sodium Arsenite Enhances the Cytotoxicity, Clastogenicity, and 6-Thioguanine-resistant Mutagenicity of Ultraviolet Light in Chinese Hamster Ovary Cells.* Mutat. Res. **148**, 83–89.

Lee, T. C., Lee, K. C., Tzeng, Y. J. Huang, R.-Y., and Jan, K. Y. (1986a), *Sodium Arsenite Potentiates the Clastogenicity and Mutagenicity of DNA Cross-linking Agents.* Environ. Mutagen. **8**, 119–128.

Lee, T. C., Wang-Wuu, S., Huang, R. Y., Lee, K. C., and Jan, K. Y. (1986b), *Differential Effects of the Pre- and Post-treatment of Sodium Arsenite on the Genotoxicity of Methylmethanesulfonate in Chinese Hamster Ovary Cells.* Cancer Res. **46**, 1854–1857.

Lee, T. C., Tanaka, N., Lamb, P. W., Gilmer, T., and Barrett, J. C. (1988), *Induction of Gene Amplification by Arsenic.* Science **241**, 79–81.

Léonard, A. (1985), *Recent Advances in Arsenic Mutagenesis and Carcinogenesis,* in: Merian, E., Frei, R. W., Härdi, W., and Schlatter, C. (eds.): *Carcinogenic and Mutagenic Metal Compounds,* pp. 443–453. Gordon & Breach, Sci. Publ., New York-London-Paris-Montreux-Tokyo.

Léonard, A., and Lauwerys, R. R. (1980a), *Carcinogenicity, Teratogenicity and Mutagenicity of Arsenic.* Mutat. Res. **75**, 49–62.

Léonard, A., and Lauwerys, R. R. (1980b), *Carcinogenicity and Mutagenicity of Chromium.* Mutat. Res. **76**, 227–239.

Léonard, A., Gerber, G. B, and Jacquet, P. (1981), *Carcinogenicity, Mutagenicity and Teratogenicity of Nickel.* Mutat. Res. **87**, 1–15.

Léonard, A., Jacquet, P., and Lauwerys, R. R. (1983), *Mutagenicity and Teratogenicity of Mercury Compounds.* Mutat. Res. **114**, 1–18.

Levan, A. (1945), *Cytological Reactions Induced by Inorganic Salt Solutions.* Nature **156**, 751–752.

Levine, B. S., Preache, M. M., and Pergament, E. (1980), *Mutagenic Potential of cis-Dichlorodiammin Platinum II in Rodents.* Toxicology **17**, 57–65.

Levis, A. G., and Bianchi, V. (1982), *Mutagenic and Cytogenetic Effects of Chromium Compounds,* in: Långard S. (ed.): *Biological and Environmental Aspects of Chromium,* pp. 171–208. Elsevier, Amsterdam.

Li, J. H., and Rossman, T. G. (1989), *Mechanism of Comutagenesis of Sodium Arsenite with n-Methyl-n-nitrosourea.* Biol. Trace Elem. Res. (in press).

Mabille, V., Roels, H., Jacquet, P. Léonard, A., and Lauwerys, R. (1984), *Cytogenetic Examination of Leucocytes of Workers Exposed to Mercury Vapour.* Int. Arch. Occup. Environ. Health **53**, 257–260.

Mäki-Paakkanen, J., Sorsa, M., and Vainio, H. (1981), *Chromosome Aberrations and Sister Chromatid Exchanges in Lead-exposed Workers.* Hereditas **94**, 269–275.

Mandel, R., and Ryser, H. J. (1984), *Mutagenicity of Cadmium and Salmonella typhimurium and its Synergism with two Nitrosamines.* Mutat. Res. **138**, 9–16.

Miller, C. A., and Costa, M. (1988), *Characterization of DNA-protein Complexes Induced in Intact Cells by the Carcinogen Chromate.* Mol. Carcinogen. **1**, 125–133.

Mitelman, F. (1988), *Catalog of Chromosome Aberrations in Cancer,* 3rd Ed. Liss, New York.

Miyaki, M, Akamatsu, N., Ono, T., and Koyama, H. (1979), *Mutagenicity of Metal Cations in Cultured Cells from Chinese Hamster.* Mutat. Res. **68**, 259–263.

Mochizuki, H., and Kada, T. (1982), *Antimutagenic Action of Cobaltous Chloride on Trp-P-1-induced Mutations in Salmonella typhimurium TA98 and TA1538.* Mutat. Res. **95**, 145–157.

Moore, M. M., and Brock, K. H. (1988), *High Concentrations of Sodium Chloride Induce a "Positive" Response at the TK Locus of L5178Y/TK$^{+/-}$ Mouse Lymphoma Cells.* Environ. Mol. Mutagen. **12**, 265–268.

Nelson, N. (1985), *Comments on the Carcinogenicity and Mutagenicity of Metals and Their Compounds,* in: Merian, E., Frei, R.W., Härdi, W., and Schlatter, C. (eds.): *Carcinogenic and Mutagenic Metal Compounds,* pp. 513–527. Gordon & Breach Sci. Publ., New York–London–Paris–Montreux–Tokyo.
Newton, M.F., and Lilly, L.J. (1986), *Tissue-specific Clastogenic Effects of Chromium and Selenium Salts in vivo. Mutat. Res. 169,* 61–69.
Nocentini, S. (1987), *Inhibition of DNA Replication and Repair by Cadmium in Mammalian Cells. Protective Interaction of Zinc. Nucleic Acids Res. 15,* 4211–4225.
Nogawa, K., Tsuritani, I., Yamada, Y., Kido, T., Honda, R., Ishuzaki, M., and Kurihara, T. (1986), *Sister-chromatid Exchanges in the Lymphocytes of People Exposed to Environmental Cadmium. Toxicol. Lett. 32,* 283–288.
Nordenson, I., and Beckman, L. (1982), *Occupational and Environmental Risks in and around a Smelter in Northern Sweden. VII. Hereditas 96,* 175–181.
Nordenson, I., and Beckman, L. (1985), *Interaction Between Some Common Clastogenic Agents,* in: Merian, E., Frei, R.W., Härdi, W., and Schlatter, C. (eds.): *Carcinogenic and Mutagenic Metal Compounds,* pp. 507–511. Gordon & Breach Sci. Publ., New York-London-Paris-Montreux-Tokyo.
Nordenson, I., Beckman, G., Beckman, L., and Nordström, S. (1978), *Occupational and Environmental Risks in and around a Smelter in Northern Sweden. II. Hereditas 88,* 47–50.
Nordenson, I., Salmonsson, S., Brun, E., and Beckman, G. (1979), *Chromosome Aberrations in Psoriatic Patients Treated with Arsenic. Human Genet. 48,* 1–6.
Nordström, S., Beckman, L., and Nordenson, I. (1979), *Occupational and Environmental Risks in and around a Smelter in Northern Sweden. VI. Hereditas 90,* 297–302.
Norppa, H., Westermark, T., Laasonen, M., Knuutila, L., and Knuutila, S. (1980), *Chromosomal Effects of Sodium Selenite in vivo. Hereditas 93,* 93–105.
Obe, G., Heller, W.D., and Vogt, H.J. (1984), *Mutagenic Activity of Tobacco Smoke,* in: Obe, G. (ed.): *Mutations in Man,* pp. 223–246. Springer, Berlin–Heidelberg–New York–Tokyo.
Oberly, T.J., Piper, C.E., and McDonald, D.S. (1982), *Mutagenicity of Metal Salts in the L5178Y Mouse Lymphoma Assay. J. Toxicol. Environ. Health 9,* 367–376.
Ogawa, H.I., Sakata, K., Inouye, T., Jyosui, S., Niyitani, Y., Kakimoto, K., Moroshita, M., Tsuruta, S., and Kato, Y. (1986), *Combined Mutagenicity of Cobalt(II) Salt and Heteroaromatic Compounds in Salmonalla typhimurium. Mutat. Res. 172,* 97–104.
Okui, T., and Fujiwara, Y. (1986), *Inhibition of Human Excision DNA Repair by Inorganic Arsenic and the Co-mutagenic Effect in V79 Chinese Hamster Cells. Mutat. Res. 172,* 69–76.
O'Riordan, M.L., Highes, E.G., and Evans, H.J. (1978), *Chromosome Studies on Blood Lymphocytes of Men Occupationally Exposed to Cadmium. Mutat. Res. 58,* 305–311.
Petres, J., Baron, D., and Hagedorn, M. (1977), *Effects of Arsenic on Cell Metabolism and Cell Proliferation: Cytogenetic and Biochemical Studies. Environ. Health Perspect. 19,* 223–227.
Pilinskaya, M.A. (1970), *Chromosome Aberrations in Persons Contacted with Ziram. Genetika 6,* 157–163.
Pilinskaya, M.A. (1974), *Results of Cytogenetic Examination of Persons Occupationally in Contact with the Fungicide Zineb. Genetika 10,* 140–146.
Plooy, A.C.M., and Lohman, P.H.M. (1980), *Platinum Compounds with Antitumor Activity. Toxicology 17,* 169–176.
Popescu, H.I., Negru, L., and Lancranjan, I. (1979), *Chromosome Aberrations by Occupational Exposure to Mercury. Arch. Environ. Health 34,* 461–463.
Ray, J.H., Altenburg, L.C., and Jacobs, M.M. (1978), *Effects of Sodium Selenite and Methylmethanesulfonate or N-hydroxy-2-acetylaminofluorene Co-exposure on Sister-chromatid Exchange Production in Human Whole Blood Cultures. Mutat. Res. 57,* 359–368.
Rivedal, E., and Sanner, T. (1980), *Synergistic Effect on Morphological Transformation of Hamster Embryo Cells by Nickel Sulphate and Benz(a)pyrene. Cancer Lett. 8,* 203–208.
Roberts, J.J., and Thomson, A.J. (1979), *The Mechanism of Action of Antitumour Platinum Compounds. Progr. Nucleic Acids Res. Mol. Biol. 22,* 71–133.

Röhr, G., and Bauchinger, M. (1976), *Chromosome Analyses in Cell Cultures of the Chinese Hamster after Application of Cadmium Sulphate.* Mutat. Res. *40*, 125 – 130.

Rossman, T. G. (1981), *Enhancement of UV-mutagenesis by Low Concentrations of Arsenite in E. coli.* Mutat. Res. *91*, 207 – 211.

Rossman, T. G., and Li, J. H. (1989), *Mechanism of Comutagenesis of Sodium Arsenite with N-methyl-N-nitrosourea.* Biol. Trace Elem. Res. (in press).

Rossman, T. G., and Molina, M. (1986), *The Genetic Toxicology of Metal Compounds. II. Enhancement of Ultraviolet Light-induced Mutagenesis in Escherichia coli WP2.* Environ. Mutagen. *8*, 263 – 271.

Rossman, T. G., Stone, D., Molina, M., and Troll, W. (1980), *Absence of Arsenite Mutagenicity in E. coli and Chinese Hamster Cells.* Environ. Mutagen. *2*, 371 – 379.

Rossman, T. G., Molina, M., and Meyer, L. W. (1984), *The Genetic Toxicology of Metal Compounds: I. Induction of Prophage in E. coli WP2s.* Environ. Mutagen. *6,* 59 – 69.

Rossman, T. G., Meyer, L. W., and Molina, M. (1986a), *Induction of Lambda Prophage as a Screen for Genotoxic Agents.* Ann. N.Y. Acad. Sci. *463,* 347 – 348.

Rossman, T. G., Molina, M., and Klein, C. B. (1986b), in: Ramel, C. (ed.): *Genetic Toxicology of Environmental Chemicals,* pp. 403 – 408. Liss, New York.

Rossman, T. G., Zelikoff, J. T., Agarwal, S., and Kneip, T. J. (1987), *Genetic Toxicology of Metal Compounds: An Examination of Appropriate Cellular Models.* Toxicol. Environ. Chem. *14*, 251 – 262.

Rossman, T. G., Rubin, L. M., and Kneip, T. J. (1989), *Effects of $CuCl_2$ on UV-mutagenesis and on DNA Damage in a Restriction Fragment of E. coli gpt.* Toxicol. Environ. Chem. *23* (1 – 4), 65 – 72.

Sandberg, A. A. (1980), *The Chromosomes in Human Cancer and Leukemia.* Elsevier, New York-Amsterdam.

Schaaper, R. M., Koplitz, R. M., Tkeshelashvili, L. K., and Loeb, L. A. (1987), *Metal-induced Lethality and Mutagenesis: Possible Role of Apurinic Intermediates.* Mutat. Res. *177*, 179 – 188.

Schardein, J. L. (1985), *Chemically Induced Birth Defects.* Dekker, New York-Basel.

Schinzel, A. (1984), *Catalogue of Unbalanced Chromosome Aberrations in Man.* De Gruyter, Berlin-New York.

Schwanitz, G., Gebhart, E., Rott, H. D., Schaller, K. H., Essing, H. G., Lauer, O., and Prestele, H. (1975), *Chromosomenuntersuchungen bei Personen mit beruflicher Bleiexposition.* Dtsch. Med. Wochenschr. *100,* 1007 – 1010.

Sen, P., and Costa, M. (1986), *Incidence and Localization of Sister Chromatid Exchanges Induced by Nickel and Chromium Compounds.* Carcinogenesis *7,* 1527 – 1533.

Shamberger, R. J. (1985), *The Genotoxicity of Selenium.* Mutat. Res. *154*, 29 – 48.

Shinkai, T., Saijo, N., et al. (1988), *Cytogenic Effect of Carboplatin on Human Lymphocytes.* Cancer Chemother. Pharmacol. *21,* 203 – 207.

Sirover, M. A., and Loeb, L. A. (1976), *Metal-induced Infidelity During DNA Synthesis.* Proc. Natl. Acad. Sci. USA *73,* 2331 – 2335.

Sirover, M. A. (1981), *Effects of Metals in in vitro Bioassays.* Environ. Health Perspect. *40,* 153 – 188.

Skerfving. S., Hansson, K., Mangs, C., Lindsten, J., and Ryman, N. (1974), *Methylmercury-induced Chromosome Damage in Man.* Environ. Res. *7,* 83 – 98.

Skilleter, D. N. (1985), *Biochemical Properties of Beryllium Potentially Relevant to its Carcinogenicity*, in: Merian, E., Frei R. W., Härdi, W., and Schlatter, C. (eds.): *Carcinogenic and Mutagenic Metal Compounds*, pp. 371 – 386. Gordon & Breach Sci. Publ., New York-London-Paris-Montreux-Tokyo.

Snyder, R. D. (1988), *Role of Oxygen Species in Metal-induced DNA Strand Breakage in Human Diploid Fibroblasts.* Mutat. Res. *193,* 237 – 246.

Snyder, R. D., Davis, G. F., and Zachmann, P. J. (1989), *Inhibition by Metals of X-ray and Ultraviolet-induced DNA Repair in Human Cells.* Biol. Trace Elem. Res. (in press).

Stich, H. F., Wei, L., and Whiting, R. F. (1979), *Enhancement of the Chromosome-damaging Action of Some Reducing Agents. Cancer Res. 39,* 4145–4151.

Stoner, G. D., Shimkin, M. B., Troxell, M. C., Thompson, T. L., and Terry, L. S. (1976), *Test for Carcinogenicity of Metallic Compounds by the Pulmonary Tumor Response in Strain A Mice. Cancer Res. 36,* 1744–1747.

Sugiyama, M., Wang, X. W., and Costa, M. (1986), *Comparison of DNA Lesions and Cytotoxicity Induced by Calcium Chromate in Human, Mouse, and Hamster Cell Lines. Cancer Res. 46,* 4547–4551.

Sunderman, F. W., Jr. (1981), *Recent Research on Nickel Carcinogenesis. Environ. Health Perspect. 40,* 131–141.

Sunderman, F. W., Jr. (1985a), *Metal Carcinogenesis,* in: Merian, E., Frei, R. W., Härdi, W., and Schlatter, C. (eds.): *Carcinogenic and Mutagenic Metal Compounds,* pp. 325–342. Gordon & Breach Sci. Publ., New York-London-Paris-Montreux.

Sunderman, F. W., Jr. (1985b), *Recent Progress in Nickel Carcinogenesis,* in: Merian, E., Frei, R. W., Härdi, W., and Schlatter, C. (eds.): *Carcinogenic and Mutagenic Metal Compounds,* pp. 325–342. Gordon & Breach Sci. Publ., New York-London-Paris-Montreux-Tokyo.

Tamaro, M., Venturini, S., Monti-Bragadin, C., Sainacich, G., Mestroni, G., and Zassinovich, G. (1979), *Effects in Bacterial Systems of Pt(II) Complexes with Antitumor Activity. Chem.-Biol. Interact. 26,* 179–184.

Tandon, P., and Sodhi, A. (1985), *cis-Dichlorodiamine Platinum(II) Induced Aberrations in Mouse Bone-marrow Chromosomes. Mutat. Res. 156,* 187–193.

Teel, R. W. (1984), *A Comparison of the Effect of Selenium on the Mutagenicity and Metabolism of Benzo(a)pyrene in Rat and Hamster Liver S9 Activation Systems. Cancer Lett. 24,* 281–289.

Tsapakos, M. J., and Wetterhahn, K. E. (1983), *The Interaction of Chromium with Nucleic Acids. Chem.-Biol. Interact. 46,* 265–277.

Turnbull, D., Popescu, N. C., Di Paolo, J. A., and Myhr, B. C. (1979), *Cis-Platinum(II) Diamine Dichloride Causes Mutation, Transformation, and Sister-chromatid Exchanges in Cultured Mammalian Cells. Mutat. Res. 66,* 267–275.

Verschaeve, V., Tassignon, J. P., Lefevre, M., De Stoop, P., and Susanne, C. (1979), *Cytogenetic Investigation on Leukocytes of Workers Exposed to Metallic Mercury. Environ. Mutagen. 1,* 259–268.

Verschaeve, L., Kirsch-Volders, M., and Susanne, C. (1984), *Mercury-induced Segregational Errors of Chromosomes in Human Lymphocytes and in Indian Muntjac Cells. Toxicol. Lett. 21,* 247–253.

Vouk, V. P., and Piver, W. T. (1983), *Metallic Elements in Fossil Fuel Combustion Products: Amounts and Form of Emissions and Evaluation of Carcinogenicity and Mutagenicity. Environ. Health Perspect. 47,* 201–225.

Waksvik, H., and Boysen, M. (1982), *Cytogenetic Analyses of Lymphocytes from Workers in a Nickel Refinery. Mutat. Res. 103,* 185–190.

Waksvik, H., Boysen, M., and Hogetveit, A. C. (1984), *Increased Incidence of Chromosomal Aberrations in Peripheral Lymphocytes of Retired Nickel Workers. Carcinogenesis 5/11,* 1525–1527.

Watanabe, T., Endo, A., Kumai, M., and Ikeda, M. (1983), *Chromosome Aberrations and Sister Chromatid Exchanges in Styrene-exposed Workers with Reference to their Smoking Habits. Environ. Mutagen. 5,* 299–309.

Whiting, R. F., Stich, H. F., and Koropatnick, D. J. (1979), *DNA Damage and Repair in Cultured Human Cells Exposed to Chromate. Chem. Biol. Interact. 26,* 267–280.

Whiting, R. F., Wei, L., and Stich, H. F. (1980a), *Unscheduled DNA Synthesis and Chromosome Aberrations Induced by Inorganic and Organic Selenium Compounds in the Presence of Glutathione. Mutat. Res. 78,* 159–169.

Whiting, R. F., Wei, L., and Stich, H. F. (1980b), *Enhancement by Transition Metals of Chromosome Aberrations Induced by Isoniazid. Biochem. Pharmacol. 29,* 842–845.

Wiencke, J. K., Cervenka, J., and Paulus, H. (1979), *Mutagenic Activity of Anti-Cancer Agent cis-Dichlorodiammine Platinum II. Mutat. Res. 68,* 68–77.

Wolman, S. R. (1986), *Cytogenetic Heterogenicity: Its Role in Tumor Evolution. Cytogenetics 19*, 129–140.
Zakour, R. A., Kunkel, T. A., and Loeb, L. A. (1981), *Metal-induced Infidelity of DNA Synthesis. Environ. Health Perspect. 409*, 173–180.
Zelikoff, J. T., Atkins, N., and Rossman, T. G. (1986), *Mutagenicity of Soluble Metal Salts Using the V79/HGPRT Mutation Assay.* (Abstr.) *Environ. Mutagen. 8*, 95.
Zelikoff, J. T., Li, J. H., Hartwig, A., Wang, X. W., Costa, M., and Rossman, T. G. (1988), *Genetic Toxicology of Lead Compounds. Carcinogenesis 9*, 1727–1732.

Additional Recommended Literature

Bianchi, V., and Levis, A. G. (1987), *Recent Advances in Chromium Genotoxicity. Toxicol. Environ. Chem. 15*, 1–24.
Christie, N. T. (1989), *The Synergistic Interaction of Nickel(II) with DNA Damaging Agents. Toxicol. Environ. Chem. 22* (1–4), 51–59.
Costa, M. (1988), *The Role of Trace Metals in Carcinogenesis (Cancerlit Data Base Referencing Articles). Oncology Overview* (September). US Department of Health and Human Services, Washington, DC.
EEMS (European Environmental Mutagen Society) (1986), *XVIth Annual Meeting, Brussels,* Proceedings; see also *Chemosphere 16*, (10/12), N24–N37 (1987).
EEMS (European Environmental Mutagen Society) (1987), *XVIIth Annual Meeting, Zürich,* Proceedings; see also *Chemosphere,* in press.
Fiskesjö, G. (1988), *The Allium Test – An Alternative in Environmental Studies: The Relative Toxicity of Metal Ions. Mutat. Res. 197,* 243–260.
Gebhart, E. (1989), *Heavy Metal Induced Chromosome Damage. Life Chem. Rep.,* in press.
Imbra, R. J., Latta, D. M., and Costa, M. (1989), *Studies on the Mechanism of Nickel-induced Heterochromatin Damage; Effect on Specific DNA-Protein Interactions. Toxicol. Environ. Chem. 22* (1–4), 167–169.
3rd International Conference on Safety Evaluation of Chemicals (1984), *Alternative Embryotoxicity and Teratogenicity Tests, Zürich,* Proceedings; see also *Chemosphere 15* (7), N25–N34 (1986).
1st International Meeting on *Molecular Mechanisms of Metal Toxicity and Carcinogenicity, Urbino* (1988), *Proceedings; Biol. Trace Elem. Res.,* in press.
3rd IUPAC Cadmium Workshop, Jülich (1985), Proceedings, *Environ. Toxin Ser. 2;* see also *Chemosphere 16* (4), N10–N16 (1987).
Kazantzis, G. (1987), *The Mutagenic and Carcinogenic Effects of Cadmium: An Update. Toxicol. Environ. Chem. 15*, 83–100.
Kazantzis, G., and Lilly, L. J. (1986), *Mutagenic and Carcinogenic Effects of Metals,* in: Friberg, L. (ed.): *Handbook on the Toxicology of Metals,* 2nd Ed. Elsevier, Amsterdam.
Léonard, A., and Lauwerys, R. (1987), *Mutagenicity, Carcinogenicity and Teratogenicity of Beryllium. Mutat. Res. 186*, 35–42.
Léonard, A., Gerber, G. B., and Léonard, F. (1986), *Mutagenicity, Carcinogenicity and Teratogenicity of Zinc. Mutat. Res. 168*, 342–353.
Merian, E., Frei, R. W., Härdi, W., and Schlatter, C. (eds.) (1985), *Carcinogenic and Mutagenic Metal Compounds,* Vol. 1. Gordon & Breach, New York–London–Paris–Montreux–Tokyo.
Merian, E., Frei, R. W., Härdi, W., and Schlatter, C. (eds.) (1988), *Carcinogenic and Mutagenic Metal Compounds,* Vol. 2. Gordon & Breach, New York–London–Paris–Montreux–Tokyo.
Merian, E., Bronzetti, G. L., and Härdi, W. (eds.) (1989), *Carcinogenic and Mutagenic Metal Compounds,* Vol. 3, in press. Gordon & Breach, New York.

Mukherjee, A., Giri, A. K., Sharma, A., and Talukder, G. (1988), *Relative Efficacy of Short-term Tests in Detecting Genotoxic Effects of Cadmium Chloride in Mice in vivo. Mutat. Res. 206*, 285–295.

Office of Technology Assessment, Congress of the United States (1988), *Identifying and Regulating Carcinogens.* Marcel Dekker, New York-Basel.

Sharma, A., .and Talukder, G. (1987), *Effects of Metals on Chromosomes of Higher Organisms. Environ. Mutagen. 9*, 191–226.

I.19 Ecogenetics

MARIKA GELDMACHER-VON MALLINCKRODT
Erlangen, Federal Republic of Germany

1 General

Ecogenetics is understood to be a genetic predisposition for an individual reaction to environmental factors (GOEDDE, 1972; PROPPING, 1978, 1980; KALOW, 1982; KALOW et al., 1986). Ecogenetic reactions are found in all living organisms – bacteria (see Chapter I.7f, Sect. 5), plants (see Chapter I.7g), insects, fishes, birds, mammals, and humans. Various reactions to pharmaceuticals, industrial toxins, pesticides, radiation, gaseous emissions, environmental toxins, foodstuffs, and also metals are known. An indication for genetically determined differences in the reactions of organisms to environmental factors is apparent whenever instead of a unimodal distribution (Fig. I.19-1 a), a bimodal (Fig. I.19-1 b) or multimodal distribution is found upon quantification. The common basis for almost all the various genetically determined reactions is the role that proteins – often as enzymes but also transport proteins – play in almost all life processes. The genetically controlled synthesis of proteins can lead to these variations. In this chapter examples will be given for the importance of genetic factors in the metabolism of metals.

2 Bacteria

A certain strain of the widely distributed *Escherichia coli* bacteria was discovered to have two different enzymes that are responsible for a genetically determined resistance to mercury and organic mercury compounds. The two enzymes have been shown to be an organomercury hydrolase that breaks carbon-mercury bonds and a mercury reductase. The latter reduces mercury(II) to metallic mercury, which escapes from the culture medium as vapor and allows the bacteria further growth. A similar resistance has been found in a *Pseudomonas* bacterial strain (SCHOTTEL, 1978).

HERMANSSON et al. (1987) reported a greater resistance of certain strains of bacteria, which live in the boundary zone between salt water and air, to certain antibiotics and metals, when compared with the same strains from deeper water areas. The resistant strains were for the most part more deeply pigmented (increased protection from sunlight). The greater resistance to mercury was especially striking, and increased tolerance to zinc, cadmium, and chromium (in not too large amounts) was

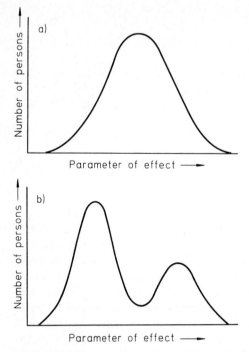

Fig. I.19-1. Unimodal (a) and bimodal (b) distribution.

also observed. The reason for these genetically dependent differences was believed to be the pressure of natural selection in this environment (greater UV radiation, high concentrations of heavy and other toxic metals) (see also Chapter I.7h).

3 Plants

Only those plant populations which are genetically tolerant to specific heavy metals can grow in soil contaminated with those metals. There are plants with resistance to lead, zinc, copper, and cadmium (MACNAIR, 1977; SIMON et al., 1977; HUTCHINSON, 1984; see also Chapters I.7f and I.7g).

4 Worms

Some types of earthworms have varying resistance to heavy metals, for example, zinc and lead (IRELAND and RICHARDS, 1977).

5 Insects

A search for duplications of the *Drosophila melanogaster* metallothionein gene (*Mtn*) yielded numerous examples of this type of chromosomal rearrangement. These duplications were widely distributed and were found in samples from four continents. They are also functional – larvae carrying *Mtn* duplications produce more *Mtn* RNA and tolerate increased cadmium and copper concentrations. It is a widely accepted hypothesis that gene duplication is one of the first steps in the evolution of new genes (MARONI et al., 1987).

6 Mammals

Certain inbred strains of mice are resistant to cadmium-induced testicular necrosis. The testicles of these mice bind less radioactive-labeled cadmium than susceptible mice. One assumes that a specific cadmium-binding protein plays a role in this. The characteristic is determined by a single autosomal recessive gene (MEISLER et al., 1979). The responsible gene is probably localized on chromosome 12 (TAYLOR, 1976). Genetically determined variations in the toxic effects of cadmium on the liver are also found in mice (TSUNOO et al., 1979).

HUANG et al. (1987) studied inbred strains of mice resistant to cadmium. Results with whole animals showed that the difference in the rate and level of metallothionein accumulation is at most twofold between sensitive and resistant strains.

BHAVE et al. (1988) investigated the methylation status, copy number, and organization of the metallothionein-I (*MT-I*) gene in hepatic and testicular DNA of mouse strains resistant (BALB/c) and susceptible (NFS) to cadmium-induced testicular toxicity. Digestion of DNA by restriction enzymes produced identical patterns for hepatic and testicular DNA of both strains, indicating that there was no apparent difference in the gross genomic organization or in copy number of the *MT-I* gene in the two types of tissues from either strain.

But differences in methylation status may account for the differential susceptibility of the two tissues to cadmium toxicity. The higher degree of *MT-I* gene methylation may result in slower or inefficient induction of *MT* in the testes, resulting in greater sensitivity to metal toxicity in testes than in liver. However, differences in methylation status alone do not seem to account for the interstrain differences in cadmium toxicity, and other factors, such as differences in genetic organization, seem to be involved in the inducibility of the *MT-I* gene in different strains.

Heavy metal induction of the synthesis of metallothioneins (MTs) provides an ideal model system for basic mechanistic studies of gene expression (see also Chapter I.12). Cell lines varying in their resistance to heavy metals have been isolated by GRADY et al. (1987) through a regimen of exposure to serially increasing levels of cadmium followed by clonal isolation. These cell lines have been used to examine the role of methylation and amplification in the Cd-resistant (Cdr) phenotype. It is suggested that regulation of expression of the *MT* genes in Cdr Chinese hamster cells is modulated at both the transcriptional and translational levels. An analysis of the *MT2* gene sequence has uncovered a potential alternative splice site in the first intron. Usage of this site would insert 3 or 12 additional amino acids between amino acids 9 and 10. Analysis of the splicing pattern of the *MT2* gene transcript in cultured cells has indicated that the second intron is preferentially removed prior to first intron excision.

Certain mice strains show varying sensitivity to lead poisoning (MYKKÄNEN et al., 1980).

Also observed in mice is the phenomenon of genetic zinc deficiency in milk (PILETZ and GANSCHOW, 1978). Nursing pups fed only the milk of homozygotic deficient animals do not survive. If the pups are given additional zinc, mortality is reduced significantly. It is assumed that the gene responsible for zinc transport from maternal blood to milk is modified. The so-called supermouse, on the other hand, accumulates more zinc from the mother during gestation than do other strains and excretes it in smaller quantities after birth. Through this it is protected against zinc deficiency (REIS and EVANS, 1977). BRUMMERSTEDT et al. (1977) reported an significant zinc deficiency among cattle (lethal trait A 46).

If manganese supplements are given to pregnant mice that are homozygotic for the "pallid" gene, the appearance of a genetically determined ataxia is prevented (OVERHOFF and FORTH, 1978).

An inheritable defect which leads to a heavy accumulation of copper in the liver has been found in Bedlington terriers. The hepatocellular lysosomes appear under the microscope as dense, insoluble, copper-rich granules. From the lysosomes, STERNLIEB (1987) isolated a copper-thionein which displays homology with other known metallothioneins. Although the primary genetic defect which underlies the pathogenesis of this form of hepatic copper overload is still unknown, circumstantial evidence suggests that sequestration of the protein by lysosomes may be temporarily cytoprotective.

Genetically determined variations in the distribution of lithium between plasma and erythrocytes have been found in sheep (SCHLESS et al., 1975).

7 Humans

Genetically determined manifestations of illness, such as defective absorption, distribution or excretion of metal ions or increased sensitivity, are also known. For example:

7.1 Aluminum

A genetically dependent disturbance of the aluminum metabolism is discussed with respect to Alzheimer's disease (AD). Increased levels of aluminum have been found in the brain (CRAPPER et al., 1976). Onset of the disease is at about age 45, first appearing as memory lapses, spatial disorientation, and lack of spontaneity. Later symptoms include speech difficulties, motor disturbances, and finally senility.

The question as to whether this disease, also known as "presenile dementia", is hereditary, has been the topic of various studies (reviews by AMADUCCI and LIPPI, 1988; SINET et al., 1988; GAGE et al., 1989). Risk factors seem to include the occurrence of Alzheimer's disease or Down's syndrome in the family, or advanced age of the mother at the time of the patient's birth. Molecular biological studies on the DNA of patients having AD showed a genetically determined polymorphism which is localized on chromosome 21. The gene that codes for the formation of β-amyloids was recently found on this same chromosome. A correlation between viral, immunological, or toxic factors and AD has not been proven, but a correlation with traumatic head injury appears to exist.

7.2 Chromium

Repeated analysis of the blood and urine of chromate workers for chromium indicated that one group (A) excreted little chromium(III) in the urine and showed relatively high concentrations of chromium in the blood, while the situation was reversed for a second group (B). From this, all persons whose chromium levels were higher in plasma than in the erythrocytes were termed strong Cr(VI) reducers (B), and correspondingly all persons whose chromium levels in the erythrocytes were the same as or greater than that of the plasma were termed weak Cr(VI) reducers (A) (LEWALTER, 1985). This difference is probably genetically determined.

7.3 Copper

Wilson's disease (hepatolenticular degeneration) is an autosomal recessive hereditary disturbance of the copper metabolism resulting in increased copper storage in the organs, for example the liver (causing cirrhosis) and brain (causing intellectual and psychical degeneration). As a consequence of this storage, a low copper level in the blood and ceruloplasmin (the transport protein for copper) is also found, the mechanisms for which remain unclear (STANBURY et al., 1972).

Genetically determined changes in the terminal amino acid frequency of human serum albumin, the so-called proalbuminemia, also exist. Since dogs, swine, and chickens possess no histidine in the third position of their serum albumin, and therefore can only bind very little copper to their albumin, TAKAHASHI et al. (1987) undertook a terminal amino acid group sequence analysis of the serum albumin of two Wilson's disease patients only to find the normal situation with histidine in the third position.

Using chromosome mapping, FRYDMAN et al. (1985) localized the genetic defect that leads to Wilson's disease on human chromosome 13 (the gene for ceruloplasmin is on chromosome 3, that for albumin on chromosome 4).

Patients with Wilson's disease or glucose-6-phosphate dehydrogenase (EC 1.1.1.49) deficiency, a hereditarily determined enzyme defect of the erythrocytes, reacted to oral administration of soluble copper(II) salts with hemolytic anemia, while healthy persons showed no effects (MOORE and CALABRESE, 1980).

A genetically determined copper metabolism disturbance is also responsible for "kinky hair" disease (Menkes' disease). The symptoms are cerebral degeneration, retarded growth, muscular hypertony, cramping of noticeably kinky hair, with lowered copper levels in the serum, liver and brain and elevated copper levels in the kidneys, spleen, skin, muscles, etc. (MENKES et al., 1962; GARNICA et al., 1977; HORN, 1984). The disease is recessive and bound to the X-chromosome. Experiments with cell cultures of skin fibroblasts and lymphoblastoid cells from patients with Menkes' disease showed that cells from patients accumulated 3–10 times as much copper as corresponding cells from healthy individuals. The copper is bound to a metallothionein in the cytosol of patients' cells (MERCER et al., 1981). There are two hypotheses for the cause of this illness:

1. The defect is an abnormal regulation of the metallothionein synthesis, which secondarily leads to a large accumulation of intracellular copper. This is not in accordance with the observed genetic inheritance in which the gene for metallothionein synthesis is localized on chromosome 16 (KARIN et al., 1984; SCHMIDT et al., 1984).
2. The disease is the result of abnormal copper transport, the mechanism of which is not yet known, which leads to increased metallothionein synthesis (see also SONE et al., 1987).

7.4 Iron

Increased iron resorption and storage, especially in the parenchymatosic organs is genetically determined. Cirrhosis of the liver, "bronze diabetes", congestive heart failure, etc. result from primary or endogenous hemochromatosis (STANBURY et al., 1972; SIMON et al., 1977).

7.5 Lead

(see also Chapter I.20a, Section 2.7)

The second enzyme in the porphyrin and heme syntheses, the porphobilinogen synthetase (EC 4.2.1.24) can show decreased activity as a result of a hereditary defect. Patients with this defect can show symptoms after exposure to low concentrations of lead (DOSS and MÜLLER, 1982; DOSS et al., 1982).

7.6 Lithium

Depressive patients react differently to lithium treatment. Heredity is probably a factor in this (MENDLEWICZ et al., 1972; PANDEY et al., 1979), as shown by investigations of twins (DORUS et al., 1974, 1975). PANDEY et al. (1977) described a family with an autosomal-dominant hereditary defect of the sodium-lithium exchange system.

7.7 Magnesium

A hereditary recessive primary hypomagnesemia that is accompanied by a low level of calcium in the blood leads to tetany. This is presumably caused by a non-functional magnesium resorption in the intestine, although in some instances a disturbed excretion of magnesium through the kidneys is assumed to be the cause (LOMBECK and BREMER, 1977).

7.8 Zinc

Acrodermatitis enteropathica is an autosomal recessive genetic zinc deficiency that results in lowered zinc levels in the plasma, skin changes, hair loss, diarrhea, and possible neurological disturbances (AKSU et al., 1980). An autosomal genetic elevated zinc level is also known, yet it does not result in clinical symptoms (SMITH et al., 1976).

In its widest sense, one can also include damage to genetic material resulting from metals in the area of ecogenetics. The mutagenic properties of metals are discussed extensively in Chapter I.18 and in sections of the chapters on the individual elements in Part II.

References

Amaducci, L., and Lippi, A. (1988), *Risk Factors and Genetic Background for Alzheimer's Disease. Acta Neurol. Scand. Suppl. 116*, 13–18.
Aksu, F., Huck, W., Sperling, G., and Mietens, C. (1980), *Acrodermatitis enteropathica: heute eine behandelbare Krankheit. Med. Klin. 75*, 485–489.
Bhave, M. R., Wilson, M. J., and Waalkes, M. P. (1988), *Methylation Status and Organization of the Metallothionein-I Gene in Livers and Testes of Strains of Mice Resistant and Susceptible to Cadmium. Toxicology 50*, 231–245.
Brummerstedt, E., Basse, A., Flagstadt, T., and Andresen, E. (1977), *Animal Model of Human Disease. Acrodermatitis enteropathica, Zinc Malabsorption. Am. J. Pathol. 87*, 725–728.
Crapper, D. R., Krishnan, S. S., and Quittkat, S. (1976), *Aluminium, Neurofibrillary Degeneration and Alzheimer's Disease. Brain 99*, 67–80.
Dorus, E., Pandey, G. N., Frazer, A., and Mendels, J. (1974), *Genetic Determination of Lithium Ion Distribution. I. An in vitro Monozygotic Twin Study. Arch. Gen. Psychiatry 31*, 463–465.
Dorus, E., Pandey, G. N., and Davis, J. M. (1975), *Genetic Determinant of Lithium Ion Distribution. An in vitro and in vivo Monozygotic-Dizygotic Twin Study. Arch. Gen. Psychiatry 32*, 1097–1102.

Doss, M., and Müller, W. A. (1982), *Acute Lead Poisoning in Inherited Porphobilinogen Synthase (δ-Aminolevulinic Acid Dehydrase) Deficiency. Blut 45*, 131–139.
Doss, M., Becker, U., Sixel, F., Geisse, S., Solcher, H., Schneider, J., Kufner, G., Schlegel, H., and Stoeppler, M. (1982), *Persistent Protoporphyrinemia in Hereditary Porphobilinogen Synthase (δ-Aminolevulinic Acid Dehydrase) Deficiency under Low Lead Exposure. A New Molecular Basis for the Pathogenesis of Lead Intoxication. Klin. Wochenschr. 60*, 599–606.
Frydman, M., Bonne-Tamir, B., Farrer, L. A., Conneally, P. M., Magazanik, A., Asbel, S., and Goldwich, Z. (1985), *Assignment of the Gene for Wilson Disease to Chromosome 13: Linkage to the Esterase D Locus. Proc. Natl. Acad. Sci. USA 82*, 1819–1821.
Gage, F. H., Privat, A., and Christen, Y. (eds.) (1989), *Neuronal Grafting and Alzheimer's Disease.* Springer Verlag, Berlin-Heidelberg-New York.
Garnica, A. D., Frias, J. L., and Rennert, O. M. (1977), *Menkes' Kinky Hair Syndrome: Is it a Treatable Disorder? Clin. Genet. 11*, 154–161.
Goedde, H. W. (1972), *Genetically Determined Variability in Response to Drugs. Pharm. Weekbl. 107*, 437–466.
Grady, D. L., Moyzis, R. K., and Hildebrand, C. E. (1987), *Molecular and Cellular Mechanisms of Cadmium Resistance in Cultured Cells. Experientia (Suppl.) 52*, 456–477.
Hermansson, M., Jones, G. W., and Kjelleberg, S. (1987), *Frequency of Antibiotic and Heavy Metal Resistance, Pigmentation, and Plasmids in Bacteria of the Marine Air-Water Interface. Appl. Environ. Microbiol. 53*, 2338–2342.
Horn, N. (1984), *Copper Metabolism in Menkes' Disease,* in: Rennert, O. M., and Chan, W. Y. (eds.): *Metabolism of Trace Metals in Man,* Vol. 2. pp. 25–53. CRC Press, Boca Raton, Florida.
Huang, P. C., Morris, S., Dinman, J., Oine, R., and Smith, B. (1987), *Role of Metallothionein in Detoxification and Tolerance to Transition Metals. Experientia (Suppl.) 52*, 439–446.
Hutchinson, T. C. (1984), *Toleranzgrenzen für Pflanzen: Auswahl geeigneter Pflanzen für metallverseuchte Böden,* in: Merian, E. (ed.) *Metalle in der Umwelt,* pp. 137–139. Verlag Chemie, Weinheim-Deerfield Beach/Florida-Basel.
Ireland, M. P., and Richards, K. S. (1977), *The Occurrence and Localisation of Heavy Metals and Glycogen in the Earthworms Lumbricus rubellus and Dendrobaena rubida from a Heavy Metal Site. Histochemistry 51*, 153–166.
Kalow, W. (1982), *Ethnic Differences in Drug Metabolism. Clin. Pharmacokin. 7*, 373–400.
Kalow, W., Goedde, H. W., and Agarwal, D. P. (1986), *Ethnic Differences in Reactions to Drugs and Xenobiotics.* Alan R. Liss, New York.
Karin, M., Eddy, R. L., Henry, W. M., Haley, L. L., Byers, M. G., and Shows, T. B. (1984), *Human Metallothionein Genes are Clustered on Chromosome 16. Proc. Natl. Acad. Sci. USA 81*, 5494–5498.
Lewalter, J. (1985), *Neue methodische Ansätze in der biologischen Überwachung Arbeitsstoff-Exponierter,* in: Bolt, H. M., Piekarski, C., and Rutenfranz, J. (eds.): *Bericht über die 25. Jahrestagung der Deutschen Gesellschaft für Arbeitsmedizin,* pp. 151–166. Gentner Verlag, Stuttgart.
Lombeck, I., and Bremer, H. J. (1977), *Primary and Secondary Disturbances in Trace Element Metabolism Connected with Genetic Metabolic Disorders. Nutr. Metab. 21*, 49–64.
MacNair, M. R. (1977), *Major Genes for Copper Tolerance in Mimulus guttatus. Nature 268*, 428–430.
Maroni, G., Wise, J., Young, J. E., and Otto, E. (1987), *Metallothionein Gene Duplications and Metal Tolerance in Natural Populations of Drosophila melanogaster. Genetics 117*, 739–744.
Meisler, M., Orlowski, C., Gross, E., and Bloor, J. H. (1979), *Cadmium Metabolism in cdm/cdm Mice. Biochem. Genet. 17*, 731–736.
Mendlewicz, J., Fieve, R. R., Stallone, F., and Fleiss, J. L. (1972), *Genetic History as a Predictor of Lithium Response in Manic-depressive Illness. Lancet 1*, 599–600.
Menkes, J. H., Alter, M., Stiegleder, G. K., Weakley, D. R., and Sung, J. H. (1962), *A Sex-linked Recessive Disorder with Retardation of Growth, Peculiar Hair, and Focal Cerebral and Cerebellar Degeneration. Pediatrics 29*, 764–779.

Mercer, J. F. B., Stevenson, T., Ladzins, I., Camakaris, J., and Danks, D. M. (1981), *Studies of Metallothionein m-RNA relevant to Menkes' Syndrome and Mottled Mice. Am. J. Hum. Genet. 33*, 50A.

Moore, G. S., and Calabrese, E. J. (1980), *G6PD-deficiency: A Potential High-risk Group to Copper and Chlorite Ingestion. J. Environ. Pathol. Toxicol. 4*, 271–279.

Mykkänen, H. M., Dickerson, J. W. T., and Lancaster, M. (1980), *Strain Differences in Lead Intoxication in Rats. Toxicol. Appl. Pharmacol. 52*, 414–421.

Overhoff, H., and Forth, W. (1978), *Biologisch essentielle Elemente ("Spurenelemente"). Dtsch. Ärztebl. 75*, 301–305.

Pandey, D. N., Ostrow, D. G., Haas, M., Dorus, E., Casper, R. C., and Davis, J. M. (1977), *Abnormal Lithium and Sodium Transport in Erythrocytes of a Manic Patient and Some Members of the Family. Proc. Natl. Acad. Sci. USA 74*, 3607–3611.

Pandey, G. N., Dorus, E., Davis, J. M., and Tosteson, D. C. (1979), *Lithium Transport in Human Red Blood Cells: Genetic and Clinical Aspects. Arch. Gen. Psychiatry 36*, 902–908.

Piletz, J. P., and Ganschow, R. E. (1978), *Zinc Deficiency in Murine Milk Underlies Expression of the Lethal Milk (lm) Mutation. Science 199*, 181–183.

Propping, P. (1978), *Pharmacogenetics. Rev. Physiol. Biochem. Pharmacol. 83*, 123–173.

Propping, P. (1980), *Neue Entwicklungen in der Pharmakogenetik. Klinikarzt 9*, 422–434.

Reis, B. L., and Evans, G. W. (1977), *Genetic Influence on Zinc Metabolism in Mice. J. Nutr. 107*, 1683–1686.

Schless, A. P., Frazer, A., Mendels, J., Pandey, G. N., and Theodorides, V. J. (1975), *Genetic Determination of Lithium Ion Metabolism. II. An in vivo Study of Lithium Ion Distribution across Erythrocyte Membranes. Arch. Gen. Psychiatry 32*, 337–340.

Schmidt, C. J., Hamer, D. H., and McBride, O. W. (1984), *Chromosomal Location of Human Metallothionein Genes: Implications for Menkes' Disease. Science 224*, 1104–1106.

Schottel, J. L. (1978), *The Mercuric and Organomercurial Detoxifying Enzymes from a Plasmid-bearing Strain of Escherichia coli. J. Biol. Chem. 253*, 4341–4349.

Simon, M., Alexandre, J.-L., Bourel, M., Le Marec, B., and Scordia, C. (1977), *Heredity of Idiopathic Haemochromatosis: A Study of 106 Families. Clin. Genet. 11*, 327–341.

Sinet, P. M., Lamour, Y., and Christen, Y. (eds.) (1988), *Genetics and Alzheimer's Disease.* Springer Verlag, Berlin-Heidelberg-New York-London-Paris-Tokyo.

Smith, J. C., Zeller, J. A., Brown, E. D., and Ong, S. C. (1976), *Elevated Plasma Zinc: A Heritable Anomaly. Science 193*, 496–498.

Sone, T., Yamaoka, Y., and Tsunoo, H. (1987), *Induction of Metallothionein Synthesis in Menkes' and Normal Lymphoblastoid Cells is Controlled by the Level of Intracellular Copper. J. Biol. Chem. 262*, 5878–5885.

Stanbury, J. B., Wyngaarden, J. B., and Frederickson, D. S. (1972), *The Metabolic Basis of Inherited Disease,* pp. 1033, 1051. McGraw-Hill, New York.

Sternlieb, I. (1987), *Hepatic Lysosomal Copper-thionein. (Suppl.) 52*, 647–653.

Takahashi, N., Takahashi, Y., and Putnam, F. W. (1987), *Structural Changes and Metal Binding by Proalbumins and Other Amino-terminal Genetic Variants of Human Serum Albumin. Proc. Natl. Acad. Sci. USA 84*, 7403–7407.

Taylor, B. A. (1976), *Linkage of the Cadmium Resistance Locus to Loci on Mouse Chromosome 12. J. Hered. 67*, 389–390.

Tsunoo, H., Nakajima, H., Hata, A., and Kimura, M. (1979), *Genetical Influence on Induction of Metallothionein and Mortality from Cadmium Intoxication. Toxicol. Lett. 4*, 253–256.

I.20a Standard Setting and Risk Assessment in Human Exposure to Metals and Their Compounds

REINIER L. ZIELHUIS, Oegstgeest, The Netherlands

1 Introduction

Humans have always been exposed to metals in and around home. As soon as man started to make tools, work-related health risks will have occurred. The Roman empire may have been broken down by lead poisoning among the leading classes because of drinking from leaded vessels. Mining activities have moved metals from ore deposits and have distributed them over the ambient and occupational environment. Around the turn of the 19th/20th century reproductive risks in female lead workers were reported, although it is difficult to separate the role of poor living and working conditions from that of exposure to metals as such. HUNTER (1975) reviewed many historical examples of occupational poisoning by metals.

In the first decennia of the 20th century regulatory measures were taken to prevent lead poisoning from the use of powdered white lead, one of the first types of regulation of chemicals.

The 20th century has brought the development of modern animal and human toxicology (see Chapter I.13b); working mechanisms and exposure effect/response relationships were elucidated. In the last decades the development of analytical methods has made it possible to assess internal exposure (see Chapter I.4a). The impact of these developments on the time lag between the first suggestive case reports and the confirmation by modern time research has been illustrated by the Workshop/Conference on the Role of Metals in Carcinogenesis (BELMAN, 1981): the time lag between the first case reports of cancer and epidemiological confirmation was for arsenic about 70 years, for chromium about 20 years. The time lag up to confirmation in animal experiments was for chromium about 40 years, for nickel 30 years. For beryllium the time lag between the first evidence and confirmation in animal studies was only about 7 years, confirmation (?) in epidemiologic studies occurred about 30 years later. Similar time lags will have existed for the elucidation of non-carcinogenic effects. The time lag has been long for the "old" metals (e.g., As, Pb, Hg, Cu, Fe, Mn) and has become smaller for the "new" metals (e.g., Be, Cd, Co, Ce, In, Ni, Te, U). Toxicology has developed from low quality research "after the fact" to high quality "preventive" research.

This development of metal toxicology made standard setting possible. Around the turn of the 19th/20th century exposure limits were proposed, at first in Germany. From the thirties onwards lists of occupational exposure limits (OEL) were issued, particularly in the USA and the USSR (ILO, 1983). The USSR was the first country to make these limits statutory. In the last 20 years biological exposure limits have been developed, after biological monitoring of exposure had become an established method (see Sect. 3.4). A few decades later than the OELs, chemical quality limits were established and enforced for, e.g., ambient air, drinking water, and food. In the course of time the criterion of accepted risks shifted from prevention of early death and disease to safeguarding health (see Sect. 2.5).

This chapter only refers to health risks in man; hazards to the ecosystem are not discussed. Nevertheless, protection of the ecosystems as such is essential for safeguarding wellbeing and health of the present and – not to forget – of the future generations. For standards and guidelines see Chapter I.20b.

2 Conceptual Aspects

2.1 Standard Setting for Exposure to Metals

There exists a "family" of chemical quality limits for metals, e.g., for workroom air, ambient air, drinking water, and food. All limits should be underpinned by the same toxicological data base (see Sect. 3.1), albeit the groups most at risk (see Sect. 2.7), the sources and the procedures (see Sects. 2.11, 4.2, 5.1, 5.2) may differ. Before treating the practical aspects of setting occupational or environmental exposure limits, several basic concepts have to be discussed.

The term "standard" used in this chapter, not necessarily means "statutory standard", but refers to occupational (see Sect. 4) and environmental exposure limits (see Sect. 5) in a general sense.

2.2 Risk Assessment and Standard Setting

Risk means "the measure of the probability of the magnitude of impairment, including death" (ESMEN, 1984) at specified levels of exposure. In the context of standard setting and enforcement one can distinguish two types of risk assessment: (1) assessment of exposure effect/response relationships in order to produce basic data for standard setting (see Sect. 2.4), and (2) assessment of whether the actual exposure exceeds exposure standards (see Sect. 3.4). The first requires a more elaborate strategy of measurements (research monitoring) than the second one (compliance monitoring). Risk management is discussed in Chapter I.20b.

Research risk assessment requires four steps: assessment of (1) the probability of a quantified exposure to specified agents in a given situation, (2) the probability of internal exposure (uptake), (3) the toxic amounts/concentrations in the critical target organ(s), and (4) the specified and quantified end point(s) of health impairment (ESMEN, 1984). This sequence yields the basic data (see Sect. 3.1) for standard setting and risk management.

2.3 Critical Concentration, Organ and Effect

One should assess the probability of the occurrence of early adverse (see Sect. 2.6) effects at specified exposure levels, in order to prevent health impairment. FRIBERG (1986) presented the following strategic concepts:

- the *critical concentration for a cell:* the concentration at which adverse functional changes, reversible or irreversible, occur in the cell;
- the *critical organ concentration:* the mean concentration in the organ at the time when any of its cells reaches the critical concentration;
- the *critical organ:* that particular organ which first attains the critical (organ) concentration under specified circumstances of exposure and for a given population;
- the *critical effect:* that effect in an individual at that point in the exposure effect/response relationship from which an adverse effect starts to occur;
- the *critical population concentration (PPC):* the response rate at specified PCCs, e.g., the PCC-50 or -20 is the concentration in the critical organ at which 50 or 20% of the population has the critical effect.

2.4 Exposure Effect/Response Relationships

Because in occupational and environmental studies one usually cannot establish the "dose" as precisely as in pharmacology, one prefers the term "exposure" instead of "dose". Humans exposed differ in actual exposure and in health effects. Therefore, it is important to distinguish: (1) Exposure-effect relationships (EER), which refer

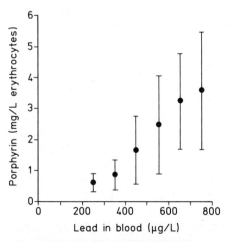

Fig. I.20a-1. Exposure(PbB)-effect(FEP) relationship in lead exposed workers.

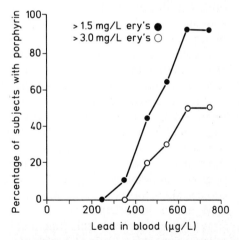

Fig. I.20a-2. Exposure(PbB)-response(FEP) relationship, based upon the same data as in Fig. I.20a-1.

either to the (semi-)quantitative relation between exposure (intensity, frequency, duration) and a specified graded (continuum of intensity) effect in an individual subject (or animal) or the average graded effect in groups of subjects (or animals). Quantal (stochastic) effects, however, either occur or do not occur, e.g., cancer, abortion. This leads to (2) exposure-response relationships (ERR), which present either the relationship between exposure and the percentage of subjects with a quantal effect or the percentage of subjects that exceeds a specified intensity of a specified graded effect.

From the viewpoint of health protection not the *average* effect in a non-existing "standard" subject is of primary importance, but the *percentage* of subjects that, at specified exposure levels, exceeds a specified graded effect or that responds with a specified quantal effect. At the same exposure level subjects usually differ in toxicokinetics and/or in coping capacity (toxico-dynamics) (see Sect. 2.7). The ERR permits to pinpoint those exposures which need correction and those (groups of) subjects who run an unaccepted risk. Fig. I.20a-1 presents an EER and Fig. I.20a-2 an ERR, based upon the same data. It is to be deplored that in the literature both terms often are used interchangeably.

2.5 The Concept of Health

Does protection of health mean prevention of early mortality, morbidity, early health impairment or even nuisance? The Dutch Public Health Council and the Dutch National MAC Commission defined health as "a non-stable condition of the human organism, of which the functional capacities leave nothing to be desired in the opinion of the general population or workers and/or according to health experts; pre-existing physical and mental capacities (e.g., determined by age or sex) have to be taken into account; the functional conditions should be comparable to those in non-exposed otherwise similar groups of the general population or workers in the same society; allowance should be made for the present state of the art, present-day objectives of health care, social acceptability and social habits" (ZIELHUIS and NOTTEN, 1978/79). This definition basically refers to epidemiological data. It is more specified than the general definition of health as given by the WHO: "optimal physical, mental, and social wellbeing". Therefore, it is better suited as a criterion for standard setting.

Present day standards are not meant to prevent merely disease or death, but to safeguard health. Moreover, a distinction between "somatic health impairment" (signs) and "nuisance" (symptoms) should be rejected, because (1) this distinction is based upon an unwarranted dichotomy between soma and psyche, whereas man functions as a whole person; (2) perceived nuisance can be assessed as objectively in groups of humans as signs of somatic impairment; (3) symptoms not necessarily precede signs; (4) both symptoms and signs may restrict "normal" functioning at work and/or in society.

The actualization of this functional concept of health may differ between countries (social habits, levels of health care, etc.) and may change in the course of time by shifts in acceptability and in the state of the art of science. Consequently, standards periodically will have to be reassessed accordingly (see Sect. 4.4).

2.6 Adverse Effects

An "effect" as such is a neutral concept: humans may respond with biological "changes". Health experts have to distinguish between "effects as such" and "adverse" effects.

The WHO (1978) defined "adverse" as biological changes (effects) which "(1) occur with intermittent or continued exposure and that result in impairment of functional capacity (as determined by anatomical, physiological, biochemical or behavioral parameters) or in a decrement of the ability to compensate additional stress, (2) are irreversible during exposure or following cessation of exposure, if such changes cause detectable decrements in the ability to compensate additional stress and (3) enhance the susceptibility of the organism to the deleterious effects of environmental influences".

SHERWIN (1983) very aptly described "adverse" as follows: "an adverse health effect is the causation, promotion, facilitation of a structural and/or functional abnormality, with the implication that the abnormality has the potential of lowering the quality of life, causing a disabling illness or leading to premature death".

Both concepts very well correspond to the concept of health as defined in Sect. 2.5.

The "adverse" character often can only be determined by health experts a posteriori, i.e., when all data are available, e.g., trends in the course of time, combination with other "effects", age, and sex.

2.7 Groups/Subjects at Extra Risk

In comparison with animal species exposed in experimental toxicology under strictly controlled conditions the individual variability of humans in actual exposure and in coping capacity at similar exposure levels is much larger (see also Chapter I.13b). This is true within (intra) and between (inter) individuals (see Sect. 2.4).

One may classify at extra risk groups/subjects as follows:
(1) Increased *external exposure* in risk situations, e.g., high arsenic (As) or lead (Pb) levels in drinking water; exposure to cadmium (Cd) at work by smoking. The work load (respiratory minute volume) increases the intake; a hot climate may increase dermal absorption;
(2) groups/subjects with *risk behavior* and other aspects of *life style*, which may lead to increased external exposure, under similar conditions of environmental or occupational exposure, e.g., toddlers with a high intake of food and drinking water per kg/b.w. and with frequent hand-mouth contact (soil, objects), workers with a poor hygienic behavior;
(3) groups/subjects with a similar external exposure (intake), but with different *toxicokinetics*, which affects the levels in the critical organs (*internal exposure*), e.g., in young children relatively more biologically available lead and deposit in bone (physiological); decreased elimination because of renal disease or impaired clearance from the airways (pathological);

(4) groups/subjects with increased *susceptibility* (coping capacity), i.e., lower critical organ levels at similar internal exposure (predisposition, preexisting diseased organs, *toxicodynamics*).

Physiological characteristics as, e.g., age, sex, pregnancy, cannot be regarded as deviant phenomena. Therefore, such groups should be regarded as *"special"* groups rather than as "at extra risk" groups.

One can broadly classify those factors which contribute to an increased risk as follows:

Work environment related factors
– work process (e.g., scouring, spraying);
– availability and use of technical or personal preventive measures; workers may carry metal dust on clothes and shoes to their home, which may lead to higher exposure at home and higher blood levels particularly in their children (KNISHKOWSKY and BAKER, 1986);
– poor hygienic behavior at work.

Life style factors, living conditions

The most important are:
– nutritional habits, maybe deviant diets;
– consumption of tobacco (contains metals, particularly Cd, and may become extra polluted at work) (ADAMSSON et al., 1979). ISKANDER et al. (1986) found 28 metal elements in cigarette tobacco, commercially manufactured in the USA. The levels in American cigarettes were much lower than in other national brands;
– consumption of alcohol: BOUDENE et al. (1975) measured in French wines on average about 400 µg/L, range up to 4 mg/L;
– consumption of coffee, tea, mineral water, non-alcoholic beverages; boilers may contaminate drinks with cadmium; pottery may contaminate acid drinks with lead; mineral waters may contain on average about 20 µg/L of arsenic (ZOETEMAN and BRINKMAN, 1976);
– personal hygiene off work;
– hobbies, "do-it-yourself" work;
– living in the vicinity of metal ore smelters, e.g., suggestive evidence or reproductive effects in Sweden in exposure to As (NORDSTRÖM et al., 1978, 1979); evidence of lung cancer risks in exposure to As (in the USA, BROWN et al., 1984; in Sweden, PERSHAGEN, 1985).

The relative importance of life style factors and of living/working conditions depends on natural, ethnic and/or cultural background, level of education, socio-economic status, etc. The above mentioned factors are in principle avoidable, although often not by means of regulatory standards. Therefore, standard setting reflects only a part of total exposure.

Host factors
– age, e.g., young children may absorb metals to a greater extent; they are more susceptible than adults to effects on the central nervous system in exposure to

alkylmercury compounds; the toxicokinetics of lead differ in comparison with adults; lower critical levels in blood with respect to heme synthesis; moreover, toddlers tend to have more polluted hand-mouth contact, which results in relatively high lead in blood levels in children 1 to 5 years of age (MAHAFFEY, 1985; BRUNEKREEF, 1985);
- sex: adult women have lower critical Pb-levels in blood with respect to an effect on the heme synthesis than adult men, which leads to relatively higher free erythrocyte porphyrin levels in blood at the same Pb-level in blood (WHO, 1980).
- pregnancy: e.g., Cd may adversely affect the placenta structure in humans (COPIUS PEEREBOOM et al., 1983);
- genetics: there may exist increased susceptibility to Pb in G6PD-deficiency, thalassemia, and sickle cell trait (CALABRESE, 1978);
- pre-existent or intercurrent predisposition or manifest disease;
- consumption of prescribed drugs, e.g., dimethyldithiocarbamate, a metabolite of disulfuram used for treatment of alcoholism, increases Ni excretion and affects the kinetics of other metals. The host factors are partly physiological, partly pathological. Standards should take into account physiological and (pre)pathological factors, which occur in relatively large groups of exposed humans; however, individual subjects with increased susceptibility (less than, e.g., 5% of those exposed) cannot be protected by standard setting.

2.8 Combined Exposure
(see also Chapter I.10)

Health and disease to a large extent reflect the impact of the *total* environment at and off work, sometimes even from before conception (ZIELHUIS, 1985a). Man usually is exposed to a mixture of chemicals, particularly at work.

One may distinguish various types of combined exposure:
(1) exposure to the *same* metal (compounds) with similar toxicokinetics and -dynamics at work, at home, and in the ambient environment, e.g., lead at work, from drinking water, food and beverages, traffic exhaust; cadmium at work, from smoking, from food and drinking water; chromium(III) and (VI) from food and from welding stainless steel;
(2) exposure to *different* compounds of the *same* metal, with different toxicokinetics and/or -dynamics (see Sect. 2.14);
(3) exposure to compounds of *different* metals and/or *other* chemicals, e.g., in the case of metal ore smelting (e.g., Cu, As, Cr, Ni, Zn);
(4) exposure to *non-chemical* workplace or environmental factors, e.g., climate, nutrition, physical exertion. An example is the job of the welder: exposure to iron oxide, Cr and Ni in welding stainless steel, welding gases, UV-radiation, hot or cold climate, high noise levels, static work posture.
Combined exposure may result in increased internal exposure, in interaction (see Chapter I.10), or in a combination of more or less independent effects, some of

which can be considered adverse. Welders may run the risks to become hard of hearing, to develop signs and symptoms of the locomotor tract, to get flashburn or arc-eye, to be overexposed to Cr, Ni, Co, infrared radiation, etc. (ZIELHUIS and WANDERS, 1986).

2.9 Scientific Evaluation and Decision Making

Setting standards requires a two step-procedure. The first step involves the evaluation of the toxicological data base (see Sect. 3.1), which permits to propose a *health-based* (HB) *recommended* exposure limit. This step has to be carried out by health experts (toxicologists, physicians, hygienists), who exclusively consider the relevance of the data for the protection of health. The second step is the decision making process, which also takes into account the socio-economic and technological constraints; this leads to the *operational* exposure limit (ZIELHUIS and NOTTEN, 1978/79).

As will be discussed in Sects. 4.2 and 5.2 such a two-step approach usually does not exist. Too often a body of experts and policy-makers is responsible for standard setting. Those "who run the risk", e.g., workers, children, general public, are not involved in this standard setting process, whereas those who "decide" upon the standards usually are not or hardly exposed themselves in their own daily life and/or at work. Moreover, those who are expected to run the risks should be fully informed both on the HB-recommended limits and on the reasons why the operational limits possibly deviate from the HB-limits. Both steps should be fully documented and published.

An explicitly documented underpinned operational standard and a clear distinction between what is "science" and what is "policy" may also help to prevent what the *British Journal of Industrial Medicine* (EDITORIAL, 1986) called unwarranted non-scientific "verbal pollution" and "paratoxicology", which particularly may be a threat in exposure to metals which are also normally widely present in the ambient environment.

HB-recommended limits are to a large extent similar for all human beings; they can be proposed by international groups of experts, convened by, e.g., the WHO (1980). However, operational exposure limits may differ between nations, because of the large differences in technological and socio-economic constraints. Therefore, operational limits have to be set by the national authorities (see Sects. 4 and 5).

2.10 The Ergonomic and the Hygienic Rule
(see also Chapter I.16)

Some metals are essential for human growth, development, and health, e.g., Co, Fe, Zn, Cu, Se, Mn. In addition there are many metals which are not essential, but xenobiotic, e.g., Hg, Pb, Cd, Sr. Intake of xenobiotic chemicals as such not necessarily causes health impairment when established threshold values are not exceeded. The intake of essential chemicals should be *optimized* (Ergonomic Rule), whereas that of non-essential xenobiotic chemicals should always be *minimized* as far as possible (Hygienic Rule). The appropriate route of entry for essential elements is by proper nutrition. Exposure to (compounds of) essential elements in, e.g., drinking water, cosmetics, at work, in ambient air, should also be minimized as far as possible. Therefore, all metals, except those intentionally added to food, should follow the Hygienic Rule, irrespective of the route of entry.

2.11 Positive versus Negative Approach

In actual practice, the decision making processes differ according to the source of exposure. The *positive* approach permits regulated chemicals to be present up to a defined level, whereas all non-regulated chemicals are forbidden. The *negative* approach permits chemicals listed to be present up to a specified limit, whereas the presence of non-listed chemicals is not regulated.

The positive approach refers to chemicals, e.g., additives, present in food; the limit is indirectly determined by the internationally agreed upon Acceptable Daily Intake (ADI), usually established jointly by the WHO and the FAO on the basis of a reasonably adequate toxicological data base. However, standards for drinking water, ambient air, workroom air, soil follow the negative approach: the toxicological data base often is highly deficient (see Sect. 3.1), because an elaborate testing program is not a priori required.

The FAO/WHO (1972) established some "provisional" weekly intakes for Pb, Hg, As, Cd; however, the documentation was and still is rather poor; for inorganic As a better underpinned lower ADI has recently been recommended (FAO/WHO, 1983).

2.12 Hazard and Toxicity

The term "hazard" may have several meanings. It may be used interchangeably with the term "risk"; however, the term risk as defined in Sect. 2.2 neither includes the probability that the agent is present in the immediate environment, including food, drinking water, etc., nor the probability that man becomes actually exposed. The probability of actual exposure depends on the positive or negative approach (see Sect. 2.11), the physico-chemical properties of the agent (e.g., volatility, water solubility, density in comparison to air), type of process, subject's behavior, the way of handling the agent at work or at home (see Sect. 2.7).

The term hazard refers to the probability that exposure occurs. Such a hazard does not necessarily imply an increased health risk, because this to a large extent also depends upon the toxic properties of the agent. The term "toxicity" can be regarded as "the ability of the agent to cause harm", the hazard as "the probability to do so". "Hazard" refers to the exposure situation, "toxicity" to the agent and "risk" to the actual probability of impaired health at a specified level of exposure, taking into account the susceptibility of exposed subjects.

A few examples may illustrate this. The mineral water in the EEC-countries may contain on average about 20 µg/L inorganic water soluble arsenic (see Sect. 2.7). The consumption per capita differs about 100-fold between France and the UK, the Netherlands, and Denmark. So, the hazard in France appears to be considerably higher than in the other countries. Whether this leads to any increased skin cancer risk can only be guessed; as far as known, it has never been studied. The daily intake of lead contaminated wine (see Sect. 2.7) generally is considerably higher in Italy and France than in other Western and Northern European countries; this may at least partly explain the upward trend in lead blood levels from Northern to Southern Europe. This should have an impact on the operational occupational exposure limit; however, the EEC Directive for protection of lead workers (EEC, 1982) did not take this into account.

2.13 Negative Data

In animal and epidemiological studies one measures a priori chosen effects with an a priori chosen method. Absence of an increased incidence or prevalence of an effect not necessarily implies that an increased health risk actually does not exist, when insensitive methods have been applied. The progress of analytical chemistry has very much decreased the detection limit and expanded the number of analytical methods. Not long ago, the detection limit for β_2-microglobulin (β_2M) in urine applied in the assessment of risks in exposure to Cd had to be counted in mg/L, whereas nowadays it is about 5 µg/L. Moreover, β_2M is pH-dependent; therefore, the less pH-dependent retinol binding protein (RBP) has been added as another indicator of renal tubular dysfunction. Recently it has been suggested that a sensitive modified SDS-PAGE electrophoresis of urinary proteins decreased the non-observed effect level for tubular lesions in exposure to Cd. This level may approach the individual acceptable health based level of Cd-U of 5 µg/g creatinine, recommended by the WHO in 1980 (VERSCHOOR et al., 1987).

In workers exposed to water soluble Cr(VI) compounds, e.g., welders and chromium platers, some authors observed evidence of early changes in renal tubular function, whereas others could not confirm this; however, the authors used different effect parameters. The location of the renal lesion also determines which parameter has to be chosen; a set of several sensitive parameters is necessary to detect early changes of tubular and/or glomerular function.

The third cause of negative data may be the small size (n) of the group of subjects examined, because n determines the level of significance.

The fourth factor is an inadequate study design that does not control for e.g. confounders. It often requires a much more elaborated study to suggest absence than an increased incidence or prevalence of effects (HERNBERG, 1981).

2.14 Metal Speciation
(see also Chapters I.3, I.4, and Part II)

In the risk assessment of exposure to metals one usually too easily assumes that measuring the concentration of metals as element in air, food, drinking water, etc., or in blood, urine, hair, saliva, yields a reliable estimate of external or internal exposure. However, man usually is not exposed to the elements as such, but to different *metal compounds* (species). Metal speciation aims at analysis of the metal species and not of the metals as such.

Compounds of the same metal may essentially differ in toxicokinetics and -dynamics and consequently in health risks. Wellknown are the differences in health risks in exposure to inorganic Pb-compounds and Pb-tetra/trimethyllead, inorganic Hg- and alkylmercury compounds. Large differences also exist between various Cr-, Ni-, As-, Se- and Sb compounds.

These differences also affect the methods of environmental and biological effect monitoring (see Sect. 3.4). For example, in biological monitoring of exposure to inorganic Hg-, alkoxyethyl- and phenylmercury compounds one may mainly rely on measurement of Hg in urine, but in the case of exposure to alkylmercury compounds one has to rely on sampling of blood and/or hair. In exposure to inorganic Pb compounds one relies on sampling blood, but in the case of tetraethyl/methyllead one has to sample urine. One cannot rely on sampling urine in the case of not or hardly water soluble Cr- or As compounds when one wants to assess the risk of cancer in nose or lungs, whereas for assessment of risks to the kidney in exposure to water soluble Cr compounds one has to sample urine. Organic As compounds from marine food are excreted unchanged with the urine, whereas inorganic As compounds are metabolized first into monomethylarsonic acid (MMAA) and subsequently into dimethylarsinic acid (DMAA), also measured in urine. Assessment of the risk of skin cancer in exposure to water soluble As compounds has to rely on measuring the excretion of the metabolites and inorganic As, after a few days abstention from marine food (ZIELHUIS and WIBOWO, 1983).

The exposure limits often refer to the element concentrations as such, and too little to the different metal compounds.

However, it should be realized that the analytical difficulties of metal speciation are considerable (see Chapter I.4).

3 Methodological Aspects of Standard Setting and Risk Assessment

3.1 The Toxicological Data Base

The basic data required for setting health based (HB-) limits (first step; see Sect. 2.9) are derived from the following four sources:

(1) *Animal studies* (in vivo, in vitro), which, e.g., elucidate working mechanisms, genotoxicity, exposure effect/response relationships (see Sect 2.4) and non-observed adverse effect/response exposure (dose) levels (NOAE/RL).

In the last decades the available set of data has considerably expanded. However, because of the negative approach for ambient and workroom air, etc. (see Sect. 2.11), there often is a lack of (sub)chronic respiratory exposure studies. An accepted method for extrapolation from oral to respiratory exposure does not yet exist. On the other hand, because human exposure to metals has a very long history in contrast to many other chemicals, standards for exposure to metals often can largely be based upon human data, at least for the widely used "old" metals, although not for recently applied "new" metals and various specified metal compounds (see Sect. 2.14 and Chapter I.13b).

(2) *Human volunteer studies*, which usually are carried out in small groups of relatively healthy young subjects and with a short duration of exposure. Because exposure to metals generally exerts health effects after rather long-term exposure, only relatively few volunteer studies on metal exposure-effect/response relationships have been carried out. One example may illustrate the limitation. COOLS et al. (1976) studied 12 subjects who ingested inorganic lead, 5 days/week for 49 days. The aim was to study the interindividual difference in the time lag before the protoporphyrin level in blood starts to increase after the PbB-level has reached about 40 µg Pb/100 mL. In this group of volunteers VERBERK (1976) did not observe any effect on the nerve conduction velocity, and BIJLSMA and DE FRANCE (1976) could not establish chromosomal aberrations after 49 days of exposure. However, this does not permit to conclude that such effects do not occur in long-term exposure to lead. Moreover, human volunteer studies cannot be used to study reproductive risks. They also never yield data which are representative for large groups of the population, e.g., children, old-age subjects, pregnant women.

Several volunteer studies have been carried out on the relationship between external exposure and parameters of internal exposure, particularly for lead (see Sect. 3.4). Such studies only present indirect evidence of health risks. Wellknown are the studies carried out by KEHOE already in the twenties and more recently by, e.g., GRIFFIN et al. (1975) and by CHAMBERLAIN et al. (1978).

(3) *Case studies*, which usually are limited to a few subjects with manifest intoxications; reliable data on exposure often are not available. Sometimes such cases may lead to useful hypotheses for further animal and human studies.

(4) *Epidemiological studies* refer to exposure-effect/response studies of groups of humans exposed to metals: groups of the general population (with emphasis on young children), ambulant workers, patients with evidence of manifest intoxication, reproduction studies; the number of studies is very large; the flow of papers is still going on.

The handbook on the toxicology of metals (FRIBERG et al., 1986) presents a recent overall review of epidemiological studies. More specified recent reviews refer either to specified metals or to specified groups of effects, e.g., BELMAN (1981) on carcinogenicity, CLARKSON et al. (1983) on reproduction, BROWN and SUNDERMAN (1985) on nickel, MAHAFFEY (1985) on lead, FOWLER (1977) on arsenic, LANGÅRD (1982) on chromium, NEEDLEMAN (1980) on effects of lead on the central nervous system, PORTER (1982) on renal effects.

In the course of time the emphasis has changed from studies of morbidity or mortality to studies of not yet clinically manifest early changes in organ functions and/or in symptoms (e.g., WILLIAMS et al., 1983) in exposure to metals.

The epidemiological data base for "old" metals is generally speaking less inadequate than for most other chemicals. Nevertheless, particularly with regard to reproductive risks and health risks in exposure to metal species the data base is still largely insufficient.

The most serious defect of many epidemiologic studies is the inadequacy of quantitative assessment of external exposure, particularly with regard to exposure in previous years (WHO, 1983). ULFVARSON (1983) demonstrated that this may lead to an underestimate of the actual health risk. This particularly decreases the validity of exposure-response relationships for effects with a long latency time (e.g., cancer).

From the health protection point of view one should not have to rely only on epidemiological studies, because ideally speaking exposure should not occur when the NOAE/RLs are not yet known from animal studies (positive approach, see Sect. 2.11). However, even when an almost complete set of animal data appears to be available, epidemiological studies still are needed to assess the validity of the extrapolation from selected inbred groups of animals exposed under strictly controlled conditions to humans with their free choice of partnership, their own life style, their own combined exposures and their own variability in coping capacity (see Sect. 2.7). Moreover, epidemiological studies often do not allow to exclude any risk (HERNBERG, 1981). The WHO (1983) published guidelines on studies in environmental epidemiology; the basic methodological methods are critically discussed. Many examples are presented of studies on exposure to metals.

3.2 Scientific Value Judgments

Even when a reasonably complete set of toxicological data is available, HB-exposure limits cannot automatically be computed. This requires several value judgments by a group of experts (see Sect. 2.9). This group first of all has to make a critical assessment of the quality of the data. It also has to make various value judgments when interpreting the health risks. This can be demonstrated from the following highly simplified exposure-effect/response relationship (see Sect. 2.4) (Fig. I.20a-3).

Exposure to one metal (compound) may induce effects A, B, and C which differ in health significance. The experts have to determine a priori which biological change should be considered as potentially adverse (*qualitative* choice) (see Sect. 2.6), and in the case of graded effects at which intensity effects should be regarded as adverse (*quantitative* choice). The observed not adverse response level ultimately determines the HB-recommended exposure limit. Moreover, it is never possible to guarantee absence of any risk to any exposed subject at any time of his life. Only when exposure does not exist at all (almost never possible for metals) can absence of risk be guaranteed. The group has to decide which percentage of subjects will probably be protected at the HB-recommended exposure limit ("nearly all"; "generally", see Sect. 4.1): 90%, 95%, 99.9%? (choice of *tolerance-persons*). When there exists a risk of, e.g., irritation of eyes and/or upper respiratory tract, does the group accept only

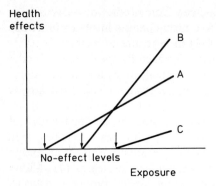

Fig. I.20a-3. Simplified exposure effect/response relationships. The curves A, B, and C represent: 1. three different effects caused by exposure to agent x; 2. effect x caused by exposure to three different agents; 3. average effect x caused by agent y in three groups of workers who differ in capacity to cope with agent y; 4. effect x caused by exposure to agent y in three periods of life, e.g., age, state of health (intercurrent disease), pregnancy.

slight irritation, e.g., two or several times per day during a period of say 5 minutes? (choice of *tolerance time*).

In addition, the expert group often may have to extrapolate from animal to man; what extrapolation model should be applied in the case of, e.g., carcinogenic risks (see also MERIAN et al., 1985, 1988, 1989)? On the basis of the same animal data the calculated risks may differ by a factor of 10^5 according to the model chosen. Does the group accept the concept of "virtually safe dose", of "operational threshold" for agents with a low carcinogenic potential? (ZIELHUIS, 1985b). The choice of the model depends on the exact working mechanism, which, however, usually is not known.

Moreover, for "new" metals an adequate set of toxicological data usually is not available. Should the experts then abstain from recommending an exposure limit? However, a recommended tentative limit, based upon an "educated guess" by a group of experts, offers better protection than no limit at all.

Even although man has been exposed to beryllium already for several decades, no consensus has been yet achieved whether it is a human carcinogen; one positive study has later on even been rejected by one of its authors (SMITH, 1981). Feeding rats with Pb-acetate induces renal cancer; on the other hand, human evidence of carcinogenicity has never conclusively been demonstrated. The international Workshop/Conference on the carcinogenicity of metals (BELMAN, 1981) concluded that cadmium may cause prostate cancer; however, DOLL (1984) afterwards confessed that for statistical reasons this conclusion was not justified. Should the experts on the basis of not yet fully conclusive evidence regard 10 or 20 µg Cd/m^3 as a "virtually safe dose" in view of cancer risks, and if so, irrespective of the type of Cd-compounds? Recently ELINDER et al. (1983) have reviewed epidemiological studies suggesting an increased lung cancer risk; a conclusive answer on the basis of epidemiological studies is not yet justified, however, see also KAZANTZIS (1989). This author concluded that prostate cancer and cerebrovascular diseases are not related to cadmium exposure, but that there is some evidence that bronchitis, emphysema, lung cancer, and stomach cancer are. Protective effects, bioavailability, smoking, interactions with arsenic, and length of follow-up have to be studied further. Cadmium chloride has recently been shown an animal pulmonary carcinogen (TAKENAKA et al., 1983; OLDIGES and GLASER, 1986).

3.3 Consensus

The experts have to make various value judgments. The uncertainties and the choices made should be fully documented and made public. This requires a consensus on the best scientific judgment available.

One may distinguish four levels of (in)consensus:
(1) There exists a consensus that a causal relation has been established between exposure and adverse effects in man.
(2) One or a few well designed studies suggest an increased risk of adverse effects at relevant exposure limits; moreover, this appears to be biologically plausible, but more research is still needed.
(3) Contradictory evidence exists.
(4) Only preliminary evidence is available or risks are extrapolated from high exposure data, maybe with non-physiological routes of administration, not corresponding to actual exposure at and off work.

Prevention of a priori chosen critical health effects (see Sect 3.2) should determine the exposure limit in the case of (1) and (2) and maybe also (3); level 4 usually can be disregarded; this often stems rather from "paratoxicology" (see Sect. 2.9) than from toxicology according to the state of the art.

3.4 Monitoring Programs
(see also Chapter I.4)

In the occupational and environmental setting the term monitoring means: "a systematic continuous or repetitive health-related activity designed to lead if necessary to corrective action" (BERLIN et al., 1984).

One distinguishes three monitoring programs with different objectives: (1) *environmental* monitoring (EM) assesses the external load by measuring the (metal) levels in, e.g., air, food, drinking water, soil; (2) *biological* monitoring (BM) assesses the internal load by measuring the (metal) levels in biological specimens; (3) *biological effect* monitoring (BEM) assesses the early effects. The *Health Surveillance* (HS) program assesses the state of health by measuring (pre)clinical, biochemical, functional, or structural a priori adverse effects. Because the ongoing development of these programs has created a confusion in terminology, ZIELHUIS and HENDERSON (1986) have recently presented a review of definitions and objectives and of the different relevance of each monitoring program with regard to risk management and the information to be given to workers and management.

For BM the most feasible biological specimens to be sampled and (bio)chemical parameters to be measured in exposure to metals are (ZIELHUIS et al., 1984; see also ANGERER and SCHALLER, 1985, 1988; MAK, 1988):

Metal	Biological Specimen
Arsenic, inorganic	
water soluble	Urine (as metabolite)
water insoluble	Urine (?)*
Cadmium	Urine, blood
Chromium	
water soluble compounds	Urine
water insoluble compounds	Urine (?)*
Cobalt	Urine
Copper	Serum or plasma
Iron	Serum or plasma
Lead, inorganic	Blood
alkyllead	Urine
Manganese	Urine (?)
Mercury, metallic and inorganic	Urine
alkylmercury	Blood, hair
ethylethoxy- and phenylmercury	Urine
Molybdenum	Urine (?)
Nickel, water soluble	Serum or plasma, urine
Selenium	Blood, serum or plasma, urine
Zinc	Serum or plasma

* May be useful when no well established specimen is known

Hair usually is exogeneously contaminated; metals may become adsorbed and even strongly bound to S-H groups; washing procedures not always remove all exogeneous metals. Therefore, BM of metal levels in hair should be regarded as indirect EM, except when exposure is exclusively by ingestion. The same is true for measuring metal levels in feces, when the fractional intestinal absorption is limited. BM of breast milk corresponds to EM for the infant.

One of the most difficult procedures is to assess the actual total intake by food and beverages; the best, but most elaborate method offers the duplicate portion technique, measuring the total metal concentration in all food and beverages ingested in 24 h (WHO, 1983).

3.5 Strategies for Compliance Risk Assessment

Assessment of risks requires a consistent strategy. A proposal for risk assessment in ambient exposure to metals has been presented by ZIELHUIS (1983). A similar strategy should – mutatis mutandis – also be applied in occupational exposure.

This strategy is a step-by-step approach to be followed both in the risk situation and in a non-exposed reference situation. It is assumed that the exposure-response relationships (see Sect. 2.4) are known. Each step determines the feasibility of the subsequent step. The steps are: (a) qualitative *identification* of pollutants by taking an inventory of origin and type of chemicals deposed, e.g., at work or at a dump site (chemical waste, harbor sludge); (b) measurement of the *concentrations* of pollutants in the relevant external environmental media; when (b) not or hardly differs between exposure and reference situation, then the following steps need not to be taken; (c) measurement/estimation of the *volumes/amounts* of the external media ingested (for toddlers ingestion of 100 – 200 mg soil/day appears to be a reasonable estimate, CLAUSING et al., 1987); (d) calculation of the total *daily intake* by multiplying (b) and (c); when (d) does not much differ between both situations, one should not go on with the following steps; (e) based upon the total intake (d) and the known exposure-response relationships and emphasizing the subgroups most at risk, one is able to estimate the *relative risk* (RR) for the exposed group; (f) only when the RR exceeds 1 in comparison to the reference group, may it be feasible to examine the health status of both index- and reference groups; however, when the RR does not or hardly differ between both groups, then there is no reason to do so; this is also the case when the risk indicates the probability of a rare disease or a disease with a long latency period.

A shortcut for the assessment of exposure may be biological monitoring of internal exposure (see Sect. 3.4); this offers relatively many possibilities in exposure to metal compounds. Then step (a) can be followed by step (d), because BM indirectly estimates the total intake (see Sect. 5.4). When exposure limits have already been set for specified pollutants in specified media (e.g., ADIs, OELs, MACs, TLVs, and biological exposure limits) then one does not need to go on, when (d) does not exceed these limits.

This strategy represents a type of compliance risk assessment. When exposure-response relationships are not known one should undertake the more elaborate research type of risk assessment (see Sect. 2.2).

It is important to maximize the data, e.g., for the amounts/volumes of media inhaled/ingested; chemicals with the highest toxicity should be regarded representative for groups of related chemicals.

4 Occupational Exposure Limits (OELs)

4.1 History and Definition

In most Western countries the American Threshold Limit Values (TLVs) have for several decades served as the basis for their national occupational exposure limits, whether or not legally enforceable, although under different names, e.g., Maximal Allowable (or Acceptable or Accepted) Concentration (MAC, in Germany MAK), in the USSR Maximal Permissible Concentration (MPC). Another important source, particularly for East-European countries, is the list of the USSR Hygienic Limits. In 1977 the International Labor Organization (ILO) adopted the generic term

"Occupational Exposure Limits" (OEL), subsequently also taken over by the WHO and the EEC. For a historical review see PAULI (1984), HENSCHLER (1984), and PAUSTENBACH (1985). Table I.20b-5 in Chapter I.20b presents an overview of OELs for metals established/proposed by various governmental departments or semi-governmental bodies of experts.

The US TLVs are intended for "use in the practice" of industrial hygiene as guidelines in the control of potential health hazards and for no other use". Each new TLV list contains information on intended changes in the near future.

In the course of time the TLVs have undergone a remarkable revolution from values which aimed at prevention of overt signs of intoxication to limits to which nearly all workers may be exposed for their working life without experiencing adverse health effects (see Sect. 4.4).

TLVs and most other OELs refer to "airborne concentrations (in work room air) of substances and represent conditions under which it is believed that nearly all workers may repeatedly be exposed day after day without adverse effect" (ACGIH, 1986/87). The limits usually represent the time-weighted arithmetic average (t.w.a.) concentration for a normal 8 h working day and a 40 h workweek for the whole working life.

The TLVs do not guarantee absence of any health risk, but "a small percentage of workers may experience discomfort at or below the limit; a smaller percentage may be affected more seriously by aggravation of a pre-existing condition or by development of an occupational illness". This statement is in contrast to that of the American Occupational Safety and Health Art, which requires that "no employee shall be affected", a goal that never can be guaranteed, unless exposure can be excluded, which is never the case for most metals.

In 1976 the Dutch government added "protection of the offspring" to the definition (ZIELHUIS and NOTTEN, 1978/79). This has also been done by the USSR authorities. Other countries have not explicitly done so, but they implicitly started to take also the protection of the offspring into account. The German MAK Commission recently explicitly added "protection of pregnant workers" (and indirectly of the fetus), but this does not cover – as in the Netherlands – occupationally induced risks to, e.g., fertility, the male and female reproductive organ systems and risks during lactation (ZIELHUIS, 1986).

The EEC is in the process of harmonizing the OELs of the 12 member countries (EEC, 1986; see Chapter I.20b). Inorganic lead and its ions have already been regulated by the EEC (1982).

4.2 The Procedures

As discussed in Sect. 2.9 standard setting requires a two-step procedure.

The following procedure has been institutionalized in the Netherlands. A Working Group of Experts (WGE) was appointed à titre personnel in 1976 for the first step; in 1977 a National MAC Commission (NMC) was convened of representatives of unions of workers, the union of the chemical industry (management), and of the government for the second step (ZIELHUIS and NOTTEN, 1978/79). A recent Dutch law on the protection of workers' health and safety (ARBO-Act) puts, in contrast to the previous Workers' Safety Act, the responsibility for workers' health protection no longer only on management, but both on management and workers. The name of the NMC has been changed per 1-1-1986 into the "Commission on OELs for Hazardous Chemicals"; however, the essential procedures remained the same. The Dutch government treats the OELs as rather strict guidelines, but they are not yet legally enforceable, although it has the authority to do so.

In some other countries more or less similar procedures, albeit with less explicitly separated steps, have been instituted in recent years. The procedure in the Netherlands still is rather unique (KLINKERT, 1984).

In Great Britain an Advisory Committee on Toxic Limits (ACTS) has been set up; members are representatives of industry, unions and local authorities and two independent experts, which undertake the "political" process of decision-making (second step) after a Working Group on the Assessment of Toxic Chemicals (WATCH, first step), consisting of experts and also some representatives of trade unions and of employers has proposed a limit (HMIB, 1984). A strict separation between the first and the second step has not been realized.

HENSCHLER (1984) discussed the procedure in the Federal Republic of Germany. The Commission for Investigation of Health Hazards of Chemical Compounds in the Work Area, instituted by the Deutsche Forschungsgemeinschaft "restricts itself to entirely scientific (medical, ethical) elements of evaluation and validation; social, economic, and technical parameters are expressly excluded". The Minister of Labor may change this or that level in taking into account socio-economic elements. The Commission does not set tolerable limits for carcinogens, but just identifies chemicals as such. For carcinogens Technical Guidance Values (Technische Richtlinienkonzentration, TRK) are established, based upon a consent of different groups, such as trade unions, employers, consumers organizations, governmental. The Commission publishes an updated list of the German MAKs every year.

The Scandinavian countries have set up a joint Nordic Expert Group for the Documentation of Occupational Exposure Limits that has already published in the Arbete och Hälsa series several reviews of toxicological data bases. Subsequently, the various Scandinavian governments establish their own operational OELs. A strict separation between both steps is not realized.

In the USA the Occupational Safety and Health Act-1970 provided the legal framework for establishing enforceable operational OELs. The government adopted the 1968-TLVs and also 22 standards developed originally by the American Standards Association (ASA), in total 440 chemicals, as "enforceable standards". Such standards can only be changed by an Act of Congress; in 1984 only 20 standards had been changed, although more than 140 changes already had been made in the TLVs since 1968. However, as discussed before, the TLVs have never meant to be used in that way (PAULI, 1984).

In the seventies the National Institute of Occupational Safety and Health (NIOSH) published a large number of extensive documentations on HB-recommended OELs, but this useful activity was stopped around 1980. This offered one to the few possibilities to compare the HB-recommended limits with the operational limits. Moreover, changes in standards are extensively documented in the Federal Register, both with regard to the HB-recommended and the operational limits; however, this covers up to now only a few metals.

A WHO Study Group proposed for inorganic lead, cadmium, mercury, and manganese HB-recommended OELs accompanied by a documentation of the toxicological data base, and agreed upon by experts from East and West (WHO, 1980). In this way the WHO succeeded in bridging the gap between the US TLVs and the USSR MPCs, at least at the level of the first step (see Sect. 4.6).

There exist quite different approaches in setting OELs for carcinogenic substances; two contrasting examples may illustrate this. The ACGIH TLV list (1986/87) presents "those substances in industrial use that have proven to be carcinogenic in man or have induced cancer in animals under appropriate experimental conditions". Some have been assigned a TLV and some not when the environmental conditions have not yet been sufficiently defined. The classification for metals is as follows:

A1a, human carcinogens: chromite ore processing (chromate); Cr(VI), certain water insoluble compounds; Ni-sulfide ore roasting, fume and dust, all with an assigned TLV.
A1b, human carcinogens, no TLV; no metals in this group.
A2, suspect of carcinogenic potential for man based on either limited epidemiological evidence or demonstration of carcinogenesis in one or more animal species by appropriate methods: antimony trioxide production (no TLV), arsenic trioxide production (no TLV), beryllium (TLV), chromates of lead and zinc (TLV, intended change).

Quite another approach is followed by the German MAK Commission. Carcinogenic substances are classified as A1 (proven human carcinogens), A2 (proven animal carcinogens, under condition which are comparable to those for possible human exposure at work), B (justifiably suspected of having carcinogenic potential). As mentioned before, the MAK Commission itself does not establish a MAK for groups A1 and A2, because no level can be established for a safe concentration range. This refers to the following metal compounds/processes: antimony trioxide, arsenic trioxide, -pentoxide, arsenous acid, arsenic acid and their salts, beryllium and its compounds, cadmium chloride, chromium(VI) compounds (as dusts/aerosols), cobalt (as dusts/aerosols), nickel (as inspirable dusts/aerosols), nickel carbonyl and zinc chromate. For group B tentative MAKs may be given, this refers to cadmium and its compounds, alkaline chromates, chromium carbonyl, and lead chromate.

Each step of the standard setting process has to be fully documented. The Dutch WGE started a series of documentations; about 60 have already been published (since 1987 in English); another 20 to 30 are in preparation. The American TLVs are also documented but in a rather summarizing fashion. No distinction has been made between HB-recommended and operational limits (ACGIH, 1986). The German MAK Commission also publishes a rather extensive regularly updated documentation (DFG, 1985–88). The USSR presented a documentation of their OELs in an English translation (SANOTSKY and ULANOVA, 1983). On the other hand, the EEC (1982) did not underpin the regulatory OEL for lead and its ions with any documentation at all (ZIELHUIS, 1985c).

4.3 OELs for Mixtures?

Workers are often exposed to mixtures of chemicals (see Sect. 2.8). LEWIS (1986) has discussed several approaches for settling OELs for mixtures. Great Britain, the Netherlands, and the EEC have not adopted a strategy to derive such limits. The German MAK Commission does not offer any practical solution in general, although this Commission is willing to provide limits for defined vapor mixtures of practical relevance. The ACGIH decided that, subject to certain conditions, a simple rule of thumb can be applied: "When two or more hazardous substances, which act on the same organ system, are present, their combined effect, rather than that of either individually, should be given primary consideration. In the absence of information to the contrary, the effects of the different hazards should be considered as additive" (ACGIH, 1986/87, but see also Chapter I.10).

Different compounds of the same metal or compounds of different metals may highly differ in toxicokinetics and/or -dynamics (see Sect. 2.14). Therefore, this rule of the thumb, which may be feasible for mixtures of solvents with predominantly narcotic action, usually does not apply to metals.

The US EPA (1986) has published official "Guidelines for the health risk assessment of chemical mixtures" in the Federal Register. Those involved in research or in practical work should consult these extensive guidelines, which deal both with addition and interaction models for mixtures of non-carcinogenic and carcinogenic chemicals.

4.4 Ongoing Reevaluation

The OELs have to be periodically reevaluated in order to adjust to new evidence and/or shift in acceptability. Two examples may demonstrate how different proposals have been made in the course of the last decades. A TLV for inorganic mercury was already established in 1946: 0.1 mg Hg/m^3, which was preceded by a similar proposal in 1943 by the American National Standards Institute, confirmed in 1971, but changed to 0.05 mg/m^3 in 1972, as recommended in 1968 by the International Symposium on Maximal Allowable Concentrations in Stockholm and similar to a recommendation by the National Institute of Occupational Safety and Health (NIOSH) in the USA in 1973. The present TLV 1988/89 still is 0.1 mg Hg/m^3 and also the official US standard under the Occupational Safety and Health Act. These levels of 0.05 – 0.1 mg Hg/m^3 are in sharp contrast to the USSR MPC of 0.01 mg/m^3 already established in 1929. The WHO (1980) recommended an HB-OEL of 0.05 mg Hg/m^3, also agreed upon by experts from the USSR and the USA.

Although already in 1933 an OEL of 0.15 mg Pb/m^3 had been recommended on the basis of a survey in American lead batteries, in that period one also used a limit of 0.5 mg/m^3. The limit of 0.15 mg Pb/m^3, at first accepted by the ACGIH, was increased to 0.20 mg/m^3 in 1957, but lowered again to 0.15 mg/m^3 in 1971, as recommended by the above mentioned symposium in 1968. However, in 1978 the NIOSH recommended a limit of 0.10 mg Pb/m^3. Nevertheless, the TLV 1988/89 still is 0.15 mg/m^3. The WHO (1980) recommended a considerably lower HB-OEL of

0.03–0.06 mg Pb/m^3, acknowledging that the lead level in blood should be regarded as the primary measure of exposure and not the Pb-level in air. In 1982 the EEC issued its Directive for the protection of workers exposed to inorganic lead and its ions; the OEL has been set at 0.15 mg/m^3, with two action levels at 0.04 and 0.075 mg Pb/m^3.

4.5 Fact and Fallacies of OELs
(see also Part II)

The OELs too often are treated in practice merely as "numbers": the same levels in mg/m^3 for different OELs too easily are treated in a similar manner, e.g., with regard to the strategy of compliance risk assessment (see Sect. 2.2). Without referring to the toxicological data base so-called "experts" even add and/or multiply OELs. However, the "number" as such does not present any indication of the actual health risks to be expected when the OEL is exceeded. Physicochemical properties, toxicokinetics and -dynamics, critical organs, severity of health effects, all determine the appropriate way of handling a situation in which compliance measurements have to be carried out. The most important rule of thumb, therefore, is: do not consider the "numbers" as such, but always do take into account the most important data of the toxicological data base for the agent.

In addition, at least the following points should be considered (ZIELHUIS, 1977a):

(1) The OELs refer to *concentrations* in air; however, except for chemicals which act directly on the surface linings of eyes, skin, and respiratory tract, e.g., V_2O_5, CdO-, ZnO fumes, and Ni-carbonyl, man does not respond to the concentration, but to the *amount* absorbed into the body per unit of time. The concept of OELs as a concentration is a scientific fallacy. The amount taken up by the respiratory tract is the product of the concentration in air×the respiratory minute volume×the percentage deposited in the lungs×the percentage absorbed×the duration of exposure; therefore, the uptake increases with the physical workload. In moderate physical activity the respiratory volume per workshift is 6 to 8 m^3; the physical workload that maximally can be sustained per workshift corresponds to a respiratory volume of about 10 m^3/8 h.

(2) The concentration is expressed as a *time weighted average* (t.w.a.) over an 8 h workshift. However, the concentrations probably often do not follow the Gaussian distribution, but a log-probability distribution, which represents a reasonably straight line. In that case the OEL should have been based on that level, sampled over a rather short period of time, that in, e.g., 90 to 97.5% of the workshift is not exceeded, particularly when the biological half-time is short and/or when the effects occur after short-term exposure. For control of metal compounds with a long biological half-life, even a t.w.a. per week may suffice, when the variation around this limit does not extremely exceed the t.w.a. limit per workshift (8 h).

(3) The OELs do not take into account exposure to other chemical and/or physical working conditions (see Sect. 2.8); they refer only to exposure to *single* chemicals.

(4) Epidemiological studies too little take into account the sometimes large *interindividual variability* in internal load and/or in response. Therefore, man is treated as a non-existent standard-man (see Sect. 2.7). The variability in toxicokinetics and/or in coping capacity within and between workers is insufficiently known. Although the definition claims that "in general" or "nearly all" workers are protected, the percentage of inadequately protected workers generally is defined.
(5) When there is no sharp distinction between the *first* and the *second step* (see Sects. 4.2), it is not possible to discover what the operational OEL actually means in terms of health protection, let alone when no complete documentation has been presented.
(6) The toxicological *data base* is often deficient (see Sect. 2.11 and 3.1).
(7) National authorities too easily adopt levels issued in other countries (notably the US TLVs and the USSR MPCs), not adjusting the limits to, e.g., climatic conditions, working hours per week, nutrition, state of health.
(8) There is no consensus how to *extrapolate*, particularly from oral exposure of animals to respiratory exposure of man at work, and for potential carcinogenic agents.
(9) The human data base largely refers to epidemiologic studies of male workers, whereas an increasing proportion of the work force nowadays consists of female subjects, many of reproductive age. Therefore, notwithstanding the increasing number of studies in the last decade, the field of human occupational *reproductive* risks still is largely terra incognita, both for female and male workers (reviews by CLARKSON et al., 1983, particularly on metals; BARLOW and SULLIVAN, 1982; ZIELHUIS et al., 1984; ZIELHUIS, 1986).
(10) As mentioned before (Sect 2.14) the OELs inadequately take into account *metal speciation*.

The German MAK Commission carried out a study of the validity of about 150 chemicals; less than 10% was based upon appropriate and sufficient animal tests and/or field experience (HENSCHLER, 1984). In 1983–1984 the Dutch Working Group of Experts (WGE) screened the complete Dutch MAC list, still to a large extent similar to the TLV list, with the following question in mind: "Are the present MACs reasonably well underpinned by the most recent documentations of the TLVs and the German MAKs, taking also into account some readily available recent papers, books, and reviews?" Reproductive risks were not taken into account, because the data base on reproductive risks still is too inadequate for the large majority of chemicals. The purpose of this screening was neither to reevaluate the total toxicological data base, nor to recommend new HB-OELs, but first of all to set priorities for those chemicals, which are widely used in industry and which need reevaluation in the near future. It should be noted that both the TLVs and MAKs and the Dutch MACs refer to operational limits, whereas the screening only took the health-based aspects into account. Therefore, when the present OELs appear to be inadequately underpinned, then one may not automatically conclude that they are too high: they may be too high, too low, or even adequate. The result of this screening was as follows (percentage of OELs that received the respective notation):

A. Insufficient data on workers with long-term exposure (60%).
B. Insufficient data on animals with long-term exposure (50%).
C. Carcinogenicity established in humans and/or animals at relevant exposure levels (3%).
D. Carcinogenicity suspected as suggested by the chemical structure or by incomplete animal data (3%).
E. Suggestive evidence that the present MAC may be too high (15%).
F. Merely based upon analogy with other agents (13%).
G. The present MAC appears to be adequately underpinned (9%).

Because various chemicals received more than one notation, the total exceeds 100%. Only about 10% appeared to be reasonably well underpinned. Moreover, 13% of the present MACs were in the process of full reevaluation and not included in the screening. Among the chemicals reviewed were about 20 metals or metal compounds still in the process of a complete reevaluation; several of these compounds will lead to a proposal for a lower HB-limit than the present MAC; of the other metals or metal compounds 38 were considered to be insufficiently underpinned.

The fact that most OELs are inadequately underpinned from the health protection point of view should not be interpreted as if the present OELs are meaningless. They serve as a highly important tool in risk management. Moreover, the present OELs have been proposed by groups of experienced occupational toxicologists, hygienists, and physicians; the present limits should be considered as highly relevant guidelines, notwithstanding the fact that the data base has to be improved. Therefore, the present state of the art makes it necessary that more research has to be carried out in order to fill up as soon as possible the sometimes wide gaps of knowledge, with priority to those chemicals to which large numbers of workers are exposed and/or the suspected health hazards may be severe. In addition, more research should be carried out on working mechanisms and on individual/group differences in toxicokinetics and/or -dynamics, in order to improve the extrapolation of animal to man and to detect subjects/groups most at risk. Furthermore, the negative approach should change into a positive approach (see Sect. 2.11).

The discussion of the facts and fallacies and the validity of the present OELs emphasizes once and again the importance of the Hygienic Rule. Moreover, OELs for xenobiotic agents are not goals to achieve, but the actual exposure at the workplace should always be minimized as far as possible (see Sect. 2.10).

The ACGIH (1988/89) explicitly stated that the TLVs should not be used for, e.g., the evaluation or control of community air pollution nuisances (see Sect. 5), in estimating the toxic potential of continuous, uninterrupted exposure or extended work periods (see Sect. 4.8), as proof or disproof of an existing disease or physical condition, or for adoption by countries, where the working conditions differ from those in the USA and where substances and processes differ. Moreover, the limits are not fine lines between safe and dangerous concentrations and should not be used by anyone untrained in the discipline of industrial hygiene.

4.6 US- versus USSR OELs

It is wellknown that the ACGIH TLVs and USSR MPCs often largely differ. The difference in approach can be summarized as follows (ZIELHUIS, 1977b):

ACGIH TLV	USSR MPC
Focused on cells or organs	Organismic emphasis upon control and integration by the central nervous system
Time weighted average	Maximum level
Compensatory mechanisms accepted	Regarded as concealed pathology
Prevention of overt health impairment	Conservation of health
Most workers protected	All workers protected
Emphasis upon epidemiological studies	Upon animal and human volunteer studies
Freedom to exploit one's own capacities	Freedom from unwanted stimuli
Practicable in the occupational setting	Medical arguments overrule feasibility; considered as goals to achieve
Reproductive risks hardly taken into account	As far as possible included

This review emphasizes the extreme points of view; in recent years much mutual exchange of information has taken place. Particularly the WHO (1980) has contributed to narrowing the gap.

4.7 Biological Exposure Limits
(see also ANGERER and SCHALLER, 1985, 1989)

The WHO (1980) proposed HB-biological exposure limits for Cd (blood and urine), Pb (blood), and metallic Hg vapor (urine). The ACGIH recently started to propose so-called Biological Exposure Indices (BEIs), e.g., for Pb and Cd, both in blood and urine. The German MAK Commission (1986) recommended "Biologische Arbeitsstofftoleranzwerte" (BAT). Table I.20b-6 (Chapter I.20b) presents an overview. It should be pointed out that particularly the BAT and the EEC limits sometimes largely exceed the HB-biological exposure limits recommended by the WHO (1980).

The EEC Directive 1982 for the protection of workers exposed to lead and its ions also sets an exposure limit for Pb in blood.

As discussed in Sect. 3.4, biological monitoring of exposure is relatively more developed for exposure to metals than for other chemicals. Therefore, the number of proposed biological exposure limits increase. Moreover, most metals and metal compounds are present as particulate dusts. This may lead to extra oral intake (primary ingestion), because of poor hygiene, and even extra respiratory intake from polluted cigarettes (see Sect. 2.8). Consequently, the biological limits better serve to estimate the total health risk. Biological assessment of exposure measures the *total* uptake, at least of water soluble compounds, irrespective of the route of intake. Particularly in the case of exposure to inorganic Pb and Cd the measurement of the levels of Pb in blood and of Cd in urine or blood offer the most adequate exposure parameters to assess the health risks, even on an individual basis. Therefore, the established exposure-effect/response levels usually often can be based upon these biological exposure indices as measure of exposure (see also Sect. 5.4).

4.8 Novel Work Schedules

Many workers work unusually long shifts (longer than 8 hours per day) and/or on other points of time (shiftwork) than in the common 8 h day shifts. Industrial toxicology has not yet started to explore this aspect of chronotoxicology, although chronopharmacology has already discovered that the toxico-kinetics and/or -dynamics of drugs may depend upon the time of intake.

In exposure to agents which act on the surface linings the risks primarily depend on the concentration in inhaled air. Extending the workshift merely extends the period of irritation; consequently, the OEL is to a large extent independent of the duration of the length of the workshift. For metal and metal compounds, which are absorbed into the body and which have a biological half-life to be counted in weeks rather than in days and which ultimately tend to achieve an equilibrium between total daily intake and excretion, the level in the critical organ is also independent of the length of the workshifts. In exposure to agents with a biological half-life of over 80 h the theoretical adjustment of the OEL to extended workshifts will be less than about 10%, which is within the confidence interval of even adequately underpinned OELs. This is the case for most metals and metal compounds. However, in exposure to chemicals with a biological half-life of 5 to 50 h the concentration in the critical organs may achieve higher levels in extended workshifts, which may exceed the critical organ level (see Sect. 2.3); in that case the OELs have to be adjusted. A general formula, applicable to all metal compounds, does not exist. The acceptable limits for each compound should be reevaluated as such.

In view of the fact that the validation of OELs particularly for several metal compounds still may be inadequate, careful monitoring of workers' internal exposure and health effects is necessary, also in the case of extended workshifts (PAUSTENBACH, 1985).

5 Environmental Exposure Limits (Env. ELs)

(see also Part II)

5.1 Differences between Env. ELs and OELs

The OELs refer to levels in workroom air. However, in the ambient environment man is exposed to pollutants by inhalation of outdoor and indoor air and by ingestion of food and drinking water with intended, incidental or accidental additives, and of home dust, street dust, etc.

The quality of ambient air, food, drinking water and beverages, and sometimes of soil may be regulated. However, only the regulation of marketed food is requested according to the positive approach (see Sect. 2.11). For example, in home grown food and privately caught fish, crabs, or seaweed the levels may exceed the regulatory limits. The limits for soil are meant to serve rather as criteria for soil sanitation than for regulating the oral intake by humans.

Table I.20b-7 (in Chapter I.20b) presents standards and guidelines for metals in drinking water, Table I.20b-8 for those in ambient air, Table I.20b-10 for Pb, Hg, and Cd in food, and Table I.20b-11 for the tolerable weekly intake of Hg, Pb, and Cd.

The population at risk in occupational exposure comprises mainly rather healthy adults, often under supervision of occupational health services, whereas the general population at risk in the ambient environment covers all ages and all subjects with various types of predisposition and even diseases without any direct supervision.

Workers are usually exposed up to 40 h/wk during up to 30–40 years; the general population may be exposed from birth to death.

Since the sixties much information on outdoor pollution has been gathered; in the last decade attention has also been paid to the indoor environment of home, schools, public buildings (see also Chapter I.6b). The indoor pollution often appears to be considerably higher than that outdoors. Moreover, people spend much more time indoors than outdoors.

The essential differences in exposure conditions and sources and in the distributions of susceptibilities among subjects most at risk may illustrate that it is not allowed to extrapolate from OELs to Env. ELs by dividing the first by, e.g., a factor of 10, as often has been done. Finally, OELs refer to much more agents than the Env. ELs.

5.2 The Procedures

The procedures for standard setting also should follow the two-step approach discussed in Sects. 2.9 and 4.2; however, this appears to be even less the case than for the OELs. Moreover, the regulation of the conditions at work usually is carried out by one governmental department, whereas ambient air, drinking water, food, cosmetics may be regulated by various governmental agencies and within these again by different subdepartments; this may easily lead to discrepancies between the various limits for the same agent (see Sect. 5.5).

5.3 An Example: Lead in Air and Blood

Particularly the level of inorganic lead in ambient air has been regulated in most industrialized countries (see Table I.20b-8). For lead the total exposure can easily be established individually by measuring the lead in blood level (PbB). One has tried to establish the relationship between the level in outdoor air (PbA) and the level in blood (PbB). On the basis of human volunteer and population studies it is assumed that with an increase of the PbA of 1 µg/m^3 the average PbB increases by 1 µg Pb/100 mL; however, this relationship usually is curvilinear: the increase of PbB per 1 µg Pb/m^3 decreases with increasing PbB. The same is true for the relation between the levels in food or drinking water with PbB.

As discussed in Sect. 2.7 young children should be regarded as the group most at risk. The increase of PbB per 1 µg Pb/m^3 (PbA) is higher in children, on the average by a factor of 2. However, as demonstrated by BRUNEKREEF (1984, 1985) for children the variation of the ratio PbB (µg/100 mL)/PbA (µg/m^3) between various studies is very large, apparently due to the extra oral intake of dust etc. which indirectly also depends on the Pb level in air. The US Environmental Protection Agency (US EPA) (1984) nevertheless proposed a PbA of 1.5 µg/m^3 on the basis of this ratio of 2.

The DUTCH PUBLIC HEALTH COUNCIL (1984) followed another approach. In 1981 it had been established in a study of 647 4–6 years old urban and rural children that in 98% the PbB did not exceed 23 µg/Pb 100 mL, in 90% not 18–19 µg/100 mL and in 50% not 130 µg/100 mL. The levels had decreased since a similar study in 1979. Both in 1979 and 1981 the PbB levels were below the 1977-EEC Guideline (see Sect. 5.4) for the general population (98% ⩽ 35 µg Pb/100 mL, 90% ⩽ 30 µg/100 mL, 50% ⩽ 20 µg/100 mL) and also below the lower Dutch Guideline adjusted for young children (98% ⩽ 30 µg Pb/100 mL, 90% ⩽ 25 µg/100 mL, and 50% ⩽ 20 µg/100 mL) (Sect. 5.4). The PbA-levels were highest in the city centers. In recent years the PbA-levels have considerably decreased; the average lead levels per year usually were below 0.5 µg/m^3 already in 1978 and have even further decreased in recent years. Therefore, instead of proposing a PbA-standard for ambient air based upon the quotient of PbB/PbA the Dutch Public Health Council advised the government to keep the annual average PbA-level as low as possible and at least below 0.5 µg Pb/m^3 with the restriction that on locations where people live regularly and for an extended period of time the 3-months average 24 h-PbA-level might exceed 1 µg Pb/m^3 at most once per two years, whereas the annual average should not exceed 0.5 µg/m^3. Under these conditions the PbB levels should not be expected to exceed the Dutch PbB-Guideline for children. This practical approach for setting an EL for lead appears to be preferred above a limit based upon questionable extrapolations and calculations which are not warranted by the facts.

5.4 Biological Monitoring and Biological Exposure Limits

As discussed in Sects. 3.4 and 4.7 and illustrated in Sect. 5.3 for lead, biological monitoring permits to assess total exposure and health risks. Consequently biological exposure limits provide the best yardstick to prevent health risks.

In 1979 and 1981 the CEC conducted nine country studies for inorganic lead, based upon the Guideline, proposed by ZIELHUIS (1974) and accepted as an EEC Directive in 1977. The PbB-levels of non-occupationally exposed adults generally were lower than in kindergarten or primary school children. Therefore, when the children do not "exceed the Guideline", then also the adults will be protected against overexposure.

In the late 1970s the total oral lead intake by 100 Dutch adults from food and beverages was median 81 µg Pb, mean 107, range 10–1460 µg Pb/24 h (duplicate portion technique), i.e., considerably lower than permitted by the – albeit poorly underpinned – provisional weekly intake of 430 µg Pb/d, 2000 µg/wk, set by the FAO/WHO Expert Committee on Food Additives (1972); the lead intake from drinking water was estimated to be av. 40 µg/24 h.

A few examples of other large-scale studies are presented: VAHTER (1982) carried out a worldwide WHO-study on the levels of Pb and Cd in blood and of Cd in the kidney cortex in 10 countries in about 200 teachers per country (about 50 for kidney cortex; cases of sudden death). The 90-percentile for PbB ranged from 92 µg Pb/L in Tokyo ($n = 92$) to 456 µg Pb/L in Bangalore ($n = 13$), the Cd levels in blood from 1.2 µg/L in Calcutta ($n = 16$) to 10.0 µg/L in Zagreb ($n = 27$). The geometric mean Cd-levels in the kidney cortex ranged from 9.0 mg/kg w.w. in Bangalore ($n = 42$) to 56.2 in Tokyo ($n = 50$).

In the USA a countrywide study was carried out: the National Health and Nutrition Examination Survey (NHANES II in 1976–80). The mean PbB-levels (±standard errors) were (ROBERTS et al., 1985):

– Children of 6 mths – 5 yrs	16.0 ± 0.42 µg Pb/100 mL	($n = 2372$)
children of 6–17 yrs	12.5 ± 0.30 µg Pb/100 mL	($n = 1720$)
– Adults of 18–74 yrs	14.2 ± 0.25 µg Pb/100 mL	($n = 5841$)
– Male	16.1 ± 0.26 µg Pb/100 mL	($n = 4945$)
– Female	11.9 ± 0.23 µg Pb/100 mL	($n = 4988$)
– Annual income < $ 6000	14.5 ± 0.40 µg Pb/100 mL	($n = 2345$)
– Annual income > $ 15000	13.5 ± 0.24 µg Pb/100 mL	($n = 3718$)
– Urban (≥ 1 million)	15.0 ± 0.37 µg Pb/100 mL	($n = 2395$)
– Urban (< 1 million)	13.9 ± 0.32 µg Pb/100 mL	($n = 3869$)
– Rural	13.0 ± 0.40 µg Pb/100 mL	($n = 3669$)

This survey shows similar trends as the EEC studies: males > female adults; children > adults; urban > rural. During the period 1976–80 the PbB-levels decreased, presumably due to the increased use of non-leaded gasoline. Both in the NHANES II and the CEC-1979 survey the PbB levels exceeded 30 µg Pb/100 mL in only 2% of the subjects examined.

Many studies have been carried out to assess the internal exposure to alkylmercury compounds. The external exposure usually is due to the oral intake of contaminated fish (Minamata incident in Japan) or bread (Iraq). The Hg-level in blood appeared to be linearly related to the daily oral intake in fishermen who were exposed for several years. In 725 subjects the relationship was y (estimated oral intake/60 kg body weight) = $0.5 x$ (level in blood in µg Hg/L) + 4 (1–10). The coefficent of x appeared to be largely independent of the daily intake. When there exclusively exists oral intake, then also monitoring the Hg-level in hair is feasible; the level in hair is on average 250 times higher than in blood. In short-term exposure one can determine a posteriori from the level in various hair segments the time and intensity of previous oral exposure; the level in blood corresponds to the level in hair about two weeks later.

The above given levels refer to total mercury; in long-term stable exposure, the levels in blood and hair reflect the intensity of daily exposure (organic + inorganic Hg); in blood the ratio inorganic/organic mercury is about 0.05, in hair 0.21; the individual variation of this ratio is large (WHO, 1976).

5.4 Discrepancies between Environmental Standards

One might expect that quality standards for ambient air, workroom air, food and drinking water should be interrelated. Based upon assumptions on the amount of air inhaled, the amount of food and drinking water ingested, duration of exposure and pulmonary absorption one might expect that a standard for one environmental source could be calculated from the standards for the other sources. However, this appears not to be the case.

ZIELHUIS (1982) compared the 1979 EEC Directive on drinking water standards with those for the ADI and the OELs for workroom air. Drinking water limits (in µg/L) were extrapolated from the tolerable weekly intake or ADI (µg/kg b.w./day) and from the TLV (mg/m^3); the extrapolated drinking water limits were compared with the limits according to the EEC Directive, expressed in µg/L (see Table I.20 b-7). The results were as follows:

Pollutant	Extrapolated from the ADI	Extrapolated from the TLV	EEC limit
Arsenic	160	5	50
Cadmium	3.3	1.25	10
Copper	1625	25	3000
Mercury	2.2	1.25	1
Lead	22	3.75	50

Even when one accepts a discrepancy of a factor of 3, surprising differences appear to exist. Since 1980, only the ADI for inorganic arsenic has considerably decreased, from 50 µg/kg b.w. to 2 µg/kg b.w., i.e., by a factor of 25; so the limit extrapolated from the ADI should now become 6.4 µg/L; the discrepancy with the EEC limit even increases. For cadmium the discrepancy is limited; however, the provisional Cd- and Pb-ADIs are poorly underpinned and probably are too high; this again increases the discrepancies. Moreover, the WHO (1980) recommended an HB-OEL of 10 µg respirable Cd/m^3; this again increases the discrepancy. The difference for Cu can readily be explained: both the ADI and the drinking water limit are based upon the organoleptic quality, whereas the TLV aims at prevention of local effects on the respiratory tract. The TLV of 0.15 mg Pb/m^3 corresponds to a PbB of about 70 µg/100 mL, which is too high to prevent adverse health effects (although accepted by the EEC and in Germany as a biological OEL). The WHO (1980) recommended an HB-OEL of 30–60 $µg/m^3$, which again increases the discrepancy between the OEL- and the EEC-limit.

It should be acknowledged that the TLVs and the EEC drinking water limits are operational limits (second step) and not recommended HB-limits; nevertheless, even when regarded as operational limits, the discrepancies are large, except for mercury.

Different governmental (sub)departments and international authorities have set the various standards; often there is no documentation; too often one only "copies" other limits from those of other countries. No official documentation is presented for the EEC Directive on drinking water limits. Techno-economic feasibility seems to have overruled a well documented health-based approach.

5.5 The Offspring as a Meeting Point of OELs and Env. ELs

Several metals and their compounds pass the placental barrier, inter alia Pb, Hg, Cr, Mn, As, Ni, Pb, and Hg, more so than Cd. In Belgium LAUWERYS et al. (1978), BUCHET et al. (1978), ROELS et al. (1978), and HUBERMONT et al. (1978) observed slightly lower or almost similar Pb and Hg levels in cord blood as compared with the levels in mothers' blood at the time of delivery; the level of Cd in cord blood, however, was considerably lower than in mothers' blood. The metal levels in breast milk probably more or less correspond to those in mothers' blood plasma.

The internal exposure to metals of the fetus and infants of occupationally exposed mothers will be considerably higher than that due to ambient exposure. Also the EEC has set different limits for the maximal PbB-level for workers (EEC, 1982) and the general population (EEC, 1977). Although the offspring cannot be regarded as a "worker in the making", the fetus is permitted to run a much higher health risk at the most susceptible period of life when the mother is occupationally exposed than when only environmentally exposed. The WHO (1980) recommended a lower limit for female workers of reproductive age than for male workers (30 µg Pb/100 mL and 40 µg Pb/100 mL, respectively); in Germany the BAT limit is 45 µg Pb/100 mL and 70 µg Pb/100 mL, respectively. However, the EEC did not adjust the acceptable limits according to sex.

Ideally speaking, OELs should be geared to the groups most at risk (see Sect. 2.7) and be similar both for male and female workers. However, policy-decision makers apparently are not yet always inclined to do so, notwithstanding laws for equal opportunity for employment of both men and women. This discrepancy in standard setting for occupational and environmental exposure, leading to different health risks to the offspring, is a matter that should be solved as soon as possible, not only for exposure to lead, but also to several other chemicals (ZIELHUIS, 1985c, 1986).

References

Adamsson, E., Piscator, M., and Nogawa, K. (1979), *Environ. Health Perspect. 28*, 219–222.
ACGIH (American Conference Governmental Industrial Hygienists) (1986), *Documentation of the Threshold Limit Values.* ACGIH, Cincinnati, Ohio.
ACGIH (American Conference Governmental Industrial Hygienists) (1988/89), *Threshold Limit Values and Biological Exposure Indices for 1988–1989.* ACGIH, Cincinnati, Ohio.
Angerer, J., and Schaller, K. H. (1985), *Analyses of Hazardous Substances in Biological Materials,* Vol 1. VCH Verlagsgesellschaft, Weinheim-Deerfield Beach/Florida-Basel.
Angerer, J., and Schaller K. H. (1988), *Analyses of Hazardous Substances in Biological Materials,* Vol. 2. VCH Verlagsgesellschaft, Weinheim-Deerfield Beach/Florida-Basel.
Barlow, S. M., and Sullivan, F. M. (1982), *Reproductive Hazards of Industrial Chemicals: An Evaluation of Animal and Human Data.* Academic Press, London.
Belman, S. (ed.) (1981), *Workshop/Conference on the Role of Metals in Carcinogenesis. Environ. Health Perspect. 40–42*, 1–252.
Berlin, A., Yodaiken, R. E., and Herman, B. A. (eds.) (1984), *Assessment of Toxic Agents at the Workplace.* Nijhoff, The Hague.
Bijlsma, J. B. and de France, H. F. (1976), *Int. Arch. Occup. Environ. Health 38*, 145–148.
Boudene, C. E., Arsac, F., and Meininger, J. (1975), in: *International Symposium on Environmental Lead Research. Arh. Hig. Rada Toksikol., Suppl. 26*, 179–189.
Brown, L. M., Pottern, L. M., and Blot, W. J. (1984), *Environ. Res. 34*, 250–261.
Brown, S. S., and Sunderman, F. W. (eds.) (1985), *Progress in Nickel Toxicology. Proceedings 3rd International Conference on Nickel Metabolism and Toxicology.* Blackwell, Oxford.
Brunekreef, B. (1984), *Sci. Total Environ. 38*, 79–123.
Brunekreef, B. (1985), *The Relationship Between Environmental Lead and Blood Lead in Children: A Study in Environmental Epidemiology.* Thesis, Agricultural University, Wageningen, The Netherlands.
Buchet, J. P., Roels, H., Hubermont, G., and Lauwerys, R. (1978), *Environ. Res. 15*, 494–503.
Calabrese, E. J. (1978), *Pollutants and High Risk Groups; the Biological Basis of Increased Human Susceptibility to Environmental and Occupational Pollutants.* Wiley, New York.
Chamberlain, A. C., Heard, M. J., et al. (1978), *Investigations into Lead from Motor Vehicles.* AERE-R9198. H. M. Stationary Office, London.
Clarkson, Th. W., Nordberg, G. F., et al. (eds.) (1983), *Reproductive and Developmental Toxicity of Metals.* Plenum Press, New York.
Clausing, P., Brunekreef, B., and van Wijnen, J. H. (1987), *Int. Arch. Occup. Environ. Health 59*, 73–82.
Cools, A., Sallé, H. J. A., et al. (1976), *Int. Arch. Occup. Environ. Health 38*, 129–139.
Copius Peereboom, J. H. J., van de Velde, W. J., and Dessing, J. M. J. (1983), *Ecotoxicol. Environ. Safe. 7*, 79–86.
DFG (Deutsche Forschungsgemeinschaft) (1985–88), *Gesundheitsschädliche Arbeitsstoffe. Toxikologisch-arbeitsmedizinische Begründung von MAK-Werten* (Henschler, D., ed.). VCH Verlagsgesellschaft, Weinheim.
Doll, R. (1984), *Ann. Occup. Hyg. 28*, 291–305.
Dutch Public Health Council (1984), *Lead, a Health Based Recommended Limit for the Quality of Outdoor Air* (in Dutch). Gezondheidsraad, 's-Gravenhage, The Hague.

Editorial (1986), *Br. J. Ind. Med. 43*, 217–219.
EEC (1977), *Directive Biological Assessment of Exposure of the General Public to Lead.* Council of Ministers, March 29.
EEC (1982), *Directive: Protection of Workers Occupationally Exposed to Lead.* Council of Ministers, July 29.
EEC (1986), *Proposal for a Directive of the Council to Change the Directive 80/1107/EEC for the Protection of Workers Against the Hazards of Exposure to Chemical, Physical and Biological Agents at Work,* June.
Elinder, C. G., Kjellström, T., et al. (1983), *Br. J. Ind. Med. 42*, 651–655.
Esmen, N. A. (1984), in: Esmen, N. A., and Mehlman, M. A. (eds.): *Occupational and Industrial Hygiene: Concepts and Methods. Adv. Mod. Environ. Toxicol. 8*, 45–74.
FAO/WHO (1972), *Tech. Rep. Ser. 505.* WHO, Geneva.
FAO/WHO (1983), *Tech. Rep. Ser. 696.* WHO, Geneva.
Fowler, B. A. (ed.) (1977), *Proceedings International Conference on Environmental Arsenic. Environ. Health Perspect. 19*, 159–242.
Friberg, L. (1986), *Risk Assessment,* in: Friberg, L., Nordberg, G. F., and Vouk, V. B. (eds.): *Handbook on the Toxicology of Metals,* Vol. I. Elsevier, Amsterdam.
Friberg, L., Nordberg, G. F., and Vouk, V. B. (eds.) (1986), *Handbook on the Toxicology of Metals,* Vols. I and II. Elsevier, Amsterdam.
Griffin, T. B., Coulston, F., and Wills, H. (1975), in: *International Symposium on Environmental Lead Research. Arh. Hig. Rada Toksikol., Suppl. 26*, 191–208.
Henschler, D. (1984), *Ann. Occup. Hyg. 28*, 79–92.
Hernberg, S. (1981), *Scand, J. Work Environ. Health 7*, Suppl. 4, 121–126.
HMIB (HM Intelligence Bulletin) (1984), *Occupational Exposure Limits.* Note EH40, HMSO, London 102, June 5.
Hubermont, G., Buchet, J. P., Roels, H., and Lauwerys, R. (1978), *Int. Arch. Occup. Environ. Health 41*, 117–124.
Hunter, D. (1975), *The Diseases of Occupations.* Engl. Univ. Press.
ILO (International Labour Office) (1983), *Encyclopaedia of Occupational Health and Safety,* Vols. I and II. ILO, Geneva.
Iskander, F. Y., Bauer, T. L., and Klein, D. E. (1986), *Analyst 3*, 107–109.
Kazantzis, G. (1989), *Is Cadmium a Human Carcinogen? Toxicol. Environ. Chem. 22* (1–4), 159–165 and 27 (1–3), 113–122.
Klinkert, H. (1984), *Paper read at the CEFIC Symposium on OELs and Harmonization and Control of OELs for the Protection of Workers. CEFIC,* Brussels.
Knishkowsky, B., and Baker, E. L. (1986), *Am. J. Ind. Med. 9*, 543–550.
Langård, S. (ed.) (1982), *Biological and Environmental Aspects of Chromium.* Elsevier, Amsterdam.
Lauwerys, R., Buchet, J. P., Roels, H., and Hubermont, G. (1978), *Environ. Res. 15*, 278–289.
Lewis, P. (1986), *Chem. Ind.* April 21, 268–271.
Mahaffey, K. R. (1985), in: Mahaffey, K. R. (ed.): *Dietary and Environmental Lead: Human Health Effects,* pp. 373–419. Elsevier, Amsterdam.
MAK (1988), *Maximum Concentrations at the Workplace, Report No. XXIII, DFG.* Commission for the Investigation of Health Hazards of Chemical Compounds in the Work Area. VCH Verlagsgesellschaft, Weinheim-Basel-Cambridge-New York.
Merian, E., et al. (1985, 1988, 1989), Vols. 1, 2 and 3 (in press): *Carcinogenic and Mutagenic Metal Compounds.* Proceedings of three IAEAC-Workshops in Geneva, Villars, and Follonica. Gordon & Breach, New York.
Needleman, H. L. (ed.) (1980), *Low Level Lead Exposure; the Clinical Application of Current Research.* Raven Press, New York.
Nordström, S., Beckman, L., and Nordensen, I. (1978), *Hereditas 88*, 43–46; 51–54.
Nordström, S., Beckman, L., and Nordensen, I. (1979), *Hereditas 40*, 291–296; 297–302.

Oldiges, H., and Glaser, U. (1986), *Paper presented at the XIth International Congress of Toxicology.* Katakyushu, Japan.
Pauli, J.R. (1984), *Am. J. Ind. Med.* 5, 227–238.
Paustenbach, D.J. (1985), in: Cralley, L.J., and Cralley, L.V. (eds.): *Patty's Industrial Hygiene and Toxicology,* Vol. 3: *Theory and Rationale of Industrial Hygiene Practice;* Vol. 3A: *The Work Environment.* Wiley Interscience Publications, New York.
Pershagen, G. (1985), *Am. J. Epidemiol.* 122, 684–694.
Porter, G.A. (ed.) (1982), *Nephrotoxic Mechanisms of Drugs and Environmental Toxins.* Plenum Medical Book Co., New York.
Roberts, J., Mahaffey, K.R., and Annest, J.L. (1985), in: Mahaffey, K.R. (ed.): *Dietary and Environmental Lead: Human Health Effects,* pp. 355–372. Elsevier, Amsterdam.
Roels, H., Hubermont, G., Buchet, J.P., and Lauwerys, R. (1978), *Environ. Res.* 15, 236–247.
Sanotsky, I.V., and Ulanova, P.P. (1983), *Hygienic and Toxicological Criteria of Harmfulness in Evaluating Hazards of Chemical Compounds.* United Nations Environment Program, Geneva.
Sherwin, R.P. (1983), *Environ. Health Perspect.* 52, 177–182.
Smith, R.J. (1981), *Science* 211, 556–557.
Takenaka, S., Oldiges, H., et al. (1983), *J. Natl. Cancer Inst. U.S.* 70, 367–373.
Ulfvarson, U. (1983), *Int. Arch. Occup. Environ. Health* 52, 285–300.
US EPA (1978/80), *National Primary and Secondary Ambient Air Quality Standards for Lead. Fed. Reg.* 43(94), 46246–46263; 45(119), 41211–41214.
US EPA (1986), *Guidelines for the Health Risk Assessment of Chemical Mixtures. Fed. Reg.* 51(165), 34014–34025.
Vahter, M. (ed.) (1982), *Assessment of Human Exposure to Lead and Cadmium Through Biological Monitoring.* National Swedish Institute of Environmental Health, Stockholm.
Verberk, M.M. (1976), *Int. Arch. Occup. Environ. Health* 38, 141–143.
Verberk, M.M. (1986), *Am. Ind. Hyg. Ass. J.* 47, 559–562.
WHO (1976), *Mercury. Environ. Health Crit. 1.* WHO, Geneva.
WHO (1978), *Principles and Methods for Evaluating the Toxicity of Chemicals,* Part I. *Environ. Health Crit. 6.* WHO, Geneva.
WHO (1980), *Recommended Health-Based Limits in Occupational Exposure to Heavy Metals. Tech. Rep. Ser. 647.* WHO, Geneva.
WHO (1983), *Guidelines on Studies in Environmental Epidemiology. Environ. Health Crit. 27.* WHO, Geneva.
Williams, M.K., Walford, J., and King, E. (1983), *Br. J. Ind. Med.* 40, 285–292.
Zielhuis. R.L. (1974), *Int. Arch. Arbeitsmed.* 32, 103–127.
Zielhuis, R.L. (1977a, 1977b), in: Grandjean, P. (ed.): *Standards Setting. Proceedings of a Course Organized by the Danish Society of Industrial Medicine. Arbejdsmiljøfondet,* etc. a, pp. 15–35; b, pp. 133–146.
Zielhuis, R.L. (1982), *Int. Arch. Occup. Environ. Health* 50, 113–130.
Zielhuis, R.L. (1983), in: *Heavy Metals in the Environment,* Vol. I, pp. 27–45. International Conference, Heidelberg. CEP Consultants Ltd., Edinburgh.
Zielhuis, R.L. (1985a), *Ann. Occup. Hyg.* 29, 463–475.
Zielhuis, R.L. (1985b), in: Merian, E., Frei, W.F., Härdi, W., and Schlatter, C. (eds.): *Carcinogenic and Mutagenic Metal Compounds. Current Topics in Environmental and Toxicological Chemistry,* pp. 477–490. Gordon & Breach Science Publishers, New York.
Zielhuis, R.L. (1985c), *Br. J. Ind. Med.* 42, 145–146.
Zielhuis, R.L. (1986), *Frauen in der Arbeitswelt: toxikologische Aspekte,* in: *Verhandlungen der Deutschen Gesellschaft für Arbeitsmedizin,* 26. Jahrestagung in Hamburg, pp. 57–71. Gentner Verlag, Stuttgart.
Zielhuis, R.L., and Henderson P.T. (1986), *Int. Arch. Occup. Environ. Health* 57, 249–257.
Zielhuis, R.L., and Notten, W.R.F. (1978/79), *Int. Arch. Occup. Environ. Health* 42, 269–281.

Zielhuis, R. L., and Wanders, S. P. (1986), *Health Effects and Medical Wastage due to Combined Exposure in Welding,* in: Stern, R. M. et al. (eds.): *Health Hazards and Biological Effects of Welding Fumes and Gases,* pp. 497–533. Excerpta Medica, Amsterdam.

Zielhuis, R. L., and Wibowo, A. A. E. (1983), *Global Environmental Monitoring System. Biological Specimen Collection in Environmental Exposure Monitoring.* UNEP-WHO, Geneva.

Zielhuis, R. L., and Wibowo, A. A. E. (1984), in: Nriagu, J. O. (ed.): *Changing Metal Cycles and Human Health.* Report of a Dahlem Workshop, Berlin 1983, pp. 323–344. Springer, Berlin.

Zoeteman, B. C. J., and Brinkman, F. J. J. (1976), in: Amavis, R., Hunter, W. J., et al. (eds.): *Hardness of Drinking Water and Public Health,* pp. 173–211. Pergamon Press, Oxford.

Additional Recommended Literature

With respect to reproductive sites:

Spermatogenesis

Wildt, K., Eliasson, R., and Berlin, M. (1983), in: Clarkson et al.: *Reproductive and Developmental Toxicity of Metals,* pp. 279–300. Plenum Press, New York.

Mental retardation in infants in relation to lead levels in cord blood

Bellinger, D., Needleman, H. L., Leviton, A., et al. (1984), *Neurobehav. Toxicol. Teratol. 6,* 387–402.

Bellinger, D., Needleman, H. L., Leviton, A., et al. (1985), *Environ. Res. 38,* 119–129.

Bellinger, D., Needleman, H. L., Leviton, A., et al. (1986), *Neurobehav. Toxicol. Teratol. 8,* 151–161.

Bellinger, D., Needleman, H. L., Leviton, A., et al. (1987), *N. Engl. J. Med. 316,* 1037–1043.

With respect to lead in blood levels:

US Agency for Toxic Substances and Disease Registry (1988), *The Nature and Extent of Lead Poisoning in the United States: A Report to Congress.* US Department of Health and Human Services, Public Health Service, Atlanta, Georgia.

With respect to interindividual variability in exposure and response:

Notton, W. F. R., Herba, R. F. M., Hunter, W. J., Monster, A. C., and Zielhuis, R. L. (eds.) (1988), *Health Surveillance of Individual Workers Exposed to Chemical Agents.* Springer Verlag, Berlin.

I.20 b Standards, Guidelines, and Legislative Regulations Concerning Metals and Their Compounds

ULRICH EWERS, Düsseldorf, Federal Republic of Germany

1 Introduction

During the last decades concern about the hazards of toxic metals and other environmental chemicals for human health and the environment has increased worldwide. In many countries effective legislative and administrative measures have been taken to reduce environmental pollution and to prevent adverse effects resulting from exposure to chemicals in the work environment and the general environment. The legislation and administrative procedures for the control of toxic chemicals are largely based on technical standards and guidelines and on environmental and occupational health standards. This chapter tries to present an overview on standards, guidelines, and other legislative regulations concerning metals, metalloids, and metal compounds that have been enforced in various countries or have been recommended by international organizations. Reviews of the current standards and regulations concerning lead have been presented recently by BERLIN and VAN DER VENNE (1987) and ZABEL (1989).

1.1 Technical Standards and Guidelines

Technical standards and guidelines are intended to be a tool in achieving the following objectives:

- to increase the safety of industrial plants and other technical installations with regard to the prevention of accidents, which are associated with the release of large amounts of hazardous substances;
- to reduce or minimize the continuous release of toxic substances arising from the normal operation of a technical or industrial process;
- to ensure the safe collection, storage, transportation, and disposal of hazardous wastes;
- to ensure the safe handling and use of chemicals by means of an appropriate classification, labelling and packaging of dangerous substances and preparations;
- to ensure the safe storage and transportation of hazardous chemicals and preparations.

1.2 Environmental and Occupational Health Standards

The aim of environmental health standards is to protect individuals, human populations, and their progeny from the adverse effects of hazardous environmental factors, including chemicals. Similarly, it is the aim of occupational health standards to protect workers' health (and that of their progeny) from the adverse effects of occupational exposure to chemical, physical, and biological factors. Basically the same principles apply to prevent adverse effects of these factors on ecosystems.

Environmental health standards for chemicals may be formulated either in terms of concentrations in environmental components (e.g., air, water, food, soil, consumer products) or in terms of amounts of substances that may be taken up into the body within a given period of time. These concentrations and amounts should be sufficiently low so that the threshold dose (if it exists and can be determined) will not be reached, and that the population of concern will not be subject to an "unacceptable" risk, even following life-time exposure (WHO, 1978).

The conceptual problems of setting environmental and occupational health standards are discussed in Chapter I.20a of this book. As stated there, standard setting should include a two-step procedure. The first step should be the derivation of a health-based exposure limit based on an evaluation of the toxicological data available, including exposure-effect and exposure-response relationships and taking into account the conditions and different routes of exposure. The second step should be the translation of the derived health-based limits into operational standards by the responsible authority or government. In reaching a decision on this, policy makers may have to consider a variety of other factors such as technical feasibility, economic impact, means of enforcement and monitoring, and other public health priorities.

The scientific basis of environmental and occupational health standards frequently is described in so-called criteria documents. Such documents have been prepared by several organizations, e.g., World Health Organization (WHO, 1980, 1984, 1987), American Conference of Governmental Industrial Hygienists (ACGIH, 1988), U.S. Environmental Protection Agency (EPA, 1986), Criteria Group of the Swedish National Board on Occupational Safety and Health, U.S. National Institute of Occupational Safety and Health (NIOSH), and the West-German MAK Commission (MAK, 1988; HENSCHLER, 1988; HENSCHLER and LEHNERT, 1986). It has to be noted, however, that many standards and guidelines lack a sound toxicological or ecotoxicological basis and documentation, and that many of the standards currently applied were primarily set for administrative purposes in order to provide guidelines for good practice in environmental and occupational health management.

2 Regulations Concerning Industrial Plants

2.1 Standards and Guidelines Regulating Emissions and Discharges from Industrial Plants

2.1.1 Emissions in Ambient Air

In most industrialized countries factories manufacturing and processing metals, metal compounds, and other chemicals require a permission to operate. In general, permission is given by the regional authorities provided that certain conditions are met. The installations concerned include ferrous and nonferrous smelters, steel and ferroalloy manufacturing plants, battery plants, metal casting, forming and molding factories, electroplating plants, and plants manufacturing chemicals. In order to obtain the required permission the operator of the factory must submit details of the proposed structure, methods of operation, and expected emissions and discharges. Usually it is prescribed that the operator must use the best available technology to reduce the emission of hazardous substances as far as possible, taking into consideration, however, economic aspects. In some countries maximum admissible emission levels have been established for a number of hazardous substances. As an example the First General Administration Regulation pursuant to section 7(1) of the West-German Federal Immission Control Law may be mentioned, which includes maximum admissible emission levels for various metals and metal compounds (Table I.20b-1). Similar or identic emission standards are applied in Switzerland persuant to the Clean Air Ordinance (1985) issued under the Environment Protection Law (1983). Emission standards for metals, metalloids, and metal compounds and standards of performance have also been set by the U.S. Environmental Protection Agency (Table I.20b-2), the Canadian Minister of the Environment (Air Pollution Control Directorate, 1981), and the West-German VDI Commission on Air Pollution Control (VDI, 1986).

Some countries also require environmental impact statements, i.e., a systematic evaluation of the effects that an installation is expected to have in a given area. For factories above a certain size a public hearing is required before a final decision is taken. Factories that were operating before regulations for emissions and discharges were issued or strengthened are usually either exempt from compliance or given a specified period of time to make necessary adjustments, sometimes with governmental grants. Alterations or extensions automatically become subject to the regulations in force.

2.1.2 Effluents into Surface- and Groundwaters

In many countries the discharge of waste water also requires permission from the responsible authorities, and detailed technical regulations have been issued to reduce the amount and concentration of toxic substances in waste waters. The U.S. Environmental Protection Agency, e.g., has promulgated a number of effluent limitation

Table I.20b-1. Emission Standards for Metals and Metal Compounds According to the 1st General Administration Regulation Issued under the West-German Federal Immission Control Law (1974, Last Amendment 1986)[a]

Category	Substance	Maximum Admissible Concentration in Stack Gas (mg/m^3)[b]
A. Toxic Metals and Metal Compounds:		
I	Cd and Cd compounds (as Cd)	0.2
	Hg and Hg compounds (as Hg)	
	Tl and Tl compounds (as Tl)	
II	As and As compounds (as As)	0.1
	Co and Co compounds (as Co)	
	Ni and Ni compounds (as Ni)	
	Se and Se compounds (as Se)	
	Te and Te compounds (as Te)	
III	Sb and Sb compounds (as Sb)	0.5
	Pb and Pb compounds (as Pb)	
	Cr and Cr compounds (as Cr)	
	Cu and Cu compounds (as Cu)	
	Mn and Mn compounds (as Mn)	
	Pt and Pt compounds (as Pt)	
	Pd and Pd compounds (as Pd)	
	Rh and Rh compounds (as Rh)	
	V and V compounds (as V)	
	Sn and Sn compounds (as Sn)	
B. Carcinogenic Metals and Metal Compounds:		
I	Be and Be compounds (as Be) as far as in respirable form	0.1
II	arsenic trioxide, arsenic pentoxide, arsenites, and arsenates (as As) as far as in respirable form	
	calcium chromate, chromium(III) chromate, strontium chromate, and zinc chromate (as Cr)	1.0
	Co and insoluble Co compounds (as Co) as far as in respirable form	
	Ni, NiS, NiO, NiCO$_3$, Ni(CO)$_4$ (as Ni) as far as in respirable form	

[a] The emission standards apply to the sum of all substances of the same category. They are only applied when the total mass stream of these substances exceeds a certain limit (Category I: >1 g/h; category II: >5 g/h, category III: >25 g/h)
[b] The maximum admissible concentrations refer to the stack gas in the so-called normal state at 1013 mbar pressure and 0°C

guidelines and standards pursuant to sections 301, 304, 306, and 307 of the Federal Water Pollution Control Act (Table I.20b-3). The regulations include limitations for existing sources and standards of performance for new installations. Standards of performance for industrial sources and municipal waste water treatment plants have been issued in the Federal Republic of Germany under section 7(a) of the Water Resource Act (see Table I.20b-4).

Table I.20b-2. U.S. Environmental Protection Agency Emission Standards and Regulations for Toxic Metal Emissions Issued under Sections 111 and 112 of the Clean Air Act (1970)

National Emission Standards for Hazardous Air Pollutants (NESHAP)	Arsenic (from glass manufacturing plants) Arsenic (from primary copper smelters) Arsenic (from arsenic trioxide and metallic arsenic production) Beryllium Mercury
Standards of Performance for New Stationary Sources (SPNSS)	Primary aluminum reduction plants Primary copper smelters Primary lead smelters Primary zinc smelters Secondary lead smelters and lead-acid battery plants Metallic mineral processing plants Surface coating of metal furniture Oxygen process steelmaking facilities

Source: Bureau of National Affairs, Inc., *Environment Reporter*, Washington, D.C.

Table I.20b-3. U.S. Environmental Protection Agency Effluent Guidelines and Standards Issued under the Federal Water Pollution Control Act (1972). List of Guidelines and Standards Concerning Metal Containing Effluents

Effluent Guidelines and Standards for Electroplating
Effluent Guidelines and Standards for Fertilizer Manufacturing
Effluent Guidelines and Standards for Iron and Steel Manufacturing
Effluent Guidelines and Standards for Nonferrous Metals
Effluent Guidelines and Standards for Ferroalloy Manufacturing
Effluent Guidelines and Standards for Metal Finishing
Effluent Guidelines and Standards for Mining and Dressing
Effluent Guidelines and Standards for Inorganic Chemicals
Effluent Guidelines and Standards for Aluminum Forming
Effluent Guidelines and Standards for Copper Forming
Effluent Guidelines and Standards for the Battery Manufacturing Point Source Category
Effluent Guidelines and Standards for Nonferrous Metals Forming and Metal Powders
Effluent Guidelines and Standards for Metal Molding and Casting

The Commission of the European Communities has adopted a number of Directives containing limit values and quality objectives for the discharge of dangerous substances into the aquatic environment (CEC Directives 76/464/EEC and 86/280/EEC). The Directives 82/176/EEC, 83/513/EEC, and 84/156/EEC specially deal with mercury discharges from the chlor-alkali electrolysis industry and cadmium and mercury containing effluents from other industries. The Directive 80/68/EEC deals with the protection of groundwater against pollution caused by certain dangerous substances. In the Annex some toxic metals and metalloids are listed, which could have a harmful effect on the groundwater.

Table I.20b-4. Minimum Requirements for Effluents from Industrial Plants Issued under Section 7a of the West-German Water Resource Act (1957, Last Amendment 1986). List of General Administration Regulations Concerning Metal Containing Effluents

General Administration Regulation Number	Industrial Plants Concerned
17	Ceramic manufacturing
24	Iron and steel manufacturing
26	Mineral processing
27	Ore mining and dressing
34	Manufacturing of barium compounds
37	Manufacturing of pigments
38	Textile manufacturing
39	Nonferrous smelters
40	Electroplating, metal finishing, metal casting, metal forming, metal molding, battery manufacturing plants
41	Glass manufacturing
42	Chloralkali electrolysis
44	Fertilizer manufacturing

2.1.3 Disposal of Dangerous Waste
(see also Chapters I.2 and I.7b)

During recent years, many countries have introduced special legislation to ensure the appropriate disposal of hazardous waste. Local authorities have commonly been made responsible for applying these regulations. According to the CEC Directive on Toxic and Dangerous Waste (78/319/EEC) the member states of the EEC must ensure that toxic and dangerous waste is kept separate from other matter, that the packaging is appropriate and that the labelling indicates its composition and origin. At the site of deposition the composition of dangerous waste must be recorded. When transported, such waste must be accompanied by an identification sheet. In the Annex, a number of toxic and dangerous substances (including antimony, arsenic, beryllium, cadmium, chromium(VI), lead, mercury, selenium, tellurium, thallium and their compounds, metal carbonyls, and soluble copper compounds) are listed, which require priority consideration. Within the framework of this Directive detailed regulations for hazardous waste management have been enforced in many EEC member states. Similar regulations were issued in many countries outside the EEC. Three CEC Directives especially deal with wastes from the titanium dioxide industry (Directives 78/176/EEC, 82/883/EEC, and 83/29/EEC).

2.2 Worker Protection

Occupational exposure limits calculated as time-weighted average concentrations of substances in the workroom air during an eight-hour shift are used in most countries to ensure the protection of workers against damage by exposure to harmful substances in the work environment. Such limits have been set either in the form of guidelines and recommendations or as legal standards. The terms generally used are "Threshold Limit Values" (TLVs), "Maximum Allowable Concentrations" (MACs), or "Occupational Exposure Limits"(OELs). For some substances, mainly strongly irritating agents, "Short-Term Exposure Limits" (STEL) and "Ceiling Limits" have also been recommended. Table I.20b-5 presents a summary of OELs for metals, metal compounds, and metalloids currently applied in various countries or recommended by WHO. In contrast to environmental health standards, which mostly were set for metals irrespective of their chemical form or valency state, a number of OELs take into account the different physical, chemical, and toxicological properties of various metal compounds (e.g., Cr(III) and Cr(VI) compounds, inorganic and organic lead and mercury compounds).

OELs generally are based on the best available information from industrial experience, from experimental human and animal studies, and, if possible, from a combination of the three (criteria documents see Sect. 1.2 above and MARKARD and STOPP, 1984). It has to be noted, however, that the amount and nature of the information for establishing OELs varies from substance to substance.

In many cases the setting of scientifically-based OELs is beyond the possibilities and scientific manpower of a country. A number of countries, therefore, has adopted OELs that have been developed and implemented in other countries. In the Western countries the TLVs set by the ACGIH and the MAC-values established by the West-German MAK Commission are playing an important role as guideline values. In the Eastern hemisphere, the MACs developed and implemented in the Soviet Union are widely used. Moreover, the health-based occupational exposure limits recommended by WHO are used as a basis of standard setting. The procedures of setting OELs practiced in various countries have been described by HENSCHLER (1981), SCHLEGEL (1981), ZIELHUIS and NOTTEN (1979), PAULL (1984), LUNDBERG and HOLMBERG (1984), TOYAMA (1985), CASTLEMAN and ZIEM (1988), and NORDBERG et al. (1988).

In addition to OELs for workroom air, biological exposure limits for metals and other chemicals have been recommended. Such biological limits refer to maximum tolerable metal concentrations in biological materials, mainly blood and urine (see, for instance, ANGERER and SCHALLER, 1985, 1988; CLARKSON et al., 1988). The advantage of determining such parameters is to provide a measure of the worker's individual overall exposure (LAUWERYS, 1983; HENSCHLER and LEHNERT, 1986). Workers are not expected to suffer any ill effects as long as the metal concentrations in blood or urine are maintained below the recommended biological exposure limits. Measurements exceeding these limits are not necessarily indicators of a disease process, but should give an indication to reduce the intensity of exposure. If deviate measurements persist for a longer time the individual should be examined by a physician to determine whether there is any health effect. Biological exposure limits for metals currently recommended by various institutions are listed in Table I.20b-6. In the case of lead, the West-German MAK Commission additionally has recommended a limit value for ALA-U as an indicator of effect.

In many countries pre-employment and periodic medical examinations of exposed workers during employment are required. In the Federal Republic of Germany,

Table I.20b-5. Occupational Exposure Limits (maximum admissible time-weighted average concentrations for a 8-hour workday and a 40-hour workweek) for Metals and Metal Compounds. Unit: mg/m³. C = classified as carcinogen

Substance	CH[a]	D[b]	I[c]	S[d]	SU[e]	USA (ACGIH)[f]	WHO[g]
Aluminum							
— metal & oxide	6	6	—	—	—	10	—
— pyro powders	—	—	—	—	—	5	—
— welding fumes	—	—	—	10/4[h]	—	5	—
— soluble salts	2	—	—	—	—	2	—
— alkyls	2	—	—	2	—	2	—
Antimony							
— metal	0.5	0.5	0.5	0.5	0.2–0.5	0.5	—
— compounds & oxides (as Sb)	0.5/C	C	0.03	0.5	—	0.5	—
— hydride (as Sb)	0.5	0.5	—	0.3	—	—	—
Arsenic							
— soluble compounds (as As)	0.2/C	C	0.25	0.05	0.3	0.2	—
— hydride	0.16	0.2	0.1	0.05	—	0.2	—
Barium							
— soluble compounds (as Ba)	0.5	0.5	—	—	0.1	0.5	—
Beryllium							
— metal & compounds (as Ba)	0.002/C	C	0.002	0.002	0.001	0.002/C	—
Cadmium							
— metal & inorganic compounds (as Cd)	0.05/C	C	0.05	0.05/0.02[h]	—	0.05/C	0.01
Calcium							
— cyanamide	0.5	1	—	—	—	0.5	—
— hydroxide	5	—	—	—	—	5	—
— oxide	2	5	0.015	2	0.1	2	—
Cesium							
— hydroxide (as Cs)	2	—	—	—	0.3	—	—
Chromium							
— metal & compounds except Cr(VI) (as Cr)	0.5	—	0.5	0.5	—	0.5	—
— Cr(VI) compounds	0.05/C	C	0.05	0.02	0.091–0.02	0.05	—
— chromyl chloride	—	—	—	—	—	0.15	—

Cobalt								
— metal	0.1/C	—	—	—	0.05	—	0.05	—
— inorganic compounds (as Co)	—	—	—	—	—	—	0.1	—
Copper								
— dust	1	C	1	—	$1/0.2^h$	0.5	1	—
— fume	0.1	C	0.1	—	—	—	0.1	—
Hafnium	0.5	—	0.5	—	—	—	0.5	—
Indium								
— metal & compounds (as In)	0.1	—	—	—	—	—	0.1	—
Iron								
— Fe(III) oxide	6	—	6	—	3.5	—	5	—
— iron pentacarbonyl (as Fe)	0.8	—	0.8	—	—	—	0.8	—
— soluble salts (as Fe)	—	—	—	—	—	—	—	—
Lead								
— inorganic compounds (as Pb)	0.1	—	0.1	0.15	$0.1/0.05^h$	0.01–0.07	0.15	0.03–0.06
— tetraethyl lead (as Pb)	0.075	—	0.075	0.10	0.05	—	0.1	—
— tetramethyl lead (as Pb)	0.075	—	0.075	0.15	0.05	—	0.15	—
Lithium								
— hydride	0.025	—	0.025	—	—	—	0.025	—
Magnesium								
— oxide fume	6	—	6	—	—	—	10	—
Manganese								
— metal & compounds (as Mn)	5	—	5	2.5	$2.5/1.0^h$	—	5	0.3
— metal (fume)	1	—	—	—	—	—	1	—
— cyclopentadienyl tricarbonyl (as Mn)	0.1	—	—	—	—	0.1	0.1	—
— tetroxide	1	—	1	—	—	0.3	—	—
Mercury								
— alkyl compounds (as Hg)	0.01	—	0.01	0.01	0.01	0.005	0.01	—
— aryl compounds (as Hg)	0.01	—	0.01	—	—	—	0.1	—
— inorganic compounds (as Hg)	0.1	—	—	0.05	0.05	0.01–0.005	0.1	0.05
— mercury vapor	0.05	—	0.1	—	0.05	—	0.05	—
Molybdenum								
— soluble compounds (as Mo)	5	—	5	—	5	—	5	—
— insoluble compounds (as Mo)	10	—	15	—	$10/5^h$	—	10	—

Table I.20b-5 (continued)

Substance	CH[a]	D[b]	I[c]	S[d]	SU[e]	USA (ACGIH)[f]	WHO[g]
Nickel							
– metal	0.5	C	—	0.5	—	1	—
– soluble compounds (as Ni)	0.05	C	—	0.1	—	0.1	—
– nickel tetracarbonyl (as Ni)	0.35/C	C	—	0.007	0.0005	0.1	—
– oxide and sulfide (as Ni)	0.5/C	C	—	0.01	—	1/C	—
Osmium							
– tetroxide (as Os)	0.002	0.002	—	—	—	0.002	—
Platinum							
– metal	1	—	—	—	—	1	—
– soluble salts (as Pt)	0.002	0.002	—	—	—	0.002	—
Rhodium							
– metal	0.1	—	—	—	—	1	—
– insoluble compounds (as Rh)	—	—	—	—	—	1	—
– soluble compounds (as Rh)	0.001	—	—	—	—	0.01	—
Selenium							
– compounds (as Se)	0.1	0.1	0.2	—	0.1	0.2	—
– hexafluoride (as Se)	0.4	—	0.4	—	—	0.2	—
– hydride	0.17	—	0.2	—	—	—	—
Silver							
– metal	0.1	0.01	—	—	—	0.1	—
– soluble compounds (as Ag)	0.1	—	—	—	—	0.01	—
Sodium							
– azide	0.2	0.2	—	—	—	0.3	—
– bisulfite	—	—	—	—	—	5	—
– fluoroacetate	0.05	0.05	—	—	—	0.05	—
– hydroxide	2	2	—	2	—	2	—
Tantalum							
– metal & compounds (as Ta)	5	5	—	—	—	5	—
Tellurium							
– metal & compounds (as Te)	0.1	0.1	—	0.1	0.01	0.1	—
– hexafluoride (as Te)	0.2	—	—	—	—	0.2	—

I.20b Standards, Guidelines and Legislative Regulations Concerning Metals

Thallium							
– soluble compounds (as Tl)	0.1	0.1	–	–	0.01	0.1	–
Tin							
– metal	–	–	2	–	–	2	–
– inorganic compounds (as Sn)	–	0.1	–	0.1	–	2	–
– organic compounds (as Sn)	–	–	0.1	–	–	0.1	–
Titanium							
– dioxide	6	6	–	–	–	10	–
Tungsten							
– metal & insoluble compounds (as W)	–	–	5	–	–	5	–
– soluble compounds (as W)	–	–	1	–	–	1	–
Uranium							
– compounds (as U)	–	0.25	–	–	–	0.2	–
Vanadium							
– dust & fume (as V_2O_5)	–	0.05	0.2	–	–	0.05	–
Yttrium							
– metal & compounds	–	5	–	–	–	1	–
Zinc							
– chloride	–	–	1	–	–	1	–
– oxide	–	5	5	–	–	5	–
Zirconium							
– compounds (as Zr)	–	5	–	–	–	5	–

[a] Schweizerische Unfallversicherungsanstalt (SUVA), Arbeitshygienische Grenzwerte 1988, Maximale Arbeitsplatzkonzentrationen gesundheitsschädlicher Stoffe (MAK-Werte), Luzern (Switzerland) 1988
[b] MAK (1988), see References
[c] Valori limite per gas e vapori tossici secondo la Commissione Ministero del Lavoro-ENPI
[d] National Swedish Board of Occupational Safety and Health, Ordinance AFS 1984 : 5 on Occupational Exposure Limit Values. Solna (Sweden) (1984)
[e] Source: KETTNER, H. (1979), *Staub Reinhalt. Luft* 39, 56–62
[f] American Conference of Governmental Industrial Hygienists (ACGIH), Threshold Limit Values for Chemical Substances in the Work Environment. ACGIH, Cincinnati (OH) (1988)
[g] Recommended Health-based Limits for Occupational Exposure to Heavy Metals, Report of a WHO Study Group. Technical Report Series No. 647. WHO, Geneva (1980)
[h] Total dust/respirable dust

Table I.20b-6. Biological Exposure Limits for Metals Recommended for Monitoring Occupational Exposures to Metals

Substance	Medium	BAT Values (FRG)[a]	BEI (ACGIH)[b]	WHO[c]	CEC[d]
Aluminum	Urine	200 µg/L	–	–	–
Arsenic and inorganic compounds	Urine	–	–	–	300 µg/L
Cadmium	Blood	15 µg/L	10 µg/L	5 µg/L	
	Urine	15 µg/L	10 µg/g creat.	5 µg/g creat.	15 µg/g creat.
Chromium and compounds	Urine	–	–	–	15 µg/g creat.
Lead	Blood	70 µg/dL[e]	50 µg/dL[f]	40 µg/dL[g]	70 µg/dL[h]
	Urine	–	150 µg/g creat.	–	–
Mercury (metal and inorganic compounds)	Blood	50 µg/L	–	–	–
	Urine	200 µg/L	–	50 µg/g creat.	–
Mercury (organic compounds)	Blood	100 µg/L	–	–	100 µg/L
Vanadium	Urine	–	–	–	50 µg/L

[a] Biological Tolerance Value (BAT). Source: MAK (1988)
[b] Biological Exposure Indices (BEI). Source: ACGIH (1988)
[c] WHO (1980)
[d] Commission of the European Communities (CEC), Occupational Health Guidelines for Chemical Risk (R. ROI, W.G. TOWN, W.G. HUNTER, L. ALESSIO). CEC, Luxembourg (1983)
[e] Women <45 years: 30 µg/dL. Additionally, for δ-aminolevulinic acid a biological limit of 15 mg/L (women <45 years: 6 mg/L) is recommended
[f] Additionally for zinc protoporphyrin in blood the following exposure indices are recommended: 250 µg Zn/100 mL erythrocytes or 100 µg/100 mL blood
[g] Women in the reproductive age: 30 µg/dL
[h] Council Directive (82/605/EEC) on the protection of workers against hazards from lead and its inorganic compounds in the workplace area

e.g., these examinations are prescribed in detail by the "Principles for preventive medical examinations in occupational health" promulgated by the West-German "Berufsgenossenschaften". With regard to metals the following principles have been developed: lead and its compounds (G 2), lead alkyls (G 3), mercury and its compounds (G 9), chromium(VI) compounds (G 15), arsenic and its compounds (G 16), cadmium and its compounds (G 32), nickel and its compounds (G 38).

3 Control of Chemical Substances and Products

3.1 Industrial Chemicals

According to the Council Directive on the Approximation of Laws, Regulations, and Administrative Provisions Relating to the Classification, Packaging and Labelling of Dangerous Substances (67/548/EEC, amended several times by subsequent Council Directives; see under 'References') the packaging of dangerous substances must be adequate, preclude loss of content and be free from liability to damage from, or combination with, the content. Substances must not be marketed unless labelling requirements are met, i.e., the name and origin of the substance, the danger symbol and special risks must all be given. Annex I lists dangerous substances classified in accordance with the degree of hazard and specific nature of risk.

The seventh amendment of the above mentioned Directive (79/831/EEC) regulates the marketing of new substances. New substances must not be marketed unless notified to the competent authority of one of the member states and packaged and labelled in accordance with this Directive. At least 45 days before marketing, manufacturers and importers are required to submit to the competent authority a notification, which includes a dossier about the possible adverse effects of the substance on humans and ecosystems, the proposed classification and labelling categories, and proposals concerning recommended precautions relating to the safe use of the substance. Methods are specified for determining physicochemical properties, toxicity, ecotoxicity, and environmental hazards. Substances marketed in quantities less than one ton per year per manufacturer need not to be notified. The same applies to substances that have already been on the Community market until September 1981. An inventory of these substances has been published.

Regulations similar to those enforced in the member states of the EEC are practiced in other industrial countries such as the United States (Toxic Substances Control Act, 1976), Japan (Chemical Substances Control Law, 1973), Sweden (Act on Products Hazardous to Health and the Environment, 1973), and Switzerland (Ordinance on Substances Hazardous to the Environment, 1986). The Organization for Economic Cooperation and Development (OECD) has adopted several decisions and recommendations regarding test guidelines, mutual acceptance of test data, and principles of good laboratory practice (GLP).

3.2 Consumer Products

Consumer products that contain chemicals present special problems with regard to control due to the widespread and disperse nature of their use. In various countries specific regulations have been issued under various laws for specified product groups such as cosmetics, paints, table and kitchenware, do-it-yourself products, cleansing and washing materials, toys, batteries, and domestic pesticides and fertilizers. In various countries the use of arsenic, cadmium, and lead compounds in paints used for toys or in cosmetic preparations or as constituents of enamels for ceramics and potteries is prohibited or restricted. Other regulations concern the use of cadmium compounds as stabilizers and pigments for plastics and the prohibition or restriction of cadmium-coated goods. In some countries the marketing of mercury and its compounds is prohibited except for specified uses and applications. Another regulation often listed under consumer product legislation concerns the lead content of fuel for motor vehicles (see Sect. 4.2). In various countries regulations have also been issued with regard to the import, production, and disposal of cadmium and mercury containing batteries. The Swiss legislation distinguishes regarding handling of toxic substances between prohibited, very strong and caustic, strong and caustic, less hazardous, and weak poisons (since 1972; EWERS and MERIAN, 1984).

4 Environmental Health Standards and Guidelines for Metals

4.1 Standards and Guidelines Concerning Metals in Drinking Water

Standards and guidelines for the chemical quality of drinking water have been issued in many countries as well as in the form of a CEC Directive, which is binding to the member states of the European Communities. These standards usually include maximum admissible concentrations for metals, metalloids and other inorganic constituents (see Table I.20b-7). The guideline values for drinking water quality recommended by WHO are intended to provide a scientific basis for the setting of standards in different countries, which, if properly implemented, will ensure the safety of drinking water supplies. The principles and the scientific basis of the recommended guideline values are described in criteria monographs (WHO, 1984). The problems and general principles of setting standards for the chemical quality of drinking water have been discussed by ZIELHUIS (1982). The U.S. approach of developing water quality criteria and standards has been described by STARA and KRIVAK (1980) and in the ENVIRONMENT REPORTER (1988).

Standards for metals in drinking water (and other environmental media) generally are set for the metals as such, irrespective of their chemical form or valency state. In some cases it seems desirable to establish standards for different metal compounds and valency states taking into account their different physicochemical and toxicological properties (ZIELHUIS, 1982).

4.2 Standards and Guidelines for Metals in Ambient Air

In many countries standards and guidelines have been set for airborne lead as the most abundant toxic metal in ambient air (see Table I.20b-8). Since the combustion of leaded gasoline represents the major source of lead in urban areas the lead content of gasoline has been limited in many countries. According to the CEC Directive 78/611/EEC the lead content of leaded gasoline must be in the range of 0.15 to 0.40 g/L. In the Federal Republic of Germany and some other member states the

Table I.20b-7. Standards and Guidelines for Metals in Drinking Water. Unit: mg/L

Metal	WHO[a]	CEC[b] GL	CEC[b] MAC	CH[c]	USA EPA[d]	USA NRC[e]	YU[f]
Aluminum	0.2	0.05	0.2	–	–	–	0.2
Antimony	–	–	0.01	–	–	–	–
Arsenic	0.05	–	0.05	0.05	0.05	<0.05	0.05
Barium	–	0.1	–	1	–	–	–
Cadmium	0.005	–	0.005	0.01	0.005	<0.01	1
Calcium	–	100	–	–	–	–	200
Chromium	0.05	–	0.05	0.05	0.02	<0.05	0.05
Copper	1	0.1	–	1	–	1	0.05
Iron	0.3	0.05	0.2	0.3	–	–	0.3
Lead	0.05	–	0.05	0.05	0.05[g]	<0.05	0.05
Magnesium	–	30	50	–	–	–	50
Manganese	0.1	0.02	0.05	0.05	–	0.05	0.05
Mercury	0.001	–	0.001	0.002	0.001	–	0.001
Molybdenum	–	–	–	–	–	–	0.5
Nickel	–	–	0.05	–	–	<0.05	–
Potassium	–	10	12	–	–	–	–
Selenium	0.01	–	0.01	0.01	0.01	–	0.01
Silver	–	–	0.01	0.05	–	–	0.05
Sodium	200	20	–	–	–	<100	–
Strontium	–	–	–	–	–	–	2
Zinc	5	0.1	–	5	–	5	5

[a] WHO (1984), Guidelines for Drinking Water Quality, Vol. 1, Recommendations
[b] Council of the European Communities, Council Directive Relating to the Quality of Water Intended for Human Consumption (80/778/EEC). GL = guide level; MAC = maximum admissible concentration
[c] Switzerland: Ordinance on Food Additives and Food Contaminants (1986)
[d] U.S. Environmental Protection Agency National Primary and Secondary Drinking Water Regulations, Maximum contaminant levels
[e] National Research Council (NRC) (1977), Drinking Water and Health. National Academy of Sciences, Washington, D.C.
[f] Yugoslavia: Ordinance on Microbiological and Physicochemical Standards of Drinking Water (1979)
[g] According to *Environment Reporter 19* (15), Current Developments (12. August 1988), p. 597, the enforcement level is reduced to 0.005 mg Pb/L, but the measurement is then taken from water leaving the treatment plant, not at the consumer's tap

maximum admissible level of lead was set at 0.15 g/L. In many countries unleaded gasoline is increasingly used (see also CONCAWE, 1990).

Air quality standards for metals other than lead have been set only in a few countries as shown in Table I.20b-8. Additionally, standards for toxic metals in dust fallout have been issued in some countries in order to prevent excessive accumulation of toxic metals in soil (Table I.20b-9). Since lead in surface dust represents an important source of lead exposure for young children, the necessity of standards for lead

Table I.20b-8. Standards for Metals in Ambient Air. Unit: $\mu g/m^3$

Metal	CEC[a]	CH[b]	D[c]	DDR[d]	I[e]	USA[f]	YU[g]
Arsenic	–	–	–	3.0	–	–	–
Cadmium	–	0.01	0.05	0.05	–	–	0.04
Lead	2.0	1.0	2.0	0.7	2.0	1.5[h]	2.0
Manganese	–	–	–	10.0[i]	–	–	–
Mercury	–	–	–	0.3	–	–	–

[a] Council Directive on a limit value for lead in the air (82/884/EEC)
[b] Switzerland: Clean Air Ordinance (1985) issued under sections 12, 13, 16, and 39 of the Environment Protection Act (1983)
[c] Federal Republic of Germany: First General Administration Regulation (1974, Last Amendment 1986) to the Federal Immission Control Law (1974, Last Amendment 1986)
[d] German Democratic Republic: 3rd Administration Regulation to the 5th Ordinance issued under the Land Cultivation Law (1982)
[e] Italy: Decret on Maximum Admissible Levels of Pollutants in Ambient Air (1983)
[f] U.S. Environmental Protection Agency, National Primary and Secondary Ambient Air Quality Standards issued under Sections 109 and 301(a) of the Clean Air Act (1976)
[g] Yugoslavia: Maximum Admissible Levels of Air Pollutants, adopted by the Council on the Human Environment and Indoor Air Quality, 5th October 1979. *Zastita Atmosfere 15/979*, 43–44
[h] Maximum arithmetic mean averaged over a calendar quarter
[i] As MnO_2

Table I.20b-9. Standards for Toxic Metals in Dust-Fall. Unit: $\mu g/m^2$ day

Substance	CH[a]	D[b]	DDR[c]	YU[d]
Cadmium	2	5	5	7.5/2.5[e]
Lead	100	250	–	500/250[e]
Thallium	2	10	–	–
Zinc	400	–	–	–

[a] Switzerland: Clean Air Ordinance (1985) issued under the Environment Protection Law (1983)
[b] Federal Republic of Germany: First General Administration Regulation (1974, Last Amendment 1986) to the Federal Immission Control Law (1974, Last Amendment 1986)
[c] German Democratic Republic: 3rd Administration Regulation to the 5th Ordinance to the Land Cultivation Law (1982)
[d] Yugoslavia: Clean Air Commission, Limit Values for Air Pollutants (1979)
[e] The lower limit values were set to protect specified ecosystems

in dust or dust fall-out has been emphasized repeatedly (DUGGAN, 1983; BRUNEKREEF et al., 1983). It has been proposed that 1000 µg/g may be an appropriate dust lead standard to prevent increased lead exposure of children (LAXEN et al., 1987; EWERS et al., 1988).

In 1987, the WHO Regional Office for Europe published "Air Quality Guidelines" for Europe, which include guidelines for several metals in air. The following guideline values are recommended: cadmium, $1-5$ ng/m^3 (1 year average for rural areas) and $10-20$ ng/m^3 (1 year average for urban areas); lead, $0.5-1.0$ µg/m^3 (1 year average); manganese, 1 µg/m^3 (1 year average); mercury (indoor air), 1 µg/m^3 (1 year average); vanadium, 1 µg/m^3 (24 h average).

4.3 Recommendations and Standards for Metals as Food Contaminants

Maximum admissible levels of metals and metalloids were issued for specified dietary products such as fish, wine and other beverages in a number of countries. In Italy, for example, the maximum admissible levels of metals in wine are: lead, 0.3 mg/L; copper, 1.0 mg/L; zinc, 5 mg/L. According to the West-German Wine Ordinance the maximum admissible concentrations of metals in wine and liqueur are: aluminum, 8 mg/L; arsenic, 0.1 mg/L; cadmium, 0.01 mg/L; copper, 5 mg/L; lead, 0.3 mg/L; tin, 1 mg/L; zinc, 5 mg/L. Maximum admissible concentrations for metals and other contaminants were also set for mineral and table water according to the West-German Ordinance on Mineral and Table Water (1984). The maximum admissible contaminant levels are equal to the MAC-values of the CEC Directive Relating to the Quality of Water Intended for Human Consumption (80/778/EEC). Other regulations concern maximum admissible levels of mercury in fish and shell fish (Italy: 0.7 ppm; Switzerland: 0.5 ppm; West Germany: 1 ppm). The Swiss Ordinance on Additives and Contaminants in Dietary Products (1986) contains maximum tolerable levels for aluminum, arsenic, cadmium, chromium, copper, lead, mercury, nickel, selenium, tin, and zinc in specified dietary products such as beverages, fruits, and vegetables. In 1979 and 1986, the West-German Federal Health Agency promulgated guideline values for cadmium, lead, and mercury in and on foodstuffs, which are listed in Table I.20b-10. These guideline values essentially represent the upper normal limits of the cadmium, lead, and mercury concentrations in dietary products presently found in West Germany and are intended to be used as guidelines for the responsible authorities and the food industry.

4.4 Recommendations Concerning the Tolerable Weekly Intake of Toxic Metals via Food

In 1972, the Joint WHO/FAO Expert Committee on Food Additives evaluated some food contaminants and additives including cadmium, mercury, and lead. The "acceptable daily intake for man" was considered to be zero. Since this is not feasible, the committee made recommendations concerning the "provisional tolerable weekly intake for man". The intake levels recommended are summarized in Table I.20b-11. Regarding lead it is emphasized that the recommended intake level does not apply to infants and children. The Joint WHO/FAO Expert Committee on Food Additives also made a recommendation concerning the tentative "maximum acceptable load of arsenic", which was set at 0.05 mg per kg body weight per day (WHO, 1973). Although widely used it must be stated that these recommendations lack a sound toxicological basis.

Table I.20b-10. Guideline Values for Cadmium, Lead and Mercury in and on Foodstuffs Recommended by the West-German Federal Health Agency. Units: mg/kg fresh weight or mg/L Relating to the Edible Constituents[a]

	Lead	Cadmium	Mercury
Milk	0.03	0.0025	0.01
Condensed milk	0.3	0.05	0.01
Cheese	–	0.05	0.01
Cheese (except hard cheese)	0.25	–	–
Hard cheese	0.5	–	–
Eggs	0.25	0.05	0.03
Beef	0.25	0.1	0.03
Veal	0.25	0.1	0.03
Pork	0.25	0.1	0.03
Ground meat	0.25	0.1	0.03
Chicken meat	0.25	0.1	0.03
Bovine liver	0.8	0.5	0.1
Calf liver	0.8	0.5	0.1
Pork liver	0.8	0.5	0.1
Bovine kidney	0.8	1.0	0.1
Calf kidney	0.8	0.5	0.1
Pork kidney	0.8	1.0	0.1
Meat products	0.25	0.1	0.05
Sausages	0.25	0.1	0.05
Fish	0.5	0.1	1.0[b]
Fish products	0.5	0.1	1.0[b]
Canned fish	1.0	0.1	1.0[b]
Wheat grains	0.3	0.1	0.03
Rye grains	0.4	0.1	0.03
Rice grains	0.4	0.1	0.03
Potatoes	0.25	0.1	0.02
Green vegetables (except kale, herbs and spinach)	0.8	0.1	0.05
Kale	2.0	–	–
Pot herbs	2.0	–	–
Spinach	–	0.5	–
Sprout vegetables	0.5	0.1	0.05
Fruit vegetables	0.25	0.1	0.05
Root vegetables	–	0.1	–
Celery	–	0.2	–
Pomaceous fruits	0.5	0.05	0.03
Fruits with stones	0.5	0.05	0.03
Berries, small fruits	0.5	0.05	0.03
Citrus fruits	0.5	0.05	0.03
Other fruits	0.5	0.05	0.03
Hard/shelled fruits	0.5	0.05	0.03
Refreshing drinks	0.2	0.05	0.01
Wine	0.3[c]	0.01[c]	0.01
Beer	0.2	0.03	0.01

[a] Source: *Bundesgesundheitsblatt 29* (1986), pp. 22–23
[b] Maximum contaminant levels according to the Ordinance on Mercury in Fish and Seafood (1975)
[c] Maximum contaminant levels according to the Wine Ordinance

Table I.20b-11. Provisional Tolerable Weekly Intake of Mercury, Lead, and Cadmium[a]

Substance	Provisional Tolerable Weekly Intake for Man		Acceptable Daily Intake for Man mg/kg body weight
	mg/person	mg/kg body weight	
Mercury			
– total mercury	0.3	0.005	None
– methylmercury (expressed as mercury)	0.2	0.0033	None
Lead[b]	3	0.05	None
Cadmium	0.4 – 0.5	0.0067 – 0.0083	None

[a] Source: Evaluation of Certain Food Additives and the Contaminants Mercury, Lead and Cadmium. 16th Report of the Joint WHO/FAO Expert Committee on Food Additives. *WHO Tech. Rep. Ser. 505*, Geneva 1972
[b] These intake levels do not apply to infants and children

Table I.20b-12. Guideline Values for the Judgement of Metal Concentrations in Human Blood, Urine, and Hair Recommended by the West-German Federal Health Agency

Metal	Material	Unit	Categories[a]		
			I	II	III
Arsenic	Urine	µg/L	< 15	15 – 40	> 40
	Hair	µg/g	< 1	> 1	
Cadmium	Blood				
	– children	µg/L	< 1	1 – 3	> 3
	– adults	µg/L	< 2	2 – 5	> 5
	Hair	µg/g	< 1.5	> 1.5	
Lead	Blood				
	– women in the reproductive age and children	µg/dL	< 15	15 – 25	> 25
	– adults	µg/dL	< 15	15 – 35	> 35
	Hair	µg/g	< 20	> 20	
Mercury	Blood	µg/L	< 3	3 – 10	> 10
	Urine	µg/L	< 5	5 – 20	> 20

[a] Category I: normal range. Category II: increased concentration; sampling and measurement should be repeated. Category III: significantly increased concentration; adverse health effects cannot be excluded, if increased exposure persists for a longer time; the source of exposure should be identified and eliminated or reduced

4.5 Biological Exposure Limits for Metals in Human Body Fluids and Tissues

Biological exposure limits concerning environmental exposure of the general population to metals primarily have been proposed for lead. The recommendations refer either to the distribution of blood lead (PbB) levels in groups of the general popula-

Table I.20b-13. Guideline Values for Tolerable Metal Concentrations in Agricultural Soil. Units: mg/kg dry weight

Metal	Guideline Values According to KLOKE[a]		Guideline Values According to the Swiss Ordinance on Soil Contaminants[b]	
	Common Concentration Range	Maximum Tolerable Concentration	Total[c]	Water Soluble[d]
Antimony	<0.1 – 0.5	5	–	–
Arsenic	2 – 20	20	–	–
Boron	5 – 30	25	–	–
Beryllium	1 – 5	10	–	–
Cadmium	0.1 – 1	3	0.8	0.03
Chromium	2 – 50	100	75	–
Cobalt	1 – 10	50	25	–
Copper	1 – 20	100	50	0.7
Gallium	<0.5 – 10	10	–	–
Lead	0.1 – 20	100	50	1.0
Mercury	0.1 – 1	2	0.8	–
Molybdenum	<1 – 5	5	5	–
Nickel	2 – 50	50	50	0.2
Selenium	0.1 – 5	10	–	–
Tin	1 – 20	50	–	–
Thallium	<0.1 – 0.5	1	1	–
Uranium	<0.1 – 1	5	–	–
Vanadium	10 – 100	50	–	–
Zinc	3 – 50	300	200	0.5
Zirconium	<10 – 300	300	–	–

[a] Source: KLOKE (1980), Mitteilungen des Verbandes Deutscher Landwirtschaftlicher Untersuchungs- und Forschungsanstalten (VDLUFA) Heft 1 – 3 (1980)
[b] Ordinance on Soil Contaminants (1986) issued under Sections 33 and 39 of the Environmental Protection Law (1983)
[c] Extractable with concentrated HNO_3
[d] Extractable with aqueous $NaNO_3$ solution

tion or to maximum tolerable PbB levels in individuals, mainly children. According to the CEC Directive on the biological screening of the population for lead (77/312/EEC) 50% of an investigated population group should have PbB-levels <20 µg/dL; 90% of the population group should have PbB levels <30 µg/dL, and 98% PbB levels <35 µg/dL. If these reference levels are exceeded, action has to be taken to trace and reduce the sources of exposure. According to a guideline of the Dutch Ministry of Public Health concerning the evaluation of blood lead surveys in pre-school children the following reference levels are recommended: 50. percentile: <20 µg/dL; 90. percentile: <25 µg/dL; 98. percentile: <30 µg/dL (BRUNEKREEF et al., 1983). Individual action levels, i.e., PbB levels, which should give rise to actions to trace and reduce the sources of exposure on an individual basis, range from 20 to 30 µg/dL in various countries.

Biological exposure limits for other metals have not been recommended so far. Recently, the West-German Federal Health Agency proposed biological limits as a basis for the judgement of metal concentrations in blood, urine, and hair of non-occupationally exposed individuals. The recommended categorical limits are shown in Table I.20b-12.

4.6 Standards and Guidelines for Metals in Soil, Sewage Sludge, Compost, and Fertilizers

Since toxic metals in soil may be taken up by plants and thus enter the food chain, limit values for maximum tolerable metal concentrations in agricultural and garden soil were set in various countries. As an example the Swiss Ordinance on Soil Contaminants (1986) issued under sections 33 and 39(1) of the Environment Protection Act (1983) are shown in Table I.20b-13. In the Federal Republic of Germany the guideline values proposed by KLOKE (1980) are widely used for the judgement of metal concentrations in agricultural and garden soil (see also Table I.20b-13 and Chapter I.7).

In some traditional industrial areas and near hazardous waste disposal sites the toxic contamination of soil and groundwater may represent a health risk for the residents of these areas. Questions arise whether these areas can be used for residential or agricultural purposes. In order to provide a basis for administrative decisions on redevelopment measures the Dutch authorities have developed guideline values for contaminant levels in soil and groundwater. The guideline values concerning toxic metals are shown in Table I.20b-14.

The use of sewage sludge and compost, which may contain high levels of certain toxic metals, has also been regulated in various countries in order to prevent undue

Table I.20b-14. Dutch Guideline Values for Metal Concentrations in Soil and Groundwater for the Judgement of Contaminant Levels in Soil[a]. A: reference levels; B: levels, which should give rise to a more detailed examination; C: levels, which should give rise to a detailed examination of the area with regard to redevelopment measures

Metal	Soil (mg/kg dry weight)			Groundwater (µg/L)		
	A	B	C	A	B	C
Arsenic	$15 + 0.4 (L+H)$[b]	30	50	10	30	100
Barium	200	400	2000	50	100	500
Cadmium	$0.4 + 0.007 (L+3H)$	5	20	1.5	2.5	10
Cobalt	20	50	300	20	50	200
Copper	$15 + 0.6 (L+H)$	100	500	15	50	200
Chromium	$50 + 2L$	250	800	1	50	200
Lead	$50 + L + H$	150	600	15	50	200
Mercury	$0.2 + 0.0017 (2L+H)$	2	10	0.05	0.5	2
Molybdenum	10	40	200	5	20	100
Nickel	$10 + L$	100	500	15	50	200
Tin	20	50	300	10	30	150
Zinc	$50 + 1.5 (2L+H)$	500	3000	150	200	800

[a] Leidraad Bodemsanering. Aflevering 4. November 1988, Staatsuitgeverij's, Gravenhage
[b] L = percentage of clay; H = percentage of organic matter in soil

Table I.20b-15. Maximum Admissible Concentrations of Toxic Metals in Compost and Sewage Sludge Used in Agriculture and in Soil, on which Compost or Sewage Sludge is Applied. Unit: mg/kg dry weight

Metal	CH[a]	CEC[b]		D[c]		I[d]	
	Compost	Sewage Sludge	Soil	Sewage Sludge	Soil	Compost	Soil
Arsenic	–	–	–	–	–	10	10
Cadmium	3	20–40	1–3	20	3	10	3
Chromium	150	–	–	1200	100	500	2000 (CrIII)
						10	15 (CrVI)
Cobalt	25	–	–	–	–	–	–
Copper	150	1000–1750	50–140	1200	100	600	3000
Lead	150	750–1200	50–300	1200	100	500	500
Mercury	3	16–25	1–1.5	25	2	10	15
Molybdenum	5	–	–	–	–	–	–
Nickel	50	300–400	30–75	200	50	200	1000
Zinc	500	2500–4000	150–300	3000	300	2500	10000

[a] Switzerland: Ordinance on Substances Hazardous to the Environment (1986) issued under the Environment Protection Law (1983) and Water Protection Law (1971)
[b] Council Directive on the protection of the environment, and in particular of the soil, when sewage sludge is used in agriculture (86/278/EEC)
[c] Federal Republic of Germany: Ordinance on Sewage Sludge (1982) issued under Section 15 of the West-German Waste Disposal Act (1977)
[d] Italy: Supplemento ordinario alla Gazetta Ufficiale No. 25, of 13. 09. 1984, Rome

accumulation of metals in the food chain. According to the West-German "Ordinance on Sewage Sludge" (1982) issued under the Waste Disposal Act (1978) a standardized supervision procedure to control the levels of specified metals in sewage sludge and soil has been established. Sewage sludge may only be applied on land used for agricultural, forestry or horticultural purposes, if the metal concentrations in the sewage sludge as well as in the soil are below the maximum admissible contaminant levels shown in Table I.20b-15. Maximum admissible concentrations of toxic metals in compost, sewage sludge and soil, on which sewage sludge is intended to be applied, were also established in some other countries such as Italy and Switzerland and, more recently, by a CEC guideline (see Chapter I.7c, Part II, and Table I.20b-15).

In some countries maximum contaminant levels have also been set for some toxic metals in fertilizers. According to the Swiss "Ordinance on Substances Hazardous to the Environment" (1986) fertilizers may not contain more than 2000 ppm of chromium and 4000 ppm of vanadium on a dry-weight basis. Fertilizers with a phosphate content of more than 1% may not contain more than 50 ppm of cadmium.

References

ACGIH (American Conference of Governmental Industrial Hygienists) (1988), *Documentation of the Threshold Limit Values for Chemical Substances and Physical Agents in the Work Environment*, 4th Ed. ACGIH, Cincinnati, Ohio.
Air Pollution Control Directorate (Canada) (1981), *The Clean Air Act – Compilation of Regulations and Guidelines*. Environment Protection Service, Report 1-AP-81-1. Toronto.
Angerer, J., and Schaller, K. H. (1985, 1988), *Analyses of Hazardous Substances in Biological Materials,* Vols. 1 and 2 (with Discussions and References). VCH Verlagsgesellschaft, Weinheim-Basel-Cambridge-New York.
Berlin, A., and van der Venne, M. T. (1987), *Regulation at the European Community Level to Eliminate or Reduce Lead-related Health Hazards.* Trace Elem. Med. *4*, 166–172.
Brunekreef, B., Noy, D., Bierstecker, K., and Boleij, J. (1983), *Blood Lead Levels of Dutch Children and their Relationship to Lead in the Environment.* J. Air Pollut. Control Assoc. *33*, 872–876.
Castleman, B. I., and Ziem, G. E. (1988), *Corporate Influence on Threshold Limit Values.* Am. J. Ind. Med. *13*, 531–559.
CEC (Commission of the European Communities)
 – *Council Directive on the Approximation of Laws, Regulations and Administrative Provisions Relating to the Classification, Packaging and Labelling of Dangerous Substances* (67/548/EEC). Amendmends by Council Directives 69/81 EEC, 70/189/EEC, 71/144/EEC, 73/146/EEC, 75/409/EEC, 76/907/EEC, 79/370/EEC, 79/831/EEC, 82/232/EEC, 83/437/EEC, 84/449/EEC, 86/431/EEC, and 87/432/EEC.
 – *Council Directive Concerning the Quality Required of Surface Water Intended for Abstraction of Drinking Water in the Member States* (75/440/EEC). OJ No. L 194 of 25.7.1975, p. 26.
 – *Council Directive on Pollution Caused by Certain Dangerous Substances Discharged into the Aquatic Environment of the Communities* (76/464/EEC). OJ No. L 129 of 18.5.1976, p. 23.
 – *Council Directive on Biological Screening of the Population for Lead* (77/312/EEC). OJ No. L 105 of 28.04.1977, p. 10.
 – *Council Directive on Waste from Titanium Dioxide Industry* of 02.02.1978 (78/176/EEC). OJ No. L 54 of 25.02.1978, p. 20.
 – *Council Directive on Toxic and Dangerous Waste* (78/319/EEC). OJ No. L 84 of 31.03.1978, p. 43.
 – *Council Directive on the Approximation of the Law of the Member States Concerning the Lead Content of Petrol* (78/611/EEC). OJ No. L 197 of 22.07.1978, p. 19.
 – *Council Directive on the Protection of Groundwater Against Pollution Caused by Certain Dangerous Substances* (80/68/EEC). OJ No. L 20 of 26.01.1980, p. 43.
 – *Council Directive Relating to the Quality of Water Intended for Human Consumption* (80/778/EEC). OJ No. L 229 of 30.8.1980, p. 11.
 – *Council Directive on the Protection of Workers from Harmful Exposure to Chemical, Physical and Biological Agents at Work* (80/1107/EEC). OJ No. L 327 of 3.12.1980, p. 8.
 – *Council Directive on Limit Values and Quality Objectives for Mercury Discharges by the Chlor-alkali Electrolysis Industry* (82/176/EEC). OJ No. L 81 of 27.03.1982, p. 29.
 – *Council Directive on the Protection of Workers Against Hazards of Lead and its Inorganic Compounds at the Workplace* (82/1107/EEC). OJ No. L 247 of 23.8.1982, p. 12.
 – *Council Directive on a Limit Value for Lead in the Air* (82/884/EEC). OJ No. L 378 of 31.12.1982, p. 15.
 – *Council Directive on the Procedure of the Surveillance and Monitoring of Environments Concerned by the Titanium Dioxide Industry* (82/883/EEC). OJ No. L 378 of 31.12.1982, p. 1.
 – *Council Directive on Limit Values and Quality Objectives for Cadmium Discharges* (83/513/EEC). OJ No. L 74 of 24.01.1983, p. 1.
 – *Council Directive Amending Directive 78/176 EEC on Waste from the Titanium Dioxide Industry* (83/29/EEC). OJ No. L 32 of 24.01.1983, p. 28.

- Council Directive on Limit Values and Quality Objectives for Mercury Discharges by Sectors other than the Chlor-alkali Electrolysis Industry (84/156/EEC). OJ No. L 74, 17.03.1984, p. 74.
- Council Directive on the Approximation of the Laws of the Member States Concerning the Lead Content of Petrol (85/210/EEC). OJ No. L 96 of 03.04.1985, p. 25.
- Council Directive on Containers of Liquids for Human Consumption (85/339/EEC). OJ No. 1 176 of 06.07.1985, p. 18.
- Council Directive on the Protection of the Environment, and in Particular of the Soil, when Sewage Sludge is Used in Agriculture (86/278/EEC). OJ No. L 181 of 0.4 07.1986, p. 6.
- Council Directive on Limit Values and Quality of Objectives for Discharges of Certain Dangerous Substances Included in List I of the Annex to Directive 76/464/EEC (86/280/EEC). OJ No. L 181 of 04.07.1986, p. 16.

Clarkson, T. M., Friberg, L., Nordberg, G. F., and Sager, P. R. (eds.) (1988), *Biological Monitoring of Toxic Metals.* Plenum Press, New York-London.

Concawe (1990), *Report No. 2/90: Motor Vehicle Emission Regulations and Fuel Specifications – 1990 Update*, prepared by J. S. McArragher et al., Brussels.

Duggan, M. J. (1983), *Contribution of Lead in Dust to Children's Blood Lead. Environ. Health Perspect. 50*, 371–381.

Environment Reporter (1988), *Current Developments 19*(15), 597 (August 12, 1988).

EPA (U.S. Environmental Protection Agency) (1986), *Air Quality Criteria for Lead*, Vols. I–IV. Environmental Criteria and Assessment Office, Research Triangle Park, North Carolina.

Ewers, U. (1988), *WHO-Guidelines for Air Quality in Europe. Öff. Gesundheitswes. 50*, 626–629.

Ewers, U., and Merian, E. (1984), *Schutzvorschriften und -richtlinien betreffend Metalle und Metallverbindungen*, in: Merian, E. (ed.): *Metalle in der Umwelt*, pp. 283–289. Verlag Chemie, Weinheim-Deerfield Beach/Florida-Basel.

Ewers, U., Freier, I., Krämer, U., Jermann, E., and Brockhaus, A. (1988), *Schwermetalle im Staubniederschlag und im Boden, ihre Bedeutung für die Schwermetallbelastung von Kindern. Staub Reinhalt. Luft 48*, 27–33.

Henschler, D. (1981), *Maximale Arbeitsplatzkonzentrationen – Grundlagen, Entwicklung, Beratungsmodell.* in: Deutsche Forschungsgemeinschaft: *Wissenschaftliche Grundlagen zum Schutz vor Gesundheitsschäden durch Chemikalien am Arbeitsplatz*, pp. 29–40. H. Boldt Verlag, Boppard, FRG.

Henschler, D. (ed.) (1988), *Gesundheitsschädliche Arbeitsstoffe. Toxikologisch-arbeitsmedizinische Begründung von MAK-Werten.* VCH Verlagsgesellschaft, Weinheim-Basel-Cambridge- New York.

Henschler, D., and Lehnert, G. (eds.) (1986), *Biologische Arbeitsplatz-Toleranz-Werte (BAT-Werte). Arbeitsmedizinisch-toxikologische Begründungen.* VCH Verlagsgesellschaft, Weinheim-Deerfield Beach/Florida-Basel.

IUAPPA (International Union of Air Pollution Prevention Associations) (1988), *Clean Air Around the World. The Law and Practice of Air Pollution Control in 14 Countries in 5 Continents.* IUAPPA, Brighton, UK.

Kloke, A. (1980), *Orientierungsdaten für tolerierbare Gesamtgehalte einiger Elemente in Kulturböden* (Richtwerte 1980). *Mitt. VDLUFA*, Heft 1–3.

Lauwerys, R. R. (1983), *Industrial Chemical Exposure: Guidelines for Biological Monitoring.* Biomedical Publications, Davis, California.

Laxen, D. P. H., Raab, G. M., and Fulton, M. (1987), *Children's Blood Lead and Exposure to Lead in Household Dust and Water – A Basis for an Environmental Standard for Lead in Dust. Sci. Total Environ. 66*, 235–244.

Lundberg, P., and Holmberg, B. (1985), *Ann. Am. Conf. Ind. Hyg. 12*, 249–252.

MAK (1988), *Maximum Concentrations at the Workplace and Biological Tolerance Values for Working Materials, DFG Report No. XXIV.* VCH Verlagsgesellschaft, Weinheim-Basel-Cambridge-New York.

Markard, C., and Stopp, M. (1984), *How to Find Informations – Criteria Documents, Monographs, and Data Banks* (in German), in: Merian, E. (ed), *Metalle in der Umwelt*, pp. 293–300. Verlag Chemie, Weinheim-Deerfield Beach/Florida-Basel.

NIOSH (National Institute for Occupational Safety and Health), *Criteria for a Recommended Standard*
- (1973) *Occupational Exposure to Chromic Acid,*
- (1973) *Occupational Exposure to Inorganic Mercury,*
- (1975) *Occupational Exposure to Chromium VI,*
- (1975) *Occupational Exposure to Inorganic Arsenic,*
- (1977) *Occupational Exposure to Inorganic Nickel,*
- (1978) *Occupational Exposure to Inorganic Lead* (Revised Criteria)
  U.S. Department of Health, Education and Welfare, Washington, D.C.

Nordberg, G. F., Frostling, H., Lundberg, P., and Westerholm, P. (1988), *Swedish Occupational Exposure Limits: Development in Scientific Evaluation and Documentation. Am. J. Ind. Med. 14,* 217–222

Paull, J. M. (1984), *The Origin and Basis of Threshold Limit Values. Am. J. Ind. Med. 5,* 227–238.

Schlegel, H. (1981), *MAK-Werte in der Schweiz.* in: Deutsche Forschungsgemeinschaft: *Wissenschaftliche Grundlagen zum Schutz vor Gesundheitsschäden durch Chemikalien am Arbeitsplatz,* pp. 106–112. H. Boldt Verlag, Boppard, FRG.

Stara, J. F., and Krivak, J. (1980), *Prog. Water Tech. 13,* 267–275.

Toyama, T. (1985), *Permissible and Control Limits of Toxic Substances at Places of Work in Japan. Am. J. Ind. Med. 8,* 87–89.

VDI (Verein Deutscher Ingenieure) (1986), *Handbuch Reinhaltung der Luft,* Vols. 1–6. Beuth Verlag, Berlin-Köln.

WHO/FAO Joint Committee on Food Additives (1972), *Evaluation of Certain Food Additives and the Contaminants Mercury, Lead, and Cadmium. WHO Tech. Rep. Ser. 505.*

WHO (1973), *Trace Elements in Human Nutrition. WHO Tech. Rep. Ser. 532.*

WHO (1978), *Principles and Methods for Evaluating the Toxicity of Chemicals. Environ. Health Crit. 6.*

WHO (1980), *Recommended Health-based Limits in Occupational Exposure to Heavy Metals. WHO Tech. Rep. Ser. 647.*

WHO (1984), *Guidelines for Drinking Water Quality,* Vols. 1–3. WHO, Geneva.

WHO (1987), *Regional Office for Europe, Air Quality Guidelines for Europe.* WHO Regional Publications. European Series No. 23, Copenhagen.

Zabel, T. F. (1989), *Current Standards and Their Relation to Environmental Behaviour and Effects – The Case of Lead. Sci. Total Environ. 78,* 187–204.

Zielhuis, R. L. (1982), *Standards for Chemical Quality of Drinking Water: A Critical Assessment. Int. Arch. Occup. Environ. Health 50,* 113–130.

Zielhuis, R. L., and Notten, W. R. F. (1979), *Permissible Levels for Occupational Exposure: Basic Concepts. Int. Arch. Occup. Environ. Health 42,* 269–281.

Part II Metals and Metalloids, and Their Ions and Compounds

II.1 Aluminum

JOHN SAVORY and MICHAEL R. WILLS, Charlottesville, Virginia, USA[1]

1 Introduction

Aluminum is a ubiquitous element in the environment, i.e., living organisms are always in contact with some of its chemical compounds. In soils mainly aluminum silicates, hydroxides, and oxides are found, which may be transformed into ionic forms during soil acidification. At a pH of 5 or below roots of intolerant plant species may be severely damaged (see also Chapter I.7 d). Aluminum toxicity for fish is also high at low pH.

Aluminum ions have been demonstrated to be toxic especially in individuals with impaired renal function leading to reduced elimination and to excess in serum and tissues. There is not much information available on the biochemistry of aluminum and the mechanisms of its toxic effects. Most of the interest in aluminum has been in the clinical setting of the hemodialysis unit. Here aluminum toxicity occurs because of contamination of dialysis solutions and treatment of the patients with aluminum containing phosphate binding gels. Aluminum has been shown to be the major contributor to the dialysis encephalopathy syndrome and dialysis osteodystrophy. Other clinical disturbances associated with aluminum toxicity are a microcytic anemia and metastatic extraskeletal calcification. Some circumstantial evidence links aluminum with such devastating neurological diseases as amyotrophic lateral sclerosis and Alzheimer's disease. Aluminum overload can be treated effectively by chelation therapy with desferrioxamine and hemodialysis.

Major problems exist with the analytical methods for measuring aluminum which result from inaccurate techniques and contamination difficulties. The most widely used analytical technique is electrothermal atomic absorption spectrometry which can provide reliable measurements in the hands of a careful analyst. To understand environmental and biological processes aluminum speciation is of primary importance.

---

[1] Prof. Dr. Bernhard Ulrich, Göttingen, has contributed valuable information from European sources.

2 Physical and Chemical Properties, and Analytical Methods

2.1 Physical and Chemical Properties

Aluminum has an atomic mass of 27.0 (only one natural isotope), atomic number 13, melting point 660.4 °C, and boiling point 2467 °C. The unique physical and chemical properties of aluminum (e.g., its low density of 2.79 g/cm^3 and its high conductivity and ductility) account for its diverse use.

One of the most interesting physical constants of aluminum is its ionic radius of 57 pm. In combination with its high oxidation potential of +1.66 V and the +3 oxidation state, the ionic radius accounts for the stability of aluminum compounds. Because of the stability of aluminum compounds, no simple chemical or physical process has been developed to remove aluminum from its ore, bauxite (Al_2O_3). On the other hand, metallic aluminum easily forms a protecting aluminum oxide film on its surface, a property which proves very useful.

Aluminum forms a variety of trivalent compounds. It can be burned in air to form solid Al_2O_3 evolving large amounts of heat. It may also be used to reduce other metal oxides. Aluminum reacts with fluorine to form a series of stable complex ions; it reacts less with the larger chlorine atom. It will combine with complex anions such as phosphates and sulfates as well as many organic species.

2.2 Analytical Methods

The key to the success in aluminum monitoring lies in the availability of accurate and precise analytical methods together with guidelines for contamination-free specimen collection. Sample contamination (e.g., from glassware, additives, and hypodermic needles) during collection and of reagents during analysis are major problems in the measurement of aluminum. SAVORY and WILLS (1986) reviewed the analytical problems associated with aluminum measurements and provided recommendations for specimen collection and control of contamination. RØISET and SULLIVAN (1987), for instance, studied the effect of filtration and centrifugation on the determination of aluminum in water. SULLIVAN et al. (1986) compared frequently used methods for the determination of aqueous aluminum. MÖLLER (1986) analyzed waters and soils for aluminum by flow-injection analysis using colorimetric methods.

Several methods are available for the determination of total aluminum in biological and other materials (see also Chapter I.4a). Chemical and physicochemical methods are in most practical situations insensitive and inaccurate; X-ray fluorescence is specific but lacks sensitivity; neutron activation analysis is complex and subject to interferences, although it is a very sensitive technique. Nuclear magnetic resonance spectroscopy is not very sensitive but useful to get information on speciation (SADLER, 1989). Electrothermal atomic absorption spectrometry is the most widely used technique and can produce reliable results, provided the matrix effects on standardization are recognized and corrected.

Many problems existed in the past with aluminum methodology and these have, for instance, been detailed in the study of VERSIECK and CORNELIS (1980) on the variability in reported normal plasma concentrations (see also Sect. 6.6). Even with reliable methods mean normal values varied between 2 and 42 µg/L. Other reports listed normal mean concentrations ranging from 72 to 1460 µg/L, although obvious interferences were present in these methods (VERSIECK and CORNELIS, 1980). Most of the values above 5–10 µg/L plasma for "normal" persons are, however, wrong because of sampling and/or analytical errors (see also Chapter I.4a and SCHALLER and VALENTIN, 1984, 1985). Above 10 µg/L it is necessary to investigate the reasons for the concentration found; the biological tolerance value is 170 µg/L (MAK, 1987). Standard methods for monitoring aluminum in body fluids have been developed by BURATTI et al. (1984), FLEISCHER and SCHALLER (1984), SCHALLER and VALENTIN (1984), VANDERVOET et al. (1985), and WELZ (1985).

BETTINELLI et al. (1985) have made a thorough evaluation of the direct measurement of aluminum in serum using EAAS with the stabilized temperature furnace with the L'vov platform. Their recommendation was to use pyrolytically coated graphite tubes and to keep the method as simple as possible with minimal sample pretreatment. SLAVIN (personal communication) has also been unable to observe major slope change problems between aqueous and serum matrix samples and recommends a procedure similar to those reported by BETTINELLI et al. (1985), LEUNG and HENDERSON (1982), and BROWN et al. (1984). EAAS procedures have been developed with the use of auto-sampling to improve precision. Pyrolytically coated graphite tubes are recommended together with a pyrolytic graphite platform. Argon is preferred over nitrogen as the purge gas since argon produces a larger and less variable signal. Some type of background correction is recommended, and the Zeeman correction system probably provides the most sensitive and reliable results. In the direct methods for serum analysis, standards and serum samples are diluted with an equal volume of an aqueous solution containing $Mg(NO_3)_2$ (2 g/L). The autosampler is programmed to deliver a 15 µL aliquot of the sample onto the platform for final analysis.

LEUNG and HENDERSON (1982) prepared standards using a serum pool containing a minimal amount of endogenous aluminum, whereas BETTINELLI et al. (1985) recommended aqueous standards. The use of simple aqueous standards should be preferred, provided that there are no matrix effects from serum (see also SCHALLER and VALENTIN, 1984). Recently a method has been developed in the authors' laboratory where the proteins are removed from the sample prior to the final atomic spectroscopic measurement (BROWN et al., 1984).

Atomic emission using inductively coupled plasma instruments has great potential although sensitivity appears to be a problem; this technique is relatively free of matrix effects.

Solid tissues must be homogenized, dried, ashed, and/or dissolved to produce a liquid sample prior to EAAS analysis. Bone samples for analysis are washed free of marrow by a strong stream of distilled water and digested in concentrated nitric acid. Soft tissue samples, such as brain, liver, muscle, etc., also must be homogenized before processing, and this can be accomplished easily by pommelling the tissue in a "Stomacher" blender. Distilled water (5 mL) is added to the bag with the tissue, the bag is sealed and placed in the blender and blended for 5 to 15 min which completely homogenizes the sample (SUNDERMAN et al., 1985). The homogenate can then be processed for analysis.

Aluminum can be localized in tissue by histochemical staining, electron-probe X-ray microanalysis, and by laser microprobe mass spectrometry. These last two are sophisticated techniques which will be powerful tools in future studies of aluminum toxicity.

Quality assurance in aluminum measurements follows established principles used in all areas of clinical chemistry. Internal quality control materials, usually pooled plasma, at two or three different concentrations should be used on a daily basis. The concentrations of aluminum in these materials should coincide with various decision levels. Interlaboratory control materials are also an important part of any quality assurance scheme. Presently, such a program is conducted by Dr. Andrew Taylor through the Robens Institute (University of Surrey, UK). The Institute mails 3 water and 3 serum samples monthly to laboratories for aluminum analysis. The data and statistical information of the 38 participating laboratories is then made available for review.

In environmental analytical chemistry (especially in soils and waters, but also in biological fluids) aluminum speciation becomes more and more important (BRULAND, 1983; SMEYERS-VERBEKE and VERBEELEN, 1984, 1987; ÖHMAN and SJÖBERG, 1988; and others) to understand processes and reactions (see also Sects. 4 through 6, Chapter I.7d, TINKER and BARRACLOUGH, 1988). COURTYN et al. (1989) have separated undissolved, dissolved non-complexed and dissolved complexed aluminum by stepwise ion exchange techniques. Fluoride and humic acid complexes can also be distinguished.

3 Sources, Production, Important Compounds, Uses, Waste Products, Recycling

3.1 Occurrence, Production

Whereas all soils contain aluminum compounds (the average earth's crust concentration is of the order of 8.13%; SAAGER, 1984), only enriched minerals are of economic interest as raw material for the production of aluminum oxide, metallic aluminum, aluminum alloys, and aluminum compounds. The term bauxite is used for sedimentary rocks that contain economically recoverable quantities of the aluminum minerals gibbsite, boehmite, and diaspore (HUDSON et al., 1985). In 1982 globally almost 70 million tons of bauxite (with 45–60% aluminum oxide) were extracted (SAAGER, 1984), about 28% of it in Guinea, about 20% in Australia, about 11% in Brazil, and 8% in Jamaica. Generally bauxite is digested, a sodium aluminate solution is formed with NaOH according to the Bayer process (developed in 1887), from which impurities are physically separated, and pure $Al(OH)_3$ is precipitated (HUDSON et al., 1985). At about 102 °C aluminum hydroxides are converted into aluminum oxide hydroxides (the metastable systems play also a role in environmental chemistry, for instance, in soils and waters), and all these compounds may be thermally transformed into aluminum oxides (alumina), such as diaspore and corundum (HUDSON et al., 1985), which is the third hardest naturally occurring substance after diamond and carborundum (SiC). Adding a trace of chromium or cobalt to alumina yields the precious gems ruby and sapphire, respectively. Alumina is the raw material for metallic aluminum, but also for absorbant and catalytic industrial aluminas and for ceramic uses (HUDSON et al., 1985). $Al(OH)_3$ is also an antacid in pharmaceuticals (see Sect. 5).

The only method now used industrially to produce primary aluminum is the Hall-Héroult process (FRANK et al., 1985). Aluminum oxide is fused and dissolved in cryolite, Na_3AlF_6 (a process which may lead to ecotoxically undesired hydrogen fluoride emissions). For the electrowinning reduction a relatively great amount of energy is consumed, partly from the carbon anode (prebaked carbon anodes or self-baking Söderberg anodes are employed) and partly from electricity (FRANK et al., 1985). Thermodynamic considerations and current efficiency are very important. Global annual primary aluminum production is fluctuating between 13 and 16 million tons (SAAGER, 1984; FRANK et al., 1985; SANDER, 1986), about 3.3 million tons are produced in the USA, about 2.4 million tons in the Soviet Union, about 1.1 million tons in Canada, about 0.7 million tons in the Federal Republic of Germany, and about 0.6 million tons in Norway. Since 30–40% of the direct costs in the energy intensive aluminum production account for electricity, there is more and more the tendency to transfer it to locations where electricity is cheap (e.g., Canada, Latin America, Norway, and Australia) and/or to recycle aluminum waste (SAAGER, 1984; FRANK et al., 1985; SANDER, 1986). These developments lead to great overcapacities (SANDER, 1986), the more so as consumption practically does not increase.

3.2 Uses, Waste Products, Recycling

In the USA in 1982 about 33% of the aluminum was consumed in the packaging industry, about 20% in the building industry, 17% in transportation, 11% in the electrical industry, 7% for consumer's goods, and 7% in machine building (SAAGER, 1984).

Aluminum alloys are produced by melting, metal treatment, casting, extrusion and forging (LYLE and GRANGER, 1985). Uses are castings and wrought products, powder metallurgic products, alloying additions to metals, coatings on metals to prevent corrosion and oxidation, and reducing agents to produce metals from their oxides and fluorides (LYLE and GRANGER, 1985). In all these processes and uses emissions are possible. Commercial alloys are usually multicomponent systems. Copper, magnesium, manganese, and silicon increase the strength of aluminum (LYLE and GRANGER, 1985). Chromium, zirconium, and titanium may influence as-cast grain size and nature of recrystallization (LYLE and GRANGER, 1985). Metallic glass of the formula $Al_{90}Fe_5Ce_5$ has a very high strength (e.g., a tensile strength of 940 MPa) and is not brittle (TRUEB, 1988a). According to TRUEB (1989) Du Pont has developed a polycrystalline aluminum oxide fiber "FP" to reinforce non-iron metallic, ceramic, and plastic bonded materials.

Important inorganic aluminum compounds are aluminum sulfate (e.g., for water treatment), potassium alum (e.g., for tanning skins, mordants, coagulating agents, styptic pencils, cosmetics, and hardening and setting acceleration in the production of marble cement and alabaster plaster), sodium and barium aluminates (e.g., for water purification, acceleration of concrete solidification, production of synthetic zeolites, in the paper industry (to increase opacity) and in the enamel industry), and the Friedel-Crafts catalyst aluminum chloride (HELMBOLDT et al., 1985).

Organoaluminum compounds are mobile liquids, low-melting waxes, or crystalline solids, and may irritate and burn skin (breathing fumes should be avoided). They are mainly used as catalysts (ZIETZ, 1985) or the alkoxides for adjusting the viscosity of varnishes, to impregnate textiles, and for antitranspirants in cosmetics (HELMBOLDT et al., 1985).

Wastes from aluminum and its compounds become more and more problematic. Fortunately, the value of aluminum scrap is increasing, because recycling allows to save electricity (SANDER, 1986). In the USA (1982) 33% of the required aluminum is produced by recycling (SAAGER, 1984), in the Federal Republic of Germany 38% in 1988. There is a certain tendency to replace aluminum by steel, titanum, magnesium, wood, paper, and/or plastics (SAAGER, 1984). Aluminum containing ashes from waste incineration have only restricted use as road construction material because the aluminum compounds tend to transformations and efflorescence (TOUSSAINT, 1989).

4 Distribution in the Environment, in Foods, and in Living Organisms

Sea water contains 0.13 to 1.1 ppb aluminum, lowest at about 1000 m depth, higher at the surface and in the deeper ocean (BRULAND, 1983). Salinity and biological processes also influence the concentration of dissolved aluminum (VAN BEUSEKOM, 1987). Some lakes, rivers, ground waters, and domestic tap water supplies contain aluminum in high concentrations, either naturally or because aluminum has been added as a flocculant in the purification process. Acidic rain markedly increases the "natural" aluminum content of water. For instance, HAINES (1988) has studied the mobilization from sediments into lakes. Solubility and interactions depend, of course, on speciation which is different for various pHs (see also NRIAGU et al., 1987). ÖHMAN and SJÖBERG (1988) have developed thermodynamic calculations for the aqueous aluminum system, also to predict identity and amount of solid phases. Aluminum oxide hydroxides form dublets and triplets, and complexes with various ligands in hydrogeochemical processes. The ubiquitous compounds (including aluminum oxides) influence many interactions and bind other metal hydroxides, pollutants, and nutrients at their surfaces (GOH et al., 1988). According to ALLARD et al. (1988) complexes of $Al(OH)SO_4$ must also be considered, when looking at the pH effects on accumulation and redistribution of other metals in polluted stream beds. GUBALA (1988) demonstrated that concentrations of dissolved aluminum ions do not vary very much because of saturation effects. Aluminum solubility of gibbsite is lowest at pH 5.7 (it increases below and above).

In soils aluminum exists almost exclusively in the form of silicates, hydroxides, and oxides. The solubility of these cmpounds is in the range of amorphous to crystalline gibbsite, $Al(OH)_3$ (BOLT and BRUGGENWERT, 1976). During soil acidification aluminum is released from silicates in ionic form. At pH values above 5.0 this aluminum becomes strongly bound as polymeric hydroxy cation at the surface of the silicate minerals. Exchangeable aluminum as well as the Al concentration in the soil solution are negligible. In the pH range from 5 to 4.2 aluminum ions occupy increasing fractions of exchangeable cations.

The concentration of aluminum in the soil solution remains still low, in agreement with the solubility of gibbsite. At pH values below 4.2 the solubility of the Al compounds is high enough so that Al ions can reach a considerable fraction of the total sum of cations in the soil solution (regarding speciation see also DRISCOLL, 1984; SKEFFINGTON, 1984; and TINKER and BARRACLOUGH, 1988).

Acid deposition has great influence on the chemistry of Al in soils. The deposition of sulfuric acid results in the formation of Al hydroxy sulfates (PRENZEL, 1983). In these reactions, the adsorption of sulfate on surface bound Al hydroxy compounds as well as precipitation may be involved. The reaction can be described by the equation

$$Al(OH)_3 + H_2SO_4 \rightarrow Al(OH)SO_4 + 2H_2O .$$

A considerable fraction of the sulfuric acid deposited since the beginning of industrialization has been stored in the soil as Al hydroxy sulfates. The accumulation of Al sulfates took place in the rooted subsoil where organic acids and thus organic Al complexes are negligible. At pH values approaching 4.2 and lower by the presence of other acids, e.g., HNO_3 (from the emission of NO_x as well as by nitrification in the soil), these sulfates dissolve again:

$$Al(OH)SO_4 + HNO_3 \rightarrow Al^{3+} + SO_4^{2-} + NO_3^- .$$

In acid seepages aluminum cations play a great role (MAYER, 1989): if soils are not able to buffer, more aluminum acidity leaves the system than acidity is introduced. There are more or less acidified subsoils next to each other.

By this process Al ion concentrations of 2 mmol/L can be reached in soil solutions. Such concentrations cause the die-back of root systems also of Al tolerant species. Since these processes occur in the rooted subsoil, where the Al sulfates have been accumulated, the deep reaching part of the root system of acid tolerant tree species such as beech and spruce is especially hit (ULRICH, 1989). The strong increase in the emission of NO_x in Central Europe since 1960 with its consequence of increased deposition of HNO_3 has thus triggered in acid forest soils the withdrawal of the active root system of the trees from the subsoil into the top soil. In the top soil of forest ecosystems, Al in soil solution is complexed by organic acids. At pH values below 4, the acid stress in top soils is due to protons and not to Al ions. By decreasing the uptake of Ca and Mg, and by changing the depth gradient of fine root distribution in soil, Al toxicity plays a central role in the present forest decline in

Central Europe. The driving force for this development is the deposition of acidity, resulting from the emission of SO_2 and NO_x (ULRICH, 1983).

Aluminum is also found in the air mainly as aluminosilicates associated with dust particles. In rural areas the atmosphere contains normally less than 0.5 µg/m^3 Al, in urban areas these concentrations are higher, and near cement plants they are often above 10 µg/m^3 (SORENSON et al., 1974). According to SEILER (1987) it seems that atmospheric concentrations in North America are higher than in Europe.

Since aluminum compounds are found everywhere, all plants (depending on the amounts available in the environment) and animals, and therefore feed and food, contain them, too. Some plants, such as orache, mustard, sweat leaf, ilex, and mosses, may accumulate aluminum compounds. However, the concentrations are not very high because of the poor solubilities of most of the aluminum compounds (see Chapters I.8 and I.13). Cereals, soil vegetables, meat, and liver contain aluminum in the range of about 5 mg/kg. Concentrations in dairy products and fish are much lower. In mammals increased concentrations have been found in bones and lungs (see also GELDMACHER-VON MALLINCKRODT, 1984, and HORNSTEIN, 1988). Minimal increases in food originate from aluminum cooking pots (see also Sect. 5). Human adults contain about 100 mg aluminum in their bodies (about 40% of it in the muscles, about 30% in the bones). Concentrations in blood plasma are discussed in Sect. 2.2. Aluminum concentrations in urine of healthy humans may be somewhat higher (mean values 7–17 µg/L, BURATTI et al., 1984; SCHALLER and VALENTIN, 1984), but urine is often contaminated.

5 Uptake, Absorption, Transport and Distribution, Metabolism and Elimination in Plants, Animals, and Humans

When Al ions enter plant cells, they can interfere with different enzymes and cause disturbances to the cell metabolism. Aluminum belongs, therefore, to the elements with high toxicity (FOY et al., 1978). Since the acidification of soils is a natural process, evolution led to the development of Al tolerance. For Al intolerant species, the very low Al concentrations in the soil solution in the pH range of 5 to 4.5 are already sufficient to cause severe root damage. Plants accustomed to acid soils possess mechanisms to prevent the influx of Al ions into root cells. This can be performed by forming organic Al complexes, or by binding or precipitating the Al ions in the cell wall of the root cortex. According to WOOD (1989) aluminum cations are, for instance, complexed to polysaccharides at algal cell surfaces, and additionally calcium ions protect against aluminum ions which were accidentally transported through the ATP-dependent Ca^{2+} channels. In the binding of Al ions, OH^- and phosphate ions may also be involved. The formation of Al hydroxy polymers is bound to pH values >5.0. If these pH values cannot be maintained, disturbances in cell wall formation and in the meristematic tissue of the root tip may occur. Increased ratios of Al to Ca and Mg ions in the soil solution decrease or even prevent the uptake of these nutrient cations by roots and their transport to the shoots. Al toxicity is the main cause for the change in species composition from neutral to acid soils. Even Al tolerant species may be severely stressed by continuous damage to the root system and unsufficient nutrient uptake. A continuous damage to the root system can be overcome by root regeneration, but it consumes photosynthases which are missing for production in the shoot. In acid tolerant tree species, fine root production can amount to half of net photosynthesis (PERSSON, 1980).

FOLKESON et al. (1987), KELSEY (1987), and STAM (1987) studied connections between locally different aluminum concentrations in soil solutions and uptake by trees and effects on roots. Especially in podzol (in which aluminum is less leached out than in brown forest soil), in soils where old vegetation was burned down, and in soils which were changed by acid precipitation, organic monomeric aluminum (in healthy soils 90–95% of the total aluminum) declines and more inorganic aluminum ions are present, relatively, which leads to changes in the rhizosphere and to decreased root growth of spruce and sugar maples.

FRICK and HERRMANN (1988) found that at low pH the major part of aluminum is deposited on/in the exuviae of lotic mayfly nymphs, and is thus left behind during moulting or emerging. But a small significant amount is internally accumulated and may be transferred to terrestrial predators, such as pied flycatchers. The uptake of aluminum by mussels has been studied, for instance, by MALLEY et al. (1988). For the uptake and distribution in fish see Sect. 6.3.

Animals and humans take up aluminum compounds by inhalation and/or gastrointestinal absorption, preterm infants eventually also from parenteral nutrition (the latter may lead to poorer renal function and to reduced aluminum excretion; HEWITT et al., 1987).

Regarding inhalation two occupational sources have been considered (GELDMACHER-VON MALLINCKRODT, 1984): the exposure to vapors and fumes during production of corundum from bauxite and pure alumina in an arc (corundum smelter disease), and to very fine aluminum dust (see, for instance, SCHALLER and VALENTIN, 1984; and HARWERTH et al., 1987). Inhalation of the latter may lead to acute metal-fume fever, chronic aluminosis, interstitial, non-nodular pulmonary fibrosis, chronic bronchitis, and eventually encephalopathy besides disturbances of blood coagulation (GELDMACHER-VON MALLINCKRODT, 1984).

There was indeed a report over twenty years ago (MCLAUGHLIN et al., 1962) of a ball-mill room worker who developed encephalopathy and pulmonary fibrosis associated with inhalation of aluminum dust. Increased aluminum concentrations in the urine of industrially exposed individuals have been documented (SJOGREN et al., 1983). A higher incidence of respiratory complaints has been reported among aluminum smelters workers (CHAN-YEUNG et al., 1983a). No significant effects have been found on musculoskeletal or hematological parameters in these workers (CHAN-YEUNG et al., 1983b). Studies of the incidence of cancers among aluminum industry workers have been inconclusive (ANDERSEN et al., 1982; ROCKETTE and ARENA, 1983). Most reviews of industrial aluminum exposure have concluded that the aluminum industry entails no greater occupational health hazard than other comparable industries (DEHAMEL, 1983; LÉONARD, 1988). A Lyon IARC working group (WHO, 1984) has also studied industrial exposure in aluminum production, but there is little evidence that the fumes (eventually some by-products from the baked carbon anodes) are carcinogenic to humans.

The impact of environmental aluminum on human health was comprehensively reviewed by SORENSON et al. (1974). These authors evaluated the total human exposure to aluminum from food, water, and atmospheric sources in a wide variety of geographical locations and concluded that the daily intake of aluminum ranges from 1 to 100 milligrams (mean: 5 mg). Absorption of aluminum from the intestinal tract and its subsequent effect on mineral metabolism has been studied in rats (BERLYNE et al., 1972; FEINROTH et al., 1982), dogs (HENRY et al., 1982), cows (KAPPEL et al., 1983), and rabbits (THORNTON et al., 1983). These animal studies, as well as human studies (GORSKY et al., 1979; GREGER and BAIER, 1983a), have demonstrated a small net positive balance of ingested aluminum in mammals. In one metabolic balance study it was reported that aluminum was retained only when large (>125 mg/day) amounts were ingested, and that there was no retention of aluminum at dietary intakes of 5 mg/day (GREGER and BAIER, 1983a). The effect of an increased dietary intake of aluminum on mineral metabolism in humans has been studied (SPENCER and LENDER, 1979; GREGER and BAIER, 1983b), but there is little agreement on the health effects of low to moderate exposure to aluminum.

Because the intestinal absorption of aluminum at high levels of intake has been documented, the propriety of aluminum cookware has been questioned (LEVICK, 1980). The aluminum content of food has been shown to increase after cooking in aluminum pots (TRAPP and CANNON, 1981; KONING, 1981; LIONE, 1984), but the contribution from this source to total dietary intake is minimal. There has been recent concern on the effect of acid rain on the environmental availability of aluminum and resulting increased human exposure (JACKSON and HUANG, 1983). There is some evidence of higher incidences of certain neurological disorders in geographical areas that are rich in environmental aluminum particularly in drinking water. Nonenvironmental sources of aluminum exposure include surgical implants, industry, and pharmaceuticals (DRUMMOND et al., 1983).

Aluminum containing pharmaceuticals are an important source of aluminum. Significant amounts of aluminum are found in antacid preparations, which may contain up to 700 mg of $Al(OH)_3$ containing 242 mg of elemental aluminum. This amount, in a single dose of antacid, represents about 50 times the average normal daily ingestion of aluminum from other sources.

Aluminum is absorbed from the gastrointestinal tract; normal subjects have a rise in serum concentration followed by excretion in the urine (MAURAS et al., 1982). It appears that less than 1 to 2 percent of an oral dose of aluminum might be absorbed. However, the chemical form (speciation) of the aluminum which is ingested appears to play an important role in this absorption process. In experimental rats the simultaneous administration of citrate has been reported to have a significant enhancing effect on the intestinal absorption of aluminum and its subsequent deposition in the cerebral cortex and bone (SLANINA et al., 1984). It could be proposed that the citrate facilitated the absorption of aluminum by the formation of a chelate complex.

The site of aluminum absorption in the gastrointestinal tract is not well defined although it could be proposed, from a comparison with other metals, that it takes place in either in the duodenum or proximal jejunum. In vitro studies using everted rat gut sac preparations have provided some evidence that the absorption of aluminum involves an energy-dependent, carrier-mediated mechanism (FEINROTH et al., 1982). Other workers (VANDERVOET and DE WOLFF, 1984) using in vivo perfusion of the rat small intestine have reported evidence that aluminum absorption may involve a two-step mechanism and proposed that it could be similar to the mechanism for iron absorption. The two-steps in the intestinal absorption of iron, which is an active process against a concentration gradient, are mucosal uptake and net transfer to either the serosal surface in vitro or to the blood stream in vivo (MANIS and SCHACHTER, 1962). It could be proposed that the intestinal absorption is through the iron transport mechanism in view of the role of transferrin in the transport of that metal in the blood compartment. It is possible that intestinal aluminum absorption may also involve the calcium transport mechanism and in support of this proposal are in vivo findings in the rat that vitamin D and its active metabolites increase aluminum absorption and result in significant retention (CHAN et al., 1986). The active intestinal absorption of calcium also involves a two-step mechanism in which 1,25-dihydroxycholecalciferol is the dominant factor (WILLS, 1973). It is, however, established that a number of other variables play important contributory roles in intestinal calcium absorption. MICHELL (1989) has discussed transduction in cells as a target for aluminum compounds which influences receptor systems.

Other factors could potentially play major roles in the intestinal absorption of aluminum in disease states. In studies on a group of eight uremic patients receiving a daily aluminum intake varying from 1.5 to 3.4 g it was reported that they absorbed 100 to 568 mg of aluminum daily (CLARKSON et al., 1972). The aluminum content of iliac bone biopsies increased in the two patients who absorbed the greatest amounts of aluminum. CAM and his colleagues (1976) proposed that the significant difference in intestinal aluminum absorption between normal subjects and patients with chronic renal failure could theoretically be due to an intrinsic abnormality of the gut wall in patients with uremia. Increased serum concentrations of parathyroid hormone are a recognized feature of the uremic state. The potential role of parathyroid hormone in the hyperaluminemia of chronic renal failure by increasing the intestinal absorption of aluminum is discussed in Sect. 6.2.

Aluminum metabolism has not been studied extensively mainly because of the lack of availability of a relatively stable radioisotope. In rats aluminum has been reported to accumulate in bone, brain, liver, and adrenal glands (ONDREICKA et al., 1966). In normal humans, the highest concentrations of aluminum are found in the lungs, bone, and muscle (SKALSKY and CARCHMAN, 1983; see also GELDMACHER-VON MALLINCKRODT, 1984). There has also been a report that brain aluminum concentrations increase with age (MCDERMOTT et al., 1979). It can be assumed that the kidneys are the predominant route of excretion in humans in view of the hyperaluminemia which occurs in oliguric or anuric patients with chronic renal failure. Parenteral administration of aluminum to dogs results in approximately one-third of an injected dose of aluminum chloride being excreted within two hours (KOVALCHIK et al., 1978). Biliary excretion is minimal.

6 Effects on Plants, Animals, and Humans
(see also SIGEL and SIGEL, 1988)

6.1 Effects on Microorganisms and Plants

Not many studies related to effects on microorganisms are known, and it rather seems unlikely that there are any (see also Sect. 6.3 and Chapter I.18); chromosome mutations were only observed in plants, insects, and mammals. Regarding plants (see also Sects. 4 and 5, and GELDMACHER-VON MALLINCKRODT, 1984) tolerance is possible, but also toxicity. The latter may be measured by comparing root lengths (PETERSON and GIRLING, 1981). Symptoms of aluminum toxicity are similar to those of calcium deficiency. Tolerance of some plant species may be inherited genetically and is related to their ability to absorb and use phosphorus (PETERSON and GIRLING, 1981). Another possible mechanism which reduces toxicity is the precipitation of soluble aluminum around roots (PETERSON and GIRLING, 1981). Despite high aluminum levels in some plants, such as the sweat leaf, no reports are known on animal toxicity because of browsing (PETERSON and GIRLING, 1981).

6.2 Miscellaneous Biochemical Effects
(see also ALFREY 1987, and Sections 5, 6.4, and 6.5)

As already indicated, aluminum speciation plays a great role in biochemical processes. CORAIN and BOMBI (1986, 1989) looked, for instance, at solubility, destabilization, and cumulative inhibition effects of various systems of aluminum compounds, depending on pH. Biologically inert inorganic aluminum ions may be coordinated to organic ligands, and perhaps abnormal accumulation is due to a carrier. JOSHI and CHO (1987) described the role of ferritin and transferrin in the transport and detoxification of metal ions. Long-term uptake of aluminum or Alzheimer's disease (see Sects. 6.4 and 6.5) influences ferritin concentrations (also in the brain) which causes dephosphorylation of proteins. Complexation of aluminum with inositol-phosphate seems to be the critical biochemical reaction (TRUEB, 1988b). HEWITT et al. (1987) studied erythrocyte aluminum and copper levels in aluminum overloaded patients and found that desferrioxamine-bound aluminum (see also Sect. 6.6) does not enter the red cells.

In patients with chronic renal failure, MAYOR et al. (1981) have proposed that parathyroid hormone may contribute to the hyperaluminemia by increasing intestinal absorption and by influencing tissue distribution. The interrelations between parathyroid hormone and the homeostasis of aluminum and calcium remain to be defined. In aluminum poisoning the secretion rate of parathyroid hormone has been reported by some workers to be either reduced (CANATA et al., 1983) or unchanged (BISWAS et al., 1982).

6.3 Acute and Chronic Effects on Animals and Humans
(see also GELDMACHER-VON MALLINCKRODT, 1984; BOURDON et al., 1986)

The ecophysiology of acid stress in aquatic organisms has been discussed by numerous experts, for instance, at a symposium in Antwerp (WITTERS and VANDERBORGHT, 1987). To understand fish toxicity (DIETRICH et al., 1988), the parameters of naturally occurring aluminum, solubility, speciation, acidification, gill morphology, physiology, and pH toxicity, must be considered. Water flows from the mouth through the gills (gill archs, gill lamellae with pavement and chloride cells (with a pical cavity for exchange of electrolytes). Below pH 5 gill permeability is high, there is a high Na^+ and Cl^- flux, and a slight activation of NH_4^+/Na^+ exchange. Aluminum toxicity has different mechanisms: below pH 5 pH toxicity is important, and extensive gill damage, mucification and circulating collapse are observed. Between pH 5 and 6.5 several mechanisms must be considered, including acidosis, but probably no mucification. Above pH 6.5 slight toxicity of polymeric $Al(OH)_3$, and rather high toxicity of aluminum ions occur. In the laboratory no effects were observed with rainbow trout at pH 5.2 in the absence of aluminum, whereas in the presence of aluminum heavy loss of electrolyte and extreme gill damage occur (DIETRICH et al., 1988). The same experts found also mitigating effects of sodium chloride on aluminum toxicity to brown trout in acidified lakes in granitic areas in the Ticino, Switzerland, (similar situations to those in Scandinavia). Critical for the fish is the loss of electrolytes which leads to reduced permeability of the gills. It seems that the ratio of $NaCl/Ca^{2+}$ concentrations in ambient water determines aluminum fish toxicity. GOOSSENAERTS et al. (1988) looked especially at effects of high aluminum concentrations on the gills of rainbow trout. They combined bulk-analysis (with ICP-ES), microanalysis (with LAMMA), and TEM (transmission electron microscopy; a TEM section may be perforated by a high-power laser beam). Aluminum is mostly found on the outer surface cells, less within the (dead) epithelial cells, and never in the inner (chloride) cells of the gills. BIRCHALL (1989) has discussed competition in the system aluminum/silicate/phosphate. Silicate binds aluminum much stronger in the gills than the complexing ligands. In fact, it protects in a more powerful way than other ions, because aluminum is kept out of the fish.

Little is known about acute and chronic effects on mammals, except in the cases after iatrogenic hemodialysis or after oral treatment with high amounts aluminum (as antacids), especially if calcium and magnesium uptake is deficient (see Sects. 6.4 and 6.5). Potential effects of aluminum inhalation are discussed in Sect. 5.

One of the first reports of poisoning which was attributed to aluminum was that of SPOFFORTH (1921). The patient, a metal worker, had memory loss, tremor, jerking movements, and in-coordination. Five years later attention was drawn to the potential health hazards of the aluminum present in city drinking water and various medicines, and possible industrial exposure (BETTS, 1926).

LÉONARD (1988) evaluated the mutagenic and carcinogenic potential of aluminum compounds. He discussed studies on animals, epidemiological studies (see also Sect. 5), mutagenicity tests in *Escherichia coli* and *Salmonella* (also with urine samples from workers exposed to aluminum), and the release of adenine by aluminum chloride, but found no indications for effects.

6.4 Dialysis Encephalopathy

There is now substantial evidence that, in patients with end-stage chronic renal failure managed by long-term intermittent hemodialysis, there is hyperaluminemia with accumulation of aluminum in various tissues. In the absence of renal function, aluminum cannot be eliminated in the urine. The excess in serum and tissue results from intestinal absorption of aluminum salts taken by mouth and from passage of aluminum across the dialysis membrane. The problem is that even if the plasma contains 90–140 μg Al per liter, only about 1% of the total body aluminum is present in the blood plasma. Even if 500 μg is transferred from plasma into the dialysate during 4 hours of hemodialysis, accumulated aluminum remains in other tissues.

The aluminum content of the dialysate depends on the content of the water with which it is made. Aluminum is normally present in raw water; the concentrations are usually low in ground waters and are almost always high in surface waters (MILLER et al., 1984).

A progressive fatal neurological syndrome in patients on long-term intermittent hemodialysis treatment for chronic renal failure was reported by ALFREY et al. (1972). The first manifestation in these patients was a speech disorder, followed by dementia, convulsions, and myoclonus. The syndrome was considered to be a metabolic disorder due to accumulation of trace metals. ALFREY and his colleagues (1976) subsequently proposed that this syndrome ("dialysis encephalopathy" or "dialysis dementia"), which arises after three to seven years' treatment, may be due to aluminum intoxication. Their proposal was based on the findings of increased aluminum content in brain, muscle, and bone tissues in affected patients. Dialysis encephalopathy was the major cause of death in one dialysis center (BURKS et al., 1976).

It was originally proposed that the increased tissue aluminum content in patients with the dialysis encephalopathy syndrome was derived from the oral aluminum hydroxide which had been administered therapeutically to control their serum phosphate. In support of this proposal was the evidence of a positive aluminum balance in patients with chronic renal failure on hemodialysis treatment who were taking oral aluminum-containing phosphate binding gels (CLARKSON et al., 1972). In one hemodialysis center the dialysis fluid was specifically shown to be the source of the aluminum (FLENDRIG et al., 1976). Most of the existing evidence is consistent with the proposal that the excess brain tissue aluminum in patients with dialysis encephalopathy stems from the dialysate, with some contribution from the gastrointestinal tract. That aluminum is the major toxic factor in the dialysis encephalopathy syndrome is now widely accepted (SIDERMAN and MANOR, 1982; MCKINNEY et al., 1982; DAVISON et al., 1982). Experts also agree that this syndrome has a common etiology with one of the osteomalacic components of dialysis osteodystrophy (ALFREY, 1978; PARKINSON et al., 1981). Hyperaluminemia and dialysis encephalopathy may occur not only in patients on hemodialysis but also in those on treatment with peritoneal dialysis (SMITH et al., 1980) and in some patients who have not been dialyzed (NATHAN and PEDERSEN, 1980; GRISWOLD et al., 1983). The non-dialyzed patients are usually children with renal failure who have been on oral treatment with aluminum hydroxide. Other workers have proposed that the etiology of dialysis encephalopathy syndrome is multifactorial (ARIEFF, 1979; PRIOR et al., 1982). Even if this is so, aluminum undoubtedly has a major role.

Nearly all patients with chronic renal failure have a normochromic normocytic anemia. In 1978 ELLIOTT and MACDOUGALL reported that anemia was associated with osteomalacic dialysis osteodystrophy and with dialysis encephalopathy. In patients with encephalopathy the hemoglobin concentration fell before the onset of neurological symptoms, and in the patients with osteodystrophy before the development of bone complications; therefore, anemia may be a useful early indication of aluminum toxicity. With regard to heme synthesis, aluminum toxicity has been linked with porphyria in laboratory animals, particularly in those with renal failure, and with hemodialysis related porphyria cutanea tarda (DISLER et al., 1982). In animals the aluminum induced porphyria is believed to be of hepatic origin. Soft-tissue or metastatic extraskeletal calcification is common in chronic renal failure. Calcification may occur in the eyes, in or around joints, in viscera, in the skin, and in arteries (WILLS, 1978). In view of the observations that aluminum can initiate the precipitation of calcium apatite (BACHRA and VAN HARSKAMP, 1970), aluminum may have a role in the pathogenesis of this complication of chronic renal failure.

In patients with normal renal function aluminum has recently been implicated as a factor in the etiology of the amyotrophic lateral sclerosis and Parkinsonism-dementia found in the indigenous (Chamorro) population of Guam (PERL et al., 1982). In brain tissue from two patients with the disease an accumulation of aluminum was demonstrated within the nuclear region and perikaryal cytoplasm of neurofibrillary tangle-bearing hippocampal neurons (PERL et al., 1982). Soil and drinking water from areas in which there was a high incidence of these disorders had a high aluminum content with low concentrations of calcium and magnesium. There is some evidence to support the hypothesis that secondary hyperparathyroidism, provoked by the chronic environmental deficiency of calcium and magnesium on Guam, may result in the increased intestinal absorption of aluminum, its subsequent deposition in the central nervous system, and the high incidence of amyotrophic lateral sclerosis and Parkinsonism-dementia (GARRUTO et al., 1984).

Aluminum has also been implicated as a neurotoxic agent in the pathogenesis of Alzheimer's disease. In brain tissue from patients with this disorder accumulations of aluminum have been identified within the nuclear region of a high percentage of neurons containing neurofibrillary tangles (PERL and BRODY, 1980; PERL et al., 1984). In the same patients, adjacent normal appearing neurons were virtually free of detectable aluminum (PERL and BRODY, 1980). The precise pathogenic role of aluminum in Alzheimer's disease is controversial and remains to be defined. It has been proposed that the accumulation of aluminum in the brain tissue of patients with Alzheimer's disease may be a secondary phenomenon rather than an etiological agent (SHORE and WYATT, 1983). In this concept the accumulation of aluminum would represent a relatively non-specific "marker" of neurons that are degenerating rather than be present as a specific etiological factor. The proposal of SHORE and WYATT (1983) is supported by the fact that the aluminum concentrations are not increased in specimens of serum and cerebrospinal fluid from patients with Alzheimer's disease. According to BIRCHALL (1989) there are possibly also interactions between silicate and aluminum in in vivo toxicity of the latter. Aluminum may bind to phosphates and inhibit yeast and brain hexokinase, but Si(OH)$_4$ reacts with aluminum ions and the solubility of aluminum silicate is lower (however, it takes time to precipitate).

6.5 Dialysis Osteodystrophy

Bone pain, as a consequence of metabolic bone disease, is a common symptom in chronic renal-failure patients on long-term intermittent hemodialysis. The progressive metabolic bone disease in these patients should be called "dialysis osteodystrophy" – a term that distinguishes it, and some aspects of its pathogenesis, from renal osteodystrophy in the undialyzed patient. The osteomalacic component of dialysis osteodystrophy is a major problem because it is associated with a high incidence of fractures. The reported incidence and rate of progression of the various components of dialysis osteodystrophy has varied not only between countries but also between dialysis centers within a country, despite apparently similar dialysis techniques (SIMPSON et al., 1973; TATLER et al., 1973). The characteristic of dialysis osteodystrophy is that it progresses or develops despite the maintenance of plasma calcium and magnesium, and their fractions, at concentrations which in a healthy person would not interfere with bone mineralization (VARGHESE et al., 1973; CHRISTOFFERSEN and CHRISTOFFERSEN, 1987; KUROKAWA et al., 1987). In the bone mineralization process the availability of phosphate, for the formation of hydroxyapatite crystals, is of crucial importance. Phosphate deficiency, with hypophosphatemia, is a recognized cause of osteomalacia; administration of aluminum containing medications can induce a state of phosphate deficiency.

The mechanism for the disordered bone formation induced by an excess of aluminum remains to be clarified. On the evidence of electron-probe X-ray microanalysis, analytical ion-microscopy, and a specific histochemical stain, the aluminum in hemodialysis patients with predominantly osteomalacia was mainly localized at the interface between the osteoid and the calcified matrix (COURNOT-WITMER et al., 1981).

6.6 Therapy of Aluminum Overload and Aluminum Monitoring
(see also Section 2.2)

The only method available for removing substantial quantities of aluminum is by chelation with desferrioxamine (DFO) (ACKRILL et al., 1980, 1982; BROWN et al., 1982; DAY et al., 1983; BERTHOLF et al., 1984; D'HAESE et al., 1988). Aluminum is bound strongly to DFO which has a molecular weight of 600 daltons, is ultrafiltrable (D'HAESE et al., 1988), and can be dialyzed readily. There is an acute rise in serum aluminum concentrations following DFO administration which is consistent with the mobilization of aluminum from tissues. It has been demonstrated that aluminum can be removed from both bone trabeculae and bone marrow by repeated DFO administration (ACKRILL et al., 1982; KUROKAWA et al., 1987; MARUMO et al., 1987).

Treatment of fracturing renal osteodystrophy by desferrioxamine provides symptomatic relief not only of osteomalacia but also of dialysis encephalopathy.

The infusion of DFO has recently been proposed as a diagnostic test for aluminum related osteodystrophy (MILLINER et al., 1984; BERLAND et al., 1985). MILLINER et al. (1984) evaluated the effect of a standard intravenous dose of DFO on plasma aluminum concentrations in 54 patients on hemodialysis treatment. Stainable bone aluminum, bone histologic findings, and bone aluminum content were studied. Baseline plasma aluminum concentrations of greater than 200 µg/L were associated with aluminum-related osteodystrophy (specificity, 95%), but concentrations of less than 200 µg/L did not exclude the diagnosis (sensitivity, 43%). After administration of DFO the increase in plasma aluminum concentration was 534 ± 260 (SD) and 214 ± 92 µg/L in patients with and without aluminum-related bone disease, respectively ($p < 0.001$), and correlated with the bone aluminum content ($r = 0.64$). An increment in plasma aluminum concentration of greater than 200 µg/L identified 35 of the 37 patients with aluminum-related osteodystrophy; sensitivity was 94% and specificity, 50%. These scientists concluded that the DFO infusion test was non-invasive, well tolerated, and of value particularly in excluding the diagnosis of aluminum-related osteodystrophy.

Recommendations of aluminum monitoring in hemodialysis programs have been made in a report of an international workshop held to study the hazards of aluminum toxicity related to renal insufficiency (SAVORY et al., 1983). The analytical reliability in risk management is, of course, very important (see, e.g., BERLIN et al., 1987; GRANDJEAN et al., 1987; NORDAL et al., 1987; SMEYERS-VERBEKE and VERBEELEN, 1987; TONGE and DAY, 1987; STEVENS et al., 1987; MAY and RAINS, 1987, who discussed interferences in aluminum determinations in indicators). It is also possible to control dialysis by computer calculation of aluminum speciation in solutions (HALL, 1989).

Reverse osmosis is the recommended method for water treatment since it provides water with a low aluminum content (< 10 µg/L), a low content of other cations, and eliminates organic contaminants which may contribute to the problems associated with hemodialysis.

Deionization of water is the next preferred mode of treatment, but it is recognized that it requires more careful monitoring since aluminum loading of the resin can occur with the danger of subsequent erratic elution.

Concentrations of water aluminum not exceeding 10 to 15 µg/L are necessary to minimize significant body uptake of aluminum during dialysis. These concentrations can be achieved using reverse osmosis and are recommended.

Between-batch differences in the aluminum concentration of dialysate concentrate exist. The recommendation is that the final concentration of the dialysate after dilution with treated water should be less than 15 µg/L and preferably less than 10 µg/L.

There is evidence that chronic ambulatory peritoneal dialysis (C.A.P.D.) treatment can remove some of the aluminum present in the plasma. The effectiveness of transfer of aluminum species during this type of treatment is of importance and needs further investigation. Because of these factors the concentration of aluminum in the C.A.P.D. fluid should be less than 10 µg/L.

In view of the problems with maintaining extremely low serum aluminum concentrations, especially with patients taking aluminum binders, the following criteria are recommended for the management of dialysis patients:

- Serum aluminum concentrations should never exceed 200 µg/L. Levels in excess of this concentration correlate with the development of dialysis encephalopathy or osteodystrophy.
- Concentrations over 100 µg/L should be viewed with concern and require careful monitoring.
- Concentrations of 60 to 100 µg/L are evidence of excessive build-up of aluminum in the body.

Hospital Hemodialysis. It is considered appropriate for the main water supply, the treated water, and the dialysis fluid to be monitored weekly. If a reverse osmosis system is used and has been shown to be reliable, then monthly monitoring should be adequate. The dialysis fluid to be monitored should be sampled immediately before it comes into contact with the membrane since contamination by metal parts of the system may occur. Serum aluminum concentration should be monitored in all dialysis patients four to six times each year.

Home Hemodialysis. It is considered appropriate for the main water supply, the treated water, the dialysis fluid, and serum aluminum concentration to be monitored when the patient comes for check-up (usually two or three times per year). If there are important fluctuations in the aluminum content of the main water supply and if reverse osmosis is not employed, then more frequent monitoring is desirable.

Peritoneal Dialysis and C.A.P.D. Serum aluminum should be monitored every two or three months.

7 Hazard Evaluation and Limiting Concentrations

Environmental contact with aluminum compounds cannot be avoided since it is the third common element in the earth's crust. However, the potential toxicity to healthy living organisms is normally low except for some plants, some aquatic animals, and for human patients with chronic renal failure (which may reduce elimination in the urine).

Drinking water for animals should contain less than 5 mg Al per L (HAPKE, 1975), and for humans less than 0.2 mg/L (E.C., 1980). Water used in hemodialysis should have aluminum concentrations lower than 30 µg/L.

In the Federal Republic of Germany the maximum concentration at the workplace is 6 mg/m^3 for aluminum metal, aluminum oxide, aluminum hydroxide, and for aluminum oxide fume as fine dust (MAK, 1987). The biological tolerance value has been determined to be 170 µg Al per liter urine, at the end of shift (MAK, 1987). As discussed in Sect. 2.2 aluminum concentrations in human blood should not exceed 10 µg/L plasma (SCHALLER and VALENTIN, 1984, 1985).

In Switzerland special attention is turned to the toxicity and the corroding effects of aluminum bromide and aluminum chloride. In the Federal Republic of Germany impairments of the respiratory tract by aluminum and its compounds are also notifiable occupational diseases.

References

Ackrill, P., Ralston, A. J., Day, J. P., and Hodge, K. C. (1980), *Successful Removal of Aluminum from a Patient with Dialysis Encephalopathy. Lancet 1,* 692–693.
Ackrill, P., Day, J. P., Garstang, F. M., Hodge, K. C., Metcalfe, P. J., Benzo, Z., Hill, K., Ralston, A. J., Ball, J., and Denton, J. (1982), *Treatment of Fracturing Renal Osteodystrophy by Desferrioxamine. Proc. Eur. Dial. Transplant Assoc. 19,* 203–207.
Alfrey, A. C. (1978), *Dialysis Encephalopathy Syndrome. Annu. Rev. Med. 29,* 93–98.
Alfrey, A. C. (1987), *Aluminum Metabolism and Toxicity in Uremia,* in: Brown, S. S., and Kodama, Y. (eds.): *Toxicology of Metals,* pp. 201–210. Ellis Horwood Ltd., Chichester, UK.
Alfrey, A. C., Mishell, J. M., Burks, J., Contiguglia, S. R., Rudolph, H., Lewin, E., and Holmes, J. H. (1972), *Syndrome of Dyspraxia and Multifocal Seizures Associated with Chronic Hemodialysis. Trans. Am. Soc. Artif. Intern. Organs 18,* 257–261.
Alfrey, A. C., LeGendre, G. R., and Kaehny, W. D. (1976), *The Dialysis Encephalopathy Syndrome. Possible Aluminum Intoxication. N. Engl. J. Med. 294,* 184–188.
Allard, B., Karlsson, S., and Hakansson, K. (1988), *Effects of pH on the Accumulation and Redistribution of Metal Compounds in a Polluted Stream Bed. Proceedings of the Conference on Trace Metals in Lakes, Hamilton (Canada). Sci. Total Environ.,* in press.
Andersen, A., Dahlberg, B. E., Magnus, K., and Wannag, A. (1982), *Risk of Cancer in the Norwegian Aluminum Industry. Int. J. Cancer 29,* 295–298.
Arieff, A. I., Cooper, J. D., Armstrong, D., and Lazarowitz, V. C. (1979), *Dementia, Renal Failure and Brain Aluminum. Ann. Intern. Med. 90,* 741–747.
Bachra, B. N., and Van Harskamp, G. A. (1970), *The Effect of Polyvalent Metal Ions on the Stability of a Buffer System for Calcification in vitro. Calcif. Tissue Res. 4,* 359–365.
Berland, Y., Chorhon, S. A., Olmer, M., and Meunier, P. S. (1985), *Predictive Value of Desferrioxamine Infusion Test for Bone Aluminum Deposits in Hemodialyzed Patients. Nephron 40,* 433–435.
Berlin A., Mattiello, G., Taylor, A., and Weber, J. P. (1987), *The Importance of Analytical Reliability in Risk Management: Aluminum and Dialysis Patients. Trace Elements in Human Health and Disease, Odense.* Symposium Report, pp. 37/38. *Environ. Health 26.* WHO Regional Office for Europe, Copenhagen.
Berlyne, G. M., Yagil, R., Ben-Ari, J., Weinberger, G., Knopf, E., and Danovitch, G. M. (1972), *Aluminium Toxicity in Rats. Lancet 1,* 564–567.
Bertholf, R. L., Roman, J. M., Brown, S., Savory, J., and Wills, M. R. (1984), *Aluminum Hydroxide-induced Osteomalacia, Encephalopathy, and Hyperaluminemia in CAPD. Treatment with Desferrioxamine. Periton. Dial. Bull. 4,* 30–32.
Bettinelli, M., Baroni, U., Fontana, F., and Poisetti, P. (1985), *Evaluation of the L'vov Platform and Matrix Modification for the Determination of Aluminum in Serum. Analyst 110,* 19–22.
Betts, C. T. (1926), *Aluminum Poisoning.* Research Publishing Co., Toledo, Ohio.
Birchall, J. D. (1989), *Aluminum Toxicity and in vivo Chemistry.* Lecture Robens Institute on Toxicity of Newer Metals, University of Surrey Guildford, UK;
see also Birchall, J. D., and Espie, A. W. (1986), *Biological Implications (via Silanol Groups) of Silicon with Metal Ions. Silicon Biochemistry, Ciba Foundation Symposium 121,* pp. 140–159, Wiley Chichester;

Birchall, J. D., and Chappell, J. S. (1987, 1988, 1989), *The Solution Chemistry of Aluminum and Silicon and its Biological Significance. Proceedings Second International Symposium on Geochemistry and Health* (ed. I. Thornton). Science Reviews Ltd., Northwood, Middlesex UK, in press;
The Chemistry of Aluminum and Silicon within the Biological Environment. Royal Society of Chemistry Monograph, UK, in press;
The Chemistry of Aluminum and Silicon in Relation to Alzheimer's Disease. Clin. Chem. 34 (2), 265–267;
Aspects of the Interaction of Silicic Acid with Aluminum in Dilute Solution and its Biological Significance. Inorg. Chim. Acta 153, 1–4;
Acute Toxicity of Aluminum to Fish Eliminated in Silicon-rich Acid Waters. Nature 338, 146–148;
Aluminum, Chemical Physiology, and Alzheimer's Disease. Lancet, October 29, 1988, 1008–1010, Lancet, April 29, 1989, 953.

Biswas, C. K., Arze, R. S., Ramos, J. M., Ward, M. K., Dewar, J. H., Kerr, D.N.S., and Kenward, D. H. (1982), *Effect of Aluminum Hydroxide on Serum Ionized Calcium, Immunoreactive Parathyroid Hormone, and Aluminum in Chronic Renal Failure. Br. Med. J. 284,* 776–778.

Bolt, G. H., and Bruggenwert, M. G. M. (1976), *Soil Chemistry a: Basic Elements.* Elsevier, Amsterdam.

Bourdon, R., Drueke, T., and Petit, L. (1986), *Aluminum: Pathologie, Biochemie, Exploration* (in French), 92 pages, *Cahiers de Toxicologie Clinique et Expérimentale No. 1.* Association Lyonnaise de Médécine légale, Editions Alexandre Lacassagne.

Brown, D. J., Dawborn, J. K., Ham, K. N., and Xipell, J. M. (1982), *Treatment of Dialysis Osteomalacia with Desferrioxamine. Lancet 2,* 343–345.

Brown, S., Bertholf, R. L., Wills, M. R., and Savory, J. (1984), *Electrothermal Atomic Absorption Spectrometric Determination of Aluminum in Serum with a New Technique for Protein Precipitation. Clin. Chem. 30,* 1216–1218.

Bruland, K. W. (1983), *Trace Elements in Seawater. Chem. Oceanogr. 8,* 177–179.

Buratti, M., Caravalli, G., Calzaferri, G., and Colombi, A. (1984), *Determination of Aluminum in Body Fluids by Solvent Extraction and Atomic Absorption Spectroscopy with Electrothermal Atomization. Clin. Chim. Acta 141,* 253–259.

Burks, J. S., Alfrey, A. C., Huddlestone, J., Norenberg, M. D., and Lewin, E. (1976), *A Fatal Encephalopathy in Chronic Haemodialysis Patients. Lancet 1,* 764–768.

Cam, J. M., Luck, V. A., Eastwood, J. B., and De Warderner, H. E. (1976), *The Effect of Aluminium Hydroxide Orally on Calcium, Phosphorus and Aluminium Metabolism in Normal Subjects. Clin. Sci. Mol. Med. 51,* 407–414.

Canata, J. B., Briggs, J. D., Junor, B.J.R., Fell, G. S., and Beastall, G. (1983), *Effect of Acute Aluminium Overload on Calcium and Parathyroid-hormone Metabolism. Lancet 1,* 501–503.

Chan, J. C. M., Jacob, M., Brown, S., Savory, J., and Wills, M. R. (1986), *Aluminum Metabolism in Rats. Effects of Vitamin D, Dihydrotachysterol, 1,25-Dihydroxyvitamin D and Phosphate-binders. Nephron,* in press.

Chan-Yeung, M., Wong, R., Maclean, L., Tan, F., Schulzer, M., Enarson, E., Martin, A., Dennis, R., and Grzybowski, S. (1983 a), *Epidemiologic Health Study of Workers in an Aluminum Smelter in British Columbia: Effects on the Respiratory System. Annu Rev. Respir. Dis. 127,* 465–469.

Chan-Yeung, M., Wong, R., Tan, F., Enarson, D., Schulzer, M., Subbarao, K., Knickerbocker, J., and Grzybowski, S. (1983 b), *Epidemiologic Health Study of Workers in an Aluminum Smelter in Kitimat, B.C. II. Effects on Musculoskeletal and Other Systems. Arch. Environ. Health 38,* 34–40.

Christoffersen, J., and Christoffersen, M. R. (1987), *Physico-chemical Aspects of Aluminum Ion Induced Bone Diseases. Symposium Abstract F1, Odense.* WHO, Regional Office for Europe, Copenhagen.

Clarkson, E. M., Luck, V. A., Hynson, W. V., Bailey, R. R., Eastwood, J. B., Woodhead, J. S., Clements, V. R., O'Riordan, J. L. H., and DeWardener, H. E. (1972), *The Effect of Aluminium Hydroxide on Calcium, Phosphorus and Aluminium Balances, the Serum Parathyroid Hormone Concentrations and the Aluminum Content of Bone in Patients with Chronic Renal Failure. Clin. Sci. 43*, 519–531.

Corain, B., and Bombi, G. G. (1986, 1989), *Aluminum Coordination Compounds in Biological Investigations: Novel Physiopathological Perspectives? Proceedings SECOTOX-Symposium, Rome. Proceedings 3rd IAEAC Workshop on Toxic Metal Compounds (Interrelation between Chemistry and Biology), Follonica (Italy). J. Toxicol. Environ. Chem. 22* (1–4), 149–157; and *Int. J. Environ. Anal. Chem. 36* (1), 13–16.

Cournot-Witmer, G., Zingraff, J., Plachot, J. J. et al. (1981), *Aluminum Localization in Bone from Hemodialyzed Patients: Relationship to Matrix Mineralization. Kidney Int. 20*, 375–385.

Courtyn, E., Vandercasteele, C., and Dams, R. (1989), *Specification of Aluminum in Acid Surface Waters. Proceedings Conference Heavy Metals in the Environement, Geneva*, Vol. 2, pp. 391–395. CEP Consultants Ltd., Edinburgh.

Davison, A. M., Walker, G. S., Ott, H., and Lewis, A. M (1982), *Water Supply Aluminium Concentration, Dialysis Dementia, and Effect of Reverse-osmosis Water Treatment. Lancet 2*, 785–787.

Day, J. P., Ackrill, P., Garstand, F. M., Hodge, K. C., and Metcalfe, P. J. (1983), *Reduction of the Body Burden of Aluminium in Renal Patients by Desferrioxamine Chelation Therapy*, in: Brown, S. S., and Savory, J. (eds.): *Chemical Toxicology and Clinical Chemistry in Metals*, pp. 353–356. Academic Press, London.

DeHamel, F. (1983), *Aluminum Smelters and Health. N. Z. Med. J. 1*, 123.

d'Haese, P. C., Lamberts, L. V., and de Broe, M. E. (1988), *Aluminum Accumulation/Toxicity in Dialysis Patients: Monitoring, Diagnosis and Therapy. Proceedings 3rd IAEAC Workshop on Toxic Metal Compounds (Interrelation between Chemistry and Biology), Follonia (Italy). J. Toxicol. Environ. Chem. 23* (1–4), 17–25.

Dietrich, D., Schlatter, C., Blau, N., and Fischer, M. (1988), *Aluminum and Acid Rain: Mitigating Effects of NaCl on Aluminum Toxicity to Brown Trout in Acid Water. Proceedings 3rd IAEAC Workshop on Toxic Metal Compounds (Interrelation between Chemistry and Biology), Follonica (Italy). J. Toxicol. Environ. Chem. 19* (1+2), 17–23.

Disler, P., Day, R., Burman, N., Bickenhorst, G., and Eales, L. (1982), *Treatment of Hemodialysis-related Porphyria Cutanea Tarda with Plasma Exchange. Am. J. Med. 72*, 989–993.

Driscoll, C. T. (1984), *Dissolution of Soil-bound aluminum in an Experimental Forest. VDI-International Conference on Origin and Effects of Acid Precipitation, Lindau*. See also *Chemosphere 13*(4), N23.

Drummond, J. L., Simon, M. R., Woodman, J. L., and Brown, S. D. (1983), *Aluminum Ion Deposition in Rat Tissues Following Implantation of a Ceramic Metal Disc. Biomater. Med. Devices Artif. Organs 11*, 147–159.

E.C. (European Community) (1980), *Council Directive* of 15 July, 1980, relating to the quality of water intended for human consumption. OJ No. L 229, 30 August, 1980, p. 11.

Elliott, H. L., and MacDougall, A. I. (1978), *Aluminium Studies in Dialysis Encephalopathy. Proc. EDTA 15*, 157–163.

Feinroth, M., Feinroth, M. V., and Berlyne, G. M. (1982), *Aluminum Absorption in the Rat Everted Gut Sac. Miner. Electrolyte Metab. 8*, 29–35.

Fleischer, M., and Schaller, K. H. (1984), *Analytical Reliable Determination of Aluminum in Serum and Plasma by Graphite Furnace AAS* (in German), in: Welz, B. (ed.): *Fortschritte in der atomspektrometrischen Spurenanalytik*, pp. 297–307. Verlag Chemie, Weinheim-Deerfield Beach/Florida-Basel.

Flendrig, J. A., Kruis, H., and Das, H. A. (1976), *Aluminium and Dialysis Dementia. Lancet 1*, 1235.

Folkeson, L., Betgkvist, B., and Olsson, K. (1987), *Metal Fluxes in Picea albis, Fagus sylvatica and Betula pendula Forest Ecosystems. Proceedings Conference Heavy Metals in the Environment, New Orleans*, Vol. 2, pp. 407–409. CEP Consultants Ltd., Edinburgh.

Foy, C. D., Chaney, R. L., and White, M. C. (1978), *The Physiology of Metal Toxicity in Plants. Annu. Rev. Plant Physiol. 29*, 511–566.

Frank, W. B., Haupin, W. E., Granger, D. A., Wei, M. W., Calhoun, K. J., and Bonney, T. B. (1985), *Aluminum,* in: *Ullmann's Encyclopedia of Industrial Chemistry,* 5th Ed., Vol. A1, pp. 459–480. VCH Verlagsgesellschaft, Weinheim-Deerfield Beach/Florida-Basel.

Frick, K. G., and Herrmann, J. (1988), *Aluminum Accumulation in a Lotic Mayfly at Low pH. Proceedings 1st European SECOTOX Conference on Ecotoxicology, Copenhagen,* in press.

Garruto, R. M., Fukatsu, R., Yanagihara, R., Gajdusek, D. C., Hook, G., and Fiori, C. E. (1984), *Imaging of Calcium and Aluminum in Neurofibrillary Tangle-bearing Neurons in Parkinsonism-dementia of Guam. Proc. Natl. Acad. Sci. USA 81*, 1875–1879.

Geldmacher-von Mallinckrodt, M. (1984), *Aluminum* (in German), in: Merian, E. (ed.): *Metalle in der Umwelt,* pp. 301–308. Verlag Chemie, Weinheim-Deerfield Beach/Florida-Basel.

Goh, T. B., Racz, G. J., and Cheslock, M. M. (1988), *Oxides as Sources and Sinks of Nutrients and Pollutants in the Soil Environment. Proceedings 1st Soil Residue Analysis Workshop, Winnipeg. Environ. Anal. Chem. 34* (3), 227–237.

Goossenaerts, C., Witters, H., Jacob, W., Vanderborght, O., and Van Grieken, R. (1988), *A Microanalytical Study of the Gills of Aluminum-exposed Rainbow Trout. Proceedings 3rd IAEAC Workshop on Toxic Metal Compounds (Interrelation between Chemistry and Biology). Int. J. Environ. Anal. Chem. 34* (3), 227–237.

Gorsky, J. E., Dietz, A. A., Spencer, H., and Osis, D. (1979), *Metabolic Balance of Aluminum Studied in Six Men. Clin. Chem. 25*, 1739–1743.

Grandjean, P., Hørder, M., and Thomassen, Y. (1987), *Indicators of Aluminum Exposure and Mineral Metabolism in Cryolite Workers. Symposium Abstract B9, Odense.* WHO Regional Office for Europe, Copenhagen.

Greger, J. L., and Baier, M. J. (1983a), *Excretion and Retention of Low or Moderate Levels of Aluminum by Human Subjects. Food Chem. Toxicol. 21*, 473–477.

Greger, J. L., and Baier, M. J. (1983b), *Effect of Dietary Aluminum on Mineral Metabolism of Adult Males. Am. J. Clin. Nutr. 38*, 411–419.

Griswold, W. R., Reznik, V., Mendoza, A., Trauner, D., and Alfrey, A. C. (1983), *Accumulation of Aluminum in a Non-dialyzed Uremic Child Receiving Aluminum Hydroxide. Pediatrics 71*, 56–58

Gubala, C. P. (1988), *The Relationships between Lead, Aluminum and Zinc in the Sediments of a Dilute Acidic Lake. Proceedings of the Conference on Trace Metals in Lakes, Hamilton (Canada). Sci. Total Environ.,* in press.

Haines, T. A. (1988), *Interrelation of Concentration of Selected Trace Metals in the Gill, Viscera, and Carcass of White Sucker from Soft Water Lakes in Maine, USA. Proceedings of the Conference on Trace Metals in Lakes, Hamilton (Canada). Sci. Total Environ.,* in press.

Hall, S. B. (1989), *The Computer-assisted Chemical Specification of Intravenous Nutrition Fluids. Lecture Robens Institute on Toxicity of Newer Metals,* University of Surrey, Guildford, UK.

Hapke, H.-J. (1975), *Toxikologie für Veterinärmediziner* (in German). Enke, Stuttgart.

Harwerth, A., Kufner, G., and Helbing, F. (1987), *Examination of Workers Occupationally Exposed to Aluminum Dust* (in German). *Arbeitsmed. Sozialmed. Präventivmed. 22*, 2–5.

Hemboldt, O., Hudson, L. K., Stark, H., and Danner, M. (1985), *Inorganic Aluminum Compounds,* in: *Ullmann's Encyclopedia of Industrial Chemistry,* Vol. A1, pp. 527–541. VCH Verlagsgesellschaft, Weinheim-Deerfield Beach/Florida-Basel.

Henry, D. A., Nudelman, R. K., DiDomenico, N. C., Stanley, T. M., Alfrey, A. C., Goodman, W. G., Slatotolsky, E., and Coburn, J. W. (1982), *Metabolic and Toxic Effects of Aluminum in the Dog. Kidney Int. 21*, 229.

Hewitt, C. D., Day, J. P., and others (1987), *Aluminum Exposure of Preterm Infants Receiving Formula Feeds and Parenteral Nutrition. Erythrocyte Aluminum and Copper Levels in Aluminum Overloaded Patients. Symposium Abstracts H10 and H11, Odense.* WHO Regional Office for Europe, Copenhagen.

Hornstein, C. (1988), *Determination of Aluminum Concentrations in Human Organs* (Thesis, in German). Friedrich Alexander-Universität, Erlangen.
Hudson, L. K., Misra, C., and Wefers, C. (1985), *Aluminum Oxide*, in: *Ullmann's Encyclopedia of Industrial Chemistry*, 5th Ed., Vol. A1, pp. 557–594. VCH Verlagsgesellschaft, Weinheim-Deerfield Beach/Florida-Basel.
Jackson, M. L., and Huang, P. M. (1983), *Aluminum of Acid Soils in the Food Chain and Senility. Sci. Total Environ. 28*, 269–276.
Joshi, Y. G., and Cho, S. (1987), *Transferrin, Ferritin and Enzymes Sensitive to Metal Ions.* Platform Presentation, Symposium Trace Elements in Human Health and Disease, Odense. *Environ. Health 20*, 51–54; *26*, 12–13. WHO Regional Office for Europe, Copenhagen.
Kappel, L. C., Youngberg, H., Ingraham, R. H., Hembry, F. G., Robinson, D. L., and Cherney, J. H. (1983), *Effects of Dietary Aluminum on Magnesium Status of Cows. Am. J. Vet. Res. 44*, 770–773.
Kelsey, P. D. (1987), *Autogenic Soil Aluminum Concentrations and Sugar Maple Establishment in Some Mature Hardwood Forests in Northern Illinois. Proceedings Conference Heavy Metals in the Environment, New Orleans*, Vol. 2, pp. 413–415. CEP Consultants Ltd., Edinburgh.
Koning, J. H. (1981), *Aluminum Pots as a Source of Dietary Aluminum. N. Engl. J. Med. 304*, 172.
Kovalchik, M. T., Kaehing, W. D., Hegg, A. P., Jackson, J. T., and Alfrey, A. C. (1978), *Aluminum Kinetics During Hemodialysis. J. Lab. Clin. Med.*, 712–720.
Kurokawa, K., Marumo, F., Ogura, Y., Ono, T., and Suzuki, M. (1987), *Aluminum-associated Bone Disease: Considerations on Pathogenesis and Effects of Desferroxamine Treatment*, in: Brown, S., and Kodama, Y. (eds.): *Toxicology of Metals*, pp. 211–217. Ellis Horwood Ltd., Chichester, UK.
Léonard, A. (1988), *Mutagenic and Carcinogenic Potential of Aluminum and Aluminum Compounds. Proceedings 3rd IAEAC Workshop on Toxic Metal Compounds (Interrelation between Chemistry and Biology). J. Toxicol. Environ. Chem. 23* (1–4), 27–31.
Leung, F. Y., and Henderson, A. R. (1982), *Improved Determination of Aluminum in Serum and Urine with Use of a Stabilized Temperature Platform Furnace. Clin. Chem. 28*, 2139–2143.
Levick, S. E. (1980), *Dementia from Aluminum Pots. N. Engl. J. Med. 303*, 164.
Lione, A. (1984), *Aluminum in Food. Nutr. Rev. 42*, 31.
Lyle, J. P., and Granger, D. A. (1985), *Aluminum Alloys*, in: *Ullmann's Encyclopedia of Industrial Chemistry*, Vol. A1, pp. 481–526. VCH Verlagsgesellschaft, Weinheim-Deerfield Beach/Florida-Basel.
MAK (1987), *Maximum Concentrations at the Workplace and Biological Tolerance Values for Working Materials, Report No. XXIII.* DFG. VCH Verlagsgesellschaft, Weinheim-Basel-Cambridge-New York.
Malley, D. F., Hueber, J. D., and Donkersloot, K. (1988), *Effects on Ionic Composition of Blood and Tissues of Bivalves of an Addition of Aluminum and Acid to a Lake. Arch. Environ. Contam. Toxicol. 17*, 479–491.
Manis, J. G., and Schachter, D. L. (1962), *Active Transport of Iron by Intestine: Features of the Two-step Mechanism. Am. J. Physiol. 203*, 73–80.
Marumo, F., Umetani, N., Sato, N., and Takahashi, Y. (1987), *Desferroxamine (DFO): The Effectiveness for Non-invasive Diagnosis of Aluminum-related Bone Disease, and for Treatment of the Disease*, in: Brown, S. S., and Kodama, Y. (eds.): *Toxicology of Metals*, pp. 219–231. Ellis Horwood Ltd., Chichester, UK.
Mauras, Y., Allain, P., and Riberi, P. (1982), *Etude de l'absorption digestive de l'hydrocarbonate d'aluminium chez l'individu sain. Therapie 37*, 593–605.
May, J. C., and Rains, T. C. (1987), *Determination of Aluminum in Source Plasma. Symposium Abstract H9, Odense.* WHO Regional Office for Europe, Copenhagen.
Mayer, R. (1989), *The Impact of Atmospheric Acid Deposition on Soil and Vegetation. Proceedings Conference Heavy Metals in the Environement, Geneva*, Vol. 2, pp. 19–27. CEP Consultants Ltd., Edinburgh.

Mayor, G. B., Sprague, S. M., and Sanchez, T. V. (1981), *Determinants of Tissue Aluminum Concentration. Am. J. Kidney Dis. 1*, 141–145.

McDermott, J. R., Smith, I. A., Iqbal, K., and Wisniewski, H. M. (1979), *Brain Aluminum in Aging and Alzheimer Disease. Neurology 29*, 809–814.

McKinney, T. D., Basinger, M., Dawson, E., and Jones, M. M. (1982), *Serum Aluminum Levels in Dialysis Dementia. Nephron 32*, 53–56.

McLaughlin, A. I. G., Kazantzis, G., King, E., Teare, D., Porter, R. J., and Owen, R. (1962), *Pulmonary Fibrosis and Encephalopathy Associated with the Inhalation of Aluminum Dust. Br. J. Ind. Med. 19*, 253–263.

Michell, R. H. (1989), *Signal Transduction in Cells as a Target for Toxic Compounds. Plenary Lecture Vth ICT '89. Brighton, UK*, Proceedings in press, Taylor & Francis, London.

Miller, R. G., Kopfler, F. C., Kelty, K. C., Stober, J. A., and Ulmer, N. S. (1984), *The Occurrence of Aluminum in Drinking Water. J. Am. Water Works Assoc. 76*, 84–91.

Milliner, D. S., Nebeker, H. G., Ott, S. M., Andress, D. L., Sherrard, D. J., Alfrey, A. C., Slatopolsky, E. A., and Coburn, J. W. (1984), *Use of the Deferrioxamine Infusion Test in the Diagnosis of Aluminum-related Osteodystrophy. Am. Intern. Med. 101*, 775–780.

Möller, J. (1986), *Determination of Aluminum in Water and Soils. Tecator AB, FIAstar Newsletter No. 2,* Analytica '86, see also *Chemosphere* 16(5), N16/17.

Nathan, E., and Pedersen, S. (1980), *Dialysis Encephalopathy in a Non-dialysed Uraemic Boy Treated with Aluminum Hydroxide Orally. Acta Paediatr. Scand. 69*, 793–796.

Nordal, K. P., Dahl, E., Thomassen, Y., Brodwall, E. K., and Halse, J. (1987), *Seasonal Variations in Serum Aluminum: Are They Determined by Water Borne Agents? Renal Aluminum Clearance During Seasonal and Experimental Increases in Serum Aluminum Concentrations. Symposium Abstracts C10 and H7, Odense.* WHO Regional Office for Europe, Copenhagen .

Nriagu, J. O., Wong, H. K. T., and LaZerte, B. D. (1987), *Aluminum Speciation in Pore Waters of Some Lake Sediments. Proceedings Conference Heavy Metals in the Environment, New Orleans,* Vol. 1, pp. 113–117. CEP Consultants Ltd., Edinburgh.

Öhman, L.-O., and Sjöberg, S. (1988), *Thermodynamic Calculations with Special Reference to the Aqueous Aluminum System,* in: Kramer, J. R., and Allen, H. E. (eds.): *Metal Speciation,* pp. 1–40. Lewis Publishers, Forest Grove, Oregon.

Ondreicka, R., Ginter, E., and Kortus, J. (1966), *Chronic Toxicity of Aluminium in Rats and Mice and its Effects on Phosphorus Metabolism. Br. J. Ind. Med. 23*, 305–312.

Parkinson, I. S., Ward, M. K., and Kerr, D. N. S. (1981), *Dialysis Encephalopathy, Bone Disease and Anemia: The Aluminium Intoxication Syndrome During Regular Haemodialysis. J. Clin. Pathol. 34*, 1285–1294.

Perl, D. P., and Brody, A. R. (1980), *Alzheimer's Disease: X-ray Spectrometric Evidence of Aluminum Accumulation in Neurofibrillary Tangle-earing Neurons. Science 208*, 297–299.

Perl, D. P., and Pendlebury, W. W. (1984), *Aluminum (Al) Accumulation in Neurofibrillary Tangle (NFT) Bearing Neurons of Senile Dementia Alzheimer's Type (SDAT); Detection by Intraneuronal X-ray Spectrometry Studies of Unstained Tissue Sections. J. Neuropathol. Exp. Pathol. 43*, 349.

Perl, D. P., Gajdusek, D. C., Garruto, R. M., Yanagihara, R. T., and Gibbs, C. J., Jr. (1982), *Intraneuronal Aluminum Accumulation in Amyotrophic Lateral Sclerosis and Parkinsonism-dementia of Guam. Science 217*, 1053–1055.

Persson, H. (1980), *Fine Root Dynamics in a Scots Pine Stand With and Without Near-optimum Nutrient and Water Regimes. Acta Phytogeogr. Suec. 68*, 101–110.

Peterson, P. J., and Girling, C. A. (1981), in: Lepp, N. W. (ed.): *Aluminum, in Effect of Heavy Metal Pollution,* Vol. 1, pp. 214–218. Applied Science Publishers, Barking, Essex, UK.

Prenzel, J. (1983), *A Mechanism for Storage and Retrieval of Acid in Acid Soils,* in: Ulrich, B., and Pankrath, J. (eds.): *Effects of Accumulation of Air Pollutants in Forest Ecosystems*, pp. 157–170. Reidel Publ. Co., Dordrecht.

Prior, J. C., Cameron, E. C., Nickerbocker, W. J., Sweeney, V. P., and Suchowersky, O. (1982), *Dialysis Encephalopathy and Osteomalacia Bone Disease. A Case-controlled Study. Am. J. Med. 72,* 33–42.
Rockette, H. E., and Arena, A. C. (1983), *Mortality Studies of Aluminum Reduction Plant Workers: Potroom and Carbon Department. J. Occup. Med. 25,* 549–557.
Røiset, O., and Sullivan, T. J. (1987), *Effects of Filtration and Centrifugation on the Determination of Aluminum in Water. Int. J. Environ. Anal. Chem. 29*(1+2), 141–149.
Saager, R. (1984), *Aluminum,* in: *Dictionary of Metallic Raw Materials* (in German), pp. 139–144. Bank von Tobel, Zürich.
Sadler, P. (1989), *Application of NMR in Studying Interactions of Metal Ions with Biological Systems. Lecture Robens Institute on Toxicity of Newer Metals,* University of Surrey, Guildford, UK.
Sander, G. (1986), *Aluminum Capacities* (in German), *SBV Der Monat 3*/86, 31–33, Basel.
Savory, J., and Wills, M. R. (1986), *Analytical Methods for Aluminum Measurement. Kidney Int. 29,* Suppl. 18, S24–S27.
Savory, J., Berlin, A., Courtoux, C., Yeoman, B., and Wills, M. R. (1983), Summary Report of an International Workshop on *The role of Biological Monitoring in the Prevention of Aluminum Toxicity in Man: Aluminum Analysis in Biological Fluids. Ann. Clin. Lab. Sci. 13,* 444–451.
Schaller, K. H., and Valentin, H. (1984), *Aluminum in Biological Indicators for the Assessment of Human Exposure to Industrial Chemicals,* EUR 8903 EN, pp. 20–27. Commission of the European Communties.
Schaller, K. H., and Valentin, H. (1985), *BAT-Werte, Arbeitsmedizinische Toxikologische Begründungen,* Vol. 1, pp. 1–13. VCH Verlagsgesellschaft, Weinheim-Deerfield Beach/Florida-Basel.
Seiler, H.-G. (1987), *The Significance of Analytical Determinations of Trace Elements in Biological Materials.* Lecture at the Basel Chemical Society.
Shore, D., and Wyatt, R. J. (1983), *Aluminum and Alzheimer's Disease. J. Nerv. Ment. Dis. 171,* 553–558.
Siderman, S., and Manor, D. (1982), *The Dialysis Dementia Syndrome and Aluminum Intoxication. Nephron 31,* 1–10.
Sigel, H., and Sigel, A. (1988), *Aluminum and Its Role in Biology,* Vol. 24 of the *Metal Ions in Biological Systems Series.* Marcel Dekker, New York-Basel.
Simpson, W., Kerr, D. N. S., Hill, A. V. L., and Siddiqui, J. Y. (1973), *Skeletal Changes in Patients on Regular Hemodialysis. Radiology 107,* 313–320.
Sjogren, B., Lundberg, I., and Lidums, V. (1983), *Aluminium in the Blood and Urine of Industrially Exposed Workers. Br. J. Ind. Med. 40,* 301–304.
Skalsky, H. L., and Carchman, R. A. (1983), *Aluminum Homeostasis in Man. J. Am. Coll. Toxicol. 2,* 405–423.
Skeffington, R. A. (1984), *Effects of Acid Precipitation on Nutrient Cycling.* VDI International Conference on Origin and Effects of Acid Precipitation, Lindau. See also *Chemosphere 13*(4), N23.
Slanina, P., Falkeborn, Y., Frech, W., and Cedergren, A. (1984), *Aluminum Concentrations in the Brain and Bone of Rats Fed Citric Acid, Aluminum Citrate and Aluminum Hydroxide. Food Chem. Toxicol. 22,* 391–397.
Smeyers-Verbeke, J., and Verbeelen, D. (1984), *Difficulties in the Determination of Aluminum in Dialysate Concentrates by Graphite Furnace AAS using the L'vov Platform.* Proceedings 10th International Symposium on Microchemical Techniques, Antwerp.
Smeyers-Verbeke, J., and Verbeelen, D. (1987), *Symposium Abstract H1, Odense.* WHO Regional Office for Europe, Copenhagen.
Smith, D. B., Lewis, J. A., Burks, J. S., and Alfrey, A. C. (1980), *Dialysis Encephalopathy in Peritoneal Dialysis. J. Am. Med. Assoc. JAMA 244,* 365–366.
Sorenson, J. R. J., Campbell, I. R., Tepper, L. B., and Lingg, R. D. (1974), *Aluminum in the Environment and Human Health. Environ. Health Perspect. 8,* 3–95.

Spencer, H., and Lender, M. (1979), *Adverse Effects of Aluminum-containing Antacids on Mineral Metabolism. Gastroenterologia 86,* 603–606.

Spofforth, J. (1921), Case of Aluminium Poisoning. *Lancet 1,* 1301.

Stam, A. C. (1987), *Acid Precipitation: Solution Chemistry of Aluminum in the Rhizosphere of Pinus taeda. Proceedings Conference Heavy Metals in the Environment, New Orleans,* Vol. 2, pp. 476–478. CEP Consultants Ltd., Edinburgh.

Stevens, B. J., Biddle, N., and Gill, R. J. (1987), *An Australasian Inter-laboratory Assessment of Serum Aluminum Measurement. Symposium Abstract H4, Odense.* WHO Regional Office for Europe, Copenhagen.

Sullivan, T. J., Seip, H. M., and Muniz, I. P. (1986), *A Comparison of Frequently Used Methods for the Determination of Aqueous Aluminum, Int. J. Environ. Anal. Chem.* 26(1), 61–75.

Sunderman, F. W., Jr., Marzouk, A., Crisostomo, M. C., and Weatherby, D. R. (1985), *Electrothermal Atomic Absorption Spectrophotometry of Nickel in Tissue Homogenates. Ann. Clin. Lab. Sci. 15,* 229–307.

Tatler, G. L. V., Baillod, R. A., Varghese, Z., Young, W. B., Farrow, S., Wills, M. R., and Moorhead, J. F. (1973), *Evolution of Bone Disease over 10 Years in 135 Patients with Terminal Renal Failure. Br. Med. J. 3,* 315–319.

Thornton, D. J., Liss, L., and Lott, J. A. (1983), *Effect of Ethanol on the Uptake and Distribution of Aluminum in Rats. Clin. Chem. 29,* 1281.

Tinker, P. B., and Barraclough, P. B. (1988), *Root-Soil Interactions,* in: Hutzinger, O. (ed.): *Handbook of Environmental Chemistry,* Vol. 2, Part D, *Reactions and Processes,* pp. 153–175. Springer Verlag, Heidelberg-New York.

Tonge, M. D., and Day, J. P. (1987), *Interferences in Aluminum Determination in Biological Fluids by Graphite Furnace AAS. Symposium Abstract H2, Odense.* WHO Regional Office for Europe, Copenhagen.

Toussaint, A. (1989), *Use of Ashes from Waste Incineration. 6th ENVITEC '89.* Summaries of Papers, Messe Düsseldorf 30.

Trapp, G. A., and Cannon, J. B. (1981), *Aluminum Pots as a Source of Dietary Aluminum. N. Engl. J. Med 304,* 172–173.

Trueb, L. (1987), *Problems with High-Tech Fibers* (in German). *Neue Zürcher Zeitung, Forschung und Technik,* No. 143, p. 67 (24 June), Zürich.

Trueb, L. (1988a), *Metallic Glass Rich in Aluminum* (in German), *Neue Zürcher Zeitung, Forschung und Technik,* No. 274, p. 66 (23 November), Zürich.

Trueb, L. (1988b), *Aluminum and Alzheimer's Disease* (in German), *Neue Zürcher Zeitung, Forschung und Technik,* No. 303, p. 51 (28 December), Zürich; see also *Lancet* (29 October 1988), 1008–1010.

Ulrich, B. (1983), *A Concept of Forest Ecosystem Stability and of Acid Deposition as Driving Force for Destabilization,* in: Ulrich, B., and Pankrath, J. (eds.): *Effects of Accumulation of Air Pollutions in Forest Ecosystems,* pp. 1–29. Reidel Publ. Co., Dordrecht.

Ulrich, B. (1989), *Effects of Acid Deposition on Forest Ecosystems in Europe. Adv. Environ. Sci. 4,* in press.

van Beusekom, J. E. E. (1987), *Seasonal Cycles of Dissolved Aluminum and Silicate in the North Sea: Contrasts between Central and Coastal Areas. Proceedings Heavy Metals in the Environment, New Orleans,* Vol. 2, pp. 239–241. CEP Consultants Ltd., Edinburgh.

Vandervoet, G. G., and de Wolff, F. A. (1984), *A Method of Studying the Intestinal Absorption of Aluminum in the Rat. Arch. Toxicol. 55,* 168–172.

Vandervoet, G. B., de Haas, J. M., and de Wolff, F. A. (1985), *Monitoring of Aluminum in Whole Blood, Plasma, Serum, and Water by a Single Procedure Using Flameless AAS. J. Anal. Toxicol. 9* (May/June) 97–100.

Varghese, Z., Moorhead, J. F., and Wills, M. R. (1973), *Plasma Calcium and Magnesium Fractions in Chronic Renal Failure Patients on Maintenance Hemodialysis. Lancet 2,* 985–988.

Versieck, J., and Cornelis, R. (1980), *Normal Levels of Trace Elements in Human Blood Plasma or Serum. Anal. Chim. Acta 116,* 217–254.

Welz, B. (1985), *Atomic Absorption Spectrometry,* 2nd Ed., pp. 267–268, 347–348, 403 ff. VCH Verlagsgesellschaft, Weinheim-Deerfield Beach/Florida-Basel.

WHO (World Health Organization) (1984), *Industrial Exposures in Aluminum Production, Coal Gasification, Core Production, and Iron and Steel Founding, Polynuclear Aromatic Compounds,* Part 3. IARC Monograph No. 34 on the Evaluation of the Carcinogenic Risk of Chemicals to Humans, Geneva.

Wills, M. R. (1973), *Intestinal Absorption of Calcium. Lancet 1,* 820–823.

Wills, M. R. (1978), *Metabolic Consequences of Chronic Renal Failure,* 2nd Ed. Harvey Miller & Medcalf, Aylesbury, UK.

Witters, H., and Vanderborght, O. (1987), *Ecophysiology of Acid Stress in Aquatic Organisms. Proceedings International Symposium, Antwerp.* Dr. Frank Fiers, K.B.I.N., Brussels.

Wood, J. M. (1989), *Transport, Bioaccumulation, and Toxicity of Elements in Microorganisms under Environmental Stress. Proceedings Conference Heavy Metals in the Environement, Geneva,* Vol. 1, pp. 1–12. CEP Consultants Ltd., Edinburgh.

Zietz, J. R. (1985), *Organic Aluminum Compounds,* in: *Ullmann's Encyclopedia of Industrial Chemistry,* Vol. A1, pp. 543–556. VCH Verlagsgesellschaft, Weinheim-Deerfield Beach/Florida-Basel.

Additional Recommended Literature

Feret, F. R., and Sokolowski, J. (1989), *Effects of Sample Surface Integrity on X-ray Fluorescence Analysis of Aluminum Alloys. Spectros. Int. 2*(1), 34–39.

Frankenfeldt, R. E., and Gilgen, P. W. (1989), *Aluminum and Ecology. IRC-Berlin Proceedings, Abfallvermeidung in der Metallindustrie,* pp. 167–172. EF-Verlag für Energie- und Umwelttechnik, Berlin; see also Report E. Merian, *Swiss Chem. 12*(5), 44 (May 1990).

Kirchner, G. (1989), *The Importance of Secondary Aluminum in Aluminum Supply. ICR-Berlin Proceedings, Abfallvermeidung in der Metallindustrie,* pp. 173–177. EF-Verlag für Energie- und Umwelttechnik, Berlin; see also Report E. Merian, *Swiss Chem. 12*(5), 44 (May 1990).

Lewis, T. E. (1989), *Environmental Chemistry and Toxicology of Aluminum.* Lewis Publishers, Chelsea, Michigan.

Milde, G., and Schleyer, R. (1990), *The Determination of Groundwater Quality Changes by Atmospheric Inputs* (in German). *VDI Proceedings on Effects of Atmospheric Pollution on Soils,* Lindau-Meeting, in press.

Reisener, J., Krüger, J., and Vest, H. (1989), *Aluminum and Tin Beverage Cans in the Material Cycle. ICR-Berlin Proceedings, Abfallvermeidung in der Metallindustrie,* pp. 189–201. EF-Verlag für Energie- und Umwelttechnik, Berlin; see also Report E. Merian, *Swiss Chem. 12*(5), 44 (May 1990).

Steinegger, A., Rickenbacher, U.-J., and Schlatter, C. (1990), *Aluminum,* in: Hutzinger, O. (ed): *Handbook of Environmental Chemistry,* Vol. 3, Part E *Anthropogenic Compounds,* pp. 155–184. Springer-Verlag Berlin-Heidelberg-New York-London-Paris-Tokyo-Hong Kong.

Trueb, L. (1990), *Aluminum-Air Batteries* (in German). *Neue Zürcher Zeitung, Forschung und Technik,* No. 7, p. 75 (10 January), Zürich.

Trueb, L. (1990), *Metallic Glasses Based on Aluminum Alloys* (in German). *Neue Zürcher Zeitung, Forschung und Technik,* No. 49, p. 87 (28 February), Zürich.

II.2 Antimony

BRUCE A. FOWLER, Baltimore, Maryland, USA
PETER L. GOERING, Rockville, Maryland, USA

1 Introduction

Antimony is a relatively common toxic trace element normally occurring together with sulfur and arsenic. Its derivatives are utilized in a number of industrial processes and in therapeutic agents against several major tropical parasitic diseases (side effects must be considered). Recently, one antimonial compound has been found to be a potentially effective reverse transcriptase inhibitor of possible utility against the AIDS virus. Other current uses for antimony are in the manufacture of high-speed III-V semiconductors. Antimony compounds show toxic properties similar to those of arsenic. Local accumulation or dissipation, also during incineration or dumping, must be controlled. Chemical speciation and solubility must always be considered (MERIAN and STEMMER, 1984). Trivalent antimonials are generally more toxic than pentavalent forms. Exposure to these agents produces cell injury in a number of organ systems such as the lungs, heart, liver, and kidney. The mechanisms of antimony toxicity are not well understood. Antimony trioxide is carcinogenic in animal experiments.

2 Physical and Chemical Properties, and Analytical Methods

Antimony is a member of Group V of the Periodic Table of the Elements, with the atomic mass of 121.8, atomic number of 51, density of 6.7 g/cm^3; melting point of 631 °C; and boiling point of about 1750 °C, and is a glittering, brittle, but relatively resistant metalloid. It may exist in oxidation states of $-$III, $+$III, ($+$IV), or $+$V. More than 3000 organic antimony compounds are known (WIEBER, 1981/82).

In sample handling contamination must be carefully avoided. Antimony analyses may be conducted by neutron activation analysis (NAA) with counting of the ^{124}Sb isotope (KATAYAMA and ISHIDA, 1987). Polarographic analyses for various antimony compounds have also been recently reported in studies involving antiparasitic antimonials (GAWARGIOUS et al., 1987). Spectrophotometric determination, e.g., in industrial waste waters with iodide and amidines may still be used (GOLWELKER et al., 1988). Candoluminescence spectrometry (CLARK and PATEL, 1986) is a new method which has been applied to antimony in urine samples. Anodic stripping voltammetry (CONSTANTINI et al., 1985) has also been used for antimony determination in several matrices.

More recently, high performance liquid chromatography systems coupled with heated graphite furnace atomic absorption spectroscopy have been found to provide a sensitive analytical approach for analysis of both alkyl and inorganic antimony species (IRGOLIC and BRINCKMAN, 1986). Various AAS techniques (including matrix modification, hydride-formation and flameless AAS) – eventually after enrichment – are especially successful (WELZ, 1985; ELINDER and FRIBERG, 1986; EMMERLING et al., 1986). With a temperature program it is possible to make use of different preatomizing stages, e.g., to separate Sb(III) and Sb(V) species in a graphite atomizer, but SbF_6^- is difficult to hydrogenate (KRIVAN, 1988). An alternative for speciation is derivatizing Sb(III) with diethyldithiophosphate and determination of the complex by reversed phase chromatography with a diode array detector; Sb(V) species are separately quantified indirectly after reduction (IRTH and FREI, 1988). Analytical chemistry has been improved especially for the control of microelectronics, including pure silicon: SIMS, energy loss electronic spectra, and synchrotron XPS allow to determine depth distribution and diffusion of antimony in silicon (GRASSERBAUER, 1989).

3 Sources, Production, Important Compounds, Uses, Waste Products, Recycling

(see also TRUEB, 1989)

Antimony is present in the earth's crust at an estimated concentration of 0.2–0.3 ppm. The most important ores of antimony are stibnite (Sb_2S_3), kermesite ($2Sb_2S_3 \cdot Sb_2O_3$), valentinite (Sb_2O_3), and cervantite (Sb_2O_4). The annual production of antimony is estimated to be about 70000 tons with Bolivia, South Africa, the Soviet Union and the People's Republic of China as major producers (MINERALS YEARBOOK, 1981; SAAGER, 1984; HERBST et al., 1985). Antimony is recovered by roasting Sb-containing ores (HERBST et al., 1985). 55% is used in the form of antimony trioxide (SAAGER, 1984; HERBST et al., 1985). Recycling is very important (40–50% of the consumption mainly from accumulators, according to SAAGER, 1984).

Antimony is used in a number of alloys for batteries and motor bearings, and up to recently in type metal. Other alloys are used in the electronics industry for the manufacture of thermocouples. Semiconductors incorporating antimony with elements such as gallium and indium (III-V semiconductors) offer great promise for the manufacture of high-speed computer chips (DUPUIS, 1984). Gallium- and indium antimonides are used in optical information memories working with laser signals (compact discs, digital optic recording, etc.; PHILIPS, 1987).

The antimony compound ammonium-5-tungsto-2-antimoniate (HPA 23) has been demonstrated to inhibit reverse transcriptase in AIDS patients (ROSENBAUM et al., 1985; BUIMOVICI-KLEIN et al., 1986) and to facilitate increases in peripheral blood lymphocytes. Other medical uses for antimony compounds include antimony tartrate. It has been extensively used as an emetic and is the treatment of choice for a number of tropical parasites such as trypanosomes. In addition, technetium-antimony sulfur colloids have been used for the detection of lymphatic involvement in patients with malignant melanomas (Doss et al., 1986). Pyroantimoniate complexes are extensively used in electron microscopy for the localization of calcium within cells (NORTHOVER, 1985).

Other uses for antimony compounds include the manufacture of fireworks, ammunition, production of ceramics and glass, and as flame retardants (antimony trioxide is substituted to some extent by organic antimony compounds; see SAAGER, 1984). Some antimony compounds are used in the vulcanization of rubber. The toxic and foul smelling gas stibine was used to produce smoke and is an intermediate, its trialkyl derivatives are catalysts (MERIAN and STEMMER, 1984; HERBST et al., 1985). Other organic antimony compounds may be bactericides or fungicides (WIEBER, 1981/82).

4 Distribution in the Environment, in Foods and Living Organisms

4.1 Emissions, Air and Water Quality

Antimony oxides are released into the environment from smelters, coal-fired power plants and volcanoes (ZOLLER, 1984; ANDREAE et al., 1984). LANTZY and MACKENZIE (1979) estimated that 3.8×10^{10} gram/year were released globally into the environment from anthropogenic activities. Combustion of fossil fuels leads to substantial emissions (in the same order as in the cases of cadmium and beryllium; MERIAN and STEMMER, 1984). Antimony is transported in the atmosphere over long distances, for instance, from Central Europe to Norway, and accumulates there in soils, plants, mosses, etc. (STEINNES et al., 1983). BRAR et al. (1970) have reported that antimony concentrations in air over Chicago ranged between $1.4-55$ ng/m^3. In the working environment of smelter operations air concentrations may increase to $1-10$ mg/m^3 (ELINDER and FRIBERG 1986). Antimony containing dusts may also be produced during manufacturing of luminous substances (ECKHARDT, 1986/87). Antimony concentrations in seawater have been reported to be about 0.2 µg/L (mainly in the form of $Sb(OH)_6^-$ ions, $HSbO_2$, methylstibonic acid, and dimethylstibinic acid) in the North Pacific, the Gulf of Mexico and in arctic fjords (SPENCER et al., 1970; BRULAND, 1983). Stibine (SbH_3) has been reported in the air of lead acid plants at a mean concentration of 0.44 mg/m^3 (JONES and GAMBLE, 1984). HURTIG (1990) has discussed contamination of soils by antimony.

4.2 Food Chain, Plants, Animals, and Humans

Antimony concentrations in plants and animal tissues are typically low in areas not receiving anthropogenic input of antimony (RAFEL et al., 1985). But higher concentrations have been reported in farm products from areas geochemically rich in antimony (FUZAILOV et al., 1985). Tobacco contains about 0.1 mg Sb/kg dry weight, 20% of it may be inhaled (ELINDER and FRIBERG, 1986). UTHE and BLIGH (1971) reported freshwater fish to contain antimony concentrations of approximately 3.0 µg/kg on a wet weight basis. SCHLEUTZ (1977) reported antimony concentrations in powdered milk and potato powder to be 3.0–8.0 µg/kg. WINCHESTER and KEATING (1980) performed antimony analyses on oysters and generally found contents of less than 0.7 µg/kg. The daily intake of humans may be in the order of 10–70 µg, which leads to blood concentrations of about 3 µg/L (serum 1 µg/L). Concentrations in skin, adrenals, and lungs are higher (ELINDER and FRIBERG, 1986).

5 Uptake, Absorption, Transport and Distribution, Metabolism and Elimination in Plants, Animals, and Humans

Some plants, mosses, lichens, and fungi are able to accumulate antimony compounds (MERIAN and STEMMER, 1984). Antimonial compounds present in food and beverages are apparently absorbed by animals and humans at a slow rate from the gastrointestinal tract (WALTZ et al., 1965) with large amounts of antimony excreted in the feces. The possibility of enterohepatic circulation was not investigated in these studies. Inhalation of antimony compounds results in extensive pulmonary deposition (FELICETTI et al., 1974a,b).

GERHARDSON et al. (1982) and studies by FELICETTI et al. (1974a,b) showed that uptake and retention of antimony by major organ systems such as the liver, skeleton, and kidney is highly dependent upon chemical form and oxidation state. The highest tissue concentrations of antimony shortly after administration are generally in the liver, kidney, and thyroid (BOYD et al., 1931; BRADY et al., 1945; WESTRICK, 1953). In other studies (BRUNE et al., 1980) antimony concentrations were measured in organs of deceased occupationally exposed workers. They found highest antimony concentrations in the lungs, followed by the liver and kidney. Similar results were reported by LEFFLER et al. (1984) who gave intratracheal instillations of Sb_2O_3 to hamsters.

Excretion of antimony compounds appears to depend upon the chemical species administered. Pentavalent antimonials are primarily excreted in the urine while trivalent forms are mainly excreted in the feces (GOODWIN and PAGE, 1943; OTTO and MAREN, 1950; EMMERLING et al., 1986; WINSHIP, 1987). Radiolabeled antimony tracer studies (WAITZ et al., 1965) demonstrated that both rats and monkeys excreted a large fraction of the administered dose in the urine within 2 hours of administration. DJURIC et al. (1962) studied the kinetics of trivalent antimony in the rat and found slightly greater fecal elimination in comparison with urine. A multiexponential elimination curve was observed. A biological half-life of 16 days has been reported by FELICETTI et al. (1974a) in hamsters following inhalation exposure to both valence states.

6 Effects on Plants, Animals, and Humans

Antimony is a non-essential element in plants, animals, and humans. Some compounds are bactericides. The toxicity of antimonials highly depends upon chemical form and oxidation state with trivalent antimonials exerting greater toxicity than pentavalent compounds. Inhalation exposure to antimonials has been reported to produce pneumonitis, fibrosis, bone marrow damages, and carcinomas (STEMMER, 1976; GROTH et al., 1986; WINSHIP 1987). Pneumoconiosis among antimony workers has also been reported by a number of investigators (COOPER et al., 1968; BROOKS, 1981). Cardiotoxicity has been reported as an important manifestation of antimony toxicity (BRIEGER, 1954). The liver and kidney are major organs for antimony toxicity (STOKINGER, 1981). Stibine affects the nervous system and the erythrocytes (STEMMER, 1976; MERIAN and STEMMER, 1984).

Antimony potassium tartrate has been demonstrated to be positive in several mutagenicity test systems of human cells in vitro (PATON and ALISON, 1972: HASHEM and SHAWKI, 1976). Both the trivalent and pentavalent chloride salts of antimony have been reported positive in the modified rec assay (KANEMATSU et al., 1980). Viral transformation of Syrian hamster embryo cells has also been found to be facilitated by trivalent antimony compounds (CASTO et al., 1979). An extensive review of the carcinogenicity and mutagenicity test results for antimony has been given by VOUK and PIVER (1983). NIOSH (1978) and STOKINGER (1981) have reviewed available data for carcinogenicity risk among antimony workers and noted a slight excess of lung cancer in this population. Smoking habits in this population were not evaluated, and it is presently not possible to reach firm conclusions regarding antimony carcinogenesis in humans.

7 Hazard Evaluation and Limiting Concentrations

As noted by ZOLLER (1984) and ANDREAE et al. (1984), antimony is one of those elements which show anthropogenic mobilization into the environment (e.g., from incineration or dumping of wastes) and which exert both local and regional effects in a variety of ecosystems. Inhalation of antimony containing dusts should be avoided (MERIAN and STEMMER, 1984). The health effects of antimony in man are presently difficult to clearly delineate since exposure to this element usually occurs in the presence of other elements such as lead, arsenic, or mercury. There are a number of industrial standards for limiting worker exposure to both antimony particulates and stibine (MERIAN and STEMMER, 1984). These values are generally in the range of 0.2 – 2 mg Sb/m^3 air. In the Federal Republic of Germany concentrations of antimony at the workplace should not exceed 0.5 mg/m^3. Antimony trioxide is considered unmistakably carcinogenic in animal experimentation under conditions comparable to human exposure (MAK, 1987). Handling thus requires extraordinary caution. For stibine a maximum concentration of 0.1 mL/m^3 or

0.5 mg/m^3 is stipulated (MAK, 1987). Caution is advised where nascent hydrogen comes into contact with antimony (e.g., during battery charging), because stibine may be produced (MERIAN and STEMMER, 1984).

References

Andreae, M. O., Assami, T., Bertine, K. K., Buat-Menard, P. E., Duce, R. A., Filip, Z., Förstner, U., Goldberg, E. D., Heinrichs, H., Jernelöv, A. B., Pacyna, J. M., Thornton, I., Toleschall, H. J., and Zoller W. H. (1984), in: Nriagu, J. O. (ed.): *Dahlem Konferenzen Life Sciences, Report 28*, pp. 359–374. Springer Verlag, Berlin.

Boyd, T. C., Napier, L. E., and Roy, A. C. (1931), *Indian J. Med. Res. 19*, 285–194.

Brady, F. S., Lawton, A. H., Cowie, D. B., Andrews, H. L., Ness, A. T., and Ogden, G. E. (1945), *Am. J. Trop. Med. 25*, 103–107.

Brar, S. S., Nelson, D. M., Kline, J. R., Gustafson, P. F., Kanabrocki, E. L., Moore, C. E., and Hablori, D. M. (1970), *J. Geophys. Res. 75*, 2939–2945.

Bruland, K. W. (1983), *Trace Elements in Seawater. Chem. Oceanogr. 8*, 187.

Brieger, H. (1954), *Ind. Med. Surg. 23*, 521–523.

Brooks, S. M. (1981), *Lung Disorders Resulting from the Inhalation of Metals. Clin. Chest Med. 2*, 235–254.

Brune, D., Nordberg, G., and Wester, P. O. (1980), *Distribution of 23 Elements in the Kidney, Liver and Lungs of Workers from a Smeltery and Refinery in North Sweden Exposed to a Number of Elements and of a Control Group. Sci. Total Environ. 16*, 13–35.

Buimovici-Klein, E., Ong, K. R., Lange, M., England, A., McKinley, G. F., Reddy, M., Grieco, M. H., and Cooper, L. Z. (1986), *Reverse Transcriptase Activity (RTA) in Lymphocyte Cultures of AIDS Patients Treated with HPA-23. AIDS Res. 2* (4), 279–283.

Casto, B. C., Meyers, J., and Paolo, A. (1979), *Enhancement of Viral Transformation for Evaluation of the Carcinogenic or Mutagenic Potential of Inorganic Metal Salts. Cancer Res. 38*, 193–198.

Chung, C., Iwamoto, E., Yamamoto, M., and Yamamoto, Y. (1984), *Spectrochim. Acta 39B*, 459–466.

Clark, E. R., and Patel, M. (1986), *Determination of Antimony in Urine by Candoluminescence Spectrometry. Analyst 111*, 415–417.

Constantini, S., Giordano, R., Rizzica, M., and Benidetti, F. (1985), *Analyst 10*, 1355–1359.

Cooper, D. A., Pendergrass, E. P., Vorwald, A. J., Mayock, R. L., and Brieger, H. (1968), *Pneumoconiosis among Workers in an Antimony Industry. Am. J. Roentgenol. Radium Ther. Nucl. Med. 103*, 495–508.

Djuric, D., Thomas, R. G., and Lie, R. (1962), *Arch. Gewerbepathol. Gewerbehyg. 19*, 529–545.

Doi, K. (1978), *Jpn. J. Ind. Health 20*, 9.

Doss, L. L., Padilla, R. S., and Hladik, W. B. (1986), *J. Dermatol. Surg. Oncol. 12*, 1280.

Dupuis, R. D. (1984), *Science 226*, 623–629.

Eckhardt, Kl. (1986/87), *Report on Emission Reduction at Osram GmbH, Schwabmünchen* (in German). Umweltbundesamt, Berlin; see also *Umwelt 3*, 111–112 (29 May 1987).

Elinder, C.-G., and Friberg, L. (1986), *Antimony*, in: Friberg, L., et al. (eds.), *Handbook on the Toxicology of Metals*, 2nd Ed., Vol. II, pp. 26–42. Elsevier, Amsterdam.

Emmerling, G., Schaller, K. H., and Valentin, H. (1986), *Actual Knowledge on Antimony, Bismuth, Gallium, Germanium, Indium, Tellurium, and their Compounds, and Feasibility of Quantitative Determination in Biological Materials in Occupational Medicine and Toxicology. Zentralbl. Arbeitsmed.* (in German) *36*, 258–265.

Felicetti, S. A., Thomas, R. G., and McCullan, R. O. (1974a), *Metabolism of Two Valence States of Inhaled Antimony in Hamsters. Am. Ind. Hyg. Assoc. J. 35*, 292–300.

Felicetti, S. A., Thomas, R. G., and McCullan, R. O. (1974b), *Health Phys. 26*, 525–531.

Fuzailov, I., Mirzakarimov, V. M., and Fuzailov, A. Iu. (1985), *Vopr. Pitan. 1*, 41–43.

Gawargious, Y. A., Tadros, N. B., Besada, A., and Ibrahim, L. F. (1987), *Polarographic Determination of Antibilharzial Organic Antimony Compounds. Analyst 112*, 545–551.

Gerhardson, L., Brune, D., Nordberg, G. F., and Webster, P. O. (1982), *Antimony in Lung, Liver and Kidney Tissue from Deceased Smelter Workers. Scand. J. Work Environ. Health 8*, 201–209.

Golwelker, A., Patel, K. S., and Mishra, R. K. (1988), *Spectrophotometric Determination of Antimony in Industrial Waste Waters with Iodide and Amidines. Int. J. Environ. Anal. Chem.*, in press.

Goodwin, L. G., and Page, J. E. (1943), *Biochem. J. 37*, 198–209.

Grasserbauer, M. (1989), *The Challenge of Microelectronics for Analytical Chemistry. Fresensius Z. Anal. Chem. 334*, 601–619.

Groth, D. H., Stetter, L. E., Burg, J. R., Busey, W. M., Grant, G. C., and Wong, L. (1986), *Carcinogenic Effects of Antimony Trioxide and Antimony Ore Concentrate in Rats. J. Toxicol. Environ. Health 18*, 607–626.

Hashem, N., and Shawki, R. (1976), *Cultured Peripheral Lymphocytes: One Biologic Indicator of Potential Drug Hazard. Afr. J. Med. Sci. 5*, 155–163.

Herbst, A., Rose, G., Hanusch, K., Schumann, H., and Wolf, H. V. (1985), *Antimony and Antimony Compounds*, in: *Ullmann's Encyclopedia of Industrial Chemistry*, 5th Ed. Vol. A3, pp. 55–76. VCH Verlagsgesellschaft, Weinheim-Deerfield Beach/Florida-Basel.

Hurtig, H.-W. (1990), *Contamination of Soils by Antimony* (in German). *Proceedings VDI Conference on Effects of Air Pollution on Soils, Lindau*, in press.

Irgolic, K. J., and Brinckman, F. E. (1986), in: Bernhard, M., Brinckman, F. E., and Sadlev, P. J. (eds.): *The Importance of Chemical Speciation in Environmental Processes*, pp. 667–684. Springer Verlag, Berlin.

Irth, H., and Frei, R. W. (1988), *New Approaches to Trace Metal Analysis by High Performance Liquid Chromatography. Proceedings 3rd IAEAC Workshop on Carcinogenic and Mutagenic Metal Compounds (Interrelation between Chemistry and Biology), Follonica, Italy*, Vol. 14. Gordon & Breach, New York, in press.

Jones, W., and Gamble, J. (1984), *Epidemiological-environmental Study of Lead Acid Battery Workers. I. Environmental Study of Five Lead Acid Battery Plants. Environ. Res. 35*, 1–10.

Kanematsu, K., Hara, M., and Kada, T. (1980), *Rec Assay and Mutagenicity Studies on Metal Compounds. Mutat. Res. 77*, 109–116.

Katayama, Y., and Ishida, N. (1987), *Determination of Antimony in Nail and Hair by Thermal Neutron Activation Analysis. Radioisotopes 36*, 103–107.

Krivan, V. (1988), *AAS Techniques for the Speciation of Metals. Proceedings 3rd IAEAC Workshop on Carcinogenic and Mutagenic Metal Compounds (Interrelation between Chemistry and Biology), Follonica, Italy*, Vol. 14. Gordon & Breach, New York, in press.

Lantzy, R. J., and MacKenzie, F. T. (1979), *Geochim. Cosmochim. Acta 43*, 511–525.

Leffler, P., Gerhardson, L., Brun, D., and Nordberg, G. F. (1984), *Lung Retention of Antimony and Arsenic in Hamsters after the Intratracheal Instillation of Industrial Dust. Scand. J. Work Environ. Health 10*, 245–253.

MAK (1987), *Maximum Concentrations at the Workplace, Report No. XXIII, DFG*. VCH Verlagsgesellschaft, Weinheim-Basel-Cambridge-New York.

Merian, E., and Stemmer, K. (1984), *Antimony* (in German), in: Merian, E. (ed.): *Metalle in der Umwelt*, pp. 309–317. Verlag Chemie, Weinheim-Deerfield Beach/Florida-Basel.

Minerals Yearbook (1981), Plunkert, P. A. (ed.), *Antimony*, p. 93. U.S. Bureau of Mines, Government Printing Office, Washington, D.C.

NIOSH (National Institute of Occupational Safety and Health) (1978) *Criteria for a Recommended Standard-Occupational Exposure to Antimony*. U.S. Department of Health, Education and Welfare, Washington, D.C.

Northover, A. M. (1985), *The Release of Membrane-associated Calcium from Rabbit Neutrophils by Fixatives. Implications for the Use of Antimonate Staining to Localize Calcium. Histochem. J. 17*, 443–452.

Otto, G. T., and Maren, T. H. (1950), *Am. J. Hyg. 51*, 370–385.

Paton, G. R., and Allison, A. C. (1972), *Chromosome Damage in Human Cell Cultures Induced by Metal Salts. Mutat. Res. 16*, 332–336.

Philips Corp. (1987) *Neue Zürcher Zeitung, Forschung und Technik*, No. 208 (9 September), Zürich.

Rafel, I. B., Popov, I. P., and Zakusilova, R. M. (1985), *Antimony Accumulation in Agricultural Food Crops. Vopr. Pitan. 5*, 65–68.

Rosenbaum, W., Dormont, D., Spire, B., Vilmer, E., Gentilini, M., Griscelli, C., Montagnier, L., Barre-Sinoussi, F., and Chermann, J. C. (1985) *Lancet 1*, 450–451.

Saager, R. (1984), *Antimony*, in: *Metallic Raw Materials Dictionary* (in German), pp. 91–94, Bank von Tobel, Zürich.

Schleutz, R. (1977), *J. Radioanal. Chem. 37*, 539–548.

Spencer, D. W., Robertson, D. E., Turkekian, K. K., and Folsom, T. R. (1970), *J. Geophys. Res. 75*, 7688–7696.

Steinnes, E., Hvatum, O. Ø., and Bølviken, B. (1983), *Heavy Metals in Norwegian Ombrotrophic Bogs. Environ. Biogeochem. Ecol. Bull.* (Stockholm) *35*, 351–356;
see also further reports by E. Steinnes and collaborators: *Contribution from Long-range Atmospheric Transport to the Deposition of Trace Metals in Southern Scandinavia*. NILU OR 29/1984, Nordic Ministry Council;
Impact of Long-range Atmospheric Transport of Heavy Metals to the Terrestrial Environment in Norway, "Lead, Mercury, Cadmium and Arsenic in the Environment" (ed. by T. C. Hutchinson and K. M. Meema, pp. 107–117) (1987), SCOPE, J. Wiley & Sons Ltd.;
Atmospaerisk Nedfall av Tungmetallen i Norge, Rapport 334/1988 Statlig Programm for Forurensningsovervåking.

Stemmer, K. L. (1976), in: Sartorelli, A. C. (ed.): *Pharmacology and Toxicology of Heavy Metals*, Part A, Vol. 1, No. 2, pp. 157–160. Pergamon Press, Oxford.

Stokinger, H. E. (1981), in: Clayton, G. D., and Clayton, F. E. (eds.): *Patty's Industrial Hygiene and Technology*, pp. 1505–1517. John Wiley & Sons, New York.

Trueb, L. (1989), *Antimony* (in German), in: *Neue Zürcher Zeitung, Forschung und Technik*, No. 129, p. 77 (7 June), Zürich.

Uthe, J. F., and Blight, E. G. (1971), *J. Fish Res. Board Can. 28*, 786–788.

Vouk, V. B., and Piver, W. T. (1983), *Metallic Elements in Fossil Fuel Combustion Products: Amounts and Form of Emissions and Evaluation of Carcinogenicity and Mutagenicity. Environ. Health Perspect. 47*, 201–225.

Waitz, J. A., Ober, R. E., Meisenhelden, J. E., and Thompson, P. E. (1965), *Bull. WHO 33*, 537–546.

Welz, B. (1985), *Atomic Absorption Spectrometry*, 2nd Ed., pp. 268–269, 380 ff. VCH Verlagsgesellschaft, Weinheim-Deerfield Beach/Florida-Basel.

Westrick, M. L. (1953), *Proc. Soc. Biol. Med. 82*, 56–60.

Wieber, M. (1981/82), *Organoantimony Compounds*, in: Gmelin, *Handbook of Inorganic Chemistry*, 8th Ed. Springer Verlag, Berlin-Heidelberg.

Winchester, R. V., and Keating, D. L. (1980), *N. Z. J. Sci. 23*, 161–169.

Winship, K. A. (1987), *Toxicity of Antimony and its Compounds. Adverse Drug React. Acute Poisoning Rev. 6*, 67–90.

Zoller, W. H. (1984), in: Nriagu, J. O. (ed.): *Changing Metal Cycles and Human Health*, pp. 27–41. Springer Verlag, Berlin.

II.3 Arsenic

ALAIN LÉONARD, Brussels, Belgium[1]

1 Introduction

Although arsenic has almost exclusively been associated with criminal poisoning for many centuries, the matter of concern today is its contribution to environmental pollution through man's use of pesticides, non-ferrous smelters and coal-fired and geothermal power plants. The long-term consequences of exposure to inorganic forms of arsenic are important because these compounds are recognized as carcinogens affecting especially the lungs, and in some countries, drinking water contaminated through natural sources was linked to skin cancer. When discussing arsenic, speciation plays an especially important role: hydrides, halogenides, oxides, sulfides, arsenites, arsenates, and organic arsenic compounds all have very different properties.

2 Physical and Chemical Properties, and Analytical Methods

Arsenic (atomic number 33, atomic mass 74.9216) belongs to the subgroup Va of the Periodic System where it is placed below phosphorus and above antimony. The mass numbers of its isotopes range from 68 to 80, but only the natural isotope 75 is stable. Elemental arsenic exists at room temperature as metallic or gray arsenic and yellow arsenic. Gray arsenic represents the common stable form. Its density is 5.73 g/cm^3, melting point 814 °C at 36.5 bar pressure, specific heat is 0.0772 at 28 °C, and the vapor pressure is 1 bar at 604 °C (sublimating). Metallic arsenic is not soluble in common solvents. Yellow arsenic has a density of 2.03 g/cm^3 at 18 °C and is unstable, being deposited when arsenic vapors are cooled suddenly to below 0 °C. It is also more volatile than the gray form. The transformation of yellow arsenic to more stable forms is accelerated by heat, light, and catalysts such as iodine and bromine. Arsenic displays variable valences (-3, $+3$ and $+5$) and has both cationic and anionic forms.

Arsenic trihydride (arsine, AsH_3) is a colorless, extremely poisonous, neutral gas with a characteristically disagreeable garlic odor. Its melting and boiling points are -117 °C and -55 °C, respectively. Arsine is a powerful reducing agent, even for fairly weak oxidizing agents. Other examples of compounds containing arsenic in the -3 oxidation state are Na_3As, Ca_3As_2, Zn_3As_2, or $AlAs$.

---

[1] Prof. Dr. Lawrence Fishbein, Washington, D.C., has contributed valuable information from American sources.

Arsenic trioxide (As_2O_3), or white arsenic, exists in several forms such as arsenolite (density 3.65 g/cm^3, melting point 275 °C), claudetite (density 4.05 g/cm^3, melting point 315 °C), amorphous (density 3.68 g/cm^3, melting point 200 °C). Arsenic oxide, the common commercial form of the element, is prepared as a by-product of the roasting of various ores and constitutes the primary material for all arsenic compounds. In the +3 oxidation state arsenic can exist in aqueous solution as a cation, as the hydroxide, as a negative oxy-ion, and as a negative sulfarsenite ion. Arsenic can form arsenious halides (AsF_3, $AsCl_3$, $AsBr_3$, AsI_3) which in aqueous solution are readily hydrolyzed to arsenious acid and the hydrogen halides. Arsenites are also trivalent compounds.

Arsenic pentoxide (As_2O_5), arsenic pentasulfide (As_2S_5), arsenic pentaselenide (As_2Se_5) and especially arsenates represent examples of compounds in the +5 oxidation state.

Inorganic and organic derivatives of arsenic are usually classified into two main categories:

Category I: R_3As, R_2AsX, $RAsX_2$, AsX_3
Category II: R_3AsX_2, R_2AsX_3, $RAsX_4$, AsX_5.

R represents hydrogen atoms or aliphatic or aromatic organic radicals and X represents an electronegative atom or radical (F, Cl, Br, I, OH, etc.). Arsenious acid is an example of the most oxidized member of Category I, in which one finds, as derivatives, tetraalkylarsonium compounds, cacodyl derivatives, and esters of arsenious acids. Arsenic acid is an example of the most oxidized members of Category II, and trimethyl arsine oxide $(CH_3)_3AsO$ is an example of one of the lowest oxidation states. Arsenobetaine and arsenocholine are found particularly in fish (BUROW and STOEPPLER, 1987).

Speciation is of critical importance for arsenic compounds because organic and inorganic forms coexist in our environment and differ largely with respect to their toxicity. Many methods have thus been developed for the determination of total arsenic and of the various methylated and inorganic forms in solids, water, air, and biological materials (LAUWERYS et al., 1979; BUCHET et al., 1980, 1981; IARC, 1982; SAVORY and WILLS, 1984; BRULAND, 1983; FISHBEIN, 1984, 1987). The methods most frequently used for the determination of arsenic include atomic absorption spectrometry (AAS) (see also WELZ, 1985), atomic emission spectrometry (AES), flameless atomic absorption spectrometry (FAAS), gas chromatography (GC), inductively coupled plasma-atomic emission spectroscopy (ICP-AES), inductively coupled plasma-mass spectrometry (ICP-MS), neutron activation analysis (NAA), spark source mass spectrometry (SSMS), ultraviolet spectrometry, X-ray fluorescence (IARC, 1980; FISHBEIN, 1984). Total arsenic content (as in food) is one determination, but inorganic content may vary considerably. For speciation studies, AAS and GC may be combined with derivatization (EBDON et al., 1988). NORIN and VAHTER (1981), APEL and STOEPPLER (1983), VAHTER and LIND (1986), and BUROW and STOEPPLER (1987) differentiated especially between inorganic and organic arsenic compounds in fish and urine. SCHÜMANN et al. (1987) looked at the stability of arsenic compounds in blood and urine. The limits of detection vary according to the methods used and the material analyzed. The concentrations usually detected by these methods are of the order of

1 µg/L for inorganic and organic arsenic in natural waters; higher than 0.2 ng/m^3 and below 0.1 ng/m^3 for inorganic and alkyl arsenic, respectively. The biological monitoring of workers exposed to inorganic arsenic is generally achieved by the determination of urinary arsenic (LAUWERYS, 1983; and others).

3 Sources, Production, Important Compounds, Uses, Waste Products

3.1 Occurrence, Production

The earth's crust and igneous rocks contain about 3 mg/kg arsenic, coal between 0.5 and 93 mg/kg with a mean value of 17.7 mg/kg, and brown coal up to 1500 mg/kg (PETERSON et al., 1981). Arsenopyrite (FeAsS) is the most abundant ore of this element which is also found as arsenolite, As_2O_3; mimetite, $Pb_5Cl(AsO_4)_3$; olivenite, Cu_2OHAsO_4; cobaltite, CoAsS; and proustite, Ag_3AsS_3; etc. Arsenic oxide is usually produced as a by-product of copper, lead, and nickel smelting. Annual world production of arsenic oxide was about 12000 metric tons in the early part of the 20th century and increased sharply during the Second World War. Progressive replacement of arsenic insecticides by organic compounds has led to a decrease in production from up to 70000 to about 40000 tons per year (MERIAN, 1984, for review; HANUSCH et al., 1985; ARNOLD, 1988). Main producers of arsenic are Sweden, Mexico, France, the United States, the Soviet Union, and Namibia. Metallic arsenic is produced by carbon reduction of As_2O_3 (HANUSCH et al., 1985). High purity arsenic may be obtained by hydrogen reduction of recrystallized As_2O_3 or from distilled $AsCl_3$ (HANUSCH et al., 1985).

3.2 Uses and Waste Products

The most frequent application of arsenic salts used to be in the preparation of insecticides, mainly as lead arsenate and less frequently as calcium arsenate and arsenite, sodium arsenite, cupric arsenite (Scheele's green, $Cu(AsO_2)_2$) and cupric acetoarsenate (Paris green, $Cu(C_2H_3O_2)_2 \cdot 3\,Cu(AsO_2)_2$). Arsenicals found other frequent applications as herbicides (weed killers for railroad and telephone posts), desiccants to facilitate mechanical cotton harvesting, fungicides, rodenticides, insecticides, algicides, sheep dips, and wood preservatives (PETERSON et al., 1981; NEWLAND, 1982). Because of occupational and environmental risks, these uses are decreasing. Malaysia still uses great amounts of arsenic based herbicides in its rubber plantations (HANUSCH et al., 1985). Arsenic is known to promote growth in animals and roxarsone (3-nitro-4-hydroxyphenylarsonic acid, $C_6H_6AsNO_6$) or arsenilic acid ($C_6H_8AsNO_3$) are used to fatten swine and poultry (BUCK, 1971). The excretion of arsenic is very rapid, and when treatment is discontinued a few days before slaughter, the amount of arsenic falls rapidly to an acceptable level of 1 ng/kg or less.

Other uses of arsenic include the glass industry (As_2O_3, As_2Se, As_2O_5, metallic arsenic), electronic applications, colors for digital watches, the textile and tanning industries, manufacture of pigments and antifouling paints, as a light filter (thin sheets of As_2O_5), in cosmetics as depilatory agents, in the ceramic industry (AsO_5), and the manufacture of fireworks (As_4S_4). Arsenic(III) chloride is a starting material for organoarsenic compounds. Gallium arsenide plays an important role in semiconductor, integrated circuit, diode, infrared detector, and laser technology (see Chapter II.10, Gallium; HANUSCH et al., 1985). In metallurgy, arsenic is used in arsenical-copper alloys for automobile radiators and other objects assembled by soldering. Arsenical-lead alloys are employed for bullets and metallic arsenic is used as a replacement for antimony as a hardening agent for lead storage battery plates (BUCK et al., 1971). Additions of 0.5% arsenic to bronze and other alloys increase hardness and corrosion resistance (HANUSCH et al., 1985; ARNOLD, 1988).

Several arsenic compounds have found applications in veterinary medicine. Arsenilic acid ($C_6H_8AsNO_3$) is used for swine dysentery and arsenamide ($C_{11}H_{12}AsNO_5S_2$) is employed for heartworm infection in dogs. Sodium arsenite and arsenate have been used successfully to alleviate selenium poisoning in pigs, dogs, chickens, and cattle (WAHLSTROM et al., 1955).

Arsenic has been used therapeutically since 2000 years. It stimulates the production of hemoglobin and influences the metabolism of arginine, zinc, and manganese. This may explain why it was prescribed as a treatment for anemia. Several natural spring waters probably owe their tonic effects, and possibly curative properties, to arsenic (EMSLEY, 1985). Arsenic compounds, administered as Fowler's solution (potassium arsenite), Donovan's solution (arsenic iodide) or de Valagin's solution (arsenic trichloride), were employed to treat rheumatism, arthritis, asthma, malaria, trypanosome infections, tuberculosis, and diabetes (EMSLEY, 1985). N-Acetylarsanilic acid ($C_8H_{10}AsNO_4$), and primarily salvarsan (arsphenamine, $C_{12}H_{14}As_2Cl_2N_2O_2$) discovered by Ehrlich in 1909, remained in use against syphilis until the advent of antibiotics in the 1940s. Some arsenicals are still used in Africa against sleeping sickness (e.g., melarsoporol, ADAMS et al., 1986).

Special provisions for storage and handling of arsenical products have to be taken (HANUSCH et al., 1985; see also Chapter I.20). As arsenic compounds usually occur in rather dilute and dissipated form, not too much is known about waste problems caused by arsenic, but local problems of dumping have been discussed.

4 Distribution in the Environment, in Foods, and in Living Organisms

4.1 Emissions

The major sources of air-borne arsenic emissions are the smelting of metals (mainly nickel-copper smelters), burning of coal, pesticide use, and volcanoes. Global man-made releases have been estimated at 24000 tons per year, compared with natural releases of about 8000 tons per year (WALSH et al., 1979; BENNETT, 1981; FISHBEIN, 1984). According to these authors, volcanoes and coal contribute 7000 and 550 tons, respectively. However, MERIAN (1984) estimated, on the basis of data provided by NEWLAND (1982) and PETERSON et al. (1981), that arsenic emissions from coal could reach 50000 tons per year. MERIAN (1984) also considered that the majority of arsenic produced each year is reintroduced into soil, water, and air in various forms. In agreement with MERIAN, calculations made by MACKENZIE et al. (1979) also suggest that the annual releases reported by BENNETT (1981) and WALSH et al. (1979) could be underestimated. According to MACKENZIE et al. (1979), anthropogenic release of arsenic into the environment from mining, industry, and the burning of fossil fuels is as high as 124000 tons per year, compared to continental and volcanic dust fluxes of 2800 tons. Arsenic is transported over long distances in the atmosphere (PETERSON et al., 1981; STEINNES, 1979) and then precipitated.

4.2 Arsenic in Soil, Water, and Air

Arsenic is mobile within the environment and may circulate many times in various forms through the atmosphere, water, and soil before finally entering its ultimate sink: the sediments (FISHBEIN, 1988; see also SAVORY and WILLS, 1984, and PETERSON et al., 1981).

The level of arsenic in soil is about 7 mg/kg, but can be as high as 1000 mg/kg in the vicinity of metal smelters and in agricultural soils where extensive use was made of pesticides, herbicides, and defoliants (PETERSON et al., 1981; MERIAN, 1984). Arsenic compounds tend to form insoluble complexes with soils and sediments (leaching can occur; FISHBEIN, 1988). The European Community recommends that arsenic levels in soil not exceed 20 mg/kg.

In air, arsenic is present mainly in particulate form as arsenic trioxide, with background levels of 1 to 10 ng/m^3 in rural areas and 20 ng/m^3 in urban areas (NAS, 1977; NRCC, 1978). Near smelters or coal-burning plants levels can reach 1000 ng/m^3 and more (FOWLER, 1977; SAVORY and WILLS, 1984). Ten to twenty percent of the arsenic present in tobacco (less than 10 µg/kg) are volatilized in smoke.

An exhaustive review on arsenic in aquatic systems has been published by MOORE and RAMAMOORTHY (1984). Fresh water normally contains 0.15 – 0.45 µg/L arsenic, mainly in inorganic form (FERGUSON and GAVIS, 1972), but some waters in Canada, Arizona, Argentina, Chile, Bosnia, New Zealand, and Taiwan contain more than 1000 µg/L (LEVANDER, 1977; SAVORY and WILLS, 1984). Mineral waters may contain up to 50 times, hot springs up to 300 times, more arsenic than normal background levels (SAVORY and WILLS, 1984). In river-borne particulates the concentrations vary from 3 to 74 mg/kg (CRECELIUS et al., 1975), and relatively high levels, up to 60 mg/kg (RUPPERT et al., 1974) or even 675 mg/kg (LOVELL and FARMER, 1983), are found in freshwater sediments. Concentrations in excess of 3000 mg/L have been reported in some lakes contaminated by mining; lakes with arsenic levels of more than 500 mg/L caused by sodium arsenite used as insecticide have also been reported (MOORE and RAMAMOORTHY, 1984, for review). In the ocean, arsenic is found at a concentration of 0.09−24 µg/L (ISHINISHI et al., 1986; MARIJANOVIC et al., 1983). BRULAND (1983) stated that arsenic concentrations are of the order of 1.1 to 1.9 µg/L seawater, mainly in the form of arsenate, arsenite, methylarsonate, and dimethylarsinate. Reduction is possible in the surface photic zone. Distribution is of the nutrient type (biological uptake and net particulate transfer to depth; see also NRCC, 1978).

4.3 Plants, Animals, Food Chain, Humans

Natural arsenic levels in plants seldom exceed 1 mg/kg, but the leaf content may be higher if arsenic pesticides have been used. For instance, tomatoes, lima beans, cabbage, and green beans contain little arsenic even when sodium arsenate concentrations were used reducing their growth by 50%, but spinach will contain 10 mg/kg dry weight and radishes even 76 mg/kg dry weight under such conditions (PETERSON et al., 1981). As mentioned previously, smoking can be a pathway for the transfer of arsenic to man, but in recent years the level of arsenic in tobacco has diminished (SAVORY and WILLS, 1984). Wine from grapes sprayed with arsenic-containing insecticides and illegally produced whiskeys may contain higher concentrations of arsenic (ISHINISHI et al., 1986; SAVORY and WILLS, 1984).

Most food contains little arsenic (0.25 mg/kg) and consumption of seafood represents the main source of daily arsenic ingestion. In marine fish, arsenic concentrations may range from 1 to 10 mg/kg with values of over 100 mg/kg in certain bottom-feeding fish and crustaceans (FOWLER et al., 1979). Examples of arsenic concentrations sometimes encountered in marine organisms are: 3.7 mg/kg in Whitstable oysters, up to 120 mg/kg in mussels, as much as 175 mg/kg in prawns, and 4 mg /kg in plaice. In aquatic species, however, arsenic is found in the form of stable, non-toxic arsenobetaine, arsenocholine, and arsenolecithin (BUROW and STOEPPLER, 1987).

Arsenic does not usually accumulate in fresh-water fish species. However, high levels of arsenic in natural waters may be transferred to plants, invertebrates (i.e., daphnia and bivalves) and fish through the food chain. Residues of 150 mg/kg have been reported in phytoplankton (KLUMPP and PETERSON, 1979), 2400 mg/kg dry weight in zooplankton (WAGEMAN et al., 1978), 48 mg/kg in invertebrates (LEATHERLAND and BURTON, 1974) and of 220 mg/kg dry weight in fish (WAGEMAN et al., 1978). Increased levels of arsenic in urine and skin cancer were observed in human populations consuming drinking water containing inorganic arsenic (SAVORY and WILLS, 1984; see also Sect. 6.7).

Depending on species and tissues analyzed, there exists a wide variation in the level of arsenic in meat products as shown by a study done in the USA (US Department of Agriculture, 1971, cited in RADER and SPAULDING, 1979) which reports levels of 0.02−0.22 mg/kg in cattle, 0.05−0.40 mg/kg in heifers, 0.01−0.30 mg/kg in steers, 0.01−6.30 mg/kg in swine, 0.01−0.05 mg/kg in lamb, and 0.01−5.5 mg/kg in poultry. In terrestrial species arsenic is mainly found in the form of dimethylarsinic acid, methylarsonic acid, and inorganic As(III) and As(V) compounds (BUROW and STOEPPLER, 1987).

Daily human intake is between about 0.01 and 0.3 mg arsenic depending on diet. Median arsenic concentration in normal human organs and body fluids is between about 0.02 and 0.06 ppm. Higher arsenic levels are found in muscle, the lungs and femurs, while skin, teeth, nails and especially hair contain even more (ISHINISHI et al., 1986). In normal persons, arsenic content in the blood is about 0.004 mg/kg. CEBRIAN (1985) found arsenic concentrations of 0.002 mg/L in blood and 0.01 mg/L in urine in unexposed Mexicans, and concentrations of 0.008 mg/L blood and 0.3 mg/L urine in Mexicans considered at risk. Arsenic levels in hair are generally higher than 0.4 mg/kg (VALENTINE et al., 1979; EMSLEY, 1985; ISHINISHI et al., 1986; ARNOLD, 1988), but it is difficult to distinguish between deposited and incorporated arsenic (FISHBEIN, 1984).

5 Uptake, Absorption, Transport and Distribution, Metabolism and Elimination in Plants, Animals, and Humans

Inorganic and organic arsenic compounds used previously as pesticides, plant defoliants, and herbicides may accumulate in agricultural and horticultural soils and in plants (PETERSON et al., 1981). Lead arsenate, calcium arsenate, copper acetoarsenite, magnesium arsenate, zinc arsenate, zinc arsenite, sodium arsenite, monosodium methanearsonate (MSMA), disodium methanearsenate (DSMA), and cacodylic acid (CA) were used especially in cotton, tobacco, and blueberry fields. However, soil texture and competing phosphates (e.g., from fertilization) have a great influence on plant uptake. Growth of rice, barley, and alfalfa in cotton fields previously treated with arsenic may be reduced (PETERSON et al., 1981). It seems that low levels of phosphates will displace arsenic from soil particles to increase uptake and phytotoxicity, while larger amounts of phosphates will compete with arsenic at root surfaces to decrease uptake and phytotoxicity (PETERSON et al., 1981).

In 1981, BENNETT, on the basis of LIEBSCHER and SMITH's data (1968), estimated the body burden of arsenic to be 1 mg for a 70 kg person, but according to EMSLEY (1985) and KIEFFER (Chapter I.9), it is about 10 mg. Ingestion of arsenic from the diet depends to a large extent on the amounts of seafood consumed. Representative daily intake is of the order of 40 µg/d in foods of terrestrial origin (and about an additional 80 µg/d in seafood; BENNETT, 1981). Non-negligible exposure can occur through ingestion of certain drinking waters or inhalation of industrial dust. According to FISHBEIN (1984) the daily intake by inhalation can be estimated at 0.4 µg/d. Dermal absorption may also occur. Resorption proceeds rapidly, but storage is limited (SAVORY and WILLS, 1984). Absorption is of course affected by the type of the arsenic compound, its solubility and physical form (FISHBEIN, 1988). Pentavalent arsenic is more readily absorbed than trivalent, and inorganic more than organic. Arsenic trioxide is not well absorbed. Ninety-five percent of ingested trivalent arsenic and eighty percent of ingested "seafood" arsenic are absorbed from the gastrointestinal tract. Skin absorption is more rapid for trivalent than pentavalent arsenic, especially if applied in a lipid soluble carrier.

Absorption is faster through skin with lesions than through normal skin (US EPA, 1981).

After absorption by the lungs and through the gastrointestinal tract, 95 to 99% of the arsenic is located in erythrocytes, bound to the globin of hemoglobin. It is transported in the blood to other parts of the body, and within 24 hours arsenic compounds are distributed in the liver, kidney, lung, spleen, and the wall of the gastrointestinal tract. Some arsenic may also be deposited in bones, hair, nails, and skin, where it may remain.

About 70% of arsenic is excreted, mainly through urine (TAM et al., 1979a, b; ISHINISHI et al., 1986; BENNETT, 1981; SAVORY and WILLS, 1984). Pentavalent arsenic is excreted more rapidly than trivalent compounds due to the binding of As(III) protein thiol groups, and organic arsenic compounds are excreted faster than inorganic ones (FISHBEIN, 1988). The conversion of trivalent compounds to pentavalent ones, as described in Sect. 6.2, protects the body to some extent. CRECELIUS (1975, 1977b) recorded biological half-time of 10 and 30 hours for inorganic arsenic and methylated forms, respectively, whereas the biological half-time of ingested, unchanged organic seafood arsenic can be estimated to be less than 20 hours. These figures are approximately correct for humans and other mammals except rats. In rats, most of the dose remains in the blood bound to hemoglobin (FISHBEIN, 1988), and the half-time for clearance is thus about 60 to 90 days.

Arsenic intoxication is usually assessed by determining the total arsenic content in urine. The average level in non-exposed persons is normally about 8 µg per g creatine (more after seafood consumption; VAHTER and LIND, 1985) or ranges from 13 to 23 µg/L (of which 30–50% may be inorganic; WEBSTER, 1941; BRAMAN and FOREBACK, 1973; ABEL and STOEPPLER, 1983). Levels as high as 820 µg/L have been found in asymptomatic chemical workers and levels of 4000 to 6000 µg/L in non-fatal poisoning cases (GINSBURG, 1965).

6 Effects on Plants, Animals, and Humans

6.1 Effects on Microorganisms and Plants

According to FISHBEIN (1988) the populations of fresh-water zooplankton and insects are reduced at levels of 2.2 to 11.1 ppm sodium arsenate. An arsenic concentration of 0.52 ppm results in a 16 percent decrease in reproduction of *Daphnia magna*. As already mentioned, bioaccumulation of arsenic is very high in zooplankton, benthic organisms, seaweed, and algae (see also NRCC, 1978, US EPA, 1980). OTTE et al. (1988) studied the uptake of arsenic by grasses, rush, reed, and nettle from the former Rhine estuary and from waters near mines. Some grasses show higher arsenic concentrations in dead leaves than in roots. Growth of reed is more inhibited than growth of nettle. Some grasses seem to be tolerant, perhaps because of a different phosphate uptake system (see also PETERSON et al., 1981). The phytotoxicity of arsenite is, according to PETERSON et al. (1981), higher than that of arsenate, and both are much more phytotoxic than MSMA and CA when present in soil (when applied foliarly CA is most phytotoxic). Rice is very sensitive to DSMA. Arsenic compounds seem to participate in many cell reactions, either by replacing phosphorus in the phosphate groups of DNA or by reacting with sulfhydryl enzymes. It is also possible that mycorrhizal association can improve arsenic tolerance (PETERSON et al., 1981). Soil texture also determines arsenic phytotoxicity. Agricultural effects have been discussed by PETERSON et al. (1981).

6.2 Miscellaneous Biochemical Effects

The fate of arsenicals in the body is closely related to their form (ZIELHUIS, 1983, 1984). The organic arsenic compounds of low solubility present in seafood are largely excreted as such in urine (CRECELIUS, 1977b; BUROW and STOEPPLER, 1987). The absorption of inorganic arsenic compounds (As-selenide, -sulfide, Cd/Cr/Co/Mn/Ni/Ti/Zn-arsenide, Pb/Ni-arsenite, Fe/Pb/Mg/Hg/Ni/Zn-arsenate), to which persons are exposed to in ore smelting or through industrial emissions, is negligible or small.

On the contrary, water-soluble inorganic compounds (As-fluoride, -pentoxide, -trioxide, Mg/K/Na-arsenite, and Na/K-arsenate) are readily absorbed ($>80\%$) in the gut and lungs, and slowly, but significantly, through the skin. It is thought that the toxicity of trivalent arsenic to animals and humans is caused by its binding to thiol anions which inhibits some enzymatic reactions. Arsenate interferes with ATP synthesis by forming unstable arsenate esters (SAVORY and WILLS, 1984). SABBIONI (1988) has discussed strong and weak interactions, different in various animal species and humans. In some kinds of cells, As(V) may be reduced to As(III). In detoxification, trivalent compounds are oxidized to pentavalent ones and are thereafter transformed by the body to monomethylarsonic acid (MMAA) and subsequently to dimethylarsinic acid (DMAA). BUCHET and LAUWERYS (1987), for instance, studied in vivo methylation in rats. Biomethylation in mammals seems to take place mainly in the liver by enzymatic transfer of the methyl groups from S-adenosylmethionine. FOA (1985) found increased amounts of arsenic, MMA, and DMA in the urine of exposed persons, but the renal functions were not modified. The marmoset monkey is not able to methylate inorganic arsenic for detoxification. Transmethylation studies allow for the identification of substances (such as PAD) impairing methylation.

The mechanism by which arsenic alleviates selenium toxicity is still disputed. Several observations suggest that arsenic modifies the detoxification of selenium in the liver and increases the excretion of selenium. According to FROST (1970), arsenic could react directly with selenium to form As^+Se^- complexes of particle size and solubility that favor biliary excretion. However, additive effects of arsenic and selenium, when added to drinking water, have also been reported (FISHBEIN, 1988).

6.3 Symptoms of Deficiency

In Sections 4.3, 5, and 6.1 it was mentioned that arsenic may impair the growth of some plants. On the other hand, SAVORY and WILLS (1984) mentioned that arsenic may be growth-promoting and life-sustaining for some animals, and, therefore, these authors recommended feed contents of 0.3 to 0.5 mg arsenic per kg (beginning with 10 to 50 mg/kg toxic symptoms are observed). ANKE et al. (1987) found that in goats a small growth depression was related to arsenic deficiency during the intrauterine life and after weaning. Deficiency also significantly impaired the success of first service and conception rate (increased abortions). Arsenic deficiency also reduced milk production (and its fat and protein content) and led to a significantly higher mortality of the goats during the second lactation (the mitochondria of cardiac muscle showed ultrastructural changes). ANKE et al. (1987) calculated that goats, minipigs, rats, and chickens require 50 µg arsenic per kilogram of feed (dry weight).

6.4 Acute Effects on Animals and Humans

6.4.1 Animal Data
(FISHBEIN, 1988)

There are many species and strain differences in the toxicity of arsenical compounds. The purity, physical form, and solubility of the compounds also influence toxicity. In a NRCC study (1978) the range of acute toxic values for fish was similar for arsenite and arsenate. Of seven tested fish species, LC_{50} values (median lethal dose) ranged from 13.3 ppm (rainbow trout) to 41.8 ppm (bluegill). The green sunfish was relatively resistant, with a toxic dose of 150 ppm. Sodium arsenate at 250 ppm was toxic to the minnow *Phoxinus phoxinus*. At 29 ppm, spottail shiners showed fin and scale damage and hemorrhage. The lowest effect level for freshwater organisms is 40 ppb, the 7-day LC_{50} for the toad. In 1931 SURBER and MEEHAN (cited in NAS, 1977) estimated that many fish food organisms could tolerate 2 ppm of arsenic trioxide. In trout fed a high-arsenic diet, 10 percent developed hepatoma, with none observed in controls. Arsenic toxicity is not related to water hardness, but is positively correlated with temperature, based upon studies of bluegill and green sunfish. Lower pH also increases toxicity due to As^{3+} formation (US EPA, 1980, 1981; NRCC, 1978; NAS, 1977).

The effects of arsenic on marine life is less well known. Acute toxicity values for salt water invertebrates ranged from 0.5 ppm (copepoda) to 7.5 ppm (American oyster) (US EPA, 1980). The toxicity level for bay scallops is 3.49 ppm and for juvenile white shrimp, 24.7 ppm. Salmon, Atlantic silverside and stickleback showed toxicity levels intermediate to these values (US EPA, 1981).

The 96 hour LD_{50} for arsenic trioxide in Sprague-Dawley albino rats is 15.1 mg/kg when given in solution, 145 mg/kg when mixed dry into feed. The 96 h LD_{50} of arsenic trioxide in Swiss mice is 39.4 mg/kg, in C_3H mice it is 25.8 mg/kg. In rats, the 48 h LD_{75} for sodium arsenite (i.p. administration) and sodium arsenate is 4.5 mg/kg and 14 – 18 mg/kg, respectively. Although much toxicity data are presented for rats, the peculiarities of rat arsenic metabolism must be considered when making extrapolations (NAS, 1977; US EPA, 1980).

The lethal oral dose in domestic animals ranges from 1 to 25 mg/kg (sodium arsenite), which is 3 – 10 times more toxic than arsenic trioxide. Arsenic affects tissues rich in oxidative enzyme systems and is a capillary poison, resulting in hypovolemia, shock, and circulatory failure. Symptoms of inorganic arsenic poisoning are usually explosive in onset, with intense abdominal pain, vomiting, diarrhea, weakness, staggering gait, hypothermia, and death. Acute poisoning may also lead to necrosis and perforation of the stomach or intestine. Cutaneous exposure may cause blistering and edema. Symptoms of subacute poisoning include depression, anorexia, anuria, partial rear-limb paralysis, stupor, and hypothermia.

6.4.2 Human Data

The fatal human dose for ingested arsenic trioxide is 70 – 180 mg. REYNOLDS reported (in 1901) 70 deaths from 6000 poisonings resulting from ingestion of 2 – 4 mg/L arsenic in beer (IARC, 1980). Acute symptoms may occur within minutes or hours of ingestion, depending upon the vehicle, solubility, and particle size. Acute effects generally seen following ingestion of inorganic arsenic compounds include constriction of the throat followed by dysphagia, epigastric pain, vomiting, and diarrhea. If the individual survives, exfoliative dermatitis and peripheral neuritis may subsequently develop.

Other symptoms include leg cramps, shock, stupor, paralysis, and coma. Cardiac abnormalities and reversible anemia and leukopenia have also been reported. Inhalation or contact exposure to arsenic trioxide cause irritation of nasal mucosa with perforation of the nasal septum, dermatitis, and conjunctivitis. Arsenicals may also act as skin contact allergens (NAS, 1977; IARC, 1980; US EPA, 1981).

Subacute poisoning may affect a variety of organ systems, mimicking other disease states. Facial edema is characteristic, occurring in 85 percent of people exposed for 2 – 3 weeks to 3 mg/day of calcium arsenate in soy sauce (MIZUTA et al., 1956, as reported in NAS, 1977). Peripheral neuropathy, with paresthesia and numbness of extremities, decreased touch, pinprick and temperature sensation, and muscle weakness and pain, may develop and persist for years, regardless of the form of arsenic (US EPA, 1980, 1981; NAS, 1971; TILSON, 1987).

WHO (1986) especially mentions degeneration in the inner ear caused by arsenicals. Should the patient recover, axonal regeneration can occur and semilunar strips, which contain a very high concentration of arsenic, will appear across the entire base of the nails ("Mees lines"; CHISOLM, 1970; CAVANAGH, 1979).

Arsine and As(III) halogenides are more toxic than other inorganic arsenic compounds and mean occupational risks. Reduction of inorganic arsenic by nascent hydrogen (produced unintentionally) may also result in arsine. These compounds are taken up by the erythrocytes, cause hemolysis, leading to arsenic acid which also damages the kidneys (SAVORY and WILLS, 1984; ARNOLD, 1988). As a result of the rapid destruction of the red blood cells, jaundice occurs, and the urine turns a reddish-violet color due to hemoglobinuria. Blockage of free hemoglobin and acute uremia may lead to death (ARNOLD, 1988). Two courses of treatment for arsenic exposure might be predicted: through the binding of arsenic to thiol groups or increasing the rate of removal from the target tissue and the body as a whole (arsenic poisoning must be verified before treatment). The former was the basis for the development of BAL (2,3-dimercaptopropanol or British antilewisite), and this drug remains the treatment of choice in acute, severe poisoning (SAVORY and WILLS, 1984; TILSON, 1987).

Subacute poisoning of infants resulting from ingestion of 3.5 mg arsenic per day (total intake 90–140 mg) produced vomiting, diarrhea, melanosis, coughing, fever and hepatomegaly. Follow-up at 20 months indicated an increased incidence of mental retardation, epilepsy, and hearing and eye damage (MASAHIKA and HIDEYASU, 1973, as cited in US EPA, 1980).

6.5 Chronic Effects on Animals and Humans

6.5.1 Animal Data
(FISHBEIN, 1988)

Few animal studies have reported neurotoxic effects due to arsenic exposure. ROZENSHTEIN (1980, as reported in US EPA, 1984) exposed rats to arsenic trioxide aerosol at a concentration of 46 µg/m^3 for three months. The rats developed central nervous system deficits characterized by altered conditional reflexes and structural damage. OSATO (1977, as reported in US EPA, 1984) observed central nervous system disorders in rats exposed to 2 to 10 mg arsenic trioxide via stomach intubation.

6.5.2 Human Data
(see also CHISOLM, 1970; ISHINISHI et al., 1986; FISHBEIN, 1988)

Chronic inhalation exposure to arsenic compounds occurs most frequently in occupational settings. The resulting symptoms have been reported to occur in three phases with the initial appearance of weakness, loss of appetite, nausea, and diarrhea. This is followed by conjunctivitis, inflammation of the mucous membranes of the nose, larynx, and respiratory passages, coryza (inflammation of the mucous membranes), mild tracheobronchitis, skin lesions, and perforation of the nasal septum. The final phase includes symptoms of peripheral neuritis, primarily sensory in nature. Depressant effects on bone marrow with disturbances of both erythropoiesis and myelopoiesis have also been reported (SITTIG, 1985).

The classic clinical symptoms of arsenic poisoning (US EPA, 1980) include numbness, tingling, and "pins-and-needles" sensations in the extremities; stocking-glove distribution of decreased touch, pain and temperature sensation; muscular tenderness, and in some cases, paralysis and atrophy of leg and hand muscles.

Degeneration of nerve cell myeling sheaths can be verified. The symptoms may persist for years (US EPA 1980, 1981; NAS 1977).

In chronic arsenic poisoning the clinical picture is dominated by hyperkeratosis of the hands and feet, symmetrical pigmentation, conjunctivitis, tracheitis, acrocyanosis, and polyneuritis (CHISOLM, 1970; US EPA, 1980, 1981; NAS, 1977). A summary of human arsenic poisoning incidents has been presented by ISHINISHI et al. (1986). Electrocardiogram abnormalities and myocarditis have been linked to chronic arsenicism. Peripheral vascular disorders such as Raynaud's syndrome and mesenteric thrombosis were reported in children exposed to arsenic in drinking water (BORGONO and GREIBER, 1972, as reported in US EPA, 1980). A gangrenous condition of the extremities referred to as "blackfoot disease" occurred in a chronically exposed Taiwanese population (TSENG et al., 1968). Vine dressers exposed to trivalent arsenic have also shown development of peripheral vascular disorders (US EPA, 1980).

Bone marrow depression, anemia, leukopenia, and basophilic stippling are associated with chronic exposure. Arsine (AsH_3) poisoning can produce widespread hemolysis (US EPA, 1981). Cirrhosis, ascites, and destruction of renal tissues have been reported. Arsine exposure may cause renal failure (US EPA, 1981).

6.6 Mutagenic Effects

(see also ISHINISHI et al., 1986; FISHBEIN, 1987; LÉONARD, 1987; ROSSMAN et al., 1987)

As(III) and As(V) inhibit DNA, RNA, and protein synthesis (SIBATAMI, 1959; NAKAMURO and SAYATO, 1981) and replace phosphate in the nucleotides during DNA synthesis (PETERS et al., 1975). However, in contrast to other metal compounds such as those of nickel and chromium, which exhibit carcinogenic properties, arsenicals do not affect the fidelity of the polymerases involved in the biosynthesis of DNA (TKESHELASHVILLI et al., 1980).

Results obtained up to 1984 (LÉONARD, 1984; JACOBSON-KRAM and MONTALBANO, 1985) suggested that arsenicals are unable to induce gene mutations in microorganisms or eukaryotes. This was confirmed recently by MARZIN and HUNG VO PHI (1985) in a reversion assay with the *Salmonella typhimurium* TA102 strain which produced positive results with nickel and chromium compounds. It should be pointed out, however, that arsenicals can cause damage to DNA as indicated by positive results obtained in Rec-assays on *Bacillus subtilis*.

Confirming the results of experiments on plants, several assays performed in vitro on mammalian cells have shown that arsenicals display evident clastogenic properties and can induce a slight increase of SCEs. As expected, trivalent compounds are more efficient in that respect than are pentavalent ones. In vivo, positive results have been reported from the micronucleus test on mice given sodium arsenite (DEKNUDT et al., 1986), but intraperitoneal injections of 5 mg/kg $NaAsO_2$ to male mice failed to induce dominant lethality. Negative findings were obtained by POMA et al. (1981) in the somatic cells and spermatogonia of male mice treated with As_2O_3. The ability of some arsenicals to produce chromosome aberrations and

SCEs could be related to their occasional incorporation into the DNA backbone in the place of phosphorus (JACOBSON-KRAM and MONTALBANO, 1985).

Conflicting results have been reported with respect to the induction of structural chromosome aberrations in peripheral blood lymphocytes from persons exposed to arsenic (LÉONARD, 1986). In industrial plants, all workers are generally exposed to a mixture of possible mutagens. Furthermore, the correlation between the frequency of aberrations and the level of arsenic was generally poor and the methodology was sometimes questionable. It is, therefore, doubtful whether the anomalies observed were induced by arsenicals.

In spite of the fact that numerous studies have demonstrated that chromosome aberrations are regularly associated with cancer, it is generally accepted that the ability of a chemical to act as initiator is associated with its capacity to induce gene mutations. The negative results obtained in the attempts to induce cancers in laboratory animals suggest that water-soluble inorganic arsenic compounds display co-carcinogenic properties through interaction with selenium (FROST, 1967, 1970; LEVANDER and BAUMAN, 1966; OSWALD and GOERTTLER, 1971) or suppression of host resistance (STONE, 1969), but more probably by inhibition of DNA repair (JUNG and TRACHSEL, 1970; JUNG, 1971; ROSSMAN et al., 1977; FONG et al., 1980; LEE et al., 1985). For example, the development of skin cancers localized mainly in those regions of the body which are exposed to the sun could be a systemic effect resulting from the lack of repair of damage done to DNA by UV radiation. Respiratory cancers produced by insoluble inorganic arsenic compounds are local effects which have possibly been considered as secondary consequences of necrosis.

6.7 Carcinogenic Effects

Induction of cancer appears to be the most striking long-term effect of chronic exposure to inorganic arsenic. Epidemiological studies have demonstrated an evident causal relationship between environmental, occupational, and medicinal exposure of man to inorganic arsenic and cancer of the skin and lungs (NEUBAUER, 1946; NAS, 1977; IARC, 1973, 1980; WILDENBERG, 1978; FOWLER, 1979; LÉONARD and LAUWERYS, 1980; PERSHAGEN, 1981; LÉONARD, 1984). Most animal experiments, however, were not able to demonstrate carcinogenicity, except for very few observations of increased incidence of leukemia and lung cancer (SUNDERMAN, 1986; FISHBEIN, 1988).

There exists a clear association between pre-cancerous dermal keratosis, epidermoid carcinoma of the skin and, to some extent, lung cancer, and exposure of humans to water-soluble inorganic arsenic through drinking water with a high natural arsenic content or through contaminated beer and wine. The risk of induction of skin cancer has been estimated to be 1×10^{-5} per 0.02 – 0.04 µg As per liter drinking water consumption over 70 years (STARA et al., 1980; WHO, 1981). According to ZIELHUIS (1983, 1984) there are reasons to believe that this risk is overestimated, but epidemiological studies in Argentina, Chile, Canada, and Taiwan suggest correlations between drinking water that contains arsenic and blackfoot disease, Bowens disease, and skin cancer (SAVORY and WILLS, 1984; FISHBEIN, 1988).

Increased incidence of respiratory cancers is characteristic of occupational exposure to inorganic arsenic in the smelting industry, in gold mines in Zimbabwe, and among persons employed in the manufacture and packaging of pesticides or spray arsenicals. On the basis of a linear extrapolation which overestimates the risk, WHO (1981) estimated an increased lung cancer incidence of 1% in workers exposed for at least 25 years to 0.250 µg/m³ non-soluble inorganic arsenic, and for the general population 0.75% per 1 µg/m³. FISHBEIN (1988) states that the probability of death from lung cancer in persons with arsenical keratosis is 5 to 10 times higher than expected, and IARC (1980) has concluded that "there is sufficient evidence that inorganic arsenic compounds are skin and lung carcinogens in humans". In epidemiological studies of occupational lung cancer it was not always made clear, however, whether the persons were exposed to arsenic compounds only, or to a multitude of carcinogens (see also KAZANTZIS, 1987). Therapy with inorganic arsenicals has also been associated with the development of precancerous skin lesions, multiple epitheliomatosis, and bronchial carcinoma.

The first experimental studies provided no conclusive evidence that arsenic is a carcinogen in animals. In spite of the fact that positive results have been obtained recently with animals exposed to dust containing arsenic or to mixtures containing different inorganic arsenic derivatives, it is probable that inorganic arsenicals must be considered promoters rather than initiators of the cancer process. This opinion is supported by the results of short-term tests performed in vitro and in vivo to assess the mutagenic properties of arsenicals (LÉONARD and LAUWERYS, 1980; BAKER, 1984; LÉONARD, 1984; JACOBSEN-KRAM and MONTALBANO, 1985).

6.8 Teratogenic Effects
(see also IARC, 1980)

The teratogenicity of inorganic arsenic is well documented and has been the subject of several reviews (FERM, 1974; FERM and HARLON, 1983). Since very few observations have been noted on humans, estimations of the teratogenic potential of arsenic rely almost entirely on experimental studies. In non-mammalian species, developmental defects and mitotic abnormalities have been observed in sea-urchins after treatment of the gametes and embryos with sodium arsenite and sodium arsenate (PAGANO et al. 1982). In chicks, inorganic arsenic elicited stunting, micromelia, abdominal edema, and dose-related death (FRANKE et al., 1936; RIDGWAY and KARNOFSKY, 1952; BIRGE and ROBERTS, 1976), and methylarsenate caused spina bifida (ANCEL, 1946).

Inorganic arsenic enters the mammalian fetus easily, as was shown by experiments on mice (GERBER et al., 1982), rats (MORRIS et al., 1938), and hamsters (FERM, 1977; HARLON and FERM, 1977), and by observations on man (LUGO et al., 1969). The fetal uptake and distribution of As(III) and As(V) do not differ greatly, except for the accumulation of As(V) in the fetal skeleton late in the gestational period (DENCKER et al., 1983). Development abnormalities, low birth weights, malformations, and fetal deaths have been observed (FISHBEIN, 1988). The organic metabolites are stored within the placenta and are slowly released into the fetal circulation (EASTMAN and BALTIMORE, 1931; UNDERHILL and AMATRUDA, 1973).

In experimental mammals, organic arsenicals do not display evident teratogenic properties (UNDERHILL and AMATRUDA, 1973), but demonstrative positive results have been obtained in most animal models with inorganic As(III) and As(V), the former being generally more toxic. The hamster is the most sensitive species to the teratogenic effects of sodium arsenate at 20–25 mg/kg (FERM et al., 1971), whereas rats are affected at 30 mg/kg (BEAUDOIN, 1974) and mice at 45 mg/kg (HOOD and BISHOP, 1972), but the period of induction of malformations is comparable in all three species (8–10 days). In the hamster, arsenate shows a site-specific effect on the developing embryo; a very high percentage of fetuses recovered from mothers treated on the 9th day of gestation displayed neural tube defects (FERM and CARPENTER, 1968). In rats, BURK and BEAUDOIN (1977) demonstrated a site-specificity of arsenate on the embryonic urogenital system, but sodium arsenate appears to be a general, rather than site-specific, teratogen in mice, affecting different systems (HOOD and BISHOP, 1972). Simultaneous administration of selenium reduces the teratogenic action of arsenic in hamsters (HOLMBERG and FERM, 1969), whereas treatment with BAL (2,3-dimercaptopropanol or British antilewisite) protects mice (HOOD and PIKE, 1972).

LUGO et al. (1969) reported a case of fetal death after a maternal human inorganic arsenic poisoning during pregnancy, but no unusual pathological changes were observed in the fetus. Offspring of smelter employees showed reductions in birth weight and malformations (IARC, 1980).

7 Hazard Evaluation and Limiting Concentrations

According to WHO (1981) drinking water should not contain more than 50 µg arsenic per liter, and daily intake through food should not exceed 50 µg/kg body weight. This value is probably much too high, because it would correspond to a daily intake of 3.5 mg arsenic for an adult of 70 kg (see also Sect. 4). WHO (1984) reported that a daily uptake of 3–6 mg arsenic may lead to intoxication, and 0.5 mg/L drinking water may lead to skin cancer (private communication by R. ANLIKER, ETAD, Basel, 1987).

In the USA the time-weighted average concentration for a normal 8-hour workday and a 40-hour workweek (TWA) is 0.2 mg/m^3 for arsenic and soluble compounds, and 0.05 ppm and 0.2 mg/m^3 for arsine (ACGIH, 1983). In West Germany the maximum allowable workplace concentration values (MAK) are 0.05 ppm and 0.2 mg/m^3 for arsine and the technical guiding concentration (TRK) is 0.2 mg/m^3 for arsenic and other compounds (MAK, 1987) in total dust (see also LAUWERYS, 1983; WICHMAN and LEHNERT, 1987; and SCHÜMANN et al., 1987, who looked at the stability of biological risk indicators).

As has already been mentioned, IARC (1980) has determined that there is sufficient evidence that inorganic arsenic compounds are human carcinogens, but that the evidence for carcinogenicity in laboratory animals is insufficient. The unit cancer risk estimate for inhalation developed by US EPA (1984; FISHBEIN, 1988) is 50 mg/kg/day, on an adsorbed dose basis, assuming that 30% of inhaled arsenic is absorbed. The unit cancer risk for oral exposure also developed by US EPA is 15 mg/kg/day. This value was based on the epidemiological study carried out by TSENG (1968, 1977).

The use of sodium arsenite as herbicide is forbidden in many countries (see, for instance, SAVORY and WILLS, 1983). Soil concentrations should not exceed 20 mg/kg (see Sect. 4.2). Problems are known because of long-range transport of emissions (STEINNES, 1979; SAVORY and WILLS, 1983; RONDA, 1986).

References

Adams (1986), *Neuropathol. Appl. Neurobiol. 12*, 81–94.
ACGIH (American Conference of Governmental Industrial Hygienists) (1983), *Threshold Limit Values for Chemical Substances in the Work Environment,* adapted by ACGIH for 1983–1984. ISBN 0-936712-45-7.
Ancel, P. (1946), *Recherche Expérimentale sur le Spina Bifida. Arch. Anat. Microsc. Morphol. Exp. 36,* 45–63.
Anke, M., Krause, U., and Groppel, B. (1987), in: Hemphill, D. D. (ed.): *The Effect of Arsenic Deficiency on Growth, Reproduction, Life Expectancy and Disease Symptoms in Animals. Proceedings Trace Substances in Environmental Health XXI,* pp. 533–550. University of Missouri, Columbia, Missouri.
Apel, M., and Stoeppler, M. (1983), *Speciation of Arsenic in Urine of Occupationally Exposed Persons. Proceedings Heavy Metals in the Environment, Heidelberg,* pp. 517–520. CEP Consultants Ltd., Edinburgh.
Arnold, W. (1988), Chapter 8 *Arsenic* in: Seiler, H. G., Sigel, H., and Sigel, A. (eds.): *Handbook on Toxicity of Inorganic Compounds,* pp. 79–93. Marcel Dekker, New York-Basel.
Baker, R. S. U. (1984), *Evaluation of Metals in in vitro Assays, Interpretation of Data and Possible Mechanisms of Action,* in: Merian, E., Frei, R. W., Härdi, W., and Schlatter, Ch. (eds.): *Carcinogenic and Mutagenic Metal Compounds,* pp. 185–206. Gordon & Breach Science Publishers, New York.
Beaudoin, A. R. (1974), *Teratogenicity of Sodium Arsenate in Rats. Teratology 10,* 153–158.
Bennett, B. G. (1981), *Exposure of Man to Environment Arsenic – an Exposure Commitment Assessment. Sci. Total Environ. 20,* 99–107.
Birge, W., and Roberts, D. W. (1976), *Toxicity of Metals to Chick Embryos. Bull. Environ. Contam. Toxicol. 16,* 319–324.
Braman, R. S., and Foreback, C. C. (1973), *Methylated Forms of Arsenic in the Environment. Science 182,* 1247–1249.
Bruland, K. W. (1983), *Trace Elements in Seawater. Chem. Oceanogr. 8,* 185–187.
Buchet, J. P., and Lauwerys, R. R. (1987), *Study of Factors Influencing the in vivo Methylation of Inorganic Arsenic in Rats. Toxicol. Appl. Pharmacol. 91,* 65–74.
Buchet, J. P., Lauwerys, R., and Roels, H. (1980), *Comparison of Several Methods for the Determination of Arsenic Compounds in Water and in Urine. Int. Arch. Occup. Environ. Health 46,* 11–29.
Buchet, J. P., Lauwerys, R., and Roels, H. (1981), *Comparison of the Urinary Excretion of Arsenic Metabolites after a Single Oral Dose of Sodium Arsenite, Monomethylarsonate, or Dimethylarsenate in Man. Int. Arch. Occup. Environ. Health 48,* 71–79.

Buck, W. B. (1971), *Toxicity of Inorganic and Aliphatic Organic Arsenicals,* in: Gehme, F. W. (ed.): *Toxicity of Heavy Metals in the Environment,* pp. 357–374. Marcel Dekker, New York-Basel.

Burk, D., and Beaudoin, A. R. (1977), *Arsenate-induced Renal Agenesis in Rats. Teratology 16,* 247–260.

Burow, M., and Stoeppler, M. (1987), *Ingestion/Excretion Experiments for Arsenic in Fish,* in: Brätter, P., and Schramel, P. (eds.): *Trace Elements, Analytical Chemistry in Medicine and Biology,* Vol. 4. Walter de Gruyter, Berlin.

Cavanagh, J. B. (1979), *The "Dying Back" Process. Arch. Pathol. Lab. Med. 103,* 659.

Cebrian, M. E. (1985), *Proceedings 5th International Conference on Heavy Metals in the Environment, Athens.* CEP Consultants Ltd., Edinburgh.

Chisolm, J. J. (1970), *Poisoning Due to Heavy Metals. Pediatr. Clin. North Am. 17,* 591–615.

Crecelius, E. A. (1975), *Chemical Changes in Arsenic Following Ingestion by Man,* in: *Proceedings 15th Annual Hanford Life Sciences Symposium on the Biological Implications of Metals in the Environment,* Richland, Washington, 29 September–1 October.

Crecelius, E. A. (1977a), *Arsenite and Arsenate Levels in Wine. Bull. Environ. Contam. Toxicol. 18,* 227–230.

Crecelius, E. A. (1977b), *Changes in the Chemical Speciation of Arsenic Following Ingestion by Man. Environ. Health Perspect. 19,* 147–150.

Crecelius, E. A., Bothner, M. H., and Carpenter, R. (1975), *Geochemistries of Arsenic, Antimony and Related Elements in Sediments of Puget Sound. Environ. Sci. Technol. 9,* 325–333.

Deknudt, Gh., Léonard, A., Arany, J., Jenar-DuBuisson, J., and Delavignette, E. (1986), *In vivo Studies in Male Mice on the Mutagenic Effects of Inorganic Arsenic. Mutagenesis 1,* 33–34.

Dencker, L., Danielsson, B., Khayat, A., and Lindgren, A. (1983), *Disposition of Metals in the Embryo and Fetus,* in: Clarkson, Th. W., Nordberg, G. F., and Sager, P. R. (eds.): *Reproductive and Developmental Toxicity of Metals,* pp. 607–631. Plenum Press, New York-London.

Eastman, N. J., and Baltimore, M. C. (1931), *The Arsenic Content of the Human Placenta Following Arsphenamine Therapy. Am. J. Obstet. Gynecol. 21,* 60–64.

Ebdon, L., Hill, S., Walton, A. P., and Ward, R. W. (1988), *Coupled Chromatography/Atomic Spectrometry for Arsenic Speciation. Analyst 113,* 1159–1165.

Emsley, J. (1985), *Whatever Happened to Arsenic? New Sci.* 19–26, December, 10–13.

Ferguson, J., and Gavis, J. (1972), *A Review of the Arsenic Cycle in Natural Waters. Water Res. 6,* 1259–1274.

Ferm, V. H. (1974), *Effects of Metal Pollutants upon Embryonic Development. Rev. Environ. Health 1,* 237–259.

Ferm, V. H., and Carpenter, S. (1968), *Malformation Induced by Sodium Arsenate. J. Reprod. Fertil. 17,* 199–201.

Ferm, V. H., and Harlon, D. P. (1983), *Metal-induced Congenital Malformations,* in: Clarkson, Th. W., Nordberg, G. F., and Sager, P. R. (eds.): *Reproductive and Developmental Toxicity of Metals,* pp. 383–398. Plenum Press, New York-London.

Ferm, V. H., Saxon, A., and Smith, B. M. (1971), *The Teratogenic Profile of Sodium Arsenate in the Golden Hamster. Arch. Environ. Health 22,* 557–560.

Fishbein, L. (1984), *Overview of Analysis of Carcinogenic and/or Mutagenic Metals in Biological and Environmental Samples: Arsenic, Beryllium, Cadmium, Chromium, and Selenium,* in: Merian, E., Frei, R. W., Härdi, W., and Schlatter, Ch. (eds.): *Carcinogenic and Mutagenic Metal Compounds,* pp. 55–112. Gordon & Breach Science Publishers, New York.

Fishbein, L. (1987), *Perspectives of Carcinogenic and Mutagenic Metals in Biological Samples. Int. J. Environ. Anal. Chem. 28,* 21–69.

Fishbein, L. (1988), private communication.

Foa, V. (1985), *Proceedings 5th International Conference on Heavy Metals in the Environment, Athens.* CEP Consultants Ltd., Edinburgh.

Fong, K., Lee, F., and Bockrath, R. (1980), *Effects of Sodium Arsenite on Single-strand DNA Break Formation and Postreplication Repair in E. coli following UV Irradiation. Mutat. Res. 70,* 151–156.

Fowler, B. A. (1977), *Toxicology of Environmental Arsenic,* in: Goyer, R. A., and Mehlman, M. A. (eds.): *Toxicology of Trace Elements,* pp. 79–122. John Wiley & Sons, New York.

Franke, K. W., Moxon, A. L., Poley, N. E., and Tully, W. C. (1936), *Monstrosities Produced by the Injection of Selenium Salts into Hen's Eggs. Anat. Rec.* 65, 15–22.

Frost, D. V. (1967), *Arsenicals in Biology – Retrospect and Prospect. Fed. Proc. Fed. Am. Soc. Exp. Biol.* 27, 194–208.

Frost, D. V. (1970), *Tolerances for Arsenic and Selenium: A Psychodynamic Problem. World Ref. Pest Control* 9, 6–28.

Gerber, G. B., Maes, J., and Eyskens, B. (1982), *Transfer of Antimony and Arsenic to the Developing Organism. Arch. Toxicol.* 49, 159–168.

Ginsburg, J. M. (1965), *Renal Mechanisms for Excretion and Transformation of Arsenic in the Dog. Am. J. Physiol.* 268, 832–840.

Harlon, D. P., and Ferm, V. H. (1977), *Placental Permeability of Arsenate Ion During Early Embryogenesis in the Hamster. Experientia* 33, 1121–1122.

Hanusch, K., Grossmann, H., Herbst, K.-A., Rose, B., and Wolf, H. U. (1985), *Arsenic and Arsenic Compounds,* in: *Ullmann's Encyclopedia of Industrial Chemistry,* 5th Ed., Vol. A3, pp. 113–141. VCH Verlagsgesellschaft, Weinheim-Deerfield Beach/Florida-Basel.

Holmberg, R. E., and Ferm, V. H. (1969), *Interrelationships of Selenium, Cadmium and Arsenic in Mammalian Teratogenesis. Arch. Environ. Health* 18, 873–877.

Hood, R., and Bishop, S. L. (1972), *Teratogenic Effects of Sodium Arsenate in Mice. Arch. Environ. Health* 24, 62–65.

Hood, R. D., and Pike, C. T. (1972), *BAL Alleviation of Arsenate-induced Teratogenesis in Mice. Teratology* 6, 235–238.

IARC Monographs (1973), *Evaluation of Carcinogenic Risk of Chemicals to Humans. Some Inorganic and Organometallic Compounds,* Vol. 2. International Agency for Research on Cancer, Lyon.

IARC Monographs (1980), *Evaluation of Carcinogenic Risk of Chemicals to Humans. Some Metals and Metallic Compounds,* Vol. 23. International Agency for Research on Cancer, Lyon.

IARC Monographs (1982), *Evaluation of Carcinogenic Risk of Chemicals to Humans,* Suppl. 4. *Chemicals, Industrial Processes and Industries Associated with Cancer in Humans.* International Agency for Research on Cancer, Lyon.

Ishinishi, N., Tsuchiya, K., Vahter, M., and Fowler, B. A. (1986), Chapter 3 *Arsenic,* in: Friberg, L., Nordberg, G. F., and Vouk, V. B. (eds.): *Handbook on the Toxicology of Metals,* 2nd Ed., Vol. II, pp. 43–83. Elsevier, Amsterdam.

Jacobson-Kram, D., and Montalbano, D. (1985), *The Reproductive Effects Assessment Group's Report on the Mut. Environ. Mutagen.* 7, 787–804.

Jung, E. G. (1971), *Molecular Biological Investigation of Chronic Arsenic Poisoning. Z. Haut Geschlechtskrankh.* 46, 35–36.

Jung, E. G., and Trachsel, B. (1970), *Molekularbiologische Untersuchungen zur Arsencarcinogenese. Arch. Klin. Exp. Dermatol.* 237, 819–826.

Kazantzis, G., (1987), *The Mutagenic and Carcinogenic Effects of Cadmium. Toxicol. Environ. Chem.* 15, 83–100.

Klumpp, D. W., and Peterson, P. J. (1979), *Arsenic and Other Trace Elements in the Waters and Organisms of an Estuary in SW England. Environ. Pollut.* 19, 11–20.

Lauwerys, R. R. (1983), *Industrial Chemical Exposure: Guidelines for Biological Monitoring,* Chapter II,3 *Arsenic,* pp. 12–15. Biomedical Publications, Davis, California.

Lauwerys, R. R., Buchet, J. P., and Roels, H. (1979), *The Determination of Trace Levels of Arsenic in Human Biological Materials. Arch. Toxicol.* 41, 239–247.

Leatherland, T. M., and Burton, J. D. (1974), *The Occurrence of Some Trace Metals in Coastal Organisms with Particular Reference to the Solent Region. J. Mar. Biol. Assoc. U.K.* 54, 457–568.

Lee, T.C., Huang, R.Y., and Jan, K.Y. (1985), *Sodium Arsenite Enhances the Cytotoxicity, Clastogenicity and 6-Thioguanine-resistant Mutagenicity of Ultraviolet Light in Chinese Hamster Ovary Cells. Mutat. Res. 148,* 83–89.

Léonard, A. (1984), *Recent Advances in Arsenic Mutagenesis and Carcinogenesis. Toxicol. Environ. Chem. 7,* 241–250.

Léonard, A. (1986), *Chromosome Damage in Individuals Exposed to Heavy Metals,* in: Sigel, H. (ed.): *Metal Ions in Biological Systems,* pp. 229–258. Marcel Dekker, New York-Basel.

Léonard, A. (1987), *Significance of in vitro Approaches, Proceedings Trace Elements in Human Health and Disease, Odense.* WHO Regional Office for Europe, Copenhagen.

Léonard, A. (1988), *Mechanisms in Metal Genotoxicity: The Significance of in vitro Approaches. Mutat. Res. 198,* 321–326.

Léonard, A., and Lauwerys, R.R. (1980), *Carcinogenicity, Teratogenicity and Mutagenicity of Arsenic. Mutat. Res. 75,* 49–62.

Levander, O.A. (1977), *Metabolic Interrelationships Between Arsenic and Selenium. Environ. Health Perspect. 19,* 159–164.

Levander, O.A., and Bauman, C.A. (1966), *Selenium Metabolism. VI. Effect of Arsenic on the Excretion of Selenium in the Bile. Toxicol. Appl. Pharm. 9,* 106–115.

Liebscher, K., and Smith, H. (1968), *Essential and Non-essential Trace Elements. Arch. Environ. Health 17,* 881–890.

Lovell, M.A., and Farmer, J.G. (1983), *The Geochemistry of Arsenic in the Freshwater Sediments of Loch Lomond,* in: *Proceedings Heavy Metals in the Environment, Heidelberg,* Vol. 2, pp. 776–779.

Lugo, G., Cassady, G., and Palmisano, P. (1969): *Acute Maternal Arsenic Intoxication with Neonatal Death. Am. J. Dis. Child. 117,* 328–330.

MacKenzie, F.T., Lantzy, R.J., and Paterson, V. (1979), *J. Int. Assoc. Math. Geol. 11,* 99–142.

MAK (1987), *Maximum Concentrations at the Workplace, Report No. XXIII, DFG.* VCH Verlagsgesellschaft, Weinheim-Basel-Cambridge-New York.

Marijanovic, P., Jaksie, M., Orlic, I., and Valkovic, V. (1983), *Trace Elements in Seawater,* in: *Proceedings Heavy Metals in the Environment, Heidelberg,* Vol. 1, pp. 237–240.

Marzin, D.R., and Hung Vo Phi (1985), *Study of the Mutagenicity of Metal Derivatives with Salmonella typhimurium TA102. Mutat. Res. 155,* 49–51.

Merian, E. (1984), *Introduction on Environmental Chemistry and Global Cycles of Carcinogenic Metals and Their Derivatives,* in: Merian, E., Frei, R.W., Härdi, W., and Schlatter, Ch. (eds.): *Carcinogenic and Mutagenic Metal Compounds,* pp. 3–32. Gordon & Breach Science Publishers, New York.

Moore, J.W., and Ramamoorthy, S. (1984), *Heavy Metals in Natural Waters.* Springer Verlag, New York.

Morris, H.P., Laug, E.P., Morris, H.J., and Grant, R.L. (1938), *The Growth and Reproduction of Rats Fed Diets Containing Lead Acetate and Arsenic Trioxide. J. Pharmacol. Exp. Ther. 64,* 420–445.

Nakamuro, K., and Sayato, Y. (1981), *Comparative Studies of Chromosomal Aberration Induced by Trivalent and Pentavalent Arsenic. Mutat. Res. 88,* 73–80.

NAS (National Academy of Sciences) (1977), *Medical and Biologic Effects of Environmental Pollutants, Arenic.* National Research Council, Washington, D.C.

NRCC (National Research Council Canada) (1978), *Effects of Arsenic in the Canadian Environment.* Report NRCC 15391, Ottawa.

Neubauer, O. (1946), *Arsenical Cancer: a Review. Br. J. Cancer 1,* 192–196.

Newland, L.W. (1982), *Arsenic, Beryllium, Selenium and Vanadium,* in: Hutzinger, O. (ed.): *Handbook of Environmental Chemistry, Anthropogenic Compounds,* Vol. 3, Part B, pp. 27–42. Springer Verlag, Berlin-Heidelberg-New York.

Norin, H., and Vahter, M. (1981), *A Rapid Method for the Selective Analysis of Total Urinary Metabolites of Inorganic Arsenic. Scand. J. Work Environ. Health 7,* 38–44.

Oswald, H., and Goerttler, K. (1971), *Arsenic-induced Leucocis in Mice after Diaplacental and Postnatal Application.* Verh. Dtsch. Ges. Pathol. *55,* 289–293.

Otte, M. L., Rozema, J., Beek, M. A., Ernst W. H. O., and Broekman, R. A. (1988), *Uptake of Arsenic by Vegetation of the Former Rhine Estuary, and Heavy Metals and Arsenic in Water, Sediment and Plants near the Jales Gold and Silver Mine in North Portugal, Proceedings of the 3rd Conference on Environmental Contamination, Venice,* pp. 152–154, 529–531. CEP Consultants Ltd. Edinburgh.

Pagano, G., Esposito, A., Bove, P., de Angelis, M., Rota, A., Vamuakinas, E., and Giacoma Giordano, G. (1982), *Arsenic-induced Developmental Defects and Mitotic Abnormalities in Seaurchin Development.* Mutat. Res. *104,* 351–354.

Pershagen, G. (1981), *The Carcinogenicity of Arsenic.* Environ. Health Perspect. *40,* 93–100.

Peters, J., Baron, D., and Enderle, J. (1975), *Zur stimulierenden Wirkung von Arsenat auf den Einbau von ^{14}C-Thymidin in die DNA Phytohämagglutinin behandelter menschlicher Lymphocyten.* Arch. Dermatol. Forsch. *251,* 301–309.

Peterson, P. J., Girling, C. A., Benson, L. M., and Zieve, R. (1981), Chapter 8: *Metalloids,* in: Lepp, N. W. (ed.): *Effect of Heavy Metal Pollution on Plants,* Vol. 1, p. 213ff, esp. 299–322. Applied Science Publishers, London.

Poma, K., Degraeve, N., Kirsch-Volders, M., and Suzanne, C. (1981), *Cytogenetic Analysis of Bone Marrow and Spermatogonia of Male Mice after in vivo Treatment with Arsenic.* Experientia *37,* 129–130.

Rader, W. A., and Spoulding, J. E. S. (1979), in: Oehme, F. W. (ed.): *Toxicity of Heavy Metals in the Environment,* pp. 669–687. Marcel Dekker, New York-Basel.

Ridgway, L. P., and Karnofsky, D. A. (1952), *The Effects of Metals on the Chick Embryo: Toxicity and Production of Abnormalities in Development.* Ann. N.Y. Acad. Sci. *55,* 203–215.

Rondia, D. (1986), *Belgian Research on Metal Cycling in the Environment.* Presses Universitaires, Liège.

Rossman, T. G., Meyn, M. S., and Troll, W. (1977), *Effects of Arsenite on DNA Repair in Escherichia coli.* Environ. Health Perspect. *19,* 229–233.

Rossman, T. G., Zelikoff, J. T., Agarwal, S., and Kneip, T. J. (1987), *Genetic Toxicology of Metal Compounds.* Toxicol. Environ. Chem. *14,* 251–262.

Ruppert, D. F., Hopke, P. K., Clute, P., Metzger, W., and Crowley, D. (1974), *Arsenic Concentrations and Distribution in Chautauqua Lake Sediments.* J. Radioanal. Chem. *23,* 159–169.

Sabbioni, E. (1988), *Metabolism and Biotransformation of Arsenic Compounds.* Toxicol. Environ. Chem., in press.

Savory, J., and Wills, M. R. (1984), *Arsenic,* in: Merian, E. (ed.): *Metalle in der Umwelt,* pp. 319–334. Verlag Chemie, Weinheim-Deerfield Beach/Florida-Basel.

Schümann, M., Flesch, D., Krause, C., and Neus, H. (1987), *Temporal Stability of Internal Load with Arsenic and Heavy Metals, Proceedings Trace Elements in Human Health and Disease, Odense.* WHO, Copenhagen.

Sibatani, A. (1959), *In vitro Incorporation of ^{32}P into Nucleic Acids of Lymphatic Cells.* Exp. Cell Res. *17,* 131–143.

Sittig, M. (1985), *Handbook of the Toxic and Hazardous Chemicals and Carcinogens,* 2nd Ed. Noyes Publications, Park Ridge, New York.

Stara, J. F., Kello, D., and Durkin, P. (1980), *Human Health Hazards Associated with Chemical Contamination of Aquatic Environment.* Environ. Health Perspect. *34,* 145–158.

Steiner, E. (1979), *Contribution from Long-range Atomspheric Transport to the Heavy Metal Pollution of Surface Soil. Proceedings International Conference on Heavy Metals in the Environment, London,* p. 271. CEP Consultants Ltd., Edinburgh.

Stone, O. J. (1969), *The Effect of Arsenic on Inflammation, Infection and Carcinogenicity.* Tex. State J. Med. *65,* 40–43.

Sunderman, F. W., Jr. (1986), *Carcinogenicity and Mutagenicity of Some Metals and Their Compounds,* in: O'Neill, I. K., Schuller, P., and Fishbein, L. (eds.): *Environmental Carcinogens – Selected Methods in Analysis,* Vol. 8, pp. 17–43. IARC, Lyon.

Tam, G. K. H., Charbonneau, S. M., Lacroix, G., and Bryce, F. (1979a), *Confirmation of Inorganic Arsenic and Dimethylarsinic Acid in Urine and Plasma of Dog by Ion-exchange and TLC.* Bull. Environ. Contam. Toxicol. 21, 371–374.

Tam, G. K. H., Charbonneau, S. M., Lacroix, G., and Bryce, F. (1979b), *In vitro Methylation of ^{74}As in Urine, Plasma and Red Blood Cells of Humans and Dog.* Bull Environ. Contam. Toxicol. 22, 69–71.

Tilson, H. A. (1987), *Neurotoxicology of Heavy Metals.* Fundam. Appl. Toxicol. 9, 599–607.

Tkeshelasvilli, L. K., Shearman, C. W., Zakour, R. A., Koplitz, R. M., and Loeb, L. A. (1980), *Effects of Arsenic, Selenium and Chromium on the Fidelity of DNA Synthesis.* Cancer Res. U.S. 40, 2455–2460.

Tseng, W. P. (1977), *Effects and Dose-Response Relationships of Skin Cancer and Blackfoot Disease with Arsenic.* Environ. Health Perspect. 19, 109–119.

Tseng, W. P., Chu, H. W., Fong, J. M., Lin, C. S., and Yeh, S. (1968), *Prevalence of Skin Cancer in an Endemic Area of Chronic Arsenicism in Taiwan.* J. Natl. Cancer Inst. U.S. 40, 453–468.

Underhill, F. P., and Amatruda, F. G. (1973), *The Transmission of Arsenic from Mother to Fetus.* J. Am. Med. Assoc. 81, 2009–2015.

US EPA (US Environmental Protection Agency) (1980), *Ambient Water Quality Criteria for Arsenic.* EPA Office of Water Regulations and Standards, EPA 440/5-80-021, Washington, D.C.

US EPA (US Environmental Protection Agency) (1981), *An Exposure and Risk Assessment for Arsenic.* EPA Office for Water Regulations and Standards, Washington, D.C.

US EPA (US Environmental Protection Agency) (1984).

Vahter, M., and Lind, B. (1985), *Concentrations of Arsenic in Urine of the General Population in Sweden.* Sci. Total Environ. 54, 1–12.

Valentine, H., Kang, H. K., and Spivey, G. (1979), *Arsenic Levels in Human Blood, Urine and Hair in Response to Exposure via Drinking Water.* Environ. Res. 20, 24–32.

Wageman, R., Snow, N. B., Rosenberg, D. M., and Lutz, A. (1978), *Arsenic in Sediments, Water and Aquatic Biota from Lakes in the Vicinity of Yellowknife, Northwest Territories, Canada.* Arch. Environ. Contam. Toxicol. 7, 169–191.

Wahlstrom, R. C., Kamstra, L. D., and Olson, O. E. (1955), *The Effect of Arsanilic Acid and 3-Nitro-4-hydroxyphenylarsonic Acid on Selenium Poisoning in the Pig.* J. Anim. Sci. 14, 105.

Walsh, P. R., Duce, R. A., and Fashing, J. L. (1979), *Considerations of the Enrichment, Sources, and Flux of Arsenic in the Troposphere.* J. Geophys. Res. 84, 1719–1726.

Webster, S. H. (1941), *Lead and Arsenic Ingestion and Excretion in Man.* Publ. Health Rep. 56, 1953–1961.

Welz, B. (1985), *Atomic Absorption Spectrometry*, 2nd Ed., pp. 44ff, 209ff, 269ff, 357ff. VCH Verlagsgesellschaft, Weinheim-Deerfield Beach/Florida-Basel.

WHO (World Health Organization) (1981), *Arsenic. Environ. Health Crit. 18: Arsenic.* International Programme on Chemical Safety, Geneva.

WHO (World Health Organization) (1984), *Guidelines for Drinking Water Quality,* Vol. 2, pp. 63–67.

WHO (World Health Organization) (1986), *Assessment of Neurotoxicity.* Environ. Health Crit. 60, 72.

Wichman, N., and Lehnert, G. (1987), *Occupational Medical Experience and TRK-Values for Arsenic* (in German). Arbeitsmed. Sozialmed. Präventivmed. 22, 18–20.

Wildenberg, J. (1978), *An Assessment of Experimental Carcinogen-detecting Systems with Special Reference to Inorganic Arsenicals.* Environ. Res. 16, 139–152.

Zielhuis, R. L. (1983), *Total Risk Assessment for Heavy Metals,* in: Proceedings Heavy Metals in the Environment, Heidelberg, Vol. 1, pp. 27–32.

Zielhuis, R. L. (1984), *Occupational and Environmental Standard Setting for Metals: More Questions Than Answers,* in: Merian, E., Frei, R. W., Härdi, W., and Schlatter, Ch. (eds.): *Carcinogenic and Mutagenic Metal Compounds,* pp. 477–490. Gordon & Breach Science Publishers, New York.

Additional Recommended Literature

Amiard-Triquet, C., Cosson-Mannevy, M. A., Cosson, R. P., Amiard, J. C., and Métayer, C. (1989), *Elemental Bioaccumulation in Hydrothermal Vent Biota on the East Pacific Rise. Organotropism and Potential Influence of Bacteria.* Proceedings 6th International Conference on Heavy Metals in the Environment, Geneva, Vol. 1, pp. 73–76. CEP Consultants Ltd., Edinburgh.

Belzile, N., de Vitre, R. R., and Tessier, A. (1989), *Chemistry of As(V)/As(III) and Iron Oxyhydroxides in Natural Lacustrine Sediments.* Proceedings 6th International Conference on Heavy Metals in the Environment, Geneva, Vol. 1, pp. 512–515. CEP Consultants Ltd., Edinburgh.

Blanck, H. (1988), *Overview on Arsenate Pollution-induced Community Tolerance (PICT).* Proceedings 1st European SECOTOX Conference on Ecotoxicology Copenhagen, in press.

Buat-Ménard, P., Peterson, P. J., Havas, M., Steinnes, E., and Turner, D. (1987), *Arsenic*, Chapter 4, in: Hutchinson, T. C., and Meema, K. M. (eds.): *Lead, Mercury, Cadmium, and Arsenic in the Environment, SCOPE*, pp. 43 ff. John Wiley & Sons, New York.

Diamond, M. L., Mackay, D., Landsberger, S., and Cheng, A. G. (1988), *Modelling Arsenic Dynamics in a Contaminated Lake.* Proceedings International Conference on Trace Metals in Lakes, Hamilton, Ontario, Canada. Sci. Total Environ., in press.

Dushenko, W., Crowder, A., and Greig, J. (1988), *Metals and their Distribution in Wetlands of the Bay of Quinte.* Proceedings International Conference on Trace Metals in Lakes, Hamilton, Ontario, Canada. Sci. Total Environ., in press.

van Elsteren, J. T., Das, H. A., de Ligny, C. L., and Agterdenbos, J. (1989), *Determination of Arsenic(III/V) in Aqueous Samples by Neutron Activation Analysis after Sequential Coprecipitation with Dibenzyldithiocarbamate.* Anal. Chim. Acta **222**, 159–167.

Fowler, B. A., and Dobrota, M. (1989), *Toxicity of Semiconductor Metal Compounds.* Robens Institute Meeting on Toxicity and Therapeutics of Newer Metals and Organometallic Compounds, Guildford, Surrey, U.K.

Galli, B. C., Bürki, P. R., Nyffeler, U. P., and Schindler, W. (1989), *Particle Size Distribution of Airborne Trace Metals.* Int. J. Environ. Anal. Chem. **35** (2), 111–118.

Irth, H., and Frei, R. W. (1988), *New Approaches to Trace Metal Analysis by High Performance Liquid Chromatography.* 3rd IAEAC Workshop on Toxic Metal Compounds (Interrelation between Chemistry and Biology), Follonica, Italy.

de Koe, T., Rozema, J., Broekman, R. A., Otte, M. L., and Ernst, W. H. O. (1988), *Heavy Metals and Arsenic in Water, Sediments and Plants near the Jales Gold and Silver Mine in North Portugal.* Proceedings 3rd International Conference on Environmental Contamination, Venice, pp. 152–154, CEP Consultants Ltd., Edinburgh.

Lithner, G. (1988), *Mobility and Bioavailability of Arsenic and Metals in Lakes Surrounding the Ronnskar Smelters (Sweden).* Proceedings International Conference on Trace Metals in Lakes, Hamilton, Ontario, Canada. Sci. Total Environ., in press.

Mahaffey, K. R. (1989), *Nutrition Essentiality and Toxicity: Comparing What's Known about Cobalt, Arsenic and Lead.* Proceedings 23rd Annual Conference on Trace Substances in Environmental Health. J. Environ. Geochem. Health, in press.

Morgan, J. J. (1989), *Principles of Metal Speciation and Transport (including As(III) Adsorption on Goethite).* Proceedings 5th EPA and ACS Workshop on Metal Speciation and Transport in Groundwater, Jekyll Island, Georgia, in press.

Notini, M. (1988), *Studies on Baltic Sea Ecosystems: Effects of Arsenate on Growth of Algae, Macrofauna and Fish.* Proceedings 1st European SECOTOX Conference on Ecotoxicology, Copenhagen, in press.

Otte, M. L., Rozema, J., Beck, M. A., Ernst, W. H. O., and Broekman, R. A. (1988), *Uptake of Arsenic by Vegetation of the Former Rhine Estuary.* Proceedings 3rd International Conference on Environmental Contamination, Venice, pp. 529–531. CEP Consultants Ltd., Edinburgh.

Schaufelberger, F. A. (1989), *Arsenic (and Arsenic Minerals) in the Canton of Valais, Swiss Alps* (in German). *Neue Zürcher Zeitung, Forschung und Technik*, No. 14, p. 77 (18 January), Zürich.

Wangberg, S.-A., and Blanck, H. (1988), *Increased Tolerance for Arsenate in Microalgal Communities that Have Experienced Arsenate Stress in Limnocorrals. Proceedings International Conference on Trace Metals in Lakes, Hamilton, Ontario, Canada. Sci. Total Environ.*, in press.

Weiler, R. R. (1988), *The Arsenic Cycle in Ontario. Proceedings International Conference on Trace Metals in Lakes, Hamilton, Ontario, Canada. Sci. Total Environ.*, in press.

II.4 Beryllium

WALLACE R. GRIFFITTS, Denver, Colorado, USA
DAVID N. SKILLETER, Carshalton, Surrey, UK

1 Introduction

Beryllium has the lowest atomic number and density of any metal that is stable in the air. Since 1932 it has been produced industrially, because small amounts of beryllium in copper produce a very fatigue-resistant alloy that is as hard as some steel. More recently the metal has been used in neutron generators and in nuclear reactors. Brittleness prevents many uses in aircraft where its low density would otherwise be very valuable. The oxide is useful as a refractory because of its high melting point, high thermal conductivity, and low electrical conductivity.

Beryllium and its compounds are scarcely toxic when taken by mouth, but inhalation of dust of the metal or of many compounds can cause serious lung disease. Embedded in the skin these materials cause dermatitis or ulcers. Many of the effects are evidently immune reactions that vary greatly between individuals. Animal studies show beryllium to be carcinogenic, although epidemiologic studies demonstrate less effect on humans (REEVES, 1986).

2 Physical and Chemical Properties, and Analytical Methods

Beryllium, with atomic number 4, is the first element in the second column of the Periodic Table, with an atomic mass of 9.013, a density of 1.85 g/cm^3, a melting point of 1287 °C, and a boiling point of 2472 °C. Beryllium is as electronegative as aluminum and, therefore, easily forms covalent compounds. Like aluminum, beryllium metal forms an oxide coating that protects it from further oxidation. Beryllium has one stable isotope, ^{9}Be, which is the only one involved in industrial use, and three unstable isotopes, ^{7}Be, ^{8}Be, and ^{10}Be, produced by cosmic rays in the upper atmosphere or produced artificially.

Beryllium-rich materials, viz., ores and industrial products, are commonly determined analytically with instruments based upon the gamma-neutron reaction of beryllium. Small concentrations, mg/g to ng/g, in rocks and biological materials are determined by emission spectrography or atomic absorption spectrometry (HARLBUT, 1978; STIEFEL et al., 1980; WELZ, 1985; ZORN et al., 1986; ZORN and STOEPPLER, 1985; BURBA et al., 1983). Beryllium-bearing organic compounds are determined by mass spectrometry or gas chromatography (NEEB, 1978; TÖLG, 1980). For isotope measurements see, for instance, WÖLFLI et al. (1989).

3 Sources, Production, Important Compounds, Uses, Waste Products, Recycling

3.1 Occurrence

The beryllium content of the earth's crust is estimated to be 2 to 3.5 ppm (parts per million). All but a very small part of this metal is contained in the common rock-forming minerals and not in beryllium-rich minerals. This dispersion is caused by the ability of beryllium to replace silicon, since its ionic radius (0.031 nm) is close to that of silicon (0.041 nm). Plagioclase feldspar probably contains most of the world's beryllium, but pyroxenes, micas, and clays are also important hosts. During crystallization of an igneous magma, beryllium becomes progressively more concentrated in the fluid as crystallization proceeds, resulting in increasing concentrations of beryllium in successively younger igneous rocks in genetically related suites. This enrichment reaches a point where beryllium minerals can form only in the presence of complexing agents, such as F^- or CO_3^{2-}, that keep an unusual amount of beryllium in the residual solutions.

Upon weathering nearly all the beryllium of the original minerals enters clays. Such clays, and quartz, form soils and then are eroded, sorted, and redeposited, ultimately to become shales and sandstones. Most shales contain 2 to 5 ppm of beryllium whereas most sandstones contain less than 1 ppm. The other common sedimentary rocks, the limestones, contain much less than 1 ppm.

ROMNEY and CHILDRESS (1965) showed, by the use of radioactive ^7Be, that beryllium is strongly adsorbed by montmorillonitic soils and by bentonite, but not by kaolinite. Once adsorbed, the beryllium was not displaced by magnesium, barium, or calcium, even at high concentrations.

In contrast to the wide distribution of trace amounts of beryllium, most minerals in which beryllium is a major component are found in few places. By far the most common mineral is beryl ($Be_3Al_2Si_6O_{18}$), which has been the most important ore, being mined in Brazil, India, Southern Africa, and in smaller amounts elsewhere. The less common bertrandite ($Be_2SiO_4 \cdot H_2O$) has become an important ore in Utah, USA, and other uncommon minerals probably will be mined elsewhere.

3.2 Production, Use and Recycling

(see also PETZOW et al., 1985)

Fewer than 4000 tons of beryl ore or 500 tons of beryllium are used annually in the world. The total amount used in the 50-year history of the industry is about 10000 tons. Beryl ores are processed in one of two ways (GRIFFITTS, 1973):

1. They may be sintered with sodium fluorosilicate and the resulting BeF_2 leached with water.
2. They can be fused, heat-treated, and $BeSO_4$ leached with sulfuric acid.

The bertrandite ore of Utah yields its beryllium to a sulfuric acid leach.

The solution resulting from any of these processes are purified, then $Be(OH)_2$ is precipitated and ignited to form BeO, which is then converted to a variety of beryllium compounds or reduced to metal. BeO is a good electrical insulator, and because of its very high heat conductivity it is used where heat dissipating insulators are needed, as in power transistors. Furthermore, it is unaffected by thermal shock. Beryllium metal and oxide are used as moderators or neutron reflectors in nuclear reactors, and the metal is used in airplanes and spacecraft where its lightness is an asset and its brittleness can be tolerated as in brake discs. Most of the beryllium metal produced is added to copper, making an alloy containing about 2 percent beryllium that is the most fatigue-resistant metal and is as hard as some steel.

Large amounts of scrap metal from manufacturing plants are reclaimed, but little beryllium is reclaimed from the final products because of their small size and, commonly, their low beryllium content. BeO based insulators for transistors and copper alloy springs for electric switches are typical products; large heat shields for space vehicles and large parts for nuclear reactors are rare.

4 Distribution in the Environment, Foods, and Living Organisms

Water contains very little beryllium because the small amount that escapes capture by clay minerals during rock weathering and soil formation is largely adsorbed by the surfaces of mineral grains. An investigation of water from several hundred places in Western United States disclosed dissolved beryllium only in three acid mine waters. Other samples had traces of beryllium in suspended clay particles. Surface water of Eastern United States and of Siberia contain 0.1 to 0.9 ppb (parts per billion) of beryllium (WEDEPOHL, 1969). BURBA et al. (1983) found, however, rather lower concentrations in European rivers. Although weathering, mobilization, and displacement are more critical than acid rain, VAN GEEN et al. (1988) and HEATH et al. (1988) found that beryllium may behave similarly to aluminum, and ionic concentrations may increase 2 to 4 times in acid lakes (pH 4.5) and in streams in Maine (up to $1-2$ µg Be per liter during acidic episodes). The content in ocean water is about 0.0002 ppb in the form of $Be(OH)^+$ and $Be(OH)_2$. Near the surface the concentrations are even lower (MEASURES and EDMOND, 1982; see also BURBA et al., 1983, and BRULAND, 1983). In the sediments of Lake Constance (and in lung tissues) increased beryllium deposits were found in the time period before 1960, probably because of the use of beryllium hydride as rocket fuel (ZORN et al., 1988). Coal combustion plays also a role in some areas (HEATH et al., 1988).

Soils collected from about 1300 localities throughout the United States contained from less than 1 to 15 ppm of beryllium, averaging about 1 ppm (SHACKLETTE, 1984). In some small areas rocks and soils may contain larger amounts. Thus some granites contain 10 ppm, as do the soils derived from them. Soils of the beryllium district in the Lost River valley, Alaska, contain from less than 1 to 300 ppm of beryllium and average about 60 ppm (SAINSBURY et al., 1968). Such areas of beryllium-rich soils are small and are in sparsely settled areas that are not important sources of food.

Plants growing in natural soil seldom contain more than trace amounts of beryllium. The air-dried tops of alfalfa plants (*Medicago*) grown in beryllium-rich culture solutions contained 27 ppm beryllium; those of peas (*Pisum*) contained 75 ppm (ROMNEY and CHILDRESS, 1965). The air-dried tops of ladino clover (*Trifolium*) contained 18 to 22 ppm beryllium when grown in kaolinitic soil to which beryllium had been added to the extent of 16 percent of its base-exchange capacity. When grown in a mixed kaolinite-montmorillonite soil containing the same amount of beryllium the clover contained 5 to 7 ppm. These investigators also found that fruit and the edible parts of barley and bean plants contained much less beryllium than other parts of the plants. Foodstuff was found to contain the following amounts of beryllium in air-dried samples: saltwater fish 0.002 to 0.2 ppm, mammal muscle about 0.00075 ppm, mammal bones about 0.003 ppm (BOWEN, 1979); polished rice 0.08 ppm, bread 0.12 ppm, potatoes 0.17 ppm, tomatoes 0.24 ppm, and head lettuce 0.33 ppm (PETZOW and ZORN, 1974). SHACKLETTE (1980) determined the metal contents of 25 kinds of fruits and vegetables from 11 important areas of commercial food production in the United States. He did not find in any ashed sample more than 4 ppm of beryllium, the minimum detectable by the analytical method used. Man is estimated to take in about 20 micrograms of beryllium daily, mainly in food.

The plant-derived organic components of coal contain beryllium. The amounts in coal ash are generally tens to hundreds of parts per million, with values in thousands of parts per million in some coal basins (ZUBOVIC et al., 1961; ZILBERMINTZ and RUSANOV, 1936). The average beryllium content of coal ash of the United States is about 45 parts per million (STADNICHENKO et al., 1956). When the coal is burned some of the beryllium bearing ash may escape into the atmosphere. STERNER and EISENBUD (1951) reported the following average beryllium contents of air of cities in the USA, in nanograms per cubic meter: Boston 0.3, New York 0.5, Brookhaven 0.7, Cleveland 1.3, and Pittsburgh 3.0. MÜLLER (1974) found that city air contained about 12 times the beryllium concentration as rural air. The solids in dusty desert air usually contain several parts per million of beryllium.

5 Uptake, Absorption, Transport and Distribution and Metabolism and Elimination in Plants, Animals, and Humans

Hickory trees (several species of the genus *Carya*) are the most effective known accumulators of beryllium and contain as much as 1 ppm in air-dried leaves (30 ppm in ash). Other broad leaved trees and shrubs contain more than conifers, which seldom contain more than 0.1 ppm (less than 1 ppm in ash). Most studies have shown that the leaves contain more beryllium than the stems, but some desert shrubs in South-western United States and Northern Mexico contain more in the stems. ROMNEY and CHILDRESS (1965) showed that the roots of plants grown in nutrient solutions contain more beryllium than the rest of the plant, but this has not been demonstrated for plants grown in soil. Tundra plants of the Seward Peninsula, Alaska, tend to concentrate beryllium from soils that contain no more than 20 ppm: their ash contains more metal than the soil. Where the soil contains more than 50 ppm of beryllium the plant ash contains less than the soil (SAINSBURY et al., 1968).

Plants can rid themselves of at least small amounts of beryllium. The metal in the leaves of deciduous plants is, of course, lost when the leaves fall. CURTIN et al. (1974) found beryllium in the fluids transpired by several species of conifer trees; the ash of the fluid contained more beryllium than the soil nearby.

As long ago as 1949 CROWLEY et al. used radioactive beryllium added to food to demonstrate that rats absorbed during digestion only 0.2 percent of an administered dose. REEVES (1965) found little beryllium outside of the gastro-intestinal tract, liver, and skeleton of his experimental animals in which about 80% of the beryllium passed into the feces unassimilated, and about 1% of the dose was excreted in urine. Ultimately the retained beryllium was deposited in bone, whether it was administered orally or parenterally, but it may reside in the liver until excreted or translocated to the bone. It was found (ZORN et al., 1988) that an exposure level of 0.002 mg/m^3 Be corresponds to beryllium values of about 7 ng/mL in urine and about 4 ng/mL in blood.

The metal is transported in the blood mainly as colloidal beryllium phosphate and hydroxide complexes weakly associated with plasma globulins (FELDMAN et al., 1953; VACHER and STONER, 1968). These forms are taken up predominantly by the liver and the reticuloendothelial system which can suffer tissue damage (HARD et al., 1977; SKILLETER and PRICE, 1978). Focal parenchymal liver necrosis may be produced (ALDRIDGE et al., 1950; GOLDBLATT et al., 1973). Under some conditions transfer of the metal ion to the spleen (DINSDALE et al., 1981) or bile (CIKRT and BENCKO, 1975) can occur, with redistribution being particularly prominent during periods of physical stress (CLARY et al., 1972).

Beryllium enters cells exclusively by endocytosis, mainly of the colloidal and particulate forms, into lysosomes (WITSCHI and ALDRIDGE, 1968; SKILLETER and PAINE, 1979), a process which is accompanied by a dose-dependent release of the metal ion and lysosomal enzymes into the cell (SKILLETER and PRICE, 1979). The post-lysosomal fate of beryllium depends on the extent of cell damage incurred and the particular cell type in question. Macrophages suffer severe cell disruption with release of the beryllium into surrounding tissues, whereas in hepatocytes the metal ion may bind to a 250 kDa cytosolic protein (KRAMPITZ et al., 1982) or the iron storage protein ferritin (PRICE and JOSHI, 1983). However, subcellular distribution studies have repeatedly found beryllium in the cell nucleus or nucleolus (VORWALD and REEVES, 1959; TRUHAUT et al., 1968). In liver increased nuclear beryllium levels can be correlated with increased acute cellular toxicity (WITSCHI and ALDRIDGE, 1968). At low accumulation levels there appears to be a gradual transfer of the metal to the cell nucleus (SKILLETER and PRICE, 1980).

6 Effects on Plants, Animals, and Humans

6.1 Effects on Plants

ROMNEY and CHILDRESS (1965) reported that increasing the beryllium content of culture solutions from 0 to 16 ppm reduced the yields of peas (*Pisum*), soybean (*Soja max*), alfalfa (*Medicago*), and lettuce (*Latuca*) by more than 50 percent. Beryllium in culture solutions causes stubby root growth and enlarged root tips on all plants studied, even at a Be(NO$_3$)$_2$ concentration of 10^{-3} mol L^{-1}. Beryllium decreased the calcium uptake of the plants and increased the phosphorus uptake. Plant uptake of sodium, potassium, iron, and manganese were not affected by beryllium. ROMNEY and CHILDRESS' results show that beryllium in water soluble compounds is toxic to plants and that soils tend to fix the metal in forms that are unavailable to plants.

The more primitive organisms are affected somewhat differently. STEINBERG (1946) reported that beryllium stimulated growth of the fungus *Aspergillus niger*, and HOAGLAND (1952) found that it stimulated growth of the green alga *Chlorella pyrenoidsa* in a culture solution containing low concentrations of magnesium at a pH above 11.2.

6.2 Oral Uptake by Animals and Humans

Ingestion of beryllium in normal food has not been found to harm animals or man, and fish seem to be especially insensitive to the presence of relatively high concentrations of beryllium (0.15 – 8.3 g/L) in the surrounding water (JUNG, 1973). However, HEATH et al. (1988) found that beryllium, acting concurrently or synergistically with aluminum, is a fish toxin (96 h TL$_{50}$ may be as low as 3 µg Be/L for common guppy fathead minnows and blue gill sun fishes).

6.3 Inhalation by Animals and Humans

Inhalation of beryllium is the most serious human hazard, and clinical problems have occurred in industrial situations rather than in the population at large, where exposure, mainly from the burning of coal and oil, appears not to have posed any major concern. Acute pulmonary beryllium disease involving bronchitis, pneumonitis, and alveolar interstitial oedema is produced following inhalation of soluble beryllium compounds where the salt acidity is the main etiological factor (FRIEMAN and HARDY, 1970). Exposure of the lungs to BeO decreases alveolar clearance rates and macrophage phagocytic capacity (SANDERS et al., 1975; HART et al., 1984), while brief inhalation of a BeSO$_4$ aerosol causes damage to alveolar macrophage membranes, proliferative responses in type II, interstitial and endothelial cells associated with macrophage and leukocyte infiltration (WITSCHI and TOYKA, 1986). Chronic pulmonary beryllium disease is characterized by fine miliary nodulation in the lungs with increasing fibrosis and histological features of granulomata, pneumosclerosis, interstitial cell infiltration, and inclusion body formation within giant cells (SHERWIN et al., 1966; LUTAI et al., 1982). The intensity of the interstitial inflammation, interestingly, allows beryllium disease to be distinguished from sarcoidosis in which the inflammatory response is less marked. It is also known, for example, that low fired (400–500 °C) BeO causes a more severe granulomatosis than high fired (1500–1700 °C) BeO, probably caused by the more porous and rough crystallite structure of the former material (REEVES, 1983).

6.4 Skin Exposure of Animals and Humans

Skin contact with beryllium can result from the many ways in which beryllium metal or beryllium compounds are handled in the various fabrication industries or the laboratory. Systemic absorption through the skin is limited, but ionic beryllium can react with tissue components to generate an immune response (BELMAN, 1969), and beryllium materials embedded in the skin frequently prevent wound healing. Essentially, three skin lesions are usually found: dermatitis (delayed hypersensitivity), ulcerations, and granulomas (VAN ORDSTRAND et al., 1945, CURTIS, 1951; EPSTEIN, 1967). Delayed contact beryllium hypersensitivity has been produced both in experimental animals (MARX and BURRELL, 1973) and humans during occupational exposure (HENDERSON et al., 1972). Diagnostic in vitro tests have been based on lymphocyte production of macrophage inhibition factor (MIF) (HENDERSON et al., 1972; PRICE et al., 1977) or lymphocyte proliferation in the presence of beryllium compounds (HANAFIN et al., 1970; JONES-WILLIAMS and WILLIAMS, 1983). However, the latter lymphocyte transformation (LT) test may only reflect actual exposure since it appears to be reversible after reduction of exposure levels (ROM et al., 1983).

Patch-testing with Be cannot safely be employed as a predictive test for hypersensitivity in humans since this in itself can induce subject sensitization (CURTIS, 1959). Studies with guinea pigs suggest that the degree of high affinity lymphocyte Be binding may indicate the potential to express delayed Be hypersensitivity, particularly in genetically predisposed individuals (SKILLETER and PRICE, 1984; SKILLETER, 1986). The appearance of granulomas frequently follows the hypersensitivity reaction. They are generally regarded as immunological granulomas since they comprise giant and epithelioid cells of mononuclear phagocyte origin associated with lymphocyte infiltration (TURK and PARKER, 1983). Consideration has been given to an adjuvant role of Be compounds (SALVAGGIO et al., 1965; UNANUE et al., 1969). Recent studies show that $Be(OH)_2$ is very active in stimulating immunoblast proliferation (HALL, 1987)). Increased serum concentrations of immunoglobulins, particularly IgG, have also been observed in cases of chronic Be disease (RESNICK et al., 1970), although the putative tissue antigens have not been clearly identified. REEVES (1983) has suggested that the proximal antigen is an adsorptive complex of electrostatically charged beryllium particles with tissue protein rather than the simple ion-bound proteinates generally considered operative for most other sensitizing metals.

6.5 Mutagenic and Carcinogenic Effects
(see also REEVES, 1986; LÉONARD and LAUWERYS, 1987)

The most extensive record of reported cases of occupational and environmental beryllium disease has been provided by the 'Massachusetts Beryllium Case Registry' (HARDY et al., 1967; SPRINCE and KAZEMI, 1980). On the basis of these data it has been concluded that beryllium exposure poses an excess risk of lung cancer (INFANTE et al., 1980), and consequently beryllium is now generally regarded as a suspect carcinogen in man (IARC, 1980; KUSCHNER, 1981). These conclusions are substantiated by many animal studies, although there are species variations in response (IARC, 1980). REEVES (1983) has raised the interesting suggestion that beryllium induced granulomatous hypersensitivity and carcinogenicity may be mutually inhibitory processes dependent on an individual's state of immunocompetence.

Beryllium, in common with other carcinogenic metals, is genotoxic in several in vitro mammalian cell asssay (SIROVER, 1981; BAKER, 1985) and decreases the fidelity of DNA synthesis in other test systems (ZAKOUR et al., 1981). In enzyme induction studies beryllium has also been found to interfere with gene transcription (WITSCHI and MARCHAND, 1971; ORD and STOCKEN, 1981; PERRY et al., 1982), and both in vivo and in cultured cells the metal ion inhibits DNA synthesis and cell division (CHEVREMONT and FIRKET, 1951; WITSCHI, 1970; ABSHER et al., 1983). A more recent examination shows that it is the G_1 phase of the cell cycle which is most sensitive to beryllium (SKILLETER et al., 1983) in contrast to most other toxic metals which cause predominantly an S phase block (COSTA et al., 1982). The detailed mechanism of these effects remains to be established, but it is known that beryllium can act as a macromolecular complexing agent between nuclei acids and proteins (MORIN et al., 1977) and that it selectively binds with high affinity to the regulatory non-histone nuclear proteins (PARKER and STEVENS, 1979). Furthermore, it has been known for some years that beryllium can inhibit certain protein or enzyme phosphorylation reactions. This property of the metal ion, together with its ability to bind to specific regulatory proteins in cells, has been suggested as being the molecular basis for many of the toxic and carcinogenic actions of beryllium (SKILLETER, 1984). Activation of oncogenes particularly of the ras family has been shown recently to be associated with cell transformations caused by a number of carcinogens (MOHR and DUNGWORTH, 1988). The latter authors discussed the reasons for the development of pulmonary tumors by beryllium and its compounds in rats and humans, and for their non-carcinogenicity in hamsters. But it is not yet known whether this mechanism operates for beryllium induced tumors.

7 Hazard Evaluation and Limiting Concentrations

Epidemiological studies are difficult because only a few percent of exposed populations develop symptoms of beryllium disease. Different compounds differ greatly in toxic effect, and there is no relation between the extent of exposure and severity of disease. Symptoms may develop many years after the last known exposure, and beryllium disease is rather difficult to diagnose. Interactions with other constituents or dusts also introduce complications. ZORN et al. (1986) have, however, shown that a short-time exposure to beryllium dust (10–20 h) does not initiate symptoms of an acute beryllium intoxication.

The first limits for exposure to beryllium were established in the United States in 1949, and current regulations recommend that

- in working areas no more than 2 micrograms of beryllium is permitted per cubic meter of air as an average over an 8 hour working day;
- no more than 25 micrograms per cubic meter is permitted in any single determination;
- in community air, no more than 0.01 microgram per cubic meter as a 1-month average should be present.

In the Federal Republic of Germany beryllium and its compounds are included in the MAK list (1987) as material clearly shown to be carcinogenic by animal experiments (the Technical Guiding Concentration is $0.002-0.005$ mg/m^3). In Switzerland it is included in the poison list as especially toxic. Despite the stringent controls, reservations about the absolute safe levels in modern industrial settings have been raised recently (CULLEN et al., 1986). In the European Community cosmetic products containing beryllium and its compounds are prohibited (ZORN et al., 1988).

References

Absher, M., Sylwester, D., and Hart, B. A. (1983), *Time Lapse Cinematographic Analysis of Beryllium-Lung Fibroblast Interactions.* Environ. Res. *30*, 34–43.

Aldridge, W. N., Barnes, J. M., and Denz, F. A. (1950), *Biochemical Changes in Acute Beryllium Poisoning.* Br. J. Exp. Pathol. *31*, 473–484.

Baker, R. S. U. (1985), *Evaluation of Metals in in vitro Assays, Interpretation of Data and Possible Mechanisms of Action,* in: Merian, E., Frei, R. W., Hardi, W., and Schlatter, Ch. (eds.): Current Topics in Environmental and Toxicological Chemistry, Vol. 8, pp. 185–206. Gordon & Breach, New York.

Belman, S. (1969), *Beryllium Binding of Epidermal Constitutents.* J. Occup. Med. *11*, 175–183.

Bowen, H. I. M. (1979), *Environmental Chemistry of the Elements.* Academic Press, London.

Bruland, K. W. (1983), *Trace Elements in Sea-Water,* Chem. Oceanogr. *8*, 176.

Burba, P., Willmer, P. G., Betz, M., and Fuchs, F. (1983), *Atomic Absorption Determination of Beryllium Traces in Natural Water.* Int. J. Environ. Anal. Chem. *13*, 177–191.

Chrevremont, M., and Firket, H. (1951), *Action of Beryllium on Cells Cultured in vitro.* Nature *167*, 772.

Cikrt, M., and Bencko, V. (1975), *Distribution and Biliary Excretion of Beryllium after i.v. Administration in Rats of BeCl$_2$.* Arch. Toxicol. *34*, 53–60.

Clary, J. J., Hopper, C. R., and Stokinger, H. E. (1972), *Altered Adrenal Function as an Inducer of Latent Chronic Beryllium Disease.* Toxicol. Appl. Pharmacol. *23*, 365–375.

Costa, M., Cantoni, O., DeMars, M, and Swartzendruber, D. E. (1982), *Toxic Metals Produce an S-phase Specific Cell Cycle Block.* Res. Commun. Chem. Pathol. Pharmacol. *38*, 405–419.

Crowley, J. F., Hamilton, J. G., and Scott, K. G. (1949), *J. Biol. Chem. 177*, 975–984.
Cullen, M. R., Cherniack, M. G., and Kominsky, J. R. (1986), *Chronic Beryllium Disease in the United States. Semin. Respir. Med. 7*, 203–209.
Curtin, G. C., King, H. D., and Mosier, E. L. (1974), *J. Geochem. Explor. 3*, 245–263.
Curtis, G. H. (1951), *Cutaneous Hypersensitivity due to Beryllium. A Study of Thirteen Cases. Arch. Dermatol. Symphilol. 64*, 470–482.
Curtis, G. H. (1959), *The Diagnosis of Beryllium Disease with Special Reference to the Patch Test. Arch. Ind. Health 19*, 150–153.
Dinsdale, D., Skilleter, D. N., and Seawright, A. A. (1981), *Selective Injury to Rat Liver Kupffer Cells Caused by Beryllium Phosphate: An Explanation of Reticulo Endothelial Blockade. Br. J. Exp. Pathol. 62*, 383–392.
Epstein, W. L. (1967), *Granlomatous Hypersensitivity. Prog. Allergy 11*, 36–88.
Feldman, I., Havill, J. R., and Newman, W. F. (1953), *The State of Beryllium in Blood Plasma. Arch. Biochem. Biophys. 46*, 443–453.
Frieman, D. G., and Hardy, H. L. (1970), *Beryllium Disease. The Relation of Pulmonary Pathology to Clinical Course and Prognosis Based on a Study of 130 Cases from the U.S. Beryllium Case Registry. Human Pathol. 1*, 25–44.
Griffitts, W. R. (1973): *United States Mineral Resources, U.S. Geological Survey Prof. Paper 820*, pp. 85–93. U.S. Government Printing Office, Washington, D.C.
Goldblatt, P. J., Lieberman, M. W., and Witschi, H. P. (1973), *Beryllium Induced Ultrastructural Changes in Intact and Regenerating Liver. Arch. Environ. Health 26*, 48–56.
Hall, J. G. (1984), *Studies on the Adjuvant Action of Beryllium. I. Effects on Individual Lymph Nodes. Immunology 53*, 105–113.
Hanafin, J. M., Epstein, W. L., and Cline, M. J. (1970), *In vitro Studies of Granulomatous Hypersensitivity to Beryllium. J. Invest. Dermatol. 55*, 284–288.
Hard, G. C., Skilleter, D. N., and Reiner, E. (1977), *Correlation of Pathology with Distribution of Be Following Administration of Beryllium Sulphate and Beryllium Sulphosalicylate Complexes to the Rat. Exp. Mol. Pathol. 27*, 197–212.
Hardy, H. L., Rabe, E. W., and Lorch, S. (1967), *United States Beryllium Case Registry (1952–1966). Review of Its Methods and Utility. J. Occup. Med. 9*, 271–276.
Harlbut, J. A. (1978), *Determination of Beryllium in Biological Tissues and Fluids by Flameless Atomic Absorption Spectroscopy. At. Absorpt. Newsl. 17*, 121–124.
Hart, B. E., Harmsen, A. G., Low, R. S., and Emmerson, R. (1984), *Biochemical Cytological and Histological Alterations in Rat Lung Following Acute Beryllium Aerosol Exposure. Toxicol. Appl. Pharmacol. 75*, 454–465.
Heath, R. C., Miller, L. W., Perry, C. M., and Norton, S. A. (1988), *Beryllium in Surface Waters: Sources, Sinks, Mobilization, and Potential Toxicity. Proceedings Conference Trace Metals in Lakes, Hamilton, Canada. Sci. Total Environ.*, in press.
Henderson, W. R., Fukuyama, K., Epstein, W. L., and Spitler, L. E. (1972), *In vivo Demonstration of Delayed Hypersensitivity in Patients with Beryllium. J. Invest. Dermatol. 58*, 5–8.
Hoagland, M. B. (1952), *Beryllium and Growth. II The Effect of Beryllium on Plant Growth. Arch. Biochem. Biophys. 35*, 249–258.
IARC Monograph (1980), *Beryllium and Beryllium Compounds*, in: *Evaluation of the Carcinogenic Risk of Chemicals to Humans. Some Metals and Metallic Compounds*, Vol. 23, pp. 173–204. IARC Publ., Lyon.
Infante, P. F., Wagoner, J. K. and Spince, N. K. (1980), *Mortality Patterns from Lung Cancer and Non-neoplastic Respiratory Disease among White Males in the Beryllium Case Registry. Environ. Res. 21*, 35–43.
Jones-Williams, S. W., and Williams, W. R. (1983), *Value of Beryllium Lymphocyte Transformation Tests in Chronic Beryllium Disease and in Potentially Exposed Workers. Thorax 38*, 41–44.
Jung, K. D. (1973), cited in: Förstner, U., and Wittmann, G. T. W. (eds.) (1979): *Metal Pollution in the Aquatic Environment*, p. 28. Springer Verlag, Berlin.

Krampitz, G., Sauerwald, N., and Zimmermann, K. (1982), *Biochemical Mechanism of Action of Beryllium. I. Beryllium Binding in the Liver of the Hen. Z. Tierphysiol. Tierernaehr. Futtermittelkd. 48*, 120–130.

Kuschner, M. (1981), *The Carcinogenicity of Beryllium. Environ. Health Perspect. 40*, 101–105.

Léonard, A., and Lauwerys, R. (1987), *Mutagenicity, Carcinogenicity and Teratogenicity of Beryllium. Mutat. Res. 186*, 35–42.

Lutai, A. V., Lutai, G. F., Kazantserva, S. V., Fedkina, L. G., and Solormina, S. W. (1982), *Features of Morphological Response of Animal Lungs to Exposure to Various Beryllium Compounds. Gig. Tr. Prof. Zabol. 1*, 23–26.

MAK (1987), *Maximum Concentrations at the Workplace, Report No. XXIII, DFG.* VCH Verlagsgesellschaft, Weinheim-Basel-Cambridge-New York.

Marx, J. J., and Burrell, R. (1973), *Delayed Hypersensitivity of Guinea Pigs to Beryllium Salts. J. Immunol. 111*, 590–598.

Measures, C. I., and Edmond, J. M. (1982), *Beryllium in the Water Column of the Central North Pacific. Nature 297*, 51–53.

Mohr, U., and Dungworth, D. L. (1988), *Relevance to Humans of Experimentally Induced Pulmonary Tumors in Rats and Hamsters. Proceedings ILSI Symposium on Inhalation Toxicology, Hannover, FRG*, pp. 209–232. Springer Verlag, New York.

Morin, N. R., Zeldin, P. E., Kubinski, Z. O., Bhattacharya, P. K., and Kubinski, M. (1977), *Macromolecular Complexes Produced by Chemical Carcinogens and Ultraviolet Radiation. Cancer Res. 37*, 3802–3814.

Müller, J. (1979), in: *Proceedings of the International Conference on Heavy Metals in the Environment London*, pp. 300–303. CEP Consultants Ltd., Edinburgh.

Neeb, R. (1978), *Proceedings Analytika, München.* Cited in: *Chem. Rundsch. 31*, No. 24, 14th June.

Ord, M. G., and Stocken, L. A. (1981), *Enyzme Induction in Rat Liver. The Effect of Be in vivo. Biosci. Rep. 1*, 217–222.

Parker, V. H., and Stevens, C. (1979), *Binding of Beryllium to Nuclear Acidic Proteins. Chem. Biol. Interact. 26*, 167–177.

Perry, S. T., Kulkarni, S. B., Lee, K.-L., and Kenney, F. T. (1982), *Selective Effect of the Metallocarcinogen Beryllium on Hormonal Regulation of Gene Expression in Cultured Cells. Cancer Res. 42*, 473–476.

Petzow, G., and Zorn, P. (1974), *Toxicology of Beryllium Containing Materials. Chem. Ztg. 98*, 236–241.

Petzow, G., Aldinger, Fr., Jönsson, S., and Preuss, O. (1985): *Beryllium and Beryllium Compounds*, in: *Ullmann's Encyclopedia of Industrial Chemistry*, 5th Ed., Vol. A 4, pp. 11–33. VCH Verlagsgesellschaft, Weinheim-Deerfied Beach/Florida-Basel.

Price, D. J., and Joshi, J. G. (1983), *Role of in vitro Tests of Hypersensitivity in Beryllium Workers. J. Biol. Chem. 258*, 10873–10880.

Price, C. D., Jones-Williams, W., Pugh, A., and Joynson, D. H. (1977), *Ferritin. Binding of Beryllium and Other Metal Ions. J. Clin. Pathol. 30*, 24–28.

Reeves, A. L. (1965), *The Absorption of Beryllium from the Gastrointestinal Tract. Arch. Environ. Health 11*, 209–214.

Reeves, A. L. (1983), *The Immunotoxicity of Beryllium*, in: Gibson, G. G., Hubbard, R., and Parke, D. V. (eds.): *Immunotoxicology*, pp. 261–282. Academic Press, New York.

Reeves, A. L. (1986), *Beryllium*, in: Friberg, L., Nordberg, N. F., and Voux, V. B. (eds.): *Handbook on the Toxicology of Metals*, 2nd. Ed., Vol. II, pp. 95–116. Elsevier, Amsterdam.

Resnick, H., Roche, M., and Morgan, W. K. C. (1970), *Immunoglobulin Concentrations in Berylliosis. Am. Rev. Respir. Dis. 101*, 504–510.

Rom, W. N., Lockey, J. E., Bang, K. M., Dewitt, C., and Jones Jr., R. E. (1983), *Reversible Beryllium Sensitization in a Prospective Study of Beryllium Workers. Arch. Environ. Health 38*, 302–307.

Romney, E. M. and Childress, J. D. (1965), *Soil Sci. 100*, 210–217.

Sainsbury, C. L., Hamilton, J. C., and Huffmann Jr., C. (1986), *Geochemical Cycle of Selected Trace Elements in the Tin-tungsten-beryllium District, Western Seward Peninsula, Alaska*, US Geol. Surv. Bull. 1242-F. US Government Printing Office, Washington D.C.

Salvaggio, J. E., Flax, M. H., and Leskowitz, A. (1965), *Studies on Immunization. II. The Use of Beryllium as a Granuloma Producing Agent in Freund's Adjuvant.* J. Immunol. 95, 846–854.

Sanders, C. L., Cannon, W. C., Powers, G. J., Adee, R. R., and Meirer, D. M. (1975), *Toxicology of High Fired Beryllium Oxide Inhaled by Rodents.* Arch. Environ. Health 30, 546–551.

Shacklette, H. T. (1980), *Elements in Fruits and Vegetables from Areas of Commercial Production in the Conterminous United States.* U.S. Geological Survey Prof. Paper 1178, p. 149. U.S. Government Printing Office, Washington, D.C.

Shacklette, H. T. (1984), *Element Concentrations in Soils and other Surficial Materials of the Conterminous United States.* U.S. Geological Survey Prof. Paper 574-D, I-105. U.S. Government Printing Office, Washington, D.C.

Sherwin, R. P., Smart, R. H., and Scarborough, G. C. (1966), *Chronic Berylliosis and Calcospherite Deposition.* Arch. Environ. Health 12, 237–245.

Sirover, M. A. (1981), *Effects of Metals in in vitro Bioassays.* Environ. Health Perspect. 40, 163–172.

Skilleter, D. N. (1984), *Biochemical Properties of Beryllium Potentially Relevant to Its Carcinogenicity.* Toxicol. Environ. Chem. 7, 213–228.

Skilleter, D. N. (1986), *Selective Cellular and Molecular Effects of Beryllium on Lymphocytes.* Toxicol. Environ. Chem. 11 (4), 301–312.

Skilleter, D. N., and Paine, A. J. (1979), *Relative Toxicities of Particulate and Soluble Forms of Beryllium to a Rat Liver Parenchymal Cell Line Culture and Possible Mechanisms of Uptake.* Chem. Biol. Interact. 24, 19–33.

Skilleter, D. N., and Price, R. J. (1978), *The Uptake and Subsequent Loss of Beryllium by Rat Liver Parenchymal and Nonparenchymal Cells after Intravenous Administration of Particulate and Soluble Forms.* Chem. Biol. Interact. 20, 383–396.

Skilleter, D. N., and Price, R. J. (1979), *The Role of Lysosomes in the Hepatic Accumulation and Release of Beryllium.* Biochem. Pharmacol. 28, 3595–3599.

Skilleter, D. N., and Price, R. J. (1980), *Apparent Two-phase Beryllium Labelling of Hepatic Cell Nuclei Isolated after Intravenous Administration of Beryllium Compounds to Rats.* Arch. Toxicol. 45, 75–80.

Skilleter, D. N., and Price, R. J. (1984), *Lymphocyte Beryllium Binding. Relationship to Development of Delayed Beryllium Hypersensitivity.* Int. Arch. Allergy Appl. Immunol. 73, 181–183.

Skilleter, D. N., Price, R. J., and Legg, R. F. (1983), *Specific G_1-S Phase Cell Cycle Block by Beryllium as Demonstrated by Cytofluorometric Analysis.* Biochem. J. 216, 773–776.

Sprince, N. L., and Kazemi, H. (1980), *U.S. Beryllium Case Registry through 1977.* Environ. Res. 21, 44–47.

Stadnichenko, T., Zubovic, P., and Sheffey, N. B. (1956), *Minor Elements in Ash of American Coal.* Abstr.: Geol. Soc. Am. Bull. 67, No. 12, Part 2, p. 1735.

Steinberg, R. A. (1946), Am. J. Bot. 33, 210–214.

Sterner, J. H., and Eisenbud, M. (1951), *Epidemiology of Beryllium Intoxication.* Arch. Ind. Hyg. 4, 123–151.

Stiefel, Th., Schulze, K., Tölg, G., and Zorn, H. (1980), *Fresenius Z. Anal. Chem.* 300, 189–190.

Tölg, G. (1980), *Proceedings 8th International Microchemical Symposium, Graz.* Cited in: Chem. Rundsch. 33, No. 43, October.

Truhaut, R., Festy, B., and LeTalaer, J. Y. (1968), *Interaction of Beryllium with DNA and Its Incidence with Some Enzymatic Systems.* C.R. Acad. Sci. Paris Ser. D 266, 1192–1195.

Turk, J. L., and Parker, D. (1983), *Immunological Aspects of Granulomas,* in: Gibson, G. G., Hubbard, R., and Parke, D. B. (eds.): *Immunotoxicology,* pp. 251–259. Academic Press, New York.

Unanue, E. R., Askonas, B. A., and Allison, A. C. (1969), *A Role of Macrophages in the Stimulation of Immune Responses by Adjuvants.* J. Immunol. 103, 71–78.

Vacher, J., and Stoner, H. B. (1968), *The Transport of Beryllium in Rat Blood. Biochem. Pharmacol. 17*, 93–107.

van Geen, A., Measures, C. I., and Boyle, E. A. (1988), *Trace Metals and pH in Lakes of the Adirondack State Park. Proceedings Conference Trace Metals in Lakes ,Hamilton, Canada. Sci. Total Environ.*, in press.

van Ordstrand, H. A., Hughes, R., DeNardi, J. M., and Carmody, M. G. (1945), *Beryllium Poisoning. J. Am. Med. Assoc. 129*, 1084–1095.

Vorwald, A. J., and Reeves, A. L. (1959), *Pathologic Changes Induced by Beryllium Compounds. Arch. Ind. Health 19*, 190–199.

Wedepohl, K. H. (1969), *Handbook of Geochemistry*, Vol. II-I, Chap. 4. Springer Verlag, New York-Heidelberg-Berlin.

Welz, B. (1985), *Atomic Absorption Spectrometry*, pp. 273–274, 352, 389ff. VCH Verlagsgesellschaft, Weinheim-Deerfield Beach/Florida-Basel.

Witschi, H. P. (1970), *Effects of Beryllium on Desoxyribonucleic Acid Synthesizing Enzymes in Regenerating Rat Liver. Biochem. J. 120*, 623–634.

Witschi, H. P., and Aldridge, W. N. (1968), *Uptake, Distribution and Binding of Beryllium to Organelles of the Rat Liver Cell. Biochem. J. 106*, 811–820.

Witschi, H. P., and Marchand, P. (1971), *Interference of Beryllium with Enzyme Induction in Rat Liver. Toxicol. Appl. Pharmacol. 20*, 565–572.

Witschi, H. P., and Toyka, A. F. (1986), *Acute Pulmonary Toxicity of Beryllium Sulphate Inhalation in Rats and Mice: Cell Kinetics and Histopathology. Toxicol. Appl. Pharmacol. 85*, 248–256.

Wölfli, W., et al. (1989), *Determination of Particle Sink Velocity (Sedimentation) in Lakes by Controlling the $^{10}Be/^{7}Be$ Ratios* (in German). *EAWAG Annual Report*, pp. 4–19/4–20.

Zakour, R. A., Kunkel, T. A., and Loeb, L. A. (1981), *Metal Induced Infidelity of DNA Synthesis. Environ. Health Perspect. 40*, 197–205.

Zilbermintz, V. A., and Rusanov, A. K. (1936), *Acad. Sci. URSS Comptes Rendus 2*, 27–31.

Zorn, H. R., and Stoeppler, M. (1985), *Beryllium in Urine*, in: Angerer, J., and Schaller K. H. (eds.): *Analysis of Hazardous Substances in Biological Materials*, Vol. 1, pp. 57–65.

Zorn, H. R., Stiefel, Th., and Porcher, H. (1986), *Toxicol. Environ. Chem. 12*, 163–171.

Zorn, H. R., Stiefel, Th., Beurs, J., and Schlegelmilch, R. (1988), *Beryllium*, Chapter 10, in: Seiler, H. G., Sigel, H., and Sigel, A. (eds.): *Handbook on Toxicity of Inorganic Compounds*, pp. 105–114. Marcel Dekker, New York.

Zubovic, P., Stadnichenko, T., and Sheffey, N. B. (1961), *Geochemistry of Minor Elements in Coals of the Northern Great Plains Coal Province*, U.S. Geological Survey Bull. 1117-A, 1–58. U.S. Government Printing Office, Washington, D.C.

II.5 Bismuth

DAVID W. THOMAS, Adelaide, Australia

1 Introduction

Bismuth is a relatively rare element, but some alloys and inorganic and organic compounds have useful properties. Because of its position in the Periodic System it shows some similarities to lead, arsenic, and antimony (EMMERLING et al., 1986). Soluble bismuth salts form insoluble basic salts when added to water, and insoluble bismuth oxychloride when added to hydrochloric acid. These properties may limit the bioavailability in the environment and the gastrointestinal absorption of bismuth when bismuth salts are ingested orally by animals and humans.

Bismuth is present in sea water, marine animals, and land plants in very low amounts. Even smaller amounts are found in land animals, probably as a result of the limited gastrointestinal absorption. Two forms of toxicity due to bismuth have been described in man: parenteral administration of various bismuth containing compounds has resulted in an "epithelial-cutaneous" form of toxicity, whereas oral ingestion of bismuth subgallate and bismuth subnitrate has led to a "neurotoxicity" in some patients with the development of a reversible metabolic encephalopathy. Patients receiving bismuth containing medications should be monitored. On the other hand, ecotoxicological, environmental, and occupational harms are not known (EMMERLING et al., 1986).

2 Physical and Chemical Properties, and Analytical Methods

Bismuth is a grayish-white lustrous metal with a relative atomic mass of 208.98. Only one natural stable isotope (atomic mass 209) is known. However, in the natural decay chains of radioactive elements and in nuclear transformation unstable isotopes have been observed (KRÜGER et al., 1985). It is the heaviest naturally occurring stable element but is easily malleable. The melting point of bismuth is 271.3 °C and its boiling point is 1560±5 °C. It has a specific gravity of 9.747. Bismuth is the most diamagnetic of all metals, and its thermal conductivity is the lowest of all metals except mercury, but it is brittle. It has a high electrical resistance and the highest increase in electrical resistance when placed in a magnetic field (Hall effect) of any metal. On solidification, bismuth expands making it suitable for the manufacture of sharp castings of objects subject to damage at high temperatures. Alloys with tin and cadmium have low melting points. It burns in air with a blue flame.

Bismuth usually forms compounds in a valence state of three; this particularly applies to bismuth containing compounds of biological significance. Its soluble salts are characterized by forming insoluble basic salts on the addition of water. In an acid environment (as occurs in the stomach of animals) insoluble bismuth oxychloride (BiOCl) forms. This compound can bind to protein which may explain the limited absorption of bismuth from the gastrointestinal tract. Bismuth can also combine with sulfhydryl groups to form thiobismuthite compounds. Toxic states associated with bismuth have been described, but are uncommon. In the +5 oxidation state bismuth is a strong oxidizing agent, e.g., $NaBiO_6$ or BiF_5 (KRÜGER et al., 1985).

Several methods have been developed for the analysis of bismuth. EIDECKER and JACKWERTH (1987) discussed multi-element preconcentration of soil and sediment samples. Spectrophotometric methods can be used to measure concentrations of bismuth in the range of 5 to 25 mg/L (24 to 120 µmol/L). A method involving the formation of a 1:1 complex with mucic acid (MuH_6) has been described for the analysis of bismuth(III) in pharmaceutical preparations (GÓNZALEZ-PORTAL et al., 1986). Electrothermal atomic absorption spectrophotometry can be used to measure concentrations of bismuth in the range of 0.1 to 10 mg/L (0.5 to 50 µmol/L). This technique has been applied to the direct measurement of bismuth in alloys (WELCHER et al., 1974). For the measurement of bismuth in biological materials, methods with even greater sensitivity are required. Electrothermal atomic absorption spectrophotometry combined with hydride generation has been described for the measurement of bismuth in blood and urine (ROONEY, 1976). An improved version of this method has been described in which the detection limit of bismuth has been lowered to 0.1 µg/L (0.5 pmol/L) (FROOMES et al., 1988). LEE (1982) and HEINRICHS and KELTSCH (1982) determined bismuth by flameless atomic absorption spectrometry with volatilization. For the determination of bismuth in food WOLNIK et al. (1981) and HAHN et al. (1982) combined a hydride generation/condensation system with inductively coupled argon plasma emission spectrometry. Further literature references are summarized at the end of this chapter (e.g., on anodic stripping voltammetry and chromatographic flow injection analysis).

3 Sources, Production, Important Compounds, Uses, Waste Products, Recycling

Bismuth occurs in the earth's crust at an average concentration of 0.96 µmol/kg (200 µg/kg) or less, making it a relatively rare element (see Chapter I.1). Its average abundance in the continental crust is 0.39 µmol/kg (80 µg/kg), which is similar to the mean concentrations in sedimentary and low to medium grade metamorphic rocks (HEINRICHS et al., 1980). It is found at even higher concentrations in certain ores, such as bismuthinite (Bi_2S_3) and bismite (Bi_2O_3), from which it may be extracted as a bismuth concentrate. The mining of such ores is not conducted on a large

Table II.5-1. Major Mining Production of Bismuth[a] Excluding the United States[b] (tons)

Country	1982	1983	1984	1985[c]	1986[d]
Australia (in concentrates)	1501 (36.5%)	1411 (35.4%)	1352 (36.1%)	1402 (29.4%)	998 (24.5%)
Japan (metal)	486 (11.8%)	573 (14.4%)	563 (15.0%)	642 (13.5%)	635 (15.6%)
Mexico (various)	606 (14.7%)	545 (13.7%)	433 (11.6%)	925 (19.4%)	898 (22.1%)
Peru (various)	760 (18.5%)	678 (17.0%)	650 (17.4%)	785 (16.5%)	680 (16.7%)
Other	755 (18.4%)	774 (19.4%)	747 (19.9%)	1008 (21.2%)	855 (21.0%)
Total	4109	3981	3745	4762	4066

[a] Adapted from *Minerals Yearbook 1986*, Bureau of Mines, U.S. Department of the Interior, Washington D.C., 1988, Vol. I, p. 166
[b] Withheld to avoid disclosing company proprietary data
[c] Preliminary figures
[d] Estimated figures

scale because of a slump in world prices for bismuth. Major sites of mining outside Australia occur in Mexico, Japan, Peru, China, and the United States. The first four countries contributed 81% of such mining in 1983 (Table II.5-1; see also SAAGER, 1984; KRÜGER et al., 1985). Relatively highest resources exist in Japan, Australia, and Bolivia.

Most bismuth is, however, produced as a by-product in the refining of lead, copper, tin, silver, and gold containing ores. Bismuth normally follows lead in the refining process. The production of crude bismuth from bismuth-rich mixed concentrates, from lead concentrates, or from copper and tin concentrates has been described by KRÜGER et al. (1985). The annual world production is in the order of 4000 t (SAAGER, 1984; EMMERLING et al., 1986).

Metallic bismuth is used as a component of a variety of low melting alloys (particularly with tin and cadmium, for instance, for electric fuses or sprinkler systems (SAAGER, 1984; KRÜGER et al., 1985)), in tempering baths for the production of steel, as a metallurgical additive to free-machining steel, in the silvering of mirrors and in dentistry. A new application may be in high-temperature superconduction, when, for instance, perovskite structural units are combined with Bi-Sr-Ca-Cu oxide layers (BEDNORZ, 1988; ZANDBERGEN et al., 1988; LÖHLE et al., 1989; see also Chapter II.8), leading to a critical temperature of 105 K.

Bismuth containing compounds are used in a variety of industrial processes and products, such as semiconductors and cathodes in batteries, catalysts in the petrochemical, heavy organic and biochemical industries and in the preparation and recycling of uranium nuclear fuels. The Merck Index (WINDHOLZ et al., 1983) lists a total of 37 bismuth compounds, 18 of which have pharmaceutical uses. In the United States, 46% of the bismuth is used in the pharmaceutical and cosmetic industry, 26% in engineering, and 27% in the metal industry (SAAGER, 1984).

Table II.5-2. Bismuth Metal Consumed in the United States[a] (tons)

Use	1985	1986
Fusible alloys	277 (23.1%)	290 (21.9%)
Metallurgical	303 (25.2%)	350 (26.4%)
Other alloys	10 (0.8%)	13 (1.0%)
Chemicals[b]	601 (50.1%)	663 (50.1%)
Other	9 (0.8%)	8 (0.6%)
Total	1200	1324

[a] Adapted from *Minerals Yearbook 1986*, Bureau of Mines, U.S. Department of the Interior, Washington D.C., 1988, Vol. I, p. 164
[b] Includes industrial and laboratory chemicals, cosmetics and pharmaceuticals

Bismuth oxychloride and other salts are used in a variety of cosmetic preparations including creams, dusting powders, hair dyes, tints and colorings, and freckle removers. Soluble salts such as bismuth subsalicylate, sodium triglycollamate and trioglycollate have been used parenterally for infectious diseases (particularly syphilis) and warts. Bismuth salicylate, subcarbonate, subcitrate, subnitrate, glycobiasol and other salts are used orally, or have been proposed for use, in the treatment of reflux oesophagitis (BORKENT and BEKER, 1988), gastritis (MCNULTY et al., 1986), duodenal ulceration (DEKKER et al., 1986), indigestion (HAILEY and NEWSOM, 1984), diarrhea and other gastrointestinal disorders. Especially CBS (bismuth subcitrate, eventually in combination with antibiotics) is very effective still in concentrations of 10–16 mg Bi/L to inactivate mucosal *Campylobacter pylori*. This microorganism is in 80 to 90 percent of the cases responsible for gastritis and is an important cofactor for stomach and duodenum ulcers (TRUEB, 1989; see also Chapter I.14b, Sect. 4.1).

Commercial usage of bismuth per annum is summarized in Table II.5-2. Usage in pharmaceuticals, including industrial and laboratory chemicals and cosmetics, accounts for half of bismuth metal consumed in the United States in 1986, with use in alloys and as metallurgical additives accounting for most of the remaining use. This latter use may be increasing as bismuth replaces lead as an additive to improve free-machining qualities of steel. If the use of bismuth increases, there may arise greater waste problems. Since application is dissipative, recycling is not important. On the other hand, substitution of bismuth compounds is often possible, for instance, in therapeutics by antibiotica and magnesium or aluminum oxides, in cosmetics by mica and fish-scales, in low melting point alloys by plastic materials, and in steel additions by selenium or tellurium (SAAGER, 1984).

4 Distribution in the Environment, in Foods, and in Living Organisms

(see also Additional Recommended Literature)

In view of the significant production and widespread use of bismuth and its compounds, some must enter the environment and food chains. Older analytical data may, however, be wrong. Bismuth is not detectable in rain water, soil solutions, or river waters. However, it is detectable in sea water in low concentration. At the surface and down to a depth of about 1000 m one finds concentrations of 0.2 to 0.1 pmol/L (0.00004 to 0.00002 ppb), at 3000 m depth 0.015 pmol/L (0.000003 ppb), a profile behavior similar to that of manganese (BRULAND, 1983; LEE et al., 1985/86). There appear to be low concentrations of bismuth in marine animals and land plants (Table II.5-3), although the mechanisms involved in bismuth accumulation and the forms in which it is present in these live forms are not known. Bismuth concentrations in land animals and mammalian tissues are considerably lower. This is probably a result of the limited gastrointestinal absorption of bismuth, because of its poor solubility and propensity to form insoluble bismuth oxychlorides in the mammalian gastrointestinal tract. In man, small amounts of bismuth are excreted in the urine, indicating some gastrointestinal absorption, with small amounts also being detectable in blood (see Table II.5-4).

There are no known biological functions for which bismuth is considered essential. The amounts of bismuth observed in various life forms are considered to be

Table II.5-3. Occurrence of Bismuth in Plants and Animals

Sample	Average Content[a]
Marine animals	0.19 – 1.44 µmol/kg d.w. (40 – 300 µg/kg)[b]
Land plants	0.29 µmol/kg d.w. (= 60 µg/kg)[c]
Land animals	0.02 µmol/kg d.w. (= 4 µg/kg)[d]
Mammalian blood	0.05 µmol/L (= 10 µg/kg)

[a] Adapted from BOWEN (1966)
[b] Molluscs may contain more (BOWEN, 1979)
[c] According to newer studies less than 20 µg/kg (BOWEN, 1979)
[d] Bovine liver contains about 10 µg/kg d.w. (WARD, 1987)

Table II.5-4. Bismuth in Blood and Urine of Normal Individuals

Sample	Number	Value[a]
Blood concentration	67	0.01 (±0.009) µmol/L
Urine excretion	64	0.06 (±0.14) µmol/day

[a] Mean (± standard deviation), author's observations

without harmful effect. However, in man quite considerable accumulations of bismuth can develop under specific circumstances, and these have resulted in two distinct forms of toxicity (see Sect. 6).

5 Uptake, Absorption, Transport and Distribution, Metabolism and Elimination in Plants, Animals, and Humans

Bismuth is not considered to be an essential element for plants and animals. However, because of its low abundance some non-significant accumulation of bismuth may occur in marine animals and land plants, and to a lesser extent in land animals including man (Tables II.5-3 and II.5-4). Although intestinal absorption is limited in man by the poor solubility of bismuth and its propensity to form insoluble oxychloride salts, some absorption must occur to produce measurable concentrations in blood and excretion in urine (Table II.5-4). This has been confirmed by several studies which have examined gastrointestinal mucosal uptake. In rats fed colloidal bismuth subcitrate, mucosal uptake of bismuth was observed in the jejunum, ileum, duodenum, and stomach (in descending order of magnitude). Subcellular fractionation studies showed brush border membrane and cytosolic localization (STIEL et al., 1985). Similar studies in man have shown bismuth uptake in gastric mucosa (LAMBERT et al., 1988). Pharmacokinetic studies in man after oral administration of colloidal bismuth subcitrate have shown that peak plasma concentrations of bismuth occurred within 15–60 minutes after ingestion (HESPE et al., 1988), and an apparent biphasic elimination pattern (McLEAN et al., 1988).

Studies with radioisotopes of bismuth have shown accumulation in kidney, liver, spleen, bone, lung, heart, and muscle (in descending order of magnitude) after parenteral administration of $^{206}BiCl_3$ or ^{206}Bi citrate in rats (VAN DEN BROECK, 1963; ERIDANI et al., 1964; RUSS et al., 1975). More recent studies have shown similar patterns of tissue distribution in rats given tracer doses of colloidal bismuth subcitrate orally (LEE et al., 1980) or radioisotopes of various bismuth salts given intraperitoneally (ZINDENBERG-CHERR et al., 1987).

These and other studies also provide evidence that bismuth enters subcellular organelles, including the nucleus, mitochondria, and membranes with specialized functions such as synapses, contracting midbodies and spreading acrosomes (ZINDENBERG-CHERR et al., 1987; LOCKE et al., 1987; WOODS and FOWLER, 1987). There is no evidence that bismuth actively participates in any physiological metabolic activities in plants or animals.

The majority of bismuth administered orally in man is unabsorbed and excreted in the feces. The absorbed bismuth fraction is predominantly excreted in the urine, consistent with the considerable renal accumulation which has been observed after the parenteral administration of bismuth. Distribution studies with radioisotopes of bismuth also demonstrate accumulation of bismuth in organs of the gastrointestinal tract and in feces (ZINDENBERG-CHERR et al., 1987) suggesting that some bismuth may be eliminated by this route as well.

Table II.5-5. Blood Concentrations, Urinary Excretions and Renal Clearances of Bismuth in Asymptomatic Patients and Patients with "Neurotoxicity" while Ingesting Bismuth Salts

Clinical State and Salt Implicated	Blood Concentration (µmol/L)	Urinary Excretion (µmol/day)	Renal Clearance (mL/min)
Asymptomatic:			
Bismuth subgallate[a]	0.18 ± 0.11 (10)[b]	1.61 ± 1.74 (10)	6.2
Bismuth subnitrate[c]	0.17 ± 0.10 (9)	0.97 ± 0.61 (9)	4.0[d]
Bismuth subcitrate[a]	0.05 ± 0.03 (8)	1.20 ± 0.61 (8)	16.7
Neurotoxic:			
Bismuth subgallate[a]	1.00 ± 0.93 (8)	1.52 ± 0.72 (8)	1.1
Bismuth subnitrate[c]	4.27 ± 3.39 (39)	7.74 ± 10.90 (21)	1.3[d]

[a] Mean ± standard deviation, author's observations (THOMAS et al., 1977)
[b] Number of patients
[c] Mean ± range, various literature sources
[d] Estimates

Studies on patients receiving bismuth subgallate, bismuth subcitrate or colloidal bismuth subcitrate orally show significant urinary excretion of bismuth (see Table II.5-5), suggesting that this is the main route of elimination of absorbed bismuth in man.

6 Effects on Plants, Animals, and Humans

BOWEN (1979) mentioned that 27 mg Bi/L are toxic to plants, and 160 mg Bi/day are lethal to rats. The oral LD_{50} is 22 g/kg for rats, and 484 mg/kg for rabbits (KRÜGER et al., 1985). The lowest published oral lethal dose for humans is 221 mg/kg (KRÜGER et al., 1985).

Bismuth appears to have an effect on microorganisms, interfering with their growth. This is probably the basis of the oral use of pharmaceutical preparations containing bismuth for a variety of gastrointestinal disorders (see Sect. 3), including reduction of fecal odor in patients with colostomies (BURNS et al., 1974) and treatment of peptic ulcer (MCNULTY et al., 1986). Parenteral administration of bismuth was also used as an early treatment for syphilis (HEYMAN, 1944; SOLLMAN, 1938).

After intramuscular injections of soluble bismuth compounds, various toxic effects have been described in man (HEYMAN, 1944). With large doses, or smaller doses repeated over longer periods of time, toxic effects involved the kidney, liver, skin, and epithelial surfaces in intimate contact with body fluids. Subjects complained of anorexia, nausea, vomiting, colicky abdominal pain and diarrhea. Involvement of the oral cavity included pigmentation of the gums and ulcerative stomatitis. The large bowel was also affected with pigmentation and colitis. Cervicovaginitis associated with vaginal pigmentation occurred in females. An exfoliative dermatitis has also been described. One of the most common toxic effects was that of renal tubular damage, extending in some cases to acute tubular necrosis (URIZAR and VERNIER, 1966; RANDALL et al., 1972). Nephrotic syndrome has also been recorded. Jaundice and various bleeding disorders have been described, with multifocal hepatic necrosis as their most likely origin. The only central nervous system effect observed in this "epithelial-cutaneous" form of toxicity was that of headache.

The parenteral administration of soluble bismuth compounds results in accumulations of bismuth at sites of excretion, particularly the kidney (SOLLMAN et al., 1938). Epithelial damage and tissue necrosis indicate a predeliction for bismuth accumulation at sites of fluid and electrolyte transport. The similarity with toxic effects of lead are striking, with the notable exceptions of the peripheral neuropathy and the encephalopathy associated with lead toxicity. There appears to be limited penetration of the "blood-brain barrier" when bismuth is administered parenterally.

In 1974 a hitherto undescribed syndrome associated with the oral ingestion of "insoluble" bismuth subgallate and bismuth subnitrate was described independently in Australia and France (BURNS et al., 1974; BUGE et al., 1974). This comprised a unique, highly characteristic and reversible form of encephalopathy consisting of confusion, hallucinations, tremulousness, clumsiness, myoclonus, and ataxia. In severe cases coma, epilepsy, and death eventually occurred. In France some subjects also developed arthropathies associated with osteonecrosis (BUGE et al., 1975). In Australia this syndrome occurred in a proportion of subjects with end-colostomies who had been ingesting bismuth subgallate to reduce colostomy odor.

In France an identical syndrome was observed almost simultaneously in a proportion of subjects ingesting bismuth subnitrate for constipation and other gastrointestinal disorders. To date, this significant "neurotoxicity" has been observed in only one other circumstance. Two patients in West Germany developed an organic brain syndrome thought to be due to bismuth absorbed from a skin cream (KRÜGER et al., 1976). These effects are similar to toxic effects associated with aluminum, either when given orally or absorbed during hemodialysis.

Not all subjects ingesting bismuth subgallate or bismuth subnitrate developed "neurotoxicity". Blood concentrations of bismuth in asymptomatic subjects were well below the currently stated upper safe limit of 0.24 µmol/L (see Table II.5-5). Those who developed "neurotoxicity" had much higher concentrations of bismuth in the blood. That only a proportion of subjects ingesting either bismuth subgallate or bismuth subnitrate developed "neurotoxicity" remains unexplained. It is possible that when bismuth compounds are taken orally, bismuth is absorbed from the gastrointestinal tract and the bismuth species so produced have a greater predeliction for penetrating the "blood-brain barrier" and less predeliction for accumulating in epithelial tissues.

The renal clearances of bismuth in normal people (no medication) and asymptomatic patients ingesting bismuth subgallate and bismuth nitrate are similar. In patients who developed "neurotoxicity" the renal clearances of bismuth were considerably reduced despite otherwise normal renal function, with the result that higher concentrations of bismuth were observed in the blood (Table II.5-5).

Such a situation could arise as a result of the formation of large molecular chelates or protein complexes involving bismuth. When bismuth subgallate is added to blood in vitro, large molecular weight complexes can form (THOMAS et al., 1983). Irrespective of the precise mechanism, reduced renal clearance of bismuth would promote its retention, and because of lipid solubility, or some other characteristics, increase the likelihood of passage across the "blood-brain barrier" producing the "neurotoxicity" observed in patients who received these compounds.

In contrast, the renal clearance of bismuth in patients ingesting colloidal bismuth subnitrate is much greater, with correspondingly lower blood concentrations of bismuth. Retention of bismuth and resulting toxicity would seem less likely with this compound. Extensive clinical experience has shown it to be free of toxicity (HAILEY and NEWSOM, 1984; DEKKER et al., 1986; MCNULTY et al., 1986; BORKENT and BEKER, 1988), except for one case report of neuropsychiatric symptoms following ingestion of this compound (WELLER, 1988).

So far there is no evidence for carcinogenicity, mutagenicity, and teratogenicity of bismuth compounds (KRÜGER et al., 1985).

7 Hazard Evaluation and Limiting Concentrations

Little information is available on safety limits of exposure to metallic bismuth or any of its compounds. The MAK Report XXIV (1988) does not list bismuth compounds. A threshold limit value (TLV) of 5 mg/m^3 has been set for bismuth telluride by the American Conference of Governmental Industrial Hygienists in 1971 (PLUNKETT, 1987).

For pharmaceutical use of bismuth containing compounds, the upper limit of safety for the concentration of bismuth in blood has been proposed at 0.24 µmol/L (50 µg/L), and the lower limit of toxicity has been proposed at 0.48 µmol/L (100 µg/L). In the author's experience, patients who had been ingesting bismuth subgallate but not showing any toxic symptoms, had a mean blood bismuth concentration of 0.18 µmol/L with an overall range of 0.10 to 0.48 µmol/L (THOMAS et al., 1977). Except for one patient, all values were less than 0.24 µmol/L. Of those patients who exhibited "neurotoxicity", the mean blood bismuth concentration was 1.43 µmol/L with an overall range of 0.26 to 3.11 µmol/L. Two of these patients had blood bismuth concentrations less than 0.48 µmol/L. Values in patients who had been ingesting bismuth subnitrate and exhibiting "neurotoxicity" were generally higher (see Table II.5-5). This suggests that the form in which bismuth is ingested is important, and indicates the tenuous nature of proposed safe and toxic values for concentrations of bismuth in blood.

Individuals regularly or intermittently receiving preparations containing bismuth, either orally or parenterally, should be assessed periodically for either of the two forms of bismuth toxicity described above. All forms of bismuth should be withheld from individuals showing signs of toxicity. Measurements of blood bismuth concentrations may also be useful in patients taking preparations containing bismuth. All forms of bismuth should be withheld when blood bismuth concentrations are greater than 0.48 µmol/L. Those individuals whose values fall between 0.24 and 0.48 µmol/L should be carefully reviewed.

References

Bednorz, J. G. (1988), *High-temperature Superconduction in Oxides*. ACHEMA Plenary Lecture, see *Rep. Swiss Chem. 10* (11), 49; see also an information on the *International Congress on Superconduction, Interlaken*, in: *Neue Zürcher Zeitung*, No. 55, p. 7 (7 March, 1988), Zürich.

Borkent, M. V., and Beker, J. A. (1988), *Treatment of Ulcerative Reflux Oesophagitis with Colloidal Bismuth Subcitrate in Combination with Cimetidine*. *Gut 29*, 385–389.

Bowen, H. J. M. (1966), *Trace Elements in Biochemistry*. Academic Press, New York.

Bowen, H. J. M. (1979), *Environmental Chemistry of the Elements*. Academic Press, London-New York-Toronto-Sydney-San Francisco.

Bruland, K. W. (1983), *Trace Elements in Sea-water*. *Chem. Oceanogr. 8*, 187–188.

Buge, A., Rancurel, G., Poisson, M., and Dechy, H. (1974), *Encephalopathics myocloniques par les sels de Bi. Six cas observés lors de traitement arousé au long cours*. *Nouv. Presse Med. 3*, 2315–2320.

Buge, A., Huboult, A., and Rancurel, G. (1975), *Les arthropathics de l'intoxication par le Bi*. *Rev. Rhum. Mal. Osteo-Articulaires 42*, 721–729.

Burns, R., Thomas, D. W., and Barron, V. J. (1974), *Reversible Encephalopathy Possibly Associated with Bi Subgallate Ingestion*. *Br. Med. J. 1*, 220–223.

Dekker, W., Dal Monte, P. R., Bianchi Porro, G., Van Bentem, N., Boeckhorst, J. C., Crowe, J. P., Robinson, T. J., Thys, O., and Van Driel, A. (1986), *An International Multi-clinic Study Comparing the Therapeutic Efficacy of Colloidal Bismuth Subcitrate Coated Tablets with Chewing Tablets in the Treatment of Duodenal Ulceration*. *Scand. J. Gastroenterol. Suppl. 122*, 46–50.

Eidecker, R., and Jackwerth, E. (1987), *Multielement Preconcentration from Iron-containing Soils and Sediments* (in German). *Fresenius Z. Anal. Chem. 328*, 469–474.

Emmerling, G., Schaller, K. H., and Valentin, H. (1986), *Actual Knowledge on Antimony, Bismuth and Other Metals, and their Compounds, and Feasibility of Quantitative Determination in Biological Materials in Occupational Medicine and Toxicology* (in German) *Zentralbl. Arbeitsmed. 36*, 258–265.

Eridani, S., Balzarini, M., Taglioretti, D., Romussi, M., and Valentini, R. (1964), *The Distribution of Radiobismuth in the Rat*. *Br. J. Radiol. 37*, 311–314.

Froomes, P. R. A., Wan, A. T., Harrison, P. M., and McLean, A. J. (1988), *Improved Assay for Bismuth in Biological Samples by Atomic Absorption Spectrophotometry with Hydride Generation*. *Clin. Chem. 34*, 382–384.

Gónzalez-Portal, A., Baluja-Santos, C., and Bermejo-Mártinez, F. (1986), *Spectrophotometric Determination of Bismuth in Pharmaceutical Preparations Using Mucic Acid*. *Analyst 111*, 547–549.

Hahn, M. H., Wolnik, K. A., Fricke, F. L., and Caruso, J. A. (1982), *Hydride Generation/Condensation System with an Inductively Coupled Argon Plasma Polychromator for Determination of Arsenic, Bismuth, Germanium, Antimony, Selenium, and Tin in Foods*. *Anal. Chem. 54*, 1048–1052.

Hailey, F. J., and Newsom, J. H. (1984), *Evaluation of Bismuth Subsalicylate in Relieving Symptoms of Indigestion*. *Arch. Intern. Med. 144*, 269–272.

Heinrichs, H., and Keltsch, H. (1982), *Determination of Arsenic, Bismuth, Cadmium, Selenium, and Thallium by Atomic Absorption Spectrometry with a Volatilization Technique*. *Anal. Chem. 54*, 1211–1214.

Heinrichs, H., Schulz-Dobrick, B., and Wedepohl, K. H. (1980), *Terrestrial Geochemistry of Cd, Bi, Tl, Pb, Zn and Rb*. *Geochim. Cosmochim. Acta 44*, 1519–1533.

Hespe, W., Staal, H. J. M., and Hall, D. W. R. (1988), *Bismuth Absorption from the Colloidal Subcitrate*. *Lancet 2*, 1258.

Heyman, A. (1944), *Systemic Manifestiations of Bi Toxicity. Observations on four Patients with Pre-existent Kidney Disease*. *Am. J. Syph. Gonorrhea Vener. Dis. 28*, 721–732.

Krüger, G., Thomas, D. J., Weindhardt, F., and Hoyer, S. (1976), *Disturbed Oxidative Metabolism in Organic Brain Syndrome Caused by Bismuth in Skin Creams. Lancet 2*, 485–487.
Krüger, J., Winkler, P., Lüderitz, E., Lück, M., and Wolf, H. U. (1985), *Bismuth, Bismuth Alloys, and Bismuth Compounds,* in: *Ullmann's Encyclopedia of Industrial Chemistry,* 5th Ed., Vol. A4, pp. 171–189. VCH Verlagsgesellschaft, Weinheim-Deerfield Beach/Florida-Basel.
Lambert, J. R., Way, D. J., King, R. G., Wan, A., and McLean, A. J. (1988), *Bismuth Levels in the Human Gastric Mucosa. Aust. N.Z.J. Med. 18*, 405.
Lee, D. S. (1982), *Determination of Bismuth in Environmental Samples by Flameless Atomic Absorption Spectrometry with Hydride Generation. Anal. Chem. 54*, 1682–1686.
Lee, D. S., Edmond, J. M., and Bruland, K. W. (1985/86), *Earth Planet. Sci. Lett. 76*, 254–262.
Lee, S. P., Lim, T. H., Pybus, J., and Clarke, A. C. (1980), *Tissue Distribution of Orally Administered Bismuth in the Rat. Clin. Exp. Pharmacol. Physiol. 7*, 319–324.
Locke, M., Nichol, H., and Ketola-Pirie, C. (1987), *Binding of Bismuth to Cell Components: Clue to Mode of Action and Side Effects. Can. Med. Assoc. J. 137*, 991–992.
Löhle, J., Mattenberger, K., and Wachter, P. (1989), *Production and Characterization of Wires of High Temperature Superconducting Materials Based on $(Bi_{1.6}Pb_{0.4}Sr_2Ca_3Cu_4O_x)$* (in German). *ETH-Z Bulletin No. 223*, pp. 21–23. Swiss Federal Institute of Technology, Zürich.
MAK (1988), *Maximum Concentrations at the Workplace and Biological Tolerance Values for Working Materials, DFG Report No. XXIV.* VCH Verlagsgesellschaft, Weinheim-Basel-Cambridge-New York.
McLean, A. J., Froomes, P., McNeil, J. J., Wan, A. T., and Harrison, P. M. (1988), *Bismuth Subcitrate Handling in Man. Clin. Pharmacol. Ther. 43*, 186.
McNulty, C. A. M., Gearty, J. C., Crump, B., Davis, M., Donovan, I. A., Melikian, V., Lister, D. M., and Wise, R. (1986), *Campylobacter pyloridis and Associated Gastritis: Investigator Blind, Placebo Controlled Trial of Bismuth Salicylate and Erythromycin Ethylsuccinate. Br. Med. J. 293*, 645–649.
Plunkett, E. R. (1987), *Handbook of Industrial Toxicology,* 3rd Ed. Edward Arnold, London.
Randall, R. E., Osheroff, R. J., Bakerman, S., and Setter, J. G. (1972), *Bismuth Nephrotoxicity. Ann. Intern. Med. 77*, 481–482.
Rooney, R. C. (1976), *Determination of Bismuth in Blood and Urine. Analyst 101*, 749–752.
Russ, G. A., Bigler, R. F., Tilbury, R. S., Woodard, H. Q., and Laughlin, J. S. (1975), *Metabolic Studies with Radiobismuth. 1. Retention and Distribution of ^{206}Bi in the Normal Rat. Radiat. Res. 63*, 443–454.
Saager, R. (1984), *Metallic Raw Materials from Antimony to Zirconium* (in German), pp. 95–98 (Bismuth). Bank von Tobel, Zürich.
Sollman, T., Cole, H. N., and Henderson, K. (1938), *Clinical Excretion of Bismuth. VII. The Autopsy Distribution of Bismuth in Patients after Clinical Bismuth Treatment. Am. J. Syph. 22*, 555–583.
Stiel, D., Murray, D. J., and Peters, T. J. (1985), *Uptake and Subcellular Localisation of Bismuth in the Gastrointestinal Mucosa of Rats after Short-term Administration of Colloidal Bismuth Subcitrate. Gut 26*, 364–368.
Thomas, D. W., Hartley, T. F., Coyle, P., and Sobecki, S. (1977), *Clinical and Laboratory Investigations of the Metabolism of Bismuth Containing Pharmaceuticals by Man and Dogs,* in: Brown, S. S. (ed.): *Clinical Chemistry and Chemical Toxicology of Metals,* pp. 293–296. Elsevier/N. Holland Biomedical Press, Amsterdam.
Thomas, D. W., Sobecki, S., Hartley, T. F., Coyle, P., and Alp, M. H. (1983), in: Brown, S. S., and Savory, J. (eds.): *Chemical Toxicology and Clinical Chemistry of Metals,* pp. 391–394. Academic Press, New York.
Trueb, L. (1989), *New Therapeutic Concepts for Gastritis and Ulcer Treatment,* in: *Neue Zürcher Zeitung, Forschung und Technik,* No. 165, p. 57 (19 July), Zürich.
Urizar, R., and Vernier, R. L. (1966), *Bismuth Nephropathy. J. Am. Med. Assoc. 198*, 207–209.
van den Broeck, C. J. H. (1963), *Autoradiography of Various Rat Tissues after the Administration of Bismuth 210. Acta Radiol. Ther. Phys. Biol. 1*, 385–396.

Ward, N.I. (1987), *The Future of Multi-ultra-trace Element Analysis in Assessing Human Health and Disease: A Comparison of NAA and ICPSMS*, pp. 118–123, in: Extended Abstracts from the Second Nordic Symposium on Trace Elements in Human Health and Disease, Odense (Denmark). *Environ. Health 20*, WHO Regional Office for Europe, Copenhagen.

Welcher, G.G., Kriege, O.H., and Marks, J.Y. (1974), *Direct Determination of Trace Quantities of Lead, Bismuth, Selenium, Tellurium, and Thallium in High-temperature Alloys by Non-flame Atomic Absorption Spectrophotometry. Anal. Chem. 46*, 1227–1231.

Weller, M.P.I. (1988), *Neuropsychiatric Symptoms Following Bismuth Intoxication. Postgrad. Med. J. 64*, 308–310.

Windholz, M., Budavari, S., Blumetti, R.F., and Otterbein, E.S. (eds.) (1983), *The Merck Index*, 10th Ed., pp. 177–181. Merck & Co. Inc., Rahway, New Jersey.

Wolnick, K.A., Fricke, F.L., Hahn, M.H., and Caruso, J.A. (1981), *Sample Introduction System for Simultaneous Determination of Volatile Element Hydrides and Other Elements in Foods by Inductively Coupled Argon Plasma Emission Spectrometry. Anal. Chem. 53*, 1030–1035.

Woods, J.S., and Fowler, B.A. (1987), *Alteration of Mitochondrial Structure and Heme Biosynthetic Parameters in Liver and Kidney Cells by Bismuth. Toxicol. Appl. Pharmacol. 90*, 274–283.

Zandbergen, H.W., Huang, Y.K., Menken, M.J.V., Kadowaki, J.N.Li.K., Menovsky, A.A., van Tendeloo, G., and Amelinckx, S. (1988), *Electron Microscopy on the $T_c = 110$ K (midpoint) Phase in the System Bi_2O_3-SrO-CaO-CuO. Nature 332*, 620–623.

Zindenberg-Cherr, S., Parks, N.J., and Keen, C.L. (1987), *Tissue and Subcellular Distribution of Bismuth Radiotracer in the Rat: Considerations of Cytotoxicity and Microdensimetry for Bismuth Radiopharmaceuticals. Radiat. Res. 111*, 119–129.

Additional Recommended Literature on the Analytical Chemistry of Bismuth

Asami, T., Kumboto, N., and Minamisawa, K. (1988), *Natural Abundance of Cadmium, Antimony, Bismuth, and Some Other Heavy Metals in Japanese Soils. Nippon Dojo Hiryogaku. Zasshi 59* (2), 197–199 (Japanese Journal of Soil Science and Plant Nutrition).

Brumsack, H.-J. (1980), *Geochemistry of Cretaceous Black Shales from the Atlantic Ocean. Chem. Geol. 31*, 1–25.

Florence, T.M. (1972), *Determination of Trace Metals in Marine Samples by Anodic Stripping Voltammetry. J. Electranal. Chem. 35*, 237–245.

Florence, T.M. (1974), *Determination of Bismuth in Marine Samples by Anodic Stripping Voltammetry* (relative high concentrations of 0.04 to 0.1 ppb at the surface were found in the Pacific, and relative low concentrations of 1 to 8 ppb in seaweed, fish muscle and oyster flesh fresh weight; biological concentration factors were thus between about 20 and 160). *J. Electranal. Chem. 49*, 255–264.

Güçer, S. (1986), *Determination of Bismuth, Cadmium and Lead in Soil Extracts from Malataya (Turkey) by Ceramic Tube Pt-loop AAS. Proceedings 10th International Symposium Microchemical Techniques, Antwerp*, in press.

Hiltenkamp, E., and Jackwerth, E. (1988), *Investigations on the Determination of Bismuth, Cadmium, Mercury, Lead and Thallium in High-purity Gallium by Graphite Furnace AAS with Atomization of Metallic Samples. Fresenius Z. Anal. Chem. 332*, 134–139.

de Kersabiec, A.M. (1980), *Determination of Arsenic, Selenium and Bismuth in Rocks and Soils by Atomic Absorption Spectrometry. Analysis 8* (3), 97–101.

Sauer, C., and Lieser, K.H. (1986), *Trace Elements in Suspended Matter, in Colloids and in Molecular-dispersed Form in a Raw Water and a Drinking Water* (fractions separated by filtration and ultrafiltration were analyzed by atomic absorption spectrometry, atomic emission spectrometry with an inductively coupled plasma, and stripping voltammetry). *Vom Wasser 66*, 285–291.

Senesi, N., Polemio, M., and Lorusso, L. (1979), *Content and Distribution of Arsenic, Bismuth, Lithium and Selenium in Mineral and Synthetic Fertilizers and their Contribution to Soil* (bismuth was determined as the gaseous hydride generated by reduction with sodium borohydride and swept by a flow of argon into an argon-hydrogen-air flame; fertilizers contain up to 0.5 ppm bismuth; highest amounts in triplape 46% – 48%, N-P-K 20-10-10, and ternape 9-7-12). *Commun. Soil Sci Plant Anal. 10* (8), 1109–1126.

Sinemus, H. W., Melcher, M., and Welz, B. (1981), *Influence of Valence State on the Determination of Antomony, Arsenic, Bismuth, Selenium and Tellurium in Lake Water Using the Hydride AA Technique* (only Bi(III) was found in the case of bismuth). *At. Spectrosc. 2* (3), 81–86.

Valcarcel, M. (1986), *Continuous Separation Techniques in Flow Injection Analysis* (in a gas-liquid system (gas diffusion, distillation, hydride generation) a gas diffusion cell, a distillation unit and a gas absorber were inserted for the determination of bismuth). *Proceedings 3rd IAEAC Symposium on Handling of Environmental and Biological Samples in Chromatography, Palma de Mallorca. J. Chromatogr.*, in press.

II.6 Cadmium

MARKUS STOEPPLER, Jülich, Federal Republic of Germany

1 Introduction

Cadmium is a relatively volatile element and is, from present knowledge, not essential for plants, animals and human beings. Higher doses of cadmium can lead to toxic effects. Because cadmium occurs together with zinc, from which it must be separated, cadmium production depends on the production of zinc. Since eight times more cadmium has been consumed in the last 40 years than in the entire history of mankind before, problems associated with cadmium have only accelerated since about 1950. Worldwide cadmium production at present is around 17 000 metric tons/year with a tendency to decrease in the future. Cadmium and its compounds are mainly used for nickel/cadmium batteries, anticorrosive coatings of metals, pigments, and stabilizers for plastics. Global emission of cadmium compounds into the atmosphere is estimated to be around 7000 metric tons/year with more than 90% coming from anthropogenic sources (the burning of coal is a significant source). In addition to long-range transport and the ubiquitous distribution of cadmium, these atmospheric emissions, together with aqueous and solid emissions, lead to local and regional pollution of soils and river sediments. Two broad source categories may be distinguished: (a) those which produce marked impact in the vicinity of the discharge source (e.g., smelters), and (b) those originating from the dissipation of consumer goods and their wastes. The results are higher cadmium levels in the terrestrial, aquatic and marine food web. Adverse effects in plants and mammals have been observed as a result of mg/kg levels in soil and food, and in a few aquatic organisms already at µg/kg concentrations.

Cadmium uptake in mammals and humans occurs by ingestion and inhalation. For the latter, the uptake rate is significantly higher. Acute effects can be seen in the respiratory and digestive tracts. Cadmium accumulates predominantly in the kidneys with a biological half-life of more than ten years. A 50-year-old occupationally non-exposed non-smoker has an average body burden of approximately 15 mg cadmium, while for a smoker of the same age the figure is about 30 mg. Heavy long-term cadmium exposure might produce irreversible adverse renal effects. In particular, cases of vitamin and protein deficiency and bone disease have been observed.

Since a potential for cadmium to cause cancer has recently been shown in animal experiments, it has been recommended that cadmium exposure be minimized and measures be taken for current biological monitoring. According to a WHO recommendation, the daily intake of cadmium should not exceed 1 µg per kg of body weight. For this reason, limits have been set on a national basis for cadmium levels in air, water and food and maximal cadmium contents in agricultural soils.

Health risks due to cadmium and its slowly but continuously rising levels in the biosphere have been extensively discussed but are also controversial. This is particularly the case if the remarkable increases over the last 40–60 years in cadmium levels in human kidneys and grain are considered. The essentiality of cadmium in animals has also been postulated. Thus, in addition to the following text and Chapters I.6, I.7, I.8, I.12, I.17a, and I.18, the following publications are included without making a judgement about given statements therein.

ELINDER, 1977; CADMIUM-77, 1978; CADMIUM-79, 1980; MENNEAR, 1979; DIEHL, 1981; PISCATOR, 1981; UMWELTBUNDESAMT, 1981; CADMIUM-81, 1982; LAUWERYS, 1982; BMI, 1982; NATIONAL SWEDISH ENVIRONMENTAL PROTECTION BOARD, 1982; HYGIÈNE ET SÉCURITÉ, 1982; COLE and VOLPE, 1983; BDI, 1982; DRASCH, 1983; SCHELENZ, 1983; COULSTON et al., 1983; CADMIUM-83, 1984; MERIAN, 1984; TAYLOR, 1984; FRIBERG et al. 1985, 1986a, b; MERIAN et al., 1985, 1988, 1989; CADMIUM TODAY, 1986; MISLIN and RAVERA, 1986; LORENZ et al., 1986; OBERDOERSTER, 1986; OLDIGES and GLASER, 1986; ANKE et al., 1984, 1987; HUTTON et al., 1987; KAZANTZIS, 1987; MORSELT et al., 1987; NRIAGU and SPRAGUE, 1987; CADMIUM-86, 1988; STOEPPLER and PISCATOR, 1988; MORGAN, 1988; GLASER et al., 1989; TUOR and KELLER, 1989.

2 Physical and Chemical Properties, and Analytical Methods

2.1 Physical and Chemical Properties

Cadmium was discovered in 1817 by STROHMEYER during an investigation of zinc carbonate. The name is derived from the Greek word "Kadmeia" for the zinc ore calamine. Cadmium has a melting point of 320.9 °C, a boiling point of 767 °C, a density of 8.64 g/cm^3 at 20 °C and is a silver-white, lustrous and ductile metal. It is relatively soft, but this can be improved by alloying, e.g., with zinc. The atomic number of cadmium is 48, its atomic mass is 112.4 and there are eight naturally occurring isotopes (listed in order of abundance): ^{114}Cd (29%), ^{112}Cd (24%), ^{111}Cd (13%), ^{110}Cd, ^{113}Cd, ^{116}Cd, ^{106}Cd, and ^{108}Cd.

Cadmium is the 67th most abundant element. It belongs, along with zinc and mercury, to the 2nd subgroup of the Periodic Table; its oxidation state is +2 in all compounds. With a normal electrochemical potential of −0.40 relative to the hydrogen electrode, it is slightly more noble than zinc. The cadmium ion ($r = 103$ pm) is very close in size to the calcium ion ($r = 106$ pm). Therefore, some similarity can be seen between cadmium and calcium (the radius of the zinc ion is 83 pm).

Cadmium is readily soluble in nitric acid, but only slowly in hydrochloric and sulfuric acids. It is insoluble in basic solution. With regard to speciation (see also Sect. 4.2.2) one must differentiate (for instance, in the case of soil or plants), between bivalent cadmium ions (e.g., CdCl$_2$), chlorocadmium complexes, cadmium bound to proteins, cadmium bound to colloidal substances, and acid cadmium complexes (see also ANGEHRN and HERTZ, 1989, who studied speciation in forest litter by thin channel ultrafiltration). For speciation in biological materials see Sects. 5.3.3, 6.2, and LORENZ (1979).

Some cadmium compounds are colored (yellow, red, brown) and others are colorless. Salts of cadmium with strong acids are readily soluble in water; less soluble are the sulfide, the carbonate, the fluoride, and the hydroxide. The latter, however, is readily soluble (i.e., forms complexes) in ammonium hydroxide. Cadmium also forms complexes with halogen ions, and cadmium complexes are more stable than zinc complexes. More details on cadmium and its compounds can be found elsewhere (NRIAGU, 1980b; SCHULTE-SCHREPPING and PISCATOR, 1985).

In the organs of humans and mammals, stored cadmium is predominantly bound to metallothionein, a cysteine-rich protein (see Chapter I.12).

2.2 Analytical Methods

Soluble cadmium compounds can be determined in biological, environmental, and technical materials by a number of different analytical methods. If digestion is required the volatility of cadmium must be considered.

Because of very low cadmium contents around and below 1 µg/kg in various materials (e.g., body fluids, sea water, inland water, foodstuffs) there are many possibilities for contamination from external sources. These sources of error, however, were only identified after the introduction of trace analytical methods with a detection power at the ng/kg level. Thus, correct cadmium levels detected using these methods are frequently much lower than those reported in earlier publications (e.g., MART, 1979, 1983; MART et al., 1983; STOEPPLER, 1985b; FRIBERG et al., 1985).

The most frequently applied analytical method for cadmium is still atomic absorption spectrometry (AAS) with flame and graphite furnace. Zeeman background correction in connection with platform techniques allows very sensitive direct determinations in body fluids and even in solids (STOEPPLER, 1986). Electrothermal atomic absorption spectrometry with the graphite furnace after extraction is also used for routine determination of cadmium in urine (ANGERER, 1988). For cadmium contents greater than 10 µg/L analyte solution, atomic emission spectrometry with plasma (mainly ICP) excitation is a convenient method with multi-element potential (e.g., SCHRAMEL et al., 1982; STOEPPLER, 1985a).

Mass spectrometry, mainly using the most reliable isotope dilution approach (HEUMANN, 1980, 1986), and neutron activation analysis (NAA), either instrumental or radiochemical, are important multi-element methods for quality control purposes (ERDTMANN and PETRI, 1986; VERSIECK et al., 1988).

In-vivo-NAA has been used for the estimation of individual cadmium exposure. In the organ of interest (liver or kidney), a neutron beam produces extremely short-lived ^{114}Cd nuclei.

γ-Spectroscopic measurements of the radiation from this isotope result in quite reliable information about the cadmium level (ELLIS et al., 1979; ROELS et al.,

1981). Recently, in vivo determination of cadmium in kidney cortex by X-ray fluorescence was reported (CHRISTOFFERSON and MATTSON, 1983; SPANG, 1988). In vivo elemental analysis was recently and comprehensively reviewed (SCOTT and CHETTLE, 1986).

X-ray fluorescence, formerly used only for mg/kg concentrations (JENKINS, 1975), can now be applied in its total reflection mode, besides its multi-element potential down to µg/kg levels, for various materials (MICHAELIS et al., 1985). Another still-used single-element method for higher concentrations is colorimetry. Electrochemical, especially voltammetric, methods are excellent for aqueous samples because of their extraordinary detection power (NÜRNBERG, 1982; MART, 1983) but can also be applied with good results to biological materials if a reliable digestion method is used (NÜRNBERG, 1983; OSTAPCZUK et al., 1986, 1987).

The evaluation of methodogical approaches for the speciation of cadmium, mainly in aqueous systems, has been reviewed recently with emphasis also on the potential of separation and electrochemical methods (ASTRUC, 1986).

In cases of cadmium exposure (humans and mammals) from a distinct dose, renal effects resulting in increased excretion of low molecular weight and/or high molecular weight proteins can occur. In order to control this, appropriate methods based on radio-immunoassays, enzyme-immunoassays, latex-immunoassays, tests using circulating antigens, measurement of the binding of Alcian Blue to red blood cell membrane, and other, e.g., chromatographic methods have been developed and applied (HERBER, 1984; FRIBERG et al., 1986a; LAUWERYS, 1987; BERNARD and LAUWERYS, 1989, see also Sect. 6.5.2).

3 Sources, Production, Important Compounds, Uses, Waste Products, and Recycling

3.1 Sources and Distribution in Nature

Pure cadmium minerals such as greenockite (hexagonal CdS), hawleyite (cubic CdS), otavite ($CdCO_3$), monteponite (CdO), and cadmoselite (CdSe) occur very rarely. Cadmium commonly occurs in isomorphic form in zinc minerals such as zinc blende (ZnS) with cadmium contents from 0.1 to 0.5%, and galmei ($ZnCO_3$) with cadmium contents up to a maximum of 5% (NRIAGU, 1980b); SCHULTE-SCHREPPING and PISCATOR, 1985; THORNTON, 1986).

Cadmium resources are linked to resources of zinc. An estimate for these resources is about 9 million metric tons (MINERAL COMMODITY SUMMARIES, 1982). The average cadmium content of the earth's crust is estimated to be about 0.1 mg/kg (HEINRICHS et al., 1980). Weathering of minerals in geological periods has led to cadmium enrichment of sediments by a factor of 2–3. Phosphates show a broad range of cadmium contents with an average of approx. 15 mg/kg (FOERSTNER, 1980a). Cadmium levels in fossil organic materials differ significantly. For pit-coal and lignite, average contents below 2 mg/kg have been reported (FOERSTNER,

1980a; CADMIUM-81, 1982), while according to recent investigations, the cadmium content of crude oil, at far less than 1 µg/kg, is negligible (NARRES et al., 1984a).

3.2 Production

Cadmium is mainly (>95%) extracted from cadmium-enriched by-products obtained from the roasting of zinc minerals. Raw separation from other elements is either performed pyrometallurgically due to the volatility of cadmium or by leaching with sulfuric acid and subsequent precipitation (cementation, formation of cadmium sponge). Very pure cadmium is obtained by electrolytic deposition or vacuum distillation at a temperature of 420–485 °C. The final products are obtained at different purities ranging from 99.5 to 99.995% (NRIAGU, 1980b; RAUHUT and WIEGAND, 1982; SCHULTE-SCHREPPING and PISCATOR, 1985).

Cadmium extraction as a by-product of zinc amounts on average to approx. 3 kg cadmium per ton of zinc (STUBBS, 1982). Different yields are due to different cadmium contents of the ores used and/or fluctuations in the addition of residues (recycling).

Table II.6-1. The Most Important Producers of Refined Cadmium in 1986 (METALLSTATISTIK, 1987)

Country	Cadmium Production (metric tons, rounded)	
USSR	2700	(2900)[a]
Japan	2490	(2170)
USA	2350	(1950)
Canada	1550	(1300)
Belgium-Luxembourg	1380	(1530)
Federal Republic of Germany	1220	(1190)
Australia	910	(1010)
Mexico	720	(860)
Poland	600	(700)
The Netherlands	560	(455)

[a] Production in 1980 in parentheses

World-wide cadmium production increased from about 16800 tons/year in 1970 to about 18700 tons/year in 1980 with some fluctuations in the following years (i.e., 12350 tons/year in 1982, SAAGER, 1984; 19700 tons/year in 1986, METALLSTATISTIK 1981, 1987). A certain decrease in consumption is expected for the next decade because of restrictions and substitutions, and will lead to an excess supply (SAAGER 1984; TÖTSCH, 1989).

The ten most important countries for cadmium production in 1986 – with 1980 production in parentheses – are given in Table II.6-1.

3.3 Uses, Recycling, Substitution

The amount of cadmium used also fluctuates with regard to the amount produced (see Table II.6-2). For instance, cadmium consumption in the Federal Republic of Germany decreased from 2100 tons/year in 1979 to 1200 tons/year in 1988 (TÖTSCH, 1989). From Table II.6-3 it can be seen that order of uses differs significantly from that of producers (Table II.6-1) (METALLSTATISTIK, 1987).

Electrodeposited cadmium has excellent properties for protecting iron and steel against corrosion – even a thickness of 0.008 mm is sufficient for protection. Another increasingly important use is in rechargeable nickel-cadmium batteries for a wide range of applications. Other uses include cadmium pigments (cadmium sulfide, cadmium selenide and mixtures of both) which are generally very stable thermally (e.g., for plastic materials in cars). Cadmium soaps made with saturated and unsaturated fatty acids play an important role as temperature and light stabilizers

Table II.6-2. World Production and Consumption of Cadmium 1984–1986

Year	Production (t)	% of Previous Year	Consumption (t)	% of Previous Year	% of Production
1984	19730	112	17210	94.7	87.2
1985	19100	96.8	17295	100.5	90.55
1986	19700	103	19330	111.8	98.12

Table II.6-3. Consumer Countries of Cadmium 1986 (METALLSTATISTIK, 1987)

Country	Cadmium Consumption (metric tons)	
USA	5140	(3670)[a]
USSR	2700	(2900)
Belgium-Luxembourg	2255	(1620)
Japan	1980	(1930)
United Kingdom	1450	(1390)
France	1170	(1085)
Federal Republic of Germany	1150	(1200)
China	450	(425)
German Democratic Republic	400	(420)
South Korea	355	(305)

[a] Consumption in 1985 in parentheses

Table II.6-4. Consumption of Cadmium (%) in the United States, Japan, the United Kingdom and the Federal Republic of Germany from 1970 to 1982 (SCHULTE-SCHREPPING and PISCATOR, 1985)

Use	1970	1973	1976	1979	1982
Plating	36.9	30.1	36.3	34.3	28.9
Batteries	8.1	14.6	21.3	22.7	28.5
Pigments	24.1	29.8	25.0	26.7	23.8
Stabilizers	23.2	17.4	12.1	12.2	12.4
Alloys etc.	7.7	8.1	5.3	4.1	6.4

mainly for PVC (e.g., for plastic window profiles). A small percentage of cadmium (cadmium sulfide-copper sulfide) is used in solar cells for direct conversion of light into electrical energy. Cadmium is also used in nuclear reactors as a neutron absorber and in various alloys with such metals as tin, copper, and aluminum (CADMIUM-79, 1980; CADMIUM-81, 1982; SCHULTE-SCHREPPING and PISCATOR, 1985). The uses of cadmium in some Western countries are shown in Table II.6-4.

Recycling of cadmium after private and industrial consumption of cadmium-containing materials is still not significant, except for the partial recycling of used batteries and some alloys (LLOYD and WISE, 1982; BMI, 1982; BDI, 1982; SAAGER, 1984; SCHULTE-SCHREPPING and PISCATOR, 1985; OHIRA, 1988; BÖHM and TÖTSCH, 1989). However, industrial recycling of wastes from cadmium production (e.g., sludge from electroplating, filter dust, scrap from batteries, etc.) is quite common(SCHULTE-SCHREPPING and PISCATOR, 1985). For the cadmium balance in the Federal Republic of Germany and the USA, recycling is included (METALLSTATISTIK, 1981, 1987), but in Germany more than 20% was given as recycled cadmium (RAUHUT and WIEGAND, 1982). Cadmium can be substituted by zinc or aluminum in many coatings, probably by tin in stabilizers (BATZER, 1983; SAAGER, 1984; TÖTSCH, 1989; BÖHM and TÖTSCH, 1989) and for pigments by chromate, iron oxide or organic compounds, although with somewhat inferior properties (MINERAL COMMODITY SUMMARIES, 1982; SAAGER, 1984). Also for batteries there is a tendency to use no or less mercury and cadmium, as vented and closed cells are critical wastes (TÖTSCH, 1989).

4 Distribution in the Environment, in Foods, and in Living Organisms
(see also HUTTON et al., 1987)

4.1 Emissions into the Environment

Cadmium has been emitted in minor amounts into the environment from the rise of industrialization, but in greatly increased quantities after World War II, in the form of dusts and aerosols into the atmosphere, effluents into rivers and lakes, and as solids from point sources (waste, slag, incineration, coal combustion, phosphate fertilizers, sewage sludge). Especially since about 1950, this has led to some global and regional redistribution as well as to a regional and local increase of cadmium levels in the human environment, as is reflected in cadmium levels of populations from more (Japan) and less (USA, Sweden) polluted areas (KJELLSTROEM, 1979). This might have caused a situation for which the occurrence of health implications for the general population cannot be completely ruled out (UMWELTBUNDESAMT, 1981); however, the situation remains under discussion (e.g., DIEHL, 1981; BMI, 1982; DRASCH, 1983; NRIAGU, 1984; MERIAN, 1984; TUOR and KELLER, 1989; see also Sects. 6 and 7).

In order to take measurements for a lowering of emissions and to obtain estimates for future cadmium emissions in several countries, balances for cadmium emissions have been created. These balances are to some extent based on dust and deposition measurements and on data for production and turnover. Table II.6-5 summarizes recent studies from the United States, Canada, and Europe. However, it must be mentioned that these data, even if more recent ones are considered, still show a wide range and frequently are incomplete or contradictory. They should therefore be considered with caution and only as preliminary estimates. They might often be based on emission factors that are too high or too low. This is due to the fact that in most cases reliable analytical data still are not available and that more effective filters etc., already in current use, have not been considered in all studies. In some cases, therefore, effective cadmium emissions might also be overestimated if the present situation is considered.

4.1.1 Atmosphere

The total annual input of cadmium into the atmosphere has been estimated to be approximately 7000 metric tons with around 10% from natural sources (HEINDRYCKX et al., 1974; NRIAGU, 1979, 1980a; NRIAGU and PACYNA, 1988). For specific estimates see Table II.6-5.

Cadmium emissions into the atmosphere are primarily attributed to the following sources:

- non-ferrous metal production and processing, including cadmium,
- fossil fuels, waste incineration,
- iron and steel production.

Table II.6-5. Estimations for Cadmium Emissions into the Atmosphere in Water and for Solid Depositions (in metric tons per year)

Country (Region)	Year(s)	Atmosphere	Waters	Soil etc. (Landfill)	Total	Reference	Remarks
USA	1968–72	300	25	1500	1825	Nriagu (1980c)	Average from 5 years
USA	1975–85	970	410	6200	7580	Yost (1980)	Prospective computer study ten years, calculating an annual increase in consumption of cadmium
USA	1980	300	15	2160	2475	Wixson (1982)	
Canada	1972	500	–	–	–	Air Pollution Control Directorate (1976)	
European Community	1975	263	132	460	855	Rauhut (1980)	
European Community	1979/80	131[a]	–	–	–	Hutton (1983)	
Western and Eastern Europe	1984	1300[b]	–	–	–	Pacyna and Muench (1987)	
United Kingdom		14	43[c]	899[d]	956	Hutton and Symon (1986)	
Federal Republic of Germany	1977/81	83.5[e]	62	124	269.5	Umweltbundesamt (1981)	
Federal Republic of Germany	1980/82	26	66	49[f]	140	Schulte-Schrepping and Piscator (1985), BMI, 1982	
Switzerland	1980/84	8	6	25	39	Tuor and Keller (1989)	

[a] There is an increase estimated resulting in 145 t for 1990 and 164 t for 2000 due to increasing coal consumption
[b] Western Europe only approx. 630 t/a from this source
[c] Only for coastal and estuarine waters
[d] Includes many sources, e.g., production and use of cadmium containing articles (217 t/a) and municipal waste disposal (525 t/a)
[e] Another study (Schladot and Nuernberg, 1982) estimated 80 t/a for atmospheric emissions only
[f] Includes only fertilizers and waste water sludges

Table II.6-6. Estimations for the Share of Different Categories of Emitters on Atmospheric Emissions from the Sources Given in Table II.6-5 (in metric tons per year)

Country (Region)	Year(s)	Primary Metals, Including Cd-Production	Cd-Processing and Consumption	Other Nonferrous Metals	Iron and Steel	Fossil Fuels[a]	Waste Incineration	Sewage Sludge	Tires and Automobile Exhaust
USA	1968/72	102	31	[b]	—	130	—	20	5
USA	1980	102	38	[b]	10.5	130	—	20	5
Canada	1972	400	—	9	5	87	—	20	—
European Community	1979/80	20	3	13	34	9	31	1	—
Western and Eastern Europe Total	1984	850	[b]	[b]	—	250	220[c]	—	—
(Western Europe)		(400)	[b]	[b]	—	(125)	(100)	(—)	(—)
		3.7			2.3	1.9	5	0.2	
FRG	1977/81	5	1.5	[b]	37	35	5	—	0.1
FRG	1980/82	7	4	—	5	<5	3.8	—	<0.5

[a] For fossil fuels there are much lower estimates from some other authors: HEINDRYCKX et al. (1974); JACKO and NEUENDORF (1977); NRIAGU (1980c); KIRSCH et al. (1982), but also from SCHULTE-SCHREPPING and PISCATOR (1985) and HUTTON and SYMON (1986)
[b] Other nonferrous metals included in the first column
[c] Waste incineration including other sources

These data are given in some detail in Table II.6-6, based mainly on data and sources from Table II.6-5.

4.1.2 Dissolved Cadmium Compounds (Aqueous Media)

Wet chemical, e.g. electrochemical, processes can result in high levels of cadmium compounds in waste water and its accumulation in sewage sludge, or the direct pollution of rivers and lakes. From waste waters containing more than 1 mg/L, cadmium can be separated by precipitation (pH value > 9.5), electrolysis, or ion exchange down to levels of 100 to 500 µg/L. Moreover, a partial sorption on sludge occurs in biological sewage plants (PIHLAR et al., 1980). The treated solutions are subsequently further diluted in aqueous effluents. This, however, can be critical if accumulation of cadmium processing plants occurs in an area where potable water must be obtained from bank filtrates (RUHRVERBAND/RUHRTALSPERRENVEREIN, 1975). Further sources of pollution, partly from ground water, are upper soil layers which are sometimes significantly enriched with cadmium (see Sects. 4.1.3 and 4.2.3).

A detailed estimate of the most important sources and amounts was given in the already-mentioned computer study for the United States which covered ten years (YOST, 1980; viz. Table II.6-5). Based on an estimated annual aqueous emission of 410 metric tons for the entire area of the USA, the most important sources are listed in Table II.6-7. Other authors, despite giving somewhat lower values, do agree about the main contributions from the production and recycling of cadmium (SCHULTE-SCHREPPING and PISCATOR, 1985; HUTTON and SYMON, 1986).

Table II.6-7. Estimations for the Most Important Sources and Amounts of Cadmium Emissions in Aqueous Media for the USA (YOST, 1980)

Sources	Metric tons/year
Lead-zinc mines in operation	140
Inactive mines	90
Sewage sludge treatment	84
Zinc-cadmium plants	56
Production of plastics	40
Steel production	50
Nickel-cadmium batteries	30
Electroplating	20

4.1.3 Solid Deposition

Deposition of cadmium-containing waste from industrial and household garbage as well as sewage sludge at even relatively safe dumping sites can lead to mobilization by leaching (see also Sect. 4.2.2), which can, via percolating waters, result in the

Table II.6-8. Estimated Cadmium Input to Landfills in the USA (YOST, 1980) and the UK (HUTTON and SYMON, 1986) (t/a)

Deposition	USA	UK
Cadmium plated items	2800[a]	–
Pigments	470	–
Deposition of phosphate fertilizers and sewage sludge on agricultural soils	460	12.7[b]
Sewage sludge disposal and sewage sludge incineration	440	11.4[c]
Iron and steel production	400	42
Municipal waste disposal	400	525
Mining of lead and zinc ores	330	–
Nickel-cadmium batteries	280	–
Production and use of articles containing cadmium	–	217
Zinc-cadmium production	180	20.8
Cement manufacture	–	21
Fossil fuel combustion	67	60

[a] From applied 3400 t/a
[b] Only production and use of phosphate fertilizers
[c] Only disposal

pollution of ground water and even surface water. Due to the chemical properties of cadmium, dumping sites represent a potential environmental hazard and should be held under current control.

Estimates for solid deposits (landfills) that have a very different potential for cadmium mobilization are to be found in recent publications (e.g., YOST, 1980; SCHULTE-SCHREPPING and PISCATOR, 1985; HUTTON and SYMON, 1986; Chapters I.7b and I.7e). There are, however, significant differences in estimated amounts, and the data for the USA appear to be more comprehensive than those from the UK (Table II.6-8).

4.2 Immissions

4.2.1 Dry and Wet Deposition from the Atmosphere

Cadmium emitted into the atmosphere from smelters, waste incineration plants and other sources occurs mainly in the form of small aerosol particles (suspended matter of $<2\,\mu m$ diameter), which are inhaled and reach the lower airways of animals and humans (NRIAGU, 1980c).

The deposition of a significant percentage of these particles occurs in the vicinity of point sources, exponentially decreasing with the distance but also dependent upon

Table II.6-9. Estimates for Cadmium Dry Deposition (Averages) for Rural, Less Polluted and Polluted Areas of the Federal Republic of Germany (JOST, 1984)

Area	Deposition ($\mu g/m^2/d$)	Percentage of Total Area in the FRG
Rural and remote areas	0.5	90
Less polluted areas	2.0	5
Polluted areas[a]	6.0	5

[a] A very small area percentage (0.1%) close to cadmium emitters shows somewhat higher deposition rates

meteorological parameters. Models for this have been published (e.g., DAVIDSON, 1980).

Apart from the deposition of larger particles close to emitters with a residence time of approximately 2 hours, residence times ranging from 0.1 to 4 days have also been observed for particles with diameters from 1 to 10 μm. A long-range transport mechanism also exists, since approximately 50% of the emitted cadmium remains in the atmosphere and distributes, predominantly in the northern hemisphere, up to a height of 10000 meters with an average residence time of approximately 10 days (STEINNES, 1989). This has led to a steady global increase in the background cadmium level (NRIAGU, 1980a).

Cadmium levels in air are usually given in ng/m^3. Depending on cadmium contents, pollution levels vary from rural to highly industrialized areas (SCHULTE-SCHREPPING, 1978; NRIAGU, 1980c; NRIAGU and DAVIDSON, 1986). Table II.6-9 shows examples for estimated average concentration levels of cadmium in dry deposition (which are very difficult to measure) for the Federal Republic of Germany (JOST, 1984). These values are significantly lower if compared with former estimates (SCHULTE-SCHREPPING, 1978), which were probably based on erroneously high analytical data.

Measurements of metal concentrations in air dust must, of course, be complemented by determinations of dry and wet deposition. The ratio of dry to wet deposition differs depending on meteorological conditions (NRIAGU, 1980b, c; GRAVENHORST et al., 1980). On the average, approximately 90% of the precipitated cadmium is in dissolved forms (NUERNBERG et al., 1983; LUM et al., 1987). In the case of dry cadmium deposition, this can be neglected. In wet deposition, cadmium concentrations range from <0.05 μg/L to approx. 5 μg/L depending on the area, e.g., the United States, Canada, and the Federal Republic of Germany, with significantly lower values in remote areas of the USA and Canada (NUERNBERG et al., 1983; VALENTA et al., 1986b; LUM et al., 1987). Rain water may contain 0.05–0.8 μg/L and fog vapor 0.3–7 μg/L, which is much more than river water with 0.01–0.15 μg/L and lake water, with even lower concentrations (TUOR and KELLER, 1989). Cadmium contents are extremely low in arctic precipitation, with typical concentrations from <0.3 to approx. 5 ng/L (MART, 1983).

Model calculations with weighted data from deposition samplers in the Federal Republic of Germany yielded a probable yearly cadmium deposition of approximately 80 tons (NUERNBERG, 1983). This only roughly agrees with estimates of cadmium emissions for the Federal Republic of Germany of around 26 tons/year (BDI, 1982; SCHULTE-SCHREPPING and PISCATOR, 1985). However, there is still some scattering in the estimate of cadmium emissions into the atmosphere (see Table II.6-5). In rural and urban areas of the Federal Republic of Germany no statistically significant differences in wet cadmium deposition could be seen from 1980 to 1983, whereas a decrease is obvious for the highly industrialized Ruhr region and some regions with metallurgical industry (VALENTA et al., 1986b).

4.2.2 Cadmium in Waters
(see also Chapter I.5 and for aquatic organisms Section 5.1)

The input of cadmium into waters is due directly to contributions from waste water (see Sect. 4.1.2) and precipitation from the atmosphere, and indirectly from washout resulting from the weathering of minerals, soils, sewage sludge deposits, waste dumps, etc. along with drain water and ground water streams (see Sect. 4.1.3).

Rivers and Lakes. Recent investigations with improved methods have shown that with the exception of particularly polluted regions, the concentrations of dissolved and particulate-bound cadmium often are very low, i.e., below 0.1 µg/L. This is significantly lower than formerly supposed (BOWEN, 1979; MARTIN and MEYBECK, 1979; MALLE, 1985; MART et al., 1983, 1985; BREDER, 1988; HUYNH-NGOC et al., 1988; DORTEN et al., 1989). The Rhine river in West Germany, which has been carefully investigated during the last decade, is a good example. Its average cadmium content is around 0.2 µg/L, with a significant tendency to decrease further (MALLE, 1985; BREDER, 1988). The annual load amounts to approx. 100 tons, which is in good agreement with indirectly computed data based on the cadmium contents of sediments (MALLE and MUELLER, 1982).

Similarly low values, i.e., values below 0.1 µg/L, have been found in inland lakes (MARTIN et al., 1980; BEAMISH et al., 1976; IHNAT et al., 1980; BREDER, 1981, 1988; SIGG et al., 1982).

Sediments (see also Chapter I.7e). Sediments are important sources for the assessment of man-made heavy metal pollution in rivers and lakes. Compared with non-polluted sediments they render possible information about the extent of pollution, and for stationary water systems also associations with time (FOERSTNER and WITTMANN, 1983). Since compared to zinc, cadmium is commonly enriched where thermal processes are concerned, the zinc/cadmium ratio can also be used to elucidate cadmium sources (FOERSTNER, 1980b).

Sediments of non-polluted rivers and lakes show cadmium contents ranging from approx. 0.04 to 0.8 mg/kg, whereas in polluted rivers levels range from approx. 30 to 400 mg/kg (FOERSTNER, 1980a, 1986; BREDER, 1988). Extremely polluted rivers can reach 800 mg cadmium/kg and even higher (FOERSTNER, 1980a, b, 1986).

A mobilization of cadmium from particulate matter in rivers is less probable under normal conditions (GUENTHER et al., 1987), but can occur in estuarine regions, due to the influence of competing ions, a decrease in pH, organic chelators, and redox reactions (FOERSTNER, 1980b; SALOMONS and FOERSTNER, 1984).

Estuaries and Oceans. Former assumptions about strongly fluctuating trace metal contents in different oceanic regions could be corrected to more uniform and much lower levels by the identification of hitherto unknown contamination sources prior to analysis. Former errors could thus be attributed to the less satisfactory accuracy and detection power of the older methods (NUERNBERG, 1979; MART, 1979; BRULAND et al., 1979; MARTIN et al., 1980; MAGNUSSON, 1981; SPERLING, 1982; MART and NUERNBERG, 1986). A critical review of cadmium in the marine environment has been published by SIMPSON (1981).

In estuarine regions, cadmium occurs predominantly in the form of inorganic chloro complexes (SIPOS et al., 1980; RASPOR, 1980). These are produced from dissolved cadmium from rivers and wet precipitation under the influence of a large excess of cadmium, magnesium, and chloride ions. Therefore, significantly increased cadmium contents ranging from approx. 0.04 to 2 µg/L are common in the mixing zone between inland and sea water (MART, 1979; SPERLING, 1982; MART et al., 1982; VALENTA et al., 1986a). The result of this increase in the aqueous phase is bioaccumulation of cadmium, e.g., in algae, bivalves, crabs, and the inner organs of teleost fish, with accumulation factors up to 10^4 (STOEPPLER and NUERNBERG, 1979; PHILLIPS, 1980). In muscle tissue of teleost fish, however, the cadmium content is extremely low, commonly well below 5 µg/kg (STOEPPLER and NUERNBERG, 1979). Net cadmium levels in plankton are also low, even when calcium and strontium concentrations are high (YAN et al., 1988).

Dumping of cadmium-containing waste and of sewage sludge should not be permitted due to accumulation in certain marine organisms (FOERSTNER, 1980a), see also Chapter I.7b. Investigations carried out during the last two decades have shown that depth profiles of cadmium are, on average, well correlated with those of the nutrients phosphate and nitrate. In surface waters nutrient levels are low, as is also the case for cadmium with a typical range from <1 to approx. 15 ng/L. In depths between 1000 and 2000 meters, however, maximal values could be observed of around 100 ng/L for the North Atlantic. Somewhat lower concentrations occur in nutrient-depleted regions, e.g., some regions of the Indian Ocean, the Baltic Sea, and the Mediterranean Sea (BRULAND, 1983; MART and NUERNBERG, 1986; YEATS, 1988).

There are, however, marine regions such as the antarctic Weddel Sea, where upwelling of nutrient-rich deep sea water has led to cadmium contents of up to 50 ng/L at the surface (MART and NUERNBERG, 1986; YEATS, 1988). Since nutrient-rich cold water areas provide favorable living conditions for phytoplankton and krill, cadmium contents in the muscle tissue of antarctic krill are somewhat elevated (STOEPPLER and BRANDT, 1979).

4.2.3 Terrestrial Ecosystems
(see also Section 3.1, Chapter I.7, and KABATA-PENDIAS and PENDIAS, 1984)

Cadmium contents in non-polluted agricultural soils range from approx. 0.01 to 0.5 mg/kg (dry weight) (LITTLE and MARTIN, 1972; COTTENIE, 1981; THORNTON, 1986). Soil samples collected and stored from the mid-1800s to the present were recently analyzed for cadmium, and increases in soil cadmium of 27–55% due to atmospheric deposition were observed. This corresponds to an increase in the cadmium concentration of the soil plough layer of between 0.7 and 1.9 µg/kg and year. The experimental data were in good agreement with theoretical assumptions based on atmospheric emission estimates (JONES et al., 1987).

Cadmium contents are especially high in soils in the vicinity of point sources, and in cases where manure and fertilizers with high cadmium levels have been used, enrichment factors up to 10^3 were found in upper layers (FULKERSON et al., 1983; FLEISCHER et al., 1974). Local situations may be very different even close together within a habitat. WOPEREIS et al. (1989) described spatial variability which is higher for cadmium than for other metal ions due to heterogeneity at a microscopic scale. Critical cadmium concentrations in top soils are thought to be 0.5 mg/kg in sand and 1.0 mg/kg in clay (DE BOO, 1989), but in the Netherlands, levels of 0.3 mg Cd/kg sand and 0.4 mg Cd/kg clay have been found. According to ANGEHRN and HERTZ (1989) and VERLOO and ECKHOUT (1989), manure and chlorides increase the solubility of cadmium compounds in soils. Quite similar contents occur in plant matter (CROESSMANN and SEIFERT, 1982; see also Sects. 5.1 and 6.1 and Chapter I.7), the reason being that cadmium is much more easily taken up by plants than other heavy metals (PAGE, 1986).

Cultivation of agricultural soils to a depth of approx. 30 cm (plough layer) results in dilution of cadmium by distribution into a layer volume. Enrichment of cadmium in upper layers is partly due to the formation of chelates with humic acids (GIESY, 1980). Additionally, cadmium is bound to clay minerals. LOGAN and CHANEY (1987) have researched sludge risk assessment and found a maximum "plateau" value of plant tissue cadmium. Perhaps some ligands outside or inside the roots reduce cadmium uptake by precipitation. Cadmium uptake in plants, influenced also by soil pH, depends not only on the total cadmium content, but also on the ratio of the cation exchange capacity of the organic and inorganic components to the total cadmium content of the soil (LORENZ, 1979). pH values around 7 lead to a diminished cadmium release from chelates and clay minerals. Higher calcium contents are also responsible for reduced cadmium uptake in plants (LORENZ, 1979). However, cadmium accumulation by plants and animals is significant, depending on the species in question.

Cadmium can also be absorbed directly from wet precipitation by above-ground parts of plants. This leads to high cadmium levels in forests, mosses, etc. and can be used to monitor cadmium contents as well (TYLER, 1972; MARTIN and CAUGHTREY, 1975; WAGNER et al., 1985). The highest cadmium concentrations are found in lichens, litter, humus, and moss (ANGEHRN and HERTZ, 1989). The main input mechanism for cadmium is interception.

Very high cadmium contents in plant garbage and in upper soil layers may lead to damage of organic material, as has been observed in areas of Sweden where local cadmium pollution is very high (TYLER, 1975). Negative influences of metal-contaminated sewage sludge on nitrogen fixation in soil organisms have been observed recently at metal concentrations close to or even below the current guidelines for soil protection (McGRATH et al., 1987). On the other hand, there are more sensitive and more resistant microorganisms and plants which synthesize protecting proteins similar to the metallothioneins (REMACLE and HAMBUCKERS-BERHIN, 1987). Cellular cadmium is probably immobilized in thicker cell envelopes. DIELS and LÉONARD (1989) were able to identify the plasmids responsible for cadmium-resistance in Gram negative bacteria. In earthworms, cadmium bioaccumulates mainly in the intestine (RAMSEIER et al., 1989). Earthworms from soil treated with sludge contain especially high cadmium concentrations (LEVINE et al., 1987).

4.3 Cadmium in the Food Chain and in Humans
(see also Sections 4.2.2, 4.2.3, and Chapter I.8)

The most important cadmium source for human beings is food. In highly polluted areas, significantly increased cadmium contents in food products could be detected (e.g., FLEISCHER et al., 1974; ELINDER, 1985a; OSTAPCZUK et al., 1987). High individual cadmium burdens, however, could be due to smoking habits and occupational exposure. The average cadmium content of cigarettes is reported to be approximately 1.2 µg/g, with contents of up to 3 µg/g in some brands (e.g., MUELLER, 1979, 1985; ELINDER et al., 1983; ELINDER, 1985a; WATANABE et al., 1987).

A relatively wide range of data were reported for cadmium contents in different food products and total diets over the last two decades. This was also the case for estimates of daily cadmium intake via food in different countries, with a range from approx. 10 to approx. 110 µg/d (FOERSTNER, 1980a; ELINDER, 1985a). These estimates doubtlessly also include erroneously high values.

In order to avoid errors, reliable determinations of trace metals in foodstuffs require extreme care and powerful methods. If one considers, however, only recent publications with improved analytical methods, current quality assurance, and contamination control, the results are much lower and show significantly smaller ranges for a variety of foodstuffs (see Table II.6-10).

Attempts were also made to investigate retrospectively whether a trend relating cadmium levels in food to the continuous overall increase in cadmium levels in the upper layers of cultivated soils exists (JONES et al., 1987). A recent retrospective study in the Federal Republic of Germany compared cadmium contents in archived wheat, rye, and barley, some samples dating back to the 19th century, to 2000 wheat samples from the last ten years, and did not indicate a trend toward continuously rising cadmium contents in bread cereals within the variations found for all investigated samples (LORENZ et al., 1986). Similar results have also been found for cadmium in a recent comparison of archived cigarettes from the beginning of this century (KJELLSTROEM, 1979), probably because of the already mentioned dilution of cadmium in the plough layer of cultivated soils.

Table II.6-10. Typical Cadmium Levels in Selected Foodstuffs from Non-polluted Areas in Various Countries. Only recent papers were chosen[a]

Typical Contents (µg/kg fresh weight)	Food Type	Remarks
>200	Some mushrooms, some cocoa powders, dark chocolate, blue poppy seeds, semifinished products from poppy seeds, marine mussels	Mushrooms differ strongly in cadmium contents; dark chocolate from particular cocoa represents only a very small percentage of total chocolate products
≤200	Kidneys and livers from swine, cattle and sheep, most mushrooms, some baked goods with blue poppy seeds	
≤40	Wheat, wheat flour, wheat bread, bran, potatoes, root and foliage vegetables, rice, shrimp	Cadmium content in rice varies considerably
≤20	Rye flour, rye bread, beans, tomatoes, fruits, eggs, fresh water fish	Fruits and fresh water fish frequently have cadmium levels up to 10 µg/kg
≤5	Meat from poultry, swine, cattle and sheep, fish fillet (sea fish), wine, beer, fruit juices	
≤1	Tap water, milk and milk products	From recent studies it is very probable that cadmium in dairy milk on average is below 0.1 µg/kg; higher cadmium contents in tap water often are due to zinc plating of water pipes[b]

Note: Cadmium contents in foodstuffs as a result of cadmium pollution have been observed in various areas of the world, typical ranges are given elsewhere, e.g., ELINDER (1986), SHERLOCK (1986); OSTAPCZUK et al. (1987)

[a] SEEGER et al. (1978), KOOPS and WESTERBEEK (1978); STOEPPLER and NUERNBERG (1979), KNEZEVIC (1979), BARUDI and BIELIG (1980), OCKER and SEIBEL (1980), CROESSMANN and SEIFERT (1982), NARRES et al. (1984b, 1985), ELINDER (1985), STOEPPLER et al. (1985), VOS et al. (1986a, b, 1988), CIUREA et al. (1986), LORENZ et al. (1986), SCHREIBER (1986), HOFFMANN and BLASENBREI (1986), DABEKA and McKENZIE (1986, 1987), OSTAPCZUK et al. (1987)

[b] Since 1978 for the FRG the cadmium content in zinc platings of water pipes is limited to 0.01% according to DIN 2444

The total daily intake of cadmium in the United States, in most European countries, and in New Zealand is thus very likely only in the 10 to 30 µg range (SCHELENZ, 1983; ELINDER, 1985a). However, this is not the case for Japan, where the average daily intake is probably in the range of 35 to 50 µg/d due to significant cadmium pollution in soils that has led to comparatively high levels in food products, e.g., rice (ELINDER, 1985a).

If cadmium contents increase in the feed of slaughter animals, then cadmium levels in the kidney and liver are also higher. Thus, in order to maintain low cadmium levels in human food, legal threshold values for cadmium in animal feed were proposed (see also Sect. 7.3.2).

The adult human body of a non-smoker contains about 15 mg cadmium. Concentrations in human organs and body fluids are given in Sects. 5.3.3, 5.3.5, and 7.1 and also in Table II.6–11. Since kidney damage by cadmium is a particular risk, it is important to know that cadmium concentrations increase with age to about 20 (individually up to 50) mg Cd/kg kidney cortex (the Cd:Zn ratios in this organ to about 0.7) in 40 to 50 year-old non-smokers; later these parameters decrease again (EWERS et al., 1989). Cadmium distribution in bones is not homogeneous. Fetal bones contain about 20 ng Cd/g dry weight (less than adults with about 40 ng Cd/g dry weight), but contain more manganese than adults (HEDRICH et al., 1988, 1989). Both elements are subject to a process of replacement during the prenatal period.

5 Uptake, Absorption, Transport and Distribution, Metabolism and Elimination in Plants, Animals, and Humans

5.1 Uptake and Distribution in Plants and Aquatic Organisms
(see also Sections 4.2.2 and 4.2.3)

Plants may take up amounts of 0.1 to 1.5 mg cadmium per kg dry weight from soil and 0.5 mg/kg d.w. from the atmosphere (TUOR and KELLER, 1989). Cadmium is especially absorbed by plants in the vital rooting zones (LAMERSDORF, 1989), while in the subvital status some desorption takes place (unlike lead, which also accumulates significantly in subvital roots).

From laboratory experiments it is known that cadmium contents, on average, decrease significantly in plants from roots to shoot. However, the extent is strongly species dependent. Commonly, the root contains ten times as much cadmium as is contained in the shoot, i.e., the root acts as a barrier (JARVIS et al., 1976).

Many species show significantly lower cadmium levels in fruits than in the shoot. In various plants, e.g., spinach, soy beans, and wheat, the barriers are only poorly effective. In rice grains, there is even a cadmium enrichment compared to the leaves (BINGHAM et al., 1975).

Bioretention for cadmium in streaming waters at average low tide ranges from 0.001 to a maximum of 0.2%. Accumulation factors for water plants range from 5×10^2 to approx. 10^4 (related to dry weight) and for fish feed (low water fauna) at approximately the same order of magnitude (WACHS, 1981). Due to this accumulation, waters used for fishing should contain no more than 0.5 µg/L cadmium (RUF, 1981).

WIENER et al. (1988) discussed fish bioavailability in seepage lakes (pH between 4.6 and 7.7). Blue gills of 4 years of age contained 0.02–0.07 µg Cd/g d.w. (whole body) in uncontaminated lakes and up to 0.61 µg Cd/g d.w. in polluted lakes. VAN

COILLIE et al. (1988) exposed salmon eggs to cadmium and found bioaccumulation factors of about 8 for the whole embryo, but of about 25 for its vitellus. Highest concentrations were observed in the mesencephalon and in the liver. Mussels accumulate cadmium mainly in the kidney and the viscera (MALLEY and CHANG, 1988). LASENBY and GROULX (1988) determined cadmium burdens of about 100 ng in phantom midge fly larvae (a food for fish) of about 100 µg (d.w.).

5.2 Intake and Metabolism in Mammals

In the course of research about the toxicology of cadmium there have been numerous experiments performed with mammals that might be considered, with caution, as being of importance for human toxicology (e.g., TSUCHIYA, 1978; MENNEAR, 1979; CADMIUM-79, 1980; CADMIUM-81, 1982; FRIBERG et al., 1985, 1986a; OBERDOERSTER, 1986, 1989; KAZANTZIS, 1987).

The most important results of these studies can be summarized as follows:
- Particulate airborne cadmium intake through the respiratory tract yields retention rates of 10 to 40% of the total inhaled cadmium. The smallest particles (<5 µm) are deposited and absorbed preferentially in the alveolar regions of the lungs (see also Sect. 5.3.1 and Chapter I.15). But it must be considered (OBERDOERSTER, 1989; OBERDOERSTER and COX, 1989) that rats, for instance, behave differently than humans; in rats, deposition in the lungs is more intensive, whereas lung deposition in humans is reduced because of breathing through the nose with higher bronchial uptake. On the other hand, the biological half-time of cadmium in the rat lung is of the order of 70–80 days; in primates, however, about 780 days. One should also know more about differences regarding the chronic activation of inflammatory cells and the significance of metallothioneins, finger-loop proteins, oncogene activation, and DNA-repair. Humans are also more exposed to synergistic and antagonistic substances. PETERS et al. (1989) found that female mice concentrate more cadmium in the lungs (up to about 60–90 µg/g) after exposure to CdO fumes or CdS than do male hamsters (which concentrate cadmium in the kidney).
- For oral intake, the absorption (uptake) depends on the species of laboratory animal and the speciation of cadmium, and ranges from approx. 0.3 to 3%. Higher absorption occurs at low calcium and protein intake, with iron deficiency and/or lower absorption at high protein intake. There seem to be two different intestinal uptake mechanisms (SCHUEMANN et al., 1989). ANDERSEN (1989) has shown that wheat-bran in the diet reduces cadmium uptake. KODAMA et al. (1989) found that female dogs take up less cadmium (up to 80 µg/L blood) than male dogs (up to 120 µg/L) from contaminated rice, and the latter therefore excrete more (up to 250 µg Cd per day via urine).
- Within the organism, cadmium is bound predominantly to metallothionein, the biosynthesis of which is induced by cadmium (see also Chapter I.12). Cadmium is deposited mainly in the kidneys and liver, and in whole blood cadmium is primarily bound to erythrocytes.

5.3 Cadmium Metabolism in Humans

5.3.1 Inhalation

The pattern and extent of deposition depends on particle diameter. Particles with an aerodynamic diameter of 10 to 20 µm are mainly deposited in the upper respiratory tract (nose and pharynx), those with diameters of about 5 to 10 µm are deposited in the tracheobronchial region, and particles smaller than 5 µm are deposited in the alveolar region. This is also the case for aerosols from cigarette fumes. The daily uptake of cadmium from smoking cigarettes can be estimated from the average cadmium content of mainstream smoke and the cadmium deposits in kidneys of smokers and non-smokers. The fumes from 20 cigarettes lead to a retention of approx. 3 µg cadmium and a resorption of approx. 35%, i.e., around 1 µg (FRIBERG et al., 1985, 1986a; see also Sect. 5.3.3).

In workplace areas where cadmium was formerly processed, high cadmium levels up to 10^4 µg/m^3 occurred. Technical improvements have now led to a drastic reduction to levels of 5 to 10 µg/m^3 and even less at most locations (FRIBERG et al., 1985).

5.3.2 Ingestion

For occupationally non-exposed persons and non-smokers, food is the main source for cadmium. From experiments with radioactive cadmium performed with volunteers, an average human resorption of 6% from food was found. This is significantly higher than that observed for experimental animals. Based on a daily intake of 10–30 µg, this would amount to a daily cadmium absorption in most Western countries of 0.6 to 1.8 µg from food (see Sect. 4.3). Calcium, protein, and iron deficiency, however, can increase this percentage (FRIBERG et al., 1985; SCHUEMANN et al., 1989).

5.3.3 Transport, Distribution, Biological Half-Time

Cadmium is absorbed in the lung and gut and is transported via the blood to body stores, the liver being first. It is likely that more than 95% of cadmium in the blood is bound to two protein fractions in the blood cells (FRIBERG et al., 1974, 1985). In the liver, cadmium is bound to metallothionein (see Chapter I.12) and the complex produced is transported to the kidneys where it diffuses through the glomerular membrane into the tubular fluid, whereafter it is almost completely absorbed at the proximal kidney tubuli (kidney cortex). Since the biological half-time of cadmium in the liver and kidneys is greater than ten years, an age-dependent increase in cadmium content occurs (FRIBERG et al., 1985).

From investigations with materials obtained from autopsies, a cadmium body store has been estimated to be around 15 mg for non-smokers and around 30 mg for smokers (FRIBERG et al., 1985). This ratio was also found by measurements with in vivo methods (ELLIS et al., 1979; ROELS et al., 1981; SPANG, 1988).

To a certain extent, cadmium does not pass the protecting placental barrier, thus the cadmium content in the umbilical cord and body of newborns is about 50% of the cadmium level in maternal blood (FRIBERG et al., 1985; communication by EWERS, 1989).

In cases of high cadmium exposure, i.e., in industry and in highly cadmium-polluted areas of Japan, cadmium contents have been found in human livers that were up to 100 times normal levels (ELINDER, 1985b). Cadmium levels in the kidneys were in some cases also very high, if kidney damage had not in fact already occurred (see also Sect. 6.5). Since cadmium-induced renal dysfunction is associated with increased cadmium excretion, kidney levels are low in cases of cadmium-induced dysfunction.

5.3.4 Distribution Models

Mathematical models for cadmium enrichment and distribution make it possible to predict concentrations in depot organs under distinct exposure conditions. Initially, single compartment models were developed for the kidneys (TSUCHIYA and SUGITA, 1971; KJELLSTROEM, 1971), followed by an eight-compartment model (KJELLSTROEM and NORDBERG, 1978) and a two-compartment model (JAERUP et al., 1983). While the simpler models are sufficient for long-term low-level exposure, the more complex models make it possible to predict the cadmium content in various organs after short-term, as well as long-term, exposure. Based on experimental data, the eight-compartment model considers the transfer of cadmium between muscle tissue, liver, and kidneys, and obtains optimal results if for each of these compartments biological half-times of between 8 and 14 years are assumed (KJELLSTROEM and NORDBERG, 1978, 1985; HERBER et al., 1988).

5.3.5 Cadmium in Body Fluids

The cadmium concentration of whole blood reflects actual exposure in industrially-exposed persons, smokers, and non-smokers. Cadmium and protein levels in the urine normally only increase significantly if kidney dysfunction commences. However, in cases of extremely high short-term cadmium exposure, high cadmium concentrations in urine have also been reported. This has been explained by a deficiency of metallothionein storage during a sudden increase of cadmium in the body (LAUWERYS et al., 1980). A high cadmium store in the body, however, also raises to some extent cadmium levels in whole blood (FRIBERG et al., 1985). Table II.6-11 presents typical cadmium levels in human materials from recent and mainly independently confirmed literature data.

Due to the biological half-time, under normal conditions only an extremely low percentage (less than 0.01 of the total amount) of cadmium is excreted through urine. Relationships between cadmium concentrations in urine and kidney cortex have been studied, e.g., by ROELS et al. (1981). Daily cadmium excretion increases on a group

Table II.6-11. Typical Cadmium Levels in Human Materials from Recent, Mainly Confirmed Information[a] (in mg/kg fresh weight)

Material	Average (approx.)	Range	Remarks
Head hair	<0.5	<0.1−2	Close to the root proximal (i.e., 0 to 5 cm) predominantly endogenous influences can be seen, while in distal parts (>5 cm) exogenous influences dominate. In cadmium exposure, values at the high mg/kg level are possible
Kidneys, smokers	≤6		In kidney cortex of exposed persons up to 500 mg/kg
Kidneys, non-smokers	≤3		Contents in mg/kidney: "critical organ" Increases with age
Liver non-smokers	≤2	0.1−3	After exposure up to 150 possible
Feces		0.1−0.4	Depending on cadmium in food
Lung tissue, unexposed	≤0.1		Exposed persons ≥0.2
Bones	≤0.1	<0.01−0.3	
Urine	≤0.0005	<0.0001−0.003	Increases with age; also some indication on smoking habits. After exposure up to 0.2 mg/kg possible
Whole blood, non-smokers	≤0.0005	<0.0002−0.003	After exposure up to 0.2 mg/kg possible
Whole blood, smokers	<0.001	≤0.0002−0.06	
Milk	≤0.0002	<0.0001−0.003	Differences still exist among averages from different studies; sampling difficult, contamination problems

[a] Liese and Simon (1984), Elinder (1985b), Dabeka et al. (1986), Gerhardsson et al. (1986), Vanoeteren et al. (1986), Mueller (1987), Clench-Aas et al. (1987), Angerer (1988), Ewers et al. (1988), Hedrich et al. (1988)

basis with age, and is proportional to the total cadmium body store (Friberg et al., 1985).

6 Effects on Plants, Animals, and Humans

6.1 Effects on Plants
(for Microorganisms see Section 4.2.3 and Chapter I.7h)

Phytotoxic effects of heavy metals in plants have been studied in vitro, but also under experimental conditions for aquatic and terrestrial plants. For terrestrial flora, effects have been studied mostly in the context of environmental contamination. Symptoms are chloroses and necroses with characteristic color changes (CHANEY et al., 1978; COX, 1986; see also Chapters I.7f and I.7g). Critical effects include decreases in productiveness (TUOR and KELLER, 1989), reduced rates of photosynthesis and transpiration, and altered enzymatic activities. Spinach and radish growing in soil with 1 ppm cadmium contain about 40% less vitamin C and vitamin B_{12} (TUOR and KELLER, 1989).

If cadmium uptake occurs from soil, adverse effects related to cadmium levels in plants could be seen earlier than effects caused by cadmium from the atmosphere. Normal cadmium levels in plants are below 0.5 mg/kg (dry weight). Reported levels in the vicinity of point sources reach several hundred mg/kg in assimilation organs (COX, 1986).

In vegetation studies, reduction in yield could be observed from approx. 5 mg Cd per kg in soil and approx. 10 mg/kg in parts of plants above ground. However, the influence of toxic metals in almost every case is a combined and mainly amplified effect (COX, 1986). In order to consider this behavior, limits for the sum of zinc, lead, and cadmium contents in soil have been proposed (KLOKE, 1977).

6.2 Biochemical Effects, Metal Interaction, Metallothioneins

Cd-cysteinate^0, Cd-cysteinate(OH)$^-$, albumin-Cd-complexes, and aquo-Cd-complexes all play a role in blood plasma (PLANAS-BOHNE and KLUG, 1988). Albumin prevents the production of low molecular weight fractions. See Sect. 5.3.3 for the binding of cadmium in blood cells; detoxification by cytosolic proteins, such as metallothionein, has been described in Sects. 4.3.2, 5.2, 5.3.3, 6.5, and Chapter I.12. FISCHER (1989) studied in vitro cadmium toxicity in Chinese hamster cells and found elongation of all cell cycle phases, the DNA-synthesis phase being especially sensitive. Zinc and manganese inhibited cellular cadmium uptake, while selenium had no measurable effect. However, all three trace substances decreased the acute toxicity of cadmium.

As is known from animal experiments, antagonistic effects of derivatives of other metals, e.g. zinc, cobalt and selenium, can also be expected in man. Most of the available information is on the zinc/cadmium antagonism. Biochemically, a substitution of zinc in enzymes containing these elements is assumed (FRIBERG et al., 1986a). Higher cadmium uptake into the body causes a redistribution of zinc. The increase of cadmium content in kidney cortex with age is accompanied by an increase in zinc content. This was attributed to an increase in metallothionein present in the kidney cortex containing equimolar amounts of cadmium and zinc (PISCATOR and LIND, 1972). Since such a redistribution of zinc is linked with a decrease in zinc in other organs, this might impair the essential functions of this element (FRIBERG et al., 1986a).

6.3 Effects on Animals
(see also Sections 4.2.2, 4.2.3, and 5.1)

Fresh water and marine organisms respond differently to cadmium ions. Cadmium levels >1 µg/L in fresh water and >7 µg/L in sea water can initiate toxic effects (see also Sects. 5.1 and 5.7). Concentrations above 2 µg/L in fresh water and 100 µg/L in sea water are lethal for certain species. An increase in salinity and/or calcium content and a decrease in temperature both decrease cadmium toxicity (PHILLIPS, 1980; TAYLOR, 1981, 1982). BODAR et al. (1988) studied cadmium resistance (acquired during a single generation) and the synthesis of metallothionein-like proteins in *Daphnia magna*. POSTHUMA (1988) looked at cadmium-induced growth retardation in *Orchesella cincta* (a 2 mm long insect living in soil) populations and the adaption of some populations.

Terrestrial species are not commonly exposed to doses that produce toxic effects. Feeding experiments with sheep, for example, using up to 30 mg cadmium per kg feed, did not result in toxic effects (GLASER et al., 1978; ANKE et al., 1984). Studies with various species have been reviewed by VAN BRUWAENE et al. (1986).

The oral LD_{50} for rats is 150 mg Cd per kg body weight, for rabbits 150 to 300 mg/kg b.w. The intravenous LD_{50} for rats is 2 mg/kg, for rabbits 5 to 10 mg/kg (UMWELTBUNDESAMT, 1977). Species-dependent acute toxicity has been observed experimentally in numerous animal species (DOYLE, 1977). More recent studies showed carcinogenic effects in rats after injection and inhalation of cadmium chloride (OLDIGES et al., 1984) and for various cadmium compounds in rats, hamsters and mice (OLDIGES and GLASER, 1986; OLDIGES et al., 1988; see especially Sect. 6.6).

6.4 Acute Effects
(see also FRIBERG et al., 1986b)

Inhalation: Shortness of breath, weakness and fever have been observed after inhalation of cadmium-containing aerosols in industry, only 24 hours after cease of exposure. In severe cases, fatal choking fits occur. A cadmium level of 5 mg/m^3 can be lethal after eight hours exposure. Under the same conditions, a cadmium level of 1 mg/m^3 still might lead to clinical symptoms (FRIBERG et al., 1986a).

Ingestion: Oral intoxications have been observed after cadmium contamination of tap water and beverages from soldering points in water pipes and water faucets, as well as from cadmium-plated heating and cooling tubes and kitchen utensils. Cadmium can be extracted by acidic juices from the colored glazes of ceramic ware. Symptoms such as nausea, vomiting, and headache occur within a few minutes after ingestion. In severe cases, diarrhea and a shock-like state have been observed. A cadmium level of 15 mg/L in liquids causes vomiting. If proteins that provide some protection are administered at the same time, however, symptoms occur at higher cadmium concentrations (FRIBERG et al., 1986a). Protein malnutrition, on the other hand, enhances adverse effects (TANDON, 1988).

6.5 Chronic Effects

(see also FRIBERG et al., 1986b)

General: In the past, chronic effects due to long-term inhalation of cadmium-containing dust were observed frequently; these consist of a syndrome that includes lung emphysema as well as renal dysfunction (FRIBERG, 1950). The type and intensity of symptoms depend on individual disposition as well as on intensity and duration of exposure. Intensive exposure for rather short times affects the lungs, whereas less intensive long-term exposure results in kidney dysfunction (FRIBERG et al., 1986a; HALLENBECK, 1986). Long-term ingestion of large amounts of cadmium has until now only been observed in Japan. This has led to kidney dysfunction, as in industrial exposure, and a severe bone disease known as Itai-itai disease. Cellular changes in the kidneys and the female reproductive tract were studied by COPIUS-PEEREBOOM-STEGEMANN (1989).

Effects on the Kidneys (see also Sect. 5.3.3): Glomerular damage causes enhanced excretion of high molecular weight proteins in urine. Tubular damage can be recognized by an increased excretion of low molecular weight proteins in urine due to the decreased reabsorption of these proteins that can pass the glomerular filter and occur in primary urine. β_2-Microglobulin (molecular weight approx. 11 800), and the more stable retinol binding protein, can be analyzed as indicator proteins (SCHALLER et al., 1980; HERBER, 1984). Glomerular and tubular kidney damage can appear separately but also simultaneously. Kidney damage always causes an enhanced cadmium excretion in urine (ROELS et al., 1981; FRIBERG et al., 1986a; HERBER et al., 1988). In cases of advanced cadmium nephropathy, the reabsorption of amino acids, phosphates, and glucose in the kidney tubuli may also decrease, which in effect leads to an increased excretion of these compounds.

Reabsorption in the kidney continues to be affected after cessation of cadmium exposure. It is unknown, however, if the first, weak changes that appear after moderate exposure are reversible. It appears that kidney disorders caused by cadmium also influence phosphorus and calcium metabolism and lead to the elimination of mineral constituents from bones (FRIBERG et al., 1986a).

Blood and Circulation: A reversible decrease in hemoglobin content was observed frequently in the blood of persons exposed to cadmium. This might be connected to inhibition of iron resorption from the diet, causing a reduction in iron supply to bone marrow (FRIBERG et al., 1986a; SCHAEFER et al., 1988). As with experimental animals, information about a possible influence of cadmium on blood pressure in humans is contradictory (MANTHEY et al., 1981; FRIBERG et al., 1986a).

Liver: Changes in liver function in persons exposed to cadmium are, compared to observed effects in the kidneys, not very significant, though from animal experiments morphological damage and impairment of enzymatic activity might be expected (FRIBERG et al., 1986a).

Bones: As mentioned already, kidney dysfunction caused by cadmium exposure can influence the metabolism of bone-forming elements such as calcium. This can promote osteoporosis in animal experiments. If the femurs contain more than 5–10 ppm cadmium, mechanical strength in young rats is significantly reduced (OGOSHI et al., 1989).

In the Toyama region of Japan, especially in the Jintzu river area, structural changes in bones (osteoporosis and osteomalacia) and renal tubular dysfunction with severe subjective symptoms were observed in post-menopausal women from the end of the 19th century to after World War II. This was named Itai-itai disease (TSUCHIYA, 1978; KJELLSTROEM, 1986). Very high cadmium contents in potable water and foodstuffs, mainly rice (daily intake up to 300 µg), were found in this region, caused by cadmium pollution of waters from zinc/cadmium mines. Since materials from autopsies (kidney, liver, other organs and tissues) also contained greatly elevated cadmium contents, it was assumed by some researchers that this was a manifestation of a chronic oral cadmium intoxication (see detailed discussion by KJELLSTROEM, 1986). This assumption, however, was not unanimously supported. Some Japanese scientists expressed the opinion that the cause for Itai-itai disease was malnutrition and vitamin D deficiency. An evaluation of all available information compared to findings from occupational exposure confirms the important role of cadmium in this disease. However, due to the particular symptoms, other factors, such as inadequate consumption of some essential elements and vitamins in the diet, have been identified as contributing factors towards the occurrence of Itai-itai disease (TSUCHIYA, 1978; KJELLSTROEM, 1986).

6.6 Mutagenic, Carcinogenic, and Teratogenic Effects
(see also Chapter I.18)

With regard to genetic effects, it seems that cadmium ions are spindle-inhibiting (FRIBERG, 1986b) and induce repair of single strand breaks (COSTA, 1989), also because of their preference to react with proteins. SUNDERMAN (1989) discussed cadmium-substitution for zinc in the finger-loop domains of gene regulating proteins as a possible mechanism for genotoxicity. Cadmium ions may change stabilities and thus oncogene expression. FISCHER (1987) studied the mutagenic effects of cadmium ions on sister chromatid exchange induction and found that selenite may interact as an antimutagen. KAZANTZIS (1987), however, is of the opinion that mutagenic effects of cadmium in animal tests are conflicting, and that there is no clear evidence for effects in humans (but see also Sect. 2.3 in Chapter I.18).

Cancer can be induced in experimental animals through injection of high doses of cadmium, but more recent studies indicate that inhalation of different cadmium compounds can also initiate cancer (HEINRICH, 1988; OLDIGES et al., 1988). Therefore, the German commission for the investigation of the health hazards of chemical compounds in the workplace stated in 1982 that an unmistakably carcinogenic potential (at that time only for cadmium chloride) existed "under conditions which are comparable to those for possible exposure of a human being at the workplace, or from which such comparability can be deduced" (MAK, 1987; HEINRICH, 1988).

OBERDOERSTER and COX (1989) concluded from various studies that oral uptake does not lead to tumors, and that significant numbers of lung tumors (inhibited by ZnO) were especially detected in rats. About 0.01 µg Cd/m^3 in inhaled air corresponds to about 10 µg/g lung and a 10% tumor incidence, while about 0.1 µg Cd/m^3 corresponds to about 60 µg/g lung and a 50% lung tumor incidence. According to OBERDOERSTER (1989), nitrosamines, asbestos, PAHs, nitrous oxide, ozone, and sulfur dioxide are synergists and increase lung cancer risk, while zinc is an antagonist. GLASER et al. (1989) and OLDIGES et al. (1989) reported on whole body exposure of rats for up to 18 months. Bronchio-alveolar adenomas and adenocarcinomas were detected in particular after inhalation of some cadmium compounds. Cadmium chloride and cadmium oxide dust pose the highest risks; cadmium oxide fumes or cadmium sulfate and cadmium sulfide aerosols are somewhat less effective and higher concentrations are needed. Exposure to cadmium oxide dust and to zinc oxide produced no lung tumor (even when cadmium was accumulated). TAKENAKA et al. (1989) could demonstrate bronchio-alveolar hyperplasia and fibrosis in rat lungs exposed to cadmium oxide dust (28% of the rates exposed to 30 µg CdO dust/m^3 and 65% of those exposed to 90 µg CdO/m^3 had tumors). HEINRICH et al. (1989) exposed Syrian golden hamsters and NMRI mice for 12 to 15 months. Survival of hamsters was reduced by 50%, but no tumors were detected in histopathological studies. Lipoproteinosis, interstitial fibrosis, and bronchio-alveolar hyperplasia were observed in exposed mice, but not an increased lung tumor incidence (spontaneous lung tumor rate varied anyway). Only cadmium oxide exposure showed effects − cadmium chloride, cadmium sulfate, and cadmium sulfide did not. According to AUFDERHEIDE et al. (1989), 216 µg and 552 µg Cd/m^3 (especially in the form of cadmium oxide) lead to proliferate lesions in hamster lungs, and 63 µg Cd/m^3 have no influence.

As far as earlier studies about the probability of prostate carcinomas are concerned, recent investigations indicate that there is no conclusive evidence for a particular risk (PISCATOR, 1981, 1988; KAZANTZIS, 1987). KAZANTZIS (1989a, b) concluded that according to epidemiological studies the situation in humans is still not clear. Prostate cancer is definitely not related to cadmium at present exposure levels (statistical evaluation is difficult since 35% of elderly men have prostate problems), but there may be an influence of cadmium on hormones. Data regarding stomach cancer are confusing. With regard to lung cancer, factors such as protective effects, bioavailability, smoking, interactions with asbestos, and length of follow-up must be studied further. KAZANTZIS (1989a, b) discussed, in addition to two other studies, two occupationally-exposed male cohorts and found a somewhat increased tendency towards bronchitis, lung cancer, emphysema, and stomach cancer whereas prostate cancer and cerebrovascular diseases were not related to cadmium exposure.

Studies on reproductive effects and teratogenicity have been done primarily through injection of relatively high doses of cadmium compounds into rodents (see, for instance, FRIBERG et al., 1986b; KAZANTZIS, 1987), and the evidence − also for humans − is therefore not clear. Reproductive toxicity in some strains of male rodents has, however, been described, and DWIVEDI (1983) suggested that cadmium compounds inhibit spermatozoan choline acetyltransferase, decrease acetylcholine synthesis, and impair spermatozoan mobility. KUTZMAN (1984) found no reproductive effects in Fischer rats exposed to 1 mg Cd/m^3 aerosols. Women exposed to cadmium in industry had no newborns with malformations (FRIBERG et al., 1986b), but birth weights were lower and a few cases of rachitis were observed.

7 Hazard Evaluation and Limiting Concentrations

7.1 Dose-Response Relations

Data available today from studies on human beings involving morphological changes and corresponding cadmium contents in kidney cortex ("critical organ concentrations") are still limited. However, a critical range can be given based on measurements in autopsy materials, and information from studies with in-vivo monitoring methods and animal experiments. For kidney cortex, it probably lies between 100 mg/kg and 300 mg/kg cadmium with a probable critical level of 200 mg/kg (CEC, 1978; ROELS et al., 1981; LAUWERYS, 1982; FRIBERG et al., 1986a). A reliable determination of exposure data was rarely possible for chronic effects observed in the past from industrial exposure. Thus, with the aid of the already mentioned single compartment model of KJELLSTROEM (Sect. 5.3.4), sufficient for this purpose, it was computed which cadmium dose, at the workplace or from diet, leads to a critical concentration in the kidney cortex (FRIBERG et al., 1974, 1986a). Assuming a somewhat too long biological half-time of 19 years, an absorption of 25% in the lungs and a cadmium content of 13 µg/m^3 in workroom air, a worker attains approx. 200 mg Cd/kg in kidney cortex after 25 years (FRIBERG et al., 1986a). Typical cadmium concentrations at the workplace today are rather in the range of 5 to 10 µg/m^3 (see Sect. 5.3.1). The present average cadmium intake via food in non-polluted regions – Japan, of course, being excluded – is, as shown in Sect. 5.3.2, less than 30 µg. Based on experiences with the more comprehensive 8-compartment model, which assumes somewhat shorter half-times, these estimates might be somewhat on the high side. A population-critical concentration should perhaps also be considered, since with lower exposure only a part of the population is at risk for kidney damage (KJELLSTROEM et al., 1984), e.g., a 10% response in cases of 180–220 mg Cd/kg kidney cortex. In any case, the order of magnitude is certainly correct.

These estimates have been confirmed in principle if data from Itai-itai patients (NOGAWA et al., 1978, 1979) and from feces studies (KJELLSTROEM, 1979) are taken into account: A total cadmium intake from diet of 240 to 480 µg/day led to a higher incidence of proteinuria and increased cadmium levels in urine, when compared with controls (LAUWERYS, 1982). In order to guarantee safe protection also for sensitive individuals and to consider other cadmium sources, such as cigarette smoke, the WHO threshold limits (see Sects. 7.3.2 and 7.3.3) have been set significantly lower.

7.2 Health Risks, Control, and Therapy

For a short-term, not excessive cadmium exposure via the respiratory and digestive tract, the prognosis is generally favorable (FRIBERG et al., 1986a, see also Sect. 5.3.1). Treatment of a relatively high single cadmium dose by administration of chelators such as D-penicillamine can be favorable if done immediately after exposure (FRIBERG et al., 1986a). It has been reported several times, however, that upon treatment of long-term cadmium exposure with chelating agents, kidney damage can occur. Thus, apart from exceptional cases, chelate therapy is not recommended (UMWELTBUNDESAMT, 1977; FRIBERG et al., 1986a).

In case of chronic inhalative or oral poisoning, the less favorable prognosis depends on effects which have already occurred. A deterioration was observed sometimes even after termination of exposure (UMWELTBUNDESAMT, 1977). In case of severe kidney damage with bone disease, a direct treatment of metabolic impairments is recommended. This can be achieved by administration of phosphates and calcium or vitamin D in the case of Itai-itai disease, combined with sufficient amounts of proteins and zinc (KJELLSTROEM, 1986).

For monitoring, the repeated determination of cadmium in whole blood is recommended (see also Sects. 2.2, 5.3.5, and 7.3.4). In general, however, elevated cadmium levels in whole blood (which may also be influenced by smoking) do not allow definitive conclusions about effects that have already begun. For this, cadmium and β_2-microglobulin determinations in urine are more appropriate (FRIBERG et al., 1985, 1986a).

A rather certain control today is theoretically possible by use of in vivo determination of cadmium levels in the kidney through the use of in vivo NAA. Due to the fact that the safety of in vivo NAA is controversial, only a few pilot studies with volunteers, and these only in especially severe cases of cadmium exposure, have been performed to date. Routine control is therefore still performed by means of cadmium and protein determinations. Cadmium levels above 5 µg/L in urine and whole blood can be the first signs of occupational or environmental exposure. In order to judge the obtained data, as already mentioned, the smoking habits of the monitored persons must be known (see Table II.6-11).

7.3 Threshold Limits

Because of the toxic properties of cadmium and its accumulation in organs, national and international threshold values have been elaborated which show a wide range and which also undergo steady corrections and changes (see Chapters I.20a and I.20b; CADMIUM TODAY, 1986; FRIBERG et al., 1986a,b; HEINRICH, 1988; TSUCHIYA, 1988; ANDERSON, 1988) so that only some examples can be given here. Hence a need for current biological monitoring is obvious.

7.3.1 Air

Emission, immission, and deposition threshold values have been determined for cadmium in air, as have maximum concentrations for air at the workplace.

Japan has set a threshold value of 1.0 mg/m^3 for industrial emissions of cadmium and cadmium compounds (TSUCHIYA, 1988). In other countries, significantly higher emission threshold levels are sometimes in force: 3 to 10 mg/m^3 in Yugoslavia, where the higher value is only acceptable for <30 minutes. In Canada (Ontario), a maximum of 10 mg/m^3 for <30 minutes is also permitted. In the Federal Republic of Germany, present regulations (TA LUFT, 1986) do allow in the most dangerous class I a sum value of 0.2 mg/m^3 for industrial emissions; in Australia this is 23 mg/m^3 and in the United Kingdom 39 mg/m^3 (UMWELTBUNDESAMT, 1977). Moreover, the Federal Republic of Germany has set a deposition threshold level for air dust of 5 μg/m^2 and day for cultivated soils (human and animal food) and a general daily threshold value of 7.5 μg/m^2. From this, the immission value should not exceed 40 ng/m^3 (TA LUFT, 1986). In Switzerland, precipitated dust should not contain more than $0.73 \text{ mg/m}^2/\text{year}$ (WILSON, 1986).

WHO has recommended to allow a short-term cadmium level in air of up to 250 μg/m^3 at the workplace. However, long-term exposure should not exceed 10 μg/m^3 (WHO, 1980). According to WHO (1987), air quality guidelines for Europe, mean levels should not exceed $10-20 \text{ ng/m}^3$ in urban areas and $1-5 \text{ ng/m}^3$ in rural areas over one year. Due to the fact that cadmium compounds have shown carcinogenic potential in animal experiments (OLDIGES et al., 1988), no maximum allowable concentrations at the workplace are given in the Federal Republic of Germany (MAK, 1987). In the USA, the National Institute of Occupational Safety and Health (NIOSH) recommended in 1983 the reduction of a previous, higher standard to 40 μg/m^3 as a 10-hour time-weighted average for 40 hours per week, with a ceiling of 200 μg/m^3 averaged over a 15-minute period. Referencing the recently stated carcinogenicity of cadmium, however, the US Environmental Protection Agency (EPA) has proposed listing cadmium as a hazardous air pollutant (ANDERSON, 1988). Threshold levels for dust in workplace air are 10 μg/m^3 in Finland, 200 μg/m^3 in the United Kingdom for metal dust and 50 μg/m^3 for oxide dust, 100 μg/m^3 in the USSR, and 200 μg/m^3 (cadmium dust) in Switzerland (FOERSTNER, 1980a). In Sweden, and recently also in Denmark, cadmium processing and use (the latter with some exceptions) have been banned (NATIONAL SWEDISH ENVIRONMENTAL PROTECTION BOARD, 1982; EDLUND, 1986; HERON, 1986).

7.3.2 Food, Pottery, Soils, Sewage Sludge, and Animal Feed
(see also Section 4.2.3 and Chapters I.7 and I.8)

Based on the assumption of a critical cadmium concentration of 200 mg/kg (see Sects. 6.5 and 7.1), the World Health Organization has set a threshold value of 50 mg/kg in the kidney cortex that is still to be considered safe (WHO, 1972). If resorption and the biological half-time of cadmium in the kidneys are taken into account, this corresponds to a daily cadmium intake of 1 µg/kg body weight. From this, a weekly cadmium dose of 400 to 500 µg can be deduced (WHO, 1972).

Because of different cadmium contents in food (see Table II.6-10) practice-oriented guidelines have been proposed by official organizations in some countries. For example, this is particularly the case for rice in Japan (0.4 mg/kg; TSUCHIYA, 1988). In the Federal Republic of Germany, upper values have been recommended for a broad selection of foods, ranging from 0.0025 mg/kg for milk to 1.0 mg/kg for beef and pork kidneys (fresh weight) (RICHTWERTE 86).

Because cadmium can enter food by dissolution from pottery and cooking utensils, test protocols have been established in some countries. The procedures are based on a treatment with 3 to 4% acetic acid at ambient temperature for a certain length of time, and then the cadmium content of this test solution is determined. Cadmium threshold limits in these solutions range from 0.1 mg/L in Sweden, 0.2 to 0.7 mg/L (depending on the material investigated) in the United Kingdom, 0.5 mg/L in Italy and the USA, and 1.0 mg/L in Denmark and South Africa (FOERSTNER 1980a).

In order to limit cadmium contents in plant material, it is necessary to limit the cadmium content of agricultural soils and of sewage sludge used as fertilizer (see Chapter I.7c). For the Federal Republic of Germany, levels of 3 mg Cd/kg in air-dried soil and 20 mg/kg (dry weight) in sewage sludge have been set (KLÄRSCHLAMMVERORDNUNG, 1982). Lower values exist in the Netherlands and some Scandinavian countries (PURVES, 1981). In other European countries and the United States, different threshold levels for sewage sludge, which also depend on the concentrations of other elements, are under discussion or in preparation. They range for cadmium from 0.1 kg/ha to 6.7 kg/ha and year (FOERSTNER, 1980a; COTTENIE, 1981; PURVES, 1981; ANDERSON, 1988).

Threshold limits in animal feed are necessary to avoid an increase of cadmium contents in meat and other food of animal origin.

7.3.3 Waters

There are various regulations in force concerning the cadmium content of waters. In the Federal Republic of Germany a limit of 6 µg/L (Verordnung über Trinkwasser und Brauchwasser, 1975) has been set and the Commission of the European Community has given a limit of 5 µg/L in their guidelines (CEC, 1975, 1980). A threshold level of 10 µg/L has been set for various countries: the United States (ANDERSON, 1988), Japan (TSUCHIYA, 1988), the Soviet Union, and Australia (FOERSTNER, 1980a). There are also attempts in some countries to reduce these values to lower ones, for example, in the United States, to 5 µg Cd/L drinking water (ANDERSON, 1988) with a final goal of 1 µg/L (VAN WAMBEKE, 1978).

To avoid fish intoxication (see also Sect. 5.1) the following threshold levels are presently set for cadmium in various countries: the threshold level for waters according to the Directory of the European Community from June 16, 1975, is 5 µg Cd/L; the United States threshold for chronic fish toxicity is 0.4 to 12 µg/L; and the threshold for tap water in the European Community, in accordance with a proposal from May 25, 1975, is 5 µg/L (RUF, 1981; see also Chapters I.20a and I.20b).

Regulations for waste water differ significantly, but there is also a tendency to lower the given threshold values. In Japan, the cadmium concentration of industrial waste water should not exceed 100 µg/L (TSUCHIYA, 1988); in the United States, a concentration of up to 40 µg/L is permitted if the water mass of the water stream into which the waste is led exceeds that of the waste by a factor of 10 (FOERSTNER, 1980a).

For the European Community, threshold limits are in force from 1986 that are different for different sources: 300 µg/L for waste water from zinc mining, lead and zinc refining and 500 µg/L for all other waste waters. From 1989 on, the limit will be set to 200 µg/L for all waste waters and then be reduced even further (CEC, 1983).

7.3.4 Humans

Analytical comparisons performed recently have shown that experienced laboratories applying proven methods for the biological monitoring of exposed populations and for trend control, provide quite reliable data for cadmium in body fluids (SUNDERMAN et al., 1982; VAHTER, 1982; STOEPPLER, 1985a; SCHALLER et al., 1987; LIND et al., 1987; VESTERBERG and ENGQVIST, 1988; ANGERER, 1988; OSTAPCZUK and STOEPPLER, 1988).

The World Health Organization has recommended a biological threshold value of 10 µg Cd/L in whole blood and 10 µg/g creatinine in urine (WHO, 1980). In contrast to these recommendations, however, still higher values are proposed by national organizations. In the Federal Republic of Germany, for instance, the Commission for the Investigation of Health Hazards of Chemical Compounds at the Workplace proposed the following biological tolerance values: cadmium in whole blood 1.5 µg/dL; cadmium in urine 15 µg/L (MAK, 1987). From the accuracy and precision reached in cadmium determination, however, trend controls for the general population with randomly selected groups also appear to be feasible now (EWERS et al., 1988; VESTERBERG and ENGQVIST, 1988).

References

Air Pollution Control Directorate (1976), *National Inventory of Sources and Emissions of Cadmium*. APDC, Environmental Protection Service, Int. Rep. APCE-76-2, Environment Canada, Ottawa.

Anderson, E. L. (1988), *Risk Assessment and Risk Management of Cadmium Exposure in the United States*. In: *Cadmium-86, Edited Proceedings Fifth International Cadmium Conference*, San Francisco, pp. 168–174. Cadmium Association, London, Cadmium Council, New York, International Lead Zinc Research Organization, Research Triangle Park, North Carolina.

Andersen, O. (1989) *Oral Cadmium Chloride Toxicity and Intestinal Cadmium Uptake: Effects of Diet and Chelators. Toxicol. Environ. Chem.* 23 (1–4), 105–120.

Angehrn, C., and Hertz, C. (1989), *Distribution of Heavy Metals in Various Litter Horizons and Forest Soils; Factors Affecting the Investigation of Heavy Metal Speciation in Forest Soils: Ultrafiltration. Int. J. Environ. Anal. Chem.* 39(1), 81–99.

Angerer, J. (1988), *Cadmium Determination in Urine*, in: *Analyses of Hazardous Substances in Biological Materials*, Vol. 2, pp. 85–96, DFG Publication. VCH Verlagsgesellschaft, Weinheim-Basel-Cambridge-New York.

Anke, M., Groppel, B., and Kronenmann, H. (1984), *Significance of Newer Essential Trace Elements. Proceedings 3rd Workshop on Trace Elements in Medicine and Biology, Neuherberg-München*, pp. 421–465. W. de Gruyter, Berlin-New York.

Anke, M., Groppel, B., and Schmidt, A. (1987), *New Results on the Essentiality of Cadmium in Ruminants*, in: Hemphill, D. D. (ed): *Proceedings Symposium Trace Substances in Environmental Health – XXI*, pp. 556–566. University of Missouri, Columbia.

Astruc, M. (1986), *Evaluation of Methods for the Speciation of Cadmium*, in: Mislin, H., and Ravera, O. (eds.): *Cadmium in the Environment*, pp. 12–17. Birkhäuser, Basel-Boston-Stuttgart.

Aufderheide, M., Tiedemann, K.-U., Mohr, U., and Heinrich, U. (1989), *Quantification of Proliferate Lesions in Hamster Lung after Chronic Inhalation of Cadmium Compounds. Toxicol. Environ. Chem.*, in press.

Barudi, W., and Bielig, H. J. (1980), *Heavy Metal Content (As, Pb, Cd, Hg) of Vegetables Which Grow Above Ground and Fruits. Z. Lebensm. Unters. Forsch. 170*, 254–257.

Batzer, H. (1983), *Use and Possibilities for Substitution of Cadmium Stabilizers. Ecotoxicol. Environ. Saf. 7*, 117–121.

BDI (Bundesverband der Deutschen Industrie) (1982), *Cadmium – Eine Dokumentation*. BDI, Köln.

Beamish, R. J., Blouw, L. M., and McFarlane, G. A. (1976), *A Fish and Chemical Study of 109 Lakes in the Experimental Lakes Area (ELA), Northwestern Ontario, with Appended Reports on Lake Whitefish Ageing Errors and the Northwestern Ontario Baitfish Industry. Techn. Rep. 607*, 106. Environment Canada, Fisheries and Marine Service.

Bernard, A., and Lauwerys, R. (1989), *Early Markers of Cadmium Nephrotoxicity: Biological Significance and Predictive Value. Toxicol. Environ. Chem.*, in press.

Bingham, F. T., Page, A. L., Mahler, R. J., and Ganje, T. J. (1975), *Growth and Cadmium Accumulation of Plants Grown on Soil Treated with Cadmium-enriched Sewage Sludge. J. Environ. Qual. 4*, 207–211.

BMI (Bundesministerium des Innern) (1982), *Protokoll der Sachverständigenanhörung zu Cadmium*. Bundesministerium des Innern und Umweltbundesamt, Berlin.

Bodar, C. W. M., Kluytmans, J. V., van Montfor, J. C. P., Voogt, P. A., and Zandee, D. I. (1988), *Cadmium Resistance and the Synthesis of Metallothionein-like Proteins in Daphnia magna. Proceedings 3rd International Conference on Environmental Contamination, Venice*, pp. 79–81. CEP Consultants Ltd., Edinburgh.

Böhm, E., and Tötsch, W. (1989), *Cadmium Substitution, Standard Perspectives*. Verlag TÜV Rheinland, Köln.

Bown, H. J. M. (1979), *Environmental Chemistry of the Elements*. Academic Press, London-New York-Toronto-Sidney-San Francisco.

Breder, R. (1981), *Die Belastung des Rheins mit toxischen Metallen*. Doctoral Thesis, Universität Bonn.

Breder, R. (1988), *Cadmium in European Inland Waters*, in: Hutzinger, O., and Safe, S. H. (eds.): *Environmental Toxins*, Vol. 2, *Cadmium*, Stoeppler, M., and Piscator, M. (Vol. eds.), pp. 159–169. Springer, Berlin-Heidelberg-New York.

Bruland, K. W. (1983), *Trace Elements in Sea Water*, Chap. 45, in: Riley, J. P., and Chester, R.: *Chemical Oceanography*, Vol. 8. Academic Press, London.

Bruland, K. W., Franks, R. P., Knauer, G. A., and Martin, J. H. (1979), *Sampling und Analytical Methods for the Determination of Copper, Cadmium, Zinc, and Nickel at the Nanogram per Liter Level in Sea Water. Anal. Chim. Acta 105*, 233–245.

Cadmium-77 (1978), *Edited Proceedings First International Cadmium Conference, San Francisco*. Metal Bulletin, London.

Cadmium-79 (1980), *Edited Proceedings Second International Cadmium Conference, Cannes*. Metal Bulletin, London.

Cadmium-81 (1982), *Edited Proceedings Third International Cadmium Conference, Miami.* Cadmium Association, London, Cadmium Council, New York, International Lead Zinc Research Organization (ILZRO), New York.
Cadmium-83 (1984), *Edited Proceedings Fourth International Cadmium Conference, Munich.* Cadmium Association, London, Cadmium Council, New York, International Lead Zinc Research Organization (ILZRO), New York.
Cadmium-86 (1988), *Edited Proceedings Fifth International Cadmium Conference, San Francisco.* Cadmium Association, London, Cadmium Council, New York, International Lead Zinc Research Organization, Research Triangle Park, North Carolina.
Cadmium Today (1986), *Proceedings of a Seminar Organized by the Cadmium Association and Held at the Palais des Congrès, Brussels,* on 26 June 1985, Cadmium Association, London.
CEC (Commission European Communities) (1975),*Council Directive Concerning the Quality Required for Surface Water Intended for Abstraction of Drinking Water in the Member States* (75/464/EEC), OJ No. L 129 of 25 July 1975, p. 26.
CEC (Commission European Communities) (1978), *Criteria (Dose/Effect Relationships)* for Cadmium. Pergamon, Oxford.
CEC (Commission European Communities) (1983), *Richtlinie des Rates vom 26.* September 1983 betreffend *Grenzwerte und Qualitätsziele für Cadmiumableitungen* (83/514/EWG), Amtsblatt der Europ. Gemeinschaft, Nr. L 291/1–8.
CEC (Commission European Communities (1980), *Council Directive Related to the Quality of Water Intended for Human Consumption* (80/778/EEC), OJ No. 229 of 30 August 1980, p. 11.
Chaney, R. L., Hundemann, P. T., Palmer, W. T., Small, R. J., White, M. C., and Decker, A. M. (1978) *Proceedings 1977 National Conference on Composting of Municipal Residues and Sludges,* pp. 87–97. Informations Transfer Inc., Pochville, Maryland.
Christoffersson, J.-O., and Mattson, S. (1983), *Polarized X-Rays in XRF-Analysis for Improved in-vivo Detectability of Cadmium in Man. Phys. Med. Biol.* **28**, 1135–1144.
Ciurea, J. C., Lipka, Y. F., and Humbert, B. E. (1986), *Trace Analysis of Mineral Elements in Raw Materials and Finished Products in the Chocolate Industry. Mitt. Geb. Lebensm. Hyg.* **77**, 509–519.
Clench-Aas, J., Thomassen, Y., Levy, F., Moseng, J., and Skaug, K. (1978), *Cadmium in Urine in Inhibitants of Oslo-Nydalen, Holmestrand and Sörumsand as a Function of Air Cadmium and Other Socio-Economic Factors.* NILU OR: 68/87, Norwegian Institute for Air Research.
Cole, J. F., and Volpe, R. (1983), *The Effect of Cadmium on the Environment. Ecotoxicol. Environ. Saf.* **7**, 151–159.
Copius Peereboom-Stegemann, J. H. J. (1989), *Cadmium Effects on the Female Reproduction Tract. Toxicol. Environ. Chem.* **23** (1–4), 91–99.
Costa, M. (1989), *Genotoxicity of Cadmium and Other Metal Ions. Toxicol. Environ. Chem.*, in press.
Cottenie, A. (1981), *Sludge Treatment and Disposal in Relation to Heavy Metals. Proceedings International Conference on Heavy Metals in the Environment,* pp. 167–175. CEP Consultants Ltd., Edinburgh.
Coulston, F., Korte, F., Rohleder, H., and Klein, W. (eds.) (1983), *Proceedings International Symposium Ecotoxicology of Cadmium,* May 6–8, 1981. *Ecotoxicol. Environ. Saf.* **7**, 1–159.
Cox, R. M. (1986), *Contamination and Effects of Cadmium in Native Plants,* in: Mislin, H., and Ravera, O. (eds.): *Cadmium in the Environment,* pp. 101–109. Birkhäuser, Basel-Boston-Stuttgart.
Croessmann, G., and Seifert, D. (1982), in: *Cadmium-81, Edited Proceedings Third International Cadmium Conference, Miami,* pp. 82–87. Cadmium Association, London, Cadmium Council, New York, ILZRO, New York.
Dabeka, R. W., and McKenzie, A. D. (1986), *Graphite Furnace Atomic Absorption Spectrometric Determination of Lead and Cadmium in Food after Nitric-Perchloric Acid Digestion and Coprecipitation with Ammonium Pyrrolidine Dithiocarbamate. Can J. Spectrosc.* **31**, 44–52.

Dabeka, R. W., and McKenzie, A. D. (1987), *Lead, Cadmium and Fluoride Levels in Market Milk and Infant Formulas in Canada. J. Assoc. Off. Anal. Chem.* **70**, 753–757.

Dabeka, R. W., Kardinski, K. F., McKenzie, A. D., and Bajdik (1986), *Survey of Lead, Cadmium and Fluoride in Human Milk and Correlation of Levels with Environmental and Food Factors. Food Chem. Toxicol.* **24**, 913–921.

Davidson, C. I. (1980), in: Nriagu, J. O.: *Cadmium in the Environment*, Part I, pp. 115–139. John Wiley & Sons, New York.

de Boo, W. (1989), *Cadmium in Soils and Crops. Toxicol. Environ. Chem.*, in press.

Diehl, J. F. (1981), *Cadmium und Umwelt (Fakten, Daten, Hintergründe). VDI-Schriftenreihe Nr. 3*, Verband der Chemischen Industrie, Frankfurt/Main.

Diels, L., and Léonard, A. (1989), *Plasmid Bound Resistances to Heavy Metals in Soil Microorganisms. Toxicol. Environ. Chem.* **23** (1–4), 79–89.

Dorten, W. S., Elbaz-Poulichet, F., Mart, L. R., and Martin, J.-M. (1989), *Reassessment of the River Input of Trace Metals into the Mediterranean Sea. Ambio*, in press.

Doyle, J. J. (1977), *Effects of Low Levels of Dietary Cadmium in Animals. A Review. J. Environ. Qual.* **6**, 111–116.

Drasch, G. A. (1983), *An Increase of Cadmium Body Burden for this Century – An Investigation on Human Tissues. Sci. Total Environ.* **26**, 111–119.

Dwivedi, C. (1983), *Contam. Toxicol.* **12**, 151–156.

Edlund, S. (1986), *Experience of the Cadmium Regulation in Sweden*, in: *Cadmium Today*, pp. 57–59. Cadmium Association, London.

Elinder, C.-G. (1977), *Cadmium Concentration in Samples of Human Kidney Cortex from the 19th Century. Ambio* **6**, 270–272.

Elinder, C.-G. (1985a), *Cadmium, Uses, Occurrence, and Intake*, in: Friberg, L., Elinder, C.-G., Kjellstroem, T., and Nordberg,, G. F. (eds.): *Cadmium and Health: A Toxicological and Epidemiological Appraisal*, Vol. I, *Exposure, Dose and Metabolism*, pp. 23–63. CRC Press, Boca Raton, Florida.

Elinder, C.-G. (1985b), *Normal Values for Cadmium in Human Tissues, Blood, and Urine in Different Countries*, in: Friberg, L., Elinder, C.-G., Kjellstroem, T., and Nordberg, G. F. (eds.): *Cadmium and Health: A Toxicological and Epidemiological Appraisal*, Vol. I, *Exposure, Dose and Metabolism*, pp. 81–102. CRC Press, Boca Raton, Florida.

Elinder, C.-G. (1986), *Cadmium: Uses, Occurrence, and Intake*, in: Friberg, L., Elinder, C.-G., Kjellström, T., and Nordberg, G. F. (eds.): *Cadmium and Health, a Toxicological and Epidemiological Appraisal*, Vol. 1, *Exposure, Dose and Metabolism*, pp. 23–63. CRC Press, Boca Raton, Florida.

Elinder, C.-G., Kjellstroem, T., Lind, B., Linnman, L., Piscator, M., and Sundstedt, K. (1983), *Cadmium Exposure from Smoking Cigarettes. Variations with Time and Country Where Purchased. Environ. Res.* **32**, 220–227.

Ellis, K. J., Vartsky, D., Zanzi, J., Cohn, S. H., and Yasamura, S. (1979), *Cadmium: In-vivo Measurements in Smokers and Nonsmokers. Science* **205**, 323–325.

Erdtmann, G., and Petri, H. (1986), *Nuclear Activation Analysis: Fundamentals and Techniques*, in: Elving, P. J. (ed.): *Treatise on Analytical Chemistry*, 2nd Ed., Part I, Vol. 14, Sect. K, pp. 419–643. John Wiley & Sons, New York.

Ewers, U., Brockhaus, A., Freier, J., Jermann, E., and Dolgner, R. (1988), *Exposure to Cadmium of the West-German Population – Results of Biological Monitoring Studies 1980–1986*, in: Hutzinger, O., and Safe, S. H.: *Environmental Toxins*, Vol. 2, *Cadmium*, Stoeppler, M., and Piscator, M. (Vol. eds.), pp. 93–113. Springer, Berlin-Heidelberg-New York.

Ewers, U., Freier, I., Turfeld, M., Jermann, E., and Brockhaus, A. (1989), *Cadmium, Copper and Zinc Relationship in Human Kidney. Toxicol. Environ. Chem.*, in press.

Fischer, A. B. (1987), *Mutagenic Effects of Cadmium Alone and in Combination with Antimutagenic Selenite*. Proceedings 6th International *Conference on Heavy Metals in the Environment, New Orleans*, Vol. 2, pp. 112–114. CEP Consultants Ltd., Edinburgh.

Fischer, A. B. (1989), *In-Vitro Studies of Cadmium Cytotoxicity and Genotoxicity. Toxicol. Environ. Chem.*, in press.

Fleischer, M., Sarofim, A. F., Fassett, D. W., Hammond, P., Shacklette, H. T., Nisbet, J. C. T., and Epstein, S. (1974), *Environmental Impact of Cadmium: A Review by the Panel on Hazardous Substances. Environ. Health Perspect. 33,* 253 – 323.

Foerstner, U. (1980a), *Cadmium,* in: Hutzinger, O. (ed.): *The Handbook of Environmental Chemistry,* Vol. 3, Part A, pp. 59 – 107. Springer, Berlin-Heidelberg-New York.

Foerstner, U. (1980b), *Cadmium in Polluted Sediments,* in: Nriagu, J. O. (ed.): *Cadmium in the Environment,* Part I, pp. 306 – 363. John Wiley & Sons, New York.

Foerstner, U. (1986), *Cadmium in Sediments,* in: Mislin, H., and Ravera, O. (eds.): *Cadmium in the Environment,* pp. 40 – 46. Birkhäuser, Basel-Boston-Stuttgart.

Foerstner, U., and Wittmann, G. T. W. (1983), *Metal Pollution in the Aquatic Environment,* 2nd Ed. Springer, Berlin-Heidelberg-New York.

Friberg, L. (1950), *Health Hazards in the Manufacture of Alkaline Accumulators with Special Reference to Chronic Cadmium Poisoning. Acta Med. Scand. 138 Suppl. 240,* 1 – 124.

Friberg, L., Piscator, M., Nordberg, G. F., and Kjellstroem, T. (1974), *Cadmium in the Environment,* 2nd Ed. CRC-Press, Cleveland, Ohio.

Friberg, L., Elinder, C.-G., Kjellstroem, T., and Nordberg, G. F. (eds.) (1985), *Cadmium and Health: A Toxicological and Epidemiological Appraisal,* Volume I, *Exposure, Dose and Metabolism.* CRC Press, Boca Raton, Florida.

Friberg, L., Elinder, C.-G., Kjellstroem, T., and Nordberg, G. F. (eds.) 1986a), *Cadmium and Health: A Toxicological and Epidemiological Appraisal,* Volume II, *Effects and Response.* CRC Press, Boca Raton, Florida.

Friberg, L., Kjellstroem, T., and Nordberg, G. F. (1986b), in: Friberg, L., Nordberg, G. F, and Vouk, V. B. (eds.): *Handbook on the Toxicology of Metals,* Vol. II, pp. 130 – 184. Elsevier, Amsterdam-New York-Oxford.

Fulkerson, W., Goeller, H. E., Gailar, J. S., and Copenhaver (eds.) (1973), *Cadmium, the Dissipated Element.* Oak Ridge Natl. Lab. (Rep.) ORNL-NSF-EP (US) 21, p. 473.

Gerhardsson, L., Brune, D., Nordberg, G. F., and Wester P. O. (1986), *Distribution of Cadmium, Lead and Zinc in Lung, Liver and Kidney in Long-term Exposed Smelter Workers. Sci. Total Environ. 50,* 65 – 85.

Giesy, J. P. Jr. (1980), *Cadmium Interaction with Natural Organic Ligands,* in: Nriagu, J. O. (ed.): *Cadmium in the Environment,* Part I, pp. 237 – 256. John Wiley & Sons, New York.

Glaser, U., Kühl, U. G., and Hapke, H. J. (1978), *Toxicological-biochemical Studies on the Diagnosis of Cadmium Health Disturbances in Sheep* (in German) *Zentralbl. Veterinärmed. Ser. A 25,* 685 – 703.

Glaser, U., et al. (1989), *Special Issue 4th IUPAC Cadmium Workshop Schmallenberg-Grafschaft; Carcinogenicity of four Cadmium Compounds Inhaled by Rats. Toxicol. Environ. Chem. 27* (1 – 3), 153 – 162.

Gravenhorst, G., Perseke, C., and Robock, E. (1980), *Untersuchung über die trockene und feuchte Deposition von Luftverunreinigungen in der Bundesrepublik Deutschland.* Umweltbundesamt, Forschungsprojekt 10402600.

Guenther, K., Henze, W., and Umland, F. (1987), *Mobilization Behaviour of Thallium and Cadmium in a River Sediment. Fresenius Z. Anal. Chem. 327,* 301 – 303.

Hallenbeck, W. H. (1986), *Human Health Effects of Exposure to Cadmium,* in: Mislin, H., and Ravera, O. (eds.): *Cadmium in the Environment,* pp. 131 – 137. Birkhäuser, Basel-Boston-Stuttgart.

Hedrich, M., Bergmann, R. L., Rösick, U., Brätter, P., and Bergmann, K. E. (1988), *Lead, Cadmium and Manganese in Fetal, Infant and Adult Rib Bones from Urban and Industrial Areas of Germany,* in: Brätter, P., and Schramel, P. (eds.): *Trace Element Analytical Chemistry in Medicine and Biology,* Vol. 5, pp. 236 – 242. Walter de Gruyter, Berlin-New York.

Hedrich, M., Rösick, U., Brätter, P., Bergmann, R. L., and Bergmann, K. E. (1989), *Determination of Cadmium, Lead and Manganese in Fetal, Infant and Adult Rib Bones*. Toxicol. Environ. Chem., in press.
Heindryckx, R., Demuynk, M., Dams, R., Janssen, M., and Rahn, K. A. (1974), *Mercury and Cadmium in Belgian Aerosols*, in: Proceedings International Symposium Problems of the Contamination of Man and his Environment by Mercury and Cadmium, Luxembourg, 3–5 July 1973, pp. 135–148. EUR-5075.
Heinrich, U. (1988), *Carcinogenicity of Cadmium – Overview of Experimental and Epidemiological Results and Their Influence on Recommendations for Maximum Concentrations in the Occupational Area*, in: Hutzinger, O., and Safe, S. H. (eds.): Environmental Toxins, Vol. 2, Cadmium, Stoeppler, M., and Piscator, M. (Vol. eds.), pp. 13–25. Springer, Berlin-Heidelberg-New York-London-Paris-Tokyo.
Heinrich, U., Peters, L., Rittinghausen, S., Ernst, H., Dasenbrock, C., and Mohr, U. (1989), *Long-term Inhalation Exposure of Syrian Golden Hamster and NMRI-mice to Various Cadmium Compounds*. Toxicol. Environ. Chem., in press.
Heinrichs, H., Schulz-Dobrick, B., and Wedepohl, K. H. (1980), *Terrestrial Geochemistry of Cadmium, Bismuth, Thallium, Zinc and Lead*. Geochim. Cosmochim. Acta 44, 1519–1533.
Herber, R. F. M. (1984), *Beta-2-microglobulin and Other Urinary Proteins as an Index of Cadmium-Nephrotoxicity*. Pure Appl. Chem. 56, 957–965.
Herber, R. F. M., Verschoor, M. A., and Wibowo, A. A. E. (1988), *A Review of the Kinetics and Kidney Effects of Cadmium – Recent Epidemiological Studies*, in: Hutzinger, O., and Safe, S. H.: Environmental Toxins, Vol. 2, Cadmium, Stoeppler, M. and Piscator, M. (Vol. eds.), pp. 115–133. Springer, Berlin-Heidelberg-New York-London-Paris-Tokyo.
Heron, H. (1986), *Danish Regulations on Cadmium Containing Products*, in: Cadmium Today, pp. 60–61. Cadmium Association, London.
Heumann, K. G. (1980), *Mass Spectrometric Isotope Dilution Analysis for Accurate Determination of Elements in Environmental Samples*. Toxicol. Environ. Chem. Rev. 3, 111–129.
Heumann, K. G. (1986), *Isotope Dilution Mass Spectrometry of Inorganic and Organic Substances*. Fresenius Z. Anal. Chem. 325, 661–666.
Hoffmann, J., and Blasenbrei, P. (1986), *Cadmium in Blue Poppy Seeds and in Poppy Seed Containing Products*. Z. Lebensm. Unters. Forsch. 182, 121–122.
Hutton, M. (1983), *A Prospective Atmospheric Emission Inventory for Cadmium – The European Community as a Study Area*. Sci. Total Environ. 29, 29–47.
Hutton, M., and Symon, C. (1986), *The Quantities of Cadmium, Lead, Mercury and Arsenic Entering the UK Environment from Human Activities*. Sci. Total Environ. 57, 129–150.
Hutton, M., Chaney, R. L., Krishna Murti, C. R., Olade, M. A., and Page, A. L. (1987), Chapter 3 *"Cadmium"*, in: Lead, Mercury, Cadmium and Arsenic in the Environment (Hutchinson, T. C., and Meema, K. M., eds.) SCOPE. John Wiley & Sons, New York.
Huynh-Ngoc, L., Whitehead, N. E., and Oregioni, B. (1988), *Cadmium in the Rhone River*. Water Res. 22, 571–576.
Hygiène et Sécurité (1982), *Cadmium*. Comité de Liaison des Industries de métaux non ferreux de la Communauté Européenne, Report EUR (49 EN, List of principal community cadmium regulations), Bruxelles.
Ihnat, M., Gordon, A. D., Gaynor, J. D., Berman, S. S., Desaulniers, A., Stoeppler, M., and Valenta, P. (1980), *Interlaboratory Analysis Natural Fresh Waters for Copper, Zinc, Cadmium and Lead*. Int. J. Environ. Anal. Chem. 8, 259–275.
Jacko, R. B., and Neuendorf, D. W. (1977), *Trace Metal Particulate Emission Test Results from a Number of Industrial and Municipal Point Sources*. J. Air. Pollut. Control Assoc. 27, 989–994.
Jaerup, L., Rogenfelt, A., Elinder, G. C., Nogawa, K., and Kjellstroem, T. (1983), *Biological Half Time of Cadmium in the Blood of Workers after Cessation of Exposure*. Scand. J. Work Environ. Health 9, 327–331.
Jarvis, S. C., Jones, L. H. P., and Hopper, M. J. (1976), *Cadmium Uptake from Solution by Plants and its Transport from Roots to Shoots*. Plant Soil 44, 179–191.

Jenkins, R. (1975), *An Introduction to X-Ray Spectrometry*. Heyden & Sons, London.
Jones, K. C., Symon, C. J., and Johnston, A. E. (1987), *Retrospective Analysis of an Archived Soil Collection. II. Cadmium. Sci. Total Environ.* 67, 75–89.
Jost, D. (1984), *Luftqualität in belasteten Gebieten und fern von Emittenten. Staub-Reinhalt. Luft* 44, 137–138.
Kabata-Pendias, A., and Pendias, H. (1984), *Trace Elements in Soils and Plants*. CRC Press, Boca Raton, Florida.
Kazantzis, G. (1987), *The Mutagenic and Carcinogenic Effects of Cadmium: An Update. J. Toxicol. Environ. Chem.* 15, 83–100.
Kazantzis, G. (1989a), *Is Cadmium a Human Carcinogen?, Toxicol. Environ. Chem.*, in press.
Kazantzis, G. (1989b), *The Mortality of Cadmium Exposed Workers. Toxicol. Environ. Chem.*, in press.
Kirsch, H., Padberg, W., Scholz, A., and Zimmermeyer, G. (1982), *Cadmium Emissions from Coal-fired Power Plants*, in: *Cadmium-81, Edited Proceedings Third International Cadmium Conference, Miami*, pp. 64–68. Cadmium Association, London, Cadmium Council, New York, ILZRO, New York.
Kjellstroem, T. (1979), *Exposure and Accumulation of Cadmium in People from Japan, USA and Sweden. Environ. Health Perspect.* 28, 169–197.
Kjellstroem, T. (1984), *Conceptual Problems in Establishing the Critical Concentration of Cadmium in Human Kidney Cortex. Environ. Res.* 33, 284–295.
Kjellstroem, T. (1986), *Itai-Itai Disease*, in: Friberg, L., Elinder, C. G., Kjellstroem, T., and Nordberg, G. F. (eds.): *Cadmium and Health: A Toxicological and Epidemiological Appraisal*, Vol. II, *Effects and Response*, pp. 257–290. CRC Press, Boca Raton, Florida.
Kjellstroem, T., and Nordberg, G. F. (1978), *A Kinetic Model of Cadmium Metabolism in the Human Being. Environ. Res.* 16, 248–269.
Kjellstroem, T., and Nordberg, G. F. (1985), *Kinetic Model of Cadmium Metabolism*, in: Friberg, L., Elinder, C.-G., Kjellstroem, T., and Nordberg, G. F. (eds.): *Cadmium and Health: A Toxicological and Epidemiological Appraisal*, Vol. I, *Exposure, Dose and Metabolism*, pp. 179–197. CRC Press, Boca Raton, Florida.
Klärschlammverordnung (1982), (Abf Klär V) *Bundesgesetzblatt 1*, 734–739.
Kloke, A. (1977), *Guidelines for Tolerable Total Concentrations of Some Elements in Cultivated Soil* (in German). *Mitt. VDLUFA*, 32–38.
Knezevic, J. (1979), *Schwermetalle in Lebensmitteln*, 1. Mitt.: *Über den Gehalt an Cadmium in Rohkakao und in Kakao-Halb- und Fertigprodukten. Dtsch. Lebensm. Rundsch.* 75, 305–309.
Kodama, Y., Matsuno, K., and Tsuchiya, K. (1989), *Cadmium Distribution in Blood and Urine of Dogs after Long-term Oral Administration of Cadmium. Toxicol. Environ. Chem.*, in press.
Koops, J., and Westerbeek, D. (1978), *Determination of Lead and Cadmium in Pasteurized Liquid Milk by Flameless Atomic Absorption Specrophotometry. Neth. Milk Dairy J.* 32, 149–169.
Kutzman, R. A. (1984), *A Study of Fischer 344 Rats Subchronically Exposed to 0, 0.3, 1 or 2 mg $CdCl_2/m^3$*. Brookhaven National Laboratory IA, 222-Y01-ES-9-0043.
Lamersdorf, N. P. (1989), *The Behavior of Lead and Cadmium in the Intensive Rooting Zone of Acid Spruce Forest Soils. Toxicol. Environ. Chem.* 18 (4), 239–247.
Lasenby, D. C., and Groulx, G. R. (1988), *Cadmium Burdens in Phantom Midge Fly Larvae from Seven Lakes in South-central Ontario. Proceedings International Conference Trace Metals in Lakes, Hamilton (Canada). Sci. Total Environ.*, in press.
Lauwerys, R. L. (1982), *The Toxicology of Cadmium*. ECSC-EEC-EAEC, Brussels, Luxembourg.
Lauwerys, R. L. (1987), *Diagnostic Methods in Nephrotoxicity. Symposium Report Trace Elements in Human Health and Disease, Odense. Environ. Health* 26, 70/71. WHO Regional Office for Europe, Copenhagen.
Lauwerys, R. L., Buchet, J. P., Roels, H., Bernard, A., Chettle, D. R., Harvey, T. C., and Al Haddad, I. K. (1980), *Biological Significance of Cadmium Concentrations in Blood and Urine and their Application in Monitoring Workers Exposed to Cadmium*, in: *Cadmium-79, Edited Proceedings Second International Conference, Cannes*, pp. 164–167. Metal Bulletin, London.

Levine, M. B., Taylor, D. H., Levine, D. A., and Barrett, G. W. (1987), *Effects of Sewage Sludge on Density, Heavy Metal Accumulation, and Locomotor Activity of Lumbricid Earthworms*. Proceedings 6th International Conference on Heavy Metals in the Environment, New Orleans, Vol. 1, pp. 390–392. CEP Consultants Ltd., Edinburgh.

Liese, T., and Simon, J. (1984), *Trace Determination of Heavy Metals in Human Bones – Problems in their Analysis and Result Interpretation*. Fresenius Z. Anal. Chem. *326*, 647–655.

Lind, B., Elinder, C.-G., Friberg, L., Nilsson, B., Svartengren, M., and Vahter, M. (1987), *Quality Control in the Analysis of Lead and Cadmium in Blood*. Fresenius Z. Anal. Chem. *326*, 647–655.

Little, K. R., and Martin, M. H. (1972), *A Survey of Zinc, Lead and Cadmium in Soil and Natural Vegetation around a Smelting Complex*. Environ. Pollut. *3*, 241–254.

Lloyd, T. B., and Wise, K. J. (1982), *Cadmium Recovery from Electroplating Wastes*, in: Cadmium-81, Edited Proceedings 3rd International Cadmium Conference, Miami, pp. 53–55.

Logan, T. J., and Chaney, R. L. (1987), *Nonlinear Rate Response and Relative Crop Uptake of Sludge Cadmium for Land Application of Sludge Risk Assessment*. Proceedings 6th International Conference on Heavy Metals in the Environment, New Orleans, Vol. 1, pp. 387–389. CEP Consultants Ltd., Edinburgh.

Lorenz, H. (1979), *Binding Forms of Toxic Heavy Metals, Mechanisms of Entrance of Heavy Metals in the Food Chain and Possible Measures to Reduce Their Level in Foodstuff*. Ecotoxicol. Environ. Saf. *3*, 47–58.

Lorenz, H., Ocker, H.-D., Brueggemann, J., Weigert, P., and Sonneborn, M. (1986), *Content of Cadmium in Cereals of the Past Compared with the Present*. Z. Lebensm. Unters. Forsch. *183*, 402–405.

Lum, K. R., Kokotich, E. A., and Schroeder, W. H. (1987), *Bioavailable Cd, Pb and Zn in Wet and Dry Deposition*. Sci. Total Environ. *63*, 161–173.

Magnusson, B. (1981), *Determination of Trace Metals in Natural Waters by Atomic Absorption Spectrometry*. Dissertation, University of Göteborg.

MAK (1987) *Maximum Concentrations at the Workplace and Biological Tolerance Values for Working Materials DVF, Report No. XXIII*. Commission for the Investigation of Health Hazards of Chemical Compounds in the Work Area. VCH Verlagsgesellschaft, Weinheim-Basel-Cambridge-New York.

Malle, K. G. (1985), *Contents of Metals and Suspended Solids in the Rhine*, Part II. Z. Wasser Abwasser Forsch. *18*, 207–209.

Malle, K.-G., and Mueller, G. (1982), *Concentrations of Metal Compounds and Suspended Particles in the River Rhine* (in German). Z. Wasser Abwasser Forsch. *15*, 11–15.

Malley, D. F., and Chang, P. S. S. (1988), *Whole Lake Addition of Cadmium and ^{109}Cadmium: Radiotracer Accumulation in the Mussel Population in the First Season*. Proceedings International Conference Trace Metals in Lakes, Hamilton (Canada). Sci. Total Environ., in press.

Manthey, J., Stoeppler, M., Morgenstern, W., Nüssel, E., Opherk, D., Weintraut, A., Wesch, H., and Kübler, W. (1981), *Magnesium and Trace Metals: Risk Factors for Coronary Heart Diseases? Associations between Blood Levels and Angiographic Findings*. Circulation *64*, 722–729.

Mart, L. (1979), *Ermittlung und Vergleich des Pegels toxischer Spurenmetalle in nordatlantischen und mediterranen Küstengewässern*. Dissertation, T.H. Aachen.

Mart, L. (1983), *Seasonal Variations of Cadmium, Lead, Copper and Nickel Levels in Snow from the Eastern Arctic Ocean*. Tellus *53 B*, 131–141.

Mart, L., and Nuernberg, H. W. (1986), *The Distribution of Cadmium in the Sea*, in: Mislin, H., and Ravera, O. (eds.): *Cadmium in the Environment*. Experientia Suppl. *50*, 28–40.

Mart, L., Ruetzel, H., Klahre, P., Sipos, L., Platzek, U., Valenta, P., and Nuernberg, H. W. (1982), *Comparative Studies on the Distribution of Heavy Metals in the Oceans and Coastal Waters*. Sci. Total Environ. *26*, 1–17.

Mart, L., Nuernberg, H. W., and Dyrssen, D. (1983), *Low Level Determination of Trace Metals in Arctic Seawater and Snow by Differential Pulse Anodic Stripping Voltammetry*, in: Wong, C. S.,

Boyle, E., Bruland, K.W., Burton, J.O., and Goldberg, E.D. (eds.): *Trace Metals in Sea Water*, pp. 113–130. Plenum Press, New York.

Mart, L., Nuernberg, H.W., and Ruetzel, H. (1985), *Levels of Heavy Metals in the Tidal Elbe and its Estuary and the Heavy Metal Input into the Sea. Sci. Total Environ.* **44**, 25–49.

Martin, M.H., and Caughtrey, P.J. (1975), *Preliminary Observations on the Levels of Cadmium in a Contaminated Environment. Chemosphere* **3**, 155–160.

Martin, J.-M., and Meybeck, M. (1979), *Elemental Mass Balance Carried by Major World Rivers. Marine Chem.* **7**, 173–206.

Martin, J.H., Knauer, G.A., and Flegal, A.R. (1980), *Cadmium in Natural Waters*, in: Nriagu, J.O. (ed.): *Cadmium in the Environment*, Part I, pp. 141–145. John Wiley & Sons, New York.

McGrath, S.P., Brookes, P.C., and Giller, K.E. (1987), *Long-term Biological Effects of Applying Metal-contaminated Sewage Sludge to Soil*, in: Lindberg, S.E., and Hutchinson, T.C. (eds.): *Proceedings International Conference Heavy Metals in the Environment*, New Orleans, Vol. 1, pp. 372–376. CEC Consultants, Edinburgh.

Mennear, J.H. (1979), *Cadmium Toxicity*. Marcel Dekker, New York.

Merian, E. (1984), *Introduction in Environmental Chemistry and Global Cycles of Chromium, Nickel, Cobalt, Beryllium, Arsenic, Cadmium and Selenium, and their Derivatives. J. Toxicol. Environ. Chem.* **8**, 9–38.

Merian, E., Bronzetti, G.L., Dürbeck, H.W., Frei, R.W., Härdi, W., and Schlatter, C. (1985, 1988, 1989), *Carcinogenic and Mutagenic Metal (and Aluminum) Compounds*, Vol. 1, Part V; Vol. 2, Part VII; Vol. 3 (in press). *Proceedings of IAEAC Workshops in Geneva, Villars-sur-Ollon and Follonica*. Gordon & Breach Science Publishers, New York.

Metallstatistik 68 (1981), *Metallstatistik 1970–1980*. Metallgesellschaft AG, Frankfurt am Main.

Metallstatistik 74 (1987), *Metallstatistik 1976–1986*. Metallgesellschaft AG, Frankfurt am Main.

Michaelis, W., Prange, A., and Knoth, J. (1985), *Applications of TXRF in Multielement Analysis*, in: Sansoni, B. (ed): *Instrumentelle Multielementanalyse*, pp. 269–289. VCH Verlagsgesellschaft, Weinheim-Deerfield Beach/Florida-Basel.

Mineral Commodity Summaries (1982), US Department of the Interior, Bureau of Mines.

Mislin, H., and Ravera, O. (eds.) (1986) *Cadmium in the Environment*. Birkhäuser, Basel-Boston-Stuttgart.

Morgan, H. (ed.) (1988), *The Shipham Report*, Special Issue. *Sci. Total Environ.* **75**, 1–143.

Morselt, A.F.W., Frederiks, W.M., Copius Peereboom-Stegemann, J.H.J., and van Veen, H.A. (1987), *Mechanism of Damage to Liver Cells after Chronic Exposure to Low Doses of Cadmium Chloride, Mechanisms and Models in Toxicology. Arch. Toxicol. Suppl.* **11**, 213–215.

Mueller, C. (1987), *Cadmium Content of Human Milk. Trace Elem. Med.* **4**, 4–7.

Mueller, G. (1979), *Heavy Metal Concentrations (Cd, Zn, Pb, Cu, Zr) in the Tobacco of often Smoked Cigarettes in the Federal Republic of Germany* (in German). *Chem. Ztg.* **103**, 133–137.

Mueller, G. (1985), *Cadmium im Tabak häufig in der BR Deutschland gerauchter Zigaretten 1978 und 1985: ein Vergleich. Chem. Ztg.* **109**, 291–292.

Narres, H.D., Mohl, C., and Stoeppler, M. (1984a), *Metal Analysis in Difficult Materials with Platform Furnace Zeeman-Atomic Absorption Spectrometry. I. Direct Determination of Cadmium in Crude Oil and Oil Products. Int. J. Environ. Anal. Chem.* **18**, 267–279.

Narres, H.-D., Valenta, P., and Nuernberg, H.W. (1984b), *Die voltammetrische Bestimmung von Schwermetallen in Fleisch und inneren Organen von Schlachtrindern. Z. Lebensm. Unters. Forsch.* **179**, 440–446.

Narres, H.-D., Mohl, C., and Stoeppler, M. (1985), *Metal Analysis in Difficult Materials with Platform Furnace Zeeman-AAS. II. Direct Determination of Cadmium and Lead in Milk. Z. Lebensm. Unters. Forsch.* **181**, 111–116.

National Swedish Environmental Protection Board (1982), *The Swedish Ban on Cadmium*. Solna (Sweden).

Nogawa, K., Tshizaki, A., and Kawano, S. (1978), *Statistical Observations of the Dose-response Relationships of Cadmium Based on Epidemiological Studies in the Kakehashi River Basin. Environ. Res.* **15**, 185.

Nogawa, K., Kobayashi, E., and Honda, R. (1979), *A Study of the Relationship between Cadmium Concentrations in Urine and Renal Effects of Cadmium*. Environ. Health Perspect. 28, 161.

Nriagu, J. O. (1979), *Global Inventory of Natural and Anthropogenic Emissions of Trace Metals to the Atmosphere*. Nature 279, 401–411.

Nriagu, J. O. (ed.) (1980a), *Global Cadmium Cycle*, in: *Cadmium in the Environment*, Part I, pp. 1–12. John Wiley & Sons, New York.

Nriagu, J. O. (ed.) (1980b), *Production, Uses and Properties of Cadmium*, in: *Cadmium in the Environment*, Part I, pp. 35–70. John Wiley & Sons, New York.

Nriagu, J. O. (ed.) (1980c), *Cadmium in the Atmosphere and in Precipitation*, in: *Cadmium in the Environment*, Part I, pp. 71–114. John Wiley & Sons, New York.

Nriagu, J. O. (ed.) (1984), *Changing Metal Cycles in Human Health*. Dahlem Konferenzen. Springer, Berlin-Heidelberg-New York-Tokyo.

Nriagu, J. O., and Davidson, C. I. (eds.) (1986), *Toxic Metals in the Atmosphere*. John Wiley & Sons, New York-Brisbane-Toronto-Singapore.

Nriagu, J. O., and Pacyna, J. M. (1988), *Quantitative Assessment of Worldwide Contamination of Air, Water and Soils by Trace Metals*. Nature 333, 134–139.

Nriagu, J. O., and Sprague, J. B. (eds.) (1987), *Cadmium in the Aquatic Environment*. John Wiley & Sons, Somerset, New Jersey.

Nuernberg, H. W. (1979), *Polarography and Voltammetry in Studies of Toxic Metals in Man and his Environment*. Sci. Total Environ. 12, 35–60.

Nuernberg, H. W. (1982), *Voltammetric Trace Analysis in Ecological Chemistry of Toxic Metals*. Pure Appl. Chem. 54, 853–878.

Nuernberg, H. W. (1983), *Potential and Applications of Voltammetry in the Analysis of Toxic Trace Metals in Body Fluids*, in: Facchetti, S. (ed.): *Analytical Techniques for Heavy Metals in Biological Fluids*, pp. 209–232. Elsevier, Amsterdam.

Nuernberg, H. W., Valenta, P., and Nguyen, V. D. (1983), *The Wet Deposition of Heavy Metals from the Atmosphere in the Federal Republic of Germany*, in: *Proceedings International Conference Heavy Metals in the Environment*, Heidelberg, Vol. I, pp. 115–123. CEP Consultants Ltd., Edinburgh.

Oberdoerster, G. (1986), *Airborne Cadmium and Carcinogenesis of the Respiratory Tract*. Scand. J. Environ. Health 12, 523–537.

Oberdoerster, G. (1989), *Assessment of Lung Tumor Risk from Inhaled Environmental Cadmium*. Toxicol. Environ. Chem. 23 (1–4), 41–51.

Oberdoerster, G., and Cox, C. (1989), *Carcinogenicity of Cadmium in Animals: what is the Significance for Man, and what is next?* Toxicol. Environ. Chem., in press.

Ocker, H.-D., and Seibel, W. (1980), *Residue Situation in Cereals and Bread, Heavy Metals Concentrations (Lead, Cadmium)* (in German). Getreide Mehl Brot 34, 118–128.

Ogoshi, K., Iwami, K., Moriyama, T., Nanzai, Y., and Dohi, Y. (1988), *The Toxic Effects of Cadmium on Rat Bones, Effect of Cadmium on Bone Formation in Cultured Rat Calvariae*. Toxicol. Environ. Chem., in press.

Ohira, Y. (1989), *Current Status Concerning the Recycling of Sealed Nickel-Cadmium Batteries in Japan*, in: *Cadmium-87, Edited Proceedings*, pp. 41–44.

Oldiges, H., and Glaser, U. (1986), *The Inhalative Toxicity of Different Cadmium Compounds in Rats*. Trace Elem. Med. 3, 72–75.

Oldiges, H., Takenaka, S., Oberdoerster, G., Hochrainer, D., and König, H. (1984), *Lung Carcinomas in Rats after Low Level Cadmium Inhalation*. Toxicol. Environ. Chem. 9, 41–51.

Oldiges, H., Heinrich, U., and Glaser, U. (1988), *Inhalation Research with Different Cadmium Compounds on Wistar Rats, Syrian Hamsters and NMRI-Mice*, in: Hutzinger, O., and Safe, S. H.: *Environmental Toxins*, Vol. 2, *Cadmium*, Stoeppler, M., and Piscator, M. (Vol. eds.), pp. 33–38. Springer, Berlin-Heidelberg-New York-London-Paris-Tokyo.

Oldiges, H., Hochrainer, D., and Glaser U. (1989), *Long-term Inhalation Study with Wistar Rats and four Cadmium Compounds*. Toxicol. Environ. Chem. 19 (3+4), 217–222.

Ostapczuk, P., and Stoeppler, M. (1988), *Rapid and Reliable Voltammetric Determination of Cadmium in Environmental and Biological Materials*, in: Hutzinger, O., and Safe, S. H.: *Environmental Toxins*, Vol. 2, *Cadmium*, Stoeppler, M., and Piscator, M. (Vol. eds.), pp. 213–226. Springer, Berlin-Heidelberg-New York-London-Paris-Tokyo.

Ostapczuk, P, Valenta, P., and Nuernberg, H. W. (1986), *Square Wave Voltammetry – A Rapid and Reliable Determination Method of Zn, Cd, Pb, Cu, Ni and Co in Biological and Environmental Samples. J. Electroanal. Chem. 214*, 51–64.

Ostapczuk, P., Valenta, P., Ruetzel, H., and Nuernberg, H. W. (1987), *Application of Differential Pulse Anodic Stripping Voltammetry to the Determination of Heavy Metals in Environmental Samples. Sci. Total Environ. 60*, 1–16.

Pacyna, J. M., and Muench, J. (1987), *Atmospheric Emissions of As, Cd, Pb and Zn from Industrial Sources in Europe. Proceedings International Conference Heavy Metals in the Environment, New Orleans*, Vol. 1, pp. 20–25.

Page, A. L. (1986), *Cadmium*, in: *Handbook of Experimental Pharmacology*, Vol. 80. Springer, Berlin-Heidelberg.

Peters, L., Heinrich, U., König, H., Fuhst, R., and Mohr, U. (1989), *The Cadmium Content in Lung, Liver and Kidney of Hamsters and Mice Exposed to Various Cadmium Compounds. Toxicol. Environ. Chem.*, in press.

Phillips, D. J. H. (1980), *Toxicity and Accumulation of Cadmium in Marine and Estuarine Biota*, in: Nriagu, J. O. (ed.): *Cadmium in the Environment*, Part I, pp. 426–569. John Wiley & Sons, New York.

Pihlar, B., Valenta, P., Golimowski, J., and Nuernberg, H. W. (1980), *Voltammetric Determination of Toxic Trace Metals in Municipal Waste Waters and in the Discharge of Biological Sewage Purification Plants* (in German). *Z. Wasser Abwasser Forsch. 13*, 130–138.

Piscator, M. (1981), *Role of Cadmium in Carcinogenesis with Special Reference to Cancer of the Prostate. Environ. Health Perspect. 40*, 107–120.

Piscator, M. (1988), *Some Views on Metabolism, Toxicity and Carcinogenicity of Cadmium*, in: Hutzinger, O., and Safe, S. H.: *Environmental Toxins*, Vol. 2, *Cadmium*, Stoeppler, M., and Piscator, M. (Vol. eds.), pp. 3–12. Springer, Berlin-Heidelberg-New York.

Piscator, M., and Lind, B. (1972), *Cadmium, Zinc, Copper and Lead in Human Renal Cortex. Arch. Environ. Health 24*, 426–431.

Planas-Bohne, F., and Klug, S. (1988), *Uptake of Cadmium into Cells in Culture. Toxicol. Environ. Chem. 18* (2+3), 229–237.

Posthuma, L. (1988), *Cadmium-induced Growth Retardation in Orchesella Soil Insects from Heavy Metal Contamination Sites. Proceedings 3rd International Conference on Environmental Contamination, Venice*, Annex. CEP Consultants Ltd., Edinburgh.

Purves, D. (1981), in: *Proceedings International Conference Heavy Metals in the Environment, Amsterdam*, pp. 176–179. CEP Consultants Ltd., Edinburgh.

Ramseier, S., Tarradellas, J., Martin, M., Haerdi, W., Cuendet, G., and Honsberger, P. (1989), *Bioaccumulation of Cadmium by Earthworms Lumbricus terrestris and its Speciation. Toxicol. Environ. Chem. 22* (1–4), 189–196.

Raspor, B. (1980), *Distribution and Speciation of Cadmium in Natural Waters*, in: Nriagu, J. O. (ed): *Cadmium in the Environment*, Part I, pp. 147–236. John Wiley & Sons, New York.

Rauhut, A. (1980), *Survey of Industrial Emissions of Cadmium in the European Economic Community.* ENV/223/74/E, Commission of the European Communities, Brussels.

Rauhut, A. and Wiegand, V. (1982) *Cadmium Wastes and Possibilities of Recycling*, in: *Cadmium-81, Proceedings 3rd International Cadmium Conference, Miami*, pp. 69–74.

Remacle, J., and Hambucker-Berhin, F. (1987), *Cadmium Accumulation by Alcaligenes eutrophicus and the Role of the Envelopes. Proceedings 6th International Conference on Heavy Metals in the Environment, New Orleans*, Vol. 2, pp. 244–246. CEP Consultants Ltd., Edinburgh.

Richtwerte 86 (1986), *Richtwerte '86 für Blei, Cadmium und Quecksilber in und auf Lebensmitteln. Bundesgesundhbl. 29*, 22–23.

Roels, H. A., Lauwerys, R. R., Buchet, J. P., Bernard, A., Chettle, D. R., Harvey, T. C., and Al-Haddad, J. K. (1981), *In-vivo Measurement of Liver and Kidney Cadmium in Workers Exposed to this Metal: Its Significance with Respect to Cadmium in Blood Urine.* Environ. Res. 26, 217–240.

Ruf, M. (1981), *Aquatic Pollution from the Point of View of Fishery* (in German), Proceedings Wasser Berlin '81, pp. 415–428. Colloquium-Verlag, O. Hess, Berlin.

Ruhrverband-Ruhrtalsperrenverein (1975), *Bericht über die Ruhrwassergüte im Wasserwirtschaftsjahr 1973*, Essen.

Saager, R. (1984), *Encyclopedia on Metallic Raw Materials* (in German), pp. 83–86. Bank von Tobel, Zürich.

Salomons, W., and Foerstner, U. (1984), *Metals in the Hydrocycle.* Springer, Berlin-Heidelberg-New York.

Schaefer, S. G., Elsenhans, B., and Forth, W. (1988), *Iron and Cadmium: What is Known about the Interactions of These Metals in the Organism,* in: Hutzinger, O., and Safe, S. H.: *Environmental Toxins,* Vol. 2, Cadmium, Stoeppler, M., and Piscator, M. (Vol. eds.), pp. 27–31. Springer, Berlin-Heidelberg-New York-London-Paris-Tokyo.

Schaller, K. H., Angerer, J., Lehnert, G., Valentin, H., and Weltle, D. (1987), *External Quality Control Programmes in the Toxicological Analysis of Biological Material in the Field of Occupational Medicine – Experiences from three Round Robins in the Federal Republic of Germany.* Fresenius Z. Anal. Chem. 326, 643–646.

Schaller, K. H., Gonzales, J., Thuerauf, J., and Schiele, R. (1980), *Detection of Early Kidney Damages in Workers Exposed to Lead, Mercury and Cadmium.* Zentralbl. Bakteriol. Hyg. I. Abt. Orig. B. 171, 320–335.

Schelenz, R. (ed.) (1983), *Essentielle und toxische Inhaltsstoffe in der täglichen Gesamtnahrung.* BFE-R-83-02. Bundesforschungsanstalt für Ernährung, Karlsruhe.

Schramel, P., Klose, B. J., and Hasse, S. (1982), *Efficiency of ICP Emission Spectroscopy for the Determination of Trace Elements in Bio-medical and Environmental Samples.* Fresenius Z. Anal. Chem. 310, 209–216.

Schreiber, W. (1986), *Levels of Heavy Metals and Organohalogenic Compounds in Fishes.* Report 3, Bundesforschungsanstalt für Fischerei.

Schuemann, K., Elsenhans, B., Schäfer, S., Kolb, K., and Forth, W. (1989), *Cadmium Absorption and its Interaction with Essential Trace Metals. An Approach to the Mechanisms of Intestinal Cadmium Adsorption.* Toxicol. Environ. Chem., in press.

Schulte-Schrepping, K. H. (1978), *Cadmium Emissions and Dust Deposition, a Contribution to Clear up Interrelations* (in German). Staub-Reinhalt. Luft 38, 172–174.

Schulte-Schrepping, K. H., and Piscator, M. (1985), *Cadmium and Cadmium Compounds,* in: *Ullmanns Encyklopädie der technischen Chemie,* 4th Ed.,, Vol. A4, pp. 499–514. VCH Verlagsgesellschaft Weinheim-Deerfield Beach/Florida-Basel.

Schladot, J. D., and Nuernberg, H. W. (1982), *Atmosphärische Belastung durch toxische Metalle in der Bundesrepublik Deutschland – Emission und Deposition.* Ber. Kernforschungsanlage Jülich, Jülich 1776.

Scott, M. C., and Chettle, D. R. (1986), *In vivo Elemental Analysis in Occupational Medicine.* Scand. J. Work Environ. Health 12, 81–96.

Seeger, R., Nuetzel, R., and Dill, U. (1978), *Cadmium in Mushrooms.* Z. Lebensm. Unters. Forsch. 166, 23–34.

Sherlock, J. C. (1986), *Cadmium in Foods and the Diet,* in: Mislin, H., and Ravera, O.: *Cadmium in the Environment,* pp. 110–114. Birkhäuser, Basel-Boston-Stuttgart.

Sigg, L., Sturm, M., Stumm, W., Mart, L., and Nuernberg, H. W. (1982), *Heavy Metals in the Lake Constance.* Naturwissenschaften 69, 546–547.

Simpson, W. R. (1981), *A Critical Review of Cadmium in the Marine Environment.* Prog. Oceanogr. 10, 1–70.

Sipos, L., Raspor, B., Nuernberg, H. W., and Pytkowicz, R. M. (1980), *Interaction of Metal Complexes with Coulombic Ion-Pairs in Aqueous Media of High Salinity.* Mar. Chem. 9, 37–47.

Spang, G. (1988), *In vivo Monitoring of Cadmium Workers*, in: *Cadmium-86, Edited Proceedings*, pp. 162–164. Cadmium Association, London, Cadmium Council, New York, ILZRO, Research Triangle Park, North Carolina.
Sperling, K.-R. (1982), *Cadmium Determination in Coastal Water Samples from the German Bight*. Vom Wasser *58*, 113–142.
Steinnes, E. (1989), *Cadmium in Terrestrial Environment: Impact of Long-range Atmospheric Transport*. Toxicol. Environ. Chem. *19* (3+4), 139–145.
Stoeppler, M. (1985a), *Cadmium-Bestimmung in biologischem und Umweltmaterial*, in: Fresenius, W., Günzler, H., Huber, W., Lüderwald, I., Tölg, G., and Wisser, H. (eds.): *Analytiker-Taschenbuch*, Vol. 5, pp. 199–216. Springer, Berlin-Heidelberg-New York-Tokyo.
Stoeppler, M. (1985b), *Analytical Aspects of the Determination and Characterization of Metallic Pollutants*, in: Nuernberg, H.W. (ed.): *Pollutants and Their Ecotoxicological Significance*, pp. 317–336. John Wiley & Sons, New York.
Stoeppler, M. (1986), *Recent Methodological Progress in Cadmium Analysis*. Int. J. Environ. Anal. Chem. *27*, 231–239.
Stoeppler, M., and Brandt, K. (1979), *Comparative Studies on Trace Metal Levels in Marine Biota, II. Trace Metals in Krill, Krill Products and Fish from the Antarctic Scotia Sea*. Z. Lebensm. Unters. Forsch. *169*, 95–98.
Stoeppler, M., and Nuernberg, H.W. (1979), *Comparative Studies on Trace Metal Levels in Marine Biota, II. Typical Levels and Accumulation of Toxic Trace Metals in Muscle Tissue and Organs of Marine Organisms from Different European Seas*. Ecotoxicol. Environ. Saf. *3*, 335–351.
Stoeppler, M., and Piscator, M. (eds.) (1988), *Cadmium*, Vol. 2, in: Safe, S., and Hutzinger, O. (eds.): *Environmental Toxin Series*. Springer, Berlin-Heidelberg-New York.
Stoeppler, M., Apel, M., Bagschik, U., May, K., Mohl, C. Ostapczuk, P., Enkelmann, R., and Eschnauer, H. (1985), *Neuere Untersuchungsergebnisse über Spurenelementgehalte in deutschen und ausländischen Weinen*. Lebensmittelchem. Gerichtl. Chem. *39*, 60–61.
Stubbs, R.L. (1982), *Cadmium-Markets and Trends*, in: *Cadmium-81, Edited Proceedings 3rd International Cadmium Conference, Miami*, pp. 3–7.
Sunderman, F.W., Jr., Brown, S.S., Stoeppler, M., and Tonks, D.B. (1982), *Interlaboratory Evaluation of Nickel and Cadmium Analyses in Body Fluids*, in: Egan, H., and West, T.S. (eds.): *IUPAC Collaborative Interlaboratory Studies in Chemical Analysis*, pp. 25–35. Pergamon, Oxford-New York.
Sunderman, F.W. Jr. (1989), *Cadmium Substitution for Zinc in Finger-loop Domains of Gene Regulating Proteins as a Possible Mechanism for Genotoxicity*. Toxicol. Environ. Chem., in press.
Takenaka, S., Glaser, U., Oldiges, H., and Mohr, U. (1989), *Morphological Effects of Cadmium Oxide Aerosols on the Rat Lungs*. Toxicol. Environ. Chem., in press.
TA LUFT (Technische Anleitung zur Reinhaltung der Luft) (1986), *Verwaltungsvorschrift zur Änderung der ersten allgemeinen Verwaltungsvorschrift zum Bundesimmissionsschutzgesetz*. In force since December 1982, in improved form since 1986.
Tandon, S.K. (1988), *Protein Malnutrition and Cadmium Intoxication*, in: Hutzinger, O., and Safe, S.H.: *Environmental Toxins*, Vol. 2, *Cadmium*, Stoeppler, M., and Piscator, M. (Vol. eds.), pp. 39–52. Springer, Berlin-Heidelberg-New York-London-Paris-Tokyo.
Taylor, D. (1981), *A Summary of the Data on the Toxicity of Various Materials to Aquatic Life*, 2nd Ed., Vol. 2, 31/A/2073 Brixham Laboratory Report.
Taylor, D. (1982) in: *Cadmium-81, Edited Proceedings Third International Cadmium Conference, Miami*, pp. 75–81. Cadmium Association, London, Cadmium Council, New York, ILZRO, New York.
Taylor, D. (1984), *Cadmium – a Case of Mistaken Identity*. Mar. Pollut. Bull. *15*, 167–170.
Thornton, J. (1986), *Geochemistry of Cadmium*, in: Mislin, H., and Ravera, O. (eds.): *Cadmium in the Environment*, pp. 7–12. Birkhäuser, Basel-Boston-Stuttgart.
Tötsch, W. (1989), *Cadmium – Uses and Possibilities of Substitution*. Toxicol. Environ. Chem. *27*(1–3), 123–130.

Tsuchiya, K. (1978), *Cadmium Studies in Japan – A Review.* Kodansha Ltd., Tokyo-Elsevier/North Holland Biomedical Press, Amsterdam-New York-Oxford.

Tsuchiya, K. (1988), *Environmental Control of Cadmium in Japan: Legislative Procedures and Effectiveness,* in: *Cadmium-86, Edited Proceedings Fifth International Cadmium Conference, San Francisco,* pp. 165–167. Cadmium Association, London, Cadmium Council, New York, International Lead Zinc Research Organization, Research Triangle Park, North Carolina.

Tsuchiya, K., and Sugita, M. (1971), *A Mathematical Model for Deriving the Biological Half-life of a Chemical. Nord. Hyg. Tidskr. 53,* 105–110.

Tuor, U., and Keller, L. (1989), *Cadmium in Switzerland* (in German), in the *Special Issue Nr. 5 on Cadmium.* Swiss Association of Environmental Research (SAGUF), Bern, Switzerland, in press.

Tyler, G. (1972), *Heavy Metals Pollute Nature, May Reduce Productivity. Ambio 1,* 52–59.

Tyler, G. (1975), *Effects of Heavy Metal Pollution on Decomposition in Forest Soil.* National Swedish Environmental Board, PB 443 E, PM 542 E.

Umweltbundesamt (1977), *Luftqualitätskriterien für Cadmium,* Berichte 4, 77.

Umweltbundesamt (1981), *Ein Beitrag zum Problem der Umweltbelastung durch nicht- oder schwer abbaubare Stoffe – dargestellt am Beispiel Cadmium.*

Vahter, M. (ed.) (1982), *Assessment of Human Exposure to Lead and Cadmium Through Biological Monitoring.* National Swedish Institute of Environmental Medicine and Department of Environmental Hygiene, Karolinska Institute, Liber Tryck, Stockholm.

Valenta, P., Duursma, E. K., Merks, A. G. A., Ruetzel, H., and Nuernberg, H. W. (1986a), *Distribution of Cd, Pb and Cu between the Dissolved and Particulate Phase in the Eastern Scheldt and Western Scheldt Estuary. Sci. Total Environ. 53,* 41–76.

Valenta, P., Nguyen, V. D., and Nuernberg, H. W. (1986b), *Acid and Heavy Metal Pollution by Wet Deposition. Sci. Total Environ. 55,* 311–320.

Van Bruwaene, R., Kirchmann, R., and Impens, R. (1986), *Cadmium Contamination in Agriculture and Zootechnology,* in: Mislin, H., and Ravera, O. (eds.): *Cadmium in the Environment,* pp. 87–96. Birkhäuser, Basel-Boston-Stuttgart.

van Coillie, R., Blaise, C., and Bermingham, N. (1988), *How Traces of Cadmium can Induce a Delayed Toxicity in the Development of Young Fish.* Proceedings International Conference on Trace Metals in Lakes, Hamilton (Canada). *Sci. Total Environ.,* in press.

van Wembeke, L. (1978), *Actions of the European Communities in Brussels in the Environmental Field with Special Reference to Cadmium,* in: *Cadmium-77, Edited Proceedings First International Cadmium Conference, San Francisco,* pp. 77–79. Metal Bulletin, London.

Vanoeteren, C., Cornelis, R., and Versieck, J. (1986), *Evaluation of Trace Elements in Human Lung Tissue, I. Concentration and Distribution. Sci. Total Environ. 54,* 217–230.

Verloo, M., and Eeckhout, M. (1989), *Metal Species Transformations in Soils, an Analytical Approach. Int. J. Environ. Anal. Chem.,* in press.

Verordnung über Trinkwasser und über Brauchwasser für Lebensmittelbetriebe (Trinkwasserverordnung) (1975), *Bundesgesetzblatt I,* p. 453, corrected: p. 679.

Versieck, J., Vanballenberghe, L., and De Kesel, A. (1988), *Determination of Cadmium in Serum and Packed Blood Cells by Neutron Activation Analysis,* in: Hutzinger, O., and Safe, S. (eds.): *Environmental Toxin Series,* Vol. 2, *Cadmium,* Stoeppler M., and Piscator, M. (Vol. eds.), pp. 195–212. Springer, Berlin-Heidelberg-New York-London-Paris-Tokyo.

Vesterberg, O., and Engqvist, A. (1988), *Comparison of Two Methods for the Determination of Cadmium in Blood,* in: Hutzinger, O., and Safe, S. H.: *Environmental Toxins,* Vol. 2, *Cadmium,* Stoeppler, M., and Piscator, M. (Vol. eds.), pp. 195–204. Springer, Heidelberg-New York-London-Paris-Tokyo.

Vos, G., Hovens, J. P. C., and Hagel, P. (1986a), *Chromium, Nickel, Copper, Zinc, Arsenic, Selenium, Cadmium, Mercury and Lead in Dutch Fishery Products 1977–1984. Sci. Total Environ. 52,* 25–40.

Vos, G., Teenwen, J. J. M. H., and van Delft, W. (1986b), *Arsenic, Cadmium, Lead and Mercury in Meat, Liver and Kidneys of Swine Slaughtered in the Netherlands During the Period 1980–1985*. Z. Lebensm. Unters. Forsch. *183*, 297–401.

Vos, G., Lammers, H., and van Delft, W. (1988), *Arsenic, Cadmium, Lead and Mercury in Meat, Livers and Kidneys of Sheep Slaughtered in the Netherlands*. Z. Lebensm. Unters. Forsch. *187*, 1–7.

Wachs, B. (1981), *Heavy Metals in Aquatic Organisms, Bioaccumulation, -magnification and -retention* (in German). Sicherh. Chem. Umwelt *1*, 113–115.

Wagner, G., Bagschik, U., Burow, M., Mohl, C., Ostapczuk, P., and Stoeppler, M. (1985), *Spruce Needles as Indicators for Heavy Metal Pollution: A Comparative Study for the Environmental Specimen Bank in Differently Polluted Areas*, in: Lekkas, T. D. (ed.): Proceedings International Conference Heavy Metals in the Environment, Vol. 1, pp. 515–517. CEP Consultants, Ltd., Edinburgh.

Watanabe, T., Kasahara, M., Nakatsuka, H., and Ikeda, M. (1987), *Cadmium and Lead Contents of Cigarettes Produced in Various Areas of the World*. Sci. Total Environ. *66*, 29–37.

WHO (1972), *Evaluation of Certain Food Additives and Contaminants Mercury, Lead and Cadmium*. 16th Report of the Joint FAO/WHO Expert Committee on Food Additives, Geneva, April 4–12. WHO Tech. Rep. Ser. 505.

WHO (1980), *Recommended Health-Based Limits in Occupational Exposure to Heavy Metals*. WHO Tech. Rep. Ser. 643.

WHO (1987), *Air Quality Guidelines for Europe*. WHO Regional Publications European Series No. 23. WHO, Regional Office for Europe, Copenhagen.

Wiener, J. G., Rada, R. R., and Schmidt, P. S. (1988), *Chemistry and Bioavailability of Cadmium in Seepage Lakes in North-central Wisconsin, USA*. International Conference Trace Metals in Lakes, Hamilton (Canada). Sci. Total Environ., in press.

Wilson, D. N. (1986), *Proposed Restriction of the Use of Cadmium in Switzerland*, in: Cadmium Today, pp. 65–67. Cadmium Association, London.

Wixson, B. G. (1982), *Cadmium in the Environment: Sources, Pathways, Levels*, in: Cadmium-81, Edited Proceedings Third International Cadmium Conference, Miami, pp. 8–9. Cadmium Association London, Cadmium Council, New York ILZRO, New York.

Wopereis, M., Gascuel-Odoux, Ch. Bourrie, G., and Soignet, G. (1989), *Spatial Variability of Heavy Metals in Soil on a One Hectare Scale*. Int. J. Environ. Anal. Chem., in press.

Yan, N. D., Mackie, G. L., and Boomer, D. (1988), *Chemical and Biological Correlates of the Levels of Essential and Non-essential Metals in the Net Plankton of Canadian Shield Lakes*. Proceedings International Conference Trace Metals in Lakes, Hamilton (Canada). Sci. Total Environ. *87/88*, 419–461.

Yeats, P. A. (1988), *The Distribution of Trace Metals in Ocean Waters*. Sci. Total Environ. *72*, 131–149.

Yost, K. J. (1980), *Environmental Exposure to Cadmium in the United States*, in: Cadmium-79, Edited Proceedings Second International Cadmium Conference, Cannes, pp. 11–20. Metal Bulletin, London.

Additional Recommended Literature

Akos, I. (1990), *New Batteries Make Chargeable Accumulators Questionable* (in German). Basler Zeitung (May/June).

Alegria, A., Barbera, R., Boluda, R., Errecalde, F., Farré, R., and Lagarda, M. J. (1990), *Environmental Cadmium, Lead and Nickel Contamination: Possible Relationship between Soil and Vegetable Content*. Fresenius Z. Anal. Chem., in press.

Baccini, P. (1989), *The Control of Heavy Metal Fluxes from the Anthroposphere to the Environment*. Proceedings International Conference Heavy Metals in the Environment, Geneva, Vol. 1, pp. 13–23. CEP Consultants Ltd., Edinburgh.

Barben, H., and Studer, C. (1990), *Environmental Pollution by Nickel/Cadmium Small Accumulators* (in German). *BUWAL-Bulletin 1/90*, 18–20.

Bellomo, G., Richelmi, P., Mirabelli, F., and Berte, F. (1989), *On the Role of Mitochondria in Cd^{2+} Toxicity in Hepatocytes. Proceedings V ICT '89* (International Congress of Toxicology), Brighton. Taylor & Francis, London-New York-Philadelphia, in press.

Bundesgesundheitsamt Berlin (1988), *Toxicological and Environmental Hygienic Evaluation of EDTA in Drinking Water* (in German), in press.

Bundesgesundheitsamt Berlin (1990), *First Results on "Federal Food Monitoring* (in German), *Including Cadmium Contents of Beef Kidney and Potatos". bga Press Release 11/1990* (4 March).

Crössmann, G. (1989), *Effects of Atmospheric Trace Element Pollution. ENVITEC '89*, Summaries of Papers, "Nowea" Messe, Düsseldorf.

den Besten, P. J., Herwig, H. J., Zandee, D. I., and Voogt, P. A. (1988), *Effects of Cadmium and PCB's on Reproduction of the Sea Star Asterias rubens: Aberrations in the Early Development. Proceedings 1st European Conference on Ecotoxicology* (H. Løkke, H. Tyle, and F. Bro-Rasmussen, eds.), pp.167–168. Technical University of Denmark, Lyngby.

Diels, L., and Mergeay, M. (1989), *Isolation and Identification of Bacteria Living in Environments Severely Contaminated with Heavy Metals, Proceedings International Conference Heavy Metals in the Environment*, Geneva, Vol. 1, pp. 61–64. CEP Consultants Ltd., Edinburgh.

Glooschenko, V., Downes, C., Frank, R., Braun, H. E., Addison, E. M., and Hickie, J. (1988), *Cadmium Levels in Ontario Moose and Deer in Relation to Soil Sensitivity to Acid Precipitation.* Sci. Total Environ. *71*, 173–186.

Grootelaar, L., van der Guchte, C., and Maas-Diepeveen, J. L. (1988), *Midge Larvae in Sediment Ecotoxicology. Proceedings 1st European Conference on Ecotoxicology* (H. Løkke, H. Tyle, and F. Bro-Rasmussen, eds.), pp. 108–114. Technical University of Denmark, Lyngby.

Hahn, R., Ewers, U., Jermann, E., Freier, I., Brockhaus, A., and Schlipköter, H.-W. (1987), *Cadmium in Kidney Cortex of Inhabitants of North-West Germany: Its Relationship to Age, Sex, Smoking and Environmental Pollution by Cadmium.* Int. Arch. Occup. Environ. Health *59*, 165–176.

Heising, M., Katz, D., and Umland, F. (1989), *An Approach to Characterization of Cadmium Species in Spinach; Distribution of Cadmium and Nickel Species in Sunflower Seeds.* Fresenius Z. Anal. Chem. *334* (7), 719.

Hemelraad, J., Holwerda, D. A., and Herwig, H. J. (1989), *Cadmium Effects and Kinetics in the Freshwater Clam Anodonta cygnea. Proceedings International Conference Heavy Metals in the Environment*, Geneva, Vol. 1, pp. 479–482. CEP Consultants Ltd., Edinburgh.

Houriet, J.-Ph. (1989), *Heavy Metals in Sediments of Swiss Rivers* (in German). *BUWAL-Bulletin 4/89*, 31–43.

Jacobs, L., van Gunten, H. R., Keil, R., and Kuslys, M. (1989), *Geochemical Changes Along a River-Groundwater Infiltration Flow Path: Glattfelden, Switzerland. Proceedings Workshop on Metal Speciation and Transport in Groundwaters* (H. E. Allen and A. W. Garrison, eds.), *Jekyll Island, Georgia.* Lewis Publishers, Chelsea, Michigan, in press.

Larjava, K., Reith, J., and Klockow, D. (1989), *Development and Laboratory Investigations of a Denuder Sampling System for the Determination of Heavy Metal Species in Flue Gases at Elevated Temperatures.* Int. J. Environ. Anal. Chem., in press.

Manzo, L., Pietra, R., Edel, J., Locatelli, C., and Sabbioni, E. (1989), *Male Reproductive Toxicity of Metals – Studies in Humans. Proceedings International Conference Heavy Metals in the Environment*, Geneva, Vol. 1, p. 308. CEP Consultants Ltd., Edinburgh.

Merian, E. (1989), *Environmental Chemistry and Biological Effects of Cadmium Compounds. Central America and Caribbean Workshop on Analytical Chemistry in Sanitary and Environmental Studies, Tegucigalpa.* Toxicol. Environ. Chem. *26*(1), 27–44.

Nielsen, J. B., and Andersen, O. (1989), *Chelation in Acute Oral Cadmium Intoxication. Proceedings V ICT '89* (International Congress of Toxicology), *Brighton.* Taylor & Francis, London-New York-Philadelphia, in press.

Nordberg, G. F., Nordberg, M., Jin, T., Leffler, P., Elinder, C.-G., and Palm, B. (1989), *The Disturbance of Calcium Metabolism in Kidney Induced by Cd-Metallothionein; Tolerance against Cadmium Toxicity – Role of Changes in Cellular Cd-Transport and Metallothionein Induction in Rats; Is Cadmium Released from Metallothionein in Rejected Human Kidneys? Proceedings V ICT '89* (International Congress of Toxicology), *Brighton.* Taylor & Francis, London-New York-Philadelphia, in press.

Pärt, P. (1988), *The Perfused Fish Gill Preparation in Studies of the Bioavailability of Chemicals. Proceedings 1st European Conference on Ecotoxicology* (H. Løkke, H. Tyle, and F. Bro-Rasmussen, eds.), pp. 115–117. Technical University of Denmark, Lyngby.

Peverly, J. H. (1989), *Decreased Biological Toxicity and Mobilization of Sediment Metals. Proceedings 23rd Annual Conference Trace Substances in Environmental Health, Cincinnati. J. Environ. Geochem. Health,* in press.

Pohland, F. G., Gould, J., and Cross, W. H. (1989), *Metal Speciation and Mobility as Influenced by Landfill Disposal Practices. Proceedings Workshop on Metal Speciation and Transport in Groundwaters* (H. E. Allen and A. W. Garrison, eds.), *Jekyll Island, Georgia.* Lewis Publishers, Chelsea, Michigan, in press.

Ragan, H. A., and Mast, T. J. (1990), *Cadmium Inhalation and Reproductive Toxicity,* in: Ware, G. W. (ed.): *Reviews of Environmental Contamination and Toxicology,* Vol. 114, pp. 1–22. Springer-Verlag, New York-Berlin-Heidelberg-London-Paris-Tokyo-Hong Kong.

Ramseier, S., Deshusses, J., and Haerdi, W. (1990), *Detection and Purification of Cadmium Binding Proteins Found in Intestine of Lumbricus terrestris Earthworms.* Fresenius Z. Anal. Chem., in press.

rivm (1989), *Concern for Tomorrow: a National Environmental Survey 1985–2010.* National Institute of Public Health and Environmental Protection, Bilthoven, The Netherlands.

Sauerbeck, D. (1989), *Soil Pollution Caused by Agricultural Activities. ENVITEC '89,* Summaries of Papers, "Nowea" Messe Düsseldorf.

Sharp, J. R. (1989), *The Influence of Exposure Duration on the Embryotoxicity of Cadmium to the Freshwater Teleost Etheostoma spectabile. Proceedings 23rd Annual Conference Trace Substances in Environmental Health, Cincinnati. J. Environ. Geochem. Health,* in press.

Streit, B., Krüger, C., Lahner, G., Kirsch, S., Hauser, G., and Diehl, B. (1990), *Uptake and Storage of Heavy Metals by Earthworms in Various Soils* (in German). UWSF Z. Umweltchem. Oekotox. **2**(1), 10–13.

Tallandini, L., Turchetto, M., Coppellotti, O., and Marcassa, C. (1989), *Response to Cadmium in Fish (Zosterisessor ophiocephalus Pall.) by Different Routes: Intercellular Interactions. Proceedings International Conference Heavy Metals in the Environment, Geneva,* Vol. 1, pp. 487–490. CEP Consultants Ltd., Edinburgh.

Urlings, L., and Vijgen, J. M. H. (1989), *Summary of an in-situ Cadmium Removal. ICR-Berlin Proceedings, Altlasten 3,* p. 509. EF-Verlag für Energie- und Umwelttechnik Berlin; see also Report E. Merian, *Swiss Chem.* **12**(5), 42 (May 1990).

van Coillie, R., Bermingham, N., Blaise, C., and Vezeau, R. (1989), *Ecotoxicology of Heavy Metals: Cytochemical Hazard Assessment for Fish Development. Proceedings International Conference Heavy Metals in the Environment, Geneva,* Vol. 1, pp. 491–494. CEP Consultants Ltd., Edinburgh.

van Eck, B. T. M., Kramer, K. J. M., Kerdijk, H. N., and van Pagee, H. (1989), *Trace Metals in Dutch Coastal Waters: Speciation and Bioaccumulation by Mussels. Proceedings International Conference Heavy Metals in the Environment, Geneva,* Vol. 1, pp. 277–280. CEP Consultants Ltd., Edinburgh.

van Strahlen, N. M., and Denneman, C. A. J. (1988), *Ecotoxicological Evaluation of Soil Quality Criteria. Proceedings 1st European Conference on Ecotoxicology* (H. Løkke, H. Tyle, and F. Bro-Rasmussen, eds.), pp. 404–405. Technical University of Denmark, Lyngby.

II.7 Chromium

JOHANNES GAUGLHOFER, St. Gallen, Switzerland
VERA BIANCHI, Padova, Italy

1 Introduction

Chromium is an element found in many minerals which are widely distributed in the earth's crust. It is considered to be essential to a part of the living organisms (e.g., as biologically active chromium, BAC, for metabolism of glucose). A deficiency of chromium in animals can produce diabetes, arteriosclerosis, growth problems, and eye cataracts (KIEFFER, 1979). Over the past several decades increased quantities of chromium compounds have been used by man and introduced into the environment. The danger of environmental contamination depends on the oxidation state of chromium. In its hexavalent form it is 100 to 1000 times more toxic than the most common trivalent compounds. Hexavalent chromium compounds can reduce plant growth and cause skin inflammation or eczemas in fish, mammals, and humans, and after a longer latent time they produce lung cancer. Because of their potential cancerogenicity special care is necessary when handling compounds such as chromium trioxide, lead chromate, calcium chromate, strontium chromate, chromium(III) chromate, and alkali chromates.

2 Physical and Chemical Properties, and Analytical Methods
(see also GMELIN, 1962)

Chromium (Cr) with the atomic number 24 and atomic mass 51.996 exists in all oxidation states from $-II$ to VI, but only the trivalent and hexavalent compounds and the metallic chromium are of practical importance. The naturally occurring isotopic mixture consists mainly of ^{52}Cr (83.76%) together with three other isotopes. ^{51}Cr has a half-life of 27.8 days and is the most stable among the radioactive isotopes.

Chromium is a silvery, shiny, malleable metal with a density of 7.2 g/cm^3, it melts at about 1860°C and boils at about 2670°C. The surface of pure chromium and its alloys (chrome steel) can be passivated by treatment with a strong oxidative agent such as nitric acid which makes it largely resistant to corrosion.

Trivalent chromium is the most stable oxidation state (for speciation studies see also OBIOLS et al., 1986, 1987). However, HUFFMAN (1973) could show with theoretical considerations that at the thermodynamic equilibrium hexavalent chromium is the most prevailing form in water at pH values above about 7 when the solution is saturated with oxygen. At ambient temperature, however, high pH values combined with oxygen saturation do not occur frequently, with the exception of sea water, and indeed it was found that chromium exists in sea water mainly as chromate (BRULAND, 1983). Most trivalent chromium compounds are soluble in water only at low pH values. At pH values above 5 to 6 generally chromium(III) hydroxide precipitates. However, stable trivalent chromium complexes can be formed, for example, with sulfite ions (SO_3^{2-}) which form a compound stable at pH 9 and above, provided a sufficient excess of sulfite is present in the solution. Since hexammine complexes $[Cr(NH_3)_6]^{3+}$ are also relatively stable it is possible to dissolve freshly prepared chromium hydroxide in concentrated ammonia solution. Complexes are also formed with amino acids and proteins. This is the reason why trivalent chromium compounds are used in leather tanning. Biologically important organic complexes, besides BAC (see Sect. 5.3), are, e.g., trioxalato chromate(III) ions (see Sect. 5.2).

In chromates and dichromates the chromium is hexavalent and is readily reduced to its trivalent form and hence is a strong oxidizing agent. Chromate is industrially produced by oxidation of chromite with atmospheric oxygen at high temperatures and is widely used as an oxidizing agent or as a component of pigments. For further information see, e.g., HOLLEMANN and WIBERG (1976) and SEEL (1979).

The determination of chromium at relatively high concentrations is not difficult. There are many methods available, e.g., iodometric titration, polarography, photometry with methylene blue, atomic absorption spectrometry with flame or graphite tube (WOLF et al., 1974; BATLEY and MATOUSEK 1980; WELZ 1985), plasma emission spectrometry, neutron activation analysis, gas chromatography of volatile complexes, e.g., with acetylacetone (2,4-pentanedione), X-ray fluorescence analysis, and other methods. When determining low levels of chromium, however, serious difficulties arise. These start with the method of sampling, digestion of the sample, further preparation, and the actual analysis, and end with the interpretation of the results. For example, HUBERT (1979) pointed out that in 1964 the "normal" concentration of chromium in human blood was considered to be about 1000 ng/mL (ppb), but in 1978 this value was supposed to be 0.1 ng/mL. Today the figure of 0.1 – 1 ng/mL is generally considered as correct (ANDERSON, 1981). Such large differences can probably be explained as a result of incorrect sampling and analysis.

For chromium analysis in biological materials, e.g., urine, electrothermal atomic absorption spectrometry with a pyrolytically coated graphite tube is usually the method of choice (VEILLON et al., 1982a, b; PING et al., 1983). This is also true for the analysis of sea water (WILLIE et al., 1983). Care must be taken to eliminate matrix effects as some compounds retain the chromium in the tube and thus lower the sensitivity of the determination, an effect that cannot be eliminated by the method of standard addition (VEILLON et al., 1980). Zeeman background compensation has been proposed by FLEISCHER (1988) for the determination in urine, but see also Sect. 5.3. A review about the determination of chromium in biological materials has been given by OTTAWAY and FELL (1986).

The oxidation state of chromium can be of great importance for analysis. Tri- and hexavalent chromium can be distinguished by methods that ask for an oxidation to the hexavalent form. Determination without the oxidation step only considers chromium that was already originally in the hexavalent state (iodometry, photometry, polarography). The amount of trivalent chromium is the difference between the values found in determinations with and without oxidation. In microanalysis, e.g., of blood, these methods cannot be used, since with the highly sensitive analytical methods (graphite tube atomic absorption and neutron activation) there is no distinction between the oxidation states. Furthermore, during the digestion, necessary for the preparation of the sample, the oxidation state of the chromium is probably altered. Polarographic methods are very useful, for instance, for waste water control (HARZDORF, 1986), because measurements do not fluctuate with pH changes.

It is said that during the digestion of the sample prior to analysis volatile compounds could be lost. HUBERT (1979) could not find any losses, while TUMAN et al. (1978) and WOLF et al. (1974) found independently that especially "biologically active" chromium escapes during digestion. Chromium can also be lost as chromyl chloride (CrO_2Cl_2) (KIEFFER, 1979). These losses depend chiefly upon the method of digestion and the composition of the specimen sample. They cause a considerable uncertainty for the interpretation of the results.

3 Sources, Production, Important Compounds, Uses, Waste Products, Recycling

(see also SAAGER, 1984; ANGER et al., 1986; DOWNING et al., 1986)

3.1 Occurrence, Production

Chromium is a relatively common element with an average concentration of 100 mg/kg (ppm). It is in the 21st position on the index of the most commonly occurring elements in the earth's crust (CRC HANDBOOK OF CHEMISTRY AND PHYSICS, 1979–1980; HAMILTON and WETTERHAHN, 1988), and it is more abundant than copper or zinc. However, since few specific mechanisms for enrichment exist, there are only a limited number of ore deposits. On the other hand, many rocks and sediments contain 70–90 mg/kg of chromium (range 5–1500 mg/kg) (BOWEN 1979). Natural chromium ores are mainly chromite ($FeCr_2O_4$) and less frequently krokoite ($PbCrO_4$). For the production of chromium chromite is used exclusively. Chromium deposits are distributed very unevenly geographically. 95.1% of all known and presently economically viable reserves of chromium ores are situated in the southern part of Africa (Table II.7-1). A similar statement is true for the total of known resources (NATIONAL MATERIALS ADVISORY BOARD, 1978; SAAGER, 1984; DOWNING et al., 1986).

The annual production of chromium ore is about 9 million tons (MAXWELL, 1985; SAAGER, 1984; ANGER et al., 1986), mainly mined in the USSR, South Africa, and Albania.

Table II.7-1. Known Chromium Deposits of the World

Land	Reserves in 10^6 t	Share in %	Resources in 10^6 t	Share in %
South Africa	1083	62.4	3200	71.1
Zimbabwe	568	32.7	1136	25.2
USSR	22	1.2	44	1.0
Philippines	5	0.3	5	0.1
Turkey	2	0.1	7	0.2
Others, together	57	3.3	110	2.4
	1737	100.0	4502	100.0

About 62% of the ore is converted to ferrochrome (FICHTE and FRANKE, 1975). The chromium ore is either reduced directly to ferrochrome with coal in an electric oven or oxidized to chromate with oxygen from the air in an alkaline melt. This is then dissolved in aqueous sodium carbonate solution at above 100 °C in an autoclave producing sodium chromate. Ferrochrome is an iron-chrome alloy containing about 60% chromium. For its production one does not require the pure metal. In the Federal Republic of Germany ferrochrome is used in amounts of about 200000 tons per annum. Low-carbon ferrochromium is produced by silicothermic reduction of chromite ores (DOWNING et al., 1986). Pure chromium (consumption in the FRG 1000 t per annum) is obtained by the reduction of chromium oxide with aluminum (thermit process), by electrolysis, or via chromium iodide (FICHTE and FRANKE, 1975; DOWNING et al., 1986). About 10% of the chromium consumed originates from recycling, mainly from scrap steel (SAAGER, 1984).

3.2 Uses, Waste Products

The metal industry uses most of the chromium produced (mostly in the form of master alloys, DOWNING et al., 1986) preferably in special steels (stainless steel). During the production of such alloys small particles of chromium can escape into the atmosphere as dust (see also DOWNING et al., 1986; HOOFTMAN, 1987; and Sects. 5.3, 6.5 and 7).

In the galvanizing industry other metals are coated with a chromium layer (chrome plating) in an electrolytic process in which chromium is deposited from a sulfuric acid-chromate solution. The metal is passivated by treatment with chromic or nitric acid. During this process large volumes of chromium containing waste waters are produced which must then be treated, e.g., by recycling of used floats and rinsing liquors or by removing the chromium from the waste water by ion exchange or other methods (TSCHERWITSCHKE, 1979). Hexavalent chromium can be reduced to trivalent chromium and precipitated as hydroxide. There is always the danger that chromium containing liquids or aerosols will escape into the environment.

Chromium compounds are used in the chemical industry in various fields (ANGER et al., 1986). Chromium(III) oxide is employed as pigment, as a catalyst, and for the production of pure chromium metal, chromium(IV) oxide (magnetic) in the audio, video, and data storage (see below), and chromates are used for the oxidation of organic compounds (see, for instance, ANGER et al., 1986). Chromium compounds are contained in chromium complex dyes, and certain pigments. Textile dyestuffs should normally contain less than 500 mg/kg free chromium (ANLIKER, 1978; BUNDESGESUNDHEITSBLATT, 1984). Sometimes, however, the metal is introduced into the dyestuff molecule only during dyeing (chromate treatment in the textile industry). The annual production of chromium pigments (zinc chromate, lead chromate, combined with molybdates to form chrome yellow and chrome orange) amounts to about 120000 tons within the OECD (SCHLIEBS, 1980). Recently attempts have been made to introduce less poisonous chromium(III) titanium pigments. From all these manufacturing processes chromium can escape into the environment via the effluents. But the largest amounts of pigments and dyestuffs escape indirectly − after a delay − via the dyed products into solid wastes.

In the tanning industry, basic chromium(III) sulfate is the most important tanning agent. Normally the uptake of chromium into the leather is not complete so that it must be removed from used tanning liquors. The diluted rinsing floats and the floats from post-tanning operations (retanning, dyeing, etc.) which also contain some chromium are quite difficult to treat. Thus, usually relatively large amounts of chromium escape into the effluent. Leather dust − also present in liquid effluent − and other leather wastes contain about 2% chromium which reaches the environment as a solid waste. All leather goods end up sooner or later in the waste (see also BRONZETTI et al., 1986a).

In wood impregnation chromate is used as a component of the CCF impregnation salts (chromium, copper, fluoride). Chromium is only loosely bound to the wood directly after the treatment and can be partially washed out by rain water. The infiltration of the chromate into surface and ground water can be of local importance in the neighborhood of the works. The chromium can be removed from the rain water running off the impregnated wood by means of ion-exchange columns and can be re-used in the impregnation floats (WÄLCHLI et al., 1979). After 20 to 50 years, the wood is no longer usable and by this way the chromium will finally reach the environment.

Further applications of chromium compounds are found in the following industries: building industry (as pigments), printing industry (photomechanical reproduction processes), oil industry (as anti-corrosives), textile industry (chromium mordant for textiles and chrome dyeing processes), match industry and fireworks (additive to the inflammable mixture) (FICHTE and FRANKE, 1975). In the cassette tape industry chromium(IV) oxide is used in a specially crystallized form (annual world production about 1200 tons). It is obtained from chromyl chloride or chromium(III) oxyhydrate (SCHLIEBS, 1980). It can also be used for magnetic storage media for computers.

Chromium from effluents is enriched in the sewage sludge in the waste water treatment plant. When such sludges are directly used in farming as fertilizer or added to the compost, chromium is introduced into the soil.

When chromium in waters does not sediment, it will eventually reach the sea (annually some 100000 tons of chromium) where it will remain in colloidal form or truly dissolved on average for about 11 000 years (BOWEN, 1979) until it eventually ends up in the sediment (see also MERIAN, 1984).

Chromium containing wastes are partially deposited as solid wastes on regular dumps and thus more or less safe against further distribution in the environment. During the incineration of chromium containing wastes the metal can escape into the air in the form of fly ash or volatile chromium compounds, (e.g., chromyl chloride). It can be oxidized partially into its hexavalent form and is then more easily leached out from the ashes than without incineration. There is agreement that wastes to be incinerated should not contain mercury, chromate and/or copper.

When compost is prepared from waste, the chromium contained in it will be deposited in the receiving soil.

4 Distribution in the Environment, in Foods, and in Living Organisms

When looking at global cycles (MERIAN, 1984) it seems that most of the emitted chromium compounds are immobilized, but locally emissions into the atmosphere or into waters may be substantial.

Generally chromium is found in soil in concentrations of 10–90 mg per kg, in sea water 0.3 µg/L mainly as chromate. The amounts increase with the depth (BRULAND, 1983). Fresh water contains 1–10 µg/L chromium and the non-industrial atmosphere about 10 ng/m^3 (BOWEN, 1979; BAETJER et al., 1974). Nowadays up to 25 µg/L chromium may be found in drinking water in exceptional cases (e.g., in a well) (beer and wine contain 300–450 µg/L, HAMILTON and WETTERHAHN, 1988), and in the atmosphere of industrial cities up to 70 ng/m^3 and in smoke from coal fires up to 2 µg/m^3 (BAETJER et al., 1974).

Plants usually contain 0.02–14 mg/kg chromium on dry weight basis. However, in some Australian trees which are capable of storing chromium, such as *Leptospermum scoparium* and *Sutera fodina* (*Pimelea suteri*), much larger quantities (2–4% in the ash) can be found (BOWEN, 1979; PETERSON and GIRLING, 1981). In lichen and mosses also large quantities of chromium have been found (PETERSON and GIRLING, 1981).

In food consumed by humans and animals it is important in which form chromium is present. Biologically active chromium can be found in unrefined sugar syrup from beet and cane, in wheat germ, in black pepper and in beer yeast, which therefore are useful to prevent chromium deficiency (KIEFFER, 1979). Fruits contain very little chromium (PETERSON and GIRLING, 1981). It must, however, be mentioned that more than half of the dietary chromium is originated from other sources (preparation, cutlery) than the foodstuffs themselves (VAN SCHOOR and DEELSTRA, 1986). Stress, such as acute exercise, may lead to a reduction of chromium uptake (ANDERSON, 1987).

The chromium content of sea fish is 0.03–2 mg/kg (dry weight). The muscles of mammals contain 0.002 to 0.8 mg/kg and mammalian bones 0.1–30 mg/kg

(BOWEN, 1979). In food derived from animals chromium is biologically available in varying doses. Highest amounts of biologically active chromium are found in liver and cheese (KIEFFER, 1979). Adult humans contain about 5–20 mg of chromium, particularly in the spleen and liver. The daily requirement of a human is ca. 0.01–0.04 mg of organically complexed chromium or about 0.1–0.3 mg chromium in an inorganic form (KIEFFER, 1979; LANGARD and NORSETH, 1986; FISHBEIN, 1987).

5 Uptake, Resorption, Transport and Distribution, Metabolism, and Elimination in Plants, Animals, and Humans

5.1 General Remarks

In vitro experiments on cellular systems have shown that the oxidation state of chromium strongly influences the rate of Cr-uptake (LEVIS and BIANCHI, 1982; BIANCHI and LEVIS, 1984). Cr(VI) can easily cross the cell membranes, where the phosphate-sulfate carrier also transports the chromate anions. Conversely Cr(III) does not utilize any specific membrane carrier and its entrance into the cell is obtained by less efficient mechanisms: simple diffusion and, in the case of animal cells, endocytosis. Furthermore, Cr(III)-uptake depends on the nature of the ligands it is complexed with, to such an extent that in some instances cell membranes are practically impermeable to Cr(III) complexes. Complexes with appropriate lipophilic ligands, however, diffuse into cells with relative ease (see also WARREN et al., 1981; GALLI et al., 1988).

As the trivalent form is usually the most stable state of Cr in nature (for exceptions see Sects. 2 and 5.2), organisms deal mostly with Cr(III) compounds which they find in soil and food. While plants can accumulate chromium, showing some species-specificity in the level of accumulation, animals excrete the excess chromium present in their diet.

The efficient uptake of Cr(VI) by organisms and the manifold toxic effects of Cr(VI) inside the cells (LEVIS and BIANCHI, 1982), make Cr(VI) contamination a serious environmental hazard in many industrialized countries. Cr(VI) is rapidly reduced to Cr(III) inside the cells, and its biological activity depends on both the process of its reduction and the subsequent trapping of Cr(III) in different cell compartments. Even in organisms exposed to Cr(VI) Cr is detectable in the reduced state. Only in a minority of studies the actual localization of the Cr associated to the organs of the treated animals has been investigated and the subcellular fraction linking Cr(III) determined (e.g., CUPO and WETTERHAHN, 1985a). Frequently this issue has been neglected, which impedes to reliably assess the biological role and the possible toxic action of Cr(III) in living organisms.

5.2 Uptake and Distribution of Chromium in Plants

The uptake of chromium(III) from the soil depends upon the species of plant, and within a plant the concentrations largely differ between the different parts of the plant (SYKES et al., 1981). CARY et al. (1977a) found a similar mechanism of transportation for iron and chromium in plants, and an increased transport of chromium was observed in the case of iron deficiency (see also PETERSON and GIRLING, 1981).

The question of the biological availability of chromium from soil for the plants is not clear. HERFELD (1974) suggested that chromium from tannery wastes becomes more and more insoluble with the time due to olation (formation of complexes with hydroxyl bridges). FENKE (1977) could not find any definite change in solubility of chromium in the soil over a period of 5–12 months using an ammonium acetate solution for extraction.

Regarding the ion radius ANDERSSON (1977) suggested that chromium may be fixed in the clay fraction of the soil after some period of time. By weathering the chromium can be set free again. The high chromium content of plants growing on millions of years old serpentine soils shows that a certain amount of chromium is always available for the plants (STICHER, 1979, 1980).

The quantities of chromium in the soil which are actually dangerous for the plants depend largely on its biological availability for them. Unfortunately, all known studies have been carried out one or two years after the addition of chromium to the soil. After such a short time equilibrium cannot be reached. For the assessment of the environmental risks long-term observations are of primary importance. Experience with serpentine soils has little significance, since in this case the damage of plants is caused by high concentrations of nickel (STICHER, 1980). Nevertheless, the existing studies provide some indications. BRAUN (1974) and SYKES et al. (1981) found that no damage was done to plants even under unfavorable conditions at chromium concentrations in the soil of up to 500 mg/kg. Calcium and phosphate reduce the sensitivity of plants to chromium. SHIVAS (1980b) reported that in rhubarb and geranium no damage occurred in soil containing 6000 mg/kg chromium and more. In general, in spite of high chromium additions to the soil only low concentrations of chromium were found in the edible parts of the plants. Spinach contained the highest quantities, up to 23 mg/kg on dry weight (KICK and BRAUN, 1977). Other plants showed much lower concentrations, e.g., the grains of winter rye had a maximum chromium content of 0.5 mg/kg under similar conditions.

A good criterium for potential hazards of chromium in the soil is its bioavailability for plants. In most investigations the uptake of chromium was measured. However, up to now no systematic study has been published on chromium uptake by at least one type of plant in different soils with different chromium contents, and different intervals between the additions of chromium and the measurement. Such experiments would be rather extensive, but would bring much more reliable results than the measurement of the "availability" for plants using various chemical extraction agents (see also Chapter I.7).

The translocation of chromium from the root through the plant to the leaves is rather slow, and it seems that the main barrier is the transport into the vessels

(PETERSON and GIRLING, 1981). Plants which take up higher concentrations of chromium, e.g., cabbage type, are able to transport trioxalatochromate(III) ions more easily. Plants damaged by hexavalent chromium do not contain increased amounts of chromium in the branches or leaves because the damage takes place already in the root system (PETERSON and GIRLING, 1981).

It cannot be excluded that chromium(III) may be oxidized to chromium(VI) and then can penetrate the cells. BARTLETT and JAMES (1979) have shown that certain soils can oxidize chromium(III) chloride solutions in a short time, whereby after 24 hours the chromium(VI) concentration decreased again slowly. They found up to 30 mg/kg of chromium(VI) in the soil. CARY et al. (1977b) suggested that the uptake of chromium by the roots of plants could take place mainly via chromium(VI). SHIVAS (1980a) found that, for instance, the oxidation of chromium(III) in chrome leather waste by means of oxygen from the air at 40–50°C and pH 7 takes place within a few weeks to produce hexavalent chromium. In chrome leather waste up to 100 mg/kg of hexavalent chromium can be found. In our own experiments no oxidation of chromium(III) in various soils rich in humus was observed, not even after the pH value had been increased with lime and an addition of manganese. It was, however, possible to achieve a concentration of 0.5–2.1 mg/kg of hexavalent chromium (on dry weight) in a mixture of chrome leather dust and expanded vermiculite (a porous magnesium aluminum silicate) at pH 8–8.5 maintained at 95% rel. humidity at 29°C. In a loamy clay soil poor in humus 0.1–5.3 mg/kg hexavalent chromium were found after 16 weeks under the above stated conditions. The yielded concentration depended on the pH value (7.4–8.0) and the addition of manganese. In clay without any humus no oxidation occurred under the same conditions (GAUGLHOFER, 1986).

5.3 Uptake, Distribution, Metabolism, and Excretion in Animals and Humans
(see also LANGÅRD and NORSETH, 1986)

The animal cell is provided with a mechanism of uptake – endocytosis – which gives it additional possibilities to internalize Cr(III) as compared to plant cells which are surrounded by rigid cell walls. However, uptake of Cr(III) by endocytosis is slow and relatively inefficient. It probably plays a relevant physiological role only in the lungs, where inhaled Cr-particles are deposited and sequestered by alveolar macrophages inside their lysosomes (JOHANSSON et al., 1986). Inhalation of Cr occurs in occupational exposure (e.g., in welding plants) and can also involve Cr(VI) (see, for instance, STERN, 1986, and HOOFTMAN, 1987). Among the cellular types present in the lungs, alveolar macrophages are the most active in reduction of Cr(VI), and represent the main defense against the carcinogenic form of chromium (PETRILLI et al., 1986). The epithelial lung fluid has also some Cr(VI) reducing ability (PETRILLI et al., 1986). Although much lower than that of alveolar macrophages, it can be important in situations of low exposure, as it produces Cr(III) extracellularly and prevents Cr from entering the lung cells in its oxidized toxic form.

The physiological exposure of humans and animals to Cr takes place in the gastro-intestinal (GI) tract, where Cr is introduced with the diet as Cr(III). Given the relatively short time of contact of the Cr-containing material with the GI epithelia, absorption is markedly influenced by the nature of the Cr(III) complexes. Inorganic Cr(III) is poorly absorbed, while Cr(III) linked to amino acids or other biomolecules is more readily taken up by diffusion across the plasma membranes (GUTHRIE, 1982; LANGÅRD, 1982). Such organic complexes — referred to as biologically active chromium (BAC) — are the most efficient supply of Cr for humans and mammals. Trace amounts of absorbed Cr (below 1 µg/day) are sufficient to maintain Cr balance in men (OFFENBACHER et al., 1986). But because of the low percent absorption (GUTHRIE, 1982; LANGÅRD, 1982; OFFENBACHER et al., 1986), the optimal dietary intake is in the range of 50–200 µg/day.

Skin contact (see also Sect. 6.5) with chromium as it occurs in a number of occupational situations can result in allergic contact dermatitis or in irritant reactions evolving in ulcers (BANG PEDERSEN, 1982). Hexavalent Cr induces corrosive reactions due to its oxidizing activity, and it is also more active than Cr(III) in causing allergic effects, even if it is Cr(III) which acts as the hapten complexing skin proteins. The permeability of membranes to Cr(VI) allows more Cr to enter and to be reduced, resulting in higher amounts of immunologically active Cr than in the cases of direct contact with Cr(III) compounds. Studies of human responses to stainless steel welding (STERN, 1986; KILBURN et al., 1987; STERN, 1987) allowed the conclusion that the potencies of the chromium and nickel species depend on bringing them into solution. Diffusion through skin increases from chromium(III) nitrate to chromium(III) sulfate to chromium(III) chloride and to chromates, $CaCrO_4$ has a relatively high water solubility (163 g/L) and is carcinogenic. The much less soluble $ZnCrO_4$, $Zn(OH)_2$, $BaCrO_4$, and $PbCrO_4$ (PbO) are practically non-carcinogenic (IARC, 1980; STERN, 1987). Welding particles may change their composition, and various chromates may go to different lung locations (STERN, 1987). While chromium ulcers are exclusively found in workers, nowadays much less frequently than in the past, Cr dermatosis is more widespread, as several materials can contain contaminating Cr(VI), e.g., cement.

Absorbed Cr is transported in the body by blood, Cr(III) is bound to plasma proteins, especially transferrin (LANGÅRD, 1982), and Cr(VI) is accumulated inside the red blood cells (RBC) where it is reduced to Cr(III). In fact, the reducing ability of erythrocytes vs. Cr(VI) represents an important mechanism of detoxification (PETRILLI and DE FLORA, 1978; KORALLUS, 1986) which can be saturated in cases of experimental exposures to high doses of Cr(VI), but is probably sufficient to eliminate oxidized Cr from the blood in situations closer to "real life". However, there are individual differences (fast and slow reducers, KORALLUS, 1986). While some chromium(III) complexes may directly be taken up through cell membranes and damage DNA by interference with DNA polymerase, chromium(V) — which can be identified with FPR — breaks DNA via radical formation (BEYERSMANN and KÖSTER, 1986; BEYERSMANN, 1987). Besides the oxidation state, water solubility influences the mobilization (i.e., absorption+transport) of Cr from the site of contact with the exposed tissues. Therefore, the behavior of a given compound can only be generalized to others of the same valence and similar solubility.

Chromium transported by blood is distributed to tissues and organs which have different retention capacity (GUTHRIE, 1982; LANGÅRD, 1982; WEBER, 1983; GREGUS and KLAASSEN, 1986). The highest levels of Cr are found in liver, kidneys, spleen, and lungs. SUNDERMAN et al. (1987) have investigated the eventual increase of chromium concentrations in body fluids with view to orthopedic prosthetics. They found that the effects are negligible.

The data on Cr distribution in the body derive from different sources, namely, non-exposed humans, occupationally exposed workers, and experimental mammals. Each set of data has its inherent limitations: the published values for Cr levels in the organs and tissues (e.g., liver or blood) of control humans vary within ranges of several orders of magnitude (GUTHRIE, 1982). Such varying data are due to analytical problems and/or sample contamination before Cr measurement (Sect. 2).

The data on exposed humans almost exclusively refer to Cr accumulation in the lungs following inhalation exposure, while quantitative data for exposure through the skin are more scarce (BANG PEDERSEN, 1982). Cr concentrations in urine and blood have been proposed for the biological monitoring of exposed workers, but a number of factors influence Cr absorption and excretion, such as level and duration of the exposure, the nature of the Cr compound(s), the life-style of the individuals (e.g., smoking habits), so that urine Cr cannot be representative of the actual exposure. A better indicator of exposure to Cr(VI) could be the level of Cr in the RBC, given their life-span in circulating blood (LANGÅRD, 1982).

The route of Cr administration to experimental animals often does not reproduce the common conditions of exposure in humans: the most frequently adopted is the i.v. injection, which may be relevant to Cr(VI) exposure only, given the slow uptake of Cr(III) at the GI level. Intra-tracheal or intrabronchial instillation of Cr compounds (WEBER, 1983; LEVY and VENITT, 1986) and inhalation of Cr(III) or Cr(VI) (GLASER et al., 1985; JOHANSSON et al., 1986) have been less widely employed, although they can better mimic human occupational exposure. Several studies have monitored the distribution of Cr to different organs after such treatments and the time course of Cr removal during many hours or days thereafter (LANGÅRD, 1982; WEBER, 1983). Liver, kidneys, and spleen are the organs which generally reach the highest and most stable content of Cr. In the liver Cr is thought to be stored linked to proteins and smaller peptides, such as glutathione, while in the spleen it accumulates with the debris of RBC. Animal studies have not detected Cr transport to the brain. Conversely, Cr has been found able to penetrate the human brain (DUCKETT, 1986), although the origin of Cr detected in perivascular pallidal deposits in three patients remained uncertain.

Placental transport of Cr was dependent on the time of administration during pregnancy: more ^{51}Cr reached the fetuses of mice when administered later rather than early during pregnancy (DIAB and SOEDERMARK, 1972). Cr(VI) induced genetic (KNUDSEN, 1980) and teratogenic (GALE and BUNCH, 1979) effects in the offspring of i.p. treated mice and i.v. treated hamsters (see also LANGÅRD and NORSETH, 1986).

The main route of Cr excretion is through the kidneys with urine, also after exposure to Cr(III). Two phases can be detected in Cr excretion, a rapid one, corresponding to the clearance of Cr from the blood, and a slower phase, representing the clearance from the tissues. Excretion with bile appears to be marginal (see also SUNDERMAN, 1986).

6 Effects on Plants, Animals, and Humans

6.1 General Remarks

It is necessary to distinguish between the various oxidation states when considering the biological activity of chromium. It is usually assumed that hexavalent Cr is about 100 to 1000 times more toxic than the trivalent Cr. BRAUN (1974) found such a ratio in winter rye grown in sandy loam, and it was not only the oxidation state of the chromium but also the state of the soil which played a significant role in the extent of damage. Additions of lime reduced the activity of trivalent chromium and increased the toxicity of Cr(VI). LEVIS and MAJONE (1979) showed that in the case of hamster fibroblasts in culture (CHO cell line) the Cr(VI) LC_{50}-value was about 1.5 mg/L, while for Cr(III) it was much higher and depended on the type of compound. The LC_{50} of chromium acetate was 150 mg/L, for chrome alum, chromium chloride, and chromium sulfate it exceeded the solubility in the culture medium which was about 300 mg/L.

6.2 Effects on Microorganisms and Plants

(see also GALLI et al., 1988)

So far it has not been possible to establish that chromium is an essential element required by plants, not even in using ultrapure solutions (HUFFMAN and ALLAWAY, 1973). On the other hand, a positive reaction on the rate of growth and on the yields of plants such as potatoes, maize, rye, wheat, or oats could be shown by adding chromium to soil which was deficient in this element (SCHARRER and SCHROPP, 1935; BERTRAND and DE WOLF, 1986). The reasons for this are not clear. HUFFMAN and ALLAWAY (1973) did not exclude the possibility that chromium could affect microorganisms which impede these plants, thus having an indirect positive effect. Chromium(VI) is toxic already in low concentrations depending on the pH of the soil. However, the time of activity of Cr(VI) is short, because it is quickly reduced to Cr(III). FENKE (1977) could show that even under unfavorable conditions (low humus, sandy soils, high pH) there was no detectable Cr(VI) present after 3 months. BRONZETTI (1986a) showed that five products used by the leather tanning industry are genetically active in yeast strains.

6.3 Toxic Effects on Water Organisms

(see also VAN STEERTEGEN and VAN DE VEL, 1986)

The toxicity of Cr(VI) compounds on fish is quite high and depends on the pH (KOEMANN et al., 1977). Probably, chromic acid is much more toxic to water organisms (due to skin damage) than its salts. According to PERES (1980) LC_{50} values for fresh water fish are 250–400 mg/L and for sea fish 170–400 mg/L. STRIK et al. (1975) quoted LC_{50} values of 17–118 mg/L for fish, 0.05 mg/L for daphnias, and 0.032–6.4 mg/L for algae. They found that trouts were badly affected in water containing 10 mg/L Cr(VI). According to JUNG (1973) crabs tolerate only 0.3–0.7 mg/L and fish 0.015–0.195 mg/L of chromate. For daphnias 24 hour LC_{50} values for potassium dichromate were established to be 0.01–0.26 mg/L (MÜLLER, 1980). For fish a tolerance limit of 42 mg/L for Cr(III) in water was given by HERFELD (1974).

6.4 Deficiency Symptoms
(see also LANGÅRD and NORSETH, 1986; FISHBEIN, 1987)

Cr deficiency is infrequent in humans and usually limited to elderly people or to rare cases of patients subjected to prolonged parental nutrition. It is associated with decreased glucose tolerance, some forms of diabetes and cardiovascular diseases. Animals fed Cr-deficient diets show reduced body weight and shorter life-span (EPA, 1984). Glucose intolerance in men is ameliorated by BAC administration. Cr(III) acts as an insulin potentiating agent, supposedly by favoring insulin binding to receptors. An early recognized source of the insulin potentiating factor was brewer's yeast. The factor was called glucose tolerance factor (GTF), and MERTZ and SCHWARZ (1959) indicated Cr(III) as the active component of GTF. A partial purification of GTF from yeast demonstrated that Cr(III) is complexed to amino acids (TOEPFER et al., 1977). More recently the presence of Cr(III) in GTF has been questioned by HAYLOCK et al. (1983) who demonstrated that during purification of yeast GTF the biological activity was cleanly separated from Cr-containing material. The question remains open, but, nevertheless, it is assessed that a number of synthetic Cr(III) complexes have GTF activity (TOEPFER et al., 1977; COOPER et al., 1984; VINSON and HSIAO, 1985).

6.5 Toxic Effects on Mammals and Humans
(see also LANGÅRD and NORSETH, 1986)

An oral dose of 0.5–1 g of potassium dichromate is fatal for man, and absorption through the skin is also very dangerous (MOESCHLIN, 1972) resulting in diarrhea, bleeding from stomach and intestine, serious liver and kidney damage, and cramps (SCHLATTER and KISSLING, 1973).

Less is known on the toxicity of trivalent chromium. LEWIS and TATKEN (1980) reported that 35–350 g of chromium sulfate had no toxic effects for man when taken orally. In the case of rats, for chromium chloride, taken orally, a LD_{50} of 1800 mg/kg and for Cr(III) nitrate a value of 3250 mg/kg were found.

It has been known for a long time that workers exposed to chromate dust for extended periods suffer perforated nose walls; also ulcers can develop (DELPECH and HILLAIRET, 1869), and the sense of smell can be lost (SEEBER et al., 1976).

If hexavalent chromium compounds come into contact with skin (see also Sect. 5.3), ulcers may develop (DEWIRTZ, 1929). Usually they do not cause pain, but heal slowly. Recovery is speeded up by application of calcium-EDTA. From this observation one could deduce a damage by trivalent chromium, since only this can be complexed by Ca-EDTA (MALOOF, 1955).

NATER (1962) supposed that Cr(VI) causes allergic skin reactions, and Cr(III) does not. FREGERT and RORSMAN (1964, 1965), however, observed allergic reactions against Cr(III) compounds with most patients sensitive to Cr(VI), when these were placed on their skin. When they injected chromium(III) chloride into the skin, even all Cr(VI) sensitive patients showed an allergic reaction. The concentrations of Cr(III) that caused an allergy were higher than in the case of Cr(VI). SAMITZ and SHRAGER (1966) suggested that these differences in sensitivity might be caused by different diffusion velocities of Cr(III) and Cr(VI) in the tissues.

Eczemas caused by contact with cement are due to allergic reactions against Cr(VI) impurities in the cement (LANGÅRD and NORSETH, 1986). In a study with 5558 patients in Scandinavia 3% of the women and 12% of the men who had skin allergy reacted positively to tests with potassium dichromate (MAGNUSSON et al., 1968). This may be due to the presence of Cr(VI) compounds in very small amounts in the environment. Asthmatic bronchitis is another allergic reaction which can be caused by Cr(VI) (LANGÅRD and NORSETH, 1986). Sensibilization takes quite a long time. A sensibilized person will have an asthmatic attack 4–8 hours after inhalation of chromate containing dust or after subcutaneous injection of chromate.

6.6 Mutagenic, Carcinogenic, and Teratogenic Effects
(see also LANGÅRD and NORSETH, 1986; ANGER et al., 1986)

6.6.1 General Remarks

Some chromium compounds are recognized as human carcinogens; epidemiological studies have shown an association between occupational exposure to chromium and lung cancer (IARC, 1980, 1982). However, until now it is not clear which Cr-containing substances are the actual etiologic agents of cancer. Exposure to complex mixtures of Cr(VI) and Cr(III) compounds takes place in industrial plants where increased incidence of cancer was reported, e.g., production of chromates, of chrome pigments, or Cr plating plants (see also Sect. 5.3).

The uncertainty about the nature of carcinogenic Cr has stimulated a great number of experimental studies, both with laboratory animals and with simpler systems for genotoxicity testing. Although in animal experimentation the activity of Cr compounds could be tested under conditions similar to those of human exposure, Cr carcinogenicity could only be demonstrated with administration routes of questionable significance to the human situation (IARC, 1980; EPA, 1984). No significant increase of lung cancer was observed in animals after inhalation of Cr(VI) or Cr(III), while intrabronchial implantation produced statistically significant bronchial carcinomas in rats only with scarcely soluble Cr(VI) (LEVY and VENITT, 1986). Consistently positive results were obtained only with sparingly soluble Cr(VI) compounds using intramuscular implantation. Thus the water solubility of Cr(VI) turned out to be an important parameter affecting its activity in vivo (see also Sect. 5.3). The accepted interpretation was that soluble Cr(VI) was rapidly released from the implantation site and, depending on the dose, either was rapidly reduced and inactivated, or it was too toxic to the surrounding tissue to allow for the carcinogenic process to be started. In contrast, scarcely soluble Cr(VI) could diffuse slowly from the site of application and cause a prolonged exposure to moderate doses of the carcinogenic agent, presumably Cr(VI) (LEVIS and BIANCHI, 1982).

Short-term tests of geno- and cytotoxicity of chromium have elucidated its mechanisms of action and substantiated some important aspects of Cr carcinogenicity. Cr experimentation has developed and expanded during the 1970s and 1980s together with the practice of short-term tests, and Cr compounds have been tested in a great number of different systems. The unusual properties of Cr(VI), as compared to other metal compounds, which make it active in bacterial cells, have favored its widespread use in mutagenicity assays with bacteria, even in the evaluation of new test strains.

The Ames test, the SCE-test, the chromosome aberration test, and the test with alveolar macrophages are, for instance, sensitive systems for in-vitro studies of welding fumes, but not the HGPRT-test (HOOFTMANN, 1987). Data published on Cr genotoxicity are so numerous that here only the conclusions thereof shall be summarized. The reader is referred to recent reviews on the subject (LEVIS and BIANCHI, 1982; BIANCHI and LEVIS, 1984, 1987, 1988). The role of chromium(V), which breaks DNA via radical formation has to be studied further (see Sect. 5.3). Chromium teratogenicity in mice, golden hamsters, and chickens was discussed by LANGÅRD and NORSETH (1986).

6.6.2 Differential Activity of Cr(VI) and Cr(III) in Short-term Tests

Cr(VI) and Cr(III) compounds strongly differ in their effects on intact cells. Cr(VI) is genotoxic and cytotoxic for bacteria and eukaryotic cells, even with short-time exposures to low doses. The active concentrations vary with the sensitivity of the cellular end point under examination. Under the same conditions Cr(III) is either inactive or active at much higher concentrations, often close to the limit of solubility of the tested compounds so that the observed effects may be due to non-specific mechanisms of toxicity.

The genetic activity of Cr(VI), as detected in cellular systems in vitro, covers the whole spectrum of genotoxicity. The use of suitable bacterial strains has demonstrated that Cr(VI) causes DNA damages, which reduce the survival of repair-deficient bacteria, and gene mutations consisting both in base-pair substitutions and frameshifts. DNA damage and repair are also induced in low eukaryotes (yeasts) and mammalian cells and result in gene mutations and chromosomal effects. Chromosomal aberrations and sister chromatid exchanges are induced by Cr(VI) not only in mammalian cells treated in vitro, but also in the lymphocytes of exposed workers and in experimental animals. Cell transformation, which is the end-point closest to neoplastic transformation detectable in vitro, is also produced by Cr(VI) in mammalian cell lines.

The activity of Cr(VI) in short-term tests in vitro is independent of the composition of the single compounds and depends on the concentration of chromate anions in solution (see, for instance, STERN, 1987). The ability of $Cr_2O_4^{2-}$ anions to be transported by the membrane anion carrier (see Sect. 5) is the basic property which distinguishes Cr(VI) from Cr(III) and all the other metals and accounts for its genetic activity in any kind of cell.

The unusual effectiveness of Cr(VI) has influenced the experimental handling of Cr(III) and has contributed to establish the view that Cr(III) is genetically inactive in cellular systems.

For a long time chromosomal aberrations were the only genotoxic effect detected in mammalian cells treated with Cr(III). Moreover, they were observed at such concentrations (10^3 times higher than the Cr(VI) effective doses) that non-specific toxicity due to the utterly non-physiological conditions of treatment could not be excluded.

On the other hand, when tested directly on DNA or subcellular fractions (e.g., isolated nuclei), Cr(III) appeared able to modify the physico-chemical properties of the genetic material and to affect the accuracy of its replication. It could be suggested that its inability to act genotoxically in vivo depended on its inability to reach the genetic targets inside the cells. In fact, cell membranes lack carriers suitable to transport Cr(III) (Sect. 5), and this form of Cr, present in solution as hexa-coordinated complexes, can only pass the membrane barrier by diffusion or enter the eukaryotic cell by endocytosis. The slow, inefficient uptake of Cr(III), associated with treatment protocols tailored on the rapid action of Cr(VI), resulted in a long repeated array of negative results in short-term tests. The occasional positive genetic effects were referred to Cr(VI) contamination of the tested Cr(III) compounds (LEVIS and BIANCHI, 1982), although only in some instances Cr(VI) presence was directly checked (LEVIS and MAJONE, 1979).

When the issue of Cr(III) passage through the cell membranes was directly approached and the experimental conditions were modified in order to favor the entry of Cr(III) into the cells, a number of genetic effects started to be reported (reviewed in BIANCHI and LEVIS, 1987, 1988). The main solutions to the problem were the use of complexes of Cr(III) with lipophilic organic ligands which could more easily diffuse across the membranes, or the prolongation of treatments in order to allow more time for Cr(III) diffusion or endocytosis (see also WARREN et al., 1981; GALLI et al., 1988). By these means Cr(III) was found able to produce gene mutations and lethal genetic damage in bacteria and chromosomal effects in mammalian cells. BRONZETTI et al. (1986b) showed that chromium(III) can directly enter yeast cells (by crossing the membranes) if phosphate buffer ions are present and if the cells contain high levels of individual P_{450} cytochromes.

6.6.3 Cellular Metabolism of Chromium and its Role in Chromium Genotoxicity

The genetic activity of Cr(VI) in whole cells contrasts with its inability to damage DNA directly in vitro or to induce appreciable chromosomal damage when applied to isolated nuclei. Only the addition of metabolizing systems (microsomes, S9 fraction) to the in vitro reaction mixture made Cr active on purified DNA, via reduction of Cr(VI) to Cr(III) (JENNETTE, 1979). The types of DNA lesions obtained under these conditions were the same as observed when cultured cells or experimental animals were treated with Cr(VI), namely DNA interstrand crosslinks, strand breaks, and, in the case of chromatin, DNA-protein cross-links. In vivo such DNA damages are removed with different kinetics, in particular DNA-protein crosslinks are the most persistent, because inefficiently repaired. They seem to be the genetic lesions more directly related to the carcinogenic activity of Cr (CUPO and WETTERHAHN, 1984).

Interestingly, the final distribution of Cr in rats treated with Cr(VI) and Cr(III) was independent of the oxidation state of Cr, without significant difference in the level of Cr bound to DNA in the organs examined. But only after Cr(VI) treatment DNA damage could be detected (CUPO and WETTERHAHN, 1985a).

The results of the in vitro assays with Cr(VI) and DNA suggest that oxidized Cr must be reduced to Cr(III) to interact with the nucleic acid. Conversely, the in vivo experiments indicate that if Cr(III) reaches the genetic material it is unable to damage DNA. The two sets of data become consistent if Cr(VI) reduction inside the cell is a mechanism of activation which produces some Cr species reactive vs. DNA.

The cytoplasmic reduction of Cr(VI) has been extensively studied, and a number of enzymatic and non-enzymatic systems have been shown to take part in it (reviewed by BIANCHI and LEVIS, 1987), e.g., glutathione, cytochrome P_{450}, and DT diaphorase. During Cr(VI) reduction, a three-electron reaction occurrs stepwise, and the intermediates Cr(V) and Cr(IV) are formed. Cr(V) is sufficiently stable to be detected by FPR both in vitro and in the cells (JENNETTE, 1982; NISHIO and UYEKI, 1985; GOODGAME and JOY, 1986; ARSLAN and BELTRAME, 1985; BEYERSMANN and KÖSTER, 1987; BEYERSMANN, 1987). Such intermediates are considered as candidates for the active genotoxic form of chromium, which would either react directly with the genetic material (CUPO and WETTERHAHN, 1985b) or cause the production of active oxygen species which damage DNA (KAWANISHI et al., 1986; BEYERSMANN, 1987).

Glutathione plays a central role in Cr(VI) reduction and influences the level and type of DNA damages induced. In chick hepatocytes treated with chromate expansion of the glutathione pool increased the level of DNA strand breaks, leaving DNA-protein crosslinks unchanged, while very little DNA damage could be detected in the case of glutathione depletion (CUPO and WETTERHAHN, 1985b).

According to these data Cr(VI) reduction corresponds to Cr activation. However, sound experimental evidence points to the opposite conclusion. The work of DE FLORA and coworkers has shown that different cells and tissues have different Cr-reducing capacity (PETRILLI et al., 1985). The lowest efficiency has been found just in those tissues – lung parenchyma and muscle – which are the specific sites of tumor formation in exposed humans and experimental animals, respectively.

The problem of Cr(VI) reduction and its role in Cr genotoxicity could find a solution in the existence of thresholds for Cr activity. It is conceivable that only a fraction of the products of Cr(VI) reduction are potentially genotoxic and that they can actually express their potential only if formed close to the genetic targets. Hence, in cells where the cytoplasmic reducing systems are very active Cr(III) can be rapidly produced and sequestered in stable complexes. However, in cells with low reducing activity oxidized Cr can get closer to the genetic material, and the genotoxic species can be generated near the relevant targets. This interpretation could be substantially correct, but it is still speculative. Very little is known about Cr speciation inside the cells, especially about the nature of the Cr(III) complexes resulting from the intracellular reduction, as opposed to those originating from the extracellular compartment, and their relative reactivity towards DNA.

It is certain that Cr(III) cannot be oxidized to Cr(VI) by cell metabolism. Thus, the recently reported genetic effects of Cr(III) in intact cells (see Sect. 6.6.2). demonstrate that Cr(III) need not be formed inside the cell to be genetically active. The characterization of Cr(III) complexes in this respect is still to be done, and it can be expected to supply information vital to the understanding of Cr genotoxicity and the prediction of the genotoxic risk of Cr(III) exposure.

7 Hazard Evaluation and Limiting Concentrations
(see also Section 3.2; DOWNING et al., 1986; ANGER et al., 1986)

Trivalent chromium compounds ingested as part of food are relatively innoxious, while hexavalent Cr compounds are poisonous. In plants damage to the root system due to Cr(VI) is well known. Skin damage is critical for water organisms and humans. In the case of inhaling Cr(VI) during occupational activity cancer of the lungs may develop after long latent periods. Rapid absorption of hexavalent chromium compounds through the skin can cause acute poisoning. In the USA the threshold limit value (TLV) for Cr(VI) is fixed as 0.1 mg/m^3 in the air (BAETJER et al., 1974). In the Federal Republic of Germany chromium trioxide and chromates are classified as cancerogenic substances (without maximum tolerable concentration at occupational exposure), and there is a "technically regulated concentration" (TRK) of 0.1 mg/m^3 for calcium-, chromium(III)-, strontium-, and zirconium chromate (HAUPTVERBAND DER GEWERBLICHEN BERUFSGENOSSENSCHAFTEN, 1987; MAK 1987). In Germany there are also restrictions to dust emissions and to Cr(VI) concentrations in these dusts (DOWNING et al., 1986). In Switzerland the maximum tolerable concentration for chromic acid and chromates at occupational exposure is 0.05 mg per m^3 air (SUVA, 1980).

In most countries (incl. USA, Japan, Federal Republic of Germany, and WHO) the recommended limit of 0.05 mg/L chromium in drinking water is applied. A maximum value of 0.02 mg/L of Cr(VI) is fixed in Switzerland (SCHWEIZERISCHES LEBENSMITTELBUCH, 1972). Trivalent chromium is not mentioned there.

In Switzerland the following maximum values are fixed. For effluents discharged into communal sewers: Cr(III), 2 mg/L; Cr(VI), 0.5 mg/L; for effluents discharged into rivers and lakes: Cr(III), 2 mg/L; Cr(VI), 0.1 mg/L; and the aim for rivers and reservoirs is: Cr(III), 0.05 mg/L; Cr(VI), 0.01 mg/L in the membrane filtrate (DER SCHWEIZERISCHE BUNDESRAT, 1975). For artificial irrigation the general limit recommended by the NAS (National Academy of Sciences, Washington) is 0.1 mg/L of chromium, under certain conditions 1 mg/L (BAETJER et al., 1974). The International Committee for Rhine Waterworks recommended the following limits for total chromium at low water level in the Rhine river (IAWR, 1973): when only natural purification processes are applied, for drinking water: 0.03 mg/L; when well established processes are used for further purification: 0.05 mg/L.

In Switzerland sewage sludge used directly or indirectly (i.e., as compost) as fertilizer should contain only 1000 g/ton of Cr (on dry wt) (DER SCHWEIZERISCHE BUNDESRAT, 1981). In the Federal Republic of Germany the recommended limits are 800 to 1200 ppm of Cr in sewage sludge (dry weight) and 200 to 300 ppm for compost (HOFFMANN, 1980; see also Chapter I.7).

For agricultural soil a limit of 100 mg/kg has been recommended (FURRER et al., 1980). In Switzerland a threshold value of 75 mg Cr/kg was fixed (DER SCHWEIZERISCHE BUNDESRAT, 1986). It should be realized that some natural soils may contain higher quantities of chromium (BOWEN, 1979). In the Federal Republic of Germany a concentration of 100 mg/kg is tolerated in soil and a loading of 4 to 6 kg chromium per hectar and year has been calculated as limit for average agriculturally used soil (HOFFMANN, 1980; see also Chapters I.7 and I.20b).

In spite of the generally recognized role of chromium as an essential trace element, it is not allowed in any country to add chromium compounds to foodstuffs or medicine (KIEFFER, 1979). Therefore, up to now no limiting values have been established. However, tentatively a dietary allowance for adults of 50–200 µg chromium per day has been recommended (NAS, 1980).

References

Anderson, R. A. (1981), *Nutritional Role of Chromium. Sci. Total Environ. 17*, 13–28.
Anderson, R. A. (1987), *Stress and Trace Metal Deficiency. Proceedings Second Nordic Symposium on Trace Elements in Human Health and Disease, Odense*, No. 26, pp. 66–69. World Health Organization (WHO), Copenhagen.
Andersson, A. (1977), *The Distribution of Heavy Metals in Soils and Soil Material as Influenced by the Ionic Radius. Swed. J. Agric. Res. 7*, 83–97.
Anger, G., Halstenberg, J., Hochgeschwender, K., Vecker, G., Korallus, V., Knopf, H., Schmidt, P., and Ohlinger, M. (1986), *Chromium Compounds*, in: *Ullmann's Encyclopedia of Industrial Chemistry*, 5th Ed., Vol. A7, pp. 67–97. VCH Verlagsgesellschaft, Weinheim-Deerfield Beach/Florida-Basel.
Anliker, R. (1978), *Proceedings Scientific Basis for the Ecotoxicological Assessment of Environmental Chemicals*, Wien. Cited in *Chem. Rundsch. 31*, No. 47.
Arslan, P., and Beltrame, M. (1985), *Intracellular Chromium Reduction. 4th International Workshop on in vitro Toxicology*, Crieff, Scotland, Sept. 1985 (Abstract 1.13).
Bang-Pedersen, N. (1982), Chapter 11: *The Effects of Chromium on the Skin*, in: Langård, S. (ed.): *Biological and Environmental Aspects of Chromium, Elsevier Biomedical 1982*, pp. 249–275. Elsevier, Amsterdam.
Baetjer, A. M., et al. (1974), *Panel on Chromium: Medical and Biological Effects of Environmental Pollutants: Chromium*. National Academy of Sciences, Washington, D.C.

Bartlett, R., and James, B. (1979), *Behavior of Chromium in Soils: III. Oxidation. J. Environ. Qual. 8* (1), 31–35.

Batley, G. E., and Matousek, J. P. (1980), *Determination of Chromium Speciation in Natural Waters by Electrodeposition on Graphite Tubes for Electrothermal Atomization. Anal. Chem. 52*, 1570–1574.

Bertrand, D., and de Wolf, A. (1986), *Nécessité de l'oligo-élément chrome pour la culture de la pomme de terre. C.R. Acad. Sci. Ser. D, Paris 266*, 1494–1495.

Beyersmann, D. (1987), *Interactions with Genetic Materials. Proceedings Second Nordic Symposium on Trace Elements in Human Heath and Disease, Odense. Environ. Health 26*, 55–60. World Health Organization (WHO), Copenhagen.

Beyersmann, D., and Köster, A. (1987), *On the Role of Trivalent Chromium in Chromium Genotoxicity. Toxicol. Environ. Chem. 14*, 11–22.

Bianchi, V., and Levis, A. G. (1984), *Mechanisms of Chromium Genotoxicity. Toxicol. Environ. Chem. 9*, 1–25.

Bianchi, V., and Levis, A. G. (1987), *Recent Advances in Chromium Genotoxicity. Toxicol. Environ. Chem. 15*, 1–24.

Bianchi, V., and Levis, A. G. (1988), *Review of Genetic Effects and Mechanisms of Action of Chromium Compounds. Sci. Total Environ. 71*, 351–355.

Bigaliev, A. B., Spak, N. C., and Smagulov, A. S. (1979), *Dokl. Akad. Nauk UdSSR 245* (5), 1234–1236.

Bowen, H. J. M. (1979), *Environmental Chemistry of the Elements.* Academic Press, London-New York-Toronto-Sydney-San Francisco.

Braun, B. (1974), *Wirkung von chromhaltigen Gerbereischlämmen auf Wachstum und Chromaufnahme bei verschiedenen Nutzpflanzen*, Dissertation, University of Bonn.

Bronzetti, G., Vellosi, R., Galli, A., Nieri, R., Corsi, C., Cundari, E., Paolini, M., and del Carratore, R. (1986a), *Mutagenicity of Complex Mixtures Used in Tannery. Toxicol. Environ. Chem. 13*, 95–101.

Bronzetti, G., Galli, A., Boccardo, P., Vellosi, R., del Carratore, R., Sabbioni, E., and Edel, J. (1986b), *Genotoxicity of Chromium in vitro on Yeast: Interaction with DNA. Toxicol. Environ. Chem. 13*, 103–111.

Bruland, K. W. (1983), *Trace Elements in Sea-Water. Chem. Oceanogr. 8*, 198–200.

Bundesgesundheitsblatt (1984) *27*, No. 3, (March 1984). BGA (Bundesgesundheitsamt) Berlin.

Cary, E. C., Allaway, W. H., and Olson, O. E. (1977a), *Control of Chromium Concentrations in Food Plants 1. Absorption and Translocation of Chromium by Plants. J. Agric. Food Chem. 25* (2), 300–304.

Cary, E. C., Allaway, W. H., and Olson, O. E. (1977b), *Control of Chromium Concentrations in Food Plants 2. Chemistry of Chromium in Soils and its Availability to Plants. J. Agric. Food Chem. 25* (2), 305–309.

Cooper, J. A., Blackwell, L. F., and Buckley, P. D. (1984), *Chromium(III) Complexes and Their Relationship to the Glucose Tolerance Factor. Part II. Structure and Biological Activity of Amino Acid Complexes. Inorg. Chim. Acta 92*, 23–31.

CRC Handbook of Chemistry and Physics (1979–1980), West, R. C., and Astle, M. J. (eds.) 60th Ed., p. F-200. CRC Press, Boca Raton, Florida.

Cupo, D. J., and Wetterhahn, K. E. (1984), *Repair of Chromate-induced DNA Damage in Chick Embryo Hepatocytes. Carcinogenesis 5*, 1705–1708.

Cupo, D. J., and Wetterhahn, K. E. (1985a), *Binding of Chromium to Chromatin and DNA from Liver and Kidney of Rats Treated with Sodium Dichromate and Chromium(III) Chloride in vivo. Cancer Res. 45*, 1146–1151.

Cupo, D. J., and Wetterhahn, K. E. (1985b), *Modification of Chromium(IV)-induced DNA Damage by Glutathione and Cytochrome P-450 in Chicken Embryo Hepatocytes. Proc. Natl. Acad. Sci. USA 82*, 6755–6759.

Delpech, M. A., and Hillairet, M. (1869), *Accidents auxquels sont soumis les ouvriers. Ann. Hyg. Publique Med. Leg. 31*, 5–30.

Der Schweizerische Bundesrat (1975), *Verordnung über Abwassereinleitungen vom 8. Dez. 1975.* Systematische Sammlung des Bundesrechts No. 814.225.21.
Der Schweizerische Bundesrat (1981), *Klärschlammverordnung vom 8. April 1981.* Systematische Sammlung des Bundesrechts No. 814.225.23.
Der Schweizerische Bundesrat (1986), *Verordnung über Schadstoffe im Boden vom 9. Juni 1986.* Systematische Sammlung des Bundesrechts No. 814.12.
Dewirtz, A. P. (1929), *Über Kaliumchromatgeschwüre. Dermatol. Wochenschr.* 89 (27), 1801–1805.
Diab, M., and Soedermark, R. (1972), *Nucl. Med.* 11, 419–429
Downing, J. H., Deeley, P. D., and Fichte, R. M. (1986), *Chromium and Chromium Alloys*, in: *Ullmann's Encyclopedia of Industrial Chemistry*, 5th Ed., Vol. A7, pp. 43–65. VCH Verlagsgesellschaft, Weinheim-Deerfield Beach/Florida-Basel.
Duckett, S. (1986), *Abnormal Deposits of Chromium in the Pathological Human Brain. J. Neurol. Neurosurg. Psychiatry 49,* 296–301.
EPA (US Environmental Protection Agency) (1984), *Health Assessment Document for Chromium,* Final Report. Washington, D.C.
Fenke, K. (1977), *Die Chromaufnahme durch Kulturpflanzen bei Verwendung chromhaltiger Düngemittel,* Dissertation, University of Bonn.
Fichte, R., and Franke, H. (1975), *Chromium and Ferrochrome* (in German), in: *Ullmanns Encyklopädie der technischen Chemie,* 4th Ed., Vol. 9, pp. 589–623. Verlag Chemie, Weinheim.
Fishbein, L. (1987), *Trace and Ultra Trace Elements in Nutrition: An Overview, I. Chromium and Nickel. Toxicol. Environ. Chem.* 14, 84–88.
Fleischer, M. (1986), *Chromium,* in: Angerer, J., and Schaller, K. H. (eds.): *DFG Analysis of Hazardous Substances in Biological Materials,* Vol. 2, pp. 97–115. VCH Verlagsgesellschaft, Weinheim-Deerfield Beach/Florida-Basel.
Fregert, S. and Rorsman, H. (1964), *Allergy to Trivalent Chromium. Arch. Dermatol.* 90, 4–6.
Fregert, S., and Rorsman, H. (1965), *Patch Test Reactions to Basic Chromium(III) Sulfate. Arch. Dermatol.* 91, 233–234.
Furrer, O. J., Keller, P., Häni, H., and Gupta, S. K. (1980), *Schadstoffgrenzwerte – Entstehung und Notwendigkeit,* EAS-Seminar *Landwirtschaftliche Verwertung von Abwässerschlämmen,* Basel, 24.–28. Sept. Eidgenössische Forschungsanstalt für Agrikulturchemie und Umwelthygiene, Liebefeld-Bern, Switzerland.
Gale, T. F., and Bunch, J. D. (1979), *Teratology 19,* 81–86.
Galli, A., Boccardo, P., and Bronzetti, G. (1988), *Nitrilotriacetic Acid Effects on the Genetic Activity Induced by Chromium Chloride and Sodium Chromate in Saccharomyces cerevisiae. Toxicol. Environ. Chem.* 17 (1), 11–17.
Gauglhofer, J., (1986), *Environmental Aspects of Tanning with Chromium. J. Soc. Leather Technol. Chem.* 70, 11–13.
Glaser, U., Hochrainer, D., Kloeppel, H., and Kuhnen, H. (1985), *Low Level Chromium(VI) Inhalation Effects on Alveolar Macrophages and Immune Functions in Wistar Rats. Arch. Toxicol.* 57, 250–256.
Gmelin (1962), *Handbook of Inorganic Chemistry,* Part A, Syst. No. 52. *Chromium.* Springer Verlag, Heidelberg.
Goodgame, D. M., and Joy, A. M. (1986), *Relatively Long-lived Chromium(V) Species are Produced by the Action of Glutathione on Carcinogenic Chromium(VI). J. Inorg. Biochem.* 26, 219–224.
Gregus, Z., and Klaassen, C. D. (1986), *Disposition of Metals in Rats: A Comparative Study of Fecal, Urinary, and Biliary Excretion and Tissue Distribution of Eighteen Metals. Toxicol. Appl. Pharmacol.* 85, 24–38.
Guthrie, B. E. (1982), Chapter 6: *The Nutritional Role of Chromium,* in: Langård, S. (ed.): *Biological and Environmental Aspects of Chromium, Elsevier Biomedical, 1982,* pp. 117–148. Elsevier, Amsterdam.

Hamilton, J. W., and Wetterhahn, K. E. (1988), *Chromium*, in: Seiler, H. G., Sigel, H., and Sigel, A. (eds): *Handbook on Toxicity of Inorganic Compounds*, pp. 239–250. Marcel Dekker, New York.

Harzdorf, A. Cl. (1987), *Analytical Chemistry of Chromium Species in the Environment, and Interpretation of Results. Int. J. Environ. Anal. Chem. 29*, 249–261.

Hauptverband der gewerblichen Berufsgenossenschaften (1987), *Betriebswacht 1987*. Universum Verlagsanstalt, Wiesbaden.

Haylock, S. J., Buckley, P. D., and Blackwell, L. F. (1983), *The Relationship of Chromium to the Glucose Tolerance Factor II. J. Inorg. Biochem. 19*, 105–117.

Herfeld, H. (1974), *Können und müssen die behördlichen Vorschriften hinsichtlich des Chromgehaltes im Abwasser eingehalten werden? Leder 25* (7), 134–141.

Hoffmann, C. (1980), *Proceedings Verwendung von Müll- und Müllklärschlammkomposten in der Landwirtschaft*, pp. 12–28. Gottlieb Duttweiler-Institut, Rüschlikon/Zürich.

Hollemann, A. F., und Wiberg, E. (1976), *Lehrbuch der anorganischen Chemie*, 81st–90th Ed., pp. 867–882. Walter de Gruyter, Berlin.

Hooftman, R. N. (1987), *Proceedings of the XVIIth Annual Meeting of the European Environmental Mutagen Society*, Zürich.

Hubert, J. (1979), in: Shapcott, C., and Hubert, J. (eds.): *Developments in Nutrition and Metabolism*, Vol. 2, *Chromium in Nutrition and Metabolism*, pp. 15–30. Elsevier/North-Holland Biomedical Press, Amsterdam-New York-Oxford.

Huffman, E. W. D. (1973), *Ph. D. Thesis*, Cornell University, Ithaka, N.Y.

Huffman, E. W. D., and Allaway, W. H. (1973), *Growth of Plants in Solution Culture Containing Low Levels of Chromium. Plant Physiol. 52*, 72–75.

IARC (International Agency for Research on Cancer) (1980), *Monographs on the Evaluation of the Carcinogenic Risk of Chemicals to Humans*, Vol. 23: *Some Metals and Metallic Compounds*. IARC, Lyon.

IARC (International Agency for Research on Cancer) (1982), *Monographs on the Evaluation of the Carcinogenic Risk of Chemicals to Humans*, Supplement 4, pp. 91–93. IARC, Lyon.

IAWR (1973), *Pollution of the River Rhine and Production of Drinking Water* (in German). *Gas Wasser Abwasser 53* (6), 201.

Jennette, K. W. (1979) *Chromate Metabolism in Liver Microsomes. Biol. Trace Element Res. 1*, 55–62.

Jennette, K. W. (1982), *Microsomal Reduction of the Carcinogen Chromate, Producing Chromium(V). J. Am. Chem. Soc. 104*, 874–875.

Johanson, A., Wiernik, A., Jarstrand, C., and Camner, P. (1986), *Rabbit Alveolar Macrophages after Inhalation of Hexa- and Trivalent Chromium. Environ. Res. 39*, 372–385.

Jung, K. D. (1973), cited by Förstner, U., and Wittmann, G. T. W. (1979), *Metal Pollution in the Aquatic Environment*. Springer Verlag, Berlin-Heidelberg.

Kawanishi, S., Inoue, S., and Sano, S. (1986), *Mechanism of DNA Cleavage Induced by Sodium Chromate(VI) in the Presence of Hydrogen Peroxide. J. Biol. Chem. 261*, 5952–5958.

Kick, H., and Braun, B. (1977), *Wirkung von chromhaltigen Gerbereischlämmen auf Wachstum und Chromaufnahme bei verschiedenen Nutzpflanzen. Landwirtsch. Forsch. 30* (2), 160–173.

Kilburn, K. H., Warshaw, R., Finklea, J. F., and Sunderman Jr., F. W. (1987), *Proceedings Second Nordic Symposium on Trace Elements in Human Health and Disease, Odense*. World Health Organization (WHO), Copenhagen.

Kieffer, F. (1979), *Spurenelemente steuern die Gesundheit. Sandoz Bull. 52*, 18–19.

Knudsen, I. (1980), *The Mammalian Spot Test and its Use for the Testing of Potential Carcinogenicity of Welding Fume Particles and Hexavalent Chromium. Acta Pharmacol. Toxicol. 47*, 66–70.

Koemann, J. H., ten Holder, V. J. H. M., and Horgendoorn, A. S. (1977), *Proceedings 2nd International Symposium on Aquatic Pollutants*, Nordwijkerhout, The Netherlands.

Korallus, U. (1986), *Chromium Compounds: Occupational Health, Toxicological and Biological Monitoring Aspects. Toxicol. Environ. Chem. 12*, 47–59.

Langård, S. (1980), Chapter 4: *Chromium*, in: Waldron, H. A. (ed.): *Metals in the Environment*, pp. 111–132. Academic Press, London.
Langård, S. (1982), Chapter 7: *Absorption, Transport and Excretion of Chromium in Man and Animals*, in: Langård, S. (ed.): *Biological and Environmental Aspects of Chromium*, Elsevier Biomedical 1982, pp. 149–169. Elsevier, Amsterdam.
Langård, S., and Norseth, T. (1986), Chapter 8: *Chromium*, in: Friberg, L., Norberg, G. F., and Vouk, V. B. (eds.): *Handbook on the Toxicology of Metals*, Vol. II, pp. 185–210. Elsevier/North-Holland Biomedical Press, Amsterdam-New York-Oxford.
Levis, A. G., and Bianchi, V. (1982), Chapter 6: *Mutagenic and Cytogenetic Effects of Chromium Compounds*, in: Langård, S. (ed.): *Biological and Environmental Aspects of Chromium*, Elsevier Biomedical 1982, pp. 171–208. Elsevier, Amsterdam.
Levis, A. G., and Majone, F. (1979), *Cytotoxic and Clastogenic Effects of Soluble Chromium Compounds on Mammalian Cell Cultures. Br. J. Cancer* **40**, 523–533.
Levy, L. S., and Venitt, S. (1986), *Carcinogenicity and Mutagenicity of Chromium Compounds: The Association between Bronchial Metaplasia and Neoplasia. Carcinogenesis* **7**, 831–835.
Lewis, R. J., and Tatken, R. L. (eds.) (1980), *Registry of Toxic Effects of Chemical Substances*, 1979 Ed., Vol. 1, p. 426. National Institute for Occupational Safety and Health, Washington, D.C.
Mangusson, B., Blohm, S. G., Freget, S., Hjorth, N., Høvding, G., Pirilä, V., and Skog, E. (1968), *Routine Patch Testing. Acta Derm. Venereol.* **48**, 110–114.
MAK (1987), *Maximum Concentrations at the Workplace*, Report No. XXIII DFG. VCH Verlagsgesellschaft, Weinheim-Basel-Cambridge-New York.
Maloof, C. C. (1955), *The Use of Edathamil Calcium in Treatment of Chromium Ulcers of Skin. Arch. Ind. Health* **11**, 123–125.
Maxwell, D. G. (1985), *Chromium, Strong Recovery Continues. Eng. Min. J.* (3), 73–79.
Merian, E. (1984), *Introduction of Environmental Chemistry and Global Cycles of Chromium, Nickel, Cobalt, Beryllium, Arsenic, Cadmium, and Selenium, and Their Derivatives. Toxicol. Environ. Chem.* **8**, 9–38.
Mertz, W. (1969), *Physiol. Rev.* **49**, 163–239.
Mertz, W., and Schwartz, K. (1959), *Relation of Glucose Tolerance Factor to Impaired Glucose Tolerance in Rats on Stock Diets. Am. J. Physiol.* **196**, 614–618.
Mertz, W., Roginski, E. E., Feldman, F. J., and Thurman, D. E. (1969), *Dependence of Chromium Transfer into the Rat Embryo on the Chemical Form. J. Nutr.* **99**, 363–367.
Moeschlin, S. (1972), *Klinik und Therapie der Vergiftungen*, 5th Ed., pp. 109–112. Georg Thieme Verlag, Stuttgart.
Müller, H. C. (1980), *Proceedings Interpretation of Oecotoxicologic Test Results*, SECOTOX, Antibes, France. Cited in *Chem. Rundsch.* **33**, No. 48.
NAS (1980), *Recommended Dietary Allowances*, 9th revised Ed. National Academy of Sciences, Washington, D.C.
Nater, J. P. (1982), *Mogelijke oorzaken van chromaateczeem. Ned. Tijdschr. Geneeskd.* **106**, 1429–1431.
National Materials Advisory Board (1978), *Contingency Plans for Chromium Utilization*. National Research Council, National Academy of Sciences, Washington, D.C.
Nishio, A., and Uyeki, E. M. (1985), *Inhibition of DNA Synthesis by Chromium Compounds. J. Toxicol. Environ. Health* **15**, 237–244.
Obiols, J., Devesa, R., Garcia-Barrio, J., and Serra, J. (1986), *Speciation of Heavy Metals in Suspended Particulates in Urban Air. Toxicol. Environ. Chem.* **13**, 121–128.
Obiols, J., Devesa, R., Garcia-Barrio, J., and Serra, J. (1987), *Speciation of Chromium in Waters by Coprecipitation-AAS. Int. Environ. Anal. Chem.* **30** (3), 197–207.
Offenbacher, E. G., Spencer, H., Dowling, H. J., and Pi-Sunyer, F. X. (1986), *Metabolic Chromium Balances in Men. Am. J. Clin. Nutr.* **44**, 77–82.
Ottaway, J. M., and Fell, G. S. (1986), *Determination of Chromium in Biological Materials. Pure Appl. Chem.* **58/12**, 1707–1720.

Peres, G. (1980), *Proceedings Interpretation of Oecotoxicologic Test Results*, SECOTOX, Antibes, France. Cited in *Chem. Rundsch. 33*, No. 48.
Peterson, P. J., and Girling, C. A. (1981), in: Lepp, N. W.: *Effect of Heavy Metal Pollution on Plants*, Vol. 1: *Effects of Trace Metals on Plant Function.* Applied Science Publ., London.
Petrilli, F., and De Flora, S. (1978), *Metabolic Deactivation of Hexavalent Chromium Mutagenicity. Mutat. Res. 54*, 139–147.
Petrilli, F., Camoirano, A. Bennicelli, C., Zanacchi, P., Astengo, M., and De Flora, S. (1985), *Specification and Inducibility of the Metabolic Reduction of Chromium(VI) Mutagenicity by Subcellular Fractions of Rat Tissues. Cancer Res. 45*, 3179–3187.
Petrilli, F. L., Rossi, A. G., Camoirano, A., Romano, M., Serra, D., Bennicelli, C., De Flora, A., and De Flora, S. (1986), *Metabolic Reduction of Chromium by Alveolar Macrophages and its Relationship to Cigarette Smoke. J. Clin. Invest. 77*, 1917–1924.
Ping, L., Matsumoto, K., and Fuwa, K. (1983), *Anal. Chim. Acta 147*, 205–212.
Saager, R. (1984), *Metallic Rawmaterials Dictionary* (in German), pp. 113–117. Bank von Tobel, Zürich, Switzerland.
Samitz, M. H., and Shrager, J. (1966), *Patch Test Reactions to Hexavalent and Trivalent Chromium Compounds. Arch. Dermatol. 94*, 304–306.
Sanderson, C. J. (1976), *The Uptake and Retention of Chromium by Cells. Transplantation 21*, 526–530.
Scharrer, K., and Schropp, W. (1935), *Die Wirkung von Chromi- und Chromat-Ion auf Kulturpflanzen. Z. Pflanzenernähr. Düng. Bodenkde. 37.* 137–149.
Schlatter, C., and Kissling, U. (1973), *Akute tödliche Vergiftung mit Bichromat. Beitr. Gerichtl. Med. 30*, 382–388.
Schliebs, R. (1980), *Technical Chemistry of Chromium* (in German). *Chem. Unserer Zeit 14*, No. 1, 13–17.
Schweizerisches Lebensmittelbuch (1972), 5th Ed., Vol. 2, Chap. 27: *Trinkwasser und Mineralwasser.*
Seeber, H., Fikentscher, R., and Roseburg, B. (1976), *Geruchs- und Geschmacksstörungen bei Chromfarbenarbeitern. Z. Gesamte Hyg. Ihre Grenzgeb. 22*, 820–822.
Seel, F. (1979), *Grundlagen der analytischen Chemie*, 7. Ed., Verlag Chemie, Weinheim-New York.
Shivas, S. A. J. (1980a), *Factors Affecting the Oxidation State of Chromium Disposed in Tannery Wastes. J. Am. Leather Chem. Assoc. (JALCA) 75*, 42–48.
Shivas, S. A. J. (1980b), *The Effects of Trivalent Chromium from Tannery Wastes on Plants. J. Am. Leather Chem. Assoc. (JALCA) 75*, 288–299.
Stern, R. H. (1986), *Occupational Exposures, Delivered Doses and Exposure Limits for Chromium and Nickel. Toxicol. Environ. Chem. 12*, 185–193.
Stern, R. H. (1987), *Role of Speciation in Metal Toxicity. Proceedings Second Nordic Symposium on Trace Elements in Human Health and Disease, Odense. Environ. Health 26,* 3–7. World Health Organization (WHO), Copenhagen.
Sticher, H. (1978), *Chrom- and Nickeldynamik in Serpentinböden. Mitt. Dtsch. Bodenkd. Ges. 27*, 239–246.
Sticher, H. (1980), *Heavy Metals in Soils and Their Role in the Food Chain* (in German). *Lebensm. Wiss. Technol. 13* (3), 3–9.
Strik, J. J. T. W. A., de Iongh, H. H., van Rijn van Alkemade, J. W. A., and Wuite, T. P. (1975), in: Koeman, J. H., and Strik, J. J. T. W. A. (eds.): *Sublethal Effects of Toxic Chemicals on Aquatic Animals*, pp. 31–41. Elsevier Scientific Publishing Co., Amsterdam-Oxford-New York.
Sunderman Jr., F. W. (1986), *Kinetics and Biotransformation of Nickel and Chromium*, in: *Health Hazards and Biological Effects of Welding Fumes and Gases*, pp. 229–247. Excerpta Medica, Amsterdam.
Sunderman Jr., F. W., et al. (1987), *Cobalt, Chromium, and Nickel Concentrations in Body Fluids of Patients with Porous-coated Knee or Hip Prostheses. J. Orthop. Res.*, in press.
SUVA (Schweizerische Unfallversicherungsanstalt, Luzern) (1980): *Zulässige Werte am Arbeitsplatz – Gesundheitsschädigende Stoffe (MAK)*, Formular 1903.d.

Sykes, R. L., Corning, D. R., and Earl, N. J. (1981), *The Effect of Soil-chromium(III) on the Growth and Chromium Absorption of Various Plants. J. Am. Leather Chem. Assoc. (JALCA)* **76**, 102–125.
Toepfer, E. W., Merz, W., Polansky, M. M., Roginsky, E. E., and Wolf, W. R. (1977), *Preparation of Chromium-containing Material of Glucose Tolerance Factor Activity from Brewer's Yeast Extracts and by Synthesis. J. Agric. Food Chem.* **25**, 162–168.
Tscherwitschke, R. (1979), *Abwasserfreier Chrom-Trakt in einem modernen Kupfer-Nickel-Chrom-Automaten. Galvanotechnik* **70**, No. 7, 620–625.
Tuman, R. W., Bibo, J. T., and Doisy, R. J. (1978), *Comparison and Effects of Natural and Synthetic Glucose Tolerance Factor in Normal and Genetically Diabetic Mice. Diabetes* **27**, 49–56.
Van Schoor, O., and Deelstra, H. (1986), *Proceedings Trace Element Analytical Chemistry in Medicine and Biology*, (Neuherberg) FRG. Walter de Gruyter, Berlin.
Van Steertegen, M., and van de Vel, A. (1986), (State University of Ghent, Belgium), *Contribution SECOTOX Conference, Rome.*
Veillon, C., Guthrie, B., and Wolf, W. (1980), *Anal. Chem.* **52**, 457.
Veillon, C., Patterson, K. Y., and Bryden, N. A. (1982a), *Anal. Chim. Acta* **136**, 233–241.
Veillon, C., Patterson, K. Y., and Bryden, N. A. (1982b), *Clin. Chem.* **28/11**, 2309–2311.
Vinson, J. A., and Hsiao, K.-H. (1985), *Comparative Effect of Various Forms of Chromium on Serum Glucose: An Assay for Biologically Active Chromium. Nutr. Rep. Int.* **32**, 1–7.
Wälchli, O., Ott, R., Hugner, R., Graf, E., and Lieberherr, B. (1979), *Selektive Chromat-Elimination im Lagerplatz-Drainage-Wasser eines Holzimprägnierwerkes. Gas Wasser Abwasser* **59**, 410–412.
Warren, G., Schultz, P. B., Croft, D., Bennet, K., Abbot, E. H., and Roget, S. (1981), *Mutat. Res.* **90**, 111.
Weber, H. (1983), *Long-term Study of the Distribution of Soluble Chromate-51 in the Rat after a Single Intra-tracheal Administration. J. Toxicol. Environ. Health*, **11**, 749–764.
Welz, B. (1985), *Atomic Absorption Spectrometry*, pp. 279–281, 348 ff. VCH Verlagsgesellschaft, Weinheim-Deerfield Beach/Florida-Basel.
Willie, S. N., Sturgeon, R. E., and Berman, S. S. (1983), *Anal. Chem.* **55**, 981.
Wolf, W., Mertz, W., and Masironi, R. (1974), *Determination of Chromium in Refined and Unrefined Sugars by Oxygen Plasma Ashing Flameless Atomic Absorption. J. Agric. Food Chem.* **22**, 1037–1042.

Additional Recommended Literature

Aijar, Y., Norges, K. M., Floyd, R. A., and Wetterhahn, K. E. (1989), *Role of Chromium(V), Glutathione Thiyl Radical and Hydroxyl Radical Intermediates in Chromium(VI)-induced DNA Damage. Toxicol. Environ. Chem.* **22** (1–4), 135–148.
Antilla, S., Pääkö, P., Kokkonen, P., and Kalliomäki, P.-L. (1989), *Cadmium and Chromium in Human Lung Tissue in Relation to Smoking and Lung Cancer. Toxicol. Environ. Chem.* **22** (1–4), 91–95.
Beyersmann, D. (1989), *Biochemical Speciation in Chromium Genotoxicity. Toxicol. Environ. Chem.* **22** (1–4), 61–67.
Bronzetti, G. L., and Galli, A. (1989), *Influence of NTA on the Chromium Genotoxicity. Toxicol. Environ. Chem.* **23** (1–4), 101–104.
Buttner, B., and Beyersmann, D. (1989), *Electrophoretic and Liquid Chromatographic Characterization of Red Cell Membrane Proteins Modified by Chromate. Toxicol. Environ. Chem.* **22** (1–4), 39–49.
Davis, J. L. (1989), *Transport of Ionic Solutes in Groundwater. Proceedings Workshop on Metal Speciation and Transport in Groundwater, Jekyll Island, Georgia*, in press.

Fish, W. (1989), *Redox Reactions in Groundwater. Proceedings Workshop on Metal Speciation and Transport in Groundwater, Jekyll Island, Georgia*, in press.

de Flora, S., Camoirano, A., Serra, D., and Bennicelli, C. (1989), *Genotoxicity and Metabolism of Chromium Compounds. Toxicol. Environ. Chem. 19* (3+4), 153–160.

Gava, C., Costa, R., Zordan, M., Venier, P., Bianchi, V., and Lewis, G. A. (1989), *Induction of Gene Mutations in Salmonella and Drosophila by Soluble Cr(VI)-compounds: Synergistic Effects of Nitrilotriacetic Acid. Toxicol. Environ. Chem. 22* (1–4), 27–38.

Johnson, C. A., and Sigg, L. (1988), *Chromium Cycling in Lakewaters. Proceedings International Conference Trace Metals in Lakes, Hamilton, Ontario Canada. Sci. Total Environ.*, in press; and *EAWAG Annual Report 1989*, pp. 4–28/4–29.

Krivan, V. (1988), *AAS Techniques for the Speciation of Metals. Proceedings 3rd IAEAC Workshop on Toxic Metal Compounds (Interrelation between Chemistry and Biology), Folonica, Italy*, in press.

Langård, S. (1989), *Basic Mechanisms of the Carcinogenic Action of Chromium: Animal and Human Data. Toxicol. Environ. Chem. 24* (1+2), 1–7.

Lewalter, J., and Korallus, U. (1989), *The Significance of Ascorbic Acid and Glutathione for Chromate Metabolism in Man. Toxicol. Environ. Chem. 24* (1+4), 25–33.

McLeod, C. W., Jian, W., and Cox, A. G. (1990), *Flow Injection as an Aid to Element Speciation (e.g., to monitor Cr(III) and Cr(VI) by ICP-AES). Fresenius Z. Anal. Chem.*, in press.

Menné, T., and Nieboer, E. (1989), *Metal Contact Dermatitis: a Common and Potentially Debilitating Disease. Endeavour New Ser. 13* (3), 117–122.

Neidhard, B., and Backes, U. (1989), *Reverse FIA for On-line Determination of Chromate in the Aqueous Environment. Proceedings Sixth International Conference on Heavy Metals in the Environment, Geneva*, Vol. 1, pp. 598–601. CEP Consultants Ltd., Edinburgh.

Neidhard, B., Herwald, S., Lippmann, C., and Straka-Emden, B. (1989), *Biosampling and GFAAS: A Possibility for Determination of Chromates in Airborne Particulates. 11th International Symposium on Microchemical Techniques, Wiesbaden. Fresenius Z. Anal. Chem. 334* (7), 660.

Pohland, C. (1988), *Metal Speciation and Mobility as Influenced by Landfill Disposal Practices. Proceedings Workshop on Metal Speciation and Transport in Groundwater, Jekyll Island, Georgia*, in press.

Rai, L. C., and Dubey, S. K. (1988), *Impact of Chromium and Tin on the Cyanobacterium Anabaena doliolum: Interaction with Ca, Mg, Mn, Ni, Co and Zn. Proceedings International Conference Trace Metals in Lakes, Hamilton, Ontario, Canada. Sci. Total Environ.*, in press.

Rai, A. K., Bhardwaj, A. K., and Mehrotra, R. C. (1988), *Hazardous Metals in Jal Mahal Lake, Jaipur. Proceedings International Conference Trace Metals in Lakes, Hamilton, Ontario, Canada. Sci. Total Environ.*, in press.

Schwarzbach, E. (1989), *Chromium(VI)-compounds in the Workplace Area: Problems of Correct Sampling and Determination in Complex Matrices. Int. J. Environ. Anal. Chem.*, in press.

Sivalingham, P. M. (1989), *Enzymatic and Cellular Level Effects of Chromium. Toxicol. Environ. Chem. 19* (3+4), 119–123.

Smith, St., Peterson, P. J., and Kwan, K. H. M. (1989), *Chromium Accumulation, Transport and Toxicity in Plants. Toxicol. Environ. Chem. 24* (4), 241–251.

Stockton, R. A., Friederich, N. J., McClendon, E. L. S., and Priebe, S. R. C. (1987), *Speciation of Chromium Compounds Using ICP as an Element-specific Detection for HPLC. Proceedings Fifth International Conference on Heavy Metals in the Environment, New Orleans*, Vol. 1, pp. 328–331. CEP Consultants Ltd., Edinburgh.

Trueb, L. (1988), *Chromium, an Essential Trace Element* (in German). *Neue Zürcher Zeitung, Forschung und Technik*, No. 21, p. 79 (27 January), Zürich.

WHO-ICPS International Programme on Chemical Safety (1988), *Chromium. Environ. Health Crit. 61*. World Health Organization, Geneva.

Wolf, T., Wiegand, H. J., and Ottenwäder, H. (1989), *Different Molecular Effects on Nucleotides by Interaction with Cr(III) and Cr(VI). Toxicol. Environ. Chem. 23* (1–4), 1–8.

II.8 Cobalt

GERHARD N. SCHRAUZER, La Jolla, California, USA

1 Introduction

Cobalt is a component of vitamin B_{12}, and in this form it is essential for all higher animals and for man. As inorganic cobalt is required for bacterial vitamin B_{12} synthesis in the rumen, adequate amounts must be present in the feed of ruminants. Although the amounts of cobalt required are very small, deficiency syndromes occur in certain regions of Australia where the soil is severely cobalt deficient. In these regions, cobalt as the oxide, carbonate, chloride, acetate, or sulfate is added to the feed, to fertilizers, salt-licks, or in the form of pellets which are placed into the rumen. Although cobalt deficiency is not yet wide-spread, it is likely to become a problem in the future, as the natural cobalt content of soils is low, and depletion of cobalt occurs through agricultural practices and natural cobalt transport processes. Cobalt is transported through the rivers to the oceans. The levels of cobalt in ocean water are low, however, as the element is largely removed by coprecipitation with other metal hydroxides. Accordingly, the average cobalt content of deep-sea clays is 74 ppm and thus more than ten times greater than that of granite or sandstone. As terrestrial supplies of cobalt minerals are not unlimited, cobalt deficiency could eventually develop on a much greater scale. Therefore, detailed studies of cobalt availability and geochemical transport of the element should be encouraged.

On the other hand, because of mining activities and its wide-spread industrial uses, cobalt also belongs to the metals posing potential dangers due to excessive exposures. At risk are primarily metal workers. Exposures to cobalt-containing dusts cause damage to lungs, heart, and skin. Cobalt also belongs to the group of occupational carcinogens and is considered dangerous under conditions that may be normally encountered at the work-place. Thus, extreme caution is required for the handling of cobalt containing metallic objects; in experimental animals, malignant tumors developed on implantation of metallic cobalt.

2 Physical and Chemical Properties, and Analytical Methods

Cobalt (atomic number 27, atomic mass 58.93) belongs to the group of transition elements. Its atomic nucleus normally contains one stable isotope, ^{59}Co. However, radioactive isotopes such as ^{60}Co, ^{57}Co, ^{58}Co have been detected in atomic fallout and in marine organisms exposed to fallout. Occasionally, ^{60}Co may be encountered even in steel produced from scrap metals, due to the industrial and medical uses of ^{60}Co as a convenient source of gamma radiation (1.17 MEV). ^{60}Co is produced from ^{59}Co on reaction with thermal neutrons. Its $t_{1/2}$ is 5.7 years; decay occurs according to the reaction: ^{60}Co → ^{60}Ni (β, γ). The half-lives of all other known artificial cobalt isotopes range from minutes to days. Because of its short half-life (270 days), ^{57}Co is preferred over ^{60}Co as a tracer in certain biological applications.

Metallic cobalt forms a lustrous, gray, strongly ferromagnetic solid with a density of 8.9 g/cm^3 (20 °C), a melting point of 1495 °C, and a boiling point of 3100 °C. It is not attacked by air or water at ambient temperature, slowly dissolved by dilute non-oxidizing acids, rapidly by concentrated nitric acid. On evaporation of the nitric acid solution, crystals of the nitrate, $Co(NO_3)_2 \cdot 6 H_2O$, are obtained. The nitrate itself decomposes on heating to yield the olive-green (Co(II) oxide, CoO. In contact with air, conversion of CoO to the black Co(II)/Co(III) oxide, Co_3O_4, occurs slowly at room temperature and more rapidly on heating to 700 °C in a stream of oxygen. At lower temperatures, CoO absorbs more oxygen to produce an oxide approaching the composition Co_2O_3. However, the pure Co(III) oxide has not been obtained thus far. Heating CoO with BaO in the presence of oxygen produces solids which formally are derivatives of the (unknown) Co(IV) oxide, CoO_2, e.g., $BaCoO_3$ ("$BaO \cdot CoO_2$").

Common oxidation states of cobalt are $+II$ and $+III$. However, compounds of Co(0), Co(+I), and Co(-I) are also known.

Aqueous solutions of Co(II) salts contain the pink hexaquo ion, $[Co(H_2O)_6]^{2+}$ in which the Co^{2+}-ion is octahedrally surrounded by 6 water molecules. In addition to the hydrates, numerous coordination complexes of Co(II) and of Co(III) are known. Hexammine-cobalt(II) chloride, $Co(NH_3)_6Cl_2$, is obtained by adding concentrated aqueous ammonia to solutions of $Co(H_2O)_6Cl_2$ under exclusion of air. Cobalt(II) salts often undergo characteristic color changes on heating or on acidification, caused by the displacement of ligands or changes of coordination geometry. The pink solution of $Co(H_2O)_6Cl_2$ reacts with hydrochloric acid to produce the blue complex ion $[CoCl_4]^{2-}$ in which the cobalt(II) ion is tetrahedrally coordinated.

Most Co(II) complexes are oxygen-sensitive. On reaction with O_2, some Co(II) compounds or -complexes initially form labile O_2-adducts which release molecular oxygen on heating, thus mimicking the behavior of iron porphyrin in hemoglobin. Oxygen-bridged, binuclear μ-peroxo- and μ-superoxo-dicobalt complexes with Co-O-O-Co linkages are also formed. Co(III) complexes are the terminal oxidation products of Co(II) complexes. These are almost invariably diamagnetic, kinetically inert, and thermodynamically more stable than those of Co(II). One such compound is $Co(NH_3)_6Cl_3$, hexammine-cobalt(III) chloride (also known as luteocobalt chloride because of its characteristic brown-yellow color). Other well-known Co(III) complexes are the yellow potassium hexacyano-cobaltate(III), $K_3[Co(CN)_6]$, and tris(ethylenediamine)cobalt(III) chloride, $[Co(en)_3]Cl_3$.

The facile formation of a yellow, sparingly soluble complex salt of the composition $K_3[Co(NO_2)_6]$ has been used as a simple qualitative test for cobalt ions in solution. A more sensitive test is based on the formation of the intensely blue cobalt aluminate, $CoO \cdot Al_2O_3$, a pigment also known as Thenard's Blue. A spot-test utilizing α-nitroso-β-naphthol allows detection of 0.05 micrograms of Co at a dilution of up to 1:1 000 000.

Methods for quantitative analysis include graphite-furnace atomic absorption spectrometry (AAS; see, for instance, SAMPSON et al. (1987) and SUNDERMAN et al. (1987)), inductively coupled plasma emission spectrometry (ICP-AES), neutron activation analysis (NAA; see, for instance, VERSIECK et al., 1978), ion chromatography (HAERDI, 1989), and electrochemical methods such as adsorption differential pulse voltammetry (ADPV; see, for instance, OSTAPCZUK et al., 1983). Older photometric methods are described in the literature (BURGER, 1973; FRIES and GETROST, 1977). For a recent comparative study of the most commonly employed methods in the analysis of biological materials see MILLER-IHLI and WOLF (1986), ANGERER and SCHALLER (1985), and Chapter I.4a. During sampling – for instance, when filtering wastewater or groundwater – there is a risk of cobalt losses, which must be avoided (McKINNEY, 1987).

3 Sources, Production, Important Compounds, Uses, Waste Products, and Recycling

Cobalt is a comparatively rare element, ranking 32nd in abundance on the earth's crust. It is found in sulfidic copper ores of Zaire, Africa, and in pyrites from Ontario, Canada. Additional deposits of cobalt containing ores are found in Marocco and Northern Simbabwe. Although deposits of cobalt containing ores are found in the USA, these are presently not utilized. Instead, imported ores or concentrates are processed, notably in Louisiana.

The most common cobalt minerals are the arsenide, $CoAs_{2-3}$ (smaltite), the arsenosulfide, CoAsS (cobaltine), and the sulfide, Co_3S_4 (linneite). Cobalt derives its name from the German "kobold", a goblin or evil spirit, as early miners were often confused and fooled by the high luster of cobalt ores.

Although compounds of cobalt were known already in antiquity and used for making ceramic glazes or colored glass, the element was first isolated and identified as such by H. BRANDT, in 1742.

Cobalt is prepared from sulfidic ores by partial roasting in the presence of fluxes. This produces a residue consisting of sulfides and arsenides of Co, Ni, and Cu. Heating of the residues in the presence of NaCl causes the metals to be converted into water-soluble chlorides from which the metals are obtained by electrolysis. Current annual world production of cobalt is in the order of 15 000 metric tons, corresponding to about 1% of the annual copper production. About 20% of the cobalt comes from recycling (SAAGER, 1984).

Cobalt is a component of the so-called superalloys used to make critical parts of jet engines, gas-turbines, and of other machines operating under stress at high temperatures. It is also a component of the so-called stellites. These alloys are composed of 50–60% cobalt, 30–40% chromium, and 8–20% tungsten and are valued for their extreme hardness, strength, and heat resistance. In addition, cobalt is a component of magnetic steels and aluminum alloys with superior ferromagnetic properties. In 1975, 75% of the total cobalt was used in alloys and for steelmaking (SIBLEY, 1977). Amorphous cobalt or iron films with terbium or gadolinium additions show interesting magnetic and magneto-optic properties and are therefore suitable for storage with high information density (TRUEB, 1987).

Combinations of the oxides of cobalt with those of aluminum and silicon are constituents of blue and green ceramic glazes and of pigments; cobalt is also used in the glass industry to impart blue colors and to mask the greenish tinge of glass or of porcellain caused by iron impurities. Cobalt salts and soaps are added in amounts of up to 0.25% to oil-, latex-, or butadiene-resin-based paints to accelerate the drying process. Cobalt salts are also added to polyester- and silicone resins to promote hardening. The average annual consumption of cobalt salts for these purposes in the USA is presently in the order of 1200 tons.

Cobalt compounds are useful chemical catalysts for the synthesis of fuels (Fischer-Tropsch process), the synthesis of alcohols and aldehydes from olefins, hydrogen and carbon monoxide at elevated temperatures and pressures ("oxo process", "hydroformylation"). They are also used in petroleum refining and the oxidation of organic compounds. In the oxo process cobalt carbonyl, $Co_2(CO)_8$, is employed or generated in situ. For the selective production of n-butanol from propylene, hydrogen and CO, an organophosphine-modified cobalt carbonyl complex is used as the catalyst. Cobalt salts are proven oxidation catalysts, e.g., for the production of terephthalic acid by the oxidation of p-xylene, and the manufacture of phenol by the oxidation of toluene.

The cobalt carbonyls typify complexes of cobalt in low oxidation states. Finely dispersed cobalt reacts with gaseous carbon monoxide under pressure to yield orange, sublimable, dinuclear dicobalt octacarbonyl, $Co_2(CO)_8$ (m.p. 51 °C, decomposition above 52 °C), a derivative of Co(0). This carbonyl is soluble in organic solvents and on gentle heating decomposes into a lower, tetranuclear carbonyl of composition $Co_4(CO)_{12}$. From $Co_2(CO)_8$ and H_2 under pressure, the highly toxic, foul-smelling, oxygen-sensitive gaseous hydridocobalt tetracarbonyl, $HCo(CO)_4$, is formed which can be condensed to a yellow liquid of m.p. -26 °C. The hydrogen atom in hydridocobalt tetracarbonyl is acidic and can be substituted by metals. Thus, $HCo(CO)_4$ can be regarded as a compound of Co($-$I). In the oxo process, cobalt carbonyls and hydridocarbonyls are catalytic intermediates in reactions with olefins, CO, and H_2.

Simple salts of Co($+$I) are unknown. However, this unusual oxidation state of cobalt is generated by the reduction of certain cobalt complexes, e.g., in derivatives of dimethylglyoximato cobalt and in vitamin B_{12}. The Co($+$I) ion in complexes of this type exhibits high nucleophilic reactivity and can react with conventional alkylating agents to yield organometallic complexes with Co-C bonds. Derivatives of vitamin B_{12} are called cobalamins. Vitamin B_{12} is produced by certain microorganisms. It is isolated from bacterial sources as the cyanide. Cyanocobalamin forms purple crystals of composition $C_{63}H_{88}CoN_{14}O_{14}P$, m.p. 210 – 220 °C (dec.). The cobalt ion is in the center of a ligand bearing some resemblance to the porphyrin system, the axial positions are occupied by coordinated cyanide and 5,6-dimethylbenzimidazole. The blue-green vitamin B_{12s} is formed on reduction of vitamin B_{12} with zinc/NH_4Cl or with other strong reducing agents. It reacts with methyl iodide to yield methylcobalamin. Methylcobalamin is also formed in vivo and is an intermediate carrier of methyl groups in bacterial methionine- and methane biosynthesis. In coenzyme B_{12} (5'-deoxyadenosylcobalamin), a 5'-deoxyadenosyl residue is attached to the central cobalt atom. Coenzyme B_{12}-dependent enzymes catalyze the conversion, e.g., of methylmalonyl-CoA to succinyl-CoA and the conversion of 1,2-diols to aldehydes. Vitamin B_{12r} is the Co(II) derivative of vitamin B_{12}. This cobalamin is yellow-brown in solution and oxygen-sensitive. In general, the chemistry of the cobalt atom in the cobalamins can be simulated with simpler cobalt complexes. The best-known vitamin B_{12} model compounds are the derivatives of bis(dimethylglyoximato)cobalt, also known as the "cobaloximes" (SCHRAUZER, 1976, 1977).

4 Distribution in the Environment, in Foods, and in Living Organisms

Traces of cobalt are found in all rocks, minerals, and soils. The average cobalt content in the earth's crust is 18 ppm. Cobalt usually occurs together with nickel and iron. The average Co content of ignous rocks is 25 ppm, of shales 19 ppm, of sandstone and limestone 0.3 and 0.1 ppm, respectively; the average Co content of soils is 8 ppm (BOWEN, 1966). Hard coal contains about 8 ppm.

The solubility of cobalt is pH-dependent. In acid soils, cobalt is more mobile than in alkaline soils. Annually, approximately 21 000 tons of cobalt are transported by rivers to the oceans and about the same amount is deposited in deep-sea sediments, whose Co content is in the order of 74 ppm (TUREKIAN and WEDEPOHL, 1961). The natural transport of cobalt is not significantly affected by mining activities and industrial uses of the element. In uncontaminated samples of fresh water, cobalt concentrations are generally low, ranging from 0.1 to 10 µg/L (FOERSTNER and WITTMANN, 1981). However, in waters of polluted rivers, high cobalt levels have been observed, e.g., 4500 µg/L, in Mineral Creek, near Big Dome, Arizona (DURUM et al., 1971). The cobalt concentration in ocean waters is normally quite low and in the order of 0.002 µg/L (KNAUER et al., 1982); higher values quoted in the older literature (YOUNG, 1979) are in part suspect due to possible sample contamination or errors in analysis or interpretation.

Exceedingly low levels of cobalt were observed in soils of some areas of Australia, New Zealand, Kenia, the USSR, Florida, and in Germany's Black Forest, giving rise to cobalt deficiency syndromes in farm animals. To protect sheep and cattle in Co-deficient regions, 1 to 2 kg of cobalt sulfate must be added per hectar every 3 to 5 years. A soil is regarded as cobalt deficient if the cobalt level is below 5 mg per kg dry matter; only about 6% of the total cobalt present is bioavailable (YOUNG, 1979). MAHAFFEY (1989) has discussed absorption, tissue distribution, nutrition essentiality, and myocardial toxicity of cobalt and its compounds in humans.

5 Uptake, Absorption, Transportation and Distribution, and Metabolism and Elimination in Plants, Animals, and Humans

Cobalt uptake by plants is species dependent. For example, cobalt is hardly detectable in green beans; it is also exceedingly low in radishes, although it has been found in the leaves of the plant (KLOKE, 1980). JAROŠIK et al., (1988) found that cobalt is transported through plants principally by the transpiration flow in the xylem; it is present in considerably lower concentrations in the plant storage organs or seeds than in the vegetative parts. It is as yet unknown whether cobalt is essential for plants. In some cases, small amounts of cobalt produce positive growth effects. However, the effects are dose dependent and may be indirect. Thus, while cobalt salts in general are toxic to alfalfa, additions of traces of cobalt promote growth by stimulating nitrogen assimilation.

In zooplankton, cobalt uptake is subject to seasonal variations (VAN AS et al., 1975). Shellfish accumulate cobalt primarily in the byssus filaments and in the liver (SHIMIZU et al., 1971).

The human dietary cobalt intakes are in the order of 20–40 µg/day, according to recent estimates (SMITH and CARSON, 1980). The nearly tenfold higher older intake estimates of SCHROEDER et al. (1967) and of SNYDER et al. (1975), are no longer considered reliable. Most of the cobalt is obtained from the ingestion of foods, only a fraction arises from airborne sources. Some cobalt may also be taken up from implants or prostheses (see, for instance, SUNDERMAN et. al., 1987; WILLIAMS, 1989). Most of the cobalt ingested is inorganic; vitamin B_{12} accounts for only a very small fraction of the total cobalt intake. Absorption of the cobalt occurs in the small intestine. The uptake of inorganic cobalt is coupled with that of iron. Anemic patients responding to iron therapy also show increased cobalt absorption. In rats, iron uptake is antagonized by cobalt; a similar absorption-antagonism exists between cobalt and manganese (THOMSON and VALBERG, 1972). Presumably, cobalt binds to iron transport proteins more strongly than iron itself. Although inorganic (non-B_{12}-) cobalt is invariably present in organs and body fluids, its physiological role, if any, is unknown. Cobalt(II) ions are bound preferentially by serum albumin, which presumably also functions as the transport protein (NANDETKAR et al., 1972). In vitro, cobalt(II) may replace zinc, magnesium, and manganese in a variety of enzymes without affecting enzymatic activity (LINDSKOG, 1970). This is due to the similar ionic sizes of these cations. Whether similar displacement reactions also occur in vivo is not known. Inhaled cobalt particles are rapidly cleared from the lung and excreted.

Experiments with ^{60}Co as the tracer revealed that the renal excretion of inhaled cobalt occurs in a biphasic fashion. Most of the cobalt is excreted with a half-time of 10 days, the remainder with a half-time of 90 days. Experiments with rats produced similar results except that the retention time was generally shorter (HEWITT and HICHS, 1972). In tungsten-carbide-workers, urinary cobalt excretion correlated directly with the degree of exposure; tungsten carbide contains up to 20% cobalt (PELLET et al., 1984; PERDRIX et al., 1983).

Table II.8-1 summarizes the observed Co levels in body fluids and tissues of unexposed and exposed humans as determined by OSTAPCZUK et al., 1983.

Table II.8-1. Cobalt in Body Fluids and Tissues of Unexposed and Exposed Subjects (OSTAPCZUK et al., 1983)

Sample	n	Mean ± SD[a]	Cobalt range (in µg/kg)
Whole blood (unexposed)	9	0.09 ± 0.02	0.08 – 0.12
Whole blood (exposed)	4	1.42 ± 0.43	0.65 – 1.27
Urine (unexposed)	9	0.25 ± 0.11	0.18 – 0.47
Urine (exposed)	4	0.88 ± 0.51	0.50 – 1.75
Saliva	5	0.28 ± 0.14	0.08 – 0.63
Human liver	15	42.3 ± 15.2	20.7 – 75.6
Fingernail	3	83.2 ± 18.1	43.2 – 151
Toenail	2	29.7 ± 5.2	24.8 – 39.1

[a] SD standard deviation

It should be noted that the concentrations in Table II.8-1 are given in µg/kg. In the older literature, much higher blood Co levels were generally reported and must now be considered erroneous.

In plasma of unexposed subjects, cobalt levels are also very low, e.g., 0.108 + 0.060 µg/L, according to VERSIECK et al. (1978). The total cobalt content of a 70 kg (unexposed) man was estimated to 1.5 mg, assuming an average Co content of 20 µg/kg fresh wt. (SMITH and CARSON, 1980). The total amount of vitamin B_{12} in the body of an adult equals to about 5 mg, corresponding to 0.25 mg of Co, of which 50 – 90% are localized in the liver.

6 Effects on Microorganisms, Plants, Animals, and Humans

Ruminants require cobalt for the bacterial biosynthesis of vitamin B_{12} in the first stomach. Cobalt-deficient sheep or cattle show diminished feed intakes and weight loss. In cows milk production declines, and the frequency of miscarriages increases. All symptoms disappear rapidly by supplying small amounts of cobalt. Sheep require 2 to 4 mg of cobalt chloride per day. Additions of 20 mg of $CoCl_2 \cdot 6 H_2O$ to the feed of cows increased the casein-, lactose-, mineral, and trace-element content of the milk as well as the total amount of dry matter in milk. Calves of 45 kg weight tolerate up to 50 mg of Co per day. Sheep may be given up to 160 mg per day for at least 8 weeks. Larger amounts, however, are toxic, and single oral doses of 3.5 grams of cobalt have caused fatalities. Exposures of this magnitude can occur only through human error (YOUNG, 1979). In naturally cobalt-rich regions, the Co-content of accumulating plants such as *Astragalus* sp. may range from 2.3 to 100 mg/kg dry matter.

Vitamin B_{12} is synthesized by intestinal microorganisms, soil bacteria, and algae and is found in particularly large amounts in sewage sludge, estuary mud, and manure. As vitamin B_{12} is neither produced nor absorbed by higher plants, monogastric animals such as horses unlike ruminants require external vitamin B_{12}. This requirement can be met by coprophagy but not by additions of inorganic cobalt salts to the feed. Although vitamin B_{12} is produced in the large intestine by normal microbiological processes, it is not utilized as its absorption occurs in the small intestine. This relatively slow process is mediated by a special glucoprotein secreted by the parietal cells of the gastric mucosa (Castle's Intrinsic Factor, IF). Hence, vitamin B_{12} absorption is further compromised if secretion of IF is inadequate, e.g., in achlorhydric patients or after gastrectomy. Vitamin B_{12} is stored mainly in the liver and is excreted primarily through the bile; 2.5 micrograms are lost per day, corresponding to 0.05% of the total vitamin B_{12} pool; its biological half-life is in the order of 400 days (HERNDON et al. 1980).

Vitamin B_{12} deficiency gives rise to pernicious anemia. However, due to the small amounts required for the maintenance of health (3.5 µg/day per adult), genuine dietary deficiencies develop rarely, e.g., in strict vegetarians. Even in subjects with diminished or absent IF production, the first symptoms of vitamin B_{12} deficiency appear only in the 4th decade, due to the long biological half-life of the vitamin. Initial symptoms of pernicious anemia are weakness and fatigue, soreness of tongue and paresthesias of hands or feet. The blood of the patients exhibits low erythrocyte counts ($\leqslant 2000000$ per mm^3), the volume of the blood cells is significantly enlarged, and corpuscular hemoglobin is elevated by 32% or more. In bone marrow, a marked hyperplasia of the rubriblasts is observed. In approximately 25% of all cases, and especially in older patients, hematological abnormalities are preceded by neural and mental symptoms. The pathogenesis of the disease is not fully understood. Atrophication of gastric mucosa causing cessation of IF production may be due to a genetic defect. However, autoimmunological processes may also play important roles as antibodies against IF, and the parietal cells of the gastric mucosa are observed in a large percentage of all patients.

Vitamin B_{12} is virtually non-toxic even at high oral or injected doses; excessive amounts are rapidly excreted. However, occasionally allergic responses to injected vitamin B_{12} occur (FISHER, 1973), and adverse reactions to the combined administration of large injected doses of vitamin B_{12} and of oral vitamin C have been reported (SCHRAUZER, 1979). Vitamin B_{12} is required for methionine biosynthesis and functions in conjunction with folic acid as the intermediate carrier of the methyl group. In its coenzyme form (5'-deoxyadenosylcobalamin) it is required for the conversion of methylmalonyl-CoA to succinyl-CoA. Bacteria require vitamin B_{12} or its coenzyme in certain dehydrases, deaminases, and in methane biosynthesis.

In healthy subjects, sufficient amounts of vitamin B_{12} are available in the liver to protect against deficiency even during long periods of dietary vitamin B_{12} deprivation. In subjects whose vitamin B_{12} absorption is completely impaired, the first symptoms of deficiency become manifest when the body pool of vitamin B_{12} drops below 10% of normal. To correct deficiencies, 100 µg of vitamin B_{12} are typically administered intramuscularly one to three times per week. Oral dosing is generally ineffective, although nasal vitamin B_{12} sprays and preparations for sublingual application are sometimes used.

Inorganic cobalt salts exhibit moderate toxicity to higher animals and man. The acute toxicity of $CoCl_2 \cdot 6H_2O$ in young rats was determined to 400 mg/kg of diet, the LD_{50} of cobalt chloride in the mouse is 80 mg/kg (CARSON et al., 1986). Symptoms of chronic cobalt toxicity in humans occurred in anemic persons undergoing therapy with iron supplements to which cobalt was added to promote iron absorption and to stimulate erythropoiesis. Depending on the degree of iron deficiency and the amount of iron supplement prescribed, patients thus treated could typically receive from 0.17 to 3.19 mg of Co/kg per day over periods of days to many months. Cobalt toxicity symptoms in these patients included: anorexia, nausea, vomiting, diarrhea, substernal aches, erythema, skin rashes, tinnitus, and neurogenic deafness. In some of the cobalt treated patients paresthesias, numbness and other neurological signs developed, others suffered optic nerve damage (HERNDON et al. 1980). Erythema and several of the other symptoms can be directly attributed to the cobalt induced stimulation of erythropoiesis, leading to polycythemia.

Cobalt inhibits cellular respiration and enzymes of the citric acid cycle and thus generates a type of systemic hypoxia against which the organism responds with an increase of erythropoietin biosynthesis. Erythropoietin, a lipoprotein produced primarily in the kidneys and the liver, in turn triggers erythropoiesis in bone marrow (BERU et al., 1986).

The erythrocytes of Co-treated anemic subjects have shorter life-spans. By binding to plasma proteins, cobalt also inhibits iodine transport and is goitrogenic (HERNDON et al., 1980). In view of this evidence, the cobalt treatment of anemic subjects has been abandoned, and cobalt containing anti-anemia iron preparations are no longer marketed in the USA.

Hundreds of cases of cobalt toxicity occurred in the mid-sixties in Canada, the USA, and Belgium among consumers of certain brands of beer to which cobalt sulfate at 1 mg/L was added as a foam stabilizer. In a large percentage of heavy drinkers consuming 12 liters or more of said brands of beer, fatal congestive heart failures occurred. The cobalt intakes in these cases were 6–8 mg/day. The fact that no fatalities apparently occurred among the anemic patients treated with much larger doses of cobalt suggests that cobalt toxicity is diminished on simultaneous administration with iron. The myocardial toxicity has been attributed to a decrease of the oxidation of pyruvate in heart muscle. The cardiotoxic effects of cobalt are enhanced by thiamine- and protein deficiency, and by ethyl alcohol. In animal experiments, selenite and vitamin E were shown to protect against myocardial toxicity of cobalt (HERNDON et al., 1980). In moderate consumers of cobalt-doped beer, thyroid dysfunction and non-fatal congestive heart disease were observed. The number of cases dropped to zero after removal of the product from the market (ALEXANDER, 1972; ACHENBACH et al., 1974; BURCH et al., 1973; MORIN and COTE, 1972).

Workers exposed to cobalt containing dusts develop progressive pulmonary fibrosis and other forms of chronic lung damage (see also TAYLOR and HAWKINS, 1987). They may in addition suffer partial or complete loss of olefactory functions, present with gastrointestinal disturbances, dyspnea, and weight loss. Congestive heart defects akin to those in the drinkers of cobalt-doped beer have also been observed (BARBORIK and DUSEK, 1971).

The toxicity of cobalt is relatively independent of the chemical form. Thus, toxic effects are elicited by the oxide as well as by the metal; they are not masked by alloying. Thus, even tungsten- or titanium carbide dusts are toxic on inhalation due to their cobalt contents (COATES and WATSON, 1971; KERFOOT et al., 1975). The pulmonary fibrosis of hard-metal grinders is also significantly attributed to the cobalt content of the respective materials. Hard metals consist primarily of carbides of tungsten, titanium, and tantalum, with cobalt serving as a binding agent. Cases of hard metal fibrosis of the lung have been increasingly observed. Although hard metal fibrosis of the lung has been recognized as an occupational disease in the Federal Republic of Germany as early as in 1961, the regulation focused primarily on the risk of dust exposure when mixing the basic and especially processing the presintered material, the early regulations did not focus on the dangers of working with the finished products. Accordingly, the causal connection between pulmonary fibrosis and occupation was later often questioned in those cases where the worker had only been exposed to grinding dust from the sintered material (HARTUNG et al., 1982). By 1983, 42 cases of hard metal fibrosis were compensated by West German occupational insurance agencies. The cobalt content in lung biopsy material of one worker who had been exposed to grinding dust for 10 years was 1010 µg/kg wet weight, the urinary cobalt concentration was 7.5 µg/L (HARTUNG et al., 1982). For a detailed review of the subject matter see HARTUNG (1986).

Comparatively little is known on the toxic effects of cobalt carbonyl. In workers acutely exposed to cobalt carbonyl vapor, headaches, weakness, irritability, changes in reflexes and electrical activity of the brain were observed (HERNDON et al., 1980).

Persons handling or wearing objects made from cobalt may become sensitized to the metal; a boy wearing a wrist watch and glasses with a metallic frame developed allergic reactions due to the cobalt content of both (GRIMM, 1971); a war-veteran became cobalt-sensitized by a shell-fragment (NOVER and HEINRICH, 1971).

Cobalt is not teratogenic in humans, as evidenced from observations on pregnant women receiving 75–100 mg of cobalt chloride per day for treatment of anemia. Among workers exposed to cobalt containing tungsten carbide dust, there have been no reports of teratogenicity on the human reproductive system (NIOSH, 1977).

Although the addition of cobalt salts to cultures of human leukocytes and of diploid fibroblasts caused a reduction of the mitotic index, no evidence of mutagenicity was obtained. Similar conclusions were drawn in studies of the effects of cobalt with bacterial test systems (CARSON et al., 1986).

Metallic cobalt is carcinogenic on implantation; in rats, rhabdomyosarcomas were thus induced (WEINZIERL and WEBB, 1972). Malignant tumors also developed in rats on implantation of cobalt(II) oxide and cobalt sulfide, the latter was a more potent carcinogen than the oxide. In mice, implantation experiments established the carcinogenicity of cobalt sulfide, but not of the Co(II) oxide (GILMAN, 1962). In an in-vitro model system, additions of cobalt to the medium increased mis-incorporation during DNA biosynthesis (LOEB et al., 1979).

7 Hazard Evaluation and Limiting Concentrations

Extreme caution and elaborate protective measures are required to protect humans against even minimal exposures, since elemental cobalt as respirable dusts or aerosols and cobalt compounds of low solubility are unmistakably carcinogenic in animals under experimental conditions which are comparable to human occupational exposures. Since cobalt and its compounds furthermore are also known to elicit allergic responses, it is no longer possible to define acceptable maximum allowable concentration (MAC- or MAK-) values. In the MAK Report No. XXIII (1987) on Maximum Concentrations at the Workplace and Biological Tolerance Values issued by the Deutsche Forschungsgemeinschaft (DFG), no MAK value is given for cobalt, meaning that the industrial handling of cobalt and of its compounds requires extraordinary caution and protective measures. There is, however, a question of limitation of the carcinogenic effectiveness as dependent on solubility (MAK, 1987). Instead of the MAK value, a "Technical Guiding Concentration" (TRK) of 0.5 mg/m^3 is given, calculated as cobalt in the total dust. The "permissible exposure limit" in the USA for cobalt metal, dust, and fume adopted by the Occupational Safety and Health Administration (OSHA) is 0.1 mg/m^3, which is a time-weighted average for a conventional workday (OSHA, 1981). The American Conference of Governmental Industrial Hygienists (ACGIH) has since 1982 recommended to set this limit to 0.05 mg/m^3. A maximum allowable concentration (MAC) of 0.5 mg/m^3 was adopted by the USSR for cobalt and cobalt oxide, but of 0.01 mg/m^3 for dicobalt octacarbonyl and cobalt hydridocarbonyl or its decomposition products (HERNDON et al., 1980). In Belgium 0.01 mg/m^3 is the limit for cobalt and cobalt oxide in air (NIOSH, 1982). For drinking water no tolerance limits have been set in the USA and most other countries, but in the USSR it is 1 mg/L, which appears high as toxic effects have occurred on exposure of rats to water containing 2 mg/L. For fish and protozoa the tolerance limits for cobalt lie between 0.01 and 0.1 mg/L (LIEBMANN, 1958; JUNG, 1979). In nutrient media, inhibition of germination of plant seeds was observed at concentrations from 0.1 to 3 mg/L (BOWEN, 1979).

References

ACGIH (American Conference of Governmental Industrial Hygienists) (1975), *Transactions of the 37th Meeting,* Minneapolis, Minnesota.

Achenbach, H., and Urbaszek, W. (1972), *Z. Inn. Med. 19,* 809–816.

Achenbach, H., Urbaszek, W., Guenther, K., Schneider, Dieter, Schneider, Dietmar, Kronenberger, H., Trankmann, H., Kiessling, J., Hurlbeck, M., and Spith, G. (1974), *Z. Inn. Med. 29,* 108.

Alexander, C. S. (1972), *Am. J. Med. 53,* 395–417.

Angerer, J., and Schaller, K. H. (1985), *Cobalt in Analysis of Hazardous Substances in Biological Materials.* Vol. 1, pp. 141–151. VCH Verlagsgesellschaft, Weinheim-Deerfield Beach/Florida-Basel.

Barborik, M., and Dusek, J. (1972), *Br. Heart J. 34,* 113–116.

Beru, N., McDonald, J., LaCombo, C., and Goldwater, E. (1986), *Mol. Cell. Biol. 6,* 2571–2575.
Bowen, H. J. M. (1966), *Trace Elements in Biochemistry.* Academic Press, New York-London.
Bowen, H. J. M. (1979), *Environmental Chemistry of the Elements.* Academic Press, New York-London.
Burch, R. E., Williams, R. V., and Sullivan, J. F. (1973), *Am. J. Clin. Nutr. 26,* 403–408.
Burger, L. (1973), *Organic Reagents in Metal Analysis.* Pergamon Press, Oxford.
Carson, B. L., Ellis III, H. V., and McCann, J. L. (1986), *Toxicology and Biological Monitoring of Metals in Humans.* Lewis Publishers, Chelsea, Michigan.
Coates, E. O., and Watson, J. H. L. (1971), *Ann. Int. Med. 75,* 709–716.
Durum, W. H., Hem, J. D., and Heidel, S. G. (1971), *Reconnaissance of Selected Minor Elements in Surface Waters of the United States,* in: *U.S. Geol. Surv. Circ. 643,* 49.
Fisher, A. A. (1973), *Contact Dermatitis,* 2nd Ed. Lee & Febinger, Philadelphia.
Foerstner, U., and Wittmann, G.T.W. (1981), *Metal Pollution in the Aquatic Environment,* pp. 71 ff. Springer Verlag, Berlin-Heidelberg-New York.
Fries, J., and Getrost, H. (1977), *Organic Reagents for Trace Analysis.* E. Merck, Darmstadt.
Gilman, J. P. W. (1962), *Cancer Res. 22,* 158–162.
Grimm, I. (1971), *Berufsdermatosen 19,* 39–42.
Haerdi, W. (1989), *Panoramic Cation Determination by Ion Chromatography.* Lecture 19th IAEAC Symposium on Environmental Analytical Chemistry, Jekyll Island, Georgia.
Hartung, M. (1986), *Lungenfibrosen bei Hartmetallschleifern, Schriftenreihe des Hauptverbandes der gewerblichen Berufsgenossenschaft e.V.* Koellen Druck, Bonn.
Hartung, M., Schaller, K. H., and Brand, E. (1982), *Int. Arch. Occup. Environ. Health 50,* 53–57.
Herndon, B. L., Jacon, R. A., and McCann, J. (1980), in: Smith, I. C., and Carson, B. L. (eds.): *Trace Metals in the Environment,* Vol. 6, pp. 925–1140. Ann Arbor Science Publishers/Butterworth, Ann Arbor, Michigan.
Hewitt, P. J., and Hichs, R. (1972), in: *Nuclear Activation Techniques in the Life Sciences Symposium Proceedings, Vienna.* International Atom Energy Agency (IAEA), Vienna.
Jarošik, J., Zvára, P., Konečný, J., and Obdržálek, M. (1988), *Dynamics of Cobalt-60 Uptake by Roots of Pea Plant. Sci. Total Environ. 71,* 225–229.
Jung, K. D. (1979), in: Foerstner, U., and Wittmann, G. T. W. (eds.): *Metal Pollution in the Aquatic Environment.* Springer Verlag, Berlin.
Kerfoot, E. J., Frederick, W. G., and Domeier, E. (1975), *Am. Ind. Hyg. Assoc. 8,* 153–196.
Kloke, A. (1980), *Lecture at GDI Institute,* Zürich. Cited in *Chem. Rundsch. 35,* No. 10, 3 (March 3, 1982); see also: *Proceedings Verwendung von Klärschlammkomposten in der Landwirtschaft,* pp. 58–87. GDI-Institut, Zürich, 1980.
Knauer, G. A., Martin, J. H., and Gordon, R. M. (1982), *Nature 297,* 49.
Liebmann, H. (1958), *Handbuch der Frischwasser- und Abwasserbiologie,* Vol. II. R. Oldenbourg, München.
Lindskog, S. (1970), *Struct. Bonding 8,* 153–196.
Loeb, L. A., Sirover, M. A., and Agrawal, S.S. (1979), *Adv. Exp. Med. Biol. 91,* 103–116.
Mahaffey, K.R. (1989), *Nutrition Essentiality and Toxicity: Comparing What's Known about Cobalt, Arsenic, and Lead.* 23rd Annual Conference on Trace Substances in Environmental Health, Cincinnati, Ohio. Proceedings, *J. Environ. Geochem. Health,* in press.
MAK (1987), *Maximum Concentrations at the Workplace, Report No. XXIII, DFG.* VCH Verlagsgesellschaft, Weinheim-Basel-Cambridge-New York.
McKinney, G. L. (1987), *Communication 17th Annual IAEAC Symposium on the Analytical Chemistry of Pollutants.* Jekyll Island, Georgia. EPA, Kansas City.
Miller-Ihli, N. J., and Wolf, W. R. (1986), *Anal. Chem. 58,* 3225.
Morin, Y., and Cote, G. (1972), *Cardiovasc. Clin. 4,* 245–267.
Nandetkar, A. K. N., Basu, P. K., and Friedberg, F. (1972), *Bioinorg. Chem. 2,* 149–157.
NIOSH (National Institute of Occupational and Health) (1982), *Occupational Hazard Assessment: Criteria for Controlling Occupational Exposure to Cobalt.* DHHA (NIOSH) Publication No. 82–107. U.S. Printing Office, Washington D.C.

Nover, A., and Heinrich, I. (1971), *Klin. Monatsbl. Augenheilk. 158*, 546–550.
OSHA (Occupational Safety and Health Administration) (1981), *Occupational Safety and Health Standards*. Subpart 2: *Toxic and Hazardous Substances*. Code of Federal Regulations 29 (Part 1910.1000), pp. 673–679, Washington DC.
Ostapczuk, P., Valenta, P., Stoeppler, M., and Nuernberg, H. W. (1983), in: *Chemical Toxicology and Clinical Chemistry of Metals*, pp. 61–64. IUPAC, Oxford.
Pellet, F., Perdrix, A., Vincent, M., and Mallison, J. M. (1984), *Arch. Mal. Prof. Med. Trav. Secur. Soc. 45* (2), 81–85.
Perdrix, A., Pellet, F., Vincent, M., DeGaudemaris, R., and Mallion, J. M. (1983), *Technocol. Eur. Res. 5* (5), 233–240.
Saager, R. (1984), *Metallic Rawmaterials Dictionary*, pp. 135–138. Bank von Tobel, Zürich.
Sampson, B., Maher, E. R., and Curtis J. R. (1987), *Proceedings of the Second Nordic Symposium on Trace Elements*, Odense. World Health Organization (WHO), Copenhagen.
Schrauzer, G. N. (1976), *Angew. Chem. 88*, 465–474; *Angew. Chem. Int. Ed. Engl. 15*, 417–426.
Schrauzer, G. N. (1977), *Angew. Chem. 89*, 239–251; *Angew. Chem. Int. Ed. Engl. 16*, 233–244.
Schrauzer, G. N. (1979), *Int. Rev. Biochem. 27*, 167–188.
Schroeder, H. A., Nason, A. P., and Tipton, I. A. (1967), *J. Chronic Dis. 20*, 869–890.
Shimizu, M., Kajihara, T., Suyama, I., and Hiyama, Y. (1971), *J. Radiat. Res. 12*, 17–28.
Sibley, S. F. (1977), *Minerals Yearbook*, pp. 485–491. Washington, D.C.
Smith, I. G., and Carson, B. L. (1980), *Trace Metals in the Environment*, Vol. 6: *Cobalt*. Ann Arbor Sciences Publ., Ann Arbor, Michigan.
Snyder, W. S., Cook, M. J., Nasset, E. S., Karhausen, L. R., Howells, G. P., and Tipton, I. H. (1975), *International Commission on Radiological Protection (ICRP), Report of the Task Group on Reference Man*. ICRP Publication No. 23, New York.
Sunderman, F. W. Jr., et al. (1987), *Cobalt, Chromium, and Nickel Concentrations in Body Fluids of Patients with Porous-coated Co/Cr Prostheses, Proceedings Second Nordic Symposium on Trace Elements*, Odense. WHO, Copenhagen.
Taylor, A., and Hawkins, L. (1987), *Effects of Occupational Exposure to Cobalt, Proceedings Second Nordic Symposium on Trace Elements*, Odense. WHO, Copenhagen.
Thomson, A. B. R., and Valberg, L. S. (1972), *Am. J. Physiol. 223* (6), 1327–1329.
Trueb, L. (1987), *Corrosion Problems in Microelectronics* (in German). *Neue Zürcher Zeitung, Forschung und Technik*, No. 58, p. 69 (11 March), Zürich.
Turekian, K. K., and Wedepohl, K. H. (1961), *Bull. Soc. Geol. Soc. Am. 67*, 1129–1132.
van As, D., Fourie, H. O., and Vlegaar, C. M. (1975), *S. Afr. J. Sci. 71*, 151–154.
Versieck, J., Moste, J., Barbier, F., Vanballenberghe, L., De Rudder, J., and Cornelis, R. (1978), *Clin. Chim. Acta 87*, 135–140.
Weinzierl, S. N., and Webb, M. (1972), *Br. J. Cancer 26*, 279–291.
Williams, D. R. (1989), *Toxicological and Therapeutic Potential of Metallic Surgical Implants. Lecture Robens Institute Meeting on Toxicity and Therapeutics of Newer Metals and Organometallic Compounds*, Guildford, Surrey, UK.
Young, R. S. (1979), *Cobalt in Biology and Biochemistry*. Academic Press, London-New York-San Francisco.

Additional Recommended Literature

Angerer, J., and Schaller, K. H. (1985), *Analyses of Hazardous Substances in Biological Materials*, Vol. 1, pp. 141–151. VCH Verlagsgesellschaft, Weinheim-Deerfield Beach/Florida-Basel.
Cole, C. J., and Carson, B. L. (1980), in: *Trace Metals in the Environment* (Smith, I. C. and Carson, B. L. eds.), Vol. 6, *Cobalt*, pp. 777–924. Ann Arbor Science Publishers/Butterworth, Ann Arbor, Michigan.

Elinder, C.-G. (1984), *Carcinogenic, Mutagenic and Teratogenic Effects of Cobalt*, Plenary Lecture, Workshop on Carcinogenic and/or Mutagenic Metal Compounds, Geneva; *J. Toxcol. Environ. Chem.* 7 (3), 251–256.

Elinder, C.-G., and Friberg, L. (1979), *Cobalt*, in: Friberg, L., Nordberg, G. F., and Vouk, V. B. (eds.): *Handbook on the Toxicology of Metals,* pp. 399–410. Elsevier/North-Holland Biochemical Press, Amsterdam.

Lauwerys, R. R., (1983), in: *Industrial Chemical Exposure. Guidelines for Biological Monitoring*, Chapter II, pp. 9–50. Medical Publications, Davis, California.

Merian, E. (1984), *Environmental Chemistry and Global Cycles of Seven Selected Metals and their Derivatives*, Plenary Lecture, Workshop on Carcinogenic and/or Mutagenic Metal Compounds, Geneva; *J. Toxicol. Environ. Chem.* 8 (1), 9–38.

NAS (National Academy of Sciences) (1980), *Recommended Dietary Allowances.* Washington, DC.

Salomons, W., and Foerstner, U. (1984), *Metals in the Hydrocycle.* Springer Verlag, Berlin-Heidelberg-New York-Tokyo.

Schumacher-Wittkopf, E. (1984), *Characterisation of Cobalt-binding Proteins with Occupational Cobalt Exhibition*, Lecture, Workshop on Carcinogenic and/or Mutagenic Metal Compounds, Geneva; *J. Toxicol. Environ. Chem.* 8 (2/3), 185–193.

Underwood, E. J. (1971), *Trace Elements in Human and Animal Nutrition,* 3rd Ed. Academic Press, New York-London.

II.9 Copper

I. HERBERT SCHEINBERG, New York, USA

1 Introduction

Copper – ubiquitously distributed – is very easily complexed and is involved in many metabolic processes in living organisms some of which involve the redox potential of Cu(I)/Cu(II). Complex formation regulates copper homeostasis in the soil and in organisms and the biosynthesis of essential copper-containing proteins and enzymes. Genetic or acquired defects in these regulating mechanisms may cause deficiency or toxic excess of copper. Copper interacts with other metals such as iron (in blood formation), molybdenum (excess Mo may cause deficiency of Cu), and zinc (excess of Cu leads to Zn-deficiency in sheep, and excess of zinc can lead to copper deficiency in man).

2 Physical and Chemical Properties, and Analytical Methods

Copper, atomic number 29 and atomic mass 63, has been known for about 10000 years and occurs in metallic form or in compounds as Cu(I) or Cu(II). The red metal has a density of 8.93 g/cm^3, a melting point of 1083 °C and a boiling point of about 2590 °C. Natural copper consists of an isotopic mixture of 69.1% ^{63}Cu and 30.9% ^{65}Cu. Copper is easily manufactured from ores. Except for silver, copper is the best common conductor of heat and electricity. Compounds of Cu(I) and Cu(II) and Cu-complexes have very different properties than the metal.

The following methods (see also Chapter I.4a) are used for the quantitative determination of copper in biological and environmental samples (collection and treatment of samples require particular attention).

a) Atomic absorption spectrophotometry is at present the most frequently used method (see, e.g., WELZ, 1985). Detection limits are rather high with flame, better results are obtained with graphite cuvettes.
b) Electrochemical methods are preferred for aqueous solutions, such as polarography, differential pulse polarography, pulse voltammetry, and pulse inverse voltammetry. The latter is the most sensitive known detection method for copper.

c) Emission spectroscopy (e.g., ICP-AES, ICP-CATS, or ICP-MS) is another interesting technique (see, e.g., MONTASER and GOLIGHTLY, 1987; WARD, 1987).
d) Spectrophotometric (colorimetric) methods, e.g., with sodium diethylthiocarbamate (NaDTC), are still reliable and cheap, but not very sensitive methods for higher concentrations of copper in biological and environmental materials.
e) Neutron activation analysis, using a sample weight of 500 mg and a thermic neutron flux of 10^{13} neutrons cm^{-1} s^{-1}, is a sensitive method (WARD, 1987).
f) Roentgen fluorescence in its wavelength-dispersive variants normally is suitable only for higher concentrations. With special techniques, such as multiple reflection, lower detection limits can be reached.
g) Mass spectrometry with flash- or thermoionic excitation also achieves very low detection limits. The isotope dilution mass spectrometry is a highly precise reference method with high reliability.
h) With panoramic high performance chromatography copper, nickel, and calcium can easily be determined simultaneously (HAERDI, 1989).
i) With PIXE (particle-induced X-ray emission analysis) or SYXFA (synchrotron radiation) it is possible to study, for instance, copper mobilization in root tips (KNÖCHEL, 1989).

3 Sources, Production, Important Compounds, Uses, Waste Products, Recycling

3.1 Occurrence, Production

Copper is almost always extracted from ores in underground or open-pit mines. Copper ores contain 5–50 g/kg, manganese knobs 10 g/kg (BEPPLER et al., 1978). The most important ores contain, besides small amounts of metallic copper, Cu_2S, CuS, $CuFeS_2$, CuO, $Cu_2CO_3(OH)_2$ (SCHEINBERG, 1971; BOWEN, 1979). The biggest stocks are located in the USA, Chile, Canada, the USSR, Zambia, and Peru (FABIAN, 1986; SAAGER, 1984). Large amounts of gold containing porphyric copper ores were found in Papua-New-Guinea (TRUEB and NUTTING, 1989). Oxides and carbonates can be leached with dilute sulfuric acid, and copper may be electrolyzed from the solutions. Sulfides are crushed and ground; concentrated by flotation, following addition of air and frothers; smelted at 1500 °C with the addition of lime and silica fluxes; and freed of sulfur and iron in a converter. The resulting blister copper, about 98% pure, is fire-refined to a purity of about 99.5% which is suitable for many purposes but not for the electrical uses to which over 75% of mined copper is put, and which requires a purity of at least 99.9%, achieved by electrolysis (SCHEINBERG, 1971; SAAGER, 1984). The annual world production increased from 0.5 million tons in 1900 to about 7 million tons (FABIAN, 1986; SAAGER, 1984), a great part of which (about 30%), however, is obtained by recycling (SAAGER, 1984; RICHARDSON, 1986). Although copper has been replaced to some extent by aluminum and glass fibers in the 1980s, there is a certain shortage, and prices in US $

have increased since 1986. However, there is a continuous tendency to economize and replace copper (TRUEB and NUTTING, 1989).

3.2 Uses

Metal (see also FABIAN, 1986). In addition to its electrical applications, copper is used in water piping, stills, roofing material, and kitchenware; for chemical and pharmaceutical equipment; as a pigment; and as a precipitant of selenium (SCHEINBERG, 1971). Copper metal has been employed as an intrauterine contraceptive (ZIPPER et al., 1968), but recent toxic reactions have diminished this use.

Alloys (see also ISBELL, 1986). Alloys of copper include those with zinc (brass), tin (bronze), nickel (monel metal), aluminum, gold, lead, cadmium, chromium, beryllium, silicon, or phosphorus (SCHEINBERG, 1971).

Compounds (see also RICHARDSON, 1986). Copper sulfate is used to supplement pastures deficient in the metal; as an algicide and molluscicide in water; with lime, as a plant fungicide; as a mordant; in electroplating; and, as a component of Fehling's solution, to estimate reducing sugars in urine. Cupric oxide has been used as a component of paint for ship bottoms, and its addition to swine (BRAUDE and RYDER, 1973) and poultry (DAVIS, 1974) feed has been proposed to promote growth. Copper chromates are pigments, catalysts for liquid phase hydrogenation, and potato fungicides. The pigment known as Scheele's green is a complex mixture of cupric oxide and arsenite, and that called verdigris is cupric oxyacetate. A solution of cupric hydroxide in excess ammonia is a solvent for cellulose and has been used in the manufacture of rayon (SCHEINBERG, 1971). Certain new crystalline compounds containing copper, calcium, oxygen, and other bivalent elements, e.g., yttrium (or lanthanum)-barium-copper oxide, or copper matrice monolithes, seem to allow "warm" superconduction (TRUEB, 1987, 1988b; LÖHLE et al., 1989). Asea Brown Boveri has developed a device with $YBa_2Cu_3O_7$, limiting current in cases of short circuits (DRESCH, 1989). Copper-indium diselenide is used in photovoltaic solar cells. It is foreseeable that some of these compounds end as waste products (also in incineration). Since copper compounds have catalytic properties in many chemical reactions, they also activate formation of chlorinated dioxins and dibenzofurans during incineration (STIEGLITZ, DICKSON, and NAIKWADI, 1988). Copper contents in wastes should therefore be reduced.

4 Distribution in the Environment, in Foods, and in Living Organisms

The approximate concentration of copper in the earth's crust is (JENKINS, 1981; BOWEN, 1979; MART et al., 1982): igneous rocks, 50–90 ppm; soil (dry), 20–30 ppm; oceans, ≤0.3 µg/L; freshwater, ≤0.01 mg/L; air (USA), 10–500 ng/

m^3 (in cities, 80–90 ng/m^3). In pristine rivers, such as the river Rhône, about 0.8 mg/L of dissolved copper was found, and about 0.07 mg/kg in suspended particulate (HUYNH NGOC et al., 1988). Also a part of the copper content of the oceans is complex-bound, only the surplus is dissolved. BRULAND (1983) reported deep-water enrichment superimposed by effects resulting from deep-water scavenging. Lowest surface concentrations of 0.08 µg/L were observed in the North Pacific central gyre.

The movement of the relatively high concentration of copper from the earth's crust into the soil depends on weathering, the process of the soil's formation, drainage, oxidation-reduction potentials, the amount of organic matter in the soil, and, perhaps most important, the pH. Copper is concentrated in the clay mineral fractions with slight enrichment in clays rich in organic carbon. Almost all of the copper carried into the ocean is precipitated accounting for its lower concentration there than in freshwater.

Acid conditions – for example, the presence of acid rain in a region – promote the solubility of copper ores or metal. The exchange of copper from humate has been studied, e.g., by HERING (1988). Alkaline conditions tend to precipitate copper and may lead to deficiency of the metal in plants. Paradoxically, if the soil is excessively acid, copper salts may be solubilized and leached out – also leading to deficiency (DAVIS, 1977). In turn, soil deficiency of copper can lead to plant deficiency which may prove lethal to susceptible animals.

Copper toxicosis in plants is less usual than deficiency – in contrast to animals where toxicosis may be induced by environmental excess of the element or with normal environmental concentrations in genetically abnormal individuals (DAWSON and PRICE, 1977). In the food chain, tolerant plants and invertebrates may accumulate copper and pose a certain risk for higher animals consuming them. Plants contain 4–20 mg Cu per kg dry weight, marine algae 2–68 mg/kg, fish 0.7–15 mg/kg, muscles of mammals ca. 10 mg/kg, bones of mammals 1–26 mg/kg (BOWEN, 1979), and the adult human body ca. 100 mg (KIEFFER, 1980). Foods with higher copper contents are organs – especially liver (see, e.g., WARD, 1987) – shellfish, cocoa, and red wine. Average daily diets contain 2 to 5 mg of copper. Blood contains ca. 1 mg/L. About half of ingested copper is found in the feces.

5 Uptake, Absorption, Transport and Distribution, Metabolism and Elimination in Plants, Animals, and Humans

Copper is biologically available as Cu(I) or Cu(II) in inorganic salts and in organic complexes. In estuarine seagrasses most of the copper is found on and in the roots, bound to the cuticula and to organic matter, and only 10% in the leaves (ROZEMA, 1988). Complexes with some soil colloids or on iron compound surfaces are, however, taken up to a lesser extent.

Varying soil qualities influence the uptake of copper by the roots of plants (LEPP, 1981). It is assumed that exchange reactions and the nitrogen content of the soil are important factors for the passive transport of copper. It accumulates in the roots and in cell walls and is transported into plants in various ways and may also be excreted. Adsorbed copper compounds may be less bioavailable, for instance, for *Mycorrhiza* fungi (RUEGG, 1989). Some mosses and plants are capable of accumulating copper (BOWEN, 1979). Some pines contain several hundred ppm of Cu, whereas 5–20 ppm is an average amount for leaves of deciduous trees.

Copper may also penetrate the fruit of French beans and radishes though not into radish leaves (KLOKE, 1980). In a comparable ecosystem, water plants take in about three times more copper than land plants. Algae contain 1.5–6 ppm Cu minimum, species of *Ascophyllum* more than 100 ppm (FÖRSTNER and WITTMANN, 1979). Grasses are relatively tolerant towards copper.

Mussels and fish are capable of taking in great amounts of copper. However, this is not a problem for human nutrition (LEATHERLAND, 1979). It has long been known that copper is part of hemocuprein (FÖRSTNER and WITTMANN, 1979), constituting the prosthetic element analogous to iron in hemoglobin. Fish absorb copper directly from water and accumulate it in greater amounts than mercury, lead, or cadmium (WACHS, 1981). Also terrestrian crustacea accumulate copper more than litter (ALIKHAN, 1989).

Fish accumulate copper mainly in the liver (LEATHERLAND, 1979). An accumulation of copper from microorganisms via the food chain seldom occurs, except if fish in ponds are fed copper-contaminated dried feed (WACHS, 1981). The presence of organic constituents, such as NTA (nitrilo triacetic acid) and EDTA (ethylene diamine tetraacetic acid), considerably reduces the intake of copper by fish (FÖRSTNER and WITTMANN, 1979). Interactions have also been observed; e.g., trouts take up copper rapidly, and as a consequence, cadmium uptake in the gills is accelerated (ANDERSON, 1988). Sublethal effects of copper and zinc mixtures on aquatic organisms were also much greater than in the case of the single toxicants (PARROT et al., 1988). DIVE et al. (1988) found also synergistic effects of copper and cadmium in ciliate protozoa in effluents of the electroplating industry.

Humans can absorb about 50% of dietary copper in the gastro-intestinal tract. The rest is excreted via the bile and feces (VENUGOPAL and LUCKEY, 1979). The excretion via the urine is minimal, normally less than 50 µg/24 h. However, KODAMA (1988) reported on the increased excretion of cadmium via the urine, observed in dogs with high copper concentrations in blood. Copper accumulates in the liver, in the brain, and in the kidneys. Organic complexes permit the transport through the cell walls. According to publications of the WHO, the adult human being needs about 2–5 mg copper per day; children somewhat more, on a weight basis.

6 Effects on Plants, Animals, and Humans

Copper is one of several heavy metals that are essential to life despite being as inherently toxic as non-essential heavy metals exemplified by lead and mercury. The evolutionary process, in selecting copper to carry out specific physiological roles, has avoided its toxic potential in three ways: (1) by developing an active process for eliminating any excess copper ingested in the diet; (2) by reducing the thermodynamic activity of copper ions virtually to zero by utilizing the metal only as a prosthetic element tightly bound to specific copper proteins; and (3) by an interaction between zinc and copper (SCHEINBERG and STERNLIEB, 1984). Copper is probably only toxic to man and animals when one of these mechanisms is defective, either because of genetic or acquired causes. Then ionic copper accumulates in excess of that which can be incorporated into normally occurring copper proteins. Copper toxicosis seems wholly or in large part to be due to free copper ions combining with new copper proteins and altering their physiological functions (see also AASETH and NORSETH, 1986).

6.1 Effects on Microorganisms and Plants

Copper is toxic, however, to many bacteria (see also Sect. 6.5 and Chapter I.7h) and viruses (OLSON, 1981). For cultures of *Aureobasidium* copper even is the most toxic element (GADD, 1981). Copper sulfate and copper(I) oxide have been used as fungicides for many decades. Freshly added copper salts in concentrations of more than 50 mg/L of soil solution reduce the growth of *Mycorrhiza* fungi which live in symbiosis with roots (RÜEGG, 1989). Adsorbed aged copper compounds are, however, less bioavailable. The Microtox® bioassay, based on reduction of bacterial luminescence, is a rapid tool for assessing interactions of copper compounds with organic ligands (MOREL et al., 1988).

For plants copper toxicity is virtually unknown, a situation very much like that obtaining in man because of the protective mechanisms mentioned above. BUCK (1977) stated that "regulatory mechanisms appear to limit the concentration of copper found in plant tissues to about 20 ppm ...".

Exceptionally copper may also be toxic to plants (COTTENIE, 1981; MACNAIR, 1977), affecting thereby mainly the growth of the roots. For instance, 25 years of *Coffea* stands in Kenya may contain more than 500 kg Cu per ha from spraying, and rehabilitation problems could arise if other crops should be grown (LEPP and DiICKINSON, 1989). Normally, however, copper is hardly available for bioprocesses since it is complex-bound in the soil. There also exist copper-tolerant plants (MCNEILLY, 1981; VERKLEIJ, 1981; HUTCHINSON, 1981; COTTENIE, 1981) that thrive in media with copper concentrations up to 20 ppm. Plants growing in soils contaminated with copper – from geological, industrial, or agricultural sources, e.g., use of fungicides – may die. In such instances only copper-tolerant grasses survive (LEPP, 1981). Algae may be copper-tolerant (HUTCHINSON, 1981), especially if they are not polyphosphate-deficient (TWISS et al., 1989).

6.2 Deficiency Symptoms

6.2.1 Essentiality to Plants

Deficiency in plants, though rare, has been reported and is usually associated with less than 5 ppm in the plant. For example, KRÄMER and PODLESAK (1987) have reported on pollen sterility in wheat due to copper deficiency. Although, as noted above, clays rich in organic matter tend to a larger copper content, soils with considerable amounts of organic matter that are brought into agricultural production for the first time frequently produce crops that are copper deficient. Placing copper salts on such soils can eliminate the deficiency. Specific signs of copper deficiency have been described in fruit trees and cereal grains (DAWSON and PRICE, 1977).

6.2.2 Essentiality to Animals
(see also MYLROIE, SMITH, and SUTTLE, 1987)

As indicated above, the essentiality of copper to animals is mediated through specific copper proteins including tyrosinase, monoamine oxidase, dopamine beta-hydroxylase, ceruloplasmin, superoxide dismutase, lysyl oxidase, tryptophan 2,3-dioxygenase and cytochrome oxidase (BUCK, 1977). Of these the last is the most important and is probably directly associated with the most dramatic form of animal copper deficiency – enzootic ataxia, or swayback, in sheep. Here deficiency of cytochrome oxidase is believed to interfere with normal myelination on newborn lambs dropped by ewes feeding on copper deficient pastures. In Australia such pastures generally consist of pale grass with bright ribbons of green grass directly under telegraph or power lines from which rain has leached sufficient copper to correct the deficiency beneath the wires.

Other effects of copper deficiency in animals – some instances of which can be associated with specific copper-protein deficiencies – include anemia, depressed growth, bone disorders, achromotrichia, abnormal wool growth, impairment of reproduction, heart failure, aortic aneurysms, and gastrointestinal disturbances (BUCK, 1977). For reasons that are not understood the addition of molybdenum to diets of animals deficient in copper worsens this deficiency.

In swine copper deficiency leads to a microcytic, hypochromic anemia that resembles that of iron deficiency except that it cannot be effectively treated by administration of iron – and, indeed, may occur in pigs with an excess of iron stores and 70–80% of ringed sideroblasts in the marrow – unless copper is administered (CARTWRIGHT et al., 1956). This form of anemia has recently been detected in human patients with Wilson's disease who have received certain chelating agents (SCHEINBERG and LAWRENCE, 1988).

6.2.3 Essentiality to Humans
(see also HOTTMEIER, 1979; MYLROIE, SMITH and SUTTLE, 1987)

Although copper deficiency in man does not occur – with rare exceptions – if there is at least 2 mg of copper in the daily diet, the disturbances due to copper deficiency can be caused by severe malabsorption, diarrhea accompanied by a copper deficient diet in infants, or in the genetic disorder of transport or utilization of copper that characterizes Menkes' disease. Except for the last, for which no treatment is effective, and which is invariably fatal, copper deficiency in man is easily and simply treated by adding about 5 mg of copper, most conveniently as the acetate, to the daily diet. Practically, it is almost never necessary to supplement any but the most abnormal diets with copper to avoid its deficiency in man (SCHEINBERG and STERNLIEB, 1977). Interactions with other elements, particularly with zinc, have also been observed. For instance, copper supplementation protects against lead intoxication provided that the protein status is normal (DE FLORA et al., 1987; for zinc antagonism see Sect. 6.5).

The administration of chelating agents, such as D-penicillamine, that are capable of promoting the excretion of copper only rarely produces clinically significant symptoms of copper deficiency even when no dietary supplements of copper are used.

In humans, as in animals, the physiological role of copper is played as a prosthetic element of perhaps a dozen specific copper proteins which are very similar to those found in animals, listed above.

6.3 Effects on Animals

According to VAN GESTEL (1989) 100 mg Cu/kg soil reduces earthworm cocoon production, but not hatchability. Copper in concentrations exceeding 0.1 mg/L water is toxic to fish (FÖRSTNER and WITTMANN, 1979). Yet some fish and crawfish survive in 0.03 – 0.8 mg/L copper (FÖRSTNER and WITTMANN, 1979).

A report of an average hepatic copper content of over 1000 ppm in 32 mute swans (*Cygnus olor*) from an area polluted with copper (and lead) indicates that excesses of copper can be environmentally induced in this avian species (CLAUSEN and WOLSTRUP, 1978) that, however, shows anyway high hepatic copper concentration.

Ruminants are quite susceptible to copper toxicity. These are the only animals in which significant, and even lethal, copper toxicosis can occur without an inherited abnormality or the addition of dietary copper supplements. Copper toxicosis may develop in sheep taking forage with a normal copper content of 8 to 10 ppm, and this is even more likely to occur if the molybdenum concentration in the diet is below 0.5 ppm. The principal effects are on the liver and blood with fatal hepatitis or hemolytic anemia. In these respects cattle are much like sheep.

Non-ruminant animals – like man – are extremely resistant to the development of copper toxicosis. In fact, to increase growth rates pigs are often fed diets containing 250 ppm copper, without ill effects discernible. But since farmers in the United States often use the same feeds for ruminant and non-ruminant animals, the Federal Food and Drug Administration has limited the concentration of copper in all farm animal feeds to 15 ppm to avoid inadvertent copper poisoning of ruminants (BUCK, 1977). An exception to this immunity is found in certain Bedlington terriers who have inherited a pair of autosomal recessive genes. This leads to copper toxicosis of the liver that resembles in several respects the hepatic disorder seen in Wilson's disease of man (TWEDT et al., 1979; JOHNSON et al., 1980).

6.4 Acute Effects on Humans

For the reasons already stated above, copper is remarkably non-toxic. Soluble salts of copper are almost only poisonous when they are ingested through misguided or suicidal intent or are used as topical medical treatment of extensively burned areas of the skin. In the former instances gram quantities of copper sulfate, also known as blue vitriol or bluestone, lead to nausea, vomiting, diarrhea, sweating, and − in severe poisoning − to convulsions, coma, and death. Milligram quantities of ingested copper, commonly the result of carbonated drinking water or citrous juices that have been in prolonged contact with copper vessels, tubing, pipes, or valves, cause only vomiting and diarrhea which usually protect the patient against such serious systemic toxic effects as hemolysis, hepatic necrosis, gastrointestinal bleeding, oliguria, and azotemia. Where copper salts have been applied to large burned areas, high concentrations of serum copper and toxic manifestations have ensued (HOLTZMANN et al., 1966). The inhalation of dusts, fumes, and mists of copper salts can cause congestion of the nasal and mucous membranes, occasionally producing perforation of the nasal septum. Fumes from volatilized metallic copper can cause nausea, gastric pain, and diarrhea (SCHEINBERG, 1971; SCHEINBERG and STERNLIEB, 1977).

Hemodialysis may introduce sufficient copper into the patient's circulation to cause hemolysis (MANZLER and SCHREINER, 1970) or febrile reactions (LYLE et al., 1976). This is derived from the content of acidic dialysate with copper tubing or semipermeable membranes that are generally fabricated with copper.

6.5 Chronic Effects on Humans

Presence of a fragment of metallic copper in the eye (chalcosis bulbi) may result in the loss of the eye or sunflower cataracts (HANNA and FRAUNFELDER, 1973; ROSEN, 1949).

Bordeaux mixture, a 1−2% solution of copper sulfate neutralized with hydrated lime, is used widely to prevent mildew on grapevines, particularly in France, Portugal, and Southern Italy. Pulmonary copper deposition and fibrosis occurred in some vineyard workers after years of exposure to such solutions (PIMENTEL and MARQUES, 1969). Their lungs may be blue suggesting the presence of excess copper. More recently, granulomas and malignant tumors have appeared in these workers' livers and lungs (PIMENTEL and MENEZES, 1975; VILLAR, 1974). In contrast, studies of Chilean copper miners have shown that liver and serum concentrations of copper are normal despite years of exposure to copper sulfide and oxide dusts (SCHEINBERG and STERNLIEB, 1977).

Wilson's disease occurs with a remarkably uniform world-wide prevalence of about 30 per million. Since it is known to be transmitted by a pair of autosomal recessive genes, the heterozygous individual, in whom the disease never develops, constitutes about 1.1% of the general population (SCHEINBERG and STERNLIEB, 1984).

Almost all patients with Wilson's disease exhibit a life-long deficiency of the plasma copper protein, ceruloplasmin, and an excess of hepatic copper. The latter finding appears to be caused, at least in part, by impairment of lysosomal excretion of hepatic copper into the bile, and is associated with diminished or absent hepatic synthesis of ceruloplasmin. This genetic defect causes the net retention of no more than 1% of the dietary intake of copper – or about 20 mg per year –, yet this is sufficient to result in progressive and fatal copper toxicosis.

The diagnosis of Wilson's disease can be unequivocally ruled in, or out, in almost every patient in whom it is suspected. In any individual over the age of one year, the simultaneous presence of less than 20 mg of ceruloplasmin per 100 mL of serum, and of more than 250 µg of copper per gram of dry liver is sufficient to confirm this diagnosis. In an individual with certain characteristic physical signs of neurological disturbance, the presence of deposits of copper in Descemet's membrane of the ocular corneas – Kayser-Fleischer rings – is also sufficient to confirm the diagnosis. Since the disease is considered to be uniformly progressive and fatal if untreated, and since treatment itself can be accompanied by serious, occasionally lethal, side effects, it is oviously absolutely critical to be certain of whether the patient has the disease or not.

During the early stages of Wilson's disease the liver is capable of binding as much as 30 to 50 times its normal concentration of copper with little, if any, overt clinical disorder. Ultimately hepatic copper is released into the bloodstream, and, if this occurs together with massive necrosis of hepatic parenchyma, there may be sufficiently large amounts of copper infused to produce a severe hemolytic anemia (DEISS et al., 1971). In most patients, however, the metal diffuses into the circulation gradually, causing the plasma concentration of free copper to rise from little more than zero to 25 to 50 µg/100 mL or more. This copper diffuses out of the vascular compartment into extracellular fluids and tissues with toxic effects in susceptible cells, principally in the liver itself and in the brain. Untreated, the patient's illness progresses to death.

In about half of the patients with Wilson's disease, the first clinical symptoms are hepatic; in almost all the remaining patients, neurologic or psychiatric disturbances, or both, are seen first and reflect copper toxicity to the brain. Blood relatives of patients, most importantly siblings, must have the diagnosis ruled in or out even if they are asymptomatic, since clinical disturbances of any sort never occur before the age of 5 years, and may not appear until the fifth decade. Such individuals with deficiency of ceruloplasmin and excess of hepatic copper of the quantitative degrees described above must be treated for life just as vigorously as clinically symptomatic patients (SCHEINBERG and STERNLIEB, 1984).

Specific treatment of Wilson's disease consists of the removal of the excess deposits of copper. This is accomplished in almost all patients by the daily and life-long administration of D-penicillamine. Such therapy can reverse most, if not all, of the clinical disturbances caused by Wilson's disease in the majority of symptomatic patients; and can maintain asymptomatic patients in that state for what appears to be a normal life-span. Patients in whom intolerance to D-penicillamine develops may be equally successfully treated with trientine (triethylene tetramine dihydrochloride) (SCHEINBERG et al., 1987). Continuous oral zinc acetate therapy seems also to be successful (COSSACK, 1987) although conclusive clinical data are still lacking.

Hepatic accumulations of copper can occur in disturbances of biliary excretion, since this is a route of excretion of significant amounts of copper. Thus patients with primary biliary cirrhosis may exhibit concentrations of hepatic copper of the same order of magnitude as seen in Wilson's disease. These excess deposits of copper are secondary to a pathogenetic process that is clearly different from that of Wilson's disease, and their removal by D-penicillamine does not appear to be beneficial to the patients.

Extremely high concentrations of hepatic copper are seen in Indian childhood cirrhosis, a progressive, fatal disease of infants of unknown cause. It is not known whether the accumulation of copper, which can reach 5000 µg per gram dry liver, precedes or follows the development of the pathologic picture that may resemble subacute viral hepatitis, fulminant necrosis, of micronodular cirrhosis (STERNLIEB, 1980). In Germany five children of different farmers, younger than one year, suffered from severe liver cirrhosis of unclear etiology. They had been formula-fed with copper containing (2.5 – 10 mg/L) acid water from similar private wells (BUNDESGE-SUNDHEITSAMT, 1988).

7 Hazard Evaluation and Limiting Concentrations

Since copper is both an essential and potentially toxic element there may be risks to living beings if there is either too little or too much in the environment. Copper concentrations in wastes should be reduced, since they may be responsible for the catalytic activation of chlorinated dioxin formation during incineration (STIEGLITZ, DICKSON, and NAIKWADI, 1988).

Agricultural soils with 100 ppm or a limit of 2 – 3 kg Cu per ha per year are considered acceptable (HOFFMANN, 1980). In England even 280 kg/ha within 30 years is not excessive (PURVES, 1981). Manure contains above 200 ppm Cu (COTTENIE, 1981). The copper content of sewage sludge dry mass is in the range of 400 – 600 ppm; of municipal waste compost, 100 – 150 ppm (HOFFMANN, 1980; PURVES, 1981). In an EC Guideline of the European Council a limiting value of 0.04 mg/L of copper has been set for waters in which there are salmon and trout (RUF, 1981).

Ruminants are sensitive to both deficiency and excess of copper in the environment, with consequences on both scores that are of economic importance particularly in the raising of sheep and cattle. Non-ruminant animals, in contrast, are resistant to both the natural and experimental induction of copper deficiency or toxicosis. This state of affairs must be the consequence, almost surely, of the existence of inherited homeostatic mechanisms buffering the effects of wide environmental concentrations of this element in non-ruminants. Compelling evidence for the existence of a genetic mechanism protecting dogs from environmentally caused copper toxicosis has been referred to in Sect. 6.3.

Except for genetically abnormal individuals in whom Menkes' or Wilson's disease develops, for individuals with grossly inadequate diets, diarrhea and/or severe malabsorption, and, perhaps, for patients with primary biliary cirrhosis, cholestatic syndromes of Indian childhood cirrhosis, man is not significantly at risk for either copper deficiency or toxicosis.

In the Federal Republic of Germany the maximal allowed concentrations at workplaces are 0.1 mg/m^3 for copper fume and 1 mg/m^3 for copper dust (MAK, 1987).

In Switzerland the regulations prescribe the following maximal concentrations of copper in food: pectines, 400 ppm; fruit juices, grape juice, vinegar, and alcoholic liquors, 5 – 30 ppm; milk, 0.05 ppm; and beer, 0.2 ppm. The limit for drinking water is 1.5 ppm Cu in Switzerland. This is a relatively high value compared to maxima of: the US Public Health Service, 1 ppm; WHO, 0.05 ppm; USSR, 0.1 ppm (FÖRSTNER and WITTMANN, 1979). Some limitations were chosen for technical reasons and are not health-based.

The recommended daily intake of copper in food is 0.5 – 0.7 mg for children in the first year of life; 2.0 – 3.0 mg (MERTZ, 1980) or up to 5.0 mg (KIEFFER, 1980) for adults.

References

Aaseth, I., and Norseth, T. (1986), Chapter 10: *Copper*, in: Friberg, L., Nordberg, G. F., and Vouk, V. B. (eds.): *Handbook on the Toxicology of Metals*, Vol. II, pp. 233 – 254. Elsevier, Amsterdam.
Alikhan, M. A. (1989), *Copper and Nickel Uptake, Accumulation and Regulation by a Terrestrial Crustacea (Porcellionidae isopoda)*, in: *Proceedings of the 6th International Conference on Heavy Metals in the Environment, Geneva*, Vol. 1, pp. 475 – 478. CEP Consultants Ltd., Edinburgh.
Anderson, P. D. (1988), *Synergism and Metal Mixtures*, in: *Proceedings of the International Conference on Trace Metals in Lakes, Hamilton, Canada. Sci. Total Environ.*, in press.
Beppler, E., Fichte, R., and Berger, A. (1978), *Mangan*, in: *Ullmanns Encyklopädie der technischen Chemie*, 4th Ed., Vol. 16, pp. 454 – 455. Verlag Chemie, Weinheim-New York.
Bowen, H. I. M. (1979), *Environmental Chemistry of the Elements*. Academic Press, London.
Braude, R., and Ryder, K. J. (1973), *Agric. Sci. Canberra 80*, 489 – 493.
Bruland, K. W. (1983), *Trace Elements in Sea Water. Chem. Oceanogr. 8*, 205 – 208.
Buck, W. B. (1977), in: *Copper*, Chap. 4. National Academy of Sciences, Washington, D.C.
Bundesgesundheitsamt (1988), *Infant Liver Cirrhosis?* (in German), BGA Press Release 09/1988 (18 March), Berlin.
Cartwright, G. E., Gubler, C. J., Bush, J. A., and Wintrobe, M. M. (1956), *Blood 11*, 143 – 153.
Clausen, B., and Wolstrup, C. (1978), *Nord. Veterinaermed. 30*, 260 – 266.
Cossack, Z. (1987), *Metal-Storage Diseases*, Symposium Report, in: *Proceedings Second Nordic Symposium on Trace Elements, Odense. Environ. Health 26*, WHO (World Health Organization), Copenhagen.
Cottenie, A. (1981), in: *Proceedings of the International Conference on Heavy Metals in the Environment, Amsterdam*. CEP Consultants Ltd., Edinburgh.
Davis, G. K. (1974), *Fed. Proc. Fed. Am. Soc. Exp. Biol. 33*, 1194 – 1196.
Davis, G. K. (1977), in: *Copper*, Chap. 2. National Academy of Sciences, Washington, D.C.
Dawson, C. R., and Price, C. A. (1977), in: *Copper*, Chap. 3. National Academy of Sciences, Washington, D.C.

de Flora, S. J. S., Sharma, R. P., Coulombe, Jr., R. A., and Tandon, S. K. (1987), *Influence of Copper on Lead Intoxication during Protein Malnutrition*. Poster C7, *Symposium Abstracts and Proceedings Second Nordic Symposium on Trace Elements in the Environment, Odense*. WHO (World Health Organization), Copenhagen.
Deiss, A., Lee, G. R., and Cartwright, G. E. (1970), *Ann. Intern. Med.* **73**, 413–418.
Dive, D. G., Gabriel, Hanssens, O., and Benga-Bengomme, A. (1988), *Interactions between Components of Electroplating Industry Wastes: Influence of Nickel and of the Calcium Richness of the Water on the Cadmium-Copper-Zinc-Chromium Interactions*, in: *Proceedings of the International Conference on Trace Metals in Lakes, Hamilton, Canada. Sci. Total Environ.* **87/88**, 355–364.
Dresch, H. (1988), *Current Limiters Using Superconductors* (in German). *Neue Zürcher Zeitung, Forschung und Technik, No. 236*, p. 83 (11 October), Zürich.
Fabian, H. (1986), *Copper*, in: *Ullmann's Encyclopedia of Industrial Chemistry*, 5th Ed., Vol. A7, pp. 471–523. VCH Verlagsgesellschaft, Weinheim-Deerfield Beach/Florida-Basel.
Förstner, U., and Wittmann, G. T. W. (1979), *Metal Pollution in the Aquatic Environment*. Springer, Berlin-Heidelberg-New York.
Gadd, G. M. (1981), in: *Proceedings of the International Conference on Heavy Metals in the Environment, Amsterdam*, p. 285. CEP Consultants Ltd., Edinburgh.
Haerdi, W. (1989), *Overview on Panoramic Cation Determination by Ion Chromatography. Lecture at the 19th Annual IAEAC Symposium on Environmental Analytical Chemistry, Jekyll Island, Georgia*.
Hanna, C., and Fraunfelder, F. T. (1973), *Ann. Ophthalmol.* **5**, 9–22.
Hering, J. G. (1988), *Kinetics of Copper Complexation in Aquatic Systems. Lecture at the EAWAG Seminar, Dübendorf, Switzerland*.
Hoffmann, G. (1980), in: *Proceedings Müll- und Müllklärschlammkomposte in der Landwirtschaft*. Rüschlikon/Zürich, Gottlieb-Duttweiler-Institut.
Holtzman, N. A., Elliot, D. A., and Heller, R. H. (1966), *N. Engl. J. Med.* **275**, 347–352.
Hottmeier, H. J. (1978), *Metall* **32**, 1157–1163.
Hutchinson, T. C. (1981), in: *Proceedings of the International Conference on Heavy Metals in the Environment, Amsterdam*. CEP Consultants Ltd., Edinburgh.
Huynh-Ngoc, L., Whitehead, N. E., and Oregioni, B. (1988), *Low Levels of Copper and Lead in a Highly Industrialized River. Toxicol. Environ. Chem.* **17**, 223–236.
Isbell, Jr., C. A. (1986), *Copper Alloys*, in: *Ullmann's Encyclopedia of Industrial Chemistry*, 5th Ed., Vol. A7, pp. 525–566. VCH Verlagsgesellschaft, Weinheim-Deerfield Beach/Florida-Basel.
Jenkins, D. W. (1981), in: *EPA-600/53-80-090*. U.S. Environmental Protection Agency, Washington, D.C.
Johnson, G. F., Sternlieb, I., Twedt, D. C., Grushoff, P. S., and Scheinberg, I. H. (1980), *Am. J. Vet. Res.* **41**, 1865–1866.
Kieffer, F. (1980), in: *Neue Zürcher Zeitung, Forschung und Technik*, Nr. 198, p. 59, Zürich.
Kloke, A. (1980), in: *Proceedings Müll- und Müllklärschlammkomposte in der Landwirtschaft*. Rüschlikon/Zürich, Gottlieb-Duttweiler-Institut.
Knöchel, A. (1989), *TXRF, PIXE, SYXFA – Principles, Critical Comparison and Applications. Abstracts 11th International Symposium on Microchemical Techniques, Wiesbaden. Fresenius Z. Anal. Chem.* **334** (7), 608–609.
Kodama, Y. (1988), *Cadmium Distribution in Blood and Urine of Dogs after Long-term Administration of Low Doses of Cadmium. Special Issue of the 4th IUPAC Cadmium Workshop, Schmallenberg-Grafschaft, FRG. Toxicol. Environ. Chem.* **27** (1–3), 73–80.
Krämer, R., and Podlesak, W. (1987), in: *Proceedings Mengen- und Spuren-Elemente*, pp. 294–299. Karl-Marx-Universität, Leipzig.
Leatherland, T. M. (1979), in: *Proceedings of the International Conference on Heavy Metals in the Environment, London*. CEP Consultants Ltd., Edinburgh.
Lepp, N. W. (1981), *Effect of Heavy Metal Pollution on Plants*. Applied Science Publishers, London.

Lepp, N. W., and Dickinson, N. M. (1988), *Long-term Accumulation Patterns of Copper in Different-aged Coffea arabica Stands. Proceedings 3rd International Conference on Environmental Contamination, Venice*, pp. 67–69. CEP Consultants Ltd., Edinburgh.
Löhle, J., Mattenberger, K., and Wachter, P. (1989), *Production and Characterization of Wires of High Temperature Superconducting Materials based on* $(Bi_{1.6}Pb_{0.4}Sr_2Ca_3Cu_4O_x)$ (in German). *ETH-Z Bulletin No. 223*, pp. 21–23. Swiss Federal Institute of Technology, Zürich.
Lyle, W. H., Payton, J. E., and Hui, I. (1976), *Lancet 2*, 1324–1325.
Macnair, M. R. (1977), *Nature 268*, 428–430.
MAK (1987), *Maximum Concentrations at the Workplace, Report No. XXIII, DFG*. VCH Verlagsgesellschaft, Weinheim-Basel-Cambridge-New York.
Manzler, A. D., and Schreiner, A. W. (1970), *Ann. Intern. Med. 73*, 409–412.
Mart, L., Rützel, H., Klahre, P., Sipos, L., Platzke, U., Valenta, P., and Nürnberg, H. W. (1982), in: Wong, C. S. (ed.): *Trace Metal in Sea Water*. Plenum Press, New York.
McNeilly, T., Farrow, S., and Putwain, P. D. (1981), in: *Proceedings of the International Conference on Heavy Metals in the Environment, Amsterdam*. CEP Consultants Ltd., Edinburgh.
Mertz, W. (1980), cited in: Merian, E., *Chem. Rundsch. 33*, Nr. 23, p. 9 and Nr. 25, p. 5 (4 and 18 June).
Montaser, A., and Golightly, D. W. (1987), *Inductively Coupled Plasmas in Analytical Atomic Spectrometry*, pp. 354 ff, 611 ff. VCH Verlagsgesellschaft, Weinheim-Basel-Cambridge-New York.
Morel, J.-L., Bitton, G., and Koopman, B. (1988), *Use of Microtox® for Assessing Copper Complexation with Organic Compounds. Arch. Environ. Contam. Toxicol. 17*, 493–496.
Mylroie, A. A., Smith, J. C., and Suttle, N. F. (1987), *Effects of Dietary Copper Deficiency on Rat Pancreas, Copper Nutrition and Cardiovascular Integrity, Copper and Disease Resistance*, in: Hemphill, D. D. (ed.), *Trace Substances in Environmental Health – XXI*, pp. 44 ff, 499 ff, 514 ff. University of Missouri, Columbia.
Parrot, J. L., Bowen, G. M., and Sprague, J. B. (1988), *Patterns of Sublethal Toxicity of Copper and Zinc Mixtures Determined by the three Rapid Assays*, in: *Proceedings of the International Conference on Trace Metals in Lakes, Hamilton, Canada. Sci. Total. Environ.*, in press.
Pimentel, J. C., and Marques, F. (1969), *Thorax 24*, 678–688.
Pimentel, J. C., and Menezes, A. P. (1975), *Am. Rev. Respir. Dis. 111*, 189–195
Purves, D. (1981), in: *Proceedings of the International Conference on Heavy Metals in the Environment, Amsterdam*. CEP Consultants Ltd., Edinburgh.
Richardson, H. W. (1986), *Copper Compounds*, in: *Ullmann's Encyclopedia of Industrial Chemistry*, 5th Ed., Vol. A7, pp. 567–593. VCH Verlagsgesellschaft, Weinheim-Deerfield Beach/Florida-Basel.
Rosen, E. (1949), *Am. J. Ophthalmol. 32*, 248–252.
Rozema, J., Otte, M. L., van Schie, C., Ernst, W. H. O., Broekman, R. A., and de Koe, T. (1988), *Foliar Uptake of Heavy Metals by Estuarine Plants in Response to Contaminated Seawater Flooding, Heavy Metals and Arsenic in Water, Sediment and Plants near the Jales Gold and Silver Mine in North Portugal*, in: *Proceedings of the 3rd International Conference on Environmental Contamination, Venice*, pp. 73–75, 152–154. CEP Consultants Ltd., Edinburgh.
Rüegg, J. (1989), *Mycorrhiza-fungi and Pollutants, a New Biotest with Root Tips. Neue Zürcher Zeitung, Forschung und Technik, No. 117*, p. 77 (24 May), Zürich.
Ruf, M. (1981), *Kongreßvorträge Wasser Berlin '81*, pp. 415–428. Colloquium Verlag Otto H. Hess, Berlin.
Saager, R. (1984), *Metallic Raw Materials Dictionary* (in German), pp. 68–72. Bank von Tobel, Zürich.
Scheinberg, I. H. (1971), in: *Encyclopaedia of Occupational Health and Safety*, Vol. I, pp. 331–333. International Labour Office, Geneva.
Scheinberg, I. H., and Lawrence, C. (1988), *Sideroblastic Anemia in a Patient with Wilson's Disease Receiving Trientine* (in preparation).

Scheinberg, I. H., and Sternlieb, I. (1977), in: *Copper*, Chap. 5. National Academy of Sciences, Washington, D.C.
Scheinberg, I. H., and Sternlieb, I. (1984), *Wilson's Disease*. W. B. Saunders Co., Philadelphia.
Scheinberg, I. H., Jaffe, M. E., and Sternlieb, I. (1987), *N. Engl. J. Med. 317*, 209–213.
Sternlieb, I. (1980), *Gastroenterology 78*, 1615–1628.
Stieglitz, L., Dickson, L. C., and Naikwadi, Kr. P. (1988), *Carbonaceous Particles in Fly Ash – a Source for the de-novo Synthesis of Organochloro Compounds, Surface-catalysed Formation of Chlorinated Dibenzodioxins and Dibenzofurans During Incineration, Effect of Temperature, Carrier Gas and Precursors' Structure on PCDD and PCDF Formed from Precursors by Catalytic Activity of MSW Incinerator Fly Ash, Prevention of PCDD Formation in MSW Incinerators by Inhibition of Catalytic Activity of Fly Ash Produced*, in: Proceedings of the 8th International Symposium on Chlorinated Dioxins and Related Compounds, Umeå, Sweden. *Chemosphere*, in press.
Trueb, L. (1987), *New Superconductors: Applications and Eventually a Technical Revolution* (in German). Neue Zürcher Zeitung, Forschung und Technik, No. 131, p. 85 (10 June), and No. 290, p. 15 (14 December), Zürich.
Trueb, L. (1988a), *Gold Rush in the South Sea* (in German). Neue Zürcher Zeitung, Forschung und Technik, No. 109, p. 87 (11 May), Zürich.
Trueb, L. (1988b), *Superconduction – a Fascinating Phenomenon* (in German). Neue Zürcher Zeitung, Forschung und Technik, No. 298, p. 57 (21 December), Zürich.
Trueb, L., and Nutting, J. (1988), *Resource-dependent and Environmental Problems in the Age of New Materials* (in German). Neue Zürcher Zeitung, Forschung und Technik, No. 236, p. 73 (11 October), Zürich.
Twedt, D. C., Sternlieb, I., and Gilbertson, S. R. (1979), *J. Am. Vet. Med. Assoc. 175*, 269–275.
Twiss, M. R., Nalewaijko, Cz., and Stokes, P. M. (1989), *The Influence of Phosphorus Nutrition on Copper Tolerance in Scenedesmus chlorophyceae Strains*, in: Proceedings of the 6th International Conference on Heavy Metals in the Environment, Geneva, Vol. 2, pp. 174–177. CEP Consultants Ltd., Edinburgh.
van Gerstel, C. A. M. (1988), *A Standardized Reproduction Test with the Earthworm Eisenia fetida andrei*, in: Proceedings of the 1st European SECOTOX Conference on Ecotoxicology, Copenhagen, in press.
Venugopal, B., and Luckey, T. D. (1979), *Metal Toxicity in Mammals*. Plenum Press, New York.
Verkleij, I. A. C., Marissen, A., and Lugtenborg, T. F. (1981), in: Proceedings of the International Conference on Heavy Metals in the Environment, Amsterdam. CEP Consultants Ltd., Edinburgh.
Villar, T. G. (1974), *Am. Res. Respir. Dis. 110*, 545–555.
Wachs, B. (1981), *Sicherh. Chem. Umwelt 1*, 113–115.
Ward, N. I. (1987), *The Future of Multi-(Ultra-Trace-)Element Analysis in Assessing Human Health and Disease: A Comparison of NAA and ICP-MS*. *Environ. Health 20*, 118–123. Extended Odense Abstract, WHO (World Health Organization), Copenhagen.
Welz, B. (1985), *Atomic Absorption Spectrometry*, pp. 202–283, 408 ff. VCH Verlagsgesellschaft, Weinheim-Deerfield Beach/Florida-Basel.
Zipper, J. A., Tatum, H. J., Pastene, L., Medel, M., and Rivera, M. (1968), *Am. J. Obstet. Gynecol. 105*, 1274–1278.

Additional Recommended Literature

Bussmann, H. (1989), *State and Development of Copper Recycling. ICR-Berlin Proceedings, Abfallvermeidung in der Metallindustrie*, pp. 159–166. EF-Verlag für Energie- und Umwelttechnik, Berlin; see also Report E. Merian, *Swiss Chem. 12* (5), 42–44.

Draxler, J., and Marr, R. (1989), *Liquid Membrane Permeation. ICR-Berlin Proceedings, Abfallvermeidung in der Metallindustrie*, pp. 139–148. EF-Verlag für Energie- und Umwelttechnik, Berlin; see also Report E. Merian, *Swiss Chem. 12* (5), 40–42.

Riss, A., Hagenmaier, H., and Rotard, W. (1990), *Effects of Dioxin Emissions on Soils* (Related to Metal Recovery), *on Plant Growth*, and *on Cow Milk Quality* (in German). *VDI-Proceedings on Effects of Atmospheric Pollution on Soils, Lindau Meeting*, in press.

Streit, B., Krüger, C., Lahner, G., Kirsch, S., Hauser, G., and Diehl, B. (1990), *Uptake and Storage of Heavy Metals by Earthworms in Various Soils* (in German). *UWSF Z. Umweltchem. Oekotoxikol. 2* (1), 10–13.

Teller, M., and Michel, E. (1989), *Treatment of Dangerous Waste from Circuit Board Production in a Combined Pyrolysis/High-turbulence Combustion System. ICR-Berlin Proceedings, Abfallvermeidung in der Metallindustrie*, pp. 225–237. EF-Verlag für Energie- und Umwelttechnik, Berlin; see also Report E. Merian, *Swiss Chem. 12* (5), 44.

II.10 Gallium

PETER L. GOERING, Rockville, Maryland, USA
BRUCE A. FOWLER, Baltimore, Maryland, USA

1 Introduction

Gallium, the name of which is derived from the Latin *gallia*, meaning France, was named to honor the discovery of this element by the French chemist DE BOISBAUDRAN in 1875. The physicochemical properties of gallium are related to iron, and thus the two elements share similar metabolic pathways in various organisms. Gallium has not been shown to have a definite biological function. Primary uses of gallium are as a chemotherapeutic and radioimaging agent in clinical settings and in the manufacture of superfast integrated circuits for semiconductor and telecommunication applications. New and developing technologies, such as manufacture and use of semiconductors, are expected to increase emissions of gallium compounds (e.g., of gallium arsenide) into the environment. The primary adverse effect associated with gallium use – varying with the mode of entry into the body – is nephrotoxicity, although other toxicities have been demonstrated in laboratory and clinical studies.

2 Physical and Chemical Properties, and Analytical Chemistry

Gallium is a silver-gray metal classified as a Group III B metal in the Periodic System of Elements, has an atomic mass of 69.72, and the atomic number 31. Gallium has an ionic radius of 0.062 nm; charge density z/r, 4.84; coordination number, 6; and electron configuration, $3d^{10}$ (GANROT, 1986). The metal is unique in that it has the widest liquid range of any metal, with melting and boiling points of 29.78 °C and 2400 °C, respectively. It is stable in dry air but tarnishes in the presence of moisture or oxygen and reacts with alkalis and acids. Numerous forms of gallium occur including sesquioxide, hydroxide, chloride, sulfate, and nitrate salts. Natural isotopes of gallium include 69 and 71, and artificial radioactive isotopes are 63–68, 70, 72–76 (MERCK INDEX, 1983). ^{67}Gallium is important as a clinical diagnostic tool in nuclear medicine and has a half-life of 72 h (GANROT, 1986). The +1 and +2 oxidation states of gallium are unstable with the most stable aqueous oxidation state being +3. The chemistry of gallium in many ways resembles that of ferric iron, sharing similar ionization potentials; ionic radii, and coordination numbers (MOERLEIN and WELCH, 1981). Gallium is a strong Lewis acid, a potentially powerful proton generating system. At neutral pH, gallium preferentially binds to thiol (SH) groups (WILLIAMS, 1984).

Trace concentrations of gallium present in tissues and biological fluids can be determined by several analytical methods. Early methodology involving the fluorometric determination of gallium utilized reagents such as lumogallion, methyl-8-hydroxyquinoline, 2-(2-pyridyl)benzimidazole, and rhodamine B. These methods are hampered by several laborious extraction steps. A recent fluorometric procedure based on the reaction of gallium with salicylaldehyde thiocarbohydrazone (URENA et al., 1985) has advantages including high sensitivity, selectivity, and simplicity. The detection limit is 2 ng/mL. Another method has been developed for the spectrofluorometric determination of 1–80 ng/mL gallium in urine and blood using pyrocatechol-1-aldehyde 2-benzothiazolylhydrazone (AFONSO et al., 1985). In general, fluorometric procedures for determining gallium concentrations in biological samples are limited by extensive tissue, reagent, and glassware preparation or are susceptible to interferences from other cations and anions. The method of choice to quantitate gallium in tissues and body fluids is flameless atomic absorption spectrophotometry, which can detect gallium at levels of 100–200 parts per billion (NEWMAN, 1978). Flame atomic absorption and flame emission spectrophotometric methods are also suitable for analyzing gallium in excreta and tissues, with detection limits in the parts per million range (WEBB et al., 1984; WELZ, 1985). Phase-selective anodic stripping voltammetry is a rapid and simple electroanalytic technique which is sensitive to less than 1 ppm gallium in ashed tissue (MOORHEAD and DOUB, 1977). A simple test to control high-purity gallium has been developed by HILTENKAMP and JACKWERTH (1988).

3 Sources, Production, Important Compounds, Uses

3.1 Occurrence, Production

Gallium is mainly extracted and concentrated from bauxite (containing 100 ppm Ga), but also from zinc sulfide by-products and from wastes produced together with phosphorus in electrothermal processes for the recycling of stove dusts (BROUHIER, 1976). Gallium consumption in the United States was reported to be 8.2 tons in 1978 and imports constituted 3.7 tons (PETKOF, 1980). Domestic consumption in 1979 had risen to 9.46 tons, and projected demand in the United States by the year 2000 is approximately 35 tons based largely on expected increased use of gallium in the electronics industry (PETKOF, 1981). In the early 1980s, the world production was of the order of 20 tons annually (BROUHIER, 1976; HAYES, 1988), mainly used in the semiconductor industry.

3.2 Important Compounds and Uses

Several gallium compounds have found utility in the electronics industry including gallium arsenide, gallium phosphide, gallium antimonide, trimethyl- and triethyl gallium (DUPUIS, 1984). These compounds have seen increased use in integrated circuits for semiconductor, laser diode, Gunn effect diode, photoelectrical materials, radio, television, and solar cell applications (KIRK-OTHMER, 1978; ANONYMOUS, 1975). Compared to the standard semiconductor silicon, gallium compounds possess superior electronic and optical properties, including electron flow velocities (DUPUIS, 1984). Gallium arsenide is the material of choice for use in high frequency microwave and millimeter wave telecommunications and ultrafast supercomputers (ROBINSON, 1983). Gallium arsenide may also be of interest in the future in light concentration solar cells (the efficiency is 27.5% compared with a theoretical maximum of 30%; ANONYMOUS, 1987). Gallium antimonide can be used for optical memories (digital optical reading, compact discs, etc.), in combination with laser signals (PHILIPS, 1987). For high-purity gallium see HILTENKAMP and JACKWERTH (1988).

Due to the fact that gallium is cytotoxic and localizes in several tumors, gallium compounds are being investigated therapeutically for use as anti-tumor agents and have undergone evaluation in Phase I and Phase II clinical trials (KRAKOFF et al., 1979; SAMSON et al., 1980; SCHWARTZ and YAGODA, 1984). Gallium nitrate effectively inhibited the growth of various experimental sarcomas and carcinomas (HART and ADAMSON, 1971; HART et al., 1971), including Walker carcinoma 256, fibrosarcoma M-89, leukemia K-1964, adenocarcinoma 755, lymphoma P 1798, and osteosarcoma 124F (FOSTER et al., 1986). Gallium nitrate has been evaluated as a treatment modality for Hodgkin's disease and non-Hodgkin's lymphoma (WEICK et al., 1983; WARRELL et al., 1983). This drug produces hypocalcemia due to its property of inhibition of calcium resorption from bone. Thus, it may also be useful in treatment of cancer-related hypercalcemia due to accelerated bone resorption (WARELL et al., 1985).

The radionuclide ^{67}gallium, primarily in the citrate form, has been an important diagnostic radiopharmaceutical for the localization of specific neoplastic lesions in humans, particularly bronchogenic carcinomas, Hodgkin's disease, and non-Hodgkin's lymphomas (SILBERSTEIN, 1976; JOHNSTON, 1981). ^{67}Gallium citrate has been found to have broader application in the diagnosis and staging of various inflammatory and infectious diseases including pulmonary lesions such as pneumonia, sarcoidosis, pneumocystis carinii, idiopathic fibrosis; renal lesions such as acute pyelonephritis; abdominal infections including peritonitis and abscesses; and bone and joint infections such as osteomyelitis (HOFFER, 1981). According to BLAIR et al. (1987) ^{67}GaNO$_3$ is also a good model for the intestinal absorption and subsequent tissue distribution of aluminum (relatively somewhat more gallium being accumulated in the spleen, somewhat more aluminum in the femur).

The melting point of gallium has also been proposed to be used as a fixed standard to calibrate thermistor probes of electronic thermometers used for clinical chemistry and experimental laboratory analyses. This standard is appropriate because: (1) the melting point of gallium (29.78 °C) falls within the region of critical importance to laboratory biological determinations (25–37 °C), and (2) the melt can be maintained accurately and constantly for several hours (MANGUM, 1977).

4 Distribution in the Environment, Food, and Living Organisms

Gallium constitutes 5×10^{-4}% of the earth's crust and is found primarily in the mineral germanite, a copper sulfide ore, and is also found in zinc blends, bauxite, aluminum clays, and ores of manganese, chromium, and iron (MERCK INDEX, 1983). It is the 32nd most abundant element in the earth's crust. According to older measurements seawater concentrations are of the order of 0.03 ppb (BRULAND, 1983). Gallium is present in part per million concentrations in coal (HILDEBRAND and CUSHMAN, 1978). In semiconductor and solar cell production indoor gallium arsenide dust emission losses are relatively high (FOWLER and DOBROTA, 1989).

Concentrations in plants are of the order of 10 ppb to 2 ppm (dry matter; BOWEN 1979). Mammalian muscle may contain 1.4 ppb gallium, marine fish 100 times more (BOWEN, 1979). WARD (1987) found 5–8 ppb gallium in bovine liver. The occurrence of gallium in healthy human tissues (in parts per billion) is: plasma, 0.1; liver, 0.7 to 1; brain, 0.6; kidney, 0.9; lung, 5; lymph node, 7; muscle, 0.3; testis, 0.9; and ovary, 2 (LANGE, 1973; HAMILTON et al., 1972).

5 Uptake, Absorption, Transport and Distribution, Metabolism and Elimination in Animals and Humans

Gallium administered in mg quantities to laboratory animals is rapidly deposited in bone primarily at the site of osteogenic activity and persists there for some time (DUDLEY and LEVINE, 1949; DUDLEY and MARRER, 1952; MAUREL et al., 1974). In rats and humans, soft tissues such as liver and kidney (as well as the spleen; see Sect. 3.2) initially concentrate gallium but levels decline rapidly (NELSON et al., 1972; MOERLEIN et al., 1981). Gastrointestinal uptake of gallium appears to be negligible (DUDLEY et al., 1950), and iron deficiency does not enhance uptake (VALBERG et al., 1981). Following i.v. injection in humans, gallium nitrate disappearance from plasma was biphasic with an initial half-life of 1.5 h and a terminal half-life of 25 h which is highly dependent upon renal function (KRAKOFF et al., 1979). Whole-body retention of gallium may involve two compartments, one with a short biological half-life and the other, which contains 80% of the dose, with a half-life of 20 to 30 days (WATSON et al., 1973). Following pulmonary exposure to gallium oxide and gallium arsenide particulates, 36–44% of the administered dose remained in the lungs 14 days after treatment (WEBB et al., 1986). Gallium has a longer half-life in the lung than arsenic, which is methylated (FOWLER and DOBROTA, 1989).

The uptake and distribution of gallium in organisms appears to be influenced by binding to at least three iron-binding molecules, whose relative affinity for gallium ranks as follows: siderophores > lactoferrin > transferrin (HOFFER, 1980). Gallium also binds to ferritin. Gallium administered i.v. binds rapidly to plasma proteins (HARTMAN and HAYES, 1969), principally transferrin. Transferrin increases uptake of gallium in various tumor cell lines (SEPHTON, 1981), and ferric iron competes with gallium for uptake (RASEY et al., 1981). In contrast, transferrin inhibited uptake while iron deficiency markedly enhanced uptake of gallium in hepatocytes (SCHEFFEL et al., 1979). In the presence of tumors and inflammation, gallium concentrates in tissues which contain tumors and inflammatory processes, and controversy exists as to the mechanism(s) involved in this preferential uptake and distribution. Localization of ^{67}Ga in tumors appears to be biphasic with (1) an early phase, during which tumor gallium can be extracted by iron-binding chelators, such as Desferal, and (2) a second phase (after 24 h) during which gallium is no longer extractable. This suggests that gallium may accumulate preferentially by an early weak binding or diffusion mechanism followed by an intracellular mechanism which binds gallium with higher affinity (HOFFER, 1980). Hyperpermeability of tumor vessels and endocytotic processes may also be partially responsible (HOFFER, 1980). The mechanism by which gallium nitrate is selectively localized in tumors may involve the increased binding of gallium-transferrin complexes to transferrin cell surface receptors, whose density is elevated in some tumors (SCHWARTZ and YAGODA, 1984; DE ABREW, 1981; LARSON et al., 1979; SEPHTON and HARRIS, 1975). Other studies suggest that uptake of gallium may be a function of tumor pH and the free ionic gallium concentration (VALLABHAJOSULA et al., 1981; HAYES and BROWN, 1975). Other factors which influence the tissue distribution of gallium include: presence of malignancy, age, sex, inflammation, tissue viability, diet, pregnancy, lactation, and exposure to ionizing radiation (HAYES, 1983). Gallium localization at sites of inflammation may be mediated by gallium binding to lactoferrin, which is heavily concentrated in leukocytes, a common cell type in inflammatory exudates, or gallium binding with high affinity to siderophores, ferric iron chelates, found in microorganisms (HOFFER, 1980; EMERY and HOFFER, 1980).

At the subcellular level, gallium is primarily associated with lysosomes in normal and tumor tissue and small lysosomal-like particles in tumor tissue (SWARTZENDRUBER et al., 1971; BROWN et al., 1973; BERRY et al., 1983). Gallium in tumor is also associated to a high degree with a $4-5 \times 10^4$ dalton glycoprotein (LAWLESS et al., 1978).

The elimination of gallium from rats is rapid with excretion occurring primarily via the urinary route during the first 24 hours following treatment. Fecal elimination appears not to result from biliary excretion but it seems to occur via passage of gallium across the gastric wall from blood into the stomach (MAUREL et al., 1974). Following i.v. injection in humans, 25 – 60% of the dose is excreted in the urine within the first 24 h. Subsequent excretion is slower and occurs through the bowel (KRAKOFF et al., 1979; HOFFER, 1980). Following injection of gallium in rabbits, 96% of the excreted gallium dose was found in the urine (DUDLEY et al., 1949; DUDLEY and MARRER, 1952). In constrast to the above results, gallium excretion was primarily limited to the fecal route following pulmonary exposure of rats to gallium arsenide (WEBB et al., 1984).

6 Effects on Animals and Humans

6.1 Effects on Aquatic Animals

In a fish study, gallium completely inhibited hatching of carp eggs (HILDEBRAND and CUSHMAN, 1978).

6.2 Miscellaneous Biochemical Effects

Administration of gallium nitrate to rats led to tubular necrosis resulting from the occlusion of the renal tubular lumina with precipitates of gallium, calcium, and phosphate (NEWMAN et al., 1979). Hypercalcuria was also evident. The nephrotoxicity and hypercalcuria could be attenuated by administration of an osmotic diuretic. In humans, bolus gallium nitrate administration produced nephrotoxicity as measured by increased blood urea nitrogen and creatinine values with a decreased creatinine clearance (KRAKOFF et al., 1979). Hypercalcuria and anemia were also observed. Lower doses infused over longer periods reduced the nephrotoxicity of gallium nitrate (WARRELL et al., 1983; LEYLAND-JONES et al., 1983).

Gallium has been shown to have effects at the molecular level. Gallium nitrate inhibited blood, liver, and kidney aminolevulinic acid (ALA) dehydratase in vitro (GOERING et al., 1988). Glutamate dehydrogenase was also inhibited by gallium in vitro (WHITE and YIELDING, 1975). Gallium chloride inhibits the formation of cerebrospinal fluid after administration to rats, by a mechanism involving OH^- depletion which reduces production of HCO_3^-, an essential aspect of cerebrospinal fluid formation (VOGH et al., 1985).

6.3 Acute Effects on Mammals (and Humans)

An important toxic effect of gallium which limits its clinical use is nephrotoxicity. Other studies have demonstrated the hepatotoxic potential of gallium nitrate. Liver injury occurred in dogs administered gallium nitrate repeatedly for five days as evidenced by sulfobromophthalein retention and increases in alkaline phosphatase and serum transaminases (RAKIETEN et al., 1973).

Symptoms of acute toxicity from gallium compounds include the following: drowsiness, metallic taste, anorexia, nausea, vomiting, folliculitis, maculopapular rash, skin edema, exfoliative dermatitis, lymphopenia, leukopenia, decreased platelets, anemia, erythrocytopenia, and bone marrow depression (BRIGGS and OWENS, 1980). ROSCINA (1983) reported on a case of petechial rash followed by a radial neuritis after a short exposure to a small amount of fume containing gallium fluoride.

Pulmonary toxicity following intratracheal instillation of gallium oxide and/or gallium arsenide to rats included increases in total lung content of lipids, protein, DNA, and lung and spleen weight (WEBB et al., 1986; GOERING et al., 1988;

FOWLER and DOBROTA, 1989). In these same studies, perturbations in heme biosynthesis were evident following gallium arsenide exposure, including inhibition of kidney and blood ALA dehydratase and elevation in urinary uroporphyrin. Pneumonitis und chronic inflammation of the alveolar septa were observed following repeated i.p. injections of gallium nitrate (HART et al., 1971).

The lethal effects of gallium are similar in male mice and female rats with corresponding LD_{50} values of 80 mg/kg and 68 mg/kg, respectively (HART et al., 1971). In the same study, gallium was shown to prolong the life-span of rats inoculated with an experimental carcinoma and inhibited the growth of 6 to 8 solid tumors in rodents. The mechanism of tumor inhibition and cytotoxicity produced following gallium administration has been suggested to involve alterations in tumor metabolism of calcium and magnesium, probably by displacing these elements from their usual molecular binding sites which impair processes critical for rapid cell proliferation (ANGHILERI, 1977, 1979).

6.4 Chronic Effects on Mammals

A chronic bioassay exposing mice to 5 ppm gallium in drinking water resulted in loss of body weight, lower survival rates, and an elevated tumor incidence, primarily lymphomas and leukemias (SCHROEDER and MTCHENER, 1971).

6.5 Mutagenic and Teratogenic Effects

Gallium has been demonstrated to produce DNA damage in the rec assay (NISHIOKA, 1975) but has been either negative or untested in other mutagenicity test systems (HANSEN and STERN, 1984).

Gallium was not teratogenic in hamsters following acute injection but an increase in fetal resorptions was observed (FERM and CARPENTER, 1970).

7 Hazard Evaluation and Limiting Concentrations

Difficulty exists in providing a risk assessment for gallium exposure because of deficiencies in data concerning geochemical cycles, toxicity, and critical exposure pathways (air, food, and water) for gallium (ANDREAE et al., 1984). The volatility of gallium may be a critical factor in the mobility of gallium beyond the workplace or factory environments where it is used (ANDREAE et al., 1984). Because of increased use of gallium compounds in new and developing technologies, gallium emissions into the biosphere are expected to increase (JAWORSKI et al., 1984). At present no levels of tolerance have been established for gallium (HAYES, 1988). Exposure to gallium arsenide rather results in systemic arsenic intoxication (HAYES, 1988), although the gallium moiety may produce toxicity (GOERING et al., 1988).

References

Afonso, A. M., Santana, J. J., and Garcia Montelongo, F. J. (1985), *Pyrocatechol-1-aldehyde 2-benzothiazolylhydrazone as Reagent for the Spectrofluorimetric Determination of Nanogram Amounts of Gallium in Urine and Blood Serum.* Anal. Lett. *18*, 1003–1012.

Andreae, M. O., Asami, T., Bertine, K. K., Buat-Menard, P. E., Duce, R. A., Filip, Z., Förstner, U., Goldberg, E. D., Heinrichs, H., Jernelöv, A. B., Pacyna, J. M., Thornton, I., Tobschall, H. J., and Zoller, W. H. (1984), in: Nriagu, J. O. (ed.): *Changing Metal Cycles and Human Health*, pp. 359–374. Springer Verlag, New York.

Anghileri, L. J. (1977), *On the Mechanism of Accumulation of ^{67}Ga by Tumors.* Oncology *34*, 74–77.

Anghileri, L. J. (1979), *Effects of Gallium and Lanthanum on Experimental Tumor Growth.* Eur. J. Cancer *15*, 1459–1462.

Anonymous (1975), *Mineral Facts and Problems*, p. 402. Bureau of Mines, U.S. Department of Interior, U.S. Government Printing Office, Washington DC.

Anonymous (1987), *Solar Cells*, (in German), *Neue Zürcher Zeitung*, Forschung und Technik, Nr. 28, p. 67, 4 February, Zürich; see also Chem. Eng. News 64/27, 34 (1986).

Berry, J. P., Escaig, F., Poupon, M. F., and Galle, P. (1983), *Localization of Gallium in Tumor Cells. Electron Microscopy, Electron Probe Microanalysis and Analytical Ion Microscopy.* Int. J. Nucl. Med. Biol. *10*, 199–204.

Blair, J. A., Farrar, G., and Morton, A. P. (1987), in: *Proceedings International Conference on Heavy Metals in the Environment*, New Orleans, Vol. 2, pp. 14–16. CEP Consultants Ltd., Edinburgh.

Bowen, H. J. M. (1979), *Environmental Chemistry of the Elements*, p. 249. Academic Press, London.

Briggs, T. M., and Owens, T. W. (1980), *NIOSH Technical Report No. 80–112.* U.S. Department of Health, Education, and Welfare, Cincinnati, Ohio.

Brouhier, O. (1976), *Gallium and Gallium Compounds* (in German), in: *Ullmanns Encyklopädie der technischen Chemie*, 4th Ed., Vol. 12, pp. 67–72. Verlag Chemie, Weinheim-New York.

Brown, D. H., Swartzendruber, D. C., Carlton, J. E., Byrd, B. L., and Hayes, R. L. (1973), *The Isolation and Characterization of Gallium-binding Granules from Soft Tissue Tumors.* Cancer Res. *33*, 2063–2067.

Bruland, K. W. (1983), *Trace Elements in Sea-water.* Chem. Oceanogr. *8*, 179.

de Abrew, S. (1981), *Assays for Transferrin and Transferrin Receptors in Tumor and Other Mouse Tissues.* Int. J. Nucl. Med. Biol. *8*, 217–221.

Dudley, H. C., and Levine, M. D. (1949), *Studies of the Toxic Action of Gallium.* J. Pharmacol. Exp. Ther. *95*, 487–493.

Dudley, H. C., and Marrer, H. H. (1952), *Studies of the Metabolism of Gallium, III. Deposition in and Clearance from Bone.* J. Pharmacol. Exp. Ther. *106*, 129–134.

Dudley, H. C., Maddox, G. E., and LaRue, H. C. (1949), *Studies of the Metabolism of Gallium.* J. Pharmacol. Exp. Ther. *96*, 135–138.

Dudley, H. C., Henry, K. E., and Lindsley, B. F. (1950), *Studies of the Toxic Action of Gallium. II.* J. Pharmacol. Exp. Ther. *98*, 409–417.

Dupuis, R. D. (1984), *Metallorganic Chemical Vapor Deposition of III-V Semiconductors.* Science *226*, 623–629.

Emery, T., and Hoffer, P. B. (1980), *Siderophore-mediated Mechanism of Gallium Uptake Demonstrated in the Microorganism Ustilago sphaerogena.* J. Nucl. Med. *21*, 935–939.

Ferm, V. H., and Carpenter, S. J. (1970), *Teratogenic and Embryopathic Effects of Indium, Gallium, and Germanium.* Toxicol. Appl. Pharmacol. *16*, 166–170.

Foster, B. J., Clagett-Carr, K., Hoth, D., and Leyland-Jones, B. (1986), *Gallium Nitrate: The Second Metal with Clinical Activity.* Cancer Treat. *70*, 1311–1319.

Fowler, B. A., and Dobrota, M. (1989), *Toxicity of Semiconductor Metal Compounds. Lecture at the Robens Institute Meeting on Toxicity and Therapeutics of Newer Metals and Organometallic Compounds*, Guildford, Surrey, U.K.

Ganrot, P. O. (1986), *Metabolism and Possible Health Effects of Aluminium. Environ. Health Perspect.* 65, 363–441.

Goering, P. L., Maronpot, R. R., and Fowler, B. A. (1988), *Effect of Intratracheal Gallium Arsenide Administration on Aminolevulinic Acid Dehydratase in Rats: Relationship to Urinary Excretion of Aminolevulinc Acid. Toxicol. Appl. Pharmacol.* 92, 179–193.

Hamilton, E. I., Minski, M. J., and Cleary, J. J. (1972/1973), *Problems Concerning Multi-element Assay in Biological Materials. Sci. Total Environ.* 1, 1–14.

Hansen, K., and Stern, R. M. (1984), *A Survey of Metal-induced Mutagenicity in vitro and in vivo. J. Am. Coll. Toxicol.* 3, 381–430.

Hart, M. M., and Adamson, R. H. (1971), *Antitumor Activity and Toxicity of Salts of Inorganic Group 3a Metals: Aluminium, Gallium, Indium, and Thallium. Proc. Natl. Acad. Sci. USA* 68, 1623–1626.

Hart, M. M., Smith, C. F., Yancey, S. T., and Adamson, R. H. (1971), *Toxicity and Antitumor Activity of Gallium Nitrate and Periodically Related Metal Salts. J. Natl. Cancer Inst.* 47, 1121–1127.

Hartman, R. E., and Hayes, R. L. (1969), *The Binding of Gallium by Blood Serum. J. Pharmacol. Exp. Ther.* 168, 193–198.

Hayes, R. L. (1983), *The Interaction of Gallium with Biological Systems. Int. J. Nucl. Med. Biol.* 10, 257–261.

Hayes, R. L. (1988), Chapter *Gallium*, in: Seiler, H. G., Sigel, H., and Sigel, A. (eds.): *Handbook on Toxicity of Inorganic Compounds*, pp. 297–300. Marcel Dekker, New York.

Hayes, R. L., and Brown, D. H. (1975), *Biokinetics of Radiogallium*, in: Pabst, H. W., Hor, G., and Schmidt, H. A. E. (eds.): *Nuklear-Medizin*, pp. 837–848. F. K. Schattauer Verlag, Stuttgart.

Hildebrand, S. G., and Cushman, R. M. (1978), *Toxicity of Gallium and Beryllium to Developing Carp Eggs (Cyprinus carpio) Utilizing Copper as a Reference. Toxicol. Lett.* 2, 91–95.

Hiltenkamp, E., and Jackwerth, E. (1988), *Investigations on the Determination of Bi, Cd, Hg, Pb and Tl in High-purity Gallium by Graphite Furnace AAS with Atomization of Metallic Samples. Fresenius Z. Anal. Chem.* 332, 134–139.

Hoffer, P. (1980), *Gallium: Mechanisms. J. Nucl. Med.* 21, 282–285.

Hoffer, P. B. (1981), *Use of Gallium-67 for Detection of Inflammatory Disease: A Brief Review of Mechanisms and Clinical Applications. Int. J. Nucl. Med. Biol.* 8, 243–247.

Jaworski, J. F., Baldi, F., Bernhard, M., Brinckman, F. E., Hecht, H. P., Kloke, A., Legovic, T., McKenzie, J. M., Nriagu, J. O., Page, A. L., Sauerbeck, D. R., and Wasserman, K. J. (1984), *Routes of Exposure to Humans and Bioavailability*, in: Nriagu, J. O. (ed.): *Changing Metal Cycles and Human Health*, pp. 375–390. Springer Verlag, New York.

Johnston, G. S. (1981), *Clinical Applications of Gallium in Oncology. Int. J. Nucl. Med. Biol.* 8, 249–255.

Kirk-Othmer: *Encyclopedia of Chemical Technology* (1978), 3rd Ed., p. 248. John Wiley & Sons, New York.

Krakoff, I. H., Newman, R. A., and Goldberg, R. S. (1979), *Clinical Toxicologic and Pharmacologic Studies of Gallium Nitrate. Cancer* 44, 1722–1727.

Lange, H. H. (1973), *The Natural Concentration of Gallium in Human Tissues. Nucl. Med.* 12, 178–185.

Larson, S. M., Rasey, J. S., Allen, D. R., and Grunbaum, Z. (1979), *A Transferrin-mediated Uptake of Galium-67 by EMT-6 Sarcoma. II. Studies in vivo (BALB/c Mice): Concise Communication. J. Nucl. Med.* 20, 843–846.

Lawless, D., Brown, D. H., Hubner, K. F., Colyer, S. P., Carlton, J. E., and Hayes, R. L. (1978), *Isolation and Partial Characterization of a 67Ga-binding Glycoprotein from Morris 5123C Rat Hepatoma. Cancer Res.* 38, 4440–4444.

Leyland-Jones, B., Bhalla, R. B., Farag, F., Williams, L., Coonley, C. J., and Warrell, Jr., R. P. (1983), *Administration of Gallium Nitrate by Continuous Infusion: lack of Chronic Nephrotoxicity Confirmed by Studies of Enzymuria and beta-2-Microglobulinuria.* Cancer Treat. Rep. 67, 941–942.

Mangum, B. W. (1977), *The Gallium Melting-point Standard: Its Role in our Temperature Measurement System.* Clin. Chem. 23, 711–718.

Maurel, E., Rouquie, A., Bonnafous, M., and Bouissou, H. (1974), *[Excretion and Localization of Gallium in the Rat].* Pathol. Biol. 22, 859–865.

Merck Index (1983), 10th Ed., pp. 621–622. Merck & Co., Rahway, New Jersey.

Moerlein, S. M., and Welch, M. J. (1981), *The Chemistry of Gallium and Indium as Related to Radiopharmaceutical Production.* Int. J. Nucl. Med. Biol. 8, 277–287.

Moerlein, S. M., Welch, M. J., Raymond, K. N., and Weitl, F. L. (1981), *Tricatecholamide Analogs of Enterobactin as Gallium- and Indium-binding Radiopharmaceuticals.* J. Nucl. Med. 22, 710–719.

Moorhead, E. D., and Doub, Jr., W. H. (1977), *Tissue-Sequestered Group III-A Metals: Dilute, High Temperature NH_4SCN for Simultaneous Measurement of In and Ga by Phase-selective Stripping Voltammetry.* Anal. Lett. 10, 673–684.

Nelson, B., Hayes, R. L., Edwards, C. L., Kniseley, R. M., and Andrews, G. A. (1972), *Distribution of Gallium in Human Tissues after Intravenous Administration.* J. Nucl. Med. 13, 92–100.

Newman, R. A. (1978), *Flameless Atomic Absorption Spectrometry Determination of Gallium in Biological Materials.* Clin. Chim. Acta 86, 195–200.

Newman, R. A., Brody, A. R., and Krakoff, I. H. (1979), *Gallium Nitrate (NSC-15200) Induced Toxicity in the Rat: A Pharmacologic, Histopathologic and Microanalytical Investigation.* Cancer 44, 1728–1740.

Nishioka, H. (1975), *Mutagenic Activities of Metal Compounds in Bacteria.* Mutation Res. 31, 185–189.

Petkof, B. (1980), in: *Minerals Yearbook*, Vol. 1, pp. 335–337. Department of the Interior, Bureau of Mines, U.S. Governmental Printing Office, Washington, DC.

Petkof, B. (1981), in: *Mineral Facts and Problems*, pp. 323–328. Bureau of Mines Bulletin 761, U.S. Department of Interior, U.S. Government Printing Office, Washington, DC.

Philips (1987), in: *Neue Zürcher Zeitung*, Forschung und Technik, Nr. 208, p. 84, 9 September, Zürich.

Rakieten, N., Cooney, D. A., Homan, E. R., and Davis, R. D. (1973), South Shore Analytical Research Laboratories, Inc., Islip, New York, US National Technical Information Service.

Rasey, J. S., Nelson, N. J., and Larson, S. M. (1981), *Relationship of Iron Metabolism to Tumor Cell Toxicity of Stable Gallium Salts.* Int. J. Nucl. Med. Biol. 8, 303–313.

Robinson, A. L. (1983), *GaAs Readied for High-speed Microcircuits.* Science 219, 275–277.

Roscina, T. A. (1983), *Gallium and Compounds*, in: Parmeggiani, L. (ed.), *Encyclopaedia of Occupational Health and Safety*, Vol. 1, pp. 938–939. International Labour Office, Geneva.

Samson, M. K., Fraile, R. J., Baker, L. H., and O'Bryan, R. (1980), *Phase I–II Clinical Trial of Gallium Nitrate (NSC-15200).* Cancer Clin. Trials 3, 131–136.

Scheffel, U., Wagner, Jr., H. N., Frazier, J. M., and Tsan, M-F. (1979), *Gallium-67 Uptake by the Liver: Studies Using Isolated Rat Hepatocytes and Perfused Livers.* J. Nucl. Med. 25, 1094–1100.

Schroeder, H. A., and Mitchener, M. (1971), *Scandium, Chromium(VI), Gallium, Yttrium, Rhodium, Palladium, Indium in Mice: Effects on Growth and Life Span.* J. Nutr. 101, 1431–1437.

Schwartz, S., and Yagoda, A. (1984), *Phase I–II Trial of Gallium Nitrate for Advanced Hypernephroma.* Anticancer Res. 4, 317–318.

Sephton, R. (1981), *Relationships Between the Metabolism of ^{67}Ga and Iron.* Int. J. Nucl. Med. Biol. 8, 323–331.

Sephton, R. G., and Harris, A. W. (1975), *Gallium-67 Citrate Uptake by Cultured Tumor Cells, Stimulated by Serum Transferrin.* J. Natl. Cancer Inst. 54, 1263–1266.

Silberstein, E. B. (1976), *Cancer Diagnosis. The Role of Tumor-imaging Radiopharmaceuticals.* Am. J. Med. 60, 226–237.

Swartzendruber, D. C., Nelson, B., and Hayes, R. L. (1971), *Gallium-67 Localization in Lysosomal-like Granules of Leukemic and Nonleukemic Murine Tissues.* J. Natl. Cancer Inst. 46, 941–952.

Urena, E., Garcia de Torres, A., Cano Pavon, J. M., and Gomez Ariza, J. L. (1985), *Determination of Traces of Gallium in Biological Materials by Fluorometry.* Anal. Chem. 57, 2309–2311.

Valberg, L. S., Flanagan, P. R., Haist, J., Frei, J. V., and Chamberlain, M. J. (1981), *Gastrointestinal Metabolism of Gallium and Indium: Effect of Iron Deficiency.* Clin. Invest. Med. 4, 103–108.

Vallabhajosula, S. R., Harwig, J. F., and Wolf, W. (1981), *The Mechanism of Tumor Localization of Gallium-67 Citrate: Role of Transferrin Binding and Effect of Tumor pH.* Int. J. Nucl. Med. Biol. 8, 363–370.

Vogh, B. P., Godman, D. R., and Maren, T. H. (1985), *Aluminium and Gallium Arrest Formation of Cerebrospinal Fluid by the Mechanism of OH-Depletion.* J. Pharmacol. Exp. Ther. 233, 715–721.

Ward, N. I. (1987), *The Future of Multielement Analysis in Assessing Human Health and Disease: A Comparison of NAA and ICPSMS.* Environ. Health Crit. 20, 118–123.

Warrell, Jr., R. P., Coonley, C. J., Straus, D. J., and Young, C. W. (1983), *Treatment of Patients with Advanced Malignant Lymphoma Using Gallium Nitrate Administered as a Seven-day Continuous Infusion.* Cancer 51, 1982–1987.

Warrell, Jr., R. P., Isaacs, M., Coonley, C. J., Alcock, N. W., and Bockman, R. S. (1985), *Metabolic Effects of Gallium Nitrate Administered by Prolonged Infusion.* Cancer Treat. Rep. 69, 653–655.

Watson, E. E., Cloutier, R. J., and Gibbs, W. D. (1973), *Whole-body Retention of ^{67}Ga-Citrate.* J. Nucl. Med. 14, 840–842.

Webb, D. R., Sipes, I. G., and Carter, D. E. (1984), *In vitro Solubility and in vivo Toxicity of Gallium Arsenide.* Toxicol. Appl. Pharmacol. 76, 96–104.

Webb, D. R., Wilson, S. E., and Carter, D. E. (1986), *Comparative Pulmonary Toxicity of Gallium Arsenide, Gallium(III) Oxide, or Arsenic(III) Oxide Intratracheally Instilled into Rats.* Toxicol. Appl. Pharmacol. 82, 405–416.

Weick, J. K., Stephens, R. L., Baker, L. H., and Jones, S. E. (1983), *Gallium Nitrate in Malignant Lymphoma: A Southwest Oncology Group Study.* Cancer Treat. Rep. 67, 823–825.

Welz, B (1985), *Atomic Absorption Spectrometry*, 2nd. Ed., pp. 283–284, 390 ff. VCH Verlagsgesellschaft, Weinheim-Deerfield Beach/Florida-Basel.

White, W. E., and Yielding, K. L. (1975), Ga^{3+} *Inhibition of Glutamate Dehydrogenase.* Ala. J. Med. Sci. 12, 244–246.

Williams, R. J. P. (1984), *Structural Aspects of Metal Toxicity,* in: Nriagu, J. O. (ed.): *Changing Metal Cycles and Human Health,* pp. 251–263. Springer Verlag, New York.

II.11 Germanium

BENJAMIN R. FISHER and PETER L. GOERING, Rockville, Maryland, USA
BRUCE A. FOWLER, Baltimore, Maryland, USA

1 Introduction

Germanium (Ge) was discovered by Clemens Winkler in 1886 and derives its name from the Latin word Germania. It is a Group IV metal in the Periodic System of Elements and shares similar properties with other elements in its group, especially silicon. Although it is not an essential mineral, it can be found in trace quantities in foods, such as fruits and vegetables. Industrially, it is refined from zinc and lead-zinc-copper ores and can be found in relatively high amounts in coal and lignites. Ge-containing compounds have found applications in medical imaging devices, solid state electronics, infrared sensors, semiconductors, phosphors in fluorescent lamps, and dental casting alloys. As in the cases of arsenic, tin, and lead, speciation must always be considered. The intermediate germanium tetrachloride is an irritant. Trialkylgermanium compounds are rather less toxic than the corresponding lead and tin alkyls (VOUK, 1986). Ge compounds have been used therapeutically as anti-tumor agents in humans, but recent reports concerning lack of efficacy and associated neurotoxicity make the future use of these compounds as antitumor agents equivocal.

2 Physical and Chemical Properties, and Analytical Methods

Germanium is a grayish-white, brittle metalloid with the following physical properties: atomic number, 32; atomic mass, 72.59; specific gravity, 5.323; melting point, 937.4 °C; boiling point about 2830 °C. It is stable in air, but can become explosive as a fine particulate mixed in air. Elemental germanium is insoluble in water, hydrochloric acid, and dilute basic hydroxides, but is readily attacked by nitric or sulfuric acids and basic peroxides, nitrates, and carbonates (MERCK INDEX, 1983). Numerous forms of germanium occur including dioxide (used for catalysts and as raw material for the metal; STANDAERT, 1976), di- and tetrachloride, and tetrahydride (germane, toxic gas of bad odor; used for germanium coating; STANDAERT, 1976). Germanium is commonly found in solution as germanic acid, $Ge(OH)_4$, and although it has a valence of +2 and +4, compounds with +4 valence are favored under normal environmental conditions (BRAMAN and

TOMPKINS, 1978). There are five naturally occurring isotopes of germanium found in the following relative amounts: 70 (21%), 72 (27%), 73 (8%), 74 (37%), 76 (8%); the radioactive forms 65–69, 71, 75, 77, and 78 are artificially produced (MERCK INDEX, 1983).

The spectrophotometric assay devised by LUKE and CAMPBELL (1956) is used to measure germanium in the range of 0.02 to 1.0 ppm and is based on the reaction of the metalloid with phenylfluorone in an acidic solution. Because of potential interference from other metals, the germanium must first be extracted from the sample with carbon tetrachloride. Spark source mass spectrometry has been used to measure germanium in human tissue samples with a detection limit of 7 µg Ge per kg of wet tissue (HAMILTON et al., 1972). Emission spectrography methods can detect germanium down to 1 µg in biological samples (GELDMACHER-VON MALLINCKRODT and POOTH, 1969), and in air and water with detection limits of 0.1–0.4 ng (BRAMAN and TOMPKINS, 1978). Atomic absorption spectrometry has been used to measure germanium content in biological samples with a detection limit of 1 ppb (MINO et al., 1980; WELZ 1985). Losses due to volatility must be avoided.

3 Sources, Production, Important Compounds, and Uses

Zinc sulfide and lead-zinc-copper sulfide ores (especially from Zaire, Namibia, and the Soviet Union; STANDAERT, 1976) are the major sources of germanium, which is refined from these ores by heating in the presence of chlorine and air, and subsequently reduced in the presence of hydrogen (STANDAERT, 1976; GERBER, 1988). Germanium can also be obtained from the ash of coal and other lignites. The world production of Ge, as of 1982, has been estimated at approximately 125 tons Ge per year (U.S. BUREAU OF MINES, 1983). With zone-refining techniques, high purities (required for semiconductors) are obtained (GERBER, 1988).

Various inorganic Ge compounds (germanium telluride) have been used in the development of semiconductors, catalysts and medical imaging devices. Silver-copper-germanium alloys are used in dental castings and in oral restorations (YOUDELIS and YOUDELIS, 1981; TOWNSEND et al., 1983). Germanium glass has a high refractive index, making it useful for camera and wide-angle lenses (STANDAERT, 1976). It is also transparent to infrared rays. The use in fiber optical systems has also increased (VOUK, 1986). Superconductive niobium-germanium alloys may also be of some interest and are described in Chapter II.23 (Nb). All these applications may lead to additional wastes. Organogermanium compounds have been used as antitumor agents and as intestinal astringents in veterinary medicine.

Biosilicification in invertebrates, such as sponges and diatoms, has been studied using ^{68}Ge, which mimics silicon in a number of biological processes (AZAM et al., 1973). This isotope is used because it has a relatively longer half-life (288 d) compared to ^{31}Si (156 min).

Bismuth germanate ($Bi_4Ge_3O_{12}$) is a scintillation material with a very high z number and a high density (7.13 g/cm^3). At low temperatures, the light yield and energy resolution of bismuth germanate makes it a superior compound for use in imaging devices such as positron and gamma cameras, compared to the detection efficiency of the standard NaI(Tl) system (CHO and FARUKHI, 1977).

Various forms of spirogermanium (SGe), such as 4,4-dialkyl-4-germacyclohexanone and 8,8-dialkyl-8-germaazaspiro(4,5)decane, were proposed as antitumor agents for human use, specifically for patients with advanced metastatic lymphomas (LEGHA et al., 1983). The major dose limiting effect of SGe treatment is acute neurologic toxicity, which appears to be reversible and schedule dependent. Although SGe treatment initially appeared to have antitumor potential, recent reports suggest that treatment with SGe at maximum tolerable doses is ineffective ($<20\%$ response rate) in treating patients with advanced breast cancer (KUEBLER et al., 1984; MEHTA and GAIN, 1984; PINNAMANENI et al., 1984), carcinoma of the cervix (BRENNER et al., 1985), non-Hodgkin's lymphomas (EISENHAUER et al., 1985a), metastatic melanomas (EISENHAUER et al., 1985b), ovarian carcinomas (BRENNER et al., 1983; KAVANAGH et al., 1985), prostatic cancer (DEXEUS et al., 1986), or advanced renal cell carcinomas (SCHULMAN et al., 1984).

4 Distribution in the Environment, in Foods, and in Living Organisms

Germanium constitutes approximately 0.0007% of the earth's crust with the highest concentrations found in zinc and lead-zinc-copper ores, coal, and lignites. The element occurs naturally as argyrodite $[(Ag_2S)_4GeS_2]$ and as germanite ($7CuSFeSGeS_2$) (MERCK INDEX, 1983). The average Ge concentration in fresh, estuarine, and saline water (Tampa, USA) was determined to be 0.016, 0.029, 0.079 µg/L, respectively (BRAMAN and TOMPKINS, 1978). BRULAND (1983) reported on the nutrient-type distribution in sea water, with extremely low concentrations of less than 0.0005 ppb at the surface and 0.009 ppb in 2000–5000 m depth. Rain and tap water concentrations were found to be 0.045 and 0.0088 µg/L.

Soil samples in the United States were shown to contain 0.6 to 1.3 mg Ge/kg (SCHROEDER and BALASSA, 1967a). Foods which contain relatively high levels (µg Ge/g wet weight) were: raw clams, 1.9–2.03; salmon, 1.23; canned tuna, 2.23; homogenized milk, 1.51; butter, 1.23; bran wheat, 1.41; canned baked beans, 4.67; and canned tomato juice, 5.76. Meat content is rather low (according to BOWEN (1979) 0.14 ppm in mammal muscle). The average range of Ge content in natural biological samples is 0.1 to 1.0 ppm, but it has been suggested that some medicinal plants contain large amounts of germanium. MINO et al. (1980) found the highest Ge content in green tea (9 ppb) and the amount of germanium in ginseng has been reported as low as 6 ppb (MINO et al., 1980) and as high as 300 ppm (ASHAI, 1973).

The mean Ge content of lung and hilar lymph node tissue of individuals in urban areas (Cincinnati, USA) was determined to be 0.40 µg/g and 1.98 µg/g, respectively (SWEET et al., 1978). NAGATA et al. (1985) have reported Ge contents (µg Ge/g wet tissue) in healthy human tissues to be: heart, 0.075; liver, 0.291; spleen, 0.108; brain, 0.189; pancreas, 0.218; and vertebra, 8.820.

5 Uptake, Absorption, Transport and Distribution, Metabolism and Elimination in Animals and Humans

The primary means of germanium uptake in humans is through dietary consumption, and a secondary source is inhalation via industrial exposure. Daily dietary intake can vary between 400 µg (high protein, low carbohydrate diet) and 3500 µg (selected ovo-vegetarian diet) depending on the food types consumed (SCHROEDER and BALASSA, 1967a). Germanium is readily absorbed and excreted in animals and humans. In rats, 73.6% of an oral dose of sodium germanate was absorbed within 4 hours and 96.4% within 8 hours (ROSENFELD, 1954). In this study, germanium was found to be distributed uniformly between red blood cells and plasma and was not bound to plasma proteins. High concentrations of germanium were found in the liver, kidney, spleen, and gastrointestinal tract of dogs and rabbits administered Ge oxide via iv injection (DUDLEY and WALLACE, 1952). Mice given a chronic germanium dietary supplement demonstrated a preferential spleenic accumulation (SCHROEDER and BALASSA, 1967b), whereas rats administered a single ip injection showed the longest retention to be in both the spleen and kidneys (MEHARD and VOLCANI, 1975). In humans, germanium is widely distributed throughout the body. In users of Ge preparations, particularly high accumulations of germanium were found in the spleen, vertebra, renal cortex, brain, and skeletal muscle (NAGATA et al., 1985), as well as in hair, nails, and toenails (MORITA et al., 1986).

Biotransformation of organo-Ge compounds is catalyzed by the liver microsomal cytochrome P-450 enzyme system, probably through a C-hydroxylation mechanism (PROUGH et al., 1981). The major excretory route for germanium is via the kidneys in both animals (DUDLEY and WALLACE, 1952; ROSENFELD, 1954) and humans (SCHROEDER and BALASSA, 1967a). In humans the half-life is about 1 1/2 days (which increases to 4–5 days for the kidneys; EMMERLING et al., 1986).

6 Effects on Animals and Humans
(see also Chapter I.13b, Sections 4 and 6)

6.1 Effects on Invertebrates and Microorganisms

At low concentrations, germanium administered in the form of germanic acid mimics the metabolic pathways of silicon. At high concentrations, it inhibits silicon transport by acting as a classical competitive inhibitor (AZAM et al., 1973). This has been shown to have detrimental effects in organisms in which silicification and/or silicon are essential. In fresh water sponges, germanic acid (1–25 µg/mL) disrupts spicule formation (ELVIN, 1972; SIMPSON et al., 1983). Germanium has also been shown to be toxic to marine gastropods (RINKEVICH, 1986). It appears to have minimal antimicrobial activity (VENUGOPAL and LUCKEY, 1978).

6.2 Miscellaneous Biochemical Effects

The basis for proposals to use SGe analogs as antitumor agents was based on evidence that these compounds suppress protein, RNA, and DNA synthesis *in vitro* (RICE et al., 1977; SCHEIN et al., 1980; HILL et al., 1982). "Quiescent" cultures appear to be less sensitive to SGe than cells in a highly proliferative growth phase (HILL et al., 1982). SGe concentrations as low as 1.0 µg/mL for 24 hours decreased cell survival in certain cell lines. Cell killing by germanium appears to be mediated by two different mechanisms. At low concentrations, germanium blocks cellular proliferation, resulting in "reproductive death". At higher concentrations, cell killing is a result of cytolysis (YANG and RAFLA, 1983). In vitro testing of SGe has demonstrated a synergistic effect in the killing of human tumor cells when combined with other antitumor drugs such as 5-fluorouracil or *cis*-platin (HILL et al., 1984).

Ge compounds act as immunomodulators by affecting a variety of immune system components. Although dibutyl-Ge-dichloride is not cytotoxic to lymphocytes up to 64 µg/mL, LI et al. (1982) found a 50% reduction in antibody production. DIMARTINO et al. (1986) have reported enhancement of interleukin 1 production and depression of interleukin 2 and 3 in mice administered SGe. The organogermanium compound Ge-132 (carboxyethyl-Ge sesquioxide) at 300 mg/kg was found to increase interferon activity and NK cell activity of spleen cells in mice 24 hours after an oral administration. Peritoneal cytotoxic macrophages were also induced 48 hours following oral administration of Ge-132 (ASO et al., 1985). Ge compounds demonstrated a strain-specific protective effect against methylcholanthrene-induced tumors in mice (KUMANO et al., 1978), and sera from mice treated with Ge-132 inhibited ascites tumors (SUZUKI et al., 1985a, b). The antitumor activity of Ge-132 is believed to be mediated via induced lymphokines and expressed through macrophages and/or T cells (SUZUKI et al., 1985a, b).

Morphine-induced analgesia in rats was enhanced by Ge-132 (500 mg/kg, po) and this compound has been administered to reduce pain in patients with advanced carcinomas (HACHISU et al., 1983). Because this effect is completely abolished by naloxone, it is hypothesized that the analgesic effect of Ge-132 may be associated with the endogenous opioid system. KOMURO et al. (1986) have proposed that Ge-132 acts by inhibiting the action of endogenous opioid catalytic enzymes, thereby prolonging the analgesic effect.

6.3 Acute Effects on Animals and Humans
(see also VOUK, 1986)

The toxicity of SGe in mice and dogs has been determined by HENRY et al. (1980). In mice, the iv LD 50 is 41 – 44.5 mg/kg (125 – 133 mg/m^2). The im LD 50 is approximately 3-fold greater, with females exhibiting higher sensitivity. In dogs, the LD 100 is the same with both routes of administration (40 mg/kg; 800 mg/m^2). Convulsive seizures occurred in dogs after administration of lethal im doses and sublethal iv doses. Necrosis and degeneration of mitotically active tissue was noted, as well as necrosis, hemorrhage, and edema at the muscle injection site. A review of LD 50 and

LD100 values for laboratory animals administered various Ge compounds can be found in VENUGOPAL and LUCKEY (1978). In humans, the chemotherapeutic use of SGe is limited by its acute neurotoxicity. LEGHA et al. (1983) have determined the maximum tolerable dose in humans to be 100 mg/m^2/day (iv bolus), but this dose can be exceeded by extending the scheduling regime. Common symptoms of acute Ge neutrotoxicity include dizziness, numbness of the mouth and extremities, blurred vision, tinnitus, auditory hallucinations, as well as impaired coordination. In addition, various degrees of anorexia, nausea, and vomiting are associated with SGe treatment. A grand mal seizure has been reported after an accidental iv administration of approximately 600 mg/m^2 (LEGHA et al., 1983). All of the side effects appear transient and reversible if dosing is decreased or SGe therapy is stopped. Renal, hematologic, gastrointestinal, and pulmonary toxicity have also been reported (BRENNER et al., 1983; DIXON et al., 1984; SCHULMAN et al., 1984).

6.4 Chronic Effects on Animals and Humans

In vivo studies have reported a decrease in median life span and longevity in male mice given continual Ge dietary supplementation. Symptoms of chronic Ge toxicity include inhibition of growth and fatty degeneration of the liver (VENUGOPAL and LUCKEY, 1978). Skin rash, myopathy, and death due to acute renal failure have also been observed in chronic users of Ge preparations (NAGATA et al., 1985).

6.5 Mutagenic, Carcinogenic, and Teratogenic Effects

LI et al. (1982) have reported that dibutylgermanium dichloride induced mutations at the HGPRT locus in CHO cells with the mutational frequency increasing proportionally with dose up to 400 µg/mL. These investigators postulate that the mutagenic effect of these lipid-soluble organo-metal compounds is mediated by penetration of biological membranes and subsequent electrophilic binding to nitrogenous bases of DNA and RNA. Although in in-vivo studies increasing accumulation of Ge in the spleen was noted with age, no increase in the rate of carcinogenesis was noted in either sex (SCHROEDER and BALASSA, 1967b; regarding antitumor activity see Sect. 6.2). Sodium germanate and germanium dioxide seem not to be teratogenic in mammals, but dimethylgermanium oxide produces malformations in chick embryos (VOUK, 1986; GERBER, 1988).

7 Hazard Evaluation and Limiting Concentrations

At the present time, there are no air or occupational standards for germanium, except for germanes (GeH_4, Ge_2H_6, Ge_3H_8, etc.) in the USA (see also EMMERLING et al., 1986: germane has effects on the cardio-vascular system, and on the liver and the kidney in animals), and for some inorganic germanium compounds ($1-2$ mg/m^3) in the Soviet Union (GERBER, 1988). A decade ago, germanium was not considered an industrial health hazard (VENUGOPAL and LUCKEY, 1978), but its increased utilization in new and developing technologies (e.g., electronics) may warrant closer monitoring. Since occupational exposure is primarily through inhalation of dust (EMMERLING et al., 1986) and fumes during processing and refining, adequate exhaust ventilation should be used in these areas to minimize Ge intake. Therapeutically, the maximum tolerable doses in humans are 100 mg/m^2/day iv bolus and 120 mg/m^2 administered over $2-3$ hours (LEGHA et al., 1983). These doses can be exceeded slightly by increasing the duration of the scheduling regime. Ecotoxicity seems to be low (GERBER, 1988).

References

Ashai, K. (1973), *Shoku no Kagaku 12*, 81–85.
Aso, H., Suzuki, F., Yamaguchi, T., Hayashi, Y., Ebina, T., and Ishida, N. (1985), *Induction of Interferon and Activation of NK Cells and Macrophages in Mice by Oral Administration of Ge-132, an Organic Germanium Compound. Microbiol. Immunol. 29*, 65–74.
Azam, F., Hemmingsen, B. B., and Volcani, B. E. (1973), *Arch. Mikrobiol. 92*, 11–20.
Bowen, H. J. M. (1979), *Environmental Chemistry of the Elements*, p. 249. Academic Press, London.
Braman, R. S., and Tompkins, M. A. (1978), *Anal. Chem. 50*, 1088–1096.
Brenner, D. E., Jones, H. W., Rosenshein, N. B., Forastiere, A., Dillon, M., Grumbine, F., Tipping, S., Burnett, L., Greco, F. A., and Wiernik, P. H. (1983), *Phase II Evaluation of Spirogermanium in Advanced Ovarian Carcinoma. Cancer Treat. Rep. 67*, 193–194.
Brenner, D. E. Rosenshein, N. B., Dillon, M., Jones, H. W., Forastiere, A., Tipping, S., Burnett, L. S., Greco, F. A., and Wiernik, P. H. (1985), *Phase II Study of Spirogermanium in Patients with Advanced Carcinoma of the Cervix. Cancer Treat. Rep. 69*, 457–458.
Bruland, K. W. (1983), *Trace Elements in Sea-water. Chem. Oceanogr. 8*, 180–182.
Cho, Z. H., and Farukhi, M. R. (1977), *Bismuth Germanate as a Potential Scintillation Detector in Positron Cameras. J. Nucl. Med. 18*, 840–844.
Dexeus, F. H., Logothetis, C., Samuels, M. L., and Hossan, B. (1986), *Phase II Study of Spirogermanium in Metastatic Prostate Cancer. Cancer Treat. Rep. 70*, 1129–1130.
DiMartino, M. J., Lee, J. C., Badger, A. M., Muirhead, K. A., Mirabelli, C. K., and Hanna, N. (1986), *Antiarthritic and Immunoregulatory Activity of Spirogermanium. J. Pharmacol. Exp. Ther. 236*, 103–110.
Dixon, C., Hagemei, F., Legha, S., and Bodey, G. (1984), *Pulmonary Toxicity Associated with Spirogermanium. Cancer Treat. Rep. 68*, 907–908.
Dudley, H. C., and Wallace, E. J. (1952), *Arch. Ind. Hyg. Occup. Med. 6*, 263–270.
Eisenhauer, E., Quirt, I., Connors, J. M., Maroun, J., and Skillings, J. (1985a), *A Phase II Study of Spirogermanium as Second Line Therapy in Patients with Poor Prognosis Lymphoma. An NCI Canada Clinical Trials Group Study. Invest. New Drugs 3*, 307–310.
Eisenhauer, E., Kerr, I., Bodurtha, A., Iscoe, N., McCulloch, P., Pritchard, K., and Quirt, I. (1985b), *A Phase II Study of Spirogermanium in Patients with Metastatic Malignant Melanoma. An NCI Canada Clinical Trials Group Study. Invest. New Drugs 3*, 303–305.
Elvin, D. W. (1972), *Effect of Germanium upon Development of Siliceous Spicules of Some Freshwater Sponges. Exp. Cell Res. 72*, 551–553.

Emmerling, G., Schaller, K. H., and Valentin, H. (1986), *Actual Knowledge on Germanium Compounds, and Feasibility of Quantitative Determination in Occupational Medicine and Toxicology.* Zbl. Arbeitsmed. *36*, 258–265.

Geldmacher-von Mallinckrodt, M., and Pooth, M. (1969), *Simultaneous Spectrographic Testing for 25 Metals and Metalloids in Biological Material.* Arch. Toxicol. *25*, 5–18.

Gerber, G. B. (1988), Chapter 24 *Germanium*, in: Seiler, H. G., Sigel, H., and Sigel, A. (eds.), *Handbook on Toxicity of Inorganic Compounds*, Marcel Dekker, New York.

Hachisu, M., Takahashi, H., Koeda, T., and Sekizawa, Y. (1983), *Analgesic Effect of Novel Organogermanium Compound, Ge-132.* J. Pharmacobio-Dyn. *6*, 814–820.

Hamilton, E. I., Minski, M. J., and Clearly, J. J. (1972), Sci. Total Environ. *1*, 341–374.

Henry, M. C., Rosen, E., Port, C. D., and Levine, B. S. (1980), *Toxicity of Spirogermanium in Mice and Dogs after iv or im Administration.* Cancer Treat. Rep. *64*, 1207–1210.

Hill, B. T., Whatley, S. A., Bellamy, A. S., Jenkins, L. Y., and Whelan, R. D. (1982), *Cytotoxic Effects and Biological Activity of 2-Aza-8-germanspiro[4,5]-decane-2-propanamine-8,8-diethyl-N,N-dimethyl Dichloride (NSC 192965; Spirogermanium) in vitro.* Cancer Res. *42*, 2852–2856.

Hill, B. T., Bellamy, A. S., Metcalfe, S., Hepburn, P. J., Masters, J. R., and Whelan, R. D. (1984), *Identification of Synergistic Combinations of Spirogermanium with 5-Fluorouracil or Cisplatin Using a Range of Human Tumour Cell Lines in vitro.* Invest. New Drugs 2, 29–33.

Kavanagh, J. J., Saul, P. B., Copeland, L. J., Gershenson, D. M., and Krakoff, I. H. (1985), *Continuous-infusion Spirogermanium for the Treatment of Refractory Carcinoma of the Ovary: A Phase II Trial.* Cancer Treat. Rep. *69*, 139–140.

Komuro, T., Kakimoto, N., Katayama, T., and Hazato, T. (1986), *Inhibitory Effects of Ge-132 (Carboxyethyl Germanium Sesquioxide) Derivatives on Enkephalin-degrading Enzymes.* Biotechnol. Appl. Biochem. *8*, 379–386.

Kuebler, J. P., Tormey, D. C., Harper, G. R., Chang, Y. C., Khandekar, J. D., and Falkson, G. (1984), *Phase II Study of Spirogermanium in Advanced Breast Cancer.* Cancer Treat. Rep. *68*, 1515–1516.

Kumano, N., Nakai, Y., Ishikawa, T., Koinumaru, S., Suzuki, S., and Konno, K. (1978), *Effect of Carboxyethylgermanium Sesquioxide on the Methylcholanthrene-induced Tumorigenesis in Mice.* Sci. Rep. Res. Inst. Tohoku Univ. (Med.) *25*, 89–95.

Legha, S. S., Ajani, J. A., and Bodey, G. P. (1983), *Phase I Study of Spirogermanium Given Daily.* J. Clin. Oncol. *1*, 331–336.

Li, A. P., Dahl, A. R., and Hill, J. O. (1982), Toxicol. Appl. Pharmacol. *64*, 482–485.

Luke, C. L., and Campbell, M. D. (1956), *Photometric Determination of Germanium with Phenylfluorone.* Anal. Chem. *28*, 1273–1276.

Mehard, C. W., and Volcani, B. E. (1975), *Similarity in Uptake and Retention of Trace Amounts of ^{31}Silicon and ^{68}Germanium in Rat Tissues and Cell Organelles.* Bioinorg. Chem. *5*, 107–124.

Mehta, C. R., and Cain, K. C. (1984), *Charts for the Early Stopping of Pilot Studies.* J. Clin. Oncol. *2*, 676–682.

Merck Index (1983), 10th Ed., p. 630. Merck and Co., Inc., Rahway, New Jersey.

Mino, Y., Oto, N., Sakao, S., and Shimomura, S. (1980), *Determination of Germanium in Medicinal Plants by Atomic Absorption Spectrometry with Electrothermal Atomization.* Chem. Pharm. Bull. (Tokyo) *28*, 2687–2691.

Morita, H., Shimomura, S., Odagawa, K., Saito, S., Sakigawa, C., and Sato, H. (1986), *Determination of Germanium and Some Other Elements in Hair, Nail, and Toenail from Persons Exposed and Unexposed to Germanium.* Sci. Total Environ. *58*, 237–242.

Nagata, N., Yoneyama, T., Yanagida, K., Ushio, K., Yanagihara, S., Matsubara, O., and Eishi, Y. (1985), *Accumulation of Germanium in the Tissues of a Long-term User of Germanium Preparation Died of Acute Renal Failure.* J. Toxicol. Sci. *10*, 333–341.

Pinnamaneni, K., Yap, H. Y., Legha, S. S., Blumenshein, G. R., and Bodey, G. P. (1984), *Phase II Study of Spirogermanium in the Treatment of Metastatic Breast Cancer.* Cancer Treat. Rep. *68*, 1197–1198.

Prough, R. A., Stalmach, M. A., Wiebkin, P., and Bridges, J. W. (1981), *The Microsomal Metabolism of the Organometallic Derivatives of the Group-IV Elements, Germanium, Tin and Lead. Biochem. J. 196*, 763–770.

Rice, L. M., Slavik, M., and Schein, P. (1977), *Clinical Brochure: Spirogermanium* (NSC-192965). National Cancer Institute, Bethesda, Maryland.

Rinkevich, B. (1986), *Does Germanium Interact with Radular Morphogenesis and Biomineralization in the Limpet Lottia gigantea? Comp. Biochem. Physiol. C 83*, 137–141.

Rosenfeld, G. (1954), *Metabolism of Germanium. Arch. Biochem. Biophys. 48*, 84–94.

Schein, P. S., Slavik, M., Smythe, T., Hoth, D., Smith, F., Macdonald, J. S., and Woolley, P. V. (1980), *Phase I Clinical Trial of Spirogermanium. Cancer Treat. Rep. 64*, 1051–1056.

Schroeder, H. A., and Balassa, J. J. (1967a), *Abnormal Trace Metals in Man: Germanium. J. Chronic Dis. 20*, 221–224.

Schroeder, H. A., and Balassa, J. J. (1967b), *Arsenic, Germanium, Tin and Vanadium in Mice: Effects on Growth, Survival and Tissue Levels. J. Nutr. 92*, 245–252.

Schulman, P., Davis, R. B., Rafla, S., Green, M., and Henderson, E. (1984), *Phase II Trial of Spirogermanium in Advanced Renal Cell Carcinoma: A Cancer and Leukemia Group B Study. Cancer Treat. Rep. 68*, 1305–1306.

Simpson, T. L., Garrone, R., and Mazzorana, M. (1983), *Interaction of Germanium (Ge) with Biosilicification in the Freshwater Sponge Ephydatia mülleri: Evidence of Localized Membrane Domains in the Silicalemma. J. Ultrastruct. Res. 85*, 159–174.

Standaert, R. (1976), *Germanium and Germanium-compounds*, in: *Ullmanns Encyclopädie der technischen Chemie* (in German), 4th Ed., Vol. 12, pp. 221–226. Verlag Chemie, Weinheim-New York.

Sweet, D. V., Crouse, W. E., and Crable, J. V. (1978), *Chemical and Statistical Studies of Contaminants in Urban Lungs. Am. Ind. Hyg. Assoc. J. 39*, 515–526.

Suzuki, F., Brutkiewicz, R. R., and Pollard, R. B. (1985a), *Ability of Sera from Mice Treated with Ge-132, an Organic Germanium Compound, to Inhibit Experimental Murine Ascites Tumours. Br. J. Cancer 52*, 757–763.

Suzuki, F., Brutkiewicz, R. R., and Pollard, R. B. (1985b), *Importance of T-Cells and Macrophages in the Antitumor Activity of Carboxyethylgermanium Sesquioxide (Ge-132). Anticancer Res. 5*, 479–483.

Townsend, J. D., Hamilton, A. I., and Sbordone, L. (1983), *Biologic Evaluation of a Silver-Copper-Germanium Dental Casting Alloy and a Gold-Germanium Coating Alloy. J. Dent. Res. 62*, 899–903.

U.S. Bureau of Mines (1983), *Mineral Commodity Summaries*. Bureau of Mines, U.S. Department of the Interior, Washington, D.C.

Venugopal, B., and Luckey, T. D. (1978), in: *Metal Toxicity in Mammals – 2*, pp. 177–180. Plenum Press, New York.

Vouk, V. (1986), Chapter 11 *Germanium*, in: Friberg, L., Nordberg, G. F., and Vouk, V. B. (eds.), *Handbook on the Toxicology of Metals*, 2nd Ed., Vol. II, pp. 255–266. Elsevier Science Publishers, Amsterdam.

Welz, B. (1985), *Atomic Absorption Spectrometry*, 2nd Ed., pp. 284–285, 408. Verlag Chemie, Weinheim-Deerfield Beach/Florida-Basel.

Yang, S. J., and Rafla, S. (1983), *Effect of Spirogermanium on V79 Chinese Hamster Cells. Am. J. Clin. Oncol. 6*, 331–337.

Youdelis, W. V., and Youdelis, R. A. (1981), *Silver-Copper-Germanium Alloys (Potential for Oral Restorations). Can. Dent. Assoc. J. 47*, 101–106.

II.12 Gold*

C. FRANK SHAW III, Milwaukee, Wisconsin, USA

1 Introduction

Gold is a noble metal which is widely dispersed in the environment, but at a very low average concentration, about 5 µg/kg in the earth's crust (see Chapter I.1). Very low concentrations occur naturally in plant and animal tissues. The quantities released to the environment by human activities are small because of economic incentives for recycling and because elemental gold is inert. These concentrations are much lower than those which give rise to toxic effects, and gold is not known to cause any environmental or industrial health problems. Detailed knowledge of its biological effects is known from the medicinal use of gold complexes.

2 Physical and Chemical Properties, and Analytical Methods

Gold (atomic number 79, atomic mass 196.97, density 19.3 g/cm^3, melting point 1064°C, boiling point 2966°C) is most commonly encountered as the lustrous, yellow, precious metal of jewelry and coinage. Although metallic gold is chemically inert, six discrete, well-characterized oxidation states are known: $-$I, 0, I, II, III, and V (PUDDEPHATT, 1978; SHAW, 1979; JONES, 1981, 1983). At present, there is no evidence that the $-$I, II, or V states exist in the environment or in biological systems. Metallic gold and the $+$I and $+$III states are biologically significant. Gold(0) occurs as the bulk metal and in colloidal suspensions and is the most stable form of gold. Heat, light, or chemical agents often decompose other oxidation states to metallic or colloidal gold.

Neither gold(I) nor gold(III) are stable as free or aquated ions. They occur as complexes with inorganic and/or organic ligands. Gold(I) is a soft Lewis acid and is stabilized by soft ligands (Lewis bases) such as phosphines, thiols, cyanide, and sometimes arsines, thioethers or aromatic nitrogen bases. The simple halides, AuX and AuX$_2^-$, tend to disproportionate into Au(0) and Au(III) in aqueous solution. With anionic ligands (thiolates, cyanide, etc.), linear, two-coordinate complexes are

---

* Contribution No. 328 from the Center for Great Lakes Studies, University of Wisconsin-Milwaukee

usually stable: $Au(CN)_2^-$, $Au(S_2O_3)^{3-}$. Compounds having a stoichiometry of 1:1 Au:thiolate form oligomers with bridging thiolates between adjacent gold(I) ions. Phosphines and other neutral ligands can form 2-, 3-, or 4-coordinate complexes, $[AuL_n^+]$, n = 2, 3, 4, which are linear, trigonal, and tetrahedral, respectively. For gold(III), square-planar, four-coordinate complexes are the most commonly encountered structures, e.g., $AuCl_4^-$, $Au(CN)_4^-$, and $(CH_3)_3AuP(CH_3)_3$. Many gold(III) complexes (especially the halides) are powerful oxidizing agents and are easily reduced by mercaptans, thioethers, phosphines, and even disulfides (PUDDEPHATT, 1978; SHAW, 1979).

Graphite furnace atomic absorption spectroscopy (GFAAS) is the most common and sensitive technique for routine determination of gold at the concentrations occurring in the environment (BROWN and SMITH 1980; WELZ, 1985). Flame atomic absorption spectroscopy is faster and less complicated, but the detection limits are much higher. Neutron activation analysis is comparable to GFAAS, but it is slow, expensive, and not readily available (BROWN and SMITH, 1980). Inductively coupled plasma (ICP) techniques, sometimes coupled to mass spectrometers, promise even greater sensitivity and freedom from interferences (see also MONTASER and GOLIGHTLY, 1987). They have the advantage that they can be interfaced directly to high pressure liquid chromatographs for insights into the nature of the ligands bound to gold (ELDER and TEPPERMAN, 1987). Colorimetric methods are slow and usually complicated by chemical interferences with other ions. Samples for colorimetric, flame AAS, and polarographic analysis must be chemically prepared (usually by acid digestion). X-ray microprobe techniques for in situ analysis of gold in plant and animal tissues are very sensitive to concentrated gold deposits; as little as 10^{-18} g can be detected. For highly dispersed gold in similar media, however, it is much less sensitive (>500 µmol/L) (SHAW, 1979).

Mössbauer spectroscopy and EXAFS/XANES (extended X-ray absorption fine structure/X-ray absorption near edge spectroscopy) are important techniques for identifying the chemical forms of gold in various systems including biological samples (ELDER et al., 1983; ELDER and EIDSNESS, 1987; PARRISH, 1982).

3 Sources, Production, Uses, Recycling

Gold occurs in nature primarily as metallic gold in alluvial deposits which contain large aggregates, in reef deposits which contain small flecks in quartz minerals, and as tellurides, Au_2Te, Au_2Te_3, and $(Au,Ag)_2Te$. The annual production of gold, primarily from Canada, the Soviet Union, and South Africa, is approximately 1500 tons. The cumulative production since historical times exceeds 100000 tons (RENNER and ADAMSON, 1976). Gold is extracted by gravitational processes ("panning"), amalgamation, and cyanide extraction. Substantial quantities are also obtained as a by-product of electrolytic copper refining.

Gold is used in speciality coinage and jewelry, in the electronics industry, for decorative purposes (in bulk or via electroplating), in dental alloys, and in anti-arthritic drugs. Small quantities are used in various heterogeneous catalysts. The amounts of gold released to the environment are detectable but very small. This may be attributed to several factors: (1) the refractory nature of gold (mp 1064 °C, bp ca. 2700 °C); (2) the chemical inertness of elemental gold; and (3) the economic incentives for efficient recovery and recycling (about 40% of the supply) of gold wastes (see also SAAGER, 1984).

4 Distribution in the Environment and in Living Organisms

Large amounts of gold are found in the oceans, but the concentrations are so small, ca. 4 ng/L (BOWEN, 1979; BRULAND, 1983), that extraction has not been feasible (PUDDEPHATT, 1978). Some plants, however, are able to concentrate gold from very dilute aqueous solutions. For example, impatiens and garden balsam accumulate $Au(CN)_4^-$ from 29 µg Au/mL solutions to the extent of 180 and 78 µg Au/g, respectively, which suggests a mechanism for mobilization of gold into the biosphere by cyanogenic plants (SHACKLETTE et al., 1970). Mechanisms of activation by microorganisms have also been documented (PARES and MARTINET, 1964; HALLBAUER, 1978). Accumulation of gold by the alga *Chlorella vulgaris* occurs at very low concentrations and can be substantial (up to 10% dry weight) at higher concentrations. It depends strongly on the oxidation state and ligation of the gold (GREENE et al., 1986; DARNALL et al., 1986; WATKINS, et al., 1986). Although mixtures of $AuCl_2^-$ and $AuCl_4^-$ can be methylated by methylcobalamin (B_{12}), the methyl groups are transferred to chloride ions forming methylchloride (AGNES et al., 1971). Thus, there is no evidence that living organisms activate gold as they do activate elemental mercury.

Evidence for the release of gold into the environment in industrial areas has been presented. Comparisons of gold in the hair of moles and hares in urban and rural Czechoslovakia show greater concentrations of gold in the urban-dwelling specimens (PAUKERT and OBRUSNIK, 1986). This finding is substantiated by detection of gold at ca. 50 ng Au/g of dust (corresponding to 5.2 ng Au/m^3 of air) in Ghent (SCHUTYSER, 1977) and by enhanced rates of long-range atmospheric deposition of gold into an Adirondack lake (Canada) during the past 30 years (GALLOWAY and LIKENS, 1979).

5 Uptake, Absorption, Transport and Distribution, Metabolism and Elimination in Animals and Humans

Although SCHROEDER (1960) classified gold as an element "not present at birth, but accumulated from the environment", more sensitive analytical techniques have detected small concentrations in fetal liver, 2.5 ± 1.3 ng Au/g (ALEXIOU et al., 1977). The ratio of gold concentrations in fetal liver to placenta are less than one, however, suggesting that it is not an essential element (ALEXIOU et al., 1977). Blood levels in man are 0.35 ng Au/mL (SMITH et al., 1973). Organ concentrations are variable, with the median values for adult liver and kidney under 10 µg Au/g ash (TIPTON, 1960). Thumbnail concentrations of gold are often greater in females (0.1 – 6.4; average, 2.6 µg/g) than in males (0.1 – 0.7; average 0.4 µg/g) (KANABROCKI et al., 1979). High concentrations of gold (90 and 70 µg/g) were found in two skin biopsy samples taken from under gold rings (BROWN et al., 1982).

Because gold and its compounds do not cause any known environmental health problems, their distribution and metabolism are known primarily from the use of gold(I) complexes as antiarthritic drugs. Definitive reviews have appeared on the biochemistry (SADLER, 1976; SHAW, 1979; BROWN and SMITH, 1980), and pharmacology (SCHATTENKIRCHNER and MÜLLER, 1982; CROOKE et al., 1986; SNYDER et al., 1987) of various gold complexes. Their metabolites concentrate primarily in the liver and kidney at short times after administration, but autopsies performed 10 to 20 years after cessation of gold therapy (1 – 2 g Au cumulative total) indicate that gold is slowly mobilized to other tissues, becoming more widely dispersed. The retention of gold in tissues and circulating after these long periods indicate that the half-life for gold excretion from deep storage compartments is very long.

The reactions of gold compounds with tissues and fluids depend greatly on the oxidation state, the ligands initially bound to gold, and on environmentally accumulated ligands. The oligomeric gold(I) thiolates (e.g., myochrysine or solganol) must be injected, but phosphine containing complexes (e.g., auranofin or Et_3PAuCl) can be absorbed orally. In the blood, the gold(I) thiolates react with proteins, primarily serum albumin, via ligand exchange reactions. In contrast, up to 50% of the gold administered as auranofin enters red and white blood cells, with the remainder bound to albumin and other proteins.

Hydrogen cyanide absorbed from tobacco smoke alters the distribution of gold compounds in vivo (GRAHAM et al., 1984). Ligand exchange reactions of gold are generally very labile, and the carrier ligands are displaced, then metabolized in vivo and/or excreted. Two general models for gold(I) biochemistry, (1) an equilibrium distribution among protein and non-protein thiols (THOMPSON, 1978) and (2) a thiol shuttle model for transport across biological membranes (SNYDER et al., 1986), have found recent support from evidence of facile inter-protein gold transfer reactions (SHAW et al., 1987).

At the subcellular level, gold deposits in lysosomes, giving them a characteristic, new morphology which led to the name "aurosomes" (GHADIALLY, 1979). Examination of these deposits by EXAFS/XANES (extended X-ray absorption fine structure/X-ray absorption near edge spectroscopy) demonstrated that they contain gold(I) ions (ELDER et al., 1983), not colloidal or metallic gold as previously supposed. A substantial fraction of the gold in mammalian cell cytosol is metallothionein bound (SHAW, 1980, and op. cit.; GLENNAS and RUGSTAD, 1985; MOGILNICKA and WEBB, 1981). Although gold does induce metallothionein (MONIA et al., 1987), the resulting aurothioneins have altered structures (LAIB et al., 1985) which may cause their rapid biological turnover (MONIA et al., 1986).

In mammals, gold(III) is rapidly and readily reduced to gold(I), so that aurosomes isolated 18 hours after administering $AuCl_4^-$ contain only gold(I). Possible reductants include protein disulfides (SHAW et al., 1980; WITKIEWICZ and SHAW, 1981), thiols (LIBENSON, 1943), and thioethers (ISAB and SADLER, 1977). The increased toxicity of gold(III) compounds compared to gold(I) compounds may be related to the additional damage caused by the oxidizing ability of gold(III).

6 Effects on Animals and Man

The most important biological effect of gold is its well established anti-arthritic activity (DAVIS and HARTH, 1979, 1982; SCHATTENKIRCHNER and MÜLLER, 1982). The mechanism(s) of chrysotherapy remain elusive despite over a half century of use. Gold may have both antiinflammatory and immunoregulatory activities (LEIBFARTH and PERSELLIN, 1981; LEWIS and WALZ, 1982). Gold toxicity is not uncommon during chrysotherapy. The most common side-effects are nephrotoxicity and various mucocutaneous reactions; thrombocytopenia and aplastic anemia are rare but life threatening. Various thiol treatments are used to mobilize and redistribute gold (the term "chelation therapy" is inappropriate because the linear coordination geometry of gold precludes chelate formation by commonly employed ligands).

The anti-cancer activity of various gold complexes depends on the cell line and the ligands (SIMON et al., 1979; MIRABELLI et al., 1985, 1986). Anti-tumor activity correlates well with potent cytotoxicity, although the converse is not true (MIRABELLI et al., 1986). Gold is also reported to have anti-pemphigus activity (PENNYS et al., 1976).

Skin reactions to gold jewelry are rare and occur most frequently among chrysotherapy patients who develop reactions to gold drugs (RASPON, 1984). In certain strains of mice and in Japanese quail, gold thioglucose induces an experimental obesity mediated by damage to the hypothalamus (SHAW, 1979; DEBBONS et al., 1977).

7 Hazard Evaluation and Toxicity Limiting Concentrations

Orally ingested gold (metallic or dust) is not significantly absorbed. The routes of administration, the oxidation state of gold, and the ligands present all significantly affect the toxicity of gold compounds (NIOSH, 1982; LIBENSON, 1943; SABIN and WARREN, 1940). For example, LD_{50} values (mg Au/kg) for gold compounds administered intravenously to mice are $NaAuCl_4$, 41; $Na_3Au(S_2O_3)_2$, 37; $[Na_2AuSTm]_n$ (myochrysine), 202 (a value of 1500 mg/kg for the LD_{50} of $AuCl_3$, quoted in Merck Index and propagated by the NIOSH RTECS (1979) and others, appears to be an error).

Gold compounds, especially gold(III) complexes, are toxic to plants and animals including man. Although the small quantities of gold mobilized into the environment from human activities significantly increase the naturally low levels of soils and waters, there are no known incidents of environmentally induced gold toxicity. Gold is conspicuously absent from major treatises on the toxicity of metals in the workplace (BROWNING, 1969) and in plants (LEPP, 1981). The side-effects of chrysotherapy are the major source of human toxicity, and they can be controlled by careful monitoring for symptoms.

References

Agnes, G., Beadle, S. Hill, H. A. O., Williams, F. R., and Williams, R. S. P. (1971),*Methylation by Methyl Vitamin B12. J. Chem. Soc. Chem. Commun.*, 850–851.

Alexiou, D., Grimanis, A. P., Grimani, M., Papaevangelou, G., Koumantakis, E., and Papadatos, C. (1977), *Trace Elements (Zn, Co, Se, Ru, Br, Au) in Human Placenta and Newborn Liver at Birth. Pediatr. Res. 11*, 646–648.

Bowen, H. J. M. (1979), *Environmental Chemistry of the Elements.* Academic Press, London.

Brown, D. H., and Smith, W. E. (1980), *The Chemistry of the Gold Drugs Used in the Treatment of Rheumatoid Arthritis. Chem. Soc. Rev. 9*, 217–239.

Brown, D. H., Smith, W. E., Fox, P., and Sturrock, R. D. (1982), *The Reactions of Gold(0) with Amino Acids and the Significance of these Reactions in the Biochemistry of Gold. Inorg. Chim. Acta 67*, 27–30.

Browning, E. (1969), *Toxicity of Industrial Metals.* Butterworths, London.

Bruland, K. W. (1983), *Trace Elements in Sea Water, Chem. Oceanogr. 8*, 209–210.

Crooke, S. T., Snyder, R. M., Butt, T. R., Ecker, D. J., Allaudeen, H. S., Brett, M., and Mirabelli, C. K. (1986), *Cellular and Molecular Pharmacology of Auranofin and Related Gold Complexes. Biochem. Pharmacol. 35*, 3421–3423.

Darnall, D. W., Greene, B., Henzl, M. T., Hosea, J. M., McPherson, R. A., Sneddon, J., and Alexander, M. D. (1986), *Selective Recovery of Gold and Other Metal Ions from an Algal Biomass. Environ. Sci. Tech. 20*, 206–208.

Davis, P., and Harth, M. (1979), *J. Rheumatol. 6, Suppl. 5*, 1–164.

Davis P., and Harth, M. (1982), *J. Rheumatol. 9 Suppl. 8*, 1–209.

Debbons, A. F., Krimsky, I., Maayan, M. L., Fani, K., and Jiminez, F. A. (1977), *Gold Thioglucose Obesity Syndrome. Fed. Proc. 36*, 143–149.

Elder, R. C., and Eidsness, M. K. (1987), *Synchroton X-Ray Studies of Metal-Based Drugs and Metabolites. Chem. Rev. 87*, 1027–1046.

Elder, R. C., and Tepperman, K. (1987), *Determination of Gold Drug Metabolites in Human Fluids. Proceedings of the International Meeting on Gold and Silver, Washington D.C.,* May 13–14. The Gold Institute, Washington, DC.

Elder, R. C., Eidsness, M. K., Heeg, M. J., Tepperman, K. G., Shaw, C. F. III, and Schaeffer, N. (1983), *Gold-based Antiarthritic Drugs and Metabolites Extended X-Ray Absorption Fine Structure (EXAFS) Spectroscopy and X-Ray Absorption Near Edge Spectroscopy (XANES). Am. Chem. Soc. Symp. Ser. 209*, 385–400.

Galloway, J. N., and Likens, G. E. (1979), *Atmospheric Enhancement of Metal Deposition in Adirondack Lake Sediments. Limnol. Oceanogr. 24*, 427–433.

Ghadially, F. N. (1979), *The Aurosome. J. Rheumatol. 6, Suppl. 5*, 45–51.

Greene, B., Hosea, M., McPherson, R., Henzl, M., Alexander, M. D., and Darnall, D. W. (1986), *Interaction of Gold(I) and Gold(III) Complexes with Algal Biomass. Environ. Sci. Technol. 20*, 627–632.

Glennas, A., and Rugstad, H. E. (1985), *Acquired Resistance to Auranofin in Cultured Human Cells. Scand. J. Rheumatol. 14*, 230–238.

Graham, G. G., Haavisto, T. M., Jones, H. M., and Champion, G. D. F. (1984), *The Effects of Cyanide on the Uptake of Gold by Red Blood Cells. Biochem. Pharmacol. 33*, 1257–1262.

Hallbauer, D. K. (1978), *Witwaterstand Gold Deposits. Gold Bull. 11*, 18–23.
Isab, A. A., and Sadler, P. J. (1977), *Reactions of Gold(III) Ions with Ribonuclease A and Methionine Derivatives in Aqueous Solution. Biochim. Biophys. Acta 492*, 322–330.
Jones, P. G. (1981), *X-Ray Structural Investigations of Gold Compounds. Gold Bull. 14*, 102–118; 159–166.
Jones, P. G. (1983), *X-Ray Structural Investigations of Gold Compounds. Gold Bull. 16*, 114–124.
Kanabrocki, E. L., Kanabrocki, J. C., Greco, J., Kaplan, E., Oester, Y. T., Brar, S. S., Gustafson, P. S., Nelson, D. M., and Moore, C. E. (1979), *Instrumental Analysis of Trace Elements in Thumbnails of Human Subjects. Sci. Total Environ. 13*, 131–140.
Laib, J. E., Shaw, C. F. III, Petering, D. H., Eidsness, M. K., Elder, R. C., and Garvey, J. S. (1985), *Formation and Characterization of Aurothioneins: Au, Zn, Cd-Thionein, Au, Cd-Thionein, and (Thiomalato-Au)$_x$-Thionein. Biochemistry 24*, 1977–1986.
Leibfahrt, J. H., and Persellin, R. H., (1981), *Review: Mechanisms of Action of Gold. Agents Actions 11*, 458–472.
Lepp, N. W. (1981), *Effect of Heavy Metal Pollution on Plants*, Vol. 2. Applied Science Publishers, London.
Lewis, A. J., and Walz, D. T. (1982), *Immunopharmacology of Gold. Prog. Med. Chem. 19*, 1–58.
Libenson, L. (1943), *Toxicity and Mode of Action of Gold Salts. Exp. Med. Surg. 3*, 146–153.
Mirabelli, C. K., Johnson, R. K., Sung, C. M., Faucette, L. F., Muirhead, K., Crooke, S. T. (1985), *Evaluation of the in vivo Antitumor Activity and in vitro Cytotoxic Properties of Auranofin, a Coordinated Gold Compound, in Murine Tumor Models. Cancer Res. 45*, 32–39.
Mirabelli, C. K., Johnson, R. K., Hill, D. T., Faucette, L. F., Girard, G. R., Kuo, G. Y., Sung, C. M., and Crooke, S. T. (1986), *Correlation of the in vitro Cytotoxic and in vivo Antitumor Activities of Gold(I) Coordination Complexes. J. Med. Chem. 29*, 218–223.
Mogilnicka, E. A., and Webb, M. (1983), *Time-Dependent Uptake and Metallothionein Binding of Gold, Copper and Zinc in the Rat Kidney. Biochem. Pharmacol. 32*, 1341–1346.
Monia, B. P., Butt, T. R., Ecker, D. J., Mirabelli, C. K., and Crooke, S. T. (1986), *Metallothionein Turnover in Mammalian Cells. J. Biol. Chem. 26*, 1097–1099.
Monia, B. P., Butt, T. R., Mirabelli, C. K., Ecker, D., Sternberg, E., and Crooke, S. T. (1987), *Induction of Metallothionein is Correlated with Resistance to Auranofin, a Gold Compound, in Chinese Hamster Ovary Cells. Am. Soc. Pharmacol. Exp. Ther. 31*, 21–26.
Montaser, A., and Golightly, D. W., (1987), *Inductively Coupled Plasmas in Analytical Atomic Spectrometry*, pp. 354, 611. VCH Verlagsgesellschaft, Weinheim-Basel-Cambridge-New York.
NIOSH (National Institute of Occupational Safety and Health) (1982), *Registry of Toxic Effects of Chemical Substances* (RTECS), U.S. Government Printing Agency, Washington, D.C.
Parès, Y., and Martinet, R. (1964), *Intervention des bacteries dans le cycle de l'or. Bull. Bur. Rech. Geol. Min. Fr. 3*, 1–29.
Parrish, R. V. (1982), *Gold and Mössbauer Spectroscopy. Gold Bull. 15*, 51–63.
Paukert, J., and Obrusnik, I. (1986), *The Hair of the Common Hare and The Common Vole as Indicators of Environmental Pollution. J. Hyg. Epidemiol. Microbiol. Immunol. 30*, 27–32.
Pennys, N. S., Eaglestein, W. E., and Frost, P. (1976), *Management of Pemphigus with Gold Compounds. Arch. Dermatol. 112*, 185–187.
Petering, H. G. (1976), *Pharmacology and Toxicology of Heavy Metals: Gold. J. Pharmacol. Exp. Ther. A1*, 119–125.
Puddephatt, R. J. (1978), *The Chemistry of Gold*. Elsevier, Amsterdam.
Raspon, W. S. (1984), *Skin Contact with Gold and Gold Alloys. Gold Bull. 17*, 102–108.
Renner, H., and Adamson, R. J. (1976), *Gold*, in: *Ullmanns Encyklopädie der technischen Chemie*, 4th Ed., Vol. 12, pp. 383–395. Verlag Chemie, Weinheim-New York.
Saager, R. (1984), *Metallic Raw Materials Dictionary* (in German), pp. 50–51. Bank von Tobel, Zürich.
Sabin, A. B., and Warren, J. (1940), *The Curative Effect of Certain Gold Compounds on Experimental, Proliferative Chronic Arthritis in Mice. J. Bacteriol. 40*, 823–856.
Sadler, P. J., (1976), *The Biological Chemistry of Gold. Struct. Bonding 29*, 171–219.

Schattenkirchner, M., and Müller, W. (1982): *Modern Aspects of Gold Therapy*, p. 229. Karger Verlag, Basel.

Schroeder, H. A. (1960), in: Seven, H. A. (ed.): *Metal Binding in Medicine*, pp. 59–67. Lippincott, Philadelphia.

Schutyser, P., Govaerts, A., Dams, R., and Hoste, J. (1977), *Neutron Activation Analysis of Platinum Metals in Airborne Particulate Matter. J. Radionanal. Chem. 37*, 651–660.

Shacklette, H. T., Lakin, H. W., Hubert, A. E., and Curtin, G. C. (1970), *Absorption of Gold by Plants. Geol. Surv. Bull.* 1314-B. U.S. Government Printing Office, Washington, DC.

Shaw, C. F. III (1979), *The Mammalian Biochemistry of Gold: An Inorganic Perspective of Chrysotherapy. Inorg. Perspect. Biol. Med.*, 287–355.

Shaw, C. F. III (1980), *The Biochemistry and Subcellular Distribution of Gold in Kidney Tissue: Implications for Chrysotherapy and Nephrotoxicity*, in: Rainsford, K. D. (ed.), *Trace Elements in the Pathogenesis and Treatment of Inflammatory Conditions*, Chap. 26. Birkhäuser, Basel.

Shaw, C. F. III, Cancro, M. P., Eldridge, J. E., and Witkiewicz, P. (1980), *Gold(III) Oxidation of Disulfides in Aqueous Solution. Inorg. Chem. 19*, 3198–3201.

Shaw, C. F. III, Coffer, M. T., Klingbeil, J., and Mirabelli, C. K. (1987), *Application of a ^{31}P NMR Chemical Shift: Gold Affinity Correlation to Hemoglobin-Gold Binding and the First Inter-Protein Gold Transfer Reaction. J. Am. Chem. Soc. 110*, 729–734.

Simon, T. M., Kunishima, D. H., Vibert, G. J., and Lorber, A. (1979), *Inhibitory Effects of a New Oral Gold Compound on Hela Cells. Cancer 44*, 1965–1975.

Smith, P. M., Smith, E. M., and Gottlieb, N. L. (1973), *Gold Distribution in Whole Blood During Chrysotherapy. J. Lab. Clin. Med. 82*, 930–937.

Snyder, R. M., Mirabelli, C. K., and Crooke, S. T. (1986), *Cellular Association, Intracellular Distribution, and Efflux of Auranofin via Sequential Ligand Exchange Reactions. Biochem. Pharmacol. 35*, 923–932.

Snyder, R. M., Mirabelli, C. K., and Crooke, S. T. (1987) *The Cellular Pharmacology of Auranofin. Semin. Arthritis Rheum. 17*, 71–80.

Thompson, H. O., Blaszak, J., Knudtson, C. J., and Shaw, C. F. III (1978), *Characterization of Gold in the Cytosol of Rat-Kidney Cortex Cells. Bioinorg. Chem. 9*, 375–388.

Tipton, I. H. (1960), in: Seven, A. A. (ed.), *Metal Binding in Medicine*, pp. 27–42. Lippincott, Philadelphia.

Venugopal, B., and Luckey, T. D. (1979), *Metal Toxicity in Mammals*, Vol. 2, pp. 35, 38. Plenum Press, New York.

Watkins, J. W. II, Elder, R. C., Greene, B., and Darnall, D. W. (1986), *Determination of Gold Binding in an Algal Biomass Using EXAFS and XANES Spectroscopies. Inorg. Chem. 26*, 1147–1151.

Welz, B. (1985), *Atomic Absorption Spectrometry*, pp. 285, 405 ff. VCH Verlagsgesellschaft, Weinheim-Deerfield Beach/Florida-Basel.

Witkiewicz, P. L., and Shaw, C. F. III (1981), *Oxidative Cleavage of Peptide and Protein Disulphide Bonds by Gold(III): A Mechanism for Gold Toxicity. J. Chem. Soc. Chem. Commun.*, 112–1114.

Additional Recommended Literature

Anonymous (1987), *Refining of Gold and Silver* (in German), *Neue Zürcher Zeitung, Forschung und Technik, No. 280*, p. 79, (2. Dec.), Zürich.

Harvie, D. I. (1989), *John Stewart MacArthur: Pioneer Gold and Radium Refiner. Endeavour New Ser. 13* (4), 179–184.

Trueb, L. (1989) *The Gold in the Blackhills of South Dakota* (in German). *Neue Zürcher Zeitung, Forschung und Technik, No. 49*, p. 77 (28 Feb.) and *No. 248*, p. 77 (25 Oct.), Zürich.

II.13 Indium

BRUCE A. FOWLER, Baltimore, Maryland, USA
PETER L. GOERING, Rockville, Maryland, USA

1 Introduction

Indium is a relatively rare non-essential metal which has been demonstrated to produce both liver and kidney injury following parenteral administration. Environmental or occupational poisonings have rarely been reported since this metal is poorly absorbed from the gastrointestinal tract and only moderately absorbed via inhalation. The mechanisms of indium toxicity are not well understood, but ultrastructural/biochemical studies at the subcellular level have suggested that this element damages both the structure and functional activities of the endoplasmic reticulum.

2 Physical and Chemical Properties, and Analytical Chemistry

Indium (In) has an atomic mass of 114.8, atomic number 49, and a density of 7.31 g/cm^3. It has a melting point of 156.6 °C and a boiling point of 2080 °C. It is a member of Group III of the Periodic Table of Elements. Indium may exist in the positive oxidation states of I, II, or III. The radionuclide ^{111}In may be used as a diagnostic agent to localize malignant and inflammatory lesions following its intravenous administration (HAYES, 1988).

There are a number of analytical methods available for indium of which atomic absorption spectroscopy (AAS) using either the flame (FOWLER et al., 1983) or heated graphite atomizer (ARMANNSSON and OVENDEN, 1980) are most commonly utilized (see also WELZ, 1985). Other methods include spectrochemical methods (KINSER et al., 1976). Neutron activation analysis (NAA) has also been applied to a number of matrices (MATTHEWS and RILEY, 1970; REY et al., 1970; KUZNECOV, 1979), as well as polarographic methods (JONES and LEE, 1979).

3 Sources, Production, Important Compounds, and Uses

Indium is a relatively rare element which is recovered as a by-product of zinc smelting (SMITH et al., 1978). Enrichment after roasting of zinc concentrates is possible by extraction, ion exchange or pyrometallurgic processes, followed by raffination (WIESE, 1977). World production of indium was about 50 tons per year in 1982 and the main producing countries are the United States, the Union of Soviet Socialist Republics, Chile, and South Africa (U.S. BUREAU OF MINES, 1983).

A major and increasing use of indium is in the semiconductor industry for the production of high-speed III–V semiconductors (DUPUIS, 1984). Copper-indium-diselenide may be valuable in photovoltaic solar cells. Indium compounds are also incorporated into various solder alloys as a hardening agent. Indium oxide is utilized in medicine for organ scanning and for treating of liver tumors (STERN et al., 1966, 1967; HART and ADAMSON, 1971; OLKKONEN et al., 1977; SMITH et al., 1978; SCHEFFEL et al., 1982). In the future some local problems may arise with wastes.

4 Distribution in the Environment, in Foods, and in Living Organisms

Indium compounds are present in relatively low concentrations in the general environment, but much higher levels may occur in local areas as a result of smelting activities. Studies of open ocean seawater from the Atlantic Ocean (MATTHEWS and RILEY, 1970) using neutron activation analysis have reported In concentrations in the range of 0.1 ng/L (see also BRULAND, 1983). Similar studies conducted on seawater in a Norwegian fjord (JOHANSON and STEINES, 1976) reported In concentrations of about 0.7 ng/L. Indium is also rare in soils and has been estimated to comprise only $10^{-5}\%$ of the earth's crust (SUNDERMAN and TOWNLEY, 1960). Indium concentrations in air and rain have also been reported as 0.3 ng/m^3 and 0.59 ng/L, respectively (PEIRSON et al., 1973). In contrast, studies conducted near a smelter in Idaho (RAGAINI et al., 1977) reported soil values for In of 2.05 mg/kg and air concentrations of 5.79 ng/m^3. Similar studies conducted in a residential area of Liege, Belgium, demonstrated an air concentration of 1.91 ng/m^3 (DAMS, 1973).

Indium concentrations in plant and animal tissues have been reported to range from below detection limits to around 10 µg/kg for beef and ham. Marine organisms collected near smelter outfalls have been reported to contain concentrations up to 15 mg In/kg (SMITH et al., 1978).

5 Uptake, Absorption, Transport and Distribution, Metabolism and Elimination in Animals and Humans

The in vivo chemical disposition of ^{114}In hydroxide or citrate was studied by SMITH et al. (1957) in rats following intratracheal instillation. These authors found that most of the administered dose was taken up into the tracheobronchial lymph nodes.

LEACH et al. (1961) reported similar findings in rats exposed to In_2O_3 via inhalation. Other investigators (MORROW et al., 1958) estimated the absorbed dose of ^{114}In sesquioxide particles at between 3 and 6% of the total dose following intratracheal instillations at 1 hour after dosing and about 18% for sequential 1-hour exposures after 4 days. Humans given $^{111}InCl_3$ or ^{111}In-diethylenetriamine pentaacetic acid by ultrasonic nebulization (ISITMAN et al., 1974) showed deposition in major airways but little alveolar uptake.

SMITH et al. (1960) reported the gastrointestinal absorption of indium to be about 0.5% of the administered dose in rats. Studies by HEADING et al. (1971) in human volunteers showed that these individuals absorbed less than 2% of a ^{113}In dose administered as a diethylenetriamine pentaacetic acid complex while COATES et al. (1973) reported no detectable absorption in human adult volunteers following administration of $^{113}InCl_3$.

Several investigators (HOSAIN et al., 1969; CASTRONOVO and WAGNER, 1973) have shown that ionic indium is transported in blood bound to transferrin. Mice given a single intravenous injection of indium were found to clear this dose from the blood within 3 days. Studies by RUSSAK et al. (1981) showed that tissue uptake of ^{113}In in mice following a single intravenous injection was markedly dependent upon thyroid function. In vitro studies by DETTMAN et al. (1981) showed that the half-life of ^{111}In-oxide in mouse spleen cells was 67 hours. The distribution of indium in body organs is highly dependent upon its chemical form. Ionic forms of indium are extensively accumulated by the kidney while colloidal indium oxide is accumulated by the liver and the reticuloendothelial system (CASTRONOVO, 1970; CASTRONOVO and WAGNER, 1971). At the subcellular level (FOWLER et al., 1983; GOERING et al., 1987), indium has been found to be primarily associated with lysosomes and bound to high molecular weight cytosolic proteins. Excretion of indium also highly depends on the chemical form administered. Indium administered in an ionic form is primarily excreted in the urine while colloidal indium complexes are primarily excreted in the feces (CASTRONOVO, 1970; CASTRONOVO and WAGNER, 1971, 1973).

6 Effects on Plants, Animals, and Humans

In the literature there are no reports published on indium toxicity to plants. Intratracheal administration of In_2O_3 (50 mg/kg) to rats has been found (PODOSINOVSKIJ, 1965) to produce a pneumonitis and initial stages of pulmonary fibrosis. Implantation of indium-treated silver discs into rabbits was not found to elicit foreign-body reactions (HARROLD, 1943). Parenteral administration of indium has been reported by a number of investigators to produce both liver and kidney damage, but the primary target is somewhat dependent upon the chemical form of indium administered. Several investigators (DOWNS et al., 1959: CASTRONOVO, 1970; CASTRONOVO and WAGNER, 1971; WOODS and FOWLER, 1982) have reported morphological damage to renal proximal tubule cells of rodents injected with $InCl_3$. Liver damage has also been reported in rats given a single intraperitoneal injection of InCl$_3$ (WOODS et al., 1979; FOWLER et al., 1983). Injection of colloidal indium has been shown to preferentially damage cells of the reticuloendothelial system in the liver and spleen (DOWNS et al., 1959; STERN et al., 1967; CASTRONOVO, 1970; CASTRONOVO and WAGNER, 1971). Reduced hemoglobin and leucopenia have also been reported in rats, mice, and rabbits following treatment with ionic indium (STEIDLE, 1933; McCORD et al., 1942; DOWNS et al., 1959; YOSHIKAWA and HASEGAWA, 1971). SCHROEDER and MITCHENER (1971) reported mild growth depression in mice given ionic indium in drinking water for prolonged time periods. Ultrastructural/biochemical studies (WOODS et al., 1979; WOODS and FOWLER, 1982; FOWLER et al., 1983) have shown marked inhibition of enzymes in the heme pathway which was associated with structural disruption of both the endoplasmic reticulum and the mitochondria in livers and kidneys.

Co-administration of ferric dextran in rats and hamsters has been shown to protect against toxicity to the liver and to embryos of these species (GABBIANI et al., 1962; FERM, 1970). EVDOKIMOFF and WAGNER (1972) also reported that thorium dioxide solutions and gelatin prevented damage to the reticuloendothelial system by colloidal indium hydroxide. EYBL et al. (1987) showed protective effects by manganese salts.

Studies concerning indium toxicity in humans are more limited. RAICIULESCU et al. (1972) reported vascular shock in 3 patients out of 770 injected with colloidal indium. ^{113}In is used for liver scanning purposes.

There are no available reports of indium carcinogenicity or mutagenicity. FERM and CARPENTER (1970) reported malformations in digits of fetuses from pregnant hamsters given a single intravenous injection of indium chloride at 1 mg In/kg. At higher dose levels (2–20 mg In/kg) embryolethality was observed.

7 Hazard Evaluation and Limiting Concentrations

Risk assessment for indium compounds is currently difficult due to a lack of baseline data concerning the cycling of indium in the environment (ANDREAE et al., 1984) and human exposure from both environmental and occupational sources. Such data will be of increasing importance in the near future due to the expected use of indium compounds in developing high-technology industries such as semiconductor manufacture. A TLV-TWA level for indium and its compounds of 0.1 mg In/m^3 of air has been estblished by the American Conference of Governmental Industrial Hygienists (HAYES, 1988).

References

Andreae, M. O., Asami, T., Bertine, K. K., Buat-Menard, P. E., Duce, R. A., Filip, Z., Forstner, U., Goldberg, E. D., Heinrichs, H., Jernelov, A. B., Pacyna, J. M., Thornton, I., Tobschall, H. J., and Zoller, W. H. (1984), in: Nriagu, J. O. (ed.), *Changing Metal Cycles and Human Health*, pp. 359–374. Springer-Verlag, New York.

Armannsson, H., and Ovenden, P. J. (1980), *Int. J. Environ. Anal. Chem.* 8, 127–136.

Bruland, K. W. (1983), *Trace Elements in Sea-water, Chem. Oceanogr.* 8, 179.

Castronovo, F. P. (1970), *Factors Affecting the Toxicity of the Element Indium.* (Doctoral Thesis), Johns Hopkins University, Baltimore, Maryland.

Castronovo, F. P., and Wagner, H. N. (1971), *Factors Affecting the Toxicity of the Element Indium. Br. J. Exp. Pathol. 52*, 543–559.

Castronovo, F. P., and Wagner, H. N. (1973), *J. Nucl. Med.* 14, 677–682.

Coates, G., Gilday, D. L., Cradduck, T. D., and Wood, D. E. (1973), *Measurement of the Rate of Stomach Emptying Using Indium-113m and a 10-Crystal Rectilinear Scanner. Can. Med. Assoc. J. 108*, 180–183.

Dams, R. (1973), *Meded. Fac. Landbouwewet. Rijksuniv. Gent 38*, 1869–1884.

Dettman, G. L., Vanderburg, B. L., Johnson, R. D., and Meyer, G. W. (1981), *Indium-111-oxide Labeled Mouse Spleen Cells. Int. J. Nucl. Med. Biol. 8*, 137–143.

Downs, W. L., Scott, J. K., Steadman, L. T., and Maynard, E. A. (1959), *The Toxicity of Indium.* The University of Rochester Atomic Energy Project Report No. UR-558. University of Rochester, Rochester, New York.

Dupuis, R. D. (1984), *Science 226*, 623–629.

Evdokimoff, V., and Wagner, H. N. (1972), *Reduction of Indium Toxicity by Blockade of the Reticuloendothelial System. J. Reticuloendothel. Soc. 11*, 599–603.

Eybl, V., Caisová, D., Kotyzová, M., Kortenská, M., Kovtenský, J., and Sykora, J. (1987), *Proceedings Second Nordic Symposium on Trace Elements, Odense.* World Health Organization (WHO), Copenhagen.

Ferm, V. H., and Carpenter, S. J. (1970), *Teratogenic and Embryopathic Effects of Indium, Gallium, and Germanium. Toxicol. Appl. Pharmacol. 16*, 166–170.

Fowler, B. A., Kardish, R. M., and Woods, J. S. (1983), *Alteration of Hepatic Microsomal Structure and Function by Indium Chloride. Ultrastructural, Morphometric, and Biochemical Studies. Lab. Invest. 48*, 471–478.

Gabbiani, G., Selye, H., and Tuchweber, B. (1962), *Br. J. Pharmacol.* 19, 508–512.

Goering, P. L., Mistry, P., and Fowler, B. A. (1987), *Mechanisms of Metal-induced Cell Injury*, in: Haley, T. J., and Berndt, W. O. (eds.): *Handbook of Toxicology*, pp. 384–425. Hemisphere Publishing Corp., New York.

Harrold, G. C., Meek, S. F., Whitman, N., and McCord, C. P. (1943), *J. Ind. Hyg. Toxicol.* 25, 233–237.

Hart, M. M., and Adamson, R. H. (1971), *Antitumor Activity and Toxicity of Salts of Inorganic Group 3a Metals: Aluminum, Gallium, Indium, and Thallium. Proc. Natl. Acad. Sci. USA 68*, 1623–1626.

Hayes, R. L. (1988), Chapter 27 *Indium*, in: Seiler, H. G., Sigel, H., and Sigel, A. (eds.), *Handbook on Toxicity of Inorganic Compounds*, pp. 323–336. Marcel Dekker, New York.

Heading, R. C., Tothill, P., Laidlow, A. J., and Maynard, E. A. (1971), *Gut* 12, 611–615.

Hosain, F., McIntyre, P. A., Poulose, K., Stern, H. S., and Wagner, N. H. (1969), *Binding of Trace Amounts of Ionic Indium-113m to Plasma Transferrin. Clin. Chim. Acta 24*, 69–75.

Isitman, A. T., Manoli, R., Schmidt, G. H., and Holmes, R. A. (1974), *An Assessment of Alveolar Deposition and Pulmonary Clearance of Radiopharmaceuticals after Nebulization. Am. Roentgenol. Radium Ther. Nucl. Med. 120*, 776–781.

Johanson, O., and Steines, E. (1976), *Int. J. Appl. Radiat. Isot.* 27, 163–167.

Jones, E. A., and Lee, A. F. (1976), *The Determination of Thallium and Indium in Sulphide Concentrates.* National Institute for Metallurgy Report No. 2022, 33 p. Randburg, South Africa.

Kinser, R. E., Keenan, R. G., and Kupel, R. E. (1976), *Am. Ind. Hyg. Assoc. J.* 26, 249–254.

Kuznecov, R. A. (1979), *J. Anal. Chem. USSR* 34, 406–408 (in Russian).

Leach, L. J., Scott, J. K., Armstrong, R. D., and Steadman, L. T. (1961), *The Inhalation Toxicity of Indium Sesquioxide in the Rat.* The University of Rochester Atomic Energy Project Report No. UR-590. University of Rochester, Rochester, New York.

Matthews, A. D., and Riley, J. P. (1970), *Anal. Chim. Acta 51*, 287–294.

McCord, C. P., Meek, S. F., Harrold, G. C., and Heussner, C. E. (1942), *J. Ind. Hyg. Toxicol. 24*, 243–254.

Morrow, P. E., Gibb, F. R., Cloutier, R., Casarett, L. J., and Scott, J. K. (1958), *Fate of Indium Sesquioxide and of Indium^{114} Trichloride Hydrolysate Following Inhalation in Rats.* University of Rochester Atomic Energy Project, Report No. UR-508. University of Rochester, Rochester, New York.

Olkkonen, H., Suonio, S., Lahtinen, T., and Penttila, I. M. (1977), *Comparison of 113mIn and 133Xe i.v. Injection Methods for Evaluation of Placental Blood Flow. Int. J. Nucl. Med. Biol. 4*, 179–183.

Peirson, D. H., Cawse, P. A., Salmon, L., and Cambray, R. S. (1973), *Nature (London) 241*, 252–256.

Podosinovskij, V. V. (1965), *Gig. Sanit. 30*, 28–34 (in Russian).

Ragaini, R. C., Ralston, H. R., and Roberto, N. (1977), *Environ. Sci. Technol. 11*, 773–781.

Raiciulescu, N., Niculescu-Zinca, D., and Stoichita-Papilan, M. (1972), *Rev. Roum. Med. Interne 9*, 55–60 (in Roumanian).

Rey, P., Wakita, H., and Schmitt, R. A. (1970), *Anal. Chim. Acta 11*, 773–781.

Russak, E. M., Vest, M. R., Born, G. S., Kessler, W. V., and Shaw, S. M. (1981), *Effect of Altered Thyroid States on Tissue Uptake of 113mIn in the Mouse. Int. J. Nucl. Med. Biol. 8*, 115–117.

Scheffel, U., Tsam, M.-F., Michell, T. G., Camargo, E. E., Braine, H., Ezekowitz, M. D., Nichfoloff, L., Hill-Zobel, R., Murphy, E., and McIntyre, P. A. (1982), *Human Platelets Labeled with In-111 8-Hydroxyquinoline: Kinetics, Distribution, and Estimates of Radiation Dose. J. Nucl. Med. 23*, 149–156.

Schroeder, H. A., and Mitchener, M. (1971), *Scandium, Chromium(VI), Gallium, Yttrium, Rhodium, Palladium, Indium in Mice: Effects on Growth and Life Span. J. Nutr. 101*, 1431–1437.

Smith, G. A., Thomas, R. G., Black, B., and Scott, J. K. (1957), *The Metabolism of Indium ^{114m} Administered to the Rat by Intratracheal Intubation.* The University of Rochester Atomic Energy Report No. UR-500. University of Rochester, Rochester, New York.

Smith, G. A., Thomas, R. G., and Scott, J. K. (1960), *Health Phys. 4*, 101–108.

Smith, I. C., Carson, B. L. and Hoffmeister, F. (1978), *Trace Elements in the Environment*, Vol. 5. *Indium.* Ann Arbor Science Publishers, Ann Arbor, Michigan.

Steidle, H. (1933), *Naunyn-Schmiedebergs Arch. Exp. Pathol. Pharmakol. 173*, 459–465 (in German).

Stern, H. S., Goodwin, D. A., Wagner, H. N., and Kramer, H. H. (1966), *Nucleonics 24*, 57–59.

Stern, H. S., Goodwin, D. A., Scheffel, U., Wagner, H. N., and Kramer, H. A. (1967), *Nucleonics 25*, 62–68.

Sunderman, D. N., and Townley, C. W. (1960), *The Radiochemistry of Indium*, p. 46. USAEC Division of Technical Information Extension, Oak Ridge, Tennessee.

U.S. Bureau of Mines (1983), *Mineral Commodity Summaries.* Bureau of Mines, Washington, D.C.

Welz, B. (1985), *Atomic Absorption Spectrometry*, pp. 286–287, 390 ff. Verlag Chemie, Weinheim-Deerfield Beach/Florida-Basel.

Wiese, U. (1977), *Indium and Indium Compounds* (in German), in: *Ullmanns Encyklopädie der technischen Chemie*, 4th Ed., Vol. 13, pp. 197–206. Verlag Chemie, Weinheim-New York.

Woods, J. S., and Fowler, B. A. (1982), *Selective Inhibition of delta-Aminolevulinic Acid Dehydratase by Indium Chloride in Rat Kidney: Biochemical and Ultrastructural Studies. Exp. Mol. Pathol. 36*, 306–315.

Woods, J. S., Carver, G. T., and Fowler, B. A. (1979), *Altered Regulation of Hepatic Heme Metabolism by Indium Chloride. Toxicol. Appl. Pharmacol. 49*, 455–461.

Yoshikawa, H., and Hasegawa, T. (1971), *Igaku to Seibutsugaku 83*, 45–48 (in Japanese).

II.14 Iron

HELMUT A. HUEBERS, Houston, Texas, USA

1 Introduction

Iron is the most abundant element in the core of the earth and one of the most abundant in the earth's crust. Besides aluminum, it is the most important metallic element in the terrestrial environment. With regard to its biological activity, iron is also the most versatile of all the elements. Life without iron is, in all likelihood, impossible since the enormous quantities of this metal in the earth's core resulted and still result in the formation of an effective shield that deflects various forms of solar and cosmic radiation. The unique properties of iron undoubtedly also led to its key role in the catalysis of metabolic processes. Because of the myriad number of important reactions in which iron participates, all organisms require a mechanism for its assimilation so as to avoid the ill effects that result from iron deficiency, which afflicts hundreds of millions of people in the world, particularly children and menstruating women. As well as being extremely useful, iron can also be highly toxic to cellular constituents when present in excess, but the problem of toxic iron overload is virtually limited to man and is far less frequent than iron deficiency.

2 Physical and Chemical Properties, and Analytical Methods

Iron (Fe, atomic number 26, atomic mass 55.8) is a shiny, white and malleable metal with a density of 7.9 g/cm^3 and a melting point of 1535 °C (its boiling point is almost 3000 °C). Naturally occurring iron is composed of the isotopes ^{56}Fe (92%), ^{54}FE (6%) and ^{57}Fe (2%), and it occurs in four allotropic modifications: a-iron is the stable modification at room temperature which is transformed to β-iron at 770 °C; at above 900 °C it is converted to γ-iron followed by a further transformation to δ-iron at above 1400 °C. a-Iron is similar to cobalt and nickel as a ferromagnetic metal. At 770 °C (the Curie point) iron loses its ferromagnetic properties and displays a paramagnetic behavior.

Iron is a reactive metal, but it is stable in dry air and water free of carbon dioxide. This stability is the result of a coating of iron oxide which prevents further oxidation. Under these conditions, iron behaves similarly to aluminum and chromium. In biological environments iron is oxidized, even by atmospheric oxygen, first to the ferrous form (Fe^{2+}) and then to the ferric form (Fe^{3+}). The compounds FeO, Fe_3O_4, and δ-Fe_2O_3 are interconvertible. The ferric state of iron is very prone to undergo hydrolysis, i.e., to form insoluble ferric hydroxide polymers with hydroxyl ions generally referred to as "rust". Because of this reaction iron is rarely found in nature in its elementary form, except in meteorites. Higher oxidation states of iron can be synthesized under harsh conditions. For example, ferrate(VI) is found in an oxo anion form which can be obtained using strong oxidizing agents in strongly alkaline media (HOLLEMANN and WIBERG, 1971). The toxic intermediate iron pentacarbonyl is an orange liquid. Iron(IV) can be obtained by oxidizing iron(III) bound to certain strongly stabilizing ligands. Aside from the possibility of forming iron(IV) peroxidases, there is no evidence for involvement of iron in oxidation states other than II or III in biochemistry.

In aqueous solutions iron occurs as iron(II) or iron(III) or as organic ferrous and ferric complexes. Under aerobic conditions and when pH approaches neutrality, the ferric form of inorganic salts is by far the most prevalent one ($3d^5$ configuration). Under these conditions all ferrous iron is oxidized and converted into the more stable ferric form (according to BRULAND, 1983, it exists in sea water mainly as $Fe(OH)_3^0$ and $Fe(OH)_2^+$). Without the presence of iron chelators, the brown precipitate of ferric hydroxide, with its poorly organized crystal structure, will form. Its solubility is as low as 0.56 µg of iron per liter in a solution of neutral pH. Ferric hydroxide is only very poorly bioavailable. Prussian Blue, $Fe_4[Fe(CN)_6]_3$, is the oldest synthetic complex compound and is a famous pigment (LUDI, 1988).

A great variety of methods are available to determine the presence of iron in inorganic material or in biological tissues. Sample contamination must carefully be avoided; organic matrices may have to be destroyed initially. The most important analytical techniques in the clinical area are colorimetric (GOODWIN et al., 1966) and atomic absorption spectroscopy (OLSON and HAMLIN, 1969; WELZ, 1985; ELINDER, 1986). In the former, iron with or without prior acid digestion of the sample is reduced to the ferrous form by the addition of ascorbic or thioglycolic acid, followed by the addition of the ferrous chromogen to produce color which may be quantified. The most extensively used chromogen to date is the disodium salt of the sulfonated derivative of bathophenanthroline (molar extinction coefficient 22 140 at $\lambda = 534$ nm), but even greater sensitivity is achieved with ferrozine, 3-(2-pyridyl)-5,6-bis(4-phenylsulfonic acid)-1,2,4 triazine disodium salt: molar extinction coefficient 28 000 at $\lambda = 562$ nm).

Atomic absorption spectroscopy does not necessarily offer an advantage in sensitivity, and in biological material its accuracy may be compromised by a matrix effect and interactions with other metals. Yet another method of quantitative iron determination is nuclear magnetic resonance which is highly specific for iron (see also WARD, 1987, for NAA and ICP-MS).

3 Sources, Production, Import Compounds, Uses, Waste Products, and Recycling

As the fourth most abundant element, iron constitutes about 4.7% of the earth's crust. It is extracted from ores, of which hematite, Fe_2O_3, magnetite, Fe_3O_4, limonite, $FeO(OH)$, pyrite, FeS_2, and bornite, Cu_5FeS_4 are the most common. The process takes place in a blast furnace which is charged by a mixture of iron ore, coke, and limestone. Air is blown into the bottom of the furnace, and carbon monoxide reduces the iron oxide to metallic iron:

$$Fe_2O_3 + 3\,CO \rightarrow 2\,Fe + 3\,CO_2.$$

The limestone combines with impurities, mainly silica, to form a "slag" of calcium silicate. When the hearth of the furnace is tapped, the iron runs into comblike molds called "pigs" to form pig iron. Pig iron is very impure and is used as a starting material for making cast iron, wrought iron, or steel. The latter is the general name for alloys or iron. Ordinary steel contains a small percentage (less than 0.5%) of carbon, but there are many special alloys such as tool steels, spring steels, and stainless steels which also contain other metals. Steel may be made in a Bessemer converter or in an open-hearth or electric furnace. The Bessemer process is quick, producing about 30 tons of steel in 15 minutes. In the open-hearth process, which takes about a day, a much larger quantity of steel is produced.

Metallic iron produced by carbon monoxide reduction of iron oxides is chemically unstable under atmospheric conditions and is slowly oxidized and converted to ferrous and ferric compounds. These compounds are found in the waste water of the steel industry. Iron oxides in particulate form are also found in the water used to clean gases from blast and steel-making furnaces.

Technically advanced processes for steel making generate more suspended solids and somewhat less dissolved ferrous sulfate. Large amounts of waste containing iron hydroxide are also created from the process used to extract aluminum from bauxite. In the Bayer process this ferric oxide is called "red mud" and is actually a mixture of iron oxide and sodium aluminum silicate. This "red mud" usually ended up in nearby lakes, but more recently has found use as a landfill, after having been filtered through sand. In addition, "red mud" can be used as a substitute for fluorspar in making steel. Iron in the form of steel is the chief metal used for making automobiles. Its chemical neighbor nickel is used in stainless steel and as a first coating in chromium plating.

In 1982 the iron and steel industry produced more than 780 million tons of iron ore (31% in the Soviet Union, 12% in Brazil, 11% in Australia, and 9% in the People's Republic of China) and about 650 million tons of raw steel (SAAGER, 1984; ELINDER, 1986). Iron and its alloys are used mainly in the construction, transportation, machine-manufacturing, and energy industries (e.g., magnets), see, for instance, SAAGER (1984). As for other basic manufacturing industries, it consumes large quantities of resources and in turn produces atmospheric, water-borne and solid wastes. Although dust extraction plants associated with basic oxygen furnaces have largely eliminated the traditional brown smoke associated with steel works, some of the highly toxic fumes produced during steel-making still escape. Sulfur dioxide is still vented from most sintering plants because the quantities of gas emitted are less than and of a lower concentration than those released from power generating plants fitted with scrubbers. At least a few countries require sulfur dioxide scrubbing at sintering plants. The steel industry is increasingly making use of scrap metal both in classical steel-making processes and in smaller mini steel mills that use electric arc furnaces. According to SAAGER (1984), 68% of the steel consumption in the United States in 1981 was produced from scrap.

Recycling of scrap metal enables important savings to be made in energy and raw materials by by-passing the coking, sintering, and blast furnace stages, but it brings its own environmental hazards, particularly for the work force. Many of these problems could be solved by using cleaner scrap, or by its pretreatment. The electric arc furnace is also inherently noisy.

4 Distribution in the Environment, in Foodstuffs, and in Living Organisms

Iron is a non-volatile, lithophilic element. The iron cycle comprises weathering of rocks, which is necessary for water-mediated resedimentation (BRUNNER and BACCINI, 1981). The human contribution to the geochemical cycle of iron is important on a global scale. It is estimated that the annual amount of iron mined exceeds the amount carried to the ocean from natural rock-weathering by a factor of eight. This includes an estimation that about one-quarter of the iron produced is destroyed by corrosion and resedimented each year. More precise figures are available for the industrialized countries. For example, in Switzerland, about 2 million tons of iron are imported each year, of this 3% is found in household garbage and industrial waste. A small percentage (0.4%) reaches the earth in the form of precipitates and a similar fraction ends up in the sewers. A mere 0.1% will eventually reach the sewage sludge (BRUNNER and BACCINI, 1981). In this fashion, iron waste is recycled back into the earth albeit in the form of very small particles. Coal also contains substantial amounts of iron (see Chapter I.1).

In lakes and rivers iron is contained in heterogeneous sediments. This type of sediment, however, is quite different from the fairly pure layers of ferric hydroxide precipitated at the same time when blue-green algae in the ancient oceans brought about the formation of oxygen, which in turn led to the oxidation and precipitation of large amounts of iron from the waters of the earth. BUFFLE (1988) differentiates in lake water between particulate iron (especially in summer) and colloidal and electroactive Fe(III) (additionally in autumn). In the sulfide layer MnO_x oxidizes Fe(II) to Fe(III). In the critical depth, iron particles consist in small amorphous, porous balls of less than 0.5 μm in diameter made up of about 50% Fe(II), about 50% Fe(III), and some $Ca_3(PO_4)_2$. Labile "FeS" complexes are in equilibrium with Fe^{2+} and S^{2-}, possibly Fe_2S_2. DE VITRE et al. (1988) described the role of colloidal iron phosphate species, formed at the oxic-anoxic interfaces in lakes. Iron/manganese oxides are carriers, sinks, and again sources for many absorbed nutrients and pollutants in waters, sediments, and soils. MOREL (1989) discussed, for instance, models of absorption on such iron oxides.

Through evolution, iron was selected from among the transition elements to form the "energy backbone" in vertebrate physiology. It is involved in a variety of essential biochemical reactions ranging from catalytic effects (e.g., cytochromes, catalases, peroxidases, and other mitochondrial enzymes) to involvement in DNA synthesis, oxygen transport, and electron transfer (ZETKIN and SCHALDACH, 1974; GUTTERIDGE, 1987). In the adult human body, there are three to four grams of functional and stored iron (Table II.14-1). Most of the former is involved in erythropoiesis and constitutes a continuous cycle of iron (Fig. II.14-1) (FINCH, 1979). The second largest fraction of iron within the body is in the form of ferritin and hemosiderin. Essential tissue iron is present in myoglobin and in cell enzymes. Only four milligrams of iron are bound to plasma transferrin, whose important function is iron transport and exchange between body tissues (Fig. II.14-1).

Biologically available forms of iron may be put into two categories. One is heme, where iron is chelated by porphyrin, a special water-insoluble ligand. To this class belong compounds such as hemoglobin, myoglobin, and cytochromes. The second iron pool, consisting of non-heme iron, is chemically very heterogeneous and consists largely of protein-bound iron as in ferritin and hemosiderin, ovotransferrin, albumin, casein, and also soluble low molecular weight iron complexes with organic ligands. The latter include ascorbic acid, citric acid, tricarboxylic acids, amines, amino acids, and sugars. In addition, iron present in an insoluble form as ferric phosphate, ferric phytate or ferric hydroxide may contribute to the non-heme iron pool by way of food. Values referring to the iron distribution in the environment,

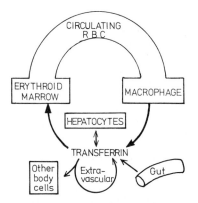

Fig. II.14-1. Schematic representation of human iron exchange. Iron is absorbed by the intestine and is bound to transferrin in the plasma. Most transferrin-iron in the body, however, results from the release of iron from the hemoglobin of senescent rest cells and its subsequent acquisition by apo-transferrin. Whether from the gut or from the reticulo-endothelial system (macrophage), where hemoglobin is degraded, transferrin-iron is largely directed to the erythroid bone marrow where it is then delivered to hemoglobin synthesizing erythroid precursor cells. In this way effective reutilization of iron is guaranteed.

Table II.14-1. Distribution of Iron

Source	Iron Content	Remarks
A. Iron in the Environment		
Iron ores	20–69%	
Soils	0.7–4.2%	Up to 55% in oxisols (laterite)
Ground water	⩽0.5–100 mg/L	Higher values found in the absence of oxygen and in the presence of organic material
Drinking water	max. 0.3 mg/L	U.S. Public Health Service drinking water standard
River water (average)	0.67 mg/L	Up to 100 mg/L in acidic mine drainage streams
Ocean water (surface)	0.01–0.14	According to BRULAND (1983); in anoxic deep waters Fe^{2+} concentrations may increase upto 60 µg/L
Atmosphere	0.9–1.2 µg/m³	As particulate ferric oxide
B. Iron in Foodstuffs (see also Chapter I.8)		
High iron	⩾2.0 mg/100 kcal	Organ meats, dried legumes, egg yolk, cocoa, shell fish
Moderate	0.7–1.9 mg iron/100 kcal	Bulk of the average diet
Low	<0.7 mg/100 kcal	Milk, milk products, refined flour, potatoes
Mixed Western diet	6 mg/1000 kcal	
Diet in developing countries	10 mg/1000 kcal	Bulk of the iron is poorly available for intestinal absorption
C. Iron in the Human Body		
Male	49 mg/kg	Percentage distribution: hemoglobin iron: 70% storage iron: 2.5% (ferritin hemosiderin), plasma iron: 0.1% remainder: 4.9% (myoglobin, cytochromes and iron containing enzymes)
Female	38 mg/kg	
D. Iron in Medicinal Preparations		
Ferrous sulfate		Therapeutic dose: 250 mg per day;
Ferrous succinate		
Ferrous lactate	~60 mg/tablet	Maximal utilizable dose for hemoglobin synthesis: 50 mg;
Ferrous fumarate		Lethal dose: 200–250 mg/kg body weight
Ferrous gluconate		
Ferrous citrate		
E. High Iron Content Alcoholic Beverages		
Kafir beer	40–80 mg/L	Causes Bantu-siderosis
Red wine	5–6 mg/L	

in food, and in the human body are depicted in Table II.14-1 (see also Chapter I.8). The liver and spleen have especially high iron concentrations (0.25 to 0.8 ppm according to ELINDER, 1986, and WARD, 1987).

5 Uptake, Absorption, Transport and Distribution, Metabolism and Elimination in Plants, Animals, and Humans

5.1 Plants

Very little is known about the molecular mechanism of iron uptake and transport in bacteria (BOWEN, 1979) and plants (WANN and HILLS, 1973) although the ability of plants to accumulate iron (up to 5.5% of dry weight in lichens) and genetic differences have been well recognized. The process of iron uptake by plants involves the uptake of iron from the soil into the root and the transport of iron from the root to the leaves. The intake of iron must be continuous as green plants grow, since iron does not move from older to newer leaves. Before uptake by the roots can occur, the very insoluble iron present in the soil (mostly as ferric hydroxide) must be solubilized. PETERSON and GIRLING (1981) distinguished iron-efficient and iron-inefficient plants (the latter may develop iron deficiency) and discussed the variety of plant species differing in their use of iron.

One mechanism for the solubilization of iron is the release of hydrogen ions by the root which lowers the pH in the root zone. This favors Fe(III) solubility; the reduction of Fe(III) to Fe(II) is achieved by special reducing agents secreted concomitantly with the hydrogen ions by the root. The root-absorbed Fe(II), which is oxidized to Fe(III) near the metaxyleme is chelated by citrate and transported in the metaxyleme to the top of the plant. Another postulated mechanism by which iron solubilization and absorption may be achieved is through the action of siderochromes present in the soil as produced by microbes or possibly excreted by plants. These iron binders could sequester the soil iron without a reduction step. Uptake of the formed iron complex into the root would then be achieved by a special receptor-mediated process similar to that described in fungi and bacterial systems (EMERY, 1978). Storage forms of iron in plants include phyto-ferritin and ferric phosphate. FARAGO (1987) demonstrated that iron is solubilized by phenolic acids from plants, a process which has implications for iron bioavailability. For instance, the leaf litter of heather (*Calluna*) produces conditions for the persistence of organic ligands of the phenolic acid type in the soil, and thus causes iron mobilization on a long-term basis.

5.2 Animals and Humans

Iron absorption in animals and man can be defined briefly as a process of translocation of iron from the luminal (mucosal) to the contraluminal (serosal) site of the intestinal epithelium (HUEBERS and RUMMEL, 1984; HUEBERS, 1986). Central to the functioning of the iron absorption system is the duodenal and proximal jejunal absorptive cell through which most iron is absorbed (ZETKIN and SCHALDACH, 1974).

The main features of mucosal iron transfer can be summarized as follows: (1) dependence on metabolic energy; (2) limited capacity, i.e., a maximum dietary iron absorption in man of 3–5 mg/day; (3) adaptation to demand; (4) preference of mucosal to serosal transport; (5) specificity for iron; (6) genetic control; and (7) dependence on protein synthesis (HUEBERS and RUMMEL, 1984).

Iron absorption involves the digestion of food, interactions between free iron and food components, motility effects, the uptake of various forms of iron by the mucosa in the upper intestine ("mucosal uptake"), mucosal processing of iron, and finally the movement of iron from the mucosal cell to the interior body ("transfer"). Iron uptake in the mucosa is more rapid than its release. Thus the transfer is the rate limiting step in the iron absorption process (FINCH and HUEBERS, 1986).

Only a small amount of the iron consumed is actually utilized. About 7 mg per 1000 kcal (equivalent to a daily adult consumption of 8–18 mg iron) is taken in with little variation among persons of different economic status. Absorption depends mainly on the body's need for iron, and recent studies have shown that the interactions of nutritional components present in the intestinal mucosa are of primary importance for the availability of iron. Available dietary iron – even in the iron-deficient subject where iron absorption is maximal – is about 3 mg/day, but may be considerably less if the diet is of poor quality (BJOERN-RASMUSSEN et al., 1974).

There are two types of absorbable iron in the diet. One is heme iron, found in meat, poultry, and fish; the second is non-heme iron, found in numerous vegetables, rice, wheat, and corn. Heme iron is of high availability in humans, and more importantly, its absorption is unaffected by other food constituents. While heme iron is highly available, it constitutes only 1–3 mg/day in the average Western diet, and even less in the diet of the world's poorer population. As a consequence much of the work in the last decade has gone into defining the reactions that affect absorption of the larger pool of non-heme iron in food.

Non-heme iron is sensitive to blocking substances, e.g., tannates, phosphates, calcium, bran, fiber, and chelates used as food preservatives such as EDTA. However, the absorption of non-heme iron is increased by certain substances such as ascorbate, meat, and certain amino acids (LAYRISSE et al., 1968; SAYERS et al., 1973, 1974). Thus the absorption of non-heme iron from any kind of diet represents the balance between food constituents that promote or inhibit the absorption of iron.

Iron from heme and non-heme sources appears to enter a common pool within the mucosal cell. Most of the absorbed iron is temporarily stored as ferritin, to be released over the next few hours. Other iron is held by ferritin until the cell is exfoliated. There is little information about the manner in which iron transverses the mucosal cell. Because of the fact that the transit of physiological amounts of iron across the mucosal cell is extremely rapid (absorbed iron appears within seconds in experimental animals regardless of the state of the iron balance) one would infer an extremely small pool of iron in transit. In more prolonged studies with larger doses of iron, a slower phase of absorption is increasingly apparent, extending over 12 to 36 hours.

Iron leaving the mucosal cell passes into the extravascular space of the lamina propria where it is complexed to plasma transferrin and rapidly appears in circulation. In all probability, the iron binding capacity of transferrin is frequently exceeded, in which case iron enters the portal system either unbound or bound nonspecifically to albumin. If iron is not bound to transferrin in the blood, it is rapidly removed by hepatocytes. The internal distribution of iron bound to transferrin is a function of the interaction between this protein and transferrin receptors on the surface of body cells (HUEBERS and FINCH, 1987). The widely varying needs of different tissues are met by variations in receptor numbers. The importance of transferrin in the internal distribution of iron is illustrated by its absence as a result of a genetic defect. Severe iron deficiency anemia occurs due to an inadequate delivery of iron to the erythron at the same time other body tissues (e.g., the liver) are overloaded with iron (HEILMEYER et al., 1961; GOYA et al., 1972).

The unique feature in human iron metabolism is the limited loss of iron from the body (10% of body iron per year or 1 mg per day in the adult male). This loss may vary from 0.5 mg in iron deficiency to 5.0 mg in iron overload. Iron is excreted via feces, urine, and dermal loss (ELINDER, 1986), and its biological half-life is of the order of 10–20 years, lung clearance being faster (ELINDER, 1986).

6 Biological Effects on Plants, Animals, and Humans

6.1 Plants

Iron is essential to the physiological processes of all living organisms. BOWEN (1979) reported on the amounts necessary for healthy growth of bacteria, fungi, algae, insects, birds, and mammals. In plants iron is required for chlorophyll synthesis (EMERY, 1978). When the amount of available iron does not meet minimal needs, plants develop chlorosis, which is manifested by the yellowing or blanching of normally green parts such as leaf tops. Availability of iron to the roots of plants is not only affected by the iron content of the soil but also by its pH and phosphate content. Other metals may compete for absorption or impair the absorption process. Iron deficiency can be corrected by the addition of iron chelates (such as iron EDTA) to the soil or nutrient solution (CHABAREK and MARTELL, 1959; HECK and BAILEY, 1950). Plants also show the ability to alter iron delivery to the top of the plant by increasing or decreasing the concentration of the specific chelating agent inside the plant, as well as that in the root exudate. Iron and manganese oxides also play an important indirect role in the soil (PETERSON and GIRLING, 1981) with implications for agriculture and plant growth – they rapidly fix other (beneficial or harmful) elements, which once fixed are no longer bioavailable to plants.

6.2 Deficiency Symptoms

Infants and menstruating or pregnant women are most vulnerable, due to their increased iron requirements (BOTHWELL et al., 1979), and the number of individuals with nutritional iron deficiencies are judged to be in the hundreds of millions. The effects of iron deficiency are due in part to the anemia which results and in part to the depletion of essential tissue enzymes (FINCH and COOK, 1984; HUEBERS and FINCH, 1982). Iron deficiency may or may not exhibit symptoms, depending on its severity. Persistent tiredness, even with adequate rest, is the most common sign. Other symptoms may include paleness, tired and aching muscles, headache, dizziness, and shortness of breath. If a low intake and absorption of iron is the cause of an iron-deficiency anemia, an iron supplement may be all that is needed to correct the problem. The usual therapeutic dose is 60 mg of ferrous iron given three times daily. The usual response in the non-bleeding patient is a 0.2 g per day increase in hemoglobin, beginning with the third or fourth day of treatment. If given with meals, the effectiveness of iron may be reduced by half, or more. Therapy is continued until the hemoglobin is normalized (BOTHWELL et al., 1979).

If iron stores are to be reconstituted, iron must be taken for a prolonged period of time, i.e., 6 to 12 months. In rare cases where oral iron is ineffective, parenteral iron may be given in the form of imferon intravenously. However, iron supplements and medicinal iron should not be taken indiscriminately, and possible toxicity should be kept in mind (see Section 7).

In iron deficiency absorption of other heavy metals is increased as well, and iron deficient individuals are therefore particularly vulnerable to their toxic effects. This has been observed with respect to cadmium and lead (SIX and GOYER, 1972; KLAUDER and PETERING, 1975; HUEBERS et al., 1987).

6.3 Effects on Animals and Humans

Excessive iron accumulation occurs in individuals with exceedingly high rates of erythropoiesis (e.g., thalassemia and sideroblastic anemia) and also in the genetic disorder idiopathic hemochromatosis. In these conditions, large amounts of iron are deposited in the hepatic parenchyma and other parenchymal tissues ultimately leading to tissue damage. In idiopathic hemochromatosis, the excessive body stores of iron can be depleted by phlebotomy. In anemic disorders associated with iron overload, treatment with an iron chelate such as desferrioxamine can reduce the iron accumulations and ameliorate tissue damage.

It is believed that one of the major reasons why iron may be toxic to cells is that it catalyzes the production of the hydroxyl radical. The hydroxyl radical is the most potent oxidizing agent that can exist in aqueous medium. Oxygen radicals are known to be toxic to living cells because they may induce peroxidation of lysosomal membranes, depolymerize hyaluronate, cause endothelial and lung damage, and cause platelet aggregation resulting in serotonin release. It has also been reported that free iron and fully saturated ferritin, which accumulate in rheumatoid synovial membranes, contribute to the development of damage that occurs in rheumatoid arthritis through the production of organic oxygen radicals and lipid peroxides.

Acute iron poisoning from ingested iron is unlikely to be encountered from any other source other than medicinal iron. Hemorrhagic gastritis, intravascular clotting, acidosis and shock may rapidly lead to death. Desferrioxamine is the drug of choice for treating acute episodes or accidental poisoning in small children. The drug is given both parenterally and by gastric tube (McENERY, 1966, 1971; MÖSCHLIN and SCHINDER, 1964).

According to ELINDER (1986) it is unlikely that ferric oxides are carcinogens. An increase in lung cancer mortality in some studies was more likely related to other simultaneous exposures (e.g., radon), but particles may serve as carriers for carcinogens (e.g., for benzo[a]pyrene).

7 Hazard Evaluation and Limiting Concentrations

The chemical characteristics of iron sharply restrict its availability to living organisms despite the vast amount of iron present in the earth. Genetic differences in plants would appear to be important in determining their geographic distribution, for plant growth is sensitive to soil changes that affect iron availability. Animals appear to be less dependent on variations in dietary iron because of the large amount of iron uptake, but artificial diets have been shown to result in iron deficiency.

Man is particularly susceptible to iron deficiency because of a very limited iron intake and availability (FINCH and HUEBERS, 1982). Changes in man's diet over the last few thousand years from one high in animal protein to one high in vegetable protein appears to be an important factor and suggests the need for dietary manipulation to increase iron intake and/or availability. The problem of iron overload is virtually limited to man and is far less frequent than iron deficiency. Iron overload is a consequence of a malfunction in the regulation of iron absorption. Since iron overload develops on a "normal" diet, the medical approach must be one of improved detection and individual treatment.

In conclusion, iron is a metal of low toxicity. Only iron concentrations exceeding 10 to 200 mg/L of nutrient solution have been found to be toxic to plants, and amounts in excess of 200 mg/day are considered toxic for man (BOWEN, 1979). Almost no information is available on the toxic effects of iron on the microorganisms commonly found in rivers and lakes. A study group of the World Health Organization determined that the maximal iron content of drinking water should not exceed 100 µg/L. The United States Public Health Service regards iron concentrations of up to 300 µg/L as acceptable.

Elemental iron of small particle size is inert but may influence the functions of the respiratory organs. Therefore, a maximal concentration of 6 mg iron oxide/m^3 has been recommended (MAK, 1987). Certain organic iron compounds (e.g., pentacarbonyl iron) are very toxic, and maximal concentrations in air should not exceed 0.1 ppm of 0.8 mg/m^3 (MAK, 1987).

References

Bjoern-Rasmussen, E., Hallberg, L., Isaksson B., and Arvidsson, B. (1974), *Food Iron Absorption in Man. Application of the Two-pool Extrinsic Tag Method to Measure Heme and Nonheme Iron Absorption from the Whole Diet. J. Clin. Invest.* 53, 247–255.

Bothwell, T. H., Charlton, R. W., Cook, J. D., and Finch, D. A. (1979), *Iron Metabolism in Man.* Blackwell Scientific Publications, Oxford.

Bowen, H. J. M. (1979), *Environmental Chemistry of the Elements.* Academic Press, New York-London.

Bruland, K. W. (1983), *Trace Elements in Seawater, Chem. Oceanogr.* 8, 203–204.

Brunner, P. H., and Baccini, P. (1981), *Forschung und Technik; Neue Zürcher Zeitung,* No. 70, p. 65 (25 March), Zürich.

Buffle, J., Geneva (1988), *Combining Field Measurements for Speciation in Non-Perturbable Water Samples. 4th Plenary Lecture,* Jekyll Island, Georgia, in: Kramer, J. R., and Allen, H. E. (eds.): *Metal Speciation,* pp. 99–124. Lewis Publishers, Chelsea, Michigan.

Chabarek, S., and Martell, A. E. (1959), *Organic Sequestering Agents,* pp. 416–504. John Wiley & Sons, New York.

de Vitre, R. R., Buffle, J., Perret, D., Leppard, G. G., and Tercier, M.-L. (1988), *Chemical Characterization of a Colloidal Iron Phosphate Species Formed at the Oxic-anoxic Interface of a Small Eutrophic Lake; Colloidal Iron Phosphate (the Sizing and Morphology of an Amorphous Species); in-situ Voltammetry in Natural Waters (Advantages and Development of a New Voltammetric System). International Conference Trace Metals in Lakes,* Hamilton, Canada, Abstracts, pp. 20, 42, 69. *Sci. Total Environ.,* in press.

Elinder, C.-G. (1986), Chapter 13 *Iron,* in: Friberg, L., Nordberg, G. F., and Vouk, V. B. (eds.), *Handbook on the Toxicology of Metals,* 2nd Ed., Vol. II, pp. 276–297. Elsevier, Amsterdam.

Emery, T. (1987), in: Sigel, H. (ed.), *Metal Ions in Biological Systems.* Marcel Dekker, New York.

Farago, M. E. (1987), *Lecture,* Jekyll Island, Georgia; quoted in *TRAC* 8 (9), XI and *Swiss Chem.* 9 (9), 52.

Finch, C. E. (1979), in: *Iron.* Subcommittee on Iron, N.A.S. Committee on Medical and Biologic Effects of Environmental Pollutants (ed.). Johns Hopkins University Park Press, Baltimore, Maryland.

Finch, C. A., and Cook, J. D. (1984), *Iron Deficiency. Am. J. Clin. Nutr.* 39, 471–477.

Finch, C. A., and Huebers, H. A. (1982), *Perspectives in Iron Metabolism. N. Engl. J. Med.* 306, 1520–1528.

Finch, C. A., and Huebers, H. A. (1986), *Iron Metabolism. Clin. Physiol. Biochem.* 4, 5–10.

Goodwin, J. F., Murphy, B., and Guillemette, M. (1966), *Direct Measurement of Serum Iron and Binding Capacity. Clin. Chem.* 12, 47–57.

Goya, N., Miyazaki, S., Kodate, S., and Ushio, B. (1972), *A Facility of Congenital Atransferrinemia. Blood* 40, 239–245.

Gutteridge, J. M. C., Potters Bar, U.K. (1987), *Lecture 2nd Nordic Symposium on Trace Elements in Human Health and Disease,* Odense, Denmark. WHO Regional Office for Europe, Copenhagen; *Environ. Health* 26, 46–48.

Heck, W. E., and Bailey, L. F. (1950), *Plant. Physiol.* 25, 573–582.

Heilmeyer, von L., Keller, W., Vivell, O., Keiderling, W., Betke, K., Wohler, F., and Schultze, H. E. (1961), *Congenital Transferrin Deficiency in a Seven-year Old Girl. Dtsch. Med. Wochenschr.* 86, 1745–1751.

Holleman, A. F., and Wiberg, E. (1971), *Lehrbuch der Anorganischen Chemie,* 4th Ed., pp. 827–844. Walter de Gruyter & Co., Berlin.

Huebers, H. A. (1986), *Iron Absorption: Molecular Aspects and Its Regulation. Acta Hematol. Jpn.* 49, 1528–1535.

Huebers, H. A., and Finch, C. A. (1982), *Introduction, Clinical Aspects of Iron Deficiency. Semin. Hematol.* 19, 3–5.

Huebers, H. A., and Finch, C. A. (1987), *The Physiology of Transferrin and Transferrin Receptors. Physiol. Rev. 67*, 520–582.

Huebers, H. A., and Rummel, W. (1984), *Protein Mediated Epithelial Iron Transfer. Handb. Exp. Pharmacol. 70* (1), 513–541.

Huebers, H., Huebers, E., and Rummel, W. (1975), *Mechanism of Iron Absorption: Iron-binding Proteins and Dependence of Iron Absorption on an Elutable Factor,* in: Kief, H. (ed.), *Iron Metabolism and Its Disorders,* pp. 13–24. Excerpta Medica, Amsterdam.

Huebers, H. A., Huebers, E., Csiba, E., Kunzler, C., Rummel, W., and Finch, C. A. (1987), *The Cadmium Effect on Iron Absorption. Am. J. Clin. Nutr. 45*, 1007–1015.

Iones, K. G. (1980), *Lecture Workshop on Environmental Effects of Chlorinated Dioxines,* Rome. Cited after: *Chem. Rundsch. 34,* No. 18, 29 April 1981.

Karlson, P. (1972), *Kurzes Lehrbuch der Biochemie für Mediziner und Naturwissenschaftler,* 8th Ed., p. 350. Georg Thieme Verlag, Stuttgart.

Klauder, D. S., and Petering, H. G. (1975), *Protective Value of Dietary Copper and Iron against Some Toxic Effects of Lead in Rats. Environ. Health Perspect. 12,* 77–80.

Layrisse, M., Martinez-Torres, C., and Rodie, M. (1968), *Effect of Interaction of Various Foods on Iron Absorption. Am. J. Clin. Nutr. 21,* 1175–1183.

Ludi, A. (1988), *Prussian Blue. Chem. Unserer Zeit 22*(4), 123–127.

MAK (1987), *Maximum Concentrations at the Workplace, Report No. XXIII, DFG.* VCH Verlagsgesellschaft, Weinheim-Basel-Cambridge-New York.

McEnery, J. T. (1971), *Clin. Toxicol. 4,* 603–613.

McEnery, J. T., and Greengard, J. (1966), *Treatment of Acute Iron Ingestion with Desferroxamine in 20 Children. J. Pediatr. 68,* 773–779.

Morel, F. (1989), *The Invisible Ligand and the Elusive Metal (Analytical Consequences of Aquatic Kinetics; Kinetics of Sorption and Exchange Reactions. 19th International Symposium on Environmental Analytical Chemistry,* Jekyll Island, Georgia. *TRAC 8* (9), 319 and *Swiss Chem. 11* (9), 46.

Möschlin, A., and Schinder, U. (1964), *Treatment of Primary and Secondary Hematochromatosis and Acute Iron Poisoning with a New Potent Iron-eliminating Agent (Desferrioxamine B),* in: Gross, F. (ed.), *Iron Metabolism,* pp. 525–550. Springer-Verlag, Berlin.

Olson, C. (1958), *C. R. Trav. Lab. Carlsberg 31,* 41–59.

Olson, A. D., and Hamlin, W. B. (1968), *A New Method for Serum Iron and Total Iron-binding Capacity by Atomic Absorption Spectrophotometry. Clin. Chem. 15,* 439, 444.

Peterson, P. J., and Girling, C. A. (1981), in: Lepp, N. W. (ed.), *Effect of Heavy Metal Pollution on Plants,* Vol. 1, pp. 213–278. Applied Science Publishers, London.

Pronk, C., Oldensiel, H., and Lequin, H. C. (1974), *A Method for Determination of Serum Iron, Total Iron Binding Capacity and Iron in Urine by Atomic Absorption Spectrophotometry with Manganese as Internal Standard. Clin. Chim. Acta 50,* 35–41.

Saager, R. (1984), *Metallic Raw Materials Dictionary* (in German), pp. 103–107. Bank von Tobel, Zürich.

Sayers, M. H., Lynch, S. R., Charlton, R. W., Bothwell, T. H., Walker, R. B., and Mayet, F. (1973), *The Effects of Ascorbic Acid Supplementation on the Absorption of Iron in Maize, Wheat, and Soya. Br. J. Haematol. 24,* 209–218.

Sayers, M. H., Lynch, S. R., Charlton, R. W., Bothwell, T. H., Walker, R. B., and Mayet, F. (1974), *Iron Absorption from Rice Meals Cooked with Fortified Salt Containing Ferrous Sulphate and Ascorbic Acid. Br. J. Nutr. 31,* 367–375.

Six, K. M., and Goyer, R. A. (1972), *The Influence of Iron Deficiency on Tissue Content and Toxicity of Ingested Lead in the Rat. J. Lab. Clin. Med. 79,* 128–136.

Wann, E. V., and Hills, W. A. (1973), *J. Hered. 64,* 370–371.

Ward, N. I. (1987), *Element Analysis: a Comparison of NAA and ICP-SMS, Environ. Health 20,* 118–123. 2nd Nordic Symposium on Trace Elements in Human Health and Disease, Odense, Denmark. WHO Regional Office for Europe, Copenhagen.

Welz, B. (1985), *Atomic Absorption Spectrometry,* 2nd Ed., pp. 289–291, 346, 357, 368ff, 400ff. VCH Verlagsgesellschaft, Weinheim-Deerfield Beach/Florida-Basel.

Zetkin, M., and Schaldach, H. (1974), in: Schaldach, H. (ed.), *dtv Wörterbuch der Medizin,* 5th Ed., Vol. 1, pp. 356. Deutscher Taschenbuch Verlag, München; Georg Thieme Verlag, Stuttgart.

Additional Recommended Literature

Allekotte, H., and von Röpenack, A. (1989), *Hematite – the Solution to a Disposal Problem – an Example from the Zinc Industry. ICR-Berlin Proceedings, Abfallvermeidung in der Metallindustrie*, pp. 43–53. EF-Verlag für Energie- und Umwelttechnik, Berlin; see also Report E. Merian, *Swiss Chem. 12* (5), 40.

Browning, E. (1961), *Toxicity of Industrial Metals*. Butterworths, London.

Epstein, E. (1972), *Mineral Nutrition of Plants – Principles and Perspectives*. John Wiley & Sons, New York.

de Vitre, R. R., Buffle, J., Belzile, N., Tercier, M. L., and Tessier, A. (1990), *In-situ Speciation Studies at the Water-Sediment Interface; Voltammetric Sensors for in-situ Measurement of Trace Elements in Water. Fresenius Z. Anal. Chem.*, in press.

Feichtinger, H., and Rennhard, C. (1990), *Metalloid-Metal Compound Materials* (in German). *Neue Zürcher Zeitung, Forschung und Technik*, No. 19, p. 77 (24 January), Zürich.

Hewitt, E. J., and Smith, T. A. (1974), *Plant Mineral Nutrition*. John Wiley & Sons, New York.

Kief, H. (1975), *Iron Metabolism and its Disorders*. Excerpta Medica, Amsterdam.

Klimmer, O. R. (1958), *Arch. Toxikol. 24*, 15.

Lee, D. H. K. (1972), *Metabolic Contaminants and Human Health*. Academic Press, New York.

Martinez-Torres, C., and Layrisse, M. (1971), *Am. J. Clin. Nutr. 24*, 531–540.

Müller, R. (1987), *To find out about Rust in Ferroconcrete. Neue Zürcher Zeitung, Forschung und Technik*, No. 46, p. 77 (25 February), Zürich.

Speidel, M. O. (1987), *Modern Nitrogen-Alloyed Steels. Neue Zürcher Zeitung, Forschung und Technik*, No. 76, p. 67 (1 April), Zürich.

Stokinger, H. E. (1962), in: Patty, F. A. (ed.), *Industrial Hygiene and Toxicology*, 2nd Ed., Vol. II. John Wiley & Sons, New York.

Trueb, L. (1989), *Iron – the Most Important Element. Neue Zürcher Zeitung, Forschung und Technik*, No. 117, p. 85 (24 May), Zürich.

Zobrist, J., et al. (1989), *The Fate of Iron in a Granitic Lake (Canton Ticino)* (in German). *EAWAG Annual Report*, pp. 4–29/4–31.

II.15 The Lanthanides

PETER L. GOERING and BENJAMIN R. FISHER, Rockville, Maryland, USA
BRUCE A. FOWLER, Baltimore, Maryland, USA

1 Introduction

The lanthanides consist of a group of elements in the Periodic Table with successive atomic numbers 57 (lanthanum) through 71 (lutetium). They are not rare, the term "rare earth elements" is misleading ("hided elements" from the Greek word for "hiddenness" would be better). The abundances of cerium, neodymium, and lanthanum in the earth's crust are higher than those of lead, tin, and cobalt. No lanthanide elements are known to be nutritionally essential in animals or humans; however, many replace calcium in a number of calcium-mediated biological processes, a fact which according to EVANS (1983) and ARVELA (1979) may be of certain risk in several functions. While the toxicity of most lanthanides is generally considered as low, new and developing technologies may increase the level of various lanthanides in the environment and exposure of workers. More detailed reviews of the chemistry, physiology, technology, and toxicology of the lanthanides are available (see VENUGOPAL and LUCKEY, 1978; EVANS, 1983; BULMAN, 1988). Absorption from oral uptake is poor, but inhalation and dermal uptake may be more significant. Speciation has an influence on effects, and there seem to be sex differences.

2 Physicochemical Properties, and Analytical Methods

The lanthanides constitute a series of 15 inner-transition elements with very similar physicochemical properties. These Group III metals usually exhibit an oxidation state of +3; however, some lanthanides also exhibit valences of +2 (La, Eu, Sm, and Yb) and +4 (Ce, Pr, and Tb). Lanthanides are thermodynamically defined as hard Lewis acids (weak electron acceptors) and therefore have a high affinity for oxygen-containing centers of amino acids and other cellular constituents (FOWLER et al., 1984).

Most lanthanides are bright, silver-gray metals which are malleable, ductile, and generally soft enough to cut with a knife. Some lanthanides, such as La, Eu, Ce, Nd, and Pr, are highly reactive, corroding and tarnishing when exposed to air. Others such as Gd, Ho, Tm, and Yb, are reactive in moist air but not dry air, while Sm, Tb, Dy, Er, and Lu are generally stable (HAMMOND, 1986). Other physicochemical properties are summarized in Table II.15-1.

Table II.15-1. Some Physicochemical Properties of the Lanthanides[a]

Element	Symbol	Atomic Number	Atomic Mass	Ionic Radius[b]	Melting Point[c]	Boiling Point[c]
Lanthanum	La	57	139	1.06	920	3454
Cerium	Ce	58	140	1.03	798	3257
Praseodymium	Pr	59	141	1.01	931	3212
Neodymium	Nd	60	144	0.99	1010	3127
Promethium	Pm	61	145	0.98	1080	2460
Samarium	Sm	62	150	0.96	1072	1778
Europium	Eu	63	152	0.95	822	1597
Gadolinium	Gd	64	157	0.94	1311	3233
Terbium	Tb	65	159	0.92	1360	3041
Dysprosium	Dy	66	162.5	0.91	1409	2335
Holmium	Ho	67	165	0.89	1470	2720
Erbium	Er	68	167	0.88	1522	2510
Thulium	Tm	69	169	0.87	1545	1727
Ytterbium	Yb	70	173	0.86	824	1193
Lutetium	Lu	71	175	0.85	1656	3315

[a] Compiled from *Handbook of Chemistry and Physics*, 67th Ed. (1986), CRC Press, Boca Raton, Florida
[b] Ionic radius ($\times 10^{-10}$ m) in the +3 oxidation state
[c] °C

A simple and rapid separation with simultaneous determination of 14 lanthanide ions by capillary tube isotachophoresis has been described using 2-hydroxyisobutyric acid as a complexing agent (NUKATSUKA et al., 1981). Ion chromatography followed by postcolumn derivatization with 4-(2-pyridylazo)resorcinol is a practical method for monitoring and screening lanthanides (RUBIN and HEBERLING, 1987). The simultaneous determination of La, Ce, and Nd in human bronchial lavage fluids by energy-dispersive X-ray fluorescence has been reported with detection limits of 1 to 10 ppm (MAIER et al., 1986). Elevated levels of La, Ce, Nd, Eu, Yb, and Lu in plasma and synovial fluid of rheumatoid arthritis patients have been determined using radiochemical neutron activation analysis with detection limits of approximately 0.1 to 0.2 ppb (ESPOSITO et al., 1986). Several nuclear properties of a number of lanthanides, such as large thermal neutron cross-sections and abundance of radionuclides with unique gamma radiation of convenient half-life, make them excellent candidates for analysis by neutron activation (VENUGOPAL and LUCKEY, 1978).

Spectrophotometric trace analysis of these elements is not practical because of tedious extraction steps and interferences from other elements. WELZ (1985), however, described AAS techniques in the presence of potassium with a nitrous oxide/acetylene flame or with a graphite furnace. A repetitive spectral subtraction method to improve spectrophotometry and ion exchange absorptiometry has been discussed by IIDA (1987) resp. YOSHIMURA and TAKETSU (1987).

3 Sources, Production, and Uses

Lanthanides are constituents of many different minerals in igneous rocks, shale, and silicates; however, the two major sources for commercial production are monazite (a phosphate) and bastnasite (a fluorocarbonate). These two minerals which may be digested with H_2SO_4 or NaOH, contain most of the lanthanides in various concentrations, including (on a percentage basis) La (25); Sm (3); Yb, Ho, and Tb (0.03–0.05); and Lu and Tm (0.003–0.01). Promethium is a man-made element and has not yet been found in the earth's crust. Abundant deposits of monazite are found on beaches in India, river sands in Brazil, Australia, Malaysia, and the United States while major deposits of bastnasite are located in Southern California (KACZMAREK, 1981; ASCHE, 1988). Annual world production of lanthanide oxides is of the order of 25000 tons (BULMAN, 1988) to 50000 tons (ASCHE, 1988). Separation and purification are very expensive.

Most lanthanides are produced by reducing the associated anhydrous halide ions with calcium metal. Advances in ion exchange and solvent extraction techniques have led to much easier isolation of the lanthanides at considerable cost savings. Rhône-Poulenc developed liquid-liquid extraction further and now separates the lanthanides industrially in La Rochelle (France), Freeport (Texas), and Pinjawa (Australia) (ASCHE, 1988). Recently, electrochemical deposition technology has been reported to be a fast, simple, and specific method for isolating and separating various lanthanide ions (HAMMOND, 1986).

The uses of lanthanide compounds are diverse and are expanding due to modern technological advances. Lanthanides are presently used to improve the mechanical properties of steels (LINEBARGER and MCCLUHAN, 1981), providing strength and corrosion resistance. Lanthanides (La, Ce, Pr, Sm) are used in the motion picture industry as a core material in carbon-arc electrodes for studio and projection lighting. Lanthanides (Ce, Pr, Nd) are utilized in the manufacture and polishing of glass products, including lenses. Many lanthanides have sharp absorption bands in the ultraviolet, near-infrared, and visible wavelengths, a property that produces brilliant pastel colors for use as colorants in glass and porcelain enamel glazes (for instance, praseodymium to dye tiles in yellow shades, whereas gadolinium and europium give red effects in television tubes; ASCHE, 1988).

Other lanthanides have been used for various phosphors in color television tubes (Eu, Gd), as laser materials (Tb, Eu, Sm, Dy), and as garnets for microwave applications (Gd, Tm). The high thermal neutron absorption cross-sections and high melting points of various lanthanides (Sm, Eu, Dy, Gd) are two properties that make them ideal for use in nuclear reactor control rod applications. Several lanthanides have been used as catalysts in various industrial and chemical processes, such as refining of petroleum, automobile emission control, and as hydrocarbon catalysts in self-cleaning ovens (Ce).

Cerium oxide (as an additive) and Pt-Ce and Ni-Ce (bimetallic supports) improve automobile exhaust-catalyst properties, because of their increased oxygen storage capacity (DUPREZ et al., 1986). Mischmetal, an alloy consisting of La, Ce, and Pr, is a pyrophoric alloy suitable for cigarette lighter flints. Cerium oxide was already used together with thorium oxide 100 years ago for the impregnation of the famous Auer glowing filaments (ASCHE, 1988).

Lanthanides have excellent magnetic properties, e.g., $Fe_{14}Nd_2B$ is very efficient (ten times more than ferrite magnets and Co/Sm magnets) in permanent magnets in electromotors, and is used especially in starters, windshield washers, and for computerized data storing, ASCHE, 1988). $SmCo_5$ has the highest resistance to demagnetization of any known substance (VENUGOPAL and LUCKEY, 1978; HAMMOND, 1986; BULMAN, 1988). Lanthanides, e.g, copper oxides with barium/lanthanum additions, may play a role in supraconduction at temperatures above the boiling point of helium in the future (TRUEB, 1987; BEDNORZ, 1988; CERUTTI, 1988) whereas barium titanate coated with lanthanum seems to be promising for warm electricity conduction because resistance decreases with temperature (ASCHE, 1988). Nickel-cadmium batteries may be replaced by lanthanum-nickel accumulators.

Lanthanides possess unique spectroscopic and chemical properties that make them powerful probes to examine various biochemical properties, such as the structure and function of calcium-binding sites on proteins, enzymes, biological membrane transporters, and calcium-coupled stimulation-excitation responses in cells (for review see EVANS, 1983). Magnetic resonance imaging of the placental circulation using Gd-DTPA (diethylene-triaminepentaacetic acid) as a paramagnetic marker has been demonstrated in the Rhesus monkey in vivo and in the perfused human placenta (PANIGEL et al., 1988).

4 Distribution in the Environment and in Living Organisms

The lanthanides are widely distributed and are found in abundance in the earth's crust over a relatively large range. The average concentrations of lanthanides in the earth's crust are shown in Table II.15-2. Lanthanides are found in North Atlantic ocean waters in very low concentrations (picomol/kg) which increase with depth (ELDERFIELD and GREAVES, 1981). According to BRULAND (1983) predominant

Table II.15-2. Average Concentration of Lanthanides in the Earth's Crust[a]

Element	ppm	Element	ppm
La	30	Tb	0.9
Ce	60	Dy	3.0
Pr	8.2	Ho	1.2
Nd	28	Er	2.8
Sm	6.0	Tm	0.5
Eu	1.2	Yb	3.0
Gd	5.4	Lu	0.5

[a] Compiled from *Handbook of Chemistry and Physics*, 67th Ed. (1986), CRC Press, Boca Raton, Florida

species are carbonato-complexes, e.g., $La(CO_3)^+$; in the case of La, Ce, and Nd, concentrations are about 0.002 – 0.005 ppb (other lanthanides 4 to 20 times less).

In areas near large deposits of lanthanides, studies of the transport pathways of four of these elements from soil to plants and farm animals demonstrated that intake and soil abundance decrease in the following order: Ce > La > Th > Sm (LINSALATA et al., 1986). WARD (1987) found about 0.1 ppm Nd, 0.02 ppm La, 0.02 ppm Ce, and 0.003 Pr in bovine liver, using NAA or ICP-SMS. Levels of the lanthanides in healthy human tissues have been reported as follows: liver, 0.005; kidney, 0.002; lung, 0.004; testes, 0.002; and bone, 0.2 – 1.0 µg elements per g ash (HAMILTON et al., 1972/73; BRUNE et al., 1980; GERHARDSSON et al., 1984).

5 Uptake, Absorption, Accumulation, and Metabolism in Animals and Humans

The lanthanides are generally absorbed very poorly across the gastrointestinal mucosa. In animals, the tissue distribution patterns are greatly influenced by which lanthanide is tested, its chemical composition, and route of administration. The lanthanide concentration in mouse liver was 4.2 times greater than in bone after subcutaneous injection of $LaCl_3$ and 9.9 times greater after intraperitoneal administration (SPODE and GENSICKE, 1958). Significant amounts were also found to accumulate in the kidney, spleen, and pancreas. Similar results have been obtained after intramuscular injection of $LaCl_3$ in rats (DURBIN et al, 1956). Inhalation studies of $^{140}LaCl_3$ in dogs showed the highest accumulation of La in the lungs, gastrointestinal tract, liver, skeleton, and bronchial lymph nodes (CUDDIHY and BOECKER, 1970).

In humans, lanthanides tend to accumulate in liver and bone, and occupational exposure may increase concentrations in specific organs. Although liver and kidney concentrations were equivalent, a 2-fold higher accumulation of lanthanides in lung was demonstrated in smelter workers compared to controls (GERHARDSSON et al., 1984). The La-concentration in these workers did not decline with time after exposure, indicating that these elements have a long biological half-life.

6 Effects on Animals and Humans

6.1 Cellular Effects

Utilizing a pulmonary macrophage assay, most lanthanides appeared to be less toxic than Cd, and only $LaCl_3$, $CeCl_3$, and Nd_2O_3 demonstrated any significant cytotoxicity (LC_{50} values of 52, 29 and 101 mmol L^{-1}, respectively) as measured by this assay (PALMER et al., 1987). La has also been shown to be less cytotoxic than Mn^{2+} and is efficient in promoting Sendai virus induced fusion of animal cells (WASSERMAN et al., 1976). Lanthanides possess several biological properties that may make them useful as therapeutic agents for diseases such as arthritis. These include inhibition of lymphocyte activation, stimulus-mediated cell secretion, and neutrophilic chemotaxis and aggregation, inhibition of collagenase, and stabilization of collagen fibrils (EVANS, 1983).

6.2 Miscellaneous Biochemical Effects

Because of its size and high charge density, it has been postulated that lanthanides should have a high electrostatic affinity for any negative Ca^{2+} binding site (LETTVIN et al., 1964). In this fashion, La could compete for, or displace, Ca^{2+} from its binding site and act as a biological antagonist. Both functional and electron microscopy studies have demonstrated that La exhibits a preferential intracellular distribution, binding to the sarcoplasmic reticulum of muscle cells and mitochondrial membranes in hepatocytes (CHEVALLIER and BUTOW, 1971; MARTINEZ-PALOMO et al., 1973; LEHNINGER and CARAFOLI, 1971; REED and BYGRAVE, 1974a; HAKSAR et al., 1976; EVANS, 1983). La-ions have been shown to mimic the action of Ca^{2+} by activating calmodulin or calmodulin-like proteins (AMELLAL and LANDRY, 1983). Other reports indicated that the biological activity of La-ions is not mediated via a Ca^{2+}-associated action (NAYLER and HARRIS, 1976; SWAMY et al., 1976). Regardless of the mechanism, lanthanum and the lanthanides inhibited the electron transport process in mitochondria (MELA, 1968, 1969; REED and BYGRAVE, 1974b), interfered with axonal activity (VAN BREEMAN and DE WEER, 1970), and inhibited contractions in both smooth and cardiac muscle (VAN BREEMAN, 1969; WEISS and GOODMAN, 1969, 1975; SANBORN and LANGER, 1970). While most lanthanides promoted actin tubule formation, Er, Tm, Yb, Lu, and La interacted with actin but prevented tubule formation, which may be attributed to the high electrostatic charge of these specific lanthanide ions (DOS REMEDIOS et al., 1980; CURMI et al., 1982). Because the binding of lanthanide ions can alter the structural integrity of biologically active molecules, it is not surprising that there are numerous reports on the ability of these ions to alter the activity of a variety of enzymes (HOLTEN et al., 1966; GOMEZ et al., 1974; KOVER et al., 1976; ARVELA, 1979; GARIEPY et al., 1983; EVANS, 1983).

Some lanthanides have been reported to exhibit antitumor activity both in vitro and in vivo. The activity of sarcomas in rats and lymphatic leukemia and lymphosarcomas in mice was inhibited by lanthanum (HISADA and ANDO, 1973; ANGHILERI, 1975). The survival of ascitic tumor and HeLa cells was reduced by La in vitro, and the antitumor activity of La was enhanced when used in conjunction with hyperthermia (ANGHILERI et al., 1983; ANGHILERI and ROBERT, 1987). The mechanisms involved in the antitumor activity may be related to La-induced changes in membrane permeability leading to disruption of the membrane potential of tumor cells and tumor metabolism of Ca^{2+} and Mg^{2+} (LEVINSON et al., 1972; ANGHILERI, 1973a, b).

6.3 Acute Effects on Animals and Humans

The LD_{50} for $LaCl_3$ in the chick embryo was 14 mg per egg with hemorrhaging being the most prevalent symptom among dead embryos (MACHLIN et al., 1952). The neurotoxicity of $LaCl_3$ in newborn chicks may be related to the inhibition of synaptosomal enzymes, such as calcium-ATPase, magnesium-ATPase, and cholinesterase (BASU et al., 1982).

Early biochemical changes observed in rats after acute intravenous injection of lanthanides included an increase in plasmafree fatty acid levels. This was followed by an increase in fatty infiltration of the liver (SNYDER et al., 1960; SNYDER and STEPHENS, 1961). At higher doses, signs of acute systemic toxicity included writhing, ataxia, labored respiration, altered posture and sedation. In this study, rats died within 48 to 96 hours after administration, and in general males exhibited a higher tolerance than females (HALEY, 1965). The acute toxicity of the lanthanides is dependent on the route of administration (BRUCE et al., 1963).

Approximate LD_{50} values for lanthanide nitrates after intraperitoneal injection ranged from 225 to 480 mg/kg in female mice and from 210 to 335 mg/kg in female rats. Oral LD_{50} values for these salts ranged from 2750 to 4200 mg/kg in rats. Lanthanide (Ce, Pr, Nd, Sm) nitrates were highly toxic to female rats after intravenous injection; LD_{50} values ranged from 4 to 9 mg/kg. Male rats were an order of magnitude less susceptible than female rats. Poor absorption from the gastrointestinal tract and peritoneal cavity was a major factor limiting the toxicity by these routes of administration. When high doses of lanthanide compounds were orally administered to rats, La ammonium nitrate was the most toxic with an LD_{50} of 830 mg/kg (COCHRAN et al., 1950). No oral toxicity was observed with doses up to 10 g/kg and 5 g/kg for the oxide and sulfate salts, respectively. A summary of LD_{50} values for various lanthanide salts in a variety of animal species has been presented by HALEY (1965) and VENUGOPAL and LUCKEY (1978).

After local injection or inhalation, the most pronounced effects of lanthanides are skin and lung granulomas. Transient pulmonary lesions indicative of acute chemical pneumonitis have been reported after intratracheal injection or prolonged inhalation of lanthanide mixtures in guinea pigs (SCHEPERS, 1955). Lanthanide salts injected into rats and guinea pigs precipitated at the injection site and initiated an inflammatory response (GRACAR et al., 1957). In these animals, intraperitoneal injection of lanthanide salts resulted in a generalized peritonitis. Of the citrate salts, Ce was the least and La the most toxic (LD_{50} range: 78–146 mg/kg); of the chlorides, La was the least and Ce the most toxic (LD_{50} range: 353–372 mg/kg). In general, guinea pigs exhibited a higher tolerance than rats.

6.4 Chronic Effects on Animals and Humans

Feeding mice and rats lanthanide salts at a dietary level of 1% resulted in growth inhibition. Mice fed varying levels of lanthanide oxides (La, Dy, Eu, Yb, Tb, and Sm) continually over three generations exhibited no differences in mortality, morbidity, morphological development, growth rate, reproductive capacity or survival between the treated groups and controls (HUTCHESON et al., 1975). Repeated injection of La, Ce, Pr, or Nd chloride salts into rabbits produced hematologic alterations while no effects were observed after repeated oral administration (OELKERS and VINCKE, 1938). Continuous inhalation of particulate lanthanide fluoride blends by guinea pigs for three years resulted in diffuse and focal cellular proliferation and accumulation in subpleural and peribronchiolar regions and subacute chemical bronchitis after four months into the study (SCHEPERS, 1955). Chronic bronchitis and emphysematous changes were evident after one year.

In humans, chronic exposure to the lanthanides occurs primarily via inhalation in occupational settings. There are reports of lanthanide pneumoconiosis occurring in photoengravers and smelter workers (HEUCK and HOSCHER, 1968; HUSAIN et al., 1980; VOCATURO et al., 1983; SULOTTO et al., 1986).

6.5 Mutagenic, Carcinogenic, and Teratogenic Effects

La and Tb ($0.1-1.0$ mmol L^{-1}) are capable of promoting the neoplasmic transformation of mouse JB6 epidermal cells. These elements may promote neoplasmic transformation via the activation of protein kinase C (SOTIROUDIS, 1986; SMITH et al., 1986).

A decrease in the number of successful pregnancies and average litter size could be induced by a single injection of $LaCl_3$ (44 mg/kg, ip) into pregnant mice (ABRAMCZUK, 1985). Although no external malformations were observed, this dose reduced litter size by as much as 75% when administered between days 4 and 6 or days 14 and 16 of gestation.

7 Hazard Evaluation

Difficulty exists in providing an adequate risk assessment for exposure to the lanthanides due to a dearth of exposure data. Values for threshold limits and maximum permissible concentrations for the lanthanide elements are also scarce. No significant exposure to humans via food, water, or air is believed to occur via anthropogenic perturbation of geochemical cycling of these metals (ANDREAE et al., 1984). Due to the low toxicity of the lanthanides, these elements are not considered occupational health hazard except in lanthanide refining and lithography industries. However, increased use of these elements in new and developing technologies may increase human exposure and environmental levels (VENUGOPAL and LUCKEY, 1978). A mode of action might be correlated to the high capacity of the lanthanides to replace Ca^{2+}-ions in vivo (ARVELA, 1979; EVANS, 1983; see also Chapter II.35 Yttrium).

References

Abramczuk, J. W. (1985), *The Effects of Lanthanum Chloride on Pregnancy in Mice and on Preimplantation of Mouse Embryos in vitro. Toxicology 34*, 315–320.

Amellal, M., and Landry, Y. (1983), *Lanthanides are Transported by Ionophore A23187 and Mimic Calcium in the Histamine Secretion Process. Br. J. Pharmacol. 80*, 365–370.

Andreae, M. O., Asami, T., Bertine, K. K., Buat-Menard, P. E., Duce, R. A., Filip, Z., Förstner, U., Goldberg, E. D., Heinrichs, H., Jernelöv, A. B., Pacyna, J. M., Thornton, I., Tobschall, H. J., and Zoller, W. H. (1984), *Changing Biogeochemical Cycles*, in: Nriagu, J. O. (ed.), *Changing Metal Cycles and Human Health*, pp. 359–374. Springer Verlag, New York.

Anghileri, L. J. (1973a) *In vivo Distribution of Calcium, Magnesium, Lanthanum and Sodium Polyphosphates. Arzneimittelforschung 23*, 1720–1721.

Anghileri, L. J. (1973b), *The Effects of Lanthanum on the Metabolism of Calcium in Tumors. Int. J. Clin. Pharmacol. 8*, 146–153.

Anghileri, L. J. (1975), *On the Antitumor Activity of Gallium and Lanthanides. Arzneimittelforschung 25*, 793–795.

Anghileri, L. J., and Robert, J. (1987), *Effects of Hyperthermia and Lanthanum on Tumor Cell Leakage. Int. J. Clin. Pharmacol. Ther. Toxicol. 25*, 374–378.

Anghileri, L. J., Crone, M. C., Marchal, C., and Robert, J. (1983), *Comparative Enhancement of Hyperthermia Lethality on Tumor Cells by Procaine and Lanthanum. Neoplasma 30*, 547–549.

Arvela, P. (1979), *Toxicity of Rare Earths, Prog. Pharmacol. 2*, 69–113.

Asche, W. (1988), *Hided (Rare) Elements, Chem. Rundsch. 41* (35), 9.

Basu, A., Chakrabarty, K., Chatterjee, G. C. (1982), *Neurotoxicity of Lanthanum Chloride in Newborn Chicks. Toxicol. Lett. 14*, 21–25.

Bednorz, G. (1988), *Supraconduction* (in German), *Neue Zürcher Zeitung, Forschung und Technik*, No. 298, pp. 57/58 (21 December), Zürich.

Bruce, D. W., Hiltbrink, B. E., and DeBois, K. P. (1963), *The Acute Mammalian Toxicity of Rare Earth Nitrates and Oxides. Toxicol. Appl. Pharmacol. 5*, 750–759.

Bruland, K. W. (1983), *Trace Elements in Seawater, Chem. Oceanogr. 8*, 210–212.

Brune, D., Nordberg, G., and Wester, P. O. (1980), *Distribution of 23 Elements in the Kidney, Liver and Lungs of Workers from a Smeltery and Refinery in North Sweden Exposed to a Number of Elements and of a Control Group. Sci. Total Environ. 16*, 13–35.

Bulman R. A. (1988), *Yttrium and Lanthanides*, in: Seiler, H. G., Sigel, H., and Sigel, A. (eds.), *Handbook on Toxicity of Inorganic Compounds*, pp. 769–785. Marcel Dekker, New York.

Cerutti, H. (1988), *Supraconduction* (in German), *Neue Zürcher Zeitung, Forschung und Technik*, No. 27, p. 65 (3 February), Zürich.

Chevallier, J., and Butow, R. A. (1971), *Calcium Binding to the Sarcoplasmic Reticulum of Rabbit Skeletal Muscle. Biochemistry 10*, 2733–2737.

Cochran, K. W., Doull, J., Mazur, M., and DuBois, K. P. (1950), *Acute Toxicity of Zirconium, Columbium, Strontium, Lanthanum, Cesium, Tantalum, and Yttrium. Arch. Ind. Hyg. Occup. Med. 1*, 637–650.

Cuddihy, R. G., and Boecker, B. B. (1970), *Kinetics of Lanthanum Retention and Tissue Distribution in the Beagle Dog Following Administration of 140-$LaCl_3$ by Inhalation, Gavage and Injection. Health Phys. 19*, 419–426.

Curmi, P. M., Barden J. A., and dos Remedios, C. G. (1982), *Conformational Studies of G-Actin Containing Bound Lanthanide. Eur. J. Biochem. 122*, 239–244.

dos Remedios, C. G., Barden, J. A., and Valois, A. A. (1980), *Crystalline Actin Tubes. II. The Effect of Various Lanthanide Ions on Actin Tube Formation. Biochim. Biophys. Acta 624*, 174–186.

Duprez, D., Le Normand, F., and Garin, F. (1986), *Catalysis and Automotive Pollution Control, Proceedings CAPOC Meeting*, Free University, Brussels, Belgium.

Durbin, P. W., Williams, M. H., Gee, M., Newman, R. H., and Hamilton, J. G. (1956), *Metabolism of the Lanthanons in the Rat. Proc. Soc. Exp. Biol. Med. 91*, 78–88.

Elderfield, H., and Greaves, M. J. (1981), *Earth Planet. Sci. Lett. 55*, 163.
Esposito, M., Oddone, M., Accardo, S., and Cutolo, M. (1986), *Concentrations of Lanthanides in Plasma and Synovial Fluid in Rheumatoid Arthritis. Clin. Chem. 32*, 1598.
Evans, C. H. (1983), *Interesting and Useful Biochemical Properties of Lanthanides. Trends Biochem. Sci. 8*, 445–449.
Fowler, B. A., Abel, J., Elinder, C.-G., Hapke, H.-J., Kogi, J. H. R., Kleiminger, J., Kojima, Y., Schoot Uiterkamp, A. J. M., Silbergeld, E. K., Silver, S., Summer, K. H., and Williams, R. J. P. (1984), *Structure, Mechanism, and Toxicity*, in: Nriagu, J. O. (ed.), *Changing Metal Cycles in Human Health*, pp. 391–406. Springer Verlag, New York.
Gariepy, J., Sykes, B. D., and Hodges, R. S. (1983), *Lanthanide-induced Peptide Folding: Variations in Lanthanide Affinity and Induced Peptide Conformation. Biochemistry 22*, 1765–1772.
Gerhardsson, L., Wester, P. O., Nordberg, G. F., and Brune, D. (1984), *Chromium, Cobalt and Lanthanum in Lung, Liver and Kidney Tissue from Deceased Smelter Workers. Sci. Total Environ. 37*, 233–246.
Gomez, J. E., Birnbaum, E. R., and Darnall, D. W. (1974), *The Metal Ion Acceleration of the Conversion of Trypsinogen to Trypsin. Lanthanide Ions as Calcium Ion Substitutes. Biochemistry 13*, 3745–3750.
Gracar, J., Gorst, E., and Lowry, W. (1957), *Comparative Toxicity of Stable Rare Earth Compounds. Arch. Ind. Health 15*, 9–14.
Haksar, A., Maudsley, D. V., Peron, F. G., and Bedigian, E. (1976), *Lanthanum: Inhibition of ACTH-stimulated Cyclic AMP and Corticosterone Synthesis in Isolated Rat Adrenocortical Cells. J. Cell. Biol. 68*, 142–153.
Haley, T. J. (1965), *Pharmacology and Toxicology of the Rare Earth Elements. J. Pharm. Sci. 54*, 663–670.
Hamilton, E. I., Minski, M. J., and Cleary, J. J. (1972/73), *Sci. Total Environ. 1*, 341–374.
Hammond, C. R. (1986), *The Elements*, in: *Handbook of Chemistry and Physics*, 67th Ed., pp. B12-B42. CRC Press, Boca Raton, Florida.
Heuck, F., and Hoscher, R. (1968), *Am. J. Radiol. 104*, 777–783.
Hisada, K., and Ando, A. (1973), *Radiolanthanides as Promising Tumor Scanning Agents. J. Nucl. Med. 14*, 615–617.
Holten, V. Z., Kyker, G. C., and Pulliam, M. (1966), *Effects of Lanthanide Chlorides on Selected Enzymes. Proc. Soc. Exp. Biol. Med. 123*, 913–919.
Husain, M. H., Dick, J. A., and Kaplan, Y. S. (1980), *Rare Earth Pneumoconoisis. J. Soc. Occup. Med. 30*, 15–19.
Hutcheson, D. P., Gray, D. H., Venugopal, B., and Luckey, T. D. (1975), *Studies of Nutritional Safety of Some Heavy Metals in Mice. J. Nutr. 105*, 670–675.
Iida, Y. (1987), *Repetitive Spectral Subtraction Method for the Spectrophotometric Determination of Rare Earth Elements, Fresenius Z. Anal. Chem. 328*, 547–552.
Kaczmarek, J. (1981), in: Gschneider, K. A. Jr. (ed.), *Industrial Applications of Rare Earth Elements, ACS Symposium 164*, pp. 135–164. American Chemical Society, Washington, D.C.
Kover, A., Szabolcs, M., and Csabai, A. (1976), *The Effect of La^{3+} on the Characteristics of Fragmented Sarcoplasmic Reticulum. Acta Biochim. Biophys. Acad. Sci. Hung 11*, 23–35.
Lehninger, A. L., and Carafoli, E. (1971), *The Interaction of La^{3+} with Mitochondria in Relation to Respiration-coupled Ca^{2+} Transport. Arch. Biochem. Biophys. 143*, 506–515.
Lettvin, J. Y., Pickark, W. F., McColloch, W. S., and Pitts, C. (1964), *Nature 202*, 1338–1339.
Levinson, C., Smith, T. C., and Mikiten, T. M. (1972), *Lanthanum-induced Alterations of Cellular Electrolytes in Ehrlich Ascites Tumor Cells: A New View. J. Cell. Physiol. 80*, 149–154.
Linebarger, H. F., and McCluhan, T. K. (1981), in: Gschneider, K. A. Jr.(ed.), *Industrial Applications of Rare Earth Elements, ACS Symposium 164,* pp. 19–42. American Chemical Society, Washington, D.C.
Linsalata, P., Eisenbud, M., and Franca, E. P. (1986), *Ingestion Estimates of Th and the Light Rare Earth Elements Based on Measurements of Human Feces. Health Phys. 50*, 163–167.
Machlin, L. J., Pearson, P. B., and Denton, C. A. (1952), *Arch. Ind. Hyg. Occup. Med. 6*, 441–444.

Maier, E. A., Dietemann-Molard, A., Rastegar, F., Heimburger, R., Ruch, C., Maier, A., Roegel, E., and Leroy, M. J. F. (1986), *Simultaneous Determination of Trace Elements in Lavage Fluids from Human Bronchial Alveoli by Energy-dispersive X-Ray Fluorescence. 2: Determination of Abnormal Lavage Contents and Verification of the Results. Clin. Chem. 32*, 664–668.

Martinez-Palomo, A., Benitez, D., and Alanis, J. (1973), *Selective Deposition of Lanthanum in Mammalian Cardiac Cell Membranes. Ultrastructural and Electrophysiological Evidence. J. Cell. Biol. 58*, 1–10.

Mela, L. (1968), *Interactions of La and Local Anesthetic Drugs with Mitochondrial Ca and Mn Uptake. Arch. Biochem. Biophys. 123*, 286–293.

Mela, L. (1969), *Reaction of Lanthanides with Mitochondrial Membranes. Ann. NY Acad. Sci. 147*, 824–828.

Nayler, W. G., and Harris, J. P. (1976), *Inhibition by Lanthanum of the $Na^+ + K^+$ Activated, Ouabain-sensitive Adenosinetriphosphatase Enzyme. J. Mol. Cell Cardiol. 8*, 811–822.

Nukatsuka, I., Toga, M., and Yoshida, H. (1981), *Separation of Lanthanides by Capillary Tube Isotachophoresis Using Complex-forming Equilibria. J. Chromatogr. 205*, 95–102.

Oelkers, E., and Vincke, E. (1938), *Arch. Exp. Pathol. Pharm. 188*, 53–61.

Palmer, R. J., Butenhoff, J. L., and Stevens, J. B. (1987), *Cytotoxicity of the Rare Earth Metals Cerium, Lanthanum, and Neodymium in vitro: Comparison with Cadmium in a Pulmonary Macrophage Primary Culture System. Environ. Res. 43*, 142–156.

Panigel, M., Coulam, C., Wolf, G., Zeleznik, A., Leone, F., and Podesta, C. (1988), *Magnetic Resonance Imaging (MRI) of the Placental Circulation Using Gadolinium-DTPA as a Paramagnetic Marker in the Rhesus Monkey in vivo and the Perfused Human Placenta in vitro. Trophoblast Res. 3*, in press.

Reed, K. C., and Bygrave, F. L. (1974a), *Accumulation of Lanthanum by Rat Liver Mitochondria. Biochem. J. 138*, 239–252.

Reed, K. C., and Bygrave, F. L. (1974b), *The Inhibition of Mitochondrial Calcium Transport by Lanthanides and Ruthenium Red. Biochem. J. 140*, 143–155.

Rubin, R. B., and Heberling, S. S. (1987), *Metal Determinations by Ion Chromatography: A Complement to Atomic Spectroscopy. Am. Lab. Boston 19*, 46–55.

Sanborn, W. G., and Langer, G. A. (1970), *Specific Uncoupling of Excitation and Concentration in Mammalian Cardiac Tissue by Lanthanum. J. Gen. Physiol. 56*, 191–217.

Schepers, G. W. H. (1955), *The Biological Action of Rare Earths. II. Arch. Ind. Health 12*, 306–316.

Smith, B. M., Gindhart, T. D., and Colburn, N. H. (1986), *Possible Involvement of a Lanthanide-sensitive Protein Kinase C Substrate in Lanthanide Promotion of Neoplastic Transformation. Carcinogenesis 7*, 1949–1956.

Snyder, F., and Stephens, N. (1961), *Proc. Soc. Exp. Biol. Med. 106*, 202–204.

Snyder, F., Cress, E. A., and Kyker, G. C. (1960), *Rare-earth Fatty Liver. Nature 185*, 480–481.

Sotiroudis, T. G. (1986), *Lanthanide Ions and Cd^{2+} are Able to Substitute for Ca^{2+} in Regulating Phosphorylase Kinase. Biochem. Int. 13*, 59–64.

Spode, E., and Gensicke, F. (1958), *Naturwissenschaften 45*, 135–136.

Sulotto, F., Romano, C., Berra, A., Botta, G. C., Rubino, G. R., Sabbinoni, E., and Pietra, R. (1986), *Rare-earth Pneumoconiosis: A New Case. Am. J. Ind. Med. 9*, 567–575.

Swamy, V. C., Triggle, C. R., and Triggle, D. J. (1976), *The Effects of Lanthanum and Thulium on the Mechanical Responses of Rat Vas deferens. J. Physiol. 254*, 55–62.

Trueb, L. (1987), *Compactation in "Warm" Supraconduction* (in German), *Neue Zürcher Zeitung, Forschung und Technik*, No. 302, p. 47 (30 December), Zürich.

Van Breeman, C. (1969), *Blockade of Membrane Calcium Fluxes by Lanthanum in Relation to Vascular Smooth Muscle Contractility. Arch. Int. Physiol. Biochim. 77*, 710–716.

Van Breeman, C., and De Weer, P. (1970), *Lanthanum Inhibition of ^{45}Ca Efflux from the Squid Giant Axon. Nature 226*, 760–761.

Venugopal, B., and Luckey, T. D. (1978), *Toxicity of Group III Metals*, in *Metal Toxicity in Mammals – 2*, pp. 135–157. Plenum Press, New York.

Vocaturo, G., Colombo, F., Zanoni, M., Rodi, F., Sabbioni, E., and Pietra, R. (1983), *Human Exposure to Heavy Metals. Rare Earth Pneumoconiosis in Occupational Workers.* Chest 83, 780–783.

Ward, N. I. (1987), *A Comparison of NAA and ICP-SMS in Assessing Human Health and Disease.* Environ. Health 20, 118–123. 2nd Nordic Symposium on Trace Elements in Human Health and Disease, Odense. W.H.O. Regional Office for Europe, Copenhagen.

Wasserman, M., Zakal, N., Loyter, A., and Kulka, R. G. (1976), *A Quantitative Study of Ultramicroinjection of Macromolecules into Animal Cells.* Cell 7, 551–556.

Weiss, G. B., and Goodman, F. R. (1969), *Effects of Lanthanum on Contraction, Calcium Distribution and Ca-45 Movements in Intestinal Smooth Muscle.* J. Pharmacol. Exp. Ther. 169, 46–55.

Weiss, G. B., and Goodman, F. R. (1975), *Interactions Between Several Rare Earth Ions and Calcium Ion in Vascular Smooth Muscle.* J. Pharmacol. Exp. Ther. 195, 557–564.

Welz, B. (1985), *Atomic Absorption Spectrometry,* 2nd Ed, pp. 291–293. VCH Verlagsgesellschaft Weinheim - Deerfield Beach/Florida - Basel.

Yoshimura, K., and Taketatsu, T. (1987), *Determination of Neodymium by Ion-exchange Absorptiometry with the f-f Electron Transition Band,* Fresenius Z. Anal. Chem. 328, 553–536.

Additional Recommended Literature

Carreira, L. A. (1990), *Metal Interaction with Humic Substances (The Eu(III) Model to Study Binding).* Fresenius Z. Anal. Chem., in press; see also Reports E. Merian in *TRAC* and *Swiss Chem.*

Evans, C. H., Pittsburgh (1989), *Biological Effects of Rare Earth Metals,* Communication at the Meeting on Toxicity and Therapeutics of Newer Metals and Organometallic Compounds, Guildford, Surrey, U. K. (24 July), in press; see also Report E. Merian in *Swiss Chem.* 12 (1–2), 25.

Lin, Y., Bailey, G. W., and Lynch, A. T., Beijing and Athens, Georgia (1988), *Metal Interactions at Sulfide Mineral Surfaces: Part II. Adsorption and Desorption of Lanthanum.* Proceedings International Symposium on Environment, Life Elements and Health, in press.

Mottram, Ph. (1989), *Bright Future for New Lanthanide Chelates in Time-resolved Fluorometry.* Eur. Clin. Lab. News (November), 4.

Owen, R. M., and Olivarez, A. M. (1988), *Geochemistry of Rare Earth Elements in Pacific Hydrothermal Sediments.* Mar. Chem. 25, 183–196.

Velthorst, N. H., Schreurs, M., and Gooijer, C. (1990), *Detection Methods in Liquid Chromatography Based on Long-living Luminiscence (using Tb(III) for Labelling).* Fresenius Z. Anal. Chem., in press; see also Reports E. Merian in *TRAC* 9(9), in press, and in *Swiss Chem.* 12, in press.

Note Added in Proof

Dr. A. Crucq, Free University B-1050 Brussels organized 11th September 1990 a discussion on CeO_2 *as Modifier for Stabilization of Automobile Exhaust Catalyst Substrate,* Proceedings in press.

II.16 Lead

ULRICH EWERS and HANS-WERNER SCHLIPKÖTER
Düsseldorf, Federal Republic of Germany

1 Introduction

In nature lead is a ubiquitous, non-essential element. Its natural concentrations are not very high; the average amounts of many other metals and their compounds are higher, and it is ranked only at about the 36th place of the elements in the earth's crust. However, in the last fifty years great amounts of lead have been extracted, concentrated, and used by man, and re-emitted into the environment. Lead concentrations are now locally and regionally much higher than they used to be. Animals and humans are thus exposed to new health risks and their body tissues and fluids may contain more lead than normal. Fortunately, in the very last years atmospheric contamination is again decreasing in industrial countries. Especially unborns, newborns, and children may still be at greater risk, if they inhale or take in orally lead compounds. Unwanted secondary effects such as the hindrance of natural body's defense mechanisms are also observed. Neurobehavioral deficits must also be considered. The concentration of lead in blood is generally accepted as an indicator of the individual's lead exposure. As regards the general population, adults should not have higher concentrations than 35 to 40 µg Pb in 100 mL blood. For the protection of special risk groups (fetus, young children) concentrations in pregnant females and in children should be below 20 µg/100 mL.

Further reduction of lead emissions is necessary because lead compounds introduced into the environment by human activities are not decomposed and accumulate locally and in biological organisms.

2 Physical and Chemical Properties, and Analytical Methods

2.1 Physical and Chemical Properties

Lead (symbol Pb) has an atomic number of 82, an atomic mass of 207.19, and is located in Group IVa of the Periodic Table. It is a bluish-white, soft metal with a density of 11.34 g/cm^3, a melting point of 327.5°C, and a boiling point of about 1740°C. Natural lead consists of 52% ^{208}Pb, 24% ^{206}Pb, 23% ^{207}Pb, and 1% ^{204}Pb. Radioactive ^{210}Pb from radon decay plays also a role, for instance, in waters and sediments (BOYLE, 1987; CORNWALL, 1987).

In the absence of air lead is not attacked by pure water. In the presence of atmospheric oxygen, however, lead becomes susceptible to attack by acids, including very weak acids such as carbonic acid and even water. The plumbosolvency of fresh water is a well known phenomenon. Plumbosolvency is considerably reduced by small amounts of carbonate and silicate in the water and, therefore, decreases with water hardness. When heated in air, metallic lead is oxidized to PbO and, upon further heating, to Pb_3O_4.

In most inorganic compounds lead is in the +2 oxidation state. The salts of Pb(II), lead oxides and lead sulfide, are not readily soluble in water, with the exception of lead acetate, lead chlorate, and, to some extent, lead chloride. Inorganic Pb(IV) compounds are unstable and strong oxidizing agents.

Because of their use as antiknock agents in gasoline, tetramethyllead and tetraethyllead are the most important organolead compounds. Both are colorless liquids with boiling points of 110 °C and 200 °C, respectively. At these temperatures or slightly below, they start to decompose.

2.2 Analytical Methods

(see also Chapter I.4a)

For the quantitative determination of lead in environmental and biological samples the following methods are applied: atomic absorption spectroscopy (AAS), optical emission spectroscopy, X-ray fluorescence (XRF), isotope dilution mass spectrometry (IDMS), colorimetric or spectrophotometric analysis using dithizone (diphenylthiocarbazone), and electrochemical methods (voltammetry). The majority of these methods is restricted to the measurement of total lead and cannot directly identify the various compounds of lead. Informations on detection limits are presented in Chapter I.4a.

For AAS, the lead atoms in the sample must be vaporized either in a precisely controlled flame or in a furnace. Furnace systems in AAS applying graphite tubes or cups offer a high sensitivity and the possibility to analyze small samples. AAS currently is the preferred method for routine lead analyses in environmental and biological specimens (WELZ, 1985; ANGERER and SCHALLER, 1985).

Optical emission spectroscopy is based on the measurement of light emitted by elements when they are excited in an appropriate energy medium. The technique has been used to determine the lead content of soils, rocks, minerals, and airborne dust at the 5 – 10 µg/g level. The primary advantage of the method is that it allows the simultaneous measurement of a large number of elements in small samples. More recent activities have focused attention on the inductively coupled plasma (ICP) system as a valuable means of excitation and analysis (MONTASER and GOLIGHTLY, 1987). The ICP system offers a higher degree of sensitivity with less analytical interference than is typical for many other emission spectroscopy systems.

X-ray fluorescence also allows the simultaneous identification of several elements using a high energy irridiation source. The method is based on the principle that an intense electron beam, which is directed on a sample, produces several forms of radiation, including X-rays, the wavelengths of which depend on the elements present in the material, and their intensities depend on the relative quantities of these elements. The technique offers the advantage that sample degradation can be kept to a minimum. It has been used, e.g., for the in vivo determination of lead in teeth and bones (SHAPIRO et al., 1978; PRICE et al., 1984).

Isotopic dilution mass spectrometry (IDMS) is the most accurate measurement technique presently known. No other technique serves more reliably as a comparative reference technique. IDMS has been used for analyses of subnanogram concentrations of lead in a variety of sample types. The isotopic composition of lead produced from ores of different origin has been used as a means of tracing the origin and distribution of anthropogenic lead (FACCHETTI and GEISS, 1982).

Colorimetric analysis of lead using dithizone as the reagent is a "classical" method which has been used for many years. Prior to the development of the IDMS method it frequently served as a reference method by which other methods were tested.

Among the electrochemical methods differential pulse polarography (DPP) and anodic stripping voltammetry (ASV) offer sufficient analytical sensitivity for most lead measurements in environmental and biological samples. Current practice with commercially available equipment allows lead analysis at subnanogram concentrations with a precision of about 5–10% on a routine basis. With stripping voltammetry at rotating mercury film electrodes it is, for instance, possible to determine really dissolved fractions in waters (MART, 1988). Determination of lead in blood and urine by inverse voltammetry (DPP) has been described by ANGERER and SCHALLER (1988). Of course, complete chemical decomposition of insoluble fractions in biological samples is necessary before electrochemical methods can be used (see also Chapter I.4a).

For analyses in biological materials (biomonitoring and bioindication) see also Sections 4.7.1, 4.7.3, 4.7.4, 6.2, and 7.7. Hematological parameters are often quantified, too.

2.3 Problems of Lead Analyses in Environmental and Biological Samples

A major source of error in the determination of lead in environmental and biological samples is the secondary contamination, which can occur during sampling and analysis. Failure to recognize the source of contamination (collecting containers, reagents, labware, laboratory atmosphere, hand contact) may result in the generation of artificially high analytical results. It is, therefore, important that emphasis is placed on a careful control of the blank problem and on extensive internal and external quality control (WHO EURO, 1981; VAHTER, 1982; KNEIP, and FRIBERG, 1986). It should be mentioned that the US National Bureau of Standards (NBS), the International Atomic Energy Commisson (IAEC) and some other organizations offer reference materials with certified metal concentrations. In part, such materials are also commercially available. These materials should be used to control the accuracy of any analytical procedure.

3 Sources, Production, Important Compounds, Uses, Waste Products, Recycling

3.1 Occurrence

Lead is a naturally occurring element, which represents an almost ever-present constituent of the earth's crust. It can be found in all environmental media (air, soil, rocks, sediments, water) and in all components of the biosphere. The average crustal abundance of lead as estimated on the basis of several reports is 16 ppm (NRIAGU, 1978a). Detailed reviews on the occurrence of lead in various environmental media have been published by NAS (1972, 1980), WHO (1977), NRIAGU (1978a, b), CHOW (1978), NRCC (1978) and US EPA (1986). A summary of environmental lead concentrations is shown in Table II.16-1 (see also Chapter I.1).

The most important lead minerals are galena (lead sulfide), cerussite (lead carbonate), and anglesite (lead sulfate). Galena occurs mostly in deposits, which also contain zinc minerals and small amounts of copper, iron, cadmium, and other metals. A review of the most important lead deposits and the world lead reserves, including economic and subeconomic resources, has been given by KESLER (1978), MELIN (1974), and SAAGER (1984). Relatively large deposits exist in the USA, Australia, the Soviet Union, and Canada (together two thirds of the global resources; SAAGER, 1984).

Table II.16-1. Summary of Environmental Lead Concentrations

Medium	Unit	Area			
		Remote	Rural	Urban	Near Smelters
Soil (total)	µg/g d.w.	5–40	5–40	10–50	20–2000
Soil (upper layer)	µg/g d.w.	5–40	5–60	50–300	300–20000
Air	µg/m^3	<0.01	<0.2	0.2–2.0	1–20
Dust fall-out	µg/m^2/day	<5	20–80	50–300	500–10000
Street dust	µg/g	5–50	50–200	500–2000	1000–20000
Surface water	µg/L		1–10	5–30	10–1000
Drinking water	µg/L		1–20	1–40	
Drinking water (plumbosolvent)	µg/L		100–1000	100–1000	

d.w. = dry weight

3.2 Production

The production of lead from lead ores involves several steps: mining and processing of the crude lead ore, roasting and sintering, reduction, and refining. Detailed descriptions of the lead smelting process have been presented by MELIN (1974), HOWE (1981), and CASSADY (1980).

The first steps of processing lead ore involve mining, crushing, milling, and concentrating. Since lead is frequently associated with zinc in the same ore body the concentrators usually consist of two flotation circuits, one for lead and one for zinc.

In the sinter plant undesirable sulfur is removed by roasting. This operation also properly prepares the mixture for the blast furnace by agglomeration (i.e., gathering the material together in a porous mass). The material that results is called sinter.

The sinter is mixed with coke and conveyed to the top of the blast furnace. Air is introduced through tuyères, or ports, near the bottom of the furnace, thus supplying oxygen to burn the coke, which in turn furnishes the heat necessary to smelt the charge into two products: a lead bullion and slag. Other metals present in the lead ore (e.g., copper, gold, silver, arsenic, antimony, and bismuth) remain in the solution with the bullion. The lead bullion collects at the bottom of the furnace and is drawn off in a continuous pour. The lighter slag floats on the top of the bullion and is tapped off intermittently.

The process of refining lead consists of several steps. First, sulfur is added to the hot blast-furnace bullion. Upon cooling the copper rises to the surface as a dross, which is skimmed. Copper-free lead from the drossing kettles is then pumped through a continuous softening process which, through oxidation, removes arsenic and antimony as a skim. An electric furnace is used to treat this skim, producing an arsenical-antimonal lead known as hard lead. The next phase of the refining process is the removal of gold and silver. This is achieved by the addition of zinc. Gold and silver combine with the zinc and, on cooling, form a dross which is skimmed off. Excessive zinc is removed by a vacuum dezincing process. The addition of sodium hydroxide removes the last traces of impurities to give a final product with a purity of more than 99.9%.

In 1982 globally about 3.7 million tons of lead were produced from mined ores, about 600000 t from each of the USA and the Soviet Union (SAAGER, 1984). Producers with about 250000 t each are the St. Joe Minerals and Asarco Inc. in the USA, Penorraya in France, Cominca in Canada, Rio Tinto Zinc Co. in the United Kingdom, and Mount Isa Mines Holdings in Australia. In the Western countries about 1.3 million tons of lead originate from recycled scrap and waste (SAAGER, 1984). In the UK even about 60% of the needed lead is produced by recycling.

3.3 Uses

In 1982, the total consumption of refined lead was about 5.2 million tons worldwide (WORLD METAL STATISTICS, 1983). About 40% of all lead consumed are used for the production of lead-acid batteries. The production of tetramethyllead (TML) and tetraethyllead (TEL), which are used as antiknock agents in gasoline, accounts for approximately 10% of the world lead consumption. In the 1970s a maximum of TML and TEL (more than 250000 t/a in the USA) was produced, since then consumption has decreased again (NRIAGU, 1989). Lead based pigments are used as a protective coating for steel structures, for paints used on highways and for other exterior uses. The use of lead for interior paint in houses has been prohibited in many countries. Lead chemicals are used in glassware and ceramics and as stabilizers in

Table II.16-2. Principal Uses of Lead in Different Countries (URBANOWICZ, 1986)

Country	Percentage of the Total Lead Consumption per Country in 1981					
	Batteries	TEL[a]	Chemicals	Cable Sheathing	Sheet and Pipe	Other Uses
Canada[a]	43		27	3		27
France	50	6	14	7	9	14
FRG	43		27	8	15	7
Italy	36	4	20	15	13	12
Japan	54		18	10	6	12
UK	29	21		7	20	23
USA	66	10	7	1	2	14
Western World[b]	51	6	14	16		13

[a] Source: NRCC (1978)
[b] Source: SAAGER (1984)

plastics. Lead sheets, cable sheathings, solder, ammunition, bearing alloys, type metal, tubes, weights and ballast, and low melting alloys together account for about 20% of the lead used. PbO_2 and lead tetraacetate are oxidizing agents. Some data on the uses of lead in different countries are summarized in Table II.16-2. In recent years lead consumption has been decreasing partially because leaded gasoline is used less, because one tries to substitute heavy batteries, and because lead tubes for many applications are not allowed any longer (SAAGER, 1987). Especially the production of TML and TEL is decreasing since about 1980 because of the detection of increased levels of lead in blood (particularly in children) and because catalysts reducing exhausts to avoid smog are destroyed in the presence of lead.

A large proportion of lead waste is recovered and recyled in secondary lead smelters. Concerning lead storage batteries the recovery is more than 90% in some countries. The recovery of lead sheets, pipes, and cable sheathings may vary between 50 and 90%. Lead from alloys is recovered to a substantially lesser degree.

For further data concerning the uses of lead, lead alloys, and lead compounds see WHO (1977), ROBINSON (1978), HOWE (1981), MCCORMACK et al. (1981), and URBANOWICZ (1986).

4 Distribution in the Environment, in Foods, and in Living Organisms

4.1 Background Information
(see also Section 4.6)

Lead in rocks is often found at concentrations of 10–20 ppm. Granitic rocks tend to have higher levels than basaltic ones. The reported average concentration of lead in normal soil range from 10–40 mg/kg dry weight (ppm). Most values are below 20 ppm. In general, soils tend to reflect the composition of their parent materials. Due to deposition of atmospheric lead the top layers of soil usually contain higher lead levels than deeper layers. Significantly increased lead concentrations in surface soils have been found particularly in inner city areas, near busy highways and near lead works.

Reviews on the lead concentration found in natural waters, including rain water, snow, and ice, surface waters, ground and spring waters, and ocean water have been given by CHOW (1978) and VALENTA et al. (1986) (see Sect. 4.3). The lead content of lake and river waters usually is in the range of 0.1–10 µg/L. It is always necessary to distinguish between dissolved lead and suspended matter. The lead levels in ground and spring waters may vary over a wide range depending on the peculiar geochemical conditions. The avarage lead concentration in surface ocean water depends on emissions and is about 0.01–0.03 µg/L (MART, 1988); in deep ocean water it ranges from about 0.001 to 0.004 µg/L (BRULAND, 1983; NÜRNBERG and MART, 1985). Concentrations at the surface have rather been decreasing in the recent years (BOYLE, 1987). The lead content of lake and river sediments was used to estimate environmental lead levels in past times. For detailed reviews of the lead levels in lake, marine, and river sediments see NRIAGU (1978a) and HUYNH-NGOC et al. (1988).

Atmospheric lead levels in remote continental areas fall in the range of 0.1–10 ng/m^3. Marine atmospheres generally contain less lead than continental air. The lowest atmospheric lead concentration measured anywhere was 0.046 ng/m^3 found in the west winds of the North Atlantic (CHESTER et al., 1974). Since the combustion of leaded gasoline represents the major source of atmospheric lead, urban areas have significantly higher air lead concentrations than rural areas. In the 1960s and 1970s the annual mean air lead levels in towns and cities in the USA, Canada, and Europe were found to be in the range of 0.4–4.0 µg/m^3. In central urban areas with high traffic density air lead levels of up to 10 µg/m^3 were recorded. The reduction of lead in fuel in the mid-1970s and the increasing use of lead-free gasoline led to a substantial decline of atmospheric lead concentrations in urban areas during the last years. The air lead levels presently found in European and North-American cities are in the range of 0.2–0.8 µg/m^3. In rural areas the air lead levels usually are in the range of 0.05–0.3 µg/m^3. Due to atmospheric transport of airborne lead, which may range over hundreds and thousands of kilometers, even remote areas show increased air lead concentrations when compared to "baseline" air lead levels.

4.2 Sources of Lead in the Environment

Lead and its compounds may enter the environment at any point during mining, smelting, processing, use, recycling, or disposal. Global emissions were about 500 000 t/a around 1919, about 2 million t/a around 1940, and about 4.5 million t/a around 1970 (NRIAGU, 1989). Estimates of the dispersal of lead emissions into the environment indicate that the atmosphere is the major initial recipient. Mobile and stationary sources of lead emissions tend to be concentrated in areas of high population density, and near smelters. From these emission sources, lead moves through the atmosphere to various components of the environment. It is deposited on soil, surface waters, and plants and thus is incorporated into the food chain of animals and man. Atmospheric lead is also an important component of street dust. Furthermore, lead is inhaled directly from the atmosphere by humans and animals.

4.2.1 Sources of Lead Emissions to the Atmosphere

The major part of lead found in the atmosphere results from the combustion of leaded gasoline. Organolead emissions may result from the production of TML and TEL and from evaporation and automotive emissions of unburned TML or TEL. Several reports indicate that automobile traffic contributes about 90% of the total atmospheric lead emissions. In the USA the annual lead emission from gasoline combustion was estimated to be 34881 tons in 1984 (US EPA, 1986). This corresponds to 89.4% of the total lead emissions in the USA. However, the use of leaded gasoline in the Western world is decreasing. Other mobile sources, including aviation use of leaded gasoline and diesel and jet fuel combustion contribute insignificantly to lead emissions into the atmosphere.

Organolead compounds, in particular tetramethyllead and tetraethyllead, which are used as antiknock additives in gasoline, occur in urban atmospheres at levels ranging from 5 to 200 ng/m^3. Since evaporation and automotive emissions of unburned TML or TEL are the main emission sources, the highest concentrations are found in parking garages, gasoline stations, and busy streets in central urban areas. The major part of atmospheric organolead is in the vapor phase. Usually, organolead makes up to 5–10% of the concentration of particulate lead in urban areas (NIELSEN, 1984).

Automotive lead emissions mainly occur as PbBrCl in fresh exhaust particles. Particles initially formed by condensation of lead compounds in the combustion gases are quite small (<0.1 µm in diameter). These particles are subject to growth by coagulation and, when airborne, can remain suspended in the atmosphere for 7–30 days and travel hundreds and thousands of kilometers from their original source. Larger particles are formed as a result of agglomeration of smaller condensation particles and have shorter atmospheric lifetimes. According to TER HAAR et al. (1972) approximately 35% of the lead in the gasoline burned by the vehicle is emitted as small particles, <0.25 µm MMAD (mass median aerodynamic diameter) and approximately 40% are emitted as larger particles (>10 µm MMAD). The remainder of the lead consumed in gasoline combustion is deposited in the engine and exhaust system.

Mining, smelting, and refining as well as the manufacturing of lead-containing compounds and goods may also give rise to substantial lead emissions. Particularly smelters of lead ores are well known to create significant pollution problems in local areas. Their influence on the surrounding air and soil depends to a large extent on the height of the stack, the emission control devices, the topography and other local features. In general, there is a zone of heavy soil pollution around the smelters caused by high levels of lead fall-out resulting from dust emissions of the smelters. The zone of high lead fall-out and soil pollution usually is extended to 5–10 km from the smelter, whereas the zone of high air lead levels is restricted to a smaller area. In the USA, the annual lead emissions from primary lead smelters was estimated to be 1150 tons in 1984 (US EPA, 1986).

Secondary smelters producing lead from scrap are comparatively small but numerous and frequently situated close to human settlements. Several studies showed that pollution in the surroundings of such smelters resulted in an increased lead intake of people living nearby (for review see US EPA, 1986).

Further stationary sources of lead emissions are coal-fired power plants, incinerators of solid waste, sewage sludge or waste oil, and the manufacturing of lead glass, storage batteries, and lead additives for gasoline. The lead emission from coal-fired power plants located in the member states of the European Community has been estimated to be in the order of 200 tons per year (SABBIONI et al., 1984). In the USA, the lead emission from coal and oil combustion has been estimated at 380 tons per year accounting for about 1% of the total anthropogenic lead emission to the atmosphere in the USA (US EPA, 1986).

4.2.2 Lead in Waste Waters

The major fraction of lead in municipal waste water results from lead-containing dust fall-out, which reaches the drainage system with rain water. Studies in the USA show that the concentration of lead, which can be present in the water in dissolved or particulate form, varies over a wide range, depending on traffic density, industrial emissions and climatic factors. The values range from ten to several hundred, in some areas even several thousand µg/L (CHOW, 1978). In the industrialized Rhône river, for instance, about 0.1 µg dissolved Pb per liter and about 50 mg solid Pb per kg suspended particulate matter or sediment have been found (HUYNH-NGOC et al., 1988).

4.2.3 Lead in Solid Wastes

On a local level special problems may arise with regard to waste and slag dumps from lead mines and primary lead smelters. These waste materials may contain high lead levels and represent a hazard for playing children and grazing animals if there are no appropriate enclosure and security measures (MAGS, 1983). According to BACCINI (1989) municipal waste contains about 0.3 mg lead per kg. In incineration lead ends up mainly in slag. From landfills lead may be leached out at higher pH values (BACCINI, 1989). Windblown dust from these deposits may also significantly contaminate the surroundings including human settlements located there.

4.3 Environmental Transport and Distribution

From a mass balance point of view, the transport and distribution of lead from stationary and mobile sources into other environmental media is mainly through the atmosphere. Large discharges may also occur directly into natural waters and onto the land, but in such cases lead tends to be localized near the points of discharge because of the low solubility of the lead compounds that are formed upon contact with soil and water.

Depending on the particle size airborne suspended particles may have a long residence time in the atmosphere. Studies from remote areas indicate that lead containing particles are transported over substantial distances, up to thousands of kilometers, by general weather systems. MUROZOMI et al. (1969) and BOUTRON (1982) have found significantly increased lead levels in polar ice and glaciers resulting from long-range transport of lead containing airborne particles. STEINNES et al. (1987) have demonstrated that lead compounds (and some other metal compounds) are enriched in mosses, in podzolic soils, in ombrotropic peat, and in other environmental targets in southernmost Norway from deposition of long-range transported atmospheric pollution (from the European continent).

Lead may be removed from the atmosphere either by dry or wet deposition. The most efficient clearing mechanism is rain (wet deposition). According to the distance of atmospheric transport the transport pattern has been classified as "near fall-out", "far fall-out", and "airborne". "Near fall-out" was defined as the deposition in the immediate surroundings of the emission sources (roadways, stationary sources). "Far fall-out" was defined as fall-out away from the emission sources, but within the geographic area (20–200 km), and "airborne" designated small particles, which were subject to long-range atmospheric transport and ultimately deposited elsewhere. According to data of HUNTZICKER et al. (1975) on the distribution of lead emissions from motor vehicles in the Los Angles basin, about 55% of the emitted lead can be attributed to the category of "near fall-out", 10% to "far fall-out", and 35% to "airborne".

In rural areas of Europe and North America the annual deposition rates are typically in the range of 20–80 µg/m^2/day. Although there are only very few data available it appears that in less industrialized areas of the world the lead deposition rates are considerably lower. In Greenland and the Antarctic values below 1 µg/m^2/day have been recorded. In large cities the deposition rates are about 1.5–10 times higher than in rural areas. Typical values presently found in West-German and British cities range from 50–300 µg/m^2/day (RUSSELL and STEPHENS, 1983; RADERMACHER et al., 1986). Near busy highways significantly higher levels have been found. The highest recorded levels of lead deposition occur near primary lead smelters.

VALENTA et al. (1986) found between about 15 and about 50 µg Pb^{2+}/L rain in West Germany, and a wet deposition of about 30 µg Pb^{2+}/m^2/day in rural regions, 50 µg Pb^{2+}/m^2/day in urban areas, and 100–150 µg Pb^{2+}/m^2/day in regions with metallurgical industry, varying with time and having in general a decreasing tendency (see also CRÖSSMANN, 1989). Regarding deposition on forests, lead compounds are not concentrated in canopies as other metal compounds (KLOCKOW and LARJAVA, 1988), interception is important, lead compounds are introduced by stem-flow, but mobility in litter is very low (HERTZ and ANGEHRN-BERTINAZZI, 1988).

Atmospheric lead deposited on soil is retained in the upper (2–5 cm) soil layers, especially in soils with at least 5% organic matter and a pH of 5 or above. A high organic humus fraction and a high soil pH are the most important factors in immobilizing lead. The concentration of anthropogenic lead in the upper 2–5 cm is, therefore, determined by the flux of atmospheric lead to the soil surface. Near roadsides, the lead concentration in surface soil decreases exponentially up to 25 m from the edge of the road. Near primary and secondary smelters, lead in soil decreases exponentially within a 5–10 km zone around the smelter complex. In the 1980s concentrations in dusts and soils, however, have been decreasing in the USA and in Western Europe (CRÖSSMANN, 1989; ELIAS, 1989; STREHLOW and BARLTROP, 1989).

4.4 Lead in the Food Chain
(see also Chapters I.8a and I.8b, Sections 5.2.2 and 5.2.3, and Table II.16-4)

Lead in food can result from various sources: lead uptake of plants growing on high-lead soils or treated with lead arsenate pesticides, surface deposition of lead on plants consumed by food-producing animals or man, inadvertent addition of lead during food processing, and leaching of lead from cans with lead soldered seams or from improperly glazed pottery used as food storage or dining utensils. Furthermore, home food preparation practices may be potential sources of additional lead residues in food. Lead in cooking water may also become part of the diet.

Numerous estimates and investigations exist of the lead content of individual foods. Large-scale market basket surveys have been conducted in the UK (MAFF, 1975, 1983), USA (US Food and Drug Administration, as quoted by US EPA, 1986), in the Federal Republic of Germany (WEIGERT et al., 1984), and several other countries. Wheat and potatoes contain on the average about 0.04 to 0.09 ppm Pb, meat about 0.01 to 0.06 ppm, and liver about 0.3 ppm.

Although there is a large variability of lead concentrations in individual foods depending on various factors including the geographic location in which the food was grown, the following generalizations can be made:

- Vegetables: the lead concentrations are highest in the roots, lower in stems and leaves, and lowest in flowers or seeds. An exception are leafy vegetables that may retain surface dust or dirt that is not readily removed by washing.
- Fruits: the highest lead levels are found in the peel and stem.
- Foods of animal origin: muscle meat, unprocessed milk, and eggs have a relatively low lead content, whereas organ meat, particularly kidney and to a lesser extent liver, is substantially higher in lead concentrations.
- Alcoholic beverages (see also Sect. 4.7.3): wine may contain substantial amounts of lead with mean values between 50 and 100 µg/L and individual samples ranging up to several hundred µg/L (ECKARD and BERTRAM, 1987) depending on the bottle sealing. High lead concentrations have also been found in illegally distilled whisky (moonshine whisky). Lead contamination of the latter may occur from lead-soldered joints of the distilling apparatus (MORRE, 1986a).

The total daily intake of lead varies widely among populations and individuals. It is difficult to measure accurately the dietary uptake, and many informations in the

Table II.16-3. Average Dietary Lead Intake of Adults and Children in Different Countries

Country	Population	Lead Intake (µg/day)	Method	Reference
Belgium	Adult males	91	Fecal excretion method	CLAEYS-THOREAU et al. (1987)
		94	Duplicate meal study	
Malta	Adult males	401	Fecal excretion method	CLAEYS-THOREAU et al. (1987)
Mexico	Adult males	177	Fecal excretion method	CLAEYS-THOREAU et al. (1987)
Sweden	Adult males	24	Fecal excretion method	CLAEYS-THOREAU et al. (1987)
UK	Adults (average)	113	Total diet study	DHHS (1980)
	Adult males	146		
	Children (2 years)	70		
	Children (3–4 years)	80		
USA	Adults (average)	33[a]		NAS (1980)
	Children	25[b]		US EPA (1986)
	Adult females	32[b]		
	Adult males	45[b]		
West Germany	Adult males	147	Total diet study	WEIGERT et al. (1984)
	Adult females	104		

[a] Estimate based on various studies
[b] US EPA estimate of baseline human lead exposure in the USA, referring to Americans living in rural areas and not occupationally exposed to lead

literature may be not comparable. Some recent estimates of the average total dietary lead intakes of adults and children in various countries are shown in Table II.16-3. The wide differences between industrial and non-industrial countries (Malta?!) need further explanation.

Within the past few years, attention has been directed to the lead content of commercial infant foods (SCHMIDT and HILDEBRANDT, 1983). A number of studies suggest that infants and young children, on a body weight basis, may have a significantly higher lead intake than adults (MAHAFFEY, 1978; US EPA, 1986). Regarding bottle fed infants the lead content of drinking water may be of significant importance.

4.5 Lead in Drinking Water

Lead in drinking water may result from contamination of the water source or from the use of lead materials in the water distribution system. Lead entry into drinking water from the latter is increased in water supplies with soft water and pH values below 6.5. Human exposure may occur through direct ingestion of the water or via food preparation in such water. The lead concentrations normally found in drinking water are of the range of 2–25 µg/L.

Numerous studies indicate that the major source of lead contamination of drinking water is the distribution system itself, particularly in older urban areas with old houses frequently equipped with lead water pipes. The highest lead levels are encountered in "first-draw" samples, i.e., water sitting in the piping system for an extended period of time.

4.6 History of Global Lead Pollution
(see also NRIAGU, 1989)

The history of global lead pollution has been assembled from chronological records of the deposition of lead in polar snow and ice strata, marine and freshwater sediments, and the annual rings of trees. These records aid in establishing natural background levels of lead in environmental and biological media and document the sudden increase of atmospheric lead emissions at the time of the industrial revolution, with a later burst during the 1920s when lead alkyls were first added to gasoline. Pond sediment analyses have shown a 20-fold increase in lead deposition during the last 150 years, documenting not only the increasing use of lead since the beginning of the industrial revolution, but also the relative fraction of natural versus anthropogenic lead inputs. Perhaps the best chronological record is that of the polar ice strata of MUROZOMI et al. (1969), which extends nearly three thousand years back in time. The authors found an about 10-fold increase between 1750 and 1940 and an accelerated increase after 1940, which was attributed to the increasing lead emissions since the beginning of the industrial revolution and the increased use of lead alkyls in gasoline in the 1920s. At the South Pole, BOUTRON (1982) found a 4-fold increase of lead in snow from 1957 to 1977 but no increase from 1927 to 1957. The author concluded that the extensive atmospheric lead pollution, which began in the 1920s and primarily was located in the Northern hemisphere did not reach the South Pole until the mid-1950s. The historical picture of global lead production since pre-historical times has been pieced together from many sources by SETTLE and PATTERSON (1980).

The history of human lead exposure has been estimated on the basis of lead analyses of ancient teeth and bones (for review see BARRY, 1978; US EPA, 1986). Overall, the available studies show a significant increase of the lead concentrations in historical and modern human bones and teeth compared with prehistorical remains of members of premetallurgical societies. It should be noted, however, that the use of results from such studies as a basis for comparison with modern findings requires special caution if erroneous conclusions are to be avoided. Most investigators agree that the potential for uptake of lead among populations of ancient civilizations was probably greater than in modern times considering that the indiscriminate use of the metal in cooking utensils, storage vessels, water conduits, and many other household purposes could have resulted in a significantly higher lead uptake than in modern populations. Apparently, the lead concentrations in human bones markedly increased with the introduction of metallurgical processes and the increase of production and utilization of lead, particularly in the Roman Empire and during medieval times (DRASCH, 1982).

4.7 Lead in Mammalian (Including Human) Tissues and Body Fluids

In individuals not occupationally exposed to lead, the lead content of soft tissues is generally below 0.5 µg/g wet weight, with higher values found in the aorta, liver, and kidney cortex. After age 20, the lead content of soft tissue does not show age-related changes (BARRY, 1978).

The lead content of bones and teeth increases throughout life, since lead becomes localized and accumulates in human calcified tissues (bones and teeth). This accumulation begins with fetal development and continues to approximately the age of 60 years. The amount of lead stored in the skeleton of men aged 60–70 years ranges up to 200 mg, while in women lower values have been measured. Various studies show that approximately 95% of the total body lead is lodged in bones of human adults. Concentrations in blood and bones have, however, been decreasing during the last years (DRASCH et al., 1987, 1988; ELIAS, 1989; STREHLOW and BARLTROP, 1989). Tooth lead levels also increase with the age at a rate proportional to exposure and roughly proportional to blood lead (STEENHOUT and POURTOIS, 1981; STEENHOUT, 1982).

4.7.1 Lead in Teeth

Due to the continuous accumulation of lead in calcified tissues the lead content of teeth increases with age. The long-term exposure to lead, therefore, can be estimated by measuring the lead concentration in deciduous teeth (children) or permanent teeth (adults). It has to be noted, however, that the lead concentration in teeth varies between different tooth tissues and between different tooth types. The highest lead concentrations are found in the inner and outer surfaces of the tooth, i.e., the outer layer of enamel and the circumpulpal dentine. The lead concentration in whole teeth decreases from the medial incisors to the premolars and molars, respectively.

The average lead concentrations of whole decidous teeth usually fall in the range of 2–8 µg/g dry weight (EWERS et al., 1982; FERGUSSON and PURCHASE, 1987; BROCKHAUS et al., 1988; PATERSON et al., 1988). The lead levels in the circumpulpal dentine usually are about 10 times higher. According to a Belgian study the lead content of permanent teeth averages about 20 µg/g dry weight (STEENHOUT and POURTOIS, 1982). Since tooth lead levels can be regarded as a measure of long-term lead exposure of an individual, differences in exposure levels are often more clearly reflected by tooth lead than by blood lead levels (BROCKHAUS et al., 1988).

4.7.2 Lead in Hair

Measurement of lead in hair allows an estimation of exposure over a period of one or several months. It has to be noted, however, that the measurement may yield inaccurate results due to external contamination of the hair surface. In children, the average lead content of head hair varies between 10 and 20 µg/g dry weight (WIBOWO et al., 1986).

4.7.3 Lead in Blood

(see also ANGERER and SCHALLER, 1985, 1988; but see also Section 6.5.2)

It is generally agreed that the level of lead in blood (PbB) as a measure of absorbed lead is the best indicator of current exposure (WHO, 1977, 1980). In interpreting results it has to be taken into account, however, that PbB reflects a dynamic equilibrium between exposure (absorption), retention, release, and elimination. Under steady state conditions, e.g., those prevailing in the general population or during long-term, unchanged occupational exposure, PbB gives a good picture of current exposure, but shortly after changes in exposure intensity PbB becomes a poorer indicator. For example, after the beginning of occupational exposure, it takes about two months for the PbB level to reach a steady state. After termination of exposure PbB decreases slowly. The half-life of lead in blood usually is about 2 – 4 weeks, but it can be significantly longer, if the lead body burden is large.

On the average, the PbB levels found in the general population usually are in the range of 5 – 20 µg/100 mL. In the literature, a great number of studies on the distribution of blood lead concentrations in the general population as well as in selected population groups (e.g., children living near smelters or in central urban areas) have been reported (for reviews see VAHTER, 1982; LEHNERT and SZADKOWSKI, 1983; HUNTER, 1986; TSUCHIYA, 1986). According to a European Community Directive (77/312/EEC) requiring member states to carry out blood lead screening campaigns separated by at least two years, large-scale blood lead surveys were conducted in several European countries in 1979 and 1981 (BERLIN et al., 1983; CLAEYS-THOREAU et al., 1983; QUINN, 1985; HUEL et al., 1986). Throughout these studies emphasis was placed on a strict quality control. An extensive population blood lead survey was also conducted within the framework of the Second National Health and Nutrition Examination Survey (NHANES II) in the USA from 1976 – 1980 (MAHAFFEY et al., 1982). International studies have been conducted within the framework of the "UNEP/WHO Pilot Project on Assessment of Human Exposure to Pollutants through Biological Monitoring" (VAHTER, 1982) and more recently by CLAEYS-THOREAU et al. (1987). Studies in remote, non-industrialized areas (PIOMELLI et al., 1980; WATANABE et al., 1985) show that baseline PbB levels are in the range of 3 – 5 µg/100 mL.

In the general population, PbB levels are influenced by a number of demographic, social, and local factors and by individual habits. Males have, on an average, higher PbB levels than females. People living in central urban areas have higher PbB levels than people from suburban and rural areas, the latter usually showing the lowest mean PbB values. Individuals, particularly children, living near lead smelters or near busy highways have higher PbB levels than otherwise comparable subjects. People living in areas with soft, highly plumbosolvent water and in aged houses with lead pipes or plumbings in the domestic water distribution system may have increased PbB levels compared with other population groups. A number of studies conducted in the USA showed elevated PbB levels in children living in aged houses with interior lead paint. Smokers have 10–15% higher PbB than levels than non-smokers. People who drink distinctly contaminated (alcoholic) beverages, particularly wine, tend to have higher PbB levels than non-drinkers. The large amount of literature on this matter has been reviewed by HUNTER (1986) and US EPA (1986); see also Sect. 4.4.

The literature on the relationship between air lead and blood lead levels has been critically evaluated by CHAMBERLAIN (1983a), BRUNEKREEF (1984) and US EPA (1986). At air lead levels below 5 µg/m^3 the increase of blood lead is in the range of 1.0–3.0 µg/100 mL per µg/m^3. At higher air lead concentrations the slope of the blood lead vs. air lead relationship is significantly smaller.

PbB levels are also affected by age. Various studies indicate that PbB reaches a peak in early childhood (1–4 years) followed by a decrease to about age 12–15 years. Thereafter, the PbB levels increase again, particularly in men, up to the age around 50 years when they begin to decline (for review see HUNTER, 1986, and US EPA, 1986).

International comparisons applying strict quality control to assure comparibility of results show that there is a considerable variation of the mean PbB levels in different countries. In the UNEP/WHO study reported by VAHTER (1982) the median PbB values ranged from 6.0 µg/100 mL in Tokyo (Japan) to 23.6 µg/100 mL in Mexico City. The subjects from Stockholm (Sweden), Baltimore (USA), and Zagreb (Yugoslavia) had median PbB values of 7.3 µg/100 mL, 7.5 µg/100 mL, and 9.0 µg/100 mL, respectively, whereas people from Brussels (Belgium) had a median of 15.2 µg/100 mL. In a recent study on human lead exposure in Belgium, Malta, Mexico, and Sweden the median PbB levels ranged from 7.7 µg/100 mL (Sweden) to 24.7 µg/100 mL (Malta) (CLAEYS-THOREAU et al., 1987). Measurements of the fecal lead excretion that were conducted throughout the study showed significant differences between the four countries suggesting that the different PbB levels are attributed to differences in oral lead intake via food and wine (see Table II.16-3; see also Section 4.4).

A number of studies consistently show a decrease of the PbB levels in the general population since the end of the 1970s. The fall of PbB was first noted in the NHANES II study for the period 1976–1980 (ANNEST et al., 1983) and also when comparing the results of the CEC blood surveys of 1979 and 1981 (BERLIN et al., 1983; CLAEYS-THOREAU et al., 1983; QUINN, 1985; HUEL et al., 1986). Other studies confirm this time trend and indicate a further fall of blood lead since 1981 (CLAEYS-THOREAU et al., 1987; ENGLERT et al., 1986; BROCKHAUS et al., 1988). The most reasonable explanation for this decline appears to be the reduction of lead in gasoline since 1976 (ANNEST, 1983; ANNEST et al., 1983). Other factors in explaining the general decrease of PbB levels are the apparent further reduction of lead in the diet (which, in part, appears to be related also to the reduction of lead in gasoline and, in part, to the reduction of lead in canned food), the decrease of water lead in areas with naturally plumbosolvent water, and the increased public awareness of lead as a health hazard.

Apart from this long-term decreasing trend of blood lead, several studies point to certain seasonal variations of PbB levels with higher values in summer and autumn compared to winter and spring. The differences are mostly pronounced in children. The possible reasons of these variations have been discussed by HUNTER (1986).

Various studies show that people, particularly children, of low socioeconomic status tend to have higher PbB levels than otherwise comparable subjects. Apparently, this difference is attributable to a combination of various factors such as living in old, deteriorated houses (which are more likely to have lead water pipes or lead plumbings and interior lead paint than modern houses) and living in polluted areas (central urban areas, lead smelter areas). As described above, these factors, probably together with less care for personal hygiene, may increase the risk of lead absorption.

4.7.4 Lead in Urine
(see also ANGERER and SCHALLER, 1988; but see also Section 6.5.2)

Increased lead concentrations in urine indicate a high lead absorption, but a normal rate of lead excretion does not reliably exclude the possibility of excessive absorption. Lead in urine is dependent on the PbB level but is also influenced by other, mostly unknown factors, so that no direct conclusions about exposure and the extent of absorption can be derived from the lead levels in urine (WHO, 1977). The average lead concentration in urine of individuals without occupational or other excessive lead exposure is in the range of $5-20\,\mu g/L$.

5 Uptake, Absorption, Transport and Distribution, Metabolism and Elimination in Plants, Animals, and Humans

5.1 Uptake of Lead in Plants

In soils with natural lead concentrations ($15-30\,\mu g/g$), only trace amounts of lead are absorbed by plants. The amount absorbed increases when the concentration of lead in soil increases or when the binding capacity of soil for lead decreases (low organic fraction and low pH). Uptake by roots does not necessarily mean that lead reaches the stem, leaves, or fruits. Rather, the process should be seen as a soil-root continuum that strongly favors rentention of lead by the soil and roots.

Most of the lead in or on plants occurs on the surfaces of the leaves and the trunk and stem, where it is deposited by dry or wet deposition and from where it can be removed to a large extent by washing. The surface concentration of lead in trees, shrubs, and grasses usually extends the internal concentration by a factor of at least five. Translocation of lead from the outer surface to the inner parts of the leaves, tubers, fruits, and seeds is very small. The major effect of surface lead at ambient concentrations seems to be on subsequent components of the grazing food chain and on the decomposer food chain following litter-fall. The effect of lead fall-out on lead in crops has been extensively discussed by CHAMBERLAIN (1983b).

A review of the mechanisms of lead uptake and transport in plants and of the lead concentrations found in different plant species has been given by PETERSON (1978).

5.2 Lead Absorption in Animals and Humans

5.2.1 Respiratory Absorption

The movement of lead from ambient air to the bloodstream is a two-part process: deposition of some fraction of inhaled air lead in the deeper part of the respiratory tract and absorption of the deposited fraction. The deposition rate mainly is determined by the particle size distribution of the inhaled particles and the ventilation rate. For adult humans, the deposition rate of particulate airborne lead occurring in rural and urban atmospheres is about 30–50%. All of the lead deposited in the lower respiratory tract appears to be absorbed so that the overall absorption rate is governed by the deposition rate. Results from autopsy results showing no lead accumulation in the lung indicate a total absorption of the deposited lead. The chemical form of the lead compounds inhaled does not seem to be a major determinant of the extent of alveolar lead absorption.

All of the available data for lead uptake via the respiratory tract in humans have been obtained with adults. Respiratory uptake of lead in children, while not fully quantifiable, seems to be comparatively greater on a body-weight basis. The smaller airway dimensions also appear to increase the relative deposition rate of lead in children (ROSEN and SORELL, 1978; US EPA, 1986).

5.2.2 Gastrointestinal Absorption

Gastrointestinal absorption of lead mainly involves lead uptake from food, drinking water, and beverages as well as lead deposited in the upper respiratory tract and swallowed. It also includes uptake of non-food materials such as dust and paint chips, which may be unwittingly ingested by young children due to their mouthing activity and pica habits (see Sect. 5.2.5).

By use of metabolic balance and isotopic studies (for reviews see HOLTZMAN, 1978, and US EPA, 1986) the gastrointestinal lead absorption in the human adult has been determined to be in the order of 10–15%. The rate can be significantly increased under fasting conditions to 45% compared to lead ingested with food. For children, the fraction absorbed is much higher than in adults, ranging up to 50% (ALEXANDER et al., 1973). Experimental animal studies also show that young animals absorb a much greater amount of lead than the adult.

Animal studies also indicate that certain dietary factors such as milk fasting, low calcium and vitamin D, and iron deficiency may enhance lead absorption from the gut. The effects of nutritional factors and other metals on the absorption and metabolism of lead have been reviewed extensively by MAHAFFEY (1980) and CHISOLM (1980). BEEBY and RICHMOND (1988) found that lead uptake in snails depends on variations in calcium metabolism (car park snails show an adaption to high ambient Pb levels). GUNN et al. (1988) informed on lead uptake by bivalve molluscs, which is correlated with 1 mol/L HCl-extractable lead in the sediment, but present iron compounds have an influence, too.

5.2.3 Percutaneous Absorption

Absorption of inorganic lead compounds through the skin is much less significant than through the respiratory and gastrointestinal tract (MOORE, 1986a; US EPA, 1986). GUNADILAKA (1988) found, however, that razorfish and eel – in contrast to other metals which accumulate in kidneys and liver – concentrate lead in gills and skin (JENSEN, 1984).

5.2.4 Transplacental Transfer of Lead
(see also Section 6.6.2)

Lead uptake by the human and animal fetus occurs readily. This uptake is apparent by the 22th week of gestation in humans and increases throughout fetal development. Cord blood contains significant amounts of lead, correlating with maternal blood lead levels (LAUWERYS et al., 1978). A further evidence for the transplacental transfer of lead is the detection of lead in fetal tissues.

5.2.5 Contribution of Different Sources of Human Lead Exposure
(see also Chapter I.6b)

In general, typical levels of human lead exposure may be attributed to four components of the human environment: food, inhaled air, drinking water, and dusts of various types. A rough estimate of a typical exposure situation regarding the intake and absorption of lead from air, food, and drinking water is shown in Table II.16-4. Food and beverages constitute the major sources of exposure to lead among adults not employed in lead-related industries and among children without pica (i.e., the ingestion of non-food items such as lead-containing paint chips).

There are many conditions, even in non-urban environments, where an individual may increase his lead exposure by choice, habit, or unavoidable circumstances. These conditions can be regarded as additive exposures being added to the "normal" human lead exposure described above. As shown in Table II.16-4 some of these additive exposures may substantially exceed exposure from food, drinking water and air. The most important additive exposures are:

– Occupational exposures: the highest and most prolonged exposures to lead are found among workers employed in primary and secondary lead smelters and in

Table II.16-4. Estimate of Lead Intake and Absorption in Children and Adults from Food, Beverages, Drinking Water, and Air and Some Possible Additional Exposures

Population	Source	Lead Intake (μg/day)	Lead Absorbed[a] (μg/day)
Children	Food and beverages	50	25
	Drinking water (10 μg/L; 1 L/day)	10	5
	Air (0.5 μg/m^3; 5 m^3/day)	2.5	1[b]
	Possible additional exposures:		
	Household and street dust (1000 μg/g; 100 mg/day)[c]	100	50
	Plumbosolvent drinking water (100 μg/L; 1 L/day)	100	50
Adults	Food and beverages	100	10
	Drinking water (10 μg/L; 2 L/day)	20	2
	Air (0.5 μg/m^3; 15 m^3/day)	7.5	3[b]
	Possible additional exposures:		
	Plumbosolvent drinking water (100 μg/L; 2 L/day)	200	20
	Smoking[d]	10	4[b]
	Wine consumption (100 μg/L; 0.25 L/day)	25	3
	Occupational exposure[e]	360	144[b]

[a] Assuming a gastrointestinal absorption of 50% in children and of 10% in adults
[b] Assuming a deposition rate of 40% in the alveolar region and a complete absorption of the deposited particles
[c] Increased lead concentrations in dust near lead smelters or busy highways, in central urban areas or houses with interior lead paint may significantly increase the intake of lead in children
[d] Assuming a lead content of 10 μg/g cigarette, an inhalation of 5% of this lead (0.5 μg/cigarette), and a consumption of 20 cigarettes/day
[e] Assuming a 8 h shift at an average concentration of 50 μg/m^3 on 5 days/week and a breathing volume of 10 m^3/8 h

lead battery plants. Due to the diversity and extent of the industrial applications of lead, exposures to lead may also occur at various other industrial workplaces (WHO, 1977; NRCC, 1978). The major route of occupational lead exposure is by inhalation.
— Secondary occupational exposures: lead workers may carry substantial amounts of lead-containing dust on their skin, hairs, shoes, and clothes to their homes and, thereby, contaminate the home environment. Various studies show that the family members of lead workers, particularly young children, have a high risk of increased lead absorption. According to one study (MORTON et al., 1982) showering and shampooing, in addition to changing clothes and shoes, are necessary to effectively reduce the lead exposure of the children.
— Ingestion of lead containing dust: children place their mouths on dustcollecting surfaces, lick non-food-items with their tongues, and put dirty fingers into their mouths. This fingersucking and mouthing activity are natural forms of behavior for young children that expose them to some of the highest concentrations of lead in their environment. A single gram of dust may contain ten times more lead than the total diet of the child. There are numerous studies showing that lead in street-

and household dust is a major determinant of lead exposure in children living in central urban areas and near smelters (ROELS et al., 1980; LANDRIGAN and BAKER, 1981; BRUNEKREEF et al., 1983; DUGGAN, 1983; REAGAN and SILBERGELD, 1989; STREHLOW and BARLTROP, 1989).
- Lead containing paints: studies conducted in the USA indicate that peeling and flaking lead paint from painted surfaces inside dwelling houses is a major source of increased lead concentrations in household dust, which may be ingested by young children. Peeling and flaking paint may also be directly ingested by children. Moreover, the weathering of lead containing paint on outer walls of houses may contribute significantly to soil contamination in the surroundings. The problem of lead paint inside homes seems to be restricted mainly to deteriorated urban areas in the USA. A great number of pediatric lead intoxications or elevated blood lead levels in children from these areas have been attributed to the unwitting ingestion of lead paint (LIN-FU, 1980).
- Consumption of home-grown vegetables and fruit from family gardens with high-lead soils or near sources of atmospheric lead (GALLACHER et al., 1984).
- Living in houses with lead water pipes or lead plumbings in the water distribution system. This may give rise to high water lead concentrations, particularly in areas with soft and/or slightly acidic water.
- Tobacco smoking: trace amounts of lead present in tobacco are inhaled with the mainstream smoke. Various studies show that smokers have slightly higher blood lead levels than non-smokers (BROCKHAUS et al., 1983; ELINDER et al., 1983).
- Drinking of wine or illegally distilled whisky containing high lead levels (ELINDER et al., 1983; see also Sects. 4.4 and 4.7.3).

Extensive reviews of the sources of human lead exposure have been presented by WHO (1977), MAHAFFEY (1978), LEHNERT and SZADKOWSKI (1983), MOORE (1986a), and US EPA (1986).

5.3 Metabolism of Lead in Humans and Animals

The absorbed lead enters the bloodstream from where it is distributed to various organs and tissues. Redistribution then occurs in relation to the relative affinity of each tissue for lead.

More than 95% of the lead in human blood is bound to erythrocytes under steady state conditions. Most of the erythrocyte lead is bound within the cell, primarily to hemoglobin. In plasma and extracellular fluids, nearly all of the lead is bound to proteins, mainly albumin and some high-molecular-weight globulins. It is unknown, which of these binding forms constitutes the active, diffusible fraction for movement of lead into tissues and cells.

According to reports of the TASK GROUP ON METAL ACCUMULATION (1973) and TASK GROUP ON METAL TOXICITY (1978) the amount of lead in the organism can be divided into three fractions: blood lead and some rapidly exchanging soft tissues with a half-life of about 19 days; soft tissues and a rapidly exchangeable bone fraction with a half-life of about 21 days; and bones and the skeleton with a half-life of about 10 to 20 years (but one should distinguish between lead in trabecular bone and lead in compact bone, with a longer half-life of the latter). For practical purposes the first two compartments can be combined. The amount of lead in organs and bones which may be potentially active toxicologically in terms of being available to biological sites of action can be estimated by the measurement of plumburesis in response to administration of a chelating agent, specifically $CaNa_2EDTA$. The chelatable amount of lead probably consists of the mobilizable fraction of lead in soft tissues and the exchangeable bone fraction.

Dietary lead in humans and animals that is not absorbed passes through the gastrointestinal tract and is eliminated with the feces. The same applies to air lead that is swallowed and not absorbed. Absorbed lead is excreted primarily in urine. The mechanism of urinary excretion appears to be essentially glomerular filtration. Other excretion routes are gastrointestinal secretion, and nails and sweat. The rate of biliary excretion in humans is not known. Lead is also excreted in human milk in concentrations up to 10 µg/L.

An extensive review of the literature on the distribution and storage of lead in human tissues has been presented by BARRY (1978). The literature on metabolic studies has been evaluated by WHO (1977), HOLTZMAN (1978), NRCC (1978), NAS (1980), and US EPA (1986).

6 Effects on Plants, Animals, and Humans

It has already been mentioned that lead compounds destroy the activity of catalysts necessary to reduce automobile exhausts.

6.1 Effects on Microorganisms and Plants

As regards the effects of lead on the terrestrial microbiota it appears that microorganisms are more sensitive to soil lead pollution than plants. Changes in the composition of bacterial populations may be an early indication of lead effects. These changes, with the more resistant organisms dominating, may be very drastic at certain locations. Delayed composition may occur at 750 µg Pb/g soil and nitrification inhibition at 1000 µg/g (US EPA, 1986). Usually, the effects are smaller in clay and peat soils compared to sandy soils (DOELMAN, 1978).

Since most of the physiologically active tissues of plants are involved in growth, maintenance, and photosynthesis, it can be expected that lead might interfere with one or more of these processes. Indeed, such interferences have been observed in laboratory experiments at lead concentrations greater than those normally found in the field, except near smelters or mines (KOEPPE, 1981). Studies of lead effects on other plant processes, especially maintenance, flowering, and hormone development, have not been conducted, and no conclusions can be drawn concerning these processes (KOEPPE, 1981).

As described in Sect. 4.3 lead in soil is strongly immobilized by the humic fraction. The uptake of lead via the root system depends on the amount of available lead in soil moisture. The literature of experimental studies reviewed by US EPA (1986) supports the conclusion that inhibition of plant growth begins at lead concentrations of less than 1 µg/g soil moisture and becomes completely inhibitory at levels between 3 and 10 µg/g. Plant populations that are genetically adapted to high-lead soils may achieve 50% of their normal root growth at lead concentrations above 3 µg/g. Plants that absorb nutrients from deeper soil layers may receive less lead. A few species of plants have the genetic capability to adapt to high lead soils. Plant communities near smelter sites, therefore, may experience a shift toward lead-tolerant plant populations (ANTONOVICS et al., 1971). According to SMITH and BRADSHAW (1972) lead-tolerant plant populations of *Festuca rubra* and *Agrostis tenuis* can be used to stabilize toxic mine wastes with lead concentrations up to 80000 µg/g.

6.2 Miscellaneous Biochemical Effects

Lead has many diverse physiological and biochemical effects, which generally are of a deleterious nature. No evidence has been presented for an essential function of lead in the metabolism of humans and animals. The general aspects of molecular mechanisms underlying the biochemical toxicology of lead have been reviewed by KACEW and SINGHAL (1980). A review of the subcellular effects of lead has been presented by the US EPA (1986). The effects of lead on energy metabolism and the neurochemical and neurophysiological correlations of lead toxicity have been reviewed by BULL (1980), HRDINA et al. (1980), COOPER and SIGWART (1980), and SILBERGELD (1983).

Lead interferes with several enzymes that participate in the heme synthesis pathway. The most important effects are (1) stimulation of the mitochondrial enzyme δ-aminolevulinic acid synthetase (ALA-S), which mediates the formation of δ-aminolevulinic acid; (2) direct inhibition of the cytosolic enzyme δ-aminolevulinic acid dehydratase (ALA-D), which catalyzes the formation of porphobilinogen from two molecules of δ-aminolevulinic acid (ALA), and (3) inhibition of ferrochelatase (heme synthetase), which mediates the insertion of iron(III) into protoporphyrin IX to form heme.

Erythrocyte ALA-D activity is the most sensitive biological indicator of lead toxicity. It is almost completely inhibited at PbB levels in excess of 70–80 µg/100 mL. The PbB threshold value, at which 10% of persons exhibit decreased ALA-D activity is about 10 µg/100 mL (PIOTROWSKI and O'BRIEN, 1980). As far as is known, ALA-D activity is without functional importance in the mature erythrocyte, in which heme synthesis no longer occurs. For this reason, inhibition of ALA-D activity is considered a subcritical effect of no direct biological importance.

Inhibition of ALA-D activity by lead is reflected by elevated levels of its substrate, ALA, in blood, urine, and soft tissues. Urinary ALA (ALA-U) is frequently employed as an indicator of excessive lead exposure in lead exposed workers. The diagnostic value of this measurement in pediatric screening is limited, however, when only spot urine samples are collected. More satisfactory results can be obtained with

24 h urine samples. Numerous studies indicate that there is a direct correlation between PbB and log ALA-U in human adults and children. The PbB threshold for increases of ALA-U is about 40 μg/100 mL.

Inhibition of ferrochelatase by lead results in an accumulation of protoporphyrin IX in erythrocytes. In lead exposure, the porphyrin acquires a zinc ion in lieu of iron, thus forming zinc protoporphyrin (ZPP). Elevation of free erythrocyte protoporphyrin (FEP) or erythrocyte ZPP is closely correlated with PbB and is generally accepted as a true critical effect of undue lead exposure. In adults, the PbB threshold for ZPP elevation is about 20–30 μg/100 mL, females being more sensitive than males (PIOTROWSKI and O'BRIEN, 1980). In children the threshold is below 20 μg/100 mL. According to one study (CAVALLIERI et al., 1981) the no-response level seems to be below 10 μg/100 mL.

Anemia is a well known manifestation of clinical lead poisoning. Although there are conflicting results in the literature it appears that there is a negative correlation between hemoglobin (Hb) and blood lead at higher PbB concentrations. The effects of lead on Hb production involve disturbances of both heme and globin synthesis. Iron deficiency as a potential confounding factor, the non-specifity of the Hb parameter as well as the great interindividual variation in its sensitivity may give rise to many difficulties in finding a significant relationship between PbB and Hb. The PbB threshold for reduced hemoglobin content appears to be about 50 μg/100 mL in adults and 40 μg/100 mL in children. The hemolytic component of lead-induced anemia appears to be attributed to a decreased erythrocyte survival time resulting from increased cell fragility and increased osmotic resistance.

Extensive reviews of the hematological effects of lead including dose-effect and dose-response relationships have been presented by ZIELHUIS (1975), WHO (1977), NRCC (1978), HERNBERG (1980), MOORE (1986a), WHO (1980), TSUCHIYA (1986), and US EPA (1986).

6.3 Effects on Domestic Animals and Wildlife

Although the incidence of lead poisoning among animals is difficult to evaluate accurately, it has been suggested that lead is one of the most frequent single sources of accidental poisoning in domestic animals (NAS, 1972). The most important sources of lead are lead containing dust fall-out on pastures near lead smelters or lead mines, lead-based paint and improperly disposed wastes such as oil wastes and storage battery casings. Fall-out from automotive exhausts has been shown to increase the lead intake, but has not yet been demonstrated to account for reported cases of lead poisoning. Various groups of wild living animals, e.g., rats, rabbits, and birds, have been studied as indicators of environmental lead pollution.

Hazards for lead poisoning may exist for grazing cattle near lead smelters or lead mines, where lead fall-out or soil lead are extremely high. Numerous fatal intoxications have been reported from such areas. The acute signs characteristic for lead poisoning include central nervous system disorders, excitement, stupor or depression, motor abnormalities, and blindness. Some animals may die without showing any of these signs. Clinical signs frequently found in lead-poisoned horses include difficulty in breathing, which leads to a characteristic roaring sound, stiffness, clumsiness, enlarged joints, facial paralysis, muscular weakness, and poor appetite. Lead intoxications in dogs have been reported from areas where lead paint is common in dwelling houses and seem to result from ingestion of peeling or flaking lead paint.

Reviews of the literature on the toxicity of lead in domestic animals and wildlife, including aquatic organisms, have been presented by FORBES and SANDERSON (1978), WONG et al. (1978), and US EPA (1986). It is also known that pellet wedges in wild birds (ducks and other water fowl) paralyze muscles, and the birds may die with starvation.

GOCHFELD and BURGER (1988) demonstrated, for instance, that lead impairs selectively behavioral functions of feeding (e.g., manipulation of fish) in young common terns.

6.4 Effects of Inorganic Lead on Experimental Animals and Humans

6.4.1 Clinical Diagnosis and Sequelae of Lead Intoxications in Humans

The clinical diagnosis of lead poisoning is not always easy. It is made partly on the basis of subjective and objective symptoms, a variety of signs, biochemical analyses indicating an accumulation or increased excretion of metabolic products, and evidence of lead exposure. Individuals may be asymptomatic or symptomatic, presenting a wide variety of symptoms and being in different states of the disease. Usually, the onset of lead poisoning, even in cases of acute intoxications, is not a sharply defined event but rather a continuum of changes from normality to illness.

In adults, the most important symptoms of severe lead poisoning are (in descending order of frequency): abdominal pain (colic), constipation, vomiting, non-abdominal pain, asthenia, paresthesia, psychological symptoms, and diarrhea. Mild symptoms and signs include: tiredness, lassitude, constipation, slight abdominal discomfort or pain, anorexia, sleep disturbances, irritability, anemia, paleness, and less frequently diarrhea and nausea (POSNER et al., 1978). The presence of a blue line in the gums and of a metallic taste are further useful indicators of increased lead absorption.

The signs and symptoms in children are somewhat different than in adults. For example, peripheral neuropathy is more common in adults, while encephalopathy is much more common in children. The most important symptoms of pediatric lead poisoning (in descending order of frequency) are: drowsiness, irritability, vomiting, gastrointestinal symptoms, ataxia, stupor, and fatigue. Further signs may include behavioral changes, speech disturbances, and intercurrent fever and dehydration (POSNER et al., 1978).

6.4.2 Effects of Lead on the Peripheral Nervous System

It is a well known fact that peripheral paresis occurs in severe lead poisoning, but such cases are extremely rare today. In recent times, attention has been directed mainly to electrophysiologically detectable functional abnormalities that occur in the peripheral nerves in the absence of clinical neurological signs. The most important neurophysiological abnormalities recorded consist of slowing of the nervous motor conduction velocity, especially that of the slower fibers, and electromyographic abnormalities, such as fibrillations and a diminished number of motor units in maximal contraction. Dose-effect relationships for some conduction velocities (maximum conduction velocity of the median and ulnar nerves, sensory conduction velocity and conduction velocity of the slower motor fibers), especially of the arm nerves, have been reported to occur in the PbB range of 30–70 µg/100 mL. PbB levels in excess of 50 µg/100 mL are associated with an increasing frequency of abnormal conduction velocities and electromyographic abnormalities among occupationally exposed workers. The frequency of such abnormalities increases at higher exposure levels. Reviews of the toxic effects of lead on the peripheral nervous system have been presented by HERNBERG (1980), COOPER and SIGWART (1980), WHO (1980), and US EPA (1986).

6.4.3 Effects of Lead on the Central Nervous System (CNS), Neurobehavioral Deficits

Numerous cases have been documented in which heavy lead exposure caused encephalopathy in children and adults. The major symptoms were dullness, restlessness, irritability, headache, muscular tremor, hallucinations, and loss of memory and ability to concentrate. The signs and symptoms may progress to delirium, mania, convulsions, paralysis, and coma. Severe encephalopathic conditions with fatal outcome have been reported to occur markedly more often in children than in adults. Prior to the introduction of chelation therapy as a standard medical practice the mortality rate for lead encephalopathy cases among children was approximately 65%. Following the introduction of chelation therapy the mortality could be reduced substantially (US EPA, 1986). Children who survive lead encephalopathy may be left with permanent neurological sequelae such as recurrent seizures and mental retardation. According to LANSDOWN (1986) it should be noted that there is an enormous variation in individual susceptibility. Some children suffer from irreversible brain damage or even death at PbB levels of about 100 µg/L, whereas others appear to show no effects at levels double this (see also Sect. 6.5.2).

In recent years the question as to whether lower intensities of exposure that do not give rise to overt signs of lead toxicity can cause impairments of CNS functions has received particular attention. To study such effects different psychological performance tests and symptom questionnaires have been applied. Several studies consistently show some impairment in the performance of different psychological functions and an excess of subjective central nervous symptoms among workers with comparatively low lead exposures (PbB levels between 60 and 80 µg/100 mL). Tests measuring visual motor functions and visual intelligence seem to be the most sensitive indicators of such effects. Impairments of a wider area of psychological functions appear to be detectable when the exposure range is somewhat higher (HERNBERG, 1980; VALCIUKAS et al., 1978).

Children as a risk group for CNS effects have received particular attention in studies dealing with lead-induced neuropsychological deficits at PbB levels between 10 and 50 µg/100 mL. The earlier work describing cognitive dysfunctions, namely IQ-deficits, impairment of eye-hand-coordination, attention deficits, and behavioral abnormalities such as hyperactivity has been critically evaluated by BORNSCHEIN et al. (1980) and RUTTER (1983). These reviewers concluded that these studies did not provide convincing evidence for cognitive deficits at PbB levels below 40 µg/100 mL. More recent studies, however, give rise to extend the range of concern down to lower PbB levels. In a critical review of these studies LANSDOWN (1986) came to the following conclusions (see also NEEDLEMAN, 1987, who considered also interactions by other pollutants):

(a) The effect of lead on intelligence and cognitive functioning, if it exists at all within the range of lead burdens currently experienced by the general population of children, is small. Most studies point to differences of up to 5 to 6 IQ points with data suggesting that lead accounts for no more than 2 or 3% of the variance in cognitive performance. This is much less than is accounted for by genetic and many other environmental factors. However, small is a relative term, and a population shift of 5 IQ points up or down is not negligible.
(b) Recent epidemiological studies conducted in Britain, West Germany, and in the USA consistently suggest that lead is causally related to deficits in cognitive functioning (see also FULTON, 1989).
(c) There is no convincing consistent evidence for an association between moderate levels of lead exposure and behavioral patterns in general.
(d) The question of a possible threshold PbB level is still open and may be difficult to answer, possibly because the notion of a clear cut-off point is questionable.

For detailed information on the results of the most recent studies the reader is referred to the proceedings of a workshop on lead exposure and child development which was held at the University of Edinburgh in September 1986 (SMITH et al., 1989).

The literature on animal studies on the behavioral toxicity of inorganic lead has been reviewed by BORNSCHEIN et al. (1980), JASON and KELLOG (1980), MICHAELSON (1980), and WINNEKE (1985, 1986). Motor activity and learning of tasks of various difficulty and complexity are behavioral dimensions which have received particular attention. These studies show that lead can be considered causative for certain neurobehavioral deficits at moderate or even low exposure levels, which are not associated with overt signs of toxicity. Some of these neurobehavioral deficits resemble cognitive deficits in man.

Generally, learning of visual discrimination has proved to be most sensitive to the effects of lead. In contrast, increased motor activity (hyperactivity) appears to be only secondarily related to lead exposure, and primarily to lead-induced undernutrition during early stages of brain development. Many of these effects seem to be irreversible, particularly if lead exposure occurs maternally and/or in early life.

6.4.4 Effects of Lead on the Kidney

Lead can produce acute and chronic nephropathies depending on the dose and duration of exposure. Two general types of effect have been described. The first one is a rather clear-cut renal tubular damage characterized by generalized aminoaciduria, hypophosphatemia with relative hyperphosphaturia, and glucosuria. The condition results from decreased tubular reabsorption and, therefore, reflects proximal tubular damage. Such effects have been observed in children with clinical lead poisoning or undue high PbB levels (WHO, 1977) and could also be demonstrated in lead exposed experimental animals (US EPA, 1986). A different kind of renal effect has been seen in workers with prolonged high-level lead exposure. This condition, which commonly is referred to as chronic lead nephropathy, is characterized by slow development of contracted kidneys with arteriosclerotic changes, interstitial fibrosis, glomerular atrophy, and hyaline degeneration of the vessels. The syndrome can develop and progress to renal failure long after undue lead exposure has terminated. All available evidence indicates, however, that a prolonged high lead exposure is necessary, even in childhood, to produce this progressive chronic nephropathy. Currently, it is only rarely encountered in occupational exposure. For reviews on the renal effects of lead see WHO (1977) and CHOIE and RICHTER (1980).

6.4.5 Effects of Lead on the Cardiovascular System

A number of studies show that lead poisoning is associated with cardiotoxic effects (electromyocardiographic abnormalities) in both children and adults (for review see WHO, 1977; US EPA, 1986).

The association between excessive lead exposure and hypertension or, more broadly, increased blood pressure has been studied since the beginning of this century. Although there are some contradictory results from various investigations the majority of studies point to a significant association between high levels of lead exposure and increased blood pressure. The causal relationship between lead exposure and increased blood pressure is supported by animal studies. Several recent studies on occupationally lead-exposed workers and some other occupational groups with elevated lead exposure show a positive relationship between blood lead and blood pressure even at moderately increased PbB levels (MOREAU et al., 1982; KIRKBY and GYNTELBERG, 1985; DE KORT et al., 1986; WEISS et al., 1986).

In addition to these studies on specific cohorts, analyses of data from two large-scale general population studies (British Regional Heart Study and US NHANES II) provide highly convincing evidence of small but statistically significant associations between PbB and increased blood pressure in adult men (POCOCK et al., 1984; PIRKLE et al., 1985). The strongest association appears to exist for males aged 40–59 years and for systolic somewhat more than for diastolic blood pressure. The lower range of PbB levels that may be associated with an increase of blood pressure remains to be determined. The above studies point toward moderately increased PbB levels ($>30\ \mu g/100\ mL$) as being associated most clearly with blood pressure increases. According to the study of MOREAU et al. (1982) there might be effects even at lower PbB levels. It should be noted, however, that the lead-blood pressure effect is small (a few mm Hg) and that age, body weight, and hereditary factors are much more important determinants of blood pressure than lead.

The biochemical and physiological mechanisms by which lead influences the cardiovascular system to induce blood pressure increases have not yet been fully evaluated. The increase of blood pressure appears to result mainly from changes in vascular reactivity and sympathetic tone, both of which may be dependent on lead-related changes of the intracellular calcium ion concentration. It has also been shown that lead (even at very low levels) produces measurable effects on the renin-angiotensin system (for a review of the literature on this matter see US EPA, 1986).

6.4.6 Effects of Lead on Other Organs and Tissues

There is no definite evidence for adverse effects of lead on the liver. Impairment of thyroid and adrenal function and gastrointestinal disorders may occur in cases of clinical lead poisoning. GRANDJEAN et al. (1989) studied delayed blood regeneration after phlebotomy in lead exposed workers from battery production. With about 40 µg lead per 100 mL blood, reserve capacity for blood formation is decreased, but there is still a bone marrow reserve to form hemoglobin.

Lead increases the susceptibility of laboratory animals to endotoxins and infectious agents. Lead induced immunosuppression is detectable in experimental animals at low exposure levels, which are not associated with overt signs of lead toxicity. The literature on the immunotoxic effects of lead has been evaluated by NRCC (1978) and US EPA (1986); see also Chapter I.18, Sect. 2.3. According to KAZANTZIS (1989) immunoglobin levels may be decreased, and possibly there is an effect on NK cell production.

6.5 Effects of Organolead Compounds on Living Organisms

Extensive reviews on the properties, uses, occurrence, and toxicology of organolead compounds have been published in a monograph edited by GRANDJEAN (1984). The most important chemicals among the organolead compounds are tetramethyllead (TML) and tetraethyllead (TEL), which are used on a large scale as antiknock additives in gasoline.

6.5.1 Effects on Plants

Tetraalkyllead compounds (R_4Pb) released into the environment undergo dealkylation caused by photolysis and by chemical degradation in the atmosphere. The primary decomposition product is trialkyllead (R_3Pb^+), which represents the most stable and most toxic derivative. The toxicity of trialkyllead compounds is largely determined by the length of the alkyl chain. In homologous series from trimethyl- to trioctyllead toxicity increases with increasing chain length. Triphenyllead is of moderate toxicity, depending on the organism used, but usually exceeding that of trimethyllead and triethyllead. The toxic effects observed at high concentrations include disturbances of fundamental processes such as photosynthesis, growth, development, mitosis, and cytokinesis. Usually, however, the atmospheric concentrations are too low to exert such adverse effects on plant species. Even massive contamination of aquatic environments with concentrated antiknock compounds seems to create toxic damages only at limited locations. The toxic effects of organolead compounds on plants and aquatic systems have been extensively reviewed by ROEDERER (1984).

6.5.2 Effects on Animals and Humans

The following sections mainly deal with TML and TEL, which are the most important organolead compounds. The toxic properties of these compounds are rather well documented, whereas data concerning the toxicity of other organolead compounds are scarce.

Since TML and TEL are volatile, absorption is usually by inhalation, but percutaneous absorption following skin contact may also be of significant importance. Due to their lipophilic properties, both TML and TEL can easily penetrate intact skin and mucous membranes (JENSEN, 1984). Inside the body the substances are rapidly cleared from the blood and degraded to the trialkyllead compounds (R_3Pb^+). Trimethyllead is metabolized faster and by a different mechanism than the higher trialkyllead compounds.

Animal experiments indicate that the toxicity of tetraalkyllead is caused by the trialkyllead compounds formed by degradation processes, mainly in the liver. Once formed, trialkyllead compounds are fairly stable in biological tissues, but there is a further degradation to dialkyllead and inorganic lead compounds. The brain is considered the critical organ in organolead intoxications. Experimental studies show that trialkyllead compounds can easily pass the blood-brain barrier (CREMER and CALLAWAY, 1961).

Organic lead compounds are mostly eliminated as inorganic lead compounds by excretion in the feces. A smaller part is excreted in the urine, mainly in the form of dialkyllead and inorganic lead compounds.

The blood lead level is not considered a reliable index of organolead exposure due to the short residence time of the organolead compounds in blood. The commonly used alternative index, lead in urine, is also of questionable value since there is no correlation between urinary lead levels and CNS effects (JENSEN, 1984).

The absorption of a sufficient quantity of TEL, whether briefly at a high rate or for a prolonged period of time at a lower rate, induces a more or less severe encephalopathic syndrome. Milder manifestations are insomnia, lassitude, and nervous excitation which reveals itself in lurid dreams and dream-like waking states of anxiety, in association with tremor, hyperreflexia, spasmodic muscular contractions, bradycardia, vascular hypotension, and hypothermia. The more severe responses include recurrent (sometimes nearly continuous) episodes of complete desorientation with hallucinations, facial contortions, and intense general somatic muscular activity with resistance to physical restraint.

Diagnosis usually is based on exposure history, a combination of symptoms and signs of the presenting illness, and the excretion of high amounts of lead in urine. Because of the rapid clearance from blood the PbB levels are only moderately increased. Usually, there are no morphological and chemical abnormalities in the hematological findings. Reviews of organolead exposures and intoxications have been presented by GRANDJEAN (1984), KEHOE (1985) and NRIAGU (1989).

6.6 Mutagenic, Carcinogenic, and Teratogenic Effects of Lead

6.6.1 Genotoxic and Carcinogenic Effects
(see also Chapter I.18, Section 2.3)

In several studies, chromosomal aberrations were found in peripheral lymphocytes of lead exposed populations whose PbB levels ranged from 10–100 µg/100 mL. Other studies could not find any significant increase in chromosomal aberrations. Reviews on the literature have been published by WHO (1977), IARC (1980), and US EPA (1986). At present, no definite conclusions can be drawn from the available studies. According to ZELIKOFF et al. (1988) insoluble lead sulfide (readily phagocytized) and moderately soluble lead nitrate and acetate may be genotoxic by an indirect mechanism disturbing important enzymatic functions (lending support to the view that lead may be a carcinogen). However, lead induced mutations seem not to be a result of direct damage to DNA, since the three compounds are mutagenic at the HPRT locus in V79 cells, but do not induce SCE and DNA single-strand breaks.

Lead acetate, lead subacetate, and lead phosphate are carcinogenic to rats and lead subacetate to mice. Following oral or parenteral administration these substances induce both benign and malignant tumors of the kidney. Gliomas could also be induced in rats by lead acetate and lead subacetate given by the oral route. The literature on this matter has been reviewed and evaluated by IARC (1980) and US EPA (1986). The International Agency for Research on Cancer (IARC, 1980) concluded that, although adequate human data are missing, it is reasonable, for practical purposes, to regard the above compounds as to present a carcinogenic risk to humans. However, in the Federal Republic of Germany only lead arsenate and lead chromate are classified as carcinogenic substances (MAK, 1988).

The literature on human epidemiological studies of the carcinogenicity of lead and its inorganic compounds has also been evaluated by IARC (1980) and US EPA (1986). There are several studies consistently showing an increased cancer mortality among lead smelter and battery plant workers, particularly with respect to cancers of the gastrointestinal tract and respiratory system (COOPER, 1976, 1985; SHEFFET et al., 1982). Another epidemiological study has noted an increased mortality from renal cancer in a group of lead smelter workers (SELEVAN et al., 1984). Some of these studies have been criticized for methodological reasons. The IARC and US EPA, therefore, stated that no final conclusions could be drawn from the available studies and that further epidemiological studies are required in which other factors that may contribute to the observed effects are well controlled for and the disease process is assessed in individuals with well documented exposure histories. KAZANT-ZIS (1989) has mentioned unusual (difficult to explain) renocancers in children, and that lead may be a tumor promotor. It seems that exposure of more than 20 years may lead to increased cancer incidence.

6.6.2 Effects of Lead on Reproduction and Development

A number of studies dating back to the 19th and early 20th century show an increased occurrence of miscarriages and stillbirths in women exposed to high levels of lead during pregnancy. Since the time of these reports pregnant women and also women in the reproductive age have largely been excluded from occupational lead exposure.

As regards possible effects of moderately increased lead exposure, two recent studies (NORDSTRÖM et al., 1978, 1979) suggest a relationship between lead exposure and an increased frequency of spontaneous abortions. Other studies (for review see US EPA, 1986) have provided evidence suggestive of a relationship between maternal lead exposure and shortened gestation and reduced fetus size and growth. Further studies in this field are needed before final conclusions can be drawn. Nevertheless, it has been stated (MAK, 1988) that damage to the developing organism cannot be excluded when pregnant women are exposed, even when MAK and BAT values are adhered to.

At present, no reliable information is available for assessing the effects of lead on human ovarian function or other factors affecting female fertility. The results of an earlier study (PANOVA, 1972, as quoted by WHO, 1977) have been questioned (ZIELHUIS and WIBOWO, 1976, as quoted by US EPA, 1986). As regards male fertility some studies show that lead at high exposure levels may exert adverse effects on the testes resulting in reduced or abnormal spermatogenesis. For a review of the literature see WHO (1977), US EPA (1986), and Chapter I.18, Sect. 4.

Reviews of experimental studies on the effects of lead on reproduction have been presented by BELL and THOMAS (1980) and SILBERGELD (1983). These studies show that lead, at high levels, is toxic to the reproductive organs in the adult and is lethal or teratogenic to the developing fetus. At lower levels of exposure, several parameters of reproduction are affected which may significantly reduce fertility. These include: mutational effects on ovaries and sperm, reduced numbers of sperm, alterations of hormone chemistry, decreases of placental function and compromised growth and development of the foetus.

The association between prenatal lead exposure and the occurrence of congenital anomalies has been investigated in three recent studies (NEEDLEMAN et al., 1984; ERNHART et al., 1986; MCMICHAEL et al., 1986). Of the three studies only NEEDLEMAN et al. (1984) reported significant effects related to lead exposure (increased occurrence of various minor malformations). Based on a critical evaluation of these studies the US EPA (1986) stated that, at present, no definite conclusion can be drawn regarding the existence of an association between commonly encountered levels of prenatal lead exposure in humans and the occurrence of congenital anomalies.

As regards the effects of prenatal lead exposure on postnatal development (6–24 months) several recent studies using the Mental Development Index (MDI) of the Bayley Scales of Infant Development consistently show significant declines in the MDI scores in relation to the children's PbB levels measured at birth in the unbilical cord. Some other investigations point to significant associations between prenatal lead exposure and growth retardation as well as deficits in neurobehavioral performance tests (SMITH et al., 1989; see also Sect. 6.4.3). The studies have been reviewed by DAVIS and SVENDSGAARD (1987) who concluded that "there now can be little doubt that exposure to lead, even at blood levels as low as 10–15 µg/100 mL and possibly lower, is linked with undesirable developmental outcomes in human fetuses and children. These effects include neurobehavioral development, reduced gestational age, lowered birth weight, and other possible effects on early development and growth". Similar conclusions were drawn by the US EPA (1986). The existence of a causal relationship between lead exposure and such effects is supported by experimental animal studies (for reviews see WINNEKE, 1985, 1986).

7 Hazard Evaluation and Limiting Concentrations

7.1 Emission Control for Stationary Sources

Primary and secondary lead smelters, battery plants, and other lead processing plants are subject to extensive regulations in most industrialized countries. Regulations include emission standards (maximum admissible concentrations of lead in stack gases) as well as standards of performance (see Chapter I.15b). In the Federal Republic of Germany and in Switzerland the maximum allowable concentration of lead in the stack gas was set at 5 mg/m^3 at a total mass flow of 25 g/h or more.

7.2 Regulations Concerning Lead in Gasoline

Since the combustion of leaded gasoline represents the major source of lead emissions into the atmosphere, the lead content of leaded gasoline has been limited in many countries. A review of national regulations of maximum lead content of motor gasoline has been presented by JENSEN and GRANDJEAN (1984).

Already in the 1970s the USA reduced the gasoline concentrations drastically after the introduction of catalysts to avoid exhausts and smog.

According to CEC Directive 78/611/EEC the lead content of leaded gasoline distributed and consumed in the member states of the European Community must be in the range of 0.15–0.40 g/L. In some European countries the maximum admissible lead content of leaded gasoline has been set at 0.15 g/L (e.g., Austria, Denmark, Norway, Sweden, Switzerland, and the Federal Republic of Germany). In Switzerland and West Germany "normal" gasoline is sold only in the unleaded form.

7.3 Regulations Concerning Lead in Ambient Air

According to CEC Directive 82/884/EEC the maximum admissible concentration of lead in ambient air was set at 2.0 µg/m^3 (measured as elemental lead and averaged over one year). In Switzerland, the limit value was set at 1.0 µg/m^3. In the USA, the National Ambient Air Quality Standard for lead is 1.5 µg/m^3 (maximum arithmetic mean averaged over a calender quarter).

Governmental standards for lead fall-out have been issued in Switzerland (100 µg/m^2/day), West Germany (250 µg/m^2/day), and Yugoslavia (500 µg/m^2/day; 250 µg/m^2/day to protect specified ecosystems).

7.4 Regulations Concerning Lead in Soil, Compost, and Sewage Sludge

Governmental regulations of this kind are scarce. According to the Swiss Ordinance on Soil Contamination (1986) lead in agricultural soil that can be extracted with concentrated nitric acid should be below 50 mg Pb/kg dry weight. The water-soluble lead content should be below 1.0 mg/kg. A value of 100 ppm has been proposed as the maximum tolerable lead concentration in agricultural soil (KLOKE, 1980). In West Germany, sewage sludge applied to agricultural soil must not contain more than 1200 mg Pb/kg dry weight, and the soil to which sewage sludge is applied must not contain more than 100 mg Pb/kg (Ordinance on Sewage Sludge, 1982). In Switzerland, the maximum admissible level of lead in compost was set at 150 mg/kg dry weight (Ordinance on Substances Hazardous to the Environment). According to REAGAN and SILBERGELD (1989) soil levels greater than 50–150 mg/kg pose a significant risk to young children when swallowed.

7.5 Regulations Concerning Lead in Drinking Water

According to CEC Directive 80/778/EEC the maximum admissible lead concentration in drinking water is 50 µg/L. The same value has been adopted by the US Environmental Protection Agency and also has been recommended by WHO (WHO, 1984). The US EPA, however, has recently suggested to limit the contamination of water supplied by water systems to 5 µg/L (BRISKIN, 1989). If water levels are above 20 µg/L after corrosion control treatment, public must be warned. In the USA 42% of first flush water samples in schools contain still more than 20 µg/L of lead (ELIAS, 1989). Education is important.

7.6 Regulations and Recommendations Concerning Lead in Food and Beverages

In 1972, the Joint WHO/FAO Committee on Food Additives proposed a "provisional tolerable weekly intake" of 3 mg per person, corresponding to 0.05 mg Pb/kg body weight (WHO, 1972). It should be emphasized that this recommendation does not apply to infants and children.

Regulations and recommendations concerning lead in individual food and dietary products are scarce. In West Germany and Italy, the maximum admissible lead content in wine has been set at 0.3 mg/L. The Federal Health Agency of West Germany has promulgated guideline values for lead, mercury, and cadmium in individual food including vegetables, fruits, meat, milk and milk products, and beverages. The guideline values do not have a toxicological basis but they represent upper normal limits of the distribution of metal concentrations in dietary products presently found in West Germany (BGA, 1979, 1986; WEIGERT et al., 1984).

7.7 Control of Lead in the Workplace Area

The maximum allowable concentrations of lead and its inorganic compounds in workroom air range from $0.1-0.15$ mg/m^3 (as time-weighted average concentration during an eight-hour shift) in various countries. A WHO working group recommended a range of $0.03-0.06$ mg/m^3 (WHO, 1980). The occupational exposure limits for tetraethyllead and tetramethyllead usually vary between 0.075 and 0.15 mg/m^3 (JENSEN and GRANDJEAN, 1984).

In addition to the above exposure limits for lead in air, biological exposure limits, relating primarily to lead in blood, have been recommended. The West-German MAK commission recommends a biological exposure limit of 70 µg/100 mL for males and females >45 years, and a limit value of 30 µg/100 mL for women <45 years (MAK, 1988). According to CEC Directive 80/1107/EEC the maximum admissible PbB level in workers is 70 µg/100 mL. Higher levels (70–80 µg/100 mL) can be tolerated if ALA-U is below 20 mg/g creatinine and ZPP is below 20 µg/g Hb. A WHO working group (WHO, 1980) recommended a health-based biological exposure limit of 40 µg Pb/100 mL for males and females over the reproductive age. In females of the reproductive age the PbB level should be kept as low as possible. In order to protect the fetus PbB should not exceed 30 µg/100 mL. The control of lead exposure may also be based on the measurement of δ-aminolevulinic acid excretion in urine (ALA-U). For this parameter the following limits have been proposed: 15 mg/L, but 6 mg/L for women <45 years (MAK, 1988); 20 mg/g creatinine (CEC Directive 80/1107/EEC).

In many industrialized countries pre-employment and periodic medical examinations are obligatory for monitoring workers with occupational lead exposure.

7.8 Miscellaneous Regulations

In West Germany and several other countries the use of paints containing more than 1% lead (as soluble Pb) is prohibited for uses inside dwelling houses. In the UK the lead content of paints and related materials has been controlled since 1927 but there is no governmental regulation. According to a voluntary agreement between the Government and the Paintmakers' Association warning labels are put on tins of paint containing more than 1% total lead in the dry film.

References

Alexander, F.W., Delves, H.T., and Clayton, B.E. (1973), in: *Environmental Health Aspects of Lead*, pp. 319–331. Commission of European Communities, Directorate General for Dissemination of Knowledge, Center for Information and Documentation (CID), Luxembourg.
Angerer, J., and Schaller, K.H. (1985), *Lead, Analysis of Hazardous Substances in Biological Materials*, Vol. 1, pp. 155–164. VCH Verlagsgesellschaft, Weinheim-Deerfield Beach/Florida-Basel.
Angerer, J., and Schaller, K.H. (1988), *Lead, Analysis of Hazardous Substances in Biological Materials*, Vol. 2, pp. 183–193. VCH Verlagsgesellschaft, Weinheim-Basel-Cambridge-New York.
Annest, J.L. (1983) *Trends in the Blood Lead of the US Population. The Second National Health and Nutrition Examination Survey* (NHANES II) 1976–1980, in: Rutter, M., and Jones, R.R. (eds.): *Lead versus Health – Sources and Effects of Low Level Lead Exposure*, pp. 33–58. J. Wiley & Sons, Chichester-New York-Brisbane-Toronto-Singapore.
Annest, J.L., Pirkle, J.L., Makuc, D., Nesse, J.W., Bayse, D.D., and Kovar, M.G. (1983), *Chronological Trend in Blood Lead Levels Between 1976 and 1980*. N. Engl. J. Med. 308, 1373–1377.
Antonovics, J. (1971), *Heavy Metal Tolerance in Plants*. Adv. Ecol. Res. 7, 1–85.
Baccini, P. (1989), *The Control of Heavy Metal Fluxes from the Anthroposphere to the Environment. Seventh International Conference on Heavy Metals in the Environment*, Geneva. Proceedings, Vol. 1, pp. 13–23. CEP Consultants Ltd., Edinburgh.
Barry, P.S.I. (1978), Distribution and Storage of Lead in Human Tissues, in: Nriagu, J.O. (ed.): *The Biogeochemistry of Lead in the Environment*, Part B, pp. 97–150. Elsevier, Amsterdam-New York-Oxford.
Beeby, A., and Richmond, L. (1988), *Calcium Metabolism in Two Populations of the Snail Helix aspersa on a High Lead Diet*. Arch. Environ. Contam. Toxicol. 17, 507–511.
Bell, J.U., and Thomas, J.A. (1980), *Effects of Lead on Mammalian Reproduction*, in: Singhal, R.L., and Thomas, J.A. (eds.): *Lead Toxicity*, pp. 169–186. Urban & Schwarzenberg, Baltimore-München.
Berlin, A., Langevin, M., Wagner, H.M., Krause, C., and Yeoman, B. (1983), *Biological Monitoring and Environmental Health*. Wiss. Umwelt 1, 124–137.
BGA (Bundesgesundheitsamt) (1979), *Richtwerte '79 für Blei, Cadmium und Quecksilber in und auf Lebensmitteln*. Bundesgesundheitsblatt 22, 282–283.
BGA (Bundesgesundheitsamt) (1986), *Richtwerte '86 für Blei, Cadmium und Quecksilber in und auf Lebensmitteln*. Bundesgesundheitsblatt 29, 22–23.
Bornschein, R., Pearson, D., and Reiter, L. (1980), *Behavioral Effects of Moderate Lead Exposure in Children and Animal Models*. Part 2, *Animal Studies*. CRC Crit. Rev. Toxicol. 8, 101–152.
Boutron, C. (1982), *Atmospheric Trace Metals in the Snow Layers Deposited at the South Pole from 1928 to 1977*. Atmos. Environ. 16, 2451–2459.
Boyle, E.A. (1987), *Evolution of Anthropogenic Lead in the Ocean. Sixth International Conference on Heavy Metals in the Environment*, New Orleans, Proceedings Vol. 1, pp. 9–11. CEP Consultants Ltd., Edinburgh.
Briskin, J. (1989), *Goals and Implications of EPA's New Drinking Water Lead Standards. 23rd Annual Conference on Trace Substances in Environmental Health*, Cincinnati. Proceedings, J. Environ. Geochem. Health, in press.
Brockhaus, A., Freier, I., Ewers, U., Jermann, E., and Dolgner, R. (1983), *Levels of Cadmium and Lead in Blood in Relation to Smoking, Sex, Occupation, and Other Factors in an Adult Population of the FRG*. Int. Arch. Occup. Environ. Health 52, 167–175.
Brockhaus, A., Collet, W., Dolgner, R., Engelke, R., Ewers, U., Freier, I., Jermann, E., Krämer, U., Manojlovic, N., Turfeld, M., and Winneke, G. (1988), *Exposure to Lead and Cadmium of Children Living in Different Areas of North-West Germany: Results of Biological Monitoring Studies 1982–1986*. Int. Arch. Occup. Environ. Health 60, 211–222.
Bruland, K.W. (1983), *Trace Elements in Sea-water*. Chem. Oceanogr. 8, 182–185.

Brunekreef, B. (1984), *The Relationship Between Air Lead and Blood Lead in Children: A Critical Review. Sci. Total Environ. 38*, 79–123.
Brunekreef, B., Veenstra, S. J., Biersteker, K., and Boteij, J. S. (1981), *The Arnhem Lead Study. Lead Uptake by 1- to 3-year-old Children Living in the Vicinity of a Secondary Lead Smelter in Arnhem, The Netherlands. Environ. Res. 25*, 441–448.
Brunekreef, B., Noy, B., Biersteker, K., and Boteij, J. (1983), *Blood Lead Levels of Dutch City Children and Their Relationship to Lead in the Environment. J. Air Pollut. Control Assoc. 33*, 872–876.
Bull, R. J. (1980), *Lead and Energy Metabolism,* in: Singhal, R. L., and Thomas, J. A. (eds.): *Lead Toxicity,* pp. 119–168. Urban & Schwarzenberg, Baltimore-München.
Cassady, M. E. (1980), *State of the Art: Historical Perspective of Smelting. Am. J. Ind. Med. 1*, 265–282.
Cavallieri, A., Baruffini, A., Minoia, C., and Bianco, L. (1981), *Biological Response of Children to Low Levels of Inorganic Lead. Environ. Res. 25*, 415–423.
Chamberlain, A. C. (1983a), *Effect of Airborne Lead on Blood Lead. Atmos. Environ. 17*, 677–692.
Chamberlain, A. C. (1983b), *Fallout of Lead and Uptake by Crops. Atmos. Environ. 17*, 693–706.
Chester, R., Aston, S. R., Stoner, H. J., and Bruty, D. (1974), *J. Res. Atmos. 10*, 777–789.
Chisolm, J. J. (1980), *Lead and Other Metals: A Hypothesis of Interaction,* in: Singhal, R. L., and Thomas, J. A. (eds.): *Lead Toxicity,* pp. 461–482. Urban & Schwarzenberg, Baltimore-München.
Choie, D. D., and Richter, G. W. (1980), *Effects of Lead on the Kidney,* in: Singhal, R. L., and Thomas, J. A. (eds.): *Lead Toxicity,* pp. 187–121. Urban & Schwarzenberg, Baltimore-München.
Chow, T. J. (1978), *Lead in Natural Waters,* in: Nriagu, J. O. (ed.): *The Biogeochemistry of Lead in the Environment,* Part A, pp. 185–218. Elsevier, Amsterdam-New York-Oxford.
Claeys-Thoreau, F., Bruaux, P., Ducoffre, G., and Lafontaine, A. (1983), *Exposure to Lead of the Belgian Population. Int. Arch. Occup. Environ. Health 53*, 109–117.
Claeys-Thoreau, F., Thiessen, L., Bruaux, P., Ducoffre, G., and Verduyn, G. (1987), *Assessment and Comparison of Human Exposure to Lead Between Belgium, Malta, Mexico and Sweden. Int. Arch. Occup. Environ. Health 59*, 31–41.
Cooper, W. C. (1976), *Cancer Mortality Patterns in the Lead Industry. Ann. N.Y. Acad. Sci. 271*, 250–259.
Cooper, W. C. (1985), *Mortality Among Employees of Lead Battery Plants and Lead-producing Plants 1947–1980. Scand. J. Work. Environ. Health 11*, 331–345.
Copper, G. S., and Sigwart, C. D. (1980), *Neurophysiological Effects of Lead,* in: Singhal, R. L., and Thomas, J. A. (eds.): *Lead Toxicity,* pp. 401–423. Urban & Schwarzenberg, Baltimore-München.
Cornwall, J. C. (1987), *Migration of Metals in Sediment Pore Waters: Problems for the Interpretation of Historical Deposition Rates.* Sixth International Conference of Heavy Metals in the Environment, New Orleans. Proceedings Vol. 2, pp. 233–235. CEP Consultants Ltd., Edinburgh.
Cremer, J. E., and Callaway, S. (1981), *Further Studies on the Toxicity of Some Tetra- and Trialkyllead Compounds. Br. J. Ind. Med. 18*, 277–287.
Crössmann, G. (1989), *Effects of Atmospheric Trace Element Pollutions.* Sixth ENVITEC'89, Summaries of Papers. Nowea Messe, Düsseldorf; see for instance *Swiss Chem. 11* (7–8), 53, and *Chemosphere*, in press.
Davis, J. M., and Svendsgaard, D. J. (1987), *Lead and Child Development. Nature 329*, 297–300.
de Kort, W. L. A. M., Verschoor, M. A., Wibowo, A. A. E., and van Hemmen, J. J. (1986), *Occupational Exposure to Lead and Blood Pressure: A Study in 105 Workers. Am. J. Ind. Med. 11*, 145–156.
Deutsche Gesellschaft für Ernährung (1984), *Ernährungsbericht 1984,* Frankfurt/M.
DHSS (Department of Health and Social Security) (1980), *Lead and Health.* Report of a DHSS Working Party on Lead in the Environment. HMSO, London.
Doelman, P. (1978), *Lead and Terrestrial Microbiota,* in: Nriagu, J. O. (ed.): *The Biogeochemistry of Lead in the Environment,* Part B, pp. 343–354. Elsevier, Amsterdam-New York-Oxford.

Drasch, G. (1982), *Lead Burden in Prehistorical, Historical and Modern Human Bones. Sci. Total Environ.* 24, 199–231.

Drasch, G.A., and Ott, J. (1988), *Lead in Human Bones, Investigations on an Occupationally Non-exposed Population in Southern Bavaria (FRG), II. Children. Sci. Total Environ.* 68, 61–69.

Drasch, G.A., von Meyer, L., and Kauert, G. (1987), *A Comparison of the Lead Concentrations of Human Bones 1974–1984. 2nd Nordic Symposium on Trace Elements in Human Health and Disease, Odense.* Abstract, p. G5. WHO Regional Office for Europe, Copenhagen; and: Drasch, G.A., Böhm, J., and Baur, C. (1987), *Lead in Human Bones, Investigations on an Occupationally Non-exposed Population in Southern Bavaria (FRG), I. Adults. Sci. Total Environ.* 64, 303–315.

Duggen, J.M. (1983), *Contribution of Lead in Dust to Children's Blood Lead. Environ. Health Perspec.* 50, 371–381.

Eckard, R., and Bertram, H.P. (1987), *Lead in Wine – Toxicological Aspects. Trace Elem. Med.* 4, 1–3.

Elias, R.W. (1989), *Monitoring Soils and Restoring them. 23rd Annual Conference on Trace Substances in Environmental Health, Cincinnati. Proceedings, J. Environ. Geochem. Health,* in press.

Elinder, C.G., Friberg, L., Lind, B., and Jawaid, M. (1983), *Lead and Cadmium Levels in Blood Samples from the General Population of Sweden. Environ. Res.* 30, 233–253.

Englert, N., Krause, C., Thron, H.L., and Wagner, M. (1986), *Untersuchungen zur Bleibelastung ausgewählter Bevölkerungsgruppen in Berlin (West). Bundesgesundheitsblatt* 29, 322–326.

Ernhart, C.B., Wolf, A.W., Kennard, M.J., Erhard, P., Filipovich, H.F., and Sokol, R.J. (1986), *Intrauterine Exposure to Low Levels of Lead: The Status of the Neonate. Arch Environ. Health* 41, 287–291.

Ewers, U., Brockhaus, A., Winneke, G., Freier, I., Jermann, E., and Krämer, U. (1982), *Lead in Deciduous Teeth of Children Living in a Non-ferrous Smelter Area and a Rural Area of the FRG. Int. Arch. Occup. Environ. Health* 50, 139–151.

Facchetti, S., and Geiss, F. (1982), *The Lead Isotopic Experiment.* Status Report, Commission of the European Communities, EUR 8352 EN. Brussels-Luxembourg.

Fergusson, J.E., and Purchase, N.G. (1987), *The Analysis and Levels of Lead in Human Teeth: A Review. Environ. Pollut.* 46, 11–44.

Forbes, R.M., and Sanderson, G.C. (1978), *Lead Toxicity in Domestic Animals and Wildlife,* in: Nriagu, J.O. (ed.): *The Biogeochemistry of Lead in the Environment,* Part B, pp. 225–278. Elsevier, Amsterdam-New York-Oxford.

Friberg, L., and Vahter, M. (1983), *Assessment of Exposure to Lead and Cadmium Through Biological Monitoring: Results of a UNEP/WHO Global Study. Environ. Res.* 30, 95–128.

Fulton, M. (1989), *Lead Exposure and Child Development – Some Methodological Issues. Seventh International Conference on Heavy Metals in the Environment, Geneva. Proceedings,* Vol. 1, pp. 94–102. CEP Consultants Ltd., Edinburgh.

Gallacher, J.E.J., Elwood, P.C., Phillips, K.M., Davies, B.E., Ginnever, R.C., Toothill, C., and Jones, D.T. (1984), *Vegetable Consumption and Blood Lead Concentrations. J. Epidemiol. Commun. Health* 38, 173–176.

Gochfeld, M., and Burger, J. (1988), *Effects of Lead on Growth and Feeding Behavior of Young Common Terns. Arch. Environ. Contam. Toxicol.* 17, 513–517.

Grandjean, P. (1984), Organolead Exposures and Intoxications, in: Grandjean, P. (ed.): *Biological Effects of Organolead Compounds,* pp. 227–242. CRC Press, Boca Raton, Florida.

Grandjean, P., Jensen, B.M., Sandø, S.H., Jørgensen, P.J., and Antonson, S. (1989), *Delayed Blood Regeneration after Phlebotomy in Lead-exposed Workers. V International Congress of Toxicology. Proceedings,* in press, Taylor & Francis Ltd., London.

Gunadilaka, A. (1988), *Accumulation from Pelagic and Benthic Foodwebs. International Conference on Trace Metals in Lakes, Hamilton (Canada). Proceedings, Sci. Total Environ.,* in press.

Gunn, A. M., Winnard, D. A., and Hunt, D. T. E. (1988), *Trace Metal Speciation in Sediments and Soils. Proceedings of a Workshop on Metal Speciation, Jekyll Island, Georgia,* pp. 261–294. Lewis Publishers, Chelsea, Michigan.

Hernberg, S. (1980), *Biochemical and Clinical Effects and Responses as Indicated by Blood Concentration,* in: Singhal, R. L., and Thomas, J. A. (eds.): *Lead Toxicity,* pp. 367–400. Urban & Schwarzenberg, Baltimore-München.

Hertz, J., and Angehrn-Bertinazzi, C. (1988), *Distribution and Speciation of Heavy Metals in Various Litter Horizons and Forest Soils. Proceedings of a IAEAC Soil Residue Analysis Workshop, Winnipeg (Canada). J. Toxicol. Environ. Chem.,* in press.

Holtzman, R. B. (1978), *Application of Radiolead to Metabolic Studies,* in: Nriagu, J. O. (ed.): *The Biogeochemistry of Lead in the Environment,* Part B, pp. 37–96. Elsevier, Amsterdam-New York-Oxford.

Howe, H. E. (1981), in: Kirk-Othmer, *Encyclopedia of Chemical Technology,* 3rd Ed., Vol. 14, pp. 98–139. J. Wiley & Sons, New York-Chichester-Brisbane-Toronto.

Hrdina, P. D., Hanin, I., and Dubas, T. C. (1980), *Neurochemical Correlates of Lead Toxicity,* in: Singhal, R. L., and Thomas, J. A. (eds.): *Lead Toxicity,* pp. 273–300. Urban & Schwarzenberg, Baltimore-München.

Huel, G., Boudene, C., Jouan, M., and Lazar, P. (1986), *Assessment of Exposure to Lead of the General Population in the French Community Through Biological Monitoring. Int. Arch. Occup. Environ. Health 58,* 131–139.

Hunter, J. (1986), in: Lansdown, R., and Yule, W. (eds.): *The Distribution of Lead. The Lead Debate, The Environment, Toxicology and Child Health,* pp. 96–130. Croom Helm, London-Sydney.

Huntzicker, J. J., Friedlander, S. K., and Davidson, C. J. (1975), *Material Balance for Automobile-emitted Lead in Los Angeles Basin. Environ. Sci. Technol. 9,* 448–457.

Huynh-Ngoc, L., Whitehead, N. E., and Oregoni, B. (1988), *Low Levels of Copper and Lead in a Highly Industrialized River. J. Toxicol Environ. Chem. 17,* 223–236.

IARC (International Agency for Research on Cancer) (1980), *IARC Monographs on the Evaluation of the Carcinogenic Risk of Chemicals to Humans,* Vol. 23, pp. 325–415, Lyon (France).

Jason, K. M., and Kellog, C. K. (1980), *Behavioral Neurotoxicity of Lead,* in: Singhal, R. L., and Thomas, J. A. (eds.): *Lead Toxicity,* pp. 241–272. Urban & Schwarzenberg, Baltimore-München.

Jensen, A. A. (1984), *Metabolism and Toxicokinetics,* in: Grandjean, P. (ed.): *Biological Effects of Organolead Compounds,* pp. 97–115. CRC Press, Boca Raton, Florida.

Jensen, A. A., and Grandjean, P. (1984), *Governmental Regulations,* in: Grandjean, P. (ed.): *Biological Effects of Organolead Compounds,* pp. 259–266. CRC Press, Boca Raton, Florida.

Kacew, S., and Singhal, R. L. (1980), *Aspects of Molecular Mechanisms Underlying the Biochemical Toxicology of Lead,* in: Singhal, R. L., and Thomas, J. A. (eds.): *Lead Toxicity,* pp. 43–78. Urban & Schwarzenberg, Baltimore-München.

Kazantzis, G. (1989), *Evaluation of the Evidence for Lead Carcinogenicity. 23rd Annual Conference on Trace Substances in Environmental Health, Cincinnati. Proceedings, J. Environ. Geochem. Health,* in press.

Kehoe, R. (1985), in: *Encyclopaedia of Occupational Health and Safety,* 3rd Ed., pp. 1197–1199. International Labour Office, Geneva.

Kesler, S. E. (1978), *Economic Lead Deposits,* in: Nriagu, J. O. (ed.): *The Biogeochemistry of Lead in the Environment,* Part A, pp. 73–98. Elsevier, Amsterdam-New York-Oxford.

Kirkby, H., and Gyntelberg, F. (1985), *Blood Pressure and Other Cardiovascular Risk Factors of Long-term Exposure to Lead. Scand. J. Work. Environ. Health 11,* 15–19.

Klockow, D., and Larjava, K. (1988), *Trace Metal Deposition to Forest Canopies. Proceedings of the 3rd IAEAC Workshop on Toxic Metal Compounds, Follonica (Italy). J. Toxicol. Environ. Chem.,* in press.

Kloke, A. (1980), *Orientierungsdaten für tolerierbare Gesamtgehalte von Metallen in Kulturböden. Mitteilungen des Verbandes Deutscher Landwirtschaftlicher Untersuchungs- und Forschungsanstalten (VDLUFA) 1–3,* 11–13.

Kneip, T. J., and Friberg, L. (1986), *Sampling and Analytical Methods,* in: Friberg, L., Nordberg, G. F., and Vouk, V. B. (eds.): *Handbook on the Toxicology of Metals,* Vol. I, pp. 36 – 67. Elsevier, Amsterdam-New York-Oxford.

Koeppe, D. E. (1981), in: Lepp, N. W. (ed.): *Effects of Heavy Metals in Plants,* Vol. I, pp. 55 – 76. Applied Science Publ., Barking (UK).

Landrigan, P. J., and Baker, E. L. (1981), *Exposure of Children to Heavy Metals from Smelters: Epidemiology and Toxic Consequences. Environ. Res. 25,* 204 – 224.

Lansdown, R. (1986), *Lead, Intelligence, Attainment and Behaviour,* in: Lansdown, R., and Yule, W. (eds.): *The Lead Debate,* pp. 235 – 270. Croom Helm, London-Sydney.

Lauwerys, R., Buchet, J. P., Roels, H., and Hubermont, G. (1978), *Placental Transfer of Lead, Mercury, Cadmium and Carbon Monoxide in Women. Environ. Res. 15,* 278 – 289.

Lehnert, G., and Szadkowski, D. (1983), *Die Bleibelastung des Menschen.* Verlag Chemie, Weinheim-Deerfield Beach/Florida-Basel.

Lin-Fu, J. (1980), *Lead Poisoning and Undue Lead Exposure in Children: History, Current Status,* in: Needleman, H. L. (ed.): *Low Level Lead Exposure – The Clinical Implications of Current Research,* pp. 5 – 16. Raven Press, New York.

MAFF (Ministry of Agriculture, Fisheries and Food) (1975), *Survey of Lead in Food.* Working Party on the Monitoring of Foodstuffs for Heavy Metals, Fifth Report. HMSO, London.

MAFF (Ministry of Agriculture, Fisheries and Food) (1983), *Food Additives and Contaminants,* Committee Report on the Review of Metals in Canned Food. HMSO, London.

MAGS (Ministerium für Arbeit, Gesundheit und Soziales des Landes Nordrhein-Westfalen) (1983), *Umweltprobleme durch Schwermetalle im Raum Stolberg 1983,* Düsseldorf.

Mahaffey, K. R. (1978), *Environmental Exposure to Lead,* in: Nriagu, J. O. (ed.): *The Biogeochemistry of Lead in the Environment,* Part B, pp. 1 – 36. Elsevier, Amsterdam-New York-Oxford.

Mahaffey, K. R. (1980), *Nutrient-Lead Interactions,* in: Singhal, R. L., and Thomas, J. A. (eds.): *Lead Toxicity,* pp. 425 – 460. Urban & Schwarzenberg, Baltimore-München.

Mahaffey, K. R., Annest, J. L., Roberts, J., and Murphy, R. S. (1982), *National Estimates of Blood Lead Levels: United States 1976 – 1980. N. Engl. J. Med. 307,* 573 – 579.

MAK (1988), *Maximum Concentrations at the Workplace, Report No. XXIV, DFG.* VCH Verlagsgesellschaft, Weinheim-Basel-Cambridge-New York.

Mart, L. (1988), *Electrochemical Ultratrace Detection of Metals in Unpolluted Water and Snow.* Proceedings of the 3rd IAEAC Workshop on Toxic Metal Compounds, Follonica (Italy). *J. Toxicol. Environ. Chem.,* in press.

McCormack, W. B., Moore, R. and Sandy, C. A. (1981), in: Kirk-Othmer: *Encyclopedia of Chemical Technology,* 3rd, Ed., Vol. 14, pp. 180 – 195. J. Wiley & Sons, New York-Chichester-Brisbane-Toronto.

McMichael, A. J., Vimpani, G. V., Robertson, E. F., Baghurst, P. A., and Clark, P. D. (1986), *The Port Pirie Cohort Study: Maternal Blood Lead and Pregnancy Outcome. J. Epidemiol. Commun. Health 40,* 18 – 25.

Melin, A. (1974), *Blei,* in: *Ullmanns Encyklopädie der technischen Chemie,* 4th Ed., Vol. 8, pp. 542 – 583. Verlag Chemie, Weinheim.

Michaelson, A. E. (1980), *An Appraisal of Rodent Studies on the Behavioral Toxicity of Lead: The Role of Nutritional Status,* in: Singhal, R. L., and Thomas, J. A. (eds.): *Lead Toxicity,* pp. 301 – 366. Urban & Schwarzenberg, Baltimore-München.

Montaser, A., and Golightly, D. W. (eds.) (1987), *Inductively Coupled Plasmas in Analytical Atomic Spectrometry.* VCH Verlagsgesellschaft, Weinheim-Basel-Cambridge-New York.

Moore, M. R. (1986a), *Lead in Humans,* in: Landsdown, R., and Yule, W. (eds.): *The Lead Debate – The Environment, Toxicology and Child Health,* pp. 54 – 95. Croom Helm, London-Sydney.

Moore, M. R. (1986b), *Lead in Soils,* in: Lansdown, R., and Yule, W. (eds.): *The Lead Debate – The Environment, Toxicology and Child Health,* pp. 131 – 189. Croom Helm, London-Sydney.

Moreau, T., Orssaud, G., and Juguet, B. (1982), *Plombémie et Pression Artérielle: Premiers Résultats d'une Enquête Transversale de 431 Sujets de Sexe Masculin* (blood lead levels and arterial pressure: initial results of a cross-sectional study of 431 male subjects). *Rev. Epidemiol. Santé Publ. 30*, 395–397.

Morton, D. E., Saah, A. J., Silberg, S. L., Owens, W. L., Roberts, M. A., and Saah, M. D. (1982), *Lead Absorption in Children of Employees in a Lead-related Industry. Am. J. Epidemiol. 115*, 549–555.

Murozomi, M., Chow, T. J., and Patterson, C. (1969), *Chemical Concentrations of Pollutant Lead Aerosols, Terrestrial Dusts and Sea Salts in Greenland and Antarctic Snow Strata. Geochim. Cosmochim. Acta 33*, 1247–1294.

NAS (National Academy of Sciences) (1972), *Biological Effects of Atmospheric Pollutants – Airborne Lead in Perspective*. Washington, DC.

NAS (National Academy of Sciences) (1980), *Lead in the Human Environment*. Washington, DC.

Needleman, H. L. (1987), *Low Level Lead Exposure and Children's Intelligence: A Quantitative and Critical Review of Modern Studies. Sixth International Conference on Heavy Metals in the Environment, New Orleans. Proceedings,* Vol. 1, pp. 1–8. CEP Consultants Ltd., Edinburgh.

Needleman, H. L., Rabinowitz, M., Leviton, A., Linn, S., and Schoenbaum, S. (1984), *The Relationship Between Prenatal Exposure to Lead and Congenital Anomalies. J. Am. Med. Assoc. 251*, 2956–2959.

Nielsen, T. (1984), *Atmospheric Occurrence of Organolead Compounds,* in: Grandjean, P. (ed.): *Biological Effects of Organolead Compounds,* pp. 43–62. CRC Press, Boca Raton, Florida.

Nordström, S., Beckman, L, and Nordenson, I. (1978), *Occupational and Environmental Risks in and around a Smelter in Northern Sweden: I. Variations in Birth Weight. Hereditas 88,* 43–46.

Nordström, S., Beckman, L., and Nordenson, I. (1979), *Occupational and Environmental Risks in and around a Smelter in Northern Sweden: V. Spontaneous Abortions among Female Employees and Decreased Birth in their Offspring. Hereditas 90,* 291–296.

NRCC (National Research Council Canada) (1978), *Effects of Lead in the Environment* –1978. NRCC/CNRC Publ. No. 16736, Ottawa.

Nriagu, J. O. (1978a), *Lead in Soils, Sediments and Major Rock Types,* in: Nriagu, J. O. (ed.): *The Biogeochemistry of Lead in the Environment,* Part A, pp. 15–72. Elsevier, Amsterdam-New York-Oxford.

Nriagu, J. O. (1978b), *Lead in the Atmosphere,* in: Nriagu, J. O. (ed.): *The Biogeochemistry of Lead in the Environment,* Part A, pp. 137–184. Elsevier, Amsterdam-New York-Oxford.

Nriagu, J. O. (1989), *The History of Leaded Gasoline. Seventh International Conference on Heavy Metals in the Environment, Geneva. Proceedings,* Vol. 2, pp. 361–365. CEP Consultants Ltd., Edinburgh.

Nürnberg, H. W., and Mart, L. (1985), *Distribution and Fate of Heavy Metals in the Ocean. Fifth International Conference on Heavy Metals in the Environment, Athens. Proceedings,* Vol. 2, pp. 340–342. CEP Consultants Ltd., Edinburgh.

Paterson, L. J., Raab, G. M., Hunter, R., Laxen, D. P. H., Fulton, N., Fell, G. S., Halls, D. J., and Sutcliffe, P. (1988), *Factors Influencing Lead Concentrations in Shed Deciduous Teeth. Sci. Total Environ. 74*, 219–233.

Peterson, P. J. (1978), *Lead and Vegetation,* in: Nriagu, J. O. (ed.): *The Biogeochemistry of Lead in the Environment,* Part B, pp. 355–384. Elsevier, Amsterdam-New York-Oxford.

Piomelli, S., Corash, L., Seaman, M. B., Mushak, P., Glover, B., and Padgett, R. (1980), *Blood Lead Concentrations in a Remote Himalayan Population. Science 210,* 1135–1137.

Piotrowski, J. K., and O'Brien, B. J. (1980), *Analysis of the Effects of Lead in Tissue upon Human Health Using Dose-Response Relationships*. MARC Reports 16–18. Monitoring and Assessment Research Centre, London.

Pirkle, J. L., Schwartz, J., Landis, J. R., and Harlan, W. R. (1985), *The Relationship between Blood Lead Levels and Blood Pressure and its Cardiovascular Risk Implication. Am. J. Epidemiol. 121,* 246–258.

Pocock, S. J., Shaper, A. G., Ashby, D., Delves, T., and Whitehead, T. P. (1984), *Blood Lead Concentration, Blood Pressure, and Renal Function. Br. Med. J. 289,* 872–874.

Posner, H. S., Damstra, T., and Nriagu, J. O. (1978), *Human Health Effects of Lead,* in: Nriagu, J. O. (ed.): *The Biogeochemistry of Lead in the Environment,* Part B, pp. 173–224. Elsevier, Amsterdam-New York-Oxford.

Price, J., Baddeley, H., Kenardy, J. A., Thomas, B. J., and Thomas, B. W. (1984), *In vivo X-Ray Fluorescence Estimation of Bone Lead Concentrations in Queensland Adults. Br. J. Radiol. 57,* 29–33.

Quinn, M. J. (1985), *Factors Affecting Blood Lead Concentrations in the UK: Results of the EEC Blood Lead Surveys, 1979–1981. Int. J. Epidemiol. 14,* 420–431.

Radermacher, L., Prinz, G., and Rudoph, H. (1986), *Bericht über die Ergebnisse der im Lande Nordrhein-Westfalen in der Zeit von Januar bis Dezember 1985 durchgeführten Staub- und Schwermetallniederschlagsmessungen. Schriftenr. Landesanst. Immissionsschutz Landes Nordrhein-Westfalen Essen 64,* 7–69.

Reagan, P. L., and Silbergeld, E. K. (1989), *Establishing a Health-based Standard for Lead in Residential Soils. 23rd Annual Conference on Trace Substances in Environmental Health, Cincinnati. Proceedings, J. Environ. Geochem. Health,* in press.

Robinson, I. M. (1978), *Lead as a Factor in the World Economy,* in: Nriagu, J. O. (ed.): *The Biogeochemistry of Lead in the Environment,* Part A, pp. 99–118. Elsevier, Amsterdam-New York-Oxford.

Röderer, G. P. (1984), *Toxic Effects in Plant Organisms,* in: Grandjean, P. (ed.): *Biological Effects of Organo-lead Compounds,* pp. 63–96. CRC Press, Boca Raton, Florida.

Roels, H. A., Buchet, J. P., Lauwerys, R. R., Bruaux, P., Claeys-Thoreau, F., Lafontaine, A., and Verduyn, G. (1980), *Exposure to Lead by the Oral and the Pulmonary Routes of Children Living in the Vicinity of a Primary Lead Smelter. Environ. Res. 22,* 81–94.

Rosen, J. F., and Sorell, M. (1978), *The Metabolism and Subclinical Effects of Lead in Children,* in: Nriagu, J. O. (ed.): *The Biogeochemistry of Lead in the Environment,* Part B, pp. 151–168. Elsevier, Amsterdam-New York-Oxford.

Russell, R. R., and Stephens, R. (1983), *The Contribution of Lead in Petrol to Human Lead Intake,* in: Rutter, M., and Jones, R. R. (eds.): *Lead versus Health – Sources and Effects of Low Level Lead Exposure,* pp. 141–177. John Wiley & Sons, Chichester-New York-Brisbane-Toronto-Singapore.

Rutter, M. (1983), *Low Level Lead Exposure: Sources, Effects and Implications,* in: Rutter, M., and Jones, R. R. (eds.): *Lead versus Health – Sources and Effects of Low Level Lead Exposure,* pp. 333–370. John Wiley & Sons, Chichester-New York-Brisbane-Toronto-Singapore.

Saager, R. (1984), *Encyclopedia of Metallic Raw Materials* (in German), pp. 73–77. Bank von Tobel, Zürich.

Sabbioni, E., Goetz, L., and Bignoli, G. (1984), *Health and Environmental Implications of Trace Metals Released from Coal-fired Power Plants: An Assessment Study of the Situation in the European Community. Sci. Total Environ. 40,* 141–154.

Schmidt, E. H. F., and Hildebrandt, A. G. (eds.) (1983) *Health Evaluation of Heavy Metals in Infant Formula and Junior Food.* Springer Verlag, Berlin-Heidelberg-New York.

Selevan, S., Landrigan, P. J., Stern, F. B., and Jones, J. H. (1984), *Mortality of Lead Smelter Workers. Am. J. Epidemiol. 122,* 673–683.

Settle, D. M., and Patterson, C. C. (1980), *Lead in Albacore: Guide to Lead Pollution in Americans. Science 207,* 1167–1176.

Shapiro, I. M., Burke, A., Mitchell, G., and Bloch, P. (1978), *X-Ray Fluorescence Analysis of Lead in Teeth of Urban Children in situ: Correlation between the Tooth Lead Level and the Concentration of Blood Lead and Free Erythroporphyrins. Environ. Res. 17,* 46–52.

Sheffet, A., Thind, I., Miller, A. M., and Louria, D. B. (1982), *Cancer Mortality in a Pigment Plant Utilizing Lead and Zinc Chromates. Arch. Environ. Health 37,* 44–52.

Silbergeld, E. K. (1983), *Experimental Studies of Lead Neurotoxicity: Implications for Mechanisms, Dose-response and Reversibility,* in: Rutter, M., and Jones, R. R. (eds.): *Lead versus Health – Sources and Effects of Low Level Lead Exposure,* pp. 191–248. John Wiley & Sons, Chichester-New York-Brisbane-Toronto-Singapore.

Smith, R. A. H., and Bradshaw, A. D. (1972), *Trans. Inst. Min. Metall. B 81,* A230–237.
Smith, M. A., Grant, L. D., and Sors, A. I. (eds.) (1989), *Lead Exposure and Child Development, an International Assessment, published for the Commission of the European Communities and the US Environmental Protection Agency.* Kluwer Academic Publishers, Dortmund-Boston-London.
Steenhout, A. (1982), *Kinetics of Lead Storage in Teeth and Bones: An Epidemiologic Approach.* Arch. Environ. Health 37, 224–231.
Steenhout, A., and Pourtois, M. (1981), *Lead Accumulation in Teeth as a Function of Age with Different Exposures.* Br. J. Ind. Med. 38, 297–303.
Steinnes, E., Bølviken, B., and Hvatum, O. Ø. (1987), *Regional Differences and Temporal Trends in Heavy Metal Deposition Studied by Analysis of Different Soil Horizons and Ombrotrophic Peat.* Sixth International Conference on Heavy Metals in the Environment, New Orleans. Proceedings Vol. 1, pp. 201–203, 291–293. CEP Consultants Ltd., Edinburgh.
Strehlow, C. D., and Barltrop, D. (1989), *Impact of Reduced Petrol Lead Emissions on Diet-Pb, Dust-Pb and Blood-Pb in the United Kingdom.* 23rd Annual Conference on Trace Substances in Environmental Health, Cincinnati. Proceedings, J. Environ. Geochem. Health, in press.
Task Group on Metal Accumulation (1973), *Accumulation of Toxic Metals with Special Reference to Their Absorption, Excretion and Half-times.* Environ. Physiol. Biochem. 3, 65–107.
Task Group of Metal Toxicity (1976), in: Nordberg, G. F. (ed.): *Effects and Dose-Response Relationships of Toxic Metals.* Elsevier, Amsterdam-New York-Oxford.
Ter Haar, G. L., Lenane, D. L., Hu, J. N., and Brandt, M. (1972), *Composition, Size and Control of Automotive Exhaust Particulates.* J. Air Pollut. Control Assoc. 22, 39–46.
Tsuchiya, K. (1986), *Lead,* in: Friberg, L., Nordberg, G. F., and Vouk, V. B. (eds.): *Handbook on the Toxicology of Metals,* Vol. II, pp. 298–353. Elsevier, Amsterdam-New York-Oxford.
Urbanowicz, M. A. (1986), *The Uses of Lead Today,* in: Landsdown, R., and Yule, W., (eds.): *The Lead Debate: The Environment, Toxicology and Child Health,* pp. 25–40. Croom Helm, London-Sidney.
US EPA (U.S. Environmental Protection Agency) (1986), *Air Quality Criteria for Lead,* Vols. I–IV. EPA-600/8–83/028, Triangle Park, North Carolina.
Vahter, M. (ed.) (1982), *Assessment of Human Exposure to Lead and Cadmium through Biological Monitoring.* National Swedish Institute of Environmental Medicine and Department of Environmental Hygiene, Karolinska Institute, Stockholm.
Valciuskas, J. A., Lilis, R., Eisinger, J., Blumberg, W. E., Fischbein, A., and Selikoff, I. J. (1978), *Behavioral Indicators of Lead Neurotoxicity: Results of a Clinical Survey.* Int. Arch. Occup. Environ. Health 41, 217–236.
Valenta, P., Nguyen, V. D., and Nürnberg, H. W. (1986), *Acid and Heavy Metal Pollution by Wet Deposition.* Sci. Total Environ. 55, 311–320; and in: Georgii, H. W. (ed.): *Atmospheric Pollutants in Forest Areas,* pp. 69–78. D. Reidel, Dordrecht.
Watanabe, T., Fujita, H., Koizumi, A., Chiba, K., Miyasaka, M., and Ikeda, M. (1985), *Baseline Level of Blood Lead Concentration among Japanese Farmers.* Arch. Environ. Health 40, 170–176.
Weigert, P., Müller, J., Klein, H., Zufelde, K. P., and Hillebrand, J. (1984), *Arsen, Blei, Cadmium und Quecksilber in und auf Lebensmitteln.* Zentrale Erfassungs- und Bewertungsstelle für Umweltchemikalien des Bundesgesundheitsamtes (ZEBS), ZEBS-Hefte 1/1984, Berlin.
Weiss, S. T., Munoz, A., Stein, A., Sparrow, D., and Speizer, F. E. (1986), *The Relationship of Blood Lead to Blood Pressure in a Longitudinal Study of Working Men.* Am. J. Epidemiol. 123, 800–808.
Welz, B. (1985), *Atomic Absorption Spectrometry,* 2nd Ed., pp. 294–295, 353 ff. Verlag Chemie, Weinheim-Deerfield Beach/Florida-Basel.
Wibowo, A. A. E., Herber, R. F. M., Das, H. A., Roeleveld, N., and Zielhuis, R. L. (1986), *Levels of Metals in Hair of Young Children as an Indicator of Environmental Pollution.* Environ. Res. 40, 346–356.

Winneke, G. (1985), *Blei in der Umwelt – Ökopsychologische und psychotoxikologische Aspekte.* Springer Verlag, Berlin-Heidelberg-New York-Tokyo.
Winneke, G. (1986), *Animals Studies,* in: Lansdown, R., and Yule, W. (eds.): *The Lead Debate – The Environment, Toxicology and Child Health,* pp. 217–234. Croom Helm, London-Sydney.
WHO (World Health Organization) (1972), *Evaluation of Certain Food Additives and the Contaminants Mercury, Lead, and Cadmium.* 16th Report of the Joint FAO/WHO Expert Committee on Food Additives. *Tech. Rep. Ser. 505,* Geneva.
WHO (World Health Organization) (1977), *Environmental Health Criteria 3: Lead.*
WHO (World Health Organization) (1980), *Recommended Health-based Limits in Occupational Exposure to Heavy Metals. Tech. Rep. Ser. 647,* Geneva.
WHO (World Health Organization) (1984), *Guidelines for Drinking Water Quality,* Vol. 1.
WHO Euro (WHO Regional Office for Europe) (1981), *Quality Control in the Occupational Toxicology Laboratory. Health Aspects of Chemical Safety.* Interim Document 4, Copenhagen.
Wong, P. T. S., Silverberg, B. A., Chau, Y. K., and Hodson, P. V. (1978), *Lead and the Aquatic Biota,* in: Nriagu, J. O. (ed.): *The Biogeochemistry of Lead in the Environment,* Part B, pp. 279–342. Elsevier, Amsterdam-New York-Oxford.
World Metal Statistics (1983), World Bureau of Metal Statistics, London.
Zelikoff, J. T., Li, H. H., Hartwig, A., Wang, X. W., Costa, M., and Rossman, T. G. (1988), *Genetic Toxicology of Lead Compounds. Carcinogenesis 9*(10), 1727–1732.
Zielhuis, R. L. (1975), *Dose-Response Relationships for Inorganic Lead. Int. Arch. Occup. Environ. Health 35,* 1–18.

Additional Recommended Literature

Alegria, A., Barbaera, R., Boluda, R., Errecalde, F., Farré, R., and Lagarda, M. J. (1990), *Environmental Cadmium, Lead and Nickel Contamination Possible Relationship between Soil and Nickel Contamination Possible Relationship between Soil and Vegetable Content. Fresenius Z. Anal. Chem.,* in press.
Chau, Y. K., and Wong, P. T. S. (1990), *Recent Developments in Speciation and Determination of Organometallic Compounds in Environmental Samples. Fresenius Z. Anal. Chem.,* in press.
van Cleuvenbergen, R. J. A., and Adams, F. C. (1990), *Organolead Compounds,* in: *Handbook of Environmental Chemistry* (ed. by O. Hutzinger), Vol. 3, Part E, *Anthropogenic Compounds,* pp. 97–153. Springer Verlag, Berlin-Heidelberg-New York-London-Paris-Tokyo-Hong Kong.
Houriet, J.-P. (1989), *Heavy Metals in the Sediments of Swiss Rivers* (in German). *BUWAL Bulletin 4/89,* 31–43.
Neite, H., and Wittig, R. (1990), *Studies on the Solubility of Heavy Metal Compounds in Forest Soils and Effects on Forest Vegetation. VDI Proceedings of the Lindau Conference on Effects of Atmospheric Pollution on Soils,* in press.
Streit, B., Krüger, C., Lahner, G., Kirsch, S., Hauser, G., and Diehl, B. (1990), *Uptake and Storage of Heavy Metal Compounds by Earthworms in Various Soils* (in German), *UWSF. Z. Umweltchem. Oekotoxikol. 2*(1), 10–13. ecomed Verlagsgesellschaft, Landsberg/Lech.
Trueb, L. (1990), *Lead from Limestone: the Lead Belt in Missouri. Neue Zürcher Zeitung, Forschung und Technik No. 31,* p. 65 (7 February), Zürich.
Yakowitz, H. (1989), *Encouraging Waste Reduction/Minimization and Recycling: A Policy Oriented Overview* (dealing particularly with Lead Recycling). *Proceedings IPRE,* Brussels.

II.17 Lithium

BARTHOLOMÉ RIBAS, Madrid, Spain

1 Introduction

The element lithium has found practical uses in medicine and, in the form of its isotope ^{6}Li, as a thermonuclear fuel in both civilian and military applications. In nature, lithium is generally found in the form of silicates, and its average concentration in the lithosphere is 30 µg/g. In biological systems the cation is very stable and cannot participate directly in redox reactions, but it does react with cyclic polyethers, polypeptides, and proteins. Lithium coordinates more water molecules than any of the other elements in Group Ia (Li, Na, K, Rb, Cs, and Fr) of the Periodic Table. Epidemiological studies have shown a correlation between the absence of lithium in drinking water and a higher incidence of mental disorders. Lithium carbonate may control neurotransmitter transport in the brain and thus causes antidepressive effects (TRUEB, 1989). In biological systems lithium interacts with sodium, potassium, and magnesium. It is not accumulated during ontogeny and does not seem to constitute an environmental problem at present.

2 Physical and Chemical Properties, Analytical Methods

Lithium (from lithos, meaning stone) is a metallic element, the smallest of the alkali ions, and was discovered in 1817 by Arfredson while working in Berzelius' laboratory. Its atomic mass is 6.94 and the distance between the two lithium nuclei in Li$_2$ is 0.267 nm. The atomic radius of lithium in Li$_2$ is 0.1340 nm, but only 0.123 nm when bound to other atomic centers; the mean ionic radius is 0.068 nm, and the electronic configuration is 1 s^2, 2 s^1. The most interesting property of lithium is its extremely low density: 0.534 g/cm^3. Lithium is the hardest and has the highest melting point of all alkali metals (m.p. 179 °C, b.p. 1340 °C). A carmine red color results when lithium is held in a flame. The two naturally occurring isotopes are ^{6}Li (8%) and ^{7}Li (92%), used in their enriched form by THELLIER et al. (1983) in lithium determination, distribution and metabolism studies. Three other nuclides have been produced artificially, but they have half-lives of less than 0.8 seconds.

The alkali metals with an external electronic configuration of s^2 p^6 are stable as cations but do not form stable complexes. They interact with macromolecular ligands such as receptor proteins, thus modifying the membrane permeability in biological systems. The Li^+ ion is in some aspects an analog of the Mg^{2+} ion (as predicted by the diagonal rule of the Periodic Table) with interesting biochemical significance such as interactions in the magnesium-dependent enzyme systems of plants and animals.

Analytical chemistry of lithium has been important since the discovery and establishment of its psychiatric effects by CADE (1949) and SCHOU (1968). No radioisotopes of lithium are known to be used in radioactive measurements, but THELLIER et al. (1983) used the enriched stable isotopes ^6Li and ^7Li as tracers. As long as these isotopes can be detected and distinguished with sufficient accuracy, ionic analyzers are suitable for this purpose (THELLIER et al., 1980).

Atomic absorption and flame spectrometry are used in routine analysis of lithium in mining, metallurgy, aqueous solutions and biological fluids (BENZO und FRAILE, 1984; MUSIL et al., 1983; WELZ, 1985), and biological samples (MUNTAU, 1984; ROBINSON, 1974; BAUMANN et al., 1983). The principal emission line used for flame analysis down to the level of a few parts per million Li^+ is at 670.8 nm. FANES (furnace-AAS-nonthermic excitation spectrometry) permits the determination of lithium and calcium (HIEFTJE, 1989).

3 Sources, Production, Important Compounds, Uses, Waste Products

3.1 Occurrence and Production

Lithium is obtained by electrolysis from lithium containing minerals. The commercially most important of these (with an annual extraction of the order of 200000 to 300000 tons) are: spondumene $LiAlSi_2O_6$, lepidolite $K_2Li_3Al_4Si_7O_{21}(OH, F)_3$, petalite $LiAlSi_4O_{10}$, and amblygonite $LiAl(F, OH)PO_4$, found in North Carolina, Canada, Australia, and Zimbabwe. Lithium carbonate and lithium chloride (which may be used for the production of lithium metal by smelt electrolysis) are important intermediates (BAUER, 1978). Lithium is used in the ceramics industry, the production of alloys (with lead for bearings, with aluminum – e.g., non-corrosive and light aluminum-copper-magnesium-lithium-alloys – for aircraft construction), as an additive to lubricants, as anodes in high-performance batteries with small self-discharge (TRUEB, 1989), in the aluminum industry, for the heat shields of spacecraft, and in the construction of hydrogen bombs (SCHOU, 1984). Strategic interest makes the estimation of lithium production difficult, but it may be of the order of 25000 tons of lithium compounds per year, mainly in the United States (BAUER, 1978; BIRCH, 1988). It is mostly produced from volcanic brine, e.g., in Nevada and northern Chile (TRUEB, 1989).

3.2 Important Compounds

A great number of lithium compounds were reviewed by HART and BEUMEL Jr. (1973) and by BAUER (1978). The physico-chemical characteristics of lithium promote the highest lattice energy among all the salts of the alkali metals. Because it is highly electropositive, lithium forms compounds with other metals such as lead, mercury, thallium, and tin. Inorganic lithium compounds of major importance in industry are lithium peroxide Li_2O_2, lithium oxide Li_2O, lithium hydroxide LiOH, lithium carbonate Li_2CO_3 (used as a flux in aluminum production (TRUEB, 1989) or for porcelain and enamel production), lithium sulfate Li_2SO_4, and the lithium halides. Lithium hydride (a convenient hydrogen source) is hazardous (BIRCH, 1988). Lithium compounds of lesser industrial importance include lithium hydrogen sulfide LiSH, lithium carbide Li_2C_2, lithium amide $LiNH_2$, and others such as lithium nitrate, borates, silicates, aluminosilicates, phosphates, and hypochlorite. Organolithium compounds are generally used for laboratory purposes (HART and BEUMEL Jr., 1973; BAUER, 1978). Lithium stearate is a lubricant, stable between -20 and $150\,°C$.

In Europe there are 29 therapeutic lithium preparations available under different tradenames and consisting of such compounds as lithium carbonate, aspartate, citrate, sulfate, gluconate, monoglutamate, succinate, acetate, and orotate.

3.3 Uses, Waste Products

Lithium plays an important role in nuclear technology as a stretegic element for civilian and military production of fusion energy. In fact, lithium is a necessary raw material in the production of tritium, which together with deuterium are the basic components of the thermonuclear fuel cycle.

Bombardment of ^6Li with neutrons leads to the formation of tritium and alpha particles, the reaction of tritium with deuterium then produces neutrons and alpha particles along with 18 MeV of energy (BARRACHINA, 1985). Lithium is also used for cooling in nuclear reactors (BAUER, 1978). In spite of its high cost (approximately $5000/g) the production of tritium from lithium may become interesting, if in the future lithium demand as energetic fuel will exceed supply capacities. ^6Li could also become a fuel of primary importance, because the reaction requires or generates less tritium into the environment than classical thermonuclear fuel (MCNALLY, 1982).

Lithium batteries (for cameras, quartz watches, remote handling devices, fire alarm systems, pocket calculators, etc.) have a high energy density and a variable capacity (TRUEB, 1989). Organic solvents, such as propylene carbonate, dimethoxyethane, and γ-butyrolactone ether, are used as electrolytes. Lithium batteries should be collected after use because of the explosion risks they present upon incineration (TRUEB, 1987).

4 Distribution in the Environment

The lithium concentration of sea water is of the order of 180 ppb (BRULAND, 1983). The lithium content of drinking water in areas of the Venezuelan Andes, where goiter is endemic, ranges from 5 to 80 µg/L (BENZO and FRAILE, 1984). BOWEN (1979) mentioned concentrations of 10 to 75 ppm in soil and of 1 ppm (dry material) in plants, but in general the lithium content of soil, flora, and fauna is not well known. Concentrations in living organisms depend on the geological origin of the soil. The top 10 cm of topsoil is often poorer in lithium, probably because it is leached into, or "diluted" by, plant roots, particularly in grassland areas. Soils high in salt content reached a concentration of greater than 200 mg lithium per kilogram of dry matter (ANKE et al., 1984). Leguminous plants show concentrations of 50 mg/kg. Those from volcanic soils were 29 mg/kg, and from more heterogeneous soils the lithium content is about 12 mg/kg dry material. In grassland plant species the lithium content ranges from 0.7 to 3.6 mg/kg dry matter, the highest concentrations being found in *Ranunculus repens*. Cow hair contains from 1.8 to 8.9 mg Li/kg dry weight. ARNOLD and ANKE (1987) have reported on the lithium status of different species of ruminants in central Europe.

5 Uptake, Absorption, Metabolism, and Elimination in Plants, Animals, and Humans

The lithium uptakes and requirements of living organisms are practically unknown, but in the body a dynamic equilibrium is established between lithium and the other ions. The human body contains about 2.2 mg lithium obtained from food and drinking water, and it is excreted principally through the kidneys.

6 Effects on Plants, Animals, and Humans

6.1 Miscellaneous Biochemical Effects

Many reports indicate that lithium enhances the accumulation of inositol-P_2, -P_3, and -P_1, whereas it blocks the transformation of inositol-1,4-P_2 to inositol-1-P. Whether this is significant in therapeutical terms remains unclear at the present (DRUMMOND et al., 1987). Brain glucose-6-phosphate, α-ketoglutarate, 6-phosphogluconate, glutamate, glycogen, and serum magnesium increase at the beginning of lithium treatment, but when the accumulation of lithium stabilizes, these modifications return to their normal physiological levels.

Lithium affects the glycolytic pathway to a larger degree than it affects the citric acid cycle (PLENGE, 1982) (see Fig. II.17-1). Enzymatic measurements of the GABA cycle after chronic lithium treatment show a marked inhibition of the GABA-transaminase activity, but no effect on succinate semialdehyde dehydrogenase and glutamate dehydrogenase behavior (RIBAS et al., 1980). This may mean that GABA (γ-aminobutyric acid) is accumulated in the brain during chronic lithium treatment and

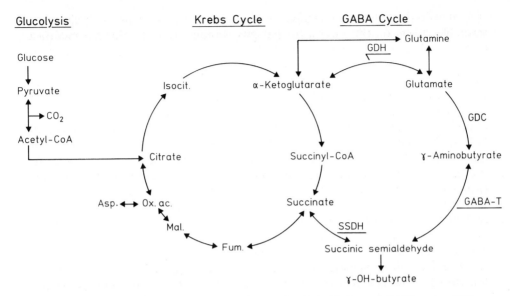

Fig. II.17-1. Primary metabolism of carbohydrates: glucolysis, Krebs and GABA cycles. Bidirectional arrows indicate the reversibility of the metabolic reactions. After chronic Li$^+$ treatment, GABA-T is inhibited; glutamate dehydrogenase and succinate semialdehyde dehydrogenase activities are found unmodified (RIBAS et al., 1980).

is responsible for the tranquilizing effect. MAO B isoenzyme is also involved in behavioral effects (RIBAS et al., 1979).

6.2 Symptoms of Deficiency

That lithium is an essential element was established by ANKE et al. (1981), who observed that goats deficient in lithium gained less weight than corresponding control goats. Long-term lithium therapy for patients with affective disorders helps in the regulation of neurotransmission. Lithium facilitates the binding of 5-hydroxytryptamine (5-HT) to receptors in the rat brain and increases 5-HT release from the hypocampus but not from the cortex (TREISER et al., 1981). Some investigators also find that lithium salts increase norepinephrine turnover in the brain and possibly enhance the deamination of this amine (SCHILDKRAUT et al., 1969), but others do not note significant changes in these reactions. Further studies will be required in order to understand the complex mechanism of lithium action.

6.3 Acute and Chronic Effects in Humans and Animals

(For teratogenic effects see Chapter I.18, Section 5, last paragraph)

Lithium therapy in cancer patients increases the tolerance versus myelosuppressive effects of systemic chemotherapy, but deserves further study. The reduction of the risk of infection and prolongation of the duration of response as well as the favorable effects on the survival of patients without electrocardiogram modifications should encourage further research (LULIARD and GELENBERG, 1982). Interesting reviews of lithium administration have been published by SCHOU (1980, 1984).

Recurrent genital herpes (HSV) infections are inhibited by topical lithium treatment. The mechanism is explained as a blocking of viral replication, inhibiting the release of arachidonic acid and formation of inflammatory mediators. Recurrent HSV (HSV 1 being herpes simplex virus type I; HSV 2 being herpes simplex virus type II) infections improve when the lithium concentration in the blood reaches 0.5 to 1 mmol/L. Polychemotherapy increases the lithium concentration in plasma. Especially recurrent affective disorders (such as psychiatric mood disorders) can be treated with lithium salts (BIRCH, 1988). Hyperactive patients require certain other compounds in combination with lithium. Haloperidol is often given concomitantly as a neuroleptic drug because of its powerful sedative effect. Some manic depressive patients may benefit from combinations of lithium with drugs such as carbamazepine, sodium valporate, and pimozide. L-Trpytophan may also be effective for manic symptoms by potentializing lithium effects when administered concomitantly (TYRER, 1985). Methyl-DOPA together with lithium induces toxical symptoms of unknown molecular mechanism. Such a combination should therefore be avoided and more research should be undertaken.

Hypothyroidism is one of the long-term side-effects of lithium treatment and can be observed by the low concentration of the thyroid hormones T_3 and T_4 in serum. Several other side-effects have been reviewed by SCHOU (1984, 1985).

Lithium intoxication can occur within hours to days after ingestion of excessive amounts of lithium. It affects the functions of the central nervous system (CNS), kidney, and heart, and causes serious disturbances in the electrolytic balance (SCHOU, 1980; FIEVE and PESELOW, 1985). CNS intoxication symptoms include ataxia, incoordination, nystagmus, and epileptiform seizures. Cardiac intoxication brings about an inversion of T-waves and depression of the ST segment, significant arrhythmia and possible congestive heart failure. Intoxicated kidney function causes significant loss of sodium and fluid, due to the diuretic and natriuretic actions of lithium. The manifestations of lithium neurointoxication can be controlled effectively by measuring the cation content in red blood cells and the red blood cell to plasma lithium ratio (ELIZUR et al., 1982). Treatment of lithium intoxication includes monitoring of the patient's emesis induced by apomorphine, syrup of ipecac, or gastric lavage. The renal clearance of lithium is increased by saline infusion of 150–300 meq NaCl per liter in the first six to ten hours (THOMSEN and SCHOU, 1968).

It can be assumed that a considerable number of people come in contact with lithium through either lithium therapy or lithium containing materials since new technological uses are constantly being developed. Studies on the lithium toxicity of medications, hospital disinfectants, foodstuffs, biocides for swimming pools, batteries (which also contain manganese dioxides; TRUEB, 1987), wristwatches, computers, electronic measuring devices, flashlights, and in special accessories chemical catalysts and reagents, and rocket fuel components have been reviewed (BRÜCKNER and HAJDU, 1983).

7 Hazard Evaluation and Limiting Concentrations

Compared to other elements such as cadmium, mercury, lead, tellurium, and beryllium, lithium has a low toxicity. Lithium hydride and lithium aluminum hydride have a MAK value of 0.025 mg/m^3 (USA) due to their causticity and properties as reducing agents. The maximum concentration of lithium hydride at the workplace is 0.025 mg/m^3 (MAK, 1987). Alkyllithium compounds are pyrophoric (BAUER, 1978).

Lithium salts are administered to prevent manic depression in humans in doses of from 600 up to 1800 mg of lithium carbonate per day. Some patients suffer from muscular weakness, gastrointestinal irritation, and a dazed feeling, but these symptoms disappear within a week. Afterwards a tremor of the hands emerges, and perhaps a goiter which shrinks or disappears when treatment is interrupted. The ingestion of more than 2 g of lithium may cause intoxication (BAUER, 1978), the warning signs of which are: sluggishness, languidness, drowsiness, a coarse tremor, muscle twitching, slurred speech, loss of appetite, vomiting and diarrhea. Lithium poisoning must be confirmed by determining the lithium level in serum, generally up to 1.5 meq/L blood (SCHOU, 1976). A level of 4 mmol/L plasma (~ 28 ppm) may be fatal (BIRCH, 1988). No specific antidote is known. The primary treatment is to maintain free respiration and prevent respiratory infection. There have been no known cases of lithium addiction and there are no withdrawal symptoms (SCHOU, 1984).

References

Anke, M., Grun, M., Groppel, B., and Kronemann, H. (1981), *The Biological Importance of Lithium* (in German), in: *Mengen- und Spurenelemente*, Vol. 1, pp. 217–239. A.W.P. Karl Marx-Universität, Leipzig, GDR.

Anke, M., Groppel, B., and Kronemann, H. (1984), *Significance of Newer Essential Trace Elements for the Nutrition of Man and Animal. Trace Elem. Anal. Chem. Med. Biol. Proc. Int. Workshop 2nd*, Vol. 3, pp. 436–440. W. de Gruyter & Co., Berlin-New York.

Arnold, W., and Anke, M. (1987), *Der Lithiumstatus verschiedener Wiederkäuerarten in Mitteleuropa. Mengen- und Spurenelemente* (Proceedings), pp. 283–288. Karl Marx-Universität, Leipzig.

Barrachina, M. (1985), *Tritium Technology, Energ. Nucl. Madrid 157*, 383–396.

Bauer, R. J. (1978), *Lithium and Lithium Compounds* (in German), in: *Ullmanns Encyklopädie der technischen Chemie*, 4th Ed., Vol. 16, pp. 261–277. Verlag Chemie, Weinheim-New York.

Baumann, W., Stadie, G., and Anke, M. (1983), *Lithium, 4. Spurenelement-Symposium*, pp. 180–185. A.W.P. der Friedrich Schiller-Universität, Jena, GDR.

Benzo, Z., and Fraile, R. (1984), *Urinary Lithium Excretion in Endemic and Non-endemic Goiter Areas in Venezuela. Trace Elem. Anal. Chem. Med. Biol. Proc. Int. Workshop 2nd*, Vol. 3, 467–474. W. de Gruyter & Co., Berlin-New York.

Birch, N. J. (1988), Chapter 32: *Lithium*, in: Seiler, H., Sigel, H., and Sigel, A. (eds.), *Handbook on Toxicity of Inorganic Compounds*, pp. 383–393. Marcel Dekker, New York.

Bowen, H. J. M. (1979), *Environmental Chemistry of the Elements*, p. 253. Academic Press, London.
Brückner, C. and Hajdu, S. (1983), *Importance of Lithium in Industrial Medicine*, in: *Lithium, 4. Spurenelement-Symposium*, pp. 370–372. A. W. P. Friedrich Schiller-Universität, Jena, GDR.
Bruland, K. W. (1983), *Trace Elements in Sea Water,* Chem. Oceanogr. *8*, 174.
Cade, J. F. (1949), *Lithium Salts in the Treatment of Psychotic Excitement.* Med. J. Aust. *36*, 349–352.
Drummond, A. H., Joels, L. A., and Hughes, P. J. (1987), *The Interaction of Lithium Ions with Inositol Lipid Signalling Systems.* Biochem. Soc. Trans. *15*, 32–35.
Elizur, A., Yeret, A., Segal, Z., and Graff, E. (1982), *Lithium and Electrolyte Plasma/RBC Ratio and Paradoxical Lithium Neurotoxicity.* Prog. Neuropsychopharmacol. Biol. Psychiatr. *6*, 235–241.
Fieve, R. R., and Peselow, E. D. (1985), *Toxicology*, in: *Lithium*, pp. 353–375. John Wiley & Sons, New York.
Hart, W. A., and Beumel, O. F., Jr. (1973), *Lithium and its Compounds*, in: *Comprehensive Inorganic Chemistry*, Vol. 1, pp. 331–367. Pergamon Press, London.
Hieffje, G. M. (1989), *Critical Comparison of Plasma Sources and Detection Methods in Atomic Spectrometric Analysis.* Fresenius Z. Anal. Chem. *334* (7), 607; see also Report E. Merian in *Swiss Chem. 11*(12), 10.
Luliard, R., and Gelenberg, A. (1982), *Hazards and Adverse Effects of Lithium.* Annu. Rev. Med. *33*, 327–352.
MAK (1987) *Maximum Concentrations at the Workplace, Report No. XXIII, DFG.* VCH Verlagsgesellschaft, Weinheim-Basel-Cambridge-New York.
McNally, J. R. (1982), *Physics of Fusion Fuel Cycles,* Nucl. Technol. Fusion *2*, 9–14.
Muntau, H. (1984), *Newer Essential Trace Elements and Analytical Reliability.* Trace Elem. Anal. Chem. Med. Biol. Proc. 2nd Int. Workshop, Vol. 3, pp. 563–580.
Musil, F., Janousek, I., and Suva, J. (1983), *Contribution to a Method for the Determination of Lithium in Biological Fluids*, in: *Lithium, 4. Spurenelement Symposium*, pp. 103–109. A. W. P. Friedrich Schiller-Universität, Jena, GDR.
Plenge, P. (1982), *Lithium Effects on Rat Brain Glucose Metabolism in vivo.* Psychopharmacology *77*, 348–355.
Ribas, B., Acobettro, R. I., Mate, C., and Santos, A. (1979), *Role of Monoamine Oxidase Isoenzymes in Rat Motor Activity, after Rubidium Chloride Treatment.* Biochem. Soc. Trans. *7*, 533–536.
Ribas, B., Gonzalez, M. P., Acobettro, R. I., and Santos, A. (1980), *4-Aminobutyrate: 2-Oxoglutarate Aminotransferase Inhibition in Rat Brain by Lithium Treatment.* Trace Elem. Anal. Chem. Med. Biol. Proc. Int. Workshop 2nd, Vol. 1, 57–65. W. de Gruyter & Co., Berlin-New York.
Robinson, J. W. (1974), *Handbook of Spectroscopy,* Vols. I and II. CRC Press, Cleveland, Ohio.
Schildkraut, J.-J., Logue, M. A., and Dodge, G. A. (1969), *The Effects of Lithium Salts on the Turnover and Metabolism of Norepinephrine in Rat Brain.* Psychopharmacologia Berlin *14*, 135–141.
Schou, M. (1968), *Lithium in Psychiatric Therapy and Prophylaxis.* J. Psychiatr. Res. *6*, 67–95.
Schou, M. (1976), *Pharmacology and Toxicology of Lithium.* Annu. Rev. Pharmacol. *16*, 231–243.
Schou, M. (1980), *The Recognition and Management of Lithium Intoxication*, in: *Handbook of Lithium Therapy*, pp. 394–402. MTP Press, Lancaster, UK.
Schou, M. (1984), *From Mine to Mind: Lithium in Psychiatry.* Intern. Med. *4*, 24–26.
Schou, M. (1985), *Lithium,* in: *Side Effects of Drugs Annual 9*, 27–32. Elsevier, Amsterdam.
Thellier, M., Hartmann, A., Lassalles, J. P., and Garrec, J. P. (1980), *A Tracer Method to Study Unidirectional Fluxes of Lithium.* Biochim. Biophys. Acta *598*, 339–344.
Thellier, M., Heurteaux, C., Garrec, J. P., Alexandre, J., and Wissocq J. C. (1983), *Use of an Ionic Probe and of the Stable ^{6}Li Isotopes to Perform Unidirectional Li-flux Measurements.*

Lithium. 4. Spurenelement-Symposium, pp. 134–138, A. W. P. Friedrich Schiller-Universität, Jena, GDR.

Thomsen, K., and Schou, M. (1968), *Renal Lithium Excretion in Man. Am. J. Physiol. 215*, 823–827.

Treiser, S. L., Cascio, C. S., O'Donohue, T. L., Thoa, N. B., Jacobowitz, D. M., and Kellar, K. J. (1981), *Lithium Increases Serotonin Release and Decreases Serotonin Receptors in the Hippocampus. Science 213*, 1529–1531.

Trueb, L. (1987), *Lithium Batteries with High Storing Capacity* (in German), *Neue Zürcher Zeitung, Forschung und Technik, No. 184*, p. 65 (12. August), Zürich.

Trueb, L. (1989), *Lithium, the Lightest of All Metals, Its Production from Brine* (in German). *Neue Zürcher Zeitung, Forschung und Technik, No. 44*, pp. 81–82 (22 February), Zürich.

Tyrer, S. P. (1985), *Lithium in the Treatment of Mania. J. Affective Disord. 8*, 251–257.

Welz, B. (1985), *Atomic Absorption Spectrometry*, pp. 295–297, 411 ff. VCH Verlagsgesellschaft, Weinheim-Deerfield Beach/Florida-Basel.

Additional Recommended Literature

Eichinger, G., and Semrau, G. (1990), *Lithium Batteries* (in German). *Chem. Unserer Zeit 24*(1, 2), 32–36, 90–96.

II.18 Magnesium

JERRY K. AIKAWA, Denver, Colorado, USA

1 Introduction

The planet Earth appeared about 4.6 billion years ago under speculative and undefined circumstances. One billion years later, life arose from non-living matter; organisms evolved from living cells in water and on land. These organisms were able to capture solar energy and use it for the synthesis of organic molecules. Thus the miracle of photosynthesis appeared to support life on Earth (AIKAWA, 1981).

Oxygen appeared in the Earth's atmosphere, primarily as a by-product of the photosynthetic process which required the presence of the magnesium atom in the green pigment chlorophyll. Chlorophyll captures the energy of the sun and converts it to chemical energy in the form of adenosine triphosphate (ATP). In the biological evolution which started about 3 billion years ago, the energy of ATP is used to reduce carbon dioxide and water to carbohydrate and oxygen. All enzyme reactions that are known to be catalyzed by ATP show an absolute requirement for magnesium compounds. It is obvious that magnesium participated in the earliest biochemical steps necessary for the evolution of life. Man has had a brief 100 000 years so far to understand the role of magnesium in biological evolution (AIKAWA, 1981).

Almost seven decades have passed since Richard Wilstätter demonstrated the central position of the magnesium atom in the chlorophyll molecule. Although much has been learned since about the photosynthetic process, the exact function of the magnesium atom in this process still eludes us. There is recent evidence suggesting that magnesium may play a central role in coordinating control of metabolism and growth in animal cells (RUBIN, 1975). Magnesium is thus an essential element, and various deficiency symptoms in plants, animals, and humans are known. On the other hand, magnesium compounds are practically non-toxic and overdoses are usually eliminated first by vomiting and/or diarrhea, although they may lead to hypermagnesemia (for which calcium compounds act as clearing antagonists). Magnesium fumes, however, can be irritating (AIKAWA, 1984).

2 Physical and Chemical Properties, and Analytical Methods

Magnesium, a member of Group IIa of the Periodic System, was discovered by Sir Humphrey Davy in 1807 and has an atomic mass of 24.312. Its atomic number is

12, its valence +2. Natural magnesium consists of the three isotopes ^{24}Mg (79%), ^{25}Mg (10%), and ^{26}Mg (11%), but does not exist in the form of the light (density 1.74 g/cm^3) metal with a melting point of 649°C and a boiling point of about 1100°C (AIKAWA, 1984). The silvery white metal is readily protected by an oxide film on the surface. At room temperature, magnesium chloride hexahydrate (MgCl$_2 \cdot$6H$_2$O) is very stable (solubility in water about 60%) and can form a double salt with potassium chloride or calcium chloride. Its solutions may hydrolyze to basic salts with corroding properties (AIKAWA, 1984; OTTO et al., 1978). Magnesium chloride makes up 17% of sea salt (OTTO et al., 1978).

Since the mid-1960s, determination of magnesium content using atomic absorption spectrometry has been improved and simplified (AIKAWA, 1984; WELZ, 1985; BIRCH, 1988), but atomic emission spectrometry is also fast and reliable (see also Chapter I.4a). FORTEZA et al. (1987) developed spectrophotometric analysis by flow injection using succinimidedioxime, a technique eliminating interference from other alkaline-earth ions, especially Ca(II).

3 Sources, Production, Important Compounds, Uses, Waste Products, and Recycling

(AIKAWA, 1981; SAAGER, 1984; TRUEB, 1989)

Magnesium constitutes about 2% of the earth's crust. It is widely distributed in nature in a variety of compounds: as carbonate in magnesite, MgCO$_3$, and dolomite, MgCO$_3 \cdot$CaCO$_3$, as the oxide in brucite Mg(OH)$_2$, and as the chloride in carnalite, MgCl$_2 \cdot$R$\cdot$6H$_2$O. Magnesium also occurs in silicate minerals of which the most important are olivine (MgFe)$_2$SiO$_4$, serpentine 3MgO:2SiO$_2 \cdot$2H$_2$O, and asbestos (CaMg)$_2$SiO$_4$. These silicate minerals are found world-wide (AIKAWA, 1981). Magnesium is also found as chloride in seawater at a concentration of 1.27 g/kg. Annual world production of magnesium is of the order of 300000 tons (AIKAWA, 1984; SAAGER, 1984). More than half of it originates from seawater; and in Freeport, Texas, alone more than a third of the world production is extracted from seawater by electrolysis (SAAGER, 1984). Besides the sedimentary dolomites, salt deposits from prehistoric seas are also worked, as are wastes from potassium mines (SAAGER, 1984). The most important magnesium producers are the United States (44%), the Soviet Union (27%), and Norway (14%) (SAAGER, 1984). The two most important processes being fusion electrolysis and thermic reduction (e.g., with ferrosilicon), but final refining is always necessary (HOY-PETERSEN, 1978).

Approximately 20% of magnesium requirements originate from secondary production and is produced through the recycling of cast magnesium and magnesium/aluminum alloys. For instance, in the United States 8000 tons of magnesium per year are produced from the caps of beverage bottles (SAAGER, 1984).

About 57% of the magnesium produced is used in aluminum alloys (for aircraft components, and for cans and containers), about 18% in other alloys, 8% as a reducing agent (e.g., in obtaining titanium, uranium, zirconium, and beryllium), 7% as corrosion protection, and about 10% is used for other purposes (SAAGER, 1984). Fine magnesium powders (obtained by atomizing), cuttings, thin foils and wires (for flash) are all very reactive (HOY-PETERSON, 1978; BIRCH, 1988).

Of the important inorganic magnesium compounds, magnesium chloride is used for metallic magnesium production and for Sorel cement. Magnesium oxide is used for fire-resistant materials, as a ceramic heat accumulator, for setting, and as an additive in fertilizers and feed (OTTO et al., 1978). Magnesium chloride and oxide were used as laxatives as well (AIKAWA, 1984), but the spasmolytic magnesium sulfate should no longer be used, especially by children and by persons with injured kidneys (AIKAWA, 1984).

Organic compounds play a role (as the so-called Grignard compounds) in the synthesis of fine chemicals and also tetramethyllead, through the oxidation of methylmagnesium chloride (JOLLY, 1978). They are normally not isolated because they decompose and are therefore not of ecotoxicological significance.

4 Distribution in the Environment, in Foods, and in Living Organisms

Rocks and minerals may contain a higher percentage of magnesium than do soils; this reflects the loss of magnesium during weathering. A major portion of soil magnesium consists of weathered primary minerals and secondary alumino-silicates in which Mg^{2+} is substituted for Al^{3+}. Mg^{2+} is held as an exchangeable cation by negatively charged silicate minerals and generally comprises from 5 to 30% of the total exchange capacity. Thus, exchangeable magnesium constitutes the major reserve of magnesium available to plants. Magnesium deficiencies usually occur when exchangeable magnesium drops below 5% of the total cation exchange capacity. This happens mainly in coarse-textured soils in areas of high rainfall (AIKAWA, 1981) especially with acid decomposition (see also Sect. 6.2). In soil magnesium may also be replaced by copper, nickel, or beryllium (LEPP, 1981). As mentioned above, magnesium ions are also important ingredients of seawater. Magnesium is widespread in all living cells (the second most common intracellular ion after potassium). Marine algae contain 6 to 20 g/kg, plants 1 to 8 g/kg, fish and mammals about 1 g/kg (dry weight). Coal and similar fuels also contain high magnesium concentrations since they are of vegetal origin (see also WEDEPOHL, Chapter I.1).

One of the greatest triumphs of early evolution was the development of a means to harness the energy of the sun, which is transmitted as light, to drive energy-requiring synthetic processes. This process occurs in higher plants in a sub-cellular organelle, the chloroplast (AIKAWA, 1978). The chloroplast is an organized set of membranes containing the pigment chlorophyll, which is the magnesium chelate of porphyrin. The magnesium atom occupies the central position in the chlorophyll molecule and appears to stabilize the structure so that it can undergo perfectly reversible one-electron oxidations. A possible reason for the role of magnesium in

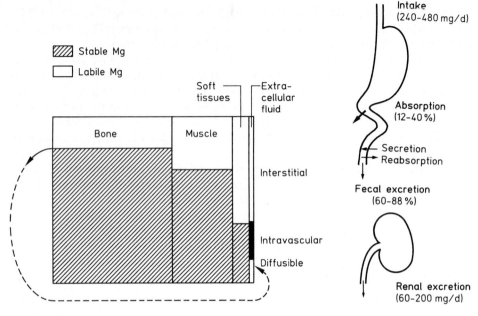

Fig. II.18-1. The anatomy of magnesium.

photosynthesis could be that it was simply available in the quantities needed for this purpose and was ideally suited for this function.

Some common foods can be ranked as follows in order of decreasing mean magnesium concentrations (mg/kg): nuts, about 2000; cereals, 800; seafood, 350; meat, 270; legumes, 240; dairy products, 180; vegetables, 170; fruits, 70; refined sugars, 60; and fats, 7 (AIKAWA, 1981). This order differs when the concentrations are ranked on the basis of the caloric values of the foods, as follows: vegetables, legumes, seafood, nuts, cereals, dairy products, fruit, meat, refined sugars, and fats. Noteworthy is the very small contribution of fats and refined sugars to the total intake of magnesium. These two, the major sources of caloric energy, are virtually devoid of magnesium.

The magnesium content of the human body ranges between 272 to 420 mg/kg of wet tissue. The total body content of magnesium for a man weighing 70 kg would be of the order of 24 g, 89% of which resides in bone and muscle. Bone contains about 60% of the total body content of magnesium at a concentration of about 1.1 g/kg wet weight. Most of the remaining magnesium is distributed equally between muscle and non-muscular soft tissues (Fig. II.18-1).

Approximately 1% of the total body content of magnesium is extracellular. The levels of magnesium in serum of healthy people are remarkably constant, remaining on average 20 mg/L and varying less than 15% from this mean value. The distribution of normal values for serum magnesium is identical in men and women and remains constant with advancing age. Approximately one-third of the extracellular magnesium is bound non-specifically to plasma proteins. The remaining 65%, which

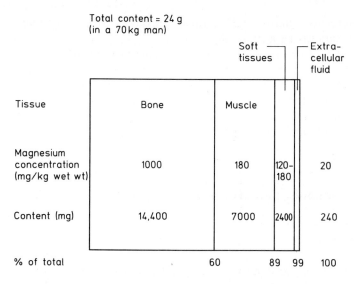

Fig. II.18-2. The physiology of magnesium.

is diffusible or ionized, appears to be the biologically active component. The magnesium content of erythrocytes varies from 53 to 12 mg/L (Fig. II.18-2). Full blood contains about 18–23 mg/L.

Regarding the intracellular/extracellular equilibrium (ratio 10:1) in the soft tissues of the body, it is not yet clear whether metabolic energy is required for the maintenance of this state. Or can this (including deficiency symptoms) be explained simply on the basis of the coordinating properties of the magnesium atom, or primarily on physicochemical adsorption?

5 Uptake, Absorption, Transport and Distribution, Metabolism and Elimination in Plants, Animals, and Humans

(see also AIKAWA, 1971)

Although it is known that magnesium is essential, not many studies have been published on the uptake of magnesium ions by plants and how natural chlorophyll is synthesized (except that the presence of iron is necessary). Some information was obtained from biomonitoring. Litter is, of course, magnesium-rich and plays a role in environmental cycling. In all organisms, interactions and equilibria were observed with calcium and phosphate ions. Endocrinological history (altered sensitivity to regulatory hormones) also seems to play a role (BIRCH, 1988). Most of the literature is related to mechanisms taking place in humans (see Fig. II.18-2 and AIKAWA, 1971). Intestinal absorption of ingested magnesium is about 12–40%, but the regulation of body content depends on kidney activity (BIRCH, 1988). The self-limiting vomiting reaction of humans against oral overdose may not work in animals (without reflex) or with narcosis (BIRCH, 1988).

Suffice it to mention here that an abnormally low magnesium level in blood of less than 13 mg/L is frequently observed in patients with congestive heart failure, cirrhosis, or renal failure after hemodialysis. Levels higher than 24 mg/L occur in patients with renal failure before therapy. The average American ingests between 240 and 480 mg of magnesium a day; magnesium intakes of 3.6 to 4.2 mg/kg/day are thought to be adequate to maintain magnesium balance in normal adults.

6 Effects on Plants, Animals, and Humans

6.1 Acute and Chronic Effects

Under normal circumstances magnesium toxicity to plants, animals, and humans is highly unlikely (BIRCH, 1988). Critical situations may occur if the ratios to calcium and/or potassium and/or zinc are not balanced (ion-antagonistic effects in plants or in patients after heart surgery; VOGEL and ANGERMANN, 1981; AIKAWA, 1984). In the event of kidney failure, it is possible for plasma magnesium levels to become elevated as a result of ingested magnesium salts or through an overloaded intravenous infusion (BIRCH, 1988). Hypermagnesemia may lead to somnolence, inhibition of autonomic function with little effect on the central nervous system, hypotonia, bradycardia, decrease in blood pressure, and eventually cardiac arrest. The LD_{50} for magnesium in dogs is about 250 mg/kg. Inhalation of magnesium oxide fumes is an industrial problem which results in metal fume fever and burns. Cutaneous reactions result from the penetration of magnesium metal through the skin (BIRCH, 1988).

6.2 Deficiency Symptoms

In the case of magnesium compounds, deficiency is a much greater risk than toxicity for all living organisms. It is obvious that magnesium deficiency results in characteristic symptoms in plants, for example chlorosis (insufficient chlorophyll production) in corn tissues, and chlorosis and loss of stomata wax in trees (AIKAWA, 1984). Magnesium deficiency is also a characteristic feature of forest soils subjected to long-term acid deposition. Exchangeable magnesium cations are leached during soil acidification. This can ultimately result in severe magnesium deficiency, which is a widespread phenomenon in Central Europe, as is seen in recent forest decline. It is the main cause for the yellowing of Norway spruce (*Picea abies*) in acid soils (ULRICH, 1989). Damage of the vascular bundles has been shown. Healthy spruce needles contain about 9.8 mg/kg potassium and 1.7 mg/kg magnesium, whereas for damaged spruce the concentrations are only 3.7 and 1.1 mg/kg, respectively (concentrations of sodium, phosphorus, and calcium are unchanged). It is not yet clear whether disorganization in the root tissue and/or translocation damage take place; there are also perhaps differences between genotypes (HUETTL, 1986).

The lack of magnesium in early spring grass may result in so-called grass tetany in animals grazing on young grass where the soil is low in available magnesium or where the nature of the soil solution limits absorption of magnesium ions by certain plants (CALVIN and BENSON, 1948). The accumulation of magnesium and phosphorus in the seed and other storage organs of the plant indicates that there is a definite relation between magnesium and phosphorus in reproduction and growth.

Magnesium is required by a wide range of metabolic enzymes involved in most fundamental body processes (BIRCH, 1988). Magnesium ions have also an influence on maturation of ova, maturation of sperm, and insemination. Animals with magnesium deficiency may develop leukocytosis and/or thymic tumors, and their mortality is increased (AIKAWA, 1981). Secondary effects are also known, since fluorosis depends not only on the fluoride concentration, but also on the magnesium concentration of drinking water (JENSEN, 1987).

The clinical symptoms of magnesium deficiency in humans (FLINK, 1956; SEELIG, 1981; WACKER, 1980) are characterized by the following features: (1) spasmophilia, gross muscular tremor, choreiform movements, ataxia, tetany, and in some instances predisposition to epileptiform convulsions; (2) hallucinations, agitation, confusion, tremulousness, delirium, depression vertigo, muscular weakness, and organic brain syndrome; (3) a low serum magnesium concentration associated with a normal serum calcium concentration and a normal blood pH; (4) low-voltage PQRS complexes, a short fixed P-R interval; (5) positive Chvostek and Trouseau signs; and (6) prompt relief of tetany when the serum magnesium concentration is restored to normal (AIKAWA, 1971). Other manifestations of clinical magnesium deficiency can occur, such as phlebothrombosis, constitutional thrombasthenia, hemolytic anemia (an allergic or osseous form of the deficiency) and oxalate lithiasis (DURLACH, 1976). Unusual symptoms and signs of magnesium deficiency are dysphagia and vertical nystagmus (HAMED and LINDEMAN, 1978).

Clinical conditions associated with depletion of magnesium include the following (AIKAWA, 1981):

(a) Fasting. Prolonged fasting is associated with a continual renal excretion of magnesium, and after two months of fasting the body deficit may amount to 20% of the total body magnesium content.
(b) Excess loss from the gastrointestinal tract. Persistent vomiting or prolonged removal of intestinal secretions by mechanical suction coupled with the administration of magnesium-free intravenous infusions can induce clinical magnesium deficiency.
(c) Surgical patients. Surgery is followed by a negative magnesium balance of several days duration, however, the magnitude of the magnesium loss is minimal and usually does not result in symptomatic magnesium deficiency.
(d) Gastrointestinal disorders. Malabsorption of magnesium occurs in conditions in which the intestinal transit is abnormally rapid or in which the major absorbing site, the small intestine, has been resected; hypomagnesemia is frequently associated with malabsorption or increased fecal loss of magnesium such as in steatorrhea.

(e) Acute alcoholism. Hypomagnesemia occurs frequently in patients with delirium tremens or chronic alcoholism (see also ZUMKLEY et al., 1986). A transient decrease in serum magnesium may occur during the withdrawal state even though the pre-withdrawal levels are normal.
(f) Cirrhosis. The magnesium content of liver tissue per unit weight is decreased in cirrhosis. Patients with cirrhosis may have clinical features consistent with magnesium deficiency in the presence of normal serum magnesium values, but with a low skeletal muscle magnesium content and normal bone and erythrocyte magnesium content.
(g) Cardiovascular disorders. Magnesium deficiency occurs in congestive heart failure (see also TSIU et al., 1986; JEPPESEN and GAM, 1987; RASMUSSEN, 1987; and SMETANA, 1987). Cardiac glycosides may induce magnesium deficiency. Magnesium deficiency is frequently associated with cardiac arrhythmias such as ventricular tachycardia and atrial or ventricular fibrillation. Geochemical studies have suggested an inverse relationship between magnesium in drinking water and cerebrovascular and cardiovascular diseases in man.
(h) Of considerable clinical interest is the recent recognition that hypokalemia may be secondary to magnesium deficiency and may be resistant to treatment unless the underlying lack of magnesium is corrected (WHANG et al., 1984). Since hypokalemia is so prevalent, the underlying magnesium deficiency is often neglected.

Diagnosis. In many patients, the clinical symptoms and signs, although non-specific, accompanied by a low serum magnesium concentration, confirm the diagnosis. However, a normal serum magnesium level does not exclude magnesium deficiency. Urinary output of magnesium has been used as an index of magnesium deficiency (HEATON, 1969), and a magnesium load test has been recommended in infants (CADDELL, 1975) and post-partum women (CADDELL et al., 1975), with the test load given either orally or intravenously (THORÈN, 1971).

Therapy. In patients with the clinical symptoms and signs of magnesium deficiency, the magnesium deficit is of the order of 10–25 mg/kg of body weight. Since less than one-half of the administered magnesium is usually retained in the body, the required therapeutic dose is 25–50 mg/kg, which can be administered parenterally over a 4-day period. In order to administer the dose safely, one should first determine the adequacy of renal function, then monitor the plasma levels of magnesium during therapy.

FLINK (1969) recommended an initial loading dose of 600 mg (6.0 g of $MgSO_4$ in 1000 mL of solution containing 5% glucose) given intravenously over a period of three hours, followed by additional doses of 600 mg every twelve hours. Another suggested regimen is the intravenous administration of 1200 mg on the first day (195 mg every two hours for three doses and then every four hours for four doses), followed by 400–600 mg/day (in divided doses) for four days.

For the treatment of arrhythmia, ISERI et al. (1975) recommended administration of 10–15 mL of a 20% magnesium sulfate solution, given intravenously over one minute, followed by a slow 4–6 hour infusion of 500 mL of 2% magnesium sulfate in 5% dextrose water. A second infusion of magnesium sulfate may be necessary should the arrhythmia recur.

7 Hazard Evaluation and Limiting Concentrations

As already mentioned, oral intoxications by magnesium compounds are exceptional and may be treated by calcium administration and/or hemodialysis. In the Federal Republic of Germany, a maximum occupational concentration of 6 mg/m^3 of magnesium oxide is recommended (an 8-hour time weight average) to prevent the impairment of respiratory organ function (MAK, 1987). In Great Britain and the United States, the general threshold limit value is 10 mg/m^3 (BIRCH, 1988).

References

Aikawa, J.K. (1963), *The Role of Magnesium in Biologic Processes.* C.C. Thomas, Springfield, Illinois.
Aikawa, J.K. (1971), *The Relationship of Magnesium to Disease in Domestic Animals and in Humans.* C.C. Thomas, Springfield, Illinois.
Aikawa, J.K. (1978), *Biochemistry and Physiology of Magnesium. World Rev. Nutr. Diet. 281,* 112.
Aikawa, J.K. (1981), *Magnesium: Its Biologic Significance.* CRC Press, Boca Raton, Florida.
Aikawa, J.K. (1984), Chapter II.13: *Magnesium*, in: Merian E. (ed.): *Metalle in der Umwelt,* pp. 465–470. Verlag Chemie, Weinheim-Deerfield Beach/Florida-Basel.
Birch, N.J. (1988), Chapter 33 *Magnesium*, in: Seiler, H.G., Sigel, H., and Sigel, A. (eds.): *Handbook on Toxicity of Inorganic Compounds,* pp. 397–403. Marcel Dekker, New York – Basel.
Caddell, J.L. (1975), *The Magnesium Load Test. I. A Design for Infants. Clin. Pediatr. Philadelphia 14,* 449.
Caddell, J.L., Sarer, F.L., and Thomason, C.A. (1975), *Parenteral Magnesium Load Test in Postpartum American Women. Am. J. Clin. Nutr. 28,* 1199.
Calvin, M., and Benson, A.A. (1948), *The Path of Carbon in Photosynthesis. Science 107,* 476.
Durlach, J. (1976), *Neurological Manifestations of Magnesium Imbalance,* in: Vinken, P.J., and Bruyn, G.W. (eds.): *Handbook of Clinical Neurology,* Vol. 28, pp. 23, 545. North-Holland, Amsterdam.
Flink, E.B. (1969), *Therapy of Magnesium Deficiency. Ann. N.Y. Acad. Sci. 162,* 901.
Flink, E.B. (1956), *Magnesium Deficiency Syndrome in Man, J. Am. Med. Assoc. 160,* 1406.
Forteza, R., Cerda, V., Maspoch, S., and Blanco, M. (1987), *Spectrophotometric Determination of Magnesium(II) by Flow-Injection Analysis Using Succinimidedioxime. Analysis 3,* 136–139.
Hamed, I.A., and Lindeman, R.D. (1978), *Dysphagia and Vertical Nystagmus in Magnesium Deficiency. Ann. Intern. Med. 89,* 222.
Heaton, F.W. (1969), *The Kidney and Magnesium Homeostasis. Ann. N.Y. Acad. Sci. 162,* 775.
Hoy-Petersen, N. (1978), *Magnesium,* in: *Ullmanns Encyklopädie der technischen Chemie,* 4th Ed., Vol. 16, pp. 319–329. Verlag Chemie, Weinheim-New York.
Huettl, R.F. (1986), *New Type of Forest Decline and Nutrient Supply. 2nd International Conference on Environmental Contamination,* Amsterdam. Proceedings, pp. 205–207. CEP Consultants Ltd., Edinburgh; see also *VDI Proceedings* (1990) of the *Lindau Conference on Effects of Atmospheric Pollutants on Soils,* in press.

Iseri, L. T., Freed, J., and Bures, A. R. (1975), *Magnesium Deficiency and Cardiac Disorders*. Am. J. Med. 58, 837.

Jensen, A. A. (1987), *Speciation of Exposure. 2nd Nordic Symposium on Trace Elements in Human Health and Disease, Odense*, Symposium Report, Environ. Health 26, 3 – 4. WHO Regional Office for Europe, Copenhagen.

Jeppesen, B. B., and Gam, J. (1987), *Magnesium Status by Acute Myocardial Infarction. 2nd Nordic Symposium on Trace Elements in Human Health and Disease, Odense*. Abstract J9. WHO Regional Office for Europe, Copenhagen.

Jolly, P. W. (1978), *Organische Magnesium-Verbindungen*, in: *Ullmanns Encyklopädie der technischen Chemie*, 4th Ed., Vol. 16, pp. 357 – 359. Verlag Chemie, Weinheim-New York.

Lepp, N. W. (1981), *Effects of Heavy Metal Pollution on Plants*, 2 Volumes. Applied Publishers, London.

Otto, W., Seeger, M., Flick, W., and Hermann, G. (1978), *Anorganische Magnesium-Verbindungen*, in: *Ullmanns Encyklopädie der technischen Chemie*, 4th Ed., Vol. 16, pp. 337 – 356. Verlag Chemie, Weinheim-New York.

Rasmussen, H. S. (1987), *Magnesium and Acute Myocardial Infarction. 2nd Nordic Symposium on Trace Elements in Human Health and Disease, Odense*. Environ. Health 20, 17 – 20; 26, 73. WHO Regional Office for Europe, Copenhagen.

Rubin, H. (1975), *Central Role of Magnesium in Coordinate Control of Metabolism and Growth in Animal Cells*. Proc. Natl. Acad. Sci. USA 72, 3551.

Saager, R. (1984), *Metallische Rohstoffe von Antimon bis Zirkonium*, pp. 145 – 148. Bank von Tobel, Zürich.

Seelig, M. S. (1981), *Magnesium Deficiency in the Pathogenesis of Disease*. Plenum Press, New York.

Smetana, R. (1987), *Diuretics-independent Hypomagnesemia in Coronary Artery Disease (CAD) and Idiopathic Dilated Cardiomyopathy (IDC). 2nd Nordic Symposium on Trace Elements in Human Health and Disease, Odense*. Abstract J10. WHO Regional Office for Europe, Copenhagen.

Thorèn, L. (1971), *Magnesium Metabolism. A Review of the Problems Related to Surgical Practice*. Prog. Surg. 9, 131.

Trueb, L. (1989), *Magnesium, the "Other" Light Metal* (in German). Neue Zürcher Zeitung, Forschung und Technik, No. 78, pp. 65 – 66 (5 April).

Tsui, J. C., Nordstrom, J. W., and Kohrs, M. B. (1986), *Relationships of Copper, Iron, and Magnesium Nutriture to Mortality in the Elderly. Trace Substances in Environmental Health – XX*. Proceedings, pp. 36 – 43, edited by Hemphill, D. D. University of Missouri, Columbia.

Ulrich, B. (1989), *Effects of Acid Deposition on Forest Ecosytems in Europe*. Adv. Environ. Sci., in press.

Vogel, G., and Angermann, H. (1981), *dtv-Atlas zur Biologie*, 2 Volumes, e.g., pp. 283ff. Deutscher Taschenbuch Verlag, Bielefeld.

Wacker, W. E. C. (1980), *Magnesium and Man*. Harvard University Press, Cambridge, Massachusetts.

Whang, R., Oei, T. O., Aikawa, J. K., Watanabe, A., Vannatta, J., Fryer, A., and Markanich, M. (1984), *Predictors of Clinical Hypomagnesemia*. Arch. Intern. Med. 144, 1794.

Welz, B. (1985), *Atomic Absorption Spectrometry*, 2nd Ed., pp. 297 – 299. Verlag Chemie, Weinheim-Deerfield Beach/Florida-Basel.

Zumkley, H., Bertram, H. P., Brandt, M., Rödig, M., and Spieker, C. (1986), *Magnesium, Aluminum, and Lead in Various Brain Areas. Trace Substances in Environmental Health XX*. Proceedings, pp. 29 – 35, edited by Hemphill, D. D., University of Missouri, Columbia.

II.19 Manganese

RAINER SCHIELE, Erlangen, Federal Republic of Germany

1 Introduction

Manganese (Mn) is in its inorganic species a ubiquitous, essential element in nature and in its occurring concentrations hardly toxic. Relatively large doses can be tolerated without injury. The manganese cycle plays a role in surface waters (interactions with the aquatic biota). Interactions with other metal compounds are also known. Environmental damage caused by this metal is not known so far. It is available in plants and animal cells in relative high concentrations in the mitochondria, where it acts as a cofactor for the activation of some enzymes.

2 Physical and Chemical Properties, and Analytical Methods

Manganese, atomic mass 54.94 and atomic number 25, is a very brittle, hard heavy metal of white-gray color; density, 7.2 up to 7.4 g/cm^3; melting point, 1244 °C; boiling point about 2000 °C. It belongs to the 7th Subgroup of the Periodic Table and is adjacent to iron. Manganese shares many mutual physico-chemical properties with iron. Manganese ores are very often found together with iron ores.

Manganese exists mostly in the oxidation state +2 in natural salts; and in stable native black manganese oxide, MnO_2, its valence is +4. Synthetic compounds are known in nearly all valence stages between −3 and +7 (COMMITTEE ON BIOLOGICAL EFFECTS OF ATMOSPHERIC POLLUTANTS, 1973). Permanganate as a strong oxidant has the valence +7. Manganese dioxide and the salts manganese carbonate, manganese sulfide, and manganese metasilicate are poorly water-soluble compounds. Important water-soluble compounds are manganese sulfate, manganese chloride, manganese nitrate, and the permanganate ion (MnO_4^-).

The electrothermal atomic absorption spectrometry (AAS) in the heated graphite tube (BUCHET et al. 1976; WELZ, 1985) and the neutron activation analysis (KASPAREK, 1979) are the most reliable analytical methods for the quantitative determination of manganese in the environment and in biological material. LAI et al. (1987) studied especially regional distribution of Mn in brain by the latter method. X-ray fluorescence analysis, emission spectroscopy (MONTASER and GOLIGHTLY, 1987, described ICP-AES techniques, WARD, 1987, ICP-MS techniques), and the classical colorimetric methods are suitable only for higher concentrations. But in recent years also very sensitive photometric techniques using the catalytic effects of manganese have been developed (MASPOCH et al., 1986). Photometric determination is, for instance, possible via a colored derivative from catalytic autoxidation of succinimide dioxime (CERDA et al., 1983, 1986). It must be noted that taking samples from body fluids or organs with steel cannulas and puncture needles may cause manganese contaminations up to the basal concentrations of the samples. This can be avoided by prerinsing with the patients' blood or still better, by the use of special cannulas made of nickel or plastic (BEHNE and BRÄTTER, 1979). With blood samples attention must be paid to possible manganese contents of the anticoagulants used.

3 Sources, Production, Important Compounds, Uses, Waste Products, Recycling

3.1 Occurrence, Production

It is presumed that the mean manganese concentration in the earth's crust amounts up to approximately 0.1%. The earth has a medium manganese content of 560 mg/kg with a widespread scattering from below 1 to over 7000 mg/kg (SHACKLETTE et al., 1971).

Elemental manganese does not exist in nature. More than 100 manganese minerals are known the most important of which is native black manganese oxide (MnO_2, e.g., pyrolusite). There are substantial workable manganese-iron deposits containing 40 to more than 50%, e.g., in the USSR, South and North Africa, South America, India, China and Australia (BEPPLER et al., 1978; SAAGER, 1984). These are exploited preponderantly in surface mining. The annual output comes to more than 20 million tons (about half in the USSR).

Important reserves for manganese, but still more for nickel, cobalt, and copper, are the so-called manganese nodules which are deposits of different metal oxides on the sea-bottom especially found in the North-East Pacific (CRONAN, 1978; SAAGER, 1984). The manganese oxides can be reduced technically by carbon monoxide, hydrogen, carbon, silicon, and aluminum (BEPPLER et al., 1978). The pure metal is preponderantly produced by electrochemical reduction of manganese sulfate. Dust can escape during the metallurgical operation (see Sect. 4.1).

3.2 Uses

The predominant portion (approximately 90%) of manganese is processed into ferro-manganese in blast furnaces and is used for alloying. This metal is necessary for binding oxygen and sulfur during the steel process to a large extent. Furthermore, it is used for the production of alloys together with aluminum, magnesium (electron), and copper (e.g., manganese bronze). Recycling does not play a role (SAAGER, 1984).

Black manganese oxide (MnO_2) is used as a depolarizer in dry-cell batteries. For various chemical reactions manganese (IV) oxide, manganese chloride, and manganese stearate are used as catalysts (for instance, manganese additives are suggested to improve the regeneration of Diesel exhaust filters, but 85% of the used manganese derivatives will be emitted (BOETTCHER, WIEDEMANN, and SCHUSTER, 1985). Manganese compounds, furthermore, are used as feed additives, fertilizers, pigments, dryers, wood preservatives, and for coating welding-rods. With reference to the organic compounds the fungicide manganese ethylene-bis(dithiocarbamate) (Maneb) and the antiknock agent methylcyclopentadienyl-manganese-tricarbonyl (MMT) are of importance.

4 Distribution in the Environment, in Foods, and in Living Organisms

4.1 Emission, Air, and Water Quality

Problems with air pollution arise during the mining, crushing, and smelting of ores as well as during steel production. Especially dust and smoke of manganese oxides (MnO_2 and Mn_3O_4) are the result. In the USA emission limits exist for ferro- and silicomanganese. In the Federal Republic of Germany emission values of 150 mg/m^3 must be kept during the production of ferromanganese in electric furnaces, in accordance with the VDI-GUIDELINE 2576 (1973) (BEPPLER et al., 1978). Generally the emission of manganese is limited to 5 mg/m^3 with mass streams of more than 25 g/L (see also MAK, 1988).

There is little proper information available referring to the immission values in the environment of manganese foundries and manganese mines.

In the USA the air concentrations in large towns are 0.01 to 10 µg/m^3, but they mostly do not exceed 0.2 µg/m^3, even in areas with specific emittents. An increase of the manganese concentration in the air from about 0.2 up to max. 1.7 µg/m^3 could be a result of replacing the antiknock agent tetraethyl lead by MMT (cf. Sect. 3.2) (COMMITTEE ON BIOLOGIC EFFECTS OF ATMOSPHERIC POLLUTANTS, 1973).

According to BRULAND (1983) manganese concentrations in the sea (especially in deeper water columns) may be of the order of 0.01 ppb (µg/L), and higher in surface waters due to atmospheric and river input. The main dissolved species are Mn^{2+} and $MnCl^+$ ions. According to KLINGHAMMER and BENDER (1980) the concentrations of manganese in seawater range from 0.03 to 0.8 µg/L. In earlier times somewhat too high values have been measured (see also Chapters I.1 and I.4a).

Manganese dioxide controls trace metal ions in natural water systems, depending on the surface properties of MnO_2 (BALIKUNGERI and HAERDI, 1985). In fact, manganese cycles (MnO_2, Mn^{4+}, Mn^{2+}) – especially at low oxygen concentrations

– regulate cellular manganese and manganese transport rates in algae (SUNDA, 1985) and production of manganese oxidizing bacteria (DE VITRE, 1986). STONE (1987) described reductive dissolution of Mn complexes and Mn(III) and Mn(II) interactions. Fe/Mn oxides in waters are carriers for many other inorganic and organic pollutants and are thus sources and sinks of them in aquatic environments incl. sediments.

4.2 Food Chain, Plants, and Animals

Almost independently of human activities, manganese is available in soils, plants, and animals. It seems that the manganese concentration in living organisms depends more on the respective species than on the concentrations in the environment. Plants contain about 1 to 700 mg/kg manganese (dry weight), sea fish 0.3 to 4.6 mg/kg, and the muscles of mammals 0.2 to 3 mg/kg (BOWEN, 1979).

Manganese does not concentrate in humans and in the majority of animals, but it accumulates in various kinds of plants such as legumes, nuts, heather and tea (BOWEN, 1979). Furthermore, some mussels and other marine invertebrates accumulate manganese up to the tenthousand-fold of the manganese content in their environment. The main manganese sources in food are grain, wholemeal bread, vegetables, nuts, and tea. Milk products and fish contain relatively low manganese concentrations. Liver may contain about 10 mg Mn per kg (WARD, 1987).

5 Uptake, Absorption, Transport and Distribution, Metabolism and Elimination in Plants, Animals, and Humans

Manganese from dust deposits and rainfall accumulates in the upper part of the soil. The enzymatic influence of microorganisms on manganese compounds in the soil and in the sea has been thoroughly investigated (EHRLICH et al., 1981).

The availability of manganese salts for plants above all depends on speciation and the pH value of the soil. Manganese cannot be absorbed from alkaline soil. The manganese concentration in plants is normally between 20 and 200 ppm (mg/kg wet weight). In a wide range neither deficiency nor symptoms of poisoning occur. Some kinds of grain and water plants are especially tolerant to concentrations up to several thousand ppm. A concentration of less than 20 ppm in leaves indicates a manganese deficiency (MATRONE et al., 1977).

Humans ingest approximately 3 mg manganese from food, about 5 µg from drinking water, and about 2 µg by inhalation per day.

Gastrointestinal resorption of soluble manganese compounds of 3% is assumed in both humans and animals (MENA et al., 1969), but it may be much higher during infancy. In the case of a concomitant iron deficiency (sideropenia), however, the resorption can increase to about 2- or 3-fold. There probably exists a reciprocal inhibition for the resorption of manganese and iron (FORTH and RUMMEL, 1973).

Ethanol increases and calcium salts may inhibit the resorption. The extent of pulmonary resorption is not known. The absorption of manganese by skin does not show a quantitative significance.

Manganese is especially stored in organs which are rich in mitochondria e.g., liver, kidney, hypophysis, and pancreas, with concentrations of about 1 mg/kg wet weight. The content in hair depends on the pigmentation (COTZIAS et al., 1971). The total manganese burden of an adult is estimated at 12 to 20 mg (COTZIAS, 1958). The normal concentration in blood comes up to 10 µg/L (BUCHET et al., 1976). Elevated manganese concentrations observed in individuals exposed to lead depend on an increase of porphyrin-bound manganese in the erythrocytes (ZIELHUIS et al., 1978). Manganese is excreted to a large extent (99%) via bile and the intestinal tract. The normal excretion amounts to less than 6 mg/100 g feces (JINDRICHOVA, 1969). Less than 1 µg/d (<0.1% of the total excretion) is excreted in urine. The half-life in the human organism has been determined to 37 ± 7 days for persons who were not exposed and 15 ± 2 days for miners working in manganese mines (COTZIAS et al., 1968). There may be an adaptation of the manganese excretion corresponding to the increased exposure. Manganese is able to permeate the placenta barrier in rats and humans and concentrates in liver and brain of the fetus (KAUR et al., 1980; TSUCHIYA et al., 1984) which seems not to lead to impairments.

6 Effects on Plants, Animals, and Humans

6.1 Deficiency Symptoms

Manganese seems to be an essential trace element for all living beings. Its significance as a cofactor for the enzyme pyruvate carboxylase is widely assumed (SCRUTTON et al., 1969). It is also assumed that manganese acts as a non-specific activator for the enzymes succinate dehydrogenase, prolidase, arginase, alkaline phosphatase, farnesyl pyrophosphatase, superoxide dismutase, glycosyl transferases, and adenosine triphosphatases. In many cases manganese can replace magnesium (MATRONE et al., 1977).

Deficiency causes growth disturbances in plants and a yellowing of the needles of conifers. A participation in the Hill reaction of photosynthesis is assumed. Birds have greater need in comparison to mammals. Manganese deficiency causes disorders in the menstrual cycle, still birth, and low birth weight as well as growth impairments in rats, mice, pigs, cows, and other animals (MATRONE et al., 1977). Disturbances of osteogenesis (formation of bone material) are traced to changes of mucopolysaccharide metabolism (LEACH, 1971) and are explained by a diminution in the activity of the glycosyl transferases. Manganese deficient diets lead to increased neonatal mortality, reduced growth, and skeletal abnormalities in offsprings of animals, and to structural defects in a number of cellular organelles (KEEN and LEACH, 1988). A lack of manganese causes ataxia in young birds, rats, and mice (disturbances especially of the gait) and loss in orientation which may have a common cause due to the missing expression of the otoliths (ERWAY et al., 1966). Furthermore, a diminished tolerance to glucose, depressed blood coagulation, decreases in triglyceride- and cholesterol levels in serum as well as structural changes of the mitochondria have been observed in animals with deficiencies of manganese. A possible manganese deficiency became obvious only in few patients, manifested by dermatitis, pigment changes, deficiencies in hair growth, hypolipidemia, and prothrombin deficiency (DOISEY, 1972). The estimated daily need of 2.5 to 5 mg for humans is normally covered by an adequate diet (MERTZ, 1979). Manganese compounds seem also to protect in acute intoxications by other substances, such as indium chloride (EYBL et al., 1987).

6.2 Effects on Plants and Animals

An excess of manganese in plants causes chlorosis. However, the tolerance of most plants is very high. In animals consuming manganese-rich food (more than 2000 ppm) an impairment of hemoglobin formation may be observed because of a reduced intake of iron into the porphyrin ring. In the case of sideropenic anemia, hemoglobin formation is reduced by a manganese concentration of 45 ppm in feed. These disturbances can be neutralized by an iron sufficient diet (400 ppm) (MATRONE et al., 1959). Some indications have been presented by BARLOW and SULLIVAN (1984) that manganese causes specific testicular damages in animals. But there is lack of data on human fertility and pregnancy after moderate exposure. Fish in acidified lakes partially show deformations of the vertebral column. This may be the result of either too high manganese- or too low calcium concentrations (FRASER and HARVEY, 1981).

6.3 Effects on Humans

Two kinds of diseases due to manganese are known in humans. These mainly concern workers occupied in manganese ore mills, smelting works, battery factories, and manganese mines. A risk exists with arc welding by the use of electrodes containing manganese. Manganese pneumonia results from acute inhalation of very high concentrations of manganese and cannot be differentiated clinically from non-occupational broncho- and lobarpneumonia. The therapeutic effect of antibiotics may be reduced. So-called manganism is a serious neuro-psychiatric disease, which is caused by inhalation of manganese in quantities of probably more than 100 mg/d during some months or years. After a short psychotic stage ("locura manganica") combined with emotional instability, hallucinations, and compulsive phenomena, a disease syndrome appears which is similar to Parkinson's disease with trembling of the hands, increased muscle tonus, rigidity of the face, and the typical impairment of the gait. Pathophysiologically it has been ascribed to disturbances of dopamine-, melanin- and probably to serotonin synthesis in the basal ganglia of the brain (MENA et al., 1970; ZIDEK, 1984). Pathologic changes may also occur in the brain cortex, the subcortical nuclei, and the spinal cord. Since not all humans will get manganism under equal conditions of exposure, a special predisposition is assumed. For instance, iron deficiency may influence the resorption and the effect of manganese. Furthermore, there seems to exist a species specificity in the sensitivity to manganese. Only in one group of primates is it possible to experimentally induce a disease which is similar to the clinical appearance of manganism in humans. Most similar are the signs of disease in chimpanzees after a single injection of manganese(III) oxide (500 mg/kg body weight) (MATRONE et al., 1977).

After termination of exposure the manganism does not progress, but neither does significant improvement occur. As in cases of Parkinson's disease due to other causalities the therapeutic administration of L-Dopa, 1,5-dioxytryptamine, and trihexylphenidyl can show improvement of some symptoms (MENA et al., 1970; ZIDEK, 1984). From 1976 to 1986, only 17 suspected cases of occupational disease caused by manganese and its compounds have been registered, and three of them have been indemnified for the first time in the Federal Republic of Germany. A mass poisoning by environmental pollution was obvious only in one case. The reason for this poisoning was spring water contaminated with manganese and zinc from dry batteries (KAWAMURA et al., 1941).

In several studies no significant relationship between manganism and manganese concentrations at the workplace could be determined, as reflected by blood, urine, and fecal levels (VALENTIN and SCHIELE, 1980). Besides the individual predisposition, the extended latent period between the exposure and the occurrence of the first symptoms may be responsible for the discrepancy. The usefulness of biological monitoring is, therefore, not yet proven for manganese.

Because of their low capability to excrete manganese, newborns and infants are suggested to be at a certain risk for a dangerous accumulation from prolonged low-level exposure to manganese (SETH and CHANDRA, 1984), inspite of the fact that the element is essential and that mother's milk may contain too little manganese in other cases.

6.4 Mutagenic, Carcinogenic, and Teratogenic Effects

Manganese salts are apparently potent mutagens in some microorganisms (LÉONARD et al., 1984) but not confirmed mutagens, teratogens (see also BARLOW and SULLIVAN, 1984) or carcinogens in mammalian systems.

7 Hazard Evaluation and Limiting Concentrations

It seems that a general risk to the environment by manganese exposure does not exist at present. Maximal immission concentrations have not been established so far. The limitation of 0.05 ppm in drinking water refers more to esthetic and technical than to toxicological reasons (MATRONE et al., 1977). The maximal allowable concentration at workplaces for an 8-hour day is up to 5 mg/m^3 in the USA (TLV), the FRG

(MAK, 1988) and in the majority of the Western countries. Because of the lower particle size and the higher content of Mn_3O_4, manganese smoke may be more dangerous than dust. A TLV (MAK) value (1988) of 1 mg/m³ is valid for manganous manganese oxide (Mn_3O_4). The TLV value for MMT amounts of 0.01 mg/m³.

So-called biological tolerance values in working materials (BAT values), esp. for blood and urine, have not been established so far. 50 µg/L urine has been suggested as a preliminary threshold value (EMPLOYMENT MEDICAL ADVISORY SERVICE, 1974). A maximal acceptable daily intake of 3 mg/kg weight can be inferred from animal experiments (ABEL and OHNESORGE, 1979). This value is 10- to 100-fold higher than the normal daily intake. Tolerance values of 0.5 up to 1 g/L have been determined for crawfish and 0.05 up to 1.2 g/L for fish. These levels are relatively high in comparison to other metals (JUNG, 1973). According to SONICH-MULLIN and VELAZQUEZ (1989) the maximum level of manganese in drinking water should not exceed 1 – 1.5 mg/L, but undesirable taste and odor are apparent already at a concentration of 0.05 mg/L (DWEL = drinking water equivalent level EPA).

References

Abel, J., and Ohnesorge, F. K. (1979), in: Gladtke, E., Heimann, G., and Eckert, I. (eds.), *Spurenelemente – Analytik, Umsatz, Bedarf, Mangel und Toxikologie*, pp. 185 – 191. Georg Thieme Verlag, Stuttgart.
Balikungeri, A., and Haerdi, W. (1985), *Surface Properties of Manganese Dioxide: The Impact of Sample Preparation and Treatment on the Maximum Exchange Capacity. Chimia 39*, 145 – 147.
Barlow, S. M., and Sullivan, F. M. (1984), *Reproductive Hazards of Industrial Chemicals*, pp. 370 – 385. Academic Press, London.
Behne, D., and Brätter, P. (1979), in Gladtke, E., Heimann, G., and Eckert, I. (eds.), *Spurenelemente – Analytik, Umsatz, Bedarf, Mangel und Toxikologie*, pp. 42 – 52. Georg Thieme Verlag, Stuttgart.
Beppler, E., Fichte, R., and Berger, A. (1978), in: *Ullmanns Encyklopädie der technischen Chemie*, 4th Ed., Vol. 16, pp. 425 – 466. Verlag Chemie, Weinheim-New York.
Boettcher, J. Wiedemann, B., and Schuster, H. D. (1985), *Proceedings Emission Reduction from Diesel Engine*. VDI, Düsseldorf.
Bowen, H. J. M. (1979), *Environmental Chemistry of the Elements*. Academic Press, New York.
Bruland, K. W. (1983), *Trace Elements in Sea-water. Chem. Oceanogr. 8*, 200 – 203.
Buchet, J. P., Lauwerys, R., and Roels, H. (1976), *Clin. Chim. Acta 73*, 481 – 486.
Cerda, V., et al. (1983), *Kinetic-catalytic Determination of Manganese(II) by Means of Succinimide-dioxime. Anal. Chim. Acta 155*, 299 – 303.
Cerda, V., et al. (1987), *Analyst 111*, 69 – 72.
Committee on Biologic Effects of Atmospheric Pollutants (1973), *Manganese*. National Academy of Sciences, Washington, D.C.
Cotzias, G. C. (1958), *Physiol. Rev. 38*, 503 – 532.
Cotzias, G. C. Horiuchi, K., Fuenzalida, F., and Mena, I. (1968), *Neurology 18*, 376 – 382.
Cotzias, G. C., Papavasiliou, P. S., Ginos, J., Steck, A., and Duby, S. (1971), *Science 176*, 410 – 412.
Cronan, D. S. (1978), *Endeavour, New Ser. 80*, 80 – 84.
De Vitre, R. (1986), *Proceedings 2nd IAEAC Workshop on Carcinogenic and/or Mutagenic Metal Compounds*, Villars, Switzerland; see also *Chemosphere 16 (4)*, N 57/58.

Doisey, E. A., Jr. (1972), in: Hemphill, D. D. (ed.), *Trace Substances in Environmental Health,* Vol. VI, pp. 193–199. University of Missouri, Columbia, Missouri.
Ehrlich, H. L., Rosson, R. A., Xenoulis, A. C., Ceccanti, B., Pakarinen, P., and Sundby, B. (1981), in: *Proceedings 5th International Symposium on Environmental Geochemistry, Stockholm.* Publishing House of the Swedish Research Councils, Stockholm.
Employment Medical Advisory Service (1974), *Biochemical Criteria in Certain Biological Media for Selected Toxic Substances,* Occasional Paper 1. Department of Employment, London.
Erway, L., Hurley, L. S., and Fraser, A. (1966), *Science 152,* 1766–1768.
Eybl, V. et al., CZ-30166 Plzeň (1987), *Influence of the Pretreatment of Manganese on the Toxic Effects of Indium in Mice. Proceedings Second Nordic Symposium on Trace Elements, Odense.* World Health Organization (WHO), Copenhagen.
Forth, W., and Rummel, W. (1973), *Physiol. Rev. 53,* 724–792.
Fraser, G., and Harvey, H. H. (1981), in: *Proceedings 5th International Symposium on Environmental Geochemistry, Stockholm.* Publishing House of the Swedish Research Councils, Stockholm.
Jindrichova, J. (1969), *Int. Arch. Gewerbepathol. Gewerbehyg. 25,* 347–359.
Jung, K. D. (1973), cited in: Förstner, U., and Wittmann, G. T. W. (1979), *Metal Pollution in the Aquatic Environment,* p. 28. Springer-Verlag, Berlin-Heidelberg-New York.
Kasparek, K. (1979), in: Gladtke, E., Heimann, G., and Eckert, I. (eds.), *Spurenelemente – Analytik, Umsatz, Bedarf, Mangel und Toxikologie,* pp. 185–191. Georg Thieme Verlag, Stuttgart.
Kaur, G., Hasan, S. K., and Srivastava, R. C. (1980), *Toxicol. Lett. 5,* 423.
Kawamura, R., Ikuta, H., Fukuzumi, S., Yamada, R., Tsubaki, S., Kodama, T., and Kurata, S. (1941), *Kitasato Arch. Exp. Med. 18,* 145–169.
Keen, C. L., and Leach, R. M. (1988), Chapter 34 *Manganese,* in: Seiler, H. G., Sigel, H., and Sigel, A. (eds.), *Handbook on Toxicity of Inorganic Compounds,* pp. 405–415. Marcel Dekker, Inc., New York.
Klinghammer, G. P., and Bender, M. L. (1980), *Earth Planet. Sci. Lett. 46,* 361.
Lai, I. C. K., Chan, A. W. K., Minski, M. J., Lim, L. and Davison, A. N. (1987), *Proceedings Second Nordic Symposium on Trace Elements, Odense.* World Health Organization, Copenhagen.
Leonard, A., Gerber, G. B., Jacquet, P., and Lauwerys, R. R. (1984), in: Kirsch-Wolders, M. (ed.): *Mutagenicity, Carcinogenicity, and Teratogenicity of Industrial Pollutants,* Chap. 2, pp. 59–126. Plenum Press, New York.
Leach, R. M., Jr. (1971), *Fed. Proc. Fed. Am. Soc. Exp. Biol. 30,* 991.
MAK (1988), *Maximum Concentrations at the Work-place. Report No. XXIV DFG.* VCH Verlagsgesellschaft, Weinheim-Basel-Cambridge-New York.
Maspoch, S., Blanco, M., and Cerda, V. (1986), *Catalytic Determination of Manganese at Ultratrace Levels by Flow Injection Analysis. Analyst 111,* 69–72.
Matrone, G., Hartmann, R. H., and Clawson, A. J. (1959), *J. Nutr. 67,* 171–175.
Matrone, G., Jenne, E. A., Kubota, J., Mena, I., and Newberne, P. M. (1977), in: *Geochemistry and the Environment,* Vol. II, pp. 29–39. National Academy of Sciences, Washington, D.C.
Mena, I., Horiuchi, K., Burke, K., and Cotzias, G. C. (1969), *Neurology 19,* 1000–1006.
Mena, I., Court, J., Fuenzalida, S., Papavasiliou, P. S., and Cotzias, G. C. (1970), *N. Engl. J. Med. 282,* 5–10.
Mertz, W. (1979), in: Gladtke, E., Heimann, G., and Eckert, I. (eds.), *Spurenelemente – Analytik, Umsatz, Bedarf, Mangel und Toxikologie,* pp. 185–191. Georg Thieme Verlag, Stuttgart.
Montaser, A., and Golightly, D. W. (1987): *Inductively Coupled Plasmas in Analytical Atomic Spectrometry,* p. 613 ff. VCH Verlagsgesellschaft, Weinheim-Basel-Cambridge-New York.
Saager, R. (1984): *Metallic Raw Materials Dictionary* (in German), pp. 108–112. Bank von Tobel, Zürich.
Scrutton, M. C., Utter, M. F., and Mildvan, A. S. (1966), *J. Biol. Chem. 241,* 3480–3487.
Seth, P. K., and Chandra, S. V. (1984), *Neurotoxicology 5,* 67–76.

Shacklette, H. T., Hamilton, J. C., Boerngen, J. G., and Bowles, J. M. (1971), *Elemental Composition of Surficial Materials in the Conterminous Unites States.* U.S. Geological Survey Paper 574-D. U.S. Government Printing Office, Washington, D.C.
Sonich-Mullin, C., and Velazquez, E. (1989), *The Risk Assessment of the Essential Element Manganese.* Proceedings 23rd Conference on Trace Substances in Environmental Health. EPA, Cincinnati; *Environ. Geochem. Health 12 Suppl.*, 379 – 386.
Stone, A. T. (1987), *Environ. Sci. Techn. and 17th Annual IAEAC Symposium on the Analytic Chemistry of Pollutants*, Jekyll Island, Georgia.
Sunda, W. (1985), *Limnol. Oceanogr. 30* (1), 71 – 80.
Tsuchiya, H., Mitani, K., Komada, K., and Nakata, T. (1984), *Arch. Environ. Health 39*, 11 – 17.
Valentin, H., and Schiele, R. (1980), *Human Biological Monitoring of Industrial Chemicals, 2. Manganese.* Commission of the European Communities, Brussels-Luxembourg.
VDI Guideline 2576 (1973), in: *Handbuch Reinhaltung der Luft*, Vol. 2. Beuth-Verlag, Berlin.
Ward, N. I. (1987), *Environmental Health* No. 20, pp. 118 – 123. WHO Regional Office for Europe, Copenhagen.
Welz, B. (1985), *Atomic Absorption Spectrometry*, pp. 299 – 300, 361 ff., 399 ff. VCH Verlagsgesellschaft, Weinheim-Deerfield Beach/Florida-Basel.
Zidek, W. (1984), in: Zumkley, H. (ed.), *Spurenelemente*, pp. 140 – 151. Georg Thieme Verlag, Stuttgart.
Zielhuis, R. L., Del Castilho, P., Herber, M. F. R., and Wibowo, E. A. A. (1978), *Environ. Health Perspect. 25*, 103 – 109.

Additional Recommended Literature

Balikungeri, A. (1989), *Acid-Base Properties of 2-Morpholino-ethanesulfonic Acid (MES), Complexation Reaction of Cu(II)-MES and Interaction of Hydrous Manganese Oxide Surface with Cu(II) in MES Buffer. Chimia 43* (1 – 2), 12 – 17.
Büchner, W., Schliebs, R., Winter, G., and Buchel, K. H. (1984), in: *Industrielle Anorganische Chemie*, Chap. 3.5, *Mangan-Verbindungen und Mangan*, pp. 285 – 297. Verlag Chemie, Weinheim-Deerfield Beach/Florida-Basel.
Giovanoli, R., and Faller, M. (1989), *Mangan(IV) oxidehydrates with Crystal Lattice Layers: Insertion of Cobalt, Nickel and Copper into Lithiophorite* (in German). *Chimia 43* (1 – 2), 11 – 12.
Hurely, L. S., Pfeiffer, C. C., Stampfl, A., Kiilunen, M., Favier, A., Rütter, J., and Neidhardt, B. (1984), *Proceedings Trace Elements, Neuherberg*, Vol. 3. Walter de Gruyter, Berlin.
Reidies, A. H. (1978), *Manganese Compounds* (in German) in: *Ullmanns Encyklopädie der technischen Chemie*, 4th Ed., Vol. 16, pp. 467 – 480. Verlag Chemie, Weinheim-New York.
Šarić, M. (1986), in: Friberg, L., Nordberg, G. F., and Vouk, V. B. (eds.), *Handbook on the Toxicology of Metals*, 2nd Ed., Vol. II, Chap. 15, pp. 354 – 386. Elsevier, Amsterdam.
Trueb, L. (1987), *Manganese* (in German) in: *Neue Zürcher Zeitung, Forschung und Technik, No. 214*, p. 95 (16. September), Zürich.
Underwood, E. J. (1977): *Trace Elements in Humans and Animal Nutrition*, 4th Ed., Chap. 7, pp. 170 – 195. Academic Press, New York.
de Vitre, R. R., Buffle, J., Tercier, M. L., Belzile, N., and Tessier, A. (1990), *In-situ Speciation Studies at the Water-Sediment Interface; Voltammetric Sensors for in-situ Measurement of Trace Elements in Water. Fresenius Z. Anal. Chem.*, in press.
WHO (1981), *Enivronmental Health Criteria 17.* World Health Organization, Geneva.
Yeats, P. A. (1988), *The Distribution of Trace Metals (especially of Nickel, Zinc, Copper, and Manganese) in Ocean Waters. Sci. Total Environ. 72*, 131 – 148.

II.20 Mercury

Rudy Von Burg, Pinole, California, USA
Michael R. Greenwood, Webster, New York, USA

1 Introduction

The poisonous properties of mercury have been known throughout recorded history and appear in ancient oriental and Roman literature (Li, 1948). Pliny the Younger was the first to call attention to a disease peculiar to slaves who worked in the mercury mines (Magos, 1975), and around 1533 Paracelsus described mercury poisoning as an occupational disease (Bidstrup et al., 1951). By the 18th century, reports of elemental mercury poisoning among miners, gilders, and mirror makers began to appear (Earles, 1964). During the 19th century, cases of mercury poisoning were reported in England, Spain, and other countries (Vroom and Greer, 1972).

The medicinal use of mercurials can also be traced back over 3000 years. They were used by Discorides Pedanius and Pliny as a treatment for syphilis and various skin disorders (Farler, 1952). During the 8th and 11th centuries, Mesue, Rhazas, and Avicenna are also reported to have used mercurial ointments as medicinals (Magos, 1975). In 1881, the antiseptic action of $HgCl_2$ was demonstrated, and between 1900 and 1920 mercurials were introduced for diuretic and chemotherapeutic purposes (Webb, 1966). However, due to an overall awareness of the toxicity of mercury, mercurials are no longer used as pharmaceutical agents. No mercury-based medicinals were listed in the "Physician's Desk Reference" of 1978 (Patty, 1981).

In the past, environmentalists and toxicologists (including clinical experts) have not always distinguished the various chemical forms (speciation) of mercury although these have very different environmental behavior, bioavailability (also for various organs), metabolism, and effects on organisms, as well as showing varying interactions. Today it is absolutely clear that metallic mercury, mercury vapor, inorganic mercury(I) and (II), alkylmercury, and phenylmercury must be distinguished. The uptake of neuro- and embryotoxic methylmercury, e.g., through contaminated aquatic organisms (Minimata disease; Tsubaki and Irukayama, 1977) is especially critical, but alkylmercury fungicides and seed treatments also lead to many poisonings (Bakir et al., 1973), particularly in developing countries.

2 Physical and Chemical Properties, and Analytical Methods

2.1 Physical and Chemical Properties

Elemental mercury has an atomic number of 80, an atomic mass of 200.59, a specific gravity of 13.55 g/cm^3, a melting point of $-39.8\,°C$ and a boiling point of 357 °C. Mercury occurs naturally as six main isotopes: ^{202}Hg (30%), ^{200}Hg (23%), ^{199}Hg (17%), ^{201}Hg (13%), ^{198}Hg (10%), and ^{204}Hg (7%), as well as trace amounts of ^{196}Hg and the relatively stable radioisotope ^{203}Hg.

Several other metals can dissolve in mercury to form various amalgams, and significant amounts of physical and chemical data are available for over 100 inorganic and organic mercury compounds (WEAST, 1978). However, the most important forms of mercury to which living organisms are exposed can be placed into three broad categories having different pharmacokinetic properties with regard to absorption, bodily distribution, accumulation, and toxic hazards. It is, therefore, necessary to discuss these categories separately.

2.1.1 Metallic Mercury

Elemental mercury is usually referred to as mercury vapor when present in the atmosphere or as metallic mercury in liquid form. This form is of considerable importance toxicologically because it has a relatively high vapor pressure and a certain water (about 20 µg/L) and lipid solubility (5–50 mg/L; BERLIN, 1986). For industrial applications, the following formula may be used to approximate the vapor pressure between 0 and 150 °C:

$$\log p = (-321.5/T) + 8.025$$

where p is the pressure in mm Hg and T the absolute temperature (PATTY, 1981). Thus a saturated atmosphere at 24 °C contains approximately 18 mg/m^3. Of equal significance is the fact that the vapor exists in a monoatomic state and upon inhalation the mercury is distributed primarily to the alveolar bed.

2.1.2 Inorganic Ions of Mercury

Mercury exists in ionic form as Hg^{2+} (mercuric salts) and Hg^+ (mercurous salts). The former readily form complexes with organic ligands, notably sulfhydryl groups. In contrast to $HgCl_2$, which is both highly soluble in water (69 g/L at 20 °C) and highly toxic, Hg_2Cl_2 is less soluble (2 mg/L at 25 °C) and correspondingly less toxic. The least soluble mercuric form is cinnabar (HgS), which has a water solubility of only 10 ng/L (WEAST, 1978).

2.1.3 Organic Mercury Compounds

Organic mercury compounds consist of diverse chemical structures in which mercury forms a covalent bond with carbon. For all practical purposes, the group is limited to alkylmercurials (methyl- and ethylmercury), arylmercurials (phenylmercury) and the family of alkoxyalkyl mercury diuretics. Organic mercury cations form salts with inorganic and organic acids, e.g., chlorides and acetates, and react readily with biologically important ligands, notably sulfhydryl groups. They also pass easily across biological membranes perhaps since the halides (e.g., H_3CHgCl, although it is not clear whether this exists in a stable form in tissues at physiological pH) and dialkylmercury are lipid soluble. The major difference among these various organomercury cations is that the stability of carbon-mercury bonds in vivo varies considerably. Thus alkylmercury compounds are more resistant to biodegradation than either arylmercury or alkoxymercury compounds (CLARKSON 1987).

2.2 Analytical Methods

With increasing concern about mercury toxicity, a whole host of techniques were developed to collect and detect submicrogram quantities of mercury in air, water, and biological samples. Many of these techniques have been described in the NIOSH "Manual of Analytical Methods" (1977). Earlier techniques used to measure mercury in biological tissue and fluids were usually colorimetric, using dithizone as the complexing agent. However, physical methods such as neutron activation and atomic absorption spectrometry are now commonly used. Gas chromatography is an important tool if the form of mercury is to be identified.

The determination of mercury in biological materials is always fraught with difficulties, and the determination of the extremely small quantities of mercury which is now possible places stringent demands on the ability of the analyst. Thus, it is best that results be checked by an independent method. Cold vapor atomic absorption (to measure total and inorganic mercury) and gas chromatography (to measure organic mercury compounds) make an ideal combination for the cross-checking of results.

2.2.1 Collection of Samples

The collection, storage, and transport of samples for mercury analysis may present problems since glass and plastic risk to be contaminated with mercury (GREENWOOD and CLARKSON, 1970; see also Chapter I.4a). Old data in the literature are therefore often too high. On the other hand, tubes used to collect air samples require appropriate precautions to prevent volatilization and loss of mercury vapor. Blood samples are conveniently collected, stored, and transported in a heparinized Vacutainer®. Urine samples should be acidified before storage or transportation, but the use of thymol as a preservative is not advised if non-destructive atomic absorption procedures are to be used.

For all biological samples in which mercury is to be analyzed, great care should be taken to avoid bacterial contamination. Bacteria not only break down methylmercury to inorganic mercury, but they are capable of reducing inorganic ions to metallic mercury which leads to loss by volatilization (MAGOS et al., 1964). Fish and water samples are best frozen to prevent biological breakdown or bacterial contamination. Water samples should be collected in glass bottles, adjusted to a pH of 2, frozen and stored in the dark (MAY et al., 1980). Also, calcium iodide can reduce mercury adsorption onto glass.

If mercury is analyzed in hair samples, care should be taken to ensure that external metal contamination has not occurred. Sources of such contamination may be occupational or shampoos and other cosmetics which may contain phenylmercury. Washing procedures generally do not result in the satisfactory removal of externally applied mercury from hair. Hair samples may be stored in paper envelopes or in polyethylene plastic bags. Correlation between blood and hair has only been demonstrated for methylmercury compounds (CLARKSON, 1977). The value of hair measurements in the case of exposure to mercury vapor has yet to be demonstrated. FRANCIS et al. (1982) found no difference in hair mercury levels between dental workers and a non-exposed control group. There was, however, a highly significant but unexplained negative correlation between the mercury concentration of hair and age.

For air sampling in an occupational setting, mercury vapor personnel samplers rely on a hopcalite filter absorber, followed by desorption and atomic absorption spectroscopy (AAS) analysis. This method requires field and reagent blanks, but has a reported detection limit in the $\mu g/m^3$ range (RATHJE and MARCERO, 1976). BREDER and FLUCHT (1984) used gold and silver adsorbers for atmospheric mercury collection and determined the electrothermally desorbed mercury by atomic absorption spectrometry.

For field or environmental sampling, NIOSH (1977a) proposed a three-stage sampler. Particulate mercury is collected on a membrane prefilter ahead of a two-section, solid phase tube. Organic mercury collects in the first section of this tube, while inorganic mercury vapor is amalgamated in the second section. Each stage is thermally desorbed separately and analyzed by AAS. Under practical circumstances, sampling, sample handling, and separation are rather delicate (see also Chapter I.4a), especially when the concentrations are very low as, for instance, with sea water (OLAFSSON, 1983; BLOOM and CRECELIUS, 1983; BREDER, 1987) and in wet deposition (AHMED et al., 1987). MAY et al. (1985) determined methylmercury in environmental samples by extraction with hydrochloric acid in the absence of light, separation in an anion exchanger, decomposition under pressure and cold vapor AAS.

Speciation analysis needs special approaches, since a total of 10 ppt mercury in rivers and lakes consists of several inorganic and organic species. HAERDI (1989) enriched aquatic samples on a microcolumn and separated labile mercury by reduction with copper. For speciation of mercury in natural waters see also SCHROEDER (1989).

2.2.2 Atomic Absorption

Cold vapor atomic absorption is the most widely used method for the determination of total mercury in biological material. Most procedures involve the digestion of the sample in an acid-oxidizing medium often involving excess permanganate and inorganic acids. Once the mercury is converted into the ionic salt, it is liberated from the digest by reduction with stannous chloride and passed through an atomic absorption unit where the absorption at 254 µm is recorded. Many variants of this technique have been published in the literature.

MAGOS and co-workers developed an atomic absorption procedure for determining total mercury in biological samples that does not necessitate sample digestion (MAGOS, 1971; GREENWOOD et al., 1977). It also has the advantage that inorganic mercury can be determined in addition to total mercury. The mathematical difference of the two results is a measurement of the organic mercury content, which in most cases of environmental exposure will be in the form of methylmercury.

Adaptations of this method have a reported detection limit of 0.5 ng Hg, which makes them suitable to determine mercury concentrations in the range of 1–10 ng/mL in blood, urine, hair, tissue, and other biological samples (SCHALLER, 1982). The technique has the advantage of high speed (each determination taking less than two minutes), high sensitivity, and the apparatus is portable and suitable for field use. A more recent modification of the Magos method (WIGFIELD and PERKING, 1982) allows the speciation of trace quantities of mercury into elemental mercury and the mercurous ion. For mercury analysis in blood and urine, ANGERER and SCHALLER (1988) use a commercially available hydride system with an amalgamation device (gold/platinum gauze) after first reducing the sample with sodium borohydride.

2.2.3 Neutron Activation

Neutron activation methods have also been used to determine total mercury content in samples (HOROWITZ et al., 1976). These methods have the advantage of high sensitivity (detection limit approximately 0.5 ng/g of sample), need no reagent blank, are independent of the chemical form of the element, and can use non-destructive methods when concentrations higher than 1 ppm are expected. The method is useful as a reference method to check the accuracy of other procedures. It has the disadvantage that it cannot be readily adapted to field use, cannot identify the form of mercury, and large numbers of samples usually require special radiation procedures and data processing. Other techniques include proton bombardment and X-ray fluorescence. These techniques have useful applications for special problems of mercury analysis such as analysis of hair sample segments (JAKLEVIC et al., 1977) or fish samples (SIPOS et al., 1979; AHMED et al., 1981).

2.2.4 Gas Chromatography

Gas chromatography with electron capture detection is the method of choice for determining the chemical form of mercury in biological samples down to 1 ng/g concentrations. WESTOO (1968) and SUMINO (1968) developed the first general procedure for gas chromatographic determinations of organic mercury compounds. Since then, adaptations of the method have been applied to a wide variety of biological samples including fish, urine, hair, and blood (VON BURG et al., 1974; CAPPON and SMITH, 1978; SCHALLER et al., 1978). All procedures involve the extraction of methylmercury with an organic halide salt, followed by a clean-up procedure prior to analysis. An important development of the gas chromatographic method was the ability to measure not only organic mercury compounds in a sample, but also inorganic mercury (CAPPON and SMITH, 1977).

2.2.5 X-Ray Fluorescence Spectroscopy

This technique has recently been verified against atomic absorption spectroscopy by MARSH et al. (1987). It was successfully used to determine the maximal level of mercury in maternal hair to assess fetal exposure. The technique utilizes X-ray scatter by atoms and has the advantage of being non-destructive (TORIBORA et al., 1982).

3 Sources, Production, Important Compounds, Uses, Waste Products, Recycling

3.1 Sources, Extraction, and Production

Mercury has a ubiquitous distribution. It is found in igneous rocks of all classes and massive quantities were undoubtedly spewed into the early Earth's atmosphere by volcanic activity (GOLDWATER, 1972). In nature, mercury occurs in a variety of physical and chemical forms. Normal soils typically contain 20–150 ppb Hg, but near known deposits the level can reach as high as 80% (WHO, 1976). Mercury is mined as cinnabar (mercury sulfide) and in some areas such as Almaden, Spain, the ore is so rich that liquid metallic mercury is also present. Generally, mercury binds strongly to the organic components in soil so that mobility by leaching is minimal and contamination of ground water is unlikely unless mercury leaches from a municipal landfill (US EPA, 1984).

The largest mercury reserves exist in Spain, the Soviet Union, and China (together about 60%; SAAGER, 1984). Normally the ores, which may contain about 1% mercury, are extracted near the surface, preconcentrated, separated by flotation, dried, and the mercury is then distilled directly from the products (ADAM et al., 1980; mercury oxide decomposes at about 350°C and mercury sulfide above 737°C). Mercury condensation and refining are important processes (ADAM et al., 1980). In

1982 about 6500 tons of mercury were mined and produced worldwide: 29% in the Soviet Union, 23% in Spain, 18% in China, 13% in the United States, and 12% in Algeria (SAAGER, 1984). Italy and Canada, also with large capacities, reduced their production for economic reasons.

Human activities have resulted in the release of a wide variety of both inorganic and organic forms of mercury. The electrical industry, chloralkali industry, and the burning of fossil fuels (coal, petroleum, etc.) release elemental mercury into the atmosphere. Metallic mercury has also been released directly to fresh water by chloralkali plants, and both phenylmercury and methylmercury compounds have been released into fresh and sea water – phenylmercury by the wood paper-pulp industry, particularly in Sweden, and methylmercury by chemical manufacturers in Japan.

3.2 Uses

(Table II.20-1; ADAM et al., 1980; SAAGER, 1984)

Important mercury compounds, which also may be released into the environment, are mercury(II) oxide, mercury(II) sulfide (cinnabar), mercury chlorides, mercury nitrates, mercury sulfates, mercury(II) thiocyanate, chloride and dithiocarbamate, borate and oleate of phenylmercury, and chloride, silicate and phosphate of alkylmercury compounds (ADAM et al., 1980).

Elemental and inorganic mercury compounds are used in the manufacture of scientific instruments (thermometers, barometers), electrical equipment (switches, rectifiers, oscillators, electrodes, batteries, meters, mercury vapor lamps, X-ray tubes, lead and tin solders), dental amalgams (see, for instance, CLARKSON et al., 1988),

Table II.20-1. Patterns of Mercury Consumption in the USA. (From EPA Criteria Document, 1979)

End Use	Annual Consumption (% total)		
	1970	1973	1985
Electric Equipment[a]	26	33	32
Caustic chloride	25	24	23
Paints	17	14	5.1
Technical instrum.	7.9	13	21.0
Dental	3.7	4.9	6.2
Catalysts	3.7	1.2	0.8
Agriculture	3.0	3.4	1.1
Laboratories	3.0	1.2	?
Pharmaceuticals	1.1	1.1	0.8
Others	9.6	4.2	9.8

[a] According to SAAGER (1984) the share of the electro-industry is probably much greater (including batteries), also because other uses are decreasing faster

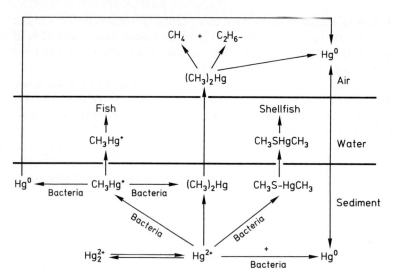

Fig. II.20-1. The mercury cycle demonstrating the bioaccumulation of mercury in fish and shellfish. NATIONAL ACADEMY OF SCIENCES (1978).

and synthetic silk. Mercury oxide batteries, however, are being replaced by manganese-alkali, manganese dioxide-zinc or nickel-cadmium batteries which contain much less mercury in the form of amalgams (SAAGER, 1984). Mercuric fulminate is no longer of importance (SAAGER, 1984). In the chemical industry mercury is used as a fluid cathode for the electrolytic production of acetic acid, chlorine, and sodium hydroxide.

In the past mercury has been used in the plating, tanning and dyeing, textile, photographic, and pharmaceutical industries. It has been used for the preparation of drugs and disinfectants, and arylmercury compounds have been used as disinfectants, fungicides (e.g., in wood preservation and paints), antiseptics, herbicides, preservatives and as a denaturant for ethyl alcohol (see also ADAM et al., 1980). Alkylmercury compounds have been used primarily as timber preservatives, disinfectants, and for treating seed against fungi (DHEW, 1977).

It has been reported that over 1200 potential disease-bearing organisms are found in or on seeds. Such organisms can reduce germination potential as well as spread disease to other crops. Mercurial seed dressings or foliar treatments gained rapid popularity during the 1960s (quoted by D'ITRI, 1972). However, the use of mercury fungicides has led to a number of large-scale poisonings involving whole communities. In the past twenty years, there have been thousands of cases and several thousand deaths due to this type of use (see Sect. 6.4). Even when used correctly, organomercurial fungicides can still contaminate soil, surface water and ground water and therefore enter the mercury cycle shown in Fig. II.20-1.

3.3 Waste Products and Recycling

A major source of mercury waste comes from the use of mercury cells in the chloralkali industry. The metal is generally recovered from collected sludges through retorting. The capacity of these units is limited although more recent developments have overcome many of the problems (OLOTKA, 1974). Unfortunately, prior to 1970 considerable quantities of such sludge were deposited in landfills or discharged into waterways in the United States and other industrialized countries. This uncontrolled disposal of mercury led to considerable accumulation in bottom sediments of rivers and lakes. In addition, the use of phenylmercury as a fungicide in Sweden stimulated considerable concern in that country when certain seed-eating bird populations began to die (JENSEN and JERNELÖV, 1967, 1969; BERGLUND et al., 1971).

Over the years, mercury cell operators have come under great pressure by governments and the general public to curb mercury losses. Diaphragm cells in the chloralkali industry are an improvement (SAAGER, 1984). Abatement efforts have limited the discharge of contaminated water effluents and restricted the dumping of contaminated sludge into landfills. Thus, mandated control has forced a recycling by industry. About 10% of today's mercury consumption originates from recycled batteries, dental material, and industrial scrap (SAAGER, 1984). At the same time, the use of mercury fungicides has been banned, thereby eliminating one source of environmental contamination.

4 Distribution in the Environment, in Foods, and in Living Organisms

4.1 Emissions, Air and Water Quality, Distribution and Biochemical Interactions in Soils

There is a natural source of mercury independent of man's actions. This is a general cycle (Fig. II.20-1) whereby the mercury is transported to surface waters by soil erosion and is circulated into the atomsphere by a natural degassing of the earth's crust and oceans. The second source of mercury is the direct or indirect result of man's activities. Total input into the atmosphere has been estimated at up to 150000 tons per year. Natural emissions (from volcanoes, wind erosion, and soil degasification) account for two-thirds of the input while man-made release (including coal and petroleum combustion) accounts for about one-third (NATIONAL ACADEMY OF SCIENCES, 1978; KORRINGA and HAGEL, 1974; BERLIN, 1986). For further information on mercury cycling in the environment the reader is referred to HUTCHINSON and MEEMA (1987). Anthropogenic emissions are actually decreasing. For example, in the Federal Republic of Germany from 1983 to 1985, emissions into water decreased from 1.1 to 0.2 tons, and emissions into the atmosphere from 5.5 to 4.2 tons (UMWELT, 1988). The use of mercury in batteries, vacuum-tube lamps and dental amalgam, however, remains about the same.

Atmospheric concentrations of 10 ng/m^3 have been reported in the Soviet Union, and 0–14 ng/m^3 from non-industrialized regions of Japan. Mercury levels of some United States cities are: Chicago, 3–39 ng/m^3; Denver, 2–5 ng/m^3; New York, 0–14 ng/m^3; San Francisco, 0.5–50 ng/m^3. The average atmospheric concentration in the United States was estimated to be in the range of 2–10 ng/m^3 (US EPA, 1984). BREDER and FLUCHT (1984) found about 1 to 8 ng Hg per m^3 in the air of the Italian Riviera, and about 10 to 20 ng/m^3 in the air of most regions of Tuscany. Lowest concentrations were found over the Mediterranean Sea. Total mercury levels in the air over mercury deposits and in some locations in Genoa have reached 1600 ng/m^3 and are related to the concentration of mercury in the ore (NRCC, 1979; BREDER and FLUCHT, 1984).

According to AHMED et al. (1987) total wet deposition of mercury is between about 10 and 100 ng per day in Stolberg-Binsfeldhammer (FRG) and between 1 and 4 ng/d in Leversbach (an unpolluted area in Germany). In the former, about 0.4% of this wet deposition is methylmercury, in the latter about 6%. This means that wet deposition of methylmercury is of the order of 1 to 3 ng per day (see also HULTBERG and HASSELROT, 1981).

Water contains mercury mainly in the form of Hg^{2+} as complex salts bound to dissolved particles. BLOOM and CRECELIUS (1983) distinguished between total mercury, easily reducible mercury, and dissolved mercury in sea water. MIERLE (1988) found about 0.3 to 2.2 ng total Hg per liter in lakes and rivers, and about 5 to 40 ng/L in brown streams and precipitation in Ontario, Canada. FITZGERALD and WATRAS (1988) measured about 1 to 2 ng per liter in lakes in Wisconsin, USA. According to BLOOM and WATRAS (1988), snow may contain 4 ng total Hg per kg and 0.05 ng methylmercury per kg, rain samples 2–5 ng total Hg per liter and 0.15 ng methylmercury per L, and lake water less than 0.6 ng methylmercury per L. BREDER et al. (1985) determined dissolved and particulate mercury in the Rhine river. Unfiltered samples contained about 5 ng/L (Lake Constance) and 50–100 ng/L (middle part of the Rhine). The Main river contained 250–400 ng/L. After filtration a minimum level of 4 ng/L was found. The mercury content of the suspended particles varied, but mercury concentrations are decreasing in the water (1972: 2.3 µg/L, 1976: 1.1 µg/L, 1986: 0.2 µg/L), and in the Rhine sediments (1977: 8 mg/kg, 1981: 3 mg/kg, 1986: 1 mg/kg) at the German-Dutch border (UMWELT, 1988). In an unpolluted trout pond, PADBERG (1986) determined about 4 ng total Hg per L and about 0.8 ng methylmercury per L. In a polluted pond the concentration of methylmercury was about the same, but the total mercury concentration is higher. Sediments normally contain about 1% of the total mercury content in the form of methylmercury. Ocean water is considered to have an average concentration of 0.4 to 2 ng Hg per L, with a range of 0.0004–0.002 ppb (BRULAND, 1983; OLAFSSON, 1983; BLOOM and CRECELIUS, 1983).

Since most of the mercury was and is produced around the Mediterranean Sea, a balance for this regional system is of special interest. VIGHI (1988) has extrapolated that annually about 236 tons (150 t from deposition, 50 t from Atlantic waters, 32 t from rivers and direct input, and 4 t in methylated form from the sediments) are introduced and the whole basin contains about 3700 tons of mercury. Fish and shellfish contain about 104 tons in total and suspended solids about 185 tons. It is thought that about 150 tons per year are volatilized again and that about 75 tons/year are sedimented. There may also be a Mediterranean outflow. Concentrations in fish are not related to levels in water (possible "hot spots" are only found in restricted areas, and different temperatures lead to different levels of methylation). Regulatory actions must take into account variances and possible local risk by fish consumption.

From the point of view of environmental exposures, methylmercury compounds present the greatest cause for concern, and the primary route of human exposure is through the food supply. Occupational exposures are generally to mercury vapor. Exposures to a wide variety of other mercury compounds also take place to a lesser extent, depending on the individual circumstances (occupational, medicinal, accidental, or environmental).

During the 1970s, a greater understanding of the role that inorganic mercury compounds play in the production of organic mercurials was achieved. In addition to direct chemical methylation of mercury, favored at high pH (WONG et al., 1981; REISINGER et al., 1981), a variety of bacterial and fungal organisms seem to have the capacity to methylate mercury, provided they are resistant to mercury toxicity. The first pathways that were described for the biological methylation of mercury involved methylcobalamin, an analog of vitamin B_{12}, which was studied in an anaerobic species of *Methanobacterium* (WOOD et al., 1968). In 1973, JERNELÖV reported the ability of *Neurospora crassa* to synthesize methylmercury and correlated this with their resistance to high concentrations of mercury in the substrate. REISINGER et al. (1984) discussed the transfer of the methylcarbenium ion from S-adenosylmethionine to inorganic mercury as a possible biological mechanism.

Dimethylmercury as well as monomethylmercury may be formed by bacteria present in sediments. Once formed, dimethylmercury can react with hydrogen sulfide to form insoluble dimethylmercury sulfide which rises to the water surface and volatilizes into the atmosphere. In the atmosphere, it undergoes decomposition back to elemental mercury and returns to land via precipitation (WOOD, 1972). JENSEN and JERNELÖV (1969) have shown that in the presence of Hg^{2+}, one molecule of dimethylmercury is converted to two molecules of monomethylmercury, e.g., in acid rain. JENSEN and JERNELÖV (1972) pointed out that conditions which promote bacterial growth also enhance the production of methylmercury.

In general, only a small fraction of the total mercury in sediment is converted to methylmercury (perhaps 1%). On an annual basis and global scale, total production of methylmercury in fresh water was estimated to be about 10 metric tons per year, and in the oceans about 480 metric tons annually. Thus, the ecological consequence of mercury release by man is a higher recirculation rate from organogenic sediments and higher concentrations of methylmercury in aquatic organisms.

Microorganisms also demethylate methylmercury compounds, splitting the carbon-mercury bond to release Hg^{2+}, then reducing Hg^{2+} to elemental mercury, which can volatilize. The organic portion can also volatilize. Phenylmercury produces benzene, methylmercury produces methane, and ethylmercury produces ethane (SILVER, 1984). Both processes are enzyme-mediated (NATIONAL ACADEMY OF SCIENCES, 1978). The presence of these enzymes is governed by plasmids and not by chromosomal genes (SILVER, 1984). Microorganisms capable of both methylation and demethylation reactions occur in aquatic sediments, soils, and human fecal material. WOOD (1972) pointed out that as a result of methylation and demethylation reactions, methylmercury concentrations will approach steady state in any given ecosystem, but the steady state concentrations will be affected by any environmental factor that influences either or both reactions.

Conversion of inorganic mercury to methylmercury is also possible from the effluents of acetaldehyde and vinylchloride manufacturing plants (TSUBAKI and IRUKAYAMA, 1977). It has also been suggested that methylmercury may be formed by the reactions of peracetic acid with inorganic mercury or absorption of mercuric chloride to acetylene resulting in chlorovinylmercury (SEBE et al., 1967) with eventual decomposition to acetaldehyde and methylmercury. Inorganic mercury ions are also available to form a variety of complex chelates. Sulfur, sulfhydryl groups, citrates, and EDTA react readily with mercury to form mercury complexes that are less bioavailable. In the following mass action equation

$$MeHgCl \rightleftarrows MeHg^+ + Cl^-$$

the presence of chloride ions shifts the reaction to the left, maintaining the uncharged species which is more readily transported across biological membranes. Decreasing the chloride ion concentration by the addition of Ca^{2+} or Mg^{2+} shifts the reaction to the right and reduces biological uptake (ROGERS and BEAMISH, 1983). The uptake of methylmercury is also favored by a low pH. At pH greater than 5, the chloride is transformed into the more soluble hydroxide which has a lower membrane penetration ability (KAWAMATA et al., 1982).

4.2 Food Chain, Plants, Animals, and Humans
(see also Chapter I.8 b)

Agricultural soils contain mercury levels between 0.06 and 0.2 mg/kg, and some edible plants such as carrots, potatoes, and mushrooms have been reported to take up mercury compounds.

MAY et al. (1985, 1987) have determined total mercury and methylmercury concentrations in living organisms and food: grass contains about 4 µg Hg per kg (0.4 µg/kg of which is methylated); brown algae contain about 10 µg/kg (2 µg/kg methylated); poplar leaves about 20 µg/kg (0.1 µg/kg methylated), earthworms about 20 µg/kg (2 µg/kg methylated); mussels about 20 µg/kg (8 µg/kg methylated); and ocean fish about 20 µg/kg (18 µg/kg methylated). Spruce needles contain about 70 µg/kg (0.5 µg/kg methylated), human liver about 70 µg/kg (14 µg/kg methylated), and carp about 70 µg/kg (67 µg/kg methylated). Mushrooms may even have 1 or more mg/kg (40 µg/kg methylated).

In rice, the use of foliar sprays can result in grain mercury levels that are 4–6 times greater than those of unsprayed controls (range 0.1–1.0 ppm). However, a survey found only a negligible amount of mercury in marketed rice with occasional concentrations of 0.015 ppm (SMART and HILL, 1968).

The concentration of methylmercury in fish is generally related to size and ecological niche. Concentrations as high as 1 mg/kg have been reported for open ocean predators such as sword fish and tuna (BERLIN, 1977). However, in industrially contaminated waters, methylmercury levels may exceed 10 mg/kg in fish muscle (BERLIN, 1977; RENZONI et al., 1979; see also Sects. 5.1 and 6.1).

Between 1956 and 1960, the Chisso Corporation, a major manufacturer of vinylchloride, discharged about 200 – 600 tons (BERTRAM et al., 1985) of mercury polluted effluents which eventually reached Minimata Bay, Japan, and accumulated in the bottom sediment at concentrations of up to several hundred micrograms per gram of dry weight. The principal mercury compounds in sediments have been reported to be mercuric oxide and mercuric sulfide. Traces of methylmercury were occasionally detected in deeper strata, but no correlation was found between the presence of aerobic or anaerobic bacteria and the presence of methylmercury. The total mercury and methylmercury contents of the seafood in Minimata Bay were higher than of seafood from control areas. The accumulation of methylmercury in fish and shellfish in Minimata Bay and subsequent consumption by humans and animals resulted in a mass outbreak of methylmercury poisoning (TSUBAKI and IRUKAYAMA, 1977). In 1964, a smaller but similar outbreak of Minimata disease was recognized in Niigata, Japan. Again, a chemical factory was identified as discharging wastes into the Argano river.

In Iraq outbreaks of mercury poisoning were also caused by contaminated food. In the 1971 – 1972 outbreak the cause was traced to homemade bread prepared from wheat seed that had been treated with a methylmercury fungicide (see Sect. 6.4).

Regardless of the source, local pollution is of concern since mercury bioaccumulates in the food chain, and fish and shellfish may be the dominant dietary sources of mercury for some human populations (US EPA, 1984; see also Sects. 5.1, 6.1, 6.6.1, and 6.6.2). The following bioconcentration factors (see also Sects. 5.1 and 5.3) have been reported:

Freshwater fish	63 000
Marine fish	10 000
Marine plants	1 000
Marine invertebrates	100 000
Freshwater plants	1 000
Freshwater invertebrates	100 000

The accumulation of mercury in fish is related to age, predation level and ecological niche. Terrestrial animals rarely have mercury levels in muscle that exceed 50 µg/kg (50 ppb). However, under certain agricultural conditions, methylmercury can reach toxic levels in game birds (BERLIN, 1977). During the Iraq epidemic, the feathers and muscle tissue of dead seed-eating birds were found to contain up to 9 – 52 ng/g (ppm). The relatively equal distribution between muscle and feathers suggests that these birds had not reached the steady state, since feathers normally concentrate mercury about eight times more than muscle (JERNELÖV, 1974). LEUMANN and BEUGGERT (1977) reported that the concentration of mercury in buzzard feathers rose from 15 – 20 ppm during the period of 1950 to 1960/65 and then declined after the prohibition of mercury-containing fungicides.

Human blood concentrations have been quoted by many groups. Since uptake from terrestrial food and drinking water is negligible, average mercury levels in the general population are less than 1 µg per liter of blood or urine (ANGERER and SCHALLER, 1988). However, it is known that marked interindividual variation can occur. Mercury concentrations in blood and urine can be affected when mercury-containing medicines are taken (ANGERER and SCHALLER, 1988) or when fish consumption is very high.

In Iraq, unexposed members of the rural population were found to have blood mercury levels of less than 5 ng/mL and some 70% of these were less than 2 ng/L (GREENWOOD, 1985). Blood samples taken from laboratory workers were found to be 10 ng Hg/L. EYL (1970) reported normal blood levels of mercury to be in the range of 5 ng/mL, although other investigators reported mean values in the range of 0.6 – 0.7 ng/mL (SCHIERLING and SCHALLER, 1981; STOEPPLER et al., 1982).

5 Uptake, Absorption, Transport and Distribution, Metabolism and Elimination in Plants, Animals, and Humans

5.1 Uptake and Absorption

Since mercury is not very phytotoxic in normally occurring concentrations (perhaps plants have a certain protection mechanism), not too much is known about mercury uptake and metabolism in plants. There is probably a root barrier, and mercury may be bound to soil. Some accumulation has been observed in mushrooms, aquatic plants, carrots, and potatoes (see also Chapter I.8a). FERRARA et al. (1989) observed that the aquatic *Posidonia oceanica* could be a biological indicator for mercury in sediments. WYTTENBACH et al. (1989) found that mercury concentrations in spruce needles increase continuously with age. BARGAGLI (1989) discussed collection, analytical procedures and data elaboration to use epiphytic lichens as biomonitors, and suggested to study only the outermost zone of the thallus. Lichens and mosses may concentrate mercury on their surfaces. HAINES (1988) found about 500 µg Hg per kg of muscle in white sucker fish from the soft water of Lost Lake in Maine, and concluded that sulfate concentration in the lake water is related to fish uptake (see also Sect. 6.1). STOKES and WREN (1987) have studied bioaccumulation of mercury by aquatic biota.

Fish accumulate mercury. This is evident from radiotracer studies (BEIJER and JERNELÖV, 1978b). However, much of this accumulation must be superficial in gills, scales, and mucous membranes. Time is probably necessary to transport it into the body, and the toxicity is low on an acute basis. Also, MITRA (1986) points out that fish kills are rare (see also Sect. 6.1).

It is difficult to estimate accurately the human dietary intake of mercury, but an estimate for different age groups for 1981 – 1982 was in the range of 3000 ng Hg per day for adults, 1000 ng/d for young children, and less than 1000 ng/d for infants (US EPA, 1984). CLARKSON (1978) attempted to calculate the maximum daily intake for the "70 kg standard adult" in the United States population (Table II.20-2). Daily intake from drinking water is generally considered to be no more than 2 ng/day, while

Table II.20-2. Estimate of Average and Maximum Daily Intakes of Mercury in the "70 kg Standard Adult" in the US Population (mercury intake μg/day/70 kg). (From EPA Mercury Criteria Document, 1979)

Medium	Average	Maximum	Predominant Form
Air	0.3	0.8	$Hg(0)$
Water	0.1	0.4	Hg^{2+}
Food	3.0	5.0	CH_3Hg^+

daily intake from air (assuming an inhalation of 20 m³/d and less than 100% retention) is considerably less than 60 ng/d. More recent estimates have been made by BERNHARD and ANDREAE (1984) and GESAMP (1987). Consequently, food provides the major portion of the daily environmental mercury intake (PATTY, 1981).

For elemental mercury, the most important route of absorption is the respiratory tract, as would be expected from its monoatomic nature, high vapor pressure, and lipid solubility. The percent retention is quite high – in the order of 80% in humans. Although confirmatory data are not available, monoalkyl mercurials (e.g., methylmercury) are probably also deposited and retained to a high degree, since they also have high vapor pressures and in some forms lipid solubility.

Gastrointestinal absorption of elemental mercury is poor (less than 0.01%) because it forms relatively large globules and the liquid may need to volatilize or ionize before crossing membranes. About 7% of inorganic mercury in food is absorbed, while organic mercury compounds are absorbed more efficiently. The extent of methylmercury absorption, even mixed with food, is about 95% in adults (CLARKSON, 1972).

In addition to populations exposed to high levels of methylmercury (MARSH et al., 1987; BERLIN, 1986), other studies examined populations with lower levels of exposure attributable to fish consumption. In Canadian Cree Indian infants (MCKEOWN-EYSSEN et al., 1983) abnormal muscle tone and deep tendon reflexes were observed in the male children when the mothers had a maximum hair concentration between 14 and 24 ppm. In a smaller New Zealand study (KJELLSTRÖM et al., 1986) where maternal hair concentrations were between 6 and 20 ppm, children performed poorly on the Denver Development Screening Test.

As with all metals, the degree of skin absorption in man is not known with any precision. Systemic absorption of alkyl mercurials may be substantial since people have been poisoned as a result of dermal application (FORBES and WHITE, 1952). In the Federal Republic of Germany (MAK, 1987) some organic mercury compounds are thought to result in cutaneous absorption and sensitization.

Absorption of elemental mercury and its inorganic salts may also occur. Certain minorities show indeed an increased mercury intake by sources other than inhalation and food. The American Conference of Governmental Industrial Hygienists (ACGIH) has, as of 1980, added the notation "skin" to their proposed Threshold Limit Value (TLV), indicating the possibility of cutaneous absorption of mercury vapors. CLARKSON et al. (1988) and THOMASSEN (1987) studied mercury vapor intake from amalgams (see also Sects. 6.3.1 and 6.6.1). In experimental animals, 5% of an aqueous solution of mercuric chloride was absorbed through the skin of guinea pigs within five hours (SKOG and WAHLBERG, 1964). In human volunteers the dermal rate of uptake was estimated to be 2.2% of the rate of uptake by the lung (HURSH et al., 1989).

5.2 Distribution

Movement of mercury from seed dressings into the new seedling has also been reported for wheat, barley, oats, and corn, but it is unlikely that high levels of mercury contamination would appear in the harvested grain. The use of foliar sprays can result in translocation and transport of mercury to crop tops, new growth, and developing fruit.

There is a good deal of interspecies variation in the distribution of the different forms of mercury. Organ distribution of inorganic and organic mercury in fish follows the general pattern of kidney, liver, brain, and muscle (STOKES and WREN, 1987). Under experimental conditions, the organ distribution of methylmercury in ducks is in decreasing order: liver, kidney, muscle, blood, brain. The level of methylmercury in eggs is slightly higher than that of the maternal blood (HEINZ, 1980), with normal values in the range of 10–20 ng/kg. As a result, the reproductive process of fish-eating birds can be jeopardized by mercury contamination, as has been seen in Scandinavia.

In general, mercury has been located in virtually every tissue or organ of the mammalian body and the distribution of elemental mercury has been shown to be affected by pretreatment with ethanol or aminotriazole (KHAYAT, 1985). However, the kidney accumulates the highest concentration followed by the liver, spleen, and brain (BIDSTRUP, 1972; NIOSH, 1978). The clinical manifestation of inorganic mercury toxicity can be partly explained by such a distribution. Despite the fact that the oxidation of metallic mercury is very rapid, it persists in the blood in sufficient amounts to reach the blood-brain and placental barriers. Upon entering tissue, elemental mercury is subject to oxidation. This results in a greater tissue concentration than does an equivalent dose of an ionic salt (CLARKSON, 1977). In red blood cells, this oxidation is thought to be catalyzed by catalase compound I. The reaction is saturable, follows Michaelis-Menten kinetics, and is stimulated by hydrogen peroxide (SICHAK et al., 1987).

Although mercury vapor and short-chained alkyl compounds such as ethyl- and methylmercury accumulate in the brain, the relative distribution to the brain is still less than in the kidney and liver (OKABE and TAKEGUCHI, 1980), so that the problem of preferential susceptibility of the nervous system, particularly to methylmercury, cannot be explained on the basis of mercury accumulation. This paradox has resulted in intensive research to correlated neurological damage with the distribution of mercury.

In the central nervous system of the rat, elemental mercury distributes more to the gray matter than to the white matter. The greatest concentrations are found in the Purkinje cells of the cerebellum and certain neurons of the spinal cord, medulla, pons and mid-brain (CASSANO et al., 1966). Studies with methylmercury show a similar distribution with the following exceptions: (1) the distribution between the gray and white matter of the cerebellum is almost the same (YOSHINO et al., 1965); (2) the Purkinje cells are spared; and (3) the highest accumulation in the cerebellum is in the visual cortex of the occipital lobe. The clinical, behavioral, and neurological manifestations of methylmercury poisoning appear to be correlated with the most severe damage to specific brain areas (EVANS et al., 1977).

An investigation (VON BURG et al., 1980) into the subcellular distribution of mercury in the rat brain found the highest concentration of methyl ^{203}HgCl in the supernatant fraction followed by the nuclear, microsomal, synaptosomal, mitochondrial, and myelin fractions. The distribution appeared to be governed by available mercury binding sites, probably sulfhydryl groups.

5.3 Metabolism and Excretion

The excretion of mercury from fish can be extremely slow. A whole-animal half-life of 1000 days has been reported. Therefore, it can be predicted that fish such as the Northern pike will accumulate mercury and not reach a steady state with respect to mercury uptake in the flesh during its life span (BEIJER and JERNELÖV, 1978b). Such a slow accumulation may eventually prove lethal for the fish.

Mercury is removed from the mammalian body by several routes. It can be volatilized from the lungs and skin, excreted in the urine, sweat, saliva, milk, intestinal mucosa, and bile, eliminated through the feces, or stored and shed in the hair, nails, or skin. The kinetics of mercury excretion is greatly influenced by the form of mercury. Mercury vapor is rapidly oxidized to divalent ionic mercury by the tissues of the body. Organomercurials such as phenylmercury and mercurial diuretics also undergo a very rapid transformation to inorganic mercury. Therefore, such labile mercurials generally show effects similar to those of inorganic mercury and the rates of elimination are consistent with their rapid conversion to the inorganic form (CLARKSON, 1977).

Short-chained alkylmercury compounds such as methylmercury do not undergo a rapid biotransformation in body tissues. The fraction of inorganic mercury formed from methylmercury appears to be less than 1% per day in humans. The unique toxic properties of methylmercury may partly be credited to the stability of the carbon-mercury bond (CLARKSON, 1977).

Methylmercury appears to follow a single compartment excretion model with first order kinetics. Reported half-times for methylmercury vary from 7–8 days in the mouse, about 70 days in humans and other primates, to 700–1000 days in some species of fish and shellfish (CLARKSON, 1972). A longer human half-life of 120 days was discovered in approximately 10% of the poisoned population in Iraq (SHAHRISTANI and SHIHAB, 1974), and a shorter half-life of 46 days was reported by GREENWOOD et al. (1978) in lactating human females. In rats, subcutaneous injected methylmercury is excreted extremely slowly until the animals are 17–18 days old. At 56 days of age, a sex difference was observed, with a more rapid excretion in females (THOMAS et al., 1982). Principal routes of removal are via the urine and predominantly the bile and feces (AASETH, 1987; ALEXANDER, 1987). Methylmercury has a bile-to-plasma ratio greater than five.

It seems that mouse, rat, and hamster females and castrates eliminate methylmercury (at higher doses) much faster than males which means that the latter are at higher risk (HIRAYAMA, YASUTAKE, DOCK, VAHTER, 1989).

In contrast to the short-chained mercurials, exposure to inorganic mercury results in a higher rate of excretion. Appoximately 80% of an inhaled dose of mercury vapor is retained. Human volunteers exposed to tracer doses of elemental Hg demonstrated first-order kinetics with a half-time of about 60 days (WHO, 1976). In experimental animals, the kinetics are more complex and most investigators recognize at least three components to the excretion pattern. An initial rapid phase, which involves about 35% of the dose, has a half-life of 2–3 days. A slower phase with a half-life of approximately 30 days accounts for 50% of the administered dose, and a final phase with a half-life of approximately 100 days accounts for the remaining 15%. The initial rapid phase has been attributed to a high fecal excretion via bile from the liver. In fish such as the skate and shark, there is an inordinately slow rate of bile formation, and this may contribute to the long biological half-life and resulting high levels of mercury in fish (BALLATORI and BOYER, 1986).

The two slower phases correspond to a clearance from the kidney, with urine accounting for almost all of the excretion from the animal (CLARKSON, 1972). In humans, variations in the urinary excretion rate determinations of inorganic mercury can be stabilized by adjusting the urinary concentration to a timed excretion and estimation of the urinary flow rate (ARAKI et al., 1986).

6 Effects on Plants, Animals, and Humans

6.1 Effects on Microorganisms, Plants, and Aquatic Organisms
(see also Sections 4.1, 5.1, 5.3, 6.6.1, and 6.6.2.1)

The antimicrobial actions of mercury salts have been known for some time, and compounds such as mercurochrome have been employed as household antiseptic agents. The toxic action on bacteria may be a result of combination with essential sulfhydryl groups. After treatment with mercuric chloride or phenylmercury nitrate, the bacteria appear dead but are easily revived by active thiol-containing agents such as thioglycolic acid, cysteine, glutathione, or even hydrogen sulfide (ALBERT, 1973). Sulfur compounds such as methionine or cysteine (where the thiol group is absent) do not show such antidotal action (SEXTON, 1963). SILVER (1984; see also Sect. 4.1) has published extensive information on the mechanism of microbial resistance to mercury and on mechanisms of controlling genes.

Organomercurials, particularly the alkyl- and arylmercury compounds, are more active as bactericides or fungicides than the inorganic salts. Phenylmercury acetate prevents the growth of a variety of fungi in vitro at concentrations of approximately 0.125 ppm. Ethylmercury salts are active at about 0.05 ppm. In contrast, mercuric acetate or chloride are active at approximately 1.0 ppm. The difference may partly be attributed to lipid solubility where penetration of a surface, such as seed coats, is required to reach a parasitic fungus (SEXTON, 1963). The toxic action of mercurials may also be related to a non-specific inhibition of a variety of intracellular enzymes. Initially, it was demonstrated that phenylmercury nitrate could inhibit oxygen consumption of a yeast culture. Subsequently, mercury salts were shown to inhibit several specific thiol-containing respiratory enzymes in vitro.

In spite of the antimicrobial action, microorganisms are known to convert inorganic mercury to organic forms and vice versa (see Sect. 4.1). For example, bacteria that methylate mercury have been isolated from the mucous material on the surface of fish, as well as from fresh and salt water sediments.

Divalent inorganic mercury can also be reduced to elemental mercury directly by widely occurring *Pseudomonas* bacteria or yeasts present in normal water supplies or the laboratory environment. Such contamination can cause the rapid loss of mercury from samples through volatilization (see Sect. 2.2.1). The alkylmercury formed by microbial action in bottom sediments (see Sect. 4.1) is bio-magnified through the food chain (BEIJER and JERNELÖV, 1978a). Bottom fauna, despite the relatively low amounts of methylmercury, may still account for a significant fraction of the gross intake by fish, although gill absorption may be more significant in some species. Despite relatively high levels of methylmercury, fish do not show any overt toxicity (BACKSTROM, 1969; see also Sect. 5.1).

At sublethal doses, mercuric chloride caused a reduction in the gross activity level as well as depressing the activity of some selected intestinal enzymes (PANIGRAHI and MISRA, 1980). However, there was a gradual recovery and overshoot in the enzyme activity rates despite continued exposure (SASTRY and GUPTA, 1978). Whole animal oxygen consumption was dramatically reduced and did not approach normal levels until the fish were placed in mercury-free water. Continual exposure to mercuric chloride for 35 days caused an increase in the mean corpuscular volume but a decrease in hemoglobin concentration, hematocrit and body weight. Blindness was observed in 60% of the test animals and according to PANIGRAHI and MISRA (1980), this observation may signify an attack on the central nervous system. Lethality could be indirect since brief exposure of goldfish to 50 ppm of methylmercury or mercuric chloride inhibited the acquisition of a shock avoidance task by 30–50% over five days of training (VON BURG, unpublished). A similar decrement in the shock avoidance task has been reported for goldfish exposed to 0.01 ppm for 24 hours (WEIR and HINE, 1970).

6.2 Miscellaneous Biochemical Effects
(see also Section 6.6)

Mercuric cations have a high affinity for sulfhydryl (-SH). In instances of inorganic mercury exposure, some forms of metal protein interaction may exert a protective effect on toxicity by binding mercury in an inactive form which may be excreted in the urine (HIRAYAMA and SHAIKH, 1978; PATTY, 1981; CHERIAN and CLARKSON, 1976). Since almost all proteins contain sulfhydryl groups or disulfide bridges, mercurials can disturb almost any function where critical or non-protected proteins are involved (CLARKSON, 1972). It is possible that a mercury ion binds to two sites of a protein molecule without deforming the chain, that it binds two neighboring chains together or that a sufficiently high concentration of mercury leads to protein precipitation. With organomercurials, the mercury atom still retains a free valency electron so salts of such compounds form a monovalent ion. If the mercury is bound to a structural protein such as keratin in the hair or nails, the disturbance will be minimal. On the other hand, if the mercury is bound to an enzyme a maximal disturbance of function can be expected (HUGHES, 1957).

PATERSON et al. (1971) reported that methylmercury induced changes in glycolytic intermediates and adenine nucleotides in the rat brain. The enzymes near the end of the glycolytic chain were the ones affected. Somewhat similar results were observed in the mouse (SALVATERRA et al., 1973). IZUNO (1976) reported an effect on the GABA shunt. However, VON BURG et al. (1979) reported that in vitro inhibition of glycolysis and the Krebs cycle, as measured by the oxygen consumption of liver slices, only occurred at methylmercury concentrations greatly in excess of that usually found in poisoned rats or human patients. YOSHINO et al. (1965) also found no change in oxygen consumption, aerobic or anaerobic glycolysis in asymptomatic animals and postulated that the development of the neurological symptoms cannot be attributed to the inhibition of respiratory enzymatic functions or a reduction in oxygen consumption.

Except for a few reports claiming a stimulatory effect on the synthesis of nucleic acids and protein in the brain of methylmercury poisoned animals, most laboratories report that protein synthesis is reduced. It is postulated that such a reduction can be attributed to the impairment of the blood-brain barrier (CHANG, 1977). JACOBS et al. (1986) showed that methylmercury chloride alters some aspects of cerebellar cell recognition through a complex mechanism initially involving depressed synthesis of specific proteins followed by alterations in cellular microtubules. IMURA et al. (1980) reported that methylmercury specifically disrupted microtubules in mouse cell culture assays. SAGER et al. (1983) found a similar effect in human fibroblasts, and methylmercury administration to newborn mice resulted in an arrest of mitosis in the cerebellum (SAGER et al., 1984). VOGEL et al. (1985) showed that methylmercury "in vitro" inhibited microtubule assembly and postulated that the mechanism for this inhibition was mediated by the binding of microtubular free sulfhydryl groups by mercury. This concept is supported by the finding that concomitant administration of glutathione (GSH) to mouse neuroblastoma cells prevented the injurious effect of methylmercury on microtubular polymerization (KROMIDAS et al., 1987).

VERITY et al. (1975) claimed that after administration of methylmercury, synaptosomes isolated from the cerebrum or cerebellum show a significant reduction in glutamate- and succinate-supported respiration during the early phase of neurotoxicity. These findings were confirmed by in vitro investigations. They concluded that these results show that methylmercury causes a specific translocation of cations across synaptosomal or mitochondrial membranes perhaps by interfering with Ca^{2+} or Na^+/K^+ ATPase systems (UZODINMA et al., 1987; AHMED-SAHIB and DESAIAH, 1987).

Mercury compounds can bind to the RNA of tobacco mosaic viruses (KATZ and SANTILLI, 1962), several synthetic polyribosomes and yeast soluble RNA (KAWADE,

1963). More rapid and drastic changes occur after inorganic mercury poisoning, perhaps owing to the fact that inorganic mercury binding to nucleosides is almost 10 times stronger compared to methylmercury (SIMPSON, 1964). However, RNA content "rebounds" after prolonged exposure to inorganic mercury and animals can recover from the neurological disturbances. These animals will also tolerate higher doses of mercury as compared to animals which have never been exposed. The "adjusting" to mercury is not understood, and it is not observed with methylmercury (CHANG, 1977). Direct and indirect evidence suggests that the reduction of RNA in nerve cells after poisoning may be attributed to either diminished RNA synthesis (CHANG et al., 1972) or an increased rate of RNA regeneration (EICHORN et al., 1970). Increased levels of DNA, RNA, and protein have also been reported (BRUBAKER et al., 1973) in the whole brain of rats, and more specifically in the anterior horn motorneurons (CHANG et al., 1972). This may indicate an increased turnover rate of such macromolecules.

Mercury has also been reported to alter the molecular structure and physicochemical properties of nucleic acids. After 11 weeks of intoxication by $HgCl_2$, there was an increase in guanine, a decrease in cytosine and a statistically significant shift in the base ratio of guanine to cytosine and purines to pyrimidines. Adenine and uracil remained unchanged. These changes could be correlated with the reactivation of RNA production following poisoning. It therefore appears that newly formed RNA was responsible for the altered base composition of the neuron, and these changes probably reflect an activation of a specific genome. This genome produces a characteristic RNA strikingly similar to the small amount of RNA (10%) normally present in the cell. Both types of RNA can be extracted at a pH of 8.3 and differ from nuclear RNA, transfer RNA, and ribosomal RNA. Therefore, the mercury-induced RNA may simply by an exaggeration of a normal constituent that allows for increased tolerance by the cell.

The various hypothetical actions or interactions of mercury compounds with the constituents of the cell have been excellently reviewed by CHANG (1977), and his summary diagram is reproduced in Fig. II.20-2. Of some additional interest is that $HgCl_2$ can block the muscarinic receptors in the rat brain, and methylmercury can block transmission of nerve impulses across myoneural junctions. VON BURG et al. (1980) hypothesized that the early manifestations of neurological impairment seen in instances of methylmercury poisoning could possibly be attributed to an early and potentially reversible block of nerve transmission. Biochemical and morphological damage would only occur at some later point in time.

The accumulation of methylmercury occurs in fetal brain, in the growing and maturing brain as well as in the adult brain. In instances of prenatal poisoning methylmercury inhibits the normal migration of nerve cells to the peripheral parts of the brain cortex thereby inhibiting normal development of the fetal brain. Upon autopsy, hypoplasia, symmetrical atrophy of the cerebrum and cerebellum and distortion of the cytoarchitecture of the cerebrum have been seen (CHOI et al., 1983).

In instances of postnatal poisoning there is general neuron degeneration and atrophy in the cerebral cortex especially in the calcarine, precentral and postcentral areas. Less pronounced changes are seen in the cerebellum (HUNTER and RUSSELL, 1954; TAKEUCHI, 1972).

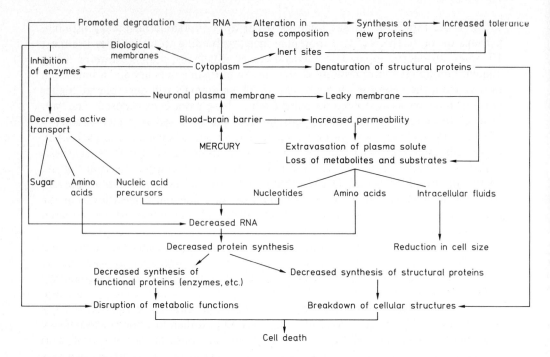

Fig. II.20-2. Hypothetical interactions of mercury compounds with the constituents of the cell (from CHANG, 1977).

In vitro cultures of fetal neurons from the human fetus or mice (SAGER et al., 1983) showed inhibition of cell mitosis.

6.3 Acute Effects on Animals and Humans

6.3.1 Elemental Mercury

All mercurials are poisonous, but metallic mercury can be considered least toxic, when compared to the other forms. Mercury vapor and liquid mercury should be distinguished and predominantly the former is inhaled. It is believed that elemental mercury converts to the ionic species to produce the toxic effect. The usual mode of exposure to metallic mercury is by skin contact, accidental oral intake, and inhalation. Intoxication from the inhaled vapor was undoubtedly common among mercury miners. The lethal concentration for humans is not known, but acute mercurialism has resulted from exposure to concentrations within the range of $1.2-8.5$ mg/m^3 (PATTY, 1981).

In dogs and rabbits, a concentration of $20-29$ mg/m^3 of air proved to be lethal in a matter of hours (FRASER et al., 1934). Signs and symptoms of acute inhalation exposure include cough, chest pain, and dyspnea, leading to bronchitis and pneumonitis (MILNES et al., 1970). Ingestion of liquid mercury results in the major portion being eliminated in the feces. SOLLMAN (1957) gave the oral LD_{10} for humans as 1429 mg/kg or approximately 100 g for a 70 kg adult.

Accidental metallic mercury poisoning is still reported in the medical literature. Such cases include inhalation toxicity among amateur gold miners that use mercury to extract gold from its ore, poisonings of neonates due to inhalation of mercury from broken thermometers, or suicide attempts by intravenous injection. Poisonings by injection, as practiced by South American boxers in an attempt to obtain a pugilistic advantage, has also been reported recently (KAMITSUKA et al., 1984). CLARKSON et al. (1988) studied possible effects of mercury vapors from dental amalgams (see also Sects. 5.1, 6.6.1, and 6.6.2.3). The BUNDESGESUNDHEITSAMT (1987) also discussed this question, since a Swedish group of experts suspected that the filling of teeth with amalgam or the removal of amalgam during pregnancy poses risks to the fetus via vapors produced. According to THOMASSEN (1987), chewing can also release minute amounts of mercury from old fillings, and mercury excreted in urine may reflect the number of dental amalgam fillings.

Elemental mercury poisoning is characterized by a metallic taste in the mouth, headache, nausea, vomiting, abdominal pain, diarrhea, and sometimes albuminuria. After a few days, there may be excess salivation or dryness of the mouth, swelling of the salivary glands, stomatitis, gingivitis, and loosening of the teeth. A dark line of HgS or dark stippling of the inflamed gums may appear. Ulcers can form on the lips and cheeks. In a young child, erythematous fingers, toes, cheeks, nose and buttocks may develop and are signs of "pink disease" (CRAVEY and BASELT, 1981). In mild cases, recovery can occur within 10–14 days or the poisoning may progress to the chronic type.

6.3.2 Ionized Inorganic Mercury

Mercury salts are corrosive to the skin and mucous membranes. They are considered to be primary skin irritants and may occasionally act as skin sensitizers (NIOSH, 1977b). Since they are relatively non-volatile, intoxication is usually by the percutaneous or oral route. The degree of toxicity appears to be related to the solubility (see Sect. 2.1) and the ionization constant of the compound. For example, the lowest reported lethal dose (LD_{10}) in humans for $HgCl_2$ is 29 mg/kg, while HgI_2, which ionizes to a lesser extent, has a reported LD_{10} of 357 mg/kg. In addition, monovalent mercury is considerably less toxic than the corresponding divalent form. The rat oral LD_{50} for HgCl (calomel) is 210 mg/kg as compared to 1 mg/kg for $HgCl_2$. Similarly, the mouse oral LD_{50} for $HgNO_3$ is 388 mg/kg versus 8 mg/kg for $Hg(NO_3)_2$ (NIOSH, 1983). Ingestion of such salts results in pharyngitis, dysphagia, nausea, vomiting, abdominal pain, bloody diarrhea, circulatory collapse, and shock. In addition to the swelling of the salivary glands, stomatitis and loosening of the teeth, nephritis, anuria, and hepatitis can occur.

6.3.3 Organic Mercury

Alkoxyalkyl- and arylmercury compounds appear to be quickly metabolized to inorganic mercury by biological systems. As a result, symptoms of intoxication are similar to those associated with inorganic mercury salts. Signs and symptoms can include loss of appetite, weight loss, diarrhea, and fatigue (D'ITRI, 1972).

Acute intoxication due to short-chain alkylmercury derivatives such as ethyl- and methylmercury produces signs and symptoms that differ significantly from intoxications due to other mercury compounds. With the exception of the tremor, signs of inorganic mercury poisoning do not occur, unless the exposure has been to both organic and inorganic mercury (BIDSTRUP, 1972). Usually several weeks are required before the symptoms of poisoning become manifest. The toxicity associated with these forms of mercury is discussed in Sect. 6.4. Table II.20-3 presents acute toxic doses for some selected mercury compounds in a comparison of their relative toxicity.

Table II.20-3. Acute Toxic Doses of Selected Mercury Compounds. (From NIOSH, 1987)

		mg/kg
HgCl	Rat, oral LD_{50}	210
	Human, oral LD_{10}	5
$HgCl_2$	Rat, oral LD_{50}	37
	Human, oral LD_{10}	29
MeHgCl	Rat, oral LD_{50}[a]	10
	Human, oral LD_{10}	5
EtHgCl	Rat, oral LD_{50}	40
	Human, oral LD_{10}	5
PhHgCl	Rat, oral Ld_{50}	60
	Human, oral LD_{10}	5

[a] Value was extracted from VON BURG et al. (1980)

6.4 Chronic Effects on Animals and Humans
(see also Sections 5.1 and 6.5)

Emotional and psychological disturbances are characteristic for clinical mercury vapor poisoning. The patient becomes excitable and irritable, particularly when criticized. Victims lose the ability to concentrate, become indecisive, fearful or depressed and may complain of headache, fatigue, and weakness. They may also demonstrate a loss of memory, insomnia, or drowsiness. The patient may exhibit a slight tremor which can interfere with fine motor movements. This tremor can affect the hands, head, lips, tongue, or jaw. The handwriting becomes imprecise with the omission of letters or becomes entirely illegible, a condition reminiscent of the "Mad Hatter" is Lewis Carroll's "Alice in Wonderland" (BISTRUP, 1972).

Other neurological disturbances caused by inorganic mercury include parathesias, neuralgias, dermographism, and affectations of taste, smell, and hearing. Signs of renal disease are common, but in the adult rat a resistance to renal damage appears to develop with repeated administration, and the neonatal rat kidney is largely insensitive to $HgCl_2$ toxicity (HALL et al., 1986; DASTON et al., 1986). Chronic nasal catarrh and epistaxis are not unusual. Many of the acute symptoms are present. Ocular lesions such as amblyopia, scotomas, browing of the iris, and discoloration of the anterior surface of the lens capsule have also been reported. Most patients will show some slow recovery when removed from exposure to inorganic mercury, but the extent of such recovery is greatly dependent upon the length of exposure and the form of the mercury.

Signs and symptoms of short-chain alkyl mercurial poisoning can include narrowing of the visual field, ataxia, dysarthria, and manifestations primarily of the nervous system (but there is a notable absence of any reported skin effects). This was dramatically seen in the 1971 epidemic in Iraq. In that country, over 100 cases of mercury poisoning were diagnosed and fourteen deaths were attributed to the use of an ethylmercury fungicide in 1956 (JALILI and ABASSI, 1961). This episode was followed in 1960 by a similar ethylmercury outbreak when 221 patients were admitted to Baghdad Hospital (DAMLUJI, 1962). In Pakistan in 1961, over 100 people developed chronic mercurial poisoning after eating treated seed (HAQ, 1963). In Alamogordo, New Mexico, (USA) in 1969, a family was stricken with classical alkylmercury poisoning after eating some hogs fed discarded waste seed that had been treated with an organomercury fungicide (CURLEY et al., 1971). In 1971–72, the most catastrophic epidemic ever recorded took place among farmers and their families in Iraq. The earliest reports on this event stated that 6530 patients were admitted to hospitals, with 459 hospital deaths (BAKIR et al., 1973). But a subsequent epidemiological study (GREENWOOD, 1985) indicated that over 2000 deaths occurred and over 60000 people were exposed. All of these episodes were due to consumption of mercury fungicide-treated seed grain that was intended for planting.

Table II.20-4. Neurological Signs in 53 Patients with Confirmed Methylmercury Poisoning. (Modified from RUSTAM and HAMDI, 1974)

Signs and Symptoms	Frequency (%)
Cerebellar disturbances	95
Sensory impairment	65
Visual disturbance	60
Speech disturbance	75
Mental abnormality	50
Pyramidal tract lesions	38
Hearing defects	19
Involuntary movement	31
Cranial nerves other than optic	19
Autonomic disorders	19
Encephalopathy	32

Table II.20-4, modified from RUSTAM and HAMDI (1974), indicates the relative frequency of neurological signs observed in 53 patients with confirmed mercury poisoning. Recovery occurred only in the mildly to moderately poisoned patients. It has been proposed (MAGOS et al., 1978) that the outcome of methylmercury intoxication depends not only on the maximum concentration of mercury but also on the duration of central nervous system exposure to the toxic concentration. Support for such a concept comes from the comparison between the Minimata and Iraqi epidemics. In Japan, the exposure was prolonged and extended over approximately 15 years. The clinical condition of these patients frequently deteriorated. In Iraq, the exposure period was not more than five months, and the number of patients that showed improvement was three times greater than the number that deteriorated (MAGOS et al., 1978). Regarding behavioral teratogenicity, WHO (1987) concluded that the mother's symptoms after methylmercury exposure usually improved. The damage to the fetal nervous system, however, seems to be permanent.

Comatose patients assume the posture seen in Fig. II.20-3. The neck is arched backward, the arms are tightly flexed with clenched fists. The legs are crossed and flexed, but to a lesser degree than the arms. Forced lower limb extension in this patient was impossible perhaps partly due to muscle "creep" (denervation shortening). Chewing and swallowing reflexes remained intact, which can explain the continued survival of many of these patients even without hospitalization. There was bowel and bladder incontinence (VON BURG, unpublished).

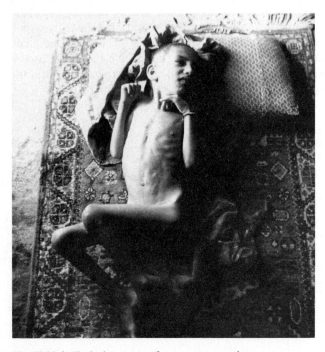

Fig. II.20-3. Typical posture of a comatose patient.

With the Iraqi patients, there was no apparent correlation between the level of mercury in the blood and the extent of observed neurological damage (RUSTAM and HAMDI, 1974). These authors suggested that individual differences in susceptibility and blood clearance times may be involved. CLARKSON et al. (1976) hypothesized that susceptibility may depend on the maximal concentration of mercury achieved in the brain.

6.5 Immunotoxicity

Methylmercury impairs both the primary and secondary immune response. Antibody production immediately after exposure to an antigen is therefore decreased (KOLLER, 1977a, b; OTTI et al., 1976). At high doses (10 ppm), methylmercury impaired the response of lymphocytes to antigen in mice. The T lymphocyte was the cell affected rather than the B lymphocyte when the secondary immune response is altered (KOLLER and ROAN, 1980a). Methylmercury also stimulated the immune system, as shown by KOLLER and ROAN (1980b) using mixed lymphocyte cultures. LINDVALL et al. (1987) demonstrated that mercury has immuno-modulatory effects in mammals. Patients with persistent active infection by the lymphotrophic Epstein-Barr virus also had abnormal amounts of mercury in their blood cells.

Mercuric chloride can induce antinuclear antibodies in mice (ROBINSON et al., 1986) and glomerular nephritis in rats (DRUET et al., 1978). The nephritis can be attributable to the induction of antiglomerular basement membrane antibodies. Some animals die during the first phase of the disease but survivors progress to the second phase which is characterized by a disappearance of porteinuria but immune complex nephritis persists. The mercury induced nephritis is only seen in the brown Norway rat and rabbits. Other rat strains tested were found to be resistant. The condition can be induced by methylmercuric chloride, ammoniated mercury, mercuresceine applied to the skin, or mercurobutol in the form of pessaries (DRUET et al., 1982).

6.6 Mutagenic, Carcinogenic, and Teratogenic Effects

6.6.1 Carcinogenesis and Mutagenesis
(see also Chapter I.18, Section 2.3)

In rats, direct intraperitoneal injection of metallic mercury can produce sarcomas. These sarcomas only developed in areas of direct contact with the metal. No tumors were observed at distant sites even though serious absorption toxicity had been noted (DRUCKREY et al., 1957). MITSUMORI et al. (1981, 1984) found renal tumors in male mice fed with methylmercury chloride (15 ppm in the diet). To the authors' present knowledge, there are no other reports that link cancer with exposure to a mercury compound. However, mercury has an affinity for thiol groups. It may therefore act as a mitotic inhibitor by interacting with the spindle in a fashion similar to colchicine and lead to aneuploidy or hyperploidy (VERSCHAEVE et al., 1979; ANDERSEN,

1986). Experimental studies on plants and animals suggest that mercury can induce chromosomal gaps and breaks.

SKERFVING et al. (1974) found a positive correlation between chromosomal aberrations and blood methylmercury levels in a fish-eating population. POPESCU et al. (1979) could not confirm an increase in aneuploidy or chromatid gaps in a population of workers exposed to either inorganic or organic mercury. However, these investigators did find a statistically greater incidence of chromosomal aberrations such as acentric fragments. VERSCHAEVE et al. (1979) suggested that dentists exposed to mercury amalgam have a greater incidence of aneuploidy while exposure to phenylmercury acetate causes hyperploidy. In two different industrial settings where the workers were exposed to mercury concentrations at or below the suggested TLV of 0.05 mg/m^3, one population suffered an increased incidence of genetic abnormalities, but the other population showed no difference from controls. The authors attributed their contradictory findings to good industrial hygiene practices. Mercury(II) chloride failed to induce chromosome aberrations in cultured mammalian cells (UMEDA and NISHIMURA, 1979) or sister chromatid exchange in cultured human lymphocytes (OGAWA, 1979). More recently, MAILHES (1983) showed that methylmercury increases the incidence of hyperploidy, but not structural aberrations, in the Syrian hamster (see also Sect. 6.6.2.1). BURKART and OGOREK (1986) studied the clastogenic effects of mercury (alone or in combination with X-ray irradiation) on the genome of V79 cells, leading to an accumulation of DNA strand breaks. It seems that mercury modulated primary damage by X-rays already at extremely low concentrations, perhaps by the stimulation of the production of sulfhydryl-carrying entities.

6.6.2 Fetal Development, Embryotoxicity, and Teratogenicity
(see also Chapter I.18, Section 4 and Sections 4 and 6.4 in this chapter)

6.6.2.1 Organic Mercury in Animals

The in utero effects of short-chain alkylmercury compounds have also been studied in experimental animal models. They are embryocidal and teratogenic in most species, but the types of malformations appear to vary with species, strain, and dose. Intrauterine death with fetal resorption or stillbirth are common even at low doses. Cleft palate, missing limbs, and brain and facial malformations are most commonly seen (FUYUTA et al., 1979). Reproductive behavior and fertility were not significantly affected in prenatally exposed mice (GATES et al., 1986). JACQUET (1984) studied the effects of methylmercuric chloride on the preimplantation mouse embryo from the two-cell stage up to implantation. The acute effects were most severe on blastocysts and unexpectedly least severe on morulae.

The findings that postnatal experimental animals have a slower rate of mercury excretion (THOMAS et al., 1982) and suckling rats have a 19-fold higher level of mercury in the brain, suggest that immature organisms may be more severely damaged by the same level of uptake than adults (JUGO, 1976). Along this line, performance deficiency in behavioral tasks after prenatal exposure to methylmercury has been reported in mice (SPYKER et al., 1972; SPYKER, 1975) and rats (ZENICK, 1974; OLSEN and BOUSH, 1975). BORNHOUSEN et al. (1980) found that as little as 0.01 mg methylmercury given daily, per os, to rats between days 6 and 9 of pregnancy produced a significant decrement in performance in the 4-month-old offspring. By the estimation of these authors, ingestion of 1 kg of fish, even that within the legal limits of methylmercury contamination, would be sufficient to produce a significant behavioral deficit in human offspring. In mice, one-tenth the contraceptive dose of phenylmercury per vaginum on day 7 of gestation resulted in fetal growth retardation, tail abnormalities, and embryonic death (KOOS and LONGO, 1976). In the hamster, the embryopathic effects of phenylmercury were similar to those seen with $HgCl_2$.

6.6.2.2 Alkylmercury in Humans

The teratogenicity of mercury compounds has been summarized by SCHARDEIN (1985). Organic mercury compounds administered to pregnant females will pass through the placenta and accumulate in the fetus. Methylmercury concentrations in the red blood cells of the newborn infant have been reported to average 28% greater than that of the mother's red blood cells (TEJNING, 1968). It has also been suggested that the presence of a fetus can partially protect the mother from the toxic symptoms of methylmercury (CLEGG, 1971; GREENWOOD et al., 1978).

After birth, the level of methylmercury in the child remains sowewhat constant owing to a delay in excretion or the continued passing of mercury from the mother by means of breast milk. Such postnatal methylmercury exposure will not only have a pathological effect on the developing nervous system, but can also produce profound degenerative changes in the epithelial cells of the proximal convoluting tubules of the developing kidney (WARE et al., 1974; CHANG and SPRECHER, 1976). Recent investigations have determined that there is a dramatic increase in the biochemical marker enzyme ornithine decarboxylase (ODC) in the brains of rat fetuses (SLOTKIN et al., 1985) or the kidneys of rat pups (BARTOLOMÉ et al., 1986) exposed to methylmercury.

Most of the information concerning the fetotoxicity of methylmercury in humans is derived from the epidemics of Minimata Bay and Iraq. Mothers usually had uneventful deliveries. Abnormalities of the skull and teeth were reported in Minimata patients, but not noted in the cases reported from Iraq. In Minimata, signs of neurological injury in the newborn began several weeks to months after birth. Lethargy, delayed movement, failure to follow visual stimuli, incoordinated sucking or swallowing, and convulsions were frequently observed. Decreased alertness, changes in emotional state, and decreased intelligence were also noted, including eight cases of idiocy. Even today, some degree of ataxia can still be seen in almost all of the surviving patients from Minimata (see also Sect. 6.4 and WHO, 1987).

In the Iraqi cases, 5 out of 15 patients suffered blindness. There were ten cases of cerebral palsy, and the infants were also below growth and expected weight for their ages (AMIN-ZAKI et al., 1974). In a five year follow-up study (AMIN-ZAKI et al.,Z 1979), the mothers' symptoms were found to have usually improved, but the damage to the fetal nervous system appeared to be permanent. Varying degrees of developmental retardation, in addition to exaggerated tendon reflexes and the pathologic extensor plantar reflex (minimal brain damage syndrome), were seen.

The persistence of such pathology is borne out by autopsy examination. The patients showed small symmetrical atrophic brains with brain weight reduced by as much as two-thirds, widespread loss of nerve cells from the cerebrum and cerebellum, and disruption of the normal cytoarchitecture (for a more detailed review, see REUHL and CHANG, 1979). In humans, the most severe methylmercury-induced congenital abnormalities to be reported are exencephaly, encephalocoele and hydrocephalus (REUHL and CHANG, 1979; see also MARSH et al., 1981, 1987, for further human data).

6.6.2.3 Inorganic Mercury

In contrast to methylmercury, inorganic mercury ions are less readily transferred across the placenta. However, embryopathic effects with $HgCl_2$ and mercuric acetate have been demonstrated in the hamster. $HgCl_2$ produced litter resorption, growth retardation, subcutaneous edema, exencephaly, and anophthalmia (KOOS and LONGO, 1976). Mercuric acetate also increased the frequency of multiple malformations (GALE and FERM, 1971). Mercuric chloride induced cataracts and fetal deaths in rats (MIYOSHI, 1959) while mercuric oxide caused a number of eye defects in both rats (RIZZO and FURST, 1972) and mice (SMITH and BERG, 1980).

The ability of maternally-administered metallic mercury to penetrate the placenta is approximately 20 times greater than of ionic mercury. In rats, up to 0.22% of the dose of inhaled mercury was present in fetal tissues within 24 hours. For comparison, 0.01% of an injected dose of mercury salt was found in the fetuses (KOOS and LONGO, 1976). Therefore, exposure to mercury vapor may be more embryopathic than an equivalent dose of inorganic mercury salt.

WANNAG and SKJAERASEN (1975) determined the blood mercury levels of 19 pregnant dental workers and a non-exposed group of women. Although the blood mercury levels were comparable between the two groups, the dental workers had an increased mercury content in the fetal membranes (see also CLARKSON et al., 1988; BUNDESGESUNDHEITSAMT, 1987; Sects. 5.1 and 6.3.1).

Despite the difference between inorganic and organic mercury in the ability to penetrate the placenta, both forms are transferred with equal facility to breast milk (MONSOUR et al., 1973). Consequently, maternal milk is a significant potential route for all forms of mercury poisoning in infants.

6.7 Treatment for Mercury Poisoning, Antagonists

The pharmacological agents chosen for the therapeutic treatment of mercury intoxication need to be directed at either inorganic or organic mercury. Knowledge of the biotransformation processes is necessary to make the correct decision on the course of action to be taken.

The agent of choice for inorganic mercury intoxication still appears to be British Anti-Lewisite (BAL) or more soluble derivatives such as 2,3-dimercaptopropane-1-sulfonic acid (Dimaval®) or its sodium salt (Unithiol®). The chemical structure of 2,3-dimercaptosuccinic acid may be sufficiently similar to BAL to also have a significant therapeutic effect. The various salts of ethylenediaminetetraacetic acid (EDTA) are not as efficient as the thiol-containing compounds.

In the case of methylmercury poisoning, BAL is contraindicated since it appears to mobilize the mercury from other tissues and increases the intracerebral penetration of the mercurial. EDTA has also been shown to have little or no therapeutic value in this case (OGAWA et al., 1976).

A variety of clinical treatments have been attempted in case of human exposure. Such treatments include gall bladder drainage or oral treatment with non-absorbable complexing resins (BAKIR et al., 1973) on the theory that biliary excreted mercury will be blocked from reabsorption. Hemodialysis with L-cysteine has also been reported to lower blood mercury levels rapidly (KOSTYNIAK et al., 1975). The efficacy of such treatments in humans remains in doubt since in Iraq both treated and untreated patients showed recovery (RUSTAM and HAMDI, 1974). However, DL-penicillamine administration to human volunteers was able to mobilize organic mercury into the urine to a greater extent than inorganic mercury (SUSUKI et al., 1976). In experimental animals, seven weeks of polythiol resin treatment resulted in a considerable mobilization of methylmercury from the brains of mice (CLARKSON et al., 1973). N-Acetyl-DL-penicillamine administered to pregnant rats reduced the maternal and fetal brain mercury levels by 40–80% (AASETH et al., 1976). In rats, treatment with dimercaptosuccinic acid reduced the body burden and brain mercury content by 60%, partially reversed the signs of intoxication and prevented the progression of cerebellar damage (MAGOS et al., 1978).

Several natural antagonists to mercury compounds in living organisms are known, for instance, selenium compounds (see Chapter II.25, Sect. 6.2, and THOMASSEN, 1987). LUNDBERG and PAULSSON (1988) added sodium selenite to lakes polluted by mercury and observed a reduction of methylmercury concentrations in perch and pike muscles (due perhaps to reduced bioavailability of mercury by binding-site competition or stimulation of demethylation).

7 Hazard Evaluation and Limiting Concentrations

Although Hg^{2+} is the predominant form of mercury present in fresh and marine water, and elemental mercury is the predominant form in the atmosphere, methylmercury is by far the most toxic form. Thus, although methylmercury represents only a small fraction of the total global mercury, much of its presence is due to the biomethylation of inorganic mercury, and it presents the greatest risk of irreversible functional damage to human and animal life.

Table II.20-5 lists the suggested maximal allowable exposures to three typical forms of mercury. The suggested occupational exposure limits for pregnant females

Table II.20-5. Maximal Allowable Exposure to Mercury. (Modified from Koos and Longo, 1976)

Mercury vapor	0.05 mg/m^3 of air
Inorganic mercury salts	0.1 mg/m^3 of air
Methylmercury	0.01 mg/100 mL blood

Table II.20-6. Occupational Exposure Limits for Pregnant Females. (Modified from Koos and Longo, 1976)

Mercury vapor	0.01 mg/m^3 of air
Aerosol inorganic mercury salts	0.02 mg/m^3 or air
Methylmercury	None

Table II.20-7. Concentrations of Total Mercury in Indicator Media and Equivalent Long-term Daily Intake of Mercury as Methylmercury[a]. (Modified from WHO Criteria Document, 1976)

Blood (ng/mL)	Hair (μg/g)	Equivalent Long-term Daily Intake (μg/kg body weight)
200 – 500	50 – 125	3 – 7

[a] Associated with the earliest effects in the most sensitive group of the adult population. The prevalence of the earliest effects could be expected to be approximately 5%.

is presented in Table II.20-6. Table II.20-7 indicates the long-term daily intake of methylmercury associated with the earliest effects on the central nervous system. This system is the most sensitive to mercury damage and can therefore be considered the target organ. Effects on the adult nervous system have been estimated to occur at blood concentrations in the range of 200 to 500 ng mercury per mL. This would correspond to a long-term daily intake of 3 to 7 μg Hg as methylmercury in the diet.

Based on much of the data presented in this chapter, a number of standards for mercury have been set. WHO (1976) and the US EPA (1979) have set a drinking water standard of 2 μg Hg/L, which would represent a daily intake of approximately 1 to 2% of the toxic level quoted in Table II.20-7. The US EPA (1985) recently issued a Health Advisory of 3 μg Hg/L for an allowable lifetime exposure in drinking water. Other drinking water standards are: Federal Republic of Germany 4 μg Hg/L; Japan 1 μg/L; Switzerland 3 μg/L.

The American Conference of Governmental Industrial Hygienists recommended an occupation exposure of 0.01 mg/m^3 for alkylmercury as a time-weighted average for an eight-hour workday (ACGIH, 1986). The corresponding standard for mercury vapor exposure is 0.05 mg/m^3. The OSHA permissible exposure level (PEL) for aryl and inorganic mercury is 0.1 mg/m^3 (NIOSH, 1985). Thus the relative toxicity of the different forms of mercurials is recognized in industrial settings. In the Federal Republic of Germany, the established Biological Working Materials Tolerance (BAT) values are 50 µg Hg per L blood or 200 µg Hg per L urine for elemental and inorganic mercury, and 100 µg Hg per L blood for organic mercury (MAK, 1987). A WHO study group recommended 50 µg Hg per L urine (SCHALLER, 1982). The MAK (1987) value is 0.1 mg/m^3 for elemental mercury and 0.01 mg/m^3 for organic mercury.

With respect to inorganic mercury the US EPA, using data for induced glomerular nephritis in the brown Norway rat (DRUET et al., 1978), proposed a LOAEL (Lowest Observed Adverse Effect Level) of 0.1 mg/kg/injection, a NOAEL (No Observable Adverse Effect Level) of 0.5 mg/kg/injection, and a safety factor of 1000 to calculate a RfD (reference dose) of 0.158 µg/kg/day. This translates to a permissible dose of 11.0 µg/day for a 70 kg man. The Lifetime Health Advisory is 1.1 µg/L for daily water consumption with a standard for drinking water set at 2 µg/L. The WHO Expert Group has recommended an international standard for mercury in drinking water at 1 µg/L (US EPA, 1987). See also Sect. 6.3.1 for information regarding a potential risk from dental amalgam.

A FAO/WHO Joint Expert Committee on food additives (1973) established a provisional tolerable weekly intake of 0.3 mg of total mercury per person, of which no more than 0.2 mg should be present as methylmercury. These amounts are equivalent to 5 µg and 3.3 µg mercury, respectively, per kilogram of body weight. Where there is evidence of an intake in excess of 0.3 mg per week, the level of methylmercury compounds should be investigated.

The identification of methylmercury accumulation in the food chain has resulted in curtailment of the use of methylmercury seed dressings and the establishment of a 10 ppm limit in the USA and a 0.5 ppm limit in Canada for mercury in fish. Over the years, several reports (BERGLUND et al., 1977; WALLACE et al., 1971) have reviewed the mercury problem and recommended an acceptable average daily intake (ADI) of 0.03 mg, or about 0.04 µg of mercury per kilogram of body weight. These levels would be approached by individuals who consume daily 500 grams of fish containing 0.6 mg Hg/kg. Such a high ingestion of fish has only been observed in specific communities (INSKIP and PIOTROWSKI, 1985). In the USA the daily consumption is only 60–100 g. The US EPA (1984) recommended a more conservative ADI of 20 µg Hg/day, which was then recommended as an AIS (maximum dose tolerable for subchronic exposure) for alkylmercury and mixed inorganic and alkylmercury. The German Health Organization (BUNDESGESUNDHEITSAMT, 1979) has established specific mercury tolerances for several principal dietary sources. For liver, 0.1 mg/kg was specified; 0.05 mg/kg for pork, 0.02 mg/kg for beef; 0.02 mg/kg for potatoes; 0.03 mg/kg for eggs; and 0.004 mg/kg for vegetables. In order to protect freshwater fish, the maximal allowable concentration of mercury in water was set between 0.05 and 0.4 µg/L (RUF, 1981), and 2 mg/kg was established for agricultural soil (HOFFMAN, 1980).

Fish-consuming birds may even be at higher risk than humans.

References

Aaseth, J. (1987), *Hepatobiliary Kinetics of Heavy Metals, as Influenced by Chelating Thiols.* Symposium Abstracts 2nd Nordic Symposium on Trace Elements in Human Health and Disease, Odense. Environ. Health 20 15–18. WHO Regional Office for Europe, Copenhagen.

Aaseth, J., Wannag, A., and Norseth, T. (1976), *The Effect of NDL-Acetylated DL-Homocysteine Thiolactone on the Mercury Distribution in Adult Rats, Rat Foetuses and Macaca Monkeys after Exposure to Methylmercuric Chloride.* Acta Pharmacol. Toxicol. 39, 302–311.

ACGIH (American Conference of Governmental Industrial Hygienists) (1980), *TLVs. Threshold Limit Values for Chemical Substances in the Work Environment Adopted by the ACGIH.* ACGIH, Cincinnati Ohio.

ACGIH (American Conference of Governmental Industrial Hygienists) (1986), *TLVs. Threshold Limit Values and Biological Exposure Indices for 1986–1987.* ACGIH, Cincinnati, Ohio.

Adam, K., Jönck, P., Gattner, H., and Zirngiebl, E. (1980), *Quecksilber* (in German), in: *Ullmanns Encyklopädie der technischen Chemie,* 4th Ed., Vol. 19, pp. 643–671. Verlag Chemie, Weinheim-Deerfield Beach/Florida-Basel.

Ahmed, R., Valenta, P, and Nürnberg, H. W. (1981), *Voltammetric Determination of Mercury Levels in Tuna Fish.* Microchim. Acta 1, 171–184.

Ahmed, R., May, K., and Stoeppler, M., (1987), *Wet Deposition of Mercury and Methylmercury from the Atmosphere.* Sci. Total Environ. 60, 249–261.

Ahmed-Sahib, K., and Desaiah, D. (1987), *Mechanism of Interaction of Cadmium and Methylmercury with Oubain-receptor Site.* Toxicologist Abstr. 290.

Albert, A. (1973), *The Covalent Bond in Selective Toxicity,* in *Selective Toxicity,* 5th Ed., p. 397. Chapman & Hall, London.

Alexander, J. (1987), *Biliary Excretion of Metals. Symposium Abstracts 2nd Nordic Symposium on Trace Elements in Human Health and Disease, Odense. Environ. Health 20,* 15–18. WHO Regional Office for Europe, Copenhagen.

Amin-Zaki, L., Elhassani, S. B., Clarkson, T. W., Doherty, R. A., and Greenwood, M. R. (1974), *Intrauterine Methylmercury Poisoning in Iraq.* Pediatrics 54(5), 587–597.

Amin-Zaki, L., Majeed, M., Elhassani, S. B., Clarkson, T. W., Greenwood, M. R., and Doherty, R. A. (1979), *Prenatal Methylmercury Poisoning: Clinical Observations over Five Years.* Am. J. Dis. Child. 133, 172–177.

Andersen, O. (1986), *Evaluation of Spindle Inhibiting Effects of Metals by Chromosome Length Measurements.* Toxicol. Environ. Chem. 12, 195–213.

Angerer, J., and Schaller, K. H. (1988), *Analyses of Hazardous Substances in Biological Materials,* Vol. 2, pp. 195–211. VCH Verlagsgesellschaft, Weinheim-Basel-Cambridge-New York.

Araki, S., Murata, K., Aomo, H., et al. (1986), *Comparison of the Effects of Urinary Flow on Adjusted and Nonadjusted Excretion of Heavy Metals and Organic Substrates in Healthy Men.* J. Appl. Toxicol. 6, 245–251.

Backstrom, J. (1969), *Distribution Studies of Mercury Pesticides in Quail and Some Fresh Water Fish.* Acta Pharmacol. 27 (Suppl. 3), 1–103.

Bakir, F., Damluji, S., Amin-Zaki, L., Murtadha, L., Khalidi, A., AlRawi, N., Tikriti, S., Dhakhir, H., Clarkson, T. W., Smith, J., and Doherty, R. (1973), *Methylmercury Poisoning in Iraq. An Interuniversity Report.* Science 181, 230–241.

Ballatori, N., and Boyer, J. (1986), *Slow Biliary Elimination of Methylmercury in the Marine Elasmobranchs Raja erinacea and Squalus acenthias.* Toxicol. Appl. Pharmacol. 85, 407–415.

Bargagli, R. (1989), *Determination of Metal Deposition Patterns by Epiphytic Lichens.* Toxicol. Environ. Chem. 18(4), 249–256.

Bartolomé, J., Alberto, G., Bartolomé, M., et al. (1985), *Postnatal Methyl Mercury Exposure: Effects on the Ontogeny of Renal and Hepatic Ornithine Decarboxylase Response to Trophic Stimuli.* Toxicol. Appl. Pharmacol. 80, 147–154.

Beijer, K., and Jernelöv, A. (1978a), *Sources, Transport and Transformation of Metals in the Environment,* in *Toxicology of Metals,* Vol. III, pp. 51–82. US Environmental Protection Agency, PB 280-115.

Beijer, K., and Jernelöv, A. (1978b), *General Aspects and Specific Data on Ecological Effects of Metals,* in *Toxicology of Metals,* Vol. III, pp. 201–222. US Environmental Protection Agency, PB 208-115.

Berglund, F., Berlin, M., Burke, G., Cederlof, R., Euler, U., Friberg, L., Holmstedt, Jonsson, E., Lung, K. G., Ramel, C., Skerfving, S., Swensson, A., and Tejning, S. (1971), *Methyl Mercury in Fish. A Toxicologic-epidemiologic Evaluation of Risks.* Report from an Expert Group. *Nord. Hyg. Tidskr. Suppl. 4.*

Berlin, M. (1977), *Mercury,* in *Toxicology of Metals,* Vol. III, pp. 301–344. US Environmental Protection Agency PB, 268-324.

Berlin, M. (1986), *Mercury,* in: Friberg, L., et al. (eds.): *Handbook on the Toxicology of Metals,* Vol. II, Chapter 16, pp. 387–435. Elsevier, Amsterdam.

Bernhard, M., and Andreae, M. (1984), *Transport of Trace Metals in Marine Food Chains,* in: Nriagu, J. O. (ed.): *Changing Metal Cycles and Human Health,* pp. 143–168. Springer-Verlag, Berlin.

Bertram, H., Kemper, F., and Zenzen, C. (1985), *Man – a Target of Ecotoxicological Influences,* in: Nürnberg, H. W. (ed.): *Pollutants and Their Ecotoxicological Significance*, Chapter 26, pp. 429–431. John Wiley & Sons, New York.

Bidstrup. P. L. (1972), *Clinical Symptoms of Mercury Poisoning in Man. Biochem. J. 130,* 59–60.

Bidstrup, P., Bonell, J., Harvey, D., and Locket, S. (1951), *Chronic Mercury Poisoning in Men Repairing Direct-current Meters. Lancet 251,* 856–861.

Bloom, N. S., and Crecelius, E. A. (1983), *Determination of Mercury in Seawater at Sub-nanogram per Liter Levels. Mar. Chem. 14,* 49.

Bloom, N. S., and Watras, J. J. (1988), *Observations on Methylmercury in Precipitation. Proceedings Conference Trace Metals in Lakes, Hamilton. Sci. Total Environ. 87/88,* 199–207.

Bornhousen, M., Musch, H., and Greim, H. (1980), *Operant Behavior Performance Changes in Rats after Prenatal Methylmercury Exposure. Toxicol. Appl. Pharmacol. 56,* 305–310.

Breder, R. (1987), *Distribution of Heavy Metals in Ligurian and Tyrrhenian Coastal Water. Sci. Total Environ. 60,* 197–212.

Breder, R., and Flucht, R. (1984), *Mercury Levels in the Atmosphere of Various Regions and Locations in Italy. Sci. Total Environ. 40,* 231–244.

Breder, R., Nürnberg, H. W., Golimowsky, J., and Stoeppler, M. (1985), *Toxic Metal Levels in the River Rhine,* in *Pollutants and Their Ecotoxicological Significance*, Chapter 15, pp. 205–225. John Wiley & Sons, New York.

Brubaker, P., Klein, R., Hermann, S. P., Lucier, G. W., Alexander, L. T., and Long, M. D. (1973), *DNA, RNA and Protein Synthesis in Brain, Liver and Kidney of Asymptomatic Methylmercury Treated Rats. Exp. Mol. Pathol. 18,* 262–282.

Bruland, K. W. (1983), Chapter 45 *Trace Elements in Sea-water, Chem. Oceanogr. 8,* 196/197.

Bundesgesundheitsamt (1979), *Bundesgesundheitsblatt 22*(15), 282.

Bundesgesundheitsamt (1987), *Amalgam-Fillings during Pregnancy* (in German). Press Release 46/1987, 5th October 1987, Berlin.

Burkart, W., and Ogorek, B. (1986), *Genotoxic Action of Cadmium and Mercury in Cell Cultures and Modulation of Radiation Effects. Toxicol. Environ. Chem. 12,* 173–183.

Cappon, C., and Smith, J. C. (1977), *Gas-chromatographic Determination of Inorganic Mercury and Organomercurials in Biological Material. Anal. Chem. 49,* 365–369.

Cappon, C., and Smith, J. C. (1978), *A Simple and Rapid Procedure for the Gas Chromatographic Determination of Methylmercury in Biological Samples. Bull. Environ. Contam. Toxicol. 19*(5), 600–667.

Cassano, G., Viola, P., Ghetti, B., and Amaducci, L. (1966), *The Distribution of Inhaled Mercury (Hg 203) Vapors in the Brain of Rats and Mice. J. Neuropathol. Exp. Neurol. 28,* 308–320.

Chang, L. W. (1977), *Neurotoxic Effects of Mercury – A Review. Environ. Res. 14*, 329–373.
Chang, L., and Sprecher, J. (1976), *Degenerative Changes in Neonatal Kidney Following in utero Exposure to Methylmercury. Environ. Res. 11*, 392–406.
Chang, L., Desnoyers, P., and Hartmann, H. (1972), *Quantitative Cytochemical Studies on RNA in Experimental Mercury Poisoning I. Changes in RNA Content. J. Neuropathol. Exp. Neurol. 331*, 389–507.
Cherian, M. G., and Clarkson, T. W. (1976), *Chem. Biol. Interact. 12,* 109–120.
Choi, B. (1983), *Effects of Prenatal Methylmercury Poisoning upon Growth and Development of Fetal Central Nervous System,* in: Clarkson, T. W., Nordberg, G. F., and Sager, P. R. (eds.): *Reproductive and Developmental Toxicity of Metals,* pp. 473–496. Plenum Press, New York.
Clarkson, T. W. (1972), *The Pharmacology of Mercury Compounds. Annu. Rev. Pharmacol. 12*, 375–406.
Clarkson, T. W. (1977), *Mercury Poisoning,* in: Brown, S. S. (ed.): *Clinical Chemistry and Chemical Toxicology of Metals,* pp. 189–200. Elsevier/North-Holland Biomedical Press, Amsterdam.
Clarkson, T. W. (1978), *Mercury Criteria Document.* US Environmental Protection Agency.
Clarkson, T. W., Small, T., and Norseth, T. (1973), *Excretion and Absorption of Methylmercury after Polythiol Resin Treatment. Arch. Environ. Health 26,* 173–176.
Clarkson, T. W., Amin-Zaki, L., and Al-Tikriti, S. (1976), *An Outbreak of Methylmercury Poisoning due to Consumption of Contaminated Grain. Fed. Proc. Fed. Am. Soc. Exp. Biol. 35*(12), 2395–2399.
Clarkson, T. W., Friberg, L., Hursh, J. B., and Nylander, M. (1988), *The Prediction of Intake of Mercury Vapor from Amalgams,* in: Clarkson, T. W.m et al. (eds.): *Biological Monitoring of Toxic Metals,* pp. 247–264. Plenum Press, New York.
Clayton, G., and Clayton, F. (eds.) (1981), *Mercury,* in: *Patty's Industrial Hygiene and Toxicology,* Vol. 2A, pp. 1769–1792. J. Wiley, New York.
Clegg, D. J. (1971), *Embryotoxicity of Mercury Compounds,* in Proceedings of the Symposium "Mercury in Man's Environment" February, 15–16. Royal Society Canada, Ottawa.
Cravey, R. H., and Baselt, R. (1981), *Pathology of Poisoning,* in *Introduction to Forensic Toxicology,* pp. 77–78. Biomedical Publications, Davis, California.
Curley, A., Sedlak, V., Girling, E., et al. (1971), *Organic Mercury Identified as the Cause of Poisoning in Humans and Hogs. Science 172,* 65.
Damluji, S. (1962), *Mercurial Poisoning with the Fungicide Granosan M. J. Fac. Med. Baghdad 4,* 83. Biol. Abstr. 41, 10589.
Daston, G., Rehnberg, B., Hall, L., and Kavlock, R. (1986), *Toxicity of Mercuric Chloride to the Developing Rat Kidney. Toxicol. Appl. Pharmacol. 85,* 39–48.
DHEW (Department of Health, Education and Welfare) (1977), *Occupational Diseases. A Guide to Their Recognition.* National Institute for Occupational Safety and Health. DHEW (NIOSH) Publication No. 77–181.
D'Itri, F. M. (1972), *The Environmental Mercury Problem.* CRC Press, Cleveland, Ohio.
Druet, P., Druet, E., Potdevin, F., and Sapin, C. (1978), *Immune Type Glomerulonephritis Induced by $HgCl_2$ in the Brown Norway Rat. Ann. Immunol. 129C,* 777–792.
Druet, P., Bellon, B., Sapin, C., Druet, E., Hirsch, F., and Fournie, G. (1982), *Nephrotoxin-induced Changes in Kidney Immunobiology with Special Reference to Mercury-induced Glomerulonephritis,* in: Bach, P. H., Bonner, F. W., Bridges, J. W., and Lock, E. A. (eds.): *Nephrotoxicity. Assessment and Pathogenesis,* pp. 206–221. John Wiley & Sons, Chichester, UK.
Druckrey, H., Hamperl, H., and Schmahl, D. (1957), *Carcinogenic Action of Metallic Mercury after Intraperitoneal Administration in Rats. Z. Krebsforsch. 61,* 511–519.
Earles, M. P. (1964), *A Case of Mass Poisoning with Mercury Vapor on Board HMS Triumph and Cadiz, 1810. Med. Hist. 8,* 281–286.
Eichorn, G., Butzow, J., Clark, P., and Shin, Y. (1970), *Studies on Metal Ions and Nucleic Acids,* in: Maniloff, J., Coleman, J. R., and Miller, M. W. (eds.): *Effects of Metals on Cells, Subcellular Elements and Macromolecules,* pp. 77–95. C. C. Thomas, Springfield, Illinois.

Evans, H. L., Garman, R. H., and Weiss, B. (1977), *Methylmercury: Exposure Duration and Regional Distribution as Determinants of Neurotoxicity in Nonhuman Primates. Toxicol. Appl. Pharmacol 41*, 15–33.

Eyl, T. (1970), *Methyl Mercury Poisoning in Fish and Human Beings. Mod. Med. 38*, 135–137.

FAO/WHO Joint Expert Committee (1973), *Evaluation of Certain Food Additives and the Contaminants of Mercury, Lead and Cadmium.* World Health Organization, Technical Report Series No 505, Geneva.

Farler, E. (1952), *The Evolution of Chemistry,* p. 21. Ronald Press, New York.

Ferrara, R., Maserti, B. E., and Paterno, P. (1989), *Mercury Distribution in Marine Sediment and its Correlation with the Posidonia oceanica Prairie in a Coastal Area by a Chlor-alkali Complex. Toxicol. Environ. Chem. 22*(1–4), 131–134.

Fitzgerald, W. F., and Watras, C. J. (1988), *Mercury in Surfacial Waters of Rural Wisconsin Lakes.* Proceedings *Conference Trace Metals in Lakes, Hamilton. Sci. Total Environ. 87/88*, 223–232.

Forbes, G., and White, J. (1952), *Chronic Mercury Poisoning in Latent Fingerprint Development. Br. Med. J. 1*, 899.

Francis, P., Birge, W., Roberts, B., and Black, J. (1982), *Mercury Content of Human Hair: A Survey of Dental Personnel. J. Toxicol. Environ. Health 10*, 667–672.

Fraser, A., Melville, K., and Stehle, R. (1934), *Mercury Laden Air: The Toxic Concentration, the Proportion Absorbed and the Urinary Excretion. J. Ind. Hyg. 16*, 77.

Fuyuta, J., Fujimoto, T., and Kiyofuji, E. (1979), *Teratogenic Effects of a Single Oral Administration of Methylmercury Chloride in Mice. Acta Anat. 104*, 356–362.

Gale, T., and Ferm, V. (1971), *Embryopathic Effects of Mercuric Salts. Life Sci. 4*, 1341–1347.

Gates, A., Doherty, R., and Cox, C. (1986), *Reproduction and Growth Following Prenatal Methylmercuric Chloride Exposure on Mice. Fundam. Appl. Toxicol. 7*, 486–493.

Gerstner, H., and Hoff, J. (1977), *Selected Case Histories and Epidemiologic Examples of Human Mercury Poisoning. Clin. Toxicol. 11*, 131–150.

GESAMP (Group of Experts on the Scientific Aspects of Marine Pollution) (1987), *Review of Potentially Harmful Substances: Arsenic, Mercury, and Selenium.* IMO/FAO/UNESCO/WHO/IAEA/UNEP, Reports and Studies No. 28. World Health Organization, Geneva.

Greenwood, M. R. (1985), *Methylmercury Poisoning on Iraq: An Epidemiological Study of the 1971–1972 Outbreak. J. Appl. Toxicol. 5*, 148–159.

Greenwood, M. R., and Clarkson, T. (1970), *Storage of Mercury at Sub-molar Concentrations. Am. Ind. Hyg. Assoc. J. 3*, 250.

Greenwood, M. R., Dhahir, P., Clarkson, T. W., Farant, J., and Chartrand, A. (1977), *Epidemiological Experience with the Magos Reagent in Determining the Different Forms of Mercury in Biological Samples by Flameless Atomic Absorption. J. Anal. Toxicol. 1*, 265–269.

Greenwood, M. R., Clarkson, T., Doherty, R. A., Gates, A. H., Amin-Zaki, L., Elhassani, S., and Majeed, M. A. (1978), *Blood Clearance Half-times in Lactating and Non-lactating Members of a Population Exposed to Methylmercury. Environ. Res. 16*, 48–54.

Haerdi, W. (Geneva) (1989), *Analysis of Heavy Metals: From the Element to Chemical Species.* Lecture at the Central American and Caribbean IAEAC Workshop on Analytical Chemistry in Sanitary and Environmental Studies, Tegucigalpa, Honduras.

Haines, T. A. (1988), *Interrelation of Concentration of Selected Trace Metals in the Gill, Viscera, and Carcass of White Sucker from Soft Water Lakes in Maine (USA).* Proceedings Conference Trace Metals in Lakes, Hamilton. *Sci. Total Environ.*, in press.

Hall, R., Wilke, W., and Fettman, M. (1986), *Renal Resistance to Mercuric Chloride Toxicity During Prolonged Exposure in Rats. Vet. Hum. Toxicol. 28*, 305–307.

Haq, I. (1963), *Agrosan Poisoning in Man. Br. Med. J. 535*, 1579–1582.

Heinz, G. (1980), *Comparison of Game Farm and Wild Strain Mallard Ducks in Accumulation of Methylmercury. J. Environ. Pathol. Toxicol. 3*, 379–386.

Hirayama, K., and Shaikh, Z. (1978), *Metabolism of Intravenous Injection of Renal Mercury Thionein in Rats. Toxicol. Appl. Pharmacol. 45*, 347.

Hirayama, K., Yasutake, A., Dock, L., and Vahter, M. (1989), *Lectures V IUTOX '89,* Brighton, U.K., Proceedings in press, Taylor & Francis Ltd.

Hoffman, G. (1980), In *Proceeding Conference Verwendung von Müll- und Müllklärschlammkomposten in der Landwirtschaft.* GDI, Rüschlikon, Switzerland.

Horowitz, P., Aronson, M., Grodzins, L., et al., (1976), *Elemental Analysis of Biological Specimens in Air with a Proton Microprobe. Science* 194, 1162–1175.

Hughes, W. L. (1957), *A Physiochemical Rationale for the Biological Activity of Mercury and its Compounds. Ann. N.Y. Acad. Sci* 65, 454–460.

Hultberg, H., and Hasselrot, B. (1981), *Mercury in the Ecosystem,* in *Project Coal Board, Health and Environment,* Vallingby. The Swedish State Board (Inf. 04-33-55).

Hunter, D., and Russell, D. (1954), *Focal Cerebral and Cerebellar Atrophy in a Human Subject Due to Organic Mercury Compounds. J. Neurol. Neurosurg. Psychiatr.* 17, 235–241.

Hursh, J., Clarkson, T., Miles, E., and Goldsmith, L. (1989), *Percutaneous Absorption of Mercury Vapor in Man. Arch. Environ. Health* 44, 120–127.

Hutchinson, T. C., and Meema, K. M. (1987), *Lead, Mercury, Cadmium and Arsenic in the Environment,* SCOPE 31. John Wiley & Sons, Chichester-New York-Brisbane-Toronto-Singapore.

Imura, N., Miura, K., Inokawa, M., and Nakada, S. (1980), *Mechanism of Methylmercury Cytotoxicity: By Biochemical and Morphological Experiments Using Cultured Cells. Toxicology* 17, 241–254.

Inskip, M., and Piotrowski, J. (1985), *Review of the Health Effects of Methylmercury. J. Appl. Toxicol.* 5, 113–133.

Izuno, Y. (1976), *Effect of Methylmercuric Chlorides on Metabolism of the GABA Shunt in Rat Brain. Kumamoto Med. J.* 29, 128–133.

Jacobs, A., Maniscalco, W., and Finkelstein, J. (1986), *Effects of Methylmercuric Chloride, Cycloheximide and Colchicine on the Reaggregation of Dissociated Mouse Cerebellar Cells. Toxicol. Appl. Pharmacol.* 86, 362–371.

Jacquet, P. (1984), *The Use of the Embryo-Culture Techniques in the Research on the Teratogenic and Mutagenic Properties of Metals. Toxicol. Environ. Chem.* 7(4), 269–285.

Jaklevic, J., French, W., Clarkson, T., and Greenwood, M. (1977), *X-Ray Fluorescence Analysis Applied to Small Samples, Adv. X-Ray Anal.* 21, 171–185.

Jalili, M., and Abbasi, A. (1961), *Poisoning by Ethyl Mercury Toluene Sulphonanilide. Br. J. Ind. Med.* 18, 303.

Jensen, S., and Jernelöv, A. (1967), *Microorganisms Capable of Methylating Mercury (II). Nordforsk Biocidinformation* 10, 4.

Jensen, S., and Jernelöv, A. (1969), *Biological Methylation of Mercury in Aquatic Organisms. Nature* 223, 753–754.

Jensen, S., and Jernelöv, A. (1972), *International Atomic Energy Report 137,* 43.

Jernelöv, A. (1973), *A New Biochemical Pathway for the Methylation of Mercury and Some Ecological Implications.* In: Miller, M., and Clarkson, T. (eds.): *Mercury, Mercurials, and Mercaptans,* pp. 315–324. C. C. Thomas Publ., Springfield, Illinois.

Jernelöv, A. (1974), *Environmental Contamination with Mercury in Iraq. Conference on Intoxication due to Alkylmercury Treated Seed, Baghdad,* November 9–13.

Jugo, S. (1976), *Retention and Distribution of 203 $HgCl_2$ in Suckling and Adult Rats. Health Phys.* 30, 240–241.

Kamitsuka, M., Robertson, W., and Fligner, C. (1984), *Metallic Mercury Poisoning. Vet. Hum. Toxicol.* 26, 5–7.

Katz, S., and Santilli, V. (1962), *The Reversible Reaction of Tobacco Mosaic Virus Ribonucleic Acid and Mercuric Chloride. Biochim. Biophys. Acta* 55, 621–626.

Kawade, Y. (1963), *The Interaction of Mercuric Chloride with Ribonucleic Acids and Polyribonucleotides. Biochem. Biophys. Res. Commun.* 10, 204–208.

Kawamata, S., Yamaura, Y., and Akagi, H. (1982), *Akagi, Nippon Kashu Elsei Zasshi* 29, 551.

Khayat, A. I. (1985), *Disposition of Metallic Mercury Vapor and Mercuric Chloride in Adult and Fetal Tissues: Influence of Pretreatment with Ethylalcohol, Aminotriazole, Selenium and Tellurium.* Acta Univ. Ups. Abstr. Uppsala Diss. Fac. Pharm. 107.

Kjellström, T., Kennedy, P., Wallis, S., et al. (1986), *Physical and Mental Development of Children with Prenatal Exposure to Mercury from Fish, Stage 1: Preliminary Tests at Age 4.* Solna National Swedish Environmental Research Board, 9. Report No. 3080.

Koller, L., and Roan, J. (1980a), *Effects of Lead, Cadmium and Methylmercury on Immunological Memory.* J. Environ. Pathol. Toxicol. *4*, 47–52.

Koller, L., and Roan, J. (1980b), *Response of Lymphocytes from Lead, Cadmium and Methylmercury Exposed Mice in the Mixed Lymphocyte Culture.* J. Environ. Pathol. Toxicol. *4*, 393–398.

Koller, L., Exon, J., and Abogast, B. (1977a), *Methylmercury: Effect on Serum Enzymes and Humoral Antibody.* J. Toxicol. Environ. Health *2*, 1115.

Koller, L., Exon, J., and Branner, J. (1977b), *Methylmercury: Decreased Antibody Formation in Mice.* Proc. Soc. Exp. Biol. Med. *155*, 602–604.

Koos, B., and Longo, L. (1976), *Mercury Toxicity in the Pregnant Woman, Fetus and Newborn Infant.* Am. J. Obstet. Gynecol. *126*, 390–409.

Korringa, P., and Hagel, P. (1974), *Proceedings International Symposia on Problems on Contamination on Man and His Environment by Mercury and Cadmium,* Luxembourg, 3–5 July, 1973. C.E.C., Luxembourg.

Kostyniak, P., Clarkson, T., Cestero, R., Freeman, R., and Abbasi, A. (1975), *An Extracorporeal Complexing Hemodialysis System for the Treatment of Methylmercury Poisoning.* J. Pharmacol. Exp. Ther. *192*, 260–269.

Kromidas, L., Jamall, I., and Trombetta, L. (1987), *Methyl Mercury Cytotoxicity: The Protective Effects of Glutathione on Microtubular Architecture.* Toxicologist Abstr. 204.

Leumann, P., and Beuggert, H. (1977), *Mercury Contamination of Aquatic Systems* (in German), Chimia *31(11)*, 447–452.

Li, C.P. (1948), *The Chemical Arts of Old China.* J. Chem. Educ., pp. 49–50. Easton, Pennsylvania.

Lindvall, A.N., Lindh, U., Grönquist, S.O., and Linde, A. (1987), *Serological Evidence of Persistent Active Epstein-Barr Virus Infection and Cellular Uptake of Mercury.* Symposium Abstracts 2nd Nordic Symposium on Trace Elements in Human Health and Disease, Odense, F4. Environ. Health 26. WHO Regional Office for Europe, Copenhagen.

Lundbergh, K., and Paulsson, K. (1988), *The Selenium Method for Treatment of Lakes Against Elevated Levels of Mercury in Fish.* Proceedings Conference Trace Metals in Lakes, Hamilton. Sci. Total Environ. *87/88*, 495–507.

Magos, L. (1971), *Selective Atomic Absorption Determination of Inorganic Mercury and Methylmercury in Undigested Biological Samples.* Analyst *96*, 847–853.

Magos, L. (1975), *Mercury and Mercurials.* Br. Med. Bull. *31,* 241–245.

Magos, L., Tuffery, A., and Clarkson, T.W. (1964), *Volatilization of Mercury by Bacteria.* Br. J. Ind. Med. *194*, 298.

Magos, L., Peristianis, G., and Snowden, R. (1978), *Post-exposure Preventive Treatment of Methylmercury Intoxication in Rats with Dimercaptosuccinic Acid.* Toxicol. Appl. Pharmacol. *45*, 463–475.

MAK (1987), *Maximum Concentrations at the Workplace and Biological Tolerance Values for Working Materials, Report No. XXIII,* DFG. VCH Verlagsgesellschaft, Weinheim-Basel-Cambridge-New York.

Marsh, D., et al. (1981), *Dose-response Relationship for Human Fetal Exposure to Methylmercury.* Clin. Toxicol. *18*, 1311–1318.

Marsh, D., Clarkson, T.W., Cox, C., Myers, G., Amin-Zaki, L., and Al-Tikriti, S. (1987), *Fetal Methylmercury Poisoning. Relationship between Concentration in Single Strands of Maternal Hair and Child Effects.* Arch. Neurol. *44*, 1017–1022.

Mailhes, J. (1983), *Methylmercury Effects on Syrian Hamster Metaphase II Oocyte Chromosomes.* Environ. Mutagen. 5, 679–686.

May, K., Reisinger, K., Flucht, R., and Stoeppler, M. (1980), *Radiochemische Untersuchungen zum Verhalten von Quecksilber- und Methylquecksilberchlorid in Süß- und Meerwasser.* Vom Wasser 55, 63–76.

May, K., Reisinger, K., Torres, B., and Stoeppler, M. (1985/1987), *Studies on the Ratio Total Mercury/Methylmercury in Biological and Environmental Materials (incl. Food Chain).* Fresenius Z. Anal. Chem. 320, 646; Proceedings *International Conference Heavy Metals in the Environment,* Athens, pp. 513–515. CEP Consultants Ltd., Edinburgh; Toxicol. Environ. Chem. 13, 153–159.

McKeown-Eyssen, G. E., Reudy, J., and Neims, A. (1983), *Methylmercury Exposure in Northern Quebec: Neurobiological Findings in Children.* Am. J. Epidemiol. 118, 470–479.

Mierle, G. (1988), *The Determination of Mercury in Lakes, Streams and Precipitation.* Proceedings Conference Trace Metals in Lakes, Hamilton. Sci. Total Environ., in press.

Milnes, J., Christopher, A., and DeSilva, P. (1970), *Acute Mercurial Pneumonitis.* Br. J. Ind. Med. 27, 334.

Mitra, S. (1986), *Mercury in the Ecosystem. Its Dispersion and Pollution Today.* Trans. Tech. Publications, Switzerland, pp. 80–84.

Mitsumori, K., Maita, K., Saito, T., Tsuda, S., and Shirasu, Y. (1981), *Carcinogenicity of Methylmercury Chloride in ICR Mice; Preliminary Note on Renal Carcinogenesis.* Cancer Lett. 12, 305–310.

Mitsumori, K., Maita, K., and Shirasu, Y. (1984), *Chronic Toxicity of Methylmercury Chloride in Rats: Pathological Study.* Jpn. J. Vet. Sci. 46, 549–557.

Miyoshi, T. (1959), *Experimental Studies on the Effects of Toxicants on Pregnancy of Rats.* J. Osaka City Med. Cent. 8, 309–318.

Monsour, M., Dryer, N., Hoffmann, H., Schulert A., and Brill, A. B. (1973), *Maternal-fetal Transfer of Organic and Inorganic Mercury via Placenta and Milk.* Environ. Res. 6, 479–484.

National Academy of Sciences (1978), *Assessment of Mercury in the Environment.* National Academy of Sciences, New York.

NIOSH (National Institute for Occupational Safety and Health) (1977a), *Manual of Analytic Methods,* 2nd Ed., Vol. 3. USDHEW No PB 276 838. Cincinnati, Ohio.

NIOSH (National Institute for Occupational Safety and Health) (1977b), *Inorganic Mercury,* in: Occupational Diseases. A Guide to Their Recognition. US DHEW. Public Health Service for Disease Control. DHEW (NIOSH) Publication No. 77-181.

NIOSH (National Institute for Occupational Safety and Health) (1983), *Registry of Toxic Effects of Chemical Substances.* National Institute for Occupational Safety and Health. US Department of Health, Education and Welfare. Cincinnati, Ohio.

NIOSH (National Institute for Occupational Safety and Health) (1985), *Pocket Guide to Chemical Hazards.* US Department of Health and Human Services, Public Health Service. Center of Disease Control, North Carolina.

Ogawa, H. (1979), *Mutagenicity of Various Metal Compounds Studied by the Induction of Sisterchromatid Exchanges in Cultured Human Lymphocytes.* Kyoto-furitsu Ika Daigaku Zasshi 88, 505–539.

Ogawa, E., Tsuzuki, H., and Yamazaki, Y. (1976), *Acceleration of Elimination of Methylmercury Chloride by Chelating Agents: A Study with $^{14}CH_3HgCl$ and $CH_3^{203}HgCl$ in Mice.* Radioisotopes 25, 21–26.

Okabe, M., and Takeuchi, T. (1980), *Distribution and Fate of Mercury in Tissues of Human Organs in Minimata Disease.* Neurotoxicology 1, 607–624.

Olafsson, J. (1983), *Mercury Concentrations in the North Atlantic in Relation to Cadmium, Aluminium and Oceanographic Parameters,* in: Wong, C. S., et al. (eds.): Trace Metals in Sea Water, pp. 475–485. Plenum Press, New York-London.

Olotka, F. T. (1974), *Can Mercury Cell Plants Continue to Comply with New Mercury Pollution Regulations?* Proceedings 1st International Congress on Mercury, Barcelona, Vol. II, pp. 329–336.

Olsen, K., and Boush, G. (1975), *Decreased Learning Capacity in Rats Exposed Prenatally and Postnatally to Low Doses of Mercury.* Bull. Environ. Contam. Toxicol. *13*, 73–79.

Otti, G., Fukuda, M., Seto, H., and Yagyo, H. (1976), *Effect of Methylmercury on Humoral Immune Responses in Mice under Conditions Simulated to Practical Conditions.* Bull. Environ. Contam. Toxicol. *15*, 175.

Padberg, S. (1986), *Mercury Balance in Natural Ecotopes in the Federal Republic of Germany* (in German). Thesis, Universität Köln.

Panigrahi, A., and Misra, B. (1980), *Toxicological Effects of Sub-lethal Concentrations of Inorganic Mercury in Freshwater Fish. Tilapia mossambica Peters.* Arch. Toxicol. *44*, 269–278.

Paterson, R., Usher, D., Biswas, R., and Sreter, J. (1971), *Acute Toxicity of Methylmercury on Glycolytic Intermediates and Adenine Nucleotides of Rat Brain.* Life Sci. *10*, 121–128.

Patty, F. A. (1981), *Mercury, Hg,* in: Clayton, S., and Clayton, F. (eds.): *Patty's Industrial Hygiene and Toxicology,* Vol 2A, pp. 1769–1792. John Wiley & Sons, New York.

Popescu, H., Negru, L., and Lancranjan, I. (1979), *Chromosome Aberrations Induced by Occupational Exposure to Mercury.* Arch. Environ. Health *34*, 461–463.

Rathje, A., and Marcero, D. (1976), *Improved Hopcalite Procedures for the Determination of Mercury Vapor in Air by Flameless Atomic Absorption.* Am. Ind. Hyg. Assoc. J. *37*, 311.

Reisinger, K., Stoeppler, M., and Nürnberg, H. W. (1981), *On the Biological Methylation of Lead, Mercury, Methylmercury and Arsenic in the Environment,* in *Proceedings International Conference on Heavy Metals in the Environment, Amsterdam,* pp. 649–652. CEP Consultants Ltd., Edinburgh.

Reisinger, K., Stoeppler, M., and Nürnberg, H. W. (1984), *Methylcarbenium Ion Transfer from S-Adenosylmethionine to Inorganic Mercury.* Toxicol. Environ. Chem. *8*(1), 45–54.

Renzoni, A., Bernhard, M., Sava, R., and Stoeppler, M. (1979), in: *Int. Comm. Scient. Expl. Medit. Sea* (IV. Journes Etude Pollution Antalya), Monaco, pp. 21–55.

Reuhl, K., and Chang, L. (1979), *Effects of Methylmercury on the Development of the Nervous System: A Review.* Neurotoxicology *1*, 21–55.

Rizzo, A., and Furst, A. (1972), *Mercury Teratogenesis in the Rat.* Proc. West. Pharmacol. Soc. *15*, 52–54.

Rogers, D., and Beamish, F. (1983), *Can. J. Fish. Aquat. Sci. 40,* 824.

Ruf, M. (1981), In *Proceedings Symposium Wassergüte, Wasser,* Berlin.

Rustam, H., and Hamdi, T. (1974), *Methyl Mercury Poisoning in Iraq.* Brain *97*, 499–510.

Saager, R. (1984), *Mercury* (in German), in: *Metallic Rawmaterials Dictionary,* pp. 99–102. Bank von Tobel, Zürich.

Sager, P., and Matheson, D. (1988), *Mechanisms of Neurotoxicity Related to Selective Disruption of Microtubules and Intermediate Filaments.* Toxicology *49*, 479–492.

Sager, P., Doherty, R., and Olmsted, J. (1983), *Interactions of Methylmercury with Microtubules in Cultured Cells and in vitro.* Exp. Cell. Res. *146*, 127–137.

Sager, P., Aschner, M., and Rodier, P. (1984), *Persistent Differential Alterations in Developing Cerebellar Cortex of Male and Female Mice after Methylmercury Exposure.* Dev. Brain Res. *12*, 1–11.

Salvaterra, P., Lown, B., Morgani, B., and Massaro, B. (1973), *Alterations in Neurochemical and Behavioral Parameters in the Mouse Induced by Low Doses of Methylmercury.* Acta Pharmacol. Toxicol. *33*, 177–190.

Sastry, K., and Gupta, P. (1978), *Chronic Mercuric Chloride Intoxication in the Digestive System of Channa punctatus.* J. Environ. Pathol. Toxicol. *2*, 443–446.

Schaller, K. H. (1972), *Staub Reinh. Luft 42,* 142–144.

Schaller, K. H., Zschiesche, W., and Gossler, K. (1978), *Fresenius Z. Anal. Chem. 290,* 113.

Schardein, J. (1985), *Mercury,* in *Chemically Induced Birth Defects,* pp. 622–632. Marcel Dekker, Inc., New York.

Schierling, P., and Schaller, K. (1981), *Arbeitsmedizin 16(3),* 57–61.

Schroeder, W. H. (1989), *Developments in the Speciation of Mercury in Natural Waters.* TRAC *(Trends in Analytical Chemistry) 8*(9), 339–342.

Sebe, K., Kitamura, S., Hayakawa, K., and Sumino, K. (1967), *Jpn. J. Pharmacol. 63*, 241.
Sexton, W. A. (1963), *Thiol and Disulphide Groups,* in *Chemical Constitution and Biological Activity,* 3rd Ed., pp. 97–98, 351. Van Nostrand Co. Inc., Princeton, New Jersey.
Shahristani, H., and Shihab, L. (1974), *Variation of Biological Half-life of Methylmercury in Man. Arch. Environ. Health 28,* 342–344.
Sichak, S., Hursh, J., and Clarkson, T. (1987), *Kinetics of Elemental Mercury Oxidation by a Suspension of Washed Erythrocytes. Toxicologist Abstr.* 154.
Silver, S. (1984), *Bacterial Transformations of and Resistance of Heavy Metals,* in: Nriagu, J. O. (ed.): *Changing Metal Cycles and Human Health,* pp. 199–223. Springer-Verlag, Berlin.
Simpson, R. (1964), *Association Constants of Methylmercury and Mercuric Ions with Nucleotides. J. Am. Chem. Soc. 86,* 2059–2065.
Sipos, L., Golinowski, J., Valenta, P., and Nürnberg, H. W. (1979), *Fresenius Z. Anal. Chem. 298,* 1–8.
Skerfving, S., Hanson, K., Mangs, C., Lindsten, J., and Tyman, N. (1974), *Methyl-mercury Induced Chromosome Damage in Man. Environ. Res. 7,* 83–98.
Skog, E., and Wahlberg, J. (1964), *A Comparative Investigation of the Percutaneous Absorption of Metal Compounds in the Guinea Pig by means of Radioisotopes.* ^{65}Zn, ^{110m}Ag, ^{115m}Cd, and ^{203}Hg. *J. Invest. Dermatol. 43,* 187–192.
Slotkin, T., Pachman, S., Kavlock, J., and Bartolomé, J. (1985), *Early Biochemical Detection of Adverse Effects of a Neurobehavioral Teratogen: Influence of Prenatal Methylmercury Exposure on Ornithine Decarboxylase in Brain and Other Tissues of Fetal and Neonatal Rats. Teratology 32,* 195–202.
Smart, N., and Hill, A. (1968), *Pesticide Residues in Foodstuffs in Great Britain, VI. Mercury Residues in Rice. J. Sci. Food Agric. 19,* 315.
Smith, B., and Berg, G. (1980), *Effects of Mercuric Oxide on Female Mice and Their Litters. Toxicol. Appl. Pharmacol. 38,* 207–216.
Sollman, T. (1957), *Manual of Pharmacology,* 8th Ed. Saunders, Philadelphia.
Spyker, J. (1975), *Assessing the Impact of Low Level Chemicals on Development: Behavioral and Latent Effects. Fed. Proc. Fed. Am. Soc. Exp. Biol. 34,* 1835–1844.
Spyker, J., Sparber, S., and Goldberg, A. (1972), *Subtle Consequences of Methylmercury Exposure: Behavioral Deviations in Offspring of Treated Mothers. Science 177,* 621–623.
Stoeppler, M., Bernhard, M., Backhaus, F., and Schulte, E. (1979), *Comparative Studies on Trace Metal Levels in Marine Biota I. Mercury in Marine Organisms from the Western Italian Coast, the Strait of Gibraltar and the North Sea. Sci. Total Environ. 13,* 209–223.
Stoeppler, M., Dürbeck, H. W., and Nürnberg, H. W. (1982), *Environmental Specimen Banking: A Challenge in Trace Analysis. Talanta 29,* 963–972.
Stokes, P. M., and Wren, C. D. (1987), *Bioaccumulation of Mercury by Aquatic Biota in Hydroelectric Reservoirs: a Review and Consideration of Mechanisms,* in: Hutchinson, T. W., and Meema, K. M. (eds.): *Lead, Cadmium and Arsenic in the Environment,* pp. 255–278. John Wiley & Sons, Chichester, UK.
Sumino, K. (1968), *Analysis of Organic Mercury Compounds by Gas Chromatography: Part I and Part II. Kobe J. Med. Sci. 14,* 115–130, 131–148.
Susuki, T., Shishido, S., and Ishihara, N. (1976), *Different Behavior of Inorganic and Organic Mercury in Renal Excretion with Reference to Effects of D-Penicillamine. Br. J. Ind. Med. 33,* 88–91.
Takeuchi, T. (1972), *Biological Reactions and Pathological Changes in Human Beings and Animals Caused by Organic Mercury Contamination,* in: Hartung, T., and Dinman, B. D. (eds.): *Environmental Mercury Contamination,* pp. 247–289. Ann Arbor Science Press, Ann Arbor, Michigan.
Tejning, S. (1968), *Report No. 68-02-20.* Department of Occupational Medicine, University Hospital, Lund, Sweden.
Thomas, D., Fisher, H., Hall, L., and Mushak, P. (1982), *Effects of Age and Sex on Retention of Mercury by Methylmercury Treated Rats. Toxicol. Appl. Pharmacol 62,* 445–454.

Thomassen, Y. (1987), *Oral Galvanism: Is Mercury the Missing Link; Selenium in Workers Exposed to Inorganic Mercury, Symposium Abstracts 2nd Nordic Symposium on Trace Elements in Human Health and Disease, Odense. Environ. Health 26,* 85–86; *20,* 152–154 (see also Round-Table Discussion No. 13), WHO Regional Office for Europe, Copenhagen.

Toribara, T., Jackson, D., French, W., et al. (1982), *Nondestructive X-Ray Fluorescence Spectrometry for Determination of Trace Elements along a Single Strand of Hair. Anal. Chem. 54,* 1844–1849.

Tsubaki, T., and Irukayama, K. (1977), *Minimata Disease.* Elsevier Scientific Publishing Co., Amsterdam.

Umeda, M., and Nishimura, M. (1979), *Inducibility of Chromosomal Aberrations by Metal Compounds in Cultured Mammalian Cells. Mutat. Res. 67,* 221–230.

Umwelt (1988), *The European River Rhine, and Balance of the Use and Fate of Mercury in the Federal Republic of Germany* (in German). Information of the German Federal Minister on Protection of the Environment and Nature, No. 9, pp. 375–378, No. 10, p. 433.

US EPA (US Environmental Protection Agency) (1979), *Assessment of Mercury.* Environmental Criteria and Assessment Office, Cincinnati, Ohio.

US EPA (US Environmental Protection Agency) (1984), *Health Effects Assessment of Mercury.* Environmental Criteria and Assessment Office. Cincinnati, Ohio.

US EPA (US Environmental Protection Agency) (1985), *Mercury Health Advisory.* Office of Drinking Water. National Technical Information Service. U.S. Department of Commerce, Springfield, Virginia.

US EPA (US Environmental Protection Agency) (1987), *Health Advisories for Legionella and Seven Inorganics.* National Technical Information Service, Washington, DC, PB 87-235586.

Uzodinma, J., Reddy, S., Jinna, R., and Desiah, D. (1987), *Comparative Studies on the Effects of Mercury on Brain ATPases of Fish and Rat. Toxicologist. Abstr. 604.*

Verity, M., Brown, W., and Cheung, M. (1975), *Organic Mercurial Encephalopathy: In vivo and in vitro Effects on Methylmercury on Synaptosomal Respiration. J. Neurochem. 25,* 759–766.

Verschaeve, L., Tassignon, J., Lefevre, M., De Stoop, P., and Susanne, C. (1979), *Cytogenetic Investigation on Leukocytes of Workers Exposed to Metallic Mercury. Environ. Mutagen. 1,* 259–268.

Vighi, M. (1988), *The Mediterranean Sea – Environmental Impact of Chemicals. 1st European Secotox Conference on Ecotoxicology, Copenhagen.* Proceedings, pp. 333–344, 496–500.

Vogel, D., Margolis, R., and Mottei, K. (1985), *The Effects of Methylmercury Binding to Microtubules. Toxicol. Appl. Pharmacol. 80,* 473–486.

Von Burg, R., Farris, F., and Smith, J.C. (1974), *Determination of Methylmercury in Blood by Gas Chromatography. J. Chromatogr. 97,* 65–70.

Von Burg, R., Lijoi, A., and Smith, C. (1979), *Oxygen Consumption of Rat Tissue Slices Exposed to Methylmercury in vitro. Neurosci. Lett. 14,* 309–314.

Von Burg, R., Northington, F., and Shamoo, A. (1980), *Methylmercury Inhibition of Rat Grain Muscarinic Receptors. Toxicol. Appl. Pharmacol. 53,* 285–292.

Vroom, F.Q., and Greer, M. (1972), *Mercury Vapor Intoxication. Brain 95,* 305–318.

Wannag, A., and Skjaerasen, J. (1975), *Mercury Accumulation in Placenta and Foetal Membranes: A Study on Dental Workers and Their Babies. Environ. Physiol. Biochem. 5,* 348–352.

Wallace, R.A., Fulkerson, W., Shults, W.D., and Lyon, W.S. (1971), *Mercury in the Environment, the Human Element, Oak Ridge.* Oak Ridge National Laboratory, ORNL NSF-EP-1.

Ware, R., Burkholder, P., and Chang, L. (1984), *Pathological Effects of Experimental Mercury Intoxication on Renal Function. Am. J. Pathol. 74,* 21a.

Weast, R.C. (1978), *Handbook of Chemistry and Physics,* 59th Ed., 1978–79. The Chemical Rubber Co., Publisher, West Palm Beach, Florida.

Webb, J. (1966), *Enzyme and Metabolic Inhibitors,* Vol. 2, p. 729. Academic Press, New York.

Weir, P., and Hine, C. (1970), *Effects of Various Metals on Behaviour of Conditioned Goldfish. Arch. Environ. Health 20,* 45–51.

Westoo, G. (1968), *Determination of Methylmercury Salts in Various Kinds of Biological Material. Acta Chem. Scand. 22,* 2277–2280.

WHO (World Health Organization) (1976), *Environmental Health Criteria 1. Mercury,* pp. 1–131. World Health Organization, Geneva.
WHO (World Health Organization) (1987), *Guidelines for the Assessment of Drugs and Other Chemicals for Behavioural Teratogenicity.* WHO Regional Office for Europe, Copenhagen.
Wigfield, D., and Perkins, S. (1982), *Speciation of Trace Quantities of Mercury Using the Double Magos Determination. J. Appl. Toxicol.* 6, 279–281.
Wong, P., Baker, M., Chau, Y. K., Mayfield, C. I., and Inniss, W. E. (1981), in *Proceedings International Conference on Heavy Metals in the Environment, Amsterdam,* pp. 645–648. CEP Consultants Ltd., Edinburgh.
Wood, J. M. (1972), *A Progress Report of Mercury. Environment* 14, 33–39.
Wood, J., Kennedy, F., and Rosen, C. (1968), *Synthesis of Methylmercury Compounds by Extracts of a Methanogenic Bacterium. Nature* 220, 173–174.
Wyttenbach, A., Tobler, L., and Bajo, S. (1989), *Na, Cl and Br in Needles of Norway Spruce and in the Aerosol Adhering to the Needles. Toxicol. Environ. Chem.* 18(4), 249–256.
Yoshino, Y., Mozai, T., and Nakao, K. (1965), *Distribution of Mercury in Brain and its Subcellular Units in Experimental Organic Mercury Poisonings. J. Neurochem.* 13, 397–406.
Zenick, H. (1974), *Behavioral and Biochemical Consequences in Methylmercury Chloride Toxicity. Pharmacol. Biochem. Behav.* 2, 709–713.

Additional Recommended Literature

Coquery, M., and Stokes, P. M. (1989), *Effect of Sediment Chemistry on the Bioavailability of Trace Metals to Aquatic Macrophytes. Proceedings 7th International Conference on Heavy Metals in the Environment, Geneva,* Vol. 2, pp. 11–14. CEP Consultants Ltd., Edinburgh.
Drauschke, S., and Birkholz, M. (1989), *Area Covering Disposal System for Refuse from Dentists. Proceedings 6th IRC Congress Berlin, Waste Reduction in the Metallurgical Industry 1,* pp. 89–105. EF-Verlag für Energie- und Umwelttechnik GmbH, Berlin.
Hirayama, N., and Gotoh, S. (1989), *Operation of the CJR Recycling Plant for Spent Dry Batteries. Proceedings 6th IRC Congress Berlin, Waste Reduction in the Metallurgical Industry 1,* pp. 265–270. EF-Verlag für Energie- und Umwelttechnik GmbH, Berlin.
Jackson, T. A. (1989), *Inhibitory and Stimulatory Effects of Cu, Cd, Zn, Hg, and Se on Microbial Production of Methyl Mercury in Sediments. Proceedings 7th International Conference on Heavy Metals in the Environment, Geneva,* Vol. 1, pp. 65–68. CEP Consultants Ltd., Edinburgh.
Matsuoka, T., Kurozu, S., Koyabu, Y., and Toita, M. (1989), *Technology for Treatment of Used Dry Batteries. Proceedings 6th IRC Congress Berlin, Waste Reduction in the Metallurgical Industry 1,* pp. 271–279. EF-Verlag für Energie- und Umwelttechnik GmbH, Berlin.
Murakawa, T. (1989), *Simultaneous Hg and NO_x Removal in Wet Scrubbers. Proceedings 6th IRC Congress Berlin, Incineration and the Environment 3,* pp. 479–487. EF-Verlag für Energie- und Umwelttechnik GmbH, Berlin.
Ogaki, Y., Fujisawa, Y., Miyachim T., and Yoshiy, J. (1989), *Mercury Removal from Flue Gas in Municipal Solid Waste Incineration Plants. Proceedings 6th IRC Congress Berlin, Incineration and the Environment 3,* pp. 489–497. EF-Verlag für Energie- und Umwelttechnik GmbH, Berlin.
Pieters, H. (1989), *Time Trends of Mercury in Pike-perch, Perch and Eel from Lake Ijssel. Proceedings 7th International Conference on Heavy Metals in the Environment, Geneva,* Vol. 1, pp. 362–365. CEP Consultants Ltd., Edinburgh.
Töpper, H. (1989), *Building and Test-principles for Amalgam Separators to Storage Amalgam Waste in Dental Practices. Proceedings 6th IRC Congress Berlin, Waste Reduction in the Metallurgical Industry 1,* pp. 75–88. EF-Verlag für Energie- und Umwelttechnik GmbH, Berlin.

II.21 Molybdenum

GEORGE K. DAVIS, Gainesville, Florida, USA

1 Introduction

Molybdenum is ubiquitous in the environment and plays a complex role in our ecosystems. The element has multiple roles metallurgically, it is a cofactor of enzymes which are essential in plants and animals, and in recent years molybdenum has been recognized as important in human health. The interactions of molybdenum with other elements have been the objects of many studies for many years. In animal metabolism the antagonism between molybdenum and copper has been extensively investigated. As molybdenum is an essential element, deficiency symptoms have now been identified. The appreciation of the importance of molybdenum is manifested by the appearance of extensive reviews (RAJAGOPALAN, 1987); CHAPPELL and PETERSEN, 1976); FRIBERG and LENER 1986; PARKER, 1986).

2 Physical and Chemical Properties, and Analytical Methods

Molybdenum with the atomic mass of 95.94 and atomic number of 42 is a silvery white metal that is very hard although softer and more ductile than tungsten. The metal does not occur natively but as molybdenite, MoS_2, wulfenite, $PbMoO_4$, and powelite, $Ca(MoW)O_4$. It has a melting point of 2617 °C, a boiling point of 5560 °C (CRC gives a lower b.p.), and a specific gravity of 10.22 at 20 °C. Valences are +2, +3, +4?, +5? and +6 (CRC HANDBOOK OF CHEMISTRY AND PHYSICS, 1985/1986; SAAGER, 1984, ELSEVIER'S PERIODIC TABLE OF ELEMENTS, 1987). SAAGER (1984) noted that when exposed to air, molybdenum is readily transformed to the volatile trioxide above 500 °C which makes processing difficult. Naturally occurring isotopes include ^{98}Mo (24%), ^{96}Mo (17%), ^{92}Mo (16%), and ^{95}Mo (16%).

Neutron activation, atomic absorption (see also WELZ, 1985), X-ray fluorescence, and polarography have largely supplanted older techniques for analysis of trace amounts of molybdenum such as thiocyanate and dithiol methods (KOSTA, 1980; LAUPRECHT, et al., 1979; PARKER, 1983, 1986; NATIONAL BUREAU OF STANDARDS, 1977; SNEDDON, et al., 1978. During sampling – for instance, when filtering waste water or groundwater – molybdenum losses must be avoided (MCKINNEY, 1987).

3 Sources, Production, Important Compounds, Uses, Waste Products, Recycling

Molybdenum is quite scarce in the earth's crust with the average crustal content of around 1.5 ppm. Basic rocks average 0.4 ppm and acid rocks about 2.3 ppm. 90 to 95 percent of the world resources of molybdenum are in disseminated porphyry deposits associated with phanerozoic orogenic belts mainly of the Cenozoic Era (SUTULOV, 1978).

The largest molybdenate deposits are in Climax, Colorado, USA with other resources in New Mexico, Arizona, California, USA; in British Colombia, Canada; in Chile, and in the Soviet Union. Molybdenum is also found in varying concentrations in shale, phosphorites, coal, and petroleum with a marked positive correlation with organic carbon (DAVIS et al., 1974). Metallurgical processing, discharge of phosphatic detergents, coal and petroleum burning, and phosphate fertilization contribute molybdenum to the environment. Copper and tungsten mining operations produce molybdenum as a by-product.

Worldwide, annual production has amounted to between 80000 and 100000 tons resulting in an overproduction with depression of prices since 1979 (SAAGER, 1984; PARKER, 1986). The ores are concentrated and transformed into molybdenum(VI) oxide and reduced (LAUPRECHT et al., 1979). Most of the molybdenum is used for the production of alloys (steel; see, for instance, LANDER, 1977; LAUPRECHT et al., 1979; SAAGER, 1984). Ferromolybdenum may also be produced directly from the sulfide by reduction with carbon in an arc, or by a metallothermic procedure. LUGSCHNEIDER et al. (1983) described the production of wear resistant materials by further processing. Molybdenum compounds are also used in some chemical products (which may lead to wastes) such as catalysts, corrosion inhibitors, flame retardants, and smoke repressants, lubricants, molybdenum blue pigments, in analytical and in electroplating techniques, and as starting material for the production of ^{99}technetium (LAUPRECHT et al., 1979; SAAGER, 1984; PARKER, 1986; DE MARIS, 1976). TRUEB (1987) described transistor diamond films on molybdenum substrates. Molybdenum is practically not recycled because its use is dissipative, and because there is an overproduction (SAAGER, 1984).

4 Distribution in the Environment and in Foods

Molybdenum is widely distributed in the environment in small amounts, compared to other trace elements. Seawater contains about 10 µg/L (BRULAND, 1983). Additions to the environment as a result of human activity can be extensive. The combustion of fossil fuels (various kinds of coal with lignite containing up to 300 ppm, and shale oil with up to 25 ppm) with fly ash containing up to 60 ppm, is a constant source (PARKER, 1986). The waste water from industrial processes, the transportation of ores and distribution of sewage can result in widespread additions of molybdenum to the environment. According to DAVIS (1989) molybdate transport in groundwater is retarded by sorption, depending on pH and phosphate and sulfate concentrations. A special source of pollution are the ignition losses from industrial activity which result in thousands of tons being exhausted into air and water. In the Tokyo Bay area and in the Black Sea, concentrations of a few hundred ppm were measured in polluted areas. The 3 billion tons of coal that are burned annually result in the intentional and unintentional release of 100000 tons of molybdenum to soils near industrial plants and can be a significant cause of the increases in the soils and plants near those establishments.

Concentrations in normal herbage may range from 0.1 to 0.3 ppm on a dry matter basis. The molybdenum is present as soluble ammonium molybdate and insoluble molybdenum oxide (MoO_3), calcium molybdate ($CaMo_4$), and molybdenum sulfide (MoS_2). Herbage values of up to 231 ppm have resulted from industrial contamination (GARDNER and HALL-PATCH, 1962). Soils, often calcareous, in the Western United States and Northwest Manitoba, Canada, have a median of 6 ppm of molybdenum with levels decreasing to a median of 0.5 ppm in the acid soils of the Eastern United States (KUBOTA, 1977; BOILA et al., 1984a,b; STONE et al., 1983).

Vegetarian diets could contain much more than average molybdenum if the vegetables were grown on alkaline soils with relatively high molybdenum content, or the diets could be very low in molybdenum if the dietary items came from areas with more acid soils with low molybdenum content. Animal products, with the exception of liver, are usually low in molybdenum (IVAN and VEIRA, 1985). An extensive listing of the molybdenum content of different foods has been published by KOIVISTOINEN (1980). Whole milk and skim milk, analyzed in Finland, contained 0.05 mg/kg Mo (wet weight), and milk powder contained 0.04 mg Mo per kg (KOIVISTOINEN, 1980). ANKE et al. (1985) found that goat milk had 116 µg Mo per liter under usual feeding conditions. Bovine and human liver contains about 3 ppm molybdenum which is relatively high compared to other metals (see WARD, 1987; WENNIG and KIRSCH, 1988). HAMILTON et al. (1972/1973) estimated that in England an adult diet that corresponded to an adult United States diet contained 128±34 µg of Mo per day (see also FRIBERG and LENER, 1986).

5 Uptake, Absorption, Transport and Distribution, Metabolism and Elimination in Plants, Animals, and Humans

The wide range of concentrations of molybdenum which have been found in plants (KUBOTA, 1977) reflects the differing concentrations in soil and plant species. Soil analyses for molybdenum are only suggestive of the possibility of a deficiency, or of an excess, in plants growing on the soil. Molybdenum in plants may vary widely. WARREN et al. (1971) noted that some vegetables have molybdenum contents that are 500 times the level of others. Individual samples of the same vegetables from different areas contain widely differing amounts of molybdenum. In a recent study, plants grown next to each other on the same soil differed in molybdenum content by 46-fold (YLARANTA and SILLANPAA, 1984). Small amounts of molybdenum appear to be essential for maximum growth, the high levels found in forages as a result of industrial contamination do not appear to depress feed production (GARDNER and HILL-PATCH, 1962).

PAPADOPOLOU and HADZISTELIOS (1988) found 400–800 ng Mo/g d.w. in marine plankton, mussels, and fish, but no indication for food chain biomagnification. The rates of absorption of molybdenum from the diet are inversely related to the dietary levels of inorganic sulfate. This relationship and the interrelationship of molybdenum, copper, and sulfate has been extensively investigated and has had importance in animal husbandry in many parts of the world (DICK, 1956a; MILLS and DAVIS, 1987). Of the various molybdenum compounds found in herbage only molybdenum sulfide (MoS_2) appears to be poorly absorbed from the intestines of animals. The biological half-life of molybdenum in mammals is comparatively short, ranging from a matter of hours to several days (FRIBERG and LENER, 1986).

CUNNINGHAM (1950) noted that molybdenum can cross the placental barrier, and high levels of molybdenum in the diet of the dam can increase the level of molybdenum in the liver of the neonate. The level of molybdenum in milk is also influenced by the level of dietary molybdenum (ANKE et al., 1985; ARCHIBALD, 1951). The molybdenum in milk is mostly in the enzyme, xanthine oxidase, but the xanthine oxidase is not increased proportionally with increases in dietary molybdenum.

The principal route of excretion of molybdenum is through the urine (see also FRIBERG and LENER, 1986) in swine (SHIRLEY et al., 1954) and in humans (TIPTON et al., 1966). Urine is not the principal excretion route in ruminants on low sulfate intakes (DICK, 1956a). Increasing the level of sulfate in the diet can markedly increase the excretion of molybdenum in the urine and reduce the level of molybdenum retained in the tissues (DICK, 1956b). Average excretion of molybdenum by humans is about 50–70 µg Mo per day (FRIBERG and LENER, 1986).

6 Effects on Plants, Animals, and Humans

Molybdenum is an essential element for several enzymes important in plant and animal metabolism. These include mammalian xanthine oxidase/xanthine dehydrogenase, aldehyde oxidase, sulfite oxidase, formate dehydrogenase, nitrate reductase, and nitrogenase.

6.1 Effects on Microorganisms and Plants

Molybdenum has a unique role in nitrogenase, an enzyme which converts molecular nitrogen into ammonia, at room temperature and normal pressure in nitrogen fixation. The ability of molybdenum to exist in a variety of oxidation states with a high affinity for many reducible substrates accounts for its remarkable catalytic effects. Molybdenum enzymes provide the key steps in the fixation of nitrogen by microorganisms and its utilization by higher plants (SCHRAUZER, 1976; POPE et al., 1980; HEWITT and NOLTON, 1980; MCKENNA, 1980).

6.2 Miscellaneous Biochemical Effects

The molybdenum enzymes, aldehyde oxidase and xanthine oxidase/xanthine dehydrogenase, are hydroxylases containing iron, sulfur, and flavin (COUGHLIN, 1980a, b) and are widely distributed. Liver and intestine are high in both enzymes which hydroxylate purines, pteridines, pyrimidines, and other heterocyclic nitrogenous compounds and aldehydes whether aliphatic, aromatic or heteroaromatic (HILLE and MASSEY, 1985).

6.3 Deficiency Symptoms

Molybdenum is necessary for plant production, even though present in plant tissue at a level much lower (0.5 ppm dry matter basis) than the critical levels for other essential elements.

Very small amounts of molybdenum in fertilizer have resulted in significant improvement in crop yields of pasture grasses and legumes, cauliflower, broccoli, lettuce, citrus, and other field crops under field conditions. Symptoms of molybdenum deficiency have been described for at least 45 different crops (JOHNSON, 1966; ALLAWAY, 1977).

A deficiency of molybdenum is most likely to occur on acid, freely drained soils rich in iron oxides. The plants with levels of molybdenum toxic for livestock most often grow on humous soils and those with impeded drainage. Toxicity has been reported on granitic soils containing as little as 1 ppm (KUBOTA and ALLAWAY, 1972; DOERGE et al., 1985; COVENTRY et al., 1985). It was the knowledge of the metabolic interrelationships of molybdenum, copper, and sulfur that emphasized its biological importance. A number of reviews have addressed aspects of this subject (MILLS and DAVIS, 1987; CLARKE and CLARKE, 1975; BREMNER, 1979; NATIONAL RESEARCH COUNCIL, 1980; SPIRO, 1985; COUGHLIN 1980a, b; ABUMRAD, 1984; MOURA and XAVIER, 1978; KAUL et al., 1985; ALLEN and GAWTHORNE, 1986; RAJAGOPALAN, 1987).

It is possible to produce a deficiency of molybdenum in rats by the inclusion of 45 or 94 mg of tungsten, as tungstate, per kilogram of diet (HIGGINS et al., 1956). The consumption of low molybdenum forage has been associated with xanthine calculi in sheep (ASKEW, 1958). Clinical symptoms of irritability leading to coma, tachycardia, tachypnea, and night blindness in a human patient receiving total parenteral nutrition were completely eliminated by supplementation with 300 µg of ammonium molybdate per day providing evidence for an essential role of molybdenum in human nutrition (ABUMRAD et al., 1981). A comparison of the molybdenum and selenium contents of the staple foods and of the soils where the foods were produced has suggested a relationship of Keshan disease with the low levels of these two elements (ZHANG, 1986). Diets low in molybdenum fed to goats (ANKE et al., 1978) and to chicks (PAYNE, 1978) resulted in detrimental effects associated with reproduction. Goats had poor conception rates and poor fetal survival. Chicks suffered high embryonic mortality and abnormal growth and development.

6.4 Acute Effects on Animals and Humans

There is convincing evidence that molybdenum is an essential element for animals and humans. The greatest need may be during the early stages of development, and inclusion of molybdenum in the diet of pregnants seems to be recommendable. Molybdenum toxicity in humans does not appear to be a problem (RAJAGOPALAN, 1987).

Calves given drinking water containing 50 ppm of molybdenum exhibited elevated blood plasma copper but non-significant ceruloplasmin changes. The uptake of copper by tissue was reduced by the molybdenum (KINCAID et al., 1986). At low concentrations of molybdenum (2 – 6 mg/kg/day) in the diet of rats, as ammonium tetrathiomolybdate, decreased ^{64}Cu absorption, plasma ceruloplasmin activity and copper concentration in the liver could be observed (MILLS et al., 1978). Ammonium thiomolybdate caused copper to accumulate with molybdenum on the same protein and in acid soluble form in both plasma and kidneys. The antagonism of tetrathiomolybdate has provided a tool in the control of copper toxicity in sheep (HUMPHRIES et al., 1986) and as an alternative to penicillamine in the treatment of Wilson's disease (WALSHE, 1986).

6.5 Chronic Effects on Animals and Humans
(see also FRIBERG and LENER, 1986)

Molybdenosis (also called "teart syndrome" or "peat scours", see FRIBERG and LENER, 1986, and WENNIG and KIRSCH, 1988) is essentially a secondary copper deficiency manifested by diarrhea, anorexia, depigmentation of hair or wool, neurological disturbances and premature death. A wide variation exists in the susceptibility of different species to molybdenum toxicity (horses are most tolerant, WENNIG and KIRSCH, 1988). Variations are also due to concurrent levels of other components of the diet including copper, zinc, sulfur, cadmium, and sulfur containing amino acids.

Dietary levels of 8 – 10 ppm of copper appear to protect cattle against molybdenum levels of 5 – 6 ppm (DICK, 1956a), but levels of molybdenum of 10 mg per day will alter copper levels in the liver of sheep, and levels of 100 mg per day for 12 weeks have caused a significant reduction of copper levels in liver (ROSS, 1970). After 20 months, sheep receiving 120 mg of molybdenum and 7.4 g of sulfate daily, had reduced hemoglobin in blood and reduced copper in wool as well as increased levels of albumin bound copper (BINGLEY, 1974). A turnover rate of ceruloplasmin in cattle higher than in sheep may account for a difference in molybdenum tolerance (MARCILESE et al., 1976). Swine are tolerant of much higher levels of molybdenum in the diet than ruminants (DAVIS, 1950).

The depressed growth of rats occurring with 300 ppm of molybdenum in the diet was obliterated by the simultaneous addition of 100 ppm of copper and 0.33 percent of sulfate (EVANS and DAVIS, 1976).

The clinical effects of low concentrations of tetrathiomolybdate in rat diets were similar to those found in cattle with molybdenosis. The protective effect of orally administered copper appeared to arise from the inhibition of intestinal absorption of the thiomolybdate. Injection of copper overcame the systemic effects of thiomolybdate when intestinal absorption of the thiomolybdate was promoted by a lower copper content of the diet (BREMNER, 1979).

Industrial releases of molybdenum have caused toxicity in cattle. These have been associated with a variety of industries such as mining of uranium, molybdenum milling, oil refining, or the use of an abandoned clay pit (CHAPPELL, 1976). The potential for molybdenum enrichment in the flora in the vicinity of coal power plants has been evaluated since molybdenum in coal has been shown to average 0.91 ± 0.52 ppm in 22 samples from 8 mines and 6.2 ± 5.4 ppm in 8 samples from a single mine. Because most of the molybdenum was retained in the bottom ash and in the precipitated fly ash, the maximum enrichment of the forage was not expected to pose a hazard to grazing livestock (KAAKINEN, 1977).

The toxicity that develops in animals appears to be mainly confined to animals grazing on humous, poorly drained granitic alluvium and black shale soils. It may be emphasized, however, that widespread stream bed sampling in the United Kingdom has indicated that molybdenum toxicity at a subclinical level may be more widespread and economically important than earlier suspected (WEBB and ATKINSON, 1965; BOILA et al., 1984a, b).

PHILLIPPO et al. (1985) found that 5 mg/kg dry mass can delay the time of onset of oestrus in cattle by at least six weeks, probably much longer. Pregnancy rate for animals receiving molybdenum was significantly lower than for those receiving adequate amounts of copper and sulfate.

In an examination of the effects of inhalation of MoS_2, at the rate of 24 mg per cubic meter as an aerosol, by rats for four months, it was noted that Unithiol (2,3-dimercapto-1-propanol Na-sulfonate) or methionine at the rate of 40 mg per kg daily would normalize blood sugar, SH groups, uric acid catalase, choline esterase and acid phosphatase activity but not xanthine oxidase (SARUKHANYAN et al., 1980). In an examination of amino acid intolerance during total parenteral nutrition, it has been noted that molybdenum reversed the amino acid intolerance (ABUMRAD et al., 1981).

6.6 Mutagenic, Carcinogenic, and Teratogenic Effects

Two epidemiological surveys conducted in 1971 in Colorado, USA, demonstrated that there were differences in the molybdenum intake but no evidence that these differences were related to cancer incidence (BRIESE, 1976). The comparatively high tolerance of non-ruminants for molybdenum and the interrelationship between copper, sulfate, and molybdenum has been well demonstrated in ruminants. Studies on carcinogenic effects provided suggestive evidence that neither very low or very high intakes of molybdenum presented mutagenic, carcinogenic, or teratogenic hazards.

7 Hazard Evaluation and Limiting Concentrations

Occupational exposure limits for soluble molybdenum compounds are prescribed at 5 mg/m^3 (ACGIH, 1986; MAK, 1987). For insoluble compounds the maximum allowances are 5 mg/m^3 (ACGIH, 1986) and 15 mg/m^3 (MAK, 1987), respectively. Guideline values of up to 5 mg/kg dry weight have been fixed for molybdenum concentrations in agricultural soils (KLOKE, 1980; SWISS ORDINANCE, 1986).

In ruminants, tolerance limits have been reported ranging from 6.2 ppm in the diet of cattle up to 1000 ppm in the diet of mule deer (NAGY et al., 1975; WARD and NAGY, 1976). Sheep appear to be more resistant than cattle and tolerate molybdenum levels in the plasma of 0.1 to 0.2 mg per liter or approximately 20 to 40 times normal plasma molybdenum levels without affecting ceruloplasmin levels, provided dietary sulfate is about 0.1 percent (DICK, 1953; SUTTLE, 1975).

Because of the copper-molybdenum interrelationship and the relationship to sulfate intake, any evaluation of possible hazard from elevated molybdenum intake in the diet must be related to the concurrent intake of sulfate and copper. In non-ruminants excessive intakes of sulfate will be a key factor in causing molybdenum-copper antagonism (ARRINGTON and DAVIS, 1953; ARTHUR, 1965; HALVERSON et al., 1960; GRAY and DANIEL, 1964).

An evaluation of the quantitative aspects of the copper-molybdenum-sulfate interrelationships required for the assessment of molybdenum dietary risk has been considered by MILLS and DAVIS (1987).

References

Abumrad, N. N. (1984), *Molybdenum – Is It an Essential Trace Metal? Bull. N.Y. Acad. Med.* **60**, 163–171.

Abumrad, N. N., Schneider, A. J., Steel, D., and Rodgers, L. S. (1981), *Amino Acid Intolerance During Prolonged Total Parenteral Nutrition Reversed by Molybdenum. Am. J. Clin. Nutr.* **34**, 2551–2559.

ACGIH (American Council of Governmental Industrial Hygiene) (1986), *Threshold Limit Values for Chemical Substances in the Work Environment.* ACGIH, Cincinnati, Ohio.

Allaway, W. H. (1977), *Perspectives on Molybdenum in Soils and Plants*, in: Chappell, W. R., and Petersen, K. K. (eds.), *Molybdenum in the Einvironment*, Vol. 2, Chap. 1. Marcel Dekker, New York.

Allen, J. D., and Gawthorne, J. M. (1986), *Involvement of Organic Molybdenum Compounds in the Interaction Between Copper, Molybdenum and Sulfur. J. Inorg. Biochem.* **27**, 95–112.

Anke, M., Grün, M., Partschefeld, M., and Groppel, B. (1978), *Molybdenum Deficiency in Ruminants*, in: Kirchgessner, M. (ed.), *Trace Element Metabolism in Man and Animals*–3,

p. 230. Institut für Ernährungsphysiologie, Technische Universität München, Freising-Weihenstephan.
Anke, M., Groppel, B., and Grün M. (1985), *Essentiality, Toxicity, Requirement and Supply of Molybdenum in Humans and Animals*, in: Mills, C. T., Bremner, I., and Chesters, J. K. (eds.), *Trace Elements in Man and Animals*, pp. 154–157. C.A.B. Farnham Royal, London.
Archibald, J. G. (1951), *Molybdenum in Cow's Milk. J. Dairy Sci. 34*, 1026–1029.
Arrington, L. R., and Davis, G. K. (1953), *Molybdenum Toxicity in the Rabbit. J. Nutr. 51*, 295–304.
Arthur, D. (1965), *Interrelationships of Molybdenum and Copper in the Diet of the Guinea Pig. J. Nutr. 87*, 69.
Askew, H. O. (1958), *Molybdenum in Relation to the Occurrence of Xanthine Calculi in Sheep. N.Z. J. Agric. Res. 1*, 447.
Bingley, J. B. (1974), *Effects of High Dose of Molybdenum and Sulfate on the Distribution of Copper in Plasma and Blood of Sheep. Aust. J. Agric. Res. 25*, 467–474.
Boila, R. J., Devlin, T. J., Drysdale, R. A., and Lillie, L. E. (1984a), *The Severity of Hypocupremia in Selected Herds of Beef Cattle in Northwestern Manitoba. Can. J. Anim. Sci. 64*, 899–918.
Boila, R. J., Devlin, T. J., Drysdale, R. A., and Lillie, L. E. (1984b), *Geographic Variation in the Copper and Molybdenum Contents of Forages Grown in Northwestern Manitoba (Canada). Can. J. Anim. Sci. 64*, 919–1936.
Bremner, I. (1979), *Toxicity of Cadmium, Zinc and Molybdenum and Their Effects on Copper Metabolism. Proc. Nutr. Soc. 38*, 235–242.
Briese, F. W. (1976), *Geographic Variation of Mortality and Cancer Morbidity in Colorado*, in: Chappell, W. R., and Petersen, K. K. (eds.), *Molybdenum in the Environment*, Vol. 1, Chap. 19. Marcel Dekker, New York.
Bruland, K. W. (1983), *Trace Elements in Seawater. Chem. Oceanogr. 8*, 208–209.
Chappell, W. R. (1976), *The Molybdenum Project – An Interdisciplinary Program*, in: Chappell, W. R., and Petersen, K. K. (eds.): *Molybdenum in the Environment*, Vol. 1, Chap. 1. Marcel Dekker, New York.
Chappell, W. R., and Petersen, K. K. (eds.) (1976), *Molybdenum in the Environment*, Vol. 1. Marcel Dekker, New York.
Clarke, E. G. C., and Clarke M. L. (1975), *Molybdenum*, in: *Veterinary Toxicology*, pp. 86–89. Williams & Wilkins, Baltimore, Maryland.
Coughlin, M. P. (1980a), *Aldehyde Oxidase, Xanthine Oxidase, and Xanthine Dehydrogenase: Hydroxylases Containing Molybdenum, Iron-Sulfur and Flavin*, in: Coughlin, M. P. (ed.), *Molybdenum and Molybdenum Enzymes*, pp. 119–185. Pergamon Press, Oxford.
Coughlin, M. P. (ed.) (1980b): *Molybdenum and Molybdenum Containing Enzymes*. Pergamon Press, Oxford.
Coventry, D. R., Hirth, J. R., Reeves, T. G., and Burnett, V. F. (1985), *Growth and Nitrogen Fixation by Subterraneum Clover in Response to Inoculation, Molybdenum Application and Soil Amendment with Lime. Soil Biol. Biochem. 17*, 791–896.
CRC Handbook of Chemistry and Physics 1985/1986, 66th Ed. CRC Press, Boca Raton, Florida.
Cunningham, I. J. (1950), *Copper and Molybdenum in Relation to Diseases of Cattle and Sheep in New Zealand*, in: McElroy, W. D., and Glass, B. (eds.), *A Symposium on Copper Metabolism*, pp. 246–273. Johns Hopkins Press, Baltimore, Maryland.
Davis, J. A. (1989), *Field Experiments. Proceedings Metal Speciation and Transport in Groundwaters, Jekyll Island, Georgia*. Lewis Publ., Chelsea, Michigan, in press.
Davis, G. K. (1950), *Influence of Copper on the Metabolism of Phosphorus and Molybdenum*, in: McElroy, W. D., and Glass, B. (eds.), *A Symposium on Copper Metabolism*, pp. 216–229. Johns Hopkins Press, Baltimore, Maryland.
Davis, G. K., Jorden, R., Kubota, J., Laitinen, A., Matrone, G., Newberne, P. M., O'Dell, B. L., and Webb, J. S. (1974), *Copper and Molybdenum*, in: *Geochemistry and the Environment*, Vol. 1, Chap. IX. National Academy of Sciences, Washington, D.C.

de Maris, J. L. (1976), *Neue Zürcher Zeitung* No. 116, p. 55. Supplement, Forschung und Technik, Zürich.
Dick, A. T. (1953), *Effects of Inorganic Sulfate on the Excretion of Molybdenum in Sheep. Aust. Vet. J. 29*, 18, 233.
Dick, A. T. (1956a), *Molybdenum and Copper Relationships in Animal Nutrition*, in: McElroy, E. D., and Glass, B. (eds.), *A Symposium on Inorganic Nitrogen Metabolism*, pp. 445–473. Johns Hopkins Press, Baltimore, Maryland.
Dick, A. T. (1956b), *Molybdenum in Animal Nutrition. Soil Sci. 81*, 229–236.
Doerge, T. A., Bottomley, P. J., and Gardner, E. H. (1985), *Molybdenum Limitations to Alfalfa Growth and Nitrogen Content on a Moderately Acid High Phosphorous Soil. Agron. J. 77*, 895–901.
Elseviers Periodic Table of the Elements (1987), compiled by P. Lof, Amsterdam.
Evans, J. L., and Davis, G. K. (1976), *Copper-Sulfur-Molybdenum Interrelationship(s) in the Growing Rat Evaluated Via the Factorial Arrangement of Diet Treatments*, in: Chappell, W. R., and Petersen, K. K. (eds.), *Molybdenum in the Environment*, Vol. 1, Chap. 10. Marcel Dekker, New York.
Friberg, L., and Lener, J. (1986), *Molybdenum* in: Friberg L. et al. (eds.), *Handbook of Toxicology of Metals*. 2nd Ed., Vol. II, pp. 446–461. Elsevier Science Publishers, Amsterdam.
Gardner, A. W., and Hall-Patch, P. K. (1962), *An Outbreak of Industrial Molybdenosis. Vet. Rec. 74*, 113–115.
Gray, L. G., and Daniel, L. J. (1964), *The Effect of Copper Status of the Rat on the Copper-Molybdenum-Sulfate Interaction. J. Nutr. 84*, 31–37.
Halverson, A. W., Pfifer, J. H., and Monty, K. J. (1960), *A Mechanism for the Copper-Molybdenum Interrelationship. J. Nutr. 71*, 95.
Hamilton, E. I., and Minski, M. J. (1972/1973), *Abundance of the Chemical Elements in Man's Diet and Possible Relations with Environmental Factors. Sci. Total Environ. 1*, 341.
Hewitt, E. J., and Nolton, B. A. (1980), *Nitrate Reductase Systems in Eukaryotic and Prokaryotic Organisms*, in: Coughlin, M. P. (ed.), *Molybdenum and Molybdenum Enzymes*, pp. 275–325. Pergamon Press, Oxford.
Higgins, E. S., Richert, D. A., and Westerfeld, W. W. (1956), *Molybdenum Deficiency and Tungsten Inhibition Studies. J. Nutr. 59*, 539.
Hille, R., and Massey, V. (1985), *Molybdenum-Containing Hydroxylases: Xanthine Oxidase, Aldehyde Oxidase and Sulfate Oxidase*, in: Spiro, T. G. (ed.), *Molybdenum Enzymes*, pp. 443–518. John Wiley & Sons, New York.
Humphries, W. R., Mills, C. F., Greig, A., Roberts, L., Inglis, D., and Halliday, G. J. (1986), *Use of Ammonium Tetrathiomolybdate in the Treatment of Copper Poisoning in Sheep. Vet. Rec. 119*, 596–598.
Ivan, M., and Veira, D. W. (1985), *Effects of Copper Sulfate Supplement on Growth, Tissue Concentration and Ruminal Solubilities of Molybdenum and Copper in Sheep Fed Low and High Molybdenum Diets. J. Dairy Sci. 68*, 891–896.
Johnson, C. M. (1966), *Molybdenum*, in: Chapman, H. D. (ed.), *Diagnostic Criteria for Plants and Soils*, pp. 286–301. University of California, Division of Agricultural Sciences, Riverside, California.
Kaakinen, J. W. (1977), *Estimating the Potential for Molybdenum Enrichment in Flora Due to Fallout from a Nearby Coal-Fired Power Plant*, in: Chappell, W. R., and Petersen, K. K., (eds.), *Molybdenum in the Environment*, Vol. 2, Chap. 12. Marcel Dekker, New York.
Kaul, Bh. B., Enemark, J. H., Merbs, S. L., and Spence, J. T. (1985), *Molybdenum(VI) dopxo, Molybdenum(V) oxo and Molybdenum(IV) oxo Complexes with 2,3:8,9 dibenzo 1,4,7,10 tetrathiodecane: Models for the Molybdenum Binding Site of the Molybdenum Cofactor. J. Am. Chem. Soc. 107*, 2885–2891.
Kincaid, R. L., Blauwiekel, R. M., and Conrath, J. D. (1986), *Supplementation of Copper as Copper Sulfate or Copper Proteinate for Growing Calves Fed Forages Containing Molybdenum. J. Dairy Sci. 69*, 160–163.
Kloke, A. (1980), *Application of Sewage Sludge in Agriculture* (in German), *Proceedings*, pp. 58–87. Gottlieb-Duttweiler-Institut, Rüschlikon-Zürich.

Koivistoninen, P. (1980), *Mineral Element Composition of Finnish Foods. Acta Agric. Scand. (Suppl. 22).*

Kosta, L., (1980), *Molybdenum*, in: *Elemental Analysis of Biological Materials*, pp. 332–333. International Atomic Energy Agency, Vienna.

Kubota, J. (1977), *Molybdenum Status of United States Soils and Plants*, in: Chappell, W. R., and Petersen, K. K. (eds.), *Molybdenum in the Environment*, Vol. 2, Chap. 6. Marcel Dekker, New York.

Kubota, J., and Allaway, W. H. (1972), *Geographic Distribution of Trace Element Problems*, in: *Micronutrients in Agriculture*, pp. 525–544. Soil Science Society of America, Madison, Wisconsin.

Lander, H. N. (1977), *Energy Related Uses in Molybdenum*, in: Chappell, W. R., and Petersen, K. K. (eds.): *Molybdenum in the Environment*, Vol. 2, pp. 773–796. Marcel Dekker, New York.

Lauprecht, W. E., Fichte, R. M., and Kuhn, M. (1979), *Molybdenum, Molybdenum Alloys and Molybdenum Compounds* (in German), in: *Ullmanns Encyklopädie der technischen Chemie*, 4th Ed., Vol. 17, pp. 23–50. Verlag Chemie, Weinheim-New York.

Lugschneider, E., Eck, R., and Ellsmayer, P. (1983), *Chromium, Molybdenum, Tungsten and Their Alloys as High Temperature and Wear Resistant Materials. Radex Rundsch. 1/2*, 52.

MAK (1987), *Maximum Concentrations at the Workplace, Report No. XXIII DFG.* VCH Verlagsgesellschaft, Weinheim-Basel-Cambridge-New York.

Marcilese, N. A., Valsecchi, R. M., and Figueiras, H. D. (1976), *Studies with ^{67}Cu and ^{64}Cu in Conditioned Copper Deficient Ruminants*, in: *Nuclear Techniques in Animal Production and Health as Related to the Soil-Plant System.* IAEA/FAO, Vienna.

McKenna, C. E. (1980), *Chemical Aspects of Nitrogenase*, in: Coughlin, M. P. (ed.): *Molybdenum and Molybdenum Enzymes*, pp. 441–461. Pergamon Press, Oxford.

McKinney, G. L. (1987), *Communication, 17th Annual IAEAC Symposium on the Analytical Chemistry of Pollutants*, Jekyll Island, Georgia.

Mills, C. F., and Davis, G. K. (1987), *Molybdenum*, in: Mertz, W. (ed.), *Trace Elements in Human and Animal Nutrition*, 5th Ed., pp. 429–463. Academic Press, New York.

Mills, C. F., Bremner, I., Ed-Gallad, T. T., Dalgarno, A. C., and Young, B. W. (1978), *Mechanisms of the Molybdenum/Sulfur Antagonism of Copper Utilization by Ruminants*, in: Kirchgessner, M. (ed.), *Trace Element Metabolism in Man and Animals-3*, pp. 150–162. Institut für Ernährungsphysiologie, Technische Universität, München, Freising-Weihenstephan.

Moura, J. J. G., and Xavier, A. V. (1978), *Molybdenum in Proteins*, in: Williams, R. J. P., and DeSilva, J. J. R. F. (eds.), *Trends in Bioinorganic Chemistry*, pp. 79–119. Academic Press, New York.

Nagy, J. G., Chappell, W. R., and Ward, G. M. (1975), *The Effects of High Molybdenum Intakes in Mule Deer. J. Anim. Sci. 41*, 412 (Abstr.).

National Bureau of Standards (1977), *Reference Material*, pp. 1571–1977. National Bureau of Standards, Washington, D.C.

National Research Council (1980), *Molybdenum*, in: *Mineral Tolerance of Domestic Animals*, pp. 328–344. National Academy of Sciences, Washington, D.C.

Papadopolou, C., and Hadzistelios, I. (1988), *Radiochemical Determination of Molybdenum in Marine Organisms from the Aegean Sea. Proceedings 3rd International Conference on Environmental Pollutants, Venice*, pp., 70–72. CEP Consultants Ltd., Edinburgh.

Parker, G. A. (1983), *Analytical Chemistry of Molybdenum.* Springer-Verlag, Berlin.

Parker, G. A. (1986), *Molybdenum* in: Hutzinger, O. (ed.), *Handbook of Environmental Chemistry*, Vol. 3, Part D, pp. 217 ff. Springer-Verlag, Berlin.

Payne, C. G. (1978), *Molybdenum Responsive Syndrome of Poultry*, in: Kirchgessner, M. (ed.), *Trace Element Metabolism in Man and Animals-3*, p. 1515. Institut für Ernährungsphysiologie Technische Universität, München, Freising-Weihenstephan.

Phillippo, M., Humphries, W. R., Bremner, I., Atkinson, T., and Henderson, G. (1985), *Molybdenum Induced Infertility in Cattle*, in: Mills, C. T., Bremner, I., and Chesters, J. K. (eds.), *Trace Elements in Man and Animals*, pp. 855–857. C.A.B. Farnham Royal, London.

Pope, M. T., Still, E. R., and Williams, R. J. P. (1980), *A Comparison Between the Chemistry and Biochemistry of Molybdenum and Related Elements*, in: Coughlin, M. P. (ed.), *Molybdenum and Molybdenum Enzymes*, pp. 3–40. Pergamon Press, Oxford.

Rajagopalan, K. R. (1987), *Molybdenum – An Essential Trace Element.Nutr. Rev. 45*, 321–328.
Ross, D. B. (1970), *The Effect of Oral Ammonium Molybdate and Sodium Sulfate Given to Lambs with High Copper Concentrations. Res. Vet. Sci. 11*, 295–297.
Saager, R. (1984), *Molybdenum*, in: *Metallic Resources Encyclopedia from Antimony to Zirconium*. Bank von Tobel, Zürich.
Sarukhanyan, Zh. G., Babyan, E. A., Mazarbtyan, R. A., Oganesyan, R. D., Dabinysan, G. B., and Kafyan, V. B. (1980), *Possibility of Using Unithiol and Methionine for Therapy of Chronic Molybdenum Poisoning in White Rats. Gig. Tr. Prof. Zabol. 3*, 38 (*Chem. Abstr. 92*, 192234).
Schrauzer, G. N. (1976), *Molybdenum in Biological Nitrogen Fixation*, in: Chappell, W. R., and Petersen, K. K. (eds.), *Molybdenum in the Environment*, Vol. 1, Chap. 17. Marcel Dekker, New York.
Shirley, R. L., Jeter, M. A., Feaster, J. P., McCall, J. T., Outler, J. C., and Davis, G. K. (1954), *Placental Transfer of ^{99}Mo and ^{45}Ca in Swine. J. Nutr. 54*, 59–64.
Sneddon, J., Ottaway, J. M., and Rowston, U. B. (1978), *Mechanism of Atomization of Molybdenum in Carbon Furnace Atomic-Absorption Spectrometry. Analyst 103*, 776–779.
Spiro, T. G. (ed.) (1985), *Molybdenum Enzymes*. John Wiley & Sons, New York.
Stone, L. R., Erdman, J. A., Fedder, G. L., and Holland, H. D. (1983), *Molybdenum in an Area Underlain with Uranium Bearing Lignites in the Northern Great Plains. J. Range Manag. 36*, 280–285.
Suttle, N. F. (1975), *The Role of Organic Sulfur in the Copper-Molybdenum-Sulfur Interrelationship in Ruminant Nutrition. Br. J. Nutr. 34*, 411.
Sutulov, A. (1978), *Resources and Production*, in: Sutulov, A. (ed.), *International Molybdenum Encyclopedia*, Vol. 1, pp. 1778–1978. Sutolov International Publications, Santiago, Chile.
Swiss Ordinance (1986), *Ordinance on Soil Contamination*, issued under Sections 33 and 39 of the Environmental Protection Law, Berne, Switzerland.
Tipton, I. H., Stewart, P. L., and Martin, P. G. (1966), *Patterns of Elemental Excretion in Long Term Balance Studies. Health Phys. 12*, 1683.
Trueb, L. (1987), *Diamond Transistors* (in German). *Neue Zürcher Zeitung, Forschung und Technik, No. 88*, p. 72, 15. April, Zürich.
Walshe, J. M. (1986), *Tetrathiomolybdate (MoS_4^{2-}) as an Anti-Copper Agent in Man*, in: Scheinberg, J. A., and Walshe, J. M. (eds.), *Orphan Diseases and Orphan Drugs*, p. 76. Manchester University Press, Manchester, U.K.
Ward, N. I. (1987), *Trace Elements in Human Health and Disease*, pp. 118–123. WHO-Report EH 20.
Ward, G. M., and Nagy, J. G. (1976), *Molybdenum and Copper in Colorado Forages. Molybdenum Toxicity in Deer and Copper Supplementation in Cattle*, in: Chappell, W. R., and Petersen, K. K. (eds.), *Molybdenum in the Environment*, Vol. 2, pp. 97–113. Marcel Dekker, New York.
Warren, H. V., Delavault, R. E., Fletcher, K., and Wilks, E. (1971), *Variations in the Copper, Zinc, Lead and Molybdenum Content of Some British Columbia Vegetables*, in: Hemphill, D. D. (ed.), *Trace Substances in Environmental Health*, Vol. IV, pp. 94–103. University of Missouri, Columbia, Missouri.
Webb, J. I., and Atkinson, W. J. (1965), *Regional Geochemical Reconaissance Applied to Some Agricultural Problems in County Limerick, Eire. Nature 208*, 1056–1059.
Welz, B. (1985), *Atomic Absorption Spectrometry*, pp. 305–306, 347 ff. VCH Verlagsgesellschaft, Weinheim-Deerfield Beach/Florida-Basel.
Wennig, R., and Kirsch, N. (1988), Chapter 36 *Molybdenum*, in: Seiler, H. G., Sigel, H., and Sigel, E. (eds.), *Handbook on Toxicity of Inorganic Compounds*, pp. 437–447. Marcel Dekker, New York.
Ylaranta, T., and Sillanpaa, M. (1984), *Micronutrient Contents of Different Plant Species Grown Side by Side. Ann. Agric. Fenn. 23*, 158–170 (*Biol. Abstr. 81*, 29435).
Zhang, X. (1986), *The Relationship Between Endemic Diseases and Trace Elements in the Natural Environment of Jilin Province of China*, in: Hemphill, D. D. (ed.), *Proceedings Trace Substances in Environmental Health XX*, p. 381. University of Missouri, Columbia, Missouri.

II.22 Nickel

F. WILLIAM SUNDERMAN, JR., Farmington, Connecticut, USA
AGNETA OSKARSSON, Uppsala, Sweden

1 Introduction

Nickel, the twenty-fourth element in order of natural abundance in the earth's crust, is widely distributed in the human environment. To understand the bioavailability and biological effects of nickel, the various classes of nickel compounds must be differentiated.

At the concentrations prevalent in natural waters, soils, and foods, *divalent nickel compounds* are relatively nontoxic for plants, fishes, birds, and mammals. In humans, adverse effects of inorganic, water-soluble nickel compounds occur after skin contact, which causes nickel dermatitis, a troubling affliction of the general population (especially women), and after inhalation, which causes respiratory tract irritation and asthma in exposed workers, such as electroplaters.

Human exposures to *inorganic, water-insoluble, nickel compounds* usually occur via inhalation of dusts or fumes (e.g., in mining, leaching, sintering, smelting, electrowinning, welding, casting, spray painting, grinding, polishing, plating, and similar industrial operations). Increased risks of respiratory tract cancers have occurred among workers in nickel refineries, usually associated with inhalation exposures to nickel sulfides and oxides. Carcinogenesis bioassays in rats have shown that certain compounds in this category (e.g., nickel subsulfide) are potent carcinogens by inhalation or parenteral routes of administration. Bioavailability of nickel compounds in this category for most plants and animals requires environmental conditions (e.g., acid rain) that promote the formation of Ni^{2+} and complexation with organic ligands (e.g., humic acids).

On account of its lipid solubility and volatility, *nickel carbonyl* is an extremely toxic vapor, which is readily absorbed upon inhalation, traversing the pulmonary alveolar membranes and blood-brain barrier. Effects of severe nickel carbonyl poisoning, as described in Sect. 6.4, include acute pneumonitis with respiratory failure and central nervous system signs (e.g., coma and convulsions), owing to cerebral edema and hemorrhage. Nickel carbonyl is carcinogenic following administration to rats by inhalation or parenteral routes.

More extensive descriptions of these effects are given later in the chapter. Readers are referred to recent monographs and reviews for detailed discussions of the environmental chemistry of nickel (NRIAGU, 1980; BENNETT, 1984) and its analysis (SCHALLER et al., 1982; STOEPPLER 1980; 1984a; 1984b; SUNDERMAN, 1984a; 1986a; SUNDERMAN et al., 1987; RAITHEL, 1987), metabolism and toxicology

(RAITHEL and SCHALLER, 1981; BENCKO, 1983; MUSHAK, 1984; SUNDERMAN, 1986b, c, d), carcinogenesis (BROWN and SUNDERMAN, 1980, 1985; SUNDERMAN, 1984b, c; COSTA and HECK, 1985; EPA, 1986; MASTROMATTEO, 1986), embryotoxicity and teratogenesis (LÉONARD et al., 1981; SUNDERMAN et al., 1983c; MAS et al., 1985), and biological monitoring (NORSETH, 1984; SUNDERMAN et al., 1986; KIILUNEN et al., 1987).

2 Physical and Chemical Properties, and Analytical Methods

2.1 Elemental Nickel and Its Alloys

Nickel is a silver-white, hard, malleable, ductile, ferromagnetic metal that maintains a high luster and is relatively resistant to corrosion. The atomic mass of nickel is 58.71, comprising a mixture of five natural isotopes with atomic masses of 58, 60, 61, 62, and 64. The melting point of metallic nickel is 1453 °C and the boiling point is 2732 °C. In the solid state, nickel has a cubic lattice at any temperature and loses ferromagnetism at 358 °C. The oxidation states of nickel include -1, 0, $+1$, $+2$, $+3$, and $+4$, but the prevalent valences are 0, as in nickel metal and its alloys, and $+2$, as discussed in the next section. Nickel atoms contain unpaired electrons in two outer 3d orbitals, and therefore can sustain changes in oxidation state involving one electron. Complexation with peptides can reduce the redox potential of the Ni^{2+}/Ni^{3+} couple from 4.2 V to 0.7–1.0 V, enabling stable Ni^{3+}-complexes to form under certain biological conditions (NIEBOER et al., 1984; CROSS et al., 1985). Valence shifts between Ni^{2+} and Ni^{3+} may be responsible for nickel-induced free-radical reactions and lipid peroxidation, as noted in Sect. 6.3.1.

2.2 Inorganic, Water-soluble Nickel Compounds

Nickel is present as Ni^{2+} in common, water-soluble nickel compounds, such as the acetate, bromide, chloride, fluoride, iodide, nitrate, sulfamate, and sulfate salts. Nickel exists in aqueous solutions primarily as the green hexaquonickel ion, $Ni(H_2O)_6^{2+}$, which is poorly absorbed by most living organisms.

2.3 Inorganic, Water-insoluble Nickel Compounds

Nickel oxides (NiO, Ni_2O_3), nickel hydroxides (Ni(OH)$_2$, NiO(OH)$_2$), nickel subsulfide (αNi_3S_2), nickel sulfides (βNiS, NiS_2), nickel arsenide (NiAs), nickel chromate ($NiCrO_4$), nickel carbonate ($NiCO_3$), nickel phosphate ($Ni_3(PO_4)_2$), nickel selenide (NiSe), and nickel titanate ($NiTiO_3$) are important compounds in this category. Nickel monoxide (NiO) exists in two major forms with different properties. Black NiO is chemically reactive and readily yields nickel salts upon contact with mineral acids, whereas green NiO is relatively refractory to solubilization.

2.4 Organic, Lipid-soluble Nickel Compounds

Nickel carbonyl, Ni(CO)$_4$, is the most important compound in this group, since it is an intermediate in the Mond process for nickel refining and is used as a catalyst in the chemical and petroleum industries. Nickel carbonyl is volatile, colorless, liquid at ambient temperatures, with a freezing point of $-25\,°C$ and a boiling point of $43\,°C$; it is considered an organic compound owing to the covalent character of the Ni-C bond. Nickel carbonyl is unstable in air, persisting under normal atmospheric conditions for only 60 s at $23\,°C$ (STEDMAN et al., 1980). However, recent studies have demonstrated traces of nickel carbonyl in the atmosphere near a busy traffic intersection, in town gas, and in cigarette smoke (FILKOVA, 1985b; FILKOVA and JAGER, 1986).

2.5 Analytical Methods

As discussed in recent reviews (SCHALLER et al., 1982; WELZ 1985; STOEPPLER, 1980, 1984a, 1984b; SUNDERMAN et al., 1986, 1988; RAITHEL et al., 1987), electrothermal atomic absorption spectrophotometry (EAAS) and differential pulse absorption voltammetry (DPAV) are practical analytical techniques that furnish the requisite sensitivity for measurements of nickel concentrations in biological samples. The detection limits for nickel determinations by EAAS analysis with Zeeman background correction are approximately 0.45 µg/L for urine, 0.1 µg/L for whole blood, 50 ng/L for serum or plasma, and 10 ng/g (dry wt) for tissues, foods, and feces (SUNDERMAN et al., 1986, 1988; ANDERSEN et al., 1986; KIILUNEN et al., 1987; ANGERER and HEINRICH-RAMM, 1989). MONTASER and GOLIGHTLY (1987) informed on application possibilities of inductively coupled plasma spectrometry (ICP-AFS and ICP-AES). Greater sensitivity can be achieved by DPAV analysis using mercury electrodes sensitized with dimethylglyoxime, furildioxime, or dithiocarbamate. For example, DPAV analyses using a dimethylglyoxime-sensitized mercury electrode provide detection limits of approximately 50 ng/L for nickel determinations in whole blood, urine, saliva, and tissue homogenates (OSTAPCZUK et al., 1983). However, present DPAV methods for nickel analysis are more cumbersome and time-consuming than the EAAS procedures. Isotope-dilution mass spectrometry provides the requisite sensitivity, specificity, and precision for nickel determinations (FASSETT et al., 1985), but has not yet been used to analyze nickel in biological samples.

Nickel carbonyl can be measured quantitatively in air and exhaled breath by gas chromatography or chemiluminescence (SUNDERMAN et al., 1968; STEDMAN et al., 1979). The chemiluminescent nickel carbonyl detector, developed by STEDMAN et al. (1979), is a portable unit for industrial use. By the action of ozone and carbon monoxide, nickel carbonyl is converted to an excited state of nickel oxide, NiO*, which emits a photon during decay to the ground state. Detection of photon emission is achieved with a specially designed chemiluminometer. Although chemiluminescent nickel carbonyl detectors are generally available for atmospheric monitoring in nickel refineries that employ the Mond process, the instruments have not yet been employed to measure nickel carbonyl concentrations in breath samples from exposed workers. An extremely sensitive technique for determination of nickel carbonyl in air has been developed by FILKOVA (1985a), based upon sampling of filtered air through a heated graphite tube to deposit nickel in the tube, and measurement of the deposited nickel by electrothermal atomic absorption spectrophotometry.

3 Production, Important Compounds, Sources, Uses, Waste Products, Recycling
(see also QUENEAU et al., 1979)

Nickel constitutes rather less than 0.008% of the earth's crust (NICHOLLS, 1973). The world's nickel is obtained primarily from sulfide ores (e.g., pentlandite and nickeliferous pyrrhotite) and, to a lesser extent, from oxide ores (e.g., laterite) (IARC; SAAGER, 1976; 1984; GRANDJEAN, 1986; MASTROMATTEO, 1986; SCHAUFELBERGER, 1987). Worldwide about 700000 to 900000 t nickel are produced annually, 26% in Canada, 19% in the USSR, and 12% in New Caledonia, smaller amounts in Cuba, Indonesia, the Philippines, and Australia (SAAGER, 1984; INCO, 1987). Nickel sulfide ores, usually mined underground, are crushed and ground, concentrated by physical methods, converted to nickel subsulfide matte, and roasted to nickel oxide. The nickel oxide may be refined electrolytically to yield nickel cathodes or refined by the Mond process, which involves reduction with hydrogen, reaction with carbon monoxide to yield nickel carbonyl, and thermal decomposition to deposit pure nickel. Nickel oxide ores, usually mined in open pits, are smelted to produce ferronickel for use in stainless steel, or reduced with sulfur to yield nickel subsulfide matte, which is refined as just described (NAS, 1975; GRANDJEAN, 1986; MASTROMATTEO, 1986).

Nickel is a constituent of more than 3000 metal alloys, (e.g., Ni-Cr-Fe alloys for cooking utensils and corrosion-resistant equipment; Ni-Cu alloys for coinage and food processing, chemical and petroleum equipment; Ni-Al alloys for magnets and aircraft parts; Ni-Cr alloys for heating elements, gas-turbines, and jet-engines). Alloys of nickel with zinc, manganese, cobalt, titanium, and/or molybdenum are used for special industrial purposes, and alloys of nickel with precious metals are used for jewelry. Nickel is also widely used in electroplating, in the manufacture of Ni-Cd batteries, in rods for arc welding, in pigments for paints and ceramics (e.g., yellow nickel titanate), in molds for ceramic and glass containers, in surgical and dental prostheses, in magnetic tapes and computer components, and in nickel catalysts. A newer application is in nickel metallized textile fibers for heating elements and reflection materials without electrostatic charge (EBNETH, 1986). Nickel catalysts are employed for organic syntheses, petroleum refining, hydrogenation of edible fats and oils (e.g., margarine), and in the final methanation step of coal gasification.

Nickel carbonyl is mainly an intermediate in the Mond process for nickel refining, but it is also used for vapor-plating in the metallurgical and electronics industries and as a catalyst for synthesis of acrylic monomers in the plastics industry.

Inadvertant formation of nickel carbonyl can occur in industrial processes that use nickel catalysts, such as coal gasification, petroleum refining, and hydrogenation of lipids. In the USA approximately 10% of nickel that is used per year undergoes recycling, worldwide even more. Nickel-bearing materials, mostly from the steel industry, are melted, refined, and used to prepare alloys similar in composition to those that entered the recycling process (NAS, 1975; SAAGER, 1984; EPA, 1986; MASTROMATTEO, 1986).

4 Distribution in the Environment, in Foods, and in Living Organisms

4.1 Emissions, Air and Water Quality, Distribution in Soils

Nickel enters groundwater and surface water from dissolution of rocks and soils, from biological cycles, from atmospheric fallout, and especially from industrial processes and waste disposal. Most nickel compounds are relatively soluble at pH values <6.5, whereas nickel exists predominantly as insoluble nickel hydroxides at pH values >6.7. Therefore, acid rain has a pronounced tendency to mobilize nickel from soil and to increase nickel concentrations in groundwaters, leading eventually to increased uptake and potential toxicity for microorganisms, plants, and animals. Nickel exists in river waters approximately half in ionic form and half as stable organic complexes (e.g., with humic acids). Nickel speciation may of course change during transport. In bottom sediments of contaminated rivers, organic nickel complexes become absorbed on silica particles, with gradual accumulation of nickel in the upper layers of mud. Nickel leached from dump sites can contribute to nickel contamination of the aquifer, with potential exotoxicity (SUNDERMAN, 1986b). Further sources for increased soil concentrations may be composts and sewage sludges (see Chapters I.7 and I.20b). Seawater contains 0.1 to 0.5 µg Ni/L (mainly in the form of Ni^{2+} cations and of chloro- and carbonato-complexes; BRULAND, 1983). Concentrations are lower at the surface (correlated with residual phosphate), and higher in greater depth (BRULAND, 1983; see also MART and NÜRNBERG, 1984; JICKELS and BURTON, 1988). Surface waters average 15 to 20 µg Ni/L, and drinking water usually contains less than 20 µg Ni/L. Drinking water samples occasionally contain much higher nickel concentrations, owing to nickel pollution of the water supply or leaching from nickel-containing pipes or nickel-plated spigots (ANDERSEN et al., 1983; SUNDERMAN et al., 1986). Arctic snow concentrations may vary between 20 and 300 µg/kg Ni (MART, 1983).

Nickel enters the atmosphere from natural sources (e.g., volcanic emissions and windblown dusts produced by weathering of rocks and soils), from combustion of fossil fuels by stationary and mobile power sources, from the emissions of nickel mining and refining operations, from metal consumption in industrial processes, and from incineration of wastes (SUNDERMAN, 1986b). In the USA, atmospheric nickel concentrations average 6 ng/m^3 for non-urban areas versus 17 ng/m^3 (in summer) and 25 ng/m^3 (in winter) for urban areas.

In industrialized regions and large cities, atmospheric nickel concentrations as high as 120 to 170 ng/m^3 have been recorded. Atmospheric concentrations of nickel are related to consumption of fossil fuels, since, for example, the nickel content of coal ranges from 4 to 24 mg/kg, whereas crude oils (especially those from Angola, Columbia and California) may contain up to 100 mg/kg (TISSOT and WELTE, 1984). Substantial atmospheric emissions of nickel derive from fly-ash that is released from coal-fired power plants; nickel derived from petroleum is released into the environment in automotive exhaust fumes. The atmospheric nickel concentration near a nickel refinery in West Virginia averaged 1.2 µg/m^3, compared to 0.04 µg/m^3 at six other sampling stations not contiguous to the nickel plant (NAS, 1975). Inhalation of nickel averages 0.4 µg/day (range from 0.2 to 1.0 µg/day) for urban dwellers and 0.2 µg/day (range from 0.1 to 0.4 µg/day) for rural dwellers (BENNETT, 1984; SUNDERMAN, 1986b). Cigarette smoking can increase inhaled nickel by as much as 4 µg per pack of cigarettes (GRANDJEAN, 1984).

4.2 Food Chain, Plants, Animals, and Humans

(see also FISHBEIN, 1987)

4.2.1 Plants

The extractable nickel content of soil affects the uptake of nickel by plant roots. Extractability of nickel from soil is influenced by physical factors (e.g., texture, temperature, and water content), chemical factors (e.g., pH, organic constituents, redox potential), and biological factors (e.g., plant species variability, microbial activity) (NAS, 1975; WALLACE et al., 1977; HEALE and ORMOND, 1982; HAZLETT et al., 1983). Extractable nickel concentrations in soils, measured by treating soil samples with solutions of potassium chloride, ammonium acetate, acetic acid, or ethylenediamine tetraacetate usually range from <0.01 to 2.6 mg/kg, and are correlated with the nickel concentrations in plant tops, which range from 0.05 to 5 mg/kg dry weight (HALSTEAD et al., 1969; NAS, 1975). In soils derived from serpentine rocks, extractable nickel can reach 70 mg/kg, which is toxic for most plants. Alkalinization of such soils decreases the uptake of nickel by plants and reduces nickel toxicity. For example, HALSTEAD (1968) showed that addition of calcium hydroxide to soil with high nickel content increased the yield of oats and reduced the nickel concentration in the plant tops.

TIFFIN (1971), using electrophoresis to fractionate nickel in plant xylem exudate, found that most of the nickel was complexed to an anionic organic ligand. THEISEN and BLINCOE (1984) used gel filtration chromatography on Sephadex G-25 to fractionate ^{63}Ni in extracts of alfalfa grown on ^{63}NiCl$_2$ treated soil. The ^{63}Ni was predominantly present in a water-soluble complex, about 2000 daltons in size. The chemical identity of this nickel-binding ligand has not been established. Nickel is an essential constituent in plant urease (DIXON et al., 1975, 1980; POLACCO, 1977). Urease-rich legumes, such as jack beans and soybeans, generally contain high nickel concentrations. Soybeans grown on nickel-deficient nutrient solutions accumulate toxic urea concentrations, which result in necrosis of leaflet tips, a characteristic of nickel deficiency (ESKEW et al., 1984). Numerous species of nickel-accumulating plants have been identified, including *Sebertia acuminata*, a tree native to New Caledonia (site of one of the world's largest nickel deposits), which attains exceptionally high concentrations of nickel (10 g/kg dry weight in leaves; 250 g/kg in latex) (JAFFRE et al., 1976, 1979; REEVES et al., 1981, 1983). Such plants usually contain elevated concentrations of citric and malic acids, which may be involved in the transport and storage of nickel (STILL and WILLIAMS, 1980).

Hundreds of foods available in the Netherlands, the United Kingdom, Finland, and Denmark have been analyzed for nickel by electrothermal atomic absorption spectrometry (ELLEN et al., 1978; EVANS et al., 1978; KOIVISTOINEN, 1980; NIELSEN and FLYVHOLM, 1984; VEIEN and ANDERSEN, 1986). In most samples, the nickel contents were <0.5 mg/kg. Nickel concentrations in nuts and cocoa ranged up to 5 and 10 mg/kg, respectively. Foods with mean nickel concentrations >1 mg/kg included oatmeal, wheat bran, dried beans, soya products, soup powder, tea leaves, hazelnuts, peanuts, lucerne seeds, sunflower seeds, licorice, spices, cocoa, and dark chocolate.

4.2.2 Aquatic Species

Shellfish and crustacea, such as oysters, mussels, lobsters, and krill, generally contain higher concentrations of nickel in their edible flesh than do the various species of fishes that have been investigated (NAS, 1975; STOEPPLER and NÜRNBERG, 1979; VEIEN and ANDERSEN, 1986). As illustrative values, the concentrations of nickel averaged 0.5 to 2.2 mg/kg dry weight in Japanese shellfish, 0.5 mg/kg dry weight in Danish mussels, and 0.13 mg/kg dry weight in krill from the Scotia Sea of the Antarctic (NAS, 1975; STOEPPLER and BRANDT, 1979; VEIEN and ANDERSEN, 1986).

Table II.22-1. Nickel Concentrations in Body Fluids and Excreta of Healthy, Non-Exposed Persons (SUNDERMAN, 1986d)

Specimens	Subjects	Nickel Concentrations mean ± SD	Range	Units
Whole blood	30 (15 m, 15 f)	0.34 ± 0.28	<0.05 – 1.05	µg/L
Serum	30 (15 m, 15 f)	0.28 ± 0.24	<0.05 – 1.08	µg/L
Urine (random collection)	34 (18 m, 16 f)	2.0 ± 1.5 2.0 ± 1.5	0.5 – 6.1 0.4 – 6.0	µg/L µg/g creatinine
Urine (24 h collection)	50 (24 m, 26 f)	2.2 ± 1.2 2.6 ± 1.4	0.7 – 5.2 0.5 – 6.4	µg/L µg/day
Feces (72 h collection)	10 (6 m, 4 f)	14.2 ± 2.7 258 ± 126	10.8 – 18.6 80 – 540	µg/g dry wt µg/day

4.2.3 Mammals

Nickel in animal food is not of great importance. Bovine meat may contain 1–13 ppb, nails a few hundred ppbs (STOEPPLER, 1984b). WARD (1987) found 0.07–0.18 ppm in bovine liver.

The estimated body burden of nickel in healthy adults averages 0.5 mg/70 kg (7.3 µg/kg body weight) (BENNETT, 1984). In postmortem tissue from adult persons without known occupational or iatrogenic exposures to nickel compounds, the highest nickel concentrations occur in bone, lung, thyroid, and adrenal, followed by kidney, heart, liver, brain, spleen, and pancreas, in diminishing order (KOLLMEIER et al. 1985; SEEMANN et al., 1985; SUNDERMAN, 1986a; REZUKE et al., 1987; RAITHEL, 1987; RAITHEL et al., 1987). Reference values for nickel concentrations in body fluids and excreta of healthy, non-exposed persons are listed in Table II.22-1.

5 Uptake, Absorption, Transport and Distribution, Metabolism and Elimination in Animals and Humans

(for Plants see Section 4.2.1)

5.1 Uptake and Absorption

In humans, oral intake of nickel is derived primarily from foods, since drinking water from public water supplies usually contains <20 µg Ni/L (SUNDERMAN et al., 1986). Estimates of the average daily intake of nickel by humans range from 0.14 to 0.6 mg/day, with an estimated upper limit of 0.9 mg/day (GRANDJEAN, 1984; FLYVHOLM et al. 1984). Nickel in Danish and Finnish diets was estimated to average 130 µg/day, mostly derived from plant products; calculations indicated that nickel intake on special occasions could reach 900 µg/day (VARO and KOIVISTOINEN, 1980; NIELSEN and FLYVHOLM, 1984; VEIEN and ANDERSEN, 1986). These findings are consistent with measurements of nickel contents of nine institutional diets in the USA, which averaged 168 ± 11 µg/day or 75 µg/10^6 calories (MYRON et al., 1978), and with the estimated dietary nickel intakes in the United Kingdom, which were between 140 and 150 µg per day (SMART and SHERLOCK, 1987). Based upon nickel concentrations in milk and infant foods, the estimated dietary intake of nickel in infants, age 1 to 12 months, ranges from 0.03 to 0.3 mg/day, which is comparatively high in relation to body weight (CLEMENTE et al., 1980).

Approximately 3 to 6% of radiolabeled Ni^{2+} is absorbed from the intestinal tract after oral administration of $^{63}NiCl_2$ or $^{57}NiCl_2$ to fasting rats or mice (HO and FURST, 1973; NIELSEN et al., 1986). Ni^{2+} is absorbed from the lumen of the perfused rat jejunum by a first-order kinetic process that is depressed by Zn^{2+} and by constituents of dried skimmed milk (FOULKES and McMULLEN, 1986). SOLOMONS et al. (1982) observed prompt and sustained elevations of plasma nickel concentrations when fasting human volunteers ingested 5 mg of nickel as an aqueous solution of nickel sulfate; however, no significant post-prandial increases of plasma nickel concentrations occurred when the same quantity of nickel sulfate was added to standard meals. The increase of plasma nickel concentrations was also suppressed when nickel sulfate was dissolved in milk, coffee, tea, or orange juice. These studies indicate that foods and beverages reduce or prevent the absorption of Ni^{2+} from the alimentary tract.

According to HOGAN and RAZNIAK (1986) accumulation and retention is age-dependent, besides variations among renal, lung, adrenal, and splenic tissues. They showed that juvenile mice are less vulnerable to nickel-induced renal dysfunction compared to adults.

In humans, approximately 35% of inhaled nickel is absorbed from the respiratory tract; the remainder is carried up the tracheobronchial mucociliary escalator and either swallowed or expectorated (BENNETT, 1984; GRANDJEAN, 1984; SUNDERMAN, 1986b). In their studies on biological monitoring (with Zeeman-corrected ETAAS) ANGERER and HEINRICH-RAMM (1989) found that about 90 µg Ni/m^3 in the working place atmosphere of welders correspond to about 5 µg Ni/L blood and to about 15 µg Ni/L urine.

5.2 Distribution

Transport of Ni^{2+} in plasma is mediated by binding to albumin and ultrafiltrable ligands (VAN SOESTBERGEN and SUNDERMAN, 1972; ASATO et al., 1975). The major nickel-binding site of serum albumin has been identified and characterized. Ni^{2+} competes with Cu^{2+} for complexation within a square planar ring that is created by (a) the terminal amino group of albumin; (b) the first two peptide nitrogen atoms at the N-terminus of the albumin molecule; and (c) the imidazole nitrogen of the histidine residue at the third position from the N-terminus (CALLAN and SUNDERMAN, 1973; GLENNON and SARKAR, 1982; LAUSSAC and SARKAR, 1980, 1984). The albumin $^{63}Ni^{2+}$-binding site is absent in persons with certain types of bisalbuminemia (FINE et al., 1983; BATHURST et al., 1987). Ultrafiltrable $^{63}Ni^{2+}$-binding ligands in plasma include amino acids (e.g., histidine) and small polypeptides (ASATO et al., 1975; SARKAR, 1984). Although a major fraction of plasma nickel is bound to nickeloplasmin, an α-macroglobulin, the nickel content of nickeloplasmin is not readily exchangeable with exogenous Ni^{2+}, and nickeloplasmin does not appear to play an important role in extracellular transport of nickel (NOMOTO et al., 1971; DECSY and SUNDERMAN, 1974). In cytosol of kidney, lung, and liver from $^{63}NiCl_2$ treated rodents, $^{63}Ni^{2+}$ is bound to several macromolecular and low-molecular weight constituents (OSKARSSON and TJÄLVE, 1979c; SUNDERMAN et al., 1981, 1983a; ABDULWAJID and SARKAR, 1983; HERLANT-PEERS et al., 1983; TEMPLETON and SARKAR, 1985, 1986).

Nickel carbonyl, $Ni(CO)_4$, is an important organic compound from the view point of toxicology. Because of its lipid solubility, nickel carbonyl can penetrate cell membranes and traverse the blood-brain barrier.

5.3 Elimination
(see also Section 6.3.2)

Urine is the predominant route for elimination of absorbed nickel in animals and humans. Nickel in urine is associated with low molecular weight complexes, which are mostly ninhydrin-positive (VAN SOESTBERGEN and SUNDERMAN, 1972; ASATO et al., 1975). Nickel concentrations in postmortem samples of human bile range from 1 to 3 µg/L, suggesting that biliary excretion of nickel may be quantitatively significant in humans (REZUKE et al., 1987). In contrast, biliary excretion of nickel accounts for only a minute fraction of total nickel elimination in rats (MARZOUK and SUNDERMAN, 1985). Other routes of nickel elimination in humans include (a) hair and dermal detritus (NECHAY and SUNDERMAN, 1973; (b) milk from lactating mothers (FEELEY, 1982; MINGORANCE and LACHIA, 1985), (c) sweat (HOHNADEL et al., 1973; COHN and EMMETT, 1978), and (d) saliva (CATTALANATTO et al., 1977). Mean concentrations of nickel in sweat from healthy adults and in milk from lactating mothers are 10 to 20 times greater than the mean concentrations of nickel in urine, suggesting that appreciable quantities of nickel may be eliminated by these routes under conditions of profuse sweating or during lactation. In humans with environmental, occupational, or iatrogenic exposures to nickel, the kinetics of nickel uptake, distribution, and elimination depend upon the physical and chemical properties of the specific nickel compounds. Hence, relevant kinetic data are considered in Sects. 6.3.1 and 6.3.2, which discuss the biological effects of the major categories of inorganic nickel compounds.

Investigations of $Ni(CO)_4$ kinetics in rodents have shown that a fraction of inhaled or injected $Ni(CO)_4$ is excreted in the exhaled breath within three hours post-exposure, and the remainder undergoes intracellular oxidation of Ni^0 to Ni^{2+}, with release of carbon monoxide, which becomes bound to hemoglobin. During 48 hours post-exposure, the carbon monoxide is exhaled and most of the residual Ni^{2+} is excreted in the urine (SUNDERMAN and SELIN, 1968; SUNDERMAN et al., 1968; KASPRZAK and SUNDERMAN, 1969; OSKARSSON and TJÄLVE, 1979b).

6 Effects on Animals and Humans
(for Plants see Section 4.2.1 and Chapters I.7f, I.7g and for animals also I.13b, Section 4)

6.1 Deficiency Symptoms

Nickel is an essential element for animal nutrition, based upon results from experimental induction of dietary nickel deficiency in chicken, rats, swine, and goats (ANKE et al., 1984; NIELSEN, 1984; SPEARS, 1984; FISHBEIN, 1987). In rats, nickel deficiency is associated with growth retardation and reduction of blood hemoglobin concentrations, hematocrit values, and erythrocyte counts, evidently mediated by impaired iron absorption (SCHNEGG and KIRCHGESSNER, 1975a, b, 1976). No specific physiological function of nickel has been established in animals, although recent studies suggest that nickel may serve as a cofactor for activation of calcineurin, a calmodulin-dependent phosphoprotein phosphatase (KING et al., 1985).

6.2 Effects on Aquatic Animals

The toxicity of nickel for adult fishes and fish cell lines is generally low (REHWOLDT et al., 1971; BABICH et al., 1986); the toxicity of nickel for embryonic and larval stages of fishes and amphibia varies widely, from LC_{50} values of 50 µg Ni/L for rainbow trout (*Salmo gardneri*) and narrow-mouthed toad (*Gastrophyrne carolinensis*) to 2.8 mg/L for goldfish (*Carassius auratus*) and 11 mg/L for Fowler's toad (*Bufo fowleri*) (BIRGE and BLACK, 1980). Reproduction of most aquatic species is not significantly impaired at nickel concentrations <10 µg/L, although this generalization excludes the narrow-mouth toad, for which the LC_{10} concentration is only 4.1 µg Ni/L (BIRGE and BLACK, 1980). Exposure of a freshwater crustacean species (*Daphnia magna* Straus) to nickel sulfate at concentrations from 5 to 10 µg Ni/L for three generations resulted in extermination, which may explain the recent disappearance of *Daphnia* from the lower Don River in the USSR (LAZAREVA, 1985).

6.3 Miscellaneous Biochemical Effects on Mammals

6.3.1 Inorganic, Water-soluble Nickel Compounds

The kinetics of soluble Ni^{2+}-compounds administered to rats and rabbits by parenteral routes conform to a two-compartment model (ONKELINX and SUNDERMAN, 1980). The half-time for diminution of plasma ^{63}Ni in $^{63}NiCl_2$ treated rats is approximately 10 hours (SUNDERMAN et al., 1976a). For comparison, the half-time for diminution of plasma nickel concentrations in human volunteers after oral intake of nickel sulfate is approximately 11 hours (CHRISTENSEN and LAGESSON, 1981). In rats that received $^{63}NiCl_2$ by i.v. injection, 93% of the ^{63}Ni-dose was eliminated during 4 days post-injection, including 90% in urine and 3% in feces (SUNDERMAN and SELIN, 1968). Studies of ^{63}Ni distribution in several rodent species at various dosages, time-intervals, and injection routes show that Ni^{2+} is accumulated most avidly in kidney and that uptake of ^{63}Ni in brain is substantially less than in other organs (PARKER and SUNDERMAN, 1974; OSKARSSON and TJÄLVE, 1979a; CARVALHO and ZIMMER, 1982; SUNDERMAN, 1986c).

Based upon recent summaries (SUNDERMAN, 1986d; SUNDERMAN et al., 1986), the following are the principal forms of biochemical reactions in nickel toxicity induced by parenteral administration of water-soluble, inorganic nickel compounds to experimental animals: (A) Peroxidative degradation of membrane lipids occurs in $NiCl_2$ treated rats, as evidenced by increased concentrations of thiobarbituric acid chromogens in liver, kidney, and lung homogenates, increased concentrations of conjugated dienes in hepatic microsomal lipids, and increased exhalation of ethane and ethylene (SUNDERMAN et al., 1985; DONSKOY et al., 1986; KASPRZAK et al., 1986;

KNIGHT et al., 1986; SUNDERMAN, 1987). (B) Heme oxygenase activity is increased in kidney and liver microsomes of $NiCl_2$ treated rats (MAINES and KAPPAS, 1977; SUNDERMAN et al., 1983b).

6.3.2 Inorganic, Water-insoluble Nickel Compounds

Certain inorganic, water-insoluble compounds (e.g., nickel subsulfide, aNi_3S_2) slowly dissolve during in vitro incubation in serum or renal cytosol (KUEHN and SUNDERMAN, 1982). Following i.m. administration of $^{63}Ni_3S_2$ to rats, cumulative excretion of ^{63}Ni during 8 weeks post-injection averaged 67% of the dose in urine and 7% in feces; residual ^{63}Ni at the injection site averaged 19% of the dose at 20 to 24 weeks post-injection and 14% at 31 weeks (SUNDERMAN et al., 1976b). Whole-body kinetic parameters for ^{63}Ni were computed by use of a three-compartment model, based upon measurements of ^{63}Ni in urine, feces, injection site, and viscera of $^{63}Ni_3S_2$ treated rats (SUNDERMAN et al., 1976b; ONKELINX and SUNDERMAN, 1980). OSKARSSON et al. (1979) demonstrated gradual mobilization of solubilized ^{63}Ni and ^{35}S from the sites of i.m. or s.c. administration of $^{63}Ni_3S_2$ or $Ni_3{}^{35}S_2$ to mice. X-ray diffractometry of insoluble nickel-containing particles that remained at the injection site did not reveal aNi_3S_2, but demonstrated Ni_7S_6 and βNiS (OSKARSSON et al., 1979). These in vivo observations are consistent with findings of KASPRZAK and SUNDERMAN (1977), who showed that aNi_3S_2 is oxidized to βNiS during in vitro incubation in rat serum and subsequently undergoes further oxidation to soluble (e.g., $NiSO_4$) and insoluble (e.g., $Ni(OH)_2$) nickel compounds. The half-time for urinary nickel elimination derived from insoluble nickel compounds inhaled by mold-makers in the glass industry was estimated to range from 30 to 50 hours (RAITHEL et al., 1982). The half-time for urinary elimination of nickel compounds inhaled by welders was estimated to be 53 hours (ZOBER et al., 1984).

6.4 Acute Effects on Animals and Humans

(A) Natural killer (NK) cell activity and T-cell mediated immune responses are inhibited in spleens of $NiCl_2$-treated mice (SMIALOWICZ et al., 1984, 1985), and acute thymic involution develops in $NiCl_2$-treated rats (KNIGHT et al., 1987). (B) Hyperglycemia occurs in rodents after administration of $NiCl_2$, apparently mediated by increased pancreatic secretion of glucagon (CLARY, 1975; HORAK et al., 1978; KASPRZAK et al., 1986). (C) Nephrotoxicity, manifested by glomerular and tubular histopathology, proteinuria, enzymuria, aminoaciduria, and increased renal metallothionein concentration, occurs in $NiCl_2$-treated rats (GITLITZ et al., 1975; SUNDERMAN and HORAK, 1981; SUNDERMAN and FRASER, 1983; KASPRZAK et al., 1986). Marked hypothermia and diminution of metabolic rate occur in mice after an i.p. injection of $NiCl_2$ (GORDON and STEAD, 1986). (D) Hepatotoxicity, manifested by microvesicular steatosis, transient depletion of hepatic glutathione, increased serum activities of alanine and aspartate aminotransferases, and diminished activity of serum alkaline phosphatase, develops in $NiCl_2$-treated rats (DONSKOY et al.,

1986). (E) Acute coronary vasoconstriction occurs following i.v. administration of $NiCl_2$ to dogs (RUBANYI et al., 1984). (F) Bronchoalveolar hyperplasia develops in rats following repeated parenteral injections of $NiCl_2$ (KNIGHT et al., 1988).

The lung is the primary target organ for acute nickel carbonyl toxicity in rats, following administration either by intravenous injection or by inhalation (HACKETT and SUNDERMAN, 1967, 1968). Hundreds of cases of nickel carbonyl poisoning have been reported in industrial workers (SUNDERMAN, 1971, 1979; VON LUDEWIGS and THEISS, 1970; VUOPALA et al., 1970; ZHICHENG, 1986; ZHICHENG et al., 1986). In the initial phase of acute nickel carbonyl poisoning, the symptoms are mild and nonspecific, including nausea, vertigo, headache, tachypnea, and chest pain. After 12 to 36 hours, severe symptoms develop with cough, dyspnea, tachycardia, cyanosis, and profound weakness. Occasionally, the onset of these symptoms is delayed as long as one week post-exposure. In fatal cases, death is attributed to diffuse interstitial pneumonitis and cerebral hemorrhage and edema (NAS, 1975). The affected organs include the lungs, brain, heart, liver, kidneys, adrenals, and spleen. Recovery from acute nickel carbonyl poisoning is often protracted, since one-third of the patients develop a neurasthenic syndrome and weakness that may persist for three to six months (ZHICHENG, 1986). Sodium diethyldithiocarbamate is generally used as the antidote in patients with acute nickel carbonyl poisoning (SUNDERMAN and SUNDERMAN, 1958; SUNDERMAN, 1971, 1979).

6.5 Chronic Effects on Animals and Humans

Following administration of inorganic, water-insoluble, nickel compounds to rodents, three principal categories of chronic toxic effects have been demonstrated: (A) Pulmonary damage occurs in rodents chronically exposed to inhalation of nickel dust, αNi_3S_2, or NiO, associated with bronchoalveolar hyperplasia in rats and proliferation of granular pneumocytes and increased phospholipid concentrations in lungs of rabbits (CAMNER et al., 1984; HORIE et al., 1985; BENSON et al., 1986). (B) Erythrocytosis occurs in rats from two weeks to six months after an intrarenal injection of αNi_3S_2, associated with increased renal production of erythropoietin (JASMIN and RIOPELLE, 1976; SUNDERMAN et al., 1982; HOPFER et al., 1984a, 1985a). (C) Disseminated arteriosclerotic lesions develop in rats at eight to eighteen weeks after an intrarenal injection of αNi_3S_2 (HOPFER et al., 1984b).

Workers (e.g., electroplaters, electrolytic refinery workers) who inhale vapors of water-soluble nickel salts occasionally develop chronic respiratory diseases, including asthma, bronchitis, and pneumoconiosis, sometimes associated with hypertrophic rhinitis, sinusitis, nasal polyposis, anosmia, and/or perforations of the nasal septum (NIOSH, 1977; IZMEROV, 1984; PETO et al., 1984).

Iatrogenic Ni^{2+}-poisoning occurred in 23 patients during extracorporeal hemodialysis, owing to leaching of nickel from nickel-plated surfaces of a central water heating-tank (WEBSTER et al., 1980). The patients developed pronounced hypernickelemia, associated with acute nausea, vomiting, weakness, headache, and palpitations; remission of symptoms occurred within a few hours after cessation of hemodialysis. Hypernickelemia occurs, to a lesser degree, in almost all patients with end-stage renal disease who are treated by extracorporeal or peritoneal dialysis (HOPFER et al., 1985b; WILLS et al., 1985; DRAZNIOWSKI et al., 1985).

Intravenous medications (e.g., albumin and radiographic contrast media) that are contaminated with nickel are potential sources of nickel toxicity and allergy (SUNDERMAN, 1983; LEACH and SUNDERMAN, 1985, 1987; FELL and MAHARAJ, 1986). Although nickel dermatitis usually is provoked by external exposures to nickel alloys in jewelry, coinage, etc., oral intake of soluble nickel salts can evoke allergic reactions in nickel-sensitive patients (CHRISTENSEN and MOLLER, 1975). Immunological toxicity and allergic effects by nickel and its compounds are in fact a rather frequent problem, since the prevalence of nickel sensitivity is about 8 – 14% for women and about 1% for men (ANDERSEN et al., 1983; GASSNER, 1986; GRANDJEAN et al., 1989; MAIBACH and MENNÉ, 1989; MENNÉ, 1987; NIEBOER, 1987; NIELSEN, 1989; NIELSEN and FLYVHOLM, 1984; ZACKER and IPPEN, 1984).

6.6 Mutagenic, Carcinogenic and Teratogenic Effects

Regarding mutagenicity testing, it is difficult to draw conclusions based on the effects of nickel compounds in bacterial systems. Mammalian cell studies are better models (ROSSMAN et al., 1987; HARTWIG and BEYERSMANN, 1987; CHRISTIE, 1989). Ni(II) ions synergistically enhance effects of X-rays, of ultraviolet light, and of chromate and/or benzo(a)pyrene (HARTWIG and BEYERSMANN, 1987; CHRISTIE, 1989). Especially DNA-protein interactions and effects on DNA-repair were observed. Nickel seems to be a comutagen and a cocarcinogen.

Two groups of investigators have shown that administration of *soluble nickel salts* to rats in drinking water promotes the carcinogenic action of nitrosamines. OU et al. (1980) studied the induction of nasopharyngeal carcinomas in rats that received a subcarcinogenic parenteral dose of dinitrosopiperazine (DNP) as the initiator, followed by oral administration of nickel sulfate for six weeks as the promotor. Carcinomas of the nasopharynx, nasal cavities, and palate developed in 5/22 rats that received DNP+Ni^{2+}, whereas no carcinomas occurred in three control groups. KUROKAWA et al. (1985) studied the induction of renal neoplasms in rats that received N-ethyl-N-hydroxyethylnitrosamine (EHEN) in drinking water for two weeks to initiate carcinogenesis, followed by nickel chloride for 25 weeks in the drinking water as a promotor. When the experiment was terminated at 27 weeks, renal cell tumors were found in 12/15 rats that received EHEN+Ni^{2+}, compared to 3/15 rats that received only EHEN. RIVEDAL and SANNER (1981, 1982) found that combined in vitro treatment of Syrian hamster embryo cells with nickel sulfate and benzo(a)pyrene resulted in a transformation frequency of 10.7%, compared to 0.5% and 0.6% for the individual substances. The mechanism of tumor promotion by soluble nickel compounds has not been elucidated, but it may possibly reflect generation of oxygen free radicals by the Ni^{2+}/Ni^{3+} redox couple (SUNDERMAN, 1987).

There is some evidence that inhalation of soluble nickel compounds contributed to mortality risks from respiratory tract cancers in workers at a Norwegian nickel refinery (PEDERSEN et al., 1973; EPA, 1986). WU et al. (1986) noted that environmental nickel contamination occurs in China in high-risk areas for nasopharyngeal carcinoma. They speculated that synergistic interaction of nickel and Epstein-Barr virus (EBV) may contribute to the development of nasopharyngeal carcinomas, based upon their finding that nickel increases in vitro cell proliferation of EBV-positive lymphoblastoid cell lines and enhances early expression of EBV antigen. ISACSON et al. (1985) noted correlation between nickel concentrations in town water supplies in Iowa, USA, and the incidences of human lung and bladder cancers. The authors emphasized that nickel may not be the causative factor, but rather an indicator of anthropogenic contamination.

Cancers develop in rodents following administration of various *water-insoluble nickel compounds* (e.g., aNi_3S_2, NiO) by inhalation or parenteral routes (IARC, 1976; SUNDERMAN, 1980, 1984b; HORIE et al., 1985; EPA, 1986).

Epidemiological studies have shown increased incidences of lung and nasal sinus cancers, and possibly laryngeal cancers, among nickel refinery workers (MORGAN, 1958; DOLL, 1958, 1984; PEDERSEN et al., 1973; EPA, 1986; GRANDJEAN, 1986; MASTROMATTEO, 1987). The highest incidences have occurred among workmen involved in roasting, smelting, and electrolysis. The identity of the nickel compounds that induce the cancers is uncertain, but nickel subsulfide (aNi_3S_2), nickel monosulfide (βNiS), and nickel oxides (NiO) are suspected to be the principal carcinogens. The extensive literature on nickel carcinogenesis in experimental animals, neoplastic transformation of cell cultures, and in vitro genotoxicity of nickel compounds has been reviewed by SUNDERMAN (1980, 1984b), COASTA and HECK (1985), and GRANDJEAN (1986).

Nickel carbonyl has been reported to induce cancers in rats following exposures by inhalation or parenteral routes (SUNDERMAN and DONNELLY, 1965; LAU et al., 1972; EPA, 1986). Exposure of pregnant rats to nickel carbonyl on days 7 and 8 of gestation caused frequent ocular anomalies in the offspring, including anophthalmia and microphthalmia (SUNDERMAN et al., 1979). Exposure of pregnant hamsters to nickel carbonyl on days 4 or 5 of gestation caused sundry fetal malformations, including exencephaly and cystic lungs (SUNDERMAN et al., 1980). Fetal *malformations* developed in rodents following administration of divalent nickel salts to pregnant dams during fetal organogenesis (SUNDERMAN et al., 1983c; MAS et al., 1985).

7 Hazard Evaluation and Limiting Concentrations
(see also Section 1)

In many countries, tolerance limits have been established for nickel concentrations in soil, compost and sewage sludge (see Sect. 4.1, and Chapters I.7 and I.20b). For hazards to aquatic species, see Sect. 6.2. Oral uptake of divalent soluble nickel compounds is seldom a significant problem for mammals, except with regard to exacerbation of nickel dermatitis (see Sect. 6.5).

According to a tabulation by GRANDJEAN (1986) of industrial regulations in eleven nations, the atmospheric limits for occupational exposures to nickel metal and sparingly soluble nickel compounds are: (a) 1 mg/m^3, Japan, Netherlands, UK, and

USA; (b) 0.5 mg/m^3, Denmark, FRG, Sweden, and USSR; (c) 0.1 mg/m^3, Norway; and (d) 0.05 mg/m^3, Czechoslovakia. The corresponding atmospheric limits for occupational exposures to soluble nickel compounds are: (a) 1 mg/m^3, Denmark, Japan, and USA; (b) 0.1 mg/m^3, Netherlands, Norway, Sweden, and UK; (c) 0.05 mg/m^3, Czechoslovakia and FRG; and (d) 0.005 mg/m^3, USSR. In Sweden the atmospheric limits for occupational exposure to nickel oxide and nickel carbonate are 0.1 mg/m^3, and the limit for nickel subsulfide is 0.01 mg/m^3. In the Federal Republic of Germany nickel compounds (in the form of respirable dusts/aerosols) are listed in MAK (1987), as working materials which have been unequivocally proven carcinogenic (no MAK limitations; handling requires extraordinary caution and protection measures; technical guiding concentration 0.05 mg/m^3).

The limit value for nickel carbonyl is 7 µg/m^3 in air for all countries, except the Federal Republic of Germany, where it is included in the list of carcinogenic materials, with a technical guiding concentration of 0.7 mg/m^3 (MAK, 1987).

In the USA the Environmental Protection Agency is currently formulating regulations for nickel concentrations in drinking water. In the USSR the maximum allowable concentrations for nickel are 0.2 mg/L in drinking water and 1 mg/L in waste water directed to biological treatment (IZMEROV, 1984).

References

Abdulwajid, A. W., and Sarkar, B. (1983), *Nickel-sequestering renal glycoprotein. Proc. Natl. Acad. Sci. USA 80*, 4509–4512.

Andersen, J. R., Gammelgaard, B., and Reimert, S. (1986), *Direct determination of nickel in human plasma by Zeeman-corrected atomic absorption spectrometry. Analyst 111*, 721–722.

Andersen, K. E., Nielsen, G. D. Flyvholm, M., Fregert, S., and Gruvberge, B. (1983), *Nickel in tap water. Contact Dermatitis 2*, 140–143.

Angerer, J., and Heinrich-Ramm, R. (1989), *Occupational exposure to cobalt and nickel: biological monitoring. Int. J. Environ. Chem. 35* (2), 81–88.

Anke, M., Groppel, B., Kronemann, H., and Grun, M. (1984), *Nickel – an essential element*, in: Sunderman, F. W., Jr. (ed.), *Nickel in the Human Environment*, pp. 339–365, IARC, Lyon.

Asato, N., Van Soestbergen, M., and Sunderman, F. W. Jr. (1975), *Binding of $^{63}Ni(II)$ to ultrafiltrable constituents of rabbit serum in vivo and in vitro. Clin. Chem. 21*, 521–527.

Babich, H., Shopsis, C., and Borenfreund, E. (1986), *In vitro cytotoxicity testing of aquatic pollutants (cadmium, copper, zinc, nickel) using established fish cell lines. Ecotoxicol. Environ. Safety 11*, 91–99.

Bathurst, I. C., Brennan, S. O., Carrell, R. W., Cousens, L. S., Brake, A. J., and Barr, P. J. (1987), *Yeast KEX2 protease has the properties of a human proalbumin converting enzyme. Science 235*, 348–350.

Bencko, V. (1983), *Nickel: A review of its occupational and environmental toxicology. J. Hyg. Epidemiol. Microbiol. Immunol. 27*, 237–247.

Bennett, B. G. (1984), *Environmental nickel pathways to man*, in: Sunderman, F. W., Jr. (ed.), *Nickel in the Human Environment*, pp. 487–495. IARC, Lyon.

Benson, J. M., Henderson, R. F., McClellan, R. O., Hanson, R. L., and Rebar, A. H. (1986), *Comparative acute toxicity of four nickel compounds to F344 rat lung. Fund. Appl. Toxicol. 7*, 340–347.

Birge, W. J., and Black, J. A. (1980), *Aquatic toxicology of nickel*, in: Nriagu, J. O. (ed.), *Nickel in the Environment*, pp. 349–366. Wiley, New York.

Brown, S. S., and Sunderman, F. W., Jr. (eds.) (1980), *Nickel Toxicology*, pp. 1–193. Academic Press, London.

Brown, S. S., and Sunderman, F. W., Jr. (eds.) (1985), *Progress in Nickel Toxicology*, pp.1–244. Blackwells, Oxford.

Bruland, K. W. (1983), *Trace elements in seawater*, Chem. Oceanogr. 8, 204–205.

Callan, W. M., and Sunderman, F. W., Jr. (1973), *Species variations in binding of $^{63}Ni(II)$ by serum albumin*. Res. Commun. Chem. Pathol. Pharmacol. 5, 459–472.

Camner, P., Casarett-Bruce, M., Curstedt, T., Jarstrand, C., Wiernik, A., Johansson, A., Lundborg, M., and Robertson, B. (1984), *Toxicology of nickel*, in: Sunderman, F. W., Jr. (ed.), *Nickel in the Human Environment*, pp. 267–276. IARC, Lyon.

Carvalho, S. M. M., and Ziemer, P. L. (1982), *Distribution and clearance of ^{63}Ni administered as $^{63}NiCl_2$ in the rat: intratracheal study*. Arch. Environ. Contam. Toxicol. 11, 245–248.

Cattalanatto, F. A., Sunderman, F. W., Jr., and Macintosh, T. R. (1977), *Nickel concentrations in human parotid saliva*. Ann. Clin. Lab. Sci. 7, 146–151.

Christensen, O. B., and Lagesson, V. (1981), *Nickel concentration of blood and urine after oral administration*. Ann. Clin. Lab. Sci. 11, 119–125.

Christensen, B., and Moller, H. (1975), *External and internal exposure to the antigen in the hand eczema of nickel allergy*. Contact Dermatitis 1, 136–141.

Christie, N. T. (1989), *The synergistic interaction of Ni(II) with DNA damaging agents*. Toxicol. Environ. Chem. 22 (1–4), 51–59.

Clary, J. J. (1975), *Nickel chloride-induced metabolic changes in the rat and the guinea pig*. Toxicol. Appl. Pharmacol. 31, 55–56.

Clemente, F. G., Cigna-Rossi, L., and Santaroni, G. P. (1980), *Nickel in foods and dietary intake of nickel*, in: Nriagu, J. O. (ed.), *Nickel in the Environment*, pp. 493–498. Wiley, New York.

Cohn, J. R., and Emmett, E. A. (1978), *The excretion of trace metals in human sweat*. Ann. Clin. Lab. Sci. 8, 270–275.

Costa, M., and Heck, J. D. (1985), *Perspectives on the mechanism of nickel carcinogenesis*. Adv. Inorg. Biochem. 6, 285–309.

Cross, J. E., Hughes, D. M., and Williams, D. R. (1985), *Critical review of the evidence for nickel(III) in animals and man*, in: Brown, S. S., and Sunderman, F. W., Jr. (eds.), *Progress in Nickel Toxicology*, pp. 109–112. Blackwells, Oxford.

Decsy, M. I., and Sunderman, F. W., Jr. (1974), *Binding of ^{63}Ni to rabbit serum a_2-macroglobulin in vivo and in vitro*. Bioinorg. Chem. 3, 87–94.

Dixon, N. E., Gazzola, C., Blakeley, R. L., and Zerner, B. (1975), *Jack bean urease. A metalloenzyme: A simple biological role for nickel?* J. Am. Chem. Soc. 97, 4131–4133.

Dixon, N. E., Blakeley, R. L., and Zerner, B. (1980), *Jack bean urease (EC 3.5.1.5). III. The involvement of active-site nickel ion in inhibition by β-mercaptoethanol, phosphoramidate, and fluoride*. Can. J. Biochem. 58, 481–488.

Doll, R. (1958), *Cancer of the lung and nose in nickel workers*. Br. J. Ind. Med. 15, 217–223.

Doll, R. (1984), *Nickel exposure: A human health hazard*, in: Sunderman, F. W., Jr. (ed.), *Nickel in the Human Environment*, pp. 3–21. IARC, Lyon.

Donskoy, E., Donskoy, M., Forouhar, F., Gillies, C. G., Marzouk, A., Reid, M. C., Zaharia, O., and Sunderman, F. W., Jr. (1986), *Hepatic toxicity of nickel chloride in rats*. Ann. Clin. Lab. Sci. 16, 108–117.

Drazniowsky, M., Parkinson, I. S., Ward, M. K., Channon, S. M. and Kerr, D. N. S. (1985), *A method for the determination of nickel in water and serum by flameless atomic absorption spectrophotometry*. Clin. Chim. Acta 145, 219–226.

Ebneth, H. (1986), *Metallic Fibers*. Textilveredlung 21, 105–108.

Ellen, G., van den Bosch-Tibbesma, G., and Douma, F. F. (1978), *Nickel content of various Dutch foodstuffs*. Z. Lebensm. Unters. Forsch. 166, 145–147.

EPA (Environmental Protection Agency) (1986), *Health Assessment Document for Nickel and Nickel Compounds*, pp. 1–383. US EPA Office of Environmental Health Assessment, Washington, D.C.

Eskew, D. L., Welch, R. M., and Norvell, W. A. (1984), *Nickel in higher plants: Further evidence for an essential role*. Plant Physiol. 76, 691–693.

Evans W. H., Read, Y. I., and Lucas, B. E. (1978), *Evaluation of a method for the determination of total cadmium, lead and nickel in foodstuffs using measurement by flame atomic absorption spectrophotometry*. Analyst 103, 580–594.

Fassett, J. D., Moore, L. J., Travis, J. C., and DeVoe, J. R. (1985), *Laser resonance ionization mass spectrometry*. Science 230, 262–267.

Feeley, R. M. (1982), *Major and Trace Element Composition of Human Milk at Early Stages of Lactation*. Ph. D. Dissertation, Athens, Georgia, pp. 1–138. University of Georgia.

Fell, G. S., and Maharaj, D. (1986), *Trace metal contamination of albumin solutions used for plasma exchange*. Lancet 2, 467–468.

Filkova, L. (1985a), *Analysis of Nickel Carbonyl in Air*. Report P-17-335-457-03. Institute of Hygiene and Epidemiology, Prague.

Filkova, L. (1985b), *The danger of contamination of the living environment with nickel carbonyl*. Cesk. Hyg. 30, 243–249.

Filkova, L., and Jager, J. (1986), *Non-occupational exposure to nickel tetracarbonyl*. Cesk. Hyg. 31, 255–259.

Fine, J. M., Abdo, Y., Rochu, D., Rousseau, J., and Dautrevaux, M. (1983), *Identification of the human albumin variant "Gainesville" with proalbumin "Christchurch"*. Blood Transfus. Immunohaematol. 26, 341–346.

Fishbein, L. (1987), *Trace and ultra trace elements in nutrition*. Toxicol. Environ. Chem. 14, 73–99.

Flyvholm, M. A., Nielsen, G. D., and Andersen, A. (1984), *Nickel content of food and estimation of dietary intake*. Z. Lebensm. Unters. Forsch. 179, 427–431.

Foulkes, E. C., and McMullen, D. M. (1986), *On the mechanism of nickel absorption in the rat jejunum*. Toxicology 38, 35–42.

Gassner, M. (1986), *Frequency of Allergic Diseases in School-children in a Rural Population* (in German). Abstract for the Meeting of the Swiss Society of Occupational Medicine.

Gitlitz, P. H., Sunderman, F. W., Jr., and Goldblatt, P. J. (1975), *Aminoaciduria and proteinuria in rats after a single intraperitoneal injection of Ni(II)*. Toxicol. Appl. Pharmacol. 34, 430–440.

Glennon, J. D., and Sarkar, B. (1982), *Nickel(II) transport in human blood serum. Studies of nickel(II) binding to human albumin and to native-sequence peptide, and ternary-complex formation with L-histidine*. Biochem. J. 203, 15–23.

Gordon, C. J., and Stead, A. G. (1986), *Effect of nickel and cadmium chloride on autonomic and behavioral thermoregulation in mice*. Neurotoxicology 1, 97–106.

Grandjean, P. (1984), *Human exposure to nickel*, in: Sunderman, F. W., Jr. (ed.), *Nickel in the Human Environment*, pp. 469–485. IARC, Lyon.

Grandjean, P. (1986), *Health Effects Document on Nickel*, pp. 1–204. Ontario Ministry of Labour, Toronto.

Grandjean, P., Nielsen, G. D., and Andersen, O. (1989), *Human nickel exposure and chemobiokinetics*, in: Maibach, H. I., and Menné, T. (eds.), *Nickel and the Skin: Immunology and Toxicology*, Chapter 2, pp. 9–34. CRC Press, Boca Raton, Florida.

Hackett, R. L., and Sunderman, F. W., Jr. (1967), *Acute pathological reactions to administration of nickel carbonyl*. Arch. Environ. Health 14, 604–613.

Hackett, R. L., and Sunderman, F. W., Jr. (1968), *Pulmonary alveolar reaction to nickel carbonyl: Ultrastructural and histochemical studies*. Arch. Environ. Health 16, 349–362.

Halstead, R. L. (1968), *Effect of different amendments on yield and composition of oats grown on a soil derived from serpentine material*. Can. J. Soil. Sci. 48, 301–305.

Halstead, R. L., Finn, B. J., and MacLean, A. J. (1986), *Extractability of nickel added to soils and its concentration in plants*. Can. J. Soil. Sci. 49, 335–342.

Hartwig, A., and Beyersmann, D. (1987), *Enhancement of UV and chromate mutagenesis by nickel ions in the Chinese hamster HGPRT assay*. Toxicol. Environ. Chem. *14*, 33–42.

Hazlett, P. W., Rutherford, G. K., and van Loon, G. W. (1983), *Metal contaminants in surface soils and vegetation as a result of nickel/copper smelting at Coniston, Ontario, Canada*. Reclamat. Reveg. Res. *2*, 123–137.

Heale, E. L., and Ormond, P. (1982), *Effects of nickel and copper on Acer rubrum, Cornus stolonifera, Lonicera tatarica, and Pinus resinosa*. Can. J. Bot. *60*, 2674–2681.

Herlant-Peers, M. C., Hildebrand, H. F., and Kerckaert, J. P. (1983), *In vitro and in vivo incorporation of $^{63}Ni(II)$ into lung and liver subcellular fractions of Balb/C mice*. Carcinogenesis *4*, 387–392.

Ho, W., and Furst, A. (1973), *Nickel excretion by rats following a single treatment*. Proc. West. Pharmacol. Soc. *16*, 245–248.

Hogan, G. R., and Razniak, S. L. (1986), *Patterns of radionickel distribution in female mice of different ages*. Trace Subst. Environ. Health 20, 57–64.

Hohnadel, D. C., Sunderman, F. W., Jr., Nechay, M. W. and McNeely, M. D. (1973), *Atomic absorption spectrometry of nickel, copper, zinc, and lead in sweat from healthy subjects during sauna bathing*. Clin. Chem. *19*, 1288–1292.

Hopfer, S. M., Sunderman, F. W., Jr., Reid, M. C., and Goldwasser, E. (1984a), *Increased immunoreactive erythropoietin in serum and kidney extracts of rats with Ni_3S_2-induced erythrocytosis*. Res. Commun. Chem. Pathol. Pharmacol. *43*, 155–170.

Hopfer, S. M., Sunderman, F. W., Jr., McCully, K. S., Reid, M. C., Liber, C., Spears, J. R., and Serur, J. (1984b), *Studies of the pathogenesis of arteriosclerosis induced in rats by intrarenal injection of a carcinogen, nickel subsulfide*. Ann. Clin. Lab. Sci. *14*, 355–365.

Hopfer, S. M., Sunderman, F. W., Jr., and Goldwasser, E. (1985a), *Effects of unilateral intrarenal administration of nickel subsulfide to rats on erythropoietin concentrations in serum and in extracts of both kidneys*, in: Brown, S. S., and Sunderman, F. W., Jr. (eds.), *Progress in Nickel Toxicology*, pp. 97–100. Blackwells, Oxford.

Hopfer, S. M., Linden, J. V., Crisostomo, M. C., Catalanatto, F. A., Galen, M., and Sunderman, F. W., Jr. (1985b), *Hypernickelemia in hemodialysis patients*. Trace Elem. Med. *2*, 68–72.

Horak, E., and Sunderman, F. W., Jr. (1973), *Fecal nickel excretion by healthy adults*. Clin. Chem. *19*, 429–430.

Horak, E., Zygowicz, E. R., Tarabishy, R., Mitchell, L. M., and Sunderman, F. W., Jr. (1978), *Effects of nickel chloride and nickel carbonyl upon glucose metabolism in rats*. Ann. Clin. Lab. Sci. *8*, 476–482.

Horie, A., Haratake, J., Tanaka, I., Kodama, Y., and Tsuchiya, K. (1985), *Electron microscopical findings with special reference to cancer in rats caused by inhalation of nickel oxide*. Biol. Trace Elem. Res. *7*, 223–239.

IARC (International Agency for Research on Cancer) (1976), *Nickel*, in: *Monographs on the Evaluation of Carcinogenic Risk of Chemicals to Man*, Vol. 11, pp. 75–104. IARC, Lyon.

INCO (International Nickel Company Ltd.) (1987), *Annual Report*, p. 7. Toronto, Ontario, Canada.

Isacson, P., Bean, J. A., Splinter, R., Olson, D. B., and Kohler, J. (1985), *Drinking water and cancer incidence in Iowa*. Am. J. Epidemiol. *121*, 856–869.

Izmerov, N. F. (1984), *Nickel and Its Compounds*, pp. 1–36. Centre of International Projects, Moscow. USSR State Committee for Science and Technology (GKNT).

Jaffre, T., Brooks, R. R., Lee, J., and Reeves, R. D. (1976), *Sebertia acuminata: A nickel-accumulating plant from New Caledonia*. Science *193*, 579–580.

Jaffre, T., Kersten, W., Brooks, R. R., and Reeves, R. D. (1979), *Nickel uptake by Flacourtiaceae of New Caledonia*. Proc. R. Soc. London Ser. B *205*, 385–394.

Jasmin, G., and Riopelle, J. L. (1976), *Renal carcinomas and erythrocytosis in rats following intrarenal injection of nickel subsulfide*. Lab. Invest. *35*, 71–78.

Jickels, T. D., and Burton, J. D. (1988), *Cobalt, copper, manganese and nickel in the Sargasso Sea*. Mar. Chem. *23*, 131–144.

Kasprzak, K. S., and Sunderman, F. W., Jr. (1969), *The metabolism of nickel carbonyl-^{14}C*. Toxicol. Appl. Pharmacol. 15, 295–303.

Kasprzak, K. S., and Sunderman, F. W., Jr. (1977), *Mechanisms of dissolution of nickel subsulfide in rats serum*. Res. Commun. Chem. Pathol. Pharmacol. 16, 95–108.

Kasprzak, K. S., Waalkes, M. P., and Poirier, L. A. (1986), *Effects of magnesium acetate on the toxicity of nickelous acetate in rats*. Toxicology 42, 57–68.

Kiilunen, M., Jarvisalo, J., Makitie, O., and Aitio, A. (1987), *Analysis, storage stability and reference values for urinary chromium and nickel*. Int. Arch. Occup. Environ. Health 59, 43–50.

King, M., Lynn, K. K., and Huang, C. (1985), *Activation of the calmodulin-dependent phosphoprotein phosphatase by nickel ions*, in: Brown, S. S., and Sunderman, F. W., Jr. (eds.), Progress in Nickel Toxicology, pp. 117–120. Blackwells, Oxford.

Knight, J. A., Hopfer, S. M., Reid, M. C., Wong, S. H. Y., and Sunderman, F. W., Jr. (1986), *Ethene (ethylene) and ethane exhalation in Ni(II)-treated rats, using an improved rebreathing apparatus*. Ann. Clin. Lab. Sci. 16, 386–394.

Knight, J. A., Rezuke, W. N., Wong, S. H. Y., Hopfer, S. M., Zaharia, O., and Sunderman, F. W., Jr. (1987), *Acute thymic involution and increased lipoperoxides in thymus of nickel chloride-treated rats*. Res. Commun. Chem. Pathol. Pharmacol. 55, 101–109.

Knight, J. A., Rezuke, W. N., Gillies, C. G., Hopfer, S. M., and Sunderman, F. W., Jr. (1988), *Bronchoalveolar hyperplasia in rats after parenteral injections of nickel chloride*. Toxicol. Pathol. 16, 350–359.

Koivistoinen, P. (ed.) (1980), *Mineral element composition of finnish foods: Na, K, Ca, Mg, P, S, Fe, Cu, Mn, Zn, Mo, Co, Ni, Cr, F, Se, Pb, Al, B, Rb, Hg, As, Cd, and ash*. Acta Agric. Scand. Suppl. 22, 1–171.

Kollmeier, H., Wittig, C., Seemann, J., Wittig, P., and Rothe, R. (1985), *Increased chromium and nickel content in lung tissue*. J. Cancer Res. Clin. Oncol. 110, 173–176.

Kuehn, K., and Sunderman, F. W., Jr. (1982), *Dissolution half-times of nickel compounds in water, rat serum, and renal cytosol*. J. Inorg. Biochem. 17, 29–39.

Kurokawa, Y., Matsushima, M., Imazawa, T., Takamura, N., Takahashi, M., and Hayashi, Y. (1985), *Promoting effect of metal compounds on rat renal tumorigenesis*. J. Am. Coll. Toxicol. 4, 321–330.

Lau, T. J., Hackett, R. L., and Sunderman, F. W., Jr. (1972), *The carcinogenicity of intravenous nickel carbonyl in rats*. Cancer Res. 32, 2253–2258.

Laussac, J. P., and Sarkar, B. (1980), *Nickel(II) binding to the NH_2-terminal peptide segment of human serum albumin: ^{13}C- and ^{1}H-nuclear magnetic resonance investigation*. Can. J. Chem. 58, 2055–2060.

Laussac, J. P., and Sarkar, B. (1984), *Characterization of the copper(II)- and nickel(II)-transport site of human serum albumin. Studies of copper(II) and nickel(II) binding to peptide 1–24 of human serum albumin by ^{13}C and ^{1}H NMR spectroscopy*. Biochemistry 23, 2832–2838.

Lazareva, L. P. (1985), *Changes of biological parameters in chronic effects of low concentrations of copper and nickel on Daphnia magna Straus*. Gidrobiol. Zh. 21 (5), 53–56.

Leach, C. N., Jr., and Sunderman, F. W., Jr. (1985), *Nickel contamination of human serum albumin solutions*. N. Engl. J. Med. 313, 1232.

Leach, C. A., Jr., and Sunderman, F. W., Jr. (1987), *Hypernickelemia following coronary arteriography, caused by nickel in the radiographic contrast medium*. Ann. Clin. Lab. Sci. 17, 137–144.

Léonard, A., Gerber, G. B., and Jacquet, P. (1981), *Carcinogenicity, mutagenicity and teratogenicity of nickel*. Mutat. Res. 87, 1–15.

Maibach, H. I., and Menné, T. (eds.) (1989), *Nickel and the Skin: Immunology and Toxicology*. CRC Press, Boca Raton, Florida.

Maines, M. D., and Kappas, A. (1977), *Metals as regulators of heme metabolism*. Science 198, 1215–1221.

MAK (1987), *Maximum Concentrations at the Workplace, Report No. XXIII, DFG*. VCH Verlagsgesellschaft, Weinheim-Basel-Cambridge-New York.

Mart, L. (1983), *Seasonal variations of Cd, Pb, Cu and Ni levels in snow from the eastern Arctic Ocean. Tellus 3B*, 131–141.
Mart, L, and Nürnberg, H. W. (1984), *Trace metal levels in the eastern Arctic Ocean. Sci. Total Environ. 39*, 1–14; see also *Proceedings Heavy Metals in the Environment, Athens*, Vol. 2, pp. 340–342 (1985). CEP-Consultants Ltd., Edinburgh.
Marzouk, A., and Sunderman, F.W., Jr. (1985), *Biliary excretion of nickel. Toxicol. Lett. 27*, 65–71.
Mas, A., Holt, D., and Webb, M. (1985), *The acute toxicity and teratogenicity of nickel in pregnant rats. Toxicology 35*, 47–57.
Mastromatteo, E. (1986), *Nickel. Am. Ind. Hyg. Assoc. J. 47*, 589–601.
Menné, T. (1987), *Nickel allergy*, in: Symposium Abstracts, *8. Immunological Effects, Trace Elements in Human Health and Disease.* Odense University, Denmark.
Mingorance, M. D., and Lachia, M. (1985), *Direct determination of some trace elements in milk by electrothermal atomic absorption spectrometry. Anal. Lett. 18*, 1519–1531.
Montaser, A., and Golightly, D. W. (1987), *Inductively Coupled Plasmas in Analytical Atomic Spectrometry*, pp. 354, 605ff. VCH Verlagsgesellschaft, Weinheim-Basel-Cambridge-New York.
Morgan, J. G. (1958), *Some observations on the incidence of respiratory cancer in nickel workers. Br. J. Ind. Med. 15*, 224–234.
Mushak, P. (1984), *Nickel metabolism in health and disease. Clin. Lab. Annu. 3*, 249–299.
Myron, D. R., Zimmerman, T. J., Shuler, T. R., Klevay, L. M., Lee, D. E., and Nielsen, F. H. (1978), *Intake of nickel and vanadium of humans. A survey of selected diets. Am. J. Clin. Nutr. 31*, 527–531.
NAS (National Academy of Sciences) (1975), *Nickel, Medical and Biologic Effects of Environmental Pollutants*, pp. 1–277. NAS Press, Washington, D.C.
Nechay, M. W., and Sunderman, F. W., Jr. (1973). *Measurements of nickel in hair by atomic absorption spectrometry. Ann. Clin. Lab. Sci. 3*, 30–35.
Nicholls, D. (1973), *Nickel*, in: Trotman-Dickensen, A. F. (ed.), *Comprehensive Inorganic Chemistry,* p. 1109. Pergamon Press, Oxford.
Nieboer, E. (1987), *Inhalation allergy of metals*, in: Symposium Abstract, *8. Immunological Effects, Trace Elements in Human Health and Disease.* Odense University, Denmark.
Nieboer, E., Stetsko, P. I., and Hin, P. Y. (1984), *Characterization of the Ni(III)/Ni(II) redox couple for the Ni(II)-complex of human serum albumin. Ann. Clin. Lab. Sci. 14*, 409.
Nielsen, F. H. (1984), *Ultratrace elements in nutrition. Ann. Res. Nutr. 4*, 22–41.
Nielsen, G. D. (1989), *Oral challenge of nickel-allergic patients with hand eczema*, in: Nieboer, E., and Aitio, A. (eds.), *Nickel and Human Health, Current Perspectives. Adv. Environ. Sci. Technol.*, in press.
Nielsen, G. D., and Flyvholm, M. (1984), *Risks of high nickel intake with diet*, in: Sunderman, F. W. Jr. (ed.), *Nickel in the Human Environment*, pp. 333–338. IARC, Lyon.
Nielsen, G. D., Andersen, O., Jensen, M., and Grandjean, P. (1986), *Gastrointestinal nickel absorption: A new experimental model using the gamma-emitting isotope ^{57}Ni. Proceedings of the UOEH International Symposium on Bio- and Toxico-kinetics of Metals.* Kitakyushu, Univ. Occup. Environ. Health.
NIOSH (National Institute of Occupational Safety and Health) (1977), *Criteria for a Recommended Standard: Occupational Exposure to Inorganic Nickel*, pp. 1–282. U.S. Department of Health, Education and Welfare, Washington, D.C.
Nomoto, S., McNeely, M. D., and Sunderman, F. W. Jr. (1971), *Isolation of a nickel a_2-macroglobulin from rabbit serum. Biochemistry 10*, 1647–1651.
Norseth, T. (1984), *Chromium and Nickel*, in: Aitio, A., Riihimaki, V., and Vainio, H. (eds.), *Biological Monitoring and Surveillance of Workers Exposed to Chemicals*, pp. 49–59. Hemisphere Press, Washington, D.C.
Nriagu, J. O. (ed.) (1980), *Nickel in the Environment*, pp. 1–833. Wiley, New York.
Onkelinx, C., and Sunderman, F. W., Jr. (1980), *Modeling of nickel metabolism*, in: Nriagu, J. O. (ed.), *Nickel in the Human Environment*, pp. 525–545. Wiley, New York.
Oskarsson, A., and Tjälve, H. (1979a), *An autoradiographic study on the distribution of $^{63}NiCl_2$ in mice. Ann. Clin. Lab. Sci. 2*, 47–59.

Oskarsson, A., and Tjälve, H. (1979b), *The distribution and metabolism of nickel carbonyl in mice.* Br. J. Ind. Med. 36, 326–335.

Oskarsson, A., and Tjälve, H. (1979c), *Binding of ^{63}Ni by cellular constituents in some tissues of mice after the administration of $^{63}NiCl_2$ and $^{63}Ni(CO)_4$.* Acta Pharmacol. Toxicol. 45, 306–314.

Oskarsson, A., Andersson, Y., and Tjälve, H. (1979), *Fate of nickel subsulfide during carcinogenesis studied by autoradiography and X-ray powder diffraction.* Cancer Res. 39, 4175–4182.

Ostapczuk, P., Valenta, P., Stoeppler, M., and Nürnberg, H. W. (1983), *Voltammetric determination of nickel and cobalt in body fluids and other biological materials*, in: Brown, S. S., and Savory, J. (eds.), *Chemical Toxicology and Clinical Chemistry of Metals*, pp. 62–64. Academic Press, London.

Ou, B., Lu, Y., Huang, X., and Feng, G. (1980), *The promoting action of nickel in the induction of nasopharyngeal carcinoma in rats.* Guang. Dong. Med. 1 (2), 32–34.

Parker, K., and Sunderman, F. W., Jr. (1974), *Distribution of ^{63}Ni in rabbit tissues following intravenous injection of $^{63}NiCl_2$.* Res. Commun. Chem. Pathol. Pharmacol. 7, 755–762.

Pedersen, E., Hogetveit, A. C., and Andersen, A. (1973), *Cancer of respiratory organs among workers at a nickel refinery in Norway.* Int. J. Cancer 12, 32–41.

Peto, J., Cuckle, H., Doll, R., Hermon, C., and Morgan, L. G. (1984), *Respiratory cancer mortality of Welsh nickel refinery workers*, in: Sunderman, F. W., Jr. (ed.), *Nickel in the Human Environment*, pp. 37–46. IARC, Lyon.

Polacco, J. C. (1977), *Is nickel a universal component of plant ureases?* Plant Sci. Lett. 10, 249–255.

Queneau, P. E., Roorda, H. J., Wassermann, G., and Blankenstein, K. (1979), *Nickel*, in: *Ullmanns Encyklopädie der technischen Chemie*, 4th Ed., Vol. 17, pp. 239–302. Verlag Chemie, Weinheim-New York.

Raithel, H. J. (1987), Research Report *Nickel* (in German), Schriftenreihe des Hauptverbandes der gewerblichen Berufsgenossenschaften e.V.

Raithel, H. J., and Schaller, K. H. (1981), *Toxicity and carcinogenicity of nickel and its compounds. A review of the current status.* Zentralbl. Bakteriol. Hyg. Abt. 1 Orig. B 173, 63–91.

Raithel, H. J., Schaller, K. H., Mohrmann, W., Mayer, P., and Henkels, U. (1982), *Untersuchungen zur Ausscheidungskinetik von Nickel bei Beschäftigten in der Glas- und Galvanischen Industrie*, in: Fliedner, T. M. (ed.), *Bericht über die 22. Jahrestagung der Deutschen Gesellschaft für Arbeitsmedizin. Kombinierte Belastungen am Arbeitsplatz*, pp. 223–228.

Raithel, H. J., Ebner, G., Schaller, K. H., Schellmann, B., and Valentin, H. (1987), *Problems in establishing norm values for nickel and chromium concentrations in human pulmonary tissue.* Am. J. Ind. Med. 12, 55–70.

Reeves, R. D., Brooks, R. R., and Macfarlane, R. M. (1981), *Nickel uptake by Californian Streptanthus and Caulanthus with particular reference to the hyperaccumulator S. polygaloides Gray (Brassicaceae).* Am. J. Bot. 68, 708–712.

Reeves, R. D., Macfarlane, R. M., and Brooks, R. R. (1983), *Accumulation of nickel and zinc by western north American genera containing serpentine-tolerant species.* Am. J. Bot. 70, 1297–1303.

Rehwoldt, R., Bida, G., and Nerrie, B. (1971), *Acute toxicity of copper, nickel, zinc ions to some Hudson River fish species.* Bull. Environ. Contam. Toxicol. 6, 445–448.

Rezuke, W. N., Knight, J. A., and Sunderman, F. W., Jr. (1987), *Reference values for nickel concentrations in human tissues and bile.* Am. J. Ind. Med. 11, 419–426.

Rivedal, E., and Sanner, T. (1981), *Metal salts as promoters of in vitro morphological transformation of hamster embryo cells initiated by benzo(a)pyrene.* Cancer Res. 41, 2950–2953.

Rivedal, E., and Sanner, T. (1982), *Evaluation of tumor promoters by the hamster embryo cell transformation assay*, in: Bartsch, H., and Bartsch, B. (eds.), *Host Factors in Human Carcinogenesis*, pp. 251–258. IARC, Lyon.

Rossman, T. G., Zelikoff, J. T., Agarwal, S., and Kneip, T. J. (1987), *Genetic toxicology of metal compounds (an examination of appropriate cellular models).* Toxicol. Environ. Chem. 14 (4), 251–262.

Rubanyi, G., Ligeti, L., Koller, A., and Kovach, A. G. B. (1984), *Nickel ions and ischemic coronary vasoconstriction. Bibl. Cardiol.* **38**, 200–208.

Saager, R. (1984), *Metallic Raw Materials Dictionary* (in German), pp. 130–134. Bank von Tobel, Zürich.

Sarkar, B. (1984), *Nickel metabolism*, in: Sunderman, F. W., Jr. (ed.), *Nickel in the Human Environment*, pp. 367–384. IARC, Lyon.

Schaller, K. H., Stoeppler, M., and Raithel, H. J. (1982), *The analytical determination of nickel in biological matrices. A summary of present knowledge and experience. Staub Reinhalt. Luft* **42**, 137–140.

Schaufelberger, F. A. (1987), *Nickel and its ores* (in German), *Neue Zürcher Zeitung*, Forschung und Technik, Nr. *103*, p. 65, 6 May, Zürich.

Schnegg, A., and Kirchgessner, M. (1975a), *Essentiality of nickel for the growth of animals. Z. Tierphysiol. Tierernähr. Futtermittelkd.* **36**, 63–74.

Schnegg, A., and Kirchgessner, M. (1975b), *Changes in the hemoglobin content, erythrocyte count and hematocrit in nickel deficiency. Nutr. Metab.* **19**, 263–278.

Schnegg, A., and Kirchgessner, M. (1976), *Absorption and metabolic efficiency of iron during nickel deficiency. Int. J. Vitamin Nutr. Res.* **46**, 96–99.

Seemann, J., Wittig, P., Kollmeier, H., and Rothe, G. (1985), *Analytical measurements of Cd, Pb, Zn, Cr and Ni in human tissues. Lab. Med.* **9**, 294–299.

Smart, G. A., and Sherlock, J. C. (1987), *Nickel in Foods and the Diet. Food Add. Contam.* **4**, 61–71.

Smialowicz, R. J., Roger, R. R., Riddle, M. M., and Stott, G. A. (1984), *Immunologic effects of nickel: I. Suppression of cellular and humoral immunity. Environ. Res.* **33**, 413–427.

Smialowicz, R. J., Rogers, R. R., Riddle, M. M., Garner, R. J., Rowe, D. G, and Luebke, R. W. (1985), *Immunological effects of nickel: II. Suppression of natural killer cell activity. Environ. Res.* **36**, 56–66.

Solomons, N. W., Viteri, F., Shuler, T. R., and Nielsen, F. H. (1982), *Bioavailability of nickel in man: Effects of foods and chemically-defined dietary constituents on the absorption of inorganic nickel. J. Nutr.* **112**, 39–50.

Spears, J. W. (1984), *Nickel as a "newer trace element" in the nutrition of domestic animals, J. Anim. Sci.* **59**, 823–834.

Stedman, D. H., Tammaro, D. A., Branch, D. K., and Pearson, R., Jr. (1979), *Chemiluminescence detector for the measurement of nickel carbonyl in air. Anal. Chem.* **51**, 2340–2342.

Stedman, D. H., Pearson, R., Jr., and Yalvac, E. D. (1980), *Nickel carbonyl: Decomposition in air and related kinetic studies. Science* **208**, 1029–1031.

Still, E. R., and Williams, R. J. P. (1980), *Potential methods for selective accumulation of nickel(II) ions by plants. J. Inorg. Biochem.* **13**, 35–40.

Stoeppler, M. (1980), *Analysis of nickel in biological materials and natural waters*, in: Nriagu, J. O., (ed.), *Nickel in the Environment*, pp. 661–822. Wiley, New York.

Stoeppler, M. (1984a), *Analytical chemistry of nickel*, in Sunderman, F. W., Jr. (ed.), *Nickel in the Human Environment*, pp. 459–468. IARC, Lyon.

Stoeppler, M. (1984b), *Recent improvements for nickel analysis in biological materials*, in: *Trace Element Analytical Chemistry*, Vol. 3, pp. 539–557. Walter de Gruyter & Co., Berlin.

Stoeppler, M., and Brandt, K. (1979), *Comparative studies on trace metal levels in marine biota. II. Trace metals in krill, krill products, and fish from the Antarctic Scotia Sea. Z. Lebensm. Unters. Forsch.* **169**, 95–98.

Stoeppler, M., and Nürnberg, H. W. (1979), *Comparative studies on trace metal levels in marine biota. III. Typical levels and accumulation of toxic trace metals in muscle tissue and organs of marine organisms from different European seas. Ecotoxicol. Environ. Safety* **3**, 335–351.

Sunderman, F. W. (1971), *The treatment of acute nickel carbonyl poisoning with sodium diethyldithiocarbamate. Ann. Clin. Res.* **3**, 182–185.

Sunderman, F. W. (1979), *Efficacy of sodium diethyldithiocarbamate (Dithiocarb) in acute nickel carbonyl poisoning. Ann. Clin. Lab. Sci.* **9**, 1–10.

Sunderman, F. W., Jr. (1980), *Recent research on nickel carcinogenesis*. Environ. Health Perspect. 40, 131–141.

Sunderman, F. W., Jr. (1983), *Potential toxicity from nickel contamination of intravenous fluids*. Ann. Clin. Lab. Sci. 3, 1–4.

Sunderman, F. W., Jr. (1984a), *Nickel*, in: Vercruysse, A. (ed.), *Hazardous Metals in Human Toxicology*, pp. 279–306. Elsevier, Amsterdam.

Sunderman, F. W., Jr. (1984b), *Recent progress in nickel carcinogenesis*. Toxicol. Environ. Chem. 8, 235–252.

Sunderman, F. W., Jr. (ed.) (1984c), *Nickel in the Human Environment*, pp. 1–529. IARC, Lyon.

Sunderman, F. W., Jr. (1986a), *Determination of nickel in body fluids, tissues, excreta, and water*, in: *IARC Monographs on Environmental Carcinogens – Selected Methods of Analysis*, Vol. 8, pp. 319–334. IARC, Lyon.

Sunderman, F. W., Jr. (1986b), *Sources of exposure and biological effects of nickel exposure*, in: *IARC Monographs on Environmental Carcinogens – Selected Methods of Analysis*, Vol. 8, pp. 79–92. IARC, Lyon.

Sunderman, F. W., Jr. (1986c), *Kinetics and biotransformation of nickel and chromium*, in: Stern, R. M., Berlin, A., Fletcher, A. C., and Jarvisalo, J. (eds.), *Health Hazards and Biological Effects of Welding Fumes and Gases*, pp. 229–247. Excerpta Medica, Amsterdam.

Sunderman, F. W., Jr. (1986d), *Nickel*, in: Seiler, H. G., Sigel, H., and Sigel, A. (eds.), *Handbook on Toxicity of Inorganic Compounds*, pp. 453–468. Marcel Dekker, New York.

Sunderman, F. W., Jr. (1987), *Lipid peroxidation as a mechanism of acute nickel toxicity*. Toxicol. Environ. Chem. 15, 59–69.

Sunderman, F. W., and Donnelly, A. J. (1965), *Studies of nickel carcinogenesis. Metastasizing pulmonary tumors in rats induced by the inhalation of nickel carbonyl*. Am. J. Pathol. 46, 1027–1041.

Sunderman, F. W., Jr., and Fraser, C. B. (1983), *Effects of nickel chloride and diethyldithiocarbamate on metallothionein in rat liver and kidney*. Ann. Clin. Lab. Sci. 13, 489–495.

Sunderman, F. W., Jr., and Horak, E. (1981), *Biochemical indices of nephrotoxicity exemplified by studies of nickel nephropathy*, in: Brown, S. S., and Davies, D. S. (eds.), *Organ-Directed Toxicity: Chemical Indices and Mechanisms*, pp, 52–64. Pergamon, London.

Sunderman, F. W., Jr., and Selin, C. E. (1968), *The metabolism of nickel-63 carbonyl*. Toxicol. Appl. Pharmacol. 1, 297–318.

Sunderman, F. W., and Sunderman, F. W., Jr. (1958), *Nickel poisoning. VIII. Dithiocarb: A new therapeutic agent for persons exposed to nickel carbonyl*. Am. J. Med. Sci. 236, 26–31.

Sunderman, F. W., Jr., Roszel, N. O., and Clark, R. J. (1968), *Gas chromatography of nickel carbonyl in blood and breath*. Arch. Environ. Health 16, 836–843.

Sunderman, F. W., Jr., Kasprzak, K. S., Horak, E., Gitlitz, P., and Onkelinx, C. (1976a), *Effects of triethylenetetramine upon the metabolism and toxicity of $^{63}NiCl_2$ in rats*. Toxicol. Appl. Pharmacol. 38, 177–188.

Sunderman, F. W., Jr., Kasprzak, K. S., Lau, T. J., Minghetti, P. P., Maenza, R. M., Becker, N. B., Onkelinx, C., and Goldblatt, P. J. (1967b), *Effects of manganese on carcinogenicity and metabolism of nickel subsulfide*. Cancer Res. 36, 1790–1800.

Sunderman, F. W., Jr., Alpass, P. R., Mitchell, J. M., and Baselt, R. C. (1979), *Eye malformations in rats: Induction by prenatal exposure to nickel carbonyl*. Science 203, 550–553.

Sunderman, F. W., Jr. Shen, S. K., Reid, M. C., and Allpass, P. R. (1980), *Teratogenicity and embryotoxicity of nickel carbonyl in Syrian hamsters*. Teratogen. Carcinogen. Mutagen. 1, 223–233.

Sunderman, F. W.,Jr., Costa, E. R., Fraser, C., Hui, G., Levine, J. L., and Tse, T. P. H. (1981), *^{63}Ni-Constituents in renal cytosol of rats after injection of $^{63}NiCl_2$*. Ann. Clin. Lab. Sci. 11, 488–496.

Sunderman, F.W., Jr., Hopfer, S.M., Reid, M.C., Shen, S.K., and Kevorkian, C.B. (1982), *Erythropoietin-mediated erythrocytosis in rodents after intrarenal injection of nickel subsulfide*. Yale J. Biol. Med. 55, 123–136.

Sunderman, F. W., Jr., Mangold, B. L., Wong, S. H. Y., Shen, S. K., Reid, M. C., and Jansson, I. (1983a), *High-performance size-exclusion chromatography of ^{63}Ni-constituents in renal cytosol and microsomes from ^{63}NiCl$_2$ treated rats.* Res. Commun. Chem. Pathol. Pharmacol. *39*, 477–492.

Sunderman, F. W., Jr., Reid, M. C., Bibeau, L. M., and Linden, J. V. (1983b), *Nickel induction of microsomal heme oxygenase activity in rodents.* Toxicol. Appl. Pharmacol. *68*, 87–95.

Sunderman, F. W., Jr., Reid, M. C., Shen, S. K., and Kevorkian, C. B. (1983c), *Embryotoxicity and teratogenicity of nickel compounds*, in: Clarkson, T. W., Nordberg, G. F., and Sager, P. R. (eds.), *Reproductive and Developmental Toxicity of Metals*, pp. 399–416. Plenum Press, New York.

Sunderman, F. W., Jr., Marzouk, A., Hopfer, S. M., Zaharia, O., and Reid, M. C. (1985), *Increased lipid peroxidation in tissues of nickel chloride-treated rats.* Ann. Clin. Lab. Sci. *15*, 229–236.

Sunderman, F. W., Jr., Aitio, A., Morgan, L. M., and Norseth, T. (1986), *Biological monitoring of nickel.* Toxicol. Ind. Health *2*, 17–78.

Sunderman, F. W., Jr., Hopfer, S. M., and Crisostoma, M. C. (1988), *Nickel analysis by atomic absorption spectrometry.* Methods Enzymol. *158*, 382–391.

Templeton, D. M., and Sarkar, B. (1985), *Peptide and carbohydrate complexes of nickel in human kidney.* Biochem. J. *230*, 35–42.

Templeton, D. M., and Sarkar, B. (1986), *Nickel binding to the C-terminal tryptic fragment of a peptide from human kidney.* Biochim. Biophys. Acta *884*, 382–386.

Theisen, M. O., and Blincoe, C. (1984), *Biochemical form of nickel in alfalfa.* J. Inorg. Biochem. *21*, 137–146.

Tiffin, L. O. (1971), *Translocation of nickel in xylem exudate of plants.* Plant Physiol. *48*, 273–277.

Tissot, B. P., and Welte, D. H. (1984), *Petroleum Formation and Occurrence*, 2nd Ed. Springer Verlag, Berlin-Heidelberg.

Van Soestbergen, M., and Sunderman, F. W., Jr. (1972), *^{63}Ni complexes in rabbit serum and urine after injection of ^{63}NiCl$_2$.* Clin. Chem. *18*, 1478–1484.

Varo, P., and Koivistoinen, P. (1980), *Mineral element composition of finnish food. XII. General discussion and nutritional evaluation*, in: Koivistoinen, P. (ed.) *Mineral Element Composition of Finnish Foods: Na, K, Ca, Mg, P, S, Fe, Cu, Mn, Zn, Mo, Co, Ni, Cr, F, Se, Si, Pb, Al, B, Br, Hg, As, Cd, and Ash.* Acta Agric. Scand. Suppl. *22*, 165–171.

Veien, N. K., and Andersen, M. R. (1986), *Nickel in Danish food.* Acta Derm. Venereol. *66*, 502–509.

von Ludewigs, H. J., and Theiss, A. M. (1970), *Arbeitsmedizinische Erkenntnisse bei der Nickelcarbonylvergiftung.* Zentralbl. Arbeitsmed. *20*, 329–339.

Vuopala, U., Huhti, E., Takkunen, J., and Huikko, M. (1970), *Nickel carbonyl poisoning. Report of 25 cases.* Ann. Clin. Res. *2*, 214–222.

Wallace, A., Romney, E. M., Cha, J. W., Soufi, S. M., and Chaudhry, F. M. (1977), *Nickel phytotoxicity in relationship to soil pH manipulation and chelating agents.* Commun. Soil Sci. Plant Anal. *8*, 757–764.

Ward, I. N. (1987), *The future of multi-(ultra-trace)element analysis in assessing human health and disease: a comparison of NAA and ICPSMS.* Environmental Health, Vol. 20, pp. 118–123. Extended Odense Abstract, WHO Copenhagen.

Webster, J. D., Parker, T. F., Alfrey, A. C., Symthe, W. R., Kubo, H., Neal, G., and Hull, A. R. (1980), *Acute nickel intoxication by dialysis.* Ann. Intern. Med. *92*, 631–633.

Welz, B. (1985), *Atomic Absorption Spectrometry*, 2nd Ed., pp. 306–308, 400 ff. VCH Verlagsgesellschaft, Weinheim-Deerfield Beach/Florida-Basel.

Wills, M. R., Brown, C. S., Bertholf, R. L., Ross, R., and Savory, J. (1985), *Serum and lymphocyte, aluminium and nickel in chronic renal failure.* Clin. Chim. Acta *145*, 193–196.

Wu, Y., Luo, H., and Johnson, D. R. (1986), *Effect of nickel sulfate on cellular proliferation and Epstein-Barr virus antigen expression in lymphoblastoid cell lines.* Cancer Lett. *32*, 171–197.

Zacher, K. D., and Ippen, H. (1984), *The nickel-eczema in joung girls* (in German). Dermatosen *32* (2), 46–54.

Zhicheng, S. (1986), *Acute nickel carbonyl poisoning: a report of 179 cases*. Br. J. Ind. Med. 43, 422–424.
Zhicheng, S., Lata, A., and Yuhua, H. (1986), *A study of serum monoamone oxidase (MAO) activity and the EEG in nickel carbonyl workers*. Br. J. Ind. Med. 43, 425–426.
Zober, A., Welte, D., and Schaller, K. H. (1984), *Untersuchungen zur Kinetik von Chrom und Nickel in biologischem Material während einwöchigen Lichtbogenschweißens mit Chrom-Nickel-haltigen Zusatzwerkstoffen*. Schweissen Schneiden 10, 461–464.

Additional Recommended Literature

Barben, H., and Studer, C. (1990), *Environmental Contamination by Nickel/Cadmium Accumulators* (in German). BUWAL Bulletin 1/90, 18–20, Berne.
Bozec, C. (1989), *Speciation and Carcinogenic Classification – Some Limits of Assessment and Prevention Procedures – the Case of Nickel*. Toxicol. Environ. Chem. 23 (1–4), 169–180.
Cambell, J. A., Whitelaw, K., Riley, J. P., Head, P. C., and Jones P. D. (1988), *Contrasting Behavior of Dissolved and Particulate Nickel and Zinc in a Polluted Estuary*. Sci. Total Environ. 71, 141–155.
Cornett, R. J., Chant, L., and Evans, R. D. (1988), *Nickel Diagenesis and Partitioning in Lake Sediments*. Abstracts Trace Metals in Lakes, Hamilton, Canada, p. 17. Proceedings Sci. Total Environ. 87/88, 157–170.
Draxler, J., and Marr, R. (1989), *Liquid Membrane Permeation (LMP)*. Proceedings 6th IRC Congress Berlin, Waste Reduction in the Metallurgical Industry 1, pp. 139–148. EF-Verlag für Energie- und Umwelttechnik GmbH, Berlin.
Grandjean, P., Andersen, O., and Nielsen, G. D. (1988), *Nickel*, in: Alessio, L., Berlin, A., Boni, M., and Roi, R. (eds.), *Biological Indicators for the Assessment of Human Exposure to Industrial Chemicals*. EUR 11478 EN, pp. 57–80. Commission of the European Communities, Brussels.
Herrmann, W. A. (1988), *100 Years Metalcarbonyls* (in German). Chem. Unserer Zeit 22 (4), 113–122.
Imbra, R. J., Latta, D. M., and Costa, M. (1989), *Studies on the Mechanisms of Nickel-induced Heterochromatin Damage; Effect on Specific DNA-Protein Interactions*. Toxicol. Environ. Chem. 22 (1–4), 167–179.
Klockow, D., and Kaiser, R.-D. (1985), *Generation of Various Types of Nickel Containing Aerosols and Detection of Nickel Species in These Aerosols*, in: Brown, S. S., and Sunderman, F. W., Jr. (eds.), *Progress in Nickel Toxicology*, p. 235ff, Blackwell, Scienfitic Publications, Boston.
Menné, T., and Nieboer, E. (1989), *Metal Contact Dermatitis: A Common and Potentially Debilitating Disease*. Endeavour New Ser. 13 (3), 117–122.
Schmid, G., and Marquardt, K. (1989), *Avoidance of Waste and Recovery in the Galvanic Industry by Application of Electrodialysis*. Proceedings 6th IRC Congress Berlin, Waste Reduction in the Metallurgical Industry 1, pp. 125–138. EF-Verlag für Energie- und Umwelttechnik GmbH, Berlin.
Sigel, H., and Sigel, A. (1988), *Nickel and its Role in Biology*, in: Metal Ions in Biological Systems, Vol. 23. Marcel Dekker, New York.
Sunderman, F. W., Jr. (1989), *Mechanisms of Nickel Carcinogenesis*. Scand. J. Work Environ. Health 15, 1–12.
Sunderman, F. W., Heinrich-Ramm, R., Hopfer, S. M., Nielsen, G. D., Hildebrand, H. F., and Lumb, G., et al. (1987), *Nickel* (Platform Presentation Trace Elements in Human Health and Disease; Extended Abstracts). Environ. Health 20, 66–89.
Yeats, P. A. (1988), *The Distribution of Trace Metals (especially of Nickel, Zinc, Copper and Manganese) in Ocean Waters*. Sci. Total Environ. 72, 131–149.

II.23 Niobium

PETER L. GOERING, Rockville, Maryland, USA
BRUCE A. FOWLER, Baltimore, Maryland, USA

1 Introduction

Niobium is a Group V element, named after the Greek mythological figure NIOBE, daughter of TANTALUS. The metal was discovered by Charles Hatchett in 1801 and isolated by Blomstrand in 1864. For many years the terms "niobium" and "columbium" were used interchangeably; however, the name "niobium" was officially adopted by the International Union of Pure and Applied Chemistry in 1950. Niobium is not a very rare element. Its concentration in the earth's crust is of the same order as those of lead and cobalt, and its production (in the form of Nb_2O_5) is rather higher than that of cobalt (SAAGER, 1984). The significant properties of niobium which make it suitable for a wide variety of technological applications (especially in alloys) are its high melting point, low neutron-capture cross-section, corrosion resistance, and excellent mechanical characteristics. However, recycling is difficult because of the wide range of applications (SAAGER, 1984). While some data are available on the metabolism of ^{95}Nb, (which has also been found in the environment as a decay product after nuclear tests; WENNIG and KIRSCH, 1988) data on the toxicity of niobium compounds are sparse.

2 Physical and Chemical Properties, and Analytical Methods

Niobium is a steel-gray, lustrous, ductile, and malleable metal classified as a Group V metal in the Periodic Table of elements, has atomic mass 92.9, and atomic number 41. Niobium has a melting point of $2468 \pm 10\,°C$ and a boiling point of approximately $4742\,°C$. The element is inert to HCl, HNO_3, or aqua regia at room temperature, but is attacked by alkali hydroxides or oxidizing agents at all temperatures. When processed at even moderate temperatures, it must be placed in a protective environment to prevent decomposition. Numerous forms of niobium occur, including pentachloride, pentafluoride, pentoxide, and potassium oxypentafluoride salts. Several oxidation states are known including +2, +3, and +4, but +5 is the most common and stable state. One natural isotope of niobium exists, ^{93}Nb, and artificial isotopes include 88 – 92 and 94 – 101 (HAWLEY, 1977; HAMMOND, 1986/87; MERCK INDEX, 1983).

Very few reports are available dealing with low-level analysis of niobium in biological tissues. The reagent morin (3, 5, 7, 2', 4'-pentahydroxyflavone) forms a red precipitate with niobium in acidic solutions, which has been used for the gravimetric determination of niobium; however, several other elements (Ti, V, Mo, Zr) interfere with the analysis. Niobium can also be determined photometrically using this reagent (BURGER, 1973). Elimination of the major interfering ions (Ti, Zr, V, and Be) in niobium analysis has been successfully achieved by spectrophotometry combined with solvent extraction (CCl_4) with the use of diethyldithiocarbamate (BURGER, 1973). Pyrogallol and 8-hydroxyquinoline have been useful reagents for the spectrophotometric determination of niobium in the 0.3 – 8 ppm range. Determination of niobium in water samples with limits of detection in the 0.1 – 0.2 ppb range has been achieved with high-dispersion spectrography, inductively-coupled plasma atomic emission spectroscopy, and X-ray emission spectroscopy (WENNIG and KIRSCH, 1988). Flame AAS in the presence of potassium or aluminum seems to be less sensitive (detection limits about 1 ppm; WELZ, 1985).

3 Sources, Production, Important Compounds and Uses

Niobium is found in niobite, niobite-tantalite, pyrochlore, carbonatites, and euxenite. These ores are found primarily in Brazil (about 86% of the world production; SAAGER, 1984), Canada, Zaire, Nigeria, and the Soviet Union. Pyrochlore represents the major source of known world reserves of niobium (BARTLETT, 1973). The annual world production of niobium (in the form of Nb_2O_5 concentrates) is about 23000 tons (FICHTE et al., 1979; SAAGER, 1984), but in the future niobium may be substituted in alloys by other metals (also by ceramics).

Extracting and refining processes for niobium consist of a series of operations, including upgrading the ores by preconcentration, disruption of the niobium-containing matrix via an ore-opening procedure, pure niobium compound preparation, reduction to metallic niobium, followed by refining, consolidation, and fabrication of the metal (FICHTE et al., 1979; SAAGER, 1984; PAYTON, 1985). Niobium is so closely associated with tantalum that they must be separated either by fractional crystallization or by solvent extraction (e.g., with ketones after treatment with hydrofluoric acid) before purification (HAWLEY, 1977; FICHTE et al., 1979). Chlorination, introduction of the products into melted sodium chloride, and fractionation of $NbCl_5$ in a rectified distilling column is another method (FICHTE et al., 1979).

The principal commercial use for niobium is in the form of ferroniobium used to alloy carbon and stainless steels, nonferrous metals, and metals used in arc-welding rods. These alloys provide greater strength, cryogenic ductility, and formability; thus, niobium-containing alloys have found utilization in pipeline construction (BARTLETT, 1973; SAAGER, 1984; HAMMOND, 1986/87). Titanium-aluminum-niobium alloys are being developed for biocompatible, high-strength surgical implants (SEMLITSCH et al., 1985). Metal-resin composites containing niobium as a filler have potential use as restorative materials in dentistry (MISRA and BOWEN, 1977). Niobium has been used as a getter to remove gases in electronic vacuum tubes. It possesses superior superconductive properties in strong magnetic fields, which may be exploited for direct large-scale electric power generation applications (MERCK INDEX, 1983; HAMMOND, 1986/87). Niobium-germanium and niobium-titanium alloys are superconductive up to 23 K (BEDNORZ, 1988). Radioactive niobium microspheres are used in experimental studies of blood flow in skin (NATHANSON and JACKSON, 1975), placenta (BUSS et al., 1975), cerebrum (WELLENS et al., 1975), and heart (NEILL et al., 1975).

4 Distribution in the Environment and in Living Organisms

Elemental niobium has an average value of 24 ppm in the earth's crust (PAYTON, 1985). Coal slags from 12 sources contained niobium with values ranging from 10–24 ppm (STETTLER et al., 1982). In the same study, similar values were obtained from mineral slag samples from copper and nickel smelters. Analysis of land plants revealed niobium levels of less than 0.4 mg/kg dry weight; however, plants located near niobium deposits demonstrated a marked capacity to accumulate and concentrate the metal to niobium levels greater than 1 mg/kg (TYUTINA et al., 1959). Marine plants, such as seaweeds, appear to concentrate niobium from seawater contaminated with nuclear fallout (YAMATO et al., 1984). Niobium occurs in seawater in the form of $Nb(OH)_6^-$ at approximately 0.005 ppb (BRULAND, 1983). Human adults are estimated to contain an average body burden of 112 mg Nb (SCHROEDER and BALASSA, 1965), but this estimation may be too high (WENNIG and KIRCH, 1988), and human hair contains approximately 2 ppm.

5 Uptake, Absorption, Transport and Distribution, Metabolism and Elimination in Plants, Animals, and Humans

As mentioned in Sect. 4 niobium may accumulate in plants. Niobium is not an essential nutrient for humans or animals. Studies on metabolism of niobium compounds are few; of these, most utilize the radionuclide ^{95}Nb, a by-product of fissionable material. Absorption of ^{95}Nb from the gastrointestinal tract varies, but an upper absorption limit of 5% of the amount ingested has been reported in a number of species (THOMAS et al., 1967; CUDDIHY, 1978). Approximately 1% of an oral dose of ^{95}Nb was absorbed in mouse, rat, monkey, and dog (FURCHNER and DRAKE, 1971). Though absorption from the stomach for most substances is usually not of the magnitude of that from the small intestine, studies of ligated rat gastrointestinal tract segments in situ demonstrated a six-fold higher absorption of ^{95}Nb from the stomach compared to duodenal and jejunal segments (EISELE and MRAZ, 1981).

Several studies have demonstrated striking age differences with regard to gastrointestinal absorption of niobium. While ^{95}Nb was poorly absorbed following oral administration to adult rats, the nuclide was absorbed to a considerable extent by suckling rats resulting in a 10- to 1000-fold higher whole-body retention than in adults (SHIRAISHI and ICHIKAWA, 1972). It was postulated that this resulted from the more active pinocytosis found in the absorptive epithelium of sucklings. Percentages of an orally administered dose of ^{95}Nb remaining in the gastrointestinal tract and its contents in rats determined 4 days after dosing decreased with increasing age at dosing through the 21st day (MRAZ and EISELE, 1977a). The absorbed dose recovered in the body ranged from 6% in the newborn to slightly less than 0.1% in weanling or adult animals. Tissue accumulation of ^{95}Nb four days after administration occurred primarily in liver, kidney, and bone. Considerably more ^{95}Nb was absorbed after oral administration in newborn sheep and swine compared to weanlings (MRAZ and EISELE, 1977b).

Inhalation exposures of dogs to ^{95}Nb oxalate and oxide aerosols resulted in 60% absorption of the total dose (CUDDIHY, 1978). In rats, exposure via inhalation to tracer ^{95}Nb and particulate ^{95}Nb resulted in absorption of 36% and 71% of the total dose (THOMAS et al., 1967).

The order of accumulation of ^{95}Nb in tissues other than lung following intravenous, intraperitoneal or inhalation exposure of rats were bone > liver > kidney (THOMAS et al., 1967; FLETCHER, 1969; FURCHNER and DRAKE, 1971; CUDDIHY, 1978). Following parenteral administration of ^{95}Nb, 50% of the dose was localized in bone and resided in that tissue with an effective half-life of 30 days (HAMILTON, 1947).

Mice exposed to 5 ppm sodium niobate (as metal) in drinking water from weanling until natural death (approximately 2 years) accumulated niobium primarily in spleen > heart > liver > lung > kidney (SCHROEDER et al., 1968). A similar study in rats resulted in a tissue accumulation in spleen > kidney > heart > lung > liver (SCHROEDER et al., 1970).

The lungs retained the highest concentration of ^{95}Nb following inhalation exposure of dogs to ^{95}Nb-oxide particles; most upper respiratory tract deposits cleared within 2 days, but ^{95}Nb in deeper lung regions was retained with a biological half-life of greater than 300 days (CUDDIHY, 1978). Rates of clearance from lung were highly dependent on the ^{95}Nb chemical form and particle size (median areodynamic diameters). Exposure of rats to a tracer ^{95}Nb aerosol resulted in high lung retention with a biological half-life of 120 days (THOMAS et al., 1967).

Pregnant rats and rabbits demonstrated different tissue distribution patterns of ^{95}Nb following intravenous injection (SCHNEIDEREIT and KRIEGEL, 1986). In rats, all maternal tissues exhibited higher concentrations compared to the correponding fetal organs; the highest fetal/maternal ratio was 0.6 for bone. In rabbits, fetal bone exhibited a 3.5-fold higher concentration of ^{95}Nb compared to maternal bone.

Retention of ^{95}Nb-oxalate in the Atlantic croaker after a single intraperitoneal injection was expressed as two exponential rate functions. A rapid and long-term phase having biological half-lives of 5 and 465 days, respectively, were evident with each phase representing retention of approximately 50% of the administered dose (BAPTIST et al., 1970).

The fecal to urinary excretion ratio of ^{95}Nb stabilized at 1:1 and 3:1 from 1 to 60 days following a single inhalation exposure of rats to tracer ^{95}Nb and tracer plus a particulate carrier, respectively (THOMAS et al., 1967). The higher fecal excretion in the niobium plus carrier animals represents a greater amount of material being removed from the respiratory tract by ciliary action with subsequent swallowing. Urinary excretion of niobium exceeded fecal excretion by a factor 3 to 9 in mice, rats, monkeys, and dogs following intravenous and intraperitoneal exposure (FURCHNER and DRAKE, 1971). Excretion of niobium in humans occurred primarily via the urinary route (SCHROEDER and BALASSA, 1965).

6 Effects on Animals

Nothing seems to be known about effects on plants and humans.

6.1 Acute Effects in Mammals

There is a dearth of toxicity data for niobium compounds. Respiratory paralysis was determined as the cause of death in cats administered 5 mg Nb per kg, iv, as $NbCl_5$. The LD_{50} values for mice and rats injected intraperitoneally with niobium chloride ($NbCl_5$) were 21 and 14 mg Nb per kg, respectively. Following intraperitoneal administration of potassium niobate ($4 K_2O \cdot 3 Nb_2O_5 \cdot 16 H_2O$) LD_{50} values were determined to be 13 and 86 mg Nb per kg for mice and rats, respectively (VENUGOPAL and LUCKEY, 1978). Niobate has been reported to inhibit the activity of succinic dehydrogenase *in vitro* and to oxidize 5-hydroxytryptophan (COCHRAN et al., 1950).

6.2 Chronic Effects in Mammals

Sodium niobate given to rats at a concentration of 5 ppm in drinking water over their life-span enhanced growth rate in males but not females (SCHROEDER et al., 1970). In the same study, longevity, defined as the mean age of the last surviving 10%, was significantly reduced in male rats (SCHROEDER et al., 1970). A 22% decrease in serum cholesterol was observed in female rats, but not in males. In a similar study using mice, a decreased median life-span and longevity associated with suppression of growth of older animals was observed. A two-fold increased incidence of fatty degeneration of the liver also was apparent in mice fed niobium (SCHROEDER et al., 1968). Niobium does not appear to be mutagenic, and has some affinity for tumors (WENNIG and KIRSCH, 1988).

7 Hazard Evaluation and Limiting Concentrations

Adequate risk assessment for human exposure to niobium compounds is hampered by the scarcity of metabolism and toxicity data. Recommended maximum allowable concentrations of niobium hydride in air were established at 6 mg/m^3 (PAYTON, 1985). There is a 10 mg/m^3 threshold limit value for niobium nitride in the U.S.S.R. and a 5 mg/m^3 value for niobium carbide in Switzerland. A limit of 0.01 mg niobium per liter has been established in the U.S.S.R. for drinking water (WENNIG and KIRSCH, 1988). The U.S. Environmental Protection Agency has published effluent limitations and wastewater pretreatment standards for primary Nb/Ta production (WENNIG and KIRSCH, 1988).

References

Baptist, J. P., Hoss, D. E., and Lewis, C. W. (1970), *Retention of ^{51}Cr, ^{59}Fe, ^{60}Co, ^{65}Zn, ^{85}Sr, ^{95}Nb, ^{141m}In and ^{131}I by the Atlantic Croaker (Micropogon undulatus)*. Health Phys. 18, 141–148.
Bartlett, E. S. (1973), *Niobium and Compounds*, in: Hampel, C. A., and Hawley, G. G. (eds.), *The Encyclopedia of Chemistry*, 3rd Ed., pp. 711–712. Van Nostrand Reinhold Co., New York.
Bednorz, J. G. (1988), *Plenary Lecture*, 22nd ACHEMA, Frankfurt am Main.
Bruland, K. W. (1983), *Trace Elements in Seawater*, Chem. Oceanogr. 8, 208.
Burger, K. (1973), in: *Organic Reagents in Metal Analysis*, pp. 111, 126, 196. Pergamon Press, New York.
Buss, D. D., Bisgard, G. E., Rawlings, C. A., and Rankin, J. H. (1975), *Uteroplacental Blood Flow During Alkalosis in the Sheep*. Am. J. Physiol. 228, 1497–1500.
Cochran, K. W., Doull, J., Mazur, M., and DuBois, K. P. (1950), *Acute Toxicity of Zirconium, Columbium, Strontium, Lanthanum, Cesium, Tantalum, and Yttrium*. Arch. Ind. Hyg. Occup. Med. 1, 637–650.
Cuddihy, R. G. (1978), *Deposition and Retention of Inhaled Niobium in Beagle Dogs*. Health Phys. 34, 167–176.
Eisele, G. R., and Mraz, F. R. (1981), *Absorption of ^{95}Nb from Ligated Segments of the Gastrointestinal Tract of the Rat*. Health Phys. 40, 235–238.
Fichte, R., Retelsdorf, H.-J., and Rothmann, H. (1979), *Niobium and Niobium Compounds*, in *Ullmanns Encyclopädie der technischen Chemie*, 4th Ed., Vol. 17, pp. 303–314. Verlag Chemie, Weinheim-New York.
Fletcher, C. R. (1969), *The Radiological Hazards of Zirconium-95 and Niobium-95*. Health Phys. 16, 209–220.
Furchner, J. E., and Drake, G. A. (1971), *Comparative Metabolism of Radionuclides in Mammals. VI. Retention of ^{95}Nb in the Mouse, Rat, Monkey, and Dog*. Health Phys. 21, 173–180.
Hamilton, J. G. (1947), *The Metabolism of the Fission Products and the Heaviest Elements*. Radiology 49, 325–333.
Hammond, C. R. (1986/87), *The Elements*, in: *Handbook of Chemistry and Physics*, 67th Ed., p. B-26. CRC Press, Boca Raton, Florida.
Hawley, G. G. (1977), in: *The Condensed Chemical Dictionary*, 9th Ed., pp. 610–611. Van Nostrand Reinhold Co., New York.
Merck Index (1983), 10th Ed., p. 6408. Merck & Co., Inc., Rahway, New Jersey.
Misra, D. N., and Bowen, R. L. (1977), *Sorption of Water by Filled-Resin Composites*. J. Dent. Res. 56, 603–612.

Mraz, F. R., and Eisele, G. R. (1977a), *Gastrointestinal Absorption of ^{95}Nb by Rats of Different Ages*. Radiat. Res. *69*, 591–593.

Mraz, F. R., and Eisele, G. R. (1977b), *Gastrointestinal Absorption, Tissue Distribution, and Excretion of ^{95}Nb in Newborn and Weanling Swine and Sheep*. Radiat. Res. *72*, 533–536.

Nathanson, S. E., and Jackson, R. T. (1975), *Blood Flow Measurements in Skin Flaps*. Arch. Otolaryngol. *101*, 354–357.

Neill, W. A., Oxendine, J., Phelps, N., and Anderson, R. P. (1975), *Subendocardial Ischemia Provoked by Tachycardia in Conscious Dogs*. Am. J. Cardiol. *35*, 30–36.

Payton, P. H. (1985), *Niobium and Niobium Compounds,* in: Kirk-Othmer Concise Encyclopedia of Chemical Technology, pp. 783–785. John Wiley & Sons, New York.

Saager, R. (1984), *Metallic Raw Materials Dictionary* (in German), pp. 157–159. Bank von Tobel, Zürich.

Schneidereit, M., and Kriegel, H. (1986), *Comparative Distribution of Niobium-95 in Maternal and Fetal Rats*. Experientia *42*, 619–620.

Schroeder, H. A., and Balassa, J. J. (1965), *Abnormal Trace Metals in Man: Niobium*. J. Chronic Dis. *18*, 229.

Schroeder, H. A., Mitchener, M., Balassa, J. J., Kanisawa, M., and Nason, A. P. (1968), *Zirconium, Niobium, Antimony and Fluorine in Mice: Effects on Growth, Survival and Tissue Levels*. J. Nutr. *95*, 95–101.

Schroeder, H. A., Mitchener, M., and Nason, A. P. (1970), *Zirconium, Niobium, Antimony, Vanadium, and Lead in Rats: Life Term Studies*. J. Nutr. *100*, 59–68.

Semlitsch, M., Staub, F., and Weber, H. (1985), *Titanium-Aluminium-Niobium Alloy, Development for Biocompatible High Strength Surgical Implants*. Biomed. Tech. *30*, 334–339.

Shiraishi, Y., and Ichikawa, R. (1972), *Absorption and Retention of ^{144}Ce and ^{95}Zr-^{95}Nb in Newborn, Juvenile and Adult Rats*. Health Phys. *22*, 373–378.

Stettler, L. E., Donaldson, H. M., and Grant, G. C. (1982), *Chemical Composition of Coal and Other Mineral Slags*. Am. Ind. Hyg. Assoc. J. *43*, 225–238.

Thomas, R. G., Thomas, R. L., and Scott, J. K. (1967), *Distribution and Excretion of Niobium Following Inhalation Exposure of Rats*. Am. Ind. Hyg. Assoc. J. *28*, 1–7.

Tyutina, N. A., Aleskovskii, V. B., and Vasilev, P. I. (1959), Geochimistry *6*, 668–671.

Venugopal, B., and Luckey, T. D. (1978), *Toxicity of Group V Metals and Metalloids,* in: Metal Toxicity in Mammals-2, pp. 227–229. Plenum Press, New York.

Wellens, D. L., Wouters, L. J., De Reese, R. J., Beirnaert, P., and Reneman, R. S. (1975), *The Cerebral Blood Distribution in Dogs and Cats. An Anatomical and Functional Study*. Brain Res. *86*, 429–438.

Welz, B. (1985), *Atomic Absorption Spectrometry*, 2nd Ed., pp. 308, 394ff. VCH Verlagsgesellschaft, Weinheim-Deerfield Beach/Florida-Basel.

Wennig, R., and Kirsch, N. (1988), Chapter 38: *Niobium,* in: Seiler, H. G., Sigel, H., and Sigel, A. (eds.), Handbook on Toxicity of Inorganic Compounds, pp. 469–473. Marcel Dekker, New York.

Yamato, A., Miyagawa, N., and Miyanaga, N. (1984), *Radioactive Nuclides in the Marine Environment – Distribution and Behaviour of ^{95}Zr, ^{95}Nb Originated from Fallout*. Radioisotopes *33*, 449–455.

Additional Recommended Literature

Trueb, L. (1990), *Niobium and Tantalum – Unequal Brothers* (in German). Neue Zürcher Zeitung, Forschung und Technik, No. 73, pp. 77–78 (28 March), Zürich.

II.24 Platinum-Group Metals

HERMANN RENNER, Frankfurt/Main, Federal Republic of Germany
GABRIELLA SCHMUCKLER, Haifa, Israel

1 Introduction

As a result of technological developments, the metals of the platinum group have gained enormously in importance, especially as catalysts (FCI, 1989; TRUEB, 1989). In the accessible areas of the environment, in foodstuff and in living organisms they are found only in minute quantities and, compared with other heavy metals, their toxicity is significant only in exceptional cases. However, inhalation of or contact with the aerosols of certain compounds may lead to allergic reactions, dermatitis, respiratory difficulties, coughing, and cyanose-like symptoms. Little is known about the biological effects caused by the platinum-group metals and their compounds. Fig. II.24-1 shows pertinent relations (KEMPER, 1987). For these reasons, particularly low MAK values (maximum allowable concentration) have been established. Even so, the quantities emitted from the catalytic converters of automobile exhausts are still more than 100 times lower than the MAK values (LAHMANN and THRON, 1987). Further study into this area seems worthy of consideration, especially in light of the appreciable quantities of the lighter platinum-group metals which result from nuclear fission in atomic power plants (BUCKOW et al., 1986).

2 Physical and Chemical Properties, and Analytical Methods

2.1 Atomic Structure

In the Periodic System of the Elements, which reflects the systematic organization of the electron shells of the atoms, the platinum-group metals (often referred to as PGMs) or platinum-group elements (PGEs) form two triads in the second and third series of transition metals. Each of these triads is structurally related to certain noble gases: the three lighter elements to krypton (Kr), the three heavier ones to xenon (Xe) (Table II.24-1). Both groups of metals, which appear below one another in the Periodic System, have energetically related valence electron orbitals and use internal valence orbitals in addition to the external ones. They thus belong to the transition metals, or more specifically to Group VIII of the Periodic System.

In nuclear power plants fed with ^{235}U and $^{239/241/243}$Pu as fissionable materials or with ^{238}U as breeder material, rubidium, palladium, and ruthenium, among

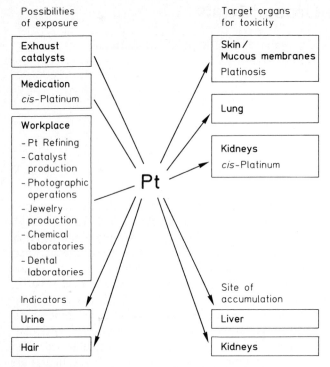

Fig. II.24-1. Schematic representation of platinum effects.

others, are formed as fission products. Rhodium appears in the form of stable, non-radioactive ^{103}Rh; palladium contains ^{107}Pd, a low-level beta-emitter; and ruthenium contains ^{106}Ru, which is a highly radioactive gamma-emitter.

2.2 Properties

Apart from the properties typical for the transition metals, such as their catalytic effects, a tendency to form complexes (HOLLECK, 1983; COTTON and WILKINSON, 1982; GREENWOOD and EARNSHAW, 1988; BRADFORD et al., 1988) and colored compounds, and the existence of a multitude of oxidation states, all platinum-group metals are characterized by being noble metals. Their melting points (ranging from 1554 °C for palladium to 3054 °C for osmium) and boiling points are high, and they are among the most chemically inert elements in the metallic state. Metals grouped in horizontal rows within the triads have similar properties; for example, Ru-Rh-Pd, which are light PGMs with densities of about 12 g/cm^3, and Os-Ir-Pt, which are heavy PGMs with densities of about 22 g/cm^3 (SAAGER, 1984). Relationships between the three vertical pairs Ru-Os, Rh-Ir, and Pd-Pt, as well as the diagonal combinations Ru-Ir and Rh-Pd(-Au) are also noteworthy (AYLWARD and FINDLAY, 1982; GMELIN, 1938–57; DEGUSSA, 1967; RENNER, 1979).

Table II.24-1. Atomic Structures Properties of the Platinum-group Metals (COOPER, 1968)

Element		Atomic Number = Nuclear Charge	Mean Atomic Mass	Density (g/cm^3)	Melting Point (°C)	Boiling Point (°C)	Naturally Occurring Isotopes
Ruthenium	Ru	44	101.07	12.2	2250	3900	96 (5.5%), 98 (1.9%), 99 (12.7%), 100 (12.6%), 101 (17.1%), 102 (31.6%), 104 (18.6%)
Rhodium	Rh	45	102.91	12.4	1965	3730	103 (100%)
Palladium	Pd	46	106.4	12.0	1552	2929	102 (1.0%), 104 (11.0%), 105 (22.2%), 106 (27.3%), 108 (26.7%), 110 (11.8%)
Osmium	Os	76	190.2	22.4	3000	5000	184 (0.02%), 186 (1.6%), 187 (1.6%), 188 (13.3%), 189 (16.1%), 190 (26.4%), 192 (41.0%)
Iridium	Ir	77	192.22	22.5	2410	4530	191 (37.3%), 193 (62.7%)
Platinum	Pt	78	195.09	21.6	1769	3830	190[a] (0.01%), 192 (0.8%), 194 (32.9%), 195 (33.8%), 196 (25.3%), 198 (7.2%)

[a] Naturally occurring radioactive isotope

2.3 Analytical Methods

Because of the similarities between platinum-group elements, their chemical analysis is among the most difficult and expensive in classical analytical chemistry (GDMB, 1961–75; BEAMISH and VAN LOON, 1977; ENSSLIN et al., 1964; KALLMAN, 1987; LÜSCHOW, 1980; WOGRINZ, 1936). Even the complete dissolution of the samples is difficult. The preliminary separation of the bulk of the noble metals from base metals, sometimes referred to as docimasy (from the Greek dokimasia, meaning examination), is often carried out by a fire-assay method involving lead alloys (ENSSLIN et al., 1964). Platinum is in most cases isolated as the sparingly soluble ammonium hexachloroplatinate(IV), $(NH_4)_2[PtCl_6]$. Palladium is separated by precipitating the dimethyl-glyoxime complex; osmium and ruthenium are separated in the form of the compounds osmium(VIII) oxide, OsO_4, and ruthenium(VIII) oxide, RuO_4. In order to obtain rhodium and iridium, selective chlorination at high temperature is employed. Physical methods of analysis, which are cheaper and faster, have been introduced in recent years (ULLMANN, 1980). These include X-ray fluorescence, atomic absorption spectroscopy (AAS) (WELZ, 1985), and emission spectra analysis. According to BRADFORD and CHASE (1988) and YANG (1989), graphite furnace AAS is the method of choice for determinations in biological fluids, for example, *cis*-platinum monitoring or measurements in dried algae. For trace analysis, mass spectroscopy and neutron activation analysis are used as well, while plasma excitation is gaining importance due to its improved detection threshold and its capability of producing rapid and relatively inexpensive simultaneous analyses (see KALLMAN, 1987; BRADFORD, CHASE, and LÉONARD 1988, for the most recent literature). GOLDBERG et al. (1986; 1988) have discussed the analytical chemistry of the PGMs (including speciation) in sea water.

3 Sources, Production, Important Compounds, Uses, and Recycling

3.1 Abundance
(CABRI, 1980; PARTHE, 1972; PARTHE and CROCKET, 1978; EDWARDS and SILK, 1987; FCI, 1989; TRUEB, 1989; NEUMÜLLER, 1979–1988)

With respect to their atomic number and atomic mass, the platinum-group metals occupy a median position. Their abundance in the universe, based on the genesis of the atomic nuclei, appears to be in the range of 10^{-4} ppm (BRESCH, 1977). Hydrogen and helium constitute almost 99.9% of all extant atoms or atomic nuclei, and their concentration in the stars, which synthesize the elements, is above average. Accordingly, the heavier elements, such as the PGMs, are found in greater amounts in the planets, and hence in the Earth (RENNER, 1979; QUIRING, 1962).

It appears that a strong concentration gradient exists in the Earth, the PGM content of the core being higher than that of the mineral crust. The order of

Table II.24-2. Distribution of Platinum-group Metals in the Earth (PARTHE, 1972; PARTHE and CROCKET, 1978)

Region	Extent (radius/thickness) (km)	Total mass (tons)	Concentration of PGMs (ppm)	Mass of PGMs (tons)
Earth, total	~6400	~$6 \cdot 10^{21}$	~30	~$2 \cdot 10^{17}$
Earth core ("Siderosphere")	~3500	~$2 \cdot 10^{21}$	~70	~$2 \cdot 10^{17}$
Earth mantle[a]	~2900	~$4 \cdot 10^{21}$	~0.05	~$5 \cdot 10^{14}$
Earth crust[b]	~16	~$3 \cdot 10^{19}$	~0.01	~$3 \cdot 10^{11}$
minable	~5	~$7 \cdot 10^{18}$	~0.01	~$7 \cdot 10^{10}$
prospected				~$6 \cdot 10^{4}$
mined				~$3 \cdot 10^{3}$
Hydrosphere	~2.7	~$1.6 \cdot 10^{18}$	$<10^{-6}$	$<10^{6}$
Biomass (dry fraction)		~10^{12} [d]	$<10^{-7}$	<10
Atmosphere	~80[c]	~$5 \cdot 10^{15}$	$<10^{-15}$	<1

[a] The subdivision of the earth according to GOLDSCHMIDT, using the terms "Chalkosphere", Sulfide-Oxide Shell, Eklogite Shell, and Lithosphere, is no longer used (SCHWAB, 1974, NEUMÜLLER, 1979–1988)
[b] Varies between about 30 and 60 km beneath the continents and between 5 and 8 km under the oceans (SCHWAB, 1974)
[c] Upper limits of homosphere and mesopause; interplanetary space begins at 60000 km altitude
[d] The vast majority on land

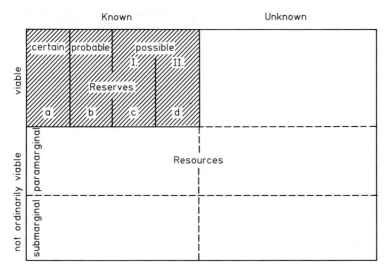

Fig. II.24-2. Classification of mineral resources. (a) Extent, form, and content of the ore body are known with a margin of error of 10%. (b) Extent, form, and content of the ore body are known with a margin of error of 20%. (c) Extent, form, and content of the ore body are known with a margin of error of 30%. (d) Extent, form, and content of the ore body are known with a margin of error of 50%.

magnitude of this gradient is about 10^4 (Table II.24-2). The massive vertical motion and the material transport involved therein cause significant quantities of PGMs to be located in regions close to the earth's surface.

The amounts of PGMs present in the upper parts of the earth's crust, down to a depth of 5 km and hence technologically attainable, are still enormous when compared with our requirements, but only a small fraction of the pertinent ores is sufficiently rich for commercial exploitation. Under the current classification of mineral deposits (Fig. II.24-2, MCKELVEY and CAMERON, 1973), there exist some 50 000 tons of PGMs in known reserves, while a further 10 000 tons have been proven as known resources. Resources of PGMs are about twice as large as those of gold (SAAGER, 1984). Included among these are deposits with a PGM content in the range of 0.05 to 1000 ppm. While the primary deposits contain chalcogenidic and similar compounds of the platinum-group elements, the secondary deposits (alluvial platinum) – which are formed by hydrothermal reactions, erosion, relocation due to water courses, etc. – are in the pure metallic state. The PGMs so far available have been found in primary deposits, i.e., they are included in those relatively unchanged magmatic formations which contain 99% of the resources (SAAGER, 1984). The secondary, or alluvial, deposits have for the most part been depleted since they are more readily reached and easily worked. In the ores, PGMs are generally found in mixtures of the following proportions:

ruthenium	23%	osmium	21%
rhodium	6%	iridium	7%
palladium	18%	platinum	25%

Sea water, with its low PGM content of less than 0.001 mg/m^3, is not a viable source, and no significant enrichment in the manganese nodules on the sea floor has been observed.

Finally, an estimate can be made of the addition of PGMs through the scattering of cosmic matter. A daily influx of a maximum of 100 tons of meteorites, including cosmic dust, is the commonly accepted figure. Even assuming a high average PGM content of 100 ppm, this would mean a daily addition of only 10 kg PGMs, almost invariably dispersed throughout the globe. A massive influx of cosmic matter with a high percentage of iridium towards the end of the Cretaceous period is thought to have occurred (ALVAREZ, 1979).

The rhodium formed in the fuel elements during nuclear fission, which after separation occurs only in the stable form ^{103}Rh, corresponds quantitatively to about one-third of the present mining production of this particularly valuable element (BUCKOW et al., 1986; RENNER and TRÖBS, 1984). The palladium arising in the same way appears to be less serviceable, and similarly derived ruthenium seems to be entirely unsuitable for future use.

3.2 Production

The mining of ores, either underground or by opencast (strip) mining, is followed by enrichment (FCI, 1989; TAFEL, 1951; RENNER, 1979; DOWSING, 1980; PAWLEK, 1983). Concentrates with a high percentage of metals can be obtained by physical methods of enrichment, such as gravity separation, e.g., in floating dredgers, but even quite simple tools and magnetic separation are capable of improving alluvial platinum. After the extraction of ferrometallic constituents by gravity separation and the flotation of non-ferrous metal sulfides (which also contain, as inclusions, noble metal minerals), the enrichment of ores from primary deposits usually requires the application of pyrometallurgical methods, during which the platinum-group metals generally associate themselves with nickel. The residues from nickel refining can therefore be starting materials for platinum separation. The palladium market especially depends on nickel production in the Soviet Union and Canada because the ores in these countries contain (relative to platinum) much more palladium than those in South Africa (SAAGER, 1984).

The wet-chemical separation of PGMs makes use of some of the most complex and expensive methods known in chemical engineering (GMELIN, 1938–57; RENNER, 1979; DOWSING, 1980, PAWLEK, 1983). The most common methods for this purpose are crystallization and solvent extraction, often in combination with redox reactions at high temperatures, as well as special distillation processes. Solubilization requires special dissolution or fusion techniques adapted to each specific metal. Ion exchange plays an important part in the recovery of small but valuable traces of metals from industrial waste water, but it can also be applied in separation and purification operations. Recycling of platinum-group metals largely employs the same wet-chemical, and occasionally pyrometallurgical, methods that are used in the production of metals from ores (RENNER, 1979, 1987). Recycling is a rapidly growing market – in 1982, 12% of the platinum and 8% of the palladium were recycled (SAAGER, 1984).

Production was about 1 ton in 1880, which increased to 7 tons annually by 1900 and was obtained mostly from Colombia and the Ural Mountain region of Russia. New deposits were found, the most prominent among them being the Merensky reef in the South African veld. Rising nickel production in Canada and Siberia provided additional quantities of the platinum-group metals, primarily palladium, which were found associated with nickel and copper. Present total annual world mining production of platinum-group metals amounts to about 200 tons. Platinum and palladium constitute some 45% each of that amount, rhodium and ruthenium about 4% each, and osmium and iridium less than 1% each (ROBSON, 1987; HARGREAVES and WILLIAMSON, 1984; RENNER and TRÖBS, 1984, 1986). The Republic of South Africa and Bophuthatswana produce approximately 50% (mainly platinum and rhodium), the Soviet Union produces about 40% (primarily palladium), and Canada a little over 5%.

3.3 Compounds

Due to their tendency to form complexes and their numerous oxidation states, the platinum-group metals form an exceptionally large number of compounds (GMELIN, 1938–57; RENNER, 1979; BRAUER, 1981; KIRK-OTHMER, 1982; FCI, 1989; NEUMÜLLER, 1979–1988). The magnetic ores contain chalcogenides, arsenides, antimonides, etc., which are compounds that also take part in the pyrometallurgical reactions applied in the industrial enrichment of PGMs to produce concentrates. In the ensuing wet-chemical separations, an important role is played by the chlorine complexes of the PGMs, which form sparingly soluble ammonium and alkali salts. Hexachloroplatinic(IV) acid, $H_2[PtCl_6]$, which in aqueous solution is a strong acid, is the starting point for most platinum compounds and preparations. More recently, preference has sometimes been given to platinum nitrate for the impregnation of catalysts. Simple binary compounds are of lesser importance. In the covalently bonded volatile tetroxides of osmium and ruthenium, the highest of all the possible oxidation states, VIII, is attained. Increasing significance is being attached to the phosphine complexes, such as monochloro-(tris)triphenylphosphine rhodium(I), which serve as homogeneous catalysts.

3.4 Uses
(GMELIN, 1986)

Only after the discovery of America, where it had been used occasionally by the natives to make jewelry and cult objects, did platinum become the object of investigation by scientists (FCI, 1989; TRUEB, 1989; NEUMÜLLER, 1979–1988). Only in the 19th century, however, did demand and production begin to increase – as coinage in Russia, in electrical and chemical engineering, in dentistry, and for jewelry during the Belle Epoque. Demand for platinum and rhodium was greatly increased as a result of the invention of ammonia oxidation by Ostwald and Brauer, in which platinum served as a catalyst. This process enabled the large-scale establishment of the fertilizer industry toward the end of the First World War, which helped to feed a growing world population (HOLZMANN, 1967).

After the Second World War it was above all the petroleum industry that became a large consumer of platinum. It was discovered that the refining of high-boiling crude oil fractions in the presence of platinum-aluminum oxide catalysts enabled the production of hydrocarbon molecules for specific purposes such as internal combustion engine fuels, heating oil, and naphtha, the basic materials of petrochemistry (FALBE and HASSEROTH, 1978). With advancing mechanization in nearly all fields, the demand for platinum-group metals has continued to grow at a fast pace in the ensuing decades. The most significant applications in this respect are in chemical catalysis, chemical laboratory, and instrumentation engineering (platinum, and in some cases iridium and rhodium), electrical engineering (palladium contacts, platinum/iridium spark plugs), measuring techniques (thermocouples made of platinum/rhodium, platinum resistance thermometers and thin film resistance elements for temperature measurement), glass technology (crucibles and spinnerets made of platinum/rhodium), the textile industry (spinnerets made of gold-platinum), jewelry-making (platinum jewelry, primarily in Japan), and surgical implants (WILLIAMS, 1989).

The past ten years have once again produced new technological developments, the most important of these being in the automobile industry, where platinum (and also palladium and rhodium) are used as exhaust catalysts that oxidize harmful combustion by-products. These include partially oxidized hydrocarbons, carbon monoxide, and nitrogen oxides, which can be reduced to about one-tenth of their original amounts (KOBERSTEIN, 1984; HARRISON et al., 1981). Three-way catalysts contain 0.9 to 2.3 g platinum, 0.2 to 0.3 g rhodium, and occasionally some palladium. According to GROENENDAHL, COOPER, and others (1987), there will be a growing rhodium demand for three-way catalysts in the 1990s; 39% of the costs of the ceramic carriers are contributed by platinum, 21% by rhodium. Interactions with supports, washcoats, cerium additives, and the crystalline structure of the PGMs adversely affect the durability and lead to the undesirable deactivation of these catalyst systems. Large-scale oxo-synthesis − building carbon compounds such as aldehydes from hydrocarbons, carbon monoxide and hydrogen − makes increasing use of homogeneous catalysts based on organic rhodium complexes (FALBE and BAHRMANN, 1981).

Successful use has been made of certain platinum compounds in cancer therapy, and clinical testing on a large scale has been carried out. These compounds are the so-called *cis*-platinum, *cis*-[PtCl$_2$(NH$_3$)$_2$] (ROSENBERG et al., 1965; LIPPARD, 1982; LIPPERT and BECK, 1983), and organically substituted compounds of this type (THOMSON, 1977a, b). More recently, carboplatinum (or *para*-platinum) has been added to this list (BARNARD, 1986; BARNARD et al., 1986), as is illustrated in Fig. II.24-3. On account of these developments, an increase in platinum consumption is predicted. Corresponding compounds of other PGMs, and of some base metals as well, with antitumor capabilities have been reported (ENGELHARD Corp., 1982/84; KÖPF and KÖPF-MAIER, 1987; MURRER, 1989).

Demand for cathodic protection against the corrosion of a large variety of steel structures in water has grown considerably and with it the demand for platinated anodes. Titanium anodes coated with ruthenium oxide are used in the industrial elec-

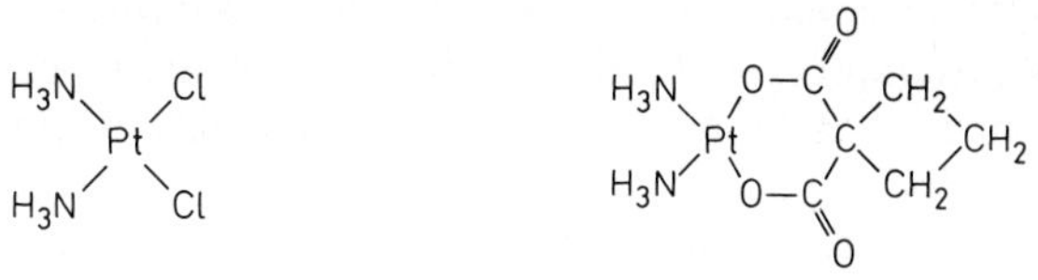

cis-Platinum
cis-Diammino-
dichloroplatinum(II)

Carboplatinum = para-platinum
diammino (1,1-cyclobutane-
dicarboxylate) platinum (II)

Fig. II.24-3. Structure of typical platinum-based anticancer drugs.

trolysis of chloro-alkalis, and in most cases the coatings also contain iridium. Iridium crucibles are employed in the growth of single crystals (RENNER and TRÖBS, 1986). If hydrogen assumes a leading role in future energy production, the uses of platinum as a catalyst in fuel cells (LUNDBLAD, 1983) may acquire much greater significance than even today (BÖLKOW, 1986/87). In electronic engineering palladium increasingly takes the place of gold.

The discrepancies between the demand for the different PGMs is evident in their prices, which as of January 8, 1988, were (in US dollars per gram): Pt – 16.66, Pd – 4.20, Rh – 42.00, Ir – 11.50, Ru – 2.50 and Os – 26.66. Principal areas of demand include automobile exhaust catalysts (Pt, Rh), jewelry (Pt), electrical engineering (Pd, Ru), electrochemical processes (Ru, Ir), and dentistry (Pd) (SAAGER, 1984). As to countries of origin, the greatest imbalance affects rhodium, since 95% of all known mineral reserves lie in Southern Africa (RENNER and TRÖBS, 1984).

4 Distribution in the Environment, in Foods, and in Living Organisms

Compared with most of the other elements, platinum-group metals have a limited influence on the biosphere, mainly because only small quantities ever reach living organisms. The amount that has reached the environment consists of what has been mined up to now (somewhat less than 3000 tons, of which about 1400 are platinum and 1400 are palladium), the magmatic and sedimentary deposits near the earth's surface, and an unspecified quantity of platinum-group metals that are dissolved in the oceans. The amount of PGMs recently introduced to rivers, lakes, and the oceans through industrial wastes is minute. Their effects, if any, are minimized by the fact that the great majority of the PGMs used in industry are in metallic form and are there almost completely inert as far as biological reactions are concerned. Disintegration over geological periods have added considerable quantities of PGMs to the oceans, but the amounts are insufficient for either extraction or physiological, let alone toxic, effects. GOLDBERG et al. (1986, 1988; YANG, 1989) found in surface sea water concentrations of about 100 to 200×10^{-6} ppb platinum (depending on depth; lower values at the surface), 1.5×10^{-6} ppb iridium (mainly in the III state), less than 5×10^{-6} ppb ruthenium (mainly in the III state) and 40×10^{-6} ppb palladium (mainly as $PdCl^+$ ions in the II state), as compared with 90×10^{-6} ppb gold (mainly in the I state). Rhenium behaves differently — it exists mainly as the inert perrhenate(VII) ion and can be enriched — as molybdenum is — to reducing forms (rhenium was possibly oxidized preferentially by the first oxygen in the atmosphere). The residence time of rhenium in sea water is of the order of 200 million years, platinum about 1 million years, gold about 500000 years, silver and ruthenium about 300000 years and palladium, nickel, and iridium less. According to BRULAND (1983) ^{107}Ru and ^{108}Ru are also found in sea water, because they are discharged as products of nuclear fission. Organic matter in the sea may quite conceivably be expected to absorb certain amounts of PGMs, since at least some of the metals are probably available in ionic form.

As to any interactions between PGMs and the environment, these could arise from compounds formed during the intermediate stages of separation or refining processes, or from such compounds that are expressly produced for further use. Reactions of highly diffuse PGMs, such as are used in catalysts, can also not be entirely disregarded. For instance, MAENHAUT et al. (1986) reported on the suspended particulate matter inside a platinum catalyst plant at a symposium on the distribution of metals in the environment. They found that platinum is associated with particles larger than 1 μm, but only less than 1 millionth of the platinum on the production line was emitted into the environment, which corresponded to a concentration of about 30 mg/m^3 at the workplace.

When automobile exhaust catalysts are installed on a large scale, as is done in most of the densely populated industrialized countries to reduce the emission of pollutants, a certain distribution of PGMs, especially platinum but also some rhenium and palladium, must be reckoned with (STOPINSKI, 1977; ROSNER and HERTEL, 1986). In this case the PGMs are transmitted into the environment as suspension materials in conjunction with $\gamma\text{-}Al_2O_3$ (which serves as a carrier for the highly dispersed metals) that eventually settle, for the most part in elemental form. In San Diego, California, the platinum content of dust is about 0.6 to 0.8 ppm (10% of it as dissolved forms; LAHMANN and THRON, 1987).

More recently, investigations into the environmental impact of PGMs have increased. Many more results of measurements have been published than was the case only a few years ago, and already a relatively well-rounded picture is available (ROSNER and HERTEL, 1986). It is today commonly accepted that an emission of 0.001–0.002 mg PGM per kilometer traveled is likely, while in undiluted exhaust fumes the PGM content amounts to 0.001 to 0.002 mg/m^3. Both figures keep within the maximum allowable concentrations set for the different elements. The same figures also correspond to the loss of PGMs in the operation of the catalysts. Such losses total 0.1–0.2 g PGMs during the 80000 kilometer service life stipulated for the catalysts by US regulations. The PGM content of air close to the road surface drops by orders of magnitude as compared with dense exhaust fumes due to the combined effect of the admixture of air and the rapid settling of the mechanically dislodged coarser particles. The fraction of particles absorbable by the lungs constitutes no more than 20% at most, and the existence of water-soluble compounds, such as platinum carbonyls, has not yet been proven and, in any case, their bioavailability would probably be small (BRADFORD and CHASE, 1988).

Nevertheless, further research needs to be done (FORTH, 1987) in view of the large amounts of PGMs emitted, and such research is proceeding. The aims of this work are the extension of analytical possibilities, chemical characterization of the PGM compounds created, laws governing the size distribution and nature of the particles contained in the aerosols, and the accumulation of PGMs in the soil and commercial crops. Special attention should be given to the mutagenic, carcinogenic, and allergenic effects, but investigations so far do not point to concrete health dangers posed by automobile exhaust catalysts. Beyond doubt, however, is the improved public health that followed the elimination of such poisonous substances as lead, nitrous oxides, hydrocarbons, and carbon monoxide due to these catalysts (LAHMANN and THRON, 1987). Some of the hypotheses recently given attention in the communication media concerning possible risks involved in the introduction of the catalysts bear no relationship to the present state of knowledge and should be regarded as speculations that in some cases border on absurdity (NIEPER, 1989).

In industrial ammonia production for the manufacture of nitric acid, large amounts of platinum and rhodium are lost through attrition of the catalysts. Depending on the process involved, the loss of platinum and rhodium (in a ratio of approximately 10 to 1) is variously estimated at between 0.05 and 0.35 grams per ton of nitric acid produced. With an annual production of about 6 million tons of nitric acid, the total platinum and rhodium loss adds up to about one ton at over 100 production sites. Nothing has been published regarding exhaust analyses with respect to platinum. However, environmental pollution by platinum-group metals as a result of the conversion of nitric oxide to nitric acid may be discounted.

Redox processes may increase the solubility of platinum in aqueous media. Very little is known about the PGM content of foodstuff, plants, and animals. Oceanic algae contain 0.001 to 0.002 ppm ruthenium and plants about 0.004 ppm (relative to dry weight; BOWEN, 1979). In the muscles of mammals, about 0.002 ppm platinum and palladium and 2×10^{-5} ppm iridium (relative to dry weight) have been found.

5 Uptake, Absorption, Transport, Distribution, Metabolism, and Elimination in Plants, Animals, and Humans

YANG (1989) has determined very high concentration factors about (5000 – 100000) for platinum, palladium, and ruthenium in four marine algal species, up to concentrations of about 1 ppb in the case of platinum. Nothing definite is known of the metabolism of the platinum-group metals in living organisms, the one exception being drugs based on *cis*-platinum. In this case platinum is discharged relatively rapidly through the kidneys, so that poisoning is largely treatable. The inhalation of PGMs, on the other hand, could be more problematic in this respect. Results are increasingly appearing in the literature (ROSNER and HERTEL, 1986; MASSARO, 1981). Elimination seems to be different for various chemical species, but biological half-lives are rather short. Bacteria can transform water soluble platinum compounds into methylplatinum compounds by demethylation of methylcobalamin (vitamin B_{12}; BOWEN, 1979; BRADFORD and CHASE, 1988).

6 Effects on Animals and Humans

Since the heavy metal compounds, of which the platinum-group metal compounds form a part, are, as a rule, more or less toxic, the PGMs have long been the subject of medical research and debates. Knowledge about the pathogenic nature of the PGMs has expanded in recent years through their pharmacological applications.

The general toxic effects of most of the heavy metal compounds stem from the physiological incompatibility with living cells. Platinum, palladium, ruthenium, and osmium are in this regard classified in close proximity to lead. Various kinds of compounds react differently, and rhodium and iridium are also part of the overall picture, even if they are relatively less harmful. The consequences of intake manifest themselves in the kidneys, liver, spleen, mucous membranes, and nervous system. However, seen in relation to lead, the abundance of which exceeds that of the PGMs by some four orders of magnitude and which may thus be termed ubiquitous, it becomes evident that the danger presented by the rare noble metals is comparatively low (BARLOW and SULLIVAN, 1984).

Results are available from widespread research into the carcinogenic and mutagenic properties of platinum-group metals (ROSNER and HERTEL, 1986; SIDERIS et al., 1986; LIPPARD, 1982; MERIAN et al., 1985, 1988). MONTI-BRAGADIN (1977), MERIAN et al. (1985), and COLUCCIA et al. (1984) have all discussed mutagenic effects of platinum and ruthenium complexes.

About half of all persons who come in regular contact with certain platinum salts are affected by allergic reactions known as platinosis, but for the most part only after exposure for a certain period of time (RENNER, 1979; STOPINSKI, 1977; CLEARE et al., 1976). Problems occur mainly in the mucous membranes and range from symptoms similar to those of the common cold to serious asthmatic attacks and fear of asphyxiation. For persons already sensitized, the amounts necessary to trigger symptoms are much lower than those liable to induce heavy metal poisoning. In practice, the phenomena are most often caused by aerosols of hexachloroplatinic(IV) acid solutions and by airborne dust particles that contain the salts of this acid. The number of persons actually affected or at risk, mainly those employed in platinum separation plants and in catalyst manufacturing, hardly exceeds a few thousand world-wide. Apart from a permanent sensitization which leads to reactions with other allergens as well, all symptoms disappear with a change in the workplace. No permanent health damage has been observed, but when symptoms appear, they are frequently disruptive both for the individual and the work environment. Advances in medicine and hygienics have brought about a clear improvement. Platinum as the pure metal does not seem to cause allergies. Platinum group metals thus play also a certain role for surgical implants (WILLIAMS, 1989). Palladium and rhodium salts are also less problematic (BRADFORD and CHASE, 1988).

The use of *cis*-platinum compounds, especially *cis*-dichlorodiamminoplatinum(II) ($PtCl_2(NH_3)_2$, Peyrone's chloride) as cytostatics, constitutes an interaction between the metals and living organisms (STOPINSKI, 1977; THOMSON, 1977a, b; RENNER, 1979; KÖPF and KÖPF-MAIER, 1981; MURRER, 1989). At the time of this writing, this is most significant in human beings. The effects of this compound as a growth and cell-division inhibitor were discovered in the 1960s. These properties presumably stem from the inhibition of DNA synthesis as a result of bifunctional involvements with the DNA double helix. The clinical application of *cis*-platinum compounds, particularly in treatment of cancers in the urogenital (testicular teratoma), head, and neck regions, and of childhood tumors, has been widely adopted. The substances are used in the form of a lyophilisate with doses of several milligrams of platinum. The side-effects caused by such amounts of platinum are considerable, although the metal is eventually excreted. Initially, however, it takes a heavy toll on the body's metabolism and impairs the well-being of the patient. The main disadvantages of *cis*-platinum are its nephrotoxicity and its suppression of bone marrow formation (myelosuppression) (AASETH, 1987; BRADFORD, 1988; SCHRAUZER, 1988). Experiments with mice and rats demonstrated carcinogenic side-effects (ROSNER and HERTEL, 1986). Selenium reduces these toxicities without reducing the cytostatic effects (AASETH, 1987; SCHRAUZER, 1988). Hydration/diuresis and antiemetics have also been recommended (MURRER, 1989). Carboplatin (see Fig. II.24-3) is a second generation anticancer drug with reduced nephrotoxicity (BRADFORD, 1988) and with successful application in small cell lung cancer (MURRER, 1989). The lethal dose of platinum has been determined to be 12 mg/kg, based on animal LD_{50} experiments. It should be noted that allergic reactions were observed in therapeutic applications.

In some cases, the handling of volatile compounds such as osmium tetroxide OsO_4 (b.p. = 131 °C) and ruthenium tetroxide RuO_4 (b.p. = 108 °C) can be toxic. The evaporation heat of these compounds is so low that even at room temperature a high saturation vapor pressure is soon reached. Both compounds are irritants of the mucous membranes. The action of osmium oxide on the mucous membranes leads to a dimming of the cornea, which is perceived as a "ring" around lights. This is reversible, but at the onset can cause considerable visual impairment.

Rhodium trichloride is lethal for rats and rabbits at doses of 200 mg/kg, probably acting on the central nervous system (LANDOLT et al., 1972). This compound, administered through drinking water to mice at concentrations of 5 ppm, leads to the development of lymphoma leukemia tumors (SCHRÖDER and MITCHENER, 1971).

7 Risk Assessment and Limiting Concentrations

The risk factors of the platinum-group metals and their compounds for living organisms are, generally speaking, very low, due to the low concentration of these elements in the environment and their low bioavailability. Specific risks due to overexposure can be nearly discounted since production of PGMs is relatively low, only 200 tons per year, 3000 tons in all until now. Much greater quantities of far more toxic metals are produced each year, for example, mercury, 7000 tons; cadmium, 18 000 tons; arsenic, 28 000 tons; chromium, 3 million tons; and lead, 6 million tons. Since the metallic form of PGMs predominates, their influence on the environment is even less than their total weight would suggest. Their high price makes it imperative to keep irretrievable losses, which would find their way into the environment, to an absolute minimum in technological processes.

Comparatively strict MAK values have been set for platinum compounds, not so much because of their toxicity, but because of their sensitizing effect (MAK, 1987; HENSCHLER, 1983). In the Federal Republic of Germany, for example, the 1987 MAK for platinum compounds was set at 0.002 mg/m^3 for metallic platinum, the value for osmium oxide at 0.0002 mg/m^3. In the USA "threshold limit values" (TLV) have been set at the same levels, with additional regulations for rhodium aerosols (0.1 mg/m^3) and salts (0.001 mg/m^3). The Federal Republic of Germany has set an upper limit of 3 mg/L (3 ppm) platinum or osmium in waste water.

References

Aaseth, J. (1987), *Proceedings 2nd Nordic Symposium on Trace Elements in Human Health and Disease, Odense.* World Health Organization (WHO), Copenhagen.
Alvarez, S. (1979), *New Sci. 82*, 788.
Aylward, G. H., and Findlay, T. J. (1982), *Datensammlung Chemie*, 2nd Ed. Verlag Chemie, Weinheim-Deerfield Beach/Florida-Basel.
Barlow, S. M., and Sullivan, F. M. (1984), *Reproductive Hazards of Industrial Chemicals.* Academic Press, London-New York-Orlando/Florida.
Barnard, C. F. J. (1986), *Scientific Symposium on Paraplatinum, Plat. Met. Rev. 30*, 116–119.
Barnard, C. F. J., Cleare, M. J., and Hydes, P. C. (1986), *Second Generation Anticancer Platinum Compounds, Chem. Br.*, Nov., 1001–1004.
Beamish, F. E., and Van Loon, J. C. (1977), *Analysis of Noble Metals.* Academic Press, New York.
Bölkow, L. (1986/87), *Energie im nächsten Jahrhundert, Scheidewege 16*, 374–382.
Bowen, H. J. M. (1979), *Environmental Chemistry of the Elements.* Academic Press, New York.
Bradford, C. W., Chase, B. J., and Léonard, A. (1988), Chapters 29, 41, 43, 45, 49, and 51 in: Seiler, H. G., Sigel, H., and Sigel, A. (eds.), *Handbook on Toxicity of Inorganic Compounds*, pp. 341ff, 501ff, 517ff, 533ff, 561ff, and 571ff. Marcel Dekker, New York.
Brauer, G. (1981), *Handbuch der präparativen anorganischen Chemie*, Vol. 3. Enke Verlag, Stuttgart.
Bresch, C. (1977), *Zwischenstufe Leben.* Piper Verlag, München.
Bruland, K. W. (1983), *Trace Elements in Seawater, Chem. Oceanogr. 8*, 209.
Buckow, G., Kohfahl, A., Mehling, O., Quillmann, H., Stahl, D., and Thomas, W. (1986), *Verwertung von Reststoffen aus einer Wiederaufarbeitungsanlage.* Gesellschaft für Reaktorsicherheit (GRS), Köln.
Cabri, L. J. (1980), *Platinum Group Elements: Mineralogy, Geology, Recovery*, CIM Special Vol. 23. Canadian Institute of Mining and Metallurgy, Ontario.
Cleare, M. C., Hughes, E., Yacoby, B., and Pepys, Y. (1976), *Clin. Allergy 6*, 183–195.
Coluccia, M. et al. (1984), *Lecture Workshop on Carcinogenic and/or Mutagenic Metal Compounds, Geneva, J. Toxicol. Environ. Chem. 8* (1), 1–9.

Cooper, D. G. (1968), *The Periodic Table*. Butterworths & Co., London; (1972), *Das Periodensystem der Elemente*. Verlag Chemie, Weinheim.

Cotton, F. A., and Wilkinson, G. (1982), *Anorganische Chemie*, 4th Ed. Verlag Chemie, Weinheim-Deerfield Beach/Florida-Basel.

DEGUSSA (1967), *Edelmetall-Taschenbuch*, Frankfurt.

Dowsing, R. J. (1980), *Spotlight on the Platinum Metals. Met. Mater.*, May, 41–48; July, 32–42.

Edwards, A. M., and Silk, M. H. (1987), *Platinum in South Africa*. Council for Mineral Technology (MINTEK), Randburg.

Engelhard Corp. (1982/84), *Ionic Bis(dicarboxylato)palladate(II) Anti-tumor Agents. Eur. Patent* 0098134, 26 June, 1982/11 January, 1984.

Ensslin, F. (1984), *Edelmetall-Analyse*. Springer Verlag, Berlin.

Falbe, F., and Bahrmann, H. (1981), *Homogene Katalyse in der Technik. Chem. Unserer Zeit 15*, 37.

Falbe, F., and Hasserodt, U. (1978), *Katalysatoren, Tenside and Mineralöladditive*. Thieme Verlag, Stuttgart.

FCI (Fonds der Chemischen Industrie) (1989), *Edelmetalle – Gewinnung, Verarbeitung, Verwendung. Folienserie No. 12*, Frankfurt/Main.

Forth, W. (1987), *Gesundheitsgefährdung durch den Katalysator? Dtsch. Ärztebl. 84*, 1603–1606.

GDMB (Gesellschaft Deutscher Metallhütten- und Bergleute) (1961–1975), *Analyse der Metalle* (Chemikerausschuß, ed.). Springer Verlag, Berlin.

Gmelin, (1938–1957), *Handbuch der Anorganischen Chemie*, System No. 68, 8th Ed. Springer Verlag, Berlin.

Gmelin, (1986), *Handbuch der Anorganischen Chemie*, System No. 68, Supplement, 8th Ed. Springer Verlag, Berlin.

Goldberg, D. (ed.) (1988), Scripps Institution of Oceanography, La Jolla, California: *Information presented at the EPA and ACS Metal Speciation Workshop, Jekyll Island (Georgia)*, May 1987; *Comparative Marine Chemistry of the Platinum Group Metals and Their Periodic Table Neighbors*, in: Kramer, J. R., and Allen, H. E. (eds.): *Metal Speciation*, pp. 201–217. Lewis Publishers, Chelsea, Michigan.

Goldberg, E. D., Hodge, V., Kay, P., Stallard, M., and Kokte, M. (1986), *Some Comparative Marine Chemistries of Platinum and Iridium. Appl. Geochem. 1*, 227–233; *Determination of Platinum and Iridium in Marine Waters, Sediments, and Organisms. Anal. Chem. 58*, 616–620.

Greenwood, N. N., and Earnshaw, A. (1988), *Die Chemie der Elemente*. VCH Verlagsgesellschaft, Weinheim-Basel-Cambridge-New York.

Groenendaal, W., Cooper, B. J., Gandhi, H. S., Taylor, K. C., Löwendahl, L. O., Kruse, N., Fisher, G. B., Hecker, W. C., Duprez, D., van Delft, V. C. M. J. M., and Garin, F. (1987), in: Crucq, A., and Frennet, A. (eds.), *Catalysis and Automotive Pollution Control, Proceedings of the CAPOC Symposium,* Brussels. Elsevier Science Publ., Amsterdam.

Hargreaves and Williamson (1984), *The Platinum Industry, Prospects in Recovery*. Shearson/American Express, London.

Harrison, B., Cooper, B. J., and Wilkins, A. J. J. (1981), *Control of Nitrogen Oxide Emissions from Automobile Engines – Development of the Rhodium/Platinum Three-Way Catalyst Systems, Plat. Met. Rev. 25*, 14–31.

Henschler, D. (1983), *Gesundheitsschädliche Arbeitsstoffe, Toxikologisch-arbeitsmedizinische Begründung der MAK-Werte*. Verlag Chemie, Weinheim-Deerfield Beach/Florida-Basel.

Holleck, H. (1983), *Die Konstitution ternärer Systeme der Übergangsmetalle der 4., 5. und 6. Gruppe mit Rhenium oder Platinmetallen und Kohlenstoff. Metall 37*, 475–485, 703–708.

Holzmann, H. (1967), *Über die katalytische Oxidation von Ammoniak bei der Salpetersäure-Herstellung. Chem. Ing. Tech. 39*, 89–95.

Kallman, S. (1987), *A Survey of the Determination of the Platinum Group Elements. Talanta 34*(8), 677–698.

Kemper, F. H. (1987), *Metalle – Belastung für den Menschen? Erzmetall 40*, 541–549.

Kirk-Othmer (1982), *Encyclopedia of Chemical Technology*, 3rd Ed., Vol. 18. Interscience Publisher, New York.

Koberstein, E. (1984), *Katalysatoren zur Reinigung von Autoabgasen, Chem. Unserer Zeit.* 18, 2, 37–45.

Köpf, H., and Köpf-Maier, P. (1981), *Metallocendihalogenide als potentielle Cytostatica. Nachr. Chem. Tech. Lab.* 29, 154–156.

Köpf, H., and Köpf-Maier, P. (1987), *Cytostatische Nicht-Platinmetall-Komplexe: neue Perspektiven für die Krebsbehandlung? Naturwissenschaften* 74, 374–382.

Lahmann, E., and Thron, J. (1987), *Platin und Katalysatoren – mögliche Auswirkungen auf die Umwelt.* Seminar des Bundesgesundheitsamtes *Schwermetalle in der Umwelt*, Berlin, January 1987. Gustav Fischer Verlag, Stuttgart.

Landolt, R. R., Berk, H. W., and Russel, T. H. (1972), *Toxicol. Appl. Pharmacol.* 21, 589.

Lippard, S. J. (1982), *Science 218*, 1075–1082; see also *Lecture on Platinum Anticancer Drugs – How Might They Work?* Basel, 8 January, 1987.

Lippert, B., and Beck, W. (1983), *Platin-Komplexe in der Krebstherapie. Chem. Unserer Zeit 17*, 190–199.

Lundblad, H. L. (1983), *Phosphoric Acid Fuel Cell Platinum Use Study*, NASA CR 168130, US Department of Energy, Morgantown.

Lüschow, H. M. (1980), *Probenahme, Theorie und Praxis.* Verlag Chemie, Weinheim-Deerfield Beach/Florida-Basel.

MAK (1987), *Maximum Concentrations at the Workplace, DFG Report No. XXIII.* VCH Verlagsgesellschaft, Weinheim-Basel-Cambridge-New York.

Maenhaut, W., Cantaert, C., and Cafmeyer, J. (1986), *Characterization of the Suspended Particulate Matter Inside a Platinum Catalyst Plant.* Belgian Research on Metal Cycling in the Environment, pp. 101–111. Presses Universitaires de Liège, Belgium.

Massaro, E. J., et al. (1981), *Sensitive Biochemical and Behavioral Indicators of Trace Substance Exposure: Part II. Platinum*, EPA-600/51-81-015, US Environment Protection Agency, Research Triangle Park, North Carolina.

McKelvey, V. E., and Cameron, E. N. (1973), *The Mineral Potential of the United States 1975–2000.* University Wisconsin Press, Madison.

Merian, E., Frei, R. W., Härdi, W., and Schlatter, C. (1985), *Carcinogenic and Mutagenic Metal Compounds*, esp. pp. 171ff, 220ff, 453ff, 467ff. 1. International Association of Environmental Analytical Chemistry, Geneva. Gordon & Breach, London.

Merian, E., Frei, R. W., Härdi, W., and Schlatter, C. (1988), *Carcinogenic and Mutagenic Metal Compounds 2*, p. 312. International Association of Environmental Analytical Chemistry, Villars. Gordon & Breach, London.

Monti-Bragadin, C. (1977), *Chem. Rundsch.* 30 (23), 17.

Murrer, B. (1989), *Metal Based Antitumor Drugs. Lecture Robens Institute Meeting on Toxicity and Therapeutics of Newer Metals and Organometallic Compounds*, Guildford, Surrey, UK, 24 July.

Neumüller, O.-A. (1979–1988), *Römpps Chemie-Lexikon*, 8th Ed. Frankh'sche Verlagsbuchhandlung, Stuttgart.

Nieper, H. A. (1989), *Der steuerbegünstigte Lungenkrebs. raum & zeit spezial 2.* Ehlers Verlag, München.

Parthe, E. (1972), in: Wedepohl, K. H. (ed.), *Handbook of Geochemistry*, Vol. II/3. Springer Verlag, Berlin.

Parthe, E., and Crocket, J. H. (1978), in: Wedepohl, K. H. (ed.), *Handbook of Geochemistry*, Vol. II/5. Springer Verlag, Berlin.

Pawlek, F. (1983), *Metallhüttenkunde.* Walter de Gruyter, Berlin.

Quiring, H. (1962), *Platinmetalle, die metallischen Rohstoffe.* Enke Verlag, Stuttgart.

Renner, H. (1979), *Platin Metals and Compounds* (in German), in: *Ullmanns Encyklopädie der technischen Chemie*, 4th Ed., Vol. 18, pp. 697–728. Verlag Chemie, Weinheim-New York.

Renner, H. (1987), *Behandlung und Verwertung von verbrauchten edelmetallhaltigen Katalysatoren*, in: Straub, H., Hösel, G., and Schenkel, W. (eds.), *Müll- und Abfallbeseitigung* (Müll-Handbuch), Kennzahl 8595, 2nd. Ed. Erich Schmidt Verlag, Berlin.
Renner, H., and Tröbs, U. (1984), *Rohstoffprofil Rhodium. Metall* **38**, 1002–1005.
Renner, H., and Tröbs, U. (1986), *Rohstoffprofil Iridium. Metall* **40**, 726–729.
Robson, G. G. (1987), *Platinum 1987*. Johnson Mathey, London.
Rosenberg, B., van Camp, L., and Krigas, T. (1965), *Inhibition of Cell Division in E. coli by Electrolysis Products from a Platinum Electrode. Nature* **205**, 698.
Rosner, G., and Hertel, R. F. (1986), *Gefährdungspotential von Platinemissionen aus Automobilabgas-Katalysatoren, Staub Reinhalt. Luft* **46**, 281–285.
Saager, R. (1984), *Metallische Rohstoffe von Antimon bis Zirkonium*. Bank von Tobel, Zürich.
Schrauzer, G. N. (1988), University of California, La Jolla, *Personal Information*; see also Imura, N., Naganuma, A., Satoh, M., and Koama, Y., *J. Univ. Environ. Occup. Health Jpn.* **9**, 223–229 (1987).
Schröder, H. A., and Mitchener, M. (1971), *Selenium and Tellurium in Rats: Effect on Growth, Survival and Tumors. J. Nutr.* **101**, 1531–1540.
Schwab, R. G. (1974), *Was wissen wir über die tieferen Schichten der Erde?, Angew. Chem.* **86**, 612–624.
Sideris, E. G., Hoffmann, R. L., Lakhanisky, T., and Villani, G. (1986), *European Environmental Mutagenic Society, Bruxelles*.
Stopinski, O. (1977), *Platinum-group Metals*. National Research Council, Washington, DC.
Tafel, V. (1951), *Lehrbuch der Metallhüttenkunde*, Vol. 1. S. Hirzel Verlag, Leipzig.
Thomson, A. J. (1977a), *The Mechanism of Action of Anti-tumor Platinum Compounds. Plat. Met. Rev.* **21**, 2–15.
Thomson, A. J. (1977b), *Platinverbindungen für die Krebstherapie. Nachr. Chem. Tech. Lab.* **50**, 20–23.
Trueb, L. (1989), *Platinmetalle – Vom Bergwerk zum Katalysator. Neue Zürcher Zeitung No. 134 (14 June), Forschung und Technik, Zürich*; see also *Das Platin und seine fünf Geschwister, zu einer Geschichte der Platinmetalle. Neue Zürcher Zeitung* No. 298 (21 December, 1983), Forschung und Technik, Zürich.
Ullmann, (1980), *Encyklopädie der technischen Chemie*, 4th Ed., Vol. 5. Verlag Chemie, Weinheim-New York.
Welz, B. (1985), *Atomic Absorption Spectrometry*, pp. 289ff, 310ff, 311ff, 314ff, 317ff, 319. VCH Verlagsgesellschaft, Weinheim-Deerfield Beach/Florida-Basel.
Williams, D. F. (1989), *Toxicological and Therapeutic Potential of Metallic Surgical Implants. Lecture Robens Institute Meeting on Toxicity and Therapeutics of Newer Metals and Organometallic Compounds, Guildford, Surrey, UK,* 24 July.
Wogrinz, A. (1936), *Analytische Chemie der Edelmetalle*. Enke Verlag, Stuttgart.
Yang, J. S. (1989), *The Comparative Chemistry of Platinum Group Metals and Their Periodic Neighbors in Marine Macrophytes. Proceedings Heavy Metals in the Environment, Geneva*, Vol. 2, pp. 1–4. CEP Consultants Ltd., Edinburgh.

II.25 Selenium

LAWRENCE FISHBEIN, Washington, DC, USA

1 Introduction

There is increasing recognition that selenium is an important metalloid with industrial, environmental, biological, and toxicological significance. Environmental selenium compounds may originate from metal smelting, coal combustion, or the disposal of wastes, but speciation must always be considered. Some plants accumulate selenium. It is an essential element in many species, including humans, in which it is a component of glutathione peroxidase which is required for the metabolism and removal of hydrogen peroxide and lipid peroxidases from cells. There is a narrow range of selenium intake that is consistent with health; outside of this range, deficiency diseases and toxicity occur.

The toxicology of selenium and its compounds, often conflicting and controversial, is of continuing importance for a variety of reasons including: (1) the long-established selenium poisoning of domestic animals foraging on seleniferous plants; (2) disorders in humans and animals resulting from selenium deficiencies and excesses; (3) the nutritional essentiality of the element; (4) the protective effect of selenium against metal toxicity and the metabolic interactions between the element and vitamin E and other antioxidants; (5) the reported carcinogenicity as well as anticarcinogenicity, antimutagenicity and anticlastogenicity; and (6) the increasingly reported and detailed inverse relationship between the dietary intake of selenium and cancer incidence and mortality.

2 Physical and Chemical Properties, and Analytical Methods

2.1 Physical and Chemical Properties

Selenium belongs to the VIa Group of the Periodic Table, located between sulfur and tellurium, and it possesses chemical and physical properties which are intermediate between metal and non-metal, although the resemblance between selenium and sulfur is more pronounced in a number of respects than that between selenium and tellurium. The atomic number and mass of the element are 34 and 78.96, respectively; it has a melting point of 217 °C and a boiling point of 685 °C. Between 74 and

217 °C, exothermic modifications of crystalline forms are observed (WIESE, 1982). Pure selenium, like sulfur, is in fact allotropic and exists in three forms, e.g., a gray or "metallic" thermodynamically stable hexagonal form, a red monoclinic form and a vitreous amorphous form. The electrical conductivity of the "metallic" form, which is a semi-conductor, increases with temperature and principally with exposure to light (NAZARENKO and ERMAKOV, 1972).

There are six natural isotopes which exist under normal conditions:

Isotope	^{74}Se	^{76}Se	^{77}Se	^{78}Se	^{80}Se	^{82}Se
Abundance (%)	0.87	9.02	7.58	23.52	49.82	9.19

Additionally, there are ten short-lived man-made isotopes of which ^{75}Se, ^{77m}Se, and ^{82m}Se are the most utilized in neutron activation and radiology (NEWLAND, 1982).

Selenium can exhibit four valence states: $-2, 0, +4$, and $+6$. Most of the organic and inorganic complexation chemistry involves the -2 oxidation state, and selenium often replaces sulfur (e.g., selenomethionine and selenocysteine; NAZARENKO and ERMAKOV, 1972; NEWLAND, 1982). Selenium chemistry is to a large extent nonmetallic with hydrogen and most metals reacting directly with selenium(-2). Metal selenides are practically insoluble. Selenium compounds in which selenium has oxidation states of $+4$ and $+6$ include: halides such as SeF_4 and SeF_6, selenium dioxide SeO_2, selenium trioxide SeO_3, and oxy acids such as $OSe(OH)_2$ and selenic acid H_2SeO_4. Many of the oxy salts (e.g., selenites and selenates) are soluble in water (GLOVER et al., 1979; NEWLAND, 1982).

2.2 Analytical Methods

Since selenium is often present in both environmental and biological samples in very low concentrations, highly sensitive analytical techniques as well as extra precautions in the collection, storage and preparation of samples are needed to prevent loss from volatilization or contamination (SHENDRIKAR, 1974; GLOVER et al., 1979; EINBRODT and MICHELS, 1984; FISHBEIN, 1985).

The methods generally employed for analyzing selenium in biological and environmental materials initially involve the destruction of organic constituents with concurrent oxidation of the element to the tetravalent or hexavalent state and its subsequent determination by a variety of techniques including: atomic absorption spectrometry, fluorimetry, spectrophotometry, neutron-activation analysis, plasma emission spectrometry, gas chromatography, and to a lesser extent X-ray fluorescence and polarography (NAZARENKO and ERMAKOV, 1972; GLOVER et al., 1979; RAPTIS et al., 1980a, b, 1983; VICKREY and BUREN, 1980; BEM, 1981; VERLINDEN et al., 1981; NÜRNBERG, 1982; FISHBEIN, 1985; ROBBERECHT and VAN GRIEKEN, 1980, 1982; COOPER, 1974; COOPER et al., 1974; OLSON et al., 1973; OLSON, 1976; EINBRODT and MICHELS, 1984; WELZ, 1985; NEVE and MOLLE, 1986a, b; NORHEIM and HAUGEN, 1986; SAEED, 1986; WELZ and VERLINDEN, 1986; WHO, 1987; THOMASSEN et al., 1988).

2.2.1 Chromatographic Methods

Tetravalent selenium and various aromatic o-diamines may react to form piazselenols (INHAT, 1974; GLOVER et al., 1979; FISHBEIN, 1985) which can be extracted into organic solvents and measured by gas chromatography and detected by electron capture or microwave emission spectrometry. The o-diamines which have been used include: 2,3-diaminonaphthalene (YOUNG and CHRISTIAN, 1977) and the 4-chloro- (MAKIDA et al., 1979), 4,5-dichloro- (STIJVE and CARDINALE, 1975), 4-nitro- (SLU and BERMAN, 1983), and 3,5-dibromo- (SHIMOISHI, 1977) derivatives of 1,2-diaminobenzene.

Selenium recovery ranges from 75–90% and was assessed by using a ^{75}Se-labeled tracer for liquid scintillation spectrometric assay. Gas chromatographic conditions allowed the detection of selenium concentrations below a ppb, and the mean deviation and relative accuracy averaged 2.3 and 3.4%, respectively. This method has been used in human population studies to assess selenium-mercury correlations and to examine the selenium content and form in specific protein functions of fish muscle.

The determination of selenium (in the form of a stable and volatile piazselenol complex) in environmental samples (biological tissues, coal, fly ash, and scrubber solutions) using gas chromatography with a microwave emission spectrometric (MES) detection system was described by TALMI and ANDREN (1974). Although hexavalent selenium (selenate) is also present in biological materials, relatively few methods are available for its determination. These include spectrophotometry (DESAI and PAUL, 1981), fluorimetry (SUGIMURA and SUZUKI, 1973) and gas chromatography (SHIMOISHI, 1977). Organoselenium and selenite are determined by digesting the sample in concentrated nitric acid, and the total selenium is determined by further treatment of the digest with hydrochloric acid. The difference between the two values obtained represents the selenate content.

In liver and other animal tissues, inorganic forms of selenium such as sodium selenite($+4$) may be reduced to hydrogen selenide(-2) through a combination of non-enzymatic and enzymatic reactions involving glutathione and glutathione reductase (GANTHER, 1979; SUNDE and HOEKSTRA, 1980). A recent procedure for the identification of hydrogen selenide and other volatile selenols (methyl selenol) by derivatization with 1-fluoro-2,4-dinitrobenzene (GANTHER and KRAUS, 1984) involved the initial generation of the volatile selenide by reducing selenious acid or dimethyl selenide with zinc dust and hydrochloric acid.

The possible metabolic interrelationships between the organic and inorganic forms of selenium mostly involve reduction and methylation. Trimethylselenonium ion is a important urinary metabolite at doses of selenite insufficient to trigger the respiratory excretion of dimethyl selenide (BLOTCKY et al., 1985). The determination of trimethyl selenonium ion (TMSe) in urine has generally been accomplished by the use of the radiotracer ^{75}Se (KUBO et al., 1977) and more recently by initially employing anion exchange-cation exchange chromatography, selectively capturing the TMSe on a cation exchange resin followed by irradiation of this fraction with neutrons and final radioassay for ^{77m}Se activity (BLOTCKY et al., 1985).

2.2.2 NAA Techniques

Because of its high sensitivity ($10^{-8}-10^{-9}$ of Se) neutron activation analysis (NAA) is used widely for the determination of selenium in biological materials as well as in environmental water, and 17.5 s ^{77m}Se, 18.6 min ^{81}Se, or 120 day ^{75}Se may be measured (IAEA, 1972; GLOVER et al., 1979; NEWLAND, 1982). The most important step in the NAA of water is the enrichment of selenium via selective extraction, non-selective extraction, or freeze-drying (ROBBERECHT and VAN GRIEKEN, 1982).

2.2.3 Atomic Absorption Spectrometry
(see also WELZ, 1985)

Routine methods for selenium determination which are applicable to the analysis of plant materials include colorimetry, fluorimetry, atomic fluorescence using hydride generation (TSUJII and KUGA, 1974; THOMPSON, 1975), flame atomic absorption and hydride generation (BROOKS et al., 1983; HAHN et al., 1981), and flameless atomic absorption (WAUCHOPE and MCWHORTER, 1977; BROOKS et al., 1983). A summary of methods for the determination of selenium in foodstuffs, the source that contributes most to the intake of selenium, has been provided by HOFSOMMER and BIELIG (1981). The detection of selenium in food by hydride generation and atomic absorption spectrometry has been accomplished with a detection limit of about 2 ng selenium (BROOKS et al., 1983). The determination, speciation, and concentration levels of selenium in all types of environmental waters have been reviewed by ROBBERECHT and VAN GRIEKEN (1982) with particular stress on the basic difficulties, the efficacy of methods, and the preconcentration steps explained in view of the sub-µg/L concentration levels. Additionally published data on speciation and concentration levels in various water samples have been critically reviewed.

A procedure for studying the chemical forms of selenium in sediments and planktonic material has been developed that uses a multistep nitric/perchloric acids digestion to solubilize total selenium and a weak sodium hydroxide treatment to release selenite and selenate. The solubilized selenium species are determined by a selective hydride generation/atomic absorption technique. The detection limit for total particulate selenium is 10 ng/g using a sample of 0.2 g (CUTTER, 1985; ALEXANDER et al., 1988).

The selenium in the gas phase has been determined unambiguously relatively infrequently in environmental samples. Total selenium vapor in ambient air was reported to span a wide range, $0.006-5$ ng/m^3 (MOSHER and DUCE, 1983), and the separation of different species of alkylated selenium compounds from air has also been reported (REAMER and ZOLLER, 1980; CHAU et al., 1976; JIANG et al., 1982, 1983; WONG and GOULDEN, 1975).

Since total airborne selenium levels are very low, an efficient technique that will preconcentrate the analysis is required. The common air sample procedure based on cryogenic trapping (WONG and GOULDEN, 1975; JIANG et al., 1982) has limited utility. A recent procedure for the determination of total selenium (as well as tellurium) in the gas phase employed an initial adsorption on gold-coated beads and charcoal. The traps were then boiled in dilute HCl, and the resulting solution was analyzed for selenium and tellurium by graphite furnace atomic absorption spectrometry. Detection limits down to about 0.1 ng/m^3 allowed the ready detection of selenium and tellurium in rural air with a precision of about ±4% and ±6%, respectively, at the nanogram level (MUANGNOICHAROEN et al., 1986).

MACPHERSON et al. (1988) have recently described a comparison of methods for the determination of selenium in biological fluids. Direct determination of selenium in biological material by graphite furnace AAS suffers from problems such as preatomization losses and spectral interferences. These problems have largely been overcome by oxygen ashing in graphite tubes and Zeeman effect background correction for an accurate direct determination of serum selenium by graphite furnace AAS in the presence of a copper-magnesium matrix modifier (NEVE and MOLLE, 1986a, b). Recent determinations of selenium in human body fluids and tissues employ HGAAS (hydride-generation atomic absorption spectrometry) (NORHEIM and HAUGEN, 1986; WELZ and VERLINDEN, 1986; ANGERER and SCHALLER, 1988) and electrothermal AAS (SAEED, 1986) have been described.

Blood selenium levels have been widely used as indices of dietary selenium intake and selenium status in both animals and humans (LEVANDER, 1972, 1982, 1986). Red cell selenium level is used to monitor longer-term selenium status because of the relatively long life-span of erythrocytes while plasma or serum Se-levels are shorter-term indicators (ROBINSON and THOMSON, 1983). It should be noted that no one method is entirely satisfactory for measuring the *bioavailability* of selenium. The more rapid assays of glutathione peroxidase activity are of limited use as markers of selenium status and are complementing rather than replacing measurement of selenium in tissues and biological fluids. Because of the growing emphasis on the interrelationships of Se with other nutrients, a whole battery of measurements is being increasingly employed for a more precise measurement of selenium status (ROBINSON and THOMAS, 1983; BURK et al., 1981; BURK, 1986; MUTANEN, 1986; LEVANDER, 1986; VAN FASSEN et al., 1986; VAN DEN HAUTE et al., 1986; NEVE and MOLLE, 1986a, b; LUTEN et al., 1986). The relationship between urinary selenium excretion (little is known concerning the chemical nature of the many selenium compounds in urine) and selenium intake has been recently reviewed (NAHAPETIAN et al., 1983; ROBBERECHT and DEELSTRA, 1984; ROBINSON and THOMSON, 1983; LEVANDER, 1986).

Hair analysis of selenium has been used in animals to diagnose both deficiency and toxicity (HIDIROGLOU et al., 1965), and in China there have been suggestions of a close relationship between hair and blood selenium levels in people living in areas either affected or unaffected by Keshan disease (CHEN et al., 1980). However, hair analyses to assess selenium status in Western countries have not been employed extensively since hair from individuals can be highly contaminated with Se-containing anti-dandruff shampoos (DAVIES, 1982; LEVANDER, 1986) or other Se-containing impurities.

3 Sources, Production, Important Compounds, Uses, Waste Products

The earth's crust contains an average selenium content of 0.05–0.09 ppm, approximating that of cadmium and antimony and ranking above molybdenum, silver, mercury, and uranium. Higher concentrations of selenium are found in volcanic rock (up to 120 ppm), sandstone, uranium deposits and carbonaceous rocks (ELKIN, 1982; FISHBEIN, 1983; EINBRODT and MICHELS, 1984; MERIAN, 1985). There are approximately 40 minerals in which selenium predominates. Selenium is recognized as a major constituent in at least 22 selenides, six sulfosalts, one oxide, four selenites, and one selenate, and as a minor constituent of 24 sulfides and tellurides. Although higher levels of selenium are found in berzelianite (Cu_2Se), tiemannite (HgSe), and naumannite (Ag_2Se), these are not considered ore material since selenium is produced from minerals mined for their other constituents (e.g., copper, lead, silver, etc.) (COOPER et al., 1974; NAZARENKO and ERMAKOV, 1972; SINDEEVA, 1964; WILBER, 1983; WIESE, 1982).

Principal sources of selenium are thus copper sulfide ores in Canada, the United States, Bolivia, and the Soviet Union (COOPER et al., 1974; EINBRODT and MICHELS, 1984). The Se/S ratio of magmatic sulfides is of the order of 1:10 (PETERSON et al., 1981; MERIAN, 1985). Selenium concentrations can be elevated in the vicinity of copper and lead smelters and refineries, and superphosphates can also contain about 20 ppm selenium (PETERSON et al., 1981; MERIAN, 1985).

More than 90% of the U.S. selenium output and more than 8% of the world's production is derived from the anode mud deposited during electrolytic refining of copper. Copper anodes contain 0.5 to 280 kg of selenium per ton of copper (AGETON, 1970), but only a fraction of the selenium can be recovered. Precipitated fly dust and slug are also used for selenium recovery (WIESE, 1982; JENNINGS and YANNOPOULOS, 1974). All commercial processes for the production of selenium may be considered as modifications or combinations of four fundamental methods, e.g., smelting with soda ash, roasting with soda ash, direct oxidation, and roasting with sulfuric acid. World production of selenium is of the order of 1300 to 1500 tons annually (HÖGBERG and ALEXANDER, 1986; NEWLAND, 1982; MERIAN, 1985; ALEXANDER et al., 1988). Secondary sources for recovery of selenium include: factory scrap generated during manufacture of selenium rectifiers, burned-out rectifiers, spent catalysts, and used xerography copying cylinders. However, applications of selenium in the chemical industry are considered largely dissipative. Important selenium compounds include selenium dioxide, selenium disulfide, cadmium selenide and sulfoselenide, sodium and other selenites, and some organic selenium compounds.

Selenium is widely employed in industrial products and processes in the glass industry (flat-glass, pressed or blown glass, and glassware, coloration and reduction of infra-red transmission) using approximately 27–30% (GLOVER et al., 1979; WIESE, 1982; MERIAN, 1985). Electronic uses of selenium in high purity form are related to its semiconductor and photoelectric characteristics and account for approximately 25–35% of the use pattern of the element. Hence, it is widely employed in semiconductors, thermoelements, photoelectric and photocells and xerographic materials.

Table II.25-1. Some Selenium Compounds and Their Uses

Selenium	Rectifiers, photoelectric cells, blasting caps; in xerography, stainless steel; as a dehydrogenation-catalyst
Tellurium-selenium alloys	Erasing optical stores (see Chapter II.28)
Sodium selenate (Na_2SeO_4)	As insecticide; in glass manufacture; in veterinary pharmaceuticals
Sodium selenite (Na_2SeO_3)	In glass manufacture; as a soil additive for selenium deficient areas
Selenium diethyldithiocarbamate	Fungicide; vulcanizing agent
Selenium monosulfide (SeS)	In veterinary medicine; dandruff removal
Selenium disulfide (SeS_2)	In veterinary medicine; dandruff removal
Selenium dioxide (SeO_2)	Catalyst for oxidation, hydrogenation or dehydrogenation of organic compounds
Selenium hexafluoride (SeF_6)	As gaseous electric insulator
Selenium oxychloride ($SeOCl_2$)	Solvent for sulfur, selenium, tellurium, rubber, bakelite, gums, resins, glue asphalt, and other materials
Aluminum selenide (Al_2Se_3)	Preparation of hydrogen selenide for semiconductors
Ammonium selenite [$(NH_4)_2SeO_3$]	Manufacture of red glass
Cadmium selenide (CdSe)	Photoconductors, photoelectric cells, rectifiers
Cupric selenate ($CuSeO_4$)	In coloring copper and copper alloys
Tungsten diselenide (WSe_2)	In lubricants

Approximately 14% of selenium is used in inorganic pigments (principally as cadmium sulfoselenide used in plastics, paints, enamels, inks, rubber, and ceramics). An additional 10–15% of selenium is used in a broad spectrum of applications including: accelerators and vulcanizing agents in rubber production, in stainless steel and as selenides of refractory metals for use in lubricants. Some of these products may end up as disposed waste. Medical uses of selenium have been restricted in the past to tropical application, e.g., for treatment of dandruff, for which the compounds are usually in the form of selenium mono- and disulfides. Selsun Blue^{TM} (1% mixture) and Selsun Red^{TM} (2.5% mixture) are the major products. Additionally, there is limited use of radioactive selenium, [^{75}Se]-selenomethionine, as a diagnostic scanning and labeling agent for various malignant tumors (NAS, 1976; NEWLAND, 1982). Table II.25-1 lists some selenium compounds and their uses.

4 Distribution in the Environment, in Food, and in Living Organisms
(Table II.25-2)

4.1 Air (Emissions and Quality)

A small fraction of selenium may exist in the gaseous state in the atmosphere (see also Sect. 2.2.3). However, the majority of atmospheric selenium is expected to be present in particulate form (NAS, 1976). Although chemical reactions in the troposphere may cause speciation of selenium, removal of atmospheric selenium

Table II.25-2. Selenium in the Environment: Summary of Representative Values (Range of Values in Parentheses)

Source	Concentrations		Source (Man)	Transfer Rates	
Atmosphere			Intake		
urban	3 ng m^{-3}	(0.01–30)	ingestion	70 µg d^{-1}	(10–220)
rural	1.3 ng m^{-3}	(0.01–3)	inhalation – urban	0.07 µg d^{-1}	(0.0002–0.7)
			– rural	0.03 µg d^{-1}	(0.0002–0.07)
Lithosphere			Absorption		
agricultural soil	0.4 µg g^{-1}	(0.1–2)	GI tract	0.8	
Hydrosphere			lungs – retention	0.35	
freshwater	0.2 µg L^{-1}	(0.02–10)	– absorption	0.6	
ocean (less at the surface)	0.2 µg L^{-1}	(0.03–0.2)			
Biosphere					
primary accumulator plants	1000 µg g^{-1}	(100–5000)			
secondary accumulator plants	100 µg g^{-1}	(10–500)			
food crops	0.1 µg g^{-1}	(0.05–1)			
Man					
tissues	0.1 µg g^{-1}	(<0.05–5)			
blood	0.07–0.15 µg mL^{-1}	(0.05–0.5)			

occurs primarily through wet and dry deposition and its association with submicrometer aerosols (MOSHER and DUCE, 1983; NAS 1976; JICKELLS et al., 1984).

Selenium is thought to be emitted naturally into the atmosphere as volatile dimethyl selenide (JIANG et al., 1983), selenium dioxide, or elemental selenium, which may be released during fossil fuel combustion (ANDREW et al., 1975), since selenium is present in inorganic forms in coal at concentrations ranging from less than 1 to about 10 ppm (NAS, 1976; PETERSON et al., 1981; EINBRODT and MICHELS, 1984; WIESE, 1982). Selenium dioxide may again be hydrated or reduced (by SO_2 present) or oxidized (by air). In rain water samples from Japan (SUZUKI et al., 1981) and California (CUTTER, 1978) selenite is the major selenium species, while rain and snow in urban Belgium contain variable quantities of both selenite and selenate (ROBBERECHT and VAN GRIEKEN, 1980).

Recent studies by CUTTER and CHURCH (1986) indicate that fossil fuel combustion enriches selenite in precipitation over the western Atlantic relative to sub-surface waters (average sub-surface selenite/selenate = 0.09). As a result, inputs of selenite and selenate through wet deposition and upwelling have different effects on parameters such as surface-water residence times for these two species (CUTTER and BRULAND, 1984).

Selenium concentrations in air are of the order of a few ng/m^3, the levels depending primarily on coal-burning power plants, copper refining plants or selenium rectifier plants in the vicinity (SALMON et al., 1978; HÖGBERG and ALEXANDER, 1986).

Table II.25-3 illustrates emissions of selenium for all sources in 28 European countries in 1979 (PACYNA et al., 1984). Approximately 420 tons of selenium were released into the atmosphere of Europe in 1979, with 373 tons of that from coal and oil combustion; refuse incinerators contributed another 32 tons, and 13 tons were

Table II.25-3. Emissions of Selenium from All Sources in Europe in 1979 (tons/year)[a]

Country	Selenium	Country	Selenium
Albania	0.5	Italy	24.0
Austria	4.5	Netherlands	7.9
Belgium	11.4	Norway	1.2
Bulgaria	9.7	Poland	37.0
Czechoslovakia	18.0	Portugal	1.4
Denmark	3.8	Romania	13.1
Finland	4.1	Spain	10.9
France	18.0	Sweden	5.4
German Democratic Republic	24.1	Switzerland	0.6
Federal Republic of Germany	46.6	Turkey	5.1
Greece	3.1	USSR	120.0
Hungary	4.6	United Kingdom	36.0
Iceland[b]	53.0	Yugoslavia	7.6
Ireland	1.0	Luxemburg	0.2

[a] Total 420 tons/year
[b] kg/year

emitted from zinc-cadmium smelters (PACYNA et al., 1984). In the United States, combustion of fossil fuels accounts for about 62% of total selenium emissions, while non-ferrous metal smelting and refining, and glass and ceramics manufacturing account for 26% and 5%, respectively (NAS, 1976; BENNETT, 1983). Recent studies have suggested a moderate enrichment factor (EF) for selenium calculated with respect to crustal abundance with similarities for EF in both fly ash from a number of coal-fired power plants and refuse incinerators in the U.K. (WADGE et al., 1986).

Additional less significant sources of atmospheric selenium may be of natural origin such as volcanic emissions and volatile releases from plants, soils, and animals. Newspaper and cardboard contain 6 to 9 ppm selenium which may be dispersed with stack emission during incineration (PETERSON et al., 1981). Other possible sources of selenium in the atmosphere include the incineration of rubber tires, which have a reported selenium concentration of between 0.7 and 2.0 ppm (HASHIMOTO et al., 1970).

The total (volatile and particulate) anthropogenic emission of selenium reported by ROSS (1985) to be 6700 to 8300 tons per year is in good agreement with the estimate of 3020 to 9625 tons per year made by NRIAGU and PACYNA (1988), who estimated that volatile selenium accounts for about 40% of the selenium released. These figures were calculated from statistics on global production or consumption of industrial goods (including combustion) and from emission factors. Another viewpoint was expressed by MOSHER and DUCE (1987) who estimated mean global emission of selenium compounds from natural sources to be 6000 to 13 000 tons per year (60–80% of marine biogenic origin). PETERSON (1987) studied the site-dependent and seasonal flux from land to atmosphere (release and re-absorption) and described the difficulties in estimating global cycling of selenium because the contribution of fires, the role of plants, reactions with microorganisms, the effect of temperature on oxidation and reduction, and the ambiguity of selenium soil profile studies are all not sufficiently understood.

The concentration of alkyl selenides (principally dimethyl selenide) in air is about 1 ng/m^3, depending on location (JIANG et al., 1983). Dimethyl selenide and dimethyl diselenide are the principal compounds released from soils, lake sediments, and sewage sludge through microbial activity (REAMER and ZOLLER, 1980; PETERSON, 1987). High concentrations of alkyl selenides have been reported near lakes, sea shores, and sewage treatment works where concentrations of over 5 ng/m^3 have been reported (REAMER, 1980). The average ratio of selenium to sulfur in air is about 10^{-4}, the same ratio as in fossil fuels (HASHIMOTO et al., 1970; BENNETT, 1983). An extrapolated minimum background level (at the zero level for sulfur) of selenium in rural European air has been estimated at 0.2 ng/m^3 (PRIEST et al., 1981; BENNETT, 1983). The residual selenium levels may be attributed to sources not associated with sulfur (e.g., volatile releases from plants, soils and microorganisms), as noted above. An additional source of selenium in air which may be of importance is cigarette smoke, where it has been found at levels of 0.001 to 0.063 µg/cigarette (JENKINS, 1985; NORMAN, 1977), originating from the tobacco, rather than the paper.

4.2 Water Quality
(Fig. II.25-1)

The major sources of selenium pollution in aquatic ecosystems, including the oceans, are domestic waste water effluents, coal-burning power plants, and non-ferrous metal smelters. Mobilization of selenium into the biosphere, as caused by man, is estimated by NRIAGU and PACYNA (1988) to be 79000 tons per year (median value of the terrestrial plus aquatic inputs, minus atmospheric emissions; soils also receive large quantities).

Although selenium levels are generally low (less than 10 ppb) in ground and surface waters (see also EINBRODT and MICHELS, 1984), it should be noted that depending on geological factors, ground water may reach much higher concentrations (up to 6000 ppb) (GLOVER et al., 1979). A survey by STRAIN et al. (1981) of 3000 first-draw water samples in the U.S. showed that about 18% of the samples (or about 540) had a selenium content that exceeded the EPA standard of 10 ppb, and the highest measured level was 450 ppb. These same samples showed elevated concentrations of elements associated with corrosion of household plumbing by soft, acidic water. The potential impact of acid precipitation on selenium was recently noted by MUSHAK (1985) who suggested that altered acidity by both soil and aquatic systems poses a risk for altered biotransformation processes. Lowered pH in soils appears to reduce selenium availability. In the presence of iron, selenium precipitates as insoluble ferric selenite in water of less than pH 7, while at high pH the selenites may be

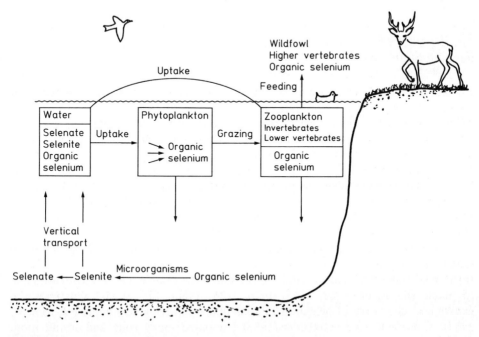

Fig. II.25-1. Biogeochemical cycling of selenium in aquatic systems.

oxidized to soluble selenates, hence increasing the concentrations in water up to 10–400 ppb (GLOVER et al., 1979; NEWLAND, 1982).

Recently, high levels of selenium (as evidenced by selenium toxicosis in waterfowl) have been reported in the drainage waters from irrigated lands in a large portion of California's San Joaquin Valley, which contains some of the major agricultural counties in the United States. It seems that drainage problems involve waterlogging and associated accumulation of excess salts in the irrigated crop-root zone (TANJI et al., 1986). According to KNIGHT et al. (1987) the agricultural drain water contains up to 1.35 ppb selenium, which may be reduced, methylated and mineralized, but also bioconcentrated (see also Sect. 6.1).

Selenium may also be found in waste water streams of coal-fired power plants, ore refineries, and industrial plants (e.g., selenium rectifier production) where selenium is utilized. Stack and quench waters in industry may contain 14 ppb Se (NEWLAND, 1982). Levels of selenium in fresh waters suggest a median concentration of 0.2 ppb with a range of 0.02 to 1 ppb (BENNETT, 1983; EINBRODT and MICHELS, 1984).

Concentrations of selenium in sea water are generally low (e.g., 0.004 ppb at the surface and 0.06 ppb in the deep ocean in the form of selenites, and in the form of selenates, 0.03 ppb at the surface and 0.12 in deep waters; BRULAND, 1983). It has been estimated that 7700 to 8000 tons of selenium are introduced into the sea annually (NEWLAND, 1982; MERIAN, 1985). BRULAND (1983) also described deselenification (comparable to denitrification) in the Pacific and decreased concentrations in anoxic bottom waters.

4.3 Soil Contamination

(see also Section 6.6.2)

Though selenium is ubiquitous, its distribution is uneven (WHO, 1987). Selenium concentrations in soil are in fact extremely variable, ranging from 0.1 µg/g in selenium-deficient areas to over 1 part per thousand in seleniferous areas. A continental-wide survey of selenium in the soil in the United States (912 samples) gave a range of 0.1 to 4.3 ppm with a geometric mean of 0.3 ppm (SHACKLETTE et al., 1974). Cultivated surface soils have on average a somewhat lower concentration of 0.5 ppm. The notable seleniferous areas, such as in Ireland and parts of the United States, and selenium-deficient areas, such as in parts of New Zealand, China, and Finland, reflect the wide geographic variations in the natural occurrence of selenium (BERROW and BURRIDGE, 1980; BENNETT, 1983).

Selenium is bound in acid soils as ferric selenite of very low solubility and hence reduces the availability of the element to vegetation (SHAMBERGER, 1986). In contrast, in alkaline soils, more selenium is present in the soluble selenate form available to plants (BENNETT, 1983; NAS, 1976; EPA, 1986). Fig. II.25-2 illustrates the generalized chemistry of selenium in soils and weathering sediments (NEWLAND, 1982). ELRASHIDI and ADRIANO (1987) compared many soils and found more SeO_3^{2-} at high pH with high redox potential, more SeO_3^{2-} at neutral pH with

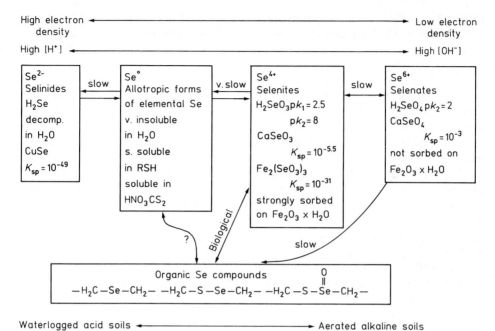

Fig. II.25-2. Generalized chemistry of selenium in soils and weathering sediments.

medium redox potential, and more Se^{2-} at low pH with low redox potential. Sandy loams contain $HSeO_3^-$ and $FeHSeO_3$.

4.4 Food Chain, Plants, Animals, and Humans
(see also Chapter I.8; Figs. II.25-1 and II.25-3)

The principal source of selenium for most individuals is the diet, and the uptake of the element from food depends largely on the chemical form of selenium present (NAS, 1979; GLOVER et al., 1979; FISHBEIN, 1983, 1986a; MEDINSKY et al., 1985; LO and SANDI, 1980; BENNETT, 1983; EINBRODT and MICHELS, 1984; LEVANDER, 1986). Various plants growing on selenium-rich soil absorb and also accumulate selenium. In plants it is found as organic compounds, including amino acids that contain selenium. Selenomethionine has been shown to be the predominant form of Se in wheat, soybeans, and selenium-enriched yeast (NAS, 1983; GLOVER et al., 1979; VERNIE, 1984; BEILSTEIN and WHANGER, 1986). The selenium content of plants differs widely among species as exemplified by accumulator plants (generally non-food plants) which can incorporate 1000 – 10000 ppm of selenium per kg of dry weight (VERNIE, 1984; GLOVER et al., 1979). Occasionally unusually high concentrations of selenium are reported in some legumes, nuts, and mushrooms. Secondary selenium absorbing plants contain 50 – 500 ppm, while most plants, grains, and grasses rarely contain more than 30 ppm, the general range being 0.05 to 1 ppm

(BENNETT, 1983). As already mentioned, tobacco contains 0.1 to 1 ppm selenium. KNIGHT et al. (1987) reported on cases of rapid selenium accumulation in top aquatic trophic levels due to bioaccumulation and translocation through aquatic food chains, leading, for instance, to a decrease of algal cell replication.

The concentration of selenium in individual food products is influenced by the type of food, its origin and processings (NAS, 1976; GLOVER et al., 1979; EINBRODT and MCHELS, 1984; VERNIE, 1984; LO and SANDI, 1980; WHO, 1970; ROBINSON and THOMSON, 1983). Most dietary intake of selenium occurs via plant foods, with the major exception being fish and certain seafoods. Recent total diet studies reported the following selenium levels: 0.07 ppm for dairy products; 0.20 ppm (0.10–0.40 ppm) for meats, fish and poultry; 0.24 ppm (0.10–0.40 ppm) for grains and cereal products, and trace amounts in the other food groups (JOHNSON and MANSKE, 1977; MAHAFFEY et al., 1975; MORRIS and LEVANDER, 1970; KARLSEN et al., 1981; LO and SANDI, 1980; ROBINSON and THOMSON, 1983).

Human spleen, lung, myocardium, skeletal muscle, and brain contain decreasing amounts of selenium, and fat contains practically none (GLOVER et al., 1979; BENNETT, 1983; GANTHER, 1986). Regional differences of selenium content in individuals are expected, reflecting variable dietary intake. Whole body selenium content has been found to be approximately 3–6 mg in a low selenium region in New Zealand (STEWART et al., 1978) and about 13 mg in places where there are sufficiently high dietary intakes of Se (ICRP, 1975). The normal average blood Se-concentration is about 70–150 µg/L; at birth it may be higher, then it drops slightly up to 20 years of age. There are, however, great individual variations ($\pm 50\%$). Control parameters are serum selenium levels, Se in erythrocytes, glutathione peroxidase in erythrocytes and serum (VON STOCKHAUSEN, 1986; EINBRODT and MICHELS, 1984; WASOWICZ and ZACHARA, 1987; ANGERER and SCHALLER, 1988). THORLING et al. (1986) have reported recently on a collaborative study on the selenium status in Europe (in a cancer preventive study), and confirmed that serum levels in healthy individuals not employed in industry, of between 20 and 65 years old, from 17 locations in 10 countries, ranged from 63 µg/L in Greece to 109 µg/L in England (London), the average being 85 µg/L. The predominant chemical form of selenium in mammalian tissues appears to be selenocysteine which may arise from the conversion of selenomethionine (BEILSTEIN and WHANGER, 1986).

Several studies on daily dietary intake of selenium in different countries suggest estimates ranging from 28–30 µg/day (New Zealand) to 326 µg/day (Venezuela) (NAS, 1976; GLOVER et al., 1979; LEVANDER, 1976, 1987). The estimated normal dietary intake of selenium for humans in most parts of the world ranges from 4–35 µg/person in infants to 60–300 µg/person in adults (LO and SANDI, 1980; KAZANTZIS, 1981). A human requirement of about 60–120 µg Se per day has been suggested to prevent deficiency diseases (FOOD AND NUTRITION BOARD, 1977; VERNIE, 1984). Based on more recent human data, LEVANDER (1987) suggested a safe and adequate daily intake of about 80 µg/day for men and 55 µg/day for women (1 g/kg/day is better). However, humans can adapt, as evidenced by people with Keshan and Kasin-Beck disease in China, where uptake in the affected person is about 7 µg/person/day and from studies in Finland (MUTANEN and KOIVISTOINEN, 1983; ALFTHAN, 1986), where daily intake varies between 21 and 56 µg/person/day (depending on the success of the domestic crop). It is now considered that the minimum daily selenium requirement is about 14–19 µg/day, while the physiological requirement, based on the saturation of plasma GSH-P_x activity may be 40 µg/day (LEVANDER, 1987). Based on a lack of definitive toxic effects on man from selenium of food origin, it has been variously estimated that 500 µg/man/day (LO and SANDI, 1980) and 1000–1500 µg/man/day (KOLLER and EXON, 1986) may be regarded as maximum tolerable levels.

5 Uptake, Absorption, Transport and Distribution, and Elimination in Plants, Animals, and Humans

Aspects of the transport behavior and transformations of selenium have been reviewed (NAS, 1976; GLOVER et al., 1979; NEWLAND, 1982; FISHBEIN, 1983; DÜRRE and ANDREESEN, 1986). There is increasing evidence which emphasizes the importance of microbial transformations of selenium in the cycling of this element (ZIEVE and PETERSON, 1984), mainly to dimethyl selenide in fresh water sediments, for instance (CHAU et al., 1976). Little is known about parameters influencing biotransformation (MUSHAK, 1985), but it was reported that the relative biomethylation of inorganic selenium by microbiota in sediments is directly proportional to pH (BAKER et al., 1981).

As noted earlier, many plants and algae convert selenium into volatile compounds such as methyl and dimethyl selenides. This may be a mechanism for reduction of selenium to tolerable non-toxic levels (NEWLAND, 1982). Some accumulation plants have a strong odor caused by methyl and dimethyl selenides, but according to ALEXANDER et al. (1988) selenium does not seem to be an essential element for plants.

Increasing soil acidity reduces the rate of biotransformation of selenium compounds to dimethyl selenide (HAMDY and GISSEL-NIELSEN, 1976; DORAN and ALEXANDER, 1976) which has the result of also reducing movement of soil selenium to plants. Foliar uptake of the volatile dimethyl selenide may also be a route by which plants assimilate selenium (ZIEVE and PETERSON, 1984).

Laboratory and field experiments have shown that selenium bioaccumulates in fish (ADAMS and JOHNSON, 1981). The bioconcentration factor for selenium in fresh water and marine fish has been reported to be about 400 (CALLAHAN et al., 1979). However, selenium concentrations in fish collected from power plant cooling tower reservoirs have been found to be thousands of times higher than those of the waters the fish inhabit. In some fish, selenium concentrations may be higher in the gonads than in the rest of the body (BAUMANN and GILLESPIE, 1986). MIKKELSEN et al. (1987) and KNIGHT et al. (1987) have reported on the extreme accumulation of selenium by mosquito fish and wildlife (which leads to a decrease in fertility and teratogenicity in birds) from agricultural drain water in California. On the other hand, LUNDBERG and PAULSON (1988) demonstrated that the addition of sodium selenite to lakes restricts the availability of mercury to aquatic organisms (perch and pike) and interacts in mercury metabolism.

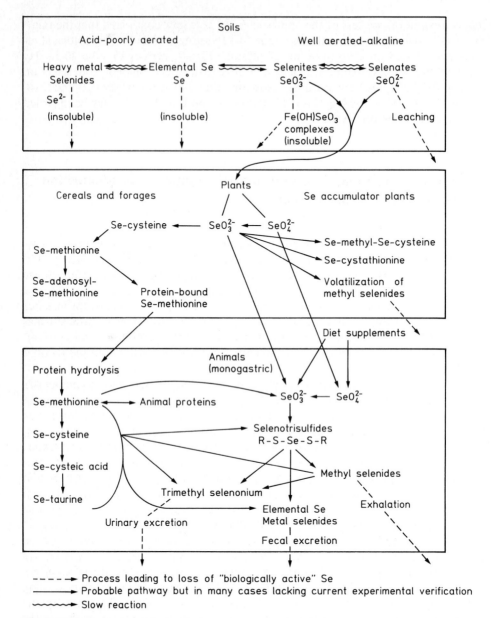

Fig. II.25-3. Chemical and biochemical changes in selenium possibly involved in its movement from soil through plants and animals.

Table II.25-4. Current Levels of Selenium in the Background Environment and in Humans

	Air	Soil	Diet	Body
Inhalation pathway	3 ng m^{-3}			0.00002 µg g^{-1}
Ingestion pathway	1 ng m^{-3}	0.4 µg g^{-1}	70 µg d^{-1}	0.10 µg g^{-1}
			Total	0.1 µg g^{-1}

The chemical and biochemical changes in selenium possibly involved in its movement from soil through plants and animals are depicted in Fig. II.25-3 (NEWLAND, 1982). In nature, all valence forms of selenium are known to exist with the specific forms dependent on the solubility and oxidation-reduction reactions which are possible in the environment (NEWLAND, 1983).

The transfer of selenium from general environmental sources to man occurs primarily through inhalation and ingestion. Selenium utilization includes: (a) its absorption; (b) transport; (c) excretion; and (d) distribution in tissue together with (e) its metabolic transformation to the biochemically active form (ROBINSON and THOMAS, 1983; YOUNG et al., 1982; LEVANDER, 1985; GANTHER, 1984, 1986). Current levels of selenium in the background environment and in man are shown in Table II.25-2 (BENNETT, 1983). From the product of transfer factors, BENNETT (1983, Table II.25-4) calculated that the air concentration values yield contributions to body selenium of 0.02 ng/g through inhalation and 0.1 µg/g through ingestion. The estimated mean concentration of selenium in the body of 0.1 µg/g corresponds to a whole body content of 7 mg.

MEDINSKY and co-workers (1985) recently developed a model to evaluate the potential uptake of selenium in body tissues through chronic inhalation of selenium at the workplace or by ingestion of selenium in food, and to predict resulting equilibrium organ concentrations. Selenium transport rates between five compartments including lung, gastrointestinal tract, blood, liver, and other tissues were estimated. Results from model simulations were compared to published tissue distribution data in rats and dogs, and the model was then modified to predict equilibrium organ concentrations of selenium in people after continual exposure through air or diet. Daily intake levels of 100 µg/day and a functional value of 0.8 were used. Although model predictions indicated that most of the total body selenium in humans is likely to arise from the diet (Se in the urban atmosphere contributes a very small part of the total body selenium) it was projected that continual inhalation of selenium at the threshold limit value (TLV, 200 µg/m^3) could contribute significantly to the total body burden. Selenium levels predicted for lung, liver, and blood after inhalation of Se at the TLV were 22000, 1200, and 400 ng/g tissue, respectively. It was also noted that predicted lung concentrations were near those that produced toxic effects in animals after ingestion of selenium. Continuous human ingestion in the diet was also simulated at levels of 50, 100, and 500 µg Se per day. In this model, higher levels of intake and high bioavailability reflect states of selenium deficiencies (MEDINSKY et al., 1985).

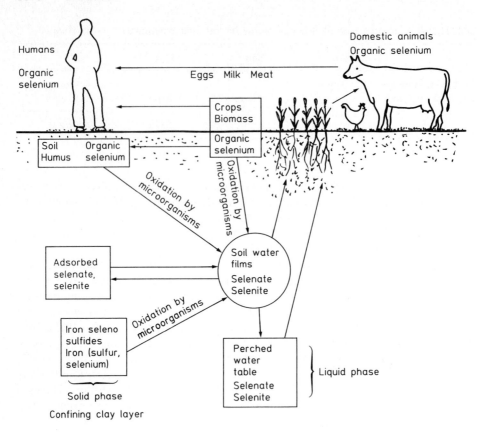

Fig. II.25-4. Biogeochemical cycling of selenium in terrestrial systems.

Ingested selenium is readily absorbed both from naturally seleniferous food and from its soluble compounds with the estimated absorption ranging up to 100% of the ingested dose depending on the chemical form of selenium, its bioavailability, and the animal species (GLOVER et al., 1979; YOUNG et al., 1982). Dietary factors include the presence or absence of promoters or inhibitors such as ascorbic acids, phytate, fiber, sugar, fats and proteins, various mineral-mineral interactions and mineral-micronutrient interactions, type and degree of food processing, and the concomitant ingestion of certain drugs. A number of physiological factors such as nutritional state, growth, pregnancy, etc., and pathological states can also influence selenium utilization.

Absorption of [^{75}Se]-selenomethionine was greater (96%) compared to either [^{75}Se]-selenite (80%) in several studies of New Zealand women (YOUNG et al., 1982). Gastrointestinal absorption of organic forms of selenium is more complete than absorption of selenium as selenite. Absorbed selenium is widely distributed by blood to organs and tissues with the highest concentrations found in the liver and kidneys in several animal species kept on adequate or high-selenium diets. Various

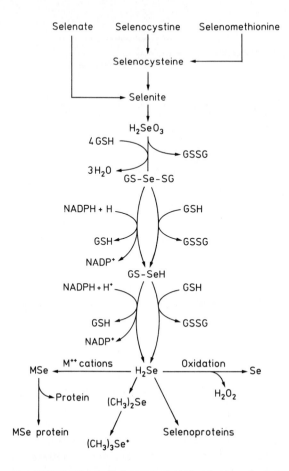

Fig. II.25-5. Metabolic interrelationships between both inorganic and organic forms of selenium.

metabolic interrelationships between both inorganic and organic forms of selenium are illustrated in Fig. II.25-5, while the relationship of ingested selenium to tissue selenium is illustrated in Fig. II.25-6 (BURK, 1986). Mammals (including humans) excrete naturally occurring selenium from foods in approximately equal proportions via urine and feces (EINBRODT and MICHELS, 1984; WHO, 1987). Very little is excreted via sweat and respiration (WHO, 1987). Overdoses of selenium compounds are excreted via urine, and eventually through expired air (WHO, 1987). Urine normally contains about 5–30 µg Se per liter (ANGERER and SCHALLER, 1988).

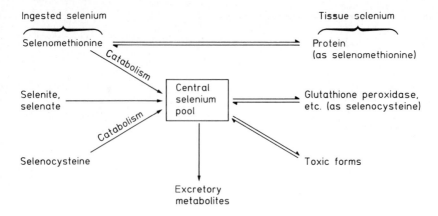

Fig. II.25-6. Relationship of ingested selenium to tissue selenium.

6 Effects on Plants, Animals, and Humans
(see also Section 6.6)

6.1 Effects on Microorganisms and Plants

Escherichia coli, Clostridium thermoaceticum, and *Clostridium sticklandii* seem to need selenium in their enzyme systems (PETERSON et al., 1981). KNIGHT (1987) described the ecotoxicological dynamics of selenium in aquatic ecosystems. He compared the influence of dissolved selenium compounds and selenium compounds containing sediments (which may be mineralized or fossilated). Selenium compounds may decrease algal cell replication (sublethal effects are observed at the 100 ppb level for *Selenastrum capriconium*, while other species are more tolerant, exhibiting effects at about 1 ppm). Water hardness, presence of sulfate, and/or temperature also play a role. In general, selenium is not essential for plants (PETERSON et al., 1981). Some plants are selenium tolerant. In non-tolerant plant species selenium compounds may impair germination and growth and lead to chlorosis (PETERSON et al., 1981). In the laboratory, it was observed that some damage to plants may be due to a replacement of sulfur in proteins by selenium (PETERSON et al., 1981).

6.2 Miscellaneous Biochemical Effects

A number of studies indicate that various selenium compounds are metabolized via two primary pathways: reduction followed by methylation or direct incorporation into or binding by proteins (DIPLOCK, 1976, 1981; GLOVER et al., 1979; GANTHER, 1986). Starting with the +6 oxidation state (selenate) there are applicable reductions to the +4 oxidation state (selenite), followed by still further reduction. Selenate may undergo enzymatic activation with ATP to yield adenosine-5′-selenophosphate which undergoes non-enzymatic reduction to selenite by means of glutathione (DILWORTH and BANDURSKI, 1977; GANTHER, 1986). The reduction-methylation pathway is responsible for the production of dimethyl selenide, which is widely regarded as a detoxified form of selenium (GANTHER, 1986). Dimethyl selenide has been considered as an intermediate metabolite exhaled only when its rate of formation exceeds the rate of further methylation to trimethyl selenonium ion − a urinary Se-metabolite in which selenium has a low biological activity both nutritionally (TSAY et al., 1970) and toxicologically (OBERMEYER et al., 1971; GLOVER et al., 1979; GANTHER, 1986). It is likely that dimethyl selenide and trimethyl selenonium arise by methylation of H_2Se or methyl selenol (GANTHER, 1986):

$$H_2Se \xrightarrow{(1)} CH_3SeH \xrightarrow{(2)} CH_3SeCH_3 \xrightarrow{(3)} (CH_3)_3Se^+.$$

Selenium in the form of selenomethionine can be incorporated directly in place of methionine, although non-ruminant animals are apparently unable to synthesize this selenoamino acid from inorganic selenium compounds (GLOVER et al., 1979). Selenocysteine, which has been identified as the form of selenium in glutathione peroxidase (Fig. II.25-5; KRAUS et al., 1983; GANTHER and KRAUS, 1984; GANTHER, 1986), may occur in other proteins in animals.

Besides glutathione peroxidase with its antioxidant properties in cells, ARTHUR et al. (1987) mentioned selenoproteins in sperm mitochondria, plasma, and muscles, which are involved in the impairment of neutrophilic functions, in the hepatic conversion of thyroxine to triiodothyronine, and in changes of the plasma pyruvate kinase activity (see also FLOHÉ et al., 1987). GSH-S-transferases show substrate specificities. KOLLER (1987) found that selenium interacts with B and T lymphocytes, macrophages, and natural killer cells as well. DWORKIN (1986) studied the effects of selenium deficiency in acquired immunodeficiency syndrome.

Selenium has strong interactions with other nutrients such as vitamins E and C. On the one hand, it functions as an antagonist to counteract to the toxicity of metals such as mercury, cadmium, arsenic, silver, lead, *cis*-platinum, and copper. (SUGAWARA and SUGAWARA, 1987, found that dietary selenium reduces cadmium absorption. According to IMURA et al., 1987, the nephrotoxicity of *cis*-platinum is reduced without changing its cytostatic activity.) On the other hand, some metals such as zinc and tellurium are antagonists to selenium and can interfere with its absorption or action (EINBRODT and MICHELS, 1984; HÖGBERG and ALEXANDER, 1986; WHO, 1987). Sulfates and phosphates can be additional factors that modify the metabolism and excretion of selenium (DIPLOCK, 1976, 1981; NAS, 1976; SCHWARZ, 1976; WHANGER, 1981; ROBINSON and THOMSON, 1983; DIPLOCK et al., 1986a, b; VAN FAASEN et al., 1986; GLOVER et al., 1979; GANTHER, 1986). Selenium has also been reported to alter cytochrome P-450-dependent drug metabolism (SHULL et al., 1979).

6.3 Deficiency Symptoms
(see also Section 4.4)

The discovery of selenium as an essential element was made almost three decades ago by SCHWARZ and FOLTZ (1957) in their classic experiments on dietary necrosis in rats, and many other selenium disorders were subsequently identified in sheep, cattle, swine, and poultry (SCHWARZ, 1976; NAS, 1976). For example, selenium deficiency diseases in several farm species include: white muscle disease in sheep, calves, and horses; hepatosis dietetica and mulberry heary in swine; and exudative diathesis in chickens. Diseases related to selenium deficiency occur in many countries of the world, and it has been estimated that selenium deficiency has caused more economic losses than selenium toxicity (VERNIE, 1984; MOXON and OLSON, 1974). These deficiency diseases are economically significant in regions where the soil contains low levels of selenium available for plant uptake. In some countries, selenium as sodium selenite or selenate is added at a level of 0.1–0.2 mg Se per kg of feed to prevent various deficiency diseases in swine and poultry (NAS, 1976). Selenite has been shown to be more effective than selenomethionine in protecting chicks from exudative diathesis, while selenomethionine is far more protective than selenite against pancreatic fibrosis, another manifestation of selenium deficiency in chicks. Recently ANKE et al. (1989) have also described the effect of selenium deficiency on reproduction, milk performance, and growth of goats.

Selenium is an essential trace element at low levels of intake and produces toxic symptoms when ingested at levels only three to five times higher than those required for adequate nutrition (NAS, 1976; DIPLOCK, 1981; YOUNG, 1981). Only extremely low levels of selenium, usually in the parts per billion range, are needed to satisfy nutritional requirements for the element, and in some cases vitamin E may partially replace selenium. A variety of diseases, including cancer, have been linked with a low selenium status (KIEFFER, 1987), and these are illustrated in Table II.25-5 (ROBINSON and THOMSON, 1983; WHO, 1987). Keshan disease is a potentially fatal cardiomyopathy – irregular pulsation, enlarged and indurated hearts, and mitochondrial swelling (crystallization) – found especially in girls, women, calves, and cows in certain areas of China where selenium levels are low in staple foods (KESHAN DIS-

Table II.25-5. Diseases Which Have Been Linked with a Low Selenium Status

Protein energy malnutrition	Cancer
Hemolytic anemia	Muscular complaints
Sickle cell anemia	Arthritis
Keshan disease	Muscular dystrophy
Hypertension	Multiple sclerosis
Ischaemic heart disease	Neuronal ceroid lipofuscinosis (NCL)
Alcoholic cardiomyopathy	Ageing
Alcoholic cirrhosis	Werner's syndrome
Pancreatitis	Muscular degeneration
Cystic fibrosis	Diabetic retinopathy
Infertility	Cataract

EASE RESEARCH GROUP, 1979a,b; YOUNG, 1981; CHEN et al., 1980; DONGXA, 1986; ZHANG, 1987; CHEN, 1987). Mean blood levels of 21 µg/L have been reported in the area where Keshan disease is found. The average dietary intake in the area was 3 – 11 µg selenium per day (YANG et al., 1983). Dietary supplementation of 0.3 – 1 mg Se per week has been used successfully in the prevention of Keshan disease (CHINESE ACADEMY OF MEDICAL SCIENCES, 1979). Current evidence would suggest that a daily human selenium consumption of about 30 µg is necessary to prevent Keshan disease, while approximately 90 µg/day/adult should be the minimum daily requirement for optimal biological performance (KOLLER and EXON, 1986).

An additional selenium deficiency disease that has been reported in China is Kaschin-Beck disease. This is characterized by collagen breakdown and joint deterioration (DONGXA, 1986) and can be alleviated by selenium supplementation.

As noted earlier, selenium is the active center for glutathione peroxidase, an enzyme that is principally involved in the removal of peroxides from cells (NAS, 1976; GLOVER et al., 1979), and the presence of this enzyme strongly indicates the significant place of selenium in human nutrition. Selenium deficiency may also lead to the loss of other important selenoproteins (ARTHUR et al., 1987).

6.4 Effects on Animals

Selenium toxicity in animals has been extensively reviewed (OLSON et al., 1973; COOPER and GLOVER, 1974; FISHBEIN, 1977; WILBER, 1983; GLOVER et al., 1979; OLSON, 1986; KOLLER and EXON 1986; WHO, 1987). Acute toxicity occurs both in livestock and laboratory animals. A variety of factors affect toxicity: form and concentration of selenium, route of administration, and animal species, age, sex, etc. Laboratory animals have exhibited various degrees of damage to internal organs, vascular edema, and hemorrhage when exposed to elevated amounts of selenium in food, water, and vapor (NAS, 1976; NEWLAND, 1982; GLOVER et al., 1979).

The range disease "blind staggers" is a form of acute selenium toxicity. The signs of acute toxicity in farm and experimental animals most commonly mentioned in the literature appear to be garlic odor of the breath, dyspnea, pulmonary edema, tachycardia, emesis, diarrhea, depression, ataxia, incoordination, paralysis and excessive salivation (OLSON et al., 1973; OLSON, 1986).

Chronic toxicity in livestock is commonly caused by ingestion of vegetation containing 3 to 20 ppm of selenium for prolonged periods and manifested by a loss of vitality, lameness, elongated and disfigured hooves, degeneration of the internal organs, and loss of hair. Alkali disease is a form of chronic selenium poisoning associated with consumption of grains containing more than 5 ppm selenium over weeks or months (NAS, 1976; GLOVER et al., 1979; NEWLAND, 1982; OLSON, 1986). Laboratory animals exhibit similar symptoms, and a dietary level of 5 to 10 ppm has been found to induce chronic toxicity in rats with the vascular system and liver being the most affected by long-term exposure (NAS, 1976; GLOVER et al., 1979; NEWLAND, 1982; OLSON, 1986). As little as 7.2 ppm dietary selenium from grain was found to be toxic to dogs. Signs included restricted food intake, reduced growth rate, decreased blood hemoglobin, ascites, and liver atrophy, necrosis and cirrhosis (RHIAN and

MOXON, 1943; OLSON, 1986). Selenites appear to be more toxic than selenates and selenides (HALVERSON et al., 1966; PALMER et al., 1983; EINBRODT and MICHELS, 1984; OLSON, 1986). The National Academy of Sciences of the U.S. has accepted 5 mg Se per kg diet as the dividing level between toxic and non-toxic feeds (NAS, 1980).

6.5 Effects on Humans

Man seems to be less sensitive to overdoses of selenium than animals (EINBRODT and MICHELS, 1984). Human toxicological data available from occupational health have been reviewed extensively (COOPER, 1967; GLOVER, 1970; IZRAELSON, 1973; GLOVER et al., 1979; OLSON, 1986; HÖGBERG and ALEXANDER, 1986; ALEXANDER et al., 1988; WHO, 1987) with exposure occurring in both the production of selenium and in various manufacturing industries utilizing it. The acute cases have usually been accidental and are, in many cases, quite clearly substantiated as resulting from excessive selenium intake. Most of the described symptoms and signs of poisoning due to inhalation of selenium fumes or Se-containing dust are non-specific and include: irritation of the eyes, nose, and throat; nausea and vomiting, indigestion and other intestinal disturbances; increased body temperature; lassitude; and ill-defined psychoneurotic symptoms such as irritability (GLOVER et al., 1979; NAS, 1976; OLSON et al., 1973; OLSON, 1986). Five cases of industrial poisoning due to selenosis have been reported resulting from the presence of less than 0.2 ppm hydrogen selenide in the air of a metal etching and buffing operation. Predominating symptoms were nausea, vomiting, metallic taste, dizziness, lassitude, and fatigue (BUCHAN, 1977; OLSON, 1986).

Episodes of chronic selenosis have been reported in workers exposed to industrial selenium (KINNIGHEIT, 1962; GLOVER, 1967; HÖGBERG and ALEXANDER, 1986; OLSON, 1986). In one study involving 62 workers in a selenium rectifier plant, 35 individuals complained of amnestic difficulties, 25 of headache. Other complaints included: sleeplessness, irritability, nervous heartbeat, unappetance, and heartburn, and nine cases of inflammation of the mucous membrane (KINNIGHETT, 1962; OLSON, 1986). The major compounds of selenium which have been implicated in acute and chronic effects in occupational exposures were hydrogen selenide, selenium dioxide, and selenium oxychloride (GLOVER et al., 1979; OLSON, 1986).

Although chronic toxicity due to inhalation or absorption of selenium is usually seen only in industry, cases of chronic poisoning in seleniferous areas (for instance, the Enshi region in China) have been studied and attributed to consumption of food contaminated with selenium (NAS, 1976; GLOVER et al., 1979; NEWLAND, 1982; OLSON, 1986; EINBRODT and MICHELS, 1984; WHO, 1987). YANG et al. (1983) found average daily selenium intakes of up to 4990 µg/day and mean blood selenium levels of up to 3200 µg/L in cases of chronic selenosis. The symptoms most often found were bad teeth, icteroid skin, dermatitis, arthritis, gastrointestinal disturbances, hair loss, and diseased nails (WHO, 1987). Garlic odor of the breath, abnormalities of the nervous system including numbness, convulsions, and paralysis may be mentioned depending on the degree of toxicosis (KOLLER and EXON, 1986). A daily uptake of 1 mg of Se per kg body weight may produce chronic toxicity in humans, and 5 ppm in foods, 0.5 ppm in water or milk, has been considered by some investigators to be potentially dangerous (NEWLAND, 1982; GLOVER et al., 1979).

6.6 Mutagenic, Carcinogenic, and Teratogenic Effects

6.6.1 Introduction

Sodium selenite per se has been reported to be a weak mutagen in *Bacillus subtilis* (the rec assay) and *Salmonella typhimurium* Ta100 (NODA et al., 1979), to induce sister chromatid exchanges (SCEs) in whole-blood cultures (RAY et al., 1978), and cause unscheduled DNA synthesis (UDS) in human fibroblasts and Chinese hamster ovary (CHO) cells (LO et al., 1978; WHITING et al., 1980). EINBRODT and MICHELS (1984) and HÖGBERG and ALEXANDER (1986) mentioned studies of the mutagenicity of selenate. WHO (1987) gives an extensive review on cytotoxicity and mutagenicity, but it seems that selenium compounds are only weak mutagens and that some effects are rather related to be general toxicity of the high doses used.

Earlier observations on the toxicity of selenium, largely centered on the possible carcinogenicity of this element based on studies of male rats fed diets containing either seleniferous grain or selenium salts (NELSON et al., 1943; SEIFERT et al., 1946; TSCHERKES et al., 1963), have generally been discounted principally based on flaws of study design and ambiguous pathological findings (SHAPIRO, 1973; IARC, 1975). According to HÖGBERG and ALEXANDER (1986) the available data seem to indicate that selenium compounds (eventually with the exception of selenium sulfide) will not act as carcinogens at low or moderate doses (see also WHO, 1987). The same authors state that teratologic and embryotoxic effects – if there are any – are doubtful, because the observations are controversial and the design of the studies does not allow clear conclusions. It seems, however, that birds are at greater risk regarding fertility and teratogenicity (see also Sect. 5 and WHO, 1987). Aquatic birds such as the American coot and the black-necked stilt have shown malformations related to overdoses of selenium compounds in feed (HOFFMAN et al., 1988). Probably selenate and/or selenomethionine (a major form in edible plants such as wheat) are the most teratogenic selenium compounds in birds (WHO, 1987; HOFFMAN et al., 1988).

6.6.2 Anticarcinogenic Properties

Increasing experimental evidence has suggested that selenium possesses antineoplastic properties with numerous studies demonstrating that dietary selenium supplementation can inhibit chemically-induced tumors (primarily those of the skin, liver, colon, and mammary glands) in rodent species (VERNIE, 1984; GRIFFIN, 1979; SHAMBERGER, 1970, 1986; WHANGER, 1983; IP, 1981a, b, 1986; FISHBEIN, 1986a, b) as well as virally induced cancers and growth of transplanted cells (SCHRAUZER et al., 1976, 1980; MEDINA and SHEPHERD, 1980; MEDINA et al.,

1981). In a broad spectrum of chemical carcinogen-induced tumorigenesis, selenium supplementation inhibits the initiation and post-initiation phases of carcinogenesis (SHAMBERGER, 1970; SHAMBERGER et al., 1973; IP, 1981a; THOMPSON and BECCI, 1979, 1980). Of 39 studies published since 1949, 34 demonstrated that selenium supplementation had an inhibitory effect, three showed no effect, and two showed that selenium enhanced tumorigenicity (MEDINA, 1985, 1986).

Most of the studies involving chemically induced tumorigenesis employed principally inorganic (selenate, selenite, or selenium dioxide) additions to the diet or drinking water at concentrations of 0.5 to 6.0 µg/g, which are of the order of 20–60 times the nutritional requirement of 0.1 µg/g (NAS, 1983). There is some disagreement as to the effectiveness of various forms of selenium as cancer-preventive agents (see also SWANSON, 1986). In one study, inorganic forms of selenium were found to be four to ten times more effective in inhibiting tumor proliferations in transplantable studies than organic forms such as selenomethionine or selenocysteine (MILNER, 1985), while in another, selenite and selenomethionine were found to be equally efficacious in protecting against mammary tumorigenesis induced by DMBA in rats (IP, 1985a).

Although selenium inhibits post-initiation events, it was noted that once selenium supplementation is withdrawn, neoplasms emerge at the same rate as seen in carcinogen-treated animals not receiving selenium (MEDINA and SHEPHERD, 1980; IP and IP, 1981). Possible mechanisms that may account for the protective action of selenium against tumor occurrence include reduction of the concentration of free radicals and hydroperoxides in tissue via the action of Se-dependent glutathione peroxidase (IP, 1985a, b; VERNIE, 1984) and the antioxidant properties of selenium compounds primarily in relation to vitamin E (HORVATH and IP, 1983; IP, 1985a, b; GRIFFIN, 1979). Alteration of the metabolism of carcinogenic compounds or their interaction with tissue macromolecules (WATTENBERG, 1978; WORTZMAN et al., 1980; DIPPLE et al., 1986), modulation of the immune response (SPALLHOLZ et al., 1973; KOLLER and EXON, 1986), inhibition of cell proliferation (MEDINA et al., 1983; MEDINA, 1985, 1986), and the inhibition of the promotional phases of carcinogenesis (IP, 1981b; THOMPSON, 1984) were discussed. It was also found (SCHRAUZER et al., 1979, 1987) that even small amounts of lead reduce the anticarcinogenic activities of selenium.

A number of case control studies, most of them cross-sectional, have suggested that cancer patients may have lower blood or plasma Se-levels than do healthy matched controls (SHAMBERGER et al., 1973; McCONNELL et al., 1975, 1980; SCHRAUZER et al., 1985). For instance, a case-control analysis of longitudinal data examined during a five year period compared 111 subjects who developed cancer to 210 matched controls and found the strongest correlation between low serum selenium levels and gastrointestinal and prostatic cancers (WILLETT et al., 1983a, b). Also, a recent prospective case-control study in Finland has linked a relative selenium deficiency to an increased risk of cancer. The relative risk was from 2 to 5.8 times higher in isolated relative Se-deficiency and up to 11.4 times higher in a combined relative deficiency of selenium and vitamin E, suggesting biochemical interaction between selenium and vitamin E and endogenous antioxidants in the protection against peroxidative cell damage (SALONEN et al., 1985).

Extensive literature, which has recently been reviewed, exists on the inverse associations between soil selenium concentrations or environmental levels of selenium and the occurrence of cancer of the gastrointestinal tract, lung, and breast, among others (FISHBEIN, 1986a, b; CLARK, 1985; CLARK et al., 1984; COMBS, 1985; COMBS and CLARK, 1985; WHO, 1987; PETERSON, 1987; LEVANDER, 1987). SHAMBERGER and FROST (1969) first suggested that cancer mortality rates in the United States were inversely associated with the geographical distribution of selenium (with plant Se-levels as mapped by KUBOTA et al., 1967).

Additional studies confirmed that an increased incidence of colorectal, breast and other cancers in humans occurred in geographic regions deficient in selenium (JANSSON et al., 1975; SHAMBERGER and WILLIS, 1971). An apparent per capita dietary selenium intake in populations in 27 countries has been correlated with age-corrected mortalities from cancer at 17 major body sites, and significant inverse correlations were found for dietary intakes and cancer of the lung, breast, ovary, colon, and rectum, for leukemia and for cancer at all sites combined (SCHRAUZER et al., 1977). Further, blood selenium levels in humans who developed cancer (primarily of the gastrointestinal tract and prostate) were significantly lower than those of matched controls (WILLETT et al., 1983a). Additional recent geographical studies of selenium status and cancer mortality in the U.S. (CLARK, 1985) and the People's Republic of China (YU et al., 1985) have confirmed the earlier geographical association between plant or blood selenium and cancer mortality.

6.6.3 Antimutagenic Properties

Selenium (generally as sodium selenite) has been shown to be effective in reducing the mutagenicity of a variety of direct-acting mutagens requiring metabolic activation in the Ames *Salmonella* microsome assay (MARTIN et al., 1981; MARTIN and SCHILLACI, 1984; ROSIN and STICH, 1979; TEEL, 1984; JACOBS et al., 1977; SPEIT, 1987).

In the inhibitory studies of selenium with 7,12-dimethylbenz(*a*)anthracene (DMBA), it was suggested that selenium may have been effective by one or more of the following modes of action: (1) selenium may react directly with the ultimate carcinogen, preventing it from interacting with DNA; (2) it may stimulate or activate enzymes involved in the detoxification of the ultimate mutagen; or (3) selenium may inhibit the activation enzymes (MARTIN and SCHILLACI, 1984).

7 Hazard Evaluation and Limiting Concentrations

The ACGIH (1980) recommended a TWA-TLV (threshold limit value) for selenium in the workplace atmosphere of 0.2 mg/m^3, primarily to protect against the irritative effects of selenium (no STEL has been recommended). The ACGIH (1983) recommended a TWA-TLV of 0.2 mg/m^3 for selenium hexafluoride, and the OSHA standard for selenium has been set at 0.2 mg/m³. The ACGIH time-weighted average for hydrogen selenide is 0.05 ppm (0.2 mg/m³) and the short-term exposure is the same. The maximum allowable concentration (MAC) set by the USSR is 0.1 mg/m³, and this maximum concentration has also been recommended in the Federal Republic of Germany (MAK, 1987) whereby one short-term peak per shift of 1 mg/m³ may be allowed.

The World Health Organization guideline for selenium in drinking water is 0.01 mg/L (WHO, 1984). Several approaches have been taken by WHO to confirm this maximum acceptable concentration. Effects considered to be due to selenium toxicity have been observed at levels of intake of 0.01–0.1 mg selenium per kg of body weight per day. For a 70 kg man this would amount to daily intakes of 0.7–7.0 mg selenium. The estimates of selenium intake used to derive the recommended guideline value range from 130 to 200 µg/day, corresponding to daily intake values published in the United States and some other countries. The maximum daily selenium intake from drinking water should not exceed 10% of the recommended daily dietary intake of 200 µg, i.e. should not exceed 20 µg/day. Assuming an intake of two liters of water daily, the concentration of selenium in drinking water should not exceed 0.01 mg/L (WHO, 1984).

As described in Sections 6.4 and 6.3, feed should not contain more than 5 mg Se per kg, but deficiency may be much more critical than overdoses under practical circumstances. Humans require about 60–120 µg Se per day to prevent deficiency diseases. Dietary supplementation may be necessary for animal and human populations. In Finland it was found that selenium-enriched fertilizers lead to an appropriate dietary intake. WHO (1987) estimated safe and adequate ranges of selenium intake to be 50–200 µg/day for persons older than 7 years, 30–120 µg/day for children 4–6 years old, and 20–80 µg/day for children 1–3 years old. According to LEVANDER (1987), a minimum daily requirement of 14–19 µg/day is sufficient when humans have adapted, but the physiological requirement may be 40 µg Se per day. Although selenium levels above 200 µg/day have been suggested to be a preventative against cancer, the evidence is still believed to be ambiguous in a number of quarters and the overall problem is still under consideration and lively debate (IP, 1986b; SWANSON, 1986; SALONEN, 1986; US Dept. HHS, 1988; see also Sect. 6.6.2). In fact, evidence to date of preventative anticarcinogenic effects of selenium are largely provided by animal studies where selenium doses generally exceed the physiological intakes 5 to 100 fold.

References

ACGIH (1980), *Documentation of the Threshold Limit Values for Substances in Workroom Air*, 4th Ed. with supplements through 1981, pp. 361–362. American Conference of Governmental Industrial Hygienists, Cincinnati, Ohio.

ACGIH (1983), *TLVs, Threshold Limit Values for Chemical Substances and Physical Agents in the Work Environment with Intended Changes for 1983–1984*. American Conference of Governmental Industrial Hygienists, Cincinnati, Ohio.

Adams, W. J., and Johnson, H. E. (1981), in: Branson, D. R., and Dickson, K. L. (eds.), *Aquatic Toxicological Hazard Assessment Fourth Conference, STP 73,* pp. 124–137. American Society for Testing and Materials, Philadelphia, Pennsylvania.

Ageton, R. W. (1970), in *Mineral Facts and Problems,* Bull. No. 65, pp. 713–721. US Department of Interior, Washington, D.C.

Alexander, J., Högberg, J., Thomassen, Y., and Aaseth, J. (1988), Chapter 53 *Selenium,* in: Seiler, H. G., Sigel, H., and Sigel, A. (eds.), *Handbook on Toxicity of Inorganic Compounds,* pp. 581–595. Marcel Dekker, New York-Basel.

Alfthan, G. (1986), *Selenium Status of Nonpregnant, Pregnant Women and Neonates. Acta Pharmacol. 59,* (Suppl. 7), 142–145.

Andrew, A. W., Klein, D. J., and Talmi, Y. (1975), *Environ. Sci. Technol. 9,* 856–858.

Angerer, J., and Schaller, K. H. (1988), *Selenium,* in: *Analyses of Hazardous Substances in Biological Materials,* Vol. 2, pp. 231–247. VCH Verlagsgesellschaft, Weinheim-Basel-Cambridge-New York.

Anke, M., Angelow, L, Groppel, B., Arnhold, W., Gruhn, K., Košla, T., and Langer, M. (1989), *The Effect of Selenium Deficiency on Reproduction and Milk Performance of Goats; The Effect of Selenium Deficiency on the Feed Consumption and Growth of Goats. Ann. Anim. Nutr. Berlin 39*(4/5), 473–481, 483–490.

Arthur, J. R., Nicol, F., Boyne, R., Allen, K. G. D., Hayes, J. D., and Beckett, G. J. (1987), *Old and New Roles for Selenium. Trace Subst. Environ. Health 21,* 487–498.

Baker, M. D., Wong, P. T. S., Chau, Y. K., et al. (1981), *Proceedings of the Third International Conference of Heavy Metals in the Environment,* pp. 645–648. CEP Consultants Ltd., Edinburgh.

Baumann, P. C., and Gillespie, R. B. (1986), *Selenium Bioaccumulation in Gonads of Large Mouth Bass and Bluegill from Three Power Plant Cooling Reservoirs. Environ. Toxicol. Chem. 5,* 695–701.

Beilstein, M. A., and Whanger, P. D. (1986), *Chemical Forms of Selenium in Rat Tissues after Administration of Selenite or Selenomethionine. J. Nutr. 116,* 1711–1719.

Bem, E. M. (1981), *Environ. Health Perspect. 37,* 183–200.

Bennett, B. G. (1983), *Exposure of Man to Environmental Selenium – an Exposure Commitment Assessment. Sci. Total Environ. 31,* 117–127.

Blotcky, A. J., Hansen, G. T., Opelanio-Buencamino, L. R., and Rack, E. P. (1985), *Anal. Chem. 57,* 1937–1941.

Berrow, M. L., and Burridge, J. C. (1980), in *Inorganic Pollution and Agriculture,* MAFF Ref. Book 326, pp. 159–183. H.M.S.O., London.

Brooks, R. R., Willis, J. A., and Little, J. R. (1983), *J. Assoc. Off. Anal. Chem. 66,* 130.

Bruce, A. (1986), *Ann. Clin. Res. 18,* 8–12.

Bruland, K. W. (1983), *Trace Elements in Sea-Water, Chem. Oceanogr. 8,* 188–190.

Buchan, R. F. (1977), *Occup. Med. 3,* 439–456.

Burk, R. F. (1986), *Selenium and Cancer: Meaning of Selenium Levels. J. Nutr. 116,* 1584–1586.

Burk, R. F., Lane, J. M., Lawrence, R. A., and Gregory, P. E. (1981), *Effect of Selenium Deficiency on Liver and Glutathione Peroxidase Activity in Guinea Pigs. J. Nutr. 111,* 690–693.

Callahan, M. A., Slimak, M. W., Gabel, W. W. et al. (1979), *Water Related Environmental Fate of 129 Priority Pollutants,* Vol. I, EPA Report 440/4-79-29a. Office of Water and Waste Management, US Environmental Protection Agency, Washington, D.C.

Chau, Y. K., Wong, P. T. S., Silverberg, B. A., Luxon, P. C., and Bengert, G. A. (1976), *Methylation of Selenium in the Environment. Science 192,* 1130–1131.

Chen, X.-S. (1987), *Studies on Selenium and Keshan Disease in China.* 2nd Nordic Symposium on Trace Elements in Human Health and Disease, Odense. Abstract Mini-Symposium No. 11 and Proceedings, pp. 72–73. *Environ. Health 20,* WHO Regional Office for Europe, Copenhagen.

Chen, X.-S., Yang, G., Chen, J., et al. (1980), *Studies on the Relations of Selenium and Keshan Disease. Biol. Trace Elem. Res. 2,* 91–107.

Chinese Academy of Medical Sciences (1979), *Chin. Med. J. 92,* 471–476.

Clark, L. C. (1985), *The Epidemiology of Selenium and Cancer. Fed. Proc. 44*, 2584–2589.
Clark, L. C., Graham, G. F., Crounse, R. G., et al. (1984), *Nutr. Cancer 6*, 13.
Combs, G. F., Jr. (1985), *Selenium and Carcinogenesis. Fed. Proc. 44*, 2561–2563.
Combs, G. F., Jr., and Clark, L. C. (1985), *Can Dietary Selenium Modify Cancer Risks? Nutr. Res. 43*, 325–331.
Cooper, W. C. (1967), *Selenium Toxicity in Man*, in: Muth, O. H. (ed.), *Selenium in Biomedicine*, pp. 185–199. AVI Publ. Co., Westport, Connecticut.
Cooper, W. C. (1974), in: Zingaro, R. A., and Cooper, W. C. (eds.), *Selenium*, pp. 615–653. Van Nostrand Reinhold, New York.
Cooper, W. C., and Glover, J. (1974), *The Toxicology of Selenium and its Compounds*, in: Zingaro, R. A., and Cooper, W. C. (eds.), *Selenium*, pp. 654–674. Van Nostrand Reinhold, New York.
Cooper, W. C., Bennett, K. G., and Croxton, F. C. (1974), *The History, Occurrence, and Properties of Selenium*, in: Zingaro, R. A., and Cooper, W. C. (eds.), *Selenium*, pp. 1–30. Van Nostrand Reinhold, New York.
Cutter, G. A. (1978), *Anal. Chim. Acta 98*, 59–66.
Cutter, G. A. (1985), *Anal. Chem. 57*, 2951–2955.
Cutter, G. A., and Bruland, K. W. (1984), *Limnol. Oceanogr. 29*, 1179–1192.
Cutter, G. A., and Church, T. M. (1986), *Selenium in Western Atlantic Precipitation. Nature 322*, 720–722.
Davies, T. S. (1982), *Lancet 2*, 935.
Desei, G. R., and Paul, J. (1981), *Microchem. J. 22*, 176.
Dilworth, G. L., and Bandurski, R. S. (1977), *Biochem. J. 163*, 521–529.
Diplock, A. T. (1976), *Metabolic Aspects of Selenium Action and Toxicity. CRC Crit. Rev. Toxicol. 4*, 271–329.
Diplock, A. T. (1981), *Metabolic and Functional Defects in Selenium Deficiency. Philos. Trans. R. Soc. London Ser. B 294*, 105–117.
Diplock, A. T., Watkins, W. J., and Hewison, M. (1986a), *Ann. Clin. Res. 18*, 22–29.
Diplock, A. T., Watkins, W. J., and Hewison, M. (1986b), *Selenium and Heavy Metals. Ann. Clin. Res. 18*, 55–60.
Dipple, A., Pigott, M. A., and Milner, J. A. (1986), *Selenium Modifies Carcinogen Metabolism by Inhibiting Enzyme Induction. Biol. Trace Elem. Res. 10*, 153–157.
Dongxa, M. (1986), in: Combs, J., et al. (eds.), *Selenium in Biology and Medicine*. AVI Publ. Co., Westport, Connecticut.
Doran, J. W., and Alexander, M. (1976), *Soil Sci. Soc. Am. J. 40*, 687–690.
Dürre, P., and Andreesen, J. R. (1986), *Die biologische Bedeutung von Selen. Biol. Unserer Zeit 16*, 12–23.
Dworkin, B. M., Rosenthal, W. S., Wormser, G. P., and Weiss, L. (1986), *Selenium Deficiency in the Acquired Immunodeficiency Syndrome. J. Parenter. Enteral Nutr. 10*, 405–407.
Einbrodt, H. J., and Michels, S. (1984), *Selenium*, in: Merian, E. (ed.), *Metalle in der Umwelt*, pp. 541–554. Verlag Chemie, Weinheim-Deerfield Beach/Florida-Basel.
Elkin, E. M. (1982), *Selenium and Selenium Compounds*, in: *Kirk-Othmer Encyclopedia of Chemical Technology*, 3rd Ed., Vol. 20, pp. 575–601. John Wiley & Sons, New York.
Elrashidi, M. A., and Adriano, D. C. (1987), *Effect of Redox Potential and pH on Chemical Speciation of Inorganic Selenium in Soils. Sixth International Conference on Heavy Metals in the Environment, New Orleans. Proceedings* Vol. 1, pp. 107–109. CEP Consultants Ltd., Edinburgh.
EPA (1986), *Health Effects Assessment for Selenium and Compounds*. Environmental Criteria and Assessment Office, US Environmental Protection Agency, Cincinnati, Ohio.
Fishbein, L. (1977), *Selenium*, in: Goyer, R. A., and Mehlmann, M. A. (eds.), *Toxicology of Trace Elements*, pp. 191–240. Hemisphere, Washington-London.
Fishbein, L. (1983), *Environmental Selenium and Its Significance. Fund. Appl. Toxicol. 3*, 411–415.
Fishbein, L. (1985), in: Merian, E., Frei, R. W., Härdi, W., and Schlatter, C. (eds.), *Carcinogenic and Mutagenic Metal Compounds*, pp. 96–112. Gordon & Breach, New York-London.

Fishbein, L. (1986a), *Perspectives in Metal Carcinogenesis. I. Selenium. Arch. Geschwulstforsch.* **56**, 53–78.

Fishbein, L. (1986b), *Perspectives on Selenium Anticarcinogenicity. Toxicol. Environ. Chem.* **12**, 1–30.

Flohé, L., Strassburger, W., and Günzler, W. G. (1987), *Selen in der enzymatischen Katalyse. Chem. Unserer Zeit* **21**(2), 44–49.

Food and Nutrition Board (1977), *Are Selenium Supplements Needed (by the General Public)? J. Am. Diet. Assoc.* **70**, 249–250.

Ganther, H. E. (1979), *Metabolism of Hydrogen Selenide and Methylated Selenides,* in: Draper, H. H. (ed.), *Advances in Nutritional Research,* Vol. 2, pp. 107–128. Plenum, New York.

Ganther, H. E. (1984), *Selenium Metabolism and Function in Man and Animals,* in: Brater, P., and Schramel, P. (eds.), *Trace Element Analytical Chemistry in Medicine and Biology,* Vol. 3, pp. 3–24. Walter de Gruyter & Co., Berlin.

Ganther, H. E. (1986), *Pathways of Selenium Metabolism Including Respiratory Excretory Products. J. Am. Coll. Toxicol.* **5**, 1–5.

Ganther, H. E., and Kraus, R. J. (1984), *Identification of Hydrogen Selenide and Other Volatile Selenols by Derivatization with 1-Fluoro-2,4-dinitrobenzene. Anal. Biochem.* **138**, 396–403.

Glover, J. R. (1970), *Selenium and its Industrial Toxicology. Ind. Med. Surg.* **39**, 50–54.

Glover, J., Levander, O., Parizek, J., and Vouk, V. (1979), *Selenium,* in: Friberg, L., Nordberg, G. F., and Vouk, V. B. (eds.), *Handbook on the Toxicology of Metals,* pp. 555–577. Elsevier/North Holland Biochemical Press, Amsterdam.

Griffin, A. C. (1979), *Role of Selenium in the Chemoprevention of Cancer. Adv. Cancer Res.* **29**, 419–422.

Hahn, M. H., Luennan, R. W., Caruso, J. A., and Fricke, F. L. (1981), *J. Agric. Food Chem.* **29**, 792.

Halverson, A. W., Palmer, I. S., and Guss, P. L. (1966), *Toxicol. Appl. Pharmacol.* **9**, 477–484.

Hamdy, A. A., and Gissel-Nielsen, E. (1976), *Z. Pflanzenernaehr. Bodenkd.* **6**, 671–678.

Hashimoto, Y., Hurang, J. Y., and Yanagisawa, S. (1970), *Environ. Sci. Technol.* **4**, 157–158.

Hidiroglou, M., Carson, R. B., and Brossard, G. A. (1965), *Can. J. Anim. Sci.* **45**, 197–202.

Hoffman, D. J., Ohlendorf, H. M., and Aldrich, T. W. (1988), *Selenium Teratogenesis in Natural Populations of Aquatic Birds in Central California. Arch. Environ. Contam. Toxicol.* **17**, 519–525.

Hofsommer, H. J., and Bielig, H. J. (1981), *Z. Lebensm. Unters. Forsch.* **172**, 32.

Högberg, J., and Alexander, J. (1986), Chapter 19 *Selenium,* in: Friberg, L., Nordberg, G. F., and Vouk, V. B. (eds.), *Handbook on the Toxicology of Metals,* 2nd Ed., Vol. II, pp. 482–520. Elsevier Science Publishers, Amsterdam.

Horvath, P. M., and Ip, C. (1983), *Synergistic Effect of Vitamin E and Selenium in the Chemoprevention of Mammary Carcinogenesis in Rats. Cancer Res.* **43**, 5335–5341.

IAEA (1972), *Proceedings of a Symposium on Nuclear Activation Techniques in the Life Sciences, Bled, Yugoslavia.* International Atomic Energy Agency, Vienna.

IARC (1975), *Monographs on the Evaluation of the Carcinogenic Risk of Chemicals to Humans,* Vol. 9, *Some Aziridines, N-, S-, and O-Mustards and Selenium,* pp. 245–260. International Agency for Research on Cancer, Lyon.

ICRP (1975), *Task Group Report on Reference Man. ICRP Publ. 23.* International Commission on Radiological Protection. Pergamon Press, Oxford.

Imura, N., Naganuma, A., Satoh, M., and Koyama, Y. (1987), *J. Environ. Occup. Health Jpn.* **9**, 223–229.

Inhat, M. (1974), *J. Assoc. Off. Anal. Chem.* **57**, 368–372.

Ip, C. (1981a), *Factors Influencing the Anticarcinogenic Efficacy of Selenium in Dimethylbenz-(a)anthracene-induced Mammary Tumorigenesis in Rats. Cancer Res.* **41**, 2683–2686.

Ip, C. (1981b), *Prophylaxis of Mammary Neoplasms by Selenium Supplement Atom in the Initiation and Promotion Phases of Carcinogenesis. Cancer Res.* **41**, 4386–4390.

Ip, C. (1985a), *Selenium Inhibitions of Chemical Carcinogenesis. Fed. Proc.* **44**, 2573–2578.

Ip, C. (1985b), *Attenuation of the Anticarcinogenic Action of Selenium by Vitamin C Deficiency.* Cancer Lett. 25, 325–331.
Ip, C. (1985c), Am. Assoc. Cancer Res. 326, 124.
Ip, C. (1986a), *The Chemopreventive Role of Selenium in Carcinogenesis.* J. Am. Coll. Toxicol. 5, 7–20.
Ip, C. (1986b), Ann. Clin. Res. 18, 22–29.
Ip, C., and Ip, M.M. (1981), *Chemoprevention of Mammary Tumorigenesis by a Combined Regimen of Selenium and Vitamin A.* Carcinogenesis 2, 915–918.
Izraelson, Z.I. (1973), in: Izraelson, Z.I., Mogievskaja, O.J., and Suvorov, S.V. (eds.), *Occupational Hygiene and Pathology of Rare Metals.* Medicina, Moscow.
Jacobs, M.M., Matney, T.S., and Griffin, A.C. (1977), *Inhibitory Effects of Selenium on the Mutagenicity of 2-Acetylaminofluorene (AAF) and AAF-derivatives.* Cancer Lett. 2, 319–322.
Jansson, B., Seibert, B., and Speer, J.F. (1975), Cancer 36, 2373–2384.
Jenkins, R.A. (1985), in: O'Neill, I.K., Schuller, P., and Fishbein, L. (eds.), *Environmental Carcinogens, Selected Methods of Analysis,* Vol. I, *Some Metals: As, Be, Cd, Cr, Ni, Pb, Se, Zn.* IARC Sci. Publ. No. 71. International Agency for Research on Cancer. Lyon.
Jennings, P.H., and Yannopoulos, J.C. (1974), in: Zingaro, R.A., and Cooper, W.C.U. (eds.), *Selenium,* pp. 31–51. Van Nostrand Reinhold, New York.
Jiang, S., De Jonghen, A., and Adams, F. (1982), Anal. Chim. Acta 136, 183–190.
Jiang, S., Robberecht, H., and Adams, F. (1983), Atmos. Environ. 17, 111–114.
Jickells, T.D., Knap, A.H., and Church, T.M. (1984), J. Geophys. Res. 89, 1423–1428.
Johnson, R.D., and Manske, D.D. (1977), Pestic. Monit. J. 11, 116–131.
Karlsen, J.T., Nordheim, G., and Froslie, A. (1981), Acta Agric. Scand. 31, 165–170.
Kazantzis, G. (1981), *Role of Cobalt, Iron, Lead, Manganese, Mercury, Platinum, Selenium, and Titanium in Carcinogenesis.* Environ. Health Perspect. 40, 143–161.
Keshan Disease Research Group (1979a), Chin. Med. J. 92, 471–476.
Keshan Disease Research Group (1979b), Chin. Med. J. 92, 477–482.
Kieffer, F. (1987), *Selen, ein medizinisch bedeutungsvolles Spurenelement,* Ars Medici Nr. 2, 59–74; and *Neue Zürcher Zeitung, Forschung und Technik,* Nr. 137, p. 87 (17 June).
Kinnigheit, G. (1962), *Arbeit.* Gleichrichter Werkszeitschrift. Gesamte Hygiene Grenzgebiete 8, 350–362.
Knight, A.W., Maier, K.J., Foe, C., Ogle, R.S., Williams, M.J., Kiffney, P., and Melton, L.A. (1987), *The Dynamics of Selenium in Aquatic Ecosystems.* Trace Subst. Environ. Health 21, 361–408.
Koller, L.D. (1987), *Immunotoxicology and Risk Assessment of Drinking Water Contaminants.* Trace Subst. Environ. Health 21, 247–252.
Koller, L.D., and Exon, J.H. (1986), Can. J. Vet. Res. 50, 297–306.
Kraus, R.J., Foster, S.J., and Ganther, H.E. (1983), Biochemistry 22, 5853–5858.
Kubo, M., Yano, T., Kobayaski, H., and Ueno, K. (1977), Talanta 24, 519–521.
Kubota, J., Allaway, W.H., Carter, D.L., et al. (1967), *Selenium in Crops in the U.S. in Relation to Selenium Responsive Disease in Animals.* J. Agric. Food Chem. 15, 448.
Levander, O.A. (1972), Ann. N.Y. Acad. Sci. 393, 70–82.
Levander, O.A. (1976), *Selenium in Foods,* in: *Selenium-Tellurium in the Environment,* pp. 26–53. Industrial Health Foundation, Pittsburgh, Pennsylvania.
Levander, O.A. (1985), Fed. Proc. 44, 2579–2583.
Levander, O.A. (1982), *Selenium: Biochemical Actions, Interactions and Some Human Implications,* in: Prasad, A.S. (ed.), *Clinical, Biochemical and Nutritional Aspects of Trace Elements,* pp. 135–163. Academic Press, New York.
Levander, O.A. (1986), *The Need for Measures of Selenium Status.* J. Am. Coll. Toxicol. 5, 37–44.
Levander, O.A. (1987), *Dietary Selenium Requirements: Minimal and Optimal Intake and Disadvantages in the Use of Supplements.* 2nd Nordic Symposium on Trace Elements in Human Health and Disease, Odense. Abstract Mini-Symposium Nr. 12 and Round-Table Discussion on

Mineral Supplementation. *Environ. Health 26*, 87–90. WHO Regional Office for Europe, Copenhagen.

Lo, M. T., and Sandi, E. (1980), *Selenium: Occurrence in Foods and its Toxicological Significance – A Review. J. Environ. Pathol. Toxicol. 4*, 193–218.

Lo, L. W., Koropatnick, J., and Stich, H. F. (1978), *Mutat. Res. 49*, 305–312.

Lundberg, K., and Paulsson, K. (1988), *The Selenium Method for Treatment of Lakes Against Elevated Levels of Mercury in Fish. International Conference on Trace Metals in Lakes, Hamilton, Ontario.* Proceedings, *J. Sci. Total Environ.*, in press.

Luten, J. B., Bouquet, W., Burggraf, M. M., and Rus, J. (1986), in *Abstracts of International Workshop Trace Element Analytical Chemistry in Medicine and Biology*; Abstr. No. 60; April 21–23, Gsf-Bericht, Neuherberg, FRG.

MacPherson, A. K., Sampson, B., and Diplock, A. T. (1988), *Analyst 113*, 281–283.

Mahaffey, K. R., Corneliussen, P. E., Jelinek, C. F., and Firoino, J. A. (1975), *Environ. Health Perspec. 12*, 63–69.

MAK (1987), *Maximum Concentrations at the Workplace, Report No. XXIII, DFG.* Commission for the Investigation of Health Hazard. VCH Verlagsgesellschaft, Weinheim-Basel-Cambridge-New York.

Makida, J., Shimoishi, Y., and Toei, K. (1975), *Analyst 100*, 648.

Martin, S. E., and Schillaci, M. (1984), *Inhibitory Effects of Selenium on Mutagenicity. J. Agric. Food Chem. 32*, 426–433.

Martin, S. E., Adams, G. H., Schillaci, M., and Milner, J. A. (1981), *Antimutagenic Effect of Selenium on Acridine Orange and 7,12-DMBA in the Ames Salmonella Microsomal System. Mutat. Res. 82*, 42.

McConnell, K. P., Broghamer, W. L., Jr., Blotcky, A. J., and Hurt, O. J. (1975), *Selenium Levels in Human Blood and Tissues in Health and Disease. J. Nutr. 105*, 1026.

McConnell, K. P., Jager, R. M., Bland, K. I., and Blotcky, A. J. (1980), *J. Surg. Oncol. 15*, 67.

Medina, D. (1985), *Selenium and Murine Mammary Tumorigenesis*, in: Reddy, B., and Cohen, L. (eds.), *Diet, Nutrition and Cancer: A Critical Evaluation*. CRC Press, Boca Raton, Florida.

Medina, D. (1986), *Mechanisms of Selenium Inhibition on Tumorigenesis. J. Am. Coll. Toxicol. 5*, 21–27.

Medina, D., and Shepherd, F. (1980), *Selenium-mediated Inhibition of Mouse Mammary Tumorigenesis. Cancer Lett. 8*, 241–245.

Medina, D., Lane, H., and Shepherd, F. (1981), *Effects of Selenium on Mouse Mammary Tumorigenesis and Glutathione Peroxidase Activity. Anticancer Res. 1*, 377–382.

Medina, D., Lane, H. W., and Tracey, C. M. (1983), *Selenium and Mouse Mammary Tumorigenesis: An Investigation of Possible Mechanisms. Cancer Res. 43*, 2460s–2462a.

Medinsky, M. A., Cuddihy, R. G., Griffith, W. C., et al. (1985), *Environ. Res. 36*, 181–192.

Merian, E. (1985), *Introduction on Environmental Chemistry and Global Cycles of Chromium, Nickel, Cobalt, Beryllium, Arsenic, Cadmium, and Selenium,* in: Merian, E., Frei, R. W., Härdi, W., and Schlatter, C. (eds.), *Carcinogenic and Mutagenic Compounds*, pp. 25–32. Gordon & Breach, London.

Mikkelsen, R. L., Page, A. L., and Bingham, F. T. (1987), *Geochemistry and Health in California; Recent Experiments with Selenium. Trace Subst. Environ. Health 20*, 413–423.

Milner, J. A. (1985), *Effect of Selenium on Virally Induced and Transplantable Tumor Models. Fed. Proc. 44*, 2568–2572.

Morris, V. C., and Levander, O. A. (1970), *J. Nutr. 100*, 1383–1388.

Mosher, B. W., and Duce, R. A. (1983), *J. Geophys. Res. 88*, 6761–6768.

Mosher, B. W., and Duce, R. A. (1987), *J. Geophys. Res. 92*, 13289–13298.

Moxon, A. L., and Olson, O. E. (1974), in: Zingaro, R. A., and Cooper, W. C. (eds.), *Selenium*, pp. 675–707. Van Nostrand Reinhold, New York.

Muangnoicharoen, S., Chiou, K. Y., and Manuel, O. K. (1986), *Anal. Chem. 58*, 2811–2813.

Mushak, P. (1985), *Potential Impact of Acid Precipitation on Arsenic and Selenium. Environ. Health Perspec. 63*, 105–113.

Mutanen, M. (1986), *Ann. Clin. Res. 18*, 48–54.
Mutanen, M., and Koivistoinen, P. (1983), *Int. J. Vitam. Nutr. Res. 113*, 55–64.
Nahapetian, A. J., Janghorbani, M., and Young, V. R. (1983), *J. Nutr. 113*, 401–411.
NAS (1976), *Medical and Biological Effects of Environmental Pollutants – Selenium.* National Academy of Sciences, Washington, D.C.
NAS (1980), *Recommended Dietary Allowances*, pp. 162–164. National Academy of Sciences, Washington, D.C.
NAS (1983), *Selenium in Nutrition.* National Academy of Sciences, Washington, D.C.
Nazarenko, I. I., and Ermakov, A. N. (1972), *Analytical Chemistry of Selenium and Tellurium.* Halsted Press, New York.
Nelson, A. A., Fitzhugh, O. G., and Calvery, H. O. (1943), *Cancer Res. 3*, 230–236.
Neve, J., and Molle, L. (1986a), in *Abstracts of International Workshop Trace Element Analytical Chemistry in Medicine and Biology;* Abstr. No. 37, April 21–23. Gsf-Bericht, Neuherberg, FRG.
Neve, J., and Molle, L. (1986b), *Acta Pharmacol. Toxicol. 59*, Suppl. 7, 606–609.
Newland, L. W. (1982), in: Hutzinger, O. (ed.), *The Handbook of Environmental Chemistry,* Vol. 3, Part B, *Anthropogenic Compounds,* pp. 45–57. Springer Verlag, Heidelberg.
Noda, M., Takano, T., and Sakurai, H. (1979), *Mutagenic Activity of Selenium Compounds. Mutat. Res. 66*, 175–179.
Norheim, G., and Haugen, A. (1986), *Acta Pharmacol. Toxicol. 59*, Suppl. 7, 620–712.
Norheim, G., and Nymoen, U. K. (1981), *8th Nordic Trace Element and Microchemical Conference, Sandefjord, Norway.*
Norman, V. (1977), *Adv. Tob. Sci. 3*, 25–58.
Nriagu, J. O., and Pacyna, J. M. (1988), *Quantitative Assessment of Worldwide Contamination of Air, Water and Soils by Trace Metals. Nature 333*, 134–139.
Nürnberg, H. W. (1982), *Pure Appl. Chem. 54*, 853–878.
Obermeyer, B. D., Palmer, I. S., Olson, O. E., and Halverson, A. W. (1971), *Toxicity of Trimethylselenium Chloride in the Rat With and Without Arsenic. Toxicol. Appl. Pharmacol. 20*, 135–146.
Olson, O. E. (1976), in *Selenium-Tellurium in the Environment.* Industrial Health Foundation, Pittsburgh, Pennsylvania.
Olson, O. E. (1986), *Selenium Toxicity in Animals with Emphasis on Man. J. Am. Coll. Toxicol. 5*, 45–70.
Olson, O. E., and Frost, D. V. (1970), *Selenium in Papers and Tobacco. Environ. Sci. Technol. 4*, 686–687.
Olson, O. G., Palmer, I. S., and Whitehead, E. I. (1973), *Methods Biochem. Anal. 21*, 39–78.
Pacyna, J. M., Semb, A., and Hansen, J. E. (1984), *Telus 36B*, 163–178.
Palmer, I. S., Thiex, N., and Olson, O. E. (1983), *Nutr. Rep. Int. 27*, 249–251.
Peterson, P. J. (1987), *Selenium Biogeochemistry: Local, Regional, and Global Processes and Problems. Trace Subst. Environ. Health 21*, 353–360.
Peterson, P. J., Girling, C. A., Benson, L. M., and Zieve, R. (1981), in: Lepp, N. W. (ed.), *Effect of Heavy Metal Pollution on Plants,* Vol. I, p. 213. Applied Science Publishers, London.
Priest, P., Navarre, J. L., and Ronneau, C. (1981), *Elemental Background Concentration in the Atmosphere of an Industrialized Country. Atmos. Environ. 15*, 1325–1330.
Raptis, S. E., Knapp, G., Meyer, A., and Tölg, G. (1980a), *Fresenius Z. Anal. Chem. 300*, 18–20.
Raptis, S. E., Wegscheider, W., Knapp, G., and Tölg, G. (1980b), *Anal. Chem. 52*, 1292–1296.
Raptis, S. E., Kaiser, G., and Tölg, G. (1983), *A Survey of Selenium in the Environment and a Critical Review of its Determination at Trace Levels. Fresenius Z. Anal. Chem. 316*, 102.
Ray, J. H., Altenburg, L. C., and Jacobs, M. M. (1978), *Sister-chromatid Exchange Induction by Sodium Selenite: Dependence on the Presence of Red Blood Cells or Red Blood Cell Lysate. Mutat. Res. 57*, 359–368.
Reamer, D. C. (1980), *Methods for the Determination of Atmospheric and Alkylselenide Fluxes Using a G.C. Microwave Plasma Detector.* Ph.D. Thesis, University of Maryland, College Park.

Reamer, D. C., and Zoller, W. H. (1980), *Selenium Biomethylation Products from Soil and Sewage Sludge. Science 208*, 500–502.
Rhian, M., and Moxon, A. L. (1943), *J. Pharmacol. Exp. Ther. 78*, 249–264.
Robberecht, H. J., and Deelstra, H. A. (1984), *Clin. Chim. Acta 136*, 107–120.
Robberecht, H., and van Grieken, R. (1980), in: Hemphill, D. D. (ed.): *Trace Substances in Environmental Health*, Vol. 14, University of Missouri, Columbia.
Robberecht, H., and van Grieken, R. (1982), *Talanta 29*, 823–844.
Robinson, M. F., and Thomson, C. D. (1983), *The Role of Selenium in the Diet. Nutr. Abstr. Rev. 53*, 3–26.
Rosin, M. P., and Stich, H. F. (1979), *Assessment of the Use of the Salmonella Mutagenesis Assay to Determine the Influence of Antioxidants on Carcinogen-induced Mutagenesis. Int. J. Cancer 23*, 722–727.
Ross, H. B. (1985), *Tellus 37B*, 78–90.
Saeed, K. (1986), *Acta Pharmacol. Toxicol. 59*, Suppl. 7, 593–597.
Salmon, L., Atkins, D. H. F., Fisher, E. M. R., Healy, C., and Law, D. V. (1978), *Retrospective Trend Analysis of the Content of U.K. Air Particulate Material 1957–1974. Sci. Total Environ. 9*, 161–200.
Salonen, J. T. (1986), Ann. Clin. Res. *18*, 18–21.
Salonen, J. T., Salonen, R., Lappetelainen, R., et al. (1985), *Risk of Cancer in Relation to Serum Concentrations of Selenium and Vitamins A and E: Matched Case Control Analysis of Prospective Data. Br. Med. J. 290*, 417–421.
Schrauzer, G. N. (1987), *Effects of Interactions of Lead, Cadmium and Other Metals with Selenium on the Genesis and Growth of Malignant Tumors: A New Aspect of Chronic Metal Toxicity. J. UEOH* (Univ. Occup. Environ. Health Jpn.) *9*, 208–215.
Schrauzer, G. N., White, D. A., McGinness, J. E., et al. (1976), *Inhibition of the Genesis of Spontaneous Mammary Tumors in C3H Mice: Effects of Selenium and of Selenium Antagonistic Elements and their Possible Role in Breast Cancer. Bioinorg. Chem. 9*, 265–270.
Schrauzer, G. N., White, D. A., and Schneider, C. J. (1977), *Cancer Mortality Correlation Studies. III. Statistical Associations with Dietary Selenium Intakes. Bioinorg. Chem. 7*, 23–24.
Schrauzer, G. N., Kuehn, K., and Hamm, D. (1979), *Effects of Dietary Selenium and Lead on the Genesis of Spontaneous Mammary Tumors in Mice. Biol. Trace Elem. Res. 3*, 185–196.
Schrauzer, G. N., McGinness, J. E., and Kuehn, K. (1980), *Effects of Temporary Selenium Supplementation on the Genesis of Spontaneous Mammary Tumors in Inbred Female C3H/Se Mice. Carcinogenesis 1*, 199–201.
Schrauzer, G. N., Molenaar, T., Mead, S., et al. (1985), *Gann 76*, 374.
Schwarz, K. (1976), in *Selenium-Tellurium in the Environment*, pp. 349–376. The Industrial Health Foundation, Pittsburgh, Pennsylvania.
Schwarz, K., and Foltz, C. M. (1957), *Selenium as an Integral Part of Factor 3 Against Dietary Necrotic Liver Degeneration. J. Am. Chem. Soc. 79*, 3292–3293.
Seifter, J., Ehrich, W. E., Hudyma, G., and Mueller, G. (1946), *Thyroid Adenomas in Rats Receiving Selenium. Science 103*, 762.
Shacklette, H. T., Boerngen, J. G., and Keith, J. R. (1974), *Survey Circular 692*. US Department of Interior, Washington, D.C.
Shamberger, R. J. (1970), *Relationship of Selenium to Cancer. I. Inhibition Effect of Selenium on Carcinogenesis. J. Natl. Cancer Inst. 44*, 931–936.
Shamberger, R. J. (1986), in *Abstracts of International Workshop Trace Element Analytical Chemistry in Medicine and Biology*. Abstract No. 19; April 21–23. Gsf-Bericht, Neuherberg, FRG.
Shamberger, R. J., and Frost, D. V. (1969), *Possible Protective Effect of Selenium Against Cancer. Can. Med. Assoc. J. 100*, 682.
Shamberger, R. J., and Willis, C. E. (1971), *Selenium Distribution and Human Cancer Mortality. CRC Crit. Rev. Clin. Lab. Stud.* , 211–216.

Shamberger, R. J., Rukovena, E., and Longfield, A. K. (1973), *Antioxidants and Cancer. I. Selenium in Blood of Normal and Cancer Patients. J. Natl. Cancer Inst. 50*, 863–870.

Shamberger, R. J., Corlett, C. L., Beamon, K. D., and Kasten, B. L. (1979), *Antioxidants Reduce the Mutagenic Effect of Malonaldehyde and beta-Propiolactone. IX. Antioxidants and Cancer. Mutat. Res. 66*, 349–355.

Shapiro, J. R. (1973), in: Klayman, D. L., and Gunther, W. H. H. (eds.): *Organic Selenium Compounds: Their Chemistry and Biology*, p. 693. John Wiley, New York.

Shendrikar, A. D. (1974), *Sci. Total Environ. 3*, 155.

Shimoishi, Y. (1977), *Analyst 101*, 298.

Shull, L. R., Buckmaster, G. W., and Cheeke, P. R. (1979), *Effect of Dietary Selenium Status on in vitro Hepatic Mixed Function Oxidase Enzymes of Rats. J. Environ. Pathol. Toxicol. 2*, 1127–1138.

Sindeeva, N. D. (1964), *Mineralogy and Types of Deposits of Selenium and Tellurium.* Interscience Publ., New York.

Slu, K. W. M., and Berman, S. S. (1983), *Anal. Chem. 55*, 1605.

Spallholz, J. E., Martin, J. L., Gerlach, M. L., et al., (1973), *Immunological Responses of Mice Fed Diets Supplemented with Selenite Selenium. Proc. Soc. Exp. Biol. Med. 143*, 685.

Speit, G. (1987), *Zur Wirkung klassischer Antimutagene* (e.g., sodium selenite and glutathione) *auf den Schwesterchromatidaustausch (SCE)*, with interesting literature, *GUM 1*(2), 3–7. GIT-Verlag, Darmstadt.

Stewart, R. D. H., Griffiths, N. M., Thompson, C. D., and Robinson, M. F. (1978), *Quantitative Selenium Metabolism in Normal New Zealand Women. Br. J. Nutr. 40*, 45–54.

Stijve, T., and Cardinale, E. (1975), *J. Chromatogr. 109*, 239.

Strain, W. H., Varnes, A. W., Paxton, C. A., Drenski, T. L., and Hill, O. A. (1981), *Trace Subst. Environ. Health 15*, 104–118.

Sugawara, N., and Sugawara, C. (1987), *Effect of Oral Selenium on Cadmium Absorption from the Gastrointestinal Tract. Trace Subst. Environ. Health 21*, 440–449.

Sugimura, Y., and Suzuki, Y. (1973), *J. Oceanogr. Soc. Jpn. 33*, 23.

Sunde, R. A., and Hoekstra, W. G. (1980), *Incorporation of Selenium from Selenite and Selenocystine into Glutathione Peroxidase in the Isolated Perfused Rat Liver. Biochem. Biophys. Res. Commun. 93*, 1181–1188.

Suzuki, Y., Sugimura, Y., and Miyake, Y. (1981), *J. Metall. Soc. Jpn. 59*, 405–409.

Swanson, C. (1986), *Dietary Selenium as a Risk Factor in Human Cancer. Nestlé Research News*, pp. 55–60. Vevey.

Talmi, Y., and Andren, A. W. (1974), *Anal. Chem. 46*, 2122.

Tanji, K., Lauchi, A., and Meyer, J. (1986), *Selenium in San Joaquin Valley. Environment 28*, 6–11, 34–39.

Teel, R. W. (1984), *A Comparison of the Effect of Selenium on the Mutagenicity and Metabolism of Benzo[a]pyrene in Rat and Hamster Liver S-9 Activation Systems. Cancer Lett. 24*, 281–289.

Thomassen, Y., Lewis, S. A., and Veillon, C. (1988), Chapter *Selenium*, in: Stoeppler, M. (ed.): *Trace Metal Analysis in Biological Specimens.* PSG Publishing Co. Inc., Littleton, Massachusetts.

Thompson, K. C. (1975), *Analyst 110*, 307.

Thompson, H. J. (1984), *J. Agric. Food Chem. 32*, 422–425.

Thompson, H. J., and Becci, P. J. (1979), *Cancer Res. 39*, 1339–1346.

Thompson, H. J., and Becci, P. J. (1980), *Selenium Inhibition of N-Methyl-N-nitrosourea-induced Mammary Carcinogenesis in the Rat. J. Natl. Cancer Inst. 65*, 1299–1301.

Thorling, E. B., Overvad, K., and Geboers, J. (1986), *Ann. Clin. Res. 18*, 3–7.

Tsay, D. T., Halverson, A. W., and Palmer, I. S. (1970), *Nutr. Rep. Int. 2*, 203–207.

Tscherkes, L. A., Volgarev, M. N., and Aptekar, S. G. (1963), *Selenium Caused Tumors. Acta Univ. Int. Contra Cancrum 19*, 632.

Tsujii, K., and Kuga, K. (1974), *Anal. Chim. Acta 72*, 85.

US Dept. HHS (1988), *The Surgeon General's Report on Nutrition and Health 1988*, DHHS (PHS) Publ. No. 88-50210, pp. 219–221. US Department of Health and Human Services, Washington, D.C.

van den Haute, J., Maenhaut, W., Rinsvelt, H. I., Hurd, R. W., and Andres, J. M. (1986), in *Abstracts of International Workshop Trace Element Analytical Chemistry in Medicine and Biology;* Abstract No. 23; April 21 – 23. Gsf-Bericht, Neuherberg, FRG.
van Faasen, A., Cardinaals, J. M., van'tVeer, P., and van der Beek, E. J. (1986), in *Abstracts of International Workshop Trace Element Analytical Chemistry in Medicine and Biology;* Abstract No. 27; April 21 – 23. Gsf-Bericht, Neuherberg, FRG.
Verlinden, M., Deelstra, H., and Adriaenssens, E. (1981), *Talanta 28*, 637.
Vernie, L. N. (1984), *Selenium in Carcinogenesis. Biochim. Biophys. Acta 738,* 203 – 217.
Vickrey, T. M., and Buren, M. S. (1980), *Anal. Lett. 13*, 1465 – 1485.
von Stockhausen, H. B. (1986), in *Abstracts of International Workshop Trace Element Analytical Chemistry in Medicine and Biology;* Abstract No. 47; April 21 – 23. Gsf-Bericht, Neuherberg, FRG.
Wadge, A., Hutton, M., and Peterson, P. J. (1986), *Sci. Total Environ. 54*, 13 – 27.
Wasowicz, W., and Zachara, B. A. (1987), *Selenium Concentrations in the Blood and Urine of a Healthy Polish Sub-population. J. Clin. Chem. Clin. Biochem. 25*, 409 – 412.
Wattenberg, L. W. (1978), *Adv. Cancer Res. 26*, 197.
Wauchope, R. D., and McWhorter, C. G. (1977), *Bull. Environ. Contam. Toxicol. 17*, 165.
Welz, B. (1985), *Atomic Absorption Spectrometry*, 2nd Ed., pp. 320 – 323, 350, 357ff, 401ff. VCH Verlagsgesellschaft Weinheim-Deerfield Beach/Florida-Basel.
Welz, B., and Verlinden, M. (1986), *Acta Pharmacol. Toxicol. 59*, Suppl. 7, 577 – 580.
Whanger, P. D. (1981), in: Spallholz, J. E., Martin, J. L., and Ganther, H. E. (eds.): *Selenium in Biology and Medicine,* pp. 230 – 255. AVI Publ., Westport, Connecticut.
Whanger, P. D. (1983), *Selenium Interactions with Carcinogens. Fund. Appl. Toxicol. 3*, 424 – 430.
Whiting, R. F., Wei, L., and Stich, H. F. (1980), *Unscheduled DNA Synthesis and Chromosome Aberrations Induced by Inorganic and Organic Selenium Compounds in the Presence of Glutathione. Mutat. Res. 78*, 159 – 169.
WHO (1973), *Trace Elements in Human Nutrition,* WHO Technical Report Series No. 532. World Health Organization, Geneva.
WHO (1984), *Guidelines for Drinking-Water Quality,* Vol. 1, *Recommendations*, pp. 57 – 58. World Health Organization, Geneva.
WHO (1987), *Selenium, Environmental Health Criteria 58*, IPCS International Program on Chemical Safety. World Health Organization, Geneva.
Wiese, U. (1982), *Selen und Selen-Verbindungen,* in: *Ullmanns Encyklopädie der technischen Chemie,* 4th Ed., Vol. 21, pp. 227 – 234. Verlag Chemie, Weinheim-Deerfield Beach/Florida-Basel.
Wilber, C. G. (1983), *Selenium, A Potential Environmental Poison and a Necessary Food Constituent.* Charles C. Thomas Publ., Springfield, Illinois.
Willett, W. C., Morris, J. S., Pressel, S., and Taylor, J. D. (1983a), *Prediagnostic Serum Selenium and Risk of Cancer.* Lancet II, 130 – 134.
Willett, W. C., Polk, B. F., and Hames, C. (1983b), Lancet II, 130.
Wong, P. T. S., and Goulden, P. D. (1975), *Anal. Chem. 47*, 2279 – 2281.
Wortzman, M. J., Besbris, H. J., and Cohen, A. M. (1980), *Cancer Res. 40*, 2670.
Yang, G. Q., Wang, S., Zhou, R., et al. (1983), *Endemic Selenium Intoxication of Humans in China. Am. J. Clin. Nutr. 37*, 872 – 881.
Young, J. N., and Christian, G. D. (1977), *Anal. Chim. Acta 65*, 127.
Young, V. R. (1981), *Selenium: A Case for its Essentiality. N. Eng. J. Med. 304*, 1228 – 1230.
Young, V. R., Nahapetian, A., and Janghorbani, M. (1982), *Selenium Bioavailability with Reference to Human Nutrition. Am. J. Clin. Nutr. 35*, 1076 – 1087.
Yu, S. Y., Chu, Y. J., Gong, X. L., et al. (1985), *Regional Variability of Cancer Mortality Incidence and its Relation to Selenium Levels in China. Biol. Trace Elem. Res. 7*, 20 – 29.
Zhang, X. (1987), *The Relationship between Endemic Diseases and Trace Elements in the Natural Environment of Jilin Province of China. Trace Subst. Environ. Health 20*, 381 – 391.
Zieve, R., and Peterson, J. (1984), *Selenium Volatilization from Plants and Soils. Sci. Total Environ. 32*, 197 – 202.

Additional Recommended Literature

Jacobs, L. W. (1989), *Selenium in Agriculture and the Environment, SSSA Special Publication 23.* Soil Science Society of America and American Society of Agronomy, Madison, Wisconsin.

Jackson, T. A. (1989), *Inhibitory and Stimulatory Effects of Copper, Cadmium, Zinc, Mercury, and Selenium on Microbial Production of Methyl Mercury in Sediments. Proceedings 7th International Conference on Heavy Metals in the Environment, Geneva,* Vol. 1, pp. 65–68. CEP Consultants Ltd., Edinburgh, UK.

Levander, O. A. (1990), *Selenium: Essentiality of Trace Elements versus Toxicity in Man. Proceedings 23rd Conference on Trace Substances in Environmental Health, Cincinnati (Ohio).* Environ. Geochem. Health, Suppl. to Vol. 12, 11–19.

Yang, F. Y., Lin, Z. H., Li, S. G., Guo, B. Q., and Yin, Y. S. (1988), *Keshan Disease – an Endemic Mitochondrial Cardiomyopathy in China.* J. Trace Elem. Electrolytes Health Dis. 2, 157–163.

II.26 Silver

H. G. PETERING and C. J. MCCLAIN, Lexington, Kentucky, USA

1 Introduction

Silver is a useful noble metal with high conductivity and ductility, which is widely disseminated in the human environment as a result of its industrial and medicinal applications. At present there is no evidence that this wide environmental distribution causes any serious health risks for human beings or special hazards for higher organisms. It may, however, be a serious problem for certain forms of aquatic life and thus disturb some ecosystems. It also may pose a health hazard for selected groups of people because of occupational exposure or medicinal use.

2 Physical and Chemical Properties, and Analytical Methods

Silver (Ag) is an element of the 1 B Group of elements and is in the second transition series of the Periodic Table, thus making it a metal with coordinating properties. It has an atomic number of 47, an atomic mass of 107.87, the electronic configuration of $4\,d^{10}\,5\,s^1$, a density of 10.5 g/cm^3, a melting point of 960°C, and a boiling point of 2170°C. Silver is a relatively soft metal, which is very malleable and has the highest known values for thermal and electrical conductivity. Silver resists oxidation and remains untarnished in dry air devoid of hydrogen sulfide. Silver can exist in three cationic forms in addition to its metallic state (Ag^0), as Ag(I), Ag(II), and Ag(III), the latter two being unstable in aqueous solutions. Univalent silver (Ag$^+$) forms linear coordination complexes with the coordination number of 2, predominantly, but complex structures with coordination numbers of 3 and 6 are also known (COTTON and WILKINSON, 1972).

A number of methods for the determination of silver are available. A dithizone colorimetric method is suitable for analytical concentrations in the range of 1 to 25 µg/mL (PRZYBYLOWICZ and ZUEHLKE, 1966). Biological and clinical samples, often small in size and concentration, are better analyzed using spectroscopic methods. The detection limits for emission and atomic absorption spectrophotometric methods are approximately 1–50 µg/L (1 to 50 ng/mL) and under special conditions these limits may be as low as 1 to 2 pg/sample; see also AAS-techniques for silver (WELZ, 1985), determination of silver in plants by flameless AAS and NAA

(JONES and PETERSON, 1985), and ICP in AAS (MONTASER and GOLIGHTLY, 1987).

The anode stripping voltammetric method for determining Ag has a sensitivity of approximately 4 ng/mL (SMITH and CARSON, 1977). Since silver has such a high melting point, ordinary methods of ashing samples pose no problem of loss. The main analytical problems are solubilizing the ash, or the silver in the ash, freeing it from interfering metals or matrices, and finally incorporating it quantitatively into a suitable solvent or medium for the analytical determination.

The problem of silver chemical speciation with respect to the proximal toxic agent in aquatic environments has not been adequately addressed experimentally. This situation occurred because of the chemical environment both in sea water and in many so-called fresh surface waters. COWAN et al. (1985), however, have published an interesting theoretical paper on the subject, based on the use of the MINTEQ geochemical model (FELMY et al., 1984). They assume that AgCl is the primary bioavailable and most toxic species and provide information about sulfide (S^{2-}), dissolved organic carbon (DOC), and pH variations of sea water and some fresh waters. Their conclusions indicate the importance of the following variables on aqueous silver speciation in decreasing order of magnitude: S^{2-}, 0.0 – 10.0 µg/L; pH, 7.0 – 8.2; salinity, 27 – 32‰; and DOC, 1 – 5 mg/L. The effect of S^{2-} even at low concentrations was found to overwhelm the effects of other variables so that it was determinative. These conclusions indicate the importance of very accurate determination of S^{2-} concentrations and the desirability of very accurate pH measurement to the sensitivity of 0.01 pH unit.

3 Sources, Production, Important Compounds, Uses, Waste Products, Recycling

Although silver occurs in nature as the free metal, the amounts of this form are very limited. It is widely distributed mainly in the forms of Ag_2S (argentite), AgCl (cerargerite or horn silver), and as Ag_3As (silver arsenide). Silver often is found as an impurity in the ores of zinc, copper, and lead. It also exists in combination with sulfides of copper, arsenic, and antimony. Almost all soils, sea water, and some fresh waters contain traces of silver. As a consequence, it is also found in many foods in small but measurable amounts.

The major silver producing countries are Mexico, Peru, the Soviet Union, Canada, Australia, and the USA. (*Silver Institute Letter*, 1986). The total world production of silver in 1985 was 10700 metric tons, a slight increase from 9200 tons produced a decade ago (STOKINGER, 1980). It occurs mostly as a by-product from the extraction of lead, zinc, copper, and tin ores, and the raw silver is refined, for instance, electrolytically (SAAGER, 1984; RENNER, 1982). Current prices of silver still make secondary sources, such as waste from photographic film production and processing, profitable to refine. It is estimated that about 31% of all silver refined in 1985 originated from secondary waste sources.

Six major uses of silver consumed 86% of the 1974 production (STOKINGER, 1980; see also RENNER, 1982; SAAGER, 1984). These were: photographic materials, 28% (rising); electrical controls and conductors, 18% (rising); coins, medallions, etc. 13% (decreasing); sterling ware, 12%; brazing alloys, 8%; and electroplated ware, 7%. Most of the remaining 14% of the 1974 production was used in jewelry, alkaline batteries, mirrors, and chemical catalyst materials (STOKINGER, 1980). $AgNO_3$ is an important compound made from silver (RENNER, 1982). Water can be completely disinfected with 5–10 ppb of silver as copper-silver alloy, or a silver salt except silver sulfide, e.g., silver nitrate (ZOBRIST, 1980; RUEGGER and DUBUIS, 1981). The use of silver in medicine and dentistry is small; in the USA this amounts to about 1% of the total consumption.

It has been estimated that the total annual loss of silver to the environment from industrial wastes can amount to more than 2500 tons (SMITH and CARSON, 1977). Of this amount, about 150 tons were found in sewage sludge and about 80 tons in surface waters. BOWEN (1966) suggested that silver may pose a potential risk as a water pollutant because of the lack of recycling of mined silver. There are processes for the recovery of silver from photographic chemical residues, from amalgam wastes, and from galvanizing machinery (equipment).

4 Distribution in the Environment, in Foods, and in Living Organisms

Silver is widely distributed in the environment. In the earth's crust it is present in 0.1 to 0.2 mg/kg (HESLOP and ROBINSON, 1967; GOLDSCHMIDT, 1937); 52% occurs as ^{107}Ag isotope and 48% as ^{109}Ag isotope. Sea water contains 0.002 (less at the surface) to more than 0.4 µg/L of silver (SIBLEY and MORGAN, 1975; MARTIN et al., 1983; BRULAND, 1983), and river waters contain 0.3 to 1.3 µg/L (BOWEN, 1979, and others).

The amount of silver contaminants in air is similar to that found for mercury. In Chicago in 1970, for example, sampling filters contained silver in amounts which indicated that the sampled air contained 0.8–7.0 µg/m^3 (BARR et al., 1970).

Silver is found in rain water, which may allow it to enter many remote places in the environment. This source could account for the finding of silver in many sewage sludges in measurable, but highly variable amounts. Sludges in California, for instance, contain a mean amount of 0.18 ppm of silver, but sludges from individual localities vary widely, some having as much as 130 ppm of the element. Sludges in the USA have been found to be high enough in silver to increase the silver in sludge-amended soils by ten-fold. Waste treatment plant effluent waters in the USA usually contain less than the permissible level of 50 ppm (SMITH and CARSON, 1977).

Silver is present in plants in the range of 0.06 to 0.28 µg/g dry weight. In fungi and bacteria silver is present in amounts of about 29 and 210 µg/g, respectively. Aquatic plants tend to concentrate silver from their environments several hundred-fold.

Trace amounts of silver occur in many foodstuffs and in most animal tissues. Flour contains about 0.3 µg/g and wheat bran about 0.9 µg/g (KENT and MCCANCE, 1941). USA milk has been reported to contain amounts of silver between 0.025 and 0.054 µg/mL (MURTHY and RHEA, 1968). Beef, pork, and mutton meats contain silver in amounts ranging from 0.004 to 0.024 µg/g dry weight (ARMOUR RESEARCH, 1952), and it is found in other animal tissues in widely varying amounts. The average content of fish is 11 ppm, whereas that of mammals is 0.006 µg/g wet weight.

The average silver values for the major organs of mammals in ppm are: brain, 0.04; liver, 0.03; kidney, 0.005; and lung, 0.005 (BOWEN, 1966, 1979). The mean content of silver in human blood is less than 0.003 ppm. The mean contents of silver for some human organs in ppm are: brain, 0.004; kidney, 0.002; liver, 0.006; and lung, 0.002 (HAMILTON et al. 1972, 1973, see also BRUNE et al. 1980). Normal concentrations may be higher (FOWLER and NORDBERG, 1986).

5 Uptake, Absorption, Transport and Distribution, and Metabolism and Elimination in Animals and Humans

Silver can be absorbed through the skin, lungs, and gastrointestinal tract. Inhalation of silver by dogs allowed pulmonary absorption of silver and subsequent redistribution from the lungs to the liver, brain, and muscles (PHALEN and MORROW, 1973). It has been shown that 10% of an oral dose of soluble silver salt was absorbed from the gastrointestinal tract, and that 90% of a dose absorbed by any route was excreted in the feces (FURCHNER et al., 1968). In agreement with these experimental findings, NEWTON and HOLMES (1966) reported that very little of the silver absorbed by a human being who accidentally had inhaled a large amount of ^{110}Ag was excreted in the urine, and that most of the dose absorbed was found in the feces (see also end of Sect. 6).

The oral intake of silver from food products eaten by the general population has been estimated to be between 27 and 88 µg/day (HAMILTON and MINSKI, 1972/1973; KEHOE et al., 1940). WESTER (1971) found a somewhat lower daily intake. From this information it may be calculated that human beings may accumulate a body burden of about 9 mg of silver over a 50 year period. The half-life of silver in the liver and skin is relatively long, about 50 days (NEWTON and HOLMES, 1966; POLACHEK et al., 1960).

The silver ion (Ag^+) has great affinity for sulfhydryl, amino, and phosphate groups. It forms complexes readily with amino acids, pyrimidines, purines, nucleosides, and nucleotides as well as with phosphotides, proteins, DNA, and RNA (EICHORN, 1973). These biochemical facts suggest that silver has the potential to be either highly toxic or easily detoxified.

6 Effects on Plants, Animals, and Humans

6.1 Effects on Microorganisms and Plants

Plants are more resistant to silver toxicity than are animals. Silver toxicity for freshwater plants occurs between 50 and 7500 µg/L, and in all cases toxicity depends on the chemical compound involved. The most toxic soluble compound tested on freshwater plants is silver nitrate ($AgNO_3$), while the least toxic compounds are silver thiosulfate ($Ag_2S_2O_3$) and soluble complexes of silver chloride (AgCl). The antibacterial effects are mentioned in Sects. 3, 6.3, and 7.

6.2 Acute Effects on Animals and Humans

Fresh-water animals show a wide tolerance in their toxic responses to contact with silver. Acute toxicity may vary from 0.25 µg/L for *Daphnia magna* to 4500 µg/L for *Gammarus pseudolinmaeus*. Fish toxicity due to silver ranges from 3.9 µg/L for fathead minnows in soft water to 280 µg/L for rainbow trout in hard water. The hardness of water and/or the presence of chloride ions reduce silver toxicity for fish. In addition, the toxicity of silver for aquatic animals is also lessened if food is present during the test. The toxicity of silver for salt-water fish ranges from 4.7 µg/L for flounder to 1400 µg/L for sheephead minnows (EPA, 1980). Silver is 300 times as toxic for *Carcinus maenas* larvae as is either cobalt or antimony (AMIARD-TRIQUET, 1980).

Assuming that soluble AgCl is the principal toxic species in aqueous environments, particularly sea water and those fresh waters with substantial chloride concentrations, any decrease in pH below 8.2, with the resultant decrease in AgOH concentration, would increase the toxicity of the silver present, according to COWAN et al. (1985) (see also Sect. 2). Similarly, any appreciable presence of sulfide (S^{2-}) would greatly reduce the toxicity of soluble silver by the formation of Ag_2S or AgSH.

The oral toxicity of silver for laboratory and domestic animals is low. A single oral dose of 2.5 g of silver nitrate (1.48 g silver) caused death in dogs, while no toxicity appears to have occurred in mice given 1.05 mg/kg of silver sulfadiazine per day for 30 days (0.31 mg Ag/kg) (EPA, 1980). PETERING (1976) indicated that oral silver toxicity is low for higher organisms and human beings partly because of the low intestinal absorption. Furthermore, occupational exposures in general present no serious health problems, providing that airborne silver is kept below 0.1 mg/m^3 and dermal contact is kept at a minimum. BALDI et al. (1987, 1988) studied the susceptibility of rat hepatocytes to low levels of silver compounds.

Acute toxicity in human beings due to ingestion of silver compounds occurs only from accidental overdoses of medicinal silver compounds or overexposure in an occupational setting. Severe intoxication and deadly poisoning due to ingestion of silver compounds are only rarely reported. REINHARDT et al. (1971), however, have discussed the possibility of severe injury to the respiratory center by ingestion of silver compounds by routes other than inhalation. Generalized argyria, which is the deposition of silver in skin, hair, and other organs producing a slate gray appearance of skin, hair, cornea, and conjunctiva, has been caused by many different silver compounds (HILL and PILLSBURY, 1939), although in most cases it was due to ingestion of medicinal preparations of silver. Even when agyrria exists there usually are no observable adverse health effects. Today argyria is a rare condition because of the limited use of silver medicinals.

6.3 Chronic and Antagonistic Effects on Animals and Humans

Soluble silver compounds exhibit definite antagonism for selenium, copper, and vitamin E when given to animals, e.g., fowl, on nutritionally adequate diets, or these silver compounds may aggravate the symptoms produced by diets deficient in the above-mentioned nutrients. Such toxic interactions have been reported to occur in rats, mice, sheep, swine, chicks, turkey poults, and ducklings (EPA, 1980, p. C-94).

SHAVER and MASON (1951) found that 1500 ppm of Ag in the drinking water of rats on a vitamin E deficient diet rapidly induced muscular dystrophy, liver necrosis, and 95% mortality, a finding which was extended by DIPLOCK et al. (1967). The latter authors observed liver necrosis in rats administered silver even when the diets were completely adequate in Se and vitamin E (cf., GRASSO et al., 1969). The liver necrosis due to dietary silver also was reported to be prevented by the inclusion of vitamin B_{12} (cobalamin) in the diet in a ratio of $3/130 = B_{12}/Ag$ (BUNYAN et al., 1968). The complexity of the effects of silver is further demonstrated by experiments which showed that glutathione peroxidase (Se-enzyme) was severely reduced in the livers of Se-deficient rats administered silver. This condition and the increased fecal excretion of ^{75}Se were not improved by vitamin E supplementation, though the liver necrosis was prevented by vitamin E (DIPLOCK et al., 1971; WAGNER et al. 1975; SWANSON et al., 1974).

The antagonism of the essential nutrient copper by silver was found to be more pronounced than that of Se or vitamin E. When 0.01% silver was given for 4 weeks to chicks fed copper-adequate and copper-deficient diets, it was found that growth was restricted in chicks on both diets, but more severely on the copper-deficient diet, the mortality being 25% in chicks on the copper-adequate diet and 60% for those on the copper-deficient diet (HILL and MATRONE, 1970; HILL et al., 1964). When silver was fed to chicks at the low level of 0.001% of the diet (10 ppm) there was a reduction of elastin content of the aortas and a depression of hemoglobin formation. PETERSON and JENSEN (1975) found similar evidence of copper antagonism by silver in turkey poults. A diet adequate in copper but containing 900 ppm of silver caused high mortality, depressed growth, enlarged heart, and reduced concentrations of copper in blood, liver, spleen, and brain. Supplementing this diet with 500 ppm copper completely prevented all these signs and symptoms except growth depression, and this was less severe. Thus, prevention of the antagonistic effect of Ag occurred at a ratio of Cu:Ag of approximately 1:1.8. Silver also has been shown to reduce ceruloplasmin biosynthesis, another manifestation of the antagonism of silver and copper (SUGAWARA and SUGAWARA, 1984).

These studies demonstrate the antagonism of silver with several essential nutrients, and they raise the important question of whether intake of silver in humans that ordinarily would not be harmful may be toxic to persons consuming diets inadequate in selenium, copper, vitamin E, or cobalamin (vitamin B_{12}).

Since one of the uses of silver that potentially may result in human toxicity is its medicinal application (*Silver Institute Letter*, 1977b), it seems appropriate to consider this in detail. Silver nitrate is used in some areas to prevent blindness in newborn infants due to birth-related infections. Topical silver sulfadiazine, a cream, is standard treatment in severely burned patients to prevent and eliminate microbial infections, especially that of *Pseudomonas aeruginosa*. Amalgams of silver are in wide use as dental fillings, and some forms of silver are also used as an antiseptic barrier in other dental procedures. The dental uses are considered very safe and without health hazard. On the other hand, the medicinal use of silver sulfadiazine has the potential for toxicity because of possible cutaneous absorption through the burn injury.

The main medicinal use of silver is in fact as silver sulfadiazine cream, a topical antibacterial preparation that is used extensively in treating burn victims. Its mechanism of action is at the cell membrane and cell wall, and its antibacterial activity is greater than that of sulfadiazine alone.

Limited information is available regarding systemic absorption of silver during topical silver sulfadiazine therapy. PETERING (1976) noted that silver binds avidly to proteins, which should impair or prevent its systemic absorption. Most studies have demonstrated little apparent absorption of silver in burned animals treated with silver sulfadiazine. SANO et al. (1982) showed that when blisters were left intact in burned rats, no absorption of silver was found. However, when the blisters were debrided, increased silver levels were observed in whole blood and liver tissue. Burned, but intact keratin has also been reported to be an effective barrier to silver penetration in the eschar of a swine burn wound (FREDELL et al., 1985). Recent human data are very limited. In the most detailed report, WANG et al., (1985) showed that 11 of 509 patients had deposits of silver in the mucosa of the lips, gingiva, and cheeks. No other adverse effects of silver toxicity were noted despite the use of 20–50% suspension of silver sulfadiazine, a more concentrated preparation than that used by most other groups.

We (BOOSALIS et al., 1987) recently evaluated longitudinally serum silver concentrations and 24 h urinary excretion of silver in 23 patients with second and third degree thermal injuries. While serum silver concentrations were only modestly elevated, urinary excretion of silver was markedly elevated, especially in those patients with most severe burns. Indeed, in patients with greater than 60% burns, mean peak urinary silver excretion was 1100 µg/24 h (normal < 1 µg/24 h). Thus, at least in these patients, 24 h urinary excretion of silver appeared to be a sensitive indicator of cutaneous silver absorption.

Interestingly, alterations of the metabolism of both selenium and serum copper also were observed in these patients. Serum selenium levels were depressed throughout the patient's hospital course (BOOSALIS et al., 1986a), an observation also noted by HUNT and coworkers (1984). In a majority of patients, depressed serum copper and ceruloplasmin levels were also noted, with the lowest levels in those patients having >40% total body surface area burns (BOOSALIS et al., 1986b). This is the opposite of what one would expect to see, i.e., copper and ceruloplasmin levels both normally increasing post-trauma as part of the acute phase response. These results emphasize the importance of the earlier animal studies showing antagonism between silver and selenium, copper, and vitamin E. Further studies are required to determine if certain subjects on silver sulfadiazine therapy may benefit from supplementation with selected nutrients such as copper, selenium, or vitamin E.

6.4 Carcinogenic Effects

The ordinary uses of and contact with silver and its compounds do not appear to present a carcinogenic hazard to human beings, although subcutaneous implantation of silver foil or metallic silver particles in rodents caused sarcomas. Since the implantation of many kinds of foils and insoluble particles causes sarcomas in rodents, this observation is not considered to be a good predictor of carcinogenesis with other types of silver contact (SMITH and CARSON, 1977; EPA, 1980).

7 Hazard Evaluation and Limiting Concentrations

Metallic silver and silver compounds present an environmental hazard only for certain aquatic organisms, especially microorganisms. The potential for toxicity with silver sulfadiazine in thermal injury patients requires further investigation.

Low-level exposure to silver compounds occurs widely because of the use of soluble silver compounds to disinfect water used for drinking and for recreational uses. In Switzerland, the Soviet Union, and other European countries, silver impregnated filters are used in water purification. Similar filters have been used in German breweries, on British ships, and on commercial airplanes of many countries (EPA, 1980; *Silver Institute Letter*, 1976). Switzerland has allowed up to 200 µg Ag/L for antimicrobial activity (*Silver Institute Letter*, 1976, 1977a). The Soviets have reported that 0.1 to 0.2 ppm of silver ion is safe, stable, and longlasting for purification of drinking water (*Silver Institute Letter*, 1973). With the silver limit for drinking water being 50 µg/L or 0.05 ppm in the USA and the indications of safety at higher levels discussed above, it would seem that there is no risk for human health insofar as silver in drinking water is concerned. However, a recent review of the interaction of silver with essential nutrients, especially selenium, copper, and vitamin E, has focused attention on the potential toxicity of silver for human beings having low dietary intake of these nutrients (EPA, 1980, p. C-94 ff).

In occupational medicine, the MAK-value (maximum allowable concentration) has been fixed at 0.01 mg/m^3 (MAK, 1987). In the USA a threshold limit value (TLV) of 0.1 mg/m^3 has been set for metallic silver, and 0.01 mg/m^3 for aerosols of soluble silver compounds for the workplace (ACGIH, 1986). The USA-EPA has set a drinking water limit of 50 µg/L. The tolerance limit for fish and shellfish varies between 3 and 100 µg/L (LIEBMANN, 1958; JUNG, 1973).

References

ACGIH (American Conference of Governmental Industrial Hygiene) (1986), *Annual Report*, Cincinnati, Ohio.
Amiard-Triquet, C. (1980), *Lecture SECOTOX, Antibes. Chem. Rundsch. 33*, No. 49 (3 December).
Armour Research Foundation (1952), *Spectrographic Study of Meats for Mineral-Element Content*. National Livestock and Meat Board, Illinois Institute of Technology, Chicago, Illinois.
Baldi, C., Minola, C., di Nucci, A., Capodaglio, E., and Manzo, L. (1987), *Proceedings Second Nordic Symposium on Trace Elements, Odense*. WHO, Copenhagen-Geneva.
Baldi, C., Minola, C., di Nucci, A., Capodaglio, E., and Manzo, L. (1988), *Effects of Silver in Isolated Rat Hepatocytes. Toxicol. Lett. 41*, 261–268.
Barr, S. S., Nelson, D. M., Kline, J. R., and Gustafson, P. F. (1970), *Instrumental Analysis of Trace Elements Present in Chicago Area Surface Air. J. Geophys. Res. 74*, 2939–2945.
Boosalis, M. G., McCall, J. T., Solem, L. D., Ahrenholz, D. H., and McClain, C. J. (1986a), *Serum Copper and Ceruloplasmin Levels and Urinary Copper Excretion in Thermal Injury. Am. J. Clin. Nutr. 44*, 899–906.
Boosalis, M. G., Solem, L. D., and Ahrenholz, D. H. (1986b), *Serum and Urinary Selenide Levels in Thermal Injury. Burns (incl. Thermal Injury) 12*, 236–240.
Boosalis, M. G., McCall, J. T., Ahrenholz, D. H., Solem, L. D., and McClain, C. J. (1987), *Serum and Urinary Levels in Thermal Injury Patients. Surgery 101*, 40–43
Bowen, H. J. M. (1966), *Trace Element Biochemistry*. Academic Press, London-New York.
Bowen, H. J. M. (1979), *Environmental Chemistry of the Elements*, p. 238. Academic Press, London.
Bruland, K. W. (1983), *Trace Elements in Sea Water, Chem. Oceanogr. 8*, 209.
Brune, D., Nordberg, G. F., and Wester, P. O. (1980), *Distribution of 23 Elements in the Kidneys, Liver, and Lungs of Workers from a Smeltery and Refinery in North Sweden Exposed to a Number of Elements and a Control Group. Sci. Total Environ. 16*, 13.
Bunyan, J., Dilock, A. T., Cawthorne, M. A., and Green, J. (1968), *Vitamin E and Stress, VIII. Nutritional Effect of Dietary Stress with Silver in Vitamin E-deficient Chicks and Rats. Br. J. Nutr. 22*, 165–182.
Cotton, F. A., and Wilkinson, G. (1972), *Silver and Gold*, in: *Advanced Inorganic Chemistry*, 3rd Ed., p. 1045. Interscience Publishers, John Wiley & Sons, New York.
Cowan, C. E., Jenne, E. A., and Crecelius, E. A. (1985), *Silver Specification in Sea Water: The Importance of Sulfide and Organic Complexation*, in: Siglio, A. C., and Hattori, A. (eds.), *Marine and Estuarine Geochemistry*, Chap. 20, pp. 285–303. Lewis Publishers, Chelsea, Michigan.
Diplock, A. T., Green, J., Bunyan, J., McHale, D., and Muthy, J. R. (1967), *Vitamin E and Stress, III. The Metabolism of D-α-Tocopherol in the Rat under Dietary Stress with Silver. Br. J. Nutr. 21*, 115–125.
Diplock, A. T., Braun, H., and Lucy, J. A. (1971), *The Effect of Vitamin E on the Oxidation State of Selenium in Rat Tissue. Biochem. J. 123*, 721.
Eichorn, G. L. (1973), *Complexes of Nucleotides and Nucleosides*, in: Eichorn, G. L. (ed.), *Inorganic Biochemistry*, Vol. 2, pp. 1191–1243. Elsevier, Amsterdam.
EPA (Environmental Protection Agency) (1980), *Ambient Water Quality Criteria for Silver*, EPA 4405-80-071. Office of Water Regulations, Washington, DC.
Felmy, A. R., Girvin, D. C., and Jenne, E. A. (1984), *MINTEQ Computer Program for Calculating Aqueous Equilibria*, NTIS-PB 84157148.
Fowler, B. A., and Nordberg, G. F. (1986), *Silver*, in: Friberg, L., Nordberg, G. F., and Vouk, V. (eds.), *Handbook on the Toxicology of Metals*, 2nd Ed. pp. 521–531, Elsevier, Amsterdam.
Fredell, P. A., Solem, L. D., Ahrenholz, D. H., Bahmer, W., Grussing, D. M., Mendenhall, H. V., Erickson, E. H., and Trancik, R. J. (1985) *Burns (Care and Rehabilitation) 6*, 55–57.

Furchner, J. E., Richmond, C. R., and Drake, G. A. (1968) *Comparative Metabolism of Radionuclides in Animals, IV. Retention of Silver-110 m in the Mouse, Rat, Monkey, and Dog.* Health Phys. 15, 505–514.

Goldschmidt, V. M. (1937), *Principles of Distribution of Chemical Elements in Minerals and Rocks.* J. Chem. Soc. Part I, 655–673.

Grasso, P., Abraham, R., Hendy, R., Diplock, A. T., Greenberg, L., and Green, V. (1969), *The Role of Dietary Silver in the Production of Liver Necroses in Vitamin E-deficient Rats.* Exp. Mol. Pathol. 11, 186–189.

Hamilton, E. I., and Minski, M. J. (1972/1973), *Abundance of the Chemical Elements in Man's Diet and Possible Relations with Environmental Factors.* Sci. Total Environ. 1, 375–394.

Hamilton, E. I., Minski, M. J., and Cleary, J. J. (1972/1973), *The Concentration and Distribution of Some Stable Elements in Healthy Tissue from the United Kingdom. An Environmental Study.* Sci. Total Environ. 1, 341–376.

Heslop, R. B., and Robinson, P. L. (1967), *Copper, Silver, and Gold*, in: *Inorganic Chemistry*, 3rd Ed., Chap. 40, pp. 726–742. Elsevier, Amsterdam.

Hill, C. H., and Matrone, G. (1970), *Chemical Parameters in the Study of in vivo and in vitro Interactions of Transition Elements.* Fed. Proc. Fed. Am. Soc. Exp. Biol. 29, 1474.

Hill, W. R., and Pillsbury, D. M. (1939), *Argyria – The Pharmacology of Silver.* Williams & Wilkins, Baltimore, Maryland.

Hill, C. H., Starcher, B., and Matrone, G. (1964), *Mercury and Silver, Interactions with Copper.* J. Nutr. 83, 107–110.

Hunt, D. R., Lane, H. W., Bessinger, D., Gallagher, K., Halligan, R., Johnston, D., and Rowlands, B. J. (1984), *Selenium Depletion in Burn Patients.* J. Parenter. Enterol. Nutr. 8, 695–699.

Jones, K. C., and Peterson, P. J. (1985), *Determination of Silver in Plants by Flameless Atomic Absorption Spectrometry and Neutron Activation Analysis.* Int. J. Environ. Anal. Chem. 21, 23–32.

Jung, K. D. (1973), cited in: Förstner, U., and Wittmann, G. T. W. (1979), *Metal Pollution in the Aquatic Environment*, p. 28. Springer Verlag, Berlin.

Kehoe, R. A., Cholar, J., and Story, R. J. (1940), *A Spectrochemical Study of the Normal Range of Concentrations of Certain Trace Metals in Biological Materials.* J. Nutr. 19, 579–592.

Kent, N. L., and McCance, R. A. (1941), *Absorption and Excretion of Minor Elements by Man, I. Silver, Gold, Lithium, Boron, and Vanadium.* Biochem. J. 35, 837–844.

Liebmann, H. (1958), *Handbuch der Frischwasser- und Abwasserbiologie*, Vol. II. R. Oldenbourg, München.

MAK (1987), *Maximum Concentrations at the Workplace, Report No. XXIII, DFG.* VCH Verlagsgesellschaft, Weinheim-Basel-Cambridge-New York.

Martin, J. H., Knauer, G. A., and Gordon, R. M. (1983), *Silver Distribution and Fluxes in North-East Pacific Waters.* Nature 305, 306.

Montaser, A., and Golightly, D. W. (1987), *Inductively Coupled Plasmas in Analytical Atomic Spectrometry,* pp. 338, 354, 606. VCH Verlagsgesellschaft, Weinheim-Basel-Cambridge-New York.

Murthy, G. K., and Rhea, U. (1968), *Cadmium and Silver Content of Market Milk.* J. Dairy Sci. 51, 610–613.

Newton, D., and Holmes, A. (1966), *A Case of Accidental Inhalation of Zinc-65 and Silver-110 m.* Radiat. Res. 29, 403–412.

Petering, H. G. (1976), *Pharmacology and Toxicology of Heavy Metals: Silver.* Pharmacol. Ther. A 1, 127–130.

Peterson, R. P., and Jensen, G. S. (1975), *Interrelationship of Dietary Silver with Copper in the Chick.* Poultry Sci. 54, 771–775.

Phalen, R. F., and Morrow, P. E. (1973), *Experimental Inhalation of Metallic Silver.* Health Phys. 24, 509–518.

Polachek, A. A., Cope, B. C., Williard, R. F., and Enns, T. (1960), *Metabolism of Radioactive Silver in a Patient with Carcinoid.* J. Lab. Clin. Med. 56, 499–505.

Przybylowicz, E. P., and Zuehlke, C. W. (1966), in: Kollthoff, I. M., and Elving, P. H. (eds.), *Treatise on Analytical Chemistry*, Vol. 4, Part II, Sect. A, pp. 1–69. Interscience, New York.

Reinhardt, G., Geldmacher-von Mallinckrodt, M., Kittel, H., and Opitz, O. (1971), *Acute Fatal Poisoning with Silver Nitrate Following an Abortion Attempt*. Arch. Kriminol. *148*, 69–78.

Renner, H. (1982), *Silver and Silver Compounds*, in: *Ullmanns Encyklopädie der technischen Chemie*, 4th Ed., Vol. 21, pp. 311–364. Verlag Chemie, Weinheim-Deerfield Beach/Florida-Basel.

Ruegger, U. P., and Dubuis, R. (1981), *Oligodynamische Wasserentkeimung mit Silber und Kupfer*. Chem. Rundsch. *34*, No. 8 (18 February).

Saager, R. (1984), *Metallic Raw Materials Dictionary* (in German), pp. 63–67. Bank von Tobel, Zürich.

Sano, S., Fujimori, R., and Takashima, M. (1982), *Absorption, Excretion, and Tissue Distribution of Silver Sulfadiazine*. Burns *8*, 278–285.

Shaver, S. E., and Mason, K. E. (1951), *Impaired Tolerance to Silver in Vitamin E-deficient Rats*. Anat. Rec. *109*, 382 (Abstract).

Sibley, T. H., and Morgan, J. J. (1975), *Equilibrium Speciation of Trace Metals in Freshwater-Seawater Mixtures*, in: *Proceedings International Conference on Heavy Metals in the Environment*, Vol. 1, pp. 319–338. National Research Council of Canada, Toronto.

Silver Institute Letter (1973), *Silver Clears Up Polluted Water*. Vol. III. No. 3.

Silver Institute Letter (1976), *Filters with Silver in Planes, Houses, and Offices*. Vol. VI, No. 2.

Silver Institute Letter (1977a), *Silver Purifies Water on Drilling Rigs*. Vol. VII, No. 1.

Silver Institute Letter (1977b), *Silver in the Healing of Burns*. Vol. VIII, No. 2.

Silver Institute Letter (1986), *World Silver Mine Production Decreasing*. Vol. XVI, No. 6.

Smith, I. M., and Carson, B. L. (1977), *Trace Metals in the Environment*, Vol. 2. *Silver*. Ann Arbor Science Publishers, Ann Arbor, Michigan.

Stokinger, H. E. (1980), *The Metals – Silver*, in: Clayton, G. D., and Clayton, F. E. (eds.), *Patty's Industrial Hygiene and Toxicology*, 3rd Ed., Vol. IIA, Chap. 29. John Wiley & Sons, New York.

Sugawara, M, and Sugawara, C. (1984), *Effect of Silver on Ceruloplasmin Synthesis in Relation to Low-molecular Weight Protein*. Toxicol. Lett. *20*, 99–104.

Swanson, A. B., Wagner, P. A., Ganther, H. E., and Hoekstra, W. G. (1974), *Antagonistic Effects of Silver Tri-o-cresyl Phosphate on Selenium and Glutathione Peroxidase in Rat Liver and Erythrocytes*. Fed. Proc. Fed. Am. Soc. Exp. Biol. *33*, 693 (Abstr. 2733).

Wagner, P. A., Hoekstra, W. G., and Ganther, H. E. (1975), *Alleviation of Silver Toxicity by Selenite in the Rat in Relation to Tissue Glutathione Peroxidase*. Proc. Soc. Exp. Biol. Med. *148*, 1106–1110.

Wang, X-W., Wang, N. Z., Zhang, O. Z., Zapata-Sirvent, R. L., and Davies, J. W. L. (1985), *Tissue Deposition of Silver Following Topical Uses of Silver Sulfadiazine in Extensive Burns (incl. Thermal Injury)*. *11*, 197–201.

Welz, B. (1985), *Atomic Absorption Spectrometry*, 2nd Ed., pp. 324–325. VCH Verlagsgesellschaft, Weinheim-Deerfield Beach/Florida-Basel.

Wester, P. O. (1971), *Trace Element Balances in Two Cases of Pancreatic Insufficiency*. Acta Med. Scand., *190*, 155–161.

Zobrist, F. (1980), *Lecture Pro Aqua-Provita, Basel*. Chem. Rundsch. *33*, No. 30 (23 July).

Additional Recommended Literature

Anonymous (1987), *Refining of Gold and Silver* (in German), *Neue Zürcher Zeitung, No. 280*, p. 79 (2. Dec.). Forschung und Technik, Zürich.

Grashoff, G. (1989), *Bacteriocidal Properties of Metals: Approaches to Studying Modes of Actions. Proceedings Workshop Toxicity of Newer Metals*, University of Guildford, in press; see, for instance, Report E. Merian, *Swiss Chem.* *12*(1–2), 25 (1990).

Nordberg, G. F., and Gerhardsson, L. (1988), *Silver*, in: Seiler, H. G., Sigel, H., and Sigel, A. (eds.), *Handbook on Toxicity of Inorganic Compounds*, pp. 619–624. Marcel Dekker, New York.

II.27 Tantalum

PETER L. GOERING, Rockville, Maryland, USA
BRUCE A. FOWLER, Baltimore, Maryland, USA

1 Introduction

Tantalum (Ta) is a Group VA element, the name of which is derived from the Greek mythological figure, Tantalos, father of Niobe, because of the almost invariable occurrence of tantalum with another Group VA element, niobium (see Chapter II.23). The element was discovered by EKEBERG in 1802 and the first relatively pure preparation was produced by VON BOLTON in 1905 (HAMMOND, 1986). The physical properties and the relative biological inertness make it ideal for a variety of medical uses and for the electronics industry (see Sect. 3). While only a small amount of tantalum is absorbed following oral administration or inhalation exposure, the element has a relatively long biological half-life (1 to 2 years). The toxicity of tantalum is generally quite low and it is not considered a health hazard.

2 Physical and Chemical Properties, and Analytical Methods

Tantalum is a gray, very hard, malleable, and ductile metal. It is classified as a Group VA metal in the Periodic System of Elements, has an atomic mass of 180.95, and the atomic number 73. Tantalum has a boiling point of 5425 °C and a melting point of 2996 °C, which is exceeded only by rhenium and tungsten (HAMMOND, 1986). The element is resistant to chemical attacks up to 150 °C by acids except for hydrofluoric acid and even more resistant against alkalies. It is more reactive at higher temperature. Numerous forms of tantalum exist, including pentachloride pentafluoride, and pentoxide salts, which are insoluble in water. The most stable oxidation state for tantalum is +5. Natural isotopes of tantalum include 181 (99.99%) and 180 (0.01%, half-life greater than 10^{12} years); artificial radioactive isotopes are 172 to 179 and 182 to 186 (MERCK INDEX, 1983). Tantalum is almost twice as heavy as steel (density: 16.6 g/cm^3) and about 90% as stiff as steel, with a modulus of elasticity value, $E = 2.7 \times 10^7$ p.s.i. (TAYLOR, 1969).

Trace concentrations of tantalum in biological tissues and environmental samples can be determined by several analytical methods. A spectrophotometric method utilizing the reagent phenylfluorone can detect tantalum at levels of 0.4 – 2.5 ppm (LUKE, 1959). Tantalum in ores has been analyzed by X-ray fluorescence spectrometry with a detection limit of 300 ppm (CAMPBELL and CARL, 1956; FICHTE and ROTHMANN, 1982). This method is especially important in technical processing, such as the determination of tantalum concentrations of 5 ppm in iron after electrolytically separating the base metal (BIRKS and BROOKS, 1950). Inductively coupled plasma atomic emission spectrometry may have potential for determining tantalum in human tissues with a detection limit of 1 to 5 ppm (MARTINSEN and THOMASSEN, 1986; MONTASER and GOLIGHTLY, 1987). When using flame atomic absorption spectrometry, interference with vanadium and titanium must be avoided (WELZ, 1985).

Neutron activation analysis has been used to detect tantalum levels less than 1 ppm in marine invertebrates (BURTON and MASSIE, 1971), and in grass (DE MEESTER, 1988), and tantalum concentrations of 33 ppm were detected in lung tissue of a worker occupationally exposed to tungsten carbide dusts (EDEL et al., 1986). Radiochemical determination of ^{182}Ta has a sensitivity of 20 ng/g (HAAS and KRIVAN, 1983).

3 Sources, Production, Important Compounds and Uses

Tantalum ores are found in Thailand (48% of the global production, SAAGER, 1984), Australia, Canada, Brazil, Mozambique, Portugal, Nigeria, and Zaire. Separation of tantalum from niobium requires several complicated procedures. About half of the annually produced 1100 t Ta_2O_5 is isolated from tin slags, and only half from ores (FICHTE and ROTHMANN, 1982; SAAGER, 1984). For alloys tantalum carbide may be used directly (FICHTE and ROTHMANN, 1982). Commercial production of the element includes methods such as electrolysis of molten potassium fluotantalate, carbothermic reduction of tantalum oxide, reduction of potassium fluotantalate with sodium, or oxidation of tantalum carbide at 800–900°C, e.g. in scrap from cemented carbides (HAMMOND, 1986; FICHTE and ROTHMAN, 1982).

Tantalum has important uses in clinical medicine. A property of tantalum exploited for diagnostic radiology is its high value of $Z = 73$, making it more visible on X-rays than iodine or barium. Thus, the element has been used as a medium for bronchography to evaluate the physiology and pathology of canine and human lung (NADEL et al., 1968; STITIK and PROCTOR, 1973), to detect occult cancer (STITIK and PROCTOR, 1975), and to study clearance times in patients with chronic obstructive pulmonary disease (GAMSU et al., 1973). The short-lived radionuclide, ^{178}Ta (half-life 9.3 min), has been suggested (WILSON et al., 1987) for use in blood pool imaging to analyze cardiac function because it allows comparable imaging with less radiation exposure compared to the standard blood pool imaging radiopharmaceutical, ^{99m}Tc (half-life 6 h). Another method for assessing cardiac function involves implantation of radiopaque tantalum markers directly in myocardial tissue (INGELS et al., 1980; SANTAMORE et al., 1982).

Several properties of tantalum, such as malleability, light weight, strength, and relative biological inertness, make the metal ideal for use in surgery. It can be easily rolled into fine wire for sutures (TAYLOR, 1969), and tantalum clips have been tested for occlusion of the vas deferens for male sterilization (KOTHARI, 1982). Tantalum is well-suited for use in cranioplasty because of an oxide coating which protects it from corrosion, even after reshaping via cutting, bending, hammering, and drilling (MCFADDEN, 1971). Tantalum mesh has been used in ophthalmic surgery to repair the floor and/or medial wall of the orbit following blow-out fractures (KOBAYASHI et al., 1986). The metal may be useful as radiopaque bone markers in Roentgen stereophotogrammetry to obtain biometric data on cranial growth (SELVIK et al., 1986), and joint kinematics, especially for rheumatoid patients (EBERHARDT and SELVIK, 1986).

The chemical industry utilizes tantalum to fabricate corrosion-resistant tantalum-tungsten-cobalt and tantalum-tungsten-molybdenum alloys for manufacture of acid-proof and high-temperature stable equipment. In the electronics industry, tantalum is used for filaments, grid wires, rectifiers, and electrolytic capacitors (VENUGOPAL and LUCKEY, 1978). The metal has also been used for aircraft and missile parts, and tantalum oxide has been used to make a special glass with a high index of refraction for camera lenses (HAMMOND, 1986).

4 Distribution in the Environment and Living Organisms

Tantalum occurs almost invariably with niobium although it is the less abundant of the two elements. Tantalum occurs with other metals in the minerals columbite-tantalite [(Fe, Mn) (Ta, Nb)$_2$O$_6$] and microlite [(Na, Ca)$_2$Ta$_2$O$_6$(O, OH, F)]. The metal is present at about 1 ppm in the earth's crust (MERCK INDEX, 1983; SAAGER, 1984). There is very little geographical variability or environmental mobilization of tantalum, possibly because it is highly insoluble (FÖRSTNER, 1984). In-stack coal fly ash has been shown to contain tantalum (HOCK and LICHTMAN, 1982).

Sea-water levels seem to be very low (lower than 0.004 ppb), and tantalum is found in the form of Ta(OH)$_5^0$ (BRULAND, 1983). Concentrations of tantalum ranging from less than 1 to greater than 400 ppm have been found in a variety of marine invertebrates of the class Ascidians, or sea squirts (KIMURA and FUKAI, 1948; KOKUBU and HIDAKA, 1965; BURTON and MASSIE, 1971).

5 Uptake, Absorption, Transport and Distribution, Metabolism and Elimination in Animals and Humans

Tantalum is not essential to animals and has not been shown to have a definite biological function. Few studies exist on tantalum metabolism; those which generally have utilized radiotracer tantalum.

The distribution of ^{182}Ta was determined in freshwater clams (*Anodonta nuttalliana*) after a 10 to 12 day exposure in aquarium water (HARRISON and QUINN,

1972). Transfer across the gut and/or exterior surfaces and subsequent circulation by body fluid was evident since ^{182}Ta was localized in the stomach/digestive gland (51% of absorbed dose), intestine/gonad (14%), body fluid (6%), and body wall (5%).

Insoluble ^{182}Ta$_2$O$_5$ administered to rats by the oral route is poorly absorbed from the gastrointestinal tract with total excretion accounted for in the feces within two days (DOULL and DUBOIS, 1949). Intramuscular injection of ^{182}Ta$_2$O$_3$ either complexed with sodium citrate or alone resulted in absorption of 70% of the ^{182}Ta dose after 30 days when complexed with citrate and 15% for non-complexed ^{182}Ta; however, the relative distributions were similar (DURBIN et al., 1956; HAMILTON et al., 1950). In another study, ^{182}Ta administered orally to rats as soluble potassium tantalate was excreted rapidly and almost exclusively via the fecal route; less than 2% of the dose was retained after 1 day (FLESHMAN et al., 1971). More than 96% of the administered dose was excreted after 3 days, and less than 0.5% was recovered in urine after 7 days, although more than 97% of the total dose had been excreted. Excretion of ^{182}Ta exhibited three phases: an early phase reflecting rapid excretion ($t_{1/2} = 0.25$ days) from the gastrointestinal tract of unabsorbed compound, a second phase ($t_{1/2} = 2$ to 5 days) which may reflect loss of ^{182}Ta that is loosely bound in tissues, and a third phase accounting for elimination of ^{182}Ta that has been absorbed and localized within tissue. This phase has a relatively long biological half-life of 62 days in males and 119 days in females. After absorption, ^{182}Ta was primarily localized in the bone compartment, which retained over 40% of the total body burden after 14 days. On a per gram basis, the highest concentrations of ^{182}Ta were found in bone and kidneys (FLESHMAN et al., 1971).

Pulmonary clearance of tantalum dust following insufflation was dependent upon particle size; a 1 µm powder was removed from the alveolar regions with a clearance half-time of 2.1 years and 5 µm- and 10 µm powders were removed with a half-time of 333 days (MORROW et al., 1976). Rapid post-insufflation uptake by the pulmonary lymph nodes was observed with up to 12% of the initial alveolar burden present in the lymph nodes at 240 days and 6% present at 816 days.

Following an accidental exposure of a human to ^{182}Ta and ^{183}Ta via the inhalation route at a nuclear reactor test site, 93% of the activity was eliminated entirely in the feces in seven days (SILL et al., 1969). The remaining radioactivity was slowly eliminated at a rate of 0.05% per day. No radioactivity was detected in the urine. In a case report of a worker suffering from hard metal pneumoconiosis after working 13 years in the tungsten carbide industry, the lung concentration of tantalum was 33 ppm 3 years after having stopped work. Bronchoalveolar lavage and blood samples contained tantalum concentrations 10 000-fold higher than control subjects (EDEL et al., 1986).

6 Effects on Animals and Humans

6.1 Miscellaneous Biochemical Effects

Tantalum oxide (Ta_2O_5) particles were ingested by rabbit alveolar macrophages in cell culture and cytotoxicity occurred as evidenced by release of lactate dehydrogenase and lysozyme into the media (MATTHAY et al., 1978).

6.2 Acute Effects on Animals

Acute toxicity following bolus administration of tantalum is low and has been summarized by VENUGOPAL and LUCKEY (1978). The rat LD_{50} values obtained after oral administration of tantalum oxide, tantalum chloride, and potassium tantalum fluoride were 8000, 1900, and 2500 mg/kg, respectively. The LD_{50} values obtained after intraperitoneal injection of tantalum chloride and potassium tantalum fluoride to rats were 75 and 375 mg/kg, respectively. The LD_{50} value obtained after the intravenous injection of tantalum fluoride to mice was 110 mg/kg.

6.3 Chronic Effects on Animals and Humans

Chronic effects of tantalum can be assessed by examining studies dealing with tantalum implants or follow-up of humans and animals exposed to tantalum dust after bronchography. Tantalum implants are presently considered to be biologically inert. In one study, tantalum stock was implanted in the subperiosteal region of the mandible, buccal mucosa, and the subcutaneous paravertebral region of the back of monkeys (MEENAGHAN et al., 1979). Tissue responses assessed after 3 weeks were variable and appeared to be a function of the surface treatment of the material prior to implantation. Increased cellularity, including fibroblasts and multinucleated giant cells, was the predominant feature in the tissues.

Eighteen months after tantalum bronchography in dogs (WELLER and KAMMLER, 1973) and six months after this procedure in cats and monkeys (MASSE et al., 1973), granulomatous changes were evident in the lungs, indicative of a delayed inflammatory reaction to this element. In another study, no pathological changes were observed in lungs of dogs two years after insufflation with tantalum dust (MORROW et al., 1976).

Toxicity data related to human exposures are limited. Patients with orbital implants covered with tantalum mesh developed pain, headache, mucopurulant discharge, diffuse conjunctival inflammation, and erosion of tissues surrounding the implant 10 to 15 years post-implantation (PRZYBYLA and LAPIANA, 1982). In a case study, an individual developed chronic urticaria ten months after surgical implantation of tantalum staples (WERMAN and RIETSCHEL, 1981). The urticaria responded favorably to disulfiram chelation therapy and resolved completely after surgical removal of the staples. A Type-1 anaphylactic hypersensitivity to tantalum was suspected.

7 Hazard Evaluation and Limiting Concentrations

The short-term-exposure limit (STEL) and threshold limit value-time weighted average (TLV-TWA) for tantalum are 5 mg/m^3 and 10 mg/m^3, respectively (ACGIH, 1980). The Occupational Safety and Health Administration TWA standard for tantalum pentoxide in air is 5 mg Ta/m^3 (FEDERAL REGISTER 1974; MAK, 1987).

References

ACGIH (American Conference of Governmental Industrial Hygienists) (1980), Vol. 4, p. 385. Cincinnati, Ohio.
Birks, L. S., and Brooks, E. J. (1950), *Anal. Chem. 22*, 1017–1020.
Bruland, K. W. (1983), *Trace Elements in Sea Water, Chem. Oceanogr. 8*, 208.
Burton, J. D., and Massie, K. S. (1971), *J. Marine Biol. Assoc. UK 51*, 679–683.
Campbell, W. J., and Carl, H. F. (1956), *Anal. Chem. 28*, 960–962.
De Meester, C. (1988), Chapter 59, *Tantalum*, in: Seiler, H. G., Sigel, H., and Sigel, A. (eds.), *Handbook on Toxicology of Inorganic Compounds*, pp. 661–663. Marcel Dekker, New York.
Doull, J., and DuBois, K. (1949), in: *Metabolism and Toxicity of Radioactive Tantalum*, Part 2, p. 12. University of Chicago Toxicology Laboratory, Chicago Quarterly Progress Report.
Durbin, P. W., Scott, K. G., and Hamilton, J. G. (1956), in: *The Distribution of Radioisotopes of Some Heavy Metals in the Rat*, pp. 17–19. Lawrence Radiation Laboratory Report UCRL-3607, Berkeley, California.
Eberhardt, K. B., and Selvik, G. (1986), *Some Aspects of Knee Joint Kinematics in Rheumatoid Arthritis as Studied with Roentgen Stereophotogrammetry. Clin. Rheumatol. 5*, 201–209.
Edel, J., Pietra, R., Sabbioni, E., Rizzato, G., and Speziali, M. (1986), *Trace Metal Lung Disease: Hard Metal Pneumoconiosis. A Case Report. Acta Pharmacol. Toxicol. 59* (Suppl. 7), 52–55.
Federal Register (1974), Vol. 39, p. 23540. U.S. Government Printing Office, Washington, D.C.
Fichte, R. M., and Rothmann, H. (1982), *Tantalum and Tantalum-Compounds* (in German) in: *Ullmanns Encyklopädie der technischen Chemie*, 4th Ed., Vol. 22, pp. 395–404. Verlag Chemie, Weinheim-Deerfield Beach/Florida-Basel.
Fleshman, D. G., Silva, A. J., and Shore, B. (1971), *The Metabolism of Tantalum in the Rat. Health Phys. 21*, 385–392.
Förstner, U. (1984), *Metal Pollution of Terrestrial Waters*, in: Nriagu, J. O. (ed.), *Changing Metal Cycles and Human Health*, pp. 71–94. Springer Verlag, New York.
Gamsu, G., Weintraub, R. M., and Nadel, J. A. (1973), *Clearance of Tantalum from Airways of Different Caliber in Man Evaluated by a Roentgenographic Method. Am. Rev. Respir. Dis. 107*, 214–224.
Haas, H. F., and Krivan, V. (1983), *Fresenius Z. Anal. Chem. 314*, 532–537.
Hamilton, J. G., Scott, K. G., Asling, C. W., Wallace, P. C., and Thilo, G. (1950), *The Metabolic Properties of Various Elements*, in: *Med. Health Physics Quarterly Report*, pp. 18–22. Lawrence Radiation Laboratory Report UGRL-1143, Berkeley, California.
Hammond, C. R. (1986), *The Elements*, in: *Handbook of Chemistry and Physics*, 67th Ed., pp. B36–B37. CRC Press, Boca Raton, Florida.
Harrison, F. L., and Quinn, D. J. (1972), *Tissue Distribution of Accumulated Radionuclides in Freshwater Clams. Health Phys. 23*, 509–517.
Hock, J. L., and Lichtman, D. (1982), *Studies of Surface Layers on Single Particles of In-stack Coal Fly Ash. Environ. Sci. Technol. 16*, 423–427.

Ingels, N. B., Jr., Daughters, G. T., Stinson, E. B., and Alderman, E. L. (1980), *Evaluation of Methods for Quantitating Left Ventricular Segmental Wall Motion in Man Using Myocardial Markers as a Standard. Circulation 61*, 966–972.
Kimura, K., and Fukai, R. (1948), *J. Chem. Soc. Jpn. 69*, 75–76.
Kobayashi, H., Hayashi, M., Kawano, H., Handa, Y., Kabuto, M., and Tsuji, T. (1986), *Treatment of Blow-out Fracture. Neurol. Res. 8*, 221–224.
Kokubu, N., and Hidaka, T. (1965), *Nature 205*, 1028–1029.
Kothari, L. K. (1982), , *Biocompatibility of the vas deferens in Relation to Contraception*, in: Williams, D. F. (ed.), *Biocompatibility in Clinical Practice*, Vol. I, pp. 135–148. CRC Press, Boca Raton, Florida.
Luke, C. L. (1959), *Anal. Chem. 31*, 904–906.
MAK (1987), *Maximum Concentrations at the Workplace, Report No. XXIII, DFG*. VCH Verlagsgesellschaft, Weinheim-Basel-Cambridge-New York.
Martinsen, I., and Thomassen, Y. (1986), *Multielement Characterization of Human Lung Tissues by ICP-AES. Acta Pharmacol. Toxicol. 59* (Suppl. 7), 620–623.
Masse, R., Ducousso, R., and Nolibe, D. (1973), *Etude Expérimentale de la Pulmonaires des Particles metalliques: Application à la Bronchographie au Tantale. Rev. Fr. Mal. Respir. 1*, 1063–1066.
Matthay, R. A., Balzer, P. A., Putman, C. E., Gee, J. B. L., Beck, G. J., and Greenspan, R. H. (1978), *Tantalum Oxide, Silica and Latex: Effects on Alveolar Macrophage Viability and Lysozyme Release. Invest. Radiol. 13*, 514–518.
McFadden, J. T. (1971), *Neurosurgical Metallic Implants. J. Neurosurg. Nursing*, 123–130.
Meenaghan, M. A., Natiella, J. R., Moresi, J. L., Flynn, H. E., Wirth, J. E., and Baier, R. E. (1979), *Tissue Response to Surface-treated Tantalum Implants: Preliminary Observations in Primates. J. Biomed. Mater. Res. 13*, 631–643.
Merck Index (1983), 10th. Ed., pp. 1301–1302. Merck and Co., Rahway, New Jersey.
Montaser, A., and Golightly, D. W. (1987), *Inductively Coupled Plasmas in Analytical Atomic Spectrometry*, pp. 467–468. VCH Verlagsgesellschaft, Weinheim-Basel-Cambridge-New York.
Morrow, P. E., Kilpper, R. W., Beiter, E. H., and Gibb, F. R. (1976), *Pulmonary Retention and Translocation of Insufflated Tantalum. Radiology 121*, 415–421.
Nadel, J. A., Wolfe, W. G., and Graf, P. D. (1968), *Powdered Tantalum as a Medium for Bronchography in Canine and Human Lungs. Invest. Radiol. 3*, 229–238.
Przybyla, V. A., Jr., and LaPiana, F. G. (1982), *Complications Associated with Use of Tantalum-mesh-covered Implants. Ophthalmology 89*, 121–123.
Saager, R. (1984), *Metallic Raw Materials Dictionary* (in German), pp. 161–164. Bank von Tobel, Zürich.
Santamore, W. P., Bove, A. A., Philips, C. M., and Monster, V. (1982), *Rapid Assessment of Ventricular Function in Acute Volume Overload Using Opaque Myocardial Markers. Cathet. Cardiovasc. Diagn. 8*, 311–317.
Selvik, G., Alberius, P., and Fahlman, M. (1986), *Roentgen Stereophotogrammetry for Analysis of Cranial Growth. Am. J. Orthod. 89*, 315–325.
Sill, C. W., Voelz, G. L., Olson, D. G., and Anderson, J. I. (1969), *Two Studies of Acute Internal Exposure to Man Involving Cerium and Tantalum Radioisotopes. Health Phys. 16*, 325–332.
Stitik, F. P., and Proctor, D. F. (1973), *Tracheography with the Experimental Contrast Agent Tantalum. Ann. Otol. Rhinol. Laryngol. 82*, 838–843.
Stitik, F. P., and Proctor, D. F. (1975), *Delayed Clearance of Tantalum by Radiologically Occult Cancer. Ann. Otol. Rhinol. Laryngol. 84*, 589–595.
Taylor, D. E. (1969), *Tantalum and Tantalum Compounds*, in: *Encyclopedia of Chemical Technology*, 2nd Ed., Vol. 19, pp. 630–652. John Wiley & Sons, New York.
Venugopal, B., and Luckey, T. D. (1978), in: *Metal Toxicity in Mammals – 2*, pp. 229–231. Plenum Press, New York.
Weller, W. E., and Kammler, E. (1973), *Long-term Effect of Tantalum Dust in Connection with Inhalation Bronchography. Respiration 30*, 430–442.

Welz, B. (1985), *Atomic Absorption Spectrometry*, 2nd Ed., p. 328. VCH Verlagsgesellschaft, Weinheim-Deerfield Beach/Florida-Basel.

Werman, B. S., and Rietschel, R. L. (1981), *Chronic Urticaria from Tantalum Staples. Arch. Dermatol. 117*, 438–439.

Wilson, R. A., Kopiwoda, S. Y., Callahan, R. J., Moore, R. H., Boucher, C. A., Manspeaker, H., Castronovo, F. P., and Strauss, H. W. (1987), *Biodistribution of Tantalum-178: A short-lived Radiopharmaceutical. Eur. J. Nucl. Med. 13*, 82–85.

Additional Recommended Literature

Trueb, L. (1990), *Niobium and Tantalum – Unequal Brothers* (in German). *Neue Zürcher Zeitung, Forschung und Technik, No. 73*, pp. 77–78 (28 March), Zürich.

II.28 Tellurium

LAWRENCE FISHBEIN, Washington, DC, USA

1 Introduction

Tellurium was discovered in 1782 by F. J. Mueller von Reichenstein who referred to it as metallum problematicum or aurum paradoxym, synonyms still found in modern texts. Klaproth in 1798 established definite evidence that this metal (frequently found in association with native gold deposits and considered by many to be an alloy of bismuth and antimony) had unique properties, and named the element tellurium, derived from the Latin tellus, meaning earth (MASCHEWSKY, 1982). Tellurium is considered a rather rare element ranking about seventy-first in the order of crustal abundance.

The use applications of tellurium and its compounds are principally in the areas of solid state thermoelectric and electronic applications as well as microalloying in ferrous and non-ferrous metallurgy, and these products may be released in the form of wastes. Tellurium concentrations in ambient air (mainly from coal combustion), water, and food are generally very low where measured, and food is the apparent main source of tellurium for the general population. The inhalation route is the major route of occupational exposure. As in the case of selenium speciation determines bioavailability and absorption. Information regarding levels of exposure and the chemical form of tellurium, as well as its toxicity in humans, is sparse. In contrast to selenium, there is no documented role of essentiality for tellurium in plants or mammals.

2 Physical and Chemical Properties, and Analytical Methods

2.1 Physical and Chemical Properties

Tellurium belongs to the VIA Group of the Periodic System located between selenium and polonium and between antimony and iodine. The atomic number and atomic mass of the element is 52 and 127.6, respectively, and it has a melting point of 450 °C and a boiling point of 990 °C. Tellurium exists in two allotropic forms, an amorphous black powder and the hexagonal crystalline form (isomorphous) with silver-white shine, and, when pure, a metallic lustre. Twenty-one isotopes of tellurium are known with atomic masses ranging from 115 to 135; natural Te consists of eight isotopes (mainly ^{130}Te, ^{128}Te, and ^{126}Te), while ^{127}Te is unstable (present to the extent of 0.87%) with a half-life of 1.2×10^{13} years.

Although tellurium is more metallic than oxygen, sulfur, and selenium, it resembles them closely in most of its chemical properties. Tellurium forms compounds in oxidation states -2, $+2$, $+4$, and $+6$. It forms ionic tellurides with active metals and covalent compounds with other elements. Some examples of the valence states include: -2 (H_2Te, Na_2Te, $CuTe$); $+4$ (TeO_2; $TeBr_4$, $TeOCl_2$), and $+6$ (TeO_3, H_6TeO_6, TeF_6) (ELKIN, 1983; GERHARDSSON et al., 1986; EINBRODT and MICHELS, 1984).

2.2 Analytical Methods

The analytical chemistry of tellurium has been reviewed by COOPER (1971), CHENG and JOHNSON (1978), WELZ (1985), GERHARDSSON et al. (1986), EMMERLING et al. (1986), and ALEXANDER et al. (1988). Sample contamination and loss of volatile compounds must be avoided, and enrichment must be controlled. Atomic absorption spectrometry (AAS) and neutron activation analysis (NAA) are the principal methods for the determination of tellurium in both environmental and biological samples. The estimated limits of detection for tellurium in biological systems by NAA are about 5 µg/kg wet weight (DIKSIC and COLE, 1977; BRUNE et al., 1980; WARD, 1987; see also Chapter I.4a).

The limit of detection of AAS employing an air hydrogen flame at 214.3 mm is about 0.015 mg/L (WELZ, 1985). More recent adaptations have employed graphite furnace (electrothermal) AAS (MUANGNOICHAROEN et al., 1986; CHUNG et al., 1984; VERLINDEN et al., 1981; ANDREAE, 1984; MAHER, 1984; YU et al., 1983), inductively coupled argon plasma emission spectrometry (WOLNIK et al., 1981; WARD, 1987), inductively coupled plasma/volatile hydride method (THOMPSON et al., 1981; BARNES, 1984), hydride generation AAS (ANDREAE, 1984; WELZ, 1985; YAMAMOTO et al., 1985), and microwave induced emission spectrometry (VAN MONFORT, et al., 1979).

A number of methods have been employed for the determination of tellurium in water with detection limits of about 50 mg/L for an inductively coupled plasma/volatile hydride method (THOMPSON et al., 1981); 20 ng/L using hydride AA (SINEMUS et al., 1981); 10 ng/L for extraction and graphite furnace AAS (CHUNG et al., 1984); 10 ng/L using hydride generation and AAS after enrichment and separation with thiol cotton (YU et al., 1983); 1 ng/L for the XAD-Bismuthiol-II system (SUGIMURA and SUZUKI, 1981); and 1 ng/L by catalytic polarographic determination after preconcentration with sulfhydryl cotton fiber (AN and ZHANG, 1983). The determination of Te^{4+} and Te^{6+} in natural waters utilizing a combination of preconcentration and hydride-graphite furnace AAS procedures has been accomplished with a limit of detection of about 60 pg/L and a precision of 10–20% (ANDREAE, 1984).

The determination of Te (as well as Se) in the gas phase using specific columns of gold-coated beads and charcoal and graphite furnace AAS has permitted detection limits down to about 0.1 ng/m^3 with a precision of about ±6% at the nanogram level of Te and about ±4% at the nanogram level of Se (MUANGNOICHAROEN et al., 1986).

The utility of a nebulizer introduction system for the simultaneous determination of the volatile hydrides of Te, As, Bi, Sb, and Se in foods by inductively coupled argon plasma (ICAP) emission spectrometry permitted an improvement of detection limits (0.5 to 3 µg/L) by at least an order of magnitude compared to those obtained with conventional pneumatic nebulizers (WOLNIK et al., 1981).

Speciation of Te(IV) and Te(VI) in urine samples has been accomplished using a poly(dithiocarbamate) resin with inductively coupled plasma atomic emission spectroscopy and hydride generation (FODOR and BARNES, 1983).

The separation of tellurium from gold(III), indium, cadmium, and other elements by cation exchange chromatography in hydrochloric acid-acetone (STRELOW, 1984); a methylisobutyl ketone amine synergistic iodine complex extraction system coupled with AAS or ICAP emission spectroscopy for the determination of 18 trace elements including tellurium in geochemical samples (CARK and VIETS, 1981) and the separation of Se and Te in copper-base alloys utilizing an initial reduction of the elements by hydroxylamine hydrochloride and ascorbic acid followed by trioctyl phosphine oxidemethylisobutyl ketone (TOPO-MIBK) extraction and AAS (BEDROSSIAN, 1984) have also been reported.

3 Sources, Production, Important Compounds, Uses

3.1 Occurrence

Tellurium is widely distributed in the earth's crust in deposits of many different types, from magmatic and pegmatic to hydrothermal, especially where the deposits are associated with epithermal gold and silver deposits. Along with Se, Pd, and Ru, tellurium ranks about seventy-first in the order of crustal abundance. The average amount of Te in crustal rocks is about 0.01 ppm. Mineralogically, tellurium is closely associated in nature with sulfur and selenium, hence, it is commonly found in pyritic ores. The most common occurrence of tellurium in nature is in the form of tellurides. Of the 39 distinct minerals that have been found, those most frequently encountered are sylvanite, hessite, nagyagite, and tetradymite. Relatively large quantities of Te are present in copper sulfide and nickel sulfide deposits and frequently in lead sulfide deposits and as the telluride of gold (calavenite). In pyrite deposits, tellurium is concentrated chiefly in pentlandite, $(FeNi)_9S_8$, chalcopyrite, $CuFeS_2$, and pyrite, FeS_2, in decreasing order. The S:Se:Te ratio varies widely among the deposits and within the same deposit. There are scant data concerning tellurium in sedimentary rocks although some shales contain 0.1–2 ppm Te, and manganese nodules from the Pacific and Indian oceans contain 0.5–125 ppm (SINDEEVA, 1964; CARAPELLA, 1971; ELKIN, 1983).

Tellurium is a main component of about 40 mineral species including 24 tellurides, two tellurates, native tellurium, and a selenium-tellurium alloy. It occurs in combination with oxygen, sulfur, and ten other elements with higher atomic numbers. Commercial grade selenium contains a minimum of 99% selenium and may contain a maximum of 0.2% tellurium while high-grade selenium contains a minimum of 99.99% selenium and about 1 ppm tellurium (see also MASCHEWSKY, 1982).

3.2 Production and Uses

The percentage of tellurium recovered from copper ore is small with about 90% being lost in flotation concentration and from 20 to 60% in each metallurgical operation such as roasting, smelting, converting, fire-refinement, and slimes treatment (ELKIN, 1983). Most tellurium is recovered commercially from electrolytic copper refinery slimes in which it is present from a trace to 8%. Nearly all the tellurium in the anodes finds its way into the anode slimes, along with Se, Ag, Au, Pb, etc. The compounds Ag_2Te, $(Ag, Au)Te_2$, and Cu_2Te have been identified in copper refinery slimes (JENNINGS, 1971).

When the copper is removed from slimes by aeration in dilute sulfuric acid, oxidative pressure-leaching, much of the tellurium is dissolved which is subsequently recovered by cementing (precipitation on metallic copper), leaching the cement mud with dilute caustic soda, and neutralizing with sulfuric acid. The precipitate from the neutralization contains tellurium as tellurous acid suitable for recovery which is accomplished principally by electrolyzing the solution with stainless-steel electrodes followed by washing, drying, and melting the cathode deposit (JENNINGS, 1971; MASCHEWSKY, 1981; ELKIN, 1983).

Tellurium has been produced commercially in the U.S.A. since 1918 and in Canada since about 1934. The United States, Canada, Peru, Japan, and the Federal Republic of Germany are the world's largest producers of tellurium with the element produced to a smaller extent in Bulgaria, Yugoslavia, the German Democratic Republic, India, and the People's Republic of China. The estimated world production of tellurium in 1978 was 350 metric tons (MASCHEWSKI, 1982; ELKIN, 1983).

Approximately 80% of the world's supply of tellurium is derived from electrolytic copper refining. The remainder is recovered from slimes and slugs produced during lead refining, and from pyrite and pyrrhotite burned in pulp and paper mills and sulfuric acid plants. Important intermediates are tellurium dioxide, tellurium tetrachloride (for organic derivatives), and the very toxic tellurium hexafluoride and hydrogen telluride (MASCHEWSKY, 1982).

The spectrum of use applications of tellurium and its compounds includes: (1) as a metallurgical additive to improve the characteristics of alloys of copper, steel, lead, and bronze; (2) in catalysts and catalytic processes; (3) in the vulcanization and curing of rubber; (4) in primary (non-rechargeable) and secondary (chargeable) batteries and fuel cells; (5) in Cd-Te and Se-Te alloys as photoreceptors, solar cells and erasable optical storage (ANONYMOUS, 1986/87; PHILIPS, 1987) in the electrophotographic industry; (6) as electrodes in photochemical cells for solar cells

Table II.28-1. Some Applications of Tellurium and Its Compounds

Compound	Use
Tellurium	Improving machinability in ferritic steel and in copper, steel and bronze alloys; in cast iron for controlling depth of chill; catalyst in petroleum refining and organic chemical industry; in glass; secondary vulcanization agent in natural rubber and in styrene-butadiene rubbers
Tellurium dioxide, tellurium oxides	Additive in bright copper electroplating; catalyst for organic chemical industry; I.R. transparent glass
Tellurium diethyldithiocarbamate	Accelerator for butyl rubber
Tellurides of Mo, Te, W, and Zr	In solid self-lubricating composites in electronics, instrumentation, and aerospace fields
Tellurides of Ag, Bi, Cu, Ge, Mn, Pb, and Sb	In thermoelectric materials and as semiconductors
Tellurides of Bi and Cd	As pigments, semiconductors, and solar cells in pocket calculators and cameras
Tellurium tetrachloride and tetrabromide	Catalysts in organic chemicals industry
Sodium tellurite	For corrosion resistance of electroplated nickel
Tellurium-radioiodinated fatty acid (TPDA)	Myocardial scanning gamma-ray liquid chromatography detector for radiopharmaceuticals

(ELKIN, 1983; GERHARDSSON, et al. 1986; SELENIUM-TELLURIUM DEVELOPMENT ASSOCIATION, 1983 a – c, 1984, 1985 a, b), and (7) as semiconductor raw materials (MASCHEWSKY, 1982). Additional uses are: in infra-red transmitting glasses, in therapeutic applications (BROWNING, 1974; OLDFIELD et al., 1974) in myocardial scanning (BIANCO et al., 1984); and as a gamma-ray liquid chromatography detector for radiopharmaceuticals (NEEDHAM and DELANEY, 1983). Asatine-211-tellurium colloid has been found effective in the treatment of experimental malignant ascites in mice and has been suggested as a radiocolloid therapy for the treatment of some human tumors (BLOOMER et al., 1981). Table II.28-1 summarizes the utility of tellurium and its compounds.

4 Distribution in the Environment, in Foods, and in Living Organisms

4.1 Emissions, Air and Water Quality

The major potential sources of tellurium in ambient air originate from coal combustion (KAUTZ, 1975) and in the vicinity of copper electrolyte plants (SELJANKINA and ALEKSEEVA, 1971; GERHARDSSON et al., 1986). It has been estimated that coal combustion in the U.S. releases about 40 tons of tellurium in fly ash (DAVIDSON and

LAKIN, 1972). Tellurium (2 µg/m^3) has been measured at a distance of 2 km from an electrolytic copper-refining plant (SELJANKINA and ALEKSEEVA, 1971).

Normally Te-levels in drinking water are below the levels of detection as measured in the U.S. and the Federal Republic of Germany (SINEMUS et al., 1981; EINBRODT and MICHELS, 1984; GERHARDSSON et al., 1986). The concentrations of Te in surface water in the Western North Pacific has been reported to be 0.3 – 0.9 ng/L while levels of 8×10^{-10} g/L in South-China sea water and 0.4 – 0.7 ng/L in East-China sea water (AN and ZHANG, 1983) and about 10 ng/L in the Japanese inland sea of Seto (CHUNG et al., 1984) have also been reported.

4.2 Food Chain, Plants, Animals, Humans

Earlier mentioned mean levels in food samples have been considered much too high, since WARD (1987) has found only 0.08 to 0.09 ppm Te even in bovine liver. No information is available on concentrations in plants and animals at the present time. The human daily intake of tellurium has been estimated to be about 100 µg (NASON and SCHROEDER, 1967). Tellurium balance of a 70 kg reference man was suggested by SNYDER et al. (1975) to be: intake from food and fluids, 600 µg/day and losses (µg/day) to be: urine, 530; feces, 100; and breath, 10, respectively. Normal human blood contains about 0.25 µg Te per L (GERHARDSSON et al., 1986). It seems, however, that it is concentrated rather more in the skeleton (SLOUKA, 1970; GERHARDSSON et al., 1986).

5 Uptake, Absorption, Transport and Distribution, Metabolism and Elimination in Microorganisms, Animals, and Humans

Tellurium can be absorbed by inhalation, ingestion, and absorption through the skin. Although there are no quantitative data on the respiratory absorption of tellurium and its compounds, it is believed that this route of absorption is likely to be efficient (GERHARDSSON et al., 1986). The gastrointestinal absorption of both metallic tellurium and tellurium dioxide is poor compared to that of ingested tellurites and tellurates which are absorbed to the extent of about 10 – 30% (HOLLINS, 1969; DEVLIN, 1975; FISHBEIN, 1977; GERHARDSSON et al., 1986).

Tellurium is used in the reprographic field to enhance the blue sensitivity to photoreceptors. In such applications, very small quantities of the photoreceptor are absorbed during normal operations. Although most of the absorbed material is contained within the equipment in filters, traps, etc., a minute portion is carried out of the machine. During such operations as well as in servicing of the equipment extremely low levels of human exposure to tellurium may take place (JOHNSON et al., 1988). The time-weighted average emissions from various reprographic devices are of the order of 10^{-5} mg/m^3. The daily human intake would be 2×10^{-6} mg/kg/day, assuming a 50 kg human female inhaling 10 m^3 of air/day and full bioavailability of all the tellurium inhaled. Since no toxicity is evident near 20 mg/kg/day of tellurium, this leads to a differential factor of 10^7 between calculated human exposure and animal toxicity data (JOHNSON et al., 1988).

Tellurium is readily soluble in tissue fluids and may be transported and deposited in all organs (COOPER, 1971). The route of administration does not apparently significantly alter the pattern of soft tissue distribution of tellurium when administered as sodium tellurite or tellurous acid in HCl solutions (HOLLINS, 1969; SLOUKA, 1970; GERHARDSSON et al., 1986). A first equilibrium is reached within 1–2 hours, while uptake of tellurium by the skeleton is much slower (HOLLINS, 1969; SLOUKA, 1970). Earlier studies with rabbits suggested that the distribution of tellurium in the organs depends to some extent on whether sodium tellurite was administered orally or by injection, with the lungs and kidneys being the most important organs affected (COOPER, 1971). In studies with hamsters, elemental tellurium was identified in deposits in the central nervous system (HAGER, 1960), and according to earlier literature living cells and intestinal bacteria seem to reduce some tellurium compounds to the metallic state (SOLLMAN, 1957). In fact some bacterial strains reduce tellurites, while others form smelling methylated products (ALEXANDER et al., 1988).

In the rat approximately 90% of the blood Te is contained in the erythrocytes, probably bound to hemoglobin (HOLLINS, 1969; SLOUKA, 1970; AGNEW and CHENG, 1971; GERHARDSSON et al., 1986), whereas tellurium deposited in various other tissues is bound to soluble proteins to the extent of 58–96% (AGNEW and CHENG, 1971).

Tellurium has been reported to readily cross the blood-brain barrier of the rat (CRAVIOTO et al., 1970) and rabbit (MIZUNO, 1969) when administered as elemental tellurium either in the diet (CRAVIOTO et al., 1970) or via intramuscular injection (MIZUNO, 1969). Transplacental uptake has been reported in rats receiving i.p. injection of radioactive tellurous acid in 0.9 mol/L HCl (AGNEW, 1972).

Tellurium is transferred to the intestine via bilary excretion (HOLLINS, 1969; SLOUKA, 1970; GERHARDSSON et al., 1986). It can be eliminated via urine, feces, sweat, mammalian milk, and breath. Parenterally administered as sodium tellurite it is excreted by the rat principally via urine (14–27% within 24 hours and 33% within one week, while fecal excretion amounts to about 6% in 24 hours and 14% in one week. In studies in female dogs, about 11–16% of i.v. administered sodium tellurite was excreted within one hour and 23% within six days (DEMEIO and HENRIQUES, 1947).

Studies by DEMEIO and HENRIQUES (1947) (employing tracer ^{129}Te) and those of VIGNOLI and DEFRETIN (1964) suggest that in spite of the fact that tellurium appears to be eliminated rapidly, repeated administration could result in an accumulation in the body, especially in the bones. Although pulmonary excretion occurs, it is not considered an important route since the amount excreted is about 0.1% presumably as dimethyltelluride producing a garlic-like odor observed in animals and humans after exposure to elemental tellurium as well as tellurium(IV) compounds (DEMEIO and HENRQUES, 1947; HOLLINS, 1969; GERHARDSSON et al., 1986).

After intraperitonal or intravenous administration of tellurous acid in HCl or sodium tellurite, retention by rats may be described by two experimental functions with biological half-times of about 19 hours (42–49% of dose) and 13–15 days (51–58%), while gavage of the above compounds resulted in a third additional exponential term (half-time of three-to-seven hours), accounting for the rapid elimination of most of the non-absorbed fraction (70–80%) (HOLLINS, 1969; SLOUKA, 1970). The estimated biological half-times for the major retention sites of tellurium in rats were estimated to be about 9 days (blood), 10 days (liver), 17 days (muscle), 23 days (kidney), and 600 days (bones; standard errors of these extrapolations are high) (HOLLINS, 1969). The whole-body retention model for man suggested by ICRP (International Commission on Radiobiological Protection) (1968) assumes a biological half-time of tellurium to be about three weeks.

6 Effects on Animals and Humans

Reviews of tellurium toxicology have been reported by BROWNING (1969), COOPER (1971), DEVLIN (1975), FISHBEIN (1977), MASCHEWSKY (1982), EINBRODT and MICHELS (1984), and GERHARDSSON et al. (1986).

6.1 Miscellaneous Biochemical Effects
(see also Section 6.2)

Potassium tellurite was found to be a potent antisickling agent in vitro that inhibited red cell sickling at concentrations less than 10 µmol/L by apparently interacting with the red cell membrane (ASAKURA et al., 1984).

Tellurium ion added in vitro at levels of 1 mmol/L to the mitochondria of rat kidney and liver selectively inhibited the oxidation of NAD-dependent substrates (SILIPRANDI et al., 1971) while administration of Te to rats at maximum tolerated doses of 0.5 mg and 0.1 mg/kg body weight resulted in significant reduction of acetylcholine esterase and monoamine oxidase in serum and brain as well as in significant decrease in hepatic glutathione and in glutathione-S-transferase and alkaline phosphatase in liver and kidney (SRIVASTAVA et al., 1983).

Inhibition of catalase in erythrocytes has been noted following both the oral administration and sodium tellurite to rats at daily doses of 0.005 mg Te per kg or more for 7 months (GERHARDSSON et al., 1986) as well as in rats exposed to tellurium and tellurium dioxide aerosols (SANDRACKAJA, 1962a, b).

Increased serum cholesterol has been observed in male but not in female rats following administration of sodium tellurite (150 µg Te per kg body weight per day) for 11–30 months to weanling rats in diet and drinking water (GERHARDSSON et al., 1986).

The feeding of swine tellurium tetrachloride (500 mg/kg) for ten weeks led to the induction of lesions of selenium-vitamin E deficiency which included necrosis of cardiac and skeletal muscles and a decreased blood glutathione-peroxidase activity over the last six weeks of the feeding period (VANVLEET et al., 1981a). Pathological alterations characteristic of selenium-vitamin E deficiency were also found in ducklings fed tellurium (500 mg/kg as tetrachloride) for a few weeks. Lesions included myopathy of gizzard intestine, skeletal muscles and heart, and hydropericardium and focal cerebral malacia have also been found (VANVLEET et al., 1981b; VANVLEET, 1982).

6.2 Effects in Animals

In animals, acute tellurium intoxication results in restlessness, tremor, reduced reflexes, paralysis, convulsions, somnolence, coma, and death. Tellurium compounds are more toxic than the metal per se, and the tellurites have been found to be more toxic than the tellurates, regardless of the route of administration (DEMEIO and JETTER, 1948; SANDRACKAJA, 1962a; COOPER, 1971). Tellurites and tellurates are apparently the most toxic compounds with the exception of hydrogen telluride and tellurium hexafluoride (DEVLIN, 1975; COOPER, 1971). The toxicity of sodium tellurite is ten times that of sodium tellurate when given by i.p. injection but only twice as toxic when fed in chronic studies. Oral toxicity is somewhat lower than parenteral presumably due to reduction to metallic tellurium in the gastrointestinal tract (COOPER, 1971).

The liver, kidney, and central nervous system are the most impacted organs and systems affected by exposure to tellurium and its compounds (COOPER, 1971; DEVLIN, 1975; FISHBEIN, 1977; EINBRODT and MICHELS, 1984; GERHARDSSON et al., 1986).

Systemic administration of tellurium results in impaired growth in rats (DUCKETT and WHITE, 1974; WALBRAN and ROBINS, 1978) and the so-called black-brain syndrome (HOLLINS, 1969; SCHROEDER et al., 1967; WALBRAN and ROBINS, 1978).

In the liver simple cellular swelling to hydrophic and fatty degeneration and cell necrosis were reported in rat studies after feeding tellurium dioxide (375–1500 mg/kg diet for 24–128 days; DEMEIO and JETTER, 1948); or after inhalation of tellurium dioxide (50 mg/m^3, two hours daily, 13–15 days; SANDRACKAJA, 1962b). Similar effects were observed in rats and rabbits fed by gavage with sodium tellurite at levels of 5 and 10 mg/kg for three months and at 0.5 mg/kg for seven months (EL'NICHNYKH and LENCENKO, 1969) and in Peking ducks fed tellurium tetrachloride in the diet at levels of 50–1000 mg/kg for two–four weeks (CARLTON and KELLY, 1967). Additionally, impairment of glycogen function, detoxifying functions, and of protein metabolism were found in studies with rats exposed to tellurium dioxide by inhalation (SANDRACKAJA, 1962b) and in subacute and chronic studies in rats and rabbits administered orally sodium tellurite, 0.5–10 mg Te per kg body weight (LENCENKO and PLOTKO, 1969).

Changes in the kidney as manifested by focal vacuolization of cells and hemorrhage in the glomeruli accompanied by albuminuria and hematuria were noted in rats exposed to tellurium dioxide by inhalation (condensation aerosol, 10 and 50 mg/m^3 two hours daily, for 13–15 weeks) (SANDRACKAJA, 1962a, b). Tellurium dioxide administered in the diet at levels of 375–1500 mg Te per kg diet for 24–128 days induced changes in rats ranging from cellular swelling to frank necrosis and was in some rats accompanied by oliguria and anuria indicating severe lesions in the proximal tubular epithelium (DeMeio and Jetter, 1948). Sodium tellurite, when chemically administered, at levels of 0.5 and 10 mg/kg and sodium tellurate at 1, 10, 25 and 50 mg/kg levels in the diet caused degenerative changes in the kidneys of rats and rabbits. In all cases, sodium tellurite was more toxic than sodium tellurate (El'nichnykh and Lencenko, 1969).

Although no significant changes were noted in peripheral blood of rats fed tellurium dioxide in the diet at levels of 375–1500 mg/kg for 24–128 days (DeMeio and Jetter, 1948; Sandrackaja, 1962a, b) noted normochromic and possibly hemolytic anemia in rats exposed to tellurium dioxide and elemental tellurium aerosol at levels of 50–100 mg/m^3, 2 h daily, for 13–15 weeks.

Tellurium induced peripheral neuropathy in rats with weakness and/or paralysis of the hindlegs has been observed after tellurium exposure (1–1.25% in the diet) (Lampert and Garret, 1971; DeMeio and Jetter, 1948; Sandrackaja, 1962a, b) with the paralysis and demyelination gradually receding and disappearing in about a week (Lampert and Garret, 1971).

A diet of normal rat food containing 1% tellurium fed to adult rats for up to 3 months caused cerebral lipofusionosis with a peripheral neuropathy being suspected but not shown (Duckett and White, 1974). Also a more recent study demonstrated tellurium induced myelinopathy in adult rats (Said and Duckett, 1981). The abnormalities in the tellurium fed adult rats indicate that Te not only damages Schwann cells during the period of most active myelogenesis as suggested in weanling rats (Duckett et al., 1979), but also induced myelin changes at a stage when Schwann cells were not involved in a myelinating process under normal conditions. It was suggested that myelinopathy or a "Schwannopathy" depends on the amount of tellurium reaching the Schwann cells, and the higher vulnerability of the young rats to Te is probably due to a higher permeability of their blood-nerve barrier to tellurium (Said and Duckett, 1981). According to Duckett et al. (1979) it seems likely that Te is transformed by bacterial action in the gut (Summers and Jacoby, 1977) and then absorbed into the blood and carried by the blood (Agnew et al., 1968) to the nerves where it crosses the endothelial barrier (Lampert and Garret, 1971) and enters the Schwann cells.

6.3 Effects in Humans

Relatively few tellurium compounds (e.g., tellurium dioxide, hydrogen telluride, potassium tellurite, and tellurium hexafluoride) are of industrial health significance. The outstanding symptom that led to the recognition that tellurium might rank as an industrial poison was the strong garlic odor of the breath of workers in contact with it. The term "mild tellurism" has been used for a syndrome occurring principally where fumes of tellurium (probably in the form of hydrogen telluride) are emitted.

Serious cases of tellurium intoxication have not been generally reported from industrial exposure to tellurium and its compounds. In the earliest report of the consequences of occupational tellurium exposure in which 14 lead refinery workers were exposed to Te-fumes when the slime from electrolytic tanks was treated to recover silver, three distinctive symptoms of Te-poisoning were noted in eight of the 13 exposed workers: suppression of sweat, dryness of mouth, and the garlic odor (SHIE and DEEDS, 1920).

Iron foundry workers exposed to concentrations between 0.01 to 0.1 mg/m^3 for 22 months complained of garlic odor of the breath and sweat, dryness of the mouth and metallic taste, somnolence, anorexia, and occasional nausea. Urinary concentrations ranged from zero to 0.06 mg/L and somnolence and metallic taste in the mouth did not appear with regularity until the level of tellurium in the urine was at least 0.01 mg/L (NIOSH, 1978; STEINBERG et al., 1942).

In two cases involving accidental occupational exposure to 50 g of tellurium hexafluoride gas, typical signs and symptoms of intoxication were observed including the pungent odor on breath and excreta and tiredness, as well as intradermal deposits of tellurium on their fingers and lesser streaks of bluish-black pigmentation on the face and neck (BLACKADDER and MANDERSON, 1975).

Three cases of accidental tellurium poisoning resulting from the mistaken administration of sodium tellurite for sodium iodide during retrograde pyelography have been reported (KEALL et al., 1946). In two of the cases, death occurred after approximately six hours. The symptoms, in order of appearance, were cyanosis, vomiting, loss of consciousness, and death with autopsy revealing congestion of the lungs, liver, spleen, and kidneys, fatty change of the liver, and deposition of black tellurium in the mucosa of the bladder and of the urether.

6.4 Mutagenic, Carcinogenic, and Teratogenic Effects

In chronic life-time studies with mice, 2 ppm tellurium in the drinking water as either the tellurite or tellurate salt did not alter the incidence of spontaneous tumors although tellurite decreased longevity of females (SCHROEDER and MITCHENER, 1972). With rats, 2 ppm tellurium or tellurite did not alter growth, survival nor spontaneous tumor incidence, although some effects on fasting glucose, cholesterol, and aortic lipids were noted (SCHROEDER and MITCHENER, 1971).

Human leukocytes treated in vitro with 1.2×10^{-8} mol/L of sodium tellurite and 2.4×10^{-7} mol/L ammonium tellurite exhibited a significant increase in chromosomal breakage (PATTON and ALLISON, 1972). Tellurium compounds ($Na_2H_4TeO_6$ and Na_2TeO_3) have been reported to be potent mutagens in reverse mutation assays with *Escherichia coli* and *Salmonella* strains (KANEMATSU et al., 1980).

Elemental tellurium administered orally or by intramuscular injection induced a high incidence of hydrocephalus in newborn rats of treated mothers (GARRO and PENTSCHEW, 1964; AGNEW et al., 1968; DUCKETT, 1970, 1972; DUCKETT and SCOTT, 1971). Hydrocephalus was found in 60–100% of litters with an incidence of 25–100%. It was suggested that tellurium acts directly on the embryo (AGNEW, 1972) with the period of teratogenic vulnerability being between day 9 and 15 of gestation (DUCKETT et al., 1979; AGNEW and CURRY, 1972).

Recently JOHNSON et al. (1988) described the developmental toxicity of tellurium in Sprague-Dawlay rats and New Zealand white rabbits. Groups of pregnant rats were fed a diet containing 0, 30, 300, 3000, or 15000 ppm of tellurium during days 6 through 15 of gestation. Signs of maternal toxicity were observed during the treatment period in a statistically significant and dose-related manner at dietary concentrations of 300 ppm and greater in rats and 1750 ppm and greater in rabbits. Exposure of these pregnant rats and rabbits to tellurium had no effect upon reproduction. Both skeletal (primarily skeletal maturational delays) and soft-tissue malformations (primarily hydrocephalus) were noted in the offspring of pregnant rats exposed to the highest levels (3000 and 15000 ppm) of tellurium. Rabbit fetuses of the highest dosage group (5250 ppm) had a slightly elevated evidence of skeletal delays and nonspecific abnormalities.

7 Hazard Evaluation and Limiting Concentrations

(see also EMMERLING et al., 1986)

Hazards depend on speciation and exposure routes. Hydrogen telluride and tellurium hexafluoride are especially toxic. Tellurites are more toxic than tellurates. Oral intake seems to show a relatively low risk. Inhalation (and possibly dermal contact), e.g., of tellurium oxide and volatile tellurium compounds, is more critical. Characteristic symptoms for tellurium intoxication are a garlic odor of the breath, suppression of sweat, and dryness in the mouth. Besides kidneys, liver, and lung, the nervous system is especially endangered. It seems that nothing is known about tellurium chronic toxicity.

According to the MAK report (1987) 0.1 mg/m^3 of tellurium and its compounds is a maximum allowable concentration per day. Exposure peaks of 0.5 mg/m^3 are limited to 30 minutes (maximum twice per shift).

For an eight hour day, the threshold limit for tellurium also in the United States and the United Kingdom is 0.1 mg/m^3, while the maximum allowable concentration in the USSR is 0.01 mg/m^3 (ACGIH, 1977; BELKES, 1981). The OSHA permissible exposure limit in the United States is 0.1 mg/m^3 (OSHA, 1981). It has been suggested by GERHARDSSON et al. (1986) that it would be prudent to keep the air concentration of tellurium below 0.01 mg/m^3 to exclude the garlic odor from the breaths of occupational workers completely, and this would necessitate the concentration of Te in the urine be kept below 1 µg/L (STEINBERG et al., 1942; LAUWERYS, 1983) as well as rotating the workers on and off with tellurium (GLOVER, 1976).

References

ACGIH (American Conference of Governmental Industrial Hygienists) (1977), *Threshold Limit Values for 1977*, Cincinnati, Ohio.

Agnew, W. F. (1972), *Teratology 6*, 331–337.

Agnew, W. F., and Cheng, J. T. (1971), *Protein Binding of Tellurium-127m by Maternal and Fetal Tissues of the Rat. Toxicol. Appl. Pharmacol. 20*, 346–356.

Agnew, W. F., and Curry, E. (1972), *Period of Teratogenic Vulnerability of Rat Embryo to Induction of Hydrocephalus by Tellurium.* Experientia 28, 1444–1445.

Agnew, W. F., Gauvre, F. M., and Pudenz, R. H. (1968), *Exp. Neurol.* 21, 120–132.

Alexander, J., Thomassen. Y., and Assethy, J. (1988), Chapter 61 *Tellurium,* in: Seiler, H. G., Sigel, H., and Sigel, A. (eds.): *Handbook of Toxicology of Inorganic Compounds,* pp. 669–674. Marcel Dekker, New York.

An, J., and Zhang, Q. (1983), *Analytical Method for Ultratrace Tellurium Determination in Sea- and Environmental Water.* Int. J. Environ. Anal. Chem. 14, 73–80.

Andreae, M. O. (1984), *Determination of Inorganic Tellurium Species in Natural Waters.* Anal. Chem. 56, 2064–2066.

Anonymous (1986/87), *Solar Cells, Chem. Eng. News* 64(27), 34; *Neue Zürcher Zeitung, Forschung und Technik,* No. 28, p. 67 (4 February), Zürich.

Asakura, T., Shibutani, Y., Reilly, M. P., and DeMeio, R. H. (1984), *Antisickling Effect of Tellurite: A Potent Membrane-acting Agent in vitro.* Blood 64, 305–307.

Barnes, R. M. (1984), *Determination of Trace Elements in Biological Materials by Inductively Coupled Plasma Spectroscopy with Novel Chelating Resins.* Biol. Trace Elem. Res. 6, 93–103.

Bedrossian, M. (1984), *Determination of Microgram Amounts of Selenium and Tellurium in Copper-base Alloys by Atomic Absorption Spectrometry.* Anal. Chem. 56, 311–312.

Belkes, R. P. (1981), in: Clayton, G. D., and Clayton, F. E. (eds.): *Patty's Industrial Hygiene and Toxicology,* 3rd Ed., Vol. II-A, pp. 2135–2139. Wiley-Intersciences, New York.

Bianco, J. A., Pape, L. A., Albert, J. S., et al. (1984), *Accumulation of Radioiodinated 15-(p-Iodophenyl)-6-tellurapentadecanoic Acid in Ischemic Myocardium During Acute Coronary Occlusion and Reperfusion.* J. Am. Coll. Cardiol. 4, 80–87.

Blackadder, E. S., and Manderson, W. G. (1975), *Br. J. Ind. Med.* 32, 59–61.

Bloomer, W. D., McLaughlin, W. H., Neirinckx, R. D., et al. (1981), *Astatine-211-Tellurium Radiocolloid Cures Experimental Malignant Ascites.* Science 212, 340–341.

Browning, E. (1969), *Toxicity of Industrial Metals,* 2nd Ed., pp. 310–316. Butterworths, London.

Brune, D., Nordberg, G., and Wester, P. O. (1980), *Distribution of 23 Elements in the Kidney, Liver, and Lungs of Workers from a Smeltery and Refinery in North Sweden Exposed to a Number of Elements and a Control Group.* Sci. Total Environ. 16, 13–35.

Carapella, S. C., Jr. (1971), in: Cooper, W. C. (ed.), *Tellurium,* pp. 1–13. Van Nostrand, New York.

Carlton, W. W., and Kelly, W. A. (1967), *Toxicol. Appl. Pharmacol.* 11, 203–214.

Cheng, K. L., and Johnson, R. A. (1978), *Determination of Selenium and Tellurium.* Chem. Anal. 8, 371–419.

Chung, C. H., Iwamoto, E., Yamamoto, M., and Yamamoto, Y. (1984), *Selective Determination of Arsenic(III, V), Antimony(III, V), Selenium(IV, VI), and Tellurium(IV, VI) by Extraction and Graphite Furnace Atomic Absorption Spectrometry.* Spectrochim. Acta B 39, 459–466.

Clark, J. R., and Viets, J. G. (1981), *Back-extraction of Trace Elements from Organometallic-halide Extracts for Determination by Flameless Atomic Absorption Spectrometry.* Anal. Chem. 53, 61–65.

Cooper, W. C. (1971), in: Cooper, W. C. (ed.), *Tellurium,* pp. 281–312. Van Nostrand Reinhold, New York.

Cravioto, H., Agnew, W. F., Carregal, E. J. A., and Pudenz, R. H. (1970), *J. Neuropathol. Exp. Neurol.* 29, 158.

Davidson, D. F., and Lakin, H. W. (1972), *Tellurium. U.S. Geol. Surv. Prof. Pap.* 820, 627–630.

DeMeio, R. H., and Henriques, F. C., Jr. (1947), *J. Biol. Chem.* 169, 609–623.

DeMeio, R. H., and Jetter, W. W. (1948), *J. Ind. Hyg. Toxicol.* 30, 53–57.

Devlin, T. (1975), *A Review of the Toxicology of Tellurium and Its Compounds,* Sandia Report No. 75-8047. Sandia Laboratories, Albuquerque, New Mexico.

Diksic, M., and Cole, T. F. (1977), *Fast Determination of Molybdenum and Tellurium by Neutron Activation Analysis.* Anal. Chim. Acta 93, 261–266.

Duckett, S. (1970), *Fetal Encephalopathy Following Ingestion of Tellurium.* Experientia 26, 1239–1241.

Duckett, S. (1972), *Teratogenesis Caused by Tellurium. Ann. N.Y. Acad. Sci. 192*, 220–226.
Duckett, S., and Scott, T. (1971), *Localization of Tellurium in Tellurium-induced Hydrocephalus. Experientia 27*, 432–434.
Duckett, S., and White, R. (1974), *Cerebral Lipofuscinosis Induced with Tellurium, Electron Dispersive X-Ray Spectrophotometry Analysis. Brain Res. 73*, 205–214.
Duckett, S., Said, G., Streletz, L. G., et al. (1979), *Tellurium-induced Neuropathy: Correlative Physiological, Morphological and Electron Microscope Studies. Neuropathol. Appl. Neurobiol. 5*, 265–279.
Einbrodt, H. J., and Michels, S. (1984), *Tellurium*, in: Merian, E. (ed.): *Metalle in der Umwelt*, pp. 562–569. Verlag Chemie, Weinheim-Deerfield Beach, Florida-Basel.
Elkin, E. M. (1983), in: Kirk-Othmer: *Encyclopedia of Chemical Technology*, 3rd Ed. Vol. 22, pp. 658–679. Wiley & Sons, New York.
El'nichnykh, L. N., and Lencenko, V. G. (1969), in: Mihailov, V. A. (ed.): *Clinical Aspects, Pathogenesis and Prophylaxis of Occupational Diseases of Chemical Etiology in Ferrous and Non-Ferrous Metallurgy*, Part II, pp. 155–160. Sverdlovsk Russian Institute for Industrial Hygiene and Occupational Diseases, Sverdlovsk.
Emmerling, G., Schaller, K. H., and Valentin, H. (1986), *Actual Knowledge on Tellurium (and Other Rare Metals) and Their Compounds, and Feasibility of Quantitative Determination in Biological Materials in Occupational Medicine and Toxicology; Zentralbl. Arbeitsmed. 36*, 258–265.
Fishbein, L. (1977), *Toxicology of Selenium and Tellurium. Adv. Mod. Toxicol. 2*, 191–240.
Fodor, P., and Barnes, R. M. (1983), *Determination of Some Hydride-forming Elements in Urine by Resin Complexation and Inductively Coupled Plasma Atomic Spectroscopy. Spectrochim. Acta 38 B*, 239.
Garro, F., and Pentschew, A. (1964), *Arch. Psychiatr. Neurol. 206*, 272–280.
Gerhardsson, L., Glover, J. R., Nordberg, G. F., and Vouk, V. (1986), in: Friberg, L., Nordberg, G. F., and Vouck, V. B.: *Handbook on the Toxicology of Metals*, 2nd Ed., Vol. II, pp. 532–548. Elsevier, Amsterdam.
Glover, J. R. (1976), in: *Proceedings of the Symposium on Selenium-Tellurium in the Environment*, pp. 279–292. Industrial Health Foundation, Pittsburgh, Pennsylvania.
Hager, H. (1960), *Arch. Psychiatr. Nervenkrankh. 201*, 53.
Hollins, J. G. (1969), *Health Phys. 17*, 497–505.
ICPR (International Commission on Radiobiological Protection) (1968), *ICRP Publ. 10*, 58–64.
Jennings, P. H. (1971), in: Cooper, W. C. (ed.), *Tellurium*, pp, 14–40. Van Nostrand, New York.
Johnson, E. M., Christians, M. S., Hoberman, A. M., De Marco, C. J., Kilpper, R., and Mermelstein, R. (1988), *Developing Toxicology Investigations of Tellurium. Fund. Appl. Toxicol. 11*, 691–702.
Kanematsu, N., Hara, M., and Kada, T. (1980), *Mutat. Res. 77*, 109–116.
Kautz, K. (1975), *VGB Kraftwerkstech. 55*, 672–676.
Keall, J. H. H., Martin, N. H., and Tunbridge, R. E. (1946), *Report of Three Cases of Accidental Poisoning by Sodium Tellurite. Br. J. Ind. Med. 3*, 175–176.
Lampert, P. W., and Garret, R. S. (1971), *Lab. Invest. 25*, 380–388.
Lauwerys, R. R. (1983), in: *Industrial Chemical Exposure, Guidelines for Biological Monitoring*. Biomedical Publications, Davis, California.
Lencenko, V. G., and Plotko, E. G. (1969), in: Mihailov, V. A. (ed.): *Clinical Aspects, Pathogenesis and Prophylaxis of Occupational Diseases of Chemical Etiology in Ferrous and Non-Ferrous Metallurgy*, Part II, pp. 137–147. Sverdlovsk Research Institute for Industrial Hygiene and Occupational Diseases, Sverdlovsk.
Maher, W. A. (1984), *Determination of Tellurium by Atomic Absorption Spectroscopy with Electrothermal Atomization after Preconcentration by Hydride Generation and Trapping. Anal. Lett. 17* (A10), 979–991.
MAK (1987), *Maximum Concentrations at the Workplace, Report No. XXIII, DFG*. VCH Verlagsgesellschaft, Weinheim-Basel-Cambridge-New York.

Maschewsky, D. (1982), *Tellur und Tellur-Verbindungen*, in: *Ullmanns Encyklopädie der technischen Chemie,* 4th Ed., Vol. 22, pp. 447–454. Verlag Chemie, Weinheim-Deerfield Beach/Florida-Basel.

Mizuno, R. (1969), *Yokohamaigaku 20,* 101–122.

Muangnoicharoen, S., Chiou, K. Y., and Manuel, O. K. (1986), *Determination of Selenium and Tellurium in the Gas Phase Using Specific Columns and Atomic Absorption Spectrometry. Anal. Chem. 58,* 2811–2813.

Nason, A. P., and Schroeder, H. A. (1967), *J. Chronic Dis. 20,* 671.

Needham, R. E., and Delaney, M. F. (1983), *Cadmium Telluride Gamma-Ray Liquid Chromatography Detection for Radiopharmaceuticals. Anal. Chem. 55,* 148–150.

NIOSH (National Institute of Occupational Safety and Health) (1978), *Occupational Health Guidelines for Tellurium and Compounds (as Tellurium),* Washington, DC.

Oldfield, J. E., Allaway, W. H., Laitinen, H. A., et al. (1974), *Geochem. Environ. 1,* 64–67.

Olson, O. E., Hilderbrand, D. C., and Matthees, P. P. (1984), *Selenium and Tellurium, Tech. Instrum. Chem. 4B,* 307–331.

OSHA (Occupational Safety and Health Administration) (1981), *Occupational Safety and Health Standards Subpart 2, Toxic and Harzadous Substances.* Code of Federal Regulations 29 (Part 1910.1000) pp. 673–679. Washington, DC.

Patton, G. R., and Allison, A. C. (1972), *Mutat. Res. 16,* 332–336.

Philips (Company) (1987), *New Materials for Optic Storage, Neue Zürcher Zeitung, Forschung und Technik,* No. 208 (12 September), Zürich.

Said, G., and Duckett, S. (1981), *Muscle Nerve 4,* 319–325.

Sandrackaja, S. E. (1962a), *Experimental Studies of the Characteristics of Tellurium as an Industrial Poison.* First Moscow Medical Institute, Moscow.

Sandrackaja, S. E. (1962b), *Gig. Tr. Prof. Zabol. 2,* 44–50.

Schroeder, H. A., Buckman, J., and Balassa, J. J. (1967), *J. Chron. Dis. 20,* 147–161.

Schroeder, H. A., and Mitchener, M. (1971), *Selenium and Tellurium in Rats. Effect on Growth, Survival and Tumors. J. Nutr. 101,* 1531–1540.

Schroeder, H. A., and Mitchener, M. (1972), *Selenium and Tellurium in Mice. Arch. Environ. Health 24,* 66–71.

Selenium-Tellurium Development Association (1983a), *Bulletin No. 23.* Darien, CT.

Selenium-Tellurium Development Association (1983b), *Bulletin No. 24.* Darien, CT.

Selenium-Tellurium Development Association (1983c), *Bulletin No. 25.* Darien, CT.

Selenium-Tellurium Development Association (1984), *Bulletin No. 27.* Darien, CT.

Selenium-Tellurium Development Association (1985a), *Bulletin No. 28.* Darien, CT.

Selenium-Tellurium Development Association (1985b), *Bulletin No. 29.* Darien, CT.

Seljankina, K. P., and Alekseeva, C. S., (1971), *Gig. Sanit. 35.,* 95–96.

Shie, M. D., and Deeds, F. E. (1920), *U.S. Public Health Serv. Health Rep. 35,* 939.

Siliprandi, D., DeMeio, R. H., Toninello, A., and Zoccarato, F. (1971), *Action of Tellurite, a Reagent for Thiol Groups, on Mitochondria Oxidative Processes. Biochem. Biophys. Res. Commun. 45,* 1071–1075.

Sindeeva, N. D. (1964), *Mineralogy and Types of Deposits of Selenium and Tellurium.* Interscience, John Wiley, New York.

Sinemus, H. W., Melcher, M., and Welz, B. (1981), *Atom. Spectrosc. 2,* 81–86.

Slouka, V. (1970), *Distribution and Excretion of Intravenous Radiotellurium in the Rat. Sb. Ved. Pr. Lek. Fak. Univ. Karlovy Hradci Kralove 47,* 3–19.

Snyder, W. S., Cook, M. J., Nasset, E. S., et al. (1975), *Report of the Task Group on Reference Man, ICRP, Publ. 23.* International Commission on Radiobiological Protection, Washington, D.C., Sept. 9–14.

Sollman, T. (1957), *Pharmacology.* Saunders, Philadelphia.

Srivastava, R. C., Srivastava, R., Srivastava, T. N., and Jain, S. P. (1983), *Effect of Organo Tellurium Compounds on the Enzymatic Alterations in Rats. Toxicol. Lett. 16,* 311–316.

Steinberg, H. H., Massari, S. C., Miner, A. C., and Rink, R. (1942), *J. Ind. Hyg. Toxicol. 24,* 183.

Strelow, F. W. E. (1984), *Separation of Tellurium from Gold(III), Indium, Cadmium, and Other Elements by Cation Exchange Chromatography in Hydrochloric Acid-Acetone. Anal. Chem. 56,* 2069–2073.

Sugimura, Y., and Suzuki, Y. (1981), *Chemical Forms of Minor Metallic Elements in the Ocean. Pap. Meteorol. Geophys. 32,* 163–165.

Summers, A. O., and Jacoby, G. A. (1977), *Plasmid-determined Resistance to Tellurium Compounds. J. Bacteriol. 129,* 276–281.

Thompson, M., Pahlvanpour, B., and Thorne, L. T. (1981), *The Simultaneous Determination of Arsenic, Antimony, Bismuth, Selenium, and Tellurium in Waters by an Inductively Coupled Plasma-Volatile Hydride Method. Water Res. 15,* 407–411.

Van Montfort, P. F. E., Agterdenbos, J., and Juette, B. A. H. G. (1979), *Determination of Antimony and Tellurium in Human Blood by Microwave Induced Emission Spectrometry. Anal. Chem. 51,* 1553–1557.

VanVleet, J. F. (1982), *Amounts of Twelve Elements Required to Induce Selenium-Vitamin E Deficiency in Ducklings. Am. J. Vet. Res. 43,* 851–857.

VanVleet, J. F., Boon, G. D., and Ferrans, V. J. (1981 a), *Evaluation of the Ability of Dietary Supplements of Silver, Copper, Cobalt, Tellurium, Cadmium, Zinc and Vanadium to Induce Lesions of Selenium-Vitamin E Deficiency in Ducklings and Swine. Am. J. Vet. Res. 42,* 789–799.

VanVleet, J. F., Boon, G. D., and Ferrans, V. J. (1981 b), *Myocardial Ultrastructural Alterations in Ducklings Fed Tellurium. Am. J. Vet. Res. 42,* 1206–1217.

Verlinden, M., Deelstra, H., and Adriaenssens, E. (1981), *The Determination of Selenium by Atomic Absorption Spectrometry. Talanta 28,* 637–646.

Vignoli, L., and Defretin, J. P. (1964), *Toxicology of Tellurium. Ann. Biol. Clin. 22,* 399–409.

Walbran, B. B., and Robins, E. (1978), *Effects of Central Nervous System Accumulation of Tellurium on Behavior in Rats. Pharmacol. Biochem. Behav. 9,* 297–300.

Ward, N. I. (1987), *Environmental Health No. 20, 2nd Nordic Symposium on Trace Elements in Human Health and Disease, Odense,* Extended Abstracts, pp. 118–123. WHO, Regional Office for Europe, Copenhagen.

Welz, B. (1985), *Atomic Absorption Spectrometry,* 2nd revised Ed., pp. 329–330, 352ff, 401ff. VCH Verlagsgesellschaft, Weinheim-Deerfield Beach/Florida-Basel.

Wolnik, K. A., Fricke, F. L., Hahn, M. H., and Caruso, H. A. (1981), *Sample Introduction System for Simultaneous Determination of Volatile Elemental Hydride and Other Elements in Foods by Inductively Coupled Argon Plasma Emission Spectrometry. Anal. Chem. 53,* 1030–1035.

Yamamoto, M., Yasuda, M., and Yamamoto, Y. (1985), *Hydride-Generation Atomic Absorption Coupled with Flow Injection Analysis. Anal. Chem. 57,* 1382–1385.

Yu, M. Q., Liu, G. Q., and Jin, Q. (1983), *Determination of Trace Tellurium(IV) and Tellurium(VI) in Water by Concentration and Separation with Sulfhydryl Cotton Fiber and Atomic Absorption Spectrophotometry Following Hydride Generation. Talanta 30,* 265–270.

II.29 Thallium

FRITZ H. KEMPER and HANS P. BERTRAM
Münster, Federal Republic of Germany

1 Introduction

Thallium is a trace element which occurs in small amounts but ubiquitously in sulfur containing ores and potassium minerals. Its economic and technical importance is negligible. The anthropogenic environmental occurrence of thallium is caused by ore smelting, handling and processing of intermediate or waste products from metallurgic industry, resulting in thallium emissions. Vegetation damages have been observed around cement plants using thallium containing pyrite smelting residues. The use of thallium containing medicaments and rodenticides is rather decreasing, but accidents with rodenticides may still occur because of the Warfarin resistance developed in rodents.

Ecotoxicological importance of thallium is derived from its high acute toxicity on living organisms, comparable to that of lead and mercury. Little is known about chronic ecotoxicology. Risk assessment is only possible in limited areas of thallium pollution. Global data are missing. Moreover, nothing is known about the effects of thallium in man (Fig. II.29-1), when it is incorporated in only small amounts but continuously by food, water, or air.

2 Physical and Chemical Properties, and Analytical Methods

2.1 Physical and Chemical Properties

Thallium (Tl) with the atomic number 81 and the atomic mass of 204.37 is placed in Group IIIA (Group 13 according to the IUPAC proposal 1986) of the Periodic System of Elements, below the two metals gallium (atomic number 31) and indium (atomic number 49).

Besides the two naturally occurring isotopes Tl-203 (30%) and Tl-205 (70%), 26 artificial isotopes are known: mass range 191–210, half-life range 2.1 ms (Tl-201m) to 3.8 years (Tl-204). With its high density of 11.85 g/cm^3 thallium is a heavy metal; melting point 303.5 °C, boiling point 1457 °C.

Thallium was discovered by CROOKES and LAMY. In 1861 CROOKES found the characteristic green line of the element at 535.05 nm in the emission spectrum of

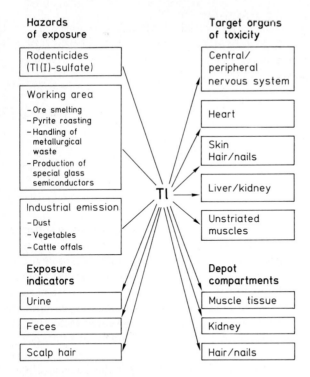

Fig. II.29-1. Toxicology of thallium (BERTRAM and KEMPER, 1983).

residues of sulfuric acid production. According to the Greek words 'thallein' (becoming green) and 'thallos' (green twig) he named the new element 'thallium'. LAMY was the first who prepared metallic thallium.

Corresponding to the place in the Periodic System of Elements, in Tl-compounds the metal occurs in the oxidation states +I and +III. In contrast to the other Group IIIA elements, the monovalent form is more stable than the trivalent. Chemical and physical properties of metallic thallium and thallium compounds are similar to those of adjacent elements, mainly to lead (atomic number 82).

2.2 Analytical Methods

Analytical methods for thallium have been reviewed by SAGER and TÖLG (1984) and SAGER (1986). TSALEV and ZAPRIANOV (1983b) reported a synopsis of the application of atomic absorption methods in determination of thallium in biological material (see also WELZ, 1985). The low detection limits of some newer specific spectroscopic methods allow in some cases the direct determination of thallium in environmental samples. Nevertheless, in many cases preconcentration procedures must be included in the analytical runs to reach sufficient accuracy.

KORKISCH and STEFFAN (1978) recommended formation of anionic thallium bromide complexes from natural waters. The extraction system sodium diethyldithiocarbamate (NaDDC)/methylisobutyl ketone (MIBK) is a widely used method for coextraction of thallium and other heavy metals from body fluids. In connection with electrothermal atomic absorption spectroscopy (ET-AAS) a detection limit of 0.1 µg/L Tl in urine or blood can be reached (for clinical application of thallium analysis, cf. BERMAN, 1980). FLEGAL and PATTERSON (1985) combined dithizone/chloroform extraction and isotope-dilution mass spectrometry.

Similar low detection limits are characteristic in neutron activation analysis (NAA). Using this method HENKE and FITZEK (1971) revealed the time course of a thallium intoxication on the basis of the longitudinal Tl-distribution of a single hair. Detection limit: 0.1 µg/kg = 0.1 ppb Tl.

Further, the following methods are used in environmental thallium analysis: Electrochemical procedures (see WANG, 1985), e.g., pulse voltammetry (WEINIG and WALZ, 1971; REINHARDT and ZINK, 1975; GEMMER-ČOLOS et al., 1981; YOU and NEEB, 1983; and ANGERER and SCHALLER, 1985, for thallium in urine); the most sensitive isotope dilution and field desorption mass spectroscopy (detection limits below 0.1 ppb, WEINIG and ZINK, 1967; ACHENBACH et al., 1980); X-ray fluorescence analysis (RETHFELD, 1980); some flame and plasma emission spectroscopic methods (e.g., WALL, 1977).

The most sensitive wavelength for emission analysis is the 535.05 nm line. Alternate lines are 276.79, 291.83, and 351.92 nm. In atomic absorption methods the Tl-wavelengths at 276.79 and 377.57 nm are used.

Photometric methods for the determination of thallium have been known since long time. Tl(I) is oxidized to Tl(III) and complexed with brilliant green, rhodamine B, or methylviolet. The complexes are extracted with organic solvents, and light absorption is measured at different wavelengths. The methods lack sensitivity for normal or environmental levels of thallium in biological material, but may be useful in the medical field for diagnosis of a severe thallium intoxication, although they may be unreliable and are thus not always recommended.

3 Sources, Production, Important Compounds, Uses, Waste Products

3.1 Occurrence

Thallium is a rare element, which occurs in the earth's crust in an estimated abundance of 0.1 to 0.5 µg/g. Its specific ionic properties (ionic radius Tl^+ 0.147 nm, electronegativity Tl^+ 2.04) are similar to those of potassium and rubidium (ionic radius K^+ 0.133 nm, Rb^+ 0.147 nm); thus, thallium occurs as trace element ubiquitously in the environment, mainly associated with K and Rb. Besides its occurrence in widespread potassium compounds such as the complex potassium silicate mica, thallium is a trace component in iron, zinc, copper, and lead minerals.

Thallium has also been found in extraterrestrial material: meteoric stones contain 0.001 to 0.2 µg/g, lunar minerals 0.0006 to 0.0024 µg/g thallium (UREY, 1952; WEDEPOHL, 1974).

In contrast to the occurrence of thallium as trace element, thallium minerals are very rare. Crookesite (from Skrikerum/Sweden) is a mixture of the selenides of copper, thallium, and silver. Similar chemical compositions have been found in berzelianite (Germany) and lorandite (Macedonia). Other minerals with high Tl-content are located in the USA and Brazil (KAZANTZIS, 1986; SCHOER, 1984).

3.2 Production

Raw material for the preparation of metallic thallium and thallium compounds are the residues of smelting copper, lead, zinc, and iron ores: fly ash, dust, sludge, residual electrolyte solutions. Thallium is enriched by precipitation of the poorly soluble Tl(I)-chloride, Tl(I)-sulfide, or Tl(I)-chromate(VI). Separation from other heavy metals is achieved using the acid solubility of Tl(I)-sulfide and the water solubility of Tl(I)-carbonate. Metallic thallium is prepared by reduction with zinc or by electrolysis (MICKE et al., 1983; SCHOER, 1984). World production of the metal is estimated to be in the range of only 10–15 tons per year.

3.3 Uses, Important Compounds

The physico-chemical properties of the stable Tl(I)-compounds are in some points similar to those of the salts of Group I elements (high water solubility of sulfate and carbonate) as well as to those of silver, mercury, and lead compounds (poor water solubility of chloride, bromide, iodide, sulfide).

A short review of inorganic and organic thallium compounds has been given by MICKE et al. (1983) or SCHOER (1984). Of the organic thallium compounds dialkylthallium ions (the dimethylthallium ion has a linear framework and is isoelectronic with dimethylmercury) are most stable (THAYER, 1988).

The most important thallium compound is Tl(I)-sulfate (Tl_2SO_4), forming colorless, tasteless, and odorless crystals with a water solubility of about 49 g/L (20 °C). Because of its high toxicity Tl(I)-sulfate is used as a rodenticide. The frequent misuse of thallium containing preparations for suicidal or criminal purposes, including accidental intoxications, led to restriction, total prohibition, or voluntary withdrawal of these products in most countries (e.g., Federal Republic of Germany, USA, Switzerland). Uses of thallium acetate as a depilatory or for treating of venereal diseases and night sweat in tuberculosis are also not practiced anymore (MANZO and SABBIONI, 1988).

Examples for the industrial use of metallic thallium and thallium compounds are:
a) deep temperature thermometers (8.5% thallium amalgam; melting point −58 °C);
b) lead/thallium alloys show good resistance against acids; multicomponent alloys such as Pb/Sb/Sn/Tl may be used as carrier material (MICKE et al., 1983);
c) ternary systems of thallium, arsenic, and selenium are used for the manufacturing of low-melting glasses with a high refractive index;
d) high permeability to infrared rays is achieved using crystals of thallium halides (combinations Tl(I)-bromide/Tl(I)-iodide and Tl(I)-chloride/Tl(I)-bromide);

e) Tl-sulfide, Tl-arsenide, Tl-selenide, and Tl-telluride are useful as semiconductor materials;
f) supraconductors such as thallium-barium-copper oxide or thallium-strontium-calcium-copper oxide have a relatively high critical temperature stability of about 90 K and more; but similar crystal structures with bismuth instead of thallium are mostly preferred to avoid toxic intermediates (BEDNORZ, 1988, and others);
g) in the medical field Tl-201 (radioactive half-life 73 h) is used for diagnostic purposes for the heart and the circulatory system (thallium scintigraphy).

3.4 Waste Products and Industrial Emissions

In contrast to the small world production of about 10–15 t thallium per year, the amount of thallium in waste material is estimated to be about 600 t/a (MICKE et al., 1983). In residues of ore smelting and metal processing thallium may be enriched. Main anthropogenic occurrence of thallium in the environment results from emissions of the highly volatile metal and its compounds from some industrial processes (SCHOER, 1984): smelting of chalcogenic ores, especially lead and zinc sulfides, yields thallium emissions.

The thallium content of some types of coal (with high sulfidic portion) may be emitted as Tl-enriched fly ash in coal burning. EWERS (1988) states that worldwide 600 t/a of thallium are emitted by coal burning power plants. Though airborne fly ash may amount to about 100 µg/g of thallium (SMITH and CARSON, 1977), this type of emission must not result in environmental contamination. SCHMIDT and DIETL (1985) found in the surroundings of a coal fired power plant normal thallium contents in soil (0.1–0.5 µg/g dry matter in non-polluted areas). Worldwide from all sources about 2000 t/a may be mobilized (EWERS, 1988).

A global risk assessment of the ecotoxicological importance of these emissions is not possible, because there are only few investigations and observations. More detailed results have been reported on the thallium flow in cement production (SCHOER, 1984). Thallium is introduced into the process by the stone (limestone, clay) and ore raw materials, but may also be added by some products used in the fabrication of special cement types.

In 1979/80 vegetation damage was observed around a cement plant in Lengerich/Westphalia (Federal Republic of Germany) and associated with thallium emissions of this factory. At this time residues of pyrite smelting with a high Tl-content (400 µg/g) were used as an additive in the production of high-class cement types. Extensive investigations in the emission area revealed elevated thallium levels in vegetables as well as in some organs of cattle. A species specific degree of thallium accumulation in plants was observed. Furthermore, there was a correlation with the thallium content in soil (LEHN and SCHOER, 1985; SCHOER, 1984). In plants of the genus *Brassica* L. (cabbage), family Cruciferae, with high sulfur content, up to 45 µg thallium per g wet weight were found (LANDESANSTALT FÜR IMMISSIONSSCHUTZ, 1980), the highest values in *Brassica oleracea* var. *sabellica* L. (kale), *Brassica oleracea* var. *capitata* L. (white- and red-heart cabbage), and *Brassica napus* L. var.

Table II.29-1. Thallium Content in Different Food Samples from the Emission Area of a Cement Plant (Own Results)

Sample	Distance from the Plant (km)	Cardinal Point	Tl-Content (µg/g wet weight)		
Pork	0.8	SW	0.284		
	1.5	NE	0.138		
	2.5	SW	0.028		
Kidney (pig)	0.8	SW	0.796		
	1.5	NE	0.244		
	2.5	SW	0.063		
Honey	0.8	SE	0.763		
	1.2	NW	0.109		
Green rape	0.4	SW	23.7		
	1.1	SE	0.904		
	10	SW	0.103		
	20	SW	0.030		
Black currants	0.15	W	0.527		
	10	SW	0.027		
			Stem	Leaves young	Leaves old
Kale	0.2	W			14.9
	1.2	NE	0.055	0.240	1.27
	1.5	NE			1.73
					0.601[a]
	2.0	NE			0.575
					0.119[a]
	2.5	NE	0.068	0.125	0.185
	4.0	SW		0.037	0.047
			Whole egg	Without shell	Egg-shell
Eggs (hen)	0.3	NW	1.26	0.394	4.94
	0.7	SE	0.631	0.149	3.53
	1.2	SE	0.787	0.248	2.14
	20	S	<0.1	<0.1	<0.1

[a] Boiled

napus (green rape). Normal Tl-content in these species ranges between 0.01 and 0.1 µg/g wet weight.

The thallium concentrations in most food samples showed a distinct relation to the distance of the cement factory (Table II.29-1) (own results).

Assessment of possible health risks for inhabitants of the emission area on the basis of chronic ingestion of small amounts of thallium is difficult (see Sect. 7). Examination of the population living in the Lengerich region and of the workers of the cement plant has been done continuously since 1979 (BERTRAM, 1980; VONGEHR, 1986; SCHURZMANN, 1986; BROCKHAUS et al., 1981). The frequency distributions of thallium content in blood, urine, and hair samples of 1980/81 showed a shift to higher values compared with those of unaffected areas. Nevertheless, the range of all results is distinctly different from the Tl-concentrations in body fluids and hair found in acute Tl-intoxications. Moreover, no individual result reached the urinary thallium level of 300 µg/L, suggested as the threshold value for factory workers (GLÖMME, 1983).

4 Distribution in the Environment, in Foods, and in Living Organisms

Normally ambient air contains less than 1 ng Tl per m^3 (EWERS, 1988). The range of thallium concentrations in sea water (in the form of Tl$^+$ ions) has been reported to be relatively constant at 0.01 to 0.016 ppb (µg/L) (BRULAND, 1983; FLEGAL and PATTERSON, 1985), in non-polluted river water 0.01 – 1 µg/L. The good water solubility of thallium compounds may result in a local contamination of river and ground water by mining, smelting, and cement waste water (FÖRSTNER and WITTMANN, 1979; SCHOER, 1984).

Only few data are available about the 'normal' content of thallium in plants. GEILMANN et al. (1960) found a concentration range of 0.021 – 0.125 µg thallium per g dry matter for some vegetables (cf. Table II.29-1) and 0.008 – 0.025 µg/g dry matter for wild meadow plants and clover. For pine trees and bilberry shrubs of the Rocky Mountains, GOUGH et al. (1979) reported a Tl-content of 2 – 15 µg/g ash, for plants of the Soviet Union 0.01 – 1.0 µg/g ash. For concentrations in the surroundings of a cement plant see Sect. 3.4.

Under environmental conditions inorganic thallium may be oxidized and stabilized in the trivalent form as dimethylthallium(III) salts by microorganisms (HUBER and KOTULLA, 1982).

The normal thallium concentration in edible organs of cattle ranges between 0.02 and about 0.1 µg/g wet weight (see also Table II.29-1).

5 Uptake, Absorption, Transport and Distribution, Metabolism and Elimination in Plants, Animals, and Humans

Thallium ions are easily taken up by plants through the roots (EWERS, 1988). The normal daily intake of thallium, mainly by food, is estimated to be about 2 µg in non-polluted areas (see also EWERS, 1988). In working environments exposure via inhalation and skin contact has also been reported (KAZANTZIS, 1986). In the surroundings of Tl-emitting plants the daily intake is elevated. SMITH and CARSON

(1977) calculated a daily thallium intake by ingestion and inhalation of 0.15 – 0.18 µg/kg body weight (10.5 – 12.6 µg/70 kg) for inhabitants of the emission region of a coal fired power plant.

Thallium is rapidly absorbed (up to 80 – 100%) by the mucous membrane after ingestion (FORTH and RUMMEL, 1975), inhalation, or contact with intact skin. A rapid distribution from blood to tissue follows absorption, to reach a steady state (TSALEV and ZAPRIANOV, 1983a). Because of the similar ionic radii of thallium and potassium (cf. Sect. 3.1), thallium is transported in the same way as potassium (GEHRING and HAMMOND, 1967) and accumulates chiefly intracellularly. Some excretory systems seem to concentrate thallium ions, as SABBIONI et al. (1980) found for salivary glands.

In animal experiments and investigations in cattle from polluted areas highest levels of thallium were found in kidney (medulla), liver, muscle tissue, and endocrine glands (testicles, thyroid gland) (HAPKE et al., 1980; KONERMANN et al., 1982; LA-MEIJER and VAN ZWIETEN, 1977; BERTRAM, 1980; RAUSCHKE, 1961; CRÖSSMANN and KONERMANN, 1987).

In cases of acute thallium intoxications in humans similar distributions were observed (WEINIG and SCHMIDT, 1966; WEINIG and WALZ, 1971; BERTRAM, 1980; ARNOLD, 1986). Specific thallium depot compartments are kidney and muscle, but no correlation between the tissue distribution and a particular sensitivity to the toxicodynamic effects of thallium could be stated (FORTH and HENNING, 1979). Brain areas densely populated with neurons have been found to accumulate thallium more than other areas (MANZO and SABBIONI, 1988). Thallium is also accumulated in testicles, which leads to reduced sperm motility (MANZO and SABBIONI, 1988).

The competitive replacement Tl^+/K^+ causes one of the most important and interesting toxicokinetic facts in thallium elimination: a direct active excretion into the intestinal lumen (LUND, 1956). In contrast to other toxic heavy metals the fecal thallium elimination is the predominant route of excretion and ranks above renal activity. There are two points of interest in this fact: a) Binding of thallium in the gastrointestinal tract by ion exchange mechanisms (Prussian blue) is the most effective treatment of acute thallium intoxications (HENNING and FORTH, 1982; FORTH, 1983). b) Active intestinal secretion results in reabsorption by enterohepatic and enterosystemic circulations, yielding a long biological half-life. According to urinary Tl-contents after accidental or suicidal ingestion of known amounts of thallium, a half-life of about 9 – 11 days was calculated in man (KEMPER, 1979). Analytical detection of orally ingested thallium in feces and urine is possible for several weeks. Besides these two excretion systems, scalp hair is a useful tool in the field of medical and forensic diagnosis and therapy control. Moreover, feces, urine, and hair samples may serve as indicator specimens in chronic exposure to small amounts of thallium (MOESCHLIN, 1980; KEMPER, 1979; HAGEDORN-GÖTZ and STOEPPLER, 1975; ANGERER and SCHALLER, 1985).

Reported values for the 'normal' thallium content in human body fluids and tissues are contradictory. In Table II.29-2 selected reference values of unexposed persons are compiled and may be used for environmental or toxicological diagnosis. Blood, however, is not considered a good indicator; concentrations in human urinary creatinine are normally below 1 µg/g (EWERS, 1988).

Table II.29-2. Normal Thallium Concentrations in Human Body Fluids and Tissue (Selected from the Reviews of Tsalev and Zaprianov, 1983b, Schoer, 1984, and Own Results)

Matrix	Reference Value	Amount of Thallium
Urine	0.05 – 1.5	µg/L (see also Angerer and Schaller, 1985; and Ewers, 1988)
Whole blood	0.5 – 2	µg/L
Hair	< 20	ng/g (see also Ewers, 1988)
Nails	< 5	ng/g
Liver	0.5 – 3	ng/g wet weight
Kidney	1 – 4	ng/g wet weight
Brain	< 1	ng/g wet weight
Feces	1	µg/day

Little is known about metabolic changes of inorganic thallium in humans. Biomethylation seems possible, as proved in a similar form for mercury (Sect. 4) (Beijer and Jernelöv, 1986).

6 Effects on Plants, Animals, and Humans

6.1 Effects on Plants

The resistance of plants to elevated thallium levels in soil and water seems to be species specific and not correlated with the ability of thallium accumulation. Plants of the genus *Brassica* (cabbage) bear high Tl-contents without any damage, whereas tobacco plants (*Nicotiana tabacum* L.) are very sensitive to thallium toxicity (Smith and Carson, 1977; for vegetation damages near cement plants, see also Sect. 3.4). At a water concentration of 2 – 3 µg thallium per gram, the activity of photosynthesis in algae is distinctly influenced, toxic effects have been observed at 7 µg/g. Förstner and Wittmann (1979) reported that the fish toxicities (salmon) of copper and thallium are similar.

6.2 Acute Effects on Animals and Humans

The mechanism of thallium toxicity in animals and humans seems to be analogous to the enzyme inhibiting effects of lead and mercury, resulting in a general histotoxicity. Damages of skin, mucous membranes, hair, and nails have been observed. The central and peripheral nervous system is the main critical organ in thallium intoxication (see also Tilson, 1987). Moeschlin (1980), Kemper (1979), Prick et al. (1955), and Forth (1983) discussed symptoms of acute thallium intoxications in humans. During the first days of an acute thallium intoxication only unspecific signs are observed, similar to those in intoxications with other metals (lead, mercury, gold, arsenic). Up to 3 days nausea, vomiting, and constipation are the initial symptoms.

From the 4th day on damage of the peripheral nervous system becomes obvious: neuralgiform pain with high intensity in the lower extremities occurs. Changes in sensitivity and hyperesthesia are located mainly in the legs and the soles of the feet. Toxicodynamic effects on the central nervous system follow: hysterical behavior, confusion, sleeplessness, and visual disturbances (see also MANZO and SABBIONI, 1988). A dark pigment precipitation in hair roots (Widy sign) from the 4th or 5th day on is an important diagnostic help.

The toxic nervous damage culminates at the end of the 2nd week. Degenerative changes in the dorsal column of the spinal cord occur (KAZANTZIS, 1986). Hyperesthesia may change to ascending paralysis. Toxic effects of thallium on heart muscle and on heart muscle metabolism result in a distinct and continuous increase in heart frequency.

The most characteristic symptom of thallium intoxication is loss of scalp, axillary, and pubic hair from the 10th to 13th day on. Lateral eyebrows are lost, too, the medial part remains. In case of recovery hair begins to regrow 1 to 2 months later. For diagnostic purposes longitudinal microscopic hair examination may be useful. Changes of hair structure and dark pigmentation will help to find the date of thallium overexposure.

Toxic damage of sebaceous and sweat glands is responsible for a generally dry and squamous skin. Deceleration of nail growth causes lunula stripes (Mees bands)

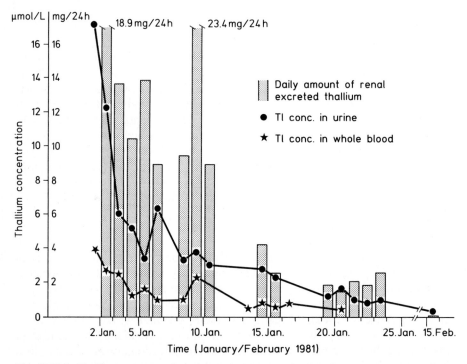

Fig. II.29-2. Thallium concentrations in whole blood and urine after oral ingestion (suicide attempt, male, 23 years) (own results).

for several months. In fatal intoxications the patient dies from central paralysis. In cases of survival a slow recovery to intact health or permanent illness (neurological and psychiatric disturbances) are possible. Fig. II.29-2 shows a six weeks control of the thallium concentration in whole blood and urine in a suicidal acute Tl-intoxication.

Fatal doses of thallium sulfate and acetate in animals have been reported (SMITH and CARSON, 1977; WEINIG and WALZ, 1971; INDUSTRIEVERBAND PFLANZENSCHUTZ, 1982): dogs 35–250 mg/kg body weight (LD_{100}), guinea pig 20–80 mg/kg b.w. (LD_{100}), rat 10–25 mg/kg b.w. (LD_{50} oral), rat 500 mg/kg b.w. (LD_{50} dermal), and mouse 50–60 mg/kg b.w. (LD_{50} oral). The fatal dose of thallium sulfate in man (orally) ranges between 0.8 and 1.0 g (10–15 mg/kg b.w.) (MOESCHLIN, 1980). BOZZA-MARUBINI et al. (1987) described iatrogenic poisoning after mistakenly implanted thallium wire into tumor mass. Since the number of workers handling with thallium or its compounds and alloys is small, only few cases of poisoning have been observed (EWERS, 1988).

SUNDERMAN, Jr. (1967) described diethyldithiocarbamate therapy of thallotoxicosis, but it seems that chelating agents are not very useful in the treatment of thallium poisoning (TILSON, 1987). Administration of potassium ions or of potassium ferric ferrocyanide(II) (Prussian blue) has instead been suggested (MICKE et al., 1983; TILSON, 1987). Symptomatic and diuretic measures have also some value (MICKE et al., 1983).

6.3 Chronic Effects on Animals and Humans, Mutagenic, Carcinogenic, and Teratogenic Effects

In animal experiments with chronic ingestion of small amounts of thallium, damage and changes in the mitochondrial system of nervous cells have also been observed. Little is known about the chronic intoxication with small amounts of thallium (KAZANTZIS, 1986).

Some investigations (CLAUSSEN, 1981) indicated that there is no thallium mutagenicity or teratogenicity (Ames and Sister chromatid exchange test); but other studies demonstrated mutagenic properties, increased frequency of DNA breaks, precancerous lesions in the female genital tract in mice, other damages of reproductive systems in animals and humans (see also Sect. 5), and chick embryo teratogenicity (results in rats, mice, and cats are conflicting; MANZO and SABBIONI, 1988). In the thallium emission area of Lengerich (cf. Sect. 3.4) no signs of mutagenicity and thallium specific embryopathy have been evident.

7 Hazard Evaluation and Limiting Concentrations

Threshold limits are important in the fields of occupational medicine and environmental control to prevent health risks. In work places slight thallium intoxications with hair loss, visual disturbances, and unspecific symptoms have been observed; but no data on the dose-response relationship are available (KAZANTZIS, 1986; FISCHER, 1974). Concentrations of up to 2 µg/L Tl in whole blood and up to 5 µg/L in urine may be considered as 'normal'. For workers a threshold limit of 300 µg/L in urine has been proposed (GLÖMME, 1983). As a result of his investigations in thallium exposed workers, MARCUS (1985) recommended a lower limit (50 µg/L). Substitution of thallium in chemical processes is an important preventional measure. WOLLMANN and FRANCK (1984) recommended the use of silver nitrate instead of thallium compounds in chemical oxidative rearrangements.

In the Federal Republic of Germany a threshold limit for thallium at work places (MAK, 1987) of 0.1 mg/m^3 (calculated for soluble Tl-compounds) has been fixed, whereby one maximum peak of 1 mg/m^3 per shift is accepted. In the USA the threshold limit value (TLV) is 0.1 mg/m^3 (ACGIH, 1984).

Some environmental levels have been proposed, but not yet fixed; in the Federal Republic of Germany: maximum thallium level in drinking water 40 µg/L; limit value for airborne thallium 10 µg/m^3.

A maximum thallium content in food has been regulated in the FRG in 1987 (BUNDESGESUNDHEITSAMT, 1987) to be 0.1 mg/kg fresh weight. To date, calculation of an acceptable daily intake value (ADI) is difficult because of missing data. Nevertheless, the EPA (Environmental Protection Agency) of the USA published in 1983 an ADI value of 37 µg/d of thallium (70 kg body weight). Other authors discuss lower values (cf. ZARTNER-NYILAS et al., 1983: 15.4 µg/d; EWERS, 1988: 14 µg/d), which are still 5–10 times higher than observed in certain situations (EWERS, 1988).

References

Achenbach, C., Hauswirth, O., Heindrichs, C., and Ziskoven, R. (1980), *Quantitative Measurement of Time-dependent Thallium Distribution in Organs of Mice by Field Desorption Mass Spectrometry. J. Toxicol. Environ. Health* 6, 519–528.
ACGIH (American Conference of Governmental Industrial Hygienists) (1984), *Threshold Limit Values for Chemical Substances and Physical Agents in the Work Environment and Biological Exposure Indices with Intended Changes for 1984–85*. 2nd Printing, Cincinnati, Ohio; cited in Friberg, L. (1986), *Risk Assessment*, in: Friberg, L., Nordberg, G. F., and Vouk, V. (eds.), *Handbook on the Toxicology of Metals*, 2nd Ed., Vol. I, pp. 269–293. Elsevier, Amsterdam.
Angerer, J., and Schaller, K. H. (1985), *Analyses of Hazardous Substances in Biological Materials*, Vol. 1, pp. 199–208. VCH Verlagsgesellschaft, Weinheim-Deerfield Beach/Florida-Basel.
Arnold, W. (1986), *Thalliumausscheidung und -verteilung bei menschlichen Vergiftungsfällen*, in: Anke, M., Baumann, W., Bräunlich, H., Brückner, C., and Groppel, G. (eds), *5. Spurenelementsymposium – Trace Elements – der Universitäten Leipzig und Jena*, pp. 1241–1253. Universität Jena.
Bednorz, J. G. (1988) (IBM CH-8803 Rüschlikon), *Plenary Lecture*, 22nd ACHEMA, Frankfurt/M., and *38th Lindau Meeting of Nobel Prize Winners* (see, e.g., *Swiss Chem.* 10(11), 39–49, and *Neue Zürcher Zeitung, Forschung und Technik*, No. 298, pp. 57–58, December 21.
Beijer, K., and Jernelöv, A. (1986), *Sources, Transport and Transformation of Metals in the Environment*, in: Friberg, L., Nordberg, G. F., and Vouk, V. B. (eds.), *Handbook on the Toxicology of Metals*, 2nd Ed., pp. 68–84. Elsevier, Amsterdam-New York-Oxford.

Berman, E. (1980), *Toxic Metals and their Analysis,* pp. 201–208. Heyden & Son, London-Philadelphia-Rheine.

Bertram, H. P. (1980), *Environmental Contamination by Thallium in the Emission Area of a Cement Factory, Human Exposure in Comparison with Acute Thallium Intoxications.* Naunyn-Schmiedeberg's Archives of Pharmacology, Supplement to 311, R22.

Bertram, H. P., and Kemper, F. H. (1983), *Thallium – Toxikologie und Ökotoxikologie,* in: Ullmanns Encyklopädie der technischen Chemie, 4th Ed., Vol. 23, pp. 110–114, Verlag Chemie, Weinheim-Deerfield Beach/Florida-Basel.

Bozza-Marrubini, M. L., Manzo, L., Sabbioni, E., and Mousty, F. (1987), in: *Proceedings of the International Conference on Heavy Metals in the Environment, New Orleans,* Vol. 2, pp. 115–117. CEP Consultants Ltd., Edinburgh.

Brockhaus, A., Dolgner, R., Ewers, U., Wiegand, H., Freier, I., Jermann, E., and Krämer, U. (1980), *Excessive Thallium Absorption Among a Population Living Near a Thallium Emitting Cement Plant,* in: Holmstedt, B., Lauwerys, R., Mercier, M., and Roberfroid, M. (eds.), *Mechanisms of Toxicity and Hazard Evaluation,* pp. 565–568. Elsevier/North-Holland Biomedical Press, Amsterdam-New York-Oxford.

Brockhaus, A., Dolgner, R., Ewers, U., Krämer, U., Soddemann, U., and Wiegand, H. (1981), *Intake and Health Effects of Thallium Among a Population Living in the Vicinity of a Cement Plant Emitting Thallium-containing Dust.* Int. Arch. Occup. Environ. Health 48 (4), 375–390.

Bruland, K. W. (1983), *Trace Elements in Seawater,* Chem. Oceanogr. 8, 173, 197.

Bundesgesundheitsamt (1987), *Bundesgesundhbl.* 30 (9), 331.

Claussen, U., Roll, R., Dolgner, R., Mattiaschk, G., Majewski, E., Stoll, B., and Röhrborn, G. (1981), *Zur Mutagenität und Teratogenität von Thallium.* Rhein. Ärztebl. 16, 469–475.

Crößmann G., and Konermann, H. (1987), *Untersuchungen über den Einfluß Thallium-kontaminierter Futtermittel auf Leistung, Gesundheit und Rückstandsbildung bei Mastbullen,* Report. Landwirtschaftskammer Westfalen-Lippe/Ministerium für Umwelt, Raumordnung und Landwirtschaft NRW; also *Zur Mobilität und Akkumulation ausgewählter anorganischer und organischer Schadstoffe bei Pflanzen und Nutztieren.* Proceedings VDI-Tagung „Wirkungen von Luftverunreinigungen auf Böden", Lindau, May 1990. VDI-Verlag, Düsseldorf, in press.

EPA (Environmental Protection Agency) (1980), *Ambient Water Quality Criteria for Thallium.* U.S. Department of Commerce, National Technical Information Service (NTIS), Washington, D.C.

Ewers, U. (1988), *Environmental Exposure to Thallium,* Sci. Total Environ. 71, 285–292.

Fischer, R. E. (1974), *Gewerbliche Vergiftungen durch Thallium und Selen und ihre Verbindungen.* Schr. Ges. Dtsch. Metallhütten-Bergleute 27, 7–72. Clausthal-Zellerfeld.

Flegal, A. R., and Patterson, C. C. (1985), *Thallium Concentrations in Seawater,* Mar. Chem. 15, 327–331.

Förstner, U., and Wittmann, G. T. W. (1979), *Metal Pollution in the Aquatic Environment,* pp. 16/17, 347 ff, 357. Springer Verlag, Berlin.

Forth, W. (1983), *Thallium-Vergiftung.* Münch. Med. Wochenschr. 125 (3), 45–50.

Forth, W., and Henning, C. H. (1979), *Thallium-Vergiftungen und ihre Behandlung.* Dtsch. Ärztebl., 2803–2807.

Forth, W., and Rummel, W. (1975), *Gastrointestinal Absorption of Heavy Metals,* in: Forth, W., and Rummel, W., (eds.), *IEPT Sect. 39 B, Pharmacology of Intestinal Absorption of Drugs,* Vol. II, pp. 599–746. Pergamon Press, Oxford.

Gehring, P. J., and Hammond, P. B. (1967), *The Interrelationship Between Thallium and Potassium in Animals.* J. Pharmacol. Exp. Ther. 155 (1), 187–201.

Geilmann, W., Beyermann, K., Neeb, K.-H., and Neeb, R. (1960), *Thallium, ein regelmäßig vorhandenes Spurenelement im tierischen und pflanzlichen Organismus.* Biochem. Z. 333, 62–70.

Gemmer-Čolos, V., Kiehnast, I., Trenner, J., and Neeb, R. (1981), *Inverse-Voltammetric Determination of Thallium.* Fresenius Z. Anal. Chem. 306, 144–149.

Glömme, J. (1983), *Thallium and Compounds,* in: Encyclopedia of Occupational Health and Safety, 3rd Ed., pp. 2170–2171. International Labor Organization, Geneva.

Gough, L. P., Shacklette, H. T., and Case, A. A. (1979), *Element Concentrations Toxic to Plants, Animals and Man. US Geol. Surv. Bull. 1466*, 49–51. US Government Printing Office, Washington, D.C.

Hagedorn-Götz, H., and Stoeppler, M. (1975), *Zum forensischen Nachweis von Thallium in menschlichen Haaren durch flammenlose Atomabsorptionsspektroskopie. Arch. Toxicol. 34*, 17–26.

Hapke, H.-J., Barke, E., and Spikermann, A. (1980), *Ansammlung von Thallium in verzehrbaren Geweben von Hammeln und Bullen in Abhängigkeit von der Thalliummenge im Futter. Dtsch. Tierärztl. Wochenschr. 87*, 376–378.

Henke, G., and Fitzek, A. (1971), *Nachweis von Thalliumvergiftungen durch Neutronenaktivierungsanalyse. Arch. Toxikol. 27*, 266–272.

Henning, C. H., and Forth, W. (1982), *The Excretion of Thallium(I) Ions Into the Gastrointestinal Tract in situ of Rats. Arch. Toxicol. 49*, 149–158.

Huber, F., and Kotulla, V. (1982), in: *Proceedings 12th Annual Symposium on the Analytical Chemistry of Pollutants, Amsterdam;* cf. *Chemosphere 11* (11), N6.

Industrieverband Pflanzenschutz (ed.) (1982), *Wirkstoffe in Pflanzenschutz- und Schädlingsbekämpfungsmitteln,* p. 329. Bintz-Verlag, Offenbach.

Kazantzis, G. (1986), *Thallium,* in: Friberg, L., Nordberg, G. F., and Vouk, V. B. (eds.), *Handbook on the Toxicology of Metals,* 2nd Ed., Vol. II, pp. 549–567. Elsevier, Amsterdam-New York-Oxford.

Kemper, F. H. (1979), *Thallium-Vergiftungen. Münch. Med. Wochenschr. 121* (42), 1357–1358.

Konermann, H., Crößmann, G., and Hoppenbrock, K. H. (1982), *Untersuchungen über den Einfluß Thallium-kontaminierter Futtermittel auf Leistung, Gesundheit und Rückstandsbildung bei Mastschweinen. Tierärztl. Umsch. 37*, 8–21.

Korkisch, J., and Steffan, I. (1978), *Determination of Thallium in Natural Waters. 8th Annual Symposium on the Analytical Chemistry of Pollutants, Geneva. Int. J. Environ. Anal. Chem. 6/2,* 111–118.

Lameijer, W., and van Zwieten, P. A. (1977), *Kinetic Behaviour of Thallium in the Rat. Arch. Toxicol. 37*, 265–273.

Landesanstalt für Immissionsschutz (1980), *Umweltbelastung durch Thallium – Untersuchungen in der Umgebung der Dyckerhoff-Zementwerke AG in Lengerich sowie anderen Thalliumemittenten im Lande Nordrhein-Westfalen,* Ministerium für Arbeit, Gesundheit und Soziales – Ministerium für Ernährung, Landwirtschaft und Forsten des Landes Nordrhein-Westfalen, Düsseldorf.

Lehn, H., and Schoer, J. (1985), *Thallium Transfer from Soils to Plants: Relations Between Chemical Forms and Plant-uptake,* in: Lekkas, T. D. (ed.), *Heavy Metals in the Environment, International Conference Athens,* Vol. II, pp. 286–288. CEP Consultants Ltd., Edinburgh.

Lund, A. (1956), *Distribution of Thallium in the Organism and its Elimination. Acta Pharmacol. Toxicol. 12*, 251–259.

MAK (1987), *Maximum Concentrations at the Workplace, Report No. XXIII, DFG.* VCH Verlagsgesellschaft, Weinheim-Basel-Cambridge-New York.

Manzo, L, and Sabbioni, E. (1988), Chapter 62 *Thallium,* in: Seiler, H. G., Sigel, H., and Sigel, A. (eds.), *Handbook on Toxicology of Inorganic Compounds,* pp. 677–688. Marcel Dekker, New York.

Marcus, R. L. (1985), *J. Soc. Occup. Med. 35*, 4–9.

Micke, H., Bertram, H. P., and Kemper, F. H. (1983), *Thallium,* in: *Ullmanns Encyklopädie der technischen Chemie,* 4th Ed., Vol. 23, pp. 103–114. Verlag Chemie, Weinheim-Deerfield Beach/Florida-Basel.

Moeschlin, R. S. (1980), *Klinik und Therapie der Vergiftungen,* 6th Ed., pp. 99–115. Thieme, Stuttgart-New York.

Prick, J. J. G., Sillevis-Smitt, W. G., and Muller, L. (1955), *Thallium Poisoning.* Elsevier Publishing Company, Amsterdam-Houston-London-New York.

Rauschke, J. (1961), *Studien über Thalliumvergiftung.* Habilitation Thesis, Universität Heidelberg.

Reinhardt, G., and Zink, P. (1975), *Analytische Probleme bei der polarographischen Thallium-Bestimmung in kleinen Organproben. Beitr. Gerichtl. Med. XXX*, 371–375.

Rethfeld, H. (1980), *Thalliumbestimmung in Pflanzenmaterial mit Hilfe der Röntgenfluoreszenzanalyse.* Z. Anal. Chem. *301*, 308.

Sabbioni, E., Marafante, E., Rade, J., di Nucci, A., Gregotti, C., and Manzo, L. (1980), *Metabolic Patterns of Low and Toxic Doses of Thallium in the Rat,* in: Holmstedt, B., Lauwerys, R., Mercier, M., and Roberfroid, M. (eds.), *Mechanisms of Toxicity and Hazard Evaluation,* pp. 559–564. Elsevier/North-Holland Biomedical Press, Amsterdam-New York-Oxford.

Sager, M. (1986), *Trace Analytical Chemistry of Thallium* (in German). Georg Thieme Verlag, Stuttgart.

Sager, M., and Tölg, G. (1984), *Spurenanalytik des Thalliums,* in: *Analytiker-Taschenbuch,* Vol. 4, pp. 443–466. Springer Verlag, Berlin-Heidelberg.

Savory, J., Roszel, N.O., Mushak, P., and Sunderman, Jr., F.W. (1968), *Measurements of Thallium in Biologic Materials by Atomic Absorption Spectrometry,* Am. J. Chem. Pathol. *50*, 505–509.

Schmidt, W., and Dietl, F. (1985), *Thallium in Fly Ashes and Soils Around a Coal Fired Power Plant,* in: Lekkas, T.D. (ed.), *Heavy Metals in the Environment, International Conference, Athens,* Vol. 1, pp. 195–197. CEP Consultants Ltd., Edinburgh.

Schoer, J. (1984), *Thallium,* in: Hutzinger, O. (ed.), *The Handbook of Environmental Chemistry,* Vol. 3, Part C, pp. 143–214. Springer Verlag, Berlin-Heidelberg-New York-Tokyo.

Schurzmann, M. (1986), *Epidemiologische Untersuchungen zum Thallium-Gehalt im Blut und Haar bei Zementarbeitern. Dissertation,* Universität Münster, FRG.

Smith, I.C., and Carson, B.L. (1977), *Trace Metals in the Environment,* Vol. 1. *Thallium.* Ann Arbor Science Publishers, Ann Arbor, Michigan.

Sunderman, Jr., F.W. (1967), *Diethyldithiocarbamate Therapy of Thallotoxicosis,* Am. J. Med. Sci. *253*(2), 209–220.

Thayer, J.S. (1988), *Organometallic Chemistry – An Overview,* pp. 52–53. VCH Verlagsgesellschaft, Weinheim-Basel-Cambridge-New York.

Tilson, H.A. (1987), *Fundam. Appl. Toxicol. 9,* 601.

Tsalev, D.L., and Zaprianov, Z.K. (1983a), *Atomic Absorption Spectrometry in Occupational and Environmental Health Significance,* Vol. I, pp. 196–199. CRC Press, Boca Raton, Florida.

Tsalev, D.L., and Zaprianov, Z.K. (1983b), *Atomic Absorption Spectrometry in Occupational and Environmental Health Significance,* Vol. II, pp. 197–201. CRC Press, Boca Raton, Florida.

Urey, H.H. (1952), *Phys. Rev. 88,* 248.

Vongehr, S. (1986), *Epidemiologische Untersuchungen zum Thallium-Gehalt im Urin und Speichel bei Zementarbeitern. Dissertation,* Universität Münster, FRG.

Wall, C.D. (1977), *The Determination of Thallium in Urine by Atomic Absorption Spectroscopy and Emission Spectrography.* Clin. Chim. Acta *76,* 259–265.

Wang, J. (1985), *Stripping Analysis.* VCH Verlagsgesellschaft, Weinheim-Deerfield Beach/Florida-Basel.

Wedepohl, K.H. (ed.) (1974), *Handbook of Geochemistry.* Springer Verlag, Berlin-Heidelberg.

Weinig, E., and Schmidt, G. (1966), *Zur Verteilung des Thalliums im Organismus bei tödlichen Thallium-Vergiftungen.* Arch. Toxikol. *21,* 199–215.

Weinig, E., and Walz, W. (1971), *Die Thallium-Verteilung in Niere und Leber bei letalen Tl-Vergiftungen.* Arch. Toxikol. *27,* 217–225.

Weinig, E., and Zink, P. (1967), *Über die quantitative massenspektrometrische Bestimmung des normalen Thallium-Gehalts im menschlichen Organismus.* Arch. Toxikol. *22,* 255–274.

Welz, B. (1985), *Atomic Absorption Spectrometry,* 2nd Ed., pp. 330ff. VCH Verlagsgesellschaft, Weinheim-Deerfield Beach/Florida-Basel.

Wollmann, T., and Franck, B. (1984), *Angew. Chem. 96*(3), 227.

You, N., and Neeb, R. (1983), *Improvements of the Inverse-Voltammetric Determination of Thallium in the Presence of Lead by Electrochemical Masking,* Fresenius Z. Anal. Chem. *314,* 394–397.

Zartner-Nyilas, G., Valentin, H., Schaller, K.-H., and Schiele, R. (1983), *Thallium – ökologische, umweltmedizinische und industrielle Bedeutung. Agrar- und Umweltforschung in Baden-Württemberg,* Vol. 3. Verlag Eugen Ulmer, Stuttgart.

II.30 Tin

Eric J. Bulten and Harry A. Meinema, Zeist, The Netherlands

1 Introduction

Tin is rather unique in the wide variety of its compounds and applications. Ever since the beginning of the bronze age the metal and its alloys have been of importance to mankind.

When considering the impact of tin and its compounds on the environment, it has to be realized that no generalization can be made as to whether tin compounds as a whole are either harmless or dangerous to life. The biological effects of a tin compound are almost exclusively determined by both the nature and the number of groups bound to tin. In contrast to, for instance, lead and mercury, which are by nature highly poisonous to living organisms, the element tin is in itself virtually non-toxic (Kirk and Othmer, 1955; Browning, 1969).

Inorganic tin salts are generally acknowledged to be of a low order of toxicity (WHO, 1980).

The toxicity of organotin compounds is largely determined by both the nature and the number of the organic groups bound to tin. The most toxic organotin chemicals belong to the class of the triorganotin compounds R_3SnX. However, whereas triethyltin acetate, Et_3SnOAc, is the most toxic of all organotin compounds (oral LD_{50}, rat, 4 mg/kg), trioctyltin chloride, Oct_3SnCl, is virtually non-toxic to animals (Smith et al., 1978).

For a proper evaluation of the impact of man-made chemicals on the environment in recent years much attention has been given to the development of more sensitive, reliable, and selective clean-up and analysis procedures for the determination of individual (organo)tin species in environmental substrates and to extended studies of the chronic toxicity and long-term effects of organotin compounds in aquatic organisms and in mammals.

2 Physical and Chemical Properties, and Analytical Methods

Metallic tin (atomic mass 118.7, mp. 232°C, bp. 2260°C) exists in two different allotropic forms. At low temperatures the stable form is gray cubic or α-tin, which changes at 18°C to β-tin, the ordinary tetragonal form. Major isotopes include ^{122}Sn (29%), ^{118}Sn (22%), ^{116}Sn (16%), and ^{119}Sn (10%).

As regards inorganic tin compounds, a great variety of both divalent and tetravalent tin derivatives have found practical application. In contrast, organotin compounds − containing by definition at least one tin-carbon bond − almost exclusively contain tetravalent tin. Depending on the number of organic groups bound to tin, these compounds are classified as mono-, di-, tri-, and tetraorganotins.

For the determination of tin as such − without making a distinction between the wide variety of tin compounds − a wealth of sophisticated *analytical techniques* is available, such as colorimetry (KIRK and POCKLINGTON, 1969; ADCOCK and HOPE, 1970; ENGBERG, 1973), emission spectrography (TIPTON et al., 1966), neutron activation (BOWEN, 1972), X-ray fluorescence (RUDOLF et al., 1973), and atomic absorption spectrometry (ENGBERG, 1973; NAKAHARA et al., 1972; EVERETT et al., 1974). Nevertheless, the information available at present does not allow one to make recommendations for an analytical technique for one particular application (HORWITZ, 1979).

Atomic absorption spectrometry (AAS; see also WELZ, 1985, and DE DONCKER et al., 1986) is most frequently used, − in particular the AAS-hydride method − the sensitivity being about 1 ppm for 1 percent absorption with a detection limit of 0.1 ppm in an aqueous solution using an air-hydrogen flame. It should be noted, however, that AAS is subject to a rather large number of interferences (BEESON, 1977). A detailed survey of the various techniques for tin analysis in divergent substrates will be published shortly by an IUPAC Task Group (CROSBY et al., 1988; see also PISCATOR, 1977).

The development of reliable methods for the quantitative extraction, separation, and determination of individual (organo)tin species present in a mixture of inorganic tin and one or more different organotin compounds, found in substrates such as organic materials, effluents, sediments, etc., has met with increasing interest in recent years. Following these techniques, organotin compounds are concentrated from dilute aqueous solutions by extraction into organic solvents or by adsorption on cation exchange material. Detection of individual organotin compounds is achieved by gas chromatography for organotin halides, organotin hydrides (after reaction with $NaBH_4$), or tetraorganotin compounds (after alkylation with Grignard reagents). Various detection methods are applied such as AAS, GC, GC-MS, LC-LEI, electrochemical and spectrofluorimetric detection and HPLC. Detection limits are in the ppb to ppt range (BLUNDEN and CHAPMAN, 1982, 1986). There is still a need for further development of these techniques for the detection of tin compounds at the ppb to ppt level in the aquatic environment, not only in the water table, but also in the sediment and the benthic life. BLAIR et al. (1986) compiled an international comparison of tributyltin measurement methods, looking at sample preparation and results of analyses. Liquid chromatography was their preferred technique. Knowledge on speciation was, for instance, further improved by PARKS et al. (1985), VALKIRS et al. (1986), RANDALL et al. (1986), DONARD et al. (1986), MACCREHAN (1987), CHAU (1987), and VAN LOON (1987).

3 Sources, Production, Important Compounds, Uses, Waste Products, Recycling

(see also PASCHEN et al., 1983; SAAGER, 1984)

Tin is dispersed in very small amounts in silicate rocks containing 2–50 ppm (HAMAGUCHI and KURODA, 1969; JOHANSEN and STEINES, 1969), the ore casserite (SnO_2) being of major commercial importance. The earth's crust contains about 2 to 3 ppm tin. The annual world production of tin amounts to about 200000 tons or somewhat more. Major tin producing countries include Malaysia (about 40%), Bolivia, Thailand, and Indonesia (KEINDL, 1970). About 70% of world total tin consumption is produced from ores, 30% being recovered from scrap metals (HEINDL, 1970; HOARE, 1974). The largest single *application of tin* is in the production of tin-plated steel, which is done either by hot-dipping or electroplating in a continuous process. Other major industrial applications include solder (tin alloys containing lead, antimony, silver, zinc, or indium), babbit, brasses, and bronzes (essentially tin-copper alloys), pewter (90–95% tin, plus 1–8% bismuth, and 0.5–3% copper), and dental amalgams (silver-tin-mercury alloys). The use of metals employing lead-based alloys containing 10–25% antimony and 3–13% tin for type is declining as a result of new printing and copying techniques.

In contrast, the application of tin compounds, particularly organometallic compounds of tin (organotins), is rapidly increasing. For detailed in-depth information on the synthesis, fundamental chemical properties, and industrial uses of inorganic and organometallic tin compounds, the reader is referred to standard reference works and recent review papers (e.g., FULLER, 1975; LUIJTEN and VAN DER KERK, 1968; NEUMANN, 1970; POLLER, 1970; LUIJTEN, 1971; BOKRANZ and PLUM, 1971; SAWYER, 1971/72; SCHUMANN and SCHUMANN, 1975; BÄHR and PAWLENKO, 1978; BLUNDEN et al., 1985; BENNET, 1983).

Table II.30-1. Major Applications of Inorganic Tin Compounds

Compound	Formula	Application
Tin(II) chloride	$SnCl_2$	Plating of steel
Tin(II) sulfate	$SnSO_4$	Plating of steel
Tin(II)-2-ethylhexoate	$Sn(C_8H_{15}O_2)_2$	Catalyst for polyurethane, production and silicon curing
Tin(II) oxalate	$Sn(C_2O_4)_2$	(Trans)esterification catalyst
Tin(II) fluoride	SnF_2	Toothpaste, dental preparations
Tin(IV) oxide	SnO_2	Ceramic opacifier, pigments
Tin(IV) chloride	$SnCl_4$	Deposition of SnO_2 films on glass (strengthening; electrically conductive films) and ceramics (decoration pigments); starting material for organotins

As regards *inorganic tin compounds*, both divalent (stannous) and tetravalent (stannic) derivatives have found important outlets in a broad spectrum of applications (cf. Table II.30-1).

Organotin compounds are remarkably versatile in their physical, chemical and biological properties, which is reflected in their rather divergent practical applications. Only very few examples of divalent organotin compounds have been reported, none of which have found practical applications.

The practically important organotin compounds all contain tetravalent tin and belong to the four classes (LUIJTEN, 1971; BOKRANZ and PLUM, 1971):

$R_4Sn \quad R_3SnX \quad R_2SnX_2 \quad RSnX_3$

in which R denotes an alkyl or aryl group bound to tin via a strongly covalent tin-carbon bond, and X stands for an anionic group attached to tin via an elec-

Table II.30-2. Major Applications of Organotins, $R_{4-n}SnX_n$ (n = 0 – 3)

Compound	Application
R_4Sn (very stable, resemble paraffins, no biocidal activity, relatively non-toxic, are, however, slowly decomposed or metabolized to toxic R_3Sn-compounds)[a]	
R = butyl, octyl, phenyl	Starting materials
R = butyl	Ziegler-Natta co-catalysts
R = phenyl	Transformer oil stabilizers
R_3SnX (some compounds are powerful fungicides and bactericides[b], depending on the nature of the organic group R)	
R = butyl (e.g., TBTO, Incidin)	Industrial biocides, e.g., in antifouling paints, wood preservatives, paint preservation; disinfectants, molluscicides
R = phenyl (e.g., Fentin, Tinmate, Brestanol)	Agricultural fungicides, antifouling paints
R = cyclohexyl, neophyl (e.g., Plictran, Bay Bue 1452, Vendex, Torque)	Miticides
R_2SnX_2 (no antifungal activity, low antibacterial and toxic activity, except for diphenyl derivatives)	
R = methyl, butyl; CH_2CH_2COOR' (e.g., Estertins)	PVC heat stabilizers, catalysts, polyurethane formation and silicon curing, heat stabilizer for food-packaging PVC
$RSnX_3$ (no biocidal activity, very low toxicity to mammals)	
R = methyl, butyl, octyl; CH_2CH_2COOR'	PVC heat stabilizers

[a] LUIJTEN, 1971; ALDRIDGE, 1978
[b] As found at TNO, Utrecht, in the early 1950s (LUIJTEN, 1971; KAARS SIJPESTEIJN et al., 1969; and references cited therein)

tronegative atom, such as halogen, $-OH$, $-OR'$, $-SH$, $-SR'$, $-OOCR'$, $-OSnR_3$, $-NR'_2$, etc.

Both the nature and the number of the organic groups R bound to tin are of decisive importance in determining the physical, chemical, and biological properties of organotins. The major applications of organotin compounds are summarized in Table II.30-2.

On the basis of scattered information from some organotin producing industries, the world consumption of organotin stabilizers and catalysts in 1986 has been estimated to be 30000–35000 tons. Together with the increasing application of triorganotin pesticides the overall world consumption of organotins in 1986 has been estimated to be of the order of 45000–50000 tons. Wastes are for obvious reasons very heterogeneous: industrial effluents, waste in incineration, dust, sewage sludge, but also dissipative burning of coal and wood, and application of organotins must be considered (see also MAGOS, 1986).

4 Distribution in the Environment, in Foods and Living Organisms

4.1 Inorganic Tin and its Derivatives

Tin is rarely detected in *air*, except near to industrial emission points (KNORR, 1975; BEESON, 1977; PISCATOR, 1977; DE DONCKER, 1986). Occasionally in Japan, at short distances from an industrial plant, levels of 3.8–4.4 µg/m^3 have been reported (ENVIRONMENT AGENCY, JAPAN, 1974). Likewise, tin concentrations in *soil* are low except in areas of tin containing minerals, or when contaminated sewage sludge is used. Out of 900 soil samples taken in the USA, only 9 samples contained more than 10 ppm of tin (maximum 20 ppm) (BEESON, 1977).

In *sea water*, fresh water, lakes, rivers, and municipal waters tin concentrations generally are well below 1 ppb (≤ 1 µg/L), viz. close to or below the analytically reliable detection limits. Older values are often wrong because PVC samples containing organotin stabilizers were used (BRULAND, 1983). For the north-eastern Atlantic (SMITH and BURTON, 1972) as well as for the Pacific (HAMAGUCHI, 1964) tin concentrations of the order of 0.01 ppb have been reported, whereas BRULAND (1983) gave newer references of the order of 0.0005 to 0.005 ppb for samples from the California coast and from the Sargasso sea. At a depth of 3000 m, concentrations are even lower (about 0.0002 ppb).

Little is known about the exact structures and forms of inorganic tin species in the environment (other than in minerals). BRULAND (1983) presented some information, also about the abundance of the dimethyltin cation. Research in this area may be of low priority in view of the low toxicity of inorganic tin in general. On the other hand, such information might be most relevant with respect to the methylation of inorganic tin resulting in the formation of potentially more toxic methyltin species. Chemical methylation of inorganic divalent tin by methylcobalamin at pH 1 has been demonstrated (DIZIKES et al., 1978). Mono-, di-, and trimethyltin species have been detected in very low concentrations in saline and/or fresh water samples (0.01–8.5 ng/L) (HODGE et al., 1979; BRAMAN and TOMPKINS, 1979). Following the Chesapeake Bay studies of BRINCKMAN (1981) some species may be transformed into more volatile methyltins such as Me_4Sn and Me_2SnH_2. Lichens concentrate tin, while forest vegetation, pasture, vegetables and cereals contain less tin than the respective soil (MAGOS, 1986).

Trace amounts of tin are present in most *natural foods* (DE GROOT et al., 1973). Estimates for the average daily intake vary between 0.2 mg (HAMILTON et al., 1972) and 1 mg (SCHROEDER et al., 1964).

Considerably larger amounts of tin may be present in *canned food* and juices. The presence of oxidizing agents (nitrates), the pH, and the storage temperature largely influence the leaching of tin into the canned material, occasionally resulting in concentrations as high as 100–500 ppm. For example, in canned tomato purée the tin concentration may rise to about 700 ppm within a period of 80 to 160 days (KNORR, 1975). Since nowadays most cans are lacquered, tin contamination by leaching is substantially reduced. In some cases stannous salts are used as additives, such as in asparagus and peas packed in glass containers (PISCATOR, 1977).

Overall it can be stated that the main exposure to inorganic tin for human beings is from food. Human organs may contain 0.1 to 1.4 mg/kg of tin, mostly in bones, liver, and lungs, the average adult body containing about 30 mg of tin.

4.2 Organotin Compounds

In recent years the environmental fate of organotin compounds has become a subject of increasing interest. Although the biocidal uses of organotin compounds comprise only about 30% of the total world consumption, they probably give rise to the largest proportion of organotin species in the environment. So far *aerial* contamination by organotins has not been reported. It is known, however, that the industrial biocide bis(tributyltin) oxide, which has a very low vapor pressure, is made volatile quite readily by co-distillation with water (MEINEMA and BULTEN, unpublished results).

Triorganotins may enter into *water* directly from industrial effluents, antifouling coatings, molluscicide formulations for the control of snails, by (aerial) spraying of agricultural biocides, and by soil leaching from fields sprayed with such biocides (BLUNDEN et al., 1984). In this respect, major attention is given to the environmental fate of tributyl- and triphenyltin compounds which form the bulk of the industrially applied organotin biocides. In harbors and areas with important pleasure craft activities, relatively high concentrations of tributyltin species, the active ingredient of antifouling paints, have been found. Tri- and dibutyltin species have been detected in the water table and in sediments of freshwater and estuarine-marine systems in France, the United Kingdom, Switzerland, Portugal, Japan, Canada, and the USA. On some occasions tributyltin levels approached or exceeded concentrations lethal for sensitive aquatic species. Toxic effects are observed at the ppb to ppt concentration levels (THOMPSON et al., 1985; MEINEMA et al., 1986). In view of the very low solubility of most commercial organotins in water (≤ 5 ppm) and their strong tendency to adhere to sediments, substantial widespread surface water contamination by organotins is unlikely.

In *soil* organotins have been detected only following spraying with triorganotin agricultural biocides. Due to the strong adsorption to soil, leaching from and transport in soil do not take place to a measurable extent (BARNES et al., 1973; SUESS and EBEN, 1973; SCHLESINGER, 1977; SCHAEFER et al., 1981).

In aqueous media organotin compounds $R_{4-n}SnX_n$ ($n = 1-3$) rapidly exchange anionic groups X for hydroxy groups resulting in the formation of the corresponding hydroxides or oxides. As the toxicity of organotins is largely determined by the nature and number of the organic groups rather than by the anionic substituents, this exchange (hydrolysis) as such does not contribute significantly to the detoxification of organotins in the environment.

The transformation of triorganotin biocides has been studied quite extensively. Both triphenyltins (BARNES et al., 1973) and tricyclohexyltin derivatives (SMITH et al., 1976) are subject to rapid stepwise photochemical deorganylation by UV light. Tributyltin species degrade at a much lower rate by UV irradiation (KLÖZER and THUST, 1976). Accordingly, sunshine accelerates the degradation of triphenyltins and tricyclohexyltins in the environment. In the dark, phenyltins in aqueous solution degrade in a similar stepwise way, albeit at a lower rate (SODERQUIST and CROSBY, 1980). In soil the half-life for the total mineralization of triphenyltin acetate has been determined to be approximately 140 days (BARNES et al., 1973). Dealkylation of bis(tributyltin) oxide in surface water proceeds rather slowly: <35% degradation after three months (KLÖTZER and THUST, 1976); half-life ≥ several months under Canadian climatic conditions (MAGUIRE and TKACZ, 1985).

The main factors limiting the persistence of tributyltin in water are sunlight and microbial degradation in water and sediment.

In the field of a variety of crops the half-life of triphenyltins has been reported to be 3–14 days, the main degradation product being inorganic tin (WHO, 1980). Food-chain studies involving the milk of cows fed with fentin containing sugar-beets showed the milk to contain about 4 ppb of fentins and some degradation products therefrom. The chance of any significant human intake was estimated to be negligible.

A potential source of organotins in *food and beverages* arises from the use of triorganotin agricultural biocides and of diorganotins as heat stabilizers in PVC packaging materials.

Residues in foods and feeds have been studied quite extensively for both triphenyltin fungicides (WHO, 1980) and tricyclohexyltin acaricides (TROMBETTI and MAINI, 1970; GETZENDANER and CORBIN, 1972; GAUER et al., 1974; WHO, 1975). The half-life of tricyclohexyltins on pears is about 2–3 weeks, on apples 5–6 weeks (TROMBETTI and MAINI, 1970). A detailed survey will be published by the IUPAC Commission on Pesticide Chemistry (CROSBY et al., 1988). From all these studies it can be concluded that foodstuffs normally do not contain any detectable triorganotin pesticides. Furthermore, triorganotin residues are largely or completely degraded during food processing (washing, sterilizing, cooking, roasting, fermentation, etc.). Diorganotin stabilizers may leach into the food contents of PVC containers, but the resulting concentration is generally less than 1 ppm (CARR, 1969).

In many of the products analyzed the "natural tin" level appeared to be substantially higher than the amount of extracted tin. It has been concluded (LAPP, 1976) that the extent of exposure to the general public and the overall environment resulting from the use of dialkyltin stabilizers in rigid PVC is small.

There is no evidence of the accumulation of organotin compounds in the environment. The compounds appear to be moderately persistent.

5 Uptake, Absorption, Transport and Distribution, Metabolism and Elimination in Microorganisms, Animals, and Humans

Taking into account the practical applications discussed earlier, major possibilities for the resorption of *inorganic* tin chemicals in plants, animals, and man are through the use of packaging materials such as tin cans, tubes, and foils. As regards *organotin* chemicals, the major environmental implications are related to the use of triorganotins as agricultural and industrial biocides and to a lesser extent to the application of $0.3-1 =$ wt.% of specific diorganotin compounds (as well as some monoorganotins) as heat stabilizers in rigid PVC. For more detailed information the reader is referred to ZUCKERMAN et al. (1978), WHO (1980), and MAGOS (1986).

5.1 Inorganic Tin

Notwithstanding an increasing number of papers so far, microbial methylation of inorganic tin has not yet been proved beyond doubt (CRAIG, 1980; see also Chapter I.7h). Transmethylation reactions, however, do take place, and microbial species have been found to be resistant to or even to accumulate tin compounds; microbial inhibition causes changes in the bacterial populations (BLAIR et al., 1981). On the other hand, it is worthy of note that both methylmercury and methyllead compounds, such as the widely used anti-knock agent tetramethyllead, are capable of methylating inorganic tin (BULTEN, 1972). On the basis of laboratory experiments with a *Pseudomonas* species, a reverse mechanism has been postulated: microbial formation of methyltin, followed by transmethylation from tin to mercury (HUEY et al., 1974).

As regards the fate of inorganic tin in animals and man it has been demonstrated that ingested inorganic tin is only poorly absorbed — tin(II): 2.85%; tin(IV): 0.64% — and is almost completely eliminated in urine. The half-life for tin(II) in rat liver and kidneys was estimated to be $10-20$ days, in bone the half-life on tin(II) and tin(IV) is about $40-100$ days (WHO, 1980; FRITSCH et al., 1977; BROWN et al., 1977).

5.2 Organotin Compounds

In recent years substantial evidence has been published demonstrating that triorganotins are quite readily degraded either by (physico-)chemical processes (see Section 4) or by metabolic systems of microorganisms and mammals (BLUNDEN and CHAPMAN, 1986).

In soil all triorganotins are subject to microbial degradation. The half-life for total dephenylation of triphenyltins in soil is about 20 weeks (BARNES et al., 1973), which is about the same period required for 50% disappearance of TBTO by microbial degradation (BARUG and VONK, 1980). A similar stepwise deorganylation in soil takes place with tricyclohexyltins (BLAIR, 1975).

A review on the fate of fentin chloride in animals has been published by BOCK (1981). In different animals, most triphenyltin (>80%) is excreted unaltered in the feces in 1–3 weeks, minor amounts of diphenyltin, monophenyltin, and inorganic tin metabolites being detectable. Likewise, tricyclohexyltins are almost exclusively excreted unchanged in the feces (>99% in nine days; BLAIR, 1975). Similarly, tributyltin acetate in mice is excreted, mainly in the feces (93% within six days), major metabolites are dibutyltin and monobutyltin species (KIMMEL et al., 1977).

With tributyltins and tricyclohexyltins hydroxylation at various positions of the (cyclo)alkyl groups presumably occurs prior to stepwise degradation. No such hydroxylation was observed with triphenyltins (KIMMEL et al., 1980).

So far no reports have been published on detailed metabolic studies of diorganotin stabilizers. Studies with dibutyltin diacetate (KIMMEL et al., 1977) suggest that in general diorganotins in animals are likewise rapidly excreted, primarily in the feces.

Biomethylation does not seem to be a significant pathway for the transformation of organotin species. In an extensive study on the occurrence of butyltin species in the aquatic environment, minor amounts of butylmethyltin species were found only infrequently (MAGUIRE et al., 1985, 1986).

6 Effects on Plants, Animals, and Humans

Tin toxicities of natural origin in plants, animals, or man are virtually unknown. According to SCHROEDER et al. (1964), tin is an essential trace element. The hormone gastrine, produced by the stomach and transferred into the blood-stream upon feeding, contains tin. On the other hand, nutrient solutions containing more than 40 mg/L of tin are toxic to plant seeds.

6.1 Effects of Inorganic Tin Compounds on Animals and Humans

Various inorganic and organometallic derivatives have been reported (SCHWARZ et al., 1970; SCHWARZ, 1974) to display a significant growth effect in rats if trace element contamination is rigidly excluded. Levels of 0.5–2 ppm of tin in the diet improved growth by 25–60 percent. Inorganic tin compounds are relatively non-toxic. Stannous chloride, orthophosphate, sulfate, oxalate, or tartrate had no toxic effects when fed to rats over a 13 week period at 450–650 ppm of tin in the diet. Stannous oxide, sulfide, and oleate showed no effect at levels three times higher (DE GROOT et al., 1974). More recently, the same group reported a ninety-day no-effect level for rats of 150 ppm of stannous chloride (DE GROOT, 1976). However, a daily intake of 2.5 mg is toxic, 150 mg is fatal (UNDERWOOD, 1977; VENUGOPAL and LUCKEY, 1978). Exposed workers accumulate tin in the lungs (stannosis).

Occasional outbreaks of food poisoning due to tin contamination in canned food have been reported, the major effects being nausea, vomiting, and diarrhea. Inorganic tin does not induce teratogenic or carcinogenic effects (PISCATOR, 1977).

6.2 Effects of Organotin Compounds on Plants and Aquatic Organisms

The toxic effects of organotin compounds on aquatic organisms, in particular tributyl- and triphenyltin compounds which are the active ingredients of antifouling ship paints, are a matter of increasing concern. Tributyl- and triphenyltin compounds appear to be highly toxic towards various aquatic biota. Toxic effects are observed at the ppm to ppt concentration levels (HALL and PINKEY, 1985). Algae, molluscs, and the larval stage of fish belong to the most sensitive organisms. In areas with little water exchange and important pleasure craft activity, unacceptably high levels of tributyltin species may occur. This matter meets with increasing attention both in Western Europe, Canada, and the USA (MEINEMA et al., 1986; THOMPSON et al., 1985). HOLM and NORRGREN (1988) have studied long-term effects (fertility, reproduction in the first and second generation) of tributyltin on fish.

6.3 Acute and Chronic Effects of Organotin Compounds on Mammals
(see also MAGOS, 1986)

Organotin compounds, with the exception of mercurials, have received more attention for their biological effects than organometallic derivatives of any other metal. In contrast to the toxicity of lead, mercury or arsenic, significant toxicity is observed only when tin is part of a quite limited number of organotin compounds, notably triorganotin derivatives.

As mentioned earlier, the toxicity of organotin compounds is largely determined by both the nature and the number of the organic groups bound to tin. The acute mammalian toxicity decreases in the order $R_3SnX > R_2SnX_2 > RSnX_3$, whereas in the alkyl series ethyl derivatives are the most toxic, the toxicity rapidly decreasing with increasing chain length of the alkyl group. Linear trialkyltins are highly phytotoxic and cannot be applied as agricultural biocides.

In Table II.30-3 some acute oral toxicity data in rats of the practically important organotins are presented (cf. POLLER, 1970; LUIJTEN, 1971; PIVER, 1973;

Table II.30-3. Acute Oral Toxicity of Commerically Important Organotin Compounds, and Some Reference Compounds, in Rats

Compound	LD_{50} (mg/kg)
$(Bu_3Sn)_2O$	150 – 234
Me_3SnOH	540
$(Neophyl_3Sn)_2O$	2630
Ph_3SnOH	125
$Me_2Sn(IOTG)_2$ [a]	620 – 1380
$Bu_2Sn(IOTG)_2$	500 – 1000
$Oct_2Sn(IOTG)_2$	1200 – 2100
$(MeOOCCH_2CH_2Sn(IOTG)_2$	1230 – 1430
$MeSn(IOTG)_3$	920 – 1700
$BuSn(IOTG)_3$	1063
$OctSn(IOTG)_3$	3400 – >4000
Me_3SnOAc	9.1
Et_3SnOAc	4.0
Pr_3SnOAc	118.3
Bu_3SnOAc	380.2
Hex_3SnOAc	1000
$Oct_3SnIOTG$	26550
Bu_4Sn	>4000
Oct_4Sn	50000

[a] IOTG isooctylthioglycolate

ALDRIDGE, 1978; SMITH et al., 1978), summarizing the results reported in hundreds of papers on this subject.

Acute toxicity studies in rats showed that short-chain trialkyltins ($C_1 - C_4$) and triphenyltins are of moderate toxicity, except for trimethyl- and triethyltin compounds, which are highly toxic. All toxic trialkyltin compounds, except for the methyl compound, produce cerebral edema, which is most pronounced in the case of triethyltin poisoning (THORACK et al., 1970; SMITH, 1973).

As to the biochemical mode of action, in vitro studies have shown that triorganotins bind to mitochondria and interfere with oxidative phosphorylation (POLLER, 1970; PIVER, 1973; ALDRIDGE, 1978). So far no effective antidote for triorganotin intoxication has been reported.

Acute and chronic toxicity studies in rats with short-chain dialkyltins showed these compounds to be markedly less toxic than trialkyltin analogs, long-chain species such as dioctyltins being virtually non-toxic by oral intake. Dialkyltins act in a manner similar to arsenicals, i.e., they inhibit α-keto acid oxidases, which contain dithiol groups. This inhibition is counteracted by 2,3-dimercaptopropanol (BAL), which accordingly can be used as an antidote (LUIJTEN, 1971).

According to the scanty information on monoorganotins these compounds are of quite low toxicity. Likewise, tetraalkyltins display very little acute oral toxicity. An important feature of tetraalkyltin poisoning, however, is the long period before toxic symptoms develop. This feature, together with the similarity of the intoxication symptoms to those of the trialkyltins, suggests a biological conversion of tetraalkyltins into the more toxic trialkyltins in the mammalian organism (LUIJTEN, 1971).

6.4 Case Studies of Human Intoxications
(see also PRESSEL, 1988)

Brief contact with dialkyl- and trialkyltin compounds causes irritation of the skin and the respiratory tract (ANGER et al., 1976). Acute intoxications due to negligence of protective measures have been reported to cause vomiting, headache, visual defects, and electroencephalographic abnormalities (ZUCKERMAN et al., 1978). In 1981 six chemical workers, although wearing protective clothing and gas masks got seriously intoxicated with dimethyl- and trimethyltin chloride while cleaning a caldron. One person died after 12 days, two had not recovered after two years, and long-term prognoses were reported to be poor (REY et al., 1986).

Widespread poisoning once occurred in France as a result of taking capsules containing diethyltin diiodide (Stalinon) for the treatment of staphylococcal skin infections. Most probably this tragic incident was due to the presence of substantial amounts of highly toxic triethyltin iodide as an impurity (LUIJTEN, 1971; ZUCKERMAN et al., 1978). At present no ethyltin derivatives are used in practice or manufactured on an industrial scale.

6.5 Carcinogenic, Teratogenic, and Immunologic Effects

There is no evidence suggestive of teratogenicity or carcinogenicity of organotin compounds (GIAVINI et al., 1980). But in 1987, in California the use of tricyclohexyltin miticides was suspended, after studies had shown that the product could cause birth defects in rabbits (ANONYMOUS, 1987). On the other hand, triphenyltin acetate was reported to hold promise as a potential tumor-growth inhibitor (BROWN, 1972). Observations by BULTEN (1980), MEINEMA et al. (1985) as well as by CROWE et al. (1980) showed that specific diorganotin compounds, such as diethyltin oxide Et_2SnO, and diphenylchlorotin hydroxide, $Ph_2ClSnOH$, display antitumor activity against P-388 lymphocytic leukemia in mice.

Immunosuppressive effects have been observed with triphenyltin acetate (VERSCHUREN et al., 1970). More recently, extensive studies have been published (SEINEN and PENNINKS, 1979; SNOEY, 1987) on reversible atrophy of thymus and thymus-dependent lymphoid tissues in rats fed dioctyltin dichloride, dibutyltin dichloride and tributyltin chloride. No such effects were observed with the more water-soluble dimethyl- and diethyltin dichloride.

7 Hazard Evaluation and Limiting Concentrations

On the basis of the available information it can be concluded that the greatest potential for human exposure to tin compounds and the effects (especially stannosis) of such exposure on health may occur with workers (manufacturers, processors). Standards for occupational exposure have been recommended in many countries (ZUCKERMAN et al., 1978). Maximum allowable concentration values (calculated as tin) have been issued for inorganic tin chemicals at 2 mg/m^3, for organotin compounds at 0.1 mg/m^3. With the latter species, special attention is drawn to resorption by the skin. To avoid local irritation momentary (5 minutes) values should not exceed 0.2 mg/m^3 (MAK, 1987). Similar threshold-limit values have been fixed in the US, except for SnO_2 (10 mg/m^3), which is regarded as fine dust only.

Once incorporated into final consumer products, the potential for tin compound exposure is not likely to pose a major human health hazard. The daily intake should not exceed 15 ng (normally 1 – 5 ng). According to JUNG (1973), fish can tolerate up to 2 mg/L.

In many countries tolerances on crops have been set, e.g., triphenyltins: 0.05 – 1 ppm, tricyclohexyltins: 0.2 – 2 ppm, trisneophyltins: 0.05 – 1 ppm (WHO, 1975; FEDERAL REGISTER, 1975 a, b).

References

Adcock, L. H., and Hope, W. G. (1970), *Analyst 95*, 868 – 874.
Aldridge, W. N. (1978), *The Biological Properties of Organogermanium, -tin and -lead Compounds*, in: Gielen, M., and Harrison, P. (eds.), *The Organometallic and Coordination Chemistry of Germanium, Tin and Lead*, pp. 1 – 30. Georgi Publ., Saphorin, Switzerland.
Anger, J. P., Anger, F., Cano, Y., Chauvel, Y., Louvet, M., van den Driessche, J., and Morin, N. (1976), *Eur. J. Toxicol. Environ. Hyg. 9* (6), 339 – 346.
Anonymous (1987), *Chem. Eng. News 15* (July 6) and *19* (October 5).
Bahr, G., and Pawlenko, S. (1978), *Methoden zur Herstellung und Umwandlung von Organozinn Verbindungen*, in: Kropf, H. (ed.), *Methoden der Organischen Chemie (Houben-Weyl)*, Vol. XIII/6. Georg Thieme Verlag, Stuttgart.
Barnes, R. D., Bull, A. T., and Poller, R. C. (1973), *Pestic. Sci. 4*, 305.
Barug, D., and Vonk, J. W. (1980), *Pestic. Sci. 11*, 77 – 82.
Beeson, K. C. (1977), *Geochemistry and the Environment*, Vol. II, pp. 88 – 92. National Academy of Sciences, Washington D.C.
Bennett, R. F. (1983), *Ind. Chem. Bull. 2* (6), 171 – 176.
Blair, E. H. (1975), *Environ. Qual. Safe., Suppl. 3*, 406.
Blair, W. R., et al., Cooney, I. I., et al., and Chan, Y. K., et al. (1981), in: *Proceedings of the International Conference on Heavy Metals in the Environmental*, Amsterdam. CEP Consultants Ltd., Edinburgh.
Blair, W. R., Olson, G. J., Brinckman, F. E., Paule, R. C., and Becker, D. A. (1986), *Report NBSIR 86-3321*. National Bureau of Standards, Gaithersburg, Maryland.
Blunden, S. J., and Chapman, A. H. (1982), *The Environmental Degradation of Organotin Compounds – a Review*, ITRI Publ. No. 626. *Environ. Technol. Lett. 3*, 267 – 272.

Blunden, S. J., and Chapman, A. H. (1986), *Organotin Compounds in the Environment*, in: Craig, P. J. (ed.), *Organometallic Compounds in the Environment*, pp. 111–159. ITRI Publ. No. 665. Longman, London.

Blunden, S. J., Hobbs, L. A., and Smith, P. J. (1984), *The Environmental Chemistry of Organotin Compounds*, in: Bowen, H. J. M. (ed.), *Environmental Chemistry*, Vol. 3, pp. 49–77. ITRI Publ. No. 640. The Royal Society of Chemistry, London.

Blunden, S. J., Cusack, P. A., and Hill, R. (1985), *The Industrial Uses of Tin Chemicals*, 346 p. The Royal Society of Chemistry, London.

Bock, R. (1981), *Triphenyltin Compounds and their Degradation Products, Residue Rev. 79*. Springer-Verlag, New York.

Bokranz, A., and Plum, H. (1971), *Fortschr. Chem. Forsch. 16*, 365.

Bowen, H. J. M. (1972), *Analyst 97*, 1003–1005.

Braman, R. S., and Tompkins, M. A. (1979), *Anal. Chem. 51*, 12.

Brinckman, F. E. (1981), *Lecture 5th International Symposium Environmental Biochemistry*, Stockholm.

Brown, M. M. (1972), Ph. D. Thesis, *Diss. Abstr. Int. B. 33* (1973) 5356.

Brown, R. A., Nazario, C. M., De Tirado, R. S., Castrillon, J., and Agard, E. T. (1977), *Environ. Res. 13*(1), 56–61.

Browning, E. (1969), Chapter *Tin*, in: *Toxicity of Industrial Metals*, 2nd Ed., pp. 323–330. Butterworths, London.

Bruland, K. W. (1983), *Trace Elements in Sea Water, Chem. Oceanogr. 8*, 182.

Bulten, E. J. (1972), unpublished results; *German Patent Application* 2 200 697.

Bulten, E. J. (1980), unpublished results; *Dutch Patent Application* 80.03160.

Carr, H. G. (1969), *Soc. Plast. Eng. 25*, 72–74.

Chau, Y. K. (1987), Contribution *Sixth International Conference on Heavy Metals in the Environment*. CEP Consultants Ltd., Edinburgh.

Craig, P. J. (1980), *Environ. Technol. Lett. 1*, 225–234.

Crosby, D. G., Engst, R., Gorbach, S., Vonk, J. W., and Woggon, H. (1988), *Pure Appl. Chem.*, in press.

Crowe, A. J., Smith, P. J., and Atassi, G. (1980), *Chem. Biol. Interact. 32*, 171–178.

De Doncker, K., Dummarey, R., Dams, R., and Hoste, J. (1986), *Anal. Chim. Acta 187*, 163–169.

De Groot, A. P. (1976), *Voeding 37* (2), 87–97; *Chem. Abstr.* (1976) *84*, 145 612.

De Groot, A. P., Feron, V. J., and Til, H. P. (1973), *Food Cosmet. Toxicol. 11*, 19–30.

Dizikes, L. J., Ridley, W. P., and Wood, J. M. (1978), *J. Am. Chem. Soc. 100*, 1010.

Donard, O. F. X., Ratsomanikis, S. and Weber, J. R. (1986), *Anal. Chem. 58*, 772–777.

Engberg, A. (1973), *Analyst 98*, 137–145.

Environment Agency Japan (1974), *Japanese Background Paper No. 2*, prepared for the *WHO Meeting on the Effects on Health on Specific Air Pollutants from Industrial Emission*, Geneva, November 4–9.

Everett, G. L., West, T. S., and Williams, R. W. (1974), *Anal. Chim. Acta 70*, 291–298.

Federal Register (1975a), Vol. 40, 33 033–33 034; *Chem. Abstr.* (1975), *83*, 191 466.

Federal Register (1975b), Vol. 40, 33 035; *Chem. Abstr.* (1975) *83*, 191 465.

Fritsch, P., de Saint Blanquat, G., and Derache, R. (1977), *Toxicology 8* (2), 165–175.

Fuller, M. J. (1975), *Tin and Its Uses 103*, 3–7.

Gauer, W. O., Seiber, J. N., and Crosby, D. G. (1974), *J. Agric. Food Chem. 22*, 252.

Getzendaner, M. E., and Corbin, H. B. (1972), *J. Agric. Food Chem. 20*, 881.

Giavini, E., Prati, M., and Vismara, C. (1980), *Bull. Environ. Contam. Toxicol. 24*, 936–939.

Hall, L. W., Jr., and Pinkney, A. E. (1985), *Acute and Sublethal Effects of Organotin Compounds on Aquatic Biota: an Interpretative Literature Evaluation, CRC Crit. Rev. Toxicol. 14* (2), 159–209.

Hamaguchi, H. (1964), *Geochim. Cosmochim, Acta 28*, 1039.

Hamaguchi, H., and Kuroda, R. (1969), in: Wedepohl, K. H. (ed.), *Biogeochemistry; Handbook of Geochemistry*, Vol. II, pp. 2–5. Springer Verlag, New York.

Hamilton, E. I., Minski, M. J., Cleary, J. J., and Halsey, V. S. (1972), *Sci. Total Environ. 1*, 205–210.

Heindl, R. A. (1970), *Tin Bur. Mines Bull.* (650), 759–771.

Hoare, W. E. (1974), *Trends in Tin Consumption: Some Technical Observations.* Fourth World Conference on Tin, Kuala Lumpur, Malaysia.

Hodge, V. F., Seidel, S. L., and Goldberg, E. D. (1979), *Anal. Chem. 51*, 1256–1259.

Holm, G., and Norrgren, L. (1988), *The Effects of Long-term Exposure of Tributyltin (BTN) on Sticklebacks and Gastropods, Proceedings 1st European SECOTOX Conference on Ecotoxicology*, University of Copenhagen, pp. 298–300.

Horwitz, W. (1979), *J. Assoc. Off. Anal. Chem. 62* (6), 1251–1264.

Huey, C., Brinckman, F. E., Grim, S., and Iverson, W. P. (1974), *Proc. Int. Conf. Transp. Persist. Chem. Aquat. Ecosyst.*

Johansen, O., and Steiner, E. (1969), *Analyst 94*, 976–978.

Jung, K. D. (1973), in: Förstner, U., and Wittmann, G. T. W. (1979), *Metal Pollution in the Aquatic Environment.* Springer-Verlag, Berlin.

Kaars Sijpesteijn, A., Luijten, J. G. A., and van der Kerk, G. J. M. (1969), in: Torgeson, D. C. (ed.), *Fungicides*, Vol. II, Chapter 7; *Organometallic Fungicides.* Academic Press, New York.

Kimmel, E. C., Fish, R. H., and Casida, J. E. (1977), *J. Agric. Food Chem. 25*, 1.

Kimmel, E. C., Casida, J. E., and Fish, R. H. (1980), *J. Agric. Food Chem. 28*, 117–122.

Kirk, R. E., and Othmer, D. F. (eds.) (1955), *Encyclopedia of Chemical Technology*, Vol. 14. Interscience Publishers, New York.

Kirk, R. S., and Pocklington, W. D. (1969), *Analyst 94*, pp. 71–74.

Klötzer, D., and Thust, U. (1976), *Chem. Tech. 28*, 614.

Knorr, D. (1975), *Lebensm. Wiss. Technol. 8*, 51–56.

Lapp, T. W. (1976), *The Manufacture and Use of Selected Alkyltin Compounds*, EPA, Office of Toxic Substances, EPA 560/6-76-011, PB-251 819.

Van Loon, J. C. (Geology, Univ. of Toronto, 1987), *Proceedings Sixth Intern. Conf. "Heavy Metals in the Environment"*, Vol. 1, pp. 259–264, CEP Consultants Ltd., Edinburgh.

Luijten, J. G. A. (1971), *Applications and Biological Effects of Organotin Compounds*, in: Sawyer, A. K. (ed.), *Organotin Compounds.* Marcel Dekker, New York.

Luijten, J. G. A., and van der Kerk, G. J. M. (1968), in: MacDiarmid, A. G. (ed.), *Organometallic Compounds of the Group IV Elements*, Vol. I, Part II, p. 91. Marcel Dekker, New York.

Mac Crehan, W. A. (1987), *Contribution at the 17th Annual IAEAC-Symposium on the Analytical Chemistry of Pollutants*, Jekyll Island, Georgia.

Magos, L. (1986), Chapter 23 Tin, in: Friberg, L., Nordberg, G. F., and Vouk, V. R. (eds.), *Handbook on Toxicology of Metals*, 2nd. Ed., Vol. II, pp. 568–593. Elsevier Science, Amsterdam.

Maguire, R. J., and Tkacz, R. J. (1985), *J. Agric. Food Chem 33*, 947–953.

Maguire, R. J., Tkacz, R. J., Chau, Y. K., Bengert, G. A. and Wong, P. T. S. (1986), *Chemosphere 15* (3), 253–274.

MAK (1987), *Maximum Concentrations at the Workplace, Report No. XXIII DFG.* VCH Verlagsgesellschaft, Weinheim-Basel-Cambridge-New York.

Meinema, H. A., Liebregts, A. M. J., Budding, H. A., and Bulten, E. J. (1985), *Rev. Silicon Germanium Tin Lead Comp. 8* (2–3), 157–168.

Meinema, H. A., van Dam-Meerbeek, T. G., and Vonk, J. W. (1986), *Evaluation of the Impact of Organotin Compounds on the Aquatic Environment.* Report to the European Economic Community, Brussels.

Nakahara, T., Munemori, M., and Musha, S. (1972), *Anal. Chim. Acta 62*, 267–278.

Neumann, W. P. (1970), *The Organic Chemistry of Tin.* Wiley Interscience, New York.

Parks, E. J., Blair, W. R., and Brinckman, F. E. (1985), *Talanta 32* (8A), 633–639.

Paschen, P., Müller, B., Pawlenko, S., and Merian, E. (1983), in: *Ullmanns Encyklopädie der technischen Chemie*, 4th Ed., Vol. 24, pp. 641–679. Verlag Chemie, Weinheim-Deerfield Beach/Florida-Basel.

Piscator, M. (1977), *Tin*, in: *Toxicology of Metals*, Vol. II, pp. 405–427. Health Effects Research Laboratory, HERL, RTP, Office of Research and Development, U.S. Environmental Protection Agency, Research Triangle Park, North Carolina.
Piver, W. T. (1973), *Environ. Health Perspect. 4*, 61–79.
Poller, R. C. (1970), *The Chemistry of Organotin Compounds*, Academic Press, New York.
Pressel, G. (1988), Chapter 64 *Tin*, in: Seiler, H. G., Sigel, H., and Sigel, A. (eds.), *Handbook on Toxicity of Inorganic Compounds*, pp. 697–703. Marcel Dekker Inc., New York.
Randall, L., Donard, O. F., and Weber J. H. (1986), *Anal. Chim. Acta 184*, 197–203.
Rey, C., Reinecke, H. J., and Besser, R. (1984), *Vet. Hum. Toxicol. 26*, 121–122.
Rudolf, H., Alfrey, A. C., and Snythe, W. R. (1973), *Trans. Am. Soc. Artif. Intern. Organs 19*, 456–461.
Saager, R. (1984), *Metallic Raw Materials Dictionary* (in German), pp. 87–90. Bank von Tobel, Zürich.
Sawyer, A. W. (ed.) (1971/1972), *Organotin Compounds*, Vols. 1–3. Marcel Dekker, New York.
Schaefer, C. H., Miura, T., Dupras Jr., E. F., and Wilder, W. H. (1981), *J. Econ. Entomol. 74* (5), 597–600.
Schlesinger, A. E. (1977), *Abstracts Marine Coatings Seminar*, Bioloxi, Massachusetts, March 1977. National Paint and Coatings Association.
Schroeder, H. A., Balassi, J. J., and Tipton, I. H. (1964), *J. Chron. Dis. 17*, 483–502.
Schumann, H., and Schumann, I. (1975), *Zinn-organische Verbindungen*, in: *Gmelin Handbuch der Anorganischen Chemie*, Vols. 1–4. Springer Verlag, Berlin.
Schwarz, K., Milne, D. B., and Vinyard, E. (1970), *Biochem. Biophys. Res. Commun. 40*, 22.
Schwarz, K. (1974), *Fed. Proc. Fed. Am. Soc. Exp. Biol. 33* (6), 1748.
Seinen, W., and Penninks, A. (1979), *Ann. NY Acad. Sci. 320*, 499–517.
Smith, M. E. (1973), *J. Neurochem. 21* (2), 357–372.
Smith, J. D., and Burton, J. D. (1972), *Geochim. Cosmochim. Acta 36*, 621.
Smith, G. N., Fischer, F. S., and Axelson, R. J. (1976), *J. Agric. Food Chem. 24*, 1225.
Smith, P. J., Luijten, J. G. A., and Klimmer, O. R. (1978), *Toxicological Data on Organotin Compounds*. ITRI Publication No. 538. International Tin Research Institute, London.
Snoey, N. J. (1987), *Triorganotin Compounds in Immunotoxicology and Biochemistry*, Ph. D. Thesis, State University, Utrecht, The Netherlands.
Soderquist, C. J., and Crosby, D. G. (1980), *J. Agric. Food Chem. 28*, 111–117.
Suess, A., and Eben, Ch. (1973), *Z. Pflanzenkrankh. Pflanzenschutz 80*, 288.
Thompson, J. A. J., Sheffer, M. G., Pierce, R. C., Chau, Y. K., Cooney, J. J., Cullen, W. R., and Maguire, R. J. (1985), *Organotin Compounds in the Aquatic Environment: Scientific Criteria for Assessing their Effects on Environmental Quality*, NRCC Rep. No. 22494, 291 p. National Research Council Canada, Associate Committee on Scientific Criteria for Environmental Quality.
Thorack, R. M., Gordon, J. S., and Prokop, J. (1970), *Int. Rev. Neurobiol. 12*, 45–86.
Tipton, I. H., Stewart, P. L., and Martin, P. G. (1966), *Health Phys. 12*, 1683–1689.
Trombetti, G., and Maini, P. (1970), *Pestic. Sci. 1*, 144.
Underwood, E. J. (1977), *Trace Elements in Human and Animal Nutrition*. Academic Press, New York.
Vakirs, A. O., et al. (1986), *Mar. Pollut. Bull. 17*, 319–324.
Venugopal, B., and Luckey, T. D. (1978), *Metal Toxicity in Mammals*, Vol. 2. Plenum Press, New York.
Verschuren, H. G., Ruitenberg, E. J., Peetoom, F., Helleman, P. W., and van Esch, G. J. (1970), *Toxicol. Appl. Pharmacol. 16* (2), 400–410.
Welz, B. (1985), *Atomic Absorption Spectrometry*, pp. 331–334, 352 ff. VCH Verlagsgesellschaft, Weinheim-Deerfield Beach/Florida-Basel.
WHO (World Health Organization) (1975), *Pestic. Residues Ser. Geneva 4*, 155.
WHO (World Health Organization) (1980), *Environmental Health Criteria 15: Tin and Organotin Compounds*, Geneva.
Zuckerman, J. J., Reisdorf, R. P., Ellis III, H. V., and Wilkinson, R. R. (1978), *ACS Symp. Ser. 82*, 388–424.

Additional Recommended Literature

de Alencastro, L. F. (1989), *Analytical Chemistry of Organotin Contaminants in the Aquatic Environment*. Proceedings First IAEAC Central American and Caribbean Workshop on Analytical Chemistry in Sanitary and Environmental Studies, Tegucigalpa, Honduras, in press.

Bayona, J. M., Tolosa, I., Albaiges, J., de Alencastro, L. F., and Tarradellas, J. (1990), *Organotin Speciation in Aquatic Matrices by CGC/FPD, ECD and MS, and LC/MS*. Proceedings 20th IAEAC Symposium on Environmental Analytical Chemistry, Strasbourg. Fresenius Z. Anal. Chem., in press (see also Reports E. Merian, *Swiss Chem.* 12(9) and *TRAC* 9(9), in press).

Blunden, S. J., and Evans, C. J. (1990), *Organotin Compounds*, in: *Handbook of Environmental Chemistry* (ed. O. Hutzinger), Vol. 3, Part E, *Anthropogenic Compounds*, pp. 1–44. Springer-Verlag, Berlin-Heidelberg-New York-London-Paris-Tokyo-Hong Kong.

Chau, Y. K., and Wong, P. T. S. (1990), *Recent Developments in the Speciation and Determination of Organometallic Compounds in Environmental Samples*. Proceedings 20th IAEAC Symposium on Environmental Analytical Chemistry, Strasbourg. Fresenius Z. Anal. Chem., in press (see also Reports E. Merian, *Swiss Chem.* 12(9) and *TRAC* 9(9), in press).

Donard, O. F. X., Astruc, M., Quevauviller, P., Durbant, F., Martin, F., Nunes, R., and Lagrange, P. (1990), *Optimization in the Determination of Inorganic and Alkylated Forms of Tin, Selenium, and Mercury by Coupled Gas Chromatography and Atomic Absorption Spectrometry; Critical Considerations on Tributyltin Determinations in the Environment*. Proceedings 20th IAEAC Symposium on Environmental Analytical Chemistry, Strasbourg. Fresenius Z. Anal. Chem., in press (see also Reports E. Merian, *Swiss Chem.* 12(9) and *TRAC* 9(9), in press).

Lund, G. (1989), *Speciation Analysis – Why and How?* Fresenius Z. Anal. Chem. 334(7), 610–611.

Omae, I. (1989), *Organotin Chemistry*. J. Organomet. Chem. Libr. 21, 356 pages.

Puddu, A., Pettine, M., La Noce, T., and Pagnotta, R. (1989), *Toxic Effects of Organotin Compounds on Marine Phytoplankton*. Proceedings 7th International Conference on Heavy Metals in the Environment, Geneva, Vol. 2, p. 166. CEP Consultants Ltd., Edinburgh, UK; see also Report E. Merian, *Swiss Chem.* 12(3), 34 (1990).

Trueb, L. (1989), *Historic Tin Mines in the Taurus Mountains* (in German), Neue Zürcher Zeitung, Forschung und Technik, No. 147, p. 84 (28 June), Zürich.

Trueb, L. (1990), *Tin from the "Tin Belt" in Southeast Asia* (in German), Neue Zürcher Zeitung, Forschung und Technik, No. 118, p. 65 (23 May), Zürich.

Vighi, M., and Bacci, E. (1989), *The Mediterranean Sea – Environmental Impact of Chemicals*. Proceedings First European Conference on Ecotoxicology (SECOTOX), Copenhagen, pp. 333–344. Laboratory of Environmental Sciences and Ecology, University of Lyngby, Denmark.

Wong, P. T. S., et al. (1989), *Occurrence of Tributyltin in Ontario Harbours* (Canada), 7th International Conference on Heavy Metals in the Environment Geneva, CEP Consultants Ltd., Edinburgh, UK; see also Report E. Merian, *Swiss Chem.* 12(3), 34 (1990).

II.31 Titanium

JACK WHITEHEAD, Maltby, Middlesbrough, Cleveland, UK

1 Introduction

Titanium in its elemental form does not occur in nature but its compounds are widely distributed. There is no evidence that it is an essential element and it is non-toxic. It is not absorbed or accumulated in the human body. The main environmental problems are associated with the disposal of waste products arising from the manufacture of titanium dioxide pigments. Titanium dioxide in its pigmentary form is classified as a nuisance dust.

2 Physical and Chemical Properties, and Analytical Methods

Titanium is a silver-gray metal and is the first member of Group 4B of the periodic classification of the elements. The metal is very resistant to corrosion and has a density of 4.5 g/cm^3, a melting point of $1670\,°C$ and a boiling point of $3260\,°C$. Titanium (IV) is the most stable valence state; the lower valence states (II) and (III) exist but are readily oxidized to the tetravalent state by air, water, and other oxidizing agents. The most important forms are white titanium dioxide (density 4.0 g/cm^3, melting point $1855\,°C$), titanium metal and its alloys, and titanium tetrachloride (freezing point $-24\,°C$, boiling point $135.8\,°C$). Naturally occurring titanium contains approximately 73% ^{48}Ti, 8% each of ^{46}Ti and ^{47}Ti, and 5% each of ^{49}Ti and ^{50}Ti.

Large amounts of titanium are determined by reduction to Ti(III) using zinc, cadmium, liquid amalgams, or aluminum. The Ti(III) is then titrated with a standard oxidizing agent. NORRIS (1984) has described the reduction of Ti(IV) in solution with chromium(II) chloride followed by automatic potentiometric titration with iron(III)-ammonium sulfate. For highest accuracy standardization must be against a sample of a pure titanium compound.

Trace amounts of titanium may be determined spectrophotometrically in acidic solution by using hydrogen peroxide or, more sensitively, Tiron (1,2-dihydroxybenzene-3,5-disulfate) with both of which titanium forms yellow complexes. X-ray fluorescence spectrometry, emission spectrography, neutron activation, and spark source mass spectrography have all been used for the determination of titanium.

In the case of AAS (atomic absorption spectrometry) a high temperature flame (nitrous oxide, acetylene) is essential and the optimum wavelengths are 364.3 nm and 365.4 nm. The sensitivity is low; with the graphite furnace a lower detection limit of approximately 0.5 µg/L can be achieved.

ICP-AES (inductively coupled plasma-atomic emission spectroscopy) is especially sensitive and is the recommended instrumental method for trace amounts (see also BROEKAERT, 1987; for the determination in urine see SCHRAMEL, 1988). The optimum wavelengths are 334.9 nm and 336.1 nm, and a detection limit of 2.0 µg/L can be achieved.

3 Sources, Production, Important Compounds, Uses, and Waste Products

(see also RÜDIGER et al., 1983; WHITEHEAD, 1983)

Titanium is not a rare metal and is widely distributed throughout the world. It is the ninth most abundant element and occurs to the extent of 0.6% in the earth's crust. Sea water contains 1–2 µg/L (see also BRULAND, 1983), probably mainly in the form of hydrated titanium oxide.

The largest deposits are of ilmenite which is an iron titanate containing 35–60% TiO_2, and the main commercial sources are Australia, Canada, Norway, South Africa, and the United States. The other commercially important mineral is rutile, containing approximately 95% TiO_2, and the major source of supply is Australia. In 1985 world production of titanium concentrates was estimated to be as follows: 4.2 million tons of ilmenite, 350000 tons of rutile and 1.15 million tons of titaniferous slag. The latter is a by-product of the manufacture of iron by the electric furnace reduction of ilmenite, and the sources of supply are Canada, Norway, and South Africa.

There are two main processes for extracting titanium from its ores. In the first, ilmenite, or titaniferous slag, is attacked with sulfuric acid to form a solution of titanium and the impurity elements. The solution is purified and the titanium separated as a hydrated titanium dioxide by a nucleation process. This is then further purified, calcined, and treated to form titanium dioxide pigments.

In the second, which usually starts with rutile although ilmenite and upgraded ilmenite can also be used, the mineral is chlorinated in the presence of a reducing agent, usually coke, to form titanium tetrachloride. After purification by distillation, the titanium tetrachloride may be reduced with sodium or magnesium to produce titanium metal or oxidized to produce titanium dioxide pigment.

Important commercial forms of titanium are the metal, dioxide, and tetrachloride. The metal and its alloys are used in applications such as aeronautics, tubings, and surgical implants (or prostheses) where strength, lightness, and resistance to corrosion are desirable. Titanium additions to chromium/nickel steel are transformed during production into titanium carbide inclusions with increased strength (ALMAR-NAESS, 1987).

Titanium dioxide because of its whiteness and high refractive index is used extensively as a white pigment. The main use for titanium dioxide pigments is in the manufacture of paint, but they are also used in plastics, rubber, paper, ceramics, fibers, printing inks, cosmetics, and foodstuffs. World production capacity in 1985 was 2.7 million tons for titanium dioxide pigments and, in 1984, 93 500 tons for titanium metal. Titanium minerals are also used in welding rod coatings, and the total market has been estimated to be in the range 100 – 110 000 tons in 1985.

Waste disposal is less a problem in the chloride process because the chlorine can be recovered and recycled. The relatively small amount of impurity chlorides produced as a by-product of the process may be neutralized with chalk or lime and surface dumped or disposed of directly in deep boreholes. Waste disposal in the sulfate process raises more serious environmental problems although the increasing use of titaniferous slag as feedstock has considerably reduced the amount of iron in the effluent. An estimate, for example, (*Umwelt*, 1981) showed that 1.7 million tons of waste acid containing 5 – 10% iron sulfate and 4 – 6% other metal sulfates were being discharged annually into the North Sea. DETHLEFSEN (1988) found that epidermal papillomas of fish are more frequent in the TiO_2 acid area of the North Sea.

The COMMISSION OF THE EUROPEAN COMMUNITIES in 1978 and 1983 produced a Directive calling for the reduction and eventual elimination of pollution from the manufacture of titanium dioxide. A further Directive in 1982 established surveillance and monitoring needs for each plant. In the UK it has been argued successfully that disposal of effluent from titanium dioxide plants into fast flowing tidal rivers ensures its rapid neutralization resulting in an acceptable level of effect on the environment (HOUSE OF LORDS SELECT COMMITTEE, 1984). An example of the care taken in monitoring such a discharge is given in the publication *Taking Care – Tioxide and the Humber Estuary* (1984). It is noted that the EEC Directive on monitoring is based on this work.

In the case of factories which are not so well situated treatment of the effluent is necessary. This involves the removal of iron and other metal sulfates, concentration and re-use of waste acid, or neutralization and dumping (SMITH and PETERSON, 1985). Useful by-products are gypsum and iron sulfate which is used in water treatment and agriculture. The treatment of the vine disease, chlorosis, is one example of the latter.

Titanium tetrachloride is also the starting point for the preparation of titanium trichloride, which is used as a catalyst in polyethylene manufacture, and a series of titanium organic compounds. Chief examples of these are isopropyl titanate and tetra-*n*-butyl titanate which are used as catalysts, crosslinking agents, and surface modifiers.

4 – 5 Distribution in Foods and in Living Organisms, Uptake, Elimination

Selected grades of titanium dioxide pigments are used as additives to impart whiteness to foodstuffs and pet foods. They are also used in edible inks, cosmetics, toothpaste and pharmaceuticals. Their use in these applications is subject to stringent regulations. Plants contain approximately 1 mg/kg of titanium based on dry weight.

It has been estimated that the daily human intake of titanium varies from 0.3 to 1 mg most of which is excreted in feces (VALENTIN and SCHALLER, 1980). It may accumulate in the lung and in the lymphatic system (NORDMAN and BERLIN, 1986). The adult body contains approximately 15 mg titanium principally in the lungs. Higher concentrations are found in persons occupationally exposed to titanium. It is mainly excreted in the feces, and the urine levels are normally below the detection limit of 1.5 µg/L (SCHRAMEL, 1988).

6 Effect on Animals and Humans

(see also NORDMAN and BERLIN, 1986)

Because of its widespread use as a pigment, the toxicology of titanium dioxide has been the subject of much investigation. The results show that titanium is non-toxic, any harmful effects with titanium compounds being associated with the anion, e.g., hydrochloric acid in the case of titanium tetrachloride. Indeed titanium dioxide has been used in the treatment of skin disorders and titanium metal has been used as a bone implant. Titanium dioxide pigments have been in use now for over fifty years, and no evidence of harmful effects has been noted. Nor is there any evidence that titanium is an essential element for either man or animals. The human body appears to have a wide range of tolerance to titanium.

Slight fibrogenic effects have been reported in animal studies using titanium hydride (BRAKHNOVA and SHURKO, 1972), titanium carbide, nitride, or boride (BRAKHNOVA, 1969; BRAKHNOVA and SAMSONOV, 1970). Potassium octatitanate fibers of specific dimension (LEE et al., 1981) and titanium phosphate fibers (GROSS et al., 1977) have caused dose related fibrosis in rats. A recent study (LEE et al., 1985) has shown that microscopic lung tumors were found in 10% of rats exposed for two years to a titanium dioxide dust concentration, 50 times the recommended Long Term Exposure Limit of 10 mg/m^3. No tumors were found in rats exposed to 10 times this concentration.

FURST and HARO (1969) and FURST (1971) showed that intramuscular injection of suspensions of titanium metal powder or titanocene in trioctanoin induced fibrosarcomas and lymphosarcomas in rats.

SCHROEDER and MITCHENER (1971) observed a reduction in the survival rate of rats and mice up to the third generation when their drinking water contained a soluble titanate equivalent to 5 mg Ti/L.

7 Hazard Evaluation and Limiting Concentrations

Titanium dioxide pigment because of its small size (approximately 20 µm) is classified as a nuisance dust and Long Term Exposure Limits of 10 mg/m^3 total dust and 5 mg/m^3 respirable dust have been specified by the UK Health and Safety Executive and the American Conference of Governmental Hygienists. In the Federal Republic of Germany, and representative for the European Continent, a MAK (1987) value of 6 mg/m^3 (maximum concentration at the workplace) has been specified. Examinations of workers with many years of service in the titanium dioxide pigment industry have shown that, although some titanium dioxide may be present in their lungs, no ill effects attributable to the pigment have been reported. A WHO (1982) report stated that titanium dioxide is biologically inert and does not possess fibrogenic characteristics. Also the results mentioned in Sect. 6 do not suggest that manufacture or usage of titanium dioxide pigments pose any risk of adverse effects on health, provided normal precautions are observed.

VALENTIN and SCHALLER (1980) have recommended that titanium should not be brought within the scope of human monitoring and stated that this is particularly true in the case of titanium dioxide. Monitoring may, however, be necessary in the case of other titanium compounds such as organic titanium derivatives.

References

ACGIH (American Conference of Governmental Industrial Hygienists) (1984), *Threshold Limit Values for Chemical Substances in the Work Environment.* ACGIH, Cincinnati, Ohio.
Almar-Naess, A. (1987), *Neue metallische Werkstoffe* (New Metallic Materials), *Neue Zürcher Zeitung No. 40*, p. 71 (16 February). *Forschung und Technik*, Zürich.
Brakhnova, I. T. (1969), *Comparative Evaluation of the Effect on Animals of Transition Metal Boride and Carbide Dusts with Respect to the Peculiarities of their Electron Structure. Gig. Tr. Prof. Zabol. 13*, 26–31.
Brakhnova, I. T., and Samsonov, G. V. (1970), *Comparison of the Effect of Nitrides of Transition Metals and Non-metals on the Body. Gig. Tr. Prof. Zabol. 13*, 26–31.
Brakhnova, I. T., and Shurko, G. A. (1972), *Hygienic Assessment of the Effect Produced on the Body by Transition Metal Hydrides with Regard to their Electronic and Crystal Structure. Gig. Sanit. 37*, 36–39.
Broekaert, J. A. C. (1987), *Trends in Optical Spectrochemical Trace Analysis with Plasma Sources. Anal. Chim. Acta 196*, 1–21.
Bruland, K. W. (1983), *Trace Elements in Sea Water. Chem. Oceanogr. 8*, 172, 198.
Commission of the European Communities (1978). *Council Directive (78/176/EEC) on Waste from the Titanium Dioxide Industry.* Official Journal of the European Communities No. L 54/19–24.
Commission of the European Communities (1982), *Council Directive (82/883/EEC) on Procedures for the Surveillance and Monitoring of Environments Concerned by Waste from the Titanium Dioxide Industry.* Official Journal of the European Communities No. L378/1–14.
Commission of the European Communities (1983), *Proposal for a Council Directive on Procedures for the Reduction and Eventual Elimination of Pollution caused by Wastes from the Titanium Dioxide Industry.* COM(83) Final. Brussels, 14 April.

Dethlefsen, V. (1988), *Marine Pollution and Fish Diseases in the North Sea.* Proceedings, University Copenhagen, pp. 362–374.

Furst, A. (1971), *Trace Elements Related to Specific Chronic Diseases. Cancer. Geol. Soc. Am. Mem. 123*, 109–110.

Furst, A., and Haro, R. T. (1969), *Survey of Metal Carcinogenesis. Prog. Exp. Tumor Res. 12*, 102–133.

Gross, P., Kociba, R., Sparschv, G. L., and Norris, J. M. (1977), *The Biological Response to Titanium Phosphate. Arch. Pathol. Lab. Med. 101*(1), 550–554.

Guidance Note from Health & Safety Executive (1986), *Occupational Exposure Limits.*

House of Lords Select Committee on the European Communities (1984), Session 1983–1984; *13th Report. Waste from the Titanium Dioxide Industry.* H.M.S.O., London.

Lee, K. P., Trochimowicz, H. J., and Reinhardt, C. F. (1985), *Pulmonary Response of Rats Exposed to Titanium Dioxide (TiO_2) by Inhalation for two Years. Toxicol. Appl. Pharmacol. 79*, 179–192.

MAK (1987), *Maximum Concentrations of the Workplace, Report No. XIII, DFG.* VCH Verlagsgesellschaft, Weinheim-Basel-Cambridge-New York.

Nordman, H., and Berlin, M. (1986), in: Friberg, L., Nordberg, G. F., and Vouk, V. (eds.), *Handbook on the Toxicology of Metals* 2nd Ed., Chap. 24, pp. 594–609. Elsevier, Amsterdam.

Norris, J. D. (1984), *Determination of Titanium in Titanium Dioxide Pigments, Paints and Other Materials by ChromiumII Chloride Reduction and Automatic Potentiometric Titration. Analyst 109*, 1475.

Rüdinger, K., Fichte, R., Feld, R., and Wolf, H. U. (1983), Chapter *Titan* (in German), in: *Ullmanns Encyklopädie der technischen Chemie,* 4th Ed., Vol. 23, pp. 267–292. Verlag Chemie, Weinheim-Deerfield Beach/Florida-Basel.

Schramel, P. (1988), *Barium, Strontium, Titanium,* in: Angerer, J., and Schaller, K. H. (eds.), *DFG Analyses of Hazardouzs Substances in Biological Materials,* Vol. 2, pp. 71–79. VCH Verlagsgesellschaft, Weinheim-Basel-Cambridge-New York.

Schroeder, H. A., and Mitchener, M. (1971), *Toxic Effects of Trace Elements on the Reproduction of Mice and Rats. Arch. Environ. Health 23*, 102–106.

Smith, I. M., and Peterson, H. C. (1985), *Sulphur Value Recovery from Waste Acid.* Paper presented at *Sulphur 85,* Nov. 10–13, sponsored by the British Sulphur Corporation Limited, London.

Taking Care – Tioxide and the Humber Estuary (1984). Tioxide Group PLC, 10 Stratton Street, London W1X 5FD.

Umwelt (1981) No. 80, 16 January; No. 81, 20 February; No. 82, 10 April. Informationen des Bundesministers des Innern zur Umweltplanung und zum Umweltschutz, Bonn.

Valentin, H., and Schaller, K. H. (1980), *Human Biological Monitoring of Chemicals, 3. Directorate-General Employment and Social Affairs.* Commission of the European Communities.

Whitehead, J. (1983), in: *Kirk-Othmer Encyclopaedia of Chemical Technology,* 3rd Ed., Vol. 23, pp. 131–176. John Wiley & Sons, New York.

WHO (World Health Organization) (1982), *Environmental Health Criteria, 24 Titanium.* WHO Geneva.

Additional Recommended Literature

Adams, R. (1984), *Titanium and Titanium Dioxide.* Financial Times Business Information Limited, 102–108 Clerkenwell Road, London.

Browning, E. (1969), *Toxicity of Industrial Metals,* 2nd Ed., Chap. 41, p. 333. Butterworths, London.

Clark, G. (1986), *Titanium Minerals. Ind. Min. London*, June, 47–54.
The Economics of Titanium (1980), 3rd Ed. Roskill Information Services Ltd., 2 Clapham Road, London.
Elwell, W. T., and Whitehead, J. (1962), in: Wilson, C. L., and Wilson, D. W. (eds.), *Comprehensive Analytical Chemistry*, Vol. 1c, Chap. 4a. Elsevier, Amsterdam.
Ferin, J., and Oberdörster, G. (1985), *Biological Effects and Toxicity Assessment of Titanium Dioxides: Anatase and Rutile. Am. Ind. Hyg. Assoc. J. 46*, 69–72.
Griffiths, J. (1985), *Minerals in Welding Fluxes – The Whys and Wherefores. Ind. Min. London*, March, 18–41.
Hygienic Guide Series (1973), *Am. Ind. Hyg. Assoc. J. 34*, 275.
Jones, A. D., Lohmann-Matthes, M.-L., and Takenaka, S. (1987), *"Nuisance" Dust Inhalation*, in: *The Design and Interpretation of Inhalation Studies* by V. Mohr (Hannover) and ILSI (Washington, D.C.). Springer Verlag, Heidelberg-Berlin-New York.
Kramer, K-H. (1983), *Titanium and Titanium Alloys. Radex-Rundsch. 75*, 1/2, 5–23.
Lee, K. P., Barras, C. E., Griffith, F. D., and Waritz, R. S. (1981), *Pulmonary Response and Transmigration of Inorganic Fibres by Inhalation Exposure. Am. J. Pathol. 102*(3), 314–323.
Lynd, Langtry, E., and Hough, R. A. (1984), in: Schrek, A. F. (ed.), *Minerals Year Book*, Vol. 1. Bureau of Mines, US Department of the Interior, Washington, D.C.
McCue, J. P. (1973), *Titanium Analysis of Human Blood. Biochem. Med. 7*, 282.
Murrer, B. (1989), *Metal-based Antitumour Drugs. Proceedings Satellite Meeting "Toxicity and Therapeutics of Newer Metals and Organometallic Compounds"*. University of Surrey, Guildford, in press; see also Report E. Merian, *Swiss Chem. 12*(1–2), 25 (1990).
Poole, W. K., and Johnston, D. R. (1968), *Estimating Population Exposure to Selected Metals – Titanium*. Research Triangle Institute, Research Triangle Park, North Carolina.
Roberts, W. T. (1983), *Titanium, Endeavour, New Ser. 7*, 4, 189–193.
Saager, R. (1984), *Metallic Raw Materials Dictionary* (in German), pp. 149–152. Bank von Tobel, Zürich.
Sandell, E. B. (1959), *Colorimetric Determination of Trace Metals*. Interscience, New York.
Scheffer, E. R. (1961), in: Kolthoff, I. M., and Elving, P. J. (eds.), *Treatise on Analytical Chemistry*, Part II, Vol. 5. Interscience Publishers, New York.
Smith, I. M., Cameron, G. M., and Peterson, H. C. (1986), *Acid Recovery Cuts Waste Output. Chem. Eng. 93*, 3, 44.
Stokinger, H. E. (1962), in: Patty, F. A. (ed.), *The Metals – Titanium*, Vol. II, p. 1154. John Wiley & Sons, New York.
Titanium & Titanium Dioxide (1986), Mineral Commodity Summaries. Bureau of Mines, US Department of the Interior, Washington, D.C.
Wennig, R., and Kirsch, N. (1988), *Titanium*, in: Seiler, H. G., Sigel, H., and Sigel, A. (eds.), *Handbook on Toxicity of Inorganic Compounds*, pp. 705–714. Marcel Dekker, New York.
Williams, D. F. (1989), *Toxicological and Therapeutic Potential of Metallic Surgical Implants*. University of Surrey, Guildford, in press; see also Report E. Merian, *Swiss Chem. 12*(1–2), 25 (1990).

II.32 Tungsten

MICHAEL HARTUNG, Erlangen, Federal Republic of Germany

1 Introduction

Tungsten is a very hard metal with enormous resistance to heat and is therefore useful for many different industrial purposes. The majority of tungsten is made into cemented tungsten carbide. Only small amounts are present in food and water and little is known about the toxicity of different forms of tungsten. Tungsten and its compounds are not, however, considered an important health hazard.

2 Physical and Chemical Properties, and Analytical Methods

Tungsten (W), with an atomic mass of 183.86 and a nuclear charge number of 74, is a shiny metal with a whitish surface and density of 19.3 g/cm^3 (STOKINGER, 1967). The powder has a dull gray appearance. Tungsten occurs in oxidation states from 2 to 6, its melting point is about 3380 °C (which is the highest melting point of all metals), and its boiling point is about 5900 °C (RÖMPP, 1966). The chemistry of this element is complex, because in addition to the large number of valencies, an interesting stereochemistry and a pronounced tendency to form polynuclear complexes are factors. Its interactions with molybdenum (in enzyme activities) and with sulfate ions have been investigated (WENNIG and KIRSCH, 1988). Tungsten belongs to Group VIb of the Periodic System and has physical and chemical properties similar to molybdenum.

The main analytical methods suitable to quantify environmental tungsten are atomic absorption spectrometry in graphite vessels (NIOSH, 1977; WELZ, 1985; WENNIG and KIRSCH, 1988), inductively coupled plasma spectrometry (MONTASER and GOLIGHTLY, 1987), and neutron activation analysis (CAWSE, 1974; WESTER, 1974). Japanese analytical chemists also use polarographic catalytic or thiocyanate absorption photometry to determine tungsten content in seawater or industrial waste water (see also SOHRIN et al. 1987).

3 Sources, Production, Important Compounds, Uses, Waste Products, Recycling

The average tungsten concentration in the earth's crust is estimated to be about 0.006% (LASSNER et al., 1983), but according to SAAGER (1984) and SOHRIN et al. (1987) only about 1 ppm (lower than vanadium and molybdenum). Tungsten normally occurs naturally as tungstate, mainly as compounds such as wolframites and scheelites, found mostly in China, the Soviet Union, and Canada. The metal is obtained from finely ground tungsten ores by means of either electrolysis or alkaline separation, which precipitates the pure metal oxide. Metallic tungsten is nonmagnetic and highly resistant to chemical treatments and therefore of special importance in the tool and drill manufacturing industry. As coal and oil extraction has declined, tungsten production has also been reduced to about 20000 tons per year. To a great extent tungsten is used as tungsten carbide or as a component of high-speed and hot-worked steels, cast hard alloys, sintered hard metals, and highly heat resistant special alloys (LASSNER et al., 1983; SAAGER, 1984). Another significant sector for the use of tungsten is the lighting industry where fine wire is employed as filament for electric bulbs (SCHMIDT, 1965). Furthermore, tungsten is an integral part of X-ray tubes, and tungsten compounds serve as pigments in dyes and inks and as catalysts, WS_2 for instance (WENNIG and KIRSCH, 1988). Tungsten oxide catalysts are used in motor fuel synthesis or in catalytic converters of automobile exhaust systems (BAKER et al., 1987). About 25% of tungsten used is recycled.

4 Distribution in the Environment

Only small concentrations of tungsten have been evident so far in the atmosphere, caused primarily by industrial emissions and nuclear fall-out. The ascertained air concentrations have been lower than 1.5 µg/m^3 according to CAWSE (1974). Rain water analyses performed in the UK by the same author showed that tungsten concentrations were mainly lower than 1 µg/L and sea water concentrations may be of the order of 0.1 µg/L (see also BRULAND, 1983). According to SOHRIN et al. (1987) the concentration in Pacific waters is much lower (0.008 ppb – about 1000 times lower than that of molybdenum), since tungsten is removed more rapidly by adsorption onto ferric hydroxide, manganese oxide, and clay minerals. It seems that the determination of the concentration in the earth's crust and in sea water needs further studies, and the information by LASSNER et al. (1983) that tungsten is as frequent as nickel and copper is questionable.

5 Uptake, Absorption, Transport and Distribution, Metabolism and Elimination in Plants, Animals, and Humans

Tungsten accumulates biologically on the leaves of *Robinia* and in *Chlorella* (WENNIG and KIRSCH, 1988). The individual animal intake from food is estimated to be up to 13.0 µg per day (WESTER, 1974). NIOSH (1977) estimates that there is evidence that in the Unites States about 30000 persons are potentially exposed to tungsten and its compounds in the workplace.

From animal experiments it has been determined that one-third of the tungsten introduced as a radioactive tracer was excreted via the urine within 24 h, and approximately two-thirds remained non-absorbed and was excreted in the feces (BALLOU, 1960). Incorporated tungsten tends to be deposited in bone and spleen, with smaller amounts going to the kidney and liver. Trace amounts were found in the lung, muscle, testis, and blood (STOKINGER, 1967). In mice, tungsten is readily transported from mother to fetus via the placenta (WIDE et al., 1986). Inhalation tests with animals showed that 60% of the inhaled activity was deposited in the respiratory tract; one-third of that was incorporated into the circulation within ten days (AAMODT, 1975). The biological half-life is low (KAZANTZIS, 1986; WENNIG and KIRSCH, 1988). Serum concentrations of healthy subjects are of the order of 6 µg/L (KAZANTZIS, 1986).

6 Effects on Plants, Animals, and Humans

Sodium tungstate solutions increase growth, yields, sugar contents, and nitrogen fixation of grapes and alfalfa (WENNIG and KIRSCH, 1988; JHA, 1969). In semistatic embryolarval bioassays, performed in fishes tungsten was the least toxic metal of *coal components* (WENNIG and KIRSCH, 1988). In embryonal cytotoxicity experiments on mice tungstate inhibited cartilage production in limb bud mesenchymal cultures at concentrations similar to those found in vivo (WIDE et al., 1986).

Occupational exposure to tungsten carbide can affect the respiratory system. Two types of respiratory diseases have been noted: bronchial asthma and progressive interstitial fibrosis. Studies in animals and humans (HARTUNG, 1986) suggest that these lung diseases are not directly caused by cemented tungsten carbide, but by the cobalt content of the material (see also LASSNER et al., 1983). SCHEPERS (1955) reported findings based on long-term tests on guinea-pigs with intratracheal and inhaled dust exposure. Nevertheless, tungsten metal and tungsten carbide are relatively inert. Only tungsten carbide in combination with cobalt caused a toxic reaction in the lung. Tungstate may interact with molybdate when fed to animals (KAZANTZIS, 1986).

7 Hazard Evaluation and Limiting Concentrations

Tungsten and its compounds are not considered important health hazards. A ranking of the toxicity of tungsten and its compounds, based on animal experiments, is as follows: sodium tungstate, tungstic oxide, ammonium paratungstate, tungsten metal (GLEASON et al., 1969). The LD_{50} of sodium tungstate for 66 day old rats was about 250 µg/kg. Acute intoxication symptoms appear along with diarrhea and respiratory arrest. KRÜGER (1912) reported on patients who had a single dose intake of between 25 and 89 g tungsten without pathogenic effects. Industrial exposure is chiefly associated with the tool-manufacturing industry where tungsten is processed together with other metals. Dust exposure to tungsten is common during milling of scheelite and wolframite.

The threshold limit values (as an 8-hour TWA) for tungsten in the work environment proposed by ACGIH (1985/1986) are 5 mg/m^3 for insoluble compounds and 1 mg/m^3 for soluble compounds. If the cobalt content of tungsten carbide dust is greater than 2% the value is recommended to be limited to 0.1 mg/m^3. MAK (1987) makes no recommendation.

References

Aamodt, R. L. (1975), *Inhalation of ^{181}W Labeled Tungstic Oxide by Six Beagle Dogs*, Health Phys. 24, 519–524.
ACGIH (American Conference of Governmental Industrial Hygienists) (1985/1986), *Threshold Limit Values and Biological Exposure Indices.*
Baker, B. G., Wolf, E. E., and Leclercq, L. (1987), *Proceedings CAPOC,* three contributions. Free University, Brussels; Elsevier, Amsterdam.
Ballou, J. E. (1960), *Metabolism of ^{185}W in the Rat.* USAEC Document HW-64112. USAEC, Tennessee.
Bruland, K. W. (1983), *Trace Elements in Sea Water.* Chem. Oceanogr. 8, 209.
Cawse, P. A. (1974), *A Survey of Atmospheric Trace Elements in the U.K. 1972-73.* Atomic Energy Research Establishment Report R 7669. Her Majesty's Stationary Office, London.
Gleason, N. M., Gosslin, R. E., Hodge, H. C., and Smith, R. P. (1969), *Clinical Toxicology of Commercial Products.* Williams & Wilkins, Baltimore, Maryland.
Hartung, M. (1986), *Lungenfibrosen bei Hartmetallschleifern – Bedeutung der Cobalteinwirkung.* Hauptverband der gewerblichen Berufsgenossenschaften, Bonn.
Jha, K. K. (1969), *Effects of Vanadium and Tungsten on Nitrogen Fixation of Medicago sativa,* J. Indian Soc. Soil. Sci. 17 (1), 11–13.
Kazantzis, G. (1986), *Tungsten,* in: Friberg, L., Nordberg, G. F., and Vouk, V. B. (eds.), *Handbook on the Toxicology of Metals,* 2nd Ed., Vol. II, pp. 610–622. Elsevier Science Publ., Amsterdam.
Krüger, R. (1912), *Kolloidales Wolfram als Ersatz für Wismut bei Röntgenaufnahmen des Magen- und Darmkanals, Münch. Med. Wochenschr.* 59, 1910.
Lassner, E., Ortner, H., Fichte, M., and Wolf, U. (1983), *Tungsten and Tungsten Compounds* (in German), in: *Ullmanns Encyklopädie der technischen Chemie,* 4th Ed., Vol. 24, pp. 457–488. Verlag Chemie, Weinheim-Deerfield Beach/Florida-Basel.

Montaser, A., and Golightly, D. W. (eds.) (1987), *Inductively Coupled Plasmas in Analytical Atomic Spectrometry*, pp. 385, 606. VCH Verlagsgesellschaft, Weinheim-Basel-Cambridge-New York.

NIOSH (National Institute of Occupational Safety and Health) (1977), *Criteria for a Recommended Standard-Occupational Exposure to Tungsten Carbide*. DHEW (NIOSH) Publication No. 77-127. Cincinnati, Ohio.

Römpp (1966), *Chemie-Lexikon*. Franckh'sche Verlagshandlung, W. Keller & Co., Stuttgart.

Saager, R. (1984), *Metallic Raw Materials Dictionary* (in German), pp. 126–129. Bank von Tobel, Zürich.

Schepers, W. G. (1955), *The Biological Action of Tungsten – Cobalt – Tantalum: Studies on Experimental Pulmonary Histopathology, Arch. Ind. Health* 12, 121–146.

Schmidt, M. (1965), *Werkstoff-Handbuch Stahl und Eisen*, 4th Ed. Stahleisen, Düsseldorf.

Sohrin, Y., Isshiki, K., Kuwamoto, T., and Nakayama, E. (1987), *Tungsten in North Pacific Waters. Mar. Chem.* 22, 95–103.

Stokinger, H. E. (1967), *Tungsten*, in: Clayton, G. B., and Clayton, F. E. (eds.), *Patty's Industrial Hygiene and Toxicology*, 3rd Ed. John Wiley & Sons, New York.

Welz, B. (1985), *Atomic Absorption Spectrometry*, 2nd Ed., pp. 260, 335–336, 405ff. VCH Verlagsgesellschaft, Weinheim-Deerfield Beach/Florida-Basel.

Wennig, R., and Kirsch, N. (1988), Chapter 67 *Tungsten*, in: Seiler, H. G., Sigel, H., and Sigel, A. (eds), *Handbook on Toxicity of Inorganic Compounds*, pp. 731–738. Marcel Dekker, New York.

Wester, P. O. (1974), *Trace Element Balances in Relation to Variations in Calcium Intake, Arteriosclerosis* 20, 207–215.

Wide, M., Danielsson, B. R. G., and Dencker, L. (1986), *Distribution of Tungstate in Pregnant Mice and Effects on Embryonic Cells in-vitro, Environ. Res.* 40, 487–498.

II.33 Uranium, Thorium, and Decay Products

WERNER BURKART, Villigen, Switzerland

1 Introduction

Uranium and thorium are ubiquitous elements found in the upper layers of the earth. They exist only as unstable, radioactive isotopes which undergo a long chain of radioactive decays to end up finally as stable isotopes of lead (SEELMANN-EGGEBERT et al., 1981). Due to their comparatively short half-lives, these decay products are present in the environment only in minute quantities. Both the chemical toxicity of the long-lived uranium and thorium parent isotopes and the radiotoxicity of all actinides and unstable decay products may cause considerable hazards after enrichment or even at environmental levels (see also TRUEB, 1989).

2 Physical and Chemical Properties, and Analytical Methods

Natural uranium, the fourth member of the actinide series, has the atomic number 92 and an atomic mass of 238.04. It is a heavy (specific gravity 18.95), silvery-white metal with a melting point of 1132°C and a boiling point of 3818°C. Finely divided uranium metal is pyrophoric. Oxidation states range from +2 to +6. In aqueous solution and in the body the oxygen-containing cation UO_6^{2+} is the most stable form.

The second actinide element thorium, atomic number 90, has an atomic mass of 232.04 and a stable oxidation state of 4. The specific gravity is 11.7 and the melting and boiling points are 1750°C and 4790°C, respectively. At higher temperatures, the powdered metal may ignite even in a pure carbon dioxide atmosphere.

The chemical properties of both uranium and thorium resemble those of the rare earth elements. Their oxides are insoluble in water and alkalis, but dissolve in acids.

The uranium or thorium content of a sample can be determined by fluorometry, alpha-spectrometry, neutron activation analysis (IGARASHI et al., 1984), X-ray microanalysis with a scanning-transmission electron microscope (LANDAS et al., 1984), mass spectrometry (for transuranic pollutants such as ^{237}Np, ^{239}Pu, etc., in the environment, see, for instance, CHASTAGNER and KANTELO, 1987), and by cathodic stripping voltammetry (VAN DEN BERG and NIMMO, 1987). In most cases, measurements of environmental levels or biological materials require preliminary

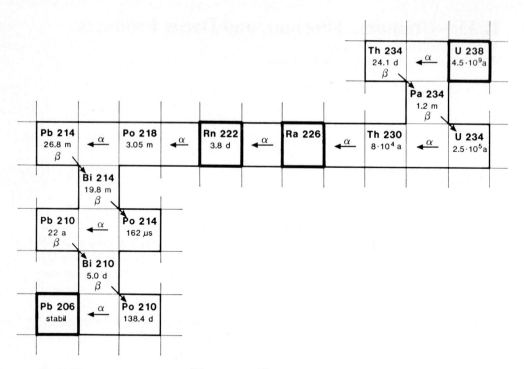

Fig. II.33-1. Decay scheme from ^{238}U to stable ^{206}Pb with half-lives and mode of decay (modified from SEELMANN et al., 1981).

sample preparations such as ashing and dissolution in acid, followed by either solvent extraction or ion exchange (BERLIN and RUDELL, 1979; SINGH et al., 1979). For uranium isotope analysis inductively coupled plasmas (ICP) may also be used (MONTASER and GOLIGHTLY, 1987). Information on the speciation of activation products has been given by BULMAN and COOPER (1986).

All decay products, with their relatively short half-lives resulting in correspondingly high specific radioactivity, i.e. high numbers of decays per second and mass unit, are best identified and measured by alpha- and gamma-spectrometry. Very low time-averaged concentrations of airborne ^{222}radon and its short-lived decay products can be determined using track etch detectors (URBAN and PIESCH, 1981).

In Fig. II.33-1, the decay chain of the natural radioactive family arising from ^{238}uranium is shown. Two other families of lesser importance for environmental exposure to radioactivity exist. Their origins are ^{232}thorium and ^{235}uranium. They finally decay to the stable lead isotopes ^{208}Pb and ^{207}Pb, respectively.

3 Sources, Production, Important Compounds, Uses, Waste Products, Recycling

3.1 Occurrence, Production

The earth's crust contains about 2.4 ppm uranium (SAAGER, 1984) and about 13 ppm thorium (VIETZKE, 1983), sea water contains about 1–3 ppb uranium (KATHREN, 1984). Typical concentrations of uranium and thorium range from 1 to 10 ppm in sandstone, shale or limestone. Granite contains up to 15 ppm uranium and up to 80 ppm thorium (KATHREN, 1984). The solubility in water and hence the migratory behavior of uranium in the lithosphere and its potential for water pollution, is strongly dependent on the oxidation state and the presence of organic chelators. Uranium(IV) which is present under anoxic conditions has a much lower solubility than U(VI) (BUNDESANSTALT FÜR GEOWISSENSCHAFTEN UND ROHSTOFFE, 1985). Higher graded uranium ores (1–6% and more) are extracted, enriched, and transformed into the hexafluoride for the selective enrichment of ^{235}U by the partial separation of the two natural uranium isotopes ^{235}U and ^{239}U (see also PLÖGER and VIETZKE, 1983). Phosphate mining and recycling of spent nuclear fuel are further sources for ^{235}U. The annual production of uranium and thorium amounts to about 40 000 t ^{235,238}U (INFCE, 1980; SAAGER, 1984) and less than 100 t ^{232}Th (MÉTIVIER, 1988).

3.2 Uses, Waste Products

Most natural uranium is mined for energy production in fission reactors (see also PLÖGER and VIETZKE, 1983; SAAGER, 1984; and FISHER, 1988). Highly enriched ^{235}uranium has military uses, either as bomb fuel for fission bombs or to ignite hydrogen bombs. Other, minor applications involve the use of natural or depleted uranium for armour-piercing shells, ship ballast and counterweights for airplanes, as negative contrast in electron microscopy, as tile glazes, and glass colors. Some actinide alloys are also candidates for electrical superconductors.

Thorium is used in incandescent mantles (so-called Welsbach mantles), the glowing part of portable gas lights, and as an alloying element with magnesium and nickel, imparting strength and creeping resistance at elevated temperatures (see also MÉTIVIER, 1988, and VIETZKE, 1983). The oxide is used to coat tungsten wires and in high quality camera lenses. It becomes a nuclear fuel by conversion to ^{233}uranium via neutron capture followed by beta-decay.

Few commerical uses exist for the radioactive decay products of uranium and thorium. The highly radiotoxic ^{226}radium was used in luminous paint on dials and in radiotherapy for the treatment of tumors. The radium daughter ^{222}radon with its half-life of 3.8 days is still used, after sealing it in minute tubes, called seeds or needles, for local irradiations in patients.

The mining and milling of uranium produce large quantities of low-level radioactive waste. The leaching of ^{226}radium by rain water into drinking water or the emanation of ^{222}radon from mill tailings may cause local contaminations. Dangerous levels of ^{222}radon and its decay products in indoor air were shown to result from the use of mine tailings as building subsoil or from gypsum produced as a by-product of phosphate mining (NCRP, 1984).

4 Distribution in the Environment, in Foods, and in Living Organisms

4.1 Emissions, Air and Water Quality, Distributions in Soils

Emissions of uranium and thorium during mining and processing are generally low. Significant increases above natural levels may occur in local aquatic systems. The decay products are of greater concern, especially the leaching of radium, and the atmospheric release of the noble gas ^{222}radon from mine tailings. The longterm control of these tailings must include the prevention of their use in construction materials for dwellings and a ban on using run-off water for irrigation. Measurable anthropogenic immissions of natural actinides and their decay products are caused by the burning of coal. Due to their extremely low solubilities, soil concentrations in the vicinity of coal-fired power plants may contain up to twice the natural concentration of about 10 ppm (WRENN et al., 1981).

Military uses (fission bombs) and accidental releases of radionuclides such as during the Chernobyl accident may lead to the contamination of large areas. However, the contribution of uranium and its decay products to the radiotoxicity of nuclear fallout is marginal (INSAG, 1986). Most of the acute and protracted dose is caused by relatively volatile and soluble fission or activation products such as ^{131}iodine, ^{137}cesium or ^{14}carbon with short, intermediate and long half-lives of 8 days, 30 years and 5730 years, respectively (EISENBUD, 1987).

4.2 Food Chains, Plants, Animals and Humans

The poor solubility and the generally low concentrations of actinides in the earth's crust lead to insignificant direct exposures and prevent an enrichment along food chains. Uptake from soil by crops is generally very low. The decreasing bioavailability from radium, uranium to thorium leads to soil-to-plant concentration ratios of $3-14 \times 10^{-3}$, 4×10^{-3} and 4×10^{-3}, respectively (GROGAN, 1985). Decay products such as radium and the volatile radon, a noble gas, may migrate more readily and lead to considerable exposures. This holds especially true for ^{238}uranium. Besides abundance, the relatively long-lived ^{226}radium and ^{222}radon isotopes of the ^{238}uranium chain are crucial factors in bringing the potential decay energy of their daughter products from the lithosphere into the biosphere. The noble gas ^{222}radon with its half-life of 3.8 days, as compared to 55 seconds for ^{220}radon arising from ^{232}thorium, may escape from subsoil, drinking water and building materials and reach critical levels in the indoor environment. Upon attachment of the radon decay products ^{214}Pb to ^{214}Po to aerosol particles, considerable fractions are deposited in the tracheobronchial and pulmonary regions of the lung and lead to a highly localized irradiation.

The actinide content of vegetables, cereals and meat is generally low. This holds also for actinide decay products with the exception of ^{210}Pb and ^{210}Po. Both these nuclides are found in the atmosphere at concentrations of up to 1 mBq/m^3 due to their indirect descendence from ^{222}Rn. Surface deposition on tobacco leaves contributes up to 50% of the ^{210}Pb and ^{210}Po body burden of smokers (NRPB, 1975). The contribution from deposited natural airborne radioactivity may also be considerable in the arctic food chain "lichen-reindeer-man".

5 Uptake, Absorption, Transport and Distribution, Metabolism and Elimination in Plants, Animals and Humans

Although uranium, thorium and all their decay products are non-essential for living organisms, the highly charged actinides form complexes with a multitude of biomolecules such as proteins, glycosaminoglycans and multivalent organic acids (citrate, malate, etc.). In view of the quite large differences in the chemical and physical properties of uranium, thorium, and their activation and decay products, it is rarely possible to use one element as a tracer for others. Since several of the actinides have more than one stable oxidation state under environmental conditions, and since these different states differ markedly in their water solubility and binding affinities to biological structures (BURKART, 1984), information on the speciation is crucial for the estimation of transfer coefficients in food webs, biokinetics in humans, and for the assessment of dose and risk (ICRP, 1979).

The transfer of uranium to the blood from the digestive tract or the lung is strongly dependent on its chemical form. Gastrointestinal transfer factors of 0.5 to 5% have been reported for the relatively soluble uranyl nitrate (HURSH, 1969). A higher value in the region of 20% is indicated for non-occupational uptake of natural uranium in the diet. For those environmental levels of intake, organically complexed uranium may be dominant (HURSH, 1973). The International Commission on Radiological Protection (ICRP) suggests a gastrointestinal transfer factor of 5% for water-soluble inorganic compounds of hexavalent uranium and 0.2% for tetravalent, relatively insoluble compounds such as UF$_4$ and UO$_2$ (ICRP, 1979). Absorption from the lung is also rapid for soluble compounds such as UF$_6$ but may take years for oxides in particulate form. The metabolic model of the ICRP (ICRP, 1979) assumes that of the uranium entering the blood, 20% and 2.3% go to mineral bone and are retained there with half-lives of 20 and 5000 days, respectively; the kidneys and all other tissues of the body together, receive 12% and 0.052% each with biological half-lives of 6 and 1000 days, respectively. The remainder of the uranium entering the blood is assumed to be excreted directly. Fig. II.33-2 shows the main pathways of soluble compounds of uranium in the human body and the biological half-lives in critical areas. Transfer

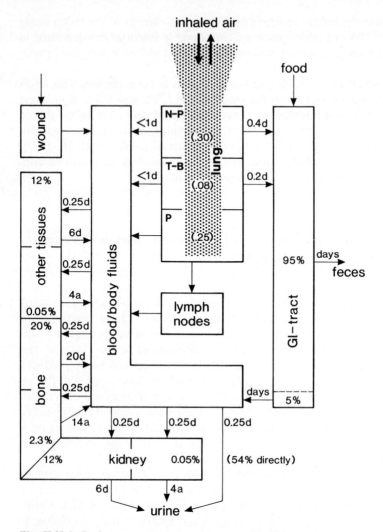

Fig. II.33-2. Body compartments, pathways, and half-lives of transfer for soluble uranium(VI). Numbers in lung air space denote fractions of inhaled activity retained in the different areas for 1 μm-diameter particles. N-P (naseopharyngeal), T-B (tracheobronchial), P (pulmonary) and lymph nodes are compartments of the respiratory tract. For the percent values in tissues, the fraction reaching the blood is set at 100%.

coefficients in the lung and in the gastrointestinal tract may vary considerably depending on the speciation, i.e. chemical form and particle size.

Thorium is poorly absorbed from both the lung and digestive tract. ICRP assumes a gastrointestinal transfer factor of 0.02%. Seventy percent of the thorium reaching the blood is translocated to the bone, 4% to the liver and 16% to all other organs and tissues of the body. The half-lives for these three compartments are 8000, 700 and 700 days, respectively (ICRP, 1979).

Table II.33-1. Daily Intake, Body Content and Resulting Dose for Uranium, Thorium and Selected Decay Products (from ICRP, 1975; ICRP, 1979; UNSCEAR, 1982, and BURKART, 1986)

	Daily Intake	Body Content	Dose in µSv/a	% of Total Dose[a]
$^{235/8}$Uranium	1.97 µg/0.023 Bq	90 µg/1.12 Bq	10	0.25
^{230,2}Thorium	3 µg/0.03 Bq	40 µg/0.45 Bq	10	0.25
^{224,6,8}Radium	2.3 pg/0.08 Bq	31 pg/1.1 Bq	20	0.5
^{222}Radon and short-lived daughter products[b]	–/900 Bq	–/30 Bq	1600	40
^{210}Lead[c]	0.44 mg/0.1 Bq	120 mg/45 Bq	70	1.75
^{210}Polonium	–/0.1 Bq	–/40 Bq	60	1.5
External radiation from U and Th series			230	5.75
Total			2000	50

[a] Based on an annual total dose of 4 mSv (400 mrem) H_{eff} from all natural and artificial radiation sources
[b] No accumulation due to half-lives shorter than 1 day
[c] Mostly in the form of stable isotopes

Radium is taken up more readily in the digestive tract due to its higher solubility as an alkaline earth element. Twenty percent is assumed to enter the blood compartment. Bone is the critical organ with a biological half-life for radium in the range of 20 years. Since the decay of radium leads to the noble gas radon with a physical half-life of 3.8 days, most of the radioactivity of the daughter product escapes from the body before further decays occur.

Upon inhalation of the aerosol-bound, short-lived ^{222}radon daughters ^{218,214}polonium, ^{214}lead and ^{214}bismuth, decay and exposure occur near to the place of deposition because the clearance mechanisms of the lung are slow compared to the physical half-lives involved (less than 30 minutes) (BURKART, 1983a, b). Table II.33-1 gives the typical intake values and the resulting body burdens and doses for the natural actinides and some decay products. In addition, the radiation field of the uranium and thorium series in the ground result in annual external doses of 90 and 140 µSv H_{eff}, respectively (UNSCEAR, 1982). H_{eff} stands for effective dose equivalent, a quantity defined by ICRP (1975) as the sum of the dose equivalents in the different organs exposed multiplied with the organ-specific weighting factors. The weighting factor represents the proportion of the detriment associated with the tissue to the total detriment when the body is irradiated uniformly. In other words, H_{eff} converts nonuniform radiation doses into the corresponding whole body dose.

6 Effects on Plants, Animals and Humans

The long-lived uranium and thorium isotopes ^{235}U, ^{238}U and ^{232}Th are both chemotoxic and radiotoxic (see also BERLIN and RUDELL, 1986; FISHER, 1988; MÉTIVIER, 1988), whereas all other isotopes and decay products with their much shorter half-lives are only critical due to the ionizing radiation emitted during their radioactive decay. This is the result of the inverse relationship between the specific radioactivity (Bq per kg) and the half-life of a radionuclide.

The dose-effect relationships for the chemical and the radiological toxicity of actinides are basically different. In general, the chemical hazard displays a steep, dose-effect function with a threshold. This also holds for acute effects from high radiation doses from therapeutic or accidental exposures which cause widespread cell death and loss of tissue functions. However, ionizing radiation may cause additional changes, so-called stochastic effects, which are based on subtle changes in the genome of cells. Experimental results and theoretical considerations suggest a linear, non-threshold dose-effect relationship for a health detriment such as cancer from low-level alpha-irradiation (BURKART, 1988).

6.1 Effects on Microorganisms and Plants

No chemotoxic or radiotoxic effects on microorganisms or plants are known from exposure to environmental radioactivity from natural and industrial sources of uranium, thorium and their decay products.

6.2 Acute Effects on Animals and Humans
(see also BEIR, 1988; UNSCEAR, 1988)

For humans and animals, uranium and its salts are highly toxic. Dermatitis, renal damage and acute arterial lesions may occur. Uranyl compounds readily complex with the phosphate containing mineral matrix of bone (RICH, 1970). The renal toxicity of uranium in animals and man is caused by the precipitation of hexavalent U in the proximal kidney tubules in the process of clearance. The resulting tissue damage leads to kidney failure and the emergence of proteins, glucose and creatinine in the urine (MORROW et al., 1982). Acute intoxication may lead to irreversible damage and to death due to renal dysfunction. Bicarbonate, which promotes the formation of an uranyl-bicarbonate complex, and chelating agents such as diethylene-triaminepentaacetic acid (DTPA) were shown to lessen the nephrotoxic effect of uranium (HODGE, 1973; CATSCH and HARMUTH, 1979).

In the context of this knowledge, acute radiation effects may only occur after the external exposure to or the ingestion of highly concentrated uranium decay products such as ^{226}radium. These include effects such as nausea, diarrhea and vomiting which occur during the first day. General weakness and epilation may follow after weeks. At acute doses above 2 Sv the so-called bone marrow syndrome becomes apparent and at even higher doses the gastrointestinal syndrome will emerge. These syndromes are caused by the arrest of proliferation of the stem cells needed to replenish the short-lived circulating lymphocytes and epithelial cells of the intestinal lining. The breakdown of the immune defense and of the barrier between body fluids and intestinal contents will lead to death in weeks or days, respectively (BURKART, 1988).

6.3 Mutagenic, Carcinogenic and Teratogenic Effects

All substances emitting ionizing radiation have to be considered mutagens and carcinogens. At higher doses, ionizing radiation is clearly teratogenic (COGGLE, 1983) and may also cause developmental defects leading to impaired brain functions, i.e., severe mental retardation (OTAKE et al., 1987). In humans, late effects from thorium and radium are known. Thorium toxicity is well known from the follow-up of more than 10000 patients injected with thorotrast, a colloidal suspension of thorium dioxide having excellent contrast properties in radiography. Increased incidence of leukemia, bone sarcomas and chromosomal aberrations are due to the radiotoxicity of thorium whereas liver diseases may by partially due to the chemotoxic action of thorium. In thorium workers effects of this actinide on peripheral lymphocytes and on liver function were also found (FARID and CONIBEAR, 1983; SERIO, 1983).

Unstable decay products of U and Th exceed their parents in their importance both in occupational and non-occupational environments as major contributors to the radiation dose. They produce only radiological effects. Three important exposure pathways exist. Radium in drinking water leads to exposure of the skeleton due to its alkaline earth behavior. In the radium-dial painting industry, several dozen cases of bone sarcomas and head carcinomas were traced to the exposure to radium at the workplace (ROWLAND et al., 1978). Radon and its short-lived decay products found ubiquitously in mining and the indoor environment, as well as polonium and lead in cigarette smoke, result primarily in the irradiation of the lung. Lung cancer due to elevated levels of radon decay products in mines is probably the best quantified occupational disease (THOMAS et al., 1985; McMICHAEL, 1989). However, repeated claims that natural radioactivity is the ultimate carcinogen of cigarette smoke remain largely unproven at this time (MARTONEN et al., 1987).

7 Risk Assessment and Limiting Concentrations

Today, radon and its decay products in indoor air are by far the most important contributors to the exposure of the public to ionizing radiation and may be responsible for as much as 10 to 20% of the lung cancer cases (JACOBI, 1984; EDLING et al., 1986; HARLEY and HARLEY, 1986; ICRP, 1987; BEIR, 1988). Synergism with smoking was found in several studies, but recent assessments of the data indicate that the two agents may act in part additively. Smoking may also reduce the latent period

Table II.33-2. Critical Effects and Occupational Exposure Limits for Uranium, Thorium and Decay Products for Different Exposure Pathways

Chemotoxicity

Element	Target Organ	Critical Effect	Exposure Limits mg/m^3
Uranium	Kidneys	Renal failure Metallocarcinogenic?	0.2 (USA: TWA)[a] 0.25 (FRG: MAK)[b]
Thorium		Metallocarcinogenic?	0.05

Radiotoxicity

Nuclide	Target Organ	Critical Effect	Exposure Limits (Bq)	
			Inhalation	Ingestion
^{238}Uranium	Bone	Sarcomas	3×10^4	5×10^5
^{232}Thorium	Bone	Sarcomas	1×10^2	3×10^4
^{226}Radium	Bone	Sarcomas	2×10^4	7×10^4
^{222}Radon[c]	Lung	Lung cancer	7×10^6	–
^{210}Lead	Skeleton, liver	Cancer	9×10^3	2×10^4
^{210}Polonium	Whole body	Cancer	2×10^4	1×10^5

[a] The short-term exposure limit (STEL) is set at 0.6 mg/m^3 (ACGIH, 1986). A STEL is defined as a 15 minute time-weighted average exposure
[b] The short-term (30 minutes) peak level is limited at 2.5 mg/m^3 (MAK, 1987)
[c] Includes dose contributions from short-lived daughters

between exposure and the first clinical signs of bronchogenic cancer (GINEVANS and MILLS, 1986; SACCOMANNO et al., 1986). Environmental levels of radium in public water supplies, in mineral waters from geological areas rich in uranium but poor in sulfates, and in some foods such as Brazil nuts may also contribute measurably to the exposure of the public to ionizing radiation (NSF, 1977; MAYNEORD et al., 1960). It has to be kept in mind, however, that with the exception of radon, the estimated health problems associated with elevated levels of natural radiation are too small to be detected even in large epidemiological studies (BEIR, 1980).

Table II.33-2 lists the occupational limits for uranium, thorium and some critical decay products. For the limit based on the chemotoxicity of uranium, American and German values for different exposure situations are given. Due to the dominance of the radiotoxicity in the case of thorium, permissible concentration limits are generally based only on the radioactivity. However, eastern block countries have set a threshold limit for thorium in workroom air (INTERNATIONAL LABOUR OFFICE, 1980). The occupational limits for exposure to radionuclides are taken from the ICRP-Publication 30 (1979). One ALI (annual limit of intake) will result in an effective dose equivalent of 50 mSv (5 rem) H_{eff}. The risk of an annual dose of 50 mSv is considered low and comparable to the risk in other safe industries. The numerical values of this philosophy of protection are at present disputed because recent developments in the dosimetry and epidemiological follow-up of atomic bomb survivors of Hiroshima and Nagasaki for cancer risk up to 1985. These led to an increase in the risk estimates for this population, which is the most important cohort for the study of the effects of ionizing radiation (BEIR, 1980, 1988; PRESTON and PIERCE, 1987; UNSCEAR, 1988; MCMICHAEL, 1989). For non-occupational settings, i.e., for the general public, the ICRP dose limit per year is set at 1 mSv or 2% of the occupational limit.

References

ACGIH (American Conference of Governmental Industrial Hygienists) (1986), *TLV's Threshold Limit Values for Chemical Substances in the Work Environment.* ACGIH, Cincinnati, Ohio.
BEIR (Committee on the Biological Effects of Ionizing Radiation) (1980), *The Effects on Populations of Exposure to Low Levels of Ionizing Radiation.* National Academy Press, Washington, D.C.
BEIR (Committee on the Biological Effects of Ionizing Radiation) (1988), *Health Risks of Radon and Other Internally Deposited Alpha-Emitters.* National Academy Press, Washington, D.C.
Berlin, M., and Rudell, B. (1979), Chapter 26: *Uranium*, in: Friberg, L., Nordberg, G. F., and Vouk, V. B. (eds.), *Handbook on the Toxicology of Metals*, pp. 647–658. Elsevier/North-Holland, Amsterdam.
Bulman, A., and Cooper, J. R. (1986), *Speciation of Fission and Activation Products in the Environment*, C.E.C. Elsevier Applied Science Publishers, London-New York.
Bundesanstalt für Geowissenschaften und Rohstoffe (1985), *Geochemischer Atlas.* E. Schweizerbart, Stuttgart.
Burkart, W. (1983a), *Assessment of Radiation Dose and Effects from Radon and its Progeny in Energy-efficient Houses. Nucl. Tech. 60*, 114–123.
Burkart, W. (1983b), *Dose Response in Radiation-induced Human Carcinogenesis: Accumulated Data do not yet Solve the Enigma. Nucl. Tech. 62*, 81–93.
Burkart, W. (1984), *Gastrointestinal Absorption of Actinides: A Review with Special Reference to Primate Data*, EIR-Report 509, CH-5303 Würenlingen.
Burkart, W. (1986), *An Estimation of Radiation Exposure and Risk from Air-tightening of Homes in the Alpine Area with Elevated Radon Source Strength. Environ. Int. 12*, 49–53.
Burkart, W. (1988), Chapter 73: *Radiotoxicity*, in: Seiler, H. G., Sigel, H. and A., *Handbook on Toxicity of Inorganic Compounds*, pp. 805–827. Marcel Dekker, Inc., New York.
Catsch, A., and Harmuth-Hoene, A. E. (1979), in: *The Chelation of Heavy Metals – International Encyclopedia of Pharmacology and Therapeutics*, p. 107. Pergamon Press, Oxford.
Chastagner, P., and Kantelo, M. V. (1987), *Determination of Transuranic Pollutants in the Environment by Thermal Ionization Mass Spectrometry.* Contribution to the 17th Annual IAEAC Symposium on the Analytical Chemistry of Pollutants, Jekyll Island, Georgia, USA.
Coggle, J. E. (1983), *Biological Effects of Radiation.* Taylor & Francis Inc., New York.
Edling, C., Wingren, G., and Axelson, O. (1986), *Quantification of the Lung Cancer Risk from Radon Daughter Exposure in Dwellings – an Epidemiological Approach. Environ. Int. 12*, 55–60.
Eisenbud, M. (1987), *Environmental Radioactivity.* Academic Press, Orlando.
Farid, I., and Conibear, S. A. (1983), *Hepatic Function in Previously Exposed Thorium Refinery Workers as Compared to Normal Controls from the Health and Nutrition Survey. Health Phys. 44*, 221–230.
Fisher, D. R. (1988), Chapter 68: *Uranium*, in: Seiler, H. G., Sigel, H. and A., *Handbook on Toxicity of Inorganic Compounds*, pp. 739–748, Marcel Dekker, Inc., New York.
Ginevans, M. E., and Mills, W. A. (1986), *Assessing the Risks of Rn Exposure: the Influence of Cigarette Smoking. Health Phys. 51*, 163–174.

Grogan, H. A. (1985), *Concentration Ratios for BIOPATH: Selection of the Soil-to-Plant Concentration Ratio Database.* EIR-Report 575, CH-5303 Würenlingen.

Harley, N. H., and Harley, J. (1986), *Risk Assessment for Environmental Exposures to Radon Daughters.* Environ. Int. 12, 39–44.

Hodge, H. C. (1973), *A History of Uranium Poisoning (1824–1942)*, in: Hodge, H. C., Stannard, J. N., and Hursh, J. B. (eds.), *Uranium, Plutonium, Transplutonium Elements*, pp. 5–68. Springer Verlag, Berlin.

Hursh, J. B., and Spoor, N. L. (1973), *Data on Man*, in: Hodge, H. C., Stannard, J. N., and Hursh, J. B. (eds.), *Uranium, Plutonium, Transplutonium Elements*, pp. 197–239. Springer Verlag, Berlin.

Hursh, J. B., Neuman, W. R., Toribara, T., Wilson, H., and Waterhouse, C. (1969), *Oral Ingestion of Uranium by Man.* Health Phys. 17, 619–621.

ICRP, International Commission on Radiological Protection (1975), *Task Group Report on Reference Man, ICRP Publ. 23.* Pergamon Press, Oxford.

ICRP, International Commission on Radiological Protection (1979), *Limits for Intakes of Radionuclides by Workers, ICRP Publ. 39*, Part 1–3. Pergamon Press, Oxford.

ICRP, International Commission on Radiological Protection (1987), *Lung Cancer Risk from Indoor Exposures to Radon Daughters, ICRP Publ. 50.* Pergamon Press, Oxford.

Igarashi, Y., Seki, R., and Ikeda, N. (1984), *Determination of Uranium in Human Tissues by the Fission Track Method.* Radioisotopes 33, 55–59.

INFCE, International Nuclear Fuel Cycle Evaluation (1980), *Fuel and Heavy Water Availability*, IAEA STI/PUB/534, Vienna.

INSAG (1986), *Summary Report: Post-Accident Review Meeting on the Chernobyl Accident*, IAEA, Vienna.

International Labour Office (1980), *Occupational Exposure Limits for Airborne Toxic Substances, Occupational Safety and Health Series No. 37*, 202–203.

Jacobi, W. (1984), *Expected Lung Cancer Risks from Radon Daughter Exposure in Dwellings*, in: Berglund, B., Lindvall, T., and Sundell, J. (eds.), *Indoor Air*, Vol. 1, pp. 31–42.

Kathren, R. L. (1984), *Radioactivity in the Environment.* Harword Academic, London.

Landas, S., Turner, J. W., Moore, K. C., and Mitros, F. A. (1984), *Demonstration of Iron and Thorium in Autopsy Tissues by its X-Ray Microanalysis.* Arch. Pathol. Lab. Med. 108, 231–233.

MAK (1987), *Report No. XXIII, DFG, Maximum Concentrations at the Workplace.* VCH Verlagsgesellschaft, Weinheim-Basel-Cambridge-New York.

Martonen, T. B., Hofmann, W., and Lowe, J. E. (1987), *Cigarette Smoke and Lung Cancer.* Health Phys. 52, 213–217.

Mayneord, W. V., Radley, J. M., and Turner, R. C. (1960), *The Origin and Metabolism of Radioactive Material in the Human Body.* Adv. Sci. 16, 363–387.

McMichael, A. J. (1989), *The Contribution of Epidemiology to Understanding the Mechanisms of Action of Carcinogens.* Proceedings Vth International Congress of Toxicology, Brighton, pp. 25–35. Taylor & Francis, London-New York-Philadelphia; see also Reports E. Merian, *EUROTOX Newslett.* 1/90, 7–8 and *Swiss Chem.* 12(1–2), 22–23 (1990).

Métivier, H. J. (1988), Chapter 63: *Thorium*, in: Seiler, H. G., Sigel, H. and A., Handbook on Toxicity of Inorganic Compounds, pp. 689–694. Marcel Dekker, Inc., New York.

Montaser, A., and Golightly, D. W. (1987), *Inductively Coupled Plasmas in Analytical Atomic Spectrometry*, pp. 255ff, 611ff. VCH Verlagsgesellschaft, Weinheim-Basel-Cambridge-New York.

Morrow, P. E., Leach, L. J., Smith, F. A., Gelein, R. M., Scott, J. B., Beiter, H. D., Amato, F. J., Picano, J. J., Yuile, C. L., and Consler, T. G. (1982), *Metabolic Fate and Evaluation of Injury in Rats and Dogs Following Exposure to the Hydrolysis Products of Uranium Hexafluoride*, NUREG/CR-2268, New York.

NCRP, National Council on Radiation Protection and Measurements (1975), *Natural Background Radiation in the United States,* NCRP Report No. 45, Washington, D.C.

NCRP, National Council on Radiation Protection and Measurements (1984), *Evaluation of Occupational and Environmental Exposures to Radon and Radon Daughters in the United States*, NCRP Report No. 78, Washington, D.C.

NSF, National Science Foundation (1977), *Drinking Water and Health*, Vol. 1, pp. 858–896. National Academy Press, Washington, D.C.

Otake, M., Yoshimaru, H., and Schull, W. J. (1987), *Severe Mental Retardation among the Prenatally Exposed Survivors of the Atomic Bombing of Hiroshima and Nagasaki: A Comparison of the Old and New Dosimetry Systems*. Radiation Effects Research Foundation RERF Technical Report 16–87, Hiroshima.

Plöger, F., and Vietzke, H. (1983), *Uranium and Uranium-Compounds* (in German), in: *Ullmanns Encyklopädie der technischen Chemie*, 4th Ed., Vol. 23, pp. 457–490. Verlag Chemie, Weinheim-Deerfield Beach/Florida-Basel.

Preston, D. L., and Pierce, D. A. (1987), *The Effects of Changes in Dosimetry on Cancer Mortality Risk Estimates in the Atomic Bomb Survivors*, Radiation Effects Research Foundation, Techn. Rep. RERF 9-87, Hiroshima.

Rich, K. E., Agarwal, R. T., and Feldman, I. (1970), *Chelation of Uranyl Ions by Adenine Nucleotides: IV. Nuclear Magnetic Resonance Investigations, Hydrogen-1 and Phosphorus-31 of the Uranyl-adenosine 5'-diphosphate and Uranyl-adenosine 5'-triphosphate Systems. J. Am. Chem. Soc.* 92, 6818.

Rowland, R. E., Stehney, A. F., and Lucas, H. F. (1978), *Dose-Response Relationships for Female Radium Dial Workers. Radiat. Res.* 76, 368–383.

Saager, R. (1984), *Uranium*, in: *Metallic Raw Materials Dictionary* (in German), pp. 165–169. Bank von Tobel, Zürich.

Saccomanno, G., Yale, C., Dixon, W., Auerbach, O., and Huth, G. C. (1986), *An Epidemiological Analysis of the Relationship Between Exposure to Radon Progeny, Smoking and Bronchogenic Carcinoma in the Uranium-mining Population of the Colorado Plateau 1960–1980. Health Phys.* 50, 605–618.

Seelmann-Eggebert, W., Pfennig, G., Münzel, H., and Klewe-Nebenius, H. (1981), *Chart of Nuclides*. Gersbach & Sohn Verlag, München.

Serio, C. S. (1983), *Measurement of Lymphoblastogenic Activity from Thorium Workers. Int. J. Radiat. Biol.* 44, 251–256.

Singh, N. P., Ibrahim, S. A., Cohen, N., and Wrenn, M. E. (1979), *Solvent Extraction Method for Determination of Thorium in Soft Tissues. Anal. Chem.* 51, 207–213.

Thomas, D. C., McNeill, K. G., and Dougherty, C. (1985), *Estimates of Lifetime Lung Cancer Risks Resulting from Radon Progeny Exposure. Health Phys.* 49, 825–846.

Trueb, L. (1989), *Two-hundred Years Uranium* (in German). *Neue Zürcher Zeitung, Forschung und Technik*, No. 224, p. 91 (27 September), Zürich.

UNSCEAR (United Nations Scientific Committee on the Effects of Atomic Radiation) (1982), *Ionizing Radiation: Sources and Biological Effects, Report to the General Assembly*. United Nations, New York.

UNSCEAR (United Nations Scientific Committee on the Effects of Atomic Radiation) (1988), *Sources, Effects and Risks of Ionizing Radiation, Report to the General Assembly*. United Nations, New York.

Urban, M., and Piesch, E. (1981), *Low Level Environmental Radon Dosimetry with a Passive Track Etch Detector Device. Radiat. Prot. Dosim.* 1, 97–109.

van den Berg, C. M. G., and Nimmo, M. (1987), *Direct Determination of Uranium in Water by Cathodic Stripping Voltammetry. Anal. Chem.* 59, 924–928.

Vietzke, H. (1983), *Thorium and Thorium Compounds* (in German), in: *Ullmanns Encyklopädie der technischen Chemie*, 4th Ed., Vol. 23, pp. 227–241. Verlag Chemie, Weinheim-Deerfield Beach/Florida-Basel.

Wrenn, M. E., Singh, N. P., Cohen, N., Ibrahim, S. A., and Saccomano, G. (1981), *Thorium in Human Tissues*, US Report, NUREG/CR-1227, New York.

II.34 Vanadium

RICHARD U. BYERRUM, East Lansing, Michigan, USA

1 Introduction

Vanadium, a commonly occurring element, has been shown to be nutritionally essential for the rat and for chicks. It is presumed therefore to be essential also for other animals including man. However, no deficiency symptoms for vanadium have been noted in man. Several organisms are known to concentrate vanadium but the role of the element in these organisms is also unknown.

2 Physical and Chemical Properties, and Analytical Methods

Vanadium (V) is element 23 in the Periodic Table of Elements and has an atomic mass of 50.94. Metallic vanadium has a density of 6.11 g/cm^3, a melting point of 1890 °C, and a boiling point of about 3000 °C. Vanadium can exist in oxidation states of II, III, IV, and V, and its biochemistry is therefore complex (see, e.g., WENNIG and KIRSCH, 1988). Naturally occurring vanadium consists of over 99% ^{51}V and the remainder as ^{50}V.

Colorimetric methods were formerly used for analysis of vanadium, e.g., the 8-hydroxyquinoline method. Presently physical methods utilizing neutron activation, emission spectrometry, mass spectrometry, atomic absorption and polarography are used (see also WELZ, 1985; MONTASER and GOLIGHTLY, 1987; SCHALLER 1988). Most recently either energy dispersive X-ray analysis or wavelength-dispersive X-ray analysis has been used particularly since the location of vanadium in a sample can be determined as well as its concentration. Also of interest is a sensitive photometric method using gallic acid for blood (KELM and SCHALLER, 1978) and for urine (HENSCHLER, 1978). Because of different behavior in the compartments, vanadium losses must be avoided during sampling, for instance, when filtering waste water or groundwater (McKINNEY, 1987). Speciation studies are even more important than analysis just for elements.

3 Sources, Production, Important Compounds, Uses, and Waste Products

(see also FICHTE et al., 1983; SAAGER, 1984; ANONYMOUS, 1987)

Vanadium compounds are widely distributed in the earth's crust and are present at an average concentration of 100 ppm of vanadium — roughly equal to the concentration of zinc and nickel. As far as is known vanadium metal does not occur naturally but the element does exist as the sulfide or in oxidized form. A few ore deposits containing vanadium are mined commercially, and in almost every case, the recovery of vanadium is associated with some other valuable products. An example is the uranium-vanadium ore carnotite ($K_2O \cdot 2UO_3 \cdot V_2O_5 \cdot 3H_2O$) obtained in Australia and the United States. In addition, roscoelite, a magnesium-iron-vanadium ore; vanadinite, containing lead and vanadium; mottramite, a lead-copper-vanadium ore; and descloizite, a lead-zinc-vanadium ore are all commercial sources of vanadium. In none of these ores is vanadium present in more than 3%. The largest resources exist in South Africa and in the Soviet Union (SAAGER, 1984).

Vanadium is recovered from the ores by pulverizing, mixing the pulverized material with sodium chloride to give 6–10% salt, and heating the mixture to about 850 °C. Under these conditions oxidized vanadium is converted to sodium metavanadate ($NaVO_3$), which is highly soluble in water. The heated mixture is then leached with hot water, and the leachate is acidified to pH 1–3 resulting in a precipitate of sodium metavanadate. This material when dried and fused with heat gives vanadium pentoxide which is the compound sold commercially for further use of vanadium in industry. About 50000 t V_2O_5 are annually produced, about 30000 t in the Western World, and about 20000 t in the Soviet Union and in China (SAAGER, 1984).

Vanadium also occurs in small amounts in fossil fuels where it exists as a porphyrin complex. Oils from Venezuela, Angola, California and Iran may contain more than 0.1‰ V (up to 1‰). Indonesian, Libyan and West African oils contain practically none (TISSOT and WELTE, 1984). The vanadium is concentrated in the ash when these fuels are burned. Such ash has also been used as a commercial source of vanadium pentoxide (V_2O_5). Pollution control equipment installed where fuel oil is burned has greatly reduced the amount of vanadium released into the air as fly ash.

A major commercial use of vanadium has been in steel production. Since vanadium is transformed into its carbide, fine perlite lamellas are produced (ANONYMOUS, 1987). Vanadium steel which contains from 0.1–3% vanadium depending on the use to which steel is put, is tough, strong, and heat resistant and withstands strain, vibration, and shock. In a conventional steel-smelting operation very little escape of vanadium compounds occurs in fume or airborne dust since the metal dissolves readily in molten iron and has negligible vapor pressure at the temperatures used. Vanadium enriched slags from steel making, however, may be deposited in the open on the ground or used as landfill and would be subject to rain and groundwater drainage. Little, however, is known about water contamination by vanadium from such sources.

Vanadium is also a major alloying element in high-strength titanium alloys ordinarily being present at a level of 4%. Treatment with aqueous solutions of strong acid and with fused salts (usually sodium hydroxide) is necessary in fabrication of titanium-vanadium alloy products. The sludge from the fused salt treatment and the spent acid are usually mixed to neutralize each other resulting in a residue containing a high concentration of vanadium halide salts. Such salts could introduce vanadium into the biosphere if care is not taken in disposal. Small amounts of vanadium have also been alloyed with copper, chromium, and aluminum to give products which are useful for special purposes.

Another important use of vanadium is as a catalyst in a variety of reactions. Vanadium pentoxide put on an inert support material is the principal catalyst used in oxidation of SO_2 to SO_3 in the production of sulfuric acid. In addition, vanadium oxychloride, tetrachloride, and triacetylacetonate are used as polymerization catalysts for soluble copolymers of ethylene and propylene. These polymers in the reaction vessel are viscous liquids which can trap the vanadium catalysts and result in a vanadium content of as much as 500 ppm in products used for packaging of food and of pharmaceuticals. Some contamination of these foods or medicinals could result but there is little information about such a possibility. Furthermore, disposal of spent catalysts also could serve as a point source for contamination of the biosphere with vanadium.

4 Distribution in the Environment, in Foods, and in Living Organisms
(see also NRIAGU and DAVIDSON, 1986)

Small amounts of vanadium compounds are found in air where there is no known anthropological contamination. Such concentrations were found to be in the range of $0.02-0.08$ ng/m^3 above the Pacific Ocean east of Hawaii and in northern Canada. About 65000 t vanadium per year naturally enter the earth's atmosphere. This vanadium is assumed to be derived from continental dust produced by wind erosion of soil and rock or from marine aerosols consisting of small particles derived from ocean spray. Over populated areas the vanadium concentration is often greater than over unpopulated regions. In the eastern United States, for instance, concentrations of vanadium ranging from about $450-1300$ ng/m^3 have been observed. In Switzerland deposition is of the order of 1.2 mg V/m^2/a (WEHRLI, 1987).

Furthermore, concentrations of vanadium are usually higher during cold periods than when the weather is warm. The increases in vanadium were assumed to be primarily from fly ash formed during burning of fossil fuels. All coal and oil has some vanadium present. Burning of these fuels during the past several years caused about 110000 t V/a to enter the atmosphere globally. Burning of other materials, such as natural gas, which contains no vanadium or wood which contains less than 2 ppm vanadium would not contribute significant amounts of vanadium to air.

The concentration of vanadium in sea water is of the order of 2 µg/L (mainly in the form of vanadates) but is about 4 times higher in the deap sea compared to the surface (BRULAND, 1983). Even though this concentration is low, the oceans would contain about 7.5×10^{12} kg of vanadium. Geographic differences in ground- and surface water concentrations would be expected because of leaching of rocks and soil or from leaching by rain water of vanadium containing materials disposed of after industrial use. A number of studies have shown vanadium concentrations ranging from 0.3 to about 200 µg/L in a variety of ground- and surface waters including those used for drinking. Additional discussion of vanadium in water has been given by FÖRSTNER and WITTMANN (1979). Weathering is an important source for natural redistribution (probably about 2400000 t V/a, WEHRLI, 1987).

Speciation of vanadium – in its different oxidation states – plays a great role in environmental chemistry, transports, and biochemistry (WEHRLI, 1987). The hydroxo complex $VO(OH)^+$ determines absorption kinetics (the two inner-sphere surface complexes $\geqslant Al-O-VO^+$ and $(\geqslant Ti-O)_2VO$ allow a quantitative description of the adsorption equilibria). Where vanadyl(IV) exists in the environment, it will show a strong tendency to be absorbed at the particulate phase. For instance, in the Lake of Zurich vanadium is mainly found in its (IV)-form, which is not very soluble, and sedimentation is of the order of 15 mg V/m^2/a. Vanadium environmental chemistry is governed by redox potentials, by hydrolysis, and by complex formation possibilities. $VO(OH)^+$ has a higher reactivity with molecular oxygen in aqueous solution than the mentioned inner-sphere surface complexes, while the half-life of the vanadyl aquion VO^{2+} is rather high (more than 8 years). Vanadyl(IV) in organic complexes is also protected against oxidation by molecular oxygen. Oxidation half-times at surfaces are only about 18 hours.

Since vanadium is widely occurring in nature it is not surprising to find it in food. Because of its varying concentration in soil, plants grown in those soils also have varying concentrations of vanadium. Usually the concentration in vegetables ranges between 0.5 to 2 ppb with an average concentration of about 1 ppb (dry weight) (ADRIANO, 1986). Among the vegetables, radish, parsley, and potato can accumulate relatively more vanadium (SOREMARK, 1967). Thirty-five plant species collected in several locations in the Rocky Mountains of the USA ranged in concentration of vanadium from none detectable to 6.6 ppm (dry weight). These same plants were growing in soils ranging from 0.65 to 98 ppm of vanadium (PARKER et al. 1978). Some plant products, such as oil rich in linoleic acid or sunflower oil, contain a much higher concentration of vanadium (0.5 to 5 mg/100 mg).

Seafood, liver, and gelatin contain relatively high quantities of vanadium, (2.4 – 44 ppb), and milk samples from several locations contained 0.4 ppb. Saltwater fish contained about 0.03 mg/kg (BOWEN, 1979). Veal and pork were found to contain 0.1 ppb in the ash of these meats. Data concerned with the vanadium content of foods is sparse at best, but several conclusions might be drawn. Vanadium is present in both animal and plant foods. The amount in a particular plant is probably related to some extent to the vanadium concentration in soil, and the vanadium content of animal produced food is probably related directly to the vanadium intake. Seafood generally is relatively high in vanadium content. The average concentration of vanadium in food consumed by man is a few ppb.

Vanadium is also found in high concentrations in at least one species of mushroom and one class of animals: The fly agaric mushroom *Amanita muscaria* contains about 100 times more vanadium than other species of mushrooms or higher plants. The vanadium level in these mushrooms seems unrelated to the vanadium concentration of the soil in which they are growing. The vanadium is present as an organic complex given the name amavanadine, and its structure has been determined (KNEIFEL and BOYER, 1973; *Chem. Unserer Zeit*, 1988). The physiological function of this vanadium compound is unknown. Vanadium is present in high concentration in a number of ascidians (sea squirts) but not in all. In some species of sea squirts the element is found in the blood along with other pigments and is in the form of a vanadium-protein complex. The yellow blood pigment tunichrome has recently been discussed in *Chem. Unserer Zeit* (1988). In other species vanadium in a trivalent state occurs in green blood cells called vanadocytes and appears to exist as a pyrrole complex. These cells also contain an unusually large amount of sulfuric acid. The biological role of vanadium is unknown, but it does not seem to be involved in oxygen transport. The vanadium concentration in these ascidians is 10000 times greater than in the sea water in which the animals are growing, and is at a level of about 2 mg/g wet weight of animal. Some species of holothurians (sea cucumbers) and at least one species of molluscs also concentrate vanadium. Some *Azotobacter* bacteria contain vanadium in nitrogenases (*Chem. Unserer Zeit*, 1988). Other invertebrates contain vanadium but at a concentration of 1 – 2 µg/g wet weight. Vertebrates also contain some vanadium at an average concentration in the range of 0.1 – 2 µg/g wet weight of tissue.

5 Uptake, Absorption, Transport and Distribution, Metabolism and Elimination in Animals and Humans

(see also XAVIER, 1986)

Vanadium compounds are absorbed in the lung and through the intestinal tract. Of course various vanadium species behave differently (WEHRLI, 1987). Vanadium included in particles entering the lung is rapidly absorbed into the blood stream. Concentrations in human blood serum in non-exposed healthy adults are between 0.02 and 0.9 µg/L (LAGERKVIST et al., 1986). On the other hand, intestinal absorption of vanadium is slight. Experiments in which rats were given radio-vanadium compounds by mouth showed intestinal absorption of only 0.5% of the amount administered. Studies in which sodium tetravanadate or ammonium vanadate were fed to man and rabbits, respectively, indicated that about 12% of the vanadium was eliminated in the urine and, therefore, was assumed to have been absorbed from the intestine. Absorbed vanadium is transported in serum mainly bound to transferrin (LAGERKVIST et al., 1986). Intracellular vanadium is most likely in the vanadyl(IV) form (LAGERKVIST et al., 1986). In cow's milk vanadium is attached to the protein lactoferrin, and this compound provides a calf with some dietary vanadium (SABBIONI and RADE, 1980).

When vanadium compounds were administered intravenously to experimental animals, about 90% was excreted in the urine and 10% in the feces. Urinary excretion is, therefore, the primary excretion route for blood vanadium. Limited experiments have been done in man in which sodium tetravanadate was administered intravenously. After a week, 81% of the vanadium injected was excreted in the urine and 9% in the feces. About 10% of the vanadium was unaccounted for and was assumed to be retained by the body. Nevertheless, vanadium seems to be relatively rapidly excreted and only a small fraction of the vanadium which enters the blood stream is deposited in the body. The human daily intake is of the order of 0.01–0.03 mg/V (WENNIG and KIRSCH, 1988).

6 Effects on Plants, Animals, and Humans

(see also WATERS, 1977; LAGERKVIST et al., 1986; regarding miscellaneous biochemical effects see also Section 4 and BRONZETTI et al., 1989)

Unfortunately vanadium species (vanadium ions, insoluble vanadium compounds, vanadyl, and vanadates) are often not distinguished in the literature. 10 to 40 mg V/L nutrient solution are toxic for plant seeds (BOWEN, 1979), whereas 100–500 µg/L stimulate biomass production of algae (WENNIG and KIRSCH, 1988). One effect that vanadium seems to have at the molecular level is to decrease the synthesis of cystine and its reduction product cysteine. Rats fed vanadium pentoxide at a dietary concentration of 25–1000 ppm had a lowered cystine content in hair. Workers exposed to V_2O_5 in the atmosphere had a lowered cystine content of fingernails.

Coenzyme A content of rat liver was appreciably reduced when the animals were fed sodium metavanadate in the diet. One of the compounds involved in the synthesis of coenzyme A is thioethanolamine which is derived from cysteine by decarboxylation. Therefore, a decrease in cysteine caused by vanadium presumably was the reason for reduced amounts of coenzyme A.

Coenzyme A is involved in a variety of metabolic reactions in which acetate is the starting compound. One of these is the synthesis of cholesterol. Although there are conflicting data in the literature, there are several studies which indicate that vanadium compounds administered to adult humans inhibit the synthesis of cholesterol, and, therefore, may affect the occurrence of atherosclerosis. Acetyl coenzyme A is also a precursor of fatty acids which would be used to synthesize triglycerides and phospholipids. Vanadium compounds administered to rats lowered liver triglycerides. Another reaction requiring acetyl coenzyme A is the synthesis of coenzyme Q, a compound involved in electron transport in mitochondria. Administration of vanadium reduced mitochondrial coenzyme Q. Vanadium also seems to affect cation transport and uptake. In the eel it inhibits (Na^+-K^+) ATPase and is a powerful vasoconstrictor (BELL et al., 1981). In mammals it was shown that vanadium is a strong inhibitor of (Na^+-K^+) ATPases (CANTLEY et al., 1977). This inhibition is caused by the substitution of vanadate for phosphate in the ATP driven reaction (EVANS et al., 1986). Recently vanadium has also been shown to inhibit a (K^+-H^+) ATPase in rabbit colon (KAUNITZ and SACKS, 1986). VO^{2+} seems also to open the K-channel of the erythrocyte membrane (WENNIG and KIRSCH, 1988).

A relationship exists between vanadium concentration in drinking water and incidence of caries in children. The greater the vanadium concentration the fewer caries develops. It was assumed that vanadium reduced enamel solubility.

The addition of sodium orthovanadate to a highly purified diet fed to growing rats enhanced their growth by 40% at a concentration of 10 µg vanadium per 100 g of diet. Interestingly, sodium pyrovanadate ($Na_4V_2O_7$) did not improve growth when added to the deficient diet. Chicks fed a diet deficient in vanadium (10 ppb) showed reduced feather growth and blood cholesterol when compared to chicks supplemented with 2 ppm NH_4VO_3 in drinking water. Vanadium is thus an essential nutrient for chicks, rats, and nitrogen-fixing soil microorganisms (WATERS, 1977). Whether or not vanadium is an essential metal for humans remains to be seen.

Chimney-sweeps who inhaled 0.7 – 14 mg vanadium pentoxide per day in soot from the cleaning operation had a urine concentration of vanadium which fluctuated between 0.15 µg/L and 13 µg/L (normal is below 3 µg/L) (VALENTIN and SCHALLER, 1976, 1977). A study was also made of metal workers exposed to vanadium. Analysis of their blood indicated a vanadium concentration of 2.5 to 54.45 µg/L (normal is under 2.5 µg/L). The concentration in urine was even higher, and the cystine content of fingernails was lowered from normal (THÜRAUF et al., 1979). WATERS (1977) reported on irritative respiratory effects in experimental animals and humans by vanadium oxides and salts (see also LAGERKVIST et al., 1986, who reported also on other symptoms). In the Federal Republic of Germany human placentas from three large industrial areas were compared for vanadium content. The placentas contained an average of about 1.3 µg vanadium. In the Ruhr district the concentration was higher (THÜRAUF et al., 1978).

Ammonium metavanadate is genotoxic (also without mice liver fractions) in yeast (BRONZETTI et al., 1989). It is able to enter cells where it destroys activating enzymes (interaction with cytochrome P450 and monooxygenase). WENNIG and KIRSCH (1988) have also reported on cytogenetic studies.

7 Hazard Evaluation and Limiting Concentrations

Vanadium in relatively high concentration can cause toxic symptoms in man. Accidental ingestion of vanadium, however, does not result in toxicity, presumably because of the low amount absorbed through the intestine. The route of entry of vanadium compounds most commonly seen in industrial exposures is through the respiratory system. Exposures are usually limited to areas when vanadium pentoxide is produced, in steel mills where V_2O_5 is used, and in cleaning boilers fired by oil containing vanadium. The effect of airborne vanadium compounds is principally on mucous membrane, resulting in irritation of eyes, nasal and throat mucosa, and lungs, accompanied by coughing. These symptoms disappeared after removal of individuals from contaminated areas. In areas in which vanadium containing particles contaminate air as a result of industrial and power plant combustion of oil, mucous membrane irritation does occur, but it is not known whether vanadium is the causitive agent. In the United States the threshold limit for vanadium pentoxide dust in air is 0.5 mg/m^3 and for fume (very small particles) 0.05 mg/m^3. Russian literature proposes even lower limits such as 0.1 mg/m^3 for V$_2$O$_5$ dust. In the Federal Republic of Germany the maximum allowable concentration at the workplace for V$_2$O$_5$ fine dust is also 0.05 mg/m^3, but short-term (30 min) peaks should not exceed 0.1 mg/m^3 (MAK, 1987). Concentration higher than the threshold limits may result in toxic symptoms (see also FAULKNER-HUDSON, 1964; PANEL ON VANADIUM, 1974, NRIAGU 1983).

References

Adriano, D. C. (1986), *Trace Elements in the Terrestrial Environment.* Springer Verlag, New York.
Anonymous (1987), *Vanadium* (in German), in: *Neue Zürcher Zeitung*, Forschung und Technik Nr. 137, p. 65. Zürich, 17. June.
Bell, M. V., Kelly, K. F., and Sargent, I. R. (1981), *Sci. Total Environ.* 19, 215–222.
Bowen, H. J. M. (1979), *Environmental Chemistry of the Elements.* Academic Press, New York.
Bronzetti, G., Morichetti, E., della Croce, C., del Carratore, R., Giromini, L., and Galli, A. (1989), *Vanadium: Genetical and Biochemical Investigations. Toxicol. Environ. Chem.*, in press.
Bruland, K. W. (1983), *Trace Elements in Seawater, Chem. Oceanogr.* 8, 198.
Cantley, L. C., Josephson, L., Warner, R., Yanagasawa, M., Lechene, C., and Guidotti, G. (1977), *Vanadate is a Potent (Na,K)-ATPase Inhibitor Found in ATP Derived from Muscle. J. Biol. Chem.* 252, 7421–7423.
Chem. Unserer Zeit (1988), *Chronik: Die Tunichrome*, Abstract by H.-J. Rudolph. *Chem. Unserer Zeit* 22 (5), 183.
Evans, J. A., Morz, G., and Gibbons, I. R. (1986), *Activation of Dynein 1 Adenosine Triphosphate by Monovalent Salts and Inhibition by Vanadate. J. Biol. Chem.* 261, 14039–14043.
Faulkner-Hudson, T. G. (1964), *Vanadium Toxicity and Biological Significance.* Elsevier, New York.
Fichte, R. M., Retelsdorf, H.-J., and Ziehl, L. (1983), in: *Ullmanns Encyklopädie der technischen Chemie*, 4th Ed., Vol. 23, pp. 491–510. Verlag Chemie, Weinheim-Deerfield Beach/Florida-Basel.
Förster, U., and Wittmann, G. T. W. (1979), *Metal Pollution in the Aquatic Environment.* Springer Verlag, New York.
Henschler, D. (1978), *Analysen in biologischem Material*, Vol. 2. Verlag Chemie, Weinheim-New York.
Kaunitz, J. D., and Sachs, G. (1986), *Identification of a Vanadate-sensitive Potassium-dependent Proton Pump from Rabbit Colon. J. Biol. Chem.* 261, 14005–14010.
Kelm, W., and Schaller, K. H. (1978), *The Quantitative Determination of Vanadium in Blood Samples of Ecologically and Occupationally Exposed Persons with a Specific and Sensitive Method. Wiss. Umwelt 1*, 34–42
Kneifel, H., and Boyer, E. (1973), *Strukturermittlung der Vanadiumverbindung des Fliegenpilzes, Amavadin. Angew. Chem.* 85, 542–543.
Lagerkvist, B., Nordberg, G. F., and Vouk, V. (1986), Chapt. 27: *Vanadium*, in: *Handbook on the Toxicology of Metals*, 2nd Ed., Vol. II, pp. 638–663. Elsevier, Amsterdam.
MAK (1987), *Maximum Concentrations at the Workplace, Report Nr. XXIII, DFG.* VCH Verlagsgesellschaft, Weinheim-Basel-Cambridge-New York.
McKinney, G. L. (1987), *Communication 17th Annual IAEAC Symposium on the Analytical Chemistry of Pollutants, Jekyll Island, Georgia.* Environmental Protection Agency, Kansas City.

Montaser, A., and Golightly, D. W. (1987), *Inductively Coupled Plasmas in Analytical Atomic Spectrometry*, pp. 605 ff. VCH Verlagsgesellschaft, Weinheim-Basel-Cambridge-New York.
Nriagu, J. O. (1983), *Changing Metal Cycles and Human Health*. Springer Verlag, New York.
Nriagu, J. O., and Davidson, C. I. (1986), *Toxic Metals in the Atmosphere*. John Wiley & Sons, New York.
Panel on Vanadium (1974), (Byerrum, R. U., Chairman), *Vanadium*. National Academy of Sciences, Washington, DC.
Parker, R. D. R., Sharma, R. P., and Miller, G. W. (1978), *Vanadium in Plants, Soil and Water in the Rocky Mountain Area and its Relationship to Industrial Operations. Trace Subst. Environ. Health 12*, 340–350.
Sabbioni, E., and Rade, I. (1980), *Relationship between Iron and Vanadium Metabolism: The Association of Vanadium with Bovine Lactoferrin. Toxicol. Lett. 5*, 381–387.
Schaller, K. H. (1988), *Vanadium*, in: M. Stoeppler, *Trace Metal Analysis in Biological Specimens*. PSG Publishing Company, Littleton, Massachusetts.
Soremark, R. (1967), *Vanadium in some Biological Specimens. J. Nutrit. 92*, 183–190.
Thürauf, J., Schaller, K. H., Syga, G., and Welte, D. (1978), *The Vanadium Concentration of the Human Placenta. Wiss. Umwelt 2*, 84–88.
Thürauf, J., Syga, G., and Schaller, K. H. (1979), *Zentralbl. Bakteriol. Parasitenkd. Infektionskr. Hyg. Abt. 1, Orig. Reihe B. 168*, 273.
Tissot, B. P., and Welte, D. H. (1984), *Petroleum Formation and Occurrence*, 2nd. Ed. Springer Verlag, Berlin-Heidelberg.
Valentin, H., and Schaller, K. H. (1976), *Jahresbericht über die 16. Jahrestagung der Deutschen Gesellschaft für Arbeitsmedizin e.V., Köln*, pp. 41–47.
Valentin H., and Schaller, K. H. (1977), in: Holzhauser, K. P., and Schaller, K. H. (eds), *Arbeitsmedizinische Untersuchungen bei Schornsteinfegern*. Georg Thieme Verlag, Stuttgart.
Waters, M. D. (1977), *Toxicology of Vanadium*, in: Goyer, R. A., and Mehlman, M. A. (eds.), *Toxicology of Trace Elements*, pp. 147–189. Halsted Press, New York.
Wehrli, B. (1987), *Vanadium in the Hydrosphere*, Thesis Nr. 8232. Swiss Federal Institute of Technology, Zürich. ADAG Administration & Druck AG; see also Lee, K. (1983), *Vanadium in the Aquatic Ecosystem*, in: Nriagu, J. O., *Aquatic Toxicology*. Wiley Interscience, New York.
Welz, B. (1985), *Atomic Absorption Spectrometry*, pp. 336–337, 408 ff. VCH Verlagsgesellschaft, Weinheim-Deerfield Beach/Florida-Basel.
Wennig, R., and Kirsch, N. (1988), Chapt. 69: *Vanadium*, in: Seiler, H. G., Sigel, H., and Sigel A. (eds.), *Handbook on the Toxicity of Inorganic Compounds*, pp. 749–765. Marcel Dekker, New York.
Xavier, A. V. (1986), *Frontiers in Bioinorganic Chemistry* (with a Section on Uptake, Essentiality and Toxicity of Vanadium). VCH Verlagsgesellschaft, Weinheim-Deerfield Beach/Florida-Basel.

II.35 Yttrium

ROGER DEUBER and THOMAS HEIM, Basel, Switzerland

1 Introduction

The name yttrium comes from the Swedish mine Ytherby. In spite of its relatively high abundance, the technical application was rare until the 1960s, due to the difficulties in separation from the lanthanides, with which it is generally found. In the last decades new techniques of separation and purification as well as new applications of yttrium alloys and compounds, e.g., as red phosphor in color television and fluorescent tubes have been developed. The production of the element is rising, and as a result the potential for increased environmental exposure to yttrium is growing. In the technical literature, yttrium is generally dealt with the lanthanides. Because of systematic reasons, above all different toxicology and differences in the metabolic behavior, this is scarcely justified. The toxicity of yttrium is generally considered as low, although there are only few data available. Inhalation may be more critical than oral uptake. A toxic mode of action may be correlated with the high capability of the Y^{3+} ion to replace Ca^{2+} ions in vivo, which are responsible for many metabolic functions. The metal is not essential for plants, animals, and humans.

2 Physical and Chemical Properties, and Analytical Methods

Yttrium (Y), a silver-gray metal with the atomic number 39 (atomic mass 88.9, density 4.47 g/cm^3, melting point 1523 °C, boiling point 3338 °C), contains 3 electrons more than the rare gas krypton. In water, only the trivalent state is stable, it is oxidized in moist air and inflammed at 500 °C, burning in a light red flame, whereby the white, in water colorless Y_2O_3 is formed. The Y(III) fluorides, oxalates, halides, carbonates, and hydroxides have low solubility (NEUMÜLLER, 1977; STOKINGER, 1981; HANDBOOK OF CHEMISTRY AND PHYSICS, 1986).

In solution, Y^{3+} hydrolizes slightly before precipitation of the hydroxide occurs above pH 6, generally producing both mononuclear and polynuclear species in small amounts ($Y(OH)^{2+}$, $pK = 7.7$; $Y(OH)_2^{2+}$, $pK = 16.4$; $Y_2(OH)_2^{4+}$, $pK = 14.23$) (BAES and BALMER, 1976).

As a strong Lewis acid, yttrium has a high affinity to strong Lewis bases such as oxygen and nitrogen compounds. This property is used in technical application.

As a biological consequence Y^{3+} forms colloids with proteins. Natural yttrium consists of the isotope 89 only, but radionuclides, such as ^{90}Y and ^{91}Y, get into the environment through radioactive fallout and as a product of nuclear fission reactions. In the USA, traces of ^{91}Y are still found in barley and wheat as a result of the overground atomic bomb tests in the 1950s (NEUMÜLLER, 1977; LUCKEY and VENUGOPAL, 1978; STOKINGER, 1981; KAZMAREK, 1981).

The determination of yttrium has formerly been made with Alizarin S, Oxin, or Arsenazo III, whereby thorium often interfered (WOOD and ADAMS, 1970). The analytical methodology has progressed through the quantitative estimation by the arc spectrographic method, using a diffraction grating of sufficient dispersion to separate the complex spectra of yttrium and the lanthanides, followed by flame photometry, and finally atomic absorption spectrometry using a nitrous oxide-acetylene flame. Yttrium, by this method for determination of air, has a range of 0.529 to 2.221 mg/m^3 with a precision (CV$_T$) of 0.054 and a sensitivity of 10 µg (at 2 µg/mL for 1 percent absorption) in 5 mL final volume (NIOSH, 1977; see also WELZ, 1985). Other possibilities of detection are X-ray fluorescence analysis (ALTIERI et al., 1981), X-ray emission spectroscopy (COWGILL, 1973), mass spectrometry (HAMILTON et al., 1973), and a large thermal neutron cross-section makes yttrium suitable for neutron activation (LUCKEY and VENUGOPAL, 1978).

3 Sources, Production, Important Compounds, Uses, and Recycling

3.1 Occurrence and Production

In general yttrium is found together with the lanthanides and occurs in minerals such as monazite, bastnasite, xenotime, yttrialite, gadonilite, yttriocrasite, and in complex minerals such as samarskite and euxenite. Monazite sand contains 3% Y and bastnasite about 0.2%. Xenotime (from Malaysia) contains about 60% Y, 10% Yb, and 10% Dy (ASCHE, 1988). The most important sources of monazite sand are in Florida, Australia, India, Malaysia, and Brazil.

The abundance of yttrium in the earth's crust is about 0.0028%, and it has place No. 32 in the Table of Elements. It is more frequent than lead (GMELIN, 1979; NEUMÜLLER, 1977).

Metallic yttrium is produced by metallothermic reduction of Y_2O_3 or YCl_3. If necessary yttrium is enriched via its slightly soluble carbonate. For separation from the lanthanides, the methods of fractioned crystallization and ion exchange- or liquid-liquid distribution chromatography are mostly used (BRUNISHOLZ and HIRSBRUNNER, 1974; VILLANI, 1980; KAZMAREK, 1981).

In 1981 KAZMAREK estimated that the world annual production of yttrium together with the lanthanides was 25 000 tons in the forms of the oxides, and approximately 60% was consumed in the USA. It can be calculated that 500 t thereof was Y_2O_3. The US Bureau of Mines estimates that the production of Y_2O_3 will increase to 2500 t per year until the year of 2000 (REINHARDT, 1982; BULMAN, 1988). In

1977 the price for highly purified Y_2O_3 was US$ 30/kg and for highly purified Y metal US$ 430/kg. Yttrium will never be a cheap metal because of the high costs for separation and purification (GMELIN, 1984).

3.2 Uses and Recycling

The main technical application of yttrium as oxide, oxidesulfide or vanadate in combination with europium as an activator is as red phosphor in color television and fluorescent tubes. Since 1965, this application has resulted in a remarkable growth of the yttrium industry. In the future, the application of yttrium as a supraconductor element may further raise its production. $YBa_2Cu_3O_7$ has, for instance, a critical temperature for supraconduction of about 92 K (PHILIPS RESEARCH, 1987; BEDNORZ, 1988), which allows cooling with liquid nitrogen.

In metallurgy yttrium metal increases the stability of heating alloys and Cu-Ni-steel against oxidation, and Y_2O_3 increases the strength of ZrO_2. In ceramic industry Y_2O_3 is used as crucible material and imparts shock resistance and low expansion in glass and ceramics. Yttrium is used in laser instruments and is a catalyst in ethylene polymerization. Several yttrium compounds are used in electronics and in data processing as store elements and modulators. Y_2O_3 forms with ThO_2 a transparent glass with high permeability ("Yttralox") and with ZrO_2 the emission source for "Nernst"-lamps which have been used for a long time in mining. Yttrium-aluminum garnet ($Y_3Al_5O_{12}$) is valued as a gem stone (VILLANI, 1980; GREINACHER, 1981; NEUMÜLLER, 1977; STOKINGER, 1981; GMELIN, 1976).

In nuclear industry Y-alloys increase the strength and resistance of reactor materials, yttrium hydride is applied as neutron moderator because of its temperature stability (GMELIN, 1976; LUCKEY and VENUGOPAL, 1978; GREINACHER, 1981).

In the literature several methods have been reported to recycle yttrium from used red phosphor (VILLANI, 1980). KORNELIUS (1988) described the recovery from South African fly ash and FeCr-dusts.

4 Distribution in the Environment and in Living Organisms

As mentioned above, the abundance of yttrium in the earth's crust is relatively high with 0.0028% or 28 ppm which is higher than the Pb, Co, and Nb concentrations. It is not known to which extent the environmental concentrations of yttrium have been increased by anthropogenic activities.

For sea water concentrations of Y^{3+} of 0.01 to 0.3 ppb are given in the literature (BRULAND, 1983; JENSEN and JORGENSEN, 1984), and it exists as YCO_3^+, $Y(OH)^{3+}$, and/or Y^{3+} ions (BRULAND, 1983). A remarkable higher concentration of 59–73 ppb has been found in the ground water in the North-East region of Switzerland (Pb: 0.3–1.8 ppb) (KEIL, 1986).

For uncultivated soils concentrations of 2.5–250 ppm have been reported (Pb: 2–200 ppm, Hg: 0.01–0.8 ppm) (JENSEN and JORGENSEN, 1984). Investigations in

Missouri (USA) of 4 different vegetation areas showed for uncultivated soils an average concentration of 23 ppm, for cultivated soils (soy beans, corn field, and pasture) an average concentration of only 15 ppm. In hickory trees, an average concentration of 46 ppm was measured (ERDMAN et al., 1976a, b). This high value could be explained by accumulation of Y^{3+} in plants. The reports in the literature on the accumulation of yttrium are contradictory. In 1958 ROBINSON et al. found in hickory trees a concentration of yttrium and the lanthanides „to a phenomenal degree" up to concentrations of 828 ppm Y^{3+}, whereas only a few ppb Y^{3+} had been detected in the ground water. COWGILL (1973) found in Pennsylvania (USA) in different water plants, among them the water lily, an accumulation of yttrium and several lanthanides of about hundred fold. URE and BACON (1978), however, could not find an evidence of accumulation in water lily gathered in Scotland. In older investigations it had been reported that algae, some microorganisms, and ferns accumulate Y^{3+} (LUCKEY and VENUGOPAL, 1978; BULMAN, 1988).

There are no data about accumulation in animals, but it seems to be likely since calcium may be replaced. WARD (1987) found by ICP-MS about 0.01 ppm in bovine liver.

The radionuclides of yttrium distribute in skin structures as follows: insoluble proteins 82–94%, lipids 6–12%, and water extracts 2–4% (HALEY, 1979). In healthy humans the following concentrations of Y^{3+} were found in England: bones 70 ppb, lymph nodes 60 ppb, lung 20 ppb, basal ganglions of the brain 20 ppb (whole brain 4 ppb), liver 10 ppb, ovary 10 ppb, kidney 6 ppb, blood 6 ppb (HAMILTON et al., 1973; GERHARDSSON et al., 1984).

5 Uptake, Absorption, Transport and Distribution, Metabolism and Elimination in Animals and Humans

In mammals the gastrointestinal absorption of soluble yttrium salts is very poor (about 0.05% of the dose taken up orally) and highly depends on the compound and its solubility (LUCKEY and VENUGOPAL, 1978). The poor absorption is owing mainly to the hydrolysis of Y-salts into precipitable, colloidal, or radiocolloidal forms of $Y(OH)_2$ or YPO_4 in the digestive tract. Soluble yttrium chelates and other chelates are more easily absorbed (STOKINGER, 1981; GMELIN, 1976; LUCKEY and VENUGOPAL, 1978).

Skin absorption of ^{90}Y by rats is affected by the pH, and radiocolloidal ^{90}Y is absorbed with difficulty by normal and injured skin (HALEY, 1979). It has been shown that deposition and retention of inhaled particulates of aerosols of yttrium are related to particle size, lung clearance mechanisms, organ and tissue distribution, and the elimination route and its magnitude.

Following inhalation, insoluble Y-salts are not fully absorbed, partly being fixed permanently in the lungs. Inhalation of ^{91}Y resulted in only 10–18% being deposited in the lungs, and the half-life in the lower respiratory tract was 19 days (GENSICKE et al., 1966; SEMENOV et al., 1966; HALEY, 1979). Y-salts are bound to gamma-globulin or albumin in colloidal forms after absorption into the blood. Y^{3+} is not present. The transport of yttrium out of the blood is rapid and follows a four-term exponential equation. The colloidal Y-protein complexes are removed by phagocytes in the reticuloendothelial system and liver. The uptake of yttrium is easy in other soft tissues, but its deposition in the bone marrow and bone is slow, and the retention of yttrium in the bone is prolonged, due to strong electrostatic bounding with carbamyl and sulfate groups. The mobilization and loss of yttrium from the bone is slow. In rats, an oral application of 936 mg YCl_3 resulted in a skeletal accumulation of up to 33 ppm (LUCKEY and VENUGOPAL, 1978; HALEY, 1979; STOKINGER, 1981).

Absorbed Y-compounds do not cross the placental membrane in pregnant rats and cattle, but are secreted into the milk. Colostrum contains relatively large amounts of yttrium when administered to the mother. Excretion of yttrium is both fecal and urinary and depends upon the form of administration. The excretion rate for injected yttrium in rats is slow; only 40% is excreted in a week, one third in the urine and the rest in feces. This slow excretion rate points to a long biological half-life of yttrium (LUCKEY and VENUGOPAL, 1978).

6 Effects on Plants, Animals, and Humans

Yttrium is not essential for plants, animals, and humans (LUCKEY and VENUGOPAL, 1978). There are no reported effects of yttrium on plants and microorganisms and no reported mechanisms for yttrium detoxication in mammals.

6.1 Miscellaneous Biochemical Effect

It is known that metals can cause changes in the cell metabolism. An attractive hypothesis relating to the possible mode of action of toxic metals would be that their primary target within the cell is a key regulatory protein. The potential consequences of the activation or deactivation of a regulatory protein could be a cascading effect with numerous biochemical processes being affected and in turn profoundly altering cellular metabolism. SUHAYDA and HAUG (1987) found that Y^{3+} ions change the activity of calmodulin, a metal-binding protein with numerous metabolic functions. The activity of calmodulin is regulated by the concentration of Ca^{2+} ions in the cell. Y^{3+} ions have a higher affinity to the binding sites of calmodulin than Ca^{2+} ions, and as a result, the activity of phosphodiesterase, which is regulated by calmodulin, was increased up to 32% when Y^{3+} ions were added. The authors suppose that there is a correlation of the high charge-to-radius of Y^{3+} and its ability to replace Ca^{2+} ions in the binding sites of calmodulin (ionic radius Ca^{2+}: 99 pm, Y^{3+}: 92 pm). Since Ca^{2+} ions are responsible for numerous metabolic functions in vivo, this effect could cause other changes in metabolism with unknown consequences. Lanthanides such as lanthanum or cerium have also been reported to be able to replace Ca^{2+} ions in several functions, e.g., affecting the cAMP dependent processes, the neuron activity, secretory processes (hormones, transmitters, and enzymes), and detoxification mechanisms, inducing calcification effects and inhibiting or activating enzyme activities (EVANS, 1983; ARVELA, 1979).

Y^{3+} ions bind to nucleoproteins, gamma-globulin, phospholipids, amino acids, ATPase and have in general a high affinity to phosphor-containing ligands. Y^{3+} ions inhibit succinic, lactic, isocitric, malic, glutamic acids and glucose-6-phosphate dehydrogenases, aldolase, and ATPase. Blood lactic acid and leukocyte counts increased in dogs following Y_2O_3 inhalation. The liver is the main organ affected by subtoxic Y-salts in rats. Due to hepatic dysfunction, liver glycogen and blood glucose levels were decreased, and blood ornithin carbamyl phosphatase activity was increased (LUCKEY and VENUGOPAL, 1978; HALEY, 1979; ARVELA, 1979; STOKINGER, 1981). Yttrium blocks the release of transmitter at the neuromuscular junction, depletes the synaptic vesicles and increases coated vesicles, membrane bound tubes and cysternae (BOWEN, 1972).

Pretreatment of thrombin with yttrium decreases its enzymic activity in transforming fibrinogen into fibrin (DUCASTING et al., 1973). Powdered Y_2O_3 and yttrium hexaboride administered intratracheally to rats decreased immunological resistance, lysozyme, complement and B-lysine in the blood (OLEFIN, 1967). Yttrium propionate increased the number of antibody producing cells in the spleen and in the serum of mice immunized with sheep erythrocytes (ZIMAKOV and ZIMAKOVA, 1971).

$Y(NO_3)_3$ injected intratesticularly causes progressive calcification of both the seminiferous tubules and interstitium in rats (SHARMA et al., 1972).

6.2 Acute Effects on Animals and Humans

Symptoms of acute yttrium toxicity in experimental animals are anorexia, asthenia, and a progressive depression of general activity. Death is due to cardiac and respiratory failure (LUCKEY and VENUGOPAL, 1978). Precise acute determination of yttrium is difficult because of its protein-precipitating capacity and the unusually great influence of the non-metallic components; differences in animal strains may be another factor (STOKINGER, 1981). As a result of poor gastrointestinal absorption, the most severe toxic effects occur after inhalation or injection of Y-compounds. The acute toxicities of Y-salts are summarized in Table II.35-1.

Compared to the lanthanides, yttrium seems to have a higher acute toxicity. The yttrium citrate-chloride complex, for example, was found by GRACA (1962) to be acutely very toxic both to guinea pigs and mice by the intraperitoneal route.

6.3 Chronic Effects on Animals and Humans

In a comparative investigation of the oxides of yttrium and the lanthanides cerium and neodymium in rats, intratracheally administered at a dose of 50 mg, Y_2O_3 showed the most pronounced changes in the lung of the three oxides tested. At eight

Table II.35-1. Acute Yttrium Toxicity (Data from LUCKEY and VENUGOPAL, 1978)

Compound	Animal	Route	Toxicity	mg/kg body weight
Yttrium oxide Y_2O_3	Rat	ip	LD_{50}	500
Yttrium oxide, dissolved in dilute HCl	Mouse	ip	LD_{100}	112
	Rat	ip	LD_{50}	57.2
	Rat	iv	MLD	5
	Rat	iv	LD_{100}	12.7
Yttrium chloride $YCl_3 \cdot 6H_2O$	Mouse	ip	LD_{50}	88
	Rat	ip	LD_{50}	45
	Guinea pig	ip	LD_{50}	85
Yttrium citrate $Y(C_6H_5O_7)$	Mouse	ip	LD_{50}	254
	Mouse	ip	LD_{50}	79
	Guinea pig	ip	LD_{50}	44
Yttrium nitrate $Y(NO_3)_3$	Mouse	sc	MLD	1660
	Mouse	ip	LD_{50}	1710
	Rat	ip	LD_{50}	350
	Rat	iv	MLD	75
	Rabbit	ip	LD_{50}	515

months, in the lung tissue in which Y_2O_3 was present characteristic granulomatous nodules developed, consisting of crystalline deposits of the oxide and cellular elements. Nodules in the peribronchial tissue compressed and deformed several bronchi, and the surrounding lung areas were emphysematous, the interalveolar walls were thin and sclerotic, and the alveolar cavities dilated. In contrast, Ce_2O_3 did not elicit serious changes and produced neither diffuse nor nodular fibrotic processes. The changes of lung tissue caused by Nd_2O_3 were only moderate. It was concluded that, regardless of the similarity of their chemical and physical properties, Y_2O_3 exhibited specific toxicologic actions which were the most severe of the three oxides studied (ISRAELSON, 1963; STOKINGER, 1981).

SCHRÖDER and MITCHENER (1971) reported growth depression in mice which were fed 5 ppm YCl_3 in their drinking water during life-time experiments. Subcutaneous implantation of metallic yttrium in mice induced growth of granulomatous tissue (TALBOT et al., 1965). Blood coagulation time was significantly longer than in control mice six months after implantation. Decreases in total leukocyte numbers and incidence of neoplasms were observed.

A 30 days exposure to Y_2O_3 particles followed by threadmill exercise showed an increase in total leukocytes and blood lactic acid and a decrease in erythrocytes. The lung showed hypertrophy, hyperplasia, desquamation of the alveolar epithelium and leukocyte infiltration (REESE et al., 1967).

6.4 Mutagenic, Carcinogenic, and Teratogenic Effects

There are contradictory reports in the literature relating to these effects of yttrium. Feeding 5 ppm of yttrium for the life-span of mice resulted in a tumor occurrence of 33%; however, the significance of the observation is difficult to assess because the control group had 26.8% tumors (SCHRÖDER and MITCHENER, 1971). No tumors were found by HUTCHESON et al. (1975) after three generations of feeding Y_2O_3.

7 Hazard Evaluation and Limiting Concentrations

The threshold limit value (TLV or MAK) in the USSR is 2 mg/m^3 and in the Federal Republic of Germany 5 mg/m^3 for all Y-compounds (MAK, 1987; one peak of 50 mg/m^3 may be accepted during one shift). In the USA, a TLV of 5 mg/m^3 was recommended for yttrium by the TLV Committee in 1960. This limit was revised downward to 1 mg/m^3 in 1964 when the TLV Committee learned from a publication from the Soviet Union that Y_2O_3 upon intratracheal administration resulted in severe lung damage. Due to poor gastrointestinal absorption, the main risk of yttrium probably occurs by inhalation of Y-containing dust. Yttrium seems to be more toxic than the lanthanides, and the mode of toxic action may be correlated with the high capacity of the metal to replace Ca^{2+} ions in vivo.

Because of the increase of production and application of yttrium, there is a high need to provide a hazard evaluation. However, on the basis of the available data this is not possible. Further investigations are required, especially upon exposure of yttrium, quantitative data on emissions in the environment by human activities and concentrations occurring in the environment, and upon chronic and biochemical effects.

References

Altieri, A., Gianello, G., Vinci, F., Porcelli, D., and Vendramin, G. (1981), *Dosaggio dell'Ittrio negli Ambienti di Lavoro per Mezzo della Fluoreszenza X. Ann. 1st. Super. Sanita, 17,* 373–378.
Arvela, P. (1979), *Toxicity of Rate-earths. Prog. Pharm. 2,* 69–113.
Asche, W. (1988), *Seltene Erden. Chem. Rundsch. 41,* No. 35, 9.
Baes, F., and Balmer, E. (1976), *The Hydrolysis of Cations,* Chapter 7: *Yttrium, the Lanthanides and Actinium,* pp. 129–138. J. Wiley & Sons, New York.
Bednorz, J.G. (1988), *Lecture ACHEMA,* see for instance *Swiss Chem. 10,* Nr. 11, 35–49, and *Chemosphere 18,* in press.
Bowen, J.M. (1972), *Effects of Rare Earths and Yttrium on Striated Muscle and the Neuromuscular Junction. Can. J. Physiol. Pharmacol. 50,* 603.
Bruland, K.W. (1983), *Trace Elements in Sea-Water. Chem. Oceanogr. 8,* 208.
Brunisholz, G., and Hirsbrunner, W. (1974), *Extraction Liquide-liquide par des Esters Alkylphosphoriques. III. Séparation de la Paire Yttrium-Gadolinum par l'Acide Di-((2-éthylhexyl)phosphorique). Helv. Chim. Acta 57,* 2483–2487.
Bulman, R.A. (1988), Chapter 70: *Yttrium and the Lanthanides,* in: Seiler, H.G., Sigel, H., and Sigel, A. (eds.), *Handbook on the Toxicity of Inorganic Compounds,* pp. 769–785. Marcel Dekker, New York.

Cowgill, U.M. (1973), *Biochemistry of the Rare-earth Elements in Aquatic Macrophytes of Linsley Pond, North Branford, Connecticut. Geochim. Cosmochim. Acta 37*, 2329–2345.

Ducasting, A., Monceyron, C., Azanza, J.L., Creach, P., and Raymond, J. (1973), *C.R. Soc. Biol. 167*, 262.

Erdman, J.A., Shacklette, H.T., and Keith, J.R. (1976a), *Elemental Composition of Corn Grains, Soybean Seeds, Pasture Grasses, and Associated Soils from Selected Areas in Missouri. Geological Survey Professional Paper 954-D*, pp. 1–17. US Governmental Printing Office, Washington, DC.

Erdman, J.A., Shacklette, H.T., and Keith, J.R. (1976b), *Elemental Composition of Selected Native Plants and Associated Soils from Major Vegetation-type Areas in Missouri. Geological Survey Professional Paper 954-C*, pp. 61–63. US Governmental Printing Office, Washington, DC.

Evans, C.H. (1983), *Interesting and Useful Biochemical Properties of Lanthanides. Trends Biochem. Sci. 8*, 445–449.

Gensicke, F., Nitschke, H.W., and Hölzer, P. (1966), *Stoffwechselverhalten von ^{91}Y, ^{144}Ce, ^{143}Pr and ^{147}Pm. Stud. Biophys. 1*, 347.

Gerhardsson, L., Wester, P.O., Nordberg, G.F., and Brune, D. (1984), *Chromium, Cobalt and Lanthanum in Lung, Liver and Kidney Tissue from Deceased Smelter Workers, Sci. Total Environ. 37*, 233–246.

Gmelin (1976), *Handbuch der Anorganischen Chemie*, Syst. Nr. 39: *Sc, Y, La and Lanthanides*, Part B2: *Separation of REE, Preparation of the Metals, Uses and Toxicology*, 8th Ed., pp. 259 and 282–283. Springer Verlag, Berlin.

Gmelin (1979), *Handbuch der Anorganischen Chemie*, Syst. Nr. 39: *Sc, Y, La and Lanthanide*, Part B7: *Reactions of Ions in Solution, Electrochemical Behaviour*, 8th Ed., pp. 16–17. Springer Verlag, Berlin.

Gmelin (1984), *Handbuch der Anorganischen Chemie*, Syst Nr. 39: *Sc, Y, La and Lanthanide*, Part A8: *Minerals, Deposits, Mineral Index*, 8th Ed., pp. 320–337. Springer Verlag, Berlin.

Graca, J.G. (1962), *Arch. Environ. Health 5*, 437.

Greinacher, E. (1981), in: Gschneidner, K.A. (ed.), *Industrial Application of Rare Earth Elements*, pp. 3–18. ACS Symposium 164, American Chemical Society, Washington, DC.

Haley, T.J. (1979), Chapter 40: *Toxicity*, in: Gschneidner, K.A., and Eyring, L. (eds.), *Handbook on the Physics and Chemistry of Rare Earths*, Vol. 4, pp. 553–585. North-Holland Publishing Comp., Amsterdam.

Hamilton, E.I., Minski, M.J., and Cleary, J.J. (1973), *The Concentration and Distribution of Some Stable Elements in Healthy Human Tissue from the United Kingdom. Sci. Total Environ. 1*, 341–374.

Handbook of Chemistry and Physics (1986), 67th Ed., CRC Press, Boca Raton, Florida.

Hutcheson, D.P., Gray, D.H., Venugopal, B., and Luckey, T.D., (1975), *Studies of Nutritional Safety of Some Heavy Metals in Mice. J. Nutr. 105*, 670.

Israelson, Z.T. (ed.) (1963), *Toxicology of Rare Metals*, pp. 195–208. Moscow.

Jensen, A., and Jorgensen, S.E. (1984), *Analytical Chemistry Applied to Metal Ions in the Environment*, in: Sigel, H. (ed.), *Metal Ions in Biological Systems*, Vol. 18, pp. 8–11. Marcel Dekker, New York.

Kazmarek, J. (1981), in: Gschneidner, K.A. (ed.), *Industrial Application of Rare Earth Elements*, pp. 135–166. ACS Symposium 164. American Chemistry Society, Washington, DC.

Keil, R. (1986), *Die chemische Zusammensetzung von EIR-Wasser, SR-Grundwasser und Aare-Wasser beim EIR. EIR-TM 44-86-12*, Paul Scherrer Institut, Würenlingen, Switzerland.

Kornelius, G. (1988), *Lecture ACHEMA*, see for instance, *Swiss Chem. 10*, No. 11 and *Chemosphere 18*, in press.

Luckey, T.D., and Venugopal, B. (1978), *Yttrium*, in: *Metal Toxicity in Mammals*, Vol. 1: *Physiological and Chemical Basis of Metal Toxicity*, pp. 131–135. Plenum Press, New York.

MAK (1987), *Maximum Concentrations at the Workplace, Report No. XXIII DFG*. VCH Verlagsgesellschaft, Weinheim-Basel-Cambridge-New York.

Neumüller, O. A. (1977), in: *Römpp's Chemielexikon*, 7th Ed., pp. 3954–3955. Franckh, Stuttgart.
NIOSH (National Institute of Occupational Safety and Health) (1977), *Manual of Analytical Methods*, 2nd Ed., Vol. 3.
Olefin, A. J. (1967), *Vrach. Delo 11*, 95.
Philips Research (1987), *Stabile Supraleiter, Neue Zürcher Zeitung, Forschung und Technik*, Nr. 292, p. 67 (16 December), Zürich.
Reese, W. O., Talbot, R. B., and Swenson, M. J. (1967), *Am. J. Vet. Res. 28*, 979.
Reinhardt, Th. (1982), in *Ullmanns Encyklopädie der technischen Chemie*, 4th Ed., Vol. 21, p. 268. Verlag Chemie, Weinheim-Deerfield Beach/Florida-Basel.
Robinson, W. O., Bastron, H., and Murata, K. J. (1958), *Biogeochemistry of the Rare-earth Elements with Particular Reference to Hickory Trees. Geochim. Cosmochim. Acta 14*, 55–67.
Sax, N. I. (ed.) (1984), *Dangerous Properties of Industrial Materials*, 6th Ed., pp. 2749–2750. Van Nostrand Reinhold, New York.
Schröder, H. A., and Mitchener, M. (1971), *Scandium, Chromium(VI), Gallium, Yttrium, Rhodium, Palladium, Indium in Mice: Effects on Growth and Life Span. J. Nutr. 101*, 1431.
Semenov, D. E., Moskalev, Yu. I., and Buldakov, L. A. (1966), *Tr. Inst. Biol. Akad. SSR Ural. Filil. 46*, 49.
Sharma, S. N., Kamboj, V. P., and Kar, A. B. (1972), *Exp. Pathol. 7*, 176.
Stokinger, H. E. (1981), Chapter 17: *The Lanthanides*, in: Clayton, G. D., and Clayton, F. E. (eds.), *Patty's Industrial Hygiene and Toxicology*, 3rd Ed., Vol. 2A, pp. 1674–1687. J. Wiley & Sons, New York.
Suhayda, C., and Haug, A. (1987), *Metal-induced Conformational Changes in Calmodulin. Bull. Environ. Contam. Toxicol. 38*, (2), 289–294.
Talbot, R. B., Davison, F. C., and Reese, W. O. (1965), USAEC. TID No. C00-1170-6, p. 23.
Ure, A. M., and Bacon, J. R. (1978), *Scandium, Yttrium and the Rare Earth Contents of Water Lily (Nuphar lutea). Geochim. Cosmochim. Acta 42*, 651–653.
Villani, F. (ed.) (1980), *Rare Earth Technology and Applications*, pp. 3–32. Noyes Data Corp., Park Ridge, New Jersey.
Ward, N. I. (1987), *Multielement Analysis, Trace Elements in Human Health and Disease. Environmental Health*, Vol. 20, pp. 118–123. WHO, Copenhagen.
Welz, B. (1985), *Atomic Absorption Spectrometry*, 2nd. Ed., pp. 337/338. VCH Verlagsgesellschaft, Weinheim-Deerfield Beach/Florida-Basel.
Wood, D. F., and Adams, M. R., (1970), *Spectrophotometric Determination of Yttrium in Chromium and Chromium-based Alloys with Arsenazo III. Analyst 95*, 556.
Zimakov, Y., and Zimakova, I. E. (1971), *Tr. 2nd Mosk. Med. Inst. 3*, 93.

II.36 Zinc

FRIEDRICH KARL OHNESORGE and MICHAEL WILHELM
Düsseldorf, Federal Republic of Germany

1 Introduction

(see also VAHRENKAMP, 1988)

Zinc has been used unwillingly for the production of brass since the 4th century A.D. As a discrete element, it was discovered in India during the 13th century and in Europe at the beginning of the 16th century by Ebener von Nürnberg; it was rediscovered in 1746 by Marggraf. Today zinc is produced in amounts of the same order of magnitude as copper, chromium, or lead.

Zinc plays an important role as an essential trace element in all living systems from bacteria to humans. The detection of the metallothioneins and their biological role gradually proved to be a substantial contribution to a better understanding of zinc metabolism and its interactions with other essential and non-essential trace metals (see Chapters I.10 and I.12).

The toxicity of zinc and most zinc-containing compounds is generally low and, with certain exceptions, of minor importance compared with the significance of zinc deficiency in plants, animals, and man. Nevertheless, industrial and household wastes sometimes contain zinc concentrations which can be harmful to the environment, although for the most part the effects of zinc-accompanying impurities, such as cadmium and lead, are much more prominent. Some hazards to aquatic organisms and to horses by zinc exposure have been observed.

2 Physical and Chemical Properties, and Analytical Methods

2.1 Physical Properties

Zinc is a bluish-white, rather soft metal, which solidifies in hexagonal crystals and belongs to the Group IIB elements of the Periodic Table. Its atomic number is 30, its atomic mass 65.39, and its density 7.14 g/cm^3. The melting point is 419.58 °C and the boiling point 907 °C. Natural zinc is composed of five stable isotopes (^{64}Zn, 48.6%; ^{66}Zn, 27.9%; ^{67}Zn, 4.1%; ^{68}Zn, 18.8%, and ^{70}Zn, 0.6%). Eighteen artificial radioactive isotopes are known, most of them with very short (milliseconds, seconds) or short (minutes) half-lives; ^{65}Zn (half-life 243.8 days) and ^{69m}Zn (half-life 13.8 hours) are mainly used in biological experiments (WEAST, 1986).

Cast zinc is brittle, but it becomes ductile at about 120 °C, and then can also be rolled.

2.2 Chemical Properties

Zinc, oxidation state +2, has a strong tendency to react with acidic, alkaline, and inorganic compounds. Because of its amphoteric properties zinc forms a variety of salts. Zinc chlorate, Zn-chloride, the sulfates and nitrates are readily soluble in water, whereas the oxide, carbonate, the phosphates and silicates, the sulfides and organic complexes are practically insoluble in water (WEAST, 1986). Because of its strong reducing properties zinc is often used in organic chemistry.

In dry air zinc oxidizes, and in wet air a basic carbonate ($2\,ZnCO_3 \times 3\,H_2O$) is formed on the surface, thereby protecting the metal from corrosion.

2.3 Analytical Methods
(see also Chapter I.4a–c)

In former times analysis was usually performed by colorimetry, mainly by determination of the zinc-dithizone complex, extracted into an organic solvent (KOCH and KOCH-DEDIC, 1974). Currently atomic flame absorption spectrophotometry is preferred with a limit of detection sufficient for zinc analysis in air, water, and most biological materials. The AAS in the electrothermal mode (furnace AAS) is too sensitive for most practical applications because of its high susceptibility to contamination (WELZ, 1985; TSALEV, 1984; ARNAUD et al., 1986; FRIEL and NGYUEN, 1986). Less commonly used (see also Chapter I.4a) are electrochemical methods, e.g., anodic stripping voltammetry (ASV) and neutron activation analysis (NAA; WARD, 1987), both usually accepted as reference methods. Only few data have been generated by inductively coupled plasma-atomic emission spectrometry (ICP; WARD, 1987), X-ray fluorescence (XRF, PIXE), and spark source or isotope dilution mass spectrometry (BERTRAM, 1983, 1984; KNEIP and FRIBERG, 1986). Recently, a fluorimetric determination of zinc in biological materials, beverages, and alloys has been described, using salicylaldehyde thiocarbazone as a complexing agent (PAVON et al., 1986). Detection limits are given in Chapter I.4a.

In general, the trace analysis of zinc is less complicated than that of most other trace elements, and it gives reliable results (HALSTED et al., 1974) provided that a strict quality control is carried out, appropriate certified reference material is used, and the sampling strategies ensure a sound statistical basis for the intended aim of the investigation (see Chapter I.4a). As usual, the neglect of these factors often impairs the authenticity of published results (BERTRAM, 1984; KNEIP and FRIBERG, 1986).

3 Sources, Production, Important Compounds, Uses, Waste Products, Recycling

3.1 Occurrence

(see Chapter I.1, and STOKINGER, 1981; JOLLY, 1982; GIESLER et al., 1983; MELIN and MICHAELIS, 1983; PÖTZSCHKE, 1983; SUESS et al., 1985)

Zinc occurs in almost all minerals in the earth's crust with a medium concentration of about 70 mg/kg. The principal ores used for production are the sulfides sphalerite (zinc blende, cubic ZnS) and wurtzite (hexagonal ZnS) and their weathering products smithsonite ($ZnSO_4$, trigonal) and hemimorphite (2 $ZnO \times SiO_2 \times H_2O$, rhombic or trigonal). Zincite (ZnO, hexagonal), zincosite ($ZnSO_4$, rhombic), and goslarite ($ZnSO_4 \times H_2O$, rhombic) are of minor importance. The most common impurities in zinc minerals are Fe, Cd, and Pb. Greenocktite consists of ZnS and CdS. Usually, Cd is about 0.05% as abundant as zinc. With the production of one ton of zinc 3 kg of cadmium are produced (SAAGER, 1984).

3.2 Production, Important Compounds and Uses

The world production of zinc was about 6×10^6 metric tons between 1980 and 1982, 20% thereof being produced in Canada, 14% in the USSR, 10% in Australia, 9% in Peru, and 5% in the USA (MELIN and MICHAELIS, 1983; SAAGER, 1984).

The mined ores are concentrated by gravity separation, floatation, magnetic and other methods with sometimes considerable loss of zinc during these processes (8–20% with the sulfide and 15–90% with other ores). The zinc concentrates are then reduced to the metal by roasting with coke (ZnS) or by an electrolytic leaching process. Refinement is achieved by a further electrolytic procedure. Lower grades of zinc contain considerable amounts of lead and cadmium; therefore, high-grade zinc is preferred today (MELIN and MICHAELIS, 1983; SAAGER, 1984).

Zinc is mainly used as a protective coating for iron and steel (by galvanizing see, for instance, TRUEB, 1988, etc.), for the production of brass and other zinc alloys (PÖTZSCHKE, 1983), of rolled zinc, zinc dust (as a pigment and reducing agent), of zinc oxide, and other zinc compounds (MELIN and MICHAELIS, 1983; SAAGER, 1984).

Zinc oxide is the most important zinc compound (GIESLER et al., 1983), as judged by the amount used and its economic value. In 1974 55–60% was used in the rubber industry as an aid in vulcanizing and as a pigment. ZnO is used in rather large amounts in the production of photocopy paper, of chemicals and paints. It is also used in floor coverings, for glasses, enamels, fabrics, plastics, lubricants, and in rayon manufacture. Zinc oxide is a long serving, well-known pharmaceutical for external application for burns and skin infections in powders and ointments.

Zinc sulfate plays some role in ore concentration and as a hardener in rayon manufacture. It can be used as a supplement for humans, animals, and plants with zinc deficiency.

Zinc chloride is used in soldering and welding fluxes, for fire proofing and as a pesticidal wood protectant, in dry batteries, in the processing of cotton, and as a cauterizing agent in medicine.

Zinc sulfide serves as a white pigment (e.g., lithopone pigment = 28% ZnS + 72% $BaSO_4$) and, highly purified, as phosphors in X-ray and TV screens, and in luminous watch faces.

Zinc soaps (stearates, palmitates, oleates) act antioxidatively, are water-repellant, and are used as lubricants as well as for the water-proofing of textiles, papers, and concrete.

Organozinc compounds, such as Ziram (zinc ethylene-1,2-bis-dithiocarbamate) and Zineb (zinc dimethyl-dithiocarbamate) are used as fungicides. In human and veterinary medicine zinc compounds are used as fungicides or antiseptics (e.g., zinc bacitracine; zinc propionate, zinc caprylate, zinc undecylate, zinc phenylsulfonate, and zinc salicylate) (FARNSWORTH et al., 1973; GIESLER et al., 1983).

3.3 Waste Disposal and Recycling

Wastage results from all stages of production and processing of zinc, leading to emissions into the atmosphere, to waste water, and to solid wastes. The latter (such as scrape zinc, filter dusts, slags, and other zinc materials) are partially recycled. The recycling quotes are given as 8% in the USA, 16% in the Federal Republic of Germany, and 22% in the Western world (MICHAELIS, 1984). Recycled zinc originates mainly from the melting of alloys (SAAGER, 1984). The waste waters are usually treated in sewage treatment plants, and the resulting sludge which contains considerable amounts of zinc in the case of industrial sources, is deposited or burned in refuse incineration plants. The total anthropogenic emissions of zinc into the atmosphere have been assessed as rather high (8400×10^3 t/a) compared with the natural emissions by soil dust, volcanic dust and emanation (360×10^3 t/a) (GALLOWAY et al., 1982). The main sources are waste waters and dust emissions resulting from the production and processing of zinc, other non-ferrous smelters, from coal power plants, and fossil combustions.

Considerable emissions can occur in the vicinity of, e.g., zinc smelters, leading to phytotoxic zinc concentrations in the soil. Sometimes zinc intoxication has been observed in domestic cattle due to high zinc forage grown on such soils, but in general, the accompanying elements such as As, Cd, Mn, and Pb are much more important from the toxicological point of view, especially for humans.

4 Distribution in the Environment, in Foods, and in Living Organisms

(see also NRIAGU and PACYNA, 1988)

4.1. Soils

The zinc concentration of non-contaminated soils ranging from 10 to 300 mg/kg are comparable with those of their rocky subsoils. On the average, levels of 20 mg/kg are found (WEDEPOHL, 1972, 1984; NRC, 1979; MACHELETT et al., 1984). The zinc content can be much higher in the vicinity of ore deposits and smelters (BURKITT et al., 1972). Like other metals, atmospheric deposition has increased zinc concentrations in surface soil from the 19th century to the present, but the exact deposition rates cannot be reconstructed, because zinc is rather mobile at the usual pH levels (GALLOWAY, 1982).

Sewage sludge, according to a governmental regulation, in the Federal Republic of Germany, may not contain more than 3000 mg zinc per kg dry weight to prevent the soil zinc content exceeding the permitted level of 300 mg/kg dry weight (ABFKLÄRV, 1982).

The zinc supply to plants mainly depends on the geological origin of the soils. If the site specific zinc contents of indicator plants (such as meadow and acre red clover, alfalfa, wheat, rye) on syenite is set at 100%, then their zinc contents on phyllite, granite, alate and peat amounts to 90–80%, on diluvial sands, gneiss, alluvial riverside soils to 80–70%, and on loess, blouder clay, new red sandstone, shell-lime, and keuper to 70 to 50% (SIEGERT et al., 1986). The zinc supply is also influenced by other factors. In lime-rich soils the availability of zinc is reduced at and above pH 7.4. The availability increases with lower pH values, but wash-out is possible under acidic conditions. Zinc deficiency in plants is also believed to be induced by carbonates and phosphates (GANIRON et al., 1969; LINDSAY, 1972; LUCAS and KNEZEK, 1972).

4.2 Waters

Seawater usually contains zinc concentrations of 0.003 to 0.6 µg/L (lower concentrations at the surface, higher concentrations at greater depth, mainly in the form of Zn^{2+} ions and of complexes; BRULAND, 1983). Freshwaters, especially rivers, are frequently contaminated by sewage and waste water and contain considerably higher zinc levels. For instance, 63% of the zinc load of waste water in Switzerland results from corrosion of galvanized drinking water pipes (BUS, 1980), and rivers in ore mining areas show zinc levels up to 21 mg/L (MINK and WILLIAMS, 1971). Zinc levels between 0.002 to 1.183 mg/L (mean 0.064 mg/L) were measured at 130 locations in the USA, and in the river Rhine (FRG) peak values of 0.4 mg/L occurred in the early 1970s, but had dropped by 1985 to a mean of 0.05 mg/L (90th percentile 0.11 mg/L) at the German-Dutch border (GOLDBERG, 1965; FÖRSTNER and WITT-

MANN, 1979; EPA, 1980; LWA, 1986). Older values in the literature are often too high (see also Chapters I.1 and I.4a).

In surface waters zinc occurs mainly bound to suspended matter, such as clay, or is precipitated with iron- or manganese oxides. Therefore, high concentrations are found in sludge. The chemistry of zinc in surface waters has been discussed by HEM (1972).

Drinking water usually contains zinc levels below 0.2 mg/L (EPA, 1980). Resident water in galvanized water pipes can contain considerably higher amounts, up to 2–5 mg/L, depending on the age of the pipes and the physico-chemical properties of the drinking water. Although unobjectionable from the toxicological point of view, the organoleptic qualities (appearance, taste) are impaired by such concentrations. Therefore, the EEC has set a guideline level of zinc in resident water of 5 mg/L (EEC, 1980). In the new drinking water regulation of the FRG the old limit value of 2 mg/L was deleted because of the absence of toxicological reasons (TRINKWV, 1986).

4.3 Atmosphere

The particle sizes of zinc dusts in ambient air are on the whole small; 52 to 70% have diameters lower than 5 µm (VDI, 1984).

According to data from the USA the zinc levels in ambient air of city areas varied between 0.1 and 1.7 µg/m^3 in 1973, with a mean annual level of lower than 1 µg/m^3 (LEE and VON LEHMDEN, 1973). These data correspond well to those from the FRG, where 0.05 µg/m^3 were measured in rural areas, 0.2 to 1.0 µg/m^3 in small towns, 0.2 to 2.0 µg/m^3 in large towns, and 0.5 to 4.0 µg/m^3 in most industrialized zones (LAHMANN, 1982).

Wet deposition predominates with 60 to 90% (GALLOWAY, 1982; PATTENDEN et al., 1982), but estimations on wet deposition from air differ considerably. According to NÜRNBERG et al. (1982, 1984) and VALENTA et al. (1986) the wet deposition in the FRG per day ranges from 10 to 70 µg/m^2 in rural, from 20 to 90 µg/m^2 in urban areas, and from 200 to more than 500 µg/m^2 in industrial areas. Data on total deposition, collected by GALLOWAY (1982), also vary by factors of about 10^2.

It was stated that about 23% of atmospheric zinc results from traffic (tire abrasion and fuel combustion). Tire stock contains 1.5% zinc, and the abrasion is estimated to be 1.2 kg Zn per 10^6 km (ONDOV et al., 1974).

4.4 Sewage Sludge and Dumping Grounds
(see also Chapter I.7c)

Treatment of wastes in sewage treatment plants leads to precipitation of zinc. The cleared effluent water normally contains zinc concentrations lower than 1 mg/L (POUND and CRITES, 1973; REED, 1972; EL BASSAM, 1982). Although the concentration seems rather low, it should be kept in mind that the resulting load is considerable because of the great volume.

The resulting native sludge is usually processed by anaerobic biological treatment, leading to a two-fold increase in zinc level by concentration. In 1982, sewage sludges in Baden-Württemberg (FRG) contained on average 1480 mg zinc per kg dry weight (SCHWEIGER, 1984). According to CHANEY (1973) and EL BASSAM (1982) the levels of zinc in composted sewage sludge vary between 500 and 5000 mg/kg dry weight, and a compilation of more than 100 data from the literature results in a mean of 2420 mg/kg, whereas the average concentration in 80 samples from sewage treatment plants in the USA was found to be 6380 mg/kg dry weight (DEAN and SMITH, 1973; KÜMMLEE, 1985). MERIAN (1982a, b) reported concentrations of 2300 mg Zn per kg dry weight in sewage sludge for Switzerland and 2000–3000 mg/kg dry weight for the FRG. According to COTTENIE (1981) manure contains 800 mg Zn per kg and sewage sludge 500 to 20000 mg/kg dry weight with a water solubility of 46% and a corresponding availability for plants.

Municipal sewage sludge is mainly dumped. Only a small part is burnt in sewage incineration plants. Another part is composted if the zinc content is relatively low and there is no risk of overloading to soil. Compliance with national regulations must be observed (see Sects. 4.1 and 7). In general, industrial sewage sludge with high zinc contents is not suitable for agricultural uses (see also Chapter I.7c).

4.5 Food Chain, Plants, Animals, Humans
(see also Sections 6.1 and 6.3)

4.5.1 Plants and Food

The zinc content of plants is species specific, dependent on the stage of vegetation as well as on zinc availability, and is influenced by the geological origin of the basic material for soil formation. Zinc supply is best from the weathering soils of syenite and phyllite (SIEGERT et al., 1986; see also Sects. 4.1 and 6.3.1).

Although the zinc requirement and absorption rate of plants is considered as low (MENGEL, 1984), some degree of zinc deficiency is quite widespread (see Sect. 6.3.1).

The zinc content in plants is also influenced by the age and vegetation state of the plant (SIEGERT et al., 1986). Usually the highest zinc content is found in young plants. During ageing the zinc concentration decreases as a result of dilution (MUGWIRA and KNEZEK, 1971). The normal zinc content of plants ranges from 15 to 100 mg/kg dry weight (EL BASSAM, 1982). Zinc deficiencies occur at levels below 20 mg/kg in leaves. Values higher than 400 mg/kg are regarded as toxic (BOAWN and RASMUSSEN, 1971; JONES, 1972). Data on the zinc content of specific plants are summarized in FIEDLER and RÖSLER (1988). French dwarf beans (*Phaseolus vulgaris*), grown in nutrient solutions with 10 mg zinc per liter, showed in the roots 5300 mg/kg, the leaves 154 mg/kg, and the fruits 131 mg/kg (FOROUGHI et al., 1982).

Good dietary sources of zinc are meat, meat products (20–60 mg/kg wet weight) and fish (about 15 mg/kg wet weight). Vegetables and cereals have about the same zinc concentrations (15 to 60 mg/kg wet weight) (UNDERWOOD, 1977; OSIS et al.,

1972), but the availability is considerably lower (see Sect. 5.2; SIEGERT et al., 1986; FSP, 1981).

4.5.2 Animals

Zinc occurs everywhere in water, both freshwater and seawater, and has also been detected in all investigated organisms. Overviews about the ecological cycle of the metal in water and its occurrence in aquatic organisms are found in RICE (1963), VINOGRADOV (1953), BERNHARD and ZATTERA (1969), PENTREATH (1971), and WOLFE (1970).

According to HENKIN et al. (1979) the average zinc content of most sea fish is between 3 and 8 ppm, with certain kinds of fish and molluscs containing up to about 25 ppm and maximum values of up to 50 ppm in individual animals. Crustaceans contain 7 to about 50 ppm of zinc with maximum values reaching a little over 100 ppm. The zinc content of oysters is about 100 to 200 ppm, with maximum values reaching about 2000 ppm. According to IKUTA (1968), however, oysters lose their high concentrations of zinc when they are transplanted into water with a low zinc content. Freshwater organisms exhibit similar zinc concentrations (UTHE and BLIGHT, 1971; HENKIN et al., 1979).

Zinc is also a constituent of snake toxins, for example, of rattlesnake poison (TU, 1977).

According to BOWEN (1979) the bones of mammals contain about 75 to 170 ppm of zinc, white muscle tissue has about 240 ppm. Protein-rich food generally exhibits increased amounts of zinc, with human milk containing about 3 mg/L (FRIBERG et al., 1979).

The zinc concentrations in organs of farm- and experimental animals (cattle, horse, goat, sheep, rat) have been reviewed quite often (HALSTED et al., 1974; NRC, 1979; RISCH and NASAROW, 1985; KOSLA et al., 1985).

4.5.3 Humans

Zinc is found in all human tissues and all body fluids. The metal is essential for growth, development, and reproduction in man. Disorders of zinc metabolism are usually due to a deficiency rather than a surplus of zinc.

The total zinc content of the human body (70 kg) is estimated to be 2300 mg (LINDEMAN and MILLS, 1980). Muscle and bone contain the major part of the total body zinc (62.2 and 28.5%, respectively), followed by the liver (1.8%). Every other organ-system contains equal to or less than 1% (NRC, 1979; see also SMITH JR. et al., 1981). The discussion on the usefulness of measuring zinc levels in special organ-systems (e.g., serum/plasma, blood, hair, nails, liver, or excretions) as indicators of an adequate zinc status in humans is still not settled (SOLOMONS, 1979; HENKIN and AAMODT, 1983; see also Sect. 6.3.3). Ribs and other bones, testes, and hair are usually regarded as best suited to reflect the zinc status because they are relatively uninvolved in the homeostasis.

The highest concentrations (in mg Zn per kg wet weight, partially recalculated from ash or dry weight) are found in prostate gland and retina (about 130), followed by nails (73 to 304), bone (53 to 117), striated muscle (38 to 70), kidney (25 to 85), pancreas (23 to 41), thyroid (24 to 37), testis (12 to 28), liver (8 to 16), and skin (6 to 19). The medium concentration in brain is given as 13 to 39, but the levels vary in different areas, and they are especially high in the nucleus amygdalae. Whole blood contains 4.8 to 9.3 mg/L, serum or plasma 0.7 to 1.8 mg/L (serum about 16% higher), and erythrocytes 7.6 to 16.0 mg/L. Na_2EDTA and NaFeEDTA (ethylenediamine tetraacetate) reduce the plasma level of zinc (SOLOMONS et al., 1979). In infants, the plasma level decreases to less than half of the birth level during the first three months of life, rising rapidly from the 3rd to 6th month, then slowly up to an age of three years to the level at adults. In preschool children zinc in hair increased significantly from 98 mg/g to 137 µg/g with age (3−7 years) (LOMBECK et al., 1988).

5 Uptake, Absorption, Transport and Distribution, Metabolism and Elimination in Plants, Animals, and Humans
(see also HENKIN, 1984; ELINDER, 1986)

5.1 Uptake and Distribution in Aquatic Plants
(see also Sections 4.5.1 and 6.3.1)

Tracer studies on the accumulation of zinc in aquatic plants (phytoplankton, periphyton, seaweed and freshwater algae) have shown that zinc is largely absorbed at the cell surface and is only passively transported (by diffusion) into the intercellular spaces (BRYAN, 1969; CUSHING and WATSON, 1968; DAVIES, 1973; HARVEY and PATRICK, 1967; ROSE and CUSHING, 1970). CUSHING and WATSON (1968) and others found that algal cells accumulated larger amounts of zinc-65 after they had been killed than they did as living cells. They attributed this difference to internal chemical changes, possibly in the pH value, which occur after respiration has ceased and which facilitate zinc absorption at certain sites within the cell.

The influence of zinc on the metabolism of aquatic plants has not been sufficiently investigated. Furthermore, a critical examination should be carried out to ascertain whether and how biological, chemical, and ecological factors in sediment and water influence the zinc content and distribution in the surrounding water and in aquatic plants.

5.2 Uptake and Distribution in Terrestrial Plants

A comprehensive overview of the absorption of zinc by plants is to be found in LINDSAY (1972) or MOORE (1972).

The daily zinc requirement (in relation to the fresh weight) amounts to 2 to 4000 mg/mol (SCHMID et al., 1965; CARROLL and LONERAGAN, 1969). HEWITT

(1966) reported that the optimal zinc concentration for plants grown in nutrient solution is 0.3 to 3.0 µmol; in contrast, CARROLL and LONERAGAN (1969) found that numerous plants exhibited good growth at 0.01 µmol and optimal growth at a zinc concentration of 0.25 µmol.

It is assumed that zinc reaches the plant roots by means of convection and predominantly by diffusion through the soil (LINDSAY, 1972). A comprehensive overview of the migration of nutrients has been given by OLSON and KEMPER (1968), and details of zinc migration have been summarized by LINDSAY (1972).

Plants growing on former cattle pastures and on farming land often suffer from zinc deficiency. It is assumed that microorganisms, which flourish in this soil, bind the available zinc (LUCAS and KNEZEK, 1972). LINDSAY (1972) reported that a serious zinc deficiency sometimes occurred in maize when sugarbeet had been previously grown on the same land.

Zinc deficiency was often observed in areas where the upper soil layer containing organic substances had been removed (GRUNES et al., 1961). LINDSAY (1972) came to the conclusion that these deficits in the uncovered subsoil are a result of the lower content of organic substances, while the pH value and the carbonate content is higher than in the upper soil layers.

Cold, wet spring weather intensifies or leads to zinc deficiency in field fruits (LUCAS and KNEZEK, 1972). The limited growth of the roots during bad weather possibly results in a restricted zone for the nutrition of the roots and a reduced microbial activity so that zinc is not released from the organic material.

Nitrogenous fertilizers have an acidifying effect and possibly enhance the cation exchange capacity of the roots. Thus, nitrogenous fertilizers are thought to lead to an increased zinc uptake in plants (VIETS et al., 1957; BOAWN et al., 1960;, DRAKE, 1964; LINDSAY, 1972).

TIFFIN (1972) investigated the transport of zinc in plants. Zinc is transported through the xylem from the roots to the apex of the plants, with a limited amount of zinc passing back from the leaves through the phloem (LINDSAY, 1972; TIFFIN, 1972).

According to LINDSAY (1972) the mobility of zinc within the plants is moderate compared with that of other trace nutrients. If the supply of the metal is normal, it can also be determined in the roots (CARROLL and LONERAGAN, 1969). When the supply is limited, the zinc is transported to the upper parts of the plants (RICEMAN and JONES, 1956; MASSEY and LOEFFEL, 1967).

Numerous aspects concerning the metabolism of zinc in plants have not yet been sufficiently investigated. Thus, the specific mechanisms of zinc uptake (also via the leaves) and transport have still to be identified at the molecular level. The physiological reasons for the large discrepancies in zinc requirements of various plants must be found, and the processes of distribution of the metal in plants, especially the migration from the growth tissues to the seeds must be clarified.

The significance of zinc for plants has only become evident in the last 50 years. SKOOG (1940) indicated the relationship between the zinc content and the auxin content in higher plants. TSUI (1948) concluded that zinc is essential for the biosynthesis of tryptophan in tomatoes and, though only indirectly, for the synthesis of auxin. NASON (1950) discovered that zinc is necessary in the synthesis of tryptophan from indole and serine. It is considered certain that the primary role of zinc is as a catalyst (SCHÜTTE, 1964); for example, carboanhydrase contains zinc in its prosthetic group.

More recently PRASK and PLOCKE (1971) found that ribosomes in the cytoplasm of *Euglena gracilis* normally contain considerable amounts of zinc, and that these organelles become extremely unstable when there is a deficiency of zinc. Confirmation that zinc is necessary for the stability of these ribosomes would also be proof that it is essential for normal growth and development (PRICE et al., 1972).

Many enzymes require the presence of zinc (HOCH and VALLEE, 1957; see also Sect. 6.2). It was shown that zinc binds pyridine nucleotides to the protein portion of yeast ADH (alcohol dehydrogenase) and simultaneously stabilizes the enzyme structure (KÄGI and VALLEE, 1960).

KESSLER and MONSELISE (1959) found that doses of zinc enhanced RNA and protein synthesis in an undernourished plant, but diminished the activity of ribonuclease in citrus leaves. In tomato plants grown under zinc deficient conditions the low zinc content is related to the carboanhydrase activity and the protein-nitrogen content (WOOD and SILBY, 1952). If zinc is supplied to microorganisms (*Neurospora*) suffering from zinc deficiency, the ADH activity is enhanced only in the presence of nitrogen, which indicates that protein synthesis processes are involved. The presence of a source of nitrogen seems to be a general characteristic of zinc sensitive functions (PRICE, 1966).

Zinc deficiency in growing organisms leads to metabolic disorders. First RNA synthesis ceases, then protein, total nitrogen and DNA values decrease (WACKER, 1962; SCHNEIDER and PRICE, 1962). In *Euglena* the absolute quantity of RNA decreases when there is a serious deficiency of zinc (PRICE, 1966). The hydrolysis of RNA increases as the supply of zinc decreases in citrus leaves (KESSLER and MONSELISE, 1959; KESSLER, 1961).

5.3 Intake and Metabolism in Aquatic Animals

HUGGETT et al. (1973) showed that there was a high positive correlation between the zinc content and the copper and cadmium values in oysters from Chesapeake Bay. It became apparent that the accumulation of heavy metals by oysters followed a certain pattern. Thus, additional pollution of the water with zinc could be determined in oysters.

Little is known about the biological availability of zinc in oysters. The zinc intake of bivalves was investigated by BRYAN (1973) (see also FÖRSTNER and WITTMANN, 1979), PHILLIPS (1976), and BLOOM and AYLING (1977). The zinc concentrations in the soft parts, the gonads, the kidneys, and especially in the important glands which produce the digestive secretions show seasonal fluctuations (according to the supply of freshwater) and are lowest at the spawning period. Temperature and the composition of their diet have an influence on their zinc intake. Interactions with iron and cobalt were observed. Mollusc shells are not good bioindicators for various reasons.

Radioactive zinc: the content of zinc-65 in marine organisms was investigated in connection with the waste water from nuclear reactors. PEARCY and OSTERBERG

(1968) monitored the γ-radiation present in the liver of tunafish (*Thunnus alalunga*) along the West coast of North America from 1962 to 1965. The zinc-65 content of the tunafish liver from Southern and Lower California lay between 10 and 100 pCi/g. Values ten times higher than these were found in samples obtained from Northern Oregon and Washington; the concentrations rose appreciably during the summer months. CAREY (1972) reported zinc-65 values in invertebrates on the sea bottom of the Oregon coast. He determined that the fauna on the bottom of the ocean accumulated zinc-65 and that the concentration fell noticeably as the depth of the water increased.

5.4 Intake in Mammals

Zinc is mainly supplied by food, the contribution by drinking water and air to the total supply can be neglected. The dietary intake for adult humans in the Western world ranges from about 7 to 15 mg per day and is quite well correlated to the intake of energy (HALSTED et al., 1974; NFCS, 1980; KAMPE, 1981; DGE, 1984). In general, women, aged persons, and those from lower socioeconomic groups have lower zinc intakes than men, younger adults, and persons of higher socioeconomic status (see also SANDSTEAD et al., 1982; JACOB et al., 1985).

Milk plays an essential role in the infant's diet. Mother's milk contains zinc levels of up to 20 mg/L in colostrum, the average being 3 mg/L for the first two months, then declining with the time of lactation. In cow's milk the mean zinc concentration is about 3.5 mg/L (range 2 to 7 mg/L), but the availability of zinc is lower than in mother's milk (see Sect. 5.5; HAMBIDGE et al., 1979). Milk formulations usually contain less than 2 mg/L, but zinc has been supplemented to the level of cow's milk since 1975. There exist some indications that preschool- and school-aged children sometimes have a marginal zinc intake (COMMNUTR, 1978; VOURI and KUITONEN, 1979; DOMINICK, 1983).

Comparison of the measured or estimated intake of zinc with the recommended dietary allowances and requirements (3 mg/day up to 0.5 years of age, 5 mg/d up to one year, 10 mg/d up to ten years; 15 mg/d up to 51 years, and 9 to 10 and 5.6 to 6.5 mg/d for men and women aged 65 years and older) (NRC, 1980; SANDSTEAD et al., 1982) leads to the conclusion that the zinc requirements for parts of the population in Western countries are only marginally catered for (SOLOMONS, 1982; HENKIN and AAMODT, 1983).

5.5 Absorption in Mammals
(see also Sections 6.3.2 and 6.3.3)

Studies in mammals and man revealed a broad range for the gastro-intestinal absorption rate which can vary between 20 and 80%. Until now the different aspects of zinc absorption and their homeostatic regulation are not fully understood. Recent summaries of the vast literature have been published by HENKIN and AAMODT (1983), SOLOMONS and COUSINS (1984), and KIRCHGESSNER and WEIGAND (1985).

The high variability of zinc absorption can be explained by several factors. Besides pathological conditions which depress the absorption rate, negative effects are recognized by interfering substances (antagonists) in the food, such as high calcium, phytate, fibers, or soybean protein (see also MILLS, 1986; O'DELL, 1987; and CHESTERS, 1987).

Furthermore, zinc absorption is affected by the zinc status, such that in zinc deficiency more zinc is absorbed, while zinc absorption is reduced when the zinc status is more than adequate (SANDSTRÖM et al., 1980; OBERLEAS and HARLAND, 1981; SNEDEKER et al., 1982; SOLOMONS et al., 1983; TURNLUND and KING, 1983; COPPEN and DAVIES, 1987).

According to a review by SOLOMONS (1982) the mean absorption rates range for drinking water from 43 to 66%, for 'meals' and 'breakfast' from 20 to 34%, for poultry and meat from 36 to 41%, for soybean and soybean meals (incl. soy portein infant formula) from 20 to 34%, for white bread from 25 to 38%, and for whole wheat bread from 14 to 17%. All these means show great variations. It has been agreed that foodstuffs of animal origin generally have a superior bioavailability compared with foodstuffs of plant origin (O'DELL et al., 1972; INGLETT, 1983).

Obviously there exist only few enhancers of zinc absorption. Red wine – but no other alcoholic beverages – is claimed to contain such a factor (MACDONALD and MARGEN, 1980). Chelating agents, such as EDTA (ethylenediamine tetraacetate) have been shown to enhance absorption in animal experiments (OBERLEAS et al., 1966; SUSO and EDWARDS, 1971). Histidine and other amino acids also increase the intestinal zinc absorption in animal experiments (GIROUX and PRAKASH, 1977; WAPNIR et al., 1983), and this may explain the better availability of zinc in protein-rich diets.

The reason for the increased availability of zinc from mother's milk compared with cow's milk remains controversial. A low molecular weight factor, either picolinic or citric acid, has been discussed for mother's milk (EVANS and JOHNSON, 1980; COUSINS, 1982; REBELLO et al., 1982; SEAL and HEATON, 1985). According to a recent report the different protein composition between mother's milk (70% whey protein and serum albumin, 20% casein) and cow's milk (20% whey protein, 80% casein) explains these differences. In rat experiments the addition of whey protein and serum albumin to a hemisynthetic diet clearly enhanced the absorption of zinc (given as sulfate, picolinate, or citrate) more than casein did (ROTH and KIRCHGESSNER, 1985).

The absorption of zinc in the intestine mainly occurs in the distal duodenum and proximal jejunum, but other parts of the small and large intestine are also involved. The absorption probably follows a saturable, carrier-mediated process (METHFESSEL and SPENCER, 1973; KOWARSKI et al., 1974; ANTONSON et al., 1979; DAVIES, 1980).

Metallothionein (see Chapter I.12) plays an essential role in the homeostatic regulation of zinc absorption in the intestine, just as in the regulation of zinc metabolism at all (SMITH and COUSINS, 1980; HAMER, 1986; PETERING and FOWLER, 1986). It can be concluded that the biology of zinc absorption is at least as complex as that of iron, and that many details still have to be elucidated (SOLOMONS and COUSINS, 1984).

5.6 Distribution in Mammals
(see also Section 4.5.3)

Absorbed zinc is predominantly transported by plasma albumin, although alpha-2-macroglobulin and possibly transferrin are also involved. Hypoalbuminemia influences therefore not only the plasma- and serum zinc levels, but also zinc absorption. There exists an exchange of zinc between its intracellular pools (high molecular weight zinc-binding proteins and metallothionein) and different organ systems as a major part of homeostasis besides absorption and excretion (SILVERMAN and RIVLIN, 1982; HENKIN and AAMODT, 1983; SOLOMONS and COUSINS, 1984). Two thirds of serum zinc is bound to albumin and one third to alpha-2-macroglobulin and to free amino acids, while 1 to 2% are ultrafiltrable.

5.7 Excretion by Mammals
(see also Section 6.3.3)

The fecal elimination of zinc varies from 5 to 10 mg/d and, therefore, represents 70 to 80% of the ingested zinc (NRC, 1979). Fecal zinc is composed of a non-absorbed dietary component and an excreted component, resulting from zinc losses by pancreatic juice, bile, serosal-luminal transport, and desquamation of the intestinal epithelia. The relative portions of the dietary and excretion component are not well known. They change from species to species, and according to the requirements of the homeostatic regulation (SPENCER et al., 1976; WEIGAND and KIRCHGESSNER, 1978, 1980; KIRCHGESSNER and WEIGAND, 1985).

Urinary zinc excretion is normally negligible with about 0.4 to 0.6 mg/d in healthy subjects (HALSTED et al., 1974; ELINDER et al., 1978; DELVES, 1981; SMITH JR. et al., 1981). Urinary zinc is mainly bound to amino acids and porphyrins, and it is not correlated to variables that determine the serum zinc levels. The major part (about 3/4) of the filtered zinc-amino acid complexes must be reabsorbed by the tubules of the kidney (NRC, 1979). Catabolic processes and the loss of muscle tissue (e.g., by starvation, injuries, surgery) increase the mobilization of zinc and its urinary excretion as well as the infusion of amino acids such as cysteine and histidine (FELL et al., 1973; SPENCER et al., 1976).

Zinc losses in sweat can be significant (up to 3 mg/d) under conditions of extreme heat, since sweat contains about 1.15 mg Zn per L (PRASAD et al., 1963a). Pregnant and lactating women lose considerable amounts of zinc by transfer into the fetus and into the milk.

Total elimination has been estimated to be of the order of 1% per day of the absorbed amount of an orally administered dose of ^{65}zinc (SULLIVAN and HEANEY, 1970; LOMBECK et al., 1975).

5.8 Biological Half-life in Mammals

Measurements in humans, following an oral or parenteral application of ^{65}zinc revealed at least a two-phased whole body elimination kinetics. 10 to 20% of the incorporated dose is eliminated with a medium half-life of 7 to 14 days while about 80% has a half-life ranging individually from 160 to 500 days. The diverse anatomical compartments (liver, muscle, bone) behave kinetically differently. For instance, in healthy volunteers the liver showed a whole body retention pattern characterized by three half-lives of 0.6 d (22%), 7.5 d (48%), and 70 d (30%). Mathematical modelling of zinc kinetics has been done by FOSTER et al. (1979), BABCOCK et al. (1982), and HENKIN and AAMODT, 1983). Less recent literature is summarized in NRC (1979). Whole body retention times decrease in experimental animals with decreasing body size (NRC, 1979).

Recently a simplified open, two-compartment calculation model has been used for the determination of the zinc status, using whole-body or fore-arm countings of injected ^{65}zinc. The normal half-life (219 ± 39 d) was prolonged from 448 to 1146 d in chronic disease states. An oral zinc loading test to detect malabsorption, deficiency, or influences on absorption was also developed (CORNELISSE, 1985; VAN DEN HAMER and CORNELISSE, 1986).

Because of the complex interrelationships of zinc in the different compartments, simple measurements in serum, whole blood, or urine, which give lowered zinc values, cannot be used as a basis of stating a zinc deficiency. Only long-term follow-up, improved methods (see above), and perhaps zinc determination in organs with a minor involvement in homeostatic regulation may provide reliable results in detecting zinc deficiency.

6 Effects on Plants, Animals, and Humans
(see also ELINDER, 1986)

6.1 Effects on Plants and Aquatic Animals

The toxicity of zinc to plants in general is low and is only observed in soils with an excessive zinc burden, for instance, on waste stockpiles, dumping grounds, or flooding areas. Some sensitive plants show signs of zinc toxicity at soil levels of about 300 mg/kg, while others are much more resistant. Growth depression in some cereals has been reported from foreland areas of the Harz mountains (FRG) at soil zinc concentrations of 300 to 320 mg/kg, whereas in maize no effects were observed at concentrations of up to 700 mg/kg (CHESNIN, 1967; MERKEL and KÖSTER, 1977). Details have been summarized in NRC (1979).

The signs of zinc toxicity in plants are not well described, and knowledge on interactions with other micronutrients is meager. Plants growing on zinc-rich soils considerably accumulate this metal, mainly in their vegetative parts, and toxic concentrations in the feed of farm animals can be reached (EL BASSAM, 1982; see Sect. 6.5).

The zinc tolerance of plants is attributed to the inactivation by complexation with malic- or citric acid and to unknown storage mechanisms in the cytoplasm or in vacuoles (TURNER and GREGORY, 1967; PETERSON, 1969; TURNER, 1970). A selection of tolerant plant species takes place on zinc-rich soils (JORDAN and LECHEVALIER, 1975).

The available data on the toxicity of zinc to aquatic organisms are mainly old, not always reliable, and must be regarded as inadequate for establishing ambient water quality criteria (EPA, 1980).

A rather susceptible species seems to be the water flea (*Daphnia magna*) with a mean lethal zinc concentration at 48 h incubation (LC_{50} 48 h) of 1.1 and 1.7 mg/L in hard and soft water, respectively (BERGLIND and DAVE, 1984). According to BIESINGER and CHRISTENSEN (1972) and BIESINGER et al. (1986) 0.14 mg zinc per liter already adversely affect the reproduction of *Daphnia magna* in a three-weeks experiment and moreover, a synergism with mercury and cadmium was observed.

In some fish species the acute zinc LC_{50} of 48 to 96 h has been reported to be in the range of 1 to 10 mg/L, depending on pH, water hardness, temperature, and other experimental circumstances. Tests were mainly performed on the rainbow trout (*Salmo gairdneri*) and the bluegill sunfish (*Lepomis macrochirus*). Other species show lesser sensitivity, for instance, the dogfish (*Scyliorhinus canicula*) (LC_{50} 48 h) 80 mg Zn per L (SANPERA et al., 1983) or the guppy (*Lebistes reticulatus*) (LC_{50} 96 h) 280 to 300 mg $ZnSO_4$ per L (SEHGAL and SAXENA, 1986).

With 0.4 to 0.9 mg/L (LC_{50} 14 d), the subacute toxicity of zinc to salmonids appears to be somewhat lower than the acute toxicity (NEHRING and GOETTL JR., 1974). Subacute and sublethal concentrations cause damage of the reproductive organs in male and female fish and impair the reproduction (UVIOVO and BEATTY, 1979; KUMAR and PANT, 1984; SEHGAL and SAXENA, 1986). With a 48 h LC_{50} of 2 to 9 mg/L the toxicity of zinc to the different stages of embryonic development seems to be of the same order of magnitude as to the parent fishes (ROMBOUGH, 1985; SHAZILI and PASCOE, 1986).

The mechanism of the toxic actions of zinc are still obscure. In vitro- but not in vivo experiments show an inhibitory effect on the respiration of the liver, kidney, and gills (TORT et al., 1984). Zinc exerts cytotoxic effects on fibroblastic cell lines of fishes which is stronger than copper or nickel but weaker than cadmium (BABICH et al., 1986). Metallothionein induction occurs obviously also in fish (ROCH and MC CARTER, 1986). In high zinc concentrations (25 mg/L) necrosis of the hepatic cells and a veil-like film formation on the gills has been observed, affecting the respiration and blood circulation (WONG et al., 1977).

Fishes accumulate zinc only moderately and mainly in the liver (normal values 20 to 60 mg/kg wet weight), whereas in shellfish (oysters) from polluted areas concentrations of 6000 to 9000 mg/kg wet weight (normal 200 to 500 mg/kg) can be reached, which act toxically to man and cause emesis (BOCKRIS, 1977).

It can be summarized that zinc constitutes a hazard to aquatic life in polluted waters (see Sects. 4.2 and 7) although other accompanying metals (cadmium, copper, lead) are much more important, particularly because they act synergistically (THOMPSON et al., 1980; ROCH and MC CARTER, 1986).

6.2 Miscellaneous Biochemical Effects
(see also Sections 5.2 and 5.5)

Evidence of zinc deficiency in the living organism is not surprising, since zinc has been found to play an important role in many biological functions such as enzyme activity, nucleic acid metabolism, protein synthesis, maintenance of membrane structure and function, as well as hormonal activity. Considering all species, zinc is a constituent of more than 200 enzymes and proteins which participate in all major metabolic processes. The confirmed mammalian zinc enzymes include carbonic anhydrase, carboxypeptidases, aminopeptidases, alkaline phosphatase, and various dehydrogenases. The stability constants of the zinc-protein complexes are the basis for their classification into zinc metalloenzymes which are defined as catalytically active metalloproteins containing stoichiometric amounts of zinc (VALLEE, 1955) and zinc-protein complexes. The large number of enzymes of this latter category loosely bind zinc and are activated by the addition of a variety of metal ions (VALLEE and WACKER, 1970). The role of zinc metalloenzymes such as alcohol dehydrogenase, superoxide dismutase, RNA and DNA polymerase, pyruvate carboxylase, and others has recently been summarized by VALLEE (1985). According to VALLEE and GALDES (1984) zinc enzymes are involved in replication, transcription, and translation of genetic material of all species. The specific role of zinc in gene transcription has been characterized by WU and WU (1987).

A number of zinc proteins have been described whose functions are not fully identified. The metallothioneins have gained special interest. These sulfur-rich proteins of low molecular weight are rich in cadmium, zinc, and possibly other bivalent metals, and it is suggested that they play a role in metal metabolism and detoxification (KÄGI and NORDBERG, 1979; FOULKER, 1982; HAMER, 1986; see also Chapter I.12 and Sect. 5.6 in this chapter).

Zinc stabilizes plasma and subcellular membranes (CHVAPIL, 1976), and it has been proposed that zinc protects membranes from free radical oxidation (BETTGER and O'DELL, 1981). The mechanism of this effect may involve the lipid component of the membrane as well as sulfhydryl groups (O'DELL, 1982).

6.3 Deficiency Symptoms in Plants, Animals, and Humans

Over 100 years ago RAULIN (1869) demonstrated that zinc was necessary for the growth of a black bread mold, *Aspergillus niger*. Subsequently, the essentiality of zinc was established for highly developed plant life (SOMMER and LIPMAN, 1926), and animals (TODD et al., 1934).

The first abnormalities of human zinc metabolism were reported by VALLEE et al. (1956). PRASAD et al. (1961) suspected zinc deficiency to occur in young men from Iran. Following studies in Egypt (PRASAD et al., 1963a; SANDSTEAD et al., 1967) confirmed this finding.

6.3.1 Zinc Deficiency in Plants

The role of zinc in plants has been summarized by the NRC (1979). Due to impaired internode growth, disorders caused by zinc deficiency (SAUCHELLI, 1969) result in a rosette condition of the growing tip and greatly distorted and unusually small leaves of apples, peach, and pecan trees. Furthermore, substantial changes in the form and growth habits of some species can be observed, resulting in shortened, stunted plants with poorly developed apical dominance. It is supposed that these symptoms are caused by an inadequate supply of the growth hormone indole acetic acid, since zinc is necessary for its synthesis. Interveinal chlorosis followed by white necrotic spots occurs in the leaves of cereals and fruit trees, suggesting a role for zinc in chlorophyll formation.

The sensitivity of plants to zinc deficiency varies: oats, wheat, barley, and rye are not very sensitive; potatoes, tomatoes, lucerne, and red clover have a medium sensitivity while maize, hop, flax, and soybeans need a high zinc supply (VIETS, 1966; see also Table II.25-3, p. 612 in HENKIN, 1984). In Asia zinc deficiency has been identified as the most widespread micronutritional disorder of wetland rice which occurs on sodic, calcareous, and poorly drained soils (AGARWALA, 1979).

In general, zinc deficiency can be overcome by application of zinc salts to the soil but, for example, the effect of zinc fertilization on the mineral nutrition of rice depends on the tolerance of zinc deficiency (CAYTON et al., 1985). Phosphate application has also been discussed to reduce zinc concentrations in plants due to interactions of both elements (ROBSON and PITMAN, 1983). As an early diagnostic tool for zinc deficiency, the determination of the activity of the zinc enzyme carbonic anhydrase has been proposed (BOUMA, 1983).

6.3.2 Zinc Deficiency in Animals

Only small amounts of the zinc stored in the body of animals are available and, therefore, it must be supplied continuously via the diet. The required amounts of zinc depend on the species, the type of diet, and its zinc content as well as on industrial emissions, and the burden of different antagonists of zinc.

In animals, only a few cases of severe zinc deficiency have been reported (MILLER, 1970). The number of marginally deficient animals, particularly in cases of insufficient supply via fodder plants, is unknown. Zinc deficiency have been studied in different species such as cattle, sheep, goats, swine, chicken, quail, rats, mice, rabbits, dogs, as well as in monkeys. The symptoms which have been described and reviewed (HALSTED et al., 1974; UNDERWOOD, 1977; NRC, 1979) have mainly been observed in young animals and include decreased growth, testicular atrophy, alopecia, and dermal lesions. Zinc deprivation in the rat reduces the food intake (WALLWORK and SANDSTEAD, 1983). In the absence of protein calorie malnutrition, zinc deficiency caused growth retardation and an impaired digestion and absorption of nutrients (PARK et al., 1985). Other authors found decreases in the packed cell volume and total leukocyte count accompanied by absolute lymphocytopenia and relative neutrophilia in guinea pigs (GUPTA et al., 1985). Several studies demonstrated that zinc deficiency affects the immune system (LUECKE, 1978; QUARTEMAN and HUMPHRIES, 1979; SANECKI et al., 1985). A maternal dietary deficiency has been reported to produce teratological manifestations in the offspring of several species (HURLEY and SCHRADER, 1972; SWENERTON and HURLEY, 1980). In general, most symptoms can be reversed by zinc supplementation.

6.3.3 Zinc Deficiency in Humans

A variety of clinical manifestations of zinc deficiency in man have been observed and reviewed several times (HALSTED et al., 1974; AGGETT and HARRIES, 1979; NRC, 1979; SOLOMONS, 1979; HAMBIDGE, 1981; PRASAD, 1983; MILLS and LINDEMANN, 1983; HENKIN and AAMODT, 1983; ZUMKLEY et al., 1983; HENKIN, 1984), but the diagnosis and effective treatment of the most common and widespread of all deficiency syndromes, the marginal zinc deficiency, still remains an outstanding problem. Because of the vast available literature, the following subsections are based on the named reviews and only few special references are cited.

A positive clinical response to zinc therapy can be regarded as the most reliable criterion for zinc deficiency. But in cases of marginal deficiencies clinical features are usually non-specific.

The use of serum or plasma as an indicator of the zinc status is limited since circulating zinc mainly reflects the exchangeable zinc that is delivered to metabolically active tissues. Furthermore, zinc values in plasma or serum greatly vary, are influenced by many factors such as pregnancy, acute or chronic infections, and it should be considered that less than 1% of the total body zinc is circulating in the blood. Because of the slow turnover, the determination of zinc in erythrocytes, leukocytes, and platelets has been discussed as a diagnosis for chronic zinc deficiency, but at present this method has not been fully established.

Similar remarks apply to hair zinc. For pharmacokinetic reasons hair zinc reflects dietary intake over a period of time which depends on the length of the analyzed sample and not the state of total-body zinc metabolism. It has to be taken into account that variations in the rate of hair growth affect the hair zinc concentrations insofar as zinc deficiency itself impairs the growth of hair. The influence of environmental contamination on hair zinc seems negligible.

In many cases of zinc deficiency in man, hypozincuria has been described. On the other hand, there are a number of clinical disorders which might be associated with zinc deficiency but where hyperzincuria is observed.

The activities of certain zinc-binding proteins may provide useful confirmation of the diagnosis of zinc deficiency. KIRCHGESSNER and ROTH (1982) suggested that the percent zinc-binding capacity of serum and the activity of alkaline phosphatase could be suitable methods to estimate the zinc supply status, especially in the marginal range. Zinc balance and kinetic studies with radioisotopes – although these techniques are at present not practicable in routine examinations – offer a very useful tool. HENKIN and AAMODT (1983) presented a new approach to the diagnosis and treatment of zinc deficiency syndromes (see also HENKIN, 1984). The authors organized zinc deficiency into the major classifications such as acute, chronic, and subacute zinc deficiency, and they defined the clinical and biochemical markers as well as the etiological factors and the symptoms for each of these states.

Zinc deficiency in man typically results from inadequate dietary intake, particularly at times of high requirement such as growth, pregnancy, and lactation (see Sect. 5.4). In addition, vegetarianism, synthetic diets, and protein calorie deficiency have been found to be related to zinc deficiency. Severe acute zinc depletion has occurred in patients receiving total parenteral nutrition without zinc supplements. The predominant wheat diet in the Middle East, which contains large quantities of phytate and fiber, reduces the availability of zinc and is thought to be the major etiological factor in cases of zinc deficiency in the rural population in Iran described by PRASAD et al. (1961).

The lack of a specific component in gastrointestinal zinc transport has been discussed as an etiological factor in the pathogenesis of acrodermatitis enteropathica – the classical most severe human zinc deficiency. Malabsorption syndromes, such as coeliac disease, pancreatic insufficiency, chronic inflammatory bowel disease, as well as immaturity of the absorptive system can be associated with zinc deficiency.

Increased body losses of zinc may occur in a wide variety of conditions such as burns, diabetes mellitus, dialysis therapy, nephrotic syndrome, excessive sweating, hepatic disease, chronic blood loss, and sickle cell anemia. Iatrogenic causes include chelating agent therapy and diuretics. Furthermore, zinc deficiency has been found in acute and chronic infectious processes and cancers.

Subacute to chronic mild zinc deficiencies in children are thought to occur more frequently than previously considered; main symptoms are the diminution of physical growth velocity and a decreased food intake (HAMBIDGE and KREBS, 1985). According to HENKIN and AAMODT (1983), the major complex of subacute zinc deficiency is related to taste and smell malfunction.

Features of acrodermatitis enteropathica (MOYNAHAN, 1974; LOMBECK et al. 1975) which develop during early infancy include skin lesions with alopecia totalis, paronechya, and bullous-pustular dermatitis characteristically distributed along the extremities, oral, anal, and genital areas. Moreover, this rare autosomal recessive disorder (see also Chapter I.19) can be associated with diarrhea, blepharitis, conjunctivitis, depressed mood, growth failure, and frequent infections.

The typical observations in zinc deficient Iranian males (PRASAD et al., 1961) included anemia, dwarfism, hypogonadism, hepatosplenomegaly, rough and dry skin, and mental lethargy.

Anorexia, growth retardation, impaired taste and dysosmia are thought to be early signs of zinc deficiency. Further clinical features are delayed wound healing, neurological disorders, night blindness, impotence, and dysarthria. The symptom appearance in children may be different from that of adults. In chronic zinc deficiency, anorexia, diarrhea, irritability and short stature may be predominant in children while in adults taste and smell malfunction, hypogonadism, and poor wound healing may appear as early signs. The main symptoms observed during an experimental zinc deficiency in male volunteers were loss of body weight and testicular hypofunction (PRASAD, 1983).

6.3.4 Treatment of Zinc Deficiency

Different zinc preparations are available for treatment of acute and severe deficiencies. In general, most cases of hypozincemia rapidly respond to zinc therapy. In cases of zinc deficiency parenteral nutrition may be necessary (LADEFOGED, 1986; PRELLWITZ, 1986).

In acrodermatitis enteropathica, zinc therapy results in a prompt clinical and biochemical remission. Without treatment, there was frequently a fatal outcome during early childhood. Zinc supplementation also increased the height-for-age scores in children (WALRAVENS et al., 1983).

The usefulness of zinc administration during pregnancy and lactation is also still discussed. Uremic neuropathy and hypogeusia in dialysis patients could be improved by dialysate zinc supplementation (SPRENGER et al., 1983). The Western populations have been said to be at risk from marginal zinc deficiency, and this has raised the question of zinc supplementation. But as long as there are no reliable screening procedures for the detection of mild zinc deficiencies and the definition of this state is not clearly established this risk remains speculative.

6.4 Acute Effects on Mammals (and Humans)

Zinc salts of strong mineral acids act as adstringents and in higher concentrations also as corrosives. The ingestion of larger doses (approx. one to several grams of $ZnCl_2$ and $ZnSO_4$) in humans can cause serious damage in the upper alimentary tract followed by severe shock symptoms. The lowest lethal dose for humans is estimated to 50 mg $ZnCl_2$ or 106 mg $ZnSO_4$ per kg body weight. Lower doses lead to nausea, emesis, stomach cramps, diarrhea, and fever. Mass intoxications have been reported from fruit juices (up to 2.2 g zinc per liter) or from meals stored in galvanized containers (BROWN et al., 1964). For drinking water, the emetic zinc concentrations range from 675 to 2280 mg/L; metallic taste is sensed at 15 mg/L, and 30 mg/L impart a milky appearance.

The oral toxicity in laboratory animals is remarkably low with LD_{50} values around 1000 to 2500 mg/kg body weight for most zinc compounds. By parenteral application the toxicity is considerably higher (RTECS, 1977; SAX, 1984).

Severe irritations in the upper and lower respiratory tract and a generalized pneumonitis can be induced by inhalative overexposure to zinc chloride smoke (but rather due to formation of hydrochloric acid) developed by smoke bombs (MARRS et al., 1983; MATARESE and MATTEWS, 1986).

Inhalation of zinc oxide fumes (particle size 0.2 to 1 µm) is the main cause of metal fume fever. The fume is formed during several procedures in the processing of zinc. The illness is characterized by sore throat, cough, hoarseness, chills, myalgias, malaise, and fever, mostly accompanied by sweating, nausea, vomiting, and sometimes also by bronchospasm, pulmonary edema, and pneumonitis. An immune complex reaction to the inhaled metal oxide appears to be the most widely accepted pathogenetic factor. A typical phenomenon of metal fume fever in workers is the quick development of the disease after exposure, its short duration, and the fast generation and loss of tolerance ('Monday morning fever') (MUELLER and SEGER, 1985). In the view of new experimental results the current threshold limit value (TLV) of 5 mg ZnO per m^3 is discussed (LAM et al., 1985; GUPTA et al., 1986).

6.5 Chronic Effects on Mammals (and Humans)

Due to the effective mechanism in homeostatic regulation, intoxications by repeated exposures (subacute to chronic) to zinc are rather rare events. In humans the therapeutic or prophylactic oral administration of high zinc doses (up to 350 mg ZnSO$_4$) usually do not cause any overt side effects with exception of mild gastrointestinal complaints in some patients. However, some warnings must be given with respect to the non-indicated supplementation with high doses of zinc. Excessive intakes interfere with copper metabolism and can aggravate marginal copper deficiency (PATTERSON et al., 1985). Interactions with calcium metabolism are also known from animal experiments (YAMAGUCHI et al., 1983; YAMAGUCHI and TAKAHASHI, 1984). Excessive intake (300 mg Zn as ZnSO$_4$ over 6 weeks to healthy adults) impairs immune responses by a reduction of lymphocyte stimulation response as well as chemotaxis and phagocytosis of granulocytes (for immunotoxicology of zinc see also NICKLIN, 1987). In renal dialysis patients dialyzed with certain disposable cuprophan-membrane coils high serum zinc levels (up to 4000 µg/L after dialysis) together with some typical symptoms of zinc toxicity (nausea, vomiting, anemia, lethargy, muscular incoordination) have been observed (BOGDEN et al., 1980).

Although farm animals are regarded as quite tolerant to high zinc levels in the diet, horses, sheep, and cattle have been intoxicated by grazing on forage in the vicinity of zinc smelters. Horses are the most zinc-sensitive farm animals and react with lameness, osteochondrosis (possibly caused by an abnormal collagen metabolism due to an inhibition of lysyl oxidase followed by a zinc provoked copper deficiency), and lymphoid hyperplasia of the spleen and lymph nodes (KOWALCZYK et al., 1986).

In ruminants typical clinical manifestations are loss of condition, diarrhea, subcutaneous edema, profound weakness and jaundice. Furthermore, pathological changes have been found in the exocrine pancreas, kidneys, liver, rumen, abomasum, small intestine, and adrenals (ALLEN et al., 1983; for older literature see NRC, 1979).

No recent reports on subchronic feeding studies in experimental animals (dogs, rats) exist; they are at least 40 years old and cannot be used to derive a non-observable effect level. Rats tolerated 0.55 and 1.0 g Zn per kg diet without symptoms; 4 to 5 g/kg diet affected the reproduction in females (reduced litter size, most offsprings born dead), and 1.0 g/kg diet caused a severe anemia, loss of weight, and finally act lethally. Dogs and cats tolerated 175 to 1000 mg ZnO per day up to 53 weeks with glucosuria in dogs and pathologic changes of the pancreas in cats. Pigs showed at 1000 mg Zn per kg diet a reduction of body weight gain, 2000 mg/kg diet caused hemorrhage in intestine and brain, swollen joints, and death. It can be concluded with prudence, that zinc doses up to 10-fold of the normal level of feed are without adverse effects.

6.6 Mutagenic, Carcinogenic, and Teratogenic Effects
(see also Chapter I.18 and LÉONARD et al., 1986)

Until now, no mutagenic effects of zinc have been observed (LÉONARD et al., 1986). Tests in vitro on *Escherichia coli* (Ames test) and in vivo on *Drosophila melanogaster* gave negative results as well as DNA-fidelity assays (NELSON, 1985). Zinc in toxic dosages (0.5% in the diet) caused in calcium deficient mice only structural chromosomal aberrations in bone marrow cells (DEKNUDT and GERBER, 1979). In workers exposed to lead and zinc a higher frequency of minor chromosomal aberrations was observed than in controls (DEKNUDT and LÉONARD, 1975), but a relationship to zinc seems to be highly questionable. A report on the formation of zinc adducts in DNA-purine nucleotides following short-term administration of low zinc doses (23 mg/L in drinking water of rats for 7 days) has to be confirmed (COOPER, 1985).

The involvement of zinc in carcinogenesis is not fully elucidated, but there is no evidence for increased mortality (ELINDER, 1986). No direct carcinogenic actions of dietary zinc deficiency or supplementation are known, but the growth rate or frequency of transplanted and chemically induced tumors is influenced by the zinc content in the diet. Both promoting and inhibiting actions have been reported depending on the experimental conditions. Zinc is needed for cellular proliferation of existing tumors, and tumor growth is retarded by zinc deficiency (LÉONARD et al., 1986). The induction of testicular tumors by direct injection of zinc salts appears to be without any practical relevance, and no conclusions can be drawn at the present time on the role of zinc in carcinogenesis from the zinc content in tumorous and normal organs of cancer patients (NRC, 1979; EPA, 1980; MILLS et al., 1981; NORDBERG and PERSHAGEN, 1985).

Like other chromates, zinc chromate is a long known human carcinogen. It causes bronchial carcinomas by inhalation of its dusts (LANGARD and VIGANDER, 1983; DAVIES, 1984; LEVY et al., 1986).

There exist no indications of adverse zinc effects on reproduction, embryotoxicity, and teratogenic actions in maternal non-toxic doses. The dithiocarbamates Ziram and Zineb already show in lower dosage adverse effects on the reproduction, which are not due to their zinc content (NRC, 1979; STOKINGER, 1981).

7 Hazard Evaluation and Limiting Concentrations

Zinc as an essential trace element can impair life functions of plants, animals, and man, either by deficiency or surplus.

To prevent deficiency in humans, recommended dietary allowances (RDAs) have been established: 10 mg/d for children and 15 mg/d for adults (for details see Sect. 5.4). The RDAs do not match with the actual intake of considerable groups of the population (see also HENKIN and AAMODT, 1983).

Recommendations and regulations have been established with regard to minimal zinc levels in the feed of farm and domestic animals, considering the requirements of the individual species (APGAR, 1979; FMV, 1981).

Overexposure to zinc by food, water, and air commonly poses no risk to the general population. Long-term administration of zinc (100 to 150 mg/d) to patients to promote wound healing usually is well tolerated but can induce anemia in cases of low copper status. This effect is readily reversible. Using a safety factor of 10 this means that an additional intake of 10 to 15 mg/d does not constitute a health hazard (EPA, 1980). The provisional maximum tolerable daily intake for man has been estimated at 0.3 to 1.0 mg zinc per kg body weight (WHO, 1983).

The aesthetic quality of drinking water is impaired by zinc concentrations higher than 5 mg/L, therefore, this level is taken by most countries as a limit value. It is not based on, but is in line with toxicological considerations (EEC, 1980; WHO, 1983). In the USA zinc is not considered in drinking water regulations at the present time, and in the Federal Republic of Germany the former limit value of 2.0 mg/L has been deleted (EPA, 1985; TRINKWV, 1986).

To protect the health of workers from occupational exposure, threshold limit values (TLV) as time-weighted averages (TWA) for some zinc compounds have been set: zinc chloride fume 1.0 mg/m^3 and zinc oxide fume 5 mg/m^3. For zinc chromate, as a known human carcinogen by its chromate content, a TLV has not been established (ACGIH, 1985/86; MAK, 1987).

Prevention of toxic actions of zinc to plants is achieved in the FRG by restricting the zinc input with composted sewage sludge to 300 mg/kg soil (dry weight) (ABF-KLÄRV, 1982, see Sect. 4.1). Other countries issued comparable regulations, partially limiting the load of the soil to 10 to 15 kg/ha per year (COTTENIE, 1981; PURVES, 1981; MERIAN, 1982b; see also Chapter I.7c).

Protection of aquatic life is ensured by a criterion of 47 µg/L as a 24 h average for total recoverable zinc in fresh water, and limit values (at any time) of 180, 320, and 570 µg/L are given, depending on the hardness of the water, e.g., at 50, 100, and 200 mg CaCO$_3$ per liter. For salt water the criterion has been established at 58 µg total recoverable Zn per liter (as a 24 h average), and the level should not exceed 170 µg/L at any time (EPA, 1980). According to an EEC guideline, fish waters should not contain more than 0.3 mg Zn per liter for salmonids and no more than 1 mg/L for cyprinids at a hardness of 100 mg CaCO$_3$ per liter (see RUF, 1981).

References

AbfKlärV (1982), *Klärschlammverordnung vom 25.6.1982*, pp. 734–739. Bundesgesetzblatt, Bonn, FRG.
ACGIH (American Conference of Governmental Industrial Hygienists) (1985/86), *Threshold Limit Values and Biological Exposure Indices*. Cincinnati, Ohio.
Agarwala, S.C. (1979), *J. Indian Bot. Soc. 58*, 297–311.
Aggett, P.J., and Harries, J.T. (1979), *Arch. Dis. Child. 54*, 909–917.
Allen, J.G., Masters, H.G., Peet, R.L., Mullins, K.R., Lewis, R.D., Skirrow, S.Z., and Fry, J. (1983), *J. Comp. Pathol. 93*, 363–377.
Antonson, D.L., Barak, A.J., and Vanderhoff, J.A. (1979), *J. Nutr. 109*, 142–147.
Apgar, J. (1979), in: NRC (1979), pp. 173–210.

Arnaud, J., Bellanger, J., Bienvenu, F., Chappuis, P., and Favier, A., (1986), *Ann. Biol. Clin. Paris* **44**, 77–87.
Babcock, A.K., Henkin, R.I., Aamodt, R.L., Foster, D.M., and Berman, M. (1982), *Metabolism* **31**, 335–347.
Babich, H., Shopsis, C., and Bodenfreund, E. (1986), *Exotoxicol. Environ. Saf.* **11**, 91–99.
Berglind, R., and Dave, G. (1984), *Bull. Environ. Contam. Toxicol.* **33**, 63–68.
Bernhardt, M., and Zattera, A. (1969), in: Nelson, D.J., and Evans, F.C. (eds.), *Symposium on Radioecology.* U.S. Atomic Energy Commission, pp. 389–398. Oak Ridge, Tennessee (CONF–670503).
Bertram, H.P. (1983), in: Zumkley, H. (ed.), *Spurenelemente,* pp. 1–11, G. Thieme Verlag, Stuttgart-New York.
Bertram, H.P. (1984), in: Zumkley, H. (ed.), *Spurenelemente in der inneren Medizin unter besonderer Berücksichtigung von Zink,* pp. 15–21. Innovations-Verlags-Gesellschaft mbH, Seeheim-Jugenheim, FRG.
Bettger, W.J., and O'Dell, B.L. (1981), *Life Sci.* **28**, 1425–1438.
Biesinger, K.E., and Christensen, G.M. (1972), *J. Fish. Res. Board Can.* **29**, 1691–1700.
Biesinger, K.E., Christensen, G.M., and Fiand, J.T. (1986), *Ecotoxicol, Environ. Saf.* **11**, 9–14.
Bloom, H., and Ayling, G.M. (1977), *Environ. Geol.* **2**, 3–22.
Boawn, L.C., and Rasmussen, P.E. (1971), *Agron. J.* **63**, 874–876.
Boawn, L.C., Nelson, C.E., Viets, F.C., Jr., and Crawford, C.L. (1960), *Wash. Agric. Exp. Stn. Tech. Bull.,* 614.
Bockris, J.O'M. (1977), *Environmental Chemistry,* pp. 461–467. Plenum Press, New York-London.
Bogden, J.D., Oleske, J.M., Weiner, B., Smith, L.G., and Najem, G.R. (1980), *Am. J. Clin. Nutr.* **33**, 1088–1095.
Bouma, D. (1983), in: Läuchli, A., and Bieleski, R.L. (eds.), *Inorganic Plant Nutrition,* p. 140. Springer Verlag, Berlin-Heidelberg-New York.
Bowen, H.J.M. (1980), *Analytica '80, München,* cited in: *Chem. Rundsch.* **33** (23) 5.
Bryan, G.W. (1969), *J. Mar. Biol. Assoc. U.K.* **49**, 225–243.
Bryan, G.W. (1973), *J. Mar. Biol. Assoc. U.K.* **53**, 145–166.
Brown, M.A., Thom, J.V., Orth, G.L., Cova, P., and Juarez, J. (1964), *Arch. Environ. Health* **8**, 657–660.
Bruland, K.W. (1983), *Trace Elements in Sea Water, Chem. Oceanogr.* **8**, 191–194.
Burkitt, A., Lester, P., and Nickles, G. (1972), *Nature* **238**, 327–328.
BUS (Bundesamt für Umweltschutz) (1980), *Mitteilung 5/80,* Bern, Switzerland.
Carroll, M.D., and Loneragan, J.F. (1969), *Aust. J. Agric. Res.* **20**, 457–463.
Cayton, M.T.C., Reyes, E.D., and Neue, H.U. (1985), *Plant Soil,* **87**, 319–327.
Chaney, R.L. (1973), *Proceedings Joint Conference Recycling of Municipal Sludges, Effluents, Land,* pp. 129–141. National Association State Universities and Land Grant Colleges, Washington, D.C.
Chesters, J.K. (1987), *Trace Substances in Environmental Health XXI, Developments in Studies of Zinc Essentiality,* Proceedings, pp. 473–486. University of Missouri; see also *Chemosphere,* in press.
Chesnin, L. (1967), *The Micronutrient Manual,* pp. 13–14. Rayonier, Inc.
Chvapil, M. (1976), *Med. Clin. North Am.* **60**, 799–812.
Commnutr (Committee on Nutrition) (1978), *Pediatrics* **62**, 408–412.
Cooper, H.K. (1985), *Toxicology* **34**, 261–270.
Coppen, D.E., and Davies, N.T. (1987), *Br. J. Nutr.* **57**, 35–44.
Cornelisse, C. (1985), *Zinc Absorption and Retention in Man.* Thesis, University of Utrecht (The Netherlands).
Cottenie, A. (1981), in: *Proceedings International Conference on Heavy Metals in the Environment Amsterdam,* pp. 167–175. CEP Consultants Ltd., Edinburgh.

Cousins, J. R. (1982), in: Prasad, A. S. (ed.), *Clinical, Biochemical and Nutritional Aspects of Trace Elements,* pp. 117–128. Alan R. Liss. Inc., New York.
Cushing, C. E., and Watson, D. G. (1968), *Oikos 19,* 143–145.
Davies, A. G. (1973), in: *Radioactive Contamination of the Culture,* pp. 403–420. International Atomic Energy Agency, Wien.
Davies, N. T. (1980), *Br. J. Nutr. 43,* 189–203.
Davies, J. M. (1984), *Br. J. Ind. Med. 41,* 158–169.
Davies, J. L., and Fell, G. S. (1974), *Clin Chim. Acta 51,* 83–92.
Dean, R. R., and Smith, Jr. J. E. (1973), *Proceedings Joint Conference Recycling of Municipal Sludges, Effluents Land,* pp. 39–43. National Association State Universities and Land Grant Colleges, Washington, D.C.
Deknudt, G., and Gerber, G. B. (1979), *Mutat. Res. 68,* 163–168.
Deknudt, G., and Léonard, A. (1975), *Environ. Physiol. Biochem. 5,* 319–327.
Delves, H. T. (1981), *Proc. Anal. Atom. Spectrosc. 4,* 1–48.
DGE (Deutsche Gesellschaft für Ernährung e.V.) (1984), *Ernährungsbericht 1984,* Frankfurt a. Main.
Dominick, H. C. (1983), *Aktuel. Ernährungsmed. 8,* 136–139.
Drake, M. (1964), in: Bear, F. E. (ed.), *Chemistry of the Soil, American Chemical Society Monographal Series 160,* pp. 395–444. Reinhold Publishing Company, New York.
EEC (European Economic Communities) (1980), *Richtlinie des Rates vom 15.7.1980 über die Qualität von Wasser für den menschlichen Gebrauch.* Amtsblatt der Europäischen Gemeinschaft vom 30.8.1980 No. L229, 11–29.
El Bassam, N. (1982), *GWF Gas Wasserfach Wasser/Abwasser 123,* 539–549.
Elinder, C.-G. (1986), Chapter 28 Zinc, in: Friberg, L., Nordberg, G. F., and Vouk, V. B. (eds.), *Handbook of the Toxicology of Metals,* 2nd Ed., Vol. II, pp. 664–679. Elsevier Science Publ., Amsterdam.
Elinder, C.-G., Kjellström, T., Linnman, L., and Pershagen, G. (1978), *Environ. Res. 15,* 473–484.
EPA (Environmental Protection Agency) (1980), *Ambient Water Quality Criteria for Zinc,* Publication 440/5-80-079. EPA, Office of Water Regulation and Standards, Criteria and Standard Division, Washington, D.C.
EPA (Environmental Protection Agency) (1985), *Proposed Rulemaking: National Primary Drinking Water Regulations; Synthetic Organic Chemicals, Inorganic Chemicals, and Microorganisms; Fed. Reg. 50* (219), 46936–47022.
Evans, G. W., and Johnson, E. C. (1980), *Pediatr. Res. 14,* 876–880.
Farnsworth, M., Kliene, C. H., and Noltes, J. G. (1973), *Zinc Chemicals.* International Lead and Zinc Research Organization Inc., London-New York.
Fell, G. S., Fleck, A., Cuthbertson, D. P., Queen, K., Morrison, C., Bessent, R. G., and Husain, S. L. (1973), *Lancet 1,* 280–282.
Fiedler, H. J., and Rösler, H. J. (1988), *Spurenelemente in der Umwelt,* pp. 97–118. F. Enke Verlag, Stuttgart.
FMV (Futtermittelverordnung) (1981), Verordnung vom 8.4.1981. Bundesgesetzblatt 1.15, Bonn, FRG.
Foroughi, M., Venter, F., and Teicher, K. (1982), *Landwirtsch. Forsch. Sonderh. 38,* 239–248.
Förstner, U., and Wittmann, C. T. W. (1979), in: *Metal Pollution in the Environment,* pp. 28, 86–102, 289–293, 297–303. Springer Verlag, Heidelberg-New York.
Foster, D. M., Aamodt, R. L., Henkin, R. I., and Berman, M. (1979), *Am. J. Physiol. 237,* R340–349.
Foulker, E. C. (ed.) (1982), *Biological Roles of Metallothionein.* Elsevier, Amsterdam.
Hamer, D. H. (1986), *Annu. Rev. Biochem. 55,* 913–951.
Friberg, L., Nordberg, C. F., and Vouk, V. B. (1979), in: *Handbook on the Toxicology on Metals,* pp. 675–685. Elsevier/North Holland, Amsterdam.
Friel, J. K., and Ngyuen, C. D. (1986), *Clin. Chem. 32,* 739–742.

FSP (Food Surveillance Paper) (1981) No. 5, *Zinc.* Ministry of Agriculture, Fisheries and Food, Her Majesty's State Office, London.
Galloway, J. N., Thornton, J. D., Norton, St. A., and Volchock, H. L. (1982), *Atmos. Environ. 16,* 1677 – 1700.
Ganiron, R. B., Adriano, D. C., Paulsen, C. M., and Murphy, L. S. (1969), *Proc. Am. Soil Sci. Soc. 33,* 306 – 309.
Giesler, E., Nehl, H., and Munk, R. (1983), in: *Ullmanns Encyklopädie der technischen Chemie,* 4th Ed., Vol. 24, pp 633 – 640. Verlag Chemie, Weinheim-Deerfield Beach/Florida-Basel.
Giroux, E., and Prakash, N. J. (1977), *J. Pharm. Sci. 66,* 391 – 395.
Goldberg, E. B. (1965), *Chem. Oceanogr. 1,* 163 – 186.
Grunes, D. L., Boawn, L. C., Carlson, C. W., and Vlets, F. G., Jr. (1961), *Agron. J. 53,* 68 – 71.
Gupta, R. P., Verma, P. C., and Gupta, R. K. P. (1985), *Br. J. Nutr. 54,* 421 – 428.
Gupta, S., Panday, S. D., Misra, V., and Viswanathan, P. N. (1986), *Toxicology 38,* 197 – 202.
Halsted, J. A., Smith, Jr. J. C., and Irwin, M. I. (1974), *J. Nutr. 104,* 345 – 378.
Hambidge, K. M. (1981), *Philos. Trans. R. Soc. London B294,* 129 – 144.
Hambidge, K. M., and Krebs, N. F. (1985), in: Gladtke, E., Heimann, G., Lombeck, I., and Eckert, I. (eds.), *Spurenelemente,* pp. 206 – 212. Georg Thieme Verlag, Stuttgart-New York.
Hambidge, K. M., Walravens, P. A., Casey, C. E., Brown, R. M., and Bender, C. (1979), *J. Pediatr. 94,* 607 – 608.
Hamer, D. H. (1986), *Annu. Rev. Biochem. 55,* 913 – 951.
Harvey, R. S., and Patrick, R. (1967), *Biotechnol. Bioeng. 9,* 449 – 456.
Hem, J. D. (1972), *Water Resour. Res. 8,* 661 – 679.
Henkin, R. I. (1984), in: Merian, E. (1st Ed.), *Metalle in der Umwelt,* pp. 597 – 629. Verlag Chemie, Weinheim-Deerfield Beach/Florida-Basel.
Henkin, R. I., and Aamodt, R. L. (1983), in: Inglett, G. E. (ed.), *Nutritional Bioavailability of Zinc,* pp. 83 – 114. ACS Symposium Series 210, American Chemical Society, Washington, D.C.
Henkin, R. I., Apgar, J., Cole, J. F., Coleman, J. E., Cotterill, C. H., Fleisher, M., Goyer, R. A., Greifer, B., Knezek, B. D., Mushak, P., Piscator, M., Stillings, B. R., Taylor, J. K., and Wolfe, D. A. (1979), in: *Zinc,* pp. 123 – 172. University Park Press, Baltimore, Maryland.
Hewitt, E. J. (1966), *Technical Communication No. 22,* pp. 541 ff. Commonwealth Agricultural Bureau, Farnham Royal, England.
Hoch, F. L., and Vallee, B. L. (1957), in: *Proceedings Trace Elements, Wooster, Ohio,* pp. 337 – 363.
Huggett, R. J., Bender, M. E., and Slone, H. D. (1973), *Water Res. 7,* 451 – 460.
Ikuta, K. (1968), *Bull. Jpn. Soc. Sci, Fish. 34,* 482 – 487.
Hurley, L. S., and Schrader, R. E. (1972), in: Pfeiffer, C. C. (ed.): *Neurobiology of the Trace Metals Zinc and Copper,* pp. 7 – 52. Academic Press, New York.
Inglett, G. E. (1983), *Nutritional Bioavailability of Zinc,* ACS Sympos. Ser., American Chemical Society, Washington, D.C.
Jacob, R. A., Russell, R. M., and Sandstead, H. H. (1985), in: Watson, R. R. (ed.): *CRC Handbook of Nutrition in the Aged,* pp. 77 – 88. CRC Press Inc., Boca Raton, Florida.
Jolly, J. H. (1982), in: *Minerals Year Book,* Vol. 1: *Metals and Minerals.* US Bureau of Mines, Dept. of the Interior, Washington, D.C.
Jones, Jr., J. B. (1972), in: Mortredt, J. J., Giordano, P. M., and Lindsay, W. L. (eds.): *Proceedings Symposium Micronutrients in Agriculture,* pp. 319 – 346. Soil Science Society of America, Madison, Wisconsin.
Jordan, M. J., and Lechevalier, E. (1975), *Ecology 56,* 78 – 91.
Kägi, J. H. R., and Nordberg, M. (eds.) (1979): *Metallothionein.* Birkhäuser Verlag, Basel-Boston-Stuttgart.
Kägi, J. H. R., and Vallee, B. L. (1960), *J. Biol. Chem. 235,* 3460 – 3465.
Kampe, W. (1981), *Gesunde Pflanz. 12,* 296 – 302.
Kessler, B. (1961), in: Reuther, W. (ed.): *Plant Analysis and Fertilizer Problems,* pp. 314 – 322. American Institute of Biological Science, Washington, D.C.
Kessler, B., and Monselise, S. P. (1959), *Physiol. Plant 12,* 1 – 7.

Kirchgessner, M., and Roth, H. P. (1982), in: Gawthorne, J. M., Howell, J. McC., and White, C. L. (eds.): *Trace Element Metabolism in Man and Animals*, pp. 327–330. Springer Verlag, Berlin-Heidelberg-New York.

Kirchgessner, M., and Weigand, E. (1985), in: Gladtke, E., Heimann, G., Lombeck, I., and Eckert, I. (eds.): *Spurenelemente*, pp. 30–46. G. Thieme Verlag, Stuttgart-New York.

Kneip, T. J., and Friberg, L. (1986), in: Friberg, L., Nordberg, G. F., and Vouk, V. B. (eds.): *Handbook on the Toxicology of Metals*, Vol. 1, pp. 36–67. Elsevier, Amsterdam-New York-Oxford.

Koch, O. G., and Koch-Dedic, G. H. (1974), in: *Handbuch der Spurenanalyse*, pp. 1328–1350. Springer Verlag, Berlin.

Kosla, T., Siegert, E., Anke, M., and Szentmihalyi, S. (1985), in: Anke, M., Brückner, C., Gürtler, H., and Grün, M. (eds.): *Mengen- und Spurenelemente*, pp. 356–366. Arbeitstagung, Karl-Marx-Universität, Leipzig.

Kowalczyk, D. F., Gunson, D. E., Shoop, C. R., and Ramberg, C. F. (1986), *Environ. Res. 40*, 285–300.

Kowarski, S., Blair-Stnec, S. C., and Schachter, D. (1974), *Am. J. Physiol. 226*, 401–407.

Kumar, S., and Pant, S. C. (1984), *Toxicol. Lett. 23*, 189–194.

Kümmlee, G. (1985), *Zum Verhalten von potentiellen Schadstoffen in Hausmüll und Hausmüllkompost, Fortschrittsber. VDI 15*, (37).

Ladefoged, K. (1986), *Proceedings Trace Element Analytical Chemistry*. de Gruyter, Berlin; see also *Chemosphere 16* (5), N11 (1987).

Lahmann, E. (1982), in: VDI-Verein Deutscher Ingenieure (ed.): *Schwermetalle in der Umwelt*, Chapter III. Ufoplan 10403186.

Lam, H. F., Conner, M. W., Rogers, A. E., Fitzgerald, S., and Amdur, M. O. (1985), *Toxicol. Appl. Pharmacol. 78*, 29–38.

Langard, S., and Vigander, T. (1983), *Br. J. Ind. Med. 40*, 71–74.

Lee, R. E. Jr., and von Lehmden, D. J. (1973), *Air Pollut, Control Assoc. 23*, 853–857.

Léonard, A., Gerber, G. B., and Léonard, F. (1986), *Mutat. Res. 168*, 343–353.

Levy, L. S., Martin, P. A., and Bidstrup, P. L. (1986), *Br. J. Ind. Med. 43*, 243–256.

Lindeman, R. D., and Mills, B. J. (1980), *Miner. Electrolyte Metab. 3*, 226–236.

Lindsay, W. (1972), *Adv. Agron. 24*, 147–186.

Lombeck, I., Schnippering, H. G., Ritzl, F., Feinendegen, L. E., and Bremer, H. J. (1975), *Lancet I*, 855.

Lombeck, I., Wilhelm, M., Hafner, D., Roloff, K., and Ohnesorge, F. K. (1988), *Eur. J. Pediatr. 147*, 179–183.

Lucas, R. E., and Knezek, B. D. (1972), in: *Proceedings Symposium Micronutrients in Agriculture, Muscle Shoals, Alabama*, pp. 265–288. Soil Science Society of America, Madison, Wisconsin.

Luecke, R. W. (1978), *Proc. Natl. Acad. Sci. USA 75*, 5660–5664.

LWA (Landesanstalt für Wasser und Abfall Nordrhein-Westfalen) (1986), *Gewässergüte-Bericht 85*. Düsseldorf, FRG.

Mac Donald, J. T., and Margen, S. (1980), *Am. J. Clin. Nutr. 33*, 1096–1102.

Machelett, B., Staiger, K., and Podlesak, W. (1984), in: Anke, M., Brückner, C., Gürtler, H., and Grün, M. (eds.): *Mengen- und Spurenelemente*, Arbeitstagung 1984, pp. 231–235. Karl-Marx-Universität, Leipzig.

MAK (1987), *Maximum Concentrations at the Workplace, Report No. XXIII, DFG*. VCH Verlagsgesellschaft, Weinheim-Basel-Cambridge-New York.

Marrs, T. C., Clifford, W. E., and Colgrave, H. F. (1983), *Toxicol. Lett. 19*, 247–252.

Massey, H. F., and Loeffel, F. A. (1967), *Agron. J. 59*, 214–217.

Matarese, S. L., and Mattews, J. I. (1986), *Chest 89*, 308–309.

Melin, A., and Michaelis, H. (1983), in: *Ullmanns Encyklopädie der technischen Chemie*, 4th Ed., Vol. 24, pp. 593–626. Verlag Chemie, Weinheim-Deerfield Beach/Florida-Basel.

Mengel, K. (1984), *Ernährung und Stoffwechsel der Pflanze*, pp. 357–359. Gustav Fischer Verlag, Stuttgart.

Merian, E. (1982a), *Schwermetalle im Klärschlamm* (Tagungsbericht GDI Rüschlikon), *Chem. Rundsch. 35* (10), 3.
Merian, E. (1982b), *Schwermetalle in der Umwelt* (Tagungsbericht Amsterdam), *Chem. Rundsch. 35* (16), 9–13.
Merkel, D., and Köster, W. (1977), *Landwirtsch. Forsch. Sonderh. 33/I,* 274–281.
Methfessel, A. H., and Spencer, H. (1973), *J. Appl. Physiol. 34,* 58–62.
Michaelis, H. (1984), in: Merian, E. (ed.), *Metalle in der Umwelt,* p. 16. Verlag Chemie, Weinheim-Deerfield Beach/Florida-Basel.
Miller W. J. (1970), *J. Dairy Sci. 53,* 1123–1135.
Mills, C. F. (1986), *Proceedings Trace Element Analytical Chemistry.* de Gruyter, Berlin; see also *Chemosphere 16* (5), N11 (1987).
Mills, B. J., and Lindemann, R. D. (1983), in: Zumkley, H. (ed.), *Spurenelemente,* pp. 197–214. Georg Thieme Verlag, Stuttgart-New York.
Mills, B. J., Broghamer, W. L., Higgins, P. J., and Lindemann, R. D. (1981), *Am. J. Clin. Nutr. 34,* 1661–1669.
Mink, L. L., and Williams, R. E. (1971), *Idaho Bur. Mines Geol. Pam. 149,* 1–30.
Moore, D. P. (1972), in: *Proceedings Symposium Micronutrients in Agriculture, Muscle Shoals, Alabama,* pp. 171–198. Soil Science Society of America, Madison, Wisconsin.
Moynahan, E. J. (1974), *Lancet II,* 399–400.
Mueller, E. J., and Seger, D. L. (1985), *J. Emerg. Med 2,* 271–274.
Mugwira, L. M., and Knezek, B. D. (1971), *Commun. Soil Sci. Plant Anal. 2,* 337–343.
Nason, A. (1950), *Science 112,* 111–112.
Nehring, R. B., and Goettl, Jr., J. P. (1974), *Bull. Environ. Contam. Toxicol. 12,* 464–469.
Nelson, N. (1985), in: Merian, E., Frei, R. W., Härdi, W., and Schlatter, C. (eds.), *Carcinogenic and Mutagenic Metal Compounds,* pp. 513–527. Gordon & Breach Sci. Publ., New York-London-Paris-Montreux-Tokyo.
NFCS (Nationwide Food Consumption Survey) (1980), *Survey 1977–1978.* USDA Science and Education Administration.
Nicklin, S. (1986), *Proceedings Second Nordic Symposium on Trace Elements in Human Health and Disease,* Odense. W.H.O., Copenhagen.
NIOSH (National Institute for Occupational Safety and Health) (1975), *Criteria for a Recommended Standard: Occupational Exposure to Zinc Oxide,* NEW-Publication No. 76, Washington, D.C.
Nordberg, F. F., and Pershagen, G. (1985), in: Merian, E., Frei, R. W., Härdi, B., and Schlatter, C. (eds.), *Carcinogenic and Mutagenic Metal Compounds,* pp. 491–506. Gordon & Breach Science Publ., New York-London-Paris-Montreux-Tokyo.
NRC (National Research Council) (1979) Subcommittee on Zinc, *Zinc.* University Park Press, Baltimore, Maryland.
NRC (National Research Council) (1980), Food and Nutrition Board, *Recommended Dietary Allowances,* 9th rev. Ed. National Academcy of Sciences, Washington, D.C.
Nriagu, J. O., and Pacyna, J. M. (1988), *Quantitative Assessment of Worldwide Contamination of Air, Water and Soils by Trace Metals. Nature 333,* 134–139.
Nürnberg, H.-W., Valenta, P., and Nguyen, V. D. (1982), in: Georgii, H. W., and Pankrath, J. (eds.), *Deposition of Atmospheric Pollutants,* pp. 143–157. M. Reidel Publ. Comp., Dordrecht-Boston.
Nürnberg, H.-W., Valenta, P., Nguyen, V. D., Gödde, M., and Urano de Carvalho, E. (1984), *Fresenius Z. Anal. Chem. 317,* 314–323.
Oberleas, D., and Harland, B. F. (1981), *J. Am. Diet. Assoc. 79,* 433–436.
Oberleas, D., Muhrer, M. E., and O'Bell, B. L. (1966), *J. Nutr. 90,* 56–62.
O'Dell, B. L. (1982), in: Gawthorne, J. M., Howell, J. McC., and White, C. L. (eds.), *Trace Element Metabolism in Man and Animals,* pp. 319–326. Springer Verlag, Berlin-Heidelberg-New York.
O'Dell, B. (1987), *Trace Substances in Environmental Health XXI,* University of Missouri; see *Chemosphere,* in press.
O'Dell, B. L., Burpo, C. E., and Savage, J. E. (1972), *J. Nutr. 102,* 653–660.

Olson, S. R., and Kemper, W. D. (1968), *Adv. Agron. 20*, 91–151.
Ondow, J. M., Zoller, W. H., and Gordow, G. E. (1974), in: *Trace Elements on Aerosols from Motor Vehicles, Proceedings 67th Annual Meeting Air Pollution Control Association*, pp. 74–197. Denver, Colorado.
Osis, D., Kramer, L., Wiatrowski, E., and Spencer, H. (1972), *Am. J. Clin. Nutr. 25*, 582–588.
Park, J. H. Y., Grandjean, C. J., Antonson, D. L., and Vanderhoof, J. A. (1985), *Pediatr. Res. 19*, 1333–1336.
Pattenden, N. J., Bransow, J. R., and Fisher, E. M. R. (1982), in: Georgii, H. W., and Pankrath, J. (eds.), *Deposition of Atmospheric Pollutants*. M. Reidel Publ. Comp., Dordrecht-Boston.
Patterson, W. P., Winkelmann, M., and Perry, M. C. (1985), *Ann. Intern. Med. 103*, 385–386.
Pavon, J. M. C., Pozo, M. E. U., and de Torres, A. G. (1986), *Anal. Chem. 58*, 1449–1451.
Pearcy, W. C., and Osterberg, C. L. (1968), *Limnol. Oceanogr. 13*, 490–498.
Pentreath, R. J. (1971), in: *Proceedings of the 2nd ENEA Seminar on Marine Radioecology*, pp. 97–126. ENEA, Paris.
Petering, T. H., and Fowler, B. A. (1986), *Environ. Health Perspect. 65*, 217–224.
Peterson, P. J. (1969), *J. Exp. Bot. 20*, 863–875.
Phillips, D. J. H. (1976), *Mar. Biol. 38*, 59–69.
Pötzschke, M. (1983), in: *Ullmanns Encyklopädie der technischen Chemie*, 4th Ed., Vol. 24, pp. 627–663. Verlag Chemie, Weinheim-Deerfield Beach/Florida-Basel.
Pound, C. E., and Crites, R. W. (1973), in: *Proceedings Joint Conference on Recycling of Municipal Sludges, Effluents, Land,* Champaign, Illinois, pp. 49–61. National Association State Universities and Land Grant Colleges, Washington, D.C.
Prasad, A. S. (1983), in: Inglett, G. E. (ed.), *Nutritional Bioavailability of Zinc*, pp. 1–14. ACS Symposium Series 210, American Chemical Society, Washington, D.C.
Prasad, A. S., Halsted, J. A., and Nadimi, M. (1961), *Am. J. Med. 31*, 532–546.
Prasad, A. S., Miale, A., Jr., Farid, Z., Schulert, A. R., and Sandstead, H. H. (1963a), *J. Lab. Clin. Med. 61*, 537–549.
Prasad, A. S., Schulert, A. R., Sandstead, H.-H., Miale, A. Jr., and Farid, Z. (1963b), *J. Lab. Clin. Med. 62*, 84–89.
Prask, J. A., and Plocke, D. J. (1971), *Plant Physiol. 48*, 150–155.
Prellwitz, W. (1986), *Proceedings Trace Element Analytical Chemistry*. de Gruyter, Berlin; see also *Chemosphere 16* (5), N11 (1987).
Price, C. A. (1966), in: Prasad, A. S. (ed.), *Zinc Metabolism*, pp. 69–89. C. C. Thomas, Springfield, Illinois.
Price, C. A., Clark, H. E., and Funkhausen, E. A. (1972), in: Mortredt, J. J., Giordano, P. M., and Lindsay, W. L. (eds.), *Proceedings Symposium Micronutrients in Agriculture, Muscle Shoals, Alabama*. Soil Science Society of America, Madison, Wisconsin.
Purves, D. (1981), in: *Proceedings Heavy Metals in the Environment Amsterdam*, pp. 176–179. CEP Consultants Ltd., Edinburgh.
Quarteman, J., and Humphries, W. R. (1979), *Life Sci. 24*, 177–184.
Raulin, J. (1869), *Ann. Sci. Nat. Bot. Biol. Veg. 11*, 93–299.
Rebello, T., Lönnerdal, B., and Hurley, L. S. (1982), *Am. J. Clin. Nutr. 35*, 1–5.
Reed, S. C. (1972), *Wastewater Management by Disposal of the Land,* Special Report 171. Cold Regions Research and Engineering Laboratory, U.S. Army Corps of Engineers, Hanover, New Hampshire.
Rice, T. R. (1963), in: Schultz, V., and Klement, A. W., Jr. (eds.), *Radioecology,* pp. 619–631. Reinhold, New York.
Riceman, D. S., and Jones, C. B. (1956), *Aust. J. Agric. Res. 7*, 495–503.
Risch, M. A., and Nasarow, S. (1985), in: Anke, M., Brückner, C., Gürtler, H., and Grün, M. (eds.), *Mengen- und Spurenelemente,* pp. 349–355. Arbeitstagung, Karl-Marx-Universität, Leipzig.
Robson, A. D., and Pitman, M. G. (1983), in: Läuchli, A., and Bieleski, R. L. (eds.), *Inorganic Plant Nutrition,* pp. 169–170. Springer Verlag, Berlin-Heidelberg-New York-Tokyo.
Roch, M., and Mc Carter, J. A. (1986), *Bull. Environ. Contam. Toxicol. 36*, 168–175.

Rombough, P. J. (1985), *Comp. Biochem. Physiol. 82C*, 115–117.
Rose, F. L., and Cushing, C. E. (1970), *Science 168*, 576–577.
Roth, H.-P., and Kirchgessner, M. (1985), *J. Nutr. 115*, 1641–1649.
RTECS (NIOSH Registry of Toxic Effects of Chemical Substances) (1977). U.S. Department of Health, Education and Welfare, National Institute for Occupational Safety and Health, Cincinnati, Ohio.
Ruf, N. (1981), in: *Proceedings Wasser Berlin*, pp. 415–428. O. H. Hess-Verlag, Berlin.
Saager, R. (1984), *Metallic Raw Materials Dictionary* (in German), pp. 78–82. Bank von Tobel, Zürich.
Sandstead, H. H., Prasad, A. S., Schulert, A. R., Farid, Z., Miale, A., Jr., Bassily, S., and Darby, W. J. (1967), *Am. J. Clin. Nutr. 20*, 422–442.
Sandstead, H. H., Henriksen, L. K., Greger, J. L., Prasad, A. J., and Good, R. A. (1982), *Am. J. Clin. Nutr. 36*, 1046–1059.
Sandström, B., Arvidsson, B., Cederblad, A., and Björn-Rasmussen, E. (1980), *Am. J. Clin. Nutr. 33*, 739–745.
Sandström, G., Cederblad, A., and Lönnerdal, B. (1983), *Am. J. Dis. Child. 137*, 726–729.
Sanecki, R. K., Corbin, J. E., and Forbes, R. M. (1985), *Am. J. Vet. Res. 46*, 2120–2123.
Sanpera, C., Vallribera, M., and Crespo, S. (1983), *Bull. Environ. Contam. Toxicol. 31*, 415–417.
Sauchelli, V. (1969), in: *Trace Elements in Agriculture*, pp. 39–57. Van Nostrand Reinhold Co., New York.
Sax, N. I. (1984), *Dangerous Properties of Industrial Materials*, 6th Ed. Van Nostrand Reinhold Comp., New York.
Schmid, W. E., Haag, H. P., and Epstein, E. (1965), *Physiol. Plant. 18*, 860–869.
Schneider, E., and Price, C. A. (1962), *Biochim. Biophys. Acta 55*, 406–408.
Schütte, K. H. (1964), *The Biology of Trace Elements. Their Role in Nutrition*, pp. 228ff. J. B. Lippincott, Philadelphia.
Schweiger, P. (1984), *Gewässerschutz Wasser Abwasser 65*, 439–449.
Seal, C. J., and Heaton, F. W. (1985), *J. Nutr. 115*, 986–993.
Sehgal, R., and Saxena, A. B. (1986), *Bull. Environ. Contam. Toxicol. 33*, 888–894.
Shazili, N. A. M., and Pascoe, D. (1986), *Bull. Environ. Contam. Toxicol. 36*, 468–474.
Siegert, E., Szentmihalyi, S., Anke, M., and Grün, M. (1985), in: Anke, M., Brückner, C., Gürtler, H., and Grün, M. (eds.), *Mengen- und Spurenelemente*, pp. 460–465. Karl-Marx-Universität, Leipzig.
Siegert, E., Anke, M., Szentmihalyi, S., Regius, A., Lokyay, D., Pavel, J., Grün, M., and Hora, K. (1986), in: Anke, M., Baumann, W., Bräunlich, H., Brückner, C., and Groppel, B. (eds.), *5. Spurenelement-Symposium*, pp. 487–493. Karl-Marx-Universität, Leipzig, Friedrich-Schiller-Universität, Jena.
Silverman, B., and Rivlin, R. S. (1982), *J. Nutr. 112*, 744–749.
Skoog, F. (1940), *Am. J. Bot. 27*, 939–951.
Smith, K. T., and Cousins, J. R. (1980), *J. Nutr. 110*, 316–323.
Smith, Jr., C. J., Andersen, R. A., Ferretti, R., Levander, O. A., Morris, E. R., Roginski, E. E., Veillon, C., Wolf, W. R., Anderson, J. B., and Mertz, W. (1981), *Fed. Proc. Fed. Am. Soc. Exp. Biol. 40*, 2120–2125.
Snedeker, S. M., Smith, S. A., and Greger, J. L. (1982), *J. Nutr. 112*, 136–143.
Solomons, N. W. (1979), *Am. J. Clin. Nutr. 32*, 856–871.
Solomons, N. W. (1982), *Am. J. Clin. Nutr. 35*, 1048–1075.
Solomons, N. W., Jacob, R. A., Pineda, O., and Viteri, F. E. (1979), *J. Nutr. 109*, 1519–1528.
Solomons, N. W., and Cousins, J. R. (1984), *Top. Nutr. Dis. 12*, 125–197.
Solomons, N. W., Guerrero, A.-M., Torun, B., Johnson, P., and Milne, D. B. (1983), *Fed. Proc. Fed. Am. Soc. Exp. Biol. 42*, 823–833.
Sommer, A. L., and Lipman, C. B. (1926), *Plant Physiol. 1*, 231–249.
Spencer, H., Osis, B., Kramer, L., and Norris, C. (1976), in: Prasad, A. S. (ed.), *Trace Elements in Human Health and Disease*, pp. 345–361. Academic Press, New York.

Sprenger, K. B. G., Bundschu, D., Lewis, K., Spohn, B., Schmitz, J., and Franz, H. E. (1983) *Kidney Int. 24*, (Suppl. 16), 315–318.

Stokinger, H. E. (1981), *Zinc,* in: Clayton, G. D., and Clayton, F. E. (eds.), *Patty's Industrial Hygienic and Toxicology,* Vol. II a, pp. 2033–2040. John Wiley & Sons, New York-Chichester-Brisbane-Toronto.

Suess, M. J., Grefen, K., and Reinisch, D. W. (eds.) (1985) *Ambient Air Pollutants from Industrial Sources,* pp. 422–433. Elsevier, Amsterdam-Oxford-New York.

Sullivan, J. F., and Heaney, R. P. (1970), *Am. J. Clin. Nutr. 23*, 170–177.

Suso, F. A, and Edwards, H. M., Jr. (1971), *Proc. Soc. Exp. Biol. Med. 36*, 211–213.

Swenerton, H., and Hurley, L. S. (1980), *J. Nutr. 110*, 575–583.

Thompson, K. W., Hendricks, A. C., and Cairns, J. C. (1980), *Bull. Environ. Contam. Toxicol. 25*, 122–129.

Tiffin, L. O. (1972), in: *Proceedings Symposium Micronutrients in Agriculture, Muscle Shoals, Alabama,* pp. 199–229. Soil Science Society of America, Madison, Wisconsin.

Todd, W. R., Elvehjem, C. A., and Hart, E. B. (1934), *Am. J. Physiol. 107*, 146–156.

Tort, L., Flos, R., and Balasch, J. (1984), *Comp. Biochem. Physiol. 77C*, 381–384.

TRINKWV (Trinkwasserverordnung) (1986), *Verordnung über Trinkwasser und über Wasser für Lebensmittelbetriebe* v. 22. Mai 1986, Bundesgesetzblatt, Teil I, 760–773, Bonn, FRG.

Trueb, L. (1988), *Hot Galvanizing* (in German). *Neue Zürcher Zeitung, Forschung und Technik, No. 232,* p. 87 (5. Oct.).

Tsalev, D. L. (1984), *Atomic Absorption Spectrometry in Occupational and Environmental Health Practice,* Vol. II, pp. 215–223. CRC Press, Boca Raton, Florida.

Tsui, C. (1948), *Am. J. Bot. 35*, 172–179.

Tu, A. T. (1977), *Venoms Chemistry and Molecular Biology.* John Wiley, New York.

Turner, R. G. (1970), *New Phytol. 69*, 725–731.

Turner, R. G., and Gregory, R. P. G. (1967), in: *Isotopes in Plant Nutrition and Physiology,* pp. 493–509. International Atomic Energy Commission, Vienna.

Turnlund, J. R., and King, J. C. (1983), *Fed. Proc. Fed. Am. Soc. Exp. Biol. 42*, 822–824.

Underwood, D. J. (1977), *Trace Elements in Human and Animal Nutrition,* 4th Ed., pp. 196–242. Academic Press, New York.

Unice, A. A., King, R. W., Jr., Kraikitpanitch, S., and Lindeman, R. D. (1978), *Am. J. Physiol. 235*, 140–145.

Uthe, J. F., and Blight, E. G. (1971), *J. Fish. Res. Board Can. 28* (5), 786–788.

Uviovo, E. J., and Beatty, D. D. (1979), *Bull. Environ. Contam. Toxicol. 23*, 650–657.

Vahrenkamp, H. (1988), *Is Zinc a Boring Element?* (in German). *Chem. Unserer Zeit 22* (3), 73–84.

Valenta, P., Nguyen, V. D., and Nürnberg, H. W. (1986), *Sci. Total Environ. 55*, 311–320.

Vallee, B. L. (1955), *Adv. Protein Chem. 10*, 317–384.

Vallee, B. L. (1985), in: Gladtke, E., Heimann, G., Lombeck, I., and Eckert, I. (eds.), *Spurenelemente,* pp. 1–22. Georg Thieme Verlag, Stuttgart-New York.

Vallee, B. L., and Galdes, A. (1984), *Adv. Enzymol. 56*, 283–430.

Vallee, B. L., and Wacker, W. E. C. (1970), in: Neurath, H., and Hill, R. (eds.), *The Proteins: Composition, Structure and Function,* Vol. 5, pp. 1–192. Academic Press, New York.

Vallee, B. L., Wacker, W. E. C., Bartholomay, A. F., and Robin, E. D. (1956), *N. Engl. J. Med. 255*, 403–408.

Van den Hamer, C. J. A., and Cornelisse, C. (1986), in: Anke, M., Baumann, E., Bräunlich, H., Brückner, C., and Groppel, B. (eds.), *5. Spurenelement-Symposium 1986,* pp. 515–521. Karl-Marx-Universität Leipzig, Friedrich-Schiller-Universität Jena, GDR.

VDI (Verein Deutscher Ingenieure) (1984), *Schwermetalle in der Umwelt,* Chapter II, III, Ufoplan 10403186.

Viets, F. G., Jr. (1966), in: Prasad, A. S. (ed.), *Zinc Metabolism,* pp. 90–128. C. C. Thomas, Springfield, Illinois.

Viets, F. G., Jr., Boawn, L. C., and Crawford, C. L. (1957), *Soil Sci. Soc. Am. Proc. 21*, 197–201.

Vinogradov, A. P. (1953), *The Elementary Composition of Marine Organisms*, pp. 647ff. Yale University, New Haven, Connecticut.
Vouri, E., and Kuitonen, P. (1979). *Acta Paediatr. Scand. 68*, 33–37.
Wacker, W. E. C. (1962), *Biochemistry 1*, 859–865.
Wallwork, J. C., and Sandstead, H. H. (1983), *J. Nutr. 113*, 47–54.
Walravens, P. A., Krebs, N. F., and Hambidge, K. M. (1983), *Am. J. Clin. Nutr. 29*, 1114–1121.
Wapnir, R. A., Khani, D. E., Bayer, M. A., and Lifshitz, F. (1983), *J. Nutr. 113*, 1346–1354.
Ward, N. I. (1987), *Environmental Health No. 20: Trace Elements in Human Health and Disease*, pp. 118–123. W.H.O. Copenhagen.
Weast, R. C. (1986), *CRC Handbook of Chemistry and Physics*, 66th Ed. CRC Press Inc., Boca Raton, Florida.
Wedepohl, K. H. (1972), in: *Handbook of Geochemistry*, Vol. II-3. Springer Verlag, Heidelberg-New York.
Wedepohl, K. H. (1984), in: Merian, E. (ed.), *Metalle in der Umwelt*, p. 4. Verlag Chemie, Weinheim-Deerfield Beach/Florida-Basel.
Weigand, E., and Kirchgessner, M. (1978), *Nutr. Metab. 22*, 101–112.
Weigand, E., and Kirchgessner, M. (1980), *J. Nutr. 110*, 469–480.
Welz, B. (1985), *Atomic Absorption Spectrometry*, 2nd Ed., pp 338/339, 346ff, 399ff. VCH Verlagsgesellschaft, Weinheim-Deerfield Beach/Florida-Basel.
WHO (World Health Organization) (1982), *Toxicological Evaluation of Certain Food Additives*, WHO Food Additives Series No. 17, pp. 320–339. World Health Organization, Geneva.
WHO (World Health Organization) (1983), *Guidelines for Drinking Water Quality*, Vol. I, *Recommendations*, p. 19. World Health Organization, Geneva.
Wolfe, D. A. (1970), *J. Fish. Res. Board Can. 27*, 47–57.
Wong, M. H., Luk, K. C., and Choi, K. Y. (1977), *Acta Anat. 99*, 450–454.
Wood, J. G., and Sibly, P. M. (1952), *Aust. J. Sci. Res. Ser. B5*, 244–255.
Wu, F. Y.-H. and Wu, C.-W. (1987), *Annu. Rev. Nutr. 7*, 251–272.
Yamaguchi, M., and Takahashi, K. (1984), *Toxicol. Lett. 22*, 175–180.
Yamaguchi, M., Takahashi, K., and Okada, S. (1983), *Toxicol. Appl. Pharmacol. 67*, 224–228.
Zumkley, H., Zidek, W., and Bertram, H. P. (1983), *Aktuel. Ernährungsmed. 8*, 116–118.

Additional Recommended Literature

Diels, L., and Mergeay, M. (1989), *Isolation and Identification of Bacteria Living in Environments Severely Contaminated with Heavy Metals. Proceedings 7th International Conference on Heavy Metals in the Environment*, Geneva, Vol. 1, pp. 61–64. CEP Consultants Ltd., Edinburgh, UK; see also Report E. Merian, *Swiss Chem. 12*(3), 32 (1990).
Draxler, J., and Marr, R. (1989), *Liquid-membrane Permeation (e.g., for Zinc Recovery in the Viscose Industry). Proceedings 6th International Recycling Congress, Berlin, Waste Reduction in the Metallurgical Industry*, pp. 139–148. EF-Verlag für Energie- und Umwelttechnik GmbH, Berlin.
von Frenckell, B. A. K., and Hutchinson, T. C. (1989), *Co-tolerance to Metals in the Grass Deschampsia cespitosa. Proceedings 7th International Conference on Heavy Metals in the Environment*, Geneva, Vol. 2, pp. 182–185. CEP Consultants Ltd., Edinburgh, UK; see also Report E. Merian, *Swiss Chem. 12*(3), 34 (1990).
Fuge, R., and Hennah, T. J. (1990), *Fluorine and Heavy Metals in the Vicinity of Brickworks. Proceedings Conference on Trace Substances in Environmental Health XXIII*, Cincinnati (Ohio); *Environ. Geochem. Health*, Suppl. to Vol. 12, 183–197.

Harmens, H., Verkleij, J. A. C., Koevoets, P., and Ernst, W. H. O. (1989), *The Role of Organic Acids and Phytochelatins in the Mechanism of Zinc Tolerance in Silene vulgaris. Proceedings 7th International Conference on Heavy Metals in the Environment*, Geneva, Vol. 2, pp. 178–181. CEP Consultants Ltd., Edinburgh, UK; see also Report E. Merian, *Swiss Chem.* *12*(3), 34 (1990).

Skoryna, S. C., Nagamachi, Y., and Dvorak, V. A. (1990), *Changing Concepts in Essentiality of Trace Elements in Life and Health. Proceedings Conference on Trace Substances in Environmental Health XXIII*, Cincinnati (Ohio), *Environ. Geochem. Health*, Suppl. to Vol. 12, 21–32.

Yakowitz, H. (1989), *Encouraging Waste Reduction/Minimization and Recycling: A Policy-oriented Overview* (especially on lead and zinc recycling). *Proceedings XXXth IPRE Symposium on Waste Reduction Management*, Haarlem, The Netherlands. IPRE, Brussels.

II.37 Zirconium

KARL-HEINZ SCHALLER, Erlangen, Federal Republic of Germany

1 Introduction

As a non-essential metal that has not been regarded as a health hazard, the effects of zirconium in the environment have received little attention. In nature, zirconium is found together with other metals in many minerals, the technologically most important being zirconia (ZrO_2, baddeleylite) and zircon ($ZrO_2 \cdot SiO_2$). As new applications for zirconium compounds are being found in the ceramics and electronics industries, the study of zirconium wastes and by-products may become more important.

2 Physical and Chemical Properties, and Analytical Methods

In the molten state, zirconium (Zr, atomic number 40, atomic mass 91.22, melting point 1852 °C) has a glossy appearance similar to that of steel, in contrast to fine zirconium powder, which is black. Zirconium belongs to the subgroup IV of the Periodic System, between the elements titanium and hafnium, two metals with which it is often found in nature. Zirconium has oxidation states ranging from II to IV, of which the tetravalent is relatively stable and abundant (VENUGOPAL and LUCKEY, 1979).

Colorimetric and fluorometric methods were formerly used to determine the zirconium content of environmental samples. A survey of the most important methods has been given by BERMAN (1981). These methods are non-specific and in general produce values that are too high. HAMILTON et al. (1972/1973) determined zirconium content by spark source mass spectrometry and obtained values that were 100 times lower than corresponding spectrophotometric investigations by SCHROEDER and BALASSA (1966). Neutron activation analysis (BROOKS, 1968), emission spectroscopy, and atomic absorption spectrometry (AAS) are the most important analytical techniques today (BERMAN, 1981; WELZ, 1985). The detection limit of flame-AAS for zirconium is 5 µg/mL, for emission spectroscopy the limits are 2 µg/mL or 5 ng/mL, using an electric arc and plasma, respectively (PINTA, 1978). An improvement in analytical methods can be expected by combining extraction techniques with graphite furnace methods, but prior to analysis zirconium compounds and -materials have to be decomposed (AYRANCI, 1989).

3 Sources, Production, Important Compounds, Uses, and Waste Products

(DRESSLER et al., 1983; SAAGER, 1984)

3.1 Occurrence and Production

The zirconium concentration of the earth's crust varies considerably, ranging from 150 to 300 mg/kg of soil (MILLER, 1965). It is therefore classified as the 20th most common element, between barium and chromium, and is consequently more abundant than nickel, tin, copper, and lead. Suitably exploitable deposits exist in Australia, South Africa, the Soviet Union, and other countries. About 650000 tons of zirconium sand are produced annually (DRESSLER et al., 1983; SAAGER, 1984). More important sources of zirconium today are the remainders left over from titanium production (SMITH and CARSON, 1978). Hafnium must be separated. Metallic zirconium can be obtained by reduction of zirconium chloride with magnesium (DRESSLER et al., 1983; SAAGER, 1984).

3.2 Uses

(see also RUBEL, 1983; DEKNUDT, 1988; TRUEB, 1990)

Pure zirconium metal is highly resistant to heat and corrosion and it imparts these properties to alloys as well. For these reasons it has become an important material in the aviation, aerospace, chemical, and surgical instrument industries, and in nuclear reactor technology. The ability of zirconium to reject neutrons is utilized for the protection of heating elements in pressurized water and hot water reactors.

Zirconium compounds can be used for water-repellent textiles, in dyes and pigments, for tanning leather and in the glass and ceramics industries (see, for instance, GAUCKLER, 1987). BAYER and WIEDEMANN (1981) reported on the mineralogy of zirconium and its uses as precious stones (zircon, phianite, and djevalithe are all very hard), mineral raw materials, foundry sand, abrasives, and corrosion resistant metals. Insoluble zirconium silicates have been used in cosmetic creams, powders, and antiperspirants. Zirconium tetrachloride is a white powder that is very moisture sensitive (DRESSLER et al., 1983).

3.3 Waste Products

Industrial by-products that contain zirconium are mainly zircon and zirconia, both insoluble in water, largely inert and of low toxicity. Water-soluble zirconium compounds are converted at pH 4–9.5 into insoluble zirconia. The only possible atmospheric emission of other zirconium compounds is that of chorinated and/or hydrolized oxychlorides from the processing of "sponge zirconium" using the Kroll process (reduction of $ZrCl_4$). Analytical data corresponding to the extent of these emissions do not exist (SMITH and CARSON, 1978), but there is no doubt that the exposure of the general population to zirconium compounds is small.

4 Distribution in the Environment, in Foods, and in Living Organisms

The average zirconium content of freshwater is between 0.002 and 0.02 ppm, and in seawater between 0.02 and 0.5 ppb (according to BRULAND, 1983, 0.01 – 0.04 ppb, mainly in the form of $Zr(OH)_4^0$ and $Zr(OH)_5^-$). It is improbable that industrial emissions of zirconium compounds lead to significant increases in these values. Zirconium does not accumulate in the food chain. Plants contain significantly lower zirconium concentrations than the soil in which they grow (SMITH and CARSON, 1978). According to BOWEN (1979), the zirconium content of organisms ranges from 0.3 to 2 ppm (dry weight) in plants and is about 0.08 ppm in mammalian muscle. The zirconium content of foodstuffs reported in the literature has for the most part been determined by methods that were analytically unreliable, as is also the case in human studies. In general, lamb, pork, eggs, dairy products, grain, and vegetables contain relatively high amounts, and daily zirconium intake is about 3.5 mg (according to SCHROEDER and BALASSA, 1966). It is remarkable that the human body contains about 300 mg zirconium (see Chapter I.9) or about 4 ppm, 67% present in the fat, 2.5% in the blood, and the rest in the skeleton, aorta, lungs, liver, brain, kidneys, and other tissues (SCHROEDER and BALASSA, 1966). Significant amounts of zirconium are present in fetuses.

5 Uptake, Absorption, Transport and Distribution, Metabolism and Elimination in Plants, Animals, and Humans

Plant uptake of zirconium from soil and fertilizers has been demonstrated. In animals, zirconium and zirconia are either absorbed by oral intake or inhalation. The greatest portion of ingested water-soluble zirconium salts is converted into zirconium oxides in the small intestine. Zirconium concentrations in the brain, kidneys, liver, lungs, and muscle of between 0.01 and 0.06 µg/g were found by HAMILTON et al. (1972/1973). In the lymph nodes concentrations vary between 0.03 and 0.06 µg/g (wet weight). These levels need to be rechecked using modern analytical methods.

Concentrations in the blood are reported to be between 0.012 and 0.028 µg/g, but zirconium could not be reliably detected in human urine. It is therefore assumed that zirconium is excreted primarily through the feces via the biliary system (DEKNUDT, 1988). It is also excreted secondarily through milk.

Inhalation exposure to water-soluble $ZrOCl_2$ indicated that the highest concentrations of zirconium occurred in the lungs and pulmonary lymph nodes. Deposition and retention in bone (femur) were greater than in the liver.

6 Effects on Plants, Animals, and Humans

SCHROEDER and BALASSA (1966) reviewed the literature on the biochemistry of zirconium and concluded that this metal is not essential for humans and animals. Investigations are not available presenting effects on plants. Zirconium salts are of low toxicity to animals. The toxicology of zirconium has been reviewed by BROWNING (1961) and SMITH and CARSON (1978). Zirconium is more toxic when administered parenterally than orally, evidently due to poor absorption from the gastrointestinal tract. The LD_{50} doses of the different zirconium salts vary from 2.5 to 10 g/kg. These salts are appreciably more toxic after intraperitoneal dosage (STOKINGER, 1967).

There is not much data on the effects of zirconium on humans. One study on 22 workers who were exposed to zirconium vapor from one to five years did not show specific health impairments (REED, 1956). Even more than 40 years later, no adverse health effects have been detected. In one Russian investigation of 120 workers who inhaled zirconium and titanium dust, headache, restlessness, and a small rise in the systolic blood pressure have been observed. In the newer literature no evidence for a systemic toxicity of zirconium could be found (SMITH and CARSON, 1978). No evidence of industrial diseases related to zirconium exposure has been documented (BROWNING, 1969). Another group of 32 males, who worked 1–17 years as handfinishers of zirconium metal reactor components and who had asked whether the dust to which they were exposed could cause chronic lung disease or cancer, or both, showed insignificant differences when compared with a control group (BRUBAKER and HADJIMICHAEL, 1981). On the other hand, DEKNUDT (1988) mentions that pulmonary granuloma have been reported in zirconium workers, as well as in rabbits following exposure to zirconium lactate.

Some zirconium compounds were formerly used as ingredients in ointments and antiperspirants. Between 1951 and 1958 more than 30 commercially obtainable deodorants (sticks and sprays) contained zirconium, and up until the 1950s zirconium compounds were also found in poison ivy and oak medications. Through the use of these deodorants and remedies, granulomas and hypersensitive reactions of the retarded type were observed in humans (EPSTEIN and ALLEN, 1964). Carcinogenic and teratogenic effects of zirconium compounds are not known.

7 Hazard Evaluation and Limiting Concentrations

With respect to the present state of knowledge, it can be assumed that naturally occurring or industrially emitted zirconium or its compounds do not pose a general risk for the environment and living organisms. Maximal emission concentrations have not been established so far. In nearly all Western countries the maximal occupational concentration for zirconium compounds in an eight hour day is 5 mg/m^3 (calculated as Zr; MAK, 1988). Harmless threshold limits in biological material cannot be evaluated so far, as cannot maximal allowable daily intakes of zirconium through food and drinking water.

References

Ayranci, B. (1989), *A Rapid Method for the Decomposition of Zirconium Silicates and Zirconium Dioxide* (in German). *Swiss Chem. 11* (6), 13–18.
Bayer, G., and Wiedemann, H.-G. (1981), *Chem. Unserer Zeit 15* (3), 88–97.
Berman, E. (1981), in: Thomas, L. C. (ed.), *Toxic Metals and their Analysis*. Heyden Int. Topics in Sciences, London-Philadelphia-Rheine.
Bowen, H. J. M. (1979), *Environmental Chemistry of the Elements*, p. 273. Academic Press, London.
Brooks, C. K. (1968), *Radiochim. Acta 9*, 157.
Browning, E. (1969), *Toxicity of Industrial Metals*. Butterworths, London.
Brubaker, R. E., and Hadjimichael, O. C. (1981), *J. Occup. Med. 23*, 543–547.
Bruland, K. W. (1983), *Trace Elements in Seawater, Chem. Oceanogr. 8*, 208.
Deknudt, G. (1988), Chapter 72 Zirconium in: Seiler, H. G., Sigel, H., and Sigel, A. (eds.), *Handbook on Toxicity of Inorganic Compounds*. Marcel Dekker, New York.
Dressler, G., Minuth, P., and Wolf, H. U. (1983), *Zirconium, Zirconium Alloys and Zirconium Compounds* (in German), in: *Ullmanns Enzyklopädie der technischen Chemie*, 4th Ed., Vol. 24, pp. 681–702. Verlag Chemie, Weinheim-Deerfield Beach/Florida- Basel.
Epstein, W. L., and Allen, J. R. (1964), *Granulomatous Hypersensitivity After Use of Zirconium-Containing Poison Oak Lotions. J. Am. Med. Assoc. 190*, 940–942.
Gauckler, L. (1987), *Tetragonal Zirconium Oxide as Raw Material for Ceramics* (in German). *Forschung und Technik, Neue Zürcher Zeitung No. 202*, p. 75 (Sept. 2), Zürich.
Hamilton, E. J., Minski, M. J., and Clearly, J. J. (1972/1973), *The Concentrations and Distributions of Some Stable Elements in Healthy Human Tissues from the United Kingdom. An Environmental Study. Sci. Total Environ. 1*, 341.
MAK (1988), *Maximum Concentrations in the Workplace, Report No. XXIV, DFG*. VCH Verlagsgesellschaft, Weinheim-Basel-Cambridge-New York.
Miller, G. L. (1965), *Zirconium*. Academic Press, New York.
Pinta, M. (1978), *Modern Methods for Trace Element Analysis*. Ann Arbor Science, Ann Arbor, Michigan.
Reed, C. E. (1956), *A Study of the Effects on the Lung of Industrial Exposure to Zirconium Dusts. Arch. Ind. Health 13*, 578.
Rubel, H. (1983), *Radex Rundsch.* (75, 1/2), 32–42.
Saager, R. (1984), *Metallic Raw Materials Dictionary* (in German), pp. 153–156. Bank von Tobel, Zürich.
Schroeder, H. A., and Balassa, J. J. (1966), *Abnormal Trace Metals in Man: Zirconium. J. Chron. Dis. 19*, 573–586.
Smith, J. C., and Carson, B. L. (1978), *Trace Metals in the Environment*, Vol. 3. Ann Arbor Science, Ann Arbor, Michigan.
Stokinger, H. E. (1967), in: Patty, F. A. (ed.), *Industrial Hygiene and Toxicology*, 2nd revised Ed., Vol. II. Interscience Publishers, New York.
Trueb, L. (1990), *Zirconium the "Golden" Element in Nuclear Applications* (discussing also Corrosion in Nuclear Reactors) (in German). *Neue Zürcher Zeitung, Forschung und Technik*, No. 85, pp. 65–66 (11 April), Zürich.
Venugopal, B., and Luckey, T. D. (eds.) (1979), in: *Metal Toxicity in Mammals*, Part 2, 2nd Ed. Plenum Press, New York-London.
Welz, B. (1985), *Atomic Absorption Spectrometry*, 2nd Ed., pp. 339–340, 400 ff. VCH Verlagsgesellschaft, Weinheim-Deerfield Beach/Florida-Basel.

Glossary

AAS atomic absorption spectrometry

achlorhydria, achlorhydric absence of hydrochloric acid from the gastric secretions (even after stimulation with histamine)

achromotrichia loss or absence of pigment from hair, canities

achylia literally, absence of chyle; lack of secretion of hydrochloric acid and enzymes by the stomach

acidophilic tolerance of, or preference for an acid environment

aciduria increased excretion of acids in the urine

acrodermatitis enteropathica rare, hereditary, inflammatory disease (Brandt's syndrome) of the skin (mainly found in infants and young children) predominantly located around the body orifices and in the folds of the large joints, accompanied by loss of hair, nutritional disturbance and diarrhea

ACTH adrenocorticotrophic hormone; secreted by the adenohypophysis of the pituitary gland, it regulates the secretion of various adrenal cortex hormones

adenine a purine base (6-aminopurine), a component of all nucleic acids (RNA and DNA)

adenocarcinoma a malignant neoplasm derived from glandular tissue or in which the tumor cells form recognizable glandular structures

ADH alcohol dehydrogenase activity

ADPV see DPAV

aerobic in the presence of atmospheric oxygen or air

AES atomic emission spectrometry; or Auger emission spectroscopy

AFS atomic fluorescence spectrometry

ahaptoglobinemia absence of haptoglobin in plasma, a common finding in hemolysis

ALAD δ-(5)aminolevulic acid dehydratase, an enzyme which is involved in the biosynthesis of the hemin-porphyrin skeleton; decrease in its activity is monitored, e.g. in the case of increased exposure to lead

albumin a group of proteins found in nearly every animal and in many vegetable tissues, e.g. serum albumin

albuminuria increased excretion of albumin in the urine; can indicate damage to the kidney

allotropic occurrence of a chemical element in different crystalline forms

alluvial, alluvium composed of eroded detrital materials, sediment, gravel, etc. deposited by running water (geologically, especially since the end of the Ice Age)

alopecia totalis complete loss of hair from the scalp, total baldness which can be congenital or acquired

alveoli air sacs of the lung

amblyopia weakness or dimness of vision without apparent defect in the eye structure (usually attributed to the optic nerve)

amenorrhea absence of menstruation

aminoaciduria more precisely: hyperaminoaciduria; pathological increase in the excretion of amino acids in urine (1–2% of the ingested amino acids are normally excreted in urine)

amnesia loss of memory

amphoteric ability of a molecule or functional group to bear a negative or a positive charge by dissociation or by addition of protons, resp.; an amphoteric substance is therefore capable of acting as either an acid or a base

amyloidosis degeneration of tissue due to deposition of paramyloid (a starch-like substance) with pathological production of immunoglobulins (antibodies)

amyotrophic, amyotrophy atrophy of muscle tissue

anaerobic in the absence of atmospheric oxygen or air

anemia decrease in the hemoglobin concentration (and also in the number of red blood cells) in the blood

aneuploidy lack of or excess of a chromosome in the normal diploid chromosome complement, e.g. trisomy 21 (mongolism caused by gene mutation)

aneurism, aneurysm a localized abnormal dilation of an artery, a lateral blood-filled sac in an artery wall

anophthalmia absence of eyes (congenital or acquired)

anorexia loss of appetite, a. nervosa: diminished urge to eat

anosmia loss of the sense of smell

anoxic oxygen deficient

antidote an agent which counteracts the effects of a poison

anuria absence of urinary excretion

APDC ammonium pyrrolidone dithiocarbamate, a complexing agent

aplastic anemia anemia resulting from failure of cell production in the bone marrow, generally unresponsive to antianemia therapy

aquo ions metal ions which only have water molecules as coordination partners

argyria deposition of silver sulfide in the skin and mucous membranes (appears as a gray-bluish discoloration), e.g. as a result of treatment with silver-containing preparations

arrhythmia irregularity of the heart beat

arthritis inflammatory deformation at the joint surfaces, e.g. a. allergica, a. rheumatica

arthropathy non-inflammatory joint disease (strictly speaking, degenerative disease, e.g. arthrosis deformans)

Ascidia sea squirts (class of marine animals)

ascites accumulation of serous fluid in the abdominal cavity, a kind of dropsy

asphyxiation suffocation, unconsciousness or death due to oxygen deprivation

asthenia weakness, absence or loss of vitality and strength

asthenospermia reduction in the vitality of sperm (or the amount of freely mobile sperm)

astringent causing contraction or shrinkage of tissue layers, thus facilitating sealing of wounds and arresting inflammation and hemorrhage

ASV anodic stripping voltammetry

asymptomatic showing no symptoms

ataxia inability to coordinate voluntary muscle movement

ATP adenosine triphosphate (coenzyme of energy supplying metabolic reactions)

atresia imperforation or closure of a natural orifice, an excretory outlet or a hollow organ of the body

atrophification, atrophy diminution of organs, muscles or cells (degeneration of cells, tissues and organs as a result of a undernourishment or circulatory malfunction)

aurosomes lysosomes enriched with gold

autosomal pertaining to any chromosome other than a sex chromosome

auxins phytohormones, substances which promote plant growth (induce elongation of the cell walls)

azoospermia absence of spermatozoa in semen

azotemia pathological increase in the amount of nitrogenous substances (waste products, urea) in the blood or serum, e.g. due to malfunction of the kidneys

BAL, British antilewisite dimercaprol, 2,3-dimercaptopropanol (antidote for poisoning caused by heavy metals; forms very stable metal complexes)

basal ganglia (also known as corpus striatum) a group of gray nuclei in the diencephalon and telencephalon, emanating from the protocephalon (forming a link in the nervous system between the associative cerebral cortex and the motor cortex)

biliary pertaining to bile or the gall bladder

bimodal showing two maxima (peaks) in a statistical series

bioavailability availability of a specific substance to living organisms under certain environmental conditions

blastocyte embryonic cell which is not yet differentiated

blepharitis inflammation of the eyelids

Bowen's disease precancerosis, usually in the skin with red, flaky, sometimes wart-like plaques which occasionally show ulcerous degeneration

brachycardia, bradycardia slow heart beat (with a rate of less than 55 beats/min)

bullous-pustular small vesicles full of pus

bypass replacement of a pathologically damaged portion of a blood vessel by implantation of a vein, artery or plastic tube

calmodulin a metal-binding protein with metabolic functions

cardiotoxic having a poisonous effect on the heart

cardiovascular concerning the heart and blood vessels

Castle's intrinsic factor a substance produced by the stomach which combines with the extrinsic factor (vitamin B_{12}) in food to yield an antianemic principle, its lack is believed to cause pernicious anemia

cation exchange capacity a measure of the capability of a substance or a (solid) material to bind cations in a reversible manner

CCSEM computer controlled scanning electron microscopy

centromedullary nail special type of nail implanted into the medullary canal of tubular bones for the fixation of fractures

cerebellum small posterior part of the brain; controls coordination of voluntary movement

cerebral pertaining or belonging to the brain

cerebrum large anterior or uppermost part of the brain (telencephalon + diencephalon + mesencephalon)

ceruloplasmin transport protein which binds about 95% of the copper concentration in blood serum. In Wilson's disease part of the copper is not bound to ceruloplasmin which leads to copper deposition in various organs

chalcogens, chalcosphere elements belonging to the VIa group of the periodic table

chemotaxis the movement of an organism or an individual cell in response to a chemical concentration gradient

chloro-complexes metal ions having at least one chloride ion as a coordination partner

chlorophyll green, magnesium-containing plant pigment which absorbs light energy for the assimilation of carbon dioxide (photosynthesis)

chlorosis a form of anemia due to iron deficiency especially in young girls; or loss of pigmentation in green plants until they appear yellow or colorless due to insufficient formation of chlorophyll

choreiform resembling chorea (a disorder characterized by irregular and involuntary, jerky movements of the face and extremities)

chromatid aberration (chromosomal aberration) abnormality of the filamentous chromosome halves (chromatids)

chrysotherapy treatment of disease using gold compounds

cilia fine, mobile hair-like processes on the surface of certain cells and organisms, eyelashes

cirrhosis damage to an organ – especially the liver – with initial inflammation, increase in the amount of connective tissue (fibrosis) followed by and shrinkage due to scarring which results in hardening and reduction in the size of the organ

clastogenic causing or inducing disruption or breakages, as of chromosomes

CMB chemical mass balance

coenzyme a nonprotein substance which, in combination with an apoenzyme, forms a complete and functional enzyme

colostomy surgical creation of an opening between the colon and the body surface

congestion, congestive concerning or resulting in abnormal accumulation of blood (in an organ)

conjunctiva membranes that line the eyelids and cover the exposed surface of the sclera

conjunctivitis inflammation of the conjunctiva

contraluminal serosal site of the intestinal epithelium

coprophagy eating dung or feces

CRM certified reference material

CSV cathodic stripping voltammetry

cyanosis reddish-blue discoloration of the skin due to insufficient oxygen saturation of the blood

cytochrome cell respiratory enzyme which transfers electrons from a suitable substrate to oxygen in the course of biological oxidation

cytoplasmic pertaining to the protoplasm of a cell, i.e. the part excluding the nucleus

cytosine 1,2-dihydro-2-oxo-4-aminopyrimidine (component of the nucleic acids DNA and RNA)

DCP direct current plasma

dementia deterioration or loss of intellectual faculties

demyelation removal of the myelin sheath of a nerve

dermatitis inflammation of the skin

dermographism over-sensitive reaction of the skin in which pressure or friction gives rise to a wheal so that a word traced is visible

Descemet's membrane a specific layer of the optic cornea in which deposits of copper are visible in patients with Wilson's disease

diabetes specifically: d. mellitus, disorder characterized by excessive blood sugar levels, glycosuria (excretion of sugar in the urine), polydipsia (abnormal thirst) and polyuria (excessive urinary output)

diagenetic concerning changes (due to pressure and temperature) occurring in sediments after their deposition. Consolidation and new mineral equilibria can result in the temperature range between that of their formation and about 200 °C

diaphorase mitochondrial flavoprotein enzyme which catalyzes the reduction of dyes, such as methylene blue

diastolic concerning the dilation and relaxation of the heart, especially of the ventricles

diathesis constitution or condition of the body which tends to make the person more than usually susceptible to certain diseases

dilatation, dilation expansion (e.g. of blood vessels)

dismutase enzyme necessary for redox reactions (dismutation)

diuresis excretion of urine

DME dropping mercury electrode

DMG dimethylglyoxime (analytical reagent)

DNA de(s)oxyribonucleic acid, carrier of genetic information

DOM dissolved organic matter

DPASV differential pulse anodic stripping voltammetry

DPAV differential pulse adsorption voltammetry

DPCSV differential pulse cathodic stripping voltammetry

DPIV differential pulse inverse voltammetry

DPP differential pulse polarography

DPSV differential pulse stripping voltammetry

drainage removal of water

dwarfism the state of being a dwarf (an abnormally undersized person)

dysarthria imperfect articulation of words due to disturbances of muscular control which result from damage to the central or peripheral nervous system

dysgeusia impairment of the sense of taste

dysosmia impairment of the sense of smell

dysphagia painful difficulty in swallowing due to disorders affecting the esophagus (gullet), e.g. stenosis (constriction), spasm (cramp)

dyspnea, dyspnoea all forms of difficulty in breathing, e.g. labored breath, shortness of breath

dystrophia, dystrophy defective nutrition, e.g. of an organ (particularly in the case of undernourishment)

EAAS electrothermal atomic absorption spectrophotometry

ECD electron capture detector

edema pathological accumulation of fluid in the tissue spaces (also known as dropsy)

EDTA ethylenediamine tetraacetic acid (chelating agent) used in analysis and therapy

ED XFA, ED XRF electron diffraction X-ray fluorescence analysis

electromyogram graphic recording of the electrical response of a muscle to electrical stimulation (measured directly with needle electrodes or indirectly via the skin)

eluate solution obtained when one material is extracted from another using a solvent, especially the outflow from a chromatographic separation column

emaciation heavy loss of body weight, growing lean and thin

emphysema abnormal distension of tissues by gases (putrefaction gases, air), especially when leading to dilatation or destruction of the alveoli

encephalocele hernia (protrusion) of the brain and its surrounding membranes through a congenital or traumatic (as a result of injury) opening in the skull

encephalopathy general term for (non-inflammatory) brain disease

endemic native to, occurring in a specific region

endocrine pertaining to internal secretions (especially of ductless hormone glands)

endocytosis uptake by a cell of material from the environment by invagination of its plasma membrane

endoprosthesis prosthetic device which is completely embedded into the body without contact to the exterior

enteritis inflammation of the intestinal tract

enterocyte an intestinal epithelial cell

enterohepatic excreted by the liver into the gall bladder, from there into the intestine, and subsequently reabsorbed through the intestinal wall

enzootic endemic occurrence of a(n) (infectious) disease in animals, e.g. enzootic ataxia – characterized by abnormal gait due to impairment of myelination of the nerves – occurs in the newborn lambs of ewes suffering from extreme copper deficiency

enzyme protein that catalyzes metabolic reactions

enzymuria increased urinary excretion of an enzyme or enzymes

epidemiology study of disease as it affects human populations rather than individuals (distribution and effects of disease in population groups) in order to establish and examine the causal relationships of disease. Statistical methods are used

epistaxis nose-bleeding

epithelium tissue which covers the external surface (skin) and lines the inner body cavities of animals and humans

EPMA electron beam microanalysis by electron probes

EPR electron paramagnetic resonance (spectroscopy)

erythema redness of the skin as a result of dilatation and suffusion of the surface blood vessels, acute, superficial inflammation of the skin

erythroblasts nucleated precursors of erythrocytes which lack nuclei (in humans)

erythrocyte red blood cell

erythrocytopenia deficiency in the number of erythrocytes

erythron the circulating erythrocytes, their precursors and all elements of the body concerned in their production

erythropoiesis formation of erythrocytes

ESCA electron spectroscopy for chemical analysis (X-ray photoelectron spectroscopy)

ET AAS electrothermal atomic absorption spectrometry

eukaryotic concerning all cells with membrane-bound, structurally discrete nuclei (or organisms composed of such cells)

exencephaly protrusion of the brain from the skull (e.g. as a result of a congenital malformation)

exudate organic material which is exuded from organisms (e.g. algae or plankton) and/or organs (e.g. blood vessels and lymphatic system) usually as a result of inflammation

FANES furnace atomic non-thermal excitation spectrometry

FD MS field desorption mass spectrometry

feces excrement, stool

ferritin most important storage protein for iron (especially in the intestinal wall) in mammals (including humans)

fetotoxic poisonous to the fetus or developing offspring

FIA flow injection analysis

fibroblast a connective tissue cell (not yet differentiated) which forms fibrous tissues in the body

fibrosis growth of (new) fibrous connective tissue in excess of that naturally present

fingerprint in analytical chemistry; signals (groups of lines in spectra, chromatograms, etc.) which are characteristic for specific elements or compounds

fluorosis poisoning due to exposure to excessive amounts of fluorine or its compounds

flux in environmental physics; flow of materials between two reservoirs per unit time

folliculitis inflammation of a follicle or follicles, usually hair follicles

Fucus seaweed, a genus of brown algae

GABA cycle, shunt an alternative metabolic pathway which can bypass part of the glycolytic pathway (GABA = γ-aminobutyric acid)

gabbro igneous rock (mainly composed of plagioclase and pyroxene) which crystallized in deep-seated layers

gastrin a hormone secreted by the pyloric mucous membrane of the stomach; it regulates the production of hydrochloric acid in the stomach as well as pancreatic activity, it also stimulates the flow of bile and gastro-intestinal motility

gastritis, gastroenteritis inflammation of the mucous membrane of the stomach and intestine

gastro-intestinal pertaining to that part of the digestive tract which includes the stomach and intestine

GC gas chromatography

genome the entire genetic material borne by the genes on a haploid (single, as in the reproductive cells: ova and sperm) set of chromosomes, or a set of genes that governs a particular trait

GF AAS graphite furnace atomic absorption spectrometry

gingivae the gums

gingivitis inflammation of the gums

glioma a tumor composed of tissue which represents the neuroglia (special supporting tissue of the central nervous system) in any one stage of its development

glomerular concerning the kidney glomeruli (bundles of blood capillaries) through which solutes are removed from the blood to maintain constant plasma levels

glucosuria, glycosuria increased excretion of sugar (glucose) in the urine

glutathione γ-glutaminylcysteinylglycine, a tripeptide, which functions as a biological redox system

glycolysis breakdown of glucose to pyruvic acid in an organism; the product is further broken down in subsequent metabolic reactions to carbon dioxide and water

glycoprotein conjugated protein consisting of a protein group and a carbohydrate group

goitrogenic producing goiter

gonads primary sex glands (testes, ovaries) which produce the reproductive germ cells

Gram negative description of microorganisms which are not stained by a certain bacterial stain (Gram's stain)

granulomata, granulomatous tissues tumor-like masses or nodules of granulation tissue with actively growing fibroblasts and capillary buds

granulomatosis tumor-like proliferation of young connective tissue which has rich blood supply, formation of multiple granulomata

grass tetany disease of cattle characterized by muscle cramps, primarily caused by the consumption of young grass which has grown too fast and is deficient in magnesium

graywacke an extremely hard, gray sandstone which, besides quartz, contains considerable quantities of feldspar, mica and other mineral fragments

guanine 2-amino-6-hydroxypurine, component of the nucleic acids DNA and RNA

half life, half time time necessary to reduce the quantity of a substance (e.g. in an organism) or of radioactive materials to half the original amount

haptene a low molecular-weight substance which reacts with a specific antibody but which by itself is unable to elicit the formation of that antibody. It is antigenic only if coupled with a carrier protein

haptoglobin a group of glycoproteins in the a_2-globulin fraction of serum, which bind free hemoglobin

hematocrit (centrifugation apparatus for) determination of the sedimentation volume of blood, i.e. the content of the cellular elements in blood in relation to the total volume

hematopoietic, hematopoesis blood-forming, specifically, concerning the formation of blood cells

hematuria red blood cells in the urine in excess of the normal physiological amount

heme iron-containing pigment part of hemoglobin

hemochromatosis disorder of iron metabolism (brown discoloration of the skin and other organs due to the destruction of red blood cells or deposition of iron pigments)

hemodialysis removal of pathological or toxic components (and waste products) from the blood with the help of an artificial kidney (diffusion through a semipermeable membrane)

hemoglobin protein which acts as the oxygen-carrying component of red blood cells and which contains four iron-containing porphyrin complexes

hemolysis, hemolytic loss of hemoglobin from the red blood cells

hemorrhagic concerning a bleeding condition

hemosiderin an insoluble iron-containing pigment found in the liver and most tissues, visible without staining methods

hemosiderosis increased deposition of iron in the organism (e.g. in cells)

heparinizing method of preventing blood coagulation, e.g. during the preparation of blood samples for biological analysis

hepatic belonging to or concerning the liver

hepatitis inflammation of the liver (diffuse or localized, generally an infectious inflammation of the vascular connective tissue of the liver)

hepatocyte a parenchymal liver cell

hepatolenticular concerning the liver and the lenticular nucleus – basal ganglia in the brain – e.g. area of degeneration in Wilson's disease

hepatosplenomegaly enlargement of the liver and spleen (due to inflammation)

heterozygous bearing two different alleles at the same gene locus

HGPRT-test hypoxanthine-guanine-phospho-ribosyl-transferase test, a mutagenicity test with mammalian cells

hippocampal concerning the hippocampus, a curved elevation in the floor of the inferior horn of the lateral ventricle

histones proteins found in the cell nucleus which are associated with the nucleic acids and play a role in structural formation (cell differentiation) by influencing gene activity

HMDE hanging mercury drop electrode

Hodgkin's disease a malignant condition characterized by progressive enlargement of lymph nodes, spleen, and lymphoid tissues

homeostasis maintenance of equilibrium with respect to the physiological functions of the body (stability of the metabolism, body temperature, pH value of the blood, blood pressure, etc.) by means of regulatory systems

homeostatic concerning homeostasis

homosphere that part of the Earth which is influenced by human activities

homozygous bearing identical alleles at a given gene locus in homologous chromosomes of an individual (the opposite of heterozygous)

HPA high pressure ashing

HPLC high performance liquid chromatography

hyaline glassy and transparent or nearly so

hydrocephalus, hydrocephaly accumulation of excess cerebrospinal fluid in the skull causing enlargement of the brain and skull

hydropericardium accumulation of water in the pericardial sac of the heart, without inflammation

hyperaluminemia increased concentration of aluminum in blood

hypercalciuria, hypercalcinuria increased excretion of calcium in urine

hyperesthesia excessive sensitivity to touch

hyperglycemia increased concentration of glucose in blood plasma

hyperkeratosis thickening of the corneous layers of skin due to excessive growth

hypermagnesemia abnormally high content of magnesium in the blood; exceeding 2.5 mg/100 mL, e.g. in the case of uremia, kidney failure or increased intake of magnesium

hypernickelemia increased concentration of nickel in blood serum or plasma

hyperparathyroidism abnormally increased activity of the parathyroid glands, which may be primary or secondary

hyperphosphaturia excessive excretion of phosphates in the urine

hyperpigmentation excess formation of pigment

hyperplasia increased number of normal cells in their normal arrangement in a tissue

hyperploidy having one or more chromosomes (or parts of chromosomes) in excess of the haploid number or a multiple of the haploid number

hypersensitivity a state of altered reactivity in which the body reacts with an exaggerated (immune) response to a foreign agent

hypertonia abnormal increase in muscle tone, e.g. also of the vegetative nervous system, certain hollow organs (stomach, gall bladder, bladder) and blood vessels

hypocalcemia reduction of calcium levels in the blood below normal

hypochromic having a reduced hemoglobin content

hypogeusia diminished sense of taste

hypogonadism underdevelopment of the sexual glands (reduced hormonal secretion and functional activity of the gonads)

hypokalemia, hypokaliemia abnormally low potassium concentration in the blood

hypomagnesemia diminished content of magnesium in blood plasma (magnesium level less than 2.0 mg/100 mL)

hypophosphatemia abnormally low amounts of phosphates in the blood

hypophysis pituitary gland situated at the base of the brain

hyposmia diminished sense of smell

hypospermia diminished number of sperm

hypothalamus the portion of the diencephalon which forms the floor and part of the lateral wall of the third ventricle

hypothermia abnormal lowering of the body temperature

hypothyroidism deficiency of thyroid activity characterized by decrease in the basal metabolic rate, tiredness and lethargy

hypotonia diminished tone of the skeletal muscles, lowered resistance to passive stretching

hypoxia reduction of oxygen supply to tissues below physiological levels despite adequate perfusion of the tissues with blood

hypozincemia abnormally decreased zinc concentration in the blood

iatrogenic concerning a condition occurring as a result of treatment by physician

ICP AES inductively coupled plasma atomic emission spectroscopy

ICP MS inductively coupled plasma mass spectrometry

idiopathic concerning a condition which is independent of other illnesses, occurring without identifiable cause

IDMS isotope dilution mass spectrometry

ileum the distal portion of the small intestine

immunoglobulins serum proteins showing antibody activity, carriers of natural defense mechanisms against infection (also known as gamma-globulins or immune globulins)

impaction wedging, separation and concentration step in the preparation of sample particles for analysis

implantation placement of an organ or tissue transplant in the body (e.g. transplantation of a tendon); also embedding of a fertilized ovum in the mucous membrane of the uterus (womb)

INAA instrumental neutron activation analysis

interstitial fluid fluid filling the intercellular spaces in tissues

interstitial space space between tissues and organs

interception process by which rainwater is caught and retained by vegetation, then evaporates so that it never reaches the ground

intestinal in or concerning the intestines or the digestive tract

intracellular within the cell

intraperitoneal being or occurring within the peritoneum or peritoneal cavity (serous lining of the abdominal cavity)

intratracheal within or through the trachea

intrauterine being or occurring within the uterus (womb)

intravascular within the blood vessels or lymph system

in vitro in a test tube, i.e. a biological reaction taking place under experimental conditions outside the organism

IPMA ion probe microanalysis (also known as ion microprobe mass analysis)

jejunum the middle section of the intestine extending from the duodenum to the ileum

keratin indigestible protein with a high cystine content, component of the skin and other parts of the body (e.g. finger nails, hair, feathers, horns)

keratosis excessive growth of the cornified epithelium, also hyperkeratosis

Keshan disease heart disease, probably caused by selenium deficiency, first observed in the Chinese province of Keshan

Krebs cycle also known as the citric acid cycle; a metabolic pathway in which energy is released and carbon dioxide and water are formed

LAFD MS laser-assisted field desorption mass spectrometry

lactation secretion of milk by the mammary glands

lactoferrin an iron-binding protein found in neutrophils and secretions (milk, tears, saliva, etc.) having bactericidal activity and acting as an inhibitor of colony formation by granulocytes and macrophages

LAMMA laser microprobe mass analysis

larynx cavity in the throat containing the vocal chords

LC_{50} lethal concentration, concentration of a substance which causes the death of 50% of the organisms in a population in a specified period of time

LD_{50} lethal dose, the single dose of a substance which causes the death of 50% of a population in a specified period of time

leukemia collective term for a group of diseases (generally of unknown cause) characterized by uncontrolled neoplastic proliferation of the leukocytopoietic system (white blood cell system) accompanied by characteristic changes in the leukocytes and their precursors

leukocytosis elevation of the white blood cell count in the blood to a level above 9000/L

leukopenia reduction of the white blood cell count in the blood to a level below 5000/L

ligands inorganic or organic molecules or ions, which form complexes with metal ions under certain conditions (e.g. CO, CO_3^{2-}, OH^- and $HO-R^1R^2-OH$)

lipid, lipidic neutral fat, e.g. the ester of fatty acids (monocarbonic acids) with glycerine; fatty

lipofuscinosis abnormal deposition of lipofuscin in tissues

lipophilic having an affinity for fat, soluble in fat

lithiasis a condition characterized by the formation of calculi and concretions in the body

lithophilic found (accumulated) in the Earth's outer crust, possessing an affinity for oxygen

lobar pneumonia inflammation of the lobes of the lung

LOD limit of detection

LOQ limit of quantification (or determination)

luminal pertaining to the lumen of a tubular structure; mucosal site of the intestinal epithelium

lymphoblastoid characterized lymphoblast cells (progenitors of lymphocytes)

lyophilisate product of freeze drying

lysosome cell organelle containing hydrolases (hydrolytic enzymes); it is active in phagocytosis, i.e. in intracellular digestion and destruction accompanied by release of the cell contents

macrophage phagocyte, participates in defense mechanisms to fight infection (amoeboid mobile cells active in phagocytosis)

maculopapular rash a skin eruption which is elevated and discolored

magmatic originating from the molten rock (magma) in the Earth's inner core

magmatite rock formed from magma (also called igneous rock)

MAK maximum allowed concentration at the work-place

malacia abnormal softening of tissues of an organ or of tissues themselves

matrix the sum of the accompanying materials in which a substance is embedded or occurs (and which must be taken into account in analytical results, matrix effects), e.g. amorphous ground substance, biological material

maxillo-fascial surgery type of oral surgery (upper jaw)

mediator transmitter substance which converts an inflammation of a tissue into functionally and morphologically recognizable symptoms

medulla central part of certain organs, e.g. bones (bone marrow) and kidneys; medulla oblongata: the upper part of the spinal cord in which important reflex centers are situated

megaloblast abnormally large, nucleated precursor of a special form of the red blood cells, the megalocytes

melanosis discoloration of the skin (abnormal deposition of melanin in the skin)

Menke's syndrome (or disease) (also known as kinky hair syndrome) a fatal malfunction of copper absorption (due to lack of copper proteins), copper transport and utilization, carried by X-chromosomes

meristematic tissues embryonal plant tissue with great growth potential

metamorphic (geological term) formed by physical and chemical changes (in rocks) at temperatures above about 200 °C

metaxylem part of the woody vascular system (which carries water and soluble, predominantly inorganic substances) in plants

methemoglobinemia increase in the concentration of methemoglobin induced by a toxic substance or due to a congenital disorder; methemoglobin cannot transport oxygen in the blood because iron(II) is oxidized to iron(III)

methionine α-amino-γ-methylthiobutyric acid; an essential amino acid, components of proteins (e.g. also of RNA)

MFE mercury film electrode

microcytic concerning abnormally small erythrocytes

micronucleus small nucleus (e.g. in viruses and protozoa)

microphthalmus congenital defect characterized by abnormally small eyes

microsomes minute granules occurring in great numbers in the cell protoplasm. They are equipped with a special set of enzymes which participate in protein synthesis

microvesicular steatosis fatty degeneration of an organ, with accumulation of minute lipid droplets

MIP microwave induced plasma

miticide an agent which kills mites

mitochondria elongated oval cell organelles in which the aerobic energy-releasing metabolism of the cell takes place (multi-enzymatic system for the respiratory chain, citric acid cycle, degradation of nutrients accompanied by the release of energy, etc.)

mitotic index the number of dividing cells per thousand cells, a measure of the rate of growth of a tissue

molluscs soft-bodied animals belonging to the phylum mollusca, such as snails, crustaceans, etc.

morula embryonic cleaved cell complex

motoricity the entire voluntary, active muscle movement

mucociliary transport of cells or particles by ciliary movement in the mucous layer, e.g. in the respiratory tract

mucocutaneous in or concerning the mucous membranes

mucopolysaccharides substances of high molecular weight (e.g. reticular substances of the mucous and connective tissues), composed of polysaccharides, uronic acids (e.g. glucuronic acid and amino sugars (e.g. glucosamine)

mucoproteins a subgroup of the glycoproteins which is linked to mucopolysaccharides (mucous substances which are secreted as protection for the skin and mucous membrane)

mucosa, mucosal, mucosa cells mucus secreting cells (especially those in the intestine)

muscle hypertonia abnormal increase in the tone of the muscles

myalgia pain in the muscles

mycelium mass of fungal hyphae which composes the vegetative body of many fungi

mycorrhiza fungi fungi living in soil in close contact with plant roots (symbiosis, nutrient exchange)

myelination, myelinization physiological process in which the nerves are ensheathed by myelin, a lipoid which is essential for normal nerve function, in the unborn and children up to the age of 4

myelinopathy any decrease of the myelin, degeneration of the white matter of the brain

myeloic concerning bone marrow

myocardial, myocardium   central muscle layer of the wall of the heart

myoclonia, myoclonus   jerky contractions of part of a muscle, an entire muscle or a group of muscles; Friedrich's syndrome (II), a rare disorder of the nervous system, which is chiefly characterized by loss of balance

myoglobin   hemoglobin-like protein which transports oxygen into muscle cells

myoneural junction   point of connection of a muscle and the nerve which enervates it

myopathia   disease of the skeletal musculature (weakness)

NAA   neutron activation analysis

Na-DDC   sodium diethyldithiocarbamate

nasopharyngeal, nasopharynx   the part of the pharynx that lies above the level of the soft palate

necrosis   death of cells, tissues or a whole organ

neoplasia, neoplastic   formation of a neoplasm (new, abnormal uncontrolled growth of a tissue)

nephritis   inflammation of the kidney

nephropathic, nephrotoxic   injurious to the kidneys

neuron   nerve cell

neuropathy   pathological changes in the peripheral nervous system; a non-inflammatory nervous disease of hereditary nature, "constitutional nervousness", accompanied by vegetative and endocrine disorders

neurotransmitter   a substance which serves to transfer the excitation of one nerve cell to another or to the target organ

normochromatic, normochromic   normally stainable or stained; in the case of red blood cells, containing the normal amount of hemoglobin

NTA   nitrilotriacetic acid, used as a substitute chelating agent instead of phosphate, also used in analysis (forms complexes with many metals)

nuclear number   (also known as mass number) sum of the particles (protons and neutrons) in an atomic nucleus

nucleogenesis   formation of atomic nuclei in the center of stars

nucleophilic sites   molecular sites (e.g. within a protein) with a free electron pair, which can interact with a positively charged site of the same or another molecule

nucleoside   component of nucleotides without the phosphate group, e.g. adenosine, guanosine; a base (such as purine, pyrimidine) is linked to a (usually) pentose sugar

nucleotide   ester of a nucleoside with phosphoric acid, forms the structural units of nucleic acids (base+monosaccharide+phosphate)

nystagmus   an involuntary rapid movement of the eyeball, which may be congenital, acquired, physiological or pathological

OAS   optical atomic spectrometry

obstipation   severe constipation

occipital lobe   posterior lobe of the cerebrum

OES   optical emission spectrometry (see also AES)

oestrus (estrus)   recurrent restricted period of sexual receptivity in female mammals other than humans

oliguria   pathologically diminished excretion of urine

ontogenesis, ontogeny   the development of the individual organism

organelles   subcellular structures having a special metabolic function (e.g. mitochondria, ribosomes, lysosomes)

osteochondrosis   different forms of degeneration of bone-cartilage tissue

osteodystrophy   defective bone formation

osteogenesis   formation of bone material

osteomalacia   weakening of the bones as a result of insufficient incorporation of mineral substances (especially calcium)

osteonecrosis   death or necrosis of bone

osteoporosis, osteoporotic   quantitative reduction in the bone tissue while the bone structure is retained

osteosarcoma   a malignant primary bone tumor composed of a malignant connective tissue stroma with evidence of malignant osteid, bone, and/or cartilage formation

osteosynthesis   type of orthopedic surgery by which fractured and fragmented bones are synthesized using plates, nails, screws and wires

otolith   small, crystalline calcareous concretion in the balance center of the inner ear (also in the outer canal in the case of chronic inflammation)

pancreatic   pertaining to the pancreas

parakeratosis   incomplete keratinization of the epidermal cells, most common cause of all forms of flaking of the skin (dandruff)

parathesia   numbness of the fingers, hands and feet mimicking a glove and stocking distribution

parenchyma   the specific cells of tissues of an organ which determine the special function of that organ (e.g. liver and nerve cells) as opposed to the supportive connective tissue

parenchymal degeneration   loss of the specialized cells accompanied by loss of specific function of that organ

parenteral   outside or not via the alimentary tract

paresis   motoric inactivation after injury to the central or peripheral nervous system

paresthesia   abnormal sensations, such as tingling, numbness of the extremities

paronychia   inflammation involving the folds of tissues around the fingernails

PCB   polychlorinated biphenyls

peak maximum, e.g. in a spectral curve or a chromatographic elution curve

pemphigus an acute bullate (with the formation of blisters) dermatosis (skin disease)

percutaneous through the skin

perfusion passage of a fluid through the vessels of an organ

pericaryal, perikaryal concerning the body as opposed to the nucleus of a cell, applied particularly to neurons

peritoneal in or concerning the peritoneum (serous lining of the abdominal cavity)

peritonitis inflammation of the peritoneum

phagocytosis process by which nutrients and foreign particles are engulfed by cells; especially by macrophages

pharmacokinetics study of the quantitative action of a drug in an organism, especially of the temporal processes of absorption, distribution, biotransformation and excretion

pharyngitis inflammation of the pharynx, including difficulty in swallowing

pharynx cavity behind the nose and mouth, throat

phenotype visible characters of an individual which can be attributed to both the genotype and the environment

phlebothrombosis presence of a clot in a vein (not associated with inflammation of the vein)

phloem conducting cells of plant vascular tissue characterized by sieve elements

phyllosphere sphere of the plant leaves

phytate salts of phytic acid (inositol-hexaphosphoric acid), which occur in plants, and thus in food

phytotoxicity injurious effect of chemical or physical agents on plants

pinocytic concerning a pinocyte (a specialized cell) or pinocytosis (ingestion of extracellular fluid into a cell by invagination and enclosure of the fluid by the cell membrane)

PIXE particle-induced X-ray emission, a method of analysis which directs ion beams onto samples (a variation of X-ray fluorescence analysis for thin surface layers)

plasmid autonomic, extrachromosomal genetic element occurring in bacteria, e.g. carrier of resistance to antibiotics

PMMA bone cement polymethylmethacrylate based cement for the fixation of endoprostheses

pneumoconiosis lung disease resulting from the deposition of inhaled particulate matter in the lungs

Pneumocystis carinii a protozoan which is the causative agent of a highly contagious, epidemic, and interstitial plasma cell pneumonia

pneumonitis collective term for inflammation of the lung

pneumothorax accumulation of air in the pleural cavity

podsol grayish-white soil horizon in mild, temperate climates which forms the lower layer of the humus-containing topsoil and contains leached particles from upper layers

polycythemia pathological increase in the number of red blood cells, thrombocytes and granulocytes in the blood

polyneuritis simultaneous inflammation of several peripheral nerves (including brain nerves)

polyposis development of multiple polyps on a mucous membrane

polyuria pathological increase in the amount of urine excreted

porphobilinogen the parent compound of the porphyrins

porphyria excretion of large amounts of different porphyrins and their precursors in the urine (indicates a dysfunction of heme synthesis)

porphyrins special respiratory pigments; e.g. components of hemoglobin and myoglobin

porphyrinuria a disorder of the porphyrin metabolism with increased (secondary) excretion of porphyrin and its precursors in the urine (not identical to porphyria)

progesterone female sex hormone (pregnancy hormone, also secreted by the corpus luteum), its functions include implantation of the fertilized ovum and immobilization of the uterus during pregnancy

prokaryotic concerning the lower, primitive cells which possess no nuclear membrane, cell division apparatus or self-duplicating organelles (e.g. mitochondria, chloroplasts)

prophage the latent stage of a phage in a lysogenic bacterium, in which the viral genome becomes inserted into a specific portion of the host chromosome and is duplicated each cell generation

proteinuria presence of excess protein in the urine

proteolytic concerning the breakdown of proteins by hydrolysis

prothrombin a glycoprotein contained in blood plasma, precursor of thrombin, an enzyme which plays an important role in blood coagulation (clotting)

protozoa primitive animals (usually unicellular)

PSA potentiometric stripping analysis

PS-MS plasma source-mass spectrometry

Purkinje cells large, differentiated cells in the cerebellar cortex

pyelography X-ray representation especially of the renal pelvis after it has been suffused by a contrast solution introduced by means of a catheter or an intravenous injection

pyelonephritis inflammation of the kidney and renal pelvis due to bacterial infection

RBC red blood cells

RCNAA radiochemical neutron activation analysis

receptor site in an organ, a cell or a macromolecule at which active substances (including toxic substances) are bound

renin-angiotensin proteolytic enzyme synthesized, stored and secreted by the kidney, a powerful vasopressor and stimulant of aldosterone

reserves the quantity of resources which can be economically exploited using the currently available processes (for the definition see U.S. Bureau of Mines, Geological Survey Circular 831 (1980), pp. 178–181)

reservoir storage container, in environmental chemistry: a natural receptacle from which the outflow of material per unit time is small compared with its total capacity

resource reserve source of supply, geological: the natural content of raw materials in the Earth's crust which are actually or potentially suitable for extraction and production (forecasts also include the marginal resources) – even under changed economic circumstances

reticuloendothelial collective term for a highly effective biological system comprising reticular and endothelial cells which function as phagocytes and storage cells and which are responsible for the body's natural detoxification and immunological processes

rhabdomyosarcoma a highly malignant tumor of the striated muscle

rhinitis inflammation of the mucous membranes of the nose

rhizosphere sphere of the root tips

rhyolite magmatic extrusive rock with a high SiO_2-content, similar to granite

RNA ribonucleic acid (structural element of the cytoplasm and cell nucleus), consisting of D-ribose sugar, the nitrogenous bases adenine, guanine, cytosine and uracil. There are three different types: transfer RNA, messenger RNA and ribosomal RNA

rubriblast pronormoblast – a nucleated precursor of the erythrocyte

SAED selected area electron diffraction

sarcoidosis a chronic granulomatous reticulosis of unknown etiology involving almost any organ or tissue

sarcoma malignant tumor arising in the connective tissue

scavenger a substance that influences the course of a chemical reaction by ready combination with free radicals; the term scavenger is also used for 1,2-dibromoethane and 1,2-dichloroethane which are added to gasoline in order to react with lead to form volatile lead halogenides, which are easily removed from the engine system

SCE, sister chromatid exchange exchange of genetic material between two chromatids of two homologous chromosomes during meiosis (reductive cell division); method of determining the genotoxic effect of substances

Schwann cell sheath cell (glia cell) that surrounds the peripheral nerves with a sheath of neurolemma and thus influences the conduction speed of the nerve processes (axons)

scintiscanning (also called radio-isotope scanning) technique in nuclear medicine to give a two-dimensional representation of organ structures after administration of a radioisotope (uses include the localization and diagnosis of disorders of the thyroid gland, kidney, liver, brain, bone, etc.)

sclerosis, sclerotic hardening of a part, especially due to inflammation and in diseases of the interstitial tissue

scotoma a blind spot or area of diminished vision in the visual field (a functional area of the retina)

SDD sodium 1,2-dihydroxy-3,5-benzenedisulfonate, a chelating agent

sedative tranquilizer, an agent which calms the nervous system

SEM scanning electron microscopy

sequestration (intracellular) separation (removal) of a (toxic) substance by a protein, e.g. metallothionein

serous concerning or consisting of serum (the liquid portion of the blood and lymph which remains after removal of the blood cells and fibrin which have formed a clot)

sickle cell anemia hereditary, genetically determined hemolytic anemia, almost exclusively found in blacks

sideroblastic concerning sideroblasts (nucleated red blood cells containing granules of iron in their cytoplasm)

siderochrome colored iron containing pigment

sideropenia iron deficiency

siderophore a macrophage containing hemosiderin, an insoluble storage form of iron which is visible microscopically

siderosis deposition of iron in tissues and organs

SIMS secondary ion mass spectrometry (for the measurement of distribution, e.g. within a surface)

sink an accumulating reservoir into which the inflow of materials considerably exceeds their outflow

sinusitis inflammation of the sinuses of the head

SMS sodium dimercaptosuccinate, a chelating agent

somatic concerning the body

somatic cell any cell of the body

somnolence prolonged drowsiness or a condition resembling trance that may continue for a number of days; sleepiness

spasmophilia a condition in which the motor nerves show abnormal sensitiveness to mechanical or electrical stimulation

speciation specifying the type of bond, compound or oxidation state of metal ions

species kind; in environmental chemistry: the various forms of chemical compounds (including complexes and salts) formed by the same element, especially a metal

spermatogenesis formation of spermatozoa (sperm)

SSMS spark source mass spectrometry

stannosis non-malignant lung disorder due to chronic inhalation of tin in the form of tin(II) chloride and tin(IV) oxide

steatosis fatty degeneration, see also microvesicular steatosis

STEM scanning transmission electron microscopy

stomatitis inflammation of the mucous membrane of the mouth

stripping in analytical chemistry: special (inverse) voltammetric process

subcutaneous under the skin

supernatant overlying clear liquid above the solid precipitate after cell fractionation and centrifugation (e.g. in an examination of the distribution of subcellular fragments from the brain)

SWV square wave voltammetry

synapse junction (site of excitation transfer) between two nerves or between a nerve and a muscle

synaptosome cell organelle containing the active substance (neurotransmitter) at a synapse

systolic blood pressure blood pressure during contraction of the heart muscle (the maximum blood pressure is measured)

tachycardia excessive activity of the heart (increase in the frequency of contraction over 100/min), e.g. as a result of strenuous exercise, but also due to certain disorders and stresses

tachypnea, tachypnoea abnormally rapid rate of breathing

TEM transmission electron microscopy

TEP total endoprosthesis, replacing a complete bone or joint

teratogenic causing malformation in an unborn child during the development of the embryo

teratospermia the presence of malformed sperm in the semen

tetany syndrome characterized by sharp flexion of wrist and ankle joints, muscle twitching, cramps and convulsions

thalassemia a form of hemolytic anemia due to a group of hereditary defects resulting in minute hypochromic erythrocytes

threshold value the minimum dose or concentration of an agent capable of producing an effect

thrombasthenia a platelet abnormality characterized by defective clot retraction and prolonged bleeding time

thrombocytopenia pathological reduction of blood platelets (under 150000/µL) in peripheral blood

tinnitus noise in the ears, e.g. ringing, buzzing

toxicokinetics study of the quantitative action of a poison in an organism over a period of time, especially the processes of absorption, distribution, biotransformation and excretion

tracheobronchial in or pertaining to the trachea (windpipe) and bronchial system

transepithelial occurring through or across an epithelium

transferrins heme-free glycoproteins in the blood serum which function as transport proteins (with a sugar component) and bind and transport absorbed nutritional iron

transmucosal through the mucous membrane (especially in the intestine)

tremor involuntary trembling; frequent, rhythmic alternate contraction of opposing groups of muscles

tryptophan essential amino acid, which participates in the biosynthesis of NAD (nicotinamide adenine dinucleotide) and is the initial precursor of the neurotransmitter serotonin

TTFA target transformation factor analysis

tubulus cells cells of the renal tubules

TXRF total X-ray reflection fluorescence

ulceration, ulcers a local defect, or excavation of the surface of an organ or tissue with an inflamed base

unimodal showing only one maximum in a statistical series

uracil 1,2,3,4-tetrahydro-2,4-dioxopyrimidine, a component of RNA

urogenital pertaining to the urinary (excretive) and genital (sex) organs

vascular concerning, belonging to, in or via (blood) vessels

vasodilatation, vasodilation increase in the diameter of peripheral blood vessels by relaxation of the smooth muscle of the vessel wall

ventricular concerning, in or via a chamber (of the heart, brain or stomach)

vesicles small bladders or sacs containing liquid; small blisters

Wilson's disease recessive autosomal, hereditary disease (if untreated, results in invalidity and death) in which toxic amounts of copper are accumulated in the liver and central nervous system

XRF X-ray fluorescence

xylem woody vascular conductive system in plants (for the transport of water and the predominantly inorganic substances dissolved in it)

ers in **boldface** refer to main entries

abandoned site, site remediation, *see* clean-up
aberration rate, *see also* chromosome   623f.
abortion, *see* fetal loss, reproduction
abrasion, rubbings, fines, grit, abrasive, grinding, *see also* tire   261, 271, 299, 1344
absorption, absorption rate, adsorption, *see also* charcoal resorption   10, 54, 68ff., 85f., 90, 92, 111, 113, 124f., 138, 140f., 148, 158f., 162, 216, 224, 255, 287f., 294, 300, 303f., 311ff., 314ff., 318, 320f., 326f., 331, 348, 353, 366, 370f., 375f., 379ff., 383, 386, 388f., 394f., 397, 419, 421f., 431, 433, 440, 449f., 453, 462, 470, 476, 485ff., 488, 493, 495ff., 498, 506, 512, 515f., 518ff., 521, 531, 533ff., 538, 541ff., 544, 549ff., 553ff., 566, 571ff., 574f., 580ff., 586, 589, 592f., 600, 602, 666, 672, 676, 680, 721ff., 724ff., 729, 735, 740f., 746, 757ff., 773, 776ff., 781, 785, 790, 793f., 796, 801, 813, 821ff., 828, 831, 862f., 865, 870, 872, 875, 880f., 883ff., 886f., 896ff., 899, 911f., 924, 933ff., 939ff., 949, 951ff., 954ff., 957, 959, 963, 965, 985, 987ff., 990, 993, 995, 998, 1000, 1011, 1013, 1018, 1028f., 1031, 1038f., 1046, 1048, 1056, 1058f., 1063, 1082, 1091f., 1094f., 1105, 1108ff., 1118, 1121, 1123, 1129f., 1132f., 1144f., 1156, 1160, 1165, 1167, 1169f., 1173, 1176, 1188, 1194f., 1197, 1200f., 1203, 1205f., 1211, 1216, 1220, 1233f., 1239, 1244, 1249f., 1261, 1270f., 1279f., 1285, 1292f., 1302ff., 1315, 1317f., 1320ff., 1323, 1326, 1328, 1333, 1345f., 1366, 1370
abundance   3, 7, 10
acceptor, *see* receptor
accident, episode   687, 777, 926, 994, 1067, 1176, 1195, 1200, 1206, 1221, 1224, 1227, 1230, 1234, 1278, 1282, 1286, 1295
accumulation, enrichment factor, *see also* depot formation, preconcentration   5, 12ff., 29, 35f., 56f., 83, 89, 107, 111, 119, 137ff., 200f., 203, 215, 219, **221ff.**, **225ff.**, 229ff., 239, 257, 261, 263f., 266, 268, 270, 273ff., 279, 282f., 292, 294, 296, 302f., 306, 311, 329, 335f., 338, 342, 358, 360ff., 370, 374ff., 377, 383, 404, 409f., 412f., 421ff., 426, 429, 433, 441, 447, 449, 453, 455, 464, 467, 469ff., 472ff., 475ff., 479, 485, 498, 501, 512ff., 515f., 518, 520, 526, 533ff., 537ff., 541f., 544, 549f., 560ff., 576, 583, 589, 592, 597f., 601, 608, 612f., 643f., 646, 701, 708, 720, 722ff., 725f., 728f., 733, 736, 738, 741, 743f., 746, 749f., 756, 758, 765, 772f., 776, 778f., 786, 793f., 796f., 799, 803, 806f., 813, 816ff., 819, 821f., 830, 839, 841ff., 845, 847, 849, 851, 855, 857, 859, 863, 876, 878, 881, 884f., 896f., 901, 903f., 906, 908, 912ff., 916, 919, 924, 926, 928, 933f., 939, 941, 954, 963, 966f., 971, 975, 980, 988, 994f., 1008, 1013, 1015f., 1018, 1027, 1031, 1038, 1041, 1046, 1048, 1052f., 1057f., 1060f., 1063, 1065, 1072f., 1077, 1081, 1086f., 1092, 1094f., 1098, 1103, 1105ff., 1109, 1111f., 1123, 1129f., 1133, 1136, 1140f., 1144f., 1156, 1160, 1162, 1164ff., 1167, 1181, 1193f., 1208f., 1212, 1217f., 1223, 1226, 1231, 1234f., 1239, 1249f., 1261, 1264, 1271, 1275, 1277f., 1289f., 1300, 1302f., 1311, 1317, 1319, 1323f., 1345, 1353, 1366, 1369, 1371
accumulator, *see* battery, heat
accuracy, *see* precision
acetaldehyde, acetylacetone, acetylene, acetone   1056, 1213, 1226, 1262, 1291
acetum, vinegar, acetic acid, acetate, *see also* peracetic acid   111, 339, 345, 358, 360, 384, 400ff., 827, 834, 860, 864, 895, 900, 904, 1017, 1047, 1052, 1106, 1237
acid, acidification, acidity, *see also* fat, gastric juice   56ff., 71, 79f., 102, 120, 157ff., 195, 245, 248, 268, 272, 276, 278, 313, 323f., 327, 342, 346, 348ff., 352, 360,

369ff., **373f.**, 376ff., 386, 393, 399f., 449f., 466, 492, 495, 587, 715, 721f., 727, 734f., 736ff., 741, 749, 777, 780, 790, 805, 827, 841, 882f., 896, 901, 903, 932, 946, 972, 991, 1013, 1027, 1030, 1040, 1049, 1091, 1093, 1095, 1163f., 1167, 1203, 1230, 1261, 1267, 1275f., 1290f., 1310, 1313, 1318, 1342, 1349
acidosis, aciduria 485, 596, 727, 954, 1349
acid rain, acid deposition, *see also* forest decline 10, 114, 191, 281, 346, 348, 405, 492, 720ff., 723f., 737, 739f., 749, 777, 850, 896, 1013, 1030, 1034, 1055, 1101, 1105, 1163, 1185
acrodermatitis enteropathica, *see* dermatitis
acrosome, *see* sperms
act(s) (in the U.S.A.), *see* regulations
ACTH, *see* adrenal gland
actinides (general), *see also* thorium, uranium 75f., 235, 243, 255, **1275**, **1278f.**, 1281, 1283, 1285f.
activated carbon, *see* carbon
activation, *see* neutron activation, radiochemical activation
adaptation, adaption 587, 1039
additive, additively, *see also* food additive, stabilizer 38, 442, 492, 659, 671, 759, 791f., 883, 961, 1016, 1027, 1037, 1149, 1159, 1174, 1180, 1183, 1214f., 1231
adenine, adenosine, ATP 423, 437, 722, 727, 759, 965, 969, 1025, 1039, 1055, 1064f., 1085, 1087, 1172, 1287, 1294, 1296, 1304, 1349, 1351, 1364, 1368
adenocarcinoma, adenoma, adenovirus 830, 911, 1187, 1349
ADH-activity, *see* alcohol dehydrogenase
ADI (acceptable daily intake), *see* daily intake
adiposis, *see* fat, body weight
administration, *see* regulation (legal)
adrenal (suprarenal) gland, adrenocorticotropic 300, 534, 725, 746, 783, 968, 999, 1108f., 1113, 1330, 1349
adults, *see also* age 209, 219, 302f., 415, 482, 485, 536, 538, 574, 579, 597, 657, 679, 705, 821, 839f., 885, 896f., 904, 934, 941, 949, 952f., 982, 984, 988ff., 994ff., 998, 1002, 1030, 1039, 1058f., 1065, 1067, 1069, 1072, 1076, 1078, 1082f., 1091, 1108f., 1111, 1119, 1129f., 1133, 1166, 1220, 1248, 1264, 1293f., 1320, 1328, 1330f.
aerobic conditions, aeration, *see also* oxidation 339, 346, 357, 433, 437, 946, 1057, 1064, 1214, 1349, 1363

Aerosol, *see* atmospheric particles, dispenser
age, *see also* annual ring, core, childbearing, reproductive age 214, 219, 225, **269ff.**, 492, 527, 531, 536, 538, 551, 555, 559, 583, 588, 594, 596, 598, 601f., 654ff., 662, 664, 738, 821, 823, 825f., 850, 913, 926, 984, 986, 999, 1013, 1028, 1034, 1048, 1057f., 1086, 1109, 1119, 1130f., 1133, 1166, 1174f., 1315, 1317, 1320
aggregation, aggregates, agglomeration 3, 27f., 34, 50, 238, 964
agriculture, agronomy, agrowaste, crops 55, 58, 90, 289, 311, 334, 343, 357ff., 361, 363ff., 367, 372, 379, 387, 399, 401, 403ff., 407f., 410, 417, 444, 449f., 459, 470, 475, 505, 706ff., 710, 746, 750, 755, 757, 804, 814, 818, 834, 838, 842, 848, 851, 857, 871f., 874, 879, 898f., 900, 903, 905, 953, 988, 1004, 1007, 1009, 1013, 1051, 1056, 1060, 1077, 1081f., 1093, 1096, 1098ff., 1145, 1160, 1164, 1167, 1170, 1181, 1190, 1240, 1248ff., 1252, 1255, 1263, 1278, 1302, 1315, 1318, 1336ff., 1340
AIDS virus, patients 743f., 748, 1182
airborne emission, *see* atmosphere
air-cell, *see* alveolus
aircraft, airplane (part, element), *see also* space 775, 777, 881, 978, 1016, 1026, 1104, 1198, 1205, 1262, 1277, 1344
air passage, *see* respiration tract
air pollution, *see also* atmosphere, contamination, occupational exposure, indoor 287, 370, 378, 410, 493, 508
air quality, *see* threshold
ALA, Δ-amino-levulinic acid and its derivatives 499, 516, 538f., 546, 600, 623, 648, 693, 914f., 917, 944, **993f.**, 1005, 1349
alabaster (plaster) 720
albinism, achromotrichia, *see also* discoloration 485, 899
albumin, albuminuria, proalbumin 514, 518, 534, 551, 577, 580, 645, 649, 826, 884, 934, 949, 953, 991, 1067, 1094, 1109, 1114, 1116ff., 1120f., 1219, 1303, 1321f., 1349
alcohol, alcoholism, *see also* ethanol, glycol, beer, wine 211, 304, 468, 485f., 656f., 703, 756, 882, 887, 904, 981, 986, 991, 1032, 1039, 1174, 1249, 1319, 1321, 1325, 1349
aldehyde, *see also* acetaldehyde 288, 485, 882, 1092f., 1097f., 1142, 1188
algae (growth) 55, 69, 95ff., 102, 223, 226, **419**, 422, 426, 428f., 442, 447, 677, 722, 758, 773f., 780, 800, 817, 865, 885, 896ff.,

907, 933, 936, 938, 948, 953, 1027, 1038, 1056, 1129, 1137, 1145, 1166f., 1172, 1235, 1252, 1294, 1302, 1317, 1319, 1356
algicide 753, 895
alkylated metallic species, *see also* organometal compounds, methylated gold, methylated and ethylated lead compounds, methylation, organotin, methylcobalamin, methylmercury 89, 432, 1246
alkylation, *see also* methylation, biomethylation 99, 153, 424, 620, 628, 633, 1244
alkyltin compounds, *see* organotin
allergic reactions, allergen, *see also* dermatitis 467, 559, 562, 591, 593, 599, **605**, 610, 614f., 761, 862, 866, 873, 886, 888f., 935, 1031, 1114, 1117f., 1121, 1135, 1145ff.
alloy, *see also* ferroalloys, steel, aluminum alloy, brass, bronze, letter 33f., 333, 557f., 560f., 563, 754, 775, 777, 789ff., 792, 800, 809, 853, 856, 873, 881f., 887, 895, 905, 921f., 929, 932, 937, 940, 958, 961, 976, 1016, 1036, 1090, 1099, 1102, 1104f., 1128f., 1137, 1159, 1193, 1205, 1211, 1213ff., 1223, 1230, 1237, 1243, 1245, 1261f., 1267, 1270, 1277, 1291, 1299, 1301, 1308, 1310ff., 1344, 1347
alluvial, alluvium 932, 1095, 1139, 1313, 1350
Alps, alpine 270, 276f., 377, 412, 1285
aluminum (general, elementary), *see also* bauxite 14, 32, **36f.**, 40f., 49f., 84, 89, 93, 95, 109, 111, 132, 145, 149, 156, 163f., 173, 175, 178, 183, 188, 197, 200, 202f., 235, 243, 292, 297, 303ff., 306, 337, 344, 373, 400, 405, 414, **451**, 467, 484, **532**, 545, 561, 572, 598, 601, **644f.**, **715ff.**, 775, 777, 780, 796, 809, 882, 894f., 911, 917, 919, 945, 947, 1017, 1034, 1036, 1084, 1125, 1128, 1261, 1291, 1301, 1359
aluminum compounds, *see also* bauxite, cryolithe, ruby, sapphire 36, 40, 67f., 85, 100, 138, 154, 240, 299f., 304, 308, 314, 319, 329, 344, 370, 373f., 403, 532, 561f., 715f., 718ff., 721, 725, 727f., 736., 792, 861, 864, 881, 947, 1016f., 1021, 1027, 1142, 1144, 1159
aluminum ions 72, 82, 97, 149, 320, 376, 628, 720ff., 723
alveolus, pulmonary alveolus, alveolar 262, 292, 304, 508, 515, 518, 520, 574ff., 582, 610f., 780, 822f., 830, 861, 867, 873f., 876, 915, 941, 943, 969, 988, 1046, 1101, 1112f., 1118, 1120, 1206f., 1209, 1305, 1324, 1350, 1355, 1363, 1371

Alzheimer's disease 304, 598, 644f., 647ff., 715, 726, **729f.**, 734, 738ff.
amalgam, amalgamation 22, 137f., 140, 557f., 561f., 932, 1046, 1048f., 1051ff., 1060, 1067, 1072, 1079f., 1088, 1193, 1197, 1230, 1245, 1261
Ames-test, *see also Salmonella* **620**, 632, 867, 1179, 1185, 1237, 1331
amines (primary, secondary), amino groups, *see also* ehtylene diamine 90, 243, 324, 430, 588, 618, 949, 1060, 1109, 1194
amino acids, *see also* GABA, protein, serine 249, 388, 420, 484, 518, 521, 523, 526, 644f., 828, 854, 862, 865, 936, 949, 952, 954, 959, 1094ff., 1112, 1165, 1194, 1304, 1321ff., 1350, 1362, 1370
amino aciduria 596, 998, 1112, 1118, 1350
ammonia 26, 76, 99, 239, 338, 369, 372, 430, 854, 880, 895, 1092, 1141, 1145, 1149
ammonium ions, ammonium compounds 73, 157, 265, 275, 338, 340f., 358, 370ff., 400ff., 426, 514, 727, 744, 805, 837, 860, 880, 918, 1091, 1093, 1098, 1100, 1106, 1141
amunition, *see* bullet, explosive
amphibian 760, 1111, 1130, 1132
amyloidosis 535, 1350
amyo..., *see* muscle
anaerobic conditions, *see* anoxic conditions
analytical error, *see* error source
anemia, anaemia 485, 494f., 535ff., 538f., 541, 544, 549, 590, 599, 646, 715, 729, 738, 754, 761, 763, 884, 886ff., 899f., 902, 906, 914, 935, 953f., 994, 1031, 1040, 1174, 1220, 1328, 1330, 1332, 1350, 1352, 1369f.
anepithymia, *see* inappetence, anorexia
aneuploidy, spindle effects 621, 829, 1071f., 1078, 1350
animals, species, *see also* animal experiment, mammal, fish, bird, invertebrate, mollusc, wild animal, amphibian 56, 225, 228, 357, 362, 406, 408f., 413, 429, 431, 442, 449, 461, 466, 470, 474, 476f., 481f., 484, 487, 491, **493f.**, 498, 501, 511, 520, 531, 533, 535f., 538, 550, 571f., 580, 588, 630, 651, 673, 722, 726f., 745ff., 753, 756ff., 759f., 762, 767, 773, 778, 780f., 789f., 793ff., 814, 818, 821, 825, 827, 833f., 838, 844, 848, 853, 859, 861f., 864, 879, 884ff., 892, 896f., 899f., 912, 914, 924ff., 927, 931ff., 935, 940f., 951, 953ff., 959, 963ff., 970f., 978f., 981, 987f., 991ff., 994, 1000, 1016, 1018, 1021, 1025, 1029ff., 1034f., 1038ff., 1044,

1056, 1058, 1060ff., 1064f., 1068ff., 1072, 1075, 1086, 1089, 1091f., 1094, 1097ff., 1101, 1105f., 1108, 1110, 1112f., 1117, 1123, 1129ff., 1145f., 1153, 1157, 1162, 1165, 1167f., 1170, 1172f., 1175, 1180, 1191, 1194ff., 1200, 1205, 1207, 1216, 1218f., 1233, 1235, 1237, 1239f., 1243, 1250f., 1258, 1264f., 1271, 1278f., 1282, 1292ff., 1302ff., 1309, 1312, 1315ff., 1321, 1323, 1325f., 1331, 1340, 1345f., 1355

animal carcinogenicity 743, 782f., 844, 858, 878f., 889, 966, 1001, 1071f., 1114f., 1147, 1177, 1198, 1221, 1264

animal experiment 145, 157, 300, 495, 498f., 501f., 504, 531, 533f., 543f., **547ff.**, **550ff.**, **553ff.**, 559, 564, 597f., 605f., 608f., 625, 630f., 655, 660ff., **663f.**, 670, 673f., 693, 724, 729, 743, 746, 764ff., 767, 772, 775, 781ff., 785, 803, 822f., 826, 828f., 831, 833, 863, 866f., 869, 879, 912, 926, 952, 988, 992, 995, 997ff., 1000, 1003, 1006, 1014, 1041f., 1060, 1062f., 1072, 1111, 1115, 1167, 1175, 1180, 1195, 1197, 1234, 1271f., 1293, 1295, 1304, 1316, 1321, 1323, 1329f.

anion (exchange) 149, 153, 156, 159, 236, 239, 245, 312, 314, 316, 327, 341, 422f., 447, 493, 495, 751, 759, 868, 910, 931, 1048, 1246, 1249

annual ring, *see also* age 113, 270, 276, 983

anode, anodizing, anodic oxidation 36, 38, 41, 137, 719, 724, 1016, 1142, 1158, 1214

anodic stripping voltammetry (ASV) 69, 137, 197, 247, 324, 743/744, 790, 800, 910, 918, 973, 1192, 1310, 1351, 1354

anophthalmia, anophthalmus microphthalmia 1074, 1115, 1350, 1363

anonexia, *see* inappetence

anoxic conditions, *see also* deaeration 55, 62, 94, 255, 324, 328, 339, 341, 346, 357, 383, 385f., 390ff., 396, 398, 433, 437f., 441, 1057, 1064, 1277, 1315, 1350

antacid, *see* drug

antagonism, antagonistic, *see also* synergism, interaction 420, 426, 505, 565, 626, 822, 826, 830, 884, 900, 964, 1025, 1030, 1074, 1089, 1094, 1096, 1099, 1173, 1178, 1187, 1190, 1196ff., 1201, 1321, 1326

anthropogenic source, anthroposphere, *see also* road, automobile exhaust, source, industrial emission, etc. 8, 10, 16, 83, 89, 102, 113, 145, 158, 196, 221, 233, 257ff., 268ff., 275, 281f., 311, 352, 360, 374, 380, 393, 405, 416, 430, 432, 443, 450, 453, 745, 747, 755, 803, 844, 849, 973, 979, 981, 983, 1006, 1053, 1055, 1115, 1138, 1162f., 1231, 1278, 1291, 1301, 1306, 1312, 1359

antibiotic, antibacterial, antiviral, *see also* penicillin, naloxone, bactericide 424, 426f., 434f., 567f., 641, 792, 799, 924, 1040, 1062f., 1195, 1197, 1366

antibody, *see* antigen

anticarcinogenicity, antitumor activity, anticlastogenic, *see also* repair mechanism, cytostatic 498, 504, 506, 527f., 568f., 626, 629, 636, 638, 911, 917, 921ff., 925f., 929, 935, 937, 943, 964, 967, 1142f., 1147ff., 1150f., 1153, 1177ff., 1180, 1182ff., 1254, 1267

anticoagulant 111, 1036

antidepressive, *see* psyche

antidote, *see* detoxification

antifouling, *see* fungicide

antigen, antipeptide, antibody, *see also* immune response 447, 605ff., 608ff., 611, 613f., 781, 806, 886, 925, 1071, 1083, 1115, 1117, 1126, 1304, 1357, 1360

antiknock (agent or compound), *see also* leaded gasoline, methylated and ethylated lead compounds, organometal compounds 972, **975**, 978, **999**, 1037, 1250

antimony, antimony compounds 21, 35, 83, 89, 130f., 135, 149, 151, 153, 159, 168, 173, 186, 193, 199, 236, 243f., 258, 335, 427, **470**, 502, **532f.**, 572, 661, 670, **743ff.**, 768f., 789, 910, 975, 1133, 1141, 1158, 1192, 1195, 1211, 1213, 1226, 1245

antimutagenicity 504, 626, 635, 829, 838, 1153, 1179, 1185, 1188

antioxidant 497, 505, 1153, 1173, 1178, 1187f., 1312

antiperspirant, *see* deodorant, perspiration

antiseptic 1045, 1052, 1062, 1197, 1312

antitranspirant, *see* deodorant, perspiration

antitumor activity, *see* anticarcinogenicity

anuria 534, 540, 725, 761, 1067, 1220, 1350

aorta, aortic aneurysm, *see also* artery 514, 899, 984, 1196, 1221, 1345

apathy, *see* tiredness

apatite, *see* calcium compounds

APDC (ammonium pyrrolidone dithiocarbamate), *see* Dithiocarb

ape, *see* monkey

appetite, *see* inappetence

apple, malic acid 485, 1107, 1249, 1279, 1304, 1324, 1326

aquatic organism, animal, life, *see also* crayfish, daphnia, fish, mollusc, oyster, plankton, shellfish 56ff., 97, 379, 436f., 452, 470, 472ff., 475ff., 479, 517, 519f., 539, 545, 691, 727, 732, 741, 760, 765, 769, 771, 784, 789, 793f., 803, 817, 821, 827, 844ff., 847, 849, 864, 870, 874, 876, 880, 890, 897, 905, 914, 940, 995, 1008, 1043, 1057f., 1062, 1078, 1082, 1086, 1088, 1099, 1107, 1111, 1115, 1117, 1123, 1130, 1163f., 1167, 1177, 1181, 1183, 1191, 1195, 1198, 1204, 1243, 1248, 1251f., 1256ff., 1259, 1309, 1316, 1319, 1323f., 1332, 1341, 1351, 1358

aquatic plant, see also algae, macrophyte 171, 226, 229, 409, 441, 691, 771f., 821, 825, 845, 847, 874, 890, 896f., 905f., 1000, 1014, 1027, 1035, 1038, 1057f., 1088, 1129, 1151, 1163, 1166, 1172, 1184, 1193, 1195, 1252, 1259, 1302, 1307f., 1317

aquo-metal complex, see hydroxocomplex

arc (radiation), see also flash, sparks 34, 132f., 209, 723, 948, 961, 1040, 1090, 1104, 1126, 1128, 1300, 1343

arctic, antarctic, see polar ice

argon 135, 148, 160, 717, 801

argyria 1196, 1350

arid area 268, 281ff., 778f.

arrhythmia 586, 589, 1020, 1032f., 1093, 1113, 1133, 1174ff., 1350, 1352, 1370

arsenate, arsenite 320, 327, 423, 427, 430, 441, 545, 555, 580, 612, 623, 626, 628, 634f., 637, 698f., 711, **751ff.**, 754, **756ff.**, 759ff., 765ff., 768, 771, 773, 895, 981, 1001

arsenic (general, elementary), arsenic compounds, see also organoarsenic compounds, arsenate, arsenide, arsine 6ff., 11ff., 15, 35, 51f., 57ff., 83, 89, 102, 119, 130ff., 135f., 140, 145, 149, 151, 153, 159, 169, 173, 176, 186, 193, 199, 201, 218, 230, 235f., 257ff., 283, 288, 294, 301ff., 307, 335, 341, 352, 355, 381, 405, 414, 427, **441ff.**, 449, **452**, 467, **469ff.**, 478, 484, 502ff., **533**, 545ff., 548, 550ff., 554ff., 571f., 579f., 583, 589, 591, 593, 596, 599f., 602, **605f.**, 612, 614f., 618, 621, 623, 625, 629, 631ff., 635f., 651, 655f., 660ff., 665, 680, 683, 698f., 703, 705, 743, 747, 749, **751ff.**, 756f., 759, 761ff., 766, 772f., 789, 840, 881, 906, 912, 915, 921, 975, 1013, 1148, 1173, 1185f., 1192, 1230, 1235, 1312

arsenic oxides 555, 589, 670, **752ff.**, 759, 760f., 770, 919

arsenide, arsenic ions, arsenic acids 12, 155f., 156, 170, 243f., 341f., 406, 501, 593, 620, 628, 670, 699, **752ff.**, 772, 881, 909f., 912ff., **915**, 917, 919, 1141, 1192, 1231

arsine, AsCl$_3$, see also hydride technique 156, 173, 343, 441, 574, 590, 751, 761, 763, 766, 931, 1213

artery, arteriosclerosis, see also aorta 218, 729, 853, 998, 1010f., 1034, 1113, 1119f., 1273, 1282, 1294, 1350, 1352

arthritis, arthropathy, see also rheumatic 567, 754, 796, 798, 927, 932/933ff., 936f., 954, 960, 964, 967, 1174, 1176, 1208, 1350f.

arthropods 223, 225

artificial silk, see rayon

asbestos 20, 23, 56, 262, 265, 277, 287f., 328, 830, 1026

ascidia, sea squirts 1205, 1293, 1351

ascites, see colic

ascorbic acid, see vitamin C

ashes, see also dry ashing, fly ash, slag, wet digestion 134, 261, 292, 296, 333, **335**, **341f.**, 344, 348, 355, 740, 858, 910, 922, 934, 963, 1095, 1120, 1192, 1233, 1290, 1292, 1317

ashing, see decomposition, dry ashing, pressure disintegration, wet digestion

asphalt 54

assimilation, see also photochemistry, nitrogen 362, 370f., 826, 884, 945, 1167, 1271f., 1293, 1352

associates, see aggregation, absorption, complex, ligands

asthma, see also dyspnea 754, 866, 1101, 1113, 1146, 1271, 1354

astringent 532, 544, 922, 1027, 1329, 1351

ataxia (ataxy), see also myoclonus 541f., 761, 796, 842, 899, 965, 994ff., 997, 1020, 1022, 1031, 1039, 1069, 1175, 1351, 1355, 1364

Atlantic, see ocean

atmophile, see also respiration 16, 87, 89f., 263, 271, 277, 280ff., 283, 287f., 292, 294ff., 303f., 307f., 555, 676, 861, 889, 913, 989, 1144, 1265

atmosphere, atmospheric emission, see also wind, cloud, mesopause 3, 14, 16, 38, 40f., 55f., 67, **87ff.**, 93, 162, 197, 256, **257ff.**, 265, 268, 273, 275f., 333, 342, 360, 371, 378, 386, 392, 405, 407, 433, 458, 460, 476, 571, 574, 582, 632, 641, 678, 680, 689ff., **700ff.**, 709, 722, 724, 738, 741, 745, 750, 754f., 773, 778, 803, 810ff., 833, 835, 840, 844, 847, 850, 880, 884, 936, 950, 971,

974, 977ff., 980, 983, 986, 991, 1004, 1007, 1011, 1025, 1033, 1037, 1042, 1050, 1053f., 1078, 1103, 1105f., 1138, 1144f., 1159ff., 1162f., 1186, 1195, 1199, 1211, 1227, 1231, 1272, 1278, 1291, 1297, 1306, 1312ff., 1340, 1344
atmospheric chemistry, see also photochemistry 265, 377
atmospheric particle, aerosol, see also flue gas, welding 12, 70, 228, 257, **260ff.**, 264, 267, **271ff.**, 277f., 280, 287f., 303, 306, 308, 377, 379, 475, 492, 506, 571, **574ff.**, 577, 582, 591, 602, 614, 762, 780, 784, 810, 814, 823, 827, 830f., 840, 847, 856, 878, 889, 938, 972, 978, 980, 1011, 1013, 1088, 1095, 1106, 1116, 1126, 1130, 1135, 1144ff., 1148, 1161f., 1198, 1218, 1220, 1279, 1286, 1290, 1302, 1338, 1340
atomic absorption spectrometry, see also electrothermal, flame AAS, flameless AAS, graphite furnace AAS, FANES 116, 119, 121ff., **124ff.**, **129ff.**, 148ff., 152f., 156, 159f., 162f., 165f., 172f., 178, 183ff., 189f., 197ff., 201ff., 209, 247, 272, 282, 717, 740, 744, 749f., 752, 768, 772, 775, 783, 787, 798, 800f., 805, 842, 854, 875, 877ff., 893, 907, 922, 928f., 937ff., 944, 946, 957, 960, 970, 972, 1010, 1013, 1016, 1023, 1026, 1034, 1043f., 1047ff., 1083, 1089, 1100, 1116, 1119, 1121, 1125, 1133, 1137, 1151, 1154, 1156, 1189, 1191f., 1200f., 1210, 1212f., 1223ff., 1226, 1228f., 1241, 1244, 1258f., 1262, 1269, 1273, 1286, 1289, 1297, 1300, 1308, 1310, 1340f., 1343, 1347, 1349, 1364
atomic emission (spectrometry) AES, see also emission spectrometry, electron spectroscopy **132ff.**, 151, 176, 187, 189, 247, 752f., 801, 805, 937, 1010, 1022, 1026, 1043, 1128, 1204, 1273, 1286, 1297, 1349, 1360, 1365
atomic fluorescence spectrometry AFS **131f.**, 148, 151, 159, 162, 175f., 179, 189f., 204, 1156, 1349
atony, weakness, see tiredness
ATP, see adenine, adenosine
atresia 535, 1351
atrophy, atrophication 762, 886, 998, 1065, 1074, 1082, 1175, 1254, 1326, 1351
aurosome, see lysosome
automation 150, 153, 186, 197, 1142
automobile, car, motor fuel 59, 257, 266, 682, 700, 754, 947, 980, 1142, 1149

automobile exhaust, automotive emission 259, 273, 276, 278, 299, 405, 571, 657, 710, 812, 961, 967, 971, 976ff., 992, 994, 1003, 1009, 1013, 1037, 1106, 1135f., 1142ff., 1145, 1149ff., 1270, 1338
automobile scrap, wrecked car, see scrap
autopsy, see forensic
autotrophic 428
auxin 1318, 1351
availability, see bioavailability
aversion to eat, see inappetence
aviation, aeronautics, see aircraft
azotemia 901

baboon, see monkey
baby, see infant, newborn
background (in analytical chemistry), see also matrix effects, Zeeman background correction 125, 133, 137, 141, 143, 147, 192, 227, 233f., 275, 657
bacteria, bacillus, see also microorganisms, Escherichia coli, Salmonella, Pseudomonas, Staphylococcus, intestinal flora 57, 86, 100, 119f., 387, 394, **419ff.**, 422ff., 425ff., 428, 431ff., 434ff., 437, 440, 446, 448, 482, 504, 509, 523, 528, 567f., 608ff., 612, 614, 619ff., 622, 632, 641, 763, 773, 792, 799, 850, 867f., 878f., 882, 885f., 888, 898, 918, 937, 951, 953, 992, 1038, 1048, 1052, 1055, 1057, 1062f., 1083, 1086, 1088, 1114, 1146, 1177, 1193, 1201, 1217, 1220, 1250, 1293, 1297, 1309, 1341, 1357, 1366f.
bactericide, see also antibiotic 745, 747, 1062f., 1201, 1246, 1361
BAL, see dimercapto compounds
balance, see material transport
Baltic sea 817
bank, shore, see coast
barium (compounds) 10, 30, 51, 57, 59, 79, 219, 561, 572, 623, 776, 862, 880, 895, 962, 1204, 1266, 1312, 1344
bark 225, 228, 376
basal ganglia, see nerve, brain
basalt 4, 6ff., 17, 976
battery, battery factory, accumulator 19, 52, 59, 343, 347, 554f., 571, 671, 689, 699f., 741, 744, 748f., 754, 791, 803, 808f., 813f., 844, 849f., 962, 975f., 979, 990, 994, 999, 1001, 1003, 1007, 1016f., 1020, 1023, 1037, 1040f., 1051ff., 1088, 1104, 1126, 1193, 1214, 1312
bauxite 20, 36, 50, 716, 718, 723, 910, 912, 947

bean, *see also* legume(n)   360, 367, 409, 420, 778, 780, 820f., 884, 897, 923, 957, 1106f., 1117, 1165, 1302, 1307, 1315, 1321, 1326
bearing metal, *see also* hard-solder   744, 976, 1016
beech, *see* deciduous tree
beef, *see also* bovine, cow   470f., 475, 834, 850, 940, 1077, 1097, 1194, 1240
beer, brewer's yeast, *see also* yeast   113, 761, 764, 820, 858, 887, 904, 950, 1198
beet, turnip, *see* cane, vegetable
behavior, behavioral, *see also* psychic disturbance, paramnesia, learning ability   109, 552, 598, 602, 655, 711, **995ff.**, 1003, 1010, 1013, 1019, 1061, 1070, 1072f., 1079, 1085f., 1088, 1118, 1226
belly-ache, *see* colic
benthic, benthos, sea-bottom   758, 1008, 1036, 1244, 1320
bentonite
benzene, toluene, xylenes, styrene   439, 638, 882, 1082, 1215
benzo(a)pyrene, *see also* polynuclear aromatic hydrocarbons   304, 499, 509, 555, 627, 636, 638, 955, 1114, 1122, 1188
berry, *see* fruit
beryllium (general, elementary)   6f., 11, 14ff., 79, 95, 145, 156, 205, 258f., 376, 502, **534**, 551, 555, 572, 580, 596, 629, 631, 637, 639, 651, 664, 670, 745, **775ff.**, **780ff.**, 783ff., 917, 1021, 1026f.
beryllium compounds, beryllium ions   593, 622f., 670, 775ff., 780f., 786, 895
beverage, *see also* drinking water, fruit (juice), well, beer, wine, alcohol   288, 462, 571, 591, 656f., 666, 677, 679, 703, 741, 746, 827, 981, 986, 988ff., 1004f., 1109, 1249, 1310, 1321
bicarbonate, *see* carbonate
bichromate, *see* chromate
bile, biliary   512, 514, 517, 519, 527, 551, 555, 581, 725, 759, 770, 779, 783, 864, 873, 886, 897, 902ff., 913, 992, 1039, 1061f., 1075, 1078, 1110, 1121f., 1217, 1322, 1345, 1351, 1355f.
bilharzial, *see* tropical medicine, intestinal flora
binding states, *see* complex, electrostatic binding, ligands, speciation, DNA damage, covalent
bioaccumulation, *see* accumulation, bioavailability
bioassay, *see* biomonitoring, sensor

bioavailability, *see also* occupational exposure   68, **95ff.**, 157, 201, 231, 255, 262, 288, **300ff.**, 311, 316, 330f., 370, 374, 379f., 386, 391, 394f., **399ff.**, 403, 405, 408ff., 412, 419ff., 422, 435, 437, 440, 455, 472, 474, 477, 479, 486, 488, 491ff., 496f., 503, 506, 511, 513ff., 516ff., 520, 531, 548, 553, 558, 565ff., 570f., 586, 589, 591, 623, 630, 651, 655, 665, 677, 685, 693, 698, 722, 727, 746, 757, 765, 771, 773f., 778, 786f., 789, 794, 818, 821f., 826, 830f., 835, 842, 845, 849, 851, 858ff., 861f., 872, 876, 879, 883ff., 890, 896ff., 902, 904, 906, 908, 912f., 916ff., 924, 933, 937, 940, 946, 951ff., 955, 963, 981, 987ff., 990, 1010, 1012, 1014, 1018, 1028f., 1031, 1038, 1045, 1055f., 1058, 1069f., 1073, 1075, 1083, 1088, 1091, 1096, 1099, 1101, 1108, 1110, 1116, 1118, 1123f., 1129, 1144f., 1147, 1157, 1160, 1164, 1166f., 1169f., 1174, 1180, 1189, 1192, 1194, 1205, 1211, 1216, 1233, 1240, 1250, 1263, 1271, 1273, 1278f., 1293, 1302, 1313, 1315ff., 1318ff., 1321, 1335, 1338, 1345, 1351, 1355
biochemical reactions, biochemical studies   484, **511ff.**, **514ff.**, 519ff., 544, 548, 559, 612, 617, 726, 759, 786, 826, 914, 925, 934, 962, 964, 993, 995, 1009, 1016, 1018f., 1025, 1033, 1043, 1053. 1063, 1065, 1073, 1086, 1088, 1093, 1099, 1111, 1124f., 1168f., 1171f., 1178, 1194, 1207, 1218, 1253, 1289, 1292, 1294, 1296f., 1303, 1306f., 1325, 1328f., 1346
biocide, *see also* antibiotic   333, 478, 1020, 1062, 1248ff., 1252
biocompatibility, *see* tolerance
biodegradation, *see also* decomposition, clean-up, purification plant, sludge, self-cleaning, degradation   339, 1047
biogeochemical cycle, *see* cycle, cycling
bioindicator, *see* biomonitoring, indicator
biological tolerance value, *see also* threshold, blood, urine, biomonitoring   675f. 679f., **698**, **705ff.**, 710, 732, 737, 739, 835, 842, 889, 1002, 1042, 1077, 1272, 1332
biomass, *see also* matrix, organic (matter) waste   96, 288, 294, 304, 371, 376, 427, 936, 938, 1170
biomethylation, *see also* methylation, organometal compounds   83, 257, **424ff.**, 432, 436, 516, 550, 759, 1055, 1075, 1082, 1085, 1167, 1187, 1235, 1250f.
biomonitoring, bioassay, *see also* indicator,

biological tolerance value   107, 219f.,
   222ff., 226, 228, 230, 246, 270f., 409f.,
   428f., 444, 521, 581, 602, 610, 612, 652,
   661, **665ff.**, 675, **679f.**, 684, 709f., 739, 749,
   769, 786, 803, 832, 835, 838, 848, 863, 890,
   892, 898, 909, 911, 973, 1006, 1008f., 1013,
   1029, 1041, 1044, 1058, 1078, 1109, 1116,
   1121, 1125, 1224
biosynthesis, biotechnology (for biotransformation *see* transformation)   421, 444, 484,
   524, 528, 587, 822, 883, 885, 893, 1162,
   1196, 1318, 1349
birch, *see* deciduous tree
bird, *see also* wild bird, poultry, hen, chicken,
   feather   223, 225, 228, 230, 435, 538, 641,
   723, 900, 953, 994f., 1008, 1039, 1053,
   1057, 1077, 1101, 1177, 1196, 1326
birth defect, *see* teratogenic effect
bismuth (general, elementary)   6ff., 11ff., 15,
   135, 145, 159, 186, 243, **789ff.**, 1245, 1272,
   1281
bismuth compounds, bismuth ions   427, 527,
   534, **568**, 570, 789f., 792, 796ff., 799, 840,
   922, 927, 1212f., 1231
bituminous (aggregate)   27f.
bivalves, *see* mollusc, shellfish
"blackfoot" disease   599, 602, 763f., 772
bladder, pemphigus   935, 1070, 1115, 1221,
   1366
blank (problems), *see also* reference material,
   sample   146, 165, 973
blast-furnace, shaft-furnace   33ff., **947f.**, 975,
   1036
bleeding, *see* hemorrhagic
blindness, *see also* vision (impaired)   540,
   901, 994, 1063, 1073, 1093, 1197, 1328
blood, *see also* erythrocyte, leukocyte, plasm,
   serum, (blood) sugar   108ff., 113, 132,
   140, 147, **156**, 164f., 168, 174, 176ff., 182,
   186, 189, 192f., 198ff., 202, **207ff.**, 213,
   216, 218f., 296, 303, 306, 474, 479, 487,
   517, 520, 546, 551, 554f., 558, 560f., 563f.,
   567, 569, 571, 574, 576f., 578ff., 581ff.,
   586, 589f., 594, 598, 600ff., 603, 609,
   612ff., 625, 636, 644f., 647, 653, 656f.,
   660ff., 666, 672, 675f., 678ff., 681, 685,
   693, 698, 705ff., 709f., 717, 722f., 725, 728,
   732, 739f., 745f., 752, 757f., 761, 764, 772,
   779, 784, 787, 790, 793ff., 796f., 799, 805,
   822ff., 825, 828, 835, 838, 840ff., 844,
   847f., 854f., 862f., 876, 884ff., 891, 896f.,
   902, 905, 910, 913ff., 916f., 934, 938, 941,
   944, 953, 971, 973, 976, 984ff., 987ff., 991,
   993f., 996ff., 999ff., 1003, 1005ff., 1008ff.,
   1011ff., 1020, 1030f., 1036, 1039, 1041f.,
   1047ff., 1050, 1057f., 1060, 1071f., 1074ff.,
   1077, 1081, 1087, 1094f., 1097, 1103,
   1107ff., 1112, 1117f., 1122ff., 1129, 1132f.,
   1137. 1157, 1160, 1166, 1169f., 1175,
   1177ff., 1188f., 1194, 1196f., 1204, 1206,
   1216ff., 1220, 1226, 1229, 1233ff., 1236ff.,
   1241, 1251, 1267, 1271, 1279ff., 1283,
   1289, 1293ff., 1296, 1302ff., 1305, 1316f.,
   1323f., 1327f., 1345, 1350ff., 1353, 1355,
   1357ff., 1360ff., 1367, 1369, 1371
blood-brain-barrier   **577ff.**, 580, 796f., 863,
   1000, 1064, 1066, 1101, 1109, 1217, 1220
blood cell (or corpuscle), *see also* erythrocyte,
   leukocyte, thrombosis   212, 571, 577f., 581,
   764, 823, 823, 826, 848, 886, 991, 1063,
   1177, 1293, 1327, 1350, 1355, 1358, 1361,
   1367, **1369f.**
blood circulation, supply of blood, *see* circulation, hemorrhagic, vascular
blood clotting (or coagulation)   499, 505,
   954, 1040, 1305, 1358, 1366f., 1369f.
blood formation, *see* heme, marrow
blood level, *see* blood
blood pigment, h(a)emochrome, *see* heme,
   porphyrine
blood pressure, *see* hypertension
boar, *see* swine
body fluids, *see* blood, urine, milk, saliva
body weight, body height, body burden, *see
   also* growth, weight   290, **482ff.**, 489, 533,
   535f., 757, 803, 825, 831, 865, 915, 999,
   1003, 1029, 1039, 1063, 1077, 1169, 1194,
   1248, 1255
bog, *see* marsh, peat
bones, skeleton, *see also* marrow, medula
   ossides, spinal   112, 144, 213, 218, 471,
   474, 482, 494, 505, 516, 534f., 538, 555,
   559, 561, 577f., 582f., 592f., 638, 655, 717,
   722ff., 725, 728ff., 731, 733ff., 737ff., 740,
   746f., 757f., 762f., 765, 771, 778f., 794,
   803, 821, 825, 828f., 839f., 842, 844, 886f.,
   896, 899, 911f., 916, 924, 963, 967, 973,
   983f., 991f., 1008, 1012ff., 1028f., 1031f.,
   1039, 1073, 1108, 1130, 1147, 1166,
   1205ff., 1209, 1216ff., 1219, 1222, 1248,
   1250, 1264, 1271, 1279ff., 1282ff., 1302f.,
   1316f., 1323, 1345, 1352, 1360, 1362ff.,
   1365f., 1368, 1370
boron, borohydride, borates   44, 48, 67,
   130f., 156, 235f., 276, 282, 341, 488, 801,
   1049, 1051, 1200, 1244, 1264f., 1304

bottle (caps)   1026
bovine, cattle, *see also* cow, ruminant,
   beef   165, 407, 412, 473, 476ff., 532, 536,
   538ff., 541ff., 544, 605, 614, 644, 754, 756,
   820, 883, 885, 900, 903, 912, 963, 987, 994,
   1091, 1094ff., 1097, 1099f., 1108, 1174,
   1216, 1228, 1231, 1233f., 1297, 1302f.,
   1316, 1318, 1326, 1330, 1357
bowel, *see* intestine
Bowen's disease   764, 1351
bradycardia   1000, 1030, 1352
brain, *see also* cerebellum, cerebrum,
   hypophysis   304, 496, 507, 516, 519, 521,
   549, 551, 577ff., 580, 593, 598, 630, 645f.,
   717, 724ff., 728ff., 733, 738f., 799, 822,
   863, 873, 888, 897, 902, 912, 923f., 966,
   997, 1000, 1015, 1018f., 1022, 1031, 1034f.,
   1039f., 1060f., 1064f., 1071ff., 1074f., 1079,
   1082, 1085ff., 1088, 1108, 1111, 1113,
   1166, 1194, 1196, 1218f., 1235, 1283, 1302,
   1317, 1345, 1351f., 1355f., 1358ff., 1363,
   1367ff., 1371
bran, *see* grain, flour, phytate
brass   895, 1245, 1309, 1311
breads and pastries, bread, *see also* flour,
   grain, nutrients   113, 467, 486, 497, 680,
   778, 819f., 844, 1038, 1057, 1321, 1366
breast cancer   923, 928, 1179, 1183ff., 1187f.
breast-feeding, *see* lactation
breast-milk, *see* milk
breathing, *see* respiration, atmophile
breathlessness, respiratory disease, *see*
   dyspnea
brewer's yeast, *see* beer, yeast
brick   8, **13**, 27f., 36, 341
bromine, bromide, hydrobromic acid   294,
   405, 733, 751f., 1228, 1230
bronchial (compartment), bronchi, *see also*
   respiration   262, 300, 574ff., 586, 611, 822,
   830, 863, 866, 875, 960, 963, 966, 969,
   1040, 1113, 1120, 1204, 1206f., 1209, 1271,
   1284, 1305, 1329, 1370
bronchial asthma, *see* asthma
bronchial carcinoma   764f., 866, 911, 1284,
   1287, 1331
bronchitis   539, 665, 723, 780, 830, 866, 966,
   1066, 1113
Brønsted acids, *see* donor
bronze   646, 895, 1037, 1214f., 1243, 1245
brown coal, *see* coal
building industry, material, edifice, *see also*
   cement, construction material   27, 365,
   719, 857, 1278

bullet, projectile, shell, missile, *see also* space
   ..., explosive   754, 888, 976, 1277
burning of refuse, wood, *see* incineration,
   firing

cabbage, *see* vegetables
cable (sheathing), *see also* electric conductor,
   pipe   976
cadmium (general, elementary)   6ff., 11ff.,
   15, 19, 35ff., 44, 50ff., 57ff., 72, 79, 89, 94,
   96, 102, 107ff., 128, 132, 136f., 140, 142,
   144, 153, 155, 158ff., 164f., 176ff., 184, 186,
   189f., 192, 198, 202, 209ff., 213f., 217ff.,
   223, 227, 229, 231, 233f., 243, 248ff., 256,
   258f., 270, 285, 288ff., 299, 301ff., 305f.,
   308, 318, 326ff., 331, 334ff., 342, 347, 354,
   357ff., 361ff., 367f., 375f., 380, 384f.,
   387f., 390ff., 394, 396, 398, 401, 407ff.,
   410, 414, 420ff., 426f., 430, 444, 449f.,
   452ff., 455, 467ff., **471ff.**, 478f., 484,
   493ff., 498, 501f., 504ff., 507f., **511f.**, 519,
   523, 526, 528f., **534f.**, 542, 545ff., 548f.,
   551ff., 554f., 571f., 576ff., 579, 581f., 594,
   596ff., 600ff., 603, **605ff.**, 608, 612ff.,
   623f., 629, 631ff., 634ff., **639**, 641ff.,
   647ff., 655ff., 660, 664f., 670, 675f., 679ff.,
   683f., 691, 698, 703ff., 708f., 711, 745, 769,
   790f., 800, **803ff.**, 807, 810ff., 814ff.,
   819ff., 822ff., 828ff., 836f., 846, 849ff.,
   877, 895, 897, 905, 954, 957, 962f., 969,
   1006, 1008, 1010, 1013f., 1021, 1052,
   1078f., 1084, 1094, 1097, 1126, 1148, 1158,
   1162, 1173, 1187f., 1190, 1200, 1213f.,
   1225f., 1309, 1311f., 1319, 1324
cadmium compounds   266, 381, 393, 432,
   440f., 502f., 508, 512, 520, 549, 552, 554f.,
   592f., 599, 602, 615, 621, 626, 629, 637,
   640, 665, 670, 698ff., 805f., **808f.**, 818, 822,
   829f., 840, 846f., 1118, 1158f., 1261
cadmium ions   82, 95f., 155, 198, 244, 248f.,
   251, 266 315, 317, 321, 324f., 340, 427,
   440f., 496ff., 499, 501ff., 523, 526ff., 592,
   618ff., 628, 805, **815ff.**, 836f., 969
caffeine, *see* coffee
calcareous, *see* lime, hypercalciuria, calcium
   compounds
calcium compounds, ions, calcite, *see also*
   hypercalciuria   8, 35, 40, 79, 82, 90f., 95,
   98f., 155, 158f., 224, 235, 248, 250, 304,
   317, 320, 363, 372, **374**, 388, 399, 408, 412,
   422, 432, 481, 484, **494ff.**, **498f.**, 503, 505,
   507f., 511f., 515f., 538, 551, 553, 566, 574,
   578f., 612, 638, 647, 715, 721f., 725ff.,

729f., 734ff., 740f., 745, 750, 753, 757, 759, 761, 776, 780, 786, 817f., 822f., 827ff., 832, 851, 853, 860, 862, 866, 870, 905, 915, 947, 952, 959, 961f., 964ff., 967ff., 970, 988f., 999, 1006, 1016, 1025f., 1029ff., 1039f., 1056, 1064, 1091, 1100, 1106, 1111, 1273, 1299, 1302ff., 1306, 1321, 1330f., 1359, 1361, 1365
calf, veal   473, 537, 540, 542, 885, 1094, 1098, 1174, 1292f.
calibration, *see* reference material
calmodulin   498f., 964, 1111, 1120, 1303, 1308, 1352
can (British: tin)   450, 459, 741, 923, 981, 986, 1005, 1026, 1248, 1250, 1252
cancer, *see* carcinogenicity
cane (sugar), beet   362, 858, 1249, 1318
canopy, *see also* deciduous tree, conifer, throughfall   169, 370ff., 374ff., 377, 980, 987, 1009, 1326, 1360
caoutchouc, *see* rubber, tire, abrasion
capillary, *see also* circulation, blood, vascular   577f., 761, 960, 969, 1048
car, *see* automobile
carbanion, methylcarbenium   436f., 1055
carbide   884, 888, 1017, 1132, 1204, 1206, 1262, 1264f., 1269ff., 1273, 1290
carboanhydrase   485, 1319, 1325f.
carbohydrates, starch, *see also* sugar   338, 481, 485f., 539, 541, 629, 722, 924, 1019, 1025, 1032, 1125, 1350, 1357, 1363
carbon, carbonaceous fractions, diamond, *see also* charcoal   7, 12, 43, 47, 76, 261, 265, 428, 439, 441, 719, 724, 753, 856, 907, 961, 1033, 1036, 1090, 1128, 1158, 1278
carbonate, carbonic acid   55, 69, 95, 99, 154, 158, 248ff., 315, 323, 338, 347, 351, 369, 371, 373, 381, 384, 390, 393, 492, 495, 541, 776, 792, 805, 856, 894, 901, 914, 921, 961, 963, 972, 974, 1015ff., 1026, 1035, 1105, 1230, 1282, 1299f., 1310, 1313, 1318
carbon oxides, carbon dioxide   4, 8, 35f., 292, 370, 428, 882, 945, 947, 1010, 1025, 1036, 1103f., 1110, 1142, 1145, 1352, 1357
carbonyl, *see* metal carbonyl
carboxylase, carboxylate, carboxyl group, carbamyl   99, 422, 949, 1039, 1073, 1078, 1086, 1303f., 1325
carboxypepsidase   485, 1325
carcinogenicity, carcinogen, cancer, *see also* tumor, neoplasm, animal carcinogenicity, co-carcinogenicity, metastase, oncogene, lung cancer, skin cancer   56, 107, 188,
217, 288, 304f., 309, 491, 495, 500ff., 503, 505ff., 508f., 521, 527, 548, 550, 555, 559, 564, 568f., 582f., 591f., 595f., 599, 601f., 608ff., 613f., 617f., 621, 626, **628f.**, 631ff., 634f., 638, 640, 651, 654, 661ff., 664, 669ff., **673f.**, 682f., 690, **694ff.**, 723f., 727, 733, 737, 743, 747ff., 750f., 764f., 767ff., 770ff., 775, 782ff., 785f., 803, 827, 829, 833, 839ff., 844f., 853, 861f., 866f., 869f., 873f., 878f., 888f., 892, 911, 915, 919, 923, 925ff., 928, 942, 955, 966, 1001f., 1007, 1009, 1012, 1014, 1019, 1041, 1043, 1071f., 1080, 1095f., 1101, 1114ff., 1117, 1119ff., 1122, 1124, 1126, 1142, 1145ff., 1148, 1150f., 1153, 1166, 1174, 1177ff., 1180, 1182ff., 1187ff., 1198, 1200, 1204, 1209, 1221, 1237, 1252, 1254, 1266, 1282ff., 1285ff., 1305, 1328, 1331f., 1337, 1346
cardiac defect, insufficiency, cardiotoxicity, *see also* heart, digitalis   747, 760, 964, 969, 998, 1020, 1034, 1084, 1133, 1174, 1189, 1204, 1218, 1304, 1352
cardiovascular, *see also* cardiac defect, circulation, heart, vascular   484, 486, 488, 589, 596, 601f., 631, 865, 906, 926, 954, 998f., 1009, 1011, 1032, 1223, 1352
caries, tooth decay, *see also* tooth   1176, 1295
carnalite, carnotite   1026, 1290
carp   542, 914, 917, 1056, 1332
carrier (protein), *see also* receptor protein, transferrin, metallothionein, finger-loop proteins   97ff., 322, 342, 379, 443, 486, 495, 498, 511, 514f., 517, 520, 551, 572, 577, 641, 725f., 862, 868, 884, 912, 924, 934, 949, 955, 991, 1028, 1038, 1131, 1230, 1304, 1352, 1357, 1360, 1362, 1370
carrot   453, 472, 1056, 1058
cartilage   482, 485, 1271, 1330, 1355
cartography   229
cascade impactor, *see* impactor
casein, *see also* cheese, protein   518, 885, 949, 1321
cassiterite   1245
casting, cast   37, 689, 921f., 929, 947, 1026, 1101, 1215, 1270
cat   532f., 542, 549, 554, 1131, 1207, 1237, 1330
catalase, catabolic   426, 485, 497, 949, 1060, 1095, 1218, 1322
catalyzed reaction, catalyst, catalysis   80, 86, 147f., 265, 337, 369, 379, 419, 425, 435, 498, 513, 516f., 590, 719f., 745, 751, 791,

856, 882, 895, 903, 907, 921f., 924f., 933, 945, 949, 954, 961, 967, 976, 992, 1003, 1036f., 1042f., 1051, 1060, 1090, 1092, 1103ff., 1135f., 1141ff., 1144ff., 1149ff., 1158f., 1183, 1193, 1212, 1214f., 1245ff., 1263, 1270, 1291, 1301, 1325, 1355

cathode, cathodic techniques, see also mercury (film) electrode 35f., 38, 137, 190, 201f., 791, 1052, 1104, 1142, 1214, 1275, 1287, 1353

cation 67, 72, 76, 79, 81, 99, 149, 154, 224, 236, 238f., 243, 248, 252, 255, 265, 311f., 314ff., **318ff.**, 325, 327ff., 347, 360, 367, 369ff., 373, 388f., 399, 404, 407, 412, 422f., 432, 440f., 447, 493, 587f., 721ff., 731, 751f., 818, 890, 910, 1015, 1027, 1030, 1063, 1105, 1155, 1213, 1226, 1244, 1247, 1294, 1306, 1318, 1352

cation exchange, see cation, ion exchange

cattle, livestock, see bovine, ruminant, cow

caustic, see corrosive

cauterization, acid burn, see corrosive

cell, cell culture, see also blood cell, hepatocyte, killer cell, mammalian cell, germ cell, somatic cell, see also cyto ... 96, 98, 421, 481, 485, 497, 501, 506f., 511, 521, 523f., 528, 533, 538, 550, 569, 578, 586, 589, 594, 605f., 611f., 617, 622, 624f., 628, 630, 634, 638, 644, 648, 653, 675, 722, 725, 727, 738, 745, 747, 750, 779ff., 783, 785ff., 799f., 819, 822, 826, 828, 843, 845, 859, 864, 867, 869f., 875f., 886, 902, 918, 926f., 929, 935f., 941, 943, 964, 969, 1025, 1035, 1064ff., 1079, 1083ff., 1087, 1111, 1115f., 1122f., 1126, 1146f., 1153, 1173, 1177, 1207, 1217, 1271, 1273, 1295, 1303f., 1317, 1324, 1331, 1355f., 1363, 1366f., 1370

cell division (cell growth), cell replication, cell cycle, cell proliferation, meristem 503, 633, 722, 782f., 786, 830, 886, 915, 1065, 1113, 1120, 1147, 1151, 1166, 1172, 1177f., 1207, 1219f., 1305, 1330, 1350f., 1359, 1363, 1367f.

cell function (metabolism), see also organelle 372, 441, 497f., 586, 589, 722, 828, 851, 859, 868, 870, 929, 945, 964, 968, 1000, 1038, 1063, 1065f., 1110, 1119, 1122f., 1219, 1303f., 1353

cell genetics, see cytogenetics

cell membrane, see cell wall, membrane

cell nucleus, see nucleus

cell physiology, cell response, cell transformation, cell proliferation, see also cytotoxicity 97f., 504, 513, 521, 533, 569, 607, 609, 619, 622f., 759, 782, 784, 786, 794, 799, 822, 851, 867, 871, 887, 913, 925, 945, 963, 993, 999, 1038, 1065, 1074, 1110, 1115, 1126, 1219, 1283, 1331, 1358

cell plasm, see cytoplasm

cell poison, see cytotoxicity

cell protoplasm, see cytoplasm

cellulose, pulp see also paper 895, 1051, 1214

cell wall, see also membrane, permeability, sclerosis 400, 406, 410, 412, 414, 422, 440, 493, 495, 531, 577, 612, 722, 806, 859, 861, 868, 897, 1109, 1197, 1305, 1317, 1366

cement, concrete, cement plant 8, 13, 27f., 54, 87, 299, 348, 477, 542, 557, 561, 571, 720, 722, 807, 814, 862, 866, 958, 1027, 1214, 1227, 1231ff., 1235, 1239ff., 1269, 1271, 1312, 1366

central nervous system CNS, see nerve, neurotoxicity

centrifugation 716, 739, 1357, 1370

ceramics 27f., 557f., 561ff., 656, 699, 719, 735, 745, 754, 800, 827, 833f., 881f., 975, 981, 1016, 1027, 1104, 1128, 1142, 1159, 1162, 1245, 1263, 1301, 1343f., 1347

cereals, see grain

cerebellum, cerebellar, see also brain, Purkinje cell, pyramidal cell 648, 1060f., 1064, 1069, 1074f., 1082, 1085, 1352, 1367

cerebrum, cerebral, cerebrovascular, cerebrospinal, see also brain 646, 648, 664, 724, 729, 830, 914, 919, 1032, 1064f., 1073ff., 1082, 1101, 1113, 1129, 1133, 1219f., 1224, 1253, 1351f., 1359, 1365

cerium, cerium compounds **959ff.**, 962f., 965f., 969f., 1133, 1142, 1209, 1303ff., 1307

certification, see reference material

ceruloplasmin 485, 514, 519, 521, 551, 577, 645, **899**, **901f.**, 1094, 1096, 1196ff., 1199, 1201, 1352

cervix, cervico-vaginal, see uterus

cesium, cesium compounds 145, 1132, 1278

chalk, chalkogen, chalkosphere 1138f., 1141, 1213, 1231, 1263, 1352

channel 29, 336, 389, 392

charcoal, animal charcoal, see also absorption 294, 1157, 1212

cheese, casein, tyrosinase 859, 899

chelating agent, chelation, chelatin, see also complex, NTA, EDTA, DTPA 44, 56, 67, 77, 82, 85, 96f., 119f., 147, 149f., 154f., 184, 189f., 194, 197f., 203, 240, 249ff.,

322f., 374, 384, 388, 400, 404, 414, 418ff., 426, 431f., 492f., 495f., 502f., 505, 511, 515f., 518ff., 525, 527, 536, 567ff., 570, 589, 618, 715, 724, 730, 735, 796, 817f., 832, 835, 851, 899f., 913, 935, 941, 949, 949, 951ff., 954, 956, 992, 996, 1027, 1056, 1078, 1084, 1116, 1125, 1207, 1223, 1237, 1277, 1282, 1285, 1287, 1302, 1321, 1328, 1342, 1364, 1368f.

chemiluminescence, see luminescence
chemotherapeutics, see drug
chest pains 1066, 1113
chicken, see also poultry 495, 520, 540, 543, 545f., 645, 754, 760, 765, 767, 769, 771, 867, 869, 872, 926, 965, 1094, 1110, 1174, 1196, 1199f., 1237, 1289, 1295, 1326
child, children, young animal, progeny, see also infant 200, 214, 217, 219f., 289, 299, 302ff., 305f., 309, 499, 547f., 551f., 555, 574, 583, 593, 598, 601, 603, 655f., 658, 662, 678f., 702f., 705f., 709f., 728, 736, 738, 763, 897, 903f., 945, 955ff., 971, 976, 979, 982, 984ff., 987ff., 990f., 994ff., 997f., 1002ff., 1006ff., 1010ff., 1013, 1027, 1059, 1073, 1083f., 1118, 1126, 1147, 1180, 1295, 1317, 1320, 1328f., 1331, 1363
child bearing (age capability), reproductive age, see also pregnancy, menopause, menstrual cycle 705, 945, 953, 1002, 1005, 1032f., 1039
chimney, see incineration, firing, organic
chloralkali production (electrolysis) 337, 423, 447, 571, 625, 691, 709f., **1051ff.**, 1081, 1143
chlorinated (organic) compounds, see also dioxins 304, 615, 846, 850, 907, 1229, 1365, 1368
chlorine, chloride, chlorination, chlorinity 12, 35, 42, 50, 76, 78, 154, 236, 242f., 248ff., 265, 340, 342, 348, 382, 387f., 430, 481, 541, 569, 589, 716, 727, 733, 751f., 754, 817f., 864, 881, 922, 932f., 961, 1035, 1047, 1052, 1106, 1128, 1137, 1195, 1214, 1218, 1230, 1244, 1262f., 1291, 1299, 1304, 1310, 1312, 1332, 1344, 1352
chlorocomplex 248f., 386, 517, 805, 817, 1026, 1105, 1352
chlorophyl(l), chloroplast, see also porphyrine 442, 953, 1025, **1027**, **1029f.**, 1326, 1352
chlorosis (greensickness, yellowing) 826, 953, 1030, 1039f., 1172, 1263, 1326, 1352
cholestatic 904

cholesterol, choline esterase, acetylcholine, choline acetyltransferase 488, 831, 965, 1040, 1095, 1131, 1218, 1221, 1294f.
chromates, chromium (VI) compounds 42ff., 52, 136, 155f., 204, 313, 315, 326, 423, 426f., 441, 455, 487, 492, 497, 502ff., 508, 513, 535, 548, 593, 599, 618ff., 621f., 625, 628f., 631ff., 634f., 638, 645, 657, 660, 670, 693, 698, 711, 809, 853ff., **856ff.**, 859, 861f., 864ff., **868ff.**, 895, 918, 1001, 1012, 1114, 1119, 1308, 1331f.
chromatid, chromatin, see also sister-chromatid exchange 625, 639, 869, 872, 1072, 1126, 1352, 1368
chromatography, see also gas-, gel-, ion (exchange)-, liquid- and thinlayer-chromatography 109, **148ff.**, 152f., 156f., 159, 190, 194f., 197, 203, 346, 744, 768, 790, 806, 1155, 1355f., 1366
chromium (general, elementary, chromium (V)), see also chromates, chromium(III) compounds or ions, ferrochrome 6ff., 11f., 15, 50f., 54, 57ff., 63, 89, 95, 107ff., 126, 138, 159, 163, 169, 171, 193, 196, 202, 207, 216, 218, 258f., 290, 294, 301ff., 307, 334, 357f., 361, 375, 381, 410, 413, **454f.**, 467, **473**, 481f., 484f., 488, 492f., 501, 505ff., 509, **513**, 520, **535f.**, 552, 562, 572, 582, 591, 596, 613, 618, 629, 631f., 634, 636f., 639, 641, 657f., 660ff., 683, 708, 719, 763, 771, **853ff.**, 877f., 881, 895, 912, 945, 947, 968, 1099, 1120, 1122, 1124, 1126, 1148, 1262, 1291, 1301, 1307ff., 1344
chromium(III) compounds (or ions) 44, 82, 136, 154, 239f., 313, 325, 426, 455, 492, 500, 503, 508ff., 513, 535, 618, 632, **645**, 651, 657, 693, 710, 853f., **857**, 859, **865**, **868ff.**, 877f., 1261, 1266
chromogen 946, 1111
chromosome, chromosome aberrations, chromosome damage, see also sister-chromatid exchange 422, 425, 430, 617, **623ff.**, 626, 629, 632ff., 637ff., 640, 643, 645f., 648f., 662, 726, 750, 763f., 770, 867f., 1001, 1055, 1072, 1078, 1084ff., 1087, 1189, 1221, 1283, 1331,1350ff., 1353, 1357, 1359, 1362, 1367f.
chronic exposure, chronic effects, delayed toxicity, see also long-term 494, 508, 532ff., 535, 537f., 542, 544ff., 547, 555, 585, **591ff.**, **595ff.**, 762f., 772, 828f., 831, 887, 901ff., 915, 926, 965f., 1019, 1030, 1068ff., 1081, 1084f., 1094f., 1113, 1120,

1124, 1131, 1169, 1175ff., 1196, 1207, 1219, 1222, 1232, 1234, 1237, 1243, 1252f., 1266, 1304, 1306, 1323, 1327f., 1346, 1365, 1369
chrysotherapy, *see also* gold **935f.**, 938, 1352
cigarette smoke, *see* tobacco smoke
ciliary, ciliated, mucociliary system, *see also* epithelium 548, 575f., 583, 897, 1109, 1131, 1319, 1352, 1363
cinnabar 437, 516f., 1046, **1050f.**
circuit, *see* electric conductor, switching
circulation, circulatory collapse, *see also* capillary, arrhythmia, vascular 533, 542, 544, 561f., 578, 593, 727, 761, 828, 915, 944, 953, 988, 1067, 1129, 1132f., 1197, 1231, 1234, 1251, 1271, 1324, 1327
cirrhosis 527, 645f., 763, 903f., 1030, 1032, 1174f., 1353
cis-Pt(NH$_3$)$_2$Cl$_2$, *see* platinum compounds
citric acid, citrate 67, 111, 419, 422, 485, 511, 515f., 518, 570, 724, 739, 792, 794f., 798f., 887, 911, 918f., 940, 949ff., 965, 1017f., 1056, 1107, 1206, 1279, 1304, 1321, 1324, 1361, 1363
citrus (plants, leaves, fruits, juice) 157, 901, 1093, 1109, 1319
city, *see* urban
clastogenic, *see* tumor
claw, talon, paw, *see* hoof, nail
clay, *see also* loam 5, 7f., 10, **13**, 58, 93, 292, 311ff., 317f., 320, 323f., 328ff., 347, 353, 370, 373, 389, 399, 405, 431, 440, 776f., 818, 860f., 879, 896, 899, 912, 992, 1095, 1231, 1270, 1313f.
clean air, clean room (technique), *see also* indoor 109, 111, 216, 546, 709f.
clean-up, clearance, *see also* biodegradation, decomposition, gas-cleaning, purification (plant), self-cleaning, sludge 156f., 349, 353, 355, 411, 450, 458, 575f., 583, 981, 987, 1044, 1050, 1130, 1243, 1249, 1254, 1282, 1295, 1302
climacteric period, *see* menopause
climate, *see* weather conditions
clonal 643
closed rooms, *see* indoor
cloth, clothes, shoes, *see also* textiles 656, 990, 1254
cloud, *see also* rain, wet deposition 266, 283, 372, 374
clover, alfalfa, lucern(e), trefoil 402, 409, 414, 757, 778, 780, 884, 1097f., 1106f., 1125, 1233, 1271f., 1313, 1326

cluster, *see* ligand, model
coagulation, coagulant, *see also* blood clotting 30, 42, 70, 261f., 720, 723, 978
coal (burning, coal fly ash, coke), *see also* bituminous, organic (matter) waste, charcoal 8, **10ff.**, 17, 33ff., 41, 54, 87, 89, 176, 184, 196, 257ff., 261, 263, 266, 275ff., 278, 282, 288, **294ff.**, 306, 333, 335, 341f., 351ff., 571, 583, 741, 745, 751, 753ff., 777f., 780, 787, 803, 806, 810, 814, 841, 856, 858, 883, 912, 921ff., 947f., 975, 979, 1012, 1027, 1051, 1053, 1082, 1090f., 1095, 1098, 1100, 1104ff., 1129, 1133, 1153, 1155, 1161ff., 1164, 1205, 1208, 1211, 1215, 1231, 1234, 1241, 1247, 1262, 1270f., 1278, 1291, 1312
coast, *see also* offshore dumping 29, 253, 256, 268, 336, 395ff., 435, 769, 813, 842, 847, 851, 961, 1079, 1081, 1091, 1162, 1247, 1263, 1320
coating, *see* protective layer
cobalamin, *see* vitamin B$_{12}$, methylcobalamin
cobalt, cobalt ions, *see also* vitamin B$_{12}$ 6ff., 11, 14f., 35, 57ff., 72, 79, 82, 89, 95, 109, 111, 132, 138, 145, 171, 175, 177, 179, 183, 195, 207, 216, 218, 224, 233, 235, 239, 258, 314, 329, 375f., 401f., 404, 409, 414, 422ff., 437, 456, 481f., 484f., 502, 529, **536**, 542, 557f., 560ff., 572, 578, 581, 599, 618, 626, 628f., 636, 658, 670, 719, 773, 826, **879ff.**, 884, 945, 959, 962, 968, 1036, 1044, 1116, 1122, 1127, 1195, **1271ff.**, 1301, 1307, 1319
cobalt compounds 503f., 507, 599, 633, 635, **880ff.**, 883f.
co-carcinogenicity 503, 629, 631, 634, 764f., 1002, 1114, 1120, 1122, 1183, 1331
co-enzyme 421f., 424, 436, 883, 886, 1294, 1351, 1353
co-factor (for enzyme activation) 1035, 1039, 1089, 1098, 1111
coffee, cocoa, chocolate 157, 203, 451f., 454ff., 459, 461f., 464, 466, 634, 656, 820, 841, 896, 898, 906, 950, 1107, 1109
coin, minting 931f., 1104, 1114, 1141, 1193
coke, *see* coal
cold, *see* cough(ing)
coli, Escherichia coli 423, 427, 437f., 620, 626, 628, 635, 637, 641, 727, 768, 771, 1151, 1172, 1221, 1331
colic, colitis, ascites, abdominal infection, peritonitis 586, 593, 761, 763, 795, 911,

925, 929, 964f., 968, 995, 1067, 1175, 1215, 1219, 1223, 1328, 1351, 1366
collapse, *see* circulation
collection, *see* deposition, precipitation, sample
colloid, colloidal, collagen, *see also* gelatin(e) 69, 119, 153, 216, 234, 245, 330, 515, 568, 589, 745, 779, 794, 797ff., 800, 805, 857, 896, 931, 934, 941f., 948, 956, 964, 1175, 1215, 1272, 1300, 1302f., 1330
colon, *see* intestine
colorimetry, *see also* iodometric titration, photometry, UV spectrometry 107, 125, 133, **147f.**, 239, 716, 743, 749, 790, 798, 806, 894, 922, 932, 939, 946, 960, 962, 968, 972f., 1022, 1026, 1033, 1036, 1047, 1128, 1154ff., 1191, 1200, 1203, 1228, 1244, 1261, 1267, 1289, 1300, 1308, 1310, 1343
colostomy, *see* intestinal hemorrhage
coma, *see* unconciousness, shock
combustion, *see* coal burning, crude oil firing, forest fire, fuel oil, incineration 95, 105, 200, 221, 233, 1061, 1280
company doctor, *see* occupational medicine
compartmentalization, *see also* distribution 95, 105, 200, 221, 233, 1061, 1280
complex, complexation, *see also* chelating agent, hydroxo complex, ligands 43f., 48, 52, 67f., 70, 76, 78ff., 85f., 90, 95f., 98ff., 102, 120, 138, 153ff., 158, 234, 236ff., **239f.**, 242, 244, **247ff.**, 255f., 316, 322ff., 328ff., 338, 342, 383f., 386, 388ff., 414, 419f., 422f., 428, 430f., 441, 457, 493, 497, 506f., 511ff., 514, 516ff., 525f., 533, 551, 556ff., 568f., 585, 588, 592, 618, 718, 721f., 724, 726f., 744f., 776, 781f., 784, 796f., 805, 826, 846, 854, 857, 859f., 862, 865, 868, 870, 872, 880, 882, 893, 896ff., 905f., 913, 931f., 934ff., 937, 941, 946, 948, 951, 953, 960, 963, 969, 1015, 1031, 1038, 1044, 1046f., 1054, 1056, 1075, 1083, 1101f., 1105f., 1109f., 1121, 1125, 1136, 1141f., 1146, 1154, 1191, 1194f., 1199, 1206, 1213, 1224, 1228f., 1261, 1269, 1279, 1282, 1292f., 1303f., 1310, 1313, 1322, 1324f., 1361, 1364, 1369
compost, *see also* litter, manure, sludge 328, 339, 348, **357**, 359ff., 365ff., 414, **707f.**, 837, 857f., 874, 890, 903, 905, 1004, 1082, 1105, 1115, 1315, 1332, 1336
compound specific, *see* speciation
computer chip, *see* electronic equipment, semiconductor, silica/silicon, tape

co-mutagen(icity) 503, 621, 624, 626, 628, 631, 633, 635ff., 1114
concentration at the workplace, *see also* biological tolerance value, occupational exposure, threshold **667ff.**, 670f., 673, 675, 681ff., **693ff.**, 701, 703, 710, 732, 737, 783, 785, 799, 831, 840, 842, 870, 875, 889f., 904, 906, 957, 1002, 1010, 1021f., 1041, 1043, 1083, 1099, 1116, 1120, 1132, 1135, 1150, 1180, 1185, 1198, 1200, 1209, 1222, 1224, 1240, 1255, 1257, 1265f., 1272, 1286, 1296, 1307, 1336, 1346f., 1362
concentration factor, *see* accumulation
concentration (lack of), *see* paramnesia, psychic disturbance, behavior, learning ability
conception, contraceptive, *see also* reproduction, fertility 494, 657, 760, 895, 1071, 1094, 1205, 1209
concrete, *see* cement
condensation 3, 261, 263, 291
conductor, conductivity, *see also* electric conductor, superconduction, thermal techniques 136, 148f., 153, 236, 716, 775, 777, 789, 906, 1159, 1191
congenital, *see* inherited syndrome
conifer, coniferous tree, *see also* needle 169, 372, 375ff., 409, 721, 723, 735, 738, 740, 778f., 841, 849, 897, 987, 1030, 1039, 1056, 1058, 1088, 1233
conjugation 499
conjunctiva, conjunctivitis 761ff., 1196, 1207, 1328, 1353
constipation, *see* obstipation
construction (building) material, *see also* building industry, material 19, 297, 947, 1278
consumption, consumer goods, *see also* household chemicals **14f.**, 271, 288, **299f.**, 303, 688, 699, 719, 792, 803, 807ff., 910, 981, 1059, 1093, 1142, 1149, 1162, 1255, 1257
contamination, contaminant, *see also* poisoning 60, 107, 109f., 112ff., 118f., 143, 150f., 153, 163, 165, 171, 212, 224, 228, 256, 270f., 279, 303, 328, 333f., 336, 341ff., 345, 349, 352ff., 367, 379f., 385, 394, 397f., 405, 409, 411, 413f., 426f., 429, 433, 450, 453, 459f., 468, 470, 475, 482, 492, 532ff., 538ff., 541ff., 544, 548, 555, 571, 585, 596, 599ff., 625, 656, 701, 703, 707f., 711, 715f., 720, 743, 749, 751, 755, 771, 773, 805, 817, 819f., 822, 826f., 837, 843, 848f., 851, 853,

859, 862f., 868, 883, 929, 949, 958, 971, 973, 985, 990f., 1014, 1036, 1048, 1050, 1052, 1057, 1060, 1063, 1075, 1080f., 1086, 1114f., 1118f., 1126, 1154, 1163f., 1186, 1193, 1200f., 1208, 1212, 1252, 1263, 1266, 1278, 1285, 1290f., 1295, 1310, 1313, 1337, 1341

contraceptive, *see* conception

control, measuring, *see also* monitoring 31ff., 36ff., 41f., 44ff., 49, 55, 68, 90, 99, 106f., 113, 129, 148, 162ff., 221ff., 229, 255, 277, 333, 343, 347, 352, 354, 363, 365f., 385, 393, 429f., 434, 511, 542, 570, 687, **699f.**, 731, 743f., 760, 783, 797, 814, 819, 831f., 835, 848f., 855, 910, 952, 961, 973, 1004, 1013, 1034, 1142, 1149, 1234, 1290, 1320

control engineering, *see also* electronic equipment 777, 1193

control material (substance), *see* reference material

cooking, cooking pots, containers 287f., 294, 300, 451, 699f., 722, 724, 737, 740, 827, 834, 895, 901, 981, 983, 1026, 1104, 1249, 1329

coordination (physicochemical), *see also* chelating agent, ligands, donor 67, 71f., 78, 84, 239, 322, 726, 735, 880, 909, 932, 937, 1015, 1352

coordination (medical), asynergia, *see also* ataxia, tremor, speech defect, spasm, hallucination 484, 509, 568, 727, 926, 954, 997, 1020, 1025, 1031, 1034, 1039, 1068, 1113, 1175, 1191, 1236, 1330, 1364f.

copier, duplicator, printer, *see* reproduction technique

copper (general, elementary) 5ff., 11f., 15, 19ff., **24**, 32f., 35, **36f.**, 39ff., 50, 52, 54, 56ff., 62ff., 89, 95, 97, 107, 126, 132, 134, 136f., 140, 153f., 158f., 164, 169, 189, 207, 211, 214, 218, 223, 233, 246, 251, 254, 258f., 268, 270, 278, 280, 284, 288, 291, 304, 307, 315f., 318, 326, 329f., 334f., 351, 357f., 361, 367, 375f., 383f., 390, 400ff., 403ff., 407, 409ff., 412, 414ff., 418, 420f., 423, 427, **455ff.**, 467f., **473**, 481f., 485, 488, 493, **495**, 504f., **512ff.**, 518f., 521, 527ff., **536f.**, 540, 544ff., 550f., 555, 567, 569f., 572, 577, 601, 605, 612, 642ff., **645f.**, 648f., 680f., 703, 719, 726, 736, 753f., 775, 777, 791, 809, 836, 838, 840, 857f., **893ff.**, 929, 932, 957, 1009, 1016, 1027, 1034, 1036f., 1044, 1048, 1089f., 1092ff., 1095ff., 1098ff., 1129, 1141, 1157ff., 1161, 1173, 1190, 1192f., 1196ff., 1199ff., 1213ff., 1216, 1223, 1235, 1245, 1290f., 1301, 1309, 1319, 1324, 1330, 1352f., 1362, 1371

copper compounds, copper complexes 240, 246, 322, 426, 428, 501, 536, 590, 637, 753, 757, 809, **893ff.**, 898, 900f., 906, 912, 921ff., 940, 962, 1158f., 1192, 1213f., 1229f.

copper deficiency 476, 484, 486, 494, 535, 574, 893, 895, **898ff.**, 903, 906, **1094**, 1096ff., **1099f.**, 1196, 1330, 1332, 1355

copper ions 68f., 71, 76, 79, 82, 86, 91f., 95ff., 101, 155, 189, 224, 235, 239, 243, 246f., 249, 313f., 316f., 321f., 324, 327, 329ff., 345, 384, 388, 420, 426, 428, 496f., 499, 501, 511, 513f., 519, 523, 567, 578, 618f., 626, 628, 646, 896, 898, 1044, 1109, 1120

cord, *see* blood, fetus, placenta

core, borehole, *see also* ice (core) 113, 391, 961, 1162, 1263

corn, maize, *see also* grain 329, 401, 408, 410, 453, 486, 864, 952, 957, 1030, 1060, 1302, 1307, 1318, 1323, 1326

cornea, keratitis, nebula 902, 1069, 1147, 1196, 1353, 1359

corpulence, *see* adiposis, body weight, fat

corrosive, corrosion, caustic, *see also* rust, oxidation 51, 344f., 532, 537, 541, 557ff., 560ff., 563f., 719, 754, 803, 808, 853, 857, 862, 891, 948, 959, 961, 1004, 1016, 1021, 1026f., 1067, 1090, 1104, 1127, 1142, 1163, 1205, 1215, 1230, 1262, 1310, 1312f., 1328, 1344, 1347

cortison, corticoid, *see also* glucose 518, 534, 1040

corundum (smelter disease) 719, 723

cosmetic(s) 216, 299, 307, 659, 677, 699, 720, 754, 783, 791f., 1048, 1263, 1244

cosmos, moon, star, *see also* meteorite 3, 945, 1140, 1229, 1364

cost-benefit, cost-effectivity, cost-competitivity 27, 38, 40f., 45ff., 49, 53, 64, 106, 118, 129, 133f., 142, 145, **160ff.**, 247, 333, 348f., 351, 688f., 719, 895, 933, 948

co-teratogen(icity) 631

cotton 153, 442, 753, 757, 1212, 1226, 1312

cough(ing) 762, 1066, 1069, 1113, 1135, 1146, 1295, 1329

coulometry, coulometric force 78, 236, 238, 846

covalent 67, 85, 314, 1141, 1246

cow, cattle, bull, *see also* bovine, beef 533, 541, 724, 737, 885, 1018, 1039, 1174, 1239f., 1249, 1293

cow's milk, *see* milk

crab, *see* crayfish

cramp, *see* spasm

crayfish, crawfish, clam, crustacea, shrimp, *see also* krill 96, 101, 194, 472, 479, 540, 542, 677, 756, 760, 817, 820, 850, 865, 897, 900, 904, 923, 1042, 1107, 1111, 1195, 1205, 1208, 1316, 1363

creatinine, creatine 111, 545, 594, 660, 758, 835, 914, 1005, 1234, 1282

criminal, *see* forensic

criteria documents 546, 688, **693**, 700, 848, 851, 1013, 1199, 1258, 1324

crops, *see* agriculture, plants

cross-link, *see also* interaction, ligands 569, 618, 620f., 635, 1263

crown, top, *see* canopy

crude oil, oil burning, *see also* fuel oil, petroleum, refinery 7, **10ff.**, 29, 52, 128, 190, 292, 807, 843, 857, 1106, 1142, 1270, 1291

crustacean, *see* crayfish

crustal..., *see* earth-crust

cryogenic, freezing, cooling 108, 112f., 131, 185, 202, 379, 385, 390, 1048, 1128, 1156, 1181, 1301, 1362

cryolite 36, 719, 736

cumulation, *see* accumulation

cutaneous tumor, *see* skin cancer

cuticle, cuticula 224, 405, 896

cyanide, *see also* ferrocyanide 23, 25f., 42, 50ff., 56, 73, 76ff., 243, 328, 430, 441, 882, 931ff., 937

cyanosis 1113, 1135, 1221, 1353

cycle, cycling, *see also* fate, life time, source, transport 3, 13, 16, 62f., 65, 67, **87ff.**, 92, 100, 102f., 222, 254, 256f., 265, 267f., 277f., 280ff., 283ff., 337f., 347, 352, 370f., 375f., 393f., 396f., 405, 408, 424f., 427f., 432f., 435f., 442, 566, 570, 602, 685, 739ff., 755, 768, 770ff., 773f., 786f., 803, 843f., 846, 849, 858, 875, 878f., 887, 892, 915, 919, 937, 942f., 948f., 966f., 969f., 981, 983, 1006f., 1011f., 1018f., 1029, 1035, 1037, 1052f., 1105, 1138f., 1162f., 1167, 1170, 1185f., 1200, 1208, 1231, 1259, 1267, 1286, 1297, 1316, 1356

cyprinid, *see* carp

cysteine, cystine 97, 430, 497, 516f., **524ff.**, 578, 592, 805, 826, 1062, 1075, 1078, 1154, 1173, 1294f., 1322, 1361

cytochromes (incl. P-450 cytochrome) 484f., 499, 513f., 580, 868f., 872, 899, 924, 949f., 1173, 1295, 1353

cytogenetics 617, 619, 623f., 628, 632ff., 635, 637, 639, 771, 868, 875, 1087, 1295

cytoplasm, *see also* ribosome 304, 372, 423, 440, 511, 586, 729, 869, 964, 1063, 1066, 1234, 1293, 1317, 1319, 1322, 1324, 1353, 1355, 1360, 1362f., 1368f.

cytosine, *see also* guanine 1065, 1353, 1368

cytosol, soluble cell fraction 439f., 501, 512, 514, 646, 779, 794, 826, 934, 938, 941, 964, 993, 1109, 1112, 1120, 1124f.

cytostatic, *see also* anticarcinogenicity 624f., 1147, 1150, 1173

cytotoxicity, dead cells, cell survival, cytolysis 435, 523, 559, 594, 606f., 743, 770, 779, 800, 839, 851, 867, 911, 915, 925, 928, 935, 937, 963, 969, 1074, 1082f., 1116, 1177f., 1207, 1271, 1282, 1324, 1364

daily intake, daily allowance, weekly intake 485f., 488, 494, 554, 595, 659f., 667, 680f., 703, **705**, 746, 757, 766, 778, 820, 823, 834, 859, 871, 884, 892, 896f., 900, 924, 927, 952, 981f., 1004, 1030, 1040, 1042, 1058f., 1076f., 1091, 1108, 1166f., 1180, 1185, 1194, 1216, 1222, 1233f., 1238, 1252, 1264, 1281, 1294, 1320, 1331f., 1347

dairy products, *see also* milk, cheese 113, 140, 165, 168, 174, 177, 471, 473, 478, 722, 746, 820, 923, 950, 1005, 1028, 1038, 1091, 1166, 1345

Daphnia 100, 756, 758, 827, 836, 865, 1111, 1120, 1195, 1324

data bank, data medium, data quality, communication, television, *see also* tape, optical information 166f., 194, 217, 283, 321, 562, 661ff., 667, 670, 672, 674, 688, 699, 710f., 857, 878, 882, 909, 911, 962, 1301, 1324

deaeration, *see also* anoxic conditions 137, 139f.

dealkylation, dearylation 433f., 475, 580, 999, 1055f., 1075, 1146, 1249, 1251

decay, *see* radioactivity

deciduous tree, *see also* canopy, leaves, citrus 267, 279f., 374, 376f., 440, 721, 723, 735, 737, 778f., 858, 897, 987, 1107, 1119, 1302, 1308, 1326

decolorization, *see* discoloration, chlorosis, necrosis

decomposition, *see also* degradation, photochemistry, atmospheric chemistry 5, 58, 108, **116ff.**, 134, 136, 138f., 144, 146, 150, 159f., 165, 170, 179, 183, 195, 201, 205, 346, 380, 388, 438, 485, 805f., 837, 848, 854f., 932, 946, 961, 972f., 976, 988, 999f., 1027, 1048f., 1055f., 1104, 1127, 1143, 1154ff., 1347

decontamination, *see* clean-up, purification

deer, roe, *see* wild animal

deficiency, *see* essentiality

defoliation, defoliant, *see* herbicide

deformation, *see* teratogen

degeneration, *see also* parenchyma 304, 533f., 539, 761, 1065, 1073, 1080, 1131, 1174, 1187, 1219f., 1351, 1365

degradation, *see also* decomposition, biodegradation 56, 338f., 344, 360, 437, 447, 527, 1111, 1249, 1251, 1255f.

dehydrogenase, dehydrase, dehydratase 427, 646, 886, 914f., 919, 1018f., 1039, 1092f., 1097, 1131, 1207, 1304, 1319, 1325, 1349

delayed toxicity, *see* chronic exposure

delta, mouth, *see* estuary

dementia, *see* paramnesia

demethylation, *see* dealkylation

dentist, dentistry, dental, dentine 557ff., 561f., 791, 921f., 929, 932, 984, 1048, **1051**, **1053**, 1067, **1072**, **1074**, 1081, **1087f.**, 1104, 1129, 1136, 1141, 1143, 1193, 1197, 1245

denuder system, *see also* diffusion 185, 194, **272**, 285, 377, 850

deodorant, *see also* odor 299f., 308, 1344, 1346

de(s)oxyribonucleic acid, *see* DNA

dephenylation, *see* dealkylation

dephosphorylation 726

deposit, *see* mineral deposit, disposal, dump, depot formation

deposition, *see also* precipitation, wet deposition, acid rain, throughfall 5, 71, 93, 114, 158, 230f., 247, 249, 260ff., **266ff.**, 277, 279ff., 283ff., 300, 311, 360, **369ff.**, 372, **373ff.**, 376ff., 379f., 393, 405, 407, 453, 458, 575f., 578, 722, 749f., 776, 786, 807, 810f., 814ff., 817f., 833, 839, 846, 861, 873, 883, 902, 933, 936, 943, 961, 969, 974, 977, 979ff., 987f., 1004, 1007, 1012, 1038, 1059, 1078, 1105, 1161, 1279, 1281, 1290f., 1294, 1302f., 1313f., 1338, 1345, 1352f., 1369

depot formation, *see also* accumulation 470, 655, 746, 902f., 912, 979, 1201, 1217, 1221, 1271, 1345, 1353

depression, *see* psyche

deratization, rat poison, *see* rodenticide

derivatization 149, 153, 246, 744, 752, 960, 1155, 1183

dermal, *see* skin

dermatitis, hypersensitivity, skin lesion, *see also* skin, allergic reactions, urogenital 527, 566, 570, 614, 647, 761ff., 764, 775, 781, 862, 878, 890, 914, 926, 935, 1020, 1040, 1069, 1101, 1114f., 1126, 1136, 1176, 1282, 1328f., 1349, 1353

dermatoma, cutaneous tumor, *see* skin cancer

Descemet's membrane, *see* cornea

desert, *see* arid area

desferrioxamine, Desferal 537, 715, 726, 730f., 733ff., 737f., 913, 954f., 957

desiccant, *see also* dry, dewatering 753

desorption, *see* mobilization

destruction, *see* decomposition

desulfurization 335, 1036

detection limit, *see* sensitivity

detector, *see also* conductor, indicator, electron capture, fluorescence 141, 148, 754, 1155, 1186, 1215, 1225, 1259, 1276, 1287

detergent, *see also* soap 389, 699, 990, 1090, 1149

determination limit, *see* sensitivity

detoxification 439, 589, 649, 832, 902, 906, 1021, 1032, 1062, 1074f., 1113, 1124, 1179, 1219, 1234, 1237, 1241, 1249, 1253, 1282, 1303f., 1325, 1350, 1368

deuterium, tritium 1017, 1021

developing countries 294f.

developmental (toxicity), *see also* underdevelopment, puberty 583, 588, 601, 615, 658, 769, 771, 1000, 1002, 1007f., 1013, 1065, 1073f., 1080f., 1083, 1085f., 1094, 1098, 1222, 1224, 1316f., 1319, 1324ff., 1328, 1365

dewatering, *see also* dry, desiccant 26f., 30, 52, 1221f., 1354

diabetes, *see also* insulin 485, 646, 754, 853, 865, 877, 1174, 1328, 1353

diagenetic 5, 12, 58, 338, 343, 391, 393, 396, 1126, 1353

diagnostic agent, *see* tracer, sensor, isotope

dialysis 53, 157, 316, 324, 327, 345, 548, 598, 715, **728**, **730ff.**, 733ff., 738f., 901, 1125f., 1328ff., 1358

diarrhea 532f., 536f., 540ff., 543f., 566, 585, 589, 593, 647, 761f., 792, 795, 827, 865, 887, 899, 901, 903, 995, 1021, 1025, 1067f., 1094, 1175, 1252, 1282, 1328ff., 1349

dichromate, *see* chromate
diet, *see* nutrients, daily intake, food
differential pulse polarography and voltammetry 69, 137, 154, 177f., 188, 842, 845, 881, 893, 973, 1103, 1349, 1354
diffusion, *see also* denuder system 247, 312, 321, 377, 423f., 512, 516, 551, 567, 571f., 575, 578f., 744, 801, 859, 862, 866, 868, 913, 1144, 1318
diffusion control(led separation), *see also* denuder system **272**, 320f.
digestibility, *see* tolerance
digestion, *see* decomposition, wet digestion, gastrointestinal, intestine, stomach
digitalis, cardiac glycoside, colchicine 587, 1032, 1071, 1082
dimercapto compounds, Dimercaprol, BAL 496, 533f., 537, 540, 543, 762, 766, 1075, 1083, 1086, 1089, 1253, 1351, 1369
dimethylglyoxime 138, 203, 882f., 1103, 1137, 1354
dimethylmercury, *see* methylmercury
diode (array detector) 149, 152, 183, 744, 911
dioxins (polychlorinated) 895, 903, 907f., 957
discharge, *see* dump, disposal
discoloration, *see also* chlorosis, necrosis, albinism 221, 1069, 1094
disinfection, *see* sterilization
disintegration, digestion, dissolution, *see* decomposition, wet digestion, solubility
dismutase 497, 514, 899, 1039, 1325, 1354
dispenser 288
dispersion (spectrography), *see also* particle 70, 392, 934, 1128, 1140
disposal, landfill 19ff., 24, **27ff.**, 36, 38, 44, 50ff., 58, 61f., 64, 255, 333f., **337ff.**, 341, 343ff., 351f., 355, 365f., 384ff., 398, 426f., 453, 533, 687, 689, 691f., 707ff., 710, 814, 837, 851, 858, 878, 947, 958, 963, 977, 979, 994, 1009, 1050, 1053, 1088, 1105, 1129, 1153, 1159, 1261, 1263, 1290ff., 1312
disposition, predisposition 641, 1041
disproportionation, *see also* transformation 432
dissipation, *see* distribution, disproportionation
dissociation 314
dissolution, *see* solubility
dissolved organic matter DOM, or carbon DOC 120f., 249, 1192, 1354
distillation, *see* volatility, pyrolysis

distribution, distribution coefficient, depletion, *see also* compartmentalization 4, 93ff., 146, 154, 158f., 195, 197, 233f., 255f., 262, 296, 302, 304, 306, 320, 328, 360, 378, 384f., 388, 390f., 396, 398f., 407, 409f., 428, 435, 440, 468, 470, 495, 505, 508, 512, 516f., 521, 531, 543, 553, 555, 558, 571, **577ff.**, 582, 586, 589, 591, 593, 603, 613, 641, 644, 647, 720, 722, 726, 743, 745ff., 754, 756f., 765, 777ff., 783f., 787, 792ff., 799f., 803, 810, 821, 823f., 826, 841, 853, 858ff., 861, 863, 871, 873, 877, 879, 883, 893, 895f., 905f., 912, 918, 923f., 933f., 938, 940f., 944, 948, 950f., 953, 962, 967, 970, 979f., 983, 986f., 991f., 1005, 1009, 1011, 1017, 1027, 1029, 1035, 1037f., 1044, 1046, 1050, 1053, 1057f., 1060f., 1078f., 1081, 1088, 1090f., 1097, 1105, 1108ff., 1111, 1117, 1119, 1121f., 1126, 1129, 1133, 1140, 1143, 1145, 1158f., 1167, 1169, 1191, 1193f., 1200f., 1205f., 1208, 1210, 1215ff., 1223, 1225, 1233f., 1238, 1240, 1247, 1250, 1262f., 1270f., 1273, 1278f., 1291ff., 1300ff., 1307, 1313, 1317f., 1322, 1345, 1369f.
disulfide, Disulfiram, dithiol groups, *see also* sulfhydryl groups 498, 657, 932, 935, 938, 1207, 1253
dithioglycerol, *see* dimercapto compounds
dithiocarb, dithiocarbamate complex 149, 657, 773, 837, 894, 1037, 1051, 1103, 1113, 1124, 1128, 1159, 1213, 1229, 1237, 1241, 1312, 1331, 1350, 1364
dithiophosphates 744
dithizone 240, 543, 972f., 1047, 1191, 1229, 1310
diuretics, diuresis 539, 914, 1020, 1034, 1045, 1047, 1061, 1147, 1237, 1328, 1354
dizziness, *see* unconciousness, coordination
DNA, DNA sequence 422, 425, 434, 439f., 497, 501, 505, 513, 569, 621f., 628f., 632, 635, 639, 643, 645, 759, 771, 786f., 862, 868f., 872, 914, 926, 1065, 1114, 1126, 1147, 1194, 1319, 1325, 1331, 1349, 1353f., 1357
DNA damage, DNA binding, *see also* chromosome, cross-link, DNA (strand) break 503, 508, 529, 620, 626, 628, 634, 637, 639, 643, 764, 771, 862, 867ff., 872, 874, 915, 1001, 1072, 1117, 1126, 1179, 1331
DNA repair system, *see also* repair mechanism 504, 506, 620f., 626, 632f.,

636f., 764, 768, 771, 822, 829, 869, 872, 914, 1114
DNA replication 628, 632, 636, 768, 868, 1147, 1325
DNA (strand) break, *see also* chromosome 497, 618, 620, 626, 628, 637, 768, 829, 869, 1001, 1237
DNA and RNA synthesis 508, 569, 619, 637, 639, 763, 772, 782, 787, 875, 888, 925, 949, 1079, 1147, 1177, 1189, 1319, 1325
dog, canine 532ff., 535ff., 538f., 541ff., 545, 551, 644f., 724f., 736, 754, 769, 772, 799, 822, 841, 897, 900, 903, 905, 914, 924f., 928, 963, 967, 1030, 1066, 1113, 1129ff., 1132f., 1169, 1175, 1194f., 1204, 1207, 1209, 1217, 1237, 1272, 1286, 1304, 1326, 1330
dolomite 1026
domestic (farm) animal, *see also* slaughter, bovine, cat, dog, sheep, swine, etc. 40, 470ff., 474ff., **531ff., 534ff., 539, 542ff.,** 545f., 761, 883, 963, 981, 987, 994f., 1008, 1033, 1099, 1123, 1153, 1170, 1174f., 1195, 1239, 1312, 1316, 1323, 1330f.
donor, donor atom 67, 71ff., 76, 79, 84, 95, 98, 234, 236f., **243ff.,** 322, 372, 421, 517, 527, 587, 910, 931, 959, 1299
dopamine (β-hydroxylase) 899, 1040f.
dose frequency 594, 1281
dose response, dose dependency 218, 287, 551, 555, 582, 590, 594, 601, **651ff.,** 661, **663f.,** 666f., 676, 688, 765, 772, 779, 782, 831, 843, 884, 994, 996, 1011ff., 1014, 1083, 1238, 1279, 1282, 1285, 1287, 1370
dosimetry, *see* monitoring
Down's syndrome, *see* Alzheimer's disease
drainage, drain, sewer, *see* pipe
dredging, dredged material 29f., 58, 61ff., 231, 333, 336, 342f., 348, 351, 354f., 379, 383ff., 387, 389, 392, 394ff., 397f., 450, 453, 461, 1140
drinking (water), *see also* beverage, well 7, 16, 109, 115, 120, 256, 301, 342, 347, 379, 386, 450, 469, 471ff., 474, 476, 478, 484, 492, 501, 533, 535, 544, 571, 591, 594ff., 599, 651f., 655, 657, 659, 661, 665, 677f., 680f., 685, **700f.,** 703, 709, 711, 724, 727, 729, 732, 735, 738, 751, 756f., 759, 763f., 766, 772, 800, 813, 829, 834, 848, 850, 858, 870f., 874, 876, 889, 901, 904, 915, 942, 950, 955, 966, 974, 982f., 988ff., 1004, 1006, 1014f., 1017f., 1031f., 1038, 1041f., 1057ff., 1076f., 1094, 1105, 1108, 1114,
1116, 1119, 1130f., 1147, 1177f., 1180, 1184, 1189, 1196, 1198, 1216, 1218, 1221, 1238, 1264, 1277f., 1287, 1292, 1295, 1305, 1313f., 1320f., 1329, 1331f., 1334, 1340f., 1347
Drosophila 643, 648, 878, 1331
drug, medicine, *see also* veterinary, chrysotherapy, platinum compounds 19, 59, 163, 299, 307, 427, 430, 445, 470, 499f., 503, 527, 532f., 539, 545, 547, **565ff.,** 568ff., 585f., 599, 613, 632f., 641, 648, 657, 684, 719, 724, 727f., 740, 743, 745, 749, 762, 790, 792, 795, 797ff., 800, 871, 909, 911, 919, 923, 926, 928, 933ff., 936f., 955, 964, 969, 1019f., 1022, 1032, 1045, 1051f., 1058, 1088, 1110, 1136, 1142f., 1145ff., 1150, 1159, 1170, 1173, 1191, 1195ff., 1203, 1210, 1215, 1225, 1227, 1264, 1267, 1277, 1291, 1311, 1327, 1346, 1352, 1366
dry ashing, drying, *see also* desiccant, dewatering, roasting 117f., 212, 290, 385, 390, 717, 778, 882, 1037, 1067, 1137, 1157, 1192, 1214, 1262, 1276, 1316, 1362
dry (cell) battery, *see* battery
dry deposition, *see* deposition
DTPA, diethylene triamine pentaacetic acid 361f., 384, 401, 408, 496, 902, 941, 962, 969, 1282
duck, *see* wild bird
dump, dumping, dump leaching, *see also* offshore dumping, disposal 5, **21ff.,** 50, 59f., **333f.,** 337f., 344f., 396, 410, 427, 743, 747, 754, 803, 811, 813f., 816f., 858, 979f., 1053, 1057, 1105, 1263, 1314f., 1323
dung, *see* manure, fertilizer
duodenum 724, 792, 794, 798, 952, 1129, 1321
duplication, *see* reproduction technique
dust, fumes, *see also* atmospheric particles 12, 25, 30, 34ff., 56, 87, 112f., 143, 259, 261, 263, 266, 273f., 278ff., 283, 285, 288, 295, **297ff.,** 302, 305, 308, 336, 344, 355, 360, 374, 408, 458, 462, 471, 473f., 477, 532, 534f., 538f., 541, 543f., 548, 554f., 571, 589ff., 598, 656, 670, 676ff., 701f., 710, 720, 722ff., 732, 738, 745, 747, 749, 755, 757, 765, 775, 778, 783, 792, 809f., 815, 822, 828, 830, 833, 846, 856ff., 861, 865f., 870, 879, 887ff., 901, 910, 912, 914, 927, 933, 935, 947f., 972, 974, 978f., 981, 989ff., 994, 1008, 1012f., 1036, 1038, 1042, 1090, 1101, 1105, 1113, 1116, 1140, 1144, 1146, 1158, 1204, 1206f., 1209, 1221,

1228, 1230, 1247, 1255, 1261, 1265, 1267, 1272, 1290f., 1296, 1301, 1306, 1311f., 1314, 1329, 1331f., 1346f.
dwarfism, nanism, *see* growth inhibition
dye, *see* paint, pigment
dysentery 471, 754
dyspnea, *see also* asthma, suffocation 537, 827, 887, 1066, 1113, 1175, 1354
dysprosium 960, 965, 1300

ear, ossicles, otoliths, tinnitus, *see also* hearing defect 761, 926, 1040, 1069, 1117, 1365, 1370
early warning (system), *see also* biomonitoring, indicator, lichen, moss, peat, sensor 222
earth crust, *see also* soil **3ff.**, 7, 36, 94, 235, 275, 294, 346, 348, 449, 452, 481, 565, 587, 605, 718, 732, 744, 753, 775, 790, 806, 853, 855, 881, 883, 895f., 912, 923, 931, 940, 945, 947, 959, 961f., 971, 974, 1026, 1036, 1053, 1090, 1101, 1104, 1127, 1129, 1138f., 1143, 1158, 1162, 1193, 1205, 1211, 1213, 1229, 1245, 1262, 1270, 1275, 1277f., 1290, 1300f., 1311, 1344, 1361, 1368
earthworm 225, 228, 230f., **331**, 529, **642**, 648,.819, 842, 845, 851, 900, 907f., 1014, 1046
E.C. Directives, *see* regulations
ecogenetics, *see also* evolution **641ff.**, 647, 726
E. coli, see coli
economy, *see* cost benefit
ecotoxicology, ecophysiology 595, 688, 699, 727, 1079, 1105, 1116f., 1172, 1181, 1184, 1227
eczema, *see also* allergy, skin 467, 566, 615, 853, 866, 875, 1117, 1126
edema, *see also* blistering, burn skin, pigmentation, pulmonary 761, 765, 925, 1011, 1113, 1175, 1253, 1354
edible oil, *see* oil
EDTA, ethylene diamine tetraacetic acid 67, 97, 111, 155, 249f., 358, 362, 384, 400ff., 420, 493, 496, 511, 516, 536f., 539, 866, 875, 897, 952f., 992, 1056, 1075, 1106, 1317, 1321, 1354
eel 475, 519, 989, 1294
effluent, *see* waste water
egg, ova, maturation of the egg, *see also* hatching, ovary 475, 535, 540, 543, 546, 769, 820, 822, 914, 917, 924, 981, 1031, 1060, 1072, 1077, 1232, 1345

Elbe (river) 184, 233f., 253, 388, 391, 472, 843
electric conductor (circuit), power line, wire 38, 754, 789, 894f., 899, 906, 908, 911, 918, 962, 1027, 1154, 1191, 1193, 1205
electric control, *see* control engineering, switching
electric equipment, industry, electricity 19, 337, 365, 719f., 775, 777, 809, 895, 948, 962, 1051, 1129, 1141ff., 1270, 1277, 1347
electric furnace, *see also* blast-furnace 34, 41, 50, 856, 947f., 975, 1037, 1262
electric fuse, *see* thermometer, melting
electro cardiogram 1020
electrochemistry, electrochemical analytical methods, *see also* polarography, voltammetry 69, 120f., **136ff.**, 140, 149, 151, 154, 157, 192, 209, 247, 255, 272, 330, 558, 587, 589, 804, 806, 813, 881, 893, 961, 972f., 1010, 1036, 1143, 1241, 1244, 1259, 1310
electrode, *see* electrolyte, cathode, anode
electrodeposition, *see* electroplating
electrodialysis, *see* dialysis
electroencephalography 1254
electrolyte, electrolytic processes, electrode, electrokinetics, *see also* chloralkali production, gold, graphite electrode, mercury electrode 35ff., 41, 53f., 62, 69, 120, 136f., 139f., 155, 236, 241f., 246f., 249, 261, 265, 324, 337, 481, 484f., 589, 727, 796, 807, 813, 856, 881, 894, 932, 961, 968, 1016, 1020, 1022, 1026, 1040, 1051f., 1104, 1113, 1115, 1143, 1151, 1158, 1192, 1204f., 1214ff., 1221, 1230, 1270, 1311, 1355
electron capture detection, ECD 148f., 1050, 1155, 1354
electron diffraction, *see* SAED
electronegativity 78, 237, 243f., 1246f.
electronic equipment, *see also* control engineering, switching 19, 59, 125, 333, 744, 749, 754, 891, 908ff., 911, 921, 927, 932, 942, 1020, 1090, 1100, 1104, 1129, 1143, 1158, 1203, 1205, 1211, 1215, 1301, 1343
electronic structure, affinity, *see also* electrons, spin 71, 587, 1135, 1230, 1265
electron impact 145
electron microscopy, *see also* SAED, STEM, PCB (particle class balance) **272ff.**, 384, 718, 727, 730, 745, 916, 1119, 1277, 1352, 1354f., 1369f.

electron pair, *see also* ion pair, spin 73, 1102, 1364
electrons, electron beam, electron flow, *see also* synchroton 67, 71ff., 79, 124, 142ff., 210, 234ff., 249, 274, 282, 312, 428, 497, 587, 909, 911, 949, 964, 972, 1064, 1129, 1355
electron spectroscopy, ERR, ESCA, *see also* atomic emission spectrometry 273, 384, 525, 558, 563f., 744, 1224, 1356
electron transfer 137, 497, 1294
electrophoresis 660, 877, 1106
electroplating, galvanization 19, 37f., 44, 49ff., 53f., 59, 61, 64, 337, 347f., 351, 358, 466, 660, 689, 700, 808f., 813f., 820, 842, 856, 866, 895, 897, 905, 932, 947, 1052, 1087, 1090, 1101, 1104f., 1113, 1122, 1126, 1193, 1215, 1245, 1311, 1313f., 1329, 1340
electrostatic binding, potential interaction, affinity 67, 82, 85, 101, 236, 241, 316, 320, 587, 781, 846, 964, 1104, 1303
electrostatic filter, precipitation 13, 22, 36, 39, 41, 134, 263
electrothermal (atomic absorption spectrometry) process 116, 147, 151, 170, 172f., 180f., 187f., 192f., 195, 198, 200f., 209, 715ff., 734, 740, 790, 854, 910, 928, 1035, 1048, 1103f., 1107, 1109, 1121, 1157, 1211f., 1215, 1224, 1229, 1310, 1354, 1356
elimination (mechanism), *see also* excretion, disposal 93, 576, 779, 985, 1084, 1088, 1110, 1145, 1159, 1240, 1250, 1263, 1293, 1302, 1323
eluent, *see* solvent
embryo (toxicity), embryo genesis, *see also* fetal, teratogenic effects 617, 622f., 629, 634, 636, 747, 765ff., 768f., 771, 822, 851, 872, 875, 916, 926, 942f., 965, 967, 1045, 1072ff., 1080ff., 1094, 1111, 1114, 1122, 1124f., 1221, 1223, 1237, 1271, 1273, 1324, 1331, 1351, 1362f., 1370
emetic, tartar emetic, antiemetic, *see also* nausea 745, 747, 1147
emission, *see* atmosphere, automobile exhaust, industrial emissions, radiation
emission spectrometry, *see also* electron spectroscopy, atomic emission, emission lines, plasma, inductively coupled plasma optical emission spectroscopy 107, 125, 132ff., 138, 209, 775, 894, 910, 922, 972, 1035, 1137, 1155, 1191, 1226f., 1229, 1244, 1261, 1289, 1343, 1364f.
emphysema, *see also* edema, pulmonary emphysema 290, 306, 665, 830, 966, 1305, 1355
enamel, *see* paint, glass, protective layer
encapsulation, *see* stabilization
encephalopathy, encephalomyocarditis 579, 598, 613, 715, 723, **728f.**, 731ff., 734f., 738f., 789, 796, 798, 995f., 1000, 1069, 1074, 1087, 1223, 1355
endocrine, endotoxin 518, 523, 528, 999, 1029, 1234, 1355, 1364
endocytosis 512, 515, 779, 859, 861, 868, 913, 1355
endothelial, *see* epithelium
energy-intensivity, energy-barrier, energy dependence 37, 40f, 45ff., 240, 422f., 427f., 725, 948, 1143, 1148, 1320
enrichment, *see* accumulation
Enshi disease 1176
enteric, enteritis, enterocyte, enteral, enterobactin 518, 531, 533f., 541, 918, 1355
enterohepatitis, *see* liver
environmental fate, *see* cycling, fate
environmental impact analysis (standard) **689**, 854f.
environmental sample, specimen, *see also* reference material, sample 108, 110, 114, 116, 119, 132, 134, 142, 144, 147, 151, 164, 170, 183, 186ff., 191, 196, 199, 205f., 230, 266, 339, 379, 390, 405, 667, 768, 773, 778, 790, 799, 816, 818, 836, 845f., 881, 883, 923, 973, 1086
environmental standard, *see* threshold, reference material, regulations, VDI Commission
enzootic 899, 1355
enzymes, enzymatic (reactions), *see also* carboxylase, catalase, dehydrogenase, glutathione lyase, oxidase, phosphatase, protease, transferase, etc., cofactor 90, 95, 99, 369, 414, 421f., 425ff., 435ff., 438ff., 441f., 466, 481, 484, 495, 497ff., 503ff., 511, 513ff., 516f., 519, 527, 533, 535ff., 538ff., 545, 587f., 593, 611, 614, 618f., 621, 628, 641, 643, 646, 648f., 722f., 730, 737, 743, 748, 759, 761, 779, 782, 785ff., 806, 826, 828, 869, 878, 884, 886f., 893, 899, 914, 918, 924f., 928, 942, 949f., 954, 962, 964ff., 968f., 993f., 999, 1001, 1016, 1018f., 1022, 1025, 1031, 1035, 1038f., 1055, 1063f., 1066, 1073, 1083, 1089, 1092f., 1097, 1099f., 1112, 1116f., 1155, 1172f., 1179, 1188, 1196, 1225, 1235, 1269, 1293,

1295, 1303f., 1319, 1325, 1349, 1353ff., 1363, 1367
epidemiology, volunteer 108, 111, 491, 502, 546, 559, 577, 582, 595, 603, 629, 631, 651, 654, 660, **662ff.**, 670, 673, 675, 678, 684, 727, 734, 739, 749, 764f., 767, 775, 782, 786, 830, 839f., 843, 866, 941, 997, 1001, 1010ff., 1015, 1059f., 1062, 1069, 1075, 1079, 1081, 1095, 1115, 1176, 1178, 1182, 1189, 1221, 1241, 1284ff., 1287, 1346, 1355
epidermis, epidermoid, see skin
epilepsy 762, 796, 1020, 1031
epithelium, transepithelial, endothelial, see also skin, ciliary 499, 513, 515, 517, 578, 727, 765, 780f., 789, 795f., 861f., 952, 954, 957, 1073, 1130, 1220, 1263, 1283, 1305, 1322, 1353, 1355, 1361, 1368, 1370
equilibrium model, see model
equine, see horse
erbium 960, 964
erica, see heather
erosion, see also weathering 4f., 29, 31, 261, 275, 311, 357, 362, 776, 1053, 1139, 1291, 1350
error (source), error evaluation 112, 163ff., 169f., 179, 182, 201f., 211, 213, 233, 280, 420, 466, 585, 633, 717, 805, 817, 819, 883, 885, 973, 983, 1310, 1314
erythema 887, 1355
erythrocyte, erythroporphyrin, erythropoiesis, see also heme 156, 178, 219, 505, 508, 513f., 516f., 520, 546, 549ff., 577, 608, 644ff., 649, 657, 726, 736, 747, 758, 761f., 772, 799, 806, 822, 862f., 877, 886f., 899, 914, 924, 934, 937, 944, 949, 953f., 991, 993f., 1012, 1020, 1022, 1029, 1032, 1039, 1060, 1073, 1086, 1110, 1113, 1119, 1123, 1125, 1157, 1166, 1186, 1201, 1217f., 1294, 1304f., 1317, 1327, 1350, 1355f., 1362ff., 1367ff., 1370
ESCA, see electron spectroscopy
Escherichia coli, see coli
essentiality, deficiency, see also nutrients, copper, iron, zinc deficiency 72, 89, 96, 141, 152, 157, 207ff., 215, 217, 288, 302f., 357, 361, 376, 406f., 428, 442, 457, 463, 466, 469f., 473, 476f., **481f.**, **484**, 487, 491ff., 494, 496, 498, 511, 513, 517, 519, 527, 531, 533, 535ff., 539ff., 544f., 550f., 553f., 566f., 572, 574, 585f., 605, 611f., 617, 644, 649, 658f., 760, 770, 773, 803, 829, 836, 846, 853, 865, 871, 874, 878f., 883ff., 886, 893, 897ff., 903, 953, 1019, 1021f., 1025,

1029ff., 1033ff., 1038ff., 1041, 1044, 1062, 1089, 1091ff., 1094, 1096ff., 1100, 1107, 1110, 1116, 1118, 1123, 1153, 1157, 1164, 1166f., 1169, 1173f., 1180ff., 1184, 1189f., 1196ff., 1200f., 1211, 1218f., 1226, 1261, 1264, 1289, 1295, 1309, 1315ff., 1318f., 1321, 1325, 1333, 1342, 1361f.
ethanol, see also alcohol, fermentation 740, 887, 1039, 1052, 1060, 1083
ethylation, see alkylation
ethylene diamine, see also EDTA 67, 78
ehtylmercury, see organomercury compounds
eukaryotic cells, eukaryotic microorganisms 422, 508, 523, 628, 763, 867f., 1098, 1317, 1356
europium 189, **959ff.**, 965, 970, 1301
evaporation, see volatility
evolution, see also ecogenetics 414ff., 427, 430, 447, 481, 587, 897, 1025, 1027
ewe, see sheep
EXAFS, see X-ray adsorption
exchange (rate, capacity), see also ion exchange, soil exchange capacity 82f., 721, 778, 896, 1042
excitability, excitation, see also irritation 140f., 247, 586, 1031, 1103, 1346
excretion, precipitation, see also elimination 471f., 476f., 493, 512, 514f., 517f., 520, 533f., 536, 539, 544, 551f., 555, 561f., 566f., 571, 576, **580**, 582f., 589, 592, 600, 644f., 647, 655, 657, 661, 722, 724f., 728, 732, 746, 753, 757ff., 767ff., 770, 772, 778f., 783, 793ff., 796, 799, 806, 821, 824, 828, 859, 861, 863f., 875, 884, 886, 896ff., 900, 902f., 910, 912f., 918, 924, 933f., 940f., 951, 953, 982, 987, 995, 1001, 1013, 1018, 1021, 1023, 1028f., 1038f., 1041, 1058, 1061ff., 1072f., 1078, 1086, 1091f., 1098, 1100, 1107f., 1110, 1112, 1117, 1119, 1122, 1124, 1129f., 1133, 1145, 1155, 1157, 1167ff., 1171ff., 1183, 1194, 1196f., 1199ff., 1205f., 1216, 1221, 1225, 1233f., 1236, 1238, 1240, 1250f., 1263f., 1271, 1279, 1293f., 1302f., 1316f., 1322, 1345, 1351, 1355ff., 1359, 1365ff., 1370
exencephaly 1074, 1115, 1356
exhalation, see respiration, excretion
exhaust, see automobile exhaust, incineration
exhaustion, see tiredness, shock
experimental animal, see animal experiment
expert system, see automation, model, statistics
explosive (blasting), see also fireworks 745, 921, 976, 1052, 1277

exposure, *see* bioavailability, occupational exposure
exposure effect response, *see* dose response
extraction, extraction scheme, fractionation, *see also* liquid extraction, solvent   25, 51, 116, 125, 149, 153, 158f., 165f., 200, 323, 341, 345, 350f., 362f., 365ff., 384f., 400f., 403f., 408, 411, 805f., 860, 894, 910, 913, 932f., 940, 947, 960f., 1004, 1026, 1048, 1050, 1106, 1118, 1143, 1155f., 1192, 1212f., 1223, 1228f., 1244, 1250, 1262, 1270, 1277, 1302, 1310, 1343, 1355, 1368
exudate, *see* perspiration, saliva, secretion
eyes, eye lid, eye irritation, optic nerve, *see also* conjunctiva, cornea, retina, vision, anophthalmia   593, 663, 672, 729, 762, 853, 887, 901f., 997, 1020, 1031, 1033, 1069, 1074, 1124, 1174, 1176, 1205, 1295, 1328, 1350f., 1353, 1364

face, *see* head
factor analysis FA, TTFA, *see also* statistics   275, 279
FANES, furnace-AAS-nonthermic excitation spectrometry   1016, 1356
fast atom bombardment MS   146, 197
fat, fatty acids, fatty degeneration, *see also* lipids, oil   338, 344, 365, 477, 481f., 485f., 497, 551, 760, 808, 926, 965, 969, 1017, 1028, 1037, 1051, 1104, 1131, 1166, 1170, 1219, 1221, 1292, 1294, 1312, 1361, 1363, 1369
fatal, *see* mortality
fate, *see also* cycle   67f., 106, 158, 290, 311, 397, 419, 429f., 560, 759, 784, 949, 1011, 1038, 1084, 1087, 1122, 1143, 1181, 1248, 1250f., 1369
fat solubility, *see* lipid, solubility
fauna, *see* animals
fear, phobia, *see* shock
feather, plume, *see also* watch spring   223, 225, 228, 230, 1057, 1295
feces, faeces, *see also* manure   112, 426, 484, 514, 519, 580, 600, 666, 746, 779, 794f., 825, 831, 873, 896f., 941, 953, 968, 982, 992, 1000, 1028, 1031, 1039, 1041, 1056, 1061f., 1067, 1107, 1111f., 1119, 1131, 1168, 1171, 1194, 1206, 1216f., 1228, 1234f., 1251, 1264, 1271, 1280, 1293f., 1303, 1322, 1345, 1353, 1356
feed, animal feed, *see* fodder
feed additive, *see* food additive
feldspar, fluorite, fluorspar   776, 947, 1357

female, *see* sex
fermentation, ferment, *see* alcohol
ferric, *see* iron ions
ferrite, ferrihydrite, *see also* iron oxides   47, 373
ferritin, *see also* transferrin   515, 519, 726, 737, 779, 785, 912, **949ff.**, 952, **954**, 1356
ferroalloy, ferromagnesium, ferrosilicon   403, 689, 1026, 1140, 1211
ferrochrome, ferromanganese, ferromolybdenum, ferronickel, ferroniobium   314, 856, 873, 1036f., 1090, 1104, 1128
ferrocyanide, *see also* Prussian blue   240, 946, 1237
ferromagnetic, *see* magnet
fertility (incl. soil fertility), *see also* reproduction, egg, sperms   65, 331, 362, 368, 386, 401, 415, 507, 533, 538, 668, 699, 760, 913, 1002, 1031, 1040, 1072, 1099, 1167, 1174, 1177, 1252, 1326, 1360, 1367
fertilizer, *see also* manure, phosphate, nitrate   19, 107, 289, 311, 359, 407, 450, 458, 461, 470, 478, 707f., 757, 801, 810, 814, 818, 834, 857, 873, 879, 1027, 1037, 1090, 1093, 1142, 1180, 1318, 1335, 1345
fetal loss, *see also* embryo (toxicity), teratogenic effects, premature birth   494, 629, 654, 760, 765f., 885, 915, 942, 1002, 1039, 1072ff., 1094, 1330
fetus, fetal   214, 498, 553, 579, 593, 597f., 601, 610, 617, 629, 631, 668, 681, 760, 765f., 768, 772, 821, 824, 839f., 863, 895, 915, 933, 942, 971, 984, 989, 1002f., 1005, 1011, 1039, 1050, 1065ff., 1070, 1072ff., 1075, 1078, 1080, 1083, 1085ff., 1094, 1115, 1124, 1130, 1133, 1200, 1222f., 1271, 1287, 1322, 1345, 1356, 1363, 1370
fever, (raised) temperature, *see also* thermometer   590, 827, 901, 964, 1000, 1030, 1112, 1197ff., 1329
fibroblast, fibrocytes, fibril, fibrosarcoma   615, 619, 637, 646, 783, 864, 888, 911, 964, 1032, 1064, 1177, 1207, 1304, 1324, 1356f., 1369
fibrogenous, fibrosis; fiber (content), fiber optics, *see also* phytate   262, 486, 496, 551, 719, 723, 738, 740, 747, 781, 830, 887f., 890, 894, 901, 911, 922, 941, 952, 998, 1170, 1174, 1263ff., 1267, 1271f., 1305, 1321, 1328, 1353, 1356
field (trial)   360ff., 367, 401, 956, 1049, 1097, 1167

field desorption MS **145**, 1229, 1356, 1361
filter, filtration, *see also* electrostatic filter, gas cleaning 36, 39, 41f., 49f., 54, 113, 119, 143, 153, 263, 266, 271f., 277, 283, 303, 324, 333, 335, 337, 348, 377, 395, 716, 739, 801, 805, 809f., 813, 870, 881, 992, 1037, 1048, 1053, 1089, 1193, 1198, 1201, 1216, 1289, 1312, 1322
financial aspects, *see* cost-benefit
fine dust, *see* dust
finger-loop protein (domains) 501, 509, 822, 847
fingerprints, *see also* multi-element analysis 107, 143, 1356
finishing, *see* metal processing
fir, *see* conifer
fireproof, fire resistance, fire protection, *see also* flame retardant 1027, 1159, 1312
fireworks, pyrotechniques, match, *see also* explosive 745, 754, 857, 961, 1021, 1137, 1299
firing, furnace, power plant, burning vegetation, *see also* blast furnace, forest fire, incineration 12, 14, 263, 277, 287, 292, 294, 300, 303f., 306ff., 335, 353f., 405, 571, 583, 745, 751, 755, 777, 841, 858, 910, 947f., 979, 1012, 1051, 1053, 1090, 1095, 1098, 1105f., 1153, 1161ff., 1164, 1167, 1181, 1215, 1231, 1234, 1241, 1278, 1290f., 1295, 1312
fish, fisherman, *see also* carp, crayfish, eel, predatory fish, salmonide, seafood, shellfish, gills 56ff., 97, 113, 188, 199, 222, 301, 343, 435f., 455, 471f., 474f., 478f., 517, 531, 534ff., 537ff., 542ff., 546, 549, 554, 597, 625, 641, 677, 680, 703f., 715, 722f., 727, 734, 736, 746, 756, 760, 768, 773, 780, 800, 817, 820f., 836, 846ff., 851, 853, 864f., 889, 896f., 900, 912, 914, 952, 989, 995, 1027, 1038, 1040, 1042, 1048f., 1052, 1054ff., 1058ff., 1061, 1063, 1072f., 1075, 1077ff., 1081, 1083, 1085, 1087f., 1092, 1101, 1107, 1111, 1116, 1122f., 1155, 1166f., 1181, 1185, 1194f., 1198, 1235, 1252, 1255, 1257, 1263, 1266, 1271, 1292, 1315f., 1324, 1339
fishery product, fish meal, fish scale 202, 471, 475f., 540, 792, 848, 1058
flame atomic absorption spectrometry 107, 116, 125f., 130, 132, 150, 153, 162, 172, 183, 189, 204, 209, 805, 854, 893, 910, 939, 960, 972, 1016, 1018, 1128, 1204, 1212, 1229, 1244, 1262, 1310

flame ionization FID, flame spectrometry 148f., 151f., 1015f., 1300
flameless atomic absorption spectrometry, *see also* graphite furnace AAS 153, 307, 395, 740, 744, 752, 784, 790, 799f., 841, 877, 910, 918, 932, 1081, 1085, 1117, 1156, 1191, 1200, 1223, 1240, 1300, 1343
flame retardant, flame proof, *see also* ignitability 745, 1090
flash smelter, flash furnace, flashlight, *see also* arc, sparks 36, 40f., 1020, 1027
flocculation, flocculant 30, 42
flora, *see* plants, flowers, intestinal flora
flotation, *see also* separation, sediment 22, 25, 349, 894, 975, 1050, 1140, 1311
flour, *see also* breads, grain, phytate 113, 486, 597, 820, 952, 1107, 1194, 1321, 1366
flow cytometry 613
flower, *see also* geranium, heather, labiate, violet 981, 992, 1231
flow injection (analysis) FIA 125, 131, 134f., 140, 150, 160f., 168, 172, 175, 177, 179, 192f., 195, 197, 716, 790, 801, 878, 1026, 1033, 1043, 1088, 1226, 1356
flue gas 40, 285, 335, 377, 850, 1088
fluidized bed, *see* incineration
fluorescence (detector), fluorescent lamp, *see also* atomic fluorescence, X-ray fluorescence 131, 434, 921, 1051, 1053, 1299
fluorine, fluorides, fluorosis, *see also* hydrofluoric acid, cryolithe, feldspar 35, 40f., 50, 73, 76, 78, 145, 207, 224, 226, 236f., 405, 716, 718, 752, 759, 776f., 805, 838, 857, 914, 961, 1031, 1117, 1133, 1154, 1159, 1179, 1214, 1219ff., 1277, 1279, 1286, 1299, 1341, 1356
fluorimetry, fluorometry 206, 786, 910, 916, 919, 970, 1154ff., 1244, 1275, 1310, 1343
flux, *see* cycle, transport, sintering
fly ash, flue ash 12, 54, 158f., 261, 263, 266, 273, 275, 277ff., 283, 296, 335f., 341f., 351ff., 355, 778, 858, 907, 1090, 1095, 1106, 1155, 1158, 1162, 1205, 1208, 1215, 1230f., 1241, 1290f., 1301
fodder, forage, feed, *see also* domestic animal 357, 361, 435, 449ff., 452ff., 460f., 463, 469, 471, 473, 475f., 533ff., 537f., 543, 545, 664, 722, 760, 821, 827, 833, 879, 885, 895, 900, 995, 1008, 1091ff., 1095, 1097f., 1100, 1153, 1174, 1177, 1180f., 1195, 1219, 1239f., 1249, 1252, 1263, 1271, 1295, 1306, 1312, 1323, 1326, 1330f.

fog, mist, *see also* moisture, smog, cloud, nebulization   80, 267, 276, 279, 377, 901
folic acid   886
food, food chain, food consumption, formula feed, *see also* seafood, nutrients   13, 56, 106, 108f., 112f., 131, 147, 157f., 163f., 174, 188, 192f., 196f., 225, 288, 290, 301, 303, 306, 357, 401, 407, 417, 419, 423f., 429, 435, **449ff.**, **452ff.**, 458, 460ff., **467ff.**, 470ff., 473ff., 479, 481f., 484ff., 487, 489, 491f., 495, 516f., 520, 555, 571, 585, 591, 594ff., 599, 652, 655, 657, 659, 661, 665f., 677ff., 680, 688, 703f., 707f., 720, 722, 724, 736f., 745, 750, 752, 754, 756, 761, 766, 777ff., 780, 790, 793, 800, 803ff., 810, 819f., 823, 831, 833f., 837f., 842, 845f., 850, 858f., 870ff., 876, 883f., 895ff., 904, 915, 921, 923f., 940, 948, 950ff., 956, 966f., 976, 978, 981f., 986ff., 989f., 1004ff., 1008, 1010, 1012f., 1018, 1020, 1027f., 1037f., 1053, 1055ff., 1059, 1077, 1084, 1086, 1090ff., 1099, 1101, 1103, 1105ff., 1108f., 1117f., 1120, 1123, 1125, 1135, 1143, 1145, 1156, 1159f., 1165ff., 1169ff., 1175ff., 1185, 1192ff., 1211, 1213, 1215f., 1226f., 1232f., 1238, 1247ff., 1252, 1263, 1269, 1278ff., 1284, 1291f., 1313, 1315f., 1320f., 1328, 1332, 1345, 1347, 1352, 1366
food additive, formula food, *see also* preserving agent   554f., 567, 659, 701, 703, 705, 711, 838, 849, 879, 887, 900, 952, 981f., 1004, 1010, 1012, 1014, 1027, 1037, 1077, 1081, 1100, 1174f., 1180, 1183f., 1187f., 1226, 1248, 1263, 1312, 1319f., 1327ff., 1330f., 1341
foot, *see* leg
forensic, *see also* poisoning   113, 213, 216, 319, 396, 751, 1074, 1080, 1230, 1234, 1240
forest (ecosystem), timber, *see also* wood   168, 225, 227f., 230, 267f., 280f., 308, 370, 372, 374, 376ff., 409, 508, 708, 721, 723, 735, 737f., 740, 805, 818, 835, 841, 848, 980, 1009, 1014, 1034, 1248
forest decline, *see also* acid rain   323, 378, 721, 1030, 1033
forest fire   258f., 275, 304
fossil fuel, *see also* coal (burning), fuel oil   10, 52, 87, 90, 100, 279, 288, 352, 405, 442, 583, 745, 750, 755, 780, 806, 810, 812, 814, 1051, 1090, 1105f., 1161f., 1290f., 1312
foundry   358, 1037
Fourier transform (spectrometry)   135, 147, 151, 177, 198, 200

fractionation, *see* separation
freckle, sunspot, *see* skin pigmentation
free radical   424, 497f., 513, 619, 632, 862, 867, 877, 1102, 1114, 1178, 1325, 1368
freezing, freeze-drying, *see* cryogenic
fruit, fruit farming, fruit juice, *see also* apple, citrus, wine, rhubarb   113, 225, 368, 407, 451ff., 455ff., 461ff., 472, 703, 757, 778, 786, 820f., 827, 836, 858, 897, 899, 904, 921, 923, 981, 987, 991, 1005, 1028, 1060, 1109, 1232f., 1248f., 1271, 1315, 1318, 1326, 1329
fuel, *see* automobile, coal, fossil fuel, fuel oil, fuel production, nuclear fuel, space (rocket), wood
fuel oil (combustion, oil firing), *see also* organic waste   41, 59, 89, 258f., 287, 309, 333, 477, 710, 843, 978f., 1142, 1161, 1290, 1295
fuel production (from waste), fuel synthesis, fuel cell   352, 365, 882, 1104f., 1143, 1148, 1150, 1214, 1270
fulvate, *see* humic substances
fumes, *see* dust
fungicide, *see also* wood preservative, Maneb, seed treatment   337, 597, 745, 753f., 895, 898, 1037, 1045, 1052f., 1057, 1062, 1069, 1080, 1159, 1246, 1248f., 1252, 1257, 1312
fungus, fungi, mushroom, *see also* mildew, parasite   223, 225, 228f., 407, 410, **419**, 422f., 428, 431, 440, 443, 449, 451f., 454ff., 459, 461, 463f., 466f., 470, 472f., 496, 746, 780, 820, 846, 897f., 906, 951, 953, 1052, 1055f., 1058, 1062f., 1165, 1193, 1292f., 1296, 1325, 1363
furnace, *see* firing, graphite furnace AAS
furuncle, *see* skin
fuse, *see* metal processing, electrolyte, explosive

GABA (γ-aminobutyric acid)   1018f., 1022, 1064, 1082, 1356
gabbro   6f., 1356
gadolinium   882, **960f.**, 969, 1306
gall bladder, *see* bile
gallic acid, gallate   789, 795ff., 798, 1289
gallium, gallium compounds   138, 175, 203, 243, 628, 744, 754, **909ff.**, 967, 1227, 1308
galvanization, *see* electroplating
gamete, *see* germ cell
ganglion (cell), *see* nerve
gangrenous, *see* "blackfoot" disease, skin pigmentation

garbage, *see* household garbage, waste
garlic (odor)   424, 751, 1175f., 1220ff.
gas, gasoline (British: petrol), *see also* leaded gasoline   171, 259, 292, 299, 547, 701, 709f., 801, 986, 1003, 1103f., 1106, 1142, 1277, 1291, 1368
gas chromatography   **148f.**, 153, 156, 159, 161, 175, 178, 191, 193, 196, 436, 752, 776, 854, 1047, 1050, 1079, 1086f., 1103, 1124, 1154f., 1186, 1244, 1259
gas cleaning, *see also* filter, wet scrubber   38f., 365
gastric hemorrhage, ulcer, gastritis, *see also* stomach   543, 568, 792, 799, 886, 901, 954, 1356
gastric juice, gastric acidity, gastric lavage   156, 487, 493, 568, 586, 1020
gastric mucous membrane, gastric mucosa   518, 535, 799, 886, 901, 913, 1349
gastrin   485, 1251, 1356
gastroenteritis   532, 535ff., 539f., 542ff., 589, 1356
gastrointestinal tract, digestion, *see also* stomach   158, 188, 499, 532f., 548, 551, 554f., 572ff., 577, 580f., 586, 589, 593, 599, 723f., 728, 746, 757f., 779, 785, 789f., 792ff., 795f., 799, 803, 832, 862, 887, 897, 899, 901, 912, 919, 924, 926, 939, 941, 952, 963, 965, 988, 992, 995, 999, 1001, 1021, 1031, 1038, 1059, 1085, 1109, 1121, 1129f., 1132f., 1160, 1169f., 1176, 1178f., 1188, 1194, 1206, 1216, 1219, 1239f., 1272, 1279ff., 1283, 1285, 1302, 1304, 1306, 1319f., 1328, 1330, 1346, 1356
gas turbine, *see* turbine
gelatin(e), *see also* colloid   942, 1292
gel (permeation liquid) chromatography, gel filtration   523, 1106, 1125
gene, genetic or genotoxic effects, genotoxic endpoints, *see also* DNA damage, cytogenetics, ecogenetics, mutagenicity   287, 411, 415, 423, 427, **430**, 433, 439, 445, 449, 458, 491, 503, 508f., 524, 528, 595, 620f., 641, 643ff., 647ff., 657, 763f., 781f., 785, 829, 847, 863f., 867f., 870, 872f., 877f., 886, 893, 898ff., 901ff., 929, 952ff., 955, 993, 997, 1055, 1088, 1296, 1357f., 1368
genetic information, gene expression, genotype, gene translation, *see also* strain   411, 446, 449f., 500, 506f., 518, 528f., 617, 619, 782, 785, 847, 896, 900, 952, 1031, 1062, 1296, 1325, 1354, 1357, 1366, 1368f.

genito-urinary cancers, *see* urogenital
genome   617, **623f.**, 643, 1065, 1072, 1282, 1357, 1367
genotoxicity, *see* mutagenicity
geranium   860
germ of a disease, *see* bacteria, viral, infection, parasite
germanium, germanium compounds, germane, germanite   19, 35, 83, 95, 141, 218, 236, 552, 554, 573, 634, 912, **921ff.**, 1129, 1255
germ cell, germination, *see also* fertility, ovary cell, sperms   415, 765, 858, 889, 899, 1015, 1052, 1167, 1172, 1181, 1319, 1357, 1359, 1365
gestation, *see* pregnancy
gibbsite   721
giddiness, staggers, *see* unconsciousness, coordination
gills   531, 544, 727, 736, 851, 897, 989, 1058, 1063, 1081, 1324
glacier, glacial ice, *see also* ice   256, 360, 400, 980
gland, *see also* secretion, thymus, thyroid, pancreas, etc.   577, 1067, 1177, 1206, 1234, 1236, 1317, 1319, 1349, 1355, 1358, 1360f.
glass (vessel), glasses, lenses, glaze, enamel   113, 115, 119, 138, 420, 557, 719, 740f., 745, 754, 881f., 888, 894, 910, 922, 961, 975, 979, 981, 984, 1017, 1048, 1104, 1112, 1122, 1142, 1158f., 1162, 1205, 1215, 1228, 1230, 1245, 1248, 1277, 1301, 1311, 1344, 1359
glass house, *see* green house
global cycle, *see* cycle
globulin, haptoglobin   549, 577, 779, 991, 1303f., 1322, 1357, 1360
glomerular   535, 580f., 660, 823, 828, 992, 998, 1071, 1077, 1080, 1112, 1219, 1357
glow discharge MS   **146**, 151, 180, 182, 192, 1277
glucose (tolerance factor), glucolysis, gluconate, glucocorticoid, glucosuria, *see also* thioglucose   426f., 432, 441, 484ff., 488, 512f., 518, 523, 528, 535, 596, 646, 828, 853, 859, 865, 872, 874f., 877, 886, 895, 998, 1017ff., 1022, 1040, 1112, 1119, 1221, 1282, 1304, 1330, 1353, 1357, 1363
gluewater color, calcimine, *see* paint
glutamate, glutamic acid   485, 914, 919, 1017ff., 1064, 1304
glutathione (peroxidase and reduction)   419, 485, 497ff., 503, 505f., 513f., 517, 519, 527,

626, 638, 863, 869, 872f., 877f., 1062, 1064, 1083, 1112, 1153, 1155, 1166f., 1171ff., 1178, 1181, 1188f., 1196, 1201, 1218, 1357
glycine, glycinate 69, 78, 97, 249
glycol, glycolysis, glycolate, glyceride 304, 516, 792, 1040, 1064, 1085, 1294, 1353, 1356f., 1359, 1361
glycoprotein, glycogen, glycosamine 514, 648, 913, 917, 1018, 1116, 1219, 1279, 1304, 1357, 1363, 1367, 1370
gneiss 6f., 1313
goat 538, 541, 545, 760, 1019, 1091, 1093f., 1110, 1174, 1181, 1316, 1326
goethite 84, 86, 100, 315f., 328, 331, 773
goiter, goitrogenic 485, 887, 1017, 1021, 1357
gold, gold coat, gold electrode, *see also* methylated gold, chrysotherapy 6f., 11, 15, 19ff., 35, 57, 83, 130f., 134, 136f., 145, 159, 254, 284, 527, 557, 578, 765, 771, 773, 791, 894f., 906f., 929, **931ff.**, 934, 937, 975, 1048f., 1067, 1139, 1142ff., 1156, 1199f., 1211ff., 1214, 1226, 1235, 1351f.
gonads, *see* testes, ovary, germ cell
good laboratory practice, *see also* interlaboratory studies, reference material 164, 595, 699
grain, cereals, *see also* corn, rice, seed 361, 367, 401, 403ff., 407ff., 410, **449**, 451ff., 455ff., 460ff., **466ff.**, 477f., 722, 757, 778, 804, 819ff., 822, 842, 844, 858, 860, 864, 899, 923, 952, 957, 981, 1028, 1038, 1056f., 1060, 1080, 1106f., 1118, 1165f., 1175, 1177, 1194, 1248, 1279, 1300, 1307, 1313, 1315, 1321, 1323, 1326, 1328, 1345
Gram negative or positive bacteria 422, 431, 439f., 448, 819, 1357
granite, granitic 5ff., 727, 778, 879, 958, 976, 1093, 1095, 1277, 1313, 1368
granulomata, granuloma, *see also* tumor 781f., 784, 786, 901, 916, 919, 965, 1113, 1207, 1305, 1330, 1346f., 1357, 1361, 1363, 1367f.
grape (juice), *see* fruit, wine
graphite electrode, *see also* carbon 137
graphite furnace (atomic absorption spectroscopy GF-AAS), *see also* flameless AAS, FANES 108f., 120, 126f., **129ff.**, 139, 142, 150f., 154, 156, 159f., 162, 169, 171ff., 175f., 180, **186ff.**, 190f., 193, 196ff., 199, 202ff., 717, 733, 735, 737, 739f., 744, 800, 805, 837, 843, 854f., 872, 878, 881, 893, 917, 932, 939, 960, 972, 1035, 1100, 1104,

1137, 1157, 1223, 1262, 1269, 1310, 1343, 1357
grass, grassland, grazing, *see also* grain, cereals, corn, meadow, clover, herb 226, 229, 372, 402, 407, 409ff., 412f., 415, 418, 441, 455, 470, 476, 478, 545, 758, 896ff., 899, 979, 987, 994, 1018, 1031, 1056, 1093, 1095, 1165, 1204, 1307, 1330, 1341f., 1357
grass tetany 1031, 1357
Great Lakes (Erie, Michigan, Ontario) 268, 278, 397, 432, 435
green fodder, *see* fodder, grass, herb
green house, *see also* test chamber 226, 362, 366, 401
green manure, *see* compost, manure
greywackes 5ff., 12
Grignard compounds **1027**, 1034, 1244
groundwater, *see also* pore water, soil solubility 4, 7, 31, 52, 55f., **57f.**, 61, 65, 89, 254f., 311f., 315, 325f., 333, 338ff., 345, 347, 351ff., 355, 379, 382f., 387, 389, 394f., 473, 689, 691, 707, 709, 720, 728, 741, 754f., 813f., 816, 850f., 857, 878, 881, 950, 977, 1050, 1052, 1089, 1091, 1105, 1163, 1233, 1289f., 1292, 1301f., 1307
growth (factor, promotion), *see also* cell division 96, 98, 214, 221, 229, 281, 399f., 405, 410, 419, 426, 430, 433, 435, 438, 441, 458, 484f., 498, 501, 507, 658, 753, 760, 770, 780, 784, 864, 872, 874, 877, 884, 895, 900, 908, 913, 916, 918, 925, 944, 953, 955, 965, 992, 1008, 1025, 1031, 1034, 1055, 1060, 1065, 1092, 1094, 1097f., 1123, 1131, 1151, 1170, 1174, 1209, 1236, 1251f., 1271, 1305, 1316, 1319, 1325, 1327f., 1331, 1351, 1359, 1362ff.
growth inhibition (disturbance, retardation), *see also* weight 226, 386, 407, 494f., 539, 541, 544, 568, 646, 648, 757, 760, 765ff., 780, 784, 827, 831, 845, 853, 898f., 915, 926, 942, 965, 993, 1000, 1002f., 1011, 1039f., 1062, 1065f., 1073f., 1080f., 1110, 1131, 1147, 1172, 1175, 1196, 1219, 1254, 1295, 1305, 1323, 1326, 1328, 1330f., 1353f., 1359
guanine, methylguanine, thioguanine, *see also* cytosine 513, 618, 621, 628, 632, 635, 770, 1065, 1357f., 1364, 1368
guidance (limiting) values, *see* threshold
guideline, *see* regulation, threshold
guinea pig 300, 308, 613, 781, 785, 965f., 1060, 1086, 1097, 1117, 1181, 1237, 1271, 1304f., 1326

gum, *see* mouth, tooth
gut, *see* intestine
gypsum, plaster   5, 54, 346, 1263, 1278

habituation, *see* tolerance
hafnium   149, 561, 1343f.
hair, hair loss, alopecia, *see also* wool   111, 143, 173, 181, 201, 206, 213f., 216ff., 219f., 223, 405, 471, 541ff., 561, 566, 594, 600, 647f., 661, 666, 680, 705, 707, 749, 757f., 772, 825, 914, 928, 933, 937, 984f., 990, 1013, 1018, 1039f., 1048ff., 1059, 1061, 1064, 1081, 1083, 1087, 1094, 1110, 1121, 1129, 1136, 1157, 1175f., 1196, 1228, 1233ff., 1236, 1238, 1240f., 1294, 1316f., 1326ff., 1349f., 1352, 1356, 1361f.
half-life, *see* life time
hallucination, confusion, *see also* coordination   796, 926, 996, 1000, 1031, 1040, 1236
hamster, hamster cells, *see also* embryo, ovary   300, 495, 550, 555, 621ff., 628, 630, 633ff., 636ff., 644, 746f., 749, 765f., 769f., 782, 785, 822, 826, 830, 836, 840, 844f., 863f., 867, 915, 929, 937, 942, 1062, 1072ff., 1084, 1114f., 1119, 1122, 1124, 1177, 1188, 1217
hand, hand-mouth contact   467, 548, 657, 762f., 886, 990, 997, 1021, 1040, 1067f., 1070, 1221, 1365
haptene   862, 1357
harbor, *see* port
hard (mineral) coal, *see* coal
hard-solder, pewter, *see also* letter   754, 775, 777, 890, 940, 975, 1245, 1269f.
hard water, *see* lime, calcium compounds
harvest, *see* agriculture
hatching, *see also* egg   544, 900, 914
hay, *see* fodder, grass, herb
hazard, *see* risk (assessment)
hazardous waste, *see* waste, household garbage
haze, *see* fog
head, headache, face, facial, *see also* brain, ear, eye, hydrocephalus, mouth, nose   552, 761, 796, 827, 888, 954, 995f., 1000, 1040, 1067f., 1072f., 1113, 1147, 1176, 1205, 1207, 1209, 1221, 1254, 1283, 1346, 1362, 1369
health injury (risk), *see also* risk (assessment)   477, 832, 835, 838f., 844, 927, 986, 1001, 1009, 1011, 1046, 1076, 1117, 1178, 1191, 1197, 1203, 1237, 1239f., 1258, 1267, 1269, 1285

heap leaching, *see* leaching
hearing, *see* ear, ossicles, hearing defect
hearing defect, deafness   658, 762, 887, 1069
heart, heart failure, vascular effects, *see also* cardiac defect, myocardium, bradycardia   646, 743, 794, 842, 879, 887, 899, 923, 1020, 1030, 1032, 1108, 1113, 1120, 1123, 1129f., 1174, 1190, 1196, 1219, 1223, 1228, 1231, 1350, 1352, 1359, 1361, 1364, 1370f.
heartbeat, *see* arrhythmia
heat, heating, hydrothermal vent, *see also* fever   4f., 8, 294, 531, 773, 777, 882, 964, 970, 1027, 1104, 1147, 1245f., 1249f., 1269f., 1290, 1344
heat conductor, *see* conductor
heather, erica   413, 951, 1038
heat shield, heat stabilizer, heat resistance, *see also* insulator   777, 1016, 1290, 1344
heaviness, ponderousness, *see* weight, ataxia
helium   134, 140, 1138
hematite   629, 947, 958
hematocrite   1063, 1110, 1123, 1357
hematological   532, 535, 723, 886, 926, 966, 973, 1001
hem(ato)opoiesis, *see* heme
hematuria, hemoxidase, hemoxigenase   1219
heme, hemoglobin, hemocuprein, *see also* erythrocyte, porphyrine   484f., 497, 499, 509, 513, 515, 534, 536ff., 539, 547, 593, 646, 657, 729, 754, 758, 762, 800, 828, 880, 893, 879, 915, 938, 942, 944, 949f., 952, 954, 956, 991, 993f., 999, 1008, 1040, 1063, 1094, 1110, 1112, 1120, 1123, 1125, 1175, 1196, 1217, 1350, 1357ff., 1364, 1367
heme synthesis, *see* heme
hemochromatosis, hemosiderin, hemosiderosis   515, 537, 646, 649, 949f., 954, 957, 1369
hemolysis, hemodialysis, hemolytic   485, 537, 549, 552, 590, 598, 646, 715, **727ff.**, **730ff.**, 733, 735, 737, 739f., 761, 763, 796, 900ff., 994, 1030f., 1033, 1075, 1083, 1113, 1119, 1174, 1220, 1349, 1358, 1369f.
hemorrhagic, hemorrhage, *see also* gastric hemorrhage, intestinal hemorrhage   533, 796, 865, 901, 925, 954, 965, 1101, 1113, 1175, 1219, 1330, 1351, 1355, 1358, 1370f.
hen, *see* chicken, poultry, layer
heparine   111, 1358
hepatic alterations, necrosis, degeneration, hepatitis, *see also* cirrhosis   494, 544, 615, 643, 779, 796, 900ff., 903, 940, 943f., 1067, 1078, 1111f., 1117, 1173f., 1177, 1187,

1196, 1200, 1218, 1304, 1324, 1328, 1358
hepatic cirrhosis, *see* cirrhosis
hepatic toxicity, hepatoma, hepatoxic, *see* liver
hepatocyte 518, 779, 849, 869, 872, 913, 949, 953, 964, 1195, 1199
hepatosplenomegal, *see* liver, spleen
herb, herbage, *see also* heather, grass 410, 413, 415f., 459, 1091, 1292
herbicide, defoliant 441, 753, 755, 757, 759, 767, 1052, 1056, 1060
herbivorous animal, *see* ruminant
hereditary factor, heritable, *see* inherited syndrome
heterotrophic 428
heterozygous 901, 1358
high performance liquid chromatography HPLC, *see* liquid chromatography
high volatility, *see* volatility, vapor pressure
histidine, histamine 243, 430, 514, 620, 645, 967, 1109, 1118, 1321f., 1349
histopathology, histology, histocompatibility, histochemical 609, 730, 918, 1112, 1235, 1273
history, *see* age
Hodgkin's disease 911, 1358
holmium 145, 960f.
homeostasis, homeostatic 300, 493, 498, 508, 511, 518ff., 521, 527, 572, 587, 590, 726, 739, 839, 903, 1316, 1320ff., 1323, 1330, 1358
homicidal, *see* murder
homogenization 108, 112f., 158, 182, 184f., 717f., 740
homosphere, *see* anthropogenic
homozygous 644
hoof, ungula, claw 542, 1175
hormone, *see also* testosterone 484, 488, 498, 501, 507, 523, 528, 534, 577, 725f., 735, 785, 830, 968, 992, 1002, 1020, 1029, 1251, 1304, 1325f., 1349, 1351, 1355f., 1359, 1367
horse, mule, equine, pony 523, 528, 532f., 535, 537ff., 540ff., 544ff., 553, 555, 886, 994, 1094, 1099, 1174, 1309, 1316, 1330
household chemicals, *see also* consumption 599, 1062
household (municipal) garbage (waste) 52, 265, 278, 333, 337, 339f., 342, 347f., 352, 355, 358, 365f., 368, 690, 707, 710, 813, 837, 845, 903, 948, 991, 1088, 1247, 1309, 1315, 1336
HPLC, *see* liquid chromatography

human milk, *see* milk
humic substances, humate, humus 69f., 90ff., 97, 102, 251, 255, 311, 315f., 319, 322f., 327ff., 346, 363, 388, 420, 450, 492, 508, 718, 818, 861, 864, 970, 980f., 993, 1095, 1101, 1105, 1366
hydrate, hydratation, *see* hydroxo complex
hydride (technique), *see also* boron, beryllium 109, 131f., 135, 149, 153, 156, 159, 173, 182f., 186, 193, 195, 203f., 744, 751, 777, 790, 798f., 801, 1017, 1021, 1049, 1132, 1156f., 1212f., 1224, 1226, 1244, 1264f.
hydrocarbon, coal gasification, *see also* methane, benzene, polynuclear (polycyclic) aromatic hydrocarbons 34, 288, 342, 499, 741, 882, 961, 1055, 1111, 1120, 1142, 1145
hydrocephalus 1074, 1221ff., 1224, 1359
hydrochloric acid 4, 34, 111, 119, 136, 156, 400, 789, 1127, 1155, 1157, 1213, 1217, 1226, 1264, 1329, 1349, 1356
hydrocycle, *see* cycle
hydrofluoric acid, hydrogen fluoride 4, 40f., 158, 405, 719, 1128, 1203
hydrogen, hydrogen bonds, hydration, hydrogen bomb 128, 148, 191f., 210, 238, 243, 327, 344, 437, 516, 587, 744, 748, 753, 761, 882, 922, 1016f., 1036, 1104f., 1138, 1142f., 1147, 1154, 1159, 1212f., 1244, 1277
hydrogen cyanide, Prussic acid 56, 934
hydrogen peroxide, *see also* peracetic acid 497f., 619, 626, 634, 874, 899, 1060, 1153, 1178, 1261
hydrogen sulfide 4, 36, 99, 370, 393, 419, 434f., 437f., 1055, 1062, 1191
hydrolysis, hydrolase **79ff.**, 245, 252, 315, 317, 439, 589, 752, 946, 1286, 1292, 1299, 1302, 1319, 1344, 1362, 1367
hydrometallurgy 35, 38, 40, 351
hydroxo complex, hydratation, hydrate 69, 79, **237ff.**, 246, 249, 311, 320ff., 345, 347f., 370ff., 589, 773, 826, 880, 931, 946, 948, 956, 1026, 1161, 1262, 1292, 1350
hydroxyl and other radicals, hydroxide (ion), hydroxylase 72, 80, 95, 99, 154, 238, 241ff., 311, 314f., 323, 369, 422, 488, 492, 495, 497, 499f., 503, 509, 529, 619, 626, 641, 719ff., 722, 728, 736, 752, 773, 779, 805, 856, 860, 877, 899, 914, 919, 924, 940, 946ff., 949, 954, 1093, 1097f., 1213, 1249, 1251, 1270, 1299
hyperactivity 996f., 1000, 1020, 1219
hyperaluminemia 725f., 728, 733, 1359

hypercalcuria, hypocalcemia, calcification 535, 729f., 733, 911, 914, 1359, 1361
hyperesthesia, *see also* psyche 1236, 1357
hyperkeratosis, *see* keratin
hypermagnesemia, hypomagnesemia 647, 1025, 1030ff., 1034, 1359
hyperpigmentation, *see* skin pigmentation
hyperplasia, *see* cell division
hypersensitiveness, oversensitivity, *see* sensitizing
hypertension, hypertony, hypotension 218, 496, 589, 597ff., 601ff., 646, 828, 998ff., 1007, 1009ff., 1013, 1030, 1174, 1346, 1358f., 1370
hypochromic 535, 899, 1236, 1359, 1370
hypogonadism, *see* growth inhibition
hypolipidemia 1040
hypophysis, hypophthalamus, hypocampus 935, 1019, 1039, 1360

ice, ice core, *see also* glacier, polar ice, snow 4, 89, 114, 238, 256, 269f., 277, 977, 983
ICP, *see* inductively coupled plasma
ignitability, *see also* flame retardant 51
ilmenite, *see* titanium dioxide
immature, *see* underdevelopment, puberty
immobilization, *see* stabilization
immune response, immunologic diseases, immunologic functions, immunotoxicity, immunosuppression 491, 494, 498, 504ff., 507f., 562, 567f., 590, 592, **605f.**, **608ff.**, 611ff., 614f., 645, 734, 775, 781f., 785f., 862, 873, 886, 900, 925, 927, 935, 937, 999, 1071, 1080, 1083, 1085, 1112, 1114, 1118ff., 1123, 1173, 1178, 1184, 1188, 1254, 1283, 1304, 1327, 1329f., 1368
immunoassay 806
immunoglobulin 781, 785, 999, 1360
impactor, impaction 264, 267, 272, 276, 280f., 283, 575, 1360
impact statement, interpretation, *see* environmental impact analysis (standard)
implant, implantation, prosthesis, restoration, *see also* surgery, transplantation **557ff.**, **560ff.**, 563f., 724, 863, 866f., 876, 879, 884, 888, 891, 898, 900, 922, 929, 941, 1072, 1104, 1128, 1133, 1142, 1146, 1151, 1198, 1204f., 1207, 1209, 1237, 1262, 1264, 1267, 1305, 1352, 1355, 1360, 1365, 1370
impotence, *see also* sperms 1002, 1328
impregnating (agent), *see also* wood preservative 720, 877, 1141, 1344

inappetence, anorexia 494, 533, 537, 541f., 544, 593, 761f., 795, 887, 914, 926, 995, 1021, 1068, 1094, 1176, 1221, 1304, 1326, 1328, 1350
incineration, combustion, *see also* automobile exhaust, firing, forest fire, ashes 38, 87, 158, 263, 265f., 278, 285, 304, 333ff., **341f.**, 344f., 348, 351ff., 355, 365, 405, 431, 442, 477, 493, 571, 583, 740, 743, 745, 747, 750, 810, 812, 814, 858, 895, 903, 907f., 977ff., 1017, 1088, 1105, 1161f., 1247, 1295, 1312, 1315
indicator, *see also* biomonitoring, early warning, sensor, tracer 107, 109, 220, **221ff.**, 228f., 285, 294, 329, 382, 412, 429, 504, 693, 731, 736, 828, 849, 937, 971, 994, 1013, 1058, 1136, 1150, 1157, 1197, 1228, 1234, 1313, 1316, 1319, 1326ff.
indium 19, 132, 175, 243, 552, 573, 744, 895, 918f., **939ff.**, 1043, 1213, 1226f., 1245, 1308
individuals, *see* risk population
indole 1319, 1326
indoor, *see also* clean room, ventilation **287ff.**, 300, 307, 539, 668, 677, 680, 693, 703, 710, 831, 912, 1162, 1278f., 1283, 1286
induction 499, 501, 509, 649
inductively coupled plasma-optical emission spectrometry ICP-ES, *see also* plasma 107ff., 120ff., 126, 132ff., **146f.**, 150, 152, 156, 160f., 167, 169, 171, 173ff., 177, 179ff., 185f., 189f., 193ff., 196ff., 200f., 203ff., 209f., 213, 256, 272, 529, 717, 727, 752, 790, 798, 800f., 846, 854, 878, 881, 894, 906f., 919, 932, 937, 957, 963, 970, 972, 1010, 1035, 1043, 1103, 1121, 1125, 1128, 1154, 1192, 1200, 1204, 1209, 1212f., 1223f., 1226, 1262, 1269, 1273, 1276, 1286, 1297, 1310, 1360
industrial (emissions, dust) 10, 87, 257, 259, 261, 263, 266, 333, 336f., 363, 365, 399, 405, 423, 430, 434, 442, 453, 459, 462, 470, 473ff., 477f., 493, 532ff., 535, 538ff., 541, 543f., 547, 571, 660f., 605, 629, 633, 641, 687, 689f., 692, 699, 721, 724, 749f., 755, 757, 759, 783, 791, 813, 815, 824, 833, 845, 866, 892, 898, 924, 927, 948, 977, 979, 982, 1030, 1091f., 1095, 1105f., 1164, 1176, 1186, 1193, 1228, 1231f., 1239, 1247f., 1256, 1263, 1270, 1278, 1292, 1295, 1297, 1306, 1312, 1314, 1326, 1340f., 1344ff., 1347

industrial toxicology, *see* occupational medicine
inert (noble) gas, *see* noble element
infant, baby, *see also* child, newborn 299, 466, 479, 574, 579, 598, 656, 662, 666f., 678, 685, 701, 703, 705, 736, 762, 838ff., 899, 904, 953, 971, 982, 990, 1003f., 1013, 1032, 1041, 1058f., 1073, 1083, 1108f., 1130, 1133, 1166, 1218, 1220, 1317, 1320, 1326, 1328, 1349, 1363
infection, *see also* bacteria, nematode, viral 606, 608f., 611f., 614, 754, 771, 911, 999, 1020, 1071, 1197, 1311, 1327f., 1355, 1360, 1367
inflammation, *see* irritation
infrared radiation, infrared spectroscopy 273, 658, 754, 921, 961, 1215, 1230
ingestion, *see* oral
inhalation, *see* respiration, atmophile, nose
inherited syndrome, heritable mutations, inheritable changes, *see also* gene, mutagenesis, ecogenetics 430, 492, 527, 550, 566, 598, 629ff., **644ff.**, **647ff.**, 726, 900, 903, 956, 999, 1002, 1011, 1074, 1349f., 1363f., 1370f.
injection (technique) 130, 549, 552, 586, 606, 970, 1041, 1061, 1067, 1071, 1077, 1081, 1110ff., 1113, 1118, 1125, 1130, 1206f., 1217, 1219, 1221, 1264, 1283, 1294, 1303f.
insect, insecticide, *see also Drosophila* 641, **643**, 723, 726, 736, 753, 755f., 758, 822, 827, 841, 845, 850, 953, 1159, 1232
insemination, *see* fertility, reproduction
insensibility, numbness, *see also* paralysis, tremor 761f.
instrumental (performance) 106f., 116, 121, 126, 129, 132f., 140ff., 1142, 1215
insulator, *see also* heat shield 777, 1159
insulin, *see also* diabetes 865
interaction, interference, interrelation, *see also* synergism, antagonism, additive, symbiosis **84ff.**, 95, 126f., 135ff., 140, 146f., 165, 176, 179, 204, 208, 218, 226, 241, 246, 255, 261, 300, 312, 314, 329f., 342, 370, 378, 380, 386, 392f., 395, 398f., 408ff., **419ff.**, 427, 430, 441, 486ff., **491ff.**, **497ff.**, **502ff.**, 505ff., 508f., 511, 516f., 531, 536f., 540, 542, 544, 553, 558, 560, 563, 566, 574, 589, 593, 613, 619f., 625, 631, 634, 636, 639, **657f.**, 663, 665, **671**, 684f., 716f., 720, 726, 730f., 739f., 759, 764, 770, 782f., 786, 826, 830, 839, 846, 872, 878, 893, 897f., 900, 905, 910, 917, 922, 929, 932, 946, 952f., 957, 960, 968, 970, 992, 997, 1007, 1010, 1015f., 1022, 1026, 1029, 1035, 1038, 1044f., 1063, 1065f., 1071, 1078, 1081, 1085, 1089, 1092f., 1095ff., 1098, 1100, 1114f., 1117, 1126, 1128, 1144, 1147, 1153, 1157, 1170f., 1173, 1178f., 1189, 1192, 1196, 1200, 1204, 1218, 1244, 1269, 1271, 1295, 1300, 1309, 1319, 1321, 1323, 1326, 1330, 1364, 1369
interception, *see* throughfall, canopy
intercomparison, *see* interlaboratory studies, interaction
interface, *see* distribution
interferon, interleukin 504, 518, 523, 528, 925, 927
interlaboratory studies, *see also* good laboratory practice 106, 108, 112, 164, 171f., 184, 186, 190, 362, 718, 840, 847
interrelation, *see* interaction
interstitial water, *see* groundwater
intestine, intestinal tract, bowel, gut, *see also* duodenum, gastrointestinal tract, mucosa 485f., 493, 495, 511f., 514ff., 518ff., 529, 532, 534, 538, 542f., 549, 566, 571f., 577, 580f., 586, 589, 634, 647, 666, 724ff., 728f., 736f., 740f., 759, 761, 794f., 819, 822f., 835, 851, 865, 884ff., 911, 922, 949, 952, 970, 988, 1029, 1031, 1039, 1061, 1063, 1070, 1081, 1092f., 1095, 1108, 1112, 1118, 1129, 1177, 1179, 1195, 1206, 1217, 1219f., 1234, 1283, 1293ff., 1296, 1321f., 1328, 1330, 1345, 1353, 1355f., 1360f., 1363, 1366, 1370
intestinal flora, *see also* bacteria, tropical medicine 749
intestinal hemorrhage, colostomy 795f., 1353
intestinal wind 795f.
intoxication, *see* poisoning
intracellular, *see* cytoplasm
intratracheal, *see* pharynx
intrauterine, *see* uterus, fetus
inverse voltammetry, *see* voltammetry
invertebrate 206, 230, 331, 377, 479, 523, 545, 756, 760, 765, 771, 819, 896, 922, 924, 1038, 1057, 1204f., 1293, 1320
"in-vitro" systems and techniques, "in-vitro" assays, *see also* short-term tests 440, 493, 505, 515, 518f., 550, 554, 561, 568, 605f., 608f., 614f., 619f., 622, 626, 632, 634, 661, 724f., 747, 765, 767, 769, 771f., 781ff., 785f., 797, 825f., 839, 859, 867ff., 871f., 884, 914, 917, 919, 928, 937, 941, 964, 967,

969, 1062ff., 1066, 1085, 1087, 1112, 1114ff., 1119, 1122, 1131, 1188, 1200, 1218, 1223, 1253, 1273, 1324, 1331, 1361
"in-vivo" systems and techniques 142, 440, 445, 493, 498, 508, 515, 519f., 524, 550, 568, 576, 580, 609, 623f., 631, 633, 661, 725, 759, 763, 765, 767f., 782, 805f., 823, 831f., 837f., 846f., 867ff., 872, 883, 888, 917, 919, 926, 934, 937, 940, 962, 964, 967, 969, 973, 1012, 1022, 1047, 1087, 1112, 1116, 1119, 1200, 1271, 1299, 1303, 1306, 1324, 1331
iodine, iodide, iodo complex, see also methyl iodide 76, 78, 145, 207, 243, 481f., 484f., 743, 749, 751f., 754, 856, 887, 1048, 1204, 1211, 1213, 1221, 1230, 1278
iodometric titration 854f.
ion exchange, ion chromatography, see also exchange (rate) 22, 25, 43, 46, 53, 69, 80, 119, 147, **149**, 153f., 157, 184f., 194f., 198, 203, **205f.**, 224, 240, 311, 315ff., **318ff.**, 326f., 330, 345, 349, 360, 367, 373, 377, 399, 404, 407, 440, 492, 523, 525, 718, 772, 813, 818, 856f., 881, 890, 894, 905, 940, 960f., 969f., 1140, 1155, 1213, 1226, 1234, 1244, 1249, 1300, 1303, 1318, 1352
ionic radiation, ionizing radiation 277, 913, 1282ff., 1285, 1287
ion microprobe, ion probe microanalysis, IPMA 145, 210, **273f.**, 730, 916, 1022, 1361
ionophore, see (carrier) protein
ion pair, see also electron pair 236, 239, 248f., 424, 846
ion selectivity 69, 324, 587
ion specific electrode, see sensor
I.Q., intelligence decrease, see learning ability, paramnesia
iridium 132, 175, 179, 243, 276, 283f., 1137, 1139ff., 1142ff., 1145f., 1149, 1151
iron (general, elementary), see also ferro ..., hematite, tinplate, steel 3, 6ff., 11f., 15, 19ff., **32ff.**, 51, 54, 57ff., 63, 72, 85, 95, 132, 158f., 179, 203, 207, 211, 218, 235, 259, 291, 294, 297, 307, 312, 344, 383, 387, 392, 400, 405, 428, 450, **457f.**, 481, 484ff., 495, 503, **514ff.**, 519, **537**, 540, 544, 551, 557f., 560, 571ff., **646**, 725, 737, 741, 779f., 798, 808, 810, 812, 814, 828, 846, 860, 882ff., 887, 893f., 909, 913, 918, **945ff.**, 962, 1029, 1034f., 1038, 1040, 1093, 1097, 1110, 1123, 1163, 1204, 1215, 1224, 1262f., 1273, 1286, 1290, 1297, 1301, 1311, 1319, 1321, 1356, 1358, 1361, 1369f.
iron complex, see also ferrocyanide, heme 528, 880, 913, 948, **951ff.**
iron compounds, see also iron oxides 56, 72, 261, 273, 313, 373, 383, 393, 396, 403, 412, 514, 546, 629, 773, 880, 896, 912, 942, **947f.**, **950f.**, 956, 989, 993f., 1140, 1163f., 1261ff., 1270
iron deficiency 504, 511, 535, 574, 822f., 860, 887, 899, 912f., 919, 945, **953ff.**, 956, 988, 994, 1038, 1041, 1352, 1369
iron ions 57, 72f., 79, 82, 86, 94f., 149, 154, 157, 181, 203f., 235, 239f., 312f., 323f., 338ff., 348, 428, 497, 499, 511, 520, 567, 626, 628, 909, 912f., **946ff.**, 951, **954**, 956, 993, 1362
iron oxides (incl. iron ores), see also ferrite, goethite, hematite 20, 23, 32ff., 55, 68, 95, 158, 313f., 319, 329, 347f., 374, 384, 657, 773, 809, 912, **945ff.**, 948f., 955, 1035f., 1038, 1093, 1229f., 1270, 1314
irradiation, see radiation, ray treatment
irreversible reaction, see also reversible reaction 597
irrigation, flooding 13, 870, 1164, 1278, 1323
irritation, stimulation, see also sensitizing, as well as the concerned organs 532f., 535ff., 538f., 541ff., 544, 559, 567, 569, 589, 664, 676, 693, 720, 762, 771, 781, 822, 853, 862, 888, 913, 915, 917, 921, 935, 938f., 965, 995f., 1020f., 1025, 1067f., 1093, 1101, 1147, 1176, 1179, 1207, 1254f., 1277, 1295, 1328f., 1349ff., 1353, 1355ff., 1358, 1361f., 1364ff., 1367ff., 1371
Irving-Williams order 79, 99
isotachophoresis 960, 969
isotope, see also tracer, radioactivity, radiochemical 3, 140, 142, 144f., 147, 174, 201, 210, 290, 325f., 409, 520, 555, 626, 725, 743, 751, 772, 775f., 779, 787, 789, 794, 798ff., 805, 823, 842, 853, 863, 879, 890, 909, 911, 916ff., 919, 922, 939, 944f., 960, 971, 988, 1008f., 1015ff., 1026, 1046, 1086, 1089, 1102, 1108, 1111f., 1127, 1129f., 1132f., 1135ff., 1144, 1154f., 1159, 1193f., 1200 1203f., 1206, 1208f., 1211f., 1227, 1231, 1243, 1271f., 1275ff., 1278, 1281f., 1284, 1286, 1289, 1293, 1300, 1302, 1309, 1319, 1327, 1368
isotope dilution (mass spectrometry) IDMS 116, 121, **144f.**, 163, 166, 169, 174, 177, 180, 183, 188f., 199, 201, 805, 840, 894, 972f., 1103, 1229, 1310, 1360

Itai-Itai patients, Itai-Itai disease 57, 62, 342, 354, 494, 553, 596, 828f., 831f., 841

jaundice, icterus, *see also* liver, hepatic ... 761, 796, 1330
jet propulsion, *see* aircraft
jewelry 19, 216, 931f., 935, 1104, 1114, 1136, 1141ff., 1193, 1301, 1344
joint (disease), synovial 557, 561, 563, 729, 911, 954, 960, 968, 995, 1175, 1205, 1208, 1330, 1349ff., 1370

kaolin, kaolinite 292, 314, 319, 440, 776, 778
keratin, kerotosis, keratogenesis 494, 535, 599, 763ff., 1064, 1197, 1361, 1365
kerosene, *see* fuel oil
Keshan disease, Kasin-Beck 1093, 1100, 1157, 1166, **1174f.**, 1181, **1184**, 1190, 1361
kidney, *see also* nephro ..., renal cancer, renal cortex 112, 142, 144, 199, 210, 212f., 217f., 290, 471ff., 474ff., 477f., 495, 507, 512f., 517f., 520, 523, 527ff., 534ff., 538f., 541, 544, 549, 551ff., 555, 577f., 580, 585f., 590, 592f., 596, 601f., 613, 646f., 661, 725, 743, 746ff., 749, 758, 761, 794ff., 798, 800, 803ff., 820ff., 823ff., 829, 831ff., 834, 838ff., 845f., 849ff., 863, 865, 872, 887, 897, 912, 914f., 924, 927, 934, 938f., 941f., 944, 963, 967f., 981, 998f., 1001, 1018, 1020, 1027, 1029f., 1033, 1039, 1060, 1062, 1069, 1073, 1079f., 1094, 1108f., 1111ff., 1119, 1124f., 1130, 1136, 1145f., 1170, 1194, 1199, 1206, 1217ff., 1220ff., 1223, 1228, 1232, 1234f., 1241, 1250, 1271, 1279f., 1282, 1284, 1302, 1307, 1317, 1319, 1324, 1330, 1345, 1349, 1351, 1357f., 1362, 1364, 1367f.
killer cell, suppressor cell, *see also* phagocytosis 504, 508, 605f., 609ff., 612, 614, 1112, 1123, 1173
killing, *see* mortality, murder
kinetics, dynamics, *see also* model 68, 79f., 95, 234, 239f., 247, **250**, 315, 320, 326f., 329f., 345, 423, 512, 514f., 520, 531, 555, 595, 600, 654f., 657, 661, 671, 673f., 676, 737, 746, 773, 840f., 850, 876, 905, 944, 957, 1009, 1013, 1042, 1046, 1060ff., 1078, 1086, 1110ff., 1118, 1122ff., 1126, 1172, 1184, 1200, 1234, 1236, 1240, 1279, 1292, 1323, 1327, 1366, 1370
kinky hair, *see* Menkes' disease
kitchen utensils, pots, *see* cooking

Krebs cycle 1019, 1064, 1361
krill 817, 847, 1107, 1123
krypton, *see* noble (inert) gas

labelling, *see* packaging, nuclear medicine
labiate 412, 415
laboratory animal, *see* animal experiment
laboratory intercomparison, *see* interlaboratory studies
lacquer, varnish, *see* paint, protective layer
lactate, lactoferrin, lactose, lactone 485, 511, 515f., 518, 885, 912f., 950, 1017, 1188, 1207, 1293, 1297, 1304f., 1346, 1361
lactation, *see also* child bearing, milk 511, 538, 551, 632, 644, 668, 760, 913, 1061, 1072f., 1081f., 1110, 1130, 1174, 1218, 1220, 1320, 1322, 1328f., 1361
lakes, *see also* Great Lakes, Lake Constance, limnology 4, 60ff., 84, 89f., 93, 100ff., 115, 191, 250, 256, 268, 270, 276, 281, 284, 342, 351, 380, 387, 394f., 397, 411, 432, 437, 443, 554, 720, 727f., 733, 736ff., 755, 771ff., 777, 787, 801, 810, 813, 815f., 821, 836, 841f., 849, 878, 933, 936, 947f., 955f., 958, 977, 1035, 1040, 1048, 1053f., 1058, 1075, 1081, 1083f., 1088, 1126, 1143, 1162, 1167, 1185, 1192f., 1247, 1259, 1297, 1339
Lake Constance, *see also* (river) Rhine 777, 846, 1054
lamb, *see also* sheep 756, 899, 1100, 1345, 1355
LAMMA, *see also* laser, mass spectrometry 145, 158, 174, 177, 196, 205, 210, 718, 727
landfill, *see* disposal, dump
land use 311, 333
lanthanides, lanthanum 72, 75f., 95, 145f., 150f., 190, 195, 200f., 203, 206, 235, 243, 255, 276, 895, 916, **959ff.**, 962ff., 965ff., 968ff., 1132, 1275, 1299f., 1302ff., 1305ff., 1308
large intestine, colon, *see* intestine
larva, grub, *see also* amphibian, fish, insect, spawning time 1111, 1195, 1252, 1271
larynx 540, 762, 1115, 1361
laser, *see also* LAMMA, field-desorption MS 130, 132, 145f., 150ff., 175, 179, 201, 718, 727, 744, 754, 911, 1118, 1301, 1361
latex, *see* rubber
laxative, aperient, purgative, *see* astringent
layer, laying hen, egg yield 543f., 546
leaching, *see also* extraction, erosion, tailing **21ff.**, 30ff., 35f., 52, 54, 57, 70, 222,

246, 266, 268, 311, 338ff., 342, **345ff.**, 349f., 352ff., 370ff., 373, 375ff., 384ff., 387, 390ff., 394, 415, 428, 723, 755, 777, 807, 813, 850, 858, 896, 899, 979, 981, 1018, 1050, 1101, 1105, 1113, 1214, 1248f., 1277f., 1290, 1292, 1311f., 1366

lead (general, elementary), *see also* organolead compounds 5ff., 11ff., 15, 19f., 32f., **34f.**, 37f., 50f., 57ff., 64f., 83, 85, 89, 93, 96, 101f., 107ff., 126, 128, 136f., 139f., 144f., 151, 153, 155, 158f., 164, 168ff., 172, 174f., 177f., 180f., 186, 189ff., 193f., 196ff., 205, 209, 213ff., 216ff., 219, 223, 233, 243, 248ff., 254, 257ff., 265, 268ff., 280ff., 283, 288, 290f., 294, 297, 299, 301f., 304ff., 308f., 318, 326, 329, 334f., 342, 348, 351, 357f., 363, 375f., 387, 390, 392, 398, 409f., 414f., 420ff., **430ff.**, 449f., 453, **458ff.**, 467, 469f., **472ff.**, 479, 484, 492f., 495, **497ff.**, 501, 504ff., 507f., 511, **515ff.**, **538f.**, 545ff., 548f., 551f., 554f., 571, 573f., 577ff., 581ff., 592f., 596f., 600ff., 603, 605, **608ff.**, 612ff., 623, 625, 629, 631ff., 635, 637, 642, 644, **646**, 648f., 651, 653, 655ff., 660, 662, 671f., 675f., 678ff., 681ff., 684f., 693, 698ff., 701ff., 704ff., 709ff., 736, 745, 747, 753f., 772f., 789, 791f., 796, 800, 813, 821, 826, 835, 837ff., 840ff., 848f., 895, 897, 900, 905, 921, 954, 957, 959, **971ff.**, 975, 983, 986, 989ff., 1016, 1021, 1034, 1039, 1051, 1100, 1127, 1133, 1137, 1145f., 1148, 1158, 1173, 1178, 1187, 1192, 1220, 1227f., 1230, 1235, 1241, 1245, 1275f., 1279, 1281, 1283f., 1290, 1300f., 1309, 1311f., 1324, 1331, 1342, 1349, 1368

lead alkyls, *see* methylated and ethylated lead

lead (inorganic) compounds, *see also* lead glance, lead halogenide 248, 266, 273, **431f.**, 503, 507f., 538, 545, 554, 588, 608, 615, 622, 629f., 633, 639, 661f., 664, 668, 670, 698, 709, 711, 749, 753, 757, 770, 814, 853, 857, 862, 921ff., 971f., **974, 976, 1000f.**, 1012, 1014, 1017, 1089, 1214, 1229f., 1368

leaded gasoline, *see also* antiknock 157, 171, 200, 276, 376, 547, 571, 700, 709f., 972, **975ff.**, 978f., 983, 986, 999, 1003, 1009, 1011ff., 1368

lead glance, galena, lead sulfide 974, 1213, 1229ff.

lead halogenide, halide 248, 266, 978

lead ions 72, 79, 95, 155, 224, 248f., 316f., 320, 324, 345, 423, 427, 440, 496f., 498f., 502, 623, 628, 668, 972f., 977, 979, 986, 1004

leak 31

learning ability, performance, asthenia, *see also* paramnesia, I.Q. 602, 901, 971, **995ff.**, 1010f., 1063, 1068, 1072f., 1085, 1283, 1287, 1328, 1350

leather (industry), *see* tanning agents

leaves, shoots, stomata, *see also* canopy, deciduous tree, litter 221, 230, 267, 360ff., 368, 372, 377, 403ff., 413, 431, 471ff., 726, 756, 758, 778f., 860f., 884, 897, 906, 951, 953, 981, 987, 1030, 1038, 1056, 1106f., 1167, 1232, 1271, 1279, 1315, 1318f., 1326, 1366

legislation, law, *see* regulation, threshold

leg, arm, extremities, limb, locomotor tract, *see also* hand 544, 568, 762f., 886, 996, 1070, 1220, 1236, 1363, 1365, 1369

legume(n), pulse, rape, *see also* bean 158, 179, 412, 415, 778, 780, 950, 1018, 1028, 1038, 1093, 1106, 1165, 1232, 1271

leishmaniasis, *see* tropical medicine

lemon (juice), *see* citrus

lense, cataract, *see* eye, photographic materials, glass

lethality, *see* mortality

lethargy, *see* tiredness

letter, type metal, *see also* alloy, hard-solder 744, 976, 1245

lettuce, *see* vegetables

leukemia, leukopenia 613, 637, 764, 911, 915, 919, 929, 942, 964, 1147, 1179, 1254, 1283, 1361

leukocyte (white blood corpuscle), leukopenia 494, 609, 611, 615, 635, 638, 761, 763, 780, 888, 914, 934, 1031, 1087, 1130, 1221, 1304f., 1326f., 1361

level, *see* blood, plasm, serum, urine

Lewis acid, *see* donor, receptor

liability (to damage) 52, 699

lichen 223ff., 226, 228ff., 409f., 746, 818, 858, 951, 1058, 1078, 1248, 1279

life style, *see also* consumption, food, tobacco smoke 288, **655f.**, 663, 863, 1006

life time, *see also* age, cycle, long-term 10, 95, 102, 111, 141, 144f., 239, 263, 321, 513, 517, 549ff., 555, 672, 676, 688, 746, 758, 803, 815, 822ff., 831, 833, 840, 848, 853, 863f., 880, 884, 886f., 901, 909, 912, 915, 918, 922, 924, 926, 934, 941, 944, 953, 960,

963, 978, 980, 985, 991f., 1000, 1013, 1015, 1039, 1061f., 1076, 1081, 1086, 1092, 1120, 1130f., 1133, 1144f., 1157, 1161, 1194, 1203f., 1206, 1218, 1221, 1231, 1234, 1249ff., 1271, 1275ff., 1278ff., 1281, 1292, 1302f., 1305, 1323, 1357

ligands, ligand exchange, ligation 67, 69, 71ff., 81ff., 86, 90ff., 95f., 98f., 138, 154f., 234, 236f., 239f., 242ff., 248ff., 254, 311, 316, 322, 348, 401, 404, 421f., 430, 440, 492, 495, 503, 509, 511, 525f., 528f., 566f., 569, 588f., 593, 726f., 818, 839, 859, 868, 880, 882, 898, 931ff., 934f., 938, 946, 949, 951, 957, 1016, 1046f., 1101, 1106, 1109, 1304, 1361

lignite, brown coal, *see* coal, bituminous, peat

lime (stone), hard water

limiting value, *see* threshold, regulation

limnology, *see also* lakes 68

lipids, lipophile, lipid peroxidation, *see also* fat, photolipid 422, 435, 470, 497, 513, 531, 541, 571f., 612, 797, 859, 914, 926, 1046f., 1059, 1063, 1101ff., 1105, 1109, 1111, 1120, 1124f., 1153, 1221, 1294, 1302, 1325, 1361, 1363

lipoprotein, lipofuchsin, myelin 515f., 519, 577, 830, 887, 1174, 1220, 1224

liquid chromatography 135, **149**, 153, 156f., 161, 174, 182ff., 192f., 523, 529, 744, 749, 773, 877f., 932, 970, 1215, 1225, 1244, 1259, 1300, 1359

liquid extraction, *see also* extraction 119, 135, 961, 1244, 1306

lithium 77, 141, 238, 545, 631, 633, 644, **646f.**, 649, **1015ff.**, 1200

lithophile, lithogenic, lithography, lithosphere 89f., 335, 948, 966, 1015, 1031, 1044, 1138, 1160, 1277f., 1312, 1361

litter (spreading, dead leaves, newborn) 223, 370f., 376, 545, 805, 818, 835, 897, 951, 966, 980, 988, 1009, 1029, 1074, 1086, 1330

liver, *see also* hepatic ..., folic acid 112, 142, 165, 199, 206, 210, 212f., 218, 290, 470ff., 473ff., 476ff., 494, 499, 501, 505f., 512ff., 517ff., 520f., 523, 525ff., 528f., 534ff., 537ff., 541, 550f., 553f., 562, 577f., 581, 586, 596, 602, 613, 638, 643ff., 646f., 649, 717, 722, 725, 729, 743, 746ff., 749, 758ff., 762, 779, 784ff., 787, 794f., 800, 805, 820ff., 823ff., 828f., 839, 843, 845f., 849, 859, 863, 865, 872, 874, 884ff., 887, 896f., 900ff., 903f., 912, 914, 917f., 924, 926, 933ff., 936, 939ff., 942, 951, 953f.,

963f., 967ff., 981, 984, 999f., 1032, 1038f., 1056, 1060, 1062, 1064, 1077, 1079, 1091ff., 1094, 1108f., 1111ff., 1119, 1124, 1130f., 1146, 1155, 1169f., 1175, 1177, 1181, 1187f., 1194, 1196f., 1199ff., 1216, 1218f., 1221ff., 1228, 1234f., 1241, 1248, 1250, 1271, 1280, 1283ff., 1292, 1294f., 1302ff., 1307, 1316f., 1320, 1323f., 1330, 1345, 1353, 1355, 1358, 1365, 1368, 1371

loam, loess 375, 403, 861, 864, 1165, 1313

lobster, *see* crayfish

long-range transport, *see also* transport mechanism 229, 231, 258, 260f., 268, 278, 282, 284f., 745, 750, 755, 771, 803, 815, 847, 933, 978, 980

long-term (behavior, effects), time lag, *see also* chronic exposure 344ff., 360, 365, 367, 386, 409, 417, 517, 547, 549, 552f., 591f., 595ff., 598f., 651, 662, 674, 728, 751, 764, 783, 824, 828, 830, 832f., 839ff., 843f., 860, 877, 906, 928, 984, 1009, 1019f., 1030, 1076, 1100, 1130, 1209, 1243, 1252, 1254, 1256, 1264f., 1271, 1323

loss (e.g. in analytical chemistry), *see also* samples, volatility, leak 109f., 805, 855, 881, 1157, 1212

loss of (eye) sight, *see* blindness, vision

loss of weight, *see* weight (loss of)

lubricant 1016f., 1090, 1159, 1311f.

lucern(e), *see* clover

luminescence (spectroscopy) 183, 195, 525, 528, 743, 748, 898, 970, 1103, 1123

luminous substance, phosphor, television 745, 921, 961, 1277, 1299, 1301, 1312

lung, *see also* pulmonary ..., pneumonia, alveolus 112, 202, 209, 212, 262, 282, 290, 292, 300, 303f., 306, 308, 493, 499, 508, 549, 554ff., 574ff., 577, 580, 582, 586, 593, 610f., 613, 615, 680, 722f., 725, 738, 743, 746f., 749, 751, 757ff., 775, 777, 780f., 783f., 787, 794, 822f., 825, 828, 830, 836, 839, 845, 847f., 861ff., 869, 877, 879, 884, 887f., 890, 901, 911f., 914, 923, 926f., 929, 941, 943, 953f., 963, 965, 967ff., 1060f., 1108f., 1111, 1113, 1115f., 1119f., 1122, 1130, 1136, 1144, 1160, 1166, 1169, 1194, 1199, 1204, 1206ff., 1209, 1217, 1221ff., 1248, 1252, 1264ff., 1267, 1271ff., 1279ff., 1283f., 1293, 1295, 1302, 1304ff., 1307, 1345f., 1350, 1361, 1366, 1369

lung cancer, pulmonary cancer, lung tumor, *see also* respiratory tract 290, 303, 305f., 502, 534, 548, 550, 555, 593, 599f., 602,

610, 632, 634, 638, 656, 661, 664f., 747, 751, 764f., 782, 784f., 830, 844, 866, 870, 877, 955, 1115, 1117, 1124, 1147, 1150, 1179, 1264, 1283ff., 1286f., 1346
lunula stripes, see Mees lines
lutetium **960f.**, 964
L'Vov platform, see graphite furnace atomic absorption spectroscopy
lyase, see also enzyme 436, 439
lymph, lymphoblast, lymphocyte, lymphoma cell, lymphatic, lymphokine 300, 534, 562, 574, 576, 605f., 608f., 612ff., 621ff., 625, 632, 636ff., 646, 649, 745, 748f., 764, 771, 781, 784, 786, 867, 911f., 914f., 919, 923, 925, 927, 940, 963f., 1001, 1071, 1083f., 1115, 1125f., 1147, 1173, 1206, 1254, 1264, 1280, 1283, 1287, 1302, 1326, 1330, 1345, 1356, 1358, 1361f., 1369
lyophilization 379, 1147, 1362
lysosome, lysyloxidase, lysate, lysozyme 512, 514, 518f., 526, 624, 649, 779, 786, 861, 899, 902, 913, 919, 934, 936, 941, 954, 1186, 1207, 1304, 1330, 1351, 1362, 1367

macrophage 292, 304, 508, 576, 582, 608ff., 611, 613f., 779ff., 786, 861, 867, 873f., 876, 925, 927, 929, 949, 963, 969, 1173, 1207, 1209, 1361f., 1366, 1369
macrophyte, see also aquatic plant 391, 1151
magma, magmatic 3ff., 7, 17, 776, 1139, 1143, 1158, 1213, 1362
magnesium, magnesium compounds, see also ferroalloy 3, 7ff., 11f., 15, 72, 76, 79, 90f., 95, 99, 132, 155, 158, 196, 218, 224, 235, 238, 248, 250, 290f., 340, **374**, 388, 399, 412, 421f., 426, 440, 481, 484, 495, 500, 503, 507f., 619, **647**, 719ff., 727, 729f., 737, 740, 757, 776, 780, 792, 817, 842, 861, 884, 915, 964, 967, 1015f., 1018, **1025ff.**, 1029f., 1034, 1036, 1039, 1056, 1120, 1157, 1262, 1277, 1290, 1344, 1352, 1357, 1359
magnesium oxide **1027**, 1033
magnet, magnetism, magnetic separation, NMR 22, 73, 127, 142, 144, 325f., 349, 384, 525ff., 529, 716, 739, 789, 857, 880, 882, 938, 945ff., 962, 969, 1104, 1129, 1140f., 1287, 1311, 1355
magnification, see accumulation
maize, see corn
MAK, see threshold, occupatinal exposure, concentration at the workplace
male, see sex

malformation, deformation, anomaly, see teratogenic effects
malic acid, see apple
malignant, see tumor
malnutrition, see essentiality, nutrients
mammal, see also monkey, dog, rabbit, cat, cow, bovine, horse, sheep, swine, ruminant, wild animal, goat, rat, mice, hamster, etc.; domestic animal, breast cancer 223, 306, 517, 520, 523f., 527, 535, 537, 543, 550, 563, 582, 605, 619, 622f., 631, 641, **643f.**, 722, 724, 726f., 758f., 766, 769, 778, 793, 803, 822, 853, 858, 862f., 865, 896, 907, 912, 914, 923, 926, 934, 953, 967, 969, 984, 1006, 1027, 1038f., 1041, 1060f., 1071, 1092, 1101, 1108, 1111, 1115, 1131f., 1145, 1166, 1171, 1177f., 1183ff., 1187f., 1194, 1211, 1243, 1251ff., 1258, 1294, 1302f., 1316, 1320, 1323, 1329, 1345, 1356, 1361, 1365
mammalian cell, see also embryo, fibroblast, germ cell, hamster, "in-vitro" systems and techniques 508, 527, 620ff., 623, 628, 632f., 636, 638, 763, 782f., 867f., 875, 934, 937, 969f., 1034, 1072, 1087, 1114, 1358
Maneb 1037
manganese, see also iron-manganese oxides, permanganate, ferrochrome, organometal compounds, antiknock 6ff., 11f., 15f., 32, 57ff., 72, 89, 95, 132, 138, 145, 159, 207, 214, 216, 218, 235, 259, 307, 312, 387, 392, 396, 403ff., 422, 428, **460**, 481f., 485, 504f., 509, **539**, 552, 571f., 574, 596, 632, 644, 670, 703, 719, 754, 780, 826, 839f., 861, 884, 904, 957, 969, **1035ff.**, 1124, 1312
manganese compounds, see also Maneb, manganese dioxide, permanganate 273, 412, 589, 622, 629, 793, 912, 942, **1035ff.**
manganese dioxide (or other oxides and ores) 10, 95, 158, 314, 321, 329, 347, 373, 384, 393, 539, 948, 953, 1020, **1035ff.**, 1038, 1040ff., 1044, 1052, 1270, 1314
manganese ions 73, 79, 82, 87, 94f., 198, 224, 239, 313, 324, 330f., 338ff., 500, 618f., 623, 626, 628, 942, 964, **1037f.**, 1042
manganese knobs, nodules 894, 1036, 1140, 1213
manganism 1040f.
manure, dung, see also compost, feces, fertilizer 292, 362, 365, 404, 818, 885, 903, 1315, 1353
marine, see ocean, seawater
marine food, see seafood

marine organism, *see* aquatic organism, aquatic plant

marrow, medulla, *see also* bones   577, 717, 730, 747, 762f., 771, 828, 886f., 899, 914, 949, 999, 1019, 1060f., 1147, 1220, 1234, 1236, 1282, 1303, 1331, 1350, 1352f., 1355, 1362f.

marsh, moor, bog, wetlands, *see also* reed   55, 58, 268, 270, 278f., 343, 352, 392, 395, 750, 773

mass balance, *see* cycle, material transport

mass spectrometry, *see also* spark source MS, isotope dilution MS, field desorption, LAMMA, SIMS, fast atom bombardment MS, glow discharge MS   **144ff.**, 152, 158, 161, 166ff., 174, 265, 436, 529, 752, 776, 800, 805, 894, 919, 957, 963, 970, 1035, 1118, 1125, 1137, 1241, 1244, 1259, 1275, 1285, 1289, 1300, 1360, 1367

mass transport, *see* material transport

match, *see* fireworks

material goods, *see also* consumption, construction, building   947, 1270, 1272f., 1308, 1339

material (or particle) transport, balance, *see also* cycle   89f., 274f., 279, 350, 365, 372, 376f., 394, 443, 735, 741, 809f., 843, 862, 875, 947, 979, 1009, 1030f., 1087, 1100, 1139, 1327, 1353

mathematical model, *see* model

matrix effects, *see also* background, biomass, organic (matter) waste   126ff., 131, 133, 135, 141, 143f., 146, 150, 157, 164f., 191, 212, 275, 384, 560, 716f., 730, 733, 744, 854, 878, 946, 1128, 1137, 1154, 1157, 1192, 1228, 1362

meadow, pasture, *see also* grass, clover   414, 478, 540, 895, 899, 994, 1018, 1093, 1233, 1248, 1302, 1307, 1313, 1318, 1330

measuring, *see* control, monitoring

measuring place, site, *see* samples

meat, flesh, *see also* muscles, beef, pork, and the starting animals, slaughter   113, 190, 202f., 455, 473, 475f., 478, 486, 722, 756, 820, 834, 843, 849, 923, 950, 952, 981, 1005, 1028, 1108, 1166, 1194, 1199, 1279, 1292, 1315, 1321

mechanisms, *see also* biochemical reactions, metabolism, repair mechanism, transformation, transport mechanism   234, 423, 436, 486, 493, 498, 501, 504, 506, 511, 514f., 518f., 521, 552, **587ff.**, 592, 610, 617, 619, 675, 725ff., 759, 767, 769, 782f., 786, 796, 822, 862, 867, 878, 893, 898, 903, 913, 915, 917, 925, 933, 964, 968, 971, 993, 1009, 1020, 1029, 1058, 1062, 1078, 1082, 1098f., 1124, 1126, 1178, 1197, 1286, 1303, 1318, 1324, 1330

medicine, medicament, *see* drug, nuclear medicine, occupational medicine, tracer

Mediterranean (sea)   175, 233, 268, 276, 393, 817, 838, 842, 1054f., 1079, 1081, **1086f.**, 1259

Mees lines, lunula stripes   761, 1236

melanosis, melanoma, melanogenesis, *see* skin cancer, skin pigmentation

melt(ing), *see also* metal processing (melting points are not indexed)   4, 719, 750, 791f., 856, 976, 1016, 1105, 1214

membrane, membrane filter, membrane permeability, *see also* cell wall   43, 95ff., 102, 119, 138, 156, 300, 372, 406, 422, 435, 440, 470, 492f., 497f., 501, 505, **511ff.**, 514ff., 517ff., 520f., 531, 572, 574, 577, 581, 586, 590, 612, 649, 728, 732, 750, 762, 780, 794, 823, 862, 868, 870, 877, 901f., 908, 926, 934, 954, 962, 964, 969, 1000, 1016, 1027, 1047f., 1056, 1058f., 1064, 1067, 1071, 1074, 1087, 1101, 1109, 1111, 1136, 1146f., 1218, 1223, 1234f., 1294f., 1304, 1325, 1330, 1341, 1355f., 1358, 1360, 1363, 1366ff., 1369f.

memory weakness, *see* paramnesia, learning ability

Menkes' disease   527, 646, 648f., **899**, 903, 1362

menopause, climacteric period, *see also* child bearing (age)   488, 583, 829, 1032f.

menstrual cycle (disturbance), *see also* child bearing (age)   1039, 1095, 1350, 1365

mental deficiency, inbecility, *see* paramnesia

mercaptane, *see also* sulfhydryl groups   72, 89, 932, 1117

mercury (general, elementary), *see also* cinnabar, methylmercury, organomercury compounds   6ff., 11ff., 15, 21, 35, 44, 51f., 56ff., 72, 89, 97, 102, 107, 113, 115, 118f., 125, 130ff., 134, 136, 140, 144, 159, 162, 168, 170, 175, 177f., 188f., 192f., 197, 223, 230, 233, 251, 257f., 265f., 279f., 296, 301ff., 334, 336, 342ff., 347, 357, 386, 392, 394ff., 397, 405, 419ff., 422f., 425ff., 428ff., **433ff.**, **436ff.**, 439, 445f., 448f., **460f.**, 467, 469, 473, **475**, 493, 495, 498, 504, **516f.**, 532, **539f.**, 546, 551f., **557f.**, 571f., 574, 580, 582f., 585f., 591, 602f.,

605, **610**, 612, 614, 625, 635, 649, 670f., 675, 680f., 684, 691, 698, 700, 703ff., 709ff., 747, 789, 840, 858, 897, 1006, 1010, 1013f., 1021, **1045ff.**, 1050f., 1055, 1059, 1062, **1066f.**, 1074, 1076f., 1079f., 1082f., 1086f., 1148, 1155, 1158, 1167, 1173, 1185, 1190, 1193, 1200, 1227, 1235, 1245, 1250, 1324
mercury (inorganic) compounds (incl. oxides), see also cinnabar 266, 344, **433ff.**, 437f., 517, 532, 539, 546, 585f., 588, 596f., 605, 610, 630, 635, 641, 661, 698, 700, 710, 1017, 1045, **1051ff.**, 1060, 1062f., 1067, **1074**, 1076f., 1080, 1083, 1158, 1230
mercury (film) electrode 120, 136ff., 202, 973, **1051ff.**, 1084, 1103, 1354, 1358, 1363
mercury ions 79, 82, 95, 99, 140, 162, 184, 188, 266, 313, 315, 323, 420, 422ff., 426, 433ff., 436ff., 439f., 447f., 497ff., 501, 519, 551, 574, 578, 580, 585, 592, 618, 628, 1045f., 1051, 1054, 1059, 1061, 1063, **1067**, **1074**, 1086
mercury vapor, see mercury (elementary)
mesopause, see also atmosphere 1138
metabolic disturbance 728
metabolism, metabolite, metabolic process, see also cell function, conjugation, mechanism, transformation 90, 106, 110, 170, 191, 194, 214, 217ff., 285, 303f., 379, 406, 419, 421f., 481f., 484f., 488f., 491f., 494, 499, 503, 509, 511ff., 514ff., 517, 519, 521, 527, 532ff., 535, 538f., 541f., 544f., 548ff., 554f., 563, 565ff., 572, 583, 586, 589, 591f., 594, 601, 617f., 626, 641, 644ff., 648f., 657, 661, 722, 724f., 733f., 736, 738, 741, 746, 749, 754, 757, 761, 766f., 770f., 778, 789, 794, 799, 821ff., 828f., 832, 841, 845, 851, 853, 859, 861f., 868, 874ff., 878, 884, 893, 896, 909, 912, 915f., 918f., 924, 929, 933f., 936, 940, 944f., 951, 953, 956, 958, 963f., 967, 987, 989, 991, 993, 995, 1000, 1006f., 1009, 1012, 1015, 1018, 1022, 1025, 1029, 1031, 1034, 1038f., 1045, 1058, 1061, 1068, 1081f., 1089, 1091ff., 1097f., 1101, 1108, 1112, 1117, 1119ff., 1122ff., 1127, 1129, 1132, 1145, 1147, 1153, 1155, 1167ff., 1171ff., 1178f., 1182f., 1188, 1194, 1197, 1200, 1205, 1208, 1216, 1219, 1233, 1235f., 1241, 1250f., 1271, 1279, 1286, 1293f., 1297, 1299, 1302, 1307, 1309, 1316ff., 1319, 1325, 1327, 1330, 1336ff., 1345, 1352, 1355ff., 1358, 1361, 1363, 1365, 1367
metal balance, see material transport

metal-carbon compounds, see organometal compounds
metal carbonyl, see also nickel (tetra)carbonyl 882, 888f., 946, 955, 1144
metalloid, see also antimony, arsenic, germanium, selenium, silicate, tellurium 72, 75, 130, 235f., 243f., 287, 302, 492, 689, 693, 700, 703, 743, 771, 922, 958, 1153
metalloprotein, see also metallothionein 174, 1117
metallothionein 96, 99, 101, 495, 501, 505, 507, 512, 514, 517ff., 520f., **523ff.**, **526ff.**, 549, 551ff., 555, 578, 581, 592, 613, 643, 646ff., 649, 805, 819, 822ff., 826f., 836, 851, 934, 937, 1081, 1095, 1100, 1112, 1124, 1309, 1321f., 1324, 1334, 1369
metallurgical production, see production
metal processing, metal smelting, see also blast furnace, flash smelter, electric furnace, foundry, pyrometallurgy **19ff.**, **31ff.**, 39ff., 52, 55, 59f., 62ff., 178, 259, 278, 281, 283, 333, **336f.**, 405, 411, 414, 431, 440, 442, 455, 544f., 571, 583, 596, 599, 601f., 625, 636, 656f., 689, 719, 723, 734f., 745, 747, 749, 751, 753ff., 759, 765f., 773, 777, 803, 810, 812, 814, 816, 833, 839, 842, 856, 888, 890, 927, 929f., 939f., 947, 963, 966f., 974, 976ff., 979ff., 983, 985, 987, 989, 991ff., 994, 1001, 1003, 1007f., 1010ff., 1016, 1036f., 1040, 1090, 1101, 1115, 1119, 1129, 1153, 1158, 1162f., 1199, 1211, 1214, 1223f., 1227, 1230f., 1233, 1272, 1277f., 1295, 1307, 1312f., 1330
metaphase 623
metastase, metastatic, see also carcinogenesis 745, 923, 927f., 1124
metavanadate, see vanadium
metaxyleme, see xyleme
meteorite, see also cosmos 3, 276, 946, 1140, 1229
meteorological, see weather conditions
methane, methanogenic 129, 204, 339, 344, 436f., 441, 883, 1055, 1088, 1104
methemoglobin(emia) 497, 534, 1362
methionine 420, 886, 937, 1062, 1085, 1154, 1168, 1170, 1362
methylarsenic compounds, see organoarsenic compounds
methylated gold 245, **932ff.**
methylated and ethylated lead compounds 153, 156, 171, 191, 194, 245, 343, 432, 547, 574, 579f., 591, 593, 612, 661, 698, 972, **975f.**, 978, 983, **999f.**, 1005, 1007, 1027, 1037, 1250

methylated tin compounds, *see* organotin (compounds)

methylation, *see also* alkylation, methylated lead compounds, methylmercury, methylcobalamin, methyl iodide, biomethylation 344, 386, 419, 422, 424, 426, 428, 430, 433f., **436ff.**, 441, 516, 550, 554, 628, 643, 647, 759, 767, 772, 886, 912, 933, 936, 1055f., 1063, 1082, 1155, 1164, 1172f., 1181, 1217, 1250

methylcholanthrene 925, 928

methylcobalamin, *see also* viamin B_{12} 424, 435ff., 441, 933, 936, 1055, 1146, 1247

methylene blue 854

methyl iodide (natural?) 432, 436, 883

methylmercury, *see also* organomercury compounds 55, 83, 95, 99, 109, 119, 130, 147, 153, 156f., 162, 165, 168, 172, 186, 188, 245, 343f., 394, 425, 433ff., 436ff., 447, 475, 498, 508, 512, 516f., 519ff., 532, 540, 547ff., 551, 554, 572, 577, 579ff., 591f., 594, 597, 600f., 610, 612, 614, 630f., 637, 657, 661, 680, 705, 1045, 1048f., 1051f., **1054ff.**, 1057, 1059ff., 1062, 1064f., **1068ff., 1071ff., 1074**, 1076, 1078ff., 1081ff., 1084ff., **1087f.**, 1190, 1230, 1250, 1259

methyl- and ethylmethane sulfonate MMS, EMS; also MES 627f., 636, 1044

methyl-vitamin B_{12}, *see* methylcobalamin

mica (schists) 7, 776, 792, 1229, 1357

mice, mouse, murine 495, 501, 504, 506ff., 520, 528f., 555, 608ff., 612ff., 615, 621f., 630, 633, 636, 638, 640, 643f., 647, 649, 738, 760, 763, 765f., 768ff., 771, 787, 822, 827, 830, 840, 844f., 863, 867, 877, 887f., 915f., 918f., 924ff., 927ff., 935, 937, 941ff., 944, 963ff., 966ff., 1001, 1039, 1043, 1061f., 1064, 1066, 1071ff., 1074f., 1079, 1081ff., 1084ff., 1109, 1112, 1118f., 1121f., 1129, 1131ff., 1147, 1185, 1187f., 1195f., 1207, 1215, 1221, 1225, 1237, 1251, 1254, 1264, 1266, 1271, 1273, 1295, 1304f., 1307f., 1326, 1331

micelle, *see* colloid, detergent, soap

microbattery, *see* battery

microbe, *see* microorganisms

microcolumn, *see* capillary

microelectronics, *see* electronic equipment

β_2-microglobulin, α_2-macroglobulin 660, 828, 832, 840, 918, 1109, 1117, 1121

micronucleus, *see* nucleus

microorganisms, unicellular organisms, protozoon, *see also* bacteria, eukaryotic and prokaryotic microorganisms, *coli, Salmonella*, etc. 10, 61, 97, 102, 311, 330f., 338, 344, 353, 367, 388, 400, 404, **419ff., 422ff., 425ff.**, 428, **430ff.**, 433, **435ff.**, 440f., **443ff.**, 448, 475, 481f., 491, 496, 501, 516, 531, 580, 701, 726, 741, 758, 763, 795, 819, 838, 864, 882, 885f., 889, 897f., 913, 916, 924, 933, 951, 955, 992, 1007, 1038, 1041, 1055, 1062f., 1082, 1092, 1105f., 1162f., 1167, 1172, 1190, 1195, 1197f., 1216, 1233, 1249ff., 1282, 1295, 1302f., 1318f., 1357, 1363, 1366f.

microphthalmy, *see* anophthalmia

microprocessor, *see* electronic equipment, semiconductor, control engineering

microscopic techniques, *see also* arc, electron microscopy, histopathology **273ff.**, 1082

microsome, microsomal 580, 618, 868, 874, 924, 929, 943, 1061, 1111f., 1125, 1179, 1185, 1363

microwave, microwave emission spectrometry 150, 1155, 1186, 1212, 1226, 1363

migration, *see* mobilization

mildew, *see also* wine 901

milk, *see also* cheese, dairy products, lactation 109f., 113, 128, 140, 147, **156**, 168, 174, 190, 301, 470f., 474ff., 478, 515f., 518, 520, 538, 567, 569, 644, 649, 666, 681, 746, 760, 820, 834, 838, 841, 843, 885, 904, 908, 923, 950, 981, 988, 992, 1005, 1041, 1061, 1073f., 1084, 1091f., 1108f., 1118, 1121, 1174, 1177, 1181, 1194, 1200, 1217, 1249, 1292f., 1303, 1316, 1320ff., 1345, 1361

millet, sorghum, *see* grain

Minamata disease, *see* Minimata disease

mineral (or ore) deposit, *see also* raw material, resource 5, 13ff., 26, 57, 70, 411f., 481, 651, 744, 753, 775f., 790f., 806, 855f., 883, 894, 912, 921ff., 932, 947, 950, 961, 963, 972, 974, 1014, 1026f., 1036, 1090, 1128f., 1139, 1141, 1143, 1151, 1158, 1181, 1188, 1200, 1204, 1211, 1213f., 1225, 1227, 1229f., 1247, 1259, 1262f., 1267, 1290, 1299f., 1311, 1313, 1343f.

mineralization 92, 411, 1164

mineral oil, *see* crude oil

mineral water, *see* groundwater, well

Minimata disease 597, 631, 680, 1045, 1057, 1070, 1073, 1084, 1087

mining, *see also* mineral deposit, ore processing 5, 7, 13f., **19ff.**, 55ff., 59f., 62f., 65, 90, 259, 333, **336**, 342, 348, 351, 354, 386f.,

411, 414f., 423, 431, 445, 596, 651, 755, 758, 765, 771, 773, 777, 790f., 813f., 829, 835, 855, 881, 883, 894, 906, 974f., 977ff., 992ff., 1016, 1022, 1026, 1036f., 1039f., 1045, 1050, 1067, 1090, 1095, 1101, 1104, 1140f., 1193, 1201, 1233, 1259, 1277f., 1283, 1290, 1299, 1301, 1311
minnow, see fish
minting, see coin
mirror   791, 1045, 1104, 1193
miscarriage, see fetal loss, reproduction
miticide, acaricide   1246, 1249, 1254, 1363
mitigating   727, 735
mitochondria   455, 484f., 760, 794, 800, 849, 942, 949, 964, 968f., 993, 1035, 1040, 1061, 1064, 1173f., 1190, 1218, 1225, 1237, 1253, 1294, 1363, 1365
mitogen, mitogenic effect   605ff., 610, 612, 614
mitosis, mitotic rate, mitotic abnormalities   617, 771, 888, 1000, 1064, 1066, 1071, 1363
MNU, see nitroso compounds
mobilization, mobility, desorption   52, 56ff., 64, 70, **87f.**, 97, 148, 158f., 231, 257, 262, 270, 276, 279, 311, 315f., 324f., 327ff., 336, **337ff.**, 342f., 345f., 348, 352, 367, 376, **379ff.**, 382, 386f., 389ff., 392ff., 395f., 408f., 419, 421, 473, 492f., 514, 519, 540, 559, 567, 578, 582f., 612, 720, 730, 747, 762, 773, 777, 784, 813f., 817, 821, 831, 839, 851, 862, 878, 883, 902, 915, 933ff., 951f., 977ff., 980, 992, 1048, 1050, 1075, 1088, 1105, 1112, 1159, 1163, 1167, 1194, 1205, 1234, 1239, 1263, 1277f., 1303, 1313, 1318, 1322, 1368
model, modelling, see also kinetics, thermodynamics, transport   68, 70, 79, 86, 91, 98, 223, 234, **241ff.**, 248ff., 254, 260ff., 274ff., 278ff., 282, 285, 291, 305, 312, 315, 321, 326, 328ff., 345, 373, 386, 393, 396, 491, 508, 519, 526, 548, 550, 554, 558, 561ff., 564, 568ff., 594, 597f., 601, **626ff.**, 664, 731, 736, 773, 815f., 824, 831, 841, 848, 883, 911, 934, 937, 1006, 1061, 1088, 1114, 1123, 1169, 1185, 1192, 1199, 1218, 1292, 1323, 1351, 1371
Moessbauer spectroscopy   273, 525, 528, 932
moisture, see also fog   443, 909, 959, 993, 1299, 1344
molecular effects, see biological reactions
mollusc, mollusk, molluscicide, see also earthworm, invertebrate, octopus, shellfish   200, 226, 230, 505, 516, 518, 850, 895, 924, 989, 1006, 1038, 1246, 1248, 1252, 1293, 1316, 1319, 1363
molybdenum, molybdenum compounds, molybdates, molybdenosis   6ff., 11f., 15, 21, 24, 89, 95, 128, 138, 201, 207, 219, 258, 304, 313, 402ff., 407, 409f., **462**, 467, **476**, 481f., 485, 536f., **540f.**, 545, 573, 628f., 857, 893, 899f., **1089ff.**, **1094ff.**, 1098, 1144, 1158, 1223, 1269ff.
molybdenum blue   1090
monitoring, testing, see also biomonitoring, control, flow cytometry   **31**, **38ff.**, 50, 129, **221ff.**, 225, 270, 277, 280, **296f.**, 306, 308, 354, 393, 400, 409, 429, 456, 468, 505, 547, 553, 583, 652, 661, **665f.**, 676, 680, 685, 688, 693, 698, 709, 716f., 730ff., 735, 739f., 753, 789, 799, 810, 818, 831f., 841, 846f., 850, 863, 874, 927, 936, 960, 1005, 1008, 1010, 1020, 1032, 1103, 1121, 1137, 1155, 1157, 1189, 1263, 1265f., 1287, 1320
monkey, ape, see also primate   201, 550, 555, 611, 746, 759, 962, 969, 1078, 1129, 1131f., 1207, 1326
monograph   710
montmorillonite   314ff., 317, 319, 327f., 440, 776, 778
moon, mooncrust, lunar, see cosmos
moor, see marsh
mordant, see pigment, tanning agent
mortality, fatal outcome, lethality, see also fetal loss   56f., 222, 343, 494, 535, 537, 540, 567f., 585, 588, 590, 598, 601f., 610, 613, 644, 649, 654, 663, 739, 760f., 767, 770, 784, 795, 827, 830, 841, 864f., 868, 885, 887, 900ff., 903, 915, 925f., 955, 963, 965, 994, 996, 1001, 1007, 1012, 1031, 1034, 1039, 1058, 1063, 1066ff., 1097, 1111, 1113f., 1131, 1151, 1153, 1179, 1187, 1189, 1196, 1201, 1207, 1219, 1221, 1237, 1241, 1248, 1252, 1256, 1264, 1272, 1287, 1305, 1324, 1329ff., 1346, 1351, 1361f., 1371
moss, see also marsh   55, 223f., **226ff.**, 229f., 282, 284f., 412, 722, 745f., 818, 858, 897, 980, 1058
mother's (breast) milk, see milk
motility, see mobilization
motor disturbance, see ataxia, psychomotoric
motor fuel, see fuel oil, gasoline
motor traffic, see automobile, road
motor vehicle exhaust, see automobile exhaust
mouse extermination, see mice, rodenticide

mouth, oral cavity, *see also* oral ..., tooth, taste, palate cleft   534, 540, 548, 569, 571, 574, 586, 795, 926, 988, 990, 1067, 1073, 1197, 1207, 1221f., 1293, 1357, 1369
mucociliary system, *see* ciliary, mucosa
mucopolysaccharide   1039, 1363
mucosa, mucus, mucous inflammation, slime, *see also* gastric mucous membrane   421, 486, 511f., 514f., 518, 520, 532f., 535, 539, 541, 543f., 571f., 574ff., 577, 581, 586, 589, 725, 727, 761f., 792, 794, 901, 935, 951ff., 963, 1000, 1058, 1061, 1063, 1067, 1109, 1136, 1146f., 1197, 1207, 1221, 1234f., 1295, 1324, 1350, 1356, 1360, 1363, 1367ff., 1370
mud, silt, *see also* marsh, sludge   36, 385, 392, 394f., 399, 885, 947, 1105, 1158, 1214, 1221
mule, *see* horse
multidentate complex, *see* chelating agent, complex
multielement analysis, *see also* AAS, ICP-ES, NAA, XRF, etc., fingerprints   125ff., 130 132f., 139, 143f., 149ff., 159ff., 162, 167, 169, 173, 176, 180ff., 187, 189, 191, 194f., 197, 199f., 203, 209, 273, 479, 843, 907, 919, 1356
municipal, *see* household garbage
murder, *see also* forensic, poisoning   600, 751
muscles (psoa), muscular, amyotrophic, *see also* meat, myoclonus   112f., 165, 212, 474, 494, 517, 535, 537, 540, 545, 563, 646, 717, 722f., 725, 728f., 757, 760ff., 778, 794, 796, 800, 817, 824, 847, 858, 867, 869, 886f., 896, 912, 923f., 954, 964, 967, 969f., 981, 995f., 1021, 1028f., 1031f., 1038, 1040, 1057ff., 1060, 1070, 1075, 1123, 1145, 1155, 1173f., 1194, 1196, 1206, 1217f., 1221, 1225, 1228, 1234, 1236, 1264, 1271, 1296, 1304, 1306, 1316f., 1322f., 1330, 1345, 1350f., 1354f., 1357, 1359f., 1363f., 1368, 1370f.
mushroom, *see* fungus
mussel, *see* shellfish, mollusc, oyster
mutagenicity, mutant, mutation, genotoxicity, *see also* DNA damage, gene, inherited syndrome, "in-vitro" systems and techniques, mammalian cell, mechanisms, reverse mutation, *Salmonella*, sister-chromatid exchange, short-term test, co-mutagen   107, 188, 287f., 303, 306, 308f., 421, 430, 435, 440, 446, **502f.**, 506, 508f., 513, 547, 583, **617ff.**, **620ff.**, 623, 626, 629, 631ff., 634ff., 637, 639f., 647, 649, 661, 683, 726f., 737, 747ff., 750, 763ff., 768ff., 771, 782, 785, 829, 837ff., 841, 847, 866ff., 869, 872, 875ff., 878, 888, 892, 915, 917f., 926, 942, 966, 1001f., 1014, 1041, 1043, 1071f., 1079, 1082, 1084, 1095f., 1114f., 1119f., 1123, 1131, 1145f., 1148, 1150, 1177f., 1183ff., 1187f., 1221, 1237, 1239, 1283, 1295, 1305, 1331, 1337, 1358, 1368
mycelium, *see* fungus
mycorrhizal, *see also* roots   759, 897f., 906, 1088, 1363
myelo ..., myeling, *see* marrow
myocardium, myocarditis, *see also* heart   763, 883, 887, 998, 1034, 1166, 1174, 1190, 1204, 1209, 1215, 1223, 1226, 1364
myoclonus, myoneural, *see also* ataxia, muscles   728, 796, 1065, 1364
myoglobins, myoglobinuria, myopathy   494, 542, 926, 949f., 1219, 1364, 1367

NADPH, *see* nicotine
nail, *see also* hoof, Mees line   112, 181, 217, 561f., 600, 749, 757f., 761, 885, 928, 934, 937, 992, 1061, 1064, 1108, 1176, 1228, 1235f., 1294f., 1316f., 1328, 1352, 1361, 1365
naloxone, nalidixic acid   427, 925
nasal, *see* nose
nausea, *see also* vomiting, emetic   552, 585, 589, 593, 762, 795, 827, 887, 901, 914, 926, 995, 1067, 1113, 1176, 1221, 1235, 1252, 1282, 1329f.
nebulization, nebulizer   134f., 147, 150f., 170, 183, 187, 191f., 195, 203f., 941, 943, 1213
neck, *see* pharynx, goiter, larynx
necrosis, *see also* hepatic alterations   362, 535, 559, 590, 643, 761, 796, 826, 902f., 914, 925, 1106, 1174f., 1187, 1196, 1200, 1218ff., 1324, 1364f.
needle, stomata, *see also* conifer   376, 849, 1030, 1036, 1039, 1056, 1058, 1088, 1277
nematode, threadworm, heartworm   754
neodymium   **959ff.**, 962f., 965f., 969f., 1304f.
neon, *see* noble (inert) gas
neoplasm, neoplastic lesion, neoplasia, *see also* tumor, carcinogenicity   578, 599, 606, 609, 628, 781, 836, 867, 875, 911, 925, 966, 969, 1113ff., 1126, 1178, 1305, 1345, 1361, 1364

nephritis, nephrosis, nephrotoxicity, nephropathy, *see also* kidney 592, 594, 598, 684, 796, 799, 828, 836, 840f., 909, 911, 914, 918, 935, 938, 998, 1067, 1071, 1077, 1080, 1112, 1124, 1147, 1173, 1282, 1364
neptunium 1275
nerve, nervous system, CNS, *see also* neuron, pyramidal cell 498, 505, 538ff., 542ff., 547, 552, 578f., 585f., 590, 592f., 656, 662, 729, 747, 762f., 796, 887, 964, 995ff., 1000, 1020, 1030, 1040, 1060, 1063ff., 1069f., 1073f., 1076, 1080, 185, 1101, 1146f., 1176, 1209, 1217, 1219f., 1222, 1225f., 1228, 1235ff., 1294, 1302, 1351, 1353ff., 1357ff., 1360, 1363ff., 1367ff., 1370f.
nerve conduction velocity, *see also* reflex 499, 662, 996, 1368
neuralgia, *see* neurotoxicity
neurochemistry, neurological examination, electrophysiology 498, 600, 993, 996, 1009, 1033, 1085
neurofibril, neurotransmitter, gliomas 304, 485, 498, 516, 729, 736, 738, 763, 996, 1001, 1015, 1019, 1065, 1304, 1364, 1370
neuron, *see also* Purkinje cell 729, 736, 738, 794, 965, 1060f., 1064ff., 1087, 1304, 1366, 1370
neurotoxicity, neurological effects, neuritis, neuropsychological deficits, *see also* behavior, psychic disturbance, paramnesia, peripheral, learning ability 83, 494, 516, 519f., 579f., 597ff., 600, 602f., 647f., 724, 728f., 761ff., 766, 771f., 789, 795ff., 800, 886ff., 902, 914, 921, 923, 926, 965, 967, 993, 995ff., 1007, 1009, 1012f., 1020, 1022, 1033, 1045, 1060f., 1064f., 1069ff., 1073, 1080f., 1084f., 1094, 1174, 1219f., 1224, 1236f., 1328f., 1364
neutron (beam, generation, absorber) 140, 151, 775, 777, 805, 809, 880, 960f., 1017, 1127, 1155, 1277, 1301, 1344, 1364
neutron activation analysis NAA 115f., 121ff., 133, **140ff.**, 151f., 154, 156, 159, 161, 173f., 177, 179, 181, 184, 202, 205, 209ff., 247, 272, 282, 296, 560, 563f., 716, 743, 749, 752, 773, 800, 805, 832, 848, 854f., 881, 894, 907, 919, 932, 938ff., 957, 960, 963, 970, 1035, 1047, 1049, 1089, 1125, 1137, 1154, 1156, 1191, 1200, 1204, 1212, 1223, 1229, 1240, 1244, 1261, 1269, 1275, 1289, 1300, 1310, 1343, 1360, 1364, 1367

neutrophil(e) 750, 964, 1326
newborn, neonate, *see also* infant, premature birth, litter 527, 579, 583, 593, 598, 608, 610, 644, 668, 681f., 766, 770, 824, 831, 863, 936, 965, 967, 971, 1003, 1008, 1011, 1039, 1041, 1064f., 1067, 1069, 1073, 1083, 1085f., 1092, 1115, 1130, 1133, 1181, 1197, 1221, 1327, 1330, 1355, 1363
nickel (general, elementary) 6ff., 11f., 15, 35, 37, 50, 52, 54, 57ff., 72, 89, 95, 107ff., 138, 157, 171f., 175, 177, 180, 193, 199f., 203, 207, 218, 233, 258, 284, 288ff., 301ff., 305ff., 328, 334, 347, 351, 358, 361, 367, 375f., 390, 392, 401, 403f., 409, 412ff., 421ff., 455, **462f.**, 467, **476**, 481f., 485, 493, 495, 501f., 507, **541f.**, 545, 552, 555, 557, 562, 571, 573, 579, 581f., 593, **610ff.**, 613ff., 625, 629, 631, 634f., 638f., 651, 657f., 662, 670, 682, 698, 740, 753f., 803, 808, 813f., 836, 844, 847, 849f., 860, 876, 883, 895, 904f., 945, 947, 961f., 1014, 1027, 1036, 1044, 1052, **1101ff.**, 1105ff., 1108ff., 1122, 1129, 1140f., 1144, 1215, 1262, 1277, 1290, 1301, 1324
nickel compounds, *see also* nickel sulfide, nickel tetracarbonyl 504, 507f., 541, 552, 593, 599, 605, 610, 621ff., 629, 636f., 661, 698, 711, 763, 850, 1101ff., **1104f.**, 1109, **1111ff., 1114ff.**, 1117f., 1125f.
nickel ions 79, 82, 224, 239, 321, 324f., 501ff., 506, 615, 618ff., 623, 628, 633, 639, 862, 1101ff., 1109ff., 1114, 1119f., 1123
nickel sulfide 502f., 507, 509, 610, 613, 622f., 629, 633, 670, **1101f.**, 1104, **1112f., 1115f.**, 1119f., 1122, 1124f., 1213
nickel (tetra)carbonyl 499, 509, 574, 610, 670, 1101, **1103ff.**, 1109f., **1113**, 1115f., 1118ff., 1122, **1124ff.**
nicotine, nicotinamide adenine dinucleotide, NADPH, *see also* tobacco 436, 439f., 1218, 1370
niobium 145, 149, 218, 557, 922, **1127ff.**, 1203ff., 1210, 1301
nitrate 73, 265, 275, 324, 341, 362, 369f., 372, 792, 795ff., 817, 880, 921, 1035, 1092, 1098, 1193, 1197, 1201, 1248, 1310, 1318
nitre, nitric acid 111, 118f., 146, 205, 224, 364, 717, 721, 837, 853, 856, 880, 921, 1004, 1127, 1145, 1149, 1155f.
nitrite, nitride 313, 364, 1264f.
nitrogen (fixation), nitrification, *see also* assimilation, donor, azotemia, steel 73, 76, 95, 98, 119f., 148, 157, 243, 347, 396,

415, 419, 421, 485, 569, 587, 721, 819, 884, 896, 914, 926, 931, 958, 992, 1092f., 1097ff., 1100, 1109, 1164, 1271f., 1293, 1295, 1299, 1301, 1318f., 1351
nitrogen oxides, nitrous oxide, *see also* smog   258, 292, 370ff., 721f., 830, 960, 1088, 1142, 1145, 1149, 1262, 1300
nitroso compounds (MNU, nitrosamines, nitrosonaphthol)   628, 830, 881, 1114
NMR, nuclear magnetic resonance, *see* magnet
noble (inert) gas, noble element, *see also* argon, helium   72, 235, 1135f., 1143, 1191, 1299
no-effect level, *see* threshold
noise, *see* background
non-histone   501
non-smoker, *see* tobacco (smoke)
North Sea   29, 103, 233, 253, 256, 268, 277, 352, 393, 396, 740, 843, 847, 1086, 1263, 1266
nose, nasal, nasopharyngeal, *see also* cough(ing)   262f., 574f., 610, 661, 761f., 822f., 865, 886, 901, 1067, 1069, 1113ff., 1117, 1122, 1176, 1280, 1295, 1355, 1368
NTA, nitrilotriacetic acid   97, 155, 249, 251, 389, 395, 397, 493, 503, 508f., 511, 513, 873, 877f., 897, 1364
nuclear charge, nuclear energy, nuclear transformation, *see also* spin   143, 181, 789, 838, 1017, 1285f., 1364
nuclear fuel, nuclear reactor, nuclear fallout   59, 140, 775, 777, 791, 809, 879f., 961, 1015, 1017, 1022, 1127, 1129, 1132f., 1135f., 1140, 1144, 1206, 1270, **1277f.**, 1284ff., 1287, 1300f., 1319, 1344, 1347
nuclear medicine   880, 909, 1159, 1225, 1277, 1368
nucleic acids, *see also* protein, DNA, ribosome, uracil   243, 481, 497, 611, 617f., 628, 632, 634, 771, 782, 785, 869, 1064f., 1080, 1194, 1199, 1304, 1325, 1349, 1353, 1357f., 1364
nucleotide, polynucleotide, nucleophilic center   513, 618, 629, 763, 878, 1064f., 1085f., 1194, 1199, 1287, 1304, 1319, 1331, 1364
nucleus, nucleation   3, 70, 235, 261, 264, 455, 779, 782, 794, 868, 1061, 1065, 1071, 1207, 1262, 1317, 1353, 1355f., 1358, 1363f., 1368
nuclide, *see* isotope
numbness, *see* paralysis

nursing mother, *see* lactation
nut, walnut, hazelnut   1028, 1038, 1107, 1165, 1326
nutrients, nutrition, *see also* autotrophic, breads, cooking, daily intake, fish, food, fungus, grass, meat, milk, seafood, vegetables, etc.   67, 94, 109, 158, 197, 207, 213, 215, 217, 219, 225, 228, 233, 270, 300, 357, 361, 374, 386, 392, 398, 407, 410, 415, 418, 453, 467, 482, 484, 486, 488f., 491ff., 496ff., 497f., 505ff., 511, 514, 518, 527, 531, 536, 538, 545, 550f., 554, 566, 598, 612, 656f., 659, 673, 683, 711, 720, 723, 736, 738ff., 756, 773, 779, 817, 828f., 835, 838, 846, 860, 865, 871, 873f., 883, 886, 889f., 892, 897, 905ff., 913, 923f., 949, 952f., 954f., 957f., 981f., 986, 988, 992f., 997, 1005f., 1010, 1013, 1021, 1033f., 1044, 1058, 1076, 1091ff., 1094ff., 1097ff., 1100, 1110, 1117f., 1121, 1123, 1125, 1129, 1153, 1157, 1165ff., 1169f., 1173ff., 1177, 1180, 1182, 1184, 1186ff., 1189, 1196ff., 1200, 1217ff., 1220, 1222, 1226, 1251ff., 1258, 1279, 1285, 1289, 1293ff., 1315, 1318ff., 1321f., 1326ff., 1329, 1336ff., 1340, 1349, 1354, 1363, 1366, 1370

oak, *see* deciduous tree
obstipation   593, 796, 995, 1235, 1365
occupational disease, *see* occupational medicine
occupational exposure, exposed workers, *see also* concentration at the workplace   107f., 111, 178, 201, 211, 218, 257, 262, 288, 303, 475, 502, 506, 527, 529, 547ff., 553ff., 571, 591, 593, 596ff., 599, 601ff., 624ff., 629, 631ff., 635, 637f., 645, 648, 651f., 654ff., 658f., 662, 664ff., 667f., 671, 674, 676f., 680ff., 683, 685, 692f., **694ff.**, 698, 709ff., 723, 727, 736, 739, 741, 745ff., 749, 753, 762, 764f., 767, 772, 781, 783, 785, 819, 823f., 829ff., 832, 840f., 846f., 861, 863, 865ff., 870, 874, 876, 878f., 884, 887ff., 891f., 906, 913, 915, 936, 942, 959, 963, 966, 970, 985, 989f., 996, 998ff., 1003, 1005ff., 1011, 1014, 1041f., 1045, 1048, 1055, 1058, 1072, 1075f., 1085, 1087, 1096, 1101, 1108ff., 1113, 1115ff., 1118, 1121f., 1126, 1136, 1169, 1176, 1179f., 1191, 1195, 1198, 1204, 1211, 1216, 1220, 1222, 1225, 1228, 1232f., 1238f., 1241, 1252, 1255, 1264ff., 1271ff., 1283ff., 1286f., 1294ff., 1331f., 1346

occupational medicine (protection), hygiene, see also risk assessment 106, 163, 284, 493, 548, 555, 594f., 651, 653, 655f., 658f., 667ff., 672ff., 675f., 681, 683f., 688, **692ff.**, 698, 710f., 723, 727, 733, 739, 747ff., 758, 761, 765, 772, 780, 782, 789, 798, 830, 833, 846, 862, 874, 884, 888, 891, 906, 923, 926, 928, 939, 967, 998, 1001, 1007f., 1012, 1014, 1022, 1040f., 1045, 1072, 1077, 1084, 1098, 1114ff., 1146, 1148, 1176, 1184, 1199, 1206, 1208, 1222ff., 1225, 1237, 1241, 1265, 1283, 1285, 1297, 1307, 1329, 1340, 1346

ocean, oceanography, see also seawater, offshore dumping 4, 8, 10, 12f., 29, 62f., 67f., 89f., 93f., 100ff., 108f., 115, 146, 155, 169, 188, 191, 233f., 241, 253, 255f., 258, 260, 268, 284f., 380, 393, 396f., 475, 517, 720, 745, 756, 773, 777, 785, 800, 817, 842, 846, 849, 879, 883, 896, 912, 924, 933, 940, 948, 950, 962, 970, 977, 1006, 1011, 1036, 1044, 1053f., 1057, 1081, 1084, 1086, 1099, 1123, 1126, 1130, 1133, 1138, 1143, 1145, 1151, 1160ff., 1163f., 1167, 1182, 1200, 1204, 1213, 1226, 1247f., 1270, 1273, 1291f., 1319f., 1341

oceanic crust 4, 6, 8, 773

octopus, cuttle fish, squid 472

ocular cornea, see cornea, eye

odor, smell, see also deodorant, garlic, intestinal wind 795f., 865, 876, 882, 921, 1041f., 1069, 1167, 1175f., 1217, 1221f., 1230, 1328, 1350, 1354, 1360

offshore dumping, waters 29, 109, 471, 1263

offspring, see newborn

oil, edible oil, see also fuel oil, crude oil 477, 486, 850, 1104, 1107, 1246, 1292

old, elderly, see age

oliguria, uremia 540, 590, 725, 733, 736, 738, 762, 901, 1220, 1329, 1365

oncogene, oncology 501, 503, 505, 509, 568, 614, 628, 782, 822, 829, 917, 919

onion, see vegetables

ontogeny, see germination

optical atomic spectroscopy, see AAS, AFS, AES

optical information, memories, lenses, properties, see also data bank 744, 911, 922, 1214, 1225

oral (uptake), ingestion, see also mouth 299f., 303, 472, 504, 514, 531ff., 534ff., 537, 542, 544, 546, 548f., 555, 559, 569, **571ff.**, 585, 588ff., 606, 646, 649, 662, 667, 673, 676ff., 679f., 724, 728, 767f., 772, 775, 779f., 784, 789, 792, 794ff., 797ff., 803, 822f., 827ff., 830, 832, 835, 841, 851, 865, 884ff., 902, 922, 924f., 927, 929, 934f., 938, 954, 957, 959, 963, 965f., 971, 986, 988, 990, 1001, 1004, 1012, 1030ff., 1033, 1066ff., 1075, 1087, 1095, 1100, 1108f., 1111, 1114f., 1117f., 1121, 1129ff., 1160, 1169ff., 1172, 1175, 1180, 1194ff., 1198, 1203, 1206f., 1216ff., 1219, 1221ff., 1234, 1236f., 1252f., 1271, 1282, 1284, 1286, 1293, 1299, 1302f., 1305, 1322f., 1326, 1328, 1345f., 1354, 1362, 1365, 1366

ordinance, decree, see regulation, threshold

ore deposit, see mineral deposit

ore processing, see also mining, metal processing **19ff.** 59, 259, 342, 596, 670, 749, 752, 759, 775, 807, 814, 856, 881, 894, 896, 912, 947, 973ff., 978, 1035, 1037, 1040, 1050, 1054, 1067, 1090, 1104, 1123, 1140f., 1164, 1192, 1203, 1214, 1227f., 1230f., 1245, 1262, 1270, 1277, 1290, 1311ff.

organ, see target (critical) organ; and the respective organs

organelle 512, 787, 794, 928, 1027, 1039, 1319, 1363, 1365, 1367, 1370

organic (matter) waste, organic layer, oil, grease, tar, see also dissolved organic matter DOM, fuel oil, matrix 41ff., 52, 311f., 314, 318, 323f., 329, 333, 338f., 361, 370ff., 376f., 395, 399, 403, 418, 428, 809, 899, 946, 979f., 994, 1037, 1144, 1154, 1200, 1228, 1292, 1295, 1297, 1318, 1354

organoaluminum compounds 720, 722f.

organoarsenic(al) compounds 99, 119, 156f., 159, 173, 175, 190, 193, 245, 424, 441, 471, 580, 661, 751f., 754, **756ff.**, 759, 765ff., 772, 1252f.

organochlorinated compounds, see chlorinated (organic) compounds

organolead compounds, see also methylated and ethylated lead compounds 157, 159, 171, 175, 431f., 474, 547, 580, 582, 608, 972, 978, **999ff.**, 1008, 1011f., **1014**, 1252, 1255

organomercury compounds, see also methylmercury, seed treatment 119, 157, 169, 186, 193, 423, 431, 433ff., **438f.**, 448, 461, 475, 517, 532, 540, 572, 585f., 597, 610, 630f., 634f., 641, 649, 657, 661, 680, 698, 1045, 1047f., **1051ff.**, 1055, 1057, 1061ff., 1064, **1068f.**, 1072, 1075, 1077, 1082, 1086ff., 1252, 1259 organometal compounds, metal-carbon compounds 83, 99, 119, 148, 153, 157, 160, 173, 182, 245, 393, 444, 507, 571f., 574, 577, 591, 743ff., 749f., 775, 910, 921f., 925ff., 928, 955, 1017, 1034, 1037, 1096, 1150, 1217, 1225, 1241, 1255, 1263, 1265, 1312

organophosphate, organophosphine 587, 882, 1201

organoselenide compounds 159, 175, 279, 424, 1154ff., 1159, 1161ff., **1165ff.**, **1168**, **1170ff.**, 1173f., 1177f., 1181, 1183, 1188f., 1259

organothallium compounds 245, **1230**, 1233, 1241

organotin compounds 153, 157, 159, 173, 176, 183, 188, 198f., 245, 410, 543, 580, 591, 1243ff., **1246ff.**, **1251ff.**, 1254ff., **1257ff.**

orthopedic(s), *see* implant

osmium, osmium tetroxide 1136f., **1139**, 1141, **1146ff.**

osmosis, osmotic 43, 47, 53, 731f., 735, 914, 994

osteoporosis, osteomalacia, osteosynthesis, osteodistrophy 488f., 494f., 534, 557, 578, 583, 596, 598, 715, **728ff.**, 731ff., 734, 739, 796, 829, 911f., 1039, 1330, 1365

ovary, ovary cell, oocyte, ovulation, *see also* germ cell, egg 535, 635, 770, 912, 923, 927f., 937, 949, 1002, 1031, 1072, 1084, 1167, 1177, 1179, 1302, 1357, 1360, 1367

oxalate, oxalic acid 67, 78, 85, 400, 420, 861, 1031, 1299

oxidase, monooxygenase, *see also* xanthine oxidase 442, 580, 626, 899, 1022, 1092f., 1097f., 1112, 1125f., 1188, 1218, 1253, 1295

oxidation, oxidizing, ozone, oxide, oxime, *see also* aerobic conditions, lipid peroxidation 34f., 42, 52, 55, 57f., 70, 80, 86, 155, 235ff., 263, 311f., 314, 318, 321, 323f., 330, 336, 344, 346, 373, 379, 384ff., 387f., 390ff., 425f., 428, 430, 443, 452, 471, 485, 492, 503, 529, 532f., 535, 551, 580, 588f., 719, 721, 736, 743, 746f., 751, 798, 804, 830, 847, 853ff., 856ff., 859, 861f., 864, 869, 872, 876, 880, 882f., 898, 901, 909, 931ff., 934f., 938f., 945ff., 948, 972, 975f., 1026f., 1035, 1038, 1049, 1060f., 1086, 1091f., 1102ff., 1110, 1112, 1127, 1136ff., 1141f., 1144, 1149, 1154, 1158ff., 1161f., 1164, 1191, 1199, 1212, 1214, 1225, 1228f., 1248f., 1253, 1261f., 1269f., 1275, 1277, 1279, 1289ff., 1292, 1300f., 1310, 1325, 1343, 1369

oxide layer, surface oxidation, *see also* protective layer 34, 558, 564, 716, 1205

oxygen, anoxemia, oxyanion, *see also* donor 3, 34, 36, 39f., 73, 76, 95, 98, 128, 137, 140, 157, 160, 236, 243, 312, 339, 344, 421, 497, 503, 506, 509, 512f., 587, 619, 626, 632, 854, 856, 880, 887, 909, 949, 961, 972, 975, 1025, 1036f., 1063f., 1087, 1114, 1144, 1157, 1212, 1214, 1292f., 1299, 1349ff., 1353, 1358, 1360ff., 1364

oyster 57, 181, 200, 472ff., 478, 746, 756, 760, 800, 1107, 1316, 1319, 1324

ozone, *see* oxidation

P-450 induction, *see* cytochrome

Pacific, *see* ocean

packaging, labelling 450, 459, 687, 692, 699, 709, 719, 1048, 1162, 1249f., 1291

PAHs, *see* polynuclear (polycyclic) aromatic hydrocarbons

paint, color, dye, *see also* pica, pigment, protective layer 38, 59, 297, 299, 699, 720, 754, 792, 806, 857, 876, 881f., 895, 961, 975, 986f., 989, 991, 995, 1005, 1051f., 1101, 1136, 1159, 1229, 1246, 1248, 1252, 1263, 1266, 1270, 1277, 1283, 1311, 1344

palate cleft 1072, 1114, 1364

palladium, palladium compounds 19, 72, 83, 128, 131, 138, 179, 181, 196, 266, 270, 279, 918, 1135ff. **1139ff.**, **1142ff.**, 1145f., 1149, 1213, 1308

pancreas, pancreatic, pancreatis 484, 494, 511, 519, 544, 581, 906, 923, 963, 1039, 1108, 1112, 1174, 1201, 1317, 1322, 1328, 1330, 1356, 1365

paper, *see also* packaging, cellulose 720, 1162, 1186, 1214, 1263, 1312

parakeratosis, *see* keratin

paralysis, paresis, dazed feeling, *see also* tremor 538, 540, 543f., 597, 761f., 887, 926, 995f., 1021, 1131, 1175f., 1219f., 1236f., 1330, 1365

paramnesia, memory weakness, I.Q. deficit, *see also* behavior, learning ability, neurotoxicity, psyche 552, 645, 685, 727f., 733, 735ff., 738f., 762, 886, **995ff.**, 1003, 1010f., 1015, 1068f., 1073, 1083, 1126, 1176, 1283, 1287, 1328, 1350, 1353

parasite, parasitic disease, *see also* germ of a disease, tropical medicine 743, 745, 1063

paraenchyma 533ff., 646, 779, 786, 869, 902, 954, 1358, 1365
paresthesia 594, 597, 761, 886f., 995, 1069, 1365
Parkinsonism, Parkinson's disease 598, 729, 736, 738, 1040f.
particle, particulate phase (in waters), particulate fraction, *see also* dispersion 12, 39, 68ff., 84, 87, 92f., 96, 101f., 107, 111, 119f., 153, 157ff., 200, 205, 228, 245, 256ff., 262, 265f., 272f., 276, 281, 283, 287, 289, 291, 294f., 297, 303f., 306ff., 353, 374, 379, 383, 387f., 395, 397ff., 405, 431, 433, 574f., 583, 586, 633, 676, 755, 757, 779, 781, 786f., 816f., 822, 842, 848, 861f., 874f., 894, 896, 907, 912f., 921, 941, 947, 955, 979, 1054, 1105, 1126, 1130f., 1144f., 1150, 1159, 1207ff., 1280, 1292f., 1295f., 1302, 1363f., 1366
particle class balance (PCB), computer controlled scanning electron microscopy CCSEM, for automated single particle microscopy 273
particle induced X-ray emission, *see* PIXE
particle size, *see also* size distribution 21, **261ff.**, 266f., 287, 292, 296, 320, 384, 405, 574ff., 589, 759, 762, 773, 822f., 955, 978, 980, 988, 1042,, 1130, 1206, 1302, 1314, 1329
partitioning, *see* compartmentalization, distribution, extraction
passivation, *see* oxide layer, protective layer
pasture, *see* meadow
pathway, *see* cycle, transport mechanism
paunch, *see* rumen
PCBs, *see* chlorinated (organic) compounds (or particle class balance)
pea, *see* legume(n)
pear, *see* fruit
peat, *see also* marsh 231, 270, 278, 280, 282, 321f., 327, 338, 412, 431, 980, 992, 1013, 1094, 1313
(anti) pemphigus activity, *see* bladder
penicillin, penicillamine, penicillinase, *see also* antibiotic 427, 430, 496, 537, 832, 900, 902f., 1075, 1086, 1094
pepper, spices 858, 1107
peptide, polypeptide, peptone 420f., 523ff., 526, 863, 938, 968, 1015, 1102, 1109, 1118, 1120, 1125, 1325, 1357
peracetic acid 1056
perch, *see* fish
perchloric acid, perchlorate 136, 837, 1156

performance, *see* behavior, learning ability
periodic table, system 72, **74ff.**, **234ff.**, 237, 243, 481, 565, 588, 743, 751, 775, 789, 804, 909, 921, 939, 959, 971, 1015f., 1025, 1035, 1127, 1135f., 1149, 1151, 1153, 1191, 1203, 1211, 1228, 1261, 1269, 1289, 1300, 1309, 1343, 1352
periphera (neuralgia, neuritis) 538, 540, 599, 761f., 995f., 1235
peritoneal bleeding, cavity, *see* hemorrhagic
peritonitis, *see* colic
permanganate 589, 1035, 1049
permeability, permease, *see also* cell wall, membrane, solubility 300, 347, 406, 423, 435, 496, 498, 517, 577f., 727, 769, 859, 862, 901, 908, 913, 964, 1016, 1039, 1066, 1126, 1220, 1230, 1301, 1341
peroxide, peroxidation, peroxidase, *see also* lipids, glutathione 390, 485, 497, 503, 513f., 519, 529, 921, 946, 949, 954, 1039, 1102, 1111, 1120, 1124, 1153, 1166, 1173, 1178, 1181, 1188, 1196, 1201, 1218, 1325
persistence, *see* accumulation, depot formation
perspiration, sweat, exudate 110, 405, 484, 720, 826, 884, 901, 953, 992, 1061, 1110, 1117, 1171, 1217, 1221f., 1230, 1236, 1322, 1328f., 1344, 1346
perturbation 89, 915
pest control, pesticide, *see also* biocide, fungicide, insecticide, miticide, mollusc, rodenticide 59, 307, 442, 447, 467, 470, 547, 641, 699, 751, 754ff., 757, 765, 981, 1078, 1086, 1240, 1247, 1249, 1312
petrol, *see* gasoline
petroleum, petrochemistry, *see also* cruide oil, fuel, oil, refinery 51, 276, 292, 337, 791, 961, 1051, 1053, 1090, 1095, 1103ff., 1106, 1125, 1142, 1215, 1297
pewter, *see* hard-solder
pH (value) 97, 159, 234, 243, 245, 311ff., 315, 318, 321ff., 325, 330, 338f., 341, 345ff., 351, 372ff., 383, 386f., 389, 393f., 403f., 410, 419, 423, 432, 453, 471, 493, 526, 589, 622, 660, 720, 722, 726f., 733, 736, 787, 855, 864, 883, 896, 913, 919, 946, 953, 981, 1038, 1164, 1167, 1182, 1192, 1248, 1263, 1302, 1318
phagocyte, phagocytosis, *see also* killer cell 561, 576, 608ff., 611, 613, 621, 623, 633, 780f., 1001, 1303, 1330, 1362, 1366, 1368
pharmaceutics, *see* drug

pharmacokinetics, *see* kinetics, mechanisms, metabolism
pharyngitis, *see* pharynx
pharynx, intratracheal (instillation), *see also* tracheobronical   540, 574f., 593, 611, 746, 749, 761, 823, 830, 863, 914, 917, 919, 940f., 944, 965, 1067, 1114f., 1122, 1147, 1176, 1271, 1280, 1295, 1304, 1306, 1329, 1354, 1360f., 1364, 1366, 1370
phenylmercury, *see* organomercury compounds
phlegm, *see* mucus, mucous membrane
phloem   1318
phobia, fear, *see* shock
phosphatase   914, 1039, 1095, 1111f., 1120, 1218, 1303f., 1325, 1327
phosphate (in function of carrier; complexing agent), *see also* dithiophosphate, organophosphate   85, 95, 158, 314, 326, 329, 421f., 426, 430, 494, 496, 513, 515f., 551, 618, 715f., 722, 726ff., 730, 734, 758f., 763, 779, 784, 828, 832, 859f., 868, 898, 948f., 951ff., 956, 967, 998, 1029, 1090f., 1173, 1194, 1282, 1294, 1304, 1310, 1313, 1326, 1360, 1364
phosphate (fertilizer, detergent, rock)   19ff., 54ff., 61, 77, 233, 259, 290, 357, 389, 407, 453, 471, 708, 757, 806, 810, 814, 817, 898, 961, 1090, 1158, 1264, 1266, 1277f.
phosphaturia, hyperphosphaturia   596, 998
phosphide, phosphine, *see also* organophosphate   910, 931f., 934, 1141, 1194
phospholipid, phosphoprotein   1111, 1113, 1120, 1294, 1304
phosphor, *see* luminous substance
phosphoric acid   146, 1150, 1364
phosphorite   1090
phosphorus (elementary, general)   93, 145, 347, 418, 430, 481, 532, 537f., 726, 735, 738, 759, 764, 771, 780, 828, 895, 907, 910, 1031, 1097f., 1304
phosphoryl group, phosphorylation   422, 782, 1253
photoacoustic spectroscopy   152
photocell, photoelectric cell, *see* photochemistry, solar cell
photochemistry, photoelectric cell, photovoltaic, photodecomposition, *see also* assimilation, semiconductor   56, 92, 223, 246, 436, 587, 723, 826, 895, 911, 931, 940, 992, 999f., 1025, 1027f., 1033, 1088, 1158f., 1214, 1216, 1235, 1249, 1352

photographic materials, photoengraving, cameras   19, 857, 966, 1017, 1020, 1052, 1136, 1192f., 1205, 1214ff., 1277
photolysis, *see* photochemistry
photometry, *see also* colorimetry   854f., 881, 928, 1036, 1128, 1229, 1269, 1289, 1300
photon, light quantum   140, 1103
photostat, photocopy, *see* reproduction technique
photosynthesis, *see* photochemistry, assimilation
phytate, *see also* fibrogenous   496, 516, 518, 566, 949, 1170, 1321, 1328, 1366
phytoplankton, *see* plankton
phytotoxicity   361f., 410f., 455, 758f., 1058, 1125, 1252, 1312, 1366
piazselenol   1155
pica, *see also* paint, hand   988f.
pigeon, *see* bird
pig, piglet, little pig, *see* swine
pigment, *see also* paint   19, 51, 333, 532, 596, 633, 641, 700, 754, 803, 808f., 814, 854, 856f., 866, 881f., 895, 946, 975, 1012, 1027, 1037, 1040, 1090, 1094, 1104, 1159, 1245, 1261ff., 1264ff., 1270, 1293, 1311f., 1344, 1349, 1352, 1358, 1367
pigmentation chromogenesis, *see* skin pigmentation, hypochromic, albinism
pike, *see* predatory fish
pine, fir, *see* conifer
pipe, water pipe, tube, drainage, tab water   56f., 63, 288, 342, 347, 391f., 399ff., 403f., 450, 492, 701, 720, 800, 816, 820, 827, 834, 873, 877, 895f., 901, 923, 976, 979, 982f., 986f., 991, 1047, 1075, 1093, 1095, 1105, 1116, 1128f., 1163f., 1167, 1250, 1262, 1277, 1290, 1299, 1304, 1313f., 1326, 1354
PIXE (particle induced X-ray emission)   143, 187, 210, 272, 894, 905, 1366
placenta, placental barrier, transplacental, cord   112, 553, **579**, 582f., 614, 657, 681, 685, 766, 768ff., 772, 824, 863, 933, 936, 944, 962, 969, 989, 1002, 1010, 1039, 1073f., 1084, 1087, 1092, 1100, 1129, 1132, 1217, 1271, 1295, 1297, 1303, 1322
plankton, *see also* microorganisms, aquatic organism   92, 95, 97, 101, 380, 410, 471, 517, 756, 758, 817, 849, 884, 1092, 1163, 1259, 1317, 1356
plants, plant uptake, *see also* algae, aquatic plant, biomass cane, canopy, conifer, clover, deciduous tree, flower, grain, grass,

legume(n), phytotoxicity, tundra, roots, vegetables, wine, etc. 56, 58, 97, 107f., 112, 158f., 180, 197, 221, 223, 226ff., 267, 278, 285, 316, 318, 324, 328, 330, 357, 360ff., 365ff., 370f., 374, 378, **399ff.**, 403f., 406f., 409f., **411ff.**, 414, 416, 429, 431, 442, 449f., 452f., 455, 458, 460ff., 463f., 467, 470f., 473ff., 476f., 481f., 484, 487f., 491, 501, 523, 527, 529, 534, 538ff., 541ff., 544, **641f.**, 648, 707, 715, 721ff., 726, 732, 737f., 745ff., 756ff., 760, 771, 773f., 778ff., 784, 789, 793ff., 803f., 818f., 821, 825f., 834, 836f., 840f., 853, 858ff., 864, 870, 872ff., 876ff., 884f., 895ff., 898, 905, 908, 912, 923, 928, 931ff., 935ff., 938, 940f., 951, 953, 955, 957f., 963, 978, 981, 987f., 992f., 999, 1006f., 1010ff., 1016, 1018, 1025, 1027, 1029ff., 1034f., 1038ff., 1056, 1058, 1062, 1069, 1072, 1089, 1091ff., 1095, 1099ff., 1105f., 1108, 1110, 1118f., 1123, 1125, 1129, 1131, 1145, 1153, 1156, 1160, 1162, 1164f., 1167f., 1172, 1174, 1179, 1186, 1189, 1191, 1193, 1195, 1200, 1211, 1216, 1227f., 1231, 1233, 1235, 1239ff., 1250ff., 1264, 1271, 1278f., 1282, 1286, 1292ff., 1297, 1302f., 1307, 1309, 1312f., 1315, 1317f., 1321, 1323, 1325f., 1331f., 1336, 1342, 1345ff., 1351f., 1362, 1366, 1371

plasm, plasma (biologic), *see also* blood, cytoplasm 109, 156, 195, 209, 212, 216, 219, 512ff., 515f., 518ff., 551, 562, 567, 571, 577f., 580f., 611, 644f., 647, 649, 666, 717, 722, 728, 730ff., 735, 740, 772, 779, 784, 794, 862, 885, 887, 901f., 912, 924, 943, 949, 953, 960, 968, 991, 1020, 1022, 1028, 1030, 1032, 1062, 1094, 1096f., 1103, 1108f., 1111, 1116, 1118, 1157, 1167, 1173, 1178, 1316f., 1322, 1325, 1327, 1349, 1357, 1359, 1366f.

plasma (physicochemical), *see also* inductively coupled plasma-optical emission spectrometry 107, 116, 132f., 135, 147f., 182, 737, 790, 805, 854, 877, 1022, 1137, 1186, 1229, 1343, 1353, 1363, 1367

plasma emission spectrometry, *see* inductively coupled plasma-optical emission spectrometry

plasma protein, *see* plasma, carrier (protein)

plasmid, plasmin 422, 424f., 427, 430f., 433f., 438ff., 447f., 508, 649, 819, 838, 1055, 1109, 1226, 1366

plaster, *see* gypsum

plastic (materials), *see also* PVC, packaging 54, 111f., 115, 138, 149f., 199, 201f., 324, 327, 334, 561f., 700, 719f., 731, 792, 803, 813, 882, 970, 976, 1036, 1048, 1075, 1080, 1104, 1128, 1132, 1159, 1223, 1245, 1263, 1291, 1301, 1311, 1366

plating, *see* electroplating

platinum, platinum metals (general, elementary) 6f., 11, 15, 19, 72, 83, 128, 138, 144, 147, 179, 202, 266, 279, 631, 938, 961, 1049, **1135ff.**, 1139ff., **1142ff.**, 1145f., 1149, 1151

platinum compounds, complex (inorganic), *cis*-platin 527f., 568ff., 593, 618, 620ff., 625, 635ff., 638, 925, 928, 1136f., **1141ff.**, 1145, 1147f., 1151, 1173

platinum compounds (organic), carboplatinum 245, 625, 1142f., 1146f., 1151

plowland, topsoil, *see* agriculture, fertilizer

plume, *see also* tailing, feather 31, 52, 233

plutonium 626, 1135, 1275, 1286

pneumonia, pneumonitis, pneumonconiosis, tuberculosis 586, 747f., 754, 780f., 911, 915, 941, 965, 969f., 1040, 1066, 1084, 1101, 1113, 1206, 1208, 1329, 1366

pneumophile, *see* atmophile

point (reverse) mutation, *see* mutagenicity

poisoning, *see also* contamination, forensic..., cell poison 107, 111, 156, 162, 304, 466, 476, 489, 527, 532ff., 536ff., 539ff., 543f., 554, 565, 582f., **585ff.**, **588ff.**, 592, 595, 597, 599ff., 641, 644, 648, 651, 685, 700, 751, 754, 758, 761ff., 768f., 772, 783, 787, 832, 839, 870, 875f., 885, 900f., 939, 955, 957, 994f., 998, 1000f., 1010, 1021, 1038, 1041, 1045, 1052, 1057, 1061, 1064ff., 1067ff., 1074f., 1078ff., 1081ff., 1085, 1088, 1098, 1100, 1113, 1124ff., 1145f., 1153, 1176, 1195, 1201, 1219ff., 1222, 1224f., 1227, 1229f., 1235, 1237ff., 1240f., 1243, 1252ff., 1256, 1272, 1282, 1304f., 1312, 1316, 1324, 1327, 1329, 1340, 1346f., 1356

poison law (Swiss Federal) 700, 783

polar ice, polar region, *see also* ice 114, 188, 233, 257f., 260f., 268ff., 274, 276ff., 279ff., 408, 745, 815, 817, 842, 847, 980, 1006, 1105, 1107, 1121

polarizability, polarization 72, 238, 587, 837

polarography, *see also* differential pulse polarography, potentiometry 107, **136ff.**, 170, 181, 188, 206, 247ff., 252, 743, 749, 844, 854f., 893, 932, 939, 1089, 1154, 1212,

1241, 1269, 1289, 1354
pollen, *see* germ cell
pollution, *see* contamination
polonium   235, 290, 1211, 1276, 1279, 1281, 1283f.
polycythemia, *see also* blood cell, erythrocyte   536, 887, 1367
polyethylene, *see* plastic (material)
polymers, *see* plastic (material)
polynuclear (polycyclic) aromatic hydrocarbons PAH, *see also* benzo(a)pyrene   285, 292ff., 303, 634, 741, 830, 882, 1093, 1178f., 1183ff.
polynucleotide, *see* nucleotide
polypeptide, *see* peptide
polyphosphate, *see* phosphate
polysaccharide, *see* carbohydrate, sugar
polyuria   534, 540, 1367
pond   26, 28, 50, 55, 341f., 897, 983, 1054
population group, *see* risk population
porcelain, China, *see also* ceramics   557, 882, 961, 1017
pore (water), *see also* groundwater, soil solubility   5, 151, 154, 341, **381ff.**, 384, 392f., 492, 577, 738, 1007
pork, *see also* swine   471, 475, 834, 940, 1077, 1194, 1232, 1292, 1345
porphyrine, porphyria, porphobilinogen   430, 499, 538f., 646, 648, 657, 662, 729, 735, 880, 882, 915, 949, 993f., 1027, 1039f., 1090, 1290, 1322, 1349, 1358, 1367
port, habor   29, 62, 336, 349f., 385, 387, 389f., 392f., 667, 1248, 1259
postnatal, *see* newborn, neonate
potable water, *see* drinking water
potassium permanganate, *see* permanganate
potato, *see* vegetables
potentiometry, potential, *see also* polarography   139f., 151, 161, 168, 178, 181f., 192, 197, 245f., 248, 587, 804, 1261, 1266, 1367
pottery, *see* ceramics
poultry, fowl, *see also* chicken   455, 471, 475, 478, 532, 534f., 537ff., 540f., 543f., 753, 756, 820, 895, 952, 1099, 1166, 1174, 1196, 1321
power plant (based on coal, oil), *see* firing
praseodymium   145, **959ff.**, 963, 965f., 1307
precipitation, precipitant, *see also* excretion, electrostatic filter, electroplating   5, 22, 42f., 45ff., 53f., 61, 80, 139, 159, 236, 240, 316, 326, 329, 341, 347f., 351, 380f., 383, 386, 388, 391, 393f., 400, 422, 424, 492, 516, 568, 575f., 622f., 721, 726, 729f., 734, 773, 813, 818, 837, 854, 856, 875, 879, 895f., 948, 1054, 1063f., 1095, 1128, 1137, 1158, 1163, 1214, 1270, 1282, 1290, 1302, 1304, 1314
precipitation, *see also* deposition, rain, snow   158, 278, 311, 315, 323, 338, 346, 359, 370f., 374, 376, 378, 386f., 749, 755, 815f., 818, 833, 948, 1004, 1007, 1054, 1079, 1084, 1105, 1161, 1182
precision   129f., 135, 143ff., 148, 163, 167, 180, 184, 190, 202, 206, 211ff., 247, 275, 835, 868, 973, 1049, 1103, 1213, 1228, 1300
preconcentration, *see also* accumulation   120, 131, 134, 143ff., 149f., 153, 184, 190, 199, 201, 206, 246, 346, 790, 798, 1050, 1128, 1156, 1212, 1224, 1228
predatory fish   475, 923, 1057, 1061f., 1075, 1078, 1320
pregnancy, pregnant women, *see also* child bearing (age)   498, 504, 544, 555, 578, 582, 632f., 644, 656ff., 662, 664, 668, 765f., 863, 888, 913, 942, 953, 966f., 971, 989, 1002f., 1010, 1040, 1067, 1073ff., 1076, 1079, 1083f., 1092, 1094f., 1115, 1121, 1130, 1170, 1181, 1221f., 1273, 1303, 1322, 1327ff., 1367
premature birth (delivery), *see also* fetal loss   494, 736
prenatal, *see* fetus
preserved food, *see* can
preserving agent, preservative, *see also* food additive   533, 952, 1047
pressure (disintegration), *see also* wet digestion   3, 5, 150f., 184, 196, 205, 1359
preterm, *see* premature birth
prevention, pretreatment, *see also* preserve...   **346ff.**, 385, 569, **594f.**, 597, 600f., 612, 665, 668, 675, 688, 698, 739, 754, 1083, 1126, 1166, 1175, 1178, 1180, 1183f., 1196f., 1278, 1332
prickly (tingling) sensation, *see* tremor
primary urine, *see* urine
primate, *see also* monkey   613, 822, 1041, 1081, 1209
printing trade, printing process, *see also* letter, reproduction technique   857, 966, 1245
production, processing, *see also* mining, ore processing, metal processing   615, 689, 718, 741, 744, 753, 776f., 790, 803, 806f., 810, 812, 855, 881, 894, 910, 922, 932, 939, 947, 961, 974f., 1016, 1026, 1036, 1050f., 1055, 1090, 1104, 1127f., 1138, 1140f.,

1144f., 1147, 1149, 1158, 1162, 1176, 1192f., 1204, 1214, 1229ff., 1245, 1249, 1257, 1262, 1270, 1277, 1290, 1300, 1306f., 1309, 1311f., 1344, 1368
projectile, *see* bullet
prokaryotic organisms   422, 523, 628, 1098, 1367
proliferation, *see* neoplasm, cell physiology
promethium   **960f.**, 1307
promoter, *see* co-carcinogen
prophage induction   619f., 637, 1367
prostate, prostate cancer   549, 664, 830, 845, 923, 927, 1178, 1316
prosthesis, *see* implant
protease   421, 1116
protection, *see* safety
protective layer, coating, *see also* oxide layer   14, 272, 333, 345, 558, 563f., 700, 716f., 719f., 775, 803, 808f., 853, 856, 945, 961, 975, 981, 1005, 1026, 1037, 1090, 1143, 1205, 1245, 1248, 1258, 1263, 1277, 1295, 1310f.
protein, protein synthesis, *see also* carrier protein, finger-loop protein, lipoprotein, metallothionein, non-histone, purine bases, sulfhydryl groups, etc.   95, 156f., 300, 365, 369, 414, 422f., 430, 435, 440f., 447f, 481, 484f., 494, 497ff., 500f., 507f., 511ff., 516f., 519f., 523ff., 527ff., 537, 540ff., 543, 545f., 551, 554, 566, 571, 574, 577ff., 581, 587, 592, 600, 609, 611, 618, 632, 634, 641, 643f., 660, 717, 726, 734, 754, 758, 760, 763, 779, 781f., 785, 790, 796, 803, 805, 819, 822ff., 826ff., 829, 832, 836, 840, 847, 851, 854, 862f., 877, 887, 892f., 898ff., 901, 905, 914, 924f., 934f., 938, 941, 949, 952f., 955, 957, 962, 964, 966, 969, 991, 998, 1015, 1063ff., 1066, 1079, 1094, 1098f., 1109, 1114, 1126, 1155, 1168, 1170, 1172f., 1194, 1197, 1201, 1217, 1219, 1222, 1279, 1282, 1293, 1300, 1302ff., 1316, 1319, 1321f., 1325ff., 1328, 1349, 1352, 1355ff., 1358, 1361ff., 1364, 1367, 1369
protein fraction   156, 158, 472, 474, 823
proteinuria   535, 555, 594, 596f., 599f., 831, 1071, 1112, 1118, 1367
prothrombin   1040
proton, protonation, *see also* donor, synchrotron   67, 72, 84, 95, 140, 143, 210, 235, 238, 244f., 321, 342, 346, 369, 372f., 401, 721, 910, 1049, 1082, 1296, 1350, 1364
protoplasm, protoporphyrin   474, 993f., 1353
protozoon, *see* microorganisms

provision, *see* prevention, preserve ...
Prussian blue, *see also* ferrocyanide   543, 581, 946, 957, 1234, 1237
Pseudomonas   432f., 438, 501, 641, 1063, 1197, 1250
psoriasis, *see* scale, skin (pigmentation)
psyche, psychic disturbance, psycho-depression, antidepressive, psychological, *see also* behavior, hyperesthesia   542, 552, 598, 602f., 631, 654, 761, 800, 902, 971, 994ff., 997, 1000, 1014ff., 1020ff., 1023, 1031, 1040, 1068, 1175f., 1236f., 1304, 1328f.
psychomotoric (disorders), psychometric   540, 590, 994ff., 997, 1039f., 1065, 1073, 1363, 1369
puberty, *see also* developmental   494, 1328
pulmonary, pulmonic, *see* atmophile, lung
pulmonary cancer, *see* lung cancer
pulmonary edema, emphysema, *see also* lung, edema, emphysema   589, 780, 828, 830, 887f., 965, 1113, 1118, 1175, 1204, 1206, 1230, 1267, 1329, 1346
pulmonary fibrosis, *see* fibrogenous
(pulse) beat, pulsation, *see* arrhythmia
pulsed amperometry, polarography, voltammetry   149, 893, 1229
purification (sewage plant), *see also* sewage, sludge, clean-up, self-cleaning   357, 365, 720, 865, 961, 981, 987, 1128, 1193, 1198, 1262f., 1299, 1301, 1312, 1314f.
purine (bases), *see also* guanine   430, 569, 1065, 1093, 1194, 1331, 1349, 1357, 1364
Purkinje cell   1060f., 1367
PVC, *see also* plastic, vinyl chloride   348, 809, 1246f., 1249f.
pyelography, pyelonephritis   911, 1221, 1367
pyramidal cell, *see also* cerebellum   1069
pyridine, pyrimidine   240, 430, 569, 628, 1065, 1093, 1194, 1319, 1353, 1364
pyrite, pyrites cinder   56f., 542, 753, 881, 947, 1213f., 1227f., 1231
pyrolysis, pyrolytic   131, 291, 342, 348, 354, 368, 717, 854, 908
pyrometallurgy   35, 807, 940, 1090, 1137, 1140f.
pyrotechnics, pyrophorics, *see* fireworks

quality control, quality assurance, *see also* reference material, interlaboratory studies   108, **163ff.**, 193, 199f., 234, 255, 262, 347, 352, 397f., 680, 700, 710f., 718, 741, 745, 772, 805, 819, 837, 846, 848, 851, 973, 985f., 1014, 1105, 1199, 1239, 1278, 1310, 1324, 1334

quality of life, *see* life style
quartz, *see also* silica(te)   113, 115, 117, 119f., 125, 131, 138, 159, 262, 776, 932, 1017, 1357

rabbit   495, 527ff., 532f., 535, 538, 549, 551, 554, 563, 608, 610, 615, 630, 724, 750, 795, 827, 913, 924, 942, 966f., 994, 1066, 1071, 1097, 1111, 1113, 1116f., 1121f., 1125, 1130, 1147, 1207, 1217, 1219ff., 1222, 1254, 1293f., 1296, 1305, 1326, 1346
rachitis, rickets   532, 831
radiation, radiator, ray treatment, *see also* arc, X-ray irradiation   115ff., 124f., 127, 129ff., 140ff., 151, 169, 211, 271, 525, 529, 629, 641, 754, 880, 894, 913, 945, 960, 1079, 1136, 1154f., 1204, 1215, 1225, 1275f., 1279, **1281ff.**, 1284ff., 1320
radical, *see* free radical
radioactivity, decay   141f., 144, 157, 211, 569, 578, 626, 776, 779, 789, 879f., 922, 971, 1127, 1133, 1136f., 1159, 1200, 1203, 1208, 1271, **1275ff.**, 1278f., **1281ff.**, 1284, 1286, 1319, 1357
radioassay, radiolabelling, radioimmunoassay, *see* isotope, monitoring, sensor, immunoassay
radiochemical activation analysis, radiochemical separation   117, 119, 121, **141f.**, 184, 805, 1084, 1099, 1121f., 1204, 1278f., 1285, 1367
radionuclide, *see* isotope
radiotracer, *see* tracer
radish   453, 456, 472, 756, 826, 884, 897, 1292
radium   59, 938, **1276ff.**, 1281ff., 1284, 1287
radon   56, 287f., 290, 955, 971, **1276ff.**, 1279, 1281, 1283ff., 1286f.
railway, railroad   50, 753
rain, rainfall, *see also* acid rain   4, 8, 50, 56, 114, 120, 144, 199, 224, 266f., 280, 289, 346, 358, 370, 372, 374, 376, 532, 793, 815, 857, 899, 923, 940, 977, 979f., 1027, 1038, 1054, 1161, 1193, 1270, 1277, 1290, 1292, 1360
Raman spectroscopy   248
rape, *see* legume(n)
rare elements, *see* lanthanides
rat   158, 300, 308, 486, 495f., 498ff., 505ff., 508, 519ff., 525f., 529, 549ff., 553ff., 576, 582, 602, 608, 613f., 630, 638, 649, 664, 724, 733f., 738ff., 746, 749, 758ff., 761f., 765ff., 770, 779, 782, 784ff., 787, 794f., 798ff., 822, 827, 830f., 841, 844, 847, 851, 865f., 869, 872f., 875f., 884, 886, 888f., 906, 913ff., 917ff., 924f., 928, 938, 940ff., 944, 957, 963ff., 968f., 994, 1001, 1019, 1022, 1039, 1060ff., 1064ff., 1067ff., 1071ff., 1074f., 1077ff., 1080, 1082ff., 1085ff., 1093ff., 1098, 1100f., 1110ff., 1113ff., 1116ff., 1119f., 1122, 1124f., 1129ff., 1132f., 1147, 1151, 1169, 1174f., 1177f., 1181, 1183, 1186ff., 1195ff., 1199ff., 1206ff., 1217ff., 1220ff., 1223, 1225f., 1237, 1240f., 1250ff., 1253f., 1264, 1266, 1272, 1286, 1289, 1293ff., 1303ff., 1316, 1326, 1330f.
rate (constant)   81f., 85, 321
raw materials, supply, *see also* resource   **10ff.**, 49, 718, 739, 741, 750, 799, 846, 876, 891, 906, 937, 948, 957, 1012, 1017, 1034, 1043, 1085, 1123, 1133, 1150f., 1198, 1201, 1209, 1230f., 1258, 1267, 1273, 1287, 1344, 1347, 1368
ray, beam, *see* radiation
Raynaud's syndrome   763
rayon, artificial silk   895, 1052, 1311f., 1341
receptor, receptor protein, *see also* carrier (protein), donor, protein, transport mechanism   71, 236, **243ff.**, 274, 276, 278ff., 501, 507, 515, 519f., 527, 593, 725, 865, 916, 953, 957, 1016, 1019, 1023, 1065, 1078, 1087, 1299, 1367
recessive   566, 643, 645ff., 900f.
recombination, *see* DNA repair, repair mechanism
rectifier   1051, 1159, 1161, 1164, 1176, 1205
recycling, recovery, regeneration   13, **26f.**, 30, 33ff., **37f.**, 42, 44, 49f., 52ff., 59, 62, 65, **335ff.**, 351f., 357, 379, 393, 398, 718ff., 741, 744, 755, 790ff., 806ff., 809, 813, 842, 844f., 855f., 881, 894, 907, 910, 931ff., 947f., 974ff., 977, 1014, 1026, 1036f., 1050, 1053, 1070, 1088, 1090, 1104f., 1126f., 1138, 1140, 1148, 1151, 1158, 1192f., 1214, 1245, 1263, 1267, 1270, 1277, 1300f., 1311f., 1333f., 1341f.
red blood cell, *see* erythrocyte
red mud, *see* mud
redox potential, redox change, redox process   71f., 86, 90, 94, 136, 140, 155, 247, 256, 311ff., 328, 338, 344f., 383f., 386, 389, 391, 393f., 412, 419, 428, 438, 485, 497, 513f., 536, 817, 878, 893, 896, 948, 956, 1015, 1102, 1106, 1114, 1121, 1140, 1145, 1164, 1169, 1182, 1292, 1354, 1357

reduction, reducing, reductase 35f., 44, 131, 312, 403, 425, 428, 436, 438f., 441, 492, 497, 645, 719, 751, 753, 761, 777, 801, 826, 856, 859, 867, 869, 871, 883, 934f., 951, 961, 974, 976, 1021, 1026, 1048f., 1063, 1090, 1098, 1104, 1128, 1155, 1161f., 1167, 1172f., 1213, 1230, 1261f., 1266, 1300, 1310f., 1344

red wine, see wine

reed, rush, see also marsh 758

reference material, reference method, reference population, see also good laboratory practice, blank, environmental sample, samples 108, 110, 112ff., 116f., 122ff., 128f., 134f., 138, 143, 146f., 159, 162, 164ff., 169f., 172, 174, 179, 186, 192ff., 196, 205, 211f., 214ff., 219, 227, 276, 283, 306, 385, 410, 488, 548, 624, 669, 676, 716f., 854, 973, 1049, 1086, 1163, 1261, 1310, 1353

refinery, refining, raffination, see also crude oil, petroleum, fuel oil, gasoline 31ff., 36f., 50f., 276, 571, 638, 748, 791, 807, 835, 857, 894, 922, 927, 932, 938, 940, 961, 966f., 974f., 978, 1026, 1050, 1095, 1101, 1103ff., 1106, 1113ff., 1122, 1128, 1136, 1142, 1144, 1158, 1161f., 1192, 1199, 1201, 1214ff., 1220, 1223, 1285, 1311

reflection, see mirror

reflex, response, see neurotoxicity

refuse, see waste

regression, retroregression, see reversible reactions, model

regulation (legal meaning), legislation, ordinance, guideline, see also environmental impact statement, poison law, VDI Commission 41f., 51f., 59, 106, 163, 227, 326, 363ff., 367f., 450, 479, 547f., 640, 651f., 656, 659, **667ff.**, 670f., 674, 677ff., 680f., **687ff.**, 690, **692f.**,, 699ff., 702ff., **706ff.**, **709ff.**, 732f., 747, 755, 766f., 772, 782f., 819, 821, 833f., 838, 840f., 843, 845, 847, 849, 870f., 873f., 889, 926, 971, 1003ff., 1009f., 1041, 1073, 1075, 1084, 1115f., 1132, 1163, 1180, 1186, 1208, 1222, 1225, 1230, 1263ff., 1278, 1313, 1315, 1331f., 1334, 1339ff.

regulation (non-legal, biological meaning; in the sense of self-regulation) 58, 504, 515, 518, 847, 893, 904, 955f., 1019, 1029, 1038, 1065, 1303

release, see solubility

remediation, see clean-up

remobilization, removal, see mobilization

renal, renal dysfunction, see also kidney 58, 186, 494, 498, 501, 506, 534f., 537, 540, 544, 546, 549, 552, 554f., 596, 598ff., 601, 655, 660, 662, 715, 723, 725f., 728ff., 731ff., 734f., 738, 740f., 759, 763, 768f., 794ff., 797, 803, 806, 824, 828f., 831, 840, 844, 884, 911f., 926, 928, 941, 998, 1011, 1020, 1023, 1028, 1030ff., 1069, 1078, 1081, 1086f., 1109, 1112, 1114, 1116, 1120, 1124f., 1236, 1282, 1284, 1330

renal cancer 664, 923, 929, 1001f., 1071

renal cortex, renal tubular 158, 472, 494, 528, 535, 555, 580f., 590, 592, 594, 596, 601f., 660, 679, 796, 806, 821, 823f., 826, 828f., 831, 833, 838, 841, 850, 914, 924, 938, 941, 964, 984, 998, 1064, 1083, 1085, 1112, 1220, 1282, 1322, 1367, 1371

repair mechanism, see also anti-carcinogenicity, DNA repair system, reversible reactions 502, 620, 628, 636, 829, 872, 1065

replacement, see substitution

reproduction, replication, see also fertility, conception, child bearing (age), layer, spawning (time) 56, 386, 413, 481, 494, 504, 583, 612, 619f., 630, 634, 651, 656, 662f., 673, 675, 682, 685, 705, 760, 767ff., 770, 828, 830f., 837, 850f., 899, 907, 917, 965, 1002, 1006, 1031, 1042, 1060, 1072, 1081, 1094, 1148, 1174, 1181, 1205, 1222, 1237, 1252, 1266, 1316, 1324, 1330f., 1365

reproduction technique, printing ink 857, 1158f., 1216, 1245, 1263, 1270, 1311

research needs 504f.

reserve, see resource

residence time, see life time

residue, see also waste 45ff., 52, 54, 342, 344, 431, 467, 478, 981, 1086, 1193, 1230, 1239f., 1249

resins, see plastic materials

resistance, see tolerance, electric conductor

resorption, see also absorption 472, 474, 589, 646, 823, 828, 859, 911, 1038f., 1041, 1250, 1255

resource, see also raw materials, mineral deposit 3, 5, **13ff.**, 51, 59, 65, 791, 806, 855f., 947, 974, 1036, 1090, 1100, 1139, 1368

respiration, respiratory tract, see also bronchial, dyspnea lung, nose, tracheobronchial 257, 261f., 271, 280, 296, 299ff., 303, 308, 367, 409, 474, 485, 492, 504, 514, 534, 538ff., 547ff., 550, 552, 555, 557, 571,

574ff., 577, 582f., 585f., 588ff., 591, 593, 599, 606, 610, 612, 614, 655, 662f., 672f., 676f., 681, 720, 723, 727, 733f., 738, 746ff., 749, 757, 761f., 764f., 767, 775, 780, 784, 786f., 803, 814, 822f., 827ff., 830, 832, 836, 840, 844, 851, 861, 863, 866, 870, 873, 884, 887ff., 901, 924, 927, 939, 941, 944, 954f., 959, 963, 965ff., 968, 971, 978, 988, 990f., 995, 1000f., 1012, 1021, 1030, 1033, 1038, 1040, 1046, 1059, 1062ff., 1066f., 1074, 1079, 1087, 1095, 1101, 1103, 1106, 1109, 1112ff., 1115f., 1119ff., 1122, 1124, 1130ff., 1133, 1135, 1145f., 1155, 1168f., 1171, 1173, 1175f., 1183, 1194f., 1200, 1203, 1206, 1208f., 1211, 1216f., 1219ff., 1222, 1233f., 1254, 1266f., 1271f., 1280, 1284, 1295, 1299, 1302, 1304, 1306, 1317, 1324, 1329, 1331, 1345, 1353f. 1363, 1366f., 1369

restriction, *see*, substitution

retention, *see* accumulation

reticuloendothelial (cells), reticulum 537, 562, 615, 779, 939, 941ff., 949, 964, 1303, 1368

retina, retinol, *see also* eye 660, 828, 1317, 1368

reverse mutation, reversion assays 621, 747, 763, 1221

reverse osmosis, *see* osmosis, membrane

reversible reactions, reversibility, *see also* irreversible reactions, repair mechanism 321, 422, 595, 748, 1012, 1250, 1254

rhenium, rhenium compounds 19, 239, 1144, 1203

rheumatic, rheumatism, anti-rheumatics, *see also* arthritis 527, 567, 754, 954, 960, 968, 1205, 1208, 1350

Rhine (river), *see also* Lake Constance 29, 57, 256, 380, 383, 388, 394, 397, 758, 771, 774, 816, 836, 842, 871, 874, 1054, 1079, 1087, 1313

rhizosphere, *see* mycorrhizal, roots

rhodamin B 910, 1229

rhodanic acid, rhodanides, thiocyanate 73, 918, 1089, 1269

rhodium, rhodium compounds 145, 918, 1136f., 1139ff., **1142f.**, 1145ff., **1148f.**, 1151, 1308

rhubarb **860**

ribosome 1064f., 1319

rice, *see also* grain, cereals 58, 451, 453, 466, 471, 494, 553, 596f., 757, 759, 778, 820ff., 829, 834, 952, 957, 1056, 1086, 1326

rime frost, *see* fog

ring-oven technique 149, 203

risk (assessment), *see also* health (injury) 19ff., 50, **51ff.**, 59f., 72, 90, 257, 287, 305, 338, 344f., 352f., 355, 470, 474, 492, 506f., 509, 535f., 548, 552, 555, 582, 599, 601, 625, 629ff., 634, 636, 647, **651f.**, 656, 659ff., **662ff.**, **665ff.**, 669, 671ff., 674ff., 679, 682f., 687ff., 691f., 699, 708ff., 731ff., 747, 749, 753, 761, 766f., 769, 772, 780, 782, 784, 797, 804, 814, 830f., 833, 835, 839, 842, 851, 859f., 870, 874, 878, 888f., 891f., 903, 915, 926f., 935f., 942, 955, 959, 966, 971, 986, 994, 1001, 1003ff., 1006, 1009, 1017, 1021f., 1030, 1033, 1041f., 1044, 1046, 1055, 1077ff., 1084, 1096, 1114ff., 1121, 1132, 1145, 1147f., 1178f., 1181, 1183f., 1187ff., 1191, 1198, 1208f., 1222, 1227f., 1231f., 1237f., 1241, 1255, 1264, 1267, 1272, 1275, 1279, 1282ff., 1285ff., 1295, 1306, 1309, 1324, 1331, 1339, 1346

risk population, individuals 208, 214, 221, 225, 289, 414f., 426, 594ff., 597f., 600ff., 607f., 612, 614, 617, 623f., 629, 631, 642, 649, 652ff., **655ff.**, 658, 667, 673f., 677f., 681f., 688, 706f., 709, 729, 747, 757, 782, 810, 829, 831, 835, 850, 886, 901, 971, 979, 985, 993, 997f., 1004, 1008, 1011, 1057, 1059, 1061f., 1072, 1074, 1081, 1101, 1114, 1146, 1191, 1232, 1267, 1285, 1320, 1331, 1355

river(s), streams, *see also* Elbe, Rhine, Scheldt, Weser 4, **8ff.**, 12ff., 25, 56ff., 60ff., 67, 84, 93, 109, 115, 159, 175f., 233f., 254ff., 268, 336, 342, 351, 354, 380, 387, 389, 392, 395ff., 398, 432, 437, 446, 450, 453, 461, 493, 720, 728, 733, 755, 777f., 793, 803, 810, 813, 815ff., 838f., 840, 843, 850, 879, 883, 896, 905, 948, 950, 955, 961, 977, 979, 1009, 1014, 1037, 1048, 1053f., 1084, 1088, 1105, 1143, 1163, 1193, 1208, 1233, 1247, 1263, 1313, 1339

river mouth, *see* estuary

RNA (ribonucleic acid), ribonuclease, *see also* DNA and RNA synthesis 500, 508, 516, 520, 618, 643, 926, 937, 1064ff., 1080, 1082, 1194, 1319, 1325, 1349, 1353, 1357, 1362, **1368**, 1371

road (construction), roadway, street, traffic, roadside 27, 266, 273f., 276, 279, 288, 297f., 305f., 372, 459, 549, 571, 677, 974f., 977f., 980f., 990, 1103, 1144

roasting, calcination 35, 40, 670, 744, 752,

807, 881, 939, 974f., 1104, 1115, 1158, 1214, 1228, 1249, 1262, 1311
robotics, *see* automation
rock (mass) 3ff., 7f., 20ff., 33, 90, 292, 296, 387, 395, 399, 405, 753, 775ff., 790, 855, 883, 948, 961, 972, 974, 976, 1011, 1027, 1050, 1090, 1105f., 1158, 1200, 1213, 1245, 1291f., 1313, 1356, 1362, 1368
rocket, *see* space . . .
rodent, *see also* guinea pig, hamster, mice, rabbit, rat, rodenticide 498, 504, 542, 576, 605, 609, 635, 786, 915, 933, 935, 937, 1010, 1109ff., 1112f., 1115, 1125, 1177, 1227
rodenticide 753, 1227f., 1230
roots, rhizosphere, *see also* mycorrhizal 361, 371, 376f., **399ff.**, 405, 407ff., 410, 412, 416, 449f., 453, 455, 458, 471f., 475, 529, 715, 721ff., 726, 738, 740, 757f., 779f., 818, 821, 840f., 860f., 870, 890, 896ff., 951, 953, 981, 987, 993, 1018, 1031, 1058, 1088, 1106, 1164, 1233, 1315, 1318, 1363, 1368
root vegetables, *see* carrot, radish, vegetables
rubber, *see also* vulcanization 299, 745, 753, 1159, 1162, 1209, 1214f., 1263
rubidium 228, 1022, 1135, 1229
ruby 719
rumen, ruminant, herbivore, *see also* bovine 475, 538ff., 541ff., 544f., 836, 879, 885f., 900, 903, 979, 1018, 1021, 1092, 1094ff., 1098ff., 1173, 1330
rust formation, rust protection, *see also* corrosive 344, **946**, 958
ruthenium 179, 969, 1135ff., **1139ff.**, 1143ff., 1146f., 1213
rutile, *see* titanium dioxide
rye, *see* grain, cereals

SAED (selected area electron diffraction) 265, 273, 1354, 1368
safety (protection) measure, *see also* threshold 31, 365, 418, 495, 497f., 500, 554f., 579, 664, 668f., 671f., 674f., 700, 709f., 783, 797, 830, 878, 900, 957, 1058, 1090, 1095, 1127, 1153, 1178, 1187, 1254, 1284, 1332, 1344
salad, *see* vegetables
salicylate, salicylaldehyde derivatives 534, 792, 798f., 910, 1310, 1312
saline, salinity, salt works, salty, brine, *see also* sea water 5, 55, 250, **386ff.**, 395, 720, 827, 846, 923, 1016, 1023, 1248
saliva, spittle 156, 581, 661, 885, 1061, 1067, 1103, 1110, 1117, 1175, 1234, 1241
Salmonella (assay), *see also* Ames-test 303, 306, 620, 632, 634f., 727, 763, 878, 1177, 1179, 1185, 1187, 1221
salmonide, salmon, trout 532, 538, 540, 544, 727, 735f., 760, 770, 822, 865, 897, 903, 923, 1054, 1111, 1195, 1235, 1324, 1332
salt deposit, *see* saline
salt water, *see* sea water
samarium 189, **959ff.**, 962f., 965
samples, sampling, sample preparation, *see also* environmental sample, denuder, tape 106, 108, **110ff.**, **115ff.**, 129, 143, 145, 150, 154f., 158, 163f., 182, 195f., 198, 201, 206, 212f., 215ff., 224, 233, 246, 266, 269, **271ff.**, 280, 294, 305, 339, 377, 379, 390, 420, 600, 667, 716ff., 727, 741, 743, 768, 778, 790, 800, 816, 836, 838, 845f., 850, 854f., 863, 878, 881, 883, 893, 923, 946, 973, 1010, 1016, 1036, 1042, 1047f., 1058, 1089, 1104, 1150, 1154, 1156, 1163, 1191ff., 1212, 1226, 1232, 1244, 1247, 1259, 1276, 1289, 1292, 1296, 1310, 1358, 1360, 1366
sand, sandy soil, sandstone 5, 28, 255, 350, 360, 363, 375, 399, 403, 431, 776, 864, 879, 883, 947, 961, 992, 1158, 1165, 1277, 1300, 1313, 1344, 1357
sapphire 719
sarcoma, sarcoidosis, *see also* carcinogenicity, fibrosarcoma, tumor 578, 781, 888, 911, 917, 964, 967, 1071, 1198, 1283f., 1365, 1368
saturation (effects) 316, 318, 515, 721
sausage, *see* meat
scab, *see* skin pigmentation
scale, psoriasis, dandruffy, *see also* skin pigmentation, fishery product 636, 1157, 1159, 1305
scandium 72, 76, 145, 235, 918, 1307f.
scavenger 497f., 527, 619, 1368
Scheldt (river) 268, 388, 848
Schwann cell 1220, 1368
scintigraphy 1231, 1368
scintillation, *see* sparks, scintigraphy
sclerosis, *see also* artery, cell wall 598, 626, 715, 729, 738, 1174, 1305, 1368
scrap 34, 37, 50, 52, 720, 777, 809, 856, 880, 948, 975, 979, 1053, 1158, 1245, 1312
scrotal tumor, *see* testes
scrubbing, *see* wet scrubber
seafood, *see also* crayfish, fish, mollusc, octopus, shellfish 156, 475, 661, 756ff., 759,

1028, 1038, 1057, 1079, 1166f., 1292, 1316
sea water, sea spray, *see also* saline, ocean
  9f. 55, 89f., 100, 114, 120, 138ff., 144, 146,
  149, 154f., 164ff., 168ff., 174f., 178, 182,
  188ff., 194f., 198f., 201f., 205, 226, 228,
  233f., 241, 243, 246ff., 205ff., 254, 258f.,
  268, 275f., 331, 374, 396, 408, 481, 488,
  517, 641, 720, 734, 745, 747, 756, 760, 767,
  770, 789, 793f., 798, 800, 805, 817, 827,
  836, 845ff., 854, 858, 872, 904, 906, 912,
  916, 923, 927, 936, 940, 943, 946, 956, 967,
  983, 1006, 1017, 1022, 1026f., 1037f., 1042,
  1048, 1051, 1054, 1057, 1063, 1075, 1079,
  1084, 1090, 1097, 1105, 1117, 1129, 1132,
  1138, 1140, 1148f., 1164, 1181, 1192f.,
  1195, 1199, 1201, 1205, 1208, 1216, 1223,
  1233, 1239, 1247f., 1258f., 1265f., 1269f.,
  1272f., 1277, 1291ff., 1296, 1301, 1306,
  1313, 1316, 1332f.
sea weed, marine alga, *see* algae
secretion, *see also* gland, perspiration,
  saliva 572f., 967, 1028, 1031, 1112,
  1303f., 1319, 1330, 1349, 1355f., 1361,
  1363, 1367
security system, *see* safety (protection) measure
sedative, sedation, tranquilizer 965, 1020,
  1369
sediment, sedimentation, *see also* benthic
  5, 7, 10, 12, 29, 55, 61, 70, 80, 89, 92ff.,
  112f., 119, 143, 147, 157ff., 176, 188, 200,
  233, 251, 255, 268, 270, 276, 281, 284, 328,
  336, 338, 348, 353ff., 360, **379**, 383ff.,
  387ff., **390ff.**, 393ff., 396ff., 409, 411,
  424ff., 427, 429, 432ff., 436f., 443, 446,
  450, 453, 461, 492f., 575, 628, 720, 736,
  738, 755, 768, 770ff., 773, 777, 787, 790,
  798, 800, 803, 806, 816, 839, 850, 855,
  857f., 883, 906, 936, 948f., 958, 970, 974,
  977, 979, 983, 989, 1007, 1009, 1011, 1014,
  1026, 1038, 1044, 1052ff., 1055ff., 1063,
  1081, 1088, 1126, 1143, 1156, 1162, 1165,
  1172, 1244, 1248f., 1292, 1314, 1317, 1350,
  1353, 1357
seed, seedling, *see also* grain, germ ...
  404f., 410, 415, 529, 539f., 597, 820, 840,
  850, 884, 889, 981, 987, 1031, 1053, 1057,
  1107, 1190, 1251, 1277, 1294, 1307, 1318
seed treatment, *see also* fungicide 435, 597,
  1045, 1052f., 1057, 1060, 1063, 1069, 1077,
  1081
seepage water, *see* groundwater, leaching, soil
  solubility

selectivity, specificity 98f., 116, 155, 159,
  321, 566ff., 587, 605, 716, 910, 1041, 1103,
  1156, 1243, 1296
selenate, selenide, selenite, selenium (inorganic) compounds, *see also* organoselenide compounds, selenium dioxide 12,
  84, 86, 131, 156, 173, 207, 423f., 442f.,
  476, 498, 503f., 517, 542, 619, 633, 636,
  638, 661, 754, 759, 806, 808, 829, 838, 887,
  895, 940, 1075, 1154ff., **1158f.**, 1161,
  1163ff., 1167, 1170ff., **1173f.**, **1176ff.**,
  1181, 1183, 1186, 1188, 1199, 1201, 1213,
  1230f.
selenium (general) 6ff., 11ff., 15, 19, 51, 57,
  59, 83, 85, 89, 100, 120, 130ff., 135ff., 151,
  153, 159, 170, 174, 176f., 182f., 185f., 190,
  193, 195, 204, 207, 211, 218, 230, 235f.,
  257f., 265, 268, 276, 278, 281, 288, 296,
  301f., 304, 307, 335, 404f., 412, 422, 424,
  442f., 461, 463f., 467, **476f.**, 481f., 485,
  489, 491, 497, 505ff., 508, 517, 520, 537,
  541f., 561, 573, 605, 612, 620, 626, 629,
  634, 754, 759, 764, 766, 769f., 772, 792,
  826, 895, 1083, 1087, 1093, 1147, 1151,
  1153ff., 1163, **1178ff.**, 1187ff., 1196ff.,
  1200f., 1211ff., 1214f., 1218f., 1223ff.,
  1226, 1230, 1239, 1361
selenium dioxide, selenium(IV) 155, 168,
  171, 443, 1154, **1158**, 1161, 1176, 1178
self-cleaning, self-purificiation, *see also* cleanup, purification 58
semiconductor, *see also* control engineering,
  electronic equipment, switching,
  photochemistry 743f., 754, 773, 791,
  909f., 911f., 916f., 921f., 940, 942, 1090,
  1100, 1154, 1158f., 1215, 1231
semimetal, *see* metalloid
seminal fluid, *see* sperms
senility, *see* paramnesia
sense (sensory) organ, *see* ear, eye, odor,
  taste
sensitivity, detection limit 116, 120f., **122f.**,
  125, 127, 130f., 134ff., 139ff., 144ff., 151,
  161, 166f., 183, 200, 221f., 225, 247, 261,
  363, 400, 415, 422, 427, 435, 438, 449, 531,
  536, 538, 542, 544, 554, 559f., 562, 622,
  632, 716, 744, 752, 819, 854f., 860, 867,
  893f., 910, 932f., 952, 972, 994, 1029, 1035,
  1041, 1049, 1103, 1114, 1128, 1137, 1156f.,
  1176, 1191, 1212f., 1216, 1228f., 1236,
  1243f., 1247, 1261f., 1264, 1296, 1300,
  1310, 1319, 1323, 1326, 1343, 1353, 1359,
  1361, 1369

sensitizing, sensitization, hypersensitivity, *see also* irritation   552, 593, 606, 609f., 781ff., 785f., 866, 888, 1059, 1067, 1103, 1146, 1207, 1346f., 1353

sensor, marker, *see also* early warning, nuclear medicine   70, 246f., 921f., 940, 942, 962, 1044, 1073, 1204f., 1209, 1328

separation, *see also* centrifugation, denuder, diffusion, filter, flotation   349ff., 354, 391, 396, 960f., 1137, 1140f., 1144, 1244, 1277, 1299ff., 1311, 1341, 1344, 1370

sequestration, *see* chelating agent

serine, serotonin, *see also* amino acids   954, 1023, 1040, 1319, 1370

serpentine   325, 328, 412, 416, 860, 876, 1026, 1106, 1118, 1122

serum, serosal, *see also* blood   109, 134, 140, **156**, 164, 169, 171f., 197, 205, 209, 212, 216, 219, 447, 488, 499, 512, 515ff., 537, 554, 561f., 577, 608, 612, 614, 645f., 649, 666, 715, 717, 724f., 728ff., 732ff., 735, 737f., 740, 746, 781, 877, 884, 901, 914, 916ff., 925, 929, 934, 951f., 956f., 1018, 1021, 1028, 1031f., 1040, 1083, 1103, 1107, 1112, 1116ff., 1119ff., 1125f., 1131, 1157, 1166, 1187, 1189, 1197, 1199, 1218, 1271, 1293, 1304, 1316f., 1321ff., 1327, 1330, 1349, 1351ff., 1357, 1360, 1369f.

sewage (plant), *see also* purification (plant), sludge, biodegradation   10, 44, 49, 182, 226, 230, 389, 424, 426, 431, 453, 690, 813, 845, 857, 948, 1162, 1312ff., 1315

sex, female, male   214, 225, 486, 492, 494, 504, 507, 552f., 633, 651, 654ff., 657, 673, 681, 705, 768, 771, 822, 837, 850, 913, 915, 925f., 934, 953, 959, 965, 982, 985, 994, 998, 1002, 1005f., 1010, 1028, 1061f., 1071, 1085f., 1131, 1166, 1174f., 1205, 1237, 1325, 1359, 1367, 1371

shale, *see also* crude oil, mica   5ff., 12f., 29, 776, 800, 883, 961, 1090, 1095, 1213, 1277

shampoo, *see* soap, detergent

shark, *see* predatory fish

SHE (Syrian hamster embryo), *see* hamster

sheep, *see also* lamb, wool   203, 472, 484, 532f., 535ff., 538, 540ff., 544ff., 644, 753, 820, 827, 839, 849, 883, 885, 893, 900, 903, 1093ff., 1096ff., 1130, 1132f., 1174, 1194, 1196, 1240, 1304, 1316, 1326, 1330

shell, *see* bullet

shellfish, mussel, *see also* crayfish, mollusc, oyster   200, 206, 223, 226, 472, 475, 479, 519, 597, 703f., 723, 737, 756, 760, 817, 820, 822, 842, 851, 884, 896f., 950, 1038, 1052, 1054, 1056f., 1061, 1092, 1107, 1198, 1319, 1324

shock, stress, *see also* unconciousness   589, 761, 827, 858, 901, 954, 1063, 1067, 1199, 1290, 1301, 1329, 1370

shoot, *see* stem

shooting, *see* bullet

shore, *see* coast, bank

shortness of breath, *see* dyspnea

short-term tests, *see also* Ames-test, *Daphnia*, *Drosophila*, mammalian cell, mutagenicity, reverse mutation, *Salmonella*, sister-chromatid exchange, yeast   502f., 595, 617f., 620f., 631, 634, 640, 693, 747, 765, 782, 784, 799, 866ff., 873f., 888, 906, 915, 1114, 1177ff., 1188

shrimp, prawn, *see* crayfish

siccative, *see* desiccant

sickle cell   657, 1174, 1218, 1223, 1328, 1369

sickness, *see* nausea

sideroblast, siderochrome, sideropenia, siderophore   432, 899, 906, 912f., 916, 951, 954, 1038, 1040, 1369

silica(te), silicon (dioxide), *see also* pneumoconiosis, quartz   3, 7f., 10, 33ff., 54, 56, 68, 84, 93, 158, 207, 235f., 292, 294, 346, 369, 373, 719, 721f., 727, **729f.**, 733f., 740, 744, 775, 777, 856, 861, 882, 894f., 921, 924, 947, 961, 972, 1015, 1017, 1026f., 1035f., 1105, 1209, 1229, 1245, 1310, 1344, 1347, 1368

silicomanganese   1037

silver, silver colloids, silver compounds   6ff., 11f., 15, 19ff., 35, 38, 51, 57, 59, 76, 89, 130, 132, 254, 268, 270, 284, 313, 336, 342, 351, 414, 420, 422, 484, 497f., 501, 573, 618, 771, 773, 791, 906, 922, 929, 975, 1048, 1144, 1158, 1173, **1191ff.**, 1194ff., 1199ff., 1213f., 1230, 1245, 1350

silver-plating, silvering   791, 1193

SIMS, secondary ion MS   145, 170, 192, 210, 273, 384, 744, 1369

simulation, *see* model, statistics

sinew, tendon (jerk?), sinus   1059, 1074, 1113, 1115, 1360

sink, *see* fate

sintering   34f., 777, 948, 974f., 1017, 1101, 1270

sister-chromatid (exchange) SCE   506, 620, 624f., 633ff., 636ff., 763f., 829, 867, 1001, 1072, 1084, 1177, 1186, 1188, 1237, 1368

size distribution, *see also* particle size 261ff., 271ff., 276, 280ff., 285, 1145
size exclusion chromatography, *see* gel (permeation liquid) chromatography
skeleton, skeletal, *see* bones
skin, *see also* allergic reactions, dermatitis, epithelium, skin cancer, wound healing, eczema, scale 112, 224, 466, 493, 531, 533, 537, 543, 548, 562, 566, 571, 577, 586, 591, 593, 609, 646f., 655, 672, 720, 729, 746, 757ff., 761f., 764, 775, 781, 783f., 795f., 853, 862f., 865f., 871, 879, 887, 901, 911, 935, 937, 953, 959, 965f., 989f., 1000, 1030, 1039, 1045, 1059ff., 1066f., 1069, 1071, 1082, 1086, 1101, 1110, 1118, 1120, 1129, 1133, 1136, 1194f., 1197, 1207, 1216, 1221f., 1228, 1230, 1233ff., 1236f., 1254f., 1264, 1302, 1305, 1317, 1328, 1349, 1353, 1355, 1358f., 1361ff., 1366, 1370
skin cancer, skin ulcer 502, 548, 550, 599, 602, 660, 745, 751, 756, 764ff., 772, 875, 966, 1177, 1263, 1351
skin cream, ointment 792, 796, 799, 1197, 1230, 1311, 1344, 1346
skin pigmentation, blistering, burnskin, edema, freckle, melanoma, *see also* wart, scale, wound healing 494, 599, 720, 745, 761ff., 792, 795, 901, 914, 923, 926f., 1030, 1039f., 1072, 1074, 1115, 1176, 1196ff., 1199, 1201, 1207, 1210, 1221, 1236, 1264, 1311, 1326, 1328, 1330, 1349ff., 1352f., 1355, 1358f., 1362, 1365f., 1371
skull, cranium, *see* bone, head
slag, slagging, cinders, coal ash 33ff., 50, 52, 134, 333, 353, 778, 810, 947, 975, 979, 1129, 1133, 1158, 1204, 1262f., 1290, 1312
slate, shale (oil), *see* shale
slaughter, meat stock, *see also* domestic animal, bovine, poultry, meat 472, 753, 821, 843, 849, 1091
sleepiness, somnolescence, disturbed sleep, *see* tiredness
slime, *see* mucous ...
sludge, sewage sludge, *see also* purification (plant), biodegradation 30, 41, 50ff., 62f., 80, 107, 112f., 143, 158f., 172, 178, 182, 189, 195, 230, 290, 311, 313, 326, 328f., 333ff., 338f., 342, 347, 349ff., 354, **357ff.**, 361, 363ff., 367f., 380, 386f., 394f., 398, 401, 404, 407ff., 410, 412, 427, 432, 441, 443f., 450, 453, 459, 461, 471, 475, 667, **707f.**, 710, 809f., 812ff., 816f., 819, 833f., 836f., 842f., 857, 871, 873f., 885, 890, 903, 905, 979, 1004, 1053, 1082, 1090, 1098, 1105, 1115, 1162, 1187, 1193, 1230, 1247, 1291, 1312ff., 1315, 1332ff., 1337
small intestine, gut, *see* intestine
smell, *see* odor
smelter(y), *see* metal processing, melt
smog, *see also* nitrogen oxides 1003
smoke, *see* dust, plume, fireworks
smoking, *see* tobacco (smoke)
snail, *see* mollusc
snake 1316, 1340
snow, *see also* ice, fog 114, 171, 188, 202, 268ff., 276f., 282, 374, 842, 977, 983, 1006, 1010, 1054, 1105, 1121, 1161
soap, shampooing, *see also* detergent, additive, fat, stabilizer 216, 808, 882, 990, 1048, 1157, 1159, 1312
sodium borohydride, *see* boron
soil, *see also* earth crust, humic substances 4, 8, 10, 12ff., 53f., 65, 80, 89, 112f., 119, 143, 157ff., 168, 178, 182, 189, 223ff., 228ff., 256, 268, 278, 285, 289f., 296, 298f., 301, 308, **311ff.**, 320, 323, 328ff., 338, 342, 347, 353, 357, **359ff.**, 363ff., 367, 371ff., 376ff., 379, 384, 387, 394f., 399ff., 402ff., 405, 408, 410ff., 413f., 419, 424f., 428f., 431, 433, 435ff., 442ff., 449f., 453, 458, 460, 462f., 470f., 473ff., 476f., 492, 505, 507, 540f., 548, 571, 648, 655, 665, 667, 677, 688, 701, **707f.**, 710, 715, 718f., 721, 729, 734, 736ff., 740, 745, 749, 755, 757, 767, 771, 776ff., 779f., 786, 790, 798, 800f., 803f., 811, 813f., 816, 818ff., 826, 833ff., 836, 838, 841ff., 848ff., 857ff., 860f., 864, 871f., 876f., 879, 883, 885, 893, 896, 898, 900, 903, 908, 923, 936, 940, 949ff., 953, 955, 963, 972, 974, 976, 978, 980f., 987, 991ff., 1004, 1008ff., 1011ff., 1014, 1018, 1027, 1030f., 1033, 1038, 1050, 1052ff., 1056, 1058, 1077, 1091, 1093, 1095ff., 1098ff., 1101, 1105f., 1115, 1118f., 1125, 1145, 1159, 1162ff., 1165, 1168f., 1179, 1182, 1186f., 1189, 1192f., 1231, 1235, 1240f., 1247ff., 1278, 1286, 1291ff., 1295, 1297, 1301f., 1307, 1312f., 1315, 1318, 1323f., 1326, 1332, 1344f., 1363, 1366
soil exchange capacity, *see also* ion exchange 311, 399, 778, 1318
soil solubility, soil solution, *see also* groundwater, pore water 159, 322ff., 327, 330, 374, 400, 410, 440, 721ff., 735, 793, 1031

solar cell, solar energy, *see also* photochemistry   809, 895, 911f., 916, 940, 1025, 1154, 1158, 1214f., 1223

solder, soldering material, plumbing, *see also* hard-solder, tinplate   450, 459, 492, 754, 827, 940, 976, 981, 986f., 991, 1051, 1245, 1312

solidification   54

solid waste, *see* waste

solubility, solubilization, solvent effects, dissolution   54ff., 80f., 158, 268, 313, 315, 326, 338, 341f., 345, 358, 360, 362f., 374, 379ff., 382ff., 386ff., 398, 400, 405, 428, 431f., 437, 442, 449, 492, 503, 512f. , 531, 541, 559, 561, 572, 574, 576, 586, 588f., 615, 659f., 676, 717, 720ff., 726f., 734f., 740, 743, 757, 759f., 776, 779, 786, 792f., 797, 804f., 815ff., 834, 840, 848, 854, 856, 858, 860, 862, 867, 874, 877, 881, 883, 894, 896, 946, 951, 972f., 977, 979, 986, 1014, 1016, 1046f., 1054, 1059, 1063, 1067, 1098, 1101ff., 1105, 1109, 1112, 1120, 1137, 1140, 1144f., 1154, 1156, 1169, 1172, 1192, 1194f., 1214, 1217, 1230, 1233, 1248, 1254, 1264, 1271, 1276ff., 1279, 1281, 1290, 1295, 1299, 1302, 1310, 1315

solvation, *see* hydroxo complex

solvent (extraction), eluent   22, 59, 147f., 154, 157, 349, 895, 961, 1017, 1128, 1140, 1155, 1159, 1192, 1229, 1287, 1310, 1355

somatic cell, soma   617, 623, 654, 763, 1000, 1369

sorption, *see* absorption

source, origin, source indication, *see also* cycle, anthropogenic source   51, 227, 257ff., 274ff., 279f., 282f., 288, 297, 299f., 305f., 308f., 311, 337, 358, 365, 379, 405, 407, 412, 428, 430, 477, 600, 718, 724, 728, 736f., 740, 744, 753ff., 772, 775, 784, 790, 805f., 810, 812ff., 816ff., 819, 831, 835, 840, 845, 855, 881, 894, 898, 907, 910, 922, 924, 932, 939, 947, 949, 961, 974, 977f., 980, 989, 991, 1003, 1016, 1026, 1036, 1038, 1050f., 1053, 1057, 1059, 1079, 1090, 1104, 1114, 1128, 1138, 1140, 1158, 1162, 1165, 1169, 1192, 1204, 1213, 1215, 1229, 1238, 1245, 1262, 1270, 1277, 1287, 1290f., 1300, 1311f., 1315, 1344

soya, soybean, *see* bean

space . . . , astronautic, rocket, missile   777, 1016, 1020, 1205, 1215, 1344

sparks, scintillation, *see also* spark source MS, arc, flash . . .   132f., 144, 209, 922, 927, 1142, 1155

spark source MS   116, 119, **144f.**, 169, 272, 296, 752, 922, 1261, 1310, 1343, 1369

spasm, cramp, convulsion, tonus, spasmolytic   590, 646, 761, 865, 901, 1000, 1027, 1031, 1040, 1101, 1176, 1219, 1329, 1354, 1357, 1363, 1369f.

spawning time   1319

speciation   65, **67ff.**, 100f., 103, 106f., 130, **152ff.**, 156ff., 170, 173f., 183, 185, 189, 191f., 194, 197, 200f., 205, 234, 245ff., 251ff., 254f., **265f.**, 273, 278, 285, 300, 316, 323ff., 327f., 331, 337f., 341f., 345, 355, 357, 377, 379, 382ff., 386, 388, **390f.**, 394, 396ff., 399, 401, 408f., 412, 416f., 419f., 424, 449, 452, 457, 467, 471, 476, 487, 492f., 503, 511, 529, 531f., 534, 565ff., 568f., 572, 580, 585, 588, 591, **661**, 663, 673, 693, 700, 715f., 718, 720, 724, 726f., 731, 735f., 738, 743f., 746f., 749, 751f., 758f., 767f., 773, 784, 790, 793, 797, 804ff., 822, 835f., 845, 848, 850f., 853, 858f., 862, 864, 869, 870, 872ff., 875ff., 878, 880, 909, 921, 931f., 935, 939, 941, 946, 956, 958f., 963, 972, 1009, 1014, 1034f., 1038, 1045, 1048ff., 1051, 1081, 1085, 1088, 1092, 1098, 1101f., 1105, 1126f., 1130, 1135f., 1138, 1141, 1144f., 1153f., 1156f., 1165f., 1169ff., 1173, 1175, 1181f., 1191f., 1201, 1211ff., 1222f., 1228, 1240, 1244, 1247f., 1259, 1261, 1269, 1275ff., 1279f., 1285, 1289, 1292ff., 1310, 1343, 1369

specificity, *see* selectivity

specimen bank, *see* environmental sample, reference material

spectrophotometry, *see* colorimetry, fluorometry, photometry, UV spectrometry

speech defect (or disorder)   728, 1021, 1069, 1328f., 1354

sperms, spermatogenesis, *see also* germ cell, testes   110, 535, 685, 763, 771, 794, 831, 1002, 1031, 1167, 1173, 1234, 1351, 1360, 1369f.

spikes   144f.

spillage, *see* leaching

spin   73

spinach, *see* vegetable(s)

spinal (or vertebral) column, cord, *see also* marrow   539, 577, 923f., 1040, 1060, 1236, 1330, 1359, 1362

spindle structure, *see* aneuploidy

spirits, liquor, *see* alcohol

spirochete, *see* syphilis

spleen 577, 606, 608, 646, 758, 779, 794, 863, 911f., 914, 923ff., 926, 941ff., 951, 963, 1060, 1108f., 1112f., 1130, 1146, 1166, 1196, 1221, 1271, 1304, 1328, 1330, 1358
sponge 35, 922, 924, 927, 929
spring, *see* watch
spring water, *see* well, groundwater
spruce, fine, fir, *see* conifer
square wave voltammetry SWV 137, 192, 845, 1370
SSMS, *see* spark source MS
stabilizer, stabilization, stability, immobilization, *see also* additive **54f.**, 61, 82, 154, 158, 236, 238, 240ff., 247ff., 321f., 328f., 342, 345, 351, 357, 367, 381, 394, 423, 484, 517, 559, 633, 700, 752, 803, 809, 819, 836, 858, 887, 975, 981, 993, 1027, 1061, 1120, 1246f., 1249ff., 1312
stack (gas) 38f., 277, 690, 1003, 1162, 1205, 1208
stag, deer, *see* wild animal
stain, mordant, *see* seed treatment, tanning agent
standard, standardization, *see* reference material
standard setting, *see* threshold
stannate, stannite, stannous, *see also* tin compounds 1049, 1246, 1248, 1252
stannosis 1252, 1255, 1369
Staphylococcus 437f., 501, 1254
statistics, statistical treatment, uncertainty, *see also* factor analysis 166f., 275, 665, 1044, 1065, 1162, 1282, 1310, 1351, 1355, 1371
steam, *see* vapor
steel (manufacture) **33f.**, 51, 111, 178, 259, 493, 544, 563f., 571, 657, 689, 720, 741, 775, 777, 791f., 808, 810, 812ff., 853, 856, 862, 880, 882, **947f.**, 958, 961, 975, 1036, 1090, 1104f., 1128, 1142, 1159, 1214f., 1245, 1262, 1270, 1273, 1290, 1295, 1301, 1311, 1343
STEM (scanning transmission electron microscopy), *see also* electron microscopy 265, **273f.**, 1275, 1369
stem (flow), shoot, *see also* xylem 279, 360f., 370, 404, 723, 779, 821, 840, 860, 980f., 987, 1232, 1283, 1318
steric factor, steric hindrance, spiral, stereoisomer 99, 242, 244, 569, 587, 618, 923, 925, 928f., 1269
sterile, sterility, sterilization, disinfection 539, 559, 899, 1020, 1052, 1193, 1198, 1201, 1205, 1246, 1249

stibine, stibnite 744f., **747f.**
stickleback, *see* fish
stimulation, *see* synergism, interaction, irritation
stomach, *see also* gastric ..., gastro ... 493, 495, 568, 571, 581, 586, 589, 761f., 790, 792, 794, 865, 885, 913, 943, 955, 1129, 1176, 1206, 1219, 1251, 1329f., 1349, 1352, 1356, 1371
stomach cancer 665, 830
stomatitis, *see* mouth
storage battery, *see* battery
store, storage, *see also* accumulation 51, 55, 158, 303, 334, **348**, 379, 399, 470, 473, 498, 514ff., 518f., 537, 551, 562, 572, 574, 578, 589, 592, 645f., 687, 738, 754, 757, 779, 805, 818, 823f., 851, 587f., 863, 882, 884, 904, 908, 934, 951f., 954, 961f., 979, 981, 983, 992, 1013, 1031, 1039, 1047, 1061, 1088, 1120, 1154, 1214, 1225, 1248, 1324, 1356, 1367ff.
stove, *see* firing
strain, strain specifity, *see also* genetic information 424, 426f., 430, 433ff., 436ff., 439f., 501, 554, 621, 628, 641, 643f., 647, 649, 760, 867, 907, 1031, 1071f., 1217, 1304
street, *see* road, automobile exhaust
stress, *see* shock
stripping technique, *see also* anodic stripping voltammetry ASV 137ff., 154, 157, 201ff., 801, 918, 1241, 1275, 1287, 1353f., 1367, 1369
strontium 79, 144, 219, 578, 817, 853, 870, 1132, 1266
styrene, *see* benzene
subcellular level, *see* cell physiology
subsidence, mitigation, *see* sedative
substance transport, *see* material transport, cycle
substitution, replacing 808f., 836, 847, 849, 895, 959, 966, 968, 1037, 1299, 1304, 1306
succinate, ethyl succinate, succinyl, succinimide dioxime 799, 886, 950, 1017ff., 1026, 1033, 1036, 1039, 1042, 1064, 1075, 1083, 1131, 1304
suck, suckling, breast-feeding, *see* lactation
suffocation, choking fit, *see also* dyspnea 827, 1354
sugar, *see also* cane, carbohydrates, diabetes, blood, urinary 486, 513, 877, 895, 949, 1028, 1095, 1112, 1170, 1249, 1271, 1353, 1357, 1363f., 1368

1432 Index

suicide, *see also* forensic ..., murder 585, 600, 901, 1067, 1230, 1234, 1236f.

sulfate, sulfite, *see also* thiosulfate 34f., 71f., 154, 248, 250, 265, 277, 279f., 304, 308, 324, 341, 369, 372, 423, 426, 513, 716, 721, 752, 854, 857, 859, 864, 894f., 898, 901, 950, 974, 1017, 1027, 1032f., 1035, 1091f., 1095ff., 1098, 1100, 1172f., 1230, 1237, 1263, 1269, 1284, 1303, 1310, 1312, 1321

sulfhydryl groups, thiols, *see also* mercaptane, dithio ..., disulfide 153, 157, 243, 414, 422, 435, 440ff., 497, 512f., 516f., 521, 526, 588, 618, 666, 758f., 762, 790, 910, 931f., 934f., 1046f., 1056, 1061ff., 1064, 1071f., 1075, 1078, 1080, 1086, 1095, 1194, 1212, 1225f., 1325

sulfides, sulfidic ores, *see also* pyrite, lead glance, nickel sulfides, cinnabar 5, 7f., 12, 22, 32ff., 39, 45, 55ff., 72, 76ff., 158, 255, 313, 323f., 336, 338, 341f., 344, 346, 370, 381ff., 384, 386, 390, 393f., 419, 428, 431f., 437, 492, 495, 516, 751, 754, 759, 805f., 808f., 830, 881, 888, 894, 901, 910, 912, 922f., 943, 947f., 970, 972, 974, 1001, 1035, 1046, 1050, 1057, 1089f., 1092, 1095, 1100, 1104, 1138, 1140, 1158f., 1192, 1195, 1199, 1230f., 1270, 1290, 1310ff.

sulfur (general, elementary), *see also* desulfurization 8, 12, 34f., 40f., 73, 76, 95, 98, 243, 284, 294, 341, 387, 405, 421, 428, 430, 443, 497, 523, 538, 587, 743, 894, 975, 1036, 1056, 1062, 1093f., 1098f., 1112, 1153f., 1162, 1212ff., 1227, 1231, 1266, 1325

sulfur dioxide, sulfur oxides 34, 36, 39ff., 64f., 67, 224, 258, 283, 292, 370ff., 394, 405, 722, 830, 948, 1161, 1291

sulfuric acid 25, 27, 34f., 54, 146, 428, 721, 777, 807, 856, 921, 961, 1158, 1214, 1228, 1262, 1291, 1293

sunflower, *see* oil

superconduction 791, 798ff., 895, 905ff., 922, 962, 967, 969, 1129, 1231, 1277, 1301, 1306, 1308

superoxide, *see* peroxide, hydrogen peroxide

superphospate, *see* phosphate, fertilizer

supervision, follow-up, *see* monitoring

supplement, *see* (food) additive

supply of blood, *see* artery, circulation

supraregional distribution, *see* long-range transport

surface water, *see* rivers, lakes

surfactant, *see* detergent

surgery, *see also* implant 557, 1110, 1204f., 1207, 1209, 1322, 1328, 1344, 1353, 1362, 1365

suspended particle, *see* dust, particle, sediment

swallowing, *see* oral

swan, *see* bird, wild bird

sweat, *see* perspiration

swine, *see also* pork 202, 471f., 474f., 478, 532, 535, 537ff., 540ff., 544, 546, 645, 753f., 756, 772, 820, 849, 895, 899f., 1039, 1069, 1080, 1092, 1100, 1110, 1130, 1133, 1174, 1196f., 1218, 1240, 1326, 1330

switching, circuit, *see also* control engineering, electric conductor 777, 895, 911, 1051

symbiosis 415

synapsis, *see* neuron

synchrotron techniques SYXFA 143f., 744, 894, 905

synergism 227, 287, 420, 426, **492**, 565, 568, 626, 636, 639, 780, 822, 830, 878, 897, 904, 925, 928, 1114f., 1117, 1119, 1183, 1190, 1213, 1283, 1324

synthetic resins, *see* plastic materials

syphilis, other venereal diseases 754, 792, 795, 1045, 1052, 1230

systemic effects, *see* circulation

tableware, dishes, *see also* cooking, porcelain, glass 699, 981

tachycardia, tachypnea, *see* arrhythmia

tailing, *see also* leaching, plume **21ff.**, 55, 61, 65, 342, 348, 351, 354, 398, 1278

tanning agents, tannates 720, 754, 854, 857, 860f., 864, 872, 874, 876, 895, 952, 1052, 1344

tantalum 14, 557, 560ff., 888, 1128, 1132f., **1203ff.**, 1207ff., 1273

tape (recording tape), also for sampling 271, 857, 882, 962, 1104

tap water, *see* pipe

target (critical) organs, organ functions, *see also* the respective organs 589f., **593ff.**, 652f., 655f., 663, 675f., 725, 738, 743, 747, 762, 1045, 1060, 1136, 1194, 1228, 1284, 1324, 1364

tars, *see* organic (matter) waste

taste 586, 876, 886, 914, 1041f., 1067f., 1176, 1230, 1314, 1328f., 1354, 1359

TCDD, *see* dioxins (polychlorinated)

tea 157, 203, 451f., 454ff., 460ff., 464, 466, 468, 656, 923, 1038, 1107, 1109

technetium 745, 1090, 1204

technical guiding concentration, *see* concentration at the workplace, regulation
teeth, *see* tooth
Teflon, *see* plastic materials
tellurium, tellurium compounds 6, 11, 15, 19, 83, 159, 190, 218, 235f., 257, 443, 573, 792, 797, 922, 932, 1021, 1083, 1151, 1153, 1156ff., 1159, 1173, 1184, 1186ff., **1211ff.**, 1214ff., **1219ff.**, **1222**, 1231
tempering bath, *see* hard-solder, steel
tendon (jerk), *see* sinew
teratogenic effects, *see also* embryo (toxicity), fetal loss, co-teratogen(icity) 495, 504, 508, 542, 547, 617, **629ff.**, 633, 635, 637, 765ff., 768ff., 785, 829ff., 863, 866f., 888, 892, 915f., 926, 942f., 966, 1001ff., 1041, 1043, 1070ff., 1073f., 1081f., 1085f., 1088, 1095f., 1114f., 1120f., 1124f., 1167, 1177, 1221ff., 1224, 1237, 1252, 1254, 1283, 1305, 1327, 1330f., 1346, 1356, 1370
terbium 145, 882, **959ff.**, 965f., 970
terrier, *see* dog
test, test battery, *see* short-term test
test chamber, *see also* green house 226
testes, testicular barrier, *see also* sperms 495, 498, 508, 538, 552f., 643, 647, 912, 963, 1002, 1040, 1147, 1234, 1271, 1304, 1316f., 1326, 1328, 1331, 1357
testosterone 488
tetany, *see also* grass tetany 647, 1031, 1370
tetramethyl lead, *see* methylated and ethylated lead, leaded gasoline
textiles (industry), *see also* cloth 720, 754, 857, 1052, 1142, 1312, 1326, 1344
thallium, thallium compounds, *see also* organothallium compounds 6ff., 11ff., 15, 19, 35f., 57, 72, 83, 95, 132, 158, 162, 178f., 186, 205, 449f., 453, 455, **464f.**, 467, 469, **477**, **542f.**, 545f., 571, 581, 588, 596, 600, 839f., 943, 1017, **1227ff.**, **1230ff.**, 1235ff.
therapeutics, therapy, *see* drug, detoxification, chrysotherapy, platinum compounds
thermal techniques, thermal conductivity detectors, thermal ionization, *see also* electrothermal, heat 135, **145f.**, 159, 749, 775, 789, 894, 910, 1026, 1090, 1104, 1118, 1137, 1211, 1213, 1215, 1285, 1300, 1360
thermodynamics, *see also* kinetics, model 68, 240ff., 325, 330, 384, 566, 719f., 738, 853, 898, 959, 1154
thermometer, thermophilic, thermocouple, *see also* fever 444, 744, 791f., 911, 918, 964, 1051, 1067, 1112, 1142, 1230, 1360
thermoplastic material, *see* plastic (materials)
thin layer chromatography, thin layer focussing 150, 437, 772
thioglucose, thioglycolic acid 935f., 946, 1062
thiol, *see* sulfhydryl groups
thionein, *see* methallothionein
thiosulfate 533
thorium 59, 138, 145, 290, 626, 942, 962f., 968, **1275ff.**, 1278, **1281ff.**, 1284ff., 1287, 1300f.
threshold, limiting value, air quality, *see also* concentration at the workplace, biological tolerance value, VDI-Commission 108, 223, 231, 260, 262, 277, 283ff., 294, 307ff., 366, 479, 505, 531, 536, 543, 548, 554f., **594f.**, **651f.**, 654, 658f., **661ff.**, 664, 667ff., **670ff.**, 673ff., **677f.**, 680, 682, **687ff.**, 690ff., **693ff.**, **700ff.**, 703, 707, 709ff., 732, 747, 749, 766f., 772, 782f., 797, 804, 821, 831ff., 834f., 837, 841, 848f., 869ff., 873, 876, 889, 903, 906, 915, 926f., 935f., 942, 955, 966, 993f., 997, 1003ff., 1013, 1021f., 1033, 1041f., 1053, 1072, 1075ff., 1078, 1083, 1096, 1099, 1115f., 1120, 1132, 1135, 1137, 1147f., 1150, 1159, 1169, 1179f., 1185, 1198, 1200, 1208, 1222, 1233, 1237f., 1240, 1252, 1255, 1257, 1264, 1266, 1272f., 1282ff., 1285f., 1295f., 1306, 1330ff., 1346, 1362, 1370
throat, *see* pharynx, brochial
thrombosis, thrombocytes, hyaluronate, *see also* prothrombin 763, 935, 954, 1031, 1040, 1304, 1367, 1370
throughfall, fallout, *see also* canopy, deposition 267, 370ff., 375ff., 479, 818, 980, 988, 994, 1007, 1360
thulium 145, **960f.**, 964, 969
thymidine, thymol 613, 621, 632, 770, 1047
thymus, thymic 1031, 1112, 1120, 1254
thyroid (gland), thyroxine, parathyroidism 484f., 725f., 729, 734f., 746, 887, 941, 999, 1020, 1108, 1173, 1187, 1234, 1317, 1360, 1368
tidal, *see* coast, estuary
time lag, *see* long-term
tin, inorganic tin compounds, *see also* cassiterite, organotin (compounds), stannate 6f., 11, 15f., 35, 79, 83, 89, 95, 130ff., 135, 139, 147, 149, 157, 159, 172f., 190, 193, 203, 207, 218, 243, 257, 301f., 422, 450, 481f., 485, **543**, 573, 741, 786,

790f., 809, 921, 959, 1017, 1049, 1051, 1192, 1204, **1243ff.**, 1247f., **1250ff.**, 1255, 1369
tinplate, tin sheet 1245
tins (British), *see* can
tire (British: tyre), tire abrasion 299, 812, 1162
tiredness, somnolence, sleepiness, lassitude, weariness 544, 585, 914, 954, 995, 1000, 1021, 1030, 1068, 1070, 1073, 1113, 1126, 1176, 1219, 1221, 1236, 1304, 1330, 1351, 1360, 1364f., 1369
titanium, titanium compounds, *see also* titanium dioxide 6f., 11f., 15, 20, 89, 218, 235, 239, 290, 314, 423, 557, 573, 622, 719f., 857, 888, 1026, 1104, 1128f., 1133, 1142, 1204, **1261ff.**, 1264ff., 1291, 1343f., 1346
titanium dioxide, ilmenite, *see also* waste acid 709f., 1261ff., **1264ff.**, 1267
TLV, *see* threshold
tobacco (smoke), *see also* nicotine, indoor 211, 216f., 287f., **289ff.**, 300, 302ff., 305ff., 308f., 548f., 576, 579, 591, 599, 624f., 634, 636, 638, 655ff., 665, 676, 746f., 755ff., 803, 819, 821, 823ff., 830ff., 838, 843, 863, 876f., 934, 961, 990f., 1006, 1064, 1082, 1103, 1106, 1162, 1166, 1186, 1235, 1279, 1283, 1286f.
α-tocopherol, *see* vitamin E
tolerance, *see also* daily intake, threshold, glucose tolerance 102, 360, 363, 365, 407, 409, **411ff.**, 414ff., 418ff., 421, 423f., 426f., 430f., 434f., 437ff., 445f., 448, 491, 501, 508, 527f., 537, 540f., 543f., 549, 559, 562, 595, 641ff., 648f., 705ff., 715, 721ff., 726, 758ff., 764, 769, 773f., 819, 827, 836, 838, 841, 851, 865, 872, 896, 898, 902, 906f., 915, 927, 937, 961, 965, 992f., 1000, 1006, 1019, 1035, 1038, 1040, 1042, 1055, 1062, 1065f., 1071, 1081, 1086, 1094ff., 1122, 1167, 1172, 1195, 1201, 1203, 1209, 1218, 1226f., 1230, 1235, 1250, 1255, 1262, 1264, 1269f., 1301, 1323f., 1326, 1330, 1332, 1341f., 1344, 1360
tomato, *see* vegetables
tongue, *see* taste
tonsil, *see* pharynx
tooth, gingiva, gum, *see also* caries 112, 177, 180, 194, 198, **213ff.**,, 216, 219, 578f., 583, 592, 598, 603, 757, 795, 973, 983f., 995, 1008, 1011ff., 1067, 1073, 1087, 1176, 1197, 1245, 1263, 1295

topsoil, *see* agriculture, soil
toxicokinetics, *see* kinetics, metabolism
toxin, *see* poisoning
tracer, *see also* isotope 119, 127, 140, 142, 151, 157, 181, 184, 195f., 261, 275f., 279, 282, 558, 568, 577, 794, 800, 842, 880, 884, 909, 921, 923, 939f., 942f., 1009, 1016, 1058, 1062, 1108, 1130f., 1155, 1189, 1204f., 1231, 1271f., 1317, 1326
tracheobronchial, tracheal cancer, tracheal injection, tracheitis, intratracheal route, *see also* pharynx, respiration 263, 574ff., 634, 762ff., 823, 863, 1109, 1209, 1279f., 1370
traffic (density), *see* road, automotive exhaust, railway, aircraft
transferrin, transferase, ferritin, *see also* carrier (protein) 514f., 519f., 551, 632, 725f., 737, 831, 862, 912f., 916ff., 919, 941, 943, 949, 953, 956f., 1039, 1112, 1218, 1293, 1322, 1358, 1370
transformation, transcription, transduction, disproportionation 92, 149, 234, 240, 256, 338, 342f., 369, 377, 390f., 419f., 422, 424, 430, 432, 434, 439, 443, 447, 487, 500, 504, 524, 579f., 610, 612, 619f., 622f., 629, 633, 644, 738, 743, 747f., 759, 769, 771, 781, 784, 789, 848, 867, 876, 924, 949, 966, 969, 1018, 1061, 1074, 1079, 1086, 1090, 1115, 1122, 1124, 1163, 1167, 1169, 1238, 1249, 1251, 1254, 1277, 1304, 1366, 1370f.
transistor, *see* electronic equipment, semiconductor
transition elements (general) 72, **77ff.**, 86, 90, 135, 204, 206, 234f., 237, 239, 243, 345, 496, 638, 648, 879, 949, 1135f., 1200, 1265
transmission TEM, *see* electron microscopy
transpiration, *see* perspiration
transplantation, *see also* implant, surgery 226, 1177f., 1185, 1331, 1360
transport mechanism, transfer, transportation, translocation, *see also* cycle, long-range transport, material transport, mobilization 4f., 8, 51, 80, 88, 94, 98, 106, 261, 265, 285, 300, 302f., 325f., 328f., 333, 337, 355, 357, 361, 370ff., 393, 397ff., 400, 410, 413, 417, 422ff., 427, 432, 437, 441, 447, 455, 467, 470, 475, 485, 493, 498, 508, **511ff.**, 514ff., 518ff., 521, 528, 548f., 551, 561f., 565, **572**, 575, **577ff.**, 582, 644, 687, 692, 719, 722f., 725f., 741, 745f., 750, 756f., 769, 771ff., 778f., 787, 794, 815, 821, 823f., 840, 851, 859f., 862f., 868, 872, 875, 878f., 883f., 896f., 899, 909, 912, 915f.,

924, 933f., 938, 940, 942, 947, 949, 951f., 957, 962f., 964, 967f., 977, 979, 987f., 1018, 1029, 1038, 1058, 1060, 1065f., 1079, 1085, 1090f., 1108f., 1116, 1118, 1120, 1125, 1129, 1145, 1162, 1166f., 1169, 1194, 1205, 1209, 1216, 1233f., 1238, 1240, 1249ff., 1267, 1271, 1279f., 1283, 1292ff., 1302ff., 1317, 1322, 1328, 1345, 1356, 1361ff., 1364, 1370
transport protein, see carrier protein
treatment, see drug
tree, see canopy, conifer, deciduous tree, plant, stem flow
tree ring, see annual ring
trembling, see tremor, coordination
tremor, see also insensibility, muscles, paralysis 541, 543, 586, 727, 796, 996, 1000, 1020f., 1031, 1040, 1068, 1219, 1352, 1370
trimethyllead ion, see methylated and ethylated lead
tropical medicine (incl. sleeping sickness), see also intestinal flora, parasite 745, 749, 754
troposphere 257, 283
trout, see salmonide
trypanosome, see tropical medicine
trypticase, tryptone, tryptophan, tryptamine 420f., 440, 620, 899, 1019f., 1041, 1125, 1131, 1318, 1370
tube, see pipe
tubili, see renal cortex
tumor, see also carcinogenicity, granulomata, killer cell, neoplasm, sarcoma 495, 504, 506, 508, 543, 559, 568, 570, 614f., 622f., 625f., 629, 634ff., 639, 745, 763, 770, 775, 781, 792, 795, 798f., 830, 862, 865f., 869, 873, 875, 879, 888, 901, 911ff., 915f., 918f., 925, 928f., 937, 939f., 964, 967f., 1001f., 1031, 1067, 1071f., 1114, 1120, 1122, 1124, 1131, 1147, 1151, 1159, 1177f., 1183, 1185, 1187f., 1215, 1221, 1237, 1254, 1264, 1277, 1305f., 1331, 1349, 1357, 1365, 1368, 1371
tuna, see predatory fish
tundra, tundra plants 779
tungsten, tungsten carbide industry, tungsten compounds 20f., 128, 192, 545, 588, 744, 786, 881, 884, 888, 1089f., 1093, 1098f., 1203f., 1206, **1269ff.**, 1272f., 1277
turbine 273, 881, 1104
type letter, see letter
tyre (British), see tire
tyrosinase, see cheese

ulceration, ulcers, see tumor, gastric hemorrhage
unborn, see fetus
unconciousness, coma, dizziness, see also shock, hallucination, coordination 590, 761, 796, 901, 926, 954, 996, 1093, 1101, 1176, 1219, 1221
underdevelopment, see also developmental, growth inhibition 608
unicellular organism, protozoon, see micro-organisms
uptake, see bioavailability, absorption, resorption
uracil, see also RNA 1065, 1368, 1371
uranium 19ff., 59, 65, 138, 202f., 422, 428, 467, 573, 626, 791, 1026, 1095, 1100, 1135, 1158, **1275ff.**, 1278, **1281ff.**, 1284, 1286f., 1290
urban (atmosphere), see also automobile, dust 159, 227, 260f., 264, 274, 281, 283, 285, 299, 304f., 308, 335, 357f., 372, 571, 678f., 700, 703, 722, 755, 778, 816, 833, 837, 839, 858, 875, 896, 929, 933, 940, 977f., 980, 983, 985, 987f., 991, 1012, 1103, 1105f., 1115, 1160f., 1169, 1199, 1291, 1314
urea, urease, uric acid, see also polyuria 485, 914, 1095, 1106, 1117, 1122
uremia, see oliguria
urine, oliguria, anuria, aciduria, etc., see also diuretics 108, 110f., 113, 119, 132, 140, **156**, 162, 164, 171f., 189, 191, 205, **207ff.**, 212f., 216, 218, 484, 529, 551, 555, 561, 577, 580f., 583, 594, 600, 602, 623, 625, 645, 660f., 666, 675f., 693, 698, 705, 707, 722ff., 727f., 732, 739, 743, 746, 748, 752f., 756ff., 759, 761, 767, 770, 772, 779, 787, 790, 793ff., 799, 805, 824, 828, 831f., 835ff., 838, 840f., 844, 847, 854, 863, 873, 885, 888, 891, 895, 897, 905, 910, 913, 915f., 941, 953, 957, 973, 987, 992ff., 1000f., 1005, 1021, 1032, 1039, 1041f., 1049f., 1058, 1061ff., 1067, 1077f., 1092, 1103, 1107ff., 1110ff., 1117, 1120, 1122, 1125, 1131, 1136f., 1157, 1168, 1171, 1173, 1189, 1194, 1197, 1199, 1206, 1213, 1216f., 1221f., 1224, 1228f., 1233ff., 1236ff., 1241, 1250, 1262, 1264, 1271, 1280, 1282, 1289, 1293ff., 1303, 1322f., 1345, 1349f., 1353ff., 1357ff., 1365, 1367
urogenital (infection, cancer) 766, 1020, 1147, 1221, 1237, 1328, 1371
uterus, womb, cervix 617, 923, 1072, 1078, 1080, 1132, 1360, 1367

UV (IR) radiation, UV light, UV oxidation 69, 120f., 139, 149, 151, 157, 239, 324, 626ff., 635, 637, 642, 657, 764, 768, 770, 785, 961, 1114, 1249
UV mutagenesis 506, 626, 633, 637, 1119
UV spectrometry, UV detector 147, 170, 200, 525, 752

vagina, sheath 795
valence state, *see* speciation, oxidation (state)
validation, *see* reference material
vanadium, vanadium compounds 6ff., 11f., 15, 87, 89, 102, 132, 138, 155, 167, 175, 202, 207, 216, 218, 258f., 270, 273, 277, 307, **465ff.**, 477, 481f., 519, **543ff.**, 573, 588, 596, 703, 708, 1121, 1133, 1200, 1204, 1270, 1272, **1289ff.**, 1293ff., 1301
vapor (pressure), *see also* volatility 153, 156, 257, 262, 342f., 574, 591, 671, 723, 916, 1046, 1059, 1068, 1101, 1156, 1175, 1290
vaporization, *see* volatility, vapor
vascular (disease), vasodilation, *see also* "blackfoot" disease, aorta, artery, cirulation 518, 544, 599, 763, 954, 969f., 998ff., 1030, 1113, 1123, 1175, 1207, 1217, 1294, 1352, 1354ff., 1358, 1361f., 1366f., 1371
VDI Commission on Air Pollution Control 689, 1037
veal, *see* calf
vegetable(s), greens, *see also* bean, carrot, legume(n), radish 361f., 367f., 401, 407, 409, 449, 451ff., 455ff., 461ff., 466, 472f., 477, 479, 532, 703, 722, 746, 756, 778, 780, 786, 820f., 826, 836, 849f., 860f., 864, 872, 895, 921, 923f., 952, 955, 981, 991, 1005, 1008, 1014, 1028, 1038, 1056, 1058, 1077, 1091, 1093, 1100, 1231ff., 1235, 1248, 1279, 1292, 1315, 1319, 1326, 1328, 1345
vegetation, *see* plants
vehicle, *see* automobile
velocity (ultrasonic measurement) 238
ventilation, *see also* indoor, wind 288, 393, 927
vertebra(l), *see* spinal
vertebrate, *see also* mammal 225, 515, 523, 949, 1163, 1293
vesicle, *see* alveolus
veterinary (medicine), *see also* drug 471, 533, 922, 1159, 1312
viability, *see* vitality
video, audio, *see* tape, data bank
vine, *see* wine
vinegar, *see* acetum

vinyl chloride, *see also* PVC 1056f.
violet, *Viola* 412f., 415f.
viral reations, and transformations, virus, *see also* AIDS virus 605, 608f., 613f., 626, 632, 645, 747f., 898, 903, 963, 1064, 1071, 1082f., 1115, 1126, 1177, 1185, 1363, 1367
viscera, *see* intestine
visibility, visible spectrometry, appearance, *see also* colorimetry 308, 961, 1314
vision (impaired or defective), *see also* eye, blindness, cornea, smog 658, 853, 901f., 926, 996, 1061, 1069, 1073, 1236, 1238, 1254, 1328, 1350, 1368
vitality, viability, *see also* fertility, reproduction, tiredness 484, 1175
vitamin A 1184, 1187
vitamin B_{12}, *see also* methylcobalamin, cobalt 424, 432, 436, 484f., 826, **879**, **882ff.**, 885f., 1055, 1146, 1196f., 1352
vitamin C, ascorbate, ascorbic acid 486, 513, 626, 826, 878, 886, 946, 949, 952, 957, 1170, 1173, 1213
vitamin D 488, 495, 499, 512, 574, 725, 734, 829, 832, 988
vitamin E 537, 887, 1153, 1173f., 1178, 1183, 1187, 1196ff., 1199ff., 1218f., 1226
vitamins (general) 481f., 489, 577, 803, 829
volatility, volatilization, vapors, *see also* thermal techniques, vapor pressure 12, 33ff., 39, 43, 46, 50, 53f., 83, 87, 89, 126f., 130f., 148f., 156f., 159, 216, 263, 294, 307, 338, 377, 379, 420, 422, 424, 426, 428, 432f., 435f., 438, 442f., 548, 574, 591, 659, 751, 755, 790, 798, 803, 805, 807, 854f., 858, 915, 922, 978, 1000, 1047ff., 1053ff., 1061, 1063, 1068, 1083, 1101, 1104, 1140f., 1147, 1154, 1162, 1167, 1183, 1189, 1212f., 1222, 1226, 1231, 1248, 1262, 1278, 1368
volcano, volcanic eruption 4, 87, 257ff., 268, 276, 284, 405, 745, 754f., 1016, 1018, 1050, 1053, 1105, 1158, 1162, 1312
voltammetry, *see also* differential pulse and anodic stripping voltammetry, potentiometry 70, 100, 108f., 115f., 118, 120ff., **136ff.**, 149, 151ff., 156, 160f., 163, 166, 170, 172, 175, 179, 183, 187, 190f., 193, 196, 201, 203, 245, **247**, 251f., 396, 806, 843ff., 893, 956, 958, 972f., 1078, 1122, 1239, 1241, 1275, 1287, 1353f., 1369
volunteer, *see* epidemiology
vomiting, *see also* nausea 532f., 535ff., 540ff., 543f., 585f., 589, 593, 761f., 795, 827, 887, 901, 914, 926, 995, 1021, 1025,

1030f., 1067, 1113, 1175f., 1221, 1235, 1252, 1254, 1282, 1324, 1329f.
vulcanization 745, 1159, 1214f.

warm-blooded animal, *see* animals, mammal, bird
wart, pustule, tubercle 792
washing agent, *see* detergent, soap
waste, refuse, *see also* household garbage, litter, organic waste, residue 21ff., **30f.**, 34ff., 41ff., 90, 222, 255, 266, **333f.**, 336, 342, 344f., 352f., 355, 358ff., 365, 384, 397, 410f., 414f., 423, 429, 442, 453, 493, 506, 533, 569, 667, 692, 707, 709f., 718ff., 740, 744, 747, 753f., 775, 790, 792, 803, 806, 809f., 812ff., 816f., 819, 835, 842, 845, 855, 857f., 861, 874, 881, 894f., 903, 905, 910, 922, 933, 940, 947f., 974ff., 979, 993f., 1014, 1016f., 1026, 1036f., 1050, 1053, 1057, 1088, 1090, 1104f., 1126, 1153, 1158f., 1161f., 1192f., 1211, 1227, 1229, 1231, 1245, 1247, 1261ff., 1265, 1270, 1277, 1290, 1292, 1309, 1311f., 1314, 1323, 1342ff., 1358
waste acid (from the titanium sulfate) process 1261, **1263**, 1265ff.
waste water, effluent 23, 25, 30, **41ff.**, 49, 51f., 61, 108, 115, 158, 205, 233, 255, 333, 341f., 347, 357f., 363, 392, 493, 532f., 535, 537, 689ff., 692, 743, 749, 810, 813, 816, 835, 837, 845, 855ff., 870, 873f., 881, 897, 947, 979, 993, 1053, 1056f., 1089f., 1116, 1140, 1148, 1163f., 1233, 1244, 1247, 1263, 1269, 1289, 1312f., 1319, 1338
watch (spring, and other parts) 754, 777, 888, 947, 1017, 1020, 1277, 1283, 1287, 1312
water hardness, *see* lime, calcium compounds
water plant, *see* aquatic plant
waterproof, water resistance 1312, 1344
water solubility, *see* solubility
weather conditions, climate, *see also* wind 38f., 226, 261, 266, 280, 308, 334, 403, 428, 449f., 655, 657, 673, 815, 979f., 1291, 1318
weathering, *see also* erosion 4, 7f., 56, 311, 322, 325, 373, 399, 401, 403, 776f., 806, 816, 860, 896, 948, 991, 1027, 1105, 1139, 1165, 1292, 1311f., 1315
weight (loss of), emaciation, *see also* bodyweight, growth inhibition 533, 538, 542ff., 553, 831, 885, 887, 915, 1019, 1039, 1068, 1073f., 1330, 1355

welding, welder, welding fume 493, 504, 508, 598, 657f., 660, 861f., 867, 874, 876, 1037, 1040, 1101, 1104, 1109, 1112, 1126, 1128, 1263, 1267, 1312
well, spring water 333, 442, 858, 876, 903, 977, 1041, 1284
Weser (river) 233f., 388
wet deposition, wet-only sampler 158, 168, 201, 266, 268, 277, 281, 283, 311, 372, 374, 377, 814f., 817f., 839, 844, 980, 987, 1013, 1048, 1054, 1078, 1161, 1314
wet digestion, wet ashing, *see also* pressure disintegration 117f., 128, 212, 272, 290, 1140f.
wet scrubber 39, 41, 263, 948, 1088, 1155
wheat (germ), *see* grain, cereals
Widy sign 1236
wild animal, wild life 112, 531, 538, 625, 638, 850, 994f., 1008, 1100, 1163, 1167, 1279
wild bird 995, 1060, 1077f., 1081, 1164, 1167, 1177, 1183, 1196, 1219, 1226
Wilson's disease 514, 527, **645f.**, 648, 899f., **901ff.**, 906f., 1094, 1352f., 1358, 1371
wind, *see also* atmosphere 4, 8, 12, 56, 1053, 1291
wine (growing), vine, viniculture, grape, *see also* mildew 113, 157, 168, 179, 199, 203, 357, 360ff., 366, 407f., 656, 660, 703f., 756, 763f., 768, 820, 847, 858, 896, 901, 904, 950, 981, 986, 990, 1005, 1008, 1263, 1271, 1321
wood, wood chips, *see also* forest 55, 259, 275f., 287f., **292**ff., 303f., 306ff., 362, 720, 1051, 1247, 1291, 1362, 1371
wood preservative, wood impregnation, *see also* fungicide, impregnating 59f., 753, 857, 877, 1037, 1052, 1246, 1312
wool, *see also* hair 899, 1094
worker, workplace, *see* occupational exposure, concentration at the workplace
worm, *see* earthworm, nematode
wound (healing), *see also* skin pigmentation 559, 781, 1196ff., 1199, 1201, 1207, 1210, 1322, 1328, 1332, 1351
wrecked car, *see* scrap
writing, *see* coordination, hand, tremor

xanthine oxidase 1092f., 1095, 1097f., 1358
xenon, *see* noble (inert) gas, noble element
xeroderma pigmentosum, *see* skin pigmentation, skin cancer

X-ray absorption, EXAFS, XANES 84, 525, 932, 934, 936, 938, 1212, 1289, 1312
X-ray damage, *see* X-ray irradiation
X-ray (powder) diffraction 265, 273, 281, 1112, 1122
X-ray (emission) spectrography, *see also* PIXE 194, 210, 738, 841, 894, 1128, 1224, 1300, 1366
X-ray (fluorescence) spectroscopy XRF, *see also* synchrotron 69f., 107ff., 116, 119ff., 133, **142ff.**, 151, 161, 163, 166, 170, 180, 182, 189, 194, 199, 206, 210f., 218, 247, 272, 384, 716, 738, 741, 752, 806, 837, 841, 843, 854, 894, 905, 960, 969, 972, 1012, 1035, 1049f., 1082, 1087, 1089, 1137, 1154, 1203, 1229, 1241, 1244, 1261, 1300, 1306, 1310, 1354, 1366, 1371
X-ray irradiation, X-ray radiation, X-rays 143, 186, 274, 277, 527, 628, 637, 837, 937, 1051, 1072, 1114, 1204, 1270, 1272, 1367
X-ray microanalysis, X-ray contrast, X-ray spectrophotogrammetry 204, 634, 718, 730, 932, 1205, 1208f., 1224, 1272, 1275, 1286, 1289
xyleme, *see also* stem (flow) 884, 951, 1106, 1125, 1318, 1362, 1371

yeast , *see also* beer 420ff., 423, 437, 440, 501, 524, 528, 730, 858, 864f., 867f., 872f., 877, 1063f., 1116, 1165, 1295, 1319
ytterbium 134, 189, **959ff.**, 964f., 1300
yttrium 72, 76, 138, 145, 203, 895, 918, 967, 1132, **1299ff.**, 1302ff., 1305ff.

Zeeman background correction **127ff.**, 132, 160, 171, 175ff., 184f., 190, 195, 198, 200, 204, 206, 717, 843, 854, 1103, 1109, 1116, 1157

zeolite 720
zinc (general, elementary) 5ff., 11ff., 15, 19ff., 32, **35ff.**, 44, 50, 52, 54, 57ff., 63ff., 72, 89, 95, 97, 107, 132, 136f., 153, 155f., 158, 175, 178, 207, 211, 214, 218, 223, 233, 249ff., 258f., 268, 270, 280, 288, 291, 294, 297, 301f., 304, 307, 315, 317, 328, 334f., 342, 347f., 351, 357f., 361ff., 367, 375f., 384, 389f., 392, 396, 401, 404f., 409ff., 413ff., 418, 423, 426, 430, 440, 444, 450, **466ff.**, 477f., 481, 484f., **495f.**, 501, 505ff., 511f., **517ff.**, 520f., 523, 527f., 536f., 540, 542, **544ff.**, 550, 553, 555, 566f., 569f., 573, 590, 605, 607f., **611ff.**, 618, 623, 625, 629, 639, 641f., 644, **647**, 649, 703, 736, 754, 803, 805, 809, 813, 816, 820, 826, 829f., 832, 835f., 838ff., 842, 847, 883f., 893, 895, 897f., 900, 906, 939, 958, 975, 1030, 1041, 1044, 1094, 1097, 1100, 1126, 1155, 1162, 1173, 1189, 1192, 1200, 1230, 1245, 1261, 1290, **1309ff.**, 1312ff., 1316ff., **1322ff.**, 1327, 1330ff., 1340ff., 1360
zinc compounds 35, 266, 393, 432, 499, 546, 568f., 670, 757, 806f., 814, 830, 857, 862, 902, 910, 912, 921ff., 974f., 994, 1012, **1229ff.**, 1309ff., **1312**, 1329f., **1332**
zinc deficiency 466, 504, 535, 545, 553, 574, 611, 644, 647, 893, 1312f., **1315**, 1318f., **1321**, 1323, **1325ff.**, 1328f.
zinc ions 79, 82, 224, 249, 266, 321, 324, 339f., 345, 420, 422, 440, 496f., 500f., 503, 523, 526f., 578, 619, 628, 994, 1313, 1316
Ziram, Zineb 636, 1312, 1331
zirconium, zirconium compounds 6f., 11f., 15, 149, 218, 299, 307, 561, 719, 870, 1026, 1032, 1132f., 1301, **1343ff.**, 1347
zooplankton, *see* plankton